Fourth Edition

Essentials of Medical Microbiology

for Undergraduate Students

As per the latest NMC Guidelines | Competency Based Medical Education (CBME) Curriculum under Graduate Medical Education Regulation

Fourth Edition

Essentials of
Medical
Microbiology

for Undergraduate Students

As per the latest NMC Guidelines | Competency Based Medical Education (CBME) Curriculum under Graduate Medical Education Regulation

Prakash Modi MD (Microbiology)
Assistant Professor
In-charge Head, Department of Microbiology
Member Secretary of Hospital Infection Control Committee
Member of College Curriculum Committee
GMERS Medical College and Hospital
Dharpur-Patan 384 265, Gujarat, India
E-mail: drprakash_md @yahoo.co.in

Formerly
Professor and Head
Department of Microbiology
PDU Govt. Medical College, Rajkot 360 001, Gujarat, India

Assistant Professor, Associate Professor, Professor and Head
Department of Microbiology
Shri MP Shah Govt. Medical College, Jamnagar 361 008, Gujarat, India

Assistant Professor
Department of Microbiology
BJ Govt. Medical College, Ahmedabad 380 004, Gujarat, India

Assistant Professor
Department of Microbiology
Govt. Medical College, Bhavnagar 364 001, Gujarat, India

Editor, *Journal of Research in Medical and Dental Science* (JRMDS)

CBS Publishers & Distributors Pvt Ltd

New Delhi • Bengaluru • Chennai • Kochi • Kolkata • Lucknow • Mumbai
Hyderabad • Jharkhand • Nagpur • Patna • Pune • Uttarakhand

CBSPD

Fourth Edition

Essentials of
Medical
Microbiology

for Undergraduate Students

As per the latest NMC Guidelines | Competency Based Medical Education (CBME) Curriculum under Graduate Medical Education Regulation

ISBN: 978-81-9798-225-5

Fourth Edition: 2025

  Reprint: 2027
Third edition: 2020
Second edition: 2017
First edition: 2015

Published by Satish Kumar Jain and produced by Varun Jain for

CBS Publishers & Distributors Pvt Ltd

4819/XI Prahlad Street, 24 Ansari Road, Daryaganj, New Delhi 110 002, India
Ph: 011-23289259, 23266838   Website: www.cbspd.com
                                    e-mail: delhi@cbspd.com

Corporate Office: 204 FIE, Industrial Area, Patparganj, Delhi 110 092, India
Ph: 011-49344934   Fax: 011-49344935   e-mail: publishing@cbspd.com; publicity@cbspd.com

Branches

- **Bengaluru:** Seema House 2975, 17th Cross, KR Road, Banasankari 2nd Stage, Bengaluru 560 070, Karnataka, India
  Ph: +91-80-26771678/79   Fax: +91-80-26771680   e-mail: bangalore@cbspd.com

- **Chennai:** 18/8B, Subbarayan Street, Shenoy Nagar, Chennai 600 030, Tamil Nadu, India
  Ph: +91-44-42032115, 45022115   e-mail: chennai@cbspd.com

- **Kochi:** 42/1325, 1326, Power House Road, Opposite KSEB, Kochi 682018, Kerala, India
  Ph: +91-484-4059061–65   Fax: +91-484-4059065   e-mail: kochi@cbspd.com

- **Kolkata:** 147, Hind Ceramics Compound, 1st Floor, Nilgunj Road, Belghoria, Kolkata 700056, West Bengal, India
  Ph: +91-33-25633055/56   e-mail: kolkata@cbspd.com

- **Lucknow:** Basement, Khushnuma Complex, 7 Meerabai Marg (behind Jawahar Bhawan), Lucknow 226001, UP, India
  Ph: +91-522-4000032   e-mail: tiwari.lucknow@cbspd.com

- **Mumbai:** PWD Shed,z Gala No. 25/26, Ramchandra Bhatt Marg, Next JJ Hospital Gate No. 2, Opp. Union Bank of India, Noorbaug, Mumbai 400009, Maharashtra, India
  Ph: +91-22-66661880/89   e-mail: mumbai@cbspd.com

Representatives

- **Gujarat**   • **Hyderabad**   • **Jharkhand**   • **Nagpur**   • **Patna**   • **Pune**   • **Uttarakhand**

For trade terms please contact customercare@cbspd.com
For general enquiries please contact info@cbspd.com

Printed at Goyal Offset Works Pvt. Ltd. Haryana, India

Preface

After the great success, appreciation and feedback of *Prakash's Notebook of Microbiology including Parasitology and Entomology* (3rd edition), I am really excited for the publication of 4th edition of this book (Name changed as *Essentials of Microbiology*). It is the end of plenty of sleepless night and restless days with the release of this 4th edition. It is designed with system-based study as per revised CBME (Competency Based Medical Education) curriculum with competency codes and numbers for topics of Microbiology and for integrated topics of other subjects. Entire curriculum is divided in **three parts** and 11 sections: Among the three parts, **Part I covers General Microbiology (Section 1), Immunology (Section 2) and Healthcare-associated Infections/HAIs (Section 3), Part II covers Systemic Microbiology (Sections 4 to 8) and Part III covers Entomology (Section 9), Practical Microbiology/DOAP/Skill Assessment (Section 10) and MCQs (Section 11).** MCQs are divided into three groups like chapter wise MCQs given at the end of every chapter, while section wise and image-based MCQs are given separately in **Chapters 125** and **126,** respectively. MCQs are the useful tools for the UG students as well as for PG aspirants.

Following latest topics/chapters are added in this edition.

- SARS CoV-2 (COVID-2019) including its variants and vaccines
- Automated culture methods
- Comments in question bank
- Precipitating factors of bacterial/viral/fungal/parasitic infections
- Laboratory acquired infections
- National Health Program and pandemic management
- Visit to Microbiology department.
- IMG, AETCOM and competencies in Microbiology as per revised CBME curriculum.
- Monkey pox outbreak 2022.
- A/H3N2 seasonal flu 2023.
- Chandipura virus encephalitis.

After the publication of 1st, 2nd and 3rd editions, I got tremendous response and feedback from entire country by students and my friends/seniors who are serving as medical teachers in different colleges, I am thankful to all of them for encouraging me to go for the 4th edition. This task could not have been possible without an eternal support from my wife Shweta, my daughters Aarya and Aadhya and youngest star of family, my son Swar. I am also appreciating and sharing my gratitude to the entire team of CBS Publisher & Distributor for their endless support and publishing such a quality book. I also appreciate the support of all book store owners and online distributors. Last but not least, again thank you to all those who were always at there for me to make this journey successful.

Microbes do not have any specific anatomical border because one organism or its species can infect multiple systems. In addition, it is also not possible to change organism-based laboratories of Microbiology department like Bacteriology, Virology, Mycology, Parasitology, etc., to system-based laboratories like CVS laboratory, CNS laboratory, etc., so as per CBME curriculum, it is difficult to transfer earlier organism-based study to strict system-based study. Strict system-based study without peer group (like *Neisseria meningitidis* with *Neisseria gonorrhoeae*, *Clostridium welchii* with *Clostridium tetani*, etc.) and without clearance of taxonomical concept of microbes, may not only hinder the understanding and the concept building of Microbiology, but it also put erroneous perceptions. Microbiology is the diagnostic branch far better a clinical branch, not the therapeutic branch (Microbiologist can guide the clinician for treatment as per laboratory tests results, but cannot treat the patient directly), which is limited to laboratory side, rarely or not to bed side. I have tried to maintain the balance between all these by describing all systemic infections as common infections followed by specific infection by specific organism. Attempt has also been made to serve the complete knowledge but author does not claim for perfection. Suggestions are welcomed from all the students and teachers about any shorts in contents that will help to improve the work and to serve more quality material to students in subsequent edition.

Thanks and regards,

Prakash Modi

Contents

SECTION 5: BACTERIAL INFECTIONS

Essentials of Medical Microbiology

• Infections of Miscellaneous Bacteria

SECTION 7: FUNGAL INFECTIONS

Index of Competencies/Code of Competencies

- MI (Microbiology).
- Other subjects (for integrated topics): BI (Biochemistry), PA (Pathology), PH (Pharmacology), CM (Community Medicine), DR (Dermatology and Venerology), DE (Dentistry), IM (General Medicine), PE (Pediatrics), SU (General Surgery), OR (Orthopedics) and CT (Respiratory Medicine).
- Competency numbers written in bold indicate the topics of practical Microbiology/skill assessment/DOAP (Demonstrate, Observe Assess and Perform).

+ve: Positive
2ME: 2-mercaptoethanol
5-HT: 5-hydroxy tryptamine
AAV: Adeno-associated virus
Ab(s): Antibody(ies)
ACE 2: Angiotensin-converting enzyme 2
Ach: Acetylcholine
ADCC: Antibody-dependent cytotoxic cells
ADS: Antidiphtheria serum
AETCOM: Attitude, ethics and communication
AFB: Acid-fast bacilli
AFLP: Amplified fragment length polymorphism
Ag(s): Antigen (s)
Ag-Ab reaction: Antigen-antibody reactions
AGN: Acute glomerulonephritis
AGS: Antigas gangrene serum
AIDS: Acquired immunodeficiency disease
ALG: Antilymphocyte globulin
ALS: Antilymphocytes serum
ALT: Alanine aminotransferase
AMA(s): Antimicrobial agent(s)
AMB: Amphotericin B
AMC: Acetyl methyl carbinol
AMI: Antibody-mediated immunity
APCs: Antigen-presenting cells
APSGN: Acute post-streptococcal glomerulonephritis
APT: Alum precipitated toxoid
APW: Alkaline peptone water
ARC: AIDS-related complex
ARDS: Acute respiratory distress syndrome
ARF: Acute rheumatic fever
ART: Antiretroviral therapy
ASO test: Antistreptolysin O test
ASOM: Acute suppurative otitis media
ASP: Antimicrobial Stewardship Program
AST: Antimicrobial sensitivity testing
ASV: Antisnake venom
ATCC: American type culture collection
ATG: Anti-thymocyte globulin
ATS: Anti-tetanus serum
ATS: Anti-thymocyte serum
B/R(s): Biochemical reaction (s)
b/w (B/W): Between
BAL: Bronchoalveolar lavage
BB: Borderline
BCDF: B-cell differentiation factor
BCG: Bacillus Calmette Guerin

BCGF: B-cell growth factor
BCR(s): B-cell receptor(s)
BCYE agar: Buffered charcoal yeast extract agar
BDBV: Bundibugyo virus
BEBOV: Bundibugyo ebola virus
BFP: Biological false-positive
BHIA: Brain heart infusion agar
BHIB: Brain heart infusion broth
BL: Borderline lepromatous
BMW: Biomedical waste
BOD incubator: Biological (biochemical) oxygen demand incubator
BP: Blood pressure
BPL: Beta propiolactone
BSE: Bovine spongiform encephalopathy
BSL: Biosafety level
BT: Blood transfusion
BT: Borderline tuberculoid
BUN: Blood urea nitogen
C/Cs: Cultural characteristics
C/F(s): Clinical feature(s)
C: Complement
CA MRSA: Community-acquired MRSA
CABSI: Catheter-associated bloodstream infection
C-activation: Complement activation
CAM: Chorioallantoic membrane
CAMP test: Christie-Atkins-Munch-Peterson test
CAUTI: Catheter-associated urinary tract infection
CBC: Complete blood count
CBNAAT: Cartridge-based nucleic acid amplification test
CCHF: Crimean Congo hemorrhagic fever
CD: Cluster of differentiation
CDC: Centres for Disease Control and Prevention
CFP-10: Culture filtrate protein -10
CFT: Complement fixation test
CFW stain: Calco-fluor-white stain
CGD: Chronic Granulomatous disease
Ch.: Chapter
CHD: Congenital heart disease
CIE: Counter immunoelectrophoresis
CIEBOV: Côte d' Ivoire ebolavirus
CJD: Creutzfeldt-Jacob disease
CL crystals: Chacot-Leyden crystals
CLED medium: Cysteine lactose electrolyte deficient medium
CMI: Cell-mediated immunity

CMN group: Corynebacteria-mycobacteria-nocardiae group

CMV: Cytomegalovirus

CNS: Central nervous system

CoNS: Coagulase-negative *Staphylococcus*

CoPS: Coagulase-positive *Staphylococcus*

COVID-19: Coronavirus disease -19

CPE(s): Cytopathic effect(s)

CPIS: Clinical pulmonary infection score

CR(s): Complement receptor(s)

CRF: Coagulase reacting factor

CRP: C-reactive protein

CSOM: Chronic suppurative otitis media

CVS: Cardiovascular system

D/D: Differential diagnosis

DCA: Deoxycholate citrate agar

DEC: Diethyl carbamazine

DGIM: Dark-ground (field) illumination microscope

DHF: Dengue hemorrhagic fever

DIC: Disseminated intravascular coagulation

DM: Diabetes mellitus

DOAP: Demonstrate, observe, access and perform

DP vaccine: Diphtheria and pertussis vaccine

DPT vaccine: Diphtheria, pertussis and tetanus vaccine

ds: double stranded

DSS: Dengue shock syndrome

DT vaccine: Diphtheria and tetanus vaccine

DTH: Delayed-type hypersensitivity

E test: Epsilomer test

e.g.: exampli gratia (for example)

EB: Elementary body

EBV: Epstein-Barr virus

EC: Exposure code

EIA: Enzyme immunoassay

ELFA: Enzyme-linked fluorescent assay

ELISA: Enzyme-linked immunosorbent assay

E-M pathway: Embden-Meyerhof pathway

EM: Electron microscope

ESAT-6: Early secretory antigen target-6

ESBL: Extended spectrum β-lactamase

ET tube: Endotracheal tube

EV: *Entero virus*

FDA: Food and Drug Administration

f-IPV: Fractional injectable polio vaccine

FITC: Fluorescein isothiocyanate

FM: Fluorescent microscope

FNAC: Fine needle aspiration cytology

FT: Formal toxoid/fluid toxoid

FTA-ABS test: Fluorescent *Treponema* antibody-ABSorption test

FTLV: Feline T-cell lymphotropic virus

G S L M: Glucose sucrose lactose mannitol

GBS: Guillain-Barre syndrome

GC: Genital chlamydiasis

G-CSF: Granulocyte-colony stimulating factor

GGMS stain: Grocott Gomori's methenamine silver stain

GI: Genital infection

GISAID: Global initiative on sharing avian influenza data

GIT: Gastrointestinal tract

GLC: Gas-liquid chromatography

GM-CSF: Granulocyte macrophage-colony stimulating factor

GNB: Gram-negative bacilli

GNC: Gram-negative cocci

GPB: Gram-positive bacilli

GPC: Gram-positive cocci

GU: Gonococcal urethritis

H & E stain: Hematoxylin and eosin stain

H/o: History of

HA MRSA: Hospital-acquired MRSA

HAI test: Hemagglutination inhibition test

HAI(s): Healthcare-associated infection(s)

HAV: Hepatitis A virus

Hb: Hemoglobin

HBLV: Human B-cell lymphotropic virus

HBV: Hepatitis B virus

HBVIg: Hepatitis B virus immunoglobulin

HCC: Hepatocellular carcinoma

HCF(s): Healthcare facility/facilities

hCG: Human chorionic gonadotropin

HCoV-EMC: Human coronavirus-Erasmus Medical Centre

HCV: Hepatitis C virus

HCW(s): Healthcare worker(s)

HDN: Hemolytic disease of newborn

HDP: Hanging drop preparation

HDV: Hepatitis D virus

HEV: Hepatitis E virus

HFMD: Hand-foot-mouth disease

HFV: Hepatitis F virus

HGE: Human granulocytic ehrlichiosis

HGV: Hepatitis G virus

HHV (1–8): Human herpes virus (1–8)

HI: Humoral immunity

Hib: *Haemophilus influenza* type b

HIV: Human immunodeficiency virus

HLA: Human leukocyte antigen

HME: Human monocytic ehrlichiosis

HMW: High molecular weight

HPLC: High performance liquid chromatography

HPV: Human papilloma virus

hr: Hour

HRCT: High-resolution CT

HSV-1/2: Herpes simplex virus-1/2

HTLV: Human T-cell lymphotropic virus

HUS: Hemolytic uremic syndrome

IARC: International Agency for Research on Cancer

IC: Inclusion conjunctivitis
IC: Intracutaneous
ICC(s): Immunocompetent cell(s)
ICMR: Indian Council of Medical Research
ICRC: Indian Cancer Research Center
ICSOL(s): Intracranial space-occupying lesion(s)
ICT: Immunochromatographic test
ICTC: Integrated Counseling and Testing Centre
ID: Intradermal
IDD(s): Immunodeficiency disease(s)
IET: Immuno enzyme test
IF test/assay: Immunofluorescent test/assay
IFN: Interferon
IFT: Immunoferritin test
Ig (s): Immunoglobulin (s)
IGRA: Interferon gamma release assay
IHA test: Indirect hemagglutination test
IM: Intramuscular
IMG: Indian medical graduate
IMN: Infectious mononucleosis
IN test: Indole nitrate/Indoso nitrate) test
IP: Incubation period
IPV: Injectable polio vaccine
Ir: Immune response
ITH: Immediate type hypersensitivity
IV: Intravenous
JC virus: John Cunningham virus
JE: Japanese Encephalitis
KB method: Kopeloff's and Beerman's method
KFD: Kyasnur forest disease
KLB: Klebs-Loeffler bacillus
KSHV: Kaposi's sarcoma-associated herpes virus
LAK cells: Lymphokine activated killer cells
LAMP: Loop-mediated isothermal amplification
LATS: Long-acting thyroid stimulator
LBW: Low-birth weight
LCB stain: Lactophenol cotton blue stain
LCM virus: Lymphocytic choriomeningitis virus
LD: Lethal dose
LF(s): Lactose fermenter(s)
LGL: Large granular lymphocyte
LGV: Lymphogranuloma venereum
LJ medium: Lowenstein Jensen medium
LL: Lepromatous leprosy
LLF(s): Late lactose fermenter(s)
LMW: Low-molecular weight
LPA: Line probe assay
LPS(s): Lipopolysaccharide(s)
LTRs: Long-terminal repeat sequences
MAC: *Mycobacterium avium*-complex
MAC-ELISA: IgM antibody capture enzyme-linked immunosorbent assay
MALDI-TOF: Matrix-assisted laser desorption ionization time of flight

MAT: Microscopic agglutination test
MCH: Mean corpuscular hemoglobin
MCHC: Mean corpuscular hemoglobin concentration
MCV: Mean corpuscular volume
MCV: Molluscum contagiosum virus
MDR: Multidrug resistance
MDR-TB: Multidrug-resistance tuberculosis
MERS-CoV: Middle-East respiratory syndrome-related coronavirus
MFS stain: Masson Fontana silver stain
MHA: Muller Hinton agar
MHC: Major histocompatibility complex
MIC: Minimum inhibitory concentration
MID: Minimum infecting dose
MIO test: Motile indole ornithine test
MIS-C: Multiorgan inflammatory syndrome in children
MLD: Minimum lethal dose
MMR vaccine: Measles-mumps-rubella vaccine
MMRV vaccine: Measles-mumps-rubella varicella vaccine
MOHFW: Ministry of Health and Family Welfare
MOI: Multiplicity of infection
MOMP: Major outer membrane protein
MR vaccine: Measles-rubella vaccine
MRSA: Methicillin-resistance *Staph. aureus*
MSP: Mycobacteria as skin pathogen
MSSA: Methicillin-susceptible *Staph. aureus*
MTC: *Mycobacterium tuberculosis* complex
MW: Molecular weight
NA: Nucleic acid
NACO: National Aids Control Organisation (New Delhi)
NANB hepatitis: Non-A non-B hepatitis
NASBA: Nucleic acid sequence base amplification
NB: Nutrient broth
NCTC: National collection of type culture
NGO(s): Non-government organization(s)
NGU: Non-gonococcal urethritis
NIH: National Institute of Health
NIV: National Institute of Virology (Pune)
NK cells: Natural killer cells
NLF(s): Non-lactose fermenter(s)
NS: Normal saline
NSF(s): Non-sucrose fermenter(s)
NTM: Nontuberculous mycobacteria
OF test: Oxidation–fermentation test
OMP: Outer membrane protein
ONPG test: O-Nitro phenyl-β-D-galactosidase test
OPV: Oral polio vaccine
PAGE: Polyacramide gel electrophoresis
PAIR: Puncture, aspiration, injection and reaspiration
PAMPs: Pathogen-associated molecular patterns
PAMs: Pulmonary alveolar macrophages
PANGOLIN: Phylogenetic Assignment of Named Global Outbreak LINeage

PAS stain: Periodic-acid Schiff stain
PBS: Peripheral blood smear
PCA: Passive cutaneous anaphylaxis
PCM: Phase-contrast microscope
PCP: *Pneumocystis pneumonia*
PCR: Polymerase chain reaction
PCV: Packed cell volume
PDAB: Para-dimethyl aminobezadehyde
PEM: Protein-energy malnutrition
PEP: Post-exposure prophylaxis
PFGE: Pulse field gel electrophoresis
PHC: Primary Health Center
PHOL stain: Pal, Hasegawa, Ono and Lee stain
PIA: *Pseudomonas*-isolation agar
PIMS: Pediatric inflammatory multisystem syndrome
PIMs: Potentially infectious materials
PJP: *Pneumocystis jirovecii* pneumonia
PK reaction: Praunitz-Kustner reaction
PML: Progressive multifocal leuko-encephalopathy
PMNs: Polymorphonuclear cells
Pn G: Penicillin G
PPA test: Phenyl pyruvic acid test
PPD: Purified protein derivative
PPE (s): Personal protective equipment (s)
PPI: Pulse polio immunization
PPLO: Pleuropneumonia like organism
ppm: parts per million
PPNG: Penicillinase-producing *Neisseria gonorrhoeae*
PPTCT: Prevention of parent-to-child transmission
PRAS: Prereduced anaerobic system
PRNT: Plaque reduction neutralization test
PRP: Polyribosyl ribitol phosphate
PRRs: Pattern recognition receptors
PT medium: Potassium tellurite medium
PVC: Polyvinylchloride
PW: Peptone water
PYG agar: Peptone yeast extract glucose agar
PYR: L- pyrrolidonyl β-naphthylamide
RA: Rheumatoid arthritis
RB: Reticulate body
RBF: Rat bite fever
RCMM: Robertson's cooked meat medium
RE system: Reticuloendothelial system
REA: Restriction endonuclease analysis
RF: Rheumatoid factor
RFLP: Restriction fragment length polymorphism
RIA: Radioimmunoassay
RNP: Ribonucleoprotein
RNTCP: Revised National Tuberculosis Control Program
RPR test: Rapid plasma reagin test
RSV: Respiratory syncytial virus
RT- PCR: Reverse transcriptase PCR
RTF: Resistance transfer factor

S/C: Subculture
SABE: Subacute bacterial endocarditis
SALT: Skin-associated lymphoid tissue
SARS-CoV (1 and 2): Severe acute respiratory syndrome related coronavirus (1 and 2)
SAT: Standard agglutination test
SC: Status code
SC: Subcutaneous
SCC: Squamous cell carcinoma
SDA: Sabouraud dextrose agar
SDA: Strand displacement amplification
SDL: Self directed learning
SDS-PAGE: Sodium dodycile sulfate-polyacramide gel electrophoresis
SGD: Small group discussion
SIM test: Sulfide indole motility test
SIV: Simian immunodeficiency virus
SLE: Systemic lupus erythematosus
SLT: Shiga-like toxin
Sp./spp.: Species
SPA: Surface protein antigen
SPE: Streptococcal pyrogenic exotoxin
SPS: Sodium polyanethol sulfonate
ss: Single stranded
SSI: Surgical site infection
SSPE: Subacute sclerosing panencephalitis
SSS: Specific soluble substance
SSSS: Staphylococcal-scalded skin syndrome
STSS: Streptococcal toxic shock syndrome
SUDV or SEBOV: Sudan ebola virus
SWG: Standard wire gauge
TAFV: Tai forest ebola virus
TATA: Tumor-associated transplant antigen
Tc cells: T cytotoxic cells (cytotoxic T cells)
TCBS agar: Thiosulfate citrate bile salt sucrose agar
TCGF: T cell growth factor
TCID: Tissue culture infective dose
TCR(s): T-cell receptor(s)
Td : Diphtheria toxoid
TF: Transfer factor
Th cells: T helper cells (CD4+)
TIG: Tetanus immunoglobulin
TIP: Translation inhibiting protein
TLRs: Toll-like receptors
TMA: Transcription-mediated amplification
TNF: Tumor necrosis factor
TPHA test: *Treponema pallidum* Hemagglutination test
TPI test: *Treponema* pallidum immobilization test
TPPA test: *Treponema pallidum* particle agglutination test
TRIC: Trachoma inclusion conjunctivitis
TSI medium: Triple sugar iron medium
TSS: Toxic shock syndrome
TSST: Toxic shock syndrome toxin
TSTA: Tumor-specific transplant antigen

TT: Tetanus toxoid
TT: Tuberculoid type
UGC: Urogenital chlamydiasis
URTI: Upper respiratory tract infection
US FDA: United State Food and Drug Administration
UV light (rays): Ultraviolet light (rays)
VAP: Ventilator-associated pneumonia
VDPV: Vaccine derived polio virus
VDRL test: Venereal Disease Research Laboratory test
–ve: Negative
VHF: Viral hemorrhagic fever
VISA: Vancomycin intermediate *Staph. aureus*

VPF: Vascular permeability factor
VRE: Vancomycin-resistant *Enterococcus*
VRSA: Vancomycin resistant *Staph. aureus*
VTM: Viral transport medium/media
VZV: Varicella zoster virus
WB test: Western blot test
WPV: Wild polio virus
XDR: Extended drug resistance
XDR-TB: Extended drug resistance tuberculosis
XLD medium: Xylose lysine deoxycholate medium
ZEBOV: Zaire ebola virus
ZN stain: Ziehl-Neelsen stain

Part I

General Microbiology, Immunology and Healthcare-associated Infections (HAIs)

SECTION 1

General Microbiology

Introduction and History of Microbiology

Chapter Outline
- Introduction
- Scientists and their Role (History)

INTRODUCTION

Microbiology

It includes the study of microorganisms like bacteria, viruses, fungi and parasites.

Branches of Microbiology

Medical microbiology: It is the branch of medical science which is related with study of microorganisms and disease produced by them (called infectious diseases) in humans. It includes diagnosis, prevention and treatment of disease and host response against microorganisms and or their products. Various branches of medical microbiology include:

- General microbiology: It includes study on general properties of microorganisms like taxonomy, morphology, physiology, etc.
- Immunology: It deals with the physiological functioning of the immune system in states of both health and diseases.
- Healthcare-associated infection (HAI): It deals with study on diagnosis, treatment and prevention of healthcare-associated infections.
- Systemic microbiology: It includes study on infections to various human body systems like
  - CVS and bloodstream infections
  - Gastrointestinal and hepatobiliary system infections
  - Skin, subcutaneous and musculoskeletal system infections
  - Central nervous system infections
  - Respiratory system infections
  - Urinary system, genital system and sexually transmitted infections
  - Miscellaneous infections like infections to eyes, ears, etc.

Food microbiology: The branch of microbiology that is involved in the study of the microorganisms that inhabit, create, enhance the flavour or contaminate the food.

Industrial microbiology: The branch of microbiology which is applied to create industrial products in mass quantities.

Soil microbiology: The study of microorganisms in soil, their functions and how they affect properties of soil or environment.

Plant microbiology: It is the study of association of plants with microorganisms.

Abiogenesis (Theory of Spontaneous Generation)

In an earliest time, people had believed that living organisms could develop from nonliving things like soil, elements, etc., **called spontaneous generation or abiogenesis.** In later part this theory was challenged by many scientists.

SCIENTISTS AND THEIR ROLE (HISTORY)

Antony Philips van Leeuwenhoek

Birth to death (1632–1723, Fig. 1.1a): He was born in **Delft, Holland**.

Fig. 1.1: (a) Antony van Leeuwenhoek and (b) Edward Jenner

1. The Dutchman was draper, who 1st prepared the single lens microscope (by own) and observed the diverse materials through it. He observed the minute organisms and other materials in rain water through an instrument (single lens microscope) with 40–300 magnification power and designated them as **animalcules.** He communicated his observation to the Royal Society of London in 1676. However, he did not realize the importance of these **animalcules.** As he worked on inventing different types of microscopes and improving the existing ones. Sure, his work helped to microbiology to achieve new heights because without a microscope, microbiology cannot exist, so he can be **called as father of microscopy.**
2. In 1678, **Robert Koch** developed a compound microscope and confirmed Leeuwenhoek's observation.
3. Almost after a century and after research of many people it was accepted that **animalcules** were the causes of many contagious disease.
4. He observed the *Giardia lamblia* in his own stool in 1681.
5. He defined the shape of bacteria as cocci, bacilli and spirochetes.

Edward Jenner

Birth to death (1749–1823, Fig. 1.1b): He was born in **England.**

Contribution in Microbiology

1. **Smallpox vaccine:** He discovered the prophylactic preparation for smallpox from cow lesion or cowpox. Such prophylactic preparation was labeled as **Vaccine (in Latin cow means Vacca)** by **Pasteur.**
2. **Father of immunology:** He is awarded as **father of immunology** for his contribution in the field of immunology.

Louis Pasteur

Birth to death (27th December 1822–28th September 1895, Fig. 1.2a): He was born in the village Dole, Jura, France on 27th December 1822. His father was a tanner.

Profession: He was trained as chemist, but his studies on fermentation led him to take interest in microorganisms.

(a) Pasteur   (b) Lister

Fig. 1.2: (a) Louis Pasteur and (b) Joseph Lister

1. **Germ theory of disease (biogenesis) (1857):** He established that putrefaction and fermentation were the result of microorganisms and their activities (Biogenesis).
2. **Disapproved abiogenesis (1860–61):** With series of classical experiments, he proved that all forms of life, even microbes, arouse only from their like and not *de novo* and disapproved abiogenesis.
3. **Contribution in sterilization technique:** He developed the steam sterilizer, hot air oven and autoclave. He invented the pasteurization (1863–65) which is useful for milk sterilization.
4. **Contribution in discovery of Pasteur pipette:** The Pasteur pipette name is given from Louis Pasteur, who used a variant of it extensively during his research. It is used to transfer small quantities of liquids in the laboratory and also to dispense small amounts of liquid medicines like eye drops.
5. **Contribution in cultivation technique:** He showed that growth medium, temperature, acidity, alkalinity and O_2 are required for successful cultivation.
6. **Studies on different diseases:** He identified the Microspora (*Nosema bombycis*) as a causative agent of **pebrine (silkworm disease)** in 1863 in France. He also studied on **anthrax, chicken cholera and hydrophobia (rabies). Pneumococci** were 1st noticed by Pasteur and Stenberg.

Mnemonic

- **Vaccines discovered by Louis Pasture:** CAR → **C**holera, **A**nthrax and **R**abies
- **Causative agents discovered by Robert Koch:** CAT → **C**holera, **A**nthrax and **T**uberculosis

7. **Coined the term vaccine:** Edward Jenner developed the prophylactic preparation for smallpox from cow **(in Latin cow means Vacca)** lesion. Pasteur termed the **word vaccine** for such prophylactic preparation.
8. **Discovery of theory of attenuation and chicken cholera vaccine:** When chicken cholera culture left on the bench for several weeks it lost its pathogenicity, but retains its ability to protect the bird against subsequent infection by it, led the discovery of theory of attenuation and live chicken cholera vaccine.
9. **Discovered live attenuated anthrax vaccine:** He attenuated the *Anthrax bacillus* by incubating at 42–43°C and proved that inoculation of such culture in animals induced specific protection against anthrax. The success of such immunization was dramatically demonstrated by an experiment on a farm at Pouilly-le-Fort in 1881 during which vaccinated sheep, cows and goats were challenged with a virulent *Anthrax bacillus* culture. All the vaccinated animals were survived while simultaneously unvaccinated animals died.
10. **Development of rabies vaccine:** He developed rabies vaccine (hydrophobia in human) in 1885.

Honor

1. **Father of microbiology:** His study on microorganisms leads the development of microbiology and he is awarded as **father of microbiology (medical microbiology, modern microbiology).**
2. **Pasteur Institute, Paris:** It is built in honour of Louis Pasteur in Paris by public contribution. Similar institutes were established in other regions for vaccine preparation and diagnosis of infectious diseases.

> **Note:** Who is the actual father of microbiology, A. Leeuwenhoek or Louis Pasteur? (link: https://www.quora.com/Who-is-the-actual-father-of-microbiology-A-Leeuwenhoek-or-Louis-Pasteur)
>
> Antony van Leeuwenhoek can be considered more specifically as the "Father of Microscopy" because of his contribution as mentioned earlier. But had said that the contribution of Louis Pasteur was far more than A. Leeuwenhoek. He single handedly explored all the fields of Microbiology; right from proving that life arises from a preexisting life to industrial microbiology. His studies and sincere work is what makes microbiology one of the most interesting subjects to study. So, Louis Pasteur can rightly be **called the "Father of Microbiology".**

Joseph Lister

Birth to death (1827–1912, Fig. 1.2b): He was born in **Scotland.**

Contribution in Microbiology

1. Pasteur's work was immediately followed by Joseph Lister in 1867 with introduction of antiseptic techniques in surgery resulting decreased morbidity and mortality due to surgical sepsis.
2. He 1st used the carbolic acid as an antiseptic agent in surgery (1865).
3. He was awarded as **father of antiseptic surgery.**
4. **Lister Institute, London:** Was built in honour of Joseph Lister in London in 1891. It works for vaccine preparation and diagnosis of infectious diseases.

Robert Koch

Birth to death (11th December 1843–27th May 1910, Fig. 1.3a): He was born in Clausthal village, Hanover, Germany, on 11th December 1843.

Profession: He was German physician and pioneering Microbiologist.

(a) Koch (b) Ehrlich

Fig. 1.3: (a) Robert Koch and (b) Paul Ehrlich

Contribution in Microbiology

1. **Contribution in microscopy:** In 1678, Robert Koch developed a compound microscope and confirmed Leeuwenhoek's observation.
2. **Contribution in staining technique:** He described the staining method for identification of bacteria in dried fixed films stained with aniline dyes.
3. **Contribution in cultivation technique:** Robert Koch invented the culture method for pure isolation of bacteria over solid media. Earliest solid medium was **cooked cut potato** used by him, later he used gelatin for solidification of media, but gelatin is not satisfactory as it has tendency to get liquefy at 24°C and also by proteolytic bacteria. **Use of agar** in solidification of media was suggested by Frau Hesse, the wife of one investigator in Koch's laboratory, who had seen her mother using agar to make jellies.
4. **Hanging drop technique:** He was the 1st to use hanging drop method to detect bacterial motility.
5. **Koch's phenomenon (1890):** Koch observed that a guinea pig already infected with tubercle bacilli gives exaggerated response when injected with tubercle bacilli or tuberculin protein. This hypersensitivity or allergic reaction **called Koch's phenomenon.**

> **Note:** Molecular Koch's postulates
>
> **Postulates:** Gene presents in microbes should encode for disease production. Gene that satisfies molecular Koch's postulates is often referred as virulence factor. The postulates were formulated by the microbiologist Stanley Falkow in 1988 and are based on Koch's postulates as follow:
> - The gene should be associated more with pathogenic species/strains than with nonpathogenic species/strains.
> - Specific inactivation of the gene associated with the suspected virulence should lead a loss in pathogenicity or virulence.
> - Replacement of the mutated gene should restore pathogenicity or virulence.
> - The gene should be expressed at some point during infection or disease process.
> - Antibodies or immune cells direct against the gene products should protect the host.
>
> **Limitation:** For many pathogenic microorganisms, it is not currently possible to apply molecular Koch's postulates, because of lack of suitable animal model for many important human diseases. Additionally, many pathogens cannot be manipulated genetically.
>
> **Difference with Koch's original postulates:** Instead of the presence or absence of a particular microorganism, they consider whether a particular virulence gene is present and active.

6. **Koch's postulates (1876)**
- **Principles:** Any organism will be accepted as causative agent of disease if it satisfied following four generalized principles called **Koch's postulates.**
  - The organism must be **present** in disease.
  - The organism must be **isolated** from the disease in pure culture.
  - Samples of the organism taken from pure culture must cause the same disease when inoculated into a healthy and susceptible **animal** in the laboratory.

- The organism must be **re-isolated** from the inoculated animal.
  - **Additional criterion in Koch's postulates:** Later additional or fifth criterion was added, which is the **detection of antibody (Ab)** from patient's serum.
- **Limitations (clinical applications) of Koch's postulates:** Following bacteria are not satisfying the Koch's postulates.
  - *M. leprae*: Not growing over artificial media.
  - *T. pallidum*: Pathogenic strains are not growing over artificial media, but non-pathogenic can grow.
  - *N. gonorrhoeae*: No animal model.
- **Molecular Koch's postulates:** Follow box.

Honour

1. He was the founder of role of bacteria in production of diseases.
2. He identified the causative agents of cholera (*Vibrio cholerae* in 1883), anthrax (*Bacillus anthracis* in 1876) and tuberculosis (*M tuberculosis* in 1882).
3. Winner of **Nobel Prize** in 1905 and known as **father of Bacteriology.**

Paul Ehrlich

Birth to death (1854–1915, Fig. 1.3b): He was born in Germany.

Profession: He was German scientist.

Contribution in Microbiology

1. He **stained** the cells and tissues to reveal their functions.
2. He reported the **acid fastness of tubercle bacilli.**
3. He introduced the method for **standardization of toxin and antitoxin and** coined the term **minimum lethal dose (MLD).**
4. **Father of chemotherapy:** He used **salvarsan** (an arsenical compound) sometimes called '**magic bullet** 'to kill spirochetes of syphilis with moderate toxic effect. He continued with his experiments till 1912 and discovered **neosalvarsan** and new branch in medicine **called chemotherapy.** Because of his extraordinary activities in medicine he is **called father of chemotherapy.**

Other Important Scientists and their Contribution in the Field of Microbiology

- Hans Christian Gram: A histologist developed the technique to identify the bacteria in tissue in 1884.
- Ziehl and Neelsen: Initially Ehrlich developed the acid fast stain in 1882 and later modified by Ziehl and Neelsen in 1882.
- Ernst Ruska and colleagues: Invented the electron microscope in 1931 for which Ernst Ruska won the Nobel Prize in Physics in 1986.
- Alexander Fleming: He discovered the penicillin from the fungus *Penicillium notatum* in 1928 and got Nobel Prize in 1945.
- von Behring Kitasato: He described antibody.
- Good Pasture: He developed the viral culture method in chick embryo in 1931.
- Kleinberger: He described the cell wall deficient form of bacteria in 1935, while studying the culture of *Streptobacillus moniliformis* in the Lister Institute, London called L-forms after the Lister Institute. He won Nobel Prize in 1941.
- Karry B Mullis: Invented the PCR technique in 1993.

Nobel Prizes

A number of scientists have been awarded with Nobel Prizes for their significant contribution and research work in the field of microbiology and are shown in Table 1.1.

Note: Year in **Table 1.1** indicates the time of Nobel Prize given.

TABLE 1.1: Nobel Laureates

Nobel Laureate	Year	Contribution
Emil A Behring	1901	Developed antitoxin of Diphtheria
Sir Ronald Ross	1902	Studied life cycle of *Plasmodium* in Mosquito
Robert Koch	1905	Invented the *M. tuberculosis*
Charles LA Laveran	1907	Studied the *Plasmodium* in unstained slide of blood
Paul Ehrlich and Elie Metchnikoff	1908	Discovered the selective theory of antibody formation
Charles Richet	1913	Discovered the anaphylaxis
Jules Bordet	1919	Role in complement and CFT
Kleinberger	1941	Defined the L-Forms
Alexander Fleming	1945	Discovered the penicillin from the *P. notatum*
F Enders, FC Robbins, TH Weller	1954	Developed the tissue culture of polio virus
JL Lederberg and EL Tatum	1958	Discovered conjugation theory in bacteria
Sir M Burnet and Sir PB Medawar	1960	Immunological tolerance
Watson and Crick	1960	Discovered the double helix DNA structure
Peyton Rous	1966	Searched viral oncogenesis
Holley, Khurana and Nirenberg	1968	Invented the genetic model
BS Blumberg	1976	Discovered the HBsAg
Barbara Mc Clintoch	1983	Discovered the transposon
George Kohler	1984	Discovered the hybridoma technique for mononclonal antibodies production
Stanley B Prusiner	1997	Discovered Prions
J Robin Warren and Barry J Marshal	2005	Discovered the *H. pylori* and its role in peptic ulcer
Luc Montagnier and F Barre Sinoussi	2008	Discovery of HIV

(Contd...)

TABLE 1.1: Nobel Laureates (*Contd...*)		
Nobel Laureate	**Year**	**Contribution**
Bruce A Beutler and Jules A Hoffmann	2011	Discovery of the theory of innate immunity
Ralph M Steinman	2011	Searched dendritic cell and its role in adaptive immunity
Sir John B Gurdon and Shin-ya Yamanaka	2012	Mature cell can be reprogrammed to become pluripotent

Common Name of the Microorganisms from the Name of Scientists

Follow Table 1.2.

TABLE 1.2: Common name of the microorganisms from the name of scientists		
Scientists	**Common name**	**Scientific name**
Victor Morax and Theodor Axenfeld	Morax-Axenfeld bacillus	*Moraxella lacunata*
Klebs and Loeffler	Klebs-Loeffler Bacillus (KLB)	*Corynebacterium diphtheriae*
Preisz and Nocard	Preisz-Nocard Bacillus	*Corynebacterium pseudotuberculosis*
Arthur Nicolair	Nicolaier's bacillus	*Clostridium tetani*
Robert Koch	Koch's bacillus	*Mycobacterium tuberculosis*
Heinrich A Johne	Johne's bacillus	*Mycobacterium paratuberculosis*
Gerard HA Hansen	Hansen's bacillus	*Mycobacterium leprae*
George Hoyt Whipple	Whipple's bacillus	*Trophyrema whipplei*
Carl Friedlander	Friedlander's bacillus	*Klebsiella pneumoniae*
Abel Rudolf	Abel's bacillus	*K. pneumoniae* subsp. *ozaenae*
Anton Von Frisch	Frisch's bacillus	*K. pneumoniae* subsp. *rhinoscleromatis*
Gaffky and Eberth	Gaffky-Eberth bacillus	*Salmonella* Typhi
Alexander Yersin	Yersin bacillus (plague bacillus)	*Yersinia pestis*
Whitmore	Whitmore's bacillus	*Burkholderia pseudomallei*
Pfeiffer	Pfeiffer's bacillus	*Haemophilus influenzae*
Koch and Weeks	Koch Weeks bacillus	*Haemophilus aegypticus*
Jules Bordet and Octave Gengou	Bordet–Gengou bacillus	*Bordetella pertussis*
Albert Doderlein	Doderlein's bacillus	*Lactobacillus acidophilus*
Eaton	Eaton agent	*Mycoplasma pneumoniae*

Scientific Name of the Microorganisms from the Name of Scientists

Follow Table 1.3.

TABLE 1.3: Scientific name of the microorganisms from the name of scientists	
Scientists	**Scientific name**
Bacteria	
Shiga and Flexner	*Shigella flexneri*
Shiga and Boyd	*Shigella boydii*
Shiga and Sonne	*Shigella sonnei*
Amedee Borrel and Bergdorfer	*Borrelia burgdorferi*
Cox and Burnet	*Coxiella burnetii*
Parasites	
Leishman and Donovan	*Leishmania donovani*
Wucherer and Bancroft	*Wuchereria bancrofti*

Special Honor to Scientists

- Father of Microscopy: Antony van Leeuwenhoek.
- Father of Microbiology (Medical Microbiology, Modern Microbiology): Louis Pasteur.
- Father of Immunology: Edward Jenner.
- Father of Antiseptic Surgery: Joseph Lister.
- Father of Bacteriology: Robert Koch.
- Father of Virology: WM Stanely.
- Father of Tumor Virology: Peyton Rous.
- Father of Mycology: Raymond Jacques Sabouraud.
- Father of Chemotherapy: Paul Ehrlich.

Causative Agents of Infectious Diseases, Methods of Detection and their Role in Health and Diseases

These include bacteria, viruses, fungi and parasites. Their detection methods and their role in health and diseases are described in respective chapter.

ACCESS YOURSELF

Short Notes

1. Robert Koch.
2. Louis Pasteur.

Short Questions for Theory/Viva Questions

1. What are Koch's postulates?
2. What are molecular Koch's postulates?
3. Name the bacteria which are not satisfying the criteria of Koch's postulates.
4. Write the common name for following bacteria.
   - *C. diphtheriae*
   - *M. tuberculosis*
   - *S.* Typhi
   - *L. acidophilus*
5. Name the following.
   - Father of microbiology
   - Father of immunology
   - Father of antiseptic surgery
   - Father of bacteriology

1. *Mycobacterium leprae* is not satisfying the Koch's postulates.

MCQs for Chapter Review

Antony Philips Van Leeuwenhoek

1. **Antony Van Leeuwenhoek is associated with:**
   a. Telescope
   b. Microscope
   c. Stains
   d. Immunization

Louis Pasteur

2. **Louis Pasteur is not associated with:**
   a. Introduction of complex media
   b. Discovery of rabies vaccine
   c. Discovery of *M. tuberculosis*
   d. Discovery of spontaneous generation theory

3. **Louis Pasteur is associated with:**
   a. Discovery of the bacillus of tuberculosis
   b. The cellular concept of immunity
   c. Introduction of anthrax vaccine
   d. Discovery of penicillin

4. **Vaccine of rabies was first discovered by:**
   a. Louis Pasteur
   b. Robert Koch
   c. Edward Jener
   d. Landsteiner

5. **Pasteur developed vaccine for:**
   a. Anthrax
   b. Rabies
   c. Chicken cholera
   d. All of the above

Robert Koch

6. **Microorganism that does not obey Koch's postulates:**
   a. *M. tuberculosis*
   b. Polio virus
   c. *M. leprae*
   d. *Streptococcus*

7. ***Vibrio cholerae* was discovered by:**
   a. Koch
   b. Metchnikoff
   c. John Snow
   d. Virchow

Other Important Scientists and their Contribution in the Field of Microbiology

8. **Electronic microscope was invented by:**
   a. Ruska
   b. Robert Koch
   c. Antony van Leeuvenhock
   d. Louis Pasteur

9. **Egg inoculation technique for cultivation of viruses was first reported by:**
   a. Louis Pasteur
   b. Ellermann and Bang
   c. Good Pasteur
   d. Lord Lister

Common Name of the Microorganisms from the Name of Scientists

10. ***C. diphtheriae* is also called as:**
    a. Klebs-Loeffler Bacilli (KLB)
    b. Roux bacilli
    c. Koch's bacilli
    d. Yersin bacilli

11. **Which of the following is called Preisz-Nocard bacillus?**
    a. *C. diphtheriae*
    b. *C. pseudotuberculosis*
    c. *M. tuberculosis*
    d. *Mycoplasma*

12. **Eaton agent is:**
    a. *Chlamydia*
    b. *Mycoplasma pneumoniae*
    c. *Klebsiella*
    d. *H. influenzae*

Answers and Explanation of MCQs

1. b
- Follow section, **Antony Philips van Leeuwenhoek** for explanation.

2. a and c
- *M tuberculosis* was discovered by Robert Koch.
- Follow section, **Louis Pasteur** for explanation of other options.

3. c
4. a } Follow section, **Louis Pasteur** for explanation.
5. d

6. c
- Follow section, **Robert Koch (limitation of Koch's postulates)** for explanation.

7. a
- **Robert Koch** discovered the causative agent of <u>C</u>holera, <u>A</u>nthrax and <u>T</u>uberculosis (**Mnemonic** CAT).

8. a
- Ernst Ruska and colleagues: Invented the electron microscope in 1931 for which Ernst Ruska won the Nobel Prize in Physics in 1986.

9. c
- Good Pasture developed the viral culture method in chick embryo in 1931.

10. a
- *C. diphtheriae* was 1st identified by **Klebs** in 1883 but 1st cultivated by **Loeffler** in 1884 hence commonly **called Klebs-Loeffler Bacillus (KLB).**
- Emile Roux contributed in the discovery of diphtheria toxin along with Alexander Yersin.
- Koch's bacilli is the common name given to *Mycobaterium tuberculosis* from the name of Robert Koch.
- Alexander Yersin who discovered the bacilli along with Kitasato in 1894 from Hong Kong at the beginning of last epidemic, from his contribution genus **called Yersinia**; however Yersin bacilli word is not used for all species of genus, but only for *Yersinia pestis* which also **called plague bacillus.**

11. b } Follow **Table 1.2** for explanation.
12. b

CHAPTER 2

Taxonomy of Microorganisms

Chapter Outline
❑ Definition and Components

DEFINITION AND COMPONENTS

Definition

Taxonomy is defined as description, identification, nomenclature and ordered classification of organisms according to their presumed natural relationships.

Components

Taxonomy includes following three components:
A. Classification/orderly arrangements.
B. Identification of unknown with known unit.
C. Nomenclature/naming of unit (bacteria).

A. Classification/orderly arrangements

Five kingdom system of classification: In 1969, RH Whittaker placed all organisms in five groups **called five kingdom system of classification.**
1. **Monera:** Prokaryote and unicellular, e.g., bacteria, blue green algae and archaebacteria.
2. **Protista:** Eukaryote and unicellular, e.g., protozoa
3. **Fungi:** Eukaryote and uni- or multicellular, e.g., fungi
4. **Plantae:** Eukaryote and multicellular, e.g., plants
5. **Animalia:** Eukaryote and multicellular, e.g., metazoa (helminths/worms), birds, animals, human, reptiles, arthropod, molluscs and coelenterates.

Modification of five kingdoms system of classification: It is modified by Margulis and Schwartz, which includes two kingdoms like **Prokaryotes and Eukaryotes** as shown in **Table 2.1** with differences.

General scheme of classification: Kingdom → Phylum (Division) → Class → Order → Family (Tribe) → Genus → Species (Specific Epithet) → Subspecies (Strain/type).

Phylogenetic classification: This hierarchical classification represents a branching tree like arrangement based on evolutionary arrangement of species.

Adsonian classification: It based on all features expressed at the time of study.

Molecular or genetic classification: Based on genetic relatedness.

Abbreviation of species: Species word is common for both singular and plural form, but abbreviation may be used as sp., and spp., for singular and plural forms, respectively.

Intraspecies classification: It is useful for epidemiological purposes. It classifies species/unit up to subspecies/strain/type by using following methods.
- **Biotypes:** Based on biochemical properties.
- **Serotypes:** Based on serological properties.
- **Bacteriophage types:** Based on susceptibility to Bacteriophage.
- **Colicin types:** Based on production of bacteriocin.

B. Identification of unknown with known unit

It is done by using morphological, biochemical, genetic and other properties.

C. Nomenclature/naming of unit/species of bacteria
- **Order:** It is labeled with suffix **"ales"**.
- **Family:** It is labeled with suffix **"aceae"**.
- **Genus name:** It is the Latin noun and starts with capital letter. Scientific name includes genus and epithet/species with Italic pattern.
- **Species/epithet name:** It starts with small letter and in Italic pattern irrespective of person or place name. It based on different properties of unit like
  - *albus* meaning white.
  - *suis* meaning pig origin.
  - *pyogenes* meaning pus.
  - *welchii* meaning person who discovered it.
  - *tetani* meaning disease produced.
  - *australis* meaning place of origin.

Essentials of Medical Microbiology

TABLE 2.1: Differences between prokaryotes and eukaryotes

Features	Prokaryotes	Eukaryotes
General features		
Meaning	Pro = primitive/immature + Karyotes = nucleus (contains immature nucleus)	Eu = true/mature + Karyotes = nucleus (contains immature nucleus)
Organism's examples	Monera (bacteria, blue green algae)	Protista, fungi, plantae and animalia
Evolutionary ancient	Yes	No
Anatomy		
Number(s) of cell	Unicellular	Uni- and multicellular
Nucleus		
• Nuclear membrane	–	+
• Nucleolus	–	+
• Chromosome	Single-circular	Multiple-linear
Cytoplasm		
• Ribosomes	+ 70S	+ 80S
• Plasmid, episomes, transposon	+	–
• Golgi complex	–	+
• Endoplasmic reticulum	–	+
• Triglyceride fats	–	+
• Mitochondria and lysosomes	–	+
Physiology		
Site of respiration	Mesosome/chondroid (It is the invagination or infolding of cytoplasmic membrane contains respiratory enzymes)	Mitochondria
Reproduction	Mostly by asexual method (binary fission)	By asexual method (budding) and/or sexual method (meiosis or mitosis)
Pinocytosis	–	+
Protoplasmic streaming	–	+
Biochemistry		
Plasma membrane		
• Sterol	– (Except in *Mycoplasma* and *Ureaplasma*)	+ (like cholesterol, ergosterol, etc.)
• Phospholipid	+	+
Cell wall		
• Muramic acid/peptidoglycan	+	–
• Diaminopimelic acid	Present in few GNB in pentapeptide bridge	Absent
• Others	Lipid and protein	Chitin, mannan, cellulose (green plants)

– (Absent), + (Present)

Notes:

Ancient: Prokaryotes are evolutionary ancient. Probably they are 1st organisms to evolve and eukaryotes evolving from prokaryotes like predecessors.

Viruses: Viruses are neither classified as prokaryotes nor eukaryotes.

Phospholipid: Present in both.

ACCESS YOURSELF

Short Notes

1. Differences between prokaryotes and eukaryotes.

Short Questions for Theory/Viva Questions

1. What is protista?

MCQs for Chapter Review

Components (Classification/Orderly Arrangements)

1. **Prokaryotes are:**
   a. Bacteria
   b. *Mycoplasma*
   c. Fungi
   d. Blue green algae
   e. Protozoa

2. **Which is an eukaryote?**
   a. *Mycoplasma*
   b. Bacteria
   c. Fungus
   d. *Chlamydia*

3. **Fungi are:**
   a. Prokaryotes
   b. Eukaryotes
   c. Plants
   d. Animals

4. **Site of respiration in prokaryote is:**
   a. Mitochondria
   b. Mesosome
   c. Endoplasmic reticulum
   d. All of above

5. Mesosomes are:
  a. Respiratory enzymes in bacteria
  b. Cytoplasmic membrane invaginations
  c. Destructive bodies
  d. Protein-forming bodies

6. Prokaryotes are characterized by:
  a. Absence of nuclear membrane
  b. Presence of microvilli on its surface
  c. Presence of smooth endoplasmic reticulum
  d. All of above

7. Which of the following is protista?
  a. Algae
  b. Fungi
  c. Protozoa
  d. Bacteria

8. Which of the following is/are bacterial taxonomy?
  a. *Chlamydia*
  b. *Rickettsia*
  c. *Mycoplasma*
  d. Prion
  e. Bacteriophage

9. True about bacteria:
  a. Mitochondria always absent
  b. Sterols always present in cell wall
  c. Divide by binary fission
  d. Can be seen only under electron microscope

10. Eukaryotes are different in causing infection because:
  a. Divide by binary fission
  b. Highly structure cell with organized cell organelles
  c. Do not have all organelles
  d. Evolutionary ancient

1. a, b, d

2. c

3. b

4. b

5. a, b

6. a

7. c

- Follow section, **classification/orderly arrangements (five kingdom system of classification and Table 2.1)** for explanation of answers of MCQs 1–7.

8. a, b and c

- *Chlamydia. Rickettsia* and *Mycoplasma* are bacteria. Prion is proteinaceous infectious virus like-particle without nucleic acid. Bacteriophage is a virus which eats bacteria.

9. a and c

- Mitochondria always absent in bacteria and respiration is possible by presence of mesosome/chondroid. Sterol is not present in all bacteria but only in *Mycoplasma.* and *Ureaplasma.* Bacteria divide by binary fission and can be seen under all types of microscopy.

10. b

- Eukaryotes are divided by meiosis or mitosis. Eukaryotes have highly structured cell with all organelles. Prokaryotes are evolutionary ancients (**Table 2.1**).

Epidemiology of Infections

Chapter Outline
- Definitions and Types of Infections
- Reservoirs, Sources and Modes of Transmission

DEFINITIONS AND TYPES OF INFECTIONS

Definitions

Infestation
- The lodgment and multiplication of infectious agents on the body surfaces or on the cloths of a host constitute infestation
- It is the lodgment and multiplication over the surfaces but no entry in body.

Contamination
- Only presence of infectious agents on surface or inside the body or inanimate object like fluid, food, clothes or soil but no multiplication
- It is the lodgment over surfaces or entry in body but no multiplication.
- Example: Bacteremia means entry of bacteria in blood and no multiplication (only detectable bacteria in blood).

Infection
- The entry and multiplication of infectious agents in the body of a host constitute infection
- It is the entry in body and multiplication but no sign-symptoms.
- Example: **Bacteremia** means entry of bacteria in blood (transient presence) and no multiplication.
- Infection is an interaction between microbes and host.

Infectious disease
- The entry, multiplication and production of clinical sign-symptoms **called infectious disease**
- It is the entry in body, multiplication and sign and symptoms.
- Examples:
  1. Septicemia: Means entry of bacteria in blood, multiplication and clinical sign-symptoms
  2. Toxemia: Means entry of bacteria in blood, multiplication and clinical sign-symptoms are

due to toxin production. It is one type of septicemia.
  3. Pyemia: Septicemia and abscess in organs like liver, spleen and other tissues.

Contagious disease: A disease that can be transmitted from one host to another by direct contact.

Communicable disease: A disease that can be transmitted from one host to another either direct or indirect mode of transmission.

Non-communicable disease: A disease that is not transmitted from one host to another.

Types of Infections

A. According to habitat/involvement of systems
Following systemic infections are described in details in respective chapters.
1. Cardio vascular system infections
2. Bloodstream infections
3. Lymphatic system infections
4. Gastrointestinal tract infections
5. Infections of abdominal viscera
6. Skin and subcutaneous tissues infections
7. Musculoskeletal infections
8. Central nervous system infections
9. Respiratory tract infections
10. Urinary tract infections (UTIs)
11. Genital tract infection
12. Eye and ear infections
13. Oral cavity, dental, periodontal and salivary gland infections.

B. According to causative agents
1. Bacterial infection
2. Viral infection
3. Fungal infection
4. Parasitic infection

5. Mixed infection: It means more than one organisms cause infection simultaneously. Following are examples:
   - Gas gangrene: **Ch. 55**
   - Fournier gangrene: **Ch. 39**
6. Co-infection or symbiotic infection: Concomitant infection by combination of microbes called co-infection. Following are examples:
   - Vincent's angina or fusospirochetosis: **Ch. 73.**
   - Adeno Associated Virus (AAV) and adeno virus: Exact infection by AAV is not known but it always required adenovirus as a helper virus to produce the infection.
   - HDV and HBV: **Ch. 88.**

C. According to nature of onset and progress

1. Acute infection: Infection characterized by sudden onset, rapid progression and often with severe symptoms.
2. Chronic infection: Infection characterized by gradual onset and slow progression.

D. According to sequence of involvement of host

1. Primary infection: Initial infection in a host by organisms.
2. Re-infection: Subsequent infection in a same host by same organisms.
3. Opportunistic infection: Infection in an immuno-deficient host by organisms.

E. According to sites/locations

1. Local or focal infection (focal sepsis): Infection that is restricted to a specific location within the body of the host.
2. Systemic or generalized infection: Infection that has been spread to several organs or areas in the body of the host.

F. According to sources of infection

1. If source is human body itself or outer source
- **Endogenous infection:** Organisms are originated from host itself. These are mostly normal flora from different parts of body. For example meningitis by *N. meningitidis* from nasopharynx.
- **Exogenous infection:** Organisms are originated from outside, mostly from environment, soil, water, human, etc.

2. If source is hospital itself or outer source
- **Nosocomial (hospital-acquired or healthcare associated) infection: Ch. 29.**
- **Community acquired infection:** Infection originated from outside the hospital environment.
- **Iatrogenic (physician induced) infection:** Infection acquired by patients from physician during diagnosis, treatment or prevention of disease.

3. If source is animal/animal products, human body or non-living objects
- **Anthroponoses/cross infections:** It means entry of infection from one human host to another human host. Word derived from anthrópos (Greek) means man and nosos (Greek) means disease. Examples include rubella, smallpox, diphtheria, gonorrhea, ringworm (*Trichophyton rubrum*), and trichomoniasis.
- **Anthropozoonoses/zoonotic infections/zoonotic diseases/zoonosis:** It means infection of human from animal or animal products. Word derived from anthrópos (Greek) means man, zoon means animal and nosos (Greek) means disease. For examples and more details follow **Ch. 47.**
- **Zooanthroponoses/reverse zoonosis:** This word is abandoned and not of much significance. It means infection of animal from human. Following are examples:
  - Tuberculosis: Both zoonotic and reverse zoonotic, with birds, cows, elephants, meerkats, mongooses, monkeys, and pigs known to have been affected.
  - Leishmaniasis: Both zoonotic and reverse zoonotic.
  - Influenza, measles, pneumonia and other pathogens are reverse zoonotic for many type of primates.
- **Sapronoses:** It means entry of infection from abiotic/nonliving substrate like soil, water, decaying plants, or animal corpses, excreta, and others. Word derived from sapros (Greek) means decaying and nosos (Greek) means disease. Following are examples:
  - Dimorphic fungi like coccidioido-mycosis and histoplasmosis
  - Monomorphic fungi like aspergillosis and cryptococcosis
  - Certain superficial mycoses like *Microsporum gypseum*, some bacterial diseases like legionellosis, and protozoan like primary amoebic meningo-encephalitis.

G. According to clinical presentation

1. **Overt infection:** Presence of clinical features of disease.
2. **Inapparent/subclinical infection:** Clinical features (C/Fs) are not apparent. It includes inapparent, missed or abortive cases. It can be detected by laboratory tests.
3. **Atypical infection:** Infection in which the typical clinical manifestation of disease are not present.
4. **Relapse:** It is the primary infection with on-off features. Following are examples:
   - Relapsing fever: It is caused by bacteria from *Borrelia* spp. Sudden onset of fever (bacteria are in blood) which lasts for 3–5 days followed by afebrile period (bacteria are in brain). After afebrile period of 3–5 days another bout of fever will occurs.
   - Relapses in malaria: In the initial infection patient feels fever during erythrocytic phase and another episode of fever during exo-erythocytic phase in case of *P. vivax*, *P. malariae* and *P. ovale*, but in *P. falciparum* there is no exo-erythocytic phase, so there is no relapse or subsequent attack of fever.

5. Latent (recrudescence): Following primary infection parasites remain in the body in silent/latent form for a very long period and again proliferating and producing the clinical disease in a later part of life when the host resistance is lowered or by specific stimulation called latent infection. For example zoster/zona/shingles by varicella zoster virus (VZV).

H. Epidemiological types

1. **Sporadic:** It means scattered about. Few cases occur irregularly, haphazardly from time to time and separated widely in space neither identifiable common sources of infection nor connected with each other. It is a starting point of epidemic when conditions are favorable for spread of infection.
2. **Endemic disease:** 'En' means in and 'demos' mean people. It defined as constant presence of a disease within a given geographic area or population group without importation from outside.
3. **Epidemic disease:** Epi means upon and demos mean people. It defined as sudden onset of disease clearly in excess of "expected occurrence."
4. **Outbreak:** Small, usually localized epidemic in the interest of minimising public alarm, unless the numbers of cases are indeed large.
5. **Pandemic:** Epidemic that spreads in many areas of the world and affecting large population.
6. **Prosodemic diseases:** Some pandemic disease (like water or air borne) spread very rapidly while some spread very slowly by person to person contact called prosodemic diseases, e.g., cerebrospinal fever.
7. **Exotic:** Disease is imported in to country in which, it do not occur otherwise.

RESERVOIRS, SOURCES AND MODES OF TRANSMISSION

Reservoirs of Infections

Definition: Persons, animals, insects, plants or inanimate objects like water, soil, etc., (or combination of all these) in which infectious agents can live, multiply and pass to susceptible host called reservoirs of infections. It includes survival, multiplication and transmission of organism.

Types of reservoirs: Reservoirs for different agents are described in respective chapters. Few are described below in two different categories.

A. According to living or non living objects which act as reservoirs

- **Living reservoirs: (1) Human reservoir:** It includes human case or carrier. For example *M. tuberculosis* (tuberculosis), *S.* Typhi (typhoid), *V. cholerae* (cholera), *B. recurrentis* (louse borne relapsing fever), *N. gonorrhoeae* (gonorrhea), *Cl tetani* (tetanus, it also has non living reservoir), *T. pallidum* (syphilis), measles, etc., **(2) Animal reservoir:** It includes zoonotic disease. For example *M. bovis* (bovine tuberculosis), rabies, yellow fever, swine flu, bird flu, etc.

- **Nonliving reservoirs:** Soil and inanimate matter can also act as reservoir. For example soil is reservoir for *Cl. tetani* (tetanus), *B. anthracis* (anthrax), *C. immitis* (coccidioido-mycosis), agents causing mycetoma, etc.

B. Other types: (1) Homologous reservoir: It means reservoir and susceptible host are from same species. For example *V. cholerae* where reservoir is man and susceptible host is man. **(2) Heterologous reservoir:** It means reservoir and susceptible host are from different species. For example, typhoid where reservoir is man, and susceptible host is bird/animal.

> **Note: Multiple reservoirs**
> *Cl. tetani* (tetanus): Human and soil reservoirs.
> *E. coli:* Human and animal reservoirs.

Sources of Infections

Definition: Persons, animals, insects, plants or inanimate objects like water or soil (or combination of all these) from which infectious agents can pass to susceptible host. It includes survival and transmission only but no multiplication.

Confusion with reservoir of infection: Sometimes it seems that reservoir and source of infection are synonyms but in fact both are different. For example;
- **Hookworm:** Reservoir of infection is man but source of infection is soil.
- **Tetanus:** Reservoir of infection and source of infection are soil.

Types of sources: Sources for different agents are described in respective chapters. Following are types (also reservoir).

A. Humans: Human is the commonest source of infections. Human may be case (patient) or carrier.

1. **Case (patient):** It is a person who harbors the pathogens with sign-symptoms of disease and serves as potential source of infection for others.
2. **Carrier:** It is a person who harbors the pathogens without any sign-symptoms of disease and serves as potential source of infection for others. Following are different types of carrier:
   - **According to clinical status: (1) Healthy carrier:** It harbors the pathogens but never suffered from the disease, e.g., polio, meningococcal meningitis, cholera, diphtheria, etc. **(2) Convalescent carrier:** It is a carrier who recovered from the disease and continue to harbors the pathogens in his body, e.g., typhoid fever, bacillary and amoebic dysentery, cholera, diphtheria, etc. **(3) Incubatory carrier:** It sheds pathogens during the incubation period of infection, e.g., measles, mumps, HBV, polio, pertussis, etc.
   - **According to duration: (1) Temporary carrier:** Carrier state lasts for <6 months, e.g., it may be healthy or convalescent or incubatory carrier. **(2) Chronic carrier:** Carrier state lasts for several

years or may for rest of life, e.g., typhoid fever, HBV, gonorrhea, etc.

- **According to portal of exit:** (1) Urinary, e.g., typhoid fever. (2) Intestinal (fecal), e.g., typhoid fever. (3) Respiratory, e.g., Influenza. (4) Nasal, e.g., Influenza. (5) Skin eruption, open wound and blood are also carriers.
- **Other types of carrier: (1) Contact carrier:** Person who acquires the pathogen from carrier. **(2) Paradoxical carrier:** Carrier who acquires pathogen from another carrier.

B. Animals and birds: These may act as case or carrier. Infectious diseases transmitted from animals to humans are called zoonoses. **(1) Cattle:** Anthrax, brucellosis and bovine tuberculosis. **(2) Goats:** Brucellosis. **(3) Sheeps:** Anthrax. **(4) Dogs:** Rabies and hydatid disease. **(5) Horses:** Glanders. **(6) Rats:** Rat bite fever, Weil's disease and plague. **(7) Pigs:** Swine flu. **(8) Birds:** Bird flu and ornithosis (psittacosis from parrot).

C. Insects: Diseases transmitted by insects are called arthropod-borne disease and that insects are called vectors. They may also act as sources of infections.

D. Soil: Spores of tetanus bacilli, fungi (*Histoplasma capsulatum*), *Nocardia asteroids*, larvae of round worm and hookworm are found on soil.

E. Water: *Vibrio cholerae*, infective hepatitis viruses (HAV) are found in water.

F. Food: Food may contain either preformed toxin of microorganisms (like *Staphylococcus aureus*, *B. cereus*, *Salmonella* species, *Cl. botulinum*, *Cl. perfrigenes*, etc.,) causing food poisoning or preexisting infectious agents (like beef tape worm in beef, pork tape worm in pork, etc.,) causing disease.

Modes of Transmission of Infections

Two categories modes of transmission of infections are as follows:

A. Direct transmission: Here mediators/vehicles for transmission of infections are not required.

1. **Droplet infections:** Droplet nuclei/particles of saliva or nasopharyngeal secretions arise during coughing, sneezing, speaking, talking or invasive procedures (bronchoscopy) are entering in to other host directly who is in close contact. Such particles are 5 µm in diameter and spread to short distance (<3 feet) can directly enter in other host. However, such larger particles can be filtered by nose. Particles 5 µm in diameter traverse to long distance and produce the air borne (indirect) infection described later in this chapter. Infection by droplet nuclei is increased in close contact, overcrowding and lack of ventilation. **Bacteria** transmitted by droplet nuclei are *Strept. pyogenes*, *N. meningitidis*, *C. diphtheriae*, *H. influenzae* type b, *B. pertussis*, *Y. pestis* and *M. pneumoniae*. **Viruses** transmitted by droplet nuclei are influenza virus, rubella virus, mumps virus, adeno virus and parvo virus B19.

2. **Infections by inoculation/injection under skin or mucosa:** Such infections are transmitted by contaminated needles or syringes like, HBV, HCV, HIV, etc., mainly in healthcare workers.

3. **Infections by contact with skin/mucosa:** Following are examples:
   - Sexually transmitted infections/diseases (STIs or STDs): **Ch. 44.**
   - Direct skin-to-skin contact transmits diseases like scabies, fungal and viral infections such as HHV-1 and 2 (HSV-1 and 2, respectively), HHV-3 (VZV) pox virus, molluscum contagiosum, etc.

4. **Infections by contact with soil**
- **Bacteria**
  - *Cl. tetani* spores are present on soil and deposition of such soil on wounded skin allows the transmission.
  - *Bacillus anthracis* spores are present in soil and in dead animal products. Animal acquired the infection by ingestion of spores present in soil, but human infection direct from soil spore is query. Human infection of anthrax occurs from animal products.
- **Parasites:** Infection occurs by walking with bare foot on contaminated soil like in *Ancyclostoma duodenale* and *Strongyloides stercoralis*.

5. **Vertical infections:** Follow **Table 3.1**.

B. Indirect transmission: Here mediators/vehicles for transmission of infections are required.

1. **Inhalation (air-borne) infections:** Particles arise during coughing, sneezing, talking or invasive procedure from patients, which are ≤5 µm in diameter and traverse to long distance can produce the air borne infection. Some droplet nuclei settle over different objects and become part of dust and cause air borne infections. **Bacterium** transmitted by air-borne route is *M. tuberculosis* (open/active pulmonary tuberculosis). **Viruses** transmitted by air borne route are measles virus, VZV, influenza virus and hemorrhagic fever viruses with pneumonia. Spores of many systemic (dimorphic) **fungi** are present in soil and they are transmitted to man by air like *Histoplasma capsulatum*, *Coccidioides immitis*, *Blastomyces dermatitidis* and *Paracoccidioides brasiliensis*.

2. **Ingestion (food- and water-borne) infections:** Like *E. coli*, *V. cholerae*, *Salmonella* spp., hepatitis A and E, poliovirus, *Rotavirus*, etc.

3. **Vector-borne infections:** For more details follow **Ch. 47.**

4. **Blood-borne infections:** By transfusion of blood or blood products following organisms are transmitted:
   - **Bacteria:** *T. pallidum*.
   - **Viruses:** HHV-4 (EBV), HHV-5 (CMV), HHV-8 (KSHV), HBV, HCV, HDV, HGV (GB/G Baker virus

Epidemiology of Infections

TABLE 3.1: Vertical infections

Time	Infections

Antenatal/before birth/congenital/ transplacental/ teratogenic
 	
Commonly called TORCH agents
	• T: Toxoplasmosis (**T**oxoplasma gondii)
	• O: **O**thers
	  – Congenital listeriosis (*Listeria monocytogenes*)
	  – Congenital syphilis (*Treponema pallidum*)
	  – Zika virus
	  – Parvo virus B19
	  – Chikungunya virus
	  – Malaria in pregnancy (*Plasmodium falciparum*)
	  – Congenital trypanosomiasis (*Trypanosoma cruzi*)
	• **R:** Congenital rubella syndrome (**R**ubella virus)
	• **C:** Congenital cytomegalovirus infection (**C**MV)
	• **H:**
	  – **H**erpesviruses like (1) congenital herpes simples (HSV) (2) Fetal varicella syndrome and congenital/ neonatal varicella by VZV
	  – **H**BV
	  – **H**IV-AIDS

Intranatal/during birth (transcervical)	Candidiasis, gonorrhea (ophthalmia neonatorum), listeriosis (neonatal listeriosis), *Strept. agalactiae* (neonatal meningitis), HSV, CMV and HPV
Postnatal/after birth by breast feeding	• Tuberculosis • CMV • HIV

C) HTLV-I, HTLV-II, HTLV-III (HIV), parvo virus B19, zika virus, etc.

- **Parasites:** *Trypanosoma cruzi, Leishmania donovani, Plasmodium* spp., *Babesia* spp., *Toxoplasma gondii,* etc.

5. **Saliva-borne infections:** Many microbes are present in saliva of humans and animals, which are deposited in tissues following bites. Following are the different examples of saliva borne infections:
   - **From human saliva (human bites): Viruses** transmitted by human saliva are HHV-4 (EBV), HHV-6 (HBLV), HHV-7 (RK virus), ECHO virus, mumps virus, etc. **Aerobic bacteria** transmitted by human bites are *Streptococcus, Staphylococcus* and *Eikenella corrodens*. **Anaerobic bacteria** transmitted by human bites are anaerobic streptococci, *Fusobacterium* and *Prevotella*. Bites by children rarely get infected while bites from adults get infected in 15–20% cases. Human bites carry aerobes (44%) and anaerobes (55%).
   - **From monkey bites:** Herpes virus simiae or B virus.

- **From dog bites:** Like rabies virus.
- **From cat bites:** Like *Bartonella henselae*.
- **Both by dog and cat bites:** *Pasteurella multocida*.
- **From rat bites (rat bite fever):** Like *Streptobacillus moniliformis* and *Spirillum minus*.

6. **Fomites-borne infections:** Contamination of fomites like towels, handkerchiefs, pens, pencils, clothes, cups, spoons, keys, etc., may transmit the infections like eye disease (trachoma), ear infection, diphtheria, dysentery, hepatitis A, etc.

7. **Infections by unclean hands and fingers:** It may transmit the infections like dysentery, trachoma, etc.

ACCESS YOURSELF

Essay/Full Question

1. Describe the epidemiological basis of common infectious diseases.

Short Notes

1. Congenital infections.

Short Questions for Theory/Viva Questions

1. Define infestation and infection.
2. What are contact carrier and paradoxical carrier?
3. Define case and carrier.
4. What are TORCH agents?

MCQs for Chapter Review

Definitions and Types

1. **Septicemia is:**
   a. Bacteria in blood
   b. Toxin in blood
   c. Pus in blood
   d. Multiplication of bacteria and toxin in blood

Reservoirs, Sources and Modes of Transmission

2. **Which of the following does not have non human reservoir**
   a. *Salmonella* Typhi
   b. *N. gonorrhoeae*
   c. *E. coli*
   d. *Clostridium tetani*
   e. *Treponema pallidum*

3. **Man is the only reservoir of:**
   a. Rabies
   b. Measles
   c. Typhoid
   d. Japanese encephalitis

4. **Which of the following is not transmitted by soil?**
   a. Coccidioidomycosis
   b. Tetanus
   c. *Brucella*
   d. Anthrax

5. **Vertically transmitted disease caused by all *except*:**
   a. *Toxoplasma*
   b. CMV
   c. HIV
   d. *Treponema pertenue*

6. **Which of the followings are transmitted by blood?**
   a. *Toxoplasma*
   b. Syphilis
   c. CMV
   d. Hepatitis B and C
   e. Hepatitis A

7. **All of the following infections may be transmitted via blood transfusion *except*:**
   a. Parvo B-19
   b. Dengue virus
   c. CMV
   d. Hepatitis G virus

8. **Most common agents responsible for human bite infections are:**

   a. Gram-negative bacilli

   b. Gram-positive bacilli

   c. Sporochetes

   d. Anaerobic streptococci

Answers and Explanation of MCQs

1. d

- Multiplication of bacteria and toxin in blood **called toxemia**, which is one type septicemia.

2. c

3. b, c

- Follow section, **reservoirs of infection** for explanation of answers of MCQs 2–3. Also follow respective chapters for more explanation.

4. c and d

- Spores of coccidioidomycosis are present in soil and transmitted by air borne route. Follow section, **modes of transmission [indirect → inhalation (air borne)]** for more explanation.
- Option b and d: Follow section, **modes of transmission (direct → contact with soil)** for explanation.
- No role of soil transmission of brucella.

5. d

- Follow section, **Table 3.1 and respective chapters** for more explanation.

6. a, b, c, d

7. b

- Follow section, **modes of transmission (indirect → blood borne)** for explanation of answers of MCQs 6–7.

8. d

- Follow section, **modes of transmission (indirect → saliva borne)** for explanation.

Normal Microbial Flora of Human Body

Chapter Outline

INTRODUCTION

Types of Microbes as per Habitats

Saprophytes

- **Meaning:** Sapros (Greek) means decayed and phyton (Greek) means plant.
- **Definition:** These are free living microbes in/on dead or decaying organic matter such as soil and water. They are incapable to multiply in living tissues except *B. subtilis*.

Commensals: (Mostly Normal Flora)

- **Meaning:** Com (Greek) means with and mensa (Greek) means table (eating at the same table).
- **Definition:** These are microbes which live in complete harmony with the host without causing any damage (living together).

Pathogens

- **Meaning:** Pathos (Greek) means suffering and gen (Greek) means to produce.
- **Definition:** These are microorganisms capable of producing disease in the host.

Host and Microbes Relationships

Normal floras are found in association with animals and also in the environment. Three types relationship between the host and microbes as follows.

Symbiosis

- **History:** This theory was given by Lynn Margulis and widely accepted.
- **Synonym:** It also called endosymbiosis or mutualism.
- **Definition:** Both microbes are dependent on each other, and both are getting benefit from each other without producing any harm to either partner.

- **Example:** Intestinal flora survive in gut produce vitamins for host.

Commensalism

- **Definition:** Association in which one partner gets benefit and other remains unharmed.
- **Example:** Flora living on skin.

Parasitism

- **Definition:** Association in which one partner (usually smaller called parasite) gets benefit with production of injuries to other partner (usually larger **called host**).
- **Example:** Microbes get benefits with production of infection to host.

> **Note: Predation**
> Association in which one animal (usually larger like tiger, lion, etc.), consumes or kills the other (usually smaller like deer, rabbit, etc.), for food. Larger animal is carnivorous called predator. Smaller is herbivorous called prey.

NORMAL FLORA

Synonym

Microbiota, microflora, microbial flora and indigenous microbial population.

Definition

Mixture of microorganisms regularly found at any anatomical site on or in the body of a healthy persons.

Advantages

1. **Immunostimulation:** They raise the immunity of body by sharing the common Ags.
2. **Endotoxin:** They liberate the endotoxin which triggers the alternative complement pathway as long as they are not produced in excessive amounts.

3. **Protection from external invaders:** Normal flora occupies body's epithelial surfaces. They prevent or interfere with colonization or invasion by other bacteria by blocking receptors (attachment) sites, competing for essential nutrients and producing anti-bacterial substances like peroxides bacteriocins, etc.
4. **Vitamins production:** Some of the intestinal flora like *E. coli* and *Bacteroides* are producing vitamin K, vitamin E and vitamin B in the gut which are available for use by host.
5. **Production of organic acids:** Like butyric acid and acetic acid which contribute in nutrition of host
6. **Digestion:** Intestinal flora degrades mucins, epithelial cells, carbohydrates and other dietary fibers and help in digestion.

Disadvantages

1. **Source of opportunistic infections,** e.g., they cause infection in immunodeficient persons.
2. **Entry in unusual sites cause infection,** e.g., intestinal flora may cause UTI.
3. **Interference with antibiotics:** By producing some enzymes like penicillinase which inactivates the penicillin and aggravates the infection.
4. **Confusion in diagnosis:** It is difficult to differentiate between normal flora and pathogens by laboratory tests.
5. **Supra-infection: Ch. 47.**

Types

Two types like transient and resident as mentioned below.

Transient flora

- **Properties:** They present for short (temporary) period in body not always. These include non-pathogenic and pathogenic groups. Among the pathogenic group some examples are like meningococcus and pneumococcus from nasopharynx. In hospital, patient may acquire the drug-resistance organisms as transient flora like MRSA from nose or skin, multi-drug resistance gram-negative bacteria like *E. coli*, *Klebsiella*, *Pseudomonas* and *Acinetobacter*, etc. Transient flora can be easily removed by improving the hygiene and taking infection control measures.
- **Transient flora in different organs:** Follow **Table 4.1**.

Resident flora

- **Properties:** They are constant flora of body or lifelong members. These include non-pathogenic groups which are harmless; infect beneficiary role.
- **Resident flora in different organs:** Described below in details.

Resident Flora of Skin

Human adult has two square meters of skin. Human skin is constantly attacked by organisms present in environment and objects come in contact, which are

TABLE 4.1: Transient flora in different organs

Organ	Flora
Skin	• **Bacteria:** *Staph. aureus, Strept. pyogenes, Strept. viridans, Enterococcus, E. coli, Proteus, Klebsiella* • **Fungi:** *Candida* spp., *M. furfur, Trychophyton*
Eyes	• **Bacteria:** *Staph. aureus, Strept. pneumoniae, Strept. viridans, Bacillus, Priopionibacterium*
Respiratory tract	• **Bacteria:** *Strept. pneumoniae, Staph. aureus, N. meningitidis, Moraxella*
Oral cavity	• **Bacteria:** *Strept. pyogenes, Lactobacillus, N. meningitidis, Moraxella, E. corrodens,* • **Fungi:** *Candida* spp. • **Viruses:** CMV, HSV
GIT	**Stomach** • **Bacteria:** *H. pylori, Streptococcus* spp., *Lactobacillus* spp. **Small intestine** • **Fungi:** *Candida* spp. • **Parasites:** *Entamoeba coli, Endolimax nana, Trichomonas hominis, Blastocystis hominis* **Large intestine** • **Bacteria:** *Corynebacterium*, MAC, *Pseudomonas* • **Fungi:** *Candida* spp. • **Parasites:** Same as small intestine
Urinary tract	• **Bacteria:** *Enterococcus, Mycoplasma, M. smegmatism, Bacteroides, Fusobacterium*
Genital tract	• **Bacteria:** *Strept. pyogenes, Enterococcus* • **Fungi:** *Candida* spp. • **Parasites:** *Trichomonas vaginalis*

loaded with microbes. Skin floras are mostly transient in nature. The populations of normal flora over the skin depend on many **factors** like differences in pH, oxygen, water, body secretions, wearing clothes, occupations, environmental conditions, etc. Following are the different **resident flora of skin.**

1. **Bacteria**
   - *Propionibacterium acne:* It is an anaerobic *Corynebacterium*. Children younger than 10 years are rarely colonized with it. It presents in areas rich in sebaceous gland.
   - Diphtheroids: These are non-pathogenic coryne-bacteria.
   - Anaerobic cocci.
   - *Staph. epidermidis*: It is the major inhabitants making up >90% of the flora.
   - *Staph. aureus:* It presents in nose, perineum and vulval skin. Its occurrence in nasal passages varies with age being greatest in newborns and less in adults. It also presents in hair follicles. Penicillin resistant *Staph. aureus* presents in hospital staff members.
   - Others: *Micrococcus*, hemolytic streptococci, *Enterococcus* spp., and non entrococci, GNB like *E. coli, Proteus* spp., etc., and nonpathogenic mycobacteria.

2. **Fungi:** *Candida* spp., *Pityrosporum ovale* and *Crypto-coccus* spp.

Resident Flora of Eyes

Conjunctiva is normally free from microbes due to constant flushing action of tears which contain lysozyme. Conjunctival floras are scanty and occasionally present. Predominant organisms of the conjunctiva are *Corynebacterium xerosis, Staphylococcus* spp., non-hemolytic *Streptococcus* spp., *Haemophilus* spp., and *Moraxella catarrhalis.*

Resident Flora of Respiratory Tract

Upper respiratory tract: It includes nose, ears, sinuses, pharynx and throat.

1. **Nose:** It includes the *Staphylococcus, Streptococcus (Strep. agalactiae), Corynebacterium, Haemophilus* and *Moraxella.*
2. **Nasopharynx:** Nasopharynx is often sterile at birth but may be contaminated within 2–3 days after birth by passage through the birth canal and by attendants. It includes the *Strept. viridans,* which established as the most prominent members of the resident flora and remain so for life. Others flora are aerobic and anaerobic bacteria such as staphylococci, pneumococci, GNC (*N. meningitidis* and *M. catarrhalis*), diphtheroids, lactobacilli, etc.
3. **Throat:** It includes the *Strept. viridans, Moraxella catarrhalis,* pneumococci and certain GNB from intestine like *E. coli, P. aeruginosa,* paracolons and *Proteus* spp.

Lower respiratory tract: It includes trachea, bronchi and lungs. Trachea contains organisms from pharynx or oral cavity. Bronchi and alveoli are normally sterile.

Resident Flora of Oral Cavity

Mouth is normally not sterile at birth. It contains same organisms as present in birth canal as mentioned below. It includes **bacteria** such as *Streptococcus mutans, Strept. sanguis, Strept. salivarius, Strept. mitior, Strept. milleri Staphylococcus* spp., *Enterococcus* spp., *Micrococcus* spp., *Lactobacillus acidophilus, Corynebacterium* spp., *Actinomyces* spp., *Propionibacterium* spp., and *Bifidobacterium* spp., *Veillonella* spp., *Neisseria* spp., *Branhmella (Moraxella)* spp., *Bacteroides* spp., *Treponema* spp., *Fusiform bacillus (Leptotricha buccalis), Actinobacillus* spp., *Capnocytophaga* spp., *Haemophilus* spp., *Eikenella* spp., *Wollinella* spp., *Selenomonas* spp., etc., **fungi** such as *Candida* spp., and *Geotrichum candidum* and **parasites** such as *Trichomonas tenax* and *Entamoeba gingivalis.*

Resident Flora of Gastrointestinal Tract (GIT)

In children: At birth the intestine is sterile but organisms are soon introduced with food. Bowels of newborns in intensive care nurseries tend to be colonized by Enterobacteriaceae like *Klebsiella, Citrobacter* and *Enterobacter.* In breast-fed children the intestine contains large numbers of streptococci and lactobacilli (*Lactobacillus bifidus* constitutes 99% of total organisms in feces). In bottle-fed children a more mixed flora exists in the bowel and lactobacilli are less prominent.

In normal adults

- **Effective factors: Diet** has a marked influence on the relative composition of the intestinal and fecal flora. Dietary habit develops toward the adult pattern which changes the bowel flora. **pH** changes also affects the intestinal flora. Esophagus contains microorganisms arriving with saliva and food. Acidic pH of stomach keeps the number of microbes at a minimum (103–105/g of contents) unless obstruction at the pylorus favors the proliferation of gram-positive cocci and bacilli. The normal acidic pH of the stomach markedly protects against infection with some enteric pathogens such as *V. cholerae.* Administration of cimetidine for peptic ulcer leads to a great increase in microbial flora of the stomach. Alkaline pH of intestine gradually allows the increment of resident flora. **Oral administration of antibiotics** can temporarily suppress the fecal flora. Neomycin plus erythromycin can suppress aerobes in 1–2 days. Metronidazole can suppress the anaerobes. The drug-susceptible microorganisms are replaced by drug-resistant ones, particularly *Staphylococcus, Enterobacter, Enterococcus, Clostridium difficile, Pseudomonas, Proteus* spp., and yeasts. The feeding of large quantities of *Lactobacillus acidophilus* may result in the temporary establishment of this organism in the gut and the concomitant partial suppression of other gut flora. **Minor traumas** during sigmoidoscopy, barium enema may induce transient bacteraemia in about 10% of procedures.
- **List of intestinal flora in adults:** More than 100 distinct types of organisms occur regularly as normal fecal floras are mentioned below.
  1. **Anaerobes:** They are mostly present in lower intestine. Anaerobes outnumber facultative aerobes organisms by 1000-fold. (1000:1). About 96–99% of the resident flora consist anaerobes like *B. fragilis* (most common), *Fusobacterium* spp., *Lactobacillus* (*L. bifidus* or bifidobacteria), *Cl. perfringens, Cl. difficile* and anaerobic gram-positive cocci (*Peptostreptococcus* spp.).
  2. **Facultative aerobes:** They are mostly present in upper intestine. About 1–4% of resident flora consist facultative aerobes like gram-negative coli-form bacteria (*E. coli*), *Enterococcus, Proteus* spp., *Pseudomonas, Lactobacillus,* etc.
  3. **Fungi:** *Candida* spp.
- **Density of flora in different intestinal parts in adults:** Follow **Table 4.2.**
- **Clinical significances**
  1. Intestinal bacteria **synthesizes** vitamin K.
  2. They have role in **conversion** of bile pigments and bile acids.

TABLE 4.2: Density of flora in different intestinal parts

Location (adult)	Bacteria/gram contents
Stomach	10^3–10^5
Duodenum	10^3–10^6
Jejunum and ileum	10^5–10^8
Cecum and transverse colon	10^8–10^{10}
Sigmoid colon and rectum	10^{11} (10–30% of fecal flora)

3. They have role in **absorption** of nutrients and breakdown products.
4. They have role in **antagonism** to microbial pathogens.
5. The intestinal flora **produce ammonia** and other breakdown products that are absorbed and can contribute to hepatic coma.
6. **Abscess formation:** The anaerobic flora of the colon, including *B. fragilis*, *Clostridium* spp., and *Peptostreptococcus* play a main role in abscess formation after bowel perforation. *Prevotella bivia* and *P. disiens* are important in abscesses of the pelvis originating in the female genital organs.
7. **Drug resistant:** *B. fragilis* are penicillin-resistant; therefore, another drug should be used.
8. Role in **superinfection or suprainfection:** Ch. 47.

Resident Flora of Urinary Tract

Upper urinary tract includes kidneys and ureters which are usually sterile.

Lower urinary tract includes bladder and urethra. **Urethra** is usually sterile or may contains same flora as skin and intestinal flora like α-hemolytic *Streptococcus*, *Enterococcus*, *Lactobacillus*, diphtheroides *Bacteroides*, *G. vaginalis*, *U. urealyticum* and *C. albicans*. These may enter in bladder can result in UTI. The urine of a healthy individual is sterile but can become contaminated due to transfer of microbes from the GIT or genital tract. Normal floras in urine are 102–104/ml.

Resident Flora of Genital Tract

Male genital organs: *M. smegmatism* can be found on the penis or genital secretion of both sexes. Its presence in urine creates confusion in diagnosis of renal tuberculosis.

Female genital organs: Norma vaginal secretion contains 10^8 bacteria/ml and also showing presence of *M. smegmatism*. Vagina has complex microbiota at different time as mentioned below.

- **At birth:** Sterile at birth, but after 24 hours acquires microbes from vagina, skin and intestine.
- **In 24 hours**: Invaded by *Micrococcus*, *Enterococcus* and diphtheroides.
- **In 2–3 days:** Maternal estrin induces deposition of glycogen in vaginal epithelium. This favors the growth of lactobacilli (also called Doderlein's bacilli) and flora similar to adult stage. Lactobacilli convert glycogen in to acid. Acidic pH prevents the vaginal colonization by foreign bacteria. After that estrin and lactobacilli are disappearing, pH becomes alkaline and vagina returns to normal flora.
- **At prepuberty:** Flora includes anaerobic cocci, *Listeria*, *Streptococcus*, Mimeae, *Mycoplasma*, *G. vaginalis*, *Neisseria*, spirochetes and *Candida* spp.
- **At puberty/adult:** *Lactobacillus* reappears, pH becomes acidic. Normal flora includes anaerobes like *Gardnerella vaginalis*, *Bacteroides*, *Mobilincus*, *Prevotella* and rare like *M. hominis*
- **At pregnancy:** Increase *Staph. epidermidis*, Doderlein's bacilli and *Candida*.
- **At menopause:** Return to prepuberty flora.

Resident Flora of Blood and Tissues

Microbes invade in blood and tissues from GIT, mouth, nasopharynx and some other body parts, but they are eliminated by defense systems of body.

PROBIOTICS

Definition: These are living microorganisms which are administered for treatment or prevention of diseases.

Properties: Probiotics are the part of normal flora of body. They are useful when normal floras are suppressed. They are available commercially in the form of capsules or sachets. Capsules or sachets contain *Bacillus coagulus*, *Bifidobacterium longum*, *Lactobacillus acidophilus*, *Saccharomyces boulardii*, etc.

Uses: Probiotics are used in following clinical conditions.
1. **Infectious diseases:** In *H. pylori* infection, in bacterial vaginosis (to restore the vaginal pH by using lactobacilli which produce the acidic pH) and in antibiotic associated diarrhea.
2. **Inflammatory diseases:** Gastroenteritis, colitis, necrotizing enteritis, irritable bowel syndromes and modulatory response in inflammation.
3. **Hypersensitivity reactions:** Modulatory response in eczema, dermatitis and other allergic disorders.
4. **Immunodeficiency diseases:** To restore the immunity
5. **High lipid profile:** Probiotics reduce the serum cholesterol by breaking the bile in the gut, thus inhibiting the reabsorption
6. **Hypertension:** Probiotics reduce the high blood pressure by producing the acetyl choline esterase (ACE) inhibitors like peptides during fermentation.

Limitations: Live organisms present in probiotics produce their actions only after their establishment in intestine, for that they have to compete with normal flora of intestine to establish them self. So, nowadays another similar preparation called prebiotics are more useful than probiotics.

PREBIOTICS

These are the preparations contain dietary fibers which when administered stimulate the growth and activity of normal flora.

ACCESS YOURSELF

Essay/Full Question

1. Normal flora of human body.

Short Notes

1. Normal flora of skin/GIT/genital tract.
2. Probiotics.

Short Questions for Theory/Viva Questions

1. Define: Saprophytes and commensals.
2. Mention the two differences between resident and transient flora of human body.
3. Mention the different types of relationship between host and microbes.
4. Write two advantages and two disadvantages of normal flora.
5. What are Doderlein's bacilli?
6. How probiotics are differentiated from probiotics.

Comments on

1. Normal floras are the useful properties of body.
2. Lactobacilli have protective role in adult vagina.
3. Prebiotics are more useful than probiotics.

MCQs for Chapter Review

Normal Flora

1. Transient colonization is caused by:
   a. HSV
   b. *Trichomonas vaginalis*
   c. *H. influenzae*
   d. *N. gonorrhoeae*
   e. *Staphylococcus aureus*

2. It is true regarding the normal microbial flora present on the skin and mucous membrane that:
   a. It cannot be eradicated by antimicrobial agents
   b. It is absent in the stomach due to the acidic pH
   c. It establishes in the body only after the neonatal period
   d. The flora in the small bronchi is similar to that of trachea

3. Common natural floras of skin are:
   a. *Streptococcus*
   b. *Staphylococcus aureus*
   c. *Candida albicans*
   d. *Bacteroides fragilis*
   e. *Propionibacterium acne*

4. Normal commensal of skin:
   a. *Staphylococcus aureus*
   b. *Candida albicans*
   c. *Bacteroides fragilis*
   d. *Propionibacterium acne*
   e. *Corynebacterium*

5. Which of the following is the main colonizer of sebaceous gland?
   a. *Propionibacterium acne*
   b. *Corynebacterium diphtheriae*
   c. *Strept. pyogenes*
   d. *Staph. aureus*
   e. *Candida*

6. In the gut, anaerobic bacteria outnumber the aerobes by ratio of:
   a. 10:1
   b. 100:1
   c. 1,000:1
   d. 10,000:1

7. Most common commensals gut flora in adult:
   a. Lactobacilli
   b. *Bacteroides*
   c. *E. coli*
   d. *Klebsiella*

8. The predominant colonic bacteria are:
   a. Largely aerobic
   b. Largely anaerobic
   c. *Bacteroides*
   d. Staphylococci

9. Bacterial count in duodenum:
   a. 10^5 per gram
   b. 10^8 per gram
   c. 10^{10} per gram
   d. 10^{12} per gram

10. Doderlein's bacilli word associated with:
   a. Lactobacilli
   b. Mimeae
   c. *Listeria*
   d. *Gardnella*

11. Normal commensal in female genital tract:
   a. *Gardnerella vaginalis*
   b. *Bifidobacterium*
   c. *Proteus*
   d. *Neisseria*

Probiotics and Prebiotics

12. Probiotics are useful for:
   a. Necrotizing enterocolitis
   b. Breast milk jaundice
   c. Hospital acquired pneumonia
   d. Neonatal seizures

13. Probiotics are differentiated from prebiotics:
   a. Probiotics contain microbes while prebiotics contains dietary fibers
   b. Probiotics contain dietary fibers while prebiotics contains microbes
   c. Probiotics are prepared from animal products while prebiotics are prepared from human products
   d. None

Answers and Explanation of MCQs

1. a, b and e
- Follow section, **transient flora** → Table 4.1 for explanation.

2. a
- Resident floras are permanent and cannot be eradicated by antimicrobial agents.

3. a, b, c, e

4. a, b, d, e

5. a
- Follow section, **resident flora of skin** for explanation of answers of MCQs 3–5.

6. c

7. b

8. b

9. a
- Follow section, **resident flora of GIT and Table 4.2** for explanation of answers of MCQs 6–9.

10. a

11. a
- Follow section, **resident flora of genital tract** → Female **genital organs** for explanation of answers of MCQs 10–11.

12. a
- Follow sections, **probiotics (uses)** for explanation.

13. a
- Follow sections, **probiotics and prebiotics (definition)** for explanation.

CHAPTER 5

Miscellaneous Topics:
World Days and Healthcare-related Symbols

Chapter Outline
- World Days
- Healthcare-related Symbols

WORLD DAYS

Follow **Table 5.1**.

TABLE 5.1: World days

Diseases	World day	Comment
Tuberculosis	24th March of every year	*M. tuberculosis* was identified and described on 24th March 1882 by Robert Koch
Leprosy	Last Sunday of January in each year	This day was chosen in commemoration of the death of Gandhi, the leader of India who understood the importance of leprosy
Polio	24th Oct. of every year	It was fixed by Rotary International over a decade ago to commemorate the birth of Jonas Salk, who led the first team to develop a vaccine against poliomyelitis.
HIV-AIDS	1st Dec. of every year	Since 1988, 1st Dec is celebrated as an AIDS day to raising awareness of the AIDS pandemic caused by HIV by Josh Lowe and mourning those who have died of the disease
Rabies	28th Sept of every year	It is the anniversary of the death of Louis Pasteur who, with his colleagues, developed the 1st effective rabies vaccine
Hepatitis	28th July of every year	It aims to raise global awareness and to encourage the prevention, diagnosis and treatment of hepatitis A, B, C, D and E
Malaria	25th April of every year	It aims the control of malaria
Health	7th April of every year	It is celebrated under the sponsorship of the WHO, as well as other related organizations

Note: World health day is one of eight official global health campaigns marked by WHO, along with World tuberculosis day, World immunization week, World malaria day, World no tobacco day, World AIDS day, World blood donor day, and World hepatitis day.

HEALTHCARE-RELATED SYMBOLS

Follow **Figs 5.1 to 5.7**.

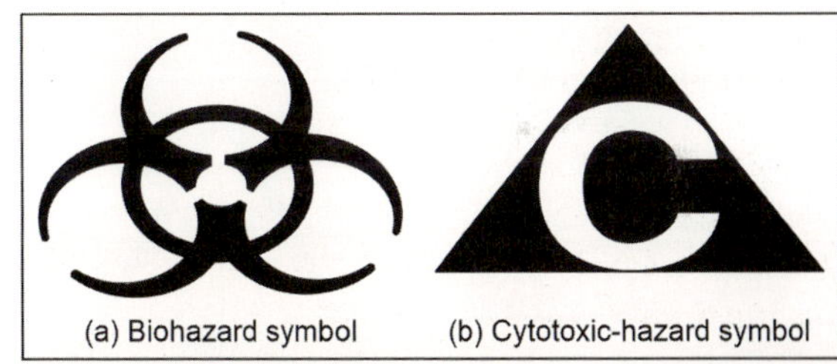

Fig. 5.1: Biohazard symbol and cytotoxic-hazard symbol

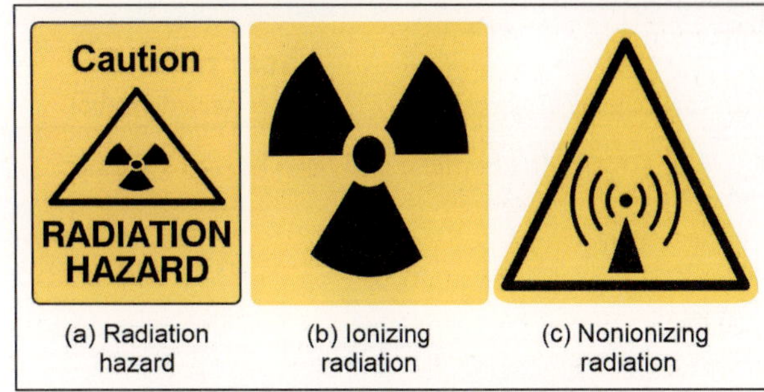

Fig. 5.2: Radiation hazard symbols

Fig. 5.3: Magnetic field symbol

Fig. 5.4: Gas hazard symbols

Fig. 5.5: Fumigation symbol

Fig. 5.6: Toxic chemical and laser hazard symbols

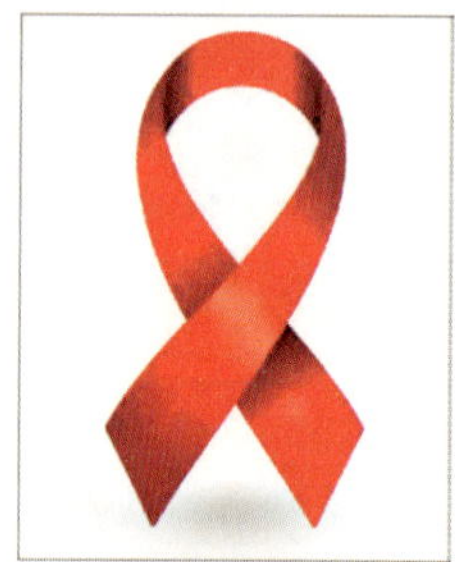

Fig. 5.7: Red ribbon symbol of HIV-AIDS

ACCESS YOURSELF

Short Questions for Theory/Viva Questions

1. Mention the day/date on which following diseases are celebrated as world day.
   - Tuberculosis
   - Leprosy
   - Rabies
   - Malaria

MCQs for Chapter Review

World Days

1. **24th March of every year is celebrated as:**
   a. World tuberculosis day
   b. World leprosy day
   c. World rabies day
   d. World malaria day

Healthcare-related Symbols

2. **Symbol shown in below image represent:**
   a. a: radiation hazard and b: cytotoxic hazard
   b. a: radiation hazard and b: biohazard
   c. a: bio medical waste and b: cytotoxic hazard
   d. a: cytotoxic and b: bio medical waste

Answers and Explanation of MCQs

1. a
- Follow section, world days (Table 5.1) for explanation.
2. c
- Follow section, healthcare-related symbols (Fig. 5.1 to 5.7) for explanation.

Morphology of Bacteria

Chapter Outline

- Shapes of Bacteria
- Size of Bacteria
- Arrangement of Bacteria
- Morphological Parts (Anatomy) of Bacteria

SHAPES OF BACTERIA

Follow **Fig. 6.1** for shapes of different bacteria.

1. Coccus: From kokkos (Greek) means berry, indicating spherical shape.
2. Bacillus: From baculus (Latin) means rod, indicating straight rod shape.
3. Coccobacillus: Indicating oval shape (intermediate between coccus and bacillus).
4. *Vibrio comma*: Vibrio from vibrare means vibrating nature and comma means curved rod like comma indicates comma shaped and motile bacillus.
5. Spirillum: Rigid spiral form.
6. Spirochete: From speira means coil and chaite means hair, flexuous spiral form.
7. *Leptospira interrogans*: Shape like interrogation mark or umbrella hook.
8. Actinomycete: Actino from actis means rays and mycete from mykes (Greek) means branching like fungi. It indicates the characteristic sun ray appearance and branching nature of bacterium.
9. *Mycoplasma*: Myco from Mykes (Greek) means branching/filamentous form and plasma means plasticity of nature. It is cell wall deficient bacteria and because of plasticity of nature assume any shape like spherical, oval, rod, balloon, filamentous, etc.

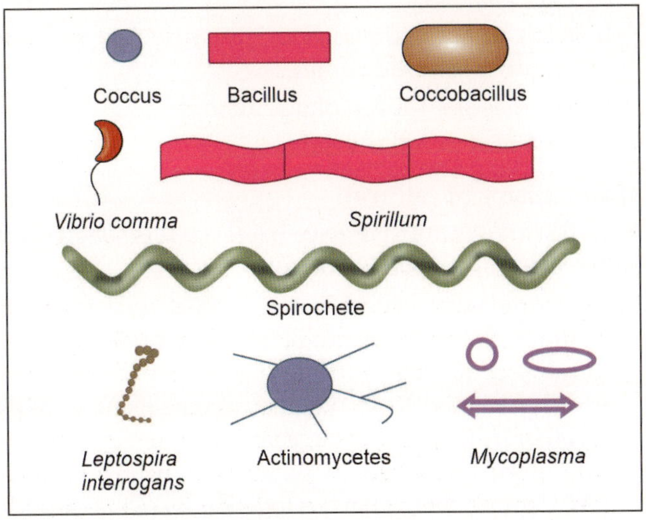

SIZES OF BACTERIA

Method to Measure the Size of Bacteria

Method to measure the size of bacteria called micrometry. Measuring unit called micrometer or micron (µm or µ).

Size of Different Bacteria

1. Coccus: 0.5–1 µm.
2. Bacillus: 0.2–0.5 µm in breadth × 1–6 µm in length.
3. Coccobacillus: 0.2–0.5 µm in breadth × 0.5–1 µm in length.
4. Spirochetes: Very large up to 10–20 µm in length.
5. Pleomorphic bacteria: Variable in size and shape.

ARRANGEMENT OF BACTERIA

Arrangement of Cocci

Follow **Fig. 6.2**.

Fig. 6.1: Shape of bacteria

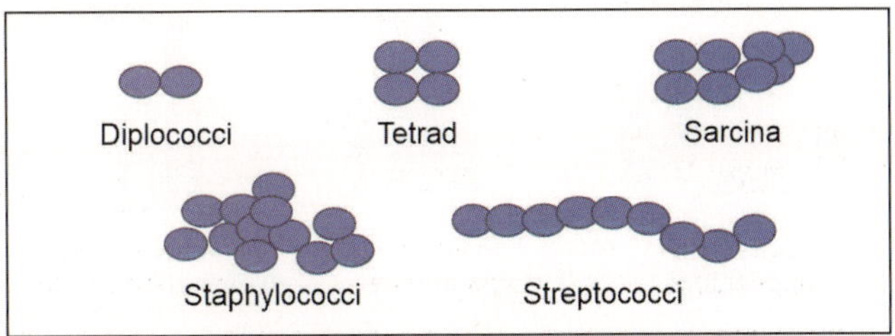

Fig. 6.2: Arrangement of cocci

1. Diplococci: Arranged in pair/group of two like pneumococcus, meningococcus, gonococcus, etc.
2. Tetrad: Arranged in group of four.
3. Sarcina: Arranged in group of eight.
4. Staphylococci: Arranged in grapes like cluster.
5. Streptococci: Arranged in chains.

Arrangement of Bacilli

Follow **Fig. 6.3**.

1. Singly: Like *E. coli*.
2. Diplobacilli: Arranged in pair like *Klebsiella pneumoniae* and *Moraxella lacunata*.
3. Streptobacilli: Arranged in chain.
4. Arranged in group or in cluster: Like *E. coli, S.* Typhi, etc.
5. S-shape: Two comma shape bacilli arranged end-to end gives S-shape appearance like in *V. cholerae* (*V comma*).
6. Spiral shape: Many comma shape bacilli arranged end-to end gives spiral-shape appearance like in *V. cholerae* (*V comma*).
7. Chinese latter or cuneiform pattern: Bacilli arranged by making an angle to each other like Chinese letter V or L in *C. diphtheriae*.
8. Palisades: Like stakes of fence in *C. diphtheriae*.
9. Fish in stream/school of fish: Bacilli arranged parallel to each other called fish in stream like in *V. cholerae* and school of fish in *H. ducreyi*.

MORPHOLOGICAL PARTS (ANATOMY) OF BACTERIA

It includes organs **(Fig. 6.4)** like bacterial cell wall, cytoplasmic (plasma) membrane, protoplasm, capsule-slime layer, flagella, fimbriae, spores, pleomorphism and involution forms. Bacterial cell wall and cytoplasmic (plasma) membrane are collectively known as cell envelop or outer layer. Capsule-slime layer, flagella and fimbriae are considered as appendages of cell wall.

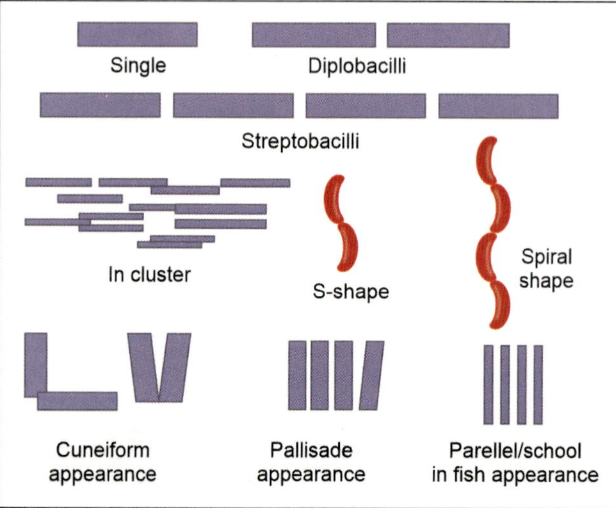

Fig. 6.3: Arrangement of bacilli

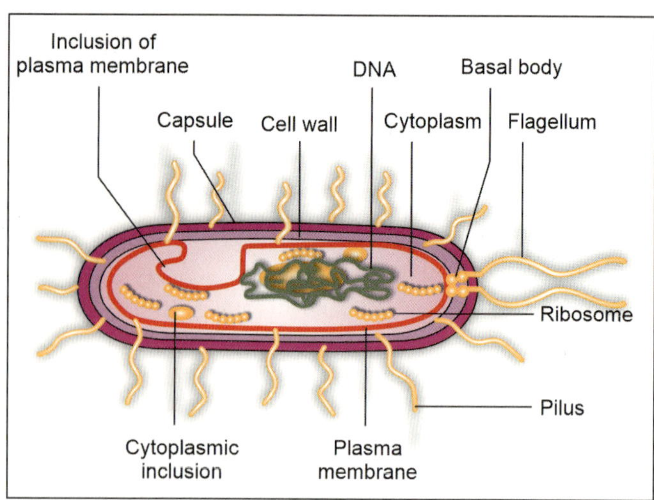

Fig. 6.4: Anatomy of bacteria

Bacterial Cell Wall

Properties: Cell wall is tough and rigid, surrounding the bacterium as like a shell. It is 10–25 nm thick and holds the 20–30% of the dry weight of the cell. It is absent in some bacteria discussed later in this chapter.

Cell wall antigen: It called somatic antigen. It is abbreviated as O-Ag, from ohne hauch (German) means without film of breath. It is opposite to hauch (German) means with film of breath for flagellar antigen (H-antigen).

Functions of cell wall

1. It provides rigidity to cell.
2. It maintains the bacterial shape.
3. It plays fundamental role in vital activity of bacteria.
4. It carries the bacterial antigen which play role in virulence and immunity.
5. It provides mechanical support to cell membrane.
6. It contains proteins (called porin proteins) that make the porin channels for diffusion of material to and fro by bacteria.
7. It helps to maintain osmotic pressure and protects cell against osmotic damages.
8. It provides site for the phage adsorption.
9. It takes part in cell division.

Biochemistry of cell wall

A. Cell wall of gram-positive bacteria: It is 80 nm thick with following structural details.

1. **Peptidoglycan layer:** It also called mucopeptide or mucoprotein or murein. Thick peptidoglycan is composed of alternating unit of N-acetyl muramic acid (NAM) and N-acetyl glucosamine (NAG) linked together by beta-1-4 linkage and a set of tetra peptide side chain attached to N-acetyle muramic acid. Tetrapeptide chain attaches together by penta-peptide bridges as shown in **Fig. 6.5**.

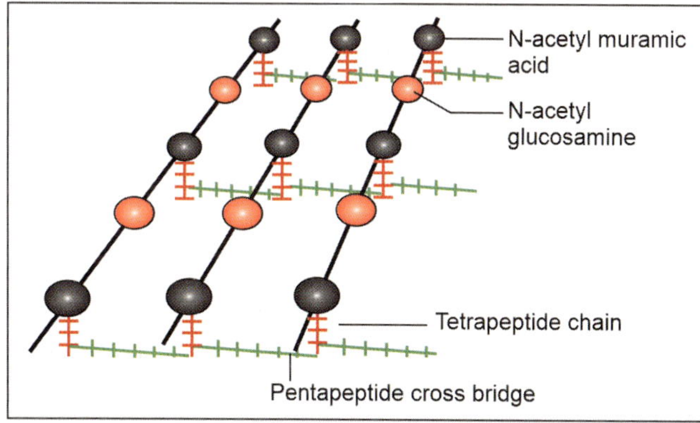

Fig. 6.5: Peptidoglycan/murein

2. Teichoic acids

- **Types of teichoic acids:** Two types.
  - Cell wall teichoic acid: Cell wall teichoic acid is polymer of ribitol (5-carbon) and covalently linked to peptidoglycan of cell wall as shown in **Fig. 6.6**.
  - Membrane teichoic acid: Membrane teichoic acid is polymer of glycerol (3-carbon) and linked to glycolipid of cytoplasmic membrane, so also called lipotechoic acid as shown in **Fig. 6.6**.
- **Functions of teichoic acids**
  - Teichoic acid helps to synthesize the peptidoglycan
  - Organ of adhesion: It helps bacteria to adhere with host cells.

3. Surface proteins:
Like M, T and R proteins are useful for grouping the bacteria (in streptococci). Surface proteins are also useful for virulence and as organs of adhesion.

B. Cell wall of gram-negative bacteria:
It is 2 nm thick. It has four layers from inner to outer are periplasmic space, thin peptidoglycan layer, outer membrane protein (OMP) and lipopolysaccharide (LPS) as shown in **Fig. 6.7** with following structural details.

1. **Periplasmic space:** It presents outer to plasma membrane contains enzymes.
2. **Peptidoglycan layer:** It is very thin layer.
3. **Outer Membrane Protein (OMP):** It is made up by phospholipids, lipoproteins and surface proteins.
   - **Phospholipids:** These are similar to plasma membrane.
   - **Lipoproteins:** OMP attaches to peptidoglycan by lipoproteins.
   - **Surface proteins:** Also called major membrane proteins or principal membrane proteins. There are following two types of surface proteins.
     - Porin protein: It makes porin channel for diffusion of material to and fro by bacteria.
     - Non-porin protein: It acts as an organ of adhesion, helps in production of exo-enzymes and also provides receptors for antibiotic agents. For example Penicillin Binding Proteins (PBPs).
4. **Lipopolysaccharide (LPS):** It has three regions as shown in **Fig. 6.8**.

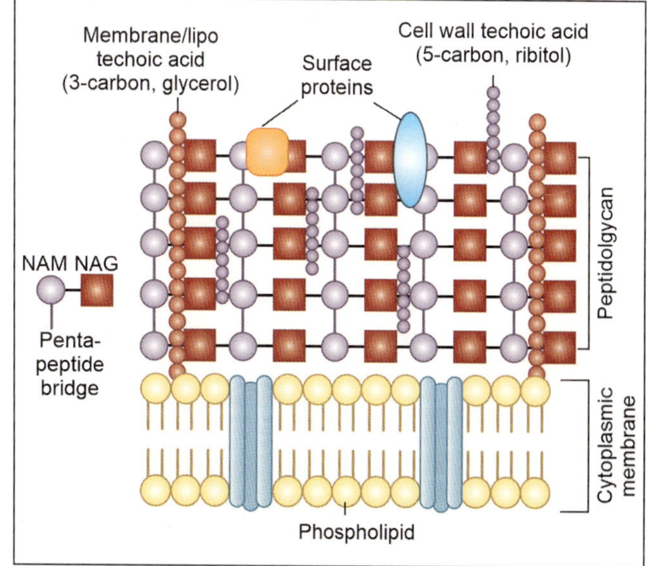

Fig. 6.6: Gram-positive cell wall

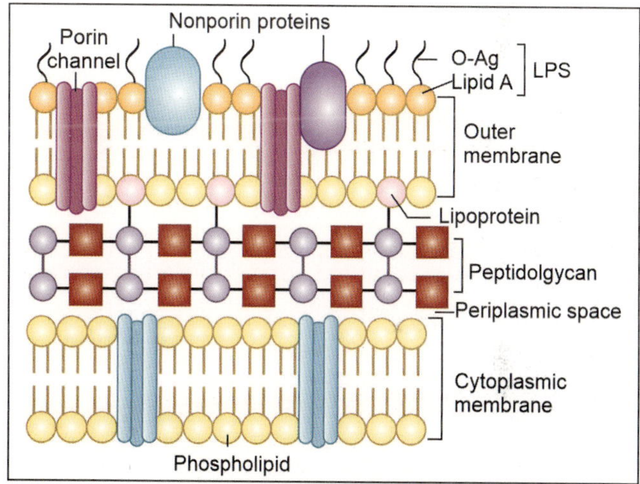

Fig. 6.7: Gram-negative cell wall

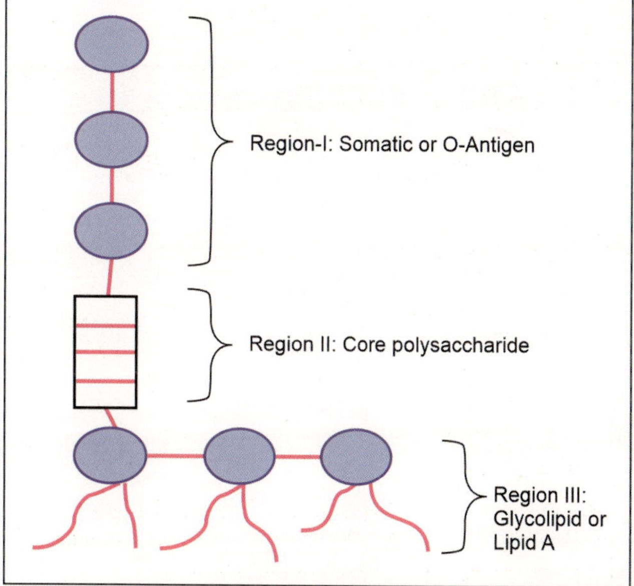

Fig. 6.8: LPS structure

1

- **Region-I:** It is the somatic/O-antigen. Formerly called Bovine-Ag. It is responsible for O-antigen specificity and typing.
- **Region-II:** It is the core polysaccharide.
- **Region-III:** It is the glycolipid or lipid A responsible for endotoxic activity like fever, shock, collapse, hemorrhage, necrosis, lethal effect, anticomplementary effect, B cell mitogenicity, antitumor activity, etc.

C. Cell wall of acid-fast bacteria: Acid fast bacteria (*Mycobacterium* spp., *Nocardia* spp., etc.) contain three layers from inner to outer are peptidoglycan, polysaccharide and fatty acids (glycolipid).

1. **Peptidoglycan:** It is similar to gram-negative bacterial cell wall and linked to polysaccharide layer by phosphodiester bond.
2. **Polysaccharide:** Layer **called arabinogalactan** and linked to mycolic acids.
3. **Fatty acid (glycolipid)**
   - It is responsible for acid fastness.
   - Glycolipid is the major part in cell wall which accounts 60% of the dry weight of cell wall.
   - It called mycolic acid in *Mycobacterium* and nocardic acid in *Nocardia* (unique mycolic/nocardic acid 6, 6′ dimycolyltrehalose called cord factor).
   - Mycolic acid or cord factor in *Mycobacterium* acts as virulence factor and responsible for cell membrane cytotoxicity, inhibition of PMNs migration, granuloma formation and activation of complement pathway.
   - Peptidoglycan-polysaccharide-mycolic acid complex forms the skeleton in cell wall of mycobacteria, so acid fastness in mycobacteria is not the property of mycolic acid alone, but depends also on the integrity of cell wall.

> **Notes: Cell wall of *Corynebacterium***
> Some workers noticed the presence of acid fast cell wall in *Corynebacterium* with three layers as mentioned above where the glycolipid called corynemycolenic acid.

Differences between gram-positive and gram-negative bacterial cell wall: Follow **Table 6.1**.

Demonstration of cell wall: Cell wall is demonstrated by following methods.
- Plasmolysis: When bacteria placed in hypertonic solution they loss the cytoplasm called plasmolysis

TABLE 6.1: Differences between gram positive and negative bacterial cell wall

Characteristics	Gram +Ve	Gram –Ve
Thickness	Thicker (80 nm)	Thin (2 nm)
Aminoacids (AA)	Few	Several
Aromatic and sulphur containing AA	–	+
Lipids	–, Or scanty	+
Techoic acid	+	–

but cell wall retains it's size and shape such cell called bacterial ghost.
- Micro dissection.
- By mechanical rupture of cell wall.
- Differential staining.
- Reaction with antibody.
- Electron microscopy.

Inhibition of cell wall synthesis: Bacteria without cell wall are called cell wall deficient bacteria.
- **Causes:** Cell wall synthesis is inhibited naturally (like in *Mycoplasma* spp., cell wall is absent naturally) or by antibiotics like penicillin or by viruses like bacteriophage or by enzymes like lysozyme or spontaneously (sudden loss of cell wall).
- **Clinical significances:** Absence of cell wall created following issues in bacterial infections.
  - Difficulty in diagnosis: Cell wall deficient bacteria are difficult to cultivate and require agar contains solid medium having right osmotic strength.
  - Difficulty in treatment: Cell wall deficient form of bacteria is ineffective to antibiotics. Sometimes they appear after treatment with penicillin.
  - Persistence of infection: Cell wall deficient bacteria are not causing the disease but causing persistence of certain chronic infection such as pyelonephritis.
  - Recurrence: They cause recurrence of infection.
- **Types of cell wall deficient bacteria:** Four types of cell wall deficient bacteria as follows:
  1. **Protoplast:** It is produced artificially by allowing lysozyme action on gram-positive bacteria in a hypertonic medium. It is spherical in shape with total loss of cell wall and cytoplasmic membrane holds the contents of cell.
  2. **Spheroplast:** It is produced artificially by allowing lysozyme action on gram-negative bacteria in a hypertonic medium. It is spherical in shape with partial loss of cell wall and some cell wall structures are retained.
  3. **L-forms:** L forms word derived from Lister institute, London. Kleinberger (Nobel Prize) while studying the culture of *Streptobacillus moniliformis* in the Lister institute, London, observed cell wall deficient, swollen, morphologically abnormal forms of bacteria and named them L-forms after the Lister institute in 1941. L forms have different types such as spontaneous (sudden development), induced (induced by some agents like penicillin), unstable (revert to original forms) and stable (irreversible). Examples of bacteria showing L-forms are *Staph. aureus*, *N. gonorrhoeae*, *M. tuberculosis* and *Streptobacillus moniliformis*.
  4. ***Mycoplasma:*** **Follow Ch. 74** for more details.

Cytoplasmic/Plasma Membrane

Properties: It presents inner to cell wall. It is elastic in nature and about 5–10 nm in size. It always presents in bacteria.

Functions

1. It is semipermeable layer acts as an osmotic barrier. It allows inflow and outflow of selective ions and molecules of the cell (called selective permeability) and prevents the others.
2. It participates in active transport of selective nutrients by enzyme permease.
3. It contains enzymes required for respiratory activity to generate energy (ATP) by electron transport and oxidative phosphorylation.
4. It participates in synthesis of cell wall components like peptidoglycan and OMP.
5. It participates in synthesis of bacterial toxins and other bacterial enzymes.
6. It provides energy for flagellar movement and chromosomal mobilization.

Biochemistry: It contains phospholipids (30–60%), protein (50–70%) and carbohydrates. Sterol (cholesterol or ergosterol) is absent in bacteria except in *Mycoplasma* spp., and *Ureaplasma urealyticum.*

Demonstration: Cytoplasmic membrane is demonstrated by Electron Microscopy (EM).

Mesosomes or chondroids (Fig. 6.4)

- **Definition:** Invaginations or infolding of the plasma membrane in to the cytoplasm called chondroids or mesosomes. They are more prominent in gram-positive bacteria.
- **Functions of mesosomes**
  1. They provide more surface area for anabolic and catabolic activities.
  2. They are the centers for respiratory enzymes and help in respiration.
  3. They act as analogs for mitochondria in cell.
  4. They take part in DNA replication and cell division or binary division.

Cytoplasm

Mixture of plasma membrane, cytoplasm, nucleus and nucleoplasm **called protoplasm**. Jelly like substance presents inner to plasma membrane but outer to nucleus **called cytoplasm**. Jelly like substance presents inner to nucleus **called nucleoplasm**. In eukaryotes nucleus is present but in prokaryotes like bacteria nucleus and nucleoplasm are absent, so we can say that cytoplasm acts as nucleoplasm which contains chromatin (chromosome, plasmid, episome and transposon). Plasma membrane is described above while cytoplasm is discussed below.

Properties of cytoplasm

- It is a colloidal system contains organic and inorganic solutes in a viscous watery solution.
- Organelles like endoplasmic reticulum, mitochondria, etc., are absent in bacteria.
- Cytoplasmic motility **called cytoplasmic (protoplasmic) streaming** is absent in bacteria unlike eukaryotic cells.

- Cytoplasm stains uniformly with basic dyes in young culture.

Intracytoplasmic inclusion/cytoplasmic matrix

1. **Ribosomes:** These are tiny granules scattered in cytoplasm. They are made up of two subunits larger (50s) and smaller (30s). They are sites for protein synthesis.
2. **Polar bodies:** They are situated at both poles of bacilli hence **called polar bodies**. They are also known as **metachromatic granules** because they had the property of changing the color of basic blue dyes used in light microscopy or **volutin granules** because they were 1st identified from *Spirillum volutans* by Meyer (1904) or **Babes-Ernst granules** because they are discovered and described by Romanian bacteriologist Babès Victor and German pathologist Paul Ernst, 1888. Biochemically they are composed of polymetaphosphate and strongly basophilic in nature. They act as energy and phosphate store house required in conditions with nutritional deficiency. They disappear when deficient nutrients are supplied. They are demonstrated by Albert stain, Loeffler's methylene blue (stain reddish violet), Neisser's stain and Ponder's stain. Bacteria contain polar bodies are *Corynebacterium diphtheriae* (causing diphtheria), *Corynebacterium xerosi, Mycobacterium leprae* (may be), *Bordetella pertussis, Gardnerella vaginalis,* and *Spirillum volutans.*
3. **Polysaccharide granules:** They stain by iodine. They act as storage products.
4. **Fat globules/lipid globules:** These are energy storage products. They contain lipid, polymetaphosphate (volition granules) and crystalline protein. They present only in vegetative stage and make confusion with spores. Stains used to differentiate spores from fat globules are mentioned in **Table 6.2**.
5. **Chromosome:** It appears as oval or elongated body. Generally it is one per cell or two or more. It is demonstrated by acid hydrolysis or by EM. It contains single-circular ds-DNA. It controls the growth, multiplication and metabolism of cell. It also controls the hereditary transmission of genetic information.
6. **Plasmid:** Extrachromosomal intracytoplasmic DNA containing material called plasmid. It serves for toxin

TABLE 6.2: Stains to differentiate spores and fat globules

Staining method	Reagents	Spores	Fat globules
Ashby's method	Malachite green	Green spores and red bacilli	Unstained
Burdon's method	Sudan black B	Unstained	Blue-black
Holbrook and Anderson method	Malachite green + Sudan black B	Green spores and red bacilli	Blue-black
Modified acid fast	0.25% H_2SO_4	Red spores and color-less bacilli	Red

production, drug resistance and other virulence purposes.

7. **Episomes:** It is a type of plasmid which is linked with chromosome.

8. **Jumping gene or transposon:** Plasmid jumps in-between chromosomal and extrachromosomal part called jumping gene or transposon.

Capsule and Slime Layer

Definitions

- **Capsule:** Many bacteria secrete the viscid materials around them that organize in a sharply defined structure called capsule **(Fig. 6.9)**.
- **Microcapsule:** Capsule is too thin to be seen under microscope called microcapsule.
- **Slime layer:** Many bacteria secrete the viscid materials around them that remains loose, unorganized as ill-defined structure called slime layer.

Capsular antigen: It abbreviated as K from Kapsel. (**German** language for capsule).

Examples of microorganisms containing capsule and slime layer

A. Capsule containing organisms

- **Capsulated fungus:** *Cryptococcus neoformans*.
- **Capsulated bacteria**
1. *Staph. aureus* (in some strains).
2. *Streptcoccus* species: These are α-**Hemolytic (Viridans group)** streptococci such as *Strept. mutans* (no Lancefield grouping) and *Strept. pneumoniae*, β-**Hemolytic** streptococci such as *Strept. pyogenes* (Lancefield Group A), *Strept. agalactiae* (Lancefield Group B) and *Strept. equisimilis* (Lancefield Group C) and *Enterococcus* spp (Lancefield Group D).
3. *Anthrax bacillus*.
4. *Clostridium welchii*.
5. *Neisseria meningitidis*.
6. *E. coli:* In some strain.
7. *Klebsiella pneumoniae, K. granulomatis* and other species.
8. *Enterobacter* spp.
9. *Yersinia enterocolitica:* Capsule occurs *in vivo* but not *in vitro*/culture.
10. *Aeromonas hydrophila*.
11. *Vibrio cholerae* O139.
12. *Haemophilus influenzae* and other species.

13. *Pasteurella multocida*.
14. *Bordetella pertussis* (no role in virulence).
15. *Bacteroides fragilis*.

B. Slime layer containing bacteria

1. *Staph. albus* (*Staph. epidermidis*).
2. *Leuconostoc*.
3. *Pseudomonas aeruginosa*.
4. *Yersinia pestis*.
5. *Rickettsia* spp.

C. Both capsule and slime layer containing bacteria

1. *Strept. salivarius*.

Biochemistry

- **Polysaccharide:** It is polysaccharide (glucan, dextran, levans) in nature in almost all bacteria.
- **Polypeptide:** In *Anthrax bacillus* capsule is polypeptide (d-glutamic acid) in nature.
- **Protein:** In *Yersinia pestis* it is protein in nature.
- **Hyaluronic acid:** In some strains of *Strept. pyogenes* (Lancefield Group A) and some strains of *Strept. equisimilis* (Lancefield Group C) capsule is hyaluronic acid in nature.
- **Glycocalyx:** Slime layer is made of glycocalyx which is the mixture of sugars (oligosaccharides) and proteins. It presents in *Staph. albus* (*Staph. epidermidis*), *P. aeruginosa*, etc.

Functions of capsule

1. **Inhibition of phagocytosis:** It enhances bacterial virulence by inhibiting phagocytosis; however capsule of *Bordetella pertussis* does not contribute in virulence.

2. **Protection:** It acts as protecting covering and protects bacteria against lysozyme, bacteriophages, colicins, antibodies, antibiotics, heavy metals, free radicals, complement action and also against desiccation.

3. **Breakdown of dietary fibers:** Refer to **Flowchart 50.2**.

4. **Abscess formation:** Capsule of certain bacteria like *Bacteroides fragilis* is toxic in nature and responsible for abscess formation.

5. **Adhesion:** It is the function of slime layer. Slime layer made up of glycocalyx has ability to binds bacteria with damaged tissues. It also helps bacteria to bind with plastic material of medical devices like catheters, suture materials, pacemakers, implants, Ryle's tube, etc., and forms the growth on such devices **called biofilm**. Uses of such devices cause diseases production. Biofilm is formed by few bacteria like *Staph. albus, P. aeruginosa*, etc.

6. **Contribution in energy and nutrition:** It attracts the nutritional materials at cell surface because of polyanionic nature.

7. **Desiccation (drying) prevention:** Capsule contains water which prevents the bacterial desiccation.

8. **Vaccine preparation:** Capsule is antigenic in nature and produces protective antibodies. This principle

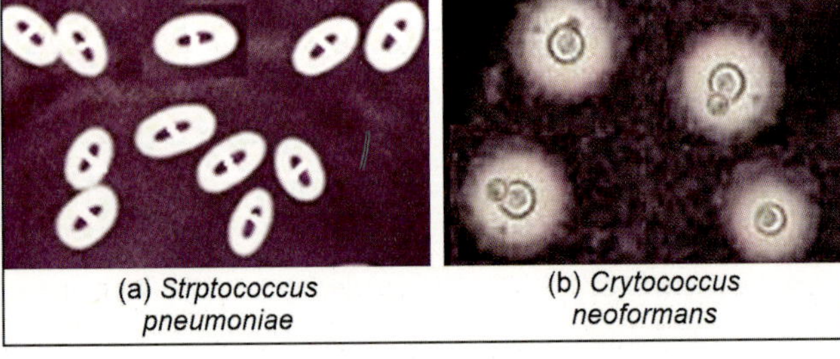

<table>
<tr><td>(a) Strptococcus pneumoniae</td><td>(b) Crytococcus neoformans</td></tr>
</table>

Fig. 6.9: Capsule under Indian ink

helps to prepare the vaccine for pneumococci, meningococci and *H. influenzae* serotype-b.

9. **Diagnosis:** It is antigenic in nature and helps in identification and typing of bacteria.

Demonstration: Following methods are used to demonstrate the capsule.

1. **Microscopy**
   - **Negative stains:** Like Indian ink (**Fig. 6.9**), nigrosin and Manveal stain.
   - **Gram's stain:** Slime layer (even capsule also) has little affinity for basic dyes, so not visible under Gram's stain. Capsule is sometimes demonstrated under Gram's stain.
   - **Special stains:** Following are the special stains used for demonstration of capsule especially for capsule of *Anthrax bacillus.*
     - Giemsa stain: Capsule appears in red color.
     - Indian ink: Capsule appears as clear halo around the bacilli.
     - Methylene blue: Capsule appears as purplish materials around the bacilli called M'Fadyean's reaction.

2. **Culture:** Mucoid colonies produced by capsulated bacteria.

3. **Serological test:** Test known as Quellung reaction or Capsular swelling, developed by Neufeld in 1902. Mix capsulated strain with specific anticapsular serum on slide and examine under microscope. Capsule becomes prominent and gets swell due to increase in refractivity.

Enhancement of capsule production

- **In *A. bacillus*** capsule production is increased in presence of 10–25% CO_2 when medium is enriched with biocarbonate and in absence of CO_2 when medium is enriched with serum, albumin, charcoal or starch.
- **In *C. neoformans*** capsule production is increased by growing the fungus on chocolate agar at 37°C in a CO_2 incubator.

Inhibition of capsule synthesis (loss of capsule)

- **Methods:** Capsule synthesis is inhibited by mutation or by repeated subculture.
- **Clinical significance:** Capsule has virulent property and loss of capsule changes virulent strain to avirulent strain.

Flagella

Definition: One or more, long, unbranched, wavy, filamentous structure for movement of organism called flagella (singular → flagellum).

Flagellar antigen: It is abbreviated as H-Ag from Hauch (German) means film of breath, as growth of motile bacteria looks like film of breath on glass surface.

Properties: It is 3–20 µm long and uniform in diameter (0.1–0.013 µm). It terminates in square tip.

Types of bacteria according to location of flagella: Follow **Flowchart 6.1** and **Fig. 6.10.**

Parts: Flagella have three parts (**Fig. 6.11**).

1. **Basal body:** It is a circular structure embedded in the cell envelope consist a central rod with following four rings:
   i. M-ring: Embedded in the cell membrane.
   ii. S-ring: Embedded in the periplasmic space.
   iii. P-ring: Embedded in the peptidoglycan.
   iv. L-ring: Embedded in the lipopolysaccharide.

2. **Hook:** It is short curved structure connecting the filament and basal body. It is broader than the filament, protein in nature (different from flagellin) and embedded in cell envelop.

3. **Filament:** It is external to the cell and made up of flagellin.

Biochemistry: Chemically flagella are composed of protein called flagellin, alike keratin and myosin.

Functions of flagella

1. **Organ of locomotion:** Flagella help in different type's movement of bacteria such as movement towards nutrients called chemotaxis, movement toward air called aerotaxis and movement toward light called phototaxis.

2. **Constitute the flagellar (H) antigen:** H antigen produces the antibody which is useful for diagnosis and typing of bacteria.

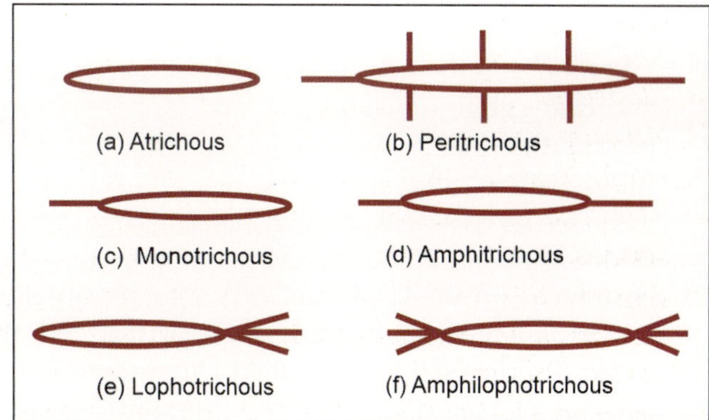

Fig. 6.10: Types of bacteria according to location of flagella

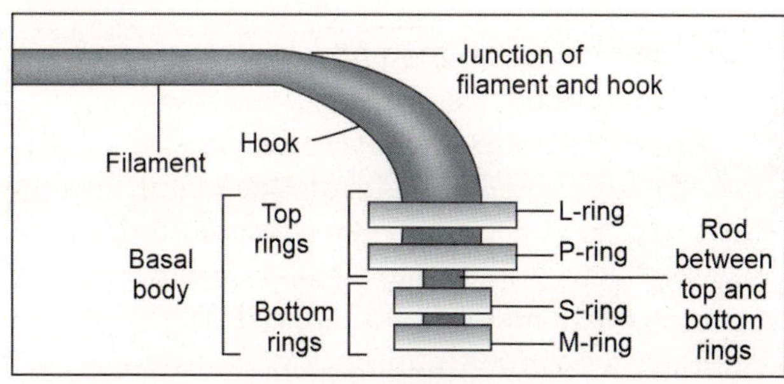

Fig. 6.11: Parts of flagella

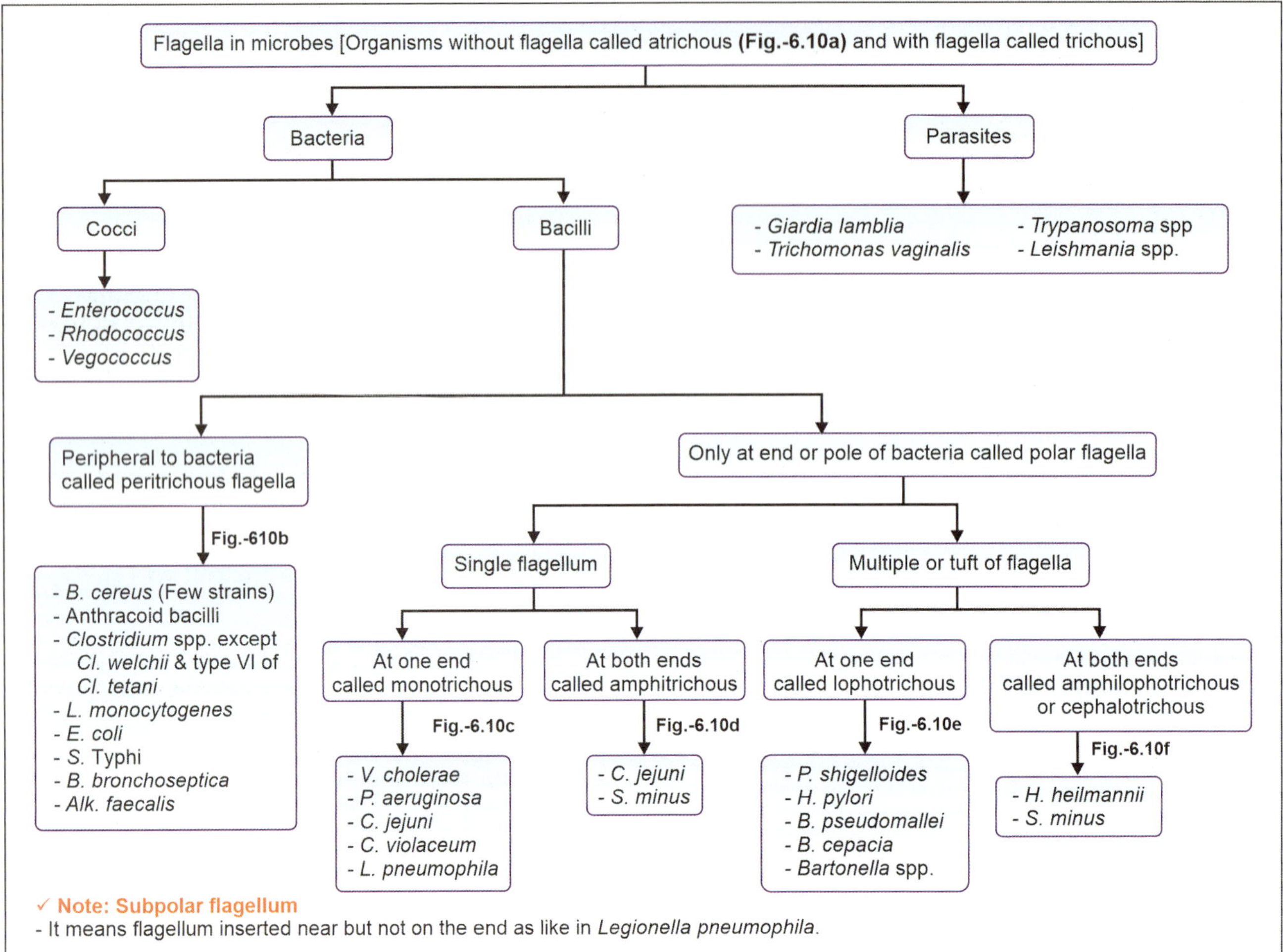

Demonstration of flagella

1. Microscopy and staining methods

- **Light microscopy**
  - Wet mount: It includes hanging drop preparation and Ryu's stain (crystal violet and tannic acid as mordant).
  - Impregnation technique: Thickness of the flagella is increased by coating them with mordants like tannic acid and potassium alum and staining them with basic fuchsin (called Gray method) or pararosaniline (called Leifson method) or silver nitrate (called West method) or crystal violet (called Difco's method). Although flagellar staining procedures are difficult to carry out, they often provide information about the presence and location of flagella which have great value in bacterial identification.
- **Other microscopical methods:** DGIM and EM.

2. Culture methods

- **By using semisolid media:** Media contain 0.2–0.5% agar. Different methods include semisolid stab agar method (**Ch. 112**), Craigie's tube technique (**Ch. 115**) and U-tube technique (**Ch. 115**).
- **By using solid media:** Media contain 1–2% agar and motile bacteria produce spreading/swarming growth on solid agar.

- **Serological test:** These are done by using specific antiserum.

Types of motility

1. **Stately motility:** It occurs in *Clostridium* spp., except. *Cl. welchii* and *Cl. tetani* type VI.
2. **Active motility:** It occurs in *E. coli*, *Proteus* spp., *S.* Typhi and *L. interrogans*.
3. **Tumbling motility:** It occurs in *L. monocytogenes* at 25°C but not at 37°C (called differential motility due to temperature dependent flagellar expression) and *C. jejuni*.
4. **Darting motility:** It occurs in *V. cholerae*, *C. jejuni* and *S. minus*.
5. **Oscillatory motility:** It occurs in *Cl. difficile*.
6. **Twitching motility:** It occurs in *Bartonella* spp.
7. **Swarming growth or swarming motility:** It occurs in *Proteus vulgaris*, *Proteus mirabilis* and *Cl. tetani*.

Phase variation

- **Definition:** Flagellar antigen (H-Ag) undergoes frequent antigenic variation called phase variation. It was 1st observed in *Salmonella* and later in other GNB.
- **Clinical significance:** It helps in bacterial typing.

Notes: Endo-flagella and bacteria showing motility without flagella

Endo-flagella: They are present in spirochetes like *T. pallidum*, *Borrelia* spp., *Leptospira* spp., etc., but not useful for motility because not protruding outside and remain in periplasmic space between peptidoglycan and plasma membrane. These bacteria show following types motility.

1. *T. pallidum:* Three types of motility like flexion—extension, translatory and cork screw.
2. *Borrelia* spp.: Lashing motility.
3. *Leptospira* spp.: Active motility with spinning or translation type of motility.

Bacteria showing motility without flagella

1. Brownian motility/movement: It is the passive movement of bacteria not due to flagella but due to air current or by fluid movement.
2. Sluggish motility: *Moraxella lacunata* which is nonflagellated but it is sluggishly motile.
3. Jerking or twitching motility: It occurs in *Eikenella corrodens*, not due to flagella but by contractile-fimbriae like filamentous appendages.
4. Gliding motility: *Capnocytophaga canimorsus* and other species of genus are lacking flagella but show gliding motility, which is very difficult to observe. *Mycoplasma pneumoniae* is without flagella but showing gliding motility.

Fimbriae

Definition: Very fine hair like surface appendages in gram-negative bacteria called fimbriae (singular → fimbria).

Synonym: Also called pili (singular → pilus) or fibrillae. The word sex pili are used only for fimbriae concerned with conjugation.

Fimbrial antigen: It is abbreviated as FiAg.

Properties: Fimbriae are shorter (0.5 µm long), thinner (0.001 µm or 10 nm) and more in numbers than flagella as shown in **Fig. 6.4**. They are unrelated to motility and are found on motile and nonmotile bacteria. They are originated from cell membrane. They are best developed in fresh culture and also in liquid media. Bacteria loose the fimbriae following subculture in solid culture.

Classification

1. **According to bacteriophage sensitivity:** Type F, type I, etc.
2. **According to their functions:** Fimbriae are named on basis of their function.
   - **Sex pili:** They found in male bacterium and help to attach with female bacterium by forming a tube like structure called conjugation tube. Transfer of genetic information from one cell to other via conjugation tube called conjugation. Fimbriae help to make conjugation tube called sex pili and protein called pilin.
   - **Fimbrial adhesin** or **pilus adhesin:** It is one type of adhesin which helps the bacterium to attach with host cells and helps in virulence. Following are different subtypes with special name.
     - Colonization Factor Antigen: It presents in ETEC and helps to adhere with intestinal epithelial cells.
     - Mannose resistant fimbria: It presents in EHEC, especially in nephritogenic strain and helps in attachment with uroepithelial cells.
     - Toxin Co-regulated Pilus (TCP): It presents in *Vibrio cholerae* and constantly binds the *Vibrio* with intestinal epithelial cells.
     - Surface agglutinogen: It favors the adhesion of *Bordetella pertussis* with respiratory epithelium.
     - Hemagglutinin adhesin (160 KDa): It favors the adhesion of *Staph. saprophyticus* with uro-epithelial cells.
   - **Contractile fimbriae:** These are filamentous appendages found in *Eikenella corrodens*. They are useful for unusual jerking or twitching motility which helps bacilli to spread and to produce corrosive effect (corroding appearance) over blood agar.
   - **Hemagglutinating pili (HA-pili):** They participate in hemagglutination reaction. They serve as organ of adhesion as mentioned above. Hemagglutination is affected by mannose, and it identified two subtypes.
     - Type-I called mannose sensitive: Inhibition of hemagglutination by bacteria after incubation with mannose.
     - Type-II called mannose resistant: No Inhibition of hemagglutination by bacteria after incubation with mannose, e.g., gonococci.

Differences between flagella and fimbriae: Follow Table 6.3.

Biochemistry: Fimbriae are made up by a protein called fimbrilin (*Neisseria gonorrhoeae*) or by techoic acid (*Staphylococcus*) or by lipotechoic acid (*Enterococcus*) or by lipotechoic acid mixed with M. protein (*Streptococcus*).

Functions: Follow classification.

Demonstration of fimbriae: Fimbriae are demonstrated by EM, liquid culture showing pellicle formation and

TABLE 6.3: Differences between flagella and fimbriae

Features	Flagella	Fimbriae
Length	Long (3–20 µm)	Short (0.5 µm)
Thickness	Thick (0.1–0.013 µm)	Thin (0.001 µm or 10 nm)
Made up by	Protein-flagellin	Protein-fimbrilin
Numbers	Less	More
Organ of adhesion	No	Yes
Organ of conjugation	No	Yes
Organ of locomotion	Yes	No

hemagglutination test which is used for demonstration of pili in *E. coli*, *K. pneumoniae*, etc.

Clinical significances

- Immune interference: Fimbriae are antigenic in nature and cross react with many related bacteria.
- Diagnostic: They help to form pellicle in liquid media by adhering the bacteria on surface. Pellicle is diagnostically useful.
- Vaccine preparation: *E. coli* diarrhea in calves and piglets and gonorrhea in humans are prevented by using vaccine prepared from fimbrial antigens.

Spores

Definition: Highly resistance, resting and metabolically inactive stage of bacterium in unfavorable conditions called spore.

Properties: Spores are resistant to heat, chemicals, cold, desiccation and other abnormal environmental conditions. Spores are formed inside the bacterial cell called endospores. Each bacterium forms one spore which on germination forms a single vegetative cell; hence it is not a method of reproduction.

Classification: Following are different spore bearing or sporing bacteria.

1. **According to O$_2$ requirement of sporing bacteria**
   - Aerobic spore bearer: *Sporosarcina* spp., (GPC), *Bacillus* spp., (GPB) and *Micromonosporum* spp., (GPB).
   - Anaerobic spore bearer: *Clostridium* spp., (GPB) and *Sporolactobacillus* spp., (GPB).

2. **According to shape and location of spore (Fig. 6.12)**
- **Central/equatorial spore:** It gives spindle shape appearance to bacilli like *Cl. bifermentans*, *Cl. sordelii* and *Bacillus* spp. Following are two different forms of central spore.
   - Spherical (non-bulging) spore: It is same in size of bacillary body like in *Bacillus* spp.
   - Oval (bulging) spore: It is larger than the size of bacillary body.
- **Subterminal spore:** It gives club shape appearance to bacilli like *Cl. perfringens* (*Cl. welchii*), *Cl. novyi, Cl. septicum, Cl. sordelii Cl. bifermentans, Cl. botulinum, Cl. sporogenes*, *Cl. histolyticum* and *Bacillus* spp. Following are two different forms of subterminal spore.

- Spherical (non-bulging) spore: It is same in size of bacillary body like in *Bacillus* spp.
- Oval (bulging) spore: It is larger than the size of bacillary body.
- **Terminal spore:** Two subtypes:
   - Spherical (non-bulging) spore: It gives drum stick appearance to bacilli like *Cl. sphenoides, Cl. tetani* and *Cl. tetanomorphum*.
   - Oval (bulging) spore: It gives tennis racket appearance to bacilli like *Cl. difficile, Cl. tertium* and *Cl. cochlearum*.

> **Notes:** Mixed spores
> - *Cl. bifermentans* and *Cl. sordelii* produce both central and subterminal spores.
> - *Bacillus* spp., produce both central or subterminal, oval or elliptical spores and same in size of bacillary body (non-bulging).

Function: Spores protect bacteria in unfavorable condition.

Sporicidal methods/agents: Methods/agents which kill the spores are called sporicidal methods/agents like

1. **Physical methods (sterilization)**
   - Autoclave (121°C at 15 lbs for 15 min.)
   - Hydroclaving
   - Hot air oven

2. **Chemical methods (disinfection)**
   - Formaldehyde
   - Glutaraldehyde (2% → cidex): Slow action in 3–10 hours
   - Iodine (1%): At higher concentration
   - Chlorine tablets and sodium hypochlorite (at higher concentration)
   - Ethylene oxide (EO)
   - H$_2$O$_2$
   - KMnO$_4$ (4%)
   - O-phthalic acid
   - Paracetic acid
   - Beta propiolactone
   - Ozone

3. **Physical + chemical methods (Chemosterilization):** Plasma sterilization.

Sporulation: Conversion of vegetative bacterium in to spore in unfavorable condition called sporulation. It begins with appearance of clear area in the protoplasm of the cell that gradually opaque with condensation of nuclear material called forespore or core. The cell membrane grows to form spore membrane or spore wall. The cell wall grow around spore wall to form cortex and multilayered tough spore coat. Some spores have extra outer covering called exosporium. Spore structure is shown in **Fig. 6.13**.

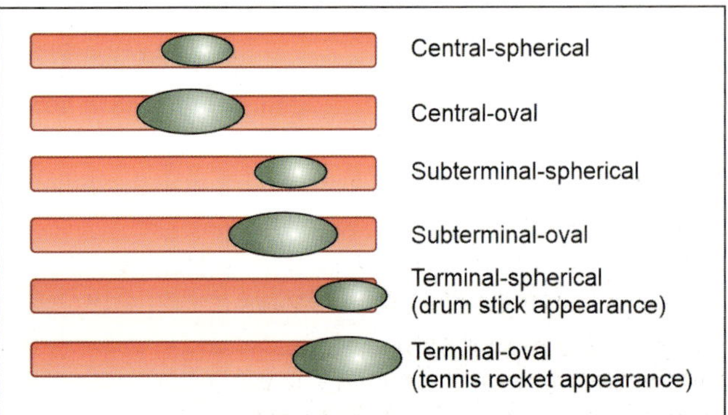

Fig. 6.12: Shape and location of spores

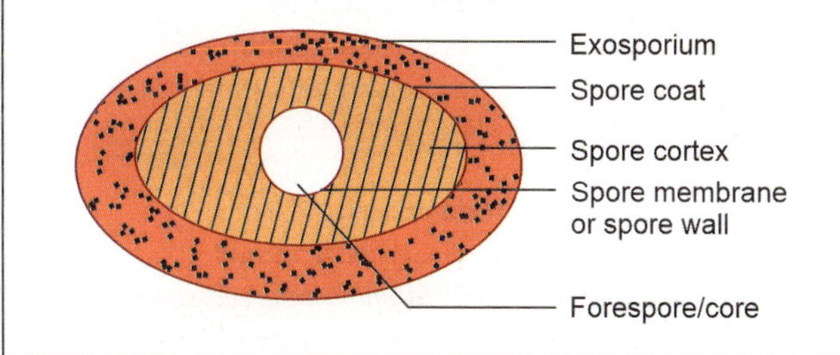

Fig. 6.13: Spore structure

Germination: It occurs in favorable conditions. Three stages are involved:

1. **Activation:** It occurs in nutritionally rich medium. Damage to spore coat is produced by heat, abrasion, acidity, etc.
2. **Initiation:** During this stage a cortex peptidoglycan and a variety of other components are degraded, water is taking up and calcium dipiclonic acid is released.
3. **Outgrowth:** A new vegetative cell with spore protoplast emerges out a period of active synthesis that terminates in cell division.

Uses of spores

1. **Biological indicator:** Spores of certain species of bacteria are used as biological indicator for different methods of sterilization as mentioned below.
   - *Bacillus (Geobacillus) stearothermophilus:* For autoclave.
   - *Cl. tetani* (non-toxigenic strain): For hot air oven.
   - *Bacillus subtilis* subspp. *niger:* For hot air oven.
2. **Bioterrorism:** Endospores of *Bacillus anthracis* were used in 2001 in anthrax bioterrorism attack on USA.

Demonstration of spores

1. **Staining and microscopy**
   - Light microscopy: By using Gram's stain (colored bacillary body with unstained spores) and modified ZN stain (pink colored spore).
   - Phase-contrast microscopy: By using wet mount
   - Other staining and differential methods of spores with fat globules: Follow **Table 6.2**.
2. **Culture**
   - Culture methods for spores of *A. bacillus*: In *A. bacillus* spore production occurs naturally only in soil but never in body. Spore production is promoted in oxalated agar by adding distilled water and 2% NaCl followed by incubation at 25–30°C in presence of O_2. Spore production is inhibited in presence of calcium chloride.
   - Culture methods for spores of *Cl. welchii*: Spores of *Cl. welchii* may develop in animal body. Media used for spore production are Phillips medium, Ellner's medium, Duncan and Strong's medium and alkaline egg medium.

Pleomorphism and Involution Forms

Definitions

- **Pleomorphism:** Certain bacteria show variation in size and shape called pleomorphism, e.g., *Cl. welchii, V. cholerae, Y. pestis, H. influenzae, B. pertussis, Mycoplasma* spp., etc.
- **Involution forms:** When variation in size and shape is present in high salt concentration medium called involution forms, e.g., *N. gonorrhoeae, Y. pestis, Cl. welchii,* etc.

Causes: Involution forms occur in aging culture especially due to high salt concentration. However, pleomorphism and involution forms also occur due to defective cell wall synthesis or by actions of autolytic enzymes.

ACCESS YOURSELF

Essay/Full Question

1. Morphology of bacteria.

Short Notes

1. Arrangement of bacteria
2. Bacterial cell wall/Cell wall appendages
3. Polar bodies
4. L-forms
5. Capsule/bacterial capsule (Answer of capsule contains bacterial portion and fungal portion)
6. Flagella/bacterial flagella (Answer of flagella contains bacterial portion and parasitic portion)
7. Spores/bacterial spores/fungal spores (Answer of spores contains bacterial portion and fungal portion)
8. Fimbriae

Short Questions for Theory/Viva Questions

1. Write the four functions of bacterial cell wall.
2. Write the four differences between gram-positive and gram-negative bacterial cell wall.
3. What is Bovine antigen?
4. What are protoplast and spheroplast?
5. Name the four bacteria showing L-forms.
6. What is selective permeability of plasma membrane?
7. Write the four functions of cytoplasmic membrane.
8. What are mesosomes or chondroides?
9. Define: Protoplasm and cytoplasm.
10. What are volutin granules?
11. Name the four stains for volutin granules.
12. Name the four bacteria contains volutin granules.
13. Write four names of cytoplasmic inclusions.
14. What are fat globules? Name the methods used to differentiate it from bacterial spores.
15. Write the four functions of capsule.
16. Write the four examples of capsulated bacteria.
17. What is Quellung reaction or capsular swelling?

18. Name the methods to increase and to decrease the capsule production in bacteria.
19. Write the four examples of bacteria with peritrichous flagella.
20. What is subpolar flagellum? Write one example.
21. Write the four examples of flagellar parasites.
22. What are endoflagella?
23. Define flagella and bacterial spores.
24. Write the four examples of sporing bacteria.
25. Write the four examples of sporicidal methods.
26. What are pleomorphism and involution forms?

Comments on

1. Endoflagella are not useful in motility.

MCQs for Chapter Review

Cell Wall

1. **Peptidoglycans are present in:**
   a. Gram-positive bacteria
   b. Gram-negative bacteria
   c. Fungi
   d. Protozoa
2. **Techoic acids:**
   a. Found in gram-positive bacteria
   b. Make the outer wall of bacteria
   c. Provide receptor for phage
   d. Influence the permeability of phage
3. **Bacterial cell wall is composed of all:**
   a. Muramic acid   b. Techoic acid
   c. Glucosamine   d. Mucopeptide
4. **The difference between gram-positive and gram-negative organisms is that gram-positive organism contains:**
   a. Techoic acid
   b Muramic acid
   c. N-acetyl neuramic acid
   d. Aromatic amino acids
5. **Acid fastness of tubercle bacilli is attributed to:**
   a. Presence of mycolic acid
   b. Integrity of cell wall
   c. Both of the above
   d. None of the above
6. **Cell wall deficient organisms are:**
   a. *Chlamydia*   b. *Mycoplasma*
   c. *Streptococcus*   d. Anaerobes
7. **Cell wall structure found in all *except*:**
   a. *Staph. aureus*
   b. *Pseudomonas aeruginosa*
   c. *Mycoplasma pneumoniae*
   d. *Corynebacterium diphtheriae*

Cytoplasmic/Plasma Membrane

8. **The cytoplasmic membrane of bacteria is responsible for:**
   a. Selective permeability   b. Motility
   c. Cell division   d. Conjugation
9. **Which is always present in bacteria?**
   a Cell wall   b. Cytoplasmic membrane
   c. Mitochondria   d. Nucleoli

Protoplasm

10. **Metachromatic granules are found in:**
    a. Diphtheria   b. *Mycoplasma*
    c. *Gardnerella vaginalis*   d. *Staphylococcus*

11. **Metachromatic granules are stained by:**
    a. Ponder's stain   b. Negative stain
    c. Gram's stain   d. Leishman's stain

Capsule and Slime Layer

12. **All are capsulated bacteria *except*:**
    a. *Neisseria*   b. *Corynebacterium*
    c. *Haemophilus*   d. *Streptococcus salivarius*
13. **In which of the following organism capsule does not act as a virulence factor?**
    a. *H. influenzae*   b. *Strept. pneumoniae*
    c. *N. meningitidis*   d. *Bordetella pertussis*
14. **Capsulated organism:**
    a. *Candida*   b. *Klebsiella*
    c. *Proteus*   d. *Cryptococcus*
    e. *Histoplasma*
15. **True about bacterial capsule is following *except*:**
    a. Prevents phagocytosis
    b. Stained by Gram's stain
    c. Protects bacteria from lytic enzymes
    d. Lost by repeated subculture
16. **Bacterial capsule is made up of:**
    a. Monosaccharide   b. Polysaccharide
    c. Long-chain fatty acid   d. Small-chain fatty acid
17. **Which of these bacteria has a capsule with polypeptide structure?**
    a. *Bacillus anthracis*   b. *Streptococcus pneumoniae*
    c. *Neisseria gonorrhoeae*   d. *Neisseria meningitidis*
18. **Quellung reaction is useful to detect:**
    a. Capsule   b. Flagella
    c. Spores   d. Fimbriae

Flagella

19. **Bacteria with tuft of flagella at one end called:**
    a. Monotrichate   b. Peritrchate
    c. Bipolar   d. lophotrichate
20. **Darting motility is shown by:**
    a. *L. monocytogenes*   b. *Proteus vulgaris*
    c. *Borrelia*   d. *V. cholerae*
21. **Darting motility occurs in *V. cholerae* also found in:**
    a. *Shigella*   b. *Campylobacter jejuni*
    c. Pneumococcus   d. *Bacillus anthracis*
    e. *Aeromonas*
22. **Following are motile organism *except*:**
    a. *Proteus*   b. *C. diphtheriae*
    c. *T. pallidum*   d. *Clostridium welchii*
23. **Tumbling motility is shown by:**
    a. *L. monocytogenes*   b. *Proteus vulgaris*
    c. *Borrelia*   d. Clostridia
24. **Bacteria are motile due to:**
    a. Flagella   b. Fimbriae
    c. Both   d. None
25. **Flagella not true:**
    a. Locomotion   b. Attachment
    c. Protein in nature   d. Antigenic
26. **Bacterium with endoflagella is:**
    a. *S.* Typhi   b. *E. coli*
    c. *T. pallidum*   d. *S. minus*
27. **Gliding motility is seen in:**
    a. *Clostridium*   b. *Listeria*
    c. *Mycoplasma*   d. *Campylobacter*

28. Example of bacterium with central spore is:
   a. *Cl. tetani*   b. *Cl. bifermentans*
   c. *Cl. welchii*   d. *Cl. tertium*

29. Subterminal spores are found in:
   a. *Clostridium sordelii*   b. *Clostridium sporogenes*
   c. *Clostridium difficile*   d. *Clostridium tertium*
   e. *Clostridium botulinum*

30. Drum-stick appearance is seen in:
   a. *Cl. sphenoides*   b. *Cl. tetani*
   b. *Cl. tetanomorphum*   d. *Clostridium tertium*

31. Spores are formed by all *except*:
   a. *E. coli*   b. *B. anthrax*
   c. *B. cereus*   d. None

32. Bacterial spores are best destroyed by:
   a. UV rays   b. Autoclaving at 121°C for 20 min.
   c. Hot air oven   d. Infrared rays

33. Sporicidal agents are:
   a. Glutaraldehyde   b. Ethylene oxide
   c. Formaldehyde   d. Bezalkonium chloride
   e. Chlorine

Answers and Explanation of MCQs

1. a and b
- Peptidoglycan is present in gram-positive (thick) and gram-negative bacteria (thin), but not in fungi and protozoa.

2. a

3. a, c, d
- Follow section, **bacterial cell wall (cell wall of gram-positive bacteria and Table 6.1 for explanation of answers of MCQs 2–3.**

4. a
- Follow **Table 6.1** for explanation.

5. c
- Peptidoglycan-polysaccharide-mycolic acids complex forms the skeleton in cell wall of mycobacteria, so acid fastness in mycobacteria is not the property of mycolic acid alone, but depends also on the integrity of cell wall.
- For more explanation follow the section bacterial cell wall (cell wall of acid-fast bacteria).

6. b

7. c
- Follow section, **bacterial cell wall (inhibition of cell wall synthesis)** for explanation of answers of MCQs 6–7.

8. a
- Motility is a function of flagella, cell division is a function of reproductive system mostly by binary fission and conjugation is due to sex pili.

9. b
- Follow section, **cytoplasmic/plasma membrane (properties)** for explanation.

10. a, c

11. a
- Follow section, **cytoplasm (polar bodies)** for explanation of answers of MCQs 10–11.

12. b

13. d

14. b, d

15. b

16. b

17. a

18. a
- Follow section, **capsule and slime layer** for explanation of answers of MCQs 12–18.

19. d
- Follow **Flowchart 6.1** for explanation.

20. d
- *Vibrio cholerae* is actively motile with single sheathed polar flagellum called darting motility.

21. b
- *Shigella*, pneumococcus and *Bacillus anthracis* are nonmotile while *Aeromonas* is motile in nature but showing motility other than darting.

22. b and d
- Follow section, **flagella (Flowchart 6.1) and notes** for explanation.

23. a

24. a

25. b
- Flagella is an organ of locomotion, it is made up by protein called flagellin and antigenic in nature. Fimbria is the organs of attachment.

26. c

27. c
- Follow section, **flagella (notes)** for explanation of answers of MCQs 26–27.

28. b

29. a, b, e

30. a, b, c

31. a
- Follow section, **bacterial spores (shape and location of spores)** for explanation of answers of MCQs 28–31.

32. b

33. a, b, c, e
- Follow section, **bacterial spores (sporicidal agents)** for explanation of answers of MCQs 32–33.

CHAPTER 7

Physiology of Bacteria

Chapter Outline

BACTERIAL NUTRITION AND GROWTH FACTORS

Definition

Nutrition is the process by which the organic and inorganic substances are obtained from surrounding environment and used for growth, metabolism and multiplication.

Nutrients or Growth Factors

Definition: Organic and inorganic substances used for nutrition are called nutrients or growth factors.

Types: Two types of nutrients:

- **Macronutrients:** These are required in relatively large quantities and play important role in cell structure and metabolism.
- **Micronutrients:** These are required in small quantities for functioning of certain enzyme systems.

Examples of nutrients required for bacteria

- **Water:** It holds 80% weight of total bacterial weight. Remaining is dry weight. It acts as vehicle to carry the nutrients to and fro from bacterial cell.
- **Carbon, nitrogen and energy:** Bacteria classified into two groups:
  1. **Lithotrophs (autotrophs):** They are able to utilize atmospheric CO_2 and N_2 for growth. They are medically less important because they survive independently in soil and water (called saprophytes). They are concerned with agriculture field and with soil fertility such as nitrogen fixing bacteria in soil. They require water, CO_2 and N_2 for growth. They have two subgroups based on energy requirement:
     - Photolithotrophs: They got energy from light.
     - Chemolithotrophs: They got energy from oxidation and reduction of chemicals like inorganic substances (Anions → phosphate, sulfate, Cations → sodium, potassium, magnesium, iron, manganese, calcium).
  2. **Organotrophs (heterotrophs):** They are not able to utilize atmospheric CO_2 and N_2 for growth. These are bacteria which are unable to synthesize their own metabolites and depend on preformed organic compounds such as carbohydrates, amino acids, nucleotides, lipids, vitamins (B_1, B_2, B_4, B_6, folic acid, B_{12}) and coenzymes for growth. Organic compound may be essential (no growth in their absence) or accessory (when they enhance growth). They are medically important to cause diseases. They have two further groups based on energy requirement:
     - Photo-organotrophs: They got energy from light.
     - Chemo-organotrophs: They got energy from oxidation and reduction of chemicals like inorganic substances.
- **Bacterial vitamins:** Certain fastidious bacteria required essential nutrients for growth in media called bacterial vitamins as shown in **Table 7.1**. They are similar to mammalian vitamins.

TABLE 7.1: Bacterial vitamins

Vitamins	Bacteria
Biotin	*Leuconostoc* spp.
B_{12}	*Lactobacillus* spp.
Folic acid	*Enterococcus faecalis*
Pantothenic acid	*Morgnella morganii*
B_6	*Lactobacillus* spp.
Niacin (nicotinic acid)	*B. abortus* and *H. influenzae*
B_2	*B. anthracis*

Carbon Dioxide

All bacteria require small amounts (1–2%) of carbon dioxide for growth. This requirement is usually met by the carbon dioxide present in the atmosphere, or produced endogenously by cellular metabolism. Energy required for utilization of CO_2 may come from light or from oxidation and reduction of chemicals like inorganic substances. However, some bacteria require higher concentration (5–10%) of CO_2 **called capnophilic bacteria** like *Brucella abortus, Neisseria* spp., *Strept. pneumoniae, Listeria monocytogenes, Helicobacter pylori, Campylobacter jejuni, Gardnerella vaginalis, Aggregatibacter* (formerly *Actinobacillus*) *actinomycetemcomitans, Eikenella corrodens, Capnocytophage* spp., *Legionella pneumophila*, etc.

Oxygen

Based on oxygen requirement, bacteria can be classified into six types:

1. **Obligate (strict) aerobes**
   - **Definition:** They grow only in presence of O_2.
   - **Examples:** *Micrococcus, Neisseria meningitidis, Moraxella lacunata, Acinetobacter baumanii, Corynebacterium jakeium, M. tuberculosis, Nocardia* spp., *Listeria monocytogenes, Serratia merscence, V. cholerae, P. aeruginosa, Bukholderia pseudomallei, Bordetella pertussis, Brucella abortus, Francisell tularensis, Alkaligenes faecalis*, etc.

2. **Obligate (strict) anaerobes**
   - **Definition:** They grow only in absence of O_2 (they die in presence of oxygen).
   - **Subtypes and examples:** Two subtypes:
     - Sporing anaerobes: *Cl. botulinum, Cl. tetani* and *Cl. novyi.*
     - Nonsporing anaerobes: For examples follow **Ch. 72.**
   - **Reasons of oxygen toxicity in anaerobes:** In presence of oxygen, hydrogen peroxide and other toxic peroxides accumulate. The enzyme catalase which splits hydrogen peroxide is present in most aerobic bacteria but absent in anaerobes. Some other enzymes like peroxidase and superoxide dismutase are also required to inactivate the oxygen products, but their deficiency in anaerobic bacteria make the oxygen toxic for anaerobes. Another reason is that obligate anaerobes possess essential enzymes that are active only in reduced state.

3. **Facultative anaerobes**
   - **Definition:** They are aerobes but can grow in absence of O_2.
   - **Examples:** *Strept. pyogenes, Staph. aureus, Strept. pneumoniae, N. gonorrhoeae, E. coli, H. influenzae*, etc.

4. **Facultative aerobes**
   - **Definition:** They are anaerobes, but can grow in presence of O_2.
   - **Examples:** *Lactobacillus* spp., etc.

5. **Aerotolerant**
   - **Definition:** They can tolerate oxygen for some times, but cannot grow in presence of O_2.
   - **Examples:** *Cl. histolyticum, Propionibacterium acne, P. avidum, P. granulosum* and *Arcobaterium* spp.

6. **Microaerophilic**
   - **Definition:** They require O_2 less than the environmental O_2 (about 5–10%).
   - **Examples:** *Clostridium welchii, Mycobacterium bovis, Listeria monocytogenes, Erysipelothrix rhusiopathiae, Campylobacter jejuni, Helicobacter pylori, Streptobacillus moniliformis, Spirillum minus, Aggregatibacter actinomycetemcomitans, Borrelia recurrentis, Borrelia bergdorferi, Leptospira interrogans,* etc.

Temperature

Temperature affects the growth and viability of bacteria.

Effect on growth

- **Definition:** For each species, there is a temperature range and growth does not occur above the maximum temperature or below the minimum temperature. The temperature at which growth occurs best is called the optimum temperature.
- **Types of bacteria:** Bacteria are classified as follows as per temperature effect on growth:
  1. Psychrophilic (cryophilic) bacteria: Bacteria which can grow best at temperature below 20°C are called psychrophilic (cryophilic) bacteria, some of them even growing at temperature as low as –7°C. For example saprophytes. They may cause spoilage of refrigerated food.
  2. Mesophilic bacteria: Bacteria which grow best at temperature of 25–40°C are called mesophilic bacteria. For example, all pathogenic bacteria. Optimum temperature for this group is 37°C.
  3. Thermophilic bacteria: Bacteria which grow best at high temperature around 55–80°C are called thermophilic bacteria. For example, *Bacillus steareothermophilus* and *Campylobacter* spp. They may cause spoilage of under processed canned food.
  4. Extremely thermophilic bacteria: Bacteria which grow at higher temperature at 250°C are called extremely thermophilic bacteria. For example, bacterial spores.

Effect on viability: Heat is an important method for destruction of microorganisms. Moist heat causes coagulation and denaturation of proteins and dry heat causes oxidation and charring. Moist heat is more lethal than dry heat. The lowest temperature that kills a bacterium under standard conditions in a given time is called the thermal death point. Under moist condition, most vegetative and hemophilic bacteria have a thermal death point is between 50°–65°C and for spores is between 100°–120°C. At low temperature

Essentials of Medical Microbiology

some species die rapidly but most survive well. Storage in the refrigerator (3°–5°C) or deep freeze cabinet (–30° to –70°C) is used for preservation of cultures. Rapid freezing as with solid carbon dioxide or the use of a stabilizer such as glycerol, minimize the death of cells on freezing.

> **Notes: Thermoduric bacteria**
>
> **Definition:** Bacteria which can survive, but do not grow to varying extents of the pasteurization process or pasteurization temperature called thermoduric bacteria.
>
> **Examples:** *Bacillus, Clostridium, Micrococcus, Streptococcus, Lactobacillus, Enterococcus* and few gram-negative rods.
>
> **Clinical significances:** The sources of contamination are poorly cleaned and sanitized utensils and equipment on farm and processing plants. These bacteria contribute to significantly higher standard plate count on pasteurized milk.
>
> **Thermoduric count:** It has been used in the dairy industry primarily as a test of care employed in utensil sanitation and as a means of detecting sources of organisms responsible for high counts in the final product.

H-ion Concentration (pH)

Range of pH: Each species has a pH range, above and below which, it cannot survive. Bacteria are sensitive to variations in pH. Strong solutions of acid or alkali (5% HCl) readily kill most bacteria, though mycobacteria are exceptionally resistant to them. pH at which, bacteria can grow best is called optimum pH.

Pathogenic bacteria: Following are different types of pathogenic bacteria as per pH effect:
1. Acidic pH: Some bacteria grow at acidic pH (<6.0) are called acidophilic bacteria. For example, *Lactobacillus* spp.
2. Neutral pH: Most of the pathogenic bacteria grow at neutral pH (7.2–7.6).
3. Alkaline pH: Some bacteria grow at alkaline pH (>7.8) like *V. cholerae*.

Moisture and Drying

Moisture is absolute requirement for growth of bacteria. The capacity to survive in dry environment varies from organism to organism. *N. gonorrhoeae* and *T. pallidum* die quickly in dry conditions. *Staph. aureus* and *M. tuberculosis* can survive drying for weeks and months. Bacterial spores can survive in drying for decades. Drying in vacuum in the cold condition is a method for the preservation of bacteria, viruses and many labile biological materials called freeze drying or lyophilization.

Light

Some bacteria required light for energy like photo-lithotrophs/photo-organotrophs. Light also required for certain bacterial functions like pigment production. Following are different types of bacteria as per requirement of light:

1. Photochromogenic bacteria: Bacteria produce the pigment only in light, not in the dark called photo-chromogenic bacteria. For examples *M. kansasii, M. marinum, M. simiae, M. asiaticum*, etc.
2. Scotochromogenic bacteria: Bacteria produce the pigment in light and in dark called scotochromogenic bacteria. For examples, *M. scrofulaceum, M. gordonae* and *M. szulgai* (scotochromogen at 37°C and photochromogen at 25°C).
3. Nonphotochromogenic (nonchromogenic) bacteria: Bacteria which do not produce pigment in presence or absence of light called nonphotochromogenic (nonchromogenic) bacteria. For examples, *M. avium, M. intracellulare, M. xenopi* and *M. ulcerans*.

Osmotic Effect

Bacteria more tolerate osmotic variation then most other cells due to the mechanical strength of their cell walls. Sudden exposure to hypertonic solutions may cause loss of water and shrinkage of cell called plasmolysis. It occurs more in gram-negative than in gram-positive bacteria. Sudden transfer to distilled water or any other hypotonic solution may cause water flow inside the cell and swelling and rupture of cell called plasmoptysis.

Mechanical and Sonic Stress

Though bacteria have tough cell walls, they may be ruptured by mechanical stress such as grinding, vigorous shaking with glass beads or ultrasonic vibration.

REPRODUCTION (MULTIPLICATION)

Methods of Reproduction

Following are different methods of reproduction.
1. **Binary fission:** Bacteria multiply by binary fission after attaining certain size and one mother cell gives two daughter cells. Nuclear division occurs at 1st followed by cell division. Cell division initiated with formation of transverse septum across the cell. Normally daughter cell completely gets separated from mother cell; however in some it attaches to mother cell due to incomplete binary fission like in *C. diphtheriae*.
2. **Budding:** Some bacteria also reproduced by budding in which a small bud (small pouch like structure) develops from the mother bacterium, which later gets separate and grows further to form a new bacterium. For example, *F. tularensis, M. pneumoniae*, etc.
3. **Filamentous formation (?),** e.g., *F. tularensis*.

Reproduction Time

Definition: The interval of time between two cell divisions or the time required for a bacterium to give rise to two daughter cells under optimum conditions is called generation time or population doubling time. One bacterium gives 10^{21} progeny in 24 hours.

Generation time of few bacteria: Follow **Table 7.2**.

TABLE 7.2: Generation time of different bacteria	
Organisms	**Generation time**
Cl welchii (GPB)	10 minutes
Coliform bacilli (lactose fermenters) like *E. coli* and *K. pneumoniae* (GNB)	20 minutes
Leptospira spp., (GNB)	4–8 hours in animals and 12–16 hours in media
M. tuberculosis (GPB)	20 hours
T. pallidum (GNB)	30–33 hours
M. leprae (GPB)	13–15 days

Effective Factors on Reproduction

1. **Nutrients and toxic products**
   - **Depletion of nutrients and accumulation of toxic products:** When bacteria grow in artificial prepared tnutritional media, multiplication is arrested after few cell divisions due to depletion of nutrients or accumulation of toxic products. Such culture called batch culture.
   - **Replacement of nutrients and removal of toxic products**
     – By using special device: A special device is used which replaces the nutrients and removes toxic products during cell division. Such culture called continuous culture. It used to maintain bacteria for industrial or research purposes.
     – In host tissue: Tissue replaces the nutrients and removes toxic products during cell division, but bacteria have to compete with host defense. It is intermediate between batch culture and continuous culture.
2. **Inhibition of cell division:** By certain drugs like penicillin.

Bacterial Count

Definition: Bacterial multiplication in number called bacterial count.

Methods: Two types of bacterial counts like total count and viable count.

1. **Total count:** Total number of cells in the sample irrespective of whether they are living or dead called total count. Total counting can be obtained by following methods:
   - Direct counting under the microscope using counting chamber
   - Counting in an electronic device as in the coulter counter
   - Using stained smear prepared by spreading a known volume of the culture over a measured area of a slide
   - Comparing relative number, in a smears of the culture mixed with known numbers of other cells

   - Opacity measurement using an absorptiometer or nephalometer
   - Separation of cells by centrifugation or filtration and measuring their wet or dry weight
   - Chemical assay of cell components like nitrogen.
2. **Viable count:** Number of living cells which are capable to multiply called viable count. Viable counting can be obtained by following methods:
   - Dilution method: It is done by using liquid media. The suspension is diluted to a point beyond which unit quantities do not yield growth when inoculated in to suitable liquid media. Several tubes are incubated with varying dilutions and the viable count calculated statistically from the number of tubes showing growth. It do not give accurate value is the drawback of this method. It is used for water culture to estimate the presumptive coliform count in drinking water.
   - Plating method: It is done by using solid media. Appropriate dilutions are inoculated on solid media, either on the surface of plates or as pour plates, followed by incubation. The numbers of colonies develop can give an estimate of the viable count. The method was employed and is described by Miles and Misra (1938).

Bacterial Growth Curve

Definition: When a bacterium is inoculated into a suitable liquid medium and incubated, it under goes the multiplication. Count the bacteria at regular interval and plotted it against the time, a growth curve is obtained called bacterial growth curve or kinetic of bacterial growth.

Phases: The curve shows the four phases during which morphological and physiological changes occur in cells are shown in **Fig. 7.1**.

1. **Lag phase**
   - It is the initial period.
   - It required to adapt the new environment, to synthesize the necessary enzymes and metabolites for multiplication.

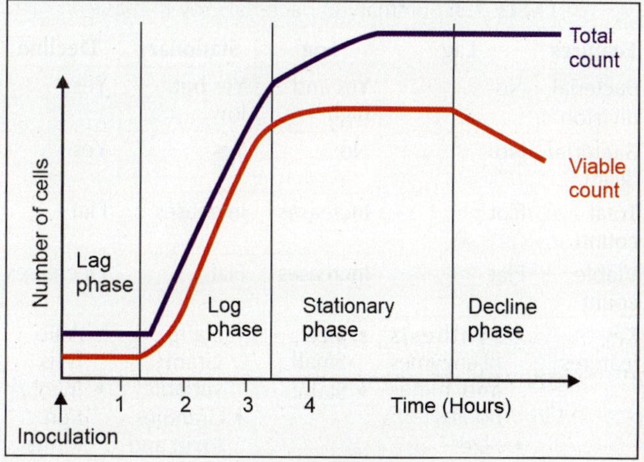

Fig. 7.1: Bacterial growth curve

- Immediately following the seeding of a culture medium there is no increase in numbers, though there may be an increase the size of a cell.
- The duration of the lag phase varies from 1–4 hours with the type of species, size of the inoculum, nature of the medium, presence of growth factors and environmental factors like temperature, O_2, CO_2, etc.

2. **Log phase/logarithmic phase/exponential phase**
   - As the time pass, cells start to divide and their numbers increase.
   - If the logarithm of the viable count is plotted against time a straight-ascending line will be obtained.
   - Smaller size and uniform staining occurs in this phase.

3. **Stationary phase**
   - After a varying period of exponential growth, cell division stops due to depletion of nutrients and accumulation of toxic products (batch culture).
   - Progeny cells formed are enough to replace the number of cells that die.
   - Viable count remains stationary as equilibrium exists between the dying and new cells.
   - Variable gram reaction, irregular staining due to storage granules and sporulation.

4. **Decline phase**
   - During this phase viable count decreases due to cell death.
   - Cell death is due to nutritional exhaustion, accumulation of toxic products (secondary metabolic products) and by autolytic enzymes.
   - Total count is runs parallel to viable count up to stationary phase, after that it continues steadily without any decline till autolysis will start.
   - After autolysis, total count also decreases.
   - Involution forms occur due to actions of autolytic enzymes, ageing culture or by defective cell wall synthesis.

Summary of bacterial growth curve: Follow **Table 7.3**.

Mnemonic
- SS: **S**tationary phase and **S**porulation
- In-In: Decl**In**e phase and **In**volution forms

METABOLISM

Steps
- Absorption of food material.
- Breakdown and utilization.
- Elimination of metabolic end products.

Metabolic Pathway

Bacteria are differentiated in their ways of metabolism.

1. **Aerobic bacteria:** They obtain their energy by oxidation (aerobic respiration). Oxygen as the ultimate hydrogen (electron) acceptor. O_2 accepts the carbon and energy to form CO_2 and H_2O. During this process energy rich phosphate bonds are released which convert ADP to ATP called oxidative phosphorylation.

2. **Anaerobic bacteria:** They obtain their energy by fermentation (anaerobic respiration). Nitrite or sulfite is hydrogen (electron) acceptor. Fermentation is an anaerobic utilization of sugar to produce acids (like lactic acid, formic acid, pyruvic acid, etc.), alcohols and gases (CO_2, H_2, etc.). During this process, energy rich phosphate bonds are released which convert ADP to ATP called substrate level phosphorylation.

3. **Facultative anaerobes:** They utilize both pathways.

Redox Potential (Oxidation–Reduction Potential or Eh)

Oxidizing agents are substances, which are capable of accepting electrons. Reducing agents are substances that are able to lose electrons. The capability of substance to accept or to lose the electron is called oxidation-reduction potential. It is abbreviated as *Eh* and measured in millivolts. It is higher in oxidized substances and lower in reducing agents. The strict anaerobes require low redox potential—less than 0.2 volts; however, the redox potential of most of the media exposed to air is +0.2 to +0.4 volts.

TABLE 7.3: Summary of bacterial growth curve

Features	Lag	Log	Stationary	Decline
Bacterial division	No	Yes and high	Yes but low	Yes
Bacterial death	No	No	Yes	Yes
Total count	Flat	Increases	Increases	Flat
Viable count	Flat	Increases	Flat	Decreases
Key features	• Synthesis of enzymes and metabolites • Size: Maximum	• Size: Small • Stain:	• Stain: Gram variable • Granules, toxin and spores	• Autolysis • Involution forms

Essay/Full Question

1. Physiology of bacteria.
2. Environmental factors affecting growth of bacteria.

Short Notes

1. Effect of oxygen on bacteria.
2. Effect of temperature on bacteria.
3. Bacterial count.
4. Bacterial growth curve.

Short Questions for Theory/Viva Questions

1. What is a bacterial vitamin?
2. What are capnophilic bacteria?
3. Write four examples of capnophilic bacteria.

4. Write four examples of strict aerobic bacteria.
5. Write four examples of strict anaerobic bacteria.
6. Write four examples of microaerophilic bacteria.
7. What are plasmolysis and plasmoptysis?
8. What is generation time?
9. Write the generation time for following bacteria: *E. coli*, *M. leprae*, *M. tuberculosis* and *T. pallidum*.
10. What are thermoduric bacteria? Write two examples.
11. What are batch culture and continuous culture?
12. Name the four phases of bacterial growth curve.
13. Name the phases of bacterial growth where sporulation and involution forms can develop.
14. Name the two methods for multiplication (reproduction) and two methods for metabolism in bacteria.
15. What is Eh or Redox potential?

Comments on

1. Oxygen acts as toxin for strict anaerobic bacteria.
2. Bacteria tolerate osmotic variation more than any other cells.
3. Bacterial multiplication is arrested after certain divisions in nutritional media.
4. Cell division stops in stationary phase of bacterial growth curve.

MCQs for Chapter Review

Bacterial Nutrition and Growth Factors

1. **What percentage of mass of bacterial cell is its dry weight?**
   a. 10–15% b. 20–25%
   c. 30–35% d. 40–45%

Environmental Factors

2. **Obligate anaerobes cannot withstand oxygen because of absence of:**
   a. Superoxide dismutase b. Catalase
   c. Peroxidase d. Cytochrome oxidase
3. **Which of the following bacterium is classified as facultative anaerobe?**
   a. *Pseudomonas* b. *Bacteroides*
   c. *Escherichia* d. *Clostridia*
4. **Strict anaerobe(s) is/are:**
   a. *Cl. novyi* b. *Cl. botulinum*
   c. a + b d. None of above
5. **Obligate anaerobes are all *except*:**
   a. *Cl. botulinum* b. *Eikinella corrodens*
   c. *Bacteroides* d. *H pylori*
6. **Which of the following is microaerophilic?**
   a. *Campylobacter* b. *Vibrio*
   c. *Bacteroides* d. *Pseudomonas*
7. **Mesophilic organisms are those that grow best at temperature of:**
   a. –20°C to –7°C b. –7°C to +20°C
   c. 25°C to 40°C d. 55°C to 80°C
   e. 90°C to 120°C

8. **Generation time for *M tuberculosis* is:**
   a. 12 days b. 5 minutes
   c. 20 hours d. 4 hours
9. **Generation time for lepra bacilli is:**
   a. 12 days b. 5 minutes
   c. 10 hours d. 24 hours
10. **Correct sequence of bacterial growth curve:**
    a. Log phase → Lag phase → Stationary phase → Decline phase
    b. Lag phase → Log phase → Stationary phase → Decline phase
    c. Stationary phase → Lag phase → Log phase → Decline phase
    d. Lag phase → Exponential phase → Log phase → Death phase
    e. Exponential phase → Lag phase → Death phase → Stationary phase
11. **True regarding lag phase:**
    a. Time taken to adopt in the new environment
    b. Growth occurs exponentially
    c. The plateau in lag phase is due to cell death
    d. It is the 2nd phase in bacterial growth curve
12. **Sporulation occurs in:**
    a. Lag phase
    b. Log phase
    c. Stationary phase
    d. Decline phase

Answers and Explanation of MCQs

1. **b**
- Follow section, **bacterial nutrition and growth factors (nutrients or growth factors → water)** for explanation.
2. **a, b, c**
- Follow section, **environmental factors (oxygen → Reasons of oxygen toxicity in anaerobic bacteria)** for explanation.
3. **c**
- *Pseudomonas* is strict aerobe while *Bacteroides* and *Clostridium* are strict anaerobes.
4. **c**
5. **b, d**
- Follow section, **environmental factors (oxygen → obligate or strict anaerobes)** for explanation of answers of MCQs 4–5.
6. **a**
- *Vibrio* and *Pseudomonas* are strict aerobes while *Bacteroides* is strict anaerobe.
7. **c**
- Follow section, **environmental factors (temperature)** for explanation.
8. **c**
9. **a**
- Generation time of different bacteria is given **Table 7.2**.
10. **b**
11. **a**
12. **c**
- Follow section, **reproduction (multiplication) → Bacterial growth curve** for explanation of answers of MCQs 10–12.

CHAPTER 8

Genetics of Bacteria

Chapter Outline

GENETIC STRUCTURE

The unit or segment of DNA for heredity is called gene. Total component of genes in the cell called genome. Study of heredity and variation called genetics.

Bacterial genome (**Fig. 8.1**) has two parts like chromosomal part and extrachromosomal part. Chromosomal part is a single (haploid), circular coiled body contains two types of nucleic acids such as DNA and RNA. Extrachromosomal part contains plasmid, episome and transposon (jumping gene).

> **Note:** Viral nucleic acid and meaning of haploid and diploid
>
> **Viral nucleic acid:** Viruses contain either DNA or RNA, but not both.
>
> **Meaning haploid and diploid:** Haploid indicates single (n) set of chromosome and diploid indicates double (2n) set of chromosome.

Chromosomal Part

Bacteria have two types of nucleic acids like DNA and RNA from chromosomal part as described below.

DNA

Full form: Deoxyribo Nucleic Acid.

History: DNA model was described by James Watson and Francis Crick (Watson and Crick) in 1953.

DNA structure

- **DNA model:** It is shown in **Fig. 8.2a.** DNA is single, spiral shape and about 1000 μm in length when straightened. It is composed of two strands (double helix) with complementary nucleotides. Length is expressed in kilobase (1 kb = 1000 base pairs). Bacterial DNA is about 4000 kb long while human DNA is about 3 million kb long. For the sake of simplicity, it

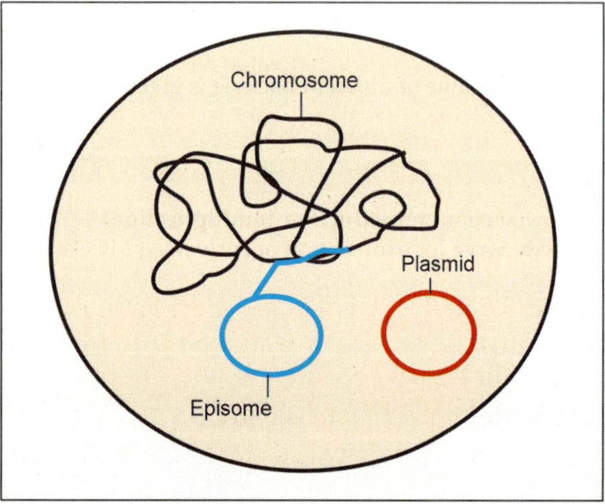

Fig. 8.1: Bacterial genome

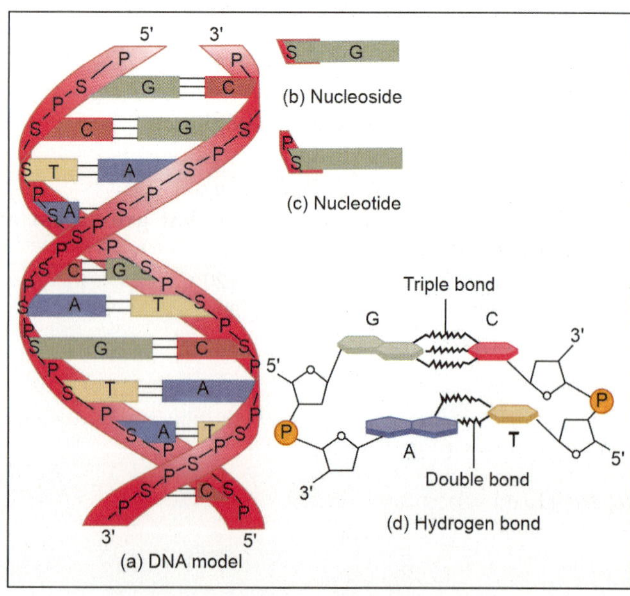

Fig. 8.2: DNA structure

is denoted as a ladder. Rungs of ladder are formed by nitrogen base united with phosphate bond. Side of ladder formed by deoxyribose sugar and phosphate bond. Bacterial DNA is available in circular/loop form without ends (closed form).

- **Nucleoside and nucleotide:** Term nucleoside and nucleotide are frequently used in the DNA model. Nucleoside **(Fig. 8.2b)** denotes sugar (s) plus nitrogen base. Nucleotide **(Fig. 8.2c)** denotes nucleoside plus phosphate bond (P).
- **Nitrogen bases:** Two types of nitrogen bases such as purine includes adenine (A) plus guanine (G) and pyrimidine includes thymine (T) plus Cytosine (C). Numbers of A are equal to that of T and also numbers of G are equal to that of C. Ration of A + T and G + C are constant for each species but vary widely from one species to another.
- **Hydrogen bond:** Both strands of DNA held together by the hydrogen bonds between the nitrogen base of opposite strand
- **Rule of bonding (Fig. 8.2d)**
  - A-T bond: Adenine (A) of one strand bound to the Thymine (T) of opposite strand by double hydrogen bonds.
  - C-G bond: Cytosine (C)) of one strand bound to the Guanine (G) of opposite strand by triple hydrogen bonds.

Functions of DNA

- **Replication:** Copying of DNA.
- **Transcription:** Synthesis of RNA from DNA.

Segment of DNA (gene): Segment of DNA called codon which is specifying for particular polypeptide chain called gene. Bacteria possess 1000–3000 genes which are coded for 1000–3000 polypeptide chains, as each gene is specify the particular polypeptide chain. Each polypeptide chain made up of 1200 nucleotides and each nucleotide consists of 400 amino acids. Chromosome contains 5×10^4 nucleotide pairs.

Introns and exons in DNA: In DNA there are several codons which do not function as gene called introns while codon which codes for particular gene are called exons. During transcription the genome is copied entirely with introns and exons. During translation the introns are excised from RNA before being translated in to the proteins.

RNA

Full form: Ribo nucleic acid.

RNA structure: Similar to DNA with minor differences as shown in **Table 8.1**.

Types: Three types:
1. mRNA: messenger RNA.
2. rRNA: ribosomal RNA.
3. tRNA: transfer RNA.

TABLE 8.1: Similarities and differences between DNA and RNA

Features	DNA	RNA
Acid	Phosphoric acid	Phosphoric acid
Purine bases	A and G	A and G
Pyrimidine bases	T and C	U and C
Sugar	D-2-Deoxyribose	D-Ribose

Function: It synthesizes the protein called translation.

Genetic Codons

Sequence of genetic information stored in the DNA called genetic codon. It was described by Nirenberg and Khorana in 1968. Each codon consists of a sequence of three nitrogen bases, so called triplet codon or triplet. Each triplet codon codes for single amino acid. For example, ACG codes for threonine. Codon is the structural terminology while code is the functional terminology. Following are the different types of codons:

1. **Sense codons:** There are total 64 codons and out of which 61 codons are code for 20 essential amino acids called sense codons.
2. **Nonsense codons (stop codons):** Out of 64 codons, 3 codons do not code for any amino acid called nonsense codons. They terminate the elongation of polypeptide chain so called stop codons. For examples UAA, UAG and UGA.
3. **Start codon:** First codon which starts the amino acid synthesis called start codon. For example, AUG which codes for methionine in eukaryotes and modified methionine [N-Formyl methionine (fMet)] in prokaryotes.
4. **Anticodon:** It is a set of three nucleotide bases present on tRNA that is complementary to the nucleotide bases of codon on mRNA.

Extrachromosomal Part

Bacteria have following three types of nucleic acids from extrachromosomal part like plasmid, episome and transpososn.

Plasmid

Definition: It is an extrachromosomal intracytoplasmic DNA material **(Fig. 8.1)**.

Properties

- **Numbers:** A bacterium can have no plasmids at all or have many plasmids (20–30) or multiple copies of a plasmid.
- **Shape:** Usually, it is closed circular molecules; however it occurs as linear molecule in *Borrelia burgdorferi*.
- **Size:** Varies from 1 kb to 400 kb.
- **Replication:** Autonomous (independent) replication.
- **Present:** It presents in bacteria and also in yeast.

1. **According to transmission**
   - **Self transmissible/conjugative plasmid:** Contains information for self transfer to other cell via conjugation, e.g., F factor
   - **Non-transmissible/non-conjugative plamid:** Does not contain information for self transfer via conjugation (can be transduced).
2. **According to functions**
   - **Fertility (F)/sex factor or sex pili:** F factor is a plasmid codes for special fimbriae or sex pili which makes a conjugation tube between two cells and for necessary enzyme of conjugation. Those bacteria that possess F factor are called F^+, such bacteria have sex pili on their surface and those are lacking called F^-. The F factor plasmid is transferred to other cells through conjugation tube. F^- cell will become F^+ when it receives the fertility factor from another F^+ cell.
   - **Resistance (R) Factor:** It is responsible for drug resistance.
   - **Colicinogenic (Col) factor:** It is responsible for bacteriocins production.
3. **According to restriction endonuclease enzyme fingerprinting.**
   - **Closely related plasmids:** They produce same or very similar fingerprinting.
   - **Unrelated plasmids:** They produce different fingerprinting.
4. **Incompatibility typing**
   - **Closely related plasmids:** They do not coexist stably in same bacterium.
   - **Unrelated plasmids:** They coexist stably in same bacterium.

Functions of plasmid: It transfers the genetic information from one cell to another cell and enables virulence properties to bacteria as mentioned below:

1. **Virulence determinant**
   - Conjugation: Fertility (F) factor is a plasmid codes for special fimbriae or sex pili. Fimbriae or sex pili make a conjugation tube between two cells and allow the transfer of genetic information from one cell to other called conjugation.
   - Iron sequestering: Plasmid carries virulence determinant genes like the Col V plasmid of *E. coli* contains genes for iron sequestering compounds.
   - Capsule and toxin in *B anthracis* are plasmid encoded. Removal of such plasmid makes the strain avirulent called Sterne strain, which used for vaccine production.
   - Invasive property in *Shigella* and EIEC is due to a special type of OMP called VMA (virulence marker antigen) which is plasmid encoded, responsible for invasion, multiplication of bacilli and destruction of epithelial cells.

2. **Drug resistance:** It codes for drug resistance to several antibiotics called resistance transfer factor (RTF). Gram-negative bacteria carry plasmids that give resistance to antibiotics such as neomycin, kanamycin, streptomycin, tetracycline, chloramphenicol, penicillins and sulfonamides.
3. **Bacteriocins production:** It codes for the production of bacteriocins called col factor.
4. **Toxins production:** It codes for the toxins production in following bacteria:
   - Enterotoxin production by *Staph. aureus*.
   - Anthrax toxin by *B anthracis*.
   - Tetanospasmin by *Cl. tetani*.
   - Labile toxin (LT) of *E. coli*.
   - Stable toxin (ST) of *E. coli*.
5. **Resistance to heavy metals:** It codes for resistance to heavy metals like Hg, Ag, Cd, Pb, etc.
6. **Resistance to UV light:** It codes for resistance to UV light (DNA repair enzymes are coded in the plasmid).
7. **Coding for enzymes:** It contains genes coded for enzymes that allow bacteria unique or unusual materials for carbon or energy sources, e.g., urease synthesis in bacteria.
8. **Metabolic plasmid:** It enables the bacteria in various metabolic activities such as nitrogen fixation, digestion of unusual substances like toluene, salicylate, camphor, etc.

Clinical applications of plasmid

1. **Used in genetic engineering as vector:** Plasmid contains many sites for artificial insertion of genes by recombinant technology. Such plasmid can be used for various purposes such as protein production, gene therapy, etc.
2. **Plasmid profiling is a useful genotyping method:** It is useful tool for epidemiological study of microbes.

Episome

Plasmid sometimes integrated with chromosomal part called 'episome' **(Fig. 8.1)** and such bacterial cells are called 'Hfr cells' (**H**igh **f**requency of **r**ecombination). Previously, it was considered as synonymous with plasmid. Jacob and Wollman coined the term episome. When it detaches with part of chromosome, it becomes free called F prime ($F^/$) factor.

Transposon (Jumping Gene)

Gene that jumps between chromosomal and extra-chromosomal part called transposon or jumping gene. It called jumping gene because it jumps between chromosomal and extrachromosomal part. It also called transposable genetic element.

BACTERIAL VARIATION

Bacteria have two types of variation like phenotypic variation and genotypic variation, as mentioned below. Differences between two types are mentioned in **Table 8.2.**

TABLE 8.2: Differences between phenotypic and genotypic variation

Phenotypic variation	Genotypic variation
Reversible	Irreversible
Temporary (unstable)	Permanent (stable)
Not heritable	Heritable
Environment effect	No environment effect

Phenotypic Variation

Meaning: Phaeno means display.

Definition: It is a physical expression of bacterial characteristics in given environment.

Examples of phenotypic variation: Such as synthesis of flagella and synthesis of enzyme described below:

- **Synthesis of flagella:** *S.* Typhi is normally flagellated, but flagella are not synthesized when it grows in phenol agar. It reversed when subcultured in broth.
- **Synthesis of enzyme:** Galactosidase is the enzymes required for *E. coli* for lactose fermentation, but it produced when *E. coli* grows in a lactose containing medium. It is not synthesized when *E. coli* grows in lactose free medium. Such enzyme which requires the presence of substance called induced enzyme and opposite which does not require the presence of substance called constitutive enzyme.

Genotypic Variation

Meaning: Geno means related to gene.

Definition: It is a genetic expression of bacterial characteristics in given environment.

Examples of genotypic variation: Such as mutation and gene transfer mechanisms/methods described below.

Mutation

History: Word "mutation (Latin)" was coined by Hugo de Vries, which means "to change".

Definitions

- **Mutation:** Change in the nucleotide sequence of DNA called mutation.
- **Wild type:** Organisms selected as reference (Normal) strains are called wild type.
- **Mutants:** Yield of mutation called mutants.
- **Mutagenesis:** The process of mutation called mutagenesis.
- **Mutagens:** Agents inducing mutations are called mutagens.

Types of mutations: Two main types:

A. According to induction

1. **Spontaneous mutation:** It occurs naturally about one in every million to one in every billion divisions. It occurs during DNA replication.
2. **Induced mutation:** It is induced by following agents:

- **Chemical agents:** Such as alkylating agents, acridine dyes (nucleoside analogs that are similar in structure to nitrogenous bases), aflatoxin, 2-amino purine, 5-bromouracil, benzpyrene (from smoke and soot) and nitrous acid (alters adenine to pair with cytosine instead of thymine).
- **Physical agents:** X-rays and gamma rays have been shown to damage DNA. UV rays are responsible for the formation of thymine dimers in which covalent links are established between the thymine molecules which change the physical shape of the DNA preventing transcription and replication.
- **Biological agents:** Such as viruses.

B. According to mechanisms: Follow **Fig. 8.3.**

1. **Point mutation:** Due to addition, deletion or substitution of single base.
   - **Substitution of a nucleotide:** It involves the changing of single base in the DNA sequence. This mistake is copied during replication to produce a permanent change. This is the most common mechanism of mutation. If a purine is replaced by a pyrimidine or *vice versa*, the substitution is **called transversion.** If one purine is replaced by the other purine or one pyrimidine is replaced by other pyrimidine, the substitution is **called transition**.
   - **Deletion or addition of a nucleotide:** Deletion or addition of a nucleotide occurs during DNA replication. When a transposon (jumping gene) inserts itself into a gene, it leads the disruption of gene **called insertional mutation**.
2. **Frame shift mutation:** Deletion or addition of a numbers of nucleotide.
3. **Multiple mutation:** It causes extensive chromosomal rearrangement.
4. **Other types**
   - **Missense mutation:** Changes in the amino acid sequence due to change in triplet codon of the particular protein called missense mutation. This could be caused by a single point mutation or a series of mutations.
   - **Nonsense mutation:** Termination of polypeptide or protein synthesis due to formation of stop codon is called a nonsense mutation. It leads incomplete or immature protein production.

A B C D E F	Normal (wild type)
A B X C D E F	Addition
A B D E F	Deletion
A B B C D E F	Duplication
A B X D E F	Substitution
A B C F E D	Inversion

Fig. 8.3: Mutation according to mechanism

- **Suppressor mutation:** It is a reversal of a mutant phenotype by another mutation at a position on the DNA distinct from that of original mutation. True reversion or back mutation results in reversion of a mutant to original form, which occurs as a result of mutation occurring at the same spot once again.
- **Lethal mutation:** Sometimes some mutations affect vital functions and the bacterial cell become nonviable. Hence, those mutations that can kill the cell are called lethal mutations.
- **Conditional lethal mutation:** Sometimes a mutation may affect an organism in such a way that the mutant can survive only in certain environmental condition called conditional lethal mutation. For example, a temperature sensitive mutant (ts mutant) can survives at permissive temperature of 35°C but not at restrictive temperature of 39°C.
- **Inversion mutation:** If a segment of DNA is removed and reinserted in a reverse direction, it is called inversion mutation.
- **Silent mutation:** Sometimes a single substitution mutation change in the DNA base sequence results in a new codon still coding for the same amino acid. Since there is no change in the product, such mutation is called silent mutation.

Clinical significance of mutation

1. **Identification of the function of gene:** Discovery of a mutation in a gene can help in identifying the function of that gene.
2. **Vaccine production:** Mutations can be induced at a desired region to create a suitable mutant, especially to produce vaccine.
3. **Drug resistant:** Spontaneous mutation can result in emergence of antibiotic resistance in bacteria like drug resistance in *M. tuberculosis.*
4. **Functional changes in bacteria:** Mutation leads functional changes like *E. coli* mutant loses the ability to ferment lactose. It can be detected on MacConkey's medium, but not in nutrient agar.
5. **Survival advantage:** Mutation affects the vital function and confers survival advantage. For example, streptomycin resistant mutant to *M. tuberculosis* develops in a patients taking treatment. It multiplies and replaces initial drug sensitive bacteria and causes survival advantage, but a patient who is not under treatment, it does not cause any survival advantage.
6. **Change in phenotypes:** Mutations can result in change in phenotypes such as
   - Appearance of novel surface antigen
   - Alternation in physiological properties
   - Change in colony morphology
   - Nutritional requirements
   - Biochemical reactions
   - Growth characteristics
   - Virulence
   - Host range

Tests to detect the mutations

1. **Fluctuation test**
   - **History:** It was developed by Luria and Delbruck in 1943.
   - **Principle:** Bacteria undergoes to spontaneous mutation when they are challenged on agar plate contains growth limiting substances like streptomycin or bacteriophage. However, the rate of mutation is wide, some bacteria mutate early while some late which leads fluctuations. Fluctuations are wide from small volume subculture (more frequent mutation) than large volume subculture (less frequent mutation).
   - **Disadvantages:** It is not widely accepted due to complicated statistical evaluation.
   - **Method:** Luria and Delbruck seeded the *E. coli* from small volume culture and single large volume culture on media contain bacteriophages. The mutant bacteria formed the colonies. Colony count was compared. There was wide fluctuation in the numbers of resistant variant in small volume culture as compare to single large volume culture.
2. **Replica plating method**
   - **History:** It was developed by Lederberg in1952.
   - **Principle:** It based on differentiation between normal strain and auxotrophic mutant in which auxotrophic mutant cannot grow in absence of particular nutrient, e.g., lysin auxotroph can grow only on media contain lysin but not on lysin deficient media.
   - **Method (Fig. 8.4):** Using velvet template, mixture of colonies (normal strain and auxotrophic mutants) of bacteria transferred from a master plate, on to two subculture plates where one plate is lacking with lysin. After incubation same colonies as master plate are obtained in subculture plates, except lysin auxotroph which do not grow on the media lacking lysin.
3. **Ames test (carcinigenicity testing)**
   - **History:** It was developed by Bruce Ames in 1970.
   - **Aim:** It is used to identify the environmental carcinogens.

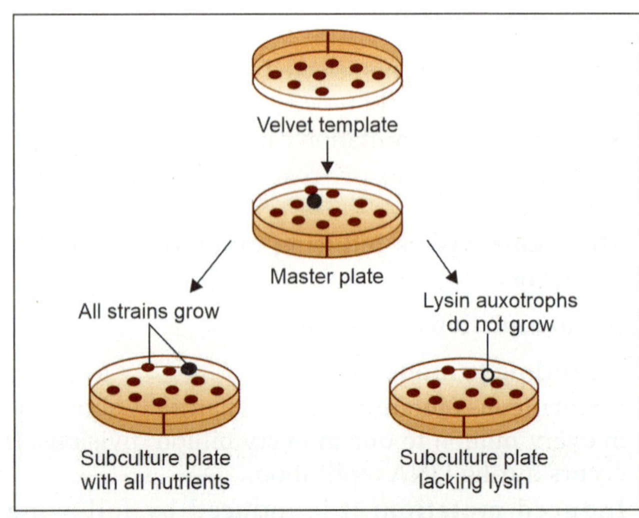

Fig. 8.4: Replica plating method

- **Principle:** It based on mutational reversion by carcinogen.
- **Method (Fig. 8.5):** Mutant strains (histidine auxotroph) of *Salmonella* which are subscribed on two agar plates contain small amount of histidine; one of the plate is added with the test mutagen. Incubate the plate at 37°C for 2–3 days. All of the histidine auxotroph will grow for the first few hours until the histidine is depleted. Once the histidine source is exhausted, mutation revertant strain can grow by their ability to synthesize the histidine. Reversed mutation may be induced by carcinogen (affect large numbers of strains) or occur spontaneously (affect only few strains). The relative mutagenicity of the carcinogen can be estimated by counting the colonies. More the numbers higher the carcinogenicity.

4. Penicilin enrichment

Gene Transfer Mechanisms/Methods

Definition: Transfer of genetic materials or genetic information from one cell to other cell called genetic recombination. Cell donating the genetic materials or genetic information called donor cell to another cell called recipient cell.

Methods of gene transfer: It includes vertical gene transfer method and horizontal gene transfer method. In vertical gene transfer method, transmission of genes occurs from parents to offsprings which is not much important. In horizontal gene transfer method transmission of genes occurs from one bacterium to other by different mechanisms such as transformation, transduction, lysogenic conversion, conjugation and transposition. Horizontal gene transfer methods are described below.

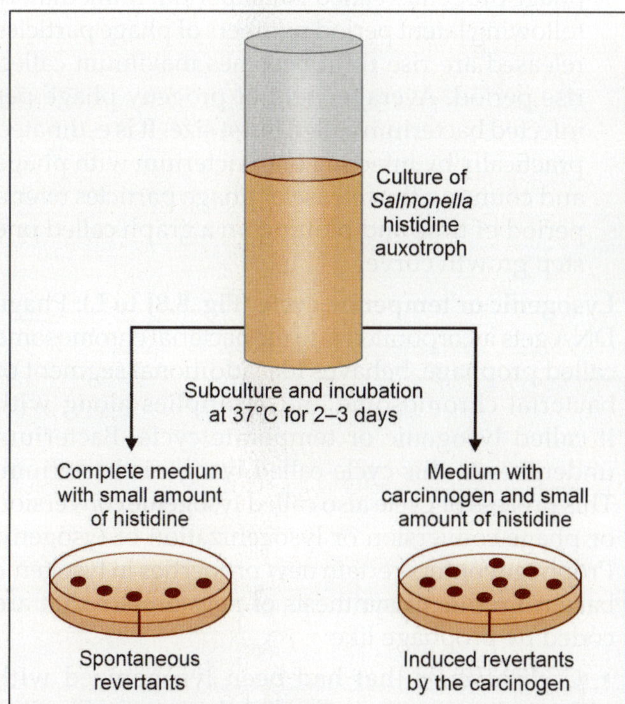

Fig. 8.5: Ames test

A. Transformation

Definition: Transfer of genetic information from one cell to other cell through agency of free/naked DNA called transformation.

Mechanism (Fig. 8.6): Transformation occurs in less than 1% of bacteria like *Bacillus*, *Haemophilus*, *Neisseria* and *Streptococcus*. When bacteria die, their DNA is released; such free DNA is referred as naked DNA. Fragments of the naked DNA are taken up by the cell wall of another bacterium and incorporated into its chromosome.

History (Fig. 8.7): This was first demonstrated in an experiment conducted by Griffith in 1928. Capsulated pneumococci give glistening, smooth (S) colonies while noncapsuled strains give rough (R) colonies. Pneumococci with a capsule (type I) are virulent and can kill a mouse while strains lacking it (type II) are harmless. Griffith found that mice died when they were injected with a mixture of live noncapsulated (R, type II) strains and heat killed capsulated (S, type I) strains. Neither of these two when injected alone could kill the mice, only the mixture of two proved fatal. Live S strains with capsule were isolated from the blood of the animal suggesting that some factor from the dead S cells converted the R strains into S type. The factor that transformed the other strain was found to be DNA by Avery, McLeod and McCarty in 1944.

Uptake of free DNA by bacteria: Some bacteria are able to take up DNA naturally. However, these bacteria only take up DNA at particular time in their growth cycle (log phase) when they produce a specific protein called a competence factor. Gram-positive bacteria take up single stranded DNA and the complementary strand is made in the recipient. Gram-negative bacteria take up double stranded DNA.

Clinical significances: Transformation occurs in nature, and it increases the virulence.

Use: It used in recombinant DNA technology.

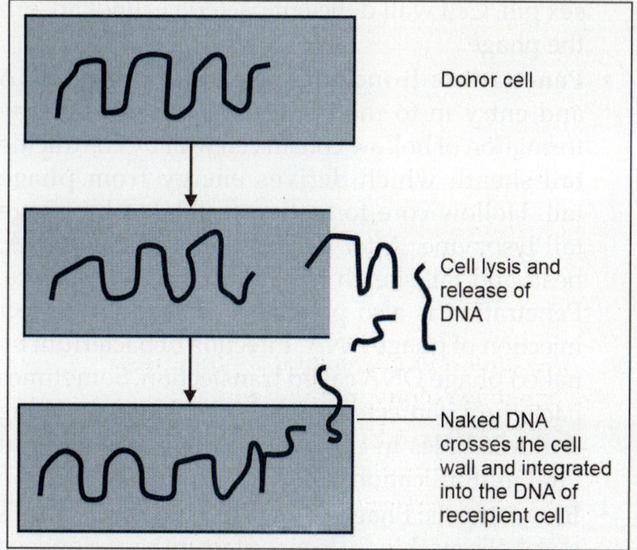

Fig. 8.6: Mechanism of transformation

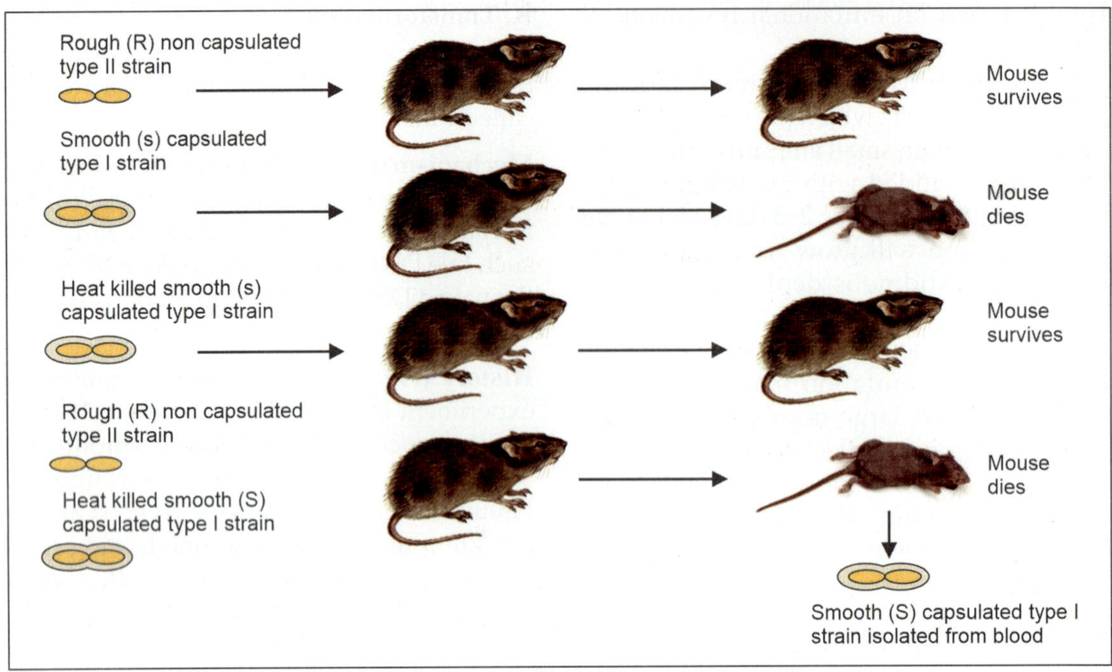

Fig. 8.7: Transformation of Griffith

B. Transduction

Definition: Transfer of genetic information from one cell to other cell through bacteriophage called transduction.

Bacteriophage: Bacteriophage is a virus that eats bacterium and uses its machinery for its own replication. To understand the mechanism of transduction it is must to study the life cycle of bacteriophage as described below:

1. **Lytic or virulent cycle (Fig. 8.8A to G):** Bacterial cell is killed with the release of mature phages called lytic cycle. Following are the stages of transduction involving a lytic cycle:

   - **Adsorption:** Bacteriophage adsorbs to a bacterium surface by its tail. It is a complementary process between phage base plate and bacterial cell wall receptors. Cell wall receptors are present at different places like Vi Ag of *S.* Typhi, on flagella or sex pili. Cell wall deficient bacteria cannot adsorb the phage.

   - **Penetration:** It includes release of phage DNA and entry in to the bacterium. It is possible by formation of hollow core in cell wall by contractile tail sheath which derives energy from phage tail. Hollow core formation facilitated by phage tail lysozyme. After penetration of phage DNA, head and tail sheath remain as ghost on surface. Penetration is also possible practically by direct injection of phage DNA. Infection of bacterium by naked phage DNA called transfection. Sometimes bacterium is infected by multiple phages resulting multiple holes in cell wall and cell lysis without viral multiplication called lysis from without.

   - **Biosynthesis:** Phage DNA directs the bacterium's metabolic machinery to manufacture bacteriophage components like head, tail and enzymes.

   - **Maturation and assembly:** Phage components synthesized in bacterium undergo maturation and assembly.

   - **Release:** After maturation bacterial cell wall gets weakened. Phage enzyme acts on weakened site and ruptures the wall with the lysis of bacterium and release of progenies of bacteriophage. The interval between entry of phage DNA and appearance of 1st intracellular phage particles called eclipse phase. It is a time required for synthesis and assembly of phage particles. The interval between infection of bacterium and 1st release of phage progenies called latent period. Immediately following latent period numbers of phage particles released are rise till it becomes maximum called rise period. Average yield of progeny phage per infected bacterium called burst size. It is estimated practically by infecting one bacterium with phage and counting the release of phage particles over a period of time and plotting on a graph called one step growth curve.

2. **Lysogenic or temperate cycle (Fig. 8.8J to L):** Phage DNA gets incorporated into the bacterial chromosome called prophage, behaves like additional segment of bacterial chromosome and multiplies along with it called lysogenic or temperate cycle. Bacterium undergoes to this cycle called lysogenic bacterium. This process or cycle also called lysogenic conversion or phage conversion or lysogenization or lysogeny. Prophage confers certain new properties in lysogenic bacterium due to synthesis of new protein that are coded by prophage like

   - *C. diphtheriae* that had been lysogenized with β-prophage produces the diphtheria toxin. Elimination of β-prophage makes the strain nontoxigenic.

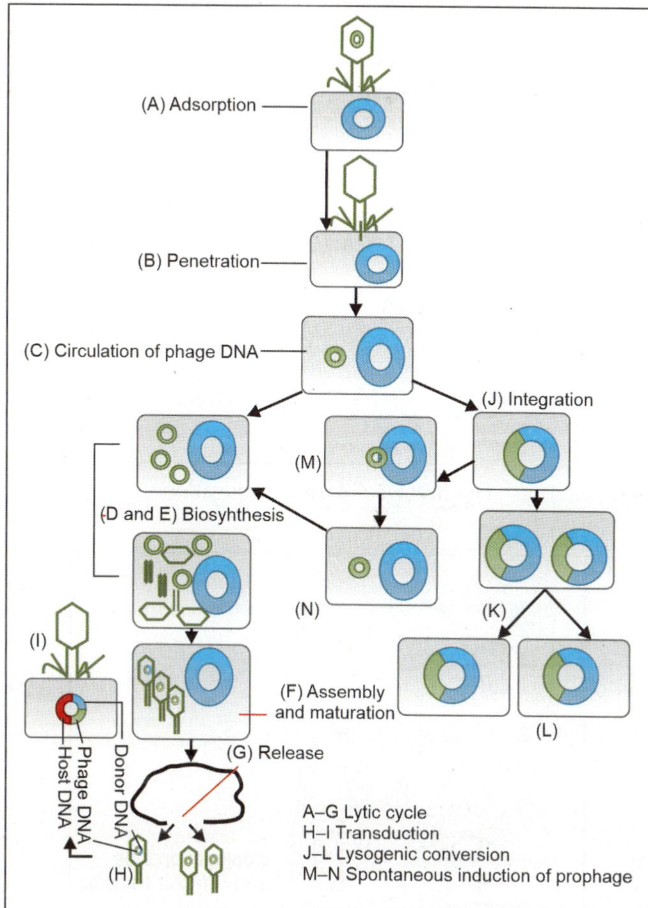

Fig. 8.8: Life cycle of bacteriophage

- Lysogenic bacteria are resist to infection by same or related phages called superinfection immunity
- It influences antigenic properties of bacterium.

3. **Variation in life cycle stage**

- **Spontaneous induction of prophage (Fig. 8.8M and N):** The prophage sometimes excised from the host chromosome during multiplication of lysogenic bacteria and starts lytic cycle with release of daughter phages called spontaneous induction of prophage. It sometimes carries along with itself a fragment of bacterial chromosome; the subsequent phage progeny may have a piece of chromosomal DNA. When such phage infects another bacterium, newer characteristics coded by that chromosomal gene are conferred.
- **Pseudolysogeny:** In certain bacteria prophage is not incorporated in host DNA but instead remains free as like plasmid called pseudolysogeny, e.g., *M. tuberculosis*.
- **Shifting of cycle:** Lysogenic cycle can be shift to lytic cycle by exposure to UV light, H_2O_2 and nitrogen mustard.
- **Infection by two phages DNA:** If a bacterium simultaneously infected by two same or related phages, it releases two types of progenies. When this occurs many of the progenies are recombinants.

Mechanism of transduction (Fig. 8.8H and I): Occasionally during the assembly of bacteriophage's components inside the host bacterium, an error occurs called packaging error that resulting the incorporation of host DNA in to bacteriophage genome. The bacteriophage carrying the donor bacterium's DNA adhere on another bacterium and inserts the donor bacterium's DNA into the recipient cell. Transduction is not confirmed to transfer the chromosomal DNA only but episome and plasmid may also be transduced.

History: Bacterial transduction was discovered by Norton Zinder and Joshua Lederberg in 1952 at the University of Wisconsin-Madison in *Salmonella*.

Types of transduction: Two types:

1. **Restricted (specialized) transduction:** It can transfer only those genes that lie adjacent to the prophage. It is best studied by lambda phage. The lambda phage that infects *E. coli* always transfers gal+ gene which is responsible for galactose fermentation.

2. **Generalized transduction:** It can transfer any bacterial gene. This process may occur with phages (lytic phages) that degrade their host DNA into pieces the size of viral genomes. If these pieces are erroneously packaged into phage particles, they can be delivered to another bacterium. Phage P22 of *S.* Typhimurium and P1 and μ of *E. coli* carry out generalized transduction.

Clinical significances of transduction

1. **Transfer of extrachromosomal nucleic acid:** Plasmid (R factor) and episome are also transferred by transduction. For example plasmid mediated penicillin resistant in *Staphylococcus* is transfer from cell to cell by transduction.

2. **Genetic mapping:** Most useful mechanism in prokaryotes for gene transfers and provides an excellent tool for genetic mapping in bacteria.

3. **Treatment of metabolic disease:** It affects the eukaryotic cell and proposed method of genetic engineering in treatment of certain inborn error of metabolism. For example, metabolic defect in fibroblast from galactosemic patients can be corrected by transduction using the lambda phage carrying the gal gene.

Differences between lytic/transduction and temperate cycle/lysogenic conversion: Follow **Table 8.3**.

TABLE 8.3: Differences between lytic and temperate cycle

Lytic cycle	Temperate cycle
Lysis of bacterium	Host bacterium is unharmed
Transfer of donor bacterium's DNA	Transfer of phage DNA
Phage acts like vehicle or carrier between donor and recipient cells	Phage DNA behaves like additional segment of bacterial chromosome and multiplies along with it and confers new properties

 C. Lysogenic conversion

Follow lysogenic/temperate cycle of bacteriophage.

D. Conjugation

Definition: Transfer of genetic information from one cell to other cell by formation of tube like structure between two cells called conjugation and such tube called conjugation tube.

History: Conjugation was 1st discovered by Joshua Lederberg and Tatum in 1946 in *E. coli* K 12 strain.

Properties of conjugation: Conjugation is a process where male/donor bacterium mates or makes physical contact with female/recipient bacterium and transfers genetic information. It is like matting in higher organisms but following conjugation female bacterium is converted in to male bacterium.

Mechanisms of conjugation: It is due to formation of conjugation tube between two cells by specialized fimbriae or sex pili present on cell surface. Such fimbriae are encoded by a plasmid called sex or fertility (F) factor. Such plasmid multiplies and acts as donor (copy of it passes to recipient cell). Such plasmid is self transmissible or conjugative. Several such plasmids were discovered and act as donors are collectively called transfer factors.

Clinical significances of conjugation: Conjugation allows transfer of genes (like plasmid, episome) or genetic information from one cell to other as below:

1. **Fertility (F)/sex factor conjugation (Fig. 8.9):** F factor is a plasmid codes for special fimbriae or sex pili and for necessary enzyme of conjugation. Fimbriae or sex pili make a conjugation tube between two cells for gene transfer. Those bacteria that possess F factor are called F^+, such bacteria have sex pili on their surface and those are lacking called F^-. It mediates its own transfer. Vertical (inheritance) or horizontal (transfer) transmission maintains plasmid. This results in the transfer of an F^+ plasmid (coded only for a sex pilus) but not chromosomal DNA from a male donor bacterium to a female recipient bacterium. A F^- female cell will become F^+ male when it receives the fertility factor from another F^+ cell and can make a sex pilus. During conjugation, no cytoplasm or cell material except plasmid passes from donor to recipient. Two cells mate and form conjugation tube through which F factor can transferred. The mating pairs can be separated by shear forces and conjugation can be interrupted. Consequently, the mating pairs remain associated for only a short time. After conjugation, the cells break apart. Following successful conjugation the recipient becomes F^+ and the donor remains F^+.

2. **R factor conjugation (Fig. 8.10):** R factor is a plasmid, which is responsible for drug resistance. It presents in gram-negative bacteria. It has two components such as RTF (resistance transfer factor) and 'r' determinant. RTF is the first part that codes for self transfer (like

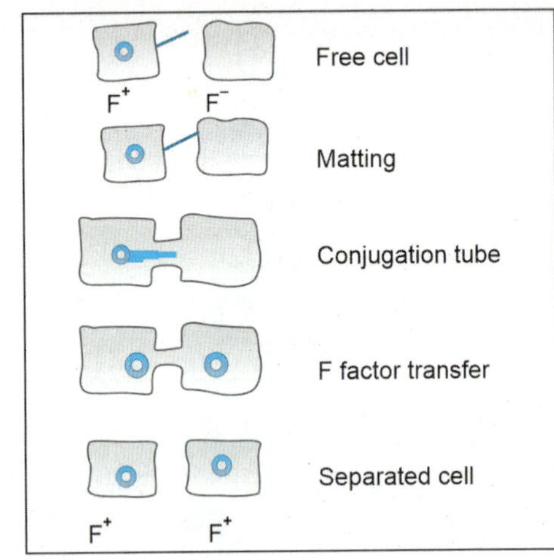

Fig. 8.9: F/Sex factor conjugation

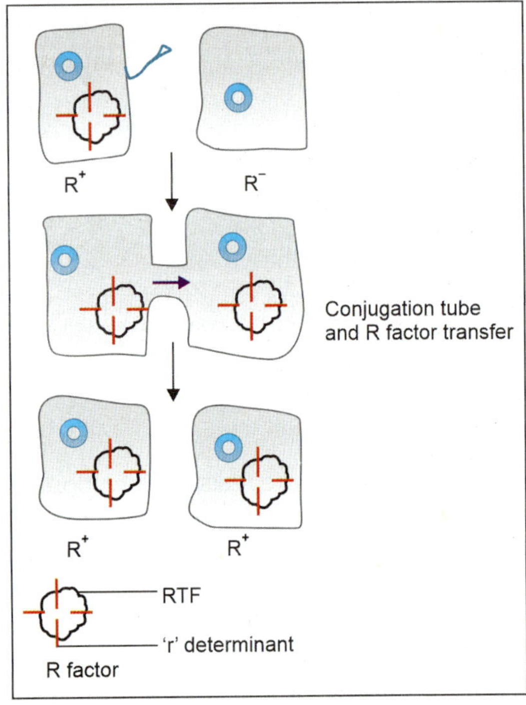

Fig. 8.10: R factor conjugation

F factor) and hemolysin-enterotoxin production in *E. coli* while r determinant is the second part that codes for antibiotic resistance. Transfer of drug resistance by R factor called transferable or episomal or infectious drug resistance. R plasmid may confer resistance to as many as eight different antibiotics at once upon the cell and by conjugation it can be rapidly transfer to bacterial population. Both F factor and R factor are self transferable but the differences are that the latter has additional genes coded for drug resistance and hemolysin-enterotoxin production. During conjugation there is transfer of resistance (R) factor from a donor bacterium to a recipient. The recipient becomes multiple antibiotic resistant and male, and is now able to transfer R-plasmid to other bacteria. When the recipient cell acquires entire R factor, it too expresses antibiotic resistance. Sometimes, RTF may disassociate from the r determinant and the two components may exist as separate entities. In such cases the host cell remains resistant to antibiotics; it cannot transfer this resistance to other cells. Transfer of multiple antibiotic resistances by conjugation has become a major problem in the treatment of certain bacterial diseases.

3. **Colicinogenic (Col) factor conjugation:** Col factor is a plasmid which is responsible for bacteriocin production. It can also transfer via conjugation.

4. **Episome conjugation/sex duction (Fig. 8.11):** It includes the transfer of episome (plasmid integrated with chromosome). Cell carrying episome can transfer gene with high frequency hence called Hfr cells. Sometimes episome detached from chromosome along with some part of chromosome and becomes free called F prime (F′) factor. When F prime (F′) factor cell mates to recipient cell (F⁻), it transfers F prime (F′) factor and host chromosome linked with it called sexduction. Following successful conjugation both cells becomes F′.

E. Transposition

Definitions: Transfer of genetic information from one cell to other cell via transposon called transposition.

History: It was discovered by Barbara Mc Clintock in plants during work in the 1940s and 1950s, for which she was awarded with Nobel Prize in 1983.

Properties of transposition

- **Random movement:** This mobile gene can move from one DNA to any DNA or even to another location on the same DNA. The movement is not totally random; there are preferred sites in a DNA molecule at which the transposable genetic element will be inserted.
- **Not capable of self replication:** It is not self replicating and depends on plasmid or chromosomal DNA for replication.
- **Transposition can be accompanied by duplication:** In many instances transposition results in removal of the element from the original site and insertion at a new site. However, in some cases the transposition event is accompanied by the duplication of the transposable genetic element. One copy remains at the original site and the other is transposed to the new site.

Structural types and significances of transposition

1. **Insertion sequences (IS)**
   - **Structure:** It is the small segment of DNA about 1–2 kb without any essential genes called IS. Such DNA is encodes for transposition.
   - **Significances**
     - Mutation: The introduction of an insertion sequence into a bacterial gene will results in inactivation of gene.
     - Selection of site for plasmid insertion in chromosome: The sites at which plasmid inserted into the bacterial chromosome are at or near insertion sequence in the chromosome.
     - Phase variation: In *Salmonella* there are two genes, which code for two antigenically different flagellar antigens. The expression of these genes is regulated by an insertion sequences.

2. **Transposons**
   - **Structure:** It is the large segment of DNA about 4–25 kb with essential genes. It carries one or more genes in the center and the two ends carrying inverted repeat sequences complementary to each (**Fig. 8.12**) other but in reverse order. Because of these, two ends contain single stranded loop and center contains double stranded stem formed by H_2 bonding between inverted repeat sequences.

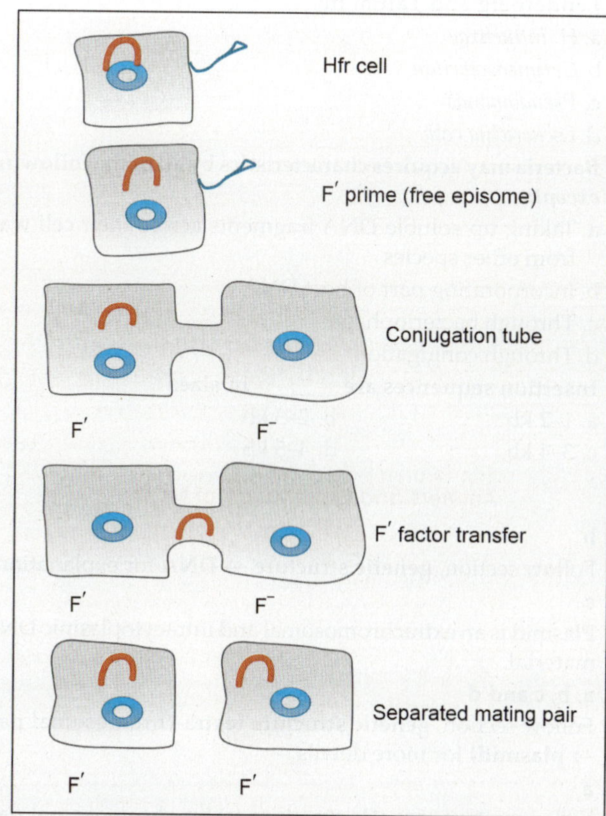

Fig. 8.11: F′ factor conjugation (sex duction)

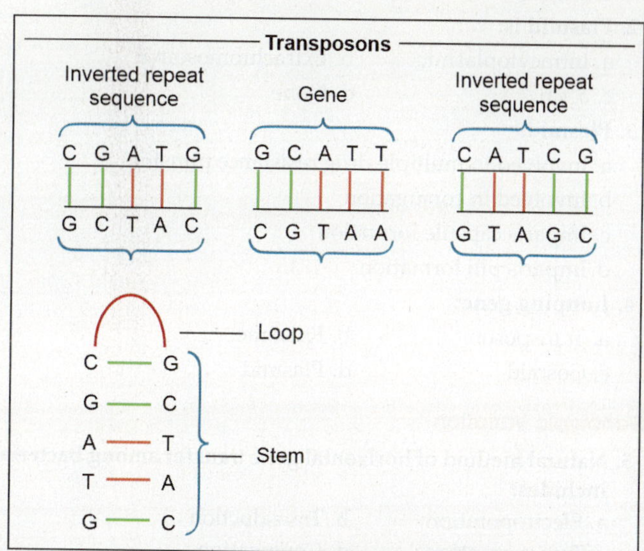

Fig. 8.12: Structure of transposon

- **Significances**
  - Drug resistance: Many antibiotic resistance genes are located on transposon. When it jumps on a transferable plasmid it carries multiple drug resistance in a bacterium and causes major medical problem.
  - Genetic engineering: It is useful in laboratory for gene manipulation.

ACCESS YOURSELF

Essay/Full Question

1. Genetic structure of bacteria.
2. Genotypic and phenotypic variation of bacteria.

Short Notes

1. Bacterial DNA
2. Plasmid
3. Mutation
4. Gene transfer mechanisms
5. Transduction
6. Bacterial conjugation.

Short Questions for Theory/Viva Questions

1. What are sense codons and non-sense codons?
2. What are introns and exons?
3. What is plasmid?
4. Name the two chromosomal and two extrachromosomal nucleic acids of bacteria.
5. What are episome and transposons?
6. Write the four differences between phenotypic and genotypic variation.

Comments on

1. Transposon is also called jumping gene.

MCQs for Chapter Review

Genetic Structure

1. **Bacterial chromosome consists of double stranded DNA, in a circular configuration the length of which is about:**
   a. 100 μm
   b. 1000 μm
   c. 10000 μm
   d. 100000 μm

2. **Plasmid is:**
   a. Intracytoplasmic
   b. Extrachromosomal
   c. a + b
   d. None

3. **Plasmid:**
   a. Involved in multiple drug resistance transfer
   b. Involved in conjugation
   c. Imparts capsule formation
   d. Imparts pili formation

4. **Jumping gene:**
   a. Transposon
   b. Episome
   c. Cosmid
   d. Plasmid

Genotypic Variation

5. **Natural method of horizontal gene transfer among bacteria includes:**
   a. Electroporation
   b. Transduction
   c. Transformation
   d. Conjugation
   e. Mutation

6. **Mechanism of direct transfer of free DNA:**
   a. Transformation
   b. Conjugation
   c. Transduction
   d. None

7. **Virus mediated transfer of host DNA from one cell to another is known as:**
   a. Transduction
   b. Transformation
   c. Transcription
   d. Integration

8. **Lysogenic conversion is:**
   a. New properties in a bacterium due to integration of phage genome
   b. Transfer of DNA from one bacterium to another by a bacteriophage
   c. Transfer of free DNA
   d. Transfer of genome during physical contact

9. **The discovery of "gene transformation" came from the study of one of the following bacteria:**
   a. *Bacillus subtilis*
   b. *Streptococcus pyogenes*
   c. *Streptococcus pneumoniae*
   d. *Escherichia coli*

10. **F factor integrated with bacterial chromosome to form:**
    a. HFr
    b. RTF+r
    c. F⁻
    d. RTF

11. **Conjugation does not involve:**
    a. Bacteriophage
    b. HFr
    c. F factor
    d. Plasmid

12. **The role of plasmid in conjugation was first described by Lenderberg and Tatum in:**
    a. *H. influenzae*
    b. *Corynebacterium*
    c. *Pseudomonas*
    d. *Escherichia coli*

13. **Bacteria may acquires characteristics by all of the following *except*:**
    a. Taking up soluble DNA fragments across their cell wall from other species
    b. Incorporating part of host DNA
    c. Through bacteriophage
    d. Through conjugation

14. **Insertion sequences are _______ in size:**
    a. 1–2 kb
    b. 2–3 kb
    c. 3–4 kb
    d. 4–5 kb

Answers and Explanation of MCQs

1. b
- Follow section, **genetic structure → DNA** for explanation.

2. c
- Plasmid is an extrachromosomal and intracytoplasmic DNA material.

3. a, b, c and d
- Follow section, **genetic structure (extra-chromosomal part → plasmid)** for more details.

4. a
- Follow section, **genetic structure (extra-chromosomal part → transposon)** for explanation

5. b, c and d
- Follow section, **bacterial variation (genotypic variation → gene transfer mechanisms/methods)** for more details.

6. a

7. a

8. a

9. c
- Follow section, **bacterial variation (genotypic variation → gene transfer mechanisms/methods → lysogenic conversion)** for explanation of answers of MCQs 6–9. Gene transformation came from the study of *Streptococcus pneumoniae* by Griffith.

10. a
- F factor is plasmid but in question its written as 'integrated with bacterial chromosome" so it considered as episome. Cells contain episome are transferring it with high frequency called HFr cells (**h**igh **f**requency of **r**ecombination), so option a is right answer.

11. a
- Bacteriophage is involved in transduction.

12. d
- Conjugation was 1st discovered by Joshua Lederberg and Tatum in 1946 in *E. coli* K 12 strain.

13. b
- Bacteria are not acquiring any properties from host cell DNA but they may acquire characteristics by
- – taking up soluble DNA fragments across their cell wall from other species called transformation.
- – through bacteriophage called transduction.
- – through conjugation tube called conjugation.

14. a
- Follow section, **bacterial variation [genotypic variation → gene transfer mechanisms → transposition → insertion sequence (IS)]** for more details.

CHAPTER 9

Precipitating Factors of Bacterial Infections

Chapter Outline

DEFINITIONS

Pathogenicity: Ability of a microbial species to produce the disease called pathogenicity.

Virulence: Ability of microbial strain to produce the disease called virulence.

Exaltation: Enhancement of virulence called exaltation. It is demonstrated by serial passage in susceptible individual.

Attenuation: Reduction of virulence called attenuation. It is demonstrated by serial passage in unfavorable host, repeated subculture, growth in high temperature, desiccation, long storage of culture or by using antiseptics like formalin.

PRECIPITATING FACTORS

Precipitating factors are also known as epidemiological determinants. There are three types of epidemiological determinants affecting the virulence of bacteria such as agent factors (virulence factors or determinants of virulence), host factors and environmental factors. All are described below.

Agent Factors (Virulence Factors or Determinants of Virulence)

Agent factors are following three types.

Intracellular or Cell Associated Factors

Bacterial adhesins: The process of attachment of pathogens with host cells called adhesion. Organs used for adhesion are called adhesins. Adhesins may account for tissue trophism and host specificity. As they used for virulence, loss of adhesins make the strain avirulent. Adhesins are usually protein and antigenic. They produce antibodies which are protective; hence, they can be used as method of prophylaxis in certain bacteria like *E. coli* diarrhea in calves and piglets and gonorrhea in humans are prevented by using vaccine prepared from fimbrial antigens. Following are the different types of adhesins:

1. **Fimbrial adhesins** or **pilus adhesins:** Follow **Ch. 6** for more details.
2. **Nonfimbrial adhesins** or **non pilus adhesins:**
   - Protein receptors in staphylococci: *Staph. aureus* has receptors for many mammalian proteins like fibronectin, fibrinogen, IgG and C1q which help in adhesion with host cells.
   - Protein F in *Streptococcus pyogenes*: It helps in attachment with pharyngeal wall.
   - Extracellular surface protein in *Enterococcus*: It helps in adhesion.
   - Slime layer (glycocalyx): It inhibits phagocytosis. Slime layer made up of glycocalyx has ability to bind bacteria with damaged tissues. It also helps bacteria to bind with plastic material of medical devices like catheters, suture materials, pacemakers, implants, Ryle's tube, etc., and forms the growth on such devices called biofilm. Use of such devices cause diseases. Biofilm is formed by few bacteria like *Staph. albus* (*Staph. epidermidis*), *P. aeruginosa*, etc.
   - Outer membrane protein (OMP) Ag: It presents in *N. gonorrhoeae*, *H. influenzae*, *Chlamydia* spp., *Rickettsia* spp. (spotted fever group), etc., which contributes in adhesion and invasion.
   - Fragment B of diphtheria toxin: Diphtheria toxin has two fragments like A and B. A is active fragment while, fragment B is for binding of toxin to cells. Antibody to fragment B protects binding of toxin to cells.

- D-galactose: It presents in *Listeria monocytogenes* and helps to bound with D-galactose receptors on macrophage's polysaccharides of Peyer's patches and epithelial cells.
- Internalins (A and B are subtypes): They present in *L. monocytogenes* and help to attach with phagocytic cells.
- Adhesin proteins in *Yersinia* spp.: Different types like MyF Ag, pH6 Ag, inv (invasive) protein, ail (attachment and invasive locus) protein and Yad A (*Yersinia* adhesin A → it binds with collagen and fibronectin to aid the invasion of tissue by organism and also it inactivates the complement).
- Filamentous hemagglutinin (FHA): It appears as filamentous structure under electron microscope, hence the name. It adheres the *B. pertussis* with cilia of respiratory epithelium. It also favors adhesion of other bacteria like *H. influenzae* and *Strept. pneumoniae* to respiratory epithelium called piracy of adhesion.
- Adhesins in *Mycoplasma pneumoniae*: Different adhesins like P1 (cytadhesin/protein Ag), bulbous enlargement and glycolipid Ag help to adhere the bacteria with RBCs (called hemadsorption), respiratory epithelium and with other cells.
- Hook of *Leptospira*: Hooks on the ends of *Leptospira* allow it to attach and latch on to host tissues.
- Cell surface lectin in *Chlamydia:* It also acts as an adhesin.

Peptidoglycan: It provides rigidity and structural integrity to cell.

Protein antigen: M protein in streptococci acts as virulence factor and inhibits the phagocytosis.

Somatic (carbohydrate) Ag: It plays role in non-suppurative lesions in streptococci.

Endotoxins: Following are different types of endotoxins as per biochemical nature.
- Lipopolysaccharides (LPSs): LPSs are present in gram-negative not in gram-positive bacteria and responsible for endotoxic activities as mentioned below. However, LPSs of *V. cholerae* have no role in the pathogenesis of disease (called cholera) in human, but intraperitoneal inoculation in mouse causes fatal effects. LPSs have different biological activities such as pyrogenicity, activation of complement, leucocytosis, macrophage inhibition, lethal action, inhibition of glucose and glycogen synthesis in liver, interferon release, depression of blood pressure, leucopenia, stimulation of B lymphocytes, induction of prostaglandin synthesis, protection of bacilli from phagocytosis (in *E. coli*), protection of bacilli from serum complement (in *Klebsiella* spp.), DIC, septic shock (endotoxic shock) and possible death.

- Lipoolygosaccharide: It presents in *H. influenzae* and acts as endotoxin like LPS as described above.
- Endotoxin in *Listeria*: An early study suggested that *L. monocytogenes* is unique among gram-positive bacteria that might possess LPS, which serves as an endotoxin. Later, LPS was found not to be true endotoxin. *Listeria* cell wall consistently contains lipoteichoic acid resembles the LPS of gram-negative bacteria in both structure and function.

Other components related to bacterial cell
- Capsular (K) Ag: Capsular (K) antigen presents in different bacteria (like *K. pneumoniae, Strept. pneumoniae,* etc.), and fungus (like *C. neoformans*) inhibits the phagocytosis; however capsule of *Bordetella pertussis* does not contribute in virulence.
- Slime layer: It inhibits the phagocytosis.
- Vi antigen: It presents in *S.* Typhi. It inhibits the phagocytosis and resists complement activity. It resists bacterial lysis by alternative pathway and by peroxidase killing. Bacilli with Vi Ag are causing more consistent disease than those are lacking.
- Antigenic cross reactivity: Like cytoplasmic membrane and vascular intima in *Streptococcus*.
- Virulence marker antigen (VMA): Invasive property in EIEC and *Shigella* is due to a special type of OMP called VMA (virulence marker antigen) which is plasmid encoded, responsible for invasion, multiplication of bacilli and destruction of epithelial cells. It is detected by VMA ELISA or by culture cells like Hela or Hep-2 cells.

Genetic factors
- Plasmid: It codes for many virulence properties of bacteria. Follow **Ch. 8** for more details.
- Tox⁺ gene: In presents in diphtheria bacilli responsible for toxin production.

Extracellular Factors
Enzymes
- **Coagulase:** It is produced by staphylococci. It forms fibrin barrier around bacteria and produces localized lesions.
- **Hyaluronidase**: It splits the components of intra-cellular connective tissues.
- **Fibrinolysin, proteases and nucleases (DNAse):** They help in initiation and spread of infection by breaking the fibrin barrier.
- **Lipase:** It helps to infect skin and subcutaneous tissues.
- **Enzymatic inactivation of the antibiotics:** Follow **Ch. 116,** for more details.
- **IgA protease:** It is produced by *N. meningitidis, N. gonorrhoeae, H. influenzae* and *Strept pneumoniae*. It destroys the IgA and reduces the local immunity.
- **Amidase:** In pneumococci, this autolytic amidase activated by surface-active agents such as bile or bile salts, cleaves bond between alanine and muramic

acid of peptidoglycan of cell wall, resulting in lysis of organisms and release of bacterial products in medium.

- **Neuraminidase (receptor destroying enzyme) in *V. cholerae*:** Formerly called 'cholera lectin'. It cleaves mucus and fibronectin and releases vibrios which are bounded to intestinal epithelium and favors their spread to other intestinal parts.
- **Other diagnostically useful enzymes are:** Catalase, phosphatase, urease, oxidase, etc.

Exotoxins: Bacteria produce two types of toxins like endotoxin **(described above)** and exotoxin. Exotoxins are described below:

- **Properties of exotoxins**
  - In contrast to endotoxins, which are integral part of bacteria; exotoxins are actively synthesized and released free in medium.
  - Exotoxins are produced by a variety of bacteria including gram-positive and gram-negative bacilli
  - They are antigenic and can be toxoided.
  - Their activity can be neutralized by antitoxins.
  - Most of the toxins have enzymatic activity.
  - Many toxins are extraordinarily powerful, small amount can be lethal.
  - Toxins can be separated from the culture broth by filtration.
  - Exotoxins are heat labile or heat stable.
    1. **Heat labile toxin producing bacteria:** *Cl. perfringens* (*Cl. welchii*), LT (labile) of ETEC, *V. cholerae*, diarrheal illness producing toxin of *B. cereus*.
    2. **Heat stable toxin producing bacteria:** *Staph. aureus*, ST (stable) of ETEC, *Y. enterocolitica*, emetic illness producing toxin of *B. cereus*.
- **Exotoxins producing bacteria:** Exotoxins are mostly produced by gram-positive bacteria and rarely by gram-negative bacteria.
  1. **Exotoxins producing gram-positive bacteria:** *Staph. aureus*, *Strept. pyogenes*, *Strept. pneumoniae*, *B. anthracis*, *B. cereus*, *Cl. welchii*, *Cl. tetani*, *Cl. botulinum*, *C. diphtheriae*, *M. ulcerans* (*M. buruli*), etc.
  2. **Exotoxins producing gram-negative bacilli:** ETEC, EIEC, EHEC, *Shigella* spp., *Y. enterocolitica*, *V. cholerae*, *P. aeruginosa*, *C. jejuni*, *H. pylori* *B. pertussis*, *F. fusiforme*, etc.
- **Nomenclature of exotoxins**
  1. **According to targeted cells/organs:** Exotoxins which attack a variety of cells are generally called cytotoxins. Further naming is done as per cell like hemolysin for RBCs, leucocidin for leucocytes, neurotoxin for nerve cells, enterotoxin for intestinal cells, etc.
  2. **According to the species, which produces them and from the disease with which they are associated:** Like cholera toxin from *Vibrio cholerae*, causes cholera and tetanus toxin from *Clostridium tetani*, causes tetanus.
  3. **According to activities:** Like adenylate cyclase or exotoxin A of *Pseudomonas aeruginosa*.
- **Biological actions of exotoxins**
  1. **Cytotoxicity:** Following toxins are causing cell lysis.
     - Hemolysin: For example, lysis of RBCs in *Staph. aureus*.
     - Leucocidin: For example, damage to polymorphonuclear leucocytes in *Staph. aureus*.
     - Verotoxin/verocytotoxin (also have enterotoxic and neurotoxic properties): It includes Shiga toxin and shiga-like toxin (SLT). Shiga toxin produced by *Shigella dysenteriae*. SLT produced by enterohemorrhagic *E. coli*, *V. cholerae*, *Aeromonas hydrophila* and *Campylobacter jejuni*. Both toxins have *in vitro* effect and *in vivo* effect. *In vitro* effect includes cytotoxicity to the cultured Vero cells (African monkey kidney cells used for cell culture are called Vero cells). For *in vivo* effect toxins have two subunits like binding (B) and active (A). Subunit A divided in to two fragments like A1 and A2. B helps in adhesion with host cells. A2 links A1 to B. A1 inactivates the host cell 60S ribosome and inhibit the protein synthesis leading to cell death. Cell death resulting discontinuity of mucosa and hemorrhage (bloody diarrhea).
     - Diphtheria toxin: It is produced by *C. diphtheriae*, *C. ulcerans* and *C. pseudotuberculosis*. It inhibits protein synthesis by inactivating elongation factor (EF-2) and causing cell death.
     - Botulinum C2: It is cytotoxic in nature while other types like A, B, C1 and D-G are neurotoxics.
     - Toxin A and B of *Clostridium difficile*.
     - Vacuolating cytotoxin produced by *H. pylori* produces injury to host cells.
     - Exotoxin A of *P. aeruginosa*: It has same actions as like diphtheria toxin.
  2. **Enterotoxicity:** Following toxins are causing outpouring of electrolytes and fluid in to intestinal lumen responsible for watery diarrhea.
     - Toxin causing diarrheal illness of *Bacillus cereus*.
     - LT and ST produced by enterotoxigenic *E. coli*.
     - Verotoxin/verocytotoxin (shiga toxin and SLT).
     - Enterotoxin of *Y. enterocolitica*.
     - Cholera toxin produced by *Vibrio cholerae*.
     - Enterotoxin of *P. aeruginosa*.
     - Enterotoxin of *C. jejuni*.
  3. **Neurotoxicity:** Following toxins act on nerve system.
     - Enterotoxin of *Staph. aureus*: It causes food poisoning features by vagal stimulation.
     - Toxin of *Bacillus cereus*: It causes emetic illness.
     - Tetanospasmin from *Cl. tetani*: It locks synaptic inhibition (presyneptic) in spinal cord by blocking the release of inhibitory neurotransmitters like glycine and gamma amino butyric acid (GABA)

→ Resulting in uncontrolled spread of impulses
→ Tonic muscle rigidity and spasm.
- Botulinum (A, B, C1 and D-G except C2) from *Cl. botulinum:* It blocks the production or release of acetylcholine and causes descending flaccid paralysis (arflexia).
- Verotoxin/verocytotoxin (shiga toxin and SLT): Vero toxin causes paralysis and death on injection in mice/rabbit. It does not act directly on CNS but on the blood vessels of CNS, neurotoxic effect is secondary.

4. **Enzyme-based actions:** Exotoxins produced by *Cl. welchii* have enzyme based actions.
5. **Connective tissues action:** Following toxins act on the extracellular matrix of connective tissue and aid in spreading the infection by breaking down extracellular matrix of connective tissue.
   - Kappa toxin (collagenase) produced by *Cl. welchii.*
   - Exfoliative toxin produced by *Staph. aureus.*
6. **Hormonal actions:** Pertussis toxin of *B. pertussis* activates the intracellular cAMP in pancreatic islets and increases the insulin secretion in animals not in humans.
7. **Immune-mediated actions**
   - Lymphocytosis produced by pertussis toxin of *B. pertussis.*
   - Superantigens: Staphylococcal enterotoxin, staphylococcal toxic shock syndrome toxin (tsst-1) and streptococcal pyrogenic exotoxin (exotoxin A and exotoxin B) are superantigens. They stimulate large numbers of T cells. Follow **Ch. 18** for more details.

- **Regulatory genes of exotoxins:** Three types as below:
  1. **Plasmid mediated exotoxins**
     - Enterotoxin production by *Staph. aureus*
     - Exfoliative toxin-B (heat labile) by *Staph. aureus*
     - Anthrax toxin by *B. anthracis*
     - Tetanospasmin by *Cl. tetani*
     - Labile toxin (LT) of *E. coli*
     - Stable toxin (ST) of *E. coli.*
  2. **Chrmosome mediated exotoxins**
     - Staphylococcal exfoliative toxin A (heat stable)
     - Streptococcal pyrogenic exotoxins B
     - Shiga toxin.
  3. **Bacteriophage mediated exotoxins**
     - Staphylococcal toxic shock syndrome toxin
     - Streptococcal pyrogenic exotoxins A and C
     - Botulinum type C and D
     - Diphtheria toxin (Tox+ gene/β-prophage)
     - Shiga like toxin (SLT)
     - Labile toxin (LT) of *V. cholerae.*
- **Detection of exotoxins**
  1. **Culture methods**
     - *In vivo:* By using laboratory animals.

- *In vitro:* By using cell/tissue culture, like Vero cell culture for verotoxin.
  2. **Serological tests:** Precipitation (like Elek's gel precipitation test for diphtheria toxin), agglutination, ELISA, RIA, etc., can be used.
  3. **Molecular tests:** DNA probe.
- **Clinical uses:** Uses and other details of different toxins are described in respective chapters.
- **Differences between two toxins:** Follow **Table 9.1**.

Biological active substances: They are released by *Cl. perfringens* in gas gangrene as listed below:
- Hemagglutinin: Active against RBCs of humans and animals in gas gangrene.
- Bursting factor: For muscle lesion in gas gangrene.
- Circulating factor: It increases sensitivity to capillary bed and also inhibits phagocytosis.
- Histamine.

Pigments: Follow **Ch. 10** for more details.

Others Factors

Invasiveness: It is defined as an ability of a pathogen to spread in the host tissues, e.g., *Streptococcus* is highly invasive producing septicemia whereas *Staphylococcus* is less invasive producing localized lesions. However, some pathogens are less invasive although produce fatal diseases, e.g., *Cl. tetani* confirmed to site of entry and produce fatal disease by elaborating the toxin.

TABLE 9.1: Differences between two exotoxin and endotoxin

Exotoxin	Endotoxin
Produced by gram-positive bacteria and also by GNB	Produced by GNB
Converted in toxoid with formalin treatment	Not converted in toxoid
Proteins	LPSs (lipoolygosaccharide in *H. influenzae*)
Heat labile (gets denatured on boiling) and few are heat stable	Heat stable (Not denatured on boiling)
MW: 50–1000 kDa	MW: 10 kDa
Actively secreted by cells and diffuses into surrounding medium	Part of cell wall and do not diffuses into surrounding medium
Separated by filtration	Separated by cell lysis
Enzymatic	Nonenzymatic
Specific pharmacological effect for each toxin	Nonspecific
Specific tissue affinity	No specific tissue affinity
Active in very minute dose (<1 µg)	Active only in very large dose (>100 µg)
Highly antigenic	Weakly antigenic
Action specifically neutralized by antibody	Neutralization by antibody is ineffective
Detected by different tests as described earlier in text	Detected by limulus lysate assay
Causes exotoxemia	Causes endotoxemia

Communicability: It is defined as an ability of microorganism to spread from one host to another host. It plays major role in development of epidemic or pandemic.

Infecting dose and lethal dose

- **Definitions**
  - Infectious dose (ID): It is the amount of pathogens (measured in numbers of microorganisms) required to cause an infection in the host.
  - Lethal dose: It is the amount of pathogens (measured in numbers of microorganisms) required to cause death in the host.
  - MID (minimum infecting dose) or MLD (minimum lethal dose): It is the minimum numbers of microorganisms required to produce clinical evidence of infection or death respectively.
  - ID_{50} or LD_{50}: It is the dose required to infect or to kill 50% animal under standard condition.
- **Effective factors over infective dose**
  - Age: Low dose requires in lower age.
  - Gastric acidity: Many microbes are susceptible to gastric acidity (like *Salmonella, Vibrio cholerae*, etc.), require high dose while many resist gastric acidity (like *Shigella*, etc.), require low dose. Factors like use of antacids, presence of local diseases like hypochlorhydria, achlorhydria, etc., can alter the gastric acidity are also effective over ID.
  - Use of antibiotics: Antibiotics reduce the competition between pathogens and normal flora, so they decrease the ID.
  - Others: Like local surgery, IDDs, etc., are also effective over ID.
- **Infective dose by different microbes:** Follow **Table 9.2**.

Route of infection: Initiation of infection depends on route of entry, e.g., *V cholerae* is infective orally only while *Streptococcus* is infective by any mode.

TABLE 9.2: Infective dose by different microbes	
Bacteria	**Infective dose (ID)**
Low infective dose	
Mycobacterium tuberculosis	<10 bacilli
Escherichia coli O157:H7	<10 bacilli
Francisella tularensis	10–50 bacilli
Shigella	10–100 bacilli
Campylobacter jejuni	$500–10^3$ bacilli
Cryptosporidium parvum	10–30 oocysts
Entamoeba coli	1 cyst
Giardia lamblia	Few cysts
High infective dose	
Bacillus anthracis	10^4 spores
Escherichia coli	$10^6–10^8$ bacilli
Salmonella Typhi	$10^3–10^6$ bacilli
Vibrio cholerae	$10^4–10^6$ bacilli

Site of infection: Bacteria differ in their site of selection in host and their ability to damage different organs, e.g., *M. tuberculosis* injected in rabbits damages kidneys and infrequently to liver and spleen while in guinea pigs lesions are mainly in liver and spleen and sparing the kidneys.

Intracellular location of microbes: Intracellular location of the organisms helps to escape the host defense and the effect of antibiotics. There are two types of microbes as per intracellular location like obligate and facultative as shown in **Table 9.3**.

Bacterial secretory system: Bacteria utilize particular strategies to release the virulence factors called bacterial secretory system. It is the important mechanism in bacterial survival and pathogenesis. There are six types of secretory system from type I to type VI.

Regulatory genes: Above all virulence factors of bacteria, responsible for pathogenicity are under control of specific genes located on the chromosome called pathogenicity islands. Removal of such genes make the bacterium avirulent. Pathogenicity islands have been detected in many bacteria like *Staphylococcus aureus, Escherichia, Shigella, Salmonella, Vibrio cholerae Helicobacter*, etc.

Host Factors

1. **Age:** Certain infections are common at particular age like *H. influenzae* B infection is common in pediatrics.
2. **Gender:** UTI is more common in female due to close proximity of genitals to anus allows the fecal contamination.

TABLE 9.3: Intracellular location of microbes	
Obligate intracellular	**Facultative intracellular**
Definition	
These microbes are not able to synthesize their own ATP and remain dependent on host cells	These microbes are able to synthesize their own ATP and remain independent on host cells and can live extracellularly
Bacteria	
M. leprae *Chlamydia* spp. *Rickettsia* spp. *C. burnetii*	*M. tuberculosis, S.* Typhi *Y. pestis, N. meningitidis* *Nocardia* spp., *Brucella* spp., *Francisella tularensis, L. monocytogenes*, etc.
Viruses	
All viruses	
Fungi	
P. jirovecii	*H. capsulatum* and *C. neoformans*
Parasites	
T. gondii, C. parvum, *Plasmodium* spp., *Leishmania* spp., *Babesia* spp., *Trypanosoma* spp.	

3. **Immune status:** Immunodeficiency status favors certain infections like *S.* Typhi, *M. tuberculosis*, etc.
4. **Blood groups:** *N. gonorrhoeae* infection is common in group O. *V. cholerae* infections is common in blood group B and least in blood group AB. Exact reasons for all these are not known.
5. **Deficiency of complement components:** Deficiency of C_5–C_9 components favor the meningococcal infection.
6. **Occupation:** Like laboratory infection in laboratory workers.
7. **Gastric acidity:** It affects over the infective dose of bacteria.

Environmental Factors

These include overcrowding, humidity and presence of reservoirs or sources or transmitting agents of infection (e.g., vectors) in environment.

ACCESS YOURSELF

Essay/Full Question

1. Epidemiological determinants of virulence of bacteria.

Short Notes

1. Bacterial adhesions.
2. Bacterial exotoxin.

Short Questions for Theory/Viva Questions

1. Write four examples of bacterial adhesions.
2. What are attenuation and exaltation?
3. Name the four bacteria producing IgA protease.
4. Write the four functions of endotoxin.
5. Name the four GNB producing the exotoxin.
6. Name the four bacteria producing the neurotoxin.
7. How following toxins are showing neurotoxicity?
8. Tetanospasmin, botulinum, shiga toxin and shiga-like toxin.
9. What is biofilm?
10. What is pathogenicity Island?

Comments on

1. IgA protease is known to reduce local immunity.
2. Removal of pathogenicity islands make the bacteria (microorganisms) avirulent.
3. UTI is more common in female.

MCQs for Chapter Review

Definitions

1. **Exaltation is:**
   a. Decreased virulence   b. Increased virulence
   c. No change   d. None

Precipitating Factors

2. **Adhesin is useful in:**
   a. Motility   b. Bacterial attachment
   c. Toxigenicity   d. Bacterial division
3. **The endotoxin which leads to endotoxic shock is actually:**
   a. Lipoprotein   b. Lipopolysaccharide
   c. Polysaccharide   d. Polyamide

4. **All are true *except*:**
   a. Exotoxin has enzymatic action
   b. Endotoxin has enzymatic action
   c. Exotoxin is highly antigenic
   d. Endotoxin is weakly antigenic
5. **Endotoxin from gram-negative organism is:**
   a. Polysaccharide
   b. Glycoprotein
   c. Lipoprotein
   d. Lipopolysaccharide
6. **True about exotoxins:**
   a. Lipopolysaccharide   b. Not antigenic
   c. Can be toxoided   d. Heat stable
7. **Exotoxins are:**
   a. Lipopolysaccharide in nature
   b. Produced by gram-negative bacilli
   c. Highly antigenic
   d. Very stable and resistant to chemical agents
8. **Septic shock is due to:**
   a. Protein   b. Lipopolysaccharide
   c. Techoic acid   d. Peptidoglycan
9. **Heat stable enterotoxin causing food poisoning is caused by all the following *except*:**
   a. *Bacillus cereus*   b. *Yersinia enterocolitica*
   c. *Staphylococcus*   d. *Clostridium perfringens*
10. **Gram-negative bacterium producing exotoxin is:**
    a. *V. cholerae*   b. *Sh. dysenteriae*
    c. *C. jejuni*   d. All of above
11. **True about mechanism of bacterial toxins:**
    a. Cholera toxin acts by inhibition of guanyl cyclase
    b. Botulinum toxin inhibits Ach release
    c. Shiga toxin of *Shigella dysenteriae* acts by inhibiting protein synthesis
    d. Diphtheria toxin acts by inhibiting protein synthesis
12. **Endotoxin of following gram-negative bacteria does not play any part in the pathogenesis of the natural disease:**
    a. *E. coli*   b. *Klebsiella*
    c. *Vibrio cholerae*   d. *Pseudomonas*
13. **Enterotoxin is produced by all *except*:**
    a. *Clostridium perfringens*
    b. *Staphylococcus aureus*
    c. *Streptococcus pyogenes*
    d. *Bacillus cereus*
14. ***Salmonella* Typhi is the causative agent of typhoid fever. The infective dose of *S.* Typhi:**
    a. One bacillus   b. 108–1010 bacilli
    c. 102–105 bacilli   d. 1–10 bacilli
15. **All of the following organisms are known to survive intracellularly *except*:**
    a. *Neisseria meningitidis*
    b. *Salmonella* Typhi
    c. *Streptococcus pyogenes*
    d. *Legionella pneumophila*
16. **Which of the following are intracellular?**
    a. Viruses   b. *Chlamydiae*
    c. *Mycoplasma*   d. *Rickettsia*
17. **Obligate intracellularly organisms is:**
    a. *Mycoplasma*   b. *Chlamydia*
    c. *Cryptococcus*   d. *H. pylori*

1. b
- Decreased virulence called attenuation and increased virulence called exaltation.

2. b
- Adhesin is the organ of adhesion and useful in bacterial attachment with host cells.

3. b
- Endotoxic shock is due to endotoxin which is lipopolysaccharide in nature while exotoxin is protein in nature.

4. b

5. d

6. c, d

7. b, c
- Follow **Table 9.1** for explanation of answers of MCQs 4–7.

8. b
- Septic shock is also called endotoxic shock, which is due to endotoxin which is lipopolysaccharide in nature.

9. d
- Follow section, **exotoxins (properties)** for explanation.

10. d
- Follow section, **"exotoxins (exotoxins producing Gram negative bacilli)"** for explanation.

11. b, c and d
- Follow section, **exotoxins (biological actions of exotoxins)** for explanation.

12. c
- Follow section, **endotoxins** (Lipopolysaccharide/LPS) for explanation.

13. c
- *Streptococcus pyogenes* produces the pyrogenic exotoxin and hemolysin (called streptolysin) but not the enterotoxin.

14. c
- Infective dose of different bacilli is mentioned in **Table 9.2**.

15. c

16. a, b, d

17. b
- Follow **Table 9.3** for explanation of answers of MCQs 15–17.

CHAPTER 10

Bacterial Growth Products

Chapter Outline

❑ Definition and Types

DEFINITION AND TYPES

Definition

During the growth, bacteria secrete the several products called bacterial growth products.

Types

Following are the different types of bacterial growth products.

Vitamins: Vitamin K, E and B are produced by intestinal flora like *Bacteroides* spp., and *E. coli.*

Chemicals/drugs

1. **Chemicals:** Acetone and butanol from *Cl. aceto-butylicum.*
2. **Toxins:** Intramuscular injection of *Cl. botulinum* toxin type A was 1st used for strabismus, is now recognized as safe and effective for many neuromuscular diseases.
3. **Enzymes**
   - **Streptokinase:** It is useful for myocardial infarction and other thomboembolic diseases.
   - **Streptodornase:** It is useful for empyema to liquefy thick pus.
4. **Antibiotics:** Following are the examples.
   - **Chloramphenicol:** Initially it was prepared from *Streptomyces venezuelae* in 1947. But nowadays, all the commercial products of chloramphenicol are prepared synthetically.
   - **Aminoglycosides:** Two categories are as follows.
     - Obtained from *Streptomyces* spp.: These drugs are labeled with suffix "mycin" such as streptomycin (1944) from *Streptomyces griseus*, kanamycin (1957) from *Streptomyces kanamyceticus*, tobramycin (1970) from *Streptomyces tenebrarius*, neomycin from *Streptomyces fradiae* and framyctin from *Streptomyces lavendulae.*
     - Obtained from *Micromonosporum* spp.: These drugs are labeled with suffix "micin" such

as gentamicin (1964) from *Micromonosporum purpurea* and sisomicin (1980) from *Micromonosporum inoyoensis.*

   - **Others**
     - Erythromycin (1952) from *Streptomyces erythreus.*
     - Mupirocin from *Pseudomonas* spp.
     - Polymyxin B (1940) from *Bacillus polymexa.*
     - Colistin (1940) from *Bacillus colistinus.*
     - Bacitracin from *Bacillus subtilis.*
     - Tyrothricin from *Bacillus bravis.*
     - Rifampicin (rifampin) from *Streptomyces mediterranei* (suffix "micin").
5. **Insectisides:** Prepared from *B. thuringenesis* and useful to prevent the food crops from diseases.

Bacteriocins

- **Definition:** Specific antibacterial substances produced by bacteria called bacteriocins.
- **History:** Bacteriocin production was 1st observed by Gratia in 1952 from *E. coli.*
- **Names of bacteriocins with producing bacteria**
  - Colicins: *E. coli* and *Sh. sonnei*
  - Diphthericin: *C. diphtheriae*
  - Megacin: *B. megaterium*
  - Proticin: *Proteus* spp.
  - Aeroginosin (pyocins): *P. aeruginosa* (*P. pyocyanea*)
  - Pesticins: *Y. pestis.*
- **Synthesis:** Bacteriocin is determined by specific plasmid called col factor. Col factor is transferred from one cell to other cell by conjugation or by transduction. Its production is stimulated by physical (UV rays) and chemical agents (nitrogen mustard).
- **Properties**
  - Bacteriocins are proteins while some are LPSs in nature.
  - They resemble like phages, e.g., pyocin appears like tail of phage under electron microscopy.

- Bacterium produces bacteriocin is immune to it, but it is susceptible to other bacteriocin.
- They adsorb on surface of susceptible cells as like phages.

- **Bacteriocin typing:** It based on ability of bacteriocin producing strain to kill standard indicator strain. Typing is useful in epidemiological typing of bacterial strain. Typing is done by plate diffusion technique as mentioned below:
  - Steps: Inoculate the test bacterium as broad streak in center of culture plates. Standard indicator strain is inoculated at right angle to original inoculum. Incubate the plate at required temperature and time.
  - Result: Pattern of inhibition of standard indicator strain represents the type of bacteriocin.

Pigments

- **Definition:** These are colored substances produced by bacteria.
- **Role of pigment in virulence:** Exact role of pigment is not known but may produce the following effects which increase the virulence of bacterium.
  - Pigment stops ciliary movement of respiratory epithelium and protects the bacterium from host defense.
  - It inhibits the growth of other bacteria and makes the bacterium dominant in mixed infections.
  - It catalyzes the production of superoxide and H_2O_2.
- **List of bacteria and pigment produced by them**
1. *Staphylococcus* **spp.:** Types and properties of pigments of staphylococci are described in **Ch. 49.**
2. *Streptococcus agalactiae* **(Group B, β-hemolytic** *Streptococcus***):** Ch. 50.
3. *Neisseria* **spp.:** Commensal *Neisseria* like *N. flavescens* (yellow pigment) and *N. flava* produce the pigment.
4. **Photochromogens:** They produce the yellow-orange pigment in light only.
5. **Scotochromogens:** They produce the yellow-orange-red pigment in light and in dark.
6. *Serratia marcescens* **(*S. prodigiosus*):** It produces red color pigment called prodigiosin. It is best produced at room/at 20°C. Its presence in sputum simulating presence of blood called pseudohemoptysis.
7. *Erwinia* **spp.:** It produces the yellow pigment.
8. *Yersinia pestis:* It absorbs the hemin and produces the dark brown pigmented colonies in blood agar and other hemin-containing media. Pigment production is essential for biofilm formation and flea blocking.
9. *Elizabethkingia* (*Flavobacterium*) *meningosepticum*: Ch. 67.
10. *Pseudomonas aeruginosa*: Ch. 67.
11. *Bordetella parapertussis:* It produces the brown diffusible pigment on nutrient agar after two days of incubation.
12. *Chromobacterium violaceum*: Ch. 71.

13. *Capnocytophaga canimorsus* **and other species:** Believed to produce yellow or orange pigments on blood agar.
14. *Legionella pneumophila:* It produces diffusible brown pigment on Feeley Gorman (FG) agar, which fluoresces in dull yellow color on UV light exposure. This may be enhanced by addition of tyrosine in medium.
15. *Prevotella melaninogenica:* It produces the hemin derived black or brown pigment.
16. *Nocardia* **spp., and** *Porphyromonas* **spp.:** These are also the pigment producing bacteria.

Exotoxins and enzymes: Follow **Ch. 9.**

Cell wall components: Like antigens, proteins, etc., are released free in medium during infection or growth of bacteria.

Others

- Gases: Like NH_3, H_2S, etc.
- Acids (by fermentation) and alkalis.

ACCESS YOURSELF

Short Notes

1. Bacterial growth products
2. Bacteriocin
3. Pigment producing bacteria.

Short Questions for Theory/Viva Questions

1. Name the four bacteria producing the bacteriocins.
2. Name the four pigment-producing bacteria.
3. Name the bacteria producing following pigments: Staphylo-xanthine, pyocyanin, prodigiosin and violacein

MCQs for Chapter Review

1. Bacterial pigment also considered as a virulence factor.

MCQs for Chapter Review

1. **Following is/are the bacteriocin(s) producing bacterium / bacteria:**
   a. *E. coli*   b. *Sh. sonnei*
   c. *C. diphtheriae*   d. *B. megaterium*
2. **Pigment-producing bacterium is:**
   a. *Streptococcus pyogenes*
   b. *Streptococcus agalactiae*
   c. *Streptococcus mutans*
   d. *Streptococcus pneumoniae*
3. **Pigment-producing colonies are seen in**
   a. *Pseudomonas*   b. Atypical *mycobacteria*
   c. *Serratia marcescens* d. All of the above

Answers and Explanation of MCQs

1. **a, b, c and d**
- Follow section, **types (bacteriocins → names of bacteriocins and producing bacteria)** for explanation.
2. **b**
3. **d**
- Follow section, **types (pigments → list of bacteria and pigments produced by them)** for explanation of answers of MCQs 2–3.

General Properties of Viruses

Chapter Outline

INTRODUCTION

Meaning of virus: Virus is a Greek word means poison. Term virus was coined by Edward Jenner in 1798.

Definitions

- **Virion:** Extracellular infectious virus particle called virion.
- **Viroid:** Virus particle without extracellular phase called viroid.

Differences between bacteria and viruses: Follow **Table 11.1**.

MORPHOLOGY OF VIRUSES

Sizes of Viruses

Viruses are very smaller than bacteria and they are determined by electron microscopy. Some larger viruses can be examined by light microscope like pox virus. Size of virus is measured in nm. They are very small, and they can pass through filter so called filterable viruses. Sizes of different viruses are mentioned in **Fig. 11.1**. Variation in size of viruses is described below:

1. Smallest (also smallest DNA) virus: *Parvovirus* [Parvo (Latin) means small] about 20 nm in size. It is as small as largest protein particle like hemocyanin.
2. Largest (also largest DNA) virus: Pox virus about 300 nm in size. It is as large as smallest bacterium like *Mycoplasma*.
3. Smallest RNA virus: Picornaviridae about 27–30 nm in size.
4. Largest RNA virus: Paramyxoviridae about 100–300 nm in size.

TABLE 11.1: Differences between bacteria and viruses

Bacteria	Viruses
Prokaryotes category	Neither prokaryotes nor eukaryotes
Larger, measured in µm	Smaller, measured in nm
Contain both DNA and RNA	Contain either DNA or RNA, but never both
Nucleic acid surrounded by plasma membrane	Nucleic acid surrounded by a capsid
Thick cell wall contains peptidoglycan and LPS	Some viruses have additional outer lipoprotein envelope
Contain organelles like mitochondria, Golgi apparatus, ribosomes, etc.	No cellular organelles
Extracellular except mycobacteria, chlamydiae, etc.	Intracellular
Contain the enzymes necessary for reproduction	Lack the enzymes necessary for reproduction
Multiply by binary fission	Multiply by replication of NA and synthesis of the viral proteins
Growth occurs on cell-free media	No growth on cell-free media
Not pass through filter except few like *M. pneumoniae*, *F. tularensis*, *C. burnetii*, *Chlamydia* spp., etc.	Pass through filter
Sensitive to antibiotics	Not sensitive to antibiotics
Few bacteria are motile	Motility absent

Viruses are very variable in shapes. Most are spherical (like parvo virus, picorna virus, etc.), and few are irregular in shape. Shapes of different viruses are mentioned in **Fig. 11.1.** Few viruses with particular shape are described below:

1. Rabies virus: Bullet shape
2. Ebola virus: Filamentous in shape
3. Pox virus: Brick shape
4. Tobacco Mosaic Virus (TMV): Rod shape
5. Bacteriophage virus: Tad pole in shape.

Structure of Viruses

Virus consists of nucleic acid, capsid, envelope, fibrils and enzymes. All are described below **(Fig. 11.2)**.

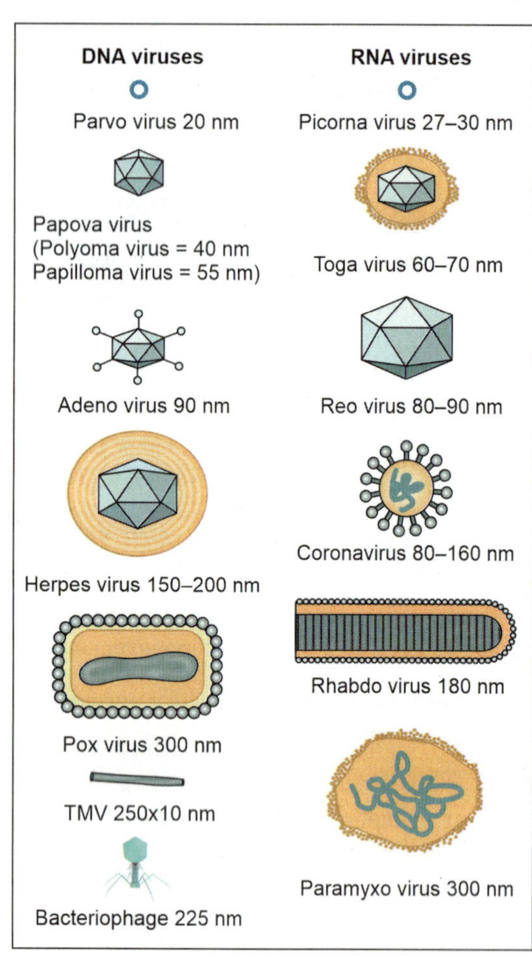

Fig. 11.1: Sizes and shapes of different viruses

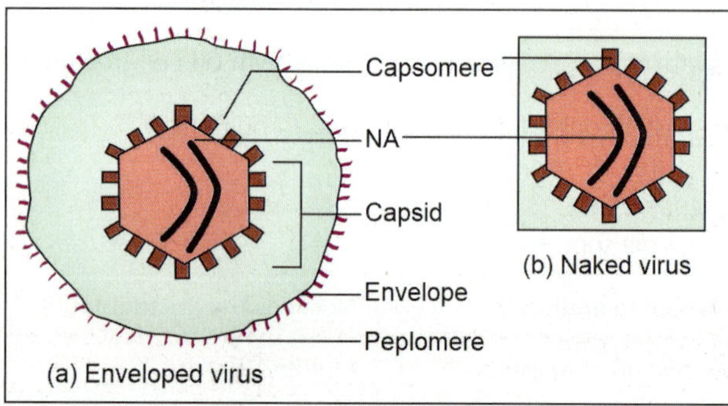

Fig. 11.2: Structure of virus

Nucleic acid (NA)/genome/core

Virus consists either DNA or RNA (never both). Protein range in NA is from about 1–50%. NA is extracted by treatment with detergent or phenol red. In some viruses like picorna virus and papova virus extracted NA is able to initiate the infection. It may be single stranded (ss) or double stranded (ds). It presents in different shapes like circular (nonsegmented) or linear (nonsegmented or segmented).

- **Circular (nonsegmented) genome:** It presents in coil form without ends in Papovaviridae and Poxviridae.
- **Linear genome**
  - Nonsegemented NA: It is available as single piece in following families of viruses
    1. **B**acteriophages (in few)
    2. **R**habdoviridae
    3. **T**ogaviridae
    4. **P**aramyxoviridae
    5. **P**neumoviridae.
  - Segmented NA: It is available in multiple pieces in following families of viruses
    1. **R**etroviridae: It consists two segments of ss-RNA (+) in *Lentivirus* (Visna virus, SIV and HIV)
    2. **O**rthomyxoviridae: It consists 8 pieces of ss-RNA (–) in influenza virus A and B while 7 pieces of ss-RNA (–) in influenza virus C.
    3. **B**unyaviridae: It consists three segments of ss-RNA (–) in *Hantavirus*, *Nairovirus* and *Phlebovirus*.
    4. **R**eoviridae: It consists 10–12 pieces of ds-RNA (+/–) in *Orthoreovirus* and *Orbivirus* while 11 pieces of ds-RNA (+/–) in *Rotavirus*.
    5. **A**renaviridae: It consists two segments of ss-RNA (–) Mammarenavirus (LCM virus, junin virus and machupo virus).

Mnemonic

- **Nonsegmented:** BTRP$_2$
- **Segmented:** ROBRA

Capsid or shell

- **Definition:** NA is surrounded by a protein coat called capsid.
- **Biochemistry:** Structural unit of capsid called capsomere which is made up by protein.
- **Functions of capsid**
  1. It protects the NA from deleterious agents like nuclease or other environmental agents.
  2. It introduces the NA in to host cell by adsorbing to cell surface.
  3. It provides antigenic property to virus.
- **Three types (symmetry) of capsid**
  1. **Icosahedral (cubical) symmetry (Fig. 11.3):** NA is surrounded by equilateral triangle of capsomeres (like soccer ball). It presents in herpes viruses, adeno viruses, etc. It composed of 12 vertices/

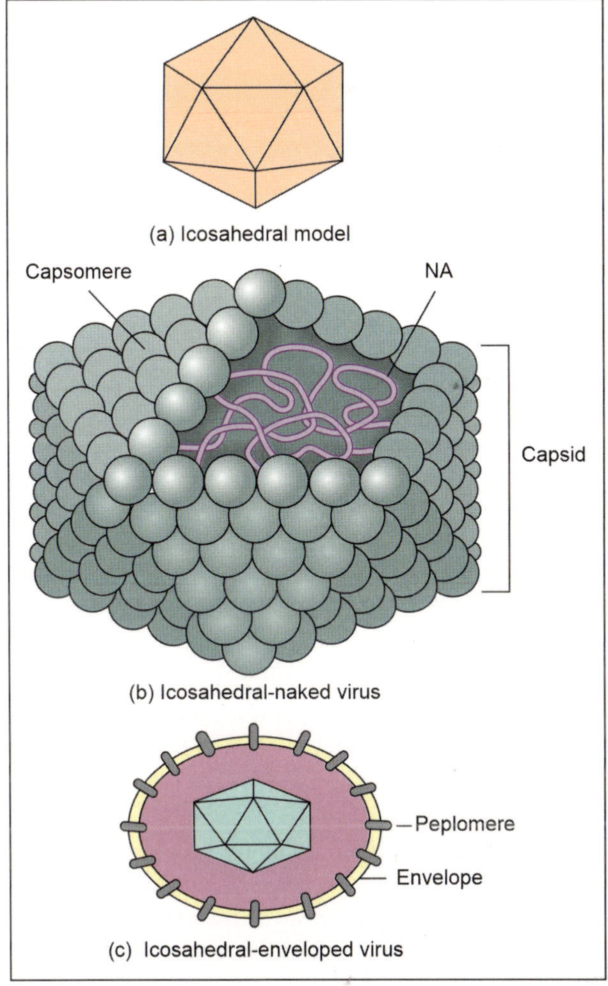

Fig. 11.3: Icosahedral symmetry of capsid

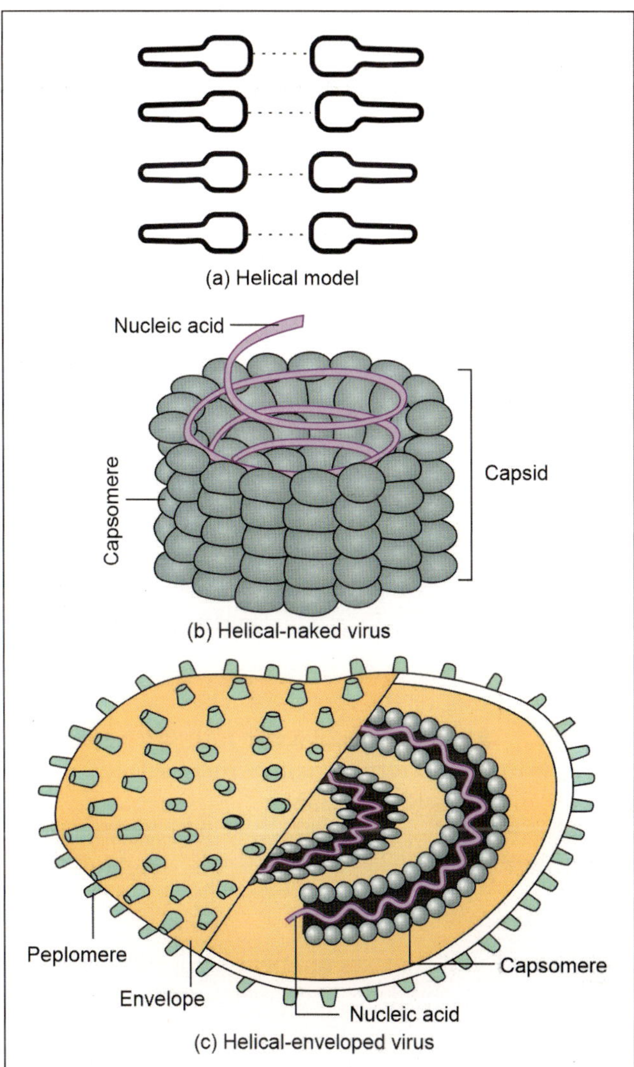

Fig. 11.4: Helical symmetry of capsid

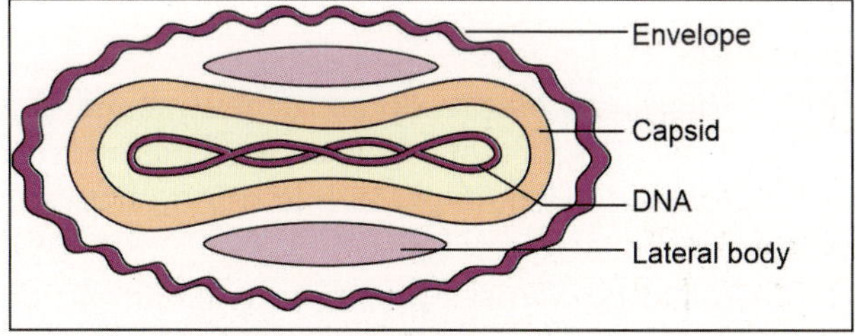

Fig. 11.5: Complex symmetry of capsid (pox virus)

corners and 20 facets/sides. Each facet has an equilateral triangle. Two types of capsomere in this type of capsid.
- Pentons: Present at vertices, always 12 in numbers
- Hexons: Present at facets, in variable numbers.

2. **Helical (spiral tube) symmetry (Fig. 11.4):** NA is surrounded by capsomeres which are arranged together in tube shape structure. The helix can be either rigid in TMV or flexible in influenza viruses.

3. **Complex symmetry (Fig. 11.5):** These viruses possess a capsid that is neither purely helical nor purely icosahedral, and that may possess extra structures such as protein tails or a complex outer wall. It presents in Poxviridae, Arenaviridae and Enterobacteriophage T4.

Envelope

- **Definitions:** Some viruses have an outer lipoprotein layer called envelope. Viruses with envelop are called enveloped viruses while some viruses without envelop are called non-enveloped (naked) viruses.
- **Biochemistry:** Envelope is made up by lipoprotein where lipid derived from the host cell plasma membrane by budding and proteins are viral encoded.

- **Functions:** It confers following properties on viruses.
  1. **Chemical:** Enveloped viruses are susceptible to lipid solvents like ether, chloroform and bile salts.
  2. **Antigenic:** Viral neutralization by specific antibodies to envelop antigens.

> **Note:**
> **Enveloped viruses (Fig. 11.2a):** NA, capsid and envelope.
> **Non-enveloped (naked) viruses (Fig. 11.2b):** NA and capsid (nucleocapsid).

1

- **Definitions:** Some envelope contains spikes like projections called peplomeres, from peplos meaning envelop.
- **Biochemistry:** Peplomeres are glycoproteins.
- **Types of peplomeres with example of influenza virus:** Influenza virus has two kinds of peplomeres.
  1. H (hemagglutinin): It allows the virus to attach to host cells (and red blood cells).
  2. N (neuraminidase): It is an enzyme that allows the mature viral particles to escape from the host cell.
- **Functions:** Peplomeres confer following properties on viruses:
  1. Antigenic: Viral neutralization by specific antibodies to peplomere antigens.
  2. Biological: Follow H and N **(described above)** peplomeres of influenza viruses.

Fibrils: It is an additional feature in some viruses such as adeno viruses. Fibrils are protruding from vertices.

Enzymes: Most viruses do not possess any enzymes for biosynthesis or energy production. They remain dependent on host cells; except influenza virus has neuraminidase and retro virus has reverse transcriptase which transcribes RNA to DNA.

RESISTANCE

Sterilization

1. **Heat and cold:** Viruses are heat labile and generally destroyed by heating at 50–60°C for 30 minute or at 4°C in a day. They are stable at low temperature. Viruses can be preserved at –70°C or for long storage. They are preserved by lyophilization (freeze drying).
2. **Radiation:** Ultraviolet, X-ray and high-energy particles inactivate the viruses.

Disinfection

1. **pH:** Viruses can be preserved at neutral pH (7.3). All viruses are killed at alkaline pH. Entero viruses are resistance to acidic pH, while rhino viruses are susceptible to acidic pH.
2. **Disinfectants:** Viruses are most resistant to disinfectants. Most common virucidal agents are $KMNO_4$, H_2O_2, hypochlorite, iodine and chlorine (but affected by organic matter present in water). Formaldehyde and BPL (beta propio lactone) are virucidal and used to prepare the killed vaccine. Enveloped viruses are susceptible to lipid solvents like ether, chloroform and bile salts. Susceptibility can be used to distinguish envelope viruses from non-envelope viruses. Naked viruses are most resistant to disinfectants. Many viruses can be stabilized by salt in concentrations of 1 mol/L, e.g., $MgCl_2$, $MgSO_4$ and Na_2SO_4.

Drug Resistance

Antibacterial agents have no effect on viruses.

VIRAL MULTIPLICATION

Steps

Total six steps (**Fig. 11.6**) described below.

Attachment or adsorption

- **Definition:** Specific binding between virus and specific receptors on the host cellular surface called adsorption.
- **Examples:**
  - HIV infects CD4 cells by specific receptors for gp120 spikes.
  - Influenza virus attaches to glycoprotein receptors on respiratory epithelium by hemagglutinin (H) spikes.
  - Polio virus can attach to lipoprotein receptors present on primate, not on rodent cell.
  - However, some viruses like picorna virus can directly inject the NA in to the host cells, which are resistant to infection by whole virus.

Penetration

- **Bacterial viruses:** Bacterial viruses called bacteriophages. They can enter in to bacteria by complex mechanisms due to tough bacterial cell wall. For more details follow **Ch. 87.**
- **Human viruses:** Human viruses can enter in host cells by mechanisms resembling phagocytosis called 'viropexis.' Human host cells do not have tough wall, so whole naked virus can enters in cell, while in enveloped virus, viral envelop fuse with plasma membrane and only nucleocapsid can enter in to cell.

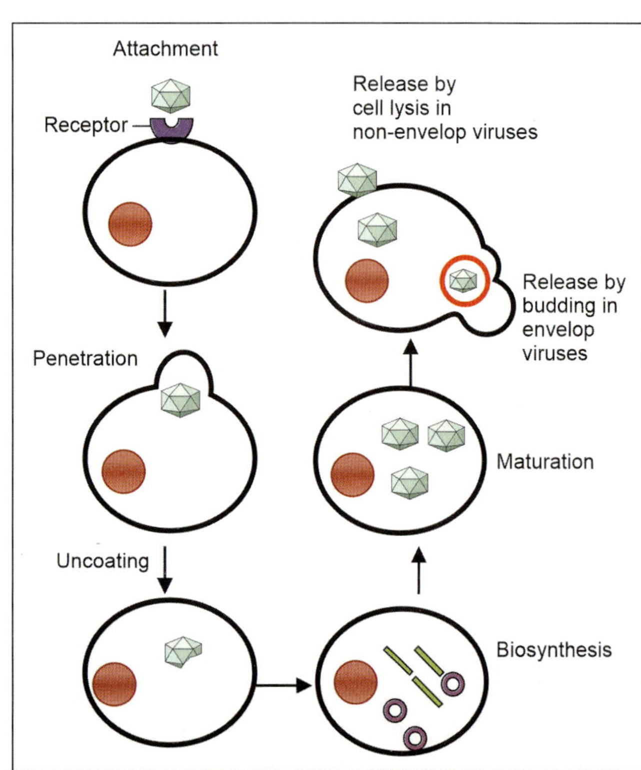

Fig. 11.6: Viral multiplication

Uncoating

- **Definition:** It is a process of stripping of envelop and capsid, so NA released in to the host cell.
- **Features:** Uncoating of virus is accomplished by lysosomal enzymes. Bacteriophage do not require uncoating because their nucleic acid is injected into the host cell. In pox virus uncoating is two steps process. In 1st step stripping of envelop by lysosomal enzymes and in 2nd step stripping of capsid by specific viral uncoating enzymes.

Biosynthesis

- **Biosynthesis of viral parts:** It includes the synthesis of NA, capsid, enzymes and regulatory proteins: Regulatory proteins help in biosynthesis by inhibiting the host cell metabolism.
- **Sites of biosynthesis:** It depends on types of particles to be synthesized and types of virus.
  1. **Biosynthesis of NA**
     - In DNA viruses: NA is synthesized in nucleus except pox virus in which it is synthesized in cytoplasm.
     - In RNA viruses: NA is synthesized in cytoplasm except orthomyxo virus, paramyxo virus and retro virus in which it is synthesized in nucleus.
  2. **Biosynthesis of viral proteins:** Take place only in cytoplasm.
- **Steps of biosynthesis**
  1. **Transcription:** It is the process of synthesis of mRNA from viral NA. Six classes are described by Baltimore in 1970.
     - **Class-1:** In ds-DNA viruses like adeno virus, herpes virus and papova virus, DNA enters in host cell nucleus and uses host cell enzymes for transcription. In partially ds-DNA virus like hepadna virus duplex is formed by DNA polymerase in cytoplasm, than mature NA enters in to host cell nucleus and uses host cell enzymes for transcription.
     - **Class-2:** It occurs in ss-DNA virus like parvo virus. DNA enters in host cell nucleus, converted into duplex and than uses host cell enzymes for transcription.
     - **Class-3:** It occurs in ds-RNA virus like reo virus where ds-RNA transcribed to mRNA by viral polymerase.
     - **Class-4:** Plus (positive) strand ss-RNA viruses like picorna virus and toga virus. RNA itself acts as mRNA and translated to viral proteins as shown in **Flowchart 11.1a.**
     - **Class-5:** Minus (negative) strand ss-RNA virus like rhabdo virus, orthomyxo virus and paramyxo virus. Viruses possess their own RNA polymerase (RNA dependent RNA polymerase) which transcribed RNA to mRNA which is translated to protein as shown in **Flowchart 11.1b.**
     - **Class-6:** Like retro viruses. Conversions of ss-RNA to ds-DNA by enzyme reverse transcriptase (RNA-dependent DNA polymerase, DNA polymerase), DNA transcribed to mRNA which is translated to protein as shown in **Flowchart 11.1c.**
  2. **Translation:** It is the process of synthesis of early or nonstructural proteins from mRNA. These are enzymes which induce and maintain the synthesis of viral components by inhibiting the host cell metabolism.
  3. **Replication:** It includes the replication of viral NA.
  4. **Synthesis of late or structural proteins:** Like capsid.

Maturation and assembly

- After synthesis of viral particles, they undergo to maturation and assembly to produce the complete virus. In adeno virus and herpes virus it takes place in nucleus. In picorna virus and pox virus it takes place in cytoplasm.
- At this stage non-enveloped virus are complete, while in enveloped virus only nucleocapsid is synthesized followed by envelope synthesis from host cell membrane by budding method.
- This envelop undergoes modification to introduce virus specific Ags like hemagglutinin (H) and neuraminidase (N) in influenza virus.

Release: Release of daughter virus is possible by following mechanisms.

- **Host cell lysis:** Viruses can be released from the host cell by lysis, a process that kills the host cell by bursting its membrane and cell wall if present. This is a feature of many viruses like bacteriophage and some animal viruses like polio virus.

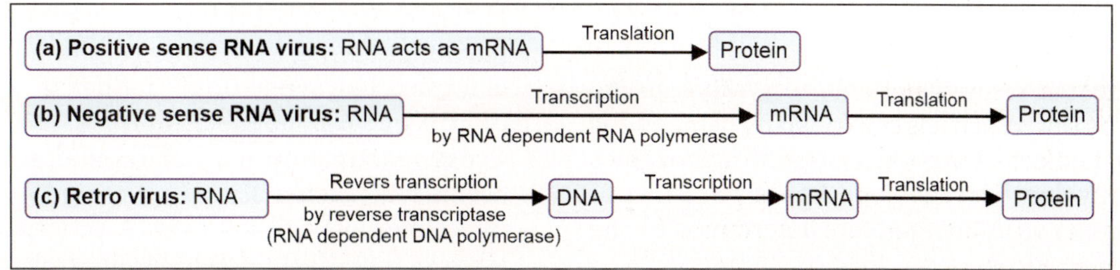

Flowchart 11.1: Mechanisms of translation in RNA viruses

- **Budding:** Enveloped viruses (e.g., HIV) are typically released from the host cell by budding. During this process the virus acquires its envelope, which is a modified piece of the host's plasma or other internal membrane.
- **Some viruses undergo a lysogenic cycle:** No release of progeny virus, but viral genome is incorporated by genetic recombination into the host's chromosome. Such viral genome "called provirus" or in the case of bacteriophage "called prophage". "Provirus" or "prophage" behaves like host genome and multiplies simultaneously with host cell's genome and confers certain new properties in host cell called lysogenic conversion. Following are examples of lysogenic conversion
  - Malignant transformation by oncogenic virus: Normal cells changed to malignant cells.
  - Conversion of nontoxigenic strain of bacteria in to toxin producing strain: Such as in *C. diphtheriae*.

Eclipse Phase

From the stage of entry to synthesis of daughter virions, virus remained underground or disappeared and cannot be demonstrated in host cells called eclipse phase. It is about 15–30 minutes in bacteriophage and 15–30 hours in human virus.

Burst Size

Release of progenies of virus per infected cell called burst size. Burst size is estimated in bacteriophages and difficult to measure in human viruses.

Abnormal Replicative Cycle

Abortive infection: Due to defect in release, virus infection in some cells does not lead the production of progenies of virus called abortive infection.

Incomplete viruses: Due to defective assembly, progenies of virus released may not be infective called incomplete viruses.

von Magnus phenomenon: Such incomplete viruses are seen when cell infected with high dose of influenza virus, where progenies of virus released have high hemagglutinin titer but low infectivity called von Magnus phenomenon. It is due to formation of incomplete viral particles lacking NA.

Defective viruses: Some viruses when infect the cells are not able to give rise fully formed progenies called defective viruses. Defective viruses can give rise fully formed progenies, only when they are coinfected by helper virus like

- Rouse sarcoma virus is not able to synthesize the envelope by self. When it is coinfected by helper virus like Avian Leukosis Virus (ALV), resulting progenies are with envelop. The envelope Ags of progenies of rouse sarcoma virus are therefore determined by the type of helper virus.
- HDV (defective virus) and HBV (helper virus)
- Adeno-associated satellite virus (defective virus) and adeno virus (helper virus).

VIRAL GENETICS

Two main mechanisms for genetic inheritance of virus like mutation and recombination. Additionally, virus also exhibits nonheritable mechanisms due to interaction between viral products.

Mutation

Some important points are highlighted at here. For more details follow **Ch. 8.**

Frequency of mutation: It is same as bacteria like 10^{-4}–10^{-8}.

Change in antigenic properties: It confers certain new properties in virus like virulence, host range, antigenicity (antigenic drift and antigenic shift in influenza virus), pock or plaque morphology and drug resistance.

Conditional lethal mutation: A mutation may affect an organism in such a way that the mutant can survive only in certain environmental condition called conditional lethal mutation. For example, a temperature sensitive mutant (ts mutant) can survive at permissive temperature of 35°C but not at restrictive temperature of 39°C. It is used in study of viral genetics and vaccine production (influenza vaccine).

Recombination

Definition: Transfer of genetic materials or genetic information from one cell to other cell in viruses (viruses are different but related) called genetic recombination.

It occurs between
- Two active (infectious) viruses.
- One active and one inactive virus.
- Two inactive (noninfectious) viruses.

Methods of viral gene transfer: Following types:
1. **Intramolecular recombination:** It occurs when two closely related viruses infect the host cell. Two viruses exchange the NA and form the hybrid that harbors the genes from both species. It occurs in ds-DNA and in some RNA viruses.
2. **Reassortment:** It occurs when different strains of same virus are growing together with different antigens. Such as one human strain, one avian strain and two pig (swine) strains of influenza virus (total four strains) growing together in pig with different H and N Ags. A hybrid may obtain H-Ag from one parent and N-Ag from other resulting a novel virus is formed like swine origin influenza virus (S-OIV) or H1N1 2009 pandemic strain.
3. **Cross reactivation/marker rescue/reactivation:** It occurs when active strain and related inactive strain infect the host cell together. A progeny virus may possess gene from active strain and one or more

genes from inactive strain. For example active strain of influenza virus (A2) grows very poorly in eggs. When A2 grows in eggs along with standard strain (A1) under activation by radiation, resulting progeny contains growth properties of A1 and antigenic properties of A2. This is useful in preparation of influenza vaccine.

4. **Multiplicity reactivation:** When cell is infected by larger dose of single virus (high MOI: Multiplicity of infection), inactivated by UV radiation, it produces changes in host cell gene and produces the live virus called multiplicity reactivation. This is useful in preparation of vaccine; however, vaccine inactivated by UV radiation is not safe.
5. **Lysogenic conversion:** Described above.
6. **Pseudovirion:** Normally viral NA is enclosed by viral capsid, but sometimes host NA is enclosed by viral capsid called pseudovirion. For example, in papova virus.
7. **Transduction:** Transfer of host NA from one cell to other cell by bacteriophage called transduction. This is useful to correct the inborn errors of metabolism.

Interaction between Viral Products/ Nonheritable Mechanisms

Phenotyping mixing: When two different viruses infect the host cell, NA of one virus is surrounded by capsid of other virus called transcapsidation. It is not a stable change; on subsequent passage original capsid will form.

Genotyping mixing: Incorporation of more than one complete genome into virus (heterozygosis). On subsequent passage, two types of viral progenies will form.

Complementation: It is the functional interaction between two viruses, in which one or both may be defective. One virus provides gene products to other virus which are defective and resulting other virus will replicate and progeny viruses are like parent virus. For example when rabbit is infected with heat inactivated virulent myxoma virus and active avirulent fibroma virus, it develops fatal myxomatosis. Heat inactivated myxoma virus is not able to produce infection due to deficiency of DNA dependent RNA polymerase. When coinfected by fibroma virus, it provides the required enzymes to produce infection.

Interference: Presence of one virus may interfere the function of other virus. Interference is produced by interferon production, by destruction of receptors required for attachment or by autointerference. In auto-interference high MOI inhibits the production of own progeny virus. Interference is useful in controlling polio outbreak by administration of live attenuated polio virus vaccine (OPV), which interfere with wild polio virus. On other hand presence of preexisting virus will interfere the vaccine virus.

Enhancement: Presence of one virus will increases the effect of (CPE) other virus called enhancement.

Definition

Viral taxonomy is the description, identification, nomenclature and ordered classification of viruses according to their presumed natural relationships.

Nomenclature

Order: It is labeled by adding the suffix 'virales'.

Family: Family is classified on the basis of morphology, genome structure and strategies of replication. Virus family is labeled by adding the suffix 'viridae'. Virus subfamily is labeled by adding the suffix 'virinae'.

Genus: Genus is classified on the basis of physicochemical/ serological differences. Genus name is given by adding the suffix *'virus'* and it is written in Italic pattern.

Species: No exact system for species nomenclature.

Classification

Properties: All microbes are grouped either in prokaryotes or in eukaryotes. Viruses are not classified in any group. Viral classification system is proposed by an International Committee on Taxonomy of Viruses in 2000.

Bases for classification: Two main types of viruses on the bases of NA such as DNA viruses and RNA viruses as mentioned in **Table 11.2.** Further types are based on following features:
- **Morphological features:** Size, shape, type of symmetry, presence or absence of envelope.
- **Genetic features:** Size of genome, strandedness (ss or ds), linear or circular, positive (+) or negative (−) sense.
- **Other features:** Such as physicochemical or serological features.

SUBVIRAL AGENTS

Viroids

Definition: Virus particles without extracellular phase called viroids.

History: Term was 1st introduced by Diener in 1971.

Morphology: Viroids contains ss-RNA, which is smaller than normal virus. HDV may resemble to viroid. Capsid and envelope are absent. They contain enzyme RNA polymerase II which instead of synthesizing RNA from DNA, uses the rolling cycle to synthesize RNA by using viroid's RNA as template.

Resistant: They are resistant to heat and sensitive to nuclease.

Pathogenicity: First viroid identified was potato spindle tuber viroid. It is a plant pathogen causing potato spindle tuber disease. Viroids also cause human and animal diseases.

TABLE 11.2: Classification of viruses

Naked or enveloped	Virus family	Genus/species	Capsid symmetry	Nucleic acid type
DNA viruses				
Enveloped	**P**oxviridae	Smallpox virus, cowpox virus, sheep pox virus, orf virus, monkeypox virus and vaccinia virus	Complex	ds
	Herpesviridae	Herpes simplex virus, varicella-zoster virus, cytomegalovirus, Epstein–Barr virus, etc.	Icosahedral	ds
	Hepadnaviridae	Hepatitis B virus	Icosahedral	Circular, partially ds
Naked	**A**denoviridae	Adeno virus	Icosahedral	ds
	Papovaviridae	HPV, SV-40 virus, JC virus and BK virus	Icosahedral	ds circular
	Parvoviridae	Parvovirus B19	Icosahedral	ss
RNA viruses				
Enveloped	**T**ogaviridae	*Rubivirus* (Rubella virus), *Pestivirus, Alphavirus*	Icosahedral	ss (+)
	Flaviviridae	Dengue virus, HCV, Yellow fever virus	Icosahedral	ss (+)
	Arenaviridae	*Mammarenavirus* (LCV, Junin virus, Machupo virus)	Complex	ss (–)
	Orthomyxoviridae	*Alphainfluenzavirus, Betainfluenzavirus, Gammainfluenzavirus, Deltainfluenzavirus*	Helical	ss (–)
	Paramyxoviridae	Mumps virus, measles virus, *henipavirus* (Nipah virus)	Helical	ss (–)
	Pneumoviridae	Human orthopneumovirus (RSV), human metapneumovirus	Helical	ss (–)
	Bunyaviridae	*Hantavirus, Nairovirus, Phlebovirus, Ukuvirus*	Helical	ss (–)
	Rhabdoviridae	*Vesicullovirus* (Vesicular stomatitis virus, Chandipura virus), *Lyssavirus* (Rabies virus)	Helical	ss (–)
	Filoviridae	Ebola virus, Marburg virus	Helical	ss (–)
	Coronaviridae	SARS-CoV1, MERS-CoV and SARS-CoV2	Helical	ss (+)
	Retroviridae	*Lentivirus* (Visna virus, SIV and HIV)	Icosahedral	ss (+)
Naked	**R**eoviridae	*Rotavirus, Orbivirus , Coltivirus, Orthoreovirus*	Icosahedral (two capsid)	ds (+/–)
	Picornaviridae	*Enterovirus, Rhinovirus*	Icosahedral	ss (+)
	Caliciviridae	*Norovirus*	Icosahedral	ss (+)
	Hepeviridae	HEV	Icosahedral	ss (+)
	Astroviridae	*Astrovirus*	Icosahedral	ss

Mnemonics

DNA viruses: All viruses have ds DNA except Parvoviridae.

- **Enveloped DNA viruses: PH2** → **P**oxviridae, **H**erpesviridae and **H**epadnaviridae.
- **Naked DNA viruses: AP2** → **A**denoviridae, **P**apovaviridae and **P**arvoviridae.
- **Icosahedral capsid:** All the DNA viruses whether enveloped or naked have icosahedral type of capsid except Poxviridae having complex type of capsid

RNA viruses: All viruses have ss RNA except Reoviridae

- **Enveloped RNA viruses: FABRha-FOT PPCR** → **F**laviviridae, **A**renaviridae, **B**unyaviridae, **Rha**bdoviridae, **F**iloviridae, **O**rthomyxoviridae, **T**ogaviridae, **P**aramyxoviridae, **P**neumoviridae, **C**oronaviridae and **R**etroviridae.
- **Naked RNA viruses: Astro-H PCR** → **A**stroviridae, **H**epeviridae, **P**icornaviridae, **C**aliciviridae and **R**eoviridae.
- **Icosahedral and helical capsid:** RNA viruses have icosahedral or helical type of capsid except Arenaviridae having complex variety of capsid.

Prions and Prion Diseases

Definition: Proteinaceous infectious virus like-particles without nucleic acid (NA) called prions.

History: Prion was identified by Stanley B Prusiner, awarded with Nobel Prize in 1997.

Morphology: Prions contain protein without NA. MW of protein is 50,000 and size is 4–6 nm.

Resistant: Prions are sensitive to protease. They are resistant to nuclease, dry heat (90°C for 3 min.) and UV-rays. Routine methods like autoclaving (120°C for 20 min.), boiling, radiation, etc., are not effective. Prions are killed by autoclaving at 134°C for 5 hours with prior treatment with acidic detergent or by treatment with 2N NaOH for several hours or by 0.5% sodium hypochlorite for 2 hours.

Pathogenicity

- **Disease name:** Disease called prion disease.
- **Incubation period:** Long incubation period may be up to 30 years, but once the disease sets, progress will be fast.
- **Sites:** CNS.
- **Pathogenesis:** Many theories are proposed for the pathogenesis of prion disease as follows:
  1. **Stanley B Prusiner theory:** It is mentioned in **Flowchart 11.2**. Because of his great work he was awarded with Noble Prize in 1997.
  2. **Abnormal protein folding:** α-helix structure of PrPc changes to structure PrP^{sc} called misfold or unfold protein. This unfolded protein is dangerous because it unfolds many other folded prion proteins. Unfold protein undergoes replication and spreads throughout tissues without using DNA or RNA to produce the disease.
- **Pathology:** Prions produce progressive vacuolation in the dendritic plus axonal process of the neurons. They cause extensive astroglial hypertrophy plus proliferation, spongiform encephalopathy, neurodegeneration in the grey matter, amyloid plaque formation in brain and gliosis. There is no inflammation and absence of immune response (non-immunogenic).
- **Clinical types:** Two types of prion diseases:
  1. **Prion diseases in animals**
     - Scrapie: It is transmitted vertically from ewe to lamb.
     - Mink encephalopathy: It spreads to mink when it feds on scrapie infected sheep's meat.
     - Mad cow disease [Bovine spongiform encephalopathy (BSE)]: It spreads to cattle when they fed on scrapie infected sheep's meat. Enzootic occurred in 1986 in Britain.

Flowchart 11.2: Pathogenesis of prion particle

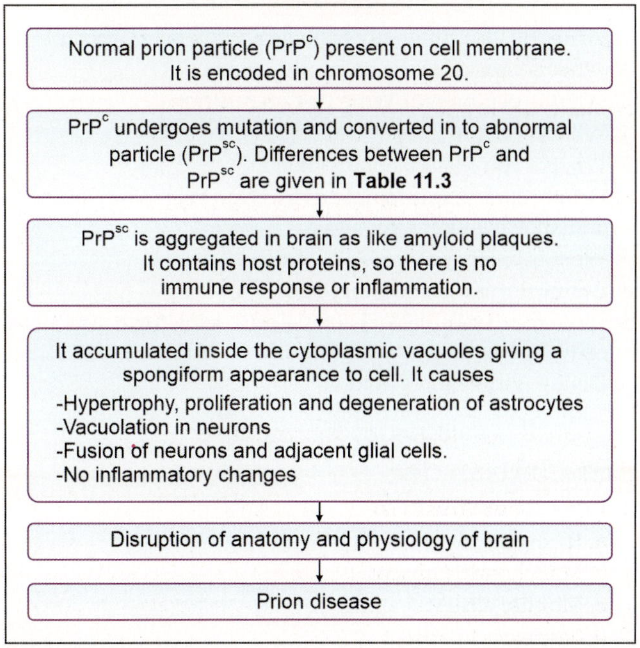

Normal prion particle (PrP^c) present on cell membrane. It is encoded in chromosome 20.

↓

PrP^c undergoes mutation and converted in to abnormal particle (PrP^{sc}). Differences between PrP^c and PrP^{sc} are given in **Table 11.3**

↓

PrP^{sc} is aggregated in brain as like amyloid plaques. It contains host proteins, so there is no immune response or inflammation.

↓

It accumulated inside the cytoplasmic vacuoles giving a spongiform appearance to cell. It causes
-Hypertrophy, proliferation and degeneration of astrocytes
-Vacuolation in neurons
-Fusion of neurons and adjacent glial cells.
-No inflammatory changes

↓

Disruption of anatomy and physiology of brain

↓

Prion disease

TABLE 11.3: Differences between PrP^c and PrP^{sc}

Features	PrP^c	PrP^{sc}
Full form	Prion particle cellular	Prion particle scrapie
Forms	Normal form	Active or pathogenic form
Structure	Elongated polypeptide with more α helix and less β structure	Globular polypeptide with less α helix and more β structure
Location	Cell membrane	Cytoplasm
Protease	Sensitive	Resistant

2. **Prion diseases in humans**
   - **CJD (Creutzfeldt-Jacob Disease):** A variety of CJD (vCJD) was occurred in persons <45 years in Britain in 1996, due to consumption of beef, infected with BSE and many cattle were slaughtered before anxiety was allayed. It occurred in sporadic due to mutation in prion. It also occurred in inherited (familiar) forms. It is **transmitted by** iatrogenic ways due to corneal transplantation, dura mater graft implantation (>160 cases have been reported), injection of pituitary growth hormone from infected person (>180 cases have been reported), electro encephalogram electrode implantation or by ingestion of beef, infected with BSE or by hereditary. **Clinical features** include senile encephalopathy, progressive incoordination, cerebellar gait, pyramidal signs, extrapyramidal dysfunctions, seizures, visual impairment, 90% myoclonus, and dementia. Death occurs within a year. It presents with defective higher cortical functions.
   - **Kuru:** Kuru means tremor, a characteristic feature of disease. **Incubation period** is about 5–10 years. It is **transmitted** due to anthropophagy or cannibalism. It is an act or practice of humans, eating the flesh or internal organs of other human beings. Word came from Caníbales, the Spanish name for the Caribs, a West Indies (New Guinea) tribe that formerly practiced cannibalism after the death of relatives due to customs. It was identified in New Guinea and disease was stopped after stopping the cannibalism. **Clinical features** include cerebellar ataxia, tremor and death. **Nobel Prize** was given to Carlton Gajdusek in 1976 for his contribution in disease.
   - **Other diseases:** Such as GSSS (Gerstmann Straussler Scheinker Syndrome) and fatal familial insomnia.

Laboratory diagnosis

- **Specimens:** Brain biopsy and CSF.
- **Testing methods**
  1. **Light microscopy:** It shows histopathological changes like hypertrophy, proliferation and

degeneration of astrocytes, vacuolation in neurons, fusion of neurons and adjacent glial cells. No inflammatory changes are present.

2. **Conformation-dependent assay:** It is specific diagnosis for measurement of PrPsc.

3. **Stress protein 14-3-3:** Elevated in CSF.

4. **Sequencing the gene:** It is important to detect the mutation to identify the familiar CJD.

5. **Abnormal EEG:** In late stage of disease, high voltage, triphasic sharp discharges are observed.

6. **MRI:** Greater than 90% cases showing the increased intensity in the basal ganglia and cortical ribboning.

Prevention: No effective measures are available.

Treatment: Several measures using drug like quinacrine and anti-PrP antibodies eliminate the prion particles from the cultured cells but failed to do so *in vivo*.

PRECIPITATING FACTORS OF VIRAL INFECTIONS

Synonym

Epidemiological determinants.

Types

Virulence of viruses is affected by following three factors.

Agent factors (virulence factors or determinants of virulence)

1. **Intracellular location:** It provides protection against the host defense system.

2. **Viral capsid:** It protects the nucleic acid against the deleterious agents.

3. **Viral adhesins:** Glycoprotein spikes present on envelope of virus like gp120 in HIV, H (hemagglutinin) protein in influenza virus, S (spike) protein in SARS-CoV2, F (fusion) protein in paramyxoviruses, etc., help in attachment with host cells **called adhesins.**

4. **Neuraminidase in influenza virus:** It is a mushroom like glycoprotein enzyme which destroys the cell receptors by splitting of N-acetyl neuraminic acid from it. It destroys the hemagglutinin (H) receptors on RBCs hence called receptor destroying enzyme (RDE) and causes reversal of hemagglutination and releases the bounded viruses to infect the other RBCs called elution.

5. **Exotoxin-producing virus:** In *Rotavirus* NSP4 (non-structural protein 4) considered as enterotoxin which increases the vascular permeability and the secretion.

Host factors

1. **Age:** Certain infections are common at particular age like *Rotavirus* common in pediatrics.

2. **Smoking:** People who smoke do have more frequent infections by rhino viruses.

3. **Immune status:** Immunodeficiency status favors certain infections like SARS-CoV2, etc.

4. **Occupation:** Like HBV, SARS-CoV2, etc., are common in laboratory workers.

Environmental factors

1. **Seasonal:** *Rhinovirus* infection occurs year round with seasonal peaks of incidence in the early fall, usually September to November and again in the spring from March to May. During these periods increased incidence up to 80% of common cold. In India, most cases of polio were reported in rainy season during June–September.

2. **Overcrowding and poor sanitation:** They also increase risk of mosquito borne viral infections.

> **Note: Important terminology**
> - Equine: Horse
> - Feline: Cat
> - Bovine: Cattle (cow/buffalo)
> - Canine: Dog
> - Murine: Rat/mice
> - Simian: Monkey
> - Swine (Porcine): Pig
> - Avian: Birds
> - Caprine: Sheep/goat

ACCESS YOURSELF

Essay/Full Question

1. Morphology of viruses.

Short Notes

1. Structure of viruses
2. Viral multiplication
3. Classification of viruses or viral taxonomy
4. Prion particle.

Short Questions for Theory/Viva Questions

1. Write the four differences between viruses and bacteria.
2. What is capsid? Write its two functions.
3. What is peplomere? Write its two functions.
4. What is eclipse phase?
5. What is von Magnus phenomenon?
6. Define: Burst size, incomplete virus, von Magnus phenomenon and abortive viral infection.
7. What is defective virus? Write two examples.
8. Define: Provirus and prophage.
9. Write four examples of bacteria producing toxin after phage conversion.
10. Define: Virion and viroid.

MCQs for Chapter Review

Introduction

1. **False about viruses is:**
   a. Ribosome absent
   b. Mitochondria absent
   c. Motility absent
   d. Nucleic acid absent

2. **Following is not the property of virus:**
   a. Contains both DNA and RNA
   b. Neither prokaryotes nor eukaryotes
   c. No growth on media
   d. Passes through filter
3. **Which of the following does not possess both DNA and RNA?**
   a. Bacteria
   b. Fungi
   c. Viruses
   d. Spirochetes
4. **All of the following are general properties of viruses** *except*:
   a. May contain both DNA and RNA
   b. Form extracellular infectious particles
   c. Heat labile
   d. Not affected by antibiotics

Morphology of Viruses

5. **Brick-shaped virus is:**
   a. Chickenpox
   b. Smallpox
   c. CMV
   d. EBV
6. **Rabies virus is in:**
   a. Bullet shape
   b. Spherical shape
   c. Brick shape
   d. Rod shape
7. **Smallest DNA virus is:**
   a. Herpes virus
   b. Adeno virus
   c. Parvo virus
   d. Pox virus
8. **The virus with smallest genome:**
   a. Reo virus
   b. Parvo virus
   c. Picorna virus
   d. HIV
9. **Segmented genome is found in all** *except*:
   a. Influenza virus
   b. Reo virus
   c. Bunya virus
   d. Rhabdo virus
10. **Segmented genome is found in:**
    a. Retro virus
    b. Rota virus
    c. Polio virus
    d. Rhabdo virus
11. **Segmented RNA is found in:**
    a. Influenza virus
    b. Rabies virus
    c. Herpes virus
    d. Molluscum contagiosum virus
12. **Segmented double stranded RNA virus is seen in:**
    a. Reo virus
    b. Myxo virus
    c. Rabies
    d. Parvo virus
13. **DNA covering material in a virus is called as:**
    a. Capsomere
    b. Capsid
    c. Nucleocapsid
    d. Envelope
14. **Symmetric protein shell which encase the nucleic acid core of virus?**
    a. Capsomere
    b. Capsid
    c. *Basidiomycetes*
    d. Fungi imperfecti

Viral Multiplication

15. **von Magnus phenomenon:**
    a. Is a normal replicative cycle
    b. Virus yield has low hemagglutination
    c. Virus has high infectivity
    d. Virus yield has high hemagglutination titer but low infectivity
16. **Which of this/these is/are an example of a defective virus?**
    a. Hepatitis C virus
    b. Delta virus
    c. Rous sarcoma virus
    d. Both b and c

Viral Genetics

17. **One virus particle prevents multiplication of second virus. This phenomenon is:**
    a. Viral interference
    b. Mutation
    c. Supervision
    d. Permutation

Viral Taxonomy

18. **Adenovirus:**
    a. ds-DNA
    b. Enveloped
    c. Complex symmetry of capsid
    d. None
19. **Human papillomavirus contains:**
    a. ds-DNA
    b. ss-DNA
    c. ds-RNA
    d. ss-RNA
20. **Lipid envelope is found in which virus?**
    a. Reo
    b. Herpes
    c. Picorna
    d. All of the above
21. **In which of the following genome has double-stranded nucleic acid?**
    a. Orthomyxo viruses
    b. Pox viruses
    c. Papova viruses
    d. Reo viruses
22. **Which of the following DNA viruses possesses a capsid with icosahedral symmetry and no lipid envelope?**
    a. Herpes virus
    b. Adeno virus
    c. Pox virus
    d. Papova virus
23. **Which is not a DNA virus?**
    a. Parvo virus
    b. Papova virus
    c. Pox virus
    d. Rhabdo virus
24. **Which of the following is not a RNA virus?**
    a. Ebola
    b. Simian-40
    c. Rabies
    d. Vesicular stomatitis virus
25. **Which is enveloped virus?**
    a. Dengue virus
    b. Norwalk virus
    c. Hepatitis A virus
    d. Adeno virus
26. **Non enveloped ss RNA virus is:**
    a. Picorna virus
    b. Pox virus
    c. Retro virus
    d. Bunya virus
27. **Which of the following virus has negative sense RNA?**
    a. Rabies virus
    b. Reo virus
    c. Corona virus
    d. Calci virus
28. **Negative sense nucleic acid genome is found in:**
    a. Polio virus
    b. Rabies virus
    c. Measles virus
    d. Picorna virus
    e. Influenza virus

Subviral Agents

29. **Which of the following infectious agent lacks RNA?**
    a. Virus
    b. Staphylococci
    c. Prions
    d. *Cryptococcus*
30. **True about viroid is:**
    a. Causes tumor in animals
    b. Lacks envelope like covering
    c. Has only genetic material
    d. Visible on light microscope
31. **True about prion protein disease is all** *except*:
    a. Myoclonus is seen in 10% of the patients
    b. Caused by infectious protein
    c. Brain biopsy is diagnostic
    d. Commonly manifests as dementia

32. Prions are best killed by:
   a. Autoclaving 121°C   b. 5% formalin
   c. Sodium hydroxide   d. Sodium hypochlorite

33. Prions are extremely resistant to dry heat and can survive dry heat at:
   a. 180° for 1 hour   b. 240° for 1 hour
   c. 360° for 1 hour   d. 90° for 3 minutes

34. Which of the following is correct about prions?
   a. Long incubation period
   b. Destroyed by autoclaving 121°C
   c. Nucleic acid present
   d. Immunogenic

35. Prions are:
   a. Made up of bacterial and viral particles
   b. Immunogenic
   c. Infectious
   d. RNA particles

36. Prions consist of:
   a. DNA and RNA   b. DNA, RNA and protein
   c. RNA and protein   d. Only proteins

37. Regarding prion protein which of the following statement is true:
   a. It is a protein product coded in viral DNA
   b. It catalyses abnormal folding of other proteins
   c. It protects disulfide bond from oxidation
   d. It cleaves normal proteins

38. True about prion is:
   a. Encoded by viral genome
   b. Associated with misfolding of protein
   c. Noninfectious
   d. Immunogenic

39. Which of the following is not a prion associated disease?
   a. Scrapie
   b. Kuru
   c. Creutzfeldt-Jacob disease
   d. Alzheimer disease

40. Which of the following is not a prion disease?
   a. Bovine spongiform encephalopathy
   b. Transmissible mink encephalopathy
   c. Scrapie
   d. Progressive multifocal leucoencephalopathy

41. Mad cow disease (Bovine spongiform encephalopathy) is similar to man in:
   a. Alzheimer disease   b. Creutzfeldt-Jacob disease
   c. Rabies   d. Polio

42. Mad cow disease is due to:
   a. Slow virus   b. *Mycoplasma*
   c. Bacteria   d. Fungi

43. Human cannabilism is associated with:
   a. Q fever   b. Sleeping sickness
   c. Trachoma   d. Kuru

44. Fatal familial insomnia is associated with:
   a. Prion disease   b. Degeneration disease
   c. Neoplastic disease   d. Vascular disease

Answers and Explanation of MCQs

1. d
2. a
3. c
• Follow **Table 11.1** for explanation of answers of MCQs 1–3.

4. a
• Follow section, **introduction, Table 11.1 and resistance (heat and cold)** for explanation.

5. b
6. a
7. c
8. b
• Follow section, **morphology of viruses (size and shapes)** for explanation of answers of MCQs 5–8.

9. d
10. a, b
11. a
12. a
• Follow section, **morphology of viruses [structure of viruses → nucleic acid (NA)/genome/core]** for explanation of answers of MCQs 9–12.

13. b
14. b
• Follow section, **morphology of viruses (structure of viruses → capsid or shell)** for explanation of answers of MCQs 13–14.

15. d
• Follow section, **viral multiplication (abnormal replicative cycle → von Magnus phenomenon)** for explanation.

16. d
• Follow section, **viral multiplication (abnormal replicative cycle → defective viruses)** for explanation.

17. a
• Follow section, **viral genetics (interaction between viral products/nonheritable mechanisms → viral interference)** for explanation.

18. a
19. a
20. b
21. b, c, d
22. b, d
23. d
24. b
25. a
26. a
27. a
28. b, c, e
• Follow section, **viral taxonomy (Table 11.2)** for explanation of answers of MCQs 18–28.

29. c
• Follow section, **subviral agents (viroids)** for explanation.

30. c
31. a
32. c, d
33. d
34. a
35. c
36. d
37. d
38. b
39. d
40. d
41. b
42. a
43. d
44. a
• Follow section, **subviral agents (prions and prion disease)** for explanation of answers of MCQs 30–44.

CHAPTER 12

Virus–Host Interactions (Viral Infections)

Chapter Outline
- Introduction
- Immune Response (Ir) to Viral Infections

INTRODUCTION

Effects on Cell by Virus

Sometimes virus causes no damage called steady-state infection or causes following types of damages.

Types of cellular damage
- Cytocidal: Cell death like in polio virus (virus causes cell degeneration which leads cell death).
- Cytolysis: Cell lysis like in bacteriophage virus (virus multiplies in cell to produce daughter viruses, so cell is not able to hold the weight and finally dies).
- Cell proliferation: For example, MCV
- Malignant transformation: For example, oncogenic virus
- Cytopathic effect (CPE): It occurs in tissue/cell culture.
- Morphological changes: For example, inclusion body formation by rabies virus, ballooning of cells in herpes virus.
- Fusion of adjacent infected cells lead polykaryocytosis or syncytium formation.
- Damage to chromosomes by measles, mumps, adeno virus, CMV, etc.
- Antigenic changes on surfaces: Such as hemagglutinin on surface of influenza infected cells, which cause hemadsorption of RBCs, tumor antigens on surface of cells infected by oncogenic virus.

Mechanisms of cellular damage
- Shutting down of protein (DNA) synthesis in cell
- Accumulation of toxic viral molecules causes distortion of cellular architecture.
- Alteration of plasma membrane permeability, which releases the lysosomal enzymes causes autolysis.

Modes of Transmission of Viruses

Direct modes: These routes do not required mediator/vehicle.

1. **Droplet nuclei:** Droplet nuclei/particles of saliva or nasopharyngeal secretion arise during coughing, sneezing, speaking, talking or invasive procedure (bronchoscopy) enter in to other host directly, who is in close contact. Such particles are ≥5 μm in diameter and spread to short distance (<3 feet) can directly enter in other host. However, such larger particles can be filtered by nose. Particles ≤5 μm in diameter are traverse to long distance and produce the air borne (indirect) infection described later in this chapter. Infection by droplet nuclei is increased in close contact, overcrowding and lack of ventilation. **Viruses** transmitted by droplet nuclei are influenza virus, mumps virus, rubella virus, adeno virus and parvo virus B19.
2. **Inoculation/injection under skin or mucosa:** Viruses transmitted by contaminated needle/syringe are HIV, HBV, HCV, etc.
3. **Contact with skin/mucosa:** It includes sexually transmitted infections, which are described in **Ch. 44** and infections transmitted by direct skin-to-skin contact such as HHV-1 plus 2, HHV-3 (VZV), poxvirus, MCV, etc.
4. **Vertical:** Follow **Table 12.1**.

Indirect modes: These routes required mediator/vehicle.

1. **Inhalation (air borne):** Particles arise during coughing, sneezing, talking or invasive procedures from patients which are ≤5 μm in diameter and traverse to long distance can produce the air borne infection. Some droplet nuclei settle over different objects and become part of dust and cause air borne infection. **Viruses** transmitted by air-borne route are influenza virus, measles virus, VZV and hemorrhagic fever viruses causing pneumonia.
2. **Ingestion (food- and water-borne):** Polio virus, *Rotavirus*, HAV, HEV, adeno virus, reo virus, etc.

TABLE 12.1: Vertical viral infections

Time	Infections
Antenatal/before birth/ congenital/transplacental/ teratogenic	• Rubella (congenital rubella syndrome) • CMV (congenital cytomegalo-virus infection) • Herpes viruses (HSV causing congenital herpes simplex while VZV causing fetal varicella syndrome and congenital/ neonatal varicella) • HBV • HIV-AIDS • Zika virus • Parvovirus B19 • Chikungunya virus
Intranatal/during birth (transcervical)	• CMV • HSV • HPV
Postnatal/after birth by breastfeeding	• CMV • HIV

3. **Blood borne:** By transfusion of blood or blood products like HHV-4 (EBV), HHV-5 (CMV), HHV-8 (KSHV), HBV, HCV, HDV, HTLV-I, HTLV-II, HTLV-III (HIV), parvo virus B19, zika virus, etc.
4. **Saliva borne**
   - From human saliva like HHV-4 (EBV), HHV-6 (HBLV), HHV-7 (RK virus), ECHO virus, mumps virus, etc.
   - From monkey bite like herpes virus simiae or B virus.
   - From dog bite like rabies virus.
5. **Vector borne**
   - **Mosquitoes-borne viruses**
     - *Aedes*: Chikungunya virus, yellow fever virus, dengue virus, zika virus, Rift Valley fever virus (also by *Culex*), Orungo virus, etc.
     - *Anopheles*: O'nyong-nyong virus.
     - *Culex*: JEV, St. Louis encephalitis virus, Ilheus virus, West Nile virus, Murray Valley encephalitis virus, Oropouche virus, Rift Valley fever virus and African horse sickness virus.
   - **Ticks-borne viruses:** RSSE virus, Central European encephalitis virus, Western Siberian encephalitis virus, Powassan encephalitis virus, Louping ill virus, Kyasanur forest disease virus, Omsk hemorrhagic fever virus, Nairobi sheep disease virus, CCHF virus and Kolarado tick fever virus.
   - **Sand fly-borne viruses**
     - *Phlebotomus*: Sand fly fever virus or three-day fever virus, Chandipura virus (also by *Sergentomyia*).
     - *Lutzomyia shannoni*: Vesicular stomatitis virus.
     - *Sergentomyia*: Chandipura virus.
6. **Fomites borne:** Contamination of fomites like towel, handkerchief, pen, pencils, clothes, cups, spoon, keys, etc., may transmit the infections like HAV, swine flu, etc.

Flowchart 12.1: Pathogenesis of viral infections

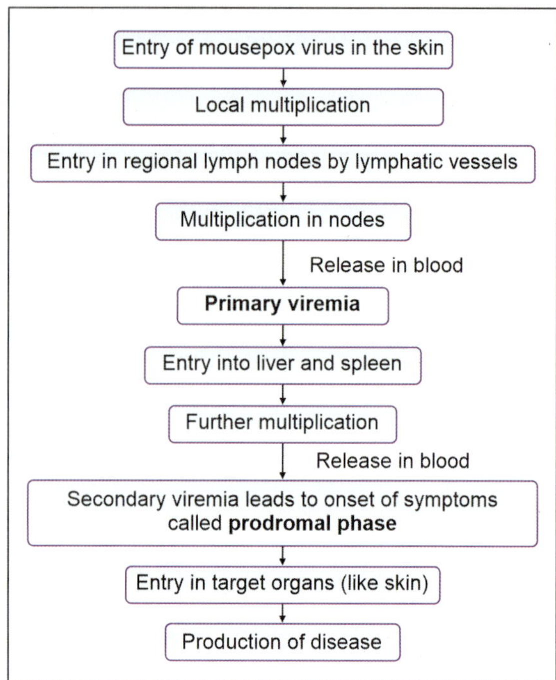

7. **Unclean hands and fingers:** Swine flu, SARS-CoV, MERS-CoV, etc.

Pathogenesis of Virus or Spread of Virus in the Body

History: Spread of viral infection was 1st studied by Fenner in 1948 by using the mousepox virus as the experimental model.

Pathogenesis: Follow **Flowchart 12.1**.

Incubation period (IP): In viral infection it depends on following factors.

- **Sites of entry, multiplication and lesion: (1)** If sites of entry and sites of lesion are same then IP is short about 1–3 days such as in *Rotavirus*. **(2)** If sites of entry and sites of lesion are distant then IP is long about 10–20 days such as in polio virus which enters by GIT and produces lesion in CNS.
- **Direct deposition of virus in blood:** Short IP about 5–6 days like in dengue or yellow fever.
- **Nature of multiplication:** IP is long about months or years, if virus multiplies slowly such as in HBV, IP is 2–6 months and in HIV after primary infection AIDS will starts approximately after 10 years.

IMMUNE RESPONSE (Ir) TO VIRAL INFECTIONS

Types of Ir of Viral Infections

Two main types of Ir like specific and nonspecific are described below.

Specific (Immunological) Ir

AMI (HI): Different classes of antibodies are formed in viral infections. IgM and IgG play major role in blood and tissue immunity. IgA provides local or mucosal immunity. Antibodies neutralize the virus by different mechanisms such as prevention of adsorption of virus to cell, enhancement of viral degradation, prevention of release of viral progenies, etc. Along with complement antibody causes damage to viral envelop and cytolysis of viral infected cells. Not all antibodies neutralize the virus, but enhance the viral infectivity and contribute to viral pathogenesis by activation of complement system followed by lysis of infected cells or by immune complex type of injury.

CMI: Cytotoxic T cells (Tc) directly kill the viral infected cells. Sensitized T cells release the cytokines, which enhance phagocytosis. Release of interferon (IFN) from T cells, which inhibits the viral multiplication and limits the infection. CMI also takes part in DTH following viral vaccination.

Nonspecific Ir

Phagocytosis: Microphages have no role in viral clearing. **Macrophages** phagocytose the virus and important in clearing virus from bloodstream.

Interferon (IFN)

- **History:** IFN was discovered in 1975 by Issacs and Lindemann, who showed that when chick CAM is treated with live or inactivated influenza virus produces antiviral substances which rendered the cell resistant to viral infection.
- **Definition:** It is a protein produced by vertebrate cells due to viral or nonviral stimulation.
- **Mechanisms of actions:** IFN has no direct action on virus, but acts on other cells of same species, rendering them refractory to viral infection. It inhibits the viral transcription in infected cells. On exposure to IFN cell produces TIP (translation-inhibiting protein), this inhibits the viral translation without affecting the cell. TIP is actually a combination of three enzymes like protein kinase, oligonucleotide synthetase and RNAase.

- **Properties of IFN: (1) Synthesis:** IFN synthesis begins within about an hour of induction and reaches in high levels in 6–12 hours. It synthesized very rapidly than Ab; hence, play primary role in defense. It is produced from host cells, not from virus so it is a host protein. IFN production increased by increasing the temperature up to 40°C. IFN production decreased by increasing the O_2 tension and by steroid therapy. **(2) Species specific:** IFN produced by one species can protect the same or related species, but not other like IFN produced by human cells can protect human cells or related species such as monkey, but not other species like mouse or chick. **(3) Cell's factors:** Cellular transcription and protein synthesis necessary for IFN production. **(4) Viral factors:** IFN is not virus specific because IFN induced by one virus can protect against any viral infection. Viruses are varying in their susceptibility to IFN. They are also varying in their capacity to induce IFN production like virulent and RNA viruses are good inducers while avirulent and DNA viruses are poor inducers. **(5) Inactivation:** IFN is inactivated by proteolytic enzyme, but not by protease or lipase. **(6) Resistant:** IFN is resistant to protease, lipase and heat at 56–60°C for 30–60 minutes. It is stable to wide range of pH (2–10). IFN is labile at pH-2. **(7) Molecular weight:** It is LMW about 17,000. **(8) Antigenic property:** It is poorly antigenic, so serological tests are not useful for its detection. **(9) Other properties:** It is nondialyzable and nonsedimentable.
- **Nomenclature of IFN:** It is abbreviated as IFN and species of origin is indicated as prefix like HuIFN for human.
- **Types of IFN:** Three types as mentioned in **Table 12.2**.
- **Clinical uses of IFN: (1) Antiviral effect (IFN-α and β):** IFN applied systemically in immunocompromised host such as in HBV and HCV infection and locally in genital warts, herpes keratitis and upper respiratory tract infection. **(2) Antibacterial effects (IFN-γ):**

Features	IFN-α	IFN-β	IFN-γ
Earlier name	Leucocytic IFN	Fibroblast IFN	Immune IFN
Type	I	II	III
Origin	Virus-infected leukocytes	Infected fibroblasts or epithelial cells	T-cells
Inducer	Viruses; dsRNA	Viruses; dsRNA	Antigen or mitogen
Biochemical nature	Non-glycosylated protein	Glycoprotein	Glycoprotein
Actions	Antiviral, anti-proliferative, ↑MHC-I expression and activates NK cells	Antiviral, anti-proliferative, ↑MHC-I expression and activates NK cells	Anti-proliferative, ↑MHC-I and II expression and immunomodulatory functions
Stability at pH 2.0	Stable	Stable	Labile
Location of gene	Chromosome 9	Chromosome 9	Chromosome 12
Receptors	α/β receptor	α/β receptor	γ receptor
Receptor gene	Located on chromosome 21	Located on chromosome 21	Located on chromosome 6
Antigenic types	16	—	—

It is used in chlamydial infection, LL, and chronic granulomatous disease. **(3) Antiparasitic effects (IFN-γ):** It is used in toxoplasmosis, leishmaniasis and malaria. **(4) Antioncogenic effects:** It is used in hairy cell leukemia, chronic myelocytic leukemia, T cell lymphoma, Kaposi's sarcoma, endocrine pancreatic neoplasm, non-Hodgkin's lymphomas, etc., **(5) Cellular effects:** It inhibits cell growth, cellular proliferation, DNA synthesis and protein synthesis. It increases expression of MHC-Ag on cell surfaces. **(6) Immunoregulatory effects:** It increases cytotoxic activity of NK, K and T cells. It activates macrophage cytocidal activity and suppressor T cells. It modulates the Ab formation. It suppresses the DTH.

- **Advantages of IFN:** It is nontoxic, non-antigenic, diffuses freely in the body and has wide antiviral activity.
- **Disadvantages of IFN: (1) Species specific:** Non-human origin IFN is not useful in human. This drawback is corrected by preparing the IFN from human cells like buffy coat leucocytes from blood bank induced by Sendai virus or Norwalk disease virus (NDV). **(2) Side effects:** It causes fever, malaise, fatigue, muscle pains and toxicity to organs.
- **Contraindications of IFN:** These are history of major depressive illness, active alcohol use, cytopenia (more than one type of blood cell deficiency), hyperthyroidism, renal transplantation, autoimmune diseases, etc.
- **IFN assay:** Estimation of IFN level called IFN assay. It is measured in IU (International Unit) per ml. It is poorly antigenic, so assay by serological method is not useful. It is done by inhibition of plaque formation by particular virus by IFN.
- **Preparation of IFN**
  - From buffy coat leukocytes: As described earlier.
  - DNA recombinant technology: Best for human use.
  - Pegylated IFN: Interferon-α linked to polyethylene glycol. This linkage results in slow absorption, decreased clearance and more sustained concentration, hence it can be administered once in a week.

Effective Factors on Ir of Viral Infections

1. **Body temperature:** Fever acts as a natural defense against viral disease as most viruses are inhibited by temperature above 39°C. However, in few cases fever causes damages like herpes simplex is reactivated by fever and produces 'fever blisters' and herpes fibrilis which is accompany with fever and other bacteria like *Strept. pneumoniae*, *Strept. pyogenes*, *Haemophilus influenza* and parasite like malaria.
2. **Age:** Most viral infections are common and more dangerous at the extremes of age.
3. **Hormones:** Administration of steroids enhance most viral infections, because of depression of immune response and inhibition of IFN synthesis. Such as

coxasackie B1 not causes disease in adult mouse, but induces fatal infection in mouse treated with cortisone. Varicella and vaccinia are lethal in patients on cortisone.

4. **Malnutrition:** Some viral infections, such as measles, produces a much higher incidence of complications and a higher case fatality rate in malnourished children than in well-fed patients.
5. **Viral infections:** Several viruses suppress the Ir, such as measles causes temporary depression of DTH, leukemia or infection by lymphocytic chorio meningitis (LCM) virus reduces the Ab synthesis and HIV destroys the CD-4 cells and reduces the CMI.

ACCESS YOURSELF

Short Notes

1. IFN.

Short Questions for Theory/Viva Questions

1. Write four examples of viruses transmitted by sexual intercourse.
2. Write four examples of viruses transmitted by placenta.
3. Write four examples of mosquito-borne viruses.
4. Write four uses of IFN.
5. Write the four differences between IFN-β and IFN-γ.

Comments on

1. Administration of steroids enhance viral infections.
2. IFN is called species specific, but not virus specific.

MCQs for Chapter Review

Introduction

1. **Which virus given below is not a teratogenic virus?**
   a. Rubella   b. Cytomegalo virus
   c. Herpes simplex   d. Measles
2. **Which of the following is/are transfusion transmitted viruses?**
   a. Hepatitis B   b. CMV
   c. HTLV-1   d. Rubella
   e. HHV-8

Immune Response (Ir) to Viral Infections

3. **True about interferon:**
   a. It is virus specific
   b. It is bacteria specific
   c. Produced from bacteria
   d. Effective against viral infection
   e. It is species specific
4. **About interferon true is:**
   a. It is synthetic viral agent
   b. Inhibits viral replication in cells
   c. It is specific for a particular virus
   d. None
5. **Interferon:**
   a. Species specific
   b. Reacts directly with virus particles to inactivate them
   c. Reacts with cells, and the affected cell becomes resistant to a number of different viruses
   d. Constitutively produced at high levels in cells, but requires an inducer for activity

6. **True about interferon is:**
   a. Host protein
   b. Viral protein
   c. Inactivated by nuclease
   d. Virus specific
7. **The temperature required for optimum production of interferon is:**
   a. 37°C   b. 30°C
   c. 40°C   d. 50°C
8. **Fibroblast in tissue culture form interferon of type:**
   a. Alpha   b. Beta
   c. Gamma   d. All of above
9. **Interferon .... in nature:**
   a. Protein   b. Lipid
   c. Polysaccharide   d. All of above

Answers and Explanation of MCQs

1. d
- Follow section, **modes of transmission and Table 12.1** for explanation.

2. a, b, c and e
- Follow section, **modes of transmission (blood borne)** for explanation.

3. d and e
- Interferon is species specific not virus or bacteria specific and effective against bacterial, viral, parasitic, malignant and other diseases.
- **Option 'c':** Follow section, **Table 12.2 (origin)** for explanation.

4. b

5. a
- Follow section, **interferon (IFN → mechanism of actions)** for explanation of answers of MCQs 4–5.

6. a

7. c
- Follow section, **interferon (IFN → properties of IFN → synthesis)** for explanation of answers of MCQs 6–7.

8. b

9. a
- Follow **Table 12.2** for explanation of answers of MCQs 8–9.

CHAPTER **13**

General Properties of Fungi

Chapter Outline

- Introduction
- Classification
- Hyphae, Pseudohyphae and Mycelium
- Precipitating Factors of Fungal Infections

INTRODUCTION

Meaning

Fungus (Latin) and Mykes (Greek) both have same meaning as mushroom (type of edible fungus), because it has mushroom like appearance.

Definition

Microorganism with mushroom like appearance is called fungus (plural; fungi). Branch of medical science related with study of fungus and disease produce by it is called Mycology.

History

Raymond Jacques Sabouraud (1964–1936) is called the father of Mycology.

Taxonomy

Fungi are eukaryotes (Eu means true and karyotes means nucleus), because they contain mature nucleus. Actually fungi are branching filamentous in nature and prefix Myco is used for some bacteria like *Mycobacterium* and *Mycoplasma,* because they have branching-filamentous form as like fungi.

Morphological Properties

- **Capsule:** *Cryptococcus neoformans* is the only capsulated fungus. Capsule aids in virulence and also helps in diagnosis of fungus.
- **Cell wall:** It contains polysaccharides like chitin, glucans, mannans, chitosan, cellulose, etc., mixed with polypeptide.
- **Plasmalemma (cytoplasmic membrane):** It presents inner to cell wall and contains glycoproteins, lipids and ergosterol.

- **Cytosol (cytoplasm):** Fungi are eukaryotic microbes and contain nucleus, nuclear membrane, mitochondria, Golgi apparatus, endoplasmic reticulum, ribosomes, etc. They have both DNA and RNA.
- **Chlorophyll:** It is absent in fungi, so fungi are not able to do photosynthesis, and they remain dependent on host cell for nutrition.
- **Reproduction:** Fungi reproduced by following methods:
  - Asexual methods: These include hyphal extension, budding or fission.
  - Sexual methods: These include meiosis and mitosis. In many fungi sexual methods of reproduction are unknown. Such fungi are called fungi imperfecti.

Differences between Bacteria and Fungi

Follow **Table 13.1**.

Useful Properties of Fungi

Food industries: Fungi are used to alter the texture, flavor, digestibility and palatability of natural and processed food. Following are the examples:

1. Mushroom of basidiomycetes is used to prepare the food.

TABLE 13.1: Differences between bacteria and fungi		
Features	**Bacteria**	**Fungi**
Taxonomy	Prokaryotes	Eukaryotes
Cell	Unicellular	Uni/multicellular
Cell wall/ envelope	Thick and contains peptidoglycan and LPS	It contains chitin, manna and other polysaccharides
Sterol in plasma membrane	Absent (except in *Myco-plasma* and *Ureaplasma*)	Present (like cholesterol, ergosterol, etc.)

2. *Candida fukuyamaensis* is used to prepare the Russian manchurian tea. It also used as medicine in many diseases.
3. *Endomyces* is used to prepare the fat.
4. *Torulopsis* is used to prepare the protein.
5. *Saccharomyces cerevisiae*: Beker's yeast is used in baking bread and bakery products. Brewer's yeast is used to prepare the beer/alcohol. It is also the common name of *S. cerevisiae*, but of different strain. It also used as vector in HBV vaccine preparation.

Drugs preparation

- *Penicillium notatum*: Penicillin was discovered from this fungus by Alexander Fleming in 1928.
- *Penicillium chrysogyneum*: Also useful to prepare the penicillin.
- *Penicillium griseofulvum*: Useful to prepare gresiofulvin.
- *Acremonium (Cephalosporium) chrysogyneum*: Useful to prepare cephalosporin.
- *Aspergillus tereus*: Useful to prepare simvastatin (cholesterol-lowering agent).
- *Claviceps purpura*: Useful to prepare ergot alkaloids like ergometrine or ergotamine.
- *Tolypocladium inflatum*: Cyclosporine an immuno-suppressive drug derived from this fungus

Vaccine preparation: Certain fungi like *Saccharomyces cerevisiae* (Brewer's yeast), *Hansenula polymorpha* and *Pichia* spp., are used as a cloning vector to prepare recombinant vaccine of HBV.

Research model: *Neurospora crasa* acts as an ideal model to study host–pathogen relationship.

Vector control

1. *Culicinomyces clavosporus*: It is useful in malaria eradication.
2. *Coelomomyces*: It is able to kill the larvae of mosquitoes.

Agriculture industries

1. *Beavaria bassiana*: It is used to control the banana root borer, sugar cane borer, sweet potato weevil and rice water weevil.
2. *Vertcilium lecani*: It is used to control the sweet potato white fly, most damaging pest in Cuba.
3. *Trichoderma*: It is used to control the soil-borne disease that attacks the tobacco, tomatoes and peppers.

Harmful Properties of Fungi

Diseases in humans

1. Mycosis (mycoses): Disease produced by fungus invasion in body called mycosis (Plural; mycoses).
2. Allergic reactions: Two types of allergic reactions like "id" reaction and systemic disease. Follow **Ch. 92** for more details.

Fungal food poisoning: These are two types like mycetism (mycetismus, muscarinism) and mycotoxicoses. Follow **Ch. 92** for more details.

Spoiling of stored food: Fungi spoil the stored grains, fruits, vegetables and foodstuff. Fungi present as bread molds in stored food. About 10% of world's foods are spoiled by fungal contamination.

Decaying of materials: Fungi cause decaying of leather, timber, fabrics, electric devices, etc.

Fungal growth on plasticize products: Fungi grow on plastic items like computer disks, videotapes, audiotapes, etc. Commonly encountered fungi are *Alternaria*, *Aspergillus*, *Epicossum*, *Paecilomyces*, *Penicillium* and *Trichoderma*.

Biological warfare

1. Yellow rain: Yellow rain was the subject of a 1981 political incident in which the US Secretary of State Alexander Haig accused the Soviet Union of supplying T-2 mycotoxin to the Communist states in Vietnam, Laos and Cambodia for use in counterinsurgency warfare. Refugees described many different forms of attacks, including a sticky yellow liquid falling from planes or helicopters, which was called yellow rain. Those exposed claimed neurological and physical symptoms including seizures, blindness and bleeding with over 10 thousand deaths. Sample analysis from the victims reported the presence of *Trichothecene* mycotoxins, including T-2 toxin, diacetoxyscirpenol (DAS) and deoxynivalenol (DON); however, presence of mycotoxin in yellow rain was challenged by later studies.
2. *Coccidioides immitis*: It is transmitted by inhalation and dangerous to work with it in laboratory enables its use in war.

CLASSIFICATION

Morphological Classification

It includes following four types.

Yeast (Fig. 13.1a): These fungi are unicellular and spherical. Cells without budding are called yeast cells or in budding form called budding yeast cells. They developed by asexual (budding) method. They are without hyphae or filamentous form. They produce the creamy or pasty colony on SDA. Examples include *Cryptococcus neoformans*, *Saccharomyces cerevesiae*, etc.

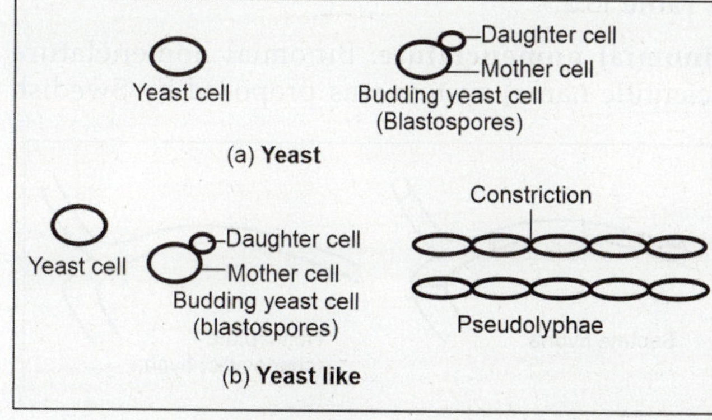

Fig. 13.1: Yeast, budding yeast cell and pseudohyphae

Yeast like (Fig. 13.1b): It is unicellular and spherical or oval. Cells without budding are called yeast cells or in budding forms called budding yeast cells. During budding process, daughter cell incompletely gets separated from mother cell and produces a chain of elongated yeast cells called pseudohyphae. It is developed by asexual (budding or binary fission) methods. It produces the creamy or pasty colony on SDA. Example includes *Candida* spp.

Mold/filamentous fungi: These fungi are multicellular. They are developed by asexual (budding/binary fission) or by sexual methods. They have filamentous hyphae **(Fig. 13.2),** which may be septate or nonseptate (coenocytic). They produce cottony, woolly, velvety, granular and pigmented colonies. Examples include dermatophytes, *Geotrichum candidum, Coccidioides immitis, Aspergillus* spp., Zygomycetes, etc.

Dimorphic fungi: They grow at two different temperatures, hence called dimorphic fungi. At 37°C they produce yeast or parasitic phase, and at 22°C they produce mold or mycelial phase. Examples include all systemic fungi like *Histoplasma capsulatum, Coccidioides immitis, Blastomyces dermatitidis* and *Paracoccidioides brasiliensis* plus *Sporothrix schenckii* and *Penicillium marneffei* (other species of *Penicillium* are not dimorphic, and they can grow at 25°C as mold).

Systemic (Taxonomical) Classification

It includes naming and typing of fungus.

Five kingdom system of classification: All organisms can be placed in five kingdoms like Monera, Protista, Fungi, Plantae and Animalia by RH Whittaker in 1969, called five kingdom system of classification. Fungi were initially classified with the plants, but in 1969 it is classified in a separate kingdom as "fungi" under the superkingdom "eukaryotes". General scheme of classification includes Superkingdom → Kingdom → Phylum (Division was used initially now replaced by phylum) → Class → Order → Family (Tribe) → Genus → Species (specific epithet) → Subspecies (subsp.)/ Variety (var.)/Forma (f.)/Forma specials (f. sp.) as mentioned in **Table 13.2**. The name of phylum, class, order and family ends by adding suffix "mycota", "mycetes", "ales" and "aceae", respectively as shown in **Table 13.2**.

Binomial nomenclature: Binomial nomenclature (scientific name) system was proposed by Swedish

TABLE 13.2: General scheme of classification of fungi

| | Superkingdom → Eukaryotes | | |
| | Kingdom → Fungi | | |
Taxon	**Suffix**	**Teleomorph (sexual)**	**Anamorph (asexual)**
Plylum	mycota	Ascomycota	Deuteromycetes
Class	mycetes	Ascomycetes	Hyphomycetes
Order	ales	Onygenales	Moniliales
Family	aceae	Onygenaceae	Moniliaceae
Genus	—	*Ajellomyces*	*Histoplasma*
Species	—	*capsulatus*	*capsulatus*
Variety	—	*capsulatus*	*capsulatus*

botanist in 1758 which includes genus name at 1st followed by species/epithet name in Italic pattern. In case of male scientist species name ends with suffix "ii", and in case of female scientist species name ends with suffix "eae". Both singular and plural are written same but in abbreviation sp., and spp., are used for singular and plural, respectively. Genus name is written in Latin noun and starts with capital letter with Italic pattern. Species/epithet name starts with small letter with Italic pattern irrespective of name of person or place. Species name is based on different properties of unit like *albicans* from white colony, *brisiliensis* from place of origin, *marneffei* from scientists, etc. For example, *Histoplasma capsulatum.*

Trinomial nomenclature: The word in the name (second epithet) might by subspecies (subsp.)/variety (var.)/forma (f.)/forma specials (f. sp.) and also italicized or underlined. For examples, *Histoplasma capsulatum* var. *capsulatum, Histoplasma capsulatum* var. *duboisii* and *Histoplasma capsulatum* var. *farciminosum.*

Taxonomical types of fungi: They are based on spore formation methods as mentioned below. This classification also used other supportive features like fatty acids analysis, zymogram pattern, DNA hybridization, RFLP, etc.

Mnemonic

- **Asexual spores:** ABC – CAS → Below explanation
- **Sexual spores:** ZABO → **Table 13.3**

1. **Asexual spores:** Fungi developed by budding, binary fission or by apical elongation.
   - **Blastospores (Fig. 13.1):** They are developed by budding method. Budding part of yeast cell called blastospore. For example, *Cryptococcus neoformans.*
   - **Chlamydospores:** They are developed by apical elongation method. Preexisting hyphal cell becomes thick, double wall and circular called chlamydospore. For examples, *Candida* spp., and *Paracoccidioides brasiliensis.* Chlamydospores produced on media like rice starch agar and cornmeal agar when incubated at 20°C. Three subtypes **(Fig. 13.3)** of chlamydospores are terminal, sessile and intercalary.

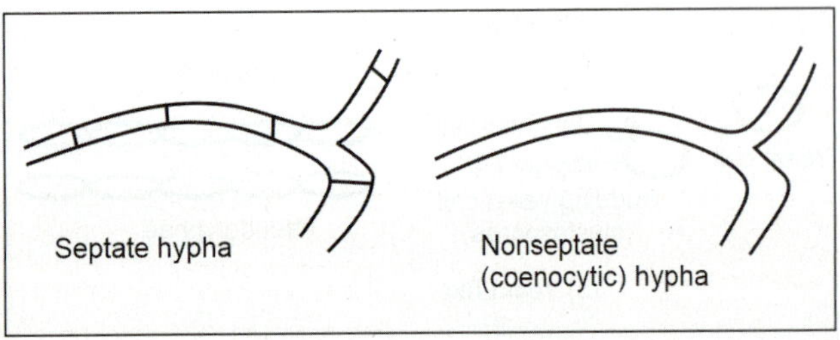

Fig. 13.2: Types of hyphae

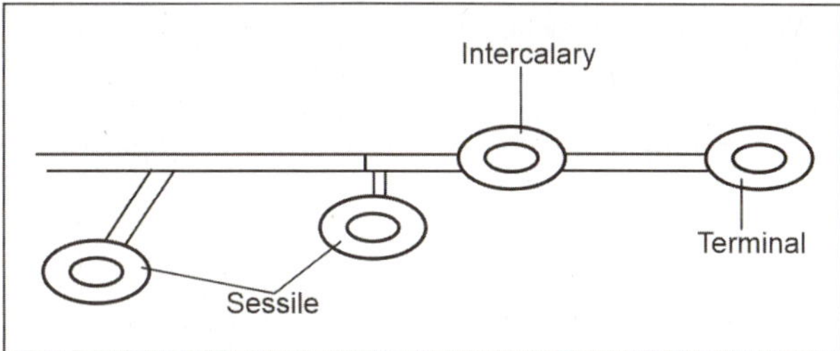

Fig. 13.3: Subtypes of chlamydospores

- **Arthrospores (Fig. 13.4):** They are developed by apical elongation method. Preexisting hyphal cells becomes thick, double wall and cuboidal or rectangular or barrel-shaped called arthrospores. They detached after fracture of supporting cell. For examples, dermatophytes, *Geotrichum candidum, Coccidioides immitis, Hortae werneckii* and *Trichosporon beigelii.*
- **Sporangiospores (Fig. 13.5):** They are developed by mitosis. These are endogenous spores. Spores are enclosed in a sac like structure called sporangium and spores called sporangiospores. For example, Zygomycetes.
- **Conidiospores or conidia (Fig. 13.6):** They are developed by mitosis. These are exogenous spores. Spores are arranged in a chain like structure and arise from a wase like structure called phialides (conodia-producing cells). For example, *Aspergillus* spp. Following are two subtypes of conidia.

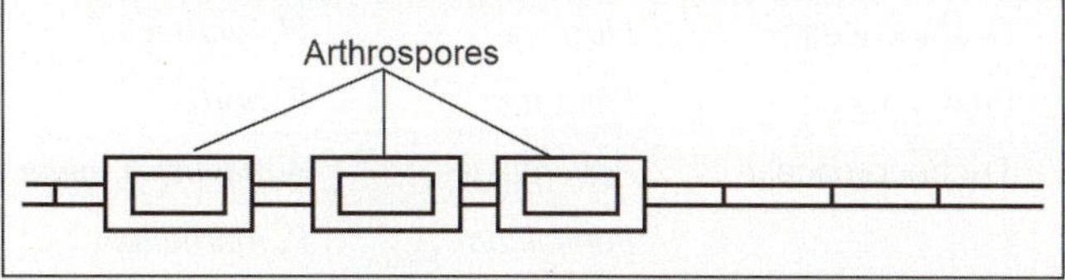

Fig. 13.4: Arthrospores

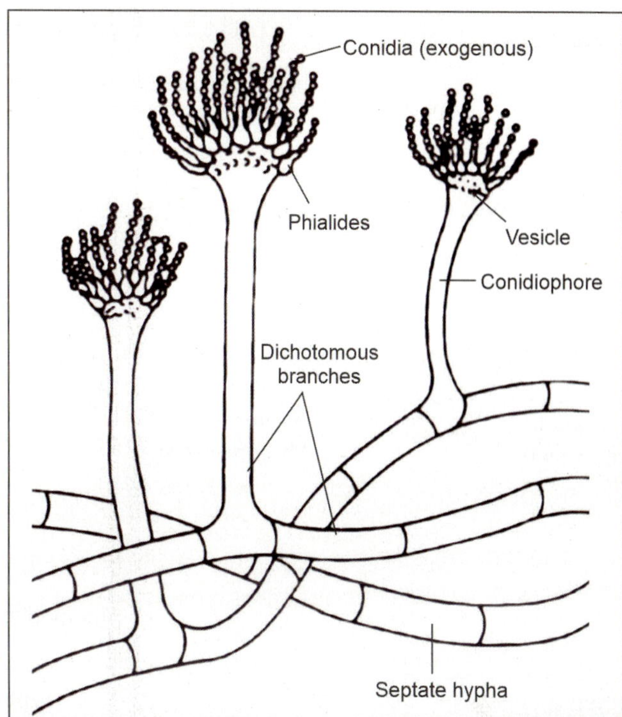

Fig. 13.6: Conidiospores

- **Microconidia:** They are further divided as basipetal and acropetal. Basipetal means youngest conidium is at base; for examples, *Penicillium, Aspergillus, Scopulariopsis brevicaulis.* Acropetal means youngest conidium is at top; for examples, *Alternaria* and *Cladosporium.*
- **Macroconidia:** They are further divided as dictyoconidia and phragmoconidia. Dictyoconidia means macroconidia with transverse and longitudinal septa; for examples, *Alternaria* and *Bipolaris.* Phragmoconidia means macroconidia with only transverse septa; for example, *Curvularia.*
- **Aleurospores (Fig. 13.7):** Macroconidia arranged the side of hyphae by a supporting cell called aleurospores. They released after detachment from cell. For examples, *Epidermophyton* and *Microsporum canis.*

2. **Sexual spores:** They are developed by miosis and mitosis. Sexual spores include zygospores, ascospores, basidiospores and oospores. All details are mentioned in **Tables 13.3, 13.4 and Fig. 13.8.**

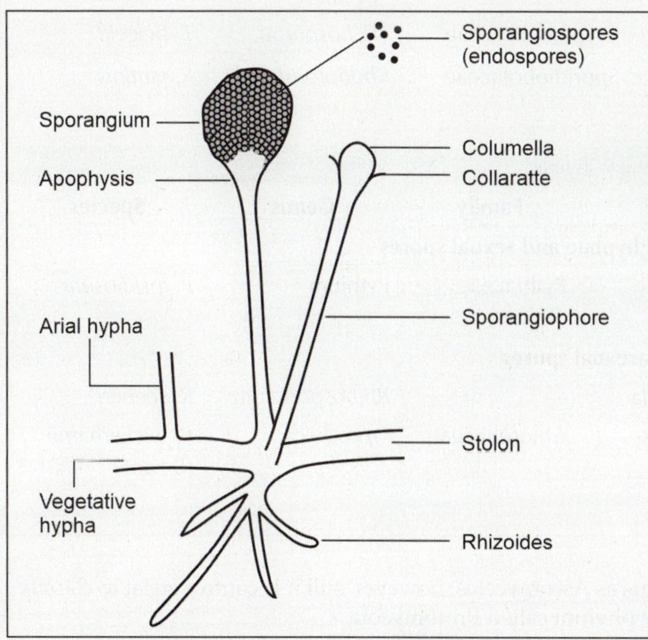

Fig. 13.5: Sporangiospores

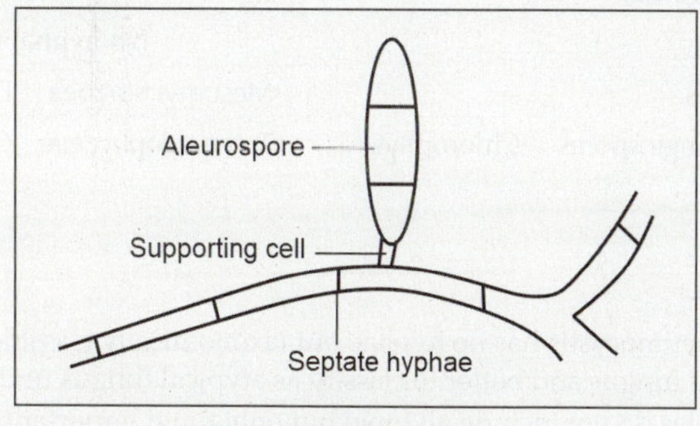

Fig. 13.7: Aleurospores

TABLE 13.3: Sexual spores with taxonomical types of fungi

Kingdom: Fungi						
Spores	**Phylum**	**Class**	**Order**	**Family**	**Genus**	**Species**
Nonseptate (aseptate/coenocytic) hyphae						
Zygospores (Fig. 13.8a)	Zygomycota	Zygomycetes (Phycomycetes)	Mucorales	Mucoraceae	*Mucor*	*M. racemosus*
					Rhizomucor	*R. pusillus*
					Rhizopus	*R. arrhizus*
						R. microspores
					Absidia	*A. corymbifera*
					Apophysomyces	*A. elegans*
				Cunninghamellaceae	*Cunninghamella*	*C. bertholletiae*
				Saksenaeaceae	*Saksenaea*	*S. vasiformis*
				Thamnidaceae	*Cokeromyces*	*C. recurvatus*
				Syncephalastraceae	*Syncephalastrum*	*S. recemosum*
			Entomophorales	Ancylistaceae	*Conidiobolus*	*C. coronatus*
				Basidiobolaceae	*Basidiobolus*	*B. ranarum*
Septate hyphae						
Ascospores (Fig. 13.8b)	Ascomycota	Ascomycetes	Pneumocystidales	Pneumocystidaceae	*Pneumocystis**	*P. jirovecii*
		Pyrenomycetes (Unitunicate)	Microascales	Microascaceae	*Pseudoallescheria*	*P. boydii*
					Scopulariopsis	*S. brevicaulis*
			Ophistomatales	Ophistomataceae	*Sporothrix*	*S. schenckii*
			Sordariales	Sordariaceae	*Neurospora*	*N. crasa*
			Hypocreales	Nectriaceae	*Fusarium*	*F. graminiarum*
				Hypocreaceae	*Acremonium*	*A. falciforme*
		Pyrenomycetes (Bitunicate)	Dothideales	Dothideaceae	*Hortaea*	*H. werneckii*
				Piedraiaceae	*Piedraia*	*P. hortae*
		Plectomycetes	Eurotiales	Trichocomaceae	*Aspergillus*	*A. flavus, A. niger*
					Penicillium	*P. marneffei*
			Onygenales	Onygenaceae	*Chrysosporium*	*C. parvum*
					Ajellomyces	*A. capsulatus*
Basidiospores (Fig.13.8c)	Basidiomycota	Basidiomycetes	Tremellales	Cryptococcaceae	*Trichosporon*	*T. beigelii*
		Pucciniomycetes	Sporidiales	Sporidiobolaceae	*Rhodotorula*	*R. glutinis*

TABLE 13.4: Parafungal agents

Spores	**Phylum**	**Class**	**Order**	**Family**	**Genus**	**Species**
Nonseptate (aseptate/coenocytic) hyphae and sexual spores						
Oospores (Fig. 13.8d)	Oomycota	Oomycetes	Pythiales	Pythiaceae	Pythium	*P. insidiosum*
No hyphal structure, asexual spores						
Spores	—	Mesomycetozoea	Dermocytida	—	*Rhinosporidium*	*R. seeberi*
Sporangiospores	Chlorophyta	Trebouxiophyceae	Chlorellales	Chlorellaceae	*Prototheca*	*P. wickerhamii* *P. zopfii*

Notes:
- *Pneumocystis has no hyphae but taxonomically classified as fungus as Ascomycetes; however, still it is controversial to classify it as fungus and better to classify as atypical fungus under the new phylum called Protomycota.
- Tables do not include all fungi but only fungi important at undergraduate level.

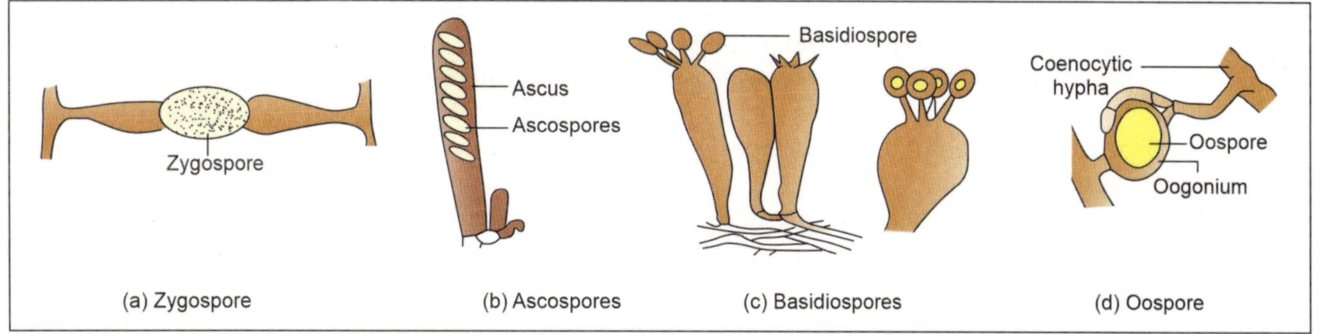

Fig. 13.8: Sexual spores

3. **Unknown method (sexual spores):** Exact method of sexual reproduction is unknown and such fungi are called miosporic fungi or mitosporic fungi or deuteromycetes or deuteromycota or hyphomyctets or fungi imperfecti.

> **Note: Parafungal agents**
> - Few pathogens like *Pythium insidiosum*, *Rhinosporidium seeberi* and *Prototheca* are producing disease similar to mycoses called parafungal agents.
> - Sexual spores: Produced by *Pythium insidiosum*.
> - Asexual spores: Produced by *Rhinosporidium seeberi* and *Prototheca*. In *Prototheca* spores are produced by internal septation and cleavage.
> - They are studied in mycology but not classified in the kingdom fungi. Their classification is given in **Table 13.4**.

Clinical Classification

Nomenclature: Fungal disease called mycosis (singular)/ mycoses (plural). Name of fungal disease are given by adding the suffix "sis" or "mycosis" to genus name; however, it is not satisfactory because same fungus causes many diseases and same disease is caused by many fungi. Some names of the fungal diseases are given by adding the suffix "i" to species name, like Histoplasmosis capsulati/duboisii/farciminosi.

Clinical types of fungi: Two types:
- **Superficial mycoses:** Ch. 90.
- **Deep mycoses:** Ch. 91.

Ecological Classification

Follow **Flowchart 13.1**.

Flowchart 13.1: Ecological types of fungus

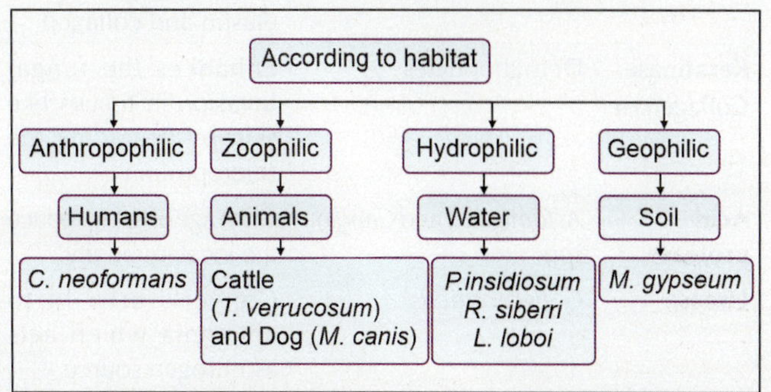

HYPHAE, PSEUDOHYPHAE AND MYCELIUM

Hypha

Meaning: Hypha derived from huphe (Greek) means Web.

Definition: It is an elongated, tubular and branching structure of fungus.

Types of hyphae

1. **According to presence or absence of septa (Fig. 13.2)**
   - Septate hyphae
   - Nonseptate hyphae (coenocytic).
2. **According to projection (Fig. 13.5)**
   - Arial hyphae: Hyphae projected in air and containing spores called arial hyphae. They are concerned with developmental function.
   - Vegetative hyphae: Hyphae submerged with surface or media called vegetative hyphae. They are concerned with nutritional function.
3. **According to color**
   - Dark (brown-black) hyphae: Brown-black pigmented hyphae occur in some fungi and such fungi called phaeoid fungi or demateceous fungi or black fungi.
   - Hyaline hyphae: Colorless hyphae occur in some fungi and such fungi called hyaline fungi.
4. **According to morphology (Fig. 13.9)**
   - Racquet hyphae: Enlargement of hyphal cell in-between the length of hyphae with one broad end and other pointed end called racquet hyphae. For examples, *T. mentagrophytes* and *E. flocossum*.

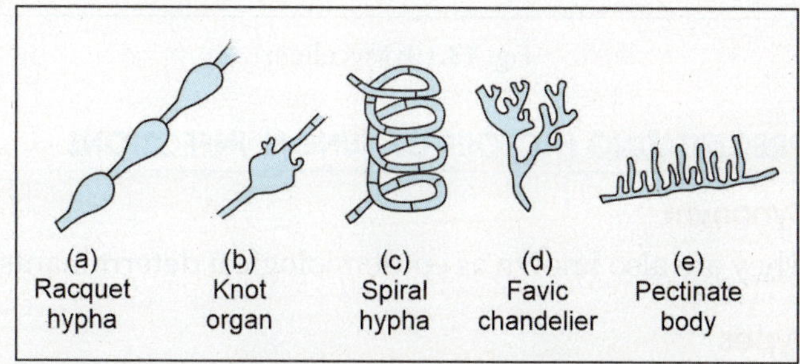

Fig. 13.9: Morphological types of hyphae

General Properties of Fungi

- Nodular hyphae or knot organ: In-between length of hyphae presence of swelling, like knot called nodular hyphae. For examples, *T. mentagrophytes*, *M. canis*, etc.
- Spiral hyphae: Hyphae present in spiral shape. For example, *T. mentagrophytes*.
- Favic chandelier: At the end of hyphae presence of multiple small projections like horn of reindeer or chandelier called favic chandelier. For examples, *T. violaceum* and *T. schoenleinii*.
- Pectinate body: One surface of hyphae is uniform while other shows some projections like broken comb called pectinate body. For example, *M. audounii*.

Pseudohyphae

Definition is described above. Differences between hyphae and pseudohyphae are given in **Table 13.5**.

TABLE 13.5: Differences between hyphae and pseudohyphae

Features	Hyphae	Pseudohyphae
Reproduction	By apical elongation	By budding
Morphology	**Fig. 13.2**	**Fig. 13.1**
Hyphal cell	Cuboidal/rectangular	Oval
Constriction in between two cells	Absent	Present
Cell wall	Uniform due to absence of constriction	Not uniform due to presence constriction
Dividing septum	Straight	Oblique/curved

Mycelium

It is an entangled mass of hyphae **(Fig. 13.10)**.

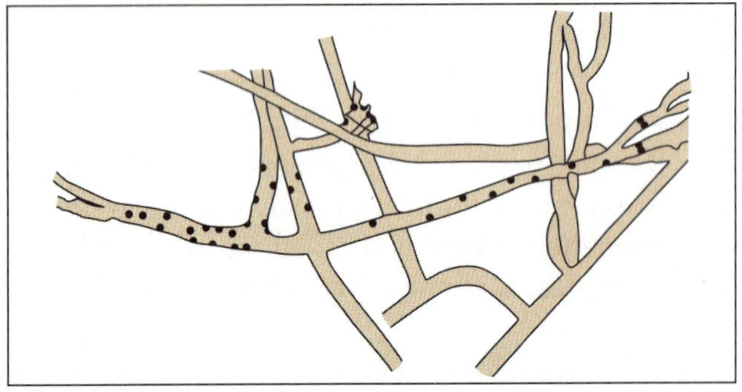

Fig. 13.10: Mycelium

PRECIPITATING FACTORS OF FUNGAL INFECTIONS

Synonym

They are also known as epidemiological determinants.

Types

Virulence of fungi is affected by following three factors.

Agent factors (virulence factors or determinants of virulence)

1. Intracellular or cell associated factors
- Fungal adhesins: Cell wall glycoprotein helps in attachment with host cells called adhesin.
- Polysaccharide and polypeptide complex: It provides rigidity and structural integrity to cell. It acts as a barrier against the effect of deleterious agents on fungi. Glucan is antiphagocytic.
- Capsular (K) Ag: Capsular (K) antigen presents in *Cryptococcus* inhibits the phagocytosis.
- Endotoxin: It is a glycoprotein presents in cell wall of *Candida*, *A. flavus* and *A. fumigates*. It is responsible for tissue necrosis, pyogenic and anaphylactic reaction.

2. Extracellular factors
- Enzymes: Follow **Table 13.6**.
- Toxins: **Ch. 92** (section → fungal food poisoning).
- Antigens: They produce allergic reaction called id reaction (candidid in *Candida* and dermatophytid in dermatophytes).
- Pigment production: Melanin pigment produced by *Cryptococcus neoformans* protects fungus against the UV radiation and effects of immune cells.
- Mannitol production: It produced by *Cryptococcus neoformans* and protects the fungus from intracellular killing by phagocytes.

3. Other factors
- Fungal dimorphism: All the dimorphic fungi (mentioned above) have two forms like mycelium and yeast phase, possess different antigenic and surface features, and require different host mechanisms to contain them.
- Thermotolerance: It is an ability of fungus to survive at 37°C. It makes fungus a potent human

TABLE 13.6: Enzymes, sources and actions in fungi

Enzyme	Sources	Actions
Elastase	*A. flavus*, *A. fumigatus* and dermatophytes	Enhances the fungal invasion in elastin containing tissues like lungs, skin, blood vessels, etc., by degrading the elastin and scleroprotein
Alkaline protease	*A. flavus*, *A. fumigatus* and *Rhizopus* spp.	Enhances the fungal invasion in tissues like lungs, by degrading the elastin and collagen
Keratinase Collagenase	Dermatophytes	Enhances the fungal invasion in tissues like skin by degrading the scleroprotein
Acid protease	*A. fumigatus* and *Candida* spp.	Cleavage of IgA to reduce the local immunity
Urease	*C. neoformans*	Converts urea in to ammonia which acts as nitrogen source

Essentials of Medical Microbiology

pathogen. *Cryptococcus neoformans* can survive at higher temperature about 41–43°C.

- Resistance to antimicrobial products released by host cells: Like yeast, spherules, etc., are showing resistant to antimicrobial products like H_2O_2 released by host cells.
- Phenotypic switching: *C. albicans* has ability to adapt the different changing conditions of host, which helps to evade the host defense mechanism and survival of fungus.
- Survival in acidic pH: Fungus grows best at acidic pH and it may tolerate the gastric acidity.

Host factors

1. **Trauma:** Local trauma may precipitate the fungal infections.
2. **Immunosuppressive diseases or conditions:** Conditions like long-term steroid therapy, antibiotic therapy, immunosuppressive therapy, renal transplantation, malignancy, DM and AIDS favor the fungal infections.
3. **Occupation:** Fungal infections are common in wooden workers like carpenters, farmers, etc.
4. **Gender:** Fungal infections are common in female due to wet work.
5. **Wet areas:** Moisture may precipitate the fungal infection in areas like inguinal, axilla and foot due to constant wearing of shoes.

Environmental factors

1. **Economical status:** Fungal infections are common in developing countries.
2. **Other factors:** Like overcrowding and poor sanitation also increase risk of fungal infections.

ACCESS YOURSELF

Essay/Full Question

1. Classification of fungi.

Short Notes

1. Useful and harmful properties of fungi
2. Morphological classification of fungi
3. Taxonomical (systemic) classification of fungi
4. Fungal spores
5. Fungal hyphae
6. Virulence factors of fungi.

Short Questions for Theory/Viva Questions

1. Write four differences between bacteria and fungi.
2. Write four differences between hyphae and pseudo-hyphae.
3. Write four examples of asexual spores of fungus.
4. Write four examples of sexual spores of fungus.
5. Define vegetative hyphae and aerial hyphae.
6. What are pectinate body and favic chandelier?
7. What is mycelium (in fungus)?

Comments on

1. Fungus is not able to do photosynthesis.

Introduction

1. **Brewer's yeast is a common name for:**
   a. *Pichia guillermondii*
   b. *Aspergillus niger*
   c. *Saccharomyces cerevisiae*
   d. *Penicillium notatum*
2. **Fungus useful for penicillin preparation is:**
   a. *Penicillium notatum*　　b. *Aspergillus tereus*
   c. *Candida fukuyamaensis* d. *Penicillium marneffei*

Classification

3. **Which of the following is only yeast?**
   a. *Candida*　　　　　　b. *Mucor*
   c. *Rhizopus*　　　　　　d. *Cryptococcus*
4. **Fungi that possess capsule is:**
   a. *Blastomyces dermatitidis*
   b. *Histoplasma*
   c. *Cryptococcus*
   d. *Coccidioides immitis*
5. **Which of the following is false regarding dimorphic fungi?**
   a. Occurs in two growth forms
   b. Can cause systemic infection
   c. *Cryptococcus* is an example.
   d. *Coccidioides* is an example.
6. **Dimorphic fungi behave like yeast at:**
   a. <10°C　　　　　　b. Body temperature
   c. >40°C　　　　　　d. *In vitro*
7. **Dimorphic fungus/fungi:**
   a. *Candida*　　　　　　b. *Cryptococcus*
   c. *Blastomycosis*　　　d. *Coccidioidomycosis*
   e. *Sporothrichosis*
8. **All are dimorphic fungi *except*:**
   a. *Blastomyces dermatitidis*
   b. *Histoplasma*
   c. *Penicillium marneffei*
   d. *Phialophora*
9. **Dimorphic fungus:**
   a. *Candida*　　　　　　b. *Histoplasma*
   c. *Rhizopus*　　　　　　d. *Mucor*
10. **The following fungi are thermally dimorphic *except*:**
    a. *Sporothrix schenckii*
    b. *Cryptococcus neoformans*
    c. *Blastomyces dermatitidis*
    d. *Histoplasma capsulatum*
11. **Budding reproduction in tissue is seen in:**
    a. *Cryptococcus, Candida*
    b. *Candida, Rhizopus*
    c. *Rhizopus, Mucor*
    d. *Histoplasma, Candida*
12. **The fungi which do not have sexual phase belong to which of the following groups?**
    a. Phycomycetes　　　　b. Fungi imperfecti
    c. Basidiomycetes　　　d. Ascomycetes
13. **Barrel-shaped spores (arthrospores) are seen with:**
    a. *Blastomyces*　　　　b. *Histoplasma*
    c. *Coccidioides*　　　　d. *Candida*
14. **A sporangium contains:**
    a. Spherules　　　　　b. Sporangiospores
    c. Chlamydospores　　d. Conidia

15. **Following is the asexual spore:**
    a. Oospore
    b. Zygospore
    c. Ascospore
    d. Blastospore
16. **Aseptate hyphae are seen in:**
    a. Phycomycetes
    b. Ascomycetes
    c. Basidiomycetes
    d. Deuteromycetes
17. **Ascospore is:**
    a. Asexual spore
    b. Sexual spore
    c. Conidia
    d. None of above
18. **Which of the following is produced sexually?**
    a. Ascospore
    b. Conidium
    c. Odium
    d. Yeast bud
19. **Fungal spores may be produced:**
    a. Singly
    b. In chains
    c. In sporangium
    d. All of above
20. **Human fungal infection is known as:**
    a. Mycosis
    b. Mycoses
    c. a + b
    d. Fungosis

Answers and Explanation of MCQs

1. **c**
   - Common names for *Saccharomyces cerevisiae* as per uses are Brewer's yeast and Baker's yeast.
2. **a**
   - Penicillin was discovered from *Penicillium notatum* by Alexander Fleming in 1928.
3. **d**
   - *Candida* is yeast like.
   - *Mucor* and *Rhizopus* are mold (filamentous) type.
4. **c**
   - In *Histoplasma capsulatum*, species name indicating the capsulated property of fungus, but it is noncapsulated.
   - *Cryptococcus neoformans* is the only capsulated fungus.
5. **c**

6. **b**
7. **c, d, e**
8. **d**
9. **b**
10. **b**
11. **a**
    - Follow section, **morphological classification** for explanation of answers of MCQs 5–11.
12. **b**
    - Phycomycetes, Basidiomycetes and Ascomycetes are showing sexual method of reproduction, while in fungi imperfecti method of reproduction is unknown.
13. **c**
    - Follow section, **classification [systemic (taxonomical) classification → asexual spores → arthrospores] and Fig. 13.4** for explanation.
14. **b**
    - Follow section, **classification [systemic (taxonomical) classification → asexual spores → sporangiospores] and Fig. 13.5** for explanation.
15. **d**
    - Oospore, zygospore and ascospore are the sexual spores.
16. **a**
17. **b**
18. **a**
    - Follow **Table 13.3** for explanation for explanation of answers of MCQs 16–18.
19. **d**
    - Follow section, **classification [systemic (taxonomical) classification]** to know about all types of fungal spores, method of production and for other details.
20. **c**
    - Mycosis is singular form for fungal infection, while mycoses is the plural form for fungal infections.

Introduction and Classification of Parasites

Chapter Outline
- Parasite
- Host
- Life Cycle

PARASITE

Meaning

Parasite word derived from parasitos (Greek) means eater at the court (meat eater).

Definition

Living organism which receives nourishment and shelter from other organism is called parasite.

Nomenclature

Parasites have Latinized name, hence written in Italics. It consists of two parts: 1st part is genus name, starts with capital letter and 2nd part is species name, starts with small letter like *Giardia lamblia*.

Classification

Different types of parasites are as follows.

According to habitat: Two types:
1. Ectoparasites: Parasites living outside on the body surface of the host called ecto-parasites. Follow **Ch. 47** and **Ch. 110** for more details.
2. Endoparasites: Parasites living inside the body of the host called endo-parasites.

According to visit to host: Following are different types.
1. Temporary parasite: It visits the host for short time.
2. Permanent parasite: Whole life passes as a parasite.
3. Facultative parasite: It lives the parasitic life when the opportunity arises.
4. Obligatory parasite: It cannot exist without parasitic life.
5. Occasional (accidental) parasite: It attacks on unusual host.
6. Wandering (aberrant) parasite: Happen to reach a place where it cannot live.

Morphological classification: Based on their cellular structure, all medically importance parasites are fall in two broad categories like protozoa and metazoa (helminths or worms) as shown in **Flowchart 14.1, Tables 14.1 and 14.2**.

Taxonomical (systemic) classification: It based on general scheme of classification such as Kingdom (sub-

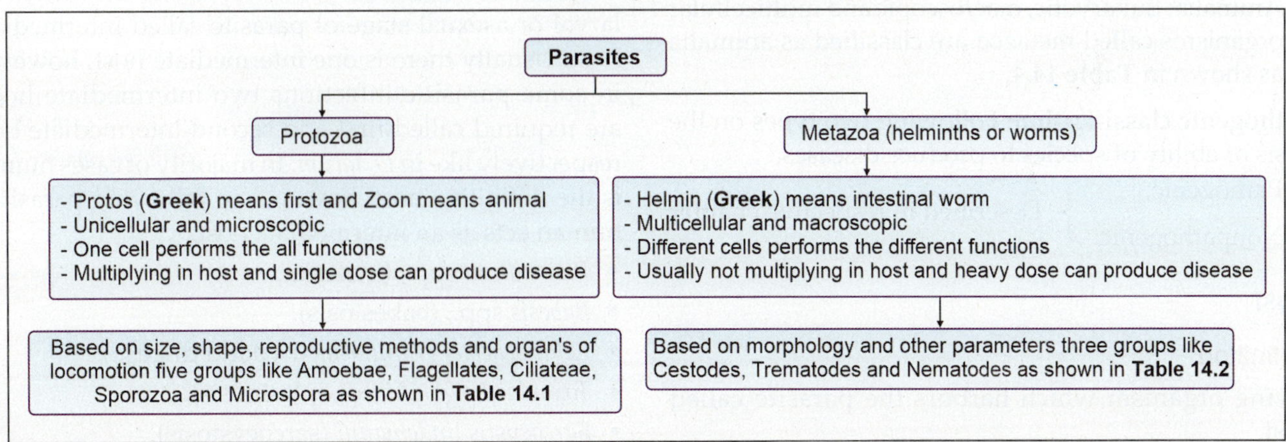

Flowchart 14.1: Morphological classification of parasites

Essentials of Medical Microbiology

TABLE 14.1: Morphological classification of protozoa

Features	Amoebae	Flagellates	Ciliateae	Sporozoa	Microspora
Organ of locomotion	Pseuopodium	Flagella	Cilia	Non-motile	Non-motile
Examples	*E. histolytica, E. coli, N. fowleri*, etc.	(1) Oral/genital flagellates: *Giardia, Trichomonas,* etc. (2) Hemo-flagellates: *Leishmania* and *Trypanosoma*	*B. coli*	(1) Blood inhabiting: *Plasmodium, Babesia* (2) GIT inhabiting: *Isospora, Toxoplasma*, etc.	*Pleistophora, Nosema, Enterocytozoon, Encephalitozon*, etc.

TABLE 14.2: Morphological classification of metazoa

Features	Cestoda	Trematoda	Nematoda
General features			
Common name	Tape worms or cestodes	Flukes or trematodes	Round worm or nematodes
Example (s)	*D. latum, T solium,* etc.	*S. haematobium, F hepatica,* etc.	*A. lumbricoides, W bancrofti,* etc.
Adult stage			
Shape	Tape like	Leaf like	Cylindrical
Segments	Segmented	Unsegmented	Unsegmented
Sexes	Not separate, hermaphrodite (monoecious)	Not separate/hermaphrodite (monoecious) except *Schistosoma*	Separate (diecious)
Head	Suckers often with hooks	Suckers, no hooks	No suckers, no hooks but well developed buccal capsule for adhesion
GIT	Absent	Present but incomplete (no anus)	Present and complete with anus
Body cavity (Coelom)	Absent	Absent	Present
Female	Oviparous	Oviparous	Oviparous/vivi (larvi) parous/ovivivi parous
Egg stage			
Opeculum	Present in Pseudophylidea and absent in Cyclophylidea	Present in monoecious trematodes and absent in diecious trematodes	Absent
Larval stage			
Larval forms	Cysticercus, cyticercoid, hydatid cyst, coenurus, coracidium, procercoid and plerocercoid	Miracidium, redia, cercaria, meta-cercaria, redia and sporocyst	Rhabditiform larvae, filariform larvae and microfilariae

kingdom) → phylum (subphylum) → class (superclass/subclass) → order (suborder) → family (super family/subfamily) → genus → species (subspecies). Medical important parasites are fall in to two kingdoms.

1. Protista: Eukaryotic, microscopic and unicellular organisms called protozoa are classified as protista as shown in **Table 14.3**.
2. Animalia: Eukaryotic, macroscopic and multicellular organisms called metazoa are classified as animalia as shown in **Table 14.4**.

Pathogenic classification: Following two types on the basis of ability of species to produce diseases.

1. Pathogenic:
2. Nonpathogenic: } Described in respective chapter

HOST

Definition

Living organism which harbors the parasite called host.

Classification

Following are different types.

Definitive (primary) host: Host which harbors the adult or sexual stage of parasite called definitive host. In majority of cases man is the definitive host.

Intermediate (secondary) host: Host which harbors the larval or asexual stage of parasite called intermediate host. Usually there is one intermediate host; however, in some parasitic infections two intermediate hosts are required called first and second intermediate host respectively like in *D. latum*. In majority of cases human is the definitive host; however, in following parasites, human acts as an intermediate host.

- *Plamodium* spp., (malaria).
- *Babesis* spp., (babesiosis).
- *Echinococcus granulosus* (hydatid disease).
- *Toxoplasma gondii* (toxoplasmosis).
- *Sarcocystis lindemanni* (sarcocystosis).

TABLE 14.3: Systemic classification of protozoa

Kingdom: → Protista
Subkingdom: → Protozoa

Phylum	Class	Subclass	Order	Suborder	Family	Genus	Species
Sarcomastigophora	**Subphylum** (Sarcodina) **Superclass** (Rhizopoda) **Class** (Lobosea/ Amoebae)		Amoebida	Tubulina	Entamoebidae	*Entamoeba*	Pathogenic and non-pathogenic species are mentioned in respective **chapters**
						Endolimax	
						Iodamoeba	
				Acanthopodina	Hartmannellidae	*Acanthamoeba*	
	Class (Heterolobosea)		Schizopyrenida		Vahlkampfiidae	*Naegleria*	
	Subphylum (Mastigophora)		Trichomonadidae		Trichomonadidae	*Trichomonas*	
						Dientamoeba	
						Retortomonas	
			Diplomonadida	Deploomonadida	Hexamitidae	*Giardia*	
				Enteromonadida		*Enteromonas*	
	Class (Zoomastigophora)		Retortamonadida			*Chilomastix*	
			Kinetoplastida	Trypanosomatina	Trypanosomatidae	*Trypanosoma*	
						Leishmania	
Ciliophora	Kinetofrgaminophorea	Vestibulifferia	Trichostomatida	Trichostomatina		*Balantidium*	
Apicomplexa	Sporozoa		Haemosporida	Haemosporina	Haemosporidae	*Plasmodium*	
						Laverania	
			Piroplasmida	Piroplasmina	Piroplamidae	*Babesia*	
		Coccidia	Eucoccidia	Eimeriina	Eimeriidae	*Cystoisospora*	
					Sarcocystidae	*Cyclospora*	
						Toxoplasma	
						Sarcocystis	
					Cryptosporidae	*Cryptosporidium*	
Microspora			Microsporida			*Enterocytozoon*	
						Encephalitozon	
						Microsporidium	
						Nosema	
						Pleistophora	

Paratenic host (carrier/transport host): Host where parasite remains viable without further development called paratenic host.

Accidental/incidental host: Host in which parasite is not found usually but enters accidentally and host will be considered as dead end host. Following are parasites where human acts as dead end host:

1. Free living amoebae: *N. fowleri*, *Acanthamoeba* spp., and *B. mandrillaris*
2. Invasive or muscular *Sarcocystis*: *S. lindamanni*
3. *Spirometra* species
4. *Echinococcus granulosus* in human
5. *Multiceps multiceps*
6. *Trichina spiralis*
7. *Angiostrongylus* species
8. *Capillaria hepatica*
9. *Gnathostoma* spp.
10. Others: Normally cycle is continuing; but when parasite enters in certain tissues, human will be act

as dead end host like in *Taenia solium* when larvae enter beyond the intestine and in *E. histolytica* when trophozoites enter beyond the intestine (extra-intestinal amoebiasis).

LIFE CYCLE

Definition

Sequential change in growth, development and multiplication of parasite called life cycle.

Types

Following are the different types:

According to reproductive methods and types of host

1. Sexual cycle/cycle in definitive host: Development and multiplication occurs by sexual methods.
2. Asexual cycle/cycle in intermediate host: Development and multiplication occurs by asexual methods.

TABLE 14.4: Systemic classification of metazoan/helminths/worms

Kingdom: → **Animalia**

Subkingdom: → **Metazoa or Helminths**

Phylum	Class	Subclass	Order	Suborder	Family	Genus	Species
Platyhe-lminths	Cestoda		Pseudophylidea		Diphylobothriidae	Diphylobothrium	Pathogenic and non-pathogenic species are mentioned in **respective chapters**
						Spirometra	
			Cyclophylidea		Taeniidae	Taenia	
						Echinococcus	
					Hymenolepidae	Hymenolepis	
					Dilepididae	Dypilidium	
	Trematoda	Digenea (So called because di (two) and genetic generation), means required two hosts for cycle	Prosostomata	Strigeata	Schistosomatidae	Schistosoma	
				Amphistomata	Fasciolidae	Fasciolopsis	
						Fasciola	
				Distomata	Heterophyidae	Heterophyes	
						Metagonimus	
					Echinostomatidae	Echinostoma	
						Paryphostomum	
					Paraphistomatidae	Watsonius	
						Gastrodiscoides	
					Opisthorchidae	Clonorchis	
						Opisthorchis	
					Triglotrematidae	Paragonimus	
Nemat-helminths	Nematoda	Aphasmidia (no caudal chemo-receptors)	Enoplida		Trichinellidae	Trichinella	
					Trichuridae	Trichuris	
						Capillaria	
		Phasmidia (Caudal Chemo-receptors)	Rhabditidia		Ascarididae	Ascaris	
					Ancyclostomatidae	Ancyclostoma	
						Necator	
					Strogyloididae	Strongyloides	
					Oxyuridae	Enterobius	
					Angiostrogylidae	Angiostrongylus	
			Spirurida		Onchocercidae (Acantho-cheilonematidae)	Wuchereria	
						Brugia	
						Onchocerca	
						Dipetalonema	
						Mansonella	
						Dirofilaria	
						Loa	
					Dracunculidae	Dracunculus	
					Gnathostomatidae	Gnathostoma	

According to numbers of host

1. Direct (simple) cycle: Parasite requires only single host to complete the cycle called direct cycle. For examples follow **Table 14.5**.
2. Indirect (complex) cycle: Parasite requires two or more host to complete the cycle called indirect cycle. Following are subtypes:
   - Requires two hosts: In this cycle one host act as definitive while other acts as intermediate host. For examples follow **Table 14.6**.

   Requires three hosts: Some parasites require three hosts, of this one act as definitive while other two acts as 1st intermediate host and 2nd intermediate host. For examples, follow **Table 14.7**.

Clinical Significances

Ideas of life cycle help to know the growth and development of parasite, to know the clinical disease produced by parasites, to diagnose the parasite in laboratory, to treat parasitic disease and to prevent the parasitic disease.

Role of Human in Parasitic Life Cycle

- **Normal (natural) life cycle:** Life cycles of many parasites normally continue between human and external environment or other host like animals, vectors, etc., all such life cycles are described with **green arrows** (↓) in respective chapters.

TABLE 14.5: Parasites with direct (simple) life cycle

Parasites	Host
Protozoa	
E. histolytica	Human
G. lamblia	Human
T. vaginalis	Human
B. coli	Pig and human
C. belli	Human
C. parvum	Human
C. cayetanensis	Human
Microspora	Human
Metazoa	
A. lumbricoides	Human
A. duodenale	Human
N. americanus	Human
S. stercoralis	Human
T. trichuria	Human

TABLE 14.6: Parasites requires two hosts

Parasites	Definitive host	Intermediate host
Protozoa		
Plamodium spp.	Mosquito	Human
Babesia spp.	Hard tick	Human
T. gondii	Cat	Human/animals
S. bovihominis	Human	Cattle
S. suihominis	Human	Pig
S. lindamanni	Cat or dog	Human
Metazoa		
T. saginata	Human	Cattle (cow)
T. solium	Human	Pig
E. granulosus	Dog	Human
H. nana (direct cycle)	Human/rat	Human/rat
H. nana (indirect cycle)	Human	Flea
D. caninum	Human	Flea
Schistosoma spp	Human	Snail
F. buski	Human	Snail
G. hominis	Human	Snail
F. hepatica	Human	Snail
A. costaricensis	Human	Slug
W. bancrofti	Human	Mosquito
B. malayi	Human	Mosquito
O. volvulus	Human	Black fly
L. loa	Human	Deer fly
Mansonella, spp.	Human	Mosquito
D. medinensis	Human	Cyclops

- **Dead end life cycle:** Many parasites once enter in human, cannot continue their life cycle and human acts as **dead end**, all such life cycles are described with

TABLE 14.7: Parasites requires three hosts

Parasites	Definitive host	1st	2nd intermediate host
Metazoa			
D latum	Human	Cyclops or diaptomus	Fish
H hetrophyes	Human	Snail	Fish
M yokogawai	Human	Snail	Fish
Echinostoma spp.	Human	Snail	Mollusc
C sinensis	Human	Snail	Fish
O felineus	Human	Snail	Fish
P westermanii	Human	Snail	Crab/cray fish
G spinigerum	Animal	Cyclops	Fish

red arrows (↓) in respective chapters. **Red arrows (↓)** also given where parasites, (especially trophozoites) are released in external environment where they get disintegrate.

- **Life cycle with autoinfection in human:** Life cycles of parasites in which autoinfection in human is possible are described with **sky blue arrows (↓)** in respective chapters.
- **Other type of life cycles:** All other type life cycles in human like exoenteric cycle in T gondii, retrograde cycle in E vermicularis, etc., are described with **black arrows (↓)** in respective chapters. These cycles allow the continuation of life cycle but these are not occurring normally or naturally.
- In certain parasitic life cycle, either naturally or accidentally human acts as both definitive host (harbors adult/sexual stage) as well as intermediate host (harbors larval/asexual stage) like in H. nana, T. solium and T. spiralis.

ACCESS YOURSELF

Short Notes

1. Differences between cestodes and trematodes/cestodes and nematodes/trematodes and nematodes.

Short Questions for Theory/Viva Questions

1. Write two differences between protozoa and metazoa.
2. Name the four parasites in which human acts as an intermediate host.
3. Name the four parasites in which human acts as dead end host.
4. Define definitive host and intermediate host.
5. Name two parasites in which human acts as both definitive host and intermediate host.

MCQs for Chapter Review

Parasite

1. **Nematodes are differentiated from other worms by:**
   a. Segmentation absent   b. Separate coelomic cavity
   c. Sexes are separate   d. They are cylindrical
   e. GIT is complete

2. **Operculated eggs are seen in:**
   a. Nematodes
   b. Cestodes
   c. Trematodes
   d. Protozoa
3. **Alimentary canal is absent in:**
   a. Cestodes
   b. Trematodes
   c. Nematodes
   d. None of above
4. **Cylindrical worms are:**
   a. Tape worms
   b. Flukes
   c. Round worms
   d. Cestodes
5. **True about trematodes:**
   a. Two hosts required
   b. Segmented
   c. Anus present
   d. Body cavity present

Host

6. **Human acts as an intermediate host in:**
   a. *L. donovani*
   b. *Plamodium* spp.
   c. *E. granulosus*
   d. b + c

Life Cycle

7. **Simple life cycle requires:**
   a. One host
   b. Two host
   c. Three host
   d. Four host
8. **Simple life cycle seen in:**
   a. *Ascaris*
   b. *T. solium*
   c. *Toxoplasma*
   d. *Giardia*
   e. *Schistosoma*
9. **Two hosts required in:**
   a. *T. solium*
   b. *E histolytica*
   c. *T. saginata*
   d. *Giardia*
   e. *Toxoplasma*
10. **The intermediate host for *T. saginata*:**
    a. Man
    b. Cow
    c. Dog
    d. Pig
11. **Which of the following parasites passes through three hosts?**
    a. *Fasciola hepatica*
    b. *Fasciola buski*
    c. *Schistosoma haematobium*
    d. *Clonerchis sinensis*
12. **Crab is the intermediate host for:**
    a. *Clonerchis sinensis*
    b. *Paragonumus westermanii*
    c. *Fasciola hepatica*
    d. *Schistosoma haematobium*
13. **Fish acts as an intermediate host in:**
    a. *D. latum*
    b. *Clonerchis sinensis*
    c. *H. diminuta*
    d. *H. nana*

Answers and Explanation of MCQs

1. **a, b, c, d, e**
2. **b, c**
3. **a**
4. **c**
- Follow **Table 14.2** for explanation of answers of MCQs 1–4.
5. **a**
- Option a: Follow **Table 14.4** for explanation (trematodes are classified under class **Digenea**).
- Option b, c and d: Follow **Table 14.2** for explanation.
6. **d**
- Follow section, **host (classification → intermediate host)** for explanation.
7. **a**
8. **a, d**
- Follow section, **life cycle (Table 14.5)** for explanation of answers of MCQs 7–8.
9. **a, c, e**
10. **b**
- Follow section, **life cycle (Table 14.6)** for explanation of answers of MCQs 9–10.
11. **d**
12. **b**
13. **a, b**
- Follow section, **life cycle (Table 14.7)** for explanation of answers of MCQs 11–13.

General Properties of Parasites

PROTOZOA

Meaning

Protozoa word derived from Protos (Greek) means first and Zoon means animal.

Variation in Species

There are about 65,000 known species of protozoa, 10,000 are parasitic and 17 are medically important.

Morphology

Size of protozoa: Protozoa are microscopic in size and visible under high power microscope. Largest protozoan is *B. coli*.

Structure of protozoa: Protozoa are unicellular and single cell performs the all functions. Structure of protozoa consists cytoplasm, nucleus and other materials.

- **Cytoplasm:** It is divided in two portions:
  - **Endoplasm (Fig. 15.1):** It is an internal, granular portion of cytoplasm. It plays role in nutrition, excretion and reproduction. It contains nucleus, contractile vacuoles, food debris, endoplasmic reticulum, Golgi bodies and RBCs in some parasite.
  - **Ectoplasm (Fig. 15.1):** It is an external, clear, hyaline portion of cytoplasm. It plays role in protection, locomotion and sensory function. Following structures are developed from ectoplasm:
    1. **Organs of locomotion:** Three types as follows:
       - **Pseudopodium (Fig. 15.1):** It is the temporary prolongation of ectoplasmic process. It helps in ingestion of food. It occurs in Rhizopodea (*E. histolytica*). Speed of locomotion by pseudopodia is 0.2–0.3 μm per seconds.
       - **Flagellum (Fig. 15.2):** It is a permanent, long, delicate and thread like filament. It occurs in Zoomastgophorea like *G. lamblia, Trichomonas*

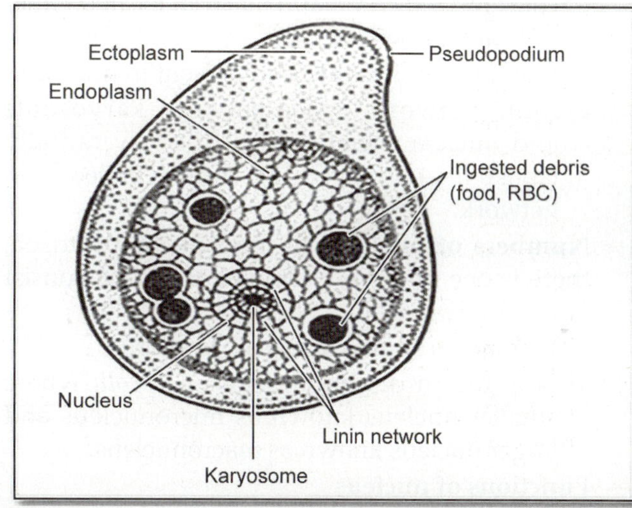

Fig. 15.1: Pseudopodium in *E. histolytica*

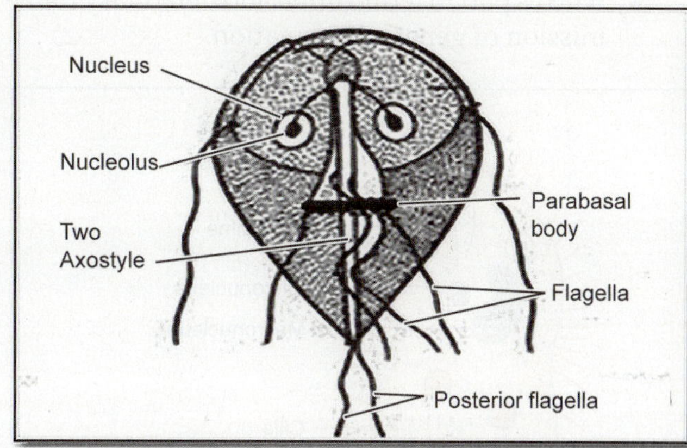

Fig. 15.2: Flagella in *G. lamblia*

spp., *Trypanosoma* spp., and *Leishmania* spp. Speed of locomotion by flagella is 15–30 μm per seconds.

- **Cilia (Fig. 15.3):** It is a permanent short, fine and needle like filament. It occurs in Ciliatea like *B. coli.* Speed of locomotion by cilias is 400–2000 μm per seconds.
2. **Rudimentary digestive organs**
   - Cytostome (cell mouth) presents in *B. coli.*
   - Cytopharynx presents in *B. coli.*
   - Cytopyge (cell anus): To excrete the food wastes.
3. **Contractile vacuoles:** It developed from ectoplasm, but situated inside the endoplasm. It concerned with excretory function and maintains the osmotic pressure.
4. **Cyst wall:** Thick resistant wall seen in cystic stage.

- **Nucleus:** It is bounded by well-defined nuclear membrane. It has following features:
  - **Contains of nucleus:** Internally it contains following materials:
    - Nucleolus (karyosomes): It is a small dot like concentrated chromatin material located either centrally or peripherally.
    - Nuclear sap: It is the fluid present in nucleus.
    - Linin network: Space between karyosome and nuclear membrane is filled by radially (spoke like) arranged fine threads called linin network.
  - **Numbers of nucleus:** In most of the protozoa, there is one nucleus. Following protozoa consist two or more nuclei:
    - Both are in similar size in *G. lamblia*
    - Both are in dissimilar size in *B. coli,* where smaller nucleus known as micronucleus and larger nucleus known as macronucleus.
  - **Functions of nucleus**
    - It controls all activities of cell.
    - It is essential for cell division, growth and replication.
    - It takes part in fertilization and hereditary transmission of genetic information.

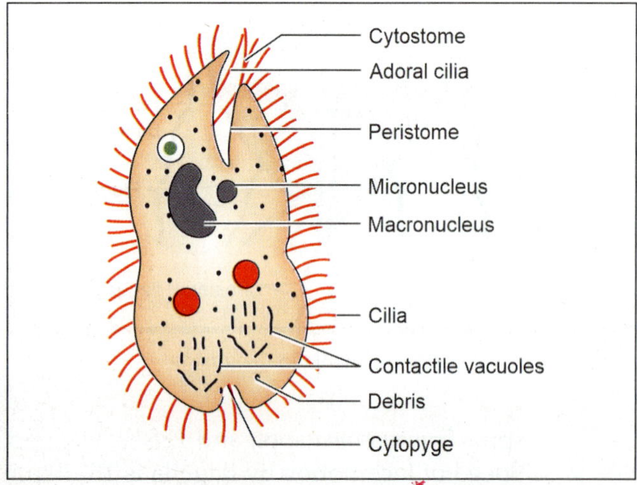

Fig. 15.3: Cilia in *B. coli*

- **Other materials**
  1. **Kinetoplast:** Nonnuclear DNA containing body called kinetoplast. It has mitochondrial structure and acts as energy store house.
  2. **Basal body:** It is a small body represents the origin of flagellum.
  3. **Chromatoid bars (chromodia):** Extranuclear chromatin materials called chromatoid bars. They present in premature cyst of *E. histolytica* and disappear in mature cyst. They stain black with iron-hematoxylin stain.
  4. **Glycogen mass:** It presents in premature cyst of *E. histolytica* and disappears in mature cyst. It stains brown with iodine.

Morphological stages of protozoa: In all protozoa mainly two morphological stages are present like trophozoite stage and cyst stage. In some protozoa intermediate stage called precyst stage also occurs.

1. **Trophozoite stage:** Word derived from tropos (Greek) means nourishment. It has following properties:
   - It is the active and feeding stage.
   - It is motile in nature.
   - Excretion of waste products from trophozoites occurs by contractile vacuoles or by osmosis.
   - Respiration takes place by osmosis.
   - It receives the nutrition from surrounding environment by different methods such as diffusion, active transport across the plasma membrane, phagocytosis through pseudopodia, mouth like structure called cytostomes or by pinocytosis for minute drop of food particles.

2. **Cyst stage:** Trophozoite transformed in to a tough wall structure called cyst. It has following properties:
   - It is an inactive, resting and resisting stage.
   - It is nonmotile in nature.
   - It is infective to human beings and transmitted by food-water, vectors or by other ways.

Methods of Reproduction

Methods of transformation: Two types:
1. Encystation: Transfer of trophozoite (active) to cyst (inactive) called encystation.
2. Excystation: Transfer of cyst (inactive) to trophozoite (active) called excystation.

Methods of multiplication: Multiplication occurs only in trophozoites stage. Following are the methods of reproduction and multiplication in protozoa:

1. **Asexual multiplication in protozoa**
   - **Binary fission:** Parasites divide in to two equal parts. Before division all the structure are duplicated. It occurs longitudinally in flagellates **(Fig. 15.4a)** or transversely in ciliates **(Fig. 15.4b)** or in *E. histolytica.*
   - **Multiple fissions (schizogony):** Parasites divides in to more than two cells. First nucleus of parent

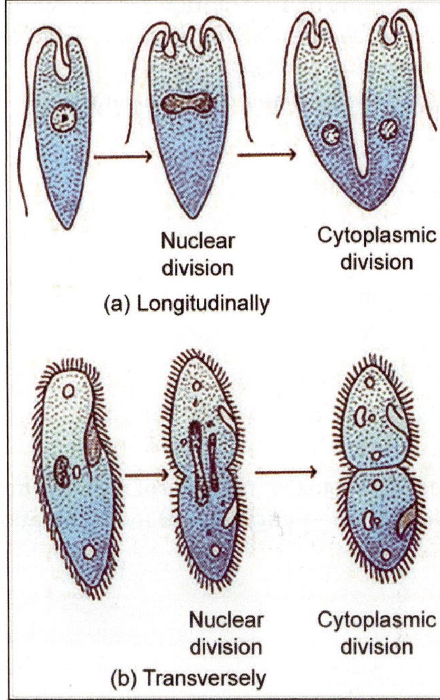

Fig. 15.4: Binary fission

cell undergoes repeated division, which are then surrounded by cytoplasm, resulting daughter cell known as merozoite. When multiplication is completed, the whole parasitic body called schizont ruptures and liberates these daughter cells, which repeat the life cycle later on. For example, *Plasmodium* spp., showing multiple fissions.

- **Endodyogeny:** Parasite undergoes the single internal budding, resulting two daughter cells are produced, e.g., *T. gondii.*

2. **Sexual multiplication in protozoa**
   - **Conjugation:** In this process two cells are united for a temporary period, during which time interchange of nuclear materials takes place. Later on, two cells will be separated. For example, *B. coli.*
   - **Gametogony (syngamy):** Two sexually differentiated cells like male and female gametocytes unite permanently to form a zygote. Which continue the life cycle later on; for example, *Plasmodium* spp.

METAZOA

Synonym

Term metazoan also called helminths or worms.

Meaning

Helminths is a Greek word derived from helmin means worm, originally referred to intestinal worms.

Morphology and Methods of Reproduction

Morphological stages, reproductive organs and reproductive methods for cestodes are described in **Ch. 100,** for trematodes in **Ch. 103** and for nematodes in **Ch. 106.**

Synonym

These factors also called epidemiological determinants.

Types

Virulence of parasites is affected by following three factors:

Agent factors (virulence factors or determinants of virulence)

1. **Intracellular or cell associated factors**
   - **Intracellular location:** It provides protection against the host defense system.
   - **Tough cyst wall in protozoa:** It provides protection against gastric acidity.
   - **Adhesins**
     - Lectin mediated adherence protein in *E histolytica*: It helps in contact with target cells.
     - Sucking disc *G lamblia*: It is the organ of adhesion. It does not invade the tissue but remains tightly adhere to the intestinal epithelium.
   - **Motility:** It helps in deeper penetration and movement toward nutrition.
   - **Spine or lateral knob in schistosomes:** It erodes the blood vessels to cause hemorrhagic manifestations.

2. **Extracellular factors**
   - **Enzymes**
     - Enzymes in *E histolytica*: Histiolysin brings the destruction and necrosis of tissues and slough formation. Organisms receive the nourishment by absorption of dissolved tissue juice. Other enzymes in *E. histolytica* like hyaluronidase, trypsin, pepsin, amylase, etc., may induce tissues destruction.
     - Enzymes in *T vaginalis* like cysteine protease which contributes in pathogenesis.
     - Enzymes in *B. coli* like hyaluronidase, which degrades the intestinal mucosa and facilitates the penetration.
   - **Toxins or metabolites production by parasites:** Liberated by trophozoites, eggs, larvae or adult stages of parasites are responsible for allergic manifestations.

Host factors

1. **Age:** Certain infections are common at particular age like thread worm in pediatrics while *T. vaginalis* infection in adults.
2. **Immune status:** Immunodeficiency status favors certain parasitic infections like *S. stercoralis, G. lamblia,* etc. Splenectomized persons are more susceptible to babesiosis.
3. **Blood group:** Group A is more prone to *G. lamblia.*
4. **Food habit:** *T. saginata* is common who eats beef while *T. solium* is common who eats pork.

1

5. Occupation: *D. medinensis* is common in bhistis (person who carries bag of water on back).

Environmental factors

1. Poor sanitation and overcrowding: They favor the mosquito-borne diseases like malaria and filaria.

2. Humidity: *Naegleria fowleri* is common in air of air cooler and live symbiotically with *Legionella* or *Listeria* and transmitted by inhalation of air of air-cooler.

ACCESS YOURSELF

Short Questions for Theory/Viva Questions

1. What are encystation and excystation?
2. Write the functions ectoplasm and endoplasm in protozoa.
3. Name the organ of locomotion in following parasites: *Entamoeba histolytica, Balantidium coli, Trypanosoma cruzi, Leishmania donvani* and *Giardia lamblia.*
4. What are monoecious and diecious parasites?

MCQ for Chapter Review

Protozoa (Methods of Reproduction)

1. Encystation is:
   a. Trophozoite → cyst
   b. Cyst → trophozoite
   c. Egg → larvae
   d. Cyst → egg

Answer and Explanation of MCQ

1. a
- Follow section, **Protozoa methods of reproduction (methods of transformation → encystation)** for explanation.

SECTION 2

Immunology

Immunity

INTRODUCTION

Meaning: Immunity word originated from immuntans (Latin) means freedom from disease.

Definition: It is the resistance exhibited by host toward injuries produced by microorganisms and/or their products.

TYPES AND EFFECTIVE FACTORS ON IMMUNITY

Types of Immunity

Innate (Native) Immunity

Definition: It is the resistance to infection that an individual possesses by virtue of his/her genetic or constitutional make-up.

Properties

- Innate immunity is the 1st line defense.
- It occurs without prior contact with microbes (already present before contact with antigen).
- Immunological memory—absent
- It is active against limited numbers of antigens.

Classification

1. **Nonspecific innate immunity:** It is in general immunity. It occurs at species, racial and at individual level.
2. **Specific innate immunity:** It is the immunity against the particular agent. It also occurs at species, racial and at individual level.

Innate immunity at the level of species, racial and individual

- **Species level:** It is the total or relative resistance to pathogens shown by all members of species. It occurs due to physiological and biochemical differences in between tissues of different species. For example, human shows resistance to plant pathogens.
- **Racial level:** It is the resistance to pathogens shown by different races of particular species. It is genetic in origin. Following are examples:
  - Anthrax resistance in Algerian sheep.
  - Peoples of African origin are more susceptible to tuberculosis than Caucasians.
  - Genetic resistance to *P. falciparum* in some parts of Africa and Mediterranean coast.
  - Genetic abnormalities in RBCs (sickle cell anemia) provide protection against malaria.
- **Individual level:** It is the resistance to pathogens shown by different individuals of particular races. It is genetic in origin. For example, homozygous twins are resistant or susceptible to lepromatous leprosy/tuberculosis. Such correlation is not seen in heterozygous twins.

Mechanisms or components of innate immunity

A. Anatomical (mechanical) barrier

1. **Skin**
   - **Mechanical barrier:** Skin acts as barrier to invasion of microbes.
   - **Chemical barrier:** Skin acts as chemical barrier and provides bactericidal activity due to presence of high salts in dry sweat, sebaceous secretions, long chain fatty acids, acidic pH (5.2–5.9, acidic pH presents in vagina, stomach, urine and skin) and soaps. Bactericidal activities of skin are detected by following ways:
     - *S.* Typhi dies in a few minutes when culture placed on skin and cultured at regular interval while *S.* Typhi remains unaffected when placed simultaneously on glass surfaces.

- Frequent washing of hands in soapy water for long periods at occupation leads pyogenic and mycotic infection.

2. **Mucosa/mucous membrane**

- **Mechanical barrier:** Mucosa can trap pathogens and prevent their entry. Mucosa secretes mucus, which is a protective barrier.

- **Chemical barrier**

  - Lysozyme: It presents in all mucosal secretions/body fluids except CSF, urine and sweats. It has powerful digestive abilities that render antigens harmless. It splits the polysaccharide components of bacterial cell wall and provides bactericidal effect. It helps in phagocytosis to form phagolysosome.

  - IgA: Mucosa also secretes IgA, which provides local immunity.

  - Respiratory mucosa: If pathogen inhaled they are then sneezed or coughed out or swallowed. Cough reflex is an important defense mechanism. Ciliary movement can propel the particles upward. Mucopolysaccharides in nasal and respiratory epithelium can combine with influenza and other viruses. If microbes reach to alveoli, they can killed by phagocytic cells called pulmonary alveolar microphages (PAMs).

  - GIT and genitourinary mucosas: Also have protective role.

3. **Hairs:** Nasal hairs can trap the pathogens.

B. Physiological mechanisms or components

1. **Eyes:** Flushing action of tears (lacrimal secretion) makes the conjunctiva free from foreign particles. Tears also contain lysozyme.

2. **Mouth:** Saliva, which constantly bathed the mouth, has inhibitory effect on microbes. It also contains lysozyme.

3. **Stomach:** Acidity destroys the pathogens.

4. **Urine:** Flushing action of urine eliminates the bacteria from urethra.

5. **Semen:** It contains antibacterial substances like spermine and zinc.

C. Biochemical: Following are biochemical substances present in blood and tissues having antimicrobial actions.

1. **Lysozyme:** Described above.

2. **B-lysin:** It is active against anthrax and related bacilli.

3. **Leukins:** It derived from leukocytes.

4. **Plakins:** It derived from platelets.

5. **Lactic acid:** It presents in muscles, tissues and at inflammatory sites.

6. **Lactoperoxidase:** It presents in milk.

7. **Properdin:** It presents in serum.

8. **Integrin:** It presents in host cells and helps to attach the host cell with extracellular matrix like

attachment of host cell with adeno virus, HFMD virus, ECHO virus and *Hantavirus*.

9. **Interferon (IFN):** It has antiviral properties. More detail is given in **Ch. 12.**

10. **Others like:** Lactoferrin, transferrin, TNF-α, fibronectin, oligosaccharides, antibodies, etc.

D. Microbiological mechanisms or components: For example, normal flora of body described in **Ch. 4.**

E. Immunological mechanisms or components

1. **Complement system:** It acts by activation of alternative pathway by bacterial endotoxin and Lectin pathway or mannose binding lectin (MBL) pathway by combination with mannose residues present on microbial surface (like *Salmonella, Neisseria, Listeria, C. neoformans* and *C. albicans*). Complement mediated lysis takes place by formation of pores on bacterial surface or by secretion of inflammatory mediators.

2. **Reticuloendothelial (RE) system/RE cells:** Ch. 22.

3. **Null cells** or **large granular lymphocytes (LGLs):** Ch. 22.

4. **Pattern recognition receptors (PRRs):** Receptors on different cells recognize the unique molecular pattern of pathogens called pathogen-associated molecular patterns (PAMPs) and such receptors called pattern recognition receptors (PRRs). Following are three types of PRPs:

   - **Toll-like receptors (TLRs):** Certain cell associated receptors involved in innate immunity have ability to recognize and to destroy the unique molecules of microbes and prevent the microbial invasion called toll-like receptors. They are present on phagocytic cells like dendritic cells and macrophages and help in phagocytosis. There are 13 TLRs.

   - **Scavenger receptors:** They bind with LPS and peptidoglycan of bacterial cell wall and also with infected, injured or apoptotic cells. These include CD-36, CD-68 and SRB-1.

   - **Mannose receptors:** They present on phagocytic cells and bind with mannose rich glycans of microbes.

F. Pathological mechanisms or components

1. **Inflammation:** Inflammation starts following biological or any other types of injuries. It provides defense by different mechanisms like outpouring of plasma which dilutes the toxic products, increases blood flow at the site of injury by vascular constriction and dilatation, formation of fibrin barrier which localizes the infection, recruiting phagocytes at the site leading to phagocytosis of microbes and acute phase response (acute phase reaction). Change in the concentration of many proteins following inflammation called acute phase proteins or acute phase reactants and this response called acute phase response or acute phase

Flowchart 16.1: Synthesis of acute phase proteins

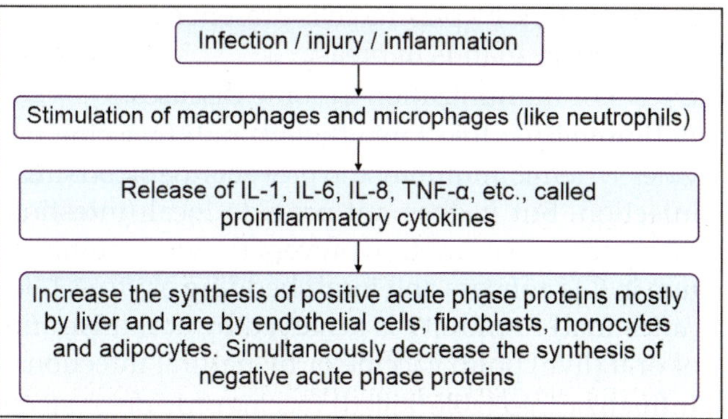

reaction. Acute phase proteins are synthesized in liver and rare by endothelial cells, fibroblasts, monocytes and adipocytes. All these acute phase proteins have antimicrobial and anti-inflammatory actions. They also have chelating actions for iron and copper making them unavailable for bacteria. Mechanisms of synthesis of acute phase proteins are highlighted in **Flowchart 16.1.** Two types of acute phase proteins like with increasing synthesis called positive acute phase proteins and decreasing synthesis called negative acute phase proteins. Examples of acute phase proteins are mentioned below.

- Positive acute phase proteins: C-reactive protein/ CRP (**Ch. 50**), procalcitonin, mannose binding protein, serum amyloid P components, α-1-acid glycoprotein, α-2 microglobulin, coagulation proteins (fibrinogen, prothrombin, von Willebrand factor and factor VIII), hepatoglobin, D-dimer protein, ferritin, ceruloplasmin, complement factors, etc.
- Negative acute phase proteins: Albumin and prealbumin, transferin, transthyretin, transcortin, antithrombin, retinol binding protein, etc.

2. **Fever:** Rise in temperature following injuries provides defense by acceleration of physiological process, antiviral action like increase interferon production and antibacterial action over certain bacteria like *T. pallidum* (therapeutic fever induction was used to treat syphilis before penicillin became available).

Acquired (Adaptive) Immunity

Definition: It is the resistance to infection that an individual acquires during life.

Properties

- Acquired (adaptive) immunity is the 2nd line defense.
- It developed after contact with antigens.
- It differentiates between two different proteins or antigens called antigenic specificity.
- Immunological memory present: It remembers the previous type of antigenic attack which helps to elicit strong immune response in subsequent antigenic attacks.

- It is active against wide range of antigens.
- Differentiation between self and non-self: It eliminates the foreign particles without disturbing the self particles. Self tolerance is the unique properties of immune system. Failure to this can cause autoimmune disease.

Mechanisms or components of acquired immunity

1. AMI/HI: By production of antibody from B cells (plasma cells).
2. CMI: By cell stimulation and production of cytokines (lymphokines) from T cells.
3. Classical complement pathway: **Ch. 21.**
4. **Antigen presenting cells (APCs): Ch. 22.**

> **Notes:**
> Many components of innate and acquired immunity like APC, complement, null cell (ADCC), etc., are interconnected in both types of immunity and make the bridge between two.

Classification: Three types of acquired immunity such as active immunity, passive immunity and combined immunity. Differences between active and passive immunity are mentioned in **Table 16.1.**

A. Active immunity: It developed due to antigenic stimulation. Two subtypes as described below:

1. **Natural active immunity:** It is developed due to infection. Following are the examples:
   - Viral infections: Some viral infections provide long lasting or lifelong immunity like polio, measles, chickenpox, etc. while some viral infections provide short lived immunity like influenza.
   - Bacterial infections: Immunity following bacterial infections is short lived unlike viral infections; however some bacteria like *S. Typhi* provides durable protection.
   - Parasitic infections: High level of Ig is found in some parasitic infections is protective and helps in recovery of diseases.

TABLE 16.1: Differences between active and passive immunity

Active immunity	Passive immunity
More effective and better protection	Less effective and less protective
Produced actively by host's immune system	Transferred passively, no active participation by host's immune system
Induced by infection or vaccination	Readymade Ab is transferred
Immunity developed after effective lag period	Immediate protection
Immunological memory present and booster effect in subsequent dose (secondary response)	No memory and subsequent dose is less effective (no secondary response)
Long lasting immunity	Transient immunity
Negative phase may occur	No negative phase
Not applicable in immuno-deficient persons	Applicable in immunodeficient persons

- Fungal infections: In most of the fungal infections protection is achieved by CMI while AMI is little or rarely useful.

2. **Artificial active immunity:** It is developed by administration of immunizing agents (immunization). Following are the examples:
   - Vaccines.
   - Toxoids.

B. Passive immunity: It is transmitted in readymade form. Two subtypes as described below.

1. **Natural passive immunity:** It is transmitted from mother to baby. Following are the examples:
   - Colostrum: It is rich in IgA and provides protection to neonate.
   - Maternal antibodies: **IgG** can transfer through placenta and provides passive protection to fetus against virus or toxin. **IgM** develops at 22nd week of gestational age but it is not able to give protection. Infant immune system starts to develop at 3rd month of life until then infant is passively protected by maternal antibodies. **Injection of Tetanus Toxoid (TT)** to mother during pregnancy minimizes the chances of neonatal tetanus due to passive transfer of antibodies. It is recommended in communities in which neonatal tetanus is common.

2. **Artificial passive immunity:** It is transmitted to recipient by administration of antibodies. Following are the examples:
   - Development of immunity after administration of antisera, antitoxin and Ig.
   - For more details: **Ch. 17.**

C. Combined immunity

- **Definition:** It is a combination of active and passive immunization.
- **Uses:** In some disease like tetanus where TIG (tetanus immunoglobulin) is injected in one arm and TT in other arm followed by full course of TT. TIG provides passive protection till the active immunity developed by TT.

Other Types of Immunity

Adoptive immunity: Passive transfer of CMI achieved by administration of viable immunological competent lymphocytes called adoptive immunity. Instead of whole immunological competent lymphocyte, only its extract is used for administration called transfer factor. For more details of transfer factor follow **Ch. 23.**

Local (mucosal) immunity

- **History:** Concept of local immunity was 1st proposed by Besredka.
- **Definition:** When particular cells or tissues are targeted by certain pathogens, they produce the immunity against such pathogens called local immunity.

- **Principle:** It based on production of IgA called secretory IgA, by plasma cells present in mucosa and by secretory glands in breast.
- **Uses:** For immunization in some diseases.
- **Poliomyelitis:** Injectable (killed) polio vaccine provides systemic immunity and prevents the blood stream infection, but it does not provide local (intestinal) immunity and not able to prevent the viral multiplication in GIT mucosa and fecal shedding of virus. Local (intestinal) immunity is achieved by administration of oral (live) polio vaccine or by natural infection.
- **Influenza:** Injectable (killed) vaccine provides systemic immunity and prevents the bloodstream infection, but it does not provide local (respiratory) immunity and not able to prevent the viral multiplication in respiratory mucosa. Local (respiratory) immunity is achieved by administration of intranasal (live) vaccine or by natural infection.

Herd immunity

- **Synonym:** Population immunity.
- **Definition:** It is the mass immunity in community to prevent the epidemic diseases.
- **Uses of herd immunity:** It is used to prevent epidemic or pandemic diseases. When herd immunity is high, spread of epidemic diseases is less and it is in mild form. Herd immunity is better understood by COVID-19 as described below. It is achieved by following two ways:
  1. **Natural infections:** Natural infection of COVID-19 can produce the antibodies for future protection; however, it is not clear how long these antibodies are protective. It provides protection to newborns or those who have compromized immune systems.
  2. **COVID-19 vaccination:** Unlike the natural infection method, vaccines create immunity without causing major illness or complications. Currently COVID-19 vaccines are not available for children, so difficult to create the herd immunity in children by vaccination.
- **Effective factors on herd immunity**
  1. Infections: Occurrence of clinical/subclinical infections in herd.
  2. Immunization to herd: It includes vaccination to large population.
  3. Herd structure: It includes population, animals, vectors and all those environmental and social factors which favor or inhibit the spread of infection.

Premunition

- **Synonym:** Infection immunity or concomitant immunity.
- **Definition:** It is the relative resistance offered by host to reinfection already harboring the microbes.
- **Properties:** Immunity to reinfection lasts as long as active infection is present, once active infection cured patients become susceptible to subsequent infection by same organisms.

- **Infection in which premunition occurs:** It occurs in syphilis, malaria and schistosomiasis.

Ring immunity

- **Synonym:** Ring vaccination.
- **Definition:** It is the vaccination of all susceptible individuals in a prescribed area around an outbreak of an infectious disease.
- **Principle:** The idea is to form a buffer of immune individuals to prevent the spread of the disease.
- **Uses:** It is used to control the outbreak of following diseases.
- **Smallpox:** Ring vaccination was used to control smallpox until the last naturally occurring case in 1977. When an infection was diagnosed, all people who were or may have been exposed were identified and vaccinated. Then, a second "ring" of people who may have been exposed to the first ring were also identified and vaccinated. Ring vaccination was also recommended by the American Academy of Pediatrics (IAP) in 2002 for smallpox, should there be another outbreak (from terrorism or whatever).
- **Foot-and-mouth disease** in livestock in the UK. Also known as surveillance and containment.
- **Polio:** When a case of polio is detected all children less than 5 years of age within a radius of 5 km are immunized within 24–48 hours. The repeat extra dose is given to the same children after one month. The ring immunization drive arrests the spread of wild viruses and creates a film of vaccine viruses in the community.

Tolerance: It is the cessation of clinical phenomenon despite infection by microbes (infection to host without any adverse effect and host become resistant to reinfection).

Effective Factors on Immunity

Age: Pediatric and old age are more susceptible to infections than adults.

- Fetus is protected by placental barrier but some pathogens (like TORCH agents) can cross this barrier and cause fetal death.
- HBV infection in newborn is usually asymptomatic, because of absence of adequate immune response, which required setting up the clinical disease.
- Prepubertal girls are sensitive to gonococcal infection due to lack of protection in vaginal epithelium.
- Poliomyelitis and chickenpox are more common in adult stage due to well developed immune system which causes hypersensitivity and more tissue damages.
- Olders are more vulnerable to infection due to waning of immunity and deformities like enlarged prostate which obstructs the urinary flow and favors the UTI.

Hormones: Endocrine dysfunctions increase the chances of infections.

- **Diabetes mellitus:** Sugar deposition in tissues favors the *Staphylococcus* infection.
- **Corticosteroids:** They decrease the immunity by antiphagocytic effect, anti-inflammatory effect, hypersensitivity, reduction of antibodies formation (however steroids neutralize the bacterial products and minimize the damages).

Nutrition: Malnutrition reduces the CMI and AMI. For example protein deficiency disease like kwashiorkor gives false-negative Mantoux test.

MEASUREMENT OF IMMUNITY

It is very difficult to measure the exact level of immunity, but some useful methods for AMI and CMI detection are described in chapter of Ir.

ACCESS YOURSELF

Essay/Full Question

1. Types and mechanisms (principles) involved in immunity.

Short Notes

1. Innate immunity.
2. Acquired immunity.

Short Questions for Theory/Viva Questions

1. Write the four differences between active and passive immunity.
2. What is local immunity/herd immunity/ring immunity?
3. What is premunition?

Comments on

1. In acquired immunity, strong immune response occurs during subsequent antigenic attacks.

MCQs for Chapter Review

Innate Immunity

1. **All of the following are part of the innate immunity *except*:**
   a. Complement   b. NK cells
   c. Macrophages   d. T cells
2. **Lysozyme is present in all following secretions of the body *except*:**
   a. Lacrimal secretion   b. CSF
   c. Saliva   d. Respiratory tract secretions
3. **All are true about innate immunity *except*:**
   a. Nonspecific
   b. First line of defense
   c. Not affected genetically
   d. Includes complement
4. **Components of innate immunity are:**
   a. T lymphocytes   b. Complement proteins
   c. B lymphocytes   d. NK cells
   e. Integrins
5. **First chemical barrier encountered by microorganisms for common exposed sites:**
   a. Lysozyme   b. Acidic pH
   c. Skin   d. Lactose
6. **Acute phase reactants in acute inflammation are:**
   a. Albumin   b. Fibrinogen
   c. Hepatoglobulin   d. Gammaglobulin

7. **Acute phase reactants are *except*:**
   a. C-reactive protein
   b. Hepatoglobulin
   c. Endothelium
   d. Fibrinogen

8. **Proinflammatory cytokines includes all *except*:**
   a. IL-1
   b. IL-2
   c. IL-6
   d. TNF-α

Acquired/Adaptive Immunity

9. **Acquired immunity also know as:**
   a. Adaptive immunity
   b. Adoptive immunity
   c. Ring immunity
   d. None of above

10. **Active immunity is not acquired by:**
    a. Infection
    b. Vaccination
    c. Ig transfer
    d. Subclinical infection

11. **Active immunity can be induced by:**
    a. Toxoides
    b. Subclinical infection
    c. Antitoxin
    d. Immunoglobulin
    e. Antigen exposure

12. **Type of immunity conferred on an individual by vaccination is:**
    a. Artificial active
    b. Artificial passive
    c. Natural active
    d. Natural passive

13. **True about active immunity:**
    a. Less effective
    b. Can be given in immunodeficient person
    c. Immunological memory present
    d. No lag period

14. **True about passive immunity:**
    a. Cannot be given with active immunity
    b. Lasts for 4–5 days
    c. It can be given before disease occurrence
    d. Can be transferred by antibodies from another host
    e. Takes long time to develop.

Answers and Explanation of MCQs

1. d
- Alternative pathway and lectin pathway of complement, NK cells (one type of null cell) and macrophages (one type of phagocytic cell) are parts of innate immunity while T cells is responsible for CMI and part of acquired immunity.

2. b
- Lysozyme is present in all body secretions/body fluids except CSF, urine and sweats.

3. c
- Innate immunity is due to genetic or constitutional make –up of an individual.

4. b, d and e
- Follow section, **innate immunity (mechanisms or components of innate immunity)** for explanation.

5. b
- Skin acts as chemical barrier and provides bactericidal activity due to presence of high salts in dry sweat, sebaceous secretion, long chain fatty acids, acidic pH (5.2–5.9, acidic pH presents in vagina, stomach, urine and skin) and soaps.

6. a, b, c

7. c
- Follow section, **innate immunity [mechanisms or components of innate immunity → inflammation)]** for explanation of answers of MCQs 6–7.

8. b
- Follow section, **acute phase response (Flowchart 16.1)** for more explanation.

9. a
- Follow section, **other types immunity (adoptive immunity and ring immunity)** for explanation of option b and c.

10. c
- Ig transfer provides an artificial passive immunity.

11. a, b and e
- Introduction of
  - Toxoids induced artificial active immunity.
  - Subclinical infection induced natural active immunity.
  - Antitoxin and immunoglobulin induced artificial passive immunity.
  - Antigen exposure like vaccine induced artificial active immunity.

12. a
- Artificial active immunity is achieved by administration of vaccine and toxoids.

13. c

14. c, d
- Follow section, **acquired immunity (Table 16.1)** for explanation of answers of MCQs 13–14.

CHAPTER 17

Immunizing Agents and Universal Immunization Schedule

IMMUNIZING AGENTS

Immunizing agents are used for artificial active immunization (immunoprophylaxis), artificial passive immunization (immunotherapy) and for combined immunization.

Used for Artificial Active Immunization (Immunoprophylaxis)

Vaccines

Definition: Vaccines are the biological products, act by reinforcing the immunological defense of the body against foreign particles.

History: Vaccine word derived from Vacca (Latin) means cow, because 1st such biological product (prophylactic preparation) was prepared by Edward Jenner for smallpox disease from pox lesion in cow. Later, Louis Pasteur coined the term vaccine for such biological product (prophylactic preparation).

Principle: After the primary immune response to the pathogens, B cells are differentiated into plasma cells and memory cells. Memory cells lie dormant in immune system; they detect the same pathogens later on and mount strong and rapid immune response. This makes the basis of vaccine development. Vaccine contains antigens/pathogens, which induces the production of memory cells and provides strong and rapid protection if same antigens/pathogens encountered later on.

Types and examples of vaccines: Following are the different types:

A. Live-attenuated vaccines

- **Preparation of live-attenuated vaccines:** They contain living microbes that have been weakened (attenuated) in the laboratory, so they cannot cause disease. Vaccines are prepared by following ways:
  - From natural virus (Jenner's smallpox vaccine).
  - By serial passage in cells/tissues (yellow fever).
  - By plaque selection (OPV).
  - From its mutant (influenza).
  - By recombination technique (influenza).
- **Advantages of live-attenuated vaccines**
  - Vaccines are the closest thing to natural infections; these vaccines are good "teachers" of the immune system. They elicit strong AMI and CMI.
  - They confer strong and lifelong immunity with only one or two doses, except OPV where three or more doses are required.
  - They administered by natural route of infection, so provide both systemic and local immunity, e.g. OPV, intranasal influenza vaccine, etc.
  - They are used for mass immunization (herd immunity).
  - They are given in combination (MMR vaccine).
- **Disadvantages of live-attenuated vaccines**
  - They are not safe for immunocompromised individuals.
  - Live-attenuated organisms may rarely mutate to a virulent form and cause diseases (disease to contact).
  - They are heat labile and inactivated if temperature is not maintained during storage and transportation.
  - They are more difficult to create for bacteria, because they have thousands of genes, thus hard to control, while easy for viruses, because viruses are simple microbes containing a small number of genes.

    – They cause immediate (local and systemic) and remote complications.
- **Contraindications of live-attenuated vaccines**
  - Person with immunodeficiency status.
  - Pregnancy.
- **Examples of live-attenuated vaccines:** Follow **Table 17.1**

B. Killed/ inactivated vaccines

- **Preparation of killed vaccines:** Microbes are inactivated by treatment with chemicals [phenol, BPL (beta propio lactone), formaldehyde or formalin], heat or radiation. However, UV radiation is unsatisfactory, because of risk of multiplicity reactivation.
- **Advantages of killed vaccines**
  - They are more stable and safe than live vaccines because the dead microbes cannot mutate back to their disease causing state.
  - They do not cause disease to contact.
  - They do not require refrigeration; can be easily stored and transported.
  - They are given in combination with polyvalent vaccine.
- **Disadvantages of killed vaccines**
  - No or poor induction of CMI.
  - Weaker immune response than live vaccines. So they require several additional doses or booster doses to maintain a person's immunity.
  - They produce short lived immunity.
  - They provide only systemic but no local immunity.
- **Contraindication:** If severe local or systemic side effect to previous doses, then subsequent doses will be contraindicated.
- **Examples of killed vaccines:** Follow **Table 17.1**
- **Differences between live and killed vaccines:** Follow **Table 17.2**.

C. Toxoid vaccines

- **Preparation of toxoids:** These vaccines are used when a bacterial toxins like *Cl. tetani, C. diphtheriae,* etc., are the causative agents. Detoxification of toxins (exotoxin) by treating with formalin called toxoid. When an immune system receives a toxoid, it learns how to fight off the natural toxin. The immune system produces antibodies that lock onto and block the toxin.
- **Advantages of toxoids:** They provide safe and effective protection.
- **Differences between toxin and toxoid:** Toxin has toxicity and antigenicity while toxoid has no toxicity but only antigenicity.
- **Examples of toxoid vaccines:** Follow **Table 17.1**.

D. Cellular fraction/subunit /acellular vaccines

- **Definition:** In subunit vaccines instead of the entire microbe (whole-agent vaccine), a fragment (subunit) of it can create an immune response.

TABLE 17.1: Types and examples of vaccines

Types	Examples
Live attenuated vaccines	**Bacterial vaccines:** BCG Typhoid oral Epidemic typhus **Viral vaccines:** Oral polio vaccine (Sabin) Measles Mumps Rubella Chickenpox Influenza (nasal spray) *Rotavirus* Yellow fever (17D)
Inactivated or killed vaccines	**Bacterial vaccines:** Typhoid Cholera Pertussis Meningococcal meningitis Plague **Viral vaccines:** Injectable polio vaccine (IPV/Salk) Hepatitis A Hepatitis B Influenza Kyasnur forest disease Japanese encephalitis (JE) Rabies (vero cell line)
Toxoid (inactivated toxin)	Tetanus toxoid (TT) Diphtheria toxoid

TABLE 17.2: Differences between live and killed vaccines

Features	Live	Killed
Numbers of dose(s)	Mostly single	Multiple
Need of adjuvant	No	Yes
Immunity	Long lived	Short lived
Protection	More	Less
Ig produced	IgA and IgG	IgG
Local (mucosal) immunity	Yes	No
CMI	Yes	Poor/no
Reversion to virulence	Yes	No
Excretion of vaccine microbes and transmission to non-immune contacts	Yes	No
Interference by other virus in host	Yes	No
Heat stability	Yes	No
Temp., maintenance during storage and transport	Yes	No

- **Preparation of subunit vaccines:** They are prepared by growing the microbes in the laboratory and then use of chemicals to break them apart and gathering of important antigens. Subunit vaccines include only the antigens or epitope (the very specific parts of the antigen) which best stimulate the immune system. Subunit vaccines may be monovalent contain one antigen (single strain) or polyvalent contain 20 or more antigens (multiple strains), of course, identifying which antigens best stimulate the immune system is a tricky and time-consuming process.

- **Advantage of subunit vaccines:** Because subunit vaccines contain only the essential antigens and not the entire microbe, the chances of adverse reactions to the vaccines are less.
- **Examples of subunit vaccines**
  - PRP Ag (capsular polysaccharide type b Ag/Hib PRP) in *H. influenzae.*
  - PT, FHA, agglutinogen 1, 2 and 3 and pertactin Ag in pertussis vaccine (combine with DPT).
  - Capsular polysaccharide Ag in pneumococcus.
  - Capsular polysaccharide Ag in meningococcus.
  - Hemagglutinin and neuraminidase subunits of the *Influenzavirus.*
  - Capsid protein in human papilloma virus (HPV).
  - All antigenic components of HBsAg (Pre-S1, Pre-S2 and S) in HBV.

E. Conjugate vaccines

- **Preparation of conjugate vaccines:** Sometimes bacterial polysaccharide is poorly immunogenic in children <2 years; its immunogenicity is increased by coupling with protein carriers like toxin, toxoid or antigen.
- **Examples of conjugate vaccines**
  - Capsular polysaccharide antigen of 7 most common serotypes of *Strept. pneumoniae,* conjugated to the CRM 197 protein (toxoid) of *C. diphtheriae,* called 7 valent conjugated vaccine.
  - PRP Ag in Hib is conjugated with diphtheria toxoid or tetanus toxoid or with meningococcal outer membrane protein (OMP) Ag.
  - Capsular polysaccharides of group A, C, W-135 and Y of *N meningitidis* are conjugated with diphtheria toxoid.

F. Recombinant (cloned) vaccines

- **Preparation of cloned vaccines:** They are prepared by cloning the desired antigen or gene in bacteria (e.g. *E. coli*) or in yeast (Baker's yeast).
- **Advantages of cloned vaccines**
  - They are not causing the disease because they do not contain the microbe, just copies of a few of its genes.
  - They produce strong antibody response.
- **Disadvantage of cloned vaccines:** They are expensive.
- **Examples of cloned vaccines**
  - Cloned vaccine currently in use is HBV vaccine, prepared by cloning the S gene in Baker's yeast.
  - Cloned vaccine also prepared in influenza virus by hybridization of antigen of newer strain and established strain.

G. Future prospects of vaccines

- **Naked DNA vaccine:** Small DNA fragment from pathogen is injected in to host cell. It integrated with host cell genome and multiplies along with to mount an immune response. It is costly but mounts stronger and long lasting protection.
- **Edible vaccine:** Gene encodes for orally active antigenic protein of pathogen is transferred to suitable plant bacteria, which are then used to infect a transgenic plant such as banana, potato, etc. Such plant starts desired antigen production in large scale. Such plant or part of plant contains antigen may be fed raw to animal or humans to produce immunity. Some examples of edible vaccines are transgenic potatoes and tomatoes against diarrheagenic pathogen. Edible banana vaccines against Norwalk virus is other example. These vaccines are low cost, heat stable, given orally, offer local immunity and produce immunity in large scale.

Administration of vaccines: The method of introduction of vaccine called vaccination while the protection offered to body after vaccination called immunization. Vaccination does not guarantee immunization. Vaccines are available and administered as plain or in mixed form, as follows:

1. **Plain/single/monovalent/univalent vaccines**
   - **Preparation:** They are designed to immunize against a single antigen (serotype) or single strain or single microorganism.
   - **Examples:** TT, BCG vaccine, *H. influenzae* type b (Hib), monovalent meningococcal vaccine contains single capsular polysaccharide of group A or C, rabies vaccine, etc.
2. **Mixed/combined/polyvalent/multivalent vaccines**
   - **Preparation:** They are designed to immunize against two or more strains of the same microorganism or two or more antigens (serotypes) of the same microorganism or against two or more microorganisms.
   - **Examples**
     - Double/bivalent vaccines: DT, DP, etc.
     - Triple/trivalent/tetravalent vaccines: DPT/DTP, MMR, etc.
     - Quadruple/quadrivalent vaccines: DPT (DTP) with Hib, DPT with HBV, MMRV, etc.
     - Pentavalent vaccines: Two types of preparations like **pentvac** contains Diphtheria, Pertussis, Tetanus, *H. influenzae* type b (Hib) and HBV and **pentaxim** contains D, P, T, Hib, and IPV.
   - **Advantages of mixed vaccine over plain vaccine:** Simple to administer, low cost and minimum hospital visit.

Adjuvants in vaccines

- **Meaning:** Adjuvant word derived from Latin word adiuvare means to help.
- **Definition:** Adjuvants are substances that are added to vaccine to increase the antigenicity/immune response.

- **Advantage:** Presence of adjuvant in vaccines reduces amount of Ag and number of doses.
- **Common adjuvants**
  1. Aluminum adjuvants: Aluminum phosphate, aluminum hydroxide and aluminum sulfate.
  2. Freund's incomplete adjuvant: Water-in-oil incorporation of protein Ag in water phase of water-in-oil emulsion.
  3. Bacteria: Like *Propionibacterium acne* and *B. pertussis*. *B. pertussis* is used in DPT (antibody response to toxoid is potentiated by *B. pertussis*).
  4. BCG.
  5. Adjuvants for laboratory animals.
     - Freund's complete adjuvant: Water-in-oil with *Nocardia* Ag or *Mycobacterium* Ag.
     - Bacterial endotoxin (LPS).
     - Vitamin A in toxic dose.
     - Fungal polysaccharide.
- **Mechanisms for immune enhancement by adjuvants:** Exact mechanisms are not known but possible mechanisms are mentioned below.
  1. Protection of Ag: It provides protection to Ag, which allows slow release of antigen for longer time (depot generation).
  2. Recruitment of APCs: By recruiting of professional antigen-presenting cells (APCs) to the site of antigen exposure, which enhances the immunogenicity leading to improve the immune response.
  3. Enhancement of innate and acquired (adaptive) immunity: Adjuvants interfere with pattern recognition receptors (PRRs), especially toll-like receptors to increase the innate immunity. Immunomodulation by local granuloma formation leads cytokine production which attracts T and B cells at the site of infection to increase the acquired (adaptive) immunity. Innate and acquired (adaptive) immunity are linked with each others, so adjuvant mediate enhancement of one can encourage the other immunity.

Adverse reactions of vaccines: Two types:
1. **Local features:** Pain, swelling/induration, ulceration, scar, etc.
2. **Systemic features:** These include **hypersensitivity symptoms** like anaphylaxis, serum sickness, etc., and **general symptoms** like fever, headache, acute flaccid paralysis by polio vaccine, anxiety, etc.

Used for Artificial Passive Immunization (Immunotherapy)

Types, preparations and examples of passive immunizing agents
1. **Antiserum**
   - Preparation: It is prepared by injecting (hyperimmune serum) the specific antigen in animals like horse or in human that leads to antibody formation.
   - Example: Rabies antiserum.

2. **Antitoxin**
   - Preparation: It is prepared by injecting (hyperimmune serum) the toxin or toxoid in animals like horse or in human that leads formation of antitoxin.
   - Examples: ATS (antitetanus serum), ADS (antidiphtheria serum), AGS (antigas gangrene serum), ASV (antisnake venom), etc.

3. **Immunoglobulins (Igs):** Igs are given by IM or IV route in 5 ml dose, if more dose than divided in 4–6 intragluteal sites. IM injection is painful, which is overcome by mixing 1 part of procaine with 10 parts of Ig. Following are two subtypes of Igs:
   - **Normal human Ig (nonspecific human Ig)**
     - Preparation: It is prepared from pooled sera of at least thousand healthy adults.
     - Example: Ig for measles and HAV.
   - **Specific human Ig**
     - Preparations: It is prepared by two ways. Those prepared from patients recovered from infectious diseases called convalescent sera and prepared from hyperimmune animals like horse or from human called hyperimmune sera.
     - Examples: Rabies Ig, HBV Ig, tetanus Ig, diphtheria Ig, Rh-D Ig, etc.

Indication of passive immunizing agents
1. Instant passive immunity: They are useful in emergencies conditions where immediate and temporary protection such as tetanus, diphtheria, snake bite, rabies, HAV, HBV, measles, etc.
2. Suppression of active immunity: For example, in Rh –ve mother with Rh +ve babies.
3. IDDs: Useful in persons with IDDs.

Disadvantages of passive immunizing agents
1. Immune elimination: This limits the passive immunization.
2. Hypersensitivity reaction: They sometimes produce serum sickness or anaphylaxis mostly with animal preparations. These are overcome by using human preparations.

Used for Combined Immunization

Combination of active (immunoprophylaxis) and passive (immunotherapy) immunization: In some diseases like diphtheria, tetanus, snake bite, rabies, HAV, HBV, measles, etc., passive immunization is prescribed with active immunization with aim to start instant immunity till the active immunity will develop.

Examples: In some disease like tetanus where TIG (tetanus immunoglobulin) is injected in one arm and TT in other arm followed by full course of TT. TIG provides passive protection till the active immunity will developed by TT.

Introduction

In May 1974 WHO launched the global immunization program called Expanded Program on Immunization (EPI) against six vaccine-preventable diseases (VPDs) like diphtheria, pertussis (whooping cough), tetanus, polio, tuberculosis and measles. Different countries employed the immunization schedule according to their priorities.

Routine Immunization

EPI was launched in India in January 1978 called Indian National Immunization Schedule as mentioned in **Table 17.3**.

TABLE 17.3: Indian National Immunization Schedule

Age	Vaccines
Birth	BCG, OPV-0, Hepatitis B
6 weeks	BCG (if not given at birth) DPT-1, Hepatitis B-1, Hib-1, OPV-1, f-IPV, *Rotavirus* 1
10 weeks	DPT-2, Hepatitis B-2, Hib-2, OPV-2, *Rotavirus* 2
14 weeks	DPT-3, Hepatitis B-3, Hib-3, OPV-3, f-IPV, *Rotavirus* 3
9 months	Measles/MR with Vita A (first doses)
16–24 months	DPT booster-1 and OPV booster, MR second dose
5–6 years	DPT booster-2 (2nd dose of DPT after 1 month if not previously immunized by DPT)
10 and 16 years	TT/Td (2nd dose of TT/Td after 1 month if not previously immunized by DPT or TT)
For pregnant women	TT1/Td1 or booster in early pregnancy TT2/Td2 one month after

Notes:

1. Interval between DPT, OPV and hepatitis B should not be less than 1 month.
2. Minor cough, fever, colds are not the contraindications of vaccine.
3. After primary immunization of DPT at 6, 10 and 14 weeks, 1st booster of DPT is given at 18–24 months followed by 2nd booster dose at 5–6 year (school entry age).
4. Severity of pertussis is decreased with age, so DPT vaccine is usually not recommended after 5 years, but only double (Td) vaccine is advised.
5. f-IPV (0.1 ml by ID rout) is given at 6th and 14th week along with bivalent OPV (since 2017).
6. First dose of vitamin A is given at 9 completed months then every 6 months up to 5 years of age (total 9 doses).
7. *Rotavirus* vaccine is given in selected states like Andhra Pradesh, Assam, Haryana, Himachal Pradesh, Jharkhand, Madhya Pradesh, Odisha, Rajasthan, Tamil Nadu, Tripura and Uttar Pradesh.
8. Pneumococcal conjugated vaccine: Given at 6 and 14 weeks followed by booster dose at 9–12 months in selected states like Bihar, Himachal Pradesh, Madhya Pradesh, Uttar Pradesh (12 districts) and Rajasthan (9 districts).
9. JE vaccine: 1st dose (0.5 ml by SC route in left upper arm) is given at 9–12 months followed by 2nd dose at 16–24 months in 231 endemic districts of Assam, Bihar, Karnataka, West Bengal and Uttar Pradesh.
10. Pentavalent: In some states pentavalent vaccine is given at 6th weeks, 10th weeks and 14th weeks. It contains Diphtheria, Pertussis, Tetanus, Hepatitis B and Hib. Other same preparation is Pentaxim, which contains diphtheria, pertussis, tetanus, IPV and Hib.

Individual Immunization

Some vaccines are available but they are not the part of immunization schedule due to high cost or others. They are prescribed after consultation with parents. For examples varicella vaccine, HAV vaccine, influenza vaccine, typhoid vaccine, COVID-19 vaccine, etc.

ACCESS YOURSELF

Essay/Full Question

1. Discuss the immunological basis of vaccines and the Universal Immunization schedule.

Short Notes

1. Immunizing agents/vaccines.
2. Universal immunization schedule.

Short Questions for Theory/Viva Questions

1. Write the two examples of each, live attenuated bacterial and viral vaccines.
2. Write four differences between live and killed vaccines.
3. What are toxin and toxoid?
4. What is adjuvant?

Short Questions for Theory/Viva Questions

1. Toxin is harmful, but not toxoid.
2. Adjuvants are responsible for immune enhancement.

MCQs for Chapter Review

Immunizing Agents

1. **Vaccination is based on the principle of:**
   a. Agglutination
   b. Phagocytosis
   c. Immunological memory
   d. Clonal detection
2. **BCG vaccine is a type of:**
   a. Killed vaccine   b. Conjugate
   c. Live attenuated   d. Recombinant
3. **Which of the following is killed vaccine?**
   a. Hepatitis B
   b. Measles
   c. Yellow fever
   d. Japanese encephalitis
4. **Which of the following is not a killed vaccine?**
   a. Yellow fever (17D)
   b. Salk (polio)
   c. Hepatitis B
   d. Human diploid cell rabies vaccine
5. **Cell fraction derived vaccine is:**
   a. Hepatitis B   b. Measles
   c. Mumps   d. Rubella

6. **The main aim of adjuvant is to increase:**
   a. Distribution
   b. Absorption
   c. Antigenicity
   d. Metabolism

7. **Role of adjuvant in vaccine is/are:**
   a. Stimulation of toll-Like receptors
   b. Activates B lymphocytes only
   c. Increases both adaptive and innate immune response
   d. Activates both B and T lymphocytes
   e. Ensures prolonged delivery of antigen

Universal Immunization Schedule

8. **According to National Immunization Schedule which of the following is recommended for a child of 5-year of age?**
   a. Pentavalent vaccine and vitamin A
   b. DT booster
   c. DT, OPV and vitamin A
   d. DPT booster and vitamin A

Answers and Explanation of MCQs

1. c
- Follow section, **vaccines (principle)** for explanation.

2. c

3. a, d

4. a
- Follow **Table 17.1** for explanation of answers of MCQs 2–4.

5. a
- Follow section, **vaccines (Cellular fraction/subunit/acellular vaccines → examples)** for explanation.

6. c
- Adjuvant is a substance that added to the vaccine to increase antigenicity or immune response.

7. a, c, d and e
- Follow section, **vaccines (adjuvant)** for explanation.

8. d
- Follow **Table 17.3 and notes** for explanation.

CHAPTER 18

Antigen (Ag)

Chapter Outline
- Antigen (Ag)
- Other Related Terms of Antigen

ANTIGEN (Ag)

Meaning

Antigen means **anti**body **gen**erator.

Definitions

Antigen: Any substance (self/foreign, protein/lipid/carbohydrate) which when enters into the body by any route produces AMI or CMI or hypersensitivity/autoimmunity or tolerance.

Immunogen: Any substance which when enters into the body by any route, must produces immune response such as AMI or CMI or hypersensitivity/autoimmunity but no tolerance.

All Immunogens are Antigens, but All Antigens are not Immunogens

Antigens are considered as particles enter in body like dust, pollen or microbes. They may induce immune response (AMI, CMI, etc.) or no immune response (tolerance). Immunogens are considered as particles which must induce an immune response but no tolerance. So immunogens contain all properties of antigens but not vice versa; hence, it is safe to say that all immunogens are antigens but all antigens are not immunogens.

Attributes of Antigenicity

Antigenicity means ability of an antigen to combine specifically with TCR, whether immunogenic or non-immunogenic. Antigenicity has following two attributes:

1. Immunogenicity: Ability of an antigen to induce an immune response called immunogenicity.
2. Immunological reactivity: Specific reaction of Ag with Ab or sensitized cell called immunological reactivity.

> **Note: Confusing terms**
> - Immunological reactivity: Few author defined it as a synonym for antigenicity, while few defined it as an attribute of antigenicity as mentioned above.
> - Immunological reaction: It is an *in vivo* Ag-Ab reaction.

Classification

Following are the different types of antigens.

A. Functional classification: Two types:

1. **Complete Ag**
   - **Synonym:** Immunogen.
   - **Definition:** Complete Ag is a substance that is capable to induce an immune response (immunogenic) by itself and also reacts specifically and in observable manner with products (like Ab) of immune response.
   - **Attributes:** It possesses both attributes of antigenicity.

2. **Incomplete Ag**
   - **Synonym:** Hapten.
   - **Meaning:** From Haptein (Greek) means to fasten.
   - **Definition:** Hapten is a substance that is incapable to induce an immune response (not immunogenic) by itself, but can reacts specifically with products (like Ab) of immune response.
   - **Attributes:** Having attribute like immunological reactivity but no immunogenicity. Haptens become immunogenic on combining with a larger molecule called carrier molecule. Hapten is generally low molecular weight lipid and carbohydrate, while carrier molecule is protein like albumin, globulin or synthetic polypeptide.
   - **Types:** Two types of haptens like simple and complex as mentioned in **Table 18.1**.
   - **Examples of haptens**
     1. Bacterial haptens: Polysaccharide capsule of pneumococci, carbohydrate–Ag of streptococci, etc.

TABLE 18.1: Types of haptens

Simple hapten	Complex hapten
Simple chemical substance	Relatively complex and large
Monovalent	Polyvalent
Can react with Ab, but unable to precipitate the reaction	Reacts with specific Ab and able to precipitate reaction

2. Drugs: Allergen causing contact dermatitis.
3. Blood group substances: Glycoproteins in ABO blood group.
4. Lipids: Cardiolipin Ag released from heart tissue in syphilis, Forssman Ag, transplantation Ag, etc.

B. According to infectious agents

1. Bacterial antigens: Such as flagellar (H) Ag, capsular (K) Ag, somatic (O) Ag, fimbrial (F) Ag, etc.
2. Viral antigens: Such as envelop Ags (gp120, gp 41, neuraminidase Ag, hemagglutinating Ag, etc.), capsid Ags (p24, p15 and p55) and nucleic acid Ag.
3. Fungal antigens: Mannan in *C. albicans*, capsular Ag in *Cryptococcus neoformans*, gp43 in *P. brasiliensis*, etc.
4. Parasitic antigens: Histidine-rich protein (HRP) and merozoite surface protein (MSP) in *Plasmodium* spp., K39 and Witebusky Klingstein Kuhn (WKK) Ag in *L. donovani*.

C. Biological classification

Antigen which requires T cell participation for antibody production from B cell is called T cell dependent antigen (TD Ag), while antigen which can directly stimulate B cell for antibody production without participation by T cell is called T cell independent antigen (TI Ag). Differences between two are mentioned in **Table 18.2**.

D. According to mode of origin

1. Exogenous antigens: These are entering in body from the outside by different routes like inhalation ingestion, injection, etc.

TABLE 18.2: Differences between TD and TI Ags

Features	TD Ag	TI Ag
T cells cooperation	Yes	No
Preliminary process by macrophages	Yes	No
Types of Ag	Soluble protein	Type I: LPS Type II: Capsular polysaccharide of pneumococci, flagellin
Degradation	Rapid	Slowly
Complement activation	No	By type II
Immunological memory	Yes	No
Polyclonal activation	No	By type-I
Ab production	Full range IgM, IgG, IgA and IgE	Limited to IgM and IgG3
Tolerance	Do not cause tolerance	High dose results in tolerance

2. Endogenous antigens: These are originated from body itself; for example, fragment of normal cell generated by metabolism or infection.

Antigenic Determinant

Synonym: Epitope.

Definition: Whole antigen is not responsible for antigenicity but smaller area on it, usually contains four or five amino acids or monosaccharide residues is responsible for antigenicity called epitope.

Properties of epitope: Epitope has MW about 400–1000, specific chemical structure, electrical charge and steric configuration. It is capable of sensitizing an immunocytes. It reacts with specific complementary site on the specific antibody called paratope or on T cell receptors. Epitope and paratope determine the specificity of immunological reactions. The presence of the same or similar epitope on the different antigens accounts for one type of antigenic cross reaction. Antigenic determinants are usually limited to those portions of the antigen that are accessible to antibody. T cell recognizes linear epitope while B cell recognizes conformational epitope.

Types of epitope: Follow **Fig 18.1.**
1. Linear/sequential: Epitope may be present as a linear segment of primary sequence.
2. Conformational: Epitope may be formed by bringing together on the surface residue from different site of peptide chain during its folding into tertiary structure.

Factors Influencing Antigenicity or Immunogenicity

Such factors called determinants of antigenicity. Following are the different factors.

Size: Antigenicity is related to molecular size. Antigen with high-molecular weight (HMW) is highly antigenic like hemocyanin. Antigen with low-molecular weight (LMW) is less antigenic. LMW substances may be rendered antigenic by adsorbing large inert particles on them like bentonite or kaolin.

Foreignness: An individual does not normally mount an immune response against his or her own normal constituent antigens. Tolerance of self antigen is conditioned by contact with them during the development of immune system. Breakdown of this

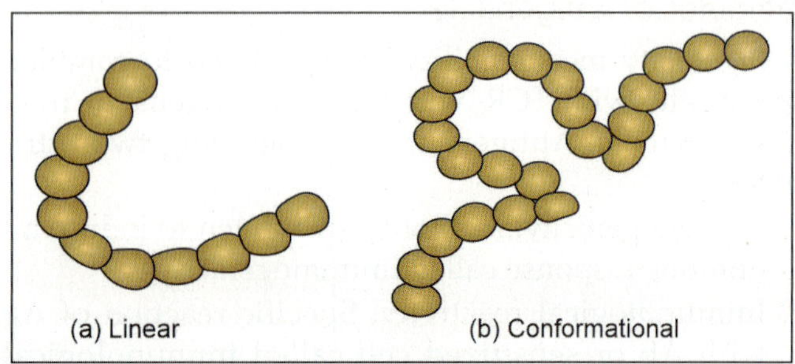

Fig. 18.1: Types of epitope

homeostatic mechanism leads autoimmunization and auto immune disease.

Biochemical nature

1. **Proteins:** The vast majority of immunogens are proteins. These may be pure proteins or they may be glycoproteins or lipoproteins. In general, proteins are usually very good immunogens, as protein made of different 20 or more amino acid residues. All proteins are not antigenic like gelatins, histones and protamines.
2. **Polysaccharides and carbohydrates:** Pure polysaccharides and lipopolysaccharides are good immunogens, but weaker than proteins and stronger than nucleic acids and lipids. Natural proteins, composed of 20 different amino acids, are better antigens than polysaccharides composed of 4–5 monosaccharide units. Carbohydrates are weaker immunogens.
3. **Nucleic acids:** Nucleic acids are usually poorly immunogenic. However, they may become immunogenic when single stranded or when complexed with proteins.
4. **Lipids:** In general lipids are non-immunogenic, although they may be haptens. Immunogenic nature of biochemical substances is proteins (except gelatins) > polysaccharides and carbohydrates > nucleic acids and lipids.

Susceptibility to tissue enzymes: Substances which are metabolized and are susceptible to the actions of tissue enzyme behave as antigens, e.g., polypeptide contains L-amino acids. Antigens introduced into the body are degraded by the host into fragments of appropriate size containing antigenic determinants. Substances which are not metabolized by tissue enzymes are not antigenic, e.g., polystyrene latex and synthetic polypeptide contains D-amino acids.

Route of administration of antigen: The route of antigen administration can also alter the nature of the response. Generally the subcutaneous route produces better Ir than the intravenous or intragastric routes.

Dose of administration of antigen: It influences the immunogenicity. There is a dose of antigen above or below which the Ir will not be optimal.

Presence of adjuvants: Substances that can enhance the immune response to an immunogen are called adjuvants. The use of adjuvants; however is often hampered by undesirable side effects such as fever and inflammation, e.g., adjuvants used in vaccines.

Host factors

1. **Genetic factors:** Some substances are immunogenic in one species but not in another. Similarly, some substances are immunogenic in one individual but not in others of same species (i.e. responders and nonresponders). The species or individuals may lack or have altered genes that code for the receptors for antigen on B cells and T cells or they may not have the appropriate genes needed for the APC to present antigen to the helper T cells.
2. **Age:** Age can also influence the immunogenicity. Usually very young and very old have a diminished ability to mount an immune response to an immunogens.

Antigenic Specificity

History: The basis of antigenic specificity is stereochemical was first demonstrated by Obermayer and Pick and later confirmed by Landsteiner.

Basis of antigenic specificity: The importance of the position of the antigenic determinant group in the antigen molecules was evidenced by the difference in specificity in compound with the group attached at the ortho, meta or para position. The influence of spatial configuration of the determinant group was shown by difference in antigenic specificity of the dextro, levo and meso isomers of substance such as tartaric acid. Antigenic specificity is not absolute and cross reactions can occur between antigens that bear stereochemical similarities.

Types: Natural tissue antigens show the following types of specificities.

1. **Species specificity**
   - **Definition:** Tissues of all individual in a species contain species-specific antigens.
   - **Clinical applications**
     - Forensic science: It has forensic applications in the identification of the species from blood and seminal stains.
     - Phylogenetic study: It has been used in tracing phylogenetic (evolutionary) relationship between species.
2. **Isospecificity**
   - **Definition:** Isoantigens are antigens found in some, but not all members of a species.
   - **Clinical applications**
     - Grouping of species: Isoantigens on human RBCs are used for blood grouping and blood transfusion (BT). Isoantigens on human WBCs are used for HLA typing and organ transplantation.
     - Paternity test: It provides valuable evidence in disputed paternity.
     - Isoimmunization: Helpful during pregnancy.
3. **Autospecificity**
   - **Definition:** Autologous or self antigens are ordinarily nonantigenic; however, some are exception **called self antigens**, as mentioned below.
   - **Clinical applications:** Some sequestrated antigens that are not normally found free in circulation or tissue fluids like eye lens protein, sperm antigen and other antigen, which are not exposed during embryonic life. When such antigens are found

free in circulation, body identifies them as foreign antigens and produces autoimmunity.

4. **Organ specificity**
   - **Definition:** Some organs such as the brain, kidneys and lens protein of different species, sharing the same antigens called organ-specific antigens.
   - **Clinical applications:** The neuroparalytic complication following anti-rabies vaccination using sheep brain vaccine is a consequence of brain-specific antigens shared by sheep and human being. The sheep brain antigen of the vaccine induces immunological response which damages the nervous tissues of human.

5. **Heterogenetic (heterophile) specificity**
   - **Definition:** The same or closely related antigens may sometimes occur in different biological species, classes and kingdoms. They are called heterogenetic or heterophile antigens.
   - **Clinical applications**
     - **Forssman antigen:** It is a lipid carbohydrate complex widely distributed among animals, birds, plants and bacteria, but absent in rabbits, so anti-Forssman antibody can be prepared in rabbits.
     - **Other heterophile antigens:** They are used in serological tests to diagnose the diseases in which antigens unrelated to etiological agents are employed. Following are examples:
       - Weil Felix reaction in typhus fever: Cross reactivity occurs between antigens of *Proteus* spp., (OX 19 and OX2 of nonmotile *Proteus vulgaris* and OX K/O antigens of nonmotile *Proteus mirabilis*) and rickettsiae spp.
       - Paul Bunnel test in infectious mononucleosis: Cross reactivity occurs between antigens of sheep (horse and cow/ox) RBCs and EBV.
       - Cold agglutinin test in primary atypical pneumonia: Cross reactivity occurs between antigens of human O blood group RBCs and *Mycoplasma pneumoniae*.
       - Streptococci MG test in primary atypical pneumonia: Cross reactivity occurs between antigens of streptococcal MG group (group F) and *Mycoplasma pneumoniae*.
       - Nonspecific CFT with WKK antigen (Witebusky, Kligenstein and Kuhn antigen) in leishmaniasis: Cross reactivity occurs between antigens of *Mycobacterium tuberculosis* and *Leishmania donovani*.

OTHER RELATED TERMS OF ANTIGEN

Superantigen

Definition: Antigen which polyclonally activates a large numbers of T cells irrespective of its antigenic specificity called superantigen.

Properties: Superantigen is medium sized protein 22–29 kDa. It is highly resistance to protease and denatured by CD4+. It produced intracellularly by bacteria and released upon infection as extracellular mature toxins. Normal antigen activates 0.0001–0.001% of the body's T-cells while superantigen activates 20% of the body's T-cells. Anti-CD3 and Anti-CD28 antibodies are also identified as highly potent superantigens and can activate up to 100% of T cells.

Mechanism of action: Conventional Ag processed by APCs before presenting to MHC –I or II molecule in the groove of αβ domain, while superantigen do not processed by APCs and directly binds laterally to variable beta (vβ) chain in TCRs and chain of MHC-II molecule on T cells (**Fig. 18.2**). It stimulates the large numbers of T cells (polyclonal activation) followed by massive releases of cytokines (IL-2). Further stimulation of T cells release more variety of cytokines. Massive release of cytokines causing polyclonal activation of B cells which lead massive immune response and multisystem damage.

Examples of superantigens

1. **Bacterial:** Staphylococcal enterotoxins, staphylococcal toxic shock syndrome toxin-1 (TSST-1), streptococcal pyrogenic exotoxin (exotoxin A and C), and antigens released by *Mycobacterium tuberculosis* and *Yersinia pseudotuberculosis*.
2. **Viral:** Nef (Negative regulatory factor) in HIV, nucleocapsid in rabies virus, mouse mammary tumor virus (retro virus) and EBV (Epstein Barr virus).
3. **Fungal:** *Malassezia furfur*.

Diseases related to superantigens: These are food poisoning, toxic shock syndrome (TSS), scalded-skin syndrome, psoriasis, acute disseminated encephalomyelitis, Kawasaki syndrome, atopic dermatitis, etc.

Mitogen

Definition: Substance that induces the division of immune cells called mitogen.

Examples of mitogens

1. Pneumococcal polysaccharide: It stimulates the B cells, causing polyclonal activation and production of large numbers of IgM.
2. Lectin glycoprotein: It binds with cells having sugar on surface and causing polyclonal activation.
3. LPS (bacterial endotoxin): It activates the B cells.

Allergen

A substance able to cause an allergic/hypersensitivity reaction called allergen like dust, pollen, ovalbumin, etc.

Tolerogen

A substance not able to produce an immune response due to LMW called tolerogen. If its molecular weight is changed, it becomes immunogen.

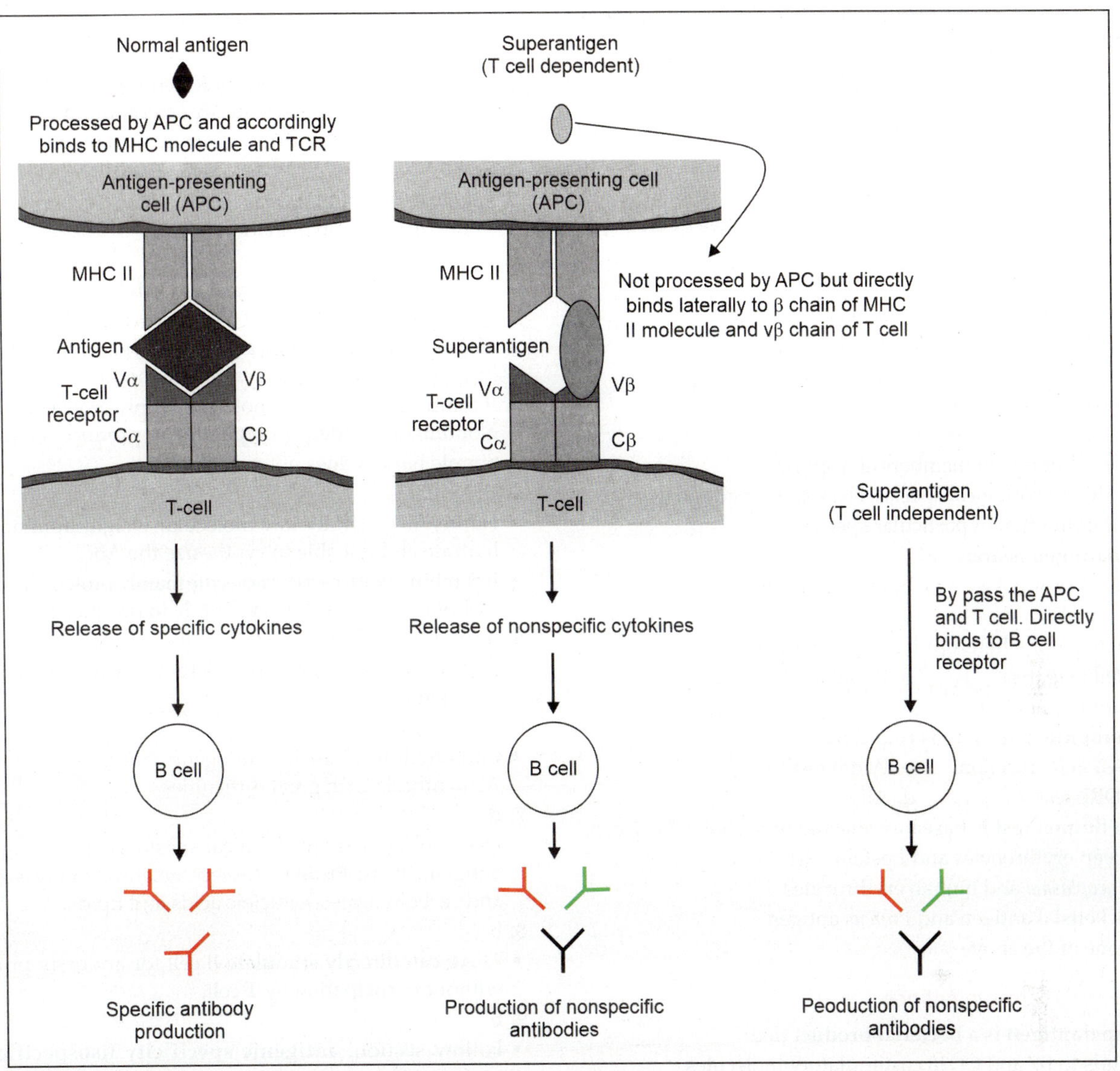

Fig. 18.2: Mechanism of action of superantigen

ACCESS YOURSELF

Essay/Full Question

1. Describe antigens and concept involved in vaccine development.

Short Notes

1. Antigenic determinants.
2. Superantigen.

Short Questions for Theory/Viva Questions

1. Define: Antigen and immunogen.
2. What is epitope?
3. What is heterophile antigen? Write two examples.
4. Name four serological tests based on the use of heterophile antigen.
5. What is superantigen? Write two examples.
6. What is mitogen? Write two examples.

Comments on

1. All immunogens are antigens but all antigens are not immunogens.

MCQs for Chapter Review

Antigen (Ag)

1. **Following are the features of antigen:**
   a. Produces antibody
   b. Produces lymphokine
   c. Produces hypersensitivity
   d. All of above

2. **Meaning of hapten is:**
   a. Fasten
   b. Antibody
   c. Toxin
   d. None of above

3. **Hapten is:**
   a. Same as epitope
   b. Small molecular weight protein
   c. Requires carrier for specific antibody production
   d. Simple hapten is precipitate

4. **Which of the following statement is true about hapten?**
   a. It induces brisk immune response
   b. It needs carrier to induce immune response
   c. It is a T independent antigen
   d. It has no association to MHC

5. **The exact part of the antigen that reacts with the immune system is called as:**
   a. Clone
   b. Epitope
   c. Idiotope
   d. Effector
6. **Which of the following is very difficult to induce antibody?**
   a. Carbohydrate
   b. Protein
   c. Ag
   d. Repeated infections
7. **Which of the following chemical nature makes a better Ag?**
   a. Lipids
   b. Nucleic acids
   c. Polysaccharides
   d. Proteins
8. **Which of the following cells acts in TI Ag?**
   a. T cells
   b. B-cells
   c. Macrophages
   d. CD8+ cells
9. **Isoantigens are found in:**
   a. All members of a species
   b. Some but not all members of a species
   c. Different biological species, class and kingdom
   d. All individual in particular species.
10. **Autoantigen is/are:**
    a. Blood group Ag
    b. Forssman Ag
    c. Both
    d. None
11. **Which is not a heterophile agglutination test?**
    a. Weil Felix test
    b. Widal test
    c. Paul Bunnel test
    d. Streptococci MG test
12. **Heterophile antibody is found in:**
    a. Weil Felix reaction
    b. Widal test
    c. VDRL test
    d. All
13. **Paul Bunnel test is based on sharing of antigens between:**
    a. Sheep erythrocytes and Epstein Barr virus
    b. *Mycoplasma* and human erythrocytes
    c. Rickettsial antigen and *Proteus* antigen
    d. None of the above

Other Related Terms to Antigen

14. **A superantigen is a bacterial product that:**
    a. Binds to B7 and CD28 costimulatory molecules
    b. Binds to the beta chain of TCR and MHC class II molecule of APC stimulating T cell activation
    c. Binds to the CD4+ molecule causing T cell activation
    d. Is presented by macrophages to a larger than normal number of T helper CD4+ T lymphocytes
15. **Superantigen causes:**
    a. Polyclonal activation of T cells
    b. Stimulation of B cells
    c. Enhancement of phagocytosis
    d. Activation of complement
16. **Superantigens are:**
    a. Erythrotoxin of *Staph. aureus*
    b. *Cl difficile* toxin
    c. Staphylococcal toxic shock syndrome toxin
    d. Cholera toxin

1. d
- Ag produces AMI/HI (Ab production) or CMI (lymphokine production) or hypersensitivity/autoimmunity or tolerance.
- Follow section, **definition** for more explanation.

2. a
- Follow section, **functional classification (incomplete antigen → meaning)** for explanation.

3. c
- Option a → Follow section, **antigenic determinant (definition)** for explanation.
- Hapten is generally low molecular weight lipid and carbohydrate and not able to synthesize the Ab.
- It combines with carrier molecule mainly protein like albumin, globulin or synthetic polypeptide to produce the Ab.
- Simple hapten does not precipitate.

4. b
- Hapten is generally low molecular weight lipid and carbohydrate and not able to synthesize the Ab.
- It combines with carrier molecule mainly protein like albumin, globulin or synthetic polypeptide to produce the Ab.

5. b
- Follow section, **antigenic determinant (definition)** for explanation.

6. a
- Carbohydrate is less immunogenic than protein and induces Ab synthesis with great difficulties.

7. d
- Descending order of chemical substance according to their antigenicity is: Proteins (except gelatin) > polysaccharides and carbohydrates > nucleic acids and lipids.

8. b
- TI Ag can directly stimulate B cell for antibody production without participation by T cell.

9. b
- Follow section, **antigenic specificity (isospecificity)** for explanation.

10. d
- Autoantigens are sperm Ag and lens Ag.

11. b

12. a

13. a
- Follow section, **antigenic specificity [heterogenetic (heterophile) specificity]** for explanation of answers of MCQs 11–13.

14. b

15. a
- Follow section, **superantigen (mechanism of action of superantigen)** for explanation of answers of MCQs 14–15.

16. c
- Follow section, **superantigen (examples of superantigen)** for explanation. Erythrotoxin is the property of *Strept pyogenes* not of *Staph. aureus*.

CHAPTER 19

Antibody (Ab)–Immunoglobulin (Ig)

Chapter Outline

- Introduction
- Immunoglobulin (Ig)
- Genes of Immunoglobulin
- Monoclonal Antibodies

INTRODUCTION

Definition: Protein (Glycoprotein) substance produced after antigenic stimulation and reacts specifically (not absolute, because of cross reactivity) with responsible antigen and in some observable manner called antibody (Ab).

Immune sera: Antisera containing antibodies are called immune sera.

Hyperimmune sera: Antisera containing high titer of antibodies are called hyperimmune sera.

Fractionation of immune sera: It is done by following methods:

1. Half saturation with ammonium sulfate separates proteins into soluble albumins and insoluble globulins. Globulins can be separated in to water soluble pseudoglobulins and insoluble euglobulins.

2. Sedimentation coefficient (S) study divides them into molecules that sediments at 7S (MW:150000) and some heavier molecules that sediments at 19S (MW:900000), designated as M or macroglobulins.

3. Electrophoretic mobilities: In 1937 plasma protein was separated by Tiselius as shown in **Flowchart 19.1**. In 1938 Tiselius and Kabat found that antibody activity was associated with γ-globulin called immunoglobulin (Ig). The confusion about the indiscriminate use of various terms was resolved when, in 1964, WHO endorsed internationally accepted term "immunoglobulins" for proteins of animal origin endowed with known antibody activity and for certain other same proteins by chemical structure.

All antibodies are immunoglobulins, but all immunoglobulins may not be antibodies: Because besides antibody globulins, Ig also includes the abnormal

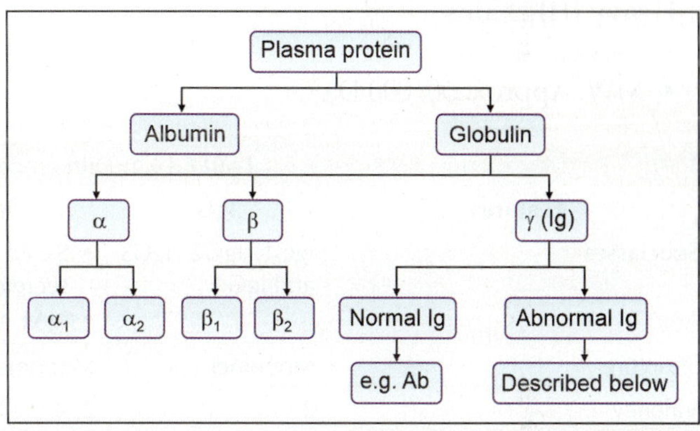

Flowchart 19.1: Parts of plasma protein

proteins found in myeloma, macroglobulinemia, cryoglobulinemia and naturally occurring subunits of Igs. Igs (abnormal Ig) are synthesized by plasma cells and to some extent by lymphocytes while Abs are synthesized by plasma cells only. Ig provides a chemical and structural concept, while the term 'Ab' is a biological and functional concept.

Properties of antibody: It constitutes 20–25% of total serum proteins. It contains proteins and sugar residues hence called glycoprotein (gp). It is heat labile and denatured at 70°C for 1 hour. Its activity is also affected by pH of medium.

IMMUNOGLOBULIN (Ig)

Structure of Ig

Polypeptides chains in Ig: Each molecule consists of two light (short) and two heavy (long) polypeptides chains (total four) of different sizes (**Fig. 19.1**). Total MW approximately is >1,50,000 Da.

1. **Light (L) chains**
   - Smaller
   - MW: Approx. 25,000 Da

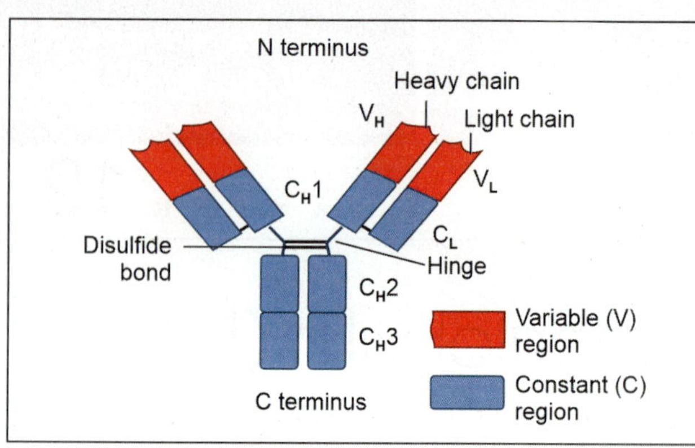

Fig. 19.1: Structure of Ig

- Light chains are similar in all classes of Ig.
- Two varieties: kappa (k from Korngold) and lambda (l from Lapari)
- A molecule of Ig may have either kappa or lambda chains, but never both together.
- In human sera 60% are kappa and 40% are lambda chains present.

2. Heavy (H) chains

- Larger
- MW: Approx. 50, 000 Da

- Heavy chains are structurally and antigenically distinct for each classes of Ig.
- Five varieties: According to H chain five varieties of Ig are designated as IgG, IgA, IgM, IgD and IgE from Greek letters like Gamma (γ), Alpha (α), Mu (μ), Delta (δ) and Epsilon (ε), respectively. This order also suggests the descending order of concentration in serum. All classes are differentiated in **Table 19.1**.

Two ends in Ig

1. Aminoterminus (N terminal/Fab fraction) end

- **Properties:** It contains both H and L chains. It is made up by amino acids where initial 110 amino acids are quite variable called variable (V) region, thus the variable light chain (V_L) and variable heavy chain (V_H) regions are defined. Sequence of amino acid beyond the variable region is relatively constant throughout the rest of molecule called constant region, thus the constant light chain (C_L) and constant heavy chain (C_H) regions are defined.
- **Function:** It acts as antigen binding site.
- **Hypervariable region:** Amino acid sequence in the variable region is sometimes hypervariable called

TABLE 19.1: Differences between each classes of Ig

Features	IgG	IgA	IgM	IgD	IgE
Subclasses	IgG1, IgG2, IgG3 and IgG4	• Serum and secretory IgA • IgA1 and IgA2	• Membrane IgM • Soluble IgM	None	None
Structure	Monomer	Monomer, Dimer	Monomer, Pentamer	Monomer	Monomer
Valency	2	2, 4	2, 10	2	2
Type of heavy chain	γ	α	μ	δ	ε
Sedimentation coefficient	7	7	19	7	8
Additional unit	–	S and J pieces	J piece	–	–
MW	150000	160000	9,00,000–10,00,000 (millionaire molecule)	180000	190000
Serum concentration in mg/ml	12 (8–16)	2 (0.6–4.2)	1.2 (0.5–2)	0.03	0.00004
Half-life (days)	23	6	5	2–8	1–5
Production (mg/kg/day)	34	24	3.3	0.4	0.0023
Intravascular distribution	45	42	80	75	50
Carbohydrate content (%)	3	8	12	13	12
Complement fixation Classical pathway Alternative pathway	++ – (Only IgG4)	– +	+++ –	– +	– –
Placental transfer	+	–	–	–	–
Present in the milk	+	+	–	–	–
Selective secretion by secretory/mucous glands	–	+	–	–	–
Heat stability (56°C in 1 hour)	Stable	Stable	Stable	Stable	Labile
Participation in agglutination	+	++	+++	–	–
Participation in precipitation	+++	Variable	+	–	–
Fixation to mast cells and basophils	–	–	+	–	–
Primary Ab response	–	–	+	–	–

hypervariable (HV) region (also called hot spot) and sometimes less variable called framework (FV) region. Hypervariable (HV) region actual makes contact with epitope called paratope or complementarity determining region (CDR).

- **Valency:** Normally antigenic determinant (epitope) binds with antigen binding site (paratope) on Ig. Monomer Ig has two antigen binding sites. Valency is defined as numbers of antigenic determinants (epitopes) bind with antigen binding sites (paratopes) on an individual Ig. Follow **Table 19.1** for numbers of valency of different classes of Ig.
- **Fd piece:** The portion of the H chain presents in the Fab fragment is called Fd piece.

2. Carboxyterminus (C terminal/Fc fraction) end

- **Properties:** It contains H chain only. It is made up by carbohydrate moieties. It linked to the constant region.
- **Functions**
  - Complement fixation: IgM > IgG3 > IgG1 and rarely IgG2 activates the classical pathway while IgG4, IgA and IgD activate the alternative pathway.
  - Skin fixation
  - Placental transfer
  - Catabolic rate
  - Formation of Ag-Ab complex by combining with Fc receptors present on host cells.

Disulfide (s-s) bonds and Ig domain: Both H chains are linked by internal disulfide bonds. H and L chains are linked together by interchain/intrachain disulfide bonds. These intrachain disulfide bonds form loops in peptide chain, and each of the loop is compactly folded to form a globular domain (Ig fold). L chain has one domain in variable region (V_L) and one domain in constant region (C_L). H chain has one domain in variable region (V_H) and 3–4 domains in constant region (C_H1, C_H2, C_H3 and C_H4) depends on Ig class. Each domain has a separate function. The variable region domains V_L and V_H are responsible for the formation of a specific antigen binding site. The C_H2 region binds C1q in the classical complement sequence. C_H3 domain mediates adherence to monocyte surface.

Hinge region in Ig: The area of the H chain in the C region between the 1st and 2nd C region domains (C_H1 and C_H2) is called hinge region. It is rich in cysteine and proline. It is more flexible and allows Ig to assume different position thus helps the Ab to reach toward Ag. Papain acts (enzymatic digestion) on hinge region to produce 1 Fc and 2 Fab fractions. It also exposed to chemicals.

Digestion or Degradation of Ig

Aim: To understand the detailed structure of Ig.

History: Porter, Edelman, Nisonoff and colleagues have digested the rabbit IgG (antibody developed against egg albumin), by papain and pepsin and they obtained the following results.

Papain digestion: It produces the three parts (**Fig. 19.2**) such as one part of Fc (crystallizable fraction) and two parts of Fab or antigen binding fraction. Fc part is an insoluble fraction which is crystallized in cold. Fab parts are soluble fractions that bind to egg albumin, but unable to precipitate with it. It's sedimentation coefficient is 3.5S.

Pepsin digestion: It produces following two parts (**Fig. 19.3**) such as one part of Fc and one part of Fab. The Fc portion is digested into smaller fragments. Both Fab fragments held together in position by disulfide bond hence called F(ab')2. It is bivalent and precipitates with the antigen. Its sedimentation coefficient is 5S.

Chemical treatment by mercaptoethanol: It cleaves the disulfide bonds in to four subunits structure (**Fig. 19.4**).

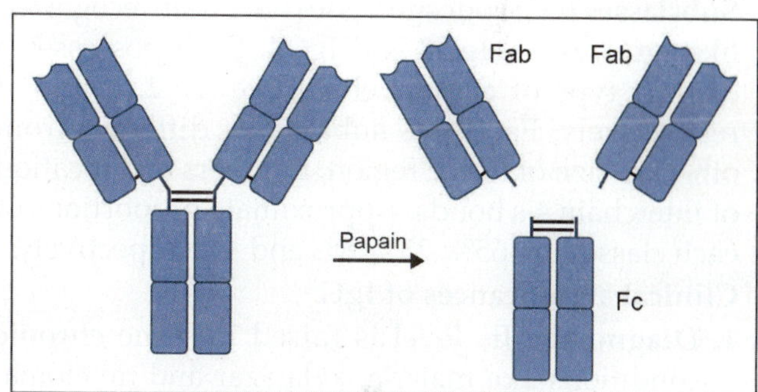

Fig. 19.2: Papain digestion

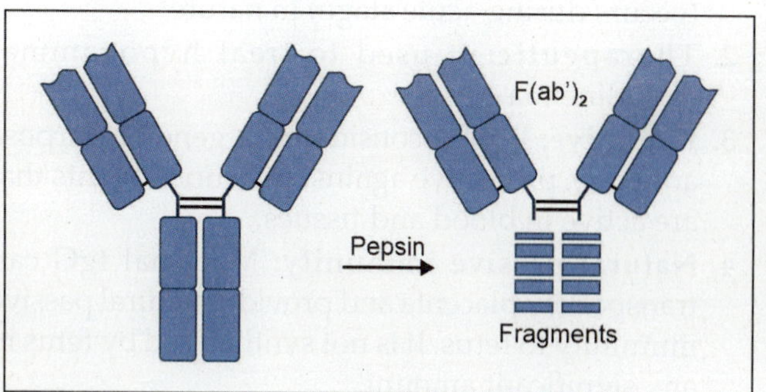

Fig. 19.3: Pepsin digestion

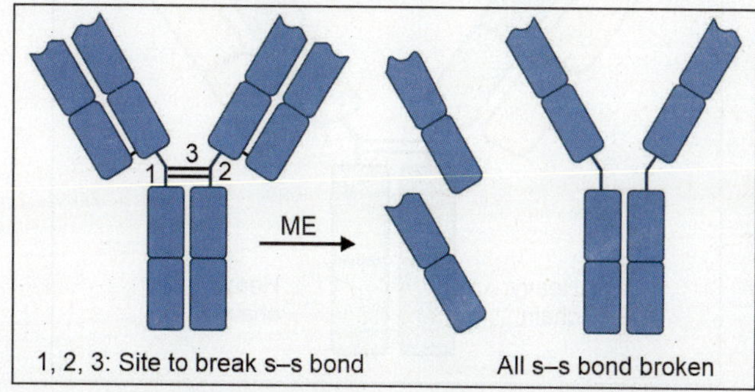

Fig. 19.4: Digestion by mercaptoethanol (ME)

Two ways to classify the Ig like structural and physical classification as below.

Structural Classification

Based on physiochemical and antigenic variation in H chain (in constant region), 5 classes of Igs have been recognized like IgG, IgA, IgM, IgD and IgE. Differences between each classes of Ig are mentioned in **Table 19.1**.

IgG

- **Properties of IgG:** It is the major serum immuno-globulin. It constitutes 70–80% of the total Ig. It is equally distributed between intravascular and extra vascular compartment. Catabolism of IgG is unique in that it varies with its serum concentration. Other properties are mentioned in **Table 19.1**.
- **Structure IgG:** It is monomer **(Fig. 19.5)** and occasionally exists in polymerized form. It contains less carbohydrate than other Ig. Its heavy chain has one V_H and three C_H domains.
- **Subclasses IgG:** Four subclasses have been recognized like IgG1, IgG2, IgG3 and IgG4. Each possesses a distinct type of gamma chain like 1, 2, 3 and 4 respectively. Each IgG subclass is different from other by size of hinge region, numbers and location of interchain s-s bonds. Approximate proportions of each classes are 65%, 23%, 8% and 4% respectively.
- **Clinical significances of IgG**
  1. **Diagnostic:** Its level is raised in some chronic conditions like malaria, kala azar and myeloma. With most antigens, IgG is a late antibody (occurs during chronic stage) and makes its appearance after the initial immune response which is IgM (occurs during acute stage) in nature.
  2. **Therapeutic:** It used to treat hypogamma-globulinemia.
  3. **Protective:** It may considered a general purpose antibody, protective against infectious agents that are active in blood and tissues.
  4. **Natural passive immunity:** Maternal IgG can transport by placenta and provides natural passive immunity to fetus. It is not synthesized by fetus in any significant amount.

5. **Isoimmunization or artificial passive immunity:** Passively administered IgG suppresses homologous antibody synthesis by feedback process. This principle is utilized in the isoimmunization of Rh –ve women who delivered a Rh +ve baby, by administration of anti-Rh (D) IgG during delivery.
6. **Phagocytosis:** IgG binds to microorganisms and enhances their phagocytosis.
7. **Extracellular killing** of the target cells coated with IgG is mediated through recognition of the surface Fc fragment by K cells bearing the appropriate receptors.
8. **Platelets function:** Interaction of IgG complexes with platelet Fc receptors probably leads aggrega-tion and release of vasoactive amines.
9. **Participation in immunological reactions:** Like
   - Complement fixation: IgG3, IgG1 and rarely IgG2 activate the classical pathway and IgG4 activates the alternative pathway.
   - Coagglutination: IgG except IgG3 binds with protein A of *Staph. aureus* (Cowan I strain) to mediate the coagglutination reaction.
   - Others: IgG participates in precipitation and neutralization of toxins and viruses.

IgA

- **Properties of IgA:** It is the 2nd most abundant class of Ig after IgG. It constitutes about 10–13% of serum Ig. Other properties are mentioned in **Table 19.1**.
- **Structure and subclasses of IgA**
  1. **According to numbers of Y shaped structure (H and L chains joined by s-s bond) and distribution:** Two subclasses.
     - **Serum IgA:** It is mostly monomer with single Y shaped structure with MW about 160,000 (7S). In humans over 80% of serum IgA is a monomer of the four chains (2 H and 2 L chains), but in most mammals it is predominantly polymeric, occurs mostly as a dimer.
     - **Secretory IgA (SIgA):** It is mostly dimer with double Y shaped structures, suggesting that the monomeric subunits are linked end-to-end by J chain at the C terminal at Ca3 regions as shown in **Fig. 19.6**. SIgA is much larger molecule than serum IgA. SIgA is the exocrine secretion (secretion on epithelial surfaces by glands/

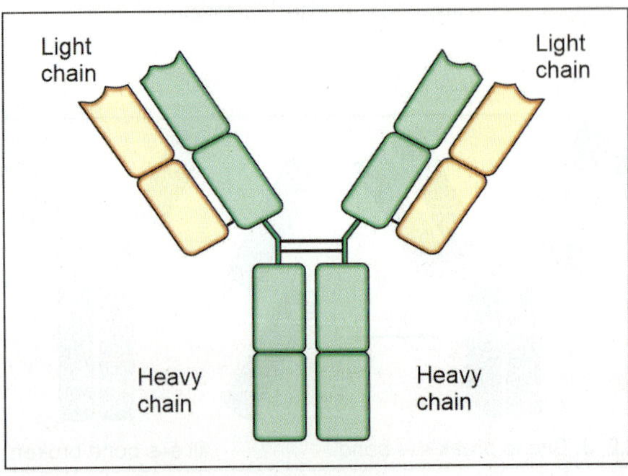

Fig. 19.5: Structure of IgG

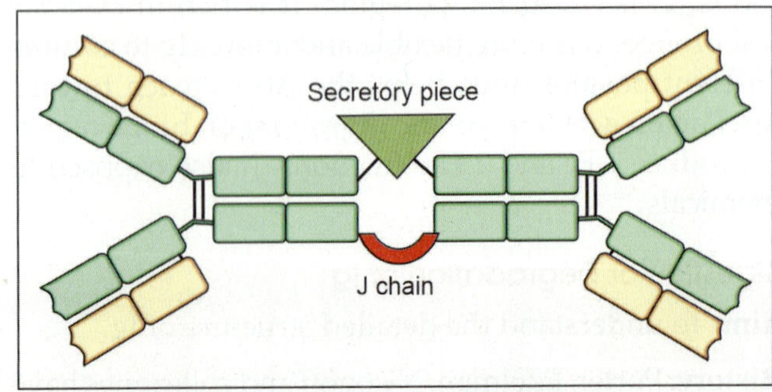

Fig. 19.6: Structure of SIgA

Essentials of Medical Microbiology

cells through duct called exocrine secretion and such glands or cells called exocrine glands/cells) and synthesized by plasma cells situated near mucosal or granular epithelium. SIgA is relatively resistant to the digestive enzymes and reducing agents. SIgA is a dimer formed by two monomer units joined together at their carboxyterminal by a glycopeptides chain called the J chain (J for joining). J chain is also produced by plasma cells situated near mucosal or granular epithelium. SIgA contains other glycine rich polypeptide called the secretory component or secretory piece. It is produced by mucosal or granular epithelial cells. Dimeric IgA binds to a receptor on the surface of the epithelial cells and is endocytosed and transported across the cell to the luminal surface. During this process, a part of the receptor remains attached to the IgA dimer, which serve secretory component. This secretory component is believed to protect IgA from denaturation by bacterial proteases.

Note:
- Origin of SIgA and J chain: Originated from plasma cells situated near mucosal or granular epithelium.
- Origin of Secretory (S) component or piece: Originated from mucosal or granular epithelial cells.

2. **According to amino acid sequences on the constant region of H chain:** Two IgA subclasses like IgA1 and IgA2.
   - **IgA1:** The hinge of IgA1 is extended and bears O-linked oligosaccharides while hinge of IgA2 is truncated. Both heavy chains bear N-linked oligosaccharides. It is dominant subclass in serum IgA (90%).
   - **IgA2:** IgA2 lacks interchain/intrachain disulfide bonds between the heavy and light chains. It holds the minor part in serum IgA (10%) but dominant subclass in secretory IgA. IgA2 ranges from 10–20% in nasal and male genital secretions, 40% in saliva, 60% in colonic and female genital secretions. Polysaccharide Ags induce more IgA2 synthesis than protein Ags.
- **Clinical significances of IgA**
  1. **Immunological reactions:** Serum IgA combines with Fc receptor on immune cells and takes part in ADCC and degranulation.
  2. **Local (mucosal) immunity:** Secretory IgA is the major Ig in colostrum, saliva, milk, tears, respiratory, intestinal and genitourinary secretion. SIgA is selectively concentrated in secretions and on the mucus surface forming a layer 'called antibody paste' and thereby preventing the entry of microbes in the body tissues and provides local immunity against respiratory and intestinal pathogens.
  3. **Complement activation:** IgA do not fix complement but it activates an alternative complement pathway.

4. **Phagocytosis:** It promotes phagocytosis and intracellular killing of microorganisms.

IgM
- **Properties of IgM:** It constitutes 5–8% of total serum Ig. Most of IgM (80%) is intravascular in distribution. Treatment of serum with 0.12 M 2-mercaptoethanol selectively destroys IgM, hence it is a simple method for differential estimation of IgG and IgM antibodies. Other properties are mentioned in **Table 19.1**.
- **Structure and subclasses of IgM:** Two physical forms.
  1. **Membrane** or **surface Ig:** It is monomer and expressed in bound form with B cell receptors (BCR) on the surface of B cells.
  2. **Soluble form:** It is a pentamer (**Fig. 19.7**) of basic five chains structure, formation of which is aided by inclusion of the J polypeptide chain of mass ~15 kDa. Each heavy chain has one V_H and four C_H domains and no hinge region. Theoretical valency is ten, but this is observed only with small haptens. With larger antigens, the effective valency falls to five, probably because of steric hindrance. It is available free in blood/plasma.
- **Clinical significances of IgM**
  1. **Phylogenetic importance:** Phylogenetically it is the oldest Ig class. It is also the earliest Ig to be synthesized by fetus, beginning by about 20 weeks of age.
  2. **Diagnostic role**
     - Diagnosis of intrauterine infection: As it is not transported by placenta and synthesized after 20 weeks of gestational age, its presence in the fetus before 20 weeks indicates intrauterine infection and its detection is useful in diagnosis of congenital infections like TORCH agents.
     - Acute (recent) infection: IgM is relatively short lived, disappearing earlier than IgG; hence it's demonstration in serum indicates acute (recent) infections.
  3. **Immune hemolysis:** A single molecule of IgM can brings about immune hemolysis, where as 1000 IgG molecules are required for the same effect.
  4. **Opsonization:** It is also 500–1000 times more effective than IgG in opsonization.
  5. **Bactericidal action:** It is 100 times more effective in bactericidal action than IgG.
  6. **Bacterial agglutination:** It is 20 times more effective in bacterial agglutination than IgG.

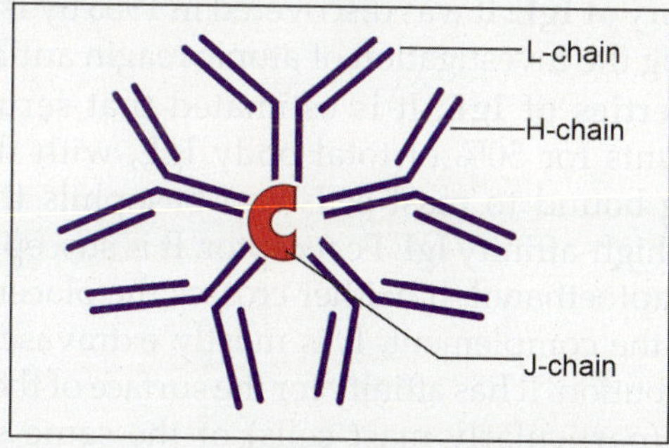

Fig. 19.7: Structure of soluble IgM

Antibody (Ab)–Immunoglobulin (Ig)

7. **Activation of classical pathway of complement:** IgM (+++) is more effective than IgG (++).

8. **Immunological reactions**
   - Neutralization: It neutralizes the toxins and viruses but less effective than IgG.
   - Precipitation: It is useful but less effective than IgG.

9. **Protective role:** It is distributed in to the intravascular space, it is believed to be responsible for protection against blood invasion by microorganisms; hence its deficiency is often associated with septicemia.

10. **Ag recognition:** Monomeric IgM is the major antibody receptor on the surface of B lymphocytes for antigen recognition.

11. **Natural Abs:** Most natural Abs are IgM type like isohemagglutinin (anti-A, anti-B), reagin Ab to syphilis, Ab to *S. Typhi* O Ag and Abs to other microbes are IgM in nature in primary or acute infections.

> **Note: IgM and IgG comparison**
> **IgG:** IgG is more effective than IgM, in neutralization, precipitation and activation of classical pathway (++).
> **IgM:** IgM is more effective than IgG, in immune hemolysis (1000 times), opsonization (500–1000 times), bactericidal action (100 times), bacterial agglutination (20 times) and activation of classical pathway (+++).

IgD

- **Properties of IgD:** It constitutes less than 1% of the serum Ig. It is mostly intravascular in distribution. It is the least important Ig in human beings. Other properties are mentioned in **Table 19.1**.
- **Structure of IgD:** Structurally, it resembles to IgG. Each heavy chain has one V_H and three C_H domains with an extended hinge region that is susceptible to proteolysis, at least on purification.
- **Clinical significances of IgD**
  1. **Antigen recognition:** It is a transmembrane monomeric form presents as a antigen-specific receptor on unstimulated mature B cells and serves as recognition receptor for Ag.
  2. **Antibody production:** Combination of IgD molecule with its corresponding antigen leads specific stimulation of the B cells either activation or cloning to produce Ab or to suppress Ab.

IgE

- **History of IgE:** It was discovered in 1966 by Ishizaka during the investigation of atopic reagin antibodies.
- **Properties of IgE:** It is estimated that serum IgE accounts for 50% of total body IgE, with the rest being bound to mast cells and basophils through their high-affinity IgE Fc receptor. It is susceptible to mercaptoethanol. It neither crosses the placenta nor fixes the complement. It is mostly extravascular in distribution. It has affinity for the surface of the tissue cells (particularly mast cells) of the same species (homocytotropism). IgE is chiefly produced in lining of the respiratory and intestinal tract. Its production increased following exposure of allergen like ovalbumin. Other properties are mentioned in **Table 19.1**.
- **Structure of IgE:** Structurally, it resembles to IgG. Each heavy chain has one V_H and four C_H domains and no hinge region.
- **Clinical significance of IgE**
  1. **Diagnostic role**
     - Allergic diseases. Its level is greatly elevated in atopic (type 1 allergic) conditions such as asthma, hay fever, eczema, etc.
     - Parasitic diseases: Children living in insanitary conditions, with high load of intestinal parasites, have high serum levels of IgE.
  2. **IgE deficiency:** It increases the susceptibility to infections.
  3. **Anaphylactic reaction:** It is responsible for anaphylactic reaction.
  4. **Prausnitz-Kustner reaction:** It mediates the Prausnitz-Kustner reaction.
  5. **Protective role:** The physiological role of IgE appears to be protection against pathogens like helminthic infections, by mast cell degranulation and release of inflammatory mediators.

Physical Classification

Two classes of Ig on the basis of availability like free form in blood or bound form with receptors on B cells.

Membrane or surface Igs: IgM and IgD (secretory IgA?) are bound on B cell receptors (BCRs) on surface hence called membrane or surface Ig. BCRs are found only on the surface of B cells and facilitate the activation of these cells. Their subsequent differentiation into either antibody producing plasma cells or memory B cells will survive in the body and remember that same antigen, so the B cells can respond faster upon future exposure.

Free or soluble Igs: IgG, IgA and IgE are secreted from the cells to be free in the blood/plasma.

Age of Ig Production

IgG: It is the only Ig which can cross the placenta. New born has high level of passively transferred IgG, immediately after birth which starts to fall after birth as shown in dark blue line in **Fig. 19.8**. Light blue line in **Fig. 19.8** suggests that immune system of newborn starts to produce the IgG in 2–6 months following birth and reaches at peak in 1 year.

IgM: It is the earliest Ig synthesized during fetal life between 3 and 6 months (approx 20 weeks); however, it is immature and at very low level. Peak level achieved after birth in 1 year.

Factors Affecting Concentration of Ig

Increase concentration: Concentration increases in diseases like leprosy, kala azar, malaria, syphilis, tuberculosis, SABE and some other chronic infections.

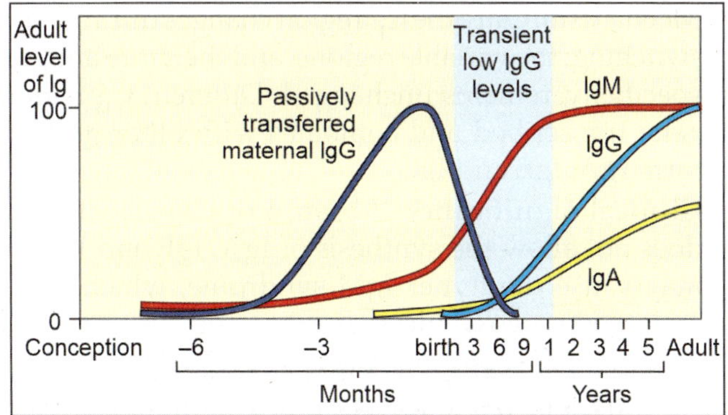

Fig. 19.8: Age of Ig production

Decrease concentration: Concentration decreases in congenital diseases, delayed production, secondary to diseases like myeloma, leukemia or drug induced. Ig is deficient in nephritic syndrome.

Ig Specificities

Idiotype specificity
- **Definition:** The specific antigenic determinants on the antigen binding sites (paratopes, variable region) are called idiotopes. The sum totals of idiotopes on an immunoglobulins molecule are called idiotype specificities.
- **Clinical significance:** It is used in vaccine production by immunizing with Fab fragments, anti-idiotype antibodies can be produced. These resemble the epitopes of the original antigen. Vaccine provides protection against the original antigen (pathogen or tumor) in experimental animals. Sequential anti-idiotypic antibody formation is the basis of Jerne's network hypothesis of immune regulation.

Isotype specificity
- **Definition:** Five classes of Ig and their subclasses called isotype specificities.
- **Clinical significance:** It is due to genetic or structural differences in constant region of H chain. All individuals of one species have same isotype or same constant region. However different species have different isotype or different constant region. When such isotype or constant region of one species is injected in to other, it will be identified as foreign, resulting Ab production.

Allotype specificity
- **Definition:** Antigenic specificities which distinguish immunoglobulins of the same classes, between different groups of individuals in the same species are called allotype specificities. It is not found in all members but some members of species called allotype.
- **Clinical significance:** It is used in paternity testing and in population genetics.

Abnormal Ig

Definition: Apart from antibodies, other structurally similar proteins are seen in serum in many pathological conditions and sometimes even in healthy persons called abnormal Ig.

Types of abnormal Ig
1. Light chains diseases
- **Bence Jones proteins:** The earliest description of an abnormal Ig was given by Bence Jones (1847), who found protein in multiple myeloma. It is the excessive production of light chains. Bence Jones proteins are the light chains of the Ig and so may occur as either K or forms in any one patient, but never both, being uniform in all other respect. This is because myeloma, a plasma cell dyscrasia in which there is unchecked proliferation of one clone of plasma cells, resulting in excessive production of the particular Ig synthesized by the clone. Such Ig called monoclonal Ig. Multiple myeloma may affect plasma cells synthesizing IgG, IgA, IgD or IgE. Because of the monoclonal nature, they have been valuable models for the understanding of immunoglobulin structure and functions. Bence Jones protein can be identified in the urine by its characteristic property of coagulation when heated to 50°C but re-dissolving at 70°C.
- **Waldenstrom's macroglobulinemia/M proteins:** Similar involvement of IgM producing cells called Waldenstrom's macroglobulinemia. In this condition there is excessive production of the respective myeloma proteins (M proteins) and of their light chains (Bence Jones proteins).

2. Heavy chain disease (HCD):
It is characterized by the overproduction of the Fc parts of the Ig heavy chain due to lymphoid neoplasia. Three types of HCD are recognized based on class of heavy chain produced by malignant cells like α, γ and μ.

3. Cryoglobulinemia:
It is a condition in which a gel or precipitate is formed on cooling the serum, which re-dissolves on warming. It may not always be associated with disease but is often found in myeloma, macroglobulinemia and autoimmune condition such as SLE. Most cryoglobulins consist of IgG, IgM or their mixed precipitates.

Important Aspects of Ig

IgG: It protects the body fluids.

IgA: It protects s the body surfaces.

IgM: It protects the bloodstream.

IgE: It mediates the hypersensitivity.

IgD: It is a recognition molecule on B cell surface. All Igs are bifunctional except IgD because IgD is found only in bound form, so its Fc region remains attached to the B cells and not available for effector function which all other Igs mediate.

GENES OF IMMUNOGLOBULIN

History

Ig chains are coded by more than one gene; this concept was 1st introduced by Dreyer and Bennett. Later it was proved by Tonegawa and Hozumi in 1975, for which they were awarded Nobel Prize in 1987.

Genes for H and L chain: Two H and two L chains of an Ig are coded by separate genes, later all they are joined.

- **H chain's genes:** H chain is coded by four genes (gene segments), of which 3 are for V region and one is for C region.
  - Variable region genes: V (variable, V_H) gene, D (Diversity, D_H) gene and J (Joining, J_H) gene.
  - Constant region genes: C (Constant, C_H) gene, which consist 9 segments like $C\alpha$, $C\beta$, $C\mu$, $C\delta$, etc.
- **L chain's genes:** L chain is coded by three genes (gene segments), of which 2 are for variable region and one is for constant region. Kappa and Lambda chains are coded by two different sets of V, J and C genes.
  - Variable region genes: V gene and J gene. D gene is absent.
  - Constant region gene: C gene.

Other genes: These include genes of enzyme, necessary for Ig formation like RAC (recombination activation) gene of recombinase enzyme.

Location of Genes

Genes for H chain, K L chain and yL chain are located not on same chromosomes but on different chromosomes as mentioned below.

- H chain gene: Located on chromosome 14.
- KL chain: Located on chromosome 2.
- yL chain: Located on chromosome 22.

Rearrangement (Recombination) of Ig Genes

Complete Ig molecule is formed after rearrangement between genes, which occurs at DNA and RNA level.

Rearrangement at DNA level: It includes the rearrangement and joining of V, D and J genes of H and L chains. C genes of H and L chains are not undergoes for rearrangement at DNA level. Rearrangement occurs 1st in H chain followed by L chain genes. In H chain V-D joining occurs 1st followed by J joining called V-D-J joining. In L chain only V-J joining occurs. Rearrangement and joining of different genes is mediated by recombinase enzyme.

Rearrangement at RNA level: V, D, J and C genes are transcribed to produce primary RNA like V region RNA, D region RNA, J region RNA and C region RNA respectively. C region RNA combines with V region RNA to form complete H and L chains. The whole C region RNA does not transcribed at a time but 1st $C\mu$ and $C\delta$ genes are transcribed and combined with V region RNA to form complete molecule of Ig like IgM, IgD, respectively.

Ig class switching

- **Synonym:** Ig isotype switching
- **Definition:** It is a biological process, which allows the synthesis of other classes or other isotypes of Ig (like IgA, IgE and IgG) after synthesis of IgM and IgD.
- **Mechanism:** Only the C_H region changes during class switching; the variable regions, and therefore antigen specificity, remains unchanged. Different C_H region gene transcribed and combines with other gene to form IgA, IgE and IgG.
- **Clinical significance:** Absence of class switching does not allow the synthesis of IgA, IgE and IgG or may cause the hyper IgM syndrome, which leads increased mucosal ulcer, diarrhea and susceptibility of skin plus respiratory tract to bacterial infections.

Differential Ig RNA processing: It is important for synthesis of membrane bound Ig (IgM and IgD) or secretory Ig.

Antibody Diversity

Definitions

- **Antibody diversity:** Human body contains enormous numbers ($\geq 10^8$) of structurally distinct antibodies, each with different specificity (paratope) called antibody diversity.
- **Antibody repertoire:** Total collection of antibodies with different specificities called antibody repertoire.

Mechanisms: Genetic diversity is due to availability of plenty of Ig genes and their genetic rearrangement by following possible mechanisms.

- **Multiple genes for each region of Ig:** Large numbers of different genes exist for each region of Ig such as $51V_H$ genes for H chain and each H chain would have one out of $51V_H$ genes. Follow **Table 19.2** for other genes of H and L chains.
- **Antigenic variation of genes:** It is also responsible for antibody diversity.
- **Multiple rearrangements (recombinations) of genes:** Multiple recombinations of genes of variable region allow the formation of multiple IgG classes. Follow **Table 19.3** for multiple recombinations of V genes.
- **Mutation/hypermutation:** After formation of Ig molecule, further changes occur in the nucleotide sequence of V region. It is due to point mutation with

TABLE 19.2: Multiple genes of H and L chains

Gene	Numbers of genes		
	H chain	K L chain	y L chain
V	51	40	30
D	27	0	0
J	6	5	4
C	9	1	4

TABLE 19.3: Multiple recombinations of V genes

Type of joining	Possible recombinations
VDJ combination in H chain	$51V_H \times 27D_H \times 6J_H = 8262$
VJ combination in K L chain	$40V_k \times 5J_k = 200$
VJ combination in Y L chain	$30V_\gamma \times 4J_\gamma = 120$
Combination in H and L chains	$8262 \times 200 \times 120 = 2.64 \times 10^6$

high frequency (10^{-3}/bp/generation then normal 10^{-8}/bp/generation), so called hypermutation.

- **Other mechanisms:** It includes junctional diversity and junctional flexibility.

MONOCLONAL ANTIBODIES

History

It was developed by Kohler and Milstein (Nobel Prize 1984) by using Hybridoma technique.

Definitions

Polyclonal antibodies: Antibodies produced in natural infections are usually against many epitopes or antigenic determinants called polyclonal antibodies. Resulting antisera contains Abs with different classes.

Monoclonal antibodies: Antibodies produced by single clone (Clone means single Ab producing cell) against single epitope or antigenic determinant called monoclonal antibodies.

Hybridomas (hybrid cells): Cells produced by fusion between mouse myeloma cells (ability to multiply indefinitely) and spleen cells (ability to produce Abs) called hybrid cells or hybridomas.

Technique of Monoclonal Antibodies

Principle: To produce the monoclonal antibody, cells have ability to produce the large number of Abs like splenic B cells and ability to multiply indefinitely like myeloma cells (cancerous plasma cells). Take these two cells and prepare a hybrid cell. Proliferate the hybrid cell in particular medium to produce large numbers of monoclonal Abs.

Steps (Fig. 19.9)

- **Preparation of hybrid cells/hybridomas:** Immunize the mouse with Ag, against which monoclonal Ab is required to produce. Cultivate the mouse myeloma cells that multiply indefinitely, do not form the Ig and deficient in enzyme HGPRTase (Hypoxanthine Guanine Phospho Ribosyl Transferase). Take the spleen cells contain HGPRTase from immunized mouse and fuse with mouse myeloma cells to produce hybrid cells hybridomas.

- **Growth of hybrid cells in HAT medium:** HAT medium contains **H**ypoxanthine, **A**minopterin and **T**hymidine kinase. To grow on HAT medium cells required to synthesize the purine which is possible by either *de novo* pathway or salvage pathway. Aminopterin blocks the *de novo* pathway while for salvage pathway cells required HGPRTase and thymidine kinase. Myeloma cells are HGPRTase deficient, so salvage pathway and ultimately purine synthesis is not possible by myeloma cells on HAT medium and they can't grow. It allows the growth of spleen cells contains HGPRTse, but it does not last longer as the cells are not immortal. It allows the growth of only hybrid cells which last longer.

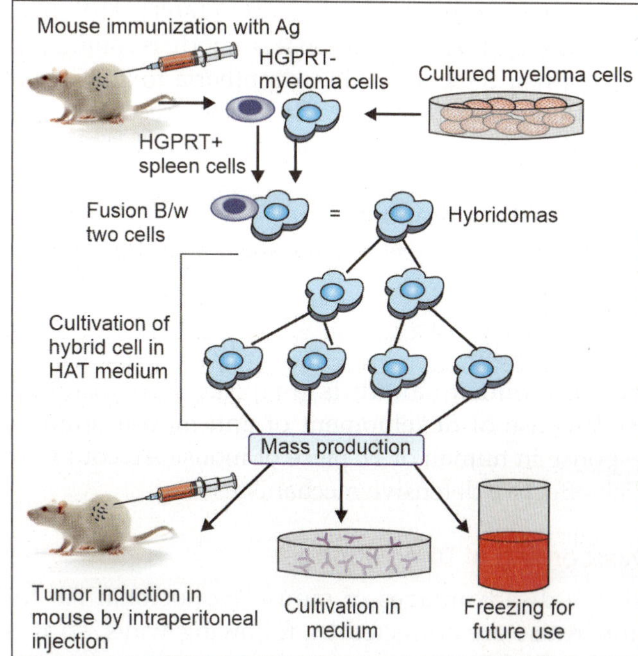

Fig. 19.9: Production of monoclonal Ab

- **Mass production of monoclonal antibodies:** These hybridomas are clone and examined for production of Abs. Such clones are selected and used for mass production of Abs by growing in culture or by growing as tumor by injecting intraperitoneally in mouse and monoclonal Abs are harvested from ascetic fluid or by freezing for future use.

Uses

1. **Prophylactic:** Monoclonal Ab is useful for vaccine preparation.

2. **Diagnostic**

   - **Isolation and identification of protein:** Monoclonal Ab is useful to isolate and to identify the interferon and coagulation factor VIII from the mixture, if present even in low concentration.

   - **Identification of cells:** Identification of CD4 and CD8 cells by using anti CD4 and anti CD8 monoclonal antibodies, respectively.

   - **Identification of antigens:** Such as identification of:
     - human chorionic gonadotropin by using anti-human chorionic gonadotropin.
     - blood group antigens by using anti-A and anti-B monoclonal antibodies.
     - tumor antigens by imaging technique like prostate specific antigens.
     - HLA by using anti-HLA monoclonal Ab.
     - HBV antigen, HIV antigen, etc.

3. **Therapeutic (immunotherapy):** Monoclonal Ab is useful:

   - for monitoring drug and protein level in serum.
   - in treatment of cancers, inflammatory and allergic diseases. Monoclonal antibody acts by suppressing immune systems, killing/inhibiting the tumor cells or by inhibiting the angiogenesis.

- as post exposure prophylaxis in rabies, HBV, etc.
- immunotoxin: Monoclonal antibody is conjugated with specific toxin (like diphtheria toxin), where it combines with specific surface receptors and allowing toxin to kill target cells such as cancer cells.

4. **Research applications**
5. **Other uses:** Abzyme is a monoclonal antibody with catalytic activity.

Disadvantages

Mouse monoclonal Ab is unsuitable for humans use because of development of anti-mouse immune response in human or Fc piece of mouse Ab could not elicit effective defensive mechanisms.

Overcoming of Disadvantages

All the disadvantages of mouse monoclonal Ab for human use are corrected by following ways, so it is suitable for human use.

1. **Coupling methods:** Coupling of Fab piece of Ig with active substances like toxins, enzymes, drugs or radiological substances.
2. **Genetic method:** (Chimeric antibodies production): By using murine variable region and human constant region.
3. **Grafting method:** Grafting of murine monoclonal Ab on CDR (Complementary Determining Region) loops on human Ig.
4. **Bacteriophage method:** Genes for monoclonal Ab production are fused in bacteriophage, which later allow infecting the bacteria and production of large quantity of Abs.

ACCESS YOURSELF

Essay/Full Question

1. Immunoglobulin (Ig).

Short Notes

1. Structure of Ig.
2. IgG/IgA/IgM.
3. Monoclonal antibody.

Short Questions for Theory/Viva Questions

1. Write the end results of digestion of Ig by papain and pepsin.
2. What is hypervariable region in Ig?
3. Write the four functions of C-terminal of Ig.
4. What is membrane Ig? Write two examples.
5. What are epitope and paratope?
6. What is the hinge region? Write its function.
7. What is the valence in IgG? Write the valence numbers for IgG, IgA, IgM, IgD and IgE.
8. Write the two names of gene, each for H and L chain of Ig.
9. What is class switching in Ig? Write its clinical significance.
10. Define: Monoclonal antibodies and polyclonal antibodies.

Comments on

1. All antibodies are immunoglobulins, but all immunoglobulins may not be antibodies.

2. IgG is useful to detect the chronic infections, while IgM is useful to detect the acute infections.
3. IgG is useful for isoimmunization.
4. IgA is known to produce local immunity.
5. 2-mercaptoethanol is a useful tool to differentiate between IgG and IgM.
6. IgM is useful to detect the intrauterine infections.
7. IgM deficiency causes septicemia.
8. All Igs are bifunctional except IgD.
9. Absence of Ig class switching causes bacterial infection.
10. Myeloma cells cannot grow in HAT medium.
11. Splenic cells can grow in HAT medium.

MCQs for Chapter Review

Introduction

1. **Electrophoretic analysis of plasma protein was done by:**
   a. Tiselius
   b. Emil A Behring
   c. Charles Richet
   d. Peyton Rous

Immunoglobulin (Ig)

2. **Papain acts on gamma globulin to form:**
   a. 2 Fc fragments
   b. 2 Fab fragments
   c. 1 Fab fragment
   d. None
3. **Portion of immunoglobulin molecule with molecular weight of 50,000:**
   a. Secretory piece
   b. H chain
   c. L chain
   d. J piece
4. **A single immunoglobulin molecule contains:**
   a. 1 light chain, 1 heavy chain
   b. 2 heavy chain, 1 light chain
   c. 2 light chain, 2 heavy chain
   d. 2 light chain, 1 heavy chain
5. **A single immunoglobulin molecule contains:**
   a. Single peptide chain
   b. Two peptide chain
   c. Non sulfur amino acid
   d. 2 long and 2 short peptide chains
6. **Numbers of variable regions on each light and heavy chain of an antibody:**
   a. 1
   b. 2
   c. 3
   d. 4
7. **Different classes of Ig are determined by:**
   a. L chain
   b. J chain
   c. H chain
   d. None
8. **Antigen binding site on antibody is:**
   a. Hinge region
   b. Constant region
   c. Variable region
   d. Hypervariable region
   e. Idiotype region
9. **Antigen combing site of antibody:**
   a. Idiotype
   b. Paratope
   c. Epitope
   d. Hapten
10. **A variable portion of antibody molecule is:**
    a. C-terminal
    b. N-terminal
    c. CHO moiety
    d. None
11. **Complement attaches to immunoglobulin at:**
    a. Aminoterminal
    b. Fab region
    c. Variable region
    d. Fc fragment
12. **The following is constitute approximately 75% of total Ig in human:**
    a. IgA
    b. IgG
    c. IgM
    d. IgD
    e. IgE

13. The most abundant Ig in human body:
    a. IgM
    b. IgG1
    c. IgG2
    d. IgG3

14. Antibody transfer from mother to fetus is:
    a. IgG
    b. IgA
    c. IgM
    d. IgD
    e. IgE

15. Which immunoglobulin is scarce in human serum?
    a. IgA
    b. IgG
    c. IgM
    d. IgD
    e. IgE

16. Heat labile immunoglobulin:
    a. IgA
    b. IgG
    c. IgE
    d. IgM

17. True about immunoglobulin:
    a. IgE has maximum concentration
    b. IgG has maximum concentration
    c. IgA has minimum concentration
    d. IgM has minimum concentration

18. True of the following is/are:
    a. IgA crosses placenta
    b. Half-life of IgG is 23 days
    c. IgD is heat stable
    d. IgE has highest carbohydrate content
    e. IgG induces leukotrienes release during inflammation

19. Which of the following Ig can cross the placenta?
    a. IgA
    b. IgM
    c. IgG
    d. IgD

20. Maximum half-life:
    a. IgA
    b. IgM
    c. IgG
    d. IgD

21. Pentamer immunoglobulin is:
    a. IgA
    b. IgG
    c. IgM
    d. IgE

22. Antibody known as millionaire molecule is:
    a. IgA
    b. IgM
    c. IgG
    d. IgE

23. Activation of classical complement pathway:
    a. IgA
    b. IgG
    c. IgM
    d. IgD

24. The most avidly complement fixing antibody is:
    a. IgA
    b. IgG
    c. IgM
    d. IgE

25. J chain is a structural part of:
    a. IgA
    b. IgM
    c. IgG
    d. a+b

26. The serum concentration of which of the following human Ig subclass is maximum?
    a. IgG1
    b. IgG2
    c. IgG3
    d. IgG4

27. The commonest IgG with maximum individual variation is:
    a. IgG1
    b. IgG2
    c. IgG3
    d. IgG4

28. Function of IgA is:
    a. Acts as mucosal barrier for infection
    b. Circulating antibody
    c. Kills virus infected cell
    d. Activates macrophages
    e. Causes delayed hypersensitivity

29. Antibody responsible for local immunity is:
    a. IgA
    b. IgM
    c. IgG
    d. IgE

30. In respiratory and GIT infections, which is the most affected immunoglobulin?
    a. IgA
    b. IgG
    c. IgM
    d. IgD

31. The secretory component of immunoglobulin molecules is:
    a. Formed by epithelial cells of lining mucosa
    b. Formed by plasma cells
    c. Formed by epithelial cells and plasma cells
    d. Secreted by bone marrow

32. Secretory piece is a structural part of:
    a. IgA
    b. IgM
    c. IgG
    d. a + b

33. Which of the following Ig is responsible for opsonization?
    a. IgA
    b. IgG
    c. IgM
    d. IgE

34. Hemagglutinis (anti-A and anti-B) are which type of antibodies?
    a. IgG
    b. IgM
    c. IgA
    d. IgE

35. Which of the immunoglobulin is associated with allergic disorders?
    a. IgG
    b. IgM
    c. IgA
    d. IgE

36. Antibody elevated in parasitic infection:
    a. IgA
    b. IgE
    c. IgG
    d. IgM

37. Ovalbumin was injected in to a rabbit. Which of the following classes of antibodies are likely to be produced initially?
    a. IgG
    b. IgM
    c. IgE
    d. IgD

38. IgE is secreted by:
    a. Mast cells
    b. Basophils
    c. Eosinophils
    d. Plasma cells
    e. Neutrophils

39. Ig are produced by:
    a. Macrophages
    b. B-cells
    c. T-cells
    d. NK cells

40. Plasma cells are derived from:
    a. T-cells
    b. B-cells
    c. Macrophages
    d. Neutrophils

41. True about immunoglobulin is:
    a. IgE fixes complement
    b. IgM fixes complement
    c. IgG found in minimum concentration
    d. IgG is elevated in primary immune response

42. Which of the following statement concerning immunoglobulin is wrong?
    a. IgM does not cross the placenta
    b. IgE increased in parasitic infection
    c. IgM increased in primary response
    d. Fetal infection is characterized by increased in IgG

43 Capacity of producing IgG starts at which age?
    a. 6th month
    c. 1st year
    d. 2nd year
    e. 3rd year

44. The earliest Ig to be synthesized by the fetus is:
    a. IgA
    b. IgG
    c. IgE
    d. IgM

45. Which immunoglobulin is least important in human being?
a. IgE
b. IgD
c. IgG
d. IgA

46. Which of the following immunoglobulin constitutes the antigen binding component of B cell receptor?
a. IgA
b. IgD
c. IgM
d. IgG

47. Antigen idiotype is related to:
a. Fc fragment
b. Hinge region
c. C-terminal
d. N-terminal

48. Ig change in variable region is:
a. Idiotype
b. Isotype
c. Allotype
d. Autotype

49. Isotype specificity in Ig is due to:
a. Changes in constant region of L chain
b. Changes in constant region of H chain
c. Changes in variable region of L chain
d. Changes in variable region of H chain

50. Isotypes refer to variations in the:
a. Heavy chain constant region
b. Heavy chain variable region
c. Light chain constant region
d. Light chain variable region

51. Which of the following statements is true about isotypic variation?
a. These result due to subtle amino acid changes resulting from allelic differences
b. These result due to changes in amino acid in heavy chain and light chain at variable region
c. Changes in heavy and light chain in constant region is responsible for class and subclass of immunoglobulin
d. These are areas in antigen that bind specifically to anti-body

52. Which precipitates at 50–60°C, but disappears on heating?
a. Heavy chain
b. Light chain
c. Both
d. None of the above

53. Bence Jones proteins are best described as:
a. α chain
b. γ chain
c. Kappa and lambda chains
d. Fibrin split products

Genes of Immunoglobulin

54. Which of the following statement is true regarding kappa, lambda and heavy chain immunoglobulin?
a. Coded on the same site of chromosome
b. Coded on the different sites of same chromosome
c. The chains are formed by genetic rearrangement after maturation
d. Different chains of same immunoglobulins are coded by different chromosomes
e. Different chains of same immunoglobulins are coded by same chromosomes

55. Class switching is required for synthesis of following immunoglobulin:
a. IgA
b. IgM
c. IgG
d. IgD

56. Synthesis of an immunoglobulin in membrane bound or secretory form is determined by:
a. One turn to two turn joining rule
b. Class switching
c. Differential RNA processing
d. Allelic exclusion

57. Antibody diversity is due to:
a. Gene rearrangement
b. Gene translocation
c. Antigenic variation
d. CD40 molecule
e. Mutation

Monoclonal Antibody

58. Theory of monoclonal antibody was developed by:
a. Theobald Smith
b. Kohler and Milstein
c. Richet
d. Paul Ehrlich

59. Hybridoma technique is used to obtain:
a. Specific antigen
b. Complement
c. Specific antibody
d. Interleukins

60. Which of these technique was used by Kohler and Milstein in 1975 to produce monoclonal antibodies?
a. Transgenic plant technology
b. Co-culture technique
c. Recombinant gene technology
d. Hybridoma technology

61. Medium for monoclonal antibody production is:
a. HAT medium
b. Kelly's medium
c. Pike's medium
d. Islam's medium

62. All of the following statements about hybridoma technology are true *except*:
a. Specific antibody producing cells are integrated with myeloma cells
b. Myeloma cells with mutation pathway grows well in HAT medium
c. Aminopterin, a folate antagonist, inhibits the *de novo* pathway
d. HGPRTase and thymidine synthetase are required for salvage pathway.

63. Use of monoclonal antibody is:
a. Immunotherapy
b. Immunological identification of cells and tissues
c. Radio immuno imaging
d. All of above

Answers and Explanation of MCQs

1. a
- Emil A Behring: Developed antitoxin to diphtheria.
- Charles Richet: Discovered the anaphylaxis.
- Peyton Rous: Discovered viral oncogenesis (father of tumor virology).

2. b
- Follow section, **digestion or degradation of Ig and Fig. 19.2** for explanation.

3. b

4. c

5. d

6. a
- Follow section, **structure of Ig** for explanation of answers of MCQs 3–6.

7. c
- Different classes/isotypes of Ig are determined by different types of H chain. Different types L chain like kappa and lambda are also available but not significant as like H chain. J chain has joining function between different Ig molecules.

8. d

9. b

10. b
- Follow section, **structure of Ig [Aminoterminus (N terminal/Fab fraction) end → hypervariable region]** for explanation of answers of MCQs 8–10.

11. d
- Follow section, **structure of Ig [Carboxyterminus (C terminal/Fc fraction) end →]** for explanation.

12. b
- IgG constitutes 70–80% of the total Ig.

13. b
- IgG constitutes 70–80% of the total Ig. There are four subclasses of IgG like IgG1, IgG2, IgG3 and IgG4 with approximate proportion is 65%, 23%, 8% and 4% respectively.

14. a
- IgG can cross the placenta and it transfers from mother to fetus.

15. e
16. c
17. b
18. b, c
19. c
20. c
21. c
22. b
23. b, c
24. c
25. d
- Follow **Table 19.1** for explanation of answers of MCQs 15–25.

26. a
27. a
- Approximate proportion of IgG1, IgG2, IgG3 and IgG4 is 65%, 23%, 8% and 4% respectively.

28. a
29. a
30. a
- Follow section, **IgA (clinical significances)** for explanation of answers of MCQs 28–30.

31. a
32. a
- Follow section, **IgA (structure and subclasses of IgA → Secretory IgA)** and **Fig. 19.6** for explanation of answers of MCQs 31–32.

33. b and c
- IgM is more effective than IgG in opsonization.

34. b
- Follow section, **IgM (clinical significances → natural Abs)** for explanation.

35. d
36. b
- IgE is elevated in allergic reactions and parasitic infections.

37. c
- IgE production increased following exposure of allergen like ovalbumin.

38. d
39. b
40. b
- After antigenic stimulation B cells are differentiated in to plasma cells and memory cells. Plasma cells are the Ab producing cells and known to produce the all kinds of Igs.

41. b
- IgE does not fix complement.
- IgM > IgG3 > IgG1 and rarely IgG2 activate the classical pathway while IgG4, IgA and IgD activate the alternative pathway.
- IgG found in maximum concentration.
- IgM is elevated in primary immune response, while IgG is elevated in secondary immune response.

42. d
- Fetal infection is characterized by increased in IgM.

43. a

44. d
- Follow section, **age of Ig production and Fig. 19.8** for explanation of answers of MCQs 43–44.

45. b
- IgD is the least important Ig in human being.

46. b and c
- IgD and IgM are the surface Igs and they are fixed with B cells by specific receptors called B cell receptor (BCR).

47. d
- Idiotype of Ig are due to change in variable region of H chain, which is present at aminoterminus end or N terminal.

48. a
49. b
50. a
- Idiotypes are due to change in variable region, isotypes are due to change in constant region of H chain. Allotypes are present in some members of species. Follow section, Ig specificities for more explanation of answers of MCQs 48–50.

51. c → Partially correct
- Isotypes of Ig are due to change in constant region of H chain only, not due to change in L chain.

52. b
- Difficult to judge the answer, but light chain is the correct answer.
- Bence Jones protein is an abnormal Ig, formed due to excessive production of L chain (either kappa or lambda forms in any one patient, but never both). It is detected from urine by coagulation (precipitation) when heated to 50°C but redissolving at 70°C.

53. c
- Follow section, **abnormal Ig (Bence Jones protein)** for explanation.

54. c and d
- Genes for H chain, K L chain and y L chain are located not on same chromosomes but on different chromosomes like number 14, 2 and 22 respectively.
- Different chains of Ig are formed by genetic rearrangement.
- Follow section, **genes of Ig** for more explanation.

55. a and c
- Follow section, **genes of Ig (Ig class switching)** for explanation.

56. c
- Follow section, **genes of Ig (Differential Ig RNA processing)** for explanation.

57. a, c and e
- Follow section, **genes of Ig (antibody diversity → mechanisms)** for explanation.

58. b
- Follow section, **monoclonal antibody (history)** for explanation.

59. c
- Hybridoma technique is used to obtain specific antibody not for specific antigen, complement or for interleukins.

60. d
- Follow section, **monoclonal antibody (definitions)** for explanation.

61. a
- Follow section, **monoclonal antibody (technique)** for explanation.

62. b
- Myeloma cells are deficient with HGPRTase so they cannot grow on HAT medium. Follow section, **monoclonal antibody** for more explanation.

63. d
- Follow section, **monoclonal antibody (uses)** for explanation.

Antigen-antibody (Ag-Ab) Reactions

INTRODUCTION

Types: Antigen and antibody combine with each other specifically and in observable manner. Different types of Ag-Ab reactions are shown in **Flowchart 20.1**.

Stages of Ag-Ab reactions: Three stages:

1. **Primary stage:** It is the initial interaction, without any visible effects. Reaction is rapid, and occurs even at low temperatures and reversible. Ag-Ab complex formation is possible by weaker intermolecular forces like ionic (electrostatic) bonds, van der Waal's forces and hydrogen bonds. Reaction is detected by estimating free and bound antigens or antibodies separately in reaction mixture by use of markers such as radioactive isotopes, fluorescent dyes or ferritin.

2. **Secondary stage:** It occurs in most instances, but not at all times. It includes demonstrable events (*in vitro*) such as precipitation, agglutination, lysis of cells, killing of live antigens, neutralization of toxins and other biologically active antigens, complement fixation, immobilization of motile organisms and enhancement of phagocytosis. Names of Ag and Ab as per reaction are mentioned in **Table 20.1**. Participation by different antibodies in different reaction is mentioned in **Table 19.1**.

3. **Tertiary stage (*In vivo*):** *In vivo* type Ag-Ab reactions initiate chain reactions that lead to tissue protection (immunity by neutralization or destruction of antigens) or lead to tissue damage (hypersensitivity or autoimmunity).

General features

1. **Reactions are specific:** Antigen combines only with its homologous antibody and *vice versa*. However, specificity is not absolute and cross reactions may occur due to antigenic similarity.

2. **Participation in reaction:** Not only fragment but entire molecule reacts. When an antigenic determinant which is present on large molecule/on 'carrier' particle reacts with its antibody, entire molecule/particle is agglutinated. Both antigens and antibodies participate in the formation of agglutinates or precipitates.

3. **No denaturation:** Denaturation of antigen or antibody do not occurs during reaction.

Flowchart 20.1: Antigen-antibody reactions

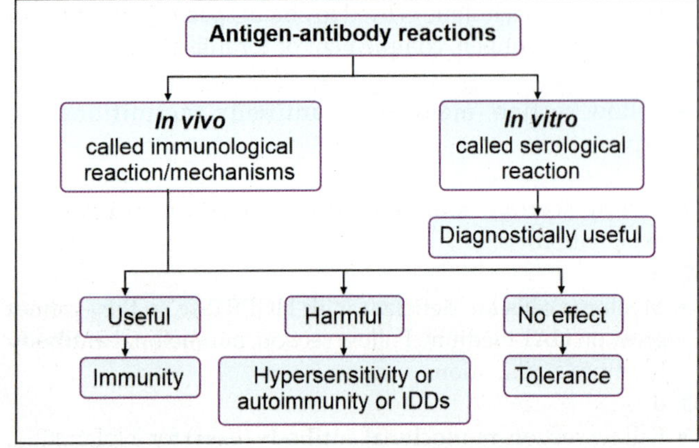

TABLE 20.1: Name of Ag and Ab as per reaction		
Reaction	**Ag**	**Ab**
Precipitation	Precipitinogen	Precipitin
Agglutination	Agglutinogen	Agglutinin

4. **Combination between Ag and Ab:** It occurs at surface. It is reversible and firm but influenced by affinity and avidity of reaction.
   - **Affinity:** It is the intensity of attraction between Ag-Ab. It depends on closeness between epitope and paratope. It is a quantitative measurement. Low affinity → weak binding → early dissociation. High affinity → strong binding → long association.
   - **Avidity:** It is the strength of bond to remain together after formation of Ag-Ab complex. IgM has low affinity than IgG, but high avidity enables it to bind very effectively with Ag.
5. **Proportions of Ag-Ab combination:** Ag-Ab can combine in varying proportions. Both antigens and antibodies are multivalent. Antibodies are generally bivalent, though IgM molecules may have five or ten combining sites. Antigens may have valences up to hundreds.

Measurement

- **Unit of measurement:** Antigen-antibody reactions are measured in mass or units or titer.
- **Titer:** It is the highest dilution of serum that shows an observable reaction in particular test. It is done for both Ag and Ab. It is influenced by nature of antigen, quantity of antigen, type of test and conditions of test.
- **Sensitivity:** It is an ability of a test to detect even very minute quantities of Ag or Ab from sample. High sensitivity of test minimizes the false-negative result.
- **Specificity:** It is an ability of test to detect reactions between homologous Ag and Ab only. High specificity of test minimizes the false-positive result.
- **Proportion:** Sensitivity and specificity of test are in inverse proportion.

SEROLOGICAL REACTIONS

Different types of serological reactions or tests are described below.

Precipitation Reaction

Definitions

Precipitation: Formation of an insoluble precipitate due to soluble Ag-Ab complex in presence of electrolytes (NaCl) at suitable temperature and pH called precipitation. Precipitation occurs at the bottom of tube in liquid media or in gels such as polyacrylamide, agar or agarose.

Flocculation: When precipitate remains suspended as floccules instead of sedimentation, reaction is called flocculation.

> **Note: Differences between precipitation and flocculation**
> **Precipitation:** Ag-Ab complex sediment at bottom of mixture. Mostly done in tube.
> **Flocculation:** Ag-Ab complex float at top of mixture. It is performed in tube and slide.

Zone Phenomenon

Definition: Amount of precipitate/agglutinate formed is influenced by relative proportions of Ags and Abs. Increasing quantities of antigens are added to same amount of antiserum in different tubes, precipitation will be found to occur most rapidly and abundantly in one of the middle tubes in which antigens and antibodies are present in optimal or equivalent proportions called zone of equivalence. In preceding tubes in which antibodies are in excess called prozone (Ab excess) phenomenon and in later tubes in which antigens are in excess called postzone (Ag excess) phenomenon, precipitation will be weak or even absent. For given antigen-antibody system, optimal or equivalent ratio will be constant, irrespective of quantity of reactants.

Clinical significance: Sera rich in antibodies may sometimes give false-negative precipitation or agglutination, unless several dilutions are tested.

Correction of prozone/post zone phenomenon: It is done by serial dilution of serum.

Mechanism of Precipitation (Lattice Hypothesis)

In 1934 Marrack proposed the lattice hypothesis to explain the mechanism of precipitation (also related to agglutination). According to him precipitate (agglutinate) results when large lattice formed between Ag and Ab. It is possible in zone of equivalence (**Fig. 20.1**) while in zone of Ab or Ag excess lattice does not enlarge.

Uses or Applications of Precipitation Reaction

1. Diagnostic: For detection of antigens, toxigenicity test in *C. diphtheriae*, VDRL testing in syphilis, etc.
2. Forensic application: In identification of blood and seminal stains.
3. Testing for food adulterants.
4. Typing or grouping of bacteria: For example, Lancefield grouping of streptococci.

Testing Methods or Procedures

Two types like qualitative test (detects the presence or absence of Ag/Ab) and quantitative test (detects the exact titer of Ag/Ab).

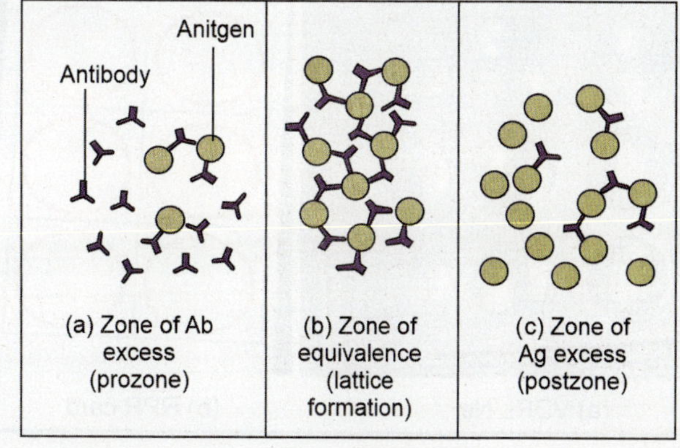

Fig. 20.1: Lattice hypothesis

Ring Test

It includes layering antigen solution over a column of antiserum in a narrow tube. A precipitate forms at junction of two liquids. Following are the examples of ring tests.

1. Ascoli's thermoprecipitin test.
2. Lancefield grouping of *Streptococcus*.

Slide Test

Drop of antigen and antiserum are placed on slide and mixed by shaking, floccules appear. Following are the examples of slide tests:

1. VDRL test
2. RPR test
3. TRUST.

VDRL test

- **History:** Initially test was developed by August Paul von Wasserman with the aid of Albert Neisser in 1906. VDRL test done today was developed by Harris, Rosenberg and Riedel in 1946.
- **Full name:** Venereal Disease Research Laboratory, New York, where the test was developed.
- **Principle:** It is the nonspecific (nontreponemal) test, based on slide flocculation reaction.
- **Antigen(s):** It is prepared from bovine heart and lipid in nature hence called cardiolipin (cardia for heart and lipin for lipid) and added with lecithin and cholesterol hence also called cholelecithin Ag. It is standardized by Pangborn in 1945. In India it is prepared by Institute of serology, Kolkata.
- **Antibody (Antibodies):** It called reagin Ab present in patient's serum which is IgG or rarely IgM in nature. IgE antibody of type I hypersensitivity (atopy) also called reagin antibody; however, there is no correlation between two.
- **Procedures:** Following are two procedures:
  1. Qualitative method: It is performed on VDRL tile/slide **(Fig. 20.2a)** contains 12 concavities/circles. Put 1 drop of heat inactivated serum (56°C for 30 minutes) in one circle. Heating is required to remove nonspecific inhibitors. Mix serum with one drop of cardiolipin antigen solution. Put negative and positive controls on separate circles. Rotate manually or on VDRL rotator at 180 rpm for 4 min. Read the results under a low power microscope.
  2. Quantitative method: All samples reactive by qualitative method are subjected to the quantitative method. Put one drop of normal saline (ns) in circle numbered as 1, 2, 3, 4… Place one drop of serum in 1st circle. Prepare the serial dilution by transferring the drop of serum from 1st to 2nd, 2nd to 3rd and so….. Add one drop of Ag in each circle and mix properly. Rotate manually or on VDRL rotator at 180 rpm for 4 min. Read the results under a low power microscope.
- **Results:** Follow **Table 20.2 and Fig. 20.3**.
- **Interpretation of results**
  - Positive: It indicates syphilis. Significant titer is 1:8.
  - Negative: It indicates absence of syphilis.
- **Uses of VDRL test**
  - Diagnostic: It is used to diagnose syphilis. Also useful to diagnose neurosyphilis. It can be performed from CSF called VDRL-CSF, where prior heating is not required.
  - Therapeutic: It is useful to monitor the treatment of syphilis where titer decline following successful therapy.
- **Advantages of VDRL test:** It is simple, rapid and most widely used test.
- **Disadvantages of VDRL test:** Chances of biological false-positive (BFP) reaction are more in VDRL test. Positive reaction with nontreponemal tests and negative result with treponemal tests in absence of past/present infection and not due to technical faults called BFP reaction. BFP occurs in 1% normal serum. BFP antibody is mainly IgM type. Cardiolipin Ag present in treponemal and also in mammalian host

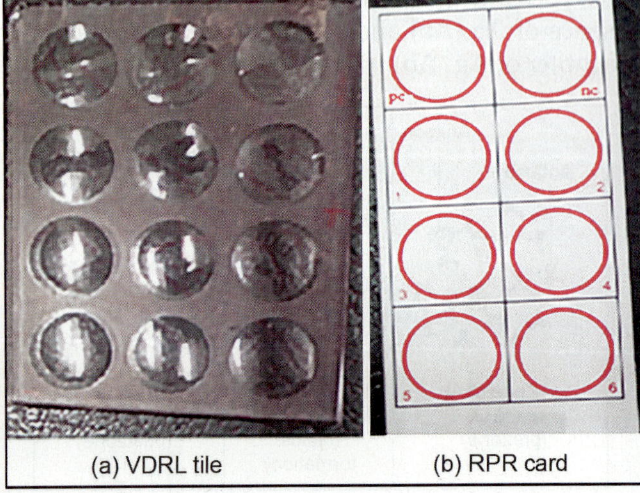

Fig. 20.2: VDRL tile and RPR card

| TABLE 20.2: Results of qualitative VDRL test ||
Results/reports	Reading
Nonreactive (NR)	No floccules
Weakly reactive (W)	Small floccules
Reactive (R)	Medium/large floccules

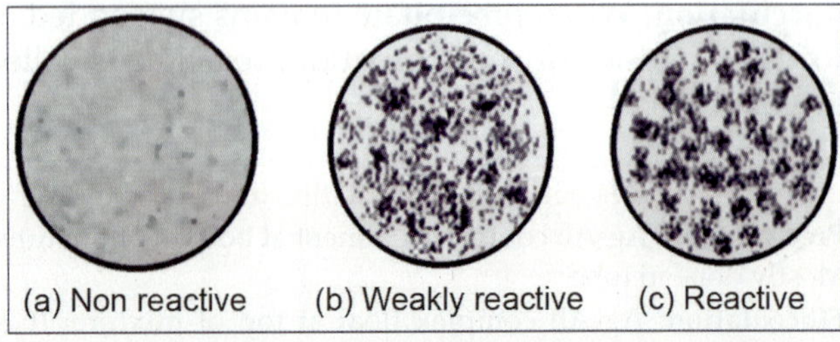

Fig. 20.3: Results of qualitative VDRL test

tissue which induces reagin Ab production. BFP occurs due to cross reaction of cardiolipin Ag with nonspecific antibodies produced by other infections. Following are two types of BFP reaction.

- Acute BFP: Antibodies last for few weeks/months like in infections, inflammation and traumas.
- Chronic BFP: Antibodies last for >6 months in chronic conditions like SLE, lepromatous leprosy (LL), hepatitis, infectious mononucleosis, malaria, tropical eosinophilia, etc.
- Others types: All serological tests of syphilis are false-positive in endemic treponematoses like yaws, pinta and endemic syphilis. BFP also occurs in HIV, genital herpes, measles, *M. pneumoniae*, relapsing fever, parenteral drugs, etc.

- **Differences between VDRL test and RPR test:** Follow **Table 20.3**.

RPR test

- **Full name:** Rapid plasma reagin test.
- **Principle:** Based on slide flocculation test.
- **Differences with VDRL:** It is similar to VDRL test, but some differences are listed in **Table 20.3**.
- **Use:** Useful to diagnose syphilis.

TRUST: It is same as like RPR test but Ag is coated with toluidine red instead of carbon particle and resulting floccules are in red color.

Tube Test

Following are the examples of tube tests:

1. Kahn test: Useful for diagnosis of syphilis.
2. Quantitative tube flocculation test: Serial dilutions of toxin/toxoid added to tubes contain fixed quantity of antitoxin. Amount of toxin or toxoid that flocculates optimally with 1 unit of antitoxin called Lf dose.

Immunodiffusion (Precipitation in Gel)

It is performed in soft (1%) agar or agarose gel. Advantages of allowing precipitation in gel than liquid medium are (1) reaction is visible as distinct band of precipitation, which is stable and can be stained for preservation. (2) Each antigen-antibody reaction gives line of precipitation, numbers of different antigens in mixture can be observed. (3) It indicates identity, nonidentity and cross reaction between different antigens. Following are the examples of immunodiffusion tests (modifications of immunodiffusion test):

1. Single diffusion in single dimension (Oudin procedure).
2. Double diffusion in single dimension (Oakley-Fulthorpe procedure).
3. Single diffusion in double dimensions (radial immunodiffusion).
4. Double diffusion in double dimensions (Ouchterlony procedure).
5. Immunoelectrophoresis.

Single diffusion in single dimension (Oudin procedure): Incorporate the Ab in agar gel in test tube. Add the Ag solution over it. Ag diffuses downward through agar gel and forms the line of precipitation as shown in **Fig. 20.4a**.

Double diffusion in single dimension (Oakley-Fulthorpe procedure): Incorporate the Ab in agar gel in test tube. Place the column of plain agar above Ab solution. Ag is layered on top of this. Ag-Ab move towards each other through column of plain agar and form band of precipitate where they meet at optimum proportion as shown in **Fig. 20.4b**.

Single diffusion in double dimensions (radial immunodiffusion): Ab is incorporated in agar gel poured on flat surface (slide or Petri dish). Ag is added to well cut on surface of gel. Ag diffuses radially from well and forms ring-shaped bands of precipitation (halos) concentrically around the well as shown in **Fig. 20.5**. Diameter of halo estimates the antigen concentration (Ag estimation). Following are the uses of this test:

- Used to estimate of Ig classes in sera.
- Used to screen sera for antibodies to influenza viruses.

TABLE 20.3: Differences between VDRL and RPR test		
Features	**VDRL test**	**RPR test**
Samples	• Heated serum • No plasma • CSF	• Unheated serum • Plasma • No CSF
Preheating of serum	Required to remove nonspecific inhibitors	Not required as choline chloride is used to remove inhibitors
Material required	Glass slide with 12 concavities/depression with 14 mm size called VDRL tile **(Fig. 20.2a)**	Plastic disposable slide with 10 circles with 18 mm size called RPR card **(Fig. 20.2b)**
Nature of Ag	Freshly made and should be used within 24 hours.	Ag can be stabilized by EDTA, stored at 4–10°C for 4–6 months and can be used long
Rotation of slide	4 minutes	8 minutes
Clumps	Clumps are small and white. Uncoated Ag	Large and black because Ag coated with carbon particles
Reading of results	By microscope	By naked eyes
Sensitivity in primary syphilis	78%	86%
Cost	Cheaper and 250 tests can be performed from 1 vial, so used as screening test	Expensive and preferred when sample size is less

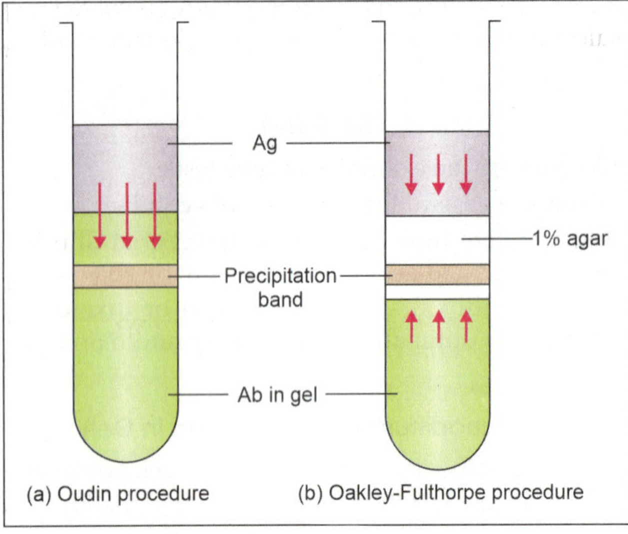

Fig. 20.4: Oudin and Oakley-Fulthorpe procedure

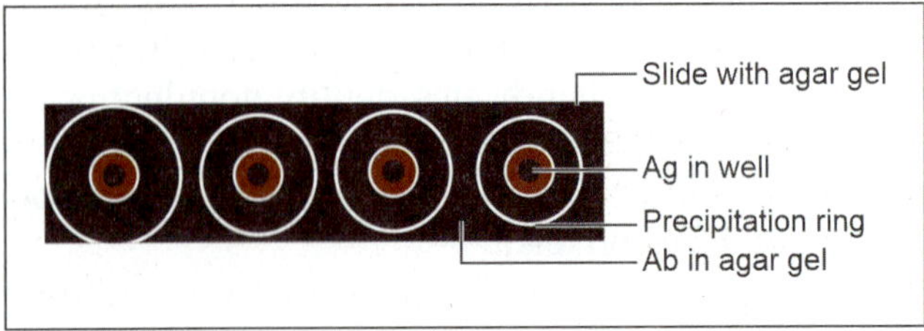

Fig. 20.5: Radial immunodiffusion

Double diffusion in double dimensions (Ouchterlony procedure): Agar gel is poured on slide and wells are cut. Ab is placed in central well and different antigens in surrounding wells. If two adjacent antigens are identical, lines of precipitate formed by them will fuse. If they are unrelated, lines will cross each other. Cross-reaction or partial identity characterized by spur formation. Different results are shown in **Fig. 20.6.** Following are the uses of this test.

- To compare different antigens and antisera directly.
- Toxigenicity testing, e.g., Elek's gel precipitations test in *C. diphtheriae.*

Immunoelectrophoresis

- **Principle:** It involves electrophoretic separation of composite antigen (serum) into its constituent proteins, followed by immunodiffusion against its antiserum, resulting in separate precipitin lines,

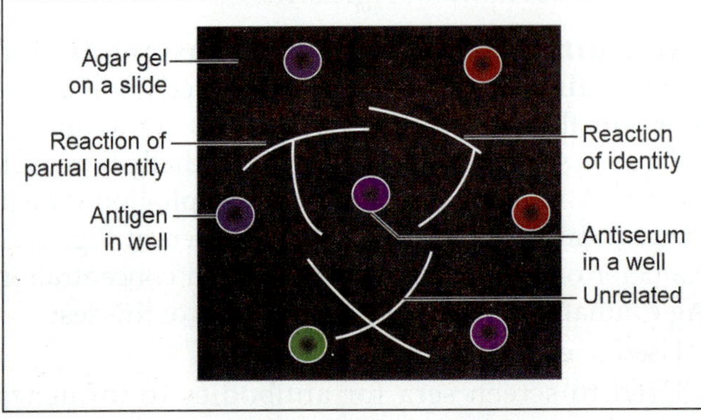

Fig. 20.6: Ouchterlony procedure

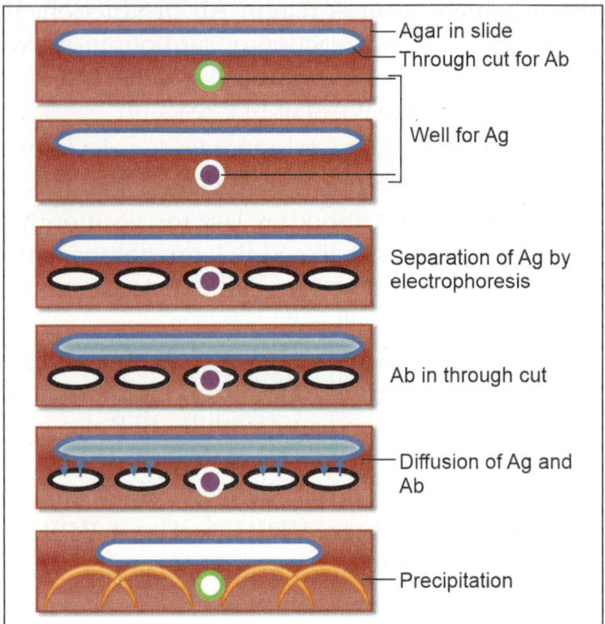

Fig. 20.7: Immunoelectrophoresis

indicating reaction between each individual protein with its antibody.

- **Steps:** Incorporate agar or agarose gel on a slide. Made an Ag well and an Ab trough cut on it. Place the antigen in well and separate by electrophoresis. Place Ab in trough cut. Allow the diffusion. Resulting formation of precipitin line (**Fig. 20.7**) can be photographed and slide dried, stained and preserved.
- **Uses:** Test is used in identification and approximate quantitation of various normal and abnormal proteins (heavy-chain diseases) of serum.

Electroimmunodiffusion

Developments of precipitin lines can be speeded up by electrically driving antigen and antibody. Following are the different types of tests based on electroimmunodiffusion tests:

1. Counter immunoelectrophoresis (CIE, countercurrent immunoelectrophoresis).
2. One-dimensional single electroimmunodiffusion (rocket electrophoresis).
3. Laurell's two-dimensional electrophoresis.

Counter immunoelectrophoresis (CIE, counter-current immunoelectrophoresis): Simultaneous electrophoresis of antigen and antibody in gel in opposite directions resulting in precipitation at mid-point between them as shown in **Fig. 20.8.** Test is used to detect the various antigens such as specific antigens of *C. neoformans* in CSF, alpha fetoprotein (AFP) in serum and Ag of *N. meningitidis* in CSF.

One-dimensional single electroimmunodiffusion (rocket electrophoresis): Place the agar gel on glass slide. Incorporate the antiserum in to agar. Antigen, in increasing concentrations, is placed in wells punched in set gel. Ag is electrophoresed into an Ab containing agarose. It forms rocket like pattern of

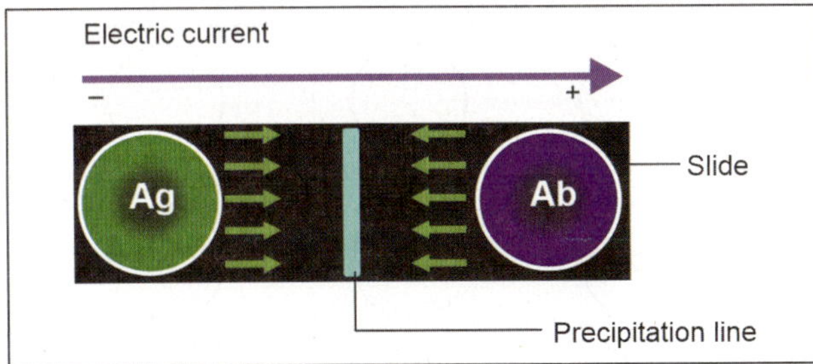
Fig. 20.8: CIE

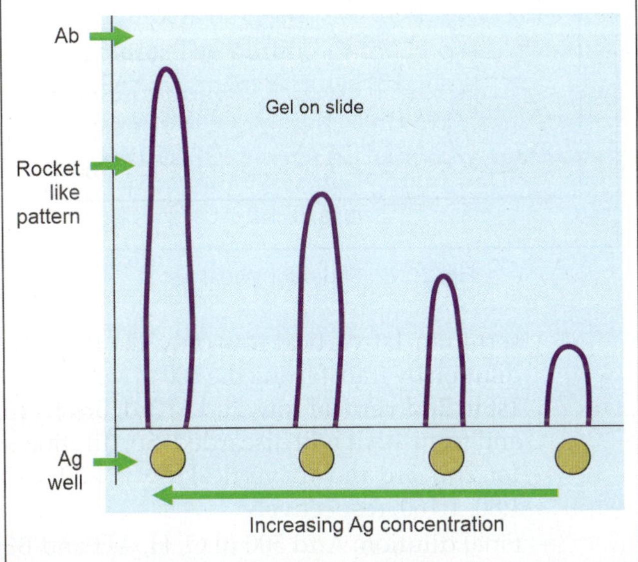
Fig. 20.9: Rocket electrophoresis

immunoprecipitation as shown in **Fig. 20.9.** Test is used for quantitative estimation of antigens.

Laurell's two-dimensional electrophoresis: Ag mixture is electrophoretically separated in direction perpendicular to that of final rocket stage as shown in **Fig. 20.10.** Test is used to quantitate each of several antigens in mixture.

Agglutination Reaction

Definition

When particular Ag is mixed with its Ab in presence of electrolytes at suitable temperature and pH, it gets clumped or agglutinated called agglutination.

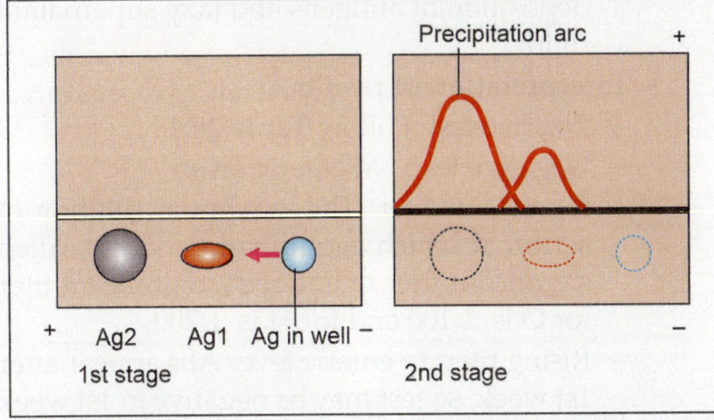
Fig. 20.10: Laurell's two-dimensional electrophoresis

Properties

It is more sensitive than precipitation for detection of antibodies. It also follows the zone phenomenon. Agglutination occurs optimally in zone of equivalence. It sediments at the bottom of tube. Incomplete or monovalent antibodies do not cause agglutination, though they combine with antigens and act as blocking antibodies, inhibiting agglutination by complete antibodies added subsequently. Blocking antibodies occur in some disease like brucellosis. Incomplete antibodies may be detected by doing the test in hypertonic (5%) saline, albumin saline or by Coombs' test.

Testing Methods or Procedures

Two types like qualitative test and quantitative test.

Types of Serological Tests Based on Agglutination

According to Method of Performance

Two types like slide agglutination method and tube agglutination method as described below.

Slide agglutination method: Mix one drop of Ab with particular Ag on slide or tile. Rotate the slide for few minutes and then observe for the clumps. Presence of clumps indicates (**Fig. 20.11a**) positive result. Clumps are visible to naked eye, may require microscopic confirmation. Absence of clumps indicates negative (**Fig. 20.11b**) result. Following are the uses of this method:

1. Identification of bacterial isolates such as serotyping of S. Typhi, *Shigella* spp. and *V. cholerae.*
2. Slide Widal test for S. Typhi and S. Paratyphi.
3. RF (Rheumatoid Factor) test for RA factor.
4. ASO test for streptococcal infection.
5. Blood grouping and cross-matching.

Tube agglutination method: It is the standard quantitative method for measurement of antibodies. Following are the uses of this method:

1. Tube Widal test for enteric fever.
2. Quantitative Ab diagnostic test for brucellosis.
3. Heterophile Ag detection by following tests.
   - Weil-Felix reaction for typhus fever.
   - Paul Bunnel test for infectious mononucleosis (**Ch. 78**).

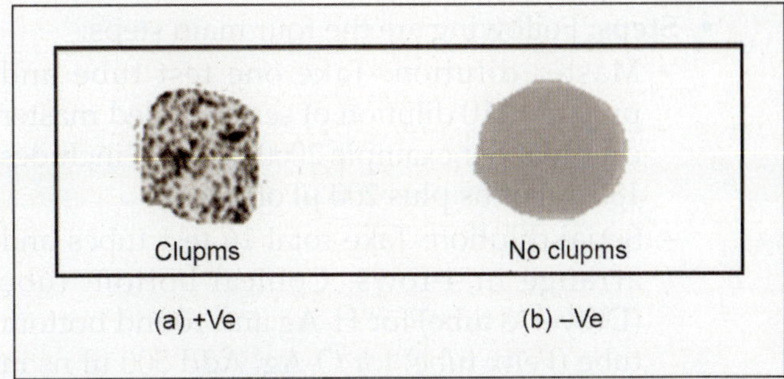
Fig. 20.11: Slide agglutination

- *Streptococcus* MG agglutination test for primary atypical pneumonia (**Ch. 74**).
- Cold agglutination test for primary atypical pneumonia (**Ch. 74**).
- Nonspecific CFT with WKK antigen (Witebusky, Kligenstein and Kuhn antigen) for leishmaniasis (**Ch. 95**).

Widal test

- **History:** Test was developed by Fernand Widal in 1896.
- **Principle:** It based on slide and tube agglutination reaction.
- **Antigens:** Somatic (O) Ag and flagellar (H) Ag of *S.* Typhi antigens. O-Ag (group specific) of *S.* Paratyphi can cross react with O-Ag of *S.* Typhi hence not included in widal test, only flagellar (species specific) Ags of *S.* Paratyphi like AH-Ag and BH-Ag are included in the Widal reaction. H-Ag is heat labile destroyed by boiling or by treatment with alcohol but not by formaldehyde. O-Ag is heat stable, alcohol stable but formaldehyde labile. Ags are prepared by following methods:
  - H-Ag prepared by treating saline suspension or broth culture of *S.* Typhi 901 H strain with 0.1% formalin.
  - O-Ag prepared by growing the *S.* Typhi 901 O strain on phenol agar (to inhibit H Ag → scrapped of in saline → mixed with 20 times its volume of absolute alcohol → heated at 40–50°C for 30 minutes → centrifuged → re-suspend the deposit in saline → add the preservative like chloroform).
- **Antibodies:** Patient serum contains antibodies.
- **Other materials:** Normal saline (ns).
- **Methods:** Two types of Widal test as follows:
  1. **Slide (rapid/screening/qualitative) Widal test**
     - **Steps:** Fig. 20.12 shows the different types of slides available commercially for slide Widal test. Mix one drop of serum and one drop of Ag (O, H, AH, BH) in respective circle on slide. Mix properly and rotate for one minute and examine for clumps. Put negative and positive controls also.
     - **Results:** Presence of clumps indicates positive result and absence of clumps indicates negative result.
  2. **Tube (quantitative) Widal test**
     - **Steps:** Following are the four main steps:
       - Master dilution: Take one test tube and prepare 1:10 dilution of serum called master dilution. For example 2000 µl (2 ml) includes 1800 µl of ns plus 200 µl of serum.
       - Serial dilution: Take total 16 test tubes and arrange in 4-rows. Conical bottom tube (Dreyer'e tube) for H-Ag and round bottom tube (Felix tube) for O-Ag. Add 500 µl ns in each test tube. Add 500 µl master diluted

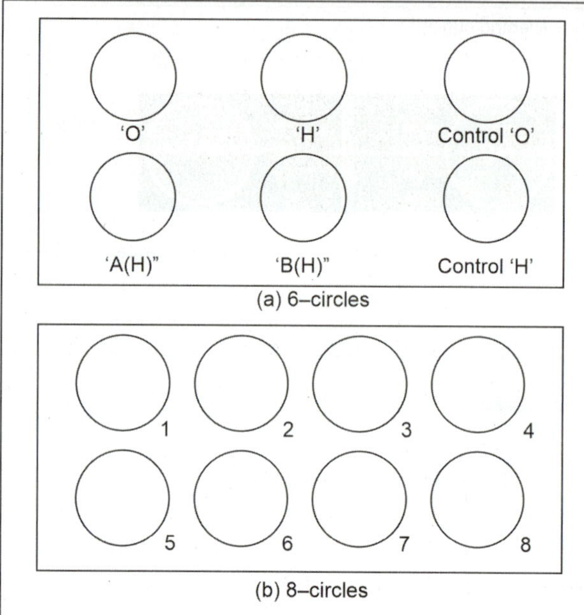

Fig. 20.12: Slides for Widal test

serum in 1st vertical row only. Do the serial dilution by transferring the 500 µl serum from 1st to 2nd vertical row, 2nd to 3rd, 3rd to 4th and from 4th it will discarded. So, dilution in 1st, 2nd, 3rd and 4th vertical row is 1:20. 1:40, 1:80, 1:160, respectively.
  - Final dilution: Add 500 µl O, H, AH and BH Ag in 1st, 2nd, 3rd and 4th horizontal row respectively. So, final dilution in 1st, 2nd, 3rd and 4th vertical row is 1:40, 1:80, 1:160 and 1:320, respectively. Put negative and positive controls also.
  - Incubation: Incubate the all tubes for 24 hours at 37°C.
- **Results**
  - Positive O-Ag agglutination: Disc like clumps or chalky granular clumps are formed at bottom of tube with clear supernatant fluid
  - Positive H-Ag agglutination: Spherical, granular, cotton woolly clumps are formed at the bottom of test tube with clear supernatant fluid
  - Negative O-Ag and H-Ag agglutination: Button formation at the bottom of tube due to deposition of antigens and lazy supernatant fluid.
- **Interpretation of results**
  - Positive test: Follow **Table 20.4**.
  - Negative test: No enteric fever
  - Significant titer: The maximum dilution of serum at which agglutination occurs called significant titer of antibody. Significant titer for O is 1:100 and for H is 1:200
  - Rising titer: In enteric fever Abs appear after 1st week, so test may be negative in 1st week and become positive in 2nd week. Titer rise

TABLE 20.4: Interpretations of tube Widal test	
Rising titer	**Interpretation**
O and H	Typhoid fever
O and AH	Paratyphoid fever A
O and BH	Paratyphoid fever B
O (Only)	Group specific (typhoidal group/enteric fever group)
H (Only)	Species specific, convalescent phase or cross reaction by other related bacteria called anamnestic reaction or false-positive result
H, AH and BH	TAB vaccination

in 3rd or 4th week (four fold rise) after which it decline gradually. If sample is taken in late stage, rising titer may not be demonstrated.

– False-positive result: Rising titer of O and H agglutinins may encounter in condition other than enteric fever called anamnestic reaction or false-positive result. It occurs in different conditions such as in normal serum, repeated subclinical infections, immunization with TAB vaccine (immunization with TAB vaccine shows high titer against *S.* Typhi. *S.* Paratyphi A and *S.* Paratyphi B, while case shows high titer only against infecting species), past history of enteric fever in healthy carrier, fimbrial Ags, tuberculosis, malaria, dengue, brucellosis, influenza, nephritic syndrome, infection with related bacterium from Enterobacticeae family, person with immunological diseases (RA and RF) and also in person from endemic area. Base line titer may be affected by endemicity of disease and antibody may be present due to inapparent infection, immunization or prior infection, especially in high endemic areas. So, distribution of agglutinin in normal sera is different in different areas.

– Differentiation between false-positive result and true case (fourfold rise): Rising titer due to anamnestic reaction may fall after 1 week while titer due to true case remains increased by four fold after 1 week, so repeated testing of paired sera at 1 week interval is more significant than single high titer.

– False-negative result: It may encounter following antibiotic treatment or by masking of O Ag by Vi Ag.

– Nonspecific Widal activity: It may occur in narcotics addicts, lipemic serum, hemolyzed samples and contaminated serum.

– No response to *Salmonella* antibodies: About 5–10% patient doesn't show any response in Widal test even though *S.* Typhi infection documented.

- **Use of Widal test:** It used to diagnose the enteric fever (typhoid and paratyphoid fever A and B).

Weil-Felix reaction

- **History:** Test was developed by Weil and Felix in 1916.
- **Principle:** It based on tube agglutination reaction.
- **Antigens:** It includes heterophile Ags like OX 19 and OX2 of nonmotile *Proteus vulgaris* and OX K (O antigens) of nonmotile *Proteus mirabilis*. Sharing of a common alkali-stable carbohydrate antigen by rickettsiae and *Proteus* makes the basis of this test. Antigens are prepared from *Proteus* strain which is isolated from urine of patient.
- **Antibodies:** Patient's serum used as antibodies. Antibodies appear rapidly during the course of the disease. They reach peak titer of up to 1:1000 or 1:5000 by the second week and declines rapidly during convalescence phase.
- **Other materials required:** Normal saline (ns).
- **Steps**
  - Master dilution: Take one test tube and prepare 1:10 dilution of serum called master dilution. For example 2000 µl (2 ml) includes 1800 µl of ns plus 200 µl of serum.
  - Serial dilution: Take total 12 test tubes and arrange in 3-rows. Add 500 µl ns in each test tube. Add 500 µl master diluted serum in 1st vertical row only. Do the serial dilution by transferring the 500 µl serum from 1st to 2nd vertical row, 2nd to 3rd, 3rd to 4th and from 4th it will discarded. So dilution in 1st, 2nd, 3rd and 4th vertical row is 1:20. 1:40, 1:80, 1:160, respectively.
  - Final dilution: Add 500 µl OX 19, OX2 and OX K Ag in 1st, 2nd and 3rd horizontal row respectively. So final dilution in 1st, 2nd, 3rd and 4th vertical row is 1:40, 1:80, 1:160, and 1: 320 respectively. Put negative and positive controls also.
  - Incubation: Incubate the all tubes for 24 hours at 37°C.
- **Results**
  - Positive test: Disc like clumps at the bottom of test tube.
  - Negative test: No clumps.
- **Interpretation of results:** Follow **Table 20.5**.
  - Positive test: Sera from epidemic and endemic typhus agglutinate OX 19 and sometimes OX 2. OX K agglutinins are found only in scrub typhus. In tick borne spotted fever both OX 19 and OX 2 are agglutinated.
  - Negative test: The test is negative or only weakly positive for OX 19 antigen in latent typhus (Brill Zinsser disease). The test is not diagnostic (negative) in rickettsial pox, ehrlichiosis, Q fever and trench fever.
  - Significant titer: It is about 1:80.
  - False-positive reaction: It may occur in some cases of urinary or other infections by *Proteus*, in typhoid fever and liver diseases.

TABLE 20.5: Interpretation of Weil-Felix reaction

Bacteria	Disease	Agglutination pattern		
		OX 19	OX 2	OX K
Rickettsia spp.	Epidemic typhus	+++	+	–
	Endemic typhus	+++	+/–	–
	Spotted fever	++	++	–
O. tsutsugamushi	Scrub typhus	–	–	+++

- **Use:** For the diagnosis and differentiation of some rickettsial diseases as shown in **Table 20.5**.

According to Free or Fixed Ag/Ab with Carrier Particles

According to free Ag/Ab: Here antiserum is mixed with whole pathogen/cell. In this type of reaction whole pathogen/cell is used as an Ag and not coated with any carrier particle, hence called active (direct) agglutination. Following are the different types of tests.

1. **Active agglutination/direct agglutination:**
   - **Principle:** When antiserum is mixed with whole pathogen it called active (direct) agglutination. In this type of reaction whole pathogen is used as an Ag and not coated with any carrier particle.
   - **Use:** Identification of bacterial isolates like serotyping of *S. Typhi*, *Shigella* spp., and *V. cholerae*.
2. **Active (direct) hemagglutination:**
   - **Principle:** When antiserum is mixed with whole RBC called active (direct) hemagglutination. In this type of reaction whole RBC is used as an Ag and not coated with any carrier particle.
   - **Uses**
     - Blood grouping test: Whole RBCs (free) are mixed with anti-A, anti-B and anti-O sera to detect the blood groups.
     - Cold agglutination test: It is positive in primary atypical pneumonia which is caused by *M. pneumoniae*. Patient's sera agglutinate to human O group RBCs at 4°C hence called cold agglutination. Clumps are dissociated at 37°C.

According to fixed Ag/Ab with carrier particle: Here Ags/Abs are not free but fixed with carrier particles hence called passive (indirect) agglutination. Different types of carrier particles used to fix Ags or Abs are mentioned in box. Following are the different types of tests according to carrier particles used:

1. Latex agglutination tests.
2. Coagglutination tests (agglutination test by using protein A of *Staph. aureus* as carrier particles).
3. Hemagglutination tests (agglutination tests by using RBCs as carrier particles).

Note: Carrier particles

1. Latex particles (latex agglutination test): They are made up from polystyrene latex, spherical in shape and 0.8–1.0 μm in size.
2. Protein A of *Staph. aureus* (coagglutination test).

---

3. RBCs (hemagglutination test).
4. Bentonite particle.
5. Carbon particle: Not for agglutination but for flocculation like RPR test (described earlier).
6. Toludine red particle: Not for agglutination but for flocculation like TRUST. It is same as like RPR test but Ag is coated with toluidine red instead of carbon particle and resulting floccules are in red color.

1. **Latex agglutination tests:** Ag/Ab is fixed with latex particle. Subtypes are as follows.
   - **Passive agglutination/indirect agglutination:** By fixing antigen with latex particle antibody is detected. It is highly sensitive. It is less specific, so false-positive result may occur. Following are the uses of this test.
     - **Modified Rose Waller test**
       - **Synonym:** It also called RA (Rheumatoid arthritis) factor test or RF (Rheumatoid factor) test.
       - **Principle:** Rheumatoid arthritis is an autoimmune diseases characterized by presence of IgM type autoantibody against Fc portion of IgG called RF / RA factor. RF is actually an anti-globulin. RF (Ab) is detected by attaching globulin (Ag) with latex particle. White clumps are visible against black background of slide. RF (Ab) is also detected by attaching globulin (Ag) with bentonite particle.
       - **Steps:** Mix one drop of serum contains Ab with one drop of reagent (Ag attached to latex particle) on a slide. Rotate slide for 1 minute and examine for the result.
       - **Results (Fig. 20.13):** Formation of white clumps against black background suggests positive result while absence of clumps suggests negative result.
       - **Interpretation of results:** Positive test indicates the RA and negative indicates absence of RA. Significant titer is 8 IU/ml.
       - **Disadvantage:** Test is non-specific because it also found positive in other connective tissue disorders.
       - **Use:** It is used to diagnose the rheumatoid arthritis.
     - **ASO (Anti-streptolysin-O) test**
       - **Principle:** It is used in acute rheumatic fever (ARF) caused by *Streptococcus pyogenes*.

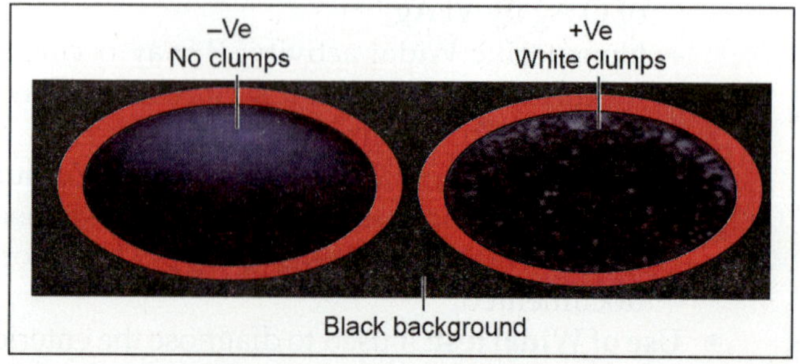

Fig. 20.13: Latex agglutination test

Essentials of Medical Microbiology

Streptococcus releases the exotoxin called streptolysin-O (Ag) which produces specific antibody called anti-streptolysin-O (ASO). ASO (Ab) is detected by attaching streptolysin-O (Ag) with latex particle. White clumps are visible against black background of slide.

- **Steps and results:** Same as modified Rose Waller test.
- **Interpretations of results:** Positive test indicates the streptococcal infection. Negative test suggests no streptococcal infection. Significant titer is 200 IU/ml.
- **Use:** Test is used to diagnose the ARF.
  - **Other tests based on passive (indirect) agglutination tests:** These include capillus test to detect HIV-Ab, detection of CRP, detection of hCG, etc.
- **Reverse passive agglutination:** Instead of antigen, antibody is adsorbed to latex particle to detect the antigen. Test is rapid, sensitive and specific. It is used for HBsAg detection in HBV infection.

2. **Coagglutination tests:** Instead of antigen, antibody is adsorbed to carrier particle like protein A of *Staph. aureus* to detect the antigen. It is similar or subtype of reverse passive agglutination, only difference is in carrier particle. Test is rapid, sensitive and specific. Following are the uses of coagglutination test.

- **S. Typhi detection:** *Staph. aureus* (Cowan I strain) which contains protein A is fixed with the Ab (IgG except IgG3) of *S. Typhi*. Mix such sensitized *Staph. aureus* with patient's serum contains Ag. Typhoid Ag combines with protein A bounded *S. Typhi* Ab and produces clumping of *Staph. aureus* called coagglutination. Test is positive in 1st week but not after that.
- **Other infections:** Co-agglutination also used for Ag detection in *Legionella pneumophilla*, *N. gonorrhoeae* and *Strept. pyogenes* infection.

3. **Hemagglutination tests:** Test called hemagglutination because carrier particle is RBC. Following are three sub-types.

- **Passive (indirect) hemagglutination:** RBC is coated with Ag from other organism to detect the Ab of that organism. Following are the examples of tests based on passive hemagglutination.

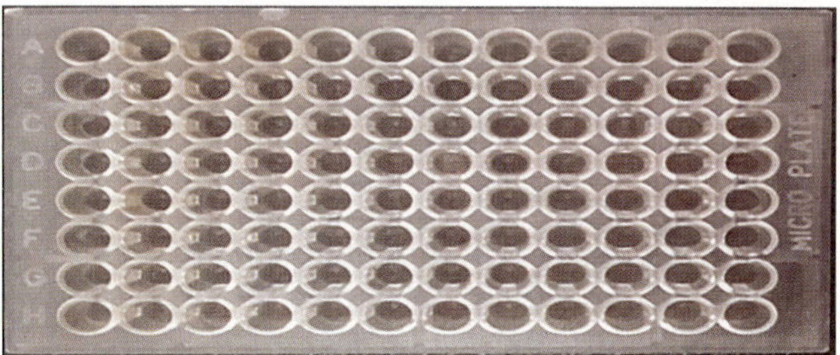

Fig. 20.14: RPHA microtiter plate

  - Rose Waller test: It is used for detection of RF or RA factor. Here, sheep RBC is coated with anti-erythrocyte Ab (amboceptor), which is used as an Ag to detect the RF (Ab).
  - TPHA (*Treponema pallidum* Hemagglutination) test: It is used to detect the specific Ab of *T. pallidum*, where RBC is coated with antigen of *T. pallidum*.
- **Reverse passive hemagglutination (RPHA) test:** Instead of antigen, RBC is coated with antibody to detect the antigen. **Figure 20.14** shows the RPHA microtiter plate with fixed wells (differentiated from ELISA microtiter plate where wells are free). RPHA test used to detect the HBsAg of HBV.

Antiglobulin Test

Synonym: Coombs' test.

History: It was invented by Coombs, Mourant and Race in 1945.

Principle: Detection of anti-Rh antibodies that do not agglutinate the Rh-positive RBCs in saline. When sera contain incomplete (blocking) anti-Rh antibodies are mixed with Rh-positive red cells, antibody globulins coat the surface of red cells, though they are not agglutinated. Such RBCs coated with Ab globulins are washed free of all unattached proteins and treated with rabbit antiserum against human γ-globulins (anti-globulin or Coombs' serum), cells are agglutinated.

Types: Coombs' test may be direct or indirect.

1. **Direct (*In vivo*) Coombs' test**

- **Steps (Fig. 20.15):** Sensitization of RBCs with incomplete antibodies takes place *in vivo* (immune mediated hemolytic anemia of newborn due to Rh incompatibility). When RBCs of case with

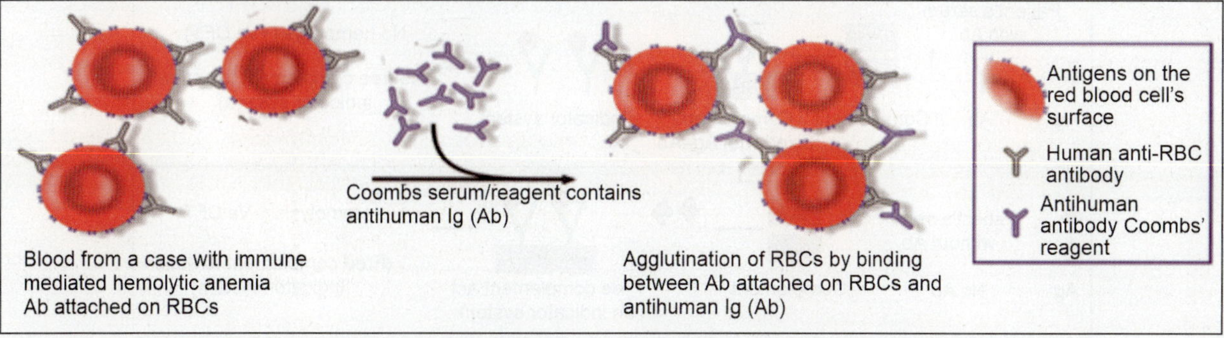

Fig. 20.15: Direct/*In vivo* Coombs' test

immune mediated hemolytic anemia are washed free of unattached proteins and mixed with drop of Coombs serum, agglutination results. Test is often negative in hemolytic disease due to ABO incompatibility.

- **Use:** To detect the immune mediated hemolytic anemia of newborn due to Rh incompatibility.

2. **Indirect (*In vitro*) Coombs' test:**
- **Steps (Fig. 20.16):** Sensitization of red cells with antibody globulin is performed *in vitro*. Other steps are same as direct test.
- **Uses:**
  - To detect any type of incomplete antibody (non-agglutinating), like in brucellosis.
  - To detect the anti-Rh Ab (free) in patient's serum.

Complement Fixation Test (CFT)

History: Bordet and Gengou (1901) described the complement fixation test by using the hemolytic indicator system.

Background: Complement is a system of different factors of serum which takes part in many immunological reactions and is absorbed during combination of antigens with antibodies.

Role of complement: It takes part in Ag-Ab reaction. In presence of appropriate antibodies, complement lyses erythrocytes, kills bacteria, immobilizes the motile organisms, promotes phagocytosis, do immune adherence and contributes to tissue damage in certain types of hypersensitivity.

Sensitivity of CFT: It is very versatile and capable of detecting as little as 0.04 mg of Ab and 0.1 mg of Ag.

Reagents of CFT: As mentioned below.
1. **Ag:** May be soluble or particulate.
2. **Ab:** Antiserum should be heated at 56°C (inactivated serum) for half an hour to destroy any complement activity if serum has and also to remove nonspecific inhibitors of complement present in serum (anti-complementary activity).
3. **Complement:** It is obtained from the guinea pig serum. As complement is heat labile, serum should be freshly drawn or lyophilized or frozen or preserve by Richardson's method.
4. **Indicator system:** It includes sensitized sheep RBC with amboceptor (amboceptor means rabbit Ab to sheep RBC).

Uses of CFT: To diagnose the various bacterial, viral, fungal and parasitic infections.

Serological tests based on CFT
1. **Wassermann reaction**
- **Principle:** It based on complement fixation.
- **Reagents:** As described above.
- **Steps:** Test consists of two steps.
  - **First step (Fig. 20.17):** Inactivated serum of patient is incubated at 37°C for 1 hour with Wassermann antigen and a fixed amount (two

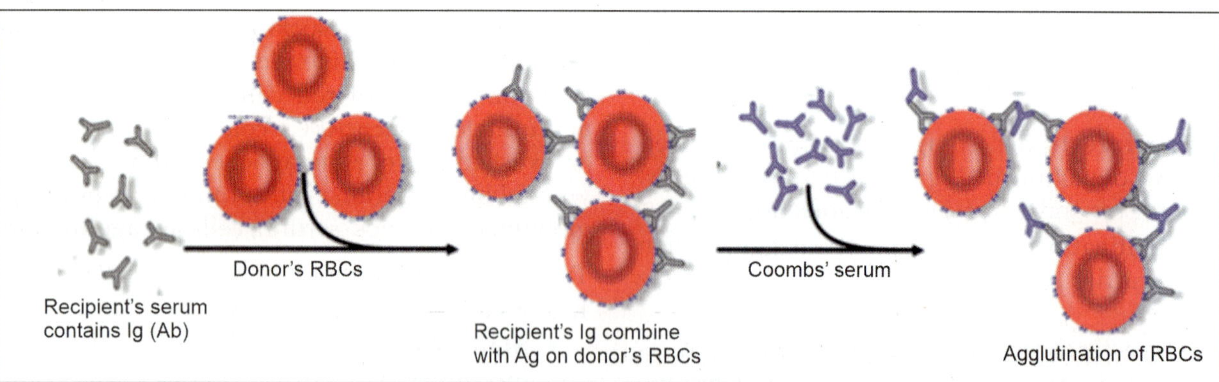

Fig. 20.16: Indirect/*In vitro* Coombs' test

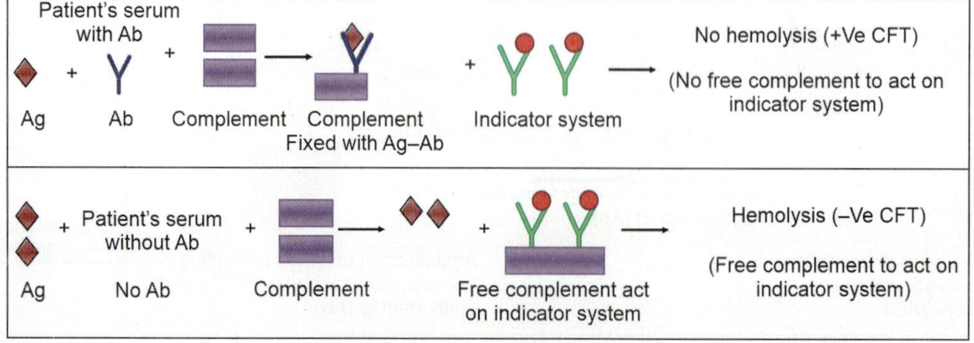

Fig. 20.17: Wassermann reaction

units) of guinea pig complement. If serum contains antibody complement will be utilized during antigen-antibody interaction. If serum does not contain the antibody, no Ag-Ab reaction occurs and complement will remain free.

- **Second step:** Add sensitized cell (sheep erythrocyte coated with amboceptor), incubate at 37°C for 30 minutes and examine the results.

- **Quality control:** Also put control tube along with the test. It has following advantages.
  - Ag and serum control: To know that they are not anticomplementary.
  - Complement control: To know that desired amount of complement is added.
  - RBCs: To know that sensitized RBCs are not lysed without complement.
- **Results**
  - Positive test: Absence of RBCs lysis indicates that complement was used up in first step and therefore, serum contains the antibody.
  - Negative test: Lysis of RBCs indicates that complement was not fixed in first step and therefore, serum did not have antibody.
- **Use:** For the serodiagnosis of syphilis.

2. **Indirect complement fixation test**
   - **Advantage:** Certain avian (duck, turkey, parrot) and mammalian (horse, cat) sera do not fix guinea pig complement.
   - **Steps:** Here the test is set up in duplicate and after the first step; the standard antiserum known to fix the complement is added to one set. If test serum contains antibody, antigen would have been used up in the first step and therefore standard antiserum added subsequently would not be able to fix complement.
   - **Result:** Hemolysis indicates a positive result.
3. **Conglutinating complement fixation test (conglutination)**
   - **Advantage:** For systems which do not fix guinea pig complement.
   - **Reagents, steps and result in brief:** Complement is obtained from horse which is nonhemolytic. Indicator system is sensitized sheep erythrocyte mixed with bovine serum. Bovine serum contains beta globulin component called conglutinin, acts as an antibody to complement. Conglutinin causes agglutination of sensitized sheep erythrocyte called conglutination if it combined with complement. If horse complement had been used up by antigen-antibody interaction in first step, agglutination of sensitized cell will not occur will indicates positive test.
4. **Immune adherence:** When some bacteria (like *V. cholerae, T. pallidum,* etc.) react with specific Abs in presence of complement and particulate materials RBCs, platelets or bacteria are aggregated and adhere to the cells called immune adherence.

5. ***Treponema pallidum* Immobilization (TPI) test:** Mix motile suspension of *T. pallidum* with specific antiserum in the presence of complement. On incubation, specific Ab inhibits the motility of *T. pallidum.*
6. **Cytolytic or cytocidal test:** When bacterium like *V. cholerae,* is mixed with its Ab in presence of complement, bacterium is killed and lysed (vibriocidal Ab test).

Neutralization Reaction

Disadvantage: Test has historical value and less useful nowadays.

Types: As follows.

A. Virus neutralization tests

- **Principle:** Neutralization of virus by specific antiserum.
- **Uses**
  - **Plaque inhibition test:** Bacteriophages are seeded in appropriate dilution on lawn culture of susceptible bacteria. They lyze the bacteria and form the lytic zone called plaques. Inhibition of plaque formation by using specific antiphage serum called plaque inhibition.
  - **Hemagglutination inhibition test:** It is used to diagnose the influenza.
  - **Other method:** Virus neutralization test can be performed in animals, chick embryo or in tissue/cell culture.

B. Toxin neutralization tests

- **Principle:** Bacterial exotoxin is good antigen and induces neutralizing antibody (antitoxin), important clinically, in protection and recovery from diseases. Toxicity of endotoxin is not neutralized by antisera.
- **Types:** Two types:
1. ***In vivo* tests**
   - Schick test: It is used to detect the immunity or susceptibility to diphtheria toxin. When diphtheria toxin is injected intradermally in patients, no reaction at injection site due to neutralization of diphtheria toxin.
   - Toxigenicity testing in *C. diphtheria* in animals: Follow **Ch. 53** for more details.
   - Toxigenicity testing in *Cl. tetani* in animals like guinea pig or mice: Follow **Ch. 55** for more details.
2. ***In vitro* tests**
   - Nagler's reaction: Used in *Cl. perfringens.* Follow **Ch. 55** for more details.
   - Toxigenicity testing in *Cl. tetani* in blood agar: Follow **Ch. 55** for more details.

Opsonization

Definitions

- **Bacteriotropin:** It is a heat stable serum factor which facilitates the phagocytosis.

- **Opsonin:** It is heat labile substance present in fresh normal serum which facilitates the phagocytosis. Term opsonin is now used to refer both these factors.

Opsonic index

- **Definition:** It is the ratio of phagocytic activity of patient's blood for a given bacterium, to phagocytic activity of normal individual's blood.
- **Measurement:** It is measured by incubating fresh citrated blood with bacterial suspension at 37°C for 15 minutes and estimating average numbers of phagocytosed bacteria per polymorphonuclear leukocytes (phagocytic index) from stained blood films.
- **Use:** It is used to study the progress of resistance of host against the progress of illness.

Radioimmunoassay (RIA)

History: It was 1st described by Berson and Yellow in 1959 and later Nobel Prize was awarded to Yallow in 1977.

Important features

- Reaction: Called binder-ligand-assay.
- Analyte or ligand: Substance (Ag) whose concentration is to be determined called analyte or ligand.
- Binder: Binding protein (antibody) which binds to the ligand called binder.
- Test permits the measurement of analytes up to picogram (10–12 g).

Methods: RIA is competitive binding assay, fixed amount of antibody and radiolabeled antigen react in presence of unlabeled antigen. Labeled and unlabeled antigens compete for limited binding sites on antibody; this competition is determined by level of unlabeled (test) antigen present in reacting system. After the reaction, antigen is separated into 'free' and 'bound' fractions and their radioactive counts measured. Concentration of test antigen calculated from ratio of bound and total antigen labels, using a standard dose response curve.

Use: It used in quantitation of hormones, drugs, tumor markers, IgE and viral antigens.

Disadvantage: It has radiation hazards.

Enzyme Immunoassay (EIA)

Advantages

EIA is versatile, highly sensitive, highly specific with the use of recombinant/synthetic Ags or with monoclonal antibodies, economical and without radiation hazards.

Disadvantages

EIA is labor intensive, requires special instruments like reader and washer and taking more time than rapid tests.

Types

Two types of EIA like homogenous and heterogeneous.

Homogeneous EIA/EMIT

No need to separate bound and free fractions in this assay. It is single step test. Example includes enzyme multiplied immunoassay technique (EMIT). EMIT is used for detection of haptens such as drugs (opiates, cocaine, barbiturates, amphetamine) and not for microbial antigens and antibodies.

Hoterogeneous EIA/ELISA

It requires separation of free and bound fractions either by centrifugation or by absorption on solid surfaces and washing. It is multistep test. Example includes ELISA which is described below.

Full form of ELISA: Enzyme-linked immunosorbent assay.

Meaning of immunosorbent: Antigens or antibodies are not free but absorbed on cellulose membrane, agarose or solid phase like tubes or microtiter plates made up from polyvinyl or polystyrene or polycarbonate.

Equipments required in ELISA

- Single channel and multichannel pipette
- Micropipette tips
- ELISA washer (**Fig. 20.18a**): It removes the unbounded materials (like Ag, Ab, chromogens, etc.) from the microtiter wells.
- ELISA reader (**Fig. 20.18b**): It detects the color by spectrophotometry. Intensity of color is directly in proportion to amount of Ag or Ab in serum.
- Other common laboratory equipments like incubator, centrifuge, timer, etc.

Reagents of ELISA: ELISA kit contains following ready to use reagents:

- Microtiter plate: With $12 \times 8 = 96$ wells
- Reagents of positive and negative controls
- Conjugate: It mostly contains AHG plus HRP means antihuman globulin (AHG) fixed with horse radish peroxidase (HRP) derived from horse. Other enzymes used in conjugate are alkaline phosphatase from *E. coli*, β-galactosidase from *E. coli* and urease.
- Substrate or color reagent or chromogen: It contains chemical like tetramethyle benzidine (TMB) or

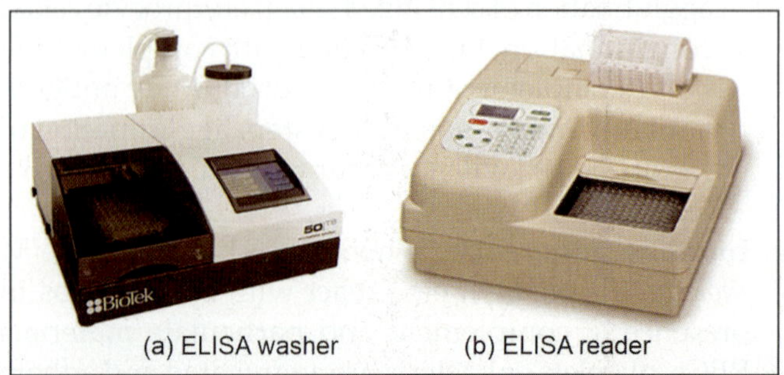

Fig. 20.18: (a) ELISA washer, (b) ELISA reader

peroxidase or paranitrophenyl phosphate (PNPP) or O-phenylene diamine dihydrochloride. It is colorless substance and when come in contact with light it produces color and gives false-positive result, so store it in dark area to avoid the contact with light.

- Stop solution: It contains acid like phosphoric acid or mineral acid, etc.

Types of ELISA: It is classified into several types on different basis:

- **On the basis of solid phase used:**
  1. Micro-ELISA: Performed in microtiter plates with 96 wells.
  2. Macro-ELISA: Performed in tubes.
- **On the basis of Ag utilized:**
  1. 1st generation: Infected cell lysates are used as an antigen.
  2. 2nd generation: Glycopeptide (recombinant Ag) used as an antigen.
  3. 3rd generation: Synthetic peptide used as an antigen.
  4. 4th generation: Simultaneously recombinant and synthetic peptide used as antigen for the detection of antibody.
- **On the basis of principle**
  1. Direct ELISA.
  2. Indirect ELISA.
  3. Sandwich ELISA.
  4. Competitive ELISA.
  5. Capture ELISA.
  6. Rapid ELISA (cylinder or cassette ELISA).
  7. ELISPOT test.
  8. IgG avidity ELISA.

1. Direct ELISA

- **Principle:** Follow **Flowchart 20.2 and Fig. 20.19a.**

Well + Ag (test serum) → Ag + Primary Ab-E (conjugate) + Substrate-chromogen → Color change

- **Use:** It is used for detection of antigen in test serum. Primary antibody (targeted against the serum antigen) is labeled with the enzyme.

- **Steps:** Remove the excess wells from microtiter plate and arrange the required numbers of wells in to plate. Wells are free and not precoated with Ags or Abs. Add the test serum (contains Ags) to the wells. Ags will fix to the wells by passive adsorption. Wash the plate to remove the unbounded Ags. Add the conjugate which contains Ab labeled with enzyme. Wash the plate to remove the unbounded conjugate. Add the substrate-chromogen or color reagent which contains colored substrate which is sensitive to enzyme of conjugate. When colored is produced take the result by ELISA reader.
- **Results:** Results are taken by two ways:
  - Naked eye: Development of color in well indicates positive test and colorless well indicates negative test.
  - By using ELISA reader: It measures the color by spectrophotometry.

2. Indirect ELISA

- **Principle:** Follow **Flowchart 20.3 and Fig. 20.19b.**

Ag in well + Primary Ab (test serum) → Ag-Ab complex + Secondary Ab-E (conjugate) + Substrate chromogen → Color

- **Use:** It is used for detection of Ab and rarely Ag in test serum.
- **Steps:** Remove the excess wells from microtiter plate and arrange the required numbers of wells in to the plate. Well contains Ag. Put negative and positive controls. Add sample (contains primary Ab specific to the Ag) to the well. Cover the plate and incubate at 37°C for few hours. Take out the plate from incubator after desired time and wash the plate to remove the unbounded Ab. Add the conjugate contains secondary Ab (anti-human immunoglobulin) labeled with enzyme. Incubate the plate at room temperature for the recommended time. Wash the plate to remove the unbounded conjugate. Add the substrate and incubate the plate in the dark area for the recommended time. Take out the plate from dark area after the recommended time and add the stop

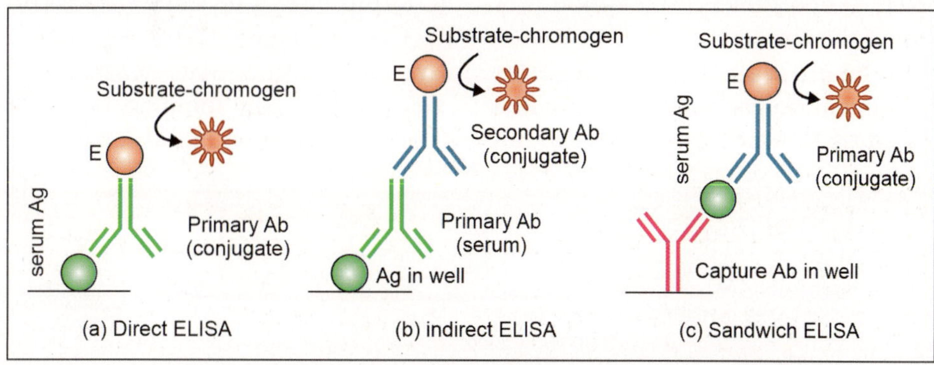

Fig. 20.19: Principle of (a) direct, (b) indirect and (c) sandwich ELISA

solution. It stops all reaction between the Ag and Ab before taking the result.

- **Results:** Results are taken by two ways:
  - Naked eye: Development of color in well (**Fig. 20.20**) indicates positive test and colorless well indicates negative test.
  - By using ELISA reader: It measures the color by spectrophotometry.

3. **Sandwich ELISA:** Called sandwich ELISA, because Ag is sandwiched on either side by fixed antibodies.
   - **Principle and steps:** Wells are fixed with monoclonal Abs. Add serum contains Ags to well. Incubate the plate, followed by washing which removes unbound Ags and Abs. Later add the conjugate contains monoclonal Ab labeled with enzyme. After sufficient time of incubation wash the plate to remove unbounded conjugate materials. Fixation of conjugate to Ag-Ab complex is identified by adding substrate, which produces color on reacting with enzyme indicates positive result. Finally, add the stop solution and read the result (**Fig. 20.19c and Flowchart 20.4**).
   - **Use:** It detects the antigen in test serum.

Flowchart 20.4: Principle of sandwich ELISA

> Capture Ab of well + Ag (test serum) + Primary Ab-enzyme + substrate–chromogen → Color

4. **Competitive ELISA:** So called because Ag in test serum competes with another Ag of the same type coated in well to bind with primary antibody.
   - **Principle and steps:** First inoculate the serum contains the Ag with primary Ab. It forms the Ag-Ab complexes while excess Abs will remain free. Add the Ag-Ab mixture to the microtiter well contains the same type of Ag. Free antibodies bound to the antigen coated on the well. More the test antigens present in the sample, lesser free antibodies will be available to bind to the antigens coated onto well. After washing (to remove free antibodies and antigens), conjugate contains secondary Ab linked with enzyme is added. After washing, a substrate–chromogen system is added. Result shows no color or less color development (**Fig. 20.21 and Flowchart 20.5**). Intensity of the color is inversely proportional to the amount of antigen present in the test serum.
   - **Use:** It detects the Ag in test serum. It also useful for Ab detection. Different formats are available like direct, indirect and sandwich ELISA. Description

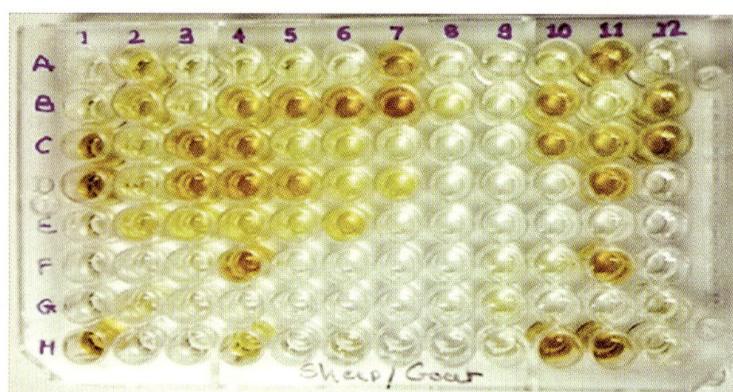

Fig. 20.20: ELISA plate with microtiter wells showing positive (colored) and negative results (colorless)

given above is indirect competitive ELISA for Ag detection.

Flowchart 20.5: Principle of competitive (indirect) ELISA

> Primary Ab + Ag (test serum) → Ag-Ab complex + free Ab → Ag in well → Ag-Ab complex + Conjugate (Secondary Ab-E) + substrate–chromogen → No color or less color

5. **Capture ELISA:** It captures particular type of Ab like IgG, IgM, etc. hence called capture ELISA. For IgM it called MAC (IgM Antibody Capture) ELISA.
   - **Principle and steps**
     - **For IgM Ab:** Add the test serum contains primary IgM Ab to a microtiter well contains antihuman-IgM Ab, followed by addition of recombinant antigen (like dengue antigen). Subsequently, add the conjugate contains secondary antibody (specific for Ag) labeled with enzyme, followed by addition of substrate-chromogen system (**Fig. 20.22a and Flowchart 20.6**). Use of avidin-biotin system helps in amplifying the signal generated between enzyme-antibody complex, thus increases the sensitivity of the assay.
     - **For IgG Ab:** Follow **Fig. 20.22b**.
   - **Uses:** It is used for Ab detection for dengue, Japanese encephalitis, West Nile virus, scrub typhus, leptospirosis and toxoplasmosis.

Flowchart 20.6: Principle of MAC ELISA

> IgM Ab (test serum) + anti-IgM Ab in well → recombinant Ag → Secondary Ab-biotin+avidin-enzyme + substrate –chromogen → color

6. **Rapid ELISA (cylinder or cassette ELISA)**
   - **Advantages:** Test is rapid (36 minutes), no need of microtiter plate, washer or reader and result is read visually. Inbuilt positive and negative controls usually provided. It separates the HIV-1 from 2.

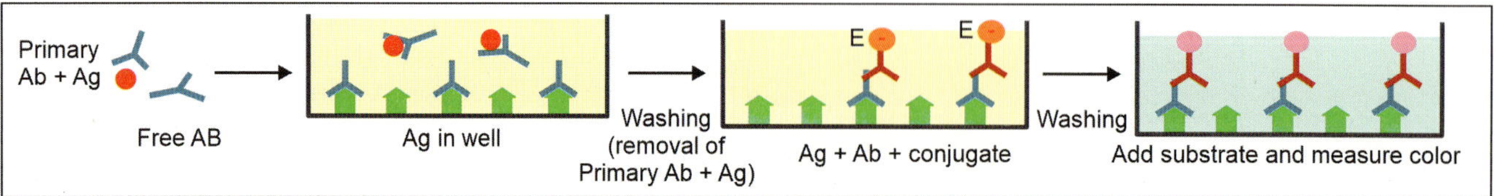

Fig. 20.21: Competitive ELISA

Essentials of Medical Microbiology

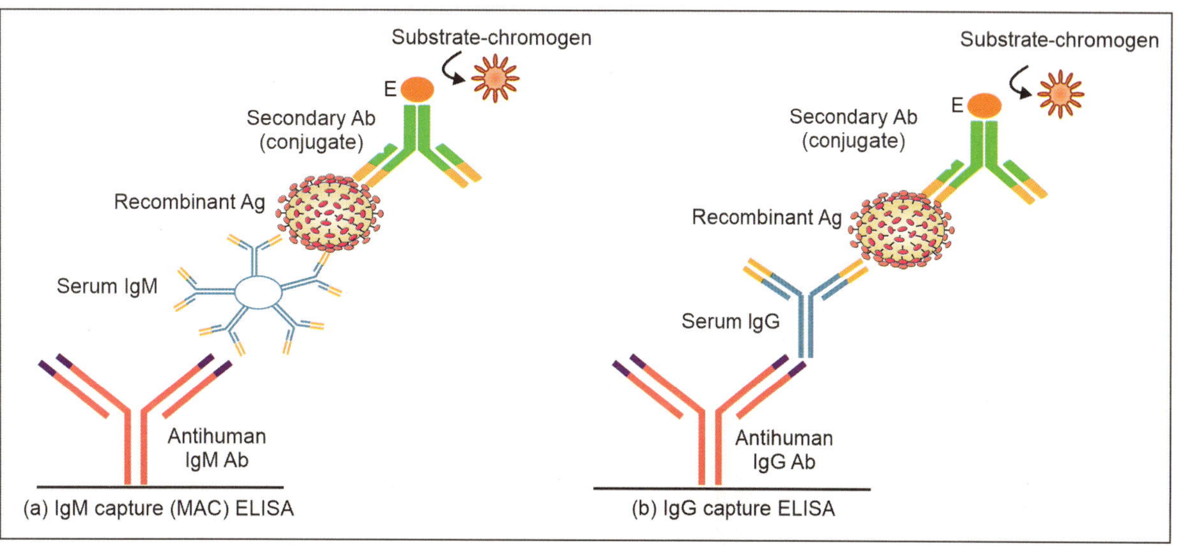

Fig. 20.22: Capture ELISA

- **Example:** Immunocomb HIV 1 and 2 bispot.
- **Reagents:** Follow **Fig. 20.23**.
- **Results:** Follow **Fig. 20.24**.

7. ELISPOT test

- **Principle:** Follow **Flowchart 20.7**.

Flowchart 20.7: Principle of ELISPOT test

Cytokine's Ab in well + cytokine (Ag) in sample → Ab-Ag complex + Anti – cytokine Ab-E + substrate → Color change

- **Steps:** Microtiter well is fixed with antibodies to cytokines. Add the sample contains cytokine or cytokine-producing cells and incubated. During this time cytokine or cytokine producing cells are attached to the antibodies fixed with microtiter well. Remove the plate from incubation and wash it to remove unbounded cytokines. Add the conjugate contains anticytokine antibody labeled with enzyme. Incubate the plate at room temperature for the recommended time. Remove the unbounded conjugate materials by washing. Add the substrate and incubate the plate in the dark area for the recommended time. Take out the plate from dark area and add the stop solution, which stops all reaction between the Ag and Ab before taking the result.
- **Results:** Results are taken by two ways:
  - Naked eye: Development of color in well indicates positive test and colorless well indicates negative test.
  - By using ELISA reader: It measures the Optical Density (OD) in each well.
- **Use:** It quantitatively measures the cells producing antibodies (plasma cells) and cytokines (macrophages). ELISPOT is currently used in IGRA (interferon gamma release assay), for diagnosis of latent tuberculosis; where the sensitized T cells capable of producing the IFN-γ are measured.

8. IgG avidity ELISA

- **Use:** It is useful to differentiate recent from past infection. Usually, recent and past infections are differentiated by presence of IgM and IgG antibodies respectively. In situations where both the antibodies are present, IgG avidity test is useful to differentiate recent from past infection especially in rubella, CMV, VZV, *Toxoplasma*, EBV, HIV, HBV, HCV, West Nile virus infection and congenital infections.
- **Principle:** Avidity of an Ab indicates how firmly it is bound with its antigen. It reflects the maturity of the antibodies, which usually increases with

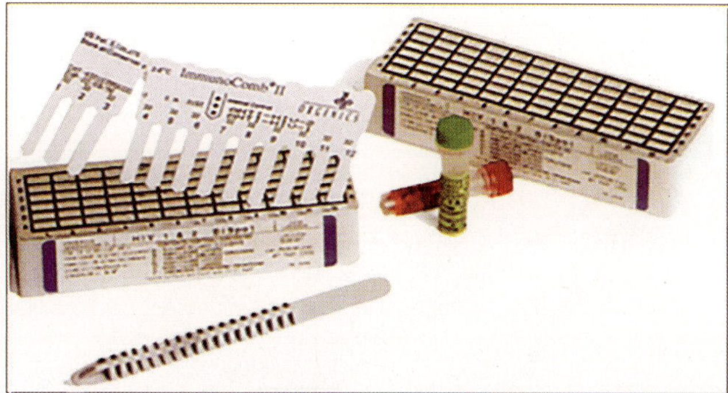

Fig. 20.23: Reagents of immunocomb HIV 1 and 2 bispot

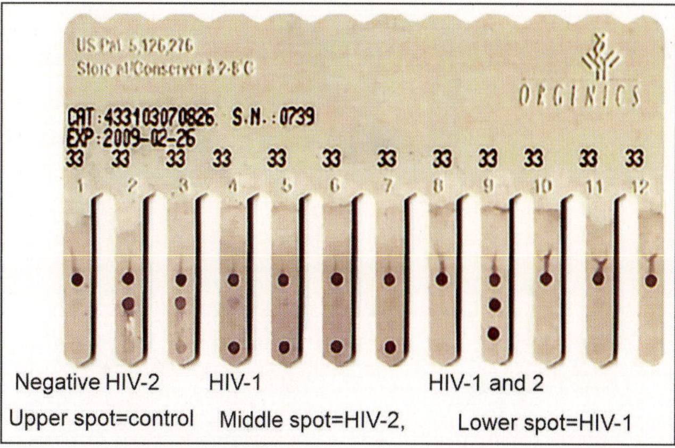

Fig. 20.24: Result of immunocomb HIV 1 and 2 bispot

time. Detection of low avidity IgG indicates recent or primary infection or IDDs. While high avidity IgG indicates past or secondary infection or reactivation or vaccination.

- **Steps:** It is performed in two parallel wells of microtiter plate. One well with untreated patient's serum and second with urea (dissociating agent) treated serum. Low avidity Abs are dissociated from the complexes and are washed away during washing steps while high avidity Abs resist urea treatment and remain bound to Ags.
- **Results:** It is expressed in percentage (%) and denotes the absorbance of the well with urea to the absorbance of the well without urea.
  - Avidity <40%: Low avidity.
  - Avidity >60%: High avidity.
  - Avidity 40–60%: Intermediate required test to repeat.

Uses of ELISA: It is universal test used for several purposes.

1. To detect Ags and Abs in infectious diseases:
   - For Ag detection: HBsAg, HBeAg, NS1 antigen for dengue, etc.
   - For Ab detection: HBV, HCV, HIV, dengue, EBV, HSV, toxoplasmosis, leishmaniasis, etc.
2. Diagnosis of human allergens: Especially by capture ELISA.
3. Food toxin: Like aflatoxin.
4. Food adulterants detection.
5. Measuring hormone level: Such as HCG (as a test for pregnancy), LH (determining the time of ovulation) and TSH, T3 and T4 (for thyroid function).
6. Measuring "rheumatoid factors" and other auto-antibody in autoimmune diseases like lupus erythematosus.
7. Detecting drugs, e.g., cocaine, opiates, etc.

Enzyme-linked Fluorescent Assay (ELFA)

Introduction: It is the modification of ELISA. VIDAS (**Fig. 20.25**) and miniVIDAS (bioMérieux) are commercially available systems based on ELFA principle. It is differ from ELISA in two ways, as it is an automated system, all steps are performed by the instrument and Ag-Ab-enzyme complex is detected by fluorometric method.

Uses

- Diagnosis of Ab: HIV-Ab, TORCH agents, measles, mumps, varicella and *H. pylori*.
- Diagnosis of Ag: HBsAg, HIV, *Cl. difficile*, rotavirus, etc.
- Detection of biomarkers, e.g., procalcitonin.
- Detection of hormones, e.g., thyroid.
- Others uses: Detection of tumor markers, cardiac markers, allergens, etc.

Steps and principle: The solid phase receptacle/SPR (**Fig. 20.26**) present in reagent strip (as like microtiter well in ELISA plate) is either coated with Ag (for Ab detection) or Ab (for Ag detection). Conjugate used here is either an antigen or antibody labeled with enzyme alkaline phosphatase and the substrate used is 4-methyl-umbelliferyl phosphate. Following Ag-Ab-enzyme conjugate complex formation; the excess of enzyme conjugate are washed out. Then the conjugate enzyme catalyzes the substrate into a fluorescent product (4-methylumbelliferone) as shown in **Fig. 20.27**. Fluorescence is measured at 450 nm (fluorometric method). Intensity of the fluorescence depends on the concentration of alkaline phosphatase present which in turn depends upon the amount of analytate (target Ag or Ab) in the sample.

Advantages: It has many advantages over ELISA like an automated system, gives quantitative results, more sensitive and specific, easy to perform, user friendly and less chances of contamination.

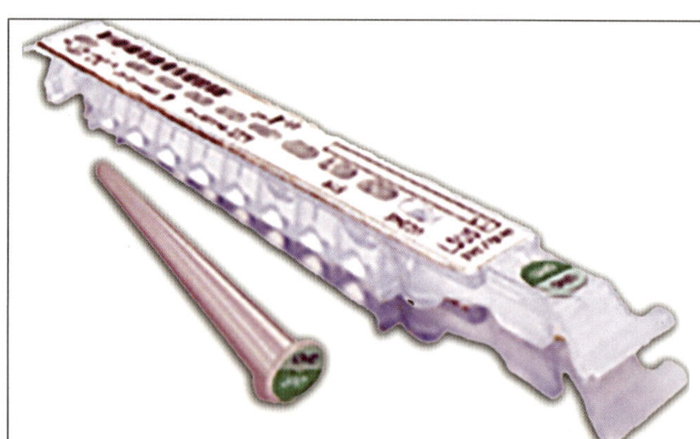

Fig. 20.26: Solid phase receptacle/SPR for ELFA (VIDAS)

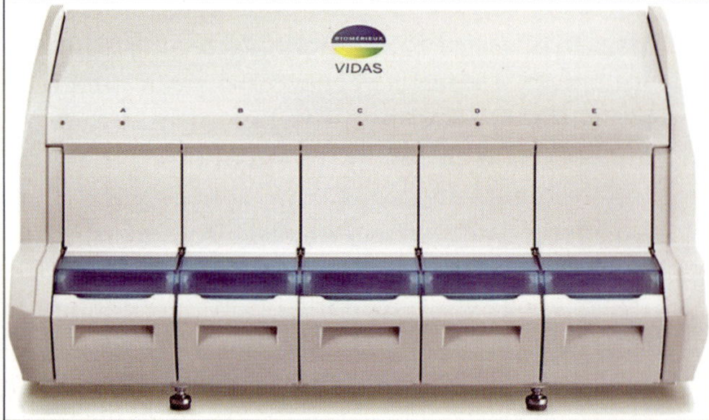

Fig. 20.25: Instrument ELFA (VIDAS)

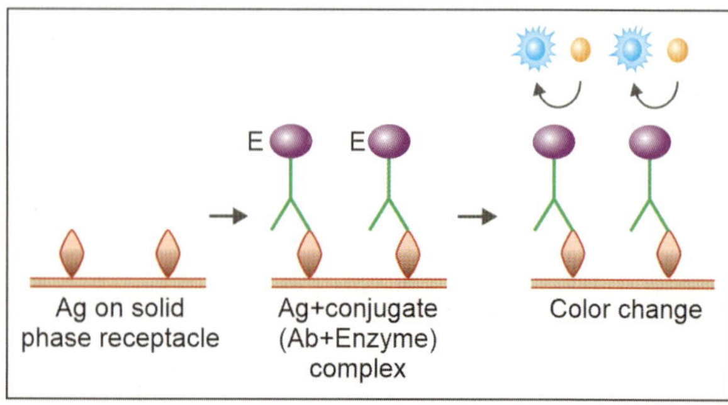

Fig. 20.27: Principle of ELFA

Disadvantages: It is expensive, can run only 12–24 numbers of tests at a time and can run 2–4 types of tests at a time.

Immunofluorescence assay (IFA)

Introduction

Fluorescent dye used is called fluorescein isothiocyanate (FITC) (blue-green fluorescence) or lissamine rhodamine (orange-red fluorescence). It absorbs the UV light and illuminates in visible light called fluorescence.

History

In 1942 Coons and his colleague conjugate the dye with Ab and such 'labeled' Ab can be used to locate and identify Ag in tissues.

Uses

For detection of auto-antibodies, hormones, tumor markers, enzymes, organs or tissues antigens and infectious agents like bacteria, viruses, fungi and parasites.

Types

Direct immunofluorescence test

- **Principle:** Ag combines with 'labeled' Ab and produces visible light on UV exposure **(Fig. 20.28a)**.
- **Disadvantage:** It separates fluorescent labeled Ab has to be prepared against each Ag to be tested.

Indirect immunofluorescence test

- **Principle:** Ag combines with sample's Ab and produces Ag-Ab complex which later combines with conjugate (dye + antiglobulin) and produces visible light on UV exposure **(Fig. 20.28b)**.
- **Advantage:** It overcomes the disadvantage of direct method by using an antiglobulin fluorescent conjugate.
- **Steps:** Steps are discussed with example of fluorescent treponemal antibody test for syphilis. Drop of test serum is placed on a slide contains the smear of *T pallidum* and incubate the smear. Slide is washed well to remove all the free serum, leaving behind only antibody globulin, if present, coated on the surface of treponemas. Smear is then treated with a fluorescent labeled Ab to human gamma globulin. Fluorescent conjugate reacts with antibody globulin bound to treponemas.
- **Results:** After washing away unbound fluorescent conjugate, slide is examined under ultraviolet illumination.
  - Positive result: Bright objects against the dark background.
  - Negative result: If sample does not contain Ab and it will not bound the conjugate and no illumination in the end.
- **Other modifications**
  - Single antihuman globulin fluorescent conjugate can be employed for detecting human antibody to any antigen.

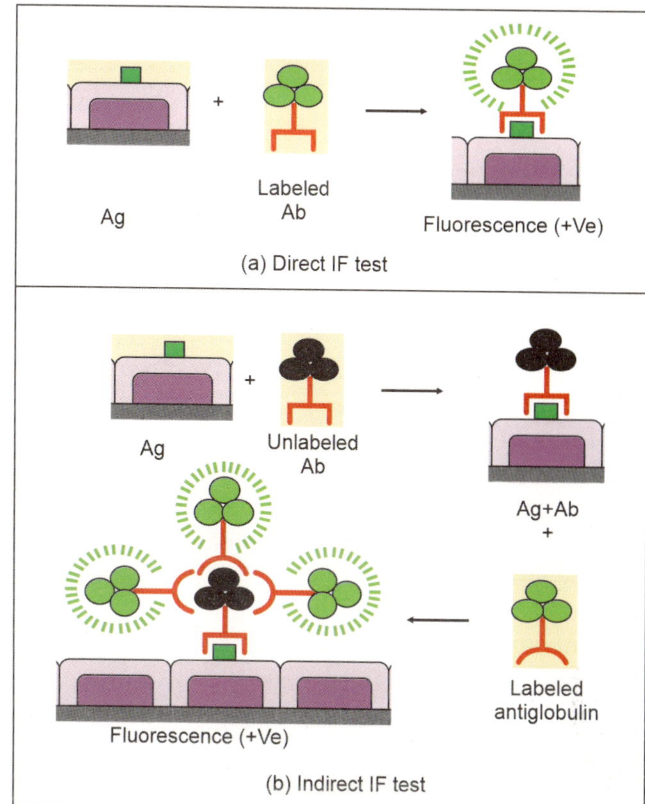

Fig. 20.28: Principle of IFA

- Fluorescent dye may also be conjugated with the complement. Labeled complement is a versatile tool and can be employed for the detection of antigen or antibody.
- Antigen also takes fluorescent labeling but not as well as antibody do.

Sandwich IF test

- **Principle:** Unlabeled Ab is allowed to react with Ag. Ag-Ab complex is then treated with fluorescent labeled Ab. Sandwich is formed, Ag is in middle while labeled and unlabeled Abs are on either side **(Fig. 20.29)**.
- **Use:** It is used for detection of antibodies.

Immunohistochemical test: It is the combined technique of serology and histology. It detects the Ag-Ab reaction in tissue (*in situ*), hence called immunohistochemical

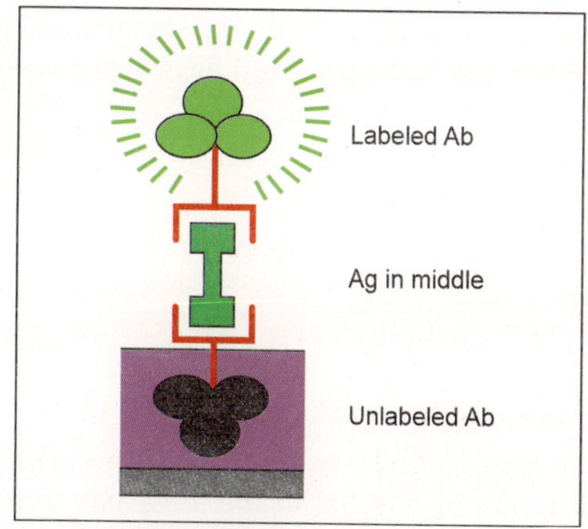

Fig. 20.29: Principle of sandwich IF test

test. Commonly used dye is FITC or rhodamine. It produces nonspecific fluorescence in tissue is the only disadvantage.

Flow cytometry

- **Principle:** Suspending the fluorescent labeled cell in a stream of fluid and passing them through an electronic detection apparatus.
- **Uses**
  - To identify cells bearing particular Ag or surface marker like CD4 and CD8 cells in HIV patients.
  - To measure the size, granularity, DNA or RNA content of cells.
  - To count the cells like differential leukocyte count.
  - For diagnosis, treatment and prognosis of cancer specially leukemia.
  - To study the cell cycle and apoptosis.
  - For research purpose.

Chemiluminescence-linked Immunoassay (CLIA)

Like RIA, EIA and IFA some chemicals are used as label to provides the signal of Ag-Ab reaction. It produces signal in the form of light which can be amplified, measured and concentration of analyte can be calculated. Commonly used chemicals are luminol or acridinium esters. It is fully automated and used in laboratories where volume of work is large.

Immunoelectroblot Technique

Synonym: Western blot (WB) test.

Advantages: It has high sensitivity like EIA and greater specificity than EIA.

Steps: Three steps:
- Separation of ligand (Ag): By PAGE.
- Blotting (fixation): Separated Ags are blotted or transferred electrophoretically from PAGE to nitro-cellulose membrane strip.
- EIA or RIA: After blotting, steps are similar to indirect ELISA. Strip contain Ag is exposed to serum contains Ab then, followed by conjugate and substrate step.

Results
- Positive result: If serum contains specific Ab to strip Ag, it produces the color band indicates the positive result.
- Negative result: No band indicates the negative result.

Use: It is the definitive or confirmatory test for detection of antibody against particular HIV-Ags like gp-120, gp-41, P-24, P-15, P-55, etc.

Immunochromatographic Test (ICT)

Synonym: It also called lateral flow assay because serum moves laterally or parallel to the long axis of strip or biscuit/card as shown in **Fig. 20.30**.

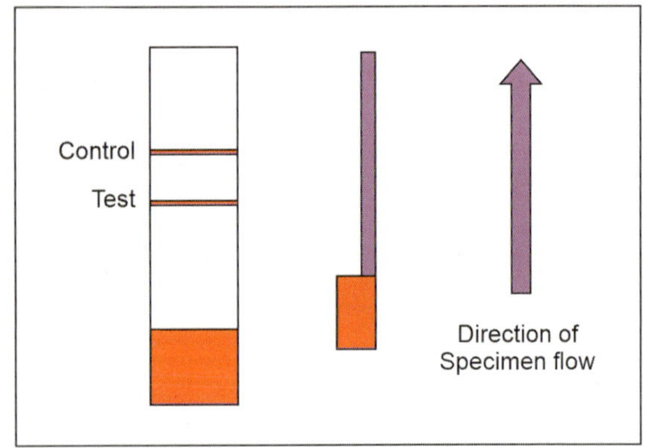

Fig. 20.30: Structure of strip and direction of specimen flow in ICT

Types: Test device is available either in the form of strip or in the form of biscuit (card) as discussed below:

1. **Strip form:** It is discussed below with example of HBsAg (virucheck) method:
   - **Structure of strip:** Test site (T) is impregnated with anti-HBsAg antibody with colloidal gold dye conjugate and control site (C) is impregnated with antiglobulin.
   - **Method:** Put the strip in test tube contains serum. Strip absorbs the serum and if serum contains HBsAg it combines with anti-HBsAg antibody-colloidal gold dye conjugate to produce visible band called test band. Control band will be produced at the control site.
   - **Results (Fig. 20.31)**
     - Positive result: Control band and test band.
     - Negative result: Control band only but no test band.
     - Invalid result: No control band with or without test band.

2. **Biscuit (card) form:** It is discussed below with example of HBsAg (SD Bioline).
   - **Structure of biscuit (card):** Biscuit has two separate windows like sample window and result window. In result window test site is impregnated with anti-HBsAg antibody with colloidal gold dye conjugate and control site is impregnated with antiglobulin.

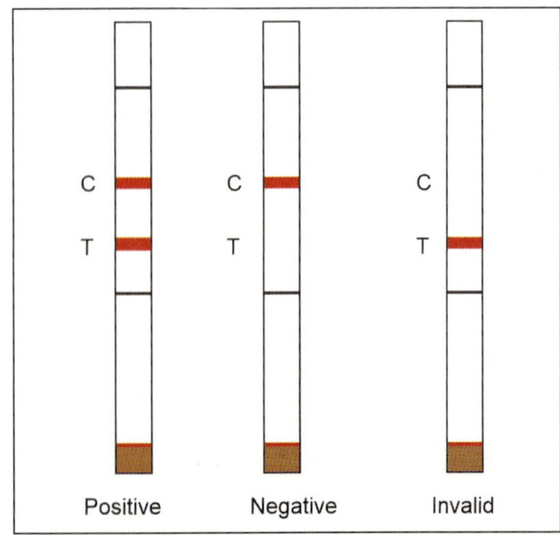

Fig. 20.31: Result of ICT (Virucheck for HBsAg) in strip form

- **Method:** Put the 2–3 drops of serum in sample window. Serum runs along the long axis of biscuit towards the result window. Read the result after 15–20 minutes. Test (T) site absorbs the serum and if serum contains HBsAg it combines with anti-HBsAg antibody-colloidal gold dye conjugate to produce visible band called test band. Control band will be produced at the control (C) site.
- **Results (Fig. 20.32):** Same as strip form method.
- **Advantages of ICT:** Test is simple, economical and nearly as sensitive and specific as EIA test. It is one step (rapid) testing because color reagent (colloidal gold) is already attached to device, so no need to add separately.
- **Disadvantages of ICT:** It is qualitative not a quantitative method.
- **Uses of ICT:** Test is used to detect HBsAg, HCV antibodies, HIV1 and 2 antibodies.

Immunoconcentration Test

Synonym: It also called vertical flow or dot blot assay or flow through assay, because serum moves vertically or perpendicular to the long axis of device **(Fig. 20.33)**.

Types: Test device is available either in the form of biscuit/card (mostly with single window) or in the form of comb. Both types are discussed as below:

1. **Biscuit/card form:** It is discussed with the example of HIV tridot for HIV-1 and 2 antibodies detection.
   - **Structure of biscuit (card):** Device is impregnated at three sites such as test site-1 with HIV-1-Ag (gp120 and gp41), test site-2 with HIV-2-Ag (gp36) and control site with antiglobulin.
   - **Method:** Add the sample contains HIV antibodies. Later add the buffer. Add the signal reagent (Protein A conjugate). Again add the buffer. Read the result within few (around 10) minutes.

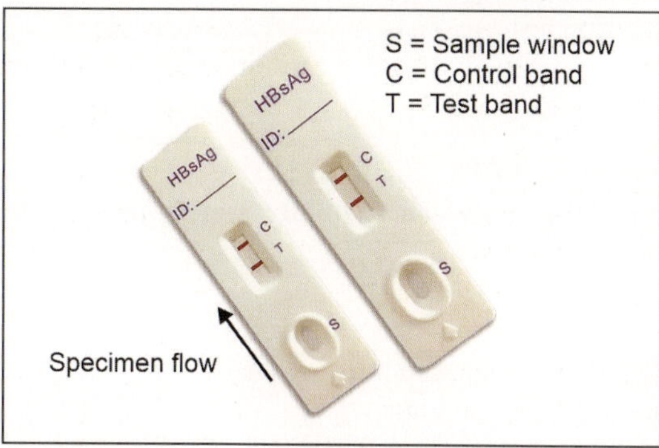

Fig. 20.32: Result of ICT (SD Bioline for HBsAg) in biscuit form

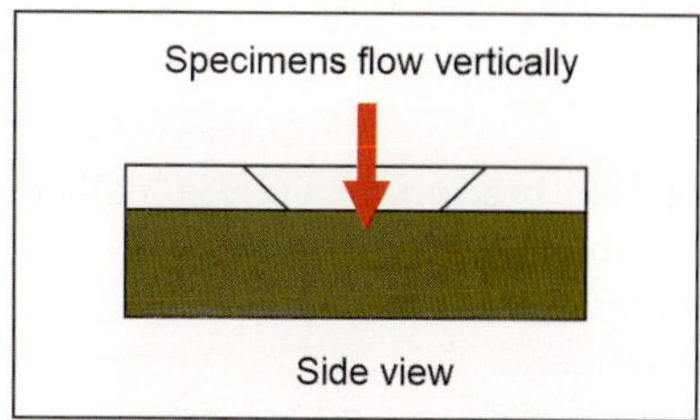

Fig. 20.33: Vertical flow

- **Results (Fig. 20.34)**
  - Positive result: Control dot and test dot. If serum contains Ab, it binds with Ag of test site (one or both) and produces colored dot indicates positive result.
  - Negative result: Control dot only but no test dot.
  - Invalid result: No control dot with or without test dot.

2. **Comb form:** It is discussed with the example of HIV comb AIDS-RS for HIV antibodies detection.
   - **Structure of comb:** Test site is impregnated with HIV-1 and 2-Ag (gp41 and gp36) and control site is impregnated with antiglobulin.
   - **Method:** Treat the comb with sample contains HIV antibodies in microtiter well as shown in **Fig. 20.35a.** Later treat the comb with buffer. Treat the comb with signal reagent (protein A conjugate). Again treat the comb with buffer. Read the result within few (around 10) minutes.
   - **Results:** Follow **Fig. 20.35b.**
     - Positive result: Control dot and test dot. If serum contains Ab, it binds with Ag of test site and produces color dot on comb.
     - Negative result: Only control dot but no test dot.
     - Invalid result: No control dot with or without test dot.

Advantage: Test is a simple, rapid, economical, qualitative, nearly as sensitive and specific as EIA.

Disadvantage: It is the two steps testing than ICT, because color reagent (protein A conjugate) is not attached to device, so need to add separately in the form of signal reagent. It is qualitative not a quantitative method.

Uses: Test is used for detection of HBsAg, HCV antibodies and HIV1 and 2 antibodies.

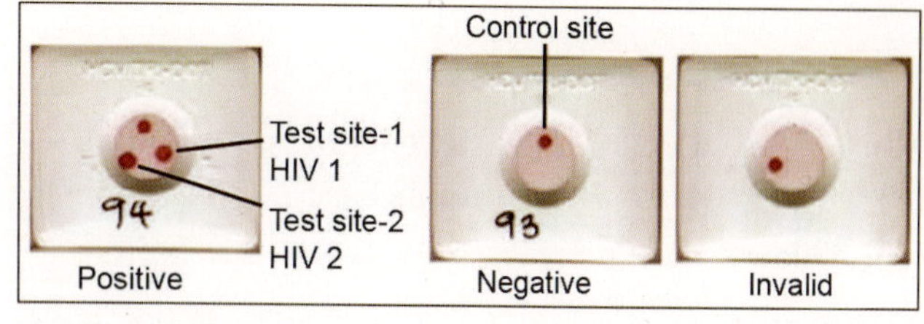

Fig. 20.34: Result of immunoconcentration test (HIV tridot) in biscuit/card form.

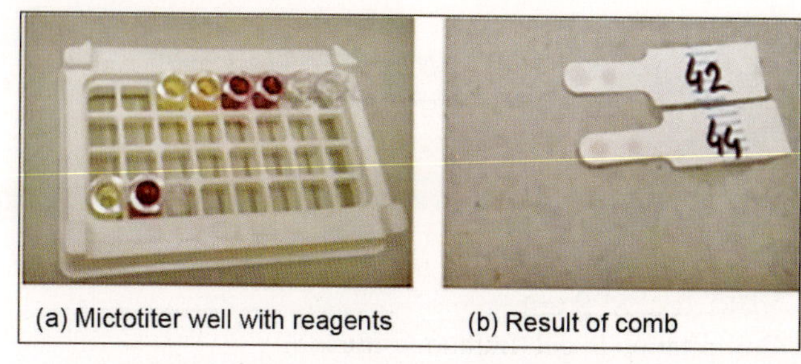

Fig. 20.35: Result of immunoconcentration test (HIV comb AIDS-RS) in biscuit/card form.

Note: Comb assay or comb type device of test

Comb type device is used in ELISA based serological test (rapid ELISA) like immunocomb HIV 1 and 2 bispot test and immuno-concentration based serological test like HIV comb AIDS-RS.

Tests by Electron Microscopy (EM)

1. **Immunoelectronmicroscopy:** Viral particles are mixed with specific antisera and observed under the Electron Microscope (EM), they are seemed to be clumped. Test is used to study virus like HAV.

2. **Immunoferritin test (IFT):** Ferritin can be conjugated with Ab and such labeled Ab reacting with an Ag can be viewed under the EM.

3. **Immunoenzyme test (IET):** Some stable enzymes like peroxidase can be conjugated with Ab. Tissue section carrying the corresponding Ag is treated with peroxidase labeled Ab. Formation of labeled Ab –Ag complex can be visualized under the EM, by microhistochemical methods. Other enzymes such as glucose oxidase, phosphatase and tyrosinase may be used.

COMPARISON OF IMMUNOASSAYS

Follow **Table 20.6.**

TABLE 20.6: Comparison of immunoassay			
Test/Assay	**Molecule used for labeling**	**Visible effect**	**Detection method**
RIA	Radioactive isotope	Emits β and γ rays	β and γ counter
ELISA	Enzyme	Color	Spectrophoto-metry by ELISA reader
ELFA	Enzyme	Color	Fluorometric
IFA	Dyes	Light	FM
CLIA	Luminol or acridinium esters	Light	Luminometer
WB	Enzyme	Color band	Naked eye
ICT	Colloidal gold	Color band	Naked eye
Vertical flow	Protein A conjugate	Color dot/ band	Naked eye
IFT	Ferritin	Black dot	EM
IET	Antibody	Dot	EM

ACCESS YOURSELF

Essay/Full Question

1. Describe different types of serological reactions.

Short Notes

1. VDRL test/RPR test
2. Precipitation in gel (Immunodiffusion)
3. Widal test/Weil-Felix test/Paul Bunnell test
4. Antiglobulin (Coombs') test.

5. Indirect (passive) agglutination test.
6. CFT/RIA/ELISA/IFA.

Short Questions for Theory/Viva Questions

1. Write the differences between immunological reaction and serological reaction.
2. What is zone phenomenon?
3. Define titer of serological reactions.
4. What are affinity and avidity of serological reactions?
5. What are sensitivity and specificity of serological reactions?
6. What is biological false-positive test in VDRL test?
7. Write two uses of CIE.
8. What is blocking (incomplete) antibody?
9. Name the four serological tests based on agglutination principle.
10. What is anamnestic reaction in Widal test?
11. What are carrier particles in serological tests?
12. Write the difference between Rose Waller and modified Rose Waller test.
13. What is amboceptor?
14. What are coagglutination and conglutination?
15. What is ELISPOT test?

Comments on

1. Precipitation in gel (immunodiffusion) is more advantageous than doing in liquid medium.
2. In Widal test O-Ag of *S. Paratyphi* is not included.
3. In ELISA test substrate or color reagent or chromogen step should be done in dark area.

MCQs for Chapter Review

Introduction

1. **All of the following forces are involved in antigen-antibody reaction *except*:**
   a. van der Waal's forces   b. Electrostatic bond
   c. Hydrogen bond   d. Covalent bond
2. **Which of the following statement is true?**
   a. Paul Bunnel test is used to diagnose measles
   b. Rose Waller test is complement fixation test
   c. Indirect hemagglutination test is less sensitive than gel diffusion test
   d. Antigen-antibody reaction cannot occur in the absence of electrolytes

Precipitation Reaction

3. **The reaction between antibody and soluble antigen is demonstrated by:**
   a. Agglutination
   b. Precipitation
   c. Complement fixation test
   d. Hemagglutination test
4. **Prozone phenomenon is due to:**
   a. Excess antigen
   b. Excess antibody
   c. Hyperimmune reaction
   d. Disproportionate antigen-antibody levels
5. **Antigen antibody precipitation is maximally seen in which of the following?**
   a. Excess of antibody
   b. Excess of antigen
   c. Equivalence of antibody and antigen
   d. Antigen hapten interaction

6. **False-positive VDRL, is seen in all *except*:**
   a. Lepromatous leprosy
   b. Infectious mononucleosis
   c. HIV
   d. Pregnancy
7. **False-positive nontreponemal serological tests for syphilis, are seen in:**
   a. HIV
   b. Collagen disorders
   c. Pediatric age group
   d. Tuberculosis
   e. Chronic liver disease
8. **True about VDRL test:**
   a. Nonspecific test
   b. Slide flocculation test
   c. Best followed for drug therapy
   d. All
9. **Reason for black color clump in RPR test is antigen coating with:**
   a. Carbon particle
   b. Latex particle
   c. Bentonite particle
   d. RBCs

Agglutination Reaction

10. **Widal test is based on:**
    a. Precipitation
    b. Flocculation
    c. Slide agglutination
    d. Tube agglutination
    e. Complement fixation test
11. **All are correct regarding Widal test, *except*:**
    a. Baseline titer differ depending upon the endemicity of the disease.
    b. High titer value in a single Widal test is not confirmative.
    c. O antibody lasts longer and hence is not indicative of recent infection.
    d. H antibody cannot differentiate between types.
12. **Coombs' test is:**
    a. Precipitation test
    b. Agglutination test
    c. CFT
    d. Neutralization test
13. **Weil-Felix reaction is based on the principle of sharing common antigen between:**
    a. *Streptococcus* and *Staphylococcus*
    b. *Rickettsia* and *Chlamydia*
    c. *Rickettsia* and *Proteus*
    d. *Rickettsia* and *Pseudomonas*
14. **A man came from Nagaland and shows positive test with OXK antigen. Diagnosis is:**
    a. Trench fever
    b. Scrub typhus
    c. Endemic typhus
    d. Epidemic typhus
15. **Rose-Waaler test is done for diagnosis of:**
    a. Brucellosis
    b. Syphilis
    c. Rheumatic fever
    d. Rheumatoid arthritis

Radioimmunoassay (RIA)

16. **Most sensitive test for antigen detection is:**
    a. RIA
    b. ELISA
    c. Immunofluorescence
    d. Passive hemagglutination
17. **Hormones are best diagnosed by:**
    a. Flow cytometry
    b. Electrophoresis
    c. ELISA
    d. RIA

Comparison of Assays

18. **The following methods of diagnosis utilize labeled antibodies *except*:**
    a. ELISA (Enzyme-linked immunosorbent assay)
    b. Hemagglutination inhibition test
    c. Radioimmunoassay
    d. Immunofluorescence

Answers and Explanation of MCQs

1. d
- Ag-Ab complex formation is possible by weaker intermolecular forces like ionic (electrostatic) bonds, van der Waal's forces and hydrogen bonds.

2. d
- Paul Bunnel test is used to diagnose infectious mononucleosis. Rose Waller test is passive hemagglutination test. Indirect hemagglutination test is more sensitive than gel diffusion test.

3. b
- In given question antigen is soluble, which is detected by precipitation.

4. b

5. c
- Follow section, **precipitation reaction (Zone phenomenon)** for explanation of answers of MCQs 4–5.

6. d

7. a, b

8. d
- Follow section, VDRL test for explanation of answers of MCQs 6–8.

9. a
- Follow **Table 20.3** for explanation.

10. c and d
- Widal test is based on agglutination principle which may be both slide as well as tube type.

11. c and d
- Base line titer of widal is different according to endemicity of areas. High titer value in a single widal test is may be due to anamnestic reaction, so it is not confirmative. It is confirmed by retesting after 1 week, so repeated testing of paired sera at 1 week interval is more meaningful than single high titer. O antigen is less immunogenic, antibody is not long lasting, so it indicates the recent infection. In Widal test, separate H antigens for *S.* Typhi, *S.* Paratyphi A and *S.* Paratyphi B are included so they can differentiate between types.

12. b
- Coombs' test fall in antiglobulin category of agglutination reaction.

13. c

14. b
- Follow section, **Weil-Felix reaction and Table 20.5** for explanation of answers of MCQs 13–14.

15. d
- Follow section, **agglutination reaction (hemagglutination tests)** for explanation.

16. a
- RIA is the most sensitive test for detection of Ag from given options. Sensitivity of different tests for Ag detection is 0.5–1 IU/ml in RIA, 1–2 mIU/ml in ELISA, 0.2 IU/ml in direct agglutination and 0.5–1 IU/ml in agglutination inhibition.

17. d
- RIA is useful in quantitation of hormones, drugs, tumor markers, IgE and viral antigens.

18. b
- In hemagglutination inhibition test, free (unlabeled) Ab is used.

Complement System

INTRODUCTION

History

Hans Ernst August Buchner (1889): Who 1st observed that bactericidal effect of serum was destroyed by heating at 55°C for 1 hour.

Pfeiffer (1894): He discovered that *V. cholerae* was lyzed when injected intraperitoneally into specifically immunized guinea pig called bacteriolysis *in vivo* or Pfeiffer's phenomenon.

Jules Bordet (1895): He observed that serum has two factors like heat stable Ab (except IgE) and heat labile called alexine.

Paul Ehrlich: He replaced the term alexine with complement, because having complementary action like Ab.

Definition

System of normal serum factors which is activated by Ag-Ab reaction and takes part in many biological activities called complement system.

Properties

Contains: It contains around 20–30 different serum proteins (glycoprotein) such as complement components from C1 to C9 (separated by electrophoresis), properdin system and regulatory proteins. It constitutes nearly 5% of the total serum proteins. All complement components come in sequence, except C4 comes after C1, but before C2. C1 is Ca^{+2} dependent and on chelation with EDTA it forms three protein subunits called C1q, r and s.

Concentration: Major component is C3 having MW of 1,95,000. C3 is in highest (1500 µg/ml) concentration while C2 (15 µg/ml) is in lowest concentration. Concentration (in µg/ml) of C1q, C1r, C1s, C4, C5, C6, C7, C8 and C9 are 150, 50, 50, 300, 80, 70, 65, 80 and 200, respectively. Their concentration will not rise by immunization.

Biosynthesis of complement components: C1 synthesized in intestinal epithelium, C2 and C4 in macrophages, C5 and C8 in spleen while C3, C6 and C9 in liver.

Temperature effect: C1 and C2 are heat labile (inactivated at 56°C in 30 minutes) while C3 and C4 are relatively heat stable. Complement as a whole is heat labile.

Binding/fixation/consumption: It binds with Ag-Ab complex, not to free Ag or Ab. Fixation is affected by Ab type such as. IgM (+++) > IgG (++, IgG3 > IgG1 > rarely IgG2, in that order) are the only immunoglobulin capable of activating C, but not IgG4, IgA, IgD or IgE. Its fixation is not affected by Ag type. It binds with Fc piece of Ig.

Presence: Complement presents in mammals, animals, birds, fish and amphibians.

COMPLEMENT PATHWAYS

Model of C Activity

It is explained by RBCs lysis with Ab. Erythrocyte (E) makes complex with antibody (A) called EA. When C components are attached to EA, it called EAC that is followed by their numbers like EAC14235 or EAC1-5. Biological (like enzymatic) activity is shown as bar ($\overline{C1}$) over components. Fragments cleaved during cascade are indicated by small letters like C3a, C3b, etc. Inactivated components are presented by using prefix 'i' like iC3b.

Complement Activation

It normally presents in the body in an inactive form. Its activity induced by Ag-Ab combination or other stimuli. C components react in a specific sequence as a cascade. Cascade is a series of reaction in which the preceding components act as enzymes on the succeeding components, cleaving in to dissimilar fragments. The larger fragments usually join the cascade. The smaller fragments which are released often possess biological effects which contribute to defense mechanisms.

Complement Pathways

Three pathways and final step is identical in all 3 pathways.

Classical pathway

- **Definition:** The chain of events in which C components react in specific sequence following activation of C1 and typically culminate in immune cytolysis is called the classical pathway.
- **Properties:** Classical pathway is initiated by formation of an Ag-Ab complex (Ab dependent). Useful antibodies are IgM (+++) > IgG (++, IgG3 > IgG1 and rarely IgG2). It also initiated by other stimuli like DNA, CRP, trypsin like enzymes or by retro viruses. It was identified at first. It is useful in active immunity.
- **Steps:** Follow **Flowchart 21.1**.
1. **C1:** Classical pathway is initiated by binding of C1 to Ag-Ab complex. It begins with the binding of C1 to the Ag-Ab complex. The recognition unit of C1 is $\overline{C1q}$, which reacts with the Fc piece of bounded IgM or IgG. C1q has six combining sites. Effective activation occurs only when C1q is attached to Ig by at least 2 of its binding sites. C1q binds in the presence of calcium ions leads to sequential activation of C1r and s ($\overline{C1qrs}$ = C) called Ag-Ab C or C1qrs complex or C1s esterase.

2. **C4:** It cleaved by $\overline{C1s}$ esterase, into two fragments like C4a which is an anaphylatoxin and C4b which binds to cell membrane along with C1qrs and forms the complex called Ag-Ab $\overline{C4b}$ or Ag–AbCqrs4b or $\overline{C4b}$.

3. **C2:** C4b in the presence of magnesium ions cleaves C2 into two fragments like C2a which remains linked to cell bound $\overline{C4b}$ and C2b which released into the fluid phase, having kinin like activity and increases vascular permeability. $\overline{C4b2a}$ has enzymatic activity and called C3 convertase.

4. **C3:** C3 convertase splits C3 into two fragments like C3a which is an anaphylatoxin and C3b which remains cell-bound along with $\overline{C4b2a}$ to form a trimolecular complex called $\overline{C4b2a3b}$ which has enzymatic activity and is called C5 convertase.

5. **C5:** Membrane attack phase starts at this level. C5 convertase cleaves C5 into two fragments like C5a which released into the medium and having anaphylatoxic and chemotactic activities (chemo-attractant). C5b is continue with cascade.

6. **C6 and C7:** They join together with C5b and form the heat trimolecular complex $\overline{C5b67}$, which has following two functions:
   - Bystander cells: Most of $\overline{C5b67}$ escape and serve to amplify the reaction by adsorbing on to unsensitized 'bystander cells' and rendering them susceptible to lysis by C8 and C9.
   - $\overline{C5b67}$ binds to cell membrane and prepare it for lysis by C8 and C9.

7. **C8 and C9:** They bind with $\overline{C5b67}$, which bounded to the cell membrane and form $\overline{C5b6\text{-}9}$ complex called membrane attack complex (MAC) causing lysis.

8. **Chemotaxis and cell lysis:** The unbound $\overline{C5b67}$ has chemotactic activity, though the effect is transient due to its rapid inactivation. The mechanism of complement mediated cytolysis is the production of holes approximately $100°A$ in diameter on the cell membrane; this disrupts the osmotic integrity of the membrane, leading to the release of the cell contents (increase permeability).

Alternative or properdin pathway

- **History:** It was 1st described by Pillemer in 1954 (antimicrobial protein present in serum).
- **Definition:** Activation of C3 without prior participation of C1, C4 and C2 is called alternative pathway.
- **Properties:** The alternate pathway bypasses participation of $\overline{C4b2a}$ (C3 convertase) and activates

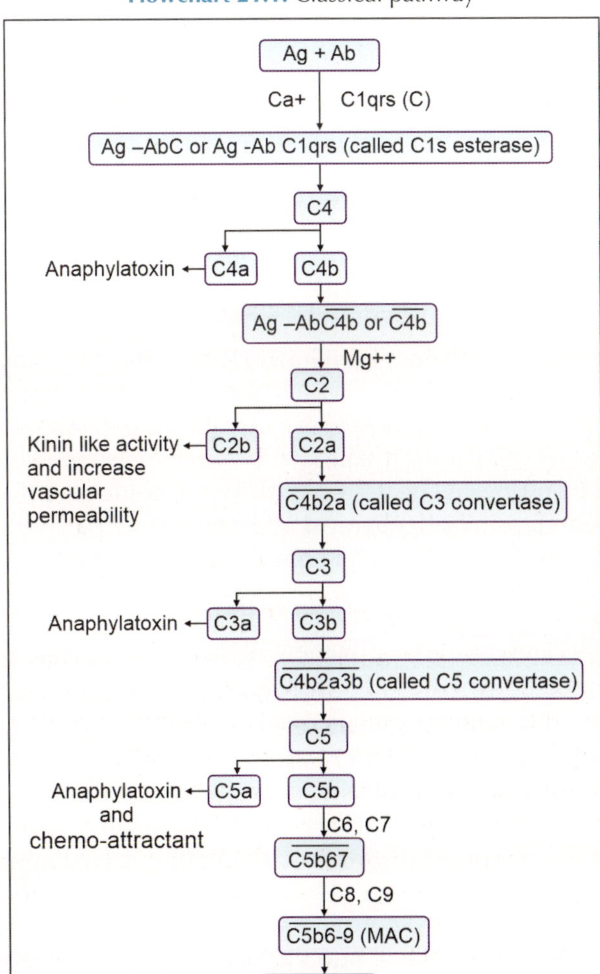

Flowchart 21.1: Classical pathway

C3 directly. It is antibody independent, but some Abs can stimulate this pathway like IgG4, IgA and IgD. It is stimulated by zymosan (yeast cell wall polysaccharide), bacterial endotoxin, IgG4, IgA, IgD, cobra venom, nephritic factor (present in patient with acute glomerulonephritis-AGN), teichoic acid from gram-positive cell walls, certain viruses, parasites, heterologous red cells, etc. It is a part of innate immunity.

- **Steps:** Follow **Flowchart 21.2**.
1. **C3b:** C3b continuously generated in serum, but free form is inactivated by factors H and I. It is protected from such inactivation by factor B also called C3B proactivator. C3b binds with factor B in presence of Mg^{++} to form C3bBb complex. Factor B is cleaved by factor D (also called C3B proactivator convertase) in to two fragments like Ba which released into the medium and Bb which binds with C3b to form C3bBb complex. C3bBb complex acts like C3 convertase (C4b2a), but it is extremely labile and loses its activity; but binding with properdin, it forms PC3bBb complex and becomes stable.
2. **Other steps:** After formation of C3 convertase further steps are like classical pathway.

Lectin or mannose-binding lectin (MBL) pathway

- **Definition:** C4 activation can be achieved by lectin without Ab and C1 participation called lectin pathway.
- **Properties:** This pathway is stimulated by combination of mannose residues present on microbial surface (like *Salmonella, Neisseria, Listeria, C. neoformans* and *C. albicans*) with MBL of serum and later combination with two specific protease like MASP1 (Mannan-binding lectin-associated serine proteases) and MASP2. It is antibody-independent and part of innate immunity. MBL is an acute phase protein, present in serum and produced in inflammatory responses.
- **Steps:** Follow **Flowchart 21.3**.
1. **C4:** In this pathway MBL binds to certain mannose residues present on many microbial surfaces and

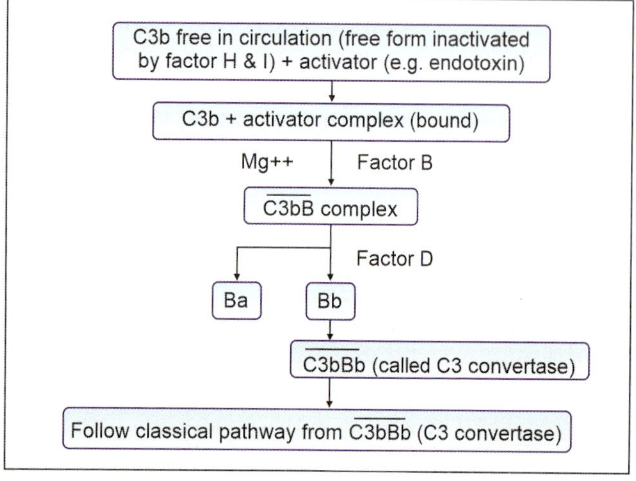

Flowchart 21.2: Alternative pathway

Flowchart 21.3: MBL pathway

subsequently interacts with MASP1 and MASP2. The Mannose residues-MBL-MASP1-MASP2 complex is similar to Ab-C1qrs complex (of classical pathway) and leads activation of C4 and cleaved it in to two fractions like C4a which is an anaphylatoxin and C4b.
2. **C2:** C4b in the presence of magnesium ions cleaves C2 into two fragments like C2a which remains linked to C4b and forms C4b2a complex and C2b which released into the fluid phase, having kinin like activity and increase vascular permeability. C4b2a has enzymatic activity and is referred to as the classical pathway C3 convertase.
3. **Other steps:** The rest steps are same as in classical pathway.

Highlights of Complement Pathways

It is mentioned in **Flowchart 21.4**. Three pathways converge at the MAC, which includes C5b-9. C5b-9 makes holes or channels in cell membrane of pathogens or target cell through which cell contents are come out and resulting cell death. C3 is the major components. It is common in both classical and alternative pathways and also acts as center for complement pathways.

Regulation of Complement Pathways

Unchecked complement activation may cause exhaustion of the complement system and serious damage to tissues hence, regulation of complement activation is necessary. Following are two types of control mechanisms:
1. **Inhibitors:** They bind to C components and halt the further functions.
   - **C1 esterase inhibitors (C1s INH):** It is heat labile and α-neuroaminoglycoprotein in nature. It does not prevent the normal progress of the complement cascade but check autolytic prolongation. It inhibits the other esterase found in serum like kininogen, plasmin and hageman factor.

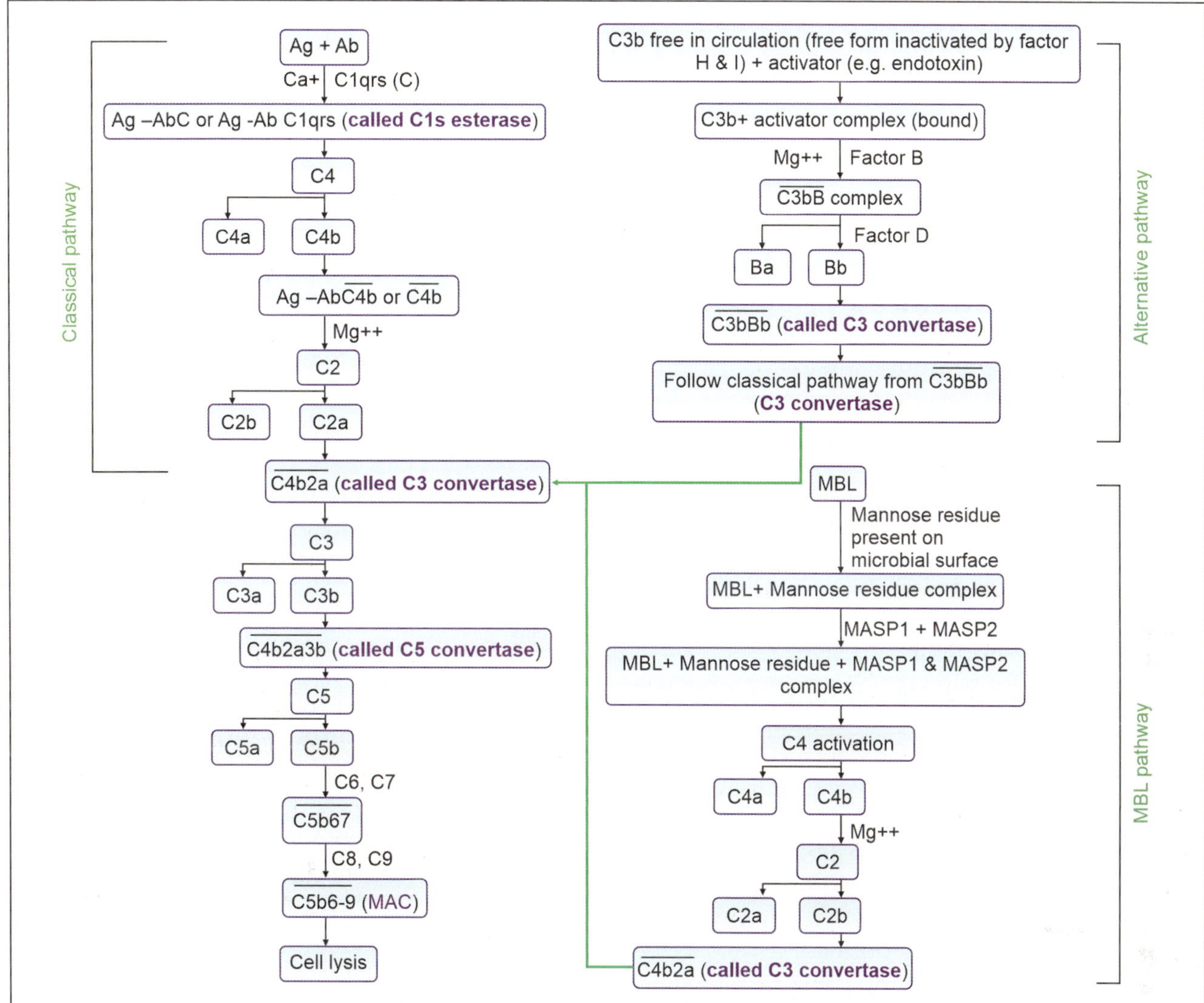

- **S protein:** It presents in serum. It binds to C67 and prevents their insertion into cell membrane modulating the cytolytic activity of the MAC.
2. **Inactivators:** These are the enzymes that destroy the complement proteins.
   - **Factor-I:** It is a serum β-globulin that control C3 activation, particularly by the alternate pathway. It also cleaves and inactivates C3b and C4b.
   - **Factor-H:** It is an another β-globulin that acts in cooperation with factor-I and modulate C3 activation
- **C4 binding protein:** It controls the activity of cell bound C4b.
- **Anaphylatoxin inactivators:** It is an α-globulin that enzymatically degrades C3a, C4a and C5a which are anaphylatoxins released during the complement cascade.
- **Other regulators:** Like decay-accelerating factor, homologous restriction factors, membrane cofactor protein, etc., have been reported to modulate the activity of the complement.

Measurement of Complement System

Measurement of complement activity: It is done by using hemolytic activity by estimating highest dilution of serum sheep RBCs lyzed by anti-RBCs antibody.

Measurement of complement components: It is done by using hemolytic activity and by radial immuno-diffusion in agar.

HEALTH AND COMPLEMENT

Biological Effects of Complement

1. **Neutralization of viruses:** It neutralizes the viruses.
2. **Anticancer activity:** It causes death of tumor cells.
3. **Hemolysis:** It lyses the RBCs by classical pathway.
4. **Bactericidal activity**
   - GNB: The complement renders GNB susceptible to lysozyme by binding holes in LPS layer which protects inner lysozyme sensitive layer of peptidoglycan and thus makes GNB sensitive to lysis (bacteriolysis).

- GPC are not susceptible to the lytic action because of their cell wall composition; however, they killed by complement without lysis.

5. **Phagocytosis:** C3b and C4b act as opsonins. CRs like C1, C4, C2, C3 and C1q present on phagocytic cells are the receptors for complement components like C3b and C4b help in opsonization.

6. **Pathogenesis of diseases:** Complement receptors-2 (CRs2/CD21on B cells) provide site for adhesion of Epstein-Barr virus (EBV).

7. **Immune adherence:** C bound to Ag-Ab complexes, Ab sensitized cells and viruses make them to adhere to cells possessing immune adherence receptors (macrophages, B-lymphocytes, primate RBCs, etc.) and increase susceptibility to phagocytosis and increase the body defense.

8. **Inflammatory:** C2b kinin released during cascade that increases the vascular permeability.

9. **Autoimmune disease:** C components are decrease in SLE and RA. They play major role in autoimmune hemolytic anemia.

10. **Hypersensitivity reaction**
    - **Anaphylatoxin activity:** C4a, C3a and C5a stimulate the release of histamine from mast cells that causes constriction of smooth muscles, increases vascular permeability and vasodilatation.
    - **Type II hypersensitivity reaction:** Such as RBCs destruction following incompatible transfusion.
    - **Type III hypersensitivity reaction:** C is required for serum sickness and arthus reaction, e.g., kidney damage in nephrotoxic nephritis.

11. **Conglutination:** Bovine serum contains β-globulin component called conglutinin (K), which causes clumping of particles or cells coated with complement called conglutination. It reacts with bound C3 in presence of Ca^{+2}. It is not Ab, but when K coated with C is injected, it produces Ab called immunoconglutinin (IK). It occurs as autoantibody in human and acts on fixed C. For conglutinating CFT follow **Ch. 20.**

12. **Endotoxic shock:** Endotoxin activates the alternative pathway. In endotoxic shock there is large amount of C-fixation and platelet adherence, resulting release of large amount of platelet factors causing DIC and thrombocytopenia, e.g., water-house Friderichsen syndrome, dengue hemorrhagic syndrome, etc.

13. Depletion of C protects against **Shwartzman reaction.**

14. **Rise in acute infection:** C components are acute phase proteins and their level rise particularly C3, C4, C5 and C6 in acute infections.

15. **Diagnostic role:** CFT is used to diagnose the different infection which is described in **Ch. 20.**

Note: Hereditary angioedema
- **Mechanism:** Reduced amount of C1 inhibitor leads to autolytic activation of C1 and uncontrolled breakdown of C4 and C2. C2b is the main mediator for edema.

- **Clinical features:** It presents with angioedema of subcutaneous tissues or mucosa of GIT or respiratory tract.
- **Treatment:** It is treated by infusion of fresh plasma as source of inhibitors. Administration of epsilon aminocaproic acid or its analogs inhibits the activation of plasma enzymes and sparing the small amount of the C1 inhibitor present.

Deficiencies of Complement Components

Disease produced due to deficiency of complement components are shown in **Table 21.1**.

TABLE 21.1: Diseases due to deficiency of complement components	
Deficiency	**Disease**
C1inhibitor	Hereditary angioedema
C1s, C4, C2	SLE and other collagen diseases
C3, regulatory proteins and C3b inactivator	Pyogenic infections
C5-8	GN-bacteremia and toxoplasmosis
C9	No particular disease
C5-9	*Neisseria* like meningococcus and gonococcus

ACCESS YOURSELF

Case Study

1. An 8-year-male child brought to skin OPD with history of pyogenic infection like boil in nose. He completed the antibiotic course and maintaining the hygiene, in spite of all these recurrence is there. Serum investigation reported the deficiency of complement component. Identify the case and answer the following:
   a. Name the complement component deficient in given case.
   b. Name the complement components and disease produced due to their deficiency.
   c. Write biological effects of complement.

Essay/Full Question

1. Complement system.

Short Notes

1. Classical pathway of complement.
2. Biological effects of complement.

Short Questions for Theory/Viva Questions

1. What is Pfeiffer's phenomenon?
2. What is conglutination?
3. Write the origin for following complement components
   C1, C2, C3 and C4
4. Name complement components and related disease produced due to their deficiency.
5. Name the C3 convertase for classical, alternate and MBL pathways of complement.

Comments on

1. Complement has direct bactericidal (lytic) effect on GNB but not on GPC.
2. Complement has ability to produce anaphylatoxin.
3. C3 acts as center for complement pathways.
4. C1 inhibitor deficiency produces angioedema.

MCQs for Chapter Review

Introduction

1. Which component of complement has highest concentration in serum?
   - a. C1
   - b. C2
   - c. C3
   - d. C9

2. Which component of complement has lowest concentration in serum?0
   - a. C1
   - b. C2
   - c. C3
   - d. C9

3. Complement formed in liver:
   - a. C2, C4
   - b. C3, C6, C9
   - c. C5, C8
   - d. C1

Complement Pathways

4. C3 convertase acts on:
   - a. C4b2b
   - b. C4b2B3a
   - c. C4b
   - d. C3

5. Which complement component is involved in both classical and alternative pathways?
   - a. C1
   - b. C2
   - c. C3
   - d. C4

6. Center for complement pathway:
   - a. C3
   - b. C1
   - c. C5
   - d. C2

7. C3 convertase in alternative complement pathway:
   - a. C4b2a
   - b. C3b
   - c. C3bBb
   - d. C3a

8. Chemoattractant is:
   - a. C5a
   - b. C1
   - c. C3
   - d. C2

9. In cell lysis by complement
   - a. It activates cyclase
   - b. Inhibits elongator factor p
   - c. Destruction of P
   - d. Increased permeability of cell membrane

10. Which of the following best denotes classical complement pathway activation in immuno-inflammatory condition?
    - a. C2, C4 and C3 decreased
    - b. C2 and C4 normal, C3 decreased
    - c. C3 normal, C2 and C4 decreased
    - d. C2, C4 and C3 all are elevated

Health and Complement

11. Deficiency of C1 inhibitor in complement system produces:
    - a. Hereditary angioedema
    - b. Toxoplasmosis
    - c. Bacteremia
    - d. Collagen disease

12. Hereditary angioneurotic edema is due to:
    - a. Deficiency of C1 inhibitors
    - b. Deficiency of NADPH oxidase
    - c. Deficiency of MPO
    - d. Deficiency of properdin

13. Which deficiency would cause *Neisseria* infection?
    - a. C5
    - b. C6
    - c. C9
    - d. C7
    - e. C8

14. Deficiency of C5–C9 complement components predispose to which infection?
    - a. Meningococci
    - b. Pneumococcal
    - c. *Pseudomonas*
    - d. All

Answers and Explanation of MCQs

1. **c**

2. **b**
   - Follow section, **introduction (properties → concentration)** for explanation of answers of MCQs 1–2.

3. **b**
   - Follow section, **introduction (properties → biosynthesis of complement components)** for the explanation.

4. **d**
   - Follow section, **complement pathways (Flowchart 21.1)** for the explanation.

5. **c**
   - Follow section, **complement pathways (Flowchart 21.1 and Flowchart 21.2)** for the explanation.

6. **a**
   - Follow section, **highlights of complement pathways (Flowchart 21.4)** for the explanation.

7. **c**
   - Follow section, **complement pathways (Flowchart 21.2)** for the explanation.

8. **a**
   - Follow section, **complement pathways (classical pathway → steps)** for the explanation.

9. **d**
   - Follow section, **complement pathways (classical pathway → steps → Chemotaxis and cell lysis)** for the explanation.

10. **a**
    - **Flowchart 21.4** shows all the pathways of complement. In classical pathway C1–C9 are utilized and decreased, while in alternative pathway all are decrease except C1, C2 and C4.

11. **a**

12. **a**

13. **a to e**

14. **a**
    - Follow **Table 21.1** for the explanation of answers of MCQs 11–14.

Structure and Functions of Immune System

Chapter Outline

- Immune System
  - Specific immune system/lymphoid system
  - Nonspecific immune system
- Major Histocompatibility Complex (MHC)

IMMUNE SYSTEM

Synonym

It also called lymphoreticular system that means it includes lymphoid system and reticuloendothelial (RE) system.

Definition

It is a complex organization of cells of diverse morphology distributed widely in different organs and tissues of the body responsible for immunity.

Parts of Immune System

There are two parts of immune system like specific immune system/lymphoid system and nonspecific immune systems as described below.

Specific Immune System/Lymphoid System

It produces the specific immune response called acquired immunity, which may be either AMI or CMI. It has two components like lymphoid organs and lymphoid cells as described below.

A. Lymphoid organs: Follow **Flowchart 22.1**.

1. Primary (central) lymphoid organs or tissues

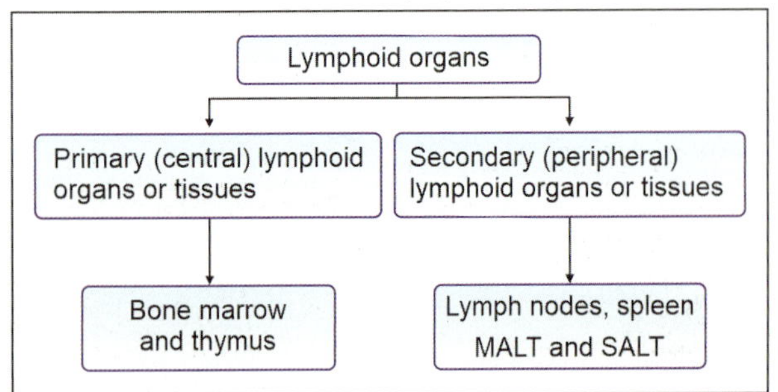

Flowchart 22.1: Lymphoid organs

Bone marrow (Bursa of fabricius in birds)

- **Origin of all blood cells:** All the blood cells are originated from the hematopoietic stem cells of bone marrow by the process called hematopoiesis. In early fetal life cells are derived from liver and yolk sac, which gradually migrate to bone marrow. By the birth, almost all stem cells occupy the position in bone marrow of large bones. With advancing age hematopoietic activity of large bones is decreased and after puberty hematopoiesis is mostly confirmed to axial bones like pelvis, vertebras, skull, sternum and ribs.

- **Hematopoiesis:** Word derived from Hemato (Greek) means blood and poiesis means to make up. It also called hemopoiesis. It is a process of formation of blood cells. Hemapoietic stem cells reside in the medulla of the bone called bone marrow and have the ability to form the all blood cells as shown in **Fig. 22.1**.

- **Maturation of cells in bone marrow:** The precursors of B-cells enter in to bone marrow from fetal liver, yolk sac and mature in B-cells (B means bursa of Fabricius in birds or bone marrow in humans).

- **Function:** Bone marrow allows the maturation of B-cells which produce AMI.

Thymus

- **Embryology:** It develops from the epithelium of the 3rd and 4th pharyngeal pouches at about 6th week of gestation. It is thus the first organ in all animal species to become predominantly lymphoid.

- **Anatomy:** It located behind the upper part of sternum in the thoracic cavity overlying heart and major blood vessels. It has two lobes surrounded by fibrous capsule. Septa arising from capsule divide the gland into an outer cortex and inner medulla as shown in

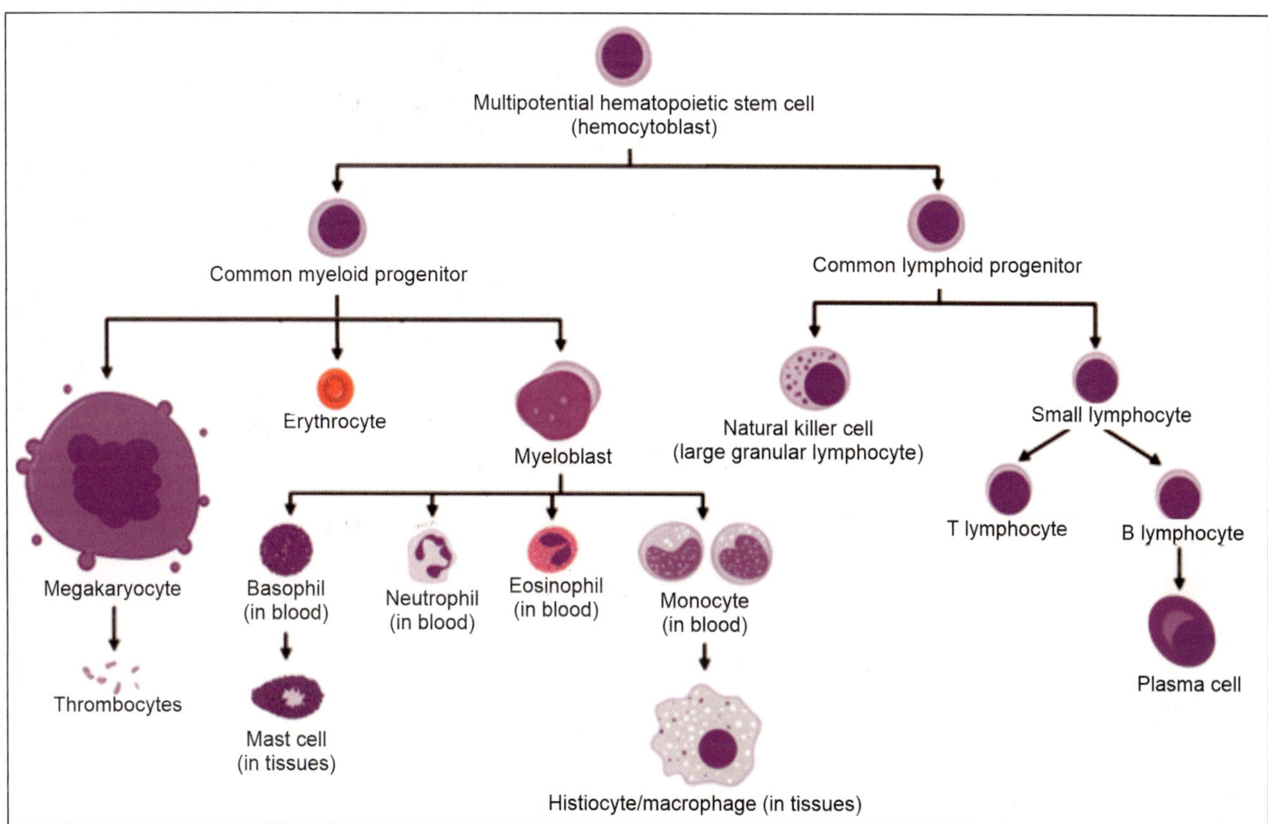

Fig. 22.1: Hematopoiesis

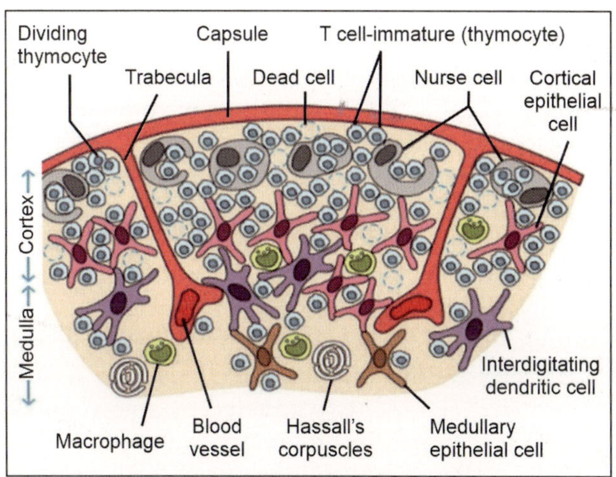

Fig. 22.2: Cross section of thymus

Fig. 22.2. Outer cortex contains plenty of immature T cells (also called thymocytes), nurse cells (special type of epithelial cells that surround the thymocytes) and cortical epithelial cells (look like star shape). Inner medulla contains T-cells (thymocytes are less in numbers but relatively mature), degenerated epithelial cells (aggregated in whorls like pattern called Hassall's corpuscles), medullary epithelial cells (look like star shape), and interdigitating dendritic cells.

- **Functions**
  - Maturation of T-cells in thymus: The precursors of lymphocytes enter into thymus from bone marrow, fetal liver, yolk sac and mature in cortex, acquire surface properties of T-lymphocytes (T means thymus dependent) and than migrate in to medulla, where they completely get mature and exit in to blood as mature T-cells. In thymus the cells become educated to mount CMI against appropriate Ags and seeded in to secondary lymphoid organs.
  - Central tolerance: About 95% cells entered in thymus are died and 5% released in circulation. The cells died are, may be not efficient to recognize MHC or believed to be self reacting in nature. Destruction of self reacting T cells prevents the development of autoimmunity. Such tolerance to self antigens mediated by thymus is called central tolerance.
  - Synthesis of hormones: Thymus hormones like thymulin, thymosin and thymopoietin are produced by epithelial cells and role in attraction of precursor T cells from bone marrow.

- **Clinical significances**
  - Di George syndrome: Congenital aplasia of the thymus leads deficiency of CMI in human called Di George syndrome and in mouse called nude mouse.
  - Runt disease: Deficiency of CMI evident from lymphopenia and deficient graft rejection.
  - Post-thymectomy effect: Thymus dependent areas in peripheral organs are grossly depleted. After originating from bone marrow and getting maturity in thymus, mature T cells enter in certain areas of peripheral lymphoid organs called thymus dependent areas. These include paracortical area in lymph nodes and white pulp in spleen. After thymectomy, mature T cells (thymus) will be

absent, so T cells (thymus) dependent areas will be depleted of T cells. Thymectomy decreases the CMI and AMI for thymus dependent Ag.

2. Secondary (peripheral) lymphoid organs or tissues

Lymph nodes

- **Anatomy:** Lymph nodes are placed along the course of lymphatic vessels. They are surrounded by a fibrous capsule. Penetrated part of capsule into the nodes called trabecula. They have three parts like outer cortex, middle between cortex and medulla called paracortical area and inner medulla as shown in **Fig. 22.3**. In the cortex, there is accumulation of lymphocytes (primary lymphoid follicles) within which germinal centers (secondary follicles) develop during antigenic stimulation. Follicles also contain dendritic macrophages, which capture and process the antigen. Paracortical area is a thymus dependent area and it contains T-lymphocytes and interdigitating cells. Inner medulla contains B-lymphocytes, plasma cells and macrophages which are arranged as elongated branching bands called medullary cords.
- **Bursa-dependent areas:** Cortical follicles and medullary cords contain B lymphocytes and constitute the bursa-dependent areas.
- **Functions of lymph nodes**
  - They act as a filter for foreign particles.
  - They phagocytose the foreign materials including microorganisms and help in development, proliferation and circulation of T and B cells.
  - They enlarge following local antigenic stimulation.

Spleen

- **Anatomy:** Spleen is the largest lymphoid organ. It is surrounded by a fibrous capsule. Penetrated part of capsule into the spleen called trabecula which divides the organ into several interconnected compartments. It has two parts like white pulp and red pulp as shown in **Fig. 22.4**. The branches of the splenic artery travel along the trabecula and on leaving them branch again to form central arterioles, which are surrounded by a sheath of the lymphoid tissue contains T-lymphocytes called white pulp. White pulp is a thymus dependent area. The periarterial lymphoid collection in the white

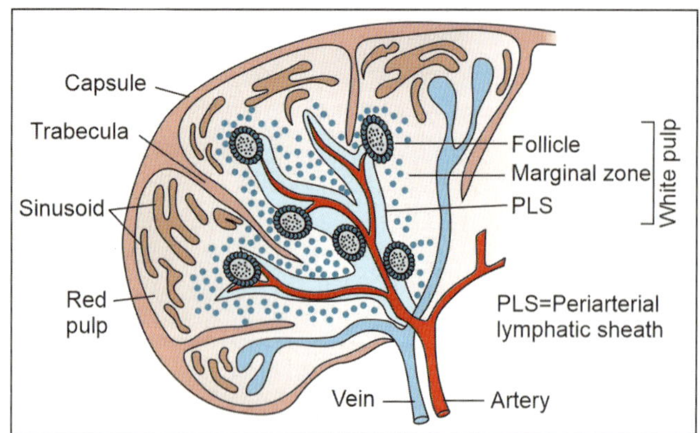

Fig. 22.4: Cross section of spleen

pulp of spleen is called malpighian corpuscles or follicles. The central arterioles proceed into the red pulp, so called because of the abundance of RBCs in it.

- **Functions:** Spleen helps to filter blood borne antigens and to synthesize the Abs against blood borne antigens.
- **Clinical significance of spleen**
  - Post-splenectomy effect: The spleen contains many macrophages (part of RE system), that phagocytose many bacteria. These macrophages are activated when bacteria are bound by IgG1 or IgG3 or C3b. These types of antibodies and complement are immune substances called opsonins molecules that bind to the surface of bacteria to facilitate phagocytosis. Immunity in forms of IgG and C3b is the human immune system response against bacterial capsules. It is done by slow passage of blood through the splenic sinuses and prolonged contact with RE cells in the cord of Billroth. When the spleen is no longer present (asplenia), IgG and C3b are still bound to bacteria, but they cannot be removed from the blood due to the loss of the splenic macrophages. Hence, the bacteria are free to cause infections. Deficiency of tuftsin, a tetrapeptide secreted by spleen plays a role in combating pneumococcal sepsis. Hence splenectomy leads to infections by encapsulated bacteria like *S. agalactiae S. pneumoniae, N. meningitidis, H. influenzae, Kleb. pneumoniae*, etc., and parasite like *Babesia* mostly in children called overwhelming post-splenectomy infection or sepsis.

Mucosa-associated Lymphoid Tissue (MALT)

- **Anatomy:** Mucosa lining the alimentary, respiratory, genitourinary and other surfaces is constantly exposed to numerous antigens. So, these areas are endowed with a rich collection of lymphoid cells or follicles which are collectively called MALT. It consist mixture of T-cells, B-cells and phagocytes. MALT includes following types.
  - **Gut-associated Lymphoid Tissue (GALT):** Lymphoid tissue lining the intestinal mucosa called GALT. It has following structure and functions.

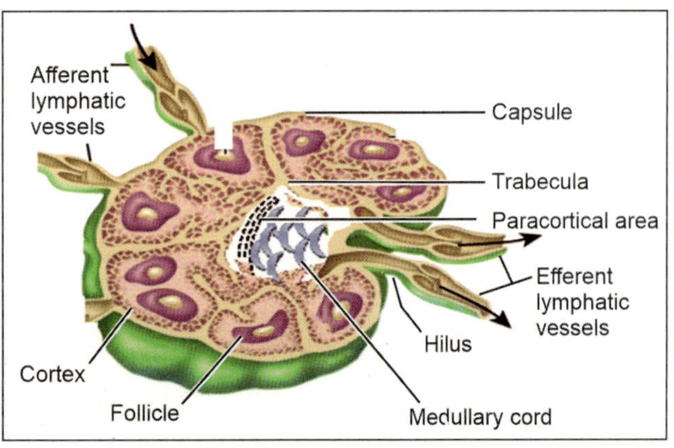

Fig. 22.3: Cross section of lymph nodes

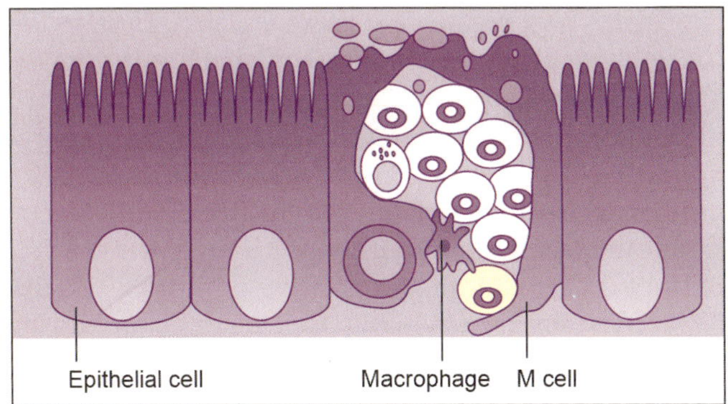

Fig. 22.5: M cells

– Epithelial layer: In includes many cells such as CD 7 2 3 γδ TCR cells (called intraepithelial lymphocytes), B cells (present in lining epithelium, lamina propria and submucosa and they secrete the secretory type of IgA which provides the local immunity), epithelial cells with microvilli and epithelial cells without microvilli called M cells. M cells do not have microvilli on surface but having pockets or invagination on basolateral side that contains T cell, B cells and macrophages as shown in **Fig. 22.5**. Invading microbes like *Vibrio, Salmonella, Shigella*, polio virus, etc., are taken up by M cell by endocytosis.
– Lamina propria: It contains B cell, Th cells and plasma cells.
– Submucosa: It contains peripheral lymphoid tissue called Peyer's patches.

■ **Tonsils:** Lingual, palatine and pharyngeal type.
■ **Bronchus-associated lymphoid tissue (BALT):** Lymphoid tissue of respiratory tract called BALT.
• **Function of MALT:** It serves for Ig production. The predominant Ig produced in the mucosa is secretory IgA and others are IgG, IgM and IgE provide local immunity.

Skin-associated lymphoid tissue (SALT): Skin also has few loose lymphocytes called Langerhans cells, act as APCs. Few authors described that Langerhans cells are the type of immature dendritic cells. (Described later in this chapter under the section of APCs).

B. Lymphoid cells: These cells provide efficient, specific and long-lasting immunity against microbes and are responsible for acquired immunity. They have following three types:
1. **Cells with T-cell receptors (TCRs):** Called lymphocytes like T cells and B cells (plasma cells).
2. **Cells without TCRs:** Called null cells or large granular lymphocytes (LGLs) because contain large azurophilic cytoplasmic granules. Examples of null cells are natural killer (NK) cells, antibody dependent cytotoxic cells (ADCCs) and lymphokine-activated killer (LAK) cells. They produce nonspecific immune response.

3. **Antigen-presenting cells (APCs):** These are two types like major cells or professional cells and minor cells or nonprofessional cells.

1. **Cells with T-cell receptors (TCRs)/lymphocytes**
• **Morphology of lymphocytes:** Lymphocytes constitutes 20–40% of WBCs and 99% cells in lymph. They are round in shape. Nucleus is round with prominent chromatin materials. Cytoplasm is thin with scattered ribosomes, but without endoplasmic reticulum, Golgi apparatus and other organelles. They are slowly motile and assume a hand mirror like appearance with nucleus in front and cytoplasmic tail.
• **Subtypes of lymphocytes**
– Based on size: This include small (5–8 μ), medium (8–12 μ) and large (12–15 μ) lymphocytes.
– Based on life span: This include short lived and long lived lymphocytes. Short lived lymphocytes last about 2 weeks and generate immune response. They perform functions of regulatory cells and effector cells also. Long lived lymphocytes last about 3 years or whole life and generate immune memory. They perform functions of memory cells.
– Based on site of maturation: This include T lymphocytes and B lymphocytes. T lymphocytes secrete lymphokines (lymphotoxins) which are useful in CMI and AMI. B lymphocytes differentiated into plasma cells to secrete antibodies that produce AMI.
– Based on functions: This include regulatory cells, effector cells and memory cells. **Regulatory cells** are involved in immune regulation. They perform the functions of helper/inducer (CD4) cells and suppressor cells (CD8). These suppressor cells do not include cytotoxicity. Regulatory cells regulate the Ir by two ways such as (1) increase functions of inducer cells and decrease functions of suppressor cells cause autoimmunity and (2) decrease functions of inducer cells and increase functions of suppressor cells cause IDDs. **Effector cells** are active cells involved in immune response. They perform the functions of cytotoxic (CD8) cells and plasma cells (CD19, B-cells). Cytotoxic (CD8) cells do not include suppression function but participate in CMI. Plasma cells (CD19, B-cells) participate in AMI. **Memory cells** are involved in immunological memory.
– Based on CD marker: CD means cluster of differentiation. Numbers of surface Ags or surface markers have been identified on lymphocytes and leukocytes by using monoclonal Abs called CD markers. When cluster of monoclonal Ab reacts against particular Ag defined as separate Ag or surface marker then separate CD number is given. CD markers or antigens are useful for differentiation of leukocytes and also to know the functional properties of cell. Around 364 CD markers have been identified. Few are mentioned in **Table 22.1**.

TABLE 22.1: CD markers

CD number	Cell type association	Former designation
CD1	• Presents on cortical thymocytes (in early stage only) and Langerhans cells • Composed on 3-polypeptide chain and β-globin	T6, Leu 6
CD2	• Receptors for sheep RBCs (SRBC) • Persists in all stage of maturation and in all mature T cells	T11, Leu 5
CD3	• Presents on all T cells, so called pan CD marker • Makes TCR-CD3 complex with TCR and transmit the signal interior to the cell following Ag binding on TCR	T3, Leu 4
CD4	Helper/inducer cells (HIV and HHV-7 receptors)	T4, Leu 3
CD8	Suppressor/cytotoxic cells	T8, Leu 2
CD19	B cells	B4, Leu 12

Normal range of CD4 in healthy person is 400–1600 cells/mm^3 and CD8 is 150–800 cells/mm^3. In normal adults and children **CD4 to CD8 ratio** is 2:1, it means CD4 constitute 65% and CD8 constitute 35% of total T cells or it means that there are about 2 CD4 cells for every CD8 cell out of total T cells. However, in newborn infants more CD4 and less CD8 cells. CD4 to CD8 ratio is 3.4–4:1, it means CD4 constitute 75–80% and CD8 constitute 20% of total T cells or it means that there are about 3.4–4 CD4 cells for every CD8 cell out of total T cells.

- **Lymphocytic recirculation:** There is constant circulation of lymphocytes in the blood, lymph, lymphatic organs and tissues to produce an immune response when Ag enters. Lymphocytes complete one cycle of recirculation in about 1–2 days. Recirculating lymphocytes can be recruited by lymphoid tissues. It is more in T cells, while B cells are more sessile (immobile).
- **Differences between T cells and B cells:** Follow **Table 22.2**.
- **Maturation of lymphocytes:** This includes maturation of T-cells, B cells and B1 (CD5) cells as described below.

Maturation of T Cells

Steps of maturation of T cells (**Fig. 22.6**) are described below.

Pro-T cells: Lymphocyte precursors called pro-T cells are developed in the fetal liver, bone marrow and yolk sac and enter in to thymus for maturation.

Pre-T cells

- Earliest identifiable lineage is CD7+ pro-T cells, which acquire CD2 on surface and CD3 in cytoplasm on entering thymus and becomes pre-T cells.
  - TCR (T-cell receptors) synthesis also takes place. TCR is a glycoprotein chain (**Fig. 22.7**). TCR occurs in two pairs of chain like αβ or γδ, which

TABLE 22.2: Differences between T-cells and B-cells

Features	T-cell	B-cell
Origin (same lineage)	All blood cells from bone marrow (also from liver and yolk sac in fetus)	
Site of maturation	Thymus	Bursa of fabricus/bone marrow
Amount	70–80% of blood lymphocytes	10–20% of blood lymphocytes
Rosette formation	Binding with sheep RBC (Erythrocytes = E) to form SRBC or E-rosette by CD2 receptor	Binding with sheep Erythrocytes (E) coated with Ab (A) and complement (C) to form EAC-rosette due to C3 receptors (CR2) on B cell surface.
CR2 (CD 21) as a receptors for EBV	–	+
Receptors for Fc piece of Ig	–	+
Thymus specific Ag	+	–
Microvilli on surface	–	+
Recognition of Ag by APCs before binding to surface receptors	Required (except in superantigen)	Not required
Function	Lymphokine (lymphotoxin) production which helps in CMI	Antibody production which helps in AMI
Lymphoblast transformation (multiplication) • Anti-CD3 • Phytohemagglutinin • Concanavalin A (Con A) • Anti-Ig • Endotoxin • *Staph. aureus* Cowan-I strain • EBV	 + + + – – – –	 – – – + + + +

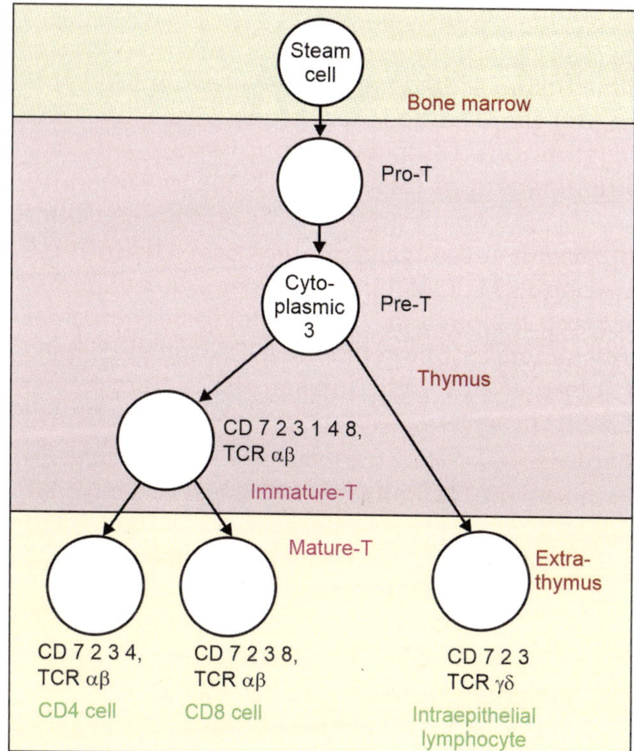

Figure 22.6: Maturation of T cells

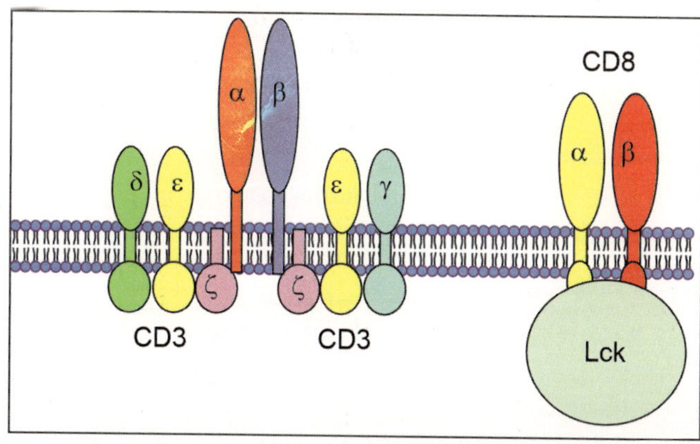

Fig. 22.7: Structure of T-cell receptor (TCR)

differentiate pre-T cells in to two lineages like αβ TCRs (present in 95% T cells) or γδ TCRs (present in 5% T cells). TCR has three domains like extracellular αβ domain (with variable and constant region in α and β chains), transmembrane domains (γε chains and δε chains) and cytoplasmic tail [CD3 molecule plus ζζ (zeta) chains]. Transmembrane domain plus cytoplasmic tail called CD3 complex while TCR with CD3 complex called TCR-CD3 complex. TCR is the site for binding of Ag. Following binding between Ag and TCR, a signal is generated that is transmitted through CD3 complex to activate the T cells. It allows the binding of only Ag recognized by APCs (except super antigen which binds directly to lateral side of β chain of TCR without earlier recognition by APCs). TCR is an analogous to Ig on B cell surface. Unlike TCR, Ag binds directly on B cell receptor without earlier recognition by APCs.

– HLA (MHC restriction) synthesis also takes place to respond to foreign Ag. Like CD4 cells react to MHC class-II molecule while CD8 cells react to MHC class-I molecule.

Immature T-cells: Immature cells in thymus contain CD 7 2 3 1 4 8 in addition to TCR and HLA.

Mature T-cells: They form outside the thymus. Immature cells differentiated into following type mature cells.

- **Naive cells:** Mostly stimulated by IL-1 of APCs like mature dendritic cells. Mature T-cell before undergoes to immune response (resting T cells) called naive cells. They help in recognition of Ag and storage of immunological memory.

- **CD 7 2 3 4 αβ TCR cells:** They called CD4 cells. Also called helper (Th) cells because help in CMI by producing lymphokines and called inducer cells because induce differentiation of B cells and proliferation of CD8 cells. They also regulate the erythropoiesis. They constitute 55–70% of the total T-cells. They contain receptors for adhesion of HIV and HHV-7. They have following subtypes.

  – Th0 cells: Original mature CD 4 cells which secrete the IL-2 having autocrine and paracrine effect called Th0 cells. IL-2 has autocrine effect and acts on Th0 cells to release IFN-γ, IL-4 and further release of IL-2. Later Th0 is differentiated in to effector cells (Th1 and Th2 cells) and memory cells. One new type Th17 also identified recently.

  – Th1 cells: They release IL-2, IFN-γ and TNF-β. Th1 cells have different functions. **(1)** Th1 cells produce **CMI** by activation of CD8 (Tc) cells by IL-2 (by endocytic pathway) which causes intracellular killing of bacteria (*M. tuberculosis, M. leprae,* etc.) and protozoa by release of free radical like NOx. **(2) Phagocytosis** by activation of macrophages. **(3)** Over activity of Th1 cells against autoantigens will cause **type 4 DTH** such as tuberculin reaction. **(4)** Over activity of Th1 cells against autoantigens will also cause **autoimmunity** such as type 1 diabetes.

  – Th2 cells: They are driven by IL-4, IL-5, IL-6, IL-10, IL-13 and other cytokines. Th2 cells have different functions. **(1)** Produce **AMI** by activation of B cells. **(2) Granuloma formation** in tuberculosis.

  – Th17 cells: These are new variants of Th0 cells, identified recently. They release IL-17 which promotes inflammation and autoimmune diseases like SLE and RA.

  – Memory cells: Concerned with recall phenomenon.

- **CD 7 2 3 8 αβ TCR cells:** They called CD8 cells. Also called cytotoxic cells, as they release cytokines which kill target cells like virus infected cells, tumor cells and allograft cells in transplanted tissues by endocytic or cytosolic pathway. Also called suppressor cells because suppress Ab synthesis by B-cells. They constitute 25–40% of the total T-cells. Some of these cells are converted into memory cells.

- **CD 7 2 3 γδ TCR cells:** They called intra epithelial lymphocytes. Their exact role is uncertain but may be useful for immune surveillance on epithelial surface (like intestinal epithelium) and defense against intracellular bacteria (innate immunity).

Maturation of B Cells

Steps of maturation of B cells (**Fig. 22.8**) are described below:

Pro-B cells: Lymphocyte precursors called pro-B cells, which are developed in the fetal liver and then in bone marrow continuous for life.

Pre-B cells: With synthesis of cytoplasmic IgM pro-B cells are converted into the pre-B cells.

Immature B-cells: Pre-B cells converted into immature B cells in bone marrow with synthesis of IgM and BCR on the cell surface. IgM molecule complexed with transmembrane molecule like Ig and Ig to form BCR as shown in **Fig. 22.9**. [In mature B cells, after combination of Ag on Fab region of Ig molecule, α and β molecule generate the signal to activate and to differentiate the B cells].

Essentials of Medical Microbiology

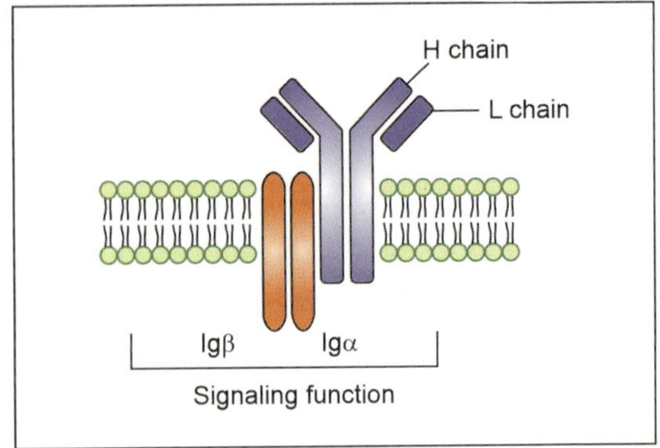

Fig. 22.9: Structure of B cell receptor (BCR)

Mature B cells: They form outside the bone marrow in peripheral organs. Immature cells converted into mature cells with synthesis of other Ig classes like IgG, IgD, IgA and IgE on surface. Immature cells differentiated into following type mature cells.

- **Naive cells:** Mature B cells before undergo to immune response (resting B cells) called naive cells. They help in recognition of Ag and storage of immunological memory.
- **Memory cells:** On contact with appropriate antigen, the mature B cells undergo clonal proliferation. Some activated B cells become long-lived memory cells responsible for the recall phenomenon. The majority of activated B cells transformed into plasma cells.
- **Plasma cells:** They are oval in shape. They are about twice the size of small lymphocytes. They have eccentrically placed oval nucleus containing large blocks of chromatin located peripherally (called cartwheel appearance). Their cytoplasm is large and contains abundant endoplasmic reticulum and a well developed Golgi apparatus. Plasma cells are end cells and have a short life span of two or three days. They secrete the antibodies and produce the AMI. Plasma cell is the best antibody producing cell, but lymphocytes, lymphoblast and transitional cells may also synthesize the Ab to some extent.

Maturation of B1 (CD5) Cells

Maturation: Separate lineages of B cells, which are predominant in fetal and early neonatal life, express the T cell markers CD5 on their surface and have been called B1 cells. Their progenitor cells move from the fetal liver to the peritoneal cavity where they multiply. They secrete low affinity polyreactive IgM antibodies, many of them are autoantibodies.

Functions: They produce T-independent 'natural' IgM antibacterial antibodies which appear in neonates seemingly without antigenic stimulus. CD5+ B cells may be relevant in the causation of autoimmune conditions.

2. Cells without TCRs/null cells/LGLs

- **Properties:** These include about 5–10% of total lymphoid cells. These are lymphocytes but produce nonspecific immune response. These are heterogeneous group of cells with difference in their functions and surface markers.

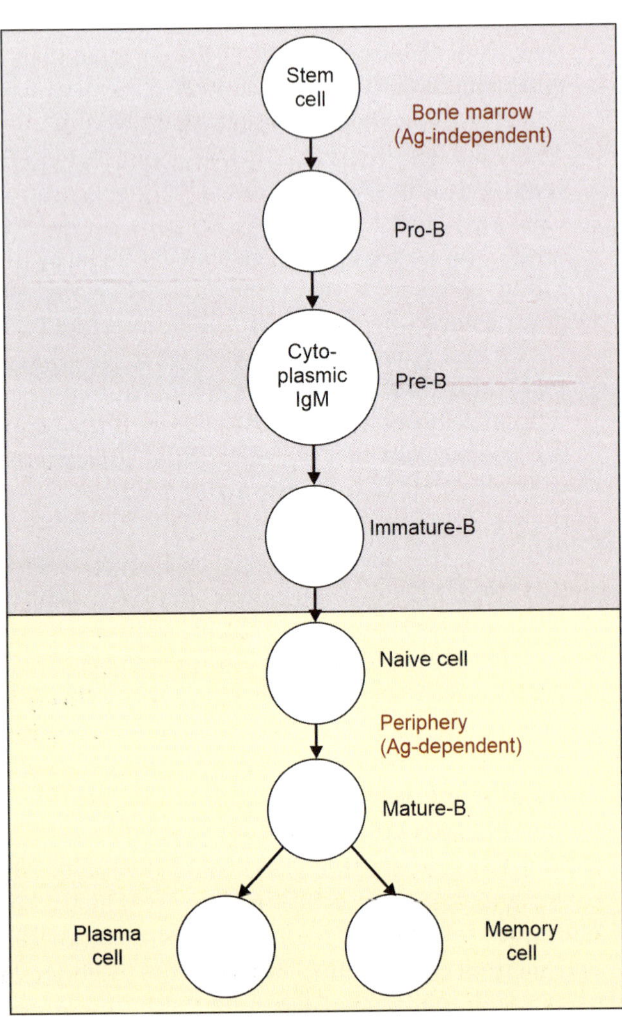

Fig. 22.8: Maturation of B cells

- **Morphology:** They are nearly double the size of the small lymphocytes. They have indented nuclei. Cytoplasm is abundant containing several azurophilic granules, mitochondria, ribosomes, endoplasmic reticulum and Golgi apparatus.
- **Types and functions of LGLs**
  - **Natural killer (NK) cells**
    - **Properties of NK cells:** They possess spontaneous cytotoxicity (cytotoxic/cytolytic/type–II hypersensitivity) towards various target cells like malignant cells and virus infected cells. Their cytotoxicity is not antibody dependent or MHC restricted. Their activity is 'natural' or 'nonimmune' and not required sensitization by prior antigenic contact and therefore they form a part of the innate immunity. They belong to a different lineage from T and B cells and are therefore normally active in 'severe combined immunodeficiency diseases', in which mature T and B cells are absent. They have CD16 and CD56 on their surface (CD3 presents on all T cells, so called pan CD marker). NK cells activity is augmented by interferon and IL-2. Ab induced proliferation does not occur in NK cells. They mature in the bone marrow instead of the thymus unlike other T cells. Their cytoplasm has granules called azurophilic granules.

> **Note: Azurophilic granules**
> - **Meaning:** Azure means blue, as the granules stain bluish in color by Romanowsky stain.
> - **Synonym:** Also called "primary granules".
> - **Contains:** Granules contain proteins such as perforin and proteases enzymes called granzymes. Granzymes include serine protease like myeloperoxidase, phospholipase A2, elastase, acid hydrolases, defensins, neutral serine proteases, bactericidal/permeability-increasing, lysozymes, proteins, cathepsin G, proteinase 3 and proteoglycans.

-
  -
    - **Differences and similarities between NK cells and Tc cells:** Tc cells and NK cells are similar in functions. NK cells act against virus infected cells and tumor cells till Tc cells are activated and carry on the functions. However, Tc cells are differ from NK cells by many ways as shown in **Table 22.3**.
    - **Functions of NK cells:** NK cells bind to the glycoprotein receptors on the surface of autologous as well as allogenic target cells and release several cytolytic factors like perforin, TNF and lymphotoxin. Perforin is a complement like substance and resembles the complement component C9. It causes pores in the target cells membrane (contain intracellular microbes or antigens) like virus infected cells, tumor cells and allograft cells in transplanted tissues, through which cytotoxic factors such as the TNF-β and granzymes enter in cells and destroy them by apoptosis.

TABLE 22.3: Differences and similarities between NK cells and Tc cells

Features	NK cells	Tc cells
Differences		
CD marker	CD16 and CD56	CD3 & CD8
Ab role	Independent	Dependent by endocytic pathway
MHC molecule	MHC restricted	MHC-1
Immune memory	No	Yes
Immunity	Innate	Acquired
Similarities		
Target cells	Virus infected and tumor cells	
Mechanisms	By producing perforins, TNF and lymphotoxin	

- **Antibody dependent cytotoxic cells (ADCCs):** These types of LGLs possess surface receptors for the Fc part of Ig. Antigen is attached at Fab region. They are capable of killing target cells sensitized with IgG like tumor cells, parasites or graft rejection. Their action is mostly IgG mediated. In certain instances IgE is useful like eosinophilic mediated killing of parasites. This antibody dependent cellular cytotoxicity is distinct from the action of cytotoxic T cells, which is independent of antibody.
- **Lymphokine activated killer (LAK) cells:** LAK cells are treated with IL-2, which are cytotoxic to tumor cells without affecting normal cells. They used in treatment of some tumors such as renal cell carcinoma.

3. Antigen-presenting cells (APCs): Any antigen presented to lymphocytes is prior presented and processed by a special type of cells called APCs. They are the following two types:
- **Major cells or professional cells**
  - **Dendritic cells:** So called because having long membranous cytoplasmic extension like dendrites of neuron. They are originated from bone marrow. They present in the peripheral blood and in the peripheral lymphoid organs, particularly in the germinal areas of the spleen and lymph nodes. Morphologically they seem highly pleomorphic, oval/spherical nucleus with small central body and many long niddle-like processes called dendritic processes as shown in **Fig. 22.10**. There are two types of dendritic cells such as (1) immature dendritic cells or Langerhans cells, present in epidermis and after making contact with Ag release the cytokines which cause loss of adhesiveness of Langerhans cells and migrate the cells free in blood called mature dendritic cells and (2) mature dendritic cells, which are known to stimulate the naïve cells. They possess MHC class 2 antigens but not Fc or sheep RBC receptors or surface Igs. Dendritic cells are involved in the presentation of antigens to T cells during the

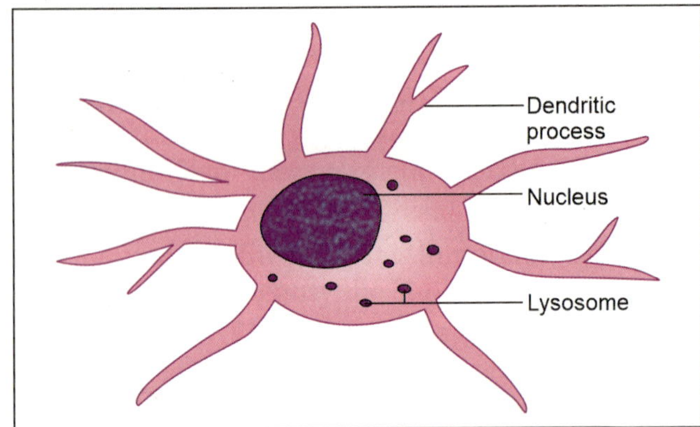

Fig. 22.10: Dendritic cell

primary immune response. They also posses the B7 and CD28 which are required for Th cells activation.

- **B-cells:** These are another antigen presenting cells, particularly during the secondary immune response.
- **Macrophages:** Any Ag presented to lymphocytes is prior presented and processed by macrophages. They trap the antigen and provide its optimal concentration to the lymphocytes. It is possible by presence of common surface Ag on both the cells called MHC-antigen. When macrophages contain different MHC-antigens, they do not participate in Ag processing and antigen will be presented to lymphocytes called MHC-restriction. If too high concentration of antigen is presented, it produces tolerance, and too low concentration may not be immunogenic.
- **Langerhans cells:** They are located in skin. They possess features of macrophages and immature dendritic cells. They process and present the Ag that reaches to the dermis.

• **Minor cells or nonprofessional cells:** These are fibroblasts (skin), thymic epithelial cells, pancreatic beta cells, vascular endothelial cells, glial cells (brain) and thyroid epithelial cells.

> **Note: Functions of lymphocytes**
> • **Types of lymphocytes according to functions:** Described above.
> • **Functions of individual lymphocytes:** These includes naive cells, CD4 (CD 7 2 3 4 $\alpha\beta$ TCR) cells, CD8 (CD 7 2 3 8 $\alpha\beta$ TCR) cells, intraepithelial lymphocytes, (CD 7 2 3 $\gamma\delta$ TCR cells), memory cells, plasma cells, B1 (CD5) cells, null cells (LGLs) and Langerhans cells. Function of all these cells are described above.

Nonspecific Immune Systems

These include reticuloendothelial (RE) system/cells, Ir by cells without TCRs, complement system and neutralization.

Reticuloendothelial (RE) System and RE Cells

Introduction: RE system produces nonspecific Ir called phagocytosis. It is an oldest defense mechanism in animals. Originating in protozoa as a combined mechanism for nutrition and defense, along the course of evolution, the phagocyte lost its nutrition function with the development of digestive enzymes. In higher organisms it specialized in the removal of foreign particles.

Definition: Cells that are responsible for engulfment and digestion of foreign particles, often with the help of Ab and C (complement) called RE cells or phagocytic cells and this mechanism of defense called phagocytosis.

History: Phagocytosis was 1st described by Metchnikoff in 1883.

Types and properties of RE cells: Two types of RE cells as mentioned below:

1. **Macrophages (mononuclear cells)**
• **Subtypes:** Follow **Flowchart 22.2**.
• **Properties of macrophages:** Blood macrophages called monocytes are about 12–15 µm in size. Tissue macrophages called histiocytes are about 15–20 µm in size. Monocytes in circulation have an approximate half-life of 3 days. Tissue macrophages survive for months. Macrophages are originating from bone marrow from precursor cells and become mature in 6 days. Multinucleated cells and epitheloid cells seen in granulomatous inflammatory lesions such as tuberculosis, originate from mononuclear macrophages. Macrophages may be activated by lymphokines, compliment components or interferon. Activated macrophages are not antigen-specific. Activated macrophages show morphological and functional changes as compared with unstimulated quiescent macrophages. They are larger, adhere better, spread faster and are more phagocytic.

2. **Microphages (polymorphonuclear cells/PMNs):**
• **Subtypes:** Follow **Flowchart 22.3**.
• **Properties of microphages:** Microphages are originating from bone marrow from precursor cells and become mature and finally released in circulation. They are short lived with half life about 2 days and few hours in tissue after penetration.

Flowchart 22.2: Subtypes of macrophages

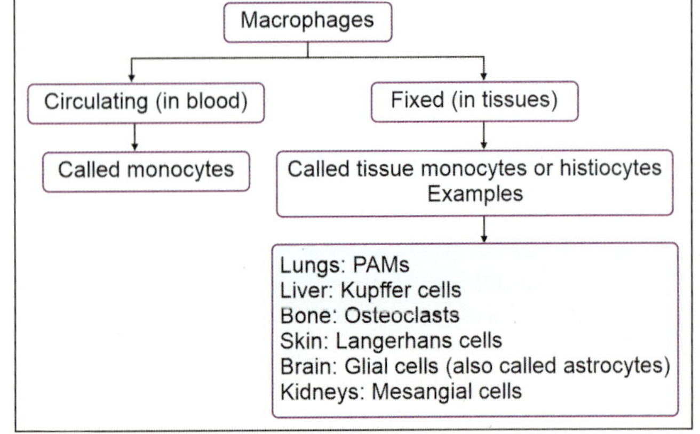

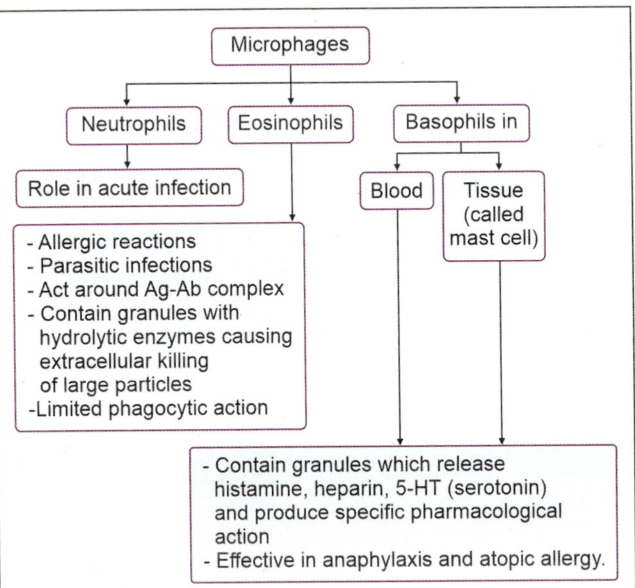

Functions of RE cells

1. **Phagocytosis:** Engulfment and digestion of foreign particles by single cell, often with the help of Ab and C called phagocytosis (Scavanger cell role). It includes following five steps:

   i. **Opsonization (Fig. 22.11a):** After entry in the body, foreign particles combine with opsonins (factors present in serum). Opsonization is defined as coating of foreign materials or antigens by substances present in serum. There are three types of opsonins such as (1) antibody (Fc opsonin) opsonins which include IgM and IgG. IgM is also 500–1000 times more effective than IgG in Opsonization. IgM/IgG binds with Fc end to the specific opsonin receptors present on the phagocytic cells, so such opsonin called Fc opsonin and receptors called Fc receptors. (2) Complement opsonins that include C3b and C4b (C-fragments) and (3) fibronectin.

   ii. **Attachment (Fig. 22.11a):** Corresponding receptors for opsonin are present on the surface of phagocytic cells, through which opsonized particles adhere with phagocytic cell. Two types of opsonin receptors like Fc receptors for IgM and IgG opsonins and C3b receptors C3b opsonin.

   iii. **Engulfment:** After an attachment, particles are ready for engulfment, which is precipitated by formation of pseudopod **(Fig. 22.11b)** around the particles. Pseudopod engulfs the particle and forms phagocytic vacuole (pseudopod and particle means phagocytic vacuole) as shown in **Fig. 22.11c**. Phagocytic vacuole remains free in cytoplasm and later combines with lysosome to form phago-lysosome (phagosome) as shown in **Fig. 22.11d**.

   iv. **Secretion or degranulation:** Secretion of metabolites from preformed stored granules (e.g. lysosomes) which kill the foreign particles. There is also new synthesis and secretion of other metabolites like IL-2, TNF, arachnoid acid metabolites (e.g. prostglandins, leukotreins and platelets activating factors) and O_2 metabolites (e.g. superoxide O_2, H_2O_2, hypochlorous acid, etc.).

   v. **Killing or degradation:** After killing of particles, they are degraded by certain metabolites, which are following 3 types.

   – O_2 dependent mechanism: It is bactericidal. O_2 dependent metabolites are HOCl, HOI, HOBr, OH, O_2, H_2O_2, etc.

   – O_2 independent mechanism: It is bactericidal. O_2 independent metabolites are lysosomal hydrolase, permeability increasing factor, defensins and cationic proteins.

   – Nitric oxide mechanism: It is fungicidal or having antiparasitic action.

2. **Act as APCs and help in specific immune response:** Described above.

3. **Antitumor activity and graft rejection:** When stimulated by cytophilic antibodies and certain lymphokines, macrophages become 'armed'. Such armed macrophages are capable of antigen-specific cytotoxicity, which is important in antitumor activity and graft rejection.

4. **M_1 marker:** Mac 1 is a protein Ag found on mouse macrophages. A similar protein on human macrophages has been named the M1 marker. This appears closely related to CR3, a cell receptor for C3 components.

5. **Release of biologically active substances:** Macrophages secrete a number of biologically active substances like hydrolytic enzymes, binding proteins (fibronectin and transferrin), TNF (cachectin), colony stimulating factor (CSF) and IL-1 (leukocyte activating factor). IL-1 acts as an endogenous pyrogen and also induces synthesis of IL-2 by T cells.

6. **Functions of microphages:** Follow **Flowchart 22.3**.

Structure and Functions of Immune System

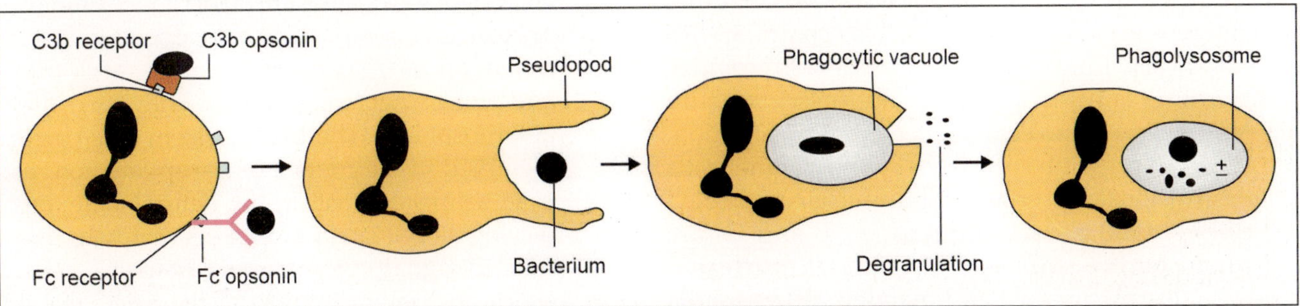

Fig. 22.11: Steps of phagocytosis

Ir by Cells without TCRs

It is described above.

Complement System

Follow **Ch. 21.**

Neutralization

Follow **Ch. 20.**

MAJOR HISTOCOMPATIBILITY COMPLEX (MHC)

Synonym: HLAs (human leukocyte antigens) complex.

Definition: It is a set of cell surface proteins (receptors) essential for the acquired immunity to recognize the foreign molecules in vertebrate, which in turn determines the histocompatibility.

History: Concept of MHC was given by Peter Gorer in 1936. George Snell identified the tissue compatibility upon transplantation and MHC locus (genes), for which he got Nobel Prize in 1980 along with Baruj Benacerraf and Jean Dausset.

Components of MHC: Two components:

1. MHC antigens or **MHC molecules** or **HLAs:**

- **Definition:** Ags present on leukocytes surface, induce immune response and responsible for allograft rejection called MHC antigens or MHC molecules or HLAs.
- **Properties:** MHC antigens are two chain glycoprotein molecules anchored on the surface membrane of cells.
- **Types:** MHC antigens have following three types:
  - **MHC class-I antigens/MHC class-I molecules/ HLAs class-I:** These consist of a heavy peptide chain (alpha chain) non-covalently linked to a much smaller peptide called beta 2-microglobulin (beta chain) as shown in **Fig. 22.12a.** The beta chain has a constant amino acid sequence and is coded by a gene on chromosome 15. The alpha chain consists of three globoid domains (alpha 1, 2 and 3) which protrude from the cell membrane and a small length of transmembrane C terminals reaching into the cytoplasm. The distal domains (alpha 1 and 2) have highly variable amino acid sequences and are folded to form a cavity or groove between them. Protein antigens processed by macrophages or dendritic cells to form small peptides are bound to this groove for presentation to CD8 T cells. The T cells will recognize the antigens only when presented as a complex with MHC class 1 molecules. T cells will not recognize the other antigens which are not presented with MHC class 1 molecule called MHC restriction. When so presented, the CD 8 cytotoxic killer cells destroy the target cells (for example virus infected cells). MHC class 1 antigens (A, B and C) are **found on the surface of virtually all nucleated cells (except sperms) and platelets.** They are the principal antigens involved in graft rejection and

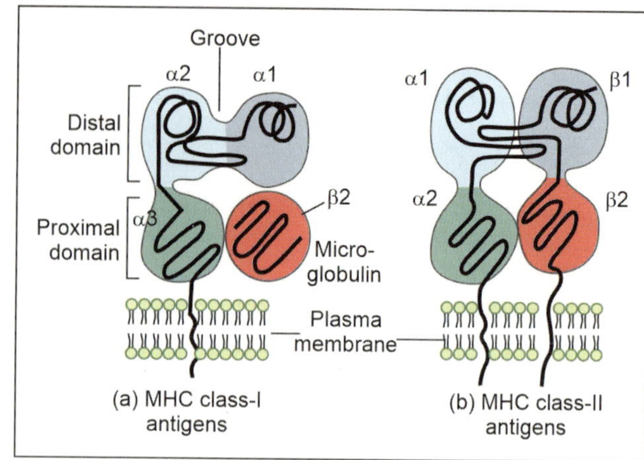

Fig. 22.12: MHC antigens/MHC-molecules/HLAs

cell mediated cytolysis. Class 1 molecules may function as components of hormone receptors.

- **MHC class-II antigens/MHC class-II molecules/ HLAs class-II:** These antigens are more restricted in distribution, being **found only on cells of immune system** like macrophages, dendritic cells activated T cells and particularly on B cells. Class 2 antigens are heterodimers, consisting of an alpha chain and a beta chain as shown in **Fig. 22.12b.** Each chain has 2 domains, the proximal domain being the constant region and the distal being the variable region. The 2 distal domains (alpha 1, beta 1) constitute the antigen binding site, for recognition by CD4 lymphocytes, in a fashion similar to the recognition of class 1 antigen peptide complex by CD8 T cells. MHC class 2 molecules are primarily responsible for the graft versus host response and mixed leukocyte reaction (MLR). Both class I and II molecules are members of the immunoglobulin gene superfamily. Differences between MHC class I and II are mentioned in **Table 22.4.** The immune response (Ir) genes which control immunological responses to specific antigens are believed to be situated in the MHC class 2 regions, probably associated with the DR locus. Ir genes have been studied extensively in mouse and located in the I region of mouse MHC. They code for Ia (I region associated) antigens consisting of 1A and 1E proteins. However, the relevance of Ir genes in human is not clear.

- **MHC class-III antigens/MHC class-III molecules/ HLAs class-III:** These are heterogenous molecules. They include complement components linked to the formation of C3 convertases, heat shock proteins and tumor necrosis factors. They also display polymorphism. The MHC system was originally identified in the context of transplantation, which is an artificial event. In the natural state, besides serving as cell surface markers that help infected cells to signal cytotoxic and helper T cells, the enormous polymorphism of the MHC maximize protection against microbial infections. By increasing

TABLE 22.4: Differences between HLA class I and II molecules

Features	MHC class I	MHC class II
Sites	All nucleated cells (except sperms) and platelets	Immune cells
Peptide Ag	Presented to CD8 T cells	Presented to CD4 T cells
Nature of peptide antigen	Endogenous or intra-cellular (viral/tumor antigen)	Exogenous
Peptide antigen (size)	8–10 amino acid long	13–18 amino acid long
Ag presentation	Cytosolic pathway	Endocytic pathway
Peptide-binding site	$\alpha1/\alpha2$ groove	$\alpha1/\beta1$ groove
CD4 or CD8 binding site	$\alpha3$ binds to CD8 molecules on T_C cells	$\beta2$ binds to CD4 on T_H cells

the specificity of self-antigens, the MHC prevents microbes with related antigen make up sneaking past host immune defenses by molecular mimicry. The primary aim of the MHC may be defense against microbes and not against the graft. MHC has been implicated in a number of nonimmunological phenomena such as individual odor, body weight in mice and egg laying in chickens.

Pleomorphism in HLA system: HLA loci are multi-allelic. HLA system is very pleomorphic; for example, at least 24 distinct alleles have been identified at HLA locus A and 50 at B. Each allele determines distinct Ag.

2. Genes of MHC or genes of HLA complex

- **Definition:** MHC Ags are regulated (encoded) by specific genes called genes of MHC or genes of HLA complex.
- **Location:** They are located on short arm of chromosome-6.
- **Types of genes:** Similar complex of genes present in different species, with following three different classes of genes **(Fig. 22.13)**.
  - Class 1 genes: They determine histocompatibility and acceptance or rejection of allografts. These include A, B and C loci.
  - Class 2 genes (D region): They regulate the Ir. These include DR, DQ and DP loci.
  - Class 3 genes (complement region): They determine the complement components C2 and C4 of the classical pathway, properdin factor B of the alternative pathway, heat shock proteins and tumor necrosis factors alpha and beta.

HLA typing

- **Definition:** Typing of HLA by using specific mono-clonal Ab (antisera) called HLA typing.
- **Preparation of antisera:** Antisera for HLA typing were obtained principally from multiparous women as they tend to have antibodies to the HLA antigens of their husbands, due to sensitization during pregnancy.
- **Typing methods**
  1. **Serological method:** It based on microcytotoxicity, in which there is complement mediated lysis of peripheral blood lymphocytes with standard set of tying sera. It is not possible for HLA DR antigens.
  2. **Mixed leukocyte reaction and primary lymphocyte typing:** For HLA DR antigen.
  3. **Genetic methods:** Typing is done by RFLP and gene sequence specific oligonucleotide probe typing.
- **Clinical significances**
  1. **Organ or tissue transplantation:** HLA typing is used primarily for testing compatibility between recipient and potential donors before transplantation.
  2. **Paternity:** It also useful in disputed paternity.
  3. **Anthropological studies:** As the prevalence of HLA types varies widely between different human races and ethnic groups, HLA typing is used in anthropological studies.
  4. **Association between HLA type and certain diseases:** An association has been observed between HLA type and certain diseases. Such diseases are generally of uncertain origin, associated with immunological abnormalities and exhibit a hereditary tendency. For examples, ankylosing spondylitis and Reiter's syndrome are associated with HLA-B27, multiple sclerosis and Goodpasture's syndrome are with HLA-DR2, myasthenia gravis, SLE, etc., are with HLA-DR3, insulin-dependent diabetes with HLA-DR3/DR4, rheumatoid arthritis with HLA-DR4 and hereditary hemochromatosis with A3/B14.

MHC restriction: It is described above.

Structure and Functions of Immune System

ACCESS YOURSELF

Essay/Full Question

1. Structure and function of immune system
2. Describe the HLA system and the immune principles involved in transplant and mechanism of transplant rejection.

Short Notes

1. Lymphoid organs
2. Types and functions of lymphocytes
3. RE system or RE cells/phagocytosis
4. MHC complex.

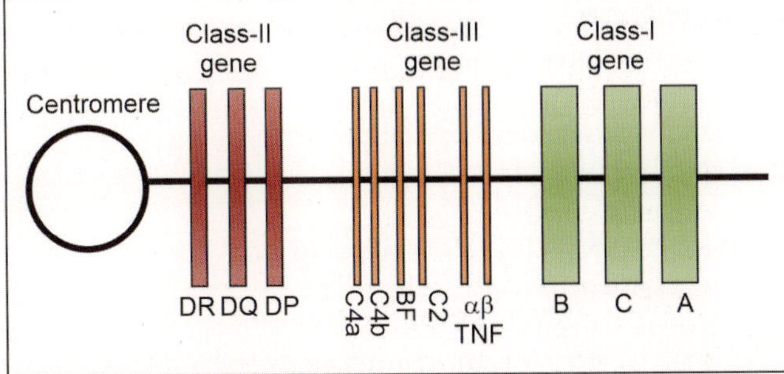

Fig. 22.13: Genes of MHC or genes of HLA complex

1. Write the two examples of each primary and secondary lymphoid organs.
2. What is thymus-dependent area? Write two examples.
3. What is overwhelming post-splenectomy infection or sepsis?
4. What is naive cell?
5. Write the functions of following Th (T helper) lymphocytes: Th0 cells, Th1 cells, Th2 cells and Th17 cells.
6. Write the functions of following immune cells. Intraepithelial lymphocytes, NK cells, astroytes and dendritic cells.
7. What is opsonization? Write two examples of opsonin.
8. Write two examples of each microphage and macrophage type phagocytic cells.
9. What is MHC restriction?

Comments on

1. Thymus dependent areas will be depleted after thymectomy.
2. Absence of spleen (asplenia) causes infections by capsulated bacteria.

MCQs for Chapter Review

Specific Immune System/Lymphoid System

1. Lymphoreticular system includes:
   - a. T-cells
   - b. B-cells
   - c. Platelets
   - d. Macrophages
   - e. Neutrophils
2. **All are peripheral lymphoid organs** *except*:
   - a. Lymph nodes
   - b. Spleen
   - c. Mucosa associated lymphoid tissue
   - d. Thymus
3. **Apart from B cell and T cell, there is a 3rd distinct type of lymphocyte. This is:**
   - a. MHC cell
   - b. NK cell
   - c. Macrophage
   - d. Neutrophil
   - e. Microglia
4. **Which of the following cell is known as large granular Lymphocyte (LGL)?**
   - a. Plasma cell
   - b. NK cell
   - c. T cell
   - d. K cell
5. **Common between B and T cells:**
   - a. Origin from same lineage
   - b. Site differentiation
   - c. Antigenic marker
   - d. Both humoral and cellular immunity
   - e. Further differentiation seen
6. **B cells maturation take place in:**
   - a. Thymus
   - b. Lymph nodes
   - c. Bone marrow
   - d. Spleen
7. **T cells mature in:**
   - a. Peyer's patch
   - b. Lymph nodes
   - c. Thymus
   - d. Bursa of fabricus
8. **Neonatal thymectomy leads to:**
   - a. Decreased size of germinal center
   - b. Decreased size of paracortical areas
   - c. Increased antibody production by B cells
   - d. Increased bone marrow production of lymphocytes
9. **T cell dependent region is:**
   - a. Cortical follicles of lymph node
   - b. Medullary cords

   - c. Mantle layer
   - d. Paracortical area
10. **Apart from T and B the other class of lymphocyte is:**
    - a. Macrophage
    - b. Astrocyte
    - c. NK cell
    - d. Langerhans cell
11. **Type of receptor present on T cells is:**
    - a. IgA
    - b. IgG
    - c. Prostaglandin
    - d. CD4
12. **Which of the following is pan T lymphocytes marker?**
    - a. CD2
    - b. CD3
    - c. CD19
    - d. CD25
13. **Group B cell lymphocyte belongs to:**
    - a. CD19
    - b. CD69
    - c. CD59
    - d. CD68
14. **The following are true for T lymphocytes, *except*:**
    - a. Constitute about 70–80% of circulating pool of lymphocytes
    - b. Release macrophage inhibition factor
    - c. Secrete specific antibodies
    - d. Release lymphotoxin
15. **Normal % of CD4 cells in a newborn:**
    - a. 35% of T cells
    - b. 45% of T cells
    - c. 55% of T cells
    - d. 65% of T cells
16. **Which cells cause rosette formation with sheep RBCs?**
    - a. T cells
    - b. NK cells
    - c. Monocytes
    - d. B cells
17. **EAC rosette formation is the property of following cells:**
    - a. T cells
    - b. B cells
    - c. Macrophages
    - d. All of the above
18. **Rosette formation with red blood cells is seen in:**
    - a. B cells
    - b. Macrophages
    - c. T cells
    - d. Monocytes
19. **What enhances multiplication of T cells in culture?**
    - a. Phytohemagglutinin
    - b. Chemotactic factor
    - c. Leukotrienes
    - d. Prostaglandin
20. **T cells are identified by:**
    - a. Rosette formation with sheep RBCs
    - b. Immunoglobulin on their surface
    - c. EAC rosette with sheep erythrocytes
    - d. Have filamentous projection on their surface
21. **Helper cells belong to:**
    - a. T cells
    - b. Macrophages
    - c. B cells
    - d. Monocytes
22. **All of the following are functions of CD4 helper cells, *except*:**
    - a. Immunological memory
    - b. Produce immunoglobulin
    - c. Activate macrophages
    - d. Activate cytotoxic cells
23. **Cellular immunity is induced by:**
    - a. NK cells
    - b. Dendritic cells
    - c. Th1 cells
    - d. Th2 cells
24. **NK cells are:**
    - a. Activated macrophages
    - b. Ab activated T cells
    - c. Null cells activated by complement
    - d. Derived from plasma cells
    - e. Independent of antibody
25. **Regarding NK cell, false statement is:**
    - a. It is activated by IL-2
    - b. It expresses CD3 receptor
    - c. It is a variant of large lymphocytes
    - d. There is antibody induced proliferation of NK cells

26. **NK cells activity is enhanced by:**
    a. IL-1
    b. TNF
    c. IL-2
    d. TGP-β

27. **All are true regarding NK cells, *except*:**
    a. CD 16 positive
    b. CD 56 positive
    c. Secrete compliment like substance
    d. Important role in viral infected cell

28. **Perforins are produced by:**
    a. Cytotoxic T cells
    b. Suppressor T cells
    c. Memory helper T cells
    d. Plasma cells
    e. NK cells

29. **NK cells provide immunity against:**
    a. Viruses
    b. Bacteria
    c. Fungi
    d. Chlamydia

30. **NK cells and cytotoxic T cells are differentiated by:**
    a. Interferon reduces NK cells activity
    b. Antibody specificity
    c. Receptor for IgG
    d. Presence in spleen

31. **Killer cells and helper cells are part of:**
    a. B cells
    b. T cells
    c. Monocytes
    d. Macrophages

32. **Most potent stimulator of naïve T cells:**
    a. Mature dendritic cells
    b. Pigment producing cells
    c. Macrophages
    d. B cells

33. **All these are antigen presenting cells (APCs), *except*:**
    a. T cells
    b. B cells
    c. Fibroblast
    d. Dendritic cells
    e. Langerhans cells

34. **Most efficient antigen presenting cell in the skin:**
    a. Dendritic cell
    b. Macrophage
    c. Langerhans cell
    d. Kupffer cell

Reticuloendothelial (RE) System and RE Cells

35. **All are mononuclear-macrophages, *except*:**
    a. Histiocytes
    b. Microglial cells
    c. Kupffer cells
    d. B cells

36. **The function common to neutrophils, monocytes and macrophages is:**
    a. Immune response
    b. Phagocytosis
    c. Liberation of histamine
    d. Destruction of old lymphocytes

37. **The process increasing the ability for phagocytosis of foreign bodies by body is called:**
    a. Cross reactivity
    b. Opsonization
    c. Immune tolerance
    d. Immune surveillance

38. **Opsonization occurs due to all, *except*:**
    a. Endotoxin
    b. Complement
    c. IgM
    d. IgD

39. **Opsonization take place through:**
    a. C3a
    b. C3b
    c. C5a
    d. C5b

40. **Bacteria coated with complement and Ig; phagocytosis is enhanced by:**
    a. Receptor mediate endocytosis
    b. Pseudopod formation
    c. Myeloperoxidase mediated destruction
    d. C3b-Fc mediated destruction

41. **Mononuclear phagocytes are produced by:**
    a. Thymus
    b. Spleen
    c. Bone marrow
    d. Liver

Major Histocompatibility Complex (MHC)

42. **HLA-I is present on:**
    a. All nucleated cells
    b. Only on cells of immune system
    c. Only on B-cells
    d. Only on T-cells

43. **MHC II are presented by:**
    a. Macrophages
    b. Dendritic cells
    c. lymphocytes
    d. Eosinophils
    e. Platelets

44. **Cell type which lacks HLA antigen is:**
    a. Monocyte
    b. Thrombocyte
    c. Neutrophil
    d. Red blood cell

45. **Peptide binding site on class I MHC molecules for presenting processed antigens to CD8 T cells is formed by:**
    a. Proximal domain of α subunits
    b. Distal domain of α subunit
    c. Proximal domains of α and β subunit
    d. Distal domains of α and β subunit

46. **T helper cell recognizes:**
    a. MHC class I
    b. MHC class II
    c. Processed peptides
    d. Surface Ig

47. **True about MHC:**
    a. Present on chromosome 4
    b. Class II comprises A, B, C loci
    c. Class III has complement
    d. Class I is involved in mixed leukocyte reaction

48. **HLA complex is on chromosome:**
    a. 6
    b. 7
    c. 8
    d. 4

49. **Gene components of HLA class I includes:**
    a. A, B, C
    b. DR
    c. DQ
    d. DP

50. **MHC class III genes encode:**
    a. Complement component C3
    b. Tumor necrosis factor
    c. Interleukin 2
    d. Beta 2 microglobulin

51. **HLA III gene codes in graft rejection:**
    a. Immunological reaction in graft rejection
    b. Complement
    c. Graft versus host reaction
    d. Immunoglobulin

52. **The role played by major histocompatibility complex-1 and 2 is to:**
    a. Transduce the signal to T cells following antigen recognition
    b. Mediate the immunogenic class switching
    c. Present antigens for recognition by T cell antigen receptors
    d. Enhance the secretion of cytokines

Answers and Explanation of MCQs

1. **a, b, d and e**
   - Lymphoreticular system includes lymphoid system and RE system. T-cells and B-cells are parts of lymphoid system while macrophages and neutrophils are parts of RE system.

2. **d**
   - Thymus is central lymphoid organ.

3. **b**

4. b
- Follow section, **specific immune system/lymphoid system (components → lymphoid cells → Cells without TCRs)** for the explanation of answers of MCQs 3–4.

5. a and e
- B cells and T cells have same origin from bone marrow. In early fetal life cells are derived from liver and yolk sac. B cells differentiation/maturation takes place in bone marrow while of T cells in thymus. Both have different antigenic markers. B cells are for humoral and T cells are for cellular immunity.

6. c

7. c
- Both B and T cells are originated from bone marrow. B cells differentiation/maturation takes place in bone marrow, while of T cells in thymus.

8. b

9. d
- Follow section, **thymus (Clinical significances → Post-thymectomy effect)** for explanation of answers of MCQs 8–9.

10. c and d
- Macrophages and astrocytes (glial cell of brain) are RE cells NK cells are lymphocytes but without TCRs. Langerhans cells are one type of lymphocytes present in skin and act like APCs.

11. d
- T cells carry CD markers, TCR and MHC receptors on the surface. CD4 presents on helper/inducer cells.

12. b

13. a
- Follow **Table 22.1** for the explanation of answers of MCQs 12–13.

14. b and c
- T cells releases macrophage activating factors not the macrophage inhibitory factors.
- For explanation of other options follow **Table 22.2**.

15. All options are wrong
- CD4 count in new born in 75–80% of total T cells.

16. a

17. b

18. a, c

19. a

20. a
- Both T and B cells can make rosette with SRBC, but T cell binds directly while B cell binds with the cell which is coated with Ab and C (complement). Follow **Table 22.2** for more explanation of answers of MCQs 17–21.

21. a
- CD4 T cells are called helper cells or inducer cells. CD8 T cells are called cytotoxic cells or suppressor cells.

22. b

23. c
- Follow section, **T cell maturation (CD 7 2 3 4 αβ TCR cells)** for the explanation of answers of MCQs 22–23.

24. e

25. d

26. c

27. a, b, c, d

28. a, e

29. a
- Follow section, **Natural killer (NK) cells** for the explanation of answers of MCQs 24–29.

30. b
- Follow **Table 22.3** for explanation.

31. b
- Both killer cells and helper cells are T cells but killer cells are without TCRs while helper cells are with TCRs.

32. a

33. a
- Follow section, **APCs (antigen-presenting cells)** for the explanation of answers of MCQs 32–33.

34. c
- From the all given options Langerhans cell is the only APC located in skin.

35. d
- Follow section, **phagocytosis (Flowchart 22.2)** for the explanation.

36. b
- Follow section, **phagocytosis (Flowchart 22.2 and Flowchart 22.3)** for the explanation.

37. b

38. a, d

39. d
- Follow section, **phagocytosis (opsonization)** for the explanation of answers of MCQs 37–39.

40. d
- C3b and Fc are the opsonins. Required for the opsonization in phagocytosis.

41. c
- Mononuclear phagocytes are called macrophages which are present in blood (called monocytes) and in tissues (called histiocytes) as shown in **Flowchart 22.2** and all they are originated from bone marrow as shown in **Fig. 22.1**.

42. a

43. a, b, c, d

44. b, d
- MHC I presents on nucleated cells (Not on RBCs, as they are non-nucleated) while MHC II presents on immune cells, which do not include platelets.

45. b
- Protein antigens processed by macrophages or dendritic cells to form small peptides are bound to the groove between distal domains like α1 and α2 of α subunit.

46. b
- T helper cells are CD4 cells which recognize the MHC class II molecules while MHC class I molecules recognized by cytotoxic T cells which are CD8 cells.

47. c
- MHC genes are present on short arm of chromosome 6.
- Class II comprises DR, DQ and DP loci.
- Class III determines the complement components C2 and C4 of the classical pathway, properdin factor B of the alternative pathway, heat shock proteins and tumor necrosis factors alpha and beta.
- Mixed leukocyte reaction is determined by DR gene of MHC class II gene.

48. a
- MHC genes are present on short arm of chromosome 6.

49. a
- Follow **Fig. 22.13** for the explanation.

50. b

51. b
- Class III gene determines the complement components C2 and C4 of the classical pathway, properdin factor B of the alternative pathway, heat shock proteins and tumor necrosis factors alpha and beta.

52. c
- After the entry of antigens, they are recognized according to MHC molecules. Those processed by MHC-I can be presented to CD8 cells and those processed by MHC-II can be presented to CD4 cells.

Immune Response (Ir)

INTRODUCTION

Definition: Specific reactivity of host to the antigenic stimulation called immune response.

Types: Follow **Flowchart 23.1**.

Flowchart 23.1: Types of Ir

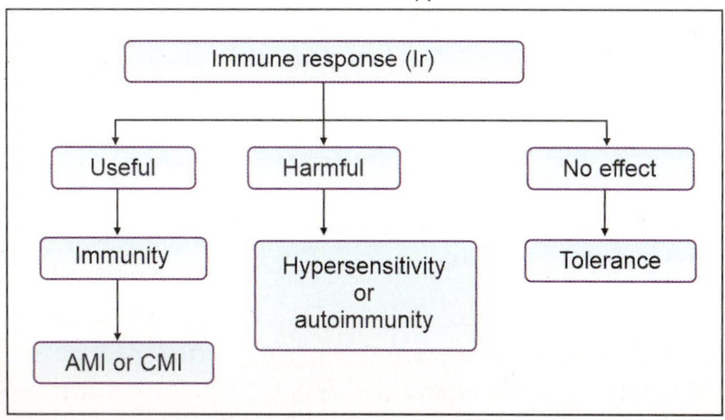

AMI and CMI: Both may work together, but differently. One may be more active than other.

AMI

Background

It is the antibody-mediated immunity. It is defined as resistance exhibited by host towards the injuries produced by microorganism and/or their products by production of antibodies from plasma cells (B cells). It also called humoral immunity (HI). Humoral word came from humor means body fluid (old term), as Ab presents free in blood and other body fluid. Ab is originated from plasma cells (B cells); however, its required participation by other immune cells.

Clinical Significances

AMI participates in infections by extra cellular pathogens like bacteria, fungi, etc., type 1, 2 and 3 (immediate) hypersensitivity reactions and in auto immune disorders.

Stages of Initiation of AMI

Contain following three stages:

1. **Afferent limb:** Entry of Ag, its distribution, fate in the tissues and its contact with appropriate immunocompetent cell.
2. **Central limb:** Processing of Ag by cell and the control of the antibody forming process.
3. **Efferent limb:** Secretion of antibody, its distribution in tissues and body fluids and manifestations due to its effects.

Pattern of Production of Ab

Contain following four patterns **(Fig. 23.1)**:

1. **Lag phase:** Stage immediately after Ag stimulation.
2. **Log phase:** Raise antibody titer.
3. **Plateau or steady phase:** Equilibrium between Ab synthesis and catabolism.

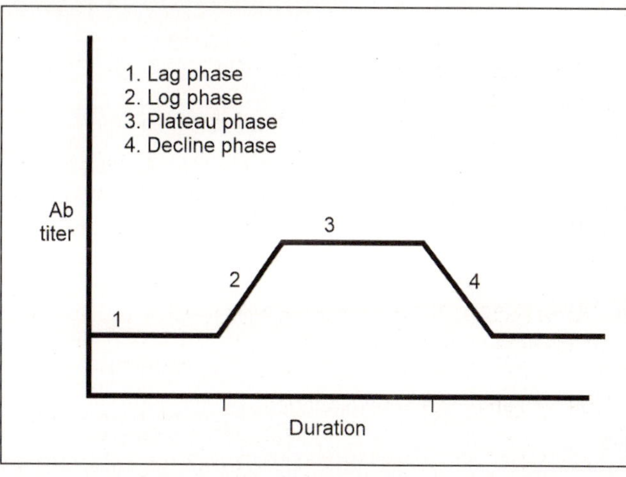

Fig. 23.1: Primary Ir

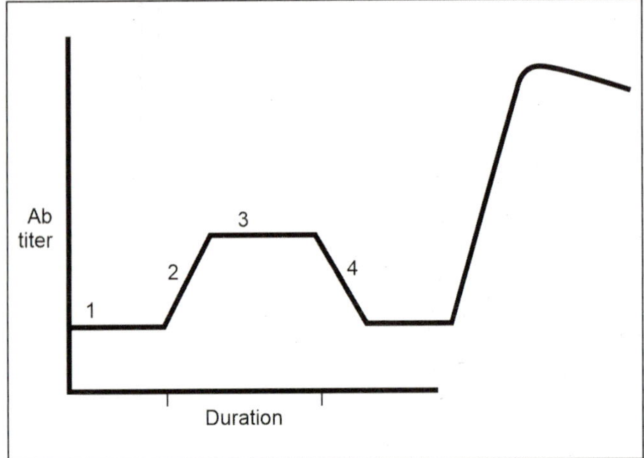

Fig. 23.2: Secondary Ir

4. **Decline phase:** Increase catabolism of Ab than production.

Antigenic Doses

1. **Priming dose:** Initial antigenic stimulation called priming dose. It is a sensitizing dose.
2. **Booster dose:** Subsequent antigenic stimulation called booster dose. It is more effective.

Primary (Fig. 23.1) and Secondary Ir (Fig. 23.2)

Ab response to initial antigenic (priming dose) stimulation called primary Ir. Ab response to subsequent antigenic (booster dose) stimulation called secondary Ir. When Ag is injected in to animals already carrying the Ab, temporary fall in Ab concentration due to combination of Ab with newly injected Ag called negative phase. Differences between primary and secondary Ir are mentioned in **Table 23.1**.

TABLE 23.1: Differences between primary and secondary Ir

Primary Ir	Secondary Ir
Slow	Prompt
Sluggish	Powerful
Short lived	Long lived
Lag phase: Long (4–10 days) and slow Ab production	Lag phase: Short (1–3 days) and rapid Ab production
Low titer of Ab	High titer of Ab
Ab lasting for short time	Ab lasting for log time
IgM type of Ab appears 1st following injuries or in primary infections	IgG type of Ab appears late or in secondary infections
Low affinity of Ab	High affinity of Ab
Includes memory B-cells: Few B-cells are converted in to Ab producing cells, majority of cells converted in to memory cells.	Includes plasma cells: B-cells are converted in to Ab producing cells (plasma cells).
Adjuvant is needed	Adjuvant is not needed
No negative phase	Negative phase occur

Fate of Antigens

The antigens are removed from circulation by two mechanisms such as by immune mechanism that includes destruction by Abs and nonimmune mechanism by phagocytosis. Following are the effective factors over fate of antigens:

1. **Physical and biochemical nature of Ag:** Proteins Ag eliminated in few days to weeks. Polysaccharide Ag slowly metabolized, persists for months or weeks, such as pneumococcal polysaccharide Ag persists for 20 years in human after single injection.
2. **Dose of Ag:** Whether induced primarily or secondarily.
3. **Routes of entry of Ag**
   - Ags introduced by IV route will rapidly localize in the spleen, liver, bone marrow, kidneys and lungs. They broken down by RE cells and excreted in urine. About 70–80% Ags are eliminated in one or two days.
   - Ags introduced by subcutaneous route are mainly localize in the draining lymph nodes only small amounts being found in the spleen.

Mechanisms of AMI (Ab Production)

Participation by cells: Three types of cells are involved in Ab production such as APCs, T lymphocytes and B lymphocytes.

Role of MHC molecules: Ag will captured and recognized by APC in accordance to MHC molecule and it is presented to T cell. Ag with MHC-II molecule is presented to CD4 (helper-Th) cell while Ag with MHC-I molecule is presented to CD8 (cytotoxic-Tc) cell. B cell also carrying Ig and MHC-II molecule, can present Ag to T-cell particularly during secondary Ir.

Steps: Steps depends on type of Ag as described below:
- **T-cell independent (TI) Ag:** Ag directly presented to B cell without processing by APC, which stimulates the mature and immature B cell to produce limited classes of antibodies (IgG3 and IgM). No memory cell formation. For more details of TI Ag follow **Ch. 18**.
- **T-cell dependent (TD) Ag:** This pathway called endocytic pathway as shown in **Fig. 23.3**. It starts with entry of exogenous Ag of extracellular microorganism or its product like toxin. Ag is processed by APC before it is presented to B cell. Ag + APC complex formation in accordance to MHC II molecule. Such complex is presented to the naïve cell of Th-cell (CD4) as it carries TCR.
   - Formation of Ag + APC + Th/CD4 (TCR) complex.
   - APC secretes IL-1, which stimulates the naïve cell of CD4 type to differentiate in to effector cell like Th0 cell. Th0 cell liberates IL-2.
   - IL-2 has autocrine effect and acts on Th0 cell to release IFN-γ, IL-4 and further release of IL-2. Later Th0 is differentiated in to effector cell (Th1 and Th2 cells) and memory cell.

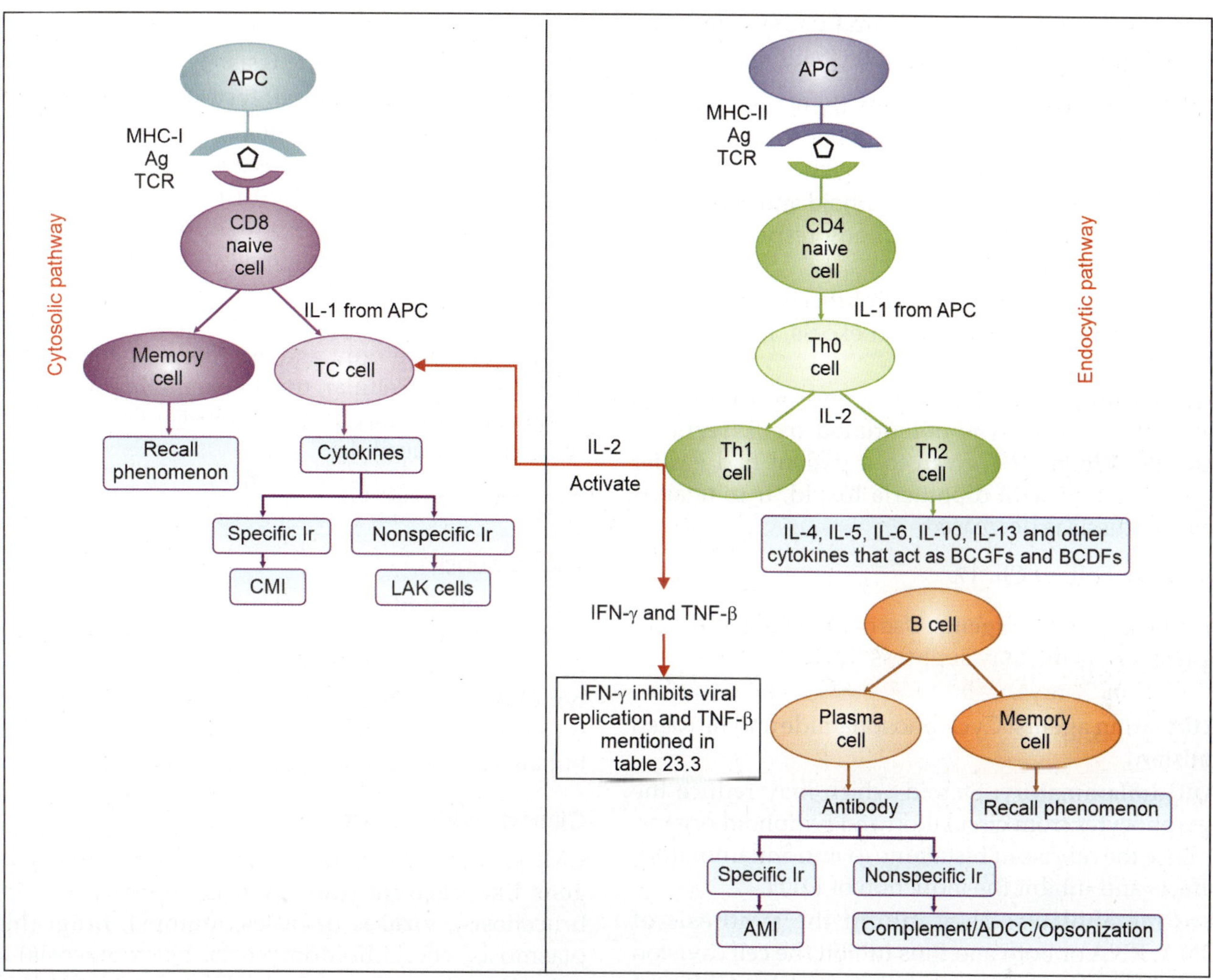

Fig. 23.3: Overview of immune response

- Th1 releases IL-2, IFN-γ and TNF-β. IL-2 is capable to activate the Tc (CD8) cell for CMI by endocytic pathway.
- Th2 cell releases IL-4, IL-5, IL-6, IL-10, IL-13 and other cytokines that act as BCGFs (B cell growth factors) and BCDFs (B cell differentiation factors) and activates the B-cell.
- Now, B-cell is differentiated in to memory cell and plasma cell.
- Memory cell is responsible for recall phenomenon and plasma cell secretes Ab which kills the Ag by opsonization (phagocytosis), complement activation, neutralization or by antibody dependent cellular cytotoxicity (ADCCs).

Factors Influencing Antibody Production

Genetic factors: Ir is under controlled of genes called Ir genes. The individual who respond to antigenic stimulation called responder and who do not called non-responder.

Age: Embryo is immunologically immature. Immunity starts with development and differentiation of lymphoid organs and cells. During the embryonic life, if any foreign Ag comes in contact with the developing lymphoid cells, it leads to **tolerance**. During the embryonic life, if any self Ag comes in contact with the developing lymphoid cells, which is released by cellular breakdown, leads to **nonantigenicity of self antigen**. Infant is protected by maternal antibodies up to 3–6 months. Immunity develops as the infant grows, mostly after 3–6 months. Full immunity developed for IgG at 5–7 years and at 10–15 years for IgA.

Nutrition: Deficiencies of vitamins, amino acids, proteins and fat reduced the both, CMI and AMI.

Routes of administration of Ag: AMI is better by parenteral route than oral or nasal route. Large particles like bacteria, RBCs, etc., are more antigenic when introduced by parenteral route. IgA produced by oral or nasal route. IgE produced by inhalation of pollen. Protein injection in to mesenteric vein or intrathymically induce tolerance.

Site of administration: HBV vaccine is more effective following injection in deltoid than gluteal injection, because of paucity of APCs in gluteal region which delaying the presentation of vaccine Ags to T and B cells.

Size and dose of Ag: Very small dose of Ag is ineffective. Ag is effective above certain level, further increase in dose also increase the Ir. However, beyond certain level Ag is ineffective and instead of increasing the Ir, it swamp and cause paralysis of the immune system called immunological paralysis by Felton (1949).

Multiple antigens: When more than two Ags are administered simultaneously, antibody response by one may affect the other called antigenic competition.

- **Two bacterial vaccines:** Like typhoid and cholera are given in mixed form, the antibody response has no effect on each other.
- **Two toxoid vaccines:** Like diphtheria and tetanus are given together with one in excess, Ir to other is inhibited.
- **Triple antigen (DPT):** Ab response to diphtheria and tetanus toxoid is potentiated by *B. pertussis* vaccine. When DPT given to a patient had earlier immunization with diphtheria toxoid, Ir to tetanus and pertussis is decreased.

Adjuvants: Follow **Ch. 17.**

Immunosuppressive agents: Use of following immuno-suppressive agents may suppress the Ir.

- **Radiation:** X-rays.
- **Alkylating agents:** Cyclophosphamide and nitrogen mustard.
- **Anti-inflammatory:** Steroid which may reduce the lymphocytes from circulation and lymphoid organs, reduce the release of histamine → anti-inflammatory effects and inhibit the induction of DTH.
- **Antimetabolites:** They inhibit the synthesis of DNA, RNA or both and thus inhibit the cell division and differentiation which is required for CMI and AMI. Commonly used antimetabolites are folic acid antagonists like methotrexate, purine analogs like 6-mercaptopurine and azathioprine, uracil analogs like 5-fluorouracil, cytosine analogs like cytosine arabinoside and alkylating agent like cyclophosphamide.
- **Cyclosporine:** It is the immunosuppressive drug derived from fungus *Tolypocladium inflatum*. It inhibits the Th cell activity. Other related drug is rapamycin.
- **Antilymphocytes serum (ALS):** It is an antiserum against lymphocytes. It acts against lymphocytes of circulation, not against the lymphocytes in lymphoid organs. Others similar antisera are antilymphocyte globulin, antithymocyte serum and antithymocyte globulin.

Effect of administration of other Ab: Ab production is inhibited by passive administration of homologous Ab. It acts by different ways like feedback mechanisms and prevention of Ag presentation to immunocompetent cells. This concept is used in treatment of many clinical conditions such as anti-Rh Ab injection in Rh-Ve women with Rh +ve child, passive immunization in diphtheria and tetanus. IV administration of Ig has immunomodulatory effect, which used in treatment of autoimmune hemolytic anemia and thrombocyto-penia.

Detection of AMI

It is detected by following tests.

1. **Non-specific (in general) tests:** General detection of Ab to Ag by agglutination test, ELISA, precipitation test, CFT and neutralization test.
2. **Specific tests:** Detection of Ab to particular Ag presents in host, e.g., Ab to diphtheria Ag (toxin) is detected by Schick test (*in vivo* and *in vitro* type).

Limitation of AMI

AMI is effective only against the antigens/microbes which are extracellular, free in circulation or bound on cell surface. It is not effective against antigens/microbes which are intracellular.

CMI

Background

It is the cell-mediated immunity. It is defined as resistance exhibited by host towards the injuries produced by microorganism and/or their products by production of lymphokines (cytokines) from sensitized T-cells. CMI is originated mostly from Tc cells (type of T lymphocytes), but also dependent on other immune cells.

Clinical Significances

CMI participates in infections by intracellular pathogens like **bacteria** (tuberculosis, leprosy, listeriosis, brucellosis), **viruses** (measles, mumps), **fungi** (histoplasmosis, cocccidioidomysosis, blastomycosis) and **parasites** (trypanosomiasis, leishmaniasis). It also participates in type 4 (delayed type) hypersensitivity, autoimmune diseases like thyroiditis, encephalomyelitis, etc., immunology in transplantation and graft versus host reaction, immunology in malignancy and immunological surveillance.

Mechanisms of CMI

Effective factors in induction of CMI: CMI is induced only by T cell dependent antigens. It is best developed by intracellular pathogens. Live vaccine is highly stimulating while killed vaccine is not very effective, but effective if contains Freund type adjuvant.

Role of MHC molecules: CMI is induced by two types of pathways (signals) as differentiated in **Table 23.2.**

Steps: Two pathways as described below.

TABLE 23.2: Differences between cytosolic pathway and endocytic pathways		
Features	**Cytosolic pathway**	**Endocytic pathway**
Type of Ag	Endogenous (intracellular) from virus/tumor cell	Exogenous (extracellular) from microorganism like toxin
MHC	Molecule I	Molecule II
Cell	Tc and Th1	Th2
Ir	CMI	AMI and CMI

1. **Cytosolic pathway:** Steps are shown in **Fig. 23.3.** It starts with entry of endogenous Ag of intracellular microorganism like virus or tumor cell. Ag is processed by APC before it is presented to Tc cell. Ag + APC complex formation in accordance to MHC I molecule. Such complex is presented to the naïve cell of cytotoxic T(Tc)/CD8 cell as it carries TCR.

   - Formation of Ag + APC + Tc/CD8 (TCR) complex.
   - APC secretes IL-1 which differentiates CD8 naïve cell into cytotoxic T (Tc) cell and memory T cell. Tc cell undergoes blast transformation and clonal proliferation.
   - Tc cell releases two types of cytokines. (1) Perforin which makes pores in the target cells membrane (contain intracellular microbes or antigens) like virus infected cells, tumor cells and allograft cells in transplanted tissues, releases their contents and destroys the cells. (2) Serine protease which induces cell death by apoptosis.
   - Memory cells are useful for recall phenomenon.
   - Tc cells effect is similar to NK cells. NK cells act against virus-infected cells and tumor cells till Tc cells are activated and carry on the function.
   - Cytokines may contribute in the effects of LAK cells.

2. **Endocytic pathway:** Steps up to stimulation of Tc (CD8) cell for CMI by endocytic pathway are described above and shown in **Fig. 23.3.** Later steps are same as described in cytosolic pathway.

Cytokines

Definitions

- Cytokines: These are biologically active substances released by activated immune cells.
- Lymhokines: These are biologically active substances released by activated T-lymphocytes. They also called lymphotoxins.
- Monokines: These are biologically active substances released by monocytes and macrophages.
- Interleukins (IL): These are biologically active substances released by leukocytes.
- Chemokines: These are biologically active substances included in chemotaxis and other leukocytes behavior.

Properties: These are soluble polypeptide (protein). They are like hormones because active in femtomolar $(10^{-15} M)$ concentration and act on cell present at distant sites. However they are unlike hormones because (1) not secreted by endocrine glands but secreted by widely distributed cells like lymphocytes, macrophages, platelets and fibroblasts. (2) Hormones are independent in actions while cytokines work together and show interaction with each other.

Actions: Cytokines have following different types of actions.

- Autocrine: Cytokines act on the cells that secrete them.
- Paracrine: Cytokines act on the surrounding cells.
- Endocrine effect: Cytokines act on cells present at distant sites.
- Pleotrophic effects: Multiple effects by same cytokine on various cells.
- Redudancy effects: Same effects by different cytokines on same cell.

Interactions: Cytokines are showing various types interactions as mentioned below:

- Synergism: Two cytokines may augment each other's actions.
- Antagonism: Two cytokines may inhibit each other's actions.
- Cascade action: It is the serial action by multiple cytokines, where one cytokine act on a target cells to release cytokines which act on subsequent cells and so on.

Types of cytokines

1. **Interleukins (IL):** Following are different types:

- **IL-1:** Initially it was described as LAF (Leukocyte activating factor) and BAF (B cell activating factor), but in 1979 it renamed as IL-I. It has two subsets like alpha and beta. It is secreted by APCs. It's production is stimulated by Ags, toxins, injuries and inflammation. Its production is inhibited by cyclosporins, corticosteiods and prostaglandins (PGs). It is useful in immunocompromized host. It has following functions:
  - Activation of CD4 cells: It stimulates the naïve cells of CD4 type to differentiate in to effector cells like Th0 cells.
  - Activation of CD8 cells: It differentiates CD8 naïve cell in to cytotoxic T (Tc) cell and memory T cell.
  - Chemotaxis: It helps neutrophils in chemotaxis.
  - Phagocytosis: It promotes phagocytosis.
  - Others: It promotes metabolic, physiological, inflammatory and hematological effects by acting on bone marrow, epithelial cells, synovial cells, osteoclasts, fibroblasts, hepatocytes, vascular endothelial cells and other target cells.
  - Role as pyrogen: It is crucial in promoting fever and so called pyrogen.
  - Role in infections: With the help of TNF it causes hematological changes in septicemia, shock and bacterial meningitis.

- **IL-2:** It is secreted by activated Th0-cells and Th1 (type of CD4 lymphocytes). It converts LGLs in to LAK cells, which can destroy NK-resistant tumor cells. This property used in treatment of tumor cells. It has following functions:
  - It modulates the immune response.
  - It is the major activator of T and B cells.
  - It stimulates cytotoxic T cells and NK cells.

- **IL-3:** It is secreted by activated T-cells. It stimulates multilineage cells of the hematopoietic system, hence called multi-colony stimulating factor (multi-CSF).

- **IL-4:** It is secreted by Th2-cells (type of CD4 lymphocytes. It acts as BCGF, BCDF, TCGF (T cell growth factor) and mast cells growth factor. It increases the action of Tc cells. It contributes in atopic hypersensitivity, as it augments IgE synthesis.
- **IL-5:** It is secreted by Th2-cells. It proliferates the activated B cells and induces maturation of eosinophils.
- **IL-6:** It is secreted by Th2-cells, B-cells, macrophages and fibroblasts. It is a pyrogen. It differentiates B cells. It induces the production of Ig. It stimulates the hepatocytes, nerve cells and hematopoietic cells. It is an inflammatory response mediator in host defense against infections.
- **Other ILs:** From IL-7 to IL-13 and IL-17 are mentioned in **Table 23.3**.
2. **Interferon (IFN): Ch. 12.**
3. **CSF (colony stimulating factor)** — Follow
4. **TNF (tumor necrosis factor)** — **Table 23.3**.
5. **Other cytokines**

Therapeutic uses of cytokines: IL1, 2, 3 and CSF are used in inflammatory diseases, infections, immuno-compromized hosts, autoimmune diseases (IL17), hematopoietic dysfunctions and neoplastic diseases.

Regulation of cytokines: Cytokines are regulated by exogenous stimuli like antigens and mitogens and endogenous stimuli like hormones (corticosteroids and endorphins), lipo-oxygenase and cyclo-oxygenase pathways. They are also regulated by each other by positive and negative feedback mechanisms.

Detection of CMI

CMI is detected by following two types of tests:

In vivo **test:** Like skin test which based on DTH. For examples, tuberculin test, lepromin test, etc.

In vitro **tests**

1. **Lymphocyte transformation test:** It based on transformation of culture sensitized T cells on contact with Ags.
2. **Target cell destruction test:** It based on destruction of target cells of culture by T cells sensitized against them.
3. **Migration inhibiting factor test**
   - **Principle:** Macrophages packed in capillary tube when placed in a tissue culture medium

TABLE 23.3: Sources and functions of cytokines

Cytokines	Sources	Functions
Other IL		
IL-7	Spleen, marrow stromal cells	BCGF and TCGF
IL-8	Macrophages and other cells	Neutrophil chemotactic factor
IL-9	Th cells	TCGF and proliferation
IL-10	Th2 cells, B cells, macrophages	Inhibits IFN production and mononuclear cell function
IL-11	Marrow stromal cells	Induces production of acute phase proteins
IL-12	T cells	Activates NK cells
IL-13	Th2 cells	Inhibits mononuclear cell function
IL-17	Th17 cells	Proinflammatory marker
CSF (Colony-stimulating factor)		
GM (Granulocyte, Mononuclear)—CSF	T cell, macrophages and fibroblasts	TCGF and macrophage growth stimulation
G (Granulocyte)—CSF	Fibroblasts and endothelium	Granulocyte growth stimulation
M (Mononuclear)—CSF	Fibroblasts and endothelium	Macrophage growth stimulation
TNF (Tumor necrosis factor): It induces hemorrhagic necrosis in certain tumor hence the name		
TNF-α	Macrophages and monocytes	Tumor cytotoxicity, wasting syndrome (cachexia, so called cachectin), lipolysis, acute phase proteins, antiviral and anti-parasitic effects, phagocytic cells activation, pyrogen, endotoxic shock
TNF-β	Th1 cells	In addition to actions of TNF-α it induces other cytokines production
Other cytokines		
TGF-β	T cells and B cells	It transforms the fibroblast, hence called TGF-β (Transforming growth factor beta), promotes wound healing, down regulation of hematological and immunological process
LIF	T cells	LIF (Leukemia inhibitory factor) helps in stem cell proliferation and eosinophil chemotaxis
IFN	Follow **Ch. 12**	

in a chamber; they migrate out, spread over the glass walls of the chamber and form lacy fan like appearance. If macrophages are from sensitized guinea pigs, addition of Ags to the culture chamber will inhibit the migration.

- **Use:** Human peripheral leukocytes packed in capillary tube and placed in a tissue culture medium in a chamber. Add the specific Ags which inhibit the migration of leukocytes.
- **Result:** By comparing with control test, a semi-quantitative assessment of migration inhibition is possible.

4. **Other tests:** These are detection of T cells by SRBC rosette test, immunofluorescence test, etc.

Passive Transfer of CMI or Adoptive Immunity

Passive transfer of CMI is achieved by administration of viable immunological competent lymphocytes called adoptive immunity. Instead of whole immunological competent lymphocytes, only its extract is used for administration called transfer factor (TF). TF is non-antigenic, dialyzable and polypeptide-polynucleotide. It is LMW about 2000–4000. It is resistant to trypsin, DNAase, RNAase and freeze thawing (stable for several years at –20°C). It is sensitive to heat at 56°C in 30 minutes. It is highly potent, 0.1 ml dose is effective. It does not induce AMI and local CMI but induces only systemic CMI. TF is used in treatment of Lepromatous Leprosy (LL), tuberculosis, mucocutaneous candidiasis, cancers such as malignant malenoma and others, autoimmune diseases like SLE, RA, sarcoidosis, multiple sclerosis, etc. It also used to restore the immunity in patients with T-cell deficiency (Wiskott-Aldrich syndrome).

INTERCONNECTIONS OF IRs

Specific Irs

Like AMI and CMI cannot work individually because

- Both required initial participation by APCs and Th cells.
- Th cells which are part of CMI secrete the IL-4, IL-5, IL-6, IL-10, IL-13 and other cytokines which allow the proliferation and differentiation of B cells in to plasma cells (Ab producing cells) and memory cells.

Nonspecific Irs

Like complement system, phagocytosis, ADCC and LAK cells are also inter connected with specific immune response as follows:

- Complement system: It acts after formation of Ag–Ab complex which required Ab synthesis by AMI.
- Phagocytosis: Opsonization in phagocytosis is dependent to IgM and IgG of AMI.
- ADCC: It uses antibodies as receptors to identify and to kill the target cells.
- LAK cells: They required lymphokines from CMI.

Overview of Irs

Follow **Fig. 23.3.**

IMMUNOLOGICAL TOLERANCE

Introduction

It also called immunological unresponsiveness. Inability of immune system to mount an immune response against particular Ag when administered subsequently called immunological tolerance.

Historical Concept

In 1949, Burnet and Fenner suggested that, if substance comes in contact with immature immunological system during embryonic life, it will be considered as self Ag and there is no Ir. Hence all body Ags are considered as self Ags and there is no Ir and all foreign Ags come in contact with immature immunological system during embryonic life are also considered as self Ags and there is no Ir. This was proved experimentally in 1953 by Medawar and colleagues by using two strains of syngenic mice. Skin graft from strain of mouse (B) is applied on another strain (A), it is rejected. When cells from strain-B are injected in strain-A in embryonic life, strain-A will not reject the graft in later part of life. Concept of self Ag is enlarged by contact with foreign Ag during embryonic life called specific immunological tolerance.

Types of Immunological Tolerance

According to quantity: It may be total or partial.

According to duration: It may be short lived or long lived.

- Examples of short lived tolerance: (1) Tolerance induced by immunosuppressive agents in adult is short lived. (2) Tolerance to bovine albumin in rabbit is also short lived which is overcome by injection of human albumin.
- Example of long lived tolerance: Tolerance by living substances like rubella and cytomegalovirus (CMV) infections, in which persistent viremia reduce Abs production called persistent tolerant infections.

Effective Factors

Induction, degree and duration of tolerance depend on following factors.

1. **Species of animal:** Tolerance is more common in rabbits and mice than chickens and guinea pigs.
2. **Immunocompetence of host:** It is very difficult to induce tolerance in host with high immuno-competence. Tolerance is induced temporary by using immunosuppressive agents.
3. **Nature of Ag:** Soluble Ags and haptens are more tolerogenic than particulate Ags. Tolerogenicity of Ags is modified by certain procedures like heat and centrifugation. Heat aggregated human Igs are more immunogenic, while disaggregated Igs are

tolerogenic. Centrifugation of serum proteins at high speed separates the tolerogenic supernatant and immunogenic sediment.

4. **Dose of Ag:** High dose causes high zone tolerance like immunological paralysis by Felton while low dose causes low zone tolerance. Intermediate dose causes no tolerance, but induces immunity.

5. **Routes of Ag administration:** Certain Ags are immunogenic in guinea pigs by intradermal route and tolerogenic by IV or oral route. Tolerance is induced best by the route which equilibrated the Ag between intra- and extravascular compartments.

Mechanisms

Following are the possible ways for tolerance.

1. **Contact of Ag in embryonic life:** Described above.

2. **Afferent mechanisms:** Contact of Ag to immune cell is interfered.

3. **Efferent mechanisms:** Ab produced is inhibited or neutralized or blocked.

4. **Central mechanisms (helplessness of B cells):** Elimination of Th cells will prevent the activation of B cells.

5. **Split tolerance:** Tolerance to either AMI or CMI but not to both, e.g., in guinea pig DTH to tuberculin Ag is blocked out interference to Ab production.

6. **Lack of genes:** Which are required for development of Ir.

Artificial Induction of Tolerance

Possible ways to induce tolerance artificially are administration of antisera or Abs, cytotoxic drugs and surgical ablation.

THEORIES OF IR

Introduction

Numbers of theories have been proposed are described below, but none of them is satisfactory.

Theories

A. Instructive theory: According to this, immunocompetent cell (ICC) can synthesize Ab of any specificity. Ag encounters the ICC and instructs it to produces the complementary Ab.

1. **Direct template theory:** According to this theory, Ag (epitope) enters in to the cell and acts as template against which Ab molecule is synthesized by cell. Such Ab is complementary to Ag (epitope).

2. **Indirect template theory:** According to this theory, not the Ag (epitope) but its genocopy (indirect template) enters in to the cell and Ab molecule is synthesized by cell. It makes the genetic changes in cell and transmitted to progeny cells. It explains

the specificity, secondary Ir and nonantigenicity to self. It becomes challenging with advance molecular biology.

B. Selective theory: According to this theory, there is selective synthesise of Ab by ICC on antigenic stimulation.

1. **Natural selection theory:** This theory was proposed by Jerne in 1955, so called Jerne's theory. According to this theory, numbers of Abs are synthesized during the embryonic life. These Abs act as receptors and combine with complementary Ags. This complex settles in cell and stimulates the cell to produce large numbers of Abs. It is not accepted because it does not explain the immunological memory.

2. **Clonal selection theory (Fig. 23.4):** It was proposed by Burnet in 1957. According to this theory numbers of ICCs were synthesized during embryonic life by somatic mutation. The cells that react with self antigens are eliminated and called forbidden clones. Their persistence and development in later life by mutation leads autoimmune process. ICCs will recognize and combine with Ags and undergo the proliferation to produce Abs. Some of the progeny cells act as memory cells. It is accepted than other theories; however, it does not explain the all features of Ir.

C. Side chain theory: It was proposed by Ehrlich in 1900. According to this theory cell has side chain (receptors) for assimilation of nutrients. Sometimes complementary Ag enters and combines with such side chains leading to interference in absorption of nutrients. By compensatory mechanisms cell starts to overproduce same receptors which circulate as Abs. It is not accepted because it does not explain the secondary Ir, tolerance and autoimmunity.

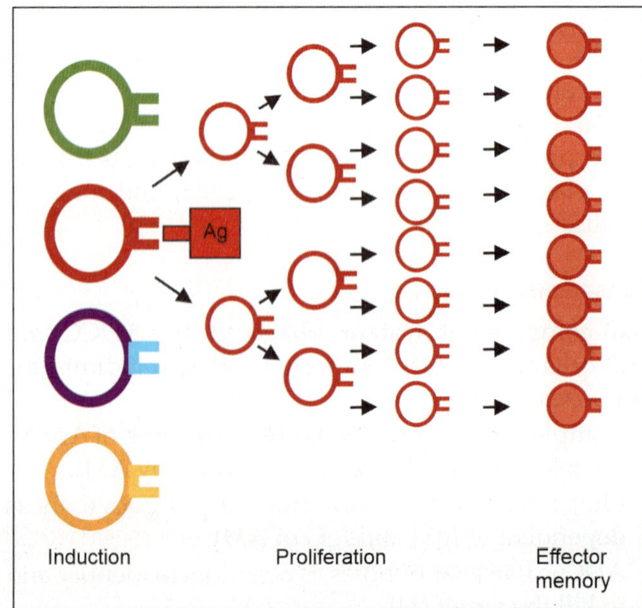

Fig. 23.4: Clonal selection model

Jerne's Network Hypothesis

Theory: It explains the mechanism of Ab response. The variable region of an immunoglobulin molecule carrying the antigen combining site is different in different antibodies. The distinct amino acid sequence at antigen combing site (paratope) and the adjacent part of the variable regions is termed as idiotype. It produces antiidotypic antibody and forms an idiotype network. The above process controls the amount of antibodies.

Nobel Prize: The above theory was given by Niels K. Jerne and he was awarded with Nobel Prize of Medicine in 1984.

Genetic Basis

Theory: Individual has capacity to produce 108 different Abs. Each Ab required separate gene. Discovery of split-gene theory, demolished the 'one gene-one protein' concept. It has important implication in biology and immunology.

Nobel Prize: The above theory was discovered by Susumu Tonegawa and he was awarded with Nobel Prize of Medicine in 1987.

ACCESS YOURSELF

Case Study

1. A 40-year-male has chronic cough and cavitary lesion in the lungs. His sputum is positive for AFB. Identify the clinical conditions and answer the following.
  a. Which type of immunity required to this patient to fight against this infection?
  b. Describe in general about the steps/mechanisms of induction of required immunity of given case and its clinical significances.
  c. How you diagnose the immunity, which patient required in given case?

Essay/Full Question

1. Immune response (Ir).

Short Notes

1. AMI/CMI
2. Cytokines.

Short Questions for Theory/Viva Questions

1. What are priming dose and booster dose of antigen?
2. What is antigenic competition?
3. Write four differences between cytosolic pathway and endocytic pathways.
4. What is adoptive immunity?
5. What is transfer factor?

Comments on

1. HBV vaccine is more effective following injection in deltoid than gluteal region.
2. Cytokines are unlike hormones and like hormones.
3. All the immune responses are interconnected.

AMI

1. **Cells involved in humoral immunity:**
   a. B cells   b. T cells
   c. Helper cells   d. Dendritic cells
2. **First immunoglobulin to appear following infection:**
   a. IgG   b. IgM
   c. IgA   d. IgE
3. **Primary immune response is mediated by:**
   a. IgE   b. IgM
   c. IgA   d. IgD
4. **Secondary immune response is mediated by:**
   a. IgG   b. IgA
   c. IgE   d. IgM
5. **True about secondary immune response is:**
   a. Long latent period
   b. Usually of low titer
   c. Antibodies appear in short time
   d. Persists for long
6. **Possible source of "second signal" to a B cell bound by specific antigen include:**
   a. EB virus
   b. Endotoxin
   c. Antigen specific T cells
   d. Plasma cells

CMI

7. **Which is concerned with cell mediated immunity?**
   a. B lymphocytes   b. T lymphocytes
   c. Eosinophils   d. Monocytes
8. **CMI is seen in:**
   a. Histoplasmosis   b. Leprosy
   c. Tetanus   d. Measles
9. **Adoptive immunity is by:**
   a. Infection
   b. Injection of antibodies
   c. Injection of lymphocytes
   d. Immunization
10. **Cell mediated immunity is transferred by:**
    a. Antibody   b. Transfer factor
    c. Interferon   d. Toxoid
11. **When transfer factor is given as treatment result in:**
    a. Natural active immunity
    b. Artificial active immunity
    c. Artificial passive immunity
    d. Adoptive immunity
12. **True about cytokine is:**
    a. It is always a polypeptide
    b. It acts on protein target
    c. It takes part in intrinsic enzymatic reactions
    d. Chemotactic
13. **IL-1 produces:**
    a. T lymphocytes activation
    b. Delayed wound healing
    c. Increased pain perception
    d. Decreased PMN release from bone marrow
14. **Interleukin I primarily acts on:**
    a. T lymphocytes   b. B lymphocytes
    c. Neutrophils   d. Macrophages

15. IL-1 is produced by:
  a. Macrophages
  b. Helper T lymphocytes
  c. B cells
  d. Cytotoxic T cells

16. IL-2 is produced by:
  a. CD4 lymphocytes
  b. CD8 cells
  c. Macrophages
  d. Neutrophils

17. Cachectin is produced by:
  a. Neutrophils
  b. Eosinophils
  c. Macrophages
  d. Basophils

18. IL-7 is produced by:
  a. Macrophages
  b. B cells
  c. T cells
  d. Dendritic cells
  e. Stromal cells

19. T cells functions are assessed by:
  a. Phagocytic index
  b. T cell count
  c. Migration inhibition test
  d. Immunoglobulin index

20. Which is not pyrogenic?
  a. IL-1
  b. TNF-α
  c. IL-4
  d. IL-6

21. Fever is caused by:
  a. IL-3
  b. IL-6
  c. IL-5
  d. IL-9

Immunological Tolerance

22. Specific immunological unresponsiveness is called tolerance. Which one of the following statements best describes immunological tolerance?
  a. Immunological maturity of the host does not play a major role.
  b. It occurs only with polysaccharide antigens.
  c. It is related to the concentration of the antibody.
  d. It is prolonged by administration of immunosuppressive drugs.

Answers and Explanation of MCQs

1. a
- Follow section, **AMI (introduction)** for explanation.

2. b

3. b

4. a

5. c, d
- Follow section, **AMI (Table 23.1)** for explanation of answers of MCQs 2–5.

6. c
- In AMI T (Th) cells participation required. Follow section, **AMI (mechanisms of AMI/Ab production → steps)** for explanation.

7. b
- Follow section, **CMI (introduction)** for explanation.

8. a, b and d
- Follow section, **CMI (clinical significances)** for explanation.

9. c

10. b

11. d
- Follow section, **CMI (Passive transfer of CMI or adoptive immunity)** for explanation of answers of MCQs 9–11.

12. a
- Cytokines are soluble polypeptides (proteins).

13. a

14. a

15. a
- Follow section, **CMI (Cytokines → IL-1)** for explanation of answers of MCQs 13–15.

16. a
- Follow section, **CMI (Cytokines → IL-2)** for explanation.

17. c

18. e
- Follow section, **CMI (Table 23.3)** for explanation of answers of MCQs 17–18.

19. c
- Follow section, **CMI (detection of CMI)** for explanation.

20. c

21. b
- IL-1, TNF-α and IL-6 are pyrogenes and they are known to induce fever.

22. d
- Follow section, **immunological tolerance (types → according to duration)** for explanation.

Hypersensitivity

INTRODUCTION

Definitions

Hypersensitivity: Excessive or exaggerated immune response in sensitized host due to contact with particular Ag leads to tissue damage, disease or even death called hypersensitivity.

Allergy: This word used by von Pirquet, and it is most confusing. Sometimes, it is used to refer the alter state of reactivity to antigen, which includes both protective (immunity) and harmful (hypersensitivity and autoimmunity) immune response. Sometimes, it is used only injurious immune response (hypersensitivity and autoimmunity). Most commonly, it is used as a synonym of hypersensitivity (no autoimmunity). Sometimes, it used to refer to only one type of hypersensitivity, namely atopy.

Allergen: A substances capable of causing an allergic or hypersensitivity reaction.

Antigenic dose: Two types. (1) Priming or sensitizing dose which means initial antigenic stimulation. (2) Booster or shocking dose which means subsequent antigenic stimulation. Subsequent dose is more effective, because it causes all manifestations of hypersensitivity.

Humoral amplification system: The pathology and clinical outcome of hypersensitivity is influenced by immune and nonimmune body mechanisms like complement, fibrinolytic system, kininogenic system, inflammation and coagulation called humoral amplification system.

Classification

According to time required by sensitized host to respond to the shocking dose of antigen: Two major types like immediate and delayed as described below with differences in **Table 24.1**.

TABLE 24.1: Differences between immediate and delayed type of hypersensitivity

Features	Immediate	Delayed
Appearance	Rapid	Slow
Duration	Short	Longer
Induction	By Ag/hapten by any route	By Ag/hapten by ID route or by skin contact or by Freund's adjuvant
Mediators	Ab mediated	Cell mediated
Passive transfer	Possible by serum	Rarely by transfer factor
Desensitization	Easy and transient	Difficult but long lasting

1. **Immediate hypersensitivity:** It starts in few minutes to few hours. It is B cell or Ab mediated. It is induced by hapten or Ag by any route. It is transferred passively by serum. Desensitization is easy, but short lived. Following are different subtypes.
   - Anaphylaxis
   - Atopy
   - Ab-mediated cell damage
   - Arthus reaction
   - Serum sickness
2. **Delayed hypersensitivity:** It starts in 24 hours, reaches peak in 48–72 hrs. It is T cell mediated without circulating Ab. It is induced by hapten or Ag or by Freund's adjuvant by intradermal route or by skin contact. It is transferred passively by Transfer Factor (TF) not by serum. Desensitization is difficult, but long lived. Following are different subtypes.
   - Infection (tuberculin) type reaction
   - Contact dermatitis type

Coombs and Gel's classification (1963): Major types are described below with comparison as mentioned in **Table 24.2**.

TABLE 24.2: Comparison between all types of hypersensitivity

Features	Type I	Type II	Type III	Type IV
Type	Immediate	Immediate	Immediate	Delayed
Response	Humoral	Humoral	Humoral	Cellular
Period between onset of symptoms and Ag contact	2–30 minutes	5–8 hours	2–8 hours	24–72 hours
Antigen	Soluble	Cell surface bound	Soluble	Soluble or cell surface bound
Mediators	IgE, histamine and pharmacological agents	IgG IgM C or phagocytic cells	IgG IgM C or Leukocytes	T cells Lymphokines Macrophages
Desensitization	Easy but short lasting	Easy but short lasting	Easy but short lasting	Difficult but long lasting
Syndrome	Anaphylaxis Atopy	Ab-mediated cell damage	Arthus reaction Serum sickness	Infection (tuberculin) type Contact dermatitis type

1. **Type I (anaphylactic, IgE or reagin dependent hypersensitivity):** Antibodies (cytotropic IgE) are fixed on surface of tissue cells in sensitized individual. The antigen combines with cell fixed antibody, leads the release of pharmacologically active substances (vasoactive amines) which produce the clinical reactions. It includes two clinical syndromes like anaphylaxis and atopy. Diseases and mediators under these syndromes are mentioned below.
   - **Anaphylaxis**
     – Examples of diseases: It include disease/reaction which occurs following exposure of allergens like honey bee bite, etc.
     – Mediators: These are IgE fixed with tissue basophils (mast cells), blood basophils (free IgE is not useful), histamine and other agents.
   - **Atopy**
     – Diseases: It includes diseases like allergic asthma, allergic conjunctivitis, allergic rhinitis (hay fever) and dermatitis.
     – Mediators: Overproduction of IgE antibodies.
2. **Type II (cytotoxic or cytolytic hypersensitivity, IgG or rarely IgM dependent hypersensitivity):** This type of reaction is initiated by IgG or rarely by IgM/IgE. Ab reacts either with the cell surface or tissue antigen. Ab produces cell/tissue damage in presence of complement or mononuclear cells. Sometimes Ab stimulates (Type-V) or inhibits the cell. Type 2 reactions are intermediate between hypersensitivity and autoimmunity. It includes only one clinical syndrome that is **Ab mediated damage**. Diseases and mediators under this syndrome are mentioned below.
   - Examples of diseases: Ab acts by four ways. (1) Lytic for cells like thrombocytopenia, agranulocytosis (agranulosis or granulopenia), autoimmune hemolytic anemia, etc. (2) Supportive to cells like ADCC. (3) Cell stimulatory (separately classified as type V hypersensitivity) like Graves' disease. (4) Cell inhibitory like pernicious anemia, myasthenia gravis, etc.
   - Mediators: Autoantibodies.

3. **Type III (immune complex or toxic complex disease):** It causes damage by Ag-Ab complex. This precipitates in and around the small blood vessels, causing damages to cell secondarily or on membrane interfering with its functions. It includes two clinical syndromes like arthus reaction and serum sickness. Diseases and mediators under these syndromes are mentioned below.
   - **Arthus reaction**
     – Examples of diseases: Like farmer's lungs.
     – Mediators: IgG, IgM, C and leucocytes.
   - **Serum sickness:**
     – Examples of diseases: Like post-streptococcal glomerulonephritis.
     – Mediators: IgG, IgM, C and leukocytes.
4. **Type IV (delayed or cell-mediated hypersensitivity):** It is not Ab mediated, but mediated by sensitized T cells. The antigen activates specifically sensitized T4 and T8 cells, leading to the secretion of lymphokines with accumulation of fluid and phagocytes. In unsensitized individual the injection of Ag provokes no response. It differs from type I hypersensitivity not only in longer interval of appearance but also in its morphology and histology. It includes two clinical syndromes like infection (tuberculin) type reaction and contact dermatitis type reaction. Diseases and mediators under these syndromes are mentioned below.
   - **Infection or tuberculin type:**
     – Examples of diseases: Infective, autoimmune conditions and allograft rejection.
     – Mediators: T cells, lymphokines and macrophages.
   - **Contact dermatitis type**
     – Examples of diseases: Dermatitis due to contact with allergens.
     – Mediators: T cells, lymphokines and macrophages.
5. **Type V hypersensitivity (stimulatory type):** It is the Ab mediated hypersensitivity. Ab causes stimulation instead of damages.

- Examples of disease: Graves' disease, Stevens-Johnson syndrome, sulfonamide-induced morbilliform rash, etc.
- Mediator: In Graves' disease, long-acting thyroid stimulator (LATS), an antibody against some determinant of thyroid cells is present. It stimulates excessive secretion of thyroid hormones leading to Graves' disease. Sometimes it is considered as a part of type II hypersensitivity. In others diseases autoantibodies like IgG or IgM mat act as mediators.

TYPE I REACTION

It is the IgE or reagin dependent hypersensitivity. Two clinical syndromes like anaphylaxis and atopy are described below.

Anaphylaxis

It is discussed with following four types.

A. Systemic anaphylaxis

History: Term anaphylaxis was given by Richet in 1902, who observed that dogs had survived with sublethal dose of toxic extracts from sea anemones, were rendered susceptible to minute dose given days or weeks later. Theobald Smith observed similar thing in guinea pigs, Ehrlich named it as Theobald Smith phenomenon.

Meaning: Ana means without and phylaxis means protection.

Definition: It is an acute and potentially fatal form of type I hypersensitivity.

Properties: It is possible only by cell fixed IgE called cytotrophic Ab but not possible by free IgE.

Agents inducing systemic anaphylaxis

- Heterologous serum therapy like ATS, AGS or ADS
- Injection of drugs like penicillin (Pn) or insulin
- Insect bites like honey bee, ant sting or wasp
- Ingestion of sea foods
- Nuts.

Factors influencing anaphylaxis

- Sensitization: It is done by injection, inhalation or contact of Ag
- Shocking Ag: It is most effective when administered by parenteral route, less by intraperitoneal or subcutaneous and least by intradermal route. Shocking Ag is identical with or related to sensitizing Ag.
- Waiting period: An interval of 2–3 weeks between sensitizing and shocking dose is required.

Mechanisms of anaphylaxis: Mechanisms for induction of anaphylaxis are described in **Flowchart 24.1 and Fig. 24.1**. Actions of different mediators are described below.

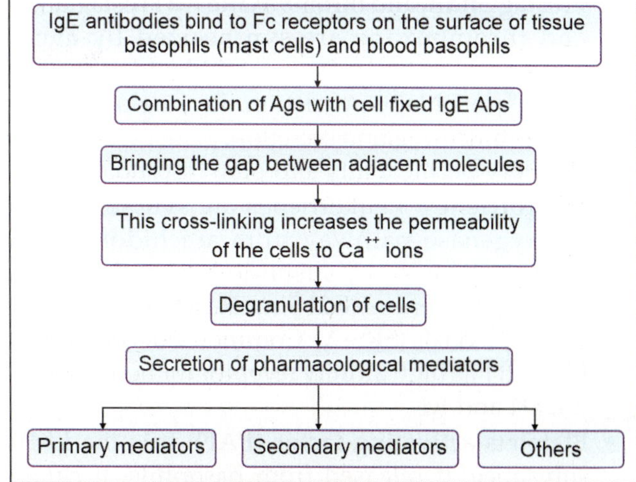

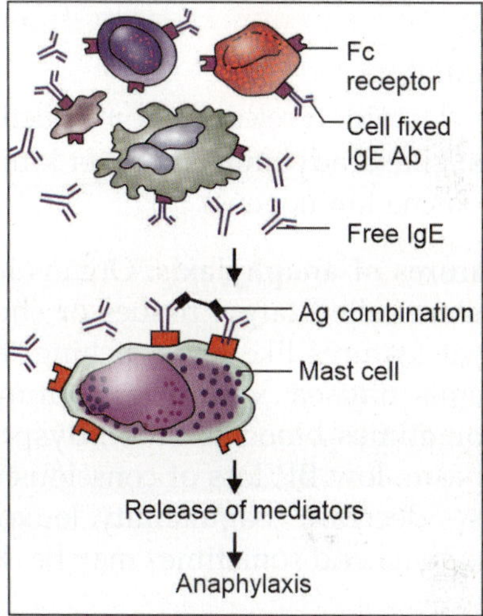

Fig. 24.1: Mechanisms of anaphylaxis

1. Primary mediators

- **Histamine:** It formed by decarboxylation of histidine from granules of mast cells, basophils and platelets. It released in to skin and stimulates the nerves to produce itching and burning sensation. It causes vasodilatation, hyperemia (flare effect) and edema by increasing the capillary permeability (wheal effect). It induces smooth muscle contraction.
- **5-HT (5-Hydroxy tryptamine)/serotonin:** It formed by decarboxylation of tryptophan. It presents in intestinal tissues, brain and platelets. It causes vasoconstriction and increases the capillary permeability. It induces smooth muscles contraction.
- **Eosinophil chemotactic factor of anaphylaxis (ECF-A):** It released from granules of mast cells. It attracts the eosinophils and causing the eosinophilia.
- **Neutrophil chemotactic factor (NCF):** It is the HMW substances. It attracts the neutrophils.
- **Enzymatic mediators:** Like protease and hydrolase are released from mast cells.

2

  2. Secondary mediators

- **Prostaglandin and thromboxane A$_2$:** Prostaglandin and thromboxane are synthesized by cyclo-oxygenase pathway from arachidonic acid. They cause bronchial constriction. Prostaglandin also affects mucous gland secretion, platelets adhesion, capillary permeability and pain threshold.
- **Leukotrienes:** Leukotrienes are synthesized by lipoxygenase pathway from arachidonic acid. They induce slow, sustained smooth muscle contraction, hence called **S**low **R**eacting **S**ubstances of **A**naphylaxis (SRS-A). Common leukotrienes are LTB4 (It has high affinity receptor for neutrophils), C4, D4 and E4.
- **Platelets-activating-factor (PAF):** It is the LMW substance. It released from basophils. It causes platelets aggregation and release of vasoactive amines.

3. Other mediators

- **Anaphylactoid:** It released from C-activation.
- **Kinins:** Like bradykinin and other kinins derived from plasma kininogens.

Clinical features of anaphylaxis: Organs involved in anaphylaxis are called target tissues or shock organs with different features like fever, itching, flushing of skin or edema, nausea, vomiting, abdominal pain, diarrhea, sometimes blood in stool, dyspnea due to bronchial spasm, low BP, loss of consciousness, blood picture shows decrease coagulability, leukopenia and thrombocytopenia and sometimes may be death.

Diagnosis of anaphylaxis

1. **Animal inoculation:** Guinea pig is the highly susceptible animal to diagnose the anaphylaxis.
2. **Radioimmunosorbent test:** It quantitatively measures the serum IgE up to nanogram level. It is highly sensitive test. In this test patient's serum contains IgE is made to reacts with paper disk/beads contains anti-IgE. After washing the disk/beads, it allows to reacts with radiolabeled anti-IgE. Radioactivity of the disk/beads is measured by gamma counter.
3. **Radioallergosorbent test:** It quantitatively measures the allergen specific serum IgE. It is highly sensitive test. In this test patient's serum contains IgE is made to reacts with paper disk/beads contains allergen, so only allergen specific IgE would bound. After washing the disk/beads it allows to reacts with radiolabeled anti-IgE. Radioactivity of the disk/beads is measured by gamma counter.

Prevention of anaphylaxis

1. **Avoidance of contact with known allergen:** This is difficult task.
2. **Desensitization:** By following two methods:
   - **Acute desensitization**
     - Method: Small amount of Ag is administered at 15 minutes interval for 1–2 hours. Ag-Ab complex is formed which releases the chemical mediators but not enough to produces the major reaction. This technique is adopted on the subsequent entry of allergen like ATS or Pn.
     - Disadvantage: It is short lasting and hypersensitivity may return after few days or months.
   - **Chronic desensitization**
     - Method: Small amount of Ag is administered at weekly interval to the hypersensitive individual. Ag stimulates the IgE blocking Ab production which prevents the contact of allergen with IgE-Ab present on mast cell.
     - Advantage: It is a long lasting procedure.

Treatment of anaphylaxis: Prompt treatment with adrenaline (0.5 ml, 1-in-1000 solution) by IM or SC route.

B. Local anaphylaxis: It has following three subtypes:

1. **Cutaneous anaphylaxis:** Intradermal introduction of small shocking dose of Ag in sensitized individual produces local wheal and flare effect called cutaneous anaphylaxis. It also developed on ingestion of allergen followed by absorption and characterized by utricaria or angioneurotic edema. Wheal effect includes central pale area due to edema by increasing the capillary permeability. Flare effect includes peripheral red (erythema) area due to hyperemia by vasodilatation. It is diagnosed by following skin tests.
   - **Intradermal injection:** 0.1 ml antigen is introduced intradermally in one forearm called test arm with normal saline in other arm called control arm. Wheal and flare response in test arm and no response in control arm suggest the positive reaction. Disadvantages of this test are like risk of anaphylaxis by test antigen and negative test does not rule out the possibility of IgE mediated hypersensitivity because it is positive in about 60% of sensitive individual. Example of test is Casoni's test in hydatid cyst.
   - **Other tests:** Like prick method and patch method.
2. **Mucosal anaphylaxis:** Entry of small shocking dose of Ag in conjunctiva, nasal mucosa or in respiratory mucosa of sensitized individual produces conjunctivitis, rhinorrhea or bronchospasm respectively. It is diagnosed by conjunctival test. In this test one drop of antigen is instilled in to the one eye called test eye with normal saline in other eye called control eye. Conjunctivitis (redness, lacrimation with itching) in test eye and no response in control eye suggest the positive test.
3. **Passive cutaneous anaphylaxis (PCA):** It was developed by Ovary in 1952. Intradermal (ID) injection of Ab, followed by IV injection of Ag fixed with dye like Evans blue, 4–24 hours afterwards will produces vasodilatation and hyperaemia at ID site. It is a wheal-flare effect. This is the extremely sensitive method for detection of antibodies. PCA can be used to detect human IgG antibody which is

heterocytotropic (capable of fixing the cells of other species) but not IgE which is homocytotropic (capable of fixing the cells of homologues species only).

C. Anaphylaxis *in vitro*: It also called Schultz-Dale phenomenon. In this reaction isolated tissue strip like intestinal or uterine muscle from sensitized guinea pig, held in bath of the Ringer's solution, will contracts on addition of the specific Ag to the bath. Reaction is specific and elicited only by Ag to which animal is sensitive.

D. Anaphylactoid reaction: IV injection of certain substances like trypsin, peptone provokes the reaction like anaphylaxis called anaphylactoid reaction. It mostly produced due to release of biological substances like C4a, C3a and C5a from complement activation.

Atopy

History: Atopy term was first coined by Coca 1923.

Meaning: Atopy means out of place or strangeness.

Definition: It is type I hypersensitivity reaction that occurs naturally, spontaneously in response to substances encountered in the environment in everyday life.

Properties: It is very difficult to induce atopy artificially. It produces local effect, but sometimes remote effect also occurs like utricaria following ingestants. Atopens are generally not good antigens when introduced parenterally. It occurs in human beings, it is not induced experimentally (artificially) in animals.

Agents causing atopy: These agents are called atopens. For examples inhalants like pollen, dust, etc., ingestants like milk, egg, etc., and contact allergen to skin and conjunctiva.

Mechanisms of atopy: About 10% populations are prone to develop sensitization to various environmental atopens such as pollen or dust. Small amount of IgE is produced by individual but atopy is due to overproduction of IgE called reagin antibody with simultaneous deficiency of IgA. When atopen enters in body it is prevented by IgA to produces damage, but due to deficiency of IgA it combines with cell fixed IgE and releases the mediators producing allergic reaction. It shows marked familial distribution and it is suspected that the sensitization is inherited probably MHC genotype. Inheritance is not sensitivity to particular Ag, but tendency to develop IgE.

> **Note: Reagin antibodies**
> Both, IgE antibody of atopy (type I hypersensitivity) and non-treponemal antibodies are called reagin antibodies; however, there is no correlation between two.

Clinical features of atopy: These depend on route of entry of atopens. **Features include** allergic asthma, allergic conjunctivitis, allergic rhinitis (hay fever) and dermatitis

Diagnosis of atopy: IgE is detected by passive agglutination test and ELISA.

Prevention of atopy: Avoidance of contact with known allergen, but difficult.

Treatment of atopy: Desensitization by injecting serum or repeated injection of Ag.

> **Note: Praunitz-Kustner reaction (PK reaction)**
> **History:** It was 1st reported by Prausnitz and Kustner in 1921.
> **Principle:** IgE is homocytotrophic, which is species specific. Only human IgE can fix to the surface of human cells. This is the basis of Prausnitz-Kustner (PK) reaction which was the original method for detecting atopic antibodies.
> **Method:** Serum is collected from Kustner (atopic hypersensitive to certain species of cooked fish) and injected in to Prausnitz by Intracutaneous (IC) route. After 24 hours small dose of cooked fish Ag is injected in to Prausnitz by intracutaneous route at same site.
> **Result:** Local wheal-flare effect at IC injection site.
> **Disadvantage:** It carries the risk of transmission of infection so no longer is used.

TYPE II REACTION

It is the cytotoxic or cytolytic hypersensitivity and IgG or rarely IgM dependent hypersensitivity. It includes only one clinical syndrome that is **Ab mediated damage** which is described below.

Types of antigens: Following are two types:
- Intrinsic Ag: "Self" antigen, part of the host cells.
- Extrinsic Ag: Adsorbed on to the cells and arise from pathogens or by drugs.

Mechanisms: Ab damages the cell by following four mechanisms:

1. **Ab mediated cell lysis**
   - **Steps:** Antigens are adsorbed on to the cells. The antibodies produced by the immune response are mostly IgG and rarely IgM bind to antigens on the patient's own cell surfaces. Such cells with Ag-Ab complex on surfaces are killed by phagocytic cells or by C-activation as shown in **Fig. 24.2**.
   - **Clinical syndromes**
     - Thrombocytopenia: Destruction of platelets by cell fixed antigens and antibodies.
     - Agranulocytosis or agranulosis or granulopenia: Destruction of WBCs (especially neutrophils → Neutropenia) by cell fixed antigens and antibodies.
     - Autoimmune hemolytic anemia: Destruction of RBCs by cell fixed antigens and antibodies.
     - Rh incompatibility (erythroblastosis fetalis): It is due to Rh incompatibility, where Rh −ve mother has anti-Rh antibodies due to earlier pregnancy with Rh +ve fetus. Such antibodies can cross the placenta and destroy the Rh +ve fetal RBCs called erythroblastosis fetalis.

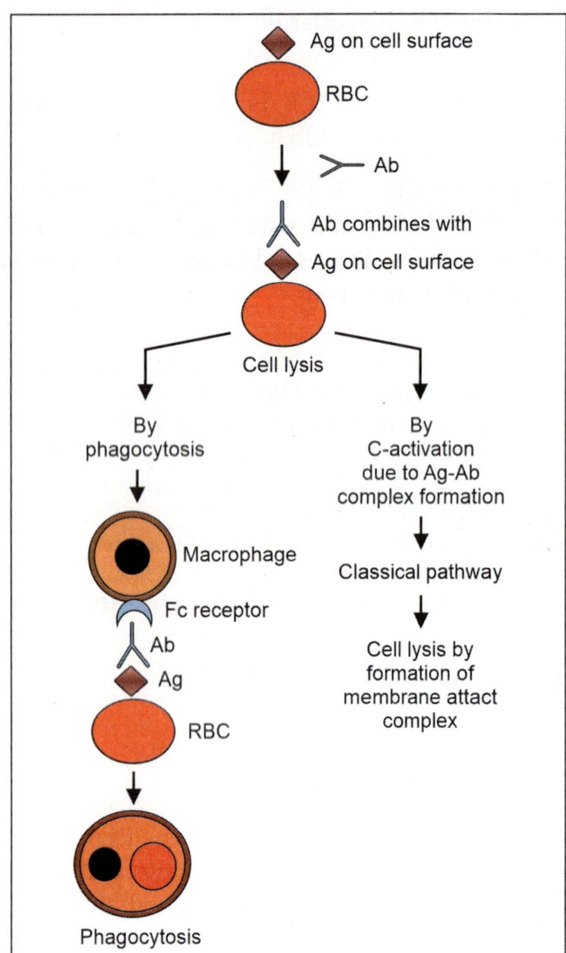

Fig. 24.2: Cell lysis by type II hypersensitivity

– ABO incompatibility (transfusion reaction): Donor's RBCs are lysed by recipient anti-RBCs antibodies due to incompatible blood transfusion.
– Drug-induced hemolytic anemia: Drug or its products, absorbed on RBC surface. Synthesis of antibody to drug or its products lyses the attached RBCs by complement activation.
– Pemphigus vulgaris: Autoantibody against desmosomal protein leads the disruption of epidermal intracellular junction.
• **Mediators:** These are IgG, IgM and C (complement) or phagocytic cells.

2. **Antibody dependent cytotoxic cells (ADCCs):** It is possible by LGLs. This type of LGLs possesses surface receptors for the Fc part of Ig. Antigen is attached at Fab region. They are capable of killing target cells sensitized with IgG like tumor cells, parasites or graft rejection. ADCC is mostly IgG mediated. In certain instances IgE is useful like eosinophilic mediated killing of parasites. This ADCC is distinct from the action of cytotoxic T cells, which is independent of antibody.

3. **Cell stimulatory:** It is separately classified as type V hypersensitivity, discussed earlier.

4. **Cell inhibitory:** Antibody produced is inhibitory in nature instead of stimulatory like in following diseases:
   • Pernicious anemia: Where antibodies against parietal cells of gastric mucosa decrease the secretion of acid leads to achlorhydria and atrophic gastritis.
   • Myasthenia gravis: Where antibody present against acetylcholine receptors at neuromuscular junctions in striated muscles. It prevents the combination of Ach with its receptors and impairs the muscle contraction.

> **Note: Cell stimulatory and cell inhibitory hypersensitivities** Both are better classified as autoimmunities. For more details **follow Ch. 26.**

TYPE III REACTION

It mediates immune complex or toxic complex disease. It includes two clinical syndromes like arthus reaction and serum sickness as described below.

Arthus Reaction

History: It was reported by Arthus in 1903. He observed that when rabbits were repeatedly injected subcutaneously with normal horse serum, the initial injections had no local effect but with later injections, there is a production of intense local edema, indurations and hemorrhagic necrosis. This is called Arthus reaction.

Definition: It is a local manifestation of type III hypersensitivity.

Passive transfer: It is transferred passively with sera containing precipitating Abs (IgG, IgM) in high titers.

Mechanism: Follow **Flowchart 24.2.**

Examples of diseases: Arthus reaction is seen in clinical conditions like farmer's lung, allergic bronchopulmonary aspergillosis, allergic fungal rhionosinusitis and bagassosis. Farmer's lung presents with hypersensitive pneumonitis due to thermophilic actinomycetes from mouldy hay or grain.

Mediators: IgG, IgM, C (complement) and leukocytes.

Flowchart 24.2: Mechanism of Arthus reaction

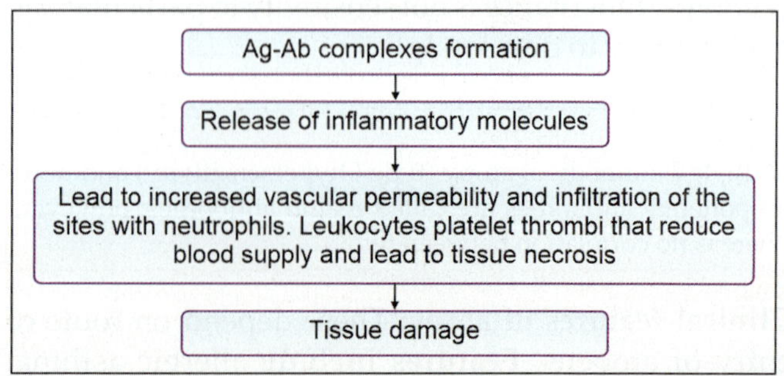

Serum Sickness

History: Originally it was described by von Pirquet and Shick in 1905.

Properties: It is a systemic manifestation of type III hypersensitivity. Single injection can serve both as the sensitizing dose and shocking dose. It appeared 7–12 days following a single injection of high concentration of foreign serum such as diphtheria antitoxin. As heterologous serum injections are not used nowadays the syndrome is more commonly seen with injection of Pn or other antibiotics.

Mechanism of serum sickness: Follow **Flowchart 24.3**.

Flowchart 24.3: Mechanism of serum sickness

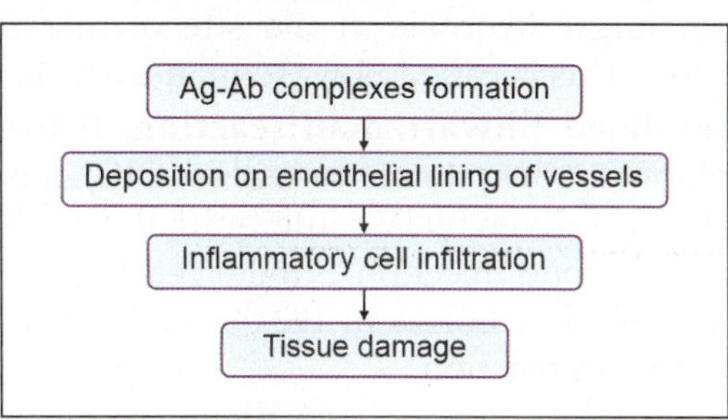

Clinical features of serum sickness: It presents with fever, lymphadenopathy, splenomegaly, arthritis, glomerulonephritis, endocarditis, vasculitis, urticarial rashes, abdominal pain, nausea, vomiting, etc. It is self limiting disease. With continued rise in Abs production the immune complex become larger and more susceptible to phagocytosis and immune elimination. When all antigens are thus eliminated and free antibodies appear the symptoms clear.

Examples of diseases

- Rheumatoid Arthritis: Autoantibody (IgM) against Fc piece of IgG to form IgM-IgG complex.
- Acute post-streptococcal glomerulonephritis (APSGN): Immune complex formation between streptococcal antigens and respective antibodies, which later deposited in renal glomeruli.
- PAN (polyarteritis nodosa): Immune complex formation but exact nature of autoantibodies is unknown.
- SLE: Autoantibodies against nuclear materials that make immune complex and deposited in PMN cells called LE cell.

TYPE IV REACTION

It is the delayed or cell mediated hypersensitivity. It includes two clinical syndromes like infection (tuberculin) type reaction and contact dermatitis type reaction as described below.

Infective (Tuberculin) Type Reaction

Causes: It is induced by following allergens.

1. **Infective conditions:** Mostly it seen in subacute or chronic infections with intracellular pathogens.

- **Bacterial infections:**
  - *M. tuberculosis* (Tuberculin type): When a small dose of tuberculin is injected intradermally in individual sensitized to tuberculoprotein by prior infection or immunization, an indurated inflammatory reaction develops within 48–72 hours. Tuberculin test provides a useful indication of the state of delayed hypersensitivity which is described in **Ch. 56**.
  - *M. leprae*: Lepromin test provides a useful indication of the state of delayed hypersensitivity which is described in **Ch. 57**.
  - Others: *L. monocytogenes, B. abortus*, etc.
- **Viral infections:** Smallpox virus, HSV (herpes simplex virus) and measles virus, etc.
- **Fungal infections:** *P. carinii, C. albicans, C. neoformans, H. capsulatum*, etc.
- **Parasitic infections:** *L. donovani*, etc.
2. **Other conditions:** Like allograft rejection and autoimmune diseases (multiple sclerosis) are also showing infective type reactions.

Mechanism: Follow **Flowchart 24.4**.

Mediators: T cells, lymphokines and macrophages.

Flowchart 24.4: Mechanism of infective type hypersensitivity

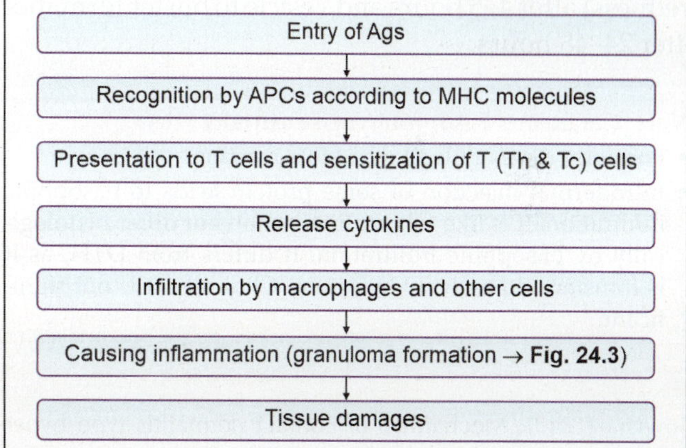

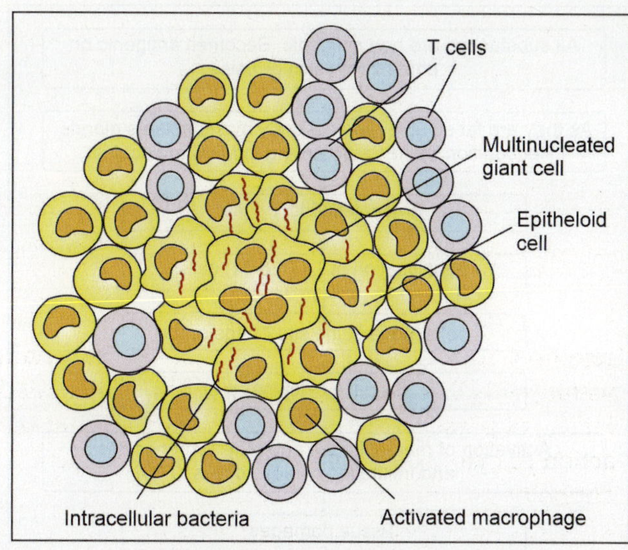

Fig. 24.3: Granuloma formation

Clinical features: It is characterized by granuloma formation in tissues.

Diagnosis: It is diagnosed by skin test like tuberculin test, lepromin test, etc.

Contact Dermatitis Type

Causes: It is induced by the following allergens:

- Chemicals: Dyes, picryl chloride and dinitrochlorobenzene.
- Metals: Like nickel, chromium, etc.
- Drugs: Penicillin.
- Plant allergens: Parthenin from parthenium.

Mechanism: Follow **Flowchart 24.5**.

Mediators: T cells, lymphokines and macrophages.

Clinical features: Contact with allergen in sensitized individual causes contact dermatitis, the lesions varying from macule and papule to vesicle that breaks down, leaving behind raw weeping area typical of acute eczematous dermatitis.

Diagnosis: It is diagnosed by patch test, in which allergen is applied to the skin under an adherent dressing. Sensitivity is detected by itching, erythema (redness) after 4–5 hours and vesicle to blister formation after 24–48 hours.

> **Note: Cutaneous basophil hypersensitivity**
> - Formerly it was called Jones-Mote reaction.
> - Intradermal injection of some protein leads to basophilic infiltration. It is like tuberculin reaction but differ histologically by basophilic infiltration. It differs from DTH, as it is transferred passively by serum. Clinically, it is not significant.

Flowchart 24.5: Mechanism of contact dermatitis type hypersensitivity

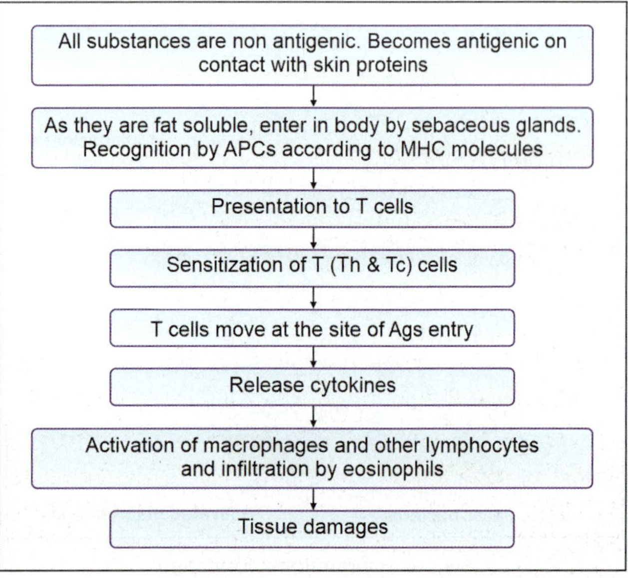

SHWARTZMAN REACTION

Synonym: Shwartzman phenomenon.

Definition: It is an overwhelming nonimmune response with multisystem involvement leading to local or systemic vasculitis.

History: Reported by Gregory Shwartzman in 1928.

Types

1. **Local Shwartzman reaction:** It based on development of local vasculitis (local hemorrhage and necrosis). Shwartzman observed that when rabbits were injected intradermally with culture filtrate (endotoxin) of *S. Typhi*, with 24 hours later IV injections by same filtrate, there is a production of intense local hemorrhagic necrosis at the site of intradermal injection. This is called Shwartzman reaction.
2. **Generalised Shwartzman reaction:** It based on development of systemic vasculitis (DIC). If both the injection are given by IV route, animal will die 12–24 hrs after second injection. Similar reaction observed by Sanarelli in Cholera in 1924 so called Sanarelli – Shwartzman reaction.

Antigenic doses

- Preparatory dose: It is an initial dose. It includes some bacterial endotoxin.
- Provocative dose: It is the subsequent (IV) dose. It includes some bacterial endotoxin, antigen antibody complex, starch, serum or kaolin.

Mechanisms

- Preparatory dose: It causes accumulation of leukocytes, which condition the site by releasing the lysosomal enzymes, interleukin (IL-1 and 6) and capillary wall damage.
- Provocative dose: It causes intravascular clotting and thrombus formation, leads to necrosis and hemorrhage (local) or DIC (generalized).

Clinical significances: Similar mechanisms are responsible in many clinical conditions like Waterhouse Friderichsen syndrome in infection by *N. meningitidis*, infection by *Staph. aureus*, etc.

Case Study

1. A 28-year-male patient brought to the medical emergency with history of honey bee bite with complains of coughing and dyspnea. Identify the hypersensitivity and answer the following:
   a. Name the types of hypersensitivity.
   b. Write mechanisms and diagnosis of given hypersensitivity.
   c. Write treatment of given hypersensitivity.
2. A 28-year-old female patient visited the skin OPD with history of skin rashes on forehead, 2 days after the application of hair dyes. A case of hypersensitivity was diagnosed. Identify the clinical type of hypersensitivity and answer the following:
   a. Name the type of hypersensitivity.

b. Name the allergic test useful to diagnose the hypersensitivity.
c. Write mechanisms of hypersensitivity in given case.
d. Write the classification of hypersensitivity.

Essay/Full Question

1. Describe the mechanisms of hypersensitivity.

Short Notes

1. Classification of hypersensitivity
2. Systemic anaphylaxis
3. Type II hypersensitivity.
4. Atopy/Arthus reaction/Serum sickness
5. Delayed type hypersensitivity.

Short Questions for Theory/Viva Questions

1. Differences between immediate and delayed type of hypersensitivity.
2. Define: Immunity and hypersensitivity.
3. What is Theobald Smith phenomenon?
4. What is anaphylactoid reaction?
5. What is Schultz-Dale phenomenon?
6. What is Shwartzman reaction?

Comments on

1. IgG is known as heterocytotropic antibody and IgE is known as homocytotropic antibody.
2. Atopy is not only due to overproduction of IgE but with simultaneous deficiency of IgA.

MCQs for Chapter Review

Type I Reaction

1. **Type I hypersensitivity is mediated by which of the following immunoglobulin?**
   a. IgA
   b. IgG
   c. IgM
   d. IgE
2. **The most important cell in type I hypersensitivity:**
   a. Macrophage
   b. Mast cell
   c. Neutrophil
   d. Lymphocyte
3. **Mast cell synthesizes and/or secretes:**
   a. Adrenaline
   b. Ach
   c. Histamine
   d. Heparin
   e. Neutrophilic chemotactic factor
4. **Example of type I hypersensitivity is:**
   a. Lepromin test
   b. Tuberculin
   c. Casoni's test
   d. Arthus reaction
5. **Type I hypersensitivity includes all of the following *except*:**
   a. Autoimmune hemolytic anemia
   b. Anaphylaxis
   c. Extrinsic asthma
   d. Hay fever
6. **Wheal and flare response is what type of hypersensitivity reaction?**
   a. Type I
   b. Type II
   c. Type III
   d. Type IV
7. **Anaphylaxis is mediated by:**
   a. 5-hydrpxytryptamine
   b. Heparin
   c. Prostaglandin
   d. Platelet activating factor
8. **Casoni's test is:**
   a. Type I hypersensitivity
   b. Type II hypersensitivity
   c. Type III hypersensitivity
   d. Type IV hypersensitivity

9. **Which leukotriene is the adhesin factor for the neutrophils on the cell surface to attach to endothelium?**
   a. B4
   b. C4
   c. D4
   d. E4
10. **Schultz-Dale phenomenon is an example of:**
    a. Type I hypersensitivity reaction
    b. Type II hypersensitivity reaction
    c. Type III hypersensitivity reaction
    d. Type IV hypersensitivity reaction
11. **PK reaction detects:**
    a. IgG
    b. IgA
    c. IgE
    d. IgM
12. **Which of the immunoglobulin shows homocytotropism?**
    a. IgG
    b. IgA
    c. IgE
    d. IgD
13. **Atopy is mediated by:**
    a. IgE
    b. IgD
    c. IgM
    d. IgA

Type II Reaction

14. **True about type II hypersensitivity reaction is:**
    a. May be complement mediated
    b. Schultz-Dale phenomenon is a type II hypersensitivity
    c. Antibody independent
    d. Role of IgE
15. **Prototype of type II hypersensitivity reaction is:**
    a. Arthus reaction
    b. SLE
    c. Autoimmune hemolytic anemia
    d. Contact dermatitis
16. **Graves' disease is an example of which type of immunological response?**
    a. Type I
    b. Type II
    c. Type III
    d. Type IV
    e. Type V
17. **Erythroblastosis fetalis is an example of which type of hypersensitivity reaction?**
    a. Type I
    b. Type II
    c. Type III
    d. Type IV
18. **Which category of hypersensitivity best describes hemolytic disease of the newborn caused by Rh incompatibility?**
    a. Type I
    b. Type II
    c. Type III
    d. Type IV
19. **Myasthenia gravis is which type of hypersensitivity?**
    a. Type I
    b. Type II
    c. Type III
    d. Type IV
20. **All are type II hypersensitivity reaction *except*:**
    a. Hemorrhagic disease of newborn
    b. Graves' disease
    c. Autoimmune disease
    d. Hemolytic anemia

Type III Reaction

21. **Type III reaction is:**
    a. Antibody mediated
    b. Immune complex mediated
    c. Cell mediated
    d. None
22. **Which is an example of type III hypersensitivity reaction?**
    a. Contact dermatitis
    b. Hemolytic anemia
    c. Serum sickness
    d. Goodpasture's syndrome
23. **Arthus phenomenon is an example of which hypersensitivity?**
    a. Type I
    b. Type II
    c. Type III
    d. Type IV

24. All of the following are immune complex disease, *except*:
   a. Serum sickness
   b. Farmer's lung
   c. SLE
   d. Graft rejection

Type IV Reaction

25. Delayed hypersensitivity involves:
   a. Neutrophils
   b. Monocytes
   c. Eosinophils
   d. Lymphocytes

26. Which of the following is type IV hypersensitivity?
   a. Arthus reaction
   b. Serum sickness
   c. Shwartzman reaction
   d. Granulomatous reaction

27. Not a delayed type hypersensitivity reactions:
   a. Arthus reaction
   b. Bronchial asthma
   c. Hemolytic anemia
   d. Multiple sclerosis

28. Contact dermatitis is:
   a. Type I hypersensitivity
   b. Type II hypersensitivity
   c. Type III hypersensitivity
   d. Type IV hypersensitivity

29. In contact dermatitis which cells play major role?
   a. T-cells
   b. B-cells
   c. Langerhans cells
   d. Macrophages

30. Skin tests are used for which hypersensitivity reactions?
   a. I
   b. II
   c. III
   d. IV
   e. V

31. Which of the following is false?
   a. Theobald Smith phenomenon is a type I hypersensitivity reaction
   b. Serum sickness is a type II hypersensitivity reaction
   c. Allograft rejection is a type IV hypersensitivity reaction
   d. Transfusion reaction is a type II hypersensitivity reaction

32. Tuberculin test is reaction of:
   a. Anaphylaxis mediated
   b. Cell mediated
   c. Antibody mediated
   d. Immune complex mediated

33. Type IV hypersensitivity reaction includes all *except*:
   a. Paul Bunnel test test
   b. Lepromin test
   c. Tuberculin test
   d. Granulomatous reaction

Answers and Explanation of MCQs

1. d
2. b
3. c, e
4. c
5. a
6. a
7. a, c, d
8. a
9. a
10. a
   - Follow section, **type I hypersensitivity** for explanation of answers of MCQs 1–10.
11. c
12. c
13. a
   - Follow section, **type I hypersensitivity (atopy)** for explanation of answers of MCQs 11–13.
14. a
   - Type II hypersensitivity mediated by autoantibody and damage to target cell occurs by phagocytes, complement, ADCC, cell stimulation or cell inhibition.
15. c
16. b, e
17. b
18. b
19. b
   - Follow section, **type II hypersensitivity** for explanation of answers of MCQs 15–19.
20. a
   - Hemorrhagic disease of newborn is due to vitamin K deficiency. Graves' disease is an autoimmune disease, where Ab is stimulatory type and it is classified as type II and separately as type V hypersensitivity. Hemolytic anemia and other autoimmune diseases are part of type II hypersensitivity.
21. b
22. c
23. c
24. d
   - Type III hypersensitivity is immune complex or toxic complex disease in nature which includes arthus reaction and serum sickness.
25. b and d
   - Type IV hypersensitivity is mediated by lymphocytes, lymphokines and macrophages (monocytes).
26. d
   - Type IV hypersensitivity, especially tuberculin type presents with granulomatous reaction.
27. a, b and c
   - Arthus reaction → Type III hypersensitivity.
   - Bronchial asthma → Type I hypersensitivity.
   - Hemolytic anemia → Type II hypersensitivity.
28. d
   - Type IV hypersensitivity includes infective (tuberculin) and contact dermatitis varieties.
29. a and d
   - Follow section, **type IV hypersensitivity (contact dermatitis)** for explanation.
30. a and d
   - Skin tests are based on immediate (type I) hypersensitivity like Casoni's test and delayed (type IV) hypersensitivity like tuberculin test.
31. b
   - Serum sickness is a type III hypersensitivity reaction.
32. b
   - Tuberculin test is type IV hypersensitivity, which is cell mediated.
33. a
   - Paul Bunnel test → based on agglutination reaction.
   - Lepromin test, tuberculin test and granulomatous reaction are based on type IV or delayed type hypersensitivity.

Immunodeficiency Diseases (IDDs)

Chapter Outline
- Primary IDDs
- Secondary IDDs

Conditions with impairment of defense mechanisms of body resulting susceptibility to infections and malignancies called immunodeficiency diseases (IDDs). Two types according to etiological factors and involvement of immune systems like primary IDDs and secondary IDDs.

PRIMARY IDDs

Immunodeficiency due to abnormality in development of immune mechanism or immune organ. It is due to congenital or genetical factors. It includes two subtypes like deficiency of specific immune system and non-specific immune system as described below.

Deficiency of Specific Immune System

It includes deficiency of AMI, CMI or both as described below.

Deficiency of AMI

X-linked agammaglobulinemia: It was 1st IDD identified by Bruton in 1952, so called Bruton's disease with defective AMI but normal CMI. Child is protected by maternal Abs up to 6 months. Symptoms start afterward. It occurs in male infants.

- **Pathogenesis:** Failure of B cells to mature beyond the pre-B cell stage in the bone marrow, because of mutations or deletions in the gene encoding B cell tyrosine kinase or Bruton's tyrosine kinase (BTK).
- **Clinical features:** It presents with atrophy of tonsils, atrophy of adenoids and recurrent serious infection by *Strept. pyogenes, N. meningitidis, H. influenzae, S. pneumoniae, P. aeruginosa,* etc. Patients respond normally to viral infections like measles and chickenpox, though patients develop poliomyelitis or progressive encephalitis following OPV or wild virus entry. Autoimmune diseases develop in ~20% of patients. Other features include hemolytic anemia, arthritis and atopic manifestations.

- **Diagnosis**
  - Blood picture: Low levels or absence of all classes of Abs in the blood. IgG level is less than tenth and IgA plus IgM less than hundredth of the normal level. Reduced or absent B cells in the peripheral blood.
  - Picture of lymph node biopsy: Depletion of bursa dependent areas. No plasma cells and germinal centers in lymph nodes even after antigenic stimulation. Maturation, numbers and functions of T cells are usually normal.
  - Other pictures: Tonsils and adenoids are atrophic.
  - Skin tests: Wheal-flare effect is not demonstrated. CMI is not affected, so all skin tests based on DTH are normal. Allograft rejection is normal.
- **Treatment:** Live microbial vaccines are contraindicated. Initial administration of 300 mg Ig/kg of body weight in three doses followed by 100 mg Ig/kg/month. Commercial preparations contain only IgA and IgM therefore whole plasma has been infused.

Transient hypergammaglobulinemia of infant: Maternal Abs catabolize at 2 months. IgG starts to develop at this age. Delay in synthesis of IgG produces this IDD. It occurs in infants of both sexes. It presents with recurrent otitis media and respiratory infection. Spontaneous recovery occurs in 18–30 months. It is transient so, treatment is not required usually.

Common variable of immunodeficiency disease (CVIDD): It occurs at 15–35 years of age so called late onset hypo γ-globulinemia. B cells normal in numbers, but fail to differentiate in to plasma cells, so reduced total Ig level. Increased Tc cell activity and decreased Th cell activity may cause the disease. It presents with recurrent pyogenic infections, malabsorption, giardiasis and increased incidence of autoimmune diseases. It is diagnosed by low levels of antibodies in blood. It is treated by administration of IM or IV Ig.

Selective Ig deficiency: Selective deficiency of one or more Abs while others are remain normal or elevated, so also called dysgammaglobulinemia. There may be presence of Ab to Ig like IgA-Ab. IgA deficiency produces rhinitis, common cold, cough, fever, steatorrhea and atopic type of allergy. IgM deficiency produces septicemia while IgG deficiency produces chronic progressive bronchiectasis. Antibiotics are used to prevent recurrent infections.

Combined deficiencies of IgA and IgG (IgG2): It includes deficiency of IgA and IgG2. Deficiency of IgG2 favors the infections by bacteria with polysaccharide capsule. Such infections are more common if associated with deficiency of IgA.

X-linked hyper-IgM syndrome: It includes deficiency of IgG, IgA, and IgE with elevated levels of IgM and normal numbers of B cells. It is both X-linked and acquired. It presents with recurrent infections, autoimmune diseases like neutropenia, thrombocytopenia, hemolytic anemia and renal lesions. It is treated by Ig by IV route.

Bloom syndrome: It also called congenital telangiectatic erythema. It occurs due to decreased IgA and IgM with or without IgG changes. It presents with telangiectatic erythema appears as macules or papules in butter fly fashion on face and other areas. It also presents with photosensitivity, repeated respiratory infections, repeated GIT infections and delay in growth, which encourage the parents to seek medical advices. There is 150–300 times increased risk of malignancy.

Nucleotidase deficiency: Ecto-5 nucleotidase deficiency is an alteration in purine metabolism that produces deficiency of B cells, which is responsible for AMI.

Transcobalamin-II deficiency: It is an inherited autosomal recessive trait. It occurs due to vitamin B12 deficiency due to defective metabolism along with immunological defect of phagocytosis, plasma cells and low Ig level. It presents with villous atrophy and megaloblastic anemia. It is treated by vitamin B12 therapy.

Deficiency of CMI

DiGeorge syndrome: It also called congenital thymic aplasia.
- **Etiopathogenesis:** In 90% cases, it occurs due to deletion of chromosome 22q11 resulting malformation of 3rd and 4th pharyngeal pouches, so all the structure developed from 3rd and 4th pharyngeal pouches like thymus, parathyroid gland (congenital aplasia or hypoplasia of thymus and parathyroid gland), portions of face and aortic arch are defective. It is not inheritance, but may be due to intrauterine infections.
- **Clinical features:**
  - Parathyroid defect: Hypocalcemic tetani is evident within 24 hours of birth due to a deficiency of parathormone or parathyroid hormone which is normally produced by the parathyroid and regulates K^+ and Ca^{+2} metabolisms.
  - Thymus defect: Defect in T cell maturation due to defect or absence of thymus, resulting depression of CMI which leads to repeated infections by bacteria, viruses and fungi (*Candida* spp., and *P. jirovecii*) and protozoa. Thymus dependant area of lymph nodes and spleen are depleted. T cells are reduced in numbers with normal B cells. DTH and graft rejection are depleted.
  - Normal AMI: AMI remains unaffected. B cells and Ig levels are normal.
  - Other features: These include Fallot's tetralogy and characteristic facial appearance.
- **Treatment:** Fetal thymus transplantation.

Chronic mucocutaneous candidiasis (CMC): It includes abnormal Ir to *C. albicans*. CMI, DTH and phagocytosis to *Candida* are defective. Circulating antibodies to *Candida* are in high titer. It presents with chronic candidiasis of skin, mucosa and nail. It is treated by transfer factor with amphotericin B.

Purine nucleoside phosphorylase (PNP) deficiency: This enzyme degrades purine to hypoxanthine and finally in to uric acid. Deficiency of enzyme is an autosomal recessive trait shows decreased CMI and recurrent or chronic infections. It presents with hypoplastic anemia, diarrhea, chronic candidiasis, pneumonia, etc. It is diagnosed as low serum uric acid. It is treated by hematopoietic stem cell transplantation to restore the immune functions.

Combined Deficiency of AMI and CMI

Nezelof syndrome: It is characterized by depressed CMI (decreased T ells) with decreased (decreased B ells), elevated or normal Ig (cellular immunodeficiency with abnormal Ig synthesis). It presents with autoimmune hemolytic anemia, diarrhea, chronic candidiasis, pneumonia and recurrent infection by bacteria, viruses, fungi and protozoa. It is diagnosed by presence of abundant plasma cells in lymph nodes, spleen, intestine and in other body tissues with thymic dysplasia. It is treated by bone marrow, thymus and TF transplantation. Antibiotics are used to control the infections.

Ataxia telangiectasia: It is an autosomal recessive disease with chromosomal abnormalities characterized by depressed CMI resulting impairment of DTH plus graft rejection, decreased IgE level, lack of serum and secretory IgA and also presence of Ab to IgA. It presents with cerebral ataxia, telangiectasia in conjunctiva, face and other parts of body usually at 5 or 6 years of life, ovarian dysgenesis and death in early life due to sinopulmonary infection or in 1st or 2nd decade of life due to malignancy. It is treated by thymus or transfer factor transplantation.

Good's syndrome (Immunodeficiency with thymoma): It occurs in adults. It presents with thymic tumor, depressed CMI, agammaglobulinemia and aplastic anemia.

Immunodeficiency with short limbed dwarfism: It is an autosomal recessive defect. It presents with short limb dwarfism, ectodermal dysplasia, thymic defect and increased susceptibility to infections.

Episodic lymphopenia with lymphocytotoxin: Lymphocytotoxin is an Ab to lymphocyte. It is a familial disease and present with lack of immunological memory and secondary Ab response is abolished.

Severe combined immunodeficiency (SCID)

- **Synonym:** It also called alymphocytosis, Glanzmann–Riniker syndrome, severe mixed immunodeficiency syndrome, thymic alymphoplasia and bubble boy disease/bubble baby disease because its victims are extremely vulnerable to infections and become famous for living in a sterile environment.

- **Definition:** It is a genetic disease characterized by the disturbed development of functional T and B cells due to mutation in different genes that result in heterogeneous clinical presentations.

- **Pathogenic subtypes:** In SCID immune system is highly compromised or considered as almost absent due to defective Ab response due to either direct involvement of B cells or improper B cells activation due to nonfunctional Th cells. Consequently, both "arms" (B and T cells) of adaptive immune system are impaired due to mutation in one of several possible genes. There are now at least nine different known genes in which mutations lead to a form SCID. Pathogenic subtypes according to mutation in different gens are mentioned in **Table 25.1**.

	TABLE 25.1: Pathogenic types of SCID
Type	**Description**
X-linked severe combined deficiency (X-SCID)	• It is the most common type occurs due to mutation in gene encodes for common γ chain. Common γ chain is a protein that is shared by receptors of different ILs like IL-2, IL-4, IL-7, IL-9, IL-15 and IL-21. Different ILs and their receptors are required for development and differentiation of T plus B cells. As common γ chain is shared by receptors of different ILs, mutation of concerned gene results in nonfunctioning of IL signaling. So, finally there is low or absent T cells plus NK cells and nonfunctional B cells which cause complete failure of the immune system. • Common γ chain is encoded by IL-2Rγ (IL-2 Receptor γ) gene which is present on X-chromosome; hence IDD due to mutation in IL-2Rγ gene called X-linked severe combined immunodeficiency. It is inherited in an X-linked recessive pattern.
Adenosine deaminase (ADA) deficiency	• After X-SCID, ADA deficiency is the second most common type of SCID. Normally ADA is required for breakdown of purines, deficiency of ADA causes accumulation of dATP, which inhibits the function of ribonucleotide reductase, enzyme required to reduce ribonucleotides in to deoxyribonucleotides. Inactivation of ribonucleotide reductase inhibits the proliferation of lymphocytes that results in low immunity. • It presents with complete absence to mild abnormal T and B cells. • It associated with chondrocyte abnormalities.
Purine nucleoside phosphorylase deficiency	• An autosomal recessive disorder involving mutations of the purine nucleoside phosphorylase (PNP) gene. PNP is a key enzyme in the purine salvage pathway. Impairment of this enzyme causes elevated dGTP levels resulting in T-cell toxicity and deficiency.
Reticular dysgenesis of De Vaal	• Inability of granulocyte precursors to form granules secondary to mitochondrial adenylate kinase 2 malfunction. • Defect at multipotent hemopoietic stem cells level resulting anemia, thrombocytopenia, neutropenia, lymphopenia, bone marrow aplasia, death in few weeks of life.
Omenn syndrome	Recombination Activating Genes (RAG-1 and 2) are encoded for recombinase enzymes, which is required for synthesis of Igs. Recombinase enzymes are involved in the first stage of V(D)J recombination, the process by which segments of a B cell or T cell's DNA are rearranged to create a new T or B cell receptor (and, in the B cell's case, the template for antibodies). Mutation in RAG-1 and 2 prevents V(D)J recombination which results in SCID.
Bare lymphocyte syndrome	**Type 1:** It is due to defect in TAP proteins, so MHC class I is not expressed on cell surface. **Type 2:** It is due to alteration in MHC class II regulatory genes, so MHC class II is not expressed on APCs. It is an autosomal recessive disease.
JAK3	It is due to mutation in gene encodes for enzyme Janus Kinase-3 (JAK3) that mediates transduction downstream of the γ chain signal.
Artemis/DCLRE1C	Although scientists have identified about a plenty of genes that cause SCID, the Navajo and Apache population has the most severe form of the disorder. It is due to the lack of an artemis gene in children. Due to this, children are unable to repair DNA or develop disease-fighting cells.
Swiss type agamma-globulinemia	Defect at lymphoid stem cell level

- **Clinical features:** SCID patients are usually affected by severe bacterial, viral, or fungal infections early in life and often present with interstitial lung disease, chronic diarrhea, and failure to thrive. Ear infections, recurrent *Pneumocystis jirovecii* pneumonia, and profuse oral candidiasis occur commonly. These babies, if untreated, usually die within one year due to severe, recurrent infections unless they have undergone successful hematopoietic stem cell transplantation.
- **Diagnosis:** (1) Real-time PCR to measure the concentration of T-cell receptor excision circles. (2) Genetic tests to detect the mutant genes.
- **Treatment:** Bone marrow transplantation and gene therapy.

Wiskott–Aldrich syndrome (WAS)

- **Pathogenesis:** It is the X-linked disease characterized by depressed CMI due to depletion of T lymphocytes in peripheral blood and also in T cell zones of lymph nodes. B cell count is normal with normal IgG and IgA, raised IgE but decreased IgM. Patient is not able to respond to polysaccharide Ag due to defect in AMI thymus showing cellular depletion.
- **Clinical features:** It presents with recurrent infections, eczema and thrombocytopenic purpura. Death of affected patient occurs in 1st decade of life due to infections, hemorrhage or lymphoreticular malignancy.
- **Treatment:** Bone marrow and TF transplantation.

Deficiency of Nonspecific Immune System

It includes deficiency of phagocytosis and complement as described below.

Deficiency of Phagocytosis

It includes two types of defects like intrinsic and extrinsic. Intrinsic means defect in cells for example enzyme deficiency and extrinsic means deficiency of opsonins, effect of drugs or presence of antineutrophil Abs. Common diseases due to deficiency of phagocytosis are described below.

Chronic granulomatous disease (CGD)

- **Pathogenesis:** It is familiar disease, X-linked (70%) in boys and autosomal recessive (30%) in girls. It is associated with defect in H_2O_2 production (due to defect in NADPH oxidase) that kills bacteria, O_2 consumption, hexose monophosphate pathway and myloperoxidase release.
- **Clinical features:** It presents with recurrent infections by catalase-positive bacteria like staphylococci, coliforms, etc., while catalase negative bacteria are handled normally. Clinical features include chronic granulomatous lesion in skin, lymph nodes, lungs and in bones with hepatosplenomegaly.
- **Diagnosis:** It is diagnosed by following tests.
  1. **Nitroblue tetrazolium (NBT) reduction test:** In normal patient phagocytic cells (microphages and macrophages) produce the enzyme NADPH oxidase which is responsible for production of H_2O_2 and other free radicals. These free radicals reduce NBT to formazan. In patient with CGD, NADPH oxidase is not produced so no production of H_2O_2 and other free radicals and no reduction of NBT to formazan.
  2. **Other tests:** Dihydrorhodamine (DHR) test and immunoblot test for NADPH production.

Myeloperoxidase deficiency: Deficiency of myeloperoxidase in leukocytes increases the chances of *C. albicans* infection.

Chediak-Higashi syndrome

- **Defective LYST gene:** It is an autosomal recessive disease due to defect/mutation in LYST gene, which is encodes for lysosomes of phagocytic cells. Lysosomes contain many enzymes required to kill the bacteria. During phagocytosis lysosome fuses with phagosome to form phagolysosome. Once phagolysosome has been formed lysosome secretes the all necessary enzymes. Due to defect in LYST gene, structure and functions of lysosome are disturbed and it is not able to fuse with phagosome to form phagolysosome, resulting impaired bacteriolysis leads to persistent infections in infancy and early childhood which may be life threatening.
- **Other defects:** Beside lysosomal fusion defect, it also linked with other following cellular defects.
  - Melanocytes: Contains large melanosome → reduced melanin production → albinism, characterized by decreased pigmentation of skin, eyes and hairs.
  - Granules in Schwann cells: Causing peripheral neuropathy.
  - Abnormal platelets: Causing bleeding disorders.
  - Giant peroxidise positive granules/inclusions in leukocytes. Peroxidise positive inclusions are result of autophagocytic activity.
  - Eyes: Photophobia and nystagmus.
  - Accelerated phase defect: WBCs divide uncontrollably and invade many organs leading to fever, bleeding disorders, overwhelming infections, organs failure and may be death.

Leukocyte G6PD deficiency: Deficiency of G6PD reduces bactericidal effect of phagocytes. It is like CGD in reduced myeloperoxidase activity and increases the chances of infections. NBT test is normal.

Job's syndrome: Exactly it is not clear but this disease is due to defect in phagocytic activity. It includes little inflammatory response and serum Ig is normal but increased IgE. It presents with multiple, recurrent staphylococcal abscess in skin and other organs, otitis media, atopic eczema and chronic nasal discharge.

Tuftsin deficiency: Tuftsin is a leukokinin discovered in Tufts University, Boston. It is tetrapeptide (Thr-Lys-

Arg). It stimulates the phagocytic activity. Deficiency of Tuftsin increases the chances of local and systemic infections.

Lazy leukocyte syndrome: Bone marrow has normal numbers of neutrophils, but due to defect in chemotaxis and mobility peripheral neutropenia occurs with poor response to inflammation and chemical stimulations. It presents with increased susceptibility to bacterial infections, otitis media, gingivitis and recurrent stomatitis.

Hyper-IgE syndrome: Here AMI and CMI are normal but increased ten times level of IgE. It presents with early onset of eczema and recurrent *Staph. aureus* and *Strept. pyogenes* infections with abscess and pneumonia.

Actin-binding protein deficiency: It presents with frequent infection and slow mobility of leukocytes.

Shwachman's disease: It presents with frequent infection, slow mobility of neutrophils, pancreatic malfunction and bone abnormalities.

Complement (C) Deficiency

Follow **Ch. 21**.

SECONDARY IDDs

Immunodeficiencies due to factors interfere with functions of immune system. They include following two subtypes.

AMI (HI) deficiency: Following are examples:
- B-cell deficiency in chronic lymphatic leukemia.
- Ig catabolism in nephritic syndrome.
- Excessive loss of serum proteins in exfoliative skin disease and protein loosing enteropathies.
- Over production of Ig in multiple myeloma.

CMI deficiency: Following are examples:
- CMI deficiency in Hodgkin's disease.
- Obstruction of lymph circulation.
- Thymus-dependent area is infiltrated with non-lymphoid cells in lepromatous leprosy (LL).
- Viral infection like measles.

Combined deficiency of AMI and CMI: It is due to different agents such as radiation, immunosuppressive drugs, AIDS, malnutrition, anti lymphocyte Ab or old age.

ACCESS YOURSELF

Case Study

1. An 11-month-old male child is brought to the hospital with history of fever and dyspnea. Mother reported attack of measles with full recovery in previous month. Lymph node biopsy revealed depleted bursa dependent areas. IgA, IgM and IgG are less than normal level. Identify the clinical conditions and answer the following:
   a. Name the clinical condition or syndrome.
   b. Write clinical features and diagnosis of clinical condition.
   c. Write treatment of clinical condition.

2. A 10-year-old girl child is brought to the pediatric OPD with history of cough, common cold and fever since a very long time. Identify the clinical conditions and answer the following:
   a. Name the Ig deficient in given clinical condition.
   b. Draw the structure of Ig deficient in given case.
   c. Describe in detail about IDDs due to deficiency of AMI.

Essay/Full Question

1. Immunodeficiency diseases (IDDs).

Short Note

1. Primary immunodeficiency diseases.

Short Questions for Theory/Viva Questions

1. Define primary immunodeficiency diseases.
2. Define secondary immunodeficiency diseases.

Comments on

1. Nitroblue tetrazolium (NBT) reduction test is negative in patient with CGD.
2. Defect/mutation in LYST gene, does not allow the formation of phagolysosome in phagocytic cells.

MCQs for Chapter Review

Deficiency of AMI

1. **The commonest primary immunodeficiency is:**
   a Common variable immunodeficiency
   b. Isolated IgA immunodeficiency
   c. Wiskott–Aldrich syndrome
   d. AIDS
2. **All are true regarding agammaglobulinemia *except*:**
   a. Loss of germinal center in lymph node
   b. Normal cortical lymphocytes
   c. Normal cortical lymphocytes in paracortex and medulla
   d. Decreased red pulp in spleen
   e. Immunodeficiency (cell mediated)
3. **A child presents with recurrent episode of sinopulmonary infection by bacteria with polysaccharide rich capsule. Deficiency of which of the following immunoglobulin subclasses should be investigated?**
   a. IgA   b. IgG1
   c. IgG2   d. IgA + IgG2
4. **True about Bloom syndrome:**
   a. Decreased IgG   b. Decreased IgM
   c. IgA absent   d. Increased IgE
5. **Nucleotidase deficiency:**
   a. Humoral immunity deficiency
   b. Acquired immunity deficiency
   c. SCIDs
   d. Cell mediated immunity deficiency
6. **Giardiasis is associated with:**
   a. Common variable immunodeficiency
   b. C1 esterase deficiency
   c. C8 deficiency
   d. Anemia

Deficiency of CMI

7. **Which is found in DiGeorge syndrome?**
   a. Tetany
   b. Eczema
   c. Mucocutaneous candidiasis
   d. Absent B and T cells
   e. Total absence of T cells

8. DiGeorge syndrome is characterized by all *except*:
   a. Congenital thymic aplasia
   b. Abnormal development of 3rd and 4th pouch
   c. Hypothyroidism
   d. Hypocalcemic tetany

9. Digeorge syndrome is characterized by:
   a. Absence or presence of very few T cells
   b. Low complement level
   c. Absence of B cells
   d. Presence of auto-antibodies

10. Which fungal infection commonly occurs in neutropenia?
   a. *Candida*
   b. *Histoplasma*
   c. *Aspergillus niger*
   d. *Aspergillus fumigatus*

11. Chronic mucocutaeneous candidiasis is primarily due to deficiency of:
   a. Cellular immunity
   b. Humoral immunity
   c. Complement components
   d. Combined humoral and cellular immunity

Combined Deficiency of AMI and CMI

12. Nezelof syndrome is characterized by:
   a. T cell defect
   b. B cell defect
   c. Deficiency of complement component C3
   d. Both T and B cell defect

13. Adenosine deaminase deficiency is seen in the following:
   a. Common variable immunodeficiency
   b. Severe combined immunodeficiency
   c. Chronic granulomatous disease
   d. Nezelof syndrome

14. Which of the following about SCID is false?
   a. Due to malfunction of adenylate kinase-2
   b. Peyer's patches are present and normal
   c. X-linked type is the most common
   d. Gene therapy is used

15. Wiskott–Aldrich not true is:
   a. Raised IgE
   b. Raised IgM
   c. Reduced IgA
   d. CD4 and CD8 defect

16. A patient presents with thrombocytopenia, eczema and recurrent infection:
   a. Wiskott–Aldrich syndrome
   b. DiGeorge syndrome
   c. Agammaglobulinemia
   d. SCID

Deficiency of Phagocytosis

17. Which of the following statement is correct regarding chronic granulomatous disease?
   a. It is an autosomal dominant disease
   b. It is characterized by abnormal bacterial phagocytosis
   c. Recurrent streptococcal infections are usual in this disease
   d. Nitroblue tetrazolium test is useful for screening

18. Chronic granulomatous disorder is due to defect in:
   a. B cell
   b. NADPH oxidase
   c. IgA
   d. T cell

19. Most common cause of chronic granulomatous disease in children is:
   a. Myeloperoxidase deficiency
   b. Defective phagocytosis
   c. Defective H_2O_2 production
   d. Job's disease

20. The NBT (Nitroblue tetrazolium) reduction assay is used to:
   a. Evaluate granulocyte function
   b. Evaluate T cell function
   c. Determine whether polymorphonuclear leukocytes can produce superoxide
   d. Stain B lymphocytes

21. *Candida* infection is common in:
   a. Chronic granulomatous disease
   b. Chediak-Higashi syndrome
   c. Myeloperoxidase deficiency
   d. Lazy leukocyte syndrome

22. Chediak-Higashi syndrome, defect is:
   a. Fusion of lysosome   b. T cells
   c. B cells   d. Complement

23. Chediak-Higashi syndrome, true is:
   a. Defect in phagocytosis
   b. Neutropenia
   c. Agammaglobulinemia
   d. IgA deficiency

24. Job's syndrome is the following type of immunodeficiency diseases?
   a. Humoral immunodeficiency
   b. Cellular immunodeficiency
   c. Disorder of complement
   d. Disorder of phagocytosis

Answers of MCQs and Explanation

1. b
- Selective IgA immunodeficiency is the most common primary IDD.

2. b, c, d and e
- Follow section, **deficiency of AMI (X-linked agammaglobulinemia)** for explanation.

3. d
- Follow section, **deficiency of AMI [Combined deficiencies of IgA and IgG (IgG2)]** for explanation.

4. b
- Follow section, **deficiency of AMI (Bloom syndrome)** for explanation

5. a
- Follow section, **deficiency of AMI (Nucleotidase deficiency)** for explanation.

6. a
- Follow section, **deficiency of AMI (Common variable immunodeficiency disease)** for explanation.

7. a, c

8. c

9. a
- Follow section, **deficiency of CMI (DiGeorge syndrome)** for explanation of answers of MCQs 7–9.

10. a

11. a
- Follow section, **deficiency of CMI (chronic mucocutaneous candidiasis)** for explanation of answers of MCQs 10–11.

12. d

- Follow section, **combined deficiency of AMI and CMI → Nezelof syndrome** for explanation.

13. b

14. b

- Follow section, **combined deficiency of AMI and CMI (SCID and Table 25.1)** for explanation of answers of MCQs 13–14.

15. b, c

16. a

- Follow section, **Combined deficiency of AMI and CMI (Wiskott–Aldrich syndrome)** for explanation of answers of MCQs 15–16.

17. d

18. b

19. c

20. c

- Follow section, **phagocytosis (chronic granulomatous disease)** for explanation of answers of MCQs 17–20.

21. c

- Deficiency of myeloperoxidase in leukocytes increases the chances of *C. albicans* infection. *Candida* infection is also common in syndrome where CMI is deficient as described in text.

22. a

23. a

- Follow section, **phagocytosis (Chediak-Higashi syndrome)** for explanation of answers of MCQs 22–23.

24. d

- Follow section, **phagocytosis (Job's syndrome)** for explanation.

Autoimmunity

Paul Ehrlich gives the concept of 'horror autotoxicus' means horror of self toxicity. Literally meaning of autoimmunity is "protection against self" but actually it is an "injury to self." Normally immune system shows tolerance to self. It is a condition in which tolerance or unresponsiveness to self is brokendown to produce autoimmune diseases.

It is defined as an abnormal immune response in which antibodies or sensitized lymphocytes are capable of reacting with self components. Following are the common features of autoimmune diseases.
- Increased level of antibodies called autoantibodies. Autoantibodies are demonstrable.
- Accumulation of Abs, plasma cells and lymphocytes at target sites like renal glomeruli.
- It is more in females, chronic and irreversible.
- It occurs with more than one lesion in patients.
- Patient got benefit after steroid therapy.

MECHANISMS/STEPS

Autoimmunity developed in two steps:
- Step I: Formation of autoantibody or sensitization of lymphocyte.
- Step II: Action of formed autoantibody or sensitized lymphocytes

Step I: Formation of Autoantibody or Sensitization of Lymphocyte

It includes following mechanisms.

Neoantigen: Native cell or tissue antigen may undergo antigenic alteration by different factors and assume a new antigenic specificity called neo-antigen or altered antigen. This altered antigen stimulates immune reaction and causes tissue injuries. Neo-antigen is formed by following factors.
- Physical: Irradiation, photosensitivity and cold-allergy.
- Chemical: Contact dermatitis occurs due to contact of chemical with skin protein. Certain drugs induce anemia, leukopenia and thrombocytopenia.
- Biological: Infectious mononucleosis in Epstein-Barr Virus. Many bacteria like *P. aeruginosa* acts on RBCs and releases the T-Ags. For more details follow **Ch. 28.**
- Genetic: Mutation.

Release of hidden or sequestered antigen
- Self Antigen: Tissue antigen which is unable to induce immune response because exposed to lymphoreticular system during embryonic life and identified as self antigen.
- Sequestered antigen: Tissue antigen anatomically confined to site which is not access by lymphoreticular system and it is unable to induce immune response but when it exposed free in circulation it is identified as foreign and induces immune response. This tissue Ag called sequestered antigen. For examples, eye lens protein, brain tissue protein, thyroglobulin and sperms.

Cross reaction with microbial antigen or molecular mimicry:
Antigenic similarities between foreign antigen (called heterophile antigen or cross reacting antigen) and self antigen are the basis of this theory. Antibody formed in response to foreign antigen may cross react with self antigen to produces tissue damages. Following are the examples of cross reacting antigens.
- Human and sheep brain antigens: Anti-rabies vaccine prepared from sheep brain (Sample vaccine) when injected in human it produces antibodies which may cause neurological injuries in humans due to antigenic cross reactivity between human brain and sheep brain antigens.
- Streptococcal antigens and human tissues: Follow **Ch. 50 (Table 50.3).**
- Cornea and HSV- I: Causing stromal keratitis.

- Others: Like joint membrane and *M. tuberculosis*, myocardium and coxsackie B virus and HLA-B27 and arthritogenic *Sh. Flexneri*.

Polyclonal B-lymphocyte activation: Instead of normal specific stimulation of B cells by Ag, there is abnormal, non-specific stimulation of B cells by antigen resulting multiple nonspecific Abs are formed. Polyclonal Ab is IgM in nature and produced by CD5+ B cells. Following are the examples of agents causing polyclonal activation of B cells.

- Chemical agents: Nystatin, 2-mercaptoethanol, etc.
- Biological agents: Virus like EBV which produces anti-sheep erythrocyte antibodies to RBCs of sheep, horse and cow. Bacteria like *Mycoplasma pneumoniae* which produces antihuman erythrocyte cold antibodies. Parasite like *Plasmodium* spp.
- Enzymes: Trypsin.

Forbidden clones: It is due to breakdown of tolerance to self Ag and production of Ir to self Ag. For example, injection of self Ag with Freund's adjuvant.

Loss of immunoregulation: It occurs in following conditions.
- Increased activity of T helper cells and decreased activity of T suppressor cells.
- Defects in development of stem cells in the thymus.
- Defect in macrophages function.
- Defect in idiotype-anti-idiotype of network.

Genetic factors: It includes following factors.
- Defect in genes of Ig or Ir.
- Association between HLA type and certain diseases: Follow **Ch. 22**.

Step II: Action of Formed Autoantibody or Sensitized Lymphocyte

Exact mechanisms are unknown, but it may cause damages by hypersensitivity reactions, as follows.

Type II (cytotoxic or cytolytic) hypersensitivity: Here Ab damages the cell by four mechanisms as described in **Ch. 24**.

Type III (immune complex or toxic complex disease) hypersensitivity: Follow **Ch. 24**.

Type IV (delayed/cell mediated) hypersensitivity: Follow **Ch. 24**.

Type V (stimulatory) hypersensitivity: Follow **Ch. 24**.

Combined effect by Ab and T-cell: Mostly seen in experimental orchitis.

CLASSIFICATION

Based on location and nature of lesion, autoimmune diseases are classified as hemocytolytic autoimmune diseases, organ specific (localized) autoimmune diseases, nonorgan-specific (systemic) autoimmune diseases and transitory autoimmune diseases as described below.

Hemocytolytic Autoimmune Diseases

Autoimmune thrombocytopenia: It is characterized by presence of Abs to platelets such as in idiopathic thrombocytopenic purpura and sedormid purpura. Later is drug induced disease.

Autoimmune leukopenia: It is characterized by presence of Abs to leukocytes such as in SLE, RA, etc.

Autoimmune hemolytic anemia: It is characterized by presence of Abs to RBCs. RBCs coated by Abs are prematurely killed by spleen and liver in hemolytic anemia. Two groups of antibodies are identified.
- **Cold antibodies:** These are complete, agglutinating IgM Abs that agglunatinate erythrocytes at 4°C but not at 37°C. These are produced in primary atypical pneumonia caused by *Mycoplasma pneumoniae*, black water fever and trypanosomiasis. These are detected by cold agglutination test.
- **Warm antibodies:** These are incomplete and non-agglutinating IgG Abs produced due to drugs like sulfonamides, α-methyldopa, etc. These are detected by direct Coomb's test.

Organ-specific (Localized) Autoimmune Diseases

Myasthenia gravis: It is characterized by presence of Abs against acetylcholine (Ach) receptors at neuro-muscular junctions in striated muscles. Abs prevent the combination of Ach with its receptor and impair the muscle contraction. Child born from affected mother shows the symptoms and gets clear with increasing age, which indicates that autoantibodies are passively acquired from mother.

Hashimoto's disease (lymhadenoid goiter): It is characterized by presence of cytotoxic autoantibodies which react mainly with thyroglobulin and also with acinar colloid, microsomal antigens and thyroid cell surface components and cause enlargement of thyroid gland. It frequently seen in females and presents with symptoms of hypothyroidism or frank myxoedema.

Graves' disease (thyrotoxicosis or hyperthyroidism): IgG type autoantibodies called long-acting thyroid stimulator (LATS), against some determinant of thyroid cells, which stimulates excessive secretion of thyroid hormones leading to Graves' disease. Sometimes, it is considered as part of type II hypersensitivity.

Pernicious anemia: It is characterized by presence of following two types of antibodies.
- Abs against parietal cells of gastric mucosa which cause achlorhydria and atrophic gastritis.
- Abs against the intrinsic factors which prevent the absorption of vitamin B12 and cause megaloblastic anemia.

Addison's disease: It is characterized by presence of antibodies against zona glomerulosa with lymphocytic infiltration of adrenal glands.

Insulin-dependent diabetes mellitus: In this disease autoreactive T cells attack on islet cell enzyme glutamic acid decarboxylase. Antibodies against various antigens of β cells are also produced but major damage by T cells that destroy the islet cells of pancreas. Coxsackie virus B4 triggers the disease because a six amino acid sequence is common between coxsackie virus B4 and glutamic acid decarboxylase.

Celiac disease: It is characterized by presence of gliadin, which is the antigen that stimulates cytotoxic T cells which attack on enterocytes, resulting in villous atrophy.

Autoimmune diseases of nervous system

1. **Multiple sclerosis:** Autoreactive T cells and activated macrophages cause demyelination of the white matter of brain. Viral infection like EBV triggers the disease mostly in people with HLA DR2 positive.
2. **Guillain-Barre syndrome:** Abs against myelin proteins are formed which result in demyelinating polyneuropathy. It occcurs in infection by *C. jejuni* and in some viral infections also.
3. **Neuroparalysis:** It occurs following rabies vaccination due to cross reaction between human and sheep brain antigens.

Autoimmune diseases of eye

1. **Phacoanaphylaxis:** Intraocular inflammation due to autoimmune response to lens proteins following cataract surgery.
2. **Perforating injury:** It involves the iris or ciliary bodies following the sympathetic opnthalmia in the opposite eye.

Autoimmune diseases of skin

1. **Pemphigus vulgaris:** Abs against desmoglein are formed which result in disruption of tight junctions between epithelial cells and formation of bullae all over the body.
2. **Bullous pephigoid:** Abs against dermal epithelial junctions has been noticed.
3. **Dermatitis herpetiformis:** Specific Ab is not identified.

Nonorgan-specific (Systemic) Autoimmune Diseases

Rheumatoid Arthritis (RA)

- **Pathogenesis:** Rheumatoid arthritis is an autoimmune diseases characterized by presence of IgM type autoantibody against Fc portion of IgG called Rheumatoid Factor (RF) or RA (Rheumatoid Arthritis) factor. Other classes of Abs like IgA, IgE and IgG are also noticed. Autoantibody (IgM) binds with circulating IgG and makes an IgM-IgG complex (Type III hypersensitivity → immune complex disease → serum sickness) which deposited in the joints and stimulates compliment system to destroy the cells and to recruit neutrophils. Antinuclear antibodies are also found frequently.
- **Clinical features:** It is more in women. It presents with polyarthritis, muscle wasting, subcutaneous nodules, serositis, myocarditis, vasculitis and other disseminated lesions.
- **Diagnosis:** It is diagnosed by Rose Waller test and modified Rose Waller test as described in **Ch. 20.**

Systemic Lupus Erythematosus (SLE)

- **Meaning:** Word systemic indicates the wide range of distribution. Lupus (Latin) means "wolf", the disease was so named in the 13th century as the rash was thought to appear like a wolf's bite. Erythematosus due to development of red rash across the nose and upper cheeks on both sides.
- **Synonym:** Simply it called lupus.
- **Etiology:** The exact cause of SLE is not clear. It is due to genetics, environmental or by following precipitating factors.
  - Age: Women of child-bearing age are affected about nine times more often than men. Most commonly it begins between the age of 15 and 45.
  - Races: African, Caribbean, and Chinese descent are at higher risk than white people.
  - Others: Female sex hormones, sunlight, smoking, vitamin D deficiency and certain infections are also believed to increase the risk.
- **Pathogenesis:** A variety of autoantibodies directed against cell nuclei (antinuclear Ab), intracytoplasmic cell constituents, thyroid, RBCs, WBCs and other unknown antigens are produced. The important immunological feature is LE (lupus erythematosus) cell phenomenon. The LE cell is a PMN leukocyte with ingested nuclear material complexed with antinuclear antibody (type III hypersensitivity → immune complex disease → serum sickness).
- **Clinical features:** Patient presents with a red rash across the nose and upper cheeks on both sides gives butterfly appearance as shown in **Fig. 26.1.** Other symptoms are painful and swollen joints, fever, chest pain, hair loss, mouth ulcers, swollen lymph

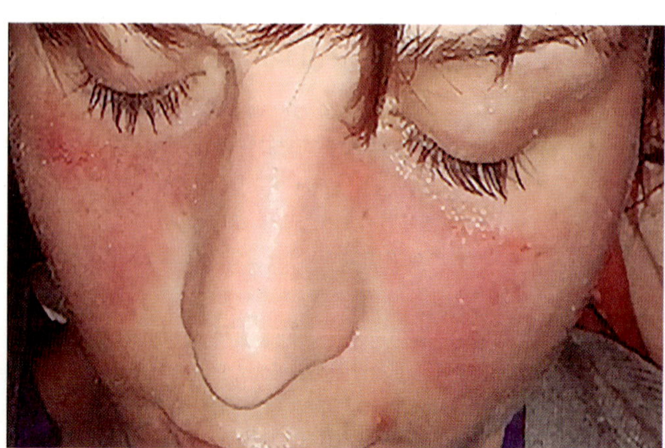

Fig. 26.1: "Butterfly rash" found in lupus

nodes and feeling tired. Often there are periods of illness, called flares, and periods of remission during which there are few symptoms. Hemolytic anemia, thrombocytopenia, leukopenia, lesion in kidneys, vascular tissues, joints, spleen and heart are seen.

- **Diagnosis**
  1. **Detection of LE cells:** It is done by incubating patient's blood or bone-marrow at 37°C and then observing for LE cells. **Giemsa stained** smear can demonstrate LE cells but less sensitive.
  2. **Detection of antinuclear Ab:** It is done by immunofluorescence test. It is sensitive but less specific.
  3. **Anti-DNA antibodies detection:** Three types of anti-DNA antibodies are seen such as Ab to ss-DNA, Ab to ds-DNA and Ab to ss and ds-DNA. They are tested by RIA or ELISA. High titer of anti-ds DNA antibody is relatively specific for SLE. Another SLE specific Ab is anti-sm antibody.
- **Treatment:** It is treated by NSAIDs, steroids and immunosuppressive agents like methotraxate and hydroxychloroquine.
- **Prognosis:** Life expectancy is lower among people with SLE. SLE significantly increases the risk of cardiovascular disease with this being the most common cause of death. With modern treatment about 80% of those affected survive more than 15 years. Women with lupus have pregnancies that are higher risk but are mostly successful.

Sjögren's syndrome: Varieties of antibodies are formed like antinuclear antibodies, RF, antibodies against salivary duct, lacrimal glands, smooth muscles, mitochondrias, thyroid gland, etc. It includes triad of kerato-conjunctivitis sicca, xerostomia with or without salivary gland enlargement and RA.

Polyarteritis nodosa (PAN): It is a component of serum sickness and other immune complex diseases (type III hypersensitivity). It is actully an autoimmune disease but autoantibody is not identified. It presents with necrotizing angitis involves the small and medium sized arteries. The disease end fatally due to coronary thrombosis, cerebral hemorrhage or gastrointestinal bleeding. It is more commonly seen in males.

Wegener's granulomatosis: It is a necrotizing granulomatous vasculitis that primarily affects the upper and lower respiratory tracts and kidneys. It is diagnosed by finding antineutrophil cytoplasmic antibodies (ANCA) in patient's serum.

Reiter's syndrome: Follow **Ch. 52** for more details.

Goodpasture's syndrome: It presents with anti-bodies against collagen in basement membranes of the kidneys and lungs. It affects young men and those with HLA-DR2 genes positive. It is diagnosed by detection of Ab and compliment C5a bound to basement membranes in fluorescent-antibody test.

Transitory Autoimmune Diseases

It occurs due to certain microbial infections or drug therapy. The infecting agent or drug induces antigenic alteration in some self-Ag initiates an Ir leading to tissue damage. It presents with anemia, thrombocytopenia or nephritis. It is transient and recovery occurs following removal of causative factors.

DIAGNOSIS

Autoimmune diseases are diagnosed by **general tests** like CRP, autoantibody titers (anti-DNA, anti-phospholipids, presence of RF) and **disease specific tests.**

TREATMENT

Autoimmune diseases are treated by following ways:
1. **T cell vaccine:** It suppresses the autoimmune cells.
2. **Synthetic blocking peptides:** They compete with autoantigens for binding to MHC molecules.
3. **Thymectomy:** Removal of thymus.
4. **Plasmapheresis:** Removes Ag-Ab complexes and reduction in symptoms for a short-term.
5. **Immunosuppressive drugs:** Mentioned above.

ACCESS YOURSELF

Case Study

1. A 40-year-female visited the orthopedic OPD with history of pain in wrist joint. Physical examination revealed subcutaneous nodule. Autoantibody is diagnosed by serological test. Identify the clinical conditions and answer the following:
   a. Name the clinical condition and one serological test useful to diagnose it.
   b. Write the mechanisms of autoimmunization.
   c. Classify the autoimmune diseases.

Essay/Full Question

1. Autoimmunity.

Short Note

1. Mechanisms of autoimmunity.

Short Question for Theory/Viva Question

1. Define autoimmunity/neoantigen/LE cell.

Comment on

1. Antirabies vaccine prepared from sheep brain (sample vaccine) causes neurological damages when used in humans.

MCQs for Chapter Review

1. **Which of the following HLA type is associated with rheumatoid arthritis?**
   a. HLA-B27
   b. HLA-A1
   c. HLA-DR4
   d. None of the above

2. **Epstein-Barr virus causes autoimmunity by:**
   a. Molecular mimicry
   b. Release of sequestrated antigen
   c. Inappropriate expression of MHC class II molecules
   d. Polyclonal B cell activation

3. **All are true about autoimmune disease *except*:**
   a. T cell recognizes the auto antigen
   b. Polyclonal B cell activation
   c. Higher incidence in male
   d. Hashimoto's thyroiditis is an example of autoimmune disease

4. **Example of sequestrated antigen is:**
   a. Eye lens protein
   b. Liver cell antigen
   c. Intestinal epithelial cell
   d. Macrophage cell

Answers and Explanation of MCQs

1. **c**
- Follow section, **mechanism/steps (genetic factors → association between HLA type and certain diseases)** for explanation.

2. **d**
- Epstein-Barr virus causes autoimmunity by polyclonal B cell activation.

3. **c**
- Autoimmune diseases are more common in female.

4. **a**
- Follow section, **mechanism/steps (release of hidden or sequestered antigen)** for explanation.

CHAPTER 27

Immunology of Transplantation and Malignancy

Chapter Outline

IMMUNOLOGY OF TRANSPLANTATION

Definitions

- Transplant or graft: Tissue or organ selected for transplantation is called transplant or graft.
- Donor: Individual from whom transplant or graft is obtained called donor.
- Recipient: Individual to whom transplant or graft is applied called recipient.

Classification of Transplants or Grafts

1. **Based on organ or tissues transplanted:** Kidney, heart, liver, skin transplant, etc.
2. **Based on anatomical site of origin and placement**
   - Orthotropic: Both sites are same like skin graft.
   - Heterotropic: Both sites are different like thyroid tissue placed in subcutaneous pocket.
3. **Based on duration after receiving the graft:** Fresh graft and stored graft.
4. **Based on viability**
   - Vital graft: Living graft like kidney, heart, liver or skin transplant.
   - Structural/static: Nonliving graft like bone or artery.
5. **Based on genetic and antigenic relationship between donor and recipient**
   - Autograft: It is the transplantation in the same individual. For example, skin graft from one site of patient to other site of same patient (donor and recipient are same).
   - Isograft: It is the transplantation between genetically identical individuals and same species. For example, twins (donor and recipient are synergic member).
   - Allograft or homograft: It is the transplantation between two genetically non-identical members but same species. For example, skin graft from cadaver (human donor) to patient (human recipient).
   - Xenograft or heterograft: It is the transplantation between genetically un-identical members and different species. For example, baboon heart (animal donor) to human (human recipient).

Allograft Rejection

When a graft is applied to genetically unrelated animal of same species, it looks healthy and seems to be accepted initially. Approximately on 4th day, graft is invaded by lymphocytes and macrophages, occlusion of blood vessels by thrombi and necrosis of graft. Graft becomes scab like and sloughs off by 10th day called 1st set rejection or reaction or response. If another graft is applied from same donor, it is rejected in accelerated fashion called 2nd set rejection or reaction or response. If another graft is applied from other donor, it is rejected by 1st set rejection. Allograft is accepted if animal is immunologically tolerant. Graft is rejected by lymphocytes and/or by antibodies with following **mechanisms.**

- **Rejection by lymphocytes:** Graft is rejected directly by Tc cells or NK cells and indirectly by releasing lymphokines or by ADCCs.
- **Rejection by antibodies:** Antibodies are formed more rapidly and abundantly during 2nd set response and play important role with CMI in 2nd set response. Antibodies formed are detected by serological tests. They deposited along the vascular endothelium and activate the complement and coagulation system which result in inflammatory cells infiltration, platelets thrombi formation, fibrin deposition and coagulative necrosis. They play role in graft rejection by ADCCs. When graft is applied to recipient with high Ab titer, **hyperacute rejection** take place. Graft becomes pale and rejected immediately in hours called white graft response. High Abs titer present

prior transplantation, pregnancy and transfusion. Hyperacute rejection takes place sometimes following kidney transplant.

Measures to Prevent Graft Rejection

MHC testing: Graft is rejected due to presence of Ags in grafted tissue, which are absent in donor. Graft is not rejected, if same Ags are present in donor and recipient. Different methods are used to match the Ags of donor and recipient are ABO blood grouping and HLA typing. HLA typing is done by microcytotoxicity testing, molecular methods like RFLP, PCR, southern blotting, etc., and tissue matching methods like MLR (mixed leukocytes reaction) or MLC (mixed leukocytes culture).

Uses of immunosuppressive agents: Follow **Ch. 23.**

Immunological enhancement: Abs produced may act opposite to CMI and prevent the graft rejection called immunological enhancement. It was 1st described by Kaliss in tumor transplant. It happens when recipient is pretreated with one or more injection of killed donor tissue and than graft is applied, the graft survival is increased. Abs produce immunological enhancement by following three possible mechanisms:

1. Afferent inhibition: Abs combine with Ags released from graft and prevent the induction of Ir.
2. Central inhibition: Abs combine with specific lymphoid cells, make them incapable to generate Ir to graft Ags.
3. Efferent inhibition: Abs coating the cells of graft and prevent their contact with sensitized lymphoid cells.

Privileged sites: These are sites where allograft are permitted to survive and safe from immunological attacks.

1. **Fetus in uterus:** Fetus is protected from maternal immunological attack by immunological barrier of placenta, relative resistance of cell membrane of trophoblastic cells to T or K cells, low density of MHC Ags on trophoblastic cells, mucopolysaccharide barrier rich in sialic acid in trophoblastic cells protect them from Tc cells, high level of α-fetoprotein which protects fetus from immunological damage by maternal leukocytes entering in fetal circulation and shedding of fetal Ags, which blocks the Abs or T cells by enhancement effect.
2. **Cartilage:** It is less accessible by immunocompetent cells (ICCs), so remain protected.
3. **Brain or hamster cheek pouch:** Absent of lymphatic drainage in these organs keep them protected.
4. **Testes:** Lymphatic drainage is ineffective, so testes remain protected
5. **Cornea:** Low vascularity prevents the cornea.

Histocompatibility Reaction

Reaction between grafted tissue and host called histocompatibility reaction. It is due to participation by Th cells and Tc cells. When antigens are presented to MHC I molecules Tc cells are involved and when antigens are presented to MHC II molecules Th cells are involved. It includes following two types:

1. **Host-versus-graft reaction or response:** Reaction of host to the grafted tissue called host-versus-graft reaction or response resulting graft is rejected.
2. **Graft-versus-host reaction or response:** Graft mounts an Ir against the Ags of host called graft versus host reaction or response. Opposite to host versus graft reaction. It occurs due to the presence of immunocompetent T cells in the graft, presence of Ags in recipient that are absent in graft, recipient is immunodeficient or recipient must not reject the graft. It is predominantly involves the mechanisms which are based on CMI. Clinical syndrome called runt disease which is characterized by retardation of growth, emaciation, diarrhea, hepatosplenomegaly, lymphoid atrophy, anemia or may be death.

Post-transplantation Infections

It occurs due to use of unrelated donor, use of immunosuppressive drugs or presence of latent infection in donor. It is common in allograft than other and bone marrow than organ transplant. Common pathogens are listed in **Table 27.1.**

IMMUNOLOGY OF MALIGNANCY

Introduction

When cell undergoes malignant transformation it loses normal Ags and acquired new Ags on surface. It makes the tumor antigenically different from normal tissues and tumor is considered as allograft and expected to induce Ir.

Clinical Evidence of Ir in Malignancy

- Spontaneous regression of malignant melanoma and neuroblastoma indicates Ir against malignancy as like Ir against infection.
- Cures by chemotherapy: Cancer like Burkitt's lymphoma and choriocarcinoma are cleared by cytotoxic drugs.
- Overcome immunity: Certain cancers are observed by autopsy findings, indicated that they are clinically controlled by body defense.
- Cellular response: Indicated by presence of lymphocytes, plasma cells and macrophages in tumor.
- Immunodeficiency state: It favors the development of cancer, like Kaposi's sarcoma, Hodgkin's lymphoma are common in AIDS.

Tumor Antigens

Following are two types of tumor Ags:

1. **Tumor-specific transplant antigen (TSTA) or tumor-associated transplant antigen (TATA):** These are the Ags present on tumor cells not on normal cells. They induce Ir when tumor is transplanted in to syngeneic

TABLE 27.1: Post-transplantation infections

Site of infection	Early (<1 month)	Middle (1–4 months)	Late (>6 months)
		After transplantation of hemapoietic stem cells	
Kidneys	—	BK virus, adeno virus	—
Brain	—	HHV-6, *T. gondii*	JC virus, *T. gondii*
Lungs	Aerobic bacteria, *Candida, Aspergillus,* HSV	Respiratory viruses (seasonal), CMV, *T. gondii, P. jiroveci*	*P. jiroveci, Nocardia,* Strept. pneumoniae
GIT	*Cl. difficile*	CMV, adeno virus	EBV, CMV
Bone marrow	—	CMV, HHV-6	CMV, HHV-6
Skin and mucosa	HSV	HHV-6	HPV and VZV
Disseminated	Aerobic bacteria	*Candida, Aspergillus,* EBV	Capsulated bacteria
		After transplantation of solid organ	
Kidneys	Bacteria and fungi by catheter	BK virus and JC virus	BK virus
CNS	—	CMV, *L. monocytogenes, T. gondii*	*L. monocytogenes, C. neoformans, Nocardia,* JC virus
Lungs	Bacterial pneumonia due to ventilation	*P. jiroveci, Aspergillus,* CMV	*P. jiroveci, Mycobacterium* spp., *Nocardia*
GIT	Peritonitis (after liver) and cholangitis	CMV (hepatitis), *Cl. difficile*	
Systemic	Bacteria and fungi by intravenous line	CMV	CMV, EBV
Donor organ	Graft infection by bacteria and fungi	CMV	EBV

animals, finally tumor is rejected. If tumor is induced by chemical Ags are tumor (cell) specific. If tumor is induced by virus Ags are virus specific.

2. **Tumor-associated Ags:** These are the Ags present on tumor cells and on normal cells. Following are the examples:
   - **Alpha-fetoprotein:** It is alpha-globulin produced by hepatocytes. It founds in embryonic life. Its level drops after birth and never found in normal adults. It increased in carcinoma of liver/hepatoma.
   - **Carcinoembryonic Ag (CEA):** It is a glycoprotein. It increased in carcinoma of colon; however, high level also found in serum in alcoholic cirrhosis; hence less diagnostic value.
   - **Prostate Specific Ag (PSA):** High level in serum indicates carcinoma of prostate.
   - **Ca-125 Ag:** It is cancer/carbohydrate 125 Ag. High level in serum indicates ovarian cancer.

Detection of Ir in Malignancy

Body produces both AMI and CMI against malignancy. CMI is protective while AMI favors the tumor growth by process of immunological enhancement. AMI is detected by Anti-TSTA Abs. CMI is detected by (1) stimulating the synthesis of lymphokines and DNA by using patient's leukocytes on exposure to tumor Ag or by (2) culture method like destruction of tumor cells by patient's Tc cells. DTH is detected by skin test by using tumor cell extract as an Ag.

Immunological Surveillance in Malignancy

Concept was given by Lewis Thomas in 1950. He postulates that CMI 'seek and destroy' the malignant cells.

Immunological surveillance includes all mechanisms that favor the growth of cancer, as mentioned below.
- IDDs: It may allow the growth of cancer.
- Larger size of tumor: It may escape the body defense.
- Tumor Ags: They act as smoke screens and cover the lymphoid cells and prevent their attack on tumor cells.
- Neutral substances of body: They cover the tumor cells and make them inaccessible by lymphoid cells.
- Immunological enhancement: AMI favors the tumor growth by suppressing the CMI.
- Cytokines formation: Like tumor growth factor-β which suppresses the CMI.
- Low level of MHC I molecules: It minimizes the chances of recognition and destruction by Tc cells.
- Low immunogenicity of tumor.

Immunotherapy in Malignancy

It is given to augment antitumor defense of body. It includes following different types:

1. **Nonspecific (active) therapy**
   - By using live agent like BCG vaccine. It augments antitumor response by activation of T lymphocytes and increasing macrophages cytotoxicity.
   - By using nonliving agents like (1) glycan (glucose polymer) from *Corynebacterium parvum* which is an immunomodulatory and (2) drugs like levamisole (anthelminthic drug) and DNC (Di Nitrochlorobenzene) which stimulate the CMI and macrophage activity.

2. **Specific therapy**
   - **Active immunization:** It is done by injecting vaccine contains tumor cell antigens (not productive or

contains tumor cell membrane antigens or tumor cells treated with neuraminidase. Later two will increase immune potential.

- **Passive immunization:** It is done by using specific antisera or monoclonal Abs. Specific antisera neutralize circulating tumor Ags and allow sensitized ICCs to act on tumor cells. Monoclonal Abs act as carrier and transport the cytotoxic and radioactive drugs to the tumor cells.

- **Adoptive immunization**
  - By using lymphocytes, transfer factor (TF) and immune RNA: All they boost up the immunity. They are prepared from persons who recovered from cancer or who is immunized against the patient's tumor.
  - LAK cells: They are useful in renal carcinoma, prevent the metastasis and prepared from NK cells treated with IL-2.
  - Thymosin: It also has antitumor activity. It is prepared from human or bovine thymus.

3. **Combined therapy:** Best result will be obtained by combining the radiotherapy, chemotherapy, immuno-therapy and surgery.

ACCESS YOURSELF

Essay/Full Question

1. Describe the immunological mechanisms of transplantation and tumor immunity/malignancy.

Case Study

1. A woman with infertility receives an ovary transplant from her sister who is an identical twin. Identify the type of graft and answer the following.
   a. Which type of graft it is?
   b. Mention the different types of grafts.
   c. Write the mechanisms and ways of prevention of allograft rejection.
   d. Describe histocompatibility reaction in details.

Short Notes

1. Graft versus host reaction.

Short Questions for Theory/Viva Questions

1. Define: Autograft, isograft, homograft and heterograft.
2. What is white graft response?
3. What is runt disease?

Comments on

1. BCG vaccine is also used in treatment of cancer.

MCQs for Chapter Review

Immunology of Transplantation

1. **Transplantation of host's own tissue is known as:**
   a. Isograft
   b. Allograft
   c. Xenograft
   d. Autograft

2. **Allograft is defined as:**
   a. Graft from self
   b. Graft from identical twin
   c. Graft from member of same species
   d. Graft from other species

3. **Types of graft, best suited for renal transplantation:**
   a. Allograft
   b. Autograft
   c. Xenograft
   d. Isograft

4. **Graft versus host reaction is caused by:**
   a. B lymphocytes
   b. T lymphocytes
   c. Macrophages
   d. Complement

5. **Type of T lymphocyte responsible for histocompatibility reaction:**
   a. Suppressor T cell
   b. Activator T cell
   c. Effector T cell
   d. Helper T cell

6. **Runt disease:**
   a. Graft rejection
   b. Graft-versus-host reaction
   c. Deficient function
   d. Complement deficiency

7. **Antibodies are most responsive to:**
   a. Recipient tissue
   b. Donor tissue
   c. Isograft
   d. Autograft

Immunology of Malignancy

8. **Alpha fetoprotein is raised in:**
   a. Hepatoma
   b. Multiple myeloma
   c. Prostrate cancer
   d. Choriocarcinoma

Answers and Explanation of MCQs

1. d

2. c

- Follow section, **classification of transplant or graft (Based on genetic and antigenic relationship between donor and recipient)** for explanation of answers of MCQs 1–2.

3. d

- Best graft is autograft, but not possible for renal transplantation, so best answer is isograft.

4. b

5. a, c, d

- Follow section, **histocompatibility reaction** for explanation of answers of MCQs 4–5.

6. b

- Follow section, **histocompatibility reaction (Graft-versus-host reaction or response)** for explanation.

7. b

- In graft rejection antibodies are formed against donor tissue.

8. a

- Follow section, **immunology of malignancy (tumor associated Ags)** for explanation.

CHAPTER 28

Immunohematology

Chapter Outline

❏ Introduction
❏ Clinical Significances

INTRODUCTION

History

Landsteiner 1st proposed the human blood group in 1900 and he was awarded the Nobel Prize late in 1930

ABO Blood Group System

Typing is done according to Ag on RBCs surface as shown in **Table 28.1**.

H Ag

RBCs of all ABO blood groups possess common H Ag, which is a precursor of A and B Ag. Amount of H Ag is related with ABO group of cells, AB group has least and O group has most amounts. Due to its universal distribution, it is less useful in blood grouping and blood transfusion (BT). Bhende et al., from Bombay defined the blood group in which there is A, B and H were absent on RBCs surface called Bombay or OH blood group. Serum of such person contains Anti-A, Anti-B and Anti-H antibodies and their sera are incompatible to all RBCs except of same group.

Rh Blood Group

In 1939, Philip Levine and Rufus Stetson demonstrated new type of Ab in serum of mother received ABO compatible blood from her husband. She had just delivered a stillbirth fetus with hemolytic disease. It indicates that woman had sensitized by Ag from fetus inherited from its father. This newer Ab called Anti-Rh (from Rhesus monkey) or anti-D Ab, which was a cause of hemolytic disease of newborn (HDN). D Ag presents on RBCs surface and corresponding Ab called anti-D or anti-Rh antibody. Person with D antigen called Rh +ve and without called Rh –ve.

Other Blood Group System

Following are other blood groups with little clinical significances.

- **Lewis blood group system:** It consists two Ags like Le^a Le^b. It differs from other blood group system, as Ags are present in plasma and saliva.
- **MN system:** Different groups are identified like M, N and MN. Later S-Ag was added to the list. Total 28 different Ags are identified.

CLINICAL SIGNIFICANCES

Clinical Significances of Blood Group

Blood transfusion (BT): Recipient's plasma should not contain any Abs which damage the donor's cells and *vice versa*. Donor RBCs should not have any Ags that are lacking in recipient. If transfused cells possess a foreign Ags, they mount Ir in recipient. Group O is transfused to any group, as it does not contain Ag like A or B, so called universal donor. Group O contain anti-A and anti-B isoantibodies which are diluted by recipient's plasma. However, sometimes donor's plasma contain high titer of isoantibodies, which damage the recipient's cells called dangerous O group. Group AB does not contain any isoantibodies, so transfused by any group called universal recipient. Rh compatibility is important when recipient is Rh negative. Following are the different complications due to BT.

1. By incompatible transfusion

- Pathogenesis: It causing clumping and intravascular hemolysis of RBCs. RBCs are coated by antibodies and phagocytosed.

TABLE 28.1: ABO blood group system		
Group	**Ag on RBCs**	**Ab in Serum**
A	A	Anti-B (mostly IgM type)
B	B	Anti-A (mostly IgM type)
AB	AB	None
O	None	Anti-A and anti-B

Essentials of Medical Microbiology

- Clinical features: Common features include shivering, tingling sensation, headache, lumbar pain, hypotension, cold-calmmy skin, cyanosis, feeble, collapse, jaundice, hematuria, oliguria, anuria, etc. Hypersensitivity reactions like utricaria, rigor, etc., occur due to presence of allergen in donor's blood. Serious reaction occurs due to hemolyzed BT.
- Diagnosis: ABO blood group substances are present in high concentration in saliva, semen, vaginal secretion and gastric juices while in low concentration in sweat, tears and urine. They are also present in tissues but not in CSF. So above mentioned samples except CSF are used for diagnosis of ABO blood group substances.

2. Infections due to blood transfusion

- Bacterial infections: *Treponema pallidum*, *Leptospira* spp., and *Pseudomonas aeruginosa*. If donor blood contains *P. aeruginosa*, it unmasks hidden Ags of RBCs called T-Ags (type of neoantigen), produce anti-T Abs which agglutinate the RBCs by all blood group sera and by normal sera also. This phenomenon called Thomsen-Freidenrich phenomenon.
- Viral infections: HBV, HCV, HDV, HIV, CJD (Creutzfeldt-Jacob Disease), CMV, etc.
- Parasitic infections: *Leishmania donovani*, *Plasmodium* spp., and *Toxoplasma gondii*.

Hemolytic Disease of Newborn (HDN)

- **Pathogenesis:** When Rh –ve women carry the Rh +ve fetus, fetal RBCs enter in to maternal circulation and produce anti Rh-Ab. This Abs are IgG in nature and can pass through placenta. However, 1st child escape the damage, during subsequent pregnancy the maternal IgG pass through placenta and damage the RBCs of fetus called erythroblastosis of fetalis.
- **Clinical features**
  - Feature in fetus: Enlarged liver, spleen or heart and fluid build up in the fetus abdomen seen via ultrasound.
  - Features in newborn: Common features are anemia, jaundice, sever edema of body, dyspnea, enlarged liver and spleen. Jaundice may be evident right after birth or after 24–48 hours after birth.
- **Diagnosis:** Anti-Rh-Ab, which is IgG in nature and incomplete. It is detected by indirect Coombs' test or by using colloid medium like 20% bovine serum albumin or by using RBCs treated with enzymes like trypsin, pepsin, ficin or bromelin.
- **Prevention**
  - Early diagnosis with identification of anti Rh-Abs by ICT at 32–34 weeks and every month thereafter.
  - Antepartum intrauterine Rh –ve blood will be transferred when disease is diagnosed in antenatal stage.
  - Introduction of RBCs in fetal peritoneal cavity. They survive and enter in to circulation.
  - Premature delivery with transfusion.
  - Isoimmunization (artificial passive immunity): Passively administered IgGs suppress homologous

antibody synthesis by feedback process. This principle is utilized in the isoimmunization of Rh –ve women who delivered Rh +ve baby by administration of anti-Rh (D) IgG during delivery.

ABO hemolytic disease

- **Pathogenesis:** It occurs in O group mother with A or B group fetus, due to presence of natural isoantibodies like anti-A and anti-B in maternal circulation.
- **Clinical features:** It is mild than HDN due to Rh incompatibility.

Blood group and susceptibility to diseases or infections: Duodenal ulcer is common in O group. Stomach cancer and giardiasis are common in group A. Gonococci are common in group B. Vibrios are common in group O, less in group AB. Exact reasons for such associations are unclear.

ACCESS YOURSELF

Short Notes

1. Clinical significance of blood grouping.
2. Enumerate and describe infections transmitted by blood transfusion.

Short Questions for Theory/Viva Questions

1. What is Bombay blood group?
2. What are universal donor and universal recipient?
3. What is Thomsen-Friedenreich phenomenon?
4. Name the bacterial infection common in following blood group: Group O and Group B.

MCQs for Chapter Review

History

1. **AB blood group antigens are known as .... factor:**
   a. Duffy
   b. Landsteiner
   c. Rhesus
   d. Lutheran
   e. Kidd

Clinical Significance of Blood Group

2. **Diagnosis of ABO incompatibility can be from all of the following *except*:**
   a. Sweat
   b. Saliva
   c. Semen
   d. CSF
3. **Thomsen-Friedenreich phenomenon is:**
   a. Red cell infection by CMV
   b. Red cell agglutination by all blood group sera
   c. Hemolysis of transfused blood
   d. Due to B antigen

Answers and Explanation of MCQs

1. **b**
- Landsteiner 1st proposed the human blood group in 1900 and he was awarded the Nobel Prize late in 1930.

2. **d**
- Follow section, **clinical significances of blood group (blood transfusion → complications → by incompatible transfusion → diagnosis)** for explanation.

3. **b**
- Follow section, **clinical significances of blood group (blood transfusion → complications → by infectious transfusion)** for explanation.

Healthcare-associated Infections (HAIs)

CHAPTER 29

Healthcare-associated Infections (HAIs): Introduction, Types, Prevention and Control

Chapter Outline

- Introduction
- Types of HAIs
- Nosocomial Infection Prevention and Control Program

INTRODUCTION

Definition

It is define as an infection acquired by a hospitalized patients, which was neither present nor in incubation at the time of admission. Word nosocomial derived from nosocomion (Greek) means hospital. It also called hospital-acquired infection.

Important Features

- Infections acquired in the hospital but clinically evident after (48 hours) discharge.
- It is due to endemic nature of organisms in institute.
- It increases morbidity, mortality and actual cost in addition to actual illness.
- It occurs in 5–10% in developed and 25% in developing countries.
- It also includes infection to hospital staff.
- Iatrogenic infection (physician induced infection): Infection acquired by patients from physician during diagnosis, treatment and prevention of disease.

Precipitating Factors (Epidemiological Determinants)

These include following three types.

Microbial (agent) factors: These include emergence of new resistant strains, increased prevalence of multidrug-resistant organisms like *Klebsiella, P. aeruginosa* and *Acinetobacter* and widespread use of antimicrobials for therapy or prophylaxis. Following are the common microbes available in hospital environment.

- **Bacteria:** CoNS, *E. coli, S. aureus*, β-hemolytic streptococci, *Klebsiella, Proteus* spp., *Pseudomonas*, etc.
- **Viruses:** HBV, HCV, HIV, RSV, *Rotavirus* spp., and *Enterovirus* spp.
- **Fungi:** *Candida albicans, Aspergillus* spp., and *Cryptococcus neoformans*.

- **Parasites**
  - Protozoa: *Toxoplasma gondii, Cryptosporidium parvum, Plasmodium* spp., *Babesia* spp., *Leishmania donovani* and *Trypanosoma* spp.
  - Helminths: *Taenia solium* and *Enterobius vermicularis*.
  - Ectoparasites: *Acarus scabies (Sarcoptes scabiei)*.

Host factors

Age: HAIs are common in infancy and old age.

Immune status: Poor immunity by AIDS, malignancies, leukemia, diabetes mellitus, renal failure, immunosuppressive therapy, radiotherapy or malnutrition may activate the normal flora and cause the endogenous infection.

Diagnostic and therapeutic interventions: Biopsies, endoscopic examinations, catheterization, intubation/ventilation, suction and surgical procedures, transfusions, dialysis, etc., may favor the HAIs.

Transfer: Frequent transfers of patients from one unit to another also increase the chances of HAIs.

Environmental factors: Crowded conditions within the hospital, poor hospital administration, poor sterilization and disinfection practices may precipitate the HAIs.

Sources of Infection

- **Endogenous:** Mostly due to activation of normal flora of body.
- **Exogenous:** Any part of hospital ecosystem like people, food, water, biomedical waste sites and hospital devices like needles, catheters, suture materials, surgical instruments, hospital beds, bed sheets, etc.

Modes of Transmission

Direct transmission

1. **Direct contact:** Infections are transmitted by contact of hands/body surfaces with other body surfaces

or with inanimate objects. Hand contact is the most common mode of transmission of nosocomial infection. Diseases which are transmitted by this route include colonization or infection with multidrug-resistant organisms (like MRSA), enteric infections and skin infections.

2. **Droplet infection:** Droplet nuclei/particles of saliva or nasopharyngeal secretion arise during coughing, sneezing, speaking or talking enter into other host directly who is in close contact. Such particles are ≥5 µm in diameter and spread to short distance (<3 feet) and directly enter in other host. However, such larger particles can be filtered by nose. Particles ≤5 µm in diameter are traverse to long distance and produce the air-borne (indirect) infections as described below. Infection by droplet nuclei is increased in close contact, overcrowding and lack of ventilation. Infections transmitted by droplet nuclei are pneumonia, pertussis, diphtheria, influenza type B, mumps, meningitis, etc.

3. **Inoculation:** Infections are transmitted by IV line.

Indirect transmission

These routes required mediators/vehicles as mentioned below.

1. **Inhalation (air-borne):** Particles arise during coughing or sneezing from patients. Particle with ≤5 µm diameter are traverse to long distance and settle over different objects and become part of dust and cause respiratory infections like open/active pulmonary tuberculosis, measles, chickenpox, pulmonary plague and hemorrhagic fever with pneumonia.

2. **Ingestion:** Infections occurs by ingestion of water or food.

TYPES OF HAIs

Overall leading agent of nosocomial infection is *Staph. aureus*. Following are the other different types of HAIs and agents.

Catheter-associated Urinary Tract Infection (CAUTI)

Definition: It is the significant bacteriuria of urine culture with or without symptoms.

Etiological agents: UTI is the most common type (33%) of nosocomial infection due to indwelling catheter. Short-term CAUTI (<30 days) occurs by *E. coli* (most common), *Klebsiella, Pseudomonas, Serratia marcescens, Enterococcus faecalis, Candida* (most common agent of UTI in ICU) spp., etc. Long-term CAUTI (≥30 days) occurs by above agents plus *Proteus, Providencia* and *Morganella*.

Pathogenicity

- **Sources of infection:** Endogenous infections occur from patient's flora while exogenous infections occur via hands of healthcare workers.
- **Modes of transmission:** If aseptic precautions are not taken during insertion or maintenance of catheter

then bacteria may spread extraluminally (2/3 cases) from above mentioned sources. Open or broken drainage bag allows intraluminal spread (1/3 cases) due to reflux of urine.

- **Precipitating factors**
  1. **Device (catheter)-related and microbial agent factors:** Catheter allows the infection by introduction of patient's flora and flora of healthcare workers by contaminated hands. Catheter causes pressure effect and irritation to the mucosa, resulting mucosal disruption and failure of mucosal secretion. All these precipitate the UTI. Presence of catheter blocks the flushing effect of urine on urethral flora which favors the colonization and infection of urethral and perineal flora to bladder. Latex catheter is more risky than silicone catheter, as it is more prone to develop urethritis, stricture formation and obstruction. Small amount of urine always remains near balloon which precipitates the UTI. Retained urine is more likely allowing the growth of bacteria.
  2. **Host factors:** UTI is more in females due to shorter urethra. Other host factors are like incomplete emptying of bladder due to pressure effect of catheter, manipulation of catheter, fecal incontinence, low immunity due to diabetes mellitus, old age or due to presence of other diseases and pregnancy which increases 20–30-fold risk of CAUTI and premature delivery.
  3. **Environmental (hospital/healthcare workers) factors:** These include poor aseptic precautions, catheterization outside the operation theater, etc.
- **Clinical features:** Cloudy urine, blood in the urine, strong urine odor, urine leakage around your catheter, pressure, pain during micturition (dysuria), frequency or urgency of urine, discomfort in lower back or stomach, chills, fever, unexplained fatigue, vomiting, etc.

Laboratory diagnosis

- **Specimen:** Collect urine from catheter port (**Fig. 29.1**) or by puncturing the catheter tube by sterile needle and syringe.
- **Testing methods**
  1. **Urine routine and microscopy:** Presence of blood cells.
  2. **Urine culture:** It identifies any bacteria or fungi in your urine. Significant bacteriuria of urine culture will be classified as $\geq 10^3$ CFU/ml with clinical features or $\geq 10^5$ CFU/ml without any clinical features.

Prevention: Insert the catheter when it is necessary and also remove once its use gets over. Use all aseptic precautions and non touch technique during insertion and maintenance. Choose the appropriate size. Clean area/skin around the catheter each day. Keep the drainage bag below bladder and above the floor. Empty the drainage bag several times per day. Keep the catheter

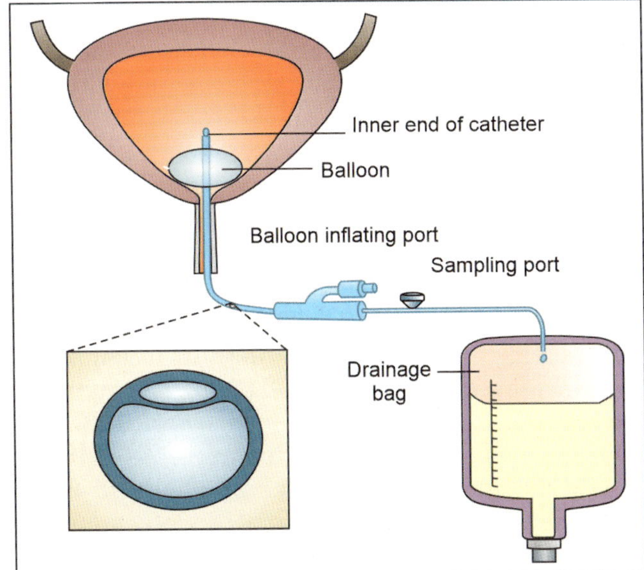

Fig. 29.1: Catheterization

tube from kinking. Wash your hands before and after touching the catheter or drainage bag. Change the catheter at least once per month. Frequent handwashing and good hygiene practices on the part of hospital staff to prevent CAUTIs.

Treatment: In symptomatic case antibiotics are given after sensitivity testing. In asymptomatic case generally treatment is not required but given when bacteriuria last for >48 hours, catheterization done in pregnant lady and urogenital surgery is required (it cause bleeding which may precipitates the bacteremia).

Ventilator-associated Pneumonia (VAP)

Definition: It is a type of lung infection that occurs after 48–72 hours in people who are on mechanical ventilation in hospitals.

Etiological agents: Pneumonia is the second most common type (15%) of nosocomial infections. Persons with VAP have increased lengths of ICU hospitalization and have up to a 20–30% death rate. There are two types of VAP like early onset occurs in ≤4 days mostly by drug-sensitive bacteria and late onset occurs in >4 days mostly by drug resistant bacteria. Common pathogens are *Pseudomonas* (24.4%), *Staph. aureus* (20.4%, of which >50% is MRSA), Enterobacteriaceae (14.1% – includes *E. coli*, *Klebsiella* spp., *Proteus* spp., *Enterobacter* spp., *Serratia* spp., *Citrobacter* spp.), *Streptococcus* species (12.1%), *Haemophilus* species (9.8%), *Acinetobacter* species (7.9%), *Neisseria* species (2.6%), *Stenotrophomonas maltophilia* (1.7%) and coagulase-negative *Staphylococcus* (1.4%). Others (4.7% – include *Corynebacterium*, *Moraxella*, *Enterococcus* and fungi). Most common group is GNB that includes *Pseudomonas* and *E. coli* as most common agents of nosocomial pneumonia.

Pathogenicity

- **Sources of infection:** Endogenous infections occur from oropharyngeal flora and gastric contents.

Exogenous infections occur from hospital environment like air, reused equipments, nebulized drugs, insertion of devices by contaminated hands, etc.

- **Modes of transmission:** Oropharyngeal flora transmitted by aspiration in unconscious patients. Ventilated patients are often kept on nasogastric tube which disrupts the lower esophageal sphincter and resulting gastric aspiration. If aseptic precautions are not taken during insertion or maintenance of devices then bacteria may inhaled from above mentioned exogenous sources.
- **Precipitating factors**
  1. **Device (ET tube) related and microbial agent factors:** ET tube allows the infection by introduction of patient's flora and flora of healthcare workers by contaminated hands. Endotracheal intubation blocks the ciliary clearance of bronchial secretions, inhibits the cough reflex and irritates the mucosa, which results in bacterial colonization. Formation of biofilms typically by gram-negative bacteria and fungal species in and outer to ET tube prevents the entry of antimicrobial agents and lowers the immunity. Collection of subglotic secretion favors the microbial growth.
  2. **Host factors:** Presence of comorbid conditions like old age, COPD, ARDS, head injury, burns, poor nutrition, earlier surgery, use of antibiotics, immobilization, unconsciousness or coma favors the VAP. Sedation decreases the natural ability to clear the respiratory secretions which increase the risk of aspiration. Supine position favors the aspiration, so patient put at 30–45° position.
  3. **Environmental (hospital/healthcare workers) factors:** These include poor aseptic precautions, use of contaminated devices, etc.
- **Clinical features:** People who are on mechanical ventilation are often sedated and are rarely able to communicate. As such, many of the typical symptoms of pneumonia will either be absent or unable to be obtained. The most important signs are high or low body temperature, purulent sputum, and hypoxia. However, these symptoms may be similar for tracheobronchitis which are differentiated by presence of hypoxia and the presence of infiltrate/consolidation on chest radiography in VAP.

Laboratory diagnosis: At the present time, there is no universally accepted, gold standard diagnostic criterion for VAP. Several clinical methods have been recommended, but none have the needed sensitivity or specificity to accurately identify VAP.

Diagnosis based on clinical pulmonary infection score (CISP): It is based on total 6 clinical, physiological, microbiological and radiographic parameters as described below and in **Table 29.1**. Score is ranged from 0 to 2. Maximum score is 12. Score ≥6 shows good correlation with the presence of VAP. Subjective variation always occurs in scoring system.

TABLE 29.1: Clinical pulmonary infection score

Parameter	Result	Score
Temperature	36.5–38.4°C	0
	38.5–38.9°C	1
	≤36 or ≥39°C	2
Leukocytes count	4,000–11,000/mm³	0
	<4,000 or >11,000/mm³	1
	≥500 band cells	2
Tracheal secretions	None	0
	Mild/nonpurulent	1
	Purulent	2
Oxygenation (PaO$_2$:FiO$_2$)	>240 or ARDS	0
	≤240 and absence of ARDS	2
Culture results (endotracheal aspirate)	No or mild growth	0
	Moderate or florid growth	1
	Moderate or florid growth and pathogen consistent with Gram's stain	2
Chest X-ray	No infiltrate	0
	Diff use/patchy infiltrate	1
	Localized infiltrate	2

1. **Clinical diagnosis:** It based on temperature, leukocytes count and tracheal secretions.
2. **Physiological diagnosis:** It based on oxygenation.
3. **Microbiological diagnosis**
   - **Specimens: (1) Blood culture:** It may reveal the microorganisms causing VAP, but often not helpful as it is positive in only 25% of clinical VAP cases. Even in cases with positive blood cultures, the bacteremia may be from a source other than the lung infection. **(2) Respiratory samples:** Different methods to collect the respiratory samples like invasive (bronchoscopy) and noninvasive (without bronchoscopy from trachea). There is no strong evidence to suggest that an invasive method to collect samples is more effective than a non-invasive method. Different samples collected and sent to laboratory without delay (not >2 hours) are endotracheal aspirate, BAL under bronchoscopic guidance, mini-BAL (mini-BAL performed blindly) without bronchoscopic guidance and protected specimen brush (PSB) which utilizes a brush at the tip of the catheter which is rubbed against the bronchial wall.
   - **Testing methods: (1) Gram's stain:** The Gram's stain provides crucial initial clues to the type and numbers of organism(s), extracelluar bacteria, presence of fibrin strands, whether the material is purulent (defined as ≥25 neutrophils and ≤10 squamous epithelial cells per low power field) or not and absence of VAP, where Gram's stain is negative. **(2) Culture:** Two types of culture reports like quantitative and semiquantitative. Quantitative

is considered significant if the diagnostic threshold is ≥10⁵ CFU/ml for endotracheal aspirate, ≥10⁴ CFU/ml for BAL and ≥10³ CFU/ml for PSB. Semi-quantitative is obtained by endotracheal sampling. It is considered positive when the growth is moderate (+++) or heavy (++++), suggests ≥10⁵ CFU/ml.

4. **Radiographic diagnosis:** Chest X-rays with presence of infiltrate(s) and/or consolidation and/or cavitation in absence CHF and ARDS.
5. **Rapid diagnosis (newer method):** In recent years, there has been a focus on rapid diagnostic methods, allowing detection of significant levels of pathogens before they become apparent on microbial cultures. These include biomarkers like IL-1β and IL-8 and molecular detection of bacteria by detecting pan-bacterial 16S gene which gives idea about bacterial load.

Prevention: VAP is prevented by following ways:
- Discontinuation of mechanical ventilation.
- Alcohol-based hand washing and sterile technique for invasive procedures.
- Limiting the amount of sedation.
- Raising the head of the bed to at least 30–45° to prevent oropharyngeal aspiration.
- Antiseptic mouthwashes with Cidex (2% chlorhexidine). Not given to those who have undergone cardiac surgery.
- Supraglottic secretion drainage: Use of special tracheal tubes with an incorporated suction. New cuff technology based on polyurethane material in combination with subglottic drainage (SealGuard Evac tracheal tube from Covidien/Mallinckrodt) showed significant delay in early and late type VAP.
- Use of silver/antibiotic-coated endotracheal tube reduces the incidence of VAP in the first 10 days of ventilation.
- Prophylactic probiotics.
- Single-dose of antibiotics within 4 hours of intubation may be effective in preventing early onset VAP in a comatose patient.
- Limiting exposure to resistant bacteria.
- Preparation of hospital antibiogram.
- Frequent educational programs to reduce unnecessary antibiotic prescription.
- Reduce reintubation rates.
- Propagate the use of noninvasive positive pressure ventilation.
- Use of oropharyngeal v/s nasopharyngeal feeding tubes.
- Endotracheal tube cuff pressure ~20 cm H$_2$O.
- Early tracheostomy.
- Small bowel feeding instead of gastric feeding.
- Preference on using heat-moisture exchangers over heater humidifiers.

- Mechanical removal of the biofilm (e.g. the mucus shaver).
- Stress/peptic ulcer prophylaxis: Intubated patients are at high risk of developing stress/peptic ulcer which causes gastrointestinal hemorrhage hence stress/peptic ulcer prophylaxis is must; however, stress/peptic ulcer prophylaxis itself causing aspiration, so only acceptable prophylaxis is sucralfate which reduces the risk of VAP.

Treatment: Once VAP has been diagnosed, starts the initial empirical therapy based on hospital antibiogram policy. It should be modified later as per drug sensitivity report. Empirical therapy given to common organisms causing VAP like *Pseudomonas* (24.4%), *Staph. aureus* and other GNB.

Surgical Wound Infection/Surgical Site Infection (SSI)

Definition: Infection that occurs at the site of surgery within 30 days (or in 90 days for some surgeries like breast, cardiac and joints including implants).

Etiological agents: Wound infection covers the 15% of total HAIs. In India, SSI occurs in 4–11 per 100 surgeries. It will be high following abdominal surgeries. Common pathogens are mentioned below as per sources.

Pathogenicity
- **Sources of infection**
  1. **Endogenous:** It includes mainly skin flora and GIT flora. Skin flora includes *Staph. aureus* and *Staph. epidermidis* (CONS). *Staph. aureus* is most common agent of surgical wound infections and surgical wound is the most common reason for establishment of nosocomial infection. GIT flora causing infections when bowel mucosa is breached (surgically most important) are *E coli, Klebsiella, Enterococcus. Bacteroides* and *Prevotella*. Flora from respiratory and genitourinary cause SSI when they are disturbed.
  2. **Exogenous:** These include hospital environment, contaminated hands and contaminated equipments. Common micobes are *Staph. aureus, Pseudomonas, Acinetobacter, Candida albicans*, etc.
- **Modes of transmission:** Body flora causes SSI when natural continuity has been breached. If aseptic precautions are not taken during surgery then infection occurs from above mentioned exogenous sources.
- **Precipitating factors**
  1. **Microbial agent factors:** These include ability of microbes to form biofilms, virulence of microbes (higher the virulence higher the SSI) and type of flora like GIT and vagina are heavily colonized (large inoculum), so surgeries at these places have more risk of SSI.
  2. **Host factors:** Presence of comorbid conditions like old age, poor nutrition, diabetes mellitus, prolonged hospitalization, obesity, etc., favor the SSI.

3. **Environmental (hospital/healthcare workers) factors:** These include poor aseptic precautions during surgery, longer time of surgery, inadequate antimicrobial prophylaxis, use of contaminated materials in surgery, etc.
- **Clinical features:** SSI typically appears 5–7 days post-procedure; however, can develop up to 3 weeks after (especially if prosthesis is inserted). The common clinical features spreading erythema, local pain, pus or discharge from the wound, wound dehiscence and persistent pyrexia. Following are clinical classes of wound.

1. **Class 1: Clean wound**
   - Uninfected and no inflammation.
   - Hollow viscus like GIT, urinary tract, respiratory tract or genital tract is not entered.
   - Rate of SSI: 2.1/1000 operations.
2. **Class 2: Clean-contaminated wound**
   - No evidence of infection or major breach in technique.
   - Surgeries included in this class are appendix, vagina and oropharynx.
   - Hollow viscus like GIT, urinary tract, respiratory tract or genital tract is entered under controlled conditions and without unusual contamination.
   - Rate of SSI: 3.3/1000 operations.
3. **Class 3: Contaminated wound**
   - Open, fresh and contaminated wound.
   - Acute nonpurulent inflammation (no purulent discharge like in dry gangrene).
   - Entry in to biliary or genitourinary tract in the presence of infected bile or urine. Gross spillage in GIT (colon surgery). Major breach in technique (open cardiac surgery).
   - Rate of SSI: 6.4/1000 operations.
4. **Class 4: Dirty/infected wound**
   - Surgery performed when active infection is already present like abdominal surgery in peritonitis or perforated viscera or abdominal abscess.
   - Old traumatic wound with retained devitalized tissue.
   - Rate of SSI: 7.1/1000 operations.

Laboratory diagnosis
Specimen: Pus/discharge/wound swab or blood.

Testing methods: These include direct microscopy, culture and tests for bio-marker like CRP, etc.

Prevention: SSI can be prevented by preoperative, intraoperative, and postoperative measures.
1. **Preoperative phase**
   - **Antibiotics prophylactic,** if indicated like mupirocin ointment for MRSA nasal carriers.
   - **Do not remove hair routinely**—if necessary do this immediately prior to surgery with an electric clipper.
   - **Shower/bathing prior to surgery with soap to reduce the bacterial load.

- Encourage weight loss, optimized nutrition (to promote wound healing), good diabetic control, and smoking cessation.

2. **Intraoperative phase**
   - Antibiotics cover: Type of drugs, time of administration and dosing schedule depend on local antibiotic policy.
   - Disinfection of surgical site by povidone-iodine or chlorhexidine.
   - Disinfection of surgical hands by alcohol or chlorhexidine.
   - Change gloves or gowns if contaminated.
   - Use an appropriate **interactive dressing** at the end of the operation to cover all surgical incisions.

3. **Postoperative phase**
   - Monitor wounds closely—the use of see-through (transparent film) dressings will limit the numbers of dressing changes, thus minimizing the chance for bacterial contamination.
   - Closely observe the wounds in difficult areas like skin creases and underneath skin folds (e.g. groin).
   - Patients may require pads to separate the wound from overlying skin or be bed bound to remove pressure on a wound.
   - Refer to a tissue viability nurse for advice on appropriate dressings for the management of surgical wounds that are healing by secondary intention.
   - Use of systemic and topical antibiotics. Meta-analysis has shown that topical antibiotics probably do prevent SSI rates when compared with no topical antibiotic or antiseptic therapy.

Treatment

- Removal of sutures or clips if present.
- Allowing for the drainage of any pus and the opportunity for wound packing if required.
- Empirical antibiotic should be started; different wounds are often caused by different organisms (e.g. a laparotomy wound infection is more likely to be caused by a coliform); however, best practice is to follow antibiotic guidelines as per culture results or local antibiotics policy.

Catheter-associated/related Bloodstream Infection (CABSI/CRBSI)

Synonym: Also called central line-related bloodstream infection (strictly used for surveillance purpose) or catheter-related sepsis.

Definition: It is defined as one or more positive blood cultures associated with systemic signs of infection such as fevers, chills, and/or hypotension in the presence of an intravenous catheter.

Etiological agents: It covers 13% of total nosocomial infections, due to central line or IV catheter. Common pathogens are *Staph. epidermidis* (30–40%), *Staph. aureus* (5–10%), *Enterococcus faecalis, Pseudomonas aeruginosa* (2–5%), *Candida* spp., etc.

Pathogenicity

- **Sources of infection and modes of transmission**
  1. **Endogenous sources:** For examples migration of skin flora along the surface of catheter with colonization of tip and migration of bacteria by blood from other foci of infection (like presence of UTI).
  2. **Exogenous sources:** For examples contamination of catheter by hands of HCWs and contamination of device/fluid at the source of production.
- **Precipitating factors**
  1. **Device (IV catheter) related and microbial agent factors**
     - Ability of microbes to form biofilms on catheter.
     - Virulence of microbes.
     - Duration of devices: Longer duration has more risk.
     - Site of IV catheter: Risk in femoral vein > jugular vein > subclavian vein.
     - Numbers of lumen: Multi-lumen catheter has higher risk than single lumen.
  2. **Host factors**
     - Foreign body reaction at the site of catheter insertion.
     - Presence of foci of infection.
     - Comorbid conditions like IDDs, burns, blood cancer, etc.
  3. **Environmental (hospital/healthcare workers) factors**
     - Use of contaminated devices.
     - Poor aseptic precautions.
     - Contaminated hands of HCWs.
- **Clinical features:** Fever or chills/rigors, hypotension, malaise, vomiting, changes in mental status, erythema, swelling, tenderness, and purulent drainage around the catheter. **As per source** there are two types of CABSI like primary which occurs due to catheter without another defined focus of infection or secondary which occurs from another defined focus of infection like urinary tract infection with subsequent bacteremia. **As per duration** there are two types of CABSI like long-term which occurs in ≥72 hours and short-term which occurs in <72 hours.
- **Complications:** Infective endocarditis (in patients with cardiac murmur), septic arthritis, osteomyelitis, spinal epidural abscess and septic emboli.

Laboratory diagnosis: Clinically diagnosed by positive blood culture with suggestive symptoms and absence of focal infection. The 'gold standard' is the combination of a positive blood culture with the same organism isolated from the catheter.

- **Specimen:** Catheter (culture from surface and/or lumen) or paired blood samples (from catheter vein and peripheral vein) during fever time or endoluminal brush sample.
- **Testing methods**
  1. **Staining technique:** The acridine orange leukocyte cytospin test (AOLC test) has been used for newborns and infants. Take catheter blood into

EDTA, lysing the red blood cells with hypotonic formol saline, pelleting the leukocytes by centrifugation, staining the cellular monolayer with acridine orange, and examining under ultraviolet light. If any bacteria are seen (in plasma or within polymorphs) the result is positive.

2. **Culture:** These include catheter culture, paired blood culture and endoluminal brush sampling.

3. **Serological tests:** ELISA has been developed for the detection of antibodies against a novel short chain lipoteichoic acid antigen of *Staph. epidermidis*.

Note: Catheter culture techniques, paired blood culture and endoluminal brush sampling

Catheter culture techniques

- **Maki roll culture technique:** Rolling the catheter tip across an agar plate and thereby inoculating those organisms adherent to the outer surface, >15 CFU being indicative of catheter colonization. Disadvantage of this method is—catheter may be contaminated by commensals as it is pulled through the skin exit site and bacteria growing on the luminal surface will not be detected.
- **Cleri flush culture technique:** Tip of the catheter to be immersed in broth and the lumen flushed through. Vortexing or sonication of the catheter tip is aimed at dislodging microorganisms embedded in the biofilm giving a higher sensitivity of culture. Disadvantage of this method is positive culture may still reflect contamination of the outer surface of the catheter.
- **Modified Cleri flush culture technique:** It involves treating the outer surface of the catheter with chlorhexidine prior to flushing the lumen to avoid surface contamination.

Paired blood culture: It spares the catheter. Catheter vein blood contains far higher numbers of organisms than peripheral vein blood. Differential colony counts of 5–10:1 (catheter: peripheral vein) is taken as diagnostic of catheter sepsis, so this method also called quantitative blood culture. Catheter blood culture becomes positive at least two hours before the peripheral culture (best detected by automated blood culture system).

Endoluminal brush sample: This technique involves passing a guidewire with a nylon brush down the catheter to its distal end, withdrawal then resulting in sampling of the catheter biofilm. The brush is vortexed with phosphate buffered saline and plated onto agar; colony counts of 100 CFU/ml are significant.

Prevention

- **Repeated quantitative blood cultures (2–3 times per week over the duration of catheterization):** It may predict some cases of CABSI (30%) prior to the development of clinical symptoms, and also help to monitor antibiotics response.
- **Selection of catheter type:** Use a single-lumen catheter unless multiple ports are essential. Use an implantable or tunneled catheter for long-term (>30 days) use. Use antimicrobial (chlorhexidene/silver sulfadiazine and minocycline/rifampin) impregnated catheter for patients at high risk. Use catheter flush solutions should contain anticoagulants. Newer device to prevent CABSI such as silver iontophoretic catheter is under evaluation.

- **Selection of catheter insertion site:** Use the subclavian route unless contraindicated. Consider the use of peripherally inserted catheter.
- **Aseptic technique during insertion:** Use optimum insertion technique including sterile gown, gloves and drapes. Clean the insertion site with alcoholic chlorhexidine gluconate solution (or alcoholic povidone iodine) and allows to dry. Disinfect the external surfaces of the catheter and connection ports with an aqueous solution of chlorhexidine gluconate or povidone iodine (unless against manufacturer's recommendations).

Treatment: Removal of the CVC is the only way to treat the infection. Empirical antibiotic therapy against likely pathogens (including resistant gram-negative organisms and methicillin-resistant *Staph. aureus*) and as per local antibiotic policy and modified later as per drug sensitivity report.

Gastroenteritis

It occurs due to contamination of hospital food and water. Common pathogens are *Staph. aureus* (enterotoxin) and *Salmonella* food poisoning, diarrhea due to *E. coli*, *Cl. difficile* and some species of *Enterovirus*.

Others HAIs

HBV, HCV and HIV infections occur due to blood transfusion, hemodialysis or by contaminated needle. Tetanus is common with poor antiseptic practice during surgery, cutting umbilical cord, injection administration, wound dressing or using materials contaminated by bacterial spores.

NOSOCOMIAL INFECTION PREVENTION AND CONTROL PROGRAM

It is implemented to manage the risk associated with the transmission of microorganisms from staff to staff, staff to patient and/or visitors, patient and/or visitors to staff, equipment and environment.

Aims and Objectives

- To interpret, uphold and implement the infection prevention and control policies and procedures in the healthcare facilities (HCFs).
- To develop written policies and procedures for standards of infection prevention and control practices, cleanliness, sanitation and asepsis in the HCFs.
- To review and analyse data on HAI in order to take corrective and preventive actions.
- To review and provide input for investigations of HCF related outbreaks.
- To develop a mechanism to supervise infection control measures in all phases of hospital activities and to promote infection control practices at all levels of the facility.
- To ensure continuing education of employees on infection prevention and control practices.

Committee: Hospital infection control programs is organized by medical superintendent, for which he/she constitutes hospital infection control committee. It includes chairman (medical superintendent), head (officer) of all clinical departments, microbiologists, epidemiologist, hospital administrators, nursing staff, pharmacist, CSSD (central sterile supplies department), mortuary, housekeeping, biomedical engineer if available and maintenance in charge of linen, laundry and kitchen.

Functions of committee: It acts as an advisory committee for superintendent to take following measures for infection control and prevention.

- **Infection surveillance:** 4 key parameters used are catheter associated UTIs, IV line associated bloodstream infections, ventilator associated pneumonia and surgical wound infections.
- **System development:** To identify, report, analyze, investigate and control the HAIs.
- **Antibiogram:** Development of guidelines on antibiotics usage, maintenance of data for sensitive or resistant drugs and follow on actions after detection of drug resistant strain.
- **Training:** Training of different categories of staff on infection prevention and control guideline.
- **Post-exposure prophylaxis:** Especially after needle prick injury like HBV vaccination, post-exposure prophylaxis of HIV to healthcare workers.
- **Outbreak management:** Identification of source and corrective measures.
- **Meeting:** All members of committee should meet regularly not less than a month and as often as required. However, in emergency (like outbreak) committee should meet promptly.

Standard Precautions

Measures taken to prevent device related infections are grouped as care bundle approach, as described earlier with device related infection. Standard precautions include following.

Hand hygiene: Hands of all healthcare workers can carry microorganisms (bacteria, viruses and fungi) that are potentially infectious to them and others; hence, hand washing before and after contact with patient is the best and inexpensive way to prevent the HAIs. Hand hygiene includes following:

- **Nail care:** Nails are the area of greatest contamination. Short nails are easier to clean and are less likely to tear gloves. Nail varnish/polish/extensions/art and acrylic nails are prohibited, regardless of color, for staff with direct patient contact or who work in direct patient care area.
- **Skin care:** Ensure the skin on your hands does not become dry or damaged. In these conditions the hands show a higher bacterial load and are difficult to remove than with intact skin. Hand lotion may be used to prevent skin damage from frequent hand washing.
- **"Five moments for hand hygiene" approach (Fig. 29.2):** It is given by WHO.
  1. Before touching a patient.
  2. Before clean or aseptic procedure.
  3. After body fluid exposure risk.
  4. After touching a patient.
  5. After touching patient's surroundings.
- **Types of hand hygiene (hand cleaning):** It includes following three types:

1. **Handwashing with soap and running water**
   - Indications: Wash hands with soap contains 4% chlorhexidine and water when visibly dirty/contaminated with blood, body fluids, etc., after using a rest room, before and after having food and after arriving and before leaving work place.
   - Effects: It removes transient microorganisms, soil, blood and other organic material from hands.
   - Steps: Follow **Fig. 29.3**.

2. **Alcohol-based hand rub**
   - Indications: Rub the hands with alcohol when hands are not visibly soiled, before having direct contact with patients, before putting on gloves, after contact with patient's intact skin (e.g. after taking pulse, BP, lifting a patient, etc.), after contact with inanimate objects (e.g. medical equipment) in the immediate vicinity of the patient, after removing gloves and hand washing with soap and water is not possible, as long as hands are not visibly soiled.
   - Effects: Rub the hands with a solution contains 70–80% alcohol and 0.5%–4% chlorhexidine for 20 seconds. It kills most transient and resident microorganisms, but ineffective if hands are visibly soiled.
   - Steps: Follow **Fig. 29.4**.

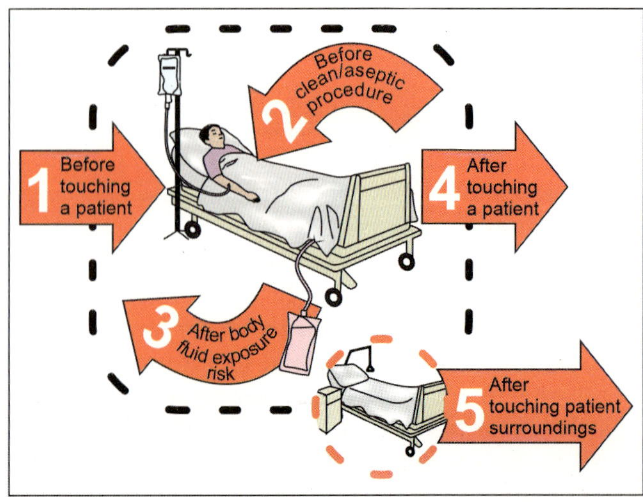

Fig. 29.2: Five moments for hand hygiene

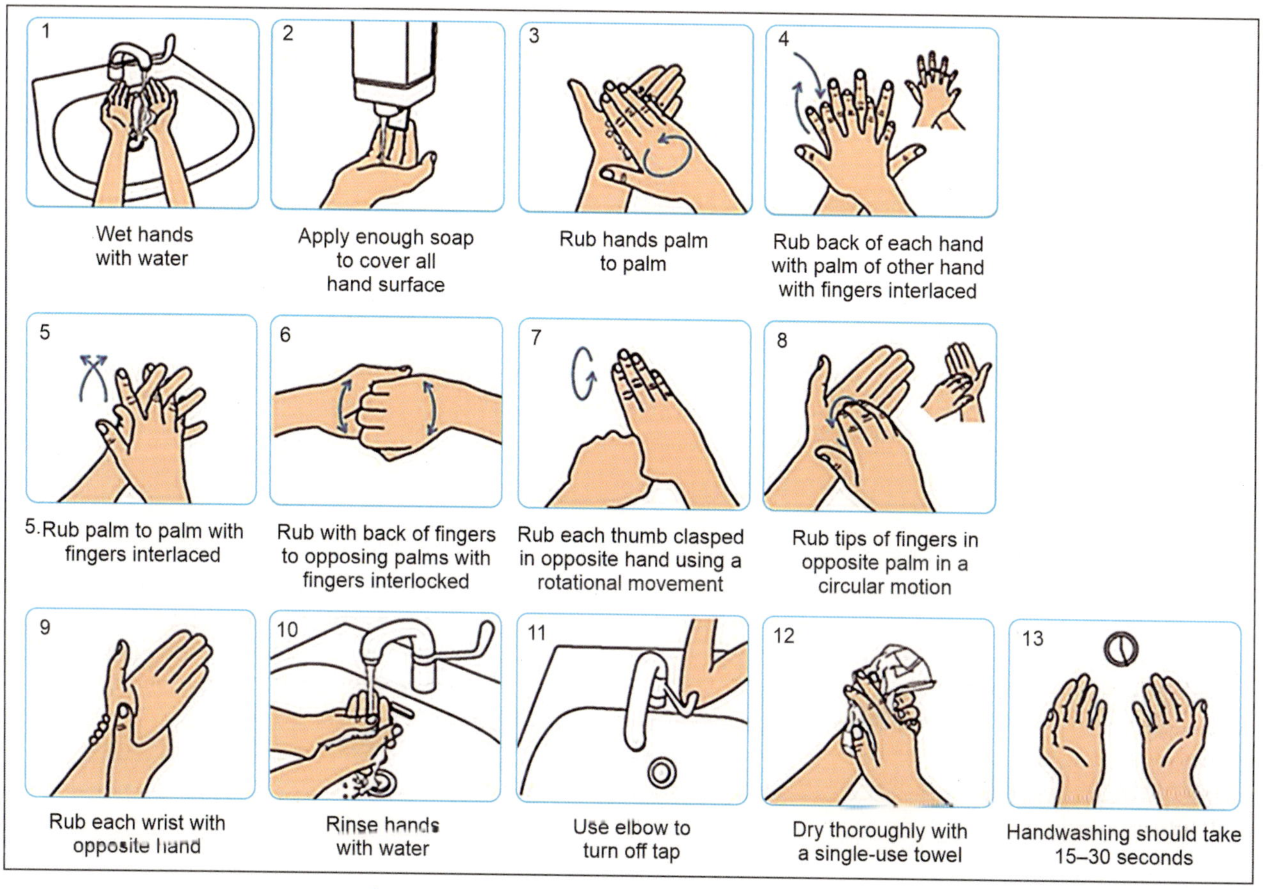

1
Wet hands with water
2
Apply enough soap to cover all hand surface
3
Rub hands palm to palm
4
Rub back of each hand with palm of other hand with fingers interlaced
5
5.Rub palm to palm with fingers interlaced
6
Rub with back of fingers to opposing palms with fingers interlocked
7
Rub each thumb clasped in opposite hand using a rotational movement
8
Rub tips of fingers in opposite palm in a circular motion
9
Rub each wrist with opposite hand
10
Rinse hands with water
11
Use elbow to turn off tap
12
Dry thoroughly with a single-use towel
13
Handwashing should take 15–30 seconds

Fig. 29.3: Handwashing with soap and running water

1a
1b
Apply a palmful of the product in cupped hand with covering of all surfaces
2
Rub hands palm to palm
3
Right palm over left dorsum with interlaced fingers and *vice versa*
4
Palm to palm with fingers interlaced
5
Back of fingers to opposing palm with fingers interlocked
6
Rotational rubbing of left thumb clasped in right palm and *vice versa*
7
Rotational rubbing, backwards and forwards with clasped fingers or right hand in left palm and *vice versa*
8
Once dry, your hands are safe

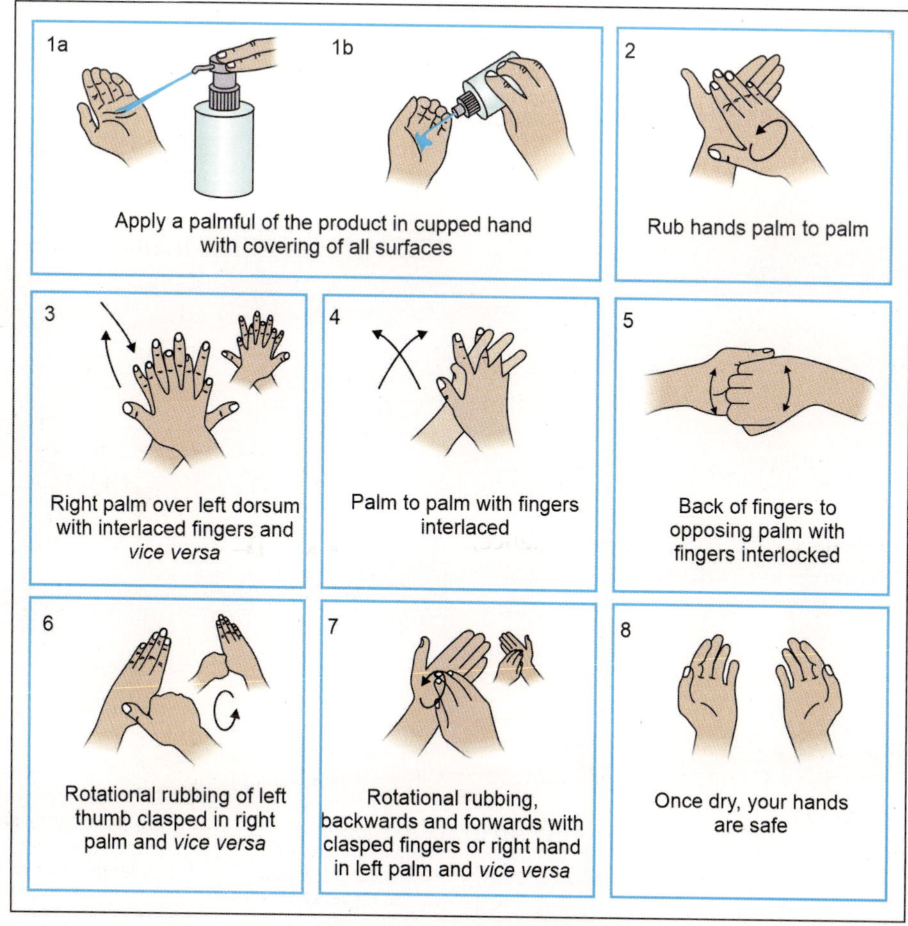

Fig. 29.4: Alcohol-based hand rub

3. Surgical hand scrub with antiseptic

- Indications: Before surgical or invasive procedure to reduce the risk of infection to patient if gloves develop tears or holes during procedure.
- Effects: It is done by 4% chlorhexidine (or alcohol based) for 3–5 min. It minimizes the microorganisms on hands under the gloves.
- Steps: Remove all jewelleries like rings, watches, bracelets. For remaining steps, follow **Fig. 29.5**.
- Points to remember: Do not dip hands into basins containing standing water even if antiseptic agents have been added. Microorganisms can survive and multiply in the solutions. Liquid soap shall be used.

Respiratory hygiene (cough etiquette): Cover the nose and/or mouth and use handkerchief/tissue paper while coughing or sneezing. Perform hand hygiene when having contact with respiratory secretions and contaminated objects/materials.

Wearing Personal Protective Equipments (PPEs)

- **Uses of PPEs:** It is used to prevent the infections to healthcare workers during different procedures like invasive procedures, sample collection like blood in CCHF virus infection, nasal/nasopharyngeal swab in SARS-CoV2 or H1N1 virus infection, etc., and collection, segregation, transport and disposal of BMWs (like heavy duty gloves and gum boot).
- **Parts of PPEs:** PPEs include gloves, gown/cover all (full body cover), shoe cover, mask/respirator, goggles, cap/hood (hood is available with cover all), face shield, etc., as shown in **Fig. 29.6**. Few parts are described below in details.

1. Gloves

- **Types of gloves used in hospital and indications**
  - Nitrile gloves **(Fig. 29.6a)**: Most resistant to puncture and used in molecular laboratory.
  - Vinyl/plastic gloves **(Fig. 29.6b)**: These are disposable gloves used for routine laboratory work.

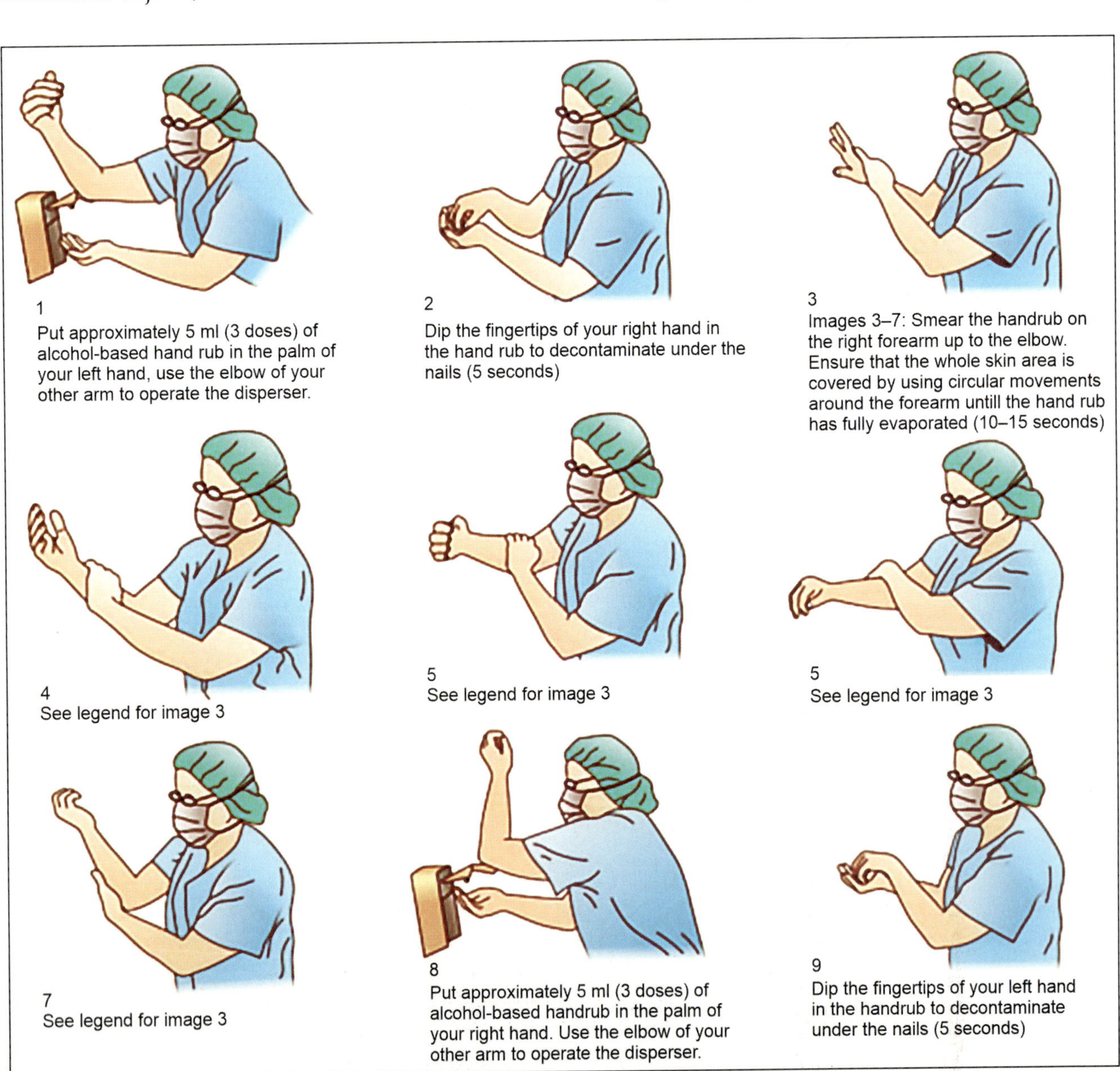

Fig. 29.5: Surgical hand scrub

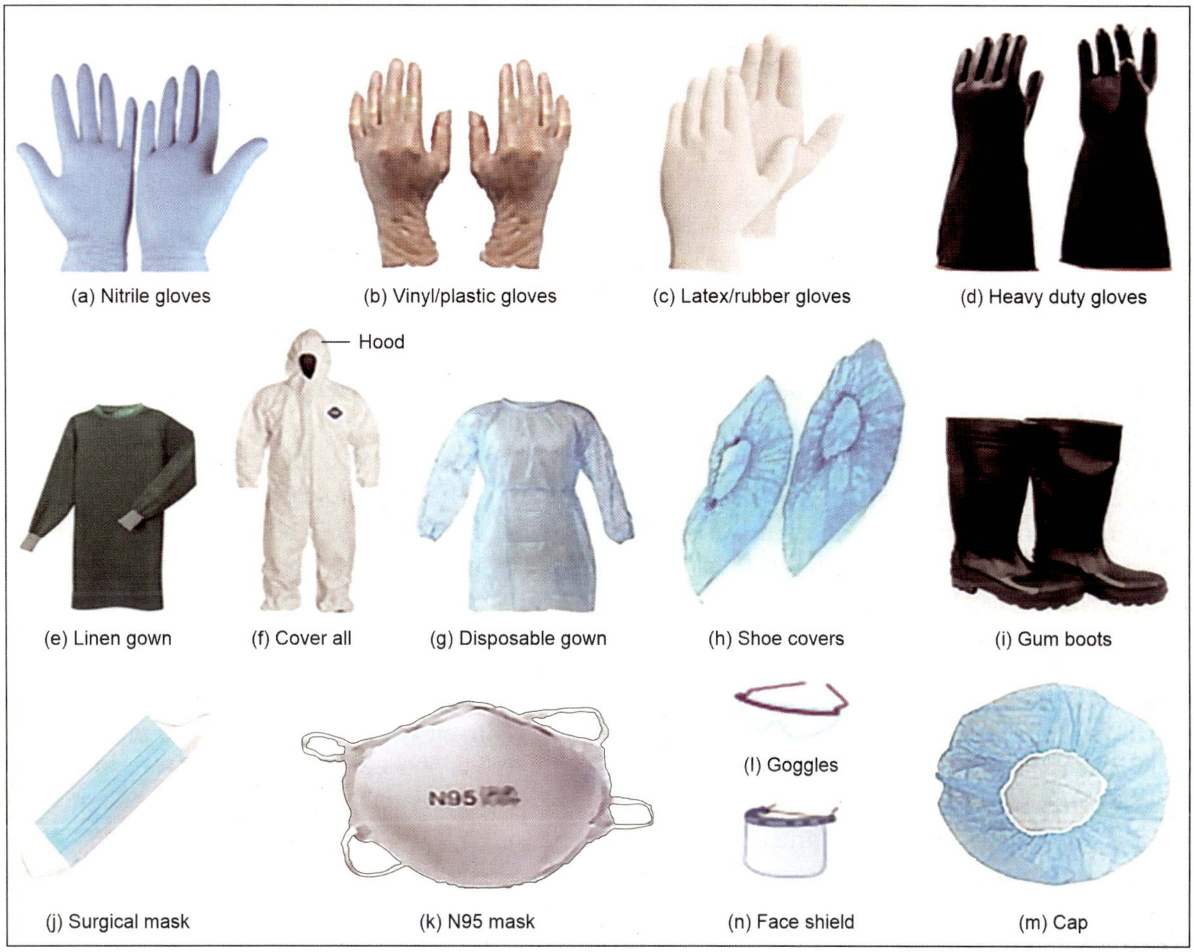

Fig. 29.6: PPE

- Latex/rubber gloves **(Fig. 29.6c)**: These are used during work with chemical or sharp items. They are coated with powder, so not useful in allergic persons.
- Heavy duty (rubber) gloves **(Fig. 29.6d)**: These are used during work with BMW management.
- **Steps of donning and doffing of gloves:** Steps are shown in **Figs 29.7 and Fig. 29.8,** respectively.

- **Hands hygiene and uses of gloves**
  - Hands hygiene before use of gloves: It is done to prevent the possible cross contamination of gloves by normal flora of body.
  - Hands hygiene after use of gloves: It is done to prevent the possible transmission of microbes

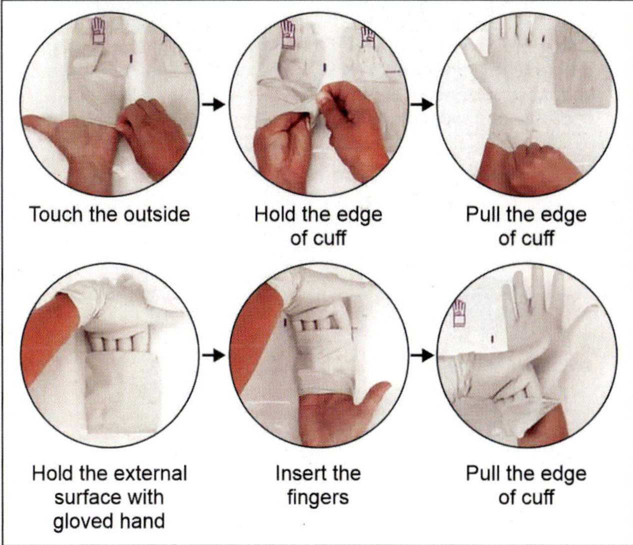

Fig. 29.7: Steps of wearing gloves (donning)

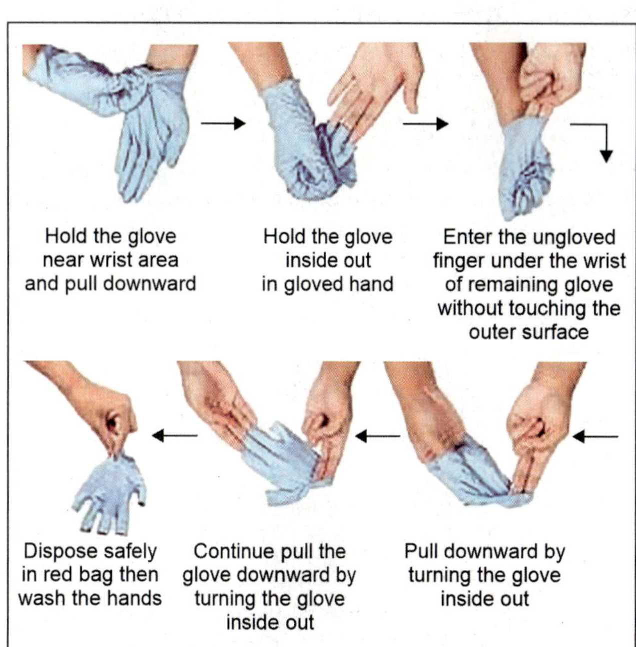

Fig. 29.8: Steps of removing gloves (doffing)

due to microtears in gloves. Gloves also create the warm, moist and occlusive environment between gloves and skin which increase the chances of infection and to prevent this, hands hygiene is important after use of gloves.

2. **Gown/cover all (full body cover)**
- **Types of gowns used in hospital and indication:** Different types are like linen gown **(Fig. 29.6e)**, cover all/full body cover **(Fig. 29.6f)** and disposable gown **(Fig. 29.6g)**. Gown is used when there is risk of cloth contamination by blood or body fluid.
- **Steps of donning and doffing of gown:** Steps of are shown in **Fig. 29.9 and Fig. 29.10,** respectively.

3. **Shoe cover**
- **Types of shoe covers used in hospital and indications:** Different types are like surgical shoes, shoe covers **(Fig. 29.6h)** and gum boots **(Fig. 29.6i)**. Shoe covers are used in laboratory or operation theater. Gum boots are used during work with BMW management.
- **Steps of donning and doffing of shoe covers:** Wear the shoe cover above the cuff of gown. Remove gently and dispose in yellow bag.

4. **Mask/respirator**
- **Types of masks used in hospital and indications:**
  - Three layers surgical mask/medical mask **(Fig. 29.6j):** It is a single use disposable mask used in hospital to protect mouth and nose from aerosols. It has three layers **(Fig. 29.11)**. Outer layer is hydrophobic, which repels the fluid like water, blood, etc. Middle layer is filter made up from melt blown material which filters the bacteria, viruses and fluid particles. Drawback of this filter is that pores size is not standardized

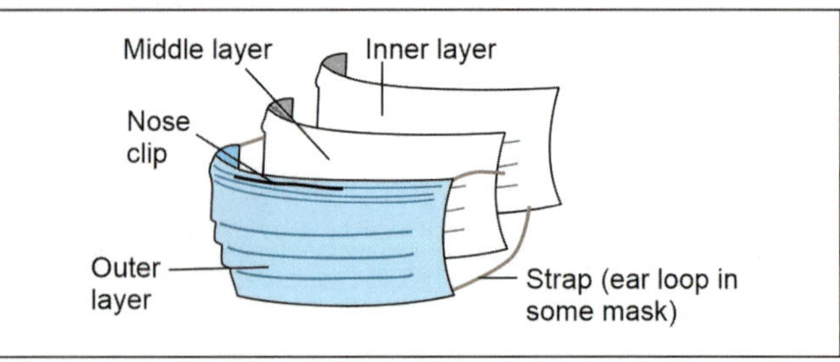

Fig. 29.11: Layers of surgical mask

as in N95 mask. Inner layer is hydrophilic which absorbs the sweat, water and spit.
  - Two layers mask: It is not used in hospital but used in food/dairy industries, restaurants, etc. It has no middle layer.
  - N95 mask or respirator **(Fig. 29.6k):** Described separately in box.
- **Steps of donning and doffing of masks:** Avoid the touching the front part of mask while wearing. Do not hang the mask below chin or on neck (called hanging mask syndrome) as contaminants from chin or neck may infect the inner layer. Surgical mask should change after 4–6 hours or earlier, if becomes wet or soiled. N95 mask should change after 8 hours or earlier, if becomes wet or soiled. Common steps of donning and doffing are shown in **Figs 29.12 and 29.13,** respectively.

5. **Goggles:** Goggles **(Fig. 29.6l)** are used to protect the mucosa of eyes.

6. **Cap/hood:** Cap is used to cover head. Sometimes cap is available as hood with cover all **(Fig. 29.6f)** or available separately **(Fig. 29.6m)**.

7. **Face shield:** Face shield **(Fig. 29.6n)** is used to protect the mucosa of eyes, nose and mouth.

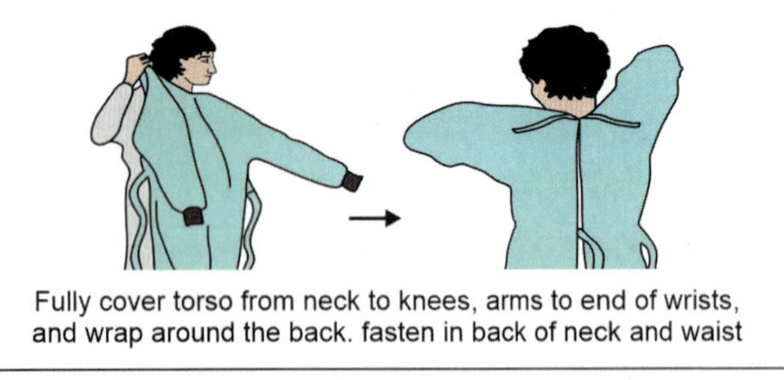

Fig. 29.9: Steps of wearing gown (donning)

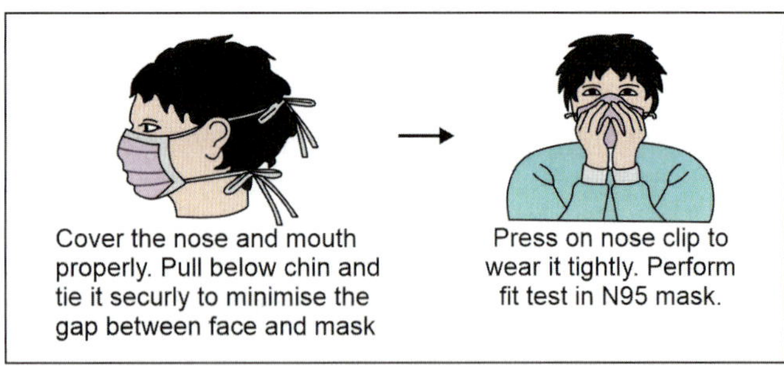

Fig. 29.12: Donning of surgical mask

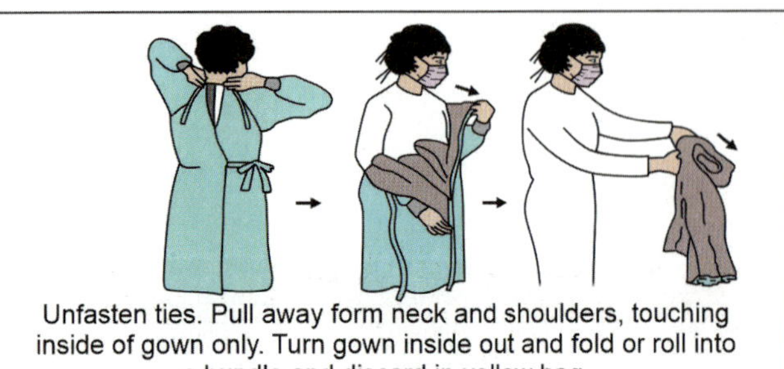

Fig. 29.10: Steps of removing gown (doffing)

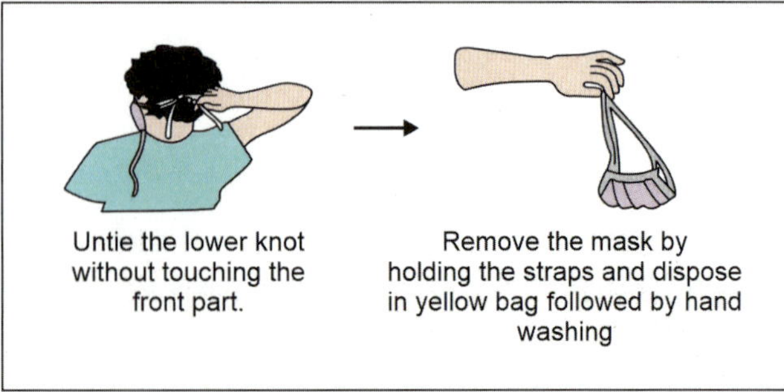

Fig. 29.13: Doffing of surgical mask

Note: N95 mask or respirator

Meaning: N suggests Respirator Rating Letter Class that means "non-oil". It is not resistant to oil hence if no oil-based particulates are present in the work environment, then you can use this mask. Other masks ratings are R (resistant to oil for 8 hours) and P (oil proof). While 95 suggest the ability to filter 95% airborne particles. Masks ending in a 99 have 99% efficiency. Masks ending in 100 are 99.97% efficient and that is the same as a HEPA quality filter. The minimum size of 0.3 micron of particulates and large droplets would not pass through the barrier, according to the Center for Disease Control and Prevention (CDC).

Material: It has total 4 layers. Outer and inner are made up of spun bond polypropylene fibers and middle two layers are of cellulose/polyester and melt blown polypropylene fibers.

Negative pressure: N95 mask is available with negative pressure. It is checked by inhalation. If there is no air leak or if it is not drawn in towards face then it should be readjusted.

Positive pressure: It is checked by exhalation. If air escape then it should be readjusted.

Fit checking: After wearing fit checking is done by compressing nasal clip, face seal and cheeks seal.

Fit testing: It is done to select the size and style for person. It is done at the time of joining and then annually.

Use and care of N95 mask: It is for single use (no re-use) and should be changed after 8 hours or earlier if teared, wet or soiled. It cannot be cleaned or disinfected.

- **Donning and doffing of PPEs:** Wearing of PPEs called donning and removing called doffing. Common precautions and steps of donning and doffing are mentioned below; however, individual variation in steps is always there as per associated risk, modes of transmission of microbes, parts available in PPE etc.

  - **Donning:** Before donning take enough food and water and urinate. It should be done in defined area. Remove all jewelleries, wrist watch, mobile and other accessories. Steps include hands hygiene with alcohol based sanitizer followed by → 1st (inner) pair of gloves → Gown or cover all → Shoe cover → N95 Mask/respirator and goggles → Cap/hood → Face shield (instead of goggles or goggles and mask) → 2nd (outer) pair of gloves → Hands hygiene.

  - **Doffing:** Do with patience and care to avoid the generation of aerosols. It should be done in defined area. Check the PPEs for any tear or contamination. Steps include hands hygiene followed by → Shoe cover → 2nd (outer) pair of gloves → Face shield → Cap/hood → Gown or cover all → N95Mask/respirator and goggles → 1st (inner) pair of gloves → Hands hygiene.

Note: Different views in donning and doffing
Few experts suggested the use of two pairs of gloves and hand sanitization in-between each steps of doffing (http://youtu.be/KXueclFu1PU).

- **Disposal of PPEs**
  - **Red bag:** All plastic parts like gloves, face shield and goggles.
  - **Yellow bag:** Shoe cover, cap/hood, gown or cover all, and N95Mask/respirator.

Prevention of sharp/splash injuries in HCWs: It requires to prevent the transmission of HBV, HCV and HIV via blood/body fluids spillage in mucosa or abraded skin or by needle prick injury. This includes use the disposable needles and syringes, post exposure prophylaxis for HBV and HIV, use of PPEs, HBV vaccination with titre documentation, clean work surface with 0.5% sodium hypochlorite, spillage management by 0.5% sodium hypochlorite, no recapping of needle, if required then use one hand scoop technique as shown **Fig. 29.14**. Needle/sharp objects should be disposed in white translucent box as per guidelines of BMW.

Preventions during surgeries: Use the sharp instruments without touching. Do not pick the needle by hands during suturing but hold it with needle holder or forceps holder. Use the double gloves technique in case of positive or suspected patients.

Environmental control: Measures to prevent infection associated with air, water and other elements.

Isolation of patients and assessment of infection risk: Diseases requiring isolation include severe influenza cases, SARS, open case of tuberculosis, anthrax, diphtheria, pertussis, chickenpox, pneumonic plague, infection by multidrug resistant pathogen and patients with low immunity.

Patient resuscitation: All staff who participate in resuscitation shall adhere to standard precautions throughout the resuscitation.

Linen: Soiled linens are the sources of infection, so required careful handling.

Sterilization and disinfection practise: It includes the autoclaving of instrumentation, disinfection of spillage, fumigation of ward, laboratory, operation theaters, other rooms, etc.

Proper collection, handling and disposal of waste: Follow **Ch. 31.**

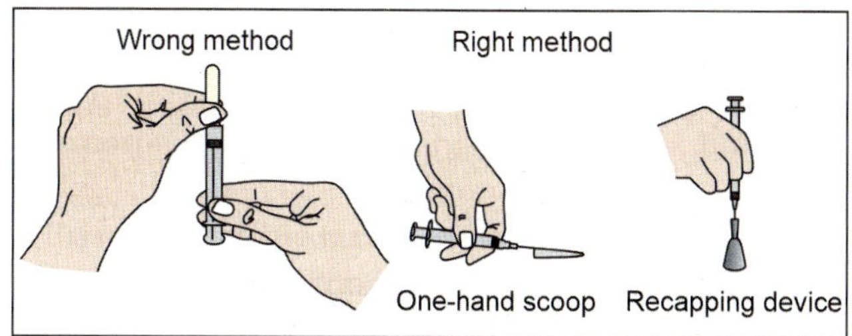

Fig. 29.14: Recapping of needle

These precautions are based on mode of transmission of pathogens with following types.

Transmission-based precautions

- **Contact precautions:** Place patient in a single room/in a room with another similarly infected patients. Consider the epidemiology of the disease and the patient population when determining patient placement. Wear clean, nonsterile gloves and gown when entering the room if substantial contacts with the patient, environmental surfaces or items in the patient's room are anticipated. Limit the movement and transport of the patient from the room for essential purposes only. If transportation is required, use precautions to minimize the risk of transmission.

- **Droplet precautions:** Place patient in a single room/in a room with similarly infected patients. Wear a mask when working within 1–2 meters of the patient. Place a mask on the patient if transport is necessary. Special air handling and ventilation are not required to prevent droplet transmission of infection.

- **Airborne precautions:** Airborne precautions are used in addition to routine practices for clients/patients/residents known or suspected of having an illness transmitted by the airborne route. Anyone who enters the room must wear appropriate PPE. N95 shall be worn and use of a negative pressure isolation room. Place patient in a single room that has a monitored negative airflow pressure and is often referred to as a "negative pressure room".

Management of an outbreak: An outbreak is defined as an unusual or unexpected increase of cases of a known HCAI of the emergence of cases of a new infection. Following are different steps of outbreak investigation.

1. Step 1: Constitution outbreak investigation team with various specialities.
2. Step 2: Verify the diagnosis, developing a case definition, define the outbreak in terms of time, person and place.
3. Step 3: Determine the magnitude of the problem and if immediate control measures are required, such as isolation or cohorting of infected cases, strict hand washing and asepsis shall be applied.
4. Step 4: Notification to concerned departments.
5. Step 5: Line listing, collection of microbiological records and patient details, epidemic curve preparation, risk factors identification, and reporting to hospital infection control committee.
6. Step 6: Identification of cause, modes of transmission, etc., by microbiological and epidemiological study.
7. Step 7: Implementation of control measures.
8. Step 8: Rule out the efficacy of control measures.

Requirements of isolation: Accommodation for the suspected or confirmed patient, in a room or area designated for infectious diseases.

Notifiable disease: Disease where steps are needed to be taken to prevent them from taking the form of epidemic or spreading from one person to another is called notifiable disease. For examples, dengue, chikungunya, Japanese encephalitis, meningococcal meningitis, typhoid fever, cholera, diphtheria, shigellosis, leptospirosis, viral hepatitis, malaria, syphilis, TB, H1N1 flu and HIV. Health care authority shall notify about the disease noticed in health care facility to concerned departments.

ACCESS YOURSELF

Essay/Full Question

1. Define healthcare-associated infections (HAIs) and enumerate the types. Discuss the factors that contribute to the development of HAIs and the methods for prevention.

Short Notes

1. CAUTI/VAP/CABSI/SSI.
2. Demonstrate infection control practices and use of PPE.

Short Questions for Theory/Viva Questions

1. Define hospital acquired infection. Write its sources.
2. Write five moments for hand hygiene.

MCQs for Chapter Review

Introduction

1. **A person gets infected in a hospital and clinical manifestations appear after he is discharged this is called:**
   a. Nosocomial infection
   b. Opportunistic infection
   c. Epizootic infection
   d. Physician induced infection
2. **All the following are true about nosocomial infection** *except*:
   a. May manifest within 48 hours of admission
   b. May develop after discharge of patient from hospital
   c. Denote new condition which is unrelated to the patients primary conditions
   d. May already present at the time of admission
3. **True common nosocomial infection** *except*:
   a. *Staph. aureus*   b. *P. aeruginosa*
   c. Enterobactericeae   d. *Mycobacterium*
4. **Most common mode of transmission of nosocomial infection:**
   a. Hand contact
   b. Droplet infection
   c. Blood and blood products
   d. Contaminated water

Types of HAIs

5. **Which parameter is not included in HAI surveillance?**
   a. CAUTI   b. CABSI/CRBSI
   c. VAP   d. Open wound infection
6. **Most common organism involved in nosocomial infection:**
   a. *Staph. aureus*   b. *E. coli*
   c. *Legionella*   d. *Strept. pneumoniae*
7. **Most common pathogen responsible for nosocomial pneumonia in ICU is:**
   a. Gram-positive organisms
   b. Gram-negative organisms
   c. *Mycoplasma*
   d. Virus infections

8. **Most common organism causing VAP:**
   a. *Legionella*
   b. *Pneumococcus*
   c. *Pseudomonas*
   d. Coagulase-negative *Staphylococcus*

9. **Clinical pulmonary infection score (CPIS) is used for:**
   a. VAP
   b. Upper respiratory tract infections
   c. Community-associated pneumonia
   d. Tuberculosis

10. **Anaerobic bacterium surgically most important:**
    a. *Bacteroides*
    b. *Staph. aureus*
    c. *Pseudomonas*
    d. Pneumococcus

11. **Most common species of *Pseudomona* causes intravascular catheter related infections is:**
    a. *P. cepacia*
    b. *P. aeruginosa*
    c. *P. maltophila*
    d. *P. mallei*

12. **Most common catheter-related bloodstream infections is:**
    a. *Candida*
    b. Gram-negative organisms
    c. CoPS
    d. CoNS

13. **Which of the following causes highest risk of nosocomial infection to a patient?**
    a. Patient admitted for elective surgery
    b. HIV patient coming in follow up OPD
    c. Patient undergoing endoscopy
    d. Patient admitted for normal delivery

Nosocomial Infection Prevention and Control Program

14. **Hand rub should not be used in which condition?**
    a. Before touching the patient
    b. After touching the patient
    c. After touching patient's surrounding
    d. Hands are visibly soiled

15. **Medical mask should be disposed in:**
    a. Red bag
    b. Yellow bag
    c. Blue cardboard box
    d. White box

Answers and Explanation of MCQs

1. **a**
2. **a, d**
- Follow section, **introduction → definition** for explanation of answers of MCQs 1–2.
3. **d**
- Follow section, **introduction → precipitating factors (microbial factor)** for explanation.
4. **a**
- Follow section, **introduction → modes of transmission** for explanation.
5. **d**
6. **a**
7. **b**
8. **c**
9. **a**
10. **a**
11. **b**
12. **d**
13. **a**
- Follow section, **types of HAIs** for explanation of answers of MCQs 5–13.
15. **d**
- Follow **Fig. 29.2** for explanation.
16. **b**
- Follow section, **prevention and control → standard precautions → Fig. 29.13** for explanation.

Sterilization and Disinfection

INTRODUCTION

History

Concept of sterilization and disinfection was practised by Pasteur, Lister and Koch.

Definitions

Sterilization: Method by which an article or surface or medium completely becomes free from all microorganisms, including pathogenic, nonpathogenic and spores and organisms are incapable to give rise the reinfection. It is done by using physical agents or methods so also called physical sterilization.

Disinfection: Method by which an article or surface or medium incompletely becomes free from all microorganisms, including pathogenic and nonpathogenic except spores and organisms are capable to give rise the reinfection. It is done by using chemical agents or methods, so also called chemical sterilization. As per definition of disinfection, spores are not killed by disinfectants; however, some disinfectants have sporicidal action.

Asepsis: It is the prevention of infection by keeping the article or surface or medium away from the source of infection. It means absence of microbes.

Antisepsis: It is the prevention of infection by inhibiting the growth of organisms.

Antiseptics: Chemical agents used to prevent the infection or growths of microbes are called antiseptics. Few authors differentiated antiseptics from disinfectants by their use on skin or mucosa (also called skin disinfectants), where later can be used for inanimate objects. Common skin and hand disinfectants are mentioned below.

- **Common antiseptics/skin disinfectants**
  - Betadine (povidone iodine), tincture iodine (iodine in 2% alcohol): Best skin disinfectant, because less irritants and causes less staining.
  - Spirit (ethyle alcohol/70% ethanol).
  - 2-propanol (isopropyl alcohol, isopropanol or propan-2-ol).
  - Hibitane (chlorhexidine).
- **Common hand disinfectants**
  1. **Alcohol based:** Commercially available preparations are sterilium, bactilium and casillium. All are same preparations with different brands. In 500 ml solution they contains 1-propanol (*n*-propanol or propan-1-ol) about 35 g, 2-propanol (isopropyl alcohol, isopropanol or propan-2-ol) about 45 g and macetronium ethyl sulfate (Ethyl-hetadecyl-dimethylammonium-ethyl sulfate) about 0.2 g. Patient care handwash is done by 3 ml (2 push) solution for 30 seconds. Surgical handwash is done by 9 ml (6 push) solution for 3 minutes for each hand.
  2. **Chlorhexidine gluconate:** About 2.5% concentrations available in hand rub and hand gel. Mode of use is same as sterilium.
  3. **Betadine (povidone-iodine) solution:** It is high level disinfectants used for surgical hand scrub.

Bacteriostatic agents: These are chemical agents capable of preventing/inhibiting multiplication of bacteria.

Bactericidal agents (germicides): These are chemical agents capable of killing the bacteria.

Sporicides or sporicidal agents: These are the chemical agents capable of killing the spores. For more details about sporicidal agent, follow **Ch. 6.**

Cleaning: It is the removal of all soil (e.g., organic and inorganic materials) and other dirt from object or surface or medium by wiping or using water with detergent called cleaning.

Decontaminations (sanitization): It is the process of keeping article or area free from contaminants.

Differences between Sterilization and Disinfection

Follow **Table 30.1**.

TABLE 30.1: Differences between sterilization and disinfection

Sterilization	Disinfection
By physical agents	By chemical agents
Kills all pathogenic, non-pathogenic microbes and spores	Kills all pathogenic and non-pathogenic microbes except spores
Complete killing of microbes, so no chances of reinfection	Incomplete killing (reduce the numbers) of microbes, so chances of reinfection

Order of Resistance of Microbes

Decreasing order of resistance of microbes to the agents used for sterilization and disinfection is mentioned in **Flowchart 30.1**.

Flowchart 30.1: Order of resistance of microbes

Prions (highest resistance) > Oocyst of *C. parvum* > bacterial spores > mycobacteria > cysts of parasites (*G. lamblia*) > small non-enveloped viruses > trophozoites > gram-negative bacteria > fungi > large non-enveloped viruses > gram-positive bacteria > enveloped viruses (lowest resistance)

Note: Terminology

Physical methods and chemical methods are used as a synonym for sterilization and disinfection, respectively, but this practice is not correct because many physical methods are giving incomplete sterilization and many chemical methods are giving complete disinfection with killing of spores.

CLASSIFICATION

Three types of methods sterilization, disinfection and chemosterilization are mentioned in **Flowchart 30.2**.

Physical Methods (Sterilization) or Physical Agents (Sterilants)

Different physical methods are described below.

Sunlight

It has bactericidal effect due to heat and UV rays but rarely useful.

Drying

About 80% of bacterial weight is due to water and drying in air cause deleterious effect on bacteria. This method is unreliable and only theoretical value.

Heat

This is the method of choice for sterilization unless it is contraindicated. Effective factors over heat are temperature of exposed heat, time of exposure, type of heat (like dry or moist heat), presence of substances like proteins, starches, fats-oils, acids-alkalis, etc., pH of medium, types of organism such as vegetative form or spores form, numbers of organisms and types of materials to be sterilized. Minimum time required to kill organisms at given temperature is called thermal death time (TDT). Two main types of heat, i.e., like dry heat and moist heat are described below.

A. Dry heat

Mechanisms of actions of dry heat: Killing effects of dry heat are due to <u>D</u>enaturation of protein, <u>O</u>xidative damage and <u>T</u>oxic effect of elevated levels of electrolytes.

Mnemonic: Mechanisms of actions

- **Dry heat mechanisms:** DOT → As above
- **Moist heat mechanisms:** DC → As below

Methods of dry heat: Following are three methods:

1. **Red heat** or **red hot** or **flaming:** It includes burning of article till it becomes red hot. Articles to be sterilized by this method are nichrome wires, nichrome loops, tips of forceps, etc.
2. **Incineration:** It includes burning of article till it becomes ash. Articles to be sterilized by this method are contaminated clothes, animal's carcasses, pathological materials, bio-medical wastes and plastic items like PVC or polythene.
3. **Hot air oven**
   - **Principle:** It includes sterilization by dry heat at 160°C for 2 hours.
   - **Holding time and temperature:** Follow **Table 30.2**. British pharmacopoeia recommended 150°C for 1 hour for oil, glycerol and dusting powder.
   - **Instrument's structure:** It is shown in **Fig. 30.1**. It is a bad conductor of heat. Heating elements are heated by electricity and kept inside the wall. It is fitted with fan for elimination of air and even distribution of air.
   - **Articles to be sterilized:** It includes articles which cannot be penetrated by steam and can tolerate high temperature. **Glass materials** like glass tubes, glass bottles, glass syringes, etc., **metal instruments** like forceps, scissors, scalpels, etc., and **pharmaceutical products** like grease, liquid paraffin, fats, dusting powders, etc., are sterilized by this method.
   - **Disadvantages:** This method required long exposure time and high temperature >185°C which may destroy the instruments or corrosive, resulting loss of hardness.

TABLE 30.2: Holding temperature and time of oven

Temperature (°C)	Time (minutes)
160	120
170	60
180	30

Physical methods

- Sunlight
- Drying
- Heat
  - Dry heat
    - Red hot or flaming
    - Incineration
    - Hot air oven
  - Moist heat
    - Below 100°C
      - Pasteurization
      - Water bath
      - Vaccine bath
      - Inspissation
    - At 100°C
      - Boiling
      - Steam sterilizer at atmospheric pressure
        - Koch or Arnold's steam sterilizer
        - Tyndallization
    - Above 100°C
      - Steam sterilizer above atmospheric pressure
        - Autoclave
        - Hydroclaving

Collectively called steam sterilizers

- Filtration
  - Liquid filter
    - Depth filter
      - Candle filter
      - Asbestos filter
      - Sintered glass filter
    - Surface filter (membrane filter)
      - Gradacol membrane filter
      - Modern membrane filter
  - Air filter
    - Surgical mask
    - HEPA filter
    - ULPA filter
- Ultrasonic and sonic vibration
- Radiation
  - Ionizing radiation
    - X-rays
    - γ-rays
    - Cosmic rays
    - Electron beams
  - Nonionizing radiation
    - UV-rays
    - Infrared radiation

Chemical methods

- Alcohols
- Aldehydes
- Dyes
- Halogens
- Phenols
- Gases
- Biguanides
- Surfactants
- Acids
- Salts (metallic salts)

Chemosterilization (physical + chemical methods)

- Gases
- Plasma sterilization (newer method)
- Duckering

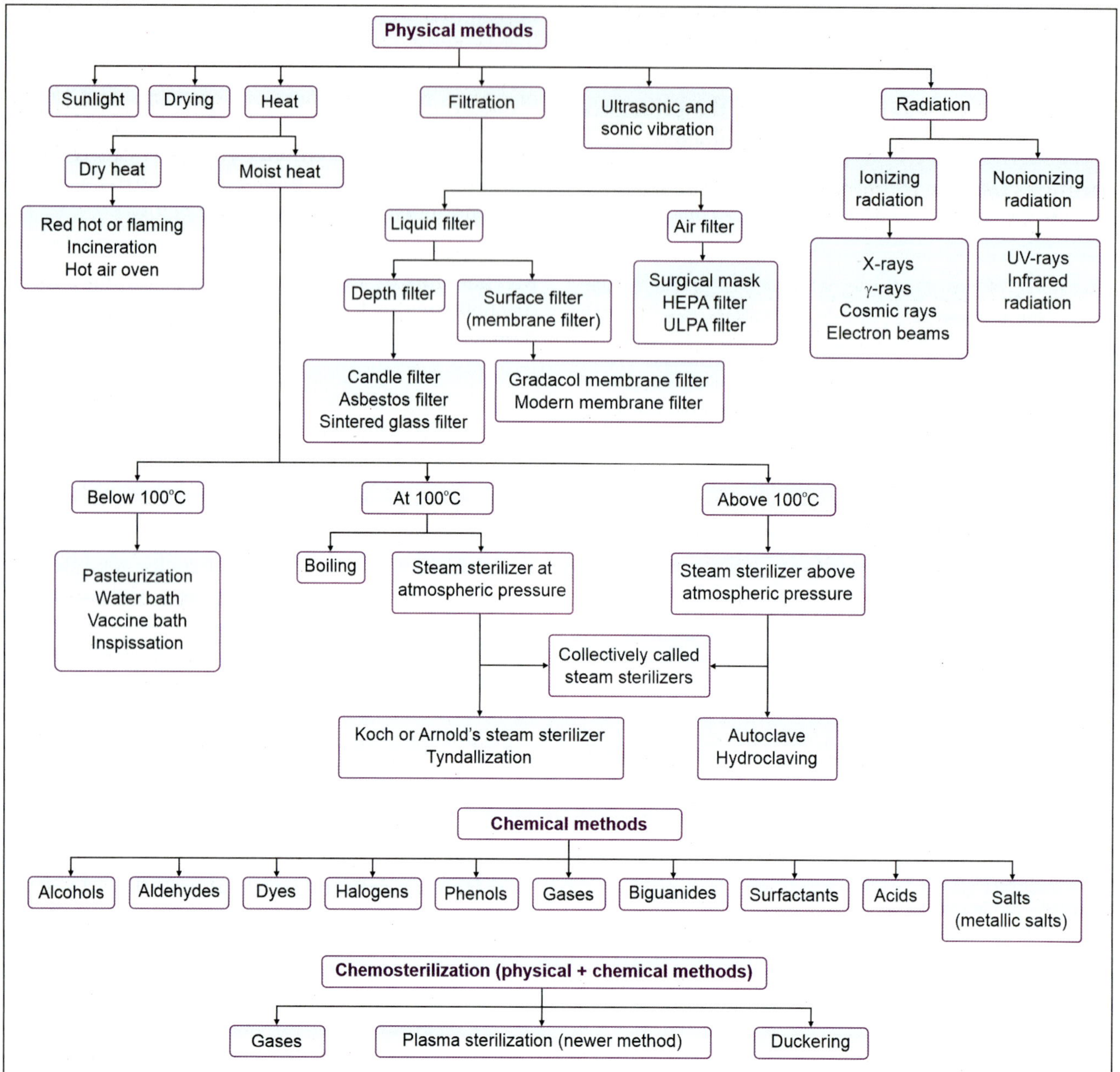

Notes
- Fraction or intermittent sterilization→both inspissation & tyndallization
- Disinfectants sensitive to organic matters: Chlorine (+++) and Phenol (– / +)

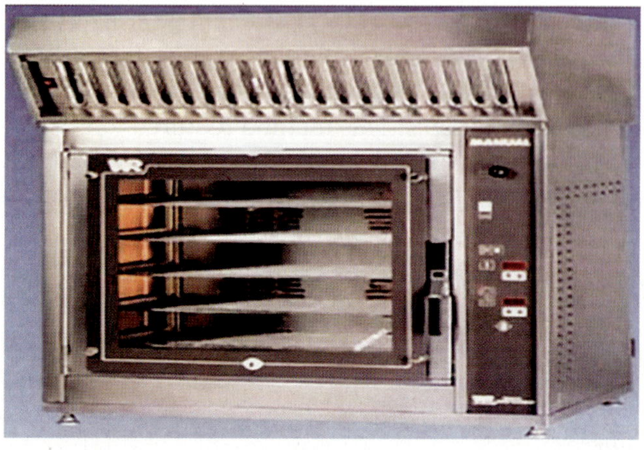

Fig. 30.1: Hot air oven

- **Precautions:** Instrument should not be overloaded. Dry the glass wares before putting in oven. Wrap the tubes or bottles in paper.
- **Method of use of hot air oven:** Arrange articles in such a way which allow free circulation of air in between the articles. Rubber except silicon rubber cannot withstand the temperature, so do not put in oven. Cotton plug may charred at 180°C, so do not put in oven. After sterilization, oven should be allowed to cool for 2 hours before opening the door to avoid the cracking of the glasswares due to sudden or uneven cooling.
- **Sterilization control:** Three types:
  - Physical control: Thermometer and thermocouple

– Chemical control: Browne's tube or green spot, which gives green color after proper sterilization.
– Biological control: A strip contains 10^6 spores of nontoxigenic strain of *Cl. tetani* or *B. subtilis* subsp. *niger* is packed in and envelope and put into an oven. After proper sterilization remove the strip and inoculate in thioglycolate broth or RCMM. Incubate anaerobically at 36°C for 5 days for sterility testing.

> **Note: Microwave oven**
> **Principle:** It is basically radio wave (with 2500 MHz or 2.5 GHz frequency) to cook the food. Radio waves are absorbed by fats, water and sugars of food to generate the heat.

B. Moist heat

Mechanisms of actions: Killing effects of moist heat are due to **D**enaturation of protein and **C**oagulation.

Moist heat is more advantageous than dry heat: Because the steam of moist condenses over the surface of spores and other organisms, which increases the temperature and water content of organisms. Later causes microbial death by hydrolysis and breaking down of bacterial proteins.

Types of moist heat: Three types of moist heat methods are mentioned in **Flowchart 30.2**.

Below 100°C

These methods kill all mesophilic-vegetative bacteria, moulds, yeasts and some viruses; however, they are not able to kill spores and some heat resistant viruses like polio virus, HBV, etc. Four methods in this category are described below.

1. Pasteurization

- **Definition:** It was defined by WHO in 1970. Heating of milk to such temperature and for such period of time as required to destroy any pathogen that may be present, while causing minimal changes in the composition, flavor and nutritive value of milk called pasteurization.
- **History:** It was invented by Louis Pasteur in 1863-65.
- **Uses:** It is used for sterilization of milk and to control the microorganisms from beverages like beer, fruit juices and vegetable juices.
- **Methods:** Two methods as described below.
  - Holder method: Milk is kept at 60°C for 30 minutes followed by rapid cooling at 5°C. It does not kill the heat-resistant bacteria like *Coxiella burnetii.*
  - Flash (HTST-High Temperature and Short Time) method: Milk is heated at 72°C for 15–20 seconds followed by rapid cooling at 13°C or lower. It is rapid and most widely used method. Very large quantity of milk per hour can be pasteurized by this method. Relatively heat resistant bacteria like *Coxiella burnetii* may survive in Holder method but killed by flash method.

- **Disadvantages:** Pasteurization does not destroy the bacterial spores, bacterial toxins and fungal toxins. It also not kills the thermoduric bacteria (**Ch. 7**)
- **Advantages:** All non sporing bacteria like *Coxiella burnetii* (Q fever, survive under Holder method), *Mycobacterium tuberculosis* (tuberculosis), *Salmonella* (slamonellosis), *Brucella* (brucellosis), etc., are killed.
- **Sterility tests of pasteurized milk**
  - **Phosphatase test:** It based on production of enzyme alkaline phosphatase by bacteria. It breaks the phenyl disodium phosphate to release phenol (phenol is benzene derivatives and used as disinfectant). Also called carbolic acid. It is performed by adding the solution contains phenyl disodium phosphate to milk in a sterile test tube. Later, add the solution contains color indicator. Incubate at 37°C for 2 hours. Change in color occurs, if free phenol is present. Phenol release is quantitated by measuring the absorbance of the color developed by a spectrophotometer. Alkaline phosphatase enzyme present in milk is destroyed by heating the milk at 60°C for 30 minutes. Presence of enzyme indicates inadequate pasteurization, or later addition of raw milk. Phosphatase test also used for *Staph. aureus* (**Ch. 49**); however, both are different.
  - **Standard plate count:** Bacteriological quality of milk is determined by culture technique called standard plate count. Most countries in the West enforce a limit of 30,000 bacterial counts per ml of pasteurized milk.
  - **Coliform count:** Test based on fermentation of lactose of milk by coliform bacteria. Coliforms ferment the lactose of milk to produce acid and gas. Acid is indicated by color change of the medium and gas is indicated by bubbles collection in the inverted Durham tube. It is performed by adding the serially diluted milk to three tubes contains MacConkey's broth and inverted Durham tube. Incubate at 37°C for 48 hours. Acid is indicated by color change of the medium and gas is indicated by bubbles collection in the inverted Durham tube. Coliform bacteria are usually completely destroyed by pasteurization; therefore their presence in pasteurized milk is an indicator of inadequate pasteurization or post pasteurization contamination. The standard in most countries is that coliform be absent in 1 ml of milk.

2. Water bath (Fig. 30.2): It is used for serum/body fluid contains coagulable proteins. Sterilization is done at 56°C for 1 hour for several successive days.

Fig. 30.2: Water bath

Fig. 30.3: Inspissator

Essentials of Medical Microbiology

3. **Vaccine bath:** It is used for vaccine of nonsporing bacteria. Sterilization is done at 60°C for 1 hour.
4. **Inspissation (Fig. 30.3):**
   - **Synonym:** Also called fractional sterilization, because it requires three successive days for sterilization of articles.
   - **Principle:** It sterilizes the article by moist heat below 100°C for three successive days. 1st exposure kills all vegetative form and in the interval between the heating remaining spores germinates into vegetative forms which are killed on subsequent heating.
   - **Uses:** It is used for sterilization of media contains serum and egg like LJ medium, Dorset egg medium and Loeffler's serum. Serum and egg generally get destroyed at high temperature.
   - **Holding time and temperature:** Sterilization can be done for three successive days as follows:
     - 1st day: 85°C for 1 hour.
     - 2nd day: 75°C for 1 hour.
     - 3rd day: 75°C for 1 hour.

At 100°C

Two methods in this category are described below.
1. **Boiling:** Heating in boiling water at 100°C for 5 minutes sufficient to kill all vegetative bacteria, HBV and some bacterial spores. Articles to be sterilized by this method are medical and surgical equipments in emergency like glass syringes. Heat labile articles and hollow or porous items where water do not penetrate in the lumen cannot be sterilized by this way. Sterilization may be promoted by adding 2% of sodium bicarbonate.

2. **Steam sterilizer at atmospheric pressure:** Two categories.
   - **Koch or Arnold's steam sterilizer:** In this method sterilization is done by moist heat at 100°C for 90 minutes at atmospheric pressure. Instrument (**Fig. 30.4**) contains copper cabinet with walls suitably lagged, conical lid with central perforation for drainage of steam, perforated tray to put article to be sterilized and water at bottom with heater. Articles to be sterilized by this methods are selective heat labile media like DCA, XLD, TCBS and selenite F broth. It does not kill the bacterial spores. It is cheap and no chances of explosion.
   - **Tyndallization:** In this method sterilization is done by moist heat at 100°C at atmospheric pressure for 20 minutes for 3 successive days. Hence, called fractional or intermittent sterilization. It kills the vegetative bacteria. Spores will germinate in favorable condition which will be killed by successive sterilization. Articles to be sterilized by this method are media containing sugar or gelatin. It kills the bacterial spores.

Above 100°C (Steam Sterilizer above Atmospheric Pressure)

Two methods in this category are described below.
1. **Autoclave**
   - **Principle:** Sterilization by moist heat above 100°C above atmospheric pressure.
   - **Instrument's structure (Fig. 30.5):** It has horizontal or vertical cabinet (cylinder) made up from stainless steel or gun metal with iron supporting sheet. Lid is available with screw clamps to make it airtight to prevent the leakage of steam, safety valve for drainage of steam, pressure gauge and temperature gauge. Perforated tray is fitted to put an article to be sterilized. Water is collected at bottom of cabinet with heater, which is operated by gas or electricity.

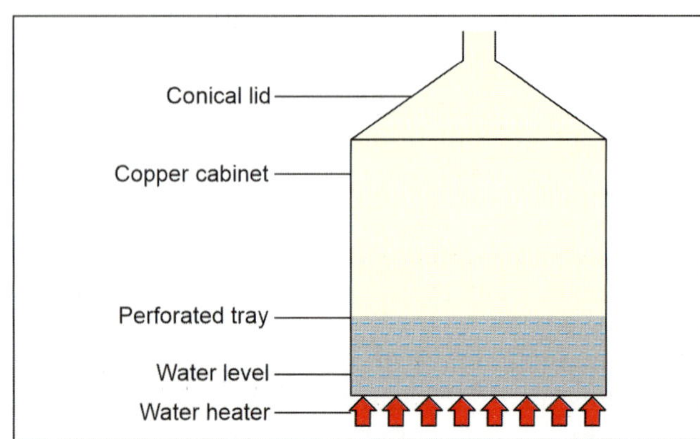

Fig. 30.4: Koch or Arnold's steam sterilizer

Fig. 30.5: Autoclave

- **Method of use of autoclave:** Add the enough amount of water. Place all articles to be sterilized on perforated tray along with sterilization control. Start the heater and close the lid. Safety valve is set to the required pressure. Allow the steam-air mixture to escape till all air has been displaced from autoclave. Make the lid airtight by screw clamps. Steam starts to collect in the cabinet and raising the pressure. When pressure reached to desired level, open the safety valve and remove excess steam. Start to count the holding time from this point. Off the heater when holding time is over. Allow the autoclave to cool till the pressure gauge indicates same pressure inside the cabinet and atmosphere.
- **Articles to be sterilized are:** Laboratory media, glass wares, hospital dressings, clothes, apron, surgical instruments, pharmacological products, fluid in sealed container, biomedical wastes before disposal, etc.
- **Holding time, temperature and pressure:** Follow **Table 30.3**.
- **Types of autoclaves**
  - Rapid cooling sterilizer like pressure cooker.
  - Gravity displacement type: Vertical type includes laboratory autoclave and horizontal type includes hospital dressing sterilizer or BMW sterilizer.
  - Negative pressure (vacuum) displacement sterilizer.
  - Positive pressure displacement sterilizer.
- **Advantages:** Autoclave is nontoxic, easy to control—monitor, inexpensive, least affected by organic/inorganic soils, bactericidal, sporicidal

TABLE 30.3: Holding temperature, pressure and time of autoclave

Temperature (°C)	Time (minutes)	Pressure (lbs)
121	15	15
126	10	20
134	3	30

and rapid cycle time. Heat penetrates medical packing, device, lumens, etc.
- **Disadvantages:** Autoclave is deleterious for heat labile instruments. There are chances of explosion due to overheating so required constant supervision during operation. No facility available for drying the load after the sterilization. No facility for air discharge, and it is very difficult to judge the complete air removal. If air is not removed completely desired temperature and pressure cannot be achieved.
- **Sterilization control:** Three types:
  - Physical control: Temperature monitoring gauge, pressure monitoring gauge and thermocouple.
  - Chemical control: Browne's tube or green spot which gives green color after proper sterilization. Autoclave tape can be used for example bowie dick tapes.
  - Biological control: A strip contains 10^6 spores of *Bacillus (Geobacillus) stearothermophilus* is packed in envelope and put into autoclave. After proper sterilization, remove the strip and inoculate in suitable medium and incubate at 55°C for 5 days for sterility testing.

2. **Hydroclaving**
  - **Principle:** Sterilization by indirect moist heat (no contact of article with steam) above 100°C under atmospheric pressure.
  - **Method of use of hydroclave:** It is an expansion of autoclave method. Keep all articles to be sterilized in inner jacket and steam is collected in outer jacket. Articles turn mechanically with the help of a series of large rotating rod, which spin continuously rupturing the waste bags and ensuring complete exposure to heat.
  - **Articles to be sterilized are:** Same as autoclave.
  - **Advantages:** It hydrolyses the organic matter. No pretreatment of waste is required. It reduces volume and weight. More heat penetrating than autoclave. It is economical.

Note: Steam sterilizer
Types: Two types as mentioned in **Table 30.4** with differences.

TABLE 30.4: Types of steam sterilizers

Features	At atmospheric pressure	Above atmospheric pressure
Examples	• Koch or Arnold's steam sterilizer • Tyndallization	Autoclave
Instrument's lid	Perforated, so no steam collection in cabinet	Not-perforated, so steam collection in cabinet
Temperature	At 100°C	>100°C
Pressure	Not rise in cabinet	Rise in cabinet due to steam collection
Supervision	Not required	Required

Principle: According to British pharmaceutical codex test, pore's size of filter prepared in such a way that retains the smallest bacteria like *S. marcescens*.

Uses: Filters are used to purify the water, separate exotoxin, separate viruses (viruses pass through the filters) from bacteria, measure the size of viruses, separate bacteriophages and to remove bacteria from heat labile liquids like sera, sugar or antibiotics solution.

Quality control: It is done by using smallest bacteria like *Serratia marcescens* and *Brevundimonas diminuta*.

Types of filters: Follow **Flowchart 30.2**.

1. Liquid filters

- **Depth filters:** These are porous filters that retain the particles throughout the depth of filter as shown in **Fig. 30.6a**. Depth filters are composed of metals, polymers and inorganic materials. In depth filters some particles still come out so not suitable for solutions contain small size bacteria. They are composed of metals that may be carcinogenic like asbestos filters. They retain the large mass of particles before becoming clogged. Flow rate of fluid in depth filter is high. They are economical and useful when fluid to be filtered contains high load of particles like filtration of fluid, beverages or chemicals in industries. Three subtypes of depth filter are described below.

  - Candle filters **(Fig. 30.7)**: They are made of special type of earth. They are useful for industrial and drinking water purification. Different types candle filters include unglazed ceramic filters like Chamberland filter, Doulton filter, etc., and diatomaceous earth filters like Berkefeld filter, Mandler filter, etc.

  - Asbestos filters: These are disposable and single-used discs. They have high adsorbing capacity. They alkalinize the filtered liquid. They are rarely useful or use with care, because of carcinogenic property. For examples Seitz filter **(Fig. 30.8)** and sterimat filter.

  - Sintered glass filter **(Fig. 30.9)**: It is made up from heat fusing finely powered glass particles. It has low absorptive properties. It is brittle and expensive.

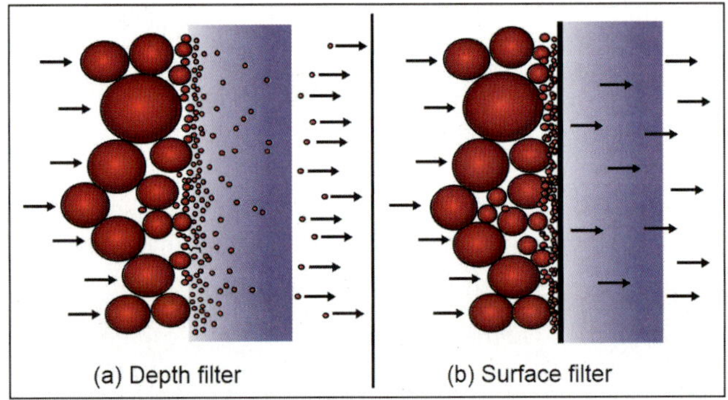

Fig. 30.6: Principle of liquid filters

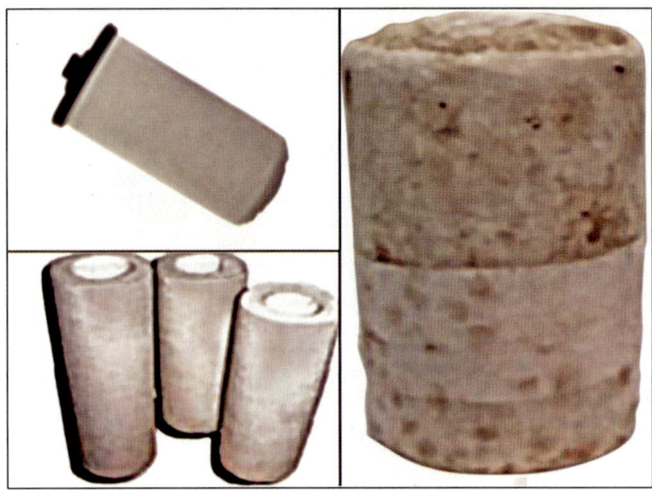

Fig. 30.7: Different types of candle filters

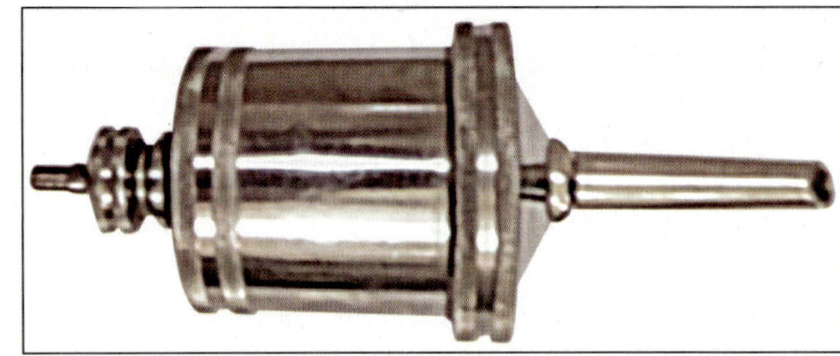

Fig. 30.8: Seitz (asbestos) filter

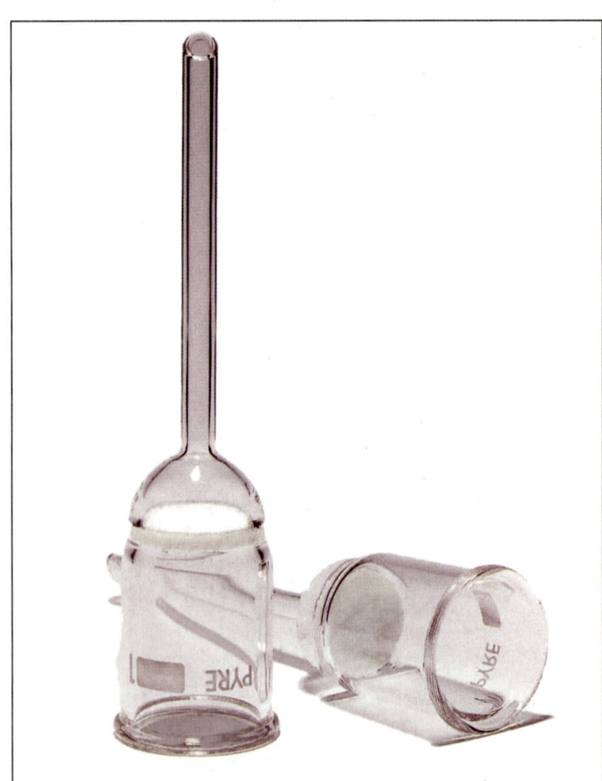

Fig. 30.9: Sintered glass filter

- **Surface filter:** It is a porous filter that retains all the particles on the surface as shown in **Fig. 30.6b**. It is made up of cellulose acetate, cellulose nitrate, polycarbonate, polyvinylidene fluoride or other synthetic materials. Average size of pores in membrane filter is 0.22 µm which remove almost all bacteria but allowing viruses to pass. Pores with 0.45 µm size are useful to retain coliform bacteria in water. Pores with 0.8 µm size are useful to remove microbes from air to prevent air borne infection and to produce the bacteria free gases. It is used for water purification and analysis, sterilization and sterility testing and for preparation of solutions for parenteral use. Examples of membrane filter are gradacol membrane filter and modern membrane filter.

2. **Air filters:** Following are the examples of air filters to deliver bacteria free air.
   - **Surgical mask:** It is the simplest example which removes 95% particles and delivers bacteria free air.
   - **HEPA filter:** It is the high-efficiency particulate air filter with pore size is ≥0.3 µm. It is used in safety cabinet to remove 99.97% of pathogens from air.
   - **ULPA filter:** It is ultra-low particulate/penetration air filter with pore size is ≥0.12 µm. It is used to remove 99.999% of dust, pollen, mold and pathogens from air.

Ultrasonic and Sonic Vibration

They have bactericidal power, but variable sensitivity to microorganisms and survivors have been found after such treatment, hence practically not useful.

Radiation

Two forms of radiation like ionizing and nonionizing. Efficacy of ionizing radiation is tested by *Bacillu pumilus*.

Ionizing radiation: It includes X-rays, γ-rays, cosmic rays and electron beams. Ionizing radiation has high penetrating power. It is lethal to DNA and other vital constitutes. It steriles the articles without increasing the temperature so called cold sterilization. Exposing object to the radiation especially to ionizing radiation called irradiation. γ-rays are best useful for this. Irradiation is done for disposable supplies like plastic syringes, bandages, catheters, gloves, etc., heat sensitive pharmaceuticals products like oils, grease, hormones, antibiotics, etc., cardboards, fabrics, metal foils, catgut sutures, bones or tissue grafts, adhesive dressings and food. Food is permitted in some countries.

Nonionizing: It includes ultraviolet/UV rays (280–200 nm) and infrared radiation. Nonionizing radiation steriles the articles with increasing the temperature so called hot air sterilization. Article sterilized by infrared radiation are disposable supplies (syringes, bandages, catheters and gloves) and by ultraviolet/UV rays (280–200 nm) are closed area like, hospital ward, laboratory, operation theater, etc.

Note: Cold sterilization
Sterilization of articles without increasing the temperature by radiation (ionising), chemicals (EO, formaldehyde, etc.), membrane or by other measures called cold sterilization.

Definition of Disinfection and Disinfectant

These are described above.

Criteria for Ideal Disinfectants

Ideal disinfectant should be effective against all micro-organisms such as bacteria, spores, viruses and fungi, active in presence of organic matter, equally effective in acid and in alkaline medium, with speedy action, with high penetrating power, compatible with other antiseptics and disinfectants, nontoxic if absorbed into circulation, inexpensive, easily available, safe, easy to use and stable. It should not corrode the metals, causes local irritation, sensitization or interferes with healing.

Effective Factors

Factors determining the potency of disinfectants are numbers of organisms, nature of organisms, time of exposure, pH of the medium, temperature, presence of extraneous materials and biofilms.

Mechanisms of Actions

Disinfectants act by different mechanisms like protein denaturation (alcohols, phenols, aldehydes and oxidants), protein coagulation (metallic salts), damage to cell membrane resulting in loss of contents (surfactant compounds and alcohols), removal of sulphhydryl (–SH) group required for enzymatic activities (EO and metallic salts), damage to RNA and DNA (aldehydes, oxidants and dyes) and substrate competition.

Types and Details of Disinfectants

Alcohols: Alcohols cause protein denaturation. They are bactericidal, fungicidal, tuberculicidal, virucidal, but not sporicidal. They have following uses.

1. **Ethanol 70% (ethyl alcohol):** Commonly called spirit (70%). It is used as skin antiseptic befor injection. It is safe to use on sites from where it can easily evaporate.
2. **2-Propanol (isopropanol) 70%:** It is used for themometer sterilization by soakig it in the solution for 10–15 minutes.
3. **1-Propanol (propanol):** Both 1-propanol and 2-propanol are often used as hand disinfectants due to their properties like high volatility, rapid effect, low toxicity and excellent bactericidal activity. Few commercially available hand disinfectants are sterillium **(Fig. 30.10a)**, casillium, etc.
4. **Methanol (methyl alcohol):** It is used for treating cabinets and incubators. Wipe inside of chamber by methanol and than put a pad moistened with methanol and water at room temperature for few hours. Methyl alcohol is toxic and inflammable. It is not useful for cut skin because it coagulates the protein under which bacteria can grow. It has no sporicidal activity.

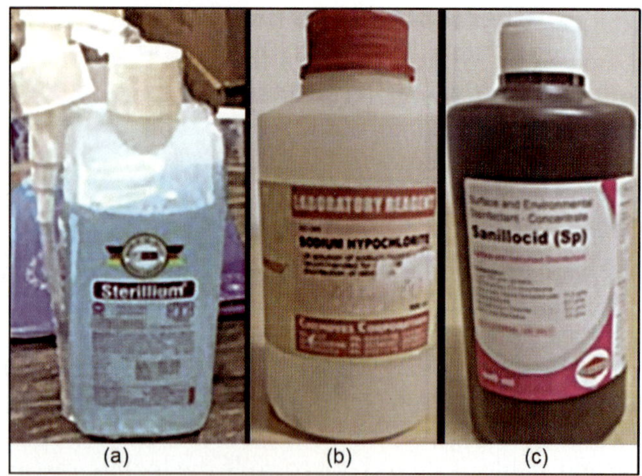

Fig. 30.10: Commercially available preparation of disinfectants commonly used in laboratory

Aldehydes: Aldehydes causing protein denaturation and also act as protoplasmic poison. They are bactericidal, fungicidal, virucidal and sporicidal.

1. **Formalin:** 37–40% aqueous solution of formaldehyde called formalin or formol. It is used to preserve biological specimens or anatomical viscera, in mortuaries for embalming, to prepare the toxoid from toxin and to destroy the spores of *Bacillus anthracis* in hairs and wools.
2. **Formaldehyde gas:** It is used for sterilization of instrument and heat sensitive catheter. It also used for fumigation as described in **Ch. 32**.
3. **Glutaraldehyde (2% solution called Cidex):** It is less irritating and more effective than formaldehyde. It is bactericidal, tuberculicidal, virucidal in 10 minutes and sporicidal in 3–10 hours (slow action). Commonly used to disinfect endoscopic instruments like bronchoscope, cystoscope, esophagoscope, etc., for 20 minutes. These instruments have lens which is heat sensitive so best sterilized by Cidex. It is used to treat corrugated rubber, anaesthetic rubber, face masks, plastic endotracheal tubes, metal instruments and polythene tubing.
4. **Ortho-phthalaldehyde:** It is available as 0.55% solution. It is used to sterile endoscopes and cystoscopes. It is more advantageous than glutaraldehyde, because it is high level disinfectant, requires low exposure at ambient temperature, has high mycobactericidal activity, does not required activation, has low vapor property, has better odor, has less toxic fumes and is more stable during storage.

Dyes: Two groups of dyes like aniline dyes and acridine dyes.

1. **Aniline dyes:** Aniline dyes are reacting with acid group of cells. They are bacteriostatic in high dilution but little bactericidal. They are more active against gram-positive (acidic pH), little or no effect on gram-negative organisms (alkaline pH) and no effect on *Mycobacterium* spp. Commonly useful aniline dyes are malachite green, brilliant green, crystal (gentian) violet, etc. Malachite green is used as selective agent in LJ medium. Aniline dyes are used as skin and wound antiseptic in boil, bed sore and Vincent's angina. It causes deep staining.

2. **Acridine dyes:** Acridine dyes impair the DNA of the organisms and thus kill the reproductive capacity of cell. They are bacteriostatic in high dilution, but little bactericidal. They are more active against gram-positive (acidic pH) organisms and little or no effect on gram-negative organisms (alkaline pH). They are not selective as like aniline dyes. Commonly useful acridine dyes are proflavine, acriflavine, euflavine, aminacrine, etc. They used as skin and wound antiseptic. They loose efficacy on light exposure, hence required to store in amber color bottle.

> **Note: Triple dye lotion (Triple DY)**
> **Contains:** 0.25% brilliant green + 0.25% crystal violet + 0.1% acriflavine.
> **Uses:** Used for burns and dressings of umbilical stumps.

Halogens

1. **Iodine:** It causes protein denaturation by oxidative effect. It is bactericidal, fungicidal, tuberculicidal, virucidal, and at higher concentration, it is sporicidal. It is more sporicidal than chlorine. At higher concentration, it causes burning and blistering of skin. In allergic individual, it causes rashes and systemic manifestations. It has following uses.
   - **Tincture iodine:** Tincture means solution prepared by dissolving drug in alcohol. It contains iodine in 2% alcohol. It used as skin disinfectant in cuts/abrasions.
   - **Iodophores (iodine and surfactant like povidone)/Povidone iodine (polyvinyl pyrrolidone and iodine/PVP-I:** Surfactant like povidone will acts as carrier and releases the free iodine. It used as disinfectant of skin, small wounds and candidal/trichomonal/nonspecific vaginitis. It also used for degerming the skin before the operation. Commercial preparations are betadine (5% cream/lotion, 7.5% scrub lotion and 200 mg vaginal pessary) and piodin (10% solution/cream and 1% mouth wash).

2. **Chlorine:** It causes protein denaturation by oxidative effect. It is bactericidal, virucidal and sporicidal (at higher concentration), but not tuberculicidal. It has few disadvantages like it is an irritant, corrosive for metals and textiles, mixer with acid will liberate toxic gases and an inactivated by organic matters. It has following uses.
   - **Chlorine:** It is used for chlorination of water by chlorine tablets:
   - **Chlorophores:** It includes two types of preparations like chlorinated lime (bleaching powder) and sodium hypochlorite **(Fig. 30.10b)**. Chlorinated lime (bleaching powder) is obtained by action of chlorine on lime. It is used for chlorination of

drinking water, swimming pools, food and dairy industries. Sodium hypochlorite is used as 1000 ppm (parts per million) for surface cleaning, 10,000 ppm for spillage (blood/body fluid) and 2500 ppm for discarding container. 1000 ppm means 0.1% (1 g/liter). Sodium hypochlorite also used for disinfection of equipments soiled with blood.

Phenols or phenol group: Phenol is obtained from distillation of coal tar between 170 and 270°C. It damages the cell membrane causing releasing of cell contents and cell lysis so called protoplasmic poison. Low concentration of phenol will precipitate the proteins. Bactericidal effect occurs by inactivation of membrane bound enzymes (dehydrogenase and oxidase). It is bacteriostatic at 0.2% and bactericidal at 1% but not sporicidal. Action does not affected by organic matters. Higher concentration of phenol causes skin burns. It has following uses:

1. **Phenol (carbolic acid):** It was 1st used by Joseph Lister as antiseptic agents in surgery (1865). It is the earliest disinfectant, seldom use now. It is very cheap and used for disinfection of urine, sputum or feces of patients. It is included in antipruritic lotion because of mild local anesthetic action.
2. **Methyl phenol or cresol:** Its commercial available preparation lysol contains 50% cresol. It is used for disinfection of utensils, excreta and handwash.
3. **Chloroxylenol:** Following are the different preparations and uses.
   - Dettol: It contains 4.8% chloroxylenol + 9% terpeniol + 13% alcohol and used as an antiseptic agent.
   - Dettolin: It contains 1% chloroxylenol and used for mouthwash.
   - 0.8% skin cream and soap
   - 1.4% obstetric cream: It used as an antiseptic during vaginal examination applied on forceps.
   - Hexacholophane: 2% is incorporated in soaps, surgical scrubs, deodorants and toilet products.
   - Orho-phenyl phenol (2- phenyl phenol or O-phenyl phenol or biphenol): 0.1% in lysol for hard and soft surface.

Gases

1. **Ethylene oxide (EO):** It alkylates the amino, carboxyl, hydroxyl and sulfhydryl groups in protein molecules. Also acts on DNA and RNA. Explosive tendency of EO is decreased by 10% CO_2/N_2 and water vapor increase its efficacy. Two types of cycles for sterilization by EO like cold cycle at 37±5°C and warm cycle at 54±5°C. In these cycles, humidity is maintained by 40–50%, and EO concentration is kept at 700 mg per liter. It is bactericidal, fungicidal, tuberculicidal, virucidal and sporicidal. Disadvantages of EO are inflammable, irritant, mutagenic, carcinogenic and explosive at concentration of >3%, which is overcome by mixing it with 10% CO_2/N_2. It is a high level disinfectant used to sterilize the heat-sensitive metal instruments (heart-lung machines, respirators, dental equipment), glasses, plastics (sutures, disposable syringe, etc.), clothes, papers (books), soil, foods and tobacco items. Water vapor increases its efficacy.
2. **Formaldehyde gas:** Described above.
3. **BPL:** Product of ketane and formaldehyde with a boiling point of 163°C. It is capable of killing all microorganisms and very active against viruses. Earlier, it was used for fumigation. It is carcinogenic.
4. **H_2O_2 fogging:** It is nontoxic and has short cycle. It is used for fumigation by fogging machine.

Biguanide: Commonly used biguanide is chlorhexidine (hibitane). It causes disruption of bacterial cell membrane and protein denaturation. It is active against gram-positive bacteria. It causes brownish discoloration of teeth. It has following uses:

- 0.5–1% tooth paste is used to prevent and to treat the gingivitis.
- 0.12–0.2% oral rinse reduces and prevents the oral infections, even in AIDS patients.
- Savlon (described above) is used as skin and obstetric antiseptic, surgical scrub, neonatal bath, mouth wash, etc.

Surface active agents: Substances that alter the energy relationship at interfaces or reduce surface or interfacial tension are called surface active agents. They are also known as surfactants, tensides or detergents. They have slow action and form film on skin and porous material (cotton, poly ethylene) under which bacteria can grow. Following are the different groups of surface active agents.

1. **Cationic (quaternary ammonium) compound:** It causes protein denaturation and change in permeability of cell membrane by interfering with phosphate group. It is bactericidal (only for gram-positive bacteria), fungicidal, tuberculicidal, sporicidal, but not virucidal and no effect on *Pseudomonas aeruginosa*. Following are the different preparations and uses of cations.
   - **Cetrimide or acetyl trimethyl ammonium bromide:** It is used alone or along with chlorhexidine. It has no effect on *Pseudomonas* and used as selective agent in cetrimide agar for isolation of *Pseudomonas*. It is used as an antiseptic for surgical instruments, gloves, utensils, baths, etc. Following are the different preparations of cetrimide.
     – Cetavlon contains 20% cetrimide.
     – Savlon liquid contains 3% cetrimide and 1.5% chlorhexidine gluconate.
     – Savlon cream contains 0.5% cetrimide and 0.1% chlorhexidine HCL.
     – Savlon hospital concentrate contains 15% cetrimide and 7.5% chlorhexidine gluconate.
   - **Benzalkonium chloride:** It is used as an antiseptic for surgical instruments. Along with glutaraldehyde and other agents (sanillocid (sp) → **Fig. 30.10c** or bacillocid) it used as surface and environmental disinfectants.

2. **Anionic/soap:** Soap can be prepared from other chemicals also. Soap prepared from saturated fatty acid (coconut oil) acts only on gram-positive bacteria. Soap prepared from unsaturated fatty acid (oleic acid) acts only on gram-negative bacteria and *Neisseria*. It has cleaning effect and more effective in warm water.

3. **Amphoteric or ampholytic or Tego compounds:** It acts on gram-positive and gram-negative bacteria and viruses, but not in general use.

Acids: They are bacteriostatic. Systemic absorption of acid causes vomiting, diarrhea, abdominal pain, visual disturbances and kidney damage. Following are the different uses of acids.

1. **Boric acid**
   - 4% solution used for irrigation of eyes, mouthwash and douche.
   - 30% boroglycerine paint used for stomatitis and glossitis.
   - 10% ointment used for cuts and abrasions.
   - It also included in prickly heat powder.

2. **Acetic acid:** 1–3% used for douche and 5% used to prevent the *Pseudomonas* infection in burns.

Metallic salts

1. **Mercury:** It reacts with sulfhydryl (-SH) group which is required for enzymatic action. It is more bacteriostatic least bactericidal and fungicidal. Mercury compounds like mercury chloride, thiomersal, phenyl mercury nitrate and mercurochrome, merthiolate are used as antiseptic agents; however, due to high toxicity and environmental hazards, they are not suggested.

2. **Silver:** It has astringent effect (precipitation of protein). It reacts with –SH, COOH, PO_4 and NH_2 groups of proteins. It is active against *N. gonorrhoeae* and *P. aeruginosa*. Tissue gets black due to deposition of reduced silver. Uses of silver compounds are like
   - 1% silver nitrate solution in ophthalmia neonatorum (caused by *N. gonorrhoeae*).
   - 1% silver nitrate touch in aphthous ulcer and hypertrophied tonsillitis.
   - 1% silver sulfadiazine in burns (prevent *P. aeruginosa* infection).
   - Water stored in silver vessels is said to become sterile, even at low silver ions concentration called oligodynamic action (very tiny amount is effective).

3. **Zinc:** It has astringent effect (precipitation of protein). Following are the different preparations and uses of zinc compound.
   - Zinc sulfate: 0.1–1%% solution used for irrigation of eyes and eye/ear drops. White lotion contains 4% zinc sulfate and 4% sulfurated potash is used for acne and impetigo.
   - Zinc chloride is used for mouthwash.
   - Zinc oxide is used as antifungal agent in paints.

4. **Copper:** Copper sulfate is used to kill algae in pools and fish tanks.

5. **Selenium:** It kills fungi plus fungal spores and prevents fungal infections. It also used in shampoos for dandruff.

6. **Calamine:** It is dermal protective and adsorbent. It is used in oily skin.

Testing of Disinfectants

Following are the various tests developed to know the efficacy of disinfectants, but none of them is satisfactory.

Rideal walker (phenol coefficient) test: Suspensions containing various numbers of *S. Typhi* are submitted to the action of varying concentration of phenol and test disinfectant. Concentration of test disinfectant which sterile the *S. Typhi* is divided by corresponding concentration of phenol called phenol coefficient (phenol = 1). This test has certain limitations like disinfectant acts directly without presence of any organic matter which does not reflect the natural condition.

Chick Martin test: It is the modification of Rideal Walker test in which disinfectant acts in presence of organic matter as like yeast or feces. It does not reflect the natural condition.

Capacity (Kelsey-Sykes) test: It tests the capacity of disinfectant to retain its activity when repeatedly used microbiologically (when microbiological load keeps increasing).

In-use (Kelsey and Maurer) test: It detects the capacity of disinfectant used in laboratory or in hospital. It detects the ability of disinfectant to inactivate the known number of standard strain of *Staphylococcus* on a given surface within a certain time.

Levels of Disinfectants

There are three levels of disinfectants like high, intermediate and low as mentioned in **Table 30.6**. High level disinfectants kill all organisms and also spore if contact time is increased. Intermediate level disinfectants kill all organisms, *Mycobacterium* spp., naked (non-enveloped) viruses, but not spores. Low level disinfectants kill vegetative bacteria, fungi and lipid enveloped viruses.

Chemosterilization (Physical and Chemical Methods)

Introduction

Sterilization by chemical agents called chemosterilization and such agents called chemosterilizers. They are useful for heat sensitive articles.

Methods

Gases: As described above.

Plasma sterilization (newer method)
- **Principle:** It consists ions, electrons or neutral particles. Radio frequency energy is applied to create an electromagnetic field. Later H_2O_2 vapor is introduced, which generates a state of plasma contains free radicals of H_2 and O_2. This state sterilizes the articles.

TABLE 30.5: Biological indicators

Methods	Control (Indicators)
Hot air oven	10^6 spores of *Cl. tetani* or *Bacillus subtilis* subsp., *niger*
Autoclave	10^6 spores of *Bacillus (Geobacillus) stearo-thermophilus*
Filtration	*Serratia marcescens* or *Brevundimonas diminuta*
Ionizing radiation	*Bacillus pumilus*
Ethylene oxide	*Bacillus atrophaeus*
Plasma sterilization	*Bacillus (Geobacillus) stearothermophilus* or *Bacillus subtilis* subsp., *niger*

- **Biological control:** Follow **Table 30.5**.
- **Uses:** Arthroscope, bronchoscope, urethroscope, etc.

Duckering: Spores of *Bacillus anthracis* are destroyed from animal products like from wool (imported in to non-endemic countries) by 2% formaldehyde at 30–40°C in 20 minutes and from hairs and bristles by 25% formaldehyde at 60°C in 6 hours. This method is called duckering.

BIOLOGICAL CONTROL OR INDICATORS

They are used to assess the efficacy of sterilization method. Examples of different indicators are mentioned in **Table 30.5**. Put the indicator with the load into the sterilizer. During this process the organism are killed. After sterilization, remove the indicator and inoculate in to different media. No growth indicates the complete sterilization.

SELECTION OF STERILIZATION AND DISINFECTION IN HEALTHCARE SET-UP

Aims

Uses of contaminated equipments considered as source of infection, so they are completely sterilized and disinfected before use.

Spaulding's Classification

Splaundig classify such equipments in three categories and appropriate methods required for sterilization and disinfection are shown in **Table 30.6**.

ACCESS YOURSELF

Essay/Full Question

1. Classify and describe the different methods of sterilization and disinfection. Choose the most appropriate method of sterilization and disinfection to be used in the laboratory, clinical and surgical practise.

Short Notes

1. Classify methods of sterilization and disinfection.
2. Disinfection.
3. Dry heat sterilization.
4. Moist heat sterilization below 100°C/at 100°C/above 100°C.
5. Pasteurization/hot air oven/fractional sterilization/autoclave/steam sterilizer/filtration.

Short Questions for Theory/Viva Questions

1. Define: Sterilization and disinfection.
2. Write the differences between sterilization and disinfection.
3. Name the articles sterilized by hot air oven.
4. Write four names of sterilization methods by moist heat below 100°C.
5. Write two examples of fractional sterilization methods.
6. Name the biological control of hot air oven, autoclave, filtration and ionizing radiation.
7. What is pasteurization? Name two methods of pasteurization.
8. What is cold sterilization? Name two agents used in cold sterilization.
9. Write four examples of high level disinfectants.

Comments on

1. Betadine (povidone-iodine) is considered as best skin disinfectant.
2. Moist heat is more advantageous than dry heat.
3. Three successive days are required for sterilization of articles by inspissation (or write the principle of inspissation).
4. Asbestos filters are rarely useful or should be use with care.
5. Sterilization by radiation (ionizing) or chemicals (disinfectants) is also known as cold sterilization.
6. Ortho-phthalaldehyde is more advantageous than glutaraldehyde.
7. Cetrimide is used as selective agent in cetrimide agar for isolation of *Pseudomonas*.

MCQs for Chapter Review

Introduction

1. 'Disinfections' kills the following:
   a. All microorganisms
   b. Pathogenic microorganisms
   c. Spores
   d. Non-pathogenic microorganisms

TABLE 30.6: Spaulding's classification

Category of materials	Tissue affected	Equipments	Level of disinfectant	Sterilization and disinfection method
Critical materials	Sterile tissues or vascular system	Cardiac catheter, implants	High level	Heat resistant articles—autoclave
			High level	Heat sensitive articles—plasma sterilization, H_2O_2 fogging, EO gas
Semicritical materials	Mucosa and non-intact skin	Endoscope	High level	2% Glutaraldehyde
		Thermometer	Intermediate level	Ethanol
Noncritical	Intact skin	BP cuffs	Low level	Cationic compounds

Essentials of Medical Microbiology

2. Asepsis means:
a. Absence of microbes
b. Disinfection of surface
c. Prevention of infection
d. Destroying all forms of microbes

3. An 'antiseptic' means:
a. An agent applied on skin to eradicate pathogenic microbes
b. Used to sterilize inanimate objects
c. An agent which kills only bacteria but not spores
d. Kills all microorganisms

4. The best skin disinfectant is:
a. Alcohol
b. Savlon
c. Betadine
d. Phenol

5. Which of the following can be readily used for hand washing?
a. Chlorhexidine
b. Isopropyl alcohol
c. Lysol
c. Cresol
d. Glutaraldehyde

6. Agent, which on addition to a colony, inhibits its growth and on a removal the colony regrows is?
a. Bacteriostatic
b. Bactericidal
c. Antibiotic
d. Antiseptic

7. Which of the following is most resistant to sterilization?
a. Cysts
b. Prions
c. Spores
c. Viruses

8. Choose the correct ones for the decreasing order of resistant to sterilization?
a. Prions, bacterial spores, bacteria
b. Bacterial spores, bacteria, prions
c. Bacteria, prions, bacterial spores
d. Prions, bacteria, bacterial spores
e. Bacterial spores, prions, bacteria

Physical Methods (Sterilization)

9. All are methods of sterilization by dry heat *except*:
a. Flaming
b. Incineration
c. Hot air oven
d. Autoclaving

10. Tyndallization is a type of:
a. Intermittent sterilization
b. Pasteurization
c. Boiling
d. Autoclaving

11. The best method of sterilization of dusting powder is:
a. Autoclaving
b. Hot air oven
c. Inspissation
d. Tyndallization

12. Glass vessels and syringes are best sterilized by:
a. Hot air oven
b. Autoclaving
c. Irradiation
d. Ethylene oxide

13. Which of the following is true about pasteurization?
a. It kills all bacteria and spores
b. It kills all bacteria except thermoduric bacteria
c. It kills 95% of microorganisms
d. All bacteria are destroyed

14. Holding temperature and time for pasteurization by Holder method is:
a. 60°C for 30 seconds
b. 72°C for 30 minutes
c. 72°C for 15–20 seconds
d. 60°C for 30 minutes

15. Vaccines are best sterilized by
a. Seitz filtration
b. Hot air oven
c. Autoclaving
c. Heat inactivation

16. Autoclaving is done in:
a. Dry air at 121°C and 15 lbs pressure
b. Steam at 100°C for 30 minutes
c. Steam at 121°C for 15 minutes
d. Dry air at 160°C for 30 minutes

17. Sterilization by autoclave is work under principle of:
a. Moist heat at 100°C
b. Moist heat above 100°C
c. Moist heat above 100°C and above atmospheric pressure
d. Moist heat below 100°C

18. Browne's tube is used for:
a. Steam sterilization
b. Radiation
c. Chemical sterilization
d. Filtration

19. Out of the following the true statement regarding sterilization is:
a. Dry heat is the best method of sterilization of liquid paraffin
b. All glass wares are best sterilized by boiling at 100°C
c. Bacterial vaccines are best sterilized by ethylene oxide
d. Pasteurization of milk by flash method is done by heating at 63°C for 30 minutes

20. Sterilization of culture media contain serum is done by:
a. Autoclaving
b. Micropore filter
c. Gamma radiation
d. Centrifugation

21. Which is a form of cold sterilization?
a. Gamma rays
b. Beta rays
c. Infrared rays
d. Autoclave

22. Irradiation can be used to sterilize A/E:
A/E = All Except
a. Bone graft
b. Suture
c. Artificial tissue graft
d. Bronchoscope

23. Sterilization method for catgut suture:
a. Steam
b. Radiation
c. Boiling
d. Burning

Chemical Methods (Disinfections)

24. False about alcohol in disinfection is:
a. Ethanol is used
b. Isopropyl alcohol is used
c. Has sporicidal activity
d. Has bactericidal activity

25. 40% formalin is used to sterilize:
a. Plastic syringe
b. All microbes + spores
c. Clothes
d. Stitches

26. Bronchoscope is sterilized by:
a. 2% Glutaraldehyde
b. Formaldehyde
c. Autoclave
d. Carbolic acid

27. Percentage of glutaraldehyde used:
a. 1%
b. 2%
c. 3%
d. 4%

28. The operating temperature in an ethylene oxide sterilization during a warm cycle is:
a. 20–35°C
b. 49–63°C
c. 68–88°C
d. 92–110°C

29. Heat labile instruments for use in surgical procedure can be best sterilized by:
a. Absolute alcohol
b. Ultraviolet rays
c. Chlorine releasing compounds
d. Ethylene oxide gas

30. **Which of the following is an important disinfectant on account of effectively destroying gram-positive and gram-negative bacteria, viruses and even spores at low pH level?**
    a. Phenol
    b. Alcohol
    c. Chlorine
    d. Hexachorophene

31. **Disinfection of sputum is done by:**
    a. Boiling
    b. Autoclaving
    c. Sunlight
    d. Burning
    e. Airing

32. **Phenolic disinfectants are:**
    a. Dettol
    b. Cresol
    c. Lysol
    d. Carbolic acid
    e. Savlon

33. **All of the sterilization methods are properly matched** *except*:
    a. Cat gut suture–Radiation
    b. Culture media–Autoclaving
    c. Bronchoscope–Autoclaving
    d. Glassware and syringes–Hot air oven

34. **Which of the following statement regarding disinfectants is not true?**
    a. Hypochlorites are bactericidal and inactivated by organic matter
    b. Glutaraldehyde is sporicidal and not inactivated by organic matter
    c. Formaldehyde is bactericidal, sporicidal and virucidal
    d. Phenol is bactericidal and readily inactivated by organic matter

35. **All are true regarding disinfectants** *except*:
    a. Glutaraldehyde is sporicidal
    b. Hypochlorites are virucidal
    c. Ethylene oxide is intermediate disinfectant
    d. Phenol usually requires organic matter to act

36. **Disposable plastic syringe is sterilized by:**
    a. Ethylene oxide
    b. Ionizing radiation
    c. Nonionizing radiation
    d. Spirit

37. **Operation theater is sterilized by:**
    a. Carbolic acid spraying
    b. Washing with soap and water
    c. Formaldehyde
    d. Ethylene oxide gas

38. **Phenol coefficient indicates:**
    a. Efficacy of a disinfectant
    b. Dilution of a disinfectant
    c. Quantity of a disinfectant
    d. Purity of a disinfectant

39. **Rideal–Walker (phenol test) coefficient is related with:**
    a. Disinfecting power
    b. Parasitic clearance
    c. Dietary equipment
    d. Statistical correlation

Biological Control or Indicators

40. **Indicator used in autoclave is:**
    a. *Clostridium tetani*
    b. *Bacillus stearothermophilus*
    c. *Bacillus pumilus*
    d. *Bacillus subtilis* var *niger*

41. **Plasma sterilization accuracy is assessed by using:**
    a. *Bacillus subtilis*
    b. *Geobacillus stearothermophilus*
    c. *Staphylococcus aureus*
    d. *Clostridium tetani*

42. **According to Splauding classification system of sterilization, following is true** *except*:
    a. "Non-critical devices" come in to contact with intact skin
    b. Semicritical equipments need low level sterilization
    c. "Semicritical devices" come in to contact with non-sterile mucous membrane or non-intact skin
    d. Cardiac catheter is critical equipment

43. **Following is/are right about high level disinfectant:**
    a. Two subtypes like heat sensitive and heat resistant
    b. Heat sensitive includes autoclave and heat resistant includes plasma sterilization, H_2O_2 fogging and EO gas
    c. Ethanol is high level disinfectant
    d. None of above

Answers and Explanation of MCQs

1. **b and d**
   - Follow section, **introduction (definitions) and Table 30.1** more explanation.
2. **a**
   - Follow section, **introduction (definitions → asepsis)** for explanation.
3. **a**
   - Antiseptic is also called skin disinfectant and can be applied on skin–mucosa to inhibit the growth of microbes.
4. **c**
5. **a, b**
   - Follow section, **introduction (definitions → antiseptics)** for explanation of answers of MCQs 4–5.
6. **a**
   - Bacteriostatic agents can inhibit the growth of microbes which is reversible and on removal organisms can regrow while bactericidal agents can kill the microbes which is irreversible and on removal organisms cannot regrow.
7. **b**
8. **a**
   - Follow section, **introduction (order of resistance of microbes) and Flowchart 30.1** for explanation of answers of MCQs 7–8.
9. **d**
10. **a**
    - Follow **Flowchart 30.2** for more explanation of answers of MCQs 9–10.
11. **b**
12. **a**
    - Follow section, **physical methods (dry heat → Hot air oven → articles to be sterilized)** for explanation of answers of MCQs 11–12.
13. **b**
    - Follow section, **physical methods (moist heat → Pasteurization)** for explanation.
14. **d**
    - Holding temperature and time for Pasteurization by Holder method is 60°C for 30 minutes.
15. **a**
    - Vaccines are best sterilized by filtration and vaccine bath (non-sporing bacterial vaccines).
16. **c**
    - Follow **Table 30.3** for more explanation.
17. **c**
    - Autoclave works under principle of sterilization by moist heat above 100°C above atmospheric pressure.

18. (a)
- Browne's tube is used for autoclave which is a type of steam sterilizer.

19. a
- All glass wares are best sterilized by hot air oven.
- Bacterial vaccines are best sterilized by filtration and vaccine bath (nonsporing bacterial vaccines).
- Pasteurization of milk by flash method is done by heating at 72°C for 15–20 seconds.

20. b
- Heat labile liquids like sera, sugar or antibiotics solution are sterilized by filtration.

21. a
- Follow section, **physical methods (radiation → ionizing radiation)** for explanation.

22. d
- Bone graft, suture and artificial tissue graft are sterilized by irradiation. Follow section, **physical methods (radiation → ionizing radiation)** for more explanation.
- Bronchoscope is sterilized by Cidex (2% glutaraldehyde).

23. b
- Cat gut suture is sterilized by irradiation. Follow section **physical methods (radiation → ionizing radiation)** for more explanation.

24. c
- Alcohol is bactericidal, fungicidal, tuberculicidal, virucidal, but not sporicidal.

25. b
- Formalin is bactericidal, fungicidal, virucidal and sporicidal.

26. a

27. b
- Follow section, **chemical methods (aldehyde → glutaraldehyde)** for explanation of answers of MCQs 26–27.

28. b

29. d
- Follow section, **chemical methods (gases → ethylene oxide)** for explanation of answers of MCQs 28–29.

30. c
- Phenol, alcohol and hexachorophene are not sporicidals while chlorine is sporicidal.

31. a, b and d
- Sputum can be initially disinfected by 5% phenol followed by boiling, autoclaving or burning (incineration). All sterile materials should follow to deep burial.

32. a, b, c and d
- Follow section, **chemical methods (Phenol)** for explanation.
- Savlon contains 3% cetrimide + 1.5% chlorhexidine gluconate.

33. c
- Bronchoscope is sterilized by Cidex (2% Glutaraldehyde).

34. d
- Phenol is bactericidal and not inactivated by organic matter.

35. c and d
- Ethylene oxide is high level disinfectant. Follow **Table 30.6** for more explanation.
- Phenol does not require organic matter to act.

36. a and b
- Disposable plastic syringe is sterilized by ethylene oxide and ionizing radiation.

37. c
- Follow section, **chemical methods (gases → formaldehyde gas)** for explanation.

38. a

39. a
- Follow section, **chemical methods (testing of disinfectant)** for explanation of answers of MCQs 38–39. Different tests are mentioned to test the efficacy/power of disinfectant.

40. b

41. b
- Follow **Table 30.5** for more explanation of answers of MCQs 40–41.

42. b
- Semi-critical equipment needs high or intermediate level sterilization. Follow **Table 30.6** for explanation.

43. a
- Follow **Table 30.6** for explanation.

Biomedical Waste (BMW) Management

Chapter Outline
- Introduction
- Types and Disposal of Waste

INTRODUCTION

Definitions

Hospital waste: This includes all waste coming out of hospital.

Biomedical waste (BMW): Solid or liquid waste including its container, generated during the diagnosis, treatment, prevention or research activities in human beings or animals by healthcare set-up or research facilities or slaughter houses or other similar settings called biomedical waste.

Objectives of Management

To prevent the injuries and accidental transmission of infection to hospital staffs (workers), patients, attendants (visitors), patient's relatives and persons working with waste disposal.

Sources of Waste

BMW generated from all hospitals like government, private, dental clinic, clinician's office, dispensaries, PHC, etc., medical research centers, training centers, blood banks, laboratories, animal houses, slaughter houses, vaccination centers, mortuaries and biotechnologies units.

Quantification of Waste

Different survey says that waste generation in India is 1/2–4 kg/bed/day in Government hospital, 1/2–2 kg/bed/day in private hospital and 1/2–1 kg/bed/day in nursing homes.

Health Hazards

BMW causes public sensitivity, infections transmission like HBV, HCV, HIV-AIDS, etc., breeding of rodents and vectors, genotoxicity, trauma from sharp items, corrosion from chemicals and radiation hazards like dizziness, vomiting, headache, etc., from X-ray and MRI plates.

BMW Regulation

Legal aspects of BMW include the Air (Control of Pollution and Prevention) Act, 1981, the Environment (Protection) Act, 1986, the Hazardous Waste (Management and Handling) Rules, 1989, the National Environmental Tribunal Act, 1995 and Bio-medical Waste (Management and Handling) Rules, 1998.

BMW rules were published vide notification number S.O. 630(E) dated 20th July. 1998, by the Government of India in the erstwhile Ministry of Environment and Forests (MoEF), provided a regulatory framework for management of biomedical waste generated in country.

Bio-medical Waste Rule (Draft), **2011**: The Ministry of Environment and Forests (MoEF) had proposed a revised draft of Bio-medical Waste Rules 2011. It is simpler and containing 8 categories of BMW; each has to be collected and segregated in different color coded bag, thus clears the confusion over the color coding of containers used for disposal of BMW rules, 1998. However, still it is under consideration and not enforced yet.

Bio-medical Waste Management Rules, **2016**: They were published in the Gazette of India, Extraordinary, Part-II, Section 3, Subsection (i), Government of India, Ministry of Environment, Forests and Climate Change as a notification on March 2016 (detail available at http://mpcb.gov.in/bomedical/pdf/BMW_Rules_2016). The major changes are in the segregation in the color-coded bags as mentioned in **Table 31.1** and **Fig. 31.1**.

TYPES AND DISPOSAL OF WASTE

Types of Waste

Nonhazardous waste: It includes 85% of total waste mostly coming from housekeeping and administrative

Essentials of Medical Microbiology

TABLE 31.1: Categories according to Bio-medical Waste Management and Handling Rules, 1998

Color of bag/container	Types of BMW	Type of bag/container	Treatment and disposal
Yellow bags /containers **(Fig. 31.1a)**	Body organs (humans + animals)	Yellow colored nonchlorinated plastic bags	Incineration /plasma pyrolysis/ deep burial
	Soiled waste (items contaminated with blood/body fluid like swab, plaster casts, dressings and bags contain residual/discarded blood /blood components)		Incineration/plasma pyrolysis/deep burial*/autoclaving or microwaving/ hydroclaving followed by shredding or mutilation or combination of sterilization and shredding
	Expired/discarded medicines: Pharmaceutical waste like antibiotics, cytotoxic drugs (all items contaminated with cytotoxic drugs along with glass or plastic ampoules, vials, etc.	Yellow colored nonchlorinated plastic bags or containers	All other discarded medicines shall be either sent back to manufacturer or incinerated at temp., >1200°C
	Chemical solid waste	Yellow colored containers/ non-chlorinated plastic bags	Incineration or plasma pyrolysis or encapsulation
	Chemical liquid waste	Separate collection system leading to effluent treatment system	Chemical liquid waste shall be pre-treated before mixing with other waste water
	Microbiology waste and other laboratory waste like blood bags and vaccines	Autoclavable safe plastic bags/ containers	Pretreat to sterilize with non-chlorinated chemicals on-site as per NACO or WHO guidelines thereafter for incineration
	Discarded linen, mattresses, beddings contaminated with blood or body fluid, mask, cap, gown and show cover	Non-chlorinated yellow plastic bags or suitable packing materials	Non-chlorinated chemical disinfection followed by incineration or plasma pyrolysis. In absence of above facilities, shredding or mutilation or combination of sterilization and shredding
Red bags /containers **(Fig. 31.1b)**	Infectious plastic waste like tubes, catheters, vacutainers, syringes without needles, plastic apron, goggles, face shield, etc.	Red colored nonchlorinated plastic bags or containers	• Autoclaving or microwaving/ hydroclaving followed by shredding or mutilation • Sterilization and shredding • Treated waste to be sent to registered / authorized recyclers or for energy recovery or plastics to diesel or fuel oil or for road making, whichever is possible • Plastic waste should not be sent to landfill sites
White box **(Fig. 31.1c)**	Sharp metal waste like needles, syringes fixed with needles, scalpels, blades, etc.	Puncture proof and leak proof box (translucent)	Autoclaving or dry heat sterilization followed by shredding or mutilation or encapsulation in metal container or cement concrete; combination of shredding cum autoclaving; and sent for final disposal to iron foundries (having consent to operate from the State Pollution Control Boards or Pollution Control Committees) or sanitary landfill or designated concrete waste sharp pit
Blue cardboard box **(Fig. 31.1d)**	• Glass items like vials, ampoules, etc., contaminated with drugs except cytotoxic drugs • Metallic body implants	Puncture proof and leak proof cardboard box	Disinfection (by soaking the washed glass waste after cleaning with detergent and sodium hypochlorite treatment) or through autoclaving or microwaving or hydroclaving and then sent for recycling

Deep burial*: Disposal by deep burial is permitted only in rural or remote area where there is no access to common biomedical waste treatment facility. This will be carried out with prior approval from the prescribed authority and as per the standards specified in schedule-III. The deep burial facility shall be located as per the provisions and guidelines issued by Central Pollution Control Board from time to time.

Note: Points to remember
- BMW rules do not specify any specific color-coded bag for general waste. It depends on hospital policy. Mostly it is collected in black bag.
- Barcode system should be introduced for segregation and disposal.

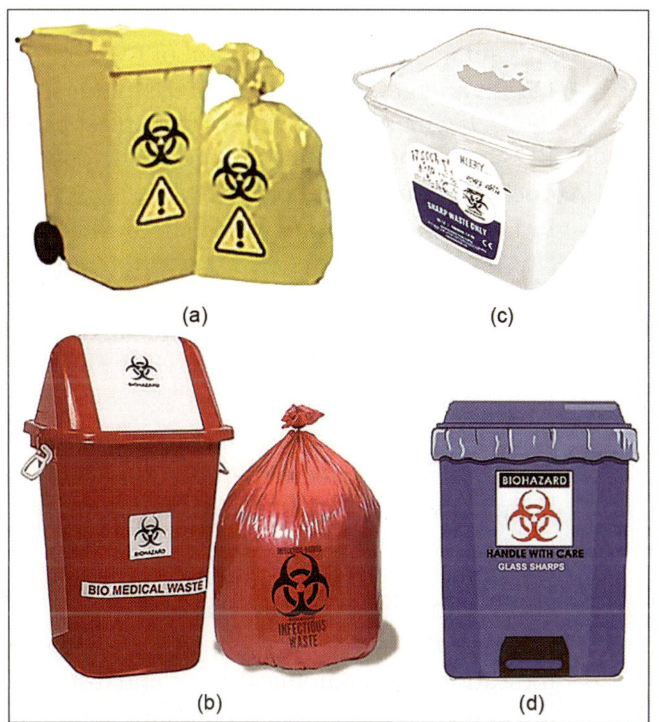

Fig. 31.1: Different color containers/bags for BMW management

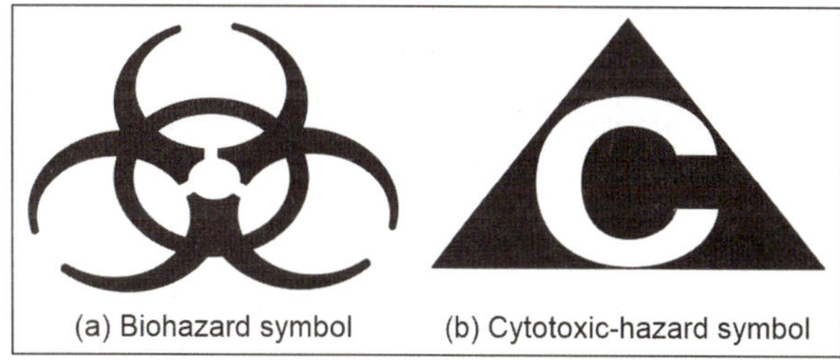

Fig. 31.2: Biohazard symbol and cytotoxic-hazard symbol

TABLE 31.2: Label for transport of biomedical waste containers/bags	
Day:__, Month ___, Year___	
Waste Category No.	Date of Generation
Waste Description: (e.g., wt in kg)	
Sender's Name and Address Phone No: Fax No.	Receiver's Name and Address Phone No. Fax No.
In case of emergency please Contact: Name and Address and Phone No.	

as shown in **Table 31.2**. Label shall be nonwashable and prominently visible.

Note: General municipal (nonhazardous) waste
Types and containers: Follow **Table 31.3 and Fig. 31.3**.

TABLE 31.3: Containers/bags and types of general municipal waste	
Color of container/bags	**Types of general waste**
Blue container (**Fig. 31.3a**)	Dry wastes like paper, foods, etc.
Green container (**Fig. 31.3b**)	Household waste

functions of hospital. For examples, general kitchen waste and office waste including food, papers, files, etc.

Hazardous waste: It includes 15% of total waste with following two categories.

1. **Noninfectious but hazardous:** It includes 5% of total waste like radioactive materials (X-ray plates), chemicals (corrosive), and pharmaceuticals materials (expired drugs), etc.
2. **Infectious and hazardous:** It includes 10% of total waste mostly generated during diagnosis, treatment or prevention of diseases. For examples, sharps (needles, syringes, glass test tubes), nonsharps (body tissues, organs, dressing materials), plastics disposables (plastic syringes, Ryle's tubes, catheters, etc), liquid waste (body fluid), etc.

Waste Disposal

Volume reduction: Reduction of volume by proper planning and using reusable items.

Collection and segregation: Biomedical waste should not be mixed with any other kind of waste. It should be collected and separated/segregated (duty of generator) at the point of generation in different color-coded bags/boxes/containers as suggested in **Table 31.1 and Fig. 31.1** before storage or transport. Bag should be sealed once filled and labeled with symbol of biohazard (**Fig. 31.2a**) or cytotoxic hazards (**Fig. 31.2b**) and with other details

Handling and transportation: There are two types of transport.
- **Intramural transport (internal):** It involves the movement of waste bag inside the hospital premises. Separate closed; cleaned and disinfected trolleys with biohazard symbol shall be used and it is not used for any other purposes. There must be a selection of waste transportation route and low activity timings (like post-OPD, post round in wards, etc.) to avoid

Fig. 31.3: Different color containers for general municipal waste

Biomedical Waste (BMW) Management

3

contact with patients. General waste shall not be transported with BMW.

- **Extramural transport (external):** It involves movement of waste for offsite treatment and/or disposal. The contractor is authorized for transport and disposal of waste. Handling and transfer lead closer contact with wastes, leading to high hazards. The transportation of clinical waste offsite shall be carried out in specially designed vehicles with a fully enclosed body and a bulk head separating the driver's compartment from the local compartment vehicle trolley. Accidental exposure chances are more during transportation, if proper care is not taken. Appropriate authorities should be informed in case of accidents occur during handling or transportation. The authorities should take the needful actions like provision of post-exposure prophylaxis.

Storage: No waste should be stored in the place of generation for more than two days, if needed one should take permission from competent authority. The storage area shall be, in a secured hospital location, limited access, cleaned, roofed, properly drained, rodent proof, insect proof with water supply, good lighting and passive ventilation. It shall not be situated near the food stores or food preparation areas or public places and marked with a biohazard symbol. This area shall be kept locked with key available to staff throughout 24 hours. Only authorized personnel are allowed to enter.

Treatment and disposal: The duration of generation of biomedical waste to the final disposal shall not exceed 48 hours including the temporary storage. Several methods are available but choice of method depends on type of waste.

1. **Liquid waste:** Following are two methods.
   - **Sewage treatment plant**
     - Indications: For liquid waste like blood, body fluids or hospital sewage.
     - Methods: Treat with disinfectants (e.g. 1–2% sodium hypochlorite) and drain in to sewer.
   - **Effluent treatment plant**
     - Indications: The liquid effluent generated during the process of washing of containers, vehicles, floors, etc.
     - Method: It is first treated by chemicals (1–2% sodium hypochlorite) and then disposed in effluent treatment plant.
2. **Solid waste:** Following are different methods.
   - **Sanitary land filling (deep burial)**
     - Indications: For sharp waste.
     - Method: It should be away from the residencies, forests and coastal waters. It is done in 2 meters depth pit. First, half fill the pit with waste followed by covering with lime within 50 cm of the surface and then remaining pit fill with soil.

- **Incineration**
  - Indications: It is useful for waste which is not reused or recycled like nonplastic infectious waste, anatomical waste, microbiological waste and disposable items. However, all such waste should be disinfected initially.
  - Method: Burning of waste till it becomes ash.
  - Advantage: Complete disposal of waste.
  - Disadvantages: Ideally it is not useful for sharp items. It generates toxic gases from plastics like PVC and expensive to maintain and to operate.
- **Autoclaves (for discard only)**
  - Indications: Infectious waste should be autoclaved before disposal. Horizontal type of autoclave is selected for this purpose.
  - Method: Same as described in **Ch. 30.**
- **Microwaving**
  - Indications: Plastic waste.
  - Method: Inactivation of microbes present in waste by electromagnetic radiation spectrum lying between 300–300,000 MHz. It is an inter-molecular heating process occurs inside the waste materials in the presence of steam. Its efficacy should be monitored regularly.
- **Plasma pyrolysis**
  - Indications: For anatomical waste and discarded linen, mattresses, beddings contaminated with blood or body fluid.
  - Method: It involves the use of ionized gas in the plasma state to convert electric energy to temperature of several thousand degrees using plasma arc torches or electrodes. The system provides high temperature combined with high UV radiation flux which destroys pathogens completely.
- **Inertization**
  - Indications: For chemical pharmaceutical waste.
  - Method: Mixing 65% pharmaceutical waste with 15% cement, 15% lime and 5% water in order to minimize the risk of toxic substances contained in waste.
- **Shredding**
  - Indications: For soiled waste, infectious plastic waste and sharp metal waste.
  - Method: It involves the cutting of waste in small pieces to make waste unrecognizable. It helps in prevention of reuse of biomedical waste and acts as identifier that the waste has been disinfected and is safe to dispose.
- **Encapsulation**
  - Indications: For chemical solid waste.
  - Method: It involves the filling of containers with waste, adding immobilizing materials and sealing the containers, to prevent the access to unscrupulous activities. The process uses cubic boxes made up of metallic drums, which are

three quarters filled with sharps or chemicals or pharmaceutical wastes and then filled with a medium such as plastic foam, cement mortar or clay materials.

Short Note

1. Define and classify hospital waste. Describe various methods of treatment of hospital waste.

Short Questions for Theory/Viva Questions

1. Draw the biohazards symbol and cytotoxic-hazard symbol.
2. Name the different types of bags for collection of biomedical waster and type of waste collected in them.

MCQs for Chapter Review

1. Microbiological waste is collected in:
   a. Yellow bag
   b. Blue bag
   c. Black bag
   d. Green container

2. Following is true method of disposal for plastic syringe without needle and plastic syringe with needle:
   a. Syringe without needle in yellow bag and syringe with needle in white puncture proof box
   b. Syringe without needle in red bag and syringe with needle in white puncture proof box
   c. Syringe without needle in black bag and syringe with needle in white puncture proof box
   d. Syringe without needle in red bag and syringe with needle in blue box

3. Sodium hypochlorite concentration for waste disposal is:
   a. 1–2%
   b. 5%
   c. 10%
   d. 15%

Answers and Explanation of MCQS

1. a

2. b

- Follow section, **waste disposal (Table 31.2)** for explanation of answers of MCQs 1–2.

3. a

- Follow section, **waste disposal (treatment and disposal → liquid waste)** for explanation.

Microbiology of Water, Milk and Air/Surfaces

Chapter Outline
- Microbiology of Water
- Microbiology of Milk
- Microbiology of Air/Surfaces

MICROBIOLOGY OF WATER

Drinking Water or Wholesome Water

Water that is fit to use for drinking, cooking, food preparation or washing without any potential of risk to human health and with following properties called drinking water or wholesome water.

- Biological properties: Free from all pathogenic micro-organisms.
- Chemical properties: Free from all chemicals like gases, metals, solvents, pesticides and hydrocarbons.
- Physical properties: Should be pleasant for taste, odor and color.

Microbiology of Water

Water flora (saprobes): These are the organisms which are normally present in water and nonpathogenic to humans. They are arising from decomposing organic matter. Their presence in water does not warrant threat. Different water floras include *Micrococcus, Pseudomonas, Serratia, Flavobacter, Alkaligenes, Acintobacter* and bacteriophages.

Waterborne pathogens: These are the organisms which are concerned with life threatening disease and their presence in water cause major threat for its use. These organisms are arising either due to sewage (feces, urine, etc.) contamination of water from animals or human source or by soil wash due to rainy water. They are causing either community acquired or hospital-acquired infections. Water pathogens with their source are mentioned in **Table 32.1**.

Indicator Organisms

Intestinal organisms which contaminate the water called indicator organisms. They are present in excess numbers, so they can be detected easily. They are more

TABLE 32.1: Waterborne pathogens	
Organism	**Disease**
Sewage source	
Escherichia coli	Diarrhea
Vibrio cholerae	Cholera
Clostridium perfrigens	Food poisoning and others
Salmonella spp.	Enteric fever and food poisoning
Enterococcus faecalis	Diarrhea
Proteus spp.	Diarrhea
Shigella spp.	Dysentery
Pseudomonas	Septicemia
Leptospira spp.	Leptospirosis
Entero viruses	Enteritis
E. histolytica	Amoebiasis
G. lamblia	Giardiasis
Cestodes like *T. saginata, T. solium,* etc.	Intestinal diseases
S. haematobium	Intestinal and urinary illness
Intestinal nematodes like *A. lumbricoides, S. stercoralis, A. duodenale, E. vermicularis,* etc.	Intestinal diseases
D. medinensis	Dracunculiasis
Soil source	
Bacillus spp.	Minor illness
Enterobacter spp.	Septicemia

resistant than other organisms to disinfectants. Their presence in water indicate that sewage contamination of water and water supplies need chlorination; however, mere presence of these indicator organisms does not mean the sewage contamination. Detection of indicator organisms gives ideas about fecal contamination of water (called fecal pollution).

Bacteriological Examinations of Water

It is done to detect the indicator organisms and water-borne pathogens (mentioned above) to rule out the fecal contamination. It is done daily to monthly depends on the size of population served.

Specimen collection: Heat sterilized screw capped bottle about 200 ml capacity should be used as container. Collect about 150 ml of water. Care should be taken to avoid the environmental or hand contamination of collecting person. Add sodium thiosulfate to neutralize the bactericidal effects of residual chlorine. Tap water is collected only after running it from tap for 2–3 minutes. To collect the water from stream or lake open the bottle after immersing it at the depth of 30 cm with mouth facing the current. Water from well is collected by bottle tide with heavy weight like stone.

Transport: Bottle should be properly labeled with date, time and source of collection and sent to the laboratory as soon as at least within 6 hours. If delay is anticipated, the bottle should be kept in ice box and protect from light.

Indicator organisms and testing methods: Following are the different tests and different microbes.

Tests to Detect the Bacteria

Plate count: Follow **Flowchart 32.1**.

Detection of coliform bacteria like *E coli*: Coliform means bacteria which ferment the lactose like *E coli*, *Klebsiella*, etc. They are detected by following tests.

1. **Presumptive coliforms count test (Fig. 32.1):** Test is called presumptive, because the final result may be due to organisms other than coliform category which further required confirmation. Most probable number (MPN) means coliforms count per 100 ml of water. It also called most probable number (MPN) test or multiple tube method. In this test water to be tested is diluted serially and inoculated in lactose broth. Coliforms if present in water ferment the lactose of medium to produce acid and gas. Acid is indicated by color change of the medium and gas is indicated by bubbles collection in the inverted Durham tube. The numbers of total coliforms are determined by counting the numbers of tubes giving

Microbiology of Water, Milk and Air/Surfaces

Flowchart 32.1: Plate count

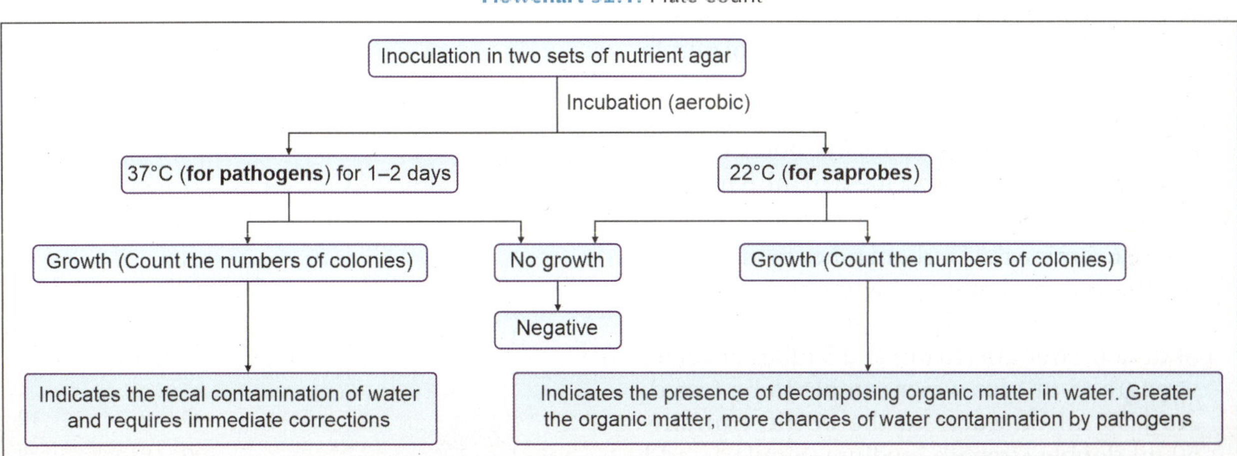

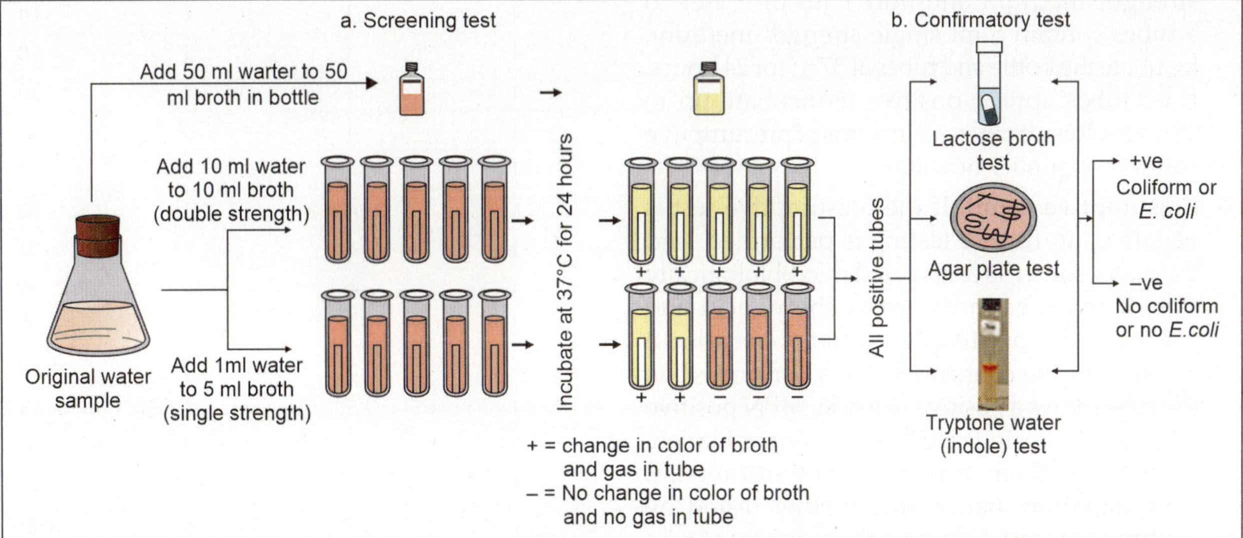

Fig. 32.1: Presumptive coliform count test

positive reaction (i.e., both color change and gas production) and comparing the pattern of positive results (the number of tubes showing growth at each dilution) with standard statistical tables. If the **presumptive test** is negative, no further testing is required, and the water source is considered microbiologically safe. If; however, any tube in the series shows acid and gas, the water is considered unsafe and then **confirmatory test** will be performed on the tube displaying a positive reaction. **Advantages** of MPN test include easy interpretation, either by observation or gas emission, dilution of sample toxins and effective analysis of highly turbid samples such as sediments, sludge, mud, etc., that cannot be analyzed by membrane filtration. **Disadvantages** of MPN test include inaccurate results, requirement of more hardware (glassware) and media and chances of false-positive results. Test is done in two steps like screening test and confirmatory test as follows.

- **Screening (presumptive) test for coliform**
  - Aims: Test is done to check whether the coliforms are present or not in water, to check the numbers of coliforms per 100 ml of water and to check the quality of water.
  - Requirements: (1) Lactose broth or MacConkey broth (with pH indicator like bromocresol purple or neutral red or others) or lauryl tryptose (lactose) broth in single and double strength concentration. (2) Glasswares like bottle (50 ml), test tubes of various capacities (10 ml, 5 ml, etc.) and Durham's tube. (3) Sterile pipette.
  - Steps: Examine the Durham tubes to make sure that they are full of liquid without air bubbles. Take 1 bottle of double strength (50 ml), 5 tubes of double strength (10 ml) and 5 tubes of single strength (5 ml) for water sample to be tested. Add 50 ml of water to the bottle contains 50 ml double strength medium. Similarly add 10 ml of water to 5 tubes contain 10 ml double strength medium and add 1 ml of water to 5 tubes contain 5 ml single strength medium. Incubate the bottle and tubes at 37°C for 24 hours. If no tubes appear positive re-incubate up to 48 hours. Results are read in terms of presumptive count and quality of water.
  - **Presumptive count:** If the presumptive test is negative, no further testing is performed, and the water source is considered microbiologically safe. If any tube in the series shows acid and gas, the test is positive and water is considered unsafe and the confirmed test is performed on the tube shows a positive reaction. MPN positive always does not indicate the fecal contamination, as some of them may be by environmental contamination; hence, it is further tested by confirmatory test. Compare the numbers of tube giving positive reaction (both color change and

gas production in Durham tube) to a standard **McCrady's probability table (Table 32.2)** and record the numbers of bacteria present in them.

TABLE 32.2: McCrady's probability table			
Numbers of tube giving appositive reaction out of			
1 bottle of 50 ml water	5 tube of 10 ml water	5 tube of 1 ml water	Bacterial count
0	0	0	0
0	0	1	1
0	0	2	2
0	1	0	1
0	1	1	2
0	1	1	2
0	1	2	3
0	2	0	2
0	2	1	3
0	2	2	4
0	3	0	3
0	3	1	5
0	4	0	5
1	0	0	1
1	0	1	3
1	0	2	4
1	0	3	6
1	1	0	3
1	1	1	5
1	1	2	7
1	1	3	9
1	2	0	5
1	2	1	7
1	2	2	10
1	2	3	12
1	3	0	8
1	3	1	11
1	3	2	14
1	3	3	18
1	3	4	20
1	4	0	13
1	4	1	17
1	4	2	20
1	4	3	30
1	4	4	35
1	4	5	40
1	5	0	25
1	5	1	35
1	5	2	50
1	5	3	90
1	5	4	160
1	5	5	180+

This is called presumptive coliform count or most probable number (MPN) of coliform present in tested water. Any further positive (positive after reincubation) test is added to the previous figures. For example, a water sample tested shows a result of 1–4 (2 × 10 ml positive, 2 × 5 ml positive) gives an MPN value of 4, means the water sample contains an estimated 4 coliforms per 100 ml.

 – **Quality of water:** The MPN is categorized as excellent, satisfactory, intermediate and unsatisfactory as mentioned in **Table 32.3**.

- **Confirmatory test or Eijkman test or differential coliform test:** Test is done to check whether the acid and gas positive screening test is due to coliform (*E. coli*) or by other bacteria. Other details are mentioned in **Table 32.4**.

2. **Membrane filtration method:** A measured volume of water is filtered through a membrane filter with a pore size of 22 µm. If bacteria present, they are retained on the surface of the filter, and later, they can grow over media on subsequent placing of membrane. Steps include filtering of water through membrane filter, which retains the water bacteria, if present. Place the membrane filter on media facing upwards. Incubate at appropriate temperature for 18 hours. Colonies that develop on the surface of the membrane are counted. After 18 hours of incubation presumptive coliform count and *E coli* count can be done.

3. **Enzyme method:** It based on detection of different enzymes of coliform and *E coli* like β-glucuronidase for fecal *E coli* and β-galactosidase for fecal coliform.

Detection of fecal streptococci: Fecal streptococci like *Enterococcus faecalis*, etc., is present in water for a very short time; hence, its detection in water suggests recent fecal contamination of water. It is detected by following methods.

1. **Glucose azide broth test:** All positive tubes of presumptive coliform test are subcultured in tubes contain 5 ml of glucose azide broth. Incubate at 45°C for 18–24 hours. Presence of *Enterococcus faecalis* is indicated by gas production within 18 hours. Confirmation is done by plating the positive tubes on MacConkey's agar or bile esculin agar. Specification is done to know the exact source of contamination like *E. faecalis* (human), *E. avium* (bird), *Strept. bovis* (cow) and *Strept. equinus* (horse).

2. **Membrane filtration method:** It is also useful for detection of fecal streptococci.

Detection of *Clostridium perfrigens*: *Cl. perfrigens* is present in water for a very long time; hence, its detection in water does not suggest recent fecal contamination of water. It is detected by litmus milk

TABLE 32.3: Quality of drinking water according to MPN

Quality of drinking water	MPN/100 ml of water	
	Coliforms/100 ml	*E. coli*/100 ml
Excellent	0	0
Satisfactory	1–3	0
Intermediate	4–9	0
Unsatisfactory	≥10	≥1

TABLE 32.4: Confirmatory tests of presumptive coliform count

Tests	Requirements	Steps	Results
Lactose broth test	3 ml lactose-broth or brilliant green lactose fermentation broth in tube	From all positive tubes (showing acid and gas) of screening test, transfer one loopful of medium to lactose-broth. Incubate the inoculated lactose-broth at 44°C (performed in thermostatically controlled water bath that do not deviate >0.5°C from 44°C) and examined after 24 ± 2 hours. If no gas production is seen, further incubate up to maximum of 48±3 hours to check gas production.	Gas production within the Durham's tube confirmed the *E. coli*.
Agar plate/slant test	Agar plate (slant)	From all positive tubes (showing acid and gas) of screening test, transfer one loopful of medium to agar plate or slant. The agar slants should be incubated at 37°C for 24± 2 hours.	Prepare the Gram's stain smear from the slants and examine it for GNB (coliform like *E. coli*). The absence of GNB constitutes a negative test (absence of coliforms in the tested sample).
Tryptone water (indole) test	3 ml tryptone water	From all positive tubes (showing acid and gas) of screening test, transfer one loopful of medium to tryptone water. Incubate the tryptone water at (44.5 ±0.2°C) for 18–24 hours. Following incubation, add approximately 0.1 ml of Kovacs reagent and mix gently.	Positive indole is indicated by a red color ring over the aqueous phase of the medium which confirmed the *E. coli*.

medium test. Inoculate the varying quantities of water in litmus milk medium. Incubate anaerobically at 37°C for 5 days. Stormy clot (stormy fermentation) indicates the presence of *Cl. perfrigens*.

Detection of other pathogens like *S.* Typhi and *V. cholerae*: Such bacteria are detected by membrane filtration method in addition to below mentioned methods.

- *S.* Typhi: Equal volume of water is added to the double strength selenite broth followed by subculturing on selective media. Growth is identified by biochemical tests and antisera.
- *V. cholerae:* Water sample is mixed with alkaline peptone water in 1:9 ratios, followed by subculturing on selective media. Growth is identified by biochemical tests and serotyping.

Tests to Detect the Viruses

Entero viruses are able to contaminate the water. There is no specific test to detect the viruses as indicator organisms. They are destroyed by chlorination with free residual chlorine at least 0.5 mg/liter, for 30 minutes (contact period) at pH <8 and turbidity is ≤1 nephelometric.

Tests to Detect the Parasites

Different parasites as mentioned in **Table 32.1** are contaminating the water. There is no specific test to detect the parasites as indicator organisms. They are resistant to chlorination.

MICROBIOLOGY OF MILK

Common Sources of Milk Contamination

Animal sources are feces, urine, infected udder, teat canal and skin. Human sources are hands of human handlers. Environment sources are water mixed with milk, air (dust) and utensils.

Microbes in Milk

According to changes done by microbes in milk: Following are different types.

1. **Acid-forming bacteria:** These are bacteria which ferment the lactose to produce acids, especially lactic acid, which leads the formation of a smooth gelatinous curd like *Strept. lactis, E. faecalis*, etc.
2. **Acid- and gas-forming bacteria:** They produce acid and gas, which lead the formation of a smooth gelatinous curd riddled with gas bubbles like coliform bacilli (responsible for ropiness in milk after formation of acid and gas), *Cl. perfrigens, Cl. butyricum*, etc.
3. **Alkali-forming bacteria:** They render the milk alkaline like aerobic spore bearers, *Alkaligenes* spp., *Achromobacter*, etc.
4. **Proteolytic bacteria:** These are bacteria, which break the protein present in milk like *B. subtilis, B. cereus. P. vulgaris*, etc.

5. **Inert bacteria:** These bacteria are not doing any changes in milk like cocci of udder, *Achromobacter* spp., and few species of pathogenic bacteria, etc.
6. **Human milk:** Human milk contains *Staph. aureus, Staph. epidermidis, Strept. mitis, Gaffkya tetragena*, etc.

According to source: Follow **Table 32.5**.

Modes of Transmission

All the organisms mentioned in **Table 32.5** are transmitted by ingestion of infected milk except like cowpox and milker's node which are transmitted by direct contact with udder during milking. *Streptobacillus moniliformis* transmitted by milk contamination from nasal secretion of rat.

Sterilization/Disinfection and Sterility Testing of Milk

Following are the different methods of sterilization/disinfection with sterility testing of milk.

Thermized milk (equal to raw milk): Raw milk that is heated at 57–68°C for 15 seconds. Sterility testing

TABLE 32.5: Different sources of milk-borne organisms

Organism	Disease
Animal source	
M. bovis	Tuberculosis
Brucella spp.	Brucellosis
Salmonella spp.	Enteric fever and food poisoning
C. burnetii	Q fever
Staph. aureus	Staphylococcal food poisoning
Streptococcus spp.	Streptococcal infections
Bacillus anthracis	Anthrax
Leptospira spp.	Leptospirosis
C. diphtheriae	Diphtheria
C. ulcerans	Throat ulcer
Campylobacter jejuni	Diarrhea/dysentery
Yersinia enterocolitica	Gastroenteritis
Streptobacillus moniliformis	Haver hill fever
Cowpox virus	Cowpox
Paravaccinia (pseudocowpox) virus	Milker's nodule (node)
Coxsackie virus	Hand-foot-mouth disease
Tick-borne encephalitis	Encephalitis
Human source	
Vibrio cholerae	Cholera
EHEC	Dysentery
Salmonella Typhi and *S.* Paratyphi	Enteric fever
Shigella spp.	Dysentery
Staph. aureus	Staphylococcal food poisoning
Streptococcus spp.	Streptococcal infections
Entero viruses	Enteritis
Hepatitis viruses	Hepatitis

is done by methylene blue reduction test. Mix 1 ml methylene blue with 10 ml of milk in a sterile test tube and incubate at 37°C in a dark place. Milk is satisfactory if no change in color in 30 minutes. If milk contains viable bacteria they reduce the methylene blue. Test is economical, simple to perform and quick. It is useful to check the quality of milk after its arrival from the producer. Sterility testing of thermized milk is also done by Resurazin test, but test is not significant.

Pasteurization: Follow **Ch. 30.**

Ultra-high temperature (UHT) method: Milk is heated very rapidly in two stages (2nd stage is under pressure) at 125°C for a few seconds followed by rapid cooling and bottling as quickly as possible. Sterility testing is done by viable count test. Inoculate the serially diluted milk in yeast extract milk agar and then incubate at 30°C or 21°C (for unopened containers) for 72 hours. Viable colony count in fixed amount of milk is done by multiplying the numbers of colonies with dilution factors. Viable colony count of UHT treated milk should be <1000/ml.

Boiling: Milk is heated at 100°C for a long time (up to 5 minutes). Sterility testing is done by turbidity test. In boiled milk all heat coagulable proteins are precipitated, so that it does not become turbid when ammonium sulfate is added. Add the ammonium sulfate to boiled milk. Absence of turbidity indicates adequate boiling.

Detection of specific pathogens

- **Tubercle bacilli:** Following centrifugation of milk at 3000 rpm, the deposit is inoculated over LJ medium or injected in guinea pig for isolation of bacilli.
- **Brucella:** Follow **Ch. 70** (section, laboratory diagnosis → animal brucellosis).

MICROBIOLOGY OF AIR/SURFACES

Introduction

Air is an important vehicle for transmission of many pathogens. A person inhale about 15 cubic meters of air per day; hence, it is must to know the microbial contents of air and its sterility.

Microbes in Air

Microbial content of air is depends on place, whether indoor air or outdoor air.

Microbes in outdoor air: They depend on many factors like density of human and animal population, nature of the soil, density of vegetation and atmospheric conditions including humidity, temperature, wind conditions, rain fall, sunlight, etc. Outdoor air mostly contains saprophytic organisms like bacterial spores or spore-bearing bacilli, *Achromobacter, Sarcina, Micrococcus,* fungal spores and fungal hyphae.

Microbes in indoor air: Indoor areas include hospital ward, operation theater, laboratory, offices, etc. Microbes in indoor air depend on density of human population, disturbances of clothes and atmospheric conditions including air conditioners, cleaning, etc. Indoor air mostly contains pathogenic microorganisms cause respiratory tract infection and transmitted either via droplet nuclei or airborne route.

Modes of Transmission

Pathogens present in air are transmitted either by droplet nuclei or by airborne as described in **Ch. 3.**

Bacteriological Examinations of Air

It is done to detect the microbes present (mentioned above) in air by following two methods.

Settle plate method: Blood agar is the most preferred medium for overall fastidious, pathogenic, saprophytes and commensals. Malt extract agar is useful for molds. Nutrient agar is also used sometimes. Open and put the Petri dish contains agar medium on surface for 30 minutes to 1 hour. Bacteria carrying large dust particles present in air are settle down by gravity on the agar. Incubate the plate at 37°C for 24 hours. Colonies are developed after incubation **(Fig. 32.2).** The numbers of the colonies formed indicate the numbers of settled particles containing bacteria.

Slit sampler technique: Special equipment called slit sampler, **(Fig. 32.3)** is required which has three parts. An area to hold the Petri dish, suction pump and slit with 0.25 mm width and outer surface with slit. Air is sucked through the equipment at a rate of 1 cubic foot (28.3 liter) per minute for 10 minutes and directed to a plate containing culture medium through the slit. The plate is rotated mechanically to even spread of organisms. Incubate the plate at 37°C for 24 hours. Colonies are developed after incubation. The numbers of the colonies formed indicate the numbers of settled particles containing bacteria. It is the most efficient and convenient method for counting the number of bacteria carrying dust particles suspended in a unit volume of air.

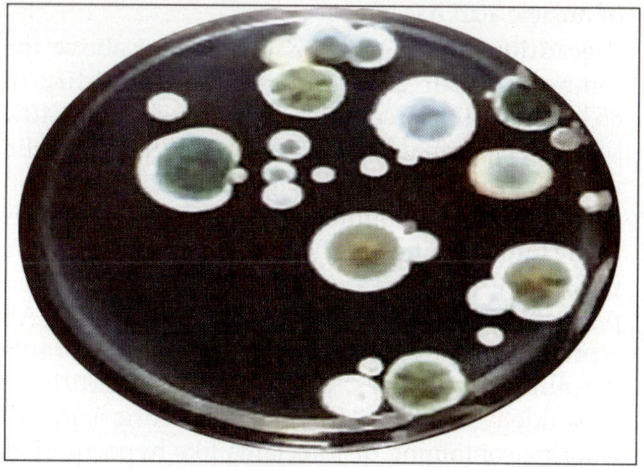

Fig. 32.2: Colonies on settle plate

Fig. 32.3: Slit sampler technique

Accepted Limit of Air Pollution

The upper limits of the bacterial count in air in various areas as follows:

- 50 per cubic feet in factories, offices and homes
- 10 per cubic feet in general operation theater
- 1 per cubic feet in operation theater for neurosurgery.

Fumigation of Indoor Air and Sterility Testing

It is done by following methods.

Formaldehyde gas fumigation

- **Legal aspect:** Formaldehyde is a schedule 1 chemical under the COSHH (Control of Substances Hazardous to health) Regulations and has a maximum exposure limit (MEL) of 2 ppm.
- **Target areas:** Hospital ward, operation theater, laboratory, offices, other rooms, books, furniture, clothes, beds, pillows, bed sheets, etc.
- **Biological (mechanism) actions of formaldehyde:** Formaldehyde is an alkylating agent, having bactericidal, virucidal, fungicidal and sporicidal properties, and it inactivates the microorganisms by reacting with carboxyl, amino, hydroxyl and sulfhydryl groups of proteins as well as amino groups of nucleic acid bases.
- **Precautions:** Fumigation is effective at above the temperature of 20°C and relative humidity of 65%. Formalin is commercially available as 40% solution of formaldehyde in water. When formalin is heated, formaldehyde vapor is generated, but concentration encountered during fumigation is many hundred times higher than this. So fumigation procedure must be carried out only by trained personnel under strictly defined conditions. All workers using formaldehyde must be aware of safe handling procedures. Under certain conditions, formaldehyde can react with hydrochloric acid and chlorine containing disinfectants like hypochlorites to form chlormethyl ether, a potent lung carcinogen.

So, hydrochloric acid and chlorine-containing disinfectants must be removed from the room before fumigation.

- **Fumigation procedure**
1. **Step 1:** It includes following **preparation**s.
   - Entire block should be thoroughly cleaned before fumigation. All apparatus such as windows, doors, floor, walls, surgery tables, all washable equipments, bulb, air conditioner, etc., should be cleaned according to manufacturer instructions.
   - Close windows and ventilators tightly. If any openings found seal it with cellophane tape or other material to avoid the leak of fume.
   - Switch off all lights and other electrical and electronical items.
   - Calculate the room size in cubic feet (Length × Breadth × Height) and calculate the required amount of formaldehyde as given in step 3.
   - Adequate care must be taken by wearing cap, mask, foot cover, spectacle, etc.
   - Formaldehyde is irritant to eye and nose; and it is also recognized as a potential carcinogen. So, the fumigating person must be provided with the personal protective equipments (PPE).
   - Paste a warning notice on the front door indicating fumigation is in progress as shown in **Fig. 32.4.**

2. **Step 2:** Following are the different **fumigation procedures**.
   - Formaldehyde fumigation with electric boiler method (recommended): For each 1000 cubic feet, 500 ml of formalin added in 1000 ml of distilled water (if not available, use tape water) in an electric boiler. Switch on the boiler, leave the room and seal the door. After 45 minutes (variable depending to volume present in the boils apparatus/its heating proficiency) switch off the boiler without entering into the room (switch off the main electric supply from outside).
   - Formaldehyde fumigation with potassium permanganate method: In this method heat is generated by an oxidizer $KMnO_4$, which

Fig. 32.4: Fumigation symbol

causes autoboiling and generates fumes from formaldehyde. Take 500 ml of formalin in 1000 ml of distilled water (or tap water) for 1000 cubic feet area, in a heat resistant container preferably in steel bucket and then add 450 g of $KMnO_4$ for 1000 cubic feet of area. Repeat the same steps in separate container for every another 1000 cubic feet until it reaches the complete area volume. Add $KMnO_4$ to all buckets simultaneously to reduce the exposure to fume (i.e., need 3–4 persons at different location). After the initiation of formaldehyde vapor, leave the room very immediately and seal it for at least 12–24 hours.

3. Step 3: Neutralization

- Before neutralization, take out the formaldehyde fumigation system from the room and then neutralize the toxicity of formaldehyde vapor with ammonia solution.
- Place a cotton ball and pour 300 ml of 10% ammonia (for each 500 ml of formaldehyde used) on the floor, at least 4 hours before (07 am) the "sterility test".
- Formaldehyde vapor reacts with ammonia gas and produces hexamine (hexamethylene-tetramine) which is considered a harmless substance.
- Switch on the air conditioner, at least 2 hours before the "sterility test".

- **Example of fumigation methods**
  - Area = L × B × H = 20 × 15 × 10 = 3000 cubic feet (Note: Make it into nearest 1000, if the volume is in fraction.
  - Formaldehyde required for fumigation: 500 ml for 1000 cubic feet so, 1500 ml of formalin required (to be diluted in 3000 ml of distilled water).
  - Ammonia required for neutralization: 300 ml of 10% ammonia for 500 ml of formalin, so 900 ml of 10% ammonia required.

- **Sterility testing:** It includes collection of air sample plate and swabs from different surfaces plus different equipments of room to be fumigated, before (prefumigation sterility samples) and after (post-fumigation sterility samples) the fumigation. Any positive culture growth from above samples indicates the repetition of fumigation.

- **Data records:** A record (log book) should be kept and properly maintained for all fumigations with following details, date and time of fumigation, date and time of neutralization, personnel involved, and the dates of "sterility test done " and their results.

- **Advantages:** Formaldehyde gas fumigation is low cost and with wide spectrum of activities like bactericidal, virucidal, fungicidal and sporicidal. Fumigants can reach the place where sprays, aerosols cannot reach. It reduces residue problems in treated areas. Fumigants are used where standard call for **"zero microbial tolerance"** in products or living environment.

- **Disadvantages:** Formalin is irritant to eye and mucosa, this effect is nullified by ammonia. It is carcinogenic for lung, nose and nasal passage.

Other methods: Above all disadvantages of formaldehyde are overcome by using following newer methods, but these are expensive.

1. Specialize air flow pattern: Only purified air is circulating, and contaminated air going out continuously.
2. Hydrogen peroxide (H_2O_2):
3. H_2O_2 with silver nitrate
   — Spread by fogging machine (**Fig. 32.5**)
4. Parasitic acid
5. Phosphine
6. 1,3 dichoropropane
7. Chloropicrin
8. Methyle isocynate
9. Hydrogene cyanide
10. Sulfuryl fluoride
11. Iodoform
12. Methyle bromide

Fig. 32.5: Fogging method

Microbiology of Water, Milk and Air/Surfaces

ACCESS YOURESLF

Essays/Full Questions

1. Describe the methods used and significance of assessing the microbial contamination of food / milk, water and air.
2. Describe the etiology and basis of waterborne diseases.

Short Note

1. Bacteriological examination of water / milk / air.

Short Questions

1. Define: Coliform bacteria and most probable number (MPN) of bacteria in water.
2. Name the four milk-borne pathogens.
3. How you fumigate the operation theater?

MCQs for Chapter Review

Microbiology of Water

1. **Which organism is considered an index of fecal pollution of drinking water supplies?**
   a. *Rotavirus*   b. *E. coli*
   c. *Salmonella* spp.   d. *Hepatitis E virus*

2. The presumptive coliform counts in water are considered unsatisfactory, if the counts exceed:

   a. 0–1/100 ml
   b. 1–3/100 ml
   c. 4–10/100 ml
   d. >10/100 ml

Microbiology of Milk

3. All are transmitted by milk *except*:

   a. Tuberculosis
   b. Brucellosis
   c. Q fever
   d. Leishmaniasis

Microbiology of Air/Surfaces

4. The width of the slit in a slit sampler used for air sampling is:

   a. 0.25 mm
   b. 0.55 mm
   c. 0.45 mm
   d. 0.35 mm

5. The upper limit of the bacterial count in air in operation theater for neurosurgery is:

   a. 50 per cubic feet
   b. 10 per cubic feet
   c. 1 per cubic feet
   d. 0 per cubic feet

Answers and Explanation of MCQs

1. b

- For index of fecal pollution coliform (lactose fermenting) bacteria are used. From given options *E. coli* is the coliform (lactose fermenting) bacterium.

2. d

- Follow section, **Microbiology of water and Table 32.3**, for explanation.

3. d

- Option a, b and c: Follow section, **Microbiology of milk and Table 32.5**, for explanation.
- Leishmaniasis: Transmitted by sand fly.

4. a

5. c

- Follow section, **Microbiology of air/surfaces → bacteriological examinations of air → slit sampler technique**, for explanation of answers of MCQs 4–5.

CHAPTER 33

Laboratory-acquired Infections

Chapter Outline

▫ Laboratory-acquired Infections

LABORATORY-ACQUIRED INFECTIONS

Definition

All infections acquired from laboratory or laboratory-related works irrelevant to symptomatic or asymptomatic in nature.

Modes of Transmission

They are transmitted by different routes like inhalation, ingestion or due to aspiration by pipette or eating in laboratory, needles stick/splash injuries, laboratory animal bites or by direct contact with contaminated articles.

Classification of Laboratory-acquired Microbes

Risk-based 4 groups are mentioned in **Table 33.1**.

Prevention

Control and preventive measures are almost same as of HAIs. Described in **Ch. 29**.

Treatment

Treatment given after exposure of laboratory microbes called post exposure prophylaxis. It includes treatment as per organism. Specific post exposure prophylaxis for important microbes like HIV (**Ch. 85**) and HBV (**Ch. 88**) are given in respective chapters.

ACCESS YOURSELF

Short Note

1. Laboratory-acquired infections.

TABLE 33.1: Classification of laboratory-acquired microbes

Group	Criteria	Bacteria	Viruses	Fungi	Parasites
1	• Nonpathogenic microbes • No human diseaset	—	—	—	—
2	• Pathogenic microbes • Cause human disease • No community spread • Treatment and prophylaxis are available	*Staphylococcus, Strepto-coccus, Clostridium* spp., *Corynebacterium diphtheriae, Enterobactericeae, Mycobacterium* (Except *M. tuberculosis*) and *Bacillus* (Except *B. anthracis*)	Adeno virus, herpes virus, influenza virus and calci virus	*Candida,* dermatophytes, *Cryptococcus* and *Aspergillus*	Clinically significant parasites
3	• Pathogenic microbes • Cause severe human disease • Community spread • Effective treatment and prophylaxis are not available	*M. tuberculosis, B. anthracis, Brucella* spp., *Coxiella burnetii* and *Francisella tularensis*	Prion particle, LCM virus, *Hantavirus,* SARS-CoV and encephalitis viruses	All systemic fungi and *Penicillium marneffei*	*Naegleria fowleri, P. falciparum, Trypanosoma* spp., *Leishmania* spp., *Echinococcus* spp., *Taenia solium*
4	• Pathogenic microbes • Cause severe human disease • Community spread • Treatment and prophylaxis are not available usually	—	Junin virus, lassa virus, machupo virus, nairo viruses, CCHF virus, ebola virus, marburg virus and variols (major and minor) virus	—	—

Part II

Systemic Microbiology

SECTION 4

Infective Syndromes or Infectious Diseases or Systemic Infections or Clinical Microbiology or Applied Microbiology

CHAPTER 34

Cardiovascular System Infections

Chapter Outline

- Anatomy of CVS
- Common Infections
  - Infections of heart
  - Infections of blood vessels
- Specific Infections

ANATOMY OF CARDIOVASCULAR SYSTEM

Cardiovascular system includes heart and blood vessels. Heart has following three layers from inner to outer:

- **Endocardium:** It is the inner most layer lining the cavities and valves.
- **Myocardium:** It is the middle layer comprising striated muscles, called cardiac muscles.
- **Pericardium:** It is the outermost layer made up of connective tissues (secretes fluid in pericardial cavity) and fat. It forms the cavity around heart called pericardial cavity.

COMMON INFECTIONS

Infections of Heart

Infective Endocarditis

Synonym: Bacterial endocarditis.

Definition: It is an infection of the inner layer (endocardium) of the heart, usually the valves.

Etiological agents: There is also a noninfective endocarditis. Following are the infective causes of endocarditis:

1. **Bacteria**
   - **Streptococci**
     - Viridans group (30–40%): Most common organism of endocarditis. Species are *Strept. sanguis, Strept. parasanguis, Strept. oralis, Strept. mitis, Strept. mutans, Strept. gordonii* and *Strept. mitior,* etc.
     - Enterococci: *Enterococcus faecalis*
     - Non-enterococci: *Strept. bovis,* etc.
   - **Staphylococci**
     - CoPS (10–27%): *Staph. aureus*
     - CoNS (1–3%): *Staph. albus (Staph. epidermidis)*
   - **GNB:** Around 1.5–13% cases of infective endocarditis are from GNB origin. *Pseudomonas aeruginosa* is the most common GNB for endocarditis. It also caused by HACEK group of microorganisms and very rarely by Enterobacticeae like *S.* Typhi and others. Less commonly reported bacteria responsible for so called "culture negative endocarditis" like *Bartonella, Chlamydia psittaci* and *Coxiella.* Such bacteria can be identified by serology, culture of the excised valve tissue, sputum, pleural fluid, and emboli, and by PCR or sequencing of bacterial 16S ribosomal RNA.
   - **Other bacteria:** *Tropheryma whipplei, Propionibacterium* sp., *Citrobacter koseri* and *Bartonella bacilliformis* were found in a patient with a bicuspid aortic valve.

2. **Fungi:** Fungal endocarditis is a fatal and one of the most serious forms of infective endocarditis. *Candida albicans* produces the biofilms around prosthetic heart valves and additionally colonizes and penetrates endothelial walls. It accounts 24–46% of fungal endocarditis with 46.6–50% mortality. Other fungal causes include *Histoplasma capsulatum* and *Aspergillus. Aspergillus* accounts for about 25% of fungal endocarditis. *Trichosporon asahii* has also been reported in a case of endocarditis.

Pathogenicity

- **Predisposing factors**
  1. **Age:** Infective endocarditis chances are increasing after 65 years of age may be due to degenerative valvular lesions.
  2. **Sex:** The male-to-female case ratio is over 2:1.
  3. **Heart diseases:** Valvular heart disease including rheumatic disease, congenital heart disease and artificial valves are the risk factors. However, 50% of all cases develop in people with no known history of valvular disease.
  4. **Medical and surgical procedures:** Viridans streptococci are the primary habitats of oral

cavity and upper respiratory tract. They enter the bloodstream usually by dental procedures (like tooth extractions) or by genitourinary manipulation. *Staphylococcus* can enter the bloodstream through procedures that cause break in the integrity of skin, like surgery, hemodialysis, during access of long-term indwelling catheters or secondary to IV injection of recreational drugs. It infects normal heart valve and having prosthetic valve. *Enterococcus* can enter the bloodstream due to defect in the gastrointestinal or genitourinary tracts. *Strept. bovis* and *Clostridium septicum* are parts of the natural flora of the bowel, associated with colorectal cancer. They cause the endocarditis due to breaking down the barrier between the gut and the blood vessels which drain the bowel bacteria. HACEK organisms are a group of bacteria that live on the dental gums, and can cause endocarditis due to poor dental hygiene or preexisting valve disease. *Candida albicans* presents in IV drug users (IV drug users tend to get their right-sided heart valves infected because the veins that are injected drain in to the right side of the heart), patients with prosthetic valves, and in person with IDDs. Electronic pacemakers can favor the infection.

5. **Other factors:** These are history of infective endocarditis, low level of white blood cells, DM, alcohol abuse, HIV/AIDS, etc.

- **Pathogenesis:** Damaged part of a heart valve allows the platelets, inflammatory cells and fibrin deposition which later allow bacteria to take hold and to form vegetations.
- **Clinical features:** Symptoms may include fever (97%), small areas of bleeding in to skin, heart murmur, feeling tired, and low RBCs count, night sweats, rigors, spleen enlargement, etc.
- **Complications:** Valvular insufficiency, heart failure, stroke, glomerulonephritis which allows blood and albumin to enter in urine, kidney failure, Oosler's nodes (painful subcutaneous lesions in the distal fingers), Roth's spot on the retina, positive serum rheumatoid factor, etc.

Classification

1. **According to duration and clinical presentation**
   - **Subacute bacterial endocarditis (SABE):** It is caused by low virulent streptococci like viridans groups. Illness is mild to moderate which progresses slowly over weeks and months (>2 weeks). It has low propensity to spread extracardiac sites.
   - **Acute bacterial endocarditis (ABE):** It is caused by high virulent *Staph. aureus*. Illness is fulminant over days to weeks (<2 weeks) or disease-producing capacity. It has high propensity to spread extra-cardiac sites. This classification is now discouraged, because the ascribed associations were not strong enough to be relied upon clinically in terms of organism and prognosis. The terms **short incubation** (meaning < about six weeks), and **long incubation** (> about 6 weeks) are preferred.

2. **According to culture results:** Infective endocarditis may also be classified as **culture-positive** or **culture-negative.** Earlier administration of antibiotics is the most common cause of a "culture-negative" endocarditis. Sometimes, microbes can take a longer time to grow in the culture media called fastidious organisms. Some examples are *Aspergillus* species, *Brucella* species, *Coxiella burnetii*, *Chlamydia* species, and HACEK group of bacteria. Due to delay in growth and identification in these organisms, patients may be falsely classified as "culture-negative" endocarditis.

3. **According to source**
   - **Nosocomial endocarditis:** It is a form of healthcare-associated endocarditis in which the infective organism is acquired during a stay in a hospital, and it is usually secondary to presence of intravenous catheters, total parenteral nutrition lines, pacemakers, etc.
   - **Community acquired:** From community and outside the hospital.

4. **According to valve type**
   - **Native valve endocarditis**
     - Right-side valve: It is caused by *Staph. aureus*. It is common in IV drug abusers (such as heroin or methamphetamine) and specially involves the tricuspid valve.
     - Left-side valve: It is infected by *Staph. aureus*, enterococci and other bacteria. Regardless of cause, left-sided endocarditis is more common in both IV drug users and nondrug users than right-sided endocarditis.
   - **Prosthetic valve endocarditis:** Two types:
     - Early onset prosthetic valve endocarditis (<1 year of valvular surgery): Nosocomial type due to intraoperative or a postoperative bacterial contamination. Mainly by *Staph. epidermidis* as it is capable of growing as a biofilm on plastic surfaces.
     - Late onset prosthetic valve endocarditis (>1 year following valvular surgery): Late prosthetic valve endocarditis is usually due to community-acquired microorganisms, mainly by viridans group of bacteria.

Diagnosis

- **Specimens:** Three blood samples every 1 hour interval.
- **Testing methods:** These are blood culture, antibiogram and echocardiography. Vegetation is present on the tricuspid valve. The transthoracic echocardiogram has 65% sensitivity and 95% specificity.
- **Modified Duke criteria for diagnosis of infective endocarditis:** Established in 1994 by the Duke Endocarditis Service and revised in 2000. It includes major and minor criteria.

1. **Major criteria**
   - Typical microorganism consistent with 2 separate blood cultures: Viridan group streptococci or *Strept. bovis* including nutritional variant strains, or HACEK group or *Staphylococcus aureus* or community acquired enterococci, in the absence of a primary focus.
   - Microorganisms consistent with persistently positive blood cultures: Two positive blood cultures drawn >12 hours apart, or all of 3 or a majority of 4 separate cultures of blood (with 1st and last sample drawn 1 hour apart). *Coxiella burnetii* detected by at least 1 positive blood culture or IgG titer for Q fever phase 1 antigen is >1:800. This was previously a minor criterion.
   - Evidence of endocarditis with positive echocardiogram: Defined as oscillating intracardiac mass on valve or supporting structures, in the path of regurgitant jets, or on implanted material in the absence of an alternative anatomic explanation, or abscess, or new partial dehiscence of prosthetic valve or new valvular regurgitation (worsening or changing of preexisting murmur not sufficient).

2. **Minor criteria**
   - Presence of precipitating factors like history of cardiac lesion, recreational drug injection, etc.
   - Presence of fever with temperature >38°C.
   - Presence of evidences of embolism: Arterial emboli, pulmonary infarct, conjunctival hemorrhage, Janeway lesions.
   - Immunological problems: Glomerulonephritis, Osler's nodes, Roth's spot and rheumatic factor.
   - Microbiologic evidence: Positive blood culture (that does not meet a major criterion) or serological evidence of infection with organism consistent with IE but not satisfying major criterion.
   - This criterion has been removed from the modified Duke criteria.
   - **Diagnostic pattern:** As per Duke criteria diagnosis of infective endocarditis can be **definite, possible, or rejected.**
     1. **Definite diagnosis**
        - One of these pathological criteria: Histology or culture of cardiac vegetation, an embolized vegetation, or intracardiac abscess from the heart finds microorganisms or active endocarditis.
        - One of these combinations of clinical criteria: 2 major clinical criteria or 1 major and 3 minor criteria or 5 minor criteria.
     2. **Possible diagnosis:** 1 major and 1 minor criterion or 3 minor criteria are fulfilled.

Prevention: Prophylactic drugs should be bactericidal rather than bacteriostatic. In some countries like the USA, high-risk patients may be given prophylactic antibiotics such as penicillin, amoxycillin (for beta lactamase bacteria) or clindamycin for penicillin allergic

patients prior to dental procedures. Such measures are not taken in certain countries like Scotland due to the fear of antibiotic resistance. All people with heart disease do not require antibiotics to prevent infective endocarditis. Heart diseases have been categorized into high, medium and low risk of developing. Those with high risk require prophylaxis before endoscopies and urinary tract procedures. Diseases listed under high risk include prior endocarditis, prosthetic heart valves, unrepaired cyanotic congenital heart diseases, completely repaired congenital heart disease in their first 6 months, incompletely repaired congenital heart diseases and cardiac transplant valvulopathy. Antibiotics were historically commonly recommended to prevent infective endocarditis in those with heart problems undergoing dental procedures. Following are the antibiotic regimens recommended by the American Heart Association.

1. **Not allergic to penicillin:** Oral amoxicillin 1 hour before the procedure or intravenous or intramuscular ampicillin 1 hour before the procedure.
2. **Allergic to penicillin:** Azithromycin or clarithromycin orally 1 hour before the procedure. Cephalexin orally 1 hour before the procedure. Clindamycin orally 1 hour before the procedure.

Treatment: Treatment is generally with intravenous antibiotics. The choice of antibiotics is based on results of blood cultures. Occasionally heart surgery is required.

Myocarditis

It is an inflammation of the myocardium. It is rarely caused by bacteria. Other etiological agents are viruses (mostly) such as coxsackie virus B (most common), adeno viruses, parvo virus B19, HHV-6 and dengue virus and parasites like *Trypanosoma cruzi* (Chagas' disease).

Pericarditis

It is an inflammation of the pericardium. Following are the etiological agents.
- Bacteria: It is rarely caused by bacteria. It occurs as as a complication of pneumonia by *Staph. aureus*, *Strept. pneumoniae*, *Haemophilus influenzae*, *Neisseria meningitidis* and of pulmonary tuberculosis by *M. tuberculosis*.
- Viruses (mostly): Coxsackie virus B (most common), ECHO virus, adeno viruses, HIV and others.

Pericardial Effusion

It is the collection of fluid in pericardial cavity. It occurs mostly as a complication of pericarditis, malignancy or by autoimmune disease of pericardium.

Infections of Blood Vessels

Aneurysm

It is an abnormal, irreversible dilatation of blood vessels. Following are the etiological agents.
- Bacteria: By *Treponema pallidum* (Tertiary syphilis).

Other bacteria are similar to endocarditis like strepto-cocci and staphylococci.

- Fungi (called mycotic aneurysm): By *Aspergillus* spp.

Infective Endarteritis

It is an inflammation of arterial wall. It occurs with or without aneurysm.

Device Related Infections of Blood Vessels

CRBSI: Ch. 29.

Infective (suppurative) thrombophlebitis: It is an inflammation of vein wall. It occurs by skin flora or by contamination from hands of HCWs. Common agents are *Staph. aureus*, members of Enterobactererales, *Candida* and *Malassezia*.

SPECIFIC INFECTIONS

ARF: Ch. 50.

Lemierre's Syndrome

It is the thrombophlebitis of internal jugular vein with bacteremia. It is caused by anaerobic bacteria like *Fusobacterium necrophorum* following oropharyngeal infection.

ACCESS YOURSELF

Essays/Full Questions

1. Describe the classification, etiology, pathogenicity, and diagnostic modalities of infective endocarditis.
2. Identify the microbial agents causing ARF and infective endocarditis.

Short Note

1. ARF.

MCQs for Chapter Review

Infective Endocarditis

1. **Common causative agent of subacute bacterial endocarditis is:**
   a. *Streptococcus faecalis*
   b. *Mycoplasma pneumoniae*
   c. Chickenpox virus
   d. *Streptococcus viridans*
2. **Which of the following is least likely to cause infective endocarditis?**
   a. *Staphylococcus albus*
   b. *Streptococcus faecalis*
   c. *Salmonella* Typhi
   d. *Pseudomonas aeruginosa*

Answers and Explanation of MCQs

1. d
2. c

- Follow section, **infective endocarditis (etiological agents)** for explanation of answers of MCQs 1–2.

CHAPTER 35

Bloodstream Infections

Chapter Outline
- Common Infections
- Specific Infections

COMMON INFECTIONS

Bacteremia

Definitions

- **Bacteremia:** It means entry/presence of bacteria in blood (transient presence) without multiplication.
- **Septicemia:** It means entry of bacteria in blood, multiplication with clinical sign-symptoms.
- **Pyemia:** It means septicemia and abscess in organs like liver, spleen and other tissues.
- **Toxemia:** Clinical sign-symptoms are due to toxin production.
- **Viremia, fungemia and parasitemia:** Similarly presence of viruses, fungi and parasite in blood called viremia, fungemia and parasitemia, respectively.

Types of bacteremia: Bacteremia may be transient, continuous or intermittent.

- **Transient bacteremia:** It occurs, when organisms, often members of the normal flora, are introduced into the blood by minimal trauma to membranes such as brushing teeth or chewing food and minor events like manipulation of infected tissues, instrumentation of contaminated mucosal surface, and surgery involving nonsterile sites (e.g., during dental procedures, after gastrointestinal biopsy, etc.).
- **Continuous bacteremia:** Bacteria are released into blood at constant rate from bacterial endocarditis or septic shock.
- **Intermittent bacteremia:** Bacteria are intermittently found in blood from undrained abscess, meningitis, pneumonia or pyogenic arthritis. Bacteria are found in blood about 45 minutes before pyrexia.

Bacterial agents causing bacteremia: These are *Staph. aureus*, CoNS, β-hemolytic *Streptococcus*, *Enterococcus*, *Strept. pneumoniae*, *E. coli*, *Klebsiella*, *Proteus*, *Salmonella*, *Pseudomonas*, *Brucella*, etc.

Pathogenicity of bacteremia

- **Modes of transmission:** Two types: **(1) Intravascular infections:** Organisms enter in blood from endocarditis, aneurysm, thrombophlebitis and by IV line. **(2) Extravascular infections:** Organisms enter in blood by lymphatic system from GIT (25%), respiratory system (20%), abscess (10%), biliary tract (5%), wound site (5%) other or uncertain sites (35%).
- **Clinical features:** Septicemia, septic shock, disseminated intravascular coagulation (DIC) and other systemic features.

Diagnosis of bacteremia

- **Specimens:** Blood.
- **Testing methods:** Ch. 118.

Pyrexia of Unknown Origin (PUO)

Synonym: It also called fever of unknown origin (FUO).

Definition: Temperature >38°C for >3 weeks and without diagnosis even after 1 week of investigations called PUO.

Etiological agents

1. **Infectious disease**
   - **Bacteria:** *Mycobacterium tuberculosis* (most common cause of PUO but remain undetected), atypical mycobacteria, *Salmonella*, *Brucella*, *Chlamydia*, *Leptospira*, *Rickettsia*, *Coxiella*, *Mycoplasma*, etc.
   - **Viruses:** CMV, arbo viruses, HIV, entero viruses, etc.
   - **Fungi:** *Candida albicans*, different systemic or opportunistic fungi.
   - **Parasites:** *Plasmodium* spp., microfilariae, *Babesia*, *Leishmania* spp., *Trypanosoma* spp., and *Toxoplasma*.
2. **Noninfectious disease:** Granulomatous disease, autoimmune disease, hypersensitivity reaction, malignancy, trauma, etc.

Diagnosis

- **Clinical diagnosis:** Based on pattern of fever, history and clinical examination.

- **Radiological diagnosis:** X-ray, USG, CT-scan, MRI, etc.
- **Laboratory diagnosis:** Different body fluid, discharge or biopsy materials are tested by following methods:
  1. **Biochemical diagnosis:** Includes liver function test, renal function test and other biochemical analysis.
  2. **Pathological diagnosis:** Includes hematological and histological diagnosis.
  3. **Microbiological diagnosis**
     - Diagnosis of bacteria: **Ch. 118.**
     - Diagnosis of viruses: **Ch. 119.**
     - Diagnosis of fungi: **Ch. 120.**
     - Diagnosis of parasites: **Ch. 121.**

Infective Anemia

Definition: Decrease in total amount of RBCs or Hb or O_2-carrying capacity of blood called anemia.

Etiological agents

1. **Iron-deficiency anemia (microcytic and hypochromic anemia due to blood loss):** Mostly caused by parasites such as *Schistosoma* spp., hookworm spp., (*Necator americanus* and *Ancylostoma duodenale*) and *Trichuris trichuria*.
2. **Megaloblastic (B12 and folic acid deficiency):** Mostly caused by parasites such as *Diphyllobothrium latum* (not in India) and hookworm spp., like *Necator americanus* and *Ancylostoma duodenale*.
3. **Dimorphic anemia (iron, B12 and folic acid deficiency):** Hookworm spp., like *Necator americanus* and *Ancylostoma duodenale*.
4. **Hemolytic anemia (destruction of RBCs):**
   - Bacteria: *Bartonella bacilliformis* and *Clostridium perfringens*.
   - Viruses: EBV and HAV.
   - Parasites: *Leishmania donovani* (autoimmune hemolytic anemia), *Plasmodium falciparum* (autoimmune hemolytic anemia) and *Babesia microti*.
5. **Aplastic anemia (normocytic normochromic type of anemia due to bone marrow dysfunction or failure):**
   - Bacteria: *Mycobacterium tuberculosis* and *Rickettsia*
   - Viruses: CMV, EBV, VZV, parvo virus B19, HIV, HHV-6 and HCV.
6. **Rare causes of anemia:** *Taenia saginata, Taenia solium* and *Fasciolopsis buski*.

Diagnosis

- Diagnosis of bacteria: **Ch. 118.**
- Diagnosis of viruses: **Ch. 119.**
- Diagnosis of fungi: **Ch. 120.**
- Diagnosis of parasites: **Ch. 121.**

Eosinophilia in Parasitic Infections

Background

- **Normal range of eosinophils:** It is 0–6% (0–4%) of total leukocyte count. Normal count is 125±100 per cubic millimeter.

TABLE 35.1: Parasites causing eosinophilia

Eosinophilia	Parasitic infections
High	Trichinosis, early stage of schistosomiasis, early stage of liver plus lung fluke infections and tropical pulmonary eosinophilia
Moderate	Ascariasis, hookworm disease, onchocerciasis, lymphatic filariasis, schistosomiasis and late stage of liver plus lung fluke infections
Low	Cysticercosis and hydatidosis

- **Distribution of eosinophils:** They are produced in bone marrow and distributed in blood and tissues. Skin, lungs and gastrointestinal tract are the major sites for eosinophils location. It is estimated that 100 eosinophils in tissue for every one eosinophil in blood.
- **Functions of eosinophils:** Eosinophils are one kind of leukocytes that are fighting with parasites and other blood invaders. They bind with parasites with the help of complement, IgE and IgG. They can protect against the parasitic infection and same time causing allergic disorders.
- **Breakdown products of eosinophils:** Called Charcot-Layden crystals (**Ch. 93**).

Definition of eosinophilia: Increase in eosinophils count of more than 5% of total leukocytes count called eosinophilia.

Causes of eosinophila

1. **Allergic disease:** Like asthma.
2. **Parasitic infections**
   - **Prozoan infection:** Like *C. belli*
   - **Helminthic infections:** Almost all parasites are causing eosinophilia. Lumen dwelling parasites like *E. vermicularis, T. solium* and *T. saginata* are not causing eosinophilia. Important parasites causing eosinophilia are given in **Table 35.1**.

Causes of pulmonary eosinophila: It is caused by larvae of *Ascaris lumbricoides* (condition called Loeffler's syndrome or eosinophilic pneumonia), *Ancyclostoma duodenale, Necator americanus* and *Strongyloides stercoralis. Toxocara canis, Toxocara cati, Paragonimus westermanii, Wuchereria bancrofti* and *Brugia malayi* are also causing pulmonary eosinophila called topical pulmonary eosinophilia or eosinophilic lungs or Weingarten's syndrome.

SPECIFIC INFECTIONS

Bacterial Infections

- *Salmonella* septicemia (*S.* Choleraesuis): **Ch. 62.**
- Brucellosis (*Brucella* spp.): **Ch. 70.**
- Relapsing fever (*Borrelia* spp.): **Ch. 73.**
- Leptospirosis (*Leptospira* spp.): **Ch. 73.**
- Infections by *Rickettsia, Orientia, Ehrlichia, Coxiella* and *Bartonella*: **Ch. 76.**

Viral Infections

- Infectious mononucleosis (EBV): **Ch. 78.**
- CMV infection: **Ch. 78.**

- AIDS (HIV I and II-retroviridae): **Ch. 85**.
- Viral hemorrhagic fever (Hemorrhagic fever viruses): **Ch. 89**.

Fungal Infections

- Candidemia (*Candida* spp.): **Ch. 90**.
- Systemic mycoses: **Ch. 91**.
- Opportunistic mycoses: **Ch. 91**.

Parasitic Infections

Protozoal bloodstream infections

- Sleeping sickness (*Trypanosoma bricei*): **Ch. 95**.
- Chaga's disease (*Trypanosoma cruzi*): **Ch. 95**.
- Kala azar or dum dum fever (visceral leishmaniasis (*Leishmania donovani*): **Ch. 95**.
- Malaria (*Plasmodium* spp.): **Ch. 97**.
- Babesiosis (*Babesia* spp.): **Ch. 97**.
- Toxoplasmosis (*Toxoplasma gondii*): **Ch. 98**.

Helminthic bloodstream infections: Microfilariae of different nematodes like *Wuchereria bancrofti, Brugia* spp., *Loa loa, Mansonella ozzardi* and *Mansonella perstans* are present in blood: **Ch. 108**.

ACCESS YOURSELF

Essays/Full Questions

1. Describe the causes and diagnosis of FUO.
2. List the common microbial agents causing anemia. Describe the morphology, mode of infection, pathogenicity, diagnosis, prevention and treatment of the common microbial agents causing anemia.
3. Describe the etiology, pathogenicity and laboratory diagnosis of kala azar, malaria, filariasis and other common parasites prevalent in India.
4. Describe the etiology, pathogenicity, opportunistic infections, diagnosis, prevention and treatment of AIDS.

Short Notes

1. Bacteremia
2. PUO
3. Eosinophilia in parasitic infections
4. CL crystals.

Short Question for Theory/Viva Question

1. Name four parasites present in peripheral blood.

MCQs for Chapter Review

Pyrexia of Unknown Origin (PUO)

1. **Fever of unknown origin is any febrile illness lasting:**
   a. 3 weeks or longer
   b. 4 weeks or longer
   c. 2 weeks or longer
   d. 1 week or longer
2. **The single most common cause of pyrexia of unknown origin:**
   a. *Mycobacterium tuberculosis*
   b. *Salmonella* Typhi
   c. *Brucella* spp.
   d. *Salmonella* Paratyphi A

3. **A veterinary doctor had pyrexia of unknown origin. His blood culture in special laboratory media was positive for gram-negative short bacillus which was oxidase positive. Which one of the following is the likely organism grown in culture?**
   a. *Pasteurella* sp.
   b. *Francisella* sp.
   c. *Batonella* sp.
   d. *Brucella* sp.
4. **The organism most likely to cause fever of unknown origin in a farmer who raises goats:**
   a. *Brucella melitensis*
   b. *Clostridium novyi*
   c. *Histoplasma capsulatum*
   d. *Mycobacterium tuberculosis*

Infective Anemia

5. **Parasite causing anemia:**
   a. Hookworm
   b. Thread worm
   c. *Ascaris*
   d. Guinea worm
6. **Microcytic hypochromic anemia found in infection of:**
   a. *Ancyclostoma*
   b. *Ascaris*
   c. *Necator*
   d. *Diphylobothrium*
7. **Megaloblastic anemia is caused by:**
   a. *Diphylobothrium latum*
   b. *Schistosoma haematobium*
   c. *Echinococcus granulosus*
   d. *Taenia solium*

Eosinophilia in Parasitic Infectionsi

8. **Which of the following is not responsible for pulmonary eosinophilia?**
   a. *Ascaris lumbricoides*
   b. *Paragonimus westermanii*
   c. *Wuchereria bancrofti*
   d. *Babesia microti*
9. **Pulmonary eosinophilia is seen in the following parasitic infections *except*:**
   a. Babesiosis
   b. Hook worm infection
   c. Strongyloidiasis
   d. Visceral larvae migrans
10. **Parasite causing pulmonary eosinophilia syndrome:**
    a. *Strongyloides*
    b. *Enterobias vermicularis*
    c. Hookworm
    d. *Trichinella*
11. **Eosinophilic pneumonia caused by *Ascaris lumbricoides* is known as?**
    a. Mafucci syndrome
    b. Loeffler's syndrome
    c. Primary pulmonary eosinophilia
    d. Sweet syndrome

Answers and Explanation of MCQs

1. a
2. a
3. d
4. a
- Follow section, **pyrexia of unknown origin (PUO) → etiological agents** and respective chapters for explanation of answers of MCQs 1–4.
5. a
6. a, c
7. a
- Follow section, **infective anemia** for explanation of answers of MCQs 5–7.
8. d
9. a
10. a, c
11. b
- Follow section, **eosinophilia in parasitic infections (causes of pulmonary eosinophilia)** for explanation of answers of MCQs 8–11.

CHAPTER 36

Lymphatic System Infections

Chapter Outline
- Common Infections
- Specific Infections

COMMON INFECTIONS

Lymphangitis

Definition: It is an inflammation of the lymphatic vessels due to infection at a site distal to the vessel.

Etiological agents: It is caused by *Strept. pyogenes* (most common in superficial lymph vessels called erysipelas), *Pasteurella multocida* and *Aeromonas hydrophila*. Nodular lymphangitis occurs due to superficial infection by *Sporothrix schenckii, Nocardia brasiliensis, Mycobacterium marinum* or *Francisella tularensis*.

Laboratory diagnosis
- Diagnosis of bacteria: **Ch. 118.**
- Diagnosis of fungi: **Ch. 120.**
- Diagnosis of parasites: **Ch. 121.**

Lymphadenitis/Lymphadenopathy

Synonym: Adenitis.

Definition: It is an inflammation (lymphadenitis) and enlargement (lymphadenopathy) of lymph node. Both words are used as synonym for each other.

Etiological agents: Follow **Table 36.1**.

Laboratory diagnosis
- **Specimens:** Blood or lymph node biopsy
- **Testing methods**
  - Diagnosis of bacteria: **Ch. 118.**
  - Diagnosis of viruses: **Ch. 119.**
  - Diagnosis of fungi: **Ch. 120.**
  - Diagnosis of parasites: **Ch. 121.**

SPECIFIC INFECTIONS

Bacterial Infections

- Tuberculosis lymphadenitis/King's evil/cold abscess/scrofula/scrophula (*M. tuberculosis*): **Ch. 56.**

TABLE 36.1: Causes of lymphadenitis

Agents	Clinical picture
Bacteria	
Salmonella spp.	Generalized
Yersinia enterocolitica	Cervical or abdominal
Francisella tularensis (tularemia)	Cervical, mediastinal, or generalized: Tender
Brucella spp.	Granulomatous nodes
Prevotella melanogenica	Less common
Rickettsia akari (rickettsial pox)	Regional
Viruses	
Cytomegalovirus	Generalized
HHV-8	Plasma cell variant of Castleman's disease
EBV (infectious mononucleosis)	Anterior cervical, mediastinal, bilateral: Discrete, firm and painless
Parvo virus and rubella virus	Posterior auricular, occipital, posterior cervical
Fungi	
Coccidioides immitis and *Histoplasma capsulatum*	Mediastinal
Parasites	
Trypanosoma cruzi	Enlarged nodes
Leishmania donovani	In African and Chinese form
Autoimmune	
SLE and RA	Generalized

- Cervical lymphadenitis/King's evil/scrof(ph)ula (NTM → *M. scrofulaceum*): **Ch. 58.**
- Bubonic plague (*Yersinia pestis*): **Ch. 64.**
- Cat scratch disease/stellate abscess (*Bartonella henselae*): **Ch. 76.**

Viral Infections

- Kaposi's sarcoma in (HHV-8): **Ch. 78.**
- Persistent generalized lymphadenopathy (PGL) in AIDS (HIV): **Ch. 85.**

Parasitic Infections

- *Toxoplasma* lymphadenopathy/Piringer–Kuchinka lymphadenopathy (*T. gondii*): **Ch. 98.**
- Winterbottom's sign (*Trypanosoma brucei*): **Ch. 95**.
- Lymphatic filariasis (*W. bancrofti*, *B. malayi* and *B. timori*): **Ch. 108**.

Essay/Full Question

1. Describe the causative agent, pathogenicity, life cycle and diagnosis of lymphatic filariasis.

Short Note

1. Lymphangitis/lymphadenitis.

MCQ for Chapter Review

Lymphangitis

1. **Most common cause of lymphangitis is**
   a. *Mycobacterium tuberculosis*
   b. *Strept. pyogenes*
   c. *Pasteurella multocida*
   d. *Aeromonas hydrophila*

Answer and Explanation of MCQ

1. **b**
- Follow section, **lymphangitis → etiological agents** for explanation.

CHAPTER **37**

Gastrointestinal Tract infections

Chapter Outline

- Introduction
- Common Infections
- Diarrheal Diseases
- Specific Infections

INTRODUCTION

Anatomy

It includes esophagus, stomach, duodenum, jejunum, ileum, appendix, cecum, ascending colon, transverse colon, descending colon, sigmoid colon, rectum and anal canal.

Normal Defense of GIT

Gastric acidity resists the growth of acid sensitive bacteria. High alkali of intestine resists the growth of certain bacteria.

Normal (Resident) Flora of GIT

Follow **Ch. 4.**

COMMON INFECTIONS

Esophagitis

It is an inflammation of the esophageal mucosa. It is caused by *Candida* spp. (most common), CMV and HSV. It occurs in persons with HIV, hematological malignancy or immunosuppressive therapy. It presents with dysphagia (difficulty or pain during swallowing) or foreign body sensation that gives a feeling that something is lodged in throat.

Gastritis

It is an inflammation of the gastric mucosa. It is caused by *Helicobacter pylori* (most common). It occurs in persons with NSAIDs use, autoimmune problems, smoking, alcohol use, cocaine use, radiation therapy or Crohn's disease. It presents with nausea, vomiting, upper abdominal pain (mostly central; however, it may occur anywhere), belching, bloating, anorexia, unexplained weight loss and heartburn.

Acid Peptic Disease

It is the disease of stomach and duodenum due to excessive presence of acid and pepsin. Etiological agents are same as gastritis. It presents with gastric and duodenal ulcer. For diagnosis and management of *H. pylori*, follow **Ch. 66.**

Enteritis

It is an inflammation of the small intestinal mucosa. It presents with fever, nausea, vomiting, abdominal pain, cramping, diarrhea, dehydration and weight loss. Following are the common etiological agents.

- Bacteria: *Campylobacter jejuni* (most common), *Staph. aureus, Bacillus cereus, Clostridium perfringens, Clostridium difficile, E. coli, Salmonella* spp., and *Shigella* spp.
- Viruses: *Rotavirus* and *Norovirus* (species → Norwalk virus).
- Other causes: Same as gastritis + celiac disease.

Colitis

It is an inflammation of the colon (large intestine). Common etiological agents are EHEC/VTEC/Shiga toxigenic *E. coli* (STEC) which causes hemorrhagic colitis (dysentery), *Shigella dysenteriae* and *Entamoeba histolytica*. It is autoimmune in ulcerative colitis and Crohn's disease. It presents with abdominal pain and tenderness, hemorrhagic diarrhea, fecal incontinence, flatulence, fatigue, anorexia, unexplained weight loss and fever.

Cecitis

It is an inflammation of the cecum. It is also known as typhlitis from typhlon (Greek) means cecum or neutropenic enterocolitis, because it is associated with neutropenia or ileocecal syndrome, because actually it involves terminal ileum and cecum. Sometimes, also it involves appendix and ascending colon. It is the polymicrobial infection caused by aerobes and anaerobes like *Enterococcus, Clostridium septicum,* members of Enterobacericeae and *Pseudomonas aeruginosa.* Use of cytotoxic drugs in cancer patients is the most common precipitating factor. It presents with nausea, vomiting, abdominal pain, tenderness, diarrhea, distended abdomen, chills and fever.

Appendicitis

It is an inflammation of the appendix. It is caused by aerobic and anaerobic GNB. It presents with nausea, vomiting, abdominal pain, tenderness, diarrhea and fever. Intestinal obstruction, perforation, peritonitis and intra-abdominal abscess are the complications.

Diverticulitis

It is an inflammation of the diverticula, which is actually the pouch develops in large intestinal wall. It is caused by aerobic and anaerobic GNB. It presents with nausea, vomiting, lower abdominal pain, diarrhea or constipation, fever or blood in stool.

Proctitis

It is an inflammation of the rectum. It is caused by microbes which are transmitted by anal intercourse like *Shigella* species, etc. It presents with itching and mucus discharge per rectum that progress to ulcer and abscess formation in rectum.

Infective Malabsorption Syndrome

Definition: Malabsorption is defined as impaired the intestinal absorption of nutrients mainly fats and other like proteins, carbohydrates, vitamins and minerals.

Etiological types: Only microbiological/infective types are discussed here.

1. **Primary:** It is due to deficiency of absorptive mucosal surface and associated enzymes.
   - **Tropical sprue:** Exact cause is not known but may be due to secretion of exotoxin by *E coli* that causes intestinal injury resulting malabsorption of B12 and folic acid producing macrocytic anemia. It presents in tropical countries like West Indies, Hong Kong, South India and Sri Lanka.
   - **Whipple's disease: Ch. 60.**
   - **Allergic and eosinophilic gastroenteritis:** It includes eosinophilic infiltration mainly in stomach followed by small and large intestine. It presents with protein loosing enteropathy and features of malabsorption.
2. **Secondary:** It is due to mucosal changes by infections, surgeries, drugs, etc., resulting impaired absorption, digestion and transport. Absorption will be impaired in *Giardia lamblia, Cyclospora cayetanensis, Cystoisospora belli, Cryptosporidium parvum, Ascaris lumbricoides* (due to antienzymatic actions) and *Capillaria phillipinensis.* Digestion and transport will be impaired in tuberculosis.

Gastroenteritis

Definitions

- **Gastroenteritis:** It is an inflammation of mucosa of stomach and intestine resulting vomiting, diarrhea, abdominal pain with or without mucus and blood in stool (dysentery) or fever or dehydration. It is better to describe the gastroenteritis with diarrhea and dysentery for easy understanding.
- **Diarrhea:** Passage of three or more liquid/semi-solid stools per day, in excess than usual habit for that person (WHO). In acute case, it lasts for <14 days.
- **Dysentery:** Presence of blood and mucus in stool often with fever, abdominal pain and tenesmus. Tenesmus means desire of defecation without resulting evacuation.

Synonym: Few authors referred gastroenteritis as an infectious diarrhea.

Etiopathogenetic classification: Clinical presentation of gastroenteritis is based on mechanisms of actions of bacterial toxin as described below. It also covers the microbes other than bacteria having tendency to cause watery diarrhea and dysentery, but may be by different mechanisms.

1. **Enterotoxicity:** Outpouring of electrolytes and fluid into intestinal lumen responsible for diarrhea called watery diarrhea or secretory diarrhea or non-invasive diarrhea. It is noninflammatory (no WBCs/pus cells in stool). It is noninvasive to epithelial cells (no cell death → no bleeding → no RBCs and mucus in stool and no systemic features like fever). Common agents for **watery diarrhea** are:
   - **Bacteria:** *Vibrio cholerae,* ETEC, EPEC, EAEC, *Staph. aureus, Bacillus cereus* with diarrheal illness, *Clostridium welchii.*
   - **Viruses:** *Rotavirus,* Norwalk virus, adeno viruses, coronavirus, *Calcivirus, Astrovirus, Enterovirus,* etc.
   - **Fungi:** *Candida albicans,* etc.
   - **Parasites:** *Giardia lamblia, Cystoisospora belli, Cryptosporidium* spp., microsporidia and *Cyclospora* spp.
2. **Cytotoxicity:** Disruption of intestinal mucosa leaving raw unprotected mucosa resulting passage of blood and mucus in stool called bloody diarrhea or dysentery or invasive diarrhea. It is inflammatory (WBCs/pus cells in stool). It is invasive to epithelial cells (cell death → bleeding → RBCs and ruptured mucosa in stool) and also extraintestinal in blood and lymph nodes (systemic features like fever). Common agents for **dysentery** are:
   - **Bacteria (bacillary dysentery):** *Clostridium difficile, Shigella* spp., EIEC (entero invasive *Escherichia coli*), EHEC (entero hamorrhagic *Escherichia coli*), *Campylobacter jejuni, Yersinia enterocolitica, Vibrio parahaemolyticus* (causing dysentery, but

nontoxigenic), *Aeromonas* spp., and may be few non-typhoidal *Salmonella* spp.

- **Parasite (parasitic dysentery):** *Entamoeba histolytica* (called amoebic dysentery), *Balantidium coli, Cystoisospora belli. Schistosoma* spp., like *S. mansoni, S. japonicum, S. intercalatum, Hetrerophyes heterophyes, Trichuris trichuria* and *Strongyloides stercoralis.*

3. **Neurotoxicity:**
   - Enterotoxin of *Staph. aureus* acts on autonomic nervous system rather than on gastrointestinal (GI) mucosa. It stimulates the vagus nerve and vomiting center. It also knows to stimulate the peristaltic activity.
   - Toxin of *Bacillus cereus*: It causes emetic illness.
   - Botulinum toxin acts on peripheral nerve system and produces the flaccid paralysis.
   - Verotoxins/verocytotoxins (shiga toxin and SLT): Verotoxins cause paralysis and death on injection in mouse/rabbit. They do not act directly on CNS, but act on the blood vessels of CNS, neurotoxic effect is secondary.

Mode of transmission: All pathogens mostly entered by ingestion.

Diagnosis of gastroenteritis
- **Specimens:** Stool, vomitus, rectal swab, water-food sample.
- **Testing**
  - Diagnosis of bacteria: **Ch. 118.**
  - Diagnosis of viruses: **Ch. 119.**
  - Diagnosis of fungi: **Ch. 120.**
  - Diagnosis of parasites: **Ch. 121.**

Food Poisoning

Synonym: Foodborne illness or foodborne disease.

Definitions
- **General definition:** Acute gastroenteritis caused by the ingestion of the food or drink contaminated with living bacteria and/or bacterial toxins, fungi and/or their toxins, viruses or parasites, inorganic chemical substances, poison delivered from plants or poison delivered from animals.
- **WHO definition:** Disease either infectious or toxic in nature, caused by agents that enter the body through the ingestion of food.
- **Intradietic toxin** or **preformed toxin:** Toxin is already present in food before ingestion called intra-dietic toxin or preformed toxin. Incubation period is short if toxin is preformed. It is present in *Staph. aureus, B. cereus* (emetic illness) and *Cl. botulinum* (foodborne botulism).

Predisposing factors
- Preparing food too far in advance
- Storing raw food at incorrect temperatures
- Incorrect cooling of food
- Incorrect heating of food
- Inadequate cooking of meats/poultry/fish
- Frozen food not fully thawed
- Cross contamination of cooked food by raw food
- Holding hot and chilled food at incorrect temperatures
- Contamination of food by food handlers.

General features: There is history of the ingestion of a common food. Similar C/Fs in majority of the cases, therefore, food poisoning must be suspected when an acute illness with gastrointestinal or neurological manifestations affects two or more persons, who have shared a meal during the previous 72 hours.

Classifications
1. **According to contains of food:** It may be of two types like infectious or intoxications as mentioned in **Table 37.1 and 37.2.**
2. **According to etiological agents**
   - **Bacterial food poisoning:** Bacteria are classified as per IP as shown in **Table 37.2.**
   - **Fungal food poisoning:** Ch. 92.
   - **Chemical food poisoning:** It occurs accidentally during insecticides spraying over food in farm or intentionally by adding flavoring agents or preservatives to food. It also occurs by products of food processing such as smoking of fleshy food or by radionucleotides.
   - **Animal food poisoning:** It occurs by fish and food from marine animals.
   - **Vegetable food poisoning:**
     - *Lathyrus sativus*/Kesari dhal (Lathyrism).
     - *Argemone maxicana*/Ujarkanta/Sialkanta/Kutila (Epidemic dropsy).
     - *Atropa beladona*/poisonous berries.
     - *Lolium temulentum*/Darnel.
     - Soya bean: It has trypsin inhibitors, which make protein unavailable for body.

TABLE 37.1: Differences between intoxication and infections

Intoxications	Infectious
Food contains preformed toxins	Food contains microbes which infect the body after consumption
Toxins (natural/preformed)	Bacterial/viral/parasitic
No invasion or multiplication	Invade and/or multiply in lining of intestine
IP: Minutes – hours	IP: Hours – days
C/Fs: Intestinal and extra-intestinal	C/Fs: Intestinal
Not communicable	Communicable – spreads from person to person
Factors: Inadequate cooking, improper handling temperatures	Factors: Inadequate cooking, poor personal hygiene, bare hand contact, cross contamination
Examples: **Table 37.2**	Examples: **Table 37.2**

TABLE 37.2: Bacterial food poisoning

Microbes	Symptoms	Sources of food
Intoxications (Preformed/intradietic toxin, so incubation period is very short)		
IP = 1–6 hours		
Staph. aureus	Nausea, vomiting and diarrhea. No fever	Meat, fish, milk and milk products
B. cereus (Emetic illness)	Nausea and vomiting. Rarely diarrhea. No fever	Rice (Chinese restaurant)
IP = 12–36 hours		
Cl. botulinum (Foodborne botulism)	Vomiting, thirst, constipation, ocular paresis, difficulty in swallowing-in speaking and in breathing	Preserved food—meat or meat products, fish, sea foods, canned vegetables
Infectious (Toxin formed inside the intestine after ingestion of organisms)		
IP = 8–16 hours		
B. cereus (Diarrheal illness)	Diarrhea, abdominal pain and fever	Vegetables and cooked meat
Cl. welchii (*Cl. perfringens*)	Abdominal pain/cramps, vomiting and diarrhea	Meat (like beef and poultry meat) and legumes
IP = More than 16 hours		
Cl. botulinum (Infant borne botulism)	Constipation, poor feeding, lethargy, weakness, pooled oral secretion, altered cry, floppiness and loss of head control	It occurs due to ingestion of food like honey contaminated by spores
Entero hemorrhagic Escherichia coli (EHEC)	Causing mild diarrhea to hemorrhagic colitis, HUS (Hemorrhagic uremic syndrome) and food poisoning in children and in elder group in developed countries	Vegetables (radish/alfalfa) or salads
Salmonella spp. (Like *S.* Typhimurium, *S.* Anatum etc.)	Cholera type diarrhea (passage of ≥1 loose stool) or dysentery type diarrhea (blood + mucus in stool) with abdominal pain, vomiting and fever	• Animal products like egg, meat, milk or milk-derivatives. • Also by ingestion of salads/vegetables which are contaminated by manure or poor handling. • Food contamination by dropping of rat, lizard and other small animals.
Infectious but nontoxigenic (No toxin formed before or after the ingestion of organisms)		
IP = More than 16 hours		
V. parahaemolyticus	Abdominal pain, nausea, vomiting, diarrhea and low grade fever.	Ingestion of sea (marine) foods like sea-fish, shrimps, crabs or mollusks (oyster). In Kolkata found in small pond fish

Notes:
- Remember food poisoning by heart because, almost in every year exam there is a MCQ of food poisoning.
- Incubation period helps to solve the MCQs based on food poisoning.
- Diarrhea is present in all except *Cl. botulinum*.

– Cabbage: Specially of Brassicaceae family has sulfar containing compound, which inhibits the thyroxine secretion.
– *Paspalam scrobiculatum*/Kodra.

Mode of transmission: All pathogens mostly entered by ingestion.

Diagnosis
- **Human samples:** Stool, vomitus or rectal swab: Tested by Gram's staining and culture methods.
- **Water, food and dairy products sample:** Are tested by special methods.

Traveler's Diarrhea

Definition: Diarrhea that develops during or shortly after, visiting the endemic areas.

Etiological agents
1. **Bacteria:**
   - ETEC (most common): Few authors defined ETEC is the most common agent for traveler's diarrhea.
   - EAEC (2nd most common): Few authors defined ETEC and EAEC are the most common agents for traveler's diarrhea.
   - Others: *Salmonella, Shigella, Vibrio cholerae, Vibro parahaemolyticus, Campylobacter jejuni* (more common in Asia), *Aeromonas hydrophila* and *Plesiomonas shigelloides*.
2. **Viruses:** *Rotavirus* (common in children) and Norwalk virus (Norwalk virus diarrhea are associated with traveling on cruise ships).
3. **Parasites:** *Entamoeba histolytica, Giardia lamblia* and *Cryptosporidium* spp., and *Cyclospora cayetanensis*.

Clinical features: Most cases begin within the first 3–5 days; characterized by a sudden onset of abdominal cramps, anorexia, and watery diarrhea. Self limited illness generally lasting for 1–5 days.

Persistent/Chronic Diarrhea

Definition: Persistent diarrhea means, which lasts for ≥14 days (usually 2–4 weeks) and chronic diarrhea means which usually lasts for >4 weeks.

Etiological agents:

- Bacteria: *Aeromonas, Campylobacter, Clostridium difficile* and *Plesiomonas*.
- Viruses: CMV common in person with IDDs.
- Parasites: *Cryptosporidium, Cyclospora, Entamoeba histolytica, Giardia,* microsporidia and *Faciola hepatica*.

SPECIFIC INFECTIONS

Bacterial Infections

- Necrotizing enteritis or enteritis necroticans or pigbel (*Clostridium perfringens*): **Ch. 55.**
- Necrotizing colitis: Not infective, but mostly occurs in colon cancer.
- Necrotizing enterocolitis: It is the fulminant disease in premature low birth weight infants leading to death. Caused by *Pseudomonas, Klebsiella,* EPEC O111:b4, *Salmonella, Clostridium perfringens,* and *Clostridium butyricum.* It presents with necrosis of bowel wall causing perforation and peritonitis.
- Pseudomembranous enterocolitis (*Clostridium difficile*): **Ch. 55.**
- Noninvasive listeriosis (*Listeria monocytogenes*): **Ch. 59.**
- Intestinal anthrax (*Bacillus anthracis*): **Ch. 54.**
- Enterobacterales infections: **Ch. 61.**
- Vibrionaceae, Aeromonadaceae and Plesiomonas infections: **Ch. 65.**
- Campylobacteriosis (*Campylobacter* spp.): **Ch. 66.**
- Helicobacter infections (*Helicobacter pylori*): **Ch. 66.**

Viral Infections

- Viral diarrhea (Diarrheal viruses): **Ch. 89.**

Parasitic Infections

Protozoal Infections

- Gastrointestinal amoebiasis (*Entamoeba histolytica*): **Ch. 93.**
- Giardiasis (*Giardia lamblia*): **Ch. 94.**
- Mega esophagus and mega colon by blood and tissue flagellate (*Trypanosoma cruzi*): **Ch. 95.**
- Infections of Ciliatea (*Balantidium coli*): **Ch. 96.**
- Infections of gastrointestinal sporozoa: **Ch. 98.**
- Infections of Microsporidia: **Ch. 99.**

GIT Infections of Cestodes

- Infections of Pseudophylladea (*Diphylobothrium latum*): **Ch. 101.**
- Infections of Cyclophylladea (*Taenia saginata, Taenia solium, Hymenolepis* spp., and *Dipylidium caninum*): **Ch. 102.**

GIT Infections of Trematodes

- **Infections of Diecious trematodes:** These are *Schistosoma mansoni* and *Schistosoma japonicum* habitat the rectal plexus and described in **Ch. 104.**
- **Infections of Monoecious trematodes:** These are monoecious trematodes of small intestine and large intestine with following examples.
  1. **Monoecious trematodes of small intestine:** These are *Fasciolopsis buski, Hetrophyes heterophyes, Matagonimus yokogawai, Echinostoma ilocanum, Echinostoma malayanum, Echinostoma revolutum, Paryphostomum supraryfex* and *Watsonius watsonis* described in **Ch. 105.**
  2. **Monoecious trematodes of large intestine:** This includes *Gastrodiscoides hominis* described in **Ch. 105.**

GIT Infections of Nematodes

1. **Small intestinal nematodes:** These are *Trichina spiralis, Ascaris lumbricoides, Ancyclostoma duodenale, Necator americanus, Trichostrongylus* species, *Strongyloides stercoralis, Strongyloides fuelleborni* and *Capillaria philippinensis* described in **Ch. 107.**
2. **Large intestinal nematodes:** These are *Trichuris trichuria* and *Enterobius vermicularis* described in **Ch. 107.**
3. **Nematodes in arteries and arterioles of ileocecal region:** This includes *Angiostrongylus costaricensis* described in **Ch. 107.**
4. **Visceral larva migrans:** This includes *Anisakis* spp., described in **Ch. 109.**

ACCESS YOURSELF

Essays/Full Questions

1. Enumerate the microbial agents causing diarrhea and dysentery. Describe the morphology, epidemiology, pathogenicity and diagnostic modalities of these agents.
2. Identify the common etiologic agents of diarrhea and dysentery.
3. Describe the etiology of acute and chronic infectious diarrhea.
4. Mention the enteric fever pathogens. Discuss the pathogenicity and the laboratory diagnosis of the diseases caused by them with appropriate test related to the duration of illness.
5. Enumerate the causative agents of food poisoning and discuss the pathogenicity and laboratory diagnosis.
6. Describe the etiopathogenicity of acid peptic disease. Discuss the diagnosis and management of the causative agent of acid peptic disease.

Short Notes

1. Infective malabsorption syndrome
2. Gastroenteritis

3. Food poisoning.
4. Traveler's diarrhea.

Short Questions for Theory/Viva Questions

1. Name four bacteria causing dysentery.
2. What is preformed toxin? Name two bacteria producing preformed toxin.

MCQs for Chapter Review

Infective Malabsorption Syndrome

1. **All cause malabsorption *except*:**
   a. Giardiasis
   b. *Ascaris lumbricoides*
   c. *Strongyloides*
   d. *Capillaria phillipinensis*
2. **Which of the following infection(s) leads to malabsorption?**
   a. *Giardia lamblia*
   b. *Ascaris lumbricoides*
   c. *Necator americanus*
   d. *Ancyclostoma duodenale*

Gastroenteritis

3. **Noninvasive diarrhea can be caused by one of following:**
   a. *Shigella*
   b. *B. cereus*
   c. *Salmonella*
   d. *Y. enterocolitica*
4. **An 18-year-old girl presents with watery diarrhea. Most likely causative agent:**
   a. *Rotavirus*
   b. *V. cholerae*
   c. *Salmonella*
   d. *Shigella*
5. **In a patient presenting with diarrhea and pus cells in stool, the causative organisms can be all *except*:**
   a. *Noncholera vibrio* 01
   b. Enterotoxigenic *E. coli*
   c. Enteroinvasive *E. coli*
   d. *Shigella dysentery* 1
   e. *Vibrio cholerae*
6. **Microorganisms invading the GIT causing gastroenteritis:**
   a. EHEC
   b. *Shigella*
   c. *Vibrio parahaemolyticus*
   d. *Campylobacter*
   e. *Salmonella*
7. **A child with fever with RBCs and pus in stool, causative organism is:**
   a. ETEC
   b. EHEC
   c. EPEC
   d. EAEC
8. **Pus cells in diarrhea seen in:**
   a. *Vibrio cholerae*
   b. EPEC
   c. *Rotavirus*
   d. *Shigella*
   e. *Campylobacter*
9. **Fecal leukocytes are absent in all the following *except*:**
   a. Giardiasis
   b. Cryptosporidiosis
   c. *Campylobacter* infection
   d. *Clostridium perfringens* infection

Food Poisoning

10. **A cook prepares sandwitches for 10 people going for picnic. Eight out of them develop severe gastroenteritis within 4–6 hours of consumption of the sandwitches. It is likely that on investigations the cook is found to be the carrier of:**
    a. *Salmonella* Typhi
    b. *Vibrio cholerae*
    c. *Entamoeba histolytica*
    d. *Staphylococcus aureus*
11. **A patient presents with vomiting. He had eaten rice 6 hours before. The most probable cause is:**
    a. *Bacillus cereus*
    b. *Staph. aureus*
    c. *Cl. difficile*
    d. All of above

12. **Kallu, a 22-year-old male had an outing with his friends and developed fever of 38.5°C, diarrhea and vomiting following eating chicken salad, 24 hours back. Two of his friends developed the same symptoms. The diagnosis is:**
    a. *S.* Enteritidis
    b. *Bacillus cereus*
    c. *Staph. aureus*
    d. *Vibrio cholerae*
13. **Most probable cause of food poisoning in a child who has eaten ice cream 16–18 hours earlier.**
    a. *Staph. aureus*
    b. *Clostridium perfringens*
    c. *Clostridium botulinum*
    d. *S.* Typhimurium
14. **Thirty eight children consumed eatables, procured from a single source at a picnic party. Twenty children developed abdominal cramps followed by vomiting and watery diarrhea 6–10 hours after the party. The most likely etiology for the outbreak is:**
    a. *Rotavirus* infection
    b. *Enterotoxigenic E. coli* infection
    c. Staphylococcal toxin
    d. *Clostridium perfringens* infection
15. **An adolescent male developed vomiting and diarrhea 1 hour after having food from a restaurant. The most likely pathogen is:**
    a. *Clostridium perfringens*
    b. *Vibrio parahaemolyticus*
    c. *Staphylococcus aureus*
    d. *S.* Typhimurium
16. **Food poisoning with diarrhea within 6 hours:**
    a. *Staph. aureus*
    b. *Clostridium perfrigens*
    c. *Clostridium botulinum*
    d. *V. cholerae*
17. **Preformed toxin is important in food poisoning due to all *except*:**
    a. *Staph. aureus*
    b. *Clostridium botulinum*
    c. ETEC
    d. *B. cereus*
18. **Toxin is implicated as the major pathogenic mechanism in all of the following bacterial diarrhea *except*:**
    a. *Vibrio cholerae*
    b. *Shigella* sp.
    c. *Vibrio parahaemolyticus*
    d. *Staphylococcus aureus*
19. **Most common organism causing food poisoning in canned food:**
    a. *Staph. aureus*
    b. *Clostridium perfringens*
    c. *Clostridium botulinum*
    d. *V. cholerae*
20. **Incubation period for *B. cereus* food poisoning following consumption of contaminated fried rice is:**
    a. 1–5 hours
    b. 8–16 hours
    c. 24 hours
    d. >24 hours

Traveler's Diarrhea

21. **Traveler's diarrhea is caused by:**
    a. *Shigella*
    b. *E. coli*
    c. *E. histolytica*
    d. Giardiasis
22. **The most common causative organism for traveler's diarrhoea is:**
    a. *E. coli*
    b. *Shigella*
    c. Norwalk virus
    d. *Rotavirus*
23. **In which of the following is true regarding the cause of traveler's diarrhea?**
    a. Giardiasis
    b. *E. coli*
    c. Amoebiasis
    d. Idiopathic without any causative organisms

1. c

2. a, b

- Follow section, **infective malabsorption syndrome** for explanation of answers of MCQs 1–2.

3. b

- Few authors referred dysentery as invasive diarrhea, which is caused by *Shigella spp.*, *Cl. difficile*, EIEC, ETEC, *C. jejuni*, *V. parahaemolyticus*, *Y. enterocolitica* and *Salmonella* spp.

4. b

- *Rotavirus* and *V. cholerae* are the agents of watery diarrhea from the given option, but *Rotavirus* causes watery diarrhea in children, so *V. cholerae* is the right answer.

5. a, b and e

- Given case is of dysentery (pus cells/RBCs in stool) which is caused by many bacteria and parasites, but in given question only bacteria are given as options.

- Follow section, **gastroenteritis (etiopathogenetic classification → cytotoxicity)** for explanation.

6. a, b, c, d, e

7. b

8. d, e

9. c

- Follow section, **gastroenteritis (etiopathogenetic classification → cytotoxicity),** for explanation of answers of MCQs 6–9.

10. d

- Given case is of food poisoning within 6 hours of consumption of sandwitches mostly caused by *B. cereus* or *Staph. aureus. B. cereus* is not given in option, so right answer is *Staph. aureus.* Follow **Table 37.1 and 37.2** for more explanation.

11. a

- Given case is of food poisoning with emetic illness of *B. cereus* which is transmitted by rice from Chinese restaurant. Follow **Table 37.1 and 37.2** for more explanation.

12. a

- Given case is of food poisoning with 24 hours incubation period due to consumption of chicken and salad. *Salmonella* Enteritidis is the right answer from the given options. Follow **Table 37.1 and 37.2** for more explanation.

13. d

- Given case is of food poisoning with 24 hours incubation period due to consumption of milk products (icecream). *Salmonella* Typhimurium is the right answer from the given options. Follow **Table 37.1 and 37.2** for more explanation.

14. d

- Given case is of food poisoning with 8–16 hours incubation period. Children are present with abdominal pain/cramps, vomiting and diarrhoea, so agent is *Cl. perfringens.* Follow **Table 37.1 and 37.2** for more explanation.

15. c

16. a

17. c

18. c

19. c

20. a

- Follow section, **gastroenteritis (food poisoning, and Table 37.2** for explanation of answers of MCQs 15–20.

21. a, b, c, d

22. a

23. b

- Follow section, **traveler's diarrhea** for explanation of answers of MCQs 21–23.

Infections of Abdominal Viscera

Chapter Outline

COMMON INFECTIONS

Peritonitis

Definition: It is an inflammation of the peritoneum.

Etiological types and agents: Two types, both are differ in their clinical presentations and etiology.

1. **Primary (spontaneous):** No apparent primary source of infection but usually it occurs in presence of other disease like cirrhosis of the liver or other conditions. It is mostly caused by GNB like *E. coli* or GPC like *Streptococcus, Enterococcus,* or *Strept. pneumoniae.*
2. **Secondary:** It results from spillage of bacteria due to rupture of adjacent organs like *E. coli* and *Bacteroides fragilis* in intestinal perforation. It also occurs in peritoneal dialysis by staphylococci due contamination of skin flora.

Clinical features

1. **Primary:** Fever (most), abdominal pain and ascites.
2. **Secondary:** Depends on primary condition like epigastric pain in ruptured gastric ulcer or right upper quadrant pain in rupture appendix.

Peritoneal Abscess

It is the pus formation in peritoneum due to peritonitis mostly by *Bacteroides fragilis* and fatal in nature.

Cholangitis

It is an inflammation of the bile duct. It presents with nausea, vomiting, right upper quadrant pain, fever, rigor, malaise and jaundice. It is caused by following agents:

1. **Bacteria:** It occurs when bacteria (polymicrobial) like *E coli, Klebsiella, Enterococcus, Enterobacter,* etc., invade the obstructed biliary tract (biliary stones).
2. **Viruses:** Hepatitis viruses and HIV.
3. **Parasites:** Following are the different mechanisms and concerned parasites causing infection or obstruction.
   - Mechanical obstruction by adult parasite of *Ascaris lumbricoides* during ectopic ascariasis.
   - Intrabiliary rupture or pressure on biliary tract in case of hydatid cyst of *Echinococcus granulosus.*
   - Mechanical obstruction, liberation of toxins, secondary bacterial infections, stone formation or cancer production in *Chlonerchis sinensis* and *Opisthorchis* spp.
   - Mechanical obstruction/inflammation in *Faciola hepatica.*

Cholecystitis

It is an inflammation of the gallbladder. Blockade of bile flow due to gallstone favors the infection by *E. coli, Klebsiella, Streptococcus, Clostridium* species, etc. It presents with nausea, vomiting, right upper quadrant pain and fever.

Splenic Abscess

It is the pus formation in spleen. It occurs rarely as primary infection but secondary by *Streptococcus* and *Staph. aureus* by blood route from infective endocarditis. It presents with enlarged spleen, pain in left upper quadrant which may spread to left shoulder, inablity to eat a large meal and feeling of fullness.

Splenomegaly

It is an enlargement of spleen. It is caused by bacteria causing splenic abscess as mentioned above, *Mycobacterium tuberculosis,* EBV, HIV, *Leishmania donovani, Plasmodium* spp., *Schistosoma mansoni* and *Schistosoma japonicum. W. bancrofti* causes splenomegaly in the stage of occult filariasis. Features of splenomegaly are almost same as splenic abscess.

Pancreatitis

Definition: It is an inflammation of the pancreas.

Etiological agents: Mostly it is noninfectious origin due to gallstones and alcohol. Only 10% of cases are infectious origin by following microbes.

1. **Bacteria:** *Mycobacterium tuberculosis, Salmonella, Campylobacter* and *Mycoplasma*.
2. **Viruses:** CMV, EBV, VZV, mumps, coxsackie viruses, echo viruses, hepatitis viruses, HIV, measles, and rubella virus.
3. **Fungi:** *Aspergillus* and *Candida*.
4. **Parasites:** It is due to migration of adult worm of *Ascaris* in and out of the duodenum (ectopic ascaris). Other parasites causing pancreatitis are *E. granulosus*, *Clonorchis* species, *P. falciparum* and *Fasciola hepatica*.

Clinical features: It presents with abdominal pain (cardinal symptom), fever, tachycardia, vomiting, anorexia and diarrhea. It occurs in different clinical forms like (1) mild interstitial pancreatitis (in 80% cases), (2) more severe/necrotizing pancreatitis (occurs in 20% cases) and (3) pancreatic abscess which is the late complication of acute necrotizing pancreatitis, occurs in >4 weeks after the initial attack.

Liver Abscess

It is the pus formation in liver occurs due to following reasons.
1. **Bacterial/pyogenic liver abscess:** In 50–80% cases liver abscess is polymicrobial mostly caused by *E coli* and *Klebsiella pneumoniae*. In more than 50% of cases exact mode of transmission is unknown. Infection spreads locally from infected organs within the peritoneal cavity like the biliary tract (most common source) or ruptured appendicitis or blood borne from distant sites. It shows the collections of pus in one or more sites. It presents with fever and right upper quadrant pain (most consistent). If left untreated, it can be fatal due to rupture of the abscess into adjacent organs or body cavities like pleural cavity, lungs and intra-abdominal (peritonitis, abscesses or sepsis). It is diagnosed by microbiological tests like blood culture or culture of the pus obtained from liver under CT guidance, pathological tests like increased WBC count with increased liver enzymes and radiological tests like CT scan to locate the abscess. It is treated by surgical drainage and systemic antimicrobial therapy as per causative agents.
2. **Fungal liver abscess:** *Candida* species (<10% cases).
3. **Parasitic/amoebic (10% cases) liver abscess:** Ch. 93.

SPECIFIC INFECTIONS IN LIVER

Bacterial Infections
- Leptospirosis (*Leptospira*): **Ch. 73.**

Viral Infections
- Viral hepatitis (hepatitis viruses): **Ch. 88.**

Parasitic Infections
Infections of abdominal viscera by Protozoa
- Hepatic amoebiasis/amoebic liver abscess (*Entamoeba histolytica*): **Ch. 93.**
- Hepatomegaly due to visceral leishmaniasis (*Leishmania donovani*): **Ch. 95.**

Infections of abdominal viscera by Cestodes
- Hydatid cyst in liver (*E. granulosus*): **Ch. 102.**
- Multilocular hydatid cyst in liver (*Echinococcus multilocularis*): **Ch. 102.**
- Polycystic hydatid cyst in liver (*E. vogeli*): **Ch. 102.**
- Unicystic hydatid cyst in liver (*E. oligarthus*): **Ch. 102.**

Infections of abdominal viscera by Trematodes
- **Infections of Diecious trematodes** (hepatomegaly by ectopic lesion in *Schistosoma mansoni* and via portal vein in *Schistosoma japonicum*): **Ch. 104.**
- **Infections of Monoecious trematodes** (hepatic trematodes/liver flukes → *Clonorchis sinensis, Opsthorchis felineus, Fasciola hepatica, Fasciola gigantica, Dicrocoelium dendriticum* and *Erytrema pancreaticum*): **Ch. 105.**

Infections of abdominal viscera by Nematodes
- Hepatic capillariasis (*Capillaria hepatica*): **Ch. 108.**
- Hepatomegaly in occult filariasis (*W bancrofti*): **Ch. 108.**
- *Toxocara canis* and *T. cati*: **Ch. 109.**

ACCESS YOURSELF

Essay/Full Question
1. Describe the etiopathogenicity and viral markers in the diagnosis of viral hepatitis. Discuss the modalities in the prevention of viral hepatitis.

Short Note
1. Liver abscess.

Short Question for Theory/Viva Question
1. Name four parasites infecting liver/biliary tract.

MCQs for Chapter Review
Cholangitis
1. **Which of the following does not cause biliary tract obstruction?**
   - a. *Ascaris lumbricoides*
   - b. *Ancyclostoma duodenale*
   - c. *Clonorchis sinensis*
   - d. *Fasciola hepatica*
2. **Which of the following may cause biliary obstruction?**
   - a. *Ancyclostoma*
   - b. *Enterobius*
   - c. *Strongyloides*
   - d. *Clonorchis*
3. **A traveler presents with conjugated hyperbilirubinemia and on investigation eggs were found in his biliary tract. Likely organism:**
   - a. *Clonorchis sinensis*
   - b. *Fasciola buski*
   - c. *Gnathostoma*
   - d. *Ascaris*

Splenic Abscess
4. **Most common cause of splenic abscess is**
   - a. Streptococci
   - b. *Klebsiella*
   - c. *E. coli*
   - d. *Salmonella*

Answers and Explanation of MCQs
1. b
2. d
3. a, (d)
- Follow section, **cholangitis** for explanation of answers of MCQs 1–3.
4. a
- Follow section, **splenic abscess** for explanation.

CHAPTER 39

Skin and Subcutaneous Tissues Infections

Chapter Outline

- Introduction
- Terminology of Skin Infections
- Common Infections
- Specific Infections

INTRODUCTIONS

Anatomy of Skin

It includes following skin layers and skin appendages.

Skin layers: From outer to inner skin layers are epidermis, dermis, subcutaneous tissues and fascia.

- **Epidermis:** It includes 5 layers, from outer to inner are stratum corneum (dead/horny layer), stratum lucidum, stratum granulosum, stratum spinosum and stratum germinativum (basal layer).
- **Dermis:** It is a connective tissue contains blood vessels, nerve endings, some superficial hair follicles and sebaceous glands.
- **Subcutaneous tissues:** It is composed of loose connective tissues, fats, deep hair follicles and sebaceous glands.
- **Fascia:** It presents below the subcutaneous tissues and contains fibrous tissues which cover the muscles, ligaments and other connective tissues.

Skin appendages: These are hairs and nails.

Functions of Skin

1. **Physical:** Skin is the body's largest and thinnest organ, serves as an anatomical barrier and protects the internal organs from the external environment.
2. **Microbiological:** Presence of normal microbial (resident) flora in skin (**Ch. 4**) and chemical substance like high salt and acidic pH prevent the colonization by many exogenous microbes.
3. **Physiological:** It controls the water excretion, body temperature, salts excretion, synthesis of important chemicals and hormones.
4. **Neurological:** Act as a sensory organ.

TERMINOLOGY OF SKIN INFECTIONS

Common Terminology

Following are the different terms used in different skin diseases.

- Scales (epidermal scales): Flakes arising from dead/horny layer of epidermis
- Crusted lesions: Scale like lesion, composed of dried blood and tissue fluid
- Ulcer: It is the breakdown of natural continuity of skin or mucosa
- Macule: Flat and nonpalpable discoloration of skin in 5 mm size
- Patch: Flat and nonpalpable discoloration of skin in >5 mm size
- Papule: Elevated palpable solid lesion in 5 mm size
- Nodule: Elevated palpable solid lesion in >5 mm size
- Plaque: Elevated palpable solid lesion in >2 cm size
- Vesicle: Fluid filled blister in 5 mm size
- Bulla: Fluid filled blister in >5 mm size
- Pustule: Pus filled lesion in 5 mm size
- Abscess: Pus filled lesion in >5 mm size
- Petechiae: Extravasation of blood into skin in <2 mm size
- Purpura: Extravasation of blood into skin in 2 mm–1 cm size
- Ecchymosis: Extravasation of blood into skin in >1 cm size.

Specific Terminology

- Eschar: Black scar formed following healing of primary wound in *Bacillus anthracis* (malignant pustule in cutaneous anthrax), *Rickettsia akari* (rickettsial pox), *Orientia tsutsugamushi* (scrub typhus) and also following burn injury.
- Rose spots: **Ch. 62** (*Salmonella* Typhi → Clinical features).

- Green nail: **Ch. 67** (*Pseudomonas aeruginosa* → Clinical features).
- Koplik's spot: **Ch. 82** (Measles virus → Clinical features).

COMMON INFECTIONS

Wound Infection

Definition: It is the injury to living tissues by living or nonliving objects in which tissue/skin has been broken.

Types

1. **Surgical wound infection/surgical site infection (SSI): Ch. 29** (Types of HAIs).
2. **Bite wound infection: Ch. 3** (Modes of transmission of infections → Saliva-borne infections).
3. **Burn wound infection**
   - **Definition:** Infection that occurs at the site of wound caused by burn or thermal injury.
   - **Etiological agents:** Initially, the infection occurs by GPC like streptococci and staphylococci from the surrounding tissue. By seven days infection occurs by bacteria like *P. aeruginosa* (most common and characterized by bluish discoloration), *Acinetobacter, Escherichia coli, Klebsiella* and *Staph. aureus,* fungi like *Candida, Mucor* and *Aspergillus* and virus like HSV.
   - **Pathogenicity**
     - Sources of infection and modes of transmission: Organisms are acquired exogenously from hospital environment or endogenously as normal flora.
     - Clinical features: Burn wound impetigo, cellulitis, purulent or exudative discharge, local pain, fever, etc.
   - **Laboratory diagnosis**
     - Specimen: Swab or pus/discharge collection from wound.
     - Testing methods: Microscopy and culture.
   - **Prevention:** Keep the wound clean.
   - **Treatment:** Do the debridement of necrosed or burned tissues. Do the wound dressing by using silver sulfadiazine cream, mafenide acetate cream, silver nitrate cream, and nanocrystalline silver. Use the systemic antibiotics as per AST report.
4. **Traumatic/accidental wound infection**
   - **Definition:** Infection that occurs at the site of wound caused by vehicle accidents or by violence.
   - **Etiological agents:** Wound may be closed or open type. Wound with infection has 5 times more mortality then noninfected wound. Early microbes causing wound infections are *Staphylococcus* spp., *Streptococcus* spp., anaerobic streptococci and GNB. *Vibrio vulnificus* and *Aeromonas hydrophila* cause wound infections on exposure of sea water. In penetrating abdominal trauma especially to colon, infections are caused by *Bacteroides fragilis*, aerobic intestinal GNB and *Enterococcus*. Traumatic/accidental wound infections also include some specific infections like local varieties are gas gangrene (*Clostridium perfringens*) and necrotizing fasciitis (polymicrobial by *Streptococcus pyogenes* and others) while systemic variety is tetanus (*Clostridium tetani*).

   - **Pathogenicity**
     - Sources of infection and modes of transmission: Organisms are acquired exogenously from environment or endogenously as normal flora.
     - Precipitating factors: Traumatic/accidental wound infection is caused by aerobes or anaerobes. Presences of comorbid conditions like old age, poor nutrition, diabetes mellitus, prolonged hospitalization, obesity, etc., are increasing the chances of infection. Other risk factors are soil contamination, contact with water, presence of foreign bodies like clothes, dung, bullet pieces, etc., also increasing the chances of infection.
     - Clinical features: Bloody, purulent or exudative discharge, local pain, fever, etc.
   - **Laboratory diagnosis:** Same as burn wound infection.
   - **Prevention:** Keep the wound clean and take injection of TT.
   - **Treatment:** Wound dressing and systemic antibiotics as per AST report.

Diabetic Foot Ulcer Infection

Definition: It is an open sore or wound occurs in about 15% cases of diabetes and mainly located on the bottom of the foot.

Etiological agents: Polymicrobial infection mostly caused by *Staph. aureus, Strept. pyogenes,* members of the Enterobacterales, *Pseudomonas aeruginosa* plus anaerobes like *Bacteroides fragilis* and others.

Pathogenicity

- **Precipitating factors**: High blood glucose impaired blood circulation in microvessels and produces peripheral motor neuropathy. Subsequent skin injury increases the risk of infections. It occurs in type 1 and 2 diabetes mellitus. It mostly occurs in foot and heals very slowly compare to healthy individual. Other risk factors are old age, smoking, foot injury, neuropathy, vascular diseases or other local/systemic disease.
- **Clinical features:** Ulcer is painless. Depending on the severity, a diabetic foot ulcer may be rated between 0–5.
  - 0: No ulcer but there is a high risk to develop
  - 1: Superficial ulcer without infection
  - 2: Deep ulcer involves ligaments, muscles, tendons and joints
  - 3: Deep ulcer with abscesses

– 4: Deep ulcer with surrounding tissues (limited to the toes and forefoot) begun to black (gangrene) due to lack of healthy blood flow to the foot.
– 5: Extensive gangrene that spread from the localized area to cover whole foot.
- **Complications:** These are sinus formation, cellulitis in surrounding skin and osteomyelitis in underlying bone. Once ulcer develops and if not treated it increases the 8-fold risk to produce lower-extremity amputation.

Laboratory diagnosis

- **Specimen:** Swab is not much useful. Pus or discharge or tissue is collected from wound.
- **Testing methods:** Microscopy and culture.

Prevention: Diabetes should be controlled to reduce neuropathic and vascular complications. Avoid all host factors as mentioned above. Trim toenails regularly. Wear the comfortable foot like diabetic shoes and socks. Do regular exercises, physiotherapy and frequent foot examination.

Treatment: Wound is treated by regular dressing and using systemic antibiotics as per AST report.

Decubitus Ulcer Infection

Synonym: Bed sore or pressure sore or pressure ulcer.

Definition: Injuries to skin and underlying tissues due to prolonged pressure on the skin and limited movement.

Etiological agents: Decubitus ulcer is mostly infected by gastrointestinal flora like *Bacteroides fragilis* or hospital bacteria like *Staph. aureus* and *P. aeruginosa*.

Pathogenicity

- **Precipitating factors:** Constant pressure against the skin limits the blood flow to the skin and increases the chances of infection. It is more common in older or patients with chronic illness (like diabetes, vascular diseases, poor nutrition and hydration) that limit their movement to change positions or cause them to spend most of their time in a bed/chair. Limited movement makes the skin most vulnerable to bedsores. It is common on bony areas like heels, ankles, back of knees, hips, tailbone/buttocks, back of body, shoulders and back or sides of the head.
- **Clinical features:** It presents with unusual changes in skin color or texture, swelling, pus-like draining, skin that feels cooler or warmer to the touch than other areas and tenderness.
- **Complications:** These are cellulitis in surrounding tissues, spread of infection deep in to muscles and bones and squamous cell carcinoma. Long-term, non-healing wound (Marjolin's ulcer) can cause squamous cell carcinoma.

Laboratory diagnosis

- **Specimen:** Swab or pus or discharge or tissues collected from ulcer.
- **Testing methods:** Microscopy and culture.

Prevention: It is prevented by shifting the position frequently to relieve pressure. Select cushions or a mattress that relieves pressure. Keep skin clean and dry. Use moisture barrier creams to protect the skin from urine and stool contamination. Change bedding and clothing frequently, if needed.

Treatment: Wound is treated by debridement of necrosed tissues, regular dressing and using systemic antibiotics as per AST report.

Fournier Gangrene

Synonym: Polymicrobial necrotizing fascitis.

Definition: It is a type of necrotizing fascitis or gangrene affecting the external genitalia (scrotum, penis and vulva). Perianal and/or perineum (area between anus and scrotum/vulva) areas.

History: It was first described by Baurienne in 1764 and is named after a French venereologist, Jean Alfred Fournier, following five cases he presented in clinical lectures in 1883.

Etiological agents: It is a polymicrobial infection with an average of 4 isolates per case like *E. coli* (most predominant aerobe) and *Bacteroides* (most predominant anaerobe). Other agents are *Staphylococcus*, *Streptococcus* (aerobic and anaerobic), *Enterococcus*, *Pseudomonas*, *Klebsiella*, *Proteus* and *Candida albicans*.

Pathogenicity

- **Precipitating factors**
  1. **Sex:** Males are affected about 40 times more than females. In women it occurs following septic abortions, vulvar or Bartholin's gland abscesses, hysterectomy and episiotomy. In men, anal intercourse may increase risk of perineal infection, either from blunt trauma to the area or by spread of rectally carried microbes.
  2. **Age:** It is common in extreme age. In children, it occurs following circumcision, omphalitis, insect bites, trauma, urethral instrumentation, perirectal abscesses, systemic infections
  3. **Comorbid conditions:** Risk is increased by HIV, long-term corticosteroid therapy, diabetes mellitus (in 60% of cases), obesity, alcoholism, vascular disease of the pelvis, malignancy, SLE, Crohn's disease and malnutrition.
  4. **No temperature effect:** Gangrene is affected by low temperature (hypothermia), but in Fournier gangrene hypothermia does not occur.
- **Clinical features:** Common features are fever, pallor, dehydration, generalized weakness, pain and swelling in the genitals or anal area, foul odor and purulent discharge from the infected tissues, crackling sound (crepitus) when touching the affected area and anemia. It begins as a subcutaneous infection; very soon necrotic patches appear in the overlying skin which later develops into gangrene.

Diagnosis: Three types.

1. **Clinical diagnosis:** Fournier gangrene is usually diagnosed clinically.
2. **Radiological diagnosis:** X-rays and ultrasounds may show the presence of gas below the surface of the skin. CT scan can be useful in determining the site of origin and extent of spread.
3. **Laboratory diagnosis**
   - **Specimens:** Different body fluid, discharge or biopsy materials, blood, etc.
   - **Testing methods**
     – Biochemical diagnosis: It includes liver function test, renal function test and other biochemical analysis.
     – Pathological diagnosis: It includes hematological and histological diagnoses.
     – Microbiological diagnosis: It includes microscopy, culture, and other tests.

Treatment: Fournier gangrene is a urological emergency which requires IV antibiotics and debridement of necrotic (dead) tissue (**Fig. 39.1a**). In addition to surgery and antibiotics, hyperbaric oxygen therapy (HBOT) may be useful and acts to inhibit the growth and to kill the anaerobic bacteria.

Paronychia

Meaning: Para (Greek) means around and onyx (Greek) means nail with suffix ia.

Synonym: It also called bar rot in the context of bartending, or run around paronychia, when it occurs around the entire nail.

Definition: It is an inflammation of skin around nail.

Etiological agents: It occurs mostly by *Staph. aureus* or *Candida albicans*. Primary inoculation of tuberculosis to the skin and nails called prosector's paronychia, named after its association with prosecutors, who prepare specimens for dissection. It could be classified as an acute or chronic as follows.

1. **Acute paronychia:** It lasts for ≤6 months, mostly caused by bacteria, affects finger nail due to biting, picking hangnails, manicure or due to pushing down cuticles too aggressively.

2. **Chronic paronychia:** It lasts for >6 months, mostly caused by *Candida* and bacteria, affects toe or finger nail due to repeated hands washing (water contact favors the *Candida* infection).

Pathogenicity

- **Precipitating factors:** It occurs due to repeated hands washing, repeated nail biting, splinter/thorn injury, picking at a hang nail, finger sucking, in-growing toe nail, manicure procedures, or chemical injuries by acids/alkalis. It mostly affects the index and middle fingers. Risk increased with presence of diabetes, drug induced immuno-suppression, pemphigus, etc.
- **Clinical features:** Common features are redness, swelling, pus filled blister (**Fig. 39.1b**), pain or discharge.
- **Complications:** These are changes in nail shape, color, or texture and detachment of nail.

Laboratory diagnosis

- **Specimen:** Swab/pus/discharge/tissues are collected from lesion.
- **Testing methods:** Microscopy and culture.

Prevention: It is prevented by avoiding above mentioned risk factors.

Treatment: It is treated by using antibacterial and antifungal drugs. Incision and drainage is done, if pus is present.

Meleney's Ulcer

It also called as progressive synergistic gangrene. It is the slowly progressive chronic infection of the subcutaneous tissue which usually begins as an ulcer following trauma or surgery and progress to subcutaneous necrosis. It is a polymicrobial infection mostly caused by *Staph. aureus*, streptococci and anaerobes. Sample should be taken from periphery of wound for better recovery of aerobic organisms.

SPECIFIC INFECTIONS

Bacterial Infections

- Skin infections of Micrococcaceae: **Ch. 49.**
- Streptococcaceae infections of skin, subcutaneous tissues, fascia and muscles including skin lesions in ARF (*Strept. pyogenes*): **Ch. 50.**
- Cutaneous diphtheria (*Corynebacterium diphtheriae*): **Ch. 53.**
- Cutaneous anthrax (*Bacillus anthrcis*): **Ch. 54.**
- Gas gangrene (*Clostridium perfringens*): **Ch. 55.**
- MSP: **Ch. 58.**
- Erysipeloid (*Erysipelothrix rhusiopathiae*): **Ch. 59.**
- Actinomycosis (*Actinomyces* spp.): **Ch. 60.**
- Nocardiosis (*Nocardia* spp.): **Ch. 60.**
- Skin infections by members of Enterobacterales (*E coli, Klebsiella, Proteus, Enterobacter, Serratia, Providentia* and *Morgnella*) like cellulitis, abscess, wound infection: **Ch. 61.**

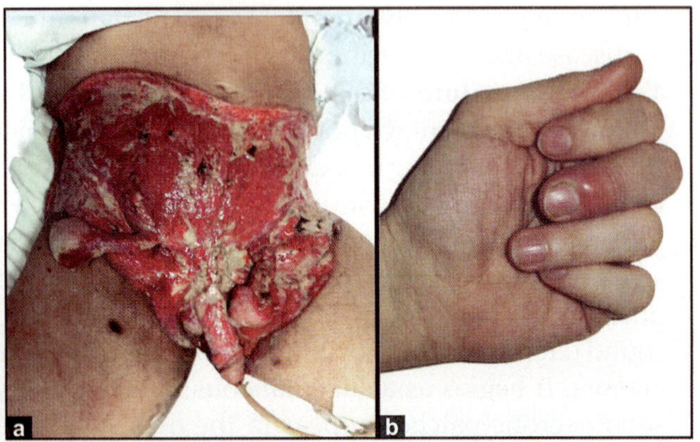

Fig. 39.1: (a) Fournier gangrene after debridement; (b) Paronychia

- Skin infections by nonfermenters like *Pseudomonas* infections, *Acinetobacter* infections, malleus (*Burkholderia mallei*) and malioidosis/Vietnam time bomb (*Burkholderia pseudomallei*): **Ch. 67.**
- Rat bite fever: **Ch. 71.**
- Nonsporing anaerobic infections: **Ch. 72.**
- Nonvenereal syphilis: **Ch. 73.**
- Endemic (nonvenereal) treponematoses: **Ch. 73.**
- Erythema migrans in Lyme disease: **Ch. 73.**
- Typhus fever and spotted fever (*Rickettsia*): **Ch. 76.**
- Verruga peruana (*Bartonella*): **Ch. 76.**

Viral Infections

- Skin infections of Pox viridae: **Ch. 77.**
- Skin infections of Herpesviridae (HHV1, 2, 3, 6, 7 and 8): **Ch. 78.**
- Exanthematous viruses (better to learn as exanthematous agents, because forgetting bacteria makes the topic incomplete and confusing): **Ch. 78.**
- Skin infections of Papovaviridae (HPV): **Ch. 79.**
- Skin infections of Parvoviridae (parvovirus B 19): **Ch. 79.**
- Skin infections of Picornaviridae (HFMD by A16, EV-71 and measles): **Ch. 80.**
- Rubella (rubella virus): **Ch. 86.**
- Skin lesions by hemorrhagic fever viruses: **Ch. 89.**

Fungal Infections

Superficial mycoses
- Surface mycoses: **Ch. 90.**
- Cutaneous mycoses: **Ch. 90.**

Deep mycosis
- Subcutaneous mycoses: **Ch. 91.**
- Skin manifestations (disseminated infections) of systemic mycoses (*Histoplasma, Blastomyces, Coccidioides* and *paracoccidioides*): **Ch. 91.**
- Disseminated skin infections in opportunistic mycoses (*Cryptococcus*): **Ch. 91.**
- Direct or disseminated skin infections in aspergillosis (*Aspergillus flavus*) and systemic zygomycosis (*Rhizopus* and *Mucor*): **Ch. 91.**

Parasitic Infections

Protozoal skin infections
- Cutaneous amoebiasis (*E. histolytica*): **Ch. 93.**
- Skin ulcer by *Acanthamoeba*: **Ch. 93.**
- Trypanosomal chancre (*Trypanosoma brucei*): **Ch. 95.**
- Chagoma (*Trypanosoma cruzi*): **Ch. 95.**
- Skin manifestation of visceral leishmaniasis and PKDL (*Leishmania donovani*): **Ch. 95.**
- Old world cutaneous leishmaniasis (*L. tropica, L. major* and *L. aethiopica*): **Ch. 95.**
- New world cutaneous leishmaniasis (*L. braziliensis, L. maxicana* and *L. amazonensis*): **Ch. 95.**

Skin infections of Cestodes: Sparganosis (*Spirometra*) mentioned in **Ch. 101.**

Skin infections of Tremaodes: Dermatitis by schistosomes (cercarial dermatitis) at the site of bite (*Schistosoma* spp.) mentioned in **Ch. 104.**

Skin infections of Nematodes

- Subcutaneous filariasis (*L. loa, O. volvulus* and *M. strptocerca*): **Ch. 108.**
- Subcutaneous (cutaneous) and serous cavity filariasis (*M. ozzardi* and *M. perstans*): **Ch. 108.**
- Diseases in subcutaneous tissues (*D. medinensis*): **Ch. 108.**
- Cutaneous larva migrans: **Ch. 109.**

> **Note: Parasites entering through skin**
> Following are the parasites which enter in body via skin but may or may not be cause diseases in skin.
> - **From water:** *Acanthamoeba* spp., like *A. culbertsoni, A. castellanii, A. hatchetti, A. polyphaga, A. rhysodes, A. astronyxis, A. divionensis* and *A. healyi, Balamuthia* spp., and *Schistosoma* spp., like *S. haematobium, S. mansoni* and *S. japonicum.*
> - **From soil:** *Hookworm* spp., like *Ancyclostoma duodenale, Ancyclostoma braziliensis, Ancyclostoma caninum* and *Necator americanus* and *Strongyloides* spp., like *S. stercoralis* and *S. fulleborne.*

ACCESS YOURSELF

Essays/Full Questions

1. Describe the etiopathogenicity of infections of skin and subcutaneous tissue. Discuss the laboratory diagnosis.
2. Describe the etiology, microbiology, pathogenicity and diagnostic features of common viral infections of the skin.

Short Note

1. Wound infections/Diabetic foot ulcer infection/Decubitus ulcer infection/Fournier gangrene/Paronychia.

Short Questions for Theory/Viva Questions

1. Name the causative agents for following skin manifestations. Malignant pustule, rose spots, green nail and Koplik's spot.
2. Name the four parasites enter through skin.

MCQs for Chapter Review

Terminology of Skin Infections

1. **Correct option for bulla:**
   a. Fluid-filled blister of size >5 mm
   b. Fluid-filled blister of size ≤5 mm
   c. Pus filled lesion of size ≤5 mm
   d. Pus filled lesion of size >5 mm

Fournier Gangrene

2. **All causes Fournier gangrene** *except*:
   a. *Staphylococcus*   b. *Streptococcus*
   c. *Clostridium*   d. *Bacteroides*

Parasites Entering through Skin

3. **Parasites enter in body by penetrating the skin are:**
   a. *Ancyclostoma duodenale* b. *Strongyloides*
   c. Round worm   d. *Trichuris trichuria*

Answers and Explanation of MCQs

1. **a**
- Follow section, **terminology of skin infections** for more explanation.

2. **c**
- Follow section, **Fournier gangrene** for more explanation.

3. **a and b**
- Follow section, **parasites entering through skin** for explanation.

CHAPTER 40

Musculoskeletal Infections

Chapter Outline

ANATOMY OF MUSCULOSKELETAL SYSTEM

It includes muscles, bones, joints and bone marrow.

COMMON INFECTIONS

Infections of Muscles

Myositis

Definition: It is an inflammation of muscles, especially skeletal muscles.

Etiological agents

1. **Bacterial:** *Staph. aureus* (most common and also causing pyomyositis), anaerobes (like *Bacteroides*), *Vibrio vulnificus*, *Aeromonas* and *Burkholderia pseudomallei*.
2. **Viral myositis:** HIV, HTLV, influenza virus, dengue virus, coxsackie viruses, and echo viruses.
3. **Fungal myositis:** *Candida* species (commonest), *Cryptococcus neoformans*, *Histoplasma capsulatum*, *Coccidioides* sp., or *Aspergillus* spp.
4. **Parasitic myositis:** Chagas' disease (*Trypanosoma cruzi*), cysticercosis (*Cysticercus cellulosae* → Larva of *Taenia solium*), trichinellosis (*Trichina spiralis*) are associated with myositis.

Clinical features: Muscular pain or swelling.

Laboratory diagnosis

- **Specimen:** Muscle biopsy.
- **Testing methods:** Microscopy and culture.

Psoas Abscess

Definition: It is the retroperitoneal collection of pus in the compartment of iliopsoas muscles.

Etiological agents: It is caused by *Staph. aureus* (80%, most common), streptococci, *E. coli* and *Mycobacterium tuberculosis*.

Pathogenicity

- **Modes of transmission:** Two types of psoas abscess as per transmission.
  1. **Primary:** It is due to hematogenous spread of microbes from distant foci in patients with diabetes, IV drug abuse, AIDS, renal failure or due to immunosuppression.
  2. **Secondary:** It is due to local spread from **gastro-intestinal disease** (most common route for secondary type) like appendicitis, diverticulitis, Crohn's disease, perforated colon carcinoma, etc., or **renal disease** (second most common) or **cold abscess** (tuberculosis) of paravertebral tissues, which penetrates the inguinal ligament to produce psoas abscess.
- **Clinical features:** Abdominal pain radiating to hip and flank, pain with thigh flexion (against resistance), limp, fever, nausea and vomiting.

Diagnosis

1. **Laboratory diagnosis**
   - Specimen: Aspiration of pus.
   - Testing methods: Microscopy and culture.
2. **Radiological methods:** USG, CT or MRI.

Treatment: It is treated by CT scan guided per-cutaneous drainage of the abscess due to the retroperitoneal location and antibiotics as per AST report.

Infections of Bones and Joints

Osteomyelitis

Definition: It is an inflammation of bones.

Etiological agents: Acute osteomyelitis is caused by *Staph. aureus* (most common), streptococci, GNB like *Salmonella*, *E. coli* and *Pseudomona aeruginosa* (in IV drug users). Subacute osteomyelitis is caused by *M. tuberculosis* or *Brucella* spp. It also caused by fungus like *Candida* spp.

Pathogenicity

- **Precipitating (host) factors:** Persons with surgery or prosthetic devices, open wounds, IV drugs users, alcohol, smoking, foci of infections or diabetes are at more risk.
- **Modes of transmission:** Bacteria enter directly in bone by trauma or presence of foreign body like prosthetic material or indirectly by blood from distant sites.
- **Sites:** Vertebras are the most common sites following hematogenous spread. In children with sickle cell disease, it is caused by *Salmonella* where sternum is the most common site. It may also occur in the long bones, pelvic bones, and clavicle.
- **Clinical features:** These are fever, local pain, swelling, redness, tenderness and difficulty in movement.
- **Complications**
  - Sinus: A deep-seated lesion like osteomyelitis occasionally develops a sinus tract to the skin surface which helps in draining the fluid and pus onto the skin.
  - Septic arthritis: Extension of bone infection in to joint.

Diagnosis: It is diagnosed radiologically by X-ray showing periosteal thickening and CT scan or MRI to reveal the extent and location of osteomyelitis. Specimens and testing methods for laboratory diagnosis include followings.

- A small piece of infected bone is collected from the most soft and necrotic area for culture.
- Blood for culture (in case of hematogenous spread).
- Bone biopsies for histopathological study and culture.

Prevention: Early identification and treatment of underlying distant infections.

Treatment: It is treated by antibiotics and surgery. Extensive debridement of necrosed tissue is done if no response to antibiotics or there is evidence of a persistent soft tissue abscess or concomitant joint infection.

Infectious Arthritis

Synonym: Suppurative/septic arthritis.

Definition: It is an inflammation/infection of joints.

Etiological agents

1. **Bacterial arthritis:** Common causes are *Staph. aureus* (commonest), *Streptococcus pyogenes* (2nd most common), CONS, *Neisseria gonorrhoeae*, diphtheroides, GNB like *Haemophilus influenzae*, *P. aeruginosa* and *E. coli* and anaerobes like *Bacteroides*. Rare causes are *Borrelia burgdorferi* (Lyme arthritis) and *Streptobacillus moniliformis* (rat bite fever → haverhill fever)
2. **Viral arthritis:** Following are different types.
   - Parvovirus B19 causes acute arthropathy: **Ch. 79.**
   - Chikungunya virus causes chronic chikungunya virus induced arthralgia: **Ch. 83.**
   - HBV and HCV cause arthritis due to immuno-complex deposition: **Ch. 88.**
3. **Fungal arthritis:** Mostly caused by *Candida* spp.
4. **Parasitic arthritis:** Occurs in *Echinococcus granulosus*.

Pathogenicity

- **Precipitating (host) factors:** Persons with osteomyelitis, joint surgery or prosthetic devices, open wounds, IV drugs users, alcohol, smoking, diabetes, foci of infection like skin infection/UTI or cancer are at more risk. Following surgery infection occurs very immediately by *Staph. aureus*, *Streptococcus* and GNB or very late after one year by CONS and diphtheroides.
- **Modes of transmission:** Infections occurs **directly** from extension of infection from the bone (osteomyelitis) or direct injection of corticosteroids into joints or insertion of prosthetic material like total hip replacement. It occurs **indirectly** where bacteria enter by blood from distant foci of infection like skin infection/UTI.
- **Sites:** It is monoarticular, but may be polyarticular in case of hematogenous spread. In bacterial arthritis knees and hips are the most commonly infected joints.
- **Clinical features:** Symptoms depends on age and the medications taken. Common symptoms are severe pain that worsens with movement, swelling of the joint, warmth and redness around the joint, fever, chills, fatigue, weakness, decreased appetite, rapid heart rate and irritability.
- **Complications:** If treatment is delayed, it causes sepsis and degeneration/permanent damage to joints.

Diagnosis

- **Specimen and testing methods for laboratory diagnosis:** Synovial fluid is collected for Gram's staining and inoculated into blood culture bottles for better diagnosis by culture. Culture can be done for mycobacteria and fungi. Blood culture is performed in case of hematogenous spread.
- **Radiology methods:** X-ray, CT scan or MRI are also useful.

Prevention: Early identification and treatment of underlying distant infections.

Treatment: It is treated by antibiotic therapy and surgery to drain the fluid.

Reiter's Syndrome or Reactive Arthritis

Follow **Ch. 52.**

Infections of Bone Marrow

It is caused by following agents.

1. **Bacteria:** Bone marrow infection occurs in brucellosis, enteric fever, tuberculosis and *Mycobacterium avium* complex infection.
2. **Viruses:** CMV in HIV patients.

3. **Fungi:** Histoplasmosis, blastomycosis and cryptococcosis.
4. **Parasites:** Leishmaniasis (amastigote form of *Leishmania* can be visualized in smears or sections made from bone marrow material) and hydatid cyst (*Echinococcus granulosus*).

SPECIFIC INFECTIONS

Bacterial Infections

- Streptococcal infections of skin, subcutaneous tissues, fascia and muscles (*Strept. pyogenes*): **Ch. 50.**
- Clostridial myositis and gas gangrene (*Clostridium perfringens*): **Ch. 55.**
- Skeletal tuberculosis (*Mycobacterium tuberculosis*): **Ch. 56.**
- Rat bite fever (by *Streptobacillus moniliformis* called haverhill fever or erythema arthriticum epidemicum or streptobacillary RBF): **Ch. 71.**
- Lyme disease (*Borrelia burgdorferi*): **Ch. 73.**

Parasitic Infections

- Tissue cyst of *T gondii*: **Ch. 98.**
- Muscular sarcocystosis (invasive sarcocystosis): **Ch. 98.**
- Sparganosis: **Ch. 101.**
- Muscular cysticercosis: **Ch. 102.**
- Trichinosis: **Ch. 107.**
- Visceral larva migrans: **Ch. 109.**

Essay/Full Question

1. Describe the etiopathogenicity and laboratory diagnosis of infections of bones and joints.
2. Discuss the etiology, pathogenicity, investigations and principles of management of following infections of bones and joints.
   - Osteomyelitis.
   - Suppurative arthritis/septic arthritis.
   - Spirochetal infection (Lyme disease).
   - Skeletal tuberculosis.

Short Notes

1. Myositis
2. Psoas abscess
3. Osteomyelitis
4. Infectious/suppurative/septic arthritis
5. Reiter's syndrome or reactive arthritis.

Short Question for Theory/Viva Question

1. Write the different modes of transmission of psoas abscess.

MCQ for Chapter Review

Myositis

1. Most common agent for myositis is:
   a. *Klebsiella pneuminiae* b. *Streptococcus pneumoniae*
   c. *Staph. aureus* d. *Listeria monocytogenes*

Answer and Explanation of MCQ

1. c
- Follow section, **myositis (etiological agents)** for more explanation.

Central Nervous System Infections

Chapter Outline

- Anatomy of CNS
- Common Infections
  - Infections of meninges
- Infections of brain
- Infections of spinal cord
- Specific Infections

ANATOMY OF CNS

CNS includes brain, spinal cord and their protective covering. Two protective coverings like outer bone and inner meninges. Meninges have three layers. From outer to inner are dura mater, arachnoid mater and pia mater. Arachnoid mater and pia mater are collectively called leptomeninges. Between and around the meninges are spaces that include the epidural, subdural, and subarachnoid spaces.

COMMON INFECTIONS

Infections of Meninges

Meningitis

Definition: It is an inflammation of leptomeninges (inner two layers of meninges).

Types and etiological agents: Two types of meningitis.

A. Septic (purulent or pyogenic) meningitis: It is characterized by presence of multinucleated cells (PMNs) in CSF with following subtypes.

1. **Acute (bacterial) meningitis:** Mostly by bacteria.
   - **<1 months/neonates (neonatal meningitis)**
     - **Most common causes:** *E. coli* (34%) > *Strept. agalactiae,* (30%) > GNB (8%) > *Listeria monocytogenes* (6% but aseptic meningitis).
     - **Other rare causes:** *Staph. aureus, Strept. pneumoniae, Neisseria meningitidis, Pseudomonas, Elizabethkingia (Flavobacterium) meningosepticum,* and *Haemophilus influenzae.*
   - **1–11 months (infants):** *Neisseria meningitidis* > *Strept. pneumoniae* > *Haemophilus influenzae.*
   - **1–6 (children) or 1–20 years:** *Neisseria meningitidis* > *Strept. pneumoniae* > *Haemophilus influenzae.*
   - **> 20 years (adults):**
     - **Most common:** *Strept. pneumoniae*
     - **Others:** *Neisseria meningitidis, Staph. aureus, Strept. pneumoniae, Listeria* and GNB like *E. coli* and *Klebsiella.*

2. **Chronic (bacterial, fungal and parasitic) meningitis:**
   - **Bacteria:** *Acinetobacter, Actinomyces, Nocardia, Mycobacterium tuberculosis, Brucella abortus, Salmonella,* etc.
   - **Fungi:** *Candida albicans, Aspergillus* spp., *Cryptococcus neoformans* and other systemic or opportunistic fungi.
   - **Parasites:** *Entamoeba histolytica, Toxoplasma gondii, Cysticercus cellulosae, Paragonimus westermanii* and *Trichinella spiralis.*

B. Aseptic meningitis: It is characterized by presence of mononuclear cells (lymphocytes) in CSF and caused by virus with exception of few bacteria.

1. **Acute (bacterial and viral) meningitis**
   - **Bacteria:** *Listeria monocytogenes.*
   - **Viruses:** Mostly caused by members of
     - Picornaviridae family (>85% cases) like Genus *Enterovirus* which includes polio virus 1, 2 and 3, coxsackie virus A7 and A9, all members of coxsackie virus B, ECHO virus serotypes like 4, 6, 9, 16, 20, 28 and 30 and entero virus 71 as described in **Ch. 80.**
     - Herpesviridae family like HSV-1 and 2, VZV, EBV and CMV as described in **Ch. 78.**
     - Arboviruses like encephalitis group as described in **Ch. 83.**
     - Robo virus like LCM virus as described in **Ch. 83.**
     - Picornaviridae family like mumps virus as described in **Ch. 80.**

2. Chronic (bacterial and viral) meningitis

- **Bacteria:** *Treponema pallidum* and *Leptospira*.
- **Viruses:** It is caused by some members of acute viral meningitis as mentioned above, Bunyaviridae (Oropouche virus) as described in **Ch. 80** and Retroviridae (HIV) as described in **Ch. 85**.

> **Note: More details about above mentioned causes of bacterial meningitis**
> 1. Neonatal meningitis by *Strept. agalactiae*: **Ch. 50.**
> 2. Pneumococcal meningitis by *Strept. pneumoniae*: **Ch. 50.**
> 3. Meningococcal meningitis (cerebrospinal fever) by *Neisseria meningitidis*: **Ch. 52.**
> 4. Meningitis by *Acinetobacter*: **Ch. 52 and 67.**
> 5. Tuberculous meningitis by *M. tuberculosis*: **Ch. 56.**
> 6. Invasive listeriosis/neonatal meningitis by *Listeria monocytogenes*: **Ch. 59.**
> 7. Meningitis by *Actinomyces*, *Nocardia* and *Tropheryma whipplei*: **Ch. 60.**
> 8. Meningitis by *E. coli* and *Klebsiella*: **Ch. 61.**
> 9. Meningitis by *Salmonella*: **Ch. 62.**
> 10. Meningitis by *Pseudomonas* and *Elizabethkingia* (*Flavobacterium*) *meningosepticum*: **Ch. 67.**
> 11. Meningitis by *Brucella abortus*: **Ch. 70.**
> 12. Meningitis by *Haemophilus influenzae*: **Ch. 68.**
> 13. Meningitis by *Treponema pallidum* (neurosyphilis) and *Leptospira* (Canicola fever and Swamp/marsh fever): **Ch. 73.**

Pathogenicity

- **Precipitating (host) factors**
  - **Low immunity:** More risk in persons with HIV-AIDS, autoimmune diseases, chemotherapy or organ transplantation.
  - **Living with close contacts:** More risk in schools, hostels, etc.
  - **Pregnancy:** It increases the risk of infection by *Listeria*. Infection can spread to the fetus.
  - **Age:** Children <5 years are at increased risk of viral meningitis and infants (<1 year) are at higher risk of bacterial meningitis.
  - **Working with animals:** Farm workers and others who work with animals have an increased risk of infection with *Listeria*.
  - **Vaccination:** Widespread vaccination against common microbes like *Neisseria meningitidis*. *Strept. pneumoniae* and *Haemophilus influenzae*, is shown to reduce the incidence of meningitis.
  - **Other factors:** Presence of CSF shunts and breach in the blood-brain barrier can precipitate the meningitis.
- **Modes of transmission**
  - Direct (continuous) spread occurs from surrounding organs or exogenously (direct entry) by trauma to skull/bone or intraneural transmission in case of rabies virus or HSV.
  - Indirect spread occurs by hematogenous route from distant organs.

- **Clinical features:** It presents with severe headache, fever, altered sensorium and following signs.
  - Neck rigidity and stiffness: Neck resistance to passive flexion.
  - Positive Kernig's sign (**Fig. 41.1a**): Inability to straighten the leg when the hip is flexed to 90 degrees due to severe stiffness of the hamstrings.
  - Positive Brudzinski's sign (**Fig. 41.1b**): Flexion of hips and knees on passive flexion of neck due to severe neck stiffness.
- **Complications:** These are seizures, hearing loss, vision loss, memory problems, arthritis, migraine, brain damage, hydrocephalus, subdural/epidural empyema and sepsis.

Laboratory diagnosis

A. Diagnosis of bacteria

- **Specimen**
  - **CSF:** It is collected by lumbar puncture under strict aseptic precautions. It is divided into three sterile containers; like 1st for cytological study, 2nd for biochemical study and 3rd for bacteriological study. It should be transported immediately to laboratory. CSF does not contain normal flora so for bacteriological examination (culture) CSF should never be refrigerated but if a delay is expected then it should be incubated at 37°C. For molecular diagnosis, freeze it.
  - **Other specimens:** Blood, nasopharyngeal swab (collected by using West's post-nasal swab in *Neisseria meningitidis*) and urine (for capsular antigen detection).
- **Testing methods**
  1. **Cytological and biochemical study to differentiate the different types of meningitis:** Follow **Table 41.1**.

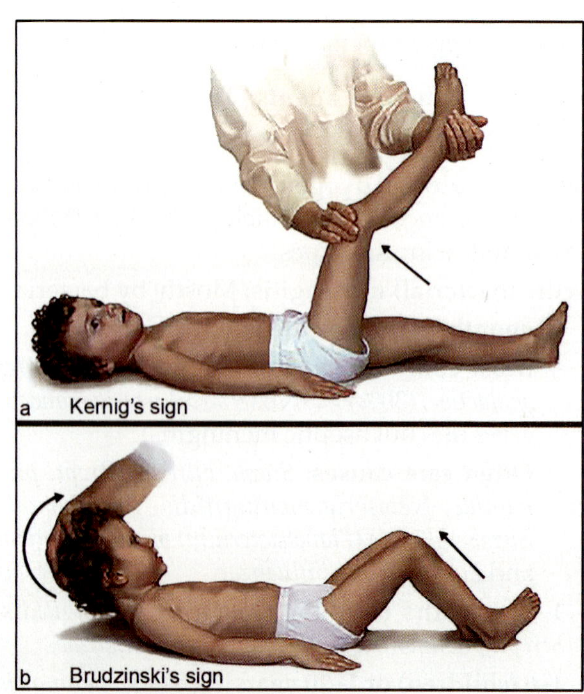

Fig. 41.1: (a) Kernig's sign; (b) Brudzinski's sign

2. **Bacteriological study done by Gram's stain**
   - **Smear preparation:** Bacterial load in CSF is very low so more concentrated smear should be prepared by two ways. (1) Centrifuge the CSF and use sediment to prepare the smear. Supernatant is used for capsular antigen detection. (2) Heap smear, which is prepared by putting the drop on smear. Put the subsequent drop on same spot of slide after allowing earlier drying.
   - **Examination:** Identification of bacteria from Gram's stain is depends on the morphological features described in respective chapters.
3. **Other methods:** Like culture (including automated Bact/Alert system), serology, molecular, etc., are as per suspected bacteria.

B. Diagnosis of viruses: Ch. 119.

C. Diagnosis of fungi: Ch. 120.

D. Diagnosis of parasites: Ch. 121.

Prevention: It is prevented by avoiding contact with sick people, using prophylactic antibiotics in case of contact with sick people and vaccinations like *Haemophilus influenzae* type B (Hib) vaccine, pneumococcal conjugate vaccine and meningococcal vaccine.

Treatment: It is treated as per causative agents.

Epidural Abscess

Definition: It is the collection of pus between dura mater and skull/spine.

Types and etiological agents: Two types:
1. **Intracranial epidural abscess:** It occurs either due to endogenous bacteria mostly from respiratory tract like viridans group of streptococci causing sinusitis or due to exogenous bacteria mostly nosocomial bacteria which enter by craniotomy.
2. **Spinal epidural abscess:** It occurs mostly by *Staph. aureus* (in 60% cases) and intestinal GNB.

Clinical features: Intracranial epidural abscess is presents with fever, headache, nausea, vomiting, malaise, etc. Spinal epidural abscess is presents with fever, back/neck pain and other neurological features.

Laboratory diagnosis
- **Specimen:** CSF and pus collection.
- **Testing methods:** Microscopy and culture.

Prevention: Early diagnosis and treatment of underlying illness like sinusitis.

Treatment: As per etiological agents.

Subdural Abscess

It is the collection of pus between the dura mater and the arachnoid mater. In 95% cases it occurs in frontal lobe. Other features are same as brain abscess.

Infections of Brain

Brain Abscess

Definition: It is the collection of pus within brain parenchyma.

Etiological agents: In 40% cases brain abscess occurs by streptococci (anaerobic, aerobic and viridans group includes *Strept. angiosus*). In 30% cases, it occurs by anaerobes like *Bacteroides fragilis* and *Fusobacterium*. In 25% cases, it occurs by members of Enterobacterales like *Proteus*, *E. coli* and *Klebsiella*. In 10% cases, it occurs by *Staphylococcus aureus* due to trauma or surgery. Few other causes are larvae of *Taenia solium* (neurocysticercosis) and *M. tuberculosis* (tuberculoma). In person with IDDs infection occurs by *Nocardia* spp., *Toxoplasma gondii*, *Aspergillus* spp., *Candida* spp., and *Cryptococcus neoformans*.

Pathogenicity
- **Modes of transmission:** Endogenous spread (45–50%) occurs from surrounding organs due to continuity of mucosa like paranasal sinuses, middle ear or mastoid process. Exogenous (10%) spread includes direct entry by trauma to skull/bone. Hematogenous (25%) spread occurs from distant organs like cyanotic CHD. In 15% cases it is cryptogenic (unknown) source.
- **Precipitating (host) factors:** It is more common in hosts with HIV-AIDS, organ transplantation, malignancy, or immunosuppressive therapy. Mostly it involves frontal-temporal lobe, followed by frontal-parietal, parietal, cerebellar, and occipital lobes.

TABLE 41.1: CSF findings in different types of meningitis

Characteristics	Normal	Pyogenic (septic) meningitis	Tuberculous (septic) meningitis	Viral/aseptic meningitis
CSF pressure (mm of water)	50–150	Severe (>180)	Moderate	Normal/ mild
TLC (per mm³)	0–5	100–10,000	10–500	25–500
Predominant cells	Lymphocytes	Neutrophils	Initially neutrophils later lymphocytes	Lymphocytes
Glucose (mg%)	40–70	<40 mg/dl (slightly decreased)	20–40 mg/dl (slightly decreased to absent)	Normal
Proteins (mg%)	15–45	>45 mg/dl (usually >250; markedly increased)	100–500 mg/dl (moderate to markedly increased)	20–80 mg/dl (normal or slightly elevated)

- **Clinical features:** It presents with fever, nausea, vomiting, severe headache (mostly unilateral, commonly on the side of the abscess), focal neurologic deficit, mental status changes due to cerebral edema, seizures, nuchal rigidity and papilloedema.
- **Complications:** Brain abscesses may rupture into the subarachnoid space causing severe meningitis with high mortality rate.

Diagnosis

- **Specimen and testing methods for laboratory diagnosis**
  - Aspiration of pus under CT guidance which is tested by direct microscopy (Gram's stain, acid-fast stain, and special fungal stains), culture (aerobic, anaerobic, fungal and mycobacterial culture)
  - Brain tissue for histopathological examinations
  - Blood: For cultures
  - Serology: Anticysticercus antibodies for neuro-cysticercosis
  - Molecular tests: PCR detecting 16S ribosomal genes
- **Radiology methods:** CT scan and MRI are also useful.

Prevention: Early identification and treatment of underlying distant infections or infections of surrounding tissues.

Treatment: It is treated by antibiotic therapy and surgery to drain the pus.

> **Note: Intracranial space-occupying lesions (ICSOLs).**
> - These are epidural abscess, subdural abscess, brain abscess, granulomatous lesions in brain like tuberculoma, space-occupying lesions by parasites like cystic lesions in hydatid cyst, neurocysticercosis, schistosomes, etc., tumor and hematoma.

Encephalitis

It is an inflammation of brain parenchyma. It mostly occurs by viruses and rare by parasites. All viruses are described in **Ch. 89.** Parasites causing encephalitis are free living amoebae (**Ch. 93**) and *Toxoplasma gondii* (**Ch. 98**). It presents with altered consciousness, behavioral changes, seizures, focal neurological deficits and sometimes, extrapyramidal signs like involuntary movements. Laboratory diagnosis is done as per etiological agents.

Encephalopathy

It is the diffuse disease of the brain that alters the brain function or structure. It is mostly non-infectious origin due to metabolic disease, hepatic encephalopathy, brain tumor, uremic encephalopathy, etc. Almost all infections causing encephalitis are causing encephalopathy; however, in some chronic infections, word 'encephalopathy' is widely used like slow virus diseases (**Ch. 89**) and cerebral malaria (**Ch. 97**). It presents with altered mental state, progressive loss of memory, personality changes, inability to concentrate, lethargy, and progressive loss of consciousness. Other symptoms are as per etiological agents. It is diagnosed by CSF examination, CT scan or MRI, electroencephalograms and few other tests are as per etiological agents.

Infections of Spinal Cord

Myelitis

Definition: It is an inflammation of spinal cord.

Etiological agents: It is caused by infections or by auto-immunization. Following are infectious causes.

1. **Viruses:** Polio virus (most common), coxsackie virus A and B, ECHO viruses, entero virus-71 (EV-71) and few members of arbo virus (JE virus, West Nile virus, tick-borne encephalitis viruses, etc.), herpes viruses, influenza virus, rare by HIV and HTLV.
2. **Bacteria:** *Mycoplasma pneumoniae*, *Mycobacterium tuberculosis*, *Treponema pallidum* and *Brucella* spp.
3. **Fungi:** *Histoplasma, Coccidioides, Blastomyces* and *Cryptococcus.*
4. **Parasites:** *Schistosoma, Echinococcus, Taenia solium, Trichinella, Plasmodium* and *Toxocara.*

Pathogenicity: It results in disruption of the connection of the brain to the rest of the body, and *vice versa.* Following are the different types as per locations:

1. **Gray matter myelitis:** It affects anterior horn of spinal cord resulting acute flaccid paralysis. It mostly occurs by polio virus (most common), coxsackie virus A and B, ECHO viruses, EV-71 and few members of arbo virus.
2. **White matter (transverse) myelitis/leukomyelitis:** It mostly occurs by herpes viruses, influenza virus, rare by HIV and HTLV.

SPECIFIC INFECTIONS

Bacterial Infections

- Tetanus (*Clostridium tetani*): **Ch. 55.**
- Botulism (*Clostridium botulinum*): **Ch. 55.**
- Lyme disease → neuroborreliosis (*Borrelia burgdorferi*): **Ch. 73.**

Viral Infections

- Myelitis (polio virus): **Ch. 80.**
- Rabies (rabies virus): **Ch. 84.**
- Slow virus diseases: **Ch. 89.**

Fungal Infections

Superficial mycoses (meningitis and encephalitis by *Candida* spp./candidiasis in person with IDDs): **Ch. 90.**

Subcutaneous mycoses (disseminated sporotrichosis in brain by *Sporothrix schenkii*): **Ch. 91.**

Systemic mycoses: Disseminated infection occurs in brain in patients with low CMI by *Histoplasma capsulatum, Blastomyces dermatitidis, Coccidioides immitis* and *Paracoccidioides brasiliensis* as described in **Ch. 91.**

Oppotunistic mycoses
- CNS cryptococcosis (*Cryptococcus neoformans*): **Ch. 91.**
- Rhinocerebral aspergillosis (*Aspergillus* spp.): **Ch. 91.**
- Rhinocerebral zygomycosis (species from class Zygomycetes): **Ch. 91.**

Parasitic Infections (Neuroparasites)

CNS infection of protozoa

- Cerebral amoebiasis (*Entamoeba histolytica*): **Ch. 93.**
- Free living amoebae like *Nagleria fowleri* (acute and purulent primary amoebic meningoencephalitis), *Acanthamoeba* spp. (granulomatous amoebic meningo-encephalitis) and *Balamuthia mandrilaris* (granulomatous amoebic meningoencephalitis): **Ch. 93.**
- Meningoencephalitis in sleeping sickness (*Trypanosoma brucei*): **Ch. 95.**
- Meningoencephalitis in Chaga's disease (*Trypanosoma cruzi*): **Ch. 95.**
- Cerebral malaria in malaria (*Plasmodium flaciparum*): **Ch. 97.**
- Tissue cyst in toxoplasmosis (*Toxoplasma gondii*): **Ch. 98.**
- Brain abscess in microsporidiosis (*Microspora* spp.): **Ch. 99.**

CNS infection of cestodes

- Sparganosis (*Spirometra* spp.): **Ch. 101.**
- Neurocysticercosis by *Cysticercus cellulosae* (larvae of *Taenia solium*): **Ch. 102.**
- Hydatid cyst by hydatid cyst (larvae of *Echinococcus granulosus*): **Ch. 102.**
- Coenurosis by coenurus (larvae of *Multiceps multiceps*): **Ch. 102.**

CNS infection of trematodes: Space-occupying lesion produced by *Schistosoma japonicum* (**Ch. 104**) and *Paragonimus westermanii* (**Ch. 105**).

CNS infection of nematodes

- Massive or hyperstrongyloidiasis (*Strongyloides stercoralis*): **Ch. 107.**
- Neuropathy, encephalopathy, meningitis, Jacksonian epilepsy (*Loa loa*): **Ch. 108.**
- Eosinophilic meningoencephalitis (*Angiostrongylus cantonensis*): **Ch. 108.**
- Visceral larva migrans (*Gnathostoma spinigerum*): **Ch. 109.**

ACCESS YOURSELF

Essays/Full Questions

1. Describe the etiopathogenicity, types, laboratory diagnosis and differentiating features/CSF findings of infectious meningitis.
2. Identify the etiology of meningitis based on given CSF parameters and discuss in details.
3. Describe the etiopathogenicity and laboratory diagnosis of encephalitis.

Short Notes

1. Bacterial meningitis/viral meningitis/viral encephalitis.
2. Brain abscess
3. ICSOL
4. Neuroparasites
5. Enumerate the indications and describe the findings in the CSF in patients with meningitis or distinguish bacterial, viral and tuberculous meningitis.

Short Questions for Theory/Viva Questions

1. What is aseptic meningitis? Write two examples of bacteria causing aseptic meningitis.
2. What is the heap smear? Write the advantage.

MCQs for Chapter Review

Meningitis

1. **Following bacteria are most often associated with acute neonatal meningitis** *except*:
   - a. *E. coli*
   - b. *Streptococcus agalactiae*
   - c. *Neisseria meningitidis*
   - d. *Listeria monocytogenes*
2. **Bacterium *not* associated with chronic meningitis is:**
   - a. *Mycobacterium tuberculosis*
   - b. *Borrelia burgdorferi*
   - c. *Strept. pneumoniae*
   - d. *Treponema pallidum*
3. **CSF analysis of pyogenic meningitis reveals all of the following, *except*:**
   - a. CSF pressure: highly elevated
   - b. Total leukocyte count: highly elevated, neutrophilic
   - c. Glucose: highly elevated
   - d. Total proteins: markedly increased

Brain Abscess

4. **Most common source/route for brain abscesses is:**
   - a. Direct inoculation
   - b. Contiguous infection
   - c. Hematogenous route
   - d. Unknown

Other Specific Infections: Parasitic Infections (Neuroparasites)

5. **Neurocysticercosis is caused by:**
   - a. *T. solium*
   - b. *T. saginata*
   - c. *D. latum*
   - d. *Ascaria lumbricides*
6. **Which of the following is *not* a neuroparasite?**
   - a. *Taenia solium*
   - b. *Acanthamoeba*
   - c. *Naegleria*
   - d. *Trichinella spiralis*
7. **Parasitic encephalitis is caused by:**
   - a. *Naegleria*
   - b. *Acanthamoeba*
   - c. *Balamuthia*
   - d. *Gnathostoma*
8. **Primary amoebic meningoencephalitis is caused by:**
   - a. *Naegleria fowleri*
   - b. *Entamoeba histolytica*
   - c. *Endolimax nana*
   - d. *Dientamoeba fragilis*
9. **Eosinophilic meningoencephalitis is caused by:**
   - a. *Gnathostoma*
   - b. *Naegleria*
   - c. *Toxocara canis*
   - d. *Angiostrongylus cantonensis*

Answers and Explanation of MCQs

1. **c**
2. **c**
- Follow section, **meningitis (types and etiological agents)** for explanation of answers of MCQs 1–2.
3. **c**
- Follow **Table 41.1** for explanation.
4. **b**
- Follow section, **brain abscess (pathogenicity → Modes of transmission)** for explanation.
5. **a**
6. **d**
7. **a, b, c, d**
8. **a**
9. **d**
- Follow section, **other specific infections → neuroparasites)** for explanation of answers of MCQs 5–9.

C H A P T E R **42**

Respiratory Tract Infections

Chapter Outline

- Introduction
- Common Infections
  - Upper respiratory tract infections
  - Lower respiratory tract infections
- Specific Infections

INTRODUCTION

Anatomy of Respiratory Tract

It has two parts like upper respiratory tract which includes nose, sinuses, pharynx (throat), epiglottis and larynx and lower respiratory tract which includes trachea, bronchi, bronchioles, lungs and pleural cavity.

Normal Defense of Respiratory Tract

It includes nasal hair, cilia, mucus, cough reflex sneez reflex, pulmonary alveolar macrophages (PAMs), secretory IgA against specific pathogens and normal (resident) flora of respiratory tract.

COMMON INFECTIONS

Upper Respiratory Tract Infections

Rhinitis (Common Cold)

It is an inflammation of nasal mucosa. It also called common cold or coryza. In 20–25% cases, it is due to viruses like rhino viruses (most common), corona-viruses, adeno viruses, influenza viruses, parainfluenza viruses, entero viruses, etc. In 10–15% cases it is due to bacteria like *Strept pyogenes*, *Mycoplasma pneumoniae*, *Chlamydia pneumoniae*, etc. Exposure to dust, pollen, etc., produce sneezing and running nose. Common features are running nose, sneezing and rarely fever.

Sinusitis

Definition: It is an inflammation of paranasal sinuses.

Etiological agents

1. **Bacteria:** *Strept pyogenes* (most common in 33% cases), *Haemophilus influenzae* (32%), *Moraxella catarrhalis* (9%), *Staph. aureus* (most common in sphenoid sinus) and anaerobes, etc.
2. **Viruses (15%):** Rhino viruses, adeno viruses, influenza viruses, parainfluenza viruses, coronaviruses, etc.

3. **Fungi:** *Aspergillus, Alternaria, Bipolaris* and *Curvularia*.
4. **Others:** Like allergic causes.

Clinical features: Frontal headache, loss of smell, facial pressure, nasal congestion and postnasal drip are common features. In bacterial sinusitis, other features like purulent nasal discharge, maxillary toothache, fever and unilateral facial pain are present. In fungal sinusitis, other features like sneezing and itchy eyes are present. It may heal by 7–10 days.

Diagnosis: It is diagnosed clinically and radiologically, but difficult to diagnosis microbiologically.

Treatment: Sinusitis is the common cause of overuse of antibiotics. Antibiotics like amoxicillin and clavulanate are given for 5–7 days in case of bacterial sinusitis when above bacterial features are present. Antibiotics are not given when symptoms suggest viral or allergic etiology; with wait and watch strategy.

Pharyngitis (Sore Throat)

Definition: It is an inflammation of pharynx (Greek) means throat.

Etiological agents

1. **Bacteria:** Most common causes are *Strept pyogenes* (streptococcal pharyngitis also called strep throat), *Corynebacterium diphtheriae*, group C and G β-hemolytic streptococci. Rare causes are *Arcanobacterium haemolyticum, Fusobacterium necrophorum, Mycoplasma pneumoniae* (usually present with cough) and *Neisseria gonorrhoeae* (due to orogenital sexual contact).
2. **Viruses:** Most common causes are influenza viruses, parainfluenza viruses, coronaviruses including SARS CoV-2 causes COVID-19, EBV, entero viruses (coxsackie A virus causes vesicular pharyngitis or herpangina), ECHO viruses and entero virus-71. Rare causes are adeno viruses, rhino viruses, HSV or HIV (as acute retroviral syndrome).

3. **Fungi:** *Candida albicans.*
4. **Noninfectious:** Pharyngitis may also be caused by mechanical, chemical, or thermal irritation, like cold air, gastroesophageal reflux disease or by drugs like pramipexole and antipsychotic drugs.

Pathogenicity

- **Precipitating (host) factors:** It is one of the most common URTI in children (presents with tonsillitis) and adults. Closed-in work, living spaces, exposure to chemical irritants or tobacco or smoking, gastro-esophageal reflux disease and low immunity are other risk factors.
- **Clinical features**
  - These include throat (pharyngeal) pain, difficulty in swallowing or drinking, runny or stuffy nose, dry cough, hoarseness, red eyes, fever, headache or body aches. Throat becomes red, swollen or ulcerative.
  - Clinical features also depend on etiological agents as mentioned below.
    1. *Strept. pyogenes:* Exudative fluid with protein, inflammatory cells and cellular debris.
    2. *Corynebacterium diphtheriae:* Pseudomembrane over tonsil (rarely in infectious mononucleosis also).
    3. EBV: Mucosal ulceration, or enlarged/swollen nasopharyngeal lymph nodes.
    4. *Candida albicans*: White patches on mucosa (oral thrush).
  - Clinically pharyngitis is classified as acute and chronic.
    1. Acute pharyngitis: It may be catarrhal, purulent or ulcerative.
    2. Chronic pharyngitis: It may be catarrhal, hyper-trophic or atrophic.
- **Complications:** Infection may extend to surrounding organs to produce tonsillitis, peritonsillar abscess, sinusitis, acute otitis media or in blood to cause sepsis.

Laboratory diagnosis

- **Specimen:** Dacron or rayon swabs are most suitable for URTI. Flocked swabs are also preferable. Cotton swabs are preferable for *Strept. pyogenes* but not for viruses (and for *B. pertussis*).
  1. **Throat swab (oropharyngeal swab):** It is an ideal specimen collected by vigorous rubbing of sterile swab over the posterior pharynx and both the tonsillar pillars. Two swabs are collected, one for direct microscopy and the other for culture.
  2. **Other specimens:** Pseudomembrane is collected if diphtheria is suspected. Nasopharyngeal swab is collected for viruses like influenza or corona-virus by inserting flexible swab through nose into posterior nasopharynx and then rotating for 5 seconds.
- **Storage:** For isolation of most bacterial pathogens swab should be processed within 4 hours. For molecular diagnosis of bacteria or viruses specimens could be stored at 4°C.
- **Transport:** By using transport media like VTM for viruses, Stuart's media/Amie's transport medium for bacteria.
- **Testing methods**
  - Direct microscopy: Albert stain for *C. diphtheriae* and Gram's stain for other bacteria.
  - Culture: Routinely used media are blood agar and chocolate agar. Other media are as per suspected organism.
  - Serological tests: Antigen detection tests are performed from throat swab for *Strept. pyogenes* and from nasopharyngeal swab for SARS-CoV2 (by ICT). Antibody detection from serum is rarely useful.
  - Molecular diagnosis: RT-PCR especially for viral causes. Biofire film array is an automated multiplex PCR which detects 22 most common pathogens (including viruses and bacteria) in 1 hour with 95% sensitivity and 99% specificity.

Prevention: Avoid smoking, sharing food, drinks, and eating utensils. Avoid contact with individuals who are sick. Wash hands often, especially before eating and after coughing or sneezing. Use alcohol based hand sanitizer.

Treatment: Drink plenty of water to prevent dehydration. Gargle with warm salty water (1 teaspoon of salt per 8 ounces of water) to ease throat pain. Drink warm liquids (tea or soup). Use cool mist vaporizer to relieve throat dryness, throat lozenges or anesthetic (lignocaine) throat sprays. Take analgesic like ibuprofen, acetaminophen (paracetamol) or aspirin (in adults only) to relieve throat pain. In oral analgesic solution the active ingredient is usually phenol. Use antibiotics to prevent complications.

> **Note:** Diagnosis of pharyngitis covers entire answer for diagnosis of URTI.

Tonsilitis

It is an inflammation of tonsils. Sometimes, pharyngitis present with inflammation of tonsil called pharyngo-tonsillitis or tonsillopharyngitis. Other details are same as pharyngitis.

Paritonsillar Abscess (Quinsy)

Definition: It is the collection of pus in connective tissues around the tonsils. It may be referred as peritonsi-litis.

Etiological agents: This deep neck infection usually results from untreated or partially treated acute tonsillitis. It is due to aerobes like *Strept. pyogenes*, *Staph. aureus*, viridans streptococci and *Haemophilus influenzae*. Most common anaerobes are *Fusobacterium necrophorum*, *Peptostreptococcus*, *Prevotella* and *Bacteroides*.

- **Precipitating (host) factors:** Usually it affects children >5 years and young adults. It does not occur in person with tonsillectomy.
- **Clinical features:** Symptoms include fever or chills, headache, throat pain usually on one side, difficulty in opening the mouth, change of voice called hot potato voice, referred ear pain and foul breath. Presence of redness and swelling in the tonsillar area of the affected side. Enlarged jugulodigastric nodes. Shifting of uvula towards the unaffected side.
- **Complications:** If not treated immediately, it can spread to adjacent tissues and causes acute hemorrhage. Other complications are respiratory blockage, aspiration pneumonia, retropharyngeal abscess, spreading of infection to throat, mouth, neck, chest and blood (sepsis) and decreased oral intake leads to dehydration.

Diagnosis

- **Specimen and testing methods for laboratory diagnosis:** Aspiration of pus from abscess which is tested by direct Gram's stain and culture (aerobic and anaerobic). Blood is collected for culture.
- **Radiology methods:** CT scan or MRI is also useful.

Prevention: Treat tonsillitis immediately. Avoid smoking and keep mouth clean.

Treatment: Fluids therapy, pain medication and antibiotic (piperacillin + tazobactam), removal of pus by incision and drainage or by needle aspiration and tonsillectomy are the answers.

Laryngitis

Definition: It is an inflammation of larynx (voice box).

Etiological types and causative agents: Two types:

1. **Acute laryngitis:** Sudden onset and symptoms lasts for <3 weeks. Most acute cases of laryngitis are caused by viruses, rare by bacteria and other causes.
   - **Viruses:** Influenza viruses, parainfluenza viruses, RSV, rhino viruses, adeno viruses, coronaviruses, and human meta-pneumo virus. In patients who have a low immunity, it is caused by other viruses such as herpes, HIV and coxsackie viruses.
   - **Bacteria:** *Strept pyogenes, C. diphtheriae, Strept. pneumoniae, Haemophilus influenzae, Moraxella catarrhalis, Bacillus anthracis, Mycobacterium tuberculosis* and *Treponema pallidum.*
   - **Fungi:** Fungus occurs in 10% of acute laryngitis cases in both immunocompetent and immunodeficient persons by *Histoplasma capsulatum, Blastomyces dermatitidis, Coccidioides immitis, Candida* spp., or *Cryptococcus neoformans.*
   - **Other causes:** These include excessive use of the vocal cords like excessive yelling, screaming, or singing and laryngeal trauma during endotracheal intubation.

2. **Chronic laryngitis:** Gradual onset and symptoms last for >3 weeks. Rarely caused by infections but by following causes.
   - **Allergy:** Like in allergic asthma.
   - **Reflux:** In gastroesophageal reflux disease.
   - **Autoimmune disorders:** In 30–75% cases of rheumatoid arthritis and 0.5–5% cases of sarcoidosis.

Pathogenicity

- **Precipitating (host) factors:** These are presence of common cold, bronchitis, sinusitis, exposure to irritating substances, such as smoking, alcoholism, stomach acid or workplace chemicals and overuse of voice, by speaking too much, speaking too loudly, shouting or singing.
- **Clinical features:** Patient presents with hoarseness and lowering/deepening of the voice. Tickling sensation, dry throat and dry cough are other features.
- **Complications:** Infective laryngitis may spread to lower respiratory tract.

Diagnosis: It is diagnosed clinically by doing laryngoscopy then by laboratory tests.

Prevention: Avoid the all risk factors mentioned above.

Treatment

- **Acute laryngitis:** It is self-limited, only symptomatic treatment is required. Antibiotics are not indicated.
- **Chronic laryngitis:** Treat the underlying causes.

Epiglottitis

Definition: It is an inflammation of epiglottis and other soft tissues above the vocal cords.

Etiological agents: Mostly it is caused by bacteria in contrary to laryngitis which is mostly caused by viruses. *Haemophilus influenzae* type b is a primary cause of epiglottitis. Other bacteria are *Strept. pyogenes, Strept. pneumoniae,* and *Staph. aureus*

Pathogenicity

- **Precipitating (host) factors:** It is common between 2–6 years of age. It is less in children who are vaccinated to *Haemophilus influenzae* type b.
- **Clinical features:** It leads significant edema and inflammation. Patient presents with fever, difficulty/pain in swallowing.
- **Complications:** Respiratory obstruction with inspiratory stridor.

Laboratory diagnosis: Swabbing of epiglottis is usually not advised because it leads respiratory obstruction. Blood culture is performed for *H. influenzae.*

Prevention: It is prevented by vaccination to common bacteria like *Haemophilus influenzae* type b.

Treatment: Antibiotics are given. Tracheostomy may be performed to avoid the respiratory obstruction.

Lower Respiratory Tract Infections

Acute Laryngotracheobronchitis (Croup)

Definition: It is an inflammation of larynx extending to trachea and bronchi.

Etiological agents: It is mostly caused by viruses same as laryngitis. Bacterial causes are divided into laryngeal diphtheria caused by *C. diphtheriae* and bacterial tracheitis, laryngotracheobronchitis and laryngotracheobronchopneumonitis are caused by *Strept. pneumoniae, H. influenzae, M. catarrhalis, Staph. aureus* and *M. pneumoniae.*

Pathogenicity

- **Precipitating (host) factors:** It is common in young children <3 years of age. It is less in children who are vaccinated to *C. diphtheriae, Haemophilus influenzae* type b and other bacteria.
- **Clinical features:** Common features are fever and runny nose. The infection leads respiratory obstruction (**Fig. 42.1**) which interferes with breathing and produces the classic symptoms of "barking/brassy nonproductive cough", inspiratory stridor and hoarseness. Symptoms are mild, moderate, or severe, often start or worse at night and normally last for 1–2 days.
- **Complications:** Pulmonary edema, pneumothorax, pneumonia, pneumomediastinum, lymphadenitis and otitis media.

Diagnosis: It is diagnosed clinically. Blood culture is useful for bacterial identification. Viruses are identified by culture methods.

Prevention: It is prevented by vaccination to common bacteria as mentioned above.

Treatment: Only symptomatic treatment is given. Antibiotics are given when bacterial infection is suspected. Steroids are given routinely. Moderate-to-severe croup may be improved temporarily with nebulized epinephrine.

Bronchitis

Definition: It is an inflammation of bronchus.

Etiological agents:

1. **Acute bronchitis:** It also called chest cold. About 20–50% cases are caused by viruses same as laryngitis. Bacterial causes are *M. pneumoniae* (5%), *Ch. pneumoniae* (5%) and *Bordetella pertussis.*
2. **Chronic bronchitis:** It is mostly caused by allergy to dust or other irritants and smoking.

Pathogenicity

- **Precipitating (host) factors:** Chronic bronchitis is common in adults, in winter and in smokers.
- **Clinical features**
  1. **Acute bronchitis:** It presents with fever, persisting cough and sputum production which is initially clear later becomes yellow (purulent) or green.
  2. **Chronic bronchitis:** It presents with cough with expectoration for at least 3 months/year during a period of 2 consecutive years. Patient feels difficulty in breathing. Presence of wheezing/rhonchi during auscultation.
- **Complications:** These are pneumonia and repeated attack may indicate chronic obstructive pulmonary disease (COPD).

Laboratory diagnosis

- **Specimen:** Sputum.
- **Testing methods:** Microscopy and culture.

Prevention: Vaccination to causative agents, washing hands, wearing mask and avoiding smoking are the useful measures.

Treatment

1. **Acute bronchitis:** Symptomatic treatment like cough syrup and bronchodilators are given. Antibiotics are given when bacterial cause is suspected.
2. **Chronic bronchitis:** Antibiotics and steroid are useful.

Bronchiolitis

Definition: It is an inflammation of bronchioles.

Etiological agents: Almost always caused by viruses same as laryngitis where RSV is most common.

Pathogenicity

- **Precipitating (host) factors:** It is common in young children, infants, in adults who smoke and in winter season. Other risk factors are low immunity, premature birth, babies have not been breast-fed, contact with other children, etc.
- **Clinical features:** Symptoms present in first few days are runny nose, stuffy nose, cough and mild fever (not always present). After week or more it presents with difficulty in breathing, expiratory wheezing (whistling noise when the child breathes out), respiratory distress with tachypnea, nasal flaring, blue lips/skin (cyanosis) due to lack of oxygen.
- **Complications:** These are otitis media, secondary bacterial infections, respiratory failure, etc.

Diagnosis: It is diagnosed clinically.

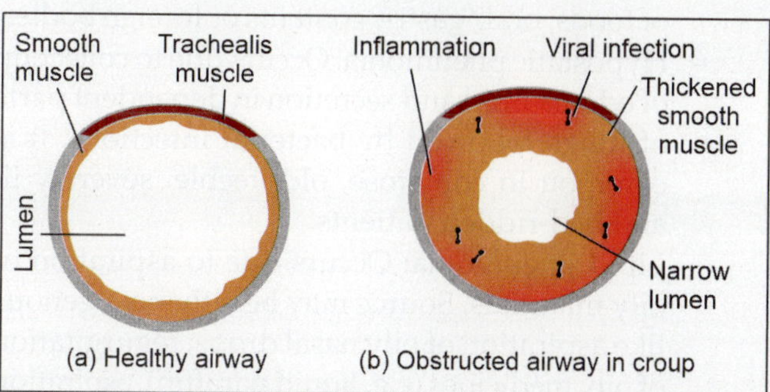

Fig. 42.1: Normal airway and obstructed airway in croup

Prevention: Avoiding contact with sick children and regular breast feeding are the useful measures.

Treatment: Symptomatic treatment is given. Antibiotics are given if secondary bacterial infections are suspected.

Pneumonia

Synonym: Pneumonitis.

Definitions

- **Pneumonia:** It is an inflammation/infection of lung parenchyma (one or both lungs) distal to terminal bronchioles which includes respiratory bronchioles, alveolar ducts, alveolar sacs and alveoli.
- **Consolidation (solidification):** It is the radiographic and macroscopic appearance of lungs in pneumonia.
- **Organization or carnification:** It is the fibrosis following pneumonia.

Etiological agents and types

1. **As per source of agents**
   - **Community acquired pneumonia:** Patients acquire infection from the community. It is caused by bacteria like *Strept. pneumoniae, H. influenzae, Mycoplasma pneumoniae, Ch. pneumoniae,* etc., and viruses like influenza, coronaviruses, RSV, para-influenza viruses, adeno viruses, etc.
   - **Hospital-acquired pneumonia:** Patients acquire infection from the healthcare facility. Causative agents are mentioned in **Ch. 29** (Ventilator-associated pneumonia → Etiological agents).
2. **Anatomical types:** It depends on involvement of parts of respiratory tract (bronchioles and/or lobes of lungs or entire lungs or interstitial space) and type of cough (productive or nonproductive).
   - **Bronchopneumonia/lobular pneumonia:** It is the infection of bronchioles that extends to alveoli. It is caused by *Staph. aureus* (most common), *Strept. pneumoniae, Klebsiella pneumoniae, E. coli, Pseudomonas aeruginosa, Haemophilus influenzae,* etc.
   - **Lobar pneumonia:** It is an infection of part of lobe, or entire single lobe or more lobes (mostly lower lobes) or one/both lungs. It is caused by *Strept. pneumoniae* (most common), *Klebsiella pneumoniae, Staph, aureus, Moraxella catarrhalis, Haemophilus influenzae,* etc.

> **Note:**
> - Distinction between pathogens of community acquired and hospital-acquired pneumonia is difficult as many pathogens present in community may be identified from hospital as MDR pathogens.
> - Bronchopneumonia (lobular) often leads to lobar pneumonia as the infection progresses. The same organism may cause one type of pneumonia in one patient, and another in a different patient.
> - Differences between bronchopneumonia (lobular) and lobar pneumonia: Follow **Table 42.1**.
> - Both bronchopneumonia (lobular) and lobar pneumonia are collectively called typical pneumonia.

- **Primary atypical (interstitial) pneumonia:** It is an infection in interstitial space (hence also called interstitial pneumonia) of lungs with non-productive (dry) cough. It is caused by following agents:
   - **Atypical pneumonia by bacteria**
     - Walking pneumonia (*Mycoplasma pneumoniae* → most common): **Ch. 74.**
     - Legionellosis (*Legionella* species): **Ch. 67.**
     - Tularemia (*Francisella tularensis*): **Ch. 71.**
     - *Chlamydia* spp.: This includes infantile pneumonia (caused by *Ch. trachomatis* serotypes D-K), psittacosis (caused by *Ch. psittaci*) and *Ch. pneumoniae* pneumonia (caused by *Ch. pneumoniae*). All are described in **Ch. 75.**
     - Scrub typhus (*Orientia tsutsugamushi*): **Ch. 76.**
     - Q fever (*Coxiella burnetii*): **Ch. 76.**
   - **Atypical pneumonia by viruses**
     - Pneumonia (Adenoviridae → serotypes 3 and 7): **Ch. 79.**
     - Lower respiratory tract infections (Pneumoviridae → RSV): **Ch. 82.**
     - COVID-19 (Coronaviridae → SARS-CoV2): **Ch. 86.**
   - **Atypical pneumonia by fungi**
     - Pneumocystosis (*Pneumocystis jiroveci*): **Ch. 91.**
   - **Atypical pneumonia by parasites**
     - Toxoplasmosis (*Toxoplasma gondii*): **Ch. 98.**
3. **Other types of pneumonia**
   - **Fatal necrotizing pneumonia (*Burkholderia cepacia*): Ch. 67.**
   - **Extensive necrotizing pneumonia (*Burkholderia pseudomallei*): Ch. 67.**
   - **Tuberculous caseous pneumonia (*Mycobacterium tuberculosis*): Ch. 56.**
   - **Pneumonia by migrating larvae of parasites:**
     - Loeffler's syndrome or Loeffler's pneumonia or eosinophilic pneumonia due to migrating larvae of *Ascaris lumbricoides*: **Ch. 107.**
     - Bronchopneumonia due to migrating larvae of *Hookworm* species: **Ch. 107.**
   - **Noninfectious pneumonia**
     - Aspiration pneumonia: Occurs due to aspiration of foods, oral/gastric contents or foreign bodies.
     - Hypostatic pneumonia: Occurs due to collection of edema fluid and secretion in dependent parts of lungs followed by bacterial infections. It is common in comatose, old, feeble, severely ill and bed-ridden patients.
     - Lipid pneumonia: Occurs due to aspiration of oily materials. Source may be either exogenous like aspiration of oily nasal drops, regurgitation of oily medicines (e.g. liquid paraffin), aspiration of oily vitamin preparations in children/severely

ill-bed ridden patients, etc., or endogenous like lipid origin following breakdown of air ways as in tuberculosis, bronchogenic cancer, bronchiectasis, etc.

Pathogenicity of pneumonia

- **Modes of transmission:** Exogenous infections occur due to entry of organisms in to respiratory tract directly by inhalation route. Endogenous infections occur by following routes:
  - Aspiration: Organisms enter in to respiratory tract by aspiration from the nasopharynx or oropharynx.
  - Hematogenous spread: Organisms enter into respiratory tract by blood from distant sites.
  - Local spread: Organisms enter into respiratory tract by local spread from surrounding organs.
- **Host (precipitating) factors:**
- **Clinical features:** } Follow **Table 42.1** for broncho and lobar pneumonia
- **Complications:**

Laboratory diagnosis

- **Specimen:** Sputum and blood.
- **Testing methods:** These are microscopy of sputum, sputum culture, blood culture, testing of biomarkers (CRP and procalcitonin are elevated) and blood cell count (leukocytosis).

Prevention: Vaccination to causative agents, washing hands, wearing mask, avoiding smoking and contact with ill persons are the useful measures.

Treatment: Follow **Table 42.1**.

Pneumatocele

Definition: It is a thin walled, air-filled cyst that develops within the lung parenchyma results from pulmonary trauma during mechanical ventilation.

Etiological agents

- **Infectious:** Most often, it occurs as a sequelae to acute pneumonia, commonly caused by *Staph. aureus*. However, pneumatocele formation also occurs with other agents, like *Strept pneumoniae, Haemophilus influenzae, Escherichia coli, Strept. pyogenes, Serratia marcescens, Klebsiella pneumoniae, Mycobacterium tuberculosis* and adeno virus. Pneumatocele is generally observed soon after the development of pneumonia but can be observed on the initial chest radiograph.
- **Noninfectious:** It includes hydrocarbon ingestion, trauma following ventilation and positive pressure ventilation.

Pathogenicity

- **Precipitating (host) factors:** One study reported that 70% of pneumatoceles were occurred in children <3 years.
- **Clinical features:** In most circumstances, pneumatocele is asymptomatic.
- **Complications:** These include secondary bacterial infections and pneumothorax.

Diagnosis: It is made by chest X-ray.

Prevention: Treatment of the underlying pneumonia with antibiotics.

Treatment: Symptomatic/supportive treatment is given.

Lung Abscess

Definition: It is a localized necrosis with suppuration within lung parenchyma.

Etiological agents

- **Bacterial/pyogeniclung abscess:** *Staph. aureus,* streptococci, *Nocardia, Legionella,* GNB (e.g. *Pseudomonas* and Enterobacterales).
- **Fungal lung abscess:** *Aspergillus,* Mucorales, *Cryptococcus,* and *Pneumocystis jirovecii.*
- **Parasitic/amoebic lung abscess:** Ch. 93 (pulmonary amoebiasis).

Pathogenicity

- **Modes of transmission**
  1. **Primary lung abscess:** It occurs in normal lungs in approximately 80% cases due to aspiration

Features	Bronchopneumonia	Lobar pneumonia
	TABLE 42.1: Differences between bronchopneumonia and lobar pneumonia	
Definition	Follow text	Follow text
Agents	Follow text	Follow text
Pathogenicity		
• Precipitating (host) factors	• Common at extremes of age (infants and old age) and with pre-existing disease like measles	• Common in adults and healthy persons
• Pathology	• Patchy consolidation with central granularity, alveolar exudation and thick septa	• It passes via stage of congestion, consolidation and resolution.
• Clinical features	• Initial 2–3 days features of acute bronchitis followed by features of lobar pneumonia	• Fever with chills, malaise, pleuritic chest pain, dyspnea, productive cough may be mucoid, purulent or bloody, cyanosis, tachycardia and tachypnea
• Complications	• Bronchiectasis and complication of lobar pneumonia	• Pleural effusion, empyema, lung abscess and fibrosis. It also spread in other organs
X-ray	Consolidation	Mottled focal opacities
Treatment	Better response to antibiotics	Variable response to antibiotics
Prognosis	Good	Poor

of contents from mouth (food and decayed teeth), nasopharynx and stomach during sleep, unconsciousness, coma, anesthesia or alcoholism. Infection occurs mostly by anaerobic bacteria (or microaerophilic streptococci).

2. **Secondary lung abscess:** It occurs in diseased lungs like pneumonia, malignancy, bronchial obstruction (distal to obstructed bronchus by tumor or foreign body), trauma, pulmonary infarct, etc., as a complication by above mentioned bacteria and fungi.

- **Clinical features:** Initially, symptoms are same as of pneumonia like fever, productive cough, and chest pain. It also presents with hemoptysis, clubbing of fingers or toes. In anaerobic bacterial infections, progression is slow and indolent with night sweats, fatigue, and occasionally foul-smelling sputum (in case of putrid lung abscess). In cases with non-anaerobic bacterial etiology like in *Staph. aureus* the course is more fulminant with rapid progression.
- **Complications:** If left untreated, it causes secondary amyloidosis, empyema due rupture into pleural cavity, lung fibrosis, etc.

Diagnosis: It is diagnosed microbiologically by blood culture, microscopy and culture of sputum and pus obtained from lungs. Radiological tests include X-ray and CT scan to locate the abscess. They are showing collections of pus in one site of >2 cm in size.

Prevention: Early diagnosis and treatment of underlying etiology.

Treatment: In primary lung abscesses clindamycin is the drug of choice. Secondary lung abscesses is treated by surgical drainage and antibiotics are given as per AST report. A prolonged course is often required until resolution of the abscess.

Pleural Effusion

Definition: It is an excess quantity of fluid collection in the pleural space. Following different terminologies are used for pleural effusion, depending on the nature of the material collected into the pleural space.
- Hydrothorax: Collection of serous fluid
- Hemothorax: Collection of blood
- Chylothorax: Collection of chyle (lymph)
- Pyothorax/empyema: Collection of pus
- Pneumothorax/collapsed lung: Collection of air (pneumo means air).

Etiological agents: Normally fluid enters in pleural space out of the parietal capillaries at a rate of 0.01 ml per kg of body weight per hour. A portion of this fluid will be absorbed into the lymphatic system, leaving only 5–15 ml of fluid inside the pleural space. Following are the conditions causing pleural effusion.

1. **Transudative:** These are congestive heart failure, cirrhosis, nephrotic syndrome, severe hypo-albuminemia, pulmonary emboli, acute atelectasis, myxoedema, peritoneal dialysis, obstructive uropathy and end-stage kidney disease.

2. **Exudative:** These are infective and noninfective causes. Infective causes are bacterial pneumonia, tuberculosis (most common cause in developing countries) and very rarely viral infections. Non-infective causes are lung cancer or metastases to the pleura from elsewhere, trauma, pulmonary infarct, autoimmune disease like RA, pancreatitis, ruptured esophagus (Boerhaave's sunfrome) and drugs.

Pathogenicity

- **Precipitating (host) factors:** It is common in elderly.
- **Clinical features:** Excess fluid can impair breathing by limiting the expansion of the lungs, chest pain, fever, weight loss, etc.
- **Complications:** These are hypoxia, lung collapse, scaring of wall called fibrothorax.

Laboratory diagnosis

- **Specimen:** Sputum, induced sputum, tracheal aspirate, BAL, protected specimen brush (PSB), lung aspirate (collected by transtracheal aspiration) and pleural fluid (collected by thoracentesis).
- **Testing methods:** Microscopy (Gram's and ZN stain) and culture.

Prevention: It is prevented by early diagnosis and treatment of underlying etiology.

Treatment: It is treated by pleural tapping (thoracentesis) and antibiotics. Specific treatment is given for underlying etiology.

SPECIFIC INFECTIONS

Bacterial Infections

Diphtheria (*Corynebacterium diphtheriae*): **Ch. 53.**

Pulmonary anthrax (*Bacillus anthracis*): **Ch. 54.**

Tuberculosis (*Mycobacterium tuberculosis*): **Ch. 56.**

Infections by NTM (nontuberculous *Mycobacterium* spp.): **Ch. 58.**

Ozena/atrophic rhinitis (*Klebsiella pneumoniae ozaenae*) **and rhinoscleroma** (*Klebsiella pneumoniae, rhino-scleromatis*): **Ch. 61.**

Pneumonic plague (*Yersinia pestis*): **Ch. 64.**

Pertusis/whooping cough (*Bordetella* spp.): **Ch. 69.**

Infection by nonfermenters

- Respiratory tract infections by *Pseudomonas* spp.: **Ch. 67.**
- Glanders/malleus (*Burkholderia mallei*): **Ch. 67.**
- Melioidosis/Vietnam time bomb (*Burkholderia pseudomallei*): **Ch. 67.**
- Respiratory tract infections by *Moraxella catarrhalis*: **Ch. 52 and 67.**
- Respiratory tract infections by *Acinetobacter* spp.: **Ch. 52 and 67.**

- Respiratory tract infections by *Alkalegenes faecalis*: **Ch. 67.**
- Respiratory tract infections by *Elizabethkingia (Flavobacter)* meningosepticum: **Ch. 67.**
- Respiratory tract infections by *Stenotophomonas maltophila* : **Ch. 67.**
- Legionellosis (*Legionella pneumophila*): **Ch. 67.**

Viral Infections

Respiratory infections of Adenoviridae: Ch. 79.
- Pharyngitis: Serotypes 1–7.
- Acute respiratory distress (ARD) in military person: Serotypes 4 and 7.
- Pharyngoconjunctival fever: Serotypes 3, 7 and 14.

Respiratory infections of Picornaviridae: Ch. 80.
- Coxsackie virus: Herpangina (vesicular pharyngitis) and other respiratory infection
- EV-68: Pneumonia
- ECHO viruses: Respiratory infections by 1, 11, 19, 20 and 22.
- Rhino viruses: Respiratory infections like common cold.

Respiratory infections of Orthomyxoviridae: Ch. 81.
- Influenza (Influenza viruses):
- A/H1N1 2009 pandemic
- Avian flu

Respiratory infections of Paramyxoviridae: Ch. 82.
- Respiratory infection by *Respirovirus* and *Rubulavirus* (Parainfluenza viruses)
- Respiratory infection by *Henipavirus* (nipah virus and hendra virus).

Respiratory infections of Pneumoviridae: Ch. 82.
- Respiratory infection by *Metapneumovirus* (human metapneumo virus).

Respiratory infections of Coronaviridae: Ch. 86.
- SARS (SARS-CoV/SARS-CoV1).
- MERS (MERS-CoV).
- COVID-19 (SARS-CoV2).

Fungal Infections

Rhinosporidiosis: Ch. 91.

Systemic mycoses
- Pulmoary histoplasmosis (*Histoplasma capsulatum*): **Ch. 91.**
- Pulmoary blastomycosis (*Blastomyces dermatitidis*): **Ch. 91.**
- Pulmoary coccidioidomycosis (*Coccidioides immitis*): **Ch. 91.**
- Pulmoary paracoccidioidomycosis (*Paracoccidioides brasiliensis*): **Ch. 91.**

Oppotunistic mycoses
- Pulmoary candidiasis (*Candida* spp.): **Ch. 90.**
- Pulmoary cryptococcosis (*Cryptococcus neoformans*): **Ch. 91.**

- Penicilliosis (*Penicillium marneffei*): **Ch. 91.**
- Aspergillosis (*Aspergillus* spp.): **Ch. 91.**
- Systemic zygomycosis (species from class → Zygomycetes): **Ch. 91.**

Parasitic Infections
- Pulmonary amoebiasis (*E. histolytica*): **Ch. 93.**
- Hydatid cyst in lungs (*Echinococcus granulosus*): **Ch. 102.**
- Paragonimiasis (*Paragonimus westermanii*): **Ch. 105.**
- Lung infection by intestinal nematodes (*Ascaris lumbricoides, Hookworm* spp., and *Strongyloides stercoralis*): **Ch. 107.**
- Tropical Pulmonary Eosinophilia (*Wuchereria bancrofti* and *Brugia malayi*): **Ch. 108.**
- Visceral larva migrans (*Toxocara canis* and *Toxocara cati*): **Ch. 109.**

ACCESS YOURSELF

Essays/Full Questions
1. Describe the etiopathogenicity, laboratory diagnosis and prevention of infections of upper and lower respiratory tract.
2. Define and describe the etiology, types, pathogenicity and complications of pneumonia.
3. Define and describe the etiology, pathogenicity and complications of lung abscess.

Short Notes
1. Pharyngitis
2. Peritonsillar abscess/quinsy
3. Pneumonia
4. Lung abscess.

Short Questions for Theory/Viva Questions
1. What is strep throat? Name the causative agent.
2. Name the most common bacterium causing following diseases. Lobular pneumonia, lobar pneumonia, atypical pneumonia and fatal necrotizing pneumonia.
3. Name the most common organism causing following diseases. Rhinitis, sinusitis, pharyngitis and pneumatocele.

MCQs for Chapter Review
Pneumatocele

1. **Pneumatoceles are seen in:**
   a. *Klebsiella pneumoniae*   b. *Streptococcus pneumoniae*
   c. *Mycoplasma pneumoniae* d. *Listeria monocytogenes*

Pneumonia

2. **Atypical pneumonia can be caused by the following microbial agents *except*:**
   a. *Mycoplasma*   b. *Legionell pneumophilla*
   c. Human coronavirus   d. *Klebsiella pneumoniae*

Answers and Explanation of MCQs

1. **a and b**
- Pneumatocele is commonly caused by *Staph. aureus*, which is not given in options. From the given options right answers are *Strept. pneumoniae* and *Klebsiella pneumoniae*.

2. **d**
- Follow section, **pneumonia (primary atypical pneumonia)** for explanation.

CHAPTER 43

Urinary Tract Infections

Chapter Outline
- Introduction
- Common Infections
- Specific Infections

INTRODUCTION

Anatomy

There are two parts of urinary tract such as upper includes kidneys and ureters while lower includes bladder and urethra.

Normal Defense of Urinary Tract

Low urine pH and high urea concentration are inhibitory for microbes. If bacteria gain access to the bladder the constant flushing action of urine eliminates bacteria. Valve like mechanisms at the junction of ureter and bladder prevents the reflux of urine from bladder to upper urinary tract. The bladder mucosal surface has antibacterial properties.

Normal (Resident) Flora of Urinary Tract

Follow **Ch. 4.**

COMMON INFECTIONS

Urinary Tract Infections (UTIs)

Etiological agents: Bacterial causes are described here with their mode of transmission. *E. coli* is the most common cause for UTI. Other causes like viruses, fungi and parasites are mentioned below as specific infections in this chapter.

Pathogenicity
- **Modes of transmission**
  1. **Ascending route:** It is most common route of infection in females from fecal contamination of perineum to urethra and kidneys. Common causes are *E. coli* (most common), *Klebsiella, Proteus, Serratia, Pseudomonas, Ch. trachomatis, Staph. saprophyticus, N. gonorrhoeae* and HSV.

  2. **Descending (hematogenous) route:** It includes spread from distant site to the kidneys via blood. This route accounts for <5% of UTIs. Common causes are *Staph. aureus* and *C. albicans.*
- **Precipitating (host) factors**
  1. **Age:** UTI is more in pediatric patients due to fecal contamination.
  2. **Sex:** More in female during sexually active stage, during pregnancy and also due to short urethra and proximity to anus.
  3. **Diseases:** Like renal stone, urethral stricture, prostate hypertrophy, tumor, vesicourethral reflux, genital prolapsed, neurogenic bladder, diabetes mellitus, local trauma, etc., increase the risk for UTI.
  4. **Instrumentation:** Like catheterization increases the risk for UTI.
- **Clinical features:** Clinically UTIs are classified as follows:
  1. **Asymptomatic bacteriuria:** It is the isolation of bacteria in urine of person without any symptoms of UTI. It is common in females and incidence increases with age about 1% in school girls and 20% in old age. It is clinically significant in pregnant women, menopause women, people undergoing prostatic surgery or any urologic procedure where bleeding is anticipated. Routine screening and treatment for asymptomatic bacteriuria is required to avoid later stage complications. It is clinically not significant in nonpregnant, premenopausal women, old age, catheterized patient, or patients with spinal injury. It does not require any screening or treatment.
  2. **Urethritis:** It is an inflammation of urethra. Patient feels discomfort or pain at the urethral meatus or a burning sensation/pain throughout the urethra during urination (dysuria).

3. **Acute urethral syndrome:** It presents in young sexually active females with pyuria. Symptoms are same as of cystitis. Bacterial count in urine is often low (10^2–10^5 CFU/ml). It is caused by usual agents of UTI plus caused by *Neisseria gonorrhoeae*, *Chlamydia*, HSV, etc.

4. **Cystitis:** It is an inflammation of bladder. It presents with dysuria, frequency, urgency, and suprapubic tenderness. Urine becomes cloudy, with bad odor, and in some cases contains blood.

5. **Ureteritis:** It is an inflammation of ureter. It presents with pain in lower abdomen or flank.

6. **Pyelonephritis:** It is an inflammation of parenchyma of kidneys, calyces and the renal pelvis. It presents with fever, chills, flank pain, vomiting, frequency, urgency and dysuria.

- **Complications:** These are stricture formation, urinary obstruction or renal/perinephric abscess.

Laboratory diagnosis: It is same as that of diagnosis of UTI caused by *E. coli* (**Ch. 61** → ***Escherichia coli*** → **UTI** → **Laboratory diagnosis of UTI**).

Prevention: It is prevented by improving the hygiene of ano-genital region with early diagnosis and treatment of underlying infections.

Treatment: Antibiotics are given as per AST report.

> **Note: Terminology**
> - Frequency: Means feeling to urinate even though there may be very little urine to pass
> - Dysuria: Pain throughout the urethra with urination
> - Pyuria: Pus in the urine or discharge from the urethra
> - Nocturia: Need to urinate during the night
> - Hematuria: Blood in urine.

Renal Abscess

Definition: It is the pus formation within parenchyma of kidney (medulla to context).

Etiological agents: Commonest bacteria causing renal abscess are *E. coli*, *Proteus* and *Klebsiella*.

Pathogenicity

- **Mode of transmission:** Infection ascends from cystitis to pyelonephritis followed by abscess.
- **Precipitating (host) factors:** It occurs due to underlying infection of urinary system.
- **Clinical features:** It presents with loin pain, fever, etc.
- **Complications:** It ruptures to produce perinephric abscess.

Laboratory diagnosis

- **Specimen:** Aspiration of pus.
- **Testing methods:** Microscopy and culture.

Prevention: It is prevented by early diagnosis and treatment of cystitis and pyelonephritis.

Treatment: Treatment includes surgical drainage and antibiotics administration as per AST report.

Perinephric Abscess

It is the pus formation around the area of kidney. It occurs due to rupture of renal abscess. Other details are same as renal abscess.

SPECIFIC INFECTIONS

Bacterial Infections

- UTI by Micrococcaceae (*Staph. aureus*, *Staph. albus*/*Staph. epidermidis* and *Staph. saprophyticus*): **Ch. 49.**
- PSGN (*Strept. pyogenes*): **Ch. 50.**
- UTI by *Enterococcus*: **Ch. 51.**
- Renal tuberculosis (*M. tuberculosis*): **Ch. 56.**
- UTI by Enterobactariceae (*E. coli*, *Citrobacter*, *Klebsiella*, *Enterobacter*, *Proteus*, *Providencia* and *Morganella*): **Ch. 61.**
- UTI by nonfermenters (*Pseudomonas* and *Acinetobacter*): **Ch. 7.**
- UTI by Mycoplasmatales (*M. hominis* and *U. urealyticum*): **Ch. 74.**
- Urogenital chlamydiasis (*Ch. trachomatis*): **Ch. 75.**

Viral Infections

- UTI causing viruses: **Ch. 89.**

Fungal Infections

- Candidal urethritis: **Ch. 90.**

Parasitic Infections

- Trichomoniasis (*Trichomonas vaginalis*): **Ch. 94.**
- Urinary schistosomiasis (*S. haematobium*): **Ch. 104.**

ACCESS YOURSELF

Essay/Full Question

1. Describe the etiopathogenicity, methods of specimen collection, and the laboratory diagnosis of UTIs.

Short Note

1. Renal abscess.

Short Question for Theory/Viva Question

1. What is asymptomatic bacteriuria? Write its clinical significances.

MCQs for Chapter Review

UTIs

1. **Most common organism implicated in the etiology of urinary tract infection in the community is:**
   a. *E. coli*
   b. *Proteus*
   c. *Pseudomonas*
   d. *Streptococcus*
2. **Most common cause of UTI in young female is:**
   a. *Staph saprophyticus*
   b. *E. coli*
   c. *Klebsiella*
   d. *Proteus*

3. **A 30-year-old male presents with urethritis. All the following can be the causative agent *except*:**
   a. *Neisseria gonorrhoeae*
   b. *Chlamydia trachomatis*
   c. *Trichomonas vaginalis*
   d. *Haemophilus ducreyi*

4. **All are common organisms causing UTI *except*:**
   a. *Streptococcus faecalis*
   b. *Escherichia coli*
   c. *Proteus mirabilis*
   d. *Haemophilus influenzae*

5. **Ascending UTI is caused by:**
   a. *Salmonella*   b. TB
   c. *E. coli*   d. *Chlamydia*
   e. *Klebsiella*

Answers and Explanation of MCQs

1. a
- *E. coli* is the most common cause of acute UTI (in 80% cases without catheter), neonatal meningitis, intra-abdominal abscess and nosocomial infection.
- *E. coli* cause 80–90% of lower UTI (cystitis) in young women.

2. b
- *E. coli* cause 80–90% of lower UTI (cystitis) in young women.

3. d

4. d
- Follow section, **urinary tract infections (etiological agents)** for explanation of answers of MCQs 3–4.

5. c and d
- Follow section, **urinary tract infections (pathogenicity → mode of transmission → ascending route)** for explanation.

Genital Tract Infections

INTRODUCTION

Anatomy of Genital Tract

Female genital tract: It includes vulva, vagina, Bartholin's gland, cervix, uterus, fallopian tube and ovary.

Male genital tract: It includes testes, epididymis, scrotum, vas deferens, urethra, prostate and penis.

Normal (Resident) Flora of Genital Tract

Follow **Ch. 4.**

COMMON INFECTIONS

Female Genital Tract Infections

Pelvic Inflammatory Disease (PID)

Definition: It is an inflammation/infection that extends from the vagina/cervix to endometrium, fallopian tube, ovary and/or peritoneum.

Etiological agents: *N. gonorrhoeae* and *Ch. trachomatis* are the most common causes of PID. Rare causes are *M. hominis, M. genitalium, U. urealyticum, Peptostreptococcus, Prevotella* species, *E. coli, Haemophilus influenzae* and *Strept. pyogenes*. It also occurs secondary to hematogenous spread in case of infections by *M. tuberculosis* or staphylococci.

Pathogenicity

- **Modes of transmission:** Primary PID occurs spontaneously due to sexually intercourse. Secondary PID occurs due to invasive intrauterine procedures or hematogenous spread from distant foci of infections like *M. tuberculosis* and *Staphylococcus*.
- **Precipitating (host) factors:** It is common in sexually active women.

- **Clinical features**
  - **Vaginitis:** It is an inflammation of the vagina presents with pain or discharge.
  - **Cervicitis:** It is an inflammation of the cervix presents with pain or discharge.
  - **Endometritis:** It is an inflammation of the endometrium presents with abdominal pain and vaginal bleeding.
  - **Salpingitis:** It is an inflammation of the fallopian tube presents with bilateral lower abdominal and pelvic pain plus swelling and fever.
  - **Oophoritis:** It is an inflammation of ovary.
  - **Peritonitis:** It is an inflammation of peritoneum.

Complications: These are tubo-ovarian abscess, infertility, irregular menstruation, perisplenitis, or pelvic abscess. In gonorrhea, sometimes peritonitis presents with perihepatic inflammation called Fitz-Hugh-Curtis syndrome. It is characterized by right upper quadrant pain, edema and erythema of liver capsule with fibrinous adhesion between liver and peritoneum.

Laboratory diagnosis

- **Specimen:** Vaginal discharge.
- **Testing methods:** Microscopy and culture.

Prevention: It is prevented by genital screening of women.

Treatment: It is treated by antibiotics and surgery in case of pelvic or tubo-ovarian abscess.

Bartholinitis

Definition: It is an inflammation of Bartholin's gland and blockade of its duct.

Etiological agents: Bartholin's gland is a mucus-producing gland presents on each side of the vaginal

TABLE 44.1: Differences between acute and chronic prostatitis

Features	Acute prostatitis	Chronic prostatitis
Agents	• <35 men: *N. gonorrhoeae* and *Ch. trachomatis* • >35 men: Enterobacterales and Enterococcus	Enterobacterales (80%), Enterococcus (15%) and rare by *Pseudomonas*
Pathogenicity • Precipitating (host) factors	• Increasing age.	• Increasing age.
• Clinical features	• Fever, chills, malaise, myalgia, dysuria, pelvic or perineal pain and cloudy urine	• Low grade fever, urinary frequency, dysuria, urgency and perineal discomfort
• Complications	• Bacteremia, epididymitis, prostatic abscess, infection to joints, or rarely chronic prostatitis	• Persistent bacteriuria and recurrent UTIs
Treatment	Single dose (IM) of ceftriaxone) then doxycycline for 10 days.	Ciprofloxacin or levofloxacin for 4 weeks.

orifice; opens through a duct on to the inner surface of the labia minora. Infection is caused by anaerobes and other bacteria originating from normal genital flora.

Male Genital Tract Infections

Orchitis

It is an inflammation of testes (testicles), mostly caused by mumps virus by the bloodborne dissemination of virus. It presents with unilateral testicular pain and swelling. Mumps orchitis is mostly unilateral so infertility is very rare.

Epididymitis

It is an inflammation of the epididymis that lasts <6 weeks, mostly presents with pain and swelling. It is caused by following agents.
• In young men: By *Ch. trachomatis* and rare by *N. gonorrhoeae*.
• In older men: It occurs following urinary tract instrumentation.
• In homosexual men: It occurs following anorectal intercourse by members of Enterobacteriaceae.

Prostatitis

It is an inflammation of prostate. Etiological agents, pathogenicity and treatment are mentioned in **Table 44.1**.

SEXUALLY TRANSMITTED INFECTIONS (STIs)

Synonym

These are also known as venereal diseases or sexually transmitted diseases (STDs).

Etiological Agents

Bacteria

1. Gonorrhea (*Neisseria gonorrhoeae*): **Ch. 52.**

2. Granuloma inguinale or donovanosis or granuloma venereum (*Klebsiella granulomatis*): **Ch. 61.**
3. Gay bowel syndrome in male homosexual (*Shigella* species): **Ch. 63.**
4. Chancroid or soft sore (*Haemophilus ducreyi*): **Ch. 68.**
5. Bacterial vaginosis (*Gardnerella vaginalis*): **Ch. 71.**
6. Syphilis (*Treponema pallidum*): **Ch. 73.**
7. Urogenital chlamydiasis and LGV (*Chlamydia trachomatis*): **Ch. 75.**
8. Others:
   • Campylobacteriasis by oral-anal sex. (*Campylobacter jejuni*): **Ch. 66.**
   • *Brucella abortus* and other *Brucella* species: **Ch. 70.**

Viruses

1. Poxviridae (MCV): **Ch. 77.**
2. Herpesviridae (HSV-1, HSV-2, CMV and KSHV): **Ch. 78.**
3. Papovaviridae (HPV most common among the viral agents): **Ch. 79.**
4. Arboviruses (zika virus): **Ch. 83.**
5. Hepatitis viruses (HBV and HCV): **Ch. 88.**
6. Retroviridae (HTLV-I, HTLV-II and HTLV-III/HIV): **Ch. 85.**

Fungi

1. Candidiasis (*Candida albicans*): **Ch. 90.**

Parasites

1. Amobiasis (*Entamoeba histolytica*): **Ch. 93.**
2. Giardiasis (*Giardia lamblia*): **Ch. 94.**
3. Trichomoniasis (*Trichomonas vaginalis*): **Ch. 94.**
4. Cystoisosporiasis/isosporiasis (*Cystoisospora/Isospora belli*) and cryptosporidiosis (*Cryptosporidium parvum*) by oral-anal sex: **Ch. 98.**

Ectoparasites

1. Scabies by itch mite (*Sarcoptes scabiei/Acarus scabiei*): **Ch. 110.**
2. Pediculosis on pubic parts by pubic louse or crab louse (*Pthirus pubis*): **Ch. 110.**

Pathogenicity

Common features of STDs: These include urethral/cervical/vaginal discharge which may be purulent watery (bacterial), thick white (candidiasis), frothy (trichomoniasis) or red (bloody), local pain, warts or swelling, lower abdominal pain, fever, malaise weakness, headache, etc.

Specific features: These include following.
• Proctitis: Inflammation of rectum following anal intercourse. It presents with itching, mucopurulent discharge, bleeding, anal pain, tenesmus.
• Anogenital warts (like in HPV): It presents on penis, vulva, vagina or in anal region.
• Genital ulcers: Painless in syphilis and painful in chancroid with discharge.
• Vulvovaginitis: Inflammation of vulva-vagina. presents with discharge with offensive smelling, itching or pain.

- Cervicitis: Inflammation of cervix. It presents with mucopurulent discharge, bleeding, edema of cervix and ulcer.
- PID: It may present as endometritis, salpingitis, oophoritis and/or peritonitis. Clinical features are described above with PID.
- Few more features are as per etiological agents and described in details in respective chapters.

Complications: These are infertility, ectopic pregnancy, premature labor, abortion, or cervical carcinoma.

Laboratory Diagnosis

- **Specimens:** Urine, fluid/discharge from genitals, swab and blood.
- **Testing methods:** Follow respective chapters for diagnosis of particular organism.

Prevention and Treatment

These depend on type of infections and discussed in detail in respective chapters.

NONGONOCOCCAL URETHRITIS (NGU)

Follow **Ch. 52.**

SPECIFIC INFECTIONS

Bacterial Infections

Genital tract tuberculosis (*M. tuberculosis*): **Ch. 56.**

Parasitic Infections

Hydrocele (*Wuchereria bancrofti, Brugia malayi, Onchocercus volvulus* and *Mansonella ozzardi*): **Ch. 108.**

ACCESS YOURSELF

Essays/Full Questions

1. Describe the etiopathogenicity and the laboratory diagnosis of infections of genitourinary system.
2. Describe the etiology, pathogenicity and diagnostic features of nonsyphilitic sexually transmitted diseases (chancroid, donovanosis and LGV).
3. Describe the etiopathogenicity and the laboratory diagnosis of sexually transmitted infections. Recommend preventive measures.

Short Note

1. PID.

Short Question for Theory/Viva Question

1. Name four sexually transmitted parasites/bacteria/viruses.

MCQs for Chapter Review

Sexually Transmitted Infections (STIs)

1. **Which of the following is/are not STD?**
   - a. HAV
   - b. HPV
   - c. HIV
   - d. Varicella zoster virus
   - e. HTLV-I

2. **Genital ulcer is seen in all *except*:**
   - a. *H. aegypticus*
   - b. *H. ducreyi*
   - c. HSV
   - d. *Chlamydia*
   - e. *T. pallidum*

3. **Which of the following is not a sexually transmitted disease?**
   - a. Hepatitis B
   - b. Amoebiasis
   - c. Bacterial vaginosis
   - d. Yaws

4. **A man presents to a STD clinic with urethritis and urethral discharge. Gram's stain shows numerous pus cells but no microorganisms. The culture is negative on the laboratory media. The most likely agent is:**
   - a. *Chlamydia trachomatis*
   - b. *Haemophilus ducreyi*
   - c. *Treponema pallidum*
   - d. *Neisseria gonorrhoeae*

5. **A young male presents with UTI, on urine examination pus cells were found but no organisms. Which method would be best used for culture?**
   - a. McCoy cell line
   - b. Thayer Martine medium
   - c. LJ medium
   - d. Levinthal medium

6. **Green frothy vaginal discharge is produced by:**
   - a. Herpes simplex
   - b. *Candida albicans*
   - c. *Trichomonas vaginalis*
   - d. Normal vaginal flora

7. **Treatment of partner is required in all infections *except*:**
   - a. *Candida*
   - b. Herpes
   - c. *Trichomonas*
   - d. *Gardnerella*

Answers and Explanation of MCQs

1. a, d
2. a
3. d
- Follow section, **sexually transmitted infections (STIs)** and respective chapters for more explanation of answers of MCQs 1–3.
4. a
- *Haemophilus ducreyi* and *Treponema pallidum* are causing lesion in external genitals.
- *Neisseria gonorrhoeae* and *Chlamydia trachomatis* are causing urethritis with pus cells under Gram's stain. *Chlamydia trachomatis* is gram-negative but not identified under Gram's stain and also not growing on routine laboratory media while *Neisseria gonorrhoeae* can.
- *H. ducreyi* can grow on routine laboratory media.
- In *T. pallidum* nonpathogenic strain can grow over media.
5. a
- In given UTI case organisms are not identified under microscopic examination, but pus cells are noticed; so, most probable organism is *Chlamydia trachomatis*, which is also not growing on routine laboratory media, but growing on cell culture media like McCoy cell line.
6. c
- Vaginal discharge is a feature of *Candida albicans* and *Trichomonas vaginalis* where it is whitish in initial one while yellowish (purulent) or white (serous) with foul smelling or green frothy in later.
7. a and d
- Exact reason is not known, but may be due to endogenous infection by these two microbes.

Ear and Eye Infections

Chapter Outline

- Introduction
- Ear Infections
  - Infection of external ear
  - Infection of middle ear
- Eye Infections
  - Bacterial infections of eyes
  - Viral infections of eyes
  - Fungal infections of eyes
  - Parasitic infections of eyes

INTRODUCTION

Anatomy of Ear

It includes external ear (that includes auricle/pinna, external auditory canal, and myringa/tympanic membrane/eardrum), middle ear, eustachian tube and inner ear.

Anatomy of Eye

It includes (**Fig. 45.1**) conjunctiva (palpebral and bulbar) sclera, pupil, cornea, uveal tract (iris, ciliary body and choroid) lens, anterior chamber with aqueous humor, posterior chamber with vitreous humor, retina, eyelid, eyelashes and lacrimal apparatus (**Fig. 45.2**).

EAR INFECTIONS

Infection of External Ear

Bacterial Infections of External Ears

Perichondritis of pinna: It is an inflammation of perichondrium and cartilage of auricle/pinna. It is mostly caused by *Staph. aureus* and *Bacillus pyocyaneus*. It presents with severe pain. It may lead to hematoma, necrosis or deformity to pinna. It is treated by antibiotics with incision and drainage.

Otitis externa

- **Definition:** It is an infection of external auditory canal.

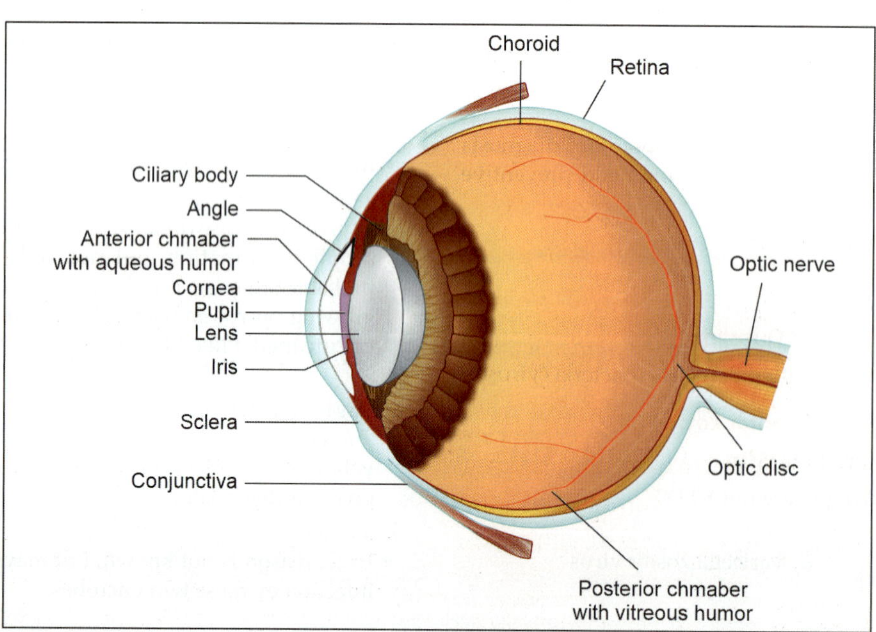

Fig. 45.1: Anatomy of eye

- **Etiological agents:** Two etiological types.
  - Localized: It presents as boil/furuncle and caused by *Staph. aureus*.
  - Diffuse/generalized: It presents as an infiltrative thickening of tissues which may be desquamative and caused by *Staph. aureus*, *Streptococcus* and GNB.
- **Pathogenicity**
  - Precipitating (host) factors: Infection is precipitated by scratching, water entry in ear and otitis media.
  - Clinical features: Both lesions produce pain, itching, white purulent ear discharge (otorrhea), tinnitus and trismus.
  - Complications: Deafness.
- **Laboratory diagnosis:** Ear discharge is tested by microscopy and culture.
- **Prevention:** It is prevented by avoiding scratching, avoiding water entry in ear, early diagnosis and treatment of otitis media.
- **Treatment:** It is treated by local plus systemic anti-biotics with incision and drainage.

Malignant necrotizing otitis externa

- **Definition:** It is the progressive necrotizing infection of external auditory canal that spreads to base of skull, parotid and temporal bone along with cranial nerves.
- **Etiological agents:** It is mostly caused by *Pseudomonas aeruginosa*. It is also caused by fungus like *Malassezia sympodialis*.
- **Pathogenicity**
  - Host (precipitating) factors: Diabetes is the precipitating factor.
  - Clinical features: It presents with granulomatous swelling in external auditory canal, with pain, ear discharge and multiple cranial nerve palsies especially facial palsy.
  - Complications: If it is not treated then produces intracranial complications.
- **Laboratory diagnosis**
  - CT scan (gallium 67 scan) is used to diagnose the bony lesions.
  - Discharge or swab is collected from lesion which is tested by microscopy and culture.
- **Prevention:** It is prevented by controlling diabetes.
- **Treatment:** It is treated by ear toilet to clean discharge, surgery to remove granuloma with antibiotics as per AST report.

Myringitis: It is an infection of external surface of myringa/tympanic membrane/eardrum. Other features are same as otitis externa.

Erysipelas: It is caused by *Strept. pyogenes*, and it is similar to skin lesion.

Viral Infections of External Ears

Herpes simplex infection: It is similar to skin lesion.

Ramsay Hunt syndrome: It is caused by herpes zoster virus presents with eruption (vesicle) in external auditory canal and paralysis of facial or other cranial nerve.

Bullous myringitis: It is caused by *Influenzavirus* and presents with hemorrhagic vesicle on external surface of eardrum, severe pain, conductive deafness and bloody ear discharge. It may complicated as otitis media.

Fungal Infections of External Ears (Otomycosis)

Definition: It is the subacute or chronic superficial fungal infection of external auditory canal.

Etiological agents: Two etiological types.

- **Molds/filamentous fungi:** *Aspergillus* is the most common cause. Common species are *Aspergillus flavus*, *Aspergillus fumigates*, *Aspergillus niger* and *Aspergillus terreus*. Other fungi are *Penicillium* species, *Pseudoallescheria boydii* and dermatophytes.
- **Yeast like:** *Candida albicans*, *C. tropicalis*, *C. krusei* and *Malassezia sympodialis* (causes malignant otitis externa).

Pathogenicity

- **Reservoirs and sources of infection:** Water and external objects entering in the ear are the reservoirs and sources of infection.
- **Modes of transmission:** Entry of water, oil or other external objects may carry the fungus in ear.
- **Precipitating factors (epidemiological determinants):**
  1. **Agent factors:** Organisms are penetrating very deep in the layers of skin; hence, long-term treatment is required.
  2. **Host factors**
     - **Age:** It is common between 2nd and 3rd decades of life.
     - **Occupation:** It is common in people exposed to external atmosphere.
     - **Entry of water:** It is common during swimming, bathing or wetting rain.
     - **Poor hygiene:** It is common in patients with poor hygiene and who are not cleaning the ear following swimming or bathing.
     - **Local problems:** Like ear diseases, application of oil, eardrops, etc., may increase the risk.
     - **Hormonal changes:** Pregnancy and menstruation may increase the risk of infection.
     - **Socioeconomic status:** It is common in people with poor socioeconomic status.
     - **Others:** IDDs, malnutrition, etc., may increase the risk.
  3. **Environmental factors:** It is common in rainy season
- **Pathogenesis:** Organisms are penetrating very deep in the layers of skin; hence, long-term treatment is required. They also invite the secondary bacterial infections.

- **Clinical features:** It presents with itching, little pain, irritation, discharge, feeling of blocked ear due to collection of discharge or fungal growth. Cotton like growth which may be identified in ear on examination which may be white (in *C. albicans*), green (in *A. fumigates*) or black (in *A. niger*). Hyperemic or edematous skin or bleeding from ear may present sometimes.
- **Complications:** These are secondary bacterial infections, deafness and local penetration in mastoid bone or even in brain.

Laboratory diagnosis

- **Specimen:** Ear discharge.
- **Testing methods:**
  1. **Microscopy:** Direct 10% KOH is used to reveal fungal hyphae or yeast cells. Gram's stain shows the presence of budding yeast cells and pseudo-hyphae in case of *Candida albicans.*
  2. **Fungal culture:** Routine culture will be done on SDA with antibiotics. Cultivation technique is mentioned in **Flowchart 120.1**. On SDA creamy white colony produced by *C. albicans*, green by *A. fumigates* and black by *A. niger*. Fungal growth on SDA is identified by LCB/PHOL stain, which shows the hyaline septate hyphae of *Aspergillus* species. Chlamydospores production by *Candida albicans* is identified on corn meal agar.

Differential diagnosis: Otomycosis should be differentiated from bacterial otitis externa, furunculosis and other local ear diseases.

Prevention: Special care should be taken during bathing and swimming to prevent the water entry in ear. Improvements in hygiene like cleaning the ear following swimming or bathing by using sterile ear buds is recommended. Avoid the application of antibacterial drops, oil or unsterile objects in ear.

Treatment: Local treatment includes antifungal drops like 1% clotrimazole and tolnaftate, 1% salicylic acid, 2% acetic acid, 1% gentian violet, hamycin and nystatin. Systemic treatment includes antibacterial drugs to control the secondary bacterial infections and analgesic to relieve the pain.

Infection of Middle Ear

Pyogenic/Suppurative and Nonspecific Otitis Media

Acute otitis media/Acute Suppurative Otitis Media (ASOM)

- **Etiological agents:** It is caused by *Strept. pneumoniae* (most common in children in 33% cases) > *Haemophilus influenzae* type b (2nd most common) > *Moraxella catarrhalis* > *Strept. pyogenes*. Few others causes are *Staph. aureus* and *Pseudomonas*.
- **Pathogenicity**
  - **Precipitating (host) factors:** It is common in infants and children of poor community. It is precipitated by underlying viral infection (like by RSV or influenza virus) and atmospheric pressure changes during flying and diving.
  - **Clinical features:** ASOM passes by following stages.
    1. **Catarrhal (congestion) stage:** It presents with occlusion of eustachian tube and congestion in middle ear. Patient complaining of fullness in ear, pain, tinnitus, autophony (self hearing), deafness and may be fever.
    2. **Exudative stage:** Exudate collected in middle ear which is initially mucoid and later becomes purulent. It puts point pressure on eardrum, which appears as yellowing nipple. Above symptoms are present with more severity.
    3. **Suppurative stage:** Exudate puts more point pressure on drum resulting perforation and purulent discharge (otorrhea).
    4. **Healing stage:** It depends on treatment taken.
  - **Complications:** If ASOM is not treated, it produces CSOM, mastoiditis or mastoid abscess.
- **Laboratory diagnosis**
  - **Specimen:** Ear discharge/swab.
  - **Testing methods:** Microscopy and aerobic plus anaerobic culture.
- **Prevention:** It is prevented by improving hygiene and treating underling viral infections.
- **Treatment:** Perform myringotomy to drain exudates/pus to relieve pressure on drum and to avoid perforation. Antibiotics (topical and systemic) are given as per AST report.

Chronic otitis media/chronic suppurative otitis media (CSOM)

- **Etiological agents:** CSOM is caused by anaerobes (most common) like *Bacteroides fragilis* and *Peptostreptococcus* plus aerobes like *Pseudomonas aeruginosa*, *E. coli*, *Staph. aureus,* and *Proteus* spp.
- **Pathogenicity**
  - **Precipitating (host) factors:** Presence of ASOM which may progress to CSOM, respiratory tract infection and trauma to ear are the risk factors.
  - **Clinical features:** Two types. **(1) Benign/safe/tubotypanic CSOM:** It presents with central ear drum perforation, intermittent ear discharge and mild-moderate conductive deafness. Polyp and choleastoma are rare. **(2) Malignant/unsafe/atticoantral:** It presents with attic or marginal ear drum perforation, continuous ear discharge and mild-moderate-sever mixed deafness. Polyp and cholesteatoma are common.
  - **Complications:** Mastoiditis or mastoid abscess.
- **Laboratory diagnosis:** It is diagnosed clinically plus microbiologically with same tests as mentioned in ASOM.
- **Treatment:** Antibiotics (topical and systemic) are given as per AST report for benign type. Surgeries like tympanoplasty, removal of polyp and cholesteatoma are required.

Malignant necrotizing otitis media (acute in nature): It is caused by *Strept. pyogenes* in children infected with measles and *Influenzavirus.* It presents with rapid destruction of whole eardrum and profuse otorrhea.

Hemorrhagic otitis media: It is caused by *Strept. haemolyticus* in person infected with *Influenzavirus* (influenza epidemic). It presents with hemorrhagic vesicle on eardrum, sever pain and hemorrhagic exudate.

Masked acute otitis media: It occurs due to inadequate antibiotics. Common cause is *Strept. pneumoniae.*

Nonpyogenic/Nonsuppurative and Specific Otitis Media

Tuberculous otitis media: Ch. 56.

Syphilitic otitis media: It is the congenital syphilis causing deafness from auditory nerve disease.

EYE INFECTIONS

Bacteria Infections of Eyes

Bacterial Conjunctivitis

Following are two types.

Acute conjunctivitis

- **Definition:** It is an acute bacterial inflammation of conjunctiva.
- **Etiological agents:** It is caused by *Staph. aureus* (most common cause) *Staph. albus* (*Staph. epidermidis*), *Strept. pneumoniae* (associated with petechial sub-conjunctival hemorrhage) and *Strept. pyogenes.*
- **Pathogenicity**
  - **Precipitating factors:** Presence of flies, hot dry climate, poor hygiene and dirty habits are the risk factors.
  - **Clinical features:** It is bilateral although one eye infected 1–2 days earlier then next eye. Common features are red eye, mild photophobia (difficult to tolerate the light), mucopurulent discharge (blennorrhea), discomfort/foreign body sensation, stickiness of lid margin during sleep, blur vision due to presence of mucus flakes in front of cornea. Conjunctiva shows congestion, swelling or petechial subconjunctival hemorrhage. Other features include matted eyelashes and swollen eyelids.
- **Laboratory diagnosis**
  - **Specimen:** Eye discharge or conjunctival scraping or eye swab moistened in sterile saline.
  - **Testing methods:** Microscopy and culture.
- **Prevention:** It is prevented by improving hygiene or habits.
- **Treatment:** Antibiotics are given in form of eye drops.

Chronic conjunctivitis

- **Definition:** It is the chronic bacterial inflammation of conjunctiva.

- **Etiological agents:** It is caused by *Staph aureus* (most common cause) *Proteus mirabilis, E. coli, Klebsiella,* etc.
- **Pathogenicity**
  - **Precipitating factors:** Alcohol, dust, smoke, eye strain due to refractive error, local irritant or chemical exposure are the risk factors.
  - **Clinical features:** Mild catarrhal inflammation of conjunctiva.
  - **Other details:** Same as acute.

Bacterial Keratitis (Corneal Ulcer)

Definition: It is the bacterial inflammation of cornea.

Etiological agents: It is caused by *Staph. aureus, Strept. pneumoniae, Pseudomonas, E. coli, Klebsiella, Neisseria, C. diphtheriae,* etc.

Pathogenicity
- **Precipitating (host) factors:** Vegetative injuries or presence of infections in conjunctiva, in eyelids or in lacrimal gland are the risk factors.
- **Clinical features:** Two types of ulcers.
  - Purulent ulcer: It presents with eye pain, photophobia, watery discharge (greenish in *Pseudomonas*), discomfort/foreign body sensation, red eye and blur vision. Ulcer presents as yellowish-white lesion with oval/irregular shape.
  - Hypopyon (means pus in anterior chamber) ulcer: It mostly occurs in *Strept. pneumoniae.*
- **Complications:** Perforation of cornea in *Pseudomonas.*

Laboratory diagnosis
- **Specimen:** Eye discharge or corneal scraping.
- **Testing methods:** Microscopy and culture.

Prevention: It is prevented by early diagnosis and treatment of eye diseases.

Treatment: Instillation of topical eye drops for gram-positive and -negative bacteria. Use cycloplegic drug (1% atropine) to relieve pain from ciliary spasm and analgesics. Use goggles to reduce photophobia, do hot fomentation and avoid dust exposure.

Bacterial Uveitis

It is the bacterial inflammation of uveal tract. Uveal tract includes iris, ciliary body and choroid. It mostly caused by viruses. Few bacteria causes uveitis are streptococci, staphylococci, pneumococci and gonococci.

Bacterial Retinitis

It is the bacterial inflammation of retina. It may be acute (in pyemia), subacute (in SABE) and chronic (in tuberculosis, leprosy, actinomycosis and syphilis). It presents with multiple retinal hemorrhages, blur vision especially in subacute type or painless loss of vision. Common complications are endophthalmitis and panophthalmitis.

Definition: It is the suppurative inflammation of inner ocular coat, adjacent structures and vitreous fluid by bacteria.

Etiological agents: It is caused by *Staph. epidermidis* (most common, 60%), *Staph. aureus* (5–10%), *Streptococcus*, *Enterococcus*, *Strept. pneumoniae*, *Pseudomonas*, *C. diphtheriae*, *Propionibacterium acne* and *Actinomyces* spp.

Pathogenicity

- **Mode of transmission:** Endogenous infection occurs in eyes via hematogenous spread from distant foci of infections or direct entry from surrounding structure like ketatitis. Exogenous infection occurs via injuries in eyes by surgeries or trauma.
- **Precipitating (host) factors:** Trauma to eye, ocular surgery and presence of IDDs.
- **Clinical features:** Classically it presents as granulomatous uveitis, diffuse retinitis and deep vitreous abscess. Common features are ocular pain, redness, lacrimation, photophobia and impairment of vision.
- **Complications:** Ocular sepsis and vision loss.

Laboratory diagnosis:

- **Specimen:** Aqueous or vitreous fluid.
- **Testing methods:** Microscopy by Gram's stain and culture.

Prevention: It is prevented by early diagnosis and treatment of underlying bacterial eye diseases.

Treatment: Administer systemic, topical and intravitreal antibacterial drugs. Vitrectomy is performed to remove bulk of lesion and to facilitate the drugs diffusion.

Bacterial Panophthalmitis

Definition: It is the bacterial inflammation of whole eyeball.

Etiological agents: Same as bacterial endophthalmitis.

Pathogenicity

- **Precipitating (host) factors:** Same as bacterial endophthalmitis.
- **Clinical features:** In addition to the features of bacterial endophthalmitis it presents with complete loss of vision, swollen eyeball and fever.
- **Complications:** Spread of infection in brain.

Laboratory diagnosis: Same as bacterial endophthalmitis.

Prevention: It is prevented by early diagnosis and treatment of underlying bacterial eye diseases.

Treatment: Administration of antibiotics with surgical removalof eyeball (evisceration) to avoid brain infection.

Infections of Lacrimal Apparatus

Following are two types like dacryocystitis and dacryoadenitis.

It is the bacterial inflammation of lacrimal sac. It may be acute or chronic as described below.

Acute dacryocystitis

- **Etiological agents:** Common bacterial causes are *Staph. aureus*, *Streptococcus*, and *Strept. pneumoniae*.
- **Pathogenicity**
  - **Precipitating (host) factors:** It is common in females. Risk increased by excessive lacrimation causing stagnation of tears, by closing of nasolacrimal duct opening due to foreign body or nasal diseases like polyp, hypertrophy, etc., by congenital defect in opening of duct and by poor hygiene.
  - **Clinical features:** It presents with painful swelling over the region of lacrimal sac (**Fig. 45.3a**) and discharge from eye.
  - **Complications:** These are abscess formation which drains externally with fistula. Other complications are conjunctivitis and chronic dacryocystitis.
- **Laboratory diagnosis**
  - **Specimen:** Eye discharge or pus from lesion.
  - **Testing methods:** Microscopy and culture.
- **Prevention:** Correction of factors mentioned above.
- **Treatment:** Do massage over area to open the duct. Do hot fomentation to relieve pain and swelling. Use topical (eye drops) and systemic antibiotics. Do lacrimal syringing to open the duct. Do incision and drainage to remove pus. Perform the surgeries like dacryocystorhinstomy or dacryocystectomy.

Chronic dacryocystitis: It occurs due to long lasting acute dacryocystitis with same causes mentioned above.

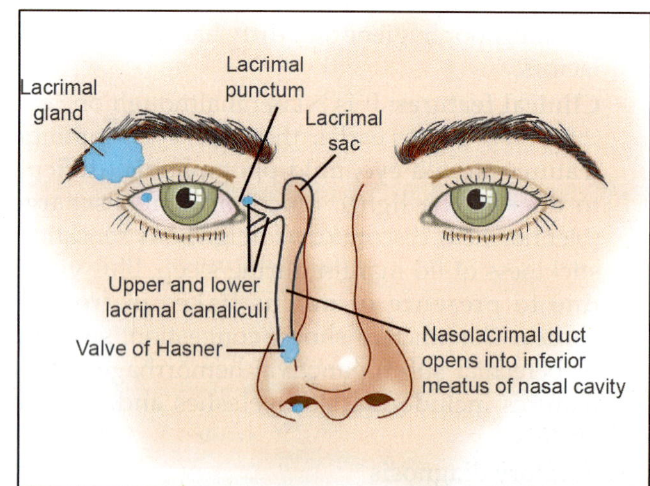

Fig. 45.2: Anatomy of lacrimal apparatus

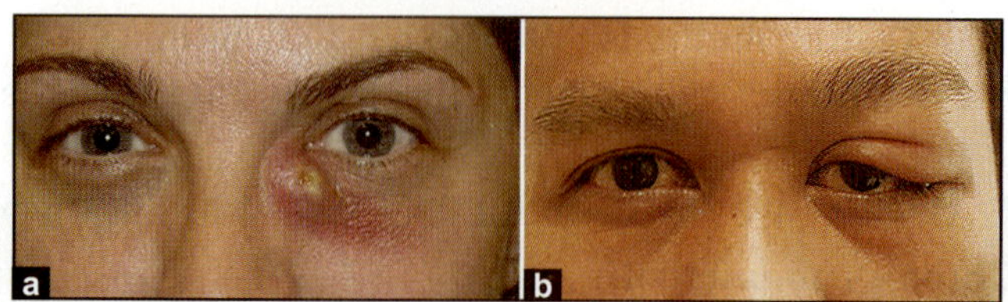

Fig. 45.3: Infections of lacrimal apparatus: (a) Dacryocystitis; (b) Dacryoadenitis

It also occurs in tuberculosis, leprosy, syphilis and occasionally in rhinooculosporidiosis.

Dacryoadenitis

It is the bacterial inflammation of lacrimal gland. It may be acute or chronic as described below.

Acute dacryoadenitis

- **Etiological agents:** It is caused by viruses (mentioned below → viral infections of eye) and bacteria like *Strept. pyogenes, Staph aureus, Strept. pneumoniae, Neisseria gonorrhoeae* and *Haemophilus influenzae.*
- **Pathogenicity:** It presents with painful swelling on the upper eyelid over the area of gland. If not treated then complicated as painful proptosis and fistula.
- **Laboratory diagnosis:** Aspirate the materials from lesion and do the microscopy and culture.
- **Treatment:** It is treated by doing hot fomentation to relieve pain and swelling, using antibiotics and doing incision and drainage to remove pus.

Chronic dacryoadenitis: It is the engorgement and hypertrophy of gland due to long lasting acute dacryoadenitis with same causes mentioned above. It also occurs in tuberculosis and syphilis. It presents with painless swelling in upper and outer part of lid associated with ptosis (**Fig. 45.3b**), displacement of eyeball and diplopia.

Infections of Eyelids

Bacterial blapharitis (chronic)

- **Definition:** It is a chronic inflammation of lid margin.
- **Etiological agents:** *Staph. aureus* is main agent.
- **Pathogenicity**
  - **Precipitating (host) factors:** It is common in childhood. Risk increased by eye strain due to muscle imbalance, refractive error or by presence of eye diseases like chronic conjunctivitis and dacryocystitis.
  - **Clinical features:** Common symptoms are lacrimation, chronic irritation, itching, gluing of cilia and photophobia, which are worse in morning.
  - **Complications:** Recurrent stye, eversion of punctum which leads epiphora (excessive lacrimation), excessive lacrimation leads eczema and ectropion (outward tuning of lid), madarosis (absence or sparseness of eyelashes), poliosis (graying of eyelashes), trichiasis (inward turning of eyelashes) and tylosis (thickening of lid margin).
- **Laboratory diagnosis:** Collect the eye discharge and test it with microscopy and culture.
- **Prevention:** It is prevented by early diagnosis and treatment of eye diseases.
- **Treatment:** Use topical (eye drops and ointment application at lid margin) and systemic antibiotics plus analgesic to reduce the inflammation.

Hordeolum externum/stye

- **Definition:** It is an acute suppurative inflammation of one of the Zeis glands.
- **Etiological agents:** *Staph. aureus* is main agent.
- **Pathogenicity**
  - **Precipitating (host) factors:** It occurs at any age but more common in young adults and children. Risk increased by eye strain due to muscle imbalance, refractive error, frequent rubbing of lid, fingering of nose to lid, diabetes (causing recurrent stye), chronic blepharitis, excessive intake of carbohydrate, alcohol, or by chronic debility.
  - **Clinical features:** It presents with swelling of lid (**Fig. 45.4a**) pain, mild watering and photophobia.
  - **Complications:** Abscess formation in stye with pus point at the lid margin/at the root of cilia.
- **Diagnosis:** It is diagnosed clinically.
- **Prevention:** Prevented by controlling risk factors.
- **Treatment:** Do hot fomentation to relieve pain and swelling. Use topical (eye drops and ointment application at lid margin) and systemic antibiotics. Use analgesic to reduce the inflammation.

Hordeolum internum

- **Definition:** It is the suppurative inflammation of meibomian gland with blockage of the duct.
- **Etiological agents:** *Staph. aureus* is main agent and also occurs due to infection in chalazion (infected chalazion).
- **Clinical features:** It has same features as like stye, but swelling (**Fig. 45.4b**), and pus point is present inside the lid appear as yellow area on everting the lid.
- **Treatment:** Same treatment as stye with vertical incision to drain the pus.

> **Note: Chalazion**
> It also called meibomian gland cyst or tarsal. It is a chronic non-infective granulomatous inflammation of meibomian gland.

Other Specific Bacterial Infections of Eye

1. Acute mucopurulent conjunctivitis (*Neisseria meningitidis*): **Ch. 52.**
2. Ocular gonorrhea in adults or acute purulent conjunctivitis or hyperacute bacterial conjunctivitis by self spread/autoinoculation/autoinfection (*Neisseria gonorrhoeae*): **Ch. 52.**

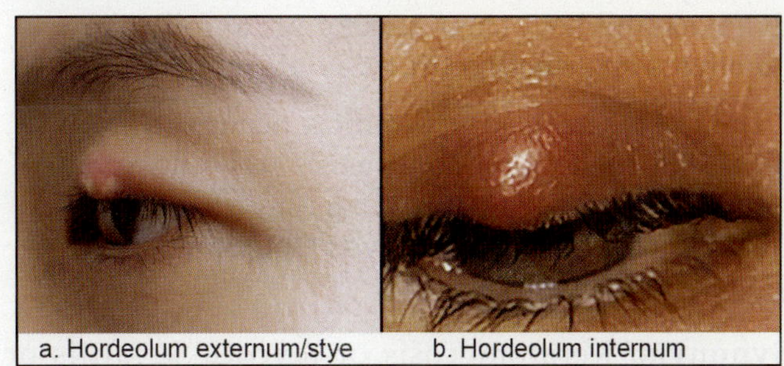

Fig. 45.4: Infections of eyelid

3. Ophthalmia neonatorum (*Neisseria gonorrhoeae*): **Ch. 52.**

4. Angular conjunctivitis or catarrhal conjunctivitis or Morax-Axenfeld conjunctivitis (*Moraxella lacunata*): **Ch. 52.**

5. Conjunctival diphtheria/pseudomembrane conjunctivitis (*Corynebacterium diphtheriae*): **Ch. 53.**

6. Pink eye (*Haemophilus aegyptius*): **Ch. 68.**

7. Trachoma (serovars A, B, Ba and C of *Chlamydia trachomatis*): **Ch. 75.**

8. Inclusion conjunctivitis (serovars D-K of *Ch. trachomatis*): **Ch. 75.**

9. Tuberculosis in eye (*M. tuberculosis*): **Ch. 56.**

10. Leprosy in eye (*M. leprae*): These include conjunctival lepromata in lepromatous leprosy, granulomatous anterior uveitis (iridocyclitis) in lepromatous leprosy, retinitis and chronic dacryocystitis.

11. Actinomycosis in eye (*Actinomyces* spp.): It presents as retinitis.

12. Tularaemia in eye (*Francisella tularensis*): It presents with ulcerative or nodular conjunctivitis.

13. Syphilis in eye (*Treponema pallidum*): It presents as gummatous conjunctivitis, gummatous anterior uveitis (iridocyclitis), diffuse choroiditis and acute plastic iritis in tertiary syphilis. Interstitial keratitis presents in congenital syphilis. Other eye features of syphilis are retinitis, chronic dacryocystitis and chronic dacryoadenitis.

14. Epidemic typhus in eye (*Ricketssia prowazekii*): It presents with uveitis as described in **Ch. 76.**

15. Scrub typhus in eye (*Orientia tsutsugamushi*): It presents with conjunctivitis and uveitis as described in **Ch. 76.**

Viral Infections of Eyes

Poxviridae: Molluscum contagiosum virus described in **Ch. 77.**

Herpesviridae: HSV-1, HSV-2, VZV (zoster) and CMV described in **Ch. 78.**

Adenoviridae: Ch. 79 (Table 79.2).

Picornaviridae: Ch. 80.
- Coxsackie A-24: Acute hemorrhagic conjunctivitis.
- Entero virus-70: Acute hemorrhagic conjunctivitis.

Paramyxoviridae: Avian avula virus 1 (new castle disease/ranikhet virus) causing acute follicular conjunctivitis described in **Ch. 82.**

Other viruses: Dacryoadenitis occurs in mumps, measles, infectious mononucleosis and influenza.

Fungal Infections of Eyes/Oculomycoses

Fungal Keratitis

Synonym: Keratomycosis or mycotic keratitis or mycotic corneal ulcer.

Definition: It is the invasive fungal infection of corneal stroma presents with ulcer.

Etiological agent
- **Molds or filamentous fungi:** Total 1/3rd cases of keratomycosis are caused by filamentous fungi, where *Aspergillus* is the most common cause followed by *Fusarium* spp., and phaeoid fungi. Different species are *Aspergillus fumigates, A. flavus, A. glucans, A. niger, Fusarium solani, F. oxysporum, Penicillium* species, *Pseudoallescheria boydii* and *Paecilomyces* species.
- **Phaeoid fungi:** It is the 3rd common cause of keratomycosis after *Aspergillus* and *Fusarium*. Different species are *Curvularia* species, *Bipolaris* species, *Alternaria* species, *Exserohilum* species, *Aureobasidium pullulans* and *Fonseca pedrosoi* var. *cladosporium*.
- **Yeast like:** These are *Candida albicans, C. tropicalis* and *C. krusei*.

Pathogenicity
- **Reservoir and source infection:** Fungi are present free in environment or on decaying vegetation.
- **Mode of transmission:** Fungi are transmitted following traumatic injuries in eyes.
- **Precipitating factors (epidemiological determinants)**
  1. **Agent factors:** *Candida* causes more localized lesion looks like collar button. *Fusarium* is more aggressive and less responsive to treatment than *Aspergillus*. Phaeoid fungi are low virulent and produce the protracted infection.
  2. **Host factors**
     - Age: It is common in middle age due to maximum activities.
     - Sex: It is common in male due to outdoor works.
     - Occupation: It is most common in agriculture workers following trauma with vegetative materials contaminated by fungi.
     - Local problems: Risk increased with earlier bacterial infection, local use of steroids for long term, local surgery like keratoplasty and wearing contact lens.
  3. **Environmental factors**
     - Season: It is common in harvesting season.
     - Distribution: *Fusarium* is most common in South East USA (Florida) and *Candida* is the most common in Northern USA. *Aspergillus* is common in India.
- **Pathogenesis:** Fungi are not able to penetrate the intact cornea; hence, trauma inoculates the fungi in cornea. Fungal spores colonize the injured tissues and develop the hypopyon corneal ulcer. Hyphae invade from the ulcer bed to deep in corneal stroma leading to lesion with edema and coagulative necrosis. Leukocytes infiltration occurs with feathery border of lesion.
- **Clinical features:** It is mostly unilateral. It presents with pain, irritation, burning or foreign body

sensation, red eye, blurred vision, photophobia and may be discharge from eye. On examination hypopyon corneal ulcer will be identified with edema in corneal stroma.

- **Complications:** It produces the blindness which is the 2nd major complication after cataract and endophthalmitis.

Differential diagnosis: Fungal keratitis should be differentiated from viral, bacterial, parasitic (*Acanthamoeba*) causes and caused by *Pthium insidiosum*.

Laboratory diagnosis

- **Specimens:** Corneal scraping or discharge.
- **Testing:** Two plates are inoculated by "C" or "S" shaped streak, to differentiate the growth from the contaminants. Other methods and findings are same as that of otomycosis.

Prevention: Avoid the application of steroid drops in eyes. Special care should be taken in wearing the contact lens and during work with vegetative materials. Wear goggles to prevent the injury.

Treatment

- **Local drops:** Use antifungal drops like 5% pimaricin (natamycin), nystatin, AMB or flucytosine lavage.
- **Systemic drugs:** Oral fluconazole.
- **Surgery:** Debridement of damaged tissues.

Fungal Uveitis

It is common in aspergillosis, candidiasis, blastomycosis and histoplasmosis. In histoplasmosis uveitis presents with bilateral mutifocal choreoretinal scars called histospots and macular lesion in retina. Histospots are due to previous granulomatous lesion. They may cause loss of central vision. However, *Histoplasma capsulatum* is not isolated from the lesions so called presumed ocular histoplasmosis syndrome.

Fungal Endophthalmitis

Definition: It is the suppurative inflammation of inner ocular coat, adjacent structures and vitreous fluid by fungi.

Etiological agents: Fungal causes are identified in 14% cases of endophthalmitis.

- **Endogenous fungi:** *Candida* is the most common cause. Common species are *C. glabrata*, *C. tropicalis* and *C. parapsilosis*. Other endogenous fungi are *Paecilomyces*, *Fusarium* species, *Curvularia lunata*, *Acremonium* species and zygomycetes species.
- **Exogenous fungi:** *Candida albicans*, *Fusarium* species, *Aspergillus* species, *Histoplasma capsulatum*, *Coccidioides immitis*, *Blastomyces dermatitidis*, *Sporothrix schenckii* and *Histoplasma capsulatum*.

Pathogenicity: Same as bacterial endophthalmitis.

Laboratory diagnosis

- **Specimen:** Aqueous or vitreous fluid.

- **Testing methods:** Two plates are inoculated by "C" or "S" shaped streak, to differentiate the growth from the contaminants. Other methods and findings are same as that of otomycosis.

Prevention: It is prevented by early diagnosis and treatment of underlying fungal eye diseases.

Treatment: Administer systemic, topical and intravitreal antifungal. Vitrectomy is performed to remove bulk of lesion and to facilitate the drugs diffusion.

Mycoses in Eyelids

It is caused by *Sporothrix schenckii*, *Blastomyces dermatitidis* and *Paracoccidioides brasiliensis* presents with chronic ulcers.

Mycoses in Eyelashes

It is caused by *Microsporum canis*, *Candida* spp., and *Malassezia furfur*.

Other Specific Ocular Mycoses

- Ocular candidiasis.
- Oculosporidiosis: **Ch. 91.**
- Ocular pneumocystosis: **Ch. 91.**
- Orbital zygomycosis/orbital mucormycosis: **Ch. 91.**

Parasitic Infections of Eyes/Oculoparasites

Infections by Protozoa

- Ulcerative keratitis/chronic amoebic keratitis (*Aanthamoeba culbertsoni*): **Ch. 93.**
- Keratitis (*Balamuthia mandrillaris*): **Ch. 93.**
- Swelling of conjunctiva at the site of entry called Roman's sign (*Trypanosoma cruzi*): **Ch. 95.**
- Ocular cerebral malaria (*Plasmodium falciparum*): **Ch. 97.**
- Anterior uveitis, focal choreoretinitis near macula/pigment ringed scar (*Toxoplasma gondii*): **Ch. 98.**
- *Microspora* eye infections like keratoconjunctivitis (*Encephalitozoon* and *Trachipleistophora*) and keratitis (*Vittaforma*): **Ch. 99.**

Infections by Metazoa or Helminths or Worms

- Granulomatous nodule/cyst in eye in spirometrosis (*Spirometra* spp.): **Ch. 101.**
- Ocular cysticercosis by larvae of *Taenia solium* (*Cysticercus cellulosae*): **Ch. 102.**
- Cyst in extraocular muscles in trichinosis (*Trichina spiralis*): **Ch. 107.**
- Lesions in eyes in loiasis (*Loa loa*): **Ch. 108.**
- Calabar swelling as like *Loa loa* in *Mansonella perstans*: **Ch. 108.**
- Lesions due to microfilariae in onchocerciasis (*Onchocercus volvulus*): **Ch. 108.**
- *Dirofilaria repens*: It is a parasite of dog. In human, it causes subcutaneous or subconjunctival nodules, so human species called *dirofilariae conjunctivae*.

ACCESS YOURSELF

Case Studies

1. A 25-year-old male, who is regularly visiting the swimming pool present in ENT OPD with greening discharge from ear. On examination of discharge by microscopy and culture, septate hyphae with exospores are identified. Identify the disease and organism and answer the following:
   a. Name the clinical condition and the most common causative agent in given case.
   b. Describe the other etiological agents of clinical condition, in given case.
   c. Write treatment of clinical condition, in given case.
2. A 30-year-old male, who is occupationally farmer visits the eye clinic with complain of discharge, red eye and photophobia. Corneal scraping is done and examined by microscopy and culture on SDA which revealed fungus with septate hyphae. Culture produces dark green colonies. Identify the disease and organism and answer the following:
   a. Name the clinical condition and the most common causative agent in given case.
   b. Describe the other etiological agents of clinical condition, in given case.
   c. Write treatment of clinical condition, in given case.

Short Notes

1. Malignant otitis externa
2. Otomycosis
3. Otitis media
4. Bacterial conjunctivitis
5. Corneal ulcer
6. Viral infections of eye
7. Oculomycosis.

Short Questions for Theory/Viva Questions

1. Name the most common bacterium as causative agent for following diseases. Malignant necrotizing otitis externa, malignant necrotizing otitis media, ASOM, CSOM, hemorrhagic otitis media and masked acute otitis media.
2. Name the four fungi causing otomycosis.
3. Name the four viruses infecting eyes with lesion(s) produced by them.
4. Name the four fungi causing mycotic keratitis.

MCQs for Chapter Review

Infections of External Ear

1. **The agent of malignant otitis externa is:**
   a. *Staphylococcus aureus*   b. *Pseudomonas* species
   c. *Streptococcus pyogenes*   d. *Candida* species
2. **White discharge from ear indicates infection with:**
   a. *C. albicans*   b. *A. fumigatus*
   c. *A. niger*   d. *M. furfur*

Infections of Middle Ears/Otitis Media

3. **A patient with history of discharge from right ear presented with severe earache. This discharge was culture and organism was found to be gram-negative cocci. The least like cause.**
   a. *Pseudomonas*   b. *Streptococcus pneumoniae*
   c. *Staphylococcus*   d. *Haemophilus influenzae*

Viral Infections of Eyes

4. **Epidemic hemorrhagic conjunctivitis is caused by:**
   a. HSV   b. VZV
   c. HIV   d. Picorna virus
5. **Which of the following causes conjunctivitis?**
   a. Adeno virus   b. Entero virus
   c. Coxsackie virus   d. Herpes virus
6. **Conjunctivitis is caused by all *except*:**
   a. CMV   b. Entero virus 70
   c. Coxsackie virus A24   d. Adeno virus

Fungal Keratitis

7. **Common fungus causing corneal ulcer:**
   a. *Aspergillus*   b. *Mucor*
   c. *Fusarium*   d. *Sporothrix*
8. **Branched septate hyphae found on corneal smear in a case of corneal ulcer is:**
   a. *Candida*   b. *Mucor*
   c. *Aspergillus*   d. *Histoplasma*
9. **In a patient, corneal ulcer scraping reveals narrow angled septate hyphae, which of the following is likely etiological agent?**
   a. *Mucor*   b. *Aspergillus*
   c. *Histoplasma*   d. *Candida*

Answers and Explanation of MCQs

1. b
- Follow section, **infections of external ear (malignant necrotizing otitis externa)** for explanation.

2. a
- Follow section, **infections of external ear (fungal infections of external ears/otomycosis → clinical features)** for explanation.

3. d
- Gram-negative cocci or coccobacilliis *H influenzae* which is the 2nd most common bacterium for acute otitis media after *Streptococcus pneumoniae*.
- Follow section, **ear infections** for more explanation.

4. d

5. a, b, c, d

6. a
- Follow section, **viral infections of eyes** for explanation of answers of MCQs 4–6.

7. a, c

8. c

9. b
- *Aspergillus* spp., *Fusarium* spp., and phaeoid fungi are common pathogens of keratitis, but *Aspergillus* is the most common etiological agent of keratitis (corneal ulcer), which shows hyaline septate hyphae with conidiophores, vesicle, phialides and chain like arrangement of conidia. Hyphae show dichotomous branches, where conidiophores and stem are equal in size (3–6 μm) and at acute/narrow angle of 450 (V-shaped).

Oral Cavity, Dental and Salivary Gland Infections

Chapter Outline

- Oral Cavity Infections
- Dental Infections
- Salivary Gland Infections

ORAL CAVITY INFECTIONS

Glositis

It is an inflammation of the tongue. It mostly caused by viruses like HSV.

Stomatitis

It is an inflammation of oral mucosa. It mostly caused by viruses like HSV. It presents with multiple painful tiny vesicular lesions on the oral mucosa and oropharynx.

Oral Thrush

It is the infection of oral mucosa by *Candida* spp., mostly in person with IDDs (like HIV-AIDS). It presents with white inflammatory exudative patch on oral mucosa which may extend to tongue, oropharynx and esophagus.

Vincent's Angina

Follow **Ch. 73.**

Ludwig's Angina

Definition: It is the diffuse cellulitis (bilateral) of the submandibular space. Submandibular space (**Fig. 46.1**) lies between the floor of the mouth and tongue on upper side, superficial layer of deep cervical facia on lower side, mandible on anterior side and hyoid bone on posterior side.

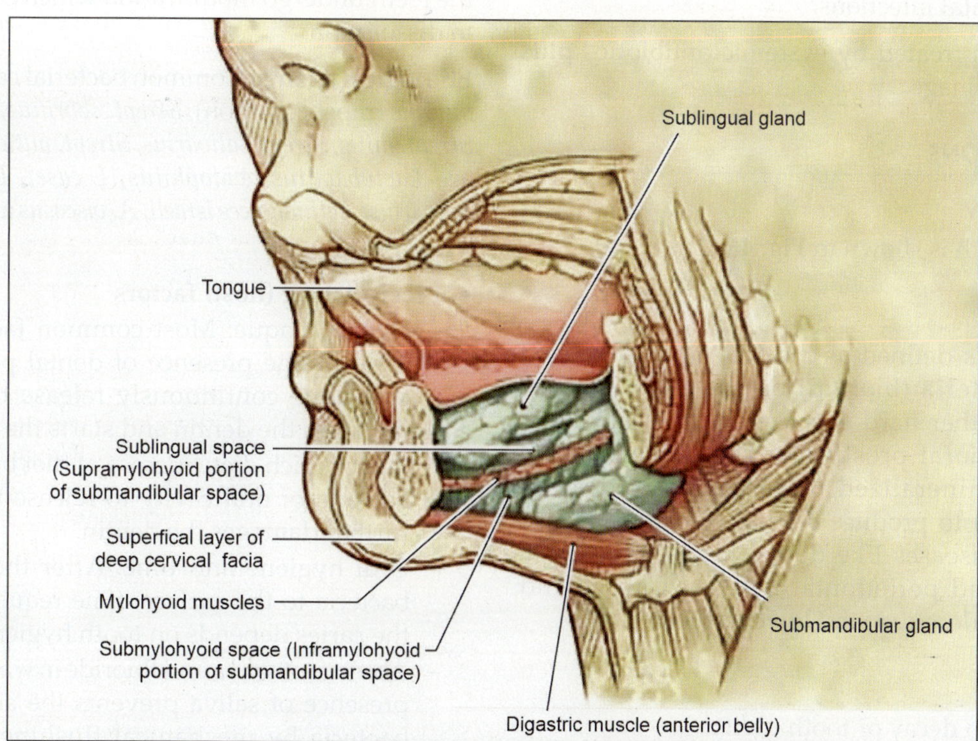

Fig. 46.1: Anatomy of submandibular space

History: It was 1st described by von Ludwig (German physician) in 1986.

Etiological agents: Infection is polymicrobial and mostly caused by anaerobes like viridans streptococci, *Peptostreptococcus*, *Prevotella*, *Porphyromonas*, *Bacteroides*, and *Fusobacterium*. Few other causes are *Staphylococcus*, *Pseudomonas*, *H influenzae* and *E coli*.

Pathogenicity

- **Sources and modes of transmission:** In 80% cases organisms are arising from dental infections, especially from an infected second or third molar tooth. Direct infection also occurs due to injuries to oral mucosa and mandible.
- **Precipitating (host) factors:** It includes preexisting dental infections.
- **Clinical features:** It is the aggressive rapidly spreading cellulitis, without lymphadenopathy. It presents with bilateral lower facial swelling around the mandible and upper neck, painful neck swelling, tooth pain, difficulty in swallowing (dysphagia) with varying degree of trismus, fever, and general malaise. Elevation of the floor of mouth, laryngeal edema and upward posterior displacement of the tongue lead the airway obstruction.
- **Complications:** These are spread of infection in parapharyngeal and retropharyngeal spaces, abscess formation, airway obstruction, septicemia and aspiration pneumonia.

Diagnosis: It is diagnosed clinically and radiologically. Microbiological diagnosis is done by microscopy and culture from aspirated pus.

Prevention: It is prevented by early diagnosis and treatment of dental infections.

Treatment: It is treated by systemic antibiotics plus incision and drainage.

DENTAL INFECTIONS

Dental Anatomy

Anatomy of tooth is shown in **Fig. 46.2**.

Dental Plaque

Dental plaque is defined as the soft nonmineralized deposit of bacteria (biofilm) on the tooth surface (**Fig. 46.3**) or other hard surfaces in the oral cavity, including intraoral prosthesis. The dental plaque which is nonmineralized undergoes calcification (mineralization) to produce calculus or tartar. It also causes other diseases like dental caries, gingivitis, periodonitis and periodontal abscess. Causes and mechanism are described in **Flowchart 50.2**.

Dental Caries

Synonym: Tooth decay or tooth cavity.

Meaning: Caries (Latin) means decay or rottenness.

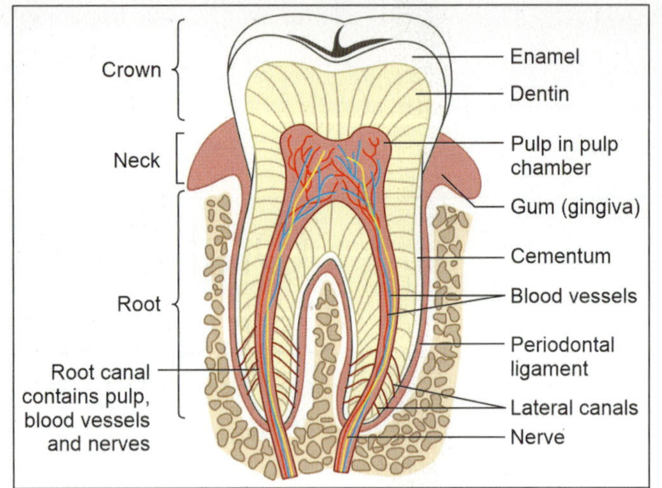

Fig. 46.2: Anatomy of tooth

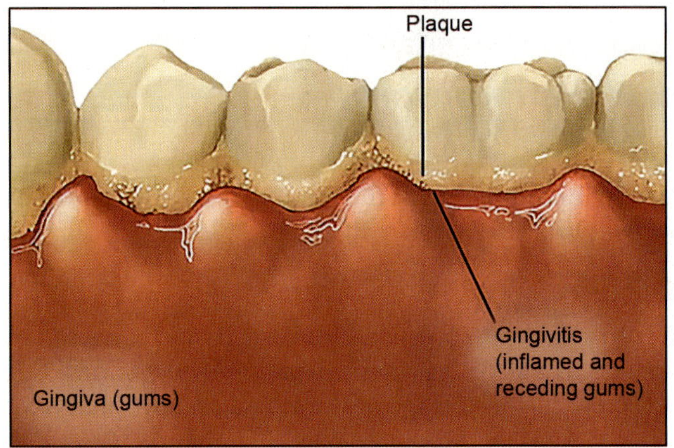

Fig. 46.3: Dental plaque

Definition: It is the disease in which calcified tissues of the teeth undergo modification which eventually leads to dissolution.

Etiological agents: Common bacterial causes are *Strept. mutans* (most common), *Strept. sobrinus*, *Strept. sanguis*, *Strept. mitis*, *Strept. salivarius*, *Strept. mitior*. Other causes are *Lactobacillus acidophilus*, *L casei*, *L fermentum*, *L salivarius*, *Actinomyces israeli*, *A. viscosus* and *A. neuslandi*.

Pathogenicity

- **Precipitating (host) factors**
  - Dental plaque: Most common factor for dental caries is the presence of dental plaque. Bacteria of plaque continuously release the acid which damages the dentin and starts the caries lesion.
  - Sucrose rich diet: Sucrose of diet broken down by bacteria of oral cavity to release the acid which further damages the dentin.
  - Oral hygiene and time: After the adherence of bacteria to the surface, time required to produce the caries depends on tooth hygiene.
  - Absence of saliva and fluoride in water: Continuous presence of saliva prevents the accumulation of bacteria by mechanical flushing effect. It also neutralizes the acid by buffering effect.

- **Clinical features:** Pain, foul smell, foul taste and appearance of small chalky-white spot on the enamel is the earlier sign followed by yellowish-brown/black (**Fig. 46.4b**) lesion then cavity formation (**Fig. 46.4c**).
- **Complications:** Infection penetrates the dentium and pulp (endodontic infections), and finally loss of tooth. Other complications are tooth abscess, swelling or pus around a tooth, chewing problems, positioning shifts of teeth after tooth loss, weight loss or nutrition problems from painful or difficult eating or chewing. It acts as focus of sepsis and causes serious or even life-threatening complications like cavernous sinus thrombosis, Ludwig's angina, endocarditis and sepsis.

Laboratory diagnosis: There are no specific micro-biological tests to diagnose the caries. However, salivary count of *Strept. mutans* (culture of saliva in mitis salivarius bacitracin agar) and *Lactobacillus* (culture of saliva in Rogosa SL agar) help to detect high-risk patients.

Prevention: Preventive measures for dental caries are reduction of oral flora by improving oral hygiene and use of antimicrobials, proper brushing and flossing to prevent the bacterial colonization on tooth, low sucrose diet, use of water with enough fluoride and topical fluoride and removal of dental plaque by scaling.

Treatment: It is treated by following ways:
- **Use of antibiotics.**
- **Fillings:** Removes the decayed material and packs this empty space with an appropriate dental filling material.
- **Crowns:** Fill the tooth with an alloy or porcelain crown covering.
- **Root canal:** Remove the damaged or dead nerve with the surrounding blood vessel tissue (pulp) and fill the area.
- **Extraction:** In some cases, the tooth may be damaged beyond repair and must be extracted if there is risk of infection spreading to the jaw bone.

Endodontic Infections

It includes the disease of dental pulp like pulpitis, necrosis etc., and periapical tissues like cellulitis, periapical abscess etc., as shown in **Fig. 46.5**. It also

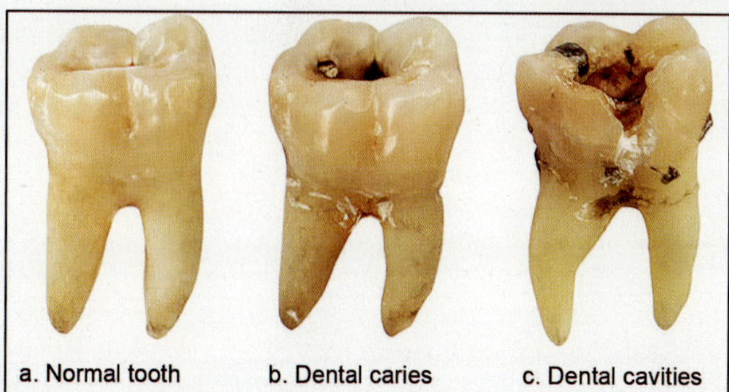

Fig. 46.4: Dental caries and cavity

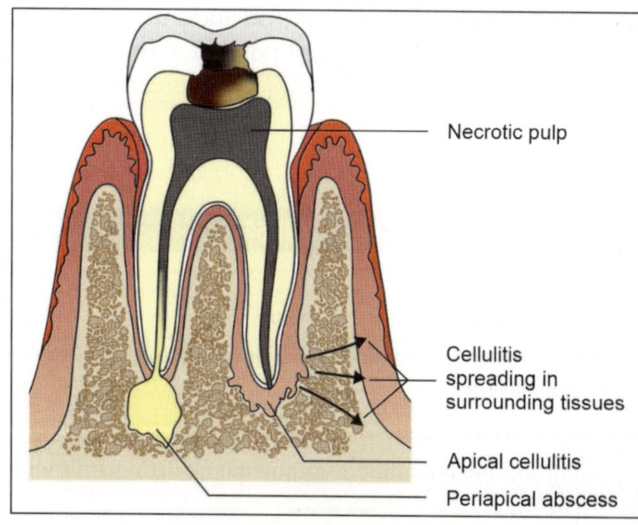

Fig. 46.5: Endodontal infections

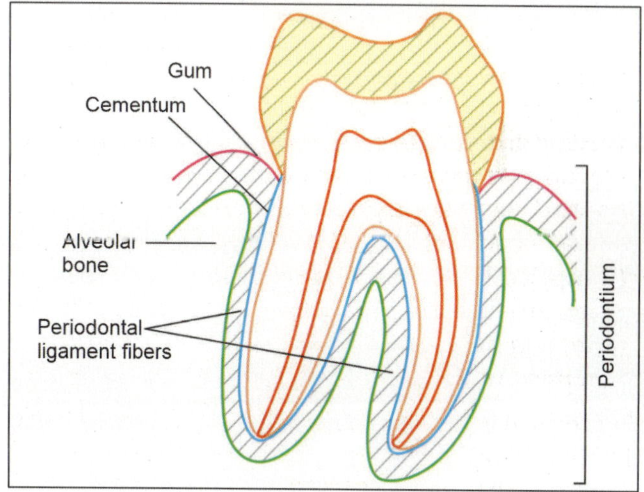

Fig. 46.6: Anatomy of periodontal tissues

presents with granuloma, cysts and with other infections. It mostly caused by bacteria of dental caries.

Periodontal Infections

Periodontal tissue contains gum, cementum, periodontal ligament and alveolar bone as shown in **Fig. 46.6**. Following are the different periodontal infections.

Gingivitis: It is an inflammation of gum. Acute gingivitis is caused by *Bacteroides, Fusobacterium, Treponema* spp., *Streptococcus, Veillonela* and *Actinomyces* spp. Chronic gingivitis is caused by *Strept. sanguis, Strept. milleri, Actinomyces israeli, A. neuslandi, Veillonela, Fusobacterium nucleatum*, etc.

Periodonitis: It is an inflammation of periodontal tissues.

SALIVARY GLAND INFECTIONS

Parotitis

It is an inflammation of parotid gland. Acute parotitis is caused by *Staphy. aureus* and members of Entero-bacterales, few other GNB and oral anaerobes. Chronic parotitis is mostly caused by *Staphy aureus*. It presents with painfull swelling over parotid and purulent drainage in mouth at the opening of parotid gland.

Mumps

Follow **Ch. 82.**

ACCESS YOURSELF

Short Notes

1. Ludwig's angina.
2. Discuss the etiopathogenicity of dental caries and its role as a focus of sepsis.

Short Question for Theory/Viva Question

1. Name the most common organism(s) causing following diseases.

Stomatitis, oral thrush, Vincent's angina, and dental caries.

MCQ for Chapter Review

Dental Infection

1. **Dental plaque and periodontal disease can be thought of as a continuum of what type of physiological process?**
   a. Biofilm formation
   b. Normal aging
   c. Exaggerated immune response
   d. Abnormal digestion

Answer and Explanation of MCQ

1. **a**
   - Dental plaque is defined as the soft non-mineralized deposit of bacteria (biofilm) on the tooth surface (**Fig. 46.3**) or other hard surfaces in the oral cavity, including intraoral prosthesis.

Miscellaneous Infections

Chapter Outline

CONGENITAL INFECTIONS

Follow **Table 3.1.**

ZOONOTIC INFECTIONS

Synonym: Zoonosis or zoonotic diseases or anthropozoonoses.

Definition: Infection transmitted from animal to human called zoonotic infections.

Meaning: Anthro (Greek) means human, zoon (Greek) means animal and nosos (Greek) means disease.

Host (predisposing factors)

- **Occupation:** Common in veterinary doctors, animal handlers, agricultural workers, butchers, hunters, nonvegetarian merchants, etc.
- **Consumer:** All those who are consuming contaminated animal products like milk, meat, eggs, etc., are at risk.
- **IDDs:** Presence of IDDs like AIDS, malignancy, etc., or other conditions like diabetes mellitus, etc., are contributing factors in development of zoonotic infections.

Etiological agents

- **Bacteria:** *Mycobacterium bovis*, *Salmonella* spp., *Brucella* spp., *Francisella tularensis*, *Leptospira*, *Rickettsia*, *Coxiella burnetii*, *Pseudomonas mallei*, *Borrelia burgdorferi*, *Bacillus anthracis*, etc.
- **Viruses:** Rabies virus, yellow fever virus, Japanese encephalitis virus, CCHF, swine flu virus, etc.
- **Fungi:** *Trichophyton* spp., *Microsporum* spp., and *Histoplasma capsulatum*.
- **Parasites:** *Taenia* spp., *Leishmania* spp., *Trypanosoma* spp., *Toxoplasma gondii*, *Plasmodium* spp., *Echinococcus granulosus*, etc.

Modes of transmission: Zoonotic infections are transmitted by different modes. Few examples are given below.

- **Direct transmission:** (1) Direct contact of abraded skin with water infected by animal's excreta and contains bacteria like *Leptospira*. (2) Direct animal bite like rabies virus.
- **Indirect transmission**
  1. **Inhalation (airborne):** Inhalation of aerosols from infected wool transmits *Bacillus anthracis*.
  2. **Ingestion (food- and waterborne)**
     - Egg ingestion contains *Salmonella*.
     - Partially cooked beef and pork in *Taenia* spp.
     - Milk of infected animal like *M. bovis*.
  3. **Vector-borne:** Blood sucking insect can transmit pathogens from animals to humans like CCHF.

Diagnosis

- **Specimens:** Human samples are collected according to system infected. Different animal samples are: (1) In anthrax: Ear sample and swab soaked in blood. (2) In rabies: Severed head. (3) In leptospirosis: Pieces of kidney. (4) In brucellosis: Pooled milk sample.
- **Testing**
  - Diagnosis of bacteria: **Ch. 118.**
  - Diagnosis of viruses: **Ch. 119.**
  - Diagnosis of fungi: **Ch. 120.**
  - Diagnosis of parasites: **Ch. 121.**

Prevention and treatment: These are depended on type of infections and discussed in details in respective chapters.

OPPORTUNISTIC INFECTIONS

Definition: Infection in immunocompromised host is called opportunistic infection.

Etiological agents: Following are the opportunistic pathogens:

- **Bacteria:** *Mycobacterium tuberculosis*, nontuberculous mycobacteria (NTM), *Salmonella* spp., *Campylobacter* spp., *Nocardia* spp., *Actinomycetes* spp., and *Legionella* spp.
- **Viruses:** VZV, HSV and CMV.
- **Fungi: Follow Ch. 91** (section opportunistic mycoses).
- **Parasites: (1) Protozoa:** *Toxoplasma gondii, Cystoisospora* (>1 month duration), *Cryptosporidium parvum* (>1 month duration), *Cyclospora, Microsporidia, Entamoeba histolytica* and *Giardia lamblia*. **(2) Metazoa:** *Strongyloides stercoralis*.

Diagnosis

- **Specimens:** According to system infected.
- **Testing:**
  - Diagnosis of bacteria: **Ch. 118.**
  - Diagnosis of viruses: **Ch. 119.**
  - Diagnosis of fungi: **Ch. 120.**
  - Diagnosis of parasites: **Ch. 121.**

Prevention and treatment: These are depended on type of infections and discussed in details in respective chapters.

SUPERINFECTION

Synonym: Suprainfection.

Definition: Appearance of new infection as a result of antimicrobial therapy called superinfection.

Mechanisms: Normal flora contributes to host defense by production of bacteriocins or nutritional competition with pathogens. Use of antibiotics may result in alteration of such flora, loss of competition and other useful functions, which allow other flora (e.g., *Candida*) to predominate, to invade and to produce superinfection. More commonly superinfection is due to use of broad spectrum AMAs like tetracycline, chloramphenicol, ampicillin, cephalosporins. It is more common with combination therapy.

Precipitating factors: Superinfection is more common in patients with immunocompromised conditions like steroid therapy, malignancies, AIDS, agranulocytosis, diabetes mellitus (DM), etc.

Sites: Superinfection is common in sites that harbor the normal flora like nasopharynx, GIT, upper respiratory tract, genital tract, urinary tract and occasionally skin.

Organisms, manifestations and treatment of super-infection: Follow **Table 47.1**.

Prevention

- **Don'ts**
  - Use of broad spectrum AMAs
  - Use of long-term antimicrobial therapy
  - Use of AMAs for self limiting disease like viral infection or for untreatable disease.

TABLE 47.1: Organisms, manifestations and treatment of superinfection

Organisms	Manifestations	Treatment
C. albicans	Diarrhea, thrush, etc.	Clotrimazole, nystatin
Staphylococcus	Enteritis	Cloxacillin
Proteus	UTI, enteritis	Cephalosporin or genta-micin
Pseudomonas	UTI, enteritis	Carbenicillin or pipericillin or gentamicin
Cl. difficile	AAC: Follow **Ch. 55**	

- **Do:** Use narrow spectrum AMAs.

> **Note:**
> Infection by HDV in patients already infected with HBV also called superinfection.

EMERGING AND REEMERGING INFECTIONS

Definitions

- **Emerging infections:** Emerging infections are those newly identified, previously unknown, spreading to new geographical areas, causing public health problems nationally or internationally and increased incidence in last 2–3 decades.
- **Reemerging infections:** Reemerging infections are those which were previously under controlled and now reappearing due to AMAs resistance or breakdown of public health measures and causing major public health problems often in epidemic proportion. For example, epidemic of chikungunya virus in 2005.

Risk factors

1. **Microbial factors:** Resistance to antibiotics and insecticides allow new strain to arise like MDR-TB, XDR-TB, MRSA, VISA, VRSA, PPNG, VRE, ESBL producers, carbapenemase producers and Amp C β-lactamase producers, etc.
2. **Host factors:** These are immunosuppressive status, malnutrition, drug abuse, illiteracy, ignorance, etc.
3. **Environmental factors:** These are high population, technology, industrial and economic development, land use, unplanned urbanization, international travel, poor or breakdown of public health measures, natural disasters, etc.

Example of emerging infections in the World since 1973: Follow **Table 47.2**.

Example of reemerging infections: These are MDR-TB, malaria, cholera, dengue, chikungunya, plague, meningococcal meningitis, etc.

Prevention: They are prevented by epidemiological surveillance, laboratory surveillance, ecological surveillance, anthropological surveillance, early diagnosis, treatment and implementation of control measures, constant monitoring and evaluation under the

TABLE 47.2: Emerging infections in the World since 1973		
Year	Microbes	Disease
1973	*Rotavirus*	Enteritis/diarrhea
1976	*Cryptosporidium parvum*	Enteritis/diarrhea
1977	*Ebolavirus*	VHF
1977	*Legionella pneumophila*	Legionnaires' disease
1977	Hantaan virus	VHF, renal failure
1977	*Campylobacter*	Enteritis/diarrhea
1980	HTLV-1	Lymphoma
1981	Toxin-producing *Staph. aureus*	Toxic shock syndrome
1982	E. coli 0157:H7	HUS
1982	HTLV-II	Leukemia
1982	*Borrelia burgdorferi*	Lyme disease
1983	HIV	AIDS
1983	*Helicobacter pylori*	Peptic ulcer
1988	Hepatitis E	Hepatitis
1989	Hepatitis C	Hepatitis
1990	Guanarito virus	VHF
1991	Encephalitozoon	Disseminated disease
1992	*Vibrio cholerae* O139	Cholera
1992	*Bartonella henselae*	Cat scratch disease
1993	Sin Nombre virus	Hanta pulmonary syndrome
1994	Sabia virus	VHF
1994	Hendra virus	Respiratory disease
1995	Hepatitis G	Hepatitis
1995	Herpes virus-8	Kaposi sarcoma
1996	vCJD prion	Variant CJD
1997	Avian influenza (H5N1)	Influenza
1999	Nipah virus	Encephalitis
1999	West Nile virus	Encephalitis
2003	SARS-CoV 1	SARS
2004	Human coronavirus NL63	
2005	Human coronavirus NKU1	
2009	H1N1 (swine flu)	Pandemic swine flu
2011	CCHF virus	Hemorrhagic fever
2012	MERS-CoV	Middle East respiratory syndrome
2013	H7N9 (bird flu)	Influenza
2016	Zika virus	Zika fever
2019	SARS-CoV2	COVID-19

supervision of global outbreak and response network to control the infection and further spread.

Emerging and reemerging infections in India

- **Bacterial diseases:** Plague, leptospirosis, brucellosis, anthrax and cholera.
- **Viral diseases:** Influenza, chikungunya, Chandipura, dengue, Japanese encephalitis, SARS, *Hantavirus* disease, Human entero virus 71 infection, Nipah virus infection and COVID-19.

Diagnosis: Follow respective chapters for diagnosis.

Prevention and treatment: These are depended on type of infections and discussed in details in respective chapters.

ONCOGENIC MICROBES

Definition: Microorganisms causing cancer are called oncogenic microbes.

Classification

Oncogenic Bacteria

Helicobacter pylori produces gastric carcinoma and lymphoma of MALT. For more details follow **Ch. 66 (section → *Helicobacter*).**

Oncogenic Viruses

Definition: Viruses which producing tumor in their natural host or in experimental animals or inducing malignant transformation of cells on culture are called oncogenic viruses.

History: Viral association with malignancy was 1st described by Peyton Rous (Father of Tumor Virology) in 1911, when he identified the fowl sarcoma caused by virus. Later, Nobel Prize was given to him in 1966.

Classification: Viruses causing cancer are listed in **Table 47.3** (human cancer) **and Table 47.4** (animal cancer).

Transformation: Conversion of normal cells into malignant cells called transformation. Following are the properties of transformed cells.

- **Alteration of morphology:** Like in fibroblast change in shape, becomes shorten, loss of parallel orientation and chromosomal aberration.
- **Alteration of metabolism:** Increased production of organic acids and acid mucopolysaccharide.
- **Alteration of growth properties:** Like change in growth rate and indefinite divisions of cells. Loss of contact inhibition, so instead of monolayer formation, there is piled up growth and microtumor formation.
- **Formation of new tumor Ags (tumor antigens):** Ch. 27.
- **Capacity to induce tumor:** They induce tumor in animals.
- **Types of viruses according to transformation**
  1. **Slow transforming viruses:** They have low oncogenic potential; do not transform the culture cells and cause malignancy of blood cells like leukemia. Example of slow transforming virus is HTLV. Replication is normal and transformation occurs after a long latent period.
  2. **Acute transforming viruses:** They have high oncogenic potential, can transform the culture cells and cause malignancy of different cells like leukemia, carcinoma, sarcoma, etc. Examples of

TABLE 47.3: Human oncogenic viruses

Family	Species	Cancer (s)
		DNA viruses
Papova-viridae	HPV	• Oropharyngeal cancer (Typ1 16) • Laryngeal or esophageal carcinoma (Typ1 16 and 18) • Genital cancers: Seen in vulva, vagina, cervix, penis, and anus [Highest risk (16, 18, 31, 45), Other high-risk (33, 35, 39, 51, 52, 56, 58, 59) • Probably high-risk (26, 53, 66, 68, 73, 82)]
Herpes-viridae	HHV-1 (HSV-1)	Lip cancer
	HHV-1 (HSV-2)	Cervical cancer
	HHV-4 (EBV)	• Nasopharyngeal carcinoma • Burkitt's lymphoma • Lymphoma (Hodgkin's and non-Hodgkin's lymphoma)
	HHV-5 (CMV)	• Carcinoma of prostate • Kaposi's sarcoma
	HHV-8 (KSHV)	Kaposi's sarcoma
Hepadna-viridae	HBV	HCC
		RNA viruses
Flavi-viridae	HCV	Hepatocellular carcinoma (HCC)
Retro-viridae	HTLV-I	• Human T cell leukemia / lymphoma • Cutaneous T cell lymphoma
	HTLV-II	T cell malignancy
	HTLV-III (HIV 1 and HIV 2)	AIDS related malignancies like • Kaposi's sarcoma (KS) • Lymphoma: Hodgkins and non-Hodgkin's type • Cervical cancer

TABLE 47.4: Animal oncogenic viruses

Family	Species	Cancer
		DNA viruses
Adeno-viridae	Type 12, 19 and 21	Sarcoma in newborn rodents
Pox-viridae	Rabbit fibroma virus	Rabbit fibroma
	Yaba virus	Benign tumor
Herpes-viridae	Marek's disease virus	Marek's disease
	Lucke's frog tuour virus	Renal adenocarcinoma in frog
	Herpes virus saimiri	Lymphoma or reticular cell carcinoma in owls, monkeys or rabbits
		RNA viruses
Retro-viridae	Rous sarcoma virus	• Avian leukosis complex (Lymphomatosis, myeloblastosis and erythroblastosis) • Sarcoma in fowls
	Murine leukosis virus	Murine leukemia and sarcoma
	Mice mammary tumor virus (Bittner's milk factor or Bittner's virus)	Breast cancer in mice
	Leukosis sarcoma virus	Leukosis and sarcoma in different animals

acute transforming viruses are rouse sarcoma virus, HBV, etc. Transformation occurs after a short latent period of weeks or months. Replication is not normal and may be following types:

– Replication defective: Viruses are not replicating normally, because they contain additional gene like viral oncogene (V-onc gene), which replaces the normal gene of virus and viruses depend on coinfection by helper virus for replication.

– Replication competent: Viruses replicate normally because they contain viral oncogene [*src* (Pronounced as 'sark') for Rous rarcoma virus] and full components of normal genes like *gag*, *pol* and *env* genes.

Viral oncogenes (V-onc/viral cancer genes): Viral genes which are encode for conversion of normal cells in to malignant cells called oncogenes (V-onc).

These genes have different properties. (1) They are not responsible for viral replication. (2) They formed by recombination between viral and cell genes. (3) Certain genes resembling V-onc are present in cancer cells called cellular oncogenes (C-onc) and in normal host cells called cellular proto-oncogenes. (4) C-onc is having all characteristics of prokaryotic cells while V-onc does not have. (5) Proto-oncogenes present in normal cells with certain functions like Oncogene-src → tyrosine specific protein kinase, Oncogene-sis → platelet derived growth factor and Oncogene-myc → NA binding protein.

Antioncogenes: Certain genes suppress the development of tumor called antioncogenes. Their absence produces tumor. For examples Rb gene and P53 gene: Absence of Rb gene produces retinoblastoma (Rb). Antioncogenes may be absent due to absence of chromosomes.

Mechanism of viral oncogenesis

• **DNA viruses:** Viral DNA is integrated in to host cell genome, which brings the transformation in host cells. It is like lysogenic conversion in bacterium.

• **RNA viruses:** They introduce oncogene into host cells, which alters preexisting cellular gene.

Oncogenic Fungi

Few fungal toxins, which are known to produce the fungal food poisoning are also having oncogenic properties concluded by IARC (International Agency for Research on Cancer). Fungal food poisoning are classified as mycotoxicoses (clinical entity due to ingestion of fungal toxin only) and mycetism/mycetismus/muscarinism or mushroom poisoning (clinical entity due to ingestion of fungal toxin along with fungus). Fungal toxins which have carcinogenic properties are described in **Ch. 92.**

Oncogenic Parasites

- *Schistosoma haematobium:* It causes the squamous cell carcinoma of urinary bladder.
- *Clonorchis sinensis:* It causes liver, bile duct (cholangiocarcinoma), gallbladder (?) and pancreas (adenocarcinoma).
- *Opisthorchis viverrini* and *Opisthorchis felineus:* Causing bile duct cancer called cholangiocarcinoma. In addition *Opisthorchis* also causes carcinoma of liver.

BIOTERRORISM

Synonym: It also called biological warfare or germ warfare.

> **Note:**
> - *Clostridium botulinum* **toxin:** Most dangerous and most likely to be used in bioterrorism is toxin of *Clostridium botulinum,* which causes botulism.
> - **Yellow rain:** ⎫
> - *C. immitis:* ⎭ Follow **Ch. 13.**

Definitions

- **Bioterrorism:** According to the CDC bioterrorism is defined as "the deliberate release of viruses, bacteria, fungi, toxins or other harmful biological agents to cause illness or death in people, animals or plants".
- **Bio-weapons or biological agents:** Agents useful for bioterrorism like viruses, bacteria, fungi or toxins are called bio-weapons or biological agents.

Classification: CDC classified the bio-weapons in 3 categories like A, B and C as shown in **Table 47.5.**

- **Category A:** These high-priority agents pose a risk to national security, can be easily transmitted and disseminated from person to person, result in high mortality, have potential major public health impact or require special action for public health preparedness.
- **Category B:** These are moderately easy to disseminate, result in moderate morbidity rates and low mortality rates; and require specific enhancements of CDC's diagnostic capacity and enhanced disease surveillance.
- **Category C:** These are emerging pathogens that could be engineered for mass dissemination in the future because of availability, ease of production and dissemination, and potential for high morbidity and mortality rates and major health impact.

Properties of bio-weapons: These are either natural agents or mutated or altered agents with increase

TABLE 47.5: Bio-weapons and disease

Bio-weapon	Disease
Category A	
Bacillus anthracis	Anthrax
Clostridium botulinum toxin	Botulism
Yersinia pestis	Plague
Variola major	Smallpox
Francisella tularensis	Tularemia
- **Flaviviruses:** Like Yellow fever virus, KFD virus, Omsk hemorrhagic fever virus - **Bunyaviruses:** Like CCHF virus, Rift valley fever virus - **Filoviruses:** Like *Ebolavirus, Marburgvirus* - **Arenaviruses:** Like Lassa virus, machupo virus, junin virus, guanarito virus and Sabia virus	Viral hemorrhagic fever
Category B	
Brucella species	Brucellosis
Epsilon toxin of *Clostridium perfringens*	Food poisoning
Salmonella species, *E. coli* O157:H7, *Shigella*	Food poisoning
Burkholderia mallei	Glanders
Burkholderia pseudomallei	Melioidosis
Chlamydia psittaci	Psittacosis
Coxiella burnetii	Q fever
Ricinus communis (castor beans)	Ricin toxin poisoning
Abrin toxin from *Abrus precatorius* (Rosary peas, Gunja, Rati, etc.)	Abrin toxin poisoning
Staphylococcal enterotoxin B	Food poisoning
Rickettsia prowazekii	Epidemic typhus
Water safety threat - *Vibrio cholerae* - *Cryptosporidium parvum*	- Cholera - Cryptosporidiosis
- **Alphaviruses:** Like Eastern equine encephalitis virus, Western equine encephalitis virus, Venezuelan equine encephalitis virus - **Flaviviruses:** Like West Nile virus, Saint Louis virus, dengue fever virus	Viral encephalitis
Category C (Emerging agents)	
Nipah virus	Encephalitis
Hantavirus	Hemorrhagic fever renal syndrome
SARS CoV1	SARS
H1N1 (a strain of influenza)	Swine flu

virulence, increase drug resistance or with increase communicability to make them more dangerous.

Advantages of using bio-weapons for terrorism: Bioterrorism is an attractive weapon because biological agents are relatively easy to obtain, inexpensive, can be easily disseminated, can cause widespread fear and panic beyond the actual physical damage and disrupt the economy.

Disadvantages of using bio-weapons for terrorism: Bio-weapons are like two-edge sword, as they damage not only to enemies, but also to friendly forces.

Mode of transmission: They spread via air, water, food or from person to person.

Preparedness before the bioterrorism: The American Red Cross, in cooperation with the CDC, has developed a detailed plan that gives ideas to people for preparedness in the event of bioterrorism. Before an attack people must store all needful items like water, food, kitchen accessories, power battery for mobile, powered radio or television, first aid kit, essential documents, stationeries, medicines, phone numbers, infants kits (includes diapers, bottles, powdered milk etc.), senior citizen kit (clothes, stick etc.) personal use kit (shampoo, deodorant, brush etc.).

Warning sign of bioterrorism: These are drug resistance to bio-weapons, multiple cases with atypical presentation and multiple deaths of animals or humans.

Steps after attack: These are collecting right information, skin cleaning, wearing clothes which fully cover the body, wearing mask, isolation of sick people, no outdoor visit, shifting to normal area, communication break to avoid person to person transmission of disease (for example, measles, influenza, avian flu, smallpox, plague and viral hemorrhagic fevers), follow messages given by public health officials, taking supportive treatment, drug distribution and taking medical advice like early diagnosis and treatment of disease.

VECTOR-BORNE INFECTIONS

Introduction: Arthropod or any living carrier (e.g., snail), which transports an infectious agent to susceptible host called vector. Vectors are responsible to transmit the bacteria, viruses and parasites from human to human, animal to animal, human to animal, animal to human or from inanimate objects to human/animals. Among the all infections >17% are vector borne. These occur mostly in tropical and subtropical countries.

Modes of transmission

1. **Method in which vectors are involved or not involved in transmission, multiplication and development**
   - **Mechanical transmission:** Parasites not enter in the body of vector, but transmitted through contamination of body parts like legs, wings, etc. Organism does not replicate or develop on the vector. Mechanical transmission occurs by crawling or flying through soiling of feet or proboscis or sometimes passes through GIT and passively excreted without development or multiplication in or on the body. For example, amoebic and bacillary dysentery by house fly which settles from one food to other.

- **Biological transmission:** Parasite enters in the body of vector and undergoes metamorphosis (development or multiplication or both) to achieve the infective stage and transmitted by vector bite, deposition of excreta on human body or by other routes. The interval between the entry of the pathogen into the vector and the vector become infective is called extrinsic incubation period. Following are the different types.
  - **Propagative:** Only multiplication, but no development of pathogens, e.g., plague bacilli in rat fleas.
  - **Cyclo-developmental:** No multiplication but only development of pathogens, e.g., micro-filariae in mosquitoes.
  - **Cyclo-propagative:** Both multiplication and development of pathogens, e.g., *Plasmodium* spp., in mosquitoes.
  - **Anterior station development (salivaria):** Organism enters in mid gut (stomach) of vector, and migrates to foregut in to the buccal cavity and finally to the salivary gland called anterior station development or salivaria. Organism multiplies here and infection occurs by releasing the organism by bites or regurgitating the contents, e.g., *T. brucei* by tse tse fly.
  - **Posterior station development (stercoraria):** Organism enters in mid gut (stomach) of vector, and migrates to hindgut in to the intestine called posterior station development or stercoraria. Organism multiplies here and infection occurs by releasing the organism in feces, e.g., *T. cruzi* by reduviid bug.

2. **According to host involvement**
   - **Invertebrate type:** (1) Diptera: Flies and mosquitoes. (2) Sphinonaptera: Fleas. (3) Orthoptera: Cockroaches. (4) Hemiptera: Bugs including kissing bugs. (5) Anoplura: Lice. (6) Acarina: Ticks and mites. (7) Copepoda: Cyclops.
   - **Vertebrate type:** Mice, rodents and bats.

3. **According to transmission channel**
   - **Man and no vertebrate host:** (1) Man-arthropod-man: Malaria. (2) Man-snail-man: *S. haematobium.*
   - **Man and another vertebrate host and non-vertebrate host:** (1) Mammal-arthropod-man: Plague. (2) Bird-arthropod-man: Encephalitis.
   - **Man and two intermediate host:** (1) Man-cyclops-fish-man: Fish tape worm. (2) Man-snail-fish-man: *C. sinensis.* (3) Man-snail-crab-man: *P. westermanii.*

4. **According to deposition of agents in host by vectors:** (1) By biting: e.g., *Plasmodium* spp., by mosquito bite. (2) By regurgitation: *T. brucei* by tse tse fly. (3) By scratching of infective feces of vectors against abraded skin: *Borrelia* in relapsing fever.

5. **Inheritance transmission:** It includes infections from vector to vector. **(1) Transovarial transmission:** Infected mother to progeny. **(2) Trans-stadial transmission:** One stage of life cycle to other stage like nymph to adult.

TABLE 47.6: Vector-borne bacteria

Bacteria	Common name of vector	Species of vector
S. Typhi (typhoid fever), *S.* Paratyphi (paratyphoid fever), *V. cholerae* (cholera), *Shigella* spp. (shigellosis), *Ch. trachomatis* (trachoma), *T. pertenue* (yaws), etc.	House fly (mechanical)	*Musca* spp.
Bartonella bacilliformis (oroya fever/carrion's disease)	Sand fly	*Lutzomyia* spp.
E. histolytica (cyst), Enterobacterales, hepatitis viruses, eggs of nematodes, polio virus, etc.	Cockroaches (mechanical)	*Periplaneta americana* and *Blatta orientalis*
R. prowazeki (epidemic typhus)	Head louse	*Pediculus capitis*
Borrelia recurrentis (louse-borne relapsing fever), *Bartonella quintana* (trench fever/five day fever/Quintana fever)	Body louse	*Pediculus corporis*
Y. pestis (plague), *R. typhi* (endemic typhus)	Rat flea	*Xenopsylla cheopis*
R. felis (endemic typhus/murine typhus/flea-borne typhus)	Cat flea	*Ctenocephalus felis*
B. duttoni, B. harmasii and *B. parkeri* (tick-borne relapsing fever)	Soft tick	*Ornithodorus lahorensis, O tholozoni* and *O crossi*
F. tularensis (tularemia), *Ehrlichia equi* (human granulocytic ehrlichiosis)	Hard (ixodid) tick	*Amblyomma* spp.
Borrelia burgdorferri (Lyme disease)	Hard (ixodid) tick	*Ixodes* spp.
Rickettsia spp. (spotted fever group)	Hard (ixodid) tick	Different species like *Dermacentor andersoni, Rhipicephalus sanguinens, Haemaphysalis leachi, Amblyoma* and *Hyalomma*
Ehrlichia chaffensis (human monocytic ehrlichiosis)	Hard (ixodid) tick	*Amblyoma*
Orientia tsutsugamushi (scrub typhus) and *Rickettsia akari* (rickettsial pox)	Trombiculid mite	*Leptotrombidium* spp.

Effective factors of vector-borne diseases: These are feeding preference, survival of vectors in environment, seasonal like malaria more in monsoon season, infectivity means ability to transmit the agent, susceptibility means ability to become infected and domesticity means degree of association with man.

Types or examples of vector-borne diseases
- **Vector-borne bacteria:** Follow **Table 47.6.**
- **Vector-borne viruses:** Follow **Ch. 83 (only arboviruses).**
- **Vector-borne parasites:** Follow **Table 47.7.**

NVBDCP
- **Introduction:** The National Vector Borne Disease Control Program (NVBDCP) is implemented in the State/UTs for prevention and control of vector borne diseases mainly malaria, filariasis, kala-azar, Japanese Encephalitis (JE), dengue and chikungunya.
- **Strategies:** Following are three strategies. Other detail about NVBDCP is beyond the scope of this book hence not given here.
  1. Early diagnosis and treatment of case with strengthening of referral services, epidemic preparedness and rapid response.
  2. Vector control to reduce the transmission which includes indoor residual spraying in selected high-risk areas, use of insecticide treated bed-nets, use of larvivorous fish, anti-larval measures in urban areas and source reduction.
  3. Supportive strategies include human resource development, research like studies on drug resistance and insecticide susceptibility, field visit, web-based management information system, vaccination against JE and annual mass drug administration against lymphatic filariasis.

ECTOPARASITIC INFESTATIONS

Definition: Parasites living outside on the body surface of the host called ectoparasites.

Examples: Following are the examples and detail description of each ectoparasite is given in **Ch. 110.**
- **Itch mite:** *Sarcoptes scabiei* (*Acarus scabiei*).
- **Louse:** Three species.
  - Head louse: *Pediculus capitis* (*Pediculus humanus capitis*).
  - Body louse: *Pediculus corporis* (*Pediculus humanus corporis / Pediculus humanus humanus*).
  - Pubis or crab louse: *Pthirus pubis.*

ACCESS YOURSELF

Essays/Full Questions
1. Enumerate the microbial agents and their vectors causing zoonotic diseases. Describe the morphology, mode of transmission, host factors, pathogenicity and discuss the clinical course, laboratory diagnosis and prevention.
2. Describe the etiology of opportunistic infections. Discuss the laboratory diagnosis.
3. Describe the etiologic agents of emerging infectious diseases. Discuss the clinical course and diagnosis.
4. Mention the oncogenic microbes. Describe the role of oncogenic viruses in the evolution of virus associated malignancy.

Essentials of Medical Microbiology

TABLE 47.7: Vector-borne parasites

Parasites	Common name of vector	Species of vector
Protozoa		
E. histolytica	House fly	Musca spp.
T. brucei	Tsetse fly	Glossina spp.
T. cruzi	Reduviid bug	Panstrongylus megistus, Triatoma infestans and Rhodnius prolixus
L. donovani	Sand fly	Lutzomyia spp. Phlebotomus spp.
L. tropica	Sand fly	Phlebotomus spp.
L. braziliensis	Sand fly	Lutzomyia spp.
Plasmodium spp.	Mosquito	Female Anopheles
Babesia spp.	Tick	Ixodid spp., and Dermacentor spp.
Helminths		
Cestodes		
D. latum*	Cyclops	—
H. nana	Rat flea	Xenopsylla cheopis
H. diminuta	Rat flea	Xenopsylla cheopis
D. caninum	Dog flea or cat flea	Ctenocephalus canis or C. felis
Trematodes		
S. haematobium*	Fresh water snail	Bulinus truncates, Planorvarius metidjensis, Ferrissia tenuis
S. mansoni*	Fresh water snail	Biomphalaria alexendrina, Australorbis glabratus
S. japonicum*	Fresh water amphibian snail	Oncomelania (Katayama or Blanfordia)
S. mekongi*	Aquatic snail	Trienla aperta
Fasciolopsis buski*	Aquatic snail	Segmentina
H. heterophyes*	Marine and brackish water snail like	Pirenella conica or Ceithidea
M. yokogawai*	Fresh water snail	Melania spp.
C. sinensis*	Planorbid snail	Helicorbis coenosus
O. felineus*	Snail	Bithynia leachi
D. dendriticum*	Land snail	—
F. hepatica*	Amphibian snail	Lymnaea truncatula
P. westermani*	Fresh water snail	Melania liberta (Semi-sulcospira liberta) or Brotia
Nematodes		
A. costaricensis	Slug	Limax maximus
W. bancrofti	Female mosquito	Culex spp. Also by Aedes and Anopheles spp.
B. malayi	Female mosquito	Mansonia spp., and Anopheles spp.
B. timori	Female mosquito	Anopheles barbirostris

L. loa	Female mango or deer fly	Chrysops spp.
O. volvulus	Female black fly	Simulium spp.
M. streptocerca	Female mosquito	Culicoides spp.
M. ozzardi		
M. perstans		
A. cantonensis	Snail or mollusc	Thelidomus aspera
G. spinigerum*	Cyclops	—
D. medinensis	Cyclops	Mesocyclops spp.

*Vector is not responsible for transmission of parasite, but required to complete the life cycle.

5. Describe the relationship between infection and cancers.
6. Describe the role of vectors in the causation of diseases and National Vector Borne Disease Control Program.
7. Describe the etiology, pathogenicity and diagnostic features of pediculosis.

Short Notes

1. Congenital infections
2. Superinfection
3. Emerging and reemerging infections
4. Bioterrorism
5. Vector-borne infections
6. Ectoparasitic infestations.

Short Questions for Theory/Viva Questions

1. Name four parasites/fungi/bacteria causing opportunistic infections.
2. Write two examples of each, DNA and RNA oncogenic viruses with cancer produced by them.
3. Name four fungal toxins causing cancer along with cancer's produced by them.
4. Name the four bio-weapons from category A.
5. What is extrinsic incubation period?
6. What is ectoparasite? Write two examples.

Comment on

1. Overuse of antibiotics cause superinfection.

MCQs for Chapter Review

Zoonotic Infections

1. **Anthropozoonosis are all *except*:**
   a. Guinea worm infection
   b. Rabies
   c. Plague
   d. Hydatid cyst
2. **Zoonotic disease are all *except*:**
   a. Typhoid   b. Anthrax
   c. Rabies   d. Q fever
3. **Zoonotic bacterial infection is/are:**
   a. Bovine tuberculosis   b. Brucellosis
   c. a + b   d. None of above

Opportunistic Infections

4. **Parasitic intestinal infestation seen in immunosuppressed patient is:**
   a. Giardisis   b. Ascariasis
   c. Liver fluke   d. Schistosomiasis
   e. *Strongyloides*

5. Humoral immunodeficiency is suspected in patient and he is under investigation. Which of the following infection(s) would be consistent with the diagnosis?
   a. Giardiasis
   b. *Pneumocystis carinii* pneumonia
   c. Recurrent sinusitis
   d. Recurrent subcutaneous abscess

Emerging and Reemerging Infections

6. Emergence or resurgence, seen in which of the following organism:
   a. Polio virus
   b. Measles virus
   c. Nipah virus
   d. West Nile virus
   e. Hepatitis B virus
7. New infectious agents are:
   a. Nipah virus
   b. *Pneumocystis jiroveci*
   c. Coronavirus
   d. SARS
   e. Prion

Oncogenic Microbes

8. Oncogenic RNA virus is:
   a. Avian leukosis virus
   b. Herpes virus
   c. Adeno virus
   d. Toga virus
9. Most common oncogenic RNA virus:
   a. Retro virus
   b. Picorna virus
   c. Orthomyxo virus
   d. Paramyxo virus
10. RNA oncogenic virus amongst the following is:
    a. HIV
    b. HTLV
    c. HBV
    d. Cytomegalo virus
11. Oncogenic virus:
    a. CMV
    b. VZV
    c. Polio virus
    d. EBV
12. Which of the following has malignant potential?
    a. HSV-1
    b. EBV
    c. CMV
    d. Varicella
13. Which of the following "oncogenic virus" is so far not shown to be oncogenic in man?
    a. Hepatitis B virus
    b. Epstein Barr virus
    c. Herpes simplex virus type 2
    d. Adeno virus
    e. Human T cell lymphotropic virus 1
14. Pancreatic Ca is caused by:
    a. *Fasciola*
    b. *Clonorchis*
    c. *Paragonimus*
    d. None
15. Cholangiocarcinoma is caused by:
    a. *Fasciola* infestation
    b. *Clonorchis* infestation
    c. *Paragonimus* infestation
    d. *Ascaris* infestation
    e. None of above
16. Cause of biliary tract carcinoma after ingesting infected fish:
    a. Gnathostoma
    b. *Angiostrongylus cantonensis*
    c. *Clonorchis sinensis*
    d. *H. diminuta*

Bioterrorism

17. True about bio-weapons category A:
    a. These agents are highest priority pathogens which pose greatest risk to national security
    b. Result in low mortality
    c. Do not cause social disruption
    d. None of the above

18. Most potent and potential agent that can be used for bioterrorism:
    a. Plague
    b. Smallpox
    c. TB
    d. *Clostridium botulinum*
19. Which of the following is not group A bioterrorism?
    a. Smallpox
    b. Hemorrhagic fever
    c. *Salmonella*
    d. Botulism
20. Category A bioterrorism agents are:
    a. Ebola
    b. *Yersinia*
    c. *Clostridium botulinum*
    d. *Rickettsia*
    e. Cholera
21. Bioterrorism group A agent:
    a. Q fever
    b. Typhus fever
    c. *Brucella*
    d. Anthrax
22. Microorganism used as weapon in biological terrorism:
    a. Smallpox virus
    b. Rabies virus
    c. Ebola virus
    d. Influenza C virus
    e. Human parvo virus
23. Which of the following is belonged to category B of bioterrorism?
    a. Cholera
    b. Anthrax
    c. Plague
    d. Botulism

Answers and Explanation of MCQs

1. a
2. a
3. c
- Follow section, **zoonotic infections** for explanation of answers of MCQs 1–3.
4. a, e
5. a, b
- Follow section, **opportunistic infections (etiological agents)** for more explanation of answers of MCQs 4–5.
6. c
7. a, c, d
- Follow section, **emerging and re-emerging infections (Table 47.2)** for explanation of answers of MCQs 6–7.
8. a
9. a
10. a, b
11. a, d
12. a, b, c
13. d
14. b
15. b
16. c
- Follow section, **oncogenic microbes** for explanation of answers of MCQs 8–16.
17. a
- Follow section, **Bioterrorism → classification (category A)** for explanation.
18. d
- *Clostridium botulinum* toxin is the most dangerous and most commonly used agent.
19. c
20. a, b, c
21. d
22. a, c
23. a
- Follow section, **Bioterrorism → classification (Table 47.5)** for explanation of answers of MCQs 19–23.

CHAPTER 48

National Health Program and Pandemic Management

Chapter Outline
- National Health Program
- Pandemic Management

NATIONAL HEALTH PROGRAM

Introduction: Government of India launched several program for rehabilitation and to prevent, to treat and to eradicate infectious/communicable and non-infectious/noncommunicable diseases. Few programs for infectious/communicable diseases are listed below.

Major health programs by Government of India

1. Revised National Tuberculosis Program (RNTCP, 1993)/National Tuberculosis Elimination program (NTEP, 2020): **Ch. 56.**
2. Poliomyelitis Eradication Program: **Ch. 80.**
3. National AIDS Control Program: Many strategies are covered in this program. Few of them like diagnosis of HIV-AIDS in high-risk group, prevention, post-exposure prophylaxis and ART to case of AIDS are described in **Ch. 85.**
4. National Vector-borne Diseases Control Program (NVBDCP) for vector-borne diseases/infections: **Ch. 47.**
5. Antimicrobial Stewardship Program (ASP): **Ch. 116.**
6. National Viral Hepatitis Control Program.
7. Integrated Disease Surveillance Program (IDSP) for epidemic prone diseases. Now IDSP is changed to Integrated Health Information Platform (IHIP).
8. National Leprosy Eradication Program.
9. National Rabies Control Program.
10. National Blindness Control Program-Trachoma.
11. Disaster Management Program in case of epidemic (like swine flu, bird flu, zika virus disease, etc.) or emergency of disease.

PANDEMIC MANAGEMENT

Introduction: To manage the sudden health crisis like COVID-19, NMC (National Medical Council) introduced the pandemic management schedule for medical students.

Measures for pandemic management

- **Administrative measures:** Like appointment of nodal officer, establishment of infection control committee, etc.
- **Preclinical measures:** These are history and origin of infection/health crisis, emerging and reemerging infection, etc.
- **Diagnostic measures**
  - Strategies for sample collection, transport, storage and microbiological testing.
  - Strategies for radiological, pathological and biochemical investigations.
- **Preventive measures**
  - General measures: Strategies for contact precautions, hand washing, sterilization and disinfection, wearing of PPEs, quarantine/isolation, contact tracing, etc.
  - Chemoprophylaxis: Strategies for use of prophylactic drugs including development of new drugs.
  - Immunoprophylaxis: Strategies for vaccine and its implementation including development of new vaccine.
- **Therapeutic measures:** Strategies for treatment including development of new drugs.
- **Others**
  - Surveillance and research.
  - Death management.
  - Biomedical waste (BMW) management.
  - Other measures include media management, financial management, etc.

Infections of Micrococcaceae

Chapter Outline

- Introduction
- Staphylococci
  - Coagulase-positive staphylococci (CoPS)
  - Coagulase-negative staphylococci (CoNS)
- Micrococci
- Stomatococci

INTRODUCTION

Differences between Micrococcaceae and Streptococcaceae

Follow **Table 49.1**.

TABLE 49.1: Differences between Micrococcaceae and Streptococcaceae

Features	Micrococcaceae	Streptococcaceae
Catalase test	+	–
Arrangement	Tetrads/cluster due to multiple planes division	Pair/chain due to single plane division

Types of Micrococcaceae

Follow **Flowchart 49.1**.

STAPHYLOCOCCI

Coagulase-positive Staphylococci (CoPS)

Staph. aureus (Staph. pyogenes)

Meaning

Staphylo word derived from Staphyle (Greek) means bunches of grapes and cocci derived from kokkos (Greek) means berry, as the bacteria are spherical in

Flowchart 49.1: Types of Micrococcaceae

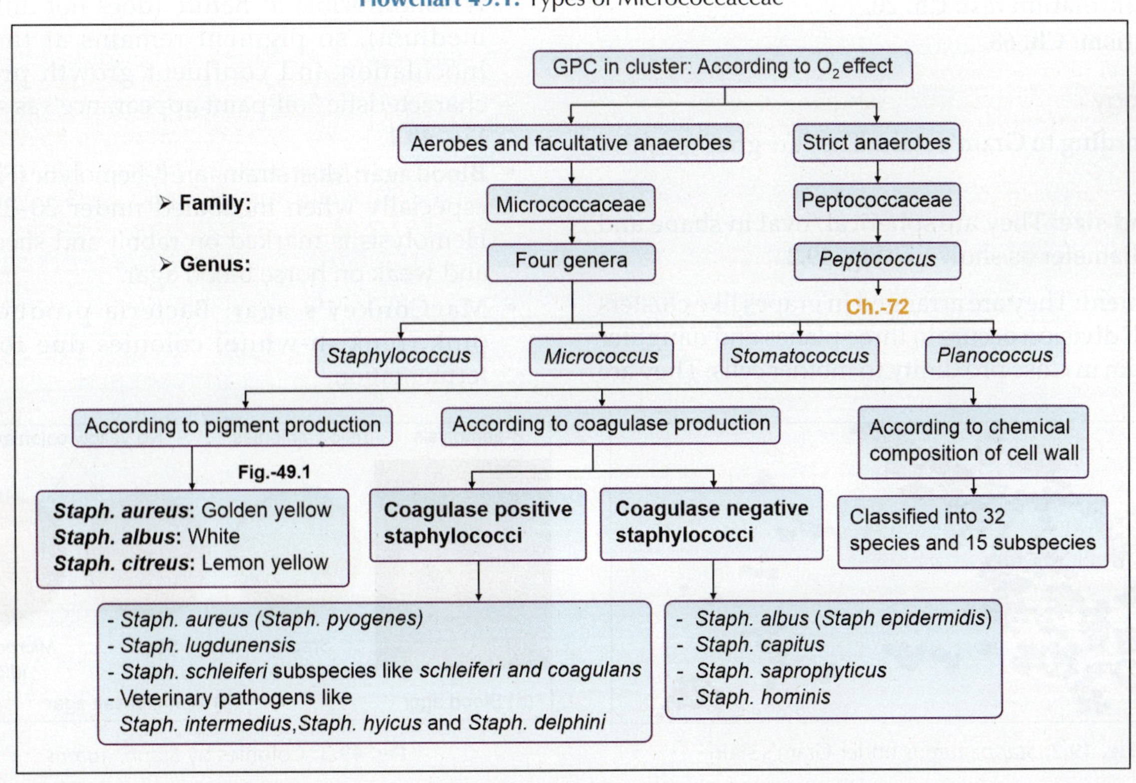

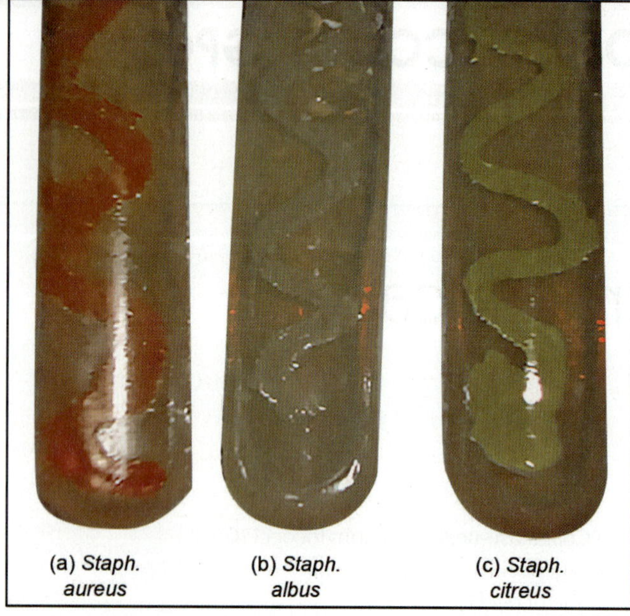

Fig. 49.1: Pigment-producing staphylococci

shape like berry and arranged in cluster like bunches of grapes hence the name is given. Aureus word derived from aurum (Latin) means gold, because of golden yellow pigment production.

Subspecies

Few author identified two subspecies like *Staph. aureus* subsp. *aureus* which is the human pathogen and *Staph. aureus* subsp. *anaerobius* which is the sheep pathogen.

Uses of *Staph. aureus*

1. **CAMP test:** Hemolytic toxin of *Staph. aureus* gives synergistic hemolysis with *Strept. agalactiae* called CAMP test. For more details follow **Ch. 50.**
2. **Coagglutination test: Ch. 20.**
3. **Satellitism: Ch. 68.**

Morphology

Type according to Gram's stain: They are gram-positive cocci.

Shape and size: They are spherical/oval in shape and 1 µm in diameter as shown in **Fig. 49.2.**

Arrangement: They are arranged in grapes like clusters due to cell division occurs in three planes and daughter cells remain in close proximity to mother cells. They are

also found in pairs, tetrads and short chains, especially when examined from liquid culture.

Capsule: Few strains contain polysaccharide capsule.

Motility and spores: They are nonmotile and non-sporing.

Fimbriae: These are the organs of adhesion made up by teichoic acid.

L-forms: Under the influence of penicillin and certain chemicals, they may change to cell wall deficient forms called L forms.

Cultural Characteristics (C/Cs)

Effective factors

- O_2/CO_2 effect: Aerobes and facultative anaerobes.
- Temperature: Range 10–42°C; optimum temperature is 37°C.
- pH: 7.4–7.6

Culture in media

1. **Liquid media:** They produce uniform turbidity in peptone water and nutrient broth.
2. **Solid media**
   - **Nutrient agar plate or slope/slant:** They produce large (2–4 mm), circular, convex, smooth, shiny, opaque, easily emulsifiable and pigmented colonies. Most strains produce golden yellow/orange pigment as shown in **Fig. 49.1a.** Golden yellow pigment of *Staph. aureus* called staphyloxanthine (Xanthine means yellow). Optimum production occurs at 22°C and in aerobic culture. Pigment production increased by incorporating 1% glycerol monoacetate or milk in medium. Chemically, pigment is lipoprotein allied to carotene. Pigment is nondiffusible in nature (does not diffuse into medium), so pigment remains at the site of inoculation and confluent growth presents a characteristic "oil-paint appearance" as shown in **Fig. 49.1.**
   - **Blood agar:** Most strains are β-hemolytic **(Fig. 49.3a),** especially when incubated under 20–25% CO_2. Hemolysis is marked on rabbit and sheep blood and weak on horse blood agar.
   - **MacConkey's agar:** Bacteria produce small pink (pinkish-white) colonies due to lactose fermentation.

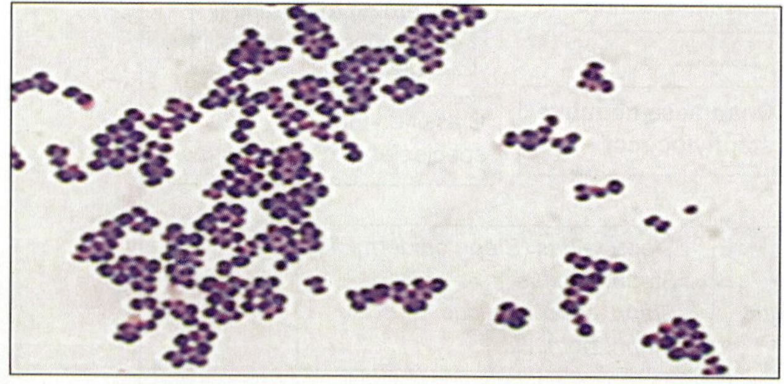

Fig. 49.2: *Staph. aureus* under Gram's stain

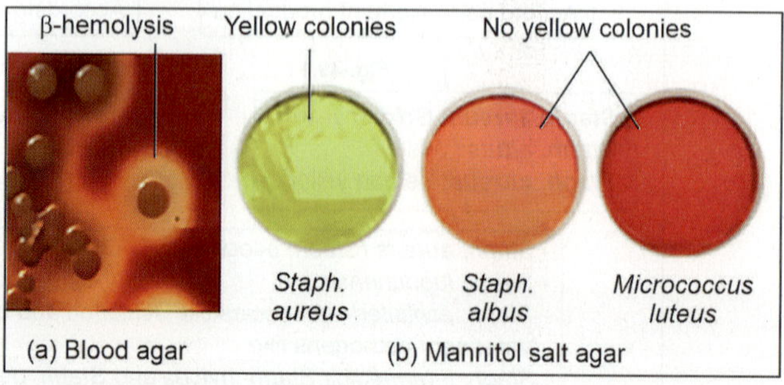

Fig. 49.3: Colonies by *Staph. aureus*

Note: Differentiation from the pigment of *Pseudomonas*
- Pigment of *Pseudomonas* is diffusible in nature, hence spread throughout the depth of nutrient agar slant and entire slant becomes pigmented.

3. Selective media

- **Liquid selective media:** These are salt broth and salt cooked meat broth. These media contain 8–10% NaCl.
- **Solid selective media**
  - Mannitol salt agar: Bacteria produce yellow colonies as shown in **Fig. 49.3b**.
  - Potassium tellurite medium: Bacteria produce black colonies.
  - Ludlam's medium: It contains lithium chloride and tellurite. Ludlam (1949) reported the effectiveness of tellurite-lithium chloride-agar in selecting coagulase-positive staphylococci. Several workers (Chapman, 1949; McDivitt and Hussemann, 1954; Baird-Parker, 1962) reported that Ludlam's medium is inhibitory to coagulase-positive cultures, and thus may fail to reveal the presence of the organism.
  - Baird-Parker medium: It is used for the selective isolation of gram-positive staphylococci. It contains lithium chloride and tellurite to inhibit the growth of microbial flora, while the included pyruvate and glycine promote the growth of staphylococci. Staphylococci produce black colonies with clear zones around them.
  - Other media: Salt milk agar and medium containing polymyxin B.

Automated culture: Like MALDI-TOF or VITEK is used to identify the species from culture.

Note: Selection of blood for blood agar
- For primary isolation, sheep blood agar is used.
- Human blood is not recommended, as it may contain antibodies and other inhibitors.

Biochemical Reaction (B/Rs)

- **Sugar fermentation tests:** *Staph. aureus* ferments numbers of sugars producing acid only. Mannitol fermented anaerobically with production of acid only which gives pink color to the medium as shown in **Fig. 49.4a**. Mannitol fermentation is diagnostically useful and helps in differentiation of species, as it is fermented by *Staph. aureus* only and not by other species of *Staphylococcus*.
- **I M Vi C tests:** – + + –
- **Catalase test:** Positive
- **Oxidase test:** Negative
- **Urease test:** Positive
- **Nitrate test:** Positive (reduces nitrate to nitrite)
- **Phosphatase test**
  - **Principle:** It is based on production of phosphatase enzyme, which breaks the phenolphthalein diphosphate to release the free phenolphthalein,

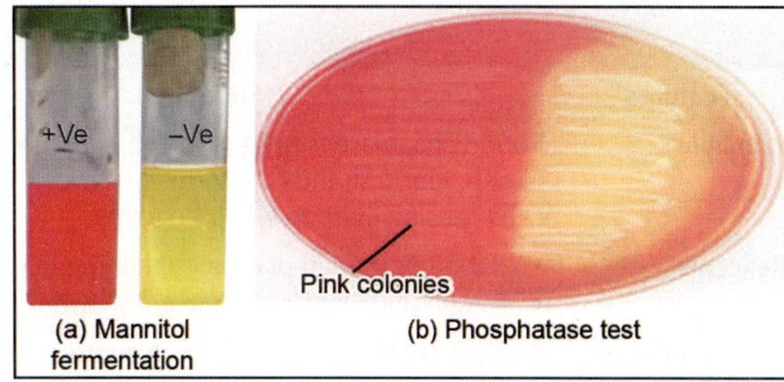

Fig. 49.4: (a) Mannitol fermentation test; (b) Phosphatase test

which changes the colonies to pink color on exposure of ammonia. Phenolphthalein is quinone derivative and used as pH indicator, which turns pink at alkaline pH by ammonia.
  - **Method:** Inoculate *Staph. aureus* in phenolphthalein diphosphate agar. After growth, expose culture to ammonia vapor and check for the result.
  - **Result:** Positive result is indicated by pink color colonies due to presence of free phenolphthalein as shown in **Fig. 49.4b**.
  - **Interpretation:** It is positive in *Staph. aureus, Staph. epidermidis* and *Staph. xylosus*.

- **Coagulase test**
  - **Background with coagulase enzyme:** Coagulase is an enzyme (protein) of unknown chemical composition. It has prothrombin like activity to convert fibrinogen in to fibrin, resulting clot formation in a suitable test system. It produces fibrin barrier at the site of *Staphylococcus* infections causing localized abscess like carbuncle, boil, etc. Two forms of enzyme like bound and free coagulase as shown in **Table 49.2**.
  - **Coagulase test:** Follow **Table 49.3 and Fig. 49.5**.
- **Lysostaphin sensitivity:** *Staph. aureus* is sensitive to lysostaphin. Lysostaphin is mixture of enzymes produced by *Staph. epidermidis*.
- **Gelatin liquefaction test:** Test is positive.
- **Lipase production test:** Most strains are lipolytic producing dense opacity when grown on egg yolk containing media due to lipase production.
- **Heat stable thermonuclease test:** Test is positive and based on production of thermostable nuclease. It is demonstrated by an ability of boiled culture to degrade DNA.

TABLE 49.2: Differences between bound and free coagulase enzyme

Bound coagulase (clumping factor)	Free coagulase
Detected by slide test	Detected by tube test
Heat stable	Heat labile
Attached to cell wall (cell surface protein)	Enzyme secreted into medium
Independent of coagulase reacting factor (CRF)	Dependent to CRF
Only 1 serotype	8 serotypes

Essentials of Medical Microbiology

TABLE 49.3: Coagulase test

Features	Slide test	Tube test
Principle	Bound coagulase is present in cell wall, not in a culture filtrate. Fibrin strands are formed between the bacterial cells causing them to clump together.	Free coagulase having prothrombin like activity. React with CRF of plasma to convert fibrinogen in to fibrin to form visible clot.
Reagents	• Rabbit or human plasma containing EDTA or heparin or oxalate as an anticoagulant. • Citrate utilizing bacteria (e.g. *Pseudomonas*) give false-positive result, so citrate plasma is not useful.	
Quality control	**Positive control:** *Staph aureus.* **Negative control:** *Staph. albus*	
Method	• Place a drop of sterile water or saline on two ends of clean and dry slide. • Emulsify the colony of test organism in one end, add one drop of plasma and mix with wooden stick. • Leave the other end as control. • Rock the slide back-forth and read for the clumps.	• Mix a 0.1 ml of organism culture suspension with 0.5 ml of human or rabbit plasma. • Incubate at 37°C for 3–4 hrs and observe for clot formation.
Results	Positive test: Clumps Negative test: No clumps } **Fig. 49.5a**	Positive test: Coagulum/clot formation Negative test: No coagulum/no clot } **Fig. 49.5b**

Interpretation	**Positive test**
	1. Only slide positive: *Staph. lugdunensis.*
	2. Only tube positive: *Staph. intermedius, Staph. hyicus* and *Staph schleferi* subspecies *coagulans.*
	3. Both slide and tube positive:
	  – **Staphylococci:** *Staph. aureus, Staph. delphini* and some strains of *Staph. schleferi* subspecies *schleferi.*
	  – **Other groups:** *E. rhusiopathiae* (tube and/or slide tests are positive with rabbit and/or bovine serum) and *Y. pestis* (exactly not defined about variation in slide and tube positivity).
	Negative test: Coagulase-negative staphylococci are mentioned in **Flowchart 49.1.**
	False-positive test: Citrate is used by many bacteria like *Pseudomonas*, so, if citrate plasma is used for testing, it will give false-positive result.
	False-negative test
	1. Positive tube test become negative after longer incubation, because of clot lysed by **fibrinolysin (staphylokinase)** produce by *staph. aureus.*
	2. Slide test may become false-negative by **capsulated strains,** because capsule masks the clumping factor.	

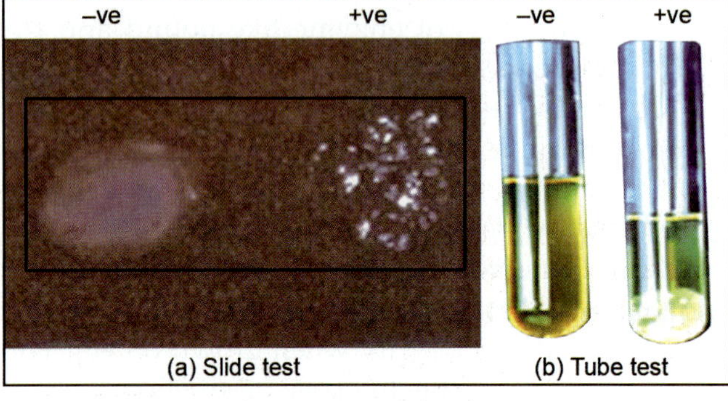

Fig. 49.5: Coagulase test

Resistance

Sterilization: Staphylococci resist drying for 3–6 months and 60°C for 30 minutes. Thermal death point is 62°C for 30 minutes.

Disinfection: Staphylococci resist 1% phenol for 15 minutes. They grow in presence of 10–15% NaCl which is important in food preservation. They are resistant to lysozymes but sensitive to lysostaphin. Lysostaphin is mixture of enzymes produced by *Staph. epidermidis.* They are killed by 1% mercury perchloride in 10 minutes. Aniline dyes like crystal violet (1 in 500,000) and brilliant green (1 in 10,000,000) are lethal.

Drug resistance: It is discussed with resistance to β-lactam antibiotics (penicillin), vancomycin and few other drugs as follows.

A. Resistance to β-lactam antibiotics (penicillin): Initially staphylococci are uniformly sensitive to penicillin, but resistant strains began to emerge, 1st in hospitals then in community. Resistant to β-lactam in staphylococci is plasmid mediated, chromosome mediated or penicillin tolerance as described below.

1. **Plasmid mediated resistance or resistance due to β-lactamase (penicillinase) production:** β-lactamase cleaves the β-lactam ring and produces the resistance. Enzyme is inducible and it's production is controlled by plasmid. Such plasmid transmitted to other bacteria by transduction (more common) or by conjugation to make them resistant to penicillin. About 90% of strains of *Staph aureus* are penicillinase producers. There are four types penicillinase enzymes from A to D. A is produced by hospital strains. β-lactamase producing bacteria are *Staph. aureus, N. gonorrheae, M. catarrhalis, Acinetobacter* spp., *E. coli, K. pneumoniae, Salmonella* spp., *P. mirabilis, Bacteroides* spp., etc.

2. **Chromosomal mediated (mutational) resistance or MRSA**
   • **Mechanisms:** Chromosomal mediated gene called mec A gene (staphylococcal cassette chromosomal mec gene) alters the penicillin binding proteins of cell wall (PBP2a) and bacterial surface receptors.

- **Definition:** Such resistance extends to cover other β-lactam antibiotics like methicillin, cloxacillin or oxacillin called methicillin resistance *Staph. aureus* (MRSA).
- **Epidemiology:** There is increasing trend of MRSA in last few decades. It varies from place to place. Around 30–40% strains of *Staph aureus* are MRSA.
- **Types:** Two types of MRSA.
  - Hospital-acquired (HA MRSA): Also called hospital strain. Sometimes, resistance extended to antibiotics like erythromycins, tetracycline, aminoglycosides, heavy metals and produces epidemic of hospital cross infection called epidemic strain or epidemic MRSA. Resistance expressed by mec A gene subtype I, II and III. It is less virulent and multidrug resistant. It produces post-operative wound infections and other hospital-acquired infections.
  - Community-acquired (CA MRSA): Resistance expressed by mec A gene subtype IV, V and VI. It is more virulent and produces the PVT (described below). It produces the soft tissues infections like necrotizing fasciitis.
- **Diagnosis of MRSA**
  - **Antibiogram:** It includes disc diffusion method and micro-dilution method. Disc diffusion method is done by using cefoxitin or oxacillin (6 µl/ml) discs. Oxacillin required 24 hours incubation at 30°C with supplementation of 2–4% NaCl in medium.
  - **Culture:** It is done by using chromogenic media for MRSA and mannitol oxacillin agar.
  - **Serological test:** It includes latex agglutination test, in which PBP2a is detected by using antibodies fixed with latex particles.
  - **Molecular method:** PCR to detect mec A gene.
- **Prevention of MRSA:** Do frequent hand washing (best method) and take other aseptic precautions during handling of instrument or working in hospital. Select the antibiotics only after antibiogram. Avoid the use of β-lactam; however, 5th generation antibiotics like ceftobiprole and ceftibuten have some role in MRSA.
- **Treatment of MRSA:** Antibiogram is necessary before treatment.
  - Local treatment: Skin lesions are treated by using 2% mupirocin oinment TDS for 5 days. Nares lesions are treated by 2% mupirocin ointment TDS for 1 week (applied at anterior part of inside of each nostrils). Skin carriers are treated by 4% chlorhexidine or 2% triclosan or 7.5% povidone iodine –daily for 1 week. Apply solution on face, nose, axilla, umbilicus, groin and perineum.
  - Systemic treatment: 1st line drugs include vancomycin (drug of choice), linezolid, deptomycin, teicolpanin, tigecycline and quinupristin/dalfopristin. 2nd line drugs rifampin with fusidic acid, cotrimoxazole, gentamicin/erythromycin or tetracycline are used in non life threatening infections

3. **Development of tolerance to penicillin:** Bacterium is only inhibited but not killed.

B. Resistance to vancomycin: Overuse of vancomycin produces two types of resistant strains like low grade and high grade.
- **Low-grade resistance:** It called VISA (Vancomycin-intermediate *Staph. aureus*). It is due to increase in cell wall thickness and common in India.
- **High-grade resistance:** It called VRSA (Vancomycin-resistant *Staph. aureus*). It is due to transfer of Van A gene by horizontal conjugation mechanism from *E. faecalis*. It is rare in India and reported from Hyderabad, Kolkata and Lucknow. VRSA is treated by linezolid, telavancin, daptomycin or quinupristin/dalfopristin.

C. Resistance to other drugs: Staphylococci also exhibit the plasmid-borne resistance to other antibiotics like erythromycins, tetracycline, aminoglycosides and almost all clinically used antibiotics.

Pathogenicity

Epidemiology: Staphylococcal outbreak is common in hospitals.

Reservoirs of infection: They are the normal flora of human skin, nose (10–30%), vagina (5–10%), perineum (10%), axilla, respiratory tract, skin glands, hairs and mucosa. Vaginal carriage is increased during menses which is responsible for TSS. Newborn colonized the bacteria in skin or nose from mother, hospital staff or environment. Umbilical stump is contaminated in hospital-borne babies which acts as carrier called shedder, which sheds the cocci in handkerchief, blankets and bed sheets for days or weeks.

Sources of infection: Case, carrier (like hospital staff), animals (like cow) or inanimate objects are the sources of staphylococcal infections.

Modes of transmission: Endogenous infection is the most common mode, and it occurs following disturbances of normal flora. Exogenous infection occurs by direct contact (from the hand of hospital staff), airborne, fomites borne.

Portal of entry: Respiratory tract, skin or mucosa.

Sites: Multisystem or organs are infected.

Precipitating factors (epidemiological determinants): Three types:

A. Agent factors (virulence factors): These are of two types like intracellular and extracellular.

1. **Intracellular (cell wall associated) factors**
   - **Peptidoglycan:** It provides rigidity and integrity to cell. It activates complement and induces release of inflammatory cytokines.
   - **Adhesins:** Fimbrial adhesin (techoic acid) acts as an organ of adhesion and protects from opsonization. Many protein receptors in staphylococci like fibronectin, fibrinogen, IgG and C1q help in adhesion with mammalian host cells.
   - **Capsular polysaccharides:** It inhibits the phagocytosis.
   - **Protein A:** It has many biological actions like chemotactic effect, antiphagocytic effect, anti-complementary effect, causing platelet damage, hypersensitivity and coagglutination. It binds to Fc portion of IgG (except IgG3) leaving Fab site free to combine with specific Ag present in samples. When protein A bearing staphylococci will mixed with IgG antiserum, they will be agglutinated called coagglutination. Coagglutination test is used for streptococcal typing, gonococcal typing, *Salmonella* diagnosis and ligand for isolation of IgG. Protein A is a B cell mitogen. It is found in 90% strain of *Staph. aureus* called Cowan 1 strain. It is absent in coagulase-negative staphylococci.
   - **Clumping factor or bound coagulase enzyme:** It is a surface protein which is responsible for "slide coagulase test". Test is routinely used for identification of *Staph. aureus*.
2. **Extracellular factors:** Two types like enzymes and toxins as mentioned below.
   - **Enzymes**
     - Free coagulase: It is described above.
     - Penicillinase (β-lactamase): As discussed above.
     - Hyaluronidase: It breaks the connective tissues and helps in spread of infection.
     - Staphylokinase (fibrinolysin), nucleases (DNAse) and proteases: All these will help in initiation and spread of infection.
     - Lipase: It helps to infect skin and subcutaneous tissues.
     - Other diagnostically useful enzymes are: Catalase, phosphatase, urease, etc.
   - **Toxins**
     - **Hemolysin (staphylolysin, cytolytic category):** Four types like α, β, γ and δ. **Alpha hemolysin** is a protein inactivated at 70°C, but reactivated paradoxically at 100°C. It lyzes rabbit RBCs, less active against sheep and human RBCs. It is leukocidal, cytotoxic, dermonecrotic after intradermal inoculation in rabbits, neurotoxic and lethal after intravenous inoculation in rabbits. It is toxic to macrophages, lysosomes, muscles, renal cortex and CVS. **Beta hemolysin** is sphingomyelinase and hemolytic for sheep RBCs but not for human and rabbit RBCs. Beta hemolysin initiates the hemolysis at

37°C but evident after chilling called hot-cold phenomenon. **Gamma hemolysin** is composed of 2 proteins both are essential for hemolytic action. **Delta hemolysin** has detergent like action on membrane of RBCs, WBCs, macrophages and platelets.

- **Leukocidin (PVT/panton-valentine toxin):** It is a cytolytic toxin. It has two components like S and F (S means slow and F means fast), similar to gamma hemolysin. Gamma hemolysin and leukocodin are bicomponent membrane active toxins grouped as synergohymenotropic toxins.
- **Enterotoxin:** There are **8 types** of enterotoxins like A, B, C1-3, D, E and H. Type A is responsible for most cases. It is heat stable (resisting 100°C for 10–40 minutes). Bacteria can grow best at 37°C, which may be the optimum temperature for toxin production. It is very potent and microgram (μg) amount can cause the illness. It is plasmid mediated, antigenic and neutralized by specific antitoxin. It is transmitted by meat, fish, milk and milk products and source of infection is usually a food handler, who is a carrier. It is a **superantigen** causing multisystem disease. It acts on autonomic nervous system (neurotoxic) rather than on gastrointestinal (GI) mucosa. It stimulates the vagus nerve and vomiting center. It also known to stimulate the peristaltic activity. It causes **food poisoning** which presents with nausea, vomiting, diarrhea, 2–6 hours after consumption of contaminated food containing preformed toxin called intradietic toxin. Fever is absent. Incubation period is very short about 2–6 hours, because it is the preformed toxin. It is a self-limiting illness; recovery occurs in a day or so. Other effects of toxin are pyrogenic, mitogenic, hypotensive, thrombocytopenic and cytotoxic. It is **detected** by latex agglutination test and ELISA.
- **Toxic shock syndrome toxin (TSST):** Initially it was known as enterotoxin F or pyrogenic exotoxin C. It has **2 types** like TSST-1 and TSST-2. TSST-1 is produced by *Staph. aureus* usually belongs to bacteriophage group I. Very rarely TSS also caused by enterotoxin B and C of *Staph. aureus*. It was 1st detected from women using highly absorbent vaginal tampons. It also reported from men and non-menstruating women due to preexisting staphylococcal infection. Now tampons related disease is rare. It is related with infections of skin, mucosa and surgical wound. It is a **superantigen** stimulates the large numbers of T cells irrespective of its antigenicity, which release cytokines like IL-2, TNF and IFN-γ that lead to the multisystem disease which may be fatal in nature. It causes toxic shock syndrome (TSS) called **staphylococcal toxic shock syndrome**

neutralized by fever, hypotension, myalgia, vomiting, diarrhea, mucosal hyperemia and an erythematous rash which desquamates subsequently. Later, it produces manifestations in liver, kidneys, GIT, CNS, etc. TSST is **detected** by latex agglutination and its gene by PCR. Clindamycin (reduces the toxin production) is the drug of choice given with semisynthetic Pn or vancomycin.

- **Exfoliative toxin/exfoliatin/epidermolytic toxin/epidermolysin:** There are **2 types** of exfoliative toxin like A and B. Exfoliative toxin produced by *Staph. aureus* usually belongs to bacteriophage group II. Exfoliative toxin A is heat stable and chromosome mediated while exfoliative toxin B is heat labile and plasmid mediated. It often occurs in newborns and infants. It causes **staphylococcal scalded skin syndrome (SSSS)** characterized by separation of outer layer of the epidermis from the underlying tissue as shown in **Fig. 49.6**. It also called **exfoliative skin disease**. Later called **Nikolsky's sign**. Exfoliatin is glutamate-specific serine proteases highly specific to the cadherin desmoglein I, an adhesion protein in the desmosomes of the stratum corneum facilitates intracellular adhesion between keratinocytes. Exfoliatin (protease) cleaves the desmoglein I resulting separation of skin layer. A very similar noninfectious condition is seen in the autoimmune skin disorder pemphigus vulgaris in which there is an IgG antibody against the cadherin desmoglein III. Clinically, it is characterized by blisters and bullae formation with separation of outer layer of the epidermis from the underlying tissue. Mucosa is spared. SSSS has two clinical forms. (1) **Severe form** occurs in newborn called Ritter's disease. It presents with lethargy, fever, poor feeding and irritability. It was 1st described in 1878 by Baron Gottfried Ritter von Rittershain,

who observed 297 cases among children in a single Czechoslovakian children's home over a 10-year period. Severe form occurs in older called toxic epidermal necrolysis. (2) **Mild form** called pemphigus neonatorum and bullous impetigo. Exfoliative toxin is **detected** by PCR, radioimmunoassay, Ouchterlony immuno-diffusion assay and reverse passive agglutination (by using latex particles) methods.

B. Host factors

1. **Age:** Human colonization of staphylococci start in newborn life in umbilical stump from hospital environment.
2. **Sex:** Vaginal carriage increase in menses which contribute in pathogenesis of TSS.

C. Environmental factors: Infections are common from hospital environment or staff.

Clinical features: *Staph. aureus* produces two types of diseases like intoxications and infections.

A. Intoxications: These include food poisoning, TSS and SSSS as described above.

B. Infections: Staphylococci produce the enzyme coagulase, which makes the fibrin barrier around the lesions and makes them localize in contrast to spreading lesions of streptococci.

1. **Skin and soft tissue infections**
   - Folliculitis: It is an infection/inflammation of hair follicle present as a pin point red spot.
   - Furuncle (boil): It is a deep folliculitis infecting hair follicle which presents with painful swelling due to collection of pus and dead tissues in hair follicle.
   - Carbuncle: It is the cluster of boils that are connected with each other beneath the skin.
   - Sycosis barbae: It is an infection of hair follicles in the beard area due to *Staph aureus*. It also called folliculitis barbae and leads to scarring and permanent hair loss. It occurs in men due to or not due to shaving and infection acquired from infected razors or towels.
   - Cellulitis, impetigo and ecthyma: **Follow Ch. 50** for more details.
   - Paronychia: It occurs around the skin of nail.
   - Sebaceous cyst: Blockage of sebaceous gland secretion resulting collection of secretion in gland which produces sebaceous cyst. Later it is infected by *Staph aureus* to produce the abscess.
   - Hidradenitis suppurativa: Hidro means water/ sweat and aden means gland. Few authors mentioned that, it is caused by *Staph. aureus*, but in fact it is neither having infective etiology

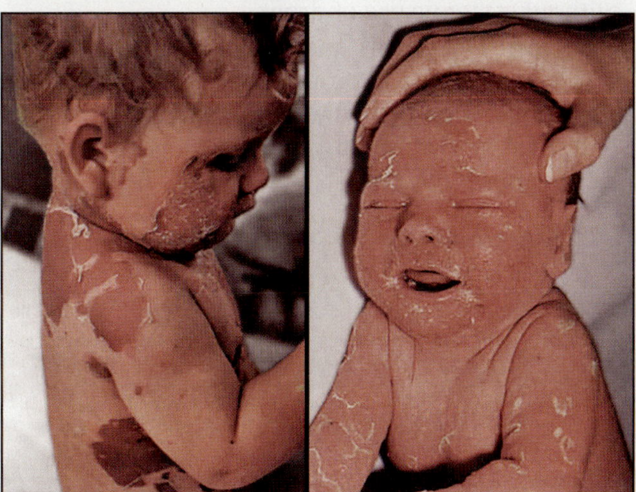

Fig. 49.6: Exfoliative skin disease

nor having contagious nature, and in contrarily, infection occurs as secondary complication.

- Others lesions: Like abscess and wound infections.

> **Note: Pseudofolliculitis barbae**
> - Inflammatory foreign body (noninfective) reaction surrounding grown facial hairs, which results from close shaving. It can also occurs on any site of body, where hair is shaved or plucked, including axilla, pubic area and legs. It is also called shaving rash or razor bumps.
> - Folliculitis barbae and pseudofolliculitis barbae can coexist.
> - It occurs mainly in person with curly hair because sharp pointed end of a recently shaved hair comes out from the skin and reenters the skin and causing a foreign body inflammatory reaction.

2. **Musculoskeletal infections:** Osteomyelitis, arthritis, botryomycosis, bursitis, myositis, pyomyositis (also called tropical pyomyositis or myositis tropicans) and abscess in skeletal muscles.
3. **Central nervous system infections:** Abscess, meningitis, encephalitis and intracranial thrombophlebitis.
4. **Respiratory system infections:** Tonsillitis, pharyngitis, sinusitis, otitis media, broncho/lobular pneumonia (most common cause), pneumatocele, lung abscess and empyema.
5. **Bloodstream infections:** Bacteremia, septicemia and pyemia.
6. **CVS infections:** Endocarditis to the native valve is caused by *Staph. aureus*. It is common in IV drug abusers and specially involves the tricuspid valve (right side valve). Few authors suggested that right side valves are commonly infected by *Staph. aureus* while left side valves are infected by *Staph. aureus, Enterococcus* and other bacteria. Endocarditis is a manifestation with prosthetic valve up to 12 months (called early onset prosthetic valve endocarditis) by *Staph. epidermidis* and after 12 months (called late onset prosthetic valve endocarditis) by viridans group of bacteria.
7. **Urinary tract infections:** It occurs due to local catheterization, implants or diabetes.
8. **Ear infections:** Perichondritis of pinna, otitis externa (localized and diffuse) and pyogenic/suppurative - nonspecific otitis media.
9. **Eyes infections:** Conjunctivitis, keratitis, corneal ulcer, uveitis, blepharitis, endophthalmitis, panophthalmitis, stye (hordeolum externum), chalazion, hordeolum internum, etc.
10. **Oral cavity and salivary gland infections:** Ludwig's angina and acute, chronic or postoperative parotitis.

Carrier

Following are the different types:
- **Newborn carrier or shedder:** Colonization of *Staphylococcus* starts early in a life like umbilical stump in newborn baby in hospital. Some carrier called shedder. It sheds the cocci in handkerchief, blankets and bedsheets for days or weeks.
- **Healthy carrier:** Healthy persons (like hospital staff) carry staphylococci in nose (10–30%), in perineum (10%) and also on the hair.
- **Vaginal carrier:** About 5–10% vaginal carrier, rises greatly during menses, a relevant factor in pathogenesis of TSS related to menstruation.

Laboratory Diagnosis

Specimen collection: Follow **Table 49.4**.

Testing methods

A. **Microscopy:** Follow morphology.
B. **Culture:** Follow C/Cs.
C. **Biochemical reactions:** Follow B/Rs.
D. **Serological test:** Anti-staphylolysin test.
E. **Molecular method:** PCR.
F. **Typing methods:** Phenotyping methods include bacteriophage typing (**Ch. 118**) and typing by antibiogram pattern. Genotyping methods include pulsed field gel electrophoresis, DNA fingerprinting, ribotyping, sequence based typing method and plasmid profile.

Prevention

Hospital infection is controlled by:
- isolation of patients with open staphylococcal lesion.
- identification of staphylococcal infection in hospital staff person and keeping them away from work till the infection will cure.
- strict aseptic precautions like hand washing and use of disinfectant to prevent the spread of hospital strains.
- diagnosis of **hospital carrier** and treated with local application of mupirocin or chlorhexidine and allow to work after complete recovery.
- selection of antibiotics only after antibiogram.

Treatment

Treatment of case

1. **Local:** Sometimes topical use of antibiotics like bacitracin, chlorhexidine or mupirocin is sufficient without systemic use of drugs.
2. **Systemic:** Drug resistance is common problem in staphylococci, the appropriate antibiotic should be chosen based on Penicillin (Pn) antibiotic sensitivity pattern as shown in **Flowchart 49.2**.

TABLE 49.4: Specimens collection	
System infected	**Specimens**
Skin, soft tissue and musculo-skeletal	Pus/discharge on swab, if available in large amount than collect by using sterile syringe
Respiratory	Sputum or BAL
CNS	CSF
Hematogenous and CVS	Blood
Urinary system	Mid stream urine samples

Flowchart 49.2: Systemic treatment of *Staph aureus*

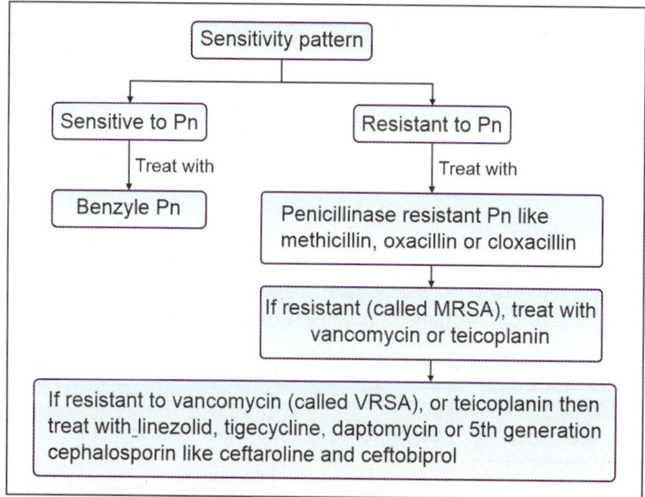

TABLE 49.5: Diagnosis of CoNS and comparison with *Staph. aureus*

Test	*Staph. aureus*	*Staph. epidermidis (albus)*	*Staph. saprophyticus*
Coagulase test	+	–	–
Thermonuclease test	+	–	–
Mannitol fermentation	Acid, no gas	–	–
Phosphatase test	+	+	–
Novobiocin (5 µg) sensitivity	S	S	R
Pigments/ colonies in NA	Golden yellow	White	No pigment production

Treatment of carrier

1. **Local:** Treat with local application of mupirocin or chlorhexidine.
2. **Systemic:** If resistant developed, than treat with rifampicin with other oral antibiotics.

Other CoPS

Only slide positive CoPS: It includes *Staph. lugdunensis* which rapidly established itself as human pathogen. It causes catheter related bacteremia, native valve endocarditis, osteomyelitis and shunt associated meningitis.

Only tube positive CoPS: It includes *Staph. intermedius, Staph hyicus* and *Staph. schleferi* subspecies *coagulans* (dog). All are animal pathogens.

Both slide and tube positive CoPS: It includes *Staph. aureus* (described above), *Staph. delphini* and *Staph. schleferi* subspecies *schleferi. Staph. delphini* causes purulent skin lesions in dolphins. Some strains of *Staph. schleferi* subspecies *schleferi* causes native valve endocarditis, osteomyelitis, nosocomial UTI and other infections associated with implanted medical devices.

Coagulase-negative Staphylococci (CoNS)

Species

Staph. capitus, Staph. hominis and *Staph. haemolyticus*: All are normal flora of skin.

Staph. albus (Staph. epidermidis): It is the most common (75–80%) isolated CoNS from clinical samples. Other details are described below.

- **Precipitating factors:** It presents as normal flora on the skin, oropharynx, vagina, lid and conjunctiva. It causes infections in person with IDDs, drug addiction and implantation of foreign bodies (indwelling prosthesis) like heart valves, shunts, intravascular catheters, etc. It colonizes different materials (medical devices like implants, catheter, etc.) called biofilm. It is due to drug resistance and extracellular polysaccharide matrix, having ability to bind with plastic material of medical devices.

- **Clinical features:** It produces following lesions.
  - Endocarditis particularly in drug addicts and having prosthetic valve.
  - Bacteremia due to its predilection for growth on implanted materials as mentioned above.
  - Others lesions: Like stitch abscess, cystitis, central line associated bloodstream infection and blapharoconjunctivitis.
- **Diagnosis:** *Staph. albus* is identified as gram-positive cocci under Gram's stain. It is mannitol nonfermenter, so no yellow color production on mannitol salt agar as shown in **Fig. 49.3b**. It is catalase-positive but coagulase-negative.
- **Treatment:** Hospital strains are multiple drug resistant. Antibiotics are given after AST report.

Staph. saprophyticus
- **Precipitating factors:** It is common in sexually active young girls and older males.
- **Clinical features:** It causes upper UTI in sexually active young girls. It is due to its enhanced capacity to adhere with uroepithelial cells by presence of 160 kDa hemagglutinin adhesin proteins.
- **Treatment:** It is sensitive to common antibiotics except nalidixic acid and novobiocin.

Laboratory Diagnosis of CoNS
Follow **Table 49.5**.

Treatment of CoNS
Treatment is same as *Staph. aureus*.

MICROCOCCI
Common species are *M. luteus* and *M. roseus*. Micrococci are GPC look like staphylococci but larger in size and gram variable. They are arranged in pairs, tetrads or irregular cluster as shown in **Fig. 49.7**. They are strict aerobes and produce the smaller colonies in culture. Differences with *Staphylococci* are mentioned in **Table 49.6**.

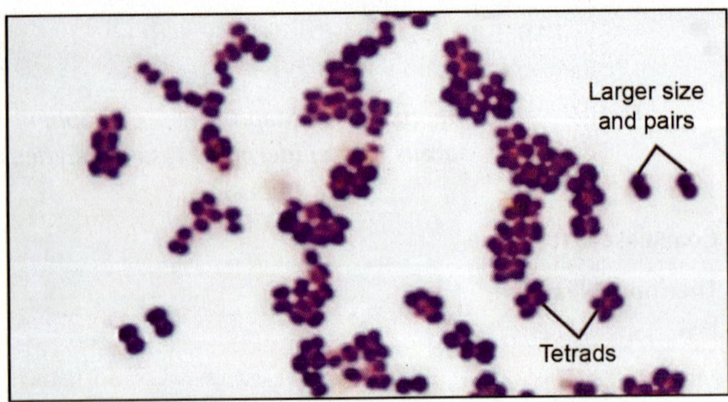

Fig. 49.7: *Micrococcus* under Gram's stain

TABLE 49.6: Differences between genera of Micrococcaceae

Features	Staphylo-coccus	Micro-coccus	Stomato-coccus
Anaerobic growth	+	−	+
O-F test (most useful)	F	O	F
Catalase	+	+	Weak
Oxidase	−	+	−
Bacitracin	R	S	S
Furazolidone	S	R	R
Lysostaphin	S	R	R
Agar adherence	−	−	+

+ = Positive, − = Negative, O − F test = Oxidation-Fermentation test, S = Sensitive, R = Resistant

Note: Common sensitivity pattern for diagnosis of bacteria
Staph. aureus: Novobiocin (5 µg) sensitive
Staph. epidermidis: Novobiocin (5 µg) sensitive.
Strept. pyogenes: Bacitracin sensitive.
Strept. pneumoniae: Optochin sensitive.
Classical *V. cholerae:* Polymixin B sensitive (50 U).
C. jejuni, C. coli and *C. curvus:* Nalidixic acid sensitive.
C. fetus: Cephalothin sensitivity.
H. pylori: Cephalothin sensitivity.

STOMATOCOCCI

Common species is *S. mucilaginosus*. It is GPC and arranged in pairs, tetrads or irregular clusters. It is aerobe and facultative anaerobe. It produces small, dome shaped, grayish brown colonies. Colonies are with rubbery consistency and adherent to agar, so string is produced when loop attached to colonies followed by its withdrawal. Differences with *staphylococci* are mentioned in **Table 49.6**.

Note: Basic steps to diagnose GPC
Follow **Flowchart 49.3**.

ACCESS YOURSELF

Case Studies

1. A 28 years female patient visited the hospital with pus discharging swelling on the hand. Gram's stain of discharge shows the presence of GPC in cluster. Culture shows the β-hemolysis on blood agar. Bacteria ferment mannitol with acid only. Identify the organism and answer the following.

Flowchart 49.3: Diagnosis of GPC

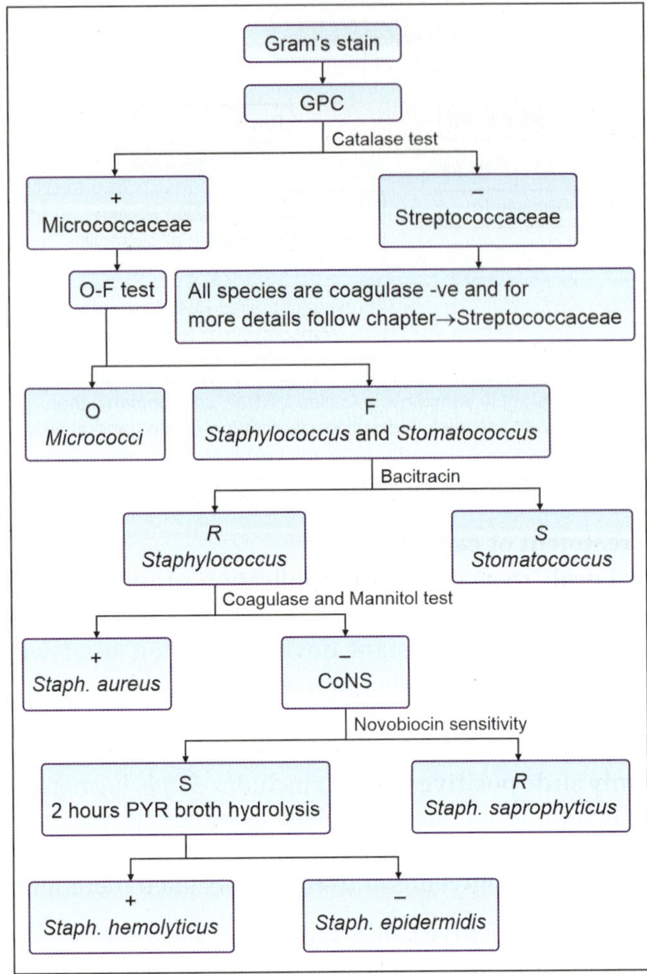

a. Name the causative agent and describe the morphology of causative agent.
b. Write the pathogenicity of causative agent.
c. Describe the lab., diagnosis of causative agent.

2. A child presents with diarrhea in 1–5 hours after consumption of food in party prepared from milk.
a. Name the causative agent. Enumerate the four toxins and four enzymes produced by causative agent.
b. Write the B/Rs used in diagnosis of causative agent.
c. Describe in detail about food poisoning (in general).

Essay/Full Question

1. *Staph aureus.*

Short Notes

1. Cultural characteristics/Biochemical reactions/Pathogenicity/Lab., diagnosis of *Staph. aureus.*
2. Toxins of *Staph. aureus*
3. Coagulase test
4. Drug resistance in *Staph. aureus*
5. MRSA
6. CoNS.

Short Questions for Theory/Viva Questions

1. How you differentiate the pigment of staphylococci from the pigment of *Pseudomonas?*
2. Write the effect of lysozyme and lysostaphin on *Staph. aureus.*
3. What is lysostaphin?

4. Write different modes of resistant to penicillin in *Staph. aureus*.
5. Name the four β-lactamase (penicillinase) producing bacteria.
6. What is hot-cold phenomenon?
7. What is coagglutination? Write its two uses.
8. Name the four enzymes produced by staphylococci.
9. Write the four differences between bound and free coagulase enzyme.
10. Name the four toxins produced by staphylococci.
11. Name the four CoPS or coagulase-positive bacteria.
12. Name the four CoNS.
13. Write the principle of phosphatase tests, used in *Staph. aureus*.

Comments on

1. Staphylococci are arranged in cluster.
2. Staphylococci produce localized lesions, while streptococci produce invasive lesions.
3. Exfoliative toxin causes separation of outer layer of skin.

MCQs for Chapter Review

Types of Micrococcaceae

1. Coagulase-positive bacterium/bacteria is/are:
a. *Staph. aureus*
b. *Staph. intermedius*
c. *Staph. hyicus*
d. All

Coagulase-positive Staphylococci (CoPS)

2. The following are characteristic features of staphylococcal food poisoning *except*:
a. Optimum temperature for toxin production is 37°C.
b. Intradietic toxin is responsible for intestinal symptoms.
c. Toxin can be destroyed by boiling for 30 minutes.
d. Incubation period is 1–6 hours.

3. All the following statements about staphylococci are true, *except*:
a. Most common source of infection is cross infection from infected people.
b. About 30% of general population is healthy nasal carrier.
c. Epidermolysin and TSS toxin are superantigen.
d. Methicilin resistance is chromosomally mediated.

4. Synergohymenotropic toxins of *Staphylococcus* consist of:
a. α-hemolysin
b. β-hemolysin
c. γ-hemolysin
d δ-hemolysin
e. Leukocidin

5. Staphylococcal toxic shock syndrome is due to:
a. Enterotoxin A
b. Enterotoxin B
c. Enterotoxin C
d. Enterotoxin D

6. Toxic shock syndrome was 1st discovered in:
a. Tampon users
b. Diabetic septicemia
c. Drug addicts
d. None

7. In *Staphylococcus* drug resistance has been transferred by:
a. Transformation
b. Transduction
c. Conjugation
d. Transfection

8. All the following statements are true regarding staphylococci, *except*:
a. A majority of infections caused by coagulase-negative staphylococci are due to *Staphylococcus epidermidis*.
b. β-lactamase production in staphylococci is under plasmid control.
c. Expression of methicillin resistance in *Staphylococcus aureus* increases, when it is incubated at 37°C on blood agar.
d. Methicillin resistance in *Staphylococcus aureus* is independent of β-lactamase production.

9. Which of the following statements is most correct regarding resistance to methicillin in MRSA?
a. Resistance is produced as a result of alteration in penicillin binding proteins (PBP).
b. Resistance is produced by production of beta-lactamase.
c. Resistance is mediated by plasmid.
d. Expression of resistance is enhanced by incubating at 37°C during susceptibility testing.

10. A diabetic patient developed cellulitis due to *Staphylococcus aureus*, which is found to be methicillin resistance on the antibiotic sensitivity testing. All of the following antibiotics are appropriate *except*:
a. Vancomycin
b. Imipenam
c. Teicoplanin
d. Linezolid

11. There is outbreak of MRSA infection in ward. What is the best way to control the infection?
a. Vancomycin given empirically to all patients
b. Fumigation of ward frequently
c. Washing hand before and after attending patient
d. Wearing mask before any invasive procedure in ICU

12. Which of the following organism is implicated in causation of botryomycosis?
a. *Staphylococcus aureus*
b. *Staphylococcus albus*
c. *Pseudomonas aeruginosa*
d. *Streptococcus pneumoniae*
e. *Streptococcus pyogenes*

13. Which of the following is not an infection of the hair follicle?
a. Folliculitis
b. Furuncle
c. Carbuncle
d. Hidradenitis

Coagulase-negative Staphylococci (CoNS)

14. *Staphylococcus aureus* differs from *Staphylococcus epidermidis* by:
a. Is coagulase-positive
b. Forms white colonies
c. A common cause of UTI
d. Causes endocarditis in drug addicts

15. A patient in ICU is on a CVP line. His blood culture shows the growth of gram-positive cocci, which are catalase-positive and coagulase-negative. The most likely agent is:
a. *Staphylococcus aureus*
b. *Staphylococcus epidermidis*
c. *Streptococcus pyogenes*
d. *Enterococcus faecalis*

16. Which of the following gram-positive organism is most common cause of UTI among sexually active women?
a. *Staphylococcus epidermidis*
b. *Staphylococcus aureus*
c. *Staphylococcus saprophyticus*
d. *Enterococcus*

17. Food poisoning is seen with:
a. *Staph. aureus*
b. *Staph. epidermidis*
c. *Staph. pyogenes*
d. *Staph. saprophyticus*

Answers and Explanation of MCQs

1. d
• Follow section, **type of Micrococcaceae (Flowchart 49.1)** for explanation.

2. c
• Bacteria can grow best at 37°C, which may be the optimum temperature for toxin production. Food poisoning occurs after consumption of contaminated food containing preformed toxins called intradietic toxin.

346

- Toxin is heat resistant.
- Symptoms start 2–6 hours after ingestion of toxin.

3. a

- It is the endogenous infection and most common source is hospital personnel. Healthy persons (like hospital staff) carry staphylococci in nose (10–30%).
- Option c is partially correct, because epidermolysin is not superantigen only TSS toxin is superantigen.
- Methicilin resistance is mediated by chromosomes while β-lactamase (penicillinase) production is plasmid mediated.

4. c and e

- γ-hemolysin is hemolytic in nature. Leukocidin is a cytolytic toxin also called Panton-Valentine toxin. γ-hemolysin and leukocidin are composed of two proteins like S and F (S means slow and F means fast), these bicomponent membrane active toxins are grouped as synergohymenotropic toxins.

5. b, c

6. a

- Follow section, **pathogenicity [Toxic shock syndrome toxins (TSST)]** for explanation of answers of MCQs 5–6.

7. b and c

- Three types of drug resistance occur in *Staphylococcus* as mentioned earlier in text, of which β-lactamase (penicillinase) production is mediated by plasmid which is transmitted from one cell to other by transduction (more common) and conjugation.

8. c

- A majority of infections caused by CoNS are due to *Staphylococcus epidermidis,* and it is the most commonly (75–80%) isolated CoNS from clinical samples.
- β-lactamase production in staphylococci is under plasmid control while MRSA is chromosome mediated.
- Expression of methicillin resistance in *Staphylococcus aureus* increases, when it is incubated at reduced temperature (<37°C, at 30°C) on blood agar.
- Methicillin resistance having no any correlation with β-lactamase production.

9. a

- MRSA is due to alteration in penicillin binding proteins (PBP) of cell membrane by chromosomes (mec A gene).

- MRSA is not beta-lactamase.
- Resistance is mediated by chromosome.
- Expression of resistance is enhanced at reduced temperature and diagnosed best by using cefoxitin or oxacillin (6 μl/ml) discs. Oxacillin required 24 hours incubation at 30°C with supplementation of 2–4% NaCl in medium.

10. b

- First-line drugs for MRSA are vancomycin (drug of choice), linezolid, deptomycin, teicolpanin, tigecycline or quinupristin/dalfopristin.
- Imipenam is useful in methicillin sensitive *Staphylococcus aureus* but not for MRSA.

11. c

- Follow section, ***Staph. aureus* (resistance → chromosomal mediated resistance or MRSA → prevention)** for explanation.

12. a

- Botryomycosis is caused by *Staph. aureus.*

13. d

- Follow section, ***Staph. aureus* (pathogenicity → clinical features → Skin and soft tissue infections)** for explanation.

14. a

- Differences between *Staphylococcus aureus* and *Staphylococcus epidermidis* are shown in **Table 49.5**.

15. b

- *Staph. aureus* is catalase and coagulase-positive.
- Following are the features of *Staph. albus* (*Staph. epidermidis*).
  - Causes central-line-associated bloodstream infection.
  - Identified as gram-positive cocci under Gram's stain and positive blood culture with catalase-positive and negative-coagulase test.
- *Streptococcus pyogenes* and *Enterococcus faecalis* are catalase and coagulase-negative.
- More points for differentiation between GPC are shown in **Table 49.5 and Flowchart 49.3**.

16. c

- Follow section, **CoNS** for explanation.

17. a and c

- *Staph. pyogenes* is the other name for *Staph. aureus.*
- Follow section, **CoPS (pathogenicity → enterotoxin)** for explanation.

Infections of Streptococcaceae

INTRODUCTION

Definition

Bacteria belong to Streptococcaceae family are catalase-negative and arranged in pair or chain due to single plane division.

Meaning

Strepto from streptos (Greek) means easily twisted/coiled/curved/bent like a chain (twisted chain) and cocci from kokkos (Greek) means berry, as the bacteria are spherical in shape like berry and arranged in chain hence the name.

Reference Centers

Reference laboratories for Lancefield grouping

1. Lady Harding Medical College, New Delhi, India.
2. Christian Medical College, Vellore, India.

Data from reference laboratories: Data showed that about 45% of hemolytic streptococcal isolates are from group A, 10–15% from group B and C each, 25% from group G, while 5% from group F.

Uses of Streptococci

1. **Streptodornase (DNAase):** Enzyme produced by *Strept. pyogenes* (group A) is used for empyema to liquefy the thick pus.
2. **Reverse CAMP test:** CAMP factor produced by *Strept. agalactiae* (group B) is used for the detection of α-toxin of *Clostridium perfringens*. More details given in **Ch. 55**.
3. **Streptokinase (fibrinolysin):** Enzyme produced by *Strept. equisimilis* (group C) is used for thrombolytic therapy in myocardial infarction and other thrombo-embolic diseases.

4. ***Streptococcus* MG (MG means minute group) strain:** It belong to *Strept. angiosus* (group F) and used for diagnosis of primary atypical pneumonia caused by *Mycoplasma pneumoniae*.

Classification of Streptococcaceae

Follow **Flowchart 50.1**.

α-HEMOLYTIC STREPTOCOCCI

α-hemolytic streptococci include viridans group and pneumoniae group as described below.

Viridans Group

Meaning

Viridans word derived from viridis means green, because bacteria from this group produce the α-hemolytic colonies which appear in green color.

Cultural Characteristics

Green color α-hemolysis occurs on blood agar **(Fig. 50.1)** due to production of H_2O_2 by bacteria which oxidizes the Hb to green color methemoglobin.

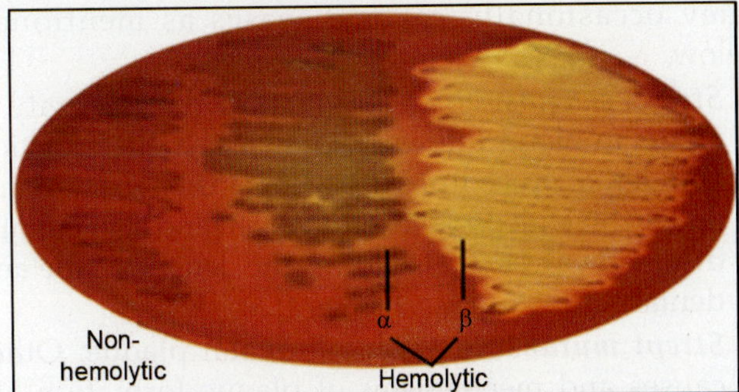

Fig. 50.1: Hemolysis on blood agar

Flowchart 50.1: Classification of streptococcaceae

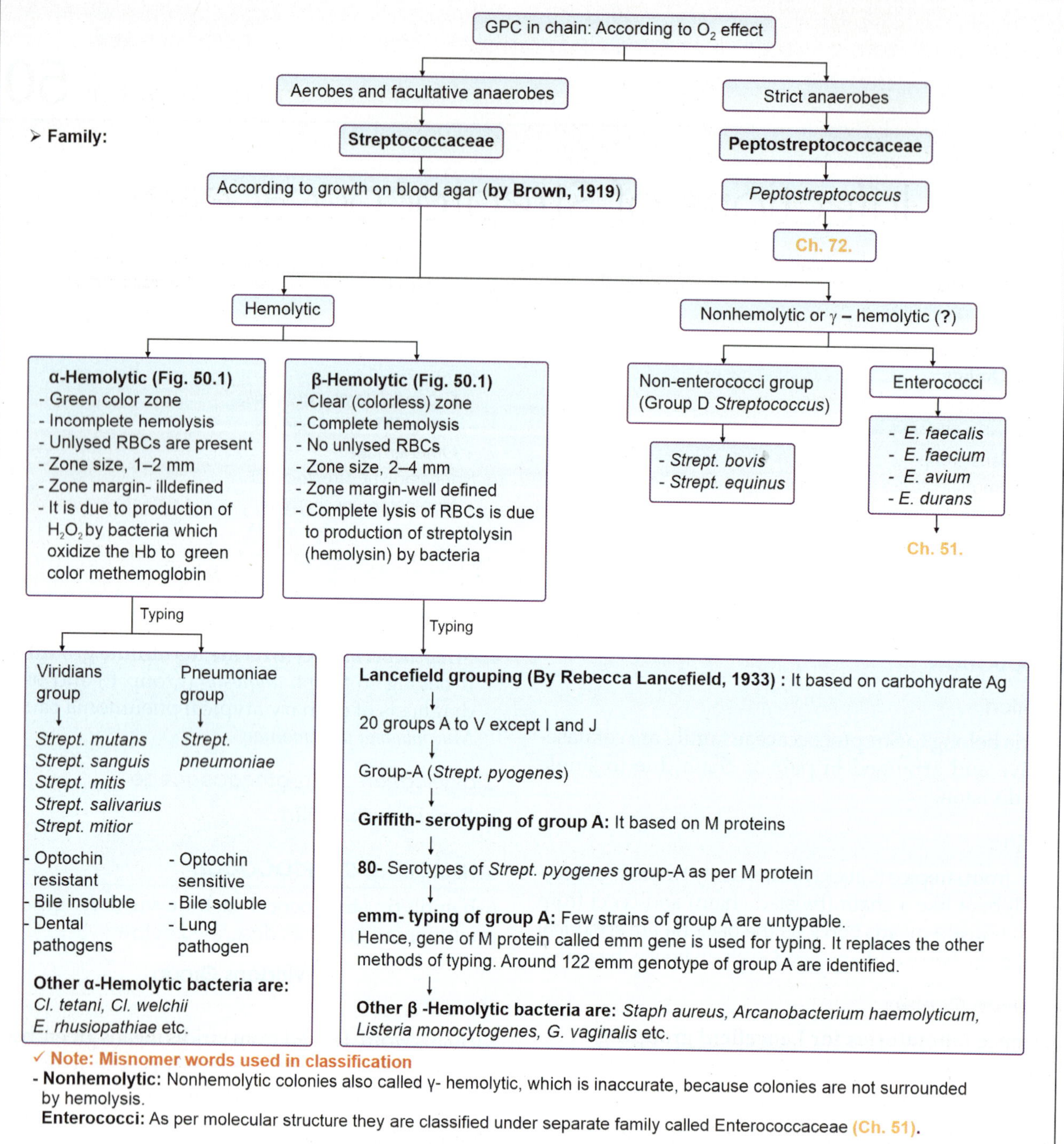

Pathogenicity

Bacteria from this group are almost nonpathogenic. They occasionally cause diseases as mentioned below.

1. *Strept. sanguis:* It is a normal flora of oral cavity. It enters in blood following chewing, tooth brushing and dental procedures to cause transient bacteremia and SABE with prosthetic, damaged and congenitally diseased valves. It also forms the dental plaque and dental caries.

2. *Strept mutans:* It produces dental plaque. Other causes and mechanisms of plaque formation are mentioned in **Flowchart 50.2.**

3. *Strept. milleri* **group:** It includes the *Strept. intermedius, Strept. angiosus* and *Strept. constellatus.* These bacteria produce brain abscess and suppurative lesions in other organs.

4. **Others species:** *Strept. mitis, Strept. salivarius, Strept. mitior,* etc.

Laboratory Diagnosis

1. **Gram's stain:** Viridans group bacteria are GPC.

2. **Culture:** Three blood samples are taken at interval of 1 hour for SABE. Bacteria produce α-hemolysis on blood agar. **Automated culture** like MALDI-TOF or VITEK is used to identify the species from culture.

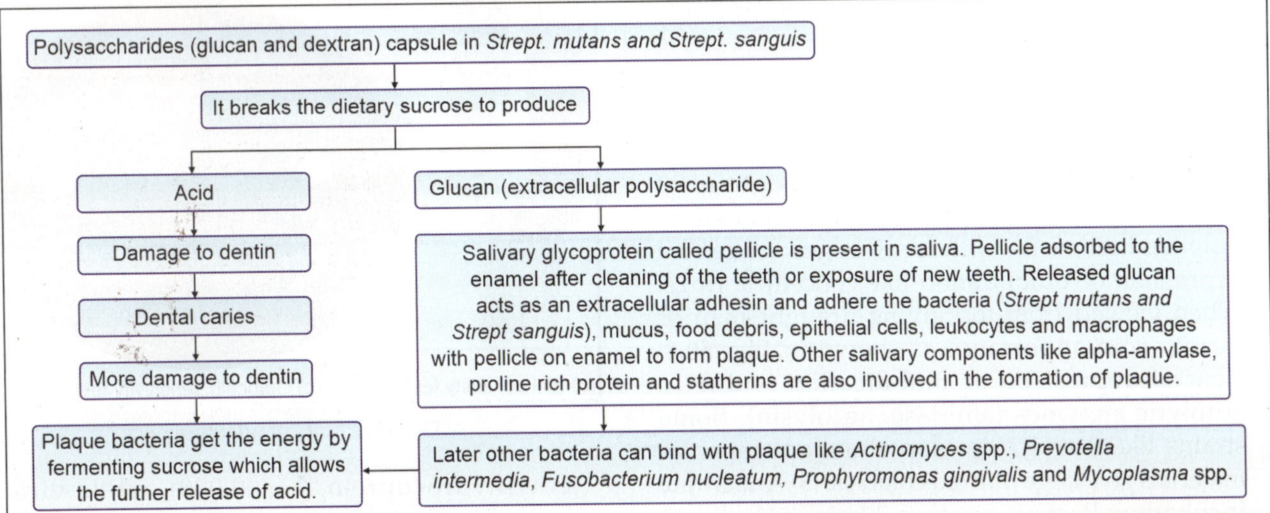

3. **Tests to differentiate pneumoniae group from viridans group:** Follow **Table 50.2**.

Pneumoniae Group (*Strept. pneumoniae*)

Synonym

It also called *Diplococcus pneumoniae* because arranged in pairs and causing pneumonia.

Morphology

Type according to Gram's stain: They are GPC.

Shape and size: They are flame or lanceolate in shape and about 1 μm in size.

Arrangement: They are arranged in pair with broad ends are apposite as shown in **Fig. 50.2**.

Motility and spores: They are nonmotile and nonsporing.

Capsule: Bacteria have polysaccharide capsule which is identified by the following tests:

1. **Gram's stain:** Capsule appears as clear halo around cocci as shown in **Fig. 50.2**.
2. **Negative stains:** Like India ink (**Fig. 50.3a**), Manveal's stain (**Fig. 50.3b**) and nigrosin stain.
3. **Quellung reaction:** It also called capsular swelling. It is described by Neufeld in 1902. Take a loopful pneumococcal suspension on a slide. Add a drop of type specific antiserum. Add loopful of methylene blue solution and observe under the microscope. Capsule appears swollen (**Fig. 50.3c**), sharply delineated, refractile and clear halo around blue

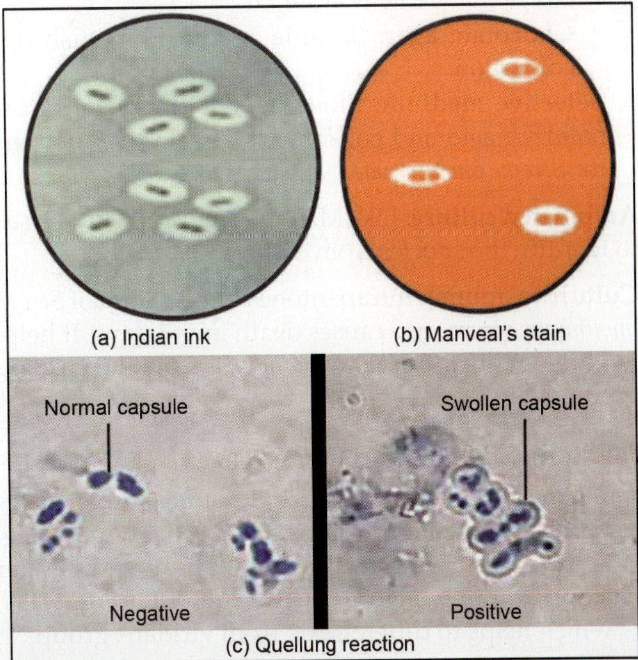

Fig. 50.3: Capsular stains (a and b) of pneumococci and Quellung reaction (c)

stained cocci. Methylene blue stains the cocci blue. It can be performed directly from sputum of acute pneumonia case. In the past it was performed bedside by using specific antiserum, used to diagnose pneumonia.

Cultural Characteristics (C/Cs)

Effective factors

- O_2 effect: They are aerobes and facultative anaerobes.
- CO_2 effect: Their growth is facilitated by 5–10 % CO_2 called capnophilic bacteria.
- Temperature: 37°C (25–42°C).
- pH: 7.8.
- They are exacting in nutritional requirements, growth occurs on media enriched with carbohydrates, blood or serum.

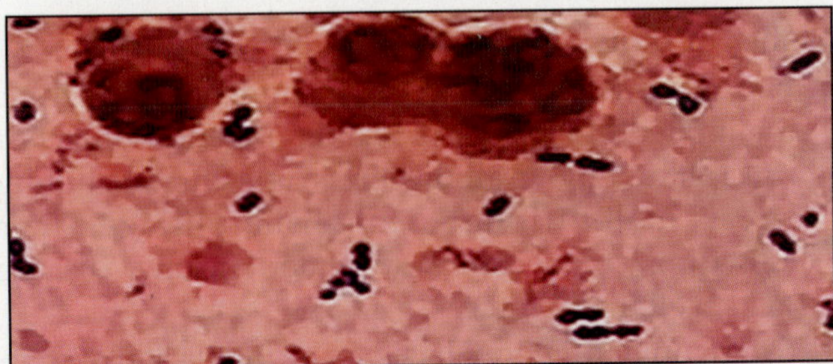

Fig. 50.2: Pneumococci under Gram's stain

5

1. **Liquid media:** Bacteria produce uniform turbidity on glucose broth.
2. **Solid media**
   - **Blood agar: (1) After 18 hours of aerobic incubation:** Colonies are small (0.5–1.0 mm), dome-shaped, glistening, α-hemolytic like viridans group of streptococci. **(2) After longer aerobic incubation:** Colonies become flat with raised edges and central depression or umbilication like concentric rings. When viewed from top giving draughtsman or carrom coin appearance as shown in **Fig. 50.4**. Central depression or umbilication is due to autolytic enzymes (amidase, autolysin). Some strains like 3 and 7 develop abundant capsular materials give large mucoid colonies. **(3) Anaerobic incubation:** Bacteria produce β-hemolysis due to oxygen labile hemolysin O.
   - **Chocolate agar:** Bacteria produce greenish discoloration.
3. **Selective medium: Blood agar with** crystal violet, nalidixic acid and colistin acts as selective medium for *Strept. pneumoniae*.

Automated culture: Like MALDI-TOF or VITEK is used to identify the species from culture.

Culture in animal: Intraperitoneal inoculation of *Strept. pneumoniae* in mouse causes death in 1–3 days. It helps to differentiate the *Strept. pneumoniae* from the viridans group where animal can survive in later.

Biochemical Reactions (B/Rs)

- **Sugar fermentation tests:** *Strept. pneumoniae* ferments numbers of sugars, producing acid only. Fermentation is tested in Hiss's serum broth/Hiss's serum agar slope. It ferments inulin with acid only **(Fig. 50.5a)** which helps to differentiate from viridans group.
- **Bile solubility test:** Follow **Table 50.1 and Fig. 50.5b**.
- **Optochin sensitivity test:** About 98% isolates of pneumococci are sensitive **(Fig. 50.5c)** to optochin

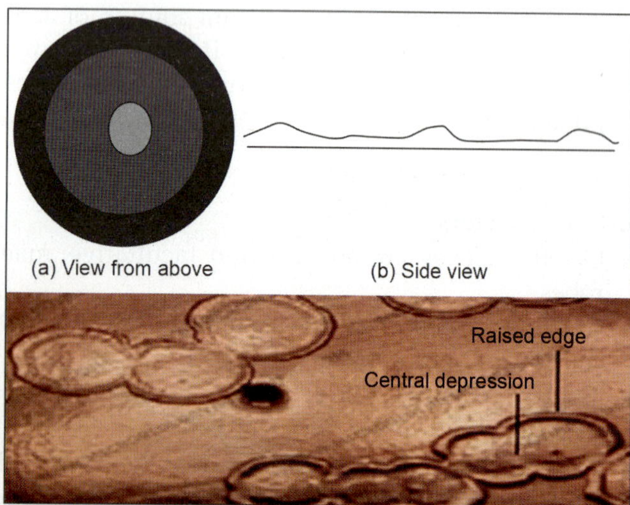

Fig. 50.4: Draughtsman appearance

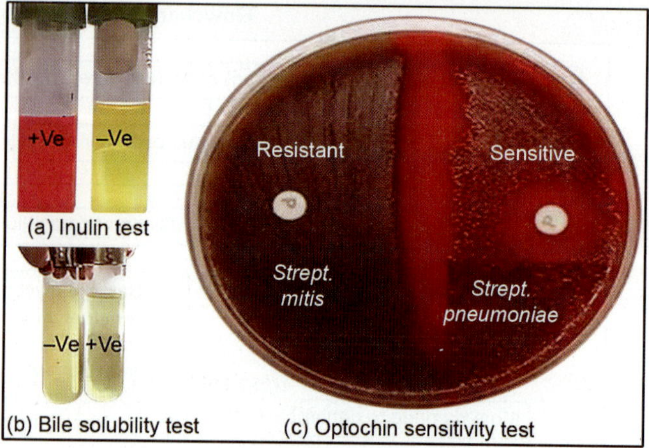

Fig. 50.5: B/Rs of pneumococci

(ethylhydrocuprein, 1/500,000)**,** while viridans group like *Strept. mitis* is resistant, which helps in differentiation of all these bacteria.

- **Catalase and oxidase test:** Both are negative.

Resistance

Sterilization: Pneumococci are sensitive to heat. Thermal death point (TDP) is 52°C for 15 minutes. They die in culture on prolonged incubation due to accumulation of toxic products. They are maintained by lyophilization or by culturing on semisolid blood agar.

Disinfection: They are destroyed by antiseptics.

Drug resistance: They are sensitive to β-lactam antibiotics; however, resistant strains start to appear since 1967. They are sensitive to optochin.

Antigenic Variation

Capsulated strains produce smooth (S) colonies and strains are virulent, on subculture they loose capsule and become rough (R), avirulent and autoagglutinable.

Pathogenicity

Disease name: Disease called pneumococcal pneumonia.

Epidemiology: In India, it occurs as sporadic or epidemic in close communities like army camps.

Reservoirs of infection: These are mostly carrier or rarely case.

Sources of infection: About 50% of human carry the pneumococci in the throat/nasopharynx (respiratory tract). Human is the source and person to person spread is possible.

Modes of transmission

- **Endogenous infection:** It occurs from nasopharynx as mentioned in **Flowchart 50.3**. Endogenous infection precipitated by viral infections, anesthesia, chilling, stress, malnutrition, alcoholism, old age, overcrowding, splenectomy, immunodeficiency disease, sickle cell disease, etc.
- **Exogenous infection:** It occurs by inhalation of nasopharyngeal/respiratory droplets.

TABLE 50.1: Bile solubility test

Features	Tube (broth) test	Culture plate test
Principle	*Strept. pneumoniae* produces autolytic enzymes (autolysin, amidase), which cleave the bond between alanine and muramic acid in peptidoglycan and cause lysis of bacteria. Bile salts (sodium deoxycholate, sodium taurocholate) activate the amidase and accelerate the lysis in culture. Test should be carried out at neutral pH, because sodium deoxycholate may precipitate at pH ≤6.5	
Reagents	• Culture suspension of *Strept. pneumoniae* in test tube. • 10% sodium deoxycholate.	• Culture of *Strept. pneumoniae* in blood agar. • 2% sodium deoxycholate.
Quality control	**Positive control:** *Strept. pneumoniae.* **Negative control:** Viridians group bacteria like *Strept. mitis.*	
Method	• Take 0.5-1 ml overnight broth culture in two test tubes. • Add 0.5 ml 10% sodium deoxycholate in one tube. • Add 0.5 ml sterile normal saline in other tube as a control. • Incubate at 37°C in incubator for 2–3 hours and read the result.	• Take two pneumococcal blood agar plates. • Add 0.5 ml 2% sodium deoxycholate in plate. • Add 0.5 ml sterile normal saline in other plate as a control. • Incubate at 37°C for 30 min in incubator (without inverting the plate) and read the results.
Results	• **Positive test:** Clearing of suspension in tube contains 10% sodium deoxycholate and no change in control tube. • **Negative test:** No change in turbidity of both tubes.	• **Positive test:** Clearing of colonies in plate contains 2% sodium deoxycholate and no change in control plate. • **Negative test:** No change in colonies in both plates.
Interpretation	**Positive test:** It indicates the *Strept. pneumoniae.* 86% strains will lyse completely while for remaining strains required additional testing like Quellung test. **Negative test:** It indicates viridans group like *Strept. mitis.*	
Use	Test helps to differentiate *Strept. pneumoniae* from viridans group.	

Flowchart 50.3: Spread of pneumococci

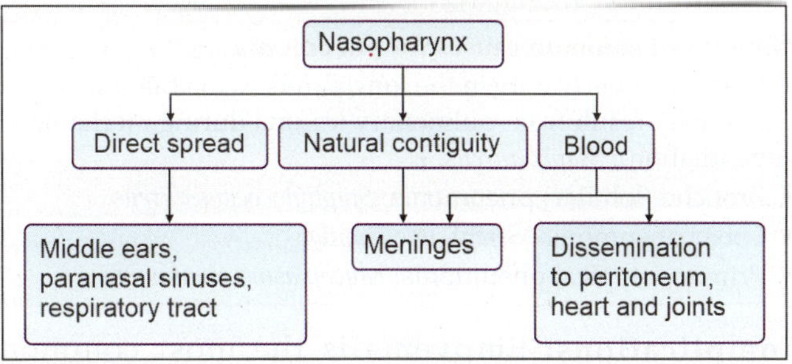

Portal of entry: Respiratory tract.

Sites: Pneumococci chiefly affect lungs but also infect other organs or tissues.

Precipitating factors (epidemiological determinants): Three types.

A. Agent factors (virulence factors): These are two types like intracellular and extracellular.

1. **Intracellular (cell wall associated) factors**
   - **Peptidoglycan:** It activates alternative pathway of complement system.
   - **Capsular Ag:** It diffused into culture medium and tissue called SSS (specific soluble substance). It inhibits the phagocytosis. It has 90 serotypes and typing is done by agglutination, precipitation or Quellung reaction (capsular swelling). Case fatality depends on virulence and serotype of bacteria. Type 3 is most virulent. Serotype 6, 14, 19F and 23F are most common in West in patient with local disease. In adults, types 1–8 are responsible for 75% cases of pneumonia with 50% fatalities. In children, types 6, 14, 19 and 23 are the major causes of disease. Anticapsular antibody is serotype specific.
   - **Techoic acid:** It is an organ of adhesion.
   - **Somatic C (carbohydrate) Ag:** It stimulates the hepatocytes to produces an abnormal protein (β-globulin), which precipitates somatic C (Carbohydrate) Ag of pneumococcus called C-reactive protein (CRP).

Flowchart 50.4: Production of CRP

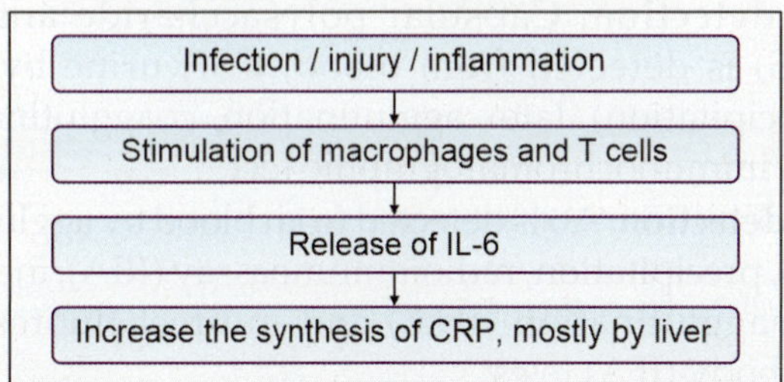

> **Note: CRP**
> - **Meaning:** So called because it precipitates somatic C (carbohydrate) Ag.
> - **History:** It was discovered by Tillett and Francis in 1930.
> - **Properties:** It is not an Ab but an abnormal, nonspecific, inflammatory protein which increased in concentration following acute inflammation, bacterial infection (like *Strept. pneumoniae*), tissue injuries, malignancy and in some other pathological conditions. Interferon-α (produced following viral infection) inhibits CRP production from liver cells which may explain the relatively low levels of CRP found during viral infections compared to bacterial infections. It is classified as positive acute phase protein or positive acute phase reactant. It is β-globulin but many workers considered it between β and γ zone. It presents in acute phase and disappears in convalescent phase. It activates the classical complement pathway. It is used as index of response in treatment of rheumatic fever and certain other conditions.
> - **Production of CRP:** Follow **Flowchart 50.4.**

- **CRP level:** Normal range of 0.2 mg/dl.
  - Mild (insignificant) increase: 0.2 – <1 mg/dl in pregnancy, heavy exercise and common cold.
  - Moderate increase: >1–10 mg/dl in pancreatitis, bronchitis, cystitis, myocardial infarction and cancer.
  - Marked increase: >10 mg/dl in acute bacterial infections, major traumas and systemic vasculitis.
- **Diagnosis:** It is one of the most common marker of acute inflammation and detected by following tests:
  - Capillary precipitation test: It is outdated method.
  - Latex (passive) agglutination test: It detects CRP up to 0.6 mg/dl and most widely used test.
  - ELISA and nephelometry: It is the most sensitive and useful to monitor the cardiovascular diseases.

2. **Extracellular factors:** Two types like enzymes and toxins as mentioned below.
   - **Enzymes**
     - **Autolytic enzymes:** Autolytic amidase activated by surface-active agents such as bile/bile salts, cleaves bond between alanine and muramic acid in peptidoglycan of cell wall, resulting lysis of organisms and release of bacterial products, which contribute in the pathogenesis of disease.
     - **IgA proteases:** It cleaves IgA and reduces the local immunity of respiratory mucosa. (other bacteria producing IgA protease are *N. miningitidis, N. gonorrheae* and *H. influenzae*).
   - **Toxins**
     - **Hemolysin:** It is O_2 labile called pneumolysin. It is weak and thiol-activated toxin. It effects on ciliary cells and PMNs (cytotoxic) and activates the classical pathway of complement by direct binding with C1q.
     - **Leukocidin:** It acts on WBCs.

B. Host factors

1. **Age:** In adults, types 1–8 are responsible for 75% cases of pneumonia with 50% fatalities. In children types 6, 14, 19 and 23 are the major cause of disease.
2. **Immunity:** Infection is precipitated by local factors like respiratory viral infections, respiratory congestion (pulmonary or mucosal), allergic disease, etc., and systemic factors like splenectomy, alcoholism, malnutrition, stress. IDDs, sickle cell disease, etc. Role of spleen in immunity is described in **Ch. 22 (section → Spleen → Clinical significance of spleen)**. Post-splenectomy patients required pneumococcal vaccine to prevent the infection.

C. Environmental factors

1. **Season:** Bacteria commonly infect in winter.
2. **Overcrowding:** Chances of infection are increased by close contact and in army camps.

Clinical features

1. **Infections in ear and paranasal sinuses:** Commonest pneumococcal infections are ASOM (most common cause in children, 33%), masked acute otitis media and sinusitis.
2. **Pneumococcal meningitis:** It causes very serious (fulminant) meningitis especially in patient of splenectomy (mortality is 20%). Most common causes for meningitis in adults are *Strept. pneumoniae* from gram-positive category and *N. meningitidis* from gram-negative category.
3. **Infections in respiratory tract:** Pneumococcus is the most common cause for lobar pneumonia and community acquired pneumonia. It also causes broncho (lobular) pneumonia for which most common cause is *Staphylococcus aureus*. Other respiratory infections produced by pneumococci are acute tracheabronchitis and acute exacerbation of chronic bronchitis which is associated with *H. influenzae*.
4. **Disseminated infections:** These are bacteremia, pericarditis, peritonitis, conjunctivitis, arthritis and endocarditis in previously damaged heart valves.
5. **Austrian syndrome:** It is a medical condition first described by Robert Austrian in 1957. It is the classical triad of pneumococcal pneumonia, endocarditis and meningitis, all caused by *Strept. pneumoniae*. It is common in alcoholic persons due to reduced spleen functions, and in males between 40 and 60 years.

Note: Most common causes for pneumonia
- Pneumatocele (cavity in the lung parenchyma filled with air that may result from pulmonary trauma during mechanical ventilation): *Staph. aureus.*
- Broncho (lobular) pneumonia: *Staphylococcus aureus.*
- Lobar pneumonia: *Strept. pneumoniae*
- Primary atypical pneumonia: *Mycoplasma pneumoniae.*

Complications: Empyema is the most common complication, characterized by collection of pus in pleural cavity. It also called pyothorax or purulent pleuritis. Mortality is 20% in pneumococcal meningitis, while 50% of survival develops acute or chronic complications like deafness, hydrocephalus and mental retardation.

Laboratory Diagnosis

Specimens: Ear discharge, laryngeal swab, sputum, CSF and blood.

Testing methods

A. Microscopy: Follow morphology.

B. Culture: Follow C/Cs.

C. Biochemical reactions: Follow B/Rs.

D. Serological tests:

1. **Quellung reaction:** Described above.
2. **Ag detection:** Capsular polysaccharide antigen (SSS) is detected from blood, CSF, urine by CIE (precipitation), latex agglutination, coagglutination and immunochromatographic test.
3. **Ab detection:** Ab is detected from blood by agglutination, precipitation, radioimmunoassay (RIA), indirect hemagglutination (IHA) and indirect fluorescent antibody (IFA) tests.

E. Molecular method: PCR.

F. Biomarkers: These are CRP and procalcitonin.

Note: Procalcitonin
- **History:** It is peptide precursor of the hormone calcitonin. It was first identified by Leonard J. Deftos and Bernard A. Roos in the 1970s.
- **Properties:** It is produced by parafollicular cells (C cells) of the thyroid and by the neuroendocrine cells of the lungs and the intestine. It is the acute phase protein (reactant).
- **Procalcitonin level:** Normal range is <0.05 ng/ml.
  - Possible local bacterial infection: 0.05–2.0 ng/ml
  - Systemic bacterial infection: 2–10 ng/ml
  - Septic shock: >10 ng/ml
- **Diagnosis:** It is diagnosed by rapid, semiquantitative immunochromatographic test from serum.

G. Tests to differentiate pneumoniae group from viridians group: Follow **Table 50.2**.

Prevention

Type specific Ab required according to capsular type. Practically, it is not possible to make complete polyvalent vaccine due to existence of 90 serotypes. Two types of pneumococcal vaccines are described below.

Polyvalent (23-valent) Polysaccharide Vaccine (PPV)
- **Preparation:** It is prepared by using capsular polysaccharide antigen of 23 most prevalent serotypes of *Strept. pneumoniae*, so also called 23 valent pneumococcal vaccine.

- **Indications:** It is not in general use but indicated to protect the high-risk individuals like absent or dysfunctional spleen, sickle cell disease, chronic renal (nephrotic syndrome), lungs and liver diseases, heart disease, coeliac disease, DM, IDDs, CSF leak due to meningeal disruption or dural tear, etc.
- **Advantages:** It provides 80–90% protection, which lasts for 5 years.
- **Contraindications:** It is contraindicated in individual with <2 year of age, lymphoreticular malignancy and immunosuppressive therapy.

Polyvalent (7-valent) conjugated vaccine
- **Preparation:** It is prepared by using capsular polysaccharide antigen of 7 most common serotypes of *Strept. pneumoniae*, which are conjugated to the CRM 197 protein (toxoid) of *C. diphtheriae*, so also called 7 valent conjugated vaccine.
- **Indication:** It used in <2 year of age; however, protection depends on serotype included in vaccine and infective serotype.
- **Contraindication:** It is contraindicated in individual <6 weeks.

Treatment

β-lactam antibiotics: *Strept. pneumoniae* is sensitive to β-lactam antibiotics. Parenteral Pn is given in severe cases, while mild cases are treated with amoxycillin. However, β-lactam resistant strains start to appear since 1967, due to alteration in penicillin binding protein.

Other antibiotics: β-lactam resistant strains are also resistant to other antibiotics like erythromycin and tetracycline. Such type of resistance strains are identified in Spain called drug resistant *Strept. pneumoniae* (DRSP) and later spread to other country. It is due to mutation or by other mechanism of gene transfer. Third-generation cephalosporin is indicated in such cases, while vancomycin is the reserved drug.

Commonly used antibiotics
- **Amoxicillin:** In otitis media, sinusitis and pneumonia.
- **Ceftriaxone and vancomycin:** In meningitis.
- **Ceftriaxone/cefotaxime and vancomycin:** In endocarditis.

β-HEMOLYTIC STREPTOCOCCI

Group A/*Strept. pyogenes*

Morphology

Type according to Gram's stain: They are GPC.

Shape and size: They are spherical or oval in shape and about 0.5–1 µm in size as shown in **Fig. 50.6**.

Arrangement: They are arranged in chains (length varies), and influenced by nature of culture medium, longer in liquid than in solid media. Chain formation is due to cocci dividing in one plane and daughter cells failing to separate completely. Other bacteria arranged

TABLE 50.2: Differences between pneumoniae group and viridans group

Features	Pneumoniae group	Viridans group
Microscopy		
• **Shape**	• Lancet/flame	• Spherical/oval
• **Arrangements**	• In pair **(Fig. 50.2)**	• In chain
• **Capsule**	• Present **(Fig. 50.3)**	• Variable
C/Cs		
• **Liquid media (Glucose broth)**	• Uniform turbidity	• Granular turbidity
• **Solid media (Blood agar)**	• Initially dome shaped and α hemolytic later carrom coin/ draughtsman appearance **(Fig. 50.4)**	• Dome shaped and α-hemolytic
• **Intraperitoneal inoculation in mouse**	• Fatal (mouse die in 1–3 days)	• Nonfatal
B /Rs: (Fig. 50.5)		
• **Inulin fermentation**	• Ferments with acid no gas	• Nonfermentative
• **Bile solubility test**	• Positive	• Negative
• **Optochin sensitivity test**	• Sensitive	• Resistance
Serological test		
Quellung test	Positive	Negative

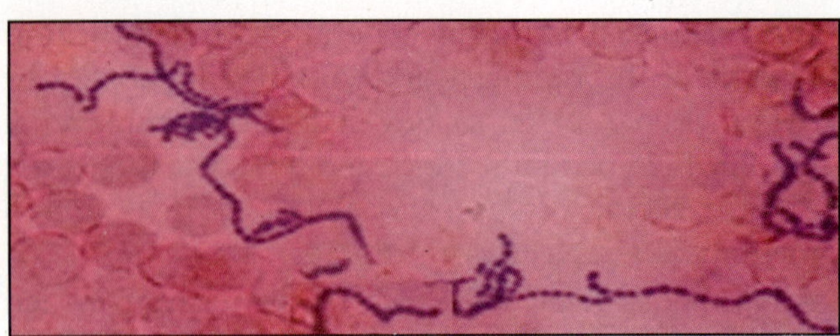

Fig. 50.6: *Strept pyogenes* under Gram's stain

in chain are *Enterococcus*, *Peptostreptococcus* and *Anthrax bacillus*.

Capsule: Group-A (*Strept. pyogenes*) and group-C (*Strept. equisimilis*) have hyaluronic acid type of capsules while groups B (*Strept. agalactiae*) and D (*Enterococcus* spp.) having polysaccharide capsules. It is best observed in young culture.

Motility and spores: They are nonmotile and non-sporing.

Fimbriae: These are the organs of adhesion consisting lipotechoic acid and mixed with M protein.

Cultural Characteristics (C/Cs)

Effective factors

- O_2/CO_2 effect: They are aerobes and facultative anaerobes. Growth and hemolysis are promoted by 10% CO_2.
- Temperature: 37°C (22–42°C).
- pH: Neutral pH
- Enriching substances: They are exacting in nutritive requirements, growth occurs only in media enriched with fermentable carbohydrates, blood or serum.

Culture in media

A. Liquid media: Granular turbidity with powdery deposit occurs in glucose broth and serum broth.

B. Solid medium: They produce β-hemolysis in blood agar **(Fig. 50.1)**.

C. Selective media
- **Blood agar with** crystal violet, nalidixic acid and colistin is selective medium for *S. pyogenes* and *S. pneumoniae*.
- **PNF medium:** Blood agar with **p**olymixin B, **n**eomycin and **f**usidic acid is also a selective medium for *S. pyogenes*.

D. Transport medium: Pike's medium is useful.

Automated culture: Like MALDI-TOF or VITEK is used to identify the species from culture.

Biochemical Reactions (B/Rs)

- **Sugar fermentation tests:** *Strept. pyogenes* ferments numbers of sugars, with acid but no gas. It does not ferment the ribose which helps to differentiate *Strept. pyogenes* from other streptococci.
- **Catalase test:** It is negative and helps to differentiate from staphylococci.

- **PYR test:** It hydrolyses the L-pyrrolidonyl β-naphthylamide (PYR) in to free β-naphthylamide by amino peptidase which produces cherry red color on addition N,N-dimethyl amino cinamaldehyde. It is useful to differentiate *Strept. pyogenes* from other streptococci.
- **Bile Solubility test:** It is negative and helps to differentiate from pneumococcus.
- **Bacitracin** (0.04 mg) **sensitivity:** It is sensitive in group A and resistant in group B, which helps in differentiation of two β-hemolytic streptococci.

Resistance

Sterilization: Streptococci are very delicate and killed by heating at 54°C in 30 minutes.

Disinfection: They are resistant to crystal violet which used as selective agent in media.

Drug resistance: They are not much resistant to antibiotics.

Pathogenicity

Epidemiology: Outbreak is common in closed communities like army camp, school boarding, etc.

Reservoirs of infection: They present as normal flora in the nose, nasopharynx and throat of human. Asymptomatic human carrier/case is the reservoir.

Sources of infection: Sources are droplets from nose, nasopharynx and throat of human. Person-to-person spread is possible.

Modes of transmission: They are transmitted by inhalation, direct contact, or fomites borne. In the tropics skin infection is due to nonbiting gnat like *Hippelates*.

Portal of entry: Respiratory tract, skin and mucosa.

Sites: Multiple systems or organs are infected.

Precipitating factors (epidemiological determinants): Three types.

A. Agent factors (virulence factors): These are two types like intracellular and extracellular.

1. **Intracellular (cell wall associated) factors**
   - **Cell wall layers:** Cell wall includes three layers from inner to outer as below **(Fig. 50.7)**.
     - **Peptidoglycan:** It provides rigidity to cell and also having pyrogenic and thrombolytic properties.
     - **Group specific carbohydrate Ag (C-Ag):** It plays important role in nonsuppurative lesions. Carbohydrate Ag is detected by capillary precipitation test or by agar gel precipitation test with specific antiserum. Organisms are grown in Todd – Hewitt broth and extracted by HCl (Lancefield's acid extraction method), by formamide (Fuller's method), by enzyme of *Streptomyces albus* (Maxted's method) or by

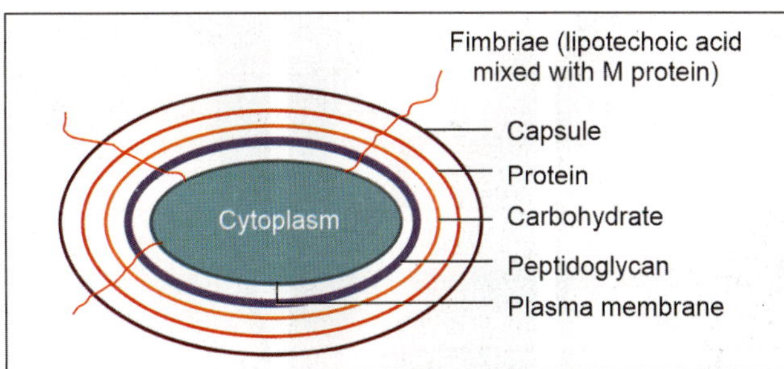

Fig. 50.7: Cell wall layers of *Streptococcus*

TABLE 50.3: Streptococcal antigenic cross reactivity	
Streptococcal Ag	**Human tissue**
Plasma membrane	Vascular intima/gomeruli
Peptidoglycan	Skin Ag
Carbohydrate (group A) Ag	Cardiac valves
M protein Ag	Myocardium
Hyaluronic acid capsule	Synovial fluid

autoclaving (Rantz and Randall's method). It helps in Lancefield grouping.

- **Protein Ags:** Three types like M, T and R proteins. **(1) M protein:** It acts as virulence factor and inhibits the phagocytosis. It is antigenic and antibody to M protein phagocytoses the cocci; hence, it is protective. It also mixed with lipotechoic acid as a structure of fimbriae. It is heat and acid stable, but sensitive to trypsin. It helps in Griffith-serotyping where M protein is extracted by Lancefield's acid extraction method and 80 serotypes of *Strept. pyogenes* group A are identified. A non-type protein associated with M protein is also identified called M associated protein. **(2) T protein:** It is acid labile and trypsin resistant. T protein is extracted by trypsin treatment to streptococci and demonstrated by slide agglutination test. It is non-virulent. **(3) R protein:** It is present in group A / *Strept. pyogenes* (2, 3, 28 and 48), group B, group C and group G. It is non-virulent.

- **Capsular Ag:** It inhibits the phagocytosis. It is not antigenic in human beings.
- **Fimbriae:** These are hair like (**Fig. 50.7**) structure consist lipotechoic acid mixed with M protein. They are helpful in attachment of streptococci with epithelial cells.
- **Other protein Ags**
  - **M associated protein:** It is described above.
  - **Protein F:** It helps in attachment with pharyngeal wall.
  - **Protein G:** It binds with Fc portion of IgG, which hinds the IgG binding site over phagocytic cells and prevents the phagocytosis.
- **Antigenic cross reactivity with human tissues:** Streptococcal antigens, which are cross reacting with human tissues are mentioned **Table 50.3.** Such cross reactivity is responsible for autoimmunity.

2. **Extracellular factors:** These are two types like enzymes and toxins as mentioned below.
- **Enzymes**
  - **Streptokinase (fibrinolysin):** It acts like plasminogen and breaks fibrin to spread the infection. It is produced by group A, C and K. It is given intravenously to treat myocardial infarction and thromboembolic disorders.
  - **Streptodornase (DNAase):** It degrades the DNA from neutrophils which is collected as pus. It helps to liquefy thick pus (serous character). This property is useful to liquefy thick pus in empyema. A preparation containing streptokinase and streptodornase is available for this purpose. It has four types like A, B, C and D. B is most common.
  - **Hyaluronidase:** It splits the hyaluronic acid which binds the connective tissues together and helps in spread of infection.
  - **Lipoproteinase (serum opacity factor/SOF):** It produces opacity when applies to agar gel contains horse or swine serum.
  - **NADase/DPNase (nicotinamide adanine dinucleotidaseddiphospho pyridine kucleotidase):** It is leukotoxic.
  - **Proteinase:** It destroys the protein.
  - **Other enzymes:** Phosphatase, esterase, amylase, N acetyl glucosaminase and neuraminidase, but their exact roles are not known.
- **Toxins**
  - **Hemolysin (streptolysin):** Two types O and S. **Streptolysin O** is O_2 labile, so active in absence of O_2 (reduced state). It is heat labile, leukotoxic, cardiotoxic, responsible for hemolysis by pour plate culture under anaerobic incubation, resembles to hemolysin of *Cl. welchii, Cl. tetani, Strept. pneumoniae* and *Cl. novyi* and detected by antistreptolysin O (ASO) test. ASO appears in serum after streptococcal infection. ASO titer >200 units is considered as significant titer. **Streptolysin S** is O_2 stable, heat stable, soluble in serum so called streptolysin S, leukocidal and responsible for hemolysis by surface (aerobic) culture.
  - **Erythrogenic toxin** or **dick toxin** or **scarletinal toxin or SPE (streptococcal pyrogenic exotoxin):** It is intradermal inoculation in susceptible individual causes erythematous reaction; hence, toxin called erythrogenic toxin and test called "dick test" (hence dick toxin). This test is used to identify the children who are susceptible to scarlet fever, so called "scarletinal toxin". It induces fever (scarlet fever) so called SPE. In addition to erythema and fever it also induces the necrotizing fasciitis and toxic shock syndrome (called streptococcal toxic shock syndrome).

It has three types like A, B and C. A and C are bacteriophage-coded, while B is chromosome-coded. Types A and C are superantigens and T cell mitogens and release large numbers of inflammatory cytokines causing fever, shock and tissue damage.

B. Host factors

1. **Age:** Infection is common between 5–8 years of age. It is less below 2 years and in adults.
2. **Immunity:** Immunity is type specific and reinfection is possible because of multiple serotypes.

C. Environmental factors

1. **Season:** Infection is common in winter in temperate countries while seasonal distribution has not been identified in tropical countries.
2. **Overcrowding:** Major factor in transmission is due to close contact. Outbreak occurs in closed communities like army camp, school boarding, etc.

Clinical features: Two types of infection like suppurative and nonsuppurative.

A. Suppurative lesions

1. **Respiratory tract infections:** Sore throat (streptococcal pharyngitis or strep throat), tonsillitis. Sometimes, acute pharyngitis presents with fever (101°F/38.3°C) and erythematous rash called scarlet fever. Streptococcal pneumonia occurs secondary to viral (like influenza) infections.
2. **Infections of skin, subcutaneous tissues, fascia and muscles**
   - **Common infections:** Bacteria produce infection of minor abrasions, burns or traumatic wounds.
   - **Cellulitis (Fig. 50.8a):** It is the infection of skin (dermis) and subcutaneous tissues. It presents with redness (generally not sharp), pain, swelling and warmth of the affected area. More common due to *Strept. pyogenes* than *Staph. aureus*.
   - **Erysipelas (Fig. 50.8b):** It is the infection of skin and superficial lymphatics. Affected skin becomes red (sharply defined from surrounding health skin), swollen and indurated. It found mainly in older patients.
   - **Lymphangitis (Fig. 50.8c):** It is the infection of lymphatic vessels. It presents with pain, swelling and redness, which run along the course of lymphatic vessels.

> **Note: Impetigo**
> It is the superficial skin infection which begins with red spot, then changes to blister (pustule), eventually breaks up, oozes fluid and develops yellowish-brown crust. It may be itchy or painful. Fever is uncommon. It mostly presents on face, neck and also on hands or legs. It founds mainly in young child. Impetigo and streptococcal lesion of scabies are the main cause of acute glomerulonephritis in children in tropics. It is further classified as nonbullous (also called contagious impetigo) and bullous impetigo as mentioned in **Table 50.4**.

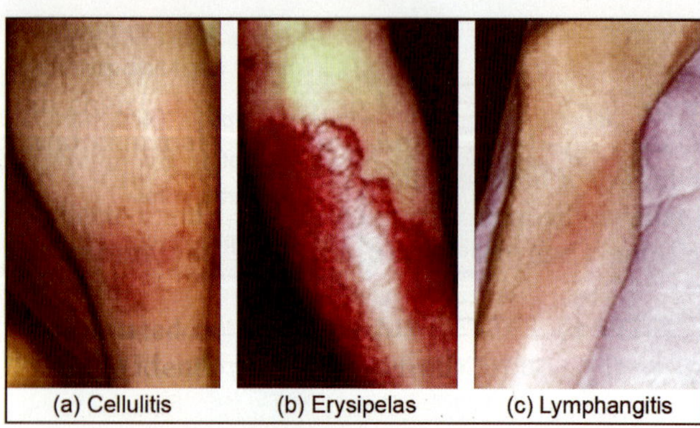

Fig. 50.8: Skin and soft tissue infections

TABLE 50.4: Types of impetigo		
Features	**Nonbullous impetigo** (Fig. 50.9a)	**Bullous impetigo** (Fig. 50.9b)
Etiology	*Staph. aureus* and *Strept. pyogenes*	*Staph. aureus*
Sites	It starts around the nose and face, but rarely affects the arms and legs	It appears in various skin areas, especially the buttocks and trunk.
Features	• It starts as small red papules like insect bites, which rapidly evolve to small blisters and then to pustules that finally scab over with a characteristic honey-colored crust. This entire process usually takes about one week. • At times, there may be nontender but swollen lymph nodes (glands) nearby.	• Bacteria produce a toxin that reduces cell to cell stickiness, causing separation between the top skin layer (epidermis) and the lower layer (dermis). This leads to the formation of a blister. (In medical terminology, blister is called bulla) • These bullae are fragile and contain a clear yellow-colored fluid. The bullae are delicate and often break with the overlying "roof" of skin lost, leaving red, raw skin with a ragged edge. A dark crust will commonly develops during the final stages of development. With healing, this crust will resolve.

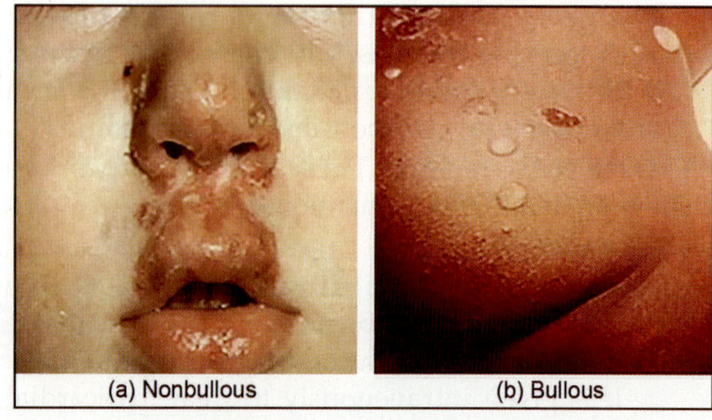

Fig. 50.9: Types of impetigo

- **Pyoderma:** Pyon (Greek) means pus and derma (Greek) means skin. It is the skin infection present with pus. It includes different clinical conditions like impetigo, folliculitis, boil (furuncle), carbuncle, tropical ulcer, pyoderma gangrenosum (autoimmune), etc., but of them impetigo is streptococcal origin. Anti-DNAase B titer and antibody to hyaluronidase are elevated, which help in diagnosis of pyoderma antecedent to acute glomerulonephritis. Moderate rise in ASO titer (not high) and do not have significant role; however, ASO test is much important in pharyngeal infection.
- **Ecthyma (deep impetigo):** It is the extension of impetigo from the epidermis to dermis. It also called deep impetigo. It presents with deeper erosion of dermis. It is caused by *Strept. pyogenes* alone or with *Staph. aureus*.

> **Note: Related terminology to ecthyma**
> - **Ecthyma diphtheriticum** or **diphtheritic whitlow:** Caused by *C. diphtheriae*.
> - **Ecthyma gangrenosum:** Caused by *P. aeruginosa*.
> - **Ecthyma contagiosum:** Caused by orf virus.
> - **Erythema marginatum:** Occurs in ARF.
> - **Erythrasma:** It is the localized infection of the stratum corneum affecting axilla and groin. It is caused by *C. minutissimum*.
> - **Erythema migrans:** Caused by *B. bergdorferi*.

- **Necrotizing fasciitis (Fig. 50.10):** It also called flesh-eating disease or flesh-eating bacteria syndrome or necrotizing soft tissue infection. It is **defined** as rapidly progressive necrosis of subcutaneous tissues, muscles and fascia. It is mixed infection **caused by** aerobic and anaerobic bacteria. M types 1 and 3 of *Strept. pyogenes* may alone be responsible for this condition which called flesh eating bacteria. Recent knowledge on the disease classified it in to following **four categories** depending on the infecting organism. **(1) Type I:** It is the most common type, caused by a mixture of bacterial types, and commonly occurs at the sites of surgery or trauma, usually in abdominal or perineal areas and accounts for 70–80% of cases. **(2) Type II:** It is caused by group A *Streptococcus* alone or often with a coinfection of *Staph. aureus*, and usually occurs on the head, neck, arm or legs. It is less often associated with predisposing risk factors such as surgery or a compromised immune system. **(3) Type III:** It is caused by *Vibrio vulnificus*, which enters the skin via puncture wounds from fish or insects in seawater. **(4) Type IV:** It is due to a fungal infection. It is **diagnosed** by isolation of *Strept. pyogenes* from the lesion, high ASO titer and high anti-DNAase B titer. It is **treated** by surgical debridement with antibiotics cover. Isolates are sensitive to penicillin *in vitro* but less effective. Combined intravenous drugs like piperacillin/tazobactam, clindamycin and vancomycin are effective.

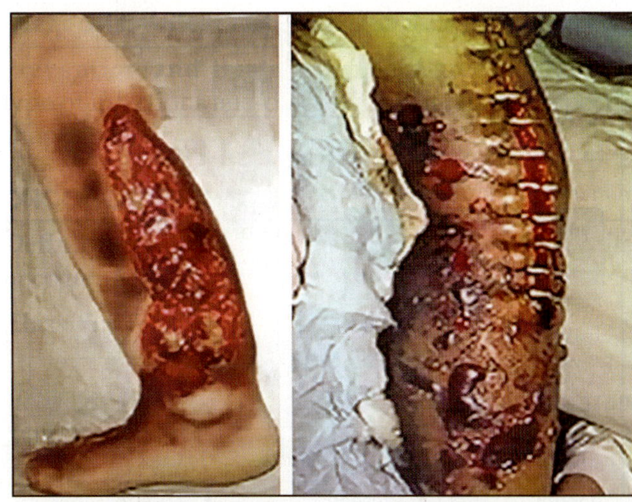

Fig. 50.10: Necrotizing fasciitis

3. **Streptococcal toxic shock syndrome:** It is characterized by DIC and multisystem failure. It is caused by strains 1, 3, 12 and 28 of M type of *Strept. pyogenes*. Sometimes, it may present with necrotizing fasciitis. It may resemble to staphylococcal TSS. Streptococcal toxic shock syndrome and necrotizing fasciitis occur in person not immune to the infecting M types.
4. **Genital infections:** These include puerperal sepsis and puerperal fever. It occurs exogenously by hospital staff and instrumentation and endogenous by anaerobic streptococci.
5. **CVS infections:** Endocarditis.
6. **Bloodstream infections:** Bacteremia, septicemia and pyemia.
7. **Eyes infections:** It produces the same lesions as *Staph. aureus* like conjunctivitis, keratitis, corneal ulcer, uveitis, blepharitis, stye (hordeolum externum), hordeolum internum, etc.
8. **Other suppurative lesions:** Abscess of brain, liver, kidneys, lungs, etc.

B. Nonsuppurative lesions: ARF and APSGN are the two nonsuppurative lesions described below with differences in **Table 50.5**.

1. **Acute rheumatic fever (ARF)**
   - **Meaning:** Rheumatic word came from rheum (Greek) means flow, as this disease flow/spread from one connective tissue to the other.
   - **Definitions:** It is an autoimmune disease of heart, skin, joints and brain characterized by fever, multiple painful joints, involuntary muscles movements and occasionally a characteristic non-itchy rash (called erythema marginatum). Basically, it is the lesion of connective tissues around the arterioles. Damage to the heart valves, in ARF called rheumatic heart disease (RHD).
   - **Onset:** It develops 2–3 weeks following repeated or persistent throat infection by any serotype of *Strept. pyogenes*.
   - **Pathogenesis:** It is an autoimmunity/type II hypersensitivity reaction due to crossreactions

TABLE 50.5: Differences between ARF and APSGN

Features	ARF	APSGN
Earlier infection	Throat	Skin or throat
Repeated attack	Common	Absent
Prior sensitization	Required	Not required
Serotypes	Any M serotype of *Strept. pyogenes*	49, 53–55, 59–61 after skin infection. 1 and 12 after pharyngitis
Pathogenesis	Type II	Type III
Immune response	Marked	Moderate
Risk factors • Age • Genetic	 • Child age • Present	 • Any age • Absent
Diagnosis • ASO titer • Anti-DNAase • Complement	 • Raised • Raised • Normal	 • Normal • Raised • Lowered
Penicillin prophylaxis	Indicated	Not indicated
Course	Progressive	Self limited
Prognosis	Poor/variable	Good

between antigens of heart, joint tissues and streptococci (M protein). Antigenic similarities between streptococci and human tissues are listed in **Table 50.3**. When antibody develops against such streptococcal antigen, it may produce the cytotoxic (type II hypersensitivity) damage to the heart and joint tissues.

- **Predisposing factors**
  - Age: It is common in children between 5 and 15 years of age. It occurs in 20% adults. It rarely presents in children below 4 years and adults above 40 years.
  - Socioeconomical factors: Poverty and over-crowding are also the contributory factors.
  - Genetic predisposition: Due to their genetics, some people are more likely to get the disease when exposed to the bacteria than others.
  - Diseases: Dental problems, bacterial endocarditis, heart transplant, artificial heart valves or congenital heart defects are also the risk factors.
- **Features (Jone's diagnostic criteria):** It was 1st published by T. Duckett Jones in 1944 in two categories like major and minor criteria.
  - **Major criteria: (1) Migratory polyarthritis (up to 75%):** It is an inflammation of large joints usually begins from legs and flow upwards. **(2) Pancarditis (up to 35%):** It is an inflammation of all three layers of heart. It is most common in myocardium called myocarditis and characterized by presence of characteristic nodules called Aschoff's nodules (described below). **(3) Subcutaneous nodules (up to 10%):** These are due to collection of collagen fibers. They present on bony prominences or tendons like back of wrists, outside the elbows and front of knees. **(4) Skin rashes (up to 10%):** Skin rash also called erythema marginatum. Red rash begins as macules on the trunk or arms spread outward and clear in the middle to form rings. It spares the face. **(5) Sydenham's chorea (up to 10%):** It is also called St. Vitus dance. It is the random, rapid and involuntary movement of the face and arms. It occurs very late in the disease mostly 3 months after the infection.
  - **Minor criteria: (1) Fever:** Temperature is 38.2–38.9°C (101–102°F). **(2) Arthralgia:** Joint pain without swelling and not the part of polyarthritis, which is the major criterion. **(3) Increased acute phase reactants** like CRP. **(4)** Leukocytosis (neutrophilia) and raised ESR. **(5) ECG features:** It suggests the 1st degree heart block, such as prolonged PR interval. This minor criterion cannot be included if major pancarditis is present. **(6) History:** Previous episode of rheumatic fever or inactive rheumatic heart disease.
- **Complications of ARF:** The heart is involved in about half of cases. Damage to the heart valves, called rheumatic heart disease (RHD), usually occurs after repeated attacks, but sometime, can occur after one. The damaged valves may result in heart failure, atrial fibrillation and infection of the valves.
- **Diagnosis of ARF:** According to revised Jone's criteria by American Heart Association, diagnosis of ARF is based on two major or one major with two minor criteria supported by laboratory evidence like raised ASO titer about >200, raised anti-DNAase titer, normal complement level and positive streptococcal culture (history of sore throat).
- **Prevention of ARF**
  - General measures: Improvement in dental health and personal hygiene.
  - Chemoprophylaxis: Penicillin is given to the patient with streptococcal sore throat infection. Long-term antibiotics are also recommended in patients with underlying disease like dental problems, bacterial endocarditis, heart transplant, artificial heart valves or congenital heart defects.

Note: Aschoff's nodule
- **Onset of Aschoff's nodule:** It develops several weeks after the onset of symptoms.
- **Three stages of Aschoff's nodule:** (1) Early/exudative/degenerative phase: It is characterized by swelling of collagen fibers around the blood vessels and lasts up to the 4th week. (2) Intermediate/proliferative/granulomatous phase: It is characterized by granulomatous reaction and occurs between 4th–13th weeks. (3) Late/fibrous/healing phase: Eventually nodule is replaced by scar.

(Contd…)

- **Macroscopy of Aschoff's nodule:** It is elliptical or fusiform in shape. It located in any layer of the heart. Occasionally presents in the pericardium. It also identified from the adventitia of the aorta.
- **Microscopy of Aschoff's nodule (Fig. 50.11):** It is composed of swollen eosinophilic collagen fibers with central fibrinoid necrosis surrounded by macrophages (called Anitschkow's cells), lymphocytes, plasma cells and multinucleated giant cells (giant cell called Aschoff's cells).
- **Aschoff's body:** Fusion of multiple Aschoff's nodules form the giant structure called Aschoff's body. It is located over bony prominences and over the tendons.
- **Anitschkow's cells (Fig. 50.12):** These are the macrophages present in heart, so called cardiac macrophages. These cells are participating in formation of Aschoff's nodules. Nuclei of these cells contain central band of clumped chromatin, that gives caterpillar like appearance in longitudinal section; hence, called caterpillar cells and owl-eye appearance in cross section.

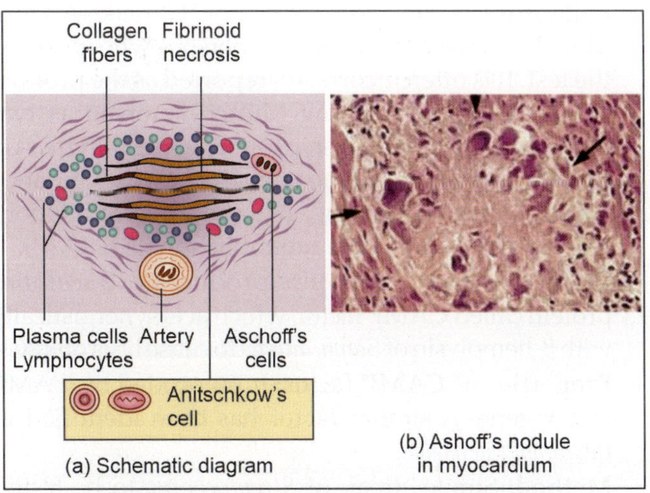

Fig. 50.11: Microscopy of Aschoff's nodule

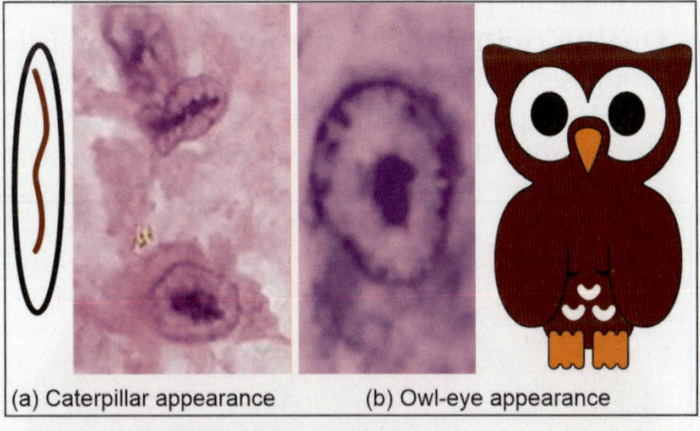

Fig. 50.12: Anitschkow's cell

- Immunoprophylaxis: No vaccines are currently available to protect against *Strept. pyogenes* infection.
- **Treatment of ARF:** Monthly injections of long-acting penicillin for five years in patients having one attack of rheumatic fever. If there is evidence of carditis, the length of therapy may be up to 40 years. It also required the continual use of low dose antibiotic like Pn, sulfadiazine or erythromycin. Anti-inflammatory like steroid or aspirin are required to reduce the inflammation.
- **Prognosis of ARF:** It is very poor.

2. Acute post-streptococcal glomerulonephritis (APSGN)

- **Onset:** It develops 2–6 weeks following skin infection (impetigo), rarely but 1–3 weeks after throat infection (pharyngitis) by M serotype of *Strept. pyogenes*. Common M serotypes are (called nephritogenic strains) 49, 53–55, 59–61 following skin plus 1 and 12 following throat infection. Prior sensitization is not necessary, it mean single infection can produce the damage.
- **Pathogenesis:** It is type III hypersensitivity reaction that occurs due to immune complex formation between streptococcal antigens and respective antibodies, which later deposited in renal glomeruli.
- **Predisposing factors**
  - Age: It occurs at any age, but common in children between 2 and 14 years of age and only 10% patients are above 40 years.
  - Socioeconomic factors: It is common in developing countries due to poverty and over-crowding.
  - Genetic predisposition: Not significant.
  - Diseases: Nothing particular.
- **Features:** It presents with hypertension, hematuria, oliguria, periorbital puffiness, anemia and flank pain.
- **Complication of APSGN:** Acute renal failure.
- **Diagnosis of APSGN:** Moderately raised or may be normal ASO titer, elevated anti-DNAase titer, elevated antibody titer to hyaluronidase, low complement level and hematuria.
- **Prevention of APSGN**
  - General measures: Improvement in personal hygiene.
  - Chemoprophylaxis: Antibiotics are given to treat skin infections.
  - Immunoprophylaxis: No vaccines are currently available to protect against *Strept. pyogenes* infection.
- **Treatment of APSGN:** It is self limited disease. Only conservative treatment is given and can heal without any permanent damage.
- **Prognosis of APSGN:** Good prognosis.

Complications: Respiratory infections produce otitis media, peritonsillar abscess, quinsy, mastoiditis, suppurative adenitis, Ludwig's angina, meningitis, cervical lymphadenitis, etc. Skin and soft tissues infections may lead to fatal septicemia.

Laboratory Diagnosis

Samples collection: Follow **Table 50.6.**

TABLE 50.6: Specimens collection

System involved	Specimens
Respiratory	Throat swab and sputum
Skin and Soft tissue	Pus/discharge by swab/syringe
Genital	Pus/discharge by swab and urine
CNS	CSF
Hematogenous and endovascular	In SABE three blood samples are collected at 1 hour interval

Testing methods

A. **Microscopy:** Follow morphology.

B. **Culture:** Follow C/Cs.

C. **Biochemical reactions:** Follow B/Rs.

D. **Serological tests**

    1. **Antistreptolysin-O (ASO) test:** Significant titer is >200 IU/ml in ARF.

    2. **Anti-DNAase B test:** Significant titer is >300 IU/ml in APSGN.

    3. **Antihyaluronidase test:** It is used for diagnosis of retrospective streptococcal infection.

    4. **Streptozyme test:** It is passive hemagglutination type test.

E. **Molecular methods:** PCR and DNA probe.

Prophylaxis

Long-term benzyl Pn is given to children to prevent rheumatic fever and reinfection. Prophylaxis is not useful for glomerulonephritis.

Treatment

Benzyl Pn is given. If allergy occurs to Pn than erythromycin or cephalexin may be used. Common modes of treatment are as follows:

- Cellulitis, erysipelas, impetigo, ecthyma and pneumonia: Local antiseptic plus systemic Pn.
- Pharyngitis: Pn
- Empyema: Pn plus drainage.
- Streptococcal toxic shock syndrome: Pn, clindamycin and intravenous Ig.

Group B/*Strept. agalactiae*

Morphology

They are GPC present in pairs or short chains and contain polysaccharide capsule.

Cultural Characteristics (C/Cs)

Blood agar: Bacteria produce larger β-hemolytic colonies about 2 mm in size with mucoid consistency. Species is identified by MALDI-TOF or VITEK.

Pigment-producing media: It was first noticed by Lancefield in 1934 in nine of twenty-four strains grown anaerobically. However, about 95–99% strains of group B can produce the pigment with modification in Islam agar. Biochemically pigment is carotenoid, produced under anaerobic incubation after 24–48 hours

incubation, bright-red in color. Pigment production is increased by cotrimoxazole. Following are the media used to detect the pigment production.

1. **Islam agar:** This medium was designed by Islam in 1977, hence the name. It is useful to detect the pigment production property of *Strept agalactiae*. Other organisms like anaerobic streptococci, *Bacteroides* and *Clostridium* species are able to grow on this medium but do not produce the pigment. Colonies of group B streptococci are 0.5–1 mm in diameter, round and pigmented (orange/red).

2. **Other media:** Columbia agar and starch serum agar.

Biochemical Reactions (B/Rs)

Hippurate hydrolysis test: Positive.

Bacitracin sensitivity test: Resistant.

PYR test: Negative

CAMP test

- **Full form:** CAMP is an acronym for "Christie–Atkins–Munch-Petersen" test, for the 3 persons who invented the test. It is often incorrectly reported as the product of four people (counting Munch-Petersen as two people). The true relationship (three people) is the reason for two en dashes and then one hyphen in "Christie–Atkins-Munch-Petersen". The name has no relationship to cyclic adenosine monophosphate (cAMP).
- **Principle:** *Strept. agalactiae* produces extracellular protein called CAMP factor, which acts synergistically with β-hemolysin of *Staph. aureus* to cause lysis of RBCs.
- **Properties of CAMP factor:** It is encoded by CAMP factor gene. A similar factor has been identified in *Bartonella henselae*.
- **Method:** Single streak of *Streptococcus* to be tested and a *Staph. aureus* are made perpendicular to each other. Left 3–5 mm distance between two streaks. Incubate the plate.
- **Quality control:** Make the control line by using group A streptococci.
- **Result:** A positive result appears as an arrowhead shaped zone (**Fig. 50.13**) of complete hemolysis.

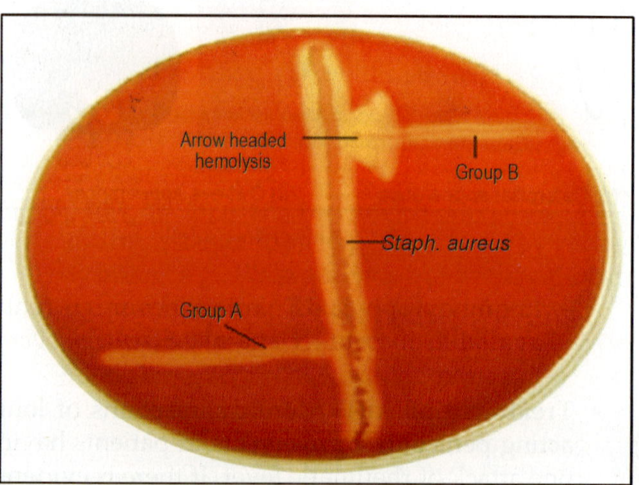

Fig. 50.13: CAMP test

- **Interpretation:** *Strept. agalactiae* is CAMP test positive, while non-group B streptococci are negative. A same test has been described for *Listeria ivanovii*, where an "arrowhead" hemolysis occurs between streaks of *Listeria ivanovii* and *Rhodococcus equi*. It also positive in *L. monocytogenes*.

Pathogenicity

Virulence factors: It has polysaccharide capsule which is virulent. Nine capsular serotypes are identified each provides the type specific protection.

Clinical features: They are the normal flora of the vagina causing following two types of infection:

- **In newborns:** It causes neonatal meningitis. It may be early or late onset type. Early onset type occurs in 1st week after birth mostly by 1a, 1b, 2, 3 and 5. It is transmitted from infected birth canal. It presents as meningitis and septicemia with 4.7% case fatality rate. Late onset type occurs between 2nd and 12th weeks of life mostly by serotype 3. It is transmitted from the environment. It presents as septicemia with 2.8% case fatality rate. Other infections in newborn are arthritis, osteomyelitis, conjunctivitis, respiratory infections, peritonitis, omphalitis, endocarditis, etc.
- **In adults:** Pneumonia and puerperal sepsis.

Diagnosis: Blood, CSF and genital discharge are collected and tested by microscopy (morphology), C/Cs and B/Rs as described earlier.

Treatment: Pn is the drug of choice for all group B infections.

Group C/*Strept. equisimilis*

It is an animal pathogen, and in human, it habitats in throat. It is the source of streptokinase, used for thrombolytic therapy. It ferments ribose and trehalose. It produces URTIs, endocarditis, osteomyelitis, brain abscess, pneumonia, puerperal sepsis and other infections. It produces streptokinase and streptolysin O, which are antigenically distinct from that produced by *Strept. pyogenes*. It is resistant to Pn, addition of gentamicin is recommended in serious cases.

Group F/*Strept. angiosus*

It grows very poorly on blood agar unless incubated under CO_2 atmosphere. They are called **minute streptococci**. One strain of this group is *Streptococcus* **MG (MG means minute group)**, used for diagnosis of primary atypical pneumonia caused by *Mycoplasma pneumoniae*. This test is called *Streptococcus* **MG test** based on detection of antibodies (agglutination reaction) to *Mycoplasma pneumoniae* by using *Streptococcus* MG strain.

Other Groups

Group G (*Strept. dysgalactiae*): It is the normal flora of throat in human, monkeys and dogs. In human, it causes UTI, endocarditis and tonsillitis.

Group H (*Strept. sanguis*) and **Group K** (*Strept. salivarius*): They may cause endocarditis.

Other includes Group M and O (*Strept. mitior*), **Group N** (*Lactococcus lactis*) and **Group R and S** (*Strept. suise*). Other *Streptococcus* species are classified as 'non-Lancefield streptococci'.

NONHEMOLYTIC STREPTOCOCCI

Two groups identified. **(1) Fecal streptococci:** Classified under a separate genus called *Enterococcus* (**Ch. 51**). **(2) Nonfecal streptococci:** Classified under streptococcal category called Group D streptococci are described below. Differences between two groups are mentioned in **Table 50.7**.

Group D/Nonenterococci Group

As per Lancefield grouping it is included in group D of β-hemolytic streptococci. *Strept. bovis* and *Strept. equines* are the species of this group. They are inhibited at high salt and 40% bile. They cause UTI and endocarditis. They are susceptible to penicillin.

COMPARISON OF STREPTOCOCCI

Follow **Table 50.8**.

TABLE 50.7: Differences between enterococci and nonenterococci group

Features	Enterococci	Nonenterococci
C/Cs		
Growth in presence of 6–5% salt, 45°C temp., high pH (9.6) and 40% bile	Yes	No
Growth on blood agar	No or α or β hemolysis	Nonhemolysis
B/Rs		
Esculin hydrolysis	Yes	No
PYR test	Yes	No
Pathogenicity		
Intestinal flora	More	Less
Pathogenicity	More	Less
Treatment		
Drug resistance	More	Less

TABLE 50.8: Comparison of streptococci

Group	Scientific name	Habitat	Colony	Tests
α-Hemolytic streptococci				
Pneumoniae group	*Strept. pneumoniae*	Lungs	α-hemolytic	Optochin: S and Bile solubility test: +ve
β-Hemolytic streptococci				
A	*Strept. pyogenes*	Throat and skin	β-hemolytic	Bacitracin: S, PYR: +ve and Trehalose: –ve
B	*Strept. agalactiae*	Genitals	β-hemolytic	Bacitracin: R, CAMP test: +ve and Hippurate hydrolysis: +ve
C	*Strept. equisimilis*	Throat	β-hemolytic	Ribose and Trehalose fermentation: +ve
Non-hemolytic streptococci				
	Enterococci **(Ch. 51)**	Colon	No or α or β hemolysis	Grow at high salt, temperature, bile and pH. PYR: +ve and Esculin hydrolysis: +ve
D	Nonenterococci	Less in colon	No hemolysis	No growth at high salt, temperature, bile and pH. PYR: –ve and Esculin hydrolysis: –ve

ACCESS YOURSELF

Case Studies

1. A female patient presents with history of fever and neck rigidity. CSF examination revealed presence of lancet shape GPC in pairs. Culture shows the α-hemolysis on blood agar. Identify the organism and answer the following:
   a. Name the causative agent and describe the morphology of causative agent.
   b. Describe the pathogenicity of causative agent.
   c. Write the laboratory diagnosis of causative agent.
2. A 10-year-old girl visited the hospital with complain of sore throat. Oral examination diagnosed the pharyngitis clinically. Throat swab examination shows the GPC in chain under Gram's stain and β-hemolysis in blood agar. Identify the organism and answer the following:
   a. Name the causative agent and describe the morphology of causative agent.
   b. Write the pathogenicity of causative agent.
   c. Describe the lab., diagnosis of causative agent.

Essays/Full Questions

1. Describe the etiology pathogenicity/pathophysiology, microscopy, criteria and complication of rheumatic fever and its diagnosis.
2. *Strept. pyogenes*/pneumococcus.

Short Notes

1. Classification of Streptococcaceae.
2. Tests to differentiate between pneumoniae group and viridians group of *Streptococcus*.
3. Bile solubility test.
4. Biochemical reactions/pathogenicity/Lab., diagnosis of *Strept. pneumoniae*.
5. Pathogenicity/Lab., diagnosis of *Strept. pyogenes*.
6. Non-suppurative complications of *Strept. pyogenes*.
7. *Strept agalactiae*.

Short Questions for Theory/Viva Questions

1. Write two uses of each, *Staph. aureus* and *S. pyogenes*.
2. Write four differences between α and β hemolysis.
3. Write four examples of α hemolytic bacteria.
4. What are Lancefield grouping and Griffith typing of streptococci?
5. Write the biochemical nature of capsule of following bacteria: *Strept. pyogenes, Strept. agalactiae, Strept. equisimilis* and *E. faecalis*
6. What is capsular swelling (Quellung reaction)?
7. What is carrom coin appearance?
8. Name the four bacteria producing the IgA protease.
9. What is Austrian syndrome?
10. Write four examples of bacteria arranged in chain.
11. Name the two bacteria (from different genera) causing toxic shock syndrome.
12. What is CAMP test?

Comments on

1. Streptococci arranged in chains.
2. α hemolytic streptococci produce green color zone, while β hemolytic streptococci produce clear zone around hemolytic colonies.
3. Longer incubation of pneumococcal culture plate produces central umbilication in colonies.
4. Adding of bile or bile salts to the pneumococcal culture suspension resulting lysis of pneumococci.

MCQs for Chapter Review

Classification of Streptococcaceae

1. **Lancefield grouping of *Streptococcus* is done by using:**
   a. M protein
   b. Group C peptidoglycan cell wall
   c. Group C carbohydrate antigen
   d. Staining properties

Viridans Group

2. **A patient of RHD developed infective endocarditis after dental extraction. Most likely organism causing this is:**
   a. *Strept viridans*
   b. *Streptococcus pneumoniae*
   c. *Streptococcus pyogenes*
   d. *Staphylococcus aureus*

Pneumoniae Group (*Strept. pneumoniae*)

3. **An infant had high grade fever and respiratory distress at the time of presentation to the emergency room. The sample collected for blood culture was subsequently positive showing growth of α-hemolytic colonies. On Gram's staining these were gram-positive cocci. In the screening test for identification, the suspected pathogen is likely to be susceptible to the following agent:**

a. Bacitracin

b. Novobiocin

c. Optochin

d. Oxacillin

4. **A patient presents with sign of pneumonia. The bacteria obtained from sputum were gram-positive cocci, which showed alpha hemolysis on sheep agar. Which of the following test will help to confirm the diagnosis?**

a. Bile solubility

b. Coagulase test

c. Bacitracin test

d. CAMP test

5. **The sputum specimen of a 7-year-old male was cultured on a 5% sheep blood agar. The culture showed α-hemolytic colonies next day. The processing of this organism is most likely to yield:**

a. Gram-positive cocci in short chains, catalase-negative and bile resistant

b. Gram-positive cocci in pairs, catalase-negative and bile soluble

c. Gram-positive cocci in clusters, catalase-positive and coagulase-positive

d. Gram-negative coccobacilli, catalase-positive and oxidase-positive

6. **True statements about pneumococcus are all *except*:**

a. Pneumolysin, a thiol-activated toxin, exerts a variety of effect on ciliary cell and PMNs

b. Autolysin may contribute to the pathogenesis of pneumococcal disease by lysing the bacteria.

c. Anticapsular antibodies are serotype specific.

d. The virulence of pneumococci depends only on the production of the capsular polysaccharide.

7. **True statements regarding pneumococcus is:**

a. Virulence is due to polysaccharide capsule.

b. Capsule is protein in nature.

c. Antibodies against capsule are not protective.

d. Resistance to antibiotics has not yet reported.

8. **Risk of pneumococcal meningitis is seen in:**

a. Post-splenectomy patient

b. Patient undergoes neurological intervention.

c. Patient following cardiac surgery

d. Patient with hypoplasia of lung

9. **In a splenectomized patient, there is increase of infection by all the organisms *except*:**

a. *Salmonella*

b. *Klebsiella*

c. *Streptococcus pneumoniae*

d. *Haemophilus influenzae*

10. **Which of the following statement about pneumococcus is false?**

a. Capsule aids in virulence.

b. Commonest cause of otitis media

c. Causes mild form of meningitis

d. Respiratory tract of carriers is the most important source of infection.

11. **Most common causative organism for lobar pneumonia is:**

a. *Staphylococcus aureus*

b. *Streptococcus pyogenes*

c. *Streptococcus pneumoniae*

d. *Haemophilus influenzae*

12. **An 8-year-old with history of pain and discharge from right ear presents with fever, neck rigidity, and a positive Kernig's sign. Discharge was stained with Gram's stain which revealed gram-positive cocci. Which of the following is most likely organism?**

a. *H. influenzae*   b. *Staphylococcus*

c. Pneumococcus   d. *Pseudomonas*

13. **C-reactive protein stands for:**

a. Capsular polysaccharide in pneumococcus

b. Concanavalin-a

c. Calretin

d. Cellular

14. **C-reactive proteins are:**

a. Alpha-globulin

b. Beta-1 globulin

c. Beta-2 globulin

d. Nonspecific inflammatory protein

15. **Following is/are true about CRP:**

a. Detected by precipitation with carbohydrate

b. Raised in acute pneumococcal infection

c. It is an antibody.

d. Detected by agglutination test

16. **C-reactive protein is:**

a. Produced by pneumococcus

b. A marker of septicemia

c. Raised in acute inflammation

d. Low in rheumatoid arthritis

17. **CRP is increased in:**

a. Acute bacterial infections

b. Malignancies

c. Major trauma

d. All of the above

18. **The 23 valent vaccine is recommended in all *except*:**

a. CSF leak

b. Chronic cardiac disease

c. Children less than 2 years

d. Nephrotic syndrome

Group A/*Strept. pyogenes*

19. ***Streptococcus pyogenes* is:**

a. GPC   b. GPB

c. GNC   d. GNB

20. **Catalase-negative β-hemolytic *Streptococcus* is:**

a. *Strept mutans*

b. *Streptococcus pneumoniae*

c. *Streptococcus pyogenes*

d. *Strept. mitior*

21. **Toxin involved in streptococcal toxic shock syndrome is:**

a. Pyrogenic exotoxin   b. Erythrogenic exotoxin

c. Hemolysin   d. Neurotoxin

22. **Which of the following streptococcal antigen cross reacts with synovial fluid?**

a. Carbohydrate (Group A)

b. Cell wall protein

c. Capsular hyaluronic acid

d. Peptidoglycan

23. ***Streptococcus*, all are true *except*:**

a. Streptodornase cleaves DNA

b. Streptolysin O is active in reduced state

c. Streptokinase is produced from serotype A, C and K

d. Pyrogenic exotoxin is plasmid-mediated.

24. The commonest organism causing cellulitis is:
a. *Streptococcus pyogenes* b. *Streptococcus faecalis*
c. *Streptococcus viridans* d. Microaerophilic streptococci

Group B/*Strept. agalactiae*

25. Scientific name for group-B, β-hemolytic *Streptococcus* is:
a. *Strept. agalactiae*   b. *Strept. pyogenes*
c. *Strept. equisimilis*   d. *Strept. angiosus*

26. A child presents with sepsis. Bacteria isolated showed beta hemolysis on blood agar, resistance to bacitracin, and a positive CAMP test. The most probable organism is:
a. *Strept. pyogenes*   b. *Strept. agalactiae*
c. *Enterococcus*   d. *Strept. pneumoniae*

Answers and Explanation of MCQs

1. c
- Follow section, **classification of Streptococcaceae (Flowchart 50.1)** for explanation.

2. a
- Viridans group includes many species like *Strept mutans, Strept. sanguis, Strept. mitis, Strept. salivarius, Strept. mitior,* etc., of which *S. sanguis* is the normal flora of oral cavity, and it enters in blood following chewing, toothbrushing and dental procedures to cause transient bacteremia and SABE (subacute bacterial endocarditis) with prosthetic, damaged and congenitally diseased valves.

3. c
- In given case, organisms are gram-positive cocci and producing alpha hemolysis on sheep agar, which include two groups like viridans group (*Strept mutans, Strept. sanguis, Strept. mitis, Strept. salivarius, Strept. mitior,* etc.) and pneumoniae group (*Streptococcus pneumoniae*). But patient presents with high grade fever and respiratory distress, so infective agent is *Strept. pneumoniae* and it is confirmed by optochin (Ethyl hydrocuprein, 1/500,000) sensitivity test. It is also confirmed by bile solubility test, but this option is not given in question.

4. a
- In given case, organisms are gram-positive cocci and producing alpha hemolysis on sheep agar, which include two groups like viridans group (*Strept. mutans, Strept. sanguis, Strept. mitis, Strept. salivarius, Strept. mitior,* etc.) and pneumoniae group (*Streptococcus pneumoniae*). But patient presents with sign of pneumonia, so infective agent is *Strept. pneumoniae,* and it is confirmed by bile solubility test. It is also confirmed by optochin (Ethylhydrocuprein, 1/500,000) sensitivity test, but this option is not given in question.

5. b
- Culture showing the α-hemolysis (greenish colonies due to incomplete hemolysis) on 5% sheep blood agar, which is given by viridans group (*Strept mutans, Strept. sanguis, Strept. mitis, Strept. salivarius, Strept. mitior,* etc.) and pneumoniae group (*Streptococcus pneumoniae*). Both are gram-positive cocci and catalase-negative but later is arranged in pairs and bile soluble.

6. d
- Pneumolysin a thiolactivated toxin. It is cytotoxic and, exerts a variety of effect on ciliary cell and PMNs. For explanation of autolysin (autolytic enzyme) follow text. Anticapsular antibodies are serotype specific. The virulence of pneumococci is not dependent only on the production of the capsular polysaccharide but also on other factors like peptidoglycan, somatic C Ag, amidase, IgA protease, hemolysin and leukocidin as described earlier in text.

7. a
- Most important virulence factor is polysaccharide capsule; however, it has other virulence factors as described earlier in text. Capsule is not in protein, but polysaccharide in nature. Antibodies against capsular antigens are protective; hence, such antigens are used in vaccine preparation. Resistance to antibiotics has been reported, as described earlier in text under the section of treatment.

8. a

9. a
- Follow section, **Pneumoniae group (*Strept. pneumoniae*) → pathogenicity (host factors)** for explanation of answers of MCQs 8–9.

10. c
- Capsule is the most important virulence factor. Commonest pneumococcal infections are otitis media and sinusitis. It causes not mild, but very serious (fulminant) meningitis. About 50% of humans carry the pneumococci in the throat/nasopharynx (respiratory tract). Respiratory tract of carriers is the most important source of infection.

11. c
- Follow section, **Pneumoniae group (*Strept. pneumoniae*) → pathogenicity (clinical features) and notes (most common causes of pneumonia)** for explanation.

12. c
- In given case, patient is suffered with otitis media and meningitis, where most likely organism is pneumococcus. Explanation is given in text under the section of clinical features.

13. All options are wrong:
- In CRP, C stands for carbohydrate antigen of cell wall of pneumococcus, neither the capsular polysaccharide Ag nor others.

14. d
- Few authors mentioned it as β-1 globulin, few as β-2 globulin, while few considered it between β and γ zone.

15. a, b and d
- CRP is detected by precipitation reaction (outdated method) with carbohydrate Ag and by agglutination test. It is not an antibody, but it is a nonspecific inflammatory protein. Its concentration raised in acute inflammation, bacterial infection (like *Strept. pneumoniae*), tissue destruction, malignancy and in some other pathological condition.

16. c
- CRP is produced by liver not by pneumococcus.

17. d
- CRP is increased in acute inflammation, bacterial infection (like *Strept pneumoniae*), tissue injuries, malignancy and in some other pathological conditions.

18. c
- Follow section, **Pneumoniae group (*Strept pneumoniae*) → prevention** for explanation.

19. a
- *Streptococcus pyogenes* is GPC arranged in chain.

20. c
- All the bacteria belong to Streptococcaceae family are catalase-negative, and from the given option all are α-hemolytic except *Streptococcus pyogenes*, which is β-hemolytic.
- Follow section, **Flowchart 50.1** and **group A/*Strept. pyogenes* (B/Rs)** for more explanation.

21. a and b
- Pyrogenic exotoxin and erythrogenic exotoxin are the synonym for one toxin. It has some other name also as

mentioned above. In addition to erythema and fever, it also induces the necrotizing fasciitis and streptococcal toxic shock syndrome.

22. c
- Follow section, **group A/*Strept. pyogenes* (pathogenicity →Table 50.3)** for explanation.

23. d
- Streptodornase (DNAase) degrades the DNA from neutrophils which is collected as pus. Streptolysin O is O_2 labile and active in absence of O_2 (reduced state). Streptokinase (fibrinolysin) is plasminogen and breaks fibrin to spread the infection, produced by group A, C, K and given IV to treat myocardial infarction and thromboembolic disorders. SPE are of 3 types like A, B and C. A and C are phage coded while B is chromosomal mediated.

24. a
- Cellulitis is mostly caused by hemolytic bacteria.

25. a
- Follow section, **comparison of streptococci (Table 50.8)** for explanation.

26. b
- *Strept. agalactiae* is causing septicemia and neonatal meningitis, giving CAMP test +ve and bacitracin resistant. Differentiation with other streptococcal species is given in **Table 50.8**.

Infections of Enterococcaceae

Chapter Outline

❑ *Enterococcus* spp.

Enterococcus spp.

Species

Family Enterococcaceae contain genus *Enterococcus*, which includes species like *E. faecalis*, *E. faecium*, *E. avium*, *E. durans*, etc.

Morphology

Enterococci are GPC, oval in shape, arranged in pair, at angle to each other or in short chain **(Fig. 51.1).**

Culture Characteristics (C/Cs)

They grow in presence of high salt concentration (6.5%), high temperature (45°C), high pH (9.6) and 40% bile. They grow on following media:

- MacConkey's agar: They produce pin point pink colonies.
- Blood agar: They are nonhemolytic. Some strains could be α or β hemolytic.
- Potassium tellurite: They produce black colonies.
- Bile esculin agar: They produce black colonies.
- Automated culture: Like MALDI-TOF or VITEK is used to identify the species of culture.

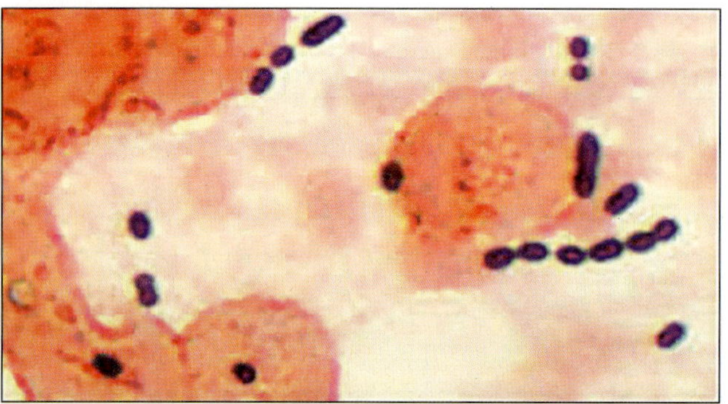

Fig. 51.1: Morphology of *Enterococcus*

Biochemical Reactions (B/Rs)

They ferment the sucrose, lactose, mannitol and sorbitol with acid only. Esculin hydrolysis and PYR tests are positive.

Pathogenicity

Virulence factors: (1) Lipotechoic acid has many functions like adhesion with host cells, regulation of autolytic enzymes of cell wall (muramidase), induction of release of cytokines (TNFα), etc. **(2) Hemolysin** lyses the RBCs in sheep and human. **(3) Extracellular surface protein** helps in adhesion. **(4) Coccolysin** is a vasoactive peptide which inactivates the endothelium. **(5) Aggregation substances** help in clumping of adjacent cells to transfer the gene like plasmid responsible for drug resistance.

Clinical features: Enterococci are the normal flora in intestine, genital tract and saliva. They cause UTI, septicemia, endocarditis, peritonitis, SABE, billiary tract infection, intra-abdominal abscesses, wound infections and suppurative infections.

Diagnosis

Specimens: Urine, blood, feces, pus/discharge from local lesions, etc.

Testing methods: Microscopy (morphology), C/Cs and B/Rs are useful test as described above.

Treatment

Enterococci are intrinsically resistant to Pn, cephalosporin and aminoglycosides. Vancomycin is the drug of choice in resistant strain; however, **vancomycin-resistant enterococci (VRE)** are emerged due to alteration/elimination of target site like D-alanyl-D-alanine chain.

Short Note

1. *Enterococcus.*

MCQ for Chapter Review

Enterococcus spp.

1. **A beta hemolytic bacterium resistant to vancomycin shows growth in 6.5% NaCl, is non bile sensitive. It is likely to be:**
   a. *Strept. agalactiae*
   b. *Strept. pneumoniae*
   c. *Enterococcus*
   d. *Strept. bovis*

Answers and Explanation of MCQ

1. c

- *Enterococcus* can grow in presence of high salt concentration (6.5%), high temperature (45°C), high pH (9.6) and 40% bile. It is nonhemolytic but some strains could be α or β hemolytic. It is intrinsically resistant to Pn, cephalosporin and aminoglycosides. Vancomycin is the drug of choice in resistant strain; however, VRE are emerged due to alteration of D-alanyl-D-alanine chain in the cell wall.

CHAPTER **52**

Infections of Neisseriae and Moraxellaceae

Chapter Outline

- Neisseriae
  - *Neisseria meningitidis*
  - *Neisseria gonorrhoeae*
  - Commensal Neisseriae
- Non-gonococcal urethritis
- Moraxellaceae
  - *Moraxella catarrhalis*
  - *Moraxella lacunata*
  - *Acinetobacter* spp.

NEISSERIAE

GNC are classified under the family Neisseriae which are catalase-positive, oxidase-positive, arranged in pairs and fastidious in growth requirements. Types of Neisseriae are mentioned in **Flowchart 52.1**.

Neisseria meningitidis

Common Name

Commonly *N. meningitidis* is known as meningococcus.

Morphology

Type according to Gram's stain: They are GNC.

Shape and size: They are half moon shape and about 0.6–0.8 µm in size.

Arrangement: They are arranged in pairs, with adjacent sides flattened as shown in **Fig. 52.1**. Cocci are

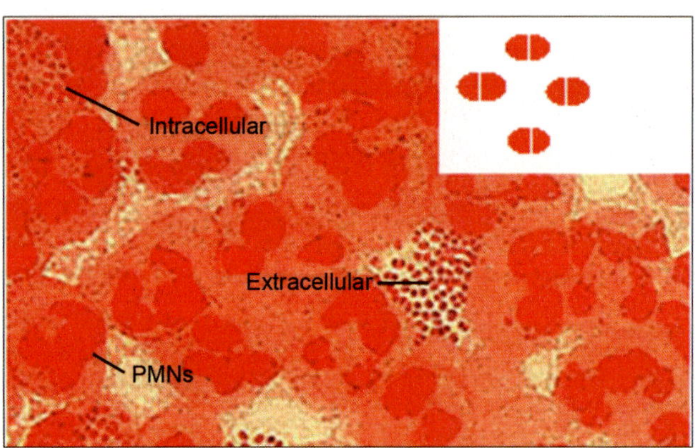

Fig. 52.1: *N. meningitidis* under Gram's stain

found intracellularly in leukocytes in CSF and often extracellularly.

Capsule: Fresh isolates are capsulated. About 60% isolates are noncapsulated.

Motility and spores: They are nonmotile and non-sporing.

Fimbriae: These are the organs of adhesion.

Culture Characteristic (C/Cs)

Effective factors

- O_2 effect: They are strict aerobes.
- CO_2 effect: Their growth is facilitated by 5–10% CO_2 called capnophilic bacteria and by high humidity.
- Temperature: 35–36°C.
- pH: 7.4–7.6.
- They are exacting in nutritional requirements. Growth occurs on media enriched with blood, serum or ascitic fluid.

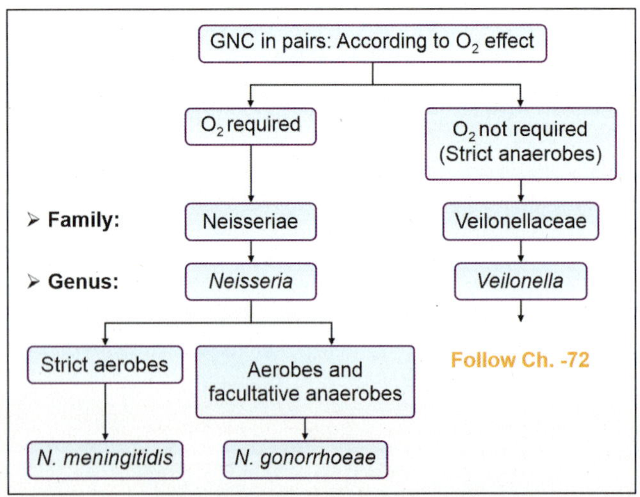

Flowchart 52.1: Types of Neisseriae

Culture in media

A. Liquid media: Poor growth occurs in liquid media, producing a granular turbidity with little or no surface growth.

B. Solid media
1. Blood agar: They produce weak hemolysis.
2. Chocolate agar and Mueller-Hinton starch casein hydrolysate agar: These are also the useful media.

C. Selective media
1. **Thayer-Martin medium:** It contains chocolate agar (heat lysed human/horse blood), vancomycin, colistin and nystatin.
2. **Modified Thayer-Martin Medium (MTMM):** It contains chocolate agar, vancomycin, colistin, nystatin and trimethoprim.
3. **Martin Lewis Medium:** It contains chocolate agar, vancomycin (higher concentration), colistin, anisomycin and trimethoprim.

Growth properties: Colonies are small (1 mm dia.) translucent, round, convex, bluish gray, with smooth glistening surface, lenticular shape, butyrous consistency and easy emulsifiable.

Automated culture: Like MALDI-TOF or VITEK is used to identify the species from culture.

Biochemical Reactions (B/Rs)

Sugar fermentation tests (Fig. 52.2): Meningococcus is oxidative, hence sugar fermentation occurs very weak, which is best tested in peptone serum agar slope contains agar and pH indicator. It ferments glucose and maltose with acid production only, but do not ferments lactose, sucrose and mannitol.

I M Vi C tests: These are not useful.

Oxidase tests: Prompt oxidase-positive by meningococcus, and gonococcus which helps in identification of these species.

Catalase test: It is positive

Resistance

Sterilization: Meningococci are easily killed by heat and drying.

Disinfection: Meningococci are killed by antiseptics and alteration in pH.

Drug resistance: They were uniformly sensitive to penicillin (Pn) and other antibiotics, but now resistant strains are common.

Pathogenicity

Disease name: Meningococci produce the infection (inflammation) of **meninges called meningococcal meningitis.**

Epidemiology: Natural infection occurs only in human. It occurs in epidemic and endemic forms.
- **World:** Infection is highest in Africa from Ethiopia to Senegal called meningitis belt of Africa. Frequent epidemics are occurring at here. Largest was in 1996, when 1,50,000 cases were reported with 15,000 deaths.
- **India:** Type A is the major cause of endemic infection in India.

Reservoir of infection: It is carrier not case. Carrier rate is 5–10% and may be high up to 90% in epidemic.

Sources of infection: Meningococci are present as normal flora in nasopharynx. Nasopharyngeal droplets of human carrier are the sources and person to person spread is possible. Case is present as a negligible source of infection.

Modes of transmission
- **Endogenous infection:** Organisms are located as normal flora in nasopharynx. Exact route for meninges involvement from nasopharynx is unknown, but possibly routes are mentioned in **Flowchart 52.2**.
- **Exogenous infection:** They entered by inhalation of nasopharyngeal/respiratory droplets or by fomites borne.

Incubation period: 3–4 days.

Portal of entry: Respiratory tract.

Sites: They mainly infect the meninges, but also infect other organs or tissues.

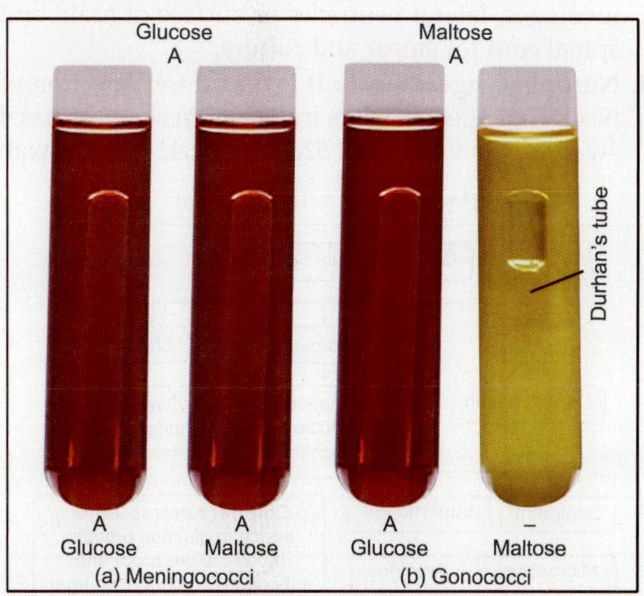

Fig. 52.2: Sugar fermentation tests

Flowchart 52.2: Spread of meningococci

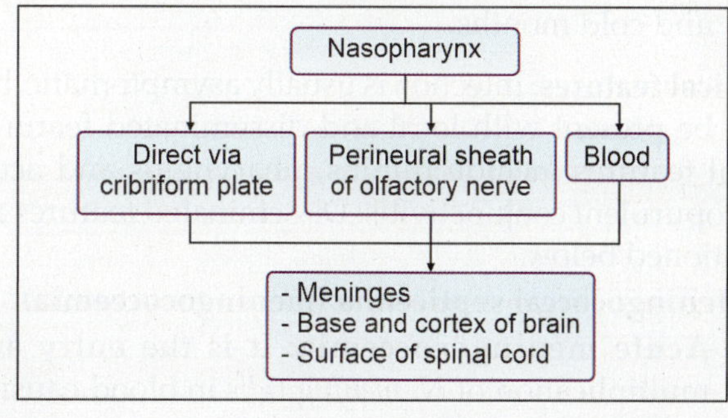

Precipitating factors (epidemiological determinants): Three types:

A. Agent factors (virulence factors): These are two types like intracellular and extracellular.

1. Intracellular (cell wall associated) factors

- **Capsular polysaccharide:** It inhibits the phagocytosis. There are total 13 serogroups like A, B, C, D, 29E, H, I, J, L, W/W-135, X, Y and Z **(A-D, H-J, X-Z, 29E, W/W-135)**. A, B, C, X, Y and W are causing invasive disease while other capsular serogroups like D, 29E, H, I, J, L and Z and noncapsular isolates are causing noninvasive disease. Group B polysaccharide is sialic acid homopolymer which is poorly immunogenic and vaccine is not available for group B *Neisseria meningitidis*. About 90% infections are due to type A (epidemic), B (epidemic and outbreaks) and C (localized outbreaks). A and W are common in developing countries, B, C and Y are common in developed countries while B is common in USA and Europe.
- **Pili (fimbriae):** They help in adhesion with host cells.
- **Lipopolysaccharide (LPS):** It is an endotoxin which is released by autolysis. It invades the cells of vascular endothelium to produce DIC and petechial hemorrhage in Waterhouse–Friderichsen syndrome. Morbidity and mortality in meningococcal meningitis and meningococcal septicemia are directly correlated with the amount of LPS.

2. Extracellular factors: It produces IgA proteases, which cleaves IgA and reduces the local immunity of respiratory mucosa.

B. Host factors

1. **Age:** It is common in children between 3 months–5 years of age.
2. **High-risk group:** They are military persons, laboratory workers, travelers to endemic areas and person with asplenia and complement components (late) C_5–C_9 deficiency.

C. Environmental factors

1. **Overcrowding:** Major factor for transmission is due to close contact especially in army camps, ships or jails.
2. **Seasonal:** Outbreak occurs more commonly in dry and cold months.

Clinical features: Infection is usually asymptomatic, but may be present with local and disseminated features. Local features include rhinitis, pharyngitis and acute mucopurulent conjunctivitis. Disseminated features are mentioned below.

1. **Meningococcal septicemia (meningococcemia)**
- **Acute meningococcemia:** It is the entry and multiplication of *N. meningitidis* in blood causing fever, malaise and petechial skin lesions. It may cause lesions in lungs, joints, adrenals and in other internal organs. It is the life-threatening septicemia occurs due to bacterial endotoxin.
- **Fulminant meningococcemia (Waterhouse-Friderichsen syndrome):** It presents with septic shock (hypotension and tachycardia), DIC and multisystem failure. It is a suppurative lesion (septicemia) and may presents with skin lesions like rash, petechiae and purpura, so also called purpura fulminans. It is mediated by LPS which stimulates the monocytes, neutrophils and endothelial cells to release the IL-1, TNF-α, IF-γ and IL-8. It is the most lethal form of septic shock and differs from other forms of septic shock by skin lesion (only meningitis presents with skin lesions) and constant development of DIC.
- **Chronic meningococcemia:** It is mostly caused by type B. It presents with fever, skin lesions and joint involvement without meningeal involvement.

2. **Meningococcal meningitis (cerebrospinal fever):** It is the suppurative lesion of meninges involving the surface of the spinal cord plus base and cortex of the brain. It presents with severe headache, rigid neck, vomiting and sensitivity to bright light accompanied with delirium, confusion and coma. Mortality is 80% in untreated cases. Epidemic is mostly caused by type A and C rare by B.

Complications and sequelae: Some cases develop chronic or recurrent meningitis. Survival may have sequelae like blindness and deafness.

Laboratory Diagnosis

Specimens

1. **CSF:** It is collected by lumbar puncture and tested as shown in **Flowchart 52.3**.
2. **Blood:** It is collected by vene puncture.
3. **Autopsy specimens:** These are collected from meninges, lateral ventricles or surface of brain and spinal cord for smear and culture.
4. **Nasopharyngeal swab:** It is useful for detection of nasopharyngeal carriers by using West's post nasal swab as shown in **Figs 52.3 and 52.4**. Collect swab

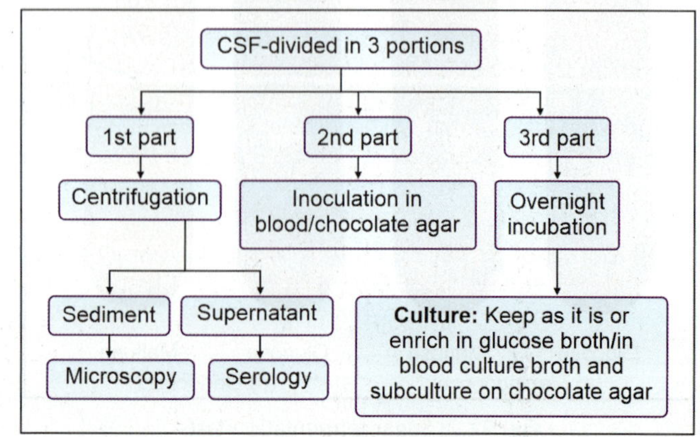

Flowchart 52.3: Testing of CSF

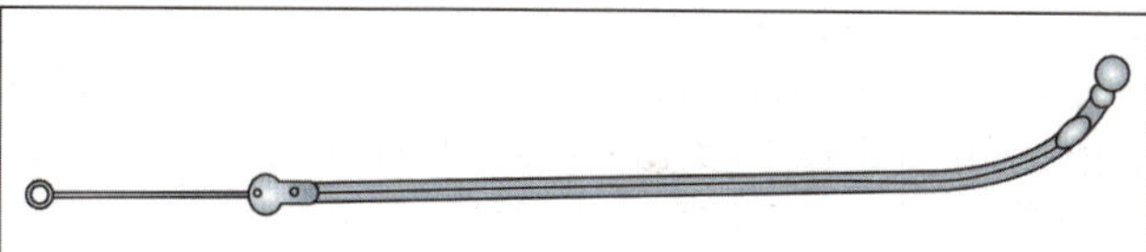

Fig. 52.3: West's postnasal swab

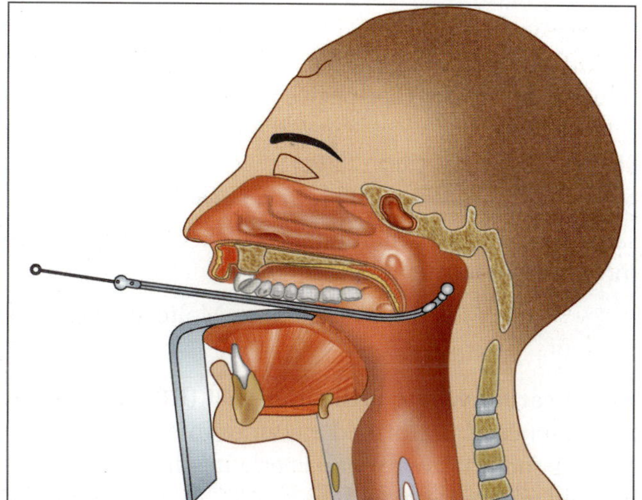

Fig. 52.4: Collection of nasopharyngeal sample by using West's postnasal swab

without contamination with saliva. Transport swab in Stuart's transport medium.

Testing methods

A. Microscopy: Follow morphology.

B. Culture: Follow C/Cs.

C. Biochemical reactions: Follow B/Rs.

D. Serological tests: Meningococcal antigens are detected by latex agglutination test and CIE.

E. Molecular method: PCR from CSF or blood.

Prophylaxis

General measures: Avoid household or close contacts with meningococcal patients.

Chemoprophylaxis: It is indicated for carriers and for persons who are in close contact to carrier. Rifampicin is the drug of choice to remove the nasopharyngeal carrier. Other drugs are ciprofloxacin, ofloxacin, azithromycin and ceftriaxone. Sulfadiazine may be used if organisms are sensitive.

Immunoprophylaxis (vaccines): No vaccine against group B because it is poorly immunogenic. Following are the different useful vaccines.

1. **Monovalent and polyvalent vaccines**
   - **Preparation:** Monovalent vaccine contains single capsular polysaccharide of group A or C and poly-valent vaccine contains capsular polysaccharide of group A, C, W-135 and Y.
   - **Indication:** For children >2 years and adults.
   - **Contraindications:** In pregnancy and in <2 years children.

   - **Disadvantages:** Immunity starts to develop after 10–14 days. It provides short-term immunity up to 3 years and not effective to nasopharyngeal carrier. It is not for journal use but useful to control the epidemic and risk groups as mentioned earlier.

2. **Conjugate vaccine**
   - **Preparation:** It contains capsular polysaccharide of group A, C, W-135 and Y are conjugated with diphtheria toxoid.
   - **Indication:** For children <2 years.
   - **Advantages:** It provides long-term immunity for >3 years. It is effective to nasopharyngeal carrier. It is available for general use and to control the epidemic and risk groups as mentioned earlier.

> **Note: Conjugate vaccines are available for**
> - *Strept. pneumoniae*
> - *N. meningitidis* (A, C, W-135 and Y)
> - *Haemophilus influenzae* type B.

Treatment

Initial therapy: It is required for case for complete recovery without sequelae. 3rd generation cephalosporin (ceftriaxone, ceftazidime, etc.) is the drug of choice for case, and it may be used for initial treatment (first 2 days).

Eradicative therapy: It is required for nasopharyngeal carrier to prevent the carrier state by eliminating the cocci. After initial therapy, eradicative therapy is given with rifampicin or ciprofloxacin.

Neisseria gonorrhoeae

Common Name

Commonly *N. gonorrhoeae* is known as gonococcus.

Morphology

Type according to Gram's stain: They are GNC.

Shape and size: They are bean or kidney-shaped and about 0.6–1 µm in size.

Arrangement: They are arranged in pairs, with adjacent sides are concave as shown in **Fig. 52.5**. They are located intracellularly in leukocytes and sometime single cell contains around hundred cocci.

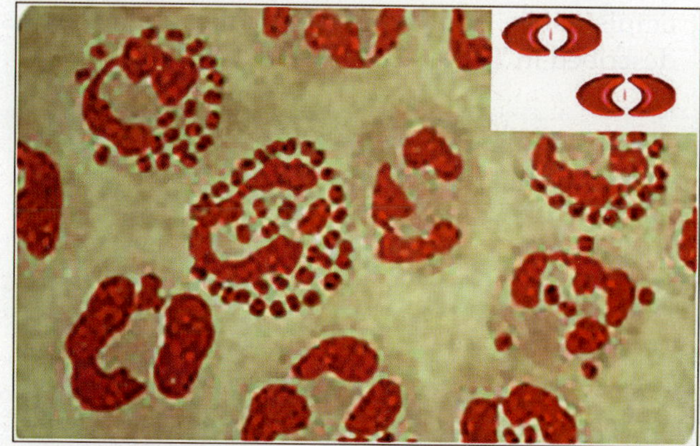

Fig. 52.5: *N. gonorrhoeae* under Gram's stain

Motility, spores and capsule: They are nonmotile, non-sporing and noncapsulated.

Fimbriae: They facilitate the adhesion of cocci to mucosal surfaces, inhibit phagocytosis and agglutinate the human RBCs called hemagglutination.

L-forms: They have cell wall deficient form, but not identified by routine tests.

Fluorescent microscopy: It is highly sensitive and specific method.

Cultural Characteristics (C/C)

Effective factors

- O_2 effect: They are aerobes and facultative anaerobes.
- CO_2 effect: Their growth is facilitated by 5–10% CO_2 called capnophilic bacteria.
- Temperature: 35–36°C.
- pH: 7.4–7.6.
- They are exacting in nutritional requirements and more difficult to grow than meningococci.

Culture in media

A. **Liquid media:** They produce poor growth in glucose broth and blood culture broth.

B. **Solid media:** They can grow on chocolate agar and Muller-Hinton agar.

C. **Selective media:**

1. **Thayer-Martin medium:** It contains chocolate agar (heat-lyzed human/horse blood), vancomycin, colistin and nystatin.
2. **Modified Thayer-Martin medium (MTMM):** It contains chocolate agar (heat-lyzed human/horse blood), vancomycin, colistin, nystatin and trimethoprim.
3. **New York city medium (NYCM):** It contains saponin lysed human/horse blood, yeast dialysate and same antibiotics as MTMM.
4. **Modified New York city medium (MNYCM):** It contains saponin lysed human/horse blood, yeast dialysate, lincomycin, colistin, amphotericin B and trimithoprim.

Growth properties: Small, round, translucent, convex or slightly umbonate colonies developed with finely granular surface, lobate margins, soft and easily emulsifiable. Four types of colonies identified as described in **Table 52.1**.

TABLE 52.1: Colonies of gonococci

T_1 and T_2	T_3 and T_4
Small	Large
Brown	Nonpigmented
Piliated cocci	Nonpiliated cocci
Autoagglutinable cocci	Nonagglutinable
Virulent cocci	Avirulent cocci
Fresh cocci	On serial subculture
T_1 called P^+ and T_2 called P^{++}	Both called P^-

D. **Transport media:** If no delay in testing (<6 hours) of specimens then it should be collected and transported in charcoal coated swab kept in Stuart's medium and charcoal containing Amie's medium. If delay in testing (>6 hours) of specimens then it should be collected and transported in commercially available transport system like JEMBEC system and Gono-Pak system.

Automated culture: Like MALDI-TOF or VITEK is used to identify the species from culture.

Biochemical Reaction (B/Rs)

Sugar fermentation tests (Fig. 52.2b): Gonococci have same properties as meningococci, except only glucose (no maltose) fermentation.

Glucose	Maltose
A	–

Rapid carbohydrate utilization test (RCUT): It based on presence of preformed enzymes in bacteria. It does not depend on growth of bacteria in sugar media. Test is rapid and more sensitive.

Catalase test: Positive.

Oxidase test: Prompt positive.

Resistance

Sterilization: Cocci die very rapidly outside the human body, so fomite-borne infection is the rare possibility. They are easily killed by heat and drying.

Disinfection: They are easily killed by antiseptics.

Drug resistance: Resistant to Pn occurs by two ways like plasmid mediated due to penicillinase production or chromosome mediated due to mutation. Resistant to fluoroquinolones occurs by two ways like alteration in DNA gyrase or by topoisomerase. Plasmid mediated resistance also occurs in tetracycline and sulfonamides.

Typing of Gonococci

1. **Serotyping:** As shown in **Flowchart 52.4**.
2. **Auxotyping:** It based on nutritional requirement of gonococci. AHU auxotype needs **a**rginine, **h**ypoxanthine and **u**racil.

Pathogenicity

Disease name: Natural disease occurs only in human (exclusive human pathogen) called gonorrhea (rrhea means discharge/flow of seeds). No natural infection occurs in animals. Mouse is infected by intracerebral inoculation and chimpanzee by urethral inoculation.

Epidemiology: Incidence of gonorrhea has been rise all over the world. WHO estimated 106 million cases globally in 2008, which was 21% higher than 2005; however, actually cases may be higher due to under-reporting and asymptomatic infections. It may be due to increase in drug resistance.

Reservoirs of infection: Asymptomatic (10% men and 50% women) carriers or cases are the reservoirs.

Sources of infection: Infected genitals or discharge from genitals act as sources of infection.

Modes of transmission: It is transmitted sexually (venereal infection) and nonvenereal infection occurs from infected birth canal to the newborn.

Incubation period: 2–8 days.

Portal of entry: Genital tract is the portal of entry in case of venereal infections and ocular organs (ophthalmia neonatorum) in case of nonvenereal infection.

Sites: It mainly affects genitals. Eyes are infected in case of nonvenereal infections.

Precipitating factors (epidemiological determinants): Three types:

A. **Agent factors (virulence factors):** These are two types like intracellular and extracellular.
  1. **Intracellular (cell wall associated) factors**
     - **Outer membrane protein (OMP):** It helps in serotyping as shown in **Flowchart 52.4**.
     - **Pili (fimbriae):** These are the organs of adhesion and made up by pilin protein.
     - **Lipopolysaccharide (LPS):** It has endotoxic activity.
  2. **Extracellular factors**
     - **IgA proteases:** It cleaves IgA, which reduces the mucosal immunity.
     - **Penicillinase (β-lactamase):** It inactivates the β-lactam antibiotics and produces the resistance to β-lactam antibiotics.

B. **Host factors**
  1. **Age:** Incidence is high between 20 and 24 years.
  2. **Sex:** Carrier rate is higher in women (50%) than men (10%).
  3. **Blood group:** It is common in blood group B, exact reason is not known.
  4. **Sexual behavior:** Unsafe sexual practise like no use of condom increases the chances of infection.
  5. **Associated disease:** Global burden of gonorrhea is increased due to increased the cases of HIV.
  6. **Others:** Like lack of health education, poor hygiene, etc., are also the contributory factors.

C. **Environmental factors:** WHO noticed highest incidence in Africa and Western pacific (China and Australia).

Pathogenesis: Follow **Flowchart 52.5**.

Clinical features: Follow **Flowchart 52.6**.

Complications
1. **Local spread:** It occurs due to natural contiguity of mucosa.
   - **Male:** In male urethritis may extends to prostate, seminal vesicles and epididymis. Chronic urethritis

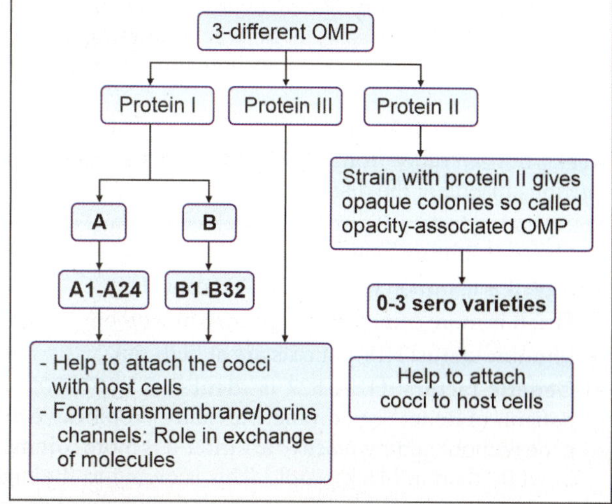

Flowchart 52.4: Outer membrane protein

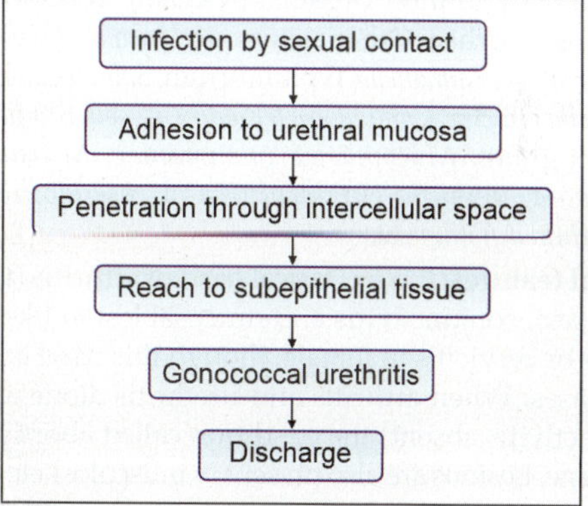

Flowchart 52.5: Pathogenesis of gonorrhea

may lead to stricture formation. It may extend to periurethral tissues, causing abscesses and multiple discharging sinuses called 'water can perineum'.
   - **Female:** In female cervicitis may extend to Bartholin's glands, endometrium and fallopian tubes. Pelvic inflammatory disease and salpingitis may lead to sterility. Sometimes, peritonitis presents with perihepatic inflammation called Fitz-Hugh-Curtis syndrome.
2. **Blood spread (metastasis):** Few strain like Por 1A, disseminated in to other organs by blood called disseminated gonococcal infections and causing arthritis, ulcerative endocarditis and rarely meningitis. It occurs in 3–5% of untreated cases.

Note: Reiter's syndrome
- **Definition:** It is an autoimmune condition characterized by classic triad of arthritis (large joints like knees), conjunctivitis and urethritis (male)/cervicitis (female) occurring after an infections, particularly from urogenital or gastrointestinal tract
- **Synonym:** It also called reactive arthritis, because arthritis in this condition is usually the result of an infection affecting another body part or called Reiter's arthritis/disease from the name of Hans Conard Julius Reiter and involvement of joints.

- **History:** It was 1st described by a German military physician and Leader of Nazi party, Hans Conard Julius Reiter in 1918. He discovered this disease, as he was examining a World War I Prussian soldier who was recuperating from a bout of diarrhea.
- **Incubation period:** Symptoms generally appear within 1–3 weeks, but can range from 4 to 35 days from the onset of the inciting episode of the disease.
- **Precipitating factors:** It can occur in epidemic form. Following are the contributory factors.
    - **Age:** It is common between 20–40 years of age.
    - **Sex:** It is more common in men than in women.
    - **Immune status:** HIV patients are at high risk.
    - **Genetic factors:** HLA-B27 is identified in 70–80% of patients of Reiter's syndrome, but many people have this gene without getting reactive arthritis. It is more common in white than in black people. This is owing to the high frequency of the HLA-B27 gene in the white population.
- **Etiological factors:** *Chlamydia trachomatis* is the most common cause of reactive arthritis following urethritis. *Shigella flexneri* is the most common organism causing reactive arthritis following diarrhea. Other agents are *Ureaplasma urealyticum*, *N. gonorrhoeae*, *Salmonella* Typhimurium, *Salmonella* Enteritidis, *Salmonella* Haldar, *Campylobacter jejuni*, *C. coli*, *Strept. pyogenes*, *viridans group*, *Mycoplasma pneumoniae*, *M. tuberculosis*, *Cyclospora*, *Yersinia enterocolitica*, *Y. pseudotuberculosis*, *Clostridium difficile*, etc.
- **Clinical features:** It is a classical triad of arthritis (large joints like knees), conjunctivitis and mucosal lesion like urethritis in male or cervicitis in female, though this triad is not found in all cases. When arthritis and urethritis alone are present (conjunctivitis absent) the condition called abortive Reiter's syndrome. Lesions are also present in musculoskeletal system, skin, nails, eyes (anterior uveitis, keratitis, scleritis, etc.), CVS, kidneys, spine (spinal inflammation called spondylo-arthropathy), etc.
- **Diagnosis:** The diagnosis is based on history and physical examination. No laboratory study or imaging finding is diagnostic. Urine, stool, joint fluid from arthrocentesis, urethral swab, throat swab or cervical swab is used for culture. Other tests are CRP, ESR, PCR for *Chlamydia*, tests to detect HLA-B27, CT and MRI.
- **Treatment:** Treatment is given as per causative agents. Symptomatic treatment includes antibiotics, NSAID, steroids, immunosuppressants and exercise to strengthen muscles and to improve joint functions.
- **Prognosis:** Prognosis is variable; 15–20% of patients may develop severe chronic sequelae.

3. **Self spread/autoinoculation/autoinfection:** Auto-inoculation by contaminated fingers in eyes in adults leads to conjunctivitis with profuse purulent discharge (blennorrhea), swollen eyelids, chemosis, corneal ulceration and rarely perforation. In adults it called ocular gonorrhea or acute purulent conjunctivitis or hyperacute bacterial conjunctivitis.

4. **Spread from pregnant lady (gonorrhea in pregnancy):** It may lead fetal loss, abortion or premature delivery. Infected birth canal may transmit the infection in to the eyes of newborn during parturition called ocular gonorrhea in neonates or called ophthalmia neonatorum. It is prevented by instillation of 1% silver nitrate solution in eyes of newborns (**Crede's method**) or by using ophthalmic preparation contains erythromycin or tetracycline. Other damages to newborn are occurring in joints and skin as mentioned in **Flowchart 52.6**.

5. **Reiter's syndrome:** Sometimes L-forms of gonococci are persist for long-time causing arthritis, conjunctivitis and urethritis/cervicitis called Reiter's syndrome.

> **Note: Reiter's syndrome, Ritter's disease and Reiter's strain**
> - **Reiter's syndrome:** Described above.
> - **Ritter's disease:** Do not get confuse with the word Ritter's disease with different spelling and different causative agent like exfoliative toxin of *Staph. aureus* (**Ch. 49**).
> - **Reiter's strain:** Ch. 73.

Laboratory Diagnosis

Specimens

1. **Acute urethritis:** Urethral discharge (50% sensitivity in culture) is collected. 1st clean the meatus with sterile gauze saline then collect the discharge directly on slide or by using loop or by using rayon/dacron swab (do not use cotton swab, because it contains inhibitory fatty acids).
2. **Chronic urethritis:** Urethral discharge is not available, so morning drop of urethral secretions or exudates following prostatic massage or urine is collected. Gonococci are identified from centrifuged deposits of urine.
3. **Asymptomatic gonorrhea:** Asymptomatic carrier of gonococcus is rare in men while clinical disease is less severe in women, many of them carry gonococci in the cervix, so swab from endocervix is the best specimens for asymptomatic gonorrhea. It has 80–90% sensitivity in culture.
4. **Eye lesions:** Purulent eye discharge is collected.
5. **Other specimens:** Cervical (endocervical) swab, vaginal swab from posterior vault, anal swab, conjunctival swab, blood and synovial fluid. Rectal swab is collected in case of anal sex.

Transport: All swabs are transported in Amies transport medium. Blood and synovial fluid are transported in trypticase soy broth.

Testing methods

A. **Microscopy:** Follow morphology.

B. **Culture:** Follow C/Cs.

C. **Biochemical reactions:** Follow B/Rs.

D. **Serological tests:** Ab is detected by using polyvalent antigens by CFT, passive (latex) agglutination test, precipitation test and RIA.

Prophylaxis

General measures: These are early detection of cases and follow-up, contact tracing, health education, good hygiene, safe sex, use of barrier contraceptives and reduction of miss use of antimicrobials.

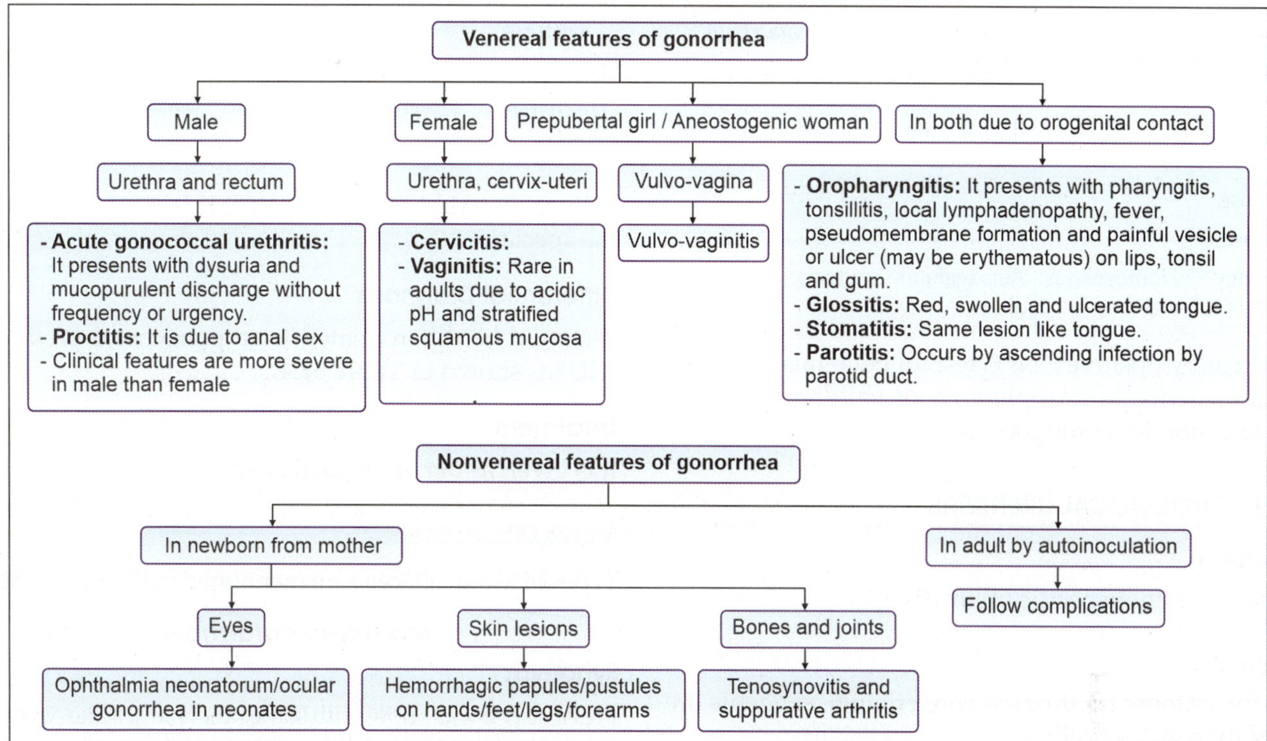

Chemoprophylaxis: Instillation of 1% silver nitrate solution in eye of newborns to prevent opthalmia neonatorum (**Crede's method**).

Immnoprophylaxis: It is ineffective and even clinical case does not provide any protection.

Treatment

Penicillin (Pn): Pn was the drug of choice since 1957, thereafter strains develop resistant to Pn. Such strains are treated with high dose, 2.4–4.8 million unit of Pn.

Regime recommended by Center for Disease Control and Prevention (CDC), USA, 1993

1. **If sensitive to ciprofloxacin:** Ceftriaxone (3rd generation cephalosporin like ceftriaxone or cefixime is the drug of choice) 125 mg single IM or ciprofloxacin 500 mg single oral dose or **ofloxacin** 400 mg single oral dose.
2. **If resistant to ciprofloxacin:** Ceftriaxone 125 mg single IM with azithromycin 1 g single oral dose or with doxycycline 100 mg twice daily for 7 days.

Differences between Meningococci and Gonococci

Follow **Table 52.2**.

Commensal Neisseriae

These are the normal flora of nasopharynx/respiratory tract and nonpathogenic, but can cause the disease. Few are rod shaped like *N. elongate* and *N. weaver,* while few are capsulated like *N. mucosa*. They can grow at lower temperature about 22°C and do not required high CO_2 concentration (not a capnophilic). They can grow on simple media like nutrient agar and cannot grow on

TABLE 52.2: Differences between meningococci and gonococci

Features	Meningococci	Gonococci
Morphology	**Fig. 52.1**	**Fig. 52.4**
• Shape	Half moon	Kidney/Bean
• Capsule	+	–
O_2 **effect**	Strict aerobes	Aerobes and facultative anaerobes
C/Cs	Easy growth	Difficult to grow
B/Rs (Fig. 2)		
Glucose	A	A
Maltose	A	–
Habitat	Nasopharynx	Genitals
Transmission	Endogenous	Exogenous
Location	Intra- and extracellular	Predominant intra-cellular
Pathogenicity	Brain and respiratory lesions	Genital lesions
Plasmid	Rarely present	Present and coded for drug resistance

selective media except *N. lactamica*. Few of them are pigmented like *N. flava* and *N. flavescens*. They ferment numbers of sugars. ONPG test is positive in *N. lactamica*. Other common features are shown in **Table 52.3**. **Pathogenic role** by few species is mentioned below.

1. *N. flavescens:* It resembles to meningococcus and reported as causative agent of meningitis. It produces yellow pigment.
2. *N. sicca:* It is avirulent and closely related to meningococcus.
3. *N. lactamica:* It is avirulent and closely related to meningococcus. It is differentiated from meningo-

Infections of Neisseriae and Moraxellaceae

TABLE 52.3: Features of commensal *Neisseria*

Species	*N. flavescen*	*N. sicca*	*M. catarrhalis*
Colonies in nutrient agar	Yellow pigment	Small, wrinkled, dry, opaque and brittle	Smooth and transparent or opaque and adherent, not easily emulsifiable
Glucose	–	A	–
Maltose	–	A	–
Serology	Homogenous Ags	Autoagglutinable	Autoagglutinable

coccus by positive ONPG test for beta-galactosidase. Its presence in nasopharynx of children provides protection to meningococcus.

NONGONOCOCCAL URETHRITIS

Synonym

It also called nonspecific urethritis.

Definition

It is the chronic urethritis where etiological agents are other than gonococci.

Etiological Agents

Bacteria: *Chlamydia trachomatis* is the most common cause in 23–55% cases. Others are *Ureaplasma urealyticum, Mycoplasma hominis, Mycoplasma genitalium, Gardnerella vaginalis* and *Acinetobacter calcoaceticus.*

Viruses: These include herpes virus, adeno virus and Cytomegalo virus (CMV).

Fungus: *Candida albicans.*

Parasite: *Trichomonas vaginalis.*

Others: NGU is caused by mechanical irritation due to catheter, cystoscope, etc., or by chemical irritation due to spermicides.

Pathogenicity

Mode of transmission: NGU is almost transmitted sexually by touching the mouth, penis, vagina or anus by penis, vagina or anus of a person who has NGU. NGU is more common in men than women.

Clinical features: Onset is longer (>1 week) and presents with dysuria, white/cloudy discharge and feeling to pass urine frequently.

Complication: Pelvic inflammatory disease.

Laboratory Diagnosis

Specimens: Urethral discharge or urine.

Testing methods: Gram's stain shows >30 pus cells/oil immersion field in urethritis. Above bacteria are difficult to identify under Gram's stain but useful to differentiate from gonococci which are stained as GNC. Other specific tests could be done as per etiological agents. Automated

culture like MALDI-TOF or VITEK is used to identify the species.

Differential Diagnosis

It should be differentiated from gonococcal urethritis (GU) as shown in **Table 52.4**.

Treatment

It is given as per etiological agents.

MORAXELLACEAE

Types of Moraxellaceae are mentioned in **Flowchart 52.7**.

Moraxella catarrhalis

Synonym

Previously it was know with some other names like *Neisseria branhmenalis* or *N. catarrhalis* or *Brahnmella catarrhalis.*

Morphology

They are GNC, oval in shape, 0.6–1 µm in size, arranged in pairs with flat adjacent surfaces as shown in **Fig. 52.6a**.

Culture Characteristics (C/Cs)

Effective factors: Their growth is facilitated by incubating under 5% CO_2 and at 18–42°C temperature.

Culture in media: They grow on nutrient agar at 18-42°C and producing nonpigmented colonies as described in **Table 52.3**. Growth also occurs on Modified Thayer-Martin medium.

Biochemical Reactions (B/Rs)

They are nonfermenters. Catalase, oxidase, tributyrin hydrolysis and DNAase tests are positive.

Pathogenicity

Modes of transmission: They are normal flora of respiratory tract and causing the endogenous infections.

Virulence factors: They secrete the β-lactamase which makes the bacteria resistant to β-lactam antibiotics.

TABLE 52.4: Differences between GU and NGU

Features	GU	NGU
Onset	48 hours	Longer, >1 week
Discharge	Purulent (flow like seeds)	White/cloudy discharge
Gram's stain	GNC in pair	Rare useful

Flowchart 52.7: Types of Moraxellaceae

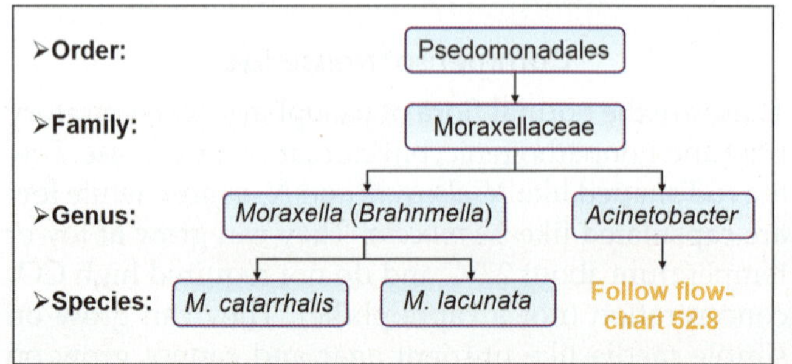

Clinical features: They are opportunistic pathogen and causing tracheobronchitis, bronchopneumonia, laryngitis, meningitis, sinusitis and otitis media. They act as common causative agent of respiratory disease in persons with COPD and old age.

Laboratory Diagnosis

Specimens: Sputum

Testing methods:
- Microscopy: Follow morphology.
- Culture: Follow C/Cs.
- Biochemical reactions: Follow B/Rs.
- Molecular method: PCR.

Treatment

Do not use β-lactam antibiotics or, if prescribed, then only in combination with sulbactam or clavulanate.

Moraxella lacunata

Synonym

Also called Morax-Axenfeld bacillus or called Morax-Axenfeld diplobacillus, because it was 1st reported by Victor Morax and Theodor Axenfeld from the case of angular conjunctivitis in 1897.

Morphology

They are GNB, rod-shaped, but becomes shorter and loose Gram's staining property if culture left out for 5 days and arranged in pairs (diplobacilli) as shown in **Fig. 52.6b**. They are nonflagellated, but sluggishly motile.

Culture Characteristics (C/Cs)

They are strict aerobes and grow on routine media like blood agar.

Biochemical Reactions (B/Rs)

They are non fermenters. Catalase, oxidase and indole tests are positive. They do not produce H₂S.

Pathogenicity

Disease name: It causes eye lesions mainly in adults, but can occur at any age called angular conjunctivitis or called catarrhal conjunctivitis or called "Morax-Axenfeld conjunctivitis".

Clinical features: Infection is characterized by chronic, mild angular blepharoconjunctivitis frequently localized on the lid at the outer canthus, typical erythema of the edges of the lids, slight maceration of the skin, most marked at the angles, especially the outer canthus, superficial infiltration of the cornea and the grayish yellow discharge, adherent to the lashes and accumulates mainly at the angles.

Laboratory Diagnosis

Specimens: Eye discharge.

Testing methods
- Microscopy: Follow morphology.
- Culture: Follow C/Cs.
- Biochemical reactions: Follow B/Rs.
- Molecular method: PCR.

Acinetobacter spp.

Classification

Initially it was grouped under tribe/family Mimeae, because mimicking GNC like Neisseriae, later classified in family Moraxellaceae. Follow **Flowchart 52.8** for more details.

Morphology

They are GNB or gram-negative coccobacilli or gram-negative diplococci like Neisseriae. They are short-stout, about 1–1.5 μm × 1.6–2.5 μm in size, capsulated and nonmotile.

Biochemical Reactions (B/Rs)

They are non fermenters, catalase-positive, oxidase negative but some strains are positive. In TSI they produce red slant/red butt (K/K or K/Nil).

Differentiating Features of Species

A baumanii: (*A. cal. anitratus, bacterium anitratum*).
- **Cultural characteristics (C/Cs):** It is strict aerobe, grow at 44°C, non-hemolytic and produces pink colony on MacConkey's medium.

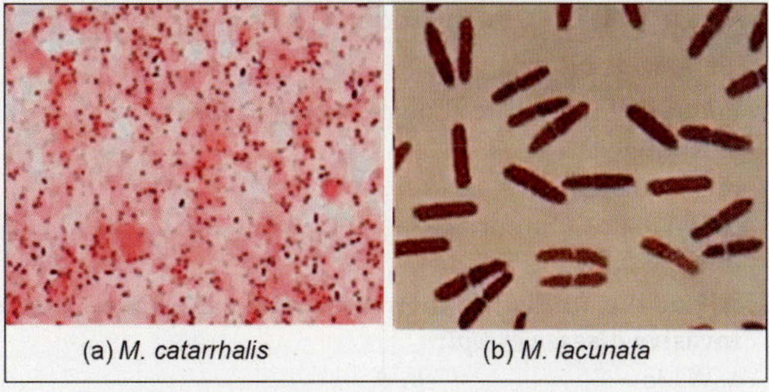

Fig. 52.6: Morphology of *Moraxella* spp.

Flowchart 52.8: Classification of *Acinetobacter*

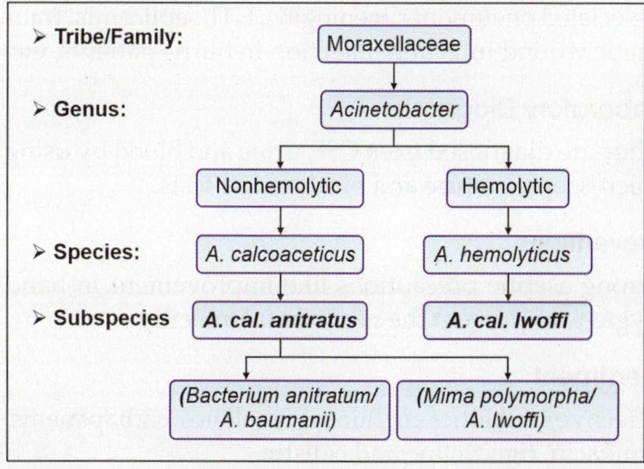

- **Biochemical reactions (B/Rs):**
  - Sugar fermentation tests: Ferments glucose, xylose, arabinose, rhamnose, 10% lactose (but not 1% lactose) with acid without gas.
  - Citrate test: Positive
  - OF test: Oxidative pattern.

A. lwoffi: (*A. cal. lwoffi, Mima polymorpha*)
- **Cultural characteristics (C/Cs):** It does not grow at 44°C, nonhemolytic and produces pale colony on MacConkey's medium.
- **Biochemical reactions (B/Rs):**
  - Sugar fermentation tests: Nonfermenter.
  - Citrate test: Negative
  - OF test: Nonfermentative pattern.

A. hemolyticus: Hemolytic colonies.

Pathogenicity

Modes of transmission

- **Endogenous infection:** These are the normal flora of skin, oral cavity and intestine. They produce infections in immunosuppressive conditions.
- **Exogenous infection:** They are also the saprophytes in soil, water and phytosphere. They are heavily available in hospital environment. They are transmitted by contact with contaminated hands of hospital staff.

Precipitating factors: Prolonged hospitalization, immunosuppressive conditions, unhygienic practice in hospital by contaminated hands of staff and warm atmosphere of hospital are the risk factors.

Virulence factors

1. **LPS:** It shows the endotoxic activities.
2. **Siderophores:** It helps in iron acquisition.
3. **Drug resistant:** It is common in hospital strains. Bacteria produce resistant to many drugs like β-lactam, aminoglycosides and quinolones. β-lactam resistance is due to production of metallo β-lactamase (MBL). Amp C β-lactamase and OXA type β-lactamase.
4. **Others:** These are ability to form biofilms and OMP antigen.

Clinical features: They produce NGU, conjunctivitis, opportunistic and nosocomial infections like ventilator-associated pneumonia, meningitis, UTI, septicemia, traumatic wound infection, infection in burns patients, etc.

Laboratory Diagnosis

They are diagnosed from CSF, urine and blood by using microscopy, culture and biochemical tests.

Prevention

Strong aseptic precautions like improvement in hand hygiene to prevent the nosocomial infections.

Treatment

Effective antibiotics are fluoroquinolones, carbapenems, amikacin, tigecycline and colistin.

ACCESS YOURSELF

Case Studies

1. An 8-year-old boy carried to emergency department with fever since 1 day along with headache, vomiting and disorientation. CSF is collected and examined by Gram's stain, which shows the capsulated gram-negative cocci in pair. Identify the organism and answer the following.
   a. Name the causative agent and describe the morphology of causative agent.
   b. Write the pathogenicity of causative agent.
   c. Describe the immunoprophylaxis of causative agent.
2. A male patient visited the STD clinic with complain of urethral discharge. History ruled out the sexual activity with many partners. Urethral discharge is collected and examined microscopically under Gram's stain which shows the gram-negative diplococci. Identify the organism and answer the following.
   a. Name the causative agent and describe the morphology of causative agent.
   b. Write the pathogenicity of causative agent.
   c. Describe the lab., diagnosis of causative agent.
3. A man presents with complain of redness, lacrimation and discharge from his left eye. Later, he developed perforation. Gram's stain from discharge revealed noncapsulated gram-negative cocci which are oxidase positive. Identify the organism and answer the following.
   a. Name the causative agent and describe the B/Rs of causative agent.
   b. Write the pathogenicity of causative agent.
   c. Mention the bacteria infecting eyes with the name of disease.

Essays/Full Questions

1. *N. meningitidis*
2. *N. gonorrhoeae.*

Short Notes

1. Pathogenicity/Lab., diagnosis of *N. meningitidis*
2. Pathogenicity/Lab., diagnosis of *N. gonorrhoeae*
3. Nongonococcal urethritis (NGU)
4. *Acinetobacter.*

Short Questions for Theory/Viva Questions

1. Write the two morphological differences between meningococcus and gonococcus.
2. What are Ritter's disease and Reiter's syndrome?
3. Name four bacteria causing Reiter's syndrome.
4. Name the four bacteria causing nongonococcal urethritis.
5. Name the four eye infecting bacteria with disease name.

Comment on

1. Vaccine is not available for group B *Neisseria meningitidis*.

MCQs for Chapter Review

Neisseria Meningitidis

1. **Following is/are true about *N. meningitidis*:**
   a. Arranged in pairs
   b. Capsulated
   c. Flat apposition surface
   d. a + b + c
2. **All of the following meningococcal serogroups cause invasive disease *except*:**
   a. W135   b. A
   c. D   d. Y

3. **Following statements about meningococcal meningitis are true *except*:**
   a. Source of infection is mainly clinical cases
   b. The disease is more common in dry and cold months of the year
   c. Chemoprophylaxis of close contact of cases is recommended
   d. The vaccine is not effective in children below 2 years of age

4. **Xavier and Yogender stay in the same hostel of same university. Xavier develops infection due to group B meningococcus. After few days Yogender develops infection due to group C meningococcus. All of the following statements are true *except*:**
   a. Educate students about meningococcal transmission and take preventive measures
   b. Chemoprophylaxis to all against both group B and group C
   c. Vaccine prophylaxis of contacts of Xavier
   d. Vaccine prophylaxis of contacts of Yogender

5. **Conjugate vaccines are available for the prevention of invasive disease caused by all of the following statements *except*:**
   a. *H. influenzae*
   b. *Strept. pneumoniae*
   c. *N. meningitidis* (group C)
   d. *N. meningitidis* (group B)

6. **Treatment of choice for meningococcal infection is:**
   a. Tetracycline
   b. Clindamycin
   c. Gentamicin
   d. Cephalosporin

7. **Young female with 3-day fever presents with headache, BP 90/60 mm Hg, heart rate of 140/min, and pinpoint spots developed distal to BP cuff. Most likely organism is:**
   a. *Burcella abortus*
   b. *Burcella suis*
   c. *N. meningitidis*
   d. *Staphylococcus aureus*

Neisseria gonorrhoeae

8. **Which of the following statement is not true about *Neisseria gonorrhoeae*?**
   a. It is an exclusive human pathogen
   b. Some strains may cause disseminated disease
   c. Acute urethritis is the most common manifestation in males
   d. All strains are highly sensitive to penicillin

9. **The virulence factors for *Neisseria gonorrhoeae* include all of the following *except*:**
   a. Outer membrane proteins
   b. IgA protease
   c. M protein
   d. Pilli

10. **Which of the following literally means "flow of seed"?**
    a. *Anthrax*
    b. *Clostridium*
    c. Gonorrhea
    d. *Proteus*

11. **True statement for gonococcal urethritis is:**
    a. Rectum and prostate are not infected
    b. Symptoms are more severe in female than male
    c. Most patients are present with symptoms of dysuria
    d. Single dose of ciprofloxacin is effective in treatment
    e. Commonly leads to arthritis

12. **The best site to obtain a swab in asymptomatic gonorrhea is:**
    a. Endocervix
    b. Urethra
    c. Lateral vaginal wall
    d. Posterior fornix

13. **For detection of gonoccocal infection in female a single swab collected from which of this site would be more useful?**
    a. Endo cervicval swab
    b. High vaginal swab
    c. Urethral swab
    d. Perineal swab

14. **Differentiation between *Neisseria gonorrhoeae* and *Neisseria meningitidis* is by:**
    a. Glucose fermentation
    b. Maltose fermentation
    c. VP reaction
    d. Indole test

Reiter's Syndrome

15. **Triad of Reiter's syndrome:**
    a. Conjunctivitis
    b. Uveitis
    c. Polyarthritis
    d. Mucosal lesions
    e. Urethritis

16. **Following is true regarding Reiter's syndrome *except*:**
    a. It is an allergic disease
    b. Most common in patient with HLA B27
    c. Most common uro-pathogen responsible is *Chlamydia trachomatis*
    d. Most common gastrointestinal pathogen responsible is *Shigella*

17. **Reactive arthritis is caused by:**
    a. *Staphylococcus*
    b. *H. influenzae*
    c. *N. gonorrhoeae*
    d. *C. trachomatis*

Nongonococcal Urethritis

18. **Nongonococcal urethritis (NGU) is caused by:**
    a. *N. gonorrhoeae*
    b. *C. albicans*
    c. *C. trachomatis*
    d. b + c

19. **Most common cause of nongonococcal urethritis is:**
    a. Menigococci
    b. *E. coli*
    c. *Chlamydia trachomatis*
    d. *Mycoplasma*

20. **A 28-year-old sexually active male presents with burning micturation. On clinical examination no ulcer in the genitals. Urine examination shows 50 WBCs/HPF, no RBCs, leukocyte esterase positive, gonococcal culture negative. What could be the most probable organism?**
    a. *Treponema pallidum*
    b. *Neisseria*
    c. *Chlamydia trachomatis*
    d. *H. ducreyi*

Answers and Explanation of MCQs

1. d
- *N. meningitidis* is capsulated, arranged in pair with flat apposition surfaces.

2. c
- A, B, C, X, Y and W are causing invasive disease while other capsular serogroups like D, 29E, H, I, J, L and Z and non-capsular isolates are causing noninvasive disease.

3. a
- Source of infection is mainly carrier not the case.
- Other options are explained in chapter.

4. c
- Xavier develops infection due to group B meningococcus which is less immunogenic and vaccine is not available.

5. d
- Conjugate vaccine is not available for group B meningococcus.

6. d
- 3rd generation cephalosporin (ceftriaxone, ceftazidime) is the drug of choice for case, and it may be used for initial treatment (first 2 days).
- Rifampicin is the drug of choice to remove the nasopharyngeal carrier and in person who is in close contact to carrier.

7. c

- It is a case of septic shock (hypotension and tachycardia) along with meningitis (headache). Patient also develop skin lesion, so organism is *N. meningitidis* because only meningococcal meningitis presents with skin lesions.

8. d

- It is an exclusive human pathogen and human disease called gonorrhea (rrhea means discharge/flow of seeds).
- Some strain like Por 1A, may cause disseminated diseases like arthritis, ulcerative endocarditis and rarely meningitis.
- Acute urethritis is the most common manifestation in males while in female the most common manifestation is cervicitis and in prepubertal girls (Aneostogenic women) the most common manifestation is vulvovaginitis or vaginitis as mentioned in **Flowchart 52.6**.
- All strains are not sensitive to penicillin and resistance has been developed to many drugs as explained in text under the section resistance.

9. c

- M protein is the virulence factor in streptococci not in gonococci.
- Follow section, *Neisseria gonorrhoeae* [pathogenicity → **agent factors (virulence factors)**] for explanation.

10. c

- Follow section, *Neisseria gonorrhoeae* (pathogenicity → **meaning)** for explanation.

11. c and d

- Proctitis (inflammation of rectum) occurs after anal sex and prostitis (inflammation of prostate) occurs due to extension from urethritis.
- Symptoms are more severe in male than female.

- Dysuria and mucopurulent discharge present without frequency or urgency.
- Single dose of ciprofloxacin or other fluoroquinolones like ofloxacin/levofloxacin is effective in treatment.
- Arthritis is not occurs commonly, but occurs in 3–5% of untreated cases.

12. a

- Follow section, *Neisseria gonorrhoeae* **(laboratory diagnosis → specimens → asymptomatic gonorrhea)** for explanation.

13. a

- Urethritis is common in male and cervicitis (vaginitis is rare follow **Flowchart 52.6** for more details) is common in female in gonococcal infection, so in female endocervical swab is more useful.

14. b

- Differentiation between *Neisseria gonorrhoeae* and *Neisseria meningitidis* is explained in **Table 52.2**.

15. a, c, d

16. a

17. c, d

- Follow section, **Reiter's syndrome** for explanation of answers of MCQs 15–17.

18. d

19. c

- Follow section, **nongonococcal urethritis (etiological agents)** for explanation of answers of MCQs 18–19.

20. c

- It is a case of NGU, because gonococcal culture is negative.
- Absence of **ulcer in the genitals is rule out the possibilities of** *Treponema pallidum* and *H. ducreyi*.
- Pus cells in urine is a feature of *Chlamydia trachomatis*.

Infections of Corynebacteria

Chapter Outline

- Introduction
- *Corynebacterium diphtheriae*
- Other Pathogenic Corynebacteriae
- Diphtheroids
- *Arcanobacterium*

INTRODUCTION

Word derived from word coryne means club as showing club shaped swelling due to presence of metachromatic granules at the ends. *Corynebacterium* contains peptido-glycan, polysaccharide, fatty acids (glycolipid) in cell wall closely related to <u>M</u>ycobacterium and <u>N</u>ocardia so called CMN group. *Corynebacterium* composed of >59 species of which 36 are medically important. They define together on genetic basis (16S rRNA, 16Sr DNA sequencing, nucleic acid hybridization). Following are the different species of *Corynebacterium*.

- **Pathogenic species:** They produce toxin. Species are *C. diphtheriae* (Kleb-Loeffler's Bacilli/KLB), *C. ulcerans*, *C. pseudotuberculosis* (*C. ovis*), *C. minutissimum*, *C. tenuis*, *C. jakeium* (formerly JK group) and *C. urealyticum*.
- **Nonpathogenic:** This group called coryneform or diphtheroids. Species are *C. pseudodiphtheriticum*, *C. xerosis* and *C. parvum*.

Corynebacterium diphtheriae

Synonyms and History

C. diphtheriae was 1st identified by Klebs in 1883, but 1st cultivated by Loeffler in 1884, hence commonly called Klebs-Loeffler Bacillus (KLB). It also called diphtheria bacilli from the name of disease.

Morphology

Type according to Gram's stain: They are GPB.

Shape and size: They are slender rod with clubbing at one or both ends and about 3–6 µm × 0.6–0.8 µm in size.

Arrangement: They are arranged in pairs, palisades (resembling stakes of a fence) or in small groups. Sometimes, they arranged end to end by making an angle to each other, resemble to letter V or L called Chinese letter appearance or Cuneiform arrangement as shown in **Fig. 53.1a**. It is due to incomplete separation of daughter cells after binary fission.

Motility, capsule and spores: Nonmotile, noncapsulated and nonsporing.

Polar bodies (Fig. 53.1b): Ch. 6.

Cultural Characteristics [C/Cs]

Effective factors

- O_2 effect: They are aerobes and facultative anaerobes.
- Temperature: 37°C (15–40°C).
- pH: 7.2.
- Growth occurs on enriched media contain blood, serum or eggs.

Culture in media

A. **Liquid media:** Follow **Table 53.1**.

B. **Solid media**

- **Loeffler's serum:** They grow very rapidly within 6-8 hours. Colonies are small, circular and white become large and yellowish on longer incubation.
- **Blood agar:** Some strains are hemolytic, while some are nonhemolytic.

C. **Selective media**

- **Potassium tellurite (PT) medium:** Medium contains heat lysed blood and 0.04% potassium

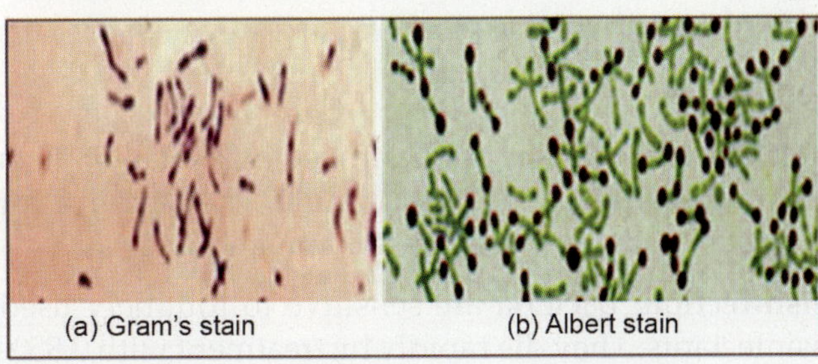

Fig. 53.1: Morphology of *C. diphtheriae*

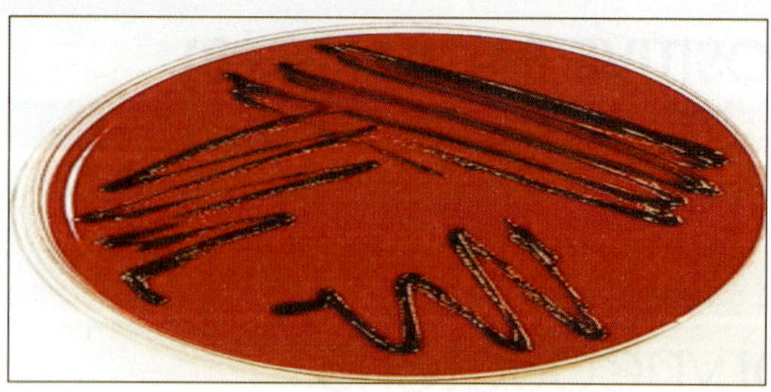

Fig. 53.2: *C. diphtheriae* in PT medium

tellurite. Tellurite inhibits growth of other bacteria and acts as selective agent. Growth is very slow (48 hours) and incubate the culture plates at least for 2 days before giving negative report. Colonies are grayish or black (**Fig. 53.2**) due to reduction of potassium tellurite to metallic tellurium. Other bacteria producing black colonies in PT medium are *Staph. aureus*, *Enterococcus faecalis* and *Erysipelothrix rhusiopathiae*.

- **McLeod's and Hoyle's medium:** It contains saponin lysed blood.
- **Cystine tellurite agar:** It contains potassium tellurite and cysteine.
- **Tinsdales medium:** It contains cystine tellurite agar and sodium thiosulfate.

Culture in animal: It is done by using the guinea pig to differentiate the toxigenic/virulent strain from no toxigenic/avirulent strain. It is broadly described in laboratory diagnosis as virulence test (toxigenicity test).

Automated culture: Like MALDI-TOF or VITEK is used to identify the species from culture.

Biochemical Reactions (B/Rs)

Sugar fermentation tests:

G	S	L	M
A	–	–	–

Other sugars fermented are galactose, maltose and dextrin with acid only. Sugar fermentation is tested in Hiss's serum water. Starch and glycogen are fermented by gravis strain only, but not by intermedius and mitis strains.

Nitrate reduction, catalase and pyrazinamidase tests: All are positive.

Oxidase and urease tests: Both are negative. Urease test helps to differentiate from *C. ulcerans* and *C. pseudo-tuberculosis*, where it is positive.

Resistance

Sterilization: Bacteria are destroyed by heat at 58°C in 10 minutes and at 100°C in 1 minute. They survive for 14 weeks on dried piece of pseudomembrane.

Disinfection: Bacteria are sensitive to routinely used disinfectants. They die rapidly by treatment with 0.85% NaCl solution.

Drug resistance: It is not a big issue in *C. diphtheriae*.

Pathogenicity

Disease name: Disease called diphtheria.

Meaning: Word derived from diphtheros (Greek) means leather, because of leathery pseudomembrane formation.

Definition: It is a toxemia and locally characterized by pseudomembrane formation.

Epidemiology: It occurs as endemic or epidemic form. The disease was 1st recognized by Bretonneau in 1826, who called it diphtherite from diphtheros (Greek) means leather.

- **In world:** Epidemic was occurred in Soviet Union in 1990 involving thousands of case with 20% deaths.
- **In India:** Condition is under controlled due to effective immunization.

Reservoirs of infection: Humans and animals like cows are the reservoirs. In humans, bacteria are present in nose or throat, and in animals, they are present in udder and transmitted via milkers by milk ingestion.

Sources of infection: Respiratory droplets arise from nose or throat of humans, and the milk (may be tissue from udder) of infected animals are the sources of infection.

Modes of transmission: Transmission occurs by air borne (inhalation), fomites borne (toys and pencils) and by milk of infected cows.

Incubation period: 3–4 days.

Portal of entry: Respiratory tract and also by skin.

Sites: Different sites like faucial (throat), nasal, conjunctival, otitic, laryngeal, genital, vulval, vaginal or prepucial and cutaneous.

Precipitating factors (epidemiological determinants): Three types

A. Agent factors (virulence factors): Bacilli do not penetrate deep in to the mucosal tissues and bacteremia do not occurs. Diphtheria is a toxemia and all clinical manifestations are due to production of powerful exotoxin called diphtheria toxin with following salient features.

History: Loeffler studied the effect of bacilli in animals and concluded that the illness is not due to bacilli itself but by its diffusible product. Emile Roux and Alexander Yersin identified the diffusible product in 1888 which called diphtheria toxin and established its pathogenic effects. Diphtheria antitoxin was discovered by von Behring in 1890.

Properties: "Park Williams 8 strain" of *C. diphtheriae* is widely used for toxin production. About 90–95% gravis and intermedius strains are toxigenic, while 80–85% mitis are toxigenic. Toxin is heat-labile and phage mediated. It is highly potent. The lethal dose for humans is about 0.1 μg of toxin per kg of body weight. Death occurs through necrosis of the heart and liver. Diphtheria toxin has also been associated with the development of myocarditis. Myocarditis secondary to diphtheria toxin is considered as one of the biggest risks to unimmunized children.

Lf unit: It is defined as Loeffler's flocculating unit. 1 Lf unit is defined as the amount of diphtheria toxin which flocculates very rapidly with one unit of antitoxin.

Other bacteria producing diphtheria toxin: *C. ulcerans* and *C. pseudotuberculosis.*

Structure: Biochemically toxin is protein in nature. Its MW is 62,000. It has two fragments like A (MW 24,000), which is active and B (MW 38,000), which is for binding of toxin to cells. Antibody to fragment B protects binding of toxin to cells.

Sites of actions: Toxin produced locally, and it spreads to distant organs by blood (exotoxemia) with special affinity for certain tissues like myocardium, adrenals glands and nerve endings.

Mechanism of action: Fragment A inhibits poly-peptide chain elongation in presence of NAD by inactivating elongation factor (EF-2). Toxin acts by inhibiting protein synthesis.

> **Note: List of toxins act by inhibiting the protein synthesis**
> - Diphtheria toxin: ⎫ By inactivation
> - Pseudomonas toxin: ⎭ of EF-2
> - Shiga like toxin (SLT): ⎫ By inactivation
> - Shiga toxin: ⎭ of 60S ribosome

Factors affecting the toxin production

- **Iron:** Minimum concentration required is 0.1 mg/l. Higher concentration (0.5 mg/l) decreases the production.
- **Tox+ phage/β prophage:** Bacteriophage infecting *Corynebacterium* is called corynephage. It introduces new nucleic acid in to bacterial chromosome called Tox⁺ phage/β prophage. Bacterium with Tox⁺ phage/β prophage is called lysogenic bacterium. Tox⁺ phage/β prophage multiplies along with bacterial chromosomes. This phenomenon is called lysogeny. In presence of new Tox⁺ phage/β prophage, bacteria acquired certain new properties [like toxin production] called phage conversion or lysogenic conversion. Elimination of Tox⁺ phage/β prophage makes the organism nontoxigenic.

Uses

- **Toxoid:** Prolonged storage, incubation at 37°C for 4–6 weeks, treatment with 0.2–0.4% formalin or acid pH converts toxin to toxoid. It has antigenicity, but no toxicity. It is capable of inducing antitoxin and reacting specifically with it. It is used in production of diphtheria vaccine.
- **Antineoplastic agent:** The drug denileukin diftitox uses diphtheria toxin as an antineoplastic

agent. Resimmune is an immunotoxin, which uses diphtheria toxin in clinical trials in cutaneous T cell lymphoma patients.

B. Host factors

1. **Age:** It is rare below one year of age due to passive protection from maternal antibodies. It is maximum between 2 and 5 years, less between 5 and 10 years and least afterward due to repeated subclinical infections.
2. **Immunity:** Single attack of disease or inapparent infection provides lifelong immunity (usually, not always), so child recovered from disease does not required active immunization. It is less in developed countries due to proper immunization while more in developing countries due to poor immunization.

C. Environmental factors

1. **Temperate regions:** Carriage is more in nose and throat. Nasal carriage harbors the bacilli for longer period.
2. **Tropical regions:** Bacilli are found more in skin and causing cutaneous diphtheria. Bacilli last for 3 years in skin. Cutaneous diphtheria may stimulate the natural immunity and also faucial diphtheria in nonimmune individual.

Pathogenesis: It is explained in **Flowchart 53.1.** It is good to remember that nontoxigenic strain can produce only local lesions (in skin/nasopharynx), while toxigenic strain can produce both local and systemic lesions. To produce the local lesion toxin production (toxigenic strain) is not necessary.

Clinical features: Following are the different clinical types.

1. **Based on sites of infection**
   - **Nasal diphtheria:** It is mildest with minimal toxemia. It presents with unilateral or bilateral nasal discharge.

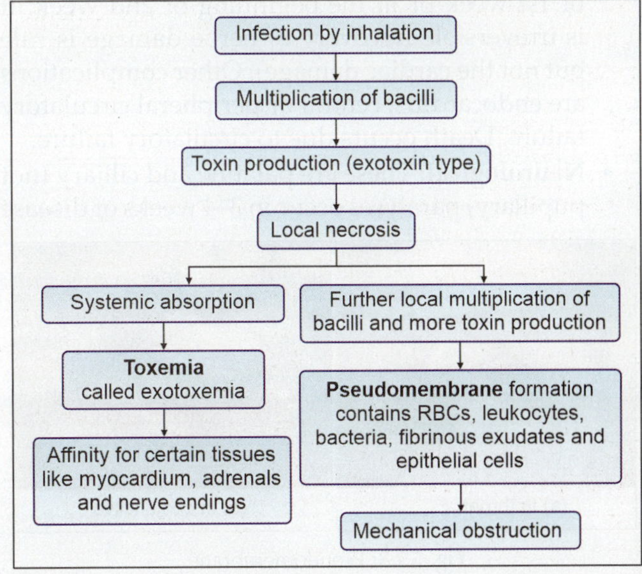

Flowchart 53.1: Pathogenesis of diphtheria

- **Faucial (tonsilopharyngeal) diphtheria:** It is the commonest variety involving throat/nasopharynx. It is more severe (moderate) than nasal diphtheria. It causes localized infection of throat, tonsils (white patch present over tonsil in child), pharynx, larynx and adjacent surface. Clinically, it presents with mild grade fever, sore throat, malaise and pseudomembrane (gray or white) in throat as show in **Fig. 53.3a.** Removal of pseudomembrane leads bleeding.
- **Laryngeal diphtheria:** It is most severe with highest mortality, so immediately required tracheostomy; however, tracheostomy is not mandatory in all cases. It presents with hoarseness and cough.
- **Conjunctival diphtheria:** It also called pseudo-membrane conjunctivitis (**Fig. 53.3b**) due to presence of pseudomembrane in conjunctiva.
- **Otitic diphtheria:** It presents with otitis media.
- **Genital diphtheria:** It involves vulva, vagina or prepuce.
- **Cutaneous diphtheria:** It occurs as an ulcer by non-toxigenic strains called ecthyma diphtheriticum or diphtheritic whitlow.

2. **Based on clinical severity**
   - **Malignant or hypertoxic:** Cervical lymph nodes are enlarged (severe adenitis) due to edema called bull neck diphtheria.
   - **Septic:** It includes ulceration, cellulitis and gangrene around pseudomembrane.
   - **Hemorrhagic:** Bleeding from the edge of pseudo-membrane, conjunctiva, nasal mucosa (epistaxis), cutaneous and generalized bleeding.

Complications

1. **Local/mechanical:** These are due to pseudomembrane which extends up in nasal mucosa and down into larynx causes suffocation and respirator obstruction called mechanical asphyxia.
2. **Systemic:** These are due to toxin production.
   - **Cardiovascular:** Myocarditis occurs at the end of 1st week or in the beginning of 2nd week. It is irreversible (recovery of nerve damage is rule but not the cardiac damage). Other complications are endocarditis, central or peripheral circulatory failure. Death occurs due to circulatory failure.
   - **Neurological:** These are palatine and cilliary (not pupillary) paralyses occur in 3–4 weeks of disease

which recovered spontaneously. It also produces neuropathy.
- **Others:** Pneumonia and degenerative changes in liver, adrenal glands and kidneys (renal failure).

Laboratory Diagnosis

Key points before moving to laboratory diagnosis: Diagnosis based on clinical ground and start treatment without waiting for laboratory report. Laboratory help needs for confirmation of disease, epidemiological purposes, research purposes, prevention of disease and for differentiation of other causes of pseudomembrane like acids/alkalis injuries, Vincent's angina and infectious mononucleosis.

Specimens: Two throat swabs are collected, one is for microscopy and 2nd is for culture. Swab also collected from nose, larynx, pseudomembrane, skin lesion or any other site where diphtheria is suspected. Collect the swab before antibiotics and antiseptic mouth wash. Rayon/dacron swab is preferred over cotton swab. Swabs are collected by using tongue depressor. Swab rubbed over affected area and pseudomembrane.

Transport: If swab cannot be inoculated promptly, it should be kept moistened with sterile serum.

Testing methods
A. **Microscopy:** Follow morphology.
B. **Culture:** Follow C/Cs.
C. **Biochemical reactions:** Follow B/Rs.
D. **Bacterial typing:** Two types of typing methods.
   1. **Phenotyping methods:**
      - **Serotyping:** Bacilli are antigenically hetero-geneous and possess three antigens like **(1) Deep-seated antigen** which is found in all *Corynebacterium* species and in *M. tuberculosis* **(2) K antigen** which is a heat labile protein and **(3) O Ag** which is a heat stable polysaccharide antigen. Typing is done by agglutination method. Gravis has 13, intermedius has 4 and mitis has 40 serotypes. Gravis type 2 is common worldwide. No connection has been established between antigenic type and other features.
      - **Biotyping:** It based on colony morphology on PT medium and other properties. Total four biotypes are identified like gravis, intermedius, mitis and belfanti by McLeod and Anderson. Initial 3 are mentioned in **Table 53.1.**
      - **Bacteriophage typing:** Ch. 118.
      - **Bacteriocin (diphthericin) typing:** Gibson and Colman demonstrated 10 pattern of bacteriocin.
   2. **Genotyping:** It is done by bacterial polypeptide analysis, DNA restriction patterns, hybridization and DNA probe.
E. **Virulence tests (toxigenicity tests):** Two types:
   1. *In vivo* **tests:** These are subcutaneous test and intra-cutaneous test as mentioned in **Flowchart 53.2** and **Flowchart 53.3,** respectively. Subcutaneous

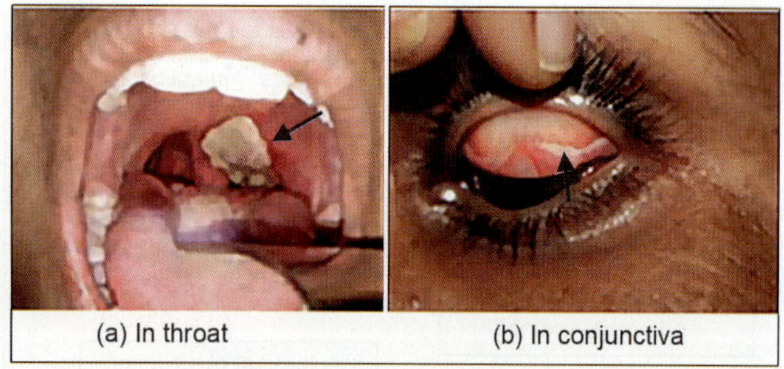

Fig. 53.3: Pseudomembrane

TABLE 53.1: Biotypes of *C. diphtheriae*

Features	Gravis	Intermedius	Mitis
Morphology			
Length	• Short rods	• Long rods	• Long-curved
Granules	• Few/no granules	• Poor granulation	• Prominent granules
Pleomorphism	• Less pleomorphic	• Very pleomorphic	• Pleomorphic
C/Cs			
Liquid media	• Surface pellicle • No turbidity • Granular deposit	• Turbid in 24 hours, clear in 48 hours • Granular deposit	• Diffuse turbidity and soft pellicle
Blood agar	Variable	Nonhemolytic	Hemolytic
PT medium/agar	Daisy head colony	Frog's egg colony	Poached egg colony
Consistency	'Cold margarine' brittle, not easily emulsifiable	Intermediate between gravis and mitis	Soft, buttery, easily emulsifiable
B/Rs			
Glycogen + starch	Fermentation	Nonfermentation	Nonfermentation
Pathogenicity			
Virulence	Severe	Moderate	Mild
Nature of disease	Epidemic	Epidemic	Endemic
Complication	Paralytic	Hemorrhagic	Obstructive

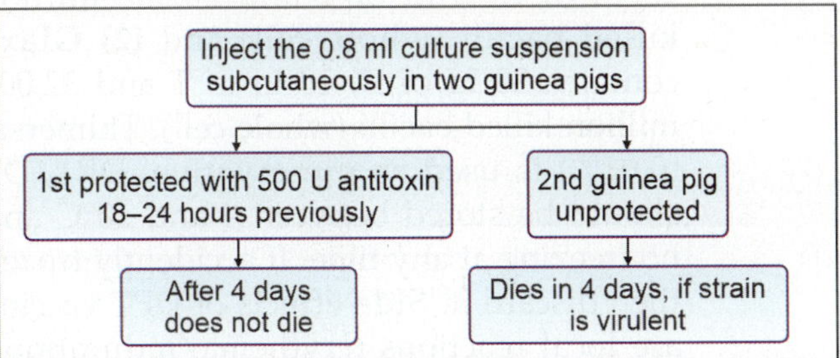

Flowchart 53.2: Subcutaneous test

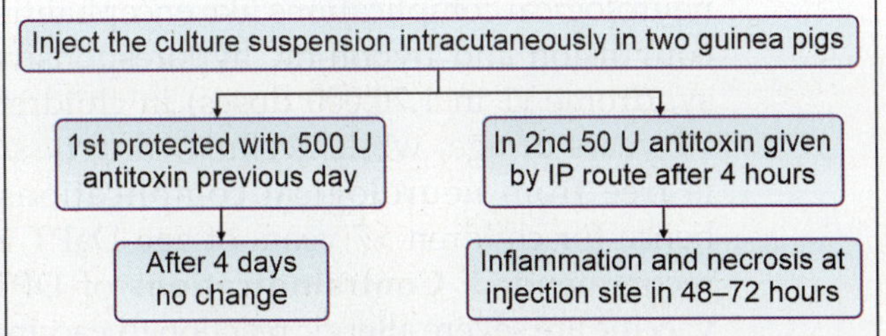

Flowchart 53.3: Intracutaneous test

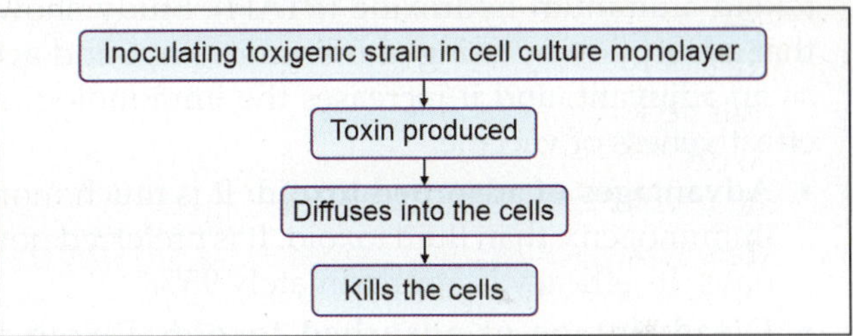

Flowchart 53.4: Tissue culture test

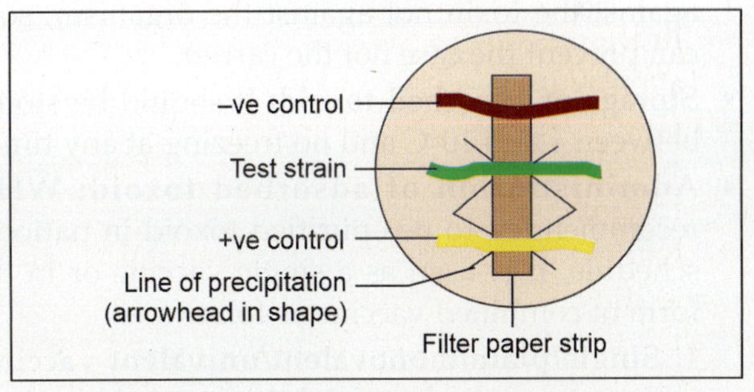

Fig. 53.4: Elek's gel precipitation test

test wastes the unprotected animal while intracutaneous test saves the unprotected animal.

***In vitro* tests:** These are
1. **Tissue culture:** Steps are mentioned in **Flowchart 53.4**.
2. **Molecular:** PCR to detect the tox$^+$ gene.
3. **Toxin detection:** By ELISA and ICT.
4. **Elek's gel precipitation test**
   - **Principle:** It based on precipitation reaction.
   - **Method:** Rectangular filter paper strip impregnated with diphtheria antitoxin (1000 u/ml) is placed in 20% normal horse serum agar. Inoculate the strain at right angle to paper strip. Also set the positive and negative controls as shown in **Fig. 53.4**. Incubate the plate at 37°C for 24–48 hours.
   - **Results**
     - Positive test: Toxins produced by bacteria will diffuse in agar and meets with antitoxin to produce the arrow head line of precipitation.
     - Negative test: No line of precipitation.
   - **Interpretation**
     - Positive test: It indicates the toxigenic strain.
     - Negative test: It indicates the nontoxigenic strain.

Note: Serological tests are not useful
- *C. diphtheriae* confirmed to the site of entry and does not invade the bloodstream, so no antibodies formation and no role of serological tests.

5

General measures: Avoid contact with carrier.

Chemoprophylaxis: Erythromycin is the drug of choice for carrier.

Immunoprophylaxis: It is based on assaying circulatory antitoxin level which is detected by passive hemagglutination test and neutralization in cell culture. In the past before doing immunization susceptibility of an individual was checked by **Schick test** which based on toxin-antitoxin neutralization. Now Schick test is obsolete. Antitoxin level of 0.01 IU or more per ml of blood is considered as index of immunity/protective titer. Immunoprophylaxis includes active, passive and combined as described below.

A. **Active immunization:** It is achieved by diphtheria toxoid. It has different types like **APT** (alum precipitated toxoid) gives reactions so not useful; **FT** (formal toxoid/fluid toxoid) or plain toxoid prepared by incubating the toxin with formalin and **purified (adsorbed) toxoid** which is adsorbed on to insoluble aluminum compound usually aluminum phosphate called purified toxoid aluminum phosphate (PTAP) or less often aluminium hydroxide called purified toxoid aluminum hydroxide (PTAH). Study shows that adsorption by using aluminum compound acts as an adjuvant, and it increases the immunological effectiveness of vaccine.

- **Advantages of adsorbed toxoid:** It is much more immunogenic than fluid toxoid. It is preferred now days. Its efficacy is approximately 95%.
- **Disadvantage of adsorbed toxoid:** Toxoid is against the toxin not against the organism, so it can prevent the case not the carrier.
- **Storage of adsorbed toxoid:** It should be stored between 4 and 10°C and no freezing at any time.
- **Administration of adsorbed toxoid:** WHO recommended to use purified toxoid in national schedule. It is used as a single vaccine or in the form of combined vaccine as follows:
  1. **Single/plain/monovalent/univalent vaccine:** For young children, diphtheria toxoid is given in a dose of 10–25 Loeffler (Lf) units. Smaller dose (1–2 Lf units) used for older children and adults.
  2. **Mixed/combined/polyvalent/multivalent vaccines**
     - Double/bivalent vaccine: Td vaccine contains **t**etanus toxoid and adult dose of **d**iphtheria toxoid (1–2 Lf units). Td vaccine is given to adults because diphtheria cases are increasing in adults. For adults, who are >18 years with completed primary immunization, Td booster is given every 10 years till 65 years. For adults, who are >18 years but not completed primary immunization, 3 doses of Td are given at 0, 1 month and 1 year.

- Triple/trivalent/tetravalent vaccine: DPT/DTP vaccine contains **d**iphtheria toxoid, whole cell **p**ertussis and **t**etanus toxoid or DaPT vaccine contains **d**iphtheria toxoid, **a**cellular **p**ertussis and **t**etanus toxoid. In DPT whole cell pertussis produces neurological complications (1 in 1,70,000 doses), hence DaPT contains **a**cellular **p**ertussis is prepared. Pertussis components are adjuvant and increase the immunogenicity of diphtheria toxoid and tetanus toxoid. In **primary vaccination,** three doses of 0.5 ml DPT are given by deep IM route in anterolateral aspect of thigh at 6th week, 10th week and 14th week of age. After primary immunization, **1st booster** of DPT is given at 18–24 months followed by **2nd booster** dose at 5–6 years (school entry age). It is not given in gluteal region, because of paucity of APCs in gluteal region delaying the presentation of vaccine Ags to T and B cells. It is also given after 5 years, but whooping cough is rare after 5 years. Two commercial preparations of DPT are available like **(1) Kasauli** contains 25 Lf of D, 5 Lf of T and 20,000 million killed bacilli (whole cell) and **(2) Glaxo** contains 30 Lf of D, 10 Lf of T and 32,000 million killed bacilli (whole cell). Thimersal (0.01%) is used as preservative. DPT/DT should be stored between 4 and 10°C and no freezing at any time. If accidently frozen then discard it. **Side effects** of DPT vaccine are local reactions (erythema/induration), exaggerated local reactions (arthus-type) and fever. Whole cell *B. petussis* causes neurological complications like encephalitis, convulsion and hypotonic hyporesponsive syndrome (1 in 1,70,000 doses) in children >7 years of age, while acellular *B. petussis* is free from neurological complications, hence for children >7 years of age DaPT is recommended. **Contraindications** of DPT vaccine are severe allergic reaction to vaccine components, or following a primary dose, >7 years of age, progressive neurological diseases or epilepsy. **Efficacy** of DPT vaccine is 70%.
- Quadruple/quadrivalent vaccine: DPT (DTP) with Hib, etc.
- Pentavalent vaccine: Two types of preparations like pentvac vaccine contains DPT, HBV and *H. influenzae* type b (Hib) or pentaxim contains DPT, IPV and Hib. In some states, instead of DPT, pentavalent vaccine is given at 6th week, 10th week and 14th week.

B. **Passive immunization:** It is used when susceptible is exposed to infection and in case of diphtheria patient. It consists of SC injection of 500–1000 U of antitoxin called antidiphtheritic serum (ADS). Adverse reaction includes hypersensitivity.

C. **Combined immunization:** It consists of administration of the first dose of adsorbed toxoid on one arm **(active immunization)** and ADS to the other arm **(passive immunization)** followed by the full course of active immunization.

Treatment

Treatment of case: Consist of ADS and antibiotic therapy. ADS is given in the dose of 20,000–1,00,000 U by SC route in serious case. Half dose is given by IV route. Antibiotic therapy includes Pn (procaine pn), which is the drug of choice. Other options are clindamycin, rifampicin or erythromycin.

Treatment of carrier: It is done with erythromycin.

Treatment of contact: Postexposure prophylaxis is given in case of close contact like house hold contact. It includes three categories:

1. **Primary immunization or booster dose received last two years:** No further treatment is required.
2. **Primary immunization or booster dose received before two years:** Only booster dose of diphtheria toxoid is given.
3. **No immunization:** It includes diphtheria vaccine, single dose of Pn G/7–10 days course of erythromycin and ADS.

OTHER PATHOGENIC CORYNEBACTERIA

C. ulcerans

Similarities with C. diphtheriae: It causes diphtheria like lesion, resembles to gravis variety of diphtheria bacillus and produces black colonies on Tinsdales medium.

Difference with C. diphtheriae: It liquefies gelatin, ferments trehalose, nitrate negative, but urease test positive.

Pathogenicity

- **Mode of transmission:** Human infection transmitted by cow's milk.
- **Toxin:** It produces 2 types of toxin one is identical to diphtheria toxin and other to toxin of *C. pseudodiphthericum* (phospholipase-D) affects sphingomyelin.
- **Clinical features:** It produces infection in cows. In human, it produces the ulcer in throat.

Treatment: Erythromycin is effective.

C. pseudotuberculosis

Synonym: Commonly, it called Preisz-Nocard bacillus.

Similarities with C. diphtheriae: It produces black colonies on Tinsdales medium and reverse CAMP test is positive.

Difference with C. diphtheriae: Urease positive.

Pathogenicity: It produces diphtheria toxin. It produces pseudotuberculosis in sheep and suppurative lymphadenitis in horses. In human, it produces the ulcer in throat.

C. minutissimum

It is lipophilic and grows in media contain 20% fetal calf serum. It produces the localized infection of the stratum corneum called erythrasma, affecting axilla and groin. On Wood's lamp examination, it gives coral red color.

C. tenuis

It produces the pigmented nodules around the axillary and pubic hair **shaft called trichomycosis axillaris.**

C. jakeium

Earlier it called JK group. It is lipophilic, strictly aerobe, pyrizinamide positive and alkaline phosphatase positive. It is an opportunistic pathogen causing bacteremia, endocarditis, meningitis and skin features. It is resistant to multiple antibiotic, responding only to vancomycin.

C. urealyticum

It is the normal flora of skin. It causes UTI like pyelonephritis, cystitis in patient with IDDs and renal transplant. It is a strong urease producer; infection of urinary tract may lead to formation of stones.

DIPHTHROIDS

Definition

Bacilli resemble to *Corynebacterium* are called diphtheroides. They are the normal flora of skin, throat, conjunctiva and other areas.

Differences between C. diphtheriae and Diphthroids

Follow **Table 53.2.**

Species

C. psudodiphthericum: Earlier it called *C. hofmannii.* It found in throat. Biochemically, it is urease and

TABLE 53.2: Differences between *C. diphtheriae* and diphtheroides

Features	C. diphtheriae	Diphtheroids
Morphology		
• Gram's stain	Weak gram-positive	Strong gram-positive
• Size	Thin	Thick
• Arrangement	Cuneiform	Pallisade
• Granules	Present	Few/absent
• Pleomorphism	More	Little
C/Cs	Growth on enriched media	Growth on ordinary media
Sucrose	Non fermenter	Fermenter
Toxin production	Toxigenic	Nontoxigenic
Virulence test	Positive	Negative

pyrazinamide positive and glucose non fermenter. It causes pharyngitis, endocarditis in patient with IDDs.

C. xerosis: It found in conjunctival sac. Biochemically, it is pyrazinamide positive, urease negative and glucose fermenter.

C. parvum: It used as an immunomodulator. It is a mixture of *Propionibacterium* spp.

Arcanobacterium

Common species is *A. haemolyticum.* Formerly it was called *C. haemolyticum.* It produces β-hemolysis on blood agar. It gives reverse CAMP test positive. It causes pharyngitis and skin ulcer.

ACCESS YOURSELF

Case Study

1. A 6-year-old patient presents with complain of throat pain and high-grade fever. Clinical examination indicates the cervical lymphadenopathy and pseudomembrane in throat. Throat swab microscopy identifies GPB under Gram's stain. Identify the organism and answer the following:
   a. Name the causative agent.
   b. Describe the pathogenicity of causative agent.
   c. Write the laboratory diagnosis/immunoprophylaxis of causative agent.

Essay/Full Question

1. *C. diphtheriae.*

Short Notes

1. Pathogenicity/Lab., diagnosis of *C. diphtheriae.*
2. Virulence test in of *C. diphtheriae.*
3. Elek's gel precipitation test.
4. DPT vaccine.

Short Questions for Theory/Viva Questions

1. Name the four bacteria producing black colonies in PT medium.
2. Write the use of Albert stain. Draw the labeled diagram of appearance of organism under Albert stain.
3. Write the four differences between *C. diphtheriae* and diphtheroids.

Essays/Full Questions

1. *C. diphtheriae* produces black colonies on PT medium.
2. Diphtheria is not a bacteremia but considered as toxemia.
3. Respiratory obstruction or mechanical asphyxia is the common complication in diphtheria.
4. In diphtheria serological tests are useful to detect the toxin not the bacteria.
5. Diphtheria toxoid is useful for case not carrier.
6. In children >7 years of age DaPT is recommended, but not the DPT.

MCQs for Chapter Review

Introduction

1. **Meaning of *Corynebacterium*:**
   a. Club shape
   b. Pigment production
   c. Leathery pseudomembrane formation
   d. Causes diphtheria

C. diphtheriae

2. **Wrong about *C. diphtheriae*:**
   a. Capsulated bacterium causes infection.
   b. Gram-negative
   c. Babes Ernst granules are seen.
   d. Chinese latter pattern
   e. Nonmotile

3. **A 12 years child presents with fever and cervical lymphadeno-pathy. Oral examination shows a gray membrane on the right tonsil extending to the anterior pillar. Which of the following medium will be ideal for the culture of the throat swab for a rapid identification of the pathogen?**
   a. Nutrient agar
   b. Blood agar
   c. Loeffler's serum slope
   d. LJ medium

4. **A child presents with white patch over the tonsils. Diagnosis is best made by culture in:**
   a. Loeffler's medium   b. LJ medium
   c. Blood agar   d. Tellurite medium

5. **Daisy head colonies are seen with following strain of *C. diphtheriae*:**
   a. Nontoxigenic strain   b. Mitis
   c. Intermedius   d. Gravis

6. **True statement about diphtheria toxin is:**
   a. Toxin is phage mediated
   b. Toxin requires for local infections
   c. Endotoxemia causes systemic manifestations
   d. Toxin acts by inhibiting synthesis of capsule

7. **Which of the following statement (s) is/are not true about diphtheria toxin?**
   a. Heat stable
   b. Toxin production is dependent on iron
   c. Inhibits the cAMP
   d. All types produce the toxin

8. **Diphtheria toxin has special affinity for:**
   a. Skeletal muscle   b. Heart muscle
   c. Lungs   d. Gut

9. **The optimum amount of iron required for toxin production in *Corynebacterium diphtheriae* is:**
   a. 1 mg/l   b. 0.1 mg/l
   c. 0.5 mg/l   d. 10 mg/l

10. **Which of the following site is most commonly affected by *C. diphtheriae*?**
    a. Skin   b. Conjunctiva
    c. Faucial   d. Kidney

11. **True about *Corynebacterium. diphtheriae* is all *except*:**
    a. Deep invasion is not seen
    b. Elek's test is done for toxigenicity
    c. Metachromatic granules are seen
    d. Toxin is mediated by chromosomal change

12. **Drug of choice for diphtheria carrier is:**
    a. Pn   b. Erythromycin
    c. Tetracycline   d. Septran

Other Pathogenic *Corynebacteria*

13. **Erythrasma is caused by:**
    a. *C. minutissimum*   b. *C. ulcerans*
    c. *C. pseudotuberculosis* b. *C. diphtheriae*

1. a
- Follow section, **introduction (meaning)** for explanation.

2. a and b
- Follow section, *Corynebacterium diphtheriae* (**morphology**) for explanation.

3. c
- Pathogen is *C. diphtheriae* which grows very rapidly within 6–8 hours in Loeffler's serum slope.
- Nutrient agar and blood agar are not useful for rapid identification.
- LJ is the selective medium for *M. tuberculosis*.

4. d
- Best diagnosis of any pathogen is made by using the selective medium. For *C. diphtheriae* selective medium is tellurite medium. Loeffler's medium is useful for the rapid diagnosis not for best diagnosis.

5. d
- C/Cs on PT medium by various biotype of *C. diphtheriae* like gravis, intermedius and mitis are described in **Table 53.1**.

6. a
- Toxin is phage (tox$^+$ phagee or β-phage) mediated.
- Toxin does not required for local infections, it is required for systemic manifestation and even non toxigenic strain can cause local lesions in skin and nasopharynx.
- Systemic manifestations are due to exotoxemia not due to endotoxemia.
- Toxin acts by inhibiting the protein synthesis.

7. a, c and d
- Toxin is heat labile.

389

- Toxin production is dependent on iron as mentioned earlier in text.
- Toxin does not inhibit the cAMP, but inhibiting the protein synthesis.
- All types produces the toxin: Partially not correct because all types means toxigenic strains, which produce the toxin and nontoxigenic strains, which do not produce the toxin, but on a contrary, all types means all biotypes, which produce the toxin.

8. b
- Follow section, *Corynebacterium diphtheriae* (**pathogenicity → agent factors → sites of action**) for explanation.

9. b
- Follow section, *Corynebacterium diphtheriae* (**pathogenicity → agent factors → factors affecting the toxin production**) for explanation.

10. c
- Follow section, *Corynebacterium diphtheriae* (**pathogenicity → clinical features → faucial diphtheria**) for explanation.

11. d
- Deep invasion is not seen by diphtheria bacilli and all systemic manifestations are due to systemic absorption of toxin not due to bacilli themselves. Elek's test is done for toxigenicity detection. Metachromatic granules are seen in bacilli. Toxin is phage mediated not chromosome mediated.

12. b
- Follow section, *Corynebacterium diphtheriae* (**treatment**) for explanation.

13. a
- Follow section, **other pathogenic corynebacteria (*C. minutissimum*)** for explanation.

Infections of Corynebacteria

- **Knisely's PLET medium:** It is a BHIA with <u>P</u>olymyxin, <u>L</u>ysozyme, <u>E</u>DTA and <u>T</u>hallous acetate.
- **E. Capsule producing media:** Described above.
- **F. Spore-producing media:** Described above.

Culture in animals: Rabbits, guinea pigs and mice are sensitive animals. Subcutaneous injection in guinea pig leads locally hemorrhagic edema and systemically enlarged, dark red spleen. Guinea pig dies within 24–72 hours.

Biochemical Reactions (B/Rs)

Sugar fermentation tests: Glucose, sucrose and maltose are fermented with acid production only.

Nitrate reduction and catalase tests: Both are positive.

Resistance

Sterilization

- **Vegetative bacilli:** Vegetative bacilli are killed by heating at 60°C in 30 minutes. Bacilli remain viable in bone marrow for a week and in skin for 2 weeks of infected animals, which have died due to anthrax. Normal heat fixation of smears may not kill the bacilli in blood films.
- **Spores:** Spores are highly resistant to physical agents like heat. They resist dry heat at 140°C for 2–3 hours and boiling for 10 minutes. Physical methods to kill the spores are described in **Ch. 6** (section → spores).

Disinfection: Spores are also highly resistant to chemical agents like 5% phenol for weeks and $HgCl_2$. Chemical methods to kill the spores are described in **Ch. 6** (section → spores).

Chemosterilization: Spores of *Bacillus anthracis* are destroyed from animal products like from wool (imported in to non-endemic countries) by 2% formaldehyde at 30–40°C in 20 minutes and from hairs and bristles by 25% formaldehyde at 60°C in 6 hours. This method is called **duckering**.

Drug resistance: Pn is the drug of choice, but resistant strains have been seen.

Pathogenicity

Disease name: It is a zoonosis of herbivores and humans are relatively resistant to infection. Disease called anthrax.

Meaning: Anthrax (Greek) means Coal, because of formation of characteristic black scar called eschar, which looks like coal.

Epidemiology:
- **World:** Disease occurs rarely.
- **India:** It is enzootic in India. Epizootic of anthrax is common in sheep in Andhra Pradesh, Tamil Nadu border causing many cutaneous and meningo-encephalitis type human infections with high mortality. Few outbreaks noticed in Karnataka and West Bengal.

Reservoirs of infection: Herbivores animals like sheeps, cattle, horses, swine, etc., are the reservoirs. Disease is common in herbivores than carnivores.

Sources of infection: Bacilli present in saliva, nasal secretions, intestinal materials, hides, bones, blood, hairs, etc., of animals that sporulate on exposure of air and serve as the sources of infection. Soil is also the source of infection, which gets contaminated from animal products. Soil is also contaminated by vulture which spread the organism from one area to another after feeding on the anthrax infected carcasses. Dried or processed hide of infected animals may harbors the spores for the years which may serve as the source of infection. Use of shaving brush made from hairs of infected animals is also the source of infection. LD50 (Lethal dose 50) means 10,000 spores are required to produce the lethal disease in 50% animals; however, sometimes 1–3 spores are enough to produce the disease.

Modes of transmission: Follow **Table 54.2**.

Incubation period: Follow **Table 54.2**.

Portal of entry: Skin, respiratory tract and GIT.

Sites: Skin (face, neck, hands, arms and back are usually infected) respiratory tract, GIT, blood, meninges, etc.

Features	Cutaneous anthrax	Pulmonary anthrax	Intestinal anthrax
Synonym	Hide porter's disease	Woolsorter's disease	—
Mode of transmission	• By direct contact by carrying hides of infected animals on back • Use of shaving brush made from hairs of infected animals • Mechanically by biting insect like *Stomoxys calcitrans*	Inhalation of spores of infected wool	Ingestion of poorly cooked meat
IP	1–3 days	1–3 days	2–7 days
Clinical features	• Most common type nowadays • Painless vesicle skin lesion called malignant pustule, which healed with black scar called eschar surrounded by non-pitting indurated edema	Hemorrhagic pneumonia	Hemorrhagic diarrhea
Prognosis	Resolved spontaneously, though 10–20% develops fatal septicemia or meningitis	100% fatal, though survival is possible with prompt treatment	High mortality

TABLE 54.2: Clinical types of anthrax

Precipitating factors (epidemiological determinants): Three types.

A. Agent factors (virulence factors): Intracellular (cell wall associated) and extracellular factors are described in **Flowchart 54.2**.

B. Host factors

1. **Age:** It is common in adults who are industrial workers.
2. **Immunity:** Single attack of disease provides life-long immunity, so second attack is very rare.
3. **Occupations:** Cutaneous anthrax is common in dock workers/abattoir workers/slaughter house workers, who are regularly touching the infected meat or carrying loads of hides and skins on their bare back. Pulmonary anthrax is common in workers of wool industry due to inhalation of dust from infected wool. Anthrax is also common in persons frequently dealing with animals like butchers, veterinarians, farmers, forest workers, etc.
4. **Community:** It is common in communities who eat the carcasses of animal dying due to anthrax.

C. Environmental factors: More than 70 species of *Bacillus* are found in soil and water. Spores are found in soil, dead animal products or hair of infected animals.

Clinical features

1. **Animal infection:** In animals, it is transmitted by ingestion of spores from soil. Animal to animal spread occurs very rarely. It is more common in herbivores than carnivores. It causes septicemia and rarely skin lesions. Bacilli are especially present in animal's spleen, which is enlarged and soft, giving rise to fever called splenic fever in ox and German name for the organism is Milzbrand bazillus (spleen destroying bacillus).
2. **Human infection:** Animal acquired the anthrax by ingestion of spores present in soil and, but human infection direct from soil spore is query. Three clinical types as shown in **Table 54.2 and Fig. 54.4**.

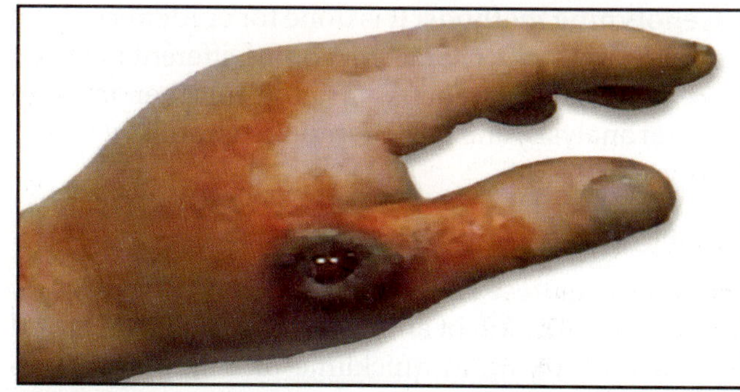

Fig. 54.4: Malignant pustule

Complications: These are septicemia and hemorrhagic meningitis.

Laboratory Diagnosis

Specimens

- **Human specimens:** These are blood, CSF, stool, vomitus, sputum, pus or discharge from pustules.
- **Animal specimens:** These are animal carcass like cut ear piece. Autopsy cannot be performed in animal died due to suspected anthrax due to risk of soil contamination.

Testing methods

A. Microscopy: Follow morphology.

B. Culture: Follow C/Cs.

C. Biochemical reactions: Follow B/Rs.

D. Serological tests:

- **Ag detection:** Ag is detected by CFT, gel diffusion, hemagglutination test, ELISA and animal inoculation technique. Animal inoculation is done by using animal samples. Apply organisms over shaven skin of guinea pig, later organisms will enter in tissues by minor abrasions. Ag is detected by Ascoli's thermoprecipitation test from tissues.
- **Ab detection:** It is detected by ELISA.
- **γ-phage sensitivity:** Spread the *Anthrax bacillus* on BHIA. Inoculate γ-phage on such plate, incubate at 37°C and observe for reaction. Clear area occurs due to lysis of bacilli.

F. Molecular method: PCR.

Flowchart 54.2: Virulence factors of *Anthrax bacillus*

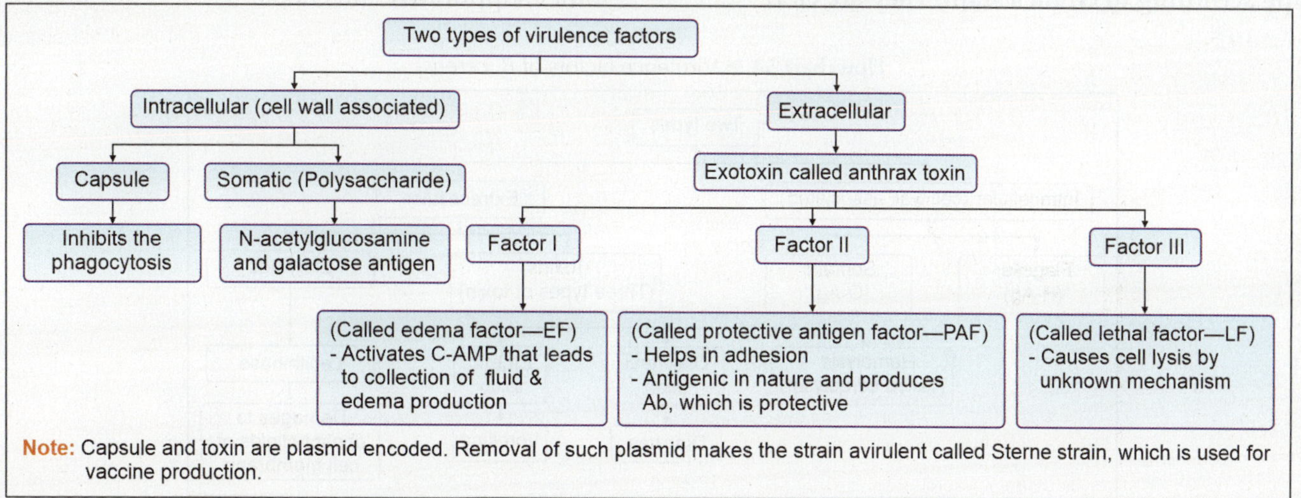

Infections of Bacillaceae (Sporing Aerobes)

G. Genotyping methods: It is done for epidemiological studies and strain characterization. Different methods are MLVA (multi locus variable number tandem repeat analysis) and AFLP (amplified fragment length polymorphism).

Prophylaxis

General measures: These are improvement of factory hygiene, sterilization of animal products like hides and wool, deep burying in quicklime of animal carcasses died due to anthrax or cremation of animals to prevent soil contamination.

Chemoprophylaxis: Doxycycline and ciprofloxacin are useful.

Immunoprophylaxis

1. **Vaccine prepared from vegetative bacilli:** It called Pasteur's anthrax vaccine prepared by attenuating anthrax bacilli at 42°C–43°C. It has only historical value.
2. **Vaccines prepared from spores:** These are used for animals and single injection can give protection for a year. These are not safe for human use. These include Sterne vaccine prepared from spores of non-capsulated, avirulent, mutant strain called Sterne strain and mazzucchi vaccine prepared from spores of stable attenuated Carbazzo strain in 2% saponin.
3. **Alum precipitated toxoid:** It is safe and effective for human and prepared from protective Ag of *Anthrax bacillus*. Three doses are given by IM route and 6 weeks interval between 1st and 2nd dose and 6 months interval between 2nd and 3rd dose. Booster dose will be given after a year, if necessary.

Treatment

Pn is the drug of choice, but resistant strains have been seen. Bacilli are susceptible to sulfonamide, erythromycin, doxycycline, ciprofloxacin and chloramphenicol. Antibiotics have no role after the production of toxin.

Bacillus cereus

Morphology

Type according to Gram's stain: They are GPB.

Motility: Bacilli are motile by peritrichous flagella, but some strains are nonmotile.

Spores: Bacilli have central and nonbulging spores.

Capsule: Bacilli are noncapsulated.

Cultural Characteristics (C/Cs)

A. Semisolid media: In gelatin medium, bacteria produce fast liquefaction in 4–7 days at 20°C.

B. Solid media
- **Nutrient agar plate:** Colonies are gray-white like *B anthracis* but less membranous consistency.
- **Blood agar:** Bacteria produce hemolytic colonies.
- **Egg-yolk agar:** Bacteria produce opacity around colonies indicating lecithinase production.
- **Chloral hydrate (0.25%) agar:** Anthrax growth is inhibited. *B. cereus* and *B. mycoides* can grow, which help to differentiate *Anthrax bacillus* from these bacilli.

C. Selective media
- **Baker's MYPA (Mannitol-egg Yolk Phenol red polymyxin Agar) medium:** Showing clear zones around pink colonies due to lecithinase production and mannitol nonfermentation. Polymyxin is selective agent.
- **Holbrook and Anderson's PEMBA (Polymyxin Egg yolk Mannitol Bromothymol blue Agar) medium:** It shows clear zones around blue colonies due to lecithinase production and mannitol non-fermentation.

Biochemical Reactions (B/Rs)

Sugar fermentation tests:

	G	S	L	M	Maltose
A	–	–	–	–

Pathogenicity

Virulence factors: Follow **Flowchart 54.3**.

Clinical features

1. **Opportunistic infections:** These are septicemia, meningitis, endocarditis, pneumonia, wound and other suppurative infections.

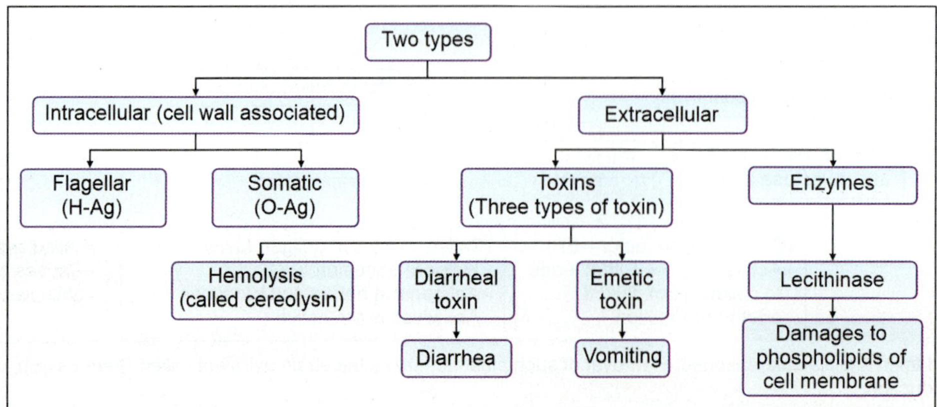

Flowchart 54.3: Virulence factors of *B. cereus*

Essentials of Medical Microbiology

TABLE 54.3: Types of food poisoning of *B. cereus*

Features	Diarrheal illness	Emetic illness
Type of food	Vegetable and cooked meat	Rice (Chinese restaurant)
Incubation period	8–16 hours after meal	1–5 hours after meal
Predominant features	Diarrhea and abdominal pain	Nausea and vomiting
Less features	Nausea, vomiting and fever	Diarrhea, abdominal pain and no fever (?)
Serotype	2, 6, 8, 9, 10 or 12	1, 3 or 5
Bacilli in stool	Less	More
Toxin nature	Heat labile and formed inside the intestine (no preformed or intradietic toxin)	Heat stable and preformed (intradietic)
Toxin resembling	Enterotoxin of *E. coli*	Enterotoxin of *Staphylococcus*
Mechanism of action of toxin	Enterotoxicity (Activation of C-AMP causes fluid collection in intestine)	Neurotoxicity (may be by vagal stimulation)

TABLE 54.4: Differences between anthracoid bacilli and *Anthrax bacillus*

Features	*Anthrax bacillus*	Anthracoid bacilli
Morphology		
Arrangement	Long chains	Short chains
Motility	Nonmotile	Motile by peritrichous flagella
Capsule	Present	Absent
M'fadyean reaction	Positive	Negative
Cultural characteristics (C/Cs)		
Growth at 45°C	Yes	No
Liquid media	Floccular deposits	Turbidity
Gelatin medium	Slow liquefaction	Fast liquefaction
Nutrient agar plate	Medusa head colony	Not typical
Blood agar	Weak hemolytic	Strong hemolytic
Penicillin (Pn) agar	No growth at 10U Pn /ml	Growth at 10U Pn /ml
0.25% Chloral hydrate agar	Growth is inhibited	Not inhibited
Animal inoculation	Pathogenic	Nonpathogenic
Others		
γ-phage sensitivity	Sensitive	Not sensitive
Salicin fermentation	Negative	Positive

2. **Food poisoning:** When food stored at warm temperature spores will germinate and produce toxin causing food poisoning. Two types of illness are produced like diarrheal and emetic, as mentioned in **Table 54.3** with differences.

Laboratory Diagnosis

Specimens: Blood, stool, vomitus and food materials.

Testing methods:

A. **Microscopy:** Follow morphology.

B. **Culture:** Follow C/Cs.

C. **Biochemical reactions:** Follow B/Rs.

ANTHRACOID BACILLI

They are also known as pseudoanthrax bacilli. Contaminants of culture resemble to *Anthrax bacillus* are called anthracoid bacilli. Differences between anthracoid bacilli and *Anthrax bacillus* are mentioned in **Table 54.4.**

ACCESS YOURSELF

Case Study

1. A female patient with history of malignant pustules with healed black scar admitted in medical ward. Identify the organism and answer the following:
   a. Name the causative agent and describe the morphology of causative agent.
   b. Describe the pathogenicity of causative agent.
   c. Write the lab., diagnosis of causative agent.

Essay/Full Question

1. *B. anthracis*

Short Notes

1. Pathogenicity/Lab., diagnosis of *B. anthracis*
2. *B. cereus*

Short Questions for Theory/Viva Questions

1. What is M'Fadyean reaction?
2. Name the four staining methods used to differentiate bacterial spores from fat globules.

3. What is duckering?

4. What are malignant pustules and eschar?

Comment on

1. Removal of plasmid from *B. anthracis* makes the bacillus avirulent.

MCQs for Chapter Review

Bacillus anthracis

1. Meaning of Anthrax:

a. Black eschar formation

b. Pseudomembrane formation

c. Susceptibility to γ phage

d. Typical colony characteristic

2. 'Bamboo stick' appearance is characteristic of:

a. *B. anthracis*

b. *Clostidium* spp.

c. *H. influenzae*

d. *Lactobacillus*

3. An abattoir worker developed pustule, which later progress to necrotic ulcer. Which of the following stain is useful for demonstration of organism from smear made from pustule?

a. Polychrome methylene blue

b. Calcofluor white

c. Giemsa

d. Modified Kinyoun's stain

4. A man, after skinning a dead animal, developed pustule on his hand. A smear prepared from the lesion showed the gram-positive bacilli in long chain which were positive for McFadyean reaction. The most likely etiological agent is:

a. *Clostridium tetani*

b. *Listeria monocytogenes*

c. *Bacillus anthracis*

d. *Actinomyces* sp.

5. A malignant pustule is a term used for:

a. An infected malignant melanoma

b. A carbuncle

c. A rapidly spreading rodent ulcer

d. Anthrax of skin

6. A malignant pustule is caused by:

a. *B. anthracis*

b. Leishmaniasis

c. Basal cell carcinoma

d. All

7. True regarding anthrax is all *except*:

a. Caused by insect bite

b. Caused by rubbing of skin

c. Cutaneous types is rare nowadays

d. Pulmonary infection occurs by inhalation

8. Which of the following statement is true regarding anthrax?

a. McFadyean reaction shows capsule.

b. Humans are usually resistant to infections.

c. Less than 100 spores can cause pulmonary infection.

d. Gram's stain shows organisms with bulging spores.

e. Sputum microscopy helps in diagnosis.

9. True about anthrax:

a. Caused by gram-positive bacilli

b. Soil reservoir

c. Spore formation takes place.

d. It is more common in carnivores than herbivores.

e. Pn is the drug of choice.

10. All of the followings are true regarding *Bacillus anthracis except*:

a. Plasmid is responsible for toxin production.

b. Cutaneous anthrax generally resolves spontaneously.

c. Capsular polysaccharide aids virulence by inhibiting phagocytosis.

d. Toxin is a complex of two fractions.

Bacillus cereus

11. Characteristic of *Bacillus cereus* food poisoning is:

a. Presence of fever

b. Presence of pain in abdomen

c. Absence of vomiting

b. Absence of diarrhea

Anthacoid bacilli

12. *Anthrax bacillus* differs from anthracoid bacilli by being:

a. Noncapsulated

b. Strict aerobe

c. Nonmotile

d. Hemolytic colonies on blood agar

Answers and Explanation of MCQs

1. a

- Follow section, *Bacillus anthracis* **(meaning)** for explanation.

2. a

- Follow section, *Bacillus anthracis* **(morphology)** for explanation.

3. a and c

- It is a case of cutaneous anthrax (hide porter's disease), where malignant pustule is the characteristic features and commonly presents in abattoir workers (slaughter house workers) due to direct contact with infected meat or carrying the infected skin on back. Organism is *Bacillus anthracis* which is capsulated in nature and stains useful to demonstrate the capsule are explained in text under the section of morphology.

- Calcofluor white: Fluorescent stain for fungi.

- Modified Kinyoun's stain: Useful for *M. tuberculosis*.

4. c

- It is a case of cutaneous anthrax (hide porter's disease), where malignant pustule is the characteristic features and commonly presents in abattoir workers (slaughter house workers) due to direct contact with infected meat or carrying the infected skin on back. Organism is *Bacillus anthracis* which is gram-positive bacilli, arranged in long chain in tissue and capsulated. Capsule appears as purplish material around bacilli under polychrome methylene blue called M'Fadyean reaction.

5. d

6. a

7. c

- Follow section, *Bacillus anthracis* **(pathogenicity and Table 54.2)** for explanation of answers of MCQs 5–7.

8. a, b, c and e

- Option a: Follow section, *Bacillus anthracis* **(morphology)** for explanation.

- Option b: Anthrax is a zoonosis and humans are relatively resistant to infection.

- Option c: Follow section, *Bacillus anthracis* [**pathogenicity → source of infection** (LD_{50})] for explanation.

- Option d: Bacilli contain spores, which are equal to the breadth of bacillary bodies, so there is no bulging. Bulging spores are present in *Clostridium* spp.
- Option e: Sputum microscopy helps in diagnosis of pulmonary anthrax.

9. a, c and e

- Soil is the source, not the reservoir.
- Anthrax is more common in herbivores than carnivores.

10. d

- Capsule and toxin are plasmid encoded. Removal of such plasmid makes the strain avirulent called Sterne strain, which used for vaccine production.

- It resolved spontaneously though 10–20% develops fatal septicemia or meningitis.
- Capsular polysaccharide aids virulence by inhibiting phagocytosis.
- Toxin is a complex of not two but three fractions.

11. b

- Follow section, *Bacillus cereus* (**pathogenicity → clinical features and Table 54.3**) for explanation.

12. c

- Follow section, **Anthacoid bacilli (Table 54.4)** for explanation.

Infections of Clostridia (Sporing Anaerobes)

Chapter Outline

INTRODUCTION

Meaning

Word clostridia derived from Kloster means spindle, because bacteria contain spores, which are larger than the bacillary body and giving bacilli a spindle shape appearance.

Uses

1. **Chemical production:** *Cl. acetobutyricum* is used for production of acetone and butanol.
2. **Biological control:** Spores of nontoxigenic strain of *Cl. tetani* are employed as biological indicator for hot air oven.
3. **Decomposition:** *Cl. perfringens* and *Cl. tetani* are the intestinal flora in humans and animals. They invade the blood and tissues after death to initiate the decomposition of body.
4. **Bioterrorism:** *Cl. botulinum* is used as category A (highest priority) pathogen for bioterrorism.
5. **Therapeutic:** Intramuscular injection of type A toxin of *Cl. botulinum* called botulinum was 1st used for strabismus, is now recognized as safe and effective for many neuromuscular diseases as shown in **Table 55.1**.

Classification

Morphological classification

- According to (A/t) shape and location of spores: **Ch. 6**.
- A/t motility: Nonmotile include *Cl. perfringens* and *Cl. tetani* type VI while other clostridia are motile by peritrichous flagella.
- A/t capsules properties: Capsulated bacilli are *Cl. perfringens* and *Cl. butyricum* while other clostridia are noncapsulated.

TABLE 55.1: Therapeutic uses of botulinum

Muscular diseases	Plastic surgery
• Myoclonus • Palatal myoclonus • Focal dystonias • Fascial spasm • Painful muscle spasm • Tremor • Parkinson's progressive supra-nuclear palsy	• Masseter hypertrophy • Facial asymmetry (post Bells) • Wrinkles • Muscles flap paralysis during healing
Ophthalmic diseases	**ENT diseases**
• Strabismus • Lower lid entropion • Acquired nystagmus • Hyperlacrimation • Thyroid ophthalmology	• Stittering • Vocal cord polyps • Hypersalivation
Genitourinary diseases	**GIT diseases**
• Detrusor sphincter dyssynergia • Vaginismus	• Achalasia • Cricopharyngeal spasm • Rectal fissure
Rehabilitation	
Painful muscle spasm, tympanomandibular related muscle spasm and focal myofascial pain	

Classification according to morphological and biochemical nature: Follow **Table 55.2**.

According to O_2 effect

- Strict anaerobes: *Cl. botulinum*, *Cl. tetani* and *Cl. novyi*.
- Aerotolerant: *Cl. histolyticum*.
- Microaerophilic: *Cl. welchii* (*Cl. perfringens*).

Clinical classification

1. **Gas gangrene**
   - Established pathogens: *Cl. perfringens*, *Cl. septicum* and *Cl. novyi*.
   - Less pathogenic: *Cl. histolyticum* and *Cl. fallax*.

Spore	Both P and S		Either P or S		Neither P nor S
	Predominant P	**Predominant S**	**Slightly P**	**Slightly S**	
Central or subterminal	*Cl. bifermentans* *Cl. botulinum* A, B and F *Cl. sporogenes* *Cl. histolyticum*	*Cl. perfringens* *Cl. septicum* *Cl. novyi*	—	*Cl. fallax* *Cl. botulinum* C, D and E	—
Terminal oval	—	*Cl. difficle*	—	*Cl. tertium*	*Cl. cochlearum*
Terminal spherical	—	—	*Cl. tetani*	*Cl. tetanomorphum*	—

TABLE 55.2: Morphological and biochemical classification of *Clostridium*

P: Proteolytic → turns meat in black color with foul smelling, **S:** Saccharolytic → turns meat in pink color

- Doubtful pathogens: *Cl. bifermentans* and *Cl. sporogenes.*

2. **Food poisoning**
   - Gastroenteritis: *Cl. perfringens* type A.
   - Necrotizing enteritis: *Cl. perfringens* type C.
   - Botulism: *Cl. botulinum.*
3. **Tetanus:** *Cl. tetani.*
4. **Acute colitis:** *Cl. difficile.*

> **Note:** Spores of *Anthrax bacillus* are never formed in animal body during life, while of *Clostridium* spp., developed in animal body during life.

Clostridium perfringens

History and Synonyms

It also called *Cl. welchii* (because it was 1st isolated and described by Welch and Nuttall in 1892 from the blood and organs of cadaver) or bacillus aerogenes capsulatus or bacillus phlegmonis emphysematosa.

Morphology

Type according to Gram's stain: They are GPB.

Shape and size: Bacilli are elongated, straight rod with rounded or truncated ends called boxcar appearance. They are 4–6 μm × 1 μm in size.

Arrangement: They are arranged singly or in chains or in small bundles as shown in **Fig. 55.1**.

Spores: Bacilli have subterminal spores, but rarely seen in culture/tissues which is the characteristic feature of bacilli.

Motility and capsule: They are nonmotile but capsulated.

Other features: Bacilli are pleomorphic, filamentous and showing involution forms.

Cultural Characteristics (C/Cs)

Effective factors

- O_2 effect: They are strict anaerobe and microaerophilic.
- Optimum temperature: 45°C (20–50°C).
- pH: 5.5–8.0.
- Generation time: 10 minutes (utilized to obtain pure culture).
- Bacilli present in intestine, so isolation from the feces except in large number is not significant. Isolation from the food is meaningful.

Culture in media

- **Robertson's cooked meat medium (RCMM):** Bacilli grow rapidly and the meat turns pink due to saccharolytic property. Subculture from this will yield pure growth.
- **Blood agar:** Bacilli produce inner narrow zone of complete hemolysis by θ toxin and a much wider outer zone of incomplete hemolysis by α toxin called target (double zone) hemolysis as shown in **Fig. 55.2**.

Infections of Clostridia (Sporing Anaerobes)

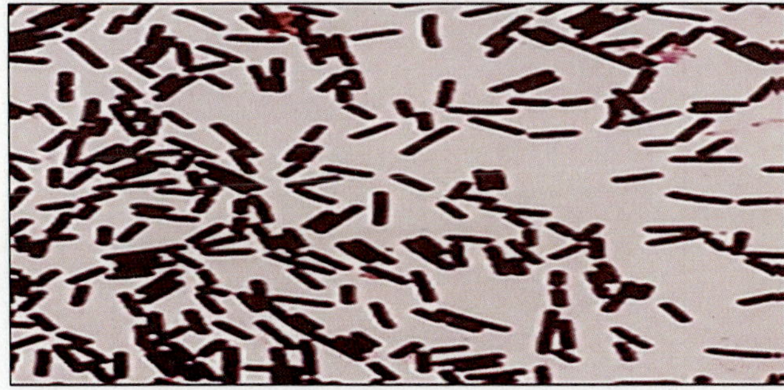

Fig. 55.1: *Cl. welchii* under Gram's stain

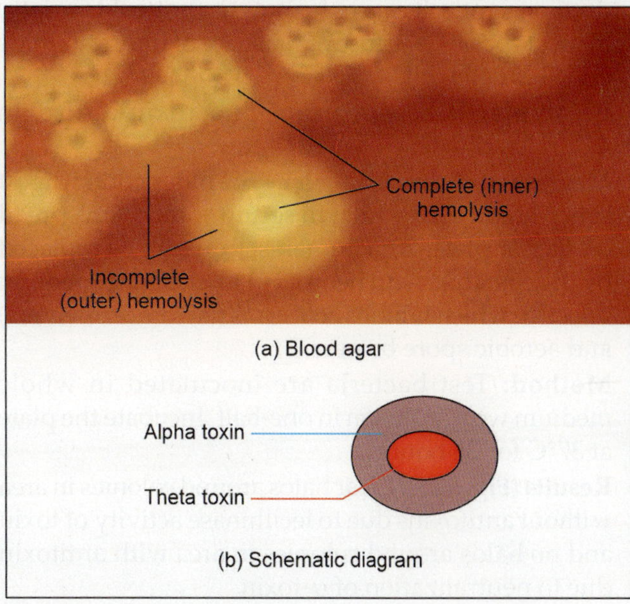

Fig. 55.2: Target hemolysis by *Cl. welchii*

- **Selective medium:** This is blood agar with neomycin.
- **Spores producing media:** These are Phillips medium, Ellner's medium, Duncan and Strong's medium and alkaline egg medium.

Automated culture: MALDI-TOF is used currently to identify the species from culture.

Culture in animals: Guinea pig is used for detection of virulence and typing.

Biochemical Reaction (B/Rs)

Saccharolytic and proteolytic properties: They are predominant saccharolytics and slightly proteolytics.

Sugar fermentation tests: Bacilli ferment glucose, sucrose, lactose and maltose with production of acid and gas.

Stormy fermentation (litmus milk test): Test is performed in litmus milk medium. Bacilli ferment lactose with acid and gas production. Acid changes litmus from blue to red and coagulates the casein (**acid clot**). Gas will disrupt the clotted milk. Paraffin plug is pushed up and shreds of clot are seen sticking to sides of the tube called "stormy fermentation".

I M Vi C tests: – + – –

Nitrate reduction test: It is positive.

H₂S test: Bacilli produce H_2S.

Nagler's reaction (Nagler effect)

- **Introduction:** Nagler reaction occurs due to α-toxin. It is produced by all types of *Cl. perfringens* (predominantly by type A). It is a lecithinase C (phospholipidase) which splits lecithin into phosphoryl choline and diglyceride in presence of Ca^{+2} and Mg^{+2}. This reaction is seen in egg yolk or serum medium and specifically neutralized by antitoxin. It is responsible for profound toxemia of gas gangrene. It is relatively heat stable, lethal and dermonecrotic. It causes hemolysis (**hot-cold variety**) in most species, because of presence of phospholipids on RBCs cell wall.
- **Principle:** It based on *in vitro* toxin-antitoxin neutralization test.
- **Medium:** It contains 6% agar + 5% fildes peptic digest sheep blood + 20% human serum + antitoxin spread on half of plate. 20% human serum will be replaced by 5% egg yolk. Add neomycin to make the medium selective which inhibits coliforms (lactose fermenters) and aerobic spore bearers.
- **Method:** Test bacteria are inoculated in whole medium with antitoxin in one-half. Incubate the plate at 37°C for 24 hours.
- **Results (Fig. 55.3):** Clear halos around colonies in area without antitoxins due to lecithinase activity of toxin and no halos around colonies in area with antitoxin due to neutralization of α-toxin.
- **Use:** It is used for rapid detection of *Cl. welchii*.

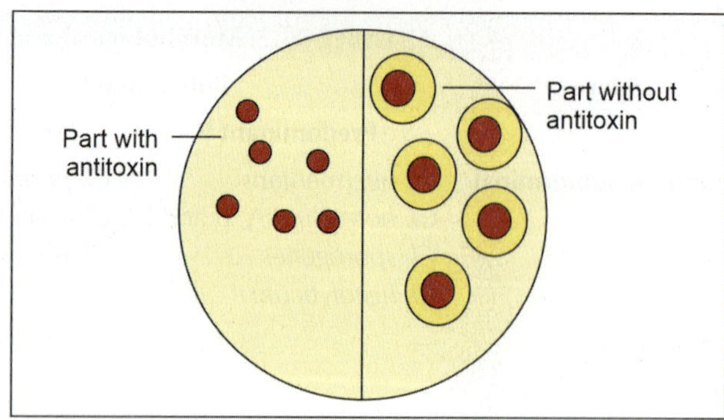

Fig. 55.3: Nagler's reaction

Interpretation: Nagler reaction may be positive by other bacteria like *Cl. novyi*, *Cl. bifermentans*, *Vibrio* and aerobic spore bearers.

Reverse CAMP test

- **Meaning:** Original CAMP test is used to detect the CAMP factor produced by *Strept. agalactiae* by using *Staph. aureus*. The test is called reverse CAMP test because CAMP factor produced by *S. agalactiae* is used for the detection of α-toxin of *Clostridium perfringens*.
- **History:** Hansen developed the test by using the synergistic relationship between the two microbes.
- **Principle:** It is based on synergistic relationship between the two microbes. Enhancement of hemolysis by α-toxin of *Clostridium perfringens* with synergistic action by CAMP factor of *Strept. agalactiae* (Group B β hemolytic *Streptococcus*).
- **Method:** A CAMP positive *Strept. agalactiae* is streaked in the center of sheep blood agar. Streak the *Cl. perfringens* at right angle to it without touching. Incubate the plate at 37°C for 24–48 hours in anaerobic conditions.
- **Result:** "Bow-tie" zone of enhanced hemolysis (**Fig. 55.4**) pointing towards *Strept. agalactiae* is seen. This is because of alpha toxin produced by *Cl. perfringens* interacts with CAMP factor and produces synergistic hemolysis.
- **Bacteria giving reverse CAMP test positive:** These are *Cl. welchii*, *C. pseudotuberculosis* and *A. haemolyticum*.

Resistance

Sterilization: Spores of gas gangrene causing strains of *Cl. perfringens* specially type A are inactivated by boiling

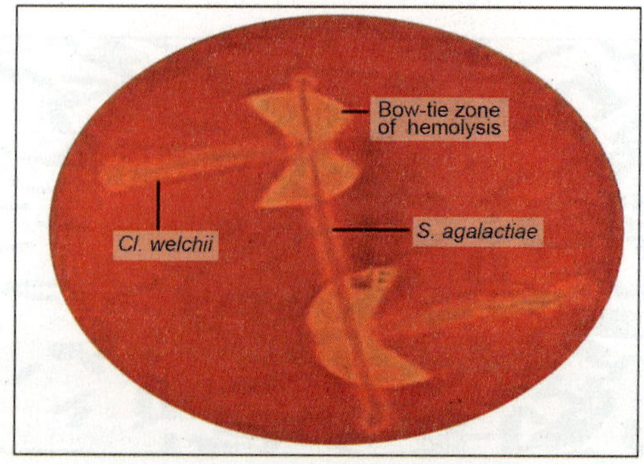

Fig. 55.4: Reverse CAMP test

in few minutes while spores of food causing strains of *Cl. perfringens* are heat resistant and can survive boiling for few hours. Physical methods to kill the spores are mentioned in **Ch. 6** (section → spores).

Disinfection: Chemical methods to kill the spores are mentioned in **Ch. 6** (section → spores).

Drug resistance: Drug resistant is not a big issue.

Pathogenicity

Disease name: *Cl. perfringens* causes different diseases like gas gangrene, food poisoning, etc., as described below.

Reservoirs of infection: *Cl. perfringens* (Type A is predominant) is a normal flora of large intestine of humans and animal. It is found in feces, and it contaminates the skin of perineum, buttocks and thigh via feces. The spores commonly found in soil, dust and air, which are the reservoirs.

Sources of infection: Spores are present in feces of humans, animals, soil, dust, clothes and hospital articles, which serve as source of infection.

Modes of transmission, incubation period, portal of entry and sites of infections: These are different in gas gangrene and food poisoning, described below.

Precipitating factors (epidemiological determinants): Three types.

A. Agent factors (virulence factors): *Cl. perfringens* is invasive; it can spread in tissues and produces following types virulence factors.

1. **Toxins:** It produces 12 distinct toxins. Major toxins are alpha toxin (described above in Nagler reaction) while other major toxins are beta, epsilon and iota, which are lethal, necrotizing and increasing the capillary permeability. Minor toxins are theta toxin, gamma, eta, delta (hemolytic for RBCs of sheep, goat and pig), kappa (collagenase), lambda (proteinase and gelatinase), mu (hyaluronidase) and nu (deoxyribonuclease). Theta (like streptolysin O) is hemolytic and antigenically related to streptolysin O. It is a thiol activated cytolysin also called perfringolysin O. Along with alpha toxin theta toxin plays an important role in gas gangrene.

2. **Enzymes:** These are neuraminidase, fibrinolysin and hyaluronidase (breaks the intracellular cement substances and allows the spread of infection).

3. **Biological active substances:** These are hemagglutinin, hemolysin (different from alpha, theta and delta), histamine, circulating factor and bursting factor (responsible for muscles damage).

B. Host factors: Infections are common following injuries during road accidents (involving crushing of large muscles), war (extensive wound with heavy contamination) or surgeries (especially amputation for vascular disease).

C. Environmental factors: Spores are distributed in soil, dust, clothes and hospital articles.

A. Gas gangrene

- **Synonym:** It also called malignant edema, anaerobic myositis, clostridial myonecrosis or clostridial myositis.
- **Definition (Fig. 55.5):** It is the rapidly spreading, edematous, myonecrosis occurs due to severe contamination of wound especially by *Cl. perfringens*.
- **History:** It was defined by Oakley in 1954.
- **Bacteriology:** Gas gangrene rarely occurs by single bacterium. It is a coinfection of clostridia with other bacteria. **Clostridial species** include *Cl. perfringens* most frequent cause (responsible for 60–80% cases of gas gangrene) followed by *Cl. novyi* and *Cl. septicum* (20–40% cases of gas gangrene by both) and less often by *Cl. histolyticum, Cl. sporogenes, Cl. fallax* and *Cl. bifermentans.* **Other bacteria** include staphylococci, anaerobic streptococci, *E. coli, Proteus* spp., etc.
- **Modes of transmission:** Endogenous infection occurs due to contamination of skin of perineum, buttock and thighs via feces. Exogenous infection occurs due to contamination of wound through soil, road dust, bits of clothing, etc. Gas gangrene may occasionally follow clean surgical procedures (amputation for vascular disease) and even injections (adrenalin).
- **Incubation period:** It is short as 7 hours or long up to 6 weeks. In *Cl. perfringens* it is about 10–48 hours (1–2 days), in *Cl. septicum* it is about 2–3 days and in *Cl. novyi* it is about 5–6 days.
- **Portal of entry:** Wounded skin.
- **Sites:** Muscles and other soft tissues.
- **Pathogenesis:** Conditions that produce low oxygen tension will favor the disease. Ca^{+2} and silicic acid in the soil allow multiplication of clostridia and causing necrosis. Alpha toxin (phospholipidase) damages the phospholipids of cell membrane and increases permeability leading to extravasations of blood and increased tension on vessels and affected muscles, which reduces the blood supply and causes further anoxic damage. Alpha toxin (phospholipidase) damages the phospholipids of cell membrane in RBCs causing hemolytic anemia and extravasated hemoglobin and myohemoglobinemia cease to act as an O_2 carrier. Abundant production of gas reduces the blood supply by pressure effect, causing further anoxic damage. Reduction of Eh (oxidation – reduction potential) and pH allow further proliferation of bacteria. Break down of carbohydrate and liberation of amino acids from

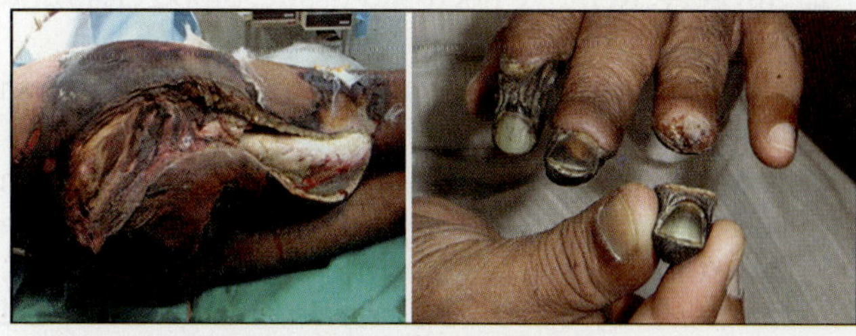

Fig. 55.5: Gas gangrene in different sites

5

protein provide ideal platform for proliferation of anaerobes. Collagenase breaks the collagen fiber and hyaluronidase breaks the intercellular substance, help clostridia to spread further.

- **Clinical types given by MacLenan**
  1. **Simple contamination of wound:** It includes small numbers of bacteria with low virulence. They do not invade underlying tissues, but causing delayed wound healing.
  2. **Anaerobic cellulitis:** It affects fascia, but not the muscles. Minimal toxin production, limited gas production and good prognosis.
  3. **Gas gangrene:** It affects the healthy muscles with abundant toxin production.
- **Clinical features:** Local features are swelling and edema of infected area, pain, tenderness, crepitation due to gas collection, foul smelling, watery discharge and rapidly spreading necrosis. General features are high grade fever, toxemia and prostration.
- **Complications:** These are shock and death due to circulatory failure.
- **Laboratory diagnosis of gas gangrene:** Diagnosis based on clinical parameters and laboratory help required for confirmation.
  - Specimens: Collect exudates or pus or discharge swab or pipette. Collect the necrotic tissues and muscle fragments. Preparation the smear directly from edge of muscle.
  - Transportation: All samples collected and transported in RCMM/thioglycollate broth.
  - Testing methods
    1. **Microscopy:** Gram's stain feature of *Cl. perfringens* are described above. In Gram's stain, *Cl. novyi* appears as large bacilli with oval subterminal spores. *Cl. septicum* appears as citron bodies and leaf-boat shaped bacilli with central/subterminal spores.
    2. **C/Cs and B/Rs:** C/Cs and B/Rs of *Cl. perfringens* are described above.
    3. **Bacterial typing:** Total five types like A, B, C, D and E are identified on the basis of *in vivo* toxin-antitoxin neutralization test. Typing is done by intracutaneous injection in guinea pig or intravenous in mouse.
    4. **Physiological method:** This includes gas liquid chromatography (GLC).
- **Prophylaxis of gas gangrene**
  1. **General measures:** Keep the wound clean.
  2. **Surgery:** Remove necrosed tissue, blood clots and foreign materials from wound.
  3. **Chemoprophylaxis:** Metronidazole, gentamicin and amoxicillin are useful.
  4. **Immunoprophylaxis:** Antigas gangrene serum (AGS) contains equine polyvalent antitoxin in a dose of 10,000 IU of *Cl. perfringens*, 10,000 IU of *Cl. novyi* and 5,000 IU of *Cl. septicum* is given by IM (in emergencies by IV) route.

- **Treatment of gas gangrene**
  1. **Surgery:** Excision of all affected parts (**amputation**) with antibiotics cover is lifesaving.
  2. **Hyperbaric O$_2$:** It also beneficial in treatment.
  3. **Antibiotics:** Metronidazole, gentamicin and amoxicillin are useful.
  4. **Passive immunization with AGS serum:** It has not yields any useful result, so it is not advised.

B. Food poisoning

- **Bacteriology:** It occurs due to enterotoxin production by some strains of type A *Cl. perfringens*. Enterotoxin is heat labile and similar to enterotoxin of ETEC and *V. cholerae*. It leads fluid accumulation in the rabbit ileal loop. Food poisoning strains also produce the alpha and theta toxins.
- **Mode of transmission:** It occurs due to ingestion of contaminated meat (like beef and poultry meat) and legumes. When contaminated meat is cooked, spores in interior may survive. During storage or warming they germinate and multiply in anaerobic condition. Large numbers of clostridia thus consumed, pass unharmed by gastric acid due to high protein in meal and reach intestines where they produce toxin (no preformed or intradietic toxin).
- **Incubation period:** It is 8–24 hours
- **Portal of entry:** GIT
- **Sites:** GIT
- **Pathogenesis:** Enterotoxin activates C-AMP, which causes collection of fluid in intestine.
- **Clinical features:** Abdominal pain/cramps, diarrhea and vomiting. These features are self-limited and recover in 24–48 hours.
- **Laboratory diagnosis of food poisoning**
  - Specimens: Stool and suspected food particles
  - Transportation: All samples collected and transported in RCMM/thioglycollate broth
  - Testing methods:
    1. Microscopy: Follow morphology (Gram's stain).
    2. Culture: Follow C/Cs.
    3. Biochemical reactions: Follow B/Rs.

C. Necrotizing enteritis or enteritis necroticans or pigbel:
It is the gas gangrene of bowel mucosa due to ingestion of pig meat with trypsin inhibitor like sweet potatoes. It occurs in susceptible host having limited proteolytic activity. It occurs due to beta toxin produced by type C of *Cl. perfringens*. Source of organism is patient own intestinal flora. It presents with intestinal features.

D. Other infections

- Necrotizing enterocolitis: It is the fulminant disease in premature low birth weight infants leading to death.
- Gangrenous appendicitis: It includes inflamed and necrosed appendix but no perforation or abscess.
- Gas gangrene of abdominal wall: It originates from intestine during abdominal surgery.

- Biliary tract infection: Two types are reported like acute emphysematous cholecystitis and post cholecystectomy septicemia.
- Urogenital infection: It occurs following nephrectomy or septic abortion.
- Thoracic infection: It is more in war due to thoracic injuries.
- Panophthalmitis: It occurs following penetrating injury of eyes.
- CNS infections: Like brain abscess and meningitis.
- CVS and BSIs: Like septicemia and endocarditis.

Clostridium tetani

Synonym

Cl tetani also known as Nicolaier's bacillus who 1st discovered the tetanus toxin.

History

Bacilli were 1st isolated by Kitasato Shibasabur in 1889 in pure culture and reproduced the disease in animals by inoculation of pure culture. He also noticed that toxin could be neutralized by specific antibodies.

Morphology

Type according to Gram's stain: They are strong GPB in young culture but in old culture they appear as Gram's variable and probably GNB.

Shape ans size: They are slender rod with straight axis, parallel sides and rounded ends and about 0.5 µm × 4–8 µm in size.

Arrangement: They are arranged singly or in chains.

Capsule: They are noncapsulated.

Motile: They are motile by peritrichate flagella except *Cl. tetani* type VI.

Spores: They have terminal and spherical spores giving drumstick appearance **(Fig. 55.6)** to bacilli.

Cultural Characteristics (C/Cs)

Effective factors
- O₂ effect: They are strict anaerobes.
- Optimum temperature: 37°C.
- pH: 7.4
- They grow on ordinary media and enriched media contain blood or serum but not glucose.

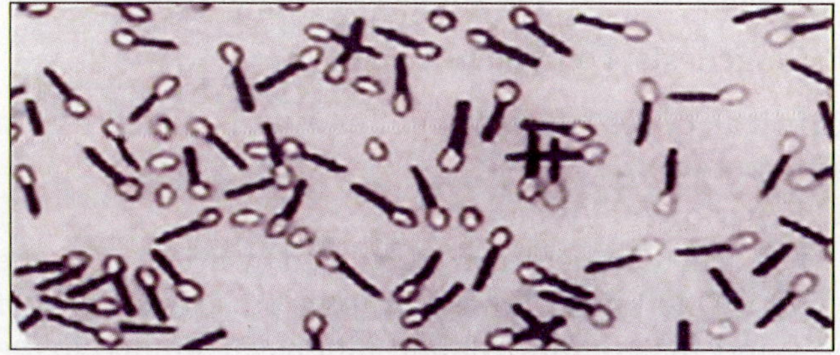

Fig. 55.6: *Cl. tetani* under Gram's stain

Culture in media

A. Liquid media
- **Thioglycollate broth:** Turbidity occurs in the bottom.
- **RCMM:** Bacteria produce gas and turbidity. Meat turned black on prolonged incubation due to slightly proteolytic activity of *Cl. tetani*. Three RCMM bottles are used, 1st bottle heated at 80°C for 15 min., 2nd bottle heated at 80°C for 5 min., and 3rd bottle left unheated. Purpose of heating is to kill vegetative forms and leaving the spores undamaged. Incubate the bottles at 37°C for 24-48 hours and do subculture on blood agar daily up to 4 days and observed for growth.

B. Semisolid medium:
Gelatin stab culture where bacteria produce fir tree growth with slow liquefaction.

C. Solid media
- **Blood agar:** Bacteria produce α-hemolysis (incomplete) initially followed by β-hemolysis (complete) due to production of hemolysin called tetanolysin. They produce swarming growth on opposite half of plate after 1–2 days anaerobic incubation.
- **MacConkey's medium:** Green flourescence is produced by bacteria.

D. Selective medium:
It is the blood agar with polymyxin B.

Culture in animals: Widely described below in laboratory diagnosis under the heading of toxigenicity testing (*in vivo*).

Biochemical Reactions (B/Rs)

Saccharolytic and proteolytic properties: They are non saccharolytics, but slightly proteolytics.

Sugar fermentation test: They **are nonferments.**

I M Vi C tests: + – – –

Nitrate reduction test: It is negative.

H₂S test: Not produced.

Resistance

Sterilization: Spores of some strains of *Cl. tetani* are destroyed by boiling in 5 minutes while spores of few strains are heat resistant and can survive boiling for 15–90 minutes. Spores can survive in soil (dry earth) for several years. Physical methods to kill the spores are mentioned in **Ch. 6** (section → spores).

Disinfection: They resist killing by 5% phenol or 0.1% mercury chloride. Chemical methods to kill the spores are mentioned in **Ch. 6** (section → spores).

Drug resistance: Drug resistant is not a big issue.

Pathogenicity

Disease name: It called tetanus.

Definition: Tetanus is toxemia occurs due to powerful neurotoxin (exotoxin) called tetanospasmin and characterized by tonic muscular spasm which is initially local followed by involvement of entire somatic muscular system.

Epidemiology: Tetanus is known from the time of Hippocrates and Aretaeus.

Reservoirs and sources of infection: Spores are ubiquitous, present in soil, intestine of humans-animals (mainly herbivorous), hospital environment, plaster of Paris, bandages, cotton, catgut, talc granules, clothes, stitch materials, syringes and needles. All these serve as reservoirs and sources of infection.

Modes of transmission: Spores are transmitted by injuries, suppurative infections like otitis media, unsterile injections, septic abortions, unhygienic practices like application of cow dung, soil or ashes on the umbilical stump, rituals such as ear boring or circumcision, etc. It is not transmitted from person to person. Spores are destroyed by phagocytes and harmless in contaminated wound. Germination and toxin production occurs in favorable conditions like reduced Eh potential, coinfection by anaerobes, necrosed and devitalized tissues and presence of foreign bodies.

Incubation period: It is 6–12 days; however, it may be short as 1 day and long as a month. It is influenced by site of wound, nature of wound, immune status of patient, dose of organism and toxigenicity of organism.

Portal of entry: Wounded skin or tissues.

Sites: Toxin targets muscles by presynaptic action.

Precipitating factors (epidemiological determinants): Three types.

A. Agent factors (virulence factors): *Cl. tetani* is little invasive, remains at the site of lodgment and produces two types of powerful exotoxins as follows.
1. **Tetanolysin (hemolysin):** It is heat labile, oxygen labile, antigenically related to hemolysin produced by *Cl. perfringens*, *Cl. novyi* and *S. pyogenes*. It is not relevant to pathogenesis.
2. **Tetanospasmin (neurotoxin)**
   - **History:** Arthur Nicolaier isolated the tetanus toxin from free-living, anaerobic soil bacteria in 1884.
   - **Properties:** It is heat labile, oxygen stable, plasmid encoded, good antigenic and neutralized by specific antitoxin. It has two chains like H-chain (MW 93,000) and L-chain (MW 52,000) joined by disulfide bond. It changes to toxoid by treatment with low concentrations of formaldehyde. Purified toxin is active in extremely small amounts. MLD is different in different species like for human MLD is 130 nanograms.

- **Effective factors over toxicity:** It is influenced by routes of administration. If entered orally, it destroyed by digestive enzymes, so no effect. Subcutaneous, intramuscular and intravenous injections are equally effective. Intraneural injection is lethal. Injection direct in CNS is more lethal.
- **Site of action:** It has strychnine like action, but acts presynaptically, while strychnine acts post-syneptically.

Note: Examples of neurotoxins
- Enterotoxin of *Staph. aureus*.
- Enterotoxin (emetic illness) of *B. cereus*
- Tetanospasmin from *Cl. tetani*
- Botulinum from *Cl. botulinum* except C2.
- Shiga like toxin: (Having cytotoxic, enterotoxic and neurotoxic properties) from *E. coli* and *V. cholerae*
- Shiga (having cytotoxic, enterotoxic and neurotoxic properties) toxin from *Shigella* spp.

B. Host factors
1. **Age:** Tetanus is the disease of 5–15 years of age, which is most vulnerable for trauma. Neonatal tetanus is common with unhygienic practices like application of cow dung, soil or ashes on the umbilical stump.
2. **Sex:** Men are more susceptible to tetanus toxin than women. In women, it is common due to septic abortion of delivery, called uterine tetanus.
3. **Occupation:** Agriculture workers are more prone to infection due to repeated contact with soil.
4. **Immunity:** No age is immune, unless protected by previous immunization. Immunity can be transferred from mother to the baby, if mother is immunized or having high immunity during pregnancy.
5. **Other factors:** Host factors which favor the transmission of spores are like injuries, suppurative infections like otitis media, unsterile injections and rituals such as ear boring or circumcision.

C. Environmental factors
1. **Distribution of spores:** Spores are ubiquitous, present everywhere as described earlier.
2. **Seasonal:** Tetanus is common in developing countries where the climate is warm in summer months and in rural areas where soil is fertile by bacteria from the animal/human feces.

Pathogenesis: Follow **Flowchart 55.1**.

Clinical types of tetanus: Tetanus is classified in different ways:
1. **Experimental types**
   - **Local tetanus:** Toxin inoculated by IM route in one of hind limb followed by tonic muscles spasms of inoculated limb. Manifestations are limited to the muscles near the wound.

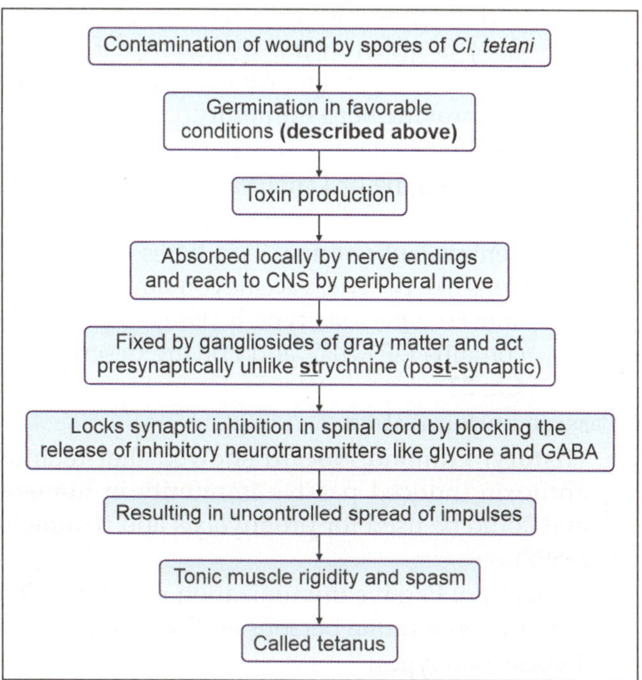

- **Ascending tetanus:** Toxin inoculated by IM route in one of hind limbs spread up into spinal cord and involved opposite hind limb, trunk and forelimbs in ascending fashion.
- **Descending tetanus:** Toxin injected intravenously; spasticity develops first in muscles of head and neck and then spreads downward like natural tetanus.

2. **According to modes of onset**
   - **Otogenic tetanus:** It develops after ear infection. Mortality rate is less.
   - **Cephalic tetanus:** It follows the head injury.
   - **Uterine tetanus:** It may be due to aseptic abortion. Mortality rate is 70–100%.
   - **Tetanus neonatorum:** It follows the unsterile treatment of the umbilical cord stump. Mortality rate is 70–100%.

Clinical features: These are increased muscle tone with generalized/neck/back rigidity/stiffness. There is tonic muscle spasm 1st local and later involving entire somatic muscular system **(Fig. 55.7a)**. Slightest

stimulation can produce generalized muscle spasm, but not seizure. Muscle spasm is very painful. It can threaten the respiration by laryngospasm or by sustained contraction of respiratory muscles. Trismus (lock jaw) is the characteristic feature **(Fig. 55.7b)** of tetanus. Back muscles are more powerful, so creating backward arc called opisthotonus, noticed by Sir Charles Bell in1809 as shown in **Fig. 55.7c**. Electromyogram shows continuous discharge of motor units and shortening/absence of silent interval normally seen after an action potential.

Complications: Mortality rate is very high about 80–90% before specific treatment and 15–50% with specific treatment.

Laboratory Diagnosis

Tetanus is diagnosed on clinical ground with history of injury and introduction of soil/feces in wound. Laboratory help is required for confirmation.

Specimens: These are wound swab, exudates or discharge from wound and necrosed tissue from the depth of wound.

Transport: All samples are collected and transported in RCMM.

Testing methods

A. **Microscopy:** Gram's stain is not reliable, because bacilli may present in wound without tetanus, and also it does not differentiate *Cl. tetani* from *Cl. tetanomorphum* and *Cl. sphenoides*.

B. **Culture:** Follow C/Cs (Most reliable).

C. **Biochemical reactions:** Follow B/Rs.

D. **Blood picture:** Leukocytosis and raised muscle's enzymes level.

E. **Bacterial typing:** Total 1–10 serotypes are identified by using flagellar Ag with agglutination reaction. Type 6 is non flagellar strain. All the types produce the same toxin which is neutralized by antitoxin of other type.

F. **Toxigenicity testing:** Two types:
   1. *In vitro* **in blood agar:** It contains 4% agar which prevents the swarming. Take one blood agar plate with antitoxin in one half and other half without antitoxin. Inoculate *Cl. tetani* on both

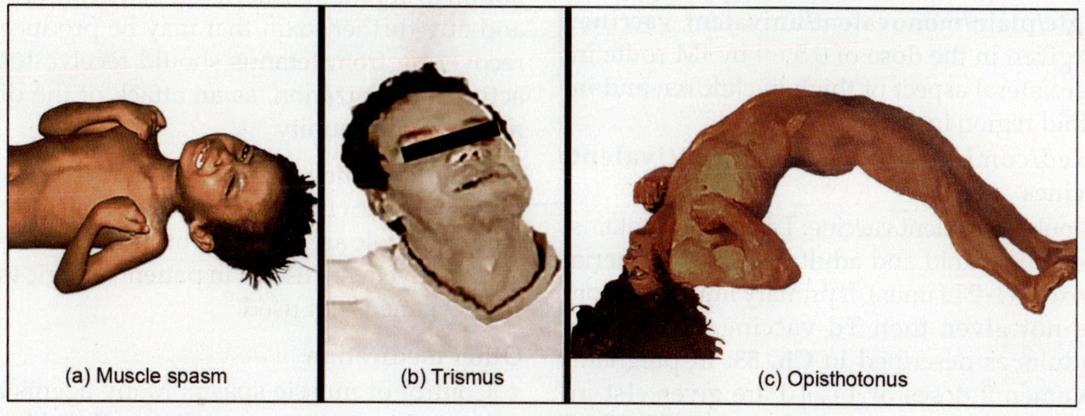

(a) Muscle spasm (b) Trismus (c) Opisthotonus

Fig. 55.7: Features in tetanus

half. Incubate the plate anaerobically for 24-48 hours and observed for hemolysis around colony. The half which contains antitoxin does not show any hemolysis. It detects only tetanolysin not tetanospasmin, which is pathogenic toxin.

2. *In vivo* **in laboratory animal:** Inoculate 0.2 ml culture in root of tail in two different mice. One is protected with 1000 U of tetanus antitoxin before 1 hour, acts as control animal. After 12–24 hours test animal shows stiffness of tail and dies within two days. Control animal does not show any features due to neutralization of toxin by antitoxin. It detects tetanospasmin.

Prophylaxis

General measures: Keep the wound clean.

Surgery: Removal of necrosed tissues, blood clots and foreign materials to prevent the development of anaerobic environment.

Chemoprophylaxis: Antibiotics are useful when given in 4 hours before toxin production. They kill the bacilli and prevent the toxin production. Pn and tetracycline are used systemically while bacitracin or neomycin used locally.

Immunoprophylaxis: Three types:

A. **Active immunization:** It is achieved by tetanus toxoid (TT) which was developed by P. Descombey in 1924. Tetanus antitoxin level of 0.01 IU or more per ml of blood is considered as index of immunity/ protective titer. TT has different types like **APT** gives reactions so not useful, **FT** or plain toxoid prepared by incubating the toxin with formalin and **purified (adsorbed) toxoid** which is adsorbed on to insoluble aluminum compound usually aluminum phosphate called purified toxoid aluminum phosphate (PTAP).

- **Advantages of adsorbed TT:** The adsorbed toxoid provides higher and long-lasting immunity than plain toxoid.
- **Adverse reaction of adsorbed TT:** Frequent injections of TT cause hypersensitivity.
- **Storage of adsorbed TT:** Adsorbed TT should be stored between 4–10°C and no freezing at any time.
- **Administration of adsorbed TT**
  1. **Single/plain/monovalent/univalent vaccine:** It is given in the dose of 0.5 ml by IM route in anterolateral aspect of thigh in children and in deltoid region in adults.
  2. **Mixed/combined/polyvalent/multivalent vaccines**
     - Double/bivalent vaccine: Td vaccine contains tetanus toxoid and adult dose of diphtheria toxoid (1–2 Lf units). If primary immunization is not given then Td vaccine is given to adults as described in **Ch. 53**. In pregnant women 2 doses of TT/Td are given, 1st in early pregnancy and 2nd one month after

or booster in early pregnancy if immunized earlier. Immunity lasts for 10 years. Booster dose is given every 10 years or if wound occurs. Avoid frequent booster doses. TT does not provide herd immunity.
- Triple/trivalent/tetravalent vaccine: DPT/ DTP vaccine or DaPT vaccine are described in **Ch. 53**.
- Pentavalent vaccine: Two types of preparations like pentvac vaccine contains DPT, HBV and *H. influenzae* type b (Hib) or pentaxim contains DPT, IPV and Hib are described in **Ch. 53**.

B. **Passive immunization**
- **History:** Edmond Nocard showed that tetanus antitoxin induced passive immunity in humans and could be used for prophylaxis and treatment (1897).
- **Indication:** Passive immunization recommended only in nonimmune persons and only once.
- **Types:** Two types:
  1. **Antitetanus serum (ATS):** 1500 IU ATS is given by SC or IM route in nonimmune persons soon after receiving any tetanus prone injury. It causes immune elimination and hypersensitivity.
  2. **Tetanus immunoglobulin (TIG) of human:** It is the preparation of choice to neutralize the toxin. 250 IU is given, which has longer half-life.

C. **Combined immunization:** TIG on one hand and TT on other hand followed by 2nd and 3rd doses of TT at one month interval. Use adsorbed toxoid, because immune response to plain toxoid may be inhibited by TIG.

D. **Integrated prophylaxis of tetanus following injury:** It depends on the type of wound and immune status of patient as shown in **Table 55.3**.

Treatment

General measures: These are isolation of patients from noise and light. Maintenance of airway, breathing and circulation.

Immunization: TIG followed by full course of active immunization. TIG may not neutralize toxin already bound to nervous tissue, it can inactivate unbound toxin and any further toxin that may be produced. Patients recovering from tetanus should receive full course of active immunization, as an attack of the disease does not confer immunity.

Antibiotics: They are used to eradicate the source of toxin by removing the vegetative cells. Penicillin and metronidazole are the drugs of choice. Clindamycin and erythromycin are useful in patient allergic to penicillin. Tetracycline is not used.

Other medication
- Control of muscle spasm: Many agents as alone or in combination are used. **Benzodizepines** (dizepam,

TABLE 55.3: Integrated prophylaxis of tetanus following injury

Category	Immune status	Wound: <6 hours, nonpenetrating, clean, no/negligible tissue damage	Wound: Contaminated
A	Completely immune in <5 years	Nothing	Nothing
B	Completely immune in >5 – <10 years	1 dose of TT	1 dose of TT
C	Completely immune in >10 years	1 dose of TT	1 dose of TT + TIG
D	Incompletely or immune status is unknown	Complete course of TT	Complete course of TT + TIG

Note: Completely immune: Patient had full course of 3 injections of toxoid

lorazepam, midazolam, etc.) are most commonly used drugs. **Barbiturates** and **chlorpromazine** are the alternative agents.
- Autonomic dysfunctions: For sympathetic over activity labetalol, esmolol or clonidine are used.

Mechanical ventilation: If spasm is unresponsive to medication, then it could be controlled by mechanical ventilation with nondepolarizing neuromuscular blocking agents. Other agents useful are propofol, dantrolene, intrathecal baclofen, succinylcholine and magnesium sulfate.

Surgery: Tracheostomy is done to maintain airway.

Clostridium botulinum

Meaning

Botulinum word derived from botulus means sausage, a type of food prepared from meat. Sausage word is used for food item in the form of cylindrical length (**Fig. 55.8a**) of minced pork or other meat.

History

Bacilli were 1st isolated by van Ermengem from the piece of ham (salted meat from upper part of leg of pig like buttocks or thighs) that caused an outbreak of botulism in 1896.

Morphology

Type according to Gram's stain: They are GPB.

Shape and size: Bacilli are straight or curve rod and 1 μm × 5 μm in size.

Arrangement: They arranged singly, in pairs or in short chains.

Motility: They are motile with peritrichate flagella.

Spores: Spores are subterminal and oval (bulging) as shown in **Fig. 55.8b**.

Capsule: They are noncapsulated.

Cultural Characteristics (C/Cs)

Effective factors
- O_2 effect: They are strict anaerobes.
- Temperature: It is 35°C, but it can grow at 1–5°C.

Culture in media

A. Routine media
- **Nutrient agar:** Bacteria produce 3–8 mm, semi-transparent colonies with a fimbriated borders.
- **RCMM:** Type A, B and F are predominant proteolytic causing blackening of meat while C, D and E are saccharolytic turn meat in to pink color.
- **Alkaline glucose gelatin media:** Spores production occurs at 20–25°C.
- **Egg yolk agar:** Bacteria produce opalescence and pearly lipolytic effect.

B. Selective medium: It is the egg yolk agar wth cycloserine, sulfamethoxazole and trimithoprim.

Automated culture: Like MALDI-TOF is used to identify the species from culture.

Culture in animals: Food or feces macerated in saline and filtered-extracted is inoculated into two mice or two guinea pigs (one as control and other as test animal) intraperitoneally. Test animal die, if toxin present. Control animal protected with antitoxin remains healthy.

Biochemical Reactions (B/Rs)

Saccharolytic and proteolytic properties: Type C, D and E are saccharolytics, while A, B and F are predominant proteolytics.

Sugar fermentation tests: It ferments glucose and maltose with production of acid and gas, but not sucrose and lactose.

H_2S test: Bacteria produce H_2S.

Resistance

Sterilization: Spores of *Cl. botulinum* are survived on boiling after 3–4 hours; even at 105°C, they are not killed completely in less than 100 minutes. Physical methods to kill the spores are mentioned in **Ch. 6** (section → spores).

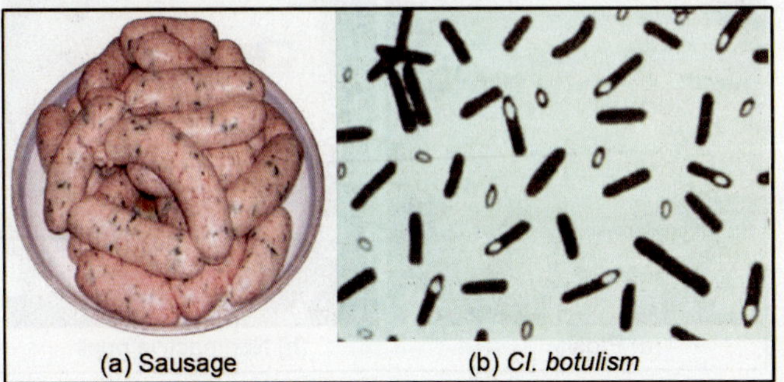

Fig. 55.8: Images of botulism

Disinfection: They resist killing by 5% phenol or 0.1% mercury chloride. Chemical methods to kill the spores are mentioned in **Ch. 6** (section → spores).

Drug resistance: Drug resistant is not a big issue.

Pathogenicity

Disease: It causes food poisoning called botulism.

Epidemiology: It is the most serious type of food poisoning, but occurs very rare. It kills two third of its victims. Infant botulism can occurs as sporadic cases, but not as an epidemic. Outbreak/epidemic of foodborne botulism has been reported from UK in 1989 with 27 cases and one death.

Reservoirs of infection: Spores are widely distributed in soil, dust and intestines of animals. Which are reservoirs of infections.

Sources of infection: Preserved food-meat or meat products, fish, sea foods and canned vegetables are the sources of infections.

Modes of transmission:
Incubation period: } Follow clinical types **(Described below).**

Portal of entry: GIT and skin (wound).

Sites: Synapse (presynaptic), neuromuscular junction, parasympathetic nerve endings and peripheral ganglia. CNS is not involved. No sensory features except blurred vision.

Precipitating factors (epidemiological determinants): Three types:

A. **Agent factors (virulence factors):** *Cl. botulinum* is noninvasive and produces powerful exotoxin in food (preformed toxin/intradietic toxin) called botulinum which is responsible for food poisoning. It produced intracellularly and released in medium only after death or autolysis of organisms not during the life of organisms. Initially, it is inactive as protoxin or progenitor toxin, converted to active form by trypsin and other proteolytic enzymes. It is heat labile (relatively heat stable), inactivated at 80°C in 30–40 minutes and at 100°C after 10 minutes. It is resistant to intestinal digestion and absorbed through small intestine in active form. MW is 70,000. It is the most lethal toxin and lethal dose for human being is 1–2 μg. Botulinum toxin once bound, then leads the permanent dysfunction of neurons. Recovery takes usually 3 months when dysfunctioned nerve terminal are replaced as a result of sprouting. It is a pure crystalline protein. It **is neurotoxin** and acts slowly, except C2, which is cytotoxic (enterotoxic). Human disease is caused by type A, B, E and F; however, few authors mentioned that all types can cause human disease. Toxin production is bacteriophage-encoded at least in types C and D. Botulinum is also produced by other species like *Cl. butyrocum* and *Cl. baratti*.

B. **Host factors:** Wound and infantile age are favoring factors.

C. **Environmental factors:** Spores are widely distributed in soil and dust.

Pathogenesis of botulism: Follow **Flowchart 55.2 and Fig. 55.9**.

Clinical types: Three main types of botulism with two newer types (total five types) as per modes of transmission.

Flowchart 55.2: Pathogenesis of botulism

Contamination of wound by *Cl. botulinum* spores and toxin production

↓

Toxin absorbed in vascular system

↓

Transported to peripheral cholinergic nerve terminals like synapse (presynaptic), neuromuscular junction, parasympathetic nerve endings and peripheral ganglias (preganglionic junction and post ganglionic nerves). CNS is not involved.

↓

Blocks the production or release of acetylcholine and causes descending flaccid paralysis (arflexia) called botulism.
- **Onset:** It starts within 18–24 hours and it is marked by diplopia, dysphagia and dysarthria
- **Other features:** It includes dyspnea, dizziness, dry mouth (thirst), ptosis **Fig. 55.9a**, blurred vision (only sensory deficit), ocular paresis, nausea, vomiting, abdominal pain, severe constipation/paralytic ileus (no diarrhea), urinary retention, upper and lower limb weakness and quadriplegia due to bilateral cranial nerve involvement.
- **Reflexes:** Following reflexes are suppressed (arflexia)
- Gag reflex may be suppressed
- Deep tendon reflex may be normal/decreased
- Pupillary reflex may be depressed or fixed called non reactive pupil as shown in **Fig. 55.9b**.
- **End stage**
- Symmetric descending paralysis is typical pattern.
- Death due respiratory paralysis or cardiac failure.

✓ **Note: Points to be remember in botulism.**
- **Botulism** presents with symmetric descending paralysis while **polio** presents with asymmetric descending paralysis
- No CNS involved
- No sensory involvement except blurred vision
- No fever. Normal or slow heart rate. Patient is responsive
- No loss of deep tendon reflex in early stage
- No pupillary reflex in **botulism** while **polio**, **diphtheria** and **porphyria** are present with pupillary reflex.

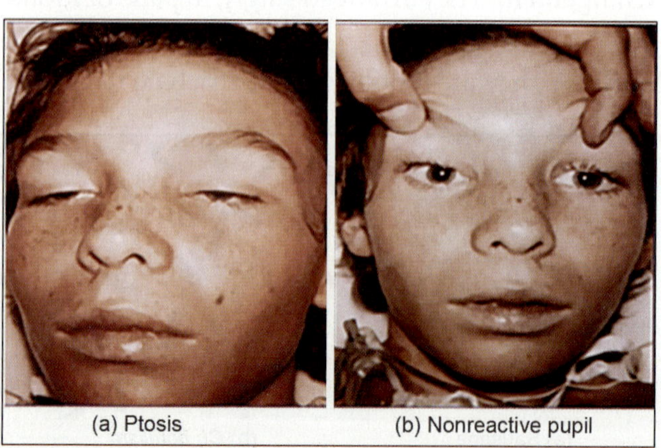

Fig. 55.9: Eye features in botulism

1. **Foodborne botulism**
   - **Modes of transmission:** It spreads by ingestion of preformed toxin in preserved meat or meat products, fish, sea foods and canned vegetables. Type F outbreak occurs especially with fish products. Spores contaminate the food, germinate and form toxin (preformed/intradietic toxin).
   - **Incubation period:** 12–36 hours following ingestion of food.
   - **Clinical features:** These are vomiting, thirst, constipation, ocular paresis, difficulty in swallowing, difficulty in speaking and difficulty in breathing are common features.
   - **Complication:** Death (20–70% cases) occurs due to respiratory failure in 1–7 days.
2. **Wound botulism**
   - **Mode of transmission:** Wound contamination by spores → germination → toxin production absorption. Most cases are due to type A.
   - **Clinical features:** Same as foodborne botulism except GIT features.
3. **Infant botulism**
   - **Modes of transmission:** It is the most common types of botulism. It also called floppy baby syndrome. It was first recognized in 1976, and is the most common form of botulism in the United States. It occurs in infants below 6 months. It occurs due to ingestion of food like honey contaminated by spores → ingestion of spores with honey in gut → germination → toxin (not a preformed/intradietic toxin). For this reason honey is not recommended for infants less than one year of age.
   - **Incubation period:** Symptoms of infant botulism begin between 3 and 30 days after an infant ingests the spores.
   - **Clinical features:** These are constipation, poor feeding, lethargy, weakness, pooled oral secretion, altered cry, floppiness and loss of head control (**Fig. 55.10**). Patient excretes toxin and spores in their feces. Most of the patient recovered with supportive therapy.
4. **Adult intestinal toxemia botulism:** It results from absorption of toxin produced *in situ* after rarely occurring intestinal colonization with toxigenic clostridia.
5. **Iatrogenic botulism:** It results from injection of botulinum.

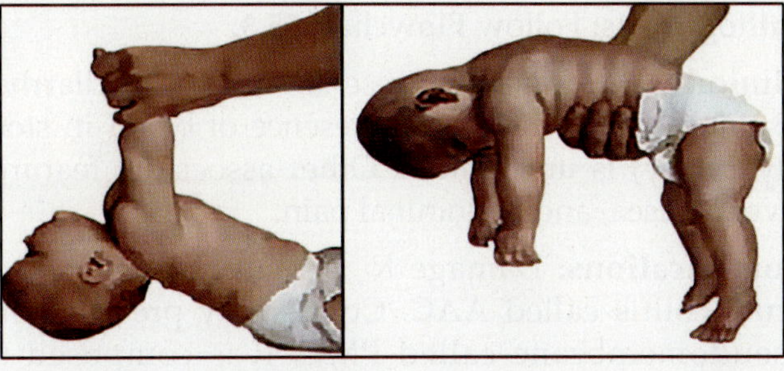

Fig. 55.10: Loss of head control in botulism

Laboratory Diagnosis

Specimens: Food, vomitus, feces or gastric fluid.

Testing methods

A. Demonstration of bacilli
   - **Microscopy:** Follow morphology.
   - **Culture:** Follow C/Cs.
   - **Biochemical reactions:** Follow B/Rs.
   - **Bacterial typing:** Eight types of *Cl. botulinum* have been identified (A, B, C1, C2, D, E, F and G) based on immunological differences in toxin produced by bacilli. C2 is cytotoxic, while all the others are neurotoxics.

B. Demonstration of toxin: Toxin is detected from food, feces, blood or liver in postmortem specimen.

C. Demonstration of antitoxin: Retrospective diagnosis can be made by detecting the antitoxin from the patient's serum, but not seen in all cases.

Prophylaxis

General measures: Proper caning and preservation of food.

Antitoxin: When an outbreak occurs, prophylactic dose of antitoxin should be given by IM route to all who consumed food.

Active immunization: Total 2 injections of aluminum sulfate adsorbed toxoid are given to laboratory workers who are at risk at interval of 10 weeks, followed by booster dose 1 year later.

Treatment

Botulism is treated by supportive therapy to maintain respiration and polyvalent antitoxin to type A, B and E may be administered immediately after diagnosis.

Clostridium difficile

Meaning

It is difficult to isolate, hence called *difficle*.

History

Bacilli were 1st isolated in 1935 from the stool of new born infant.

Morphology

Type according to Gram's stain: They are GPB but GNB in old culture.

Shape and size: They are rod shape and about 0.5–1 μm × 4–8 μm in size.

Motile: Oscillating motility occurs by peritrichate flagella.

Spores: They have terminal and oval spores.

Cultural Characteristics (C/C)

Effective factors:
   - O_2 effect: They are strict anaerobes.

- Temperature: 37°C.
- Diagnosis of *Cl. difficile* by culture is the most sensitive method.

Culture in media: (1) Blood agar: Bacteria produce non hemolytic colonies. **(2) Selective media:** These are cefoxitine cycloserine fructose agar (CCFA) and cefoxitine cycloserine egg yolk agar (CCEA).

Culture in cells: Hep-2 and human diploid cells are used for toxin demonstration. It is the most specific method for diagnosis of *Cl. difficile.*

Biochemical Reaction (B/Rs)

- They are predominant saccharolytics and weakly proteolytics.
- They ferment glucose, fructose and mannitol.
- Aesculin hydrolysis and gelatin liquefaction tests are positive.

Pathogenicity

Disease name: Disease called antibiotic-associated diarrhea (AAD) or *Clostridium difficile*-associated diarrhea or antibiotic-associated colitis (AAC) or pseudomembranous colitis (PMC).

Epidemiology: Outbreak with plenty of deaths has been reported from the different parts of world. Pseudo-membranous colitis was 1st described as a complication of *Cl. difficile* infection in 1978.

Reservoirs of infection: Spores are present as normal flora in the colon of neonate during initial 6 months. Spores are excreted in feces. Disease is present in 2–5% of the adult population, but studies have shown that colonize individuals have a less risk to disease (no endogenous infection) and never become sick, though they may still spread the infection.

Sources of infection: These are infected hands or food.

Modes of transmission: Spores present in the feces, which may contaminate the food or surface. Infection is exogenous, occurs by swallowing with infected hands, especially by healthcare workers, who get infected from the surface or food contact. It is transmitted from person to person by the feco-oral route. Once spores are ingested, their acid-resistance allows them to pass through the stomach unscathed. Upon exposure to bile acids, they germinate and multiply into vegetative cells in the colon. It is the most common bacterial cause of nosocomial diarrhea.

Incubation period: Clinical features usually develop within 5–10 days after starting antibiotics, but may occur as soon as the 1st day or up to 2 months later.

Portal of entry: GIT.

Site: Colon.

Precipitating factors (epidemiological determinants): Three types.

A. Agent factors (virulence factors): It produces two types of toxins that are glucosyltransferases that target and inactivate the Rho family of GTPase.
- **Toxin A:** It is histotoxic causes tissue damage and enterotoxic allows accumulation of fluid in colon to produce the bloody diarrhea.
- **Toxin B:** It is cytotoxin which damages the gut mucosa causing acute colitis with/without pseudo-membrane formation.

B. Host factors
1. **Age and sex:** Older age, women are at more risk.
2. **Drugs**
   - **Antibiotics:** Use of antibiotics like ampicillin/amoxicillin, tetracycline, chloramphenicol, lincomycin, clindamycin, cephalosporin (cefotaxime, ceftriaxone, cefuroxime and ceftazidime), etc., will disturb the normal flora and reduce the nutritional competition with other remaining bacteria, which allow the colonization of *Cl. difficile.* Pipericillin/tazobactam and ticarcillin/clavulanate are posses less risk for AAD.
   - **Antacids or proton-pump inhibitors:** H_2-receptor antagonists increase the risk 1.5-fold and proton-pump inhibitors like pantoprazole by 1.7 with once-daily use and 2.4 with more than once-daily use.
3. **Prolonged hospitalization:** It mostly presents in hospital environment and prolongs hospital stay and increases the chances of infection.
4. **Malignancy:** Acute myeloid leukemia and acute lymphocytic leukemia are the risk factors following chemotherapy.
5. **Other health problems:** These are use of electric rectal thermometer, central tube feeding, gastro-intestinal surgery and presence of other chronic diseases.

C. Environmental factors: The organisms form spores that are resistant to heat and alcohol-based hand cleansers or routine surface cleaners. Thus, they survive in clinical environments for long periods and they may be cultured from almost any surface. People are most often infected from the hospitals although infection outside medical settings is increasing. The rate of *Cl. difficile* acquisition is estimated to be 13% in patients with hospital stays of up to two weeks and 50% with stay longer than four weeks.

Pathogenesis: Follow **Flowchart 55.3.**

Clinical feature: Patients presents with watery diarrhea ≥3 times/day for ≥2 days. Presence of blood in stool (dysentery) is uncommon. Other associated features fever, nausea, and abdominal pain.

Complications: Damage to the gut mucosa causes acute colitis called AAC. Colitis may present with pseudomembrane called PMC. It is composed of inflammatory cells, fibrin, mucus and necrotic cells. It

Flowchart 55.3: Pathogenesis of AAC

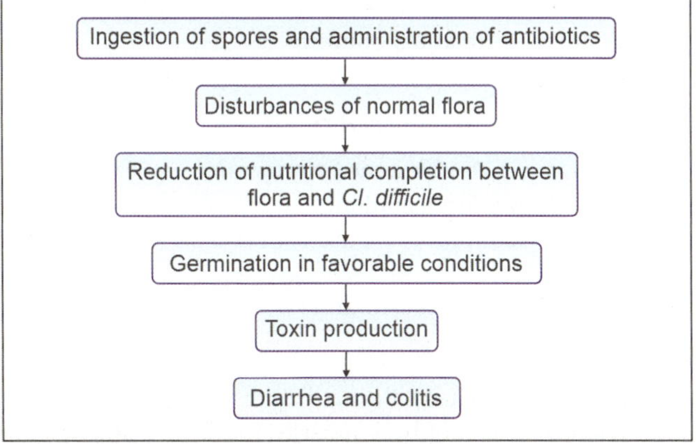

attaches to the underlying mucosa. It appears as whitish yellow plaque under the endoscopic examination (**Fig. 55.11**) and ranging from 1–2 mm to large enough to cover entire colonic mucosa. It causes problems in treatment against the effect of antibiotics. Other complications are toxic megacolon, perforation of colon and sepsis.

Laboratory Diagnosis

Specimens: Feces.

Testing methods

A. Demonstration of bacilli
- **Microscopy:** Follow morphology.
- **Culture:** Follow C/Cs.
- **Biochemical reaction:** Follow B/Rs.
- **Gas liquid chromatography (GLC)**

B. Toxin demonstration: It is demonstrated from feces by cell culture by using Hep-2 and human diploid cells and by ELISA.

C. Toxin neutralization: By antitoxin of *Cl. sordelli*.

D. Histological findings: In the earliest stage of disease tiny superficial encryptal erosions may be found called summit lesions.

E. Sigmoidoscopy: It is the most specific method along with toxin demonstration by cell culture.

F. Molecular testing: Real-time PCR aids in the diagnosis.

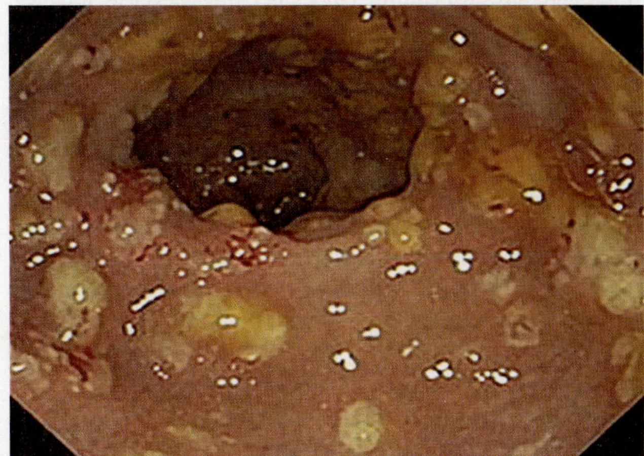

Fig. 55.11: Endoscopic image of pseudomembranous colitis, with yellow plaque seen on the wall of the sigmoid colon

Prevention

It is prevented by limiting antibiotics use, improvement of hand hygiene and cleaning in hospital. No specific vaccine is available.

Treatment

Discontinuation of antibiotics: It may result in resolution of symptoms within three days in about 20% of those infected.

Antibiotics: Antibiotic treatment may be difficult, due to resistance and physiological factors of the bacteria like spore formation plus protective effects of the pseudomembrane. Metronidazole is the drug of choice. Vancomycin, fidaxomi and bacitracin are other options. Relapse after treatment is common and seen in 15–30% cases.

Probiotics: These are less significant but may decrease the risk of relapse.

Fecal microbiota transplantation or fecal bacteriotherapy or stool transplant: It is approximately 85–90% effective in those for whom antibiotics have not worked. It involves infusion of bacterial flora acquired from the feces of a healthy donor to reverse the bacterial imbalance responsible for the recurring nature of the infection. Procedure replaces normal healthy colonic flora that had been wiped out by antibiotics and re-establishes resistance to colonization by *Cl. difficile*. There is evidence that looks hopeful that fecal transplant can be delivered in the form of a pill. They are available in the United States, but are not FDA-approved as of 2015.

Surgery: Colectomy may improve the outcomes in colitis patients.

OTHER CLOSTRIDIA

Clostridium septicum

It was first described by Pasteur and Joubert in 1887 and called *vibrion septique*. They are boat or leaf shaped pleomorphic GPB called citron bodies (**Fig. 55.12**). They are 3–8 × 0.6 µm in size, having oval, central or subterminal spores, motile by peritrichate flagella. Growth occurs anaerobically on ordinary media. Colonies are irregular and transparent initially, turning opaque on long time incubation. Hemolysis occurs on horse blood agar. It is saccharolytic and produces the

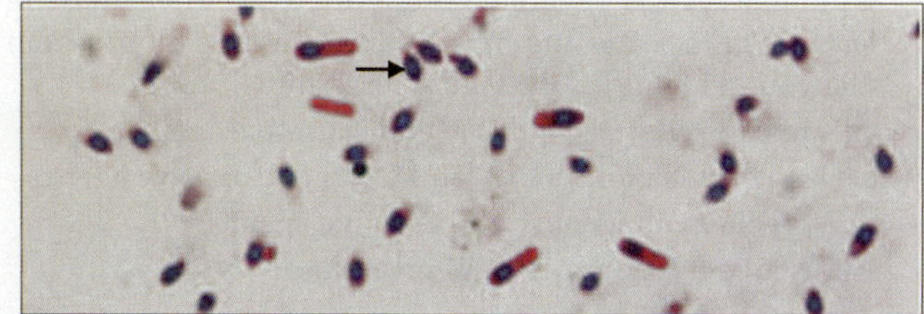

Fig. 55.12: Citron bodies (marked by arrow) of *Clostridium septicum*

gas. It has six types based on flagellar antigen. It found in soil and in animal intestine. It produces four toxins like **(1) α-toxin** which is hemolytic, dermonecrotic and lethal, **(2) β-toxin** which is leukotoxic and deoxyribonuclease, **(3) γ-toxin** which is hyaluronidase and **(4) δ-toxin** which is oxygen labile hemolysin. It also produces fibrinolysin. It is associated with gas gangrene in human being, usually with other clostridia.

Clostridium novyi

They are pleomorphic GPB, with large, oval and sub-terminal spores and motile by peritrichate flagella. They are strict anaerobes and killed on exposure of O_2. It is predominantly saccharolytic. It has four types based on toxins production. Only type A causes gas gangrene. It is widely distributed in soil, produces the four different types of toxins and associated with gas gangrene (mostly by type A) characterized by high mortality and large amount of edema fluid with little or no gas.

Clostridium histolyticum

They are GPB with subterminal bulging spores and motile by peritrichate flagella. It is aerotolerant and may grow in aerobic culture. It is predominantly proteolytic. It produces the five different types of toxins and infrequently associated with gas gangrene.

Clostridium tertium

They are GPB; often gram variable with terminal-oval spore which gives tennis racket appearance. It does not produce the exotoxin. It causes bacteremia, meningitis, septic arthritis, enterocolitis, spontaneous bacterial peritonitis, post-traumatic brain abscess, etc.

ACCESS YOURSELF

Case Studies

1. A truck driver admitted in hospital with accidental wound on his left lower limb after 5 days of accident. Wound presents with edema, severely contaminated with blood and soil and with watery discharge. Local examination of wound revealed tenderness on palpation and crepitation on auscultation. Discharge is collected and examined under Gram's stain, which noticed the GPB mixed with GPC. Identify the organism and answer the following:
   a. Name the clinical disease and describe the morphology of causative agent.
   b. Write the pathogenicity of causative agent.
   c. Describe the lab., diagnosis of causative agent.
2. An adult patient admitted in hospital following traumatic wound which is severely contaminated. Patient having features of lock jaw and muscle spasm. Tissue biopsy from the wound identified the GPB with terminal spores.
   a. Name the clinical condition and causative agent.
   b. Describe the morphology and C/Cs of causative agent.
   c. Write the pathogenicity of causative agent.
   d. Describe the prophylaxis of given case.

Essay/Full Question

1. *Cl. welchii / Cl. tetani.*

Short Notes

1. Nagler reaction/CAMP test and reverse CAMP test.
2. Gas gangrene/Tetanus/Botulism.
3. Antibiotic associated diarrhea/*(Cl. difficile).*

Short Questions for Theory/Viva Questions

1. What is target/double zone hemolysis?
2. What is stormy fermentation?
3. Write four examples of neurotoxin and bacteria producing it.
4. How does infant botulism differ from foodborne botulism?

Comments on

1. Gram's stain has less value in diagnosis of tetanus.
2. In tetanus, spasm of back muscles (backward arc/opisthotonus) is more common.
3. Overuse of antibiotics may precipitate the AAD/AAC.

MCQs for Chapter Review

Classification

1. Saccharolytic species of *Clostridium* are:
   a. *Cl. tertium* b. *Cl. cochlearum*
   c. *Cl. tetani* d. *Cl. tetanomorphum*

Clostridium perfringens

2. Boxcar appearance is the feature of:
   a. *Cl. tertium* b. *Cl. cochlearum*
   c. *Cl. tetani* d. *Cl. perfringens*
3. Gas gangrene is due to:
   a. Alpha toxin b. Theta toxin
   c. Beta toxin d. Delta toxin
   e. Epsilon toxin
4. Regarding gas gangrene one of the following is correct:
   a. It is due to *Cl. botulinum* infection
   b. *Clostridium* species are gram-negative spore forming anaerobes.
   c. The clinical features are due to release of protein endotoxin.
   d. Gas is invariably presents in muscles compartments.
5. Following statements are true regarding *Clostridium perfringens except*:
   a. It is the commonest cause of gas gangrene
   b. It is normally presents in human feces
   c. The principal toxin of the *Cl perfringens* is alpha toxin.
   d. Gas gangrene producing strains of *Cl. perfringens* produce heat resistant spores.
6. Opacity around colonies of *Cl. perfringens* is due to:
   a. Theta toxin
   b. Lecithinase
   c. Desmolase
   d. Cytokinin
7. Incubation period of gas gangrene is:
   a. 1–3 days b. 4–6 days
   c. 7–10 days d. 10–15 days
8. Gas gangrene is/are caused by all *except*:
   a. *Cl. novyi* b. *Cl. septicum*
   c. *Cl. histolyticum* d. *Cl. perfringens*
   e. *Cl. tetani*
9. True about gas gangrene:
   a. Underlying skin and muscles are normal
   b. Caused by tetanospasmin toxin
   c. Most common organism implicated is *Cl. perfringens*
   d. Passive immunization does not help

Clostridium tetani

10. All are true regarding tetanus *except*:
a. Transmitted through contaminated wound and injuries
b. Most common in winter and dry weather
c. Reservoir in soil and intestine of humans and animals
d. No herd immunity or lifelong immunity

11. True about tetanus:
a. Gram-negative spore-forming organism
b. Produces tetanolysin and tetanospasmin
c. Trismus and neck stiffness are early sign.
d. Generalized tonic-clonic seizures occurs on hyperstimulation.
e. Wound debridement is necessary.

12. Mechanism of action of tetanospasmin:
a. Inhibition of release of GABA
b. Inhibition of C-AMP
c. Inactivation of Ach receptors
d. Inhibition of c-GMP

13. Site of action of tetanus toxin:
a. Presynaptic terminal of spinal cord
b. Postsynaptic terminal of spinal cord
c. Neuromuscular junction
d. Muscle fibers

14. Which of the followings are true regarding tetanus?
a. Gram-negativee spore-forming aerobe
b. Because of tetanospasmin
c. Muscle enzymes are normal.
d. Vegetative cells will die with tetracycline.
e. Muscle spasm seen.

15. Toxigenicity of *Clostridium tetani* is tested in:
a. Rabbit
b. Horse
c. Mouse
d. Guinea pig

16. A person received complete immunization against tetanus 10 years ago. Now he presents with a clean wound without any laceration from an injury sustained 2.5 hours ago. He should now be given:
a. Full course of tetanus toxoid
b. Single dose of tetanus toxoid
c. Human tetanus globulin
b. Human tetanus globulin and single dose of tetanus toxoid

17. A 10-year old boy following to the road traffic accident presents to the casualty with a contaminated wound over left leg. He has received his complete primary immunization before preschool age and received a boosted dose of DT at school entry age. All of the following can be done *except*:
a. Injection of TT
b. Injection of human antiserum
c. Broad spectrum antibiotics
d. Wound debridement and cleaning

18. True statements regarding *Cl. tetani except*:
a. Spores are resistant to heat
b. Primary immunization consists three doses
c. Incubation period is 6–10 days
d. Man-to-man transmission is seen

Clostridium botulinum

19. Botulism is a disease of:
a. Neural transmission caused by the toxin of the bacterium *Clostridium botulinum*
b. Muscular transmission caused by the toxin of the bacterium *Clostridium botulinum*
c. Neuromuscular transmission caused by the toxin of the bacterium *Clostridium botulinum*
d. Non-neuromuscular transmission caused by the toxin of the bacterium *Clostridium botulinum*

20. Botulinum affects all *except*:
a. Neuromuscular junction
b. Preganglionic junction
c. Postganglionic nerves
d. CNS

21. Among the toxin produced by *Clostridium botulinum*, the non-neurotoxic one is:
a. A
b. B
c. C1
d. C2
e. D

22. True regarding botulism *except*:
a. Infant botulism is caused by preformed toxin
b. *Clostridium botulinum* A, B, C and F causes human disease
c. Gene for botulinum toxin is encoded by a bacteriophage
d. *Clostridium baratti* may cause botulism

23. Botulism is most commonly due to:
a. Egg
b. Milk
c. Meat
d. Dysarthria

24. All occurs in botulism *except*:
a. Diplopia
b. Diarrhea
c. Dysphagia
d. Dysarthria

25. Botulinum causes:
a. Descending flaccid paralysis
b. Ascending flaccid paralysis
c. Ascending paralysis
d. Ascending spastic paralysis

26. True regarding botulism *except*:
a. Botulism is caused by endotoxin
b. Honey ingestion causes infant botulism
c. Constipation is seen
d. Detection of antitoxin in the serum can aid in the diagnosis

27. A 18-year-old male presented with acute onset descending paralysis of 3 days duration. There is also history of blurring of vision for the same duration. On examination, the patient quadriparesis with areflexia. Both the pupils are non-reactive. The most probable diagnosis is:
a. Poliomyelitis
b. Botulism
c. Diphtheria
d. Porphyria

28. Not true about botulinum toxin:
a. Short life span
b. Increased acetylcholine release
c. Used for the treatment of blepharospasm, static and dynamic wrinkles
d. Irreversible decrease acetylcholine in neuromuscular junction

Clostridium difficile

29. *Clostridium difficile* infection occurs after:
a. Prolong antibiotic therapy
b. Pantoprazole increases the risk.
c. Associated with rectal thermometer
d. Increased with proportion of hospital stay

30. Pseudomembranous colitis, all are true *except*:
a. Toxin A is responsible for clinical manifestation.
b. Toxin B is responsible for clinical manifestation.
c. Blood in stool is common feature.
d. Summit lesion is early histopathological finding.

31. True regarding pseudomembranous colitis are all, *except*:
  a. It is caused by *Clostridium difficile*.
  b. Organism is normal commensal of gut.
  c. It is due to production of phospholipsae A.
  d. It is treated by vancomycin.

Clostridium tertium

32. True about *Clostridium tertium*:
  a. Gram's variable
  b. Terminal spores
  c. Produces the exotoxin
  d. Causes septic arthritis

Answers and Explanation of MCQs

1. a and d
- Follow section, **classification (Table 55.2)** for explanation.

2. d
- Follow section, *Cl. perfringens* **(morphology)** for explanation.

3. a and b
- Along with alpha toxin, theta toxin plays an important role in gas gangrene.

4. d
- Gas gangrene is not due to *Clostridium botulinum* infection, but due to *Clostridium perfringens*. *Clostridium* species are not gram-negative spore forming anaerobes, but they are gram-positive spore-forming anaerobes. The clinical features are not due to release of protein endotoxin, but due to release of exotoxin. Gas is invariably presents in muscles compartments.

5. d
- *Clostridium perfringens* is causing gas gangrene in 60–80% cases. It is the normal flora of human and animal intestines and normally present in human feces. *Clostridium perfringens* produces 12 distinct toxins, but the principal toxin is alpha toxin.
- Option d: Follow section, *Clostridium perfringens* **(resistance)** for explanation.

6. b
- It is due to alpha toxin, which is lecithinase.

7. a and b
- IP of bacteria causing gas gangrene is 10–48 hours (1–2 days) in *Cl. perfringens*, 2–3 days in *Cl. septicum* and 5–6 days in *Cl. novyi*. If one option is required to choose, then go with 10–48 hours/1–2 days (option a), because 60–80% of gas gangrene is caused by *Cl. perfringens*.

8. e
- Follow section, *Clostridium perfringens* **(pathogenicity → gas gangrene → bacteriology)** for explanation.

9. c and d
- Underlying skin and muscles are also infected in gas gangrene
- Caused by alpha and theta toxins of *Clostridium perfringens*, while tetanospasmin is produced by *Cl. tetani*, which is responsible for tetanus.
- Passive immunization does not help, and it is not advised.

10. b
- Tetanus is common in developing countries, where the climate is warm in summer months and in rural areas where soil is fertile by bacteria from the animal/human feces. Other options are already explained in text.

11. b, c and e
- *Cl. tetani* is gram-positive spore-forming anaerobe. Slightest stimulation can produce generalized muscle spasm, but not seizure.

12. a

13. a
- Follow section, *Clostridium tetani* **(pathogenicity → pathogenesis and Flowchart 55.1)** for explanation of answers of MCQs 12–13.

14. b and e
- *Cl. tetani* is gram-positive spore-forming anaerobe. Muscle enzymes are elevated. Vegetative cells will die with metronidazole (drug of choice).

15. c
- Follow section, *Clostridium tetani* **[laboratory diagnosis → toxigenicity testing (*in vivo*)]** for explanation.

16. b
- Follow section, *Clostridium tetani* **(prophylaxis and Table 55.3)** for explanation.

17. d
- Patient is of category B. Follow section, *Clostridium tetani* **(prophylaxis and Table 55.3)** for explanation.

18. d
- Spores of some strains of *Cl. tetani* are destroyed by boiling in 5 minutes, while spores of few strains are heat resistant and can survive boiling for 15–90 minutes. Spores can survive in soil (dry earth) for several years.
- After primary immunization by DPT at 6, 10 and 14 weeks, 1st booster of DPT is given at 18–24 months followed by 2nd booster dose (DT only) at 5–6 years (school entry age).

19. c

20. d
- Follow section, *Cl. botulinum* (pathogenesis and **Flowchart 55.2**) for explanation of answers of MCQs 19–20.

21. d
- Follow section, *Clostridium botulinum* **(pathogenicity → virulence factors)** for explanation.

22. a
- Infant botulism is not caused by preformed toxin, because spores 1st enter in gut by food especially via honey, germinate to vegetative forms in gut to produce the toxin, so toxin is not preformed.
- Other options: Follow section, *Clostridium botulinum* **(pathogenicity → virulence factors)** for explanation.

23. c
- Botulism occurs by ingestion of preformed toxin in preserved food—meat or meat products, fish, sea foods, canned vegetables. Type F outbreak occurs especially with fish products.

24. b

25. a
- Other options: Follow section, *Clostridium botulinum* **(pathogenicity → pathogenesis and Flowchart 55.2)** for explanation of answers of MCQs 24–25.

26. a
- Botulism is caused by exotoxin not by endotoxin.
- For explanation of option b follow infant botulism.
- For explanation of option c follow **Flowchart 55.2.**
- Retrospective diagnosis of botulism can be made by detecting the antitoxin from the patient's serum, but not seen in all cases.

27. b
- Descending paralysis, blurring of vision, quadriparesis with areflexia and non-reactive pupils are features of botulism due to bilateral cranial nerve involvement.
- For more explanation follow **Flowchart 55.2.**
- In poliomyelitis, diphtheria and porphyria, pupil is reactive.

Essentials of Medical Microbiology

28. a and b
- It blocks the release of acetylcholine.
- Follow **Table 55.1** for uses of toxin.
- Botulinum toxin once bound, then leads to the permanent dysfunction of neurons. Recovery takes usually 3 months when dysfunctioned nerve terminal are replaced as a result of sprouting.

29. a, b, c and d
- Follow section, *Clostridium difficile* (**pathogenicity → host factors**) for explanation.

30. c
- In pseudomembranous colitis, presence of blood in stool is uncommon feature

31. c
- Pseudomembranous colitis is caused by *Clostridium difficile*.
- Organism is normal commensal in gut of infant <6 months and 2–5% of the adult population, but studies have shown that infection is not endogenous, but exogenous.
- It is due to production of toxin with two subunits like A and B. Both these toxins are glucosyltransferases, but not the phospholipsae A.
- Metronidazole is the drug of choice and vancomycin is the other option.

32. a, b and d
- Follow section, *Clostridium tertium* for explanation.

CHAPTER 56

Infections of Mycobacterium tuberculosis

Chapter Outline

- Introduction of Mycobacteria
- *Mycobacterium tuberculosis*

INTRODUCTION OF MYCOBACTERIA

Meaning of Mycobacteria

Myco word derived from mykes (Greek) means mushroom (types of edible fungus), because this bacterium has branching or filamentous pattern as like fungus.

Synonym of Mycobacteria

All mycobacteria called acid fast bacilli (AFB), because once stained they resist decolorization with acid also.

Classification of Mycobacteria

Mycobacterium tuberculosis **complex/tubercle bacilli:** All species of this group are know to cause tuberculosis in human or in animals.

1. Human: *M. tuberculosis*
2. Bovine: *M. bovis* (in cattle)
3. Human in West Africa: *M. africanum*. It is intermediate between human and bovine type. It causes tuberculosis in West Africans by airborne route. Pathogenicity is lower than *M. tuberculosis*.
4. Avian: *M. avium* (in birds)
5. Cold blooded (fish, frog, lizard, etc.): *M. marinum*
6. Murine: *M. microti* (in vole, which is mouse like animal)
7. Other cattle pathogen: *M. caprae*
8. Seals pathogen: *M. pinnipedii*
9. Others: *M. canetti* similar to *M. africanum*

Lepra bacilli : All species of this group are known to cause leprosy in humans or in animals.

1. Human leprosy: *M. leprae*
2. Rat leprosy: *M. leprae murium*

Atypical mycobacteria: (By Runyon, 1959): Species of this group are not causing tuberculosis, but tuberculosis like lesion (hence called atypical mycobacteria).

1. Group-I: Called photochromogens
2. Group-II: Called scotochromogens
3. Group-III: Called nonphotochromogens (non-chromogens)
4. Group-IV: Called rapid growers.

Saprophytic mycobacteria: These are nonpathogenic and living free in water, soil or over decaying materials.

1. In smegma: *M. smegmatism*.
2. In butter: *M. butyricum*.
3. In grass: *M. phlei*.
4. In dung: *M. stercoris*.

Johne's bacillus: *M. paratuberculosis* (chronic specific enteritis, paratuberculosis or Johne's disease).

Causing skin ulcer (skin pathogens)

1. *M. ulcerans* (*M buruli*).
2. *M. balnei* (*M. marinum*).

Mycobacterium tuberculosis

History

M. tuberculosis was identified and described on 24th March, 1882 by Robert Koch, for this reason the **world tuberculosis day** is fixed on 24th March. Robert Koch received the Nobel Prize in physiology or medicine in 1905 for this discovery.

Synonym (Common Name)

It also called Koch's bacilli from the name of Dr Robert Koch who discovered the bacilli in 1882.

Morphology

Staining properties: Following are the different stains used to stain the AFB.

- **Gram's stain:** Mycobacteria are GPB, but strictly not correct, because after primary staining with methyl violet they resist decolorization with alcohol without iodine (no mean to use iodine and not following complete principle or all steps of Gram's stain, hence strictly not considered as GPB).

- **Acid-fast stains:** These are ZN/hot **(Fig. 56.1)** stain and Kinyoun's/cold stain as mentioned in **Ch. 112.**
- **Fluorescent stain**
  - **Principle:** It uses the auramine phenol (mixture of auramine-O and phenol) stain as primary stain, acid alcohol as decolorizer and potassium permanganate as counter stain.
  - **Steps:** Smear is prepared just like that for ZN staining. Stain it with auramine-phenol for 20 minutes (other dye used is rhodamine), then rinse it with water. Decolorize in acid alcohol, rinse with water and then counter stain with 0.1% potassium permanganate for 30 seconds. Rinse and air dry. Examine the smear under fluorescent microscope.
  - **Result:** Bacilli stain yellow rods in darkfield as shown in **Fig. 56.2**.
  - **Advantages:** It is more sensitive and rapid than acid fast stains.
  - **Disadvantages:** Dye is toxic and more expensive.

Shape and size: Bacilli are straight or slightly curved club and branching shape with 3 µm × 0.3 µm in size.

Arrangement: They are arranged singly, in pairs or in groups.

Motility, spores and capsule: They are nonmotile, non-sporing and noncapsulated.

Cell wall: Follow **Ch. 6.**

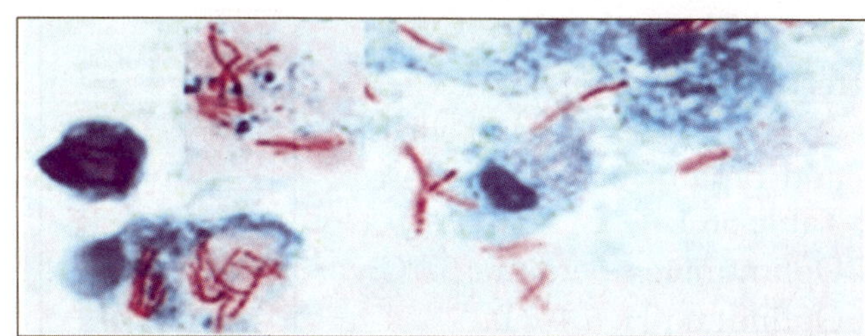

Fig. 56.1: *M. tuberculosis* under ZN stain

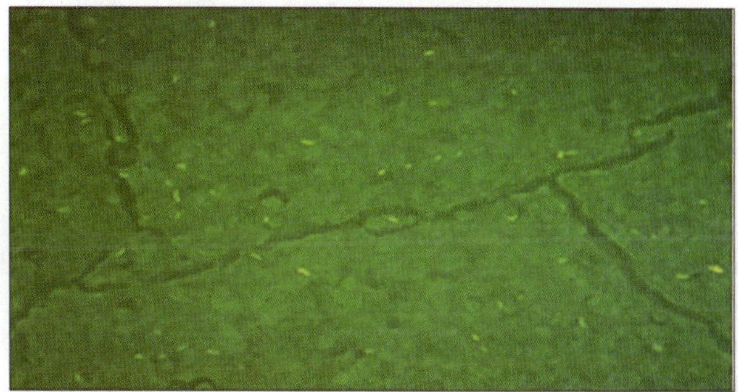

Fig. 56.2: *M. tuberculosis* under fluorescent stain

Other features: They form protoplasts in presence of lysozyme and L-forms are also present.

Morphological differences with *M. bovis*: Follow **Table 56.1**.

	TABLE 56.1: Differences between *M. tuberculosis* and *M. bovis*		
Features		*M. tuberculosis*	*M. bovis*
Morphology under ZN stain	**Length**	Long	Short
	Shape	Curved	Straight
	Strength	Beaded or barred	Stout
	Staining property	No uniform stain. Beaded or barred forms can be seen	Uniform stain
	Fast to	Acid (25%) and alcohol (95%)	Only acid
C/Cs	**O$_2$ effect**	Obligate aerobe.	Microaerophilic on primary isolation, become aerobic on subculture.
	0.5% glycerol	Improves the growth	No effect or impairs the growth
	Type of growth	Grow luxuriously on media hence called 'eugonic'	Grow sparsely on media hence called 'dysgonic'
	C/Cs on solid media	Dry, rough, raised, irregular, creamy- white colonies **(Fig. 56.3)** become buff/yellow on longer incubation	Moist, smooth, flat, regular, white colonies which easily break-up on touching
B/Rs	**Nitrate reduction**	Positive	Negative
	Niacin test	Positive	Negative
	TCH/T$_2$H test	Resistance	Sensitive
	Tween 80 hydrolysis	Variable	Always negative
Mode of transmission		Inhalation, ingestion, inoculation and congenital (by placenta) routes	Ingestion of raw milk from infected cattle
Reservoirs of infection		Humans	Animals like cattle
Sources of infection		Respiratory droplets	Infected milk of cattle
Animal pathogenicity		Pathogenic to guinea pig, but not to rabbit	Pathogenic to guinea pig and rabbit
Infectivity to human		Both are equally pathogenic for human	

Essentials of Medical Microbiology

Effective factors

- O_2 effect, 0.5% glycerol effect and type of growth: Follow **Table 56.1**.
- Optimum temperature: 37°C.
- Optimum pH: 6.4–7.0.
- Generation time: Grow very slowly in 2–8 weeks, and generation time is 20 hours.

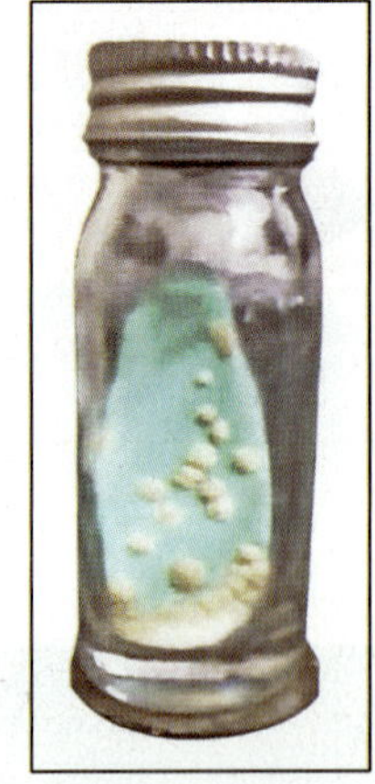

Fig. 56.3: *M. tuberculosis* in LJ medium

Culture in media: It is more sensitive than microscopy, as it can detect up to 10–100 bacilli per ml of sputum, while for microscopy to be positive, sputum should contain 10^4 bacilli/ml. It maintains the viability of microbes. However, it is a time-consuming method and required 8–12 weeks incubation before giving negative report. Three types of media are used like liquid, solid and selective.

A. Liquid media

- **Names of liquid media:** Beck's medium, Dubo's medium, Middlebrook 7H9 and 7H12 media, Proskauer medium, Sula's medium and Sauton's medium.
- **Uses of liquid media:** Liquid media are used for drug sensitivity testing, biochemical reactions, antigen preparation and vaccine preparation.
- **C/Cs in liquid media:** Follow **Flowchart 56.1**.

B. Solid media

- **Names of solid media:** Egg based media like LJ medium, Petragnini medium and dorset egg medium, blood-based medium like Tarshis medium, potato-based medium like Pawlowsky medium, serum-based medium like Loeffler medium and agar-based media like Middlebrook 7H10 medium and Middlebrook 7H11 medium.
- **Use of solid media:** LJ medium is most useful and selective medium and recommended by IUAT (International Union against Tuberculosis) for culture.
- **Cultivation technique in solid media:** Follow **Flowchart 56.2**.
- **C/Cs in solid media:** Follow **Table 56.1**.

C. Selective media:
They are prepared by adding antibiotics in liquid/solid media like polymyxin B, amphotericin B, nalidixic acid, trimethoprim and azlocillin.

Automated (newer) culture techniques: Following are different automated methods which are using middlebrook7H9 broth.

1. **MB/BacT system** (as like BacT/Alert): It based on colorimetric detection of pH change due to CO_2 production by *M. tuberculosis.* When mycobacteria multiply they produce CO_2 which increases the pH which changes the color of a blue-green sensor

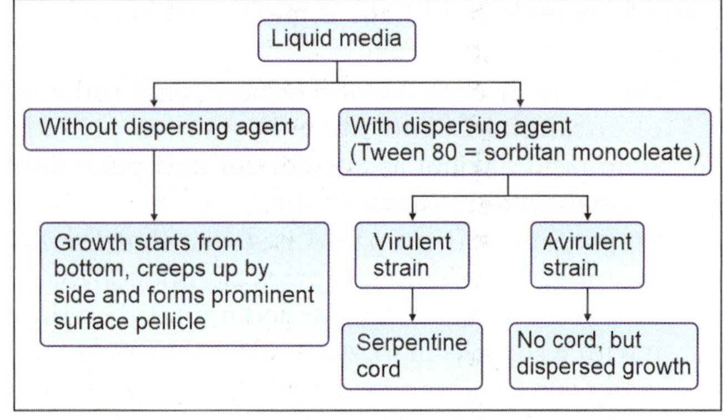

Flowchart 56.1: C/Cs in liquid media

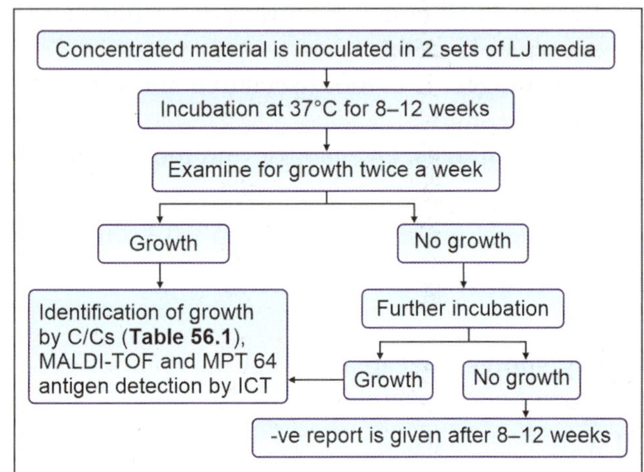

Flowchart 56.2: Cultivation technique in solid media

present at the bottom of the bottle to yellow that is detected by colorimetric system.

2. **BACTEC system:** Different principles are mentioned in **Ch. 115**.

3. **MGIT** (mycobacteria growth indicator tube): In culture tube large amount of dissolved O_2 is present which quenches the fluorescent dyes (O_2 quenched fluorescent dyes). Mycobacteria utilize the O_2 within the tube and remove the quenching effect to make the fluorescent dye free. Growth is detected by detecting the free fluorescent dye.

4. **ESP (extrasensing power) myco system:** When bacteria grow they produce the gas which is detected by gas pressure in a headspace of sealed culture bottle.

5. **Other methods:** Like MALDI-TOF and Septi-Check AFB system.

Culture in animals

1. **Guinea pig:** Both *M. tuberculosis* and *M. bovis* are pathogenic for guinea pig. Inoculate concentrated material intramuscularly into thigh of 2 guinea pigs (12 weeks age). The animals are weighed before inoculation, and, at intervals thereafter, progressive loss of weight and positive tuberculin test indicate the infection. One animal is killed after 4 weeks and autopsied. If it shows no evidence of tuberculosis, another is autopsied after 8 weeks. Autopsy findings of animal are showing enlarged draining lymph nodes and presence of tubercles in peritoneum, lungs,

at site of inoculation, but not in kidneys. Autopsy lesions have to be confirmed by acid-fast staining.
2. **Rabbit:** *M. bovis* is pathogenic to rabbit but not the *M. tuberculosis*.

Biochemical Reaction (B/Rs)

Tests to differentiate *M. tuberculosis* from *M. bovis*: Nitrate reduction test, niacin test, TCH/T2H (Thiopen2 Carboxylic acid Hydrazine) test and Tween 80 hydrolysis test are mentioned in **Table 56.1**.

Tests to differentiate *M. tuberculosis* and atypical mycobacteria: Follow **Table 56.2**.

Resistance

Sterilization: Bacteria remain viable in sputum for 20–30 hours, in droplet nuclei for 8–10 days and in culture for 6–8 months. They are killed by heat at 60°C in 15–20 minutes and sunlight after 2 hours exposure.

Disinfection: They are relatively resistant to disinfectant like 5% phenol, 15% H_2SO_4, 3% nitric acid, 5% oxalic acid and 4% NaOH. They are killed by 80% ethanol in 2–10 min, tincture iodine in 5 minutes, formaldehyde and glutaraldehyde.

Drug resistance: Resistance to antituberculous drugs is due to point mutation in regulatory genes. It occurs at a rate of once in 10^8 cell divisions. Drugs and related genes responsible for resistance are mentioned in **Table 56.3**. Different types of drug resistance are described in section of treatment (described below).

TABLE 56.2: Tests to differentiate *M. tuberculosis* from atypical mycobacteria

Tests	*M. tuberculosis*	Atypical mycobacteria
Catalase	Weak	Strong positive
Peroxidase test	Strong positive	Weak positive
Amidase test	Negative	Positive
Aryl sulfatase test	Negative	Positive

TABLE 56.3: Drug resistance genes in *M. tuberculosis*

Drugs	Genes
Isoniazide (H)	**Kat G** (Catalase and oxidase) **inhA** (Enoyl ACP reductase) **Ahpc** (Alkyl hydroperoxide reductase)
Rifampicin (R)	**rpo B** (RNA polymerase subunit B)
Pyrazinamide (Z)	**pncA** (pyrazinamidase)
Ethambutol (E)	**rpsL** (Ribosomal protein subunit 12)
Streptomycin (S)	**rpsL** (Ribosomal protein subunit 12) **rrs** (16s ribosomal RNA) **strA** (Aminoglycoside phosphotransferase gene)
Fluoroquinolones	**gyr A and B** (DNA gyrase)

Pathogenicity

Disease: It also called tuberculosis/Koch's disease. About 1/3 of world's population is died due to tuberculosis, so called 'the captain of the men death'. It has been termed as white plague.

Epidemiology: Tuberculosis has been present in humans since the time of Hippocrates. About one-third of the world's population has been infected with *M. tuberculosis* of which 90–95% remains asymptomatic while 5–10% progress to clinical disease. India carries 25% cases out of total global cases. In India, 2 deaths/ minute, >1000 deaths/day and 0.37 million deaths/year occur due to tuberculosis.

Reservoirs and sources of infection: Follow **Table 56.1**.

Modes of transmission: Follow **Table 56.1**.

Incubation period: 3–8 weeks.

Portal of entry: Respiratory system.

Sites: Pulmonary and extrapulmonary organs.

Precipitating factors (epidemiological determinants): Three types as follows:

A. Agent factors (virulence factors)
1. **Intracellular location:** Ability of *Mycobacterium* to survive and to multiply in the macrophage helps them to escape the effect of antibiotics and host immunity. Lipoarabinogalactan presents in cell wall, facilitates the intracellular survival of *M. tuberculosis*.
2. **Antigens:** Following three types of antigens are identified in *M. tuberculosis*.
   - Protein Ag: It is group specific Ag, responsible for tuberculin reaction and induces DTH.
   - Polysaccharide (arabinogalactan): It is type specific Ag. Exact role in pathogenesis is uncertain, but it induces ITH. Peptidoglycan-polysaccharide-mycolic acid complex forms the skeleton of mycobacterial cell wall.
   - Cord factor (glycolipid derivative or mycolic acid): Mycobacterial cell wall is rich in long chain fatty acid called mycolic acid. Unique mycolic acid 6,6 dimycolyltrehalose called cord factor, which is responsible for cell membrane cytotoxicity, activation of complement pathway, inhibition of migration of PMNs, which leads to granuloma formation, inhibition of phagolysosome formation, which inhibits the lysis by macrophages, development of immunity and bacilli to grow in serpentine cord (parallel arrangement of bacilli).
3. **Regulatory genes**
   - Drug resistance genes: Follow **Table 56.3**.
   - rpo V: The main sigma factor initiating the transcription of many genes.
   - erp gene: It encodes for production of proteins necessary for bacterial multiplication.
   - RD-1 (Region of difference-1): It encodes for early secretory antigen target-6 (ESAT-6) and culture filtrate protein-0 (CFP-10).

- Other genes: Leu D, Pan CD, Sig C (Sigma factor C), Sig H (Sigma factor H) and Car D genes.

B. Host factors

1. **Age:** Tuberculosis is common in all ages.
2. **Sex:** It is common in males than females.
3. **Heredity:** Tuberculosis is not a hereditary disease but study shows that inherited susceptibility is a risk factor.
4. **Nutrition:** It is widely prevalent in malnourished individual.
5. **Immunity:** IDDs like HIV-AIDS is the major risk factor for increasing the incidence and prevalence of tuberculosis. Association between tuberculosis and HIV is described below in details.
6. **Others:** Smoking, drug abuse, poor hygiene, lack of education, etc., are the other risk factors.

C. Socioeconomical factors: Overcrowding, poverty, large family, population exposure, etc., favors the tuberculosis.

Types and pathological aspects of tuberculosis: Follow Flowchart 56.3.

Pulmonary Tuberculosis

A. Primary tuberculosis/Ghon's complex/childhood tuberculosis: Initial infection in host who has not been previously infected/sensitized called primary

tuberculosis. It mostly presents in lower lobe/lower part of upper lobe, tonsils, cervical and mesenteric lymph nodes and small intestine. Pathogenesis of primary tuberculosis is described below.

1. **Pathogenesis of *M. tuberculosis*:** Generally tubercle bacilli survive in the macrophages. They lyse the macrophages and later infect the other macrophages. Humoral immunity has no role but CMI plays the major role to interact with the infected macrophages. CD4+ helper cells differentiate in Th1 cells and Th2 cells. Th1 releases the cytokines which activates the pulmonary alveolar macrophages to kill the intracellular bacilli of macrophages while Th2 releases the cytokines that are responsible for DTH and progression of tuberculosis as shown in **Flowchart 56.4 and Fig. 56.4**.
2. **Pathogenesis of *M. bovis*:** Follow **Flowchart 56.5**.

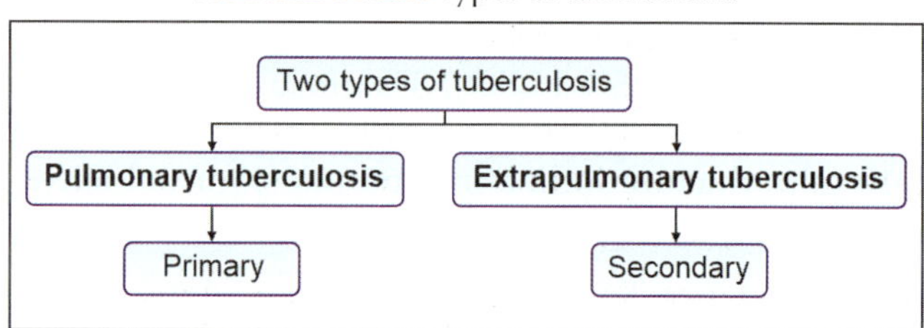

Flowchart 56.3: Types of tuberculosis

Flowchart 56.4: Pathogenesis of *M. tuberculosis*

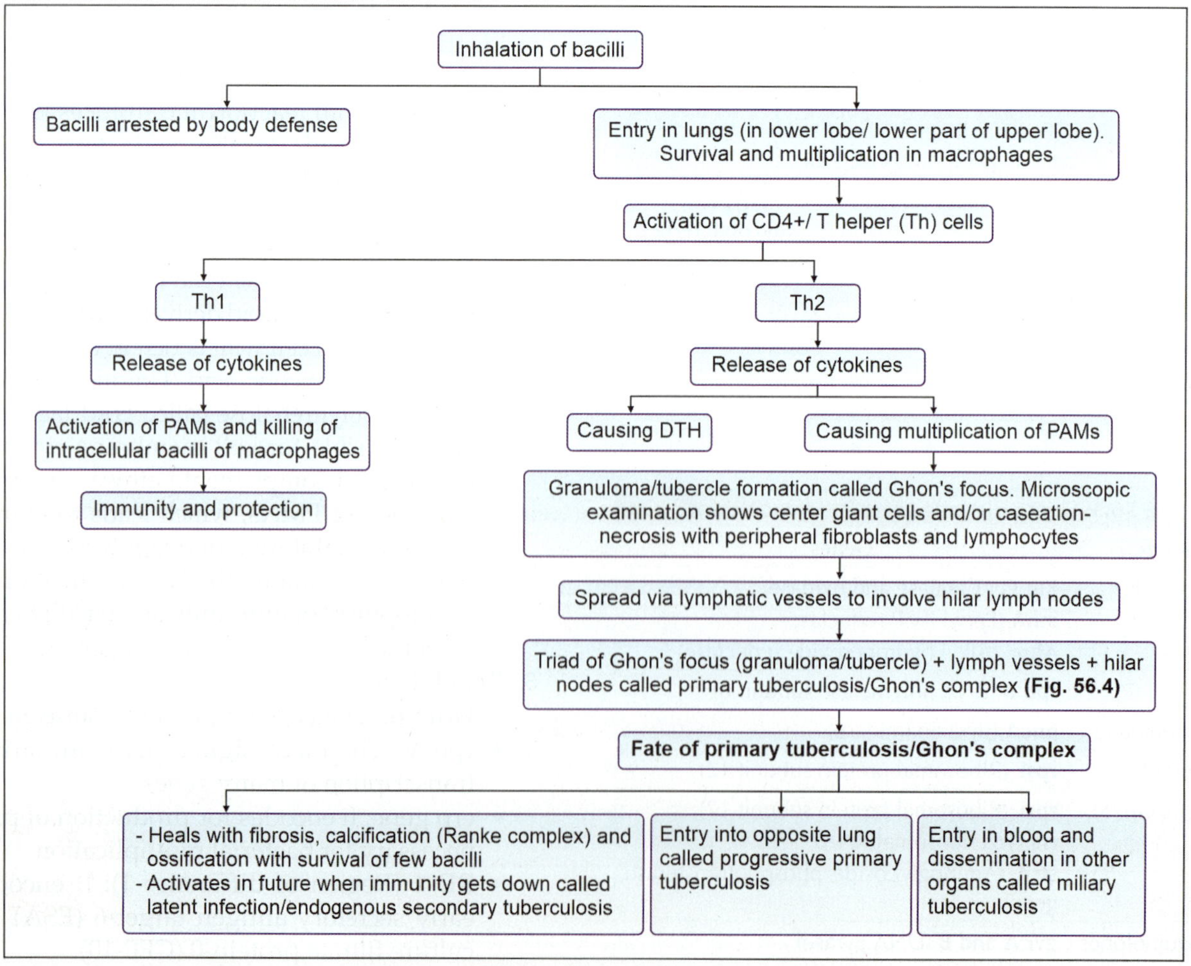

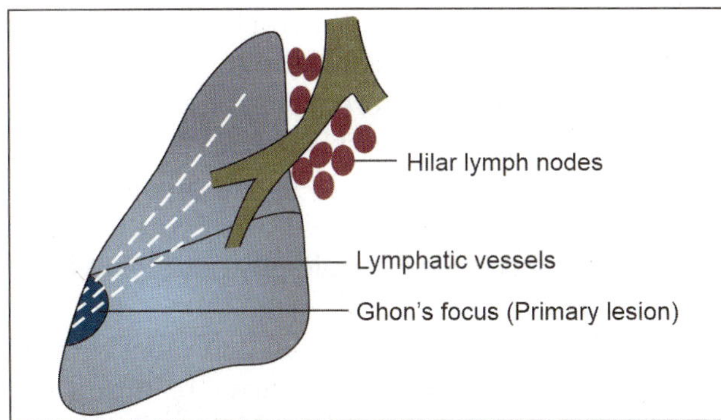

Fig. 56.4: Ghon's complex

Flowchart 56.5: Pathogenesis of *M. bovis*

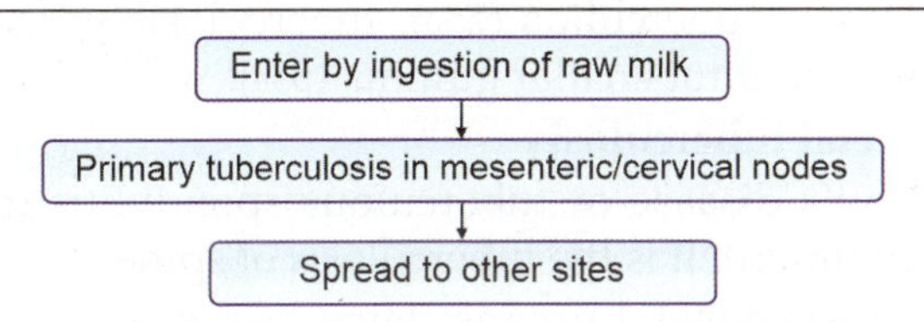

B. Secondary tuberculosis/reinfection/chronic tuberculosis/adult type tuberculosis: Infection in host who has been previously infected/sensitized called secondary tuberculosis. It mostly presents in apical lungs and other sites like tonsils, pharynx, larynx, small intestine and skin. Pathogenesis of secondary tuberculosis is described in **Flowchart 56.6.**

Differences between primary and secondary tuberculosis: Follow **Table 56.4**.

Extrapulmonary Tuberculosis

Extrapulmonary tuberculosis occurs in 15–20% of all cases of pulmonary tuberculosis. In HIV patient it is

TABLE 56.4: Differences between primary and secondary tuberculosis

Features	Primary tuberculosis	Secondary tuberculosis
Other name	Ghon's complex/ Childhood tuberculosis	Reinfection /chronic/ adult type tuberculosis
Past infection	No	Yes
Age	In children	In adults
Mode of entry	Exogenous	Exogenous and endogenous
Sites	Affects middle and lower lobe/lower part of upper lobe	Affects apical lungs
Lesion	Small	Large
Lymph node involvement	Common (hilar nodes)	Rare
Cavity formation	Rare	Common
Local spread	Rare	Common
Tuberculin test	Negative in initial stage	Positive
Clinical features	• **Respiratory:** – Asymptomatic – Productive cough with or without hemoptysis – Pleural effusion – Dyspnea and – Orthopnea • General: Low grade evening rise fever, weight loss, fatigue, anorexia and night sweat	Symptoms are same, but more pronounced

Flowchart 56.6: Pathogenesis of secondary tuberculosis

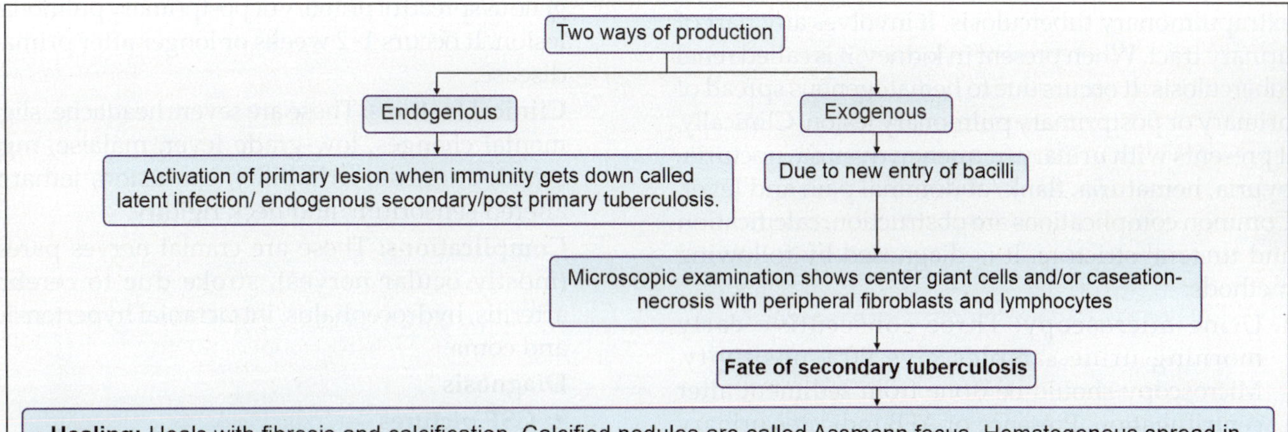

- **Healing:** Heals with fibrosis and calcification. Calcified nodules are called Assmann focus. Hematogenous spread in apical lungs called Simon's focus.
- **In immunocompetent:** Cavity formation
- Break down in to bronchus to form a cavity and leads to bacteria loaded sputum called cavitary tuberculosis. Sometimes *Aspergillus fumigatus* may produces the infection in this cavity called chronic cavitary aspergillosis or post TB aspergillosis. It presents as fungal ball (aspergilloma) causing hemoptysis, weight loss or respiratory impairment.
- Aneurysm of neighbor artery causing hemoptysis.
- Break down into pleural cavity to cause fistula and empyema
- **In immunodeficient:** No cavity formation, but instead following lesions are present.
- Widespread dissemination in whole lungs called tuberculous caseous pneumonia.
- Lymphohematogenous spread in other organs with formation of yellowish 1–2 mm size granulomatous lesions resembling millet seeds, hence called miliary tuberculosis. This is common in HIV patients.

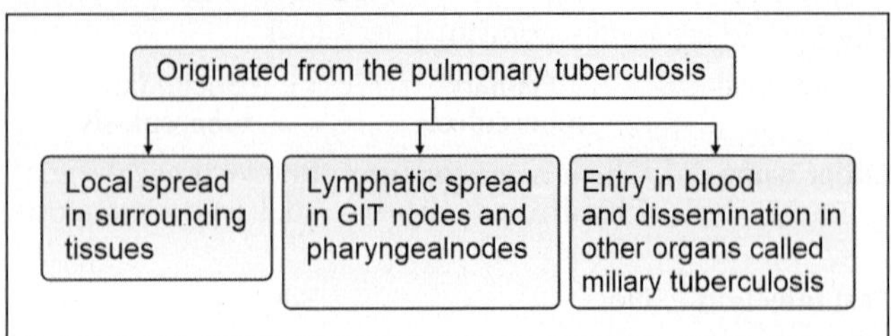

20–50%. Origin and spread of extrapulmonary tuberculosis are mentioned in **Flowchart 56.7**. It can affect any organs but clinically significant sites are mentioned below.

A. **Tuberculosis lymphadenitis:** Lymph nodes (35%) are the most common sites for extrapulmonary tuberculosis in children. It mostly affects the cervical and supraclavicular lymphnodes called scrofula/scrophula/King's evil. Scrofula is caused by *M. tuberculosis* as well as by atypical mycobacteria. It presents as painless swelling without warmth (so called cold abscess) or color changes. The nodes are usually discrete in early stage, but inflamed and produce fistula or sinus, which drains the caseous materials.

B. **Upper respiratory tuberculosis:** It involves larynx, pharynx and epiglottis. It presents with hoarseness of voice, dysphonia, and chronic productive cough.

C. **Pleural cavity tuberculosis:** It occurs in 20% cases and presents with pleural effusion. It may form the bronchoplural fistula.

D. **Pericardium tuberculosis:** It occurs due to direct spread from adjacent nodes or by hematogenous spread. It is common in older age group.

E. **Urinary system tuberculosis:** It covers 10–15% of all extrapulmonary tuberculosis. It involves any part of urinary tract. When present in kidney, it is called renal tuberculosis. It occurs due to hematogenous spread of primary or postprimary pulmonary lesion. Clinically, it presents with urinary frequency, dysuria, nocturia, pyuria, hematuria, flank/abdominal pain and fever. Common complications are obstruction, calcification and ureteral stricture. It is diagnosed by following methods:

1. Urine microscopy: Three consecutive early morning urine samples give 90% positivity. Microscopy should be done from sediment after centrifugation. Presence of AFB indicates urinary tuberculosis or presence of pus cells with negative urine culture suspect tuberculosis.

2. Urine culture: It is done manually in LJ medium or by MGIT method.

3. Imaging studies: Like CT and MRI are used to diagnose the complications.

F. **Genital tract tuberculosis:** In male, it presents with epididymitis (tender mass drain externally with fistula), orchitis and prostitis. In female, it affects fallopian tube and endometrium with formation of granuloma. It causes infertility, menstrual abnormality and adnexal swelling.

G. **Skin tuberculosis:** Different verities of cutaneous tuberculosis are mentioned below.

1. Lupus vulgaris (55%): It is the most common variety of cutaneous tuberculosis. It presents with caseation necrosis and apple jelly nodules over face in women.

2. Scrofuloderma (27%): It is the 2nd most common variety of cutaneous tuberculosis resulting from extension of scrofula which involves the cervical and supraclavicular lymph nodes with skin. It also called tuberculosis cutis colliquativa.

3. Other: Tuberculids (7%), tuberculosis verucosis (6%) and tubercular gumma (5%).

H. **Skeletal tuberculosis**

1. Pott's disease or tuberculous spondylitis (most common): It is the tuberculosis of spine.

2. Other skeletal organs: Hips and knee are also infected with advanced disease, collapse of vertebral bodies result in kyphosis (gibbus) and a paravertebral cold abscess may also form.

I. **Muscular tuberculosis (psoas abscess):** Cold abscess (tuberculosis) of paravertebral tissues penetrates the inguinal ligament and enters in the compartment of iliopsoas muscles to produce psoas abscess.

J. **Tuberculosis meningitis:** Presents in two types:

1. **Tuberculous meningitis**

   Host (precipitating) factors: It occurs approximately in 5% cases of extrapulmonary tuberculosis. It is common in young children and in adults with HIV.

   Mode of transmission: It occurs due to hematogenous spread of primary or postprimary pulmonary lesion. It occurs 1–2 weeks or longer after primary disease.

   Clinical features: These are severe headache, slight mental changes, low-grade fever, malaise, night sweat, anorexia, irritability, confusion, lethargy, altered sensorium, and neck rigidity.

   Complications: These are cranial nerves paresis (mostly ocular nerves), stroke due to cerebral arteritis, hydrocephalus, intracranial hypertension and coma.

   Diagnosis

   - **CSF pictures**
     - Cytological and biochemical studies are performed to differentiate the different types of meningitis: Follow **Table 41.1**.
     - Pathological study: When CSF is kept in a tube for 12 hours it forms coagulum called cobweb coagulum due to higher fibrin content.
     - Bacteriological study is done by acid-fast staining: It shows long slender beaded forms of AFB. It has low sensitivity (10–40%) which requires repeated lumbar puncture.

- **CSF culture:** It is the gold standard test, done manually in LJ medium (takes 4–8 weeks) or by automated liquid culture method like MGIT (takes 2–3 weeks).
- **GeneXpert assay:** Test is automated real-time PCR, with 80% sensitivity.
- **Imaging studies:** Like CT and MRI.

Prevention: Early diagnosis and treatment for pulmonary tuberculosis.

Treatment: Start treatment immediately after getting positive result with GeneXpert. Negative result does not exclude tuberculosis and requires further diagnostic tests. It responds well to anti-tubercular therapy, if started early. Adjunctive gluco-corticoids are used to reduce the CSF pressure, resulting in faster resolution.

2. **Tuberculoma (granulomatous form):** It is an uncommon manifestation of tubercular infection of CNS; presents as ICSOL. Pathological study reveals firm nodule (granulomatous lesion) with central caseous necrosis. It causes seizures and focal signs. CT or MRI shows contrast-enhanced ring lesions. Biopsy is necessary to establish the diagnosis and to differentiate it from malignancies.

K. GIT tuberculosis: It is caused by both *M. tuberculosis* and *M. bovis*. *M. bovis* is transmitted by ingestion of milk of infected cow (in developing countries). *M. tuberculosis* is transmitted by swallowing of sputum (common in children) or by hematogenous spread. Terminal ileum and cecum are the commonest sites to be infected. It also causes peritonitis.

L. Middle ear tuberculosis (tuberculous otitis media): It presents with granuloma in middle ear and multiple perforations.

M. Tuberculosis in eye: It presents with granulomatous conjunctivitis, granulomatous anterior uveitis (iridocyclitis), posterior uveitis (choroiditis) with multiple miliary tubercles, retinitis, chronic dacryocystitis and chronic dacryoadenitis.

N. Oral cavity tuberculosis: Irregular, painful tuberculous ulcer mostly presents on tongue followed by palate, buccal mucosa, lips, gingival and floor of oral cavity. Sometimes, diffuse inflammatory lesions or fissures are also present. Hematogenous spread from oral lesions causes tuberculous osteomyelitis in mandible and maxilla in later stage of disease. Bacilli are diagnosed from biopsy lesion; however, numbers are very less so tough to diagnose by microscopy while culture gives good results.

Complications of tuberculosis: Death usually occurs due to pulmonary insufficiency, pulmonary hemorrhage, secondary amyloidosis, cor pulmonale or sepsis by miliary tuberculosis.

Laboratory Diagnosis

It is discussed separately for pulmonary and extra-pulmonary cases as below.

Specimens: Two sputum samples are collected one is spot and next day early morning before meal. Bronchial washing and laryngeal swab are collected when sputum is not available. Bacilli are more in pleural wall so ideal specimen for pleural tuberculosis is pleural biopsy not the pleural fluid. Gastric aspiration is collected in child and in adult patients who swallow the sputum.

Transport of specimens: If delay about more than 2 hours is expected then refrigerated the samples to prevent multiplication of flora. If more delay is expected then equal volume of CPC (cetyl pyridium chloride)-NaCl solution is added to sputum. It liquefies the sputum, prevents the growth of other bacteria and also maintains the viability up to 8 days. Pleural fluid should be collected in citrated bulb.

Preparation of specimens: Methods used to prepare the specimens called concentration/homogenization/digestion-decontamination methods, which are of following two types:

1. **Preparation of specimens for microscopy:** It required killed bacilli without alteration of morphology or staining reaction. Treat the specimens with sodium carbonate or sodium hypochlorite or antiformin or detergents like tergitol or autoclave method or flotation methods by using hydrocarbons.
2. **Preparation of specimens for culture and animal inoculation:** They required live bacilli. Following are the different methods.
   - **Petroff's method:** Mix sputum + equal volume of 4% NAOH → Incubate at 37°C and frequent shaking for 20 min → Centrifuge at 3000 rpm, 20 min. → discard the supertant and neutralize the sediment with 0.1 N HCL and use for culture and animal inoculation.
   - **Cetrimonium bromide method:** Mix sputum with equal volume of sterile solution (contains 20 g cetrimonium bromide + 40 g of NaOH per liter of distilled water) → mix with a cotton swab; stand for 5 min. → 0.2 ml of mixture is smeared firmly with swab over LJ medium. Same swab is used for inoculating second slope after stirring contents again.
   - **Other homogenization methods:** These are treating specimen with N-acetyl L-cysteine (NALC) or pancreatin or cetrimide or dilute acids (6% H_2SO_4, 3% HCl or 5% oxalic acid).

Testing methods

A. **Microscopy:** Follow morphology.

B. **Culture:** Follow C/Cs.

C. **Biochemical reactions:** Follow B/Rs.

D. **Serological tests:** Three types of tests.

   Detection of Ag: Various antigens like lipo-arabinomannan and Ag-5 are detected by ELISA, dip stick and latex agglutination tests from clinical samples like sputum, urine, etc. Tests are less useful because of low sensitivity (40–50%). MPT 64 Ag is detected by ICT from culture growth. It is specific for *M. tuberculosis* complex (*M. tuberculosis, M. bovis* and *M. africanum*) and negative for NTM.

Detection of Ab: It is less useful, because of cross reactivity by atypical mycobacteria and also by other antigens. In 2013 WHO banned IgM ELISA because of high rate of false-positivity.

Interferon gamma release assay (IGRA): It is based on detection of γ-interferon released by sensitized T lymphocytes. Test is highly specific, and there is no chance of false-positive result. Sensitized T lymphocytes are collected from suspected case and exposed to ESAT-6 (early secretory antigen target-6) and CFP-10 (culture filtrate protein-10), which leads to the release of γ-interferon from the sensitized T lymphocytes. This γ-interferon is detected by ELISA, which is classified in following three categories:

- 1st generation: Called quantiferon-TB (QFT) assay. It responds to one tuberculoprotein like PPD (purified protein derivative).
- 2nd generation: Called Quantiferon-TB gold (QFT-G) assay. It responds to two tuberculo-proteins like ESAT-6 and CFP-10. Both are secreted by *M. tuberculosis, M. bovis, M. kansasi, M. szulgai* and *M. marinum*, hence test cannot differentiate between these organisms.
- 3rd generation: Called quantiferon-TB gold in tube (QFT-GIT) assay. It responds to three tuberculo proteins like ESAT-6, CFP-10 and TB 7.7.

E. Molecular methods: These methods have few advantages over others like they are highly sensitive than culture, take less time than culture, best useful for extrapulmonary sites, which are mostly paucibacillary types, useful for epidemiological typing and useful to detect drug sensitivity especially by detecting drug resistance genes. Following are the different types of molecular methods.

1. **PCR:** Nested PCR targeting IS6110 was the most common test used earlier. Other genes which were targeted by PCR include MPT64, 86 kDa and 38 kDa.

2. **Automated real-time PCR (qPCR):** It includes two nucleic acid amplification tests (NAATs) like cartridge based and chip based.

   - **Cartridge based:** Test called cartridge based nucleic acid amplification test (CBNAAT). Two equipments are based on this principle like GeneXpert and Xpert Ultra.
     - **GeneXpert:** It is the rapid method gives result in 2 hours. It detects up to 131 bacilli/ml. It detects *M. tuberculosis* complex DNA. It also detects rifampicin resistance gene like rpoB gene. It uses five probes targeting different sequences of rpoB gene. It has 88% sensitivity for diagnosis of TB and 95% sensitivity for rifampicin resistance detection. It has 99% specificity for diagnosis of TB and 98% specificity for rifampicin resistance detection. It is very expensive and cannot differentiate the species of *M. tuberculosis* complex.
     - **Xpert ultra:** It detects up to 16 bacilli/ml. It targets IS6110 and IS1081. It is more sensitive and specific than GeneXpert.

   - **Chip Based:** Test called chip based nucleic acid amplification test. It is a rapid method gives result in 1 hour. It is very expensive and cannot differentiate the species of *M. tuberculosis* complex. It has limited testing capacity about 1–4 samples at a time.

3. **LPA:** It identifies and differentiates between the species of *M. tuberculosis* complex (*M. tuberculosis, M. bovis* and *M. africanum*) and NTM. It detects the resistance to 1st line and 2nd-line antituberculous drugs. It is performed for smear or culture positive specimens, but nor for smear negative specimens.

4. **Others methods:** LCR, TMA, SDA, NASBA, etc.

F. Epidemiological typing

1. **Phenotyping methods:** These are bacteriophage typing **(Ch. 118)** and bacteriocin typing.
2. **Genotyping methods:** These are RFLP (done by using insertion sequence 6110), spoligotyping (it is the spacer oligotyping, which based on polymorphism in the direct repeat locus), PFGE and DNA sequencing.

G. Chromatography: GLC and HPLC for mycolic acid detection.

H. Allergic tests

History: It called tuberculin test which was discovered by von Pirquet in 1907.

Principle: It is based on DTH.

Antigens: Initially OT (old tuberculin) was used as an antigen. It was a crude preparation of tubercle bacilli. It was described by Robert Koch. OT is replaced by PPD. PPD is prepared by Siebert in 1941 by growing the human strain of *M. tuberculosis* (PPD-S) in semisynthetic medium. It also prepared from other mycobacteria like Batty bacilli (PPD-B), *M. kansasi* (PPD-Y), *M. scrofulaceum* (scrofulin), etc. However, they may cross react. PPD is standardized in terms of tuberculin unit (TU). International standardization is maintained by WHO. One TU is equal to 0.01 ml of OT or 0.00002 mg PPD. The WHO advocates a PPD tuberculin known as PPD-RT-23 with tween 80.

Methods: Tuberculin test is performed by following methods.

1. **Single puncture method/Mantoux test**
   - **Steps:** Inject 0.1 ml PPD, intradermally (in-between the layers of skin not subcutaneously) on flexor aspect of forearm with tuberculin syringe **(Fig. 56.5)**. Tuberculin syringe is calibrated in 0.01 increments with large markings at zero

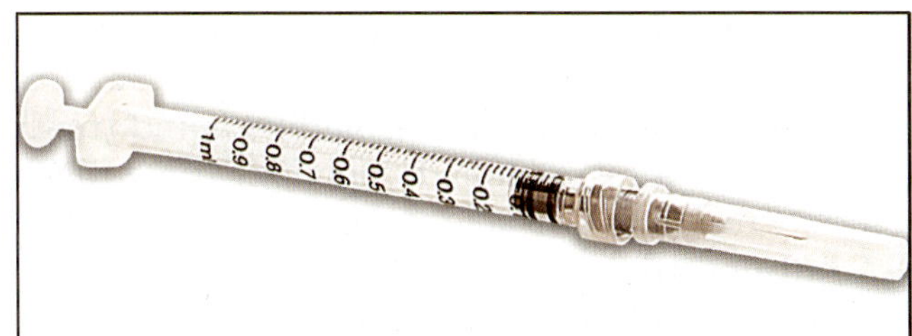

Fig. 56.5: Tuberculin syringe

(0) and each 0.05 ml, e.g., 0.05, 0.1, 0.15, 0.2, 0.25, 0.3, etc. A tuberculin syringe is available in two sizes depending on total capacity like 0.5 ml and 1 ml. Tuberculin syringes cannot be used in place of insulin syringe because the units of measure are different and the needle of a tuberculin syringe is shorter than that of an insulin syringe. Insulin is administered "subcutaneously" (under the skin) so the correct size needle is imperative. Site of injection should be examined after 48–72 hours.

- **Results:** Do not consider the erythema as it is difficult to measure. Measure only induration at its widest point transversely to long axis of forearm by ruler or caliper (in terms of length not breadth).
  - Negative: ≤5 mm.
  - Doubtful/equivocal (false-positive?): 6–9 mm.
  - Positive: ≥10 mm.
  - Strong reactor: ≥20 mm.
- **Interpretation:**
  - False-positive test (doubtful/equivocal): Infections by related mycobacteria (atypical mycobacteria).
  - False-negative test: Miliary tuberculosis, early or advanced tuberculosis, IDDs, inactive PPD preparations or faulty technique.
  - Negative test: Never contacted with tubercle bacilli and at more risk to develop tuberculosis than 6–9 mm group.
  - Positive test: Positive result in adult (type IV hypersensitivity to tuberculoprotein) indicates the exposure of bacilli either recently or in past due to infection or vaccination with or without active tuberculosis, so adults who had never contacted to tubercle bacilli give negative result. Positive result in children below 2 years is an indirect evidence of active tuberculosis, even if, it is not manifested. Test becomes positive 4–6 weeks after infection and 8–14 weeks after immunization. Positive result may revert to negative result (occasionally) on isoniazide treatment.
  - Strong reactor: At greater risk to develop tuberculosis ≥20 mm group.

- **Uses:** Mantoux test is useful:
  - to diagnose the infection in infants and young children.
  - to detect the successful vaccination.
  - to know the prevalence of tuberculosis, this is calculated by counting all tuberculin reactors in a community.
  - to know the incidence of tuberculosis, this is calculated by counting new converter to tuberculin test in a community.
  - to know the prognosis: Strong reactors (≥20 mm) are at greater risk to develop tuberculosis than ≥10 mm group, while tuberculin negative (≤5 mm) have more risk of developing tuberculosis than 6–9 mm group. In short, ≤5 mm and ≥20 mm are at high risk to develop tuberculosis. Studies indicate that among the new cases, around 92% are strong reactors. These findings justify the prognostic significance of the test.
- **Repeated (two steps) tuberculin testing:** It will not cause positive reaction in noninfected person, but may enhances the intensity **(called booster effect)** of response in reactive individuals, so another tuberculin test following 1–2 weeks later will give strong reactor.
- **Tuberculin conversion:** It is a condition at where negative result (≤5 mm) turns to positive result (≥10 mm) within a 24 months period. It indicates the latent tuberculosis, which may turn to active disease.
- **Limitations**
  - Validity of test: It is variable due to faulty technique, errors in reading the result, test materials used, etc.
  - Lack of specificity and sensitivity: Due to cross reaction by other mycobacteria and also by booster effect.
  - DTH: DTH measured by tuberculin test is irrelevant to combat the disease. It does not indicate whether the person is able to mount an immune response to tubercle bacilli or not (as in lepromin test).
  - Susceptibility: It does not indicate the resistance or susceptibility to tubercle bacilli (as in Schick test).

2. **Multiple puncture methods**
  - Heaf test: Used for screening and survey in large group, but not for diagnosis. It is quick, cheap, reliable and easy to perform.
  - Tine test: Used for individual testing, but not reliable so not recommended.

I. Drug sensitivity tests

Absolute concentration method: Media contain serial concentrations of drugs are inoculated and minimum inhibitory concentrations (MIC) is calculated.

Resistance ratio method: Two sets of media contain graded concentrations of drugs are inoculated. 1st with test strain and 2nd with standard strain of known sensitivity.

Proportion method: Numbers of colonies growing on LJ media with or without drug are compared. Strain is resistant, if more than 1% of the bacteria grow on LJ with drug.

Automated method by using BACTEC MGIT

Molecular methods: PCR-based assay, GeneXpert, line probe assay and DNA microarray.

J. **Other tests:** Detection of tuberculo-stearic acid and adenosine deaminase (ADA) are also useful methods.

> **Note: Following are the methods for diagnosis of latent tuberculosis. Other methods mentioned above are for active tuberculosis.**
> - Tuberculin test and IGRA: Both are described above.

Diagnosis of Extrapulmonary Tuberculosis

Specimens

- CSF in tuberculous meningitis: CSF shows the cobweb coagulum on standing, elevated CSF pressure, raised proteins, raised chloride, increased lymphocytes count and decreased glucose level.
- Urine in renal tuberculosis: Bacilli are shedding very intermittently in urine, hence 3–6 consecutive early morning urine samples are collected, centrifuge and use sediment for processing. Acid alcohol is used as decolorizer.
- Other specimens are blood, pleural fluid, pus in cold abscess, bone and joint fluid, lymph node aspirate, etc.

Testing methods

A. **Microscopy:** Extrapulmonary specimens are containing less numbers of bacilli so less useful.

B. **Culture:** Also a useful method.

C. **Serological test:** IGRA is best method.

D. **Molecular methods:** Best useful method.

E. **Other method:** Adenosine deaminase (ADA) detection.

Prevention

General measures: Adequate nutrition, good housing and health education.

Immunoprophylaxis: Active immunization is done by using BCG vaccine.

- **Full name:** Bacille Calmette Guerin vaccine.
- **History:** Developed by Calmette and Guerin in 1921 in France.
- **Preparation:** It is live attenuated vaccine prepared from strain of *M. bovis* attenuated by 239 serial subcultures in glycerine-bile-potato medium over period of 13 years. It is available in 0.5–1.0 mg dry powder form in ampules containing $1–2.5 \times 10^7$ colony forming units (CFU). In India, WHO recommended **Danish 1331** strain of BCG for preparation of vaccine.

It is prepared in BCG laboratory, Guindy, Chennai, India.

- **Types:** Two forms are available like liquid (fresh) form which is less stable and lyophilized (freeze dries) form which is most stable and widely used form.
- **Administration and dose:** Lyophilized form is suspended in 1 ml sterile water. Do not use distilled water, as it is irritant. Once reconstituted, it should be used in 1 hour, not later. Do not sterile skin by using the alcohol. 0.05 ml in neonate and 1 ml in older children, injected intradermally just above the insertion of left deltoid by using 26 G tuberculin syringe. It is given immediately after birth or as early as possible before 12 months. If not given than vaccine can be given maximum up to 2 years of age. Direct BCG given to new born immediately after birth in developing countries including India. Indirect BCG is given after tuberculin testing.
- **Changes following proper BCG vaccination:** Follow Flowchart 56.8.
- **Efficacy:** It has variable efficacy from 0–80%. Immunity may last for 15–20 years.
- **Complications:** Three types
  - Local: Abscess, indolent ulcer, keloid, tuberculides, confluent lesions, lupoid lesions or lupus vulgaris.
  - Regional: Enlargement and suppuration of draining lymph nodes.
  - General: Fever, mediastinal adenitis, erythema nodosum, progressive tuberculosis, disseminated BCG infection (BCGitis) or rarely nonfatal meningitis.
- **Uses (indications)**
  - BCG vaccine does not protect the person from TB, but gives protection to infant and children against serious type of infection like disseminated tuberculosis and tubercular meningitis.
  - It provides protection against leprosy and leukemia by nonspecific stimulation.
  - Adjuvant therapy: Multiple injection of BCG vaccine (called Onco TICE strain of BCG) has been tried in bladder cancer and in other diseases as an adjuvant therapy.
  - Some research suggests that it is superior for tuberculin testing than PPD.
- **Contraindications:** BCG vaccine is contraindicated in pregnancy, infant and children with AIDS, baby

Flowchart 56.8: Changes following proper BCG vaccination

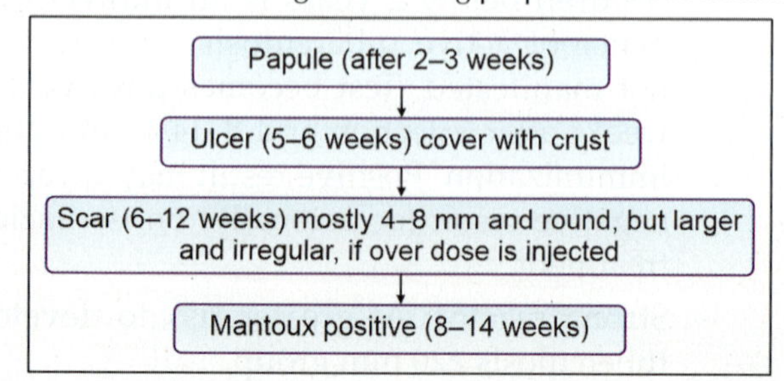

born by AFB positive mother and tuberculin test positive person.

Chemoprophylaxis

- **Aim:** It is given to prevent the progression of latent tuberculosis to active disease.
- **Indications**
  - Persons with latent tuberculosis.
  - Persons at high risk.
  - Neonate of tubercular mother.
  - Child with Mantoux positive and TB patient in family.
  - Patients with leukemia, diabetes, silicosis, AIDS, who are Mantoux positive.
  - Patients with old inactive disease who received inadequate therapy.
- **Regime (monoherapy):** Isoniazide (H) is given in the dose of 10 mg/kg for 6–12 months. If patient is resistant to H then H and rifampicin (R) are given in the dose of 10 mg/kg for 6 months.
- **Drawbacks of monotherapy:** It develops resistance.

"Stop TB strategies" of WHO: WHO launched stop TB strategies in 2006 which aims

- **By 2005:** >70% of sputum smear positive cases will be diagnosed with >85% cure rate.
- **By 2015:** Global burden of tuberculosis including prevalence and death rate will be reduced by 50%.
- **By 2050:** Global incidence of tuberculosis will be <1 million population/year.

Treatment

A. Chemotherapy

Classification of drugs: Antituberculous drugs are classified as follows:

1. **1st line drugs:** These drugs have high antitubercular efficacy, low toxicity and used routinely.
- **Bactericidal:** Isoniazide (H), rifampicin (R), pyrazinamide (Z) and streptomycin (S).
- **Bacteriostatic:** Ethambutol (E)
2. **2nd line drugs:** These drugs have low antitubercular efficacy, high toxicity or both, used in special cases and bacteriostatic only like thiacetazone (Tzn), para-aminosalicylic acid (PAS), ethionamide (Etm), cycloserine (Cys), kanamycin (Kmc), amikacin (Am), capreomycin (Cpr).
3. **Newer drugs:** Ciprofloxacin, ofloxacin, clarithromycin, azithromycin, rifabutin, pretomanid, bedaquiline, and linezolid.

Revised National Tuberculosis Control Program (RNTCP): It was introduced in 1993 in India in collaboration with WHO and World Bank. Now RNTCP is renamed as National Tuberculosis Elimination Program in 2020.

- **Strategies of RNTCP**
  - Diagnosis of >70% estimated cases by sputum smear microscopy.

- Treatment called DOTS. It is done under supervision.
- Implementing DOTS plus for MDR-TB cases.
- Cure rate not <85%.
- Involvement of NGOs (nongovernment organizations).
- **Diagnosis of tuberculosis under RNTCP:** Follow Flowchart 56.9. Do HIV counseling and testing for all suspected TB cases; however, diagnostic work for TB should not be delayed.
- **Treatment (DOTS):** Full form of DOTS is **D**irect **O**bserved **T**reatment, **S**hort course chemotherapy, so called because whole course of therapy is supervised to improve the result. It is recommended by RNTCP and WHO. Treatment response is monitored by periodic sputum microscopy. DOTS rely on 1st-line drugs isoniazide and rifampicin. It includes two phases as follows.
  - Initial/intensive phase: It is given for 2–3 months by using four (HRZE) 1st-line drugs with aim of rapid killing of bacilli to make the patient smear negative.
  - Continuous phase: It is given for 4–5 months by using two (HR) or three (HRE) 1st-line drugs with aim of killing dormant bacilli to prevent relapse.
  - Categories: It includes two categories based on sputum AFB microscopy (positive or negative), history of past treatment, sites and severity of disease as mentioned in **Table 56.5**.

Flowchart 56.9: Diagnosis of pulmonary tuberculosis under RNTCP

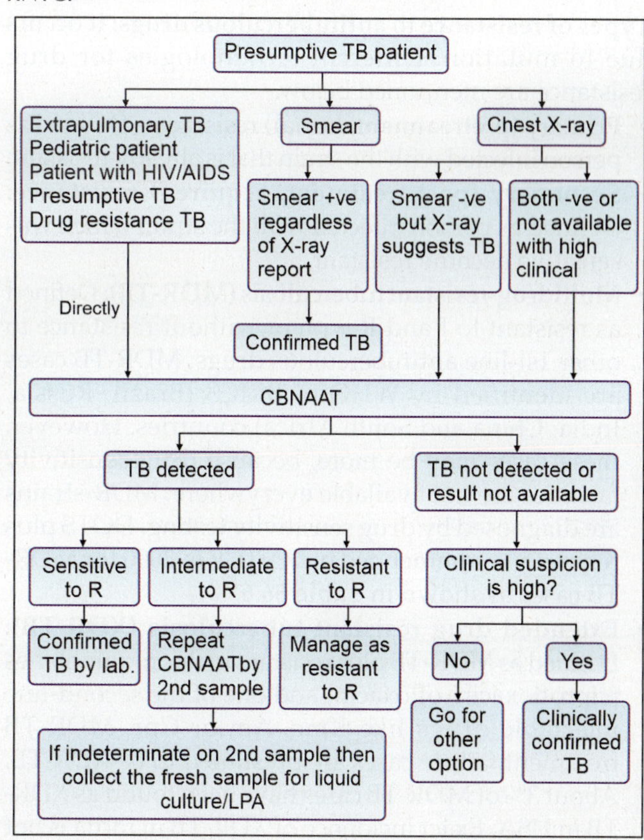

Infections of Mycobacterium tuberculosis

TABLE 56.5: Categories of tuberculosis as per RNTCP and WHO guideline, 2010

Category	Indications	Intensive phase	Continuous phase	Duration
I (New patient)	• Sputum smear positive • Sputum smear negative • New extrapulmonary	2HRZE	4HR	6 months
II (Old patient)	• Previously treated • Sputum smear positive like relapse, treatment failure or return after default • Awaiting drug sensitivity report	2HRZES + 1 HRZE	5HRE	8 months (with low risk of MDR-TB)
		Empirical MDR regime	Empirical MDR regime	18–24 months or tills DST results (with high risk of MDR-TB)

– Dosing schedule:
1. **Category I:** Drugs are given either daily or thrice weekly in one or both phases. Daily treatment in both phases is advisable, because it helps to prevent the resistance even in patient who started the treatment with primary H resistance. However, keeping in mind about drugs administration and cost reduction thrice weekly therapy is accepted in continuous phase with supervision. If constraints are still pressing then thrice weekly therapy in intensive phase is also recommended with keeping in mind about HIV coinfection or risk of HIV infection during therapy. WHO suggest inclusion of E with H and R in areas where primary H resistance is present.
2. **Category II:** 5 drugs (HRZES) are given daily for 2 months and 4 drugs (HRZE) are given daily for 1 month in intensive phase followed by continuous phase with 3 drugs (HRE) for 5 months. Outcome of regime is monitored by clinical status, microscopy and culture report.

Types of resistance to antituberculous drugs: It occurs due to mutation. Different terminologies for drug resistance are mentioned below.

- **Primary (pretreatment, initial) resistance:** Defined as person infected with the strain that is already resistant.
- **Secondary (post-treatment, acquired) resistance:** Defined as person infected with the strain which was sensitive become resistant.
- **Multidrug-resistant tuberculosis (MDR-TB):** Defined as resistant to I and R with or without resistance to other 1st-line antituberculous drugs. MDR-TB cases are identified by WHO in BRICS (Brazil, Russia, India, China and South Africa) countries. However, these cases may be more, because drug sensitivity facilities are not available everywhere. MDR-strains are diagnosed by drug sensitivity testing. DOTS plus regime is recommended by RNTCP in 2000 for MDR-TB cases as shown in **Table 56.6**.
- **Extended drug resistant tuberculosis (XDR-TB):** Defined as MDR-TB plus resistant to fluoroquinolones (ciprofloxacin, ofloxacin) and one of the second-line injectable drugs like Kmc, Am or Cpr. MDR-TB treatment failure cases are presumed to be XDR-TB. About 3% of MDR-TB cases have been found as XDR-TB in USA. Exact incidence of XDR-TB in India is not

TABLE 56.6: DOTS plus regime for MDR-TB

Intensive phase (6 drugs for 6–9 months)	Continuous phase (4 drugs for 18 months)
Kanamycin	Ofloxacin/levofloxacin
Ofloxacin/levofloxacin	Ethionamide
Ethionamide	Cycloserine
Cycloserine	Ethambutol
Pyrazinamide	
Ethambutol	

found. XDR-strains are diagnosed by drug sensitivity testing. XDR-TB cases are difficult to treat. They have a rapid progress of disease with high mortality. A 2018 meta-analysis of 12,030 patients from 25 countries in 50 studies has demonstrated that treatment success increases and mortality decreases when treatment includes bedaquiline, delamanid, later generation fluoroquinolones and linezolid. One regimen for XDR-TB called Nix-TB, a combination of pretomanid, bedaquiline, and linezolid has shown promise in early clinical trials.

Genes of resistance to antituberculous drugs: Follow **Table 56.3**.

B. Treatment for restoration of CMI

Transfer factor: It helps for recovery in immunodeficient persons.

Vaccine: It contains heat killed *M. vaccae*, an environmental *Mycobacterium*. It acts as an immunomodulator for stimulation of Th-1 cells, which promotes the protective immunity.

> **Note: Tuberculosis and HIV**
> **Introduction:** Deficiency of CMI is a major contributory factor for infection. Tuberculosis is most commonly present opportunistic infection in HIV. It present in 70-80% cases of HIV.
>
> **Mode of onset:** Following are possible mode of onset of tuberculosis in HIV patients.
> 1. **Primary infection:** It is the infection in HIV patient who is not previously infected by tuberculosis. Risk is high in immunodeficient person than immunocompetent.
> 2. **Secondary infection:** Two subtypes.
>   - Post primary infection/endogenous reinfection/reactivation of latent infection: Bacilli remain in latent form, after healing of primary infection. They are under control by immune system but once immunity down by HIV, latent infection reactivated to develop the tuberculosis.

(Contd…)

> - Exogenous reinfection: New infection in HIV patient who is previously infected by tuberculosis.
>
> **Pathology:** Patient is with CMI deficiency, so no granuloma formation and no cavitary lesions in **lungs**, but widespread dissemination of bacilli in whole lungs called tuberculous caseous pneumonia. Lymph nodes are showing poor or no granuloma in HIV positive patient, but bacillary load is more than HIV negative individual.
>
> **Clinical features:** Features are same as of tuberculosis without HIV.
>
> **Diagnosis**
> 1. **Microscopy:** No cavity formation, so bacilli are less in sputum and there is high frequency of negative sputum smear.
> 2. **Culture:** It is the best useful method as described earlier. It detects the bacilli up to 10–100 (10^1–10^2)/ml of sputum.
> 3. **Tuberculin test:** False-negative, because of poor or no CMI.
> 4. **Chest X-ray:** No cavitation, but instead widespread dissemination in lungs. It is less useful.
>
> **Treatment:** Tuberculosis can be cure, if diagnosed early and prompt treatment had been given. DOTS can control and extend the life of patient with HIV.

World Tuberculosis Day

Follow history.

ACCESS YOURSELF

Case Study

1. A 40-year-old male patient visited the medical OPD with history of low-grade evening rise fever, weight loss and productive coughing since 1 month. Sputum examination revealed AFB. Identify the organism and answer the following:
   a. Name the causative agent.
   b. Describe the morphology of causative agent.
   c. Write the pathogenicity of causative agent.
   d. Write preventive measures of clinical condition.

Essay/Full Question

1. Define and describe the etiology, types, pathogenicity, morphology/microscopic appearance and complications of tuberculosis.

Short Notes

1. Classification of mycobacteria.
2. C/Cs or pathogenicity or lab., diagnosis of *M. tuberculosis*.
3. Tuberculous meningitis.
4. Newer laboratory methods for diagnosis of *M. tuberculosis*.
5. Tuberculin test.
6. Tuberculosis and HIV.
7. BCG vaccine.

Short Questions for Theory/Viva Questions

1. Name four automated methods for cultivation of *M. tuberculosis.*
2. What are Ghon's focus and Ghon's complex?
3. Define primary and secondary tuberculosis.
4. How you diagnose the pulmonary tuberculosis under RNTCP?
5. What are MDR-TB and XDR-TB?

Comments on

1. Strictly it is not correct to classify mycobacteria as GPB.
2. For diagnosis of *M. tuberculosis* culture is more useful than microscopy.
3. *M. tuberculosis* is able to evade effects of antibiotics and host immune system.

MCQs for Chapter Review

Classification

1. **Tuberculosis complex includes all *except*:**
   a. *M. tuberculosis*   b. *M. microti*
   c. *M. kansasi*   d. *M. bovis*
2. **John's bacillus is:**
   a. *M. tuberculosis*   b. *M. paratuberculosis*
   c. *M. kansasi*   d. *M. bovis*

M. tuberculosis

3. ***Mycobacterium tuberculosis* was discovered by:**
   a. Robert Koch   b. Louis Pastuer
   c. Alexander   d. Virchow
4. **Rapid examination of tubercle bacilli is possible with:**
   a. Ziehl-Neelsen stain
   b. Kinyoun's stain
   c. Auramin–rhodamine stain
   d. Giemsa stain
5. **True regarding *Mycobacterium tuberculosis* is:**
   a. Produces visible colonies in 1 week time on Lowenstein-Jensen medium
   b. Decolorize by 20% sulfuric acid
   c. Facultative aerobe
   d. Niacin positive
6. **Which mycobacteria will show eugonic growth when grown in presence of glycerol?**
   a. *Mycobacterium bovis*
   b. *Mycobacterium ulcerans*
   c. *Mycobacterium tuberculosis*
   d. *Mycobacterium kansasii*
7. **Which of the following uses an oxygen sensitive fluorescent compound to detect presence of mycobactera?**
   a. MGIT   b. RT PCR
   c. ESP culture system   d. Conventional PCR
8. **Drug resistance in tuberculosis is due to:**
   a. Transformation   b. Transduction
   c. Conjugation   d. Mutation
9. **Reactivation tuberculosis is almost exclusively a disease of:**
   a. Lungs   b. Bones
   c. Brain   d. Lymph nodes
10. **Cavitation is most often seen in:**
    a. *Mycoplasma* pneumonia
    b. Tuberculous pneumonia
    c. Streptococcal pneumonia
    d. Staphylococcal pneumonia
11. **In which types of cutaneous tuberculosis caseation necrosis is present?**
    a. Tuberculosis verucosis
    b. Scrofuloderma
    c. Tuberculids
    d. Lupus vulgaris
12. **The most common focus of scrofuloderma is:**
    a. Lung   b. Lymph node
    c. Larynx   d. Skin

13. **Collection of a urine sample of a patient of renal tuberculosis is done:**
    a. 24 hours urine
    b. 12 hours urine
    c. In early morning
    d. Any time
14. **Which is used in digestion and decontamination of sputum in smear preparation?**
    a. NaOH
    b. KOH
    c. NaCl
    d. KCl
    e. N-acetyl L cysteine
15. **Spoligotyping typing used for:**
    a. *S.* Typhi
    b. *M. tuberculosis*
    c. *Staph. aureus*
    d. *V. cholerae*
16. **Tuberculin test is positive if induration is:**
    a. >2 mm
    b. >5 mm
    c. >7 mm
    d. 10 mm
17. **Tuberculin test positive is dependent on:**
    a. Erythema
    b. Nodule formation
    c. Induration
    d. Ulcerative change
18. **True about Mantoux is:**
    a. False-negative in fulminant disease
    b. If once done, next time it is always positive
    c. Results are given in terms of negative and positive
    d. Induration given in terms of length and breadth
    e. Always indicate active TB infection
19. **True about Mantoux is:**
    a. < 5 cm always positive
    b. Usually negative after treatment
    c. Positive reaction in children < 2 years is not important than adults
    d. Usually read after 48–72 hours
    e. False +ve in post measles state
20. **Which of the following statements regarding gamma – release – assay for diagnosis of tuberculosis is true:**
    a. First generation Quantiferon – TB assay used ESAT-6
    b. Second generation Quantiferon – TB (Gold) assay used ESAT and CFP-10
    c. These tests can distinguish between *M. tuberculosis* and *M. bovis*
    d. None of the mycobacteria gives a positive reaction with the test.
21. **All of the following is true about tuberculosis *except*:**
    a. For sputum to be positive bacilli should be >10^4/ml
    b. Niacin test differentiates *M. tuberculosis* and *M. bovis*
    c. Pathogenicity to rabbit differentiates *M. tuberculosis* and *M. bovis*
    d. Culture technique has low sensitivity.
22. **In tuberculosis, immunity is provided by:**
    a. CD4+
    b. CD8+
    c. IgG
    d. IgM
23. **True about BCG vaccine is:**
    a. Killed vaccine
    b. Subcutaneously given
    c. Given in positive tuberculin patient
    d. Live vaccine
24. **BCG vaccine in HIV positive newborn is:**
    a. Contraindicated
    b. Double dilution
    c. Half dilution
    d. Dose double
25. **In a patient, the lymph node shows necrosis with poor granuloma formation with plenty of AFB. Diagnosis suggests:**
    a. Tuberculosis in immunocompetent patient
    b. Tuberculosis in HIV patient

c. Sarcoidosis
d. Infection by *M. bovis*

Answers and Explanation of MCQs

1. **c**
2. **b**
   - Follow section, **classification of mycobacteria** for explanation of answers of MCQs 1–2.
3. **a**
   - Follow section, *Mycobacterium tuberculosis* **(history)** for explanation.
4. **c**
   - Follow section, *Mycobacterium tuberculosis* **(morphology)** for explanation.
5. **d**
   - *M. tuberculosis* produces colonies in 8–12 weeks on Lowenstein Jenson medium. Initially 20% sulfuric acid was used for decolorization, but nowadays, it is 25%. It is not facultative aerobe, but strict aerobe. It is niacin positive.
6. **c**
   - Follow **Table 56.1** for explanation.
7. **a**
   - Follow section, *Mycobacterium tuberculosis* **[culture characteristics → automated (newer) culture techniques]** for explanation.
8. **d**
   - Follow section, **resistance → drug resisatance** for explanation.
9. **a**
   - Reactivation tuberculosis is present in apical lungs and other sites are like tonsils, pharynx, larynx, small intestine and skin.
10. **b**
    - Pathological lesions produced in lungs by different microbes are: Cavitary lesion by *Mycobaterium tuberculosis,* lobar pneumonia by *Strept. pneumoniae* and interstial pneumonia (primary atypical pneumonia) by *Mycoplasma pneumoniae, Pneumocystic jirovecii,* etc.
11. **d**
    - Lupus vulgaris presents with caseation necrosis.
12. **b and d**
    - Scrofula/scrophula/King's evil affects the cervical and supraclavicular lymph nodes.
    - Scrofuloderma: It is the 2nd most common variety of cutaneous tuberculosis resulting from extension of scrofula which involves the cervical and supraclavicular lymph nodes with skin. It also called tuberculosis cutis colliquativa.
13. **c**
    - Bacilli are shedding very intermittently in urine, hence 3–6 consecutive early morning urine samples are collected, centrifuged and use sediment for processing. Acid alcohol is used as decolorizer.
14. **a and e**
    - Follow section, **diagnosis of pulmonary tuberculosis (preparation of specimens)** for explanation.
15. **b**
    - Follow section, **diagnosis of pulmonary tuberculosis (epidemiological typing → genotypic methods)** for explanation.
16. **d**
17. **c**
    - Follow section, **diagnosis of pulmonary tuberculosis [tuberculin test/Mantoux (result)]** for explanation of answers of MCQs 16–17.
18. **a**
    - False-negative occurs in overwhelming (fulminant) tuberculosis including miliary tuberculosis, convalescence from

some viral infections (like measles, chickenpox, glandular fever), lymphoreticular malignancy, sarcoidosis, malnutrition, immunosuppressive disease or therapy (steroids), impaired CMI, inactive PPD preparations or faulty technique.

- If once done, next time it is not positive; however, booster effect can occurs following (two steps) testing.
- Results are given in terms of negative; positive, doubtful (equivocal) and strong reactors.
- Induration given in terms of length not breadth. It is measured at its widest point transversely to long axis of forearm by ruler or caliper.
- It indicates active TB infection, latent infection, immunization, etc.

19. d

- Induration <5 cm is always negative. Positive result may revert to negative result (occasionally) on isoniazide treatment. Positive reaction in children <2 years is important and indicates the active infection. Read the result usually after 48–72 hours. False-negative result occurs in post measles state not false-positive.

20. b

- First generation Quantiferon-TB assay used PPD, but not ESAT-6.

- Second generation Quantiferon-TB (Gold) assay used ESAT-6 and CFP-10. ESAT-6 and CFP-10 are produced by *M. tuberculosis* and *M. bovis*, so the tests cannot distinguish between the two.
- Follow section, **interferon gamma release assay (IGRA)** for more explanation.

21. d

- Culture is more sensitive than microscopy as it can detect up to 10–100 bacilli per ml of sputum while for microscopy to be positive, sputum should contains 10^4 bacilli/ml.
- Niacin test is positive in *M. tuberculosis* and negative in *M. bovis*, so useful for differentiation *M. tuberculosis* and *M. bovis*.
- *M. bovis* is pathogenic to rabbit, but not the *M. tuberculosis*, so useful for differentiation between two.

22. a

- In tuberculosis immunity is cell mediated specially by CD4 cells, and humoral immunity (IgG and IgM) is not significant.

23. d

24. a

- Follow section, **immunoprophylaxis (BCG vaccine)** for explanation of answers of MCQs 23–24.

25. b

- Follow section, **tuberculosis in HIV** for explanation.

Infections of *Mycobacterium leprae*

Chapter Outline
- *Mycobacterium leprae*
- *Mycobacterium leprae murium*

Lepra bacilli are following two types:
1. **Causing human leprosy:** *Mycobacterium leprae.*
2. **Causing rat leprosy:** *Mycobacterium leprae murium.*

Mycobacterium leprae

Synonym (Common Name) and History

Hansen's bacilli: From the name of Gerard H Armauer Hansen, who 1st discovered the bacilli in 1868.

Lepra bacilli: Because producing the disease called leprosy.

Morphology

Staining properties: They are GPB and stain more readily than *M. tuberculosis*. They are less acid fast than *M. tuberculosis*, need 5% sulfuric acid. Differences between *M. tuberculosis* and *M. leprae* are shown in **Table 57.1**.

Morphological index (MI): Live bacilli in smear stained more uniformly than dead bacilli. Dead bacilli in smear stained less uniformly and may give granular appearance. The percentage of uniformly stained bacilli in tissue called morphological index (MI). It helps to monitor the response of antibiotics. Continued fall in MI indicates response to treatment, while rise indicates drug resistance. Percentage of solid fragmented granular bacillus is counted separately, as it gives more ideas about treatment response and better useful than MI.

Bacteriological index (BI): It is the total numbers of pluses in all smears divided by numbers of smear. Count the total numbers of bacilli (live or dead) and grade the smear as shown in **Table 57.2**. For calculation minimum 4 skin smears, 1 nasal smear and smears from both ear lobules should be examined.

Shape and size: They are straight rod, slightly curved, clubbed, lateral buds or branching forms with 1–8 μm × 0.2-0.5 μm in size (length is five times of breadth).

Arrangement: They are located extra- or intracellular in tissue monocytes (histiocytes). They are also lying extracellularly. Intracellular bacilli arranged parallel and bound together by lipid like substance called glia and giving globi or cigar in bundle appearance as shown in **Figs 57.1a and b**. Histiocytes containing globi are called foamy cells/Virchow's cells/lepra cells.

Motility, spores and capsule: They are nonmotile, non-sporing and noncapsulated.

Polar bodies or other intracellular elements: May be present.

TABLE 57.1: Differences between *M. tuberculosis* and *M. leprae*		
Features	***M. tuberculosis***	***M. leprae***
Staining technique		
Acid fastness	25%	5%
Alcohol fast	Yes	No
Culture		
Generation time	20 hours	13–15 days
Growth on media	Yes	No

TABLE 57.2: Grading of *M. leprae*	
BI/grade	**Number of AFB/oil immersion field**
1+	1–10/100 fields
2+	1–10/10 fields
3+	1–10/field
4+	10–100/field
5+	100–1000/field
6+	≥1000, globi/field

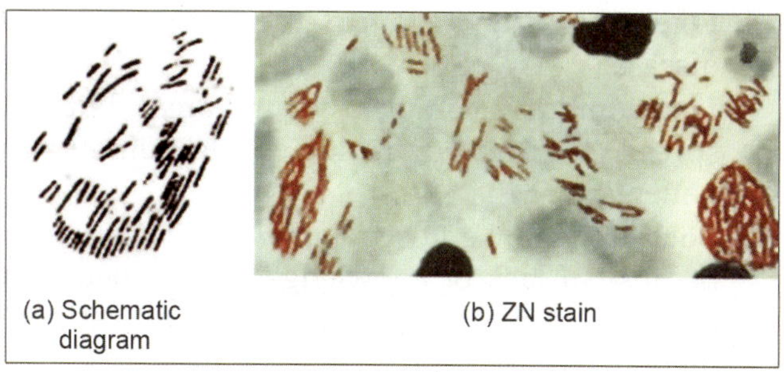

Figure 57.1: Morphology of *M. leprae*

Cultural Characteristics (C/Cs)

Effective factors: Bacilli are not cultivable in media and cultivation technique was developed by Shepard in 1960 in footpad of mouse at 20°C. Generation time in laboratory animal is 13–15 days.

Culture in media: It is not growing on artificial prepared media; however, several successful report of growth of lepra bacilli (called ICRC bacilli) from leprosy patient on human fetal spinal ganglion cell culture had been given by **ICRC [Indian Cancer Research Centre],** Bombay in 1962. ICRC bacilli are growing on LJ medium. Its relation to lepra bacilli is uncertain.

Animal culture: It is 10 times more sensitive than microscopy. Animals are **used** to isolate the bacilli, to prepare the antigen, to prepare the vaccine and to see the sensitivity pattern to antileprosy drugs. **Problems** with the uses of animals are ethical issues like requirement of permission from the animal ethical committee of institute and time-consuming technique about 6–9 months. It is done by using following animals.

Mouse: This technique was discovered by Shepard in 1960 as shown in **Fig. 57.2.** Inoculate the lepra bacilli intradermally in footpad of mouse, and keep it at low temperature about 20°C. Results are mentioned in Flowchart 57.1. Bacilli replicate very slowly in mouse (10^6 bacilli/gram of tissue after 6–8 hours), so it is the best animal for culture and for determination of sensitivity of antileprosy drugs.

Nine-banded armadillo: (*Dasypus novemcinctus*): Following inoculation of lepra bacilli in nine-banded armadillo **(Fig. 57.3)** generalized LL is developed. Natural disease occurs in some wild armadillo. Bacilli replicate very extensively in armadillo (10^{10} bacilli/gram of tissue), so it provides sufficient bacilli for study purpose/experimental work and for preparation of skin test reagent (lepromin A).

Resistance

Sterilization: Bacilli remain viable in a warm humid environment for 9–16 days and in moist soil for 46 days. They resist direct sunlight for 2 hours and UV light for 30 minutes.

Disinfection: Bacteria are resistant to many disinfectants.

Drug resistance: Resistance to dapsone is noticed.

Pathogenicity

Disease name: Disease called leprosy.

Synonym: It also called Hansen's disease, because bacilli 1st observed by Armauer Hansen (1868).

Definition: It is a chronic granulomatous disease of human primarily affecting skin, peripheral nerves and nasal mucosa, but capable to affect any tissue or organ.

Epidemiology: Leprosy is a disease of antiquity and bacilli are 1st identified by Hansen in 1868. It is recognized since the time of Vedic in **India (called Kustha Roga,** in Sushruta Samhita, 600 BC), the time of Biblic in Middle East and the time of Hipocrates, 460 BC. It is distributed worldwide, but maximum in India. It presents in all states with highest prevalence in Odisha

Infections of Mycobacterium leprae

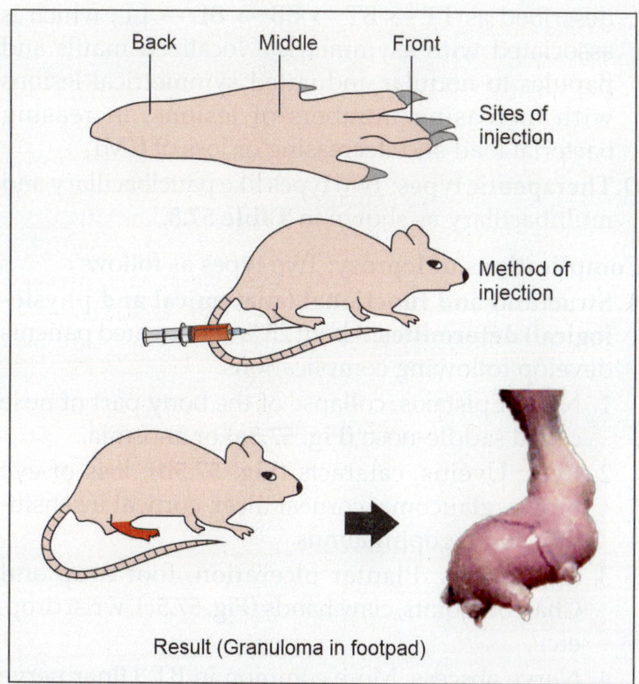

Fig. 57.2: Inoculation of *M. leprae* in footpad of mouse

Flowchart 57.1: Mouse culture

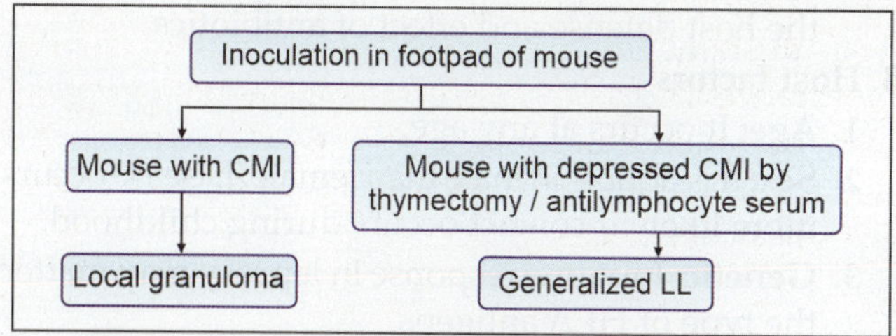

Fig. 57.3: Nine-banded armadillo

 (>5 per 1000 population) and least in Haryana (<0.1 per 1000 population).

Reservoir of infection: It is exclusive human disease, only source is human. Paucibacillary leprosy contains less numbers of bacilli, so having less risk of transmission and can transmit the infection after longer contact. Multibacillary leprosy contains large numbers of bacilli and must be considered infectious.

Source of infection: Nasal secretion is the main source. A sneeze from an untreated LL patient may contain >10^{10} bacilli.

Mode of transmission: Exactly it is not known but may be by following routes:
- Inhalation/droplet nuclei: Very large numbers of bacilli are discharged into nasal secretion.
- Skin to skin/skin to mucosa contact.
- Contact with soil or with fomites like clothes, linen, etc.
- Congenital via placenta.
- Via milk of leprosy patient to infant.
- Other rare possible routes are by tattooing with infected needles or by insect vectors.

Incubation period: It is very long from few months to 30 years (2–5 years). It is long, because of slow generation of bacilli. It is longer in LL than TT.

Portal of entry: Nasal mucosa or skin.

Sites: Bacilli can grow better in cooler tissues like skin, peripheral nerves, anterior chamber of eye, upper respiratory tract and testes and sparing the warms areas like axilla, groin, scalp and midline of back. Bacilli have affinity for Schwann cells and RE cells. Median nerve and ovaries are never involved in TT.

Precipitating factors (epidemiological determinants): Three types as follows:

A. Agent factors (virulence factors)
1. **Phenolic glycolipid-1:** It acts as virulence factor.
2. **Intracellular location:** It helps the bacilli to evade the host defense and effect of antibiotics.

B. Host factors
1. **Age:** It occurs at any age.
2. **Sex:** It is double in male than female. Disease occurs more likely if contact occurs during childhood.
3. **Genetic:** Immune response in leprosy may matter the type of HLA antigens.
4. **Immunity:** CMI is needed to develop the resistance to lepra bacilli.
5. **Close contact:** It is not highly communicable. Required longer contact for transmission and only 5% of spouse living with leprosy patient can acquired the infection.

C. Environmental and socioeconomical factors
1. **Humidity:** It favors the survival of bacilli. Bacilli remain viable for 9 days in dry condition, while for 46 days in moist area.

2. **Overcrowding:** It favors the transmission of disease.
3. **Location:** LL is common in rural areas, but now spread to urban population due to migration of patients.
4. **Social stigma:** Leprosy remains as a social stigma because of lack of knowledge, fear, superstition beliefs, complications, deformities and disfigurement of disease.

Clinical types of leprosy

A. Madrid classification/Polar classification, 1953: Four types:
1. TT (Tuberculoid type)
2. LL (Lepromatous leprosy/lepromata) } Table 57.3
3. BL [borderline (dimorphous) leprosy]: Lesions possessing characteristics of both tuberculoid and lepromatous types. It may shift to LL or TT depending on response to chemotherapy or alterations in host resistance.
4. Intermediate leprosy: It is the early unstable tissue reaction (macular lesion) which is neither characteristic of lepromatous nor tuberculoid type. Lesions may undergo spontaneous healing or may progress to tuberculoid or lepromatous types. It is bacteriologically sterile.

B. Indian classification: It includes four Madrid's types and one more type (total five types) called Indian (I) type/Pure neuritic. It is the extra type in Indian classification which is bacteriologically sterile and shows nerve involvement without skin lesion.

C. Ridley and Jopling classification, 1966: It includes five types like TT (tuberculoid type), BT (borderline tuberculoid), BB (borderline), BL (borderline lepromatous) and LL (lepromatous leprosy). All are differentiated in **Table 57.4**. Spectrum of disease is described as TT → BT → BB → BL → LL, which is associated with asymmetrical localized mauls and papules to nodular, indurated symmetrical lesions with increasing numbers of lesions, increasing bacterial load and decreasing or loss of CMI.

D. Therapeutic types: Two types like paucibacillary and multibacillary as shown in **Table 57.5**.

Complications of leprosy: Two types as follow:

A. Structural and functional (anatomical and physiological) deformities: About 25% of untreated patients develop following complications:
1. Nose: Epistaxis, collapse of the bony part of nose called saddle-nose **(Fig. 57.5a)** or anosmia.
2. Eyes: Uveitis, cataracts **(Fig. 57.5b)**, loss of eye brows, glaucoma, corneal ulcer, corneal insensitivity and lagophthalmus.
3. Extremities: Plantar ulceration, foot drop and Charcot's joints, claw hands **(Fig. 57.5c)**, wrist drop, etc.
4. Nerve abscess: More common in BT. Ulnar nerve is most frequently involved.

TABLE 57.3: Differences between TT and LL		
Features	**TT**	**LL (Lepromata)**
Bacteriology		
Bacterial load	Paucibacillary (bacilli are less or absent), bacilli are in destroyed nerve as beaded or granular forms	Multibacillary (bacilli are multiple), bacilli are inside the lepra cells as globi
Bacterial index	0–1+	4–6+
Infectivity	Low	High, because of more bacilli in lesions
Pathology		
Sites	Skin and peripheral nerves like ulnar, post auricular, peroneal and posterior tibial	All organs except lungs and CNS
Macroscopic pathology	Granulomatous lesion with predominance of CD4+ cells, IL-2, IF-γ and IL-12. Granulomatous lesion in dermis erodes the basal layer of epidermis (no clear zone)	No granuloma with predominance of CD8+ cells, IL-4, IL-5 and IL-10. Collection of lepra cells in dermis separated from epidermis (clear zone)
Microscopic pathology Granuloma formation Plasma cell infiltration Lymphocyte infiltration	 Common Poor Common	 Absent Common Not present
Clinical features		
Skin lesions	• Asymmetrical • Single/few lesion • Sharp margin • Dry, scaly hypopigmented macular lesion (**Fig. 57.4a**)	• Symmetrical • Multiple lesions • Irregular margin • Shiny maculopapular/nodular (**Fig. 57.4b**). Lesions are coalesce together to give leonine facies appearance
Nerve involvement	• Due to external compression by granuloma • Early, more sensory loss (early hypoesthesia or anesthesia)	• Due to direct bacillary invasion • Late, little or no loss of sensation (late hyposthesia)
Immunology		
Host resistance	High	Low
CMI	Good	Deficient/absent
Lepromin test	Positive	Negative
AMI	Normal	Exaggerated, with hyper γ-globulinemia and reverse albumin/globulin ratio
Ab to PGL-1	60%	95%
Autoimmunity	Autoantibodies are rare	Autoantibodies are more
Type II lepra reaction	Negative	Positive
Phagocytic cells	Mature epithelioid	Macrophage
Giant cells	Absent	Langhans type
HLA association	HLA DR-2	HLA MT-1 and HLA DQ-1
VDRL test	Negative	Biological false-positive
Lymphocyte transformation test	Positive	Negative
Cytology		
CD4+ : CD8+ ratio	Normal (2:1)	Reversed (1:2)
Others		
Prognosis	Good	Poor

TABLE 57.4: Ridley and Jopling classification

Features	TT	BT	BB	BL	LL
Bacilli seen in skin	–	–/+	+	++	+++
Bacilli in nasal secretion	–	–	–	+	++
Granuloma formation	+++	++	+	–	–
CMI	+++	++	+	+/–	–
Lepromin test	+++	+	+/–	–	–
Antibodies to lepra bacilli	+/–	+/–	+	++	+++
Phagocytic cells	Mature epithelioid	Immature epithelioid	Immature epithelioid	Macrophage	Macrophage
Type 1 reaction	–	+	+	+	+
Type 2 reaction	–	–	–	+/–	++

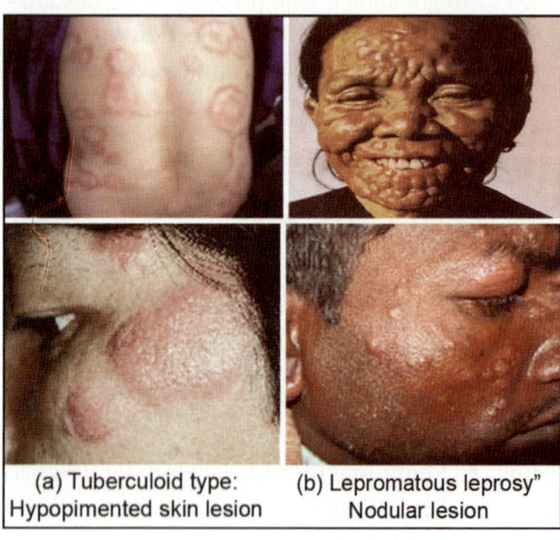

Fig. 57.4: Skin lesions in leprosy

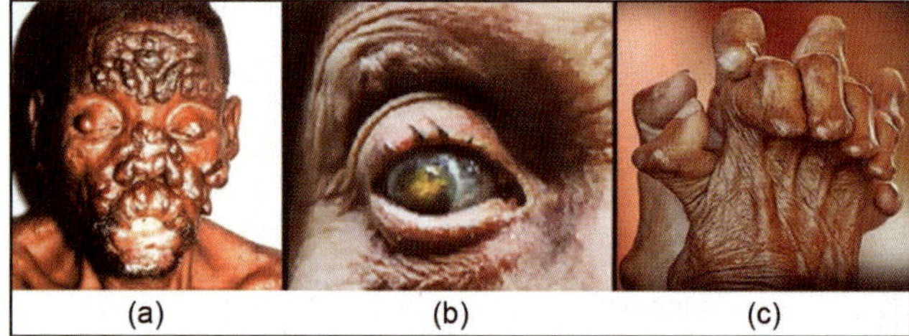

Fig. 57.5: (a) Saddle nose; (b) Cataract, (c) Claw hands

5. Testes: Orchitis followed by impotence.
6. Oral lesions: Small tumor like masses are presents on tongue, lips or hard palate called lepromas. They tend to ulcerate. Premaxillary bone recession presents frequently with or without tooth loss.

B. Allergic reaction: It called lepra reaction, defined as an acute exacerbation/allergic response during course of leprosy. Two types of such reaction:

1. **Type I (Reversal or borderline) reaction:** It occurs in initial 6 months. It presents with macules, papules or plaques without systemic features. It is the examples of type IV hypersensitivity reaction where predominant response is by Th1 with increased level of IL-2 and IFN-γ. It seen in BT, BB and in BL and treated with steroids. Borderline groups are unstable and they can shift to upgrade or downgrade direction.

TABLE 57.5: Differences between paucibacillay and multibacillary leprosy

Features	Paucibacillary	Multibacillary
Pathology		
Leprosy	TT, BT and I-Indian types	LL, BL and BB types
Nerve involvement	1	≥2
Skin lesions	1–5	≥6
Diagnosis		
Microscopy	Negative	Positive
Treatment (Multidrug therapy)		
Indication	To avoid the drug resistance in monotherapy	
Course	6 months	For 1 year or until skin smear becomes negative
Drugs	• Rifampicin (highly bactericidal) 600 mg once a month under supervision and • Dapson 100 mg daily self administered	• Rifampicin: 600 mg once a month, • Dapson 100 mg daily and • Clofazimine 50 mg daily
Follow-up	Annually at least for 4 years	Annually at least for 8 years
Other drugs	• Ethionamide or prothionamide may be added (as substitute of clofazamine). • Quinolones, minocycline, clarithromycin, rifapentine and moxifloxacin are other useful drugs • Under leprosy eradication program single lesion is managed by rifampicin, ofloxacin and minocycline	

- Upgrade direction: It seen in BL and in treated patients, increased CMI and shift to TT. It is a CMI response with influx of lymphocytes in lesions. Lesions are swollen and erythematous along with pain and tenderness.
- Downgrade direction: It seen in BT and in untreated or pregnant women, decreased CMI and shift to LL. It includes influx of bacilli in lesion.

2. **Type II (ENL, erythema nodosum leprosum):** It rarely occurs in initial 6 months. It presents with painful erythematous papules with systemic

features like subcutaneous nodules, fever, arthralgia, lymphadenopathy, neuritis, uveitis, orchitis and glomerulonephritis. It is an example of type III hypersensitivity (Arthus reaction) and main cytokine involved in the reaction is TNF-α. Predominant response is by Th2 with increase level of IL-6 and IL-8. It is seen in LL and in BL. It develops when patient is under the treatment, because of release of antigen from dead bacilli. It includes deposition of PMNs, IgG and complement in lesions. It is treated by steroids, antipyretics, thalidomide, chloroquine and clofazimine.

Laboratory Diagnosis

Specimens

1. **Nasal mucosal scraping:** It is done by using blunt scalpel. Erode the nasal septum, remove the piece of mucus membrane and take over the slide. Teased in uniform smear. It is not recommended routinely.
2. **Nasal secretion:** Early morning sample is advised. Mucous secretion is collected by blowing the nose on a clean cellophane sheet.
3. **Skin scraping (slit skin smear):** Total 5–6 different areas of skin should be sampled like forehead, buttock, chin, cheeks and ear lobules by using scalpel with blade. Edge of the lesion is the preferred site. Clean the lesion with spirit and pinched tightly to minimize the bleeding. Made a 5 mm long incision to reach up to infiltrated layers. Wipe out the blood or secretion; rotate the blade transversely to obtain the tissue pulp below epidermis, over the sterile slide. Smear (8 mm size of smear) it uniformly over slide.
4. **Biopsy:** It is taken from nodular lesions, thick nerves or from lymph nodes.

Testing method

A. **Microscopy:** Follow morphology.
B. **Culture:** Follow C/Cs.
C. **Serological tests:** Ab detecting tests have some value.
   - Tests to detect IgM Ab against PGL-1: It based on latex agglutination test. Sensitivity of ELISA is 95% in untreated LL and 60% in TT cases. Titer decreases with effective therapy. It may be positive in nonleprosy cases, so rarely useful.
   - Tests to detect specific Ab of *M leprae*: Widely used test in this category is fluorescent leprosy antibody absorption (FLA-ABS) test. It is useful to detect the early stage of disease and also to diagnose the subclinical cases. It is 92% sensitive and 100% specific.
D. **Molecular method:** PCR is in progress.
E. **Test to determine the CMI:**
   - **History:** Test called lepromin test which was discovered in 1919 by Mitsuda.
   - **Principle:** It based on DTH.
   - **Antigens:** Following three types of antigens are standardized as per bacilli contents.

   - Lepromin Ag: It gives early reaction and contains 40 million bacilli/ml. It is derived from armadillo (Lepromin A), which replaced the human lepromin (Lepromin H).
   - Mitsuda Ag: It gives early and late reaction and contains 160 million bacilli/ml.
   - Dharmendra Ag: It gives early and late reaction and contains 10 million bacilli/ml.
- **Effective factors:** It is negative in initial six months of life and may become positive up to 1 year of age. Negative result may turn to positive due to BCG vaccination.
- **Method:** Inject 0.1 ml Ag intradermally on flexor aspect of forearm and examined after 48–72 hours.
- **Results and interpretation:** Two types of reactions occur like early and late as shown in **Table 57.6**.
- **Uses:** It is not used for diagnosis of leprosy. Possible uses are mentioned in **Table 57.7**.

Prevention

Early diagnosis and treatment is the best strategies for prevention.

General measures: Isolation of patients.

Chemoprophylaxis: Long-term chemoprophylaxis by using dapsone is useful in LL.

Immunoprophylaxis: A killed leprosy vaccine had been developed in India in 2018 from *Mycobacterium indicus pranii* called MIP vaccine. Trials are given to close contacts in last three years which bring down the cases by 60%.

TABLE 57.6: Result of lepromin test

Early result	Late result
Fernandez reaction	Mitsuda reaction
Read result after 48 hours	Read result at 21st day/3rd week
Erythema and induration in 24–48 hours which disappear in 3–5 days	Nodule formation (by infiltration of lymphocytes, giant cell and epithelioid cells) after 3–4 weeks gets ulcerated and disappears in few weeks
Positive result if size of erythema and induration is >10 mm	Positive result if size of nodule is >5 mm
It is the DTH to soluble components of lepra bacilli due to infection or exposure by other ways	It indicates the status of CMI to antigens of lepra bacilli. Positive indicates good CMI and negative indicates poor CMI
Less useful	More useful

TABLE 57.7: Uses of lepromin test

Uses	+ve result	−ve result
To classify the leprosy	TT	LL
To assess the prognosis and treatment response	Good	Poor
To assess the host immunity against bacilli	Good	Poor

Chemotherapy: Monotherapy by using dapsone is given, but due to development of resistance, multi-drug therapy (MDT) is recommended as mentioned in **Table 57.5**.

Immunotherapy: Immunotherapeutic vaccine is developed from *Mycobacterium W* by National Institute of Immunology (NII), New Delhi, which is able to increase to effect of MDT.

World Leprosy Day

This day was chosen on last Sunday of January in each year in commemoration of the death of Mahatma Gandhi, the leader of India, who understood the importance of leprosy.

Mycobacterium leprae murium

It produces leprosy in rat characterized by subcutaneous induration, lymphadenopathy, emaciation, ulceration and loss of hairs. Bacilli are present large numbers in lesions. Bacilli resemble to *M. leprae,* but distinct from *M. leprae* by DNA studies.

ACCESS YOURSELF

Short Notes

1. Pathogenicity or laboratory diagnosis of *M. leprae*
2. Classification/different types of leprosy
3. Lepromin test
4. Lepra reaction.

Short Questions For Theory/Viva Questions

1. What is globi?
2. What are morphological and bacteriological index in *M. lepare?*
3. How lepromin test is useful in management of leprosy.

Comments on

1. Leprosy has long incubation period.
2. LL is more infectious than TT.
3. Lepromin test is negative in LL.

MCQs for Chapter Review

Mycobacterium leprae

1. **Which of the following is true regarding globi in a patient with lepromatous leprosy?**
   a. Consist of lipid laden macrophages
   b. Consist of macrophages filled with AFB
   c. Consist of neutrophils filled with bacteria
   d. Consist of activated macrophages
2. **For experimental work lepra bacilli are best cultured in:**
   a. Armadillos   b. Mouse foot pad
   c. Guinea pig   d. Rabbit testes
3. *Mycobacterium leprae* **can be grown in:**
   a. LJ medium   b. Roberson cooked meat medium
   c. Footpad of mouse   d. Sabouraud's agar
4. **Lepra bacilli can survive outside the human body up to:**
   a. 7 days   b. 12 days
   c. Zero days   d. 5 days

5. **Leprosy affects all of the following *except*:**
   a. Testes   b. Ovaries
   c. Eyes   d. Nerves
6. *Mycobacterium leprae* **is spread by:**
   a. Skin to skin contact   b. Blood transfusion
   c. Droplet spread   d. Ingestion
7. **The characteristic finding in a case of leprosy is:**
   a. Culture test is positive in 2–3 months in LJ medium
   b. Long contact with tuberculoid leprosy can transmit the disease
   c. CMI is seen in lepromatous leprosy
   d. Macule lesion heals spontaneously
8. **Neurological involvement is pronounced in which type of leprosy:**
   a. Tuberculoid   b. Lepromatous
   c. Borderline   d. Lucio leprosy
9. **Single skin lesion is seen in which type of leprosy:**
   a. LL   b. TT
   c. BL   d. BT
10. **Following test is not used for diagnosis of leprosy:**
    a. Lepromin test
    b. Slit skin smear
    c. Fine needle aspiration cytology
    d. Skin biopsy
11. **Mitsuda reaction is read after:**
    a. 3 days   b. 3 hours
    c. 3 weeks   d. 3 months
12. **In the management of leprosy, lepromin test is most useful for:**
    a. Herd immunity
    b. Prognosis
    c. Treatment
    d. Epidemiological investigation
13. **Exacerbation of lesions in patients of borderline leprosy is seen in:**
    a. Erythema nodosum leprosum
    b. Lepra reaction type-1
    c. Jarisch-Herxheimer reaction
    d. Resolving leprosy
14. **Characteristics of type II lepra reaction:**
    a. Erythema and edema
    b. ENL
    c. Lymphadenopathy
    d. Uveitis
    e. Trophic ulcer
15. **Main cytokine involved in erythema nodosum leprosum (ENL) reaction is:**
    a. Interleukin-02
    b. Interferon-gamma
    c. Tumor necrosis factor-alpha
    d. Macrophage colony stimulating factor
16. **ENL is seen in:**
    a. Lepromatous leprosy   b. Tuberculoid type
    c. Intermediate leprosy   d. Pure neuritic
17. **Following drug is used for the treatment of type II lepra reaction, *except*:**
    a. Chloroquine   b. Thalidomide
    c. Cyclosporine   d. Corticosteroid
18. **Fastest microbicidal agent against *M. leprae*:**
    a. Clofazimine   b. Dapsone
    c. Rifampicin   d. Minocycline

19. **Under Leprosy eradication program, the management of single lesion is:**
   a. Single dose of rifampicin and dapsone
   b. Rifampicin and dapsone for 6 months
   c. Rifampicin, ofloxacin and minocycline single dose
   d. Rifampicin and minocycline for 6 months

20. **True about lepra bacilli is:**
   a. INH inhibits their growth
   b. Antileprosy vaccine can give lifelong protection
   c. *Mycobacterium leprae* can be grown in foot pad of mouse
   d. Incubation period is 3–4 months

Answers and Explanation of MCQs

1. b
- Follow section, *Mycobacterium leprae* **(morphology)** for explanation.

2. a

3. c
- Follow section, *Mycobacterium leprae* **(culture characteristics)** for explanation of answers of MCQs 2–3.

4. b
- Follow section, *Mycobacterium leprae* **(resistance)** for explanation.

5. b
- Median nerve and ovaries are never involved in TT.
- Other organs involved in the leprosy are explained under the heading of pathogenicity (sites).

6. a and c
- Follow section, *Mycobacterium leprae* **(pathogenicity → modes of transmission)** for explanation.

7. b and d
- Culture is not possible in LJ or any other medium. Lepra bacilli can grow only in animals as described in culture characteristics
- Paucibacillary leprosy (TT) contains less numbers of bacilli, so having less risk of transmission and can transmit the infection after longer contact. Multibacillary leprosy (LL) contains large numbers of bacilli and must be considered infectious.
- CMI is absent in lepromatous leprosy.
- Macule lesion heals spontaneously in indeterminate type.

8. a
- Neurological involvement occurs early in TT due to external compression by granuloma, more sensory loss (early hyposthesia or anaesthesia).

- For more explanation follow **Table 57.3 (nerve involvement)**.

9. b
- Lesion is single/few TT and increasing from TT to LL as per spectrum of disease with increasing bacterial load and also loss or absence of CMI.

10. a
- Lepromin test is not used for diagnosis but for other purpose. Uses of lepromin test are explained in **Table 57.6**.

11. c
- Follow section, *Mycobacterium leprae* **(laboratory diagnosis → Lepromin test and Table 57.6)** for explanation.

12. b
- Positive lepromin test indicates good CMI and good prognosis while negative result indicates the poor or absence of CMI and poor prognosis, so turning of negative to positive result following antileprotic drugs, will indicates response to therapy and helps in management of leprosy.

13. b

14. b

15. c

16. a
- Follow section, *Mycobacterium leprae* **(pathogenicity → complications → lepra reaction)** for explanation of answers of MCQs 13–16.

17. c
- Lepra reaction can occur in LL, where CMI is absent or low. Cyclosporin is immunosuppressive agent and can suppress more immunity, if given so not useful.

18. c
- Rifampicin is the only bactericidal drug against *M leprae*.

19. c
- Follow section, *Mycobacterium leprae* **(pathogenicity → therapeutic types and Table 57.5)** for explanation.

20. c
- INH is not the antileprotic drugs.
- No any useful anti leprosy vaccine available which can give lifelong protection.
- *Mycobacterium leprae* can be grown in foot pad of mouse as describe in culture characteristics.
- Incubation period is very long from few months to as long as 30 years (2–5 years).

CHAPTER 58

Infections of NTM and MSP

Chapter Outline
- NTM (Non-tuberculous Mycobacteria)
- MSP (Mycobacteria as Skin Pathogen)

NTM (NONTUBERCULOUS MYCOBACTERIA)

Meaning

Mycobacteria other than typical tubercle bacilli, occasionally causing human disease, resemble to tuberculosis are called NTM.

Synonym

They are also known as anonymous mycobacteria, unclassified mycobacteria, atypical mycobacteria, MOTT (mycobacteria other than tubercle bacilli), environmental mycobacteria (as environment like water, soil, etc., are the common sources), opportunistic mycobacteria (because producing infection in person with IDDs), paratubercle bacilli or tuberculoid bacilli.

Properties

They are saprophytes and natural habitat in water and soil, but unlike saprophytic mycobacteria, they can infect humans/animals and causing opportunistic infections. Infection does not occur in person with normal immunity. They are also present as normal flora in skin, GIT and respiratory tract. They are widely present in soil, water and air. Infection with them is quite common due to direct repeated environmental exposure. Such repeated infection will decreases the efficacy of BCG vaccine by cross contamination. They are common where tuberculosis is rare and rare where tuberculosis is endemic. They are low virulent compare to *M. tuberculosis*. They are causing disease other than tuberculosis except *M. kansasi*. Overt disease present in immunocompromised and subclinical infection occurs in immunocompetent persons. Person-to-person infection does not occur. Infection is mainly asymptomatic, which cause weak positive Mantoux reaction. Some species like *M. avium*, *M. kansasi* and *M. xenopi* occur as laboratory contaminants and mistaken as tubercle bacilli in smear. They are acid as well as alcohol fast. Most NTM are strict aerobes and grow best at acidic pH. They can grow at 25°C, 37°C and 45°C on LJ medium.

Classification

It is given by Runyon in 1959 on the basis of pigment production and growth rate.

Group-I: Called photochromogens
Group-II: Called scotochromogens
Group-III: Called nonchromogens (nonphotochromogens)

(Slow grower in 4–12 weeks)

Group-IV: Called rapid growers. Grow in 7 days.

Group-I: Photochromogens

They produce yellow-orange pigment in light when young culture is exposed to light for 1 hour in presence of air and re incubated for 24–28 hours. Important species are mentioned in **Table 58.1**.

Group-II: Scotochromogens

They produce yellow-orange-red pigment in light and even in dark. Important species are mentioned in **Table 58.2**.

Group-III: Nonchromogens

These are also called non photochromogens. They do not produce pigment either in light or in dark. Important species are mentioned below.

M. malmoense: It is slow grower and causing pulmonary infection and lymphadenitis.

M. avium: It causes natural tuberculosis in birds and lymphadenopathy in pigs. It is an opportunistic human pathogen.

M. intracellularae: It also called Battey bacillus because 1st identified in Battey state hospital of tuberculosis, Georgia, USA. It is closely related *M. avium*, so collectively called MAC (*Mycobacterium avium* complex). MAC is the most common cause of pulmonary disease

TABLE 58.1: Species of photochromogens

Photochromogens (Mnemonic: ka ma si ge as)

Species	Biochemical tests	Pathogenic lesions	Other remarks
M. kansasii	Nitrate reduction, Tween 80 hydrolysis, urease and catalase tests: Positive Pyrazinamide test: Negative	Chronic pulmonary disease resembling tuberculosis	Growth at 25°C and 37°C in LJ medium
M. marinum (*M. balnei*)	Nitrate reduction test: Negative Urease test and pyrazinamide tests: Positive	Skin wart (swimming pool granuloma or fish tank granuloma), tender nodules (spread in a sporotrichoid pattern) tendonitis	Growth at 25°C and variable at 37°C
M. simiae	Niacin test: Positive (so, confused with *M. tuberculosis*)	Pulmonary disease	
M. genavense		Infection in HIV patient	
M. asiaticum	Niacin test: Negative	Pulmonary disease and bursitis	

TABLE 58.2: Species of scotochromogens

Scotochromogens (Mnemonic: s g s ce flo)

Features	Biochemical tests	C/Fs	Other features
M. scrofulaceum	Tween 80 hydrolysis, pyrazinamide and nicotinamide tests: Negative Urease test: Positive	Cervical lymphadenitis (scrofula) in children	Growth at 25°C and 37°C
M. gordonae	Tween 80 hydrolysis, pyrazinamide and nicotinamide tests: Positive but urease negative	Present in tap water and causes pulmonary infection	
M. szulgai	It is scotochromogen at 37°C and photochromogen 25°C. It produces pulmonary infection, lymphadenitis and cutaneous-subcutaneous bursitis		
M. celatum	It produces pulmonary infection		
M. flovescens	Intermediate growth rate in 7–10 days at 25–37°C with butyrous colony and yellow orange pigment. Nonpathogenic		

from atypical category. MAC is the opportunistic pathogen in HIV positive patient when CD4 count is <50/mm^3. It presents with fever, weight loss, abdominal pain, diarrhea, night sweats and lymphadenopathy. AFB found in stool, sputum and blood.

M. xenopi: It is nonchromogen, but may form scotochroogenic yellow colonies. It causes pulmonary lesion especially in HIV positive patient and epididymitis. It was originally isolated from toads. It is isolated from hospital water sources and associated with nosocomial outbreak. It is compared with MAC in **Table 58.3**.

M. paratuberculosis: It also called Johne's bacillus. It is a cattle pathogen. In human it causes Crohn's disease but its association is still under question.

M. ulcerans (*M. burulu*): Follow **Table 58.5**.

Other species: *M. nonchromogenicum, M. gastri, M. terrae, M. shimoidei,* etc.

Group-IV: Rapid Growers

They are not producing pigment and growth occur within 7 days at 25°C and 37°C. These are mostly saprophytic bacilli. *M. smegmeatism* and *M. phlei* produce the pigment. *M. fortuitum, M. chelonae* and *M abscessus* do not produce the pigment. Comparison between *M. fortuitum* and *M. chelonae* is done in **Table 58.4**. Other rapid growers are *M. vaccae, M. genevense, M. confluentis* and *M. intermedium.*

TABLE 58.3: Comparison between MAC and *M. xenopi*

Nonchromogens (Mnemonic: m a i x u p)

Features	MAC	*M. xenopi*
Urease test	Negative	Negative
Pyrazinamide test	Positive	Positive
Iron uptake	Negative	Negative
Catalase test at 68°C	Negative	Positive
Aryl sulfatase	Negative	Positive
Growth at 45°C	Positive	Positive

TABLE 58.4: Species of rapid growers

Rapid growers (Mnemonic: s p f c a)

Features	*M. fortuitum*	*M. chelonae*
Aryl sulfatase	Negative	Negative
Growth by 5% NaCl	Negative	Positive
Iron uptake	Negative	Positive
Nitrate reduction test	Positive	Positive
Growth on MacConkey's agar and at 25°C and 37°C	Positive	
Pathogenic lesion	Pulmonary infection, wound infection, abscess following injection or vaccination	

<table>
<tr><th colspan="3" style="text-align:center">TABLE 58.5: Differences between M. ulcerans and M. balnei</th></tr>
<tr><th>Features</th><th>M. marinum (M. balnei)</th><th>M. ulcerans (M. buruli)</th></tr>
<tr><td>History and synonym</td><td>1st identified in Sweden from swimming pool water called M. balnei (from balneum means bath) later identified from European and American country called M. marinum</td><td>1st identified from Australia called M. ulcerans and than from Buruli district, Uganda with large epidemic called M. buruli</td></tr>
<tr><td>Prevalence</td><td>In temperate climate</td><td>In tropical climate</td></tr>
<tr><td>Mode of transmission</td><td>Swimming pool water</td><td>Traumatic injuries</td></tr>
<tr><td>Sites</td><td>Prominence like elbows, ankles, nose etc.</td><td>Legs or arms</td></tr>
<tr><td>Incubation period</td><td>—</td><td>Few weeks</td></tr>
<tr><td>Virulence factors</td><td>—</td><td>Only mycobacteria produce exotoxin called mycolactone</td></tr>
<tr><td>Clinical features</td><td>Swimming pool/fish tank granuloma: Papular lesion gets ulcerated with spontaneous healing (self limited)</td><td>Buruli ulcer: Initially nodule formation gets ulcerated and heals with disfiguring scar (chronic progressive ulcer)</td></tr>
<tr><td>Lab., diagnosis
• Samples
• ZN stain
• LJ medium
• Pigment in light
• Footpad of mice inoculation
• Guinea pig inoculation</td><td>• Smear from edge of ulcer
• Scanty bacilli
• Grow in 1–2 weeks at 25–35°C
• Produced
• Local inflammation followed by purulent ulcer
• Non-pathogenic</td><td>• Smear from edge of ulcer
• Plenty of bacilli
• Grow in 4–8 weeks at 30–33°C
• Not produced
• Edema of limb but ulceration is infrequent
• Inflammation and necrosis</td></tr>
</table>

Flowchart 58.1: Laboratory diagnosis of atypical mycobacteria

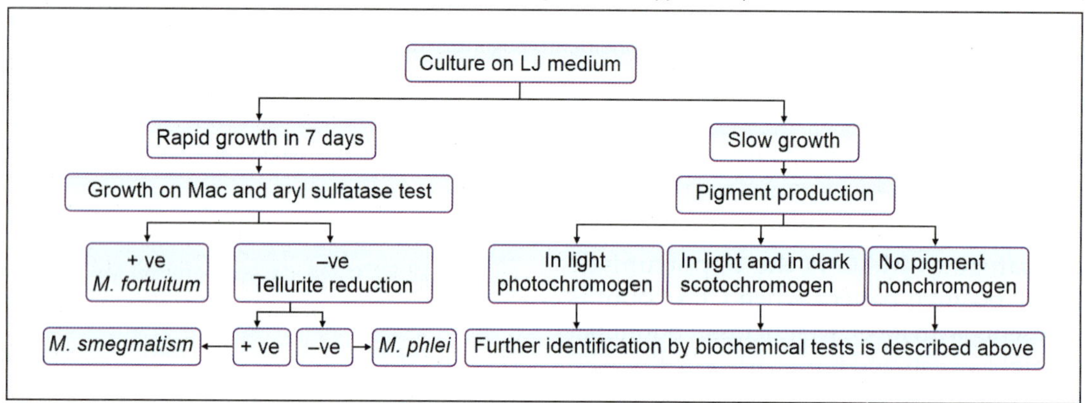

Laboratory Diagnosis

Specimens: Sputum, pus/exudates, nodular biopsy, etc.

Testing methods

A. **Microscopy:** By ZN stain.

B. **Culture and biochemical tests:** Follow **Flowchart 58.1**. Automated culture like MALDI-TOF is used to identify the species from culture.

C. **Animal pathogenicity:** Nonpathogenic for guinea pig but pathogenic for mouse.

Treatment

Anti tuberculous drugs are useful. Most atypical mycobacteria are resistant to antituberculous drugs, so drug are selected after antibiogram.

MSP (MYCOBACTERIA AS SKIN PATHOGEN)

M. tuberculosis: It produces lupus vulgaris (55%) and scrofuloderma (27%) as described in **Ch. 56**.

M. leprae: Described in **Ch. 57**.

NTM: Some **NTM** producing following skin lesions.

1. Abscesses following injection or vaccination: Abscess produced by *M. fortuitum* and *M. chelonei* as described above.

2. Swimming pool granuloma produced by *M. marinum* as described in **Table 58.5**.

3. Buruli ulcer produced by *M. ulcerans* as described in **Table 58.5**.

4. Skin lesion: *M. haemophylum*, required hemin for growth and grow at 32°C in 2–4 weeks.

> **Note: Swimming pool pathogens and diseases**
> 1. **Bacteria**
>   - *M. marinum:* Follow Table 58.5.
>   - *Chlamydia trachomatis:* Swimming pool conjunctivitis/ Adult inclusion conjunctivitis.
> 2. **Viruses:** These are serotypes 3, 4, 7 and 11 of adenoviruses causing acute follicular (swimming pool) and few serotypes of ECHO viruses.
> 3. **Fungi:** *Trichosporon beigelii* causes white piedra.
> 4. **Parasites:** *Naegleria fowleri* causes primary amoebic meningoencephalitis.

Essentials of Medical Microbiology

ACCESS YOURSELF

Short Notes

1. Atypical mycobacteria.
2. Mycobacteria as skin pathogens.

Short Questions for Theory/Viva Questions

1. Name the four pigment producing bacteria.
2. Write four examples of swimming pool pathogens.

MCQs for Chapter Review

NTM

1. **True about mycobacteria other than tuberculosis is:**
   a. Cause disseminated infection
   b. Occur in person with normal immunity
   c. Cause decreased efficacy of BCG due to cross immunity
   d. Person-to-person transmission seen
2. **Which of the following are photochromogens?**
   a. *M. fortuitum*
   b. *M. marinum*
   c. *M. simiae*
   d. *M. kansasi*
3. **Pigment producing atypical mycobacteria:**
   a. *M. fortuitum* and *M. chelonae*
   b. *M. xenopi* and MAC
   c. *M. gordonae* and *M. szulgai*
   d. *M. ulcerans*
4. **Which of the following is not a rapidly growing atypical *Mycobacterium* causing lung infections?**
   a. *M. chelonae*
   b. *M. fortuitum*
   c. *M. abscessus*
   d. *M. kansasi*
5. **Battey bacillus is:**
   a. Photochromogen
   b. Scotochromogen
   c. Nonchromogen
   d. Rapid grower
6. **A patient with diarrhea with AFB positive organism in stool. The most likely organism is:**
   a. *Mycobacterium avium intracellularae*
   b. *Mycobacterium tuberculosis*
   c. *Mycobacterium leprae*
   d. *Mycoplasma*
7. **Scotochromogens are:**
   a. *M. scrofulaceum*
   b. *M. avium*
   c. *M. szulgai*
   d. *M. gordonae*
8. **Scotochromogen produces pigment in:**
   a. Light
   b. Dark
   c. Absence of light
   d. Light and dark

9. **Which of the following is slow grower?**
   a. *M. kansasii*
   b. *M. chelonae*
   c. *M. fortuitum*
   d. *M. abscessus*

MSP

10. **Cutaneous lesion is produced by following bacteria *except*:**
    a. *M. tuberculosis*
    b. *M. leprae*
    c. *M. marinum*
    d. *M. intracellulare*
11. **Fish tank granuloma is seen in:**
    a. *M. fortuitum*
    b. *M. leprae*
    c. *M. marinum*
    d. *M. kansasi*

Answers and Explanation of MCQs

1. **c**
   - Follow section, **NTM (properties)** for explanation.
2. **b, c and d**
   - Follow section, **NTM (Group-I: photochromogens and Table 58.1)** for explanation.
3. **c**
   - Pigment producing atypical mycobacteria are photochrmogens and scotochrmogens. *M. fortuitum* and *M. chelonae* are rapid growers and do not produce the pigment. *M. xenopi*, MAC and *M. ulcerans* are nonchromogens and do not produce the pigment. *M. gordonae* and *M. szulgai* are scotochrmogens and known to produce the pigment.
4. **d**
   - *M. chelonae*, *M. fortuitum* and *M. abscessus* are rapid growers while *M. kansasi* is photochrmogen, causing tuberculosis like lesion in lungs.
5. **c** ⎱ Follow section, **NTM (group-III: nonchromogenes**
6. **a** ⎰ → *M. intracellularae*) for explanation.
7. **a, c and d**
   - Follow section, **NTM (Group-II: scotochromogens and Table 58.2** for explanation.
8. **d**
   - Photochromogens produce pigment in light, scotochrmogens produce pigment in light plus dark and nonchromogens do not produce pigment.
9. **a**
   - Photochromogens, scotochromogens and nonchromogens are the slow growers. *M. kansasii* is the photochromogen.
   - All other options are belong to rapid growers.
10. **d**
11. **c**
    - Follow section, **MSP and Table 58.5** for explanation of answers of MCQs 10–11.

Infections of Listerias and *Erysipelothrix*

Chapter Outline

LISTERIAS

Listeria monocytogenes

Classification

Order: Bacillales

Family: Listeriaceae

Genus: *Listeria*

Species: It contains 17 species like *L. aquatica, L. booriae, L. cornellensis, L. fleischmannii, L. floridensis, L. grandensis, L. grayi, L. innocua, L. ivanovii, L. marthii, L. monocytogenes, L. newyorkensis, L. riparia, L. rocourtiae, L. seeligeri, L. welshimeri* and *L. weihenstephanensis*. Out of all these species only *L. monocytogenes* is consistently associated with human illness.

Meaning and History

Genus name *Listeria* is given in honor of Joseph Lister (father of antiseptic surgery). It causes monocytosis in humans also in rabbit after experimental inoculation, hence the species name is *monocytogenes*.

Morphology

Type according to Gram's stain: They are GPB.

Shape and size: They are rod shape, about 2–3 µm × 0.5 µm in size and pleomorphic from small to medium coccobacilli as shown in **Fig. 59.1a**. Filamentous forms observed in old culture.

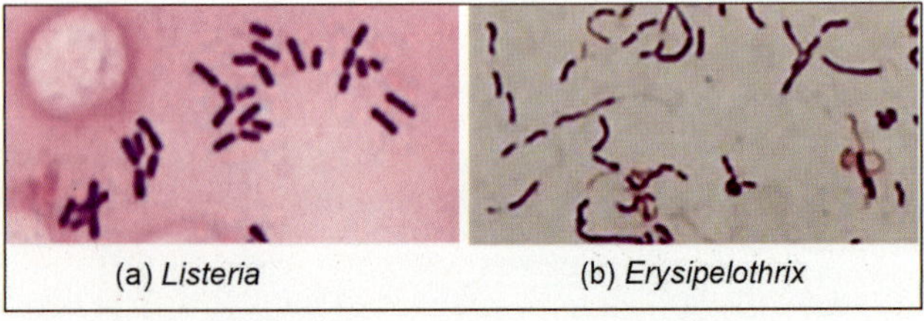

Fig. 59.1: GPB under Gram's stain

Arrangement: They arranged singly or in pairs or with angle to each other mistaken as diphtheria. Also occur in short chains.

Motility: They are showing tumbling motility (end over end) with peritrichous flagella at 25°C (room temperature) but not at 37°C (body temperature). Motility is due to temperature dependent flagella expression, called differential motility. Other bacteria showing differential motility are *Yersinia pseudotuberculosis* and *Yersinia enterocolitica*.

Spores and capsule: They are nonsporing and non-capsulated.

Cultural Characteristics (C/Cs)

Effective factors

- O_2 effect: They are facultative anaerobes.
- CO_2 effect: They grow better under 5–10% CO_2.
- Optimum temperature: 37°C (range 1–45°C).
- pH: Slightly alkaline about 9.6.
- They grow on ordinary media. Growth is improved by adding carbohydrates, blood or liver extracts in media. Bacteria can grow in presence of 0.1% tellurite and 10% salt.

Culture in media

A. Liquid media: Enrichment media like tryptose phosphate broth and thioglycollate broth are useful. Collect the sample in trypticase soy broth, incubate at 4°C and then do subculture on solid media every week for 1–6 months called cold enrichment.

B. Solid media/plating media

- **Blood agar:** They produce β-hemolysis may confuse with β-hemolytic streptococci [*L. ivanovii* gives inner zone of β-hemolysis surrounded by α-hemolysis called target (double zone) hemolysis].
- **Tryptose phosphate agar:** Colonies are 0.5–1 mm, smooth, translucent and easily emulsifiable.

C. Selective media

- **MacConkey's medium:** Bacteria grow poorly.
- **PALCAM agar:** Make the medium selective by adding the <u>P</u>olymyxin, <u>A</u>criflavine, <u>L</u>ithium chloride, <u>C</u>eftazidime, <u>A</u>esculin and <u>M</u>annitol.

Culture in animal: Instillation of culture in to the eyes of guinea pig/rabbit produces keratoconjunctivitis within 24 hours called Anton test.

Biochemical Reactions (B/Rs)

Sugar fermentation: They ferment glucose, maltose, rhamnose, salicin and α-methyl D-mannoside with acid production without gas.

I M Vi C tests: − + + −

Catalase, CAMP and aesculin hydrolysis tests: Positive

Oxidase and urease tests: Negative

Immunity

Bacilli are intracellular so antibodies are not protective. CMI is useful.

Pathogenicity

Disease name: It called listeriosis.

Epidemiology: *L. monocytogenes* serotype 4b is responsible for 5–33% of sporadic human cases worldwide and for all major foodborne outbreaks in Europe and North America since the 1980s. It is one of the most virulent foodborne pathogens, linked to the consumption of coleslaw containing cabbage with case-fatality rate is around 20–30%. It ranked third in total number of deaths among foodborne bacterial pathogens, with fatality rates exceeding even *Salmonella* (1%) and *Cl. botulinum*.

Reservoirs of infection: Common reservoirs are mammals, fish, birds, animals, ticks and crustacean. It also present as saprophytes in decaying materials, water and soil. Study suggests that 10% of human GIT are colonized by *L. monocytogenes*.

Sources of infection: *Listeria* has tendency to grow at wide range of temperature (from 1 to 45°C, refrigerator to body temperature) hence, it is present in all hot and cold (refrigerated) items. Common sources are milk of infected animals, vegetable salads, milk products like soft cheese, butter, etc., meat of turkey, undercooked chickens, hot dogs, coleslaw, pate, water, respiratory droplets, contaminated hospital devices, etc.

Modes of transmission

- **Direct**
  1. **Contact:** *Listeria* is also a common veterinary pathogen, causing abortion and encephalitis in sheeps and cattle. Occupational **contact** with animals, birds in veterinarian, animal handlers and poultry workers are the modes for human entry.
  2. **Vertical:** By placenta or by infected birth canal to neonates.
- **Indirect (vehicle borne)**
  1. **Air borne:** By inhalation of dust.
  2. **Food borne:** Infection occurs by unpasteurized (raw) milk of infected animals or cabbage contaminated with sheep manure.
  3. **Fomites-borne:** Infection occurs to neonates via contaminated hospital equipments or from contaminated hands-fingers of nurses.

Incubation period: It is variable from 3 to 70 days. The onset of gastrointestinal symptoms is unknown but probably exceeds 12 hours.

Portal of entry: Skin, GIT, respiratory tract, etc.

Sites: Meninges, brain, spinal cord, blood, uterus and also other organs.

Precipitating factors (epidemiological determinants): Three types:

A. Agent factors (virulence factors)

1. **Antigens:** Bacteria have 14 O-Ags which are heat stable and 4 H-Ags which are heat labile. Based on agglutination reaction several serovars like 1/2a, 1/2b, 1/2c, 3a, 3b, 3c, 4a, 4b, 4ab, 4c, 4d, 4e and 7 are identified. About 90% of listeriosis is caused by serovar-1/2a, 1/2b and 4b.
2. **Toxin:** Bacteria produce hemolysin-O called listeriolysin-O which is antigenically similar to streptolysin-O and pneumolysin.
3. **Intracellular location:** Intracellular nature (ability to survive within mononuclear phagocytes and host epithelial cells) of bacilli helps to evade the host defense and the effects of antibiotics.
4. **Motility:** It helps in spread of infection from one cell to other cell.
5. **Adhesins (organs of adhesion):** Two types like **(1) D-galactose in their teichoic acids** helps to bound with D-galactose receptors on macrophage polysaccharides of payer's patches and epithelial cells. **(2) Internalins A and B** for attachment with phagocytic cells. Internalin A binds to E-cadherin, while internalin B binds to the cell's Met receptors. If both of these receptors have a high enough affinity to *Listeria*'s internalin A and B, then *Listeria* will be able to invade the cell via an indirect mechanism called zipper mechanism. Corresponding receptors of adhesions are also found in blood-brain barrier and placenta, which provide the channel for manifestation in brain and fetus in uterus.
6. **Ability to grow at wide range of temperature:** Ability of bacteria to grow at temperatures as low as 0°C permits exponential multiplication in refrigerated foods. At refrigeration temperature, such as 4°C, the amount of ferric iron can affect the growth of *L. monocytogenes*.

7. **Infective dose:** Fewer than 1,000 total organisms may cause disease.
8. **Feric iron:** Bacteria produce siderophore and are able to obtain the iron from transferring.
9. **Endotoxin (LPS):** An early study suggested that *L. monocytogenes* is unique among gram-positive bacteria in that it might possess LPS, which serves as an endotoxin. Later, it was found not to be a true endotoxin. *Listeria* cell wall consistently contains lipoteichoic acids resemble the LPS of gram-negative bacteria in both structure and function.

B. Host factors

1. **Age:** Neonates and older are at more risk.
2. **Sex:** No sex bar, but disease is common in pregnancy because the organism has the ability to penetrate the endothelial layer of the placenta. Pregnancy encounters 27% of total cases mostly during 3rd trimester.
3. **Patients with immunosuppressive condition:** It is common in persons with IDDs otherwise only 7 per 1,000,000 healthy people are infected each year.

C. Environmental factors: Bacilli are distributed in mammals, fish, birds, animals, ticks and crustacean. They also present as saprophytes in decaying materials, water and soil.

Pathogenesis: Follow **Flowchart 59.1**.

Clinical features: Listeriosis is a foodborne illness and defined in two categories:

1. **Noninvasive listeriosis:** Present as febrill gastroenteritis with local GIT features.
2. **Invasive listeriosis:** Following three subtypes:
   - **Pregnancy associated listeriosis:** It causes abortion, still birth and premature delivery.
   - **Neonatal listeriosis (neonatal meningitis):** Characterized as neonatal meningitis. It is the 4th (3rd by few authors) most common cause of meningitis in newborns. Most common bacteria for meningitis in neonats (<1 months) are *E. coli* (34%) > Group B β-hemolytic streptococci (*Strept. agalactiae*, 30%) > GNB (8%) > *Listeria monocytogenes* (6%). It presents in two forms **(1) Early infection:** It starts in <5 days (mean age = 1.5 days), acquired via placenta, associated with premature delivery, low birth weight and neonatal sepsis. Common complication is granulomatous infantiseptica. Mortality rate is very high about >30%. Nosocomial outbreak is absent. **(2) Late infection:** It starts in >5 days (mean age = 14.2 days), acquired from infected birth canal or from environment or from contaminated hospital equipments or from contaminated hands-fingers of nurses, associated with neonatal meningitis. Granulomatous infantiseptica is not seen. Mortality rate is very low about >30 %. Nosocomial outbreak is present.

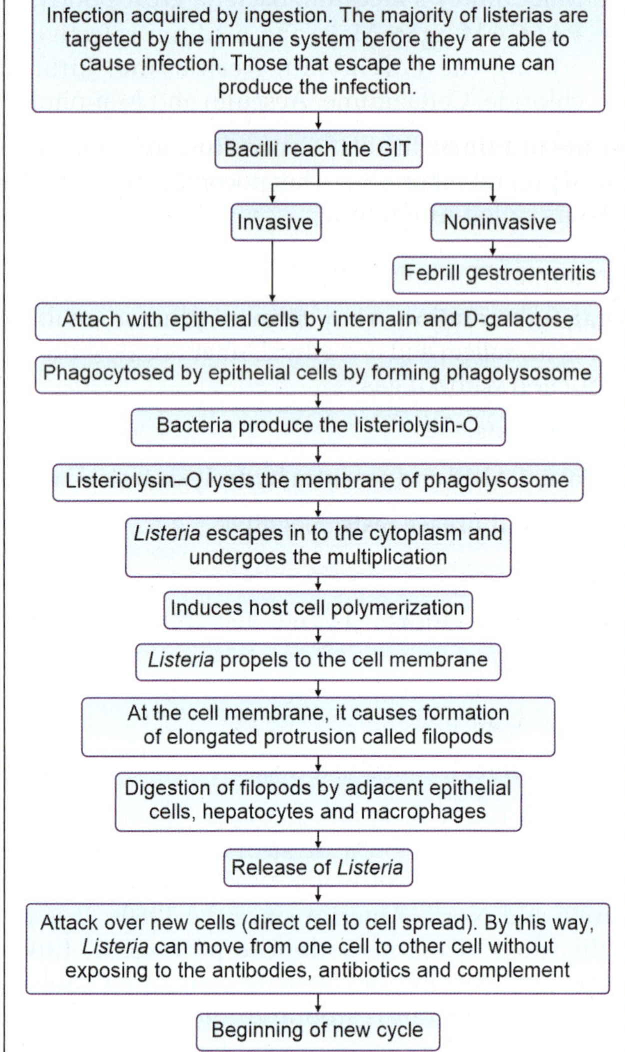

Flowchart 59.1: Pathogenesis of *L. monocytogenes*

- **Listeriosis associated with immunosuppression:** It is common in older group and presents with meningitis, meningoencephalitis, conjunctivitis, abscess, urethritis, pneumonia, endocarditis, septicemia, etc.

Complications: Surviving neonates may suffer with granulomatous infantiseptica—pyogenic granulomas distributed over the whole body and may suffer from physical retardation. Infertility occurs in adult female.

Laboratory Diagnosis

A. Human samples

Specimens: CSF, blood, sputum, vaginal discharge, placental tissue, lochia, meconium, nasopharyngeal aspirates, gastric aspirates and eye-ear-skin swab.

Collection: Follow **cold enrichment** as mentioned above.

Testing methods: These are microscopy (follow morphology), culture (follow C/Cs) and biochemical reactions (follow B/Rs).

B. Food samples

Specimens: Suspected food materials.

Collection: Follow **cold enrichment** as mentioned above.

Testing method: DNA probe

Prevention

Preventive measures are early diagnosis and treatment of infection in pregnancy, proper preparation of food by washing of vegetables and use of pasteurized milk.

Treatment

- Pregnant patient: Ampicillin or Pn G is effective. Erythromycin is given in patient allergic to Pn.
- Nonpregnant patient: Ampicillin or Pn G is effective. Cotrimoxazole and gentamicin are given in patient allergic to Pn.
- Neonate: Ampicillin.
- Cephalosporin is not effective.

Listeria ivanovii

L. ivanovii gives inner zone of β-hemolysis surrounded by α-hemolysis called target (double zone) hemolysis. It is a pathogen of mammals, specifically ruminants and rarely infects humans. It also gives CAMP test positive.

Erysipelothrix rhusiopathiae

Synonym

It also called *Erysipelothrix insidiosa.*

Morphology

They are GPB, straight-rod, slightly curve or filamentous in shape (**Fig. 59.1b**) with 1–2 μm × 0.2–0.4 μm in size. They are nonsporing, noncapsulated and nonmotile.

Cultural Characteristics (C/Cs)

Effective factors

- O_2 effect: They are microaerophilic in primary culture but aerobes and facultative anaerobes on subculture.
- Optimum temperature: 30–37°C, no growth at 4°C.
- They grow on ordinary media and growth is improved by adding glucose or serum.
- Shows **S-R** forms. **S (Smooth) forms** appear at 33°C and in alkaline pH while **R (Rough) forms** appear at 37°C in neutral pH. **R Form** suggests long filamentous form or chain.

Media

A. Liquid media: S forms produce uniform turbidity and **R forms** produce flocculent or matted growth in glucose broth and serum broth.

B. Solid media/plating media

- **Nutrient agar:** Growth is improved by adding glucose or serum.
- **Blood agar:** Bacteria produce **S-R** form on blood agar. **S form** suggests smooth colonies with circular entire edge, 0.5–1 mm, low convex, translucent and glistening surrounded by α-hemolysis, at 33°C and in alkaline pH. **R form** suggests rough colonies at 37°C in neutral pH.

C. Selective media

- **Liquid selective media:** Glucose broth or serum broth plus 0.04% kanamycin and 0.0025% vancomycin.
- **Solid selective media:** Blood agar plus 0.1% sodium azide and 0.001% crystal violet.
- **PT medium:** Black colonies.

Biochemical Reactions (B/Rs)

Sugar fermentation tests:

	G	S	L	M
	A	–	A	–

(Other sugars fermented are galactose, fructose with production of acid only.)

Coagulase test: Tube and/or slide tests are positive with rabbit and/or bovine serum.

I M Vi C tests: – – – –

H_2S: Produced in TSI

Catalase, oxidase, urease and nitrate tests: Negative

Pathogenicity

Reservoirs of infection: Animals like fish, pig, etc.

Modes of transmission: Bacteria are transmitted rarely by ingestion, but mainly by contact of wounded skin with infected animal tissues or animal products in animal handler.

Clinical features

1. **Animal infection:** Erysipelas in pig, sheep and fish.
2. **Human infection:** Bacteria produce typical lesion called erysipeloid or seal finger or whale finger. It presents as painful, purplish swelling without pus (pus present in staphylococcal and streptococcal erysipelas) on hands and on fingers due to direct contact.

Complications

1. **Local:** Arthritis, lymphangitis and lymphadenitis.
2. **Systemic:** Septicemia and endocarditis occur directly by ingestion or due to blood invasion from erysipeloid.

Laboratory Diagnosis

Specimens: Tissue or biopsy from local lesion. Blood is collected in systemic infections.

Transport: Samples are collected in liquid selective media, sent to laboratory followed by subculture over plating media.

Testing methods: These are microscopy (follow morphology), culture (follow C/Cs) and biochemical reactions (follow B/Rs).

Prevention

Educate the risk group while handling the animals.

Pn, erythromycin and other antibiotics are effective. It is intrinsically resistant to vancomycin.

ACCESS YOURSELF

Short Notes

1. Listeriosis.
2. *E. rhusiopathiae.*

Short Questions

1. What is cold enrichment?
2. What is differential motility? Name two bacteria producing it.
3. What is target (double zone) hemolysis? Name two bacteria producing it.
4. What are Anton test and Sereny test?

Comments On

1. Motility in *Listeria* is considered as differential motility.

MCQs for Chapter Review

Liateria monocytogenes

1. All of the following are true about *Listeria except*:
   a. Transmitted by contaminated milk
   b. Gram-negative bacterium
   c. Causes abortion in pregnancy
   d. Causes meningitis in neonates

2. True statement about *Listeria*:
   a. Gram-negative bacillus
   b. Motile by peritrichous flagella
   c. Commonest cause of community acquired meningitis
   d. Only one serovar is known

3. *Listeria* culture media:
   a. Baker
   b. Korthoff
   c. Tinsdale
   d. Blood agar

4. Anton test is used for:
   a. *Listeria monocytogenes*
   b. *Legionella*
   c. *Brucella*
   d. *Bordetella*

5. Major step in pathogenesis of listeriosis is:
   a. The formation of antigen-antibody complexes with resultant complement activation and tissues damage
   b. The release of hyaluronidase by *L. monocytogenes*, which contributes to its dissemination from local sites
   c. The antiphagocytic activity of the *L. monocytogenes* capsule
   d. The survival and multiplication of *L. monocytogenes*, within mononuclear phagocytes and host epithelial cells

6. Most of the cases of *Listeria* are due to:
   a. 1
   b. 4a
   c. 4b
   d. 6

7. Early onset neonatal disease caused by *Listeria monocytogenes* is characterized by all *except*:
   a. Acquired from maternal genital flora
   b. Presents as neonatal sepsis
   c. Mortality rate is less than 10%
   d. Does not cause nosocomial outbreaks

8. In a patient of *Listeria* meningitis who is allergic to penicillin, the treatment of choice is:
   a. Vancomycin
   b. Gentamicin
   c. Trimithoprim-sulfamethaxozole
   d. Ceftriaxone

Erysipelothrix rhusiopathiae

9. Route of infection in erysipelod is:
   a. Ingestion
   b. Direct inoculation
   c. Inhalation
   d. Congenital

10. Following is the lesion not containing pus:
   a. Erysipeloid
   b. Impetigo
   c. Boil
   d. Erysipelas

11. Seal fingers and whale fingers are associated with:
   a. *Listeria*
   b. *Erysipelothrix*
   c. *Corynebacterium*
   d. *Treponema*

Answers and Explanation of MCQs

1. b
- *Listeria* is a gram-positive bacterium not gram-negative.

2. b
- Gram-positive bacillus.
- Motile by peritrichous flagella.
- Commonest cause of nosocomial meningitis.
- Many serovars as mentioned with respective section.

3. d

4. a
- Follow section, *Listeria monocytogenes* (culture characteristics) for explanation of answers of MCQs 3–4.

5. d
- Intracellular nature (ability to survive within mononuclear phagocytes and host epithelial cells) of *Listeria* helps to evades the host defense like action of complement and antibodies and also the antibiotic effects.

6. c
- *L. monocytogenes* serotype 4b strains are responsible for 5–33 of sporadic human cases worldwide and for all major food borne outbreaks in Europe and North America since the 1980s.

7. c
- Mortality is >30% in early onset neonatal disease caused by *Listeria monocytogenes*.

8. b and c
- Follow section, *Listeria monocytogenes* (treatment) for explanation.

9. b
- Follow section, *Erysipelothrix rhusiopathiae* (pathogenicity → modes of transmission) for explanation.

10. a
- Erysipeloid is caused by *E. rhusiopathiae* not containing pus.
- Other lesions are caused by streptococci and staphylococci, are containing pus.

11. b
- Follow section, *Erysipelothrix rhusiopathiae* (pathogenicity → clinical features) for explanation.

CHAPTER 60

Infections of Actinomycetales

Chapter Outline

- Introduction
- *Actinomyces*
- *Nocardia*
- *Tropheryma*

Meaning

Actino word derived from actis means rays like appearance in granules and mycetes from mykes means branching or filamentous shape like fungi.

Uses

Antibiotics produced from *Streptomyces* spp., are labeled with suffix "mycin" (except rifampicin). Following are the examples.

- Streptomycin (1944): From *Streptomyces griseus*
- Kanamycin (1957): From *Streptomyces kanamyceticus*
- Tobramycin (1970): From *Streptomyces tenebrarius*
- Neomycin: From *Streptomyces fradiae*
- Framyctin: From *Streptomyces lavendulae*
- Erythromycin (1952): From *Streptomyces erythreus*
- Rifampicin (Rifampin): From *Streptomyces mediterranei*.

Classification

Follow **Flowchart 60.1**.

Actinomyces

Species

These are *A. israelii, A. bovis, A. naeslundii, A. viscosus, A. odontolyticum, A. gerencsonei* and *A. meyeri*.

Morphology

Type according to Gram's stain: They are gram-positive filamentous bacilli or cocco-bacilli as shown in **Fig. 60.1a**. They are non-acid fast.

Shape and size: They are curved or filamentous in form like fungi and variable in size.

Motility, spores and capsule: They are nonmotile, non-sporing and noncapsulated.

Flowchart 60.1: Classification of Actimomycetales

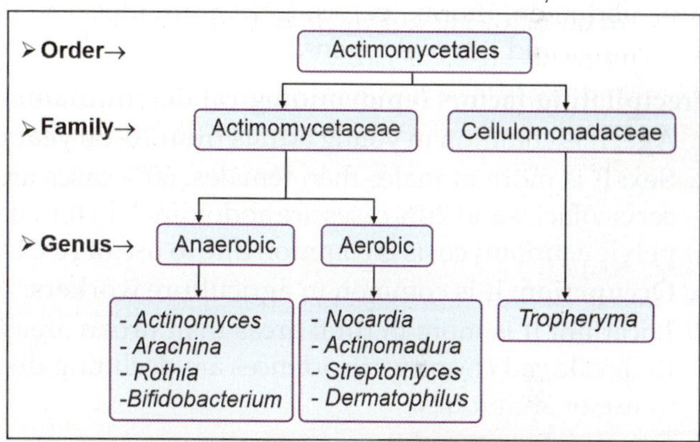

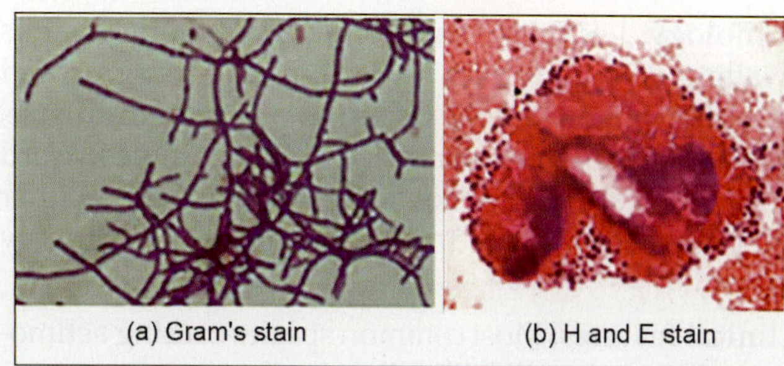

Fig. 60.1: Staining of *Actinomyces* and sulfur granules

Cultural Characteristics (C/Cs)

Effective factors

- O_2 effect: They are strict anaerobes.
- Temperature: 37°C.

Culture in media

A. **Liquid medium:** *A. israelii* produces fluffy ball like growth and *A. bovis* produces general turbidity in thioglycollate broth.

B. **Solid medium:** Bacteria produce molar tooth or spider or bread crumb like colonies after 48–72 hours in BHIA.

Automated culture: Like MALDI-TOF.

Biochemical Reactions (B/Rs)

All species ferment glucose and catalase is negative.

Pathogenicity

Disease name: It called actinomycosis. It is the chronic granulomatous lesion of connective tissues with multiples sinuses discharging sulfur granules.

Reservoir and source of infection: They are the normal flora of human mouth, intestine and vagina, so human acts as reservoir and source of infection.

Modes of transmission: Endogenous infection occurs following disturbances of normal flora by trauma, foreign body insertion like intrauterine contraceptive devices (IUCD), poor oral hygiene, etc. Actinomycosis is a coinfection associated with other bacteria like staphylococci, anaerobic streptococci, *Bacteroides*, *Fusobacterium*, *Bifidobacterium dentium*, *Aggregatibacter aphrophilus*, *Eikenella corrodens*, etc.

Sites: Most common sites are cervicofacial (angle of jaw), abdomen, thorax, pelvis, CNS, musculoskeletal, liver, lungs and internal organs.

Precipitating factors (epidemiological determinants)

1. **Age:** It is common in young adults from 10–30 years.
2. **Sex:** It is more in males then females. 60% cases are cervicofacial and 20% cases are abdominal. In female pelvic actinomycosis is common due to use of IUCD.
3. **Occupation:** It is common in agriculture workers.
4. **Location:** It is more in rural areas than urban areas. In developed countries incidences are declining due to use of antibiotics.
5. **Other risk factors:** Like poor hygiene.

Pathology: Lesion is characterized by granulomatous swelling with multiples sinuses discharging pus and granules (yellow color). Granules are <5 mm in size, hard and yellow in color called sulfur granules. Word sulfur indicates the yellow color of granules, neither the presence of sulfur in granules nor related to any contents.

Clinical features: Most common species causing actimomycosis is *A. israelii* with following features.

- **Lumpy jaw:** It presents on cheeks and submaxillary region. It is the most common type and hold about 60% cases of total actinomycosis. Most common site is angle of jaw (cervicofacial) and lesion called lumpy jaw **(Fig. 60.2)** characterized by painless, hard mass, slow growing, large swelling/abscess that grow on the head and neck. It may be due to penetration by oral flora by local traumas during dental procedures, poor oral hygiene, periodontal disease, radiation therapy or accidental injuries (broken jaw). It can also affect swine, horses, dogs, sheep and less often wild animals. Other cervicofacial lesions are otitis, sinusitis and canaliculitis.
- **Other lesions:** Actinomycosis also presents in abdomen (moslty around cecum), in thorax (in lungs,

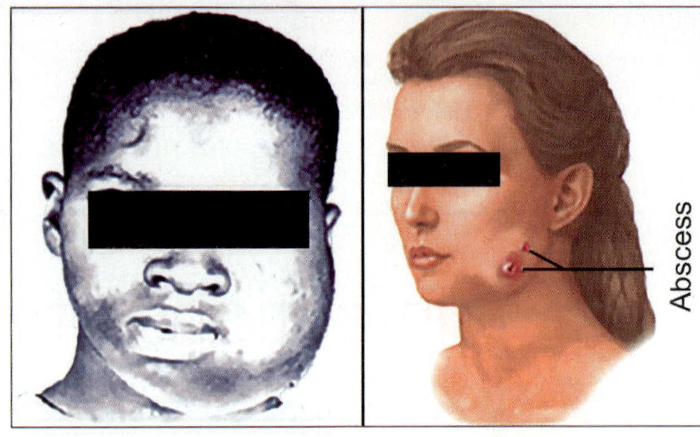

Fig. 60.2: Lumpy jaw in human

pleura and pericardium), in pelvis (mostly in IUCD users) and in subcutaneous tissues (called mycetoma). Bacteria also causing meningitis, retinitis, gingivitis and periodonitis.

Laboratory Diagnosis

Specimens: These are pus or discharge from sinus, tissue biopsy, sputum and granules. Granules are collected in test tubes containing sterile saline. Washed several times in sterile saline and slide is prepared by crushing the granules between two slides. Hold the crushed granules by sterile forceps and rubbed over medium for culture.

Testing methods

A. Microscopy

- **Gram's stain:** Bacteria appear as GPB as shown in **Fig. 60.1a**. Sulfur granules are yellowish in color. Granules appear centrally as gram-positive filaments (due to bacterial colonies) and peripheral as gram-negative (due to Ag-Ab complex by host defense mechanism) giving club shape or sun ray appearance.
- **Hematoxylin and eosin stain:** Centrally bacilli and inflammatory cells present peripherally as shown in **Fig. 60.1b**.
- **Modified ZN stain:** It is useful to differentiate from *Nocardia*.
- **Fluorescent staining:** It is also a useful method.

B. Culture: Follow C/Cs.

C. Biochemical reactions: Follow B/Rs.

D. GLC: It detects the products of glucose metabolism.

E. Molecular methods: PCR and RFLP are useful.

Treatment

Pn is the drug of choice, continue for several months. Tetracycline or erythromycin is useful in patients allergic to Pn. Do surgery, if necessary.

Nocardia

Species

These are *N. asteroids*, *N. brasiliensis*, *N. caviae* (*N. otitidiscaviarum*), etc.

Meaning and History

The genus was named for Edmond Nocard, a 19th-century veterinarian and biologist.

Morphology

Type according to Gram's stain: They are gram-positive filamentous bacilli/cocci-bacilli.

Acid fastness: They are acid fast by 1% sulfuric acid (**Fig. 60.3**) and also weakly acid fast by Kinyoun's staining. Acid fastness is due to presence of intermediate length of glycolipid called nocardic acid (as like mycolic acid in mycobacteria).

Shape and size: They are curved or filamentous in form like fungi and variable in size.

Motility, spores and capsule: They are nonmotile, nonsporing and noncapsulated.

Cultural Characteristics (C/Cs)

Effective factors

- O_2 effect: They are strict aerobes.
- Temperature: Normal temperature is 37°C where growth occurs in 2 days to 2 weeks, but can grow at 45°C in 3 days.

Culture in Media

A. Solid media: Bacteria produce dry, glabrous, wrinkled/folded or may be pigmented (yellow to orange/red color) and adherent (to the medium) colonies on Sabouraud dextrose agar without antibiotics, blood agar and BHIA.

B. Selective media

- **Sabouraud dextrose agar with chloramphenicol:** Chloramphenicol prevents the bacterial contamination by other bacteria.
- **Buffered yeast extract agar contains polymyxin and vancomycin:** Polymyxin and vancomycin are selective agents.
- **Medium for paraffin bait technique:** Media contain paraffin (sole source of carbon), are useful to identify the *Nocardia* from soil and clinical samples.
- **Lowenstein Jensen medium:** Bacteria produce moist glabrous colonies.

Automated culture: Like MALDI-TOF.

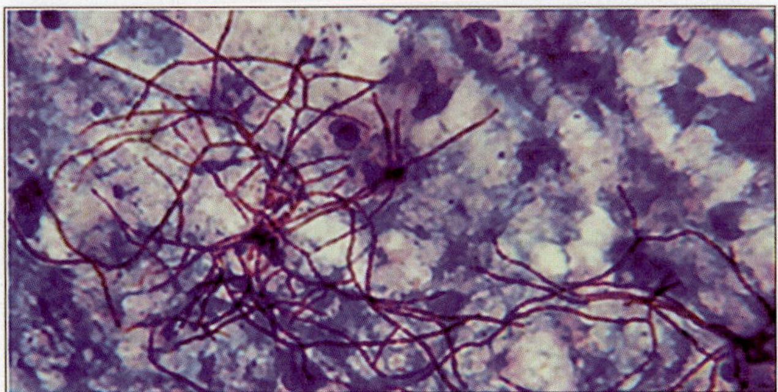

Fig. 60.3: *Nocardia* under ZN stain

Biochemical Reactions (B/Rs)

Sugar fermentation tests: They are nonfermenters, but utilize (assimilation) the numbers of sugars oxidatively.

Catalase, urease (by all species) and acetamide utilisation tests: Positive

Hydrolysis of casein, hypoxanthine and tyrosine: Positive.

Growth in presence of lysozyme: Positive.

Pathogenicity

Disease name: Disease called nocardiosis.

Epidemiology: They are normally found in soil and causing occasional sporadic disease in humans and animals throughout the world. High mortality occurs exceeds 80%, especially with brain infection; in other forms, mortality is 50%, even with appropriate therapy.

Reservoir and source of infection: These are soil and vegetative materials.

Modes of transmission: Exogenous infection occurs from soil via inhalation (pulmonary nocardiosis) or by traumas (skin and subcutaneous nocardiosis).

Portal of entry: Skin and respirator system.

Sites: Lungs, skin, subcutaneous tissues, brain, etc.

Precipitating factors (epidemiological determinants)

1. **Cord factor (glycolipid derivative or nocardic acid):** Cell wall of *Nocardia* contains intermediate length of glycolipid called nocardic acid. Unique nocardic acid 6, 6 dimycolyltrehalose called cord factor which inhibits the phagocytosis.
2. **Enzymes:** Bacteria produce super oxide dismutase and catalase, which contribute in pathogenicity.
3. **IDDs:** It is an opportunistic pathogen present commonly in person with AIDS, tuberculosis, steroid therapy and organ transplantation.

Pathology: Pulmonary nocardiosis is characterized by granuloma with neutrophilic infiltration, necrosis and abscess. Skin and subcutaneous nocardiosis is characterized by granulomatous swelling with multiple sinuses discharging pus and granules. Granules are 0.5–2 mm in size, soft, lobulated and yellowish-white in color.

Clinical Features

1. **Pulmonary nocardiosis:** Pneumonia, abscess and granulomatous lesion simulating tuberculosis.
2. **Cutaneous nocardiosis:** Abscess, cellulitis, etc.
3. **Subcutaneous nocardiosis:** Mycetoma (actinomycetoma).
4. **Systemic nocardiosis:** These include meningitis, brain abscess (most common by *N. asteroides*), keratitis, metastatic complications in kidneys and in other organs in persons with IDDs.

Specimens: Same as actinomycosis.

Testing methods: These are microscopy (follow morphology), culture (follow C/Cs) and biochemical reactions (follow B/Rs).

Treatment

Antibiotics: Pn is resistant. Sulfonamide is the drug of choice. Cotrimoxazole or minocycline will continue for several months. In immunocompromized persons, addition of amikacin and cefotaxime are advisable. Do surgery, if necessary.

Differences between *Actinomyces* and *Nocardia*

Follow **Table 60.1**.

Tropheryma

Introduction: Common species is *Tropheryma whipplei*. Previously it was known as *Trophyrema whipplei*. Disease called Whipple's disease, from the name of George Hoyt Whipple who diagnosed the disease in 1907. However, the causative agent identified later in 2003 in Johns Hopkins hospital called Whipple's bacilli.

Morphology: These are intracellular pathogens present in macrophages of lamina propria and mesenteric lymph nodes. They are GPB; however, in laboratory they can be stained as GNB or intermediate between GPB and GNB. Intracellular location is better stained by periodic acid-Schiff (PAS) stain. Culture method is not useful.

Pathogenicity: Disease called Whipple's disease or intestinal lipodystrophy. It is common at 4th–5th decades of life and in male. It is the multisystem disease. Malabsorption presents atypically with intestinal features like abdominal pain, diarrhea, weight loss, mesenteric lymphadenitis, etc., and extraintestinal features like endocarditis, meningitis, neurological disturbances, hyperpigmentation of skin, polyarthritis and also damage to kidneys, lungs, spleen and liver.

Laboratory diagnosis: Intestinal or mesenteric lymph node biopsy are tested by following methods.

1. **PAS stain** identifies the intracellular bacilli.
2. **PCR** targeting 16S ribosomal RNA is useful.
3. **D-xylose absorption test:** In normal individuals, a 25 g oral dose of D-xylose will be absorbed and excreted in the urine at approximately 4.5 g in 5 hours. A decreased urinary excretion of D-xylose is seen in conditions involving the GI mucosa, like small intestinal bacterial overgrowth and Whipple's disease. In cases of bacterial overgrowth, the values of D-xylose absorption return to normal after treatment with antibiotics. In contrast, if the D-xylose urinary excretion is not normal after a course of antibiotics, then the problem must be due to a non-infectious cause of malabsorption (like celiac disease).

Treatment: Pn, ampicillin, tetracycline or cotrimoxazole is given for 1–2 years. Hydroxochloroquine is given for 12–18 months. Chances of relapse of disease are 40%, if treatment taken for less than one year.

TABLE 60.1: Differences between *Actinomyces* and *Nocardia*

Features	*Actinomyces*	*Nocardia*
Staining		
Acid fastness	Non-acid fast	Acid fast
Culture characteristics		
O₂ effect	Anaerobic	Aerobic
Temperature for growth	37°C	37°C, but can grow at 45°C
Colonies	Nonpigmented	Pigmented
Biochemical reactions		
Sugar reaction	Fermenter	Assimilation
Catalase	Negative	Positive
Pathology and treatment		
Natural habitat	Mouth, GIT and vagina	Soil
Mode of infection	Endogenous	Exogenous
Infection occurs in	Immunocompetent	Immunodeficient
Granules Size Consistency Color	 >5 mm Hard Yellow	 0.5–2 mm Soft Yellowish white
Treatment	Pn	Sulfonamide, cotrimoxazole

Short Note

1. Acinomycosis/*Nocardia*.

Short Questions for Theory/Viva Questions

1. Write four differences between *Actinomyces* and *Nocardia*.
2. What is botryomycosis?

MCQs for Chapter Review

Actinomyces

1. **Most common site for actinomycetes is:**
   a. Cervicofacial   b. Thorax
   c. Abdomen   d. Brain
2. **Which of the following is the most predominant constituent of sulfur granules of actinimycosis?**
   a. Organisms
   b. Neutrophils and monocytes
   c. Monocytes and lymphocytes
   d. Eosinophils
3. **Color of granules of actinomycetes is:**
   a. Black
   b. Yellow
   c. Red
   d. Brown
4. **Sulfur granules are composed of:**
   a. Organisms   b. Leukocytes
   c. Erythrocytes   d. Keratinocytes

5. *Nocardia* **can be differentiated from the** *Actinomyces* **by:**
   a. ZN staining
   b. Fontana's stain
   c. Gram's stain
   d. O_2 requirement

6. *Nocardia* **resembles to** *Actinomyces* **morphologically but:**
   a. Anaerobic
   b. Facultative anaerobic
   c. Aerobic
   d. Requires CO_2 for growth

7. A characteristic infection of *Nocardia asteroides* **is:**
   a. Diarrhea
   b. Secondary dissemination to liver
   c. Brain abscess
   d. Colonic diverticulosis

8. A clinical specimen was obtained from the wound of a patient diagnosed as nocardiosis. For the selective isolation of *Nocardia* **species which of the following would be the best method?**
   a. Paraffin bait technique
   b. Castaneda's culture method
   c. Craig's culture method
   d. Hair bait technique

1. a
- Follow section, *Actimomyces* (**pathogenicity** → **sites**) for explanation.

2. a

3. b

4. a
- Sulfur granules are yellowish in color. Granules appear centrally as gram-positive filaments (due to bacterial colonies) and peripheral as gram-negative (due to Ag-Ab complex by host defense mechanism) giving club shape or sun ray appearance.

5. a, d

6. c
- Follow **Table 60.1** for explanation of answers of MCQs 5–6.

7. c
- Follow section, *Nocardia* (**pathogenicity** → **clinical features**) for explanation.

8. a
- Paraffin is the sole source of carbon and useful for selective isolation of *Nocardia* species. Castaneda's culture method is the biphasic culture method useful in many bacteria like *S.* Typhi, *Brucella*, etc. Craig's culture method is useful to differentiate motile from nonmotile bacteria. Hair bait technique is useful to diagnose dermatophytes.

Infections of Actinomycetales

C H A P T E R 61

Infections of Enterobacterales-I (*E.coli, Klebsiella,* etc.)

Chapter Outline

- Introduction
- *Escherichia coli*
- *Edwardsiella* spp.
- *Citrobacter* spp.
- *Klebsiella* spp.
- *Enterobacter* spp.
- *Hafnia* sp.
- *Serratia* spp.
- *Pantoea* spp.
- Morgnellaceae
- Erwinia sp.

INTRODUCTION

Definition of Enterobacterales

Any bacterium will be classified as a member of Enterobacterales (Enterobacteriales) order, if it has following criteria.

1. Normal habitat of intestine
2. Gram-negative bacillus
3. Non-acid fast
4. Nonsporing
5. May be capsulated
6. May be motile by peritrichate flagella (except *Shigella* and *Klebsiella*, which are nonmotile) not by polar flagella
7. Aerobe and facultative anaerobe
8. Grows on ordinary media (nonfastidious)
9. Ferments glucose with production of A/AG
10. Reduces nitrate to nitrite
11. Catalase positive (except *Sh. dysenteriae* type 1)
12. Oxidase negative (except *Plesiomonas shigelloides*)

Classification of Enterobacterales

Based on lactose fermentation in MacConkey's medium

1. **Lactose fermenters (LFs):** They are also called coliform bacilli. LFs produce pink colonies on MacConkey's medium. For examples, *Escherichia* and *Klebsiella.*
2. **Late lactose fermenters (LLFs):** They are also called paracolon bacilli. LLFs produce pale colonies earlier and later pink colonies on MacConkey's medium. For examples *Shigella sonnei* and *Citrobacter.*
3. **Nonlactose fermenters (NLFs):** NLFs produce pale/colorless colonies on MacConkey's medium. For

TABLE 61.1: Systemic classification of Enterobacterales order

Family	Genus (Genera)
Budviciaceae	*Budvicia*
Enterobacericeae	*Escherichia, Citrobacter, Klebsiella, Enterobacter, Salmonella* (**Ch. 62**), *Shigella* (**Ch. 63**), *Cedecea, Plesiomonas* (**Ch. 65**)
Hafniaceae	*Hafnia, Edwardsiella*
Morgnellaceae	*Proteus, Morganella, Providencia*
Yersiniaceae	*Serratia, Yersinia* (**Ch. 64**), *Ewingella*
Erwiniaceae	*Pantoea, Erwinia*
Pectobacteriaceae	Mostly plant pathogens

examples, *Proteus, Providentia, Morgnella, Salmonella, Shigella* (except *Sh. sonnei*) and *Yersinia.*

Systemic classification: It based on morphological, biochemical, genetical, taxonomical properties and use of three systems like Bergey's manual (1984), Edward-Ewing classification (1986) and Farmer-Kelly classification (1991). Systemic classification is given in **Table 61.1.**

Color Reaction by Enterobacterales on Different Media

Follow **Table 61.2.**

Escherichia coli

Meaning and History

In 1885, the German-Austrian pediatrician Theodor Escherich discovered this organism in the feces of healthy individuals.

TABLE 61.2: Color reaction by Enterobacterales on different media

Medium	Sugar	Indicator	Acidic pH	Alkline pH
Sugar media	G, S, L, M	Andrade's	Pink	Colorless
MacConkey's	L	Neutral red	Pink	Pale
DCA	L	Neutral red	Pink	Pale
SS agar	L	Neutral red	Pink	Pale
XLD medium	L, S	Phenol red	Yellow	pink
TSI	G, S, L	Phenol red	Yellow	pink
TCBS	S	Bromothymol blue	Yellow	Blue/green
CLED	L	Bromothymol blue	Yellow	Blue/green

Other Species of Genus

These are *E. fergusoni, E. hermanii, E. vulneris, E. blattae,* etc.

Uses

1. Conjugation was 1st discovered by Joshua Lederberg and Tatum in 1946 in *E. coli* K 12 strain.
2. *E. coli* is a very versatile host (vector) for the production of heterologous proteins like insulin, enzymes and vaccines by using recombinant technology.

Morphology

Type according to Gram's stain: They are GNB.

Shape and size: They are straight road (**Fig. 61.1a**) and 1–3 µm × 0.4–0.7 µm in size.

Arrangement: They arranged singly, in pairs or in small groups.

Capsule: It found in some strains.

Motility: They are motile by peritrichate flagella.

Spores: They are nonsporing.

Cultural Characteristics (C/Cs)

Effective factors

- O_2 effect: They are aerobes and facultative anaerobes.
- Temperature: 37°C.
- pH: 7.6

Culture in media

A. **Liquid media:** They produce uniform turbidity and heavy deposit at bottom in peptone water and glucose broth.

B. **Semisolid stab agar:** They produce 'Fan type' growth (outward from stab line).

C. Solid media

- **Nutrient agar:** Colonies are large, thick, grayish white, moist, smooth opaque or partially translucent discs. Initially smooth (S) colonies become rough (R) on subculture called S—R variation. It is due to loss of cell wall antigens and virulence.
- **Blood agar:** They produce hemolytic colonies.

D. Selective media

- **MacConkey's agar:** Bright pink spreading colonies occur due to LF as shown in **Fig. 61.2a**.
- **XLD (xylose lysine deoxycholate) medium and:** Yellow colony.
- **CLED medium (cysteine lactose electrolyte deficient):** Yellow colony.
- **DCA (deoxycholate citrate agar), SS (Salmonella-Shigella) agar and W-B (Wilson-Blair) medium:** Growth is inhibited.

Automated culture: Like MALDI-TOF or VITEK is used to identify the species from culture.

Culture in animals: Use of animals for cultivation of *E. coli* is described below with diarrheagenic strains of *E. coli*.

Biochemical Reactions (B/Rs)

Sugar fermentation tests (Fig. 61.3a): *E. coli* ferments maltose and many other sugars (mentioned below) with acid and gas.

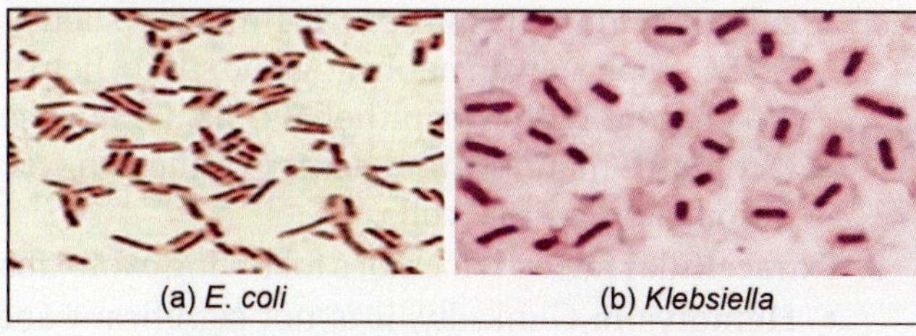

Fig. 61.1: GNB under Gram's stain

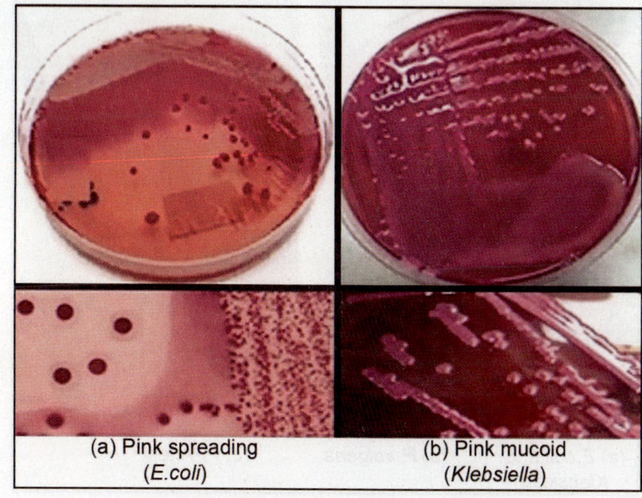

Fig. 61.2: Pink colonies on MacConkey's agar

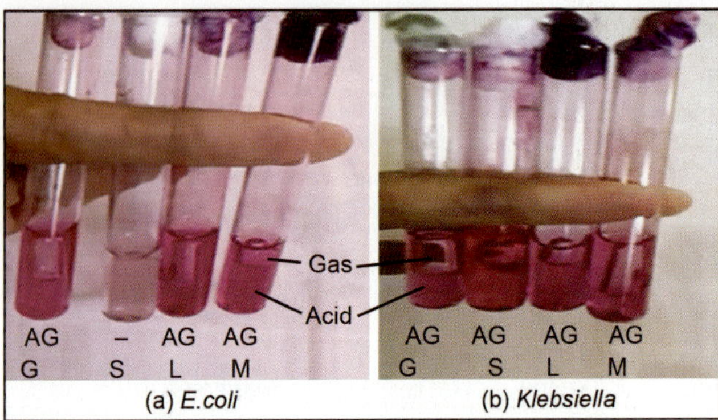

Fig. 61.3: Sugar fermentation tests

G	S	L	M
AG	V	AG	AG

I M Vi C tests (Fig. 61.4a): + + – –

Catalase and nitrate reduction tests: Positive

Oxidase and urease tests: Negative

TSI test (Fig. 61.5a): A / AG-H2S (A = Acid, AG = Acid + Gas).

Resistance

Sterilization: Bacteria survive for several days in water, soil, dust and air and they can be easily killed by moist heat at 60°C in 30 minutes.

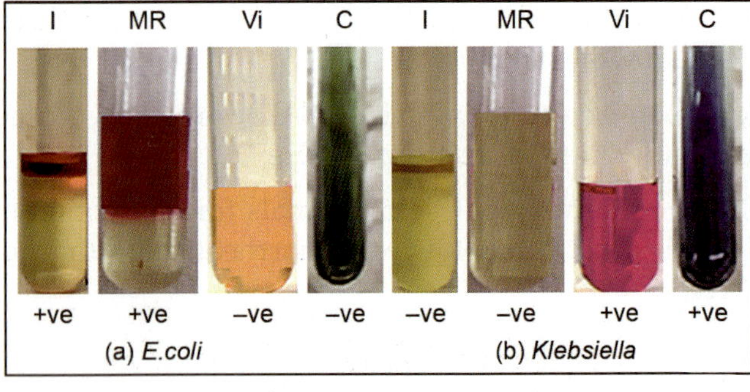

Fig. 61.4: IMViC tests

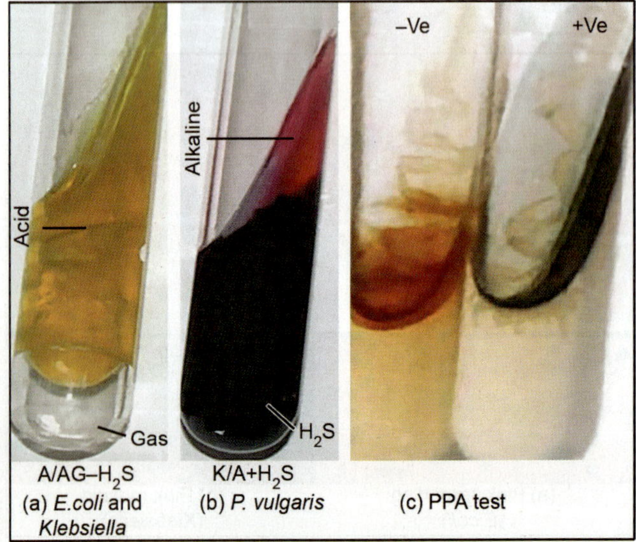

Fig. 61.5: TSI (a and b) and PPA (c) tests

Disinfection: *E. coli* also killed by 0.5–1 ppm chlorine in water.

Drug resistance: Drug resistance strains are also developed.

Pathogenicity

Disease name: *E. coli* produces UTI, diarrhea and other infections.

Reservoirs of infection: Humans and animals are the reservoirs contain *E. coli* as normal flora in the GIT.

Sources of infection: Water or food. Bacilli excreted in feces and remain viable for some days in the environment; hence detection of bacilli in water, especially a variant called thermovariant *E. coli* survives at 44°C and it is taken as recent contamination from feces of humans or animals.

Modes of transmission: *E coli* transmitted by ingestion of contaminated food and water.

Incubation period: It is about 3–5 days, may be short as 1 day and long as 10 days.

Portal of entry: GIT

Sites: Intestinal and extraintestinal.

Precipitating factors (Epidemiological determinants): Following are the factors.

A. **Agent factors (virulence factors):** These are two types like intracellular and extracellular.

1. **Intracellular (cell wall associated) factors**
   - **Capsular (K) antigen:** It is polysaccharide in nature, includes about 100 types and inhibits the phagocytosis. It masks the O-Ag and prevents the agglutination by O antiserum.
   - **Flagellar (H) antigen:** It is the organ of locomotion and includes about 75 types.
   - **Somatic antigen (O Ag or LPS or Endotoxin):** It is lipopolysaccharide exhibits endotoxic activities, includes about 170 types and inhibits the phagocytosis. It has bactericidal effects on complement.
   - **Fimbrial (Fi) antigen:** It is an organ of adhesion. Different types H Ags are present in different types of *E. coli.* **(1) Colonization factor antigen (CFA):** It presents in ETEC and helps to adhere with intestinal epithelial cells. **(2) Mannose resistant fimbria:** So, named because it is not inhibited by mannose. Different examples are Dr, P, M, S and F1C. P fimbriae help in attachment of *E. coli* with RBCs surface of P blood group called hemagglutination. P fimbriae seen in EHEC, especially in nephritogenic strain, help in attachment with uroepithelial cells. **(3) Fucose:** It contains mannose and helps in adhesion with human and animal cells.

2. **Extracellular factor:** It includes following exotoxins.
   - **Hemolysin:** Clinically, it is not significant and produced more by virulent than avirulent strains.

- **Enterotoxin:** It has three types. **(1) Heat labile toxin (LT):** It was discovered by De and colleagues from the case of adult diarrhea. It is same as cholera toxin, but less potent and plasmid encoded. Different components and actions of LT are described in **Flowchart 61.1.** Diagnostic tests are described in **Table 61.3. (2) Heat stable toxin (ST):** It is plasmid encoded. Different components and actions of ST are described in **Flowchart 61.2.** Diagnostic tests are described in **Table 61.3. (3) Verotoxin (VT)/ verocytotoxin:** Subtype of VT produced by *E. coli* is shiga like toxin (SLT). It called verotoxin, because it is cytotoxic on cultured vero cells, derived from African green monkey kidney. There are two main of categories of VT according

Flowchart 61.1: Heat labile enterotoxin

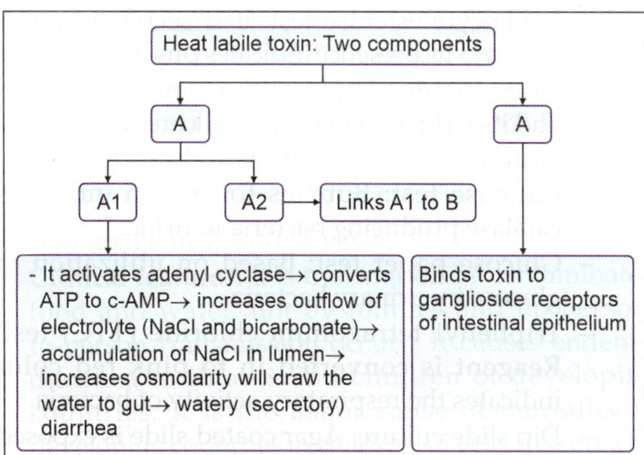

TABLE 61.3: Diagnostic tests of LT and ST

Features	LT	ST
***In vivo* tests**		
• Ligated rabbit ileal loop		
– Read at 6 hours	±	±
– Read at 18 hours	+	–
– Infant rabbit bowel	+	+
– Infant mouse intragastric (4 hrs)	–	+
• Adult rabbit/guinea pig skin test (skin blueing test): To detect the vascular permeability factor (VPF)	+	–
***In vitro* tests**		
Tissue culture test		
• Rounding of Y1 mouse adrenal cells	+	–
• Elongation of Chinese hamster ovary cells due to intracellular increase c-AMP	+	–
Serological tests		
ELISA with monoclonal Ab	+	+
Passive agglutination test	+	–
Passive immune hemolysis/precipitation/ Eiken test	+	–
Molecular tests		
DNA probe	+	+

Flowchart 61.2: Heat stable enterotoxin

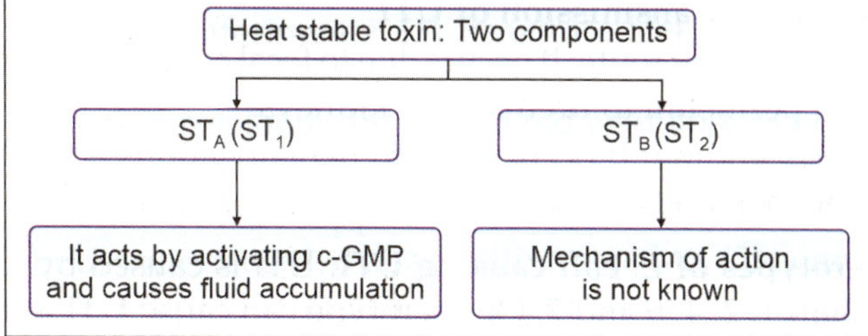

to origin of bacteria like **shiga toxin** produced by *Shigella dysenteriae* type 1 (**Ch. 63**) and **shiga like toxin (SLT)** produced by bacteria other than *Shigella* like entero hemorrhagic *E. coli* (EHEC), *V. cholerae, Aeromonas hydrophila* and *Campylobacter jejuni*. SLT is antigenically, biochemically and physically similar to shiga toxin, so called Shiga-like toxin (SLT). SLT is bacteriophage coded. According to neutralization by shiga antitoxin, SLT has two subtypes like VT1 and VT2. VT1 is neutralized by shiga antitoxin while VT2 do not neutralized by shiga antitoxin. SLT is detected by cytopathic changes in cultured Vero or HeLa cells, by demonstrating VT antibody in sera by latex agglutination test or ELISA and by molecular methods like DNA probe of VT1 and VT2 gene. SLT has following three types of biological actions.

- **Neurotoxicity:** It is detected by paralysis and death on injection in to mice or rabbits. It is neurotoxic, but this property is secondary because it does not act directly on CNS but to the blood vessels supplying the nerve.
- **Cytotoxicity:** It is detected by *in vitro* method like cytopathic changes in cultured Vero or HeLa cells and *in vivo* method. *In vivo* **method** described as toxin consists two subunits like binding (B) and active (A). Subunit A divided in to two fragments like A1 and A2. B helps in adhesion with host cells. A2 links A1 to B. A1 inactivates the host cell 60S ribosome and inhibits the protein synthesis leading to cell death. Cell death resulting discontinuity of mucosa and hemorrhage (bloody diarrhea).
- **Enterotoxicity:** It allows the collection of fluid in gut (in ligated rabbit ileal loop).

B. Host factors

1. **Age:** Younger and older adults are at a higher risk.
2. **Food habits:** Consuming unpasteurized dairy products, eating undercooked meat, drinking impure water, etc., are increasing risk of *E coli* infections.
3. **Immunity:** It is more in person with IDDs.
4. **Others:** Working around livestock.

Clinical features: *E coli* causes 3 diseases like UTI, diarrhea and other infections as defined below.

- **Clinical features:** It causes diarrhea in children between 2–6 years.
- **Virulence factors:** Showing diffuse adherence fimbriae contribute to the pathogenesis.

Other Infections

E. coli is the most common cause of acute UTI, neonatal meningitis, intra-abdominal abscess (like hepatic abscess or abscess in soft tissue) and nosocomial infection (like ventilator associated pneumonia). Other infections of *E. coli* are mentioned bellow.

- **Peritonitis:** Primary from exogenous entry of bacilli or secondary due to intestinal perforation which allows the distant spread of bacilli.
- **Wound infections:** Common in diabetic patients.
- **CNS infection:** Especially neonatal meningitis.
- **CVS infections:** Endocarditis, shock and SIRS (systemic inflammatory response syndrome).
- **Bacteremia and septicemia:** Septicemia presents with fever, hypotension and DIC usually in debilitated patients with high mortality.

Edwardsiella spp.

Edwardseilla tarda: Tarda word came from tardy means weak, because of weak sugar fermentation. They are GNB, nonmotile and noncapsulated. They ferment only glucose and maltose, indole, citrate, H_2S, lysine and ornithine decaboxylation tests are positive. They are normal intestinal flora in cold-blooded animals like snakes and other. Bacteria cause septic shock, liver abscess and diarrhea in humans. Bacteria are diagnosed from pus, blood, urine and CSF.

Edwardseilla ictaluri: It causes human septicemia.

Edwardseilla horshinae: Human pathogenicity is unknown.

Citrobacter spp.

Common species are *Citro. freundii*, *Citro. Koseri* and *Citro. amalonaticus*. They are GNB, motile, late lactose fermenters, indole variable, MR negative, VP positive, citrate positive and H_2S producer. *Citro. freundii* formerly known as Ballerup-Bathesda group. Some strain **(Bhatanagar strain)** of *Citro. freundii* shares common Ag **(like Vi Ag)** with *S.* Typhi and *S.* Paratyphi C and confusing the diagnosis. They are the intestinal flora may produce UTI, gall bladder infection, otitis media and meningitis.

Klebsiella spp.

Genus named after German microbiologist Edwin Kleb (1834–1913). It has more than 80 serotypes on the bases of capsular antigen. Many species are identified on the bases of biochemical properties like *Kleb. pneumoniae*, *Kleb. granulomatis*, *Kleb. planticola*, *Kleb. ornitholytica*, *Kleb. oxytoca* and *Kleb. terrrigena*. Phage typing is also useful in these bacteria.

Kleb. pneumoniae

Synonyms: Bacillus is also known as by other names like Friedlander's bacillus (because 1st discovered by Friedlander), bacillus mucosus capsulatus and pneumo-bacillus.

Subspecies: *K. pneumoniae pneumoniae*, *K. pneumoniae aerogenes*, *K. pneumoniae ozaenae* (Abel's bacillus) and *K. pneumoniae rhinoscleromatis* (Frisch's bacillus).

Morphology: They are GNB or cocco-bacilli, short, plump, straight road and 1–2 µm × 0.5–0.8 µm in size as shown in **Fig. 61.1b**. They arranged singly, in pairs or in small groups. They are capsuled, nonmotile and nonsporing.

Cultural characteristics (C/Cs)

- **Effective factors: (1)** O_2 effect: They are aerobes and facultative anaerobes, **(2)** Temperature: 37°C, **(3)** pH: 7.6.
- Culture in media: **(1)** Nutrient agar: Bacilli form large, dome-shaped, mucoid and sticky colonies. **(2)** MacConkey's medium: Colonies are bright pink-mucoid due to lactose fermentation as shown in **Fig. 61.2b**.
- Automated culture: Like MALDI-TOF or VITEK is used to identify the species from culture.

Biochemical reactions (B/Rs)

Sugar fermentation tests **(Fig. 61.3b)**

G	S	L	M
AG	AG	AG	AG

I M Vi C tests (Fig. 61.4b): – – + +

Catalase, nitrate reduction and urease tests: Positive

Oxidase test: Negative

TSI test **(Fig. 61.5a)**: A/AG-H_2S

Pathogenicity

1. ***Klebsiella pneumoniae pneumoniae:*** Capsule of bacteria inhibits the phagocytosis. It produces toxin similar to ST toxin of *E. coli* causes diarrhea. It produces pneumonia, UTI, diarrhea (due to entero-toxin), pyogenic infections (abscesses, meningitis and septicemia). Pneumonia is common in persons with middle or older age, diabetes mellitus, alcohol use and chronic bronchopulmonary disease. Common sero-types causing pneumonia are 1, 2 and 3. Pneumonia presents with massive mucoid inflammatory exudates of lobar or lobular distribution, involves one or more lobes of lung. Fever and cough are common features. Common complications are necrosis and abscess formation.

2. ***Klebsiella pneumoniae aerogenes:*** Earlier it was classified as *Enterobacter aerogenes*. It produces pneumonia, UTI, diarrhea, pyogenic infections (abscesses and septicemia) and nosocomial infection. It produces gas from glycerol. Aesculin hydrolysis and lysine decarboxylase tests are positive. Arginine dihydrolase test is negative.

3. *Klebsiella pneumoniae ozaenae*: It produces chronic lesion characterised by foul smelling mucopurulent nasal discharge called ozena (atrophic rhinitis) by capsular types 3–6.

4. *Klebsiella pneumoniae rhinoscleromatis*: It produces chronic granulomatous hypertrophy of nose with intracellular bacilli called rhinoscleroma by capsular type 3. It presents as tumor like growth, which causes nasal obstruction which may spread to sinuses, larynx and pharynx.

Laboratory diagnosis: Common specimens are urine, pus, blood, stool, nasal biopsy, etc., which are tested by microscopy (follow morphology), culture (follow C/Cs) and biochemical reactions (follow B/Rs).

Treatment: Klebsiellas are generally resistant to ampicillin, amoxycillin and carbenicillin. They are sensitive to cephalosporin, nitrofurantoin, co-amoxiclav and gentamicin.

Kleb. granulomatis

Reason of reclassification: Previously it was not the part of Enterobactariceae family, but later, it was identified that morphologically and antigenically it is similar to *Klebsiella*, hence reclassified as *Kleb. granulomatis*.

Synonym: Bacillus is also known by other names like *Donovania granulomatis* (Genus name given from Donovan (1905), who 1st described the intracellular bacilli from genital ulcer and species name given from granulomatous forms of disease) and *Calymmatobacterium granulomatis*.

Morphology: They are gram-negative cocco-bacilli, but better stained by Wright, Giemsa **(Fig. 61.6)** or Leishman's stain as an intracellular bacilli (called Donovan's intracellular bodies/Donovan's bodies) in mononuclear cells with bipolar staining which gives 'safety-pin or telephone handle appearance' and pinkish capsule. Bacilli are 1–2 µm in size, capsulated, nonmotile and nonsporing.

> **Note: Bacteria with safety pin appearance**
> - *Kleb. granulomatis*, *Yersinia pestis*, *Pasturella multocida*, *Francisella tularensis* *Burkholderia pseudomallei*, *Haemophilus ducreyi*, *Brucella abortus*, *Chromobacterium violaceum*, etc.

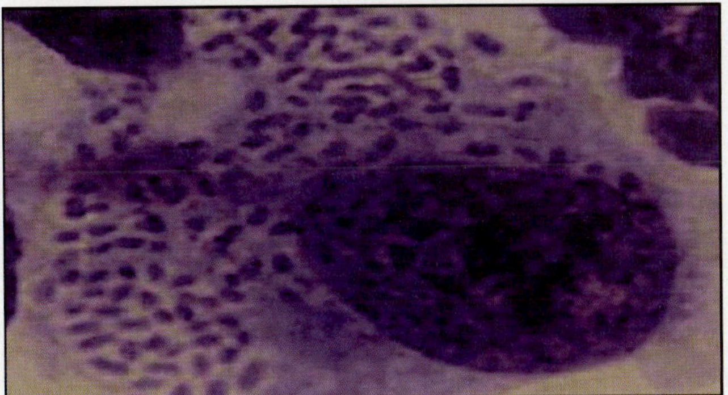

Fig. 61.6: Donovan's bodies under Giemsa stain

Cultural characteristics (C/Cs): Bacilli grow in modified levinthal agar, egg yolk medium, yolk-sac off chick embryo and HEP-2 cell lines.

Pathogenicity

- **Disease name:** Disease called donovanosis or granuloma venereum or granuloma inguinale.
- **Mode of transmission:** Sexually intercourse.
- **Incubation period:** 1–12 weeks.
- **Portal of entry:** Skin/mucosa.
- **Sites (Fig. 61.7a and Fig. 61.7b):** Genitals in 90% cases, particularly of labias are common.
- **Clinical features:** It presents with single or multiple, local, subcutaneous nodule that erodes through skin to produce granulomatous, sharply defined painless ulcer **(Fig. 61.7a and Fig. 61.7b)**. The granulomatous swelling, may heaped up via subcutaneous tissue to inguinal area which simulates the lymphadenopathy called pseudolymphadenopathy (pseudo-bubo). In fact absence of true lymphadenopathy is the hall mark of disease.
- **Complications:** It causes genital elephantiasis (labial swelling) called pseudoelephantiasis, phimosis and paraphimosis.

Laboratory diagnosis: Clean the ulcer with saline and collect the bits of tissue or biopsy materials. Bacilli are identified by microscopy and culture methods as described above. Multiplex PCR developed for simultaneous detection of sexually transmitted microbes like *T. pallidum*, *H. ducreyi*, *K. granulomatis*, *C. trachomatis* and HSV.

Prevention: Safe sex is the best preventive measure.

Treatment: Azithromycin is the drug of choice. Doxycycline (2nd choice) for 3 weeks is curative. Cotrimoxazole, chloramphenicol, gentamicin, fluoroquinolones and newer macrolides are also useful.

Enterobacter spp.

Formerly it is called *Aerobacter*. Enterobacters are GNB, motile, capsulated, lactose fermenters, indole negative, MR negative, VP positive and citrate positive. *E cloaceae* do not ferment glycerol, aesculin hydrolysis and lysine decarboxylase tests are negative while arginine dihydrolase is positive. Bacteria produce UTI, respiratory infection, wound infection, septicemia, and meningitis.

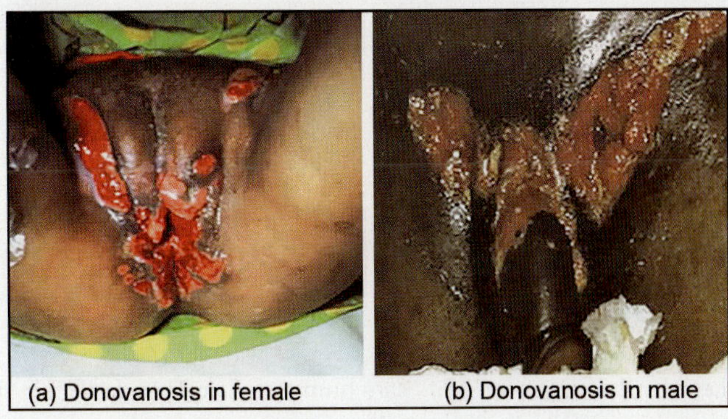

Fig. 61.7: Donovanosis in female and male

Only one species in this group is *H. alvei* which is GNB and motile. Biochemical reactions are best seen at 30°C. *H. alvei* is NLF, indole-MR negative and VP-citrate positive (best at 22°C). Lysin and other decarboxylation tests are positive. Bacteria produce opportunistic infections and isolated from blood, urine, pus and sputum.

Serratia spp.

Three species like *S. marcescens* (*S. prodigiosus*), *S. rubidaea* and *S. liquifaciens*. Bacteria are gram-negative bacilli or cocco-bacilli, motile, strict aerobes and slow lactose fermenters. Most important species is *S. marcescens* (*S. prodigiosus*), which produces red color pigment called prodigiosin. Pigment is best produced at room/low temperature about 20°C. It produces respiratory infection and presence of red pigment in sputum simulating presence of blood called pseudohemoptysis. Nosocomial infection is produced by resistant strain. It also causes meningitis, endocarditis, septicemia, peritonitis, etc.

Pantoea spp.

Pantoea (Greek) means all sources, as it has been isolated from geographical and ecological sources. Bacteria ferment lactose with production of acid and gas. They are VP positive. Arginine, lysine and ornithine decarboxylation tests are negative. *P. agglomerans* produces septicemia and salicin positive. *P. dispersa* is salicin negative.

MORGNELLACEAE

General Features

All the bacteria of family Morgnellaceae are called *Proteus* bacilli. Proteus word came from Greek god Proteus who was able to assume any shape, as bacteria are pleomorphic and variable in size and shape. They are resistant to KCN, degrade tyrosine, lactose, dulcitol and malonate are not fermented, MR positive, VP negative, arginine/lysine decarboxylation negative and PPA positive. They produce the enzyme phenyl alanine deaminase which split the phenyl alanine to phenyl pyruvic acid (PPA). PPA gives acidic pH to medium and detected by PPA test. Positive PPA test indicates green color **(Fig. 61.5c)** on adding of 10% ferric chloride within 4–5 minutes. Negative will not change the color of phenyl alanine agar slant. This is the single test which distinguishes Morgnellaceae from all other members of Enterobacterales order.

Classification

Follow **Table 61.5**.

General Scheme to differentiate the *Proteus* Bacilli in Laboratory by using Routine Tests: Follow **Flowchart 61.3**.

Flowchart 61.3: Differentiation of *Proteus* bacilli

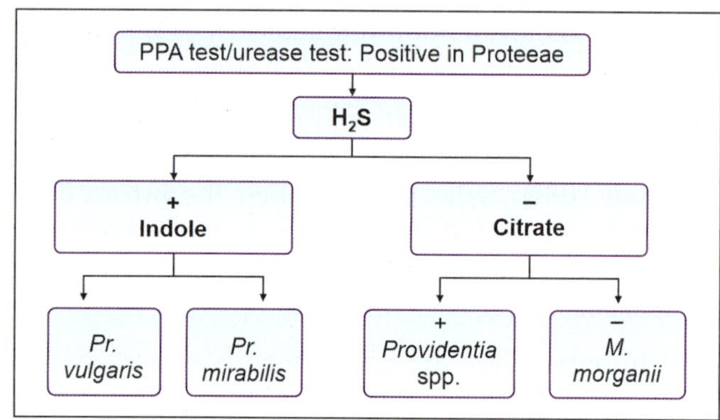

Family	Morgnellaceae					
Genus	*Proteus*		*Morgnella*		*Providencia*	
Species	*Pr. vulgaris*	*Pr. mirabilis*	*M. morganii*	*P. alcalifaciens*	*P. stuartii*	*P. rettgeri*
Features						
Adonitol fermentation	−	−	−	+	+ −	+ −
Trehalose fermentation	+ −	+	+ −	−	+	−
Indole	+	−	+	+	+	+
Methyle red	+	+	+	+	+	+
VP test	−	−	−	−	−	−
Citrate test	+	+	−	+	+	+
PPA test	+	+	+	+	+	+
Urease test	+	+	+	−	+ −	+
H₂S	+	+	−	−	−	−
Ornithin decarboxylation	−	+	+	−	−	−
Swarming	+	+	−	−	−	−
Lesions	Rare human pathogen	UTI and NI	UTI and NI	Diarrhea	UTI, burns infection	NI

Morphology

Proteus bacilli are GNB, pleomorphic with 1–3 µm × 0.5 µm in size, noncapsulated, actively motile (nonmotile strains are also observed by **Weil-Felix**) by peritrichate flagella.

Proteus vulgaris and *Pr mirabilis*

Cultural characteristics (C/Cs): They are aerobes and facultative anaerobes. C/Cs in different media are described below.

1. **Blood agar:** Cultures have putrefacient odor described as 'fishy' or 'seminal'. Bacilli produce **swarming,** defined as rapid (2–10 µm/s) and coordinated translocation of a bacterial population present as thin filmy layer in concentric circles on solid or semisolid surfaces. Swarming motility was first reported by Jorgen Henrichsen. Exact mechanisms of swarming are unknown, but may be due to multiple, thick and strong flagella which allow the actively motile cells to move away from the site of inoculation in search of nutrients. Swarming creates problem to isolate the pure growth. It is prevented by different methods like **(1) Physical methods:** These are agar overlays, poured plates, increasing agar concentration (3–4% or 6%), etc. **(2) Chemical method:** Incorporation of chloral hydrate (1:500) or sodium azide (1:500) or alcohol (5–6%) or sulfonamide or neomycin or surface active agents or boric acid (1:1000) or p-nitrophenylglycerol (0.2–0.4 nM in plates require 24 hours incubation and ≤1.0 nM in plates require 72 hours incubation) in medium. **(3) Other methods:** These are incorporation of H-antiserum in medium and uses of some media. MacConkey's, DCA and XLD are inhibiting the swarming due to presence of bile salts, W and B medium is due to bismuth sulfite and CLED medium due to electrolytes deficiency. Following are the different types and examples of swarming.

- **Continuous swarming:** It shows uniform film of growth on solid media (**Fig. 61.8a**) and it is mostly featureless. For example, *B. subtilis*.
- **Discontinuous swarming:** It may be concentric or dendritic.
  - In concentric swarming, series of concentric circles occur around site of inoculation (**Fig. 61.8b**) and give appearance like 'bull's eye', for example *Pr. mirabilis*.
  - Dendritic swarming appears in branching pattern (**Fig. 61.8c**), for example *P. aeruginosa*.

- **Other types and examples:** Also occurs in *Pr. vulgaris*, *Cl. tetani*, etc.

2. **Diene's phenomenon:** Inoculate **two identical** *Proteus* strains at different point on culture plate without anti swarming agents, resulting swarming growth coalesce without line of demarcation. Inoculate **two nonidentical** *Proteus* strains at different point on culture plate without anti-swarming agents, resulting swarming growth fails to coalesce with line of demarcation. This phenomenon called Diene's phenomenon (**Fig. 61.9a**) and line of demarcation called Diene's line. It is used to determine the identical and non-identical *Proteus* strains.

3. **MacConkey's medium:** No swarming occurs. Smooth, pale/colonies colorless (because *Proteus* bacilli are NLFs) develop as shown in **Fig. 61.9b**.

4. **Automated culture:** Like MALDI-TOF or VITEK is used to identify the species from culture.

Antigenic structure: *Proteus* bacilli have different antigens like somatic (O) Ag, flagellar (H) Ag and heterophile antigens. Weil and Felix observed that certain nonmotile strains of *Pr. vulgaris*, called 'X strains', were agglutinated by sera from typhus fever patients. This heterophilic agglutination is due to sharing of common Ag (carbohydrate) between *Proteus* and *Rickettsia*. Three nonmotile strains, OX2 and 0X19 of *Proteus vulgaris* and OXK of *Proteus mirabilis* are observed. They form basis of Weil-Felix reaction for diagnosis of some rickettsial infections.

Pathogenicity: They are the saprophytes and exogenous infections occur from soil, sewage and animal materials. They present as normal flora in moist areas like skin and intestine in human and animals responsible for endogenous infections. They produce opportunistic and nosocomial infections like UTI, wound infection, septicemia and soft tissue infections. Enzyme urease breaks urea to ammonia which makes the urine alkaline and damages the renal epithelium. Alkaline urine allows the deposition of phosphate to form the bladder stone (phosphate stone).

Laboratory diagnosis: Urine, pus and blood are tested by microscopy (follow morphology), culture (follow C/Cs), biochemical reactions (**Table 61.5**) and bacterial typing (like bacteriophage typing, ribotyping and bacteriocin typing).

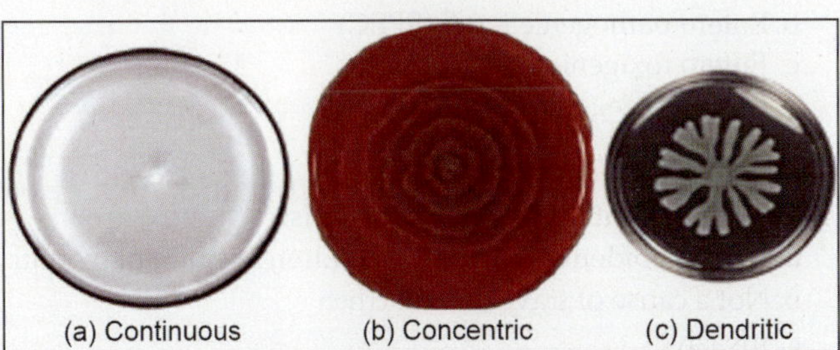

Fig. 61.8: Types of swarming

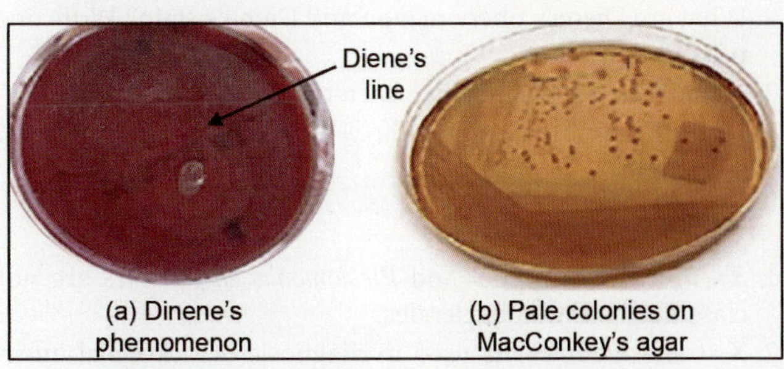

Fig. 61.9: Cultural properties of *Proteus* spp.

Note: Diene's stain
- It is useful in identification of *Mycoplasma*.

Providencia spp.

Three species as mentioned in **Table 61.5**. They are motile but do not swarm.

Morgnella morganii

All important features are mentioned in **Table 61.5**. It is motile but do not swarm. It found in human and animal feces.

Erwinia sp.

Common species is *E. herbicola* which is GNB, motile and produces yellow pigment. It is isolated from plant. It produces UTI and respiratory infection.

ACCESS YOURSELF

Case Study

1. A female patient visited the hospital with complain of fever and micturition with pain, frequency and urgency. Mid stream urine sample is collected. Direct microscopy suggests the actively motile GNB. Culture on MacConkey's medium gives pink colonies. Identify the most common pathogen and answer the following:
   a. Name the clinical condition and causative agent.
   b. Write the different mode of transmission of clinical condition.
   c. Describe the laboratory diagnosis of clinical condition.

Essay/Full Question

1. *E. coli*.

Short Notes

1. Diarrheagenic strains of *E. coli*.
2. *K pneumoniae*.
3. *K granulomatis* (Donovanosis/granuloma inguinale).
4. *Proteus*/Bacterial swarming.

Short Questions for Theory/Viva Questions

1. Define: Enterobacterales.
2. What are coliform bacilli? Write two examples.
3. What are paracolon bacilli? Write two examples.
4. Name four SLT producing bacteria.
5. What is modified MacConkey's medium? Write its use.
6. What is Congo red dye test?
7. Name the four bacteria showing safety pin appearance.
8. What are spreading and swarming growth of bacteria?
9. What are Diene's phenomenon and Diene's stain? Write one use of each.
10. Name four VP test positive bacteria.
11. Write four names of media which inhibit the bacterial swarming.

Comments on

1. *Sh. dysenteriae* type 1 and *Plesiomonas shigelloides* are not classified as Enterobacterales.
2. X strain of *Proteus* is used in diagnosis of rickettsial infections.

3. CLED medium can inhibit the *Proteus* swarming.
4. Infection by *Proteus* spp., favors the formation of bladder (phosphate) stone.

MCQ for Chapter Review

Definition of Enterobacterales

1. **Following is not the property of bacterium from Enterobacterales order:**
   a. Catalase positive   b. Oxidase positive
   c. Nitrate positive   d. Sugar fermentation
2. **Which is true of Enterobacterales?**
   a. All are catalase positive except *Shigella dysenteriae* type 1
   b. Nitrate reduction negative
   c. Glucose not fermented by all
   d. Motility by polar flagellum

Classification of Enterobacterales

3. **Coliform bacilli are defined as:**
   a. Lactose fermenters   b. Nonlactose fermenters
   c. Sucrose fermenters   d. a + b
4. ***E. coli* gives pink color with:**
   a. Chocolate agar   b. LJ medium
   c. MacConkey's medium d. Saline broth
5. **Enterobacterales are all *except*:**
   a. *Pseudomonas*   b. *Klebsiella*
   c. *V. cholerae*   d. *Proteus*
   e. *E. coli*

Escherichia coli

6. **In a case of significant bacteriuria the bacterial counts in urine are:**
   a. 100–1000 organisms/ml of urine
   b. 1000–10000 organisms/ml of urine
   c. 10000–100000 organisms/ml of urine
   d. >100000 organisms/ml of urine
7. **A 20-year-old male had pain in abdomen and mild fever followed by gastroenteritis. The stool examination showed presence of pus cells and RBCs on microscopy. The etiological agent responsible is most likely to be:**
   a. Entero invasive *E coli*
   b. Entero toxigenic *E coli*
   c. Entero pathogenic *E coli*
   d. Entero aggregative *E coli*
8. **Watery diarrhea in children is caused by:**
   a. EIEC   b. EPEC
   c. ETEC   d. EAEC
9. **Which of the following is true about EPEC?**
   a. Causes diarrhea in infants
   b. Acts by invasion of intestinal epithelial cells
   c. Adults are mostly affected
   d. Affects immunocompromised host
10. **Most common strain of *E. coli* gives traveler's diarrhea:**
    a. Entero invasive *E coli* (EIEC)
    b. Entero pathogenic *E coli* (EPEC)
    c. Entero toxigenic *E coli* (ETEC)
    d. Entero aggregative *E coli* (EAEC)

Aggegative E. coli (EAEC)

11. **True about entero toxigenic *E. coli* is:**
    a. Causes epidemic diarrhea in children in epidemic country
    b. Not a cause of traveler's diarrhea
    c. Invasive
    d. Spreads by contaminated water

12. **Regarding ETEC true is:**
    a. Invades submucosa
    b. Most common in children of developing countries
    c. Fomite borne and spreads person to person
    d. Not a common cause of traveler's diarrhea
13. **Most common cause of diarrhea in children of developing country is:**
    a. EIEC   b. EPEC
    c. ETEC   d. EAEC
14. **Congo red dye test is done for detection of:**
    a. Entero aggregative strains of *Escherichia coli*
    b. Entero hemorrhagic strains of *Escherichia coli*
    c. Entero pathogenic strains of *Escherichia coli*
    d. Entero invasive strains of *Escherichia coli*
15. **All are true about entero invasive *E. coli* is:**
    a. Sereny test positive
    b. May cause diarrhea
    c. Can cause hemolytic-uremic syndrome
    d. Produces verocytotoxin
16. **All of the followings are true about HUS *except*:**
    a. Infection may be transmitted by food
    b. HUS is caused by Verotoxin producing *Escherichia coli*
    c. HUS is more common in children
    d. HUS is rarely associated with hemorrhagic colitis
17. **A 20-year-old man presented with hemorrhagic colitis. The stool sample grew *Escherichia coli* in pure culture. The following serotype is likely to be the causative agent:**
    a. O157:H7   b. O159:H7
    c. O107:H7   d. O55:H7
18. **Hemolytic uremic syndrome is caused by:**
    a. Entero invasive *Eschericha coli*
    b. Entero toxigenic *Eschericha coli*
    c. *Eschericha coli* sero group O157:H7
    d. *Eschericha coli* sero group O127
19. **Which of these are true of *E. coli*:**
    a. The LT (labile toxin) in ETEC acts via c-AMP
    b. In type causing UTI the organism attaches via pilli
    c. The ST (stable toxin) in ETEC is responsible for hemolytic uremic syndrome (HUS)
    d. EIEC invasiveness is under plasmid control
    e. In EPEC the toxin helps in invasion by the bacteria
20. **Culture media used for diagnosis of EHEC O157:H7 is:**
    a. O culture   b. Sorbitol MacConkey's medium
    c. XLD agar   d. Deoxycholate media
21. **In *E. coli* true is:**
    a. ETEC is invasive
    b. EPEC acts via cAMP
    c. Pilli present in uropathogenic type
    d. ETEC causes HUS
22. **Labile toxin of *E. coli* can be detected by following methods *except*:**
    a. Into infant rabbit bowel
    b. Into adult rabbit skin
    c. Intragastrically into infant mouse
    d. Into tissue culture of Chinese hamster ovary cells
    e. Into Y1 mouse adrenal cells
23. **Eiken test is done for detection of:**
    a. Shiga like toxin
    b. Heat labile (LT) of *Eschericia coli*
    c. Heat stable (ST) of *Eschericia coli*
    d. Cholera toxin

24. **Which toxin is not mediated by cAMP?**
    a. *V. cholerae* 01
    b. Heat stable *E. coli* toxin
    c. Heat labile *E. coli* toxin
    d. *V. cholerae* 0137
25. **Toxin acting on cGMP:**
    a. Heat stable *E. coli* toxin
    b. Heat labile *E. coli* toxin
    c. Cholera toxin
    d. Shiga toxin
26. ***E. coli* gets attached to the surface with the help of:**
    a. Fucose   b. Concavatin
    c. Phytohemagglutinin d. Lactin
27. **True about *E. coli* is:**
    a. Sereny test is done to diagnose entero toxigenic *E. coli* (ETEC).
    b. Entero pathogenic *E. coli* (EPEC) is also invasive.
    c. UTI is caused by single serotype at a time only.
    d. *E. coli* is the most common organism to infect burn wound.
28. **Most common cause of liver abscess:**
    a. Streptococci   b. *Staph. aureus*
    c. *E. coli*   d. *Staph. pyogenes*
29. **A microbiologist wants to develop the vaccine for prevention of attachment of diarrheagenic *E. coli* to the specific receptors in the gastrointestinal tract. All of the following fimbrial adhesion would be appropriate vaccine candidates *except*:**
    a. CFA-1   b. Pi-Pilli
    c. CS-2   d. K88
30. **Sereny test is useful to detect:**
    a. Entero invasive *E. coli*
    b. *V. cholerae*
    c. Entero toxigenic *E. coli*
    d. Entero pathogenic *E. coli*

Klebsiella spp.

31. **Friedlander's bacillus is:**
    a. *Klebsiella pneumoniae*   b. *Clostridium welchii*
    c. *Chlamydia*   d. *E. coli*
32. **Which of the following is associated with extremely foul smelling infection?**
    a. *Klebsiella rhinoscleromatis*
    b. *Klebsiella seeberi*
    c. *Klebsiella zygomaticus*
    d. *Klebsiella ozaenae*
33. **A male patient presents with granulomatous penile ulcer. On Wright Giemsa stain tiny organism of 2 microns within macrophages are seen. What is the causative organism?**
    a. LGV
    b. *Calymmatobacterium granulomatis*
    c. *Neisseria*
    d. *Staph aureus*
34. **In donovanosis:**
    a. Pseudolymphadenopathy
    b. Penicillin is used for treatment
    c. Painful ulcer
    d. Suppuative lymphadenopathy
35. **'Safety pin appearance' is characteristic of:**
    a. *Yersinia pestis*
    b. *Anthrax bacillus*
    c. *Berkholderia pseudomallei*
    d. a + c

36. Diene's phenomenon is seen with:
a. *Proteus mirabilis*　　b. *Klebsiella*
c. *Proteus vulgaris*　　d. *Providentia*
e. *Morgnella*

37. *Proteus* antigen cross reacts with:
a. *Klebsiella*　　b. *Rickettsia*
c. *Chlamydia*　　d. *E coli*

Answers and Explanation of MCQs

1. b

2. a
- Follow section, **definition of Enterobacterales** for explanation of answers of MCQs 1–2.

3. a

4. c
- Follow section, **classification of Enterobacterales** for explanation of answers of MCQs 3–4.

5. a and c
- Because *Pseudomonas* and *V. cholerae* are oxidase positive
- Follow section, **definition and classification of Enterobacterales** for more explanation.

6. d
- As per Kass concept bacterial count in urine has three categories like significant bacteriuria when count is $>10^5$ bacteria/ml, doubtful/equivocal when count is between 10^4–10^5/ml and nonsignificant when count is $<10^4$/ml.

7. a
- Because entero invasive *E. coli* causing dysentery, pus cells and RBCs are present in stool. Even EHEC also causing hemorrhagic diarrhea (hemorrhagic colitis) but that option is not given.

8. b

9. a
- EPEC causing watery diarrhea in infants and children, occurs as institutional outbreaks/sporadic cases.

10. c
- ETEC and EAEC both are known to cause traveler's diarrhea, but most common by ETEC.

11. d
- ETEC causes traveler's diarrhea in children of developing countries. It is non invasive and effects are due to initial binding by CFA followed by toxin production. It is transmitted by contaminated food and water but person to person spread do to not occurs.

12. b

13. c
- Follow section, *Escherichia coli* (diarrhea → ETEC) for explanation of answers of MCQs 12–13.

14. d
- Follow section, *Escherichia coli* (diarrhea → EIEC) for explanation.

15. a
- Follow section, *Escherichia coli* (diarrhea → EHEC) for explanation.

16. d
- HUS is a rare complication following hemorrhagic colitis. Most cases of HUS are associated hemorrhagic colitis but all cases of hemorrhagic colitis may not progress to HUS.
- Follow section, *Escherichia coli* (diarrhea → EHEC) for more explanation.

17. a

18. c
- Hemorrhagic colitis or HUS is caused by EHEC, where O157:H7 is the most common serotype.

19. a, b and d
- The ST (stable toxin) in ETEC acts via cGMP.
- EPEC is nontoxigenic.

20. b
- Follow section, *Escherichia coli* (diarrhea → EHEC → detection of bacilli) for explanation.

21. c
- ETEC is noninvasive. EPEC does not produce the toxin and no activation of cAMP, which is the property of labile toxin.
- HUS is caused by EHEC.

22. c

23. b
- Follow section, *Escherichia coli* [pathogenicity → agent factors (virulence factors) → Table 61.3] for diagnostic tests of LT and ST.

24. b
- Heat labile *E. coli* toxin and cholera toxin of *V cholerae* 0137 and O1 are structurally and antigenically same, but cholera toxin is hundred times more potent than labile toxin. Both act via cAMP, while heat stable *E. coli* toxin acts via cGMP

25. a
- Already explained.

26. a
- Follow section, *Escherichia coli* [pathogenicity → agent factors (virulence factors) → fimbriae] for explanation.

27. c
- Sereny test is done to diagnose EIEC. EPEC is also non-invasive. UTI is caused by single serotype at a time only; however, recurrence can occur by same or other serotypes. Most common organism to infect burn wound is *P. aeruginosa* followed by *Staph. aureus*.

28. c
- *E. coli* is the most common cause of acute UTI (in 80% cases without catheter), neonatal meningitis, intra-abdominal abscess and nosocomial infection.

29. d
- K is the capsular antigen. Other three are fimbriae/pilli or types of pilli and known as organs of adhesion. CFA is found in entero toxigenic *E. coli*. It has many components. CS-II is the component of CFA.

30. a
- Follow section, *Escherichia coli* (pathogenicity → diarrhea → entero invasive *E. coli*) for explanation.

31. a
- Follow section, *Klebsiella* spp. (*K pneumoniae* → synonyms) for explanation.

32. d
- *Klebsiella ozaenae* causes foul smelling nasal discharge.

33. b

34. a
- Follow section, *Klebsiella* spp. (*K granulomatis*) for explanation of answers of MCQs 33–34.

35. d
- Follow section, *Klebsiella* spp. (notes → **Bacteria with safety pin appearance**) for explanation.

36. a and c
- Follow section, *Proteus vulgaris* and *Proteus mirabilis* (**cultural characteristics**) for explanation.

37. b
- Follow section, *Proteus vulgaris* and *Proteus mirabilis* (**antigenic structure**) for explanation.

Infections of Enterobacterales-II (Salmonellas)

Chapter Outline

- Enteric Fever/Typhoidal Group
- Nontyphoidal Group

- *Salmonella* food poisoning
- *Salmonella* septicemia

ENTERIC FEVER/TYPHOIDAL GROUP

History

Salmonella Typhi was 1st observed by Eberth in 1880 in the mesenteric nodes and spleen of fatal case and isolated by Gaffky in 1884, so previously called Eberth-Gaffky bacillus or *Eberthella* Typhi. Salmon and Smith described bacilli, hence the genus called *Salmonella* with removal of earlier genus name *Eberthella*.

Morphology

Type according to Gram's stain: They are GNB.

Shape and size: They are rod shaped and about 1–3 μm × 0.5 μm in size.

Arrangement: They arranged singly, in pair or in small group as shown in **Fig. 62.1**.

Motility: They are motile with peritrichate flagella (except *S.* Gallinarum and *S.* Pullorum).

Spores and capsule: They are nonsporing and non-capsulated.

Fimbriae: May be present.

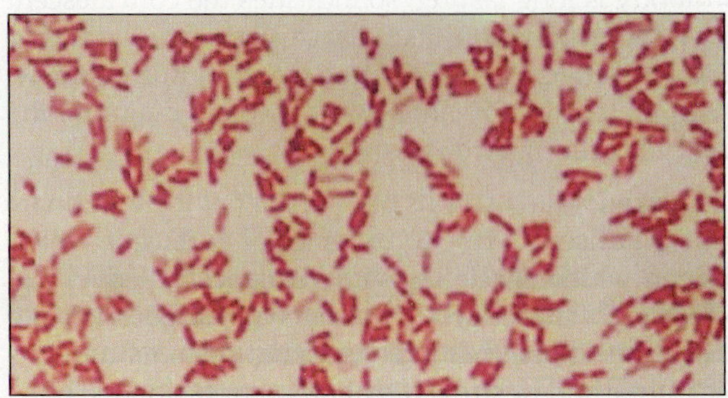

Fig. 62.1: *S.* Typhi under Gram's stain

Cultural Characteristics (C/C)

Effective factors

- O_2 effect: They are aerobes and facultative anaerobes.
- Optimum temperature: 37°C (15–41°C).
- pH: 6–8.
- *S.* Typhi and few other species of *Salmonella* required tryptophan as a growth factor.

Culture in media

A. Liquid media

1. **Enrichment media**
   - **For feces:** Selenite-F broth (selenite broth), tetrathionate broth and *Salmonella-Shigella* broth are useful for *Shigella* and *Salmonella*, but better for *Salmonella*.
   - **For blood:** Taurocholate broth is useful with dilution ration is 1:10.
2. **Transport media**
   - **Sach's glycerol buffered saline (gram-negative broth or glycerol saline transport medium):** It preserves *Salmonella* and *Shigella* from fecal specimens.
   - **Gram-negative broth:** It is a transport and selective medium for *Salmonella* and *Shigella* from clinical (feces, urine, blood, etc.) and non clinical specimens (water, food, etc.). It contains sodium citrate and sodium deoxycholate which inhibit the growth of gram-positive and coliforms. Collect the specimen; incubate for 6–8 hours followed by subculture on solid media. It is not good for rectal swab.
   - **Cary Blair medium:** It is used for transportation of EHEC (O157:H7), *Salmonella*, *Shigella* and *V. cholerae*.

B. Solid media: *Salmonella* produces 2–3 mm, circular, convex and translucent colonies on nutrient agar.

- **MacConkey's agar (Fig. 62.2a), DCA and S-S agar:** *Salmonella* is NLF and produces pale colonies.
- **W and B medium (Fig. 62.2b):** Jet black colonies produced due H_2S production by all *Salmonella* species except *S.* Paratyphi A, *S.* Cholerasuis and few other species that do not form H_2S and produce green colonies.
- **XLD agar:** *Salmonella* produces pink colonies with black center due to H_2S production.
- **Hektoen enteric agar:** It is a selective and differential medium, primarily used to recover *Salmonella* and *Shigella* from samples. It contains indicators of lactose fermentation and H_2S production; plus inhibitors to prevent the growth of gram-positive bacteria. It contains bile salt as inhibitory agent and some dyes. Bacteria produce green color colonies with color fading to the periphery.

Automated culture: Like MALDI-TOF or VITEK is used to identify the species from culture.

Biochemical Reactions (B/Rs)

Sugar fermentation tests: *S.* Typhi is anaerogenic, means does not produce gas. Following are the sugar reactions for *S.* Typhi and *S.* Paratyphi

- *S.* Typhi (**Fig. 62.3a**)

G	S	L	M	Maltose	Salicin
A	–	–	A	–	–

- *S.* Paratyphi (**Fig. 62.3b**)

G	S	L	M
AG	–	–	AG

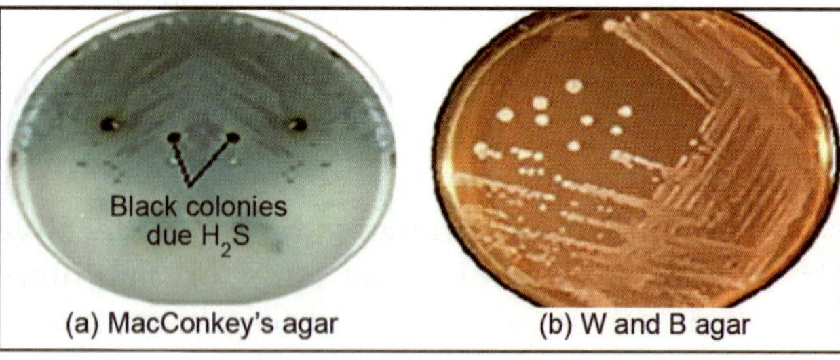

Fig. 62.2: Colonies of *S.* Typhi

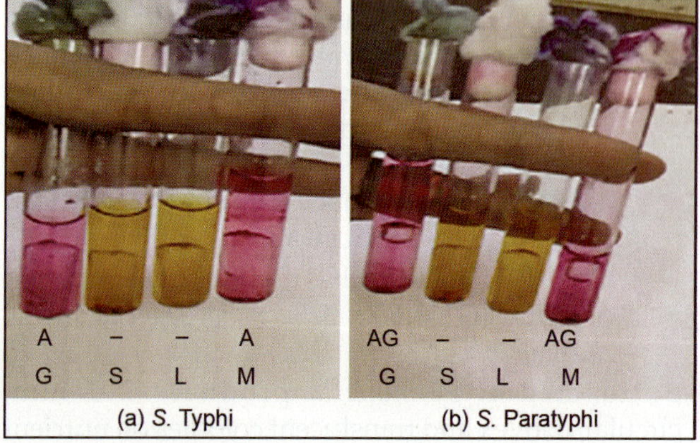

Fig. 62.3: Sugar fermentation tests

I M Vi C tests: – + – V (*S.* Typhi and few other salmonellas do not grow on Simmon's citrate medium as they required tryptophan as a growth factor).

Catalase and nitrate reduction tests: Positive

Oxidase and urease tests: Negative

TSI test: H_2S is produced by *Salmonella* spp., except *S.* Paratyphi A and *S.* Choleraesuis

- *S.* Typhi = K/A + H_2S
- *S.* Paratyphi A = K/AG – H_2S
- *S.* Paratyphi B = K/AG + H_2S

Resistance

Sterilization: Bacilli are killed by heat at 55°C in 1 hour/60°C in 15 minutes, boiling and pasteurization of milk.

Disinfection: Bacilli are killed by chlorination of water, 5% phenol and mercuric chloride in 5 minutes.

Drug resistance: Drug resistance to *Salmonella* is described below with treatment.

Antigenic Structure

Somatic antigen (O Ag or LPS or endotoxin): It is a phospholipid-protein-polysaccharide complex forms an integral part of cell wall. It is identical with endotoxin and extracted from cell by trichloracetic acid called Bovine Ag. It is heat stable, not-destroyed by boiling or by alcohol or by acids. It is group specific, so antibody produced is not discriminate between typhoid and paratyphoid fever. It is less immunogenic and induces antibody formation slowly and in low titer following infection or immunization. Antibody is not long lasting, so indicates the recent infection. Agglutination takes place more slowly and at higher temperature (50–55°C) with compact, chalky, granular clumps when mixed with O antisera. It is responsible for endotoxic activities. It helps to diagnose the typhoid fever by Widal test.

Flagellar (H) antigen: It is made up by protein called flagellin. It is heat labile and destroyed by boiling or by treatment with alcohol but not by formaldehyde. It is species specific, so antibody produced is discriminate between typhoid and paratyphoid fever. It is strongly immunogenic and induces antibody formation rapidly and in high titer following infection or immunization. Antibody is long lasting, so indicates the convalescent phase. Agglutination takes place very rapidly and at 37°C with large, loose and cotton wooly or fluffy clumps, when mixed with H-antisera. It helps to diagnose the typhoid fever by Widal test.

Vi antigen: Many strains of *S.* Typhi fail to agglutinate with O antiserum due to presence of surface/envelop antigen which is related with virulence called Vi Ag. Felix and Pitt described this antigen (analogous to K Ag of coliforms bacilli). It is polysaccharide in nature. It is heat labile and destroyed by boiling or heating at 60°C for 1 hour or by treatment with N HCl and 0.5 N NaOH, but not by alcohol or by 0.2% formalin. Vi Ag prepared by treating suspension with glycerol and alcohol. Vi antigen lost on serial subculture. Bacilli in-agglutinable with O antiserum become agglutinable after loss of Vi Ag. It presents in *S.* Typhi, *S* Paratyhi C, *S.* Dublin and in some strain of *Citrobacter freundii* (**Ballerup-Bethesda group**). It is less immunogenic and induces antibody formation in low titer following infection or by alcoholized vaccine but no antibody formation following phenolized vaccine. It inhibits phagocytosis, resists complement activity, resist bacterial lysis by alternative pathway and by peroxidase killing. Bacilli with Vi Ag are causing more consistent disease than those are lacking. Vi Ag is poorly immunogenic and antibody titer is very low, so not helpful in diagnosis of typhoid fever hence not included in Widal test. It is detected by agglutination test. Total absence of Vi antibody indicates poor prognosis (disappears in convalescence phase), and it persistence indicates carrier state. It is used for epidemiological typing of *S.* Typhi by using Vi bacteriophage. It is also useful for vaccination.

Fimbrial (Fi) Ag: It is not important in diagnosis instead causing confusion in diagnosis due to nonspecific nature and common sharing with other Enterobacterales.

Mucoid (M) antigen: It is polysaccharide in nature and responsible for mucoid colony. It presents in *S.* Paratyphi B and some strain of *E. coli*. It resembles to Vi Ag and prevents agglutination by O antiserum. Heating at 100°C for 2.5 hrs makes the strain agglutinable for O antiserum.

Antigenic Variations

H–O variation: It based on loss of H Ag. There two types of H–O variation. (1) **Phenotypic variation** where loss of H Ag occurs when *Salmonella* grows on phenol (1:800) agar. It is temporary and H Ag reappears, when it grows on media free of phenol. (2) **Genotypic variation,** where loss of H Ag occurs due to mutation. It is stable form and H Ag does not reappear. H–O variation is useful to prepare specific H and O antiserum. O antiserum is prepared by using stable mutant (genotypic variation-no H Ag) like *S.* Typhi 901-O strain. H antiserum is prepared by obtaining a population of motile cells rich in H Ag by Craigie's tube or U-tube (**Ch. 115**).

Phase variation: It presents in H Ag. Some strains posses both phases called diphasic strains and strain possesses only one phase is called monophasic strain. It occurs in two phases. Phase-I is species specific. Large numbers of antigens are found in phase-I and designated as a to z and then z1 to z68. Phase-II is group/genus specific. Antigens in this phase are designated by Arabic numerals (1–12).

V–W variation: Many strains of *S.* Typhi fail to agglutinate with O antiserum due to presence of surface/envelop antigen and agglutinate with Vi antiserum this called 'V form'. After several subcultures Vi Ag will lost and strain fails to agglutinate with Vi antiserum, but agglutinates with O antiserum, called 'W form'. Intermediate stage when strain agglutinates with both antiserums **called 'VW form'**.

S–R variation: It indicates the change in the colonies from smooth (S) to rough (R). Initially strain produces smooth (S) colony due to presence of O antigen and virulence. Strain undergoes mutational changes producing rough (R) and autoagglutinable colony with loss of O Ag and virulence. R form is common in laboratory due to repeated subculture. S–R variation can be prevented by maintaining culture in Dorset egg medium and lyophilization.

O-Ag variation: Structural changes in O Ag can be induced by bacteriophage (lysogenization, which results in new serotypes of *S.* Anatum to *S.* Newington and later to *S.* Minneapolis as shown in **Flowchart 62.1**.

Classification

Serological (Kauffmann-White) classification

1. **O-Ag (group specific):** Many serogroups were identified initially and all were labeled as A, B, C… As the numbers increased, they labeled as 1, 2, 3…, where 2 is older group A, 4 is older group B, 9 is older group D and so on. Total 67 serogroups are identified by using group specific O-Ag.
2. **H-Ag (species specific/serotype specific):** Around 2300 serotypes are identified by using species specific H-Ag (phase I and phase II). Follow **Table 62.1** for some important serotypes.

Biochemical or Kauffmann classification or biotypes: Four subgenera or subgroups like I, II, III and IV are identified as shown in **Table 62.2**.

Molecular classification: DNA hybridization study shows that there are two main species in the genus *Salmonella*. Each species further classified into subspecies and subspecies into serotypes or serovars by earlier Kauffmann-White's classification.

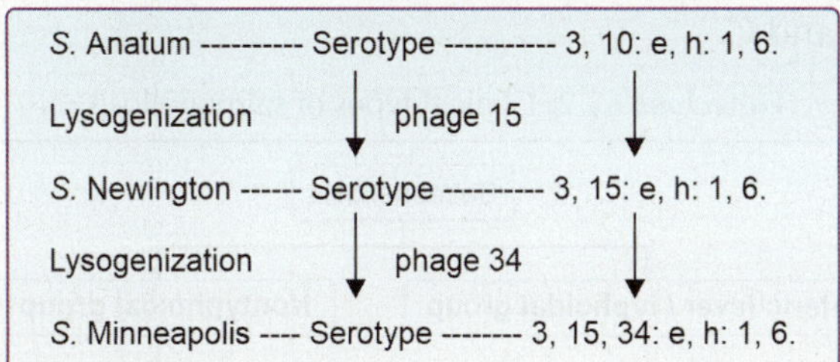

Flowchart 62.1: O-Ag variation by phage

TABLE 62.1: Kauffmann-White classification

Sero-group	Serotype	O Ag	Phase-I H-Ag	Phase-II H-Ag
2-A	*S.* Paratyphi A	1, 2, 12	a	–
4-B	*S.* Paratyphi B	1, 4, 5, 12	b	1, 2
7-C1	*S.* Paratyphi C	6, 7 (Vi)	c	1, 5
9-D	*S.* Typhi	9, 12 (Vi)	d	–

TABLE 62.2: Biotypes of *Salmonella*

Test	I	II	III	IV
Lactose	–	–	+	–
Dulcitol	+	+	–	–
d-Tartrate	+	–	–	–
Malonate	–	+	+	–
Salicin	–	–	–	+
KCN	–	–	–	+

- **S. enterica:** Six subspecies as follows. Most of the human infections are caused by subspecies *arizonae* and *enterica*.
    1. **S. enterica** subsp. *enterica* has many serotypes that include typhoidal and nontyphoidal groups.
    2. **S. enterica** subsp. *salamae*.
    3a. **S. enterica** subsp. *arizonae*.
    3b. **S. enterica** subsp. *diarizonae*.
    4. **S. enterica** subsp. *houtenae*.
    6. **S. enterica** subsp. *indica*.
- **S. bongori:** Earlier it has 5 subspecies.

> **Note: Nomenclature of *Salmonella***
> - Nomenclature of *Salmonella* is very complicated like genus *Salmonella*, species *enterica*, subspecies *enterica* and serotype Typhi/Paratyphi/Enteritidis, etc.
> - For routine use nomenclature system used is, Genus followed by serotype like *Salmonella* Typhi (*S.* Typhi)/*Salmonella* Paratyphi (*S.* Paratyphi)/*Salmonella* Enteritidis (*S.* Enteritidis), etc.

Clinical classification: Whole spectrum of disease called salmonellosis, which includes two clinical groups like enteric fever/typhoidal group and nontyphoidal group as mentioned in **Flowchart 62.2**. Both are described below in details.

Pathogenicity

Bacilli of enteric fever/typhoidal group produce a disease called enteric fever. Enteric fever is general terminology, which includes typhoid fever caused by *S.* Typhi and paratyphoid fever caused by *S.* Paratyphi A, B and C.

Flowchart 62.2: Clinical types of salmonellosis

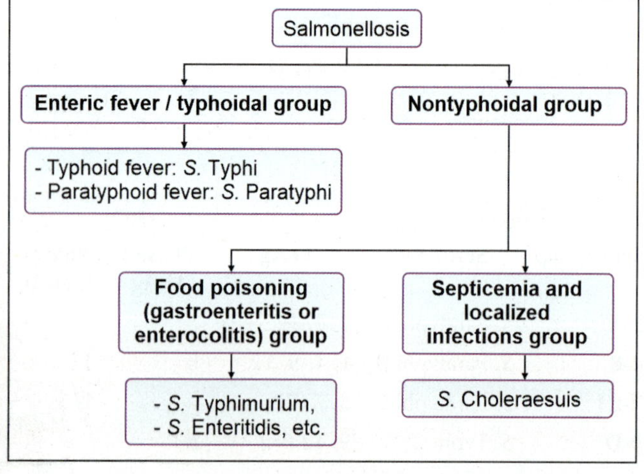

Typhoid Fever

Epidemiology: Typhoid fever is distributed worldwide. Endemic occurs throughout the year. Epidemic occurs suddenly by contamination of food, water or milk. About 22 million cases are reported annually with 6,00,000 deaths.

Reservoirs of infection: *Salmonella* Typhi is an obligate pathogen for human. Human cases or carriers of *Salmonella* are the reservoir hosts.

Sources of infection: Water or food

Modes of transmission: Ingestion

Incubation period: 7–14 days

Portal of entry: GIT

Sites: Intestinal and extraintestinal (complications).

Precipitating factors (epidemiological determinants): Following are the factors.

A. **Agent factors (virulence factors):** It includes intracellular (cell wall associated) factors like O Ag, H Ag, Vi Ag, Fi Ag and M Ag as described above under the heading of antigenic structure. Other factor is infective dose, which is very high about 10^3–10^6 bacilli can initiate the disease.

B. **Host factors**
- **Age:** It is common in 5–20 years age.
- **Local GIT factors: (1) Low gastric acidity:** Bacilli are sensitive to gastric acid. Low acidity due to achlorhydria and use of antacids can enhance the infection. **(2) Other factors**: *H. pylori* infection, inflammatory bowel disease, gastrointestinal surgery, suppression of intestinal flora by antibiotics, etc., can promote the typhoid.
- **Hygiene:** *Salmonella* infection is common in people with poor hygiene.

C. **Environmental factors:** *Salmonella* infection is more common in rural than urban. Typhoid is eliminated (decreased incidence) from developed countries due to improved water supply and also in hygiene.

Pathogenesis: Follow **Flowchart 62.3**.

Clinical features: Very mild to fatal disease.

1. **Local features:** These are abdominal discomfort with either constipation or diarrhea. Stool is khakhi-green, slimy called pea-soop stool, typically occurs in 3rd week of typhoid fever. At this time patient is in toxic stage and greater risk to develop perforation and hemorrhage. Similar stool occurs in EPEC in infants.
2. **Systemic features:** Onset is usually gradual with headache, malaise, anorexia and coated tongue. Presence of undifferentiated continuous fever (step-ladder pyrexia) with relative bradycardia and toxemia, hepatomegaly, soft and palpable spleen, **Rose spots** are the red macules, 2–4 mm in size and

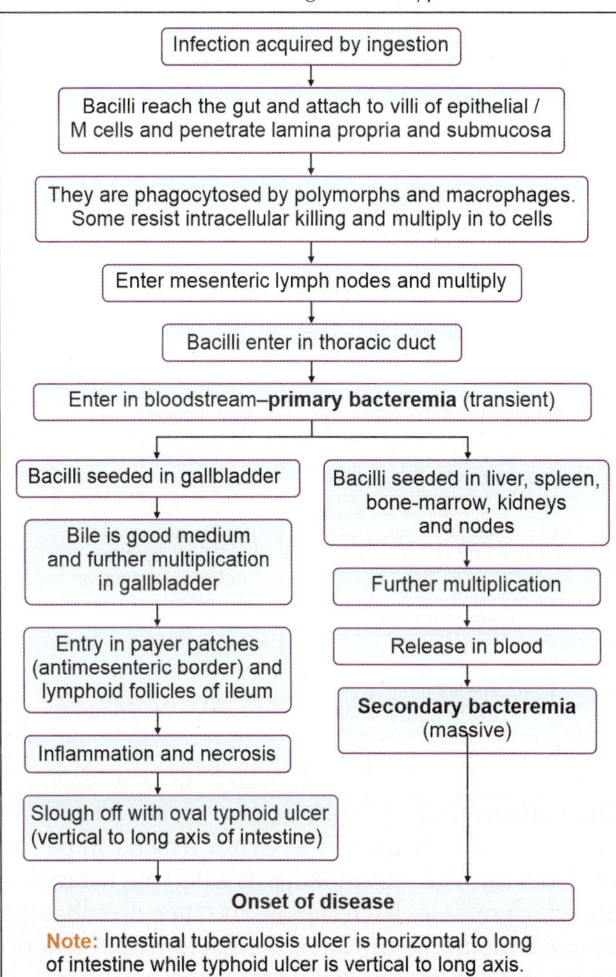

Note: Intestinal tuberculosis ulcer is horizontal to long of intestine while typhoid ulcer is vertical to long axis.

5–10 in numbers occur on lower chest and upper abdomen during 2nd or 3rd week in untreated cases of enteric fever (both in typhoid fever caused by *Salmonella* Typhi and paratyphoid fever caused by *Salmonella* Paratyphi), nontyphoidal salmonellosis and shigellosis. It fade on pressure and rarely noticeable in dark-skinned patients. These are actually bacterial emboli to the skin, occur in 1/3 cases of typhoid fever and numerous in paratyphoid fever and last for 3–4 days.

Complications

1. **Local complications:** These are intestinal perforation and hemorrhage. Stricture is not seen after typhoid but common after tuberculosis.
2. **Systemic complications:** These are circulatory collapse, bronchitis or bronchopneumonia, deafness, cholecystitis, arthritis, abscesses, periosteitis, nephritis, HUS, hemolytic anemia, venous thrombosis, meningitis, peripheral neuritis, cerebellar ataxia, muttering delirium, coma vigil, paranoid psychoses, hysteria, delirium and aggressive behavior. Case fatality rate is <1%.

Sequel: Osteomyelitis is the rare sequel.

Relapse: Slow convalescence and relapse occurs in 5–10% cases.

Epidemiology: Paratyphoid fever has variable distribution. *S.* Paratyphi A is prevalent in India and other Asian countries, Eastern Europe and South America. *S.* Paratyphi B is prevalent in Western Europe and North America. *S.* Paratyphi C is prevalent in Eastern Europe and Guyana. About 6 million cases are reported annually. Proportion of typhoid fever to paratyphoid fever A is 10:1 with B is rare and C is least.

Clinical features: *S.* Paratyphi A and B cause paratyphoid fever which resembles to typhoid fever, but generally milder. *S.* Paratyphi C causes paratyphoid fever presents as septicemia with suppurative complications.

Laboratory Diagnosis of Enteric Fever

Specimen types

1. **For detection of case:** Blood, bone marrow, bile, pus from suppurative lesion, CSF, sputum and autopsy specimens from liver, spleen mesenteric lymph node, etc.
2. **For detection of carrier:** Urine and bile.
3. **For detection of case and carrier:** Stool.

Collection of specimens

- **Blood for culture:** Make the site sterile by using appropriate disinfectant. About 5–10 ml blood is collected in adult and 2–3 ml in pediatric patient by venepuncture/phlebotomy. Blood is collected in 50–100 ml of taurocholate broth (1:10 dilution) for adult. Incubate the broth at 37°C for 24 hours followed by subculture (S/C) on solid media and selective media which are mentioned above. **Advantages** of blood culture are: **(1)** It rapidly becomes positive with 90% in 1st week, 75% in 2nd week, 60% in 3rd week and 25% after as shown in **Fig. 62.4** and **(2)** because of cross reactivity by different antigens in Widal test, blood culture is considered as gold standard for diagnosis of enteric fever almost at any time. **Disadvantages** of blood culture are **(1)** it becomes negative after antibiotic treatment and **(2)** blood contains inhibitory or bactericidal substances which are obviated by fourfold dilution of blood or by adding liquoid (SPS—sodium polyanethol sulfonate).
- **Stool for culture:** Bacilli shed in feces throughout the course of illness and even in convalescence. Collect freshly passed stool in enrichment broth like selenite-F broth or tetrathionate broth. Incubate the broth at 37°C for 24 hours followed by subculture (S/C) on solid media and selective media. **Advantages** of stool culture are **(1)** it is useful in patient taking antibiotic drugs, which are not interfering with intestinal bacilli and **(2)** used to detect the case and fecal carrier. **Disadvantages** of stool culture are **(1)** bacilli are shedding very infrequently in feces so repeated cultures are required and **(2)** it will not differentiate between case and carrier.

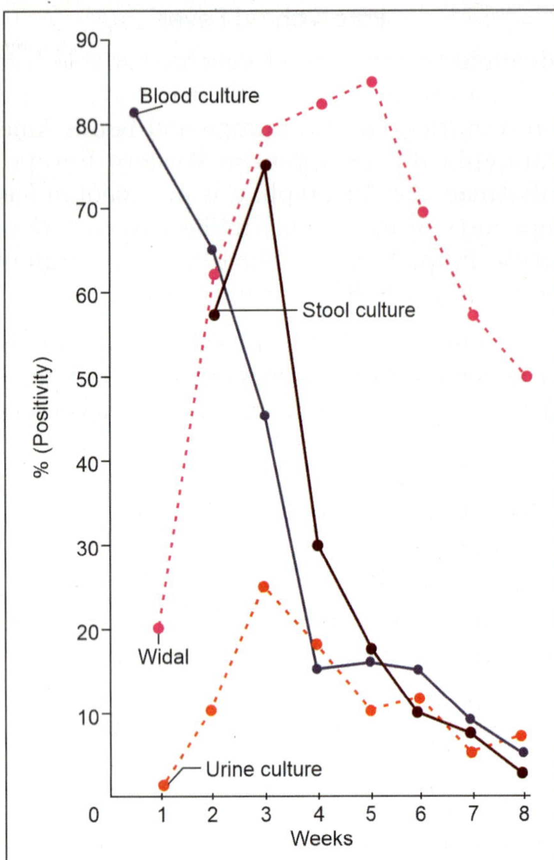

Fig. 62.4: Percentages of positivity of different tests

- **Urine for culture:** Collect the mid-stream urine sample in sterile container. Centrifuge it and take sediment in enrichment broth. Incubate the broth at 37°C for 24 hours followed by subculture (S/C) on solid and selective media. **Advantage** of urine culture is that it detects urinary carrier. **Disadvantages** of urine culture are **(1)** bacilli shed infrequently and irregularly in urine, so repeated cultures are required and **(2)** it is positive only during 2nd and 3rd week and then 25% positivity rate.
- **Bone marrow:** It is positive, even if blood culture is negative.
- **Bile:** It is collected by duodenal aspiration for detection of fecal carrier.
- **Other specimens:** Like pus from rose spot, sputum and CSF are rarely useful.

Testing methods

A. Microscopy: Follow morphology.

B. Culture: Different techniques are available.

Culture on solid media: Follow **Flowchart 62.4**.

Castaneda biphasic medium culture method (Liquid broth + solid slant phase): It contains 50–100 ml BHIB and BHIA. It avoids the contamination during subculture. Blood or bone marrow is inoculated in broth and incubated in upright position. For subculture bottle is tilted so that broth runs over the slant and then bottle is incubated in upright position. Next day observed slant for colonies.

Flowchart 62.4: Cultivation techniques and identification of growth on solid media

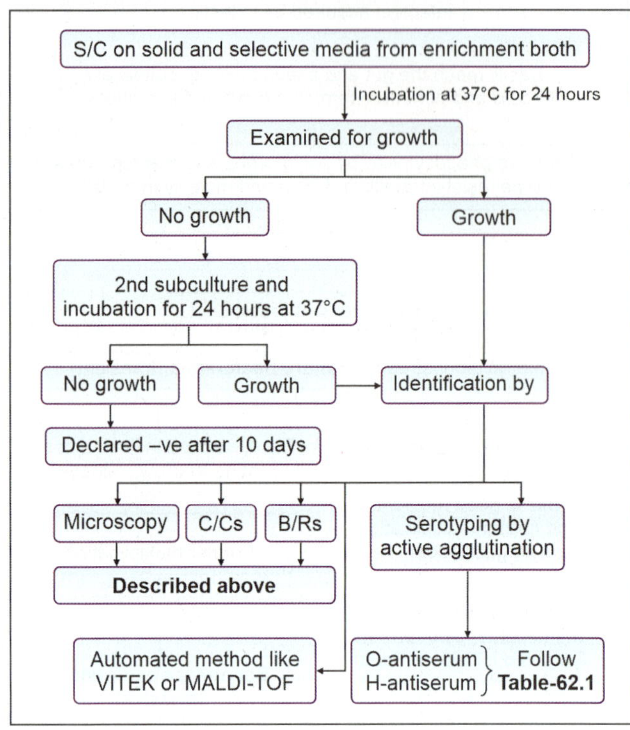

Clot culture: Collect 5 ml blood in sterile test tube and allowed to clot. Separate the serum which can be used for Widal test. Add streptokinase (100U/ml) to break the clot which releases bacteria free and then do subculture on solid media. It obviates the bactericidal action of serum and serum is available for serological tests.

> **Note: National *Salmonella* Reference Centers**
>
> **For phage typing:** National salmonella Vi phage typing center Lady Harding Medical College, New Delhi, India.
>
> **For serotyping:** For identification of uncommon serotypes, culture should be sent to following National *Salmonella* Reference Centers.
>
> 1. **Central Research Institute, Kasauli, India:** For *Salmonella* of human origin.
> 2. **Indian Veterinary Research Institute, Izzatnagar, India:** For *Salmonella* of animal origin.

C. Serological tests: Following are two types:

1. **Ab detection tests:** Because of cross reaction by related antibodies, these tests have limited value. Following are useful tests.
   - Widal test: **Ch. 20**.
   - Typhidot test: It detects IgG and IgM antibodies to outer membrane proteins antigen.
   - IDL tubex test: It detect the IgM antibodies against O9 antigen of *S.* Typhi.
   - ELISA and IgM dip stick test: Both detect the IgM antibody against LPS-Ag.
   - Dot blot assay: It detects the IgG antibodies against H antigen.
   - Other tests: Indirect hemagglutination (IHA) test, CIE, etc.

2. **Ag detection tests:** Ag is detected from blood (early phase) and urine (late phase) by using monoclonal Ab by following tests.
   - ELISA:
   - Coagglutination test: *Staph. aureus* (Cowan I strain) which contains protein A is fixed with Ab of *S.* Typhi. Mix such sensitized *Staph. aureus* with patient serum contain typhoid Ag (urine also contains typhoid Ag). Typhoid Ag combines with Ab of *S.* Typhi and produces clumping of *Staph aureus* called coagglutination. This test is rapid, sensitive and specific. Test is positive in 1st week, but not after that.

D. Pathological tests: Blood picture shows leukopenia with lymphocytosis.

E. Molecular method: PCR based tests are sensitive.

F. Salmonella typing methods
   1. **Phenotyping methods:** Like bacteriophage typing (Ch. 118), bacteriocin typing and biotyping. Biotyping for *S.* Typhi is described earlier. It is also possible in *S.* Typhimurium. Kristensen biotyping by xylose and arabinose fermentation is also useful. It also done by production of tetrathionate reductase.
   2. **Genotyping methods:** These are PFGE, multi locus enzyme electrophoresis, IS-200 profiling and random amplified polymorphic DNA typing. They are useful for the epidemiological purpose and have more discriminating power.

G. AST: Culture report should be given with AST. It can be performed by disc diffusion method or by automated method like VITEK.

Carrier

Definition: Person who carries the organism without any sign-symptoms of disease called carrier. It never presents in normal population and occurs only after infection.

S. **Typhi carrier:** Two types.
1. **Fecal carrier:** Bacilli persist in gallbladder and shed in feces, but for a different period. Antibiotics do not eliminate intestinal bacilli rapidly unlike bacilli of blood, so fecal culture is useful after antibiotics course. Fecal carrier is common, but less risky.
   - Convalescent carrier: It sheds bacilli in feces from 3 weeks to 3 months after clinical cure.
   - Temporary carrier: It sheds bacilli in feces >3 months to <1 year.
   - Chronic carrier: It sheds bacilli in feces >1 year. Example of chronic carrier is Mary Mallon/Typhoid marry (New York Food handlers or cook) who over a period of 15 years caused at least seven outbreaks affecting over 200 persons.
2. **Urinary carrier:** Bacilli persist in kidneys and shed in urine. Urinary carrier is rare, but more dangerous, because it is associated with anomalies like calculi or schistosomiasis.

S. **Paratyphi carrier:** It also occurs in paratyphoid bacilli.

Laboratory diagnosis of carrier
- **Diagnosis of fecal carrier:** It is done by repeated bile and stool culture. Collection after purgative gives more chance of isolation. Duodenal aspiration may be useful.
- **Diagnosis of urinary carrier:** It is done by repeated urine culture.
- **Diagnosis of recent carrier:** It is done by detecting Vi antigen for screening purpose.
- **Diagnosis of carrier in cities**
  - Sewer swab technique: Gauze pad left in sewers are cultured on highly selective media like Wilson and Blair medium. Positive swab one may reach to house harboring carrier.
  - Filtration technique: Filtration of sewage through millipore membrane followed by culture of membrane on highly selective media.

General Scheme to Diagnose Case and Carrier in Enteric Fever

Follow **Flowchart 62.5**.

Prevention

General measures: Improvements in sanitation and provision of protected water.

Immunoprophylaxis: Following types of vaccines are developed for *Salmonella* spp.
1. **TAB vaccine:** It is developed and tested successfully by Almroth Wright in Africa. It is **prepared** by heat

Flowchart 62.5: General scheme to diagnose case and carrier in enteric fever

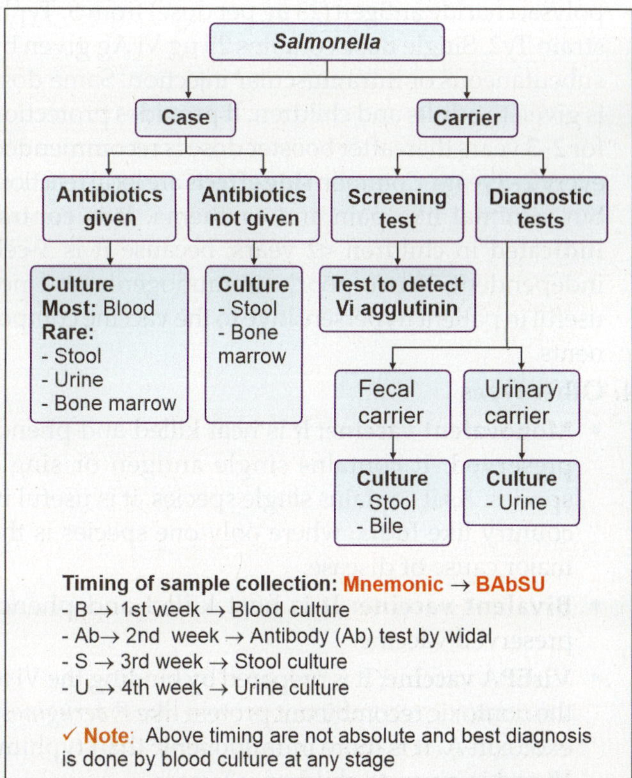

inactivation of typhoid bacillus at 50–60°C. It contains *S*. Typhi-1000 million and *S*. Paratyphi A and B, 750 million each per ml. It is **stored** at 0.5 % phenol. Two **doses** of 0.5 ml are given subcutaneously at interval of 4–6 weeks. Booster dose is given after 2–3 years. Common **side effects** are local tenderness, fever, malaise, etc. Reactions are overcome by injecting 0.1 ml intradermally. It is **used** in nonendemic areas for troops, medical and paramedical personnel and in endemic areas for children.

2. **Typhoral (live typhoid oral) vaccine:** It is **prepared** from stable mutant of *S*. Typhi strain Ty21a, lacking enzyme UDP-galactose-4-epimerase (Gal E mutant). This enzyme is responsible for production of lipopolysaccharide 'O' Ag which makes the strain virulent. Vaccine is enteric coated capsule contain 10^9 viable lyophilized mutant bacilli. **Dose** includes administration of one capsule orally before food with a glass of water or milk on day 1, 3, 5 and 7. No antibiotic should be taken during this period. On ingestion it initiates infection but 'self-destructs' due to lack of Gal E enzyme after four or five cell divisions, and therefore, cannot induce any illness. It starts protection after 1 week of last dose and immunity will last for 3–4 years. It provides 67–90% protection. Revaccination is given every 5 years with all four doses. Common **side effects** are diarrhea, abdominal pain or rashes. **Advantages** of typhoral vaccine are given by oral route, obviate the side effects by injection route, provide local and systemic immunity, safe, long acting and more convenient. It is **contraindicated** in kids with <6 years of age.

3. **Vi vaccine (typhim-Vi):** It is **prepared** by purified Vi polysaccharide antigen (25 µg per dose) from *S*. Typhi strain Ty2. Single **dose** contains 25 µg Vi Ag given by subcutaneous or intramuscular injection. Same dose is given to adults and children. It provides protection for 2–3 years thereafter booster dose is recommended every 2–3 years. Common **side effects** are local reaction, but minimal like pain and erythema. It is **contra-indicated** in children <2 years, because it is T-cell independent, hence poorly immunogenic and not useful in patient hypersensitive to the vaccine compo-nents.

4. **Other types**
   - **Monovalent vaccine:** It is heat killed and phenol preserved. It contains single antigen or single species. As it contains single species, it is useful in country like India, where only one species is the major cause of disease.
   - **Bivalent vaccine:** It is heat killed and phenol preserved vaccine.
   - **VirEPA vaccine:** It is prepared by binding the Vi to the nontoxic recombinant protein like *P. aeruginosa* exotoxin A. It is more immunogenic than typhim-Vi and is given to children <2 years.

Treatment

Treatment of case
- **MDR *Salmonella*:** Initially bacilli were sensitive to chloramphenicol, but plasmid mediated resistant strains were identified in 1972 from Mexico and Kerala (India). Resistant strain identified in Calicut, Kerala and later spread in other part of country. Ampicillin, amoxicillin, cotrimoxazole and furazolidone were sensitive to many drugs like ampicillin, amoxicillin, cotrimoxazole and furazolidone. Resistant develop to these drugs in late 1980 called multidrug resistant (MDR) *Salmonella*. Following drugs are used currently to treat MDR *Salmonella*.
  - Fluoroquinolones: Ciprofloxacin (drug of choice), ofloxacin or pefloxacin.
  - 3rd generation cephalosporins: Ceftriaxone, ceftazidime or cefotaxime.
  - Azithromycin: It is an alternative drug for empirical therapy.
- **Nalidixic acid resistant *Salmonella*:** Resistant also starts to develop against ciprofloxacin, which is detected by using nalidixic acid called nalidixic acid resistant *Salmonella*. It is either plasmid or chromosome mediated.
- **Resistant to ceftriaxone and ampicillin:** Resistant also developed against ceftriaxone, due to production of ESBL and also against ampicillin due to AmpC β-lactamase production.
- **Reversal of sensitivity:** *Salmonella* sensitive to newer drugs are now becomes resistant and instead becomes sensitive to older drugs.

Treatment of carrier: Combination of drug like ampicillin/amoxicillin and vaccine are given. They eliminate the carrier state and prevent the relapse. Surgery like cholecystectomy, pyelolithotomy or nephrectomy may be performed.

NONTYPHOIDAL GROUP

Salmonella Food Poisoning

Species: It is zoonosis, also called *Salmonella* gastro-enteritis or *Salmonella* enterocolitis caused by non-typhoidal *Salmonella* like *S*. Typhimurium (most common in the world), *S*. Enteritidis, *S*. Haldar, *S*. Anatum, *S*. Newport, *S*. Seftenberg, *S*. Heidelberg, *S*. Agona, *S*. Virchow, *S*. Indiana, etc.

Pathogenicity
- **Epidemiology:** It occurs in developed countries while enteric fever is a problem of developing countries.
- **Reservoirs of infection:** Humans and animals are the reservoir hosts for nontyphoidal group, unlike *S*. Typhi where only human cases/carriers are the reservoirs hosts. However, role of human carrier is minimal.
- **Sources of infection:** These are animal products like egg, meat, milk or milk-derivatives.

- **Modes of transmission:** Infection occurs due to ingestion of animal products especially egg and egg products. *Salmonella* can enter through the shell, if eggs left on contaminated chicken feed or feces, and grow inside. Infection also occurs due to contamination of salad/vegetables by manure or poor handling, food contamination by dropping of rat, lizard and other small animals and without food as cross infection in hospital.
- **Incubation period:** 16–24 hours.
- **Portal of entry and sites:** GIT.
- **Pathogenesis:** Nontyphoidal *Salmonella* has invasive property and not released any toxin.
- **Clinical features:** It presents with acute cholerae type diarrhea with passage of 1–2 loose stools, abdominal pain and vomiting. Acute stage subsides in 2–4 days or progresses to dysentery with presence of blood and mucus in stool. Typhoid or septicemic kind of fever develops in few days.

Laboratory diagnosis of food poisoning: It is based on isolation of bacilli from foods or feces.

Prevention: It is prevented by preventing food contamination during production, packaging, storage and marketing.

Treatment: In noninvasive cases symptoms are subsides in 2–4 days and antibiotics are not necessary. **Invasive cases** required antibiotics.

> **Note: DT 104 strain of *S. Typhimurium***
> Multidrug-resistant strain emerged in early 1990s and associated with an increased risk of bloodstream infection and hospitalization. It acquired by raw or partially cooked meat products.

Salmonella Septicemia

Species: *Salmonella* septicemia is caused by *S. Choleraesuis*.

Clinical features: Suppurative lesions such as endocarditis, meningitis, pneumonia, deep abscess, osteomyelitis and antecedent gastroenteritis may or may not present.

Laboratory diagnosis: It based on isolation of bacilli from pus, blood or feces.

ACCESS YOURSELF

Case Study

1. A pediatric patient visited the hospital with complain of step ladder type of fever since 10 days. Blood is collected for culture and serology. Culture report suggests GNB. Widal test is positive with high titer for H-Ag and O-Ag. Identify the organism and answer the following.
   a. Name the causative agent.
   b. Write the pathogenicity of causative agent.
   c. Describe the laboratory diagnosis of clinical condition.
   d. Write about prevention and treatment of clinical condition.

Essay/Full Question

1. *S.* Typhi.

Short Notes

1. Pathogenicity/Lab., diagnosis of enteric fever.
2. *Salmonella* food poisoning.
3. Widat test.

Short Questions for Theory/Viva Questions

1. What is typhoid merry?
2. Name the four salmonellas causing food poisoning.
3. Mention the reference centers for *Salmonella* phage typing and serotyping of human origin.

Comments on

1. Use of antacids favors the infection by *Salmonella*.

MCQs for Chapter Review

Enteric Fever/Typhoidal Group

1. **All of the following salmonellas are motile *except*:**
   a. *S.* Typhi
   b. *S.* Enteritidis
   c. *S.* Gallinarum
   d. *S.* Typhimurium
2. **Following are gas producing salmonellas *except*:**
   a. *S.* Typhi
   b. *S.* Enteritidis
   c. *S.* Choleraesuis
   d. *S.* Typhimurium
3. **Growth factor needed for *Salmonella*:**
   a. Tryptophan
   b. Niacin
   c. B-12
   d. Citrate
4. **There has been an outbreak of foodborne *Salmonella* gastroenteritis in the community and the stool samples have been received in the laboratory. Which is the enrichment medium of choice?**
   a. Cary Blair medium
   b. VR medium
   c. Selenite F medium
   d. Thioglycolate medium
5. **Agglutination with O antigen of *S.* Typhi is inhibited by:**
   a. Vi antigen
   b. Pilli antigen
   c. Flagellar antigen
   d. All of the above
6. **True statement about Widal test in typhoid is:**
   a. O-antigen titer remains positive for several months and reaction to it is rapid
   b. H-antigen titer remains positive for several months and reaction to it is rapid
   c. Both remain positive for several months and reaction to both is rapid
   d. None
7. **Most immunogenic in typhoid:**
   a. O antigen
   b. H antigen
   c. Vi antigen
   d. Somatic antigen
8. **Vi antigen found in:**
   a. *S.* Paratyphi A
   b. *S.* Paratyphi C
   c. *S.* Dublin
   d. *Klebsiella pneumoniae*
   e. *Citrobacter freundii*
9. **A 24-year-old cook in a hostel mess suffered from enteric fever 2 years back. The chronic carrier state in this patient can be diagnosed by:**
   a. Vi agglutination test
   b. Blood culture in brain heart infusion broth
   c. Widal test
   d. C-reactive protein

Infections of Enterobacterales-II (Salmonellas)

10. Which is true about Widal reaction?
   a. Antibody to H-Ag appears first and persists
   b. Antibody to O-Ag appears first and persists
   c. Antibodies to O and H-Ag appear simultaneously and persist
   d. None of the above

11. Pea-soop stool is characteristically seen in:
   a. Cholera
   b. Typhoid
   c. Botulism
   d. Polio in immunocompromised host

12. All of the followings are true regarding typhoid *except*:
   a. Urinary carriers are more dangerous
   b. Vi Ab is used for detecting carrier
   c. Vi is seen in normal population
   d. Urine carrier is associated with anomalies

13. Diagnosis of typhoid in 1st week is done by:
   a. Widal test   b. Stool culture
   c. Blood culture   d. Urine culture

14. In a patient with typhoid, diagnosis after 15 days of onset of fever is best done by:
   a. Blood culture   b. Widal
   c. Stool culture   d. Urine culture

15. Drugs commonly used against enteric fever are all *except*:
   a. Amikacin   b. Ciprofloxacin
   c. Ceftriaxone   d. Azithromycin

16. True about salmonellosis:
   a. Decreased incidence in developed countries
   b. Antacid and prolonged antibiotic administration promote infection
   c. Always fatal
   d. Foodborne to man and animal

17. True about typhoid:
   a. It is caused by food poisoning.
   b. Water can transmits the disease.
   c. Ty21a is an oral vaccine.
   d. Chronic carrier called when transmits up to 6 months.
   e. Widal test positive in 1st week

18. True about *Salmonella* Typhi infection in intestine are:
   a. Affects Peyer's patches
   b. Common in mesenteric border
   c. Erythrophagocytosis is the characteristics
   d. Strictures are common
   e. Typhoid ulcer always bleed very common

19. True about maximum isolation period of enteric fever:
   a. Till three consecutive negative urine/stool culture samples are obtained from patient
   b. After chloramphenicol treatment for 72 hours
   c. Disappearance of fever
   d. Widal test negative

20. For typhoid endemic country like India, Immunization of choice is:
   a. TAB vaccine
   b. Typhod 21A oral vaccine
   c. Monovalent vaccine
   d. Any of these

21. About Vi polysaccharide vaccine true is:
   a. Can be given in patient with yellow fever and hepatitis B
   b. Has many contraindications
   c. Has many serious systemic side effects
   d. Has many serious local side effects

22. All of the following statements about non-typhoid *Salmonella* are true *except*:
   a. Humans are the only reservoirs.
   b. Transmission is most commonly associated with eggs, poultry and undercooked meat.
   c. Common in immunocompromised individual
   d. Resistance to fluoroquinolones has emerged.

23. Typhoid fever is caused by:
   a. *S.* Typhi   b. *S.* Paratyphi
   c. a + c   d. None of above

Nontyphoidal Group

24. Food poisoning is produced by:
   a. *S.* Typhi   b. *S.* Paratyphi
   c. a + b   d. *S.* Typhimurium

25. *Salmonella* gastroenteritis is:
   a. Mainly diagnosed by serology
   b. Blood and mucus present in stool
   c. Caused by animal products
   d. Symptoms appear by 4–48 hours
   e. Features are due to exotoxin released

26. DT 104 strain is belong to which of the following bacteria?
   a. *Salmonella* Gallinarum   b. *Salmonella* Typhi
   c. *Salmonella* Enteritidis d. *Salmonella* Paratyphi A
   e. *Salmonella* Typhimurium

27. Microorganisms that can enter freshly laid eggs are:
   a. *Salmonella*   b. *Brucella*
   c. *Shigella*   d. *Vibrio cholerae*

28. Prolonged *Salmonella* septicemia is caused by:
   a. *S.* Enteritidis   b. *S.* Cholerasuis
   c. *S.* Typhimurium   d. *S.* Typhi

Answers and Explanation of MCQs

1. c
- All salmonellas are motile with peritrichate flagella, *except* S. Gallinarum and S. Pullorum.

2. a
- *S.* Typhi is anaerogenic, it ferments the sugars with acid only while other can ferment acid with gas.

3. a
- *S.* Typhi and few other salmonellae do not grow on Simmon's citrate medium as they required tryptophan as a growth factor.

4. c
- Follow section, **culture characteristics (C/Cs)** for explanation.

5. a

6. b

7. b

8. b, c, e

9. a

10. a
- Follow section, **Enteric fever/typhoidal group → antigenic structure** for explanation of answers of MCQs 5–10.

11. b
- Follow section, **Enteric fever/typhoidal group (typhoid fever → clinical features)** for explanation.

12. c
- *Salmonella* carrier never presents in normal population and occurs only after infection.
- Fecal carrier is common but less risky.
- Urinary carrier is rare but more dangerous because it is associated with anomalies like calculi or schistosomiasis.

13. c

14. c
- Follow section, **enteric fever/typhoidal group (General scheme to diagnose case and carrier in enteric fever → Flowchart 62.5)** for explanation of answers of MCQs 13–14.

15. a
- Follow section, **enteric fever/typhoidal group (treatment → treatment of case)** for explanation.

16. a, b and d
- Salmonellosis is more common in rural areas than urban. Typhoid fever is eliminated (decreased incidence) from the developed countries due to improved water supply and also improvement in hygiene. Bacilli are sensitive to gastric acid. Low acidity due to achlorhydria and use of antacids can enhance the infection. Suppression of intestinal flora by antibiotics can promote the infection. It is not always fatal, case fatality rate is <1. It is transmitted by contaminated food-water to humans and animals.

17. b and c
- *S.* Typhi is transmitted by contaminated food and water to humans and animals but food poisoning species are not causing typhoid.
- Ty21a strain used to prepare the typhoral (live typhoid oral) vaccine of *S* Typhi.
- Chronic carrier called when transmits the infection for >1 year.
- Widal test positive in 2nd week.

18. a
- *S.* Typhi affects Peyer's patches which lie along the antimesenteric border. Erythrophagocytosis is not seen in typhoid. It is the characteristics of *E. histolytica*.
- Strictures are uncommon, but common in tuberculosis.
- Typhoid ulcer bleed very rare.

19. a
- Maximum isolation period of enteric fever: Till three consecutive negative urine/stool culture samples collected on three separate days.

20. a
- Typhoid endemic country like India, single species like *S.* Typhi is responsible for disease, so monovalent vaccine is indicated.

21. a
- Vi polysaccharide vaccine is contraindicated in children <5 years of age and in allergic patient. No other contraindication so can be given in patient with yellow fever and hepatitis B.

22. a
- Humans and animals are the reservoir hosts for non typhoidal group, unlike *S.* Typhi where only human cases/carriers are the reservoir hosts.

23. a
- Enteric fever/typhoidal group include typhoid fever caused by *S.* Typhi and paratyphoid fever caused by *S.* Paratyphi A, B and C.

24. d

25. b, c

26. e
- Follow section, **nontyphoidal group (*Salmonella* food poisoning)** for explanation of answers of MCQs 24–26.

27. a
- Follow section, **nontyphoidal group (*Salmonella* food poisoning → mode of transmission)** for explanation.

28. b
- Follow section, **nontyphoidal group (*Salmonella* septicemia)** for explanation.

A. Agent factors (virulence factors): These are two types like intracellular (cell wall associated) and extracellular.

1. **Intracellular (cell wall associated) factors**
   - **Capsular antigen:** It presents in some strains. It masks the O-Ag and inhibits agglutination by antiserum. It inhibits the phagocytosis.
   - **Somatic antigen (O Ag or LPS or endotoxin):** It presents in large number, causes diarrhea and ulcer. LPS is responsible for endotoxic activities.
   - **Fimbriae:** These are the organs of adhesion.
   - **Invasive property:** Invasive property is due to a special type of OMP called VMA (virulence marker antigen), which is plasmid encoded, responsible for invasion, multiplication of bacilli and destruction of epithelial cells. It is detected by VMA ELISA or by culture cells like Hela or Hep-2 cells.
   - **Cross reactive antigens:** There is antigenic sharing between *Shigella*, *Salmonella* and *E. coli*.
   - **Infective dose:** *Shigella* resists gastric acidity and few bacilli can initiate the disease. Infective dose is very low, about 10–100 bacilli.
2. **Extracellular factors:** *Shigella* produces following two types of exotoxins called enterotoxins.
   - **Verotoxin/verocytotoxin (sub type → Shiga toxin):** It is produced by *Sh. dysenteriae* type-1. It is less important in pathogenesis, because nontoxigenic strain can invade the mucosa and multiply. It is the earliest exotoxin identified from GNB and chromosome mediated. Methods of detection and biological actions (neurotoxicity, cytotoxicity and enterotoxicity) of Shiga toxin are same as that of shiga like toxin (SLT) and described in **Ch. 61** under the section of agent (virulence) factors of *E coli*.
   - ***Shigella* enterocytotoxin (ShET):** It has following two forms. Both types allow the collection of fluid in gut (in ligated rabbit ileal loop).
     - *Shigella* enterocytotoxin 1 (ShET 1): It is secreted from *Sh. flexneri* 2a, structurally similar to cholera toxin and chromosome mediated.
     - *Shigella* enterocytotoxin 2 (ShET 2): It is secreted from all strains of *Sh. flexneri*. It helps in iron uptake and it is plasmid mediated.

B. Host factors

1. **Age:** It is common in young children due to poor hygiene; however, endemic shigellosis found in all ages.
2. **Immune status:** Bacilli last longer in patients with malnutrition and IDDs like AIDS but disappeared within few weeks from the feces of normal individual.
3. **Poverty, lack of sanitation and mental status:** Also favor the shigellosis.

C. Environmental factors: It is common in developing and poor countries, because of poor sanitation. It is common in mental hospitals.

Pathogenesis: Follow **Flowchart 63.1**.

Clinical types: Whole spectrum of disease called shigellosis, which includes following clinical conditions.

1. **Bacillary dysentery (invasive diarrhea/bloody diarrhea):** It presents with increase frequency of stool contains blood and mucus along with abdominal cramps and tenesmus. General features like fever and vomiting are also present. Other bacteria causing dysentery are *Cl. difficile*, EIEC, EHEC, *Y enterocolitica*, *C. jejuni*, *V. parahaemolyticus*, *Aeromonas* spp., and may by few nontyphoidal *Salmonella* spp.
2. **Bacteremia:** Bacilli enter in to the blood from the intestinal ulcer and causing bacteremia.
3. **Gay bowel syndrome:** It occurs in male homosexual.

Complications

1. **By *Sh. dysentriae* type 1:** It produces arthritis, toxic neuritis, conjunctivitis, parotitis, in children intussusception and in severe cases, it causes hemolytic uremic syndrome. Other agents causing hemolytic uremic syndrome are EHEC, *Salmonella*, *Aeromonas* spp., and *C. jejuni* and few drugs like cyclosporine, quinine and clopidogrel.
2. **Autoimmune reaction (Reiter's syndrome):** It mostly seen by *Sh. flexneri* and occurs in 3% cases. It develops in patient showing HLA B27. It is the most common organism causing reactive arthritis following diarrhea. More details of Reiter's syndrome is given in **Ch. 52**.
3. **Ekiri syndrome:** It is a toxic encephalopathy characterized by cerebral edema, seizures, altered consciousness and abnormal posture.

Laboratory Diagnosis

Specimens: Fresh stool is collected in transport media as mentioned above and sent to laboratory for testing.

Testing methods

A. Microscopy: Follow morphology.
B. Culture: Follow C/Cs.

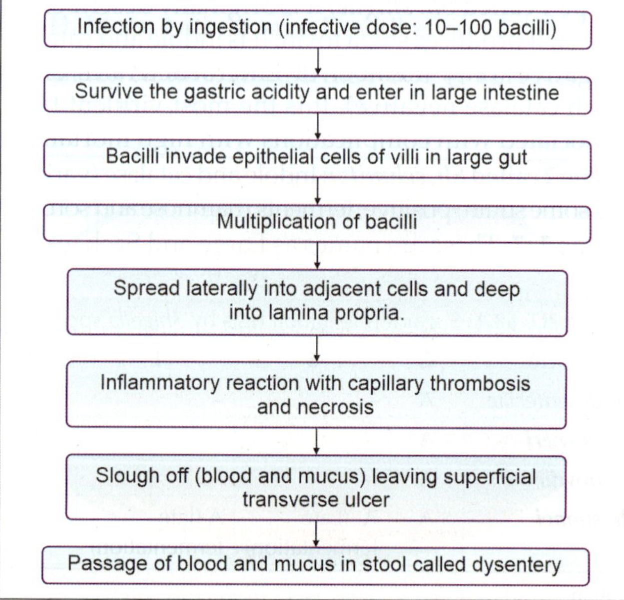

Flowchart 63.1: Pathogenesis of *Shigella*

C. Biochemical reaction: Follow B/Rs.

D. Serological tests: Agglutination by using polyvalent and monovalent antisera.

E. Bacteriocin typing: It is also useful.

Prevention

It is prevented by improving personal and environmental sanitation. Chemoprophylaxis and immunoprophylaxis have no role.

Treatment

Uncomplicated cases are self-limited but dehydration has to be corrected by ORS (Oral Rehydration Salt) therapy. Multiple plasmid mediated drug resistant *Shigella* are widely prevalent. Many strains are sensitive to nalidixic acid, norfloxacin and other flouroquinolones.

ACCESS YOURSELF

Case Study

1. A 10-year-old child admitted in hospital with complains of fever, abdominal cramps and diarrhea with presence of blood and mucus in stool. Stool culture identified the NLF bacteria on MacConkey's medium. Identify the disease and answer the following.
   a. Name the causative agent.
   b. Write the pathogenicity of causative agent.
   c. Describe the laboratory diagnosis of causative agent.

Short Note

1. Shigellosis.

Short Questions for Theory/Viva Questions

1. Name the species of *Shigella*.
2. Write the sugar fermentation tests in *Shigella* species.
3. Name the four bacteria causing dysentery.
4. What is gay bowel syndrome?
5. Name the four bacteria that cause hemolytic uremic syndrome.

MCQs for Chapter Review

1. ***Shigella* can be differentiated from *E. coli* by all of the following features *except*:**
   a. *Shigella* does not produce gas from glucose
   b. *Shigella* does not ferment lactose
   c. *Shigella* does not ferment mannitol
   d. *Shigella* has no flagella, is nonmotile
2. ***Shigella* can be divided into subgroup on the basis of ability to ferment:**
   a. Lactose
   b. Maltose
   c. Fructose
   d. Mannitol
3. **Nonlactose fermenters include all of the following *except*:**
   a. *Sh. dysenteriae*
   b. *Sh. sonnei*
   c. *Sh. flexneri*
   d. *Sh. boydii*
4. **Which of the following *Shigella* species is mannitol non-fermenter?**
   a. *Sh. sonnei*
   b. *Sh. boydii*
   c. *Sh. dysenteriae*
   d. *Sh. flexneri*
5. **Most virulent species of *Shigella* is:**
   a. *Sh. dysenteriae*
   b. *Sh. sonnei*
   c. *Sh. flexneri*
   d. *Sh. boydii*
6. **All are true about *Shigella* *except*:**
   a. Large dose is required for infection
   b. Associated with hemolytic uremic syndrome
   c. Causes bloody diarrhea with mucus
   d. Gut pathology is due to toxin
7. **Which of the following statement regarding *Shigella dysenteriae* types I is true?**
   a. It can lead to hemolytic uremic syndrome
   b. It produces an invasive enterotoxin
   c. It is facultative aerobe
   d. It is MR negative
8. **Bacteria causing bacillary dysentery is/are:**
   a. *Shigella* spp.
   b. *C. jejuni*
   c. *V. parahaemolyticus*
   d. All of above
9. **Shigellosis is best diagnosed by:**
   a. Stool examination
   b. Stool culture
   c. Sigmoidoscopy
   d. Enzyme
10. **All of the following cause hemolytic uremic syndrome:**
    a. *Shigella*
    b. *Campylobacter*
    c. EHEC
    d. *Vibrio cholerae*
11. **Hemolytic uremic syndrome is caused by:**
    a. EIEC
    b. *Shigella*
    c. *Salmonella*
    d. *Vibrio cholerae*
    e. *Klebsiella*

Answers and Explanation of MCQs

1. c
- Mannitol fermentation is useful in classification of *Shigella* as shown in **Table 63.1**. *Sh. dysenteriae* does not ferment mannitol, while remaining three species can ferment the mannitol with acid without gas production.
- *E. coli* is motile by peritrichous flagella and ferments glucose and lactose with acid and gas. *Shigella* is nonmotile and all species can ferment glucose with acid only, while lactose is fermented late by *Sh. sonnei* with acid only.

2. d
- Mannitol fermentation is useful in classification of *Shigella* as shown in **Table 63.1**. *Sh. dysenteriae* does not ferments mannitol while remaining three species can ferment the mannitol with acid without gas production.

3. b
- All *Shigella* species are nonlactose fermenters except *Sh. sonnei* which is late fermenter.

4. c
- Follow **Table 63.1** for explanation.

5. a
- *Sh. dysenteriae* is the most virulent, associated with complications (vide infra) and high mortality.

6. a and d
- Large dose is not required for infection, 10–100 bacilli (called infective dose) are enough to elicit the infection.
- Associated with hemolytic uremic syndrome and can cause bloody diarrhea with mucus.
- Toxin is not required for gut pathology, even nontoxigenic strain can invade and multiply in the enterocytes.

7. a
- *Sh. dysenteriae* types I can lead to hemolytic uremic syndrome. It produces an enterotoxin, which is noninvasive. It is aerobe and facultative aerobe. It is MR positive.

8. d
- Follow section, **pathogenicity (clinical types → bacillary dysentery)** for explanation.

9. b
- Follow section, **culture characteristics** for explanation.

10. d

11. b, c
- Follow section, **pathogenicity (complications)** for explanation of answers of MCQs 10–11.

Infections of Enterobacterales-IV (Yersinias)

INTRODUCTION

Meaning

Genus named from Alexander Yersin who discovered the bacilli along with Kitasato in 1894 from Hong Kong at the beginning of last epidemic; however, Yersin bacilli word is not used for all species of genus, but only for *Yersinia pestis*.

Classification

Follow **Flowchart 64.1**.

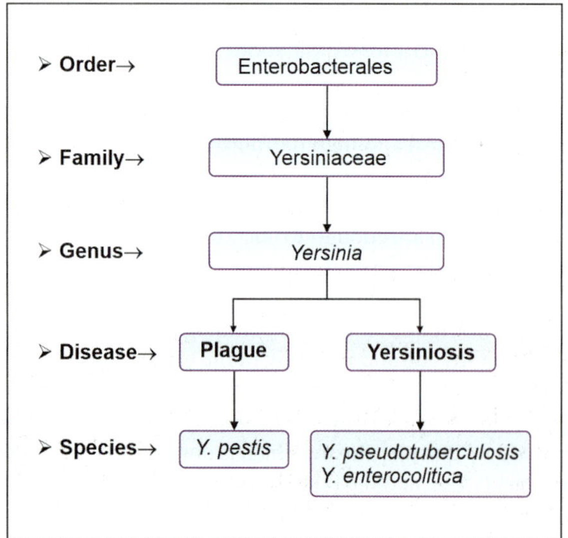

Flowchart 64.1: Classification of *Yersinia*

Yersinia pestis

Common Name

Yersinia pestis is also known as plague bacillus, as the disease produced by it or Yersin bacillus, from the Alexander Yersin.

Use

Yersinia pestis is the agent to be used in bioterrorism, especially in nonendemic regions.

Morphology

Type according to Gram's stain: They are GNB or coccobacilli and also showing bipolar staining (**Fig. 64.1**).

Giemsa stain or methylene blue or Wayson stain (Fig. 64.2) or Wright's stain: They are showing bipolar staining called safety pin or telephone handle appearance with the two ends densely stained and the clear central area. It is more in smear of tissue than smear of culture.

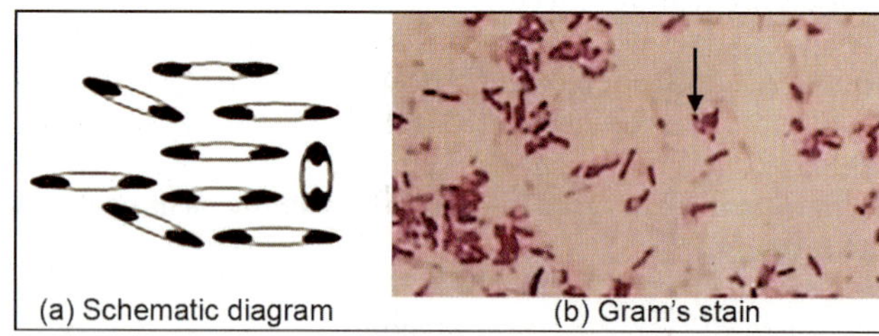

Fig. 64.1: Bipolar staining (safety pin appearance) of *Y. pestis* under Gram's stain

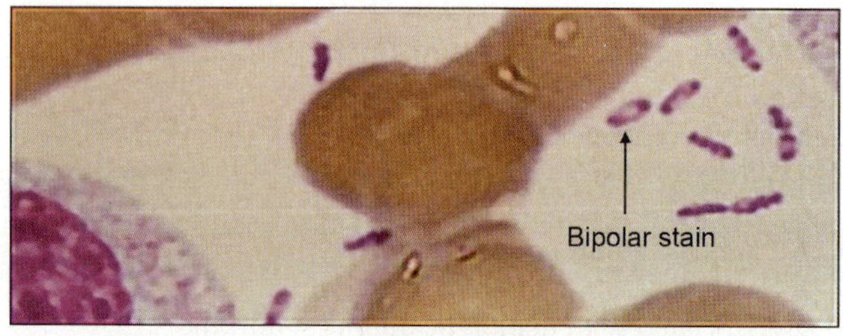

Fig. 64.2: Bipolar staining (safety pin appearance) of *Y. pestis* under Wayson stain

Shape and size: They are short, plump and oval with rounded ends and convex sides with 1.5 µm × 0.7 µm in size.

Arrangement: They arranged singly, in small groups and in short chains in liquid medium.

Slime layer (envelop): It is best developed at 37°C, but rare at optimum temperature of bacterium which is 27°C.

Motility and spores: They are nonmotile and non-sporing.

Pleomorphism and involution form: Aging culture shows great variation in size and shape like enlarged, elongated, irregular, pear-shaped or globular like yeast cells called pleomorphism. It is marked in 3% NaCl called involution form and used for identification of bacilli. It is due to defective cell wall synthesis or autolysis.

Cultural Characteristics (C/Cs)

Effective factors

- O_2 effect: They are aerobes and facultative anaerobes, but sensitive to O_2, which is overcome by anaerobiosis.
- Temperature: Temperature range is 4–45°C, optimum temp., is 27°C and slime layer develops at 37°C.
- pH: pH range is 5.0–9.6 and optimum pH is 7.2.

Culture in media

A. Liquid media

- **Nutrient broth:** Flocculent growth occurs at the bottom and along the sides of the tube, with little or no turbidity. A delicate pellicle may form later.
- **Oil/ghee broth in flask:** Medium contains oil/ghee at top. A characteristic growth occurs which hangs down into broth from the surface called 'stalactite growth' or called 'stalactites' as shown in **Fig. 64.3**.

B. Solid media

- **Nutrient agar:** Colonies are small, delicate, transparent discs, becoming opaque and increase in size on further incubation.

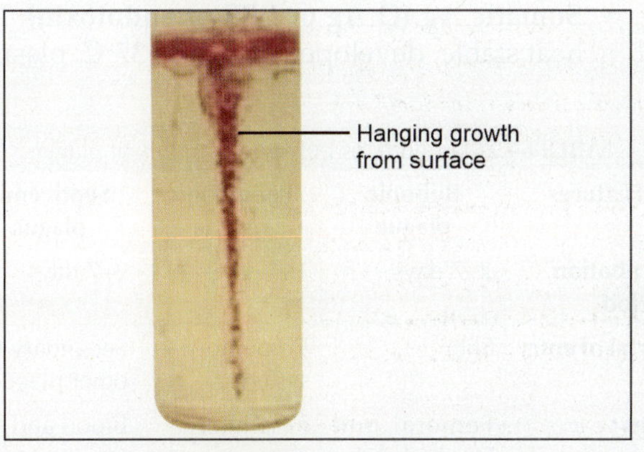

Fig. 64.3: Stalactite growth of *Y. pestis*

- **MacConkey's agar:** Pale/colorless colonies develop due to NLF that disappear after longer time due to autolysis.
- **Blood agar:** Nonhemolytic dark brown colored colonies develop due to absorption of hemin.
- **Hemin containing media:** Bacilli absorb the hemin and produce the dark brown pigmented colonies in blood agar and other hemin containing media. Pigment production is essential for biofilm formation and flea blocking.

C. **Selective media:** *Yersinia*-specific CIN agar is a selective medium for sputum and contains **c**efsulodin, **i**rgasan and **n**ovobiocin.

- **Automated culture:** Like MALDI-TOF is used to identify the species from culture.
- **Culture in animals:** Inoculate the exudates from bubo or culture suspension in guinea pig or in white rat. Development of edema, necrosis at injection site, enlargement of regional nodes, spleen and animal dies in 2–5 days. Take samples from local sites, nodes, spleen or heart blood and test samples with microscopy, culture or by B/Rs.

Biochemical Reactions [B/Rs]

Sugar fermentation tests

G	S	L	M	Maltose	Rhamnose
A	–	–	A	A	–

I M Vi C tests: – + – –

Catalase, nitrate reduction, aesculin hydrolysis, ONPG (β-galactosidase) and ornithine decarboxylation tests: Positive

Oxidase, urease, gelatin liquefaction and H_2S production tests: Negative

Biotyping: It is given by **Devignat**. There are three biotypes of *Y. pestis* based on glycerol fermentation and nitrate reduction as mentioned in **Table 64.1**.

Resistance

Sterilization: *Y. pestis* is killed by heat (at 55°C), sunlight and drying.

Disinfection: Bacilli are killed by 0.5% phenol in 15 minutes.

Drug resistance: Bacilli produce plasmid mediated resistance to many antibiotics.

Pathogenicity

Disease name: It is a zoonosis called 'plague'.

Table 64.1: Biotypes of *Y. pestis*		
Biotype	**Glycerol fermentation**	**Nitrate reduction**
Y. pestis var *antigua*	+	+
Y. pestis var *medievalis*	+	–
Y. pestis var *orientalis*	–	+

Epidemiology: It occurs in endemic, localized outbreak, epidemic or in pandemic forms. Plague is very ancient human disease. Central Asia or Himalaya is called the home land of human plague, from where it spread and caused epidemic and pandemic.

- **Plague scenario in the world:** Three pandemic reported (600 years interval) in the history with different biotypes. **(1) The first pandemic:** It called Justinian plague caused by biotypes *Y. pestis* var *antigua*. It was occurred between AD 541 and 544 (6th century) in Mediterranean countries. It began in Egypt and spread to Middle East and Mediterranean Europe. It killed 100 million people. After that 2nd-11th epidemic occurred from AD 558 to 664 in 8–12 years cycles. Finally, all known world affected by plague. **(2) The second pandemic:** It called black death, because it was present with gangrene of skin-fingers-penis and killed quarters of its victims. It was caused by biotypes: *Y. pestis* var *medievalis*. It occurred between AD 1347 and 1351 (14th century) in Europe. It began in Black sea and spread to Europe and West Russia. It killed 17–28 million people. After that epidemic continue from AD 1361 to 1480 in 2–5 years cycles. It also affected the all known world. **(3) The third pandemic:** It called modern plague. It is caused by biotypes *Y. pestis* var *orientalis*. It occurred between AD 1855 and 1918 (19th century) in Asia and in other countries. It began in Yunnan, a South Western province of China in 1855 and spread to Hong Kong in May 1894, Mumbai 1896, Kolkata 1898, Madagaskar 1898; Egypt, Portugal, Japan, Paraguay and Eastern Africa in 1899; Manila, Glasgow, Sydney and San Francisco in 1900. Local epidemics occurred subsequently throughout the world with plenty of victims. Finally whole world affected by plague and disease receded by control of rodents and vectors.
- **Plague scenario in India: (1) China pandemic (1855–1918):** It entered in India in 1896 (Bombay) and spread all over the country with 10 million death by 1918. It gradually declines thereafter; however, few scattered cases were reported till 1967. No cases reported in India from 1967 to 1994. **(2) Maharashtra outbreak (1994) of bubonic plague:** It occurred in Beed-Latur district of Maharashtra in August 1994. **(3) Surat epidemic (1994) of pneumonic plague:** It occurred in Surat and adjoining regions of Gujarat and Maharashtra in September in 1994. Around 6000 cases were reported with 60 deaths over a period of 2 months. **(4) Shimla outbreak (2002):** It occurred near Rohru of Simla (Himachal Pradesh) in February 2002 with 4 deaths. **(5) Uttaranchal outbreak (2004) of bubonic plague:** It occurred in Dangud village of Uttarkashi district of Uttaranchal in 2004 with 8 cases and four deaths. **(6) Indian foci of plague:** Total four foci of plague are known in India. **1st** is region near Kolar at the trijunction of Tamil Nadu, Karnataka and Andhra Pradesh, **2nd** is Beed-Latur

district of Maharashtra from where Surat epidemic was originated, **3rd** is Rohru in Himachal Pradesh and **4th** is Dangud village of Uttaranchal.

Reservoirs of infection: Different types of rodents are the main reservoirs as described below.

Sources of infection: These are fleas. Man-to-man transmission can occur, especially in pneumonic plague.

Modes of transmission: Direct transmission occurs in bubonic or septicemic plague due to direct contact with infected animals (like rodents) during skinning and handling. It also acquired by contact of wounded skin (produced by flea bite) with feces of infected fleas. **Indirect (vehicle borne) transmission** occurs in bubonic or pneumonic plague. In bubonic plague it occurs via flea bite (vector borne). Blocked flea cannot suck the blood because the bacterial masses block the way mechanically, but instead regurgitated the contents of proventriculus (blood mixed bacteria) into the bite wound thus transmitting the infection. In pneumonic plague transmission occurs via airborne route.

Incubation period: Extrinsic incubation period is usually 2 weeks. Intrinsic incubation period (incubation period) is mentioned in **Table 64.2**.

Portal of entry and sites: Follow **Table 64.2**.

Precipitating factors (epidemiological determinants): Following are the five types.

A. **Agent factors (virulence factors):** Plague bacilli are antigenically homogeneous, so it is very difficult to discriminate between virulent and avirulent strain and to do serotyping. Antigenic structure is complex. 20 antigens have been detected by gel diffusion and biochemical analysis. Agent factors are of two types like intra cellular (cell wall associated) and extracellular as described below.

1. **Intracellular (cell wall associated) factors**
   - **pH6 adhesin (ph6 antigen):** It helps in adhesion.
   - **Slime layer/capsular/envelope Ag (fraction 1 or F1 Ag):** It is heat labile, protein in nature, developed at 37°C, plasmid encoded, inhibits phagocytosis, necessary for full virulence, necessary for effective vaccine production and Ab to this antigen is protective in mice.
   - **Somatic Ag (O Ag or LPS or endotoxin):** It is heat stable, developed at 20 and 37°C, plasmid

TABLE 64.2: Differences between varieties of plague			
Features	**Bubonic plague**	**Pneumonic plague**	**Septicemic plague**
Incubation period	2–7 days	1–3 days	2–7 days
Portal of entry	Skin	Respiratory system	Secondary to other plague
Sites	Femoral and inguinal nodes	Respiratory system	Blood and meninges

encoded, inhibits phagocytosis and intracellular killing of bacilli. It has two subtypes like V and W, both are always produced together.

- **Unidentified surface components:** They absorb the hemin and aromatic dyes in medium to produce the pigmented colonies which help in biofilm formation and flea blocking.
- **Purin synthesis:** It also contributes in virulence.

2. **Extracellular factors**

- **Bacteriocin:** Virulent strains produce a pesticin-I which inhibits the strains of *Y. pseudotuberculosis*, *Y. enterocolitica* and *E. coli*.
- **Enzymes:** *Y. pestis* produces protease, coagulase and fibrinolysin. Protease degrades the complement, activates the mammalian plasminogen and encoded by *pla* gene. It adheres to the extracellular matrix component laminin, thus promoting the dissemination of bacteria.
- **Murine toxin:** So called because it is active in rat/mouse, but not in guinea pig or rabbit. It is heat labile protein and possessing properties of both exotoxins and endotoxins. It is toxoided, but do not diffuse free into the medium and released after cell lysis. On injection into experimental animals, it produces local edema and necrosis with systemic effects on peripheral vascular system and liver. Its role in disease in human beings is not known.

3. **Other factors**

- **Siderophore:** It helps in acquisition of iron.
- **Secretory system III:** It is the adhesin and also injects the F1 Ag in to host cells.

B. Host factors

- **Occupational factors:** Laboratory/hospital workers and persons concerned with skinning and handling of animal carcasses are at high risk.
- **Movement of people:** Risk increase to people who are moving to plague foci/region.

C. Environmental factors: Cool and humid season increase the multiplication of flea (high flea index) and rate of disease production.

D. Rodent factors: Death of diseased rat called rat fall. When diseased rat dies, flea leaves the carcasses of died rat and in the absence of other rat, it bites to human. Following different types of wild rodents are the main reservoirs for human infections.

- Desert rodent (Gerbils): *Tatera indica*
- Sewer rodent: *Rattus novergicus*
- Forest rodents (Bandicoot): *Bandicota indica*
- Domestic rodent: *Rattus rattus*. It is responsible only when numbers of other rodents are dwindled.

E. Vector (flea) factors

- **Species:** Human infection occurs mainly by rat fleas like *Xenopsylla cheopis* (in North India and more efficient vector), *X. astia* (in South India and less efficient vector) or *Ceratophyllus fasciatus*, etc., and rarely by human flea like *Pulex irritants*.

- **Blocked flea:** Plague bacilli are the natural parasites of rodents. Rodents are infected by bite of rat fleas and fleas acquired the infection from rodents during bite for blood meal (5000 bacilli per 0.5 ml blood). In the flea the bacilli multiply in stomach to such an extent that they block the proventriculus called blocked flea. Completely blocked flea dies, as it cannot obtain the blood meal while partially blocked flea called enzootic foci can survive longer up to 4 years inside the burrow. Bacilli continue multiply in flea and can infect the new rodents which may visit such burrow and eventually responsible for re-emergence of plague, hence partially blocked flea is more dangerous.
- **Flea index or cheopis index:** Average numbers of flea per rat called flea index. Plague outbreak likely occurs in place where flea index is more than 1.

Natural cycles of plague: Two cycles like urban (domestic) cycle which is continue between humans, rat fleas and domestic rodents and wild (sylvatic) cycle which occurs in nature between wild rodents and independent to human.

Clinical types and features of plague: Three major forms of plague like bubonic, pneumonic and septicemic.

1. **Bubonic plague:** Bubo means groin, because inguinal lymph nodes are commonly involved after the fleas bite. It is the common variety of plague. Depending upon the site of flea bite any node can be infected like cervical, axillary, submaxillary or femoral. Cervical or submaxillary nodes are commonly involved in children. It is characterized by intense painful swelling. Nodes become enlarged and suppurate as shown in **Fig. 64.4a**.

2. **Pneumonic plague:** It is characterized by hemorrhagic pneumonia. Cyanosis is very prominent. The bloody mucoid sputum that is coughed out contains bacilli in large numbers. It is highly infectious and in untreated patients, almost invariably fatal (30–100%).

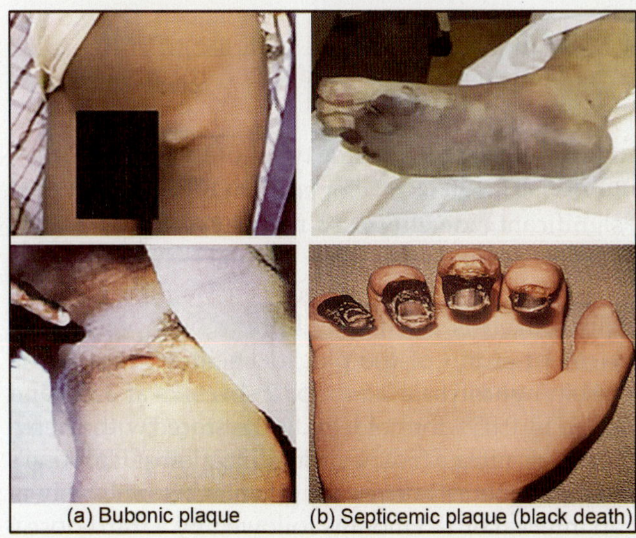

Fig. 64.4: Features of plague

3. Septicemic plague: It occurs secondary to bubonic or pneumonic plague or primarily by direct contact from laboratory infection. The bacilli enter the bloodstream from the bubo and produce septicemia, cutaneous and mucosal hemorrhage, DIC and gangrene of skin-fingers-penis as shown in **Fig. 64.4b** (hence the called black death to the pandemic of 14th century, killed a quarter of all mankind). Meningitis may occur rarely.

Laboratory Diagnosis

Diagnosis in Human

Specimens: Discharge from bubo, sputum or blood (three blood samples are collected over a period of 45 minutes).

Testing methods

A. **Microscopy:** Follow morphology.

B. **Culture:** Follow C/Cs.

C. **Biochemical reactions:** Follow B/Rs.

D. **Serological tests**
- **Passive hemagglutination tests:** Antibody to the F-I antigen may be detected by passive hemagglutination test. Rise in titer of antibodies in paired sera or titer of 128 or above in a single serum sample can be considered positive.
- **ELISA:** Tests developed for IgG and IgM.

E. **Molecular method:** PCR is the rapid and sensitive method for accurate diagnosis.

F. **Bacterial typing:** This includes bacteriophage typing and biotyping (described in B/Rs).

Diagnosis in Animals

Precautions: If rat is died of plague (rat fall) may carry infected fleas and it should be handled with care. Apply kerosene to remove ectoparasite. In laboratory carcass should be dipped in 3% lysol.

Specimens: From lymph nodes, spleen, heart blood or bone marrow.

Testing methods: These are same as human methods.

Prevention

General measures: These are control of fleas and rodents.

Immunoprophylaxis: Three types of vaccines.

1. **Live vaccine:** It is developed from EV76 strain. It has significant side effects, so not useful; however, it is used in Soviet Union.

2. **Killed vaccine:** It is **prepared** at the Haffkine institute, Mumbai. It is a whole culture antigen vaccine. A virulent strain of the plague bacillus is grown in casein hydrolysate broth for 2–4 weeks at 32°C and killed by 0.05% formaldehyde. It is **stored** with phenyl mercuric nitrate (Sokhey's modification of Haffkine's vaccine). Total 2 **doses** are given subcutaneously at an interval of 1–3 months followed by a third dose 6 months later. It gives some protection against bubonic

plague but not against pneumonic plague. Immunity does not last for more than 6 months. In contrast, an attack of plague provides more lasting immunity. Vaccine is **used** to occupationally risky persons, such as workers of plague laboratory or hospital. It has no value in plague outbreaks and mass vaccination is not advised. It is **contraindicated** in children <6 months.

3. **Subunit recombinant F1 (rF1):** Under trial.

Chemoprophylaxis: Tetracycline is the drug of choice given for 5 days. Cotrimoxazole is also useful with 5 days course.

Treatment

Streptomycin (drug of choice in normal form), doxycycline and chloramphenicol (drugs of choice in pneumonic plague, septicemic plague or meningitis) are effective. Patient should be isolated for 24 hours or until pneumonia should be ruled out by effective antibiotic therapy. Early treatment with antibiotics has reduced plague mortality from 30–100% to 5–10%.

YERSINIOSIS

It is a zoonotic infection with yersiniae other than *Y. pestis* like *Y. pseudotuberculosis* and *Y. enterocolitica*.

Yersinia pseudotuberculosis

Introduction: It causes tuberculosis like lesion in guinea pigs, rabbits and rodents hence the name pseudotuberculosis. Formerly, it was called *Pasteurella pseudotuberculosis*.

Morphology: They are GNB, oval in shape, small in size, noncapsulated and motile at 22°C (differential motility) but not at 37°C.

Cultural characteristics (C/Cs): They are aerobes and facultative anaerobes and optimum temperature for growth is 29°C. Following are different useful media.

1. **Nutrient agar:** Colonies are raised/umbonate, granular, 1 mm in size and translucent.
2. **MacConkey's agar:** Bacteria grow poorly.
3. **Blood agar:** Bacteria produce nonhemolytic colonies.
4. **Automated culture:** Like MALDI-TOF is used to identify the species from culture.

Biochemical reactions (B/Rs): Rhamnose and melibiose are fermented with acid production only. Catalase, nitrate reduction, ONPG (β-galactosidase) and urease tests are positive. Oxidase and ornithine decarboxylation tests are negative.

Pathogenicity

- **Virulence factors: (1) Adhesin proteins:** These are Inv (invasive) protein, ail (attachment and invasive locus) protein and Yad A (*Yersinia* adhesin A → it binds with collagen and fibronectin to aid the invasion of tissue by organism and also it inactivate the complement). **(2) Super Ag:** Some strains of *Y. pseudotuberculosis* express a superantigen, which causes scarlet like fever in Russia, similar condition in Japan called Izumi fever and idiopathic acute systemic vasculitis in children called Kawasaki's disease. **(3) Common serotypes:** Serotyping is done by Thal and Knapp, 1971. Antigenically, it is heterogeneous. Total 6 serotypes are identified based on heat stable O-Ag and heat labile H-Ag. Type-1 is most common human pathogen. Antigenic cross reaction occurs with *Y. pestis*. Serotypes 2 and 4 are cross react with *Salmonella's* Kauffmann-White O group B (Now 4 = *S.* Paratyphi B) and D (now 9 = *S.* Typhi).

- **Modes of transmission:** It is transmitted by direct contact of skin with water and by ingestion of food and vegetables contaminated by animal feces.
- **Clinical features: (1) In animals:** It causes tuberculosis like nodule in liver, spleen, lungs, etc. **(2) In human:** It causes acute mesenteric lymphadenitis simulating acute/subacute appendicitis (called pseudoappendicular syndrome) with typhoid like fever, gastroenteritis, hepatosplenomegaly and erythema nodosum.

Laboratory diagnosis: Tissues from mesenteric nodes, blood, etc., are tested by microscopy (follow morphology), culture (follow C/Cs) and biochemical reactions (follow B/Rs). Other tests are tube agglutination test and intradermal skin test which is similar to tuberculin or brucellin test.

Yersinia enterocolitica

Morphology: They are gram-negative coccobacilli with pleomorphism in old culture. They produce capsule *in vivo* but not in culture (*in vitro*). They are motile by peritrichous flagella at 22°C but not at 37°C (differential motility).

Cultural characteristics (C/Cs): They are aerobes and facultative anaerobes and optimum temperature for growth is 22–29°C. They can grow at 4°C (cold enrichment) in refrigerated food may cause food poisoning. Following are different useful media.

1. **Nutrient agar:** Colonies are smooth and translucent.
2. **MacConkey's agar:** Pin point pink colonies.
3. **Blood agar:** Nonhemolytic colonies.
4. **Selective medium:** Schiemann CIN medium is useful to isolate the bacilli from feces.
5. **Automated culture:** Like MALDI-TOF is used to identify the species from culture.

Biochemical reactions (B/Rs): They ferment sucrose and cellobiose with acid only. Indole, VP, catalase, nitrate reduction, ONPG (β-galactosidase), ornithine decarboxylation and urease tests are positive.

Pathogenicity

- **Virulence factor: (1) Enterotoxin:** It produces heat stable enterotoxin in refrigerated food (<30°C) causing food poisoning. **(2) Adhesin proteins:** These are different types like MyF Ag, pH6 Ag, inv (invasive) protein, ail (**a**ttachment and **i**nvasive **l**ocus)

protein and Yad A (*Yersinia* adhesin A → it binds with collagen and fibronectin to aid the invasion of tissue by organism and also it inactivates the complement). **(3) Common serotypes:** Total 60 serotypes are identified on the bases of O and H Ag. O3, O8, O9 are common human pathogens.

- **Sources of infection:** These are humans, animals, water or soil, so not a true zoonotic.
- **Modes of transmission:** Same as like *Y. pseudotuberculosis*.
- **Clinical features:** Bacteria produce 3 types of diseases in humans. **(1) In young children:** It causes self-limited gastroenteritis with (dysentery) or without blood. **(2) In older children:** It causes mesenteric lymphadenitis and terminal ileitis mimic appendicitis called pseudoappendicular syndrome, but more severe than previous one. **(3) In adults:** It causes septicemia with high fatality rate.
- **Complications:** Adult type causes polyarthritis, erythema nodosum and Reiter's syndrome.

Laboratory diagnosis: Tissues from mesenteric nodes, blood, stool, food water, soil, etc., are tested by microscopy (follow morphology), culture (follow C/Cs) and biochemical reactions (follow B/Rs). It has 6 biotypes and 10 bacteriophage types from I-X. Based on biotypes new species are noticed like *Y. frederikseni*, *Y. intermedia* and *Y. kristenseni*.

ACCESS YOURSELF

Essay/Full Question

1. *Y. pestis*.

Short Note

1. Pathogenicity or laboratory diagnosis of *Y. pestis*/Yersiniosis.

Short Questions for Theory/Viva Questions

1. What is stalactites?
2. Name the four foci of plague in India.
3. What are blocked flea and enzootic focus of plague?

Comments on

1. Partially blocked flea is more dangerous than completely blocked flea in plague.
2. Second pandemic of plague in 14th century was called black death.

MCQs for Chapter Review

Y. pestis

1. **True about Y. *pestis*:**
   a. Gram +ve   b. Gram –ve
   c. Motile   d. Nonmotile
   e. It is coccobacillus
2. **A farmer presents to the emergency department with painful inguinal lymphadenopathy and history of fever and flu like symptoms. Clinical examination reveals an ulcer in the leg. Which of the following stain should be used to detect suspected bipolar stained organisms?**
   a. Albert's stain   b. Wayson's stain
   c. Ziehl Neelsen stain   d. Mc Fadyean stain

3. **True statement about** *Y. pestis* **is/are:**
   a. Gram-positive
   b. Nonmotile
   c. Benzyl penicillin is given in prophylaxis
   d. Patients are kept isolated till 48 hours of treatment
   e. Repeated blood culture is diagnostic

4. **The drug of choice for chemoprophylaxis in contacts of patient of pneumonic plague:**
   a. Penicillin
   b. Rifampicin
   c. Erythromycine
   d. Tetracycline

5. **Which of the following drug(s) is/are used in treatment of plague?**
   a. Streptomycin
   b. Tetracycline
   c. Ciprofloxacin
   d. Chloramphenicol
   e. Cotrimoxazole

Yersiniosis

6. **True about yersiniosis:**
   a. Zoonosis
   b. Caused by *Y. pestis*
   c. By *Y. enterocolitica*
   d. By *Y. pseudotuberculosis*

Answers and Explanation of MCQs

1. b, d and e
- Follow section, *Yersinia pestis* **(morphology)** for explanation.

2. b
- Bipolar stained is done by Wayson's stain from the given options.

3. b, d and e
- *Y. pestis* is gram-negative and tetracycline is given in prophylaxis.

4. d
- Tetracycline is the drug of choice for chemoprophylaxis given for 5 days.

5. a and d
- Follow section, *Yersinia pestis* **(treatment)** for explanation.

6. a, c and d
- Follow section, **Yersiniosis** for explanation.

CHAPTER 65

Infections of Vibrionaceae, Aeromonadaceae and *Plesiomonas*

Chapter Outline

- Vibrionaceae
  - Non-halophilic vibrios
  - Halophilic vibrios
- Aeromonadaceae
- *Plesiomonas shigelloides*

VIBRIONACEAE

Vibrio word came from vibrare means vibrate, because of motile nature of bacteria. Different species of this family are mentioned in **Flowchart 65.1**.

Nonhalophilic Vibrios

Vibrio cholerae

Introduction

It also called *Vibrio comma* because of comma-shaped bacilli. It was 1st isolated by Robert Koch in 1883 in Egypt from cholera patient.

Morphology

Type according to Gram's stain: They are GNB (**Fig. 65.1a**).

Shape and size: They are short, curved, cylindrical rod with rounded or pointed ends typically comma shaped and 1.5 µm × 0.2–0.4 µm in size.

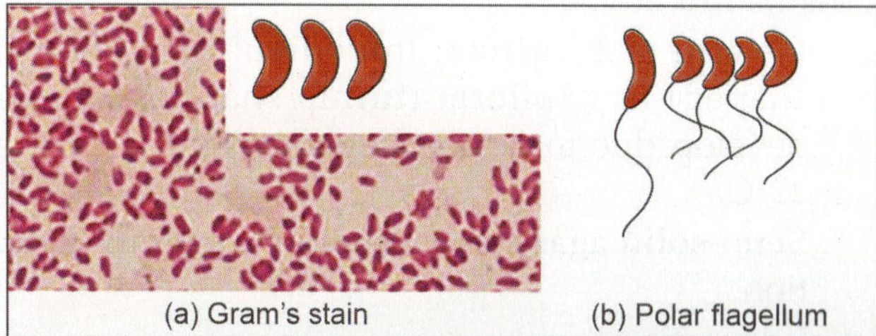

Fig. 65.1: Morphology of *V. cholerae*

Arrangement: They arranged singly or in group or parallel to each other giving 'fish in stream appearance' or end to end giving S-shape or spiral form appearance.

Motility: They are actively motile with single sheathed polar flagellum at one pole called monotrichate flagellum (**Fig. 65.1b**). Bacteria are showing darting motility under hanging drop preparation. Microscopy from young culture or from acute cholera stool reveals actively motile vibrios suggest 'swarm of gnats'.

Spores and capsule: They are nonsporing and non-capsulated. Bengal strain (O139) is capsulated.

Fimbria: It is the special organ for adhesion called 'toxin coregulated pilus (TCP)'. It colonizes the vibrios throughout the course of infection without damage or invasion to host cells.

Cultural Characteristics (C/Cs)

Effective factors

- O_2 effect: They are strict aerobes also grow anaerobically, but growth is slow and scanty.
- Temperature: Optimum temperature for growth is 37°C (16–40°C).
- pH: Optimum pH for growth is alkaline about 8.2 (6.4–9.6).
- 0.5–1% NaCl is required for optimal growth. High concentrations of NaCl (>6%) is inhibitory.

Flowchart 65.1: Classification of Vibrionaceae

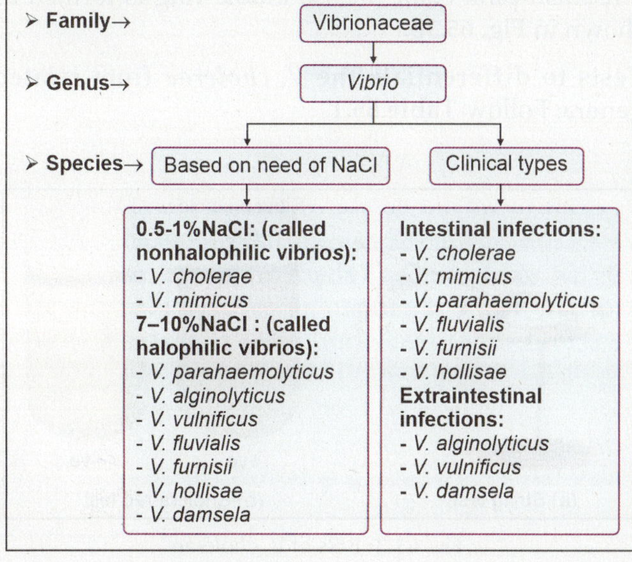

A. Liquid media
- **Ordinary/basal/simple media:** Surface pellicle occurs in peptone water within 6 hrs which breaks on shaking in membranous pieces. Turbidity and powdery deposit develop on continued incubation.
- **Enrichment media:** These are alkaline peptone water (APW) (pH–8.6) and Monsur's taurocholate tellurite peptone water (pH–9.2). They also serve as transport media.
- **Transport/holding media:** They maintain the viability of bacteria during transport, but do not allow the multiplication. Different media are Venkatraman-Ramakrishnan (VR) medium (pH 8.6–8.8), Cary-Blair transport medium (pH–8.4) and autoclaved sea water. VR medium prevents the overgrowth of commensals and bacteria remain viable for several weeks in VR medium. Cary-Blair transport medium is also useful for *Salmonella*, *Shigella* and *C. jejuni*.

B. Semisolid media
- **Gelatin stab culture:** Infundibuliform (funnel-shaped) or napiform (turnip-shaped) colonies develop due to gelatin liquefaction in 3 days at 22°C.
- **Semi-solid agar stab culture:** For motility detection.

C. Solid media
- **Nutrient agar:** Colonies are moist, translucent, 1–2 mm, round discs with bluish tinge in transmitted light, have distinctive odor.
- **Blood agar:** Colonies are α-hemolytic (green zone) initially, later becomes β-hemolytic (clear zone).

D. Selective media
- **Alkaline bile salt agar:** pH of medium is 8.2 and colonies are like nutrient agar.
- **MacConkey's agar:** Pale/colorless colony produced due to NLF **(Fig. 62.2a)**.
- **TCBS medium (thiosulphate citrate bile salts sucrose) medium:** Bromothymol blue is the pH indicator gives yellow colonies **(Fig. 65.2b)** at acidic pH due to sucrose fermentation.
- **Monsur's gelatin taurocholate trypticase tellurite medium:** Small translucent colonies occur with black center (black due to tellurite reduction), become large after 48 hours.

Automated culture: Like MALDI-TOF or VITEK is used to identify the species from culture.

Culture in animal: Numbers of animal models have been used to study the pathogenic mechanisms of cholera. The first of these was the rabbit ileal loop model of De and Chatterjee in 1953. Injection of ligated ileal loop caused fluid accumulation and ballooning.

Biochemical Reactions (B/Rs)

Sugar fermentation tests: Vibrios are oxidative and fermentative. They ferment maltose and rhamnose with acid only, but no inositol or arabinose. They may split lactose very slowly.

G	S	L	M
A	A	–	A

I M Vi C tests: + – V –

Catalase, nitrate reduction, oxidase, gelatin liquefaction, lysine decarboxylation and ornithine decarboxylation tests: Positive.

Urease and arginine decarboxylation tests: Negative.

TSI test: A/A – H_2S.

String test: Mix a loopful suspension of *V. cholerae* and a drop of 0.5% sodium deoxycholate in saline on a slide. Suspension loses its turbidity, becomes mucoid and forms 'string' between loop and mixture when loop is withdrawn slowly away from mixture as shown in **Fig. 65.3a.**

Hemolytic reaction: Mixed equal volume of broth culture and 1% sheep RBCs. Incubate for 2 hours at 37°C, than kept overnight in refrigerator at 4°C. Examine for hemolysis. Classical vibrios are nonhemolytic, while El Tor vibrios are hemolytic.

Cholera red reaction: Indole and nitrates tests are positive by *V. cholerae*. Add few drops of concentrated H_2SO_4 to 24-hour peptone water culture. With *V. cholerae* a reddish pink color nitroso-indole ring is formed as shown in **Fig. 65.3b.**

Tests to differentiate the *V. cholerae* from related genera: Follow **Table 65.1.**

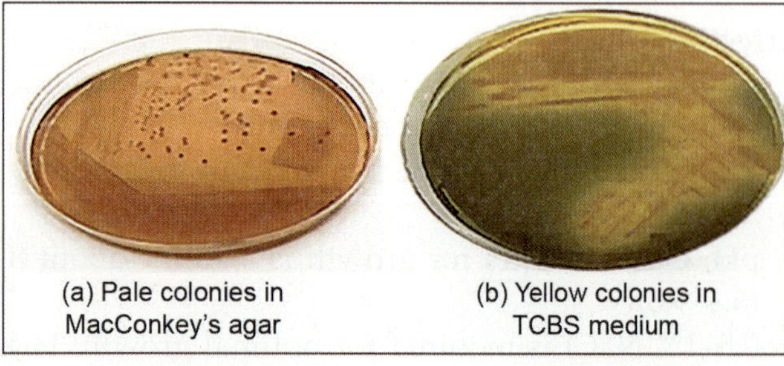

Fig. 65.2: C/Cs of *V. cholerae*

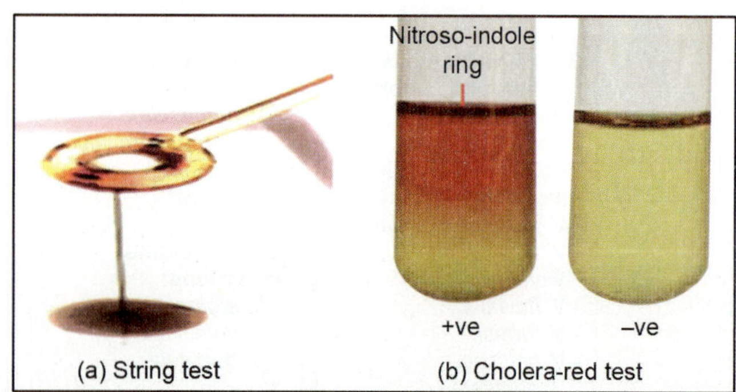

Fig. 65.3: B/Rs of *V. cholerae*

TABLE 65.1: Tests to differentiate the *V. cholerae* from related genera

| Genus | Hugh-Leifson/O-F test | | Amino acid decarboxylation test | | | String test |
	Oxidation (O)	Fermentation (F)	Lysine	Arginine	Ornithine	
Vibrio	+	A	+	–	+	+
Aeromonas	+	A/AG	–	+	–	V
Pseudomonas	+	–	V	V	V	–
Plesiomonas	+	+	+	+	+	–

A: Acid, AG: Acid + gas, V: Variable

Resistance

Sterilization: *V. cholerae* is killed by heating at 55°C in 15 minutes, by boiling in few seconds and by drying. It grows at low temperature and survives for 2–4 weeks in ice cold water and for 4–6 weeks or longer in ice. Its survival in water depends on pH, temperature, salinity and presence of other materials. In clean tape water, it survives for 30 days. Bacteria cannot survive in polluted water of Ganges due to presence of large numbers of vibriophages. It can survive for 1–2 days in a food left at room temperature but survives more than 2 weeks in food stored in cold temperature.

Disinfection: *V. cholerae* is sensitive to acids but resist high alkalinity. It is killed by normal gastric acidity, so required high infective dose and survive for 24 hours in achlorhydric patients. It is killed by chlorination of water. Chlorination does not kill the spores, sporing bacteria and viruses like polio, hepatitis, etc.

Drug resistance: It is not a major problem.

Typing of *V. cholerae*

Heiberg classification based on fermentation reaction: Follow **Table 65.2**.

Gardner and Venkatraman's serological classification: Follow **Flowchart 65.2** and **Table 65.3**.

Other methods: Phage typing **(Ch. 118)** and ribo-typing.

Pathogenicity

Disease name: Disease called cholera.

Flowchart 65.2: Gardner and Venkatraman's serological classification

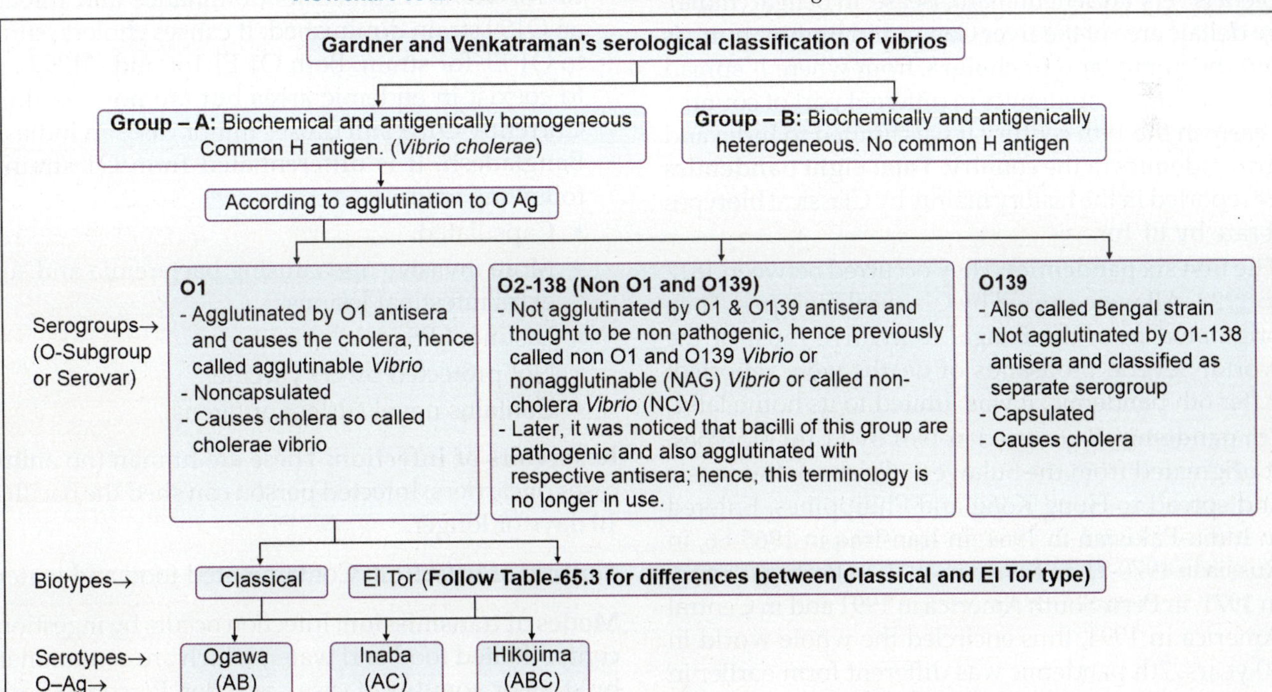

TABLE 65.2: Heiberg classification of *V. cholerae*

Group	Mannose	Sucrose	Arabinose
I	A	A	–
II	–	A	–
III	A	A	A
IV	–	A	A
V	A	–	–
VI	–	–	–
VII	A	–	A
VIII	–	–	A

TABLE 65.3: Differences between classical and El Tor type

Test	Classical	El Tor
Hemolysis of sheep RBCs	–	+
Hemagglutination of chick RBCs	–	+
Vi (VP) test	–	+
El Tor phage V sensitivity	R	S
Mukharjee phage IV sensitivity	S	R
Polymixin B sensitivity (50 U)	S	R
Cholera toxin gene	Type 1	Type 2

S = Sensitive, R = Resistant

Epidemiology: It can occur as sporadic, endemic, localized outbreak, epidemic or in pandemic form. Cholera is very ancient human disease. In Bengal (India), large deltaic area of the river Ganga and Brahmaputra is called the **home land of cholera**, from where it spread and caused many epidemics in different part of country. Till early in the 19th century it was limited to India and many epidemics in the country. **Total eight pandemics** were reported in the history mainly by Classical biotypes and rare by El Tor.

1. **The first six pandemics:** They occurred between 1817 to 1923. All were caused by Classical biotypes. They originated from the Bengal to involve most of the world. Several thousands of deaths were reported. After 6th pandemic, it was limited to its home land.

2. **7th pandemic:** It occurred in 1961 by El Tor biotypes. It originated from the Sulawesi (Celebes), Indonesia and spread to Hong Kong and Philippines. Entered in India-Pakistan in 1964, in Iran-Iraq in 1965-66, in Russia in 1970-71, in Kenya in 1971, in Portugal-Spain in 1971, in Peru-South America in 1991 and in Central America in 1994, thus encircled the whole world in 30 years. 7th pandemic was different form earlier in following ways:
   - Caused by El Tor biotypes
   - Origin was not India, but Indonesia
   - Low mortality
   - Less severe with mild to asymptomatic infections More cases were subclinical
   - High carrier rate, so remained endemic in many areas and causing periodic epidemics. Secondary

attack rate is low, fewer secondary cases in affected families (**secondary attack rate:** It is defined as the probability that infection occurs among susceptible persons within a reasonable incubation period following known contact with an infectious person or an infectious source).
- El Tor biotypes is much hardier than the Classical and survive much longer in the environment, thus involved entire world including countries of Central America, South America, Australia and other affluent countries which were never affected before.
- 7th pandemic which was caused by El Tor biotypes replaced the Classical biotypes, thus in India, Classical biotypes was hardly encountered after the El Tor biotypes took root but in Bangladesh Classical biotypes staged comeback in 1982.

3. **8th pandemic (O-139 or Bengal strain):** It occurred in October 1992 in Madras (Chennai), India and later similar outbreak occurred in other part of India. It is not agglutinable by O1-138 and identified as newer strain called O139. It was limited to coastal areas of Bay of Bengal up to West Bengal, India and adjacent areas of Bangladesh hence called Bengal strain. It replaced the El Tor *Vibrio* (of 7th pandemic). By January 1993, strain produced epidemic in Bangladesh and rapidly spread in almost 11 Asian countries and threatened to cause the next pandemic. But surprisingly in 1994 El Tor *Vibrio* regained its dominance and threat of an O139 strain diminished. It causes cholera, similar to O1 El Tor strain. Both O1 El Tor and O139 began to coexist in endemic areas but are now declining. Currently O139 still causes minor cases in India and Bangladesh. It is differentiated from O1 strain by following ways:
   - Capsulated.
   - More invasive and causing bacteremia and some extra intestinal lesions.
   - Distinct LPS.
   - Not protected by O1 vaccine.
   - Contains novel surface antigens.

Reservoirs of infection: These are human (no animal) cases or carriers. Infected person can shed the bacilli for 10 days or longer.

Sources of infection: Contaminated food and water.

Modes of transmission: Infection occurs by ingestion of contaminated food and water, which are contaminated by stool or vomitus of case or carrier. Person to person transmission is possible in household contact or in close community contact only by supplying the drinking water with contaminated hands called domestic spread. Vegetables washed with contaminated water can lead epidemic.

Incubation period: <24 hours to 5 days.

Portal of entry: GIT.

Sites: Intestine.

Precipitating factors (epidemiological determinants): Following are the types.

A. **Agent factors (virulence factors):** These are two types like intracellular and extracellular.

1. **Intracellular (cell wall associated) factors**
   - **Endotoxin (LPS):** It has no role in pathogenesis in humans, but intraperitoneal inoculation in mouse causes fatal effect.
   - **Flagella:** These are organs of locomotion and help to reach the epithelial cells.
   - **Fimbriae (TCP):** Special organ for adhesion called 'toxin coregulated pilus (TCP)'. *Vibrio* colonizes throughout the course of infection without damage or invasion to host cell. TCP constantly binds the *Vibrio* with host cell.
   - **Infective dose:** Bacilli are sensitive to acids (acid proves effective barrier against cholera), but resist high alkalinity. They are killed by normal gastric acidity so infective dose is very high about 10^4–10^6 bacilli. It can survive for 24 hours in achorhydric (absence of hydrochloric acid in gastric secretion) patients. 10^6 pathogenic *Vibrio* will not cause infection without food in normal person but same dose produces infection with food in normal person.
   - **Siderophore:** It is required for iron uptake.
   - **ToxR gene/Tox R protein:** It is a regulatory gene which regulates cholera toxin (CT), TCP and other virulence factors of *V cholerae*.

2. **Extracellular factors**
   - **Exotoxin (enterotoxin):** It has three types. **(1) Heat labile toxin (LT):** It also called cholera toxin (CT) or cholera enterotoxin (CTX) or choleragen. Heat labile *E. coli* toxin (LT) and cholera toxin of *V cholerae* 0139 and O1 are structurally and antigenically same, but cholera toxin is hundred times more potent than *E. coli*

toxin. Both act via cAMP. CT is bacteriophage coded. It can also replicate as a plasmid which can be transmitted to non toxigenic strain, rendering them toxigenic. Molecular weight (MW) is 84,000. It is toxigenic, can induce neutralizing antitoxin. It can be toxoided. It can affects only intestinal epithelial cells, no effect on any other cells. Different components and actions of CT are described in **Flowchart 65.3**. **(2) Verotoxin (VT)/verocytotoxin:** Subtype of VT produced by *V cholerae* is shiga like toxin (SLT) described in **Ch. 61. (3) Zona occludens toxin:** It is responsible for disruption of tight junction between mucosal cells.

   - **Enzymes: (1) Mucinase:** It helps to cross the protective layer of mucin. **(2) Neuraminidase (hemagglutinin protease):** Formerly, it called 'cholera lectin'. It cleaves mucus and fibronectin and releases vibrios which are bounded to intestinal epithelium and favors their spread to other intestinal parts. **(3) Other enzymes:** Like collagenase, elastase, chitinase, nucleotidase, lipase, etc., are also help in spread and virulence of *Vibrio*.

B. **Host factors**

1. **Age:** There is no age bar, but during epidemic, it targets more to children.

2. **Immunity:** People with low immunity and lack of preexisting immunity are infected more. Single attack of *Vibrio* can provide 6–9 months immunity, reinfection is possible afterward. Immunity may be local or systemic. Local immunity includes presence of antibodies like IgA, IgG and IgM in feces called coproantibodies (Copro means related to dung or feces). Systemic immunity includes vibriocidal antibodies in the serum of patients have been associated with protection against colonization and disease. Prevalence of Vibrio is measured by measuring the titer of vibriocidal antibodies.

Flowchart 65.3: Heat labile enterotoxin

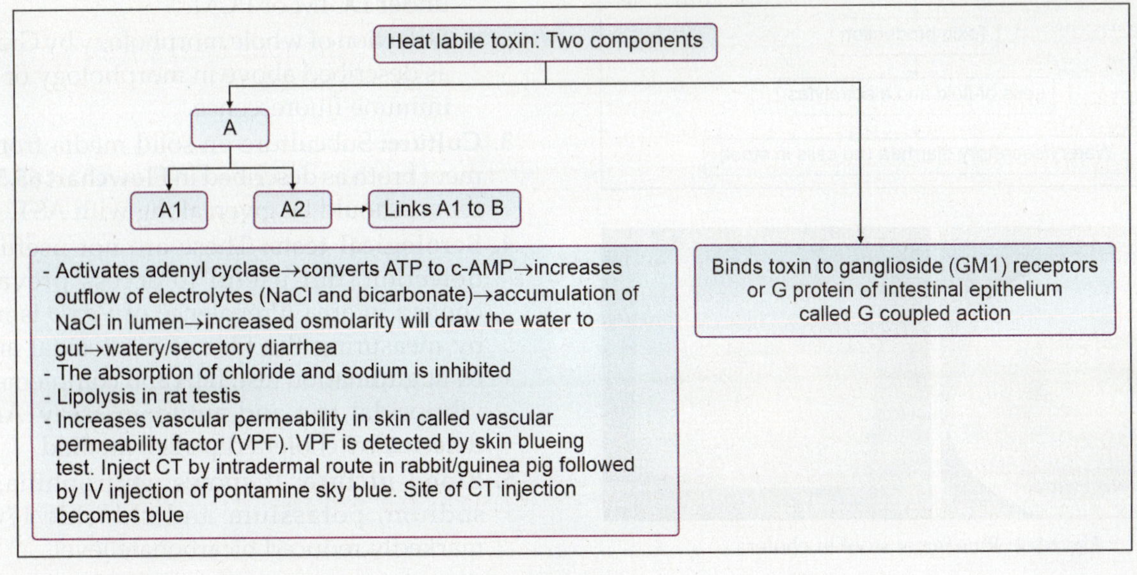

3. **Malnutrition:** It also favors the disease.
4. **Gastric acidity:** Discussed earlier under the heading of infective dose.
5. **Blood group:** It is most in O group and least in AB group. Reason is not known.

C. Environmental factors

1. **Seasons:** Cholera occurs throughout the year, but major cases occur in summer (high temperature), heavy rain fall (flooding).
2. **Habitat:** *V. cholerae* can survive extracellularly. Commonly, it habitat in brackish estuaries or coastal sew water particularly in small crustacean like crab, copepods or in plankton.
3. **Other factors:** Poor sanitation, overcrowding, and population mobility in pilgrimages, marriages, fairs, festivals, etc., favor the infection.

Pathogenesis: Follow **Flowchart 65.4**.

Clinical features: Clinical severity of cholera is variable from asymptomatic case to fatal. It begins with the sudden onset of painless **watery/secretory diarrhea** (no cells in stool) with effortless vomiting. Passage of **'rice water stool' (Fig. 65.4)** has fishy, inoffensive sweetish odor may contain mucus flakes, but no blood or pus cells. Fever is usually absent. Hypovolemic shock may cause death in <24 hours.

Complications: Severe dehydration (loss of fluid and electrolytes) causes anuria, hemoconcentration, hypovolamic shock, hypokalemia, acidosis (due to loss of bicarbonate), muscular cramps, renal failure, cardiac arrhythmia, pulmonary edema, paralytic ileus, etc.

Carrier: Four types of carriers.
1. **Incubatory carrier:** It sheds the bacilli in feces for 1–5 days.
2. **Convalescent carrier:** It excretes the bacilli for 2–3 weeks.
3. **Healthy/contact carrier:** It includes subclinical infection and sheds the bacilli for 10 days.
4. **Chronic carrier:** It sheds the bacilli for months or years (up to 10 years), more in El Tor than Classical. Persistence gallbladder infection is the reason for chronic carrier.

Laboratory Diagnosis

Detection of Case

A. Detection of case by identifying the bacilli

Specimens: (1) Stool: It is collected before antibiotics, by introducing a catheter in rectum and letting the watery stool flow in sterile screw-capped container. **(2) Rectal swab:** Good quality swab is used which absorb 0.1–0.2 ml fluid. In convalescence phase dip the swab in enrichment broth and than collect the samples.

Transport: If specimens can reach to laboratory in few hours than transfer in enrichment media. If long period before reach to laboratory then transfer in transport media. If transport media are not available, then filter paper strip may be soaked in stool and transfer in plastic envelop.

Testing methods

1. **Macroscopic examination:** 'Rice water stool', inoffensive sweetish odor, may contain mucus flakes.
2. **Microscopy**
   - Detection of darting motility of *V cholerae* by direct hanging drop preparation.
   - Inhibition of motility by specific antiserum under DGIM or PCM.
   - Detection of whole morphology by Gram's stain as described above in morphology or by direct immune fluorescence.
3. **Culture:** Subculture on solid media from enrichment broth as described in **Flowchart 65.5**. Culture report should be given along with AST.
4. **Serological tests:** These are not useful in case detection, but useful to access prevalence of cholera in area. Prevalence of *Vibrio* is measured by measuring the titer of vibriocidal antibodies by agglutination test, IHA and complement based vibriocidal test and antitoxin assay. Antigen is detected by cholera dip stick method.
5. **Blood picture:** It shows neutrophilia, normal sodium, potassium and chloride level with markedly reduced bicarbonate level.

Flowchart 65.4: Pathogenesis of cholera

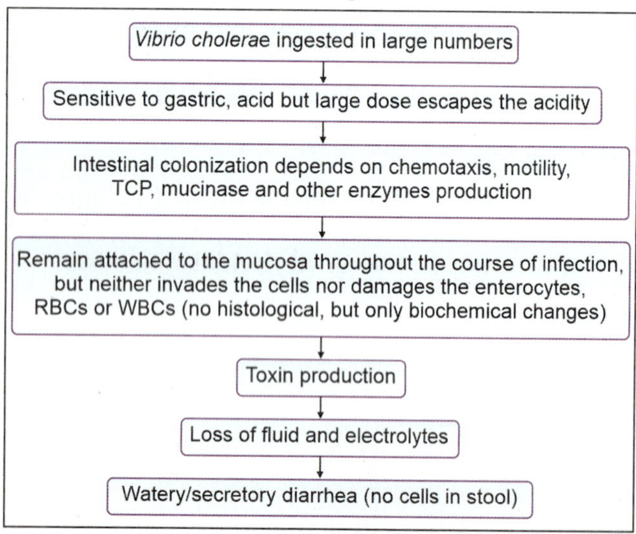

Fig. 65.4: Rice water stool in cholera

Flowchart 65.5: Cultivation techniques and identification of growth on solid media

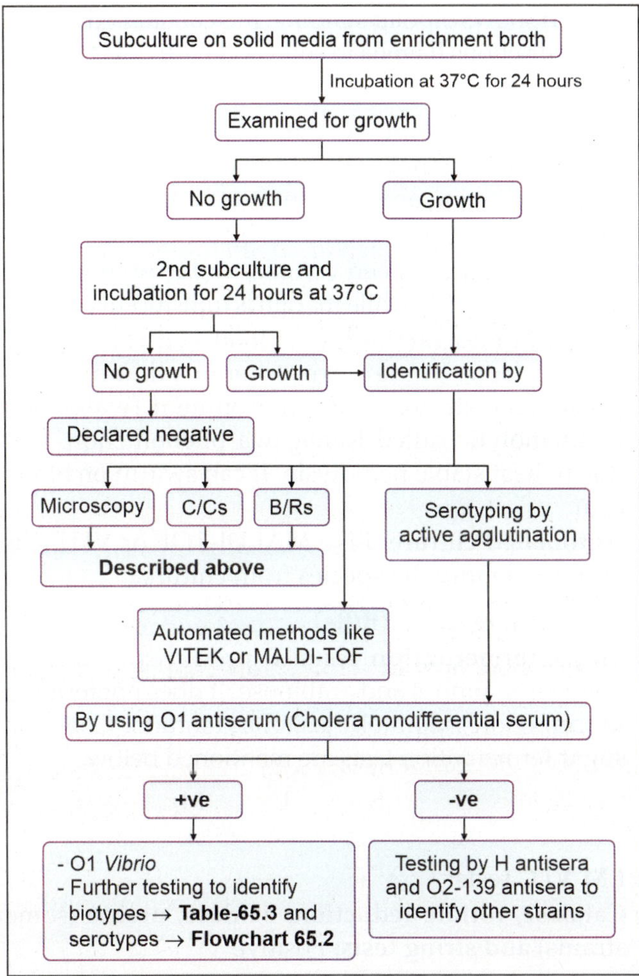

6. **Bacterial typing:** Serogrouping, serotyping and biotyping are described in **Flowchart 65.2**.
7. **Molecular method:** Multiplex PCR is performed to detect the common diarrheal pathogens.

B. Detection of case by identifying the LT: Ch. 61 (section heat labile toxin → **Table 61.3**).

Detection of Carrier

Repeated enrichment and subcultures are required for stool and bile. Stool collected by normal defecation or by purgation with mannitol (30 g) or with $MgSO_4$ (15–30 g). Bile collected by duodenal intubation.

Detection of Source

Enrichment technique: Collect 900 ml water → enrichment in 100 ml APW (pH-9.2) → incubate at 37°C for 6–8 hours → subculture on solid media.

Filtration technique: Filtration through millipore membrane → culture the membrane on selective media.

Sewage → dilute in saline → filtered through gauze → treated as water.

Prevention

General measures: These include hand hygiene, chlorination of water, improvement of environmental sanitation and food sanitation. Early detect the outbreak and take necessary steps to stop the spread.

Chemoprophylaxis: Tetracycline is the drug of choice. It is indicated to household contact or close community contact. It is not indicated for total community called mass prophylaxis, as it is not able to prevent the spread.

Immunoprophylaxis: Ideal vaccine is yet to found. Following are some useful vaccines.

- **Injectable killed vaccines:** Three different types of vaccines like **(1)** vaccine with 8000 million of *V. cholerae* per ml with equal numbers of Ogawa and Inaba serotypes, **(2)** vaccine with equal numbers of Classical and El Tor *Vibrio* and **(3)** vaccine with O-139 strain. They are given by SC or IM route. These vaccines provide only 50–60% immunity which lasts for 3–6 months. They do not provide local immunity.
- **Oral vaccines:** These are available in different types **(1)** Killed oral whole cell vaccine of O1 and O139 of *V cholerae*: Two preparations are available commercially like Shanchol from India and Euvichol from South Korea. Total 2 doses are given orally with minimum gap of 2 weeks to all groups with >1 year of age. Immunity lasts for 3 years. **(2)** Killed oral whole cell vaccine with inclusion of recombinant B subunit of cholera toxin (WC/rBS): Commercially available preparation is Dukoral. Total 2 doses are given orally with minimum gap of 1 week to all groups with >2 year of age. 3rd dose is given to children of 2–5 years. Immunity lasts for 2 years. **(3)** Oral live attenuated cholera vaccine: It is prepared by using mutant strain which lacks the gene of cholera toxin. Commercially available preparation is vaxchora. Single dose is given orally to age group between 18 and 64 years. It is indicated for persons traveling to area with active cholera cases. It is also useful to control the spread of cholera during outbreaks. It provides 90% immunity which starts after 10 days of vaccination and lasts for 3–6 months. **(4)** Live oral vaccine contains Classical *Vibrio*, El Tor *Vibrio* and O-139 strains.

WHO road map to end cholera: In 2017 WHO, launched strategy to reduce cholera deaths by 90% and to eliminate the cholera from as many as countries by 2030.

Treatment

Start **oral rehydration therapy** immediately to compensate the loss of fluid and electrolytes. Following **antibiotics** are useful but they have secondary role.

- For adult: Doxicycline or tetracycline is the drug of choice. Ciprofloxacin, 3rd generation cephalosporin and erythromycin are other options.
- For children: Furazolidine is the drug of choice, but in India cotrimoxazole is the drug of choice
- For pregnant women: Furazolidine (drug of choice).

O2-138 (Non O1 and O139)

General Features

Morphologically and biochemically, similar to O1 and O139 vibrios, but not agglutinated by O1 and O139 antisera and thought to be nonpathogenic, hence previously called non O1 and O139 *Vibrio* or nonagglutinable (NAG) *Vibrio* or called noncholera *Vibrio* (NCV). Later, it was noticed that these group are pathogenic and also agglutinated with respective antisera, hence these terminology is no longer in use.

Pathogenicity

They are **transmitted** by ingestion of sea food like raw oyster. In Calcutta they found in small pond fish. **Intestinal features** are not causing cholera, but causing gastroenteritis with abdominal pain, nausea, vomiting, diarrhea and fever. **Extraintestinal features** are otitis media, wound infection and bacteremia.

Treatment

It is same as cholera like fluid replacement and antibiotics (tetracycline, ciprofloxacin and 3rd generation cephalosporin).

Vibrio mimicus

So, called because biochemically similar to *V. cholerae*. It grows at 0.5–1% NaCl. It is sucrose nonfermenter and causes diarrheal disease in Gulf-coast of USA by ingestion of seafood (oyster) similar to *V. parahaemolyticus*.

Halophilic Vibrios

Vibrios, which grow at high concentration (7–10%) of NaCl, are called halophilic vibrios. Common species causing gastroenteritis are *V. parahaemolyticus, V. fluvialis, V. furnisii, V. hollisae*, etc., and causing extraintestinal infections are *V. alginolyticus, V. vulnificus, V. damsel*, etc. Differences between *Vibrio comma* and halophilic vibrios are mentioned in **Table 65.4**.

V. parahaemolyticus

History: It was isolated from Japan in 1951 as a causative agent of food poisoning due to ingestion of sea fish.

Morphology: Same as like *V. cholerae* except, it is pleomorphic, when grow on 3% salt agar and in old culture. It shows bipolar staining. It is capsulated, motile with peritrichous flagella in solid and with polar flagella in liquid medium.

Cultural characteristics (C/Cs): Bacteria grow in media with 8% NaCl, but not with 10% NaCl. Optimum NaCl is 2–4%. Following are different useful media.

1. **Peptone water:** It contains 8% NaCl.
2. **TCBS medium:** Green colonies occur with opaque and raised center, while margin is translucent and flat.
3. **Wagat-Suma agar (high salt blood agar):** Pathogenic strains from human are β-hemolytic and non-pathogenic strains from environment (water) are nonhemolytic called Kanagawa phenomenon. It is due to heat stable hemolysin. It can swarm on blood agar.
4. **Automated culture:** Like MALDI-TOF or VITEK is used to identify the species from culture.

Biochemical reaction (B/Rs)

1. **Sugar fermentation tests:** It produces acid from maltose, mannose and arabinose. It does not ferment salicin, xylose, adonitol, inositol and sorbitol. Common sugar fermentation tests are mentioned below.

G	S	L	M
A	–	–	A

2. **I M Vi C tests:** + – V +
3. **Catalase, nitrate reduction, oxidase, urease** (some strains) **and string tests:** Positive

Resistance: It is killed by heating at 60°C in 15 minutes. It does not grow at 4°C, but can survive refrigerator and freezing temperature. It is destroyed by drying and putting in distilled water or vinegar.

Pathogenicity:

- **Virulence factors**
  - **Antigens:** These are O-Ag (12 O-groups), K-Ag (59 types) and H-Ag.
  - **Hemolysin:** It produces the heat stable hemolysin which is not significant in virulence, but used in laboratory to test for Kanagawa phenomenon and pathogenicity. Kanagawa phenomenon

TABLE 65.4: Differences between *Vibrio comma* and halophilic vibrios

Test	Vibrio comma	V. parahaemolyticus	V. alginolyticus	V. vulnificus
Indole	+	+	+	+
VP	V	–	+	–
Nitrate	+	+	+	+
Urease	–	Some strain are +	–	–
Lactose	–	–	–	Acid
Sucrose	+	–	+	–
Swarming	–	– / +	+	–
NaCl	0.5–1%	8%	10%	<10%

positive strains are pathogenic while Kanagawa phenomenon negative strains are nonpathogenic.

- **Type III secretion system:** It is capable of injecting virulence proteins into host cells to disrupt host cell functions or to cause cell death by apoptosis.
- **Mode of transmission:** It occurs by ingestion of sea (marine) food like sea-fish, shrimps, crabs or molluscs (oyster). In Kolkata, it found in small pond fish.
- **Incubation period:** About 24 hours.
- **Clinical features: (1) Food poisoning:** It is the important cause of food poisoning throughout the world. Not all strains are pathogenic, only Kanagawa phenomenon positive strains cause food poisoning which presents with abdominal pain, nausea, vomiting, diarrhea and low-grade fever. **(2) Dysentery:** It is less common and occurs in India and Bangladesh.
- **Complications:** Moderate degree of dehydration.

Laboratory diagnosis: Stool is tested by macroscopic examination (stool with cellular exudates and often blood in case of dysentery), microscopy (follow morphology), culture (follow C/Cs) and biochemical reactions (follow B/Rs).

V. alginolyticus

It is similar to *V. parahaemolyticus* except it tolerates 10% NaCl (most salt tolerating species), VP positive and ferments sucrose. It found in sea fish and sea-water and on exposure produces eye, ear and wound infections in humans.

V. vulnificus

Previously it called L⁺ *Vibrio* or *Beneckea vulnifica*. It tolerates <10% NaCl, VP negative and ferments lactose but not sucrose. Infection occurs due to ingestion of oysters. It produces wound infection on exposure to sea water and myositis. Ingestion in patient with liver disease, it crosses the GIT without any GIT symptoms enters in blood and causing septicemia.

AEROMONADACEAE

Classification

Follow **Flowchart 65.6**.

Morphology

They are GNB, capsulated, nonsporing and motile with single polar flagellum. Some strains are nonmotile. In

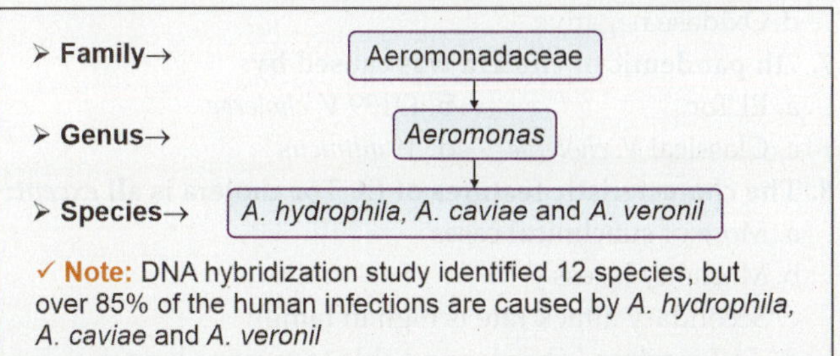

Flowchart 65.6: Classification of Aeromonadaceae

some strains lateral flagella are found. Fimbriae are also present.

Cultural Characteristics (C/Cs)

Optimum temperature is 32°C (4–42°C) and pH range is 4.5–9.0. Growth is not supported by NaCl. 90% strains produce β-hemolysis, while *A. caviae* gives non-hemolytic colonies on **blood agar**. Pale colonies are produced on **MacConkey's agar** due to NLF.

Biochemical Reactions (B/Rs)

Follow **Table 65.1**.

Pathogenicity

- Virulence factors: (1) Fimbriae: These are the organs of adhesion. (2) Capsule: It inhibits the phagocytosis. (3) LPS: It has endotoxic actions. (4) Exotoxins: These are hemolysin, aerolysin and enterotoxin (Shiga like toxin/SLT).
- Reservoir of infection: These are fresh water or brackish water or marine water or polluted water of drained pipes or in sink traps.
- Source of infection: Food and water.
- Mode of transmission: By ingestion of contaminated food and water.
- Clinical features: (1) Animal disease: *A. hydrophila* produces the red leg disease in frog and also pathogenic for other cold blooded animals like fish and reptiles. (2) Human disease: Intestinal features are gastroenteritis with abdominal pain, nausea, vomiting, watery diarrhea/dysentery and fever. Extraintestinal features are peritonitis, myositis, wound infection, bacteremia in persons with IDDs, pneumonia, pharyngitis, epiglotitis and HUS due to production of SLT.

Laboratory Diagnosis

Stool, vomitus, urine, sputum, etc., are tested by microscopy (follow morphology), culture (follow C/Cs) and biochemical reactions (follow B/Rs).

Treatment

Fluid therapy and antibiotics are the options.

Plesiomonas shigelloides

Meaning

Plesiomonas (Greek) means neighbor because closely associated with *Aeromonas*, while *shigelloide*, indicates antigenic relation to *Sh. sonnei*.

Classification

Follow **Flowchart 65.7**.

General Features

Initially it was listed under Vibrionaceae family, but DNA hybridization study suggests that it is more related

Flowchart 65.7: Classification of *Plesiomonas*

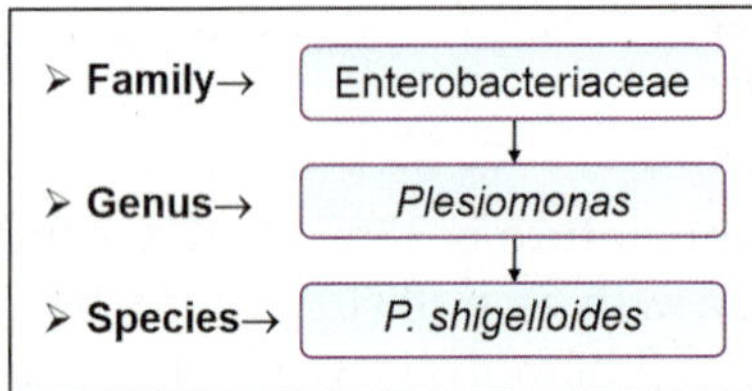

to *Proteus* spp., hence shifted to Enterobacterales order. It is the only genus in Enterobacterales order, which gives oxidase test positive. Growth is not stimulated by NaCl.

Morphology

They are GNB and motile with lophotrichous flagella.

Cultural Characteristics (C/Cs)

Optimum temperature is 37°C, but better growth occurs at 30°C. Growth is not stimulated by NaCl. It is differentiated from *Vibrio* and *Aeromonas* by tests mentioned in **Table 65.1** and also by ability to produce resistance to compound O/129. Some strains are LFs give pink colonies while some are NLFs give pale/colorless colonies on **MacConkey's agar or DCA**.

Biochemical Reactions (B/Rs)

Follow **Table 65.1**.

Pathogenicity

- Reservoirs of infection: Bacteria are saprophytes in water and soil and commensals in animal intestine (like fish) and rare in human intestine. So, water, soil and feces are the reservoirs.
- Source of infection: These are food and water.
- Mode of transmission: Ingestion of contaminated food and water.
- Clinical features: It causes gastroenteritis more severe in person with IDDs, arthritis, meningitis, septicemia and cellulitis.

Laboratory Diagnosis

Stool, blood, etc., are tested by microscopy (follow morphology), culture (follow C/Cs) and biochemical reactions (follow B/Rs).

Treatment

Fluid therapy and antibiotics are the options.

ACCESS YOURSELF

Case Study

1. A pediatric patient visited the hospital with complains of abdominal pain and diarrhea. Stool examination revealed rice water appearance and bacteria with darting motility. Identify the organism and answer the following.
   a. Name the clinical condition and causative agent.
   b. Describe the morphology and culture characteristics of causative agent.
   c. Write the classification of causative agent.
   d. Write the pathogenicity of causative agent.

Essay/Full Question

1. *V. cholerae.*

Short Notes

1. Morphology or C/Cs or B/Rs or classification of *V. cholerae.*
2. O139 (Bengal strain) vibrio.
3. Halophilic vibrios.

Short Questions for Theory/Viva Questions

1. Name two transport and two enrichment media for *V. cholerae.*
2. Write the four differences between Classical and El Tor *Vibrio.*
3. Write the two examples of each, nonhalophilic and halophilic vibrios.
4. What is Kawagawa phenomenon?

Comments On

1. Infective dose is low in shigellas while high in vibrios.

MCQs for Chapter Review

Vibrionaceae

1. **Halophilic vibrios are *except*:**
   a. *V. cholerae*   b. *V. vulnificus*
   c. *V. parahaemolyticus*   d. *V. mimicus*
   e. *V. alginolyticus*

Vibrio cholerae

2. **True about *Vibrio cholerae* is:**
   a. Can tolerate wide range of alkaline pH
   b. Nonmotile bacilli
   c. Cannot be grow in media
   d. NaCl stimulates growth
3. **All of the following are true regarding *V. cholerae*, *except*:**
   a. Transported in acidic medium
   b. Gram-negative
   c. Aerobic organism
   d. Ferments glucose
4. **The following are true of *Vibrio cholerae*, *except*:**
   a. Produces indole and reduces nitrate
   b. Synthesizes neuraminidase
   c. Dies rapidly at low temperature
   d. Vaccine confers long immunity
5. ***Vibrio cholerae* true is:**
   a. Very resistant to alkaline pH
   b. Nutritionally fastidious
   c. Best growth at 24°C
   d. Rod-shaped bacilli
6. **Which of the following statement is true about *Vibrio cholerae*?**
   a. There is no natural reservoir
   b. Transported in alkaline peptone water medium
   c. Halophilic
   d. Oxidase negative
7. **7th pandemic of cholera was caused by:**
   a. El Tor   b. O139 *V. cholerae*
   c. Classical *V. cholerae*   d. *V. mimicus*
8. **The characteristic features of EL Tor cholera is all *except*:**
   a. More of subclinical cases
   b. Mortality is less
   c. Secondary attack rate is high in family
   d. El Tor vibrio is harder and able to survive longer

9. **Strain of *Vibrio cholerae* in Bengal:**
    a. O139   b. O137
    c. O17   d. O40
10. **Cholera is caused by:**
    a. *Vibrio cholerae*-01   b. *Vibrio cholerae*-O139
    c. *Vibrio parahaemolyticus* d. NAG vibrio
11. **Not true about EL Tor vibrio O1:**
    a. Animals are the only reservoirs
    b. Epidemiologically indistinguishable from *V. cholerae*-O139
    c. Human acts as vehicle for spread
    d. Efficacy of vaccine against E1 Tor vibrio is great
12. **True regarding cholera is:**
    a. Toxin acts on GM1 receptors
    b. Toxin action is cAMP mediated
    c. Peritrichate flagella
    d. Utilizes arginine and lysine
13. **Which of the following toxin can stimulate adenylate cyclase with G protein coupled action?**
    a. Shiga toxin   b. Cholera toxin
    c. Diphtheria toxin   d. *Pseudomonas* toxin
14. **Cholera toxin effects are mediated by stimulation of which of the following second messanger?**
    a. cAMP   b. cGMP
    c. Ca^{++}-calmodulim   d. IP3/DAG
15. **Which toxin acts by ADP ribosylation?**
    a. Botulinum toxin   b. *Shigella* toxin
    c. *Vibrio cholerae*   d. Diphtheria toxin
    e. Pertussis toxin
16. **Cholera toxin is due to:**
    a. Chromosome   b. Plasmid
    c. Phage   d. Transposon
17. ***Vibrio cholerae* is able to stay in GIT because of:**
    a. Acid resistance   b. Bile resistance
    c. Motility   d. Binds to specific receptors
    e. Anaerobic potential
18. **Which of the following is true of cholera?**
    a. Recent epidemic was due to classical type
    b. Causes secretory diarrhea
    c. Caused by endotoxin
    d. Vibriocidal antibodies correspond to susceptibility
19. **True about cholera is:**
    a. Gram-negative rod
    b. Associated with fever
    c. Causes painful watery diarrhea
    d. It is an achlorhydria, which renders an individual susceptible to disease
20. **In a patient presenting with diarrhea due to *Vibrio cholerae*, which of the following will be present?**
    a. Abdominal pain
    b. Presence of leukocytes in stool
    c. Fever
    d. Neutrophilia
    e. Occurrences of many cases in same locality
21. **Antibiotic treatment of choice for treating cholera in an adult is a single dose of:**
    a. Tetracycline   b. Cotrimoxazole
    c. Doxycycline   d. Furazolidone
22. **Drug of choice for treating cholera in pregnant women is:**
    a. Tetracycline   b. Doxycycline
    c. Furazolidone   d. Cotrimoxazole

23. **Which of the following is the drug of choice for chemoprophylaxis of cholera?**
    a. Tetracycline   b. Doxycycline
    c. Furazolidone   d. Cotrimoxazole
24. **True about epidemiology of cholera:**
    a. Chemoprophylaxis is not effective
    b. Boiling cannot destroy the organism
    c. Food can transmit the disease
    d. Vaccine can give 90% protection
25. **Which is not true about *Vibrio cholerae*:**
    a. It is not halophilic
    b. Grows on simple media
    c. Man is the only natural host
    d. Cannot survive in extracellular environment
26. **True about *Vibrio cholerae* O139 all *except*:**
    a. Clinical manifestation are similar to O1 El Tor strain
    b. First discovered in Chennai
    c. Epidemiologically indistinguishable from O1 El Tor strain
    d. Produces O1 polysaccharide
27. **Which of the following about cholera is true?**
    a. Invasive
    b. Endotoxin is released
    c. Recent infection in India are of classical type
    d. Vibriocidal antibodies titer measures prevalence

Halophilic vibrios

28. ***V. parahaemolyticus* food poisoning is caused by ingestion of:**
    a. Eggs and poultry products
    b. Raw vegetable
    c. Catfish, shellfish, sea food
    d. Milk products
29. **Which of the following 'vibrios' is most commonly associated with ear infection?**
    a. *V. alginolyticus*   b. *V. parahaemolyticus*
    c. *V. vulnificus*   d. *V. fluvialius*

Aeromonanadaceae

30. **Red leg disease is caused by:**
    a. *Pseudomonas*   b. Mouldy sugar cane fiber
    c. Conidiosporum   d. *Aeromonas*

Answers and Explanation of MCQs

1. **a and d**
   - Follow section, **Vibrionaceae (classification → Flowchart 65.1)** for explanation.
2. **a and d**
   - *V. cholerae* can tolerate wide range of alkaline pH from 6.4–9.6. It is motile by monotrichous flagellum with darting type of motility, can grow in media and 0.5–1% NaCl stimulates growth.
3. **a**
   - *V. cholerae* can tolerate wide range of alkaline pH from 6.4–9.6, so transported in alkaline medium. For more explanation of transport media and their pH, follow section, C/Cs (transport media/holding media). For explanation of other options, follow section, morphology, C/Cs and B/Rs.
4. **c and d**
   - Indole and nitrate tests are positive by *V. cholerae.*, tested by cholera red reaction as explained in section B/Rs (cholera red reaction). For option b, follow section, *V. cholerae* [pathogenicity → agent factors (virulence factor) → enzymes] for

explanation. It does not die at low temperature, but it grows at low temperature (16–40°C) and survives for 2–4 weeks in ice cold water and for 4–6 weeks or longer in ice. Vaccine confers immunity lasts for 3–6 months.

5. a
- *V. cholerae* can tolerate wide range of alkaline pH from 6.4 to 9.6. Nutritionally it is not fastidious, because can grow in simple media. Best grow at 37°C (optimum temperature). It is comma shaped bacilli.

6. b
- Human is the only natural reservoir. Samples for cholera are transported in alkaline peptone water and other media as described in text. *V. cholerae* is nonhalophilic and oxidase positive.

7. a

8. c
- Follow section, *Vibrio cholerae* (**pathogenicity → 7th pandemic**) for explanation of answers of MCQs 7–8.

9. a
- Follow section, *Vibrio cholerae* [**pathogenicity → 8th pandemic (O-139 or Bengal strain)**] for explanation.

10. a and b
- *Vibrio parahaemolyticus:* Causes food poisoning and dysentery
- NAG vibrios (non 01–139 strain) had been known to produce sporadic outbreak of diarrhea and extraintestinal infections but never produced cholera epidemic so far.
- Follow section, *Vibrio cholerae* (**classification of *V. cholerae* → Flowchart 65.2**) for explanation.

11. a, b and d
- No animals but humans are the only reservoirs. Epidemiologically El Tor is distinguished from *V. cholerae*-0139 by presence of capsule and also by distinct polysaccharide called O139 LPS. Human acts as vehicle for spread by stool and vomitus. Efficacy of vaccine against El Tor vibrio is not great.

12. a and b
- *V. cholerae* is motile by monotrichate flagellum. It decarboxylase the lysine and ornithine, but not arginine. Follow section, *Vibrio cholerae* [**pathogenicity → agent factors (virulence factors) and Flowchart 65.3**] for more explanation.

13. b

14. a

15. c
- Follow section, *Vibrio cholerae* [**pathogenicity → agent factors (virulence factors) and Flowchart 65.3**] for explanation of answers of MCQs 13–15.

16. c
- Follow section, *Vibrio cholerae* [**pathogenicity → agent factors (virulence factors) → exotoxin → heat labile toxin (LT)**] for explanation.

17. c
- Follow section, *Vibrio cholerae* (**pathogenicity → pathogenesis and Flowchart 65.4** for explanation.

18. b
- Recent epidemic in 1992 was due to Bengal strain. *Vibrio* causes watery/secretory diarrhea. Endotoxin is not associated with pathogenesis. Vibriocidal antibodies present in serum are not corresponding to susceptibility, but provide protection against colonization and disease.

19. (a) and (d)
- Vibrios are gram-negative bacilli with comma shape. For options b and c follow section, *Vibrio cholerae* (**pathogenicity → clinical features**). Bacilli are sensitive to gastric acid, so achlorhydria (absence of hydrochloric acid in gastric secretion) which renders an individual susceptible to disease.

20. d and e
- Cholera present with painless watery/secretory diarrhea (no cells in stool) with effortless vomiting without fever. It presents as localized outbreak, epidemic or in pandemic forms so many cases will occur at a time in same locality. Blood picture suggests neutrophilia.

21. c

22. c
- Follow section, *Vibrio cholerae* (**treatment**) for explanation of answers of MCQs 21–22.

23. a
- Follow section, *Vibrio cholerae* (**prevention → chemoprophylaxis**) for explanation.

24. c
- Chemoprophylaxis is effective for closed contacts or house hold contact, but not for mass prophylaxis. For explanation of option b follow section, *Vibrio cholerae* (**resistance**). Food and water can transmit the disease. Vaccine can give 50–60% protection.

25. d
- *V. cholerae* can survive extracellularly in brackish estuaries or coastal sew water particularly in small crustacean like crab, copepods or in plankton.

26. d
- Options a, b and c: Follow section, *Vibrio cholerae* [**pathogenicity → 8th pandemic (O-139 or Bengal strain)**] for explanation.
- It does not produce O1 LPS but O 139 LPS.

27. b and d
- It can affect only intestinal epithelial cells, no effect on any other cells, so it is noninvasive. It released endotoxin but not contributing in pathogenesis. Recent infections in India are of El Tor type. For explanation of option d, follow section, laboratory diagnosis (serological tests).

28. c
- Follow section, **halophilic vibrios (*V. parahaemolyticus*)** for explanation.

29. a
- Follow section, **halophilic vibrios (*V. alginolyticus*)** for explanation.

30. d
- Follow section, **Aeromonadaceae (pathogenicity → Clinical features)** for explanation.

CHAPTER 66

Infections of Campylobacterales

Chapter Outline

INTRODUCTION

Campylo word derived from campylos (Greek) means curved and bacter from baktron (Greek) means rod because of curve shape of bacillus. Campylobacterales are classified in **Flowchart 66.1**.

Campylobacter jejuni

History

In 1886 a pediatrician, Theodor Escherich, observed *Campylobacter* from diarrhea samples of children. The first isolation of *C. jejuni* was done in Brussels, Belgium, from stool of diarrheal patient.

Morphology

Type according to Gram's stain: They are GNB **(Fig. 66.1).**

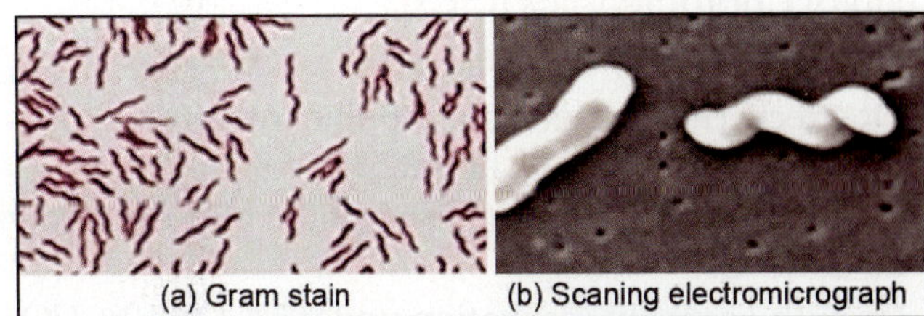

Fig. 66.1: Morphology of *C. jejuni*

Shape and size: Bacilli are curved, typically comma shaped, S-shaped or spiral (gull-wing) shaped and 0.2–0.4 μm × 1.5–5 μm in size. Under light microscopy, *C. jejuni* has a characteristic "sea-gull" shape appearance as a consequence of its helical form. When exposed to atmospheric O_2, it changes to a coccal form.

Motility: Bacilli showing darting/tumbling motility with single unsheathed polar flagellum at one end

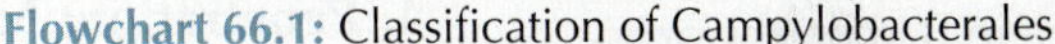

Flowchart 66.1: Classification of Campylobacterales

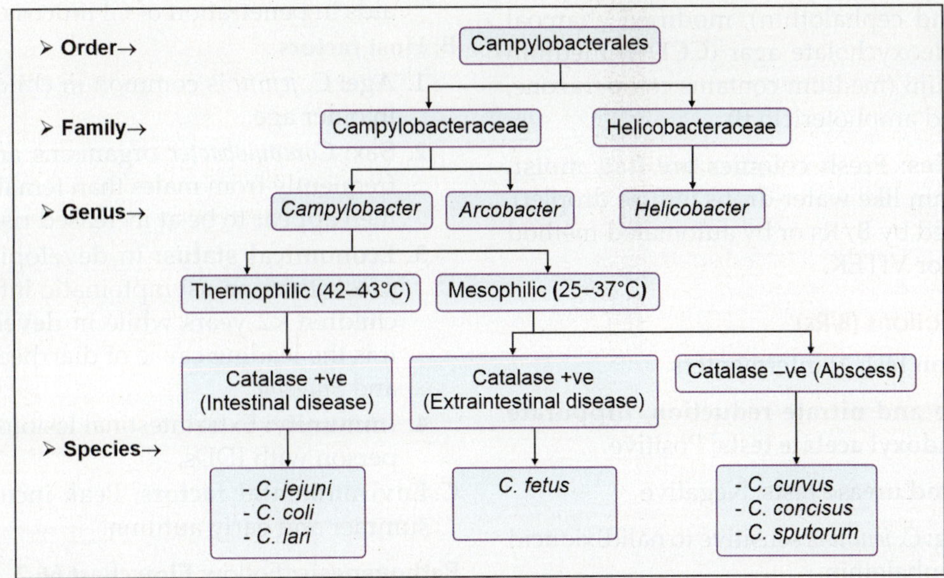

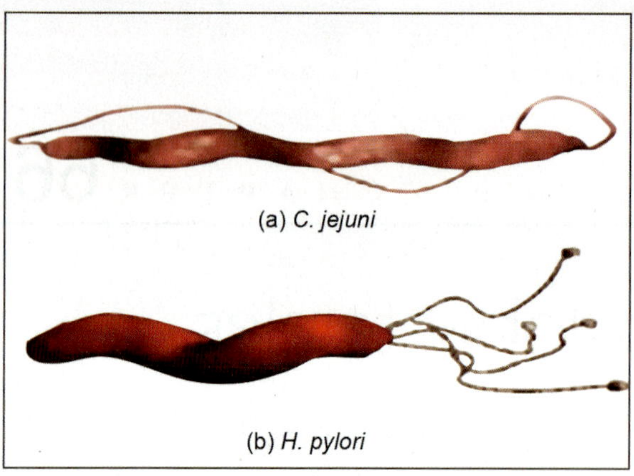

Fig. 66.2: Arrangement of flagella in *C. jejuni* and in *H. pylori*

called monotrichous or both end called amphitrichous **(Fig. 66.2a)**. Motility is best seen by DGIM or PCM.

Capsule and spores: They are nonsporing and non-capsulated.

Cultural Characteristics (C/Cs)

Effective factors

- O_2 effect: They are microaerophilic (required O_2 less than environmental O_2 about 5%).
- Temperature: They are thermophilic and can grow at 42–43°C.
- Growth occurs in incubation under 5% O_2, 10% CO_2 and 85% nitrogen.

Culture in media

A. **Liquid media:** Bacteria survive in Cary Blair transport medium for 1–2 weeks at 4°C.

B. **Solid media:** Bacteria produce nonhemolytic colonies on ***Campylobacter* blood agar**. Blood provides extra-nutrition and also inhibits toxic materials during growth of bacteria. Charcoal is best substitute for blood.

C. **Selective media:** These are Skirrow's medium, Butzler's medium, Campy BAP medium (lyzed bood agar with polymyxin B, amphotericin B, trimithoprim, vancomycin and cephalothin), modified charcoal cefoperazone deoxycholate agar (CCDA) medium and CVA medium (medium contains cefoperazone, vancomycin and amphotericin B).

Growth properties: Fresh colonies are flat, moist, translucent, 1–3 mm like water-drops (effuse droplet). Species is identified by B/Rs or by automated method like MALDI-TOF or VITEK.

Biochemical Reactions (B/Rs)

Sugar fermentation test: Nonfermenters.

Catalase, oxidase and nitrate reduction, hippurate hydrolysis and indoxyl acetate tests: Positive

H_2S production and urease tests: Negative

Sensitivity testing: *C. jejuni* is sensitive to nalidixic acid and resistant to cephalothin.

Pathogenicity

Disease name: Whole spectrum of disease called campylobacteriasis, most common cause is *C. jejuni*.

Epidemiology: *C. jejuni* are distributed worldwide, although exact figures are not available. New Zealand reported the highest national campylobacteriosis rate.

Reservoirs of infection: Animals like poultry, cattle, sheep, swine, dogs, cats and some birds are the reservoirs. There is no human reservoir.

Sources of infection: These are food and water contaminated by animal excreta and infected animal products like unpasteurized milk, raw chicken, etc.

Modes of transmission: It is a zoonosis, human infection acquired by **(1)** ingestion of raw milk or raw chicken of infected animals or food/water contaminated by animal feces as bacilli are constantly shed in feces, **(2)** direct contact with infected animals or by **(3)** oral-anal sex.

Incubation period: 1–7 days.

Portal of entry: GIT.

Sites: Mostly intestine (jejunum, ileum and colon). Extraintestinal lesions are seen in person with IDDs.

Precipitating factors (epidemiological determinants): Following three types.

A. Agent factors (virulence factors)

1. **Enterotoxin:** It is same as heat labile toxin of *V. cholerae* enterotoxin.
2. **Cytoledal distending toxin/cytotoxin:** It has no significant role, may be responsible for dysentery.
3. **Infective dose:** *Campylobacter* species are sensitive to acid, as a result, the infective dose is relatively high, and the bacteria rarely cause illness when infective dose is $<10^3$ organisms. However, for *C. jejuni* infective dose is very less about 500 bacilli.
4. **Penetration:** Motility and spiral shape of bacilli aids in penetration of GI mucosa.

B. Host factors

1. **Age:** *C. jejuni* is common in children and *C. fetus* in older age.
2. **Sex:** *Campylobacter* organisms are isolated more frequently from males than females. Homosexual men appear to be at increased risk.
3. **Economical status:** In developing countries, it presents as an asymptomatic infection except in children <2 years while in developed countries, it is the leading cause of diarrhea than *Salmonella* and *Shigella*.
4. **Immunity:** Extraintestinal lesions are common in person with IDDs.

C. Environmental factors: Peak incidence occurs in summer and early autumn.

Pathogenesis: Follow **Flowchart 66.2**.

Essentials of Medical Microbiology

Flowchart 66.2: Pathogenesis of *C. jejuni*

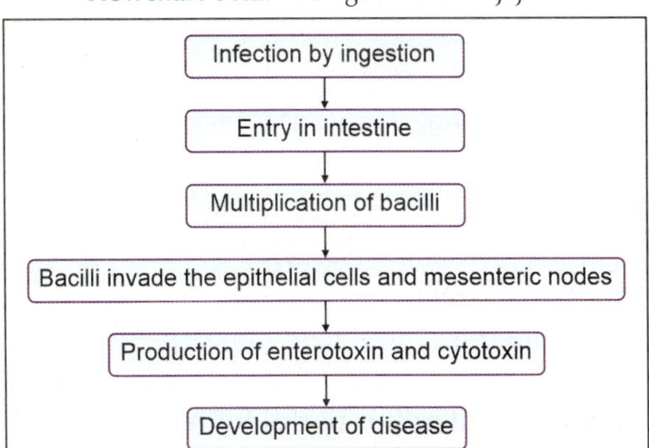

Clinical features

1. **Intestinal features:** Illness starts with fever, abdominal pain and watery-diarrhea with blood (WBC). It lasts for 1–2 weeks and self limiting.
2. **Extraintestinal features:** Very rare, but may produce the bacteremia, septicemia, meningitis and vascular infections like endocarditis, aneurysm and thrombophlebitis in immunocompromized hosts.

Complications

1. **GBS (Guillain-Barre syndrome):** It is an acute paralytic disease of peripheral nervous system mainly caused by serotype O19. It appears after 1–2 weeks of infection. It develops in 0.3 per 1000 infection. 20–40% GBS cases are due to *C. jejuni.*
2. **Alpha-chain disease:** Lymphoma of MALT.
3. **Others:** Septic abortion, Reiter's syndrome, GI perforation, HUS, thrombocytopenic purpura and periodonitis.

Laboratory Diagnosis

Specimens: Stool or rectal swab are collected in Cary-Blair medium and transported immediately to laboratory followed by subculture on solid and selective media.

Testing methods

A. **Microscopy:** Follow morphology.
B. **Culture:** Follow C/Cs.
C. **Biochemical reaction:** Follow B/Rs.
D. **Molecular method:** Multiplexed PCR like BioFire Film Array is useful. It targets 22 different gastrointestinal pathogens from stool.

E. Differentiation of *C. jejuni* from related species:

Follow **Table 66.1.**

Prevention

It is prevented by frequent hand washing before and after touching the animal products like chicken. Do not drink unpasteurized milk or untreated surface water. All poultry should be cooked to reach a minimum internal temperature of 165°F (74°C).

Treatment

Oral rehydration therapy is given to replace fluid and electrolytes loss. Erythromycin or azithromycin is the drug of choice. Other drugs are ciprofloxacin or doxycycline.

OTHER *Campylobacter* spp.

C. coli: It causes the same illness as like *C. jejuni* and differentiated by negative hippurate hydrolysis test.

C. lari: It produces the same illness as like *C. jejuni* and differentiated from *C. jejuni* and *C. coli* by resistance to nalidixic acid.

C. fetus: It is transmitted by feco-oral route. It causes bacteremia, septicemia and meningitis in hosts with IDDs.

C. curvus, C. concisus, C. sputorum and *C. mucosalis:* They cause root canal infection and abscess in tissues.

Arcobacter

Common species are *A. cryophilis, A. butzleri* and *A. nitrofigilis.* They are GNB, spiral shape, motile by unsheathed polar flagella. They are aerotolerant, grow at 15–25°C, catalase, nitrate and indoxyl acetate hydrolysis positive, hippurate hydrolysis negative, sensitive to nalidixic acid and resistant to cephalothin. They produce diarrheal disease similar to *C. jejuni.*

Helicobacter pylori

Introduction

Formerly called *Campylobacter pylori,* but in 1989, Goodwin et al., published sufficient reasons to justify the new genus name *Helicobacter.* Genus name (*Helicobacter*)

TABLE 66.1: Tests to differentiate between *Campylobacter* species

Test	C. jejuni	C. coli	C. lari	C. fetus	C. curvus
Growth at 25°C	–	–	–	+	–
Growth at 42°C	+	+	+	–	+/–
Catalase	+	+	+	+	–
H$_2$S	–	–	–	–	+
Hippurate hydrolysis	+	–	–	–	–
Indoxyl acetate hydrolysis	+	+	–	–	+
Nalidixic acid sensitivity	S	S	R	R	S
Cephalothin sensitivity	R	R	R	S	Not useful

suggests helical shape and species name, suggests genitive (close association) to pylorus (pylorus means circular opening of stomach in duodenum).

History

It was identified in 1982 by Australian scientists Barry Marshall and Robin Warren, from the person with chronic gastritis and gastric ulcers.

Morphology

Type according to Gram's stain: They are GNB under Gram's stain (**Fig. 66.3a**). They also stained by Giemsa stain or H and E stain, basic fuchsin stain (**Fig. 66.3b**), Warthin–Starry silver stain (**Fig. 66.3c**), acridine orange stain and phase-contrast microscopy.

Shape and size: They are spiral/helical shaped with 0.5–0.9 µm × 2.5–3 µm in size.

Motility: They are motile with tuft (4–8 in numbers) of sheathed flagella at one pole called lophotrichous (**Fig. 66.2b**), best seen in liquid medium.

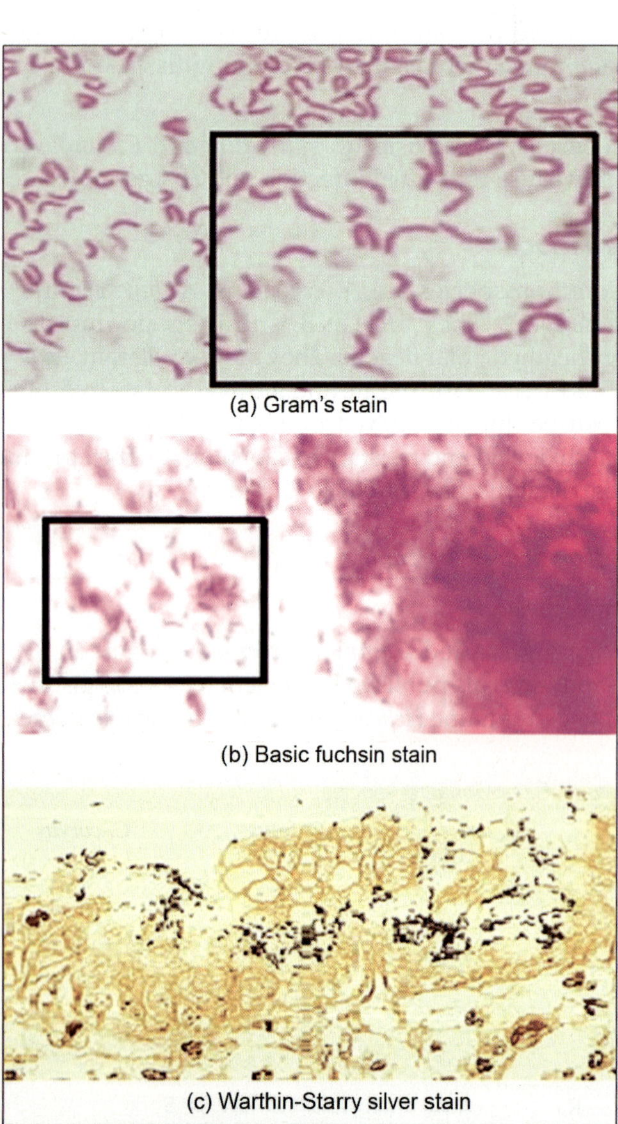

Fig. 66.3: Morphology of *H. pylori*

Cultural Characteristics (C/Cs)

Effective factors

- O_2 effect: They are microaerophilic.
- Optimum temperature is 37°C, but some strains can grow at 42–43°C.
- pH: 6.7
- Growth occurs in incubation under 5% O_2, 10% CO_2 and 85% nitrogen.

Culture in media

A. Liquid media:
- **Peptone water:** It is useful for motility detection.
- ***Brucella* broth:** It contains 1–10% horse or fetal calf serum.

B. Solid media: These are blood agar, chocolate agar, BHIA (5% horse blood), tryptose soya agar (5% sheep blood) and *Brucella* agar (5% sheep blood).

C. Selective media: Make above solid media selective by adding vancomycin, nalidixic acid and AMB. *Campylobacter* selective medium like Skirrow's medium is also useful.
- **Growth properties:** Small colonies develop in 3–7 days. Species is identified by B/Rs or by automated method like MALDI-TOF or VITEK.

Biochemical Reaction (B/Rs)

1. **Sugar fermentation tests:** Nonfermenter
2. **Phosphatase, catalase, oxidase and H_2S production tests:** Positive
3. **Rapid urease test and selective urease test:** *H. pylori* gives rapid (strong) positive urease from gastric mucosa. For more details **follow Ch. 118.**
4. **Nitrate reduction test:** Negative
5. **Nalidixic acid:** Resistance
6. **Cephalothin:** Sensitive

Pathogenicity

Reservoirs of infection: Human is the only reservoir. Only animal infected is monkey. Mucosal colonization by *H. pylori* has protective role because it minimise the occurrence of gastroesophageal reflux disease, Barrett's esophagus, adenocarcinoma of esophagus and allergic disease like asthma.

Sources of infection: These are water contaminated from the feces of case, saliva and gastric mucus.

Modes of transmission: It colonizes about 50% of human gastric antrum. *H. pylori* are contagious. Exact route is not known, person-to-person transmission occurs by either the oral–oral, orogastic (by gastric mucus) or feco-oral route. Consistent with these transmission routes, the bacteria have been isolated from feces, saliva, and dental plaque of some infected people. Transmission occurs mainly within families in developed nations, yet can also be transmitted orally by means of fecal matter through the ingestion of waste-tainted water in developing countries.

Incubation period: Few days.

Portal of entry: GIT.

Sites: It colonizes any part of stomach but gastric antrum is the most common site of infection. It colonizes the part of esophagus and duodenum commonly exposed to acid.

Precipitating factors (epidemiological determinants): These are following three types:

A. Agent factors (virulence factors)

1. **Motility:** To avoid the acidic environment of the stomach *H. pylori* uses its flagella to burrow into the mucus lining of the stomach to reach beneath the epithelial cells, where it is less acidic.
2. **Adhesin:** It is an organ of adhesion.
3. **Toxin:** Vacuolating cytotoxin produces the injury to host cells and responsible for ulcer. It is regulated by vacuolating cytotoxin gene.
4. **Urease production:** It splits urea to ammonia which buffers the gastric acidity and provides protection to the bacilli. It raises the pH from about 2 to a range of 6–7. Urease is 100 times more active than produced by *Pr. vulgaris*. Antiurease antibodies are not formed.
5. **Amidase and arginase:** They may contribute in production of ammonia.
6. **Molecular mimicry:** Cross reactivity between antigen of gastric mucosa and organism allows the formation of autoantibody responsible for chronic active gastritis.
7. **Cag Pathogenicity Island:** It is a group of genes (called cytokine associated genes) which helps in colonization of bacteria. It encodes for a specific protein called Cag A protein. Cag A protein injected by *H. pylori* in to the gastric epithelial cells which releases the cytokines (IL-1 is the major cytokine), causes cytoskeletal changes and proliferation thereby enabling *H. pylori* to colonize the gastric epithelium. Cag A gene is highly immunogenic, associated with peptic ulcer disease or gastric adenocarcinoma and patient with peptic ulcer disease or gastric adenocarcinoma have antibodies to Cag A gene. *H. pylori* and *H. influenzae* are the bacteria whose complete genome has been mapped.

B. Host factors

1. **Age:** *H. pylori* is common in adults. Childhood is a risk factor for *H. pylori* colonization, but immunity does not develop and prevalence increase with age. People infected in early age are likely to develop more intense inflammation that may progress to atrophic gastritis, gastric ulcer, gastric cancer or both. Acquisition at an older age may lead to duodenal ulcer.
2. **Diet:** Preserved food and diet rich in salt increase the risk of cancer while diet rich in vitamin C and antioxidants are protective.
3. **Smoking:** It increases risk of ulcer and cancer.
4. **Immunity:** Infection induces the AMI (local and systemic antibodies) and CMI, but not protective.

C. Environmental factors

1. **Economical status:** Over 80% of individuals infected with the bacterium are asymptomatic. More than 50% of the world's population harbor *H. pylori* in upper GIT. Infection is more common in developing countries than Western countries. Low prevalence (30% in adult) in developed countries than developing countries (80% in adult).
2. **Other risk factors:** These are low socioeconomic status, crowding, poor hygiene, etc.

Pathology: Bacteria present over the surface, but do not invade the cells. They produce chronic superficial gastritis due to primary colonization not due to re-infection. Reinfection can cause chronic gastritis but rate is very low about 0.5% per year. It also called type B gastritis (type A gastritis is due to pernicious anemia) or chronic active gastritis. Antral gastritis is linked with duodenal ulcer while pangastritis is linked with stomach ulcer (peptic ulcer) and adenocarcinoma of stomach. Infection remain lifelong, if remain untreated.

Clinical features: These are dyspepsia, stomach (abdominal) pain, nausea, bloating, belching and sometimes vomiting or black stool. Pain usually occurs on empty stomach in mid night or in early morning. Pain relieved little after meal. *H. heilmanii* causes human and animal infection. *H. cinnaedi* and *H. fennelliae* are opportunistic pathogens in HIV patient.

Complications: (1) Peptic ulcer and duodenal ulcer: *H. pylori* detected from 66 to 77% cases of peptic ulcer and duodenal ulcer, respectively. **(2) Chronic atrophic gastritis. (3) Hypergastrinemia:** Gastrin is the hormone released by stomach wall in the blood. It stimulates the secretion of gastric juice. Release of gastrin is controlled by somatostatin. Antral colonization of *H. pylori*, diminishes the somatostatin producing cells, called D cells → Decrease release of somatostatin → Loss of inhibitory control over gastrin release → Increase gastrin release called hypergastrinemia. **(4) Malignancies:** These are gastric carcinoma and MALT lymphoma. Risk is lowered with successful treatment of *H. pylori*.

Laboratory Diagnosis

Specimens: (1) For culture: It highly specific, but less sensitive, because of difficulties in isolation of *H. pylori*. Gastric mucosa and vomitus are collected. **(2) For serology:** Stool and blood. **(3) For molecular tests:** Stool, dental plaque, water and gastric juice.

Testing methods: Following two types of tests.

1. **Invasive:** Gastric mucosa collected by endoscopy and tested as mentioned in **Flowchart 66.3**.
2. **Noninvasive**
   - **Urea breath test:** Patient drink [13]Carbon or [14]Carbon labeled urea solution and then blow in a tube. If

Flowchart 66.3: Testing of gastric mucosa

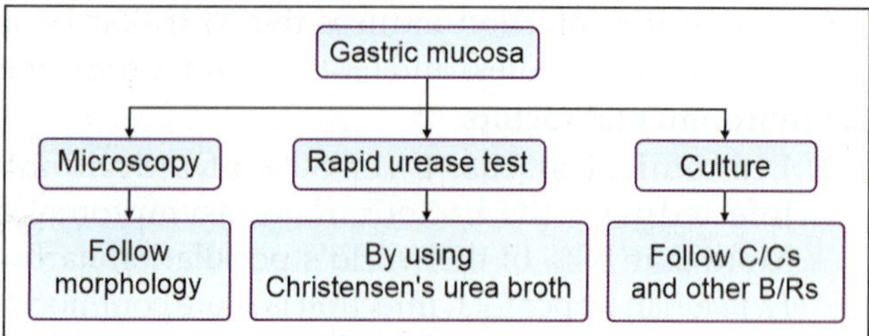

Flowchart 66.4: Treatment of *H. pylori*

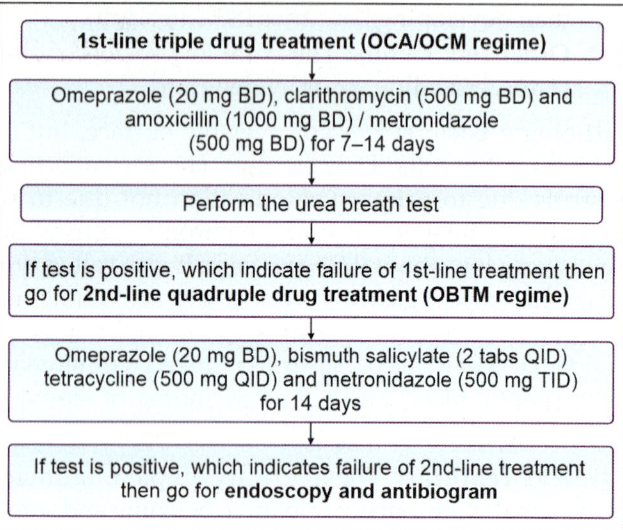

urease is present, the urea will be hydrolyses which can be detected in breath film by spectroscopy. It is the most accurate (>90% sensitivity and >90% specificity), reliable, consistent and used to monitor the treatment. It is useful in follow up after treatment and test becomes negative after improvement following antibiotics. It will be positive with massive infection.

- **Antigen detection test from stool (coproantigen):** It is done by ELISA. It is less accurate than urea breath test.
- **Antibody detection from blood:** Detection of IgG by ELISA or by immunoblot assay, but less useful.
- **Molecular method:** PCR from stool, dental plaque, water and gastric juice.

Prevention

It is prevented by improving the hygiene. Vaccine is under trial to prevent the cancer and other complications.

Treatment

Treatment is not given in asymptomatic person or as a prophylaxis. It is given for duodenal, peptic ulcer and low grade lymphoma. Treatment given for *H. pylori* can cause regression of gastric lymphoma. Regimens are mentioned in **Flowchart 66.4**.

ACCESS YOURSELF

Short Notes

1. Pathogenicity or Lab., diagnosis of *C. jejuni/H. pylori*.

Short Questions For Theory/Viva Questions

1. Name two selective media, each for *C. jejuni* and *H. pylori*.
2. Describe in short: Rapid urease test and selective urease test.

Comments On

1. Selective urease test is more advantageous than rapid urease test to detect *H. pylori*.
2. Mucosal colonization by *H. pylori* has protective role.
3. Early age or childhood colonization by *H. pylori* increases the risk of gastric cancer.
4. *H. pylori* causes hypergastrinemia.

MCQs for Chapter Review

C. jejuni

1. **Which of this is a selective medium for *C jejuni*?**
   a. Wilson and Blair medium
   b. Bile salt agar
   c. Butzler's medium
   d. TCBS medium

2. ***Campylobacter* culture media are:**
   a. Schadlar's agar
   b. CVA medium
   c. Regan Lowe medium
   d. Skirrow's medium
   e. *Campylobacter* blood agar

3. **All are true statement about *Campylobacter jejuni*, *except*:**
   a. Human is the only reservoir.
   b. Can cause GB syndrome
   c. Poultry is the source of infection.
   d. Common cause of campylobacteriasis

H. pylori

4. **True about *H. pylori*:**
   a. Vacuolating cytotoxin
   b. Antiurease antibody detection specific for type I
   c. Found in jejunum
   d. Treatment is A1(OH) 2

5. **True about *H. pylori* is all *except*:**
   a. It splits urea and produces ammonia to survive
   b. Produces gastric carcinoma
   c. Gram-negative curved rods
   d. Cag A gene is not associated with risk of duodenal ulcer

6. **Seven sheathed flagella is seen in:**
   a. *V. cholerae*
   b. *H. pylori*
   c. *P. aeruginosa*
   d. Spirochetes

7. **All are true regarding *H. pylori*, *except*:**
   a. Less prevalent in developing countries
   b. Ulcer is caused by toxigenic strains
   c. Urea breath +ve test is a rapid method of detection
   d. Gram –ve organism

8. **True about *H. pylori*:**
   a. Seen in 8–90% cases of gastric ulcer
   b. Seen in 20–25% cases of duodenal ulcer
   c. Transmitted from man-to-man, feco-orally and by orogastric route
   d. Common in adults of developing countries

9. **All of the following are true about *H. pylori*, *except*:**
   a. About 50% of world population affected
   b. 85% of population is affected in some developing countries
   c. All children in developing countries have immunity by five years of age
   d. Infection is common in low socioeconomic status

10. **True about *Helicobacter pylori* is:**
    a. Culture and Gram's staining of biopsy is the gold standard investigation.
    b. Controlled urea breath is negative with massive infection.
    c. Anti urease antibody are produced only by invasive strains.
    d. Urease activity provides protective environment to the bacilli.

11. **Noninvasive test for *H. pylori*:**
    a. Rapid urease test
    b. Urea breath test
    c. Stool antigen assay
    d. Stomach aspiration culture
    e. Biopsy

12. ***H. pylori* true about:**
    a. Gram-positive spiral organism
    b. It is a protozoon.
    c. Causes chronic gastritis in adults due to reinfection
    d. Treatment causes regression of gastric lymphoma.
    e. Duodenal mucosa normal

13. **Regarding *H. pylori* all are true, *except*:**
    a. Gram-negative bacillus
    b. Strongly associated with duodenal ulcer
    c. Associated with lymphoma
    d. C-14 urea breath test is used in diagnosis.
    e. It should be eradicated in all cases, whenever detected

14. **True about *H pylori* is all *except*:**
    a. Urea Breath test is diagnostic.
    b. Gram-negative flagellate bacilli
    c. Risk factor for development of adenocarcinoma
    d. It provides lifelong immunity.

15. **Which test is most accurate for diagnosing *H. pylori* infection?**
    a. Urea breath test   b. Serology
    c. Stool antigen   d. a + b

Answers and Explanation of MCQs

1. c

2. b, d, e
- Follow section, *Campylobacter jejuni* [**Cultural characteristics (C/Cs)** of *C. jejuni*] for explanation of answers of MCQs 1–2.

3. a
- Campylobacteriasis is the zoonosis where animals like poultry, cattle, sheep, swine, dogs, cats and some birds are the reservoirs and no human reservoir. Common complication is GB syndrome.
- Source is raw milk or raw chicken. *C. jejuni* is the most common cause of campylobacteriasis.

4. a
- In *H. pylori* vacuolating cytotoxin produces the injury to host cells and responsible for ulcer. Antiurease antibody is not formed. For option c follow section, *H. pylori* (**pathogenicity → sites**) for explanation. A1 (OH) 2 is not used in treatment.

5. d
- Cag A gene is highly immunogenic, associated with peptic ulcer disease or gastric adenocarcinoma and patient with peptic ulcer disease or gastric adenocarcinoma have antibodies to Cag A gene.

6. b
- *H. pylori* is motile by lophotrichous flagella. *V. cholerae* and *P. aeruginosa* are motile by monotrichous flagella. Spirochetes have endoflagella.

7. a
- Low prevalence (30% in adult) of *H. pylori* in developed countries than developing countries (80% in adult).

8. c
- *Helicobacter pylori* detected from 66% and 77% cases of peptic ulcer and duodenal ulcer, respectively. For option c follow section, *Helicobacter pylori* (**pathogenicity → mode of transmission**) and for option d follow section, *Helicobacter pylori* (**pathogenicity → host factors**) for explanation.

9. c
- Follow section, *Helicobacter pylori* (**pathogenicity → host factors and environmental factors**) for explanation.

10. d
- Culture is highly specific but less sensitive because of difficulties in isolation of *H. pylori*, so its not the gold standard. Controlled urea breath test is positive with massive infection. Anti urease antibody are not produced by invasive strains. Urease splits urea to ammonia which buffers the gastric acidity and provides protection to the bacilli.

11. b and c
- Follow section, *Helicobacter pylori* (**laboratory diagnosis**) for explanation.

12. d
- It is gram-negative helical/curved bacteria, not protozoon. Chronic superficial gastritis occurs due to primary colonization not due to reinfection. Reinfection can cause chronic gastritis but rate is very low about 0.5% per year. MALT is caused by *H. pylori* and treatment given for *H. pylori* can cause regression of gastric lymphoma. Duodenal mucosa is not normal because *H. pylori* can damage duodenal mucosa.

13. e

14. d

15. a
- Follow section, *Helicobacter pylori* for explanation of answers of MCQs 13–15.

Infections of Nonfermenters

Chapter Outline

INTRODUCTION

Nonfermenters are a taxonomically heterogeneous group of bacteria that cannot ferment glucose. This does not necessarily exclude that species can catabolize other sugars or have anaerobiosis. Their classification is given in **Flowchart 67.1**.

Pseudomonas aeruginoa

Synonym

It also called *Pseudomonas pyocyanea* or *Bacillus pyocyaneus*.

Morphology

Type according to Gram's stain: They are GNB.

Shape and size: They are short rods about 0.5 µm × 1.5–3 µm in size.

Arrangement: They arranged singly or in pairs or in small group as shown in **Fig. 67.1**.

Motility: They are motile by monotrichous flagellum.

Capsule: They are noncapsulated, but they have slime layer made up of glycocalyx.

Flowchart 67.1: Classification of nonfermenters

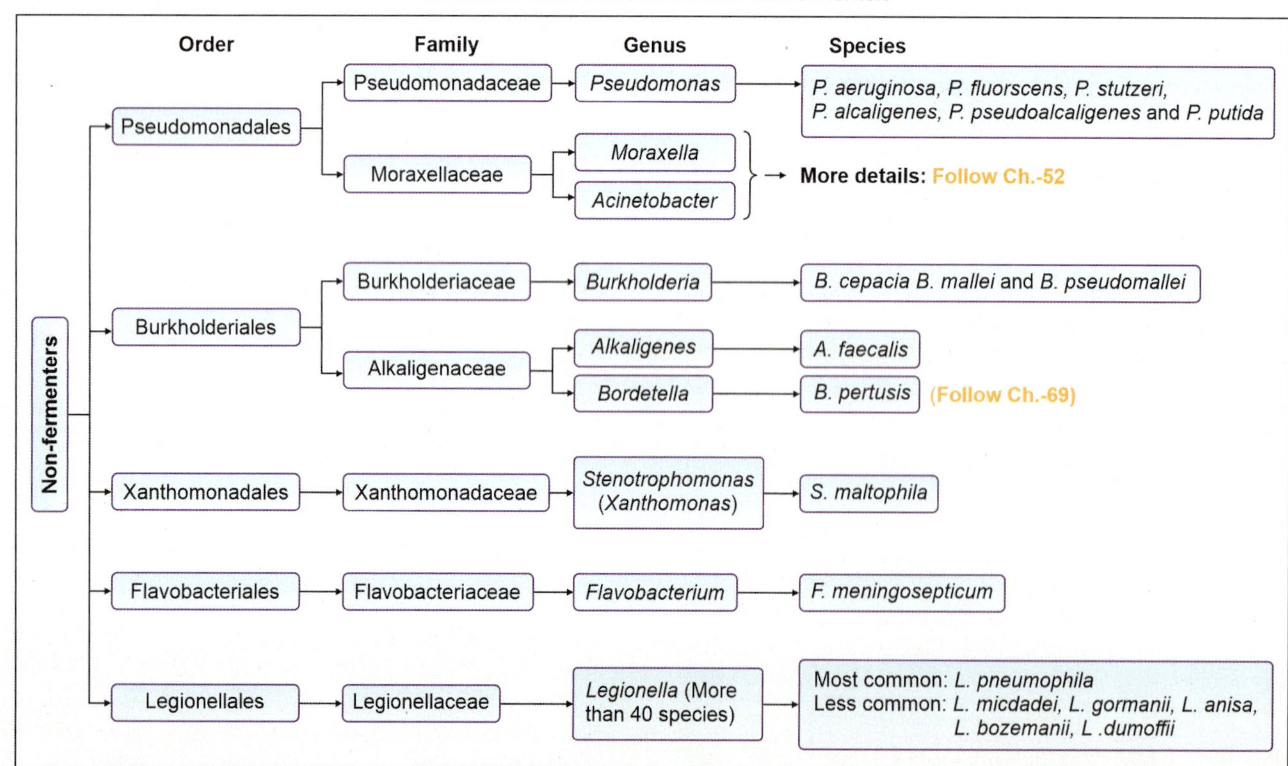

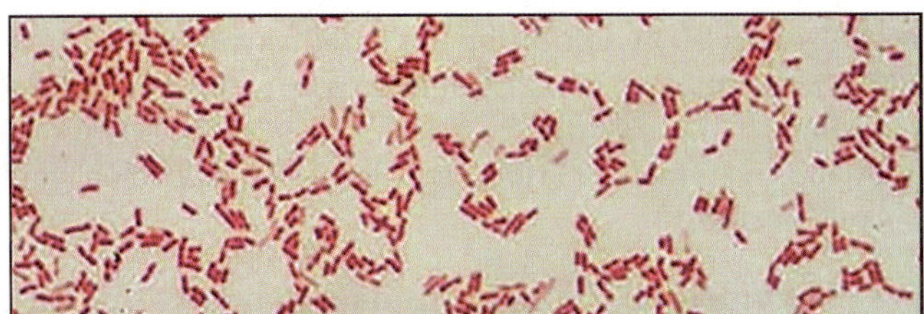

Fig. 67.1: Morphology of *P. aerugoinosa*

Spores: They are nonsporing.

Fimbriae: They present in some clinical isolates.

Cultural Characteristics (C/Cs)

Effective factors

- O₂ effect: They are strict aerobes can grow anaerobically if nitrate is available.
- Temperature: Optimum temperature for growth is 37°C (6–42°C).
- pH: They required neutral pH.

Culture in media

A. Liquid media

- **Peptone water:** Bacteria produce dense turbidity with surface pellicle.
- **Enrichment medium:** It is *Pseudomonas* enrichment broth.

B. Solid media

- **Nutrient agar plate/slant:** Colonies are large, pigmented, opaque and irregular with distinctive musty, mawkish, earthy or sweet grapes like odor. *Pseudomonas* produces **pigment** on nutrient agar which is diffusible in nature (unlike *Staphylococcus* where pigment is nondiffusible) and spreads in whole medium from the site of inoculation. About 10% strains are nonpigmented. Pigments are following four types:

1. **Pyocyanin:** It is a bluish green phenazine pigment (**Fig. 67.2a**) soluble in water and chloroform produced only by *P. aeruginosa*. It inhibits the growth of many other bacteria and makes the organism dominant in mixed infections.
2. **Pyoverdin (fluorescein):** It is a greenish yellow pigment (**Fig. 67.2b**) soluble in water, but not in chloroform produced by other species also. In old culture, it oxidized to yellowish brown pigment.

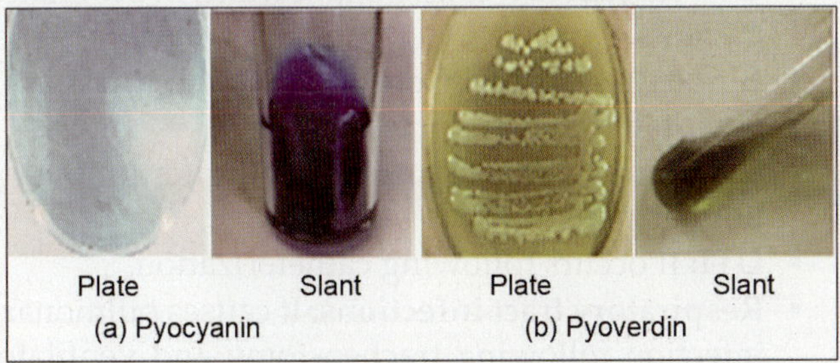

Plate | Slant | Plate | Slant
(a) Pyocyanin | (b) Pyoverdin

Fig. 67.2: Pigments of *P. aerugoinosa*

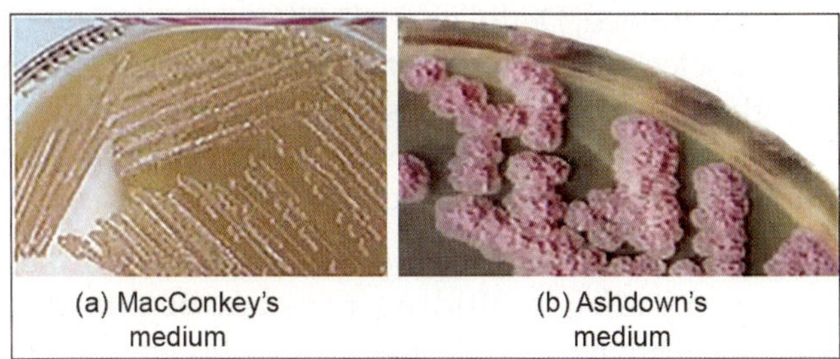

Fig. 67.3: (a) Pale colonies on MacConkey's agar by *P. aeruginosa;* (b) Wrinkled purple colonies on Ashdown's medium by *B. pseudomallei*

3. **Pyorubin:** It is red in color.
4. **Pyomelanin:** It is a brown in color.

- **Blood agar:** Some strains are hemolytic.

C. Selective media

- **MacConkey's agar (Fig. 67.3a) and DCA:** Pale colonies produced by *P. aeruginosa* due to NLF.
- **Cetrimide agar:** It contains 0.03% cetrimide.
- ***Pseudomonas* Isolation Agar (PIA):** It contains irgasan as selective agent.

D. Media for production and identification of pigments

- **For pyocyanin:** Medium P and PIA (selective) are useful for pigment identification.
- **For pyoverdin (fluorescein):** King's medium B.
- **For pyorubin:** 1% DL-Glutamate medium.
- **For pyomelanin:** Furunculosis agar, Davis-Mingioli's mineral salt medium (1% tyrosine).

Automated culture: Like MALDI-TOF or VITEK is used to identify the species from culture.

Biochemical Reactions (B/Rs)

Sugar fermentation tests: *P. aerugoinosa* is nonfermentative, but oxidative. It oxidized the sugar and produces acid only. Sugar is the only carbon source. Peptone water is unsuitable to detect weak acid production, so Hugh and Leifson's (**oxidative-fermentation test**) medium with glucose is useful to detect weak acid production from sugar.

I M Vi C tests: – – – +

Catalase, oxidase, nitrate reduction, arginine decarboxylase tests: Positive

H₂S production and urease tests: Negative

TSI: K/Nil or K/K (no change in butt, because nonfermentive and alkaline in slant).

Resistance

Sterilization: *P. aeruginosa* is killed by heating at 55°C in 1 hour.

Disinfection: It is resistant to common antiseptics and disinfectants such as quaternary ammonium compounds, chloroxylenol and hexachlorophene. It can grow in bottles of antiseptics like Dettol, iodine, cetrimide lotions or soap lotion, etc., kept for use in hospitals. Indeed, Dettol or cetrimide is incorporated in a media selective for *P aeruginosa*. It is sensitive to acids, beta glutaraldehyde, silver salts and strong phenolic disinfectants. It is susceptibility to silver and silver sulfonamide used as topical cream in burns.

Drug resistance: *P. aeruginosa* is resistant to many antibiotics called multidrug resistant or pan drug resistant. Drug resistant is due to alteration in cell wall and porin channel. Resistance to common antibiotics and antiseptics help to establish itself widely in hospitals.

Pathogenicity

Reservoirs and sources of infection: *P. aeruginosa* is widely available in hospital environment. It can survive and multiply even with minimal nutrients, if moisture is available. **Equipments** like respirators, endoscopes, stitch materials, etc., **articles** like hospital bed, cloths, etc., **medicines** like lotions, creams, eye-ear drops, etc., **stocks** of distilled water, plants and flowers may be frequently contaminated and act as reservoirs and sources of infection.

Modes of transmission: It occurs directly from environment or contaminated hospital materials.

Portal of entry: Skin, respiratory system, etc.

Sites: Skin, soft tissues, GIT, blood, respiratory system, etc.

Precipitating factors (epidemiological determinants)

A. Agent factors (virulence factors): These are two types like intracellular and extracellular.

1. **Intracellular (cell wall associated) factors**
   - **Antigens:** O-Ag (LPS/endotoxin) shows endotoxicity, flagellar Ag used as an organ of locomotion and fimbrial Ag is used as an organ of adhesion.
   - **Slime layer:** It inhibits phagocytosis. Slime layer made up by glycocalyx has ability to binds bacteria with damaged tissues. It also helps bacteria to bind with plastic material of medical devices like catheters, suture materials, pacemakers, implants, Ryle's tube, etc., called biofilm. Uses of such devices cause diseases.

2. **Extracellular factors**
   - **Pyocyanin pigment:** It stops ciliary movement of respiratory epithelium and prevents bacteria from host defense. It inhibits growth of other bacteria and makes the bacterium dominant in mixed infection. It catalyzes the production of superoxide and H_2O_2.
   - **Toxins: (1) Exotoxins:** It produces exotoxin A, S, T, U and Y. A act as NADase. It acts like diphtheria toxin and inhibits the protein synthesis by inhibiting the elongation factor-2 (EF-2). **(2) Enterotoxin:** It causes collection of fluid in ligated rabbit illeal loop. **(3) Hemolysin:** It is rarely useful.

> **Note: Toxins act by inhibiting the protein synthesis**
> - By inactivation of EF-2: Diphtheria toxin and *Pseudomonas* toxin
> - By inactivation of 60S ribosome: Shiga like toxin (SLT) and Shiga toxin

   - **Enzymes: (1) Phospholipidase:** It breaks lipid and lecithin, and leads tissue destruction. **(2) Elastase:** It degrades elastin, damages to lung parenchyma and causes hemorrhagic lesions. **(3) Protease:** It helps to spread in tissues.

B. Host factors

1. **Immunity:** It is the common cause of life threatening fever in person with IDDs.
2. **Long hospital stay:** It favors the infection.

C. Environmental factors: *P. aeruginosa* is common in hospital environmental as mentioned earlier.

Clinical features: Two types of infections.

1. **Community acquired infections (outside hospital in community)**
   - **Ear infections:** *Pseudomonas* is the most common cause of malignant otitis externa. It also causes ASOM (most common cause for ASOM is *Strept. pneumoniae* in children in 33% cases) and CSOM outside the hospital (most common cause for CSOM is anaerobes).
   - **Intestinal infection:** It causes infantile diarrhea due to heat labile enterotoxin.
   - **Respiratory tract infection:** It is the most common cause of respiratory infection in patient of cystic fibrosis.
   - **Shanghai fever:** Self limited febrile illness resembling typhoid fever.

2. **Hospital-acquired (nosocomial) infections**
   - **Localized infections of skin (blue pus → fruity odor):** These are traumatic wound infection; burn wound infection (most common agent) and bed sore. It causes ecthyma gangrenosum and many other skin lesions in patients with leukemia and malignancy. Ecthyma gangrenosum occurs singly or in small numbers on the perineum, buttocks and extremities. It causes nail bed infection following excessive exposure of hands to detergents and water. Nail gets separated from nail bed, becomes green with foul odor called green nail.
   - **Localized infections in other organs:** These are eye infections and brain abscess.
   - **CVS infection:** It causes endocarditis in debilitated patients.
   - **Bloodstream infection:** It causes septicemia in debilitated patients.
   - **CNS infection:** It causes meningitis following lumbar puncture.
   - **UTI:** It occurs following catheterization.
   - **Respiratory tract infections:** It causes pulmonary infection following tracheostomy and ventilator associated pneumonia following ventilation.

Laboratory Diagnosis

Specimens: Urine, pus, blood, sputum, CSF, swab, etc., are collected in enrichment broth and than do subculture on solid media.

Testing methods

A. Microscopy: Follow morphology.

B. Culture: Follow C/Cs.

C. Biochemical reactions: Follow B/Rs.

D. Bacterial typing: (1) Genotyping methods: PFGE is best useful, because it has more discriminatory power. **(2) Phenotyping methods:** These are following.

- Antibiogram typing: It is more useful methods.
- Serotyping: It is less useful because less discriminatory power. On the bases of O Ag 27 serotypes are identified.
- Bacteriophage typing: Less useful because less discriminatory power.
- Bacteriocin typing: Bacteriocin produced by *P. aeruginosa* is called pyocin or called aeruginosin with three types like R, F and S. Total 105 types are identified based on growth inhibition of indicator strain by pyocin. It is less useful because less discriminatory power.

Prevention

Strict aseptic prequations are required to control the infections.

Treatment

Pseudomonas is multidrug resistant. Antibiotics are given after antibiogram.

Burkholderia spp.

Bukholderia cepacia

Synonym: Previously, it was called *Pseudomonas cepacia*.

Morphology: They are GNB and motile by lophotrichous flagella.

Cultural characteristics (C/Cs): They are aerobes and optimum temperature for growth is 25–35°C. Growth occurs on nutrient agar in 48 hours, but on prolonged incubation, colonies become reddish purple. Species is identified by B/Rs or by automated method like MALDI-TOF or VITEK.

B/Rs: They acidify mannitol, sorbitol and sucrose and gives oxidase positive.

Genotyping: Based on DNA hybridization method 9 genomovars or groups are identified of which II and III are causing infection in patients of cystic fibrosis.

Pathogenicity

- **Virulence factors:** These are pili (organs of adhesion), LPS (endotoxic activities) and elastase.

- **Clinical features:**
  1. **Plant pathogen:** It is nutritionally very versatile and causes onion rot [Cepia (Latin) means onion].
  2. **Human pathogen:** It is an opportunistic pathogen and causes fatal necrotizing pneumonia in a patient with cystic fibrosis or chronic granulomatous disease. It also causes UTI, respiratory infections, wound infections, peritonitis, endocarditis and septicemia.

Treatment: It can grow in many disinfectants and even use of Pn G acts as sole source of carbon. Inherently, it is multidrug resistant.

Bukholderia mallei

Synonym: Previously, it was called *Pseudomonas mallei*.

History: It was 1st discovered by Loeffler and Schultz in 1882.

Morphology: They are GNB, but stain irregularly with beaded form, slender rod in shape, 0.5 µm × 2–5 µm in size and only nonmotile species of genus.

Cultural characteristics (C/Cs): They are aerobes and facultative anaerobes and grow at wide range of temperature.

- **Culture in media**
  1. **Nutrient agar:** Young colonies are small and translucent and old colonies are yellowish opaque.
  2. **MacConkey's agar:** No growth occurs.
  3. **Medium contains potato:** Bacteria produce amber or honey like growth.
- **Culture in animal:** Intraperitoneal injection into male guinea pig produces swelling of the testis, inflammation of tunica vaginalis and scrotal skin ulcer called straus reaction. Test is not useful. Tunica reaction also caused by *Brucella*, *C pseudotuberculosis* (Preisz-Nocard bacillus), *Actinobacillus* and *B. pseudomallei*.

Biochemical reactions (B/Rs): It is biochemically inactive, attack on glucose only and oxidase negative.

Pathogenicity

- **Disease name:** Disease called glanders.
- **Synonym:** In Latin called malleus.
- **Reservoirs and sources of infection:** These are equine animals like horses, mules and asses.
- **Modes of transmission:** It is a zoonosis transmitted by direct skin contact with infected animals in persons who handle the horses. It is the most dangerous bacteria transmitted accidentally by inhalation in laboratory workers.
- **Clinical features: (1) Animal infection:** It occurs in two forms like **glanders** presents as respiratory nodules with profuse nasal discharge and **farcy** presents as skin plus subcutaneous lymphatic nodules. **(2) Human infection:** It occurs in two forms like **acute** characterized by febrile illness with nasal discharge and **chronic** characterized by skin plus respiratory nodules or abscess.

Laboratory diagnosis: Blood, sputum, discharge from lesion, etc., are tested by microscopy (follow morphology) and culture (follow C/Cs). Skin test called mallein test is performed by SC, IC or by conjunctival route which based on DTH.

Prevention: Avoid contact with infected animals.

Treatment: Bacteria are multidrug resistant. Antibiotics are given after antibiogram. Ceftazidime is drug of choice and given with cotrimoxazole, tetracycline, amoxicillin, clavulanate or chloramphenicol.

Bukholderia pseudomallei

Synonym: Whitmore's bacilli, *Actinobacillus whitmori*, *Malleomyces pseudomallei* or *Loeffrella pseudomallei*.

History: It was discovered by Whitmore in 1913, hence called Whitmore's bacilli. Human disease was 1st described by Whitmore and Krishnaswami in 1913.

Morphology: They are GNB, showing typical 'bipolar safety pin' appearance in methylene blue stain and motile by lophotrichous type of flagella.

Cultural characteristics (C/Cs): They are strict aerobes and grow at 42°C.

1. **Solid media:** Rough corrugated colonies develop on nutrient agar, MacConkey's agar and blood agar. Species is identified by B/Rs or by automated method like MALDI-TOF or VITEK.
2. **Selective media:** Production of wrinkled purple colonies on Ashdown's medium as shown in **Fig. 67.3b**.

Biochemical reaction (B/Rs): It forms acid from several sugars, liquefies gelatine, decarboxylates arginine and causes intracellular accumulation of poly-β hydroxyl butyrate.

Pathogenicity

- **Disease name:** Disease called melioidosis.
- **Synonym:** Initially limited in Asia and North Australia, later spread to USA military persons who return from Vietnam war with exacerbation called 'Vietnam time bomb'.
- **Meaning:** Word derived from melis means disease of asses and eidos means resemblance, as it resemble to malleus (glander).
- **Epidemiology:** It presents in South East Asia, India and North Australia.
- **Reservoirs of infection:** These are rodents.
- **Sources of infection:** Bacteria present as saprophytes in soil and water in endemic area, which may act as sources of infection.
- **Modes of transmission:** It is a zoonosis transmitted by direct contact of abraded skin with contaminated water or soil. It also transmitted by inhalation route, and it is the dangerous agent to be used in bioterrorism.
- **Precipitating factors:** It produces two heat labile toxins one is lethal and second is necrotizing. DM increases the risk.

- **Clinical features: (1) Acute infection:** It produces acute pulmonary infection characterized by extensive necrotizing pneumonia may mimics tuberculosis. X-ray shows the upper lobe infiltration. Acute stage has high case fatality. **(2) Subacute infection:** Typhoid like illness. **(3) Chronic infection:** Formation of abscess/suppurative foci in skin, subcutaneous tissues, bones, muscles (myositis) and in other organs. **(4) Latent infection:** Bacilli survive in RE-system and reactivation can occur.
- **Complication:** Acute stage produces fatal septicemia.

Laboratory diagnosis: Sputum, pus, blood, urine, etc., are tested by microscopy (follow morphology) and culture (follow C/Cs) and biochemical reactions (follow B/Rs). Serological tests used in diagnosis are ELISA for IgG and IgM, latex agglutination test and indirect hemagglutination (IHA) test.

Prevention: Take the strict aseptic precautions to control the infection.

Treatment: Same as *B. mallei*.

Alcaligenes faecalis

Meaning: Genus name is given, because it is non-fermenters and gives alkaline reaction in litmus milk test. Species name is given, because it was identified from feces.

Morphology: They are GNB, short rod shaped and motile by peritrichous flagella.

Cultural characteristics (C/Cs): They are strict aerobes. Following are useful media.

1. **Blood agar:** Some strains produce the greenish colonies with fruity odor (green apple odor). Species is identified by B/Rs or by automated method like MALDI-TOF or VITEK.
2. **MacConkey's agar:** Most strains produce the thin and spreading colonies with irregular edge.

Biochemical reactions (B/Rs): It is nonfermentative but oxidative. It is citrate, oxidase and nitrate positive, but urease negative.

Pathogenicity: It presents as soil and water saprophyte, hospital contaminant in respirators, nebulizer, IV fluid, etc., and commensals in human and animal intestine. It causes nosocomial infections like UTI. It also causes gastroenteritis, typhoid like fever, local sepsis and opportunistic infections like otitis media, meningitis and pneumonia.

Laboratory diagnosis: Feces, blood, urine, pus, etc., are tested by microscopy, culture and by B/Rs as described above.

Treatment: Fluoroquinolones and nalidixic acid are the sensitive drugs.

Stenotrophomonas maltophila

Synonym: Previously it called *Pseudomonas maltophila*.

Morphology: They are GNB, 0.7–1.8 × 0.4–0.7 µm in size and motile due to polar flagella.

C/Cs: They are aerobes and grow well on MacConkey's agar producing pigmented colonies. Species is identified by B/Rs or by automated method like MALDI-TOF or VITEK.

B/Rs: It is nonfermentative but utilizes the sugar oxidatively, catalase and lysin decarboxylation are positive, but oxidase-negative (nonoxidative).

Pathogenicity

- **Reservoirs and sources of infection:** It is ubiquitous in aqueous environments, soil, and plants. It is a rhizosphere (present in soil surrounding the plant root).
- **Precipitating factors**
  - **Virulence factors: (1)** It produces LPS responsible for endotoxicities. **(2)** It produces two inducible chromosomal metallo-β-lactamases (designated L1 and L2) responsible for resistant carbapenems. **(3)** Efflux pump mechanism produces resistant to many broad-spectrum antibiotics.
  - **Host factors:** It causes high morbidity and mortality in persons with IDDs *S. maltophilia* colonizes more in patient with cystic fibrosis.
- **Clinical features:** It causes HAIs like VAP, UTI, septicemia and wound infections. It is an opportunistic pathogen causes ecthyma gangrenosum in neutropenic patients.

Laboratory diagnosis: Bacteria are diagnosed from blood, urine, sputum samples by microscopy, culture and biochemical tests which are described above.

Treatment: Multidrug resistant and useful antibiotics are tigecycline and polymyxin B. Antibiogram requires nonstandard culture methods (incubation at 30°C). Testing at the wrong temperature results in isolates being incorrectly reported as being susceptible when they are, in fact, resistant. Disc diffusion method is unreliable, so agar dilution method should be used.

Flavobacterium meningosepticum

Meaning: Genus name derived from flavus (Latin) means yellow and bacterium means related to bacteria, because of yellow color pigment production by bacteria. Species name is associated with meningitis and septicemia produced by it.

History and synonym: *Flavobacterium meningosepticum,* King, 1959), *Chryseobacterium meningosepticum,* Vandamme et al. (1994) and *Elizabethkingia meningoseptica,* Kim et al. (2005).

Morphology: They are GNB, curved rod and nonmotile.

Cultural characteristics (C/Cs): They aerobes and grow better at 40°C. Following are useful media.

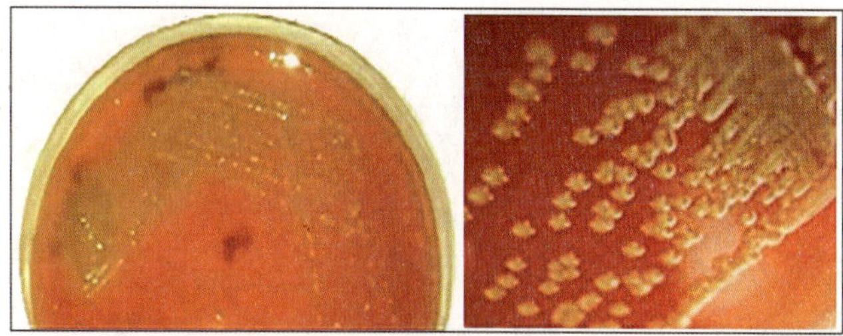

Fig. 67.4: Pigment of *F. meningosepticum* on blood agar

1. **Blood agar:** Produces pale yellow pigment after 24 hours **(Fig. 67.4)** and better at 40°C called zeaxanthine (xanthine means yellow). Species is identified by B/Rs or by automated method like MALDI-TOF or VITEK.
2. **Chocolate agar:** Also useful medium.
3. **MacConkey's agar:** Poor growth.

Biochemical reactions (B/Rs): It is weakly fermentative (oxidative) plus proteolytic, catalase, oxidase and indole positive, but urease negative.

Pathogenicity: They are ubiquitous saprophytes producing nosocomial infections (neonatal meningitis and septicemia associated with nursery outbreak) and opportunistic infections.

Laboratory diagnosis: CSF or blood is tested by microscopy, culture and by B/Rs which are described above.

Treatment: Fluoroquinolones and nalidixic acid are the sensitive drugs.

Legionella pneumophila

Meaning and history: Genus named because 1st identified in member of American legion (group of soldiers) who attended a convention in Philadelphia in 1976.

Classification: Follow **Flowchart 67.2.**

Morphology:

- **Staining properties:** They are GNB or gram-negative coccobacilli, but difficult to examine under Gram's stain from sputum. Smear contains numerous leukocytes. They are better stained by silver impregnation or fluorescent stain. Fat globules stain by Sudan black. Cell wall of *L. micdadei* is acid fast in nature.
- **Shape and size:** They are short rod, coccobacilli with pointed ends, filamentous forms **(Fig. 67.5)** and about 0.2–0.3 µm × 1–3 µm in size.

Flowchart 67.2: Classification of *Legionella* spp.

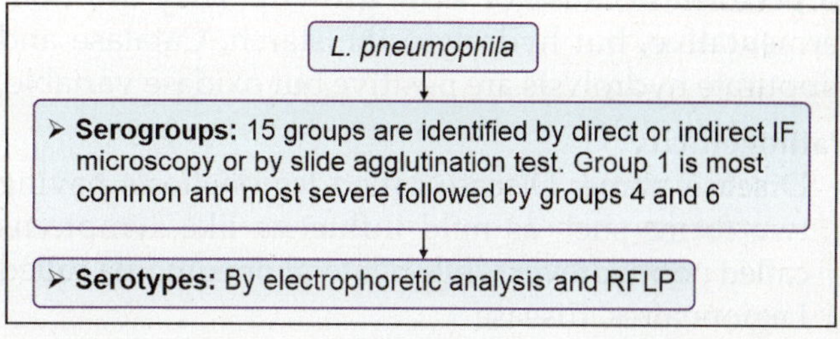

Infections of Nonfermenters

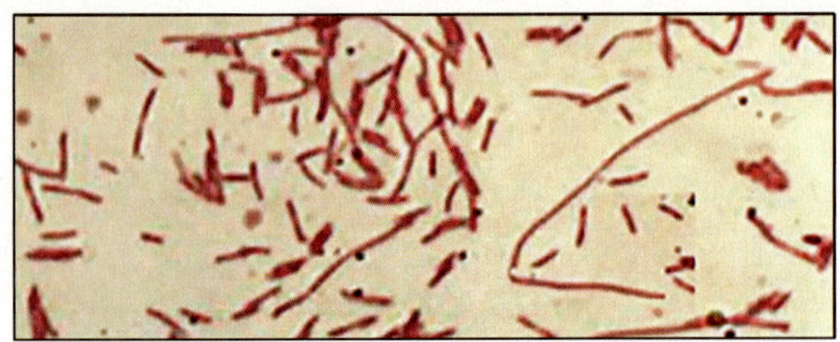

Fig. 67.5: *Morphology of L. pneumophila*

- **Motility:** They are motile by single polar or subpolar (subpolar means inserted near but not on the end) flagellum at one end.
- **Capsule and spore:** They are noncapsulated and nonsporing.

Cultural characteristics (C/Cs): Culture is highly sensitive (80–90%) and specific (100%) in diagnosis. They are strict aerobes and grow better under 5–10% CO_2 and high humidity (90%). Optimum temperature is 29–40°C and pH is 6.9. They are highly fastidious bacteria, grow better in media supplemented with L-cysteine and iron. Slow growing colonies may appear after 3–6 days.

- **Culture in media**
  1. **Liquid media:** Enrichment broth prepared from yeast extract, liver extract, L-cysteine and iron. Samples are collected in enrichment broth and after appropriate incubation, they will be subcultured on solid media.
  2. **Solid/plating media:**
     - **BCYE (buffered charcoal yeast extract) agar with L-cysteine:** It is a complex medium producing opal colonies (opal means stone with milky appearance).
     - **Feeley Gorman (FG) agar:** Diffusible brown pigment is produced which fluoresces dull yellow on UV light exposure. This may be enhanced by addition of tyrosine in medium.
     - **Egg yolk medium:** Opacity around colonies occurs due to lecithinase production; however, it is not used for diagnostic purpose.
  3. **Selective medium:** This is BCYE agar with vancomycin, cycloheximide and polymyxin-B.
- **Automated culture:** Like MALDI-TOF or VITEK is used to identify the species from culture.
- **Culture in animal:** Inoculate the organisms in guinea pig by intraperitoneal or by respiratory route and then identify from peritoneal exudates, local abscess and spleen materials.
- **Culture in hen's egg:** Bacilli grow in yolk sac.

Biochemical reactions (B/Rs): B/Rs are difficult to perform because of slow growth. They are non-fermentative, but hydrolyse the starch. Catalase and hippurate hydrolysis are positive but oxidase variable.

Pathogenicity

- **Disease name:** Disease called legionellosis having two forms such as mild influenza like symptoms called pontiac fever while bilateral pneumonia called Legionnaires' disease.

- **Nature of disease:** Legionnaires' disease occurs in epidemic or sporadic form, while pontiac fever occurs as localized outbreak with high attack rate.
- **Reservoirs of infection:** There are no carriers or animal reservoirs, but aquatic reservoirs like rivers, mud, stream, lakes, air conditioners or water coolers, etc. Bacteria survive in free living amoebae and in certain protozoa.
- **Sources of infection:** These are aerosols arise from air-conditioner or from shower heads or from infected aquatic reservoirs.
- **Modes of transmission: (1) Aspiration:** It is the common mode and transmission occurs during drinking of contaminated water or aspiration from nasopharyngeal colonization. **(2) Inhalation:** Human infection occurs by inhalation of aerosols arise from air-conditioner, shower heads, nebulizer, humidifier, device filled with tap water or from infected aquatic reservoir. Aerosolized *Legionella* can survive long and traversed to long distance may lead to epidemic. **(2) Direct instillation into the lungs:** It occurs during respiratory tract manipulation. Human-to-human transmission does not occur.
- **Incubation period:** 24–48 hours for pontiac fever and 2–10 days for Legionnaires' disease.
- **Portal of entry:** Respiratory system.
- **Sites:** Respiratory system and disseminated in other organs by local, blood or lymphatic routes.
- **Precipitating factors (epidemiological determinants):** These are following three types:
  1. **Agent factors (virulence factors):** They produce drug resistance due to **beta-lactamase** and **intracellular location of bacilli helps** to escape host defense and antibiotic effects. CMI is primary mechanism, but not protective.
  2. **Host factors:** Chances of infections are more in men, elders and person with IDDs.
  3. **Environmental factors:** It is common in developed countries (about 1–3% community acquired pneumonia and 10–30% are hospital acquired pneumonia are due to *Legionella*). Nutritional requirement of bacteria is filled by some type of algae present in aquatic reservoir.

Pathogenesis: Follow **Flowchart 67.3**.

Flowchart 67.3: Pathogenesis of legionellosis

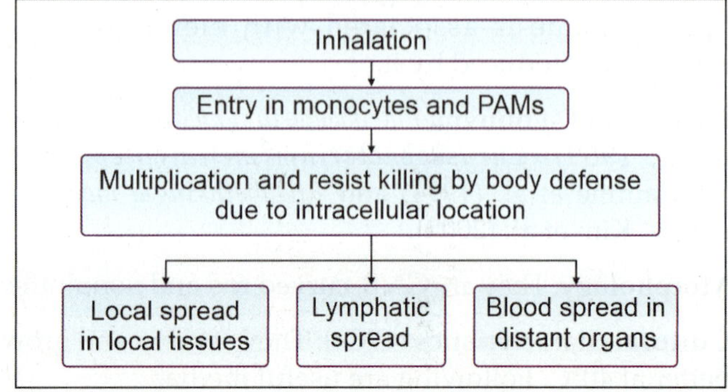

Clinical features: Two clinical forms. **(1) Pontiac fever:** It is the mild influenza like illness, presents with fever, chills, headache and myalgia. It is self limited. **(2) Legionnaires' disease:** It is a bilateral pneumonia presents with fever, nonproductive cough (atypical or interstitial pneumonia) and dyspnea may progress to diarrhea and encephalopathy (confusion). Case fatality is 15–20% due to respiratory failure or shock.

Complications: Myocarditis, pericarditis and endocarditis.

Laboratory diagnosis: Sputum, respiratory aspirates, lung tissues and blood are tested by microscopy (follow morphology), culture (follow C/Cs), and by B/Rs. **Ag is detected** from urine by ELISA and latex agglutination test. **Ab is detected (after 12 weeks)** from serum by ELISA and IIF test. Significant titer is 1:256. **Molecular method** includes BioFire Film Array (automated multiplex PCR) which targets 22 respiratory pathogens. High content of fat gives typical profile on **GLC**.

Prevention: Disinfect the water by silver ionization, commercial copper, superheat and flush methods to remove the biofilms of *L. pneumophila* from water.

Treatment: Levofloxacin, and azithromycin are drugs of choice.

ACCESS YOURSELF

Case Study

1. A 40 year male presented to medical OPD with complain of fever and productive cough. Clinically, he is diagnosed as cystic fibrosis. Gram's stain of sputum revealed gram-negative short rods, culture on MacConkey's gives NLF, oxidase and citrate are positive. Identify the case and answer the following.
   a. Name the causative agent of given condition.
   b. Write the pathogenicity of commonest agent in given case.
   c. Describe the morphology, culture characteristics and biochemical reactions of commonest agent in given case.

Essay/Full Question

1. Nonfermenters.

Short Notes

1. Pathogenicity or laboratory diagnosis of *P. aeruginosa*.
2. Glanders/melioidosis/legionellosis.

Short Questions for Theory/Viva Questions

1. Name the four non-fermentative bacteria.
2. Name four media used to detect the pigments of *Pseudomonas*.
3. What are Strauss reaction and Tunica reaction?
4. What is Vietnam time bomb?
5. Write the etiological agents for glanders and glandular fever (infectious mononucleosis).
6. Write the etiological agents for malleus and melioidosis.

Comments on

1. Pigment produced by *P. aeruginosa* spread in entire medium, but not the pigment of *Staphylococcus*.
2. Pigment produced by *P. aeruginosa* contributes in virulence.

Pseudomonas

1. **An organism grown on agar shows green colored colonies; likely organism is:**
   a. *Staphylococccus*   b. *E. coli*
   c. *Pseudomonas*   d. *Peptostreptococcus*

2. **Which of the following tests are employed to identify *Pseudomonas aeruginosa*?**
   a. Growth at 42°C   b. Positive urease test
   c. Positive indole test   d. Positive nitrate test

3. ***Pseudomonas* is which type of bacterium?**
   a. Anaerobic   b. Microaerophilic
   c. Strict aerobe   d. Obligate anaerobe

4. **Selective medium to detect the pyoverdin pigment produced by *P. aeruginosa* is:**
   a. Mannitol salt agar
   b. Furunculosis agar
   c. Thayer Martin medium
   d. King's medium B

5. **Which bacterium acts by inhibiting protein synthesis?**
   a. *Pseudomonas*   b. *Staphylococccus*
   c. *Streptococccus*   d. *Klebsiella*

6. **Ecthyma gangrenosum is caused by:**
   a. *Pseudomonas*   b. *Streptococccus*
   c. *Staphylococccus*   d. *H. influenzae*

7. ***Pseudomonas* infection is not cleaned by:**
   a. Dettol   b. Hypochlorite
   c. Chlorine   d. None

Bukholderia pseudomallei

8. **All of the following causes melioidosis:**
   a. *Bukholderia pseudomallei*
   b. *Bukholderia mallei*
   c. *Bukholderia cepacia*
   d. None

9. **The following statement is true regarding melioidosis *except*:**
   a. It is caused by *Bukholderia mallei*
   b. The agent is gram-negative aerobic bacteria
   c. Bipolar staining of etiological agent is seen by methylene blue stain
   d. The most common form of melioidosis is pulmonary infection

Legionella pneumophila

10. **All of the following are correct regarding *Legionella* except:**
    a. *Legionella* can be grown on complex media
    b. *L. pneumophila* sero group 1 is the most common serogroup
    c. *Legionella* is communicable from infected patients to others
    d. *L. pneumophila* is not effectively killed by polymorphonuclear leukocytes

11. **Most common mode of transmission of *L pneumophila* is:**
    a. Aspiration   b. Ingestion
    c. Insect bite   d. Blood

12. **Devi a 28-year-old female, has diarrhea, confusion and high-grade fever with bilateral pneumonitis. The diagnosis is:**
    a. *Legionella*
    b. *Neisseria meningitidis*
    c. *Streptococus pneumoniae*
    d. *H. influenzae*

13. Which of the following is a good medium to use for diagnosis of Legionnaires' disease?
   a. Thayer Martin medium
   b. BCYE agar
   c. Bordet Gengu medium
   d. Chocolate agar

14. True about *Legionella*:
   a. Epidemic (+)
   b. Splenomegaly
   c. Easily seen in sputum
   d. Scanty neutrophils with fed organisms
   e. Purulent sputum common

Answers and Explanation of MCQs

1. c
- *Pseudomonas* is the only organism producing the green color pigment which gives green colonies over agar. *Staphylococccus* also produces pigment, but colors are white (*Staph. albus*), lemon yellow (*Staph. citrus*) or orange (*Staph. aureus*). *E. coli* and *Peptostreptococcus* are nonpigmented.

2. a and d
- Follow section, *Pseudomonas aeruginosa* (**culture characteristics → effective factors and biochemical reactions**) for explanation.

3. c
- Follow section, *Pseudomonas aeruginosa* (**culture characteristics → effective factors**) for explanation.

4. d
- Follow section, *Pseudomonas aeruginosa* (**culture characteristics → media for production and identification of pigments**) for explanation.

5. a
- Following section, **Notes → toxins act by inhibiting the protein synthesis** for explanation.

6. a
- Follow section, *Pseudomonas aeruginosa* (**pathogenicity → clinical features**) for explanation.

7. a
- *Pseudomonas* can grow in bottles of antiseptics like Dettol, iodine, cetrimide lotions or soap lotion, etc., kept for use in hospitals. Indeed, Dettol or cetrimide is incorporated in a media selective for *P. aeruginosa*.

8. a
- *Bukholderia mallei* causes glander and *B. cepacia* causes fatal necrotizing pneumonia and infection in other systems.

9. a
- Melioidosis is caused by *Bukholderia pseudomallei*.

10. c
- *Legionella* can be grown on complex medium like BCYE agar. Sero group 1 is the most common and most sever sero group isolated from humans. *Legionella* is not communicable from infected patients to others (no man-to-man transmission). *L. pneumophila* is not effectively killed by polymorphonuclear leukocytes, because of intracellular location, it escape host defense and antibiotic effects.

11. a
- Follow section, *Legionella pneumophila* (**pathogenicity → mode of transmission**) for explanation.

12. a
- Bilateral pneumonitis is caused by *Legionella*.

13. b
- Follow section, *Legionella pneumophila* (**culture characteristic → solid/plating media**) for explanation.

14. a
- Legionnaires' disease presents as epidemic or sporadic form, without splenomegaly. For options c and d follow section, *Legionella pneumophila* (**morphology**). Sputum is non-productive.

Infections of Pasteurellales

Chapter Outline

INTRODUCTION

These are GNB live on mucosal surfaces of birds and mammals, especially in the upper respiratory tract. They are classified in **Flowchart 68.1**.

Haemophilus

Haemophilus influenzae

Meaning

Haemophilus word includes haemo or hemo means blood and philus means friendly to blood or blood loving because bacteria required factors X and V (one/both) which are present in blood.

Synonym

It also called influenza bacillus or Pfeiffer's bacillus. Pfeiffer's bacillus because of Pfeiffer (1892) identified it in sputum of patients of 1889–90 influenza pandemic as a causative agent of human influenza. But later it proved by Smith, Andrew and Laidlaw in 1933 that human influenza is due to influenza virus, hence it remained as *H. influenzae*.

Morphology

Type according to Gram's stain: They are gram-negative coccobacilli, but better stained by using dilute carbol fuchsin or safranin as counter stain. Better result obtained with Loeffler's methylene blue stain.

Shape and size: They are pleomorphic. Coccobacillary form occurs in sputum and in young culture while filamentous form occurs (**Fig. 68.1**) in CSF. They are about 0.3 μm × 1 μm in size.

Capsule: They are capsulated, which is detected by quelling reaction or by negative staining.

Motility and spores: They are nonmotile and non-sporing.

Fimbriae: They are present.

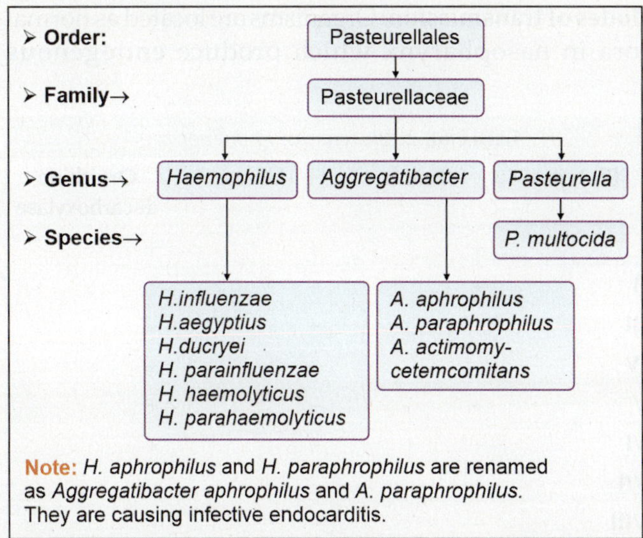

Flowchart 68.1: Classification of Pasteurellales

- **Order:** Pasteurellales
- **Family→** Pasteurellaceae
- **Genus→** Haemophilus | Aggregatibacter | Pasteurella
- **Species→** P. multocida

H.influenzae	A. aphrophilus
H.aegyptius	A. paraphrophilus
H.ducryei	A. actimomy-cetemcomitans
H. parainfluenzae	
H. haemolyticus	
H. parahaemolyticus	

Note: *H. aphrophilus* and *H. paraphrophilus* are renamed as *Aggregatibacter aphrophilus* and *A. paraphrophilus*. They are causing infective endocarditis.

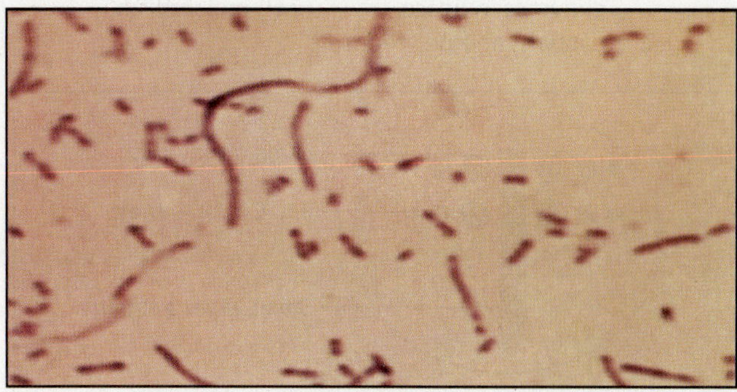

Fig, 68.1: Morphology of *H. influenzae*

Effective factors

- O$_2$ effect: They are aerobes, but grow anaerobically (facultative anaerobes).
- CO$_2$ effect: They require 5–10% CO$_2$ (capnophilic).
- Temperature: Optimum temperature is 37°C.
- They are fastidious in nature and required X and V factors for growth as mentioned in **Table 68.1**.

Culture in media

A. Liquid media

- **Phenol red broth (contains NAD):** It is used for sugar fermentation test.
- **Filde's peptic digest – blood broth:** It is the nutrient broth with peptic digest of blood.
- **Levinthal's medium:** It contains mixture of boiled-filtered blood and nutrient broth.

B. Solid media

- **Filde's peptic digest – blood agar:** It is the nutrient agar contains peptic digest of blood.
- **Blood agar:** V factor available inside the unlyzed RBCs, and it is not available free in medium, hence bacteria produce scanty growth or no growth.
- **Blood agar with *Staph. aureus* streaking:** *Staph. aureus* produces V factor which favors the growth of *H. influenzae*. Colonies are large near the streaking line, because more V factor is available, while small colonies away from streaking line called satellitism as shown in **Fig. 68.2**.
- **Chocolate agar:** It is superior than blood agar, because both X and V factors are available freely.
- **Levinthal's agar:** Development of translucent colonies with iridescence by capsulated strain.

C. Selective medium: Blood agar with bacitracin, sucrose, hemin X factor, V factor disc and phenol red. Sucrose is fermented by *H. parainfluenzae* gives yellow colonies while not fermented by *H. influenzae* gives colorless colonies. Hemin X factor is required as growth factor for *H. influenzae. H. influenzae* will identified as satellite growth around V factor disc.

Automated culture: Like MALDI-TOF or VITEK is used to identify the species from culture.

Biochemical Reactions (B/Rs)

Sugar fermentation tests: Glucose and xylose are fermented with acid production only, but not lactose, sucrose and mannitol.

Catalase, nitrate reduction and oxidase tests: Positive.

X and V factors disc test: *Haemophilus* species required factors X and/or V for their growth which is useful in identification of species. Inoculate the isolate on medium without X and V factors then place the X, V or XV discs.

- Growth around XV disc only: *H. influenzae, H aegypticus* and *H. haemolyticus*.
- Growth around X and XV disc: *H. ducryei* and *A. aphrophilus*.
- Growth around V and XV disc: *H. parainfluenzae, H. parahaemolyticus* and *A. paraphrophilus*.

Biotyping: Eight biotypes are identified on the basis of indole, urease and ornithine decarboxylation tests as shown in **Table 68.2**.

Resistance

Sterilization: *H. influenzae* is delicate organism easily killed by heat at 55°C in 30 minutes. Culture is maintained for long time by lyophillization.

Disinfection: *H. influenzae* is delicate organism easily killed by disinfectants.

Drug resistance: Plasmid-born resistance is common to many antibiotics.

Pathogenicity

Reservoirs of infection: These are human cases.

Sources of infection: These are nasopharyngeal or respiratory droplets.

Modes of transmission: Organisms are located as normal flora in nasopharynx which produce **endogenous**

TABLE 68.1: Factors required for growth of *H. influenzae*

Features	X factor	V factor
Heat	Stable	Labile at 120°C in few min.
Act as	Hemin or hematin	Bacterial vitamin or coenzymes-NAD/NADP
Function	For synthesis of heme enzymes–oxidase, catalase and peroxidase	Hydrogen acceptor in cell metabolism

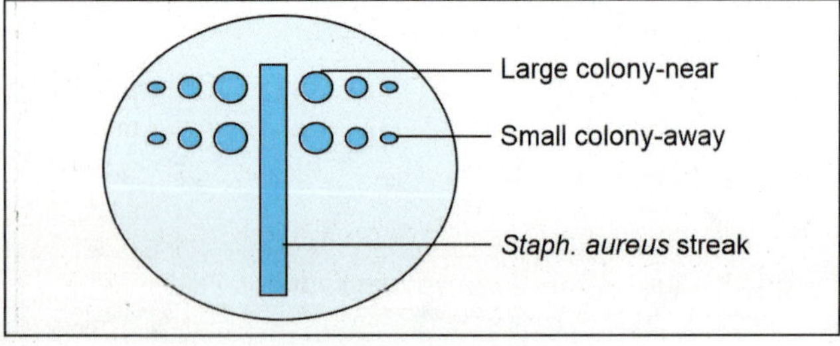

Fig. 68.2: Satellitism

TABLE 68.2: Biotypes of *H. influenzae*

Biotypes	Indole	Urease	Ornithine decarboxylase
I	+	+	+
II	+	+	–
III	–	+	–
IV	–	+	+
V	+	–	+
VI	–	–	+
VII	+	–	–
VIII	–	–	–

Flowchart 68.2: Spread of *H. influenzae*

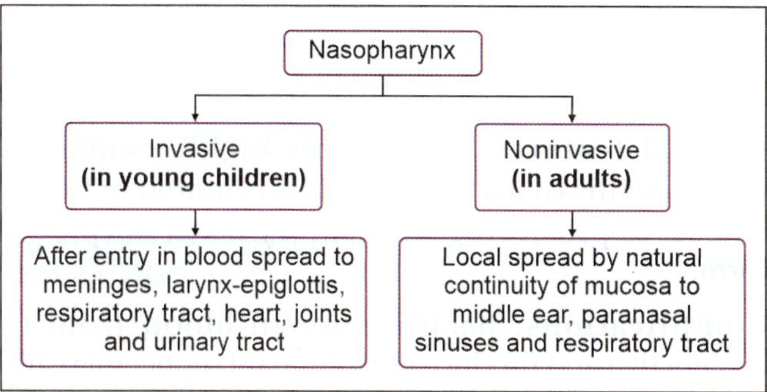

infection as mentioned in **Flowchart 68.2. Exogenous infection** occurs by inhalation of nasopharyngeal or respiratory droplets.

Incubation period: Unknown, probably short about 2–4 days.

Portal of entry: Respiratory tract in case of exogenous infection.

Sites: Meninges and respiratory system.

Precipitating factors (epidemiological determinants)

A. Agent factors (virulence factors)

1. **Plasmid:** It encodes for drug resistance.
2. **Fimbriae:** These are organs of adhesion.
3. **IgA protease:** It decreases the local immunity.
4. **Capsular polysaccharide:** It is a major antigenic determinant. Six capsular types are identified called **Pittman typing** from a–f, where b is the most common called *Haemophilus influenzae* type b (Hib). Hib is responsible for 95% of *Haemophilus influenzae* meningitis. It is invasive type. Typing is based on agglutination, Quellung reactions, precipitation, coagglutination or ELISA. Capsule inhibits the phagocytosis. It does not induce the alternative complement pathway. All the types contain hexose and hexosamine sugars, while type b contains ribose and ribitol called PRP (polyribosyl ribitol phosphate) antigen and produces IgG, IgM and IgA which are bactericidal and protective, hence this PRP-Ag is included in vaccine. Noncapsular strains cannot be typed and are called **untypable strains** and produce noninvasive or local diseases. Most infective strains are Hib, followed by untypable strains.
5. **Outer membrane proteins (OMP) Ag:** It contributes in adhesion and invasion. Total 13 serotypes of Hib are identified on the bases of OMP Ag. Antigen produces the antibody which is protective.
6. **Lipo-oligosaccharide (LOS):** OMP and LOS subtyping have an epidemiological value. LOS has endotoxic action but differ from LPS of other gram-negative bacteria.

B. Host factors

1. **Age:** It is common in young children. Young house hold contacts with patient of systemic *H. influenzae* are at increased risk of infection.

2. **Immunity:** Maternal antibodies against type b can transfer from mother to child and provide protection to the child up to the age of 6 months. Waning of such antibodies titer (after 6 months) increase the risk of infants to the type b *H. influenzae*.

Clinical features: Two types of infections.

A. Invasive infections: Around 95% of invasive infections are due to type b with following features.

- ***H. influenzae* meningitis:** It is more in young children (2 months–3 years) with 90% case fatality rate and less in old children, because of development of immunity by subclinical infection. It occurs by hematogenous spread from nasopharynx.
- **Laryngoepiglottitis:** These include laryngitis, epiglottitis and uvulitis producing laryngeal or respiratory obstruction. Children with >2 years are most vulnerable. It can be fatal in 2 hours and required urgent tracheostomy.
- **Pneumonia:** It includes lobar pneumonia or bronchopneumonia with empyema in older children and adults.
- **Bronchitis:** Hib causes acute exacerbation of chronic bronchitis and bronchiectasis.
- **UTI:** It occurs in children with urinary tract abnormalities.
- **Suppurative lesions:** These are brain abscess, arthritis, endocarditis, pericarditis and cellulitis in buccal or periorbital parts.

B. Noninvasive infections: These are mostly caused by untypable strains. Bacteria spread locally from nasopharynx to cause acute otitis media (2nd most common cause), sinusitis and exacerbation of chronic bronchitis, bronchiectasis or COPD. They cause puerperal sepsis and neonatal bacteremia from female genital tract colonization. They cause invasive illnesses in countries where Hib vaccine is used. Differences between Hib strain and untypable strains are given in **Table 68.3**.

Laboratory Diagnosis

Specimens: CSF, blood, sputum, throat or nasopharyngeal swab, pus from lesion, urine, aspirates from joint-pleural-pericardial-bronchus, etc.

Testing methods

A. Microscopy: Follow morphology.

B. Culture: Follow C/Cs.

TABLE 68.3: Differences between Hib and untypable strains of *H. influenzae*

Hib	Untypable strains
Contains PRP capsule	Noncapsulated
Spreads by blood route	Spread locally
Infects children	Infect adults
Produces invasive illnesses	Produce noninvasive illnesses
Hib vaccine is used	Vaccine not available

8. **All are true statements regarding pertussis** *except*:
   a. Secondary attack rate is average 90% in unimmunized contacts
   b. Incubation period is around 14 days
   c. Erythromycin is the drug of choice
   d. Can affect people of any age
   e. Main source of infection is chronic carrier

Answers and Explanation of MCQs

1. d
- Follow section, *Bordetella pertussis* [cultural characteristics (C/Cs)] for explanation.

2. a
- Culture of nasopharyngeal secretion is the gold standard for diagnosis, so it is the best specimens.

3. a

4. a
- Follow section, *Bordetella pertussis* (pathogenicity → incubation period) for explanation.

5. d
- *B. pertussis* adheres to normal mucosa by FHA. After adhesion it destroys cilia by tracheal cytotoxin. After ciliary damage inflammation extends into lungs produces local diseases like bronchiectasis, diffuse bronchopneumonia and lung collapse (atelactesis).

6. a
- Follow section, *Bordetella pertussis* (pathogenicity → clinical features) for explanation.

7. d
- Follow section, *Bordetella pertussis* (prevention → immunoprophylaxis → acellular vaccine) for explanation.

8. e
- Human case is the only reservoir, no animal reservoir. There is no evidence of chronic carrier or subclinical infection in pertussis.

CHAPTER 70

Infections of Brucellas

Chapter Outline
- *Brucella* spp.

Brucella spp.

Species

Human pathogens

1. *Brucella melitensis*: **Genus name** given from army doctor David Bruce (1886), who isolated organism from the spleen and **species name** given from the place Melita (Roman name, actual name Malta), from where it was isolated. It is a natural pathogen for sheep, goat and camel, but it can cause infection in human by raw milk.
2. *B. abortus*: It causes abortion in cattle (cow and buffalo), hence the species name is abortus. It was identified by Bang in 1897.
3. *B. suis*: It is a swine (pig) pathogen, hence the species name is suis. It was isolated by Traum from pig in USA in 1914.
4. *B. canis*: It is a canine (dog) pathogen, hence the species name is canis. It was isolated from the case of canine abortion and occasionally causes human disease.

Animal pathogens: *B. ovis*, which causes abortion in sheep and *B. neotomae,* which was identified from desert wood rat.

Morphology

Type according to Gram's stain (Fig. 70.1): They are gram-negative coccobacilli also showing bipolar staining.

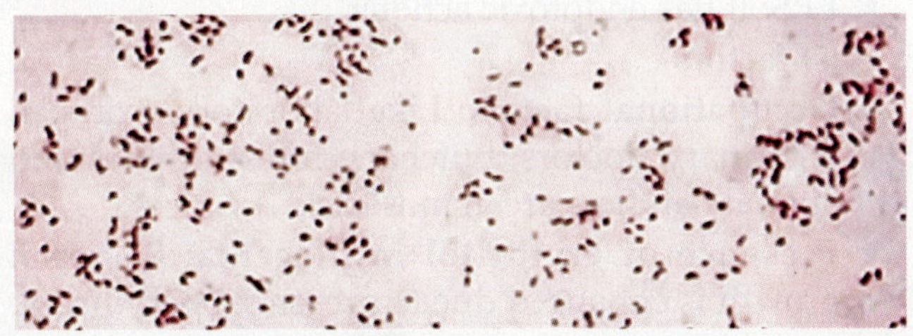
Fig. 70.1: Morphology of *B. abortus*

Shape and size: They are short rod like coccobacilli and small, about 0.5–1.5 μm × 0.5–0.7 μm in size. They are mistaken as cocci as was done by Bruce and called *Micrococcus melitensis.* Irregular forms are observed in culture. They are nonmotile, nonsporing and non-capsulated.

Arrangement: They are arranged singly or in pairs or in short chains or in clusters.

Cultural Characteristics (C/Cs)

Effective factors

- O_2 effect: They are strict aerobes.
- CO_2 effect: Better growth occurs under 5–10% CO_2 (capnophilic).
- Temperature: Optimum temperature is 37°C.
- pH: Optimum pH is 6.6–7.4.
- Slow and scanty growth occurs on ordinary media, and growth is improved by adding serum or liver extracts.

Culture in media

A. **Liquid media:** These are nutrient broth, liver infusion broth, serum dextrose broth and trypticase soy broth. Uniform growth occurs with powdery or viscous deposit at bottom in old culture.

B. **Solid media:** These are nutrient agar, liver infusion agar, serum dextrose agar, serum potato infusion agar, trypticase soy agar and tryptose agar. Small, moist, translucent and glistening colonies are developed. Colonies change to smooth, rough or mucoid forms with antigenic and virulent changes. Culture should not be declared negative before 6–8 weeks.

C. **Selective medium:** Bacitracin, polymyxin and cyclo-heximide are added in above media to make them selective.

D. **Castaneda biphasic medium culture method (liquid/broth phase plus solid/slant phase):** It avoids the contamination during subculture. Blood or bone marrow is inoculated in broth and incubated in upright position. For subculture bottle is tilted, so that broth runs over the slant and then bottle is incubated in upright position. Next day observed slant for colonies.

Automated culture: Like MALDI-TOF or VITEK is used to identify the species from culture.

Culture in animals: Guinea pig is the most susceptible laboratory animal. Tunica reaction is elicited in male guinea pig after intraperitoneal inoculation. Inject the bacilli in thigh of guinea pig. Animal will die after 6–8 weeks. Collect the samples from regional lymph nodes or from spleen and proceed for culture. Blood is collected for Ab detection.

Biochemical Reactions (B/Rs)

Glucose fermentation tests: Nonfermenter.

I M Vi C test: – – – –

Catalase, nitrate, oxidase and urease tests: Positive.

Biotyping: It based on CO_2 requirement, H_2S production, tolerance to bacteriostatic dyes (like basic fuchsin and thionin), agglutination to monospecific sera, phage lysis [Tblisi (Tb) phage] and biochemical tests with amino acids and carbohydrates. **B melitensis** has 3 biotypes (1–3), **B. abortus** has 7 biotypes (1–6 and 9) and **B. suis** has 4 biotypes (1–4). Biotypes of **B. suis** on the basis of H_2S production are American strain which is H_2S positive and Danish strain which is H_2S negative. **Reference center** for Biotyping of *Brucella* is Central Veterinary Laboratory, New Haw, UK.

Resistance

B. abortus is killed by heating at 60°C in 10 minutes, sunlight, pasteurization, acid and 1% phenol in 15 minutes.

Antigenic Structure

Somatic antigen: Somatic antigen of *Brucella* contains two epitopes like A and M. They present in different species in different amount. *B melitensis* contains 20 times as much A as M. *B. abortus* contains 20 times as much M as A. *B. suis* has intermediate antigenic pattern. They are helpful for serotyping and identification of species; however, serotyping is not straightforward, because species diagnosed serologically as *B. abortus* may be behave biochemically as *B. melitensis* and *vice versa*.

Cross-reactive antigens: Antigenic cross reaction may occurs with following bacteria.
- *V. cholerae*: Cholera vaccine produces the *Brucella* agglutinin, which lasts for 3 years.
- *E. coli*: 0:116 and 0:157.
- *S. Typhi*: 0:30 or N antigen of Kauffman and White. Also Vi Ag behaves like L antigen of *Brucella*.
- Others: *S. maltophila, Y. enterocolitica, F. tularensis,* etc.

Pathogenicity

Disease name: It called brucellosis or Mediterranean fever or Malta fever or undulant fever. Most common and most virulent human pathogen is *B. melitensis*.

Epidemiology: It is endemic in areas with availability of large numbers of animals like Mediterranean zone, Eastern Europe, Central Asia, Mexico and South America.

Reservoirs of infection: These are different animals as mentioned above with respective species. *B. abortus, B. suis* and *B. canis* are rare or intermediate human pathogens.

Sources of infection: These are animal products like milk, dairy products prepared from infected milk (cheese, ice cream, etc.), meat, urine or feces and human products like vaginal discharge. Most common animals responsible for human infection are sheep and goats.

Modes of transmission: It is a zoonosis. Human-to-human transmission can occur possibly by sex, breastfeeding or via placenta. Common routes are mentioned below.
1. **Direct transmission**
   - **Direct contact:** These include contact during sexual intercourse, occupational contact with infected animals/animal tissues in butcher, veterinarians, animal handlers, etc., or contact of abraded skin with infective materials.
   - **Inoculation:** It occurs accidentally in laboratory in conjunctiva or in mucosa.
   - **Vertical:** It occurs by placenta and breastfeeding.
2. **Indirect transmission (vehicle-borne)**
   - **Airborne:** It occurs by inhalation of dust particles arise from dried animal materials like wool.
   - **Water- and food-borne:** It occurs by ingestion of infected milk (noticed by Maltese bacteriologist Zammit), infected animal products, infected water or raw vegetables by urine, feces or carcass of infected animals.
   - **Vector-borne:** It occurs in animals by blood sucking arthropod (ticks).

Incubation period: 10–30 days.

Portal of entry: Skin, GIT, respiratory system, etc.

Sites: Blood, RE system, placenta, muscles, etc.

Precipitating factors (epidemiological determinants)
A. Agent factors (virulence factors)
   - **Intracellular location of bacilli:** It helps in survival against antibodies and drugs.
   - **LPS:** It has endotoxic actions.

B. Host factors
   - **Occupational factors:** Like laboratory workers, veterinary doctors, butchers, forest officers and animal handlers are at most risk.
   - **Presence of erythritol in placenta:** Bacterial growth is enhanced due to presence of erythritol.

Pathogenesis: Follow **Flowchart 70.1**.

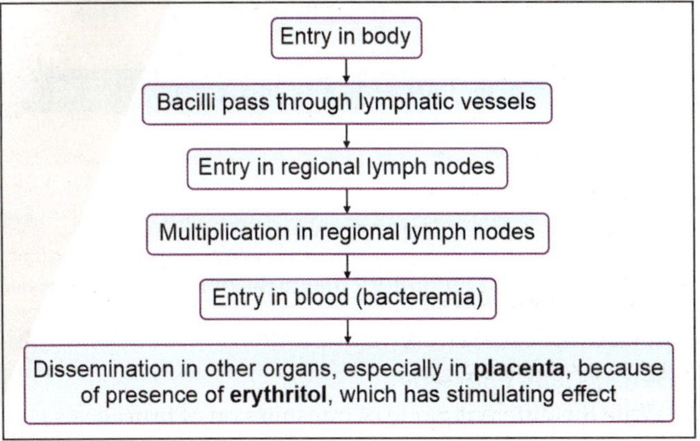

Clinical features: Brucellosis is present with nonspecific sign and symptoms.

1. **Acute/subacute brucellosis:** It is associated with prolonged bacteremia. It presents as irregular fever (undulating/on-off fever in some cases), chills, joint-muscle pain, asthmatic attack, sweating, exhaustion, anorexia, constipation, nervous irritability, etc.
2. **Chronic brucellosis:** It is nonbacteremic, low grade infection with periodic exacerbation of minor symptoms like sweating, joint pain, lassitude, minimal or no fever. It is the stage of hypersensitivity with minimal or no fever. Immunity is mostly CMI. Th1 tries to eliminate the intracellular bacilli via releasing of TNF-α, TNF-β, IL-1 and IL-12. Tissue reaction to *Brucella* consists granuloma formation with epithelial cells, giant cells, lymphocytes and plasma cells, which heals with fibrosis and sometimes with calcification. Granuloma occurs in RE system like spleen, lymph node (lymphadenopathy), liver (hepatomegaly), etc.
3. **Latent brucellosis:** Bacilli survive in RE-system and produce late infection. It is asymptomatic, but diagnosed with positive serological tests.

Complications: They are associated with bones, joints, nerve system (meningitis) or with visceral organs.

> **Note: Variation in presentation as per species**
> 1. *Brucella melitensis:* Acute and aggressive features
> 2. *B. abortus:* Chronic and insidious features
> 3. *B. suis:* Focal abscess
> 4. *B. canis:* Acute gastrointestinal symptoms

Laboratory Diagnosis

Diagnosis of Human Brucellosis

Specimens: Blood, bone marrow, urine, aspiration from nodes, sputum, CSF, breast milk, vaginal discharge, seminal fluid, etc.

Collection: Collect in trypticase broth, incubate at 37°C under 5–10% CO_2 and than s/c on solid media.

Testing methods

A.**Microscopy:** Follow morphology.
B. **Culture:** Follow C/Cs.
C. **Biochemical reactions:** Follow B/Rs.

D.Serological tests: Both IgM and IgG are appearing 7–10 days after infection. As the disease progress IgM will decline with same or rising titer of IgG. IgM is positive in acute stage while negative or weakly positive in chronic stage, hence the result will be read carefully and a negative result may not exclude the brucellosis. In chronic stage IgG can be detected. Following tests are used to detect the antibodies.

1. **Standard agglutination test (SAT)**
   - **Principle:** It based on tube agglutination.
   - **Ag:** LPS of *B. abortus* will be used as an Ag.
   - **Steps:** Mix the equal volume of serially diluted patient's serum with antigen suspension. Incubate at 37°C for 24 hours or for 50°C for 18 hours.
   - **Significant titer:** Significant titer in endemic area or following occupational exposure is ≥1:320. Rising titer is confirmed by repeating the test after 2–4 weeks. In nonendemic area titer ≥1:160 along with compatible clinical features is considered as significant.
   - **Positive test:** SAT positive indicates total antibodies (IgM and IgG) means brucellosis only, it cannot differentiate acute brucellosis (IgM) and chronic brucellosis (IgG).
   - **False-positive test:** It occurs due to immunization or due to cross reactive antigens from different bacteria as mentioned above. Cholera-induced agglutinin may be differentiated by the agglutination absorption test or treatment of serum with 2-mercaptoethanol (2ME).
   - **False-negative test:** It occurs due to prozone phenomenon (Ab excess) which is eliminated by dilution method by using 4% saline or due to effect of blocking antibodies, which are eliminated by prior heating of serum at 55°C for 30 minutes or by dilution of patient's serum by using 4% saline or detection of blocking antibodies by Coombs test. **Blocking antibodies** IgG, IgA or nonagglutinating Abs.
2. **2-mercaptoethanol (2ME) SAT:** It destroys the IgM (by breaking the disulfide bond of IgM) from the samples contain both IgG and IgM, so it detects only IgG called 2ME resistant IgG. So, only 2ME SAT positive indicates chronic brucellosis while 2ME SAT negative, but positive SAT indicates acute brucellosis. It is superior than SAT. 2ME resistant IgG can fall after adequate therapy, so reducing titer of IgG can indicates the adequate therapy and no further progress of disease.
3. **ELISA:** It is sensitive and specific test; however, result has to be confirmed by SAT. It detects IgM and IgG separately, so useful to differentiate between acute and chronic stage. It also used to detect neurobrucellosis.
4. **Dip stick test:** It is a rapid test for IgM, but less sensitive.

5. **CFT:** It is more useful to detect IgG/chronic stage.
6. **Rose Bengal card test**

E. **Brucellin test:** It based on DTH, like tuberculin test.
F. **Molecular test:** PCR is performed from blood or tissues samples. BioFire Film Array is an automated multiplex PCR which targets around 16 pathogens of bioterrorism including *Brucella*.
G. **Bacterial typing:**
1. Biotyping: Described above
2. Bacteriophage typing: **Follow Ch. 118.**

Diagnosis of Animal Brucellosis

Milk (cream of the milk) is tested by same tests as like human brucellosis. Some rapid tests are mentioned below.

1. **Milk ring test:** Mix the milk with *Brucella* Ags (prepared from killed *Brucella* and stained with hematoxylin). Incubate in water bath at 70°C for 40–50 minutes and read the result. Positive test indicates blue ring at top, leaving milk unstained if milk contains Abs. Negative test indicates no ring at top, leaving milk in blue color if no Abs in milk.
2. **Other tests:** Rapid plate agglutination test, Rose Bengal test and Whey agglutination test.

Prevention

General measures: Prevention of consumption of milk and other products of infected animals. Milk-borne infection is controlled by pasteurization. Identification and slaughtering of infected animals and development of certified *Brucella* free herds.

Immunoprophylaxis: Two types of vaccine.
1. **Animal vaccine:** Prepared from strain 19 of *B. abortus* and protective in cattle.
2. **Human vaccine:** Live attenuated vaccine is prepared from strain 19 *B. abortus* but not suitable for human use.

Chemoprophylaxis: Doxycycline plus rifampicin is given after exposure.

Treatment

For children and pregnant women: Cotrimoxazole with rifampicin or gentamicin are effective.

For adults
- Gold standard: Doxycycline daily for 45 days (6 weeks) with streptomycin by IM injection for 2 weeks.
- WHO regime: Rifampicin for 6 weeks plus doxycycline daily for 6 weeks. Relapse or treatment failure can occur in 5–10% cases.

For CNS involvement: Cotrimoxazole is added to the regime with continuation for 6 months.

ACCESS YOURSELF

Essay/Full Question
1. *Brucella.*

Short Note
1. Pathogenicity or laboratory diagnosis of *Brucella melitensis.*

Short Questions for Theory/Viva Questions
1. What is milk ring test?
2. Write the different mode of transmission of brucellosis.

MCQs for Chapter Review

1. ***Brucella melitensis* is commonly found in (animal):**
   a. Pig
   b. Camel
   c. Sheep
   d. Goat
   e. Reindeer
2. **A farmer presented with fever off and on for the past 4 years was diagnosed to be suffering from chronic brucellosis. All of the following serological tests would be helpful in the diagnosis at this state, *except*:**
   a. Standard agglutination test
   b. 2 mercaptoethanol test
   c. Complement fixation test
   d. Coombs' test
3. **All are true about *Brucella except*:**
   a. *Brucella abortus* is capnophilic.
   b. Transmission by aerosol route can occur occasionally.
   c. Pasteurization destroys it.
   d. 2ME is used to detect IgA.
4. **Rose Bengal card test is done for diagnosis of:**
   a. Brucellosis
   b. Rheumatoid arthritis
   c. Rheumatic fever
   d. Syphilis
5. **Treatment of brucellosis:**
   a. Doxycycline
   b. Streptomycin
   c. Erythromycin
   d. Pencillin
   e. Rifampin

Answers and Explanation of MCQs

1. **b, c and d**
- Follow section, **species** for explanation.
2. **a**
- SAT cannot differentiate acute brucellosis (IgM) and chronic brucellosis (IgG).
3. **d**
- 2ME is used to detect IgG.
4. **a**
- Follow section, **diagnosis (diagnosis of animal brucellosis)** for explanation.
5. **a, b and e**
- Follow section, **treatment** for explanation.

CHAPTER 71

Infections of Miscellaneous GNB

Chapter Outline

- *Francisella tularensis*
- *Gardnerella vaginalis*
- Rat-bite Fever (RBF)
- *Capnocytophaga* spp.
- *Chromobacterium violaceum*
- HACEK Group

Francisella tularensis

Classification: Follow **Flowchart 71.1**.

Meaning: Genus named after Francis for pioneering studies on tularemia a disease of rabbits and other rodents. **Species** was 1st described in Tulare country, California, hence disease is tularemia and species is tularensis.

Morphology: They are gram-negative coccobacilli with 10% carbol fuchsin showing bipolar staining. They are pleomorphic with bacillary or filamentous forms and about 0.3–0.7 μm × 0.2 μm in size. They are capsulated, nonmotile and nonsporings. *F. tularensis* resembles to *Mycoplasma* being filterable, multiplies by budding and filamentous formation besides binary fission.

Cultural characteristics (C/Cs): They are strict aerobes, optimum temperature for growth is 37°C and fastidious in growth requirements.

- **Culture in media**
  1. **Liquid media:** Contain casein hydrosylate, thiamine and cystine.
  2. **Solid media:** Minute droplet like transparent colonies appears after 3–5 days on Francis' blood dextrose cystine agar.
- **Automated culture:** VITEK is used to identify the species from culture.

Flowchart 71.1: Classification of *F. tularensis*

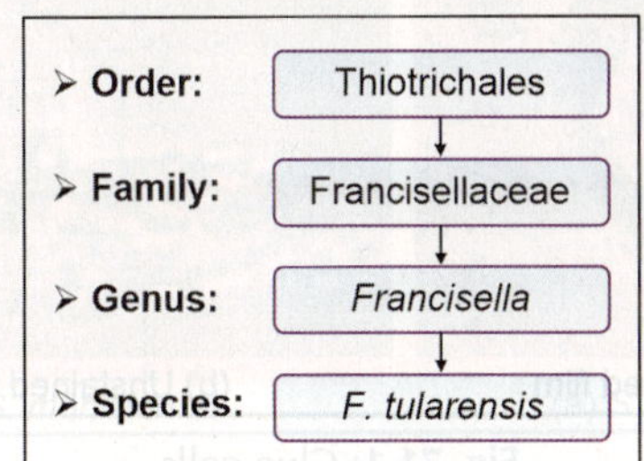

Biochemical reactions (B/Rs)

- **Sugar fermentation tests:** Glucose and maltose are fermented with acid production only.
- **Catalase test:** Variable
- **Indole, urease and ornithine decarboxylation tests:** Negative
- **Biotyping:** It is done on the basis of virulence and epidemiological nature. It is highly virulent in North America and low virulent in Europe and in Asia.

Pathogenicity: Disease called tularemia (plague like disease). Modes of transmission and clinical features are mentioned below.

1. **Direct contact:** Infection occurs by direct contact with animal tissue in butchers.
2. **Tick-bite like *Ixodes*:** Local ulcer and adenitis.
3. **Deer fly bite:** There is evidence that deer flies in the western US are involved in the transmission of tularemia, so also called deer fly fever or rabbit fever. Compared to ticks, deer flies are minor vectors of tularemia. Bite produces pain, swelling and itching.
4. **Accidental transmission in laboratory:** Ocular features like ulcerative or nodular conjunctivitis and other symptoms.
5. **Ingestion of meat or water contaminated by animal excreta:** Typhoid like fever. It also called lemming fever in Norway, because of water contamination by excreta of lemmings (water rats).
6. **Inhalation:** Causing primary atypical pneumonia.

Laboratory diagnosis: Material from local lesion or blood is tested by microscopy (follow morphology), culture (follow C/Cs) and biochemical reactions (follow B/Rs). **Serological tests like** agglutination, CFT, Coombs test and hemagglutination test are useful to detect the antibodies.

are rigid spiral in shape with tight coils at regular interval of 1 µm as shown in **Fig. 71.2b** and 2–5–10 µm × 0.2–0.5 µm in size. They show darting motility by single flagella at both ends (amphitrichous) or by tufts (2–7) of flagella at both ends (amphilophotrichous) examined by DGIM.

Cultural characteristics (C/Cs): They are microaerophilic and cannot grow on cell free media. Animal culture is done in mouse or guinea pig. In mouse, there is no sign of illness, but after 5–14 days bacilli identified in blood or peritoneal fluid.

Pathogenicity

- **Disease name:** Called sodoku.
- **Synonym:** Also called spirillary RBF.
- **Reservoirs and sources of infection:** These are rats or stagnant fresh water.
- **Mode of transmission:** Transmitted by rat bite.
- **Incubation period:** 1–4 weeks.
- **Clinical features: Animal** features include slow progressive septicemia and death of animal in few weeks. **Human** features include inflammation at the site of rat bite, enlargement of regional lymph nodes with fever and skin rashes.
- **Complication:** 10% mortality due to endocarditis.

Laboratory diagnosis: Exudates from local lesion, lymph node aspirates or blood is tested by microscopy (follow morphology) and culture (follow C/Cs).

Prevention: Rat control and use of doxycycline or oral penicillin after rat bite is effective.

Treatment: It responds to tetracycline, erythromycin and penicillin.

Capnocytophaga spp.

This genus includes 8 different species.

- **Isolated from oral cavity of humans:** *Capnocytophaga ochracea, C. gingivalis, C. granulosa, C. haemolytica, C. sputigena* and *C. leadbetteri.*
- **Isolated from oral cavity of animals like dog and cat:** *C. canimorsus* and *C. cynodegmi.*

C. canimorsus

Meaning: Genus name came from "Capno" for its dependency on CO_2 and "cytophaga" for its flexibility and gliding motility. Species name came from canine means dog, because transmitted by dog bite.

History: It was 1st observed in 1976 by Bobo and Newton from the case with meningitis and septicemia bitten by 2 different dogs on 2 consecutive days.

Morphology: They are GNB. They are rod shaped and after growth on agar plates, longer rods with curved shape appear as shown in **Fig. 71.3**. They are 1–3 µm in length and non flagellates, but showing gliding motion, although this can be difficult to see.

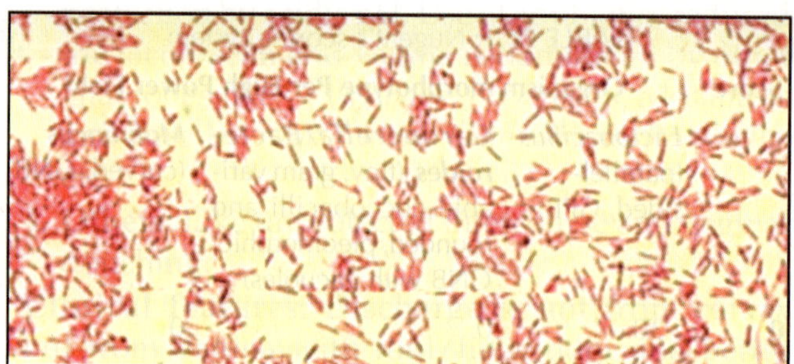

Fig. 71.3: Morphology of *C. canimorsus*

Cultural characteristics (C/Cs)

- **Effective factors:** They are anaerobes and better growth occurs under 5–10% CO_2 (capnophilic). They are fastidious can grow on media enriched with blood. Slow growing colonies may appear after 18–24 hours or longer (48–72 hours).
- **Media:** These are blood agar (with 5% sheep or rabbit blood), BHIA and chocolate agar.
- **Growth properties:** At 18 hours, colonies are usually less than 0.5 mm in diameter, spotty and convex. At 24 hours, colonies may be up to 1 mm in diameter. After 48 hours, colonies are narrow, flat and smooth with spreading edges. At this time purple, pink, orange or yellow color pigmented colonies may appear, but once they are scraped from the agar plate, they are always yellow.

Biochemical reactions (B/Rs): Difficult to perform because of slow growth.

1. **Sugar fermentation tests:** They ferment maltose and lactose.
2. **Catalase, oxidase and arginine dihydrolase tests:** Positive
3. **Nitrate reduction, urease and H_2S production test:** Negative

Pathogenicity

- **Reservoirs and sources of infection:** Bacteria present as a normal oral flora in canine and feline species and in human also.
- **Incubation period:** Symptoms appear after 2–3 days of exposure or after 1 weeks.
- **Modes of transmission: Endogenous infection** includes periodontal infection due to activation of oral flora in individual with IDDs. **Exogenous infection** (zoonosis) occurs from animals through bites, licks or even close proximity with animals.
- **Precipitating factors (epidemiological determinants)**
  1. **Agent factors (virulence factors):** *C. canimorsus* produces beta-lactamase responsible for drug resistance.
  2. **Host factors**
     - Age: Middle-aged and elder persons are at greater risk. More than 60% of sufferers are 50 years of age or older.
     - Immunity: *C. canimorsus* generally has low virulence in healthy individuals, but causes

severe illness in persons with preexisting conditions like splenectomy, alcoholism, immunosuppression due to the use of steroids (glucocorticoids) in SLE patients, beta thalassemia or smoking.

- Occupational: Individuals who spend a greater portion of their time with canines and felines are at high risk like veterinarians, breeders, pet owners, and animal keepers.
- Type of animal bite: Chances of infection after dog bites are 3–20% while after cat bites; they may increase up to 50%.

- **Clinical features**
1. **Periodontal infections:** It is due to activation of oral flora in individual with IDDs.
2. **Mild flu like symptoms:** It includes fever, vomiting, diarrhea, malaise, abdominal pain, myalgia, confusion, dyspnea, headaches and skin rashes such as exanthema.
3. **More severe case:** It includes fulminant septicemia, endocarditis, DIC and meningitis. Prior treatment with methylprednisolone has been shown to prolong bacteremia in these infections, which enables the progression of endocarditis.

Laboratory diagnosis: Because of slow growth, bacteria are difficult to diagnose by laboratory methods. Clinical history of dog or cat bite helps in diagnosis. Such history enable microbiologist for longer incubation of culture in suspected cases. Blood or CSF is tested by microscopy (follow morphology) and culture (follow C/Cs). Molecular method like PCR is also useful.

Treatment: Pn G is the drug of choice, but bacteria are able to produce beta-lactamase, so imipenem/cilastatin, clindamycin or combinations having beta-lactamase inhibitor are indicated.

Chromobacterium violaceum

Morphology: They are GNB, showing bipolar staining, long road in shape, 1.5–3-4 µm × 0.6–0.9 µm in size, motile by single flagellum at one end (monotrichous) or at both ends (amphitrichous), nonsporing and non-capsulated.

Cultural characteristics (C/Cs): They are aerobes or facultative anaerobes and grow at 35–37°C. Bacteria produce violet pigment called violacein (hence the species name is *violaceum*) on **nutrient agar** as shown in **Fig. 71.4**. Pigment is soluble in ethanol and insoluble in water and in chloroform. Bacteria produce pale colonies on **MacConkey's agar** due to NLF.

Biochemical reactions (B/Rs)
1. **Sugar fermentation tests:** They ferment glucose with acid production only.
2. **Catalase and oxidase tests:** Positive
3. **Urease test:** Negative

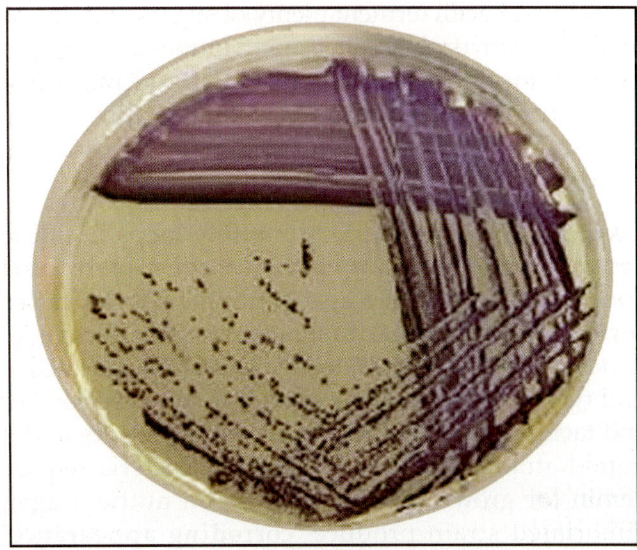
Fig. 71.4: Pigment of *C. violaceum* on nutrient agar

Pathogenicity
- **Reservoirs of infection:** These are soil and water.
- **Source of infection:** Water.
- **Modes of transmission:** It occurs by contact with wound or by ingestion of contaminated water.
- **Clinical features:** Skin lesion, local abscess or septicemia in immunosuppressed persons.

Laboratory diagnosis: Pus swab or blood tested by microscopy and culture as described above.

HACEK GROUP

Name is given by using initial of 5 genera like *Haemophilus parainfluenzae*, *Aggregatibacter* spp. (Like *A. aphrophilus*, *A. paraaphrophilus* and *A actinomycetemcomitans*) *Cardiobacterium hominis*, *Eikenella corrodens* and *Kingella kingae*.

Haemophilus parainfluenzae: Follow **Ch. 68**.

Aggregatibacter aphrophilus and *A. paraaphrophilus*: Follow **Ch. 68**.

Aggregatibacter actinomycetemcomitans: Formely it called *Actinobacillus actinomycetemcomitans*. Bacilli are GNB/GN coccobacilli, nonmotile, microaerophilic and best grow under 5–10% CO_2. Bacilli produce star shaped colonies on nutrient agar, blood agar and chocolate agar. Bacilli ferment glucose with acid production only. Catalase and oxidase are positive. They are the normal flora of mouth causes **periodontal infection**. In 30% cases of mycetoma, it is associated with *A. israelii*. Bacilli also causes endocarditis (most common pathogen from HACEK group), pericarditis, meningitis, pneumonia, UTI, abscess, empyema, etc.

Cardiobacterium hominis: It produces disease in person with preexisting heart disease, hence the name. Bacilli are GNB, pleomorphic, nonmotile, facultative anaerobes and best grow under 3–5% CO_2 and high humidity. Nutrient agar, blood agar and yeast extract agar are used

538 for culture. Bacilli ferment plenty of sugars, indole and oxidase positive while catalase and nitrate are negative. They are the normal flora of nose and throat may cause endocarditis specially in preexisting cardiovascular disease. Streptomycin and penicillin are the drugs of choice.

Eikenella corrodens: Jerking motility helps bacilli to spread and to produce corrosive effect (corroding appearance) over blood agar hence the **species called corrodens.** They are GNB, showing unusual jerking or twitching motility not due flagella, but by contractile-fimbriae like filamentous appendages. They are aerobes and facultative anaerobes. Good growth occurs under humid atmosphere and 5–10% CO_2. Bacteria require hemin for growth. Growth occurs on nutrient agar. Fimbriated strain produce **'corroding appearance'** and nonfimbriated strain produce **'noncorroding appearance'** on blood agar. Nutrient agar contains hemin, clindamycin and KNO_3 is the selective medium. They do not-ferment sugars, indole and catalase are negative while oxidase positive. They are normal flora of mouth, respiratory tract and GIT. They produce opportunistic infections like endocarditis and mixed infections with other bacteria like periodontal disease, wound infection, subcutaneous-brain-liver-skin abscess, meningitis, osteomyelitis, pneumonia, etc. Tetracycline and penicillin are effective drugs.

Kingella kingae: They are GN bacilli or coccobacilli, and sometimes, gram-negative diplococci like *Neisseria*. They grow on Thayer-Martin medium, catalase negative and oxidase positive. They are normal flora of mouth responsible for endocarditis and bone-joint-tendon infections.

ACCESS YOURSELF

Short Notes

1. Tularemia
2. Rat bite fever
3. HACEK group
4. *Capnocytophaga canimorsus.*

Short Question for Theory/Viva Question

1. Name the causative agent for following diseases: Sanghai fever, Sodoku, Haverhill fever and Pontiac fever.

MCQs for Chapter Review

Gardnerella vaginalis

1. **A-40-year-old woman presented to the gynecologist with complain of profuse vaginal discharge. There was no discharge from the cervical os on the speculum examination. The diagnosis of bacterial vaginosis was made based upon all of the following findings on microscopy *except*:**
   a. Abudance of gram variable coccobacilli
   b. Absence of lactobacilli
   c. Abundance of polymorph
   d. Presence of clue cells

Rat Bite Fever

2. **Rat bite fever may be caused by:**
   a. *Leptospira canicola*
   b. *Streptobacillus moniliformis*
   c. *Borrelia recurrentis*
   d. *Spirillum minus*

HACEK Group

3. **HACEK group includes all *except*:**
   a. *Haemophilus aphrophilus*
   b. *Acinetobacter baumanni*
   c. *E. corrodens*
   d. *Cardiobacterium hominis*
4. **True about HACEK group of bacteria is:**
   a. Anaerobes
   b. Includes *Coxiella burnetii*
   c. Gram-positive
   d. Required CO_2 for growth

Answers and Explanation of MCQs

1. c
- Follow section, *Gardnerella vaginalis* (laboratory diagnosis) for explanation.

2. b and d
- Follow section, **rat bite fever** for explanation.

3. b
- Follow section, **HACEK group (meaning)** for explanation.

4. d
- HACEK group bacteria are not anaerobes, but microaerophilic. HACEK group does not include *Coxiella burnetii*. All are gram-negative and required 5–10% CO_2 for growth.

Infections of Nonsporing Anaerobes

Chapter Outline
- Nonsporing Anaerobes

NONSPORING ANAEROBES

Definition

Bacteria which do not have spores and survive in absence of oxygen are called nonsporing anaerobes.

Clinical Significances

1. They present as normal flora in intestine and produce the vitamins.
2. *Lactobacillus acidophilus* presents in adults vagina (specially called Doderlein's bacillus) converts glycogen in to lactic acid. Such acidic pH of vagina provides local protection.
3. They produce toxins and other metabolites to elicit the local plus systemic diseases.

Classifications

Follow **Table 72.1**.

Important Features of Anaerobes

Nonsporing anaerobes are normal flora of human body as shown in **Table 72.1** and produce **endogenous infections** in person with poor immunity. Anaerobic infections are generally **polymicrobial**. Anaerobes are **coinfected** with aerobic bacteria. Exact mechanisms of **pathogenesis** are unknown but may be due to reduced blood supply resulting reduction in Eh potential and anoxic damage to tissues. **Clinical feature favors the anaerobic infections** are foul-nauseating odor, gas production gives crepitation on clinical examination, black pus, prolonged cellulitis, ulcer, abscess, resistant to treatment and no bacterial growth on aerobic media. Specific features of particular anaerobes are described below.

S. ventricularis: They are GPC arranged in group of eight.

P. niger: It belongs to family Peptococcaceae. They are GPC arranged singly, in pairs or in clusters but never in chains. They produced black colonies on blood agar and also produce the H_2S.

Peptostreptococcus **spp.:** It belongs to family Peptostreptococcaceae. They are GPC arranged in pairs or in chains.

V. parvula: It belongs to family Veilonellaceae. They are GNC arranged in pairs or in short chains.

Eubacterium **spp.:** They are GPB present as dental pathogens (periodonitis).

Propionibacterium **spp.:** They are GPB present as normal flora of skin. They are anaerobes and aerotolerant. *P acne* isolated from acne lesion. Pathogenic role of *P. avidum* and *P. granulosum* is unknown.

Lactobacillus **spp.:** They are GPB, showing bipolar staining, ferment the materials like milk or cheese with production of acid, acidophilic and grow best at pH <5.

Mobilincus **spp.:** They are gram-variable bacilli, curved shape and motile.

Bifidobacterium **spp.:** They present in branching pattern like Y-shape hence the name. They are nonmotile and pleomorphic.

Bacteroides fragilis

- **Morphology:** They are gram-negative bacilli or coccobacilli, rod shape, 1–4 µm × 0.4–0.8 µm in size, non motile, capsulated and nonsporing.
- **Culture characteristics (C/Cs):** They are strict anaerobes and grow best under 10% CO_2 at 37°C. Under 10% CO_2 and in anaerobic condition it produces smooth, circular, convex, 0.2–0.5 mm, translucent or opaque, light gray colonies on blood agar. Colonies are non hemolytic in 24 hours, but becomes β-hemolytic after longer incubation.
- **Biochemical reactions (B/Rs):** Bacilli are strongly saccharolytic (sugar assimilation), ferment sugars and aesculin hydrolysis positive.

Essentials of Medical Microbiology

TABLE 72.1: Types, sites as normal flora and pathogenic lesions of non-sporing anaerobes

Group	Anaerobes	Normal flora in				Pathogenic lesions
		Skin	Mouth	GIT	Vagina	
GPC	**Sarcina** *S. ventricularis*					Nonpathogenic
	Peptococcus *P. niger*	–	+	+	–	Abscess formation in soft tissues
	Peptostreptococcus *Pst. anaerobius* *Pst. asaccharolytic* *Pst. tetradius* *Pst. prevoti*	–	+	+	+	Puerperal sepsis, wound infection, gangrenous appendicitis, UTI, abscess in brain and in other internal organs, osteomyelitis
GNC	*Veillonella parvula*	–	+	+	+	Pathogenic role is uncertain.
GPB	**Eubacterium** *E. brachy* *E. timidum* *E. nodatum*	–	+	+	–	Periodonitis
	Propionibacterium *P. acne* *P. avidum* *P. granulosum*	+	–	–	–	Acne, endocarditis Pathogenic role is uncertain.
	Lactobacillus *L. salivarius* *L. acidophilus* *L. casei* *L. odontolyticum* *L. catenaforme*	– – – –	+ – + +	+ + – –	– + – –	Pathogenic role is uncertain. Causing bronchopneumonia
	Mobilincus *M. mulieris* *M. curtisii*	– –	– –	– –	+ +	Bacterial vaginosis. Clue cells (epithelial cells with surface covered by bacilli) are seen in stained/unstained films
	Bifidobacterium *B. dentium*	–	+	+	–	Dental caries
	Actinomyces	–	+	+	+	More details: Follow **Ch. 60**
GNB	**Bacteroides** *B. fragilis*		–	+	+	Brain and lung abscess, pneumonia, wound infection, peritonitis, hepatic abscess and other abdominal infection
	Porphyromonas *P. gingivalis* *P. endodontalis*	– –	+ +	– –	– –	Periodontal disease Dental root canal infection
	Prevotella *P. melaninogenica*	–	+	+	+	Liver abscess, lung abscess, brain abscess, mastoiditis, abdominal, gum and oral infections
	Fusobacterium *F. nucleatum* *F. necrophorum*	– –	+ –	– +	– –	Oral and pleuropulmonary infections Liver and abdominal infection, thrombophlebitis called Lemierre's syndrome
	Leptotrichia *L. buccalis*	–	+	–	–	Initially called *F. fusiforme*, causing Vincent angina (More details: Follow **Ch. 73**)
Spirochaetes	**T. pallidum** **B. vincenti**	More details: Follow **Ch. 73**				

- **Resistance:** They produce the β-lactamase (penicillinase) which is responsible for drug resistance among β-lactam group.
- **Pathogenicity:** It is the most common anaerobe isolated from the bowel and other samples. Normal stool contains 10^{11} *Bacteroides*/g than 10^8 facultative anaerobes/g. Endogenous infection occurs due to disturbances of this flora. It commonly infects GIT, lungs, brain and soft tissues. Agent (virulence) factors, host factors and clinical features are described below.
  - **Agent (virulence) factors: (1) Polysaccharide capsule** releases the IL-7, IL-8 and TNF responsible for intra-abdominal abscess. **(2) Endotoxin (LPS)** causes endotoxic activities. Structurally and functionally, it is different from other endotoxin of aerobic GNB by its low toxicity for fever and shock, which is actually from inflammation following infection rather than direct infection and low frequency for DIC and purpura. **(3) Aggresins,** which include the enzymatic activities like protease, neuraminidase, DNAase, heparinize, etc.
  - **Host factors:** Trauma, tissue necrosis, diabetes mellitus, IDDs, chronic steroid therapy, antibiotic therapy, presence of foreign body, malnutrition, cancer, etc., favor the bacterial invasion from respective surfaces.
  - **Clinical features:** Pneumonia, wound infection, peritonitis, myositis, abscess in brain, lung, liver, peritoneum and other abdominal infections occur mostly following surgery.
- **Laboratory diagnosis:** As like other anaerobes described below.
- **Treatment:** Metronidazole is effective, resistance is rare but reported.

Porphyromonas **spp.:** It differs from *B. fragilis* by ability to produce the pigment and asaccharolytic property.

Prevotella melaninigenica: Initially it called *Bacteroides melaninogenica*. It produces the hemin, derived black or brown pigment. It is moderately saccharolytic. Colonies produce the brick-red fluorescence when exposed to UV light.

Fusobacterium **spp.:** It is the long thin spindle-shaped gram-negative bacillus. *Fusobacterium necrophorum* produces toxin.

Laboratory Diagnosis of Anaerobes

Specimen collection and transport: Anaerobes are normal flora of human body, collect the sample in such a way to avoid mixture of resident flora and their presence will not prove casual role. Anaerobes are very sensitive to O_2; care is required for minimum air exposure. Airtight container or anaerobic media (RCMM, thioglycollate broth, etc.) are preferable. Also transport specimen in **PRAS (pre-reduced anaerobic sterilized) medium.** PRAS is a tube contains nitrogen fitted tightly with butyl stopper. Stuart's transport medium is also useful. Swabs are generally not ideal, but if collected then sent in to Stuart's medium. Sputum is not ideal. Lung abscess or tracheobronchial aspiration is recommended. Type of sample and method depends on site of infection. Some accepted methods are thoracocentesis, aspiration by needle-syringe, surgical excision, etc. Aspirate the abscess/discharge by needle-syringe and then inject in airtight bottle. Blood is collected in BHIA or in anaerobic media.

Testing methods

A. **Microscopy:** Gram's stain.
B. **Culture:** Inoculate the specimen on selective medium like blood agar contains neomycin, yeast extract, hemin, and vitamin K and incubate at 37°C with 5–10% CO_2 in anaerobic (gas pack) jar. Inoculate one extra plate for aerobic incubation, as control and to identify coinfection by aerobes.
C. **Biochemical reactions and serological tests:** Performed as per suspected bacteria.
D. **Ultraviolet light (Wood's lamp) exposure:** For example, exposure of culture or even wound dressing gives bright red fluorescence in *P. melaninogenica*.
E. **Other methods:** Like GLC is also useful.
F. **AST:** It is done by using gentamicin and metronidazole disc.

Treatment of Anaerobes

Drug resistance is a common problem in anaerobic infections. Drug selection depends on sites of infections, types of anaerobes and sensitivity patterns of anaerobes. Commonly useful antibiotics are penicillin, 2nd generation cephalosporin, chloramphenicol, fucidin, trimithoprim, clindamycin, tetracycline, imipenam, and aztreonam. Bacteroides are sensitive to metronidazole.

ACCESS YOURSELF

Essay/Full Question

1. Nonsporing anaerobes.
2. Enumerate the microbial agents causing anaerobic infections. Describe the pathogenicity and discuss the laboratory diagnosis of anaerobic infections.

Short Questions for Theory/Viva Questions

1. Write the two clinical significances of nonsporing anaerobes.
2. What are Doderlein's bacilli?

MCQs for Chapter Review

Clinical Significances

1. **Doderlein's bacillus word is associated with:**
   a. *Clostridium*   b. *Lactobacillus*
   c. *Treponema*   d. *Borrelia*
2. **True about anaerobic infection:**
   a. Causes toxemia
   b. Causes systemic infections
   c. Both
   d. None

3. **One of the following infection is caused by anaerobic gram-positive cocci:**
   a. Puerperal infection   b. Food poisoning
   c. Endocarditis   d. Septicemia
4. **With reference to *Bacteroides fragilis* following statements are true *except*:**
   a. *B. fragilis* is the same frequent anaerobe isolated from clinical samples
   b. *B. fragilis* is not uniformly sensitive to metronidazole
   c. The LPS formed by *B. fragilis* is structurally and functionally different from the other conventional endotoxin
   d. Shock and DIC are common in *Bacteroides*.

Pathogenicity

5. **Characteristic of anaerobic bacteria is:**
   a. Foul smelling discharge
   b. Fail to grow in aerobic media

c. Gas in tissue (crepitus)
d. All of the above

Answers and Explanation of MCQs

1. **b** } Follow section, **clinical significances** for explanation.
2. **c**
3. **a**
- Lesions produced by anaerobic gram-positive cocci are listed in **Table 72.1**. Common lesion by anaerobic gram-positive cocci from the given options is puerperal infection.
4. **d**
- *B fragilis* is the most common anaerobe isolated from the bowel and other samples. Metronidazole is the drug of choice but resistant strain has been reported. For options C and D, follow section, ***Bacteroides fragilis*** for explanation.
5. **d**
- Follow section, **important features of anaerobes** for explanation.

Infections of Spirochetales

Chapter Outline

- Introduction
- *Treponema pallidum*
- Endemic (Nonvenereal) Treponematoses
- Relapsing Fever
- Vincent's Angina
- Lyme Disease
- *Leptospira*

INTRODUCTION

Word spira derived from speira means coil and chaita means hair indicating flexuous spiral forms. **Classification of spirochetales** is given in **Flowchart 73.1**.

Treponema pallidum

Meaning

Word derived from trepos means turn, nema means thread and pallidum means pale staining.

History

T pallidum was 1st identified by Fritz Schaudinn and Erich Hoffmann in 1905. The first effective treatment (Salvarsan) was developed in 1910 by Sahachiro Hata in the laboratory of Paul Ehrlich which was followed by the introduction of penicillin in 1943.

Morphology

Staining properties: They are GNB, but better stained by following methods.

Flowchart 73.1: Classification of spirochetales

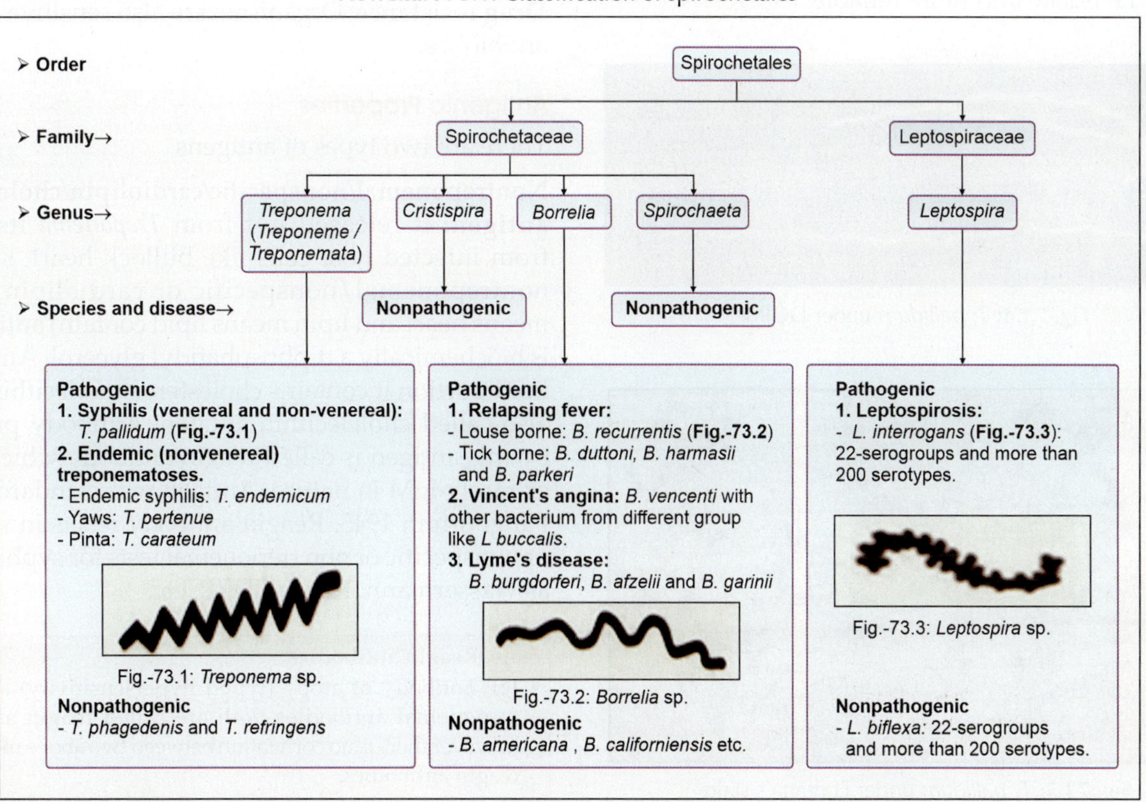

A. **Vital stains:** These are wet mounts and maintain the viability of bacteria and useful for motility detection.

- **Types of motility:** Three types of motility. **(1) Flexion and extension:** During movement secondary curve appears which disappears after some time, but primary coils remain unchanged. **(2) Translatory motility:** Whole organism move in one direction. **(3) Cork screw motility:** It is the movement around long axis.
- **Endoflagella:** Having endoflagella but not useful for motility and they remain in periplasmic space between peptidoglycan and outer membrane.
- **Detection of motility:** It is detected by two microscopic techniques. **(1) Dark ground illumination microscopy (DGIM):** Prepare the wet film from exudates, apply thin cover slip and examine under the DGIM (**Fig. 73.4**). It has low sensitivity, so required 10^4 treponemas/ml of exudate for test to be positive. **(2) Phase contrast microscopy (PCM).**

B. **Supravital stains:** They do not maintain the viability of bacteria. These are of total four types. **(1) Silver impregnation stains:** Like Fontana's stain (**Fig. 73.5**) from culture and Levaditi's stain from tissue section. **(2) Histopathological stain:** Like Giemsa stain. **(3) Negative stains:** Like Nigrosin/Indian ink stain. **(4) Direct fluorescent antibody test for *T. pallidum* (DFA-TP):** Prepare the smear from exudates and fix with acetone. Stain with reagent contains fluorescent tagged anti-*T pallidum* antiserum (specific monoclonal Ab) and examine under the fluorescent microscope. **DFA-TP** is safe and more reliable.

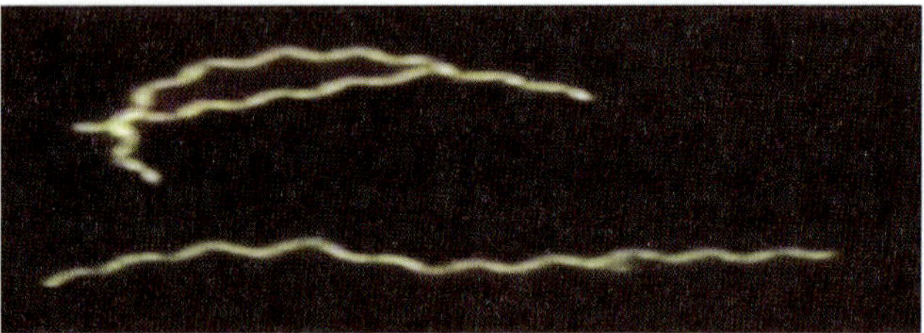

Fig. 73.4: *T. pallidum* under DGIM

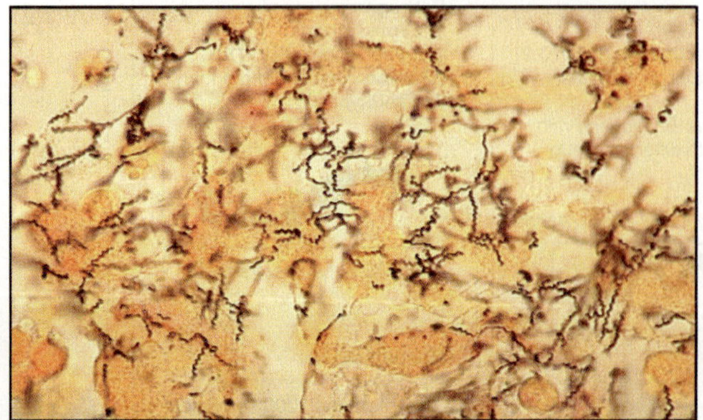

Fig. 73.5: *T. pallidum* under Fontana's stain

Shape and size: They are regular spiral in shape with sharp ends with 10 coils at interval of 1 µm. They are 10 µm long and 0.2 µm in breadth.

Spores and capsule: They are nonsporing and noncapsulated.

Cultural Characteristics (C/Cs)

Effective factors

- O_2 effect: They are strict anaerobes and killed by O_2 exposures.
- Generation time: 30–33 hours.

Media and cultivation techniques

1. **Pathogenic treponemas (treponemes):** They do not grow in artificial culture media. They are maintained for many decades by serial testicular passage in rabbits, e.g., Nichol's strain of *T pallidum*. Nichol's strain is useful to prepare the antigen for species specific treponemal tests.
2. **Nonpathogenic treponemas:** They grow in artificial culture media. *T. refringens* (Noguchi's strain) and *T. phagedenis* (Kazan's strain and Reiter's strain) can grow in **thioglycollate medium containing serum.** Reiter's strain is useful to prepare the antigen for group specific treponemal tests.

Resistance

Sterilization: Organisms are killed by heating at 41–42°C in 1 hr, refrigerating at 0–4°C in 1–3 days, drying and distilled water.

Disinfection: Organisms are killed by O_2 exposure and other disinfectants like soap, arsenic, mercury and bismuth.

Drug resistance: Organisms are also sensitive to many antibiotics.

Antigenic Properties

There are two types of antigens.

Nontreponemal/nonspecific/cardiolipin/cholelecithin antigen: It released not from *Treponema* itself, but from infected host cells like bullock heart, so called nontreponemal/nonspecific or cardiolipin (cardia means heart and lipin means lipid contain) antigen and is biochemically a diphosphatidyl glycerol. Among the lipid portion it contains cholesterol and lecithin, hence also called cholelecithin antigen. Antibody produced by this antigen is called reagin antibody which is IgG or rarely IgM in nature. Antigen was standardized by Pangborn in 1945. Reagin antibody reacts in standard or nonspecific or non treponemal tests for syphilis, such as Wassermann, Kahn, VDRL, etc.

> **Note: Reagin antibodies**
> - IgE antibody of atopy (type-I hypersensitivity) and nontreponemal antibodies both are called reagin antibody; however there is no correlation between two above mentioned reagin antibodies.

Treponemal/specific antigen: Two subtypes.

1. **Group specific antigen:** Biochemically, it is lipopoly-saccharide-protein in nature, found in pathogenic and in nonpathogenic treponemes, group specific and antibody to this antigen is detected by preparing the antigen from nonpathogenic strain like **Reiter's strain.**
2. **Species specific antigen:** It is polysaccharide in nature, found in pathogenic treponemes, species specific and antibody to this antigen is detected by preparing the antigen from pathogenic strain like **Nichol's strain.**

Immunity in Syphilis

AMI is not effective. **CMI** is more relevant. **Phagocytosis** is predominant in early syphilis. Primary infection provides some immunity to reinfection.

Premunition (concomitant infection, infection immunity): Reinfection does not appear in person already having active infection called premunition. It occurs in some parasitic infections also. Patient becomes susceptible to re-infection only when his original infection is cured.

Syphilis and AIDS

- Transmission: Abraded skin of syphilis patient allows the more chances of transmission of HIV.
- Clinical features: Syphilis progress more rapidly in AIDS patients. Concurrent infection may lead to early development of late syphilis and neurosyphilis even after treatment of primary or secondary stage.
- Diagnosis: Poor serological response due to absence of immunity and high titer in STS perhaps due to B cell activation, which may not fall even after treatment. Treponemal tests are negative with advancement of disease due to immune paralysis in end stage.

Pathogenicity, Laboratory Diagnosis, Prevention and Treatment

Disease name: Whole spectrum of disease caused by genus *Treponema* called treponematoses. Disease caused by *Treponema pallidum* called syphilis.

Meaning of syphilis: Syphilis word derived from the person name **syphilus** who died due to disease.

> **Note: Classification of treponematoses**
> 1. **Venereal treponematoses**: Like venereal syphilis.
> 2. **Nonvenereal (endemic) treponematoses**: Following four conditions.
>   - **Nonvenereal syphilis**: By *T pallidum* (**Flowchart 73.1**).
>   - **Endemic syphilis/nonvenereal syphilis/Bejel**: By *T endemicum* (**Flowchart 73.1**).
>   - **Yaws**: Follow **Table 73.2**.
>   - **Pinta**: Follow **Table 73.2**.

Epidemiology: It is distributed worldwide. There are two hypotheses know for the origin of syphilis. 1st proposes that syphilis was carried to Europe from the America by the crew of Christopher Columbus as a by product of the Columbia exchange called **Columbian theory**, while the other proposes that syphilis previously existed in Europe but went unrecognized called **pre-Columbian theory**. The 1st written records of an outbreak of syphilis in Europe occurred in 1494/1495 in Naples, Italy, during a French invasion. Because it was spread by returning French troops, the disease was "called French disease", and it was not until 1530 than the term "syphilis" was first applied by the Italian physician and poet Girolamo Fracastoro. With the discovery of penicillin and also no any extra-human reservoir, it is expected to eradicate the disease. Incidences have been increase with advancing life style, customs and habits. Clinically syphilis is similar to other diseases hence called "**the great pretender**".

Classification of syphilis: Follow **Flowchart 73.2**.

Venereal Syphilis

Reservoir of infection: Only human.

Source of infection: These are human body fluid like exudates from the lesion.

> **Note: Soft sore/chancroid/soft chancre and cold sore/fever blister/herpes labialis**
> - Soft sore: Caused by *H. ducreyi* and pain full (for more details, **follow Ch. 68**).
> - Cold sore: Caused by herpes simplex virus (for more details, **follow Ch. 78**).

Flowchart 73.2: Classification of syphilis

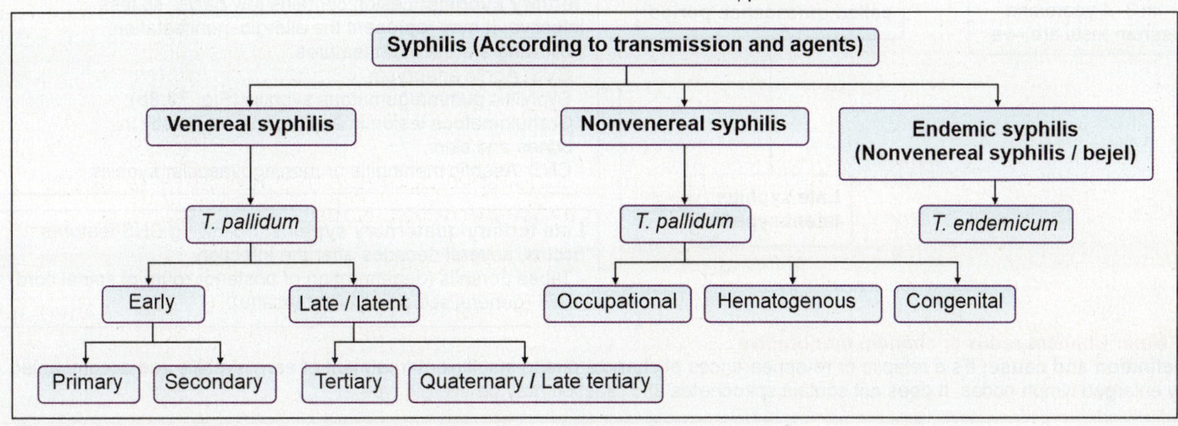

546 **Mode of transmission:** Approximately 30% persons are infected after unprotected sexual intercourse with infected person. Spirochetes enter through minor abrasions in skin and mucosa.

Incubation period: 10–90 days.

Infective dose: Infectivity is maximum in first two years of disease (primary, secondary and tertiary). After 5 years the risk is minimal. Infective dose is small about 60 treponemes are capable of infecting 50 human volunteers.

Portal of entry: Skin/mucosa.

Sites: Skin, genital organs, liver, bones, brain, CVS, CNS, etc.

Pathogenesis, stage of disease and clinical features of venereal syphilis: Follow **Flowchart 73.3 and Figs 73.6 to 73.8.**

Laboratory diagnosis of venereal syphilis

A. Microscopy: It is applicable in primary, secondary and congenital syphilis with superficial lesions. Common specimens are blood, lymph node aspiration materials, Exudates or biopsy from lesions. **To collect the exudates** clean the lesion with a gauze piece soaked in warm saline and margins gently scraped so that superficial epithelium is abraded. Apply gentle pressure at base of lesion. Collect the exudates without mixing the blood. **Biopsy** is collected from chancre, condylomata, syphilitic gumma, etc. Condylomata lata is most infectious. **For microscopical findings** follow morphology.

B. Culture: Follow C/Cs.

C. Serological tests: These are performed from serum, plasma and CSF. Sensitivity and specificity of serological tests in different stages of syphilis are mentioned **Table 73.1.** Two main types of tests as mentioned below.

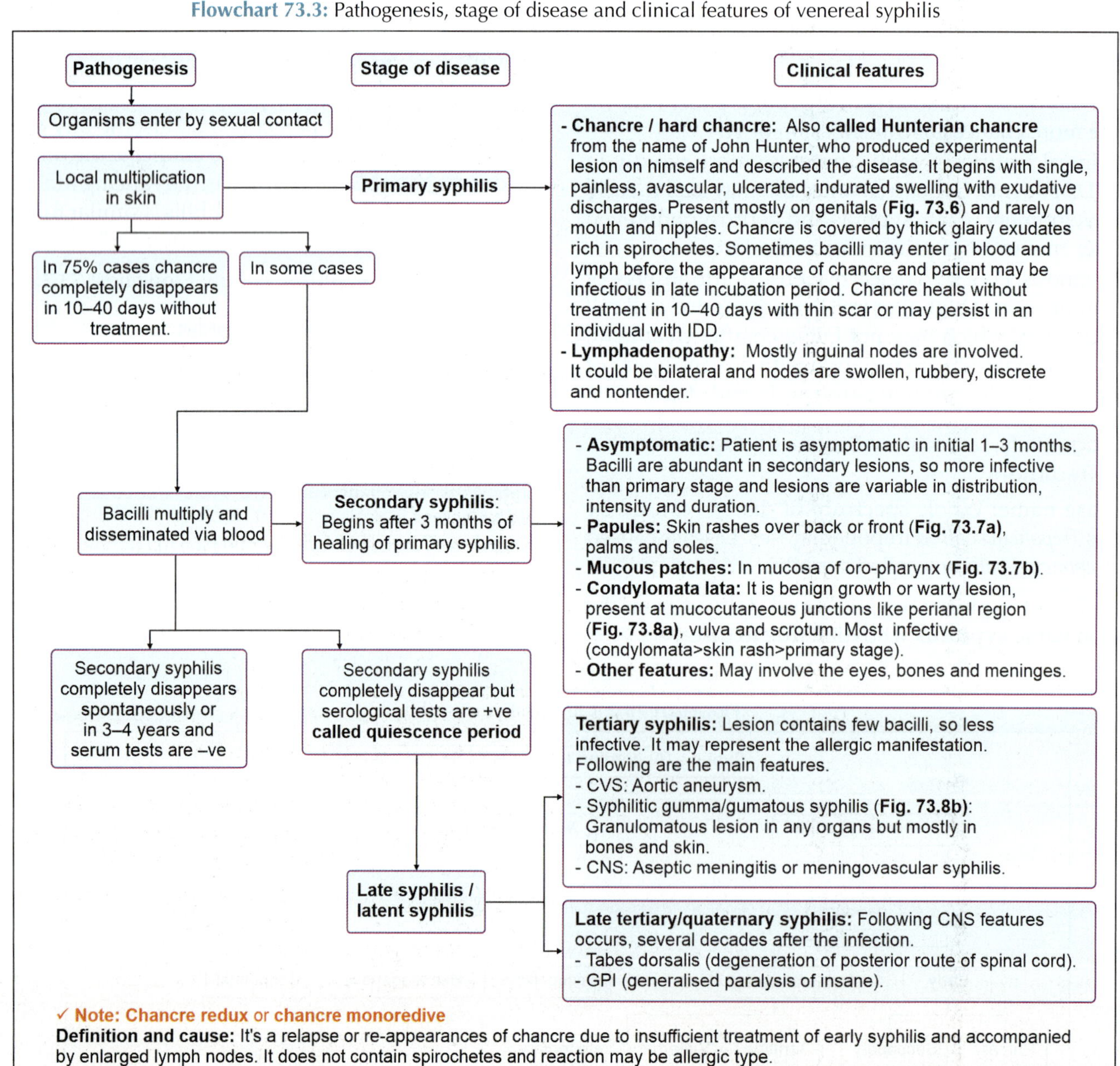

Flowchart 73.3: Pathogenesis, stage of disease and clinical features of venereal syphilis

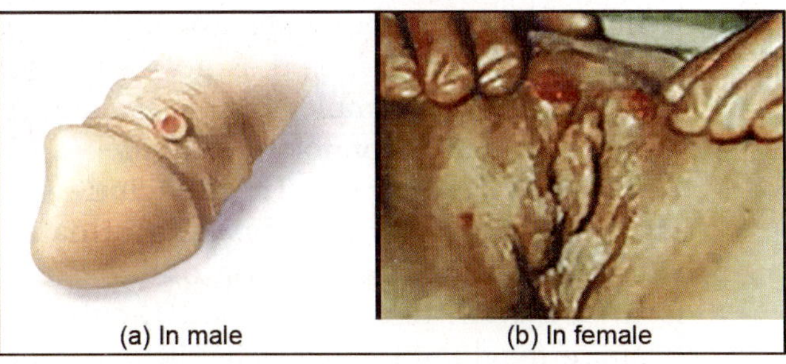

| (a) In male | (b) In female |

Fig. 73.6: Chancre in primary syphilis

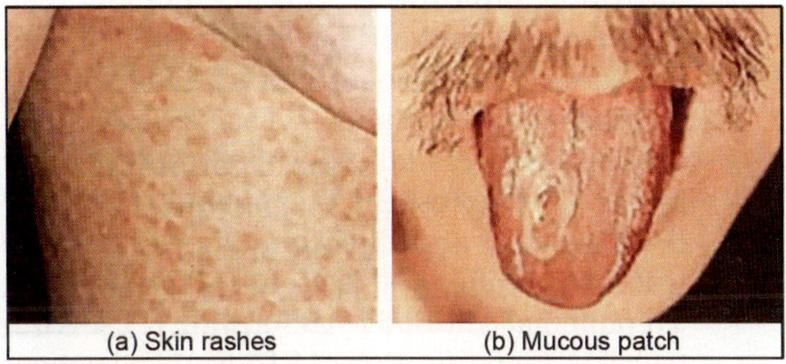

| (a) Skin rashes | (b) Mucous patch |

Fig. 73.7: Skin rashes and mucus patch in secondary syphilis

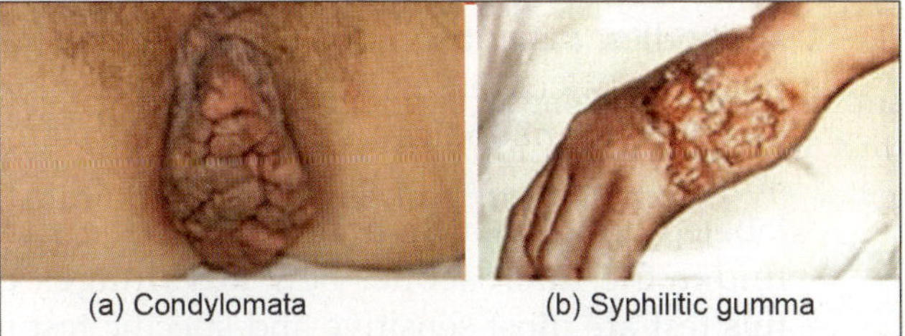

| (a) Condylomata | (b) Syphilitic gumma |

Fig. 73.8: Condylomata in secondary syphilis and syphilitic gumma in tertiary syphilis

Note: Neurosyphilis

- **Cause:** Caused by *T. pallidum*.
- **Clinical features**: It usually occurs in people who have had chronic, untreated syphilis, usually about 10–20 years after first infection and develops in about 25–40% of persons who are not treated. CDC advises that neurosyphilis can occur at any stage of a syphilis infection. It may be asymptomatic. Commonly nerve system is infected in tertiary and late tertiary syphilis with following features.
  - Tertiary syphilis: Meningitis/meningovascular syphilis.
  - Late tertiary syphilis: Features occurs several decades after the infection like tabes dorsalis (degeneration of posterior route of spinal cord) and GPI (generalized paralysis of insane).
- **Diagnosis:** Tests available for neurosyphilis are not 100% accurate. Diagnosis based on clinical findings, CSF picture and nontreponemal tests. CSF is tested by following tests.
  - CSF picture: Lymphocytic pleocytosis, normal glucose, but high (moderate) protein.
  - Nontreponemal tests: (1) VDRL-CSF: Invasion of bacilli stimulates the production of nontreponemal (nonspecific) antibodies in CSF, so it is the standard test for diagnosis of neurosyphilis. However, VDRL has 90% specificity and 30% sensitivity in neurosyphilis. It becomes negative after 1 year of treatment, so useful to monitor the neurosyphilis. (2) RPR: It is not done from CSF. It shows only 50% of seroconversion after treatment, so not useful to monitor the treatment.

  - Treponemal tests: Like FTA-ABS, TPHA and TPPA are less useful, because the treponemal (specific) antibodies have tendency to cross the CSF, so enough antibodies are already present in CSF before the development of neurosyphilis. Such tests are positive from CSF even in absence of neurosyphilis.
- **Treatment and monitoring:** Penicillin is used to treat neurosyphilis; however, early diagnosis and treatment is critical. Two examples of penicillin therapies include (1) Aqueous penicillin G 3–4 million units every four hours for 10–14 days. (2) One daily IM injection with oral probenecid four times daily, both for 10–14 days.
- **Follow-up blood tests:** They are generally performed at 3, 6, 12, 24, and 36 months to make sure the infection is gone. CSF analysis is generally performed every 6 months.
- **Effect of HIV on neurosyphilis:** Neurosyphilis remain unheard after penicillin therapy. However, concurrent infection of *T. pallidum* with HIV has been found to affect the course of syphilis. Syphilis can lie dormant for 10–20 years before progressing to neurosyphilis, but HIV may accelerate the rate of the progress. Also HIV has been found to cause penicillin therapy to fail more often. Therefore, neurosyphilis has once again been prevalent in societies with high HIV rates and limited access to penicillin.

TABLE 73.1: Sensitivity and specificity of serological tests in different stages of syphilis

Stage	Primary (%)	Secondary (%)	Late (%)	Specificity (%)
VDRL	78	100	71	98
RPR	86	100	73	98
FTA-ABST	84	100	96	97
TPHA	76	100	94	99
TPPA	88	100	*	98
EIA	90	100	*	99
WB	90	100	*	98

*Sensitivity not known

1. Non-treponemal tests

- **Other names:** They are also known as nonspecific tests, reagin antibody dependent tests, STS (standard tests for syphilis).
- These tests **based** on detection of reagin antibody of cardiolipin antigen.
- **Sensitivity of tests:** Tests are reactive 3–5 weeks after acquiring the primary infection or 7–10 days after appearance of chancre. Sensitivity is 60–70% in primary syphilis and titer is low about 1:8. Sensitivity is 100% in secondary syphilis and titer is 16–128 or more. Sensitivity is 60–70% in latent/late syphilis.
- **Clinical applications**
  - High titer and higher positivity rate also noticed in congenital syphilis.
  - They are also useful to detect neurosyphilis especially by using VDRL test.
  - Quantitative tests are useful to monitor patient's response against treatment, to indicate stage of

disease, to detect reinfection and to eliminate the effect of prozone phenomenon. Reagin tests are preferred to monitor the efficacy of antibacterial therapy, because they usually become negative or decreasing in titer following treatment.

- If treatment started in early syphilis reagin antibody test becomes negative within 6–18 months, but if treatment is started late, it remains positive in low titer.
- All serological tests are false-positive in endemic treponematoses like yaws, pinta and endemic syphilis and cannot differentiate the endemic treponematoses.

- **Tests**
  - Wasserman test/CFT: Not used now.
  - Kahn test: Tube flocculation test, not used now.
  - VDRL test: Follow **Ch. 20**.
  - RPR test: Follow **Ch. 20**.
  - Automated RPR: Large-scale testing.
  - VDRL-ELISA: Large-scale testing.
  - Unheated serum reagin test (USRT): It is similar to VDRL test except followings. EDTA is used as an Ag stabilizer and daily preparation of antigen is not required. Preheating of serum does not require as choline chloride is used to remove the inhibitors.
  - TRUST: Toluidine red unheated serum test. Ag is more stable, so it can be stored at room temperature at 26–31°C. TRUST is performed by using unheated serum. It is same as like RPR test but Ag is coated with toludine red instead of carbon particle and resulting floccules are in red color.

2. **Treponemal tests (specific tests):** Two types of tests like group specific and species specific.
   - **Group specific tests:** Antigen prepared by using nonpathogenic *Treponema* like Reiter strain. They avoid biological false-positive (BFP). Their sensitivity and specificity are lower than species specific test. Only one test in this category is Reiter protein complement fixation test (RPCFT).
   - **Species specific tests:** Antigen prepared by using pathogenic *Treponema* like Nichol's strain. They distinguish positive STS from BFP. Their sensitivity and specificity are higher in late/latent syphilis. Following are three subtypes of tests in species specific tests.
     - **By using live strain:** Test in this category is **TPI (*Treponema pallidum* immobilization) test.** Test serum is incubated with complement and *T. pallidum* maintained in complex medium anaerobically. Motility examined under DGIM. If antibodies are present in test serum, treponemes are immobilized. Positive test indicates ≥50% of treponemes are immobilized. Negative test indicates ≤20% of treponemes are immobilized. Intermediate (inconclusive) result indicates

if immobilized treponemes are in-between 20–50%. Test is considered as gold standard, but because of its complexity, it is available in highly equipped laboratory.

- **By using killed strain:** These include four tests. **(1)** *Treponema pallidum* **agglutination (TPA) test:** Killed suspension of *T. pallidum* is mixed with test serum and incubated. Examine the clumps under DGIM. It gives false-positive results, so not in use. **(2)** *Treponema pallidum* **immune adherence (TPIA) test:** Killed suspension of *T. pallidum* is mixed with test serum, complement, RBCs and incubated. In presence of antibodies *T. pallidum* adhere on RBCs. It gives false-positive results, so not in use. **(3) Fluorescent Treponemal Antibody (FTA) test:** It is an indirect immune fluorescence test. It's modified test is useful. **(4) Fluorescent Treponemal Antibody-ABSorption (FTA-ABS) test:** It is a modification of FTA test. The patient serum is 1st diluted with non pathogenic Reiter's strain to remove group specific treponemal antibodies. Patient serum is layered on a slide previously coated with killed *T. pallidum*. Treponemal antibodies present in patient serum are bound with *T. pallidum*. Add the fluorescent labeled antihuman Ig and examine the slide under the microscope. Few **advantage**s of this test are most sensitive and specific test in all stages of syphilis, 1st serological test to be positive following infection, useful to detect the neurosyphilis from CSF and useful to detect the congenital syphilis by detecting the IgM from fetal serum called IgM FTA-ABS test. Few **disadvantage**s of this test are available in well equipped laboratory and false-positive result may occur in Lyme disease which is corrected by VDRL test with negative result.
- **By using treponemal extracts:** These include four tests. **(1)** *T. pallidum* **Hemagglutination (TPHA) test**: Treponemal antigen is coated on RBCs. Mix such antigen with test sera containing antibody. Hemagglutination (clumping of RBCs) occurs. Test is employed as an automated microhemagglutination test (MHA-TP). Few **advantages** of this test are highly specific, highly sensitive as like FTA-ABS test except in primary stage (65–85%), simple, economical, no special equipment required and commercially available. These advantages make TPHA a standard confirmatory test. **(2) Latex agglutination test or** *Treponema pallidum* **particle agglutination (TPPA) test:** *T. pallidum* antigens are coated with latex/gelatin particles. It is as specific as TPHA, and most sensitive in all stages of syphilis. **(3) Enzyme immunoassays (EIA):** *T. pallidum* antigens are coated in microtiter well. It detects IgG and IgM separately. **(4) Western blot (WB)**

assay: It is highly sensitive and specific for detection of IgG and IgM antibodies separately.

D. Molecular method: PCR based techniques are useful. Multiplex PCR is developed which targets genital ulcer causing microbes like HSV, *H. ducreyi* and *T. pallidum.*

Prevention of venereal syphilis: (1) General measures include avoid sexual contact with infected person, use physical barrier like condoms and use antiseptics like potassium permanganate. **(2) Chemoprophylaxis:** Penicillin is useful **(3) Immunoprophylaxis:** No effective vaccine is available.

Treatment of venereal syphilis: (1) Patient not allergic to Pn G: Early syphilis is treated by single injection of 2.4 million units of PnG. Adverse reaction includes Jarisch-Herxheimer reaction which is frequent but harmless. It is due to hypersensitivity or due to liberation of toxic products like tumor necrosis factors form killed treponemes. Late syphilis is treated by injecting the 2.4 million units of Pn G every week for 3 weeks. Adverse reaction includes Jarisch-Herxheimer reaction which is rare but dangerous. **(2) Patient allergic to Pn G:** Doxycycline/tetracycline is effective.

Monitoring of venereal syphilis
- **Treponemal tests:** Antibodies titer by treponemal tests like FTA-ABS test or TPPA test or TPHA test, remain high even after clinical improvement, so treponemal tests have little or no value in management of syphilis.
- **Nontreponemal tests:** Antibodies titer by non-treponemal tests like VDRL test and RPR test starts to decline or become negative after treatment, so it is better to use non-treponemal tests for management of syphilis following treatment. VDRL should be done at 3 months interval for at least 1 year. Sometimes patient have low antibodies titer to non-treponemal, tests in spite of treatment called **serofast reaction,** which may be due to treatment failure.

- **For early syphilis:** Following successful treatment there should be at least fourfold decline in the titer at the end of 3rd or 4th month and eightfold decline in the titer at the end of 6th or 8th month.
- **For late/latent syphilis:** Titer decline gradually following successful treatment and low titer may persist for a years.

Nonvenereal Syphilis

A. Occupational syphilis: It occurs in doctors and nurses. Primary chancre presents on hands/fingers.

B. Hematogenous syphilis: It is transmitted by blood transfusion. Primary chancre does not occur. Transfusion syphilis is prevented by refrigerating the blood at 0–4°C for 1–3 days before transfusion.

C. Congenital syphilis

Mode of transmission of congenital syphilis: Transmission can take place through placenta at any stage of pregnancy or may be at birth from infected mother's genitals, but fetal lesions develop after 4 months of gestation, when fetal immune system starts to develop. A woman with early syphilis can infects fetus in 75–90% cases. A woman with late or 2 years old syphilis can infect fetus in 35% **cases.**

Clinical features of congenital syphilis

1. **Maternal features:** Increase risk to pregnancy. It causes abortion, live birth, stillbirth and stigmata of syphilis.
2. **Features in newborn**
   - **Early congenital syphilis:** Features develop in 0–2 years. It may be asymptomatic or symptomatic. If there is no stillbirth, then child born premature, with an enlarged lymph nodes, liver and spleen. Skeletal abnormalities (like osteochondritis, periosteitis and osteitis), pneumonia, anemia, seizure, thrombocytopenia, leukocytosis, jaundice and a bullous skin disease called pemphigus syphiliticus. Newborns will typically not develop a primary syphilitic chancre, but may present with signs of secondary syphilis (which involves skin and mucosa) like generalized skin rashes, syphilitic rhinitis/snuffles" **(Fig. 73.9a);** the mucus, which is laden with the *T pallidum* bacterium, and therefore highly infectious. Rarely, the symptoms of syphilis go unseen in infants, so that they develop the symptoms of late congenital syphilis.
   - **Late congenital syphilis:** Features develop ≥2 years. Subclinical infection occurs in 60% cases. Common features are interstitial keratitis, deafness from auditory nerve disease, Clutton's joints (bilateral swollen knees or knee effusion), gummatous periosteitis, destruction/perforation of bony and cartilage part of nasal septum, residual stigmata (with damage to bones, teeth, eyes, ears and brain), **Hutchinson's triad** (triad consisting of deafness due to auditory nerve

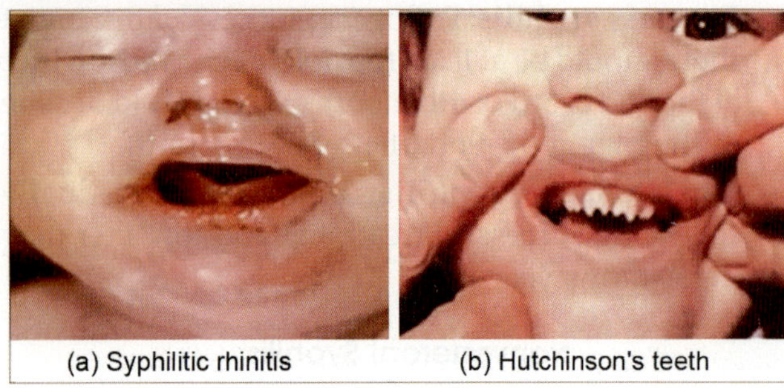

(a) Syphilitic rhinitis | (b) Hutchinson's teeth

Fig. 73.9

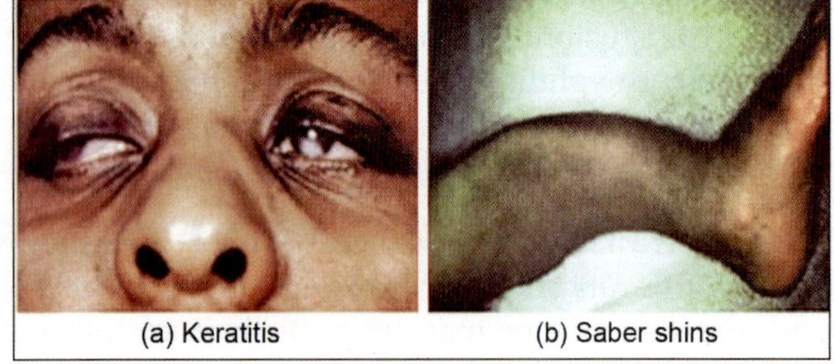

(a) Keratitis | (b) Saber shins

Figs 73.9 and 73.10: Congenital syphilis

disease, Hutchinson's teeth characterized by centrally notched or blunted, widely spaced peg-shaped upper central incisors as shown in **Fig. 73.9b**, and interstitial keratitis characterized by inflammation of the cornea as shown in **Fig. 73.10a**, which can lead to corneal scarring and potentially blindness), mulberry molars (permanent first molars with multiple poorly developed cusps), frontal bossing (prominence of the brow ridge), saddle nose (collapse of the bony part of nose), hard palate defect, swollen knees, Du Bois sign (narrowing of the little finger, poorly developed maxillae, protruding mandible, petechiae, saber shins (malformation of tibia with sharp anterior bowing as shown in **Fig. 73.10b**), enlarged nodes, jaundice, pseudoparalysis, Rhagades (linear scars at the angles of the mouth and nose due to bacterial infection of skin lesions), Higoumenakis sign, enlargement of the sternal end of clavicle in late congenital syphilis and death from congenital syphilis is usually due to bleeding into the lungs.

Diagnosis of congenital syphilis: Specimens like placenta, umbilical cord, nasal discharge, skin lesions, etc., are tested by following tests.

1. **Microscopy:** DGIM to demonstrate *T. pallidum*.
2. **Serological tests:**
   - **IgM detection:** 1st Ab appears after infection is IgM which is detected after 2nd week of infection. IgM does not cross the placenta, its presence in neonatal serum **called syphalotoxemia,** which indicates congenital syphilis. It helps to differentiate it from passively transferred maternal antibodies. IgM detected by treponemal test like IgM FTA-ABST test, TPHA test and EIA (syphilis Captia M test).

IgM also detected by VDRL test from whole sera. Commercially available tests are not recommended for IgM detection.

   - **Parallel testing of maternal and neonatal sera:** When above tests are not available, parallel testing of maternal and neonatal sera is done in which neonatal sera show the high titer than maternal sera.
   - **Serial testing of neonatal sera:** It is done to detect the presence of passively transferred maternal antibodies in neonatal serum by using VDRL test. Such titer decreases rapidly and VDRL test become negative by 3 months.
   - **Molecular method:** PCR base technique.
   - **Radiological method:** Abnormal X-rays.

Prevention of congenital syphilis: It is prevented if pregnant mother is diagnosed and treated before 4 months (16 weeks). Every pregnant woman should test with nontreponemal tests at her 1st antenatal visit. If she is on high risk again retested at 3rd trimester and at the time of delivery.

Treatment of congenital syphilis: A sick child can be treated by using antibiotics like an adult; however, any developmental symptoms are likely to be permanent. Kassowitz's law is an empirical observation used in context of congenital syphilis stating that the greater the duration between the infection of the mother and conception, the better is the outcome for the infant. Features of a better outcome include less chance of stillbirth and of developing congenital syphilis. The CDC recommends treating symptomatic or babies born to infected mother with unknown treatment status with procaine penicillin G, 50,000 U/kg dose IM a day in a single dose for 10 days. Treatment for these babies can vary on a case by case basis. Treatment cannot reverse any deformities, brain or permanent tissue damage that has already occurred.

Endemic Syphilis (Nonvenereal Syphilis/Bejel)

Follow **Table 73.2.**

ENDEMIC (NONVENEREAL) TREPONEMATOSES

Follow **Table 73.2.**

RELAPSING FEVER

Meaning

It occurs with 5–10 relapses, hence the name relapsing fever. Relapsing fever has been known since the time of Hippocrates and has occurred in epidemic, endemic and sporadic forms.

Etiological Agents

Two types as mentioned in **Table 73.3.**

Morphology

- **Staining properties:** They are GNB, but better stained by Giemsa, Leishman or dilute carbol fuchsin.

Features	Endemic syphilis (bejel)	Yaws	Pinta
Synonym	Siti, dichuchwa, njovera, skerljevo, etc.	Pian, framboesia, bouba, etc.	Mal del pinto, carate, azul, purupuru, etc.
Agent	*T. endemicum*	*T. pertenue*	*T. carateum*
Transmission	In young children via person-to-person contact by use of contaminated utensils for drinking and eating purpose	• Direct contact • Sometimes, flies may act as mechanical vector	Direct contact
Age	Early childhood	Early childhood	Late childhood
Primary syphilis	Not seen but present on nipples of mother infected by her child (oral lesion)	Extragenital papules (limbs) which break down to form ulcerative granuloma	Extragenital papules (limbs, face) without ulcer, but develops in lichenoid/psoriatic patch
Secondary/tertiary syphilis	Same as syphilis	Same as syphilis but CVS (heart) and CNS (nerve) features are rare	Secondary lesion are characterized by hypo- or hyperpigmentation
Relapse	Unknown	Common	None
Lab., diagnosis	Same as syphilis		
Treatment	All endemic treponematoses are treated by benzathine penicillin		

TABLE 73.2: Ttypes of endemic (nonvenereal) treponematoses

- **Shape and size:** Bacilli are irregular spiral shape **(Fig. 73.11a)** with 3–5 wide open coils and pointed ends. They are about 8–20 µm × 0.2–0.5 µm in size.
- **Motility:** Lashing motility examined by wet mount under DGIM or under PCM.

Cultural Characteristics (C/Cs)

- **Effective factors:** They are microaerophilic and grow at 33°C.
- **Culture in media:** They grow with difficulties in Noguchi's medium.
- **Culture in animal:** Inoculate 1–2 ml of patient's blood intraperitoneally in white mouse. Bacteria multiply in animal and appear in large number in peripheral blood within two days. Prepare smear from tail vein of animal after 2 days and examined daily for 2 weeks.
- **Culture in egg:** Growth occurs in chorioallantoic membrane of chick embryo.

Antigenic structure: One of the reasons behind occurrence of relapse in relapsing fever is due to antigenic variation in "vmp gene" of *Borrellia*. Vmp gene is located in linear plasmid which shows the DNA rearrangement. Due to antigenic variation newer strain is not protected by antibodies of earlier strain. Ultimate protection to the disease occurs after multiple attacks due to development of immunity to all strains.

Pathogenicity

- **Reservoir and source of infection:** Follow **Table 73.3**.
- **Mode of transmission:**
  - **Louse borne:** Follow **Table 73.3**.

Types	Louse borne	Tick borne
Etiological agent	*Borrelia recurrentis*	*B. duttonii, B. harmasii, B. parkeri*
Nature	Epidemic	Endemic or sporadic called place disease
Distribution	East Africa (Sudan and Ethiopia)	North America, Central Asia and Africa
Vector	*Pediculus humanus corporis*	*Ornithodorus lahorensis, O. tholozoni, O. crossi* and fowl tick like *Aargus persicus*
Reservoirs	Humans. Louse is not the reservoir because transovarial transmission does not occur in louse	Humans, rodents and ticks, because transovarial transmission occurs in ticks
Sources	Body lice	Ticks and their excreta
Modes of transmission	*Borrelia* is present in blood not in saliva or excreta of vector, so transmitted by rubbing the vector against the abraded akin but not by bite or by rubbing the excreta	*Borrelia* are present in all parts of vector and transmitted by bite or by rubbing the excreta/vector against abraded akin
Severity	More severe and highly fatal	Less severe and less fatal
Jaundice and hemorrhage	More common	Less common
Relapse	Less	More

TABLE 73.3: Types of relapsing fever

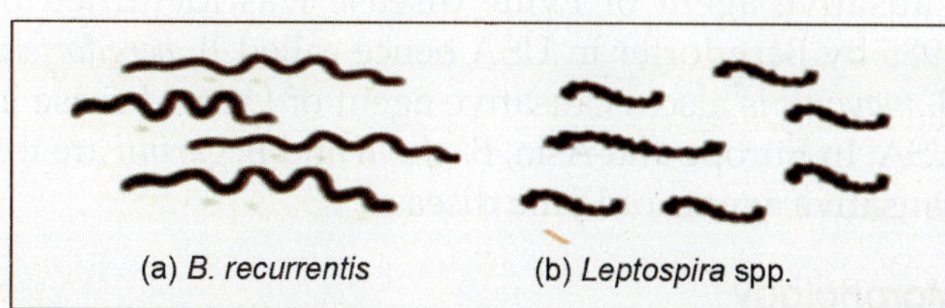

Fig. 73.11: Morphology or *B. recurrentis* and *Leptospira* spp.

– **Tick borne:** Follow **Table 73.3**.
– **Congenital:** By placenta.
– **Occupational:** By accidental contact to laboratory workers.

- **Incubation period:** 1–10 days.
- **Portal of entry:** Abraded skin.
- **Sites:** Bacteria present in blood during febrile period and in brain during afebrile period.
- **Clinical features (Table 73.3):** Sudden onset of fever (bacteria are in blood) which lasts for 3–5 days followed by afebrile period (bacteria are in brain). After afebrile period of 3–5 days another bout of fever occurs. Fever subsides after 5–10 relapses.

Laboratory Diagnosis

Blood is tested with **microscopy** (follow morphology) and **culture** (follow C/Cs). **Serological test** like Weil-Felix reaction is developed. It is rarely useful because giving false-positive reaction. *Borrelia* antibody produces agglutinin with *Proteus* OX-K Ag. **Molecular method** like multiplex PCR is developed, which targets the various *Borrelia* spp., causing relapsing fever.

Prevention

Preventive measures are identification and eradication of vectors. No vaccine is available.

Treatment

Tetracycline, chloramphenicol, penicillin and erythromycin are effective.

VINCENT'S ANGINA

Synonyms

It also called fusospirochetosis or trench mouth.

Etiological Agents

It is a symbiotic infection, caused by *Borrelia vincenti* and *Leptotrichia buccalis*.

Morphology

Follow **Table 73.4 and Fig. 73.12**.

Pathogenicity

Bacteria are the normal flora of oral cavity in humans. So, humans are the reservoirs and sources of infection. Endogenous infection occurs by activation of flora in malnutrition and viral infections. It is ulcerative gingivo-stomatitis or oropharyngitis. It also identified from other clinical conditions like lung abscess, phagedenous skin, ulcer, gangrenous balanitis, choleric diarrhea, dysentery, etc., but significance is not known.

Laboratory Diagnosis

Collect the exudates from lesion of oral cavity which is tested by **microscopy, culture and biochemical reaction** as mentioned in **Table 73.4**.

TABLE 73.4: Differences between *B. vincenti* and *L. buccalis*

Features	*B. vincenti*	*L. buccalis*
Synonym	*Treponema vincenti*	Earlier name was *Fusiform bacillus* or *Fusobacterium fusiforme*
Microscopy	Gram-negative but better stained by Giemsa/dilute carbol fuchsin	Gram-positive bacilli and stained by Gram's stain
Size	5–20 μm × 0.2–0.6 μm	3–10 μm × 1 μm
Shape	Irregular spiral with 3–5 wide open coils and pointed end	Spindle with pointed end
Motility	Motile	Nonmotile
Cultural characteristics	Heartly's broth enrich with ascitic fluid (strict anaerobes)	In blood agar contains vancomycin and neomycin, it produces gray color nonhemolytic colonies
Biochemical test	Not useful	Indole test: Positive

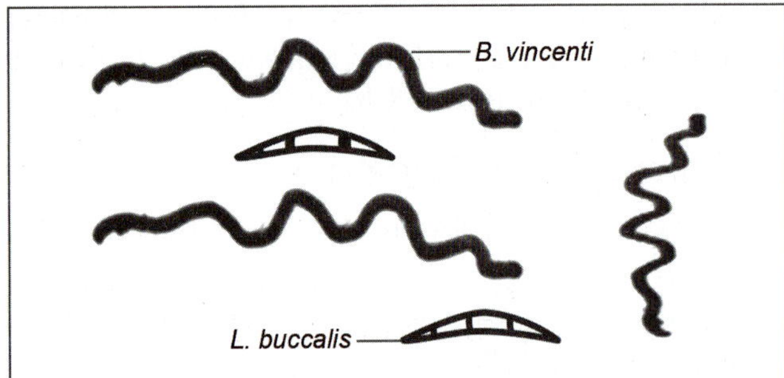

Fig. 73.12: Morphology of *B. vincenti* and *L. buccalis*

Treatment

Penicillin and metronidazole are the effective drugs.

LYME DISEASE

Synonym

It also called Lyme arthritis and Lyme borreliosis.

Meaning

New spirochetal infection identified in 1975 while studying a cluster of juvenile arthritis in **Lyme**, Connicticut, USA, hence called Lyme disease. However, later it was identified from other part of world.

Etiological Agent

Causative agent of Lyme disease was identified in 1985 by Bergdorfer in USA hence called *B. bergdorferi*. *B. mayonii* is also a causative agent of Lyme disease in USA. In Europe and Asia, *B. afzelii* and *B. garinii* are the causative agents of Lyme disease.

Morphology

They are GNB, 4–30 μm × 0.2 μm in size and motile.

17. a
- FTA-ABST is species specific test remains positive after treatment, highly positive in all stage of syphilis and false-positive in Lyme disease.

18. a
- *Treponema* is immobilized by specific antiserum in presence of complement. Such immobilization is examined under DGIM called TPI (*Treponema pallidum* immobilization) test.

19. d
- Nichol's strain is useful to prepare the antigen for species specific treponemal tests. From the given options treponemal test is TPI.

20. d
- Follow section, *Treponema pallidum* (**venereal syphilis → laboratory diagnosis of venereal syphilis**) for explanation.

21. a
- Follow section, *Treponema pallidum* (**neurosyphilis → diagnosis**) for explanation.

22. a
- Given case is of neurosyphilis, where best diagnostic test is VDRL. Follow section, Neurosyphilis (**diagnosis**) for more explanation about other options.

23. b, c and d
- Snuffle is an early manifestation in congenital syphilis.

24. a
- Follow section, *Treponema pallidum* (**diagnosis of congenital syphilis → serological tests → IgM detection**) for explanation

25. a, d and e
- Follow section, *Treponema pallidum* (**clinical features of congenital syphilis → late congenital syphilis**) for explanation.

26. a and b
- Follow section, **endemic (nonvenereal) treponematoses (classification and Table 73.2)** for explanation.

27. b

28. d

29. b
- Follow **Table 73.2** for explanation of answers of MCQs 27–29.

30. b

31. d
- *Borrelia recurrentis* causes relapsing fever. It undergoes the antigenic variation and escapes the host defense.

32. a
- Antigenic variation in *Borrelia recurrentis* is due to Vmp gene which is located in linear plasmid.

33. b

34. b, d
- Follow section, **relapsing fever (Table 73.3)** for explanation of answers of MCQs 33–34.

35. c
- *Borrelia recurrentis* causes relapsing fever, which is transmitted by vector (louse). Follow section, relapsing fever for more explanation.

36. a and b
- It is a symbiotic infection caused by *Borrelia vincenti* and *Fusobacterium fusiforme* (earlier name of *L buccalis*).

37. c
- *Borrelia recurrentis* causes relapsing fever while Lyme disease is caused by *Borrelia burgdorferi*.

38. c, d, e

39. b
- Follow section, **Lyme disease (Flowchart 73.4)** for explanation of answers of MCQs 38–39.

40. b
- Kelly's medium is used in the isolation of *Borrelia burgdorferi,* which is the causative agent of Lyme disease.

41. c
- Meningial involvement in Lyme disease suggests mononuclear lymphocytosis not the polymorphonuclear lymphocytosis. Follow section, Lyme disease for explanation of other options.

42. a
- Erythema chronicum migrans is a feature of Lyme disease, which is caused by *Borrelia bergdorferi.*

43. a
- Follow section, *Leptospira* (**note → "3Rs"**) for explanation.

44. c
- Rat is the reservoir for Weil's disease. For other types and respective reservoir, follow section, *Leptospira* (**classification and Table 73.5**).

45. d
- Weil Felix reaction is useful for *Rickettsia.*

46. a
- Follow section, *Leptospira* (**culture characteristics**) for explanation.

47. e
- Tetracycline is not the drug of choice. Follow section, *Leptospira* (**treatment**) selection of drug of leptospirosis.

48. c
- Antibodies (IgM) appear after 1st week, detectable titer presents in 2nd week, reaches peak level in 3rd-4th (1:10,000 titer) week, then decline slowly and becomes undetectable in a 6 months.

49. a
- Rat is the reservoir for Weil's disease. For other types and respective reservoir, follow section classification (**Table 73.5**).
- Pn/doxicycline is the drug of choice as per severity.
- Person to person transmission does not occur.
- Hepatorenal syndrome may occur in up to 10% of patients.

50. d
- Given case is of leptospirosis, where drug of choice is Penicillin G from the given options.

51. a
- *Leptospira*: It is a zoonosis transmitted by direct contact of abraded skin with contaminated water or mud from urine of infected animals like rat.
- *Listeria*: It is a zoonosis transmitted directly by contact or vertical route and indirectly by airborne, waterborne, foodborne and fomite-borne.
- *Legionella*: Transmitted by inhalation of aerosols arise from air-conditioner, shower heads, nebulizer, humidifier, device filled with tap water or from infected aquatic reservoir.
- *Mycoplasma*: Transmitted by inhalation route and person to person transmission is possible by close contact.

52. a
- Option a: Follow section, *Leptospira* (**pathogenicity → mode of transmission**) for explanation.
- Follow **respective chapters** for explanation of other options.

53. a
- Weil's disease is a one type of leptospirosis caused by Icterohemorrhagiae serovar of *Leptospira*.

54. a
- *Leptospira* canicola presents as an influenza like illness or aseptic meningitis.

CHAPTER 74

Infections of Mycoplasmatales

Chapter Outline

INTRODUCTION

Meaning

Myco word derived from Mykes (Greek) means branching/filamentous form and plasma means plasticity of nature.

Classification

Follow **Flowchart 74.1 and Table 74.1**.

Synonym and History

Pleuropneumonia like organism (PPLO): Because it causes pleuropneumonia in bovine. The first member of this group *Mycoplasma mycoides* subsp. *Mycoides* was isolated by Roux and Nocard in 1898 from cattle with pleuropneumonia.

Eaton agent: Eaton 1st isolated the bacterium in hamster and cotton rat. He also transmitted the infection in chick embryo by amniotic inoculation. Because it is smaller and can pass through filter as like virus, initially, it was called **Eaton agent** and grouped as virus, but later grouped as bacterium because multiplies by binary fission, sensitive to antibiotics, having both DNA and RNA and can grow on cell free media.

General Properties

- They are **cell wall deficient,** because of lacks of genes required for synthesis of peptidoglycan.
- They are **pleomorphic,** because they assume any size or shape due to cell wall deficiency.
- They are the smallest prokaryotes.
- They can **grow in cell free media**.
- They live free in the environment, but in humans, they **habitat** in respiratory system or the genitals.
- Around 16 human species (**Flowchart 74.1 and Table 74.1**) are known of *Mycoplasma*.

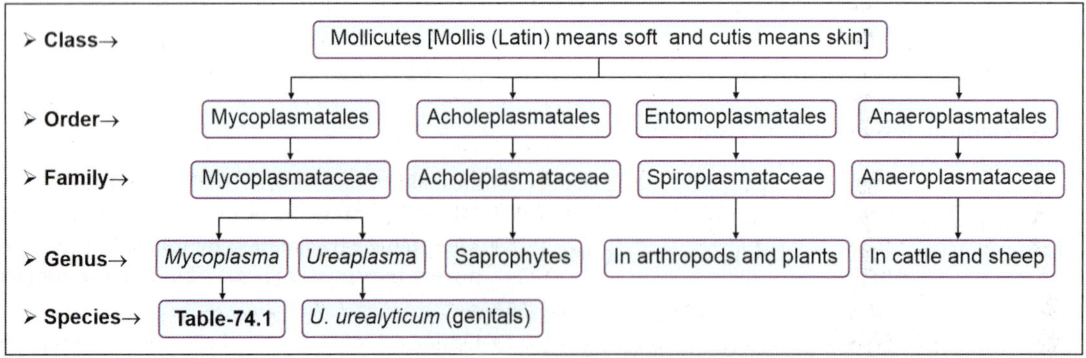

Flowchart 74.1: Classification of Mollicutes

TABLE 74.1: *Mycoplasma spp.*

Pathological status	Habitat	*Mycoplasma spp.*
Established pathogen	Respiratory system	M. pneumoniae
Presumed pathogen	Genitals	M. hominis
Nonpathogens	Mouth	M. orale, M. buccale, M. salivarium and M. faucium
	Genitals	M. fermentans, M. genitalium, M. penetrans, M. spermatophilum and M. primatum

TABLE 74.2: Differences between *Mycoplasma* and L-forms

Features	*Mycoplasma*	L-forms
Type	Stable (natural) type cell wall deficiency	Stable or unstable type cell wall deficiency
Filterable	Yes	No
Peptidoglycan precursors in cell wall	Absent	May be present
Cell membrane	Sterol present	Sterol absent
Disease	Can cause disease	Cannot cause disease, but may cause persistence of infection

- *Mycoplasma* species are the common **contaminants in viral cell culture** and interfering not in cytopathic effect, but in growth of viruses. They also interfere with reading of serological testing. They originated from the workers or from animals sera or from trypsin used in cell culture. It is very difficult to eliminate the *Mycoplasma* as contaminant, and change of cell line is required.
- *Mycoplasma* is an **opportunistic pathogen** can cause more sever and prolonged infection in persons with IDDs. AIDS associated mycoplasmas are *M. incognitus*, *M. fermentans* **(urine)**, *M. penetrans* **(urine)**, *M. pirum* **(blood)**, etc.
- The **persistence of infections** of *M. pneumoniae* even after treatment is associated with its ability to mimic host cell surface composition.
- *Mycoplasma* **and L-forms:** *Mycoplasma* shows stable type of cell wall deficiency, but genetic, biochemical and antigenic evidences are against the possibilities. Differences between *Mycoplasma* and L-forms are shown in **Table 74.2**. For more details of L-forms follow **Ch. 6**.

Mycoplasma pneumoniae

Morphology

Staining properties: They are gram-negative, but better stained by Giemsa stain. Colony on solid media is stained by Diene's stain. Fluorescent staining is also useful.

Shape: They are highly pleomorphic without fixed shape. They present in different forms like coccoid, balloon, disc, ring or star or granules or filamentous forms as shown in **Fig. 74.1a (helical → *Spiroplasma*)**.

Size: They are smallest free living microorganisms. They are highly pleomorphic without fixed size and about 125–250 nm in size. Because of smaller size and plasticity they can pass through bacterial filters. Other bacteria which can pass through filter are *F. tularensis*, *Coxiella burnetii* and *Chlamydia*.

Motility: No flagella but showing gliding motility.

Cell wall: They are devoid of cell wall, bounded by a single trilaminar unit membrane containing sterols.

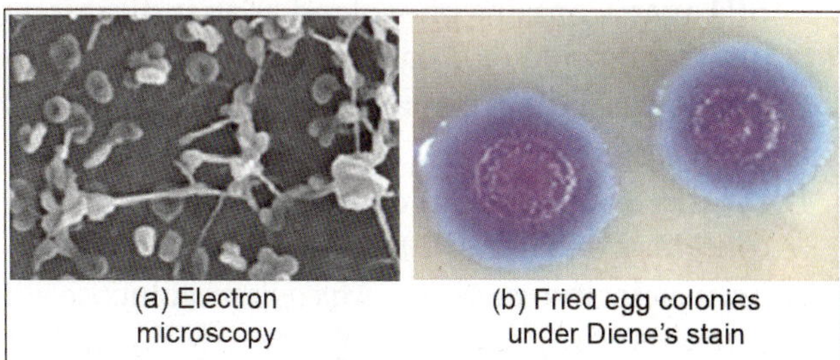

Fig. 74.1: Morphology and C/Cs of *M. pneumoniae*

Sterol presents in membrane maintains the rigidity of bacterial cell. *M. pneumoniae* is the only bacterium which possesses cholesterol in its cell membrane (obtained from the host) and possesses more genes that encode for membrane lipoprotein variations than other mycoplasmas.

Multiplication: They multiply by binary fissions, budding or beaded forms are also noticed.

Organs for adhesion: They do not have fimbriae, but show bulbous enlargement with which they attach to neuraminic acid receptors holding host cells (like respiratory epithelium) and also on RBCs called hemadsorption.

L-forms: Present and stable type.

Other features: They are nonsporing, nonflagellated and nonfimbriated.

Cultural Characteristics (C/Cs)

> **Note: Diene's phenomenon**
> - It is useful to identify the identical and non-identical *Proteus* strains (swarming).

Effective factors

- O_2 effect: They are aerobes and facultative anaerobes.
- Temperature: Optimum growth occurs at 35–37°C.
- pH: 7.8.
- Growth factors: They require sterol and cholesterol for growth. However, *Acholeplasma* of family Acholeplasmataceae do not depends on sterol for growth. Media are enriched with 20% horse/human serum (source of cholesterol) and yeast extract. 30% human ascetic fluid is also useful instead of serum for enrichment.

Culture in media

A. Liquid medium/enrichment medium/transport medium: For examples *Mycoplasma*/PPLO broth, which is actually BHIB contains peptone water, yeast extracts, horse/human serum, phenol red and glucose. *Mycoplasma* produces turbidity and color change.

B. Solid media

1. ***Mycoplasma* glucose agar:** *Mycoplasma* broth contains agar.

Essentials of Medical Microbiology

2. SP4 agar: Contains phenol red as an indicator and gives yellow color colonies.

C. Biphasic medium: It contains liquid and solid phase.

D. Selective media

1. **Liquid selective medium:** *Mycoplasma* broth contains Penicillin (Pn), polymyxin B and AMB.
2. **Solid selective medium:** *Mycoplasma* glucose agar contains Pn, polymyxin B and AMB. Thallous acetate is inhibitory for *U. urealyticum* and *M. genitalium*, so rarely useful as selective agent.

Cultivation techniques: Follow **Flowchart 74.2**.

Automated culture: Like MALDI-TOF is used to identify the species from culture.

Culture in egg: It can grow in chicken embryo.

Biochemical Reactions (B/Rs)

Glucose fermentation: *Mycoplasma* ferments glucose with acid production only which is tested in *Mycoplasma* broth.

Urease and arginine hydrolysis tests: Negative.

Resistance

Sterilization: They are destroyed by heat at 45°C in 15 minutes.

Disinfection: *M. pneumoniae* can grow in presence of 0.0002% methylene blue, while other species are inhibited. They are sensitive to surface active agents like taurocholate and digitonin.

Drug resistance: They are cell wall deficient and completely resistant to cell wall dependent drugs like β-lactam. They are sensitive to some antibiotics like tetracycline and erythromycin.

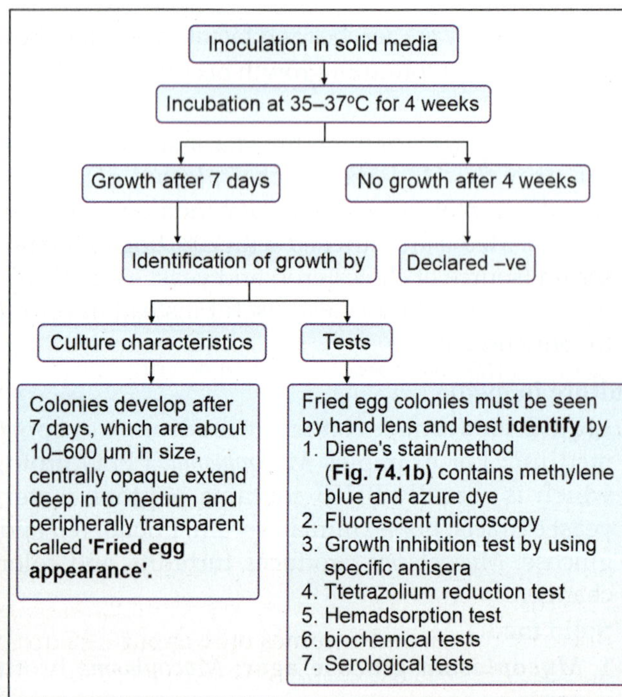

Flowchart 74.2: Cultivation techniques of *M. pneumoniae*

Pathogenicity

Disease name: It called primary atypical pneumonia. *M. pneumoniae* is the most common cause.

Meaning: It called primary because it develops independently of other diseases and called atypical because it is different from typical pneumonia with interstitial space involvement.

Pathology and clinical presentation: It presents in interstitial space but not in alveoli. In addition, this form of pneumonia is atypical in presentation (less pulmonary and more systemic symptoms) with moderate amounts of sputum, no consolidation, only small increases in WBC counts and no alveolar exudate.

Synonym: Affected person not too sick, so called **"walking pneumonia"**. Patchy inflammatory changes are present in alveolar septa and in interstitial space, so called **interstitial pneumonia.**

Epidemiology: It is distributed worldwide. Reinfection and epidemic cycling are thought to be results of P1 adhesin subtype variation. About 40% of community acquired pneumonia is due to *M. pneumoniae*.

Reservoirs of infection: Humans.

Sources of infection: Respiratory droplets.

Modes of transmission: It is transmitted by inhalation of respiratory or nasopharyngeal droplets. Person to person transmission occurs by close contact, typically in military persons.

Incubation period: 1–3 weeks.

Portal of entry: Respiratory system.

Sites: Pulmonary and extrapulmonary sites.

Precipitating factors (epidemiological determinants)

A. Agent factors (virulence factors)

- **Nuclease enzymes:** Degrades the host cell nucleic acid to generate the precursor for synthesis of their own nucleic acid.
- **Antigens: (1) Protein Ag:** It called P1 or cytadhesin which helps in adhesion to respiratory epithelium and blocks the ciliary movement called ciliostasis. It is detected by ELISA. **(2) Glycolipid Ag:** It helps in adhesion with RBCs called hemadsorption and causes hemolytic anemia. It forms the basis for detection of heterophile antibodies by cold agglutination test. It is identified by CFT. **(3) Hydrogen peroxide:** It injured the host cells by reducing glutathione, damaging lipid membranes and causes protein denaturation. **(4) Cytotoxin:** It has ADP ribosylating and vacuolating properties as like pertussis toxin. **(5) Lipoproteins:** It presents in cell membrane and appears to induce inflammation.

B. Host factors: (1) Age: It is more in children and elder group. **(2) Other risk groups:** Outbreak of *M. pneumoniae* infections occurs within groups of people in close and prolonged proximity, including schools, institutions, military bases and households.

C. Environmental factors: Infection occurs more frequently during the summer and fall months when other respiratory pathogens are less prevalent.

Pathology: **Macroscopic** finding are patchy inflammatory changes in the interstitial areas. **Microscopic** finding are no signs and symptoms of lobar consolidation, meaning that the infection is restricted to small areas, rather than involving a whole lobe. As the disease progresses the look can tend to be lobar pneumonia. Alveolar septa are widened and edematous infiltration by lymphocytes, macrophages and occasionally by plasma cells. There is absence of leukocytosis and lack of alveolar exudates, so moderate amount of sputum or no sputum at all (nonproductive).

Clinical features: It presents in interstitial space, but not in alveoli so, sputum production is minimal or not at all and manifested with some atypical features like fever, malaise, headache, sweating, myalgia, sore throat, etc. Dry irritating cough followed later by a productive cough with radiographs showing consolidation. It also causes pharyngitis, tracheobronchitis and sinusitis. Symptoms may be clear without antibiotic therapy. Consolidation in lower lobe can be observed by X-ray.

Complications: Otitis, myringitis, rashes, meningitis, encephalitis, hemolytic anemia, etc.

Laboratory Diagnosis

Specimens: Throat swab, sputum, respiratory secretions, etc.

Testing methods

A. Microscopy: Follow morphology.

B. Culture: Follow C/Cs.

C. Biochemical reactions: Follow B/Rs.

D. Blood picture: No leukocytosis, but raised ESR in 50% of patients

E. Serological tests

- **Antigen detection tests:** Antigens can be detected in respiratory specimens by direct immuno-fluorescence test, CIE, ELISA and immunoblot test.

- **Specific antibody detection tests:** Tests called specific because they are performed by using antigens from *Mycoplasma*. Tests are 55–100% sensitive. Specific antibodies developed after 1 week of infection, peak in 3–4 weeks and then gradually decline. In children IgM is increased while in adults IgA is increased. For example immunofluorescence test, latex agglutination test and ELISA (by using P1 Ag).

- **Nonspecific antibody detection tests:** Tests called nonspecific because performed by using antigens from other cells. Tests are 30–50% sensitive so less useful. **(1)** *Streptococcus* **MG test:** Do the serial dilution of patient serum. Add the heat killed suspension of *Streptococcus* MG (Group F β hemolytic *Streptococcus*/*Strept. angiosus*). Observe

for the agglutination after overnight incubation at 37°C. Significant titer is 1:20. **(2) Cold agglutination test:** In patient with primary atypical pneumonia high proportion of antibodies are produced that agglutinate human O cells at low temperature. Do the serial dilution of patient serum. Add the 0.2 % human O group cells. Do the overnight incubation at 4°C and observe for the clumps. Clumps will dissociate at 37°C. Significant titer is 1:32 or above. **(3) Indirect Coombs test:** It is useful in some cases.

F. Molecular methods: (1) Real time PCR is used for quantitative detection of *M. pneumoniae* **(2) BioFire Film Array** is an automated multiplex PCR which targets around 22 respiratory pathogens including *M. pneumoniae*. **(3) Multiplex PCR:** It targets agents causing atypical pneumonia like *M. pneumoniae*, *Legionella pneumophila* and *Chlamydia pneumoniae*.

Treatment

Macrolides like azithromycin (500 mg orally on 1st day followed by 250 mg/day for 5 days), erythromycin and clarithromycin are the drugs of choice. Other drugs like tetracycline/doxycycline or fluoroquinolones like levofloxacin, moxifloxacin and gemifloxacin (no levofloxacin) are effective.

Mycoplasma hominis

Morphology

It is same as *M. pneumoniae*.

Cultural Characteristics (C/Cs)

All media of *M. pneumoniae* are supplemented with arginine instead of glucose called **H-broth** and **H-agar** produce large fried egg colonies identified by Diene's stain.

Biochemical Reactions (B/Rs)

It does not ferment glucose but arginine hydrolysis test is positive.

Resistance

It is resistant to penicillin and rifampicin, so difficult to eliminate from the cell culture, animal host and human host.

Pathogenicity

- **Modes of transmission:** It is transmitted by vaginal intercourse, by oral to genital contact, by placenta, from infected birth canal to newborn or by tissues transplantation.

- **Precipitating factors: (1) Age:** These are inhabitant of genitourinary tract particularly in the sexually active adults. Neonates specially girls are colonized during birth from infected birth canal, but this colonization does not persist and becomes only about 10% up to age of prepuberty. After puberty colonization increase about 15% in men and women due to sexual activity.

Essentials of Medical Microbiology

(2) Race: It is more common in African Americans than in Caucasians. **(3) IDDs:** It occurs more in person with IDDs.

- **Clinical features: Urogenital** features are NGU, PID, cervicitis, urethritis, cystitis, endometritis and chorioamnionitis. **Extragenitals** features are pyelonephritis, infectious arthritis, surgical and non-surgical wound infections, septic arthritis, bacteremia, pneumonia, meningitis and salpingitis.

Laboratory Diagnosis

Same as *M. pneumoniae* but differentiated by growth on **H-broth** and **H-agar** and ability to hydrolyze arginine.

Treatment

Tetracyclines, macrolides, erythromycin, macrolides, ketolides and quinolones are useful drugs.

Ureaplasma urealyticum

Morphology

Similar to *M. pneumoniae.*

Cultural Characteristics (C/Cs)

All media of *M. pneumoniae* are supplemented with urea instead of glucose called **U-broth** and **U-agar.** It produces very tiny colonies on U agar hence called **T strain** or **T forms of *Mycoplasma.***

Biochemical Reactions

Glucose fermentation and arginine hydrolysis are negative, but urease is positive.

Pathogenicity

- **Modes of transmission:** It is transmitted by vaginal intercourse, by oral to genital contact, by placenta or from infected birth canal to new born.
- **Precipitating factors:** These are inhabitant of genitourinary tract particularly in the sexually active adults. Neonates specially girls are colonized during birth from infected birth canal, but this colonization does not persist and becomes only about 10% up to age of pre-puberty. After puberty colonization increase about 45–75% in men and women due to sexual activity.

Clinical Features

In men it causes NGU, epididymitis, proctitis, Reiter's syndrome (urethritis, conjunctivitis and arthritis), etc. **In women** it causes acute urethral syndrome, PID, salpingitis, Reiter's syndrome (cervicitis/urethritis, conjunctivitis and arthritis), bartholinitis, endometritis, and Fitz–Hugh Curtis syndrome (peritonitis and peri-hepatic inflammation).

Complications

These are infertility, abortion, ectopic pregnancy, pre-mature delivery, antenatal morbidity, puerperal sepsis, post partum fever, LBW of infant and chorioamnionitis.

Laboratory Diagnosis

It is diagnosed from genital discharges by same methods as like *M. pneumoniae* but differentiated by growth on **U-broth** and **U-agar** and urease positive.

Treatment

Doxycyclin is the drug of choice.

ACCESS YOURSELF

Short Notes

1. Primary atypical pneumonia (interstitial pneumonia).

Short Questions for Theory/Viva Questions

1. What is Eaton agent?
2. Write the four differences between *Mycoplasma* and L-forms.
3. Name four bacteria causing primary atypical pneumonia.
4. What is T-strain or tiny strain of *Mycoplasma*?
5. What is Diene's stain/method and Diene's phenomenon?

Comments On

1. Initially *Mycoplasma* was classified as virus but later grouped as bacterium.
2. Mycoplasmas are resistant to β-lactam antibiotics.

MCQs for Chapter Review

Introduction

1. **The following statements are true with references to mycoplasmas,** *except*:
    a. They are the smallest prokaryotic organisms that can grow in cell free culture media
    b. They are obligate intracellular
    c. They lack a cell wall
    d. They are resistant to β-lactam drugs

Mycoplasma pneumoniae

2. **Which of the following needs cholesterol and other lipids for growth?**
    a. *Y pestis* b. *Pseudomonas*
    c. *Proteus* d. *Mycoplasma*
3. **Which of the mycoplasmas do not require sterol for growth?**
    a. *Mycoplasma pneumoniae*
    b. *Mycoplasma hominis*
    c. *Ureaplasma*
    d. *Acholeplasma*
4. **Diene's method is used for:**
    a. *Mycoplasma* b. *Chlamydia*
    c. Plague d. Diphtheria
5. **Fried egg colony is seen in culture of:**
    a. *Mycoplasma* b. *Legionella*
    c. *Trachoma* d. *Haemophilus*
6. **In references to mycoplasmas, the following are true** *except*:
    a. They are inhibited by Pn
    b. They can reproduce in cell free media
    c. They have an affinity for mammalian cell membrane
    d. They can pass through filters of 450 nm pore size
7. **True about mycoplasmas, are all** *except*:
    a. They are L forms
    b. Sterol enhances the growth
    c. They can grow in cell free media
    d. When they grow in liquid medium, do not produces turbidity

8. **True about *Mycoplasma* is:**
   a. Causes lung infection
   b. Penicillin is the drug of choice
   c. Thick cell wall
   d. Thallium acetate inhibits the growth
9. ***M. pneumoniae* is characterized by all *except*:**
   a. Diagnosed by serum cold antibody
   b. Treatment is erythromycin
   c. Cannot be cultured from sputum
   d. Raised ESR

Ureaplasma urealytocum

10. ***Ureaplasma* naturally resistant to:**
    a. Erythromycin   b. Tetracycline
    c. Chloramphenicol   d. Cephalosporin

Answers and Explanation of MCQs

1. b

- Option b: *Mycoplasma* is not the obligate intracellular organism, because it can grow in cell free media, and it also living free in the environment. For explanation of option a and c, follow section, **introduction → general properties**. It is cell wall deficient and completely resistant to cell wall dependent drugs like β-lactams which include penicillin and cephalosporin.

2. d

3. d

- Follow section, *Mycoplasma pneumoniae* (culture characteristics → effective factors) for explanation of answers of MCQs 2–3.

4. a

5. a

- Follow section, *Mycoplasma pneumoniae* (culture characteristics → cultivation technique and Flowchart 74.2 for explanation of answers of MCQs 4–5.

6. a

- Option a and b are already explained. For explanation of option c and d follow section, **synonym and history (Eaton agent)**.

7. a, d

- Mycoplasmas are naturally deficient to cell wall, which is stable, while in L-form cell wall deficiency is due to effect of penicillin, which induced artificially and may be stable or unstable type. Sterol and cholesterol are required for growth of bacteria. *Mycoplasma* can grow in cell free media like PPLO broth, etc., and produces turbidity in liquid medium.

8. a

- *Mycoplasma* causes lung infection called primary atypical pneumonia. Penicillin is not effective because no cell wall. Thallium acetate does not inhibit the growth.

9. c

- Antigens from *Mycoplasma* produce antibodies that agglutinate the human O cell at 4°C. It can be cultured from sputum and ESR increased in 50% of cases.

10. d

- Ureaplasmas are cell wall deficient and completely resistant to cell wall dependent drugs like β-lactams, which include penicillin and cephalosporin.

CHAPTER 75

Infections of Chlamydiales

Chapter Outline

INTRODUCTION

Meaning

Word *Chlamydia* derived from chlamys means helmet (mentle), because inclusion body encloses within the cells.

Synonyms

PLT agent/PLT virus: It causes **p**sittacosis-**l**ympho-granuloma–**t**rachoma in human, so called PLT agent/PLT virus.

Energy parasites: They lack enzymes of electron transport channel (ETC) and dependent on the host cells for energy and nutrition, so called energy parasites.

Bedsonia: Because of pioneering work by **Sir Samuel Bedson** on organism causing psittacosis, this group also called Bedsonia.

Basophilic viruses: Chlamydias produce basophilic intracytoplasmic inclusion body, hence sometimes called basophilic viruses. Viruses produce eosinophilic inclusion body.

Taxonomy

Were considered as viruses: Because many features are similar to viruses like (1) obligate intracellular organism in humans, animals or bird's cells with tropism for squamous epithelial cells, macrophages of GIT and respiratory system (2) pass through filter (3) failure to grow in cell free media and (4) producing the inclusion body.

Now accepted as bacteria: Because many features are similar to bacteria like (1) have both DNA and RNA (2) have a cell wall, but lack of peptidoglycan (3) have ribosomes (4) replicate by binary fission without an "eclipse phase" and (5) sensitive to antibiotics.

Classification

It based on antigenic properties as mentioned in **Flowchart 75.1**.

Flowchart 75.1: Classification of Chlamydiales

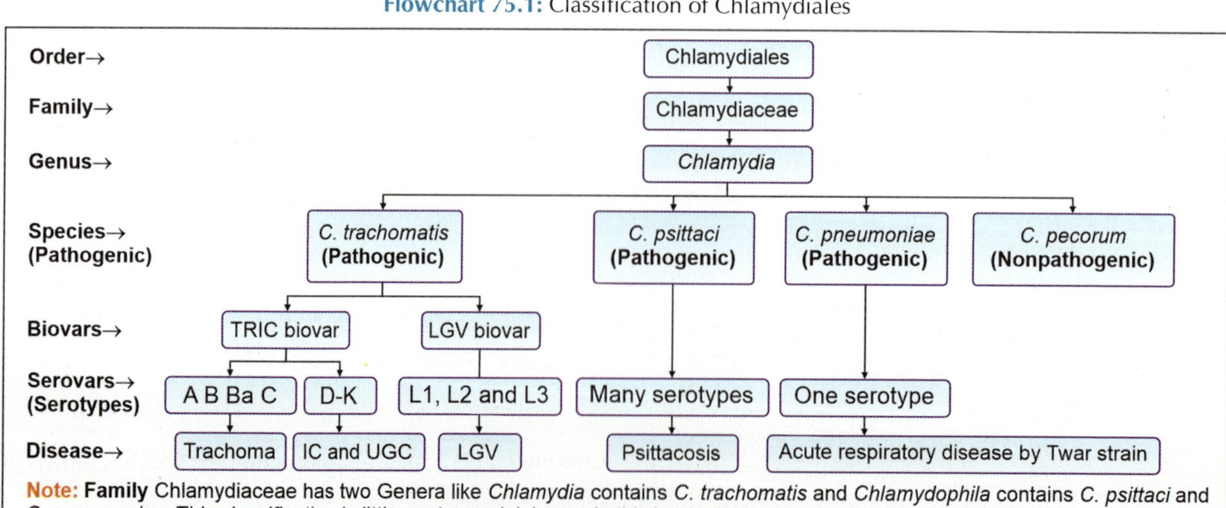

Note: Family Chlamydiaceae has two Genera like *Chlamydia* contains *C. trachomatis* and *Chlamydophila* contains *C. psittaci* and *C. pneumoniae*. This classification is little controversial, hence in this book all species are described under the genus *Chlamydia*.

S. No.	Type of antigen	Bio-chemistry	Properties	Identification
		TABLE 75.1: Antigenic properties of Chlamydiales		
1.	Genus/Group specific	Lipopolysaccharide	Like LPS of GNB	CFT
2.	Species specific	Protein (for classification in to species)	Present at envelop surface	—
3.	Serotype/intraspecies specific	Protein (for serotying of *Chlamydia* spp.)	MOMP	Micro-IF

Antigenic Structure

It has three types of antigen as shown in **Table 75.1**.

Resistance

Chlamydias are killed by heat at 56°C in a minute, phenol, formalin, ethanol and ether. Infectivity is maintained for several days at 4°C. It can be preserved by lyophillization or at –70°C.

Chlamydia trachomatis

History

Characteristic inclusion body called Halberstaedter–Prowazek body, because it was detected by Halberstaedter–Prowazek in 1907 from conjunctival smear from orangutans.

Biovars and Serovars

Follow **Flowchart 75.1**.

Life Cycle

Follow **Fig. 75.1 and Flowchart 75.2**.

Morphology

Cell wall structure: Outer cell wall of *C. trachomatis* resemble to gram-negative bacteria. It has high lipid contents, but devoid of peptidoglycan and N acetyle muramic acid. Perhaps, it contains tetrapeptide linked matrix.

Staining and microscopy

1. **Light microscopy:** It identifies both the inclusion (HP) body and EB. EB also called Miyagawa's granulocorpuscles. Chlamydias are gram-negative,

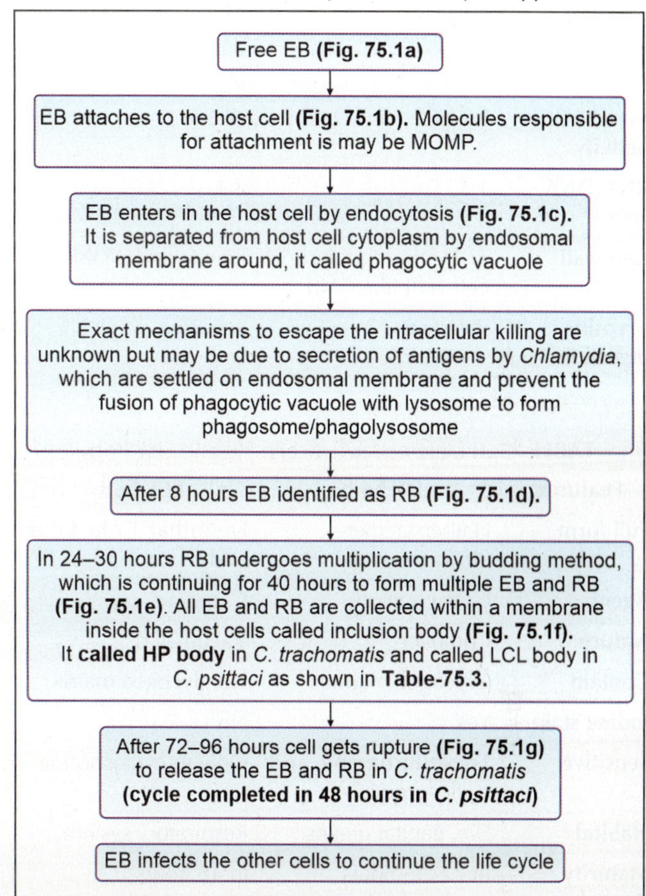

Flowchart 75.2: Life cycle of *Chlamydia* spp.

but better stained by Lugol's iodine. Inclusion body of *C. trachomatis* contains glycogen matrix, so stained with iodine and sensitive to sulfadiazine and cyclosporine. Inclusion body of *C. psittaci* does not contain glycogen matrix, so not stained with iodine. Other useful stains are like Giemsa stain, Castaneda stain, Machiavello stain and Giminea stain.

2. **Direct immunofluorescent microscopy:** It is done by using monoclonal Ab. It identifies both inclusion body and EB in conjunctival/cervical/urethral specimens. It is more sensitive and specific.

Two morphological forms: It is biphasic with two morphological forms as explained in **Table 75.2**.

Inclusion bodies: Follow **Table 75.3**.

Cultural Characteristics (C/Cs)

Chlamydia does not grow in cell free media. Cultivation is done by using following methods.

Culture in animal: Mouse was useful in past

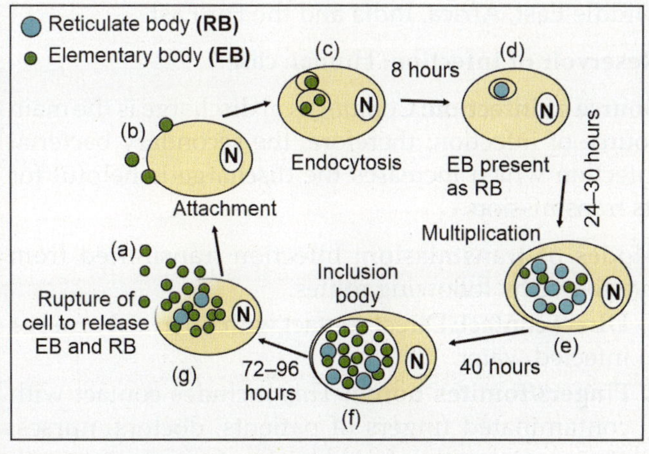

Fig. 75.1: Life cycle of *Chlamydia* spp.

5

TABLE 75.2: Morphological forms of Chlamydiales

Features	Elementary body (EB)	Reticulate body (RB)
Synonym	Miyagawa granulo-corpuscle	Initial body
Location	Extracellular	Intracellular
Size	200–300 nm	500–1000 nm
Shape	Spherical shape	Pleomorphic
Projection	Few	More
Function	Infective	Metabolic and reproductive by binary fission
Metabolic activity	Inactive	Active
RNA:DNA ratio	1:1	3:1
Cell wall	Trilaminar and rigid cell wall like GNB	Fragile and pliable
Trypsin digestion	Resistant	Sensitive

TABLE 75.3: Differences between HP and LCL body

Features	HP body	LCL body
Full form	Halberstaedter-Prowazek body	Levinthal Cole Lillie body
Agent	*C. trachomatis*	*C. psittaci*
Nature	Compact	Vacuolated
Contain	Glycogen matrix	No glycogen matrix
Iodine stain	Yes	No
Sensitive	To sulfadiazine and cyclosporine	Only to cyclosporine
Habitat	Eye, genital organs	Respiratory system
Maturity time	In 72–96 hours	In 48 hours
Released from host cell	Released with scarring of host cell	Released with lysis of host cell

Culture in eggs: It grows in yolk sac of 6–8 days old chick embryo. This method is obsolete now.

Culture in cells/tissues

- **Cells used are:** McCoy cells and Hela cells are used for *C. trachomatis*. Hep2 or human fibroblast cell line is used for *C. pneumoniae*. Other useful cells are BHK-21 and Afrcian green monkey kidney cells (Vero cells). Cells are nonreplicating (stationary phase).
- **Pretreatment of cell culture:** It is required to enhance the chlamydial replication and to increase the detection of inclusion bodies. It is done with pre-treatment of cell culture with irradiation or cycloheximide or 5-iodo-2-deoxyuridine or with diethyaminoethanol (DEAE) dextran.
- **Incubation:** Culture is incubated under 10% CO_2 for 48–72 hours. Shorter for *C. trachomatis* and longer for others.

- **Growth:** LGV biovar is more infective than TRIC biovar. Growth is enhanced by centrifugation of cells after inoculation which promotes the contact between the cell monolayer and chlamydial particles. Growth is identified by detecting the inclusion bodies.
- **Advantage:** Highly specific but less sensitive (90%). Tissue culture was considered as gold standard but less sensitive than molecular tests.
- **Disadvantages:** These are risk of laboratory infection/contamination, time consuming, labor intensive and required expertise.

Pathogenicity, Diagnosis, Prevention and Treatment

Agent factors (virulence factors): (1) LPS: LPS accumulates on host cell surface during intracellular multiplication and induces inflammatory and immunological responses which are responsible for production of disease. **(2) Host cell destruction:** When inclusion body gets mature, host cell is not able to hold the weight and finally gets rupture to release the contents. **(3) Intracellular location:** Escape the host defense and effects of antibiotics. **(4) MOMP:** It may help in attachment with host cells.

Diseases name: *C. trachomatis* produces trachoma, inclusion conjunctivitis (IC), urogenital chlamydiasis (UGC) and lymphogranuloma venereum (LGV). Later two are collectively called genital infections (GI).

Trachoma

Meaning: Name is given from trakhus (Greek) means rough, because of roughness of conjunctiva in disease.

Definition: It is a chronic keratoconjunctivitis characterized by follicular hypertrophy, papillary hyperplasia, pannus formation and cicatrization.

Synonym: It also called Egyptian ophthamia or blinding trachoma or endemic blinding trachoma or endemic trachoma or granular conjunctivitis.

Cause: Trachoma is caused by serovars/serotypes A, B, Ba and C of *C. trachomatis*.

Distribution: It is distributed worldwide and 500 million people estimated to be affected. It is endemic in Middle East, Africa, India and the Far East.

Reservoir of infection: Human case.

Source of infection: Conjunctival discharge is the main source of infection; therefore, the secondary bacterial infection which increases the discharge is helpful for its transmission.

Modes of transmission: Infection transmitted from eye-to-eye by following routes.

1. **Direct contact:** Direct contact with dust (airborne) or infected water.
2. **Fingers/fomites borne:** This includes contact with contaminated fingers of patients, doctors, nurses, etc., contact with infected tonometer or contact with

common towel, handkerchief, bedding and surma-rods.

3. **Vector-borne:** Mechanical transmission by flies.

Incubation period: It is variable about 5–12 days and influenced by the dose of infection.

Portal of entry: Conjunctiva.

Sites: Conjunctiva and cornea.

Predisposing factors

1. **Age:** It is common in infants and in early childhood otherwise no age bar.
2. **Sex:** It is more in females in numbers and in severity.
3. **Race:** No race bar but common in Jews and less in Negros.
4. **Climate:** It is common in dry and dusty weather.
5. **Occupation:** Dust, smoke and sunlight are increasing the risk therefore it is common in outdoor workers.
6. **Socioeconomic status:** Common in developing countries due to poor hygiene, overcrowding, abundantly flies population, common sharing of towel, handkerchief, bedding, surma-rods, etc., and lacks of education, especially about spread of disease.

Pathology: Following two phases:

A. Active trachoma: It has insidious (subacute/gradual but with harmful effect) onset.

- **Conjunctival pathology (Fig. 75.2):** It includes congestion, follicles, necrosis and papillary hyperplasia. **Congestion** is characterized by red eye. **Follicles** are look like boiled sago grains, present on upper palpebral (tarsal) conjunctiva but may be on upper fornix, lower fornix and sometimes in bulbar (ocular) conjunctiva. Microscopically follicles contain mononuclear histiocytes, multinucleated cells called Leber cells and few lymphocytes centrally while peripherally follicles contain lymphocytes and blood vessels. **Necrosis** present in later stage. Necrosis and Leber cells help to differentiate the trachoma from other forms of follicular conjunctivitis. **Papillary hyperplasia** contains blood vessels centrally surrounded by lymphocytes. It is a raised area, which gives red appearance to conjunctiva.
- **Corneal pathology:** It includes superficial keratitis, Herbert's follicles, pannus formation and corneal ulcer. **Keratitis** includes inflammation of cornea. **Herbert's follicles** are histologically similar to conjunctival follicles. **Pannus formation** is characterized by central blood vessels surrounded by lymphocytes. Pannus formation may be progressive and regressive. Progressive pannus **(Fig. 75.3a)** means cellular infiltration is ahead to vascularization. It presents in active trachoma. Regressive pannus **(Fig. 75.3b)** means vascularization is ahead to cellular infiltration. It presents in cicatricial trachoma. **Corneal ulcer** presents at the advancing edge of pannus. It is

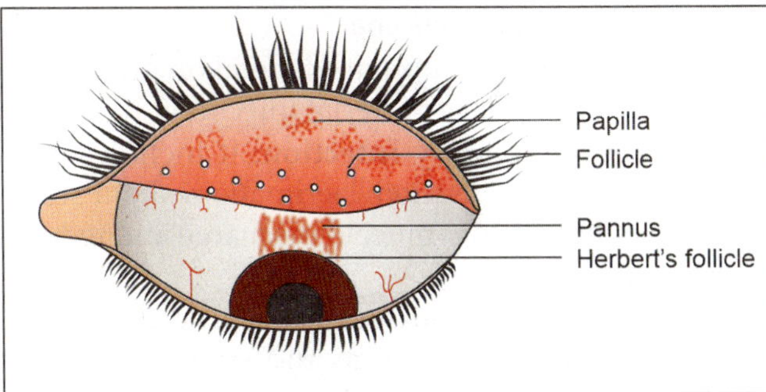

Fig. 75.2: Pathology of trachoma

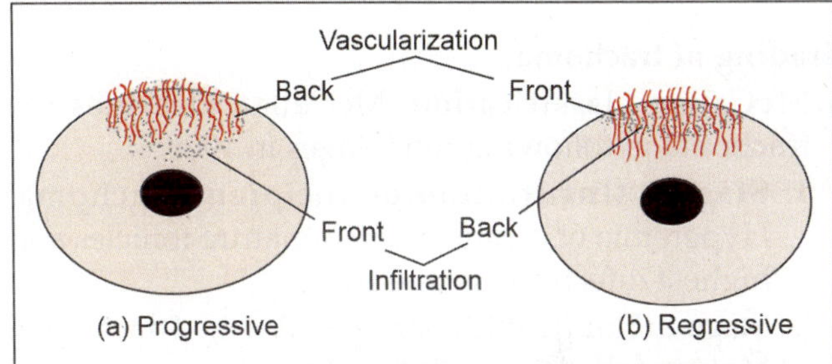

Fig. 75.3: Pannus formation

chronic in nature, which may become chronic or indolent.

B. Cicatricial trachoma

- **Conjunctival pathology:** It includes scarring and concretion. **Scarring** is irregular, star shaped or linear. Linear scar present in the sulcus subtarsalis called Arlt's line. **Concretion** is not a calcification, but it is the accumulation of dead epithelial cells with mucous deposition called gland of Henle. It is pin point to 2 mm in size. **Others pathological features** are pseudocyst and symblepharon.
- **Corneal pathology:** It includes regressive pannus (described above) and Herbal pits. **Herbal pits** are oval or circular pitted scars formed due to healing of Herbert's follicles. **Others pathological features** are corneal opacity and total corneal pannus.
- **Lid pathology:** It includes trichiasis (rubbing of eyeball by eyelashes), tylosis (thickening of lid margin), ptosis (droping of eye lid), madarosis (loss of eyelashes), ankyloblepharon (fusion of part or all the margin of eyelids) and entropion (inward folding of eyelids usually lower lids).
- **Lacrimal apparatus pathology:** It includes chronic dacryocystitis and chronic dacryoadenitis.

Pathogenesis

A. Active trachoma: Subacute inflammatory response with congestion, follicles, hyperplasia and pannus formation.

B. Cicatricial trachoma: Scar formation is due to continued mild chronic inflammation. However, scar formation is due to type-IV hypersensitivity reaction against the continued presence of chlamydial antigen.

A. Active trachoma: Symptoms depend on presence or absence of secondary bacterial infections.

- **Without secondary infection:** It includes mild foreign body sensation, lacrimation, slight stickiness of lids, scanty mucoid discharge and redness of eye.
- **With secondary infection:** It includes high mucopurulent discharge and stickiness of eye lids.

B. Cicatricial trachoma: It includes conjunctival xerosis, corneal xerosis and blindness (blinding trachoma).

Grading of trachoma

A. McCallan classification: McCallan classifies the trachoma in following four stages in 1908:

1. **Stage-1 (Infiltration or incipient trachoma:** Hyperemia of conjunctiva, immature follicles and highest infectivity is present.
2. **Stage-2 (Florid infiltration or established trachoma):** Mature follicle, papillae and progressive corneal pannus.
3. **Stage-3 (Scarring or cicatrizing trachoma):** Scar formation in palpebral conjunctiva.
4. **Stage-4 (Stage of sequelae or healed trachoma):** Features due to cicatrization. This stage is non-infectious.

B. WHO classification: WHO classifies the trachoma in following five stages in 1987:

1. **TF (Trachomatous inflammation—follicular):** Follicles are present and minimum 5 follicles are required to diagnose this stage with size of 0.5 mm.
2. **TI (Trachomatous inflammation—intense):** Inflammatory thickening of upper tarsal conjunctiva, which obscures more than half of the normal deep tarsal vessels.
3. **TS (Trachomatous scarring):** White scar formation in palpebral (tarsal) conjunctiva.
4. **TT (Trachomatous trichiasis):** At least one eye lash rubbing the eyeball.
5. **CO (Corneal opacity):** Easily visible opacity presents over pupil and causes significant visual loss (less than 6/18).

Diagnosis of trachoma: Conjunctival scraping or swab is the sample tested by microscopy (follow morphology) and culture methods (follow C/Cs).

Prevention of trachoma: Improvement in hand hygiene, health education about transmission of trachoma and control of flies are effective measures.

Treatment of trachoma: Local and oral administration of antibiotics is advised for trachoma. Doxycycline is the drug of choice. Erythromycin is the drug of choice for pregnant women. Single dose of azithromycin can gives good effect.

Inclusion Conjunctivitis (IC)

Cause of IC: It is caused by serovars/serotypes D, E, F, G, H, I, J and K.

Subtypes of IC: Two subtypes are as follows:

1. **In newborns**
   - **Disease name:** It called inclusion blenorrhea or ophthalmia neonatorum.
   - **Mode of transmission:** From the infected birth canal to the newborns.
   - **Incubation period:** 5–12 days after birth.
   - **Clinical features:** It is benign, self limited, presents with micro-pannus and conjunctival scar with chances of recurrences.
   - **Complication:** It cause infantile pneumonia at 4–16 weeks of age and presents with cough, wheezing and fever.
   - **Diagnosis:** Same as trachoma.
   - **Prevention:** Control of mother's infection.
   - **Treatment:** Azithromycin is effective.

2. **In adults**
   - **Disease name:** It is called swimming pool conjunctivitis or adult inclusion conjunctivitis.
   - **Mode of transmission:** By contact with contaminated swimming pool water from genital secretion of others/self.
   - **Clinical features:** Follicular hypertrophy and mucopurulent discharge.
   - **Diagnosis:** Same as trachoma.
   - **Prevention:** Avoiding contact with contaminated swimming pool water.
   - **Treatment:** Tetracycline or azithromycin may be given orally.

Urogenital Chlamydiasis (UGC)

Cause of UGC: It is caused by serovars/serotypes D, E, F, G, H, I, J and K.

Mode of transmission: By sexual intercourse.

Clinical features of UGC: Mostly, it is asymptomatic. **In men,** it presents with NGU, epididymitis, proctitis, Reiter's syndrome (urethritis, conjunctivitis and arthritis), etc. **In women,** it presents with acute urethral syndrome, bartholinitis, endometritis, salpingitis, PID, Reiter's syndrome (cervicitis/urethritis, conjunctivitis and arthritis) and Fitz – Hugh Curtis syndrome (peritonitis with perihepatic inflammation).

Complications of UGC: These are infertility, abortion, ectopic pregnancy, premature delivery, antenatal morbidity, puerperal sepsis, post partum fever, LBW of infant and chorioamnionitis.

Diagnosis of UGC: Mucosa scraping or swab from urethra, cervix, vagina or anus or urine is tested by following methods.

A. Microscopy: Follow morphology.

B. Culture methods: Follow C/Cs.

C. Serological tests: (1) Antigen detecting tests: Like micro-IF and ELISA. **(2) Antibody detecting tests:** Like CFT (group specific) and micro-IF (type specific). Serological tests are not useful in noninvasive conditions like trachoma and IC.

D. Molecular methods: These are highly sensitive and specific methods, replace the gold standard method like cell culture. They are used to diagnose the case, to diagnose the asymptomatic carrier and to separate the biovars and serovars. **(1)** Real-time PCR is used for quantitative detection of *C. trachomatis* **(2) BioFire Film Array** is an automated multiplex PCR which targets urogenital pathogens including *C. trachomatis*.

Prevention of UGC: Safe sex.

Treatment of UGC: Tetracycline is effective.

Lymphogranuloma Venereum (LGV)

Synonyms: It also called lymphogranuloma inguinale, poradenitis, climatic bubo and tropical bubo.

Cause of LGV: By serovars/serotypes L1, L2 and L3.

Mode of transmission: Sexual intercourse.

Incubation period: 3 days–5 weeks.

Clinical features of LGV: Three stages of disease.
1. **Primary stage:** It presents as painless papulovesicular (ulcerative) lesion on extragenital sites.
2. **Secondary stage:** It presents as suppurative enlargement of inguinal lymph nodes in male and intrapelvic-pararectal lymph nodes in female. Secondary stage (lymphadenopathy) lesion is bilateral and tender while primary stage (papulovesicular) lesion is painless. It breakdown to form pus discharging sinus.
3. **Tertiary stage:** It presents as scarring due to healing of sinus and lymphatic blockage.

Complications of LGV: Infection spreads to eyes, joints and meninges. In female, it leads rectal stricture. It also causes the elephantiasis of vulva called genital eliphantiasis or esthiomene.

> **Note: Other cause of genital eliphantiasis**
> - Filariasis: More in *W. bancrofti* and less in *B. malayi*.
> - Donovanosis (K granulomatis): Actually causing pseudo-eliphantiasis (labial swelling).

Diagnosis of LGV: Microscopy, culture, serological and molecular tests are same as UGC. Allergic test (**Frei's test**) based on DTH is useful. Antigen for Frei's test is originally prepared from bubo pus and later from mouse brain or yolk sac culture called lygranum. Antigen introduced intradermally in one forearm and other arm as control. Observe for the reaction. Induration of 7 mm in 2–5 days indicates positive test. It is not useful in trachoma and IC. False-positive results are very frequent.

Prevention of LGV: Safe sex.

Treatment of LGV: Tetracycline for three weeks.

Chlamydia psittaci

Meaning: Psittaci word derived from psittcos means parrot, because disease spreads from parrot to human.

Serovars: Follow **Flowchart 75.1**.

Morphology: It produces the inclusion body called LCL body in 24 hours. More details: Follow **Tables 75.2–75.3, Fig. 75.1 and Flowchart 75.2**.

Cultural characteristics (C/Cs): They do not grow in cell free media. Cultivation is done by using following methods.

Animal culture: Mouse was used in past.

Eggs culture: It grows in yolk sac of old chick embryo. It was used in past.

Tissue culture: Grow well in cell culture media but because of risk of lab., infection it is not advised.

Pathogenicity
- **Disease name:** Disease called psittacosis if acquired from parrot or called ornithosis if acquired from non-psittacine bird (bird other than parrot). It is a type of atypical pneumonia.
- **Reservoirs of infection:** Birds like parrots.
- **Sources of infection:** Bird excreta and droppings or nasal discharge and aerosols from human case.
- **Modes of transmission:** It is transmitted by inhalation of particles from bird excreta and parrot (bird) bites. Case to case transmission may occur as chlamydias may shed in nasal discharge, droppings and aerosols from the patients.
- **Incubation period:** 10 days
- **Portal of entry:** Respiratory tract
- **Sites:** Respiratory system, GIT, etc.
- **Precipitating factors: (1) Type of strain:** Strains from the parrots and turkeys are more virulent than other sources. **(2) Occupational:** Disease is more common in laboratory workers, poultry workers, pigeon farmers, pet shop owner, bird fanciers, veterinarians, etc. **(3) Others:** Overcrowding and caging can favor the risk of disease.
- **Clinical features:** These are mild influenza like symptoms, including diarrhea, mucopurulent discharge and emaciation to fatal pneumonia.
- **Complications:** These are septicemia, meningo-encephalitis, endocarditis, pericarditis, arthritis, typhoid like syndrome, etc.

Laboratory diagnosis
- **Specimens:** Blood, sputum and lung tissues cells like PAMs (pulmonary alveolar macrophages).
- **Testing methods**
  1. **Light microscopy:** It is useful to detect the LCL body. *C. psittaci* is gram-negative, but better stained by Giemsa stain, Castaneda stain, Machiavello stain and Giminea stain. LCL body is devoid of glycogen matrix so not stained with Lugol's

iodine and only sensitive to cyclosporine but not to sulphadiazine.

2. **Culture:** Follow C/Cs.
3. **Serological tests:** Like CFT (group specific) and micro IF (type specific).

Treatment: Tetracycline is the effective drugs.

Chlamydia pneumoniae

History: It was 1st reported by Grayston et al., from the adult patient with acute respiratory disease in Taiwan, called **TWAR strain** (from Taiwan) of *C. psittaci* due to sharing of common group specific antigen. Later, identified as separate species by species specific antigen, DNA hybridization, REA (restriction enzyme analysis), ability to grow in cell culture and only human source is available without availability of animal or avian sources.

Cultural characteristics (C/Cs): It grows very poorly in cell culture and can be isolated by using HEP-2 or human fibroblast cell lines.

Pathogenicity

- **Disease name:** It called *Ch. pneumoniae* pneumonia.
- **Modes of transmission:** Transmitted from human to human by inhalation.
- **Incubation period:** 1–3 weeks
- **Portal of entry:** Respiratory tract
- **Sites:** Respiratory system, sinuses, etc.
- **Precipitating factors:** Infection is common in young children and due to overcrowding.
- **Clinical features: (1) Respiratory features:** It causes primary atypical pneumonia (walking pneumonia) like *M. pneumoniae*, sinusitis, pharyngitis, bronchitis and adult onset of asthma. **(2) CVS features:** These include atherosclerosis of coronary, carotid and cerebral arteries.

Laboratory diagnosis: Blood and sputum are tested by **culture methods** (follow C/Cs), **serological tests** like CFT, ELISA, micro-IF, etc., and **molecular methods like (1)** Real-time PCR is used for quantitative detection of *C. pneumoniae*. **(2) BioFire Film Array** is an automated multiplex PCR, which targets around 22 respiratory pathogens including *C. pneumoniae*. **(3) Multiplex PCR:** It targets agents causing atypical pneumonia like *M. pneumoniae*, *Legionella pneumophila* and *Chlamydia pneumoniae*.

Treatment: Azithromycin or clarithromycin is the effective drug.

ACCESS YOURSELF

Case Study

1. A 5-year-old child visited the eye clinic with mucopurulent discharge and red eye. Clinical examination of eye shows the follicle in conjunctiva. Conjunctival scraping revealed inclusion body stained with iodine. Identify the case and answer the following.

a. Name the causative agent.
b. Write the pathogenicity of causative agent.
c. Write the laboratory diagnosis of causative agent.

Essay/Full question

1. *C. trachomatis*.

Short Notes

1. Morphology and life cycle of *C. trachomatis*.
2. Trachoma/Psitacosis.

Short Questions for Theory/Viva Questions

1. Write the differences between elementary body and reticulate body of *C. trachomatis*.
2. Write the differences between HP and LCL body.
3. Write the importance of Lugol's iodine in identification of *C. trachomatis* and *C. psittaci*.

Comments on

1. Chlamydias are known as energy parasites and basophilic viruses.
2. Chlamydias were initially considered as viruses now accepted as bacteria.

MCQs for Chapter Review

Introduction

1. **Taxonomically *Chlamydia* is a:**
   a. Bacteria
   b. Virus
   c. Fungus
   d. Nematode
2. **Which of the following has one serotype?**
   a. *C. psittaci*
   b. *C. pneumoniae*
   c. *C. trachomatis*
   d. None

Chlamydia trachomatis

3. **The following statements are true regarding chlamydiae, *except*:**
   a. Erythromycin is effective for therapy of chlamydial infections
   b. Their cell wall lack a peptidoglycan layer
   c. They can grow in cell free media
   d. They are obligate intracellular bacteria
4. *Chlamydia trachomatis*, **false is:**
   a. Elementary body is metabolically active
   b. It is biphasic
   c. Reticulate body divides by binary fission
   d. Inside the host cell, it evades phagolysosome
5. **All of the following are true about *Chlamydia*, *except*:**
   a. Gram-positive
   b. Causes trachoma
   c. Causative organism of psittacosis
   d. It also called basophilic virus
6. *Chlamydia trachomatis*, **serovars D-K causes:**
   a. Arteriosclerosis
   b. Trachoma
   c. Lymphogranuloma venereum
   d. Urethritis
7. *Chlamydia* **causes all the following, *except*:**
   a. NGU
   b. Pneumonia
   c. Trachoma
   d. Parotitis
8. *C. trachomatis* **is associated with following, *except*:**
   a. Endemic trachoma
   b. Inclusion conjunctivitis
   c. Lymphogranuloma venereum
   d. Community-acquired pneumonia

Essentials of Medical Microbiology

9. **The following is not a method of isolation of *Chlamydia* from clinical specimens:**
   a. Yolk sac inoculation
   b. Enzyme immune assay
   c. Tissue culture using irradiated McCoy cells
   d. Tissue culture using irradiated BHK cells

10. ***Chlamydia* in asymptomatic carrier, the most sensitive test is:**
   a. Tissue culture
   b. Nucleic acid amplification test
   c. Serology
   d. Serum electrophoresis

11. **Most sensitive method to detect cervical *C. trachomatis* is:**
   a. Direct fluorescent antibody test
   b. Enzyme immunoassay
   c. PCR
   d. Culture on irradiated McCoy cells

12. **Frei's test is useful for diagnosis of:**
   a. *Mycoplasma*
   b. *Rickettsia*
   c. Sarcoidosis
   d. *Chlamydia*

13. **Genital elephantiasis is seen in:**
   a. Donovanosis
   b. Lymph granuloma venereum
   c. Congenital syphilis
   d. Herpes simplex

Chlamydia psittaci

14. ***Chlamydia psittaci*, all are true *except*:**
   a. Acquired from bird's droppings
   b. Causes urethritis
   c. Causes pneumonia
   d. Treatment is tetracycline

15. **Levinthal Cole Lillie bodies are seen in:**
   a. LGV
   b. Psittacosis
   c. Kala azar
   d. Chickenpox

Chlamydia pneumoniae

16. **Which of the following statement is true regarding *Chlamydia pneumoniae*?**
   a. 15 serovars have been identified as human pathogen
   b. Mode of transmission is by bird excreta
   c. The cytoplasmic inclusions present in the sputum specimen are rich in glycogen
   d. Group specific antigen is responsible for the production of complement fixing antibodies

Answers and Explanation of MCQs

1. a
- Follow section, **introduction (taxonomy)** for explanation.

2. b
- Follow section, **introduction (classification)** for explanation.

3. c
- *Chlamydia* cannot grow in cell free media. For explanation of option a follow section, **treatment**, for option b follow section, **morphology** (cell wall) and for option d follow section, **taxonomy**.

4. a
- For explanation of option a, b and c follow **morphology (phases and Table 75.2)** and for option d follow section, **life cycle**.

5. a
- *Chlamydia* is gram-negative, causes trachoma and psittacosis
- Option d: Follow section, **synonyms** for explanation.

6. d
- Serovars D-K cause genital chlamydiasis, which presents with urethritis and with other features. Arteriosclerosis is caused by *C. pneumoniae*. Trachoma is caused by serovars A, B, Ba and C, while lymphogranuloma venereum is caused by serovars L1, L2 and L3 of *C. trachomatis*.

7. d

8. d
- Follow section, **pathogenicity of *Chlamydia trachomatis, Chlamydia psittaci* and *Chlamydia pneumoniae*** for explanation.

9. b
- Enzyme immune assay is the serological method used for detection of Ag or Ab not for isolation of bacteria.

10. b

11. c
- Tissue culture was considered as gold standard but less sensitive (90%). It is replaced by molecular tests like nucleic acid amplification test, PCR, etc. These are highly sensitive, specific and now considered as gold standard.

12. d
- Frei's test is the allergic test done to diagnose the LGV caused by *Chlamydia trachomatis* by serovars L1, L2 and L3.

13. a and b
- Follow section, *Chlamydia trachomatis* [pathogenicity → **Lymphogranuloma venereum and notes**] for explanation.

14. b
- Follow section, *Chlamydia pneumoniae* for explanation.

15. b
- Follow **Table 75.3** for explanation.

16. d
- Follow **Table 75.1** for explanation.

CHAPTER 76

Infections of Rickettsiales, *Coxiella* and *Bartonella*

Chapter Outline

- ❏ Rickettsiales
  - – Rickettsia spp.
  - – Orientia tsutsugamushi
  - – *Ehrlichia* spp.
- ❏ *Coxiella burnetii*
- ❏ *Bartonella* spp.

RICKETTSIALES

Meaning and History

Order, family and genus named after Howard Taylor Ricketts, who discovered the spotted fever rickettsia in 1911 and died due to illness (typhus fever) acquired during his studies. Species *R. prowazekii* is named by Da Rocha Lima in honor of von Prowazek. Charles Nicolle noticed the role of lice in transmission of epidemic typhus.

General Properties

They are obligate intracellular bacilli/coccobacilli, gram-negative in nature and transmitted by arthropod.

Taxonomy

Were considered as viruses: Because many features are similar to viruses like (1) obligate intracellular microbes and (2) grow only in cell containing media.

Now accepted as bacteria: Because many features are similar to bacteria like (1) having both DNA and RNA (2) have a cell wall which poorly stains with Gram's stain and they are mostly gram-negative bacilli/coccobacilli (3) contain ribosomes (4) replicate by binary fission and (4) sensitive to antibiotics.

Classification

Follow **Flowchart 76.1**.

Flowchart 76.1: Classification of Rickettsiales

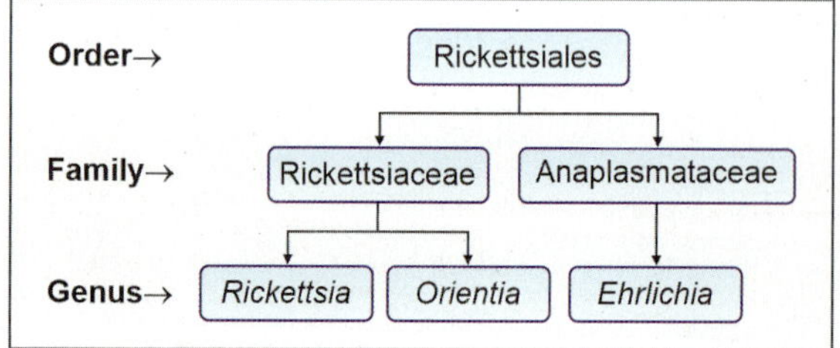

> **Note: Reasons for exclusion of *Coxiella* and *Bartonella* from Rickettsiales**
> - ***Coxiella*:** It is excluded from order Rickettsiales, because not the arthropod borne.
> - ***Bartonella*:** It is excluded from order Rickettsiales, because not the obligate intracellular, can grow in cell free media and differ in genetic properties.

Rickettsia spp.

Morphology

Staining properties: They are gram-negative bacilli/coccobacilli. Stain bluish purple with Giemsa or Castaneda stain and red with Gimenez or Machiavello stain.

Shape and size: They are pleomorphic with 0.8–2 µm × 0.3–0.6 µm in size.

Motility, spores and capsule: They are nonmotile, nonsporing and noncapsulated.

Layers: Under electron microscope, three layers are seen like from outer slime layer, cell wall and inner plasma membrane.

Cultural Characteristics (C/Cs)

Effective factors

- Growth occurs only in cell containing media (obligate intracellular). Typhus group grow in the cytoplasm of infected cells while spotted fever group grow in nucleus and cytoplasm.
- Optimum temperature: 32–35°C.

Culture in animals: Guinea pig and mouse are useful animals. When male guinea pig is inoculated intraperitoneally with *R. typhi*, it develops fever, scrotal enlargement; inflammatory adhesions between the layers of the tunica vaginalis so testis cannot be pushed back into the abdomen called tunica reaction or Neil-Mooser reaction.

Culture in eggs: They grow in the yolk sac of chick embryo. This technique was discovered by Cox.

Culture in cells: Rickettsiae can grow in detroit 6, HeLa, HEP-2, mouse fibroblast and other continuous cell lines. This is not a satisfactory method.

Resistance

Sterilization: They are destroyed at 56°C and at room temperature

Disinfection: They are preserved in skimmed milk or in SPG (**S**ucrose **P**otassium phosphate and **G**lutamate) medium.

Drug resistance: Drug resistance is not a big issue.

Antigenic Structure

Typhus fever group contains surface protein antigen (SPA). It is species specific and cross reacts with B-antigen of spotted fever group. **Spotted fever group** contains OMP-Ag as shown in **Flowchart 76.2**.

Pathogenicity

Pathogenesis: Follow **Flowchart 76.3**.

Pathological types: Two types like **typhus fever group** and **spotted fever group** are described in **Table 76.1** with subtypes, causative agents, vectors and reservoirs.

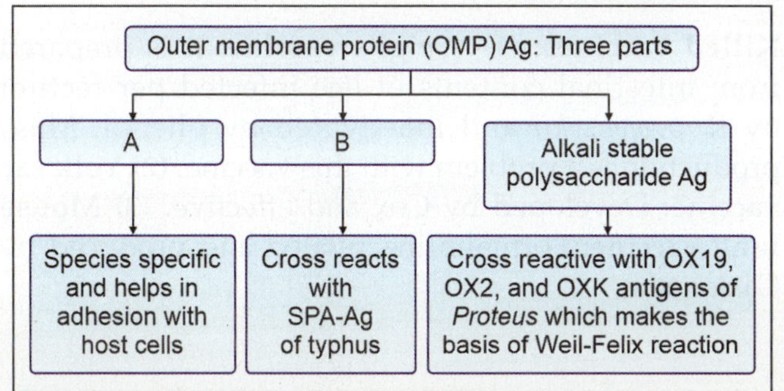
Flowchart 76.2: Antigens of spotted fever group

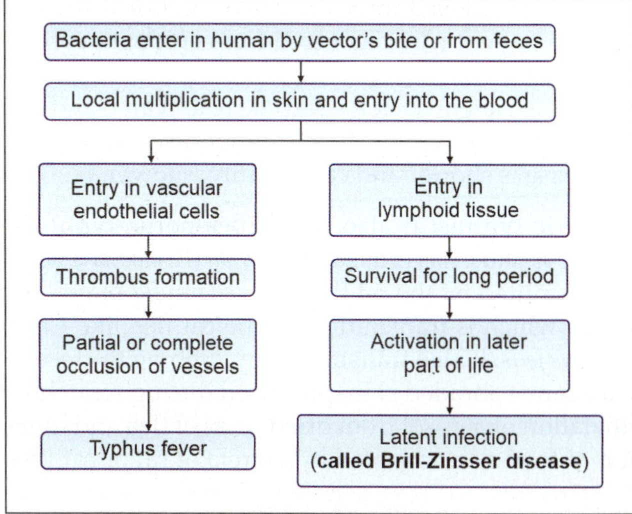
Flowchart 76.3: Pathogenesis of *Rickettsia*

Typhus Fever Group

Bacteria present in cytoplasm with following subtypes.

Epidemic typhus: Typhus **means** cloud/smoke, because of cloudy state of consciousness. It **also called** louse-borne typhus, classical typhus and Gaol fever. It is **caused** by *R. prowazekii*. It is **transmitted** by bite of head louse *Pediculus humanus corporis*, by rubbing the feces of louse against abraded skin produced during scratching, by inhalation of aerosols from dried louse feces and by instillation in conjunctiva. Incubation period is 5–15 days. **Clinical features** include onset of disease with chills and fever. Rashes appear on the 4th–5th day on the trunk and spreading over the limbs except soles, palms and face. It also causes uveitis in eye. Patient progresses to stupurous and delirium. Death occurs in 40% cases.

Brill Zinsser disease: Brill identified typhus like disease in New York, in Jewish immigrants from Europe. **Zinsser** identified the bacilli from such case and proved that it was a recrudescences of infection acquired before

TABLE 76.1: Pathological types of *Rickettsia*

Diseases	Species of *Rickettsia*	Insect vector	Reservoirs
Typhus fever group: In cytoplasm			
Epidemic typhus (Mnemonic: pp)	R. **p**rowazekii	Head louse	Humans
Recrudescent/latent typhus (Brill Zinsser disease)	R. prowazekii	Head louse	Humans
Endemic typhus (murine typhus/flea borne typhus)	R. typhi (R. mooseri)	Rat flea	Rodents
	R. felis (feline rickettsiae)	Cat flea	Cats
Spotted fever group: In nucleus and in cytoplasm			
Indian/South African/Kenyan – tick typhus Mediterranean/Israeli/ Astrakhan-spotted fever Boutonneuse fever	R. conorii	Tick	Rodents
Rocky mountain spotted fever: Lungs are infected in 7–16% cases	R. rickettsii	Tick	Rodents, dogs
Siberian tick typhus	R. siberica	Tick	Cattle
Sub-Saharan African tick typhus	R. africae	Tick	?
Queensland tick typhus	R. australis	Tick	Bush rodents
Oriental spotted fever	R. japonica	Tick	?
Rickettsial pox	R. akari	Mite	Mouse

many years. It **also called** latent typhus or recrudescent typhus. It is **caused** by *R. prowazekii*. **Clinical features** occur in patient who recovered from epidemic typhus. Bacteria remain latent in lymphatic tissue or organs for years. Such latent bacteria become reactivated and lead to recrudescent typhus. It is a milder disease, duration of disease is shorter and case fatality is lower.

Endemic typhus: It **also called** flea-borne typhus or murine typhus. It is **caused** by *R. typhi* (*R. mooseri*) which is transmitted by the rat flea like *xenopsylla cheopis* and *R. felis* which is transmitted by the cat flea like *Ctenocephalus felis*. It **also transmitted** by rubbing the feces of flea against abraded skin, produced during scratching, inhalation of aerosol from dried feces of flea and ingestion of food contaminated by excreta of rat or cat flea.

Spotted Fever Group

Bacteria present in nucleus and in cytoplasm. All subtypes with species, vector and reservoirs are mentioned in **Table 76.1**. Other details are given below.

Mode of transmission of spotted fever group: All infections are transmitted by tick bite like *Dermacentor andersoni, Rhipicephalus sanguinens, Haemaphysalis leachi, Amblyoma* and *Hyalomma*. Transovarial transmission of *Rickettsia* is present in ticks and mites, but not in lice and fleas.

Rickettsial pox: It called pox because it resembles to chickenpox. It is **also called** vesicular or varicelliform rickettsiosis. It is **caused** by *R. akari* (from akari meaning mite). It is **transmitted** by bite of mite *Liponyssoides sanguineus*. **Clinical features** include self limited-neonatal-vesicular rash (exanthema) like chickenpox. It healed with black eschar. Other features are like regional lymphadenopathy.

These pathogens are very dangerous to work in laboratory.

Specimens: Blood.

Testing methods: Follow **Flowchart 76.4**.

A. **Microscopy:** Follow morphology.
B. **Culture:** Follow culture characteristics.
C. **Serological tests:** Weil-Felix reaction (**Ch. 20**), indirect immunofluorescence test (positive after 2nd week, gold standard and significant titer is ≥1:64), IgM capture ELISA in 1st week, CFT, agglutination of rickettsial suspensions, passive hemagglutination test, radioisotope precipitation test and RIA.
D. **Molecular method:** PCR and real-time PCR are performed in 1st week from whole blood, lymph node biopsy, tissue biopsy, skin rashes biopsy and buffy coat fraction.

Prevention of *Rickettsia*

General measures: These are control of vectors, control of animal reservoirs, sterilization of contaminated clothes, beds, pillows and wearing of protective clothes before entering in endemic area.

Immunoprophylaxis: Different types like killed and live vaccines are developed for epidemic typhus, but none of them is satisfactory.

Killed vaccines: (1) Weigl's vaccine: It is prepared from intestinal contents of lice infected per rectum by *R. prowazekii* and inactivated by phenol. Mass production is a problem with this vaccine. **(2) Yolk sac vaccine:** Developed by Cox and effective. **(3) Mouse lung vaccine:** Formalin inactivated and produced by Castaneda.

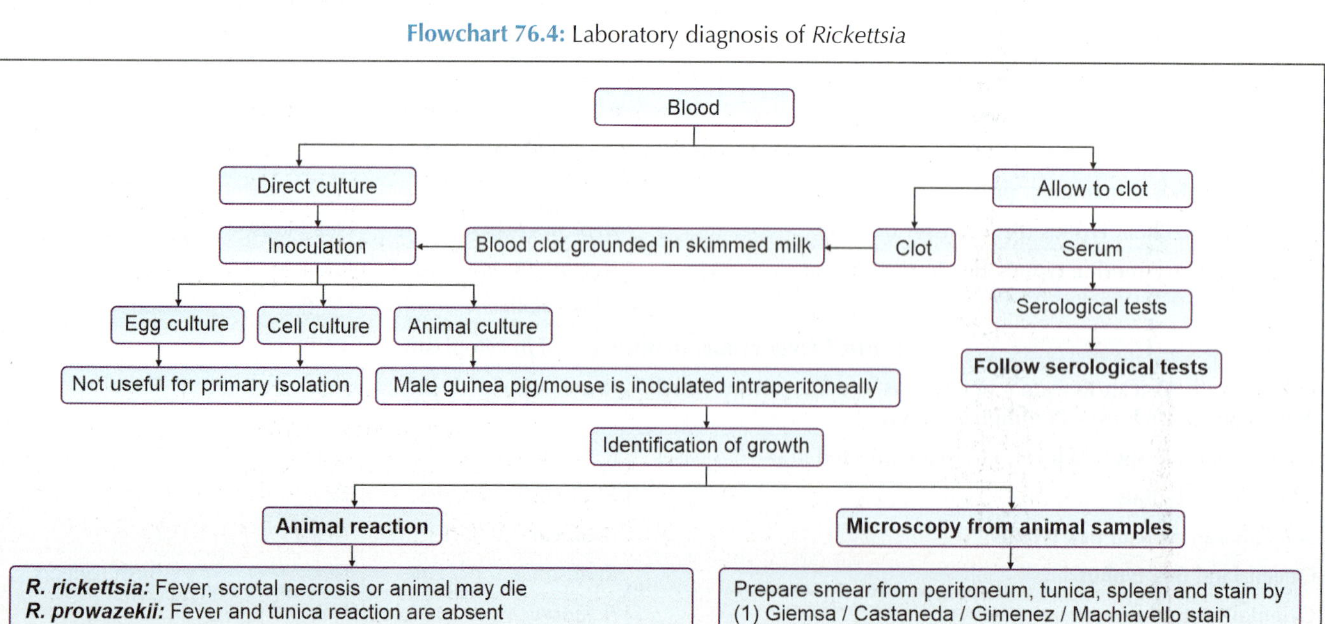

Flowchart 76.4: *Laboratory diagnosis of Rickettsia*

Live vaccine: It is prepared from the attenuated strain and highly immunogenic. It also prepared against Rocky Mountain spotted fever. It developed mild disease after administration.

Treatment

Tetracycline, chloramphenicol or ciprofloxacin is the drug of choice.

Orientia tsutsugamushi

Introduction: Tsutsuga means dangerous and Mushi means insect or mite borne. Formerly, it was called *R. tsutsugamushi* or *R. orientalis*. It was 1st observed in Japan, where it was found to be transmitted by mites.

Morphology: They are gram-negative coccobacilli, but better stained with Gimenez stain. They are pleomorphic coccobacilli with 1.2–3.0 μm × 0.6 μm in size. They are nonmotile, nonsporing and noncapsulated.

Cultural characteristics (C/Cs): They are unable to grow in cell-free media and can grow only in cell monolayers.

Classification: It is antigenically variable. Major antigenic or serotypes identified are **Karp** (accounts 50% of all infections), **Gilliam** (accounts 25% of all infections), **Kato** (accounts <10% of all infections) and **Kawasaki.** More serotypes are continued to be reported.

Pathogenicity

- **Disease name:** Disease called **scrub typhus,** because originally, it was found in scrub jungle.
- **Synonym:** It also called **chigger-borne typhus** as transmitted by Mite's larvae called **chiggers** (only stage fed on the host). Chiggers feed on serum of warm-blooded animals while adult mites fed on plants. It also called **Japanese river disease.**
- **Epidemiology:** It occurs along East Asia, from Korea to Indonesia and in the pacific Islands including Australia.
- **Reservoirs of infection:** These are rodents and birds.
- **Source of infection:** It is mites. Transovarial transmission occurs in mites.
- **Modes of transmission:** It is transmitted by bite of trombiculoid mite's larvae called **chiggers.** Mite inhabits in sharply demarcated area in soil called **mite islands.** When human visit such **mite islands** bacteria enter by the bite of mite's larvae (chiggers). Different species of mites transmit it like in India *Leptotrombidium deliensis* and in Japan *Leptotrombidium akamushi.*
- **Incubation period:** 1–3 weeks.
- **Portal of entry:** Skin.
- **Sites:** Skin, lymph nodes, lungs, meninges, etc.
- **Precipitating factors: (1) Zoonotic tetrad:** Four factors are essential for development of disease like *Orientia tsutsugamushi*, chiggers, rats and secondary or transitional form of vegetation called zoonotic tetrad. **(2) Occupation:** It is a serious problem in military persons, especially during jungle warfare. It was recognized in Indo-Burmese theater in the 2nd World war. **(3) Environmental:** It is common in scrub jungle, sandy beaches, mountain deserts and equatorial rain forests, where plenty of moisture and scrub vegetations are available.

- **Clinical features:** Eschar occurs (in <50% of western) at the site of vector's bite, maculopapular rash, regional lymphadenopathy, fever, headache, conjunctivitis, uveitis, etc. Rash develop on 4–6 days and present in <40% patients.
- **Complications:** These are atypical (interstitial) pneumonia and encephalitis. Case fatality rate is 7% in untreated cases.

Laboratory diagnosis: It is diagnosed by **serological tests** like Weil-Felix test which is positive with OX-K antigen, indirect immunofluorescence test, which is considered as gold standard and ELISA. **Molecular method** like PCR is more sensitive and specific.

Prevention

- **General measures:** Organism is highly virulent and should be handled in laboratory with BSL-3.
- **Immunoprophylaxis:** Because of constant antigenic variation, it is tough to develop the protective vaccine and vaccine developed for one locality may not be effective for other locality.

Treatment: β-Lactam is not effective, because of lack of peptidoglycan in cell wall. Aminoglycosides are not effective, because of intracellular location of bacteria. Doxycycline, rifampicin and azithromycin are effective.

Ehrlichia

Morphology: They are small in size and gram-negative, but better stained by Giemsa or Wright stain or IF-stain. They are obligatory intracellular bacteria having an affinity towards blood cells-neutrophils, lymphocytes and monocytes. In the cytoplasm of infected phagocytes, they grow within phagosomes in a cluster as like mulberry called **morula (morula means mulberry).** Morula stained by Giemsa stain.

Cultural characteristics (C/Cs): Relative success achieved by culturing *Ehrlichia* over DH82 cell line which is derived from canine (dog) histiocytoma.

Pathogenicity

- **Disease name:** It called **ehrlichiosis.**
- **Types and other details:** Three types of disease as mentioned in **Table 76.2.**

Laboratory diagnosis

- **Specimens:** Blood, CSF, etc.
- **Testing methods: (1) Microscopy:** Prepare smear from blood and CSF → Stain by Giemsa / Wright stain or by IF-stain → examined by microscope → positive test indicates the presence of morula in blood and CSF monocytes. **(2) Culture:** Follow C/Cs. **(3) Molecular:** Nested PCR.

TABLE 76.2: Types of ehrlichiosis

Features	Glandular fever	Human monocytic ehrlichiosis (HME)	Human granulocytic ehrlichiosis (HGE)
Species	*E. sennetsu*	*E. chaffensis*	*E. equi (E. phago-cytophila)*
M/T	Ingestion of fish	Tick bite like *Amblyoma*	Tick bite like *Ixodes*
C/F	Lymphoid hyperplasia, atypical lymphocytosis	Leukopenia, multisystem involvement, thrombo-cytopenia	Leukopenia, thrombocyto-penia

Prevention: Vector control is effective.

Treatment: Doxycycline and tetracycline are the effective drugs.

Coxiella burnetii

Synonym and meaning: Earlier the etiological agent of the disease was unknown hence it was called **'Query' or Q fever**. In Australia, Burnet identified the causative agent as a rickettsia, it was named **R. burnetii.** Same time same agent was identified from ticks in USA, by Cox called **R. diaporica** (Diapor means ability to pass through pores of filter).

Taxonomy: As the Q fever agent is not truly an arthropod borne and differed from other Rickettsiales member in many features, it has been excluded from Rickettsiales and renamed as *Coxiella burnetii.*

Morphology: They are pleomorphic, small gram-negative rods but better stained by Giminez and other rickettsial stains. Sometimes, they are spherical shape with 0.5–0.3 μm size or rod shape with 0.4–1 μm × 0.2–0.4 μm size. They are able to pass through filter.

Cultural characteristics (C/Cs): They grow in cytoplasmic vacuole (*Coxiella*). Following are different culture methods.

1. **Animal culture:** Guinea pig, hamster and mouse are useful for animals.
2. **Egg culture:** Bacteria grow well in yolk sac of chick embryo.
3. **Cell culture:** Bacteria are very fastidious and grow in human embryo lung fibroblast cell culture. They can be isolated from buffy-coat blood samples or tissues by shell vial method, but requires BSL-3 laboratory.

Resistance: In dried feces it survives for a year or more at 4°C and in meat at least for 1 month. In milk *Coxiella burnetii*, may survive pasteurization by Holder method but killed by Flash method. It is not completely inactivated at 60°C in 1 hour (heat resistant) or by 1% phenol in 1 hour.

Antigenic or phase variation: *Coxiella* is not showing antigenic cross reaction with *Proteus, Rickettsia, Orientia* or *Ehrlichia.* It shows two types of phase variations.

1. **Phase-I:** It occurs in fresh isolates present in guinea pig. It is autoagglutinable, phagocytosed in absence of antibody and strong immunogenic elicits good antibody response against both phases.
2. **Phase-II:** Phase-I will convert in to Phase-II on repeated passage in yolk sac. Phase-II antibody is more suitable for CFT.

Pathogenicity

- **Disease name:** It is a zoonosis called **'Query' or 'Q fever'**.
- **Epidemiology:** Derrick investigated an outbreak of typhoid like fever in abattoir workers in Brisbane, Australia in 1935 by inoculating the blood from the patient to know the causative agent. It is distributed worldwide.
- **Reservoir of infection:** Tick like ixodid ticks.
- **Source of infection:** Animal or animal products.
- **Modes of transmission**
  - **Animal transmission:** Transmitted to cattle, sheep and poultry by ixodid ticks. Transovarial transmission occurs in ticks, but tick is not responsible for human transmission.
  - **Human transmission:** Person-to-person transmission is very rare. Human transmission may or may not occur by ticks but occurs by following routes.
    1. **Contact:** It includes occupational contact during handling of wool, hides, meat or other animal products contaminated with the organisms especially in abattoir workers. Direct contact with abraded skin will favors the infection.
    2. **Ingestion:** Ingestion of milk from infected animals
    3. **Inhalation:** Inhalation of aerosols arises from dried tick feces.
- **Incubation period:** 3–30 days.
- **Portal of entry:** Skin, GIT and respiratory system.
- **Sites:** Lungs, liver, brain, meninges, endocardium, etc.
- **Agent factors (virulence factors): (1) Intracellular nature:** It is an obligate intracellular pathogen in monocytes-macrophages and evades the host defense. It shares many features of rickettsiae like intracellular location; cell wall contains peptidoglycan, presence of DNA and RNA and susceptibility to antibacterial agents but differs by heat resistant nature and no arthropod-borne transmission. **(2) Dismutase:** It is resistant to the acidic environment of phagolysosome by producing superoxide dismutase.
- **Pathogenesis:** It induces the autoantibodies particular to cardiac and smooth muscles.
- **Clinical features:** Clinical picture is variable and sometimes present as asymptomatic infection. It presents as acute and chronic form. **Acute Q fever** is characterized by primary atypical pneumonia/interstitial pneumonia. **Chronic Q fever** characterized by hepatitis, meningoencephalitis or endocarditis. Spontaneous recovery is usual.

Complications: (1) Post Q fever fatigue syndrome: It presents with fever, headache, myalgia, sweating, arthralgia and muscle fasciculation. Spontaneous recovery may occur. **(2) Latent infection:** Coxiella may remain latent in the tissues of patients for 2–3 years.

Laboratory diagnosis

- **Specimens:** Sputum, buffy-coat blood samples or tissues, milk/tissue of infected animals, etc.
- **Testing methods**
  1. **Microscopy:** Follow morphology.
  2. **Culture:** Follow C/Cs.
  3. **Serological tests:** Indirect IF test detects the Abs against phase-I and II antigens of LPS. In acute case phase II Abs are high while in chronic case phase I Abs are high.
  4. **Molecular method:** PCR.

Prevention: *C. burnetii* may survive pasteurization by Holder method but killed by Flash method. Formalin killed whole cell vaccine (licensed in Australia) and trichloracetic acid extract vaccine are developed.

Treatment: Tetracycline, rifampicin, erythromycin, clarithromycin, or ciprofloxacin is effective drug.

Bartonella spp.

Properties: It is not obligate intracellular (also extracellular) parasite hence excluded from Rickettsiaceae family. It is transmitted by arthropods. It invades blood cells and endothelial cells.

Morphology: They are GNB, but better stained by Wharthin-Starry technique. They are pleomorphic, rod shape, 1–3 μm × 0.5 μm in size and showing twitching motility with lophotrichous flagella.

Cultural characteristics (C/Cs): They grow in cell free media at 35–37°C and under 5% CO_2. Following are different useful culture methods.

1. **Liquid media:** BHIB supplemented with 10% hemin is a useful medium. Growth occurs at 35–37°C after 12–15 days. BHIB is useful for biochemical reactions.
2. **Solid media:** These are BHIA, trypticase soy agar and rabbit, sheep, human blood agar. Bacteria produce agar adherent colonies.
3. **Cell culture:** Vero cells are useful for cultivation.

Biochemical reactions (B/Rs): Tests are performed by using BHIB. Bacteria oxidize ribose. Gelatin liquefaction, hippurate hydrolysis and aesculin hydrolysis tests are positive

Pathogenicity: Three different species and disease are mentioned in **Table 76.3**.

Laboratory diagnosis

- **Specimens:** Blood, lymph node/skin biopsy, etc.

TABLE 76.3: Pathogenicity of *Bartonella* spp.

Species	*B. bacilliformis*	*B. (Rochalimaea) quintana*	*B. henselae*
Reservoir	Human only	Human only	Cats and other feline
Mode of transmission	By sandfly bite (*Lutzomyia*)	By human body louse (*Pediculus humanus corporis*)	• By cat scratch or cat bite • Also by cat flea (*Ctenocephalides felis*)
Clinical features	Peruvian medical students Daniel Carrion inoculated (1885) himself with material from verruga and developped Oroya fever, hence also called **Carrion's disease.** 1. **Oroya fever (Carrion's disease):** It presents with fever and progressive anemia due to bacterial invasion of RBCs. High mortality in untreated cases. 2. **Verruga peruana:** It is a late sequel. It presents with nodular ulcerating skin lesion. Bacilli seen inside the RBCs and in the skin lesions	1. **Trench fever/five-day fever/ Quintana fever:** It presents with fever, which lasts for 5 days, followed by remission and recurrence after 5 days with about 12 recurrences. It is non fatal disease. Slow course and prolonged convalescence lead to loss of manpower. Disease named trench fever as it was occurred in soldiers fighting in the trench in Europe The word quintana means fifth, referring to five day fever 2. **Chronic infection (latent infection):** Recrudescence may occur after 20 years	1. **Cat scratch disease/stellate abscess:** It also caused by *Afipia feli*. It presents with fever and single (discrete) lymphadenopathy at scratch site which is mobile and painless. Node contains granulomatous lesion with central star shaped necrosis (called **stellate necrosis**) surrounded by histiocytes, B cells and giant **cells called stellate abscess.** 2. **Lesions in patients with HIV/ IDDs:** • **Bacillary angiomatosis:** Vascular nodules or tumors appear on the skin, mucosa and other locations • **Bacillary peliosis:** It occurs in liver and spleen 3. **Other:** Bacteremia and endocarditis

Note: In *B. quintana* no vertical transmission occurs in lice. Feces become infectious after 5–10 days and lice remains infectious for life. Autoinoculation occurs due to scratching.

Testing methods:

1. **Microscopy:** Follow morphology.
2. **Culture:** Follow C/Cs
3. **Biochemical reactions:** Follow B/Rs.
4. **Serological tests:** Indirect immunofluorescence assay and EIA detect the antibodies against *B. quintana* and *B. henselae*
5. **Molecular methods:** They are further classified with protein profile by SDS-PAGE, RFLP and PCR.

ACCESS YOURSELF

Short Note

1. *Rickettsia/Orientia/Ehrlichia/Cat scratch disease/Q –fever.*

Short Questions for Theory/Viva Questions

1. What is tunica reaction or Neil-Mooser reaction?
2. Name the causative agent for following disease: Q-Fever, Oroya fever, Trench fever and cat scratch disease.

Comments on

1. Rickettsias were initially considered as viruses now accepted as bacteria.
2. *Coxiella* and *Bartonella* are excluded from the Rickettsiales.

MCQs for Chapter Review

Rickettsiales

1. **Tick is the vector for following rickettsial infections,** *except*:
   a. *R. rickettsii*
   b. *R. akari*
   c. *R. africae*
   d. *R. conorii*
2. **Rickettsiae belong to spotted fever group,** *except*:
   a. *R. rickettsii*
   b. *R. conorii*
   c. *R. typhi*
   d. *R. akari*
3. **Rickettsiae:**
   a. Multiply only within living cell
   b. Produce typhus fever of epidemic type only
   c. Transmitted by arthropod vectors
   d. Respond to tetracycline therapy
4. **All are true about rickettsiae,** *except*:
   a. Obligate intracellular
   b. Gram-positive bacilli
   c. Arthropods are vectors
   d. Weil-Felix test is diagnostic
5. **Neil–Mooser reaction is used to diagnose:**
   a. *Rickettsia*
   b. *Chlamydia*
   c. *Mycoplasma*
   d. Herpes
6. **A man presents with fever and chills 2 weeks after a louse bite and maculopapular rash on the trunk which spreads peripherally. The causative agent can be:**
   a. Scrub typhus
   b. Endemic typhus
   c. Rickettsial pox
   d. Epidemic typhus
7. **Brill-Zinsser disease is:**
   a. Epidemic louse-borne typhus
   b. Endemic louse-borne typhus
   c. Q fever
   d. Rocky mounted spotted fever
8. **It is true regarding endemic typhus that:**
   a. Man is the only reservoir.
   b. Flea is a vector for the disease.
   c. Rash developed in to eschar is the characteristic presentation
   d. Tissue culture is the diagnostic.
9. **True statement about endemic typhus:**
   a. It is caused by *R. rickettsii*
   b. It is transmitted by bite of fleas.
   c. Has no mammalian reservoir
   d. Can be culture in chemically defined media
10. **Following is the etiological agent of rocky mounted spotted fever:**
    a. *R. rickettsii*
    b. *R. quintana*
    c. *R. tsutsugamushi*
    d. *C. burnetii*
11. **Vascular endothelial infection is caused by:**
    a. *Rickettsia*
    b. *Mycoplasma*
    c. *Chlamydia*
    d. None
12. **Primary site of multiplication of rickettsial organisms is:**
    a. Parenchymal cells of the liver
    b. Endothelial cells of small vessels
    c. Media of arteries
    d. Advantitia of all blood vessels
13. **Which of the following used for *Rickettsia*?**
    a. Weil-Felix reaction
    b. Rose Waller test
    c. Paul-Bunnel test
    d. VDRL
14. **Transovarial transmission is a features of:**
    a. Scrub fever
    b. Epidemic typhus
    c. Endemic typhus
    d. Trench fever
15. **True about scrub typhus is:**
    a. Transmitted by larvae of trombiculid mite
    b. Incubation period is 3–4 days
    c. Eschar is diagnostic
    d. Caused by *Rickettsia typhi*
16. **All are true regarding scrub typhus.** *except*:
    a. Caused by *R. tsutsugamushi*
    b. Spreads by mites
    c. Caused by bite of adult mite
    d. Tetracycline is effective
17. **Which is not transmitted by arthropod?**
    a. *R. prowazekii*
    b. *C. burnetii*
    c. *R. akari*
    d. *R. rickettsii*
18. ***E. chaffensis* is causative agent of:**
    a. HME
    b. HGE
    c. Glandular fever
    d. None

Coxiella burnetii

19. **True statement regarding Q fever,** *except*:
    a. It is a zoonotic disease
    b. Human disease is characterized by an interstitial pneumonia
    c. No rash is seen
    d. Weil–Felix reaction is very useful for diagnosis

Bartonella spp.

20. **All are true about *B. quintana*, *except*:**
    a. Causes trench fever
    b. Not detected by Weil-Felix reaction
    c. Recurrence is common
    d. Tick is the vector
21. ***B henselae* causes all, *except*:**
    a. Oroya fever
    b. Cat scratch disease
    c. Bacillary angiomatosis
    d. SABE

22. Most common cause of bacillary angiomatosis is:
   a. *B. quintana*   b. *B. bacilliformis*
   c. *B. henselae*   d. *B. elizabethi*

23. Cat scratch disease is:
   a. Associated with positive Frei's skin test
   b. Caused by DNA virus
   c. Associated with a pathognomic histological picture
   d. Associated with regional lymphadenopathy
   e. Associated with Hanger-rose test

Answers and Explanation of MCQs

1. b

2. c
- Follow **Table 76.1** for explanation of answers of MCQs 1–2.

3. a, c and d
- Rickettsiae can multiply only within living cell (obligate intracellular) *in vivo* (like nucleus of vascular endothelial cells) and *in vitro* (like detroit 6, HeLa, HEP-2, mouse fibroblast and other continuous cell lines). They produce typhus fever of epidemic and endemic type. They are transmitted by bite or feces of insect as mentioned in **Table 76.1** and respond to tetracycline therapy.

4. b
- Rickettsiae are gram-negative. Follow section, **Rickettsiales → *Rickettsia*** for explanation of other options.

5. a
- Follow section, *Rickettsia* **(culture characteristics)** for explanation.

6. d
- From the given option louse borne infection is epidemic typhus. Scrub typhus and rickettsial pox are mite-borne. Endemic typhus is flea borne.

7. a
- Follow section, **Rickettsiales → *Rickettsia* (pathogenicity → Brill-Zinsser disease)** for explanation.

8. b
- Rodent and cat are the only reservoirs as mentioned in **Table 76.1**. Flea is a vector and no eschar formed in disease. Tissue culture is available but not satisfactory, so diagnostic methods are molecular (PCR) and serological.

9. b
- Endemic typhus is caused by *R. typhi*. Other options are already explained.

10. a
- Follow section, *Rickettsia* **(pathogenicity → pathological and Table 76.1** for explanation.

11. a

12. b
- Follow section, *Rickettsia* **(pathogenicity → pathogenesis and Flowchart 76.3)** for explanation of answers of MCQs 11–12.

13. a
- Rose Waller test is used for RF or RA factor, Paul- Bunnel test for infectious mononucleosis and VDRL for syphilis detection.

14. a
- Transovarial transmission to offspring is present in ticks and mites but not in louse and fleas. Scrub fever is mite borne, epidemic typhus and trench fever are louse borne, while endemic typhus is flea borne.

15. a
- Scrub typhus is caused by *O. tsutsugamushi* and transmitted by larvae (**called chiggers**) of trombiculid mite but not by adult stage. Incubation period is 6–21 days. Eschar is not diagnostic because it absent in < 50% of Westerners.

16. c
- Scrub typhus is transmitted by larvae (**called chiggers**) of trombiculid mite but not by adult stage.

17. b
- Follow section, **Table 76.1 and *Coxiella*** for explanation.

18. a
- Follow section, *Ehrlichia* **(pathogenicity → types and other details and Table 76.2)** for explanation.

19. d
- The Weil-Felix reaction is not diagnostic (negative) in rickettsial pox, trench fever and Q fever.

20. d

21. a

22. c
- Follow section, *Bartonella* **spp. (Table 76.3)** for explanation of answers of MCQs 20–22.

23. c and d
- Frei's skin test is useful for lymphogranuloma venereum caused by *Chlamydia trachomatis*. Cat scratch disease is caused by bacterium not by virus, associated with a pathognomic picture like stellate abscess and regional lymphadenopathy, but not with Hanger-rose test.

Viral Infections

Infections of DNA Viruses

Infections of RNA Viruses

Infections of Miscellaneous Viruses

Infections of Poxviridae

INTRODUCTION

Classification

Family: Poxviridae.

Two subfamilies: Like Chordopoxvirinae and Entomopoxvirinae. Entomopoxvirinae does not infect the mammalian. Different genera and species of Chordopoxvirinae are described below.

Genera of subfamily Chordopoxvirinae

- *Orthopoxvirus:* Pox virus infecting mammalian is called *Orthopoxvirus*. Common species are variola virus (natural virus causes smallpox in human), vaccinia virus (artificial virus and used for vaccine preparation), cowpox virus, buffalopox virus, monkeypox virus, cantagalo and arcatuba virus.
- *Avipoxvirus:* Pox virus infecting birds is called *Avipoxvirus*. Common species are turkypox virus, canarypox virus, fowlpox virus (inclusion body called Bollinger body) and pigeonpox virus.
- *Capripoxvirus:* Pox virus infecting goat and sheep is called *Capripoxvirus*. Common species are sheeppox virus, goatpox virus and lumpy skin disease virus.
- *Leporipoxvirus:* Pox virus infecting leporids like rabbits, squirrel or hares is called *Leporipoxvirus*. Common species are myxomavirus and fibroma virus.
- *Parapoxvirus:* Poxvirus infecting ungulates is called *Parapoxvirus*. Common species are human orf virus, paravaccinia (pseudocowpox) virus, bovine papular stomatitis virus, deerpox virus and sealpox virus.
- *Suipoxvirus:* Pox virus infecting swine (pig) is called *Suipoxvirus*. Common species is swinepox virus.
- *Molluscipoxvirus:* Common species is molluscum contagiosum virus (MCV).
- *Yatapoxvirus:* Common species are yabapox virus (yaba monkey tumor virus) and tanapox virus.

Antigenic Structure

All pox viruses have common nucleoprotein (NP) antigen. Antigenic cross reactivity is present in-between the different pox viruses. Twenty different antigens like heat labile L, heat stable S, LS (complex of L and S), agglutinogen and hemagglutinin are present in different pox viruses.

HUMAN DISEASES

Four genera of *Poxviridae* with respective species causing human disease are listed in **Table 77.1**.

Variola Virus

History: Microscopically, it was identified by Buist in 1887.

Use: It is useful as biological weapon.

Morphology

- **Shape and size:** It is brick shaped (**Fig. 77.1**), largest of all animal viruses about 400×230 nm in size, and only virus visible under the light microscope.
- **Genome and capsid:** It is ds-DNA virus surrounded by double layer capsid (complex variety of capsid).
- **Envelope:** It is lipoprotein in nature.
- **Lateral bodies:** On either side of capsid (between capsid and envelop) a lens-shaped bodies are present called lateral bodies. Function is unknown.
- **Inclusion body:** It called **Paschen body.** It is intra-cytoplasmic, identified by Paschen in 1906 from smallpox lesion and detected under light microscopy.
- **Multiplication:** Virus replicates in cytoplasm of infected cells.
- **Electron microscopy:** It is used for rapid diagnosis.

Culture characteristics (C/Cs)

- **Egg culture:** It grows in CAM of 11–13 days old hen's egg and form 'pocks' in 48–72 hours. Highest

TABLE 77.1: Pox viruses causing human diseases

Species	Reservoir	Distributiont	Human disease
Orthopoxvirus			
Variola virus	Humans	Eradicated	Smallpox
Vaccinia virus	Humans	—	Local skin lesion (artificial virus) and used for vaccination
Monkeypox virus	Rodents and monkeys	Africa	Rare, skin or systemic lesions
Cowpox virus	Cows	Europe	Rare, skin or systemic lesions
Buffalopox virus	Water buffalo	Indian subcontinent	Rare, skin lesions
Cantagalo and Arcatuba virus	Cattle	South America	Rare, skin or systemic lesions
Parapoxvirus			
Orf virus	Sheep and goats	Worldwide	Rare, local skin lesions called contagious pustular dermatitis
Paravaccinia (pseudocowpox) virus	Cattle	Worldwide	Rare, local skin lesions called milker's nodule (node)
Bovine papular stomatitis virus	Cattle	Worldwide	Rare, local skin lesions
Deerpox virus	Deers	Deer herds	Local pox skin lesions
Sealpox virus	Seals	Seal colonies	Local pox skin lesions
Molluscipoxvirus			
Molluscum contagiosum virus (MCV)	Human	Worldwide	Benign skin nodule called molluscum contagiosum
Yatapoxvirus			
Yabapox virus	Monkeys	Unknown	Local skin tumor
Tanapox virus	Monkeys	Africa (Zaire)	Rare, local skin lesions

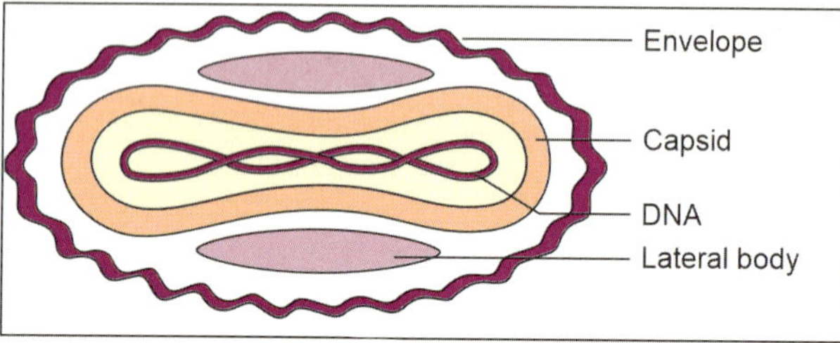

Fig. 77.1: Morphology of variola virus (complex symmetry)

temperature above which pocks are not produced called **ceiling temperature**. **Variola** pocks are small, shiny, white, convex, non-necrotic and non-hemorrhagic. Ceiling temperature for variola major is 38°C and for variola minor is 37.5°C. **Vaccinia** pocks are large, irregular, grayish, flat, necrotic and hemorrhagic. Ceiling temperature is 41°C.

- **Tissue culture:** Commonly used cells are Hela cells, Vero cells and chick embryo cells. Paschen bodies (eosinophilic inclusion bodies) will be identified in stained smear. Vaccinia, not variola produces plaque in chick embryo cell culture.
- **Animal culture:** Monkey, rabbit, calves and sheep are useful.

Resistance: Virus is susceptible to UV light, susceptible to formalin and oxidizing disinfectants. It is resistant to 50% glycerol and 1% phenol. It is enveloped, even though not inactivated by ether.

Pathogenicity

- **Disease name:** Human pox is caused by variola virus is called **smallpox.**
- **Epidemiology:** It was an ancient scourge and killed many millions of people. **Last variola major** was detected in Saiban Bibi, a Bangldeshi woman at Karimganj Railway station, Assam in 24th may 1975. **Last variola minor** was detected in Merca, Somalia in October, 1977. Natural small pox was ceased in 1977, but **Last small outbreak** was occurred in August 1978 in Birmingham in medical school following accidental spread from laboratory. It was promptly identified and controlled, but incident gave message regarding hazards of keeping stock. After that all stocks were destroyed by direction from WHO. Large stock was maintained at CDC, Atlanta, USA and at Centre for Research on Virology and Bio-technology Koltsova, Russia but they were destroyed in June 30 1999. Large stock of small pox vaccine is maintained by WHO for rapid deployment in case of bioterrorism or re-emerging of disease. It was globally eradicated on 8th may, 1980. Factors contributed in eradication were use of freeze dried vaccine and multiple punctured technique of vaccination by using bifurcated needle, which was simple, effective and economical.
- **Reservoirs of infection:** No animals/vectors reservoirs but only human reservoirs.
- **Sources of infection:** Respiratory droplets.

- **Modes of transmission:** It is transmitted via droplets/aerosols in early phase of disease or via close contact.
- **Portal of entry:** Respiratory tract.
- **Site:** Skin.
- **Incubation period:** 12 days.
- **Pathogenesis:** Follow **Flowchart 77.1**.

Clinical features: Clinically there are two types of smallpox like variola major with high mortality plus classical smallpox and variola minor with low mortality plus alastrim. **Prodromal phase** presents with non specific febrile and prostrating flu-like features lasts for 3–5 days. **Eruptive phase** has characteristic rash—single crop of centrifugal exanthems passed through macular, vesicular and pustular stages (**Fig. 77.2**) before scabbing and heal by scar formation in 2–4 weeks.

Laboratory diagnosis

- **Prodromal phase:** Blood.
- **Eruptive phase:** Fluid, scraping or scab from lesion.
- **Testing methods:** Samples are tested by **microscopy** (follow morphology), **culture** (follow C/Cs), **serological** and **molecular methods.**

Flowchart 77.1: Pathogenesis of pox (variola) virus

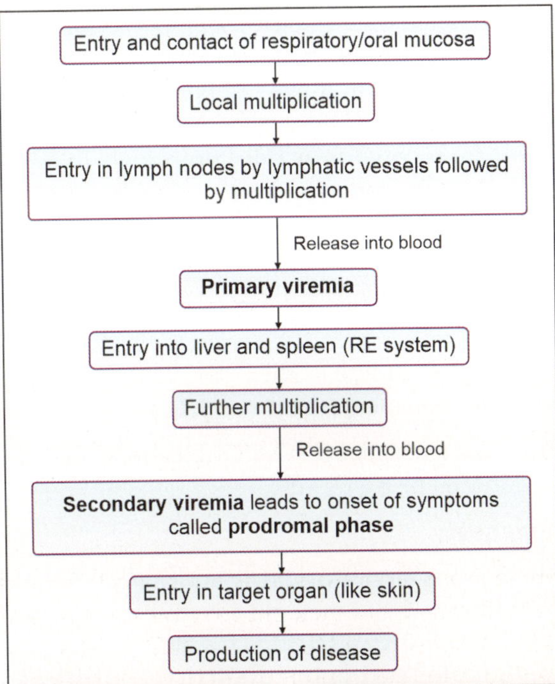

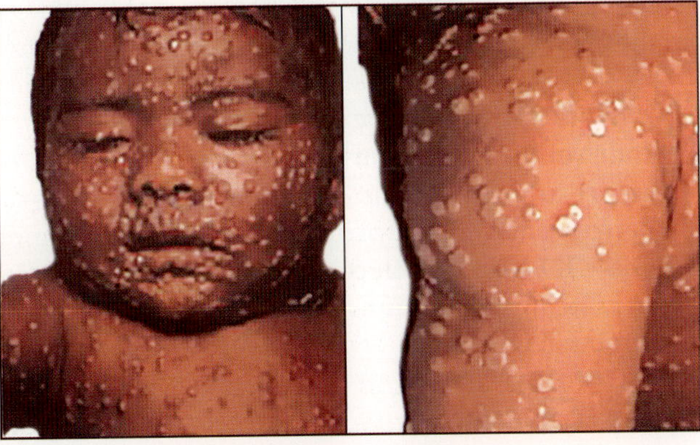

Fig. 77.2: Smallpox lesions

Prevention: It is eradicated from the world, and routine vaccination is now stopped. 1st vaccine of smallpox was developed from cowpox lesion by Edward Jenner. Freeze dried vaccine was used for control. It was injected by multiple puncture with bifurcated needle in upper arm.

Vaccinia Virus

It is similar to variola virus and both can be considered together. Initial cowpox vaccine virus was maintained by serial arm to arm passages in human, it underwent some permanent changes and becomes different from cowpox and smallpox called **vaccinia virus.** As it is not occurred in nature, so it is called **artificial virus.** Inclusion body produced by vaccinia virus called **Guarnieri body.** It is nonpathogenic, but may cause local skin lesions. **Differences between variola and vaccinia viruses** are mentioned in **Table 77.2**. It is **used** (1) for preparation of smallpox vaccine. (2) As a vector to prepare recombinant vaccine like HBV, HIV and rabies. (3) To prepare pharmacological products like neuropeptide.

TABLE 77.2: Differences between variola and vaccinia viruses

Features	Variola	Vaccinia
Type of virus	Natural	Artificial
Ceiling temp.	37.5°/38°C	41°C
Pock morphology	Small, shiny, white, convex, non-necrotic and non hemorrhagic	Large, irregular grayish, flat, necrotic and hemorrhagic
Plaque on chick embryo tissue culture	No	Yes
Inclusion body	Paschen body	Guarnieri body
Pathogenicity	Smallpox	Local skin lesion
Use for vaccine preparation	No	Yes

Monkeypox Virus

Clinically, it is same as smallpox. Rodents are the primary reservoirs followed by monkeys. It is transmitted by direct contact with animals. Person-to-person transmission occurs rarely. It is distributed in Africa. First outbreak was reported in USA in 2003 with more than 70 cases. It was transmitted from pet dogs that got infection from rodents. It produces local lesion like small pox but mild in nature and systemic illness like fever and lymphadenopathy.

> **Note: Monkeypox Outbreak 2022**
> - **Morphology, C/Cs and resistance of virus:** It was caused by monkeypox virus. Morphology, C/Cs, and resistance of virus are same as variola virus as described above.
> - **Clades:** It has two genetic clades. (1) Central African clade which produces more severe disease, more transmissible and seen in Congo basin. (2) West African clade.

- **Pathogenicity**
  - **Epidemiology:** Outbreak of human monkeypox was confirmed in May 2022, beginning with a cluster of cases found in the United Kingdom. The first recognized case was confirmed on 6th May 2022 in an individual with travel links to Nigeria (where the disease is endemic), but it has been suggested that cases were already spreading in Europe in the earlier months. From 18 May onwards, cases were reported from an increasing number of countries and regions, predominantly in Europe, but also in North and South America, Asia, Africa, and Australia.
  - **Reservoirs of infection:** These are monkeys, Gambian pouched rats (*Cricetomys gambianus*), dormice (*Graphiurus* spp.) and African squirrels (*Helioschiurus* and *Funisciurus*).
  - **Sources of infection:** These are infected animal/human products like meat, body fluid or respiratory droplets.
  - **Modes of transmission: (1) Direct:** It may spread from handling bushmeat, animal bites or scratches, body fluids, contaminated objects, or other close contact with an infected person. It also transferred vertically. **(2) Indirect:** Spread can occur by droplets and possibly the airborne route. People can spread the virus from the onset of symptoms until all the lesions have scabbed and fallen off; with some evidence of spread for more than a week after lesions have crusted.
  - **Incubation period:** 5–21 days. Generally, symptoms last for 2–4 weeks.
  - **Portal of entry:** Skin or respiratory route.
  - **Sites:** Skin, lymph nodes and respiratory system.
  - **Precipitating factors:** Cases may be severe, especially in children, pregnant women or people with suppressed immune systems. Cases may be milder in people vaccinated against smallpox in childhood.
  - **Clinical features:** Clinical features include fever, muscle pain, swollen lymph nodes, and a rash that forms blisters and then crusts over **(Fig. 77.3)**. Many cases in the 2022 monkeypox outbreak presented with genital and perianal lesions and pain when swallowing.
  - **Complications:** Secondary infections, pneumonia, sepsis, encephalitis, and loss of vision if severe eye infection. If infection occurs during pregnancy, stillbirth or birth defects may occur.
- **Laboratory diagnosis**
  - **Specimens:** Take the material from the skin lesions.
  - **Testing method:** PCR from skin material. PCR from blood is usually inconclusive, because the virus does not remain very long in the blood.
- **Prevention**
  - **General measures:** Avoid contact with suspected cases and animal products. Frequent hand washing, wearing mask and distance from case/animal/animal products are the other useful measures.
  - **Chemoprophylaxis:** Specific drug is not available.
  - **Immunoprophylaxis:** The smallpox vaccine was found to provide around 85% protection and reduces the severity of the disease.
- **Treatment:** Tecovirimat and brincidofovir are the 1st-line drugs used during outbreaks. Vaccinia Ig was also used during outbreaks.

Cowpox Virus

Rodents or cats are the primary reservoirs followed by cows. Cow lesions are present on udder or teats and enter in human during milking. It is distributed in Britain and Europe. Fatal outbreak was reported in wild animals like panthers or leopards and elephants. Human lesions appear on hands or fingers and with rare systemic illness like fever. It is distinguished from variola and vaccinia virus by hemorrhagic lesion in CAM and in rabbit skin and also by restriction endonuclease analysis.

Buffalopox Virus

It was identified from cattle in India in 1934. Lesions appear on hands of persons in contact of infected animal. Epizootic was occurred in buffaloes. Clinically, it is as like smallpox and distinguished by laboratory methods.

Orfpox Virus

It is a disease of sheep and goats and transmitted to human by animal contact. Single papulovesicular lesion with central ulcer occurs on hand, forearm or face called contagious pustular dermatitis or infectious labial dermatitis or ecthyma contagiosum or thistle disease or scabby mouth. Virus is different from variola and vaccinia virus and morphologically similar to paravaccinia virus.

Paravaccinia (Pseudocowpox) Virus

Morphologically, it is similar to orf virus. Cow lesions are entered in human during milking. Human lesion is ulcerative nodule on skin called **Milker's nodule.** It is different from cowpox as does not grows in eggs, but grows in bovine kidney culture.

Molluscum Contagiosum Virus (MCV)

History: MCV was first described, and later, it was named by Bateman in the early nineteenth century.

Morphology: Same as other pox virus as described above under variola virus **(Fig. 77.1)**.

Culture characteristics (C/Cs): Virus cannot be cultivated in egg, animal or tissue culture.

Pathogenicity

- **Disease name:** Disease called **molluscum contagiosum.**
- **Synonym:** Water wart.
- **Epidemiology:** It is distributed worldwide. It has four types from MCV-1 to MCV-4. MCV-1 is the most common and most prevalent (75–90%) and MCV-2 is common in adults.
- **Reservoirs and sources of infection:** Humans.
- **Modes of transmission:** It is transmitted by **(1) direct contact** to infected skin or contaminated clothes or toys (common use of towels by barbers or swimming pools), **(2) sexual intercourse** and **(3) autoinfection** by scratching the lesion or by contact of normal skin with infected skin.
- **Portal of entry:** Skin.
- **Sites:** Lesions are most commonly found on the face, arms, legs, torso and armpits in children (except palm

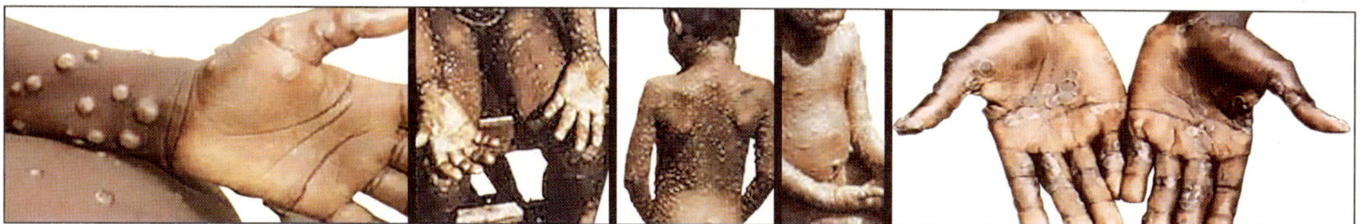

Fig. 77.3: Monkeypox lesions

and sole). Adults typically have molluscum lesions in the genital region and this is considered to be a sexually transmitted infection; however, if genital lesions are found on a child, sexual abuse should be suspected. Rarely, it affects eyelids.

- **Precipitating factors:** It is common in children between 4 and 11 years, sexually active adults and in AIDS patients.
- **Clinical features**
  - Lesions are flesh/pink-colored, dome-shaped, pearly in appearance, 1–5 mm in diameter and with a dimpled center called **umbilicated nodules** (**Fig. 77.4**), present on genitals and eyelids.
  - Lesions are generally not painful, but they may itch or become irritated. Eye lesions may complicate as chronic follicular conjunctivitis and superficial keratitis may occur. It is a self limited condition and disappears in 6 months to 5 years without treatment, but lasts longer in person with IDDs.
- **Complications:** Picking or scratching the bumps may lead the spread of viral infection, bacterial infection, scarring or eczema in 10% cases.

Laboratory diagnosis

- **Specimens:** Skin scraping or excision biopsy.
- **Testing methods: (1) Light microscopy:** It is performed by using H and E stain. Very large (20–30 µm), hyaline, intracytoplasmic and eosinophilic inclusion bodies detected in epithelial cells called molluscum bodies or called Henderson-Petterson bodies. **(2) Electron microscopy:** It is used for virus detection. **(3) Culture:** Follow C/Cs. **(4) Molecular:** PCR.

Treatment: (1) Medications: For mild cases, salicylic acid may be used which shorten duration. Other useful local agents are tretinoin cream and cantharidin. Oral cimetidine and placebo-controlled trials have demonstrated **(2) Imiquimod:** A form of immunotherapy is useful. **(3) Surgery:** Liquid nitrogen (cryosurgery) is used to freeze and to destroy lesions. Scraping of

lesions with curette can be performed. **(4) Laser:** Pulsed dye laser therapy is a safe and effective treatment for molluscum contagiosum and is generally well-tolerated by children.

Yabapox Virus

It is a monkey tumor virus related to pox virus.

Tanapox Virus

It is identified from the epidemic of febrile illness along the Tana river, Kenya hence the name. Monkeys are the susceptible animals. Patient had single pock like lesion on the upper part of body. Antigenically, it is unrelated to other pox virus and can grow in human and monkey tissue culture, but not in eggs.

ACCESS YOURSELF

Short Notes

1. Monkeypox outbreak 2022.
2. MCV.

Short Questions for Theory/Viva Questions

1. Who discovered the smallpox vaccine? What was the source?
2. Name the inclusion body of following viruses: Variola virus, vaccinia virus, fowpox virus and MCV.
3. Write the four differences between variola and vaccinia virus.

MCQs for Chapter Review

Classification

1. **Smallpox belongs to which genus of poxviruses?**
   a. *Parapoxvirus*   b. *Capripoxvirus*
   c. *Laporipoxvirus*   d. *Orthopoxvirus*
2. **Which of the following is not a pox virus?**
   a. Cowpox
   b. Molluscum contagiosum
   c. Smallpox
   d. Chickenpox
3. **Following virus is of pox virus:**
   a. Variola   b. Coxsackie
   c. ECHO   d. HSV

Antigenic Structure

4. **The protection against smallpox by previous infection with cowpox represents:**
   a. Antigenic cross reactivity
   b. Antigenic specificity
   c. Passive immunity
   d. Innate immunity

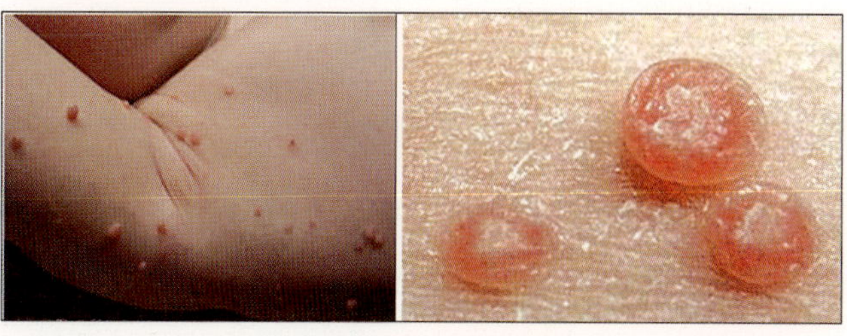

Fig. 77.4: Lesions of MCV

5. **Which year did WHO declared global eradication of smallpox?**
   a. 8th may 1975
   b. 8th may 1977
   c. 8th may 1980
   d. 8th may 1985

6. **Vaccine preparation requires which virus as vector:**
   a. Rhino virus
   b. Vaccinia virus
   c. Adeno virus
   d. Ebola virus
   e. Hepatitis B

7. **Ceiling temperature for vaccinia virus is:**
   a. 41°C
   b. 38°C
   c. 39°C
   d. 40°C

8. **Which virus cannot be cultivated?**
   a. Vaccinia
   b. Variola
   c. MCV
   d. Cowpox

9. **Which virus would not grow in egg, animal cells?**
   a. Cowpox
   b. Vaccinia
   c. Variola
   d. Molluscum

10. **Most common molluscum virus:**
    a. 1
    b. 2
    c. 3
    d. 4

11. **Umbilicated nodules are produced by:**
    a. Pox virus
    b. Entero virus
    c. Rhino virus
    d. Myxo virus

1. d

2. d

3. a

- Follow section, **classification** for explanation of answers of MCQs 1–3.

4. a

- Small pox and cowpox viruses are belong to same family and protection is due to antigenic cross reactivity. Follow section, **antigenic structure** for more explanation.

5. c

- Follow section, **human diseases (variola virus → pathogenicity → Epidemiology)** for explanation.

6. b

7. a

- Follow section, **human diseases (vaccinia virus and Table 77.2)** for explanation of answers of MCQs 6–7.

8. c

9. d

10. a

11. a

- Follow section, **human diseases (molluscum contagiosum virus)** for explanation of answers of MCQs 7–10.

Infections of Herpesviridae

Chapter Outline

- Introduction
- HHV-1 and 2 (HSV-1 and 2)
- HHV-3 (VZV)
- HHV-4 (EBV)
- HHV-5 (CMV)
- HHV-6 (HBLV)
- HHV-7 (RKV)
- HHV-8 (KSHV)
- Ceropethacine Herpes Virus-1

INTRODUCTION

Meaning

The word herpes is taken from the herpein (Greek) means to creep or crawl. This refers the spreading nature of skin lesion.

Classification

Follow **Table 78.1**.

Morphology

Shape and size: It is spherical or oval (**Fig. 78.1a**). Enveloped virus is 300 nm in size, while naked virus is 100 nm in size.

Genome and capsid: Virus contains ds-DNA surrounded by icosahedral variety of capsid.

Envelope: It presents with spikes about 8 nm long.

TABLE 78.1: Classification of Herpesviridae

Order: Herpesvirales, Family: Herpesviridae

Sub family	Genus	Species	Common name	Target cell type	Latency	Transmission
α-herpesvirinae	Simplexvirus	HHV-1	Herpes simplex virus-1 (HSV-1)	Mucoepithelial	Neuron	Contact, aerosols
α-herpesvirinae		HHV-2	Herpes simplex virus -2 (HSV-2)	Mucoepithelial	Neuron	Close contact usually sexual
α-herpesvirinae	Varicillovirus	HHV-3	Varicella zoster virus (VZV)	Mucoepithelial	Neuron	Contact or respiratory route
γ-herpesvirinae	Lymphocrypto-virus1	HHV-4	Epstein-Barr virus (EBV)	B-cell, epithelial	B-cell	Kissing, transfusions transplantation,
β-herpesvirinae	Cytomegalo-virus	HHV-5	Cytomegalo virus (CMV)	Epithelial, mono-cytes, lymphocytes	Monocytes, lympho-cytes and possibly others	Contact, blood trans-fusions, transplanta-tion, congenital
β-herpesvirinae	Roseolovirus	HHV-6	Human B-cell lymphotropic virus (HBLV)	T lymphocytes and others	T-lymphocytes and others	Contact, respiratory route
β-herpesvirinae		HHV-7	RK virus	T lymphocytes and others	T-lymphocytes and others	Unknown
γ-herpesvirinae	Rhadinovirus	HHV-8	Kaposi's sarcoma associated herpes virus (KSHV)	Endothelial cells	Unknown	Exchange of body fluids?

Teguments: Space between capsid and envelope is taken up by amorphous protein materials called teguments.

Multiplication: Virus replicates in nucleus of infected cells.

Inclusion body: It produces **Lipschutz body,** which is intranuclear Cowdry type-A type and detected under light microscopy by Giemsa stain.

Tzanck test: It was developed by Arnault Tzanck, 1947. Multinucleated giant cells called **Tzanck cells** are formed due to fusion of acanthoytic keratinocytes in pemphigus vulgaris, herpes simplex, herpes zoster and chickenpox. Tzanck test includes smear preparation and staining for identification of Tzanck cells. Scrape the base of vesicle and prepare the smear called **Tzanck smear.** Smear can be stained by 1% aqueous solution of toludine blue-O (for 15 seconds), Giemsa stain, H and E stain, methylene blue or by Papanicolaou stain and examine under microscope. Tzanck cells **(Fig. 78.1b)** are seen with faceted nuclei and homogeneously stained ground glass chromatin. Ballooning of infected cells, margination of chromatin and Lipschutz body are also identified. Tzanck test is rapid and cheap. It cannot differentiate between HSV-1 and 2. Sensitivity of stain is very low about <30% for mucosal swab.

Electron microscopy of vesicle fluid: It produces rapid result but cannot distinguish between HSV and VZV.

Immunofluorescent (IF) microscopy of skin scrapings: It can distinguish between HSV and VZV.

Resistance

Sterilization: All herpes viruses are susceptible to heat and can be stored at –70°C.

Disinfection: All herpes viruses are enveloped viruses, and they are inactivated by fat solvents like ether, chloroform, bile salts and alcohol.

Antigenic Structure

There is no common group antigen. Different species do not show the significant cross reacting antigens except HSV-1 and 2.

HHV-1 AND 2 (HSV-1 AND 2)

History

Virus was identified by histopathological study due to its association with giant cell formation in 1919. Latency of HSV was identified in 1930.

Morphology (Fig. 78.1a)
Resistance

} Described above

Differences between HHV-1 and 2

Follow **Table 78.2**.

TABLE 78.2: Differences between HHV-1 and 2		
Features	**HHV-1**	**HHV-2**
Culture characteristics (C/Cs)		
Pock in CAM	Smaller	Larger
Chick embryo fibroblast culture	Poor replication	Good replication
Neurovirulent property in animal	Less	More
Resistance		
Temp., sensitivity	Less	More
Antiviral agent sensitivity	Less resistant	More resistant to IUDR and cytarabine
Immunity		
Abs formation	70–90%	20%
Protection by Ab	Both Abs are less protective. HSV-2 provides protection against HSV-1, but not vice versa	
Pathogenicity		
Age	Young children	Young adults
Transmission		
• **Congenital**	• Less	• More (75% cases of total)
• **Contact**	• More	• Less
Site of infection	Above belt	Below belt
Encephalitis	More	Less
Cancer	Of lips	Of cervix
Common lesions	Oro-facial, mucosal and ocular	Genital
Site of latency	Trigeminal ganglia	Sacral ganglia
Recurrence	Less (1 times)	More (4 times)
More differentiation		
Antigenic structure and differentiation	>80% antigenic homology occurs between HSV-1 and 2. More differences are done by using monoclonal Ab	
Molecular	>50% genomic homology occurs between HSV-1 and 2. More differences are done by using restriction endonuclease analysis, electrophoretic analysis of viral DNA or NA hybridization	

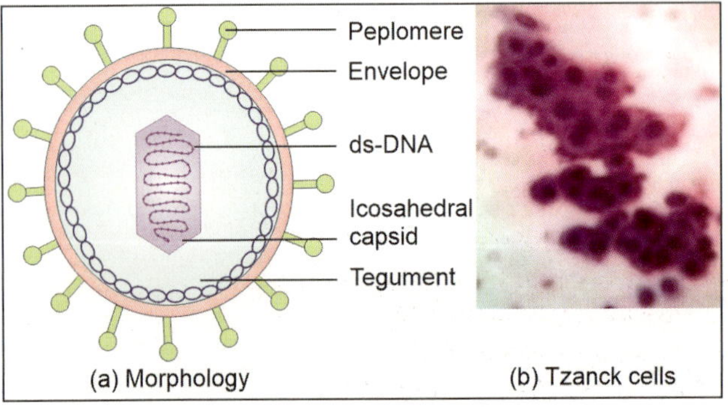

Fig. 78.1: Images of herpes virus

Culture Characteristics (C/Cs)

Egg culture: Poor growth occurs in CAM of chick embryo. Pocks are small, shiny, white and non- necrotic.

Animal culture: Poor growth occurs in mouse.

Tissue culture: Commonly used cells are McCoy cells, Hela cells, Vero cells, rabbit kidney and human amnion or human diploid fibroblasts. Typical CPE is observed in 1–2 days, but culture declared negative after 2 weeks. CPE may be in the form of giant cell formation or syncytium formation or heaped up cells or rounding or ballooning of infected cells. Shell vial method can be used to reduce detection time by <24 hours. Viral antigens are detected by neutralization test or by IF staining with specific antiserum. Cell culture is also useful for drug sensitivity testing.

Antigenic Structure

Follow **Table 78.2**.

Immunity

AMI: Antibodies may not prevent the recurrences but may reduce the severity of disease. They form in 70-90% cases in HSV-1 and 20% cases in HSV-2. Both Abs are less protective. HSV-2 provides protection against HSV-1, but not *vice versa*.

CMI: It is more important in resistance and recovery of infection. HSV infection is more common and more sever in person with IDDs.

Pathogenicity

Epidemiology: Both viruses are distributed worldwide. Nature of disease depends on age, site of infection, immune status and antigenic type of virus.

Reservoir of infection: Asymptomatic human carrier is the natural reservoir and most important in HSV-2.

Sources of infection: Carrier's saliva, skin lesion and aerosols are the main sources of infection. Period of infectivity (communicability) is 1–2 days before the rash and 4–5 days after the development of rash. Isolation of case is done of 6 days after the onset of rash. Secondary attack rate is about 70–90%.

Modes of transmission: Common mode is **inhalation** of respiratory droplets/aerosols. **HSV-1 transmitted** by close contact like mucosal (oropharyngeal) contact or direct contact with abraded skin. **HSV-2 transmitted** by sexual route or vertically via placenta, during birth (due to contact with vaginal secretion) or after birth.

Incubation period: 1–26 days (median, 6–8 days).

Portal of entry: Skin or mucosa or respiratory tract.

Sites: HHV-1 produces the lesions mostly above west (on cheeks, chin, around mouth and on the forehead) while **HHV-2** below waist (on genitals like penis and urethra in male and cervix, vagina, vulva and perineum in female); however, it is not absolute.

(1) Age: Primary infection occurs in childhood and latent infection in adults. HSV-1 is more in young children and HSV-2 is more in young adults. **(2) Sex and race:** Antibodies synthesis in HSV-2 is more in black women than white men. **(3) Immunity:** Single attack provides the lifelong immunity. **(4) IDDs:** Infections are common in HIV infected person and in other with low immunity like malnutrition. **(5) Occupation:** Infections are common in doctors or nurses on hands or fingers due to frequent contact with patients. **(6) Erythema multiforme:** It is an acute, self limited and recurring skin condition based on type-IV hypersensitivity. It is associated with certain infections, medications and other triggers. HSV is commonly associated with this and herpes antigens can be detected from skin biopsies or serum. **(7) Trauma:** Infect the cornea due to trauma leading to blindness.

Pathogenesis

- **Mechanisms of primary infection:** Follow Flowchart 78.1.
  - **Mechanisms of reactivation or latent infection:** After childhood infection, viruses remain latent in ganglia (HSV-1 in trigeminal and HSV-2 in sacral) for many years and reactivated during adult stage and produce latent infection (recurrence). Many triggers like physical (irradiation/sunlight), infective (pneumococcal, meningococcal, etc.), pathological (fever), physiological (menstruation) and psychological (stress), can provoke a recurrence (reactivation of latent infection). Primary infection is wide spread, while latent infection is localized.

Clinical features

1. **Cutaneous infections:** Herpes lesion is thin-walled, umbilicated-vesicle, which may break down to form ulcer and heals without scar. It presents mostly on cheeks, chin, around mouth and on the forehead.

Flowchart 78.1: Pathogenesis of primary infection of HHV-1 and HHV-2

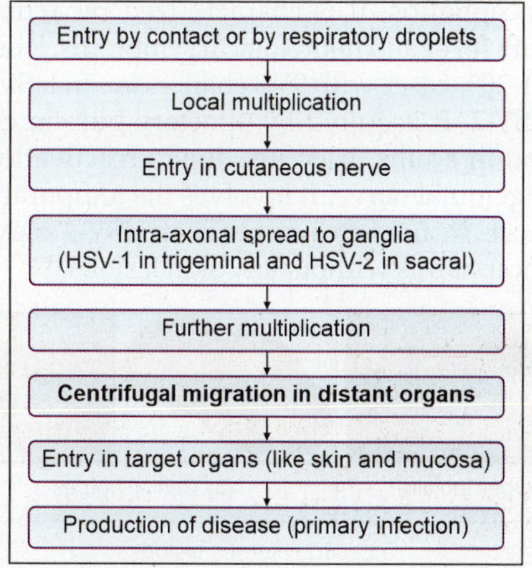

Other cutaneous varieties of herpes are mentioned below.

- **Napkin rash:** It occurs in infant on the buttock.
- **Fever blister or herpes labialis (Fig. 78.2a) or cold sore:** Fever by any cause reactivates the virus and blister formation called fever blister or herpes labialis.
- **Eczema herpeticum:** It is the generalized eruption in children with eczema by herpes.
- **Herpetic whitlow:** It is the occupational herpes on fingers or hands in doctors or nurses characterized by vesicle with ulcer.
- **Herpes gladiotorum:** It is the mucocutaneous lesion presents on the body of wrestlers.
- **Kaposi's varicelliform eruption:** Vaccinia virus also produces similar lesion, which clinically indistinguishable from herpes. Both are collectively called Kaposi's varicelliform eruption.

2. **Oral cavity infections**
   - Primary infection: It is common in buccal mucosa and presents as an acute gingivostomatitis, pharyngitis, tonsilitis and ulcerative stomatitis.
   - Recurrent infection: Herpes labialis occurs as recurrent infection and presents with painful vesicle near lips.
3. **GIT features:** These are esophagitis, hepatitis, rectal and perianal lesions (in homosexual).
4. **Respiratory features:** These are pharyngitis, tracheobronchitis and pneumonitis.
5. **Ocular features:** They are more by HSV-1 and rare by HSV-2 in adults via oculogenital contact.
   - Conjunctival features: These are follicular conjunctivitis with vesicles on lids and face with lymphadenopathy.
   - Corneal features: These are acute keratoconjunctivitis and dendritic ulcer which may heal with scar or with corneal blindness.
   - Uveal tract and retinal features: These are non-granulomatous iridocyclitis, choreoretinitis and acute necrotizing retinitis.
6. **CNS infections**
   - Encephalitis: It is characterized by acute onset with fever and neurological symptoms. It occurs in 10–20% cases with 95% chances are in HSV-1 than HSV-2. It acquired by olfactory bulb in children and in adults. It occurs due to reactivation from trigeminal nerve. It involves the temporal lobe at most. In neonates sometimes HSV-2 may cause encephalitis without any skin lesions.

- Meningitis: It is self-limiting disease and healed in weeks without sequelae. Recurrent condition called **Mollaret's meningitis.**
7. **Ear infections:** It is similar to skin lesion.
8. **Urogenital infections**
   - Primary infection/genital ulcers **(Fig. 78.2b):** Multiple, sever painful, vesiculoulcerative (bilateral and widely spaced) lesions present on external genitalia, while latent infection is mild. Ulcer may present with fever, headache, malaise, myalgia, itching, dysuria, vaginal and urethral discharge.
   - Inguinal lymphadenopathy: Enlarged, tender, firm nodes are present, often bilaterally.
   - Other urogenital infections: Urethritis, vulvovaginitis, cervicitis, endometritis and salpingitis, proctitis and perianal infections occurs following rectal intercourse.
   - Recurrent infection: It is usually milder and recovers faster than primary genital herpes.
9. **Neonatal (congenital) herpes:** It is transmitted *in utero*, during birth or after birth. Risk is more with HSV-2 (75% of total cases) than HSV-1 and with primary infection than recurrent infection in mother. **Localized lesions** occur over skin, eyes or mouth. **Generalized lesions occur** in multiple organs including encephalitis with high mortality and survival with neurological impairment.
10. **Visceral infections:** Disseminated infections occur in patients with IDDs, malnutrition and burns.
11. **Oncogenic (not proved):** HSV-1 associated with carcinoma of lips and HSV-2 associated with carcinoma of cervix.

Complications
- Skin vesicle gets ulcerated and can predispose to secondary bacterial infection. Sacral lesion (autonomous nerve system involvement).
- CNS features may produce transverse myelitis, Guillain-Barre syndrome (GBS) and Bell's palsy (peripheral nerve system involvement).

Laboratory Diagnosis
Specimens: Vesicle fluid, scraping from lesion, saliva, corneal scraping, CSF or brain biopsy.

Testing methods
A. **Microscopy:** Follow morphology.
B. **Culture:** Follow C/Cs.
C. **Serological tests**
   - **Antigen detection test:** Ag is detected by direct IF test. It is more sensitive, specific and can differentiate between HSV-1 and 2.
   - **Antibody detection test:** High Ab titer presents in primary infection and low in latent infection. Abs synthesized after 4–7 days and peak in 2–4 weeks. 1st Ab appears is IgM replaced by IgG, which lasts for life.

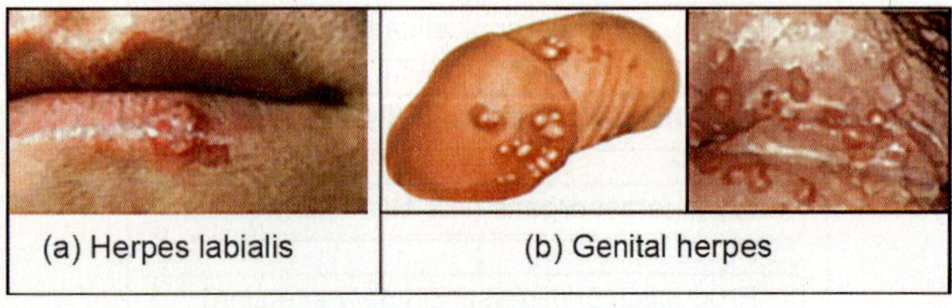

| (a) Herpes labialis | (b) Genital herpes |

Fig. 78.2: Herpes lesions

- ELISA: Specific antibodies detection by ELISA against glycoprotein G antigens like Ig1 and Ig2 can differentiate between HSV-1 and 2.
- WB test: It has 98% sensitivity and high specificity.
- Other tests are: Indirect IF test, CFT or neutralization based test.

D. Molecular methods

- PCR, real-time PCR or DNA hybridization is the useful methods for diagnosis and for differentiation between HSV-1 and 2.
- BioFire Film Array is an automated multiplex PCR which targets around multiple pathogens causing meningitis or encephalitis including HSV-1 and 2.

Prevention

No effective vaccine is developed. Recombinant HSV-2 glycoprotein vaccine is under trial. It is prevented by avoiding contact with patients and also by safe sex. Neonatal infections are prevented by giving acyclovir to mother during 3rd trimester or delivery.

Treatment

- Idoxuridine: Used topically for eye infection.
- Acyclovir and vidarabin: Drugs of choice for deep and systemic infection. Early treatment with IV acyclovir improves the outcome of encephalitis.
- Valacyclovir and famciclovir are effective orally.
- When resistance develops to these drugs, foscarnet may be useful.

HHV-3 (VZV)

History

Weller and Stoddard isolated the virus from both, chickenpox and zoster lesions, but they were not cleared that both lesions were due to same virus. This confusion was cleared by Von Bokay in 1888 and he noticed that both lesions were due to same virus. Differences between two lesions were defined by Heberden in late 18th century.

Morphology (Fig. 78.1a)

Resistance

} Described above

Culture Characteristics (C/Cs)

No growth occurs in egg culture and animal culture. It grows in **tissue culture.** Commonly used cells are Hela cells, Vero cells, human embryonic cell culture, human amnion and human fibroblasts. It produces rounding or ballooning of infected cells.

Antigenic Structure

Only 1 antigenic type is identified, and it can be differentiated from HSV-1 and 2 by specific antisera.

Immunity

Single attack confers long lasting immunity.

Clinical Types

It produces two types of disease (Suggested by von Bokay in 1889) like **primary infection** called **varicella or chickenpox** and **latent infection** called **zoster or herpes zoster or zona or shingles** hence virus **called varicella-zoster virus (VZV).**

Varicella (Chickenpox)

Pathogenicity

- **Epidemiology:** It is distributed worldwide. There are five clades (genotypes) of VZV as follows.
  - Clade 1 and 3: European/North American strain
  - Clade 2: Asian strain, especially from Japan
  - Clade 4: Strain from Europe, but its geographic origin needs further clarification
  - Clade 5: Appears in India.
- **Reservoirs of infection:** Humans.
- **Sources of infection:** Skin lesions, buccal mucosa lesions, vesicular fluid and aerosols. Infectivity is maximum in early stage, decreases with advancing of lesion and scabs are generally noninfectious.
- **Modes of transmission:** It is transmitted by droplets/aerosols, by close contact or vertically from mother to child.
- **Incubation period:** 14–21 days.
- **Portal of entry:** Skin or mucosa (conjunctiva) or respiratory tract.
- **Sites:** Mainly on trunk and spread to face, arms and legs.
- **Precipitating factors (epidemiological determinants):** (1) **Age:** It is common in children, but mild, while in adults and in pregnancy, it is rare, but more severe (2) **IDDs:** It is more common in person with IDDs like AIDS.
- **Pathogenesis:** Follow **Flowchart 78.2.**

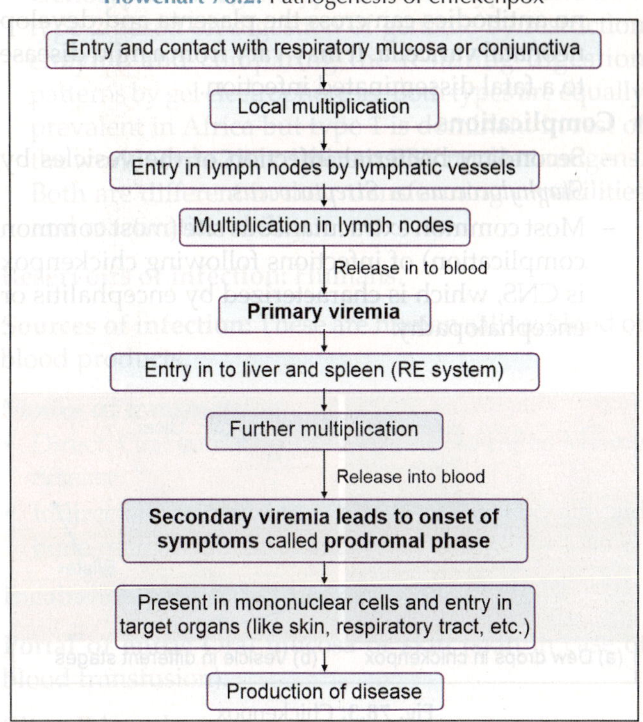

Flowchart 78.2: Pathogenesis of chickenpox

Infections of Herpesviridae

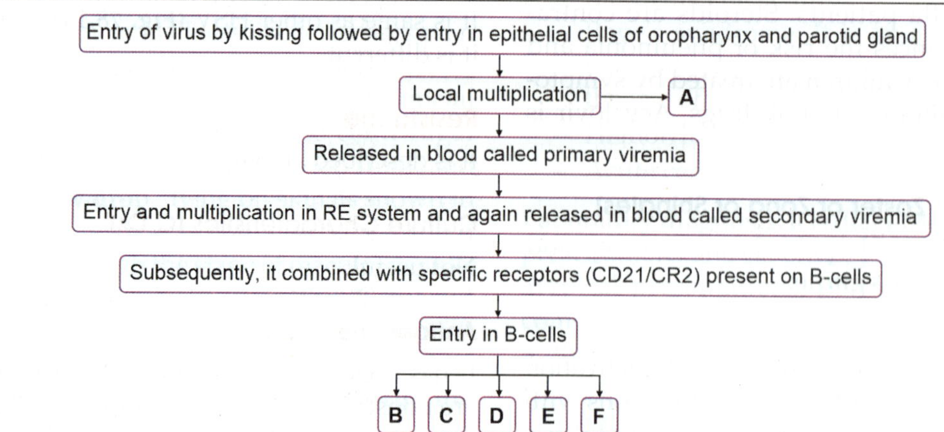

A. Shedding and malignancies: Virus sheds in saliva for months or long time and produces **nasopharyngeal carcinoma**.

B. Cell lysis: Lysis of B-cells and release of progeny viruses.

C. Long time survival in B-cells: It transforms (immortalizes) the B-cells.

D. Reactivation: In **immunocompetent persons** virus checked by T-cell and no reactivation. In **immunodeficient persons** EBV produces following types of malignancies.

1. **In patients with malaria:** Hyperendemic malaria prevalent in **Africa** is impairing the immune system of children and coinfection with EBV produces the **Burkitt's lymphoma**. Memory B cells are the reservoirs of EBV and Burkitt's lymphoma is the tumor of B cells. Immunosuppressive effect of malarial parasite allows the viral proliferation and interference with immune mechanisms.

2. **In patients with AIDS:** Reactivation of virus in persons with AIDS produces the **lymphomas (Hodgkin's and non-Hodgkin's lymphoma)**. EBV-DNA regularly found in tumor cells.

3. **In patient with AIDS and organ transplantation: Oral hairy leukoplakia** characterized by white patch on the side of the tongue **(Fig. 78.4)** with a corrugated or hairy appearance. This white lesion cannot be scraped off; it is benign and does not require any treatment, although its appearance may have diagnostic and prognostic implications for the underlying condition like AIDS. The white appearance is created by hyperkeratosis (overproduction of keratin) and epithelial hyperplasia.

E. Polyclonal activation of B-cells (autoimmunization): It produces heterophile-Abs which are detected by **Paul-Bunnel test**.

F. Infectious mononucleosis/glandular fever/kissing disease: Formation of Ag (neoantigen) on B-cell surface, which induces blast-transformation in T-cell. Multiple atypical T-cells (lymphoblasts) are examined in peripheral smear called infectious mononucleosis. Usually, it is a self-limited disease and resolved in 2–4 weeks. It consists of fever, sore throat (most common symptom), lymphadenopathy (most common sign), splenomegaly and mild transient rash. Jaundice may be seen due to hepatitis. Oral lesions include petechiae seen at the junction of soft and hard palate, sometimes presence of white pseudomembrane on tonsils or other parts of oral cavity and oral ulcers. Presence of >20% atypical lymphocytes (lymphocytosis not monocytosis) in the blood. Atypical lymphocytes are characterized by round or irregular nuclei with abundant flowing cytoplasm taking dark staining at periphery.

Note: Other causes of infectious mononucleosis
- **Infectious:** CMV, *Toxoplasma gondii* and HHV-6 (variant - B).
- **Non-infectious:** Nonspecific stimuli.

Precipitating factors (epidemiological determinants): **(1) Age:** EBV is common in infants and children of developing countries and usually asymptomatic. EBV is common in adolescence and in early adulthood patients of affluent countries and develops IM. **(2) IDDs:** Patients with AIDS or malaria or organ transplantation are at more risk. **(3) Genetic factors:** EBV infection is associated with X-linked lymphoproliferative (XLP or Duncan syndrome) disease. **(4) Environmental and racial factor:** Hyperendemic malaria prevalent in Africa is impairing the immune system of children and coinfection with EBV produces Burkitt's lymphoma. Nasopharyngeal carcinoma is common in men of Chinese origin. **(5) Coinfection:** EBV infection is also associated with CMV infection.

Pathogenesis and clinical features: Follow **Flowchart 78.3** and **Fig. 78.4**.

Complications: Most common complications are meningitis/meningoencephalitis. **Others** are splenic rupture, cardiac, pulmonary and hematologic (auto-immune hemolytic anemia may be due to formation of cold agglutinin) complications.

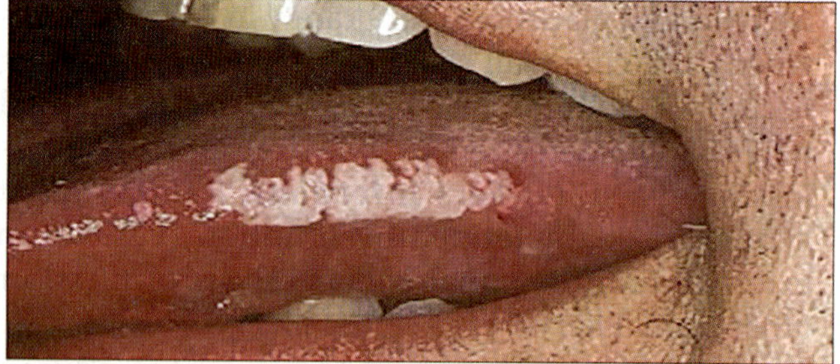

Fig. 78.4: Oral hairy leukoplakia

Laboratory Diagnosis

Specimens: Blood, saliva, etc.

Testing methods

A. Microscopy: Follow morphology.

B. Blood picture: Initially leukopenia occurs due to reduction in polymorph count, but later leucocytosis due to appearance of abnormal mononuclear cells (atypical lymphocytes/blast T cells). More than 20% lymphocytes are atypical in peripheral blood smear.

C. Cell culture: Not useful.

D. Serological tests: Best method.

1. **Tests for detection of Ags:** Like viral capsid-Ag is detected by ELISA, CFT and gel diffusion method.
2. **Tests for detection of heterophile (nonspecific) Abs:** This is Paul Bunnell test including differential absorption test and monospot or mononuclear spot test.

Paul Bunnell Test

History: It was developed by John R Paul (American physician, 1893–1971) and Walls W Bunnell (American physician, 1902–1966).

Principle: It is based on detection of heterophile antibodies by agglutination reaction. Heterophile antibodies are agglutinated with RBCs of sheep, horse and cow/ox but not with guinea pig. The Paul-Bunnell test uses sheep RBCs.

Antigen(s): Sheep RBCs.

Antibody (Antibodies): Patient serum contains heterophile antibodies. In infectious mononucleosis, IgM heterophile antibodies appear early in 1st week in 40% cases and in 3rd week in 80–90% cases and remain detectable for the first 3 months of infection and then disappear. Heterophile antibodies are also present in normal individual or after serum therapy (injection of serum) and make the confusion in diagnosis of infectious mononucleosis, which are differentiated by absorption test in which heterophile antibodies of IgM are not absorbed by guinea pig kidney cells but absorbed by ox RBCs.

Method: Inactivate the patient serum by heating at 56°C for 30 minutes. Do the serial dilution of patient serum in tube and mix the equal volume of 1% sheep RBCs. Incubate the tube at 37°C for 4 hours.

Results: Positive test is indicated by agglutination of RBCs. Test is negative if there is no agglutination.

Interpretation

- Positive test: It is positive in infectious mononucleosis in children and young adults. Agglutination titer of >256 is considered significant.
- Negative test: Test is negative in children <5 years, elderly and in patients without symptoms of infectious mononucleosis.

- False-positive test: It is due to presence of heterophile antibodies in normal individual or after serum therapy, which are differentiated by differential absorption test as follow.

Differential Absorption Test

Principle (Table 78.3): Differential absorption test based on absorption of heterophile antibodies of normal individual and due to serum therapy by guinea-pig kidney cells, but not by ox RBCs, while heterophile antibodies of infectious mononucleosis are not absorbed by guinea-pig kidney cells but absorbed by ox RBCs.

Results: Follow **Table 78.3**.

Use: For diagnosis of heterophile antibodies.

Monospot or Mononuclear Spot Test

Principle: This rapid test based on detection of heterophile Ab by slide agglutination reaction by using horse RBCs. Test serum is initially treated with guinea-pig kidney cells and ox red blood cells.

Advantages: It is rapid test, simple to perform, high sensitivity about 75%, high specificity (90%), and it replaced differential absorption test.

Disadvantages: False-positive result is seen in connective tissue disease, lymphoma, viral hepatitis and malaria.

3. **Tests for detection of specific EBV antibodies:** ELISA and indirect IF tests are used to detect following types of specific EBV antibodies.
   - Ab to viral capsid-Ag: It is IgM (recent infection) and IgG types (past infection).
   - Ab to nuclear-Ag: It suggests past infection, but fourfold rises in titer suggests the current infection.
   - Ab to early-Ag (called EA-D Ab): Elevated Ab to early Ag in acute infection and Burkitt's lymphoma is distributed in nucleus and cytoplasm of infected cells. Elevated Ab to early Ag in nasopharyngeal carcinoma is restricted to the cytoplasm of infected cells.

E. Molecular methods: Real-time PCR is useful for quantitative detection of viral gene.

TABLE 78.3: Differential absorption test

Features	Paul Bunnell test (performed 1st)	Differential absorption test		Paul Bunnell test (Repeat)
		Serum treated with guinea pig kidney cells	Serum treated with ox RBCs	
Infectious mono-nucleosis	Positive	No absorption	Absorption	Positive = Pretreatment with guinea pig kidney cells Negative = Pretreatment with ox RBCs
Serum therapy	Positive	Absorption	No Absorption	Negative for both sera
Normal serum	Positive	Absorption	No absorption	Positive = Pretreatment with ox RBCs Negative = Pretreatment with guinea pig kidney cells

Prevention and Treatment

No specific vaccine and treatment are available. A vaccine contains EBV glycoprotein is tried but found ineffective. Acyclovir is effective.

HHV-5 (CMV)

Synonym

Previously it called salivary gland virus.

Meaning and History

Word derived from Cyto (Greek) means cell and Megalic (Greek) means large because infected cell gets enlarged. Name was given by Weller et al in 1960. They subsequently isolated CMV from the urine of infants with generalized disease. Inclusion bodies and infected cells were first shown by Ribbert in 1881. Goodpasture and Talbert in 1921 were the first to suggest the "cytomegalic" nature of infected cells. The introduction of exfoliative cytology methods allowed identification of characteristic cells in the urine of infected infants. Smith in 1956, Rowe et al in 1956, and Weller et al in 1957 independently isolated human CMV strains. In 1960, Weller and coworkers proposed the term "Cytomegalo virus" and

Morphology

It is largest among the all herpes viruses about 150–200 nm. It produces prominent intranuclear Cowdry type-B inclusion body called **"owl's eye" (Fig. 78.5)**. Other morphology is same as like HSV and VZV.

Culture Characteristics (C/Cs)

CPE developed in human fibroblasts after 50 days. CPE includes inclusion body formation or multinucleated giant cell formation. Growth is identified by IF microscopy. Shell vial technique can be useful for early growth detection in 1–2 days.

Resistance

It is described above.

Antigenic Structure

Human CMV is unrelated antigenically to animal CMV except simian CMV with which it shows the minor antigenic cross reactivity. Human isolates may carry minor genetic and antigenic differences, but clinically not significant.

Immunity

Antibodies are formed, but not protective.

Pathogenicity

Epidemiology: Virus produces the periodic reactivation, latent infection and lifelong persistence. It is distributed in USA and also in some developing countries. Up to 80% asymptomatic persons containing the antibodies show the high prevalence.

Reservoirs of infection: Humans.

Sources of infection: These are saliva, blood, urine, milk, semen and cervical secretion. Congenitally infected children had high viruria for 4–5 years and they are highly infectious in early infancy.

Modes of transmission

- Direct: (1) Close contact: By body fluid and during coitus. (2) Vertical: By placenta, by contact with maternal body fluid during delivery and by breast feeding.
- Indirect: It occurs by blood, blood products (post-transfusion mononucleosis) and organ transplantation.

Portal of entry: Skin or mucosa.

Sites: Respiratory system and fetal organs.

Precipitating factors (epidemiological determinants): **(1) Age and IDDs:** It is asymptomatic in normal children and immunocompetent persons. Disseminated lesions occur in neonates and persons with IDDs. It is the most common pathogen that causes infection following organ transplantation. **(2) Socioeconomical status:** It is more in persons with low socioeconomical status and poor hygiene. About 90% patients are sero-positive in developing countries while 40–70% people are sero-positive in developed countries.

Pathogenesis: Glycoprotein receptors (spikes or peplomere) present on viral surface combine with Fc piece of irrelevant Ig, resulting masking of virus and prevent access to specific anti-CMV Ab. Once infected, the person carries the virus for life, which may be activated from time to time. It multiplies in epithelium of respiratory system, salivary glands, kidneys and cervix and sheds in the urine, saliva, tears, semen, breast milk and cervical secretion.

Clinical features

- **Primary infection:** It is asymptomatic in older children and adults with prolonged latency and occasional reactivation. Reactivation can also lead to the vertical transmission. It is common following transfusion called **post transfusion mononucleosis.** Primary infection sometimes presents with **infectious mononucleosis like syndrome** and characterized by

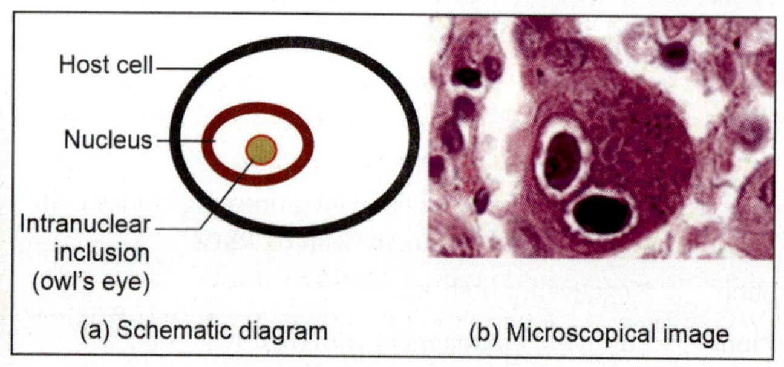

Fig. 78.5: Owl's eye

malaise, myalgia, fever, liver function abnormalities and lymphocytosis **without heterophile Ab** in serum. Primary infection in persons with IDDs is more severe and presents with pneumonia, hepatitis or generalized disease or even fatal infection.

- **Vertical infection: (1) Antenatal (congenital):** Transmission by placenta called **cytomegalic inclusion disease**. It is the most common virus causing intrauterine infection. It is associated with hepatosplenomegaly (most features), jaundice, thrombocytopenic purpura, hemolytic anemia, microcephaly, choreoretinitis, cerebral calcification resembling toxoplasmosis, often death or survival shows mental retardation. **(2) Intranatal infection:** Usually asymptomatic. **(3) Postnatal infection:** Usually asymptomatic.
- **Oncogenic lesions:** These are carcinoma of prostate and Kaposi's sarcoma.
- **Ocular infections in AIDS patients:** Most common ocular features by CMV in AIDS patients are hemorrhagic retinitis and granular retinitis, when $CD4^+$ count falls $<50/mm^3$. These are usually bilateral and present with retinal exudates and perivascular hemorrhage. They are complicated as retinal detachment, retinal atrophy or optic nerve disease.
- **Infections following organ transplantation:** CMV is the most common pathogen causing infection following organ transplantation and presents with fever, leukopenia, hepatitis, pneumonitis, esophagitis, gastritis, colitis, retinitis and meningoencephalitis.

Laboratory Diagnosis

Specimens: Urine, saliva, tears, semen, breast milk and cervical secretion and blood (buffy coat).

Testing methods

A. **Microscopy:** Follow morphology.

B. **Cell culture:** Follow C/Cs.

C. **Serological tests:** (1) Ag detection test: CMV specific pp65 (phosphoprotein 65) Ag is detected by indirect IF test but preferred very rarely, because it is time consuming and technically more demanding. (2) Ab detection test: ELISA is useful to detect the Abs against different Ags of CMV like pp150, pp65, gpgB (glycoprotein gB), gpgH and major DNA binding protein. 1st Ab appears is IgM (recent infection) in 1–2 weeks after infection. IgM detection in pregnancy helps in diagnosis of congenital infection. IgG (past infection) appears few days after IgM and persists for life. It is detected by IgG avidity test to differentiate the recent and past infection.

D. **Molecular methods:** Real time PCR is useful for quantitative detection of viral gene.

Prevention

General measures: These are screening and matching the CMV status of the donor and recipient and use of CMV negative blood for transfusions.

Chemoprophylaxis: Acyclovir and ganciclovir are used for prophylaxis.

Immunoprophylaxis

- Active immunization: Live attenuated vaccine (Towne 125 and AD 169 strain) and purified CMV polypeptide vaccine are developed, but not effective.
- Passive immunization: CMV Ig is used for prevention of congenital infection and into sero-negative recipients prior to transplant.

Treatment

Ganciclovir and foscarnet are effective in persons with IDDs.

HHV-6 (HBLV)

It was 1st isolated in 1986 from blood of patient with lymphoproliferative disease. It has two variants like A and B. Variant-B is ubiquitous, spread by saliva and produces **pediatric disease** called sixth disease or roseola infantum or exanthema subitum, which is characterized by high-grade fever and skin rashes. **In adults** it causes infectious mononucleosis like syndrome and focal encephalitis. **In persons with IDDs,** it causes pneumonia and disseminated diseases.

HHV-7 (RKV)

It called R K virus but this name is no longer in use. It shares 30–50% DNA homology with HHV-6. It was 1st isolated in 1990 from CD4 cells of healthy person. It is transmitted by saliva and produces disease called **exanthema subitum** mainly in children characterized by fever, seizure, ptyriasis rosea, respiratory and GIT features. Virus combines with specific receptors present on CD4 cells and causes further depletion of CD4 cells in HIV patients.

HHV-8 (KSHV)

History: It was 1st isolated in 1994 from tissue of Kaposi's sarcoma from HIV patient; however, its causal relationship with Kaposi's sarcoma is still to prove hence called **Kaposi's sarcoma associated Herpes Virus (KSHV).** It also identified from Kaposi's sarcoma without HIV.

Pathogenicity: It is highly prevalent in Africa (endemic) and transmitted by oral secretion. Low prevalence occurs in Asia, North Europe and North America in homosexual men. It also transmitted by blood, blood products, contaminated needles and syringes in drug abuse and by organ transplantation. **In immunocompetent** person, it causes fever and rash. **In person with organ transplantation and IDDs like HIV/AIDS,** it causes Kaposi's sarcoma, primary effusion lymphoma (body cavity lymphomas) and Castleman's disease (lymphoproliferative disease of B cells).

Diagnosis: Saliva, blood, lymphoid tissues, etc., are tested by serological tests like ELISA, indirect IF test, WB test, etc., and molecular method like PCR.

Note: Kaposi's sarcoma

It is the most common cancer of AIDS patient. It is a cancer of lymphatic endothelium and forms vascular channels that fill with blood cells, giving the tumor its characteristic bruise-like appearance. The most typical feature of Kaposi sarcoma is the presence of spindle cells forming slits containing RBCs. The tumor is highly vascular, containing abnormally dense and irregular blood vessels, which leak RBCs into the surrounding tissue and give dark color to tumor. Inflammation around the tumor may produce swelling and pain. The spindle cells of Kaposi sarcoma differentiate toward endothelial cells, probably of lymph vessels rather than blood vessels nature.

Note: Exanthema (exanthem) versus enanthema (exanthem)

Exanthema (exanthem): These are the skin eruptions or skin rashes. Sometimes it is considered as exanthematous fever also. Following are the causative agents.

- **Old list of 6 organisms: (1) 1st disease:** Measles (rubeola) caused by measles virus. **(2) 2nd disease:** Scarlet fever caused by *Strept pyogenes*. **(3) 3rd disease:** German measles (rubella) caused by rubella virus. **(4) 4th disease:** Caused by *Staph. aureus*. It was described in 1900, but its existence is not widely accepted today. **(5) 5th disease:** Erythema infectiosum or slapped cheek disease caused by parvo virus B19. **(6) 6th disease:** HHV-6 (roseola infantum or exanthema subitum) and HHV-7 (exanthema subitum).
- **Other organisms:** In addition to above 6 diseases few other viral diseases are considered as exanthematous diseases like varicella zoster (chickenpox and zona caused by VZV), mumps (caused by mumps virus), disease caused by *Rhinovirus*, Rocky mounted spotted fever (caused by *R. rickettsia*), tick-borne diseases and viral hemorrhagic fever.

Enanthema (enanthem): These are mucosal eruptions. It occurs with viral exanthema. Common enanthematous conditions are smallpox (caused by variola virus), measles/rubeola (caused by measles virus), chickenpox (caused by VZV), roseola infantum (caused by HHV-6) and hypersensitivity.

CEROPETHACINE HERPES VIRUS-1

It also called herpes virus simiae or B Virus. From initial B of the name of patient virus called **B virus.** It was 1st isolated in 1934 by Sabin and Wright from brain of laboratory worker who developed fatal ascending myelitis after being bitten by healthy monkey. It is **transmitted by** monkey bite or by contact with monkey's tissue infected with virus. It **produces vesicle** in buccal mucosa which gets ulcerated and sheds the virus to infect the contact. Disease is fatal (60% mortality) and survival shows the neurological sequelae. Antigenically, it is same as HHV but HHV-Ab does not neutralize the B virus. Formolized vaccine is tried for laboratory worker who are at risk.

Case Studies

1. An 8-year-old child is brought to the hospital with history of fever and vesicle on the lips. Mother had given the history of same illness to 2–3 colleagues of his school. Scraping is done from the vesicular lesion for Tzanck test showing Tzanck cells under microscopy. Identify the case and answer the following.
   a. Name the clinical condition and causative agent.
   b. Write pathogenicity of clinical condition.
   c. Write the diagnosis of clinical condition.
2. A 20-year-old male visited the medical OPD with history of fever and sore throat since 5 days. Blood is collected for cell count which reported the lymphocytosis with atypical lymphocytes in peripheral blood. Identify the case and answer the following:
   a. Name the clinical condition and causative agent.
   b. Write pathogenicity of clinical condition.
   c. Write the diagnosis of clinical condition.
3. A 40-year-old man with kidney transplantation before 2 months visited the medical OPD with history of fever and features suggestive of bilateral diffuse interstitial pneumonia. The laboratory report suggested the ds-DNA virus with owl's eye inclusion body. Identify the case and answer the following.
   a. Name the most common agent in given condition.
   b. Write pathogenicity of clinical condition.
   c. Write the diagnosis of clinical condition.

Essay/Full Question

1. Herpesviridae.

Short Notes

1. HHV-1 and 2 (HSV-1 and 2)/HHV-3 (VZV), HHV-4 (EBV), HHV-5 (CMV).
2. Chicken pox/infectious mononucleosis.
3. Paul Bunnell test.
4. Exanthema versus enanthema.

Short Questions for Theory/Viva Questions

1. Write four differences between HSV-1 and HSV-2.
2. What is Tzanck test?
3. Name the four viruses transmitted by saliva.
4. Name the four cancer causing herpes viruses with cancer produced by them.
5. What are exanthema and enanthema?
6. Write four examples of viral exanthematous fever.
7. Write 1 pathogenic disease produced by each RK and BK viruses.

Comments on

1. EBV is responsible for autoimmunity.

MCQs for Chapter Review

Introduction

1. **Varicella is classified under:**
   a. Entero virus   b. Retro virus
   c. Pox virus   d. Herpes virus
2. **Herpes simplex virus is:**
   a. Single-stranded DNA
   b. Double-stranded DNA
   c. Single-stranded RNA
   d. Double-stranded RNA

HHV-1 and 2 (HSV-1 and 2)

3. Herpes simplex virus causes all, *except*:
   a. Encephalitis
   b. Pharyngitis
   c. IMN (infectious mononucleosis)
   d. Whitlow

4. Encephalitis is caused by:
   a. HSV-1
   b. EBV
   c. Infectious mononucleosis
   d. CMV

5. A neonate develops encephalitis without any skin lesion, most probable causative organism is:
   a. HSV-1
   b. HSV-II
   c. Meningococci
   d. Streptococci

6. Regarding HSV-2 infection:
   a. Primary infection is usually wide spread
   b. Recurrent attacks are due to reactivation of latent infection
   c. Encephalitis can be caused by HSV-2
   d. Newborn may acquire infection via the birth canal at the time of labor
   e. Treatment is with acyclovir

7. Following is true for HSV-1:
   a. Small pock
   b. Poor replication in chick embryo fibroblast culture
   c. Above belt infection
   d. a + b + c

8. True about herpes virus:
   a. HSV encephalopathy is treated with acyclovir
   b. Oropharyngeal involvement is common in HSV-1
   c. Recurrent genital involvement is common in HSV-1
   d. Recurrence is rare in HSV-1

HHV-3 (VZV)

9. Herpes zoster is caused by:
   a. Herpes simplex type I
   b. Herpes simplex type II
   c. Epstein Barr virus
   d. Varicella

10. Infectivity of chicken pox lasts for:
   a. Till the last scab falls off
   b. 6 days after onset of rash
   c. 3 days after onset of rash
   d. Till the fever subsides

11. Rash pattern in chickenpox is:
   a. Centripetal
   b. Centrifugal
   c. Localized
   d. All

12. Most common extra skin manifestation of varicella is involvement of:
   a. CNS
   b. Lungs
   c. Kidneys
   d. CVS

13. Tests for chicken pox:
   a. Fluorescent Ab test
   b. ELISA
   c. Widal
   d. PCR

14. Chickenpox:
   a. Is commonly seen in a congenital form
   b. May be severe in a newborn child infected by the mother in late pregnancy
   c. Affects limbs more that the trunk
   d. May cause pneumonitis
   e. Should be treated with intravenous vidarabine, when there is evidence of hepatitis.

15. Shingles are seen in:
   a. IMN
   b. Herpes zoster
   c. Chickenpox
   d. Smallpox

16. Zoster recurrence occurs after infection with:
   a. HSV 1
   b. HSV 2
   c. Varicella
   d. Small pox

17. Varicella zoster remains latent in:
   a. Lymphocytes
   b. Monocytes
   c. Trigeminal ganglion
   d. Plasma cells

18. Zoster ophthalmicus is due to:
   a. Primary herpes infection of eye
   b. Herpes reactivation in optic nerve
   c. Herpes reactivation in Gasserian gaglion
   d. Herpes infection of eye in immunocompromized patient

19. Treatment of herpes zoster:
   a. Zidovudin
   b. Valcyclovir
   c. Ribavarin
   d. Nevirapine

HHV-4 (EBV)

20. Oral hairy leukoplakia caused by:
   a. Epstein-Barr virus
   b. CMV
   c. HIV
   d. HZV

21. African Butkitt's lymphoma is caused by:
   a. Epstein-Barr virus
   b. CMV
   c. HIV
   d. HZV

22. Epstein-Barr virus causes autoimmunity by:
   a. Molecular mimicry
   b. Inducing inappropriate expression of class II MHC
   c. Release of sequestere antigens
   d. Polyclonal activation of B cells

23. All are associated with EBV *except*:
   a. Infectious mononucleosis
   b. Nasopharyngeal carcinoma
   c. Oral hairy leukoplakia
   d. Epidermodysplasia

24. EB virus causes all *except*:
   a. Infectious mononucleosis
   b. Nasopharyngeal carcinoma
   c. Burkitt's lymphoma
   d. Carcinoma of cervix

25. EB virus has been implicated in the following malignancies *except*:
   a. Hodgkin's disease
   b. Non-Hodgkin's disease
   c. Nasopharyngeal carcinoma
   d. Multiple myeloma

26. True about infectious mononucleosis is:
   a. Associated with heterophile antibodies
   b. Monocytosis
   c. Associated with cold agglutinin
   d. Associated with CMV infection
   e. Self limited disease

27. A patient with sore throat has positive Paul Bunnell test. The causative organism is:
   a. EBV
   b. Herpes virus
   c. Adeno virus
   d. CMV

28. Paul Bunnell antibodies are reactive in all, *except*:
   a. Ox
   b. Sheep
   c. Dog
   d. Horse

29. **Most sensitive test for diagnosis of infectious mononucleosis:**
    a. Monospot test
    b. Paul Bunnell test
    c. Lymphocytosis in peripheral smear
    d. Culture of the virus

30. **Burkitt's virus is:**
    a. EBV
    b. HPV
    c. HIV
    d. HAV

31. **IMN true is all, *except*:**
    a. Caused by EBV
    b. Also called kissing disease
    c. Diagnosed by Paul Bunnel test
    d. RNA virus

HHV-5 (CMV)

32. **The most common presentation of congenital CMV infection is:**
    a. Hepatosplenomegaly
    b. Microcephaly
    c. Cerebral calcification
    d. Chorioretinitis

33. **Which of the following does not establish a diagnosis of congenital CMV infection in a neonate?**
    a. Urine culture of CMV
    b. IgG CMV antibodies in blood
    c. Intranuclear inclusion bodies in hepatocytes
    d. CMV viral DNA in blood by polymerase chain reaction

34. **The most sensitive and rapid test for diagnosis of CMV retinitis is:**
    a. Viral isolation from the intraocular fluid
    b. Nucleic acid detection from the intraocular fluid
    c. Viral antigen detection in vitreous
    d. Viral antibody detection in the blood by ELISA

35. **All are true regarding cytomegalo virus, *except*:**
    a. It is a DNA virus.
    b. Most commonly infected in last trimester
    c. Diagnosed by increased IgA in fetal blood
    d. Most common cause of congenital viral infection

HHV-6 (HBLV)

36. **HHV-6b causes:**
    a. Carcinoma of cervix
    b. Carcinoma of endometrium
    c. Clear cell carcinoma
    d. Focal encephalitis

HHV-8 (KSHV)

37. **Kaposi sarcoma is caused by:**
    a. Human herpes virus-2
    b. Human herpes virus-4
    c. Human herpes virus-6
    d. Human herpes virus-8

38. **HHV-8 causes:**
    a. Burkitt's lymphoma
    b. Nasopharyngeal carcinoma
    c. Kaposi sarcoma
    d. Hepatic carcinoma

39. **The tissue of origin of Kaposi sarcoma is:**
    a. Lymphoid
    b. Vascular
    c. Neural
    d. Muscular

40. **Multifocal tumor of vascular origin in a patient of AIDS:**
    a. Kaposi sarcoma
    b. Astrocytoma
    c. Gastric carcinoma
    d. Primary CNS lymphoma

Mixed MCQs of Herpesviridae

41. **All of the following are true about herpes group virus *except*:**
    a. Ether sensitive
    b. May cause malignancy
    c. HSV-II involves below diaphragm
    d. Burkitt's lymphoma involves T cells

42. **True about herpes virus is:**
    a. HSV-1 causes encephalitis
    b. EBV affects T lymphocytes
    c. CMV is always symptomatic
    d. Herpes zoster is not reactivated

Answers and Explanation of MCQs

1. d
- Varicella is actually a varicella zoster virus that belongs to family Herpesviridae. Follow section **introduction → classification** for more explanation.

2. b
- Follow section, **introduction → morphology (genome and capsid)** for more explanation.

3. c
- Follow **Flowchart 78.3** for causes of IMN (infectious mononucleosis).
- Follow section, **HHV-1 and 2 (HSV-1 and 2 → pathogenicity → clinical features)** for pathogenic lesions caused by herpes simplex virus.

4. a, Follow section, **HHV-1 and 2 (HSV-1and2)**

5. b

6. a, b, c, d, e

7. d
- Follow **pathogenicity → clinical features and Table 78.2** for explanation of answers of MCQs 5–7.

8. d
- Follow section, **HHV-1 and 2 (HSV-1 and 2) → pathogenicity and treatment) and Table 78.2** for more explanation.

9. d

10. b

11. a

12. a

13. a, b, d

14. b, d

15. b

16. c

17. c

18. c

19. b
- Follow section, **HHV-3 (VZV)** for explanation of answers of MCQs **9–19**.

20. a

21. a

22. d

23. d

24. d

25. d

26. a, d, e

27. a

28. c

29. a

30. a

31. d

- Follow section, **HHV-4 (EBV) and Flowchart 78.3** for explanation of answers of MCQs **20–31**.

32. a

33. b

34. b

35. c

- Follow section, **HHV-5 (CMV)** for explanation of answers of MCQs **32–35**.

36. d

- Follow section, **HHV-6 (HBLV)** for more explanation.

37. d

38. c

39. b

40. a

- Follow section, **HHV-8 (KSHV)** for explanation of answers of MCQs **37–40**.

41. c and d

- Herpes virus is an enveloped virus, which is made up by lipoprotein. This envelope is sensitive to lipid solvents like ether, chloroform, etc. Herpes viruses like HHV-1, 2, 4, 5 and 8 are causing malignancy. HSV-II infects below the west not below the diaphragm. Burkitt's lymphoma involves B cells not the T cells.

42. a

- EBV affects B lymphocytes. CMV is asymptomatic in older children and adults with prolonged latency and occasional reactivation. Herpes zoster is reactivated from trigeminal gaglion.

CHAPTER 79

Infections of Other DNA Viruses (Hepadnaviridae, Adenoviridae, Papovaviridae and Parvoviridae)

Chapter Outline

- Hepadnaviridae
- Adenoviridae
- Papovaviridae
- Parvoviridae

HEPADNAVIRIDAE

Virus is responsible for hepatitis in human and it is better to discuss it with other hepatitis viruses. Follow **Ch. 88** for more details.

ADENOVIRIDAE

Meaning

Name is given because; it was grown and identified from the adenoid tissue.

Use

It carries inserted DNA up to 7 kb and acts as vector in gene therapy.

History

It was 1st identified by Rowe and colleagues in 1953 by growing the surgically removed adenoid tissue on plasma clot culture.

Classification and Antigenic Structure

Family: Adenoviridae.

Genera: All adenoviruses share common complement fixing antigens. It has two genera like *Aviadenovirus* which infects the birds and *Mastadenovirus* which infects the mammals. Human species of *Mastadenovirus* are divided into six **subgenera or subgroups** (A to F) based on certain properties like hemagglutination, fiber length, DNA analysis and oncogenic potential as shown in **Table 79.1**. Group specific Ag presents on **hexons**, can be detected by ELISA or by IF test. Each group is divided in many **serotypes** by type specific Ags present on **pentons and fibres**. More than 50 serotypes are identified by neutralization test (targeting capsid Ags) and by hemagglutination, test (targeting hemagglutinin Ags) as shown in **Table 79.1**.

TABLE 79.1: Subgroup and serotypes of human adeno viruses

Sub-genus	Serotype	Hemagglutination	
		Red cells	**Pattern**
A	12,18,31	Rat	Partial
B	3, 7, 11, 14, 16, 21, 34, 35	Monkey	Complete
C	1, 2, 5, 6	Rat	Partial
D	8, 9, 10, 13, 15, 17, 19, 20, 22–30, 32, 33, 36-39, 42–47	Rat	Complete
E	4	Rat	Partial
F	40,41	Rat	Partial

Morphology

Shape and size: It appears like space vehicle (satellite) as shown in **Fig. 79.1**. It is nonenveloped (naked virus) about 70–75 nm in size.

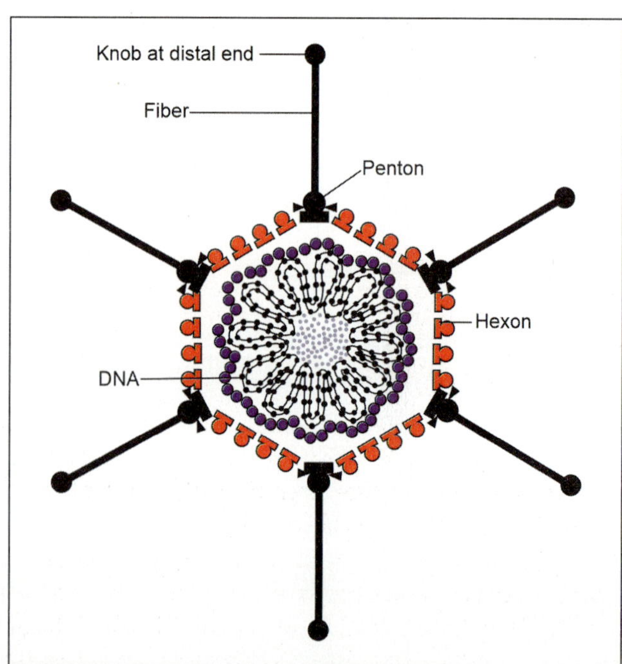

Fig. 79.1: Morphology of adeno virus

Genome and capsid: It contains ds-DNA surrounded by icosahedral variety of capsid.

Capsomeres: They are total 252 in numbers, arranged with 20 triangular facets and 12 vertices. 240 capsomeres have six neighbors called **hexons** while 12 capsomeres have five neighbors called **pentons**. Penton consists penton base anchored in capsid and rod like projection or fiber with knob at distal end.

Envelope: It is nonenveloped (naked) virus.

Electron microscopy (EM): It is used for diagnosis of serotype 40 and 41 from feces.

Culture Characteristics (C/Cs)

Tissue culture: It grows only in A549 cell line, Hela cells, HEP-2 and human embryonic kidney cells. Growth is identified by intranuclear inclusion bodies, rounding of cells or aggregation in grapes like clusters. Viral antigen from culture is detected by direct IF test.

Animal culture: Animals are not susceptible.

Resistance

It is inactivated at 50°C and nonenveloped so resist to ether and bile salts.

Pathogenicity

Epidemiology: It is distributed worldwide. One serotype can produce many diseases and same disease can be produced by many serotypes. Different serotypes are producing endemic, epidemic or outbreak.

Reservoirs of infection: Humans.

Sources of infection: Respiratory droplets, feces, eye discharge, urine, etc.

Modes of transmission: It is transmitted by droplets in early phase of disease, close contact or by feco-oral route.

Portal of entry: Respiratory tract, GIT, conjunctiva, etc.

Sites: Respiratory tract, eyes, bladder, intestine, etc.

Clinical features: Follow **Table 79.2**.

Laboratory Diagnosis

Specimens: Serum, nasopharyngeal/throat samples, eye sample like conjunctival scraping, stool and urine are collected.

Testing methods:

A. **Microscopy:** Follow morphology.

B. **Culture:** Follow C/Cs.

C. **Serological test:** **(1) ELISA** is used for diagnosis of serotype 40 and 41 from feces. **(2) Direct IF test** is used to detect the Ags from nasopharyngeal/throat samples or conjunctival scraping.

D. **Molecular method:** **Real-time PCR** is useful for quantitative detection of viral gene. **BioFire Film Array** is an automated multiplex PCR, which targets around multiple pathogens causing respiratory and GIT lesions including adeno viruses.

TABLE 79.2: Clinical features of adeno viruses	
Features	**Serotypes**
Respiratory	
Pharyngitis	1-7
Pneumonia (may be atypical)	3, 7
Acute respiratory distress (ARD) in military person	4, 7
Eyes	
Shipyard eye/epidemic keratoconjunctivitis (conjunctivitis, keratitis and lymphadenopathy in adults)	8, 19, 37
Acute follicular (swimming pool) conjunctivitis	3, 4, 7, 11
Respiratory and eyes	
Pharyngoconjunctival fever (pharyngitis, fever, conjunctivitis and lymphadenopathy in adults)	3, 7, 14
GIT	
Diarrhea (in children)	40, 41
Others	
Acute hemorrhagic cystitis	11, 21
Generalized exanthema, mesenteric adenitis, intussusception and cancer (not for human)	
In transplant recipient	
Pneumonia, hepatitis, nephritis, colitis, encephalitis and hemorrhagic cystitis	34, 35
Tumor production (noticed by Huebner in 1962)	
Sarcoma in baby hamster	12, 18
Tumors in animals	12, 18 and 31
Transformation of cell culture	All types
Human cancer	Not produced

Prevention

General measures: These are frequent hand washing, use of paper towels than clothes which easily get dry, chlorination of water to prevent the water related transmission and strict aseptic techniques during eye examination.

Immunoprophylaxis: It is given to prevent the epidemic in communities and in military persons. Killed and live vaccines are used for prevention of ARD. No vaccine is available for general use. Live vaccine available as gelatin coated capsule, contains type 4 and 7, used orally in military persons. It was highly effective but due to manufacturer issues it is not in use since 1999.

Treatment

Only symptomatic treatment is given. In case of severe pneumonia cidofovir is given.

PAPOVAVIRIDAE

Meaning

Family name is given from the initial of its genera like *Papillomavirus* and *Polyomavirus*.

Follow **Flowchart 79.1**.

Morphology

It is 50–55 nm in size, naked and contains circular ds-DNA surrounded by icosahedral variety of capsid.

Resistance

The virus is resistant to drying and heating, but killed by 100°C and UV radiation. The virus is relatively hard and immune to many common disinfectants. Exposure to 90% ethanol for at least 1 minute, 2% glutaraldehyde, 30% Savlon, and/or 1% sodium hypochlorite can disinfect the pathogen.

HPV

Morphology: General morphological features are described above.

Genomic structure: Two types of genes. **(1) Early region genes:** These are E1–E7 codes for nonstructural proteins. Products of these genes have oncogenic potential. **(2) Late region genes:** These are L1 (for major capsid protein) and L2 (for minor capsid protein) codes for structural proteins like capsid. Over 170 types are identified based on DNA sequences of L1 gene. They are labeled as 1, 2, 3, etc., and disease produced by particular types is described in pathogenicity.

Culture characteristics (C/Cs): HeLa cells used for culture are containing HPV-18 DNA.

Immunity: Some HPV types like 16, 18, 31 and 45 are not controlled by immune system resulting lifelong infection and development of cancer.

Pathogenicity

- **Disease name:** It produces precancerous lesions called **warts** or **verucca** or **human papilloma** or **human papillomatosis,** which are cauliflower like growths raised beyond the surface and cancer like squamous cell carcinoma.
- **Epidemiology:** It is distributed worldwide.
- **Reservoirs and sources of infection:** Humans.
- **Modes of transmission:** It is transmitted by direct skin-to-skin contact, vaginal sex, anal sex and occasionally by placenta. It is not spread via common items like toilet seats.
- **Portal of entry:** Skin or mucosa.
- **Sites:** Skin of arms, face, forehead or feet, soles, finger nail regions, genitals like cervix, vulva, vagina, penis and anus, larynx and mouth.
- **Precipitating factors:** It is common in children and adults and epidermodysplasia verruciformis occurs in immunocompromized individuals.
- **Clinical features:** Follow **Fig. 79.2 and Table 79.3**.
- **Complication:** Recurrence is the common problem after treatment.

Diagnosis: Biopsy from warts is tested by following methods.

1. **Naked eye:** Most lesions are visible naked eye. 5% acetic acid can be applied to improve the visibility.
2. **Microscopy:** Follow morphology.
3. **Culture:** Follow C/Cs.
4. **Serological test:** Ab detection is not much useful.
5. **Molecular methods:** PCR or hybrid capture assay used to detect the DNA and specific type.
6. **Cytological study:** Virus is detected by papanicolaou smear (better called **Pap smear**) prepared from cervical or anal scraping and histopathological staining of biopsies.

Prevention: (1) General measure includes safe sex by using condom is the best way of prevention. **(2) Immunoprophylaxis** includes subunit vaccines **prepared** in yeast by DNA recombinant technology from major capsid protein like L1. Recombinant vaccines are available which are highly effective and reduce the rate of all infections including cervical cancer. **Efficacy** is >90%. **Side effect** includes local reactions at the site of injection. These are **use**ful to reduce cervical cancer and male diseases, so given to all adolescent boys and girls at ages 11–12 years. They are **contraindicated** in pregnancy. **Three types, administration and doses of** recombinant vaccines are described below.

- **Nine valent vaccine:** It contains two noncancer causing types like 6 and 11 plus seven cancer causing types like 16, 18, 31, 33, 45, 52 and 58. Two doses are given by IM route at 11–12 years. After the 1st

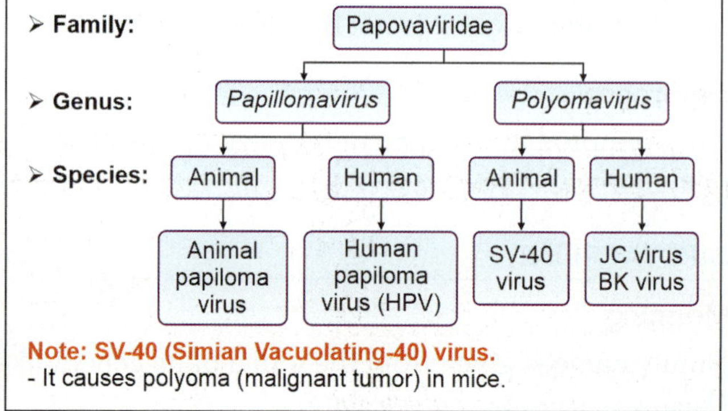

Flowchart 79.1: Classification of Papovaviridae

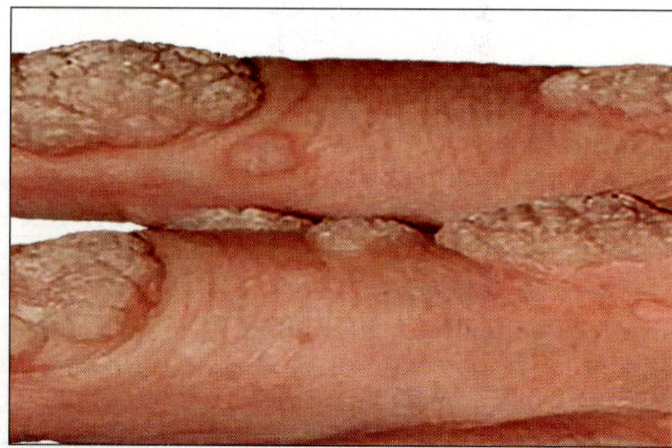

Fig. 79.2: Skin wart (verruca vulgaris) caused by HPV

TABLE 79.3: Clinical features of HPV

Disease	HPV type
Warts (verucca or papilloma or papillomatosis)	
Common warts (Verruca vulgaris): Most common type of wart. Seen commonly on hands, feet and rarely on knees or elbows in young children.	2, 7, 22
Plantar warts (Verruca plantaris): Found on the soles; grow inward, generally causing pain when walking. Seen in adolescent and young adult.	1, 2, 4, 63
Flat warts (Verruca plana): Found on the arms, face or forehead. Seen in young children and teens.	3, 10, 8
Anogenital warts (Condyloma acuminate): Seen on ano-genital parts and most common in sexually active adults.	6, 11, 42, 44 and others
Anal dysplasia (lesions)	6, 16, 18, 31, 53, 58
Epidermodysplasia verruciformis: Occurs in person with IDDs and tendency to change to SCC particularly in sun exposed areas.	More than 15 types (5, 8)
Focal epithelial hyperplasia (mouth)	13, 32
Mouth papillomas	6, 7, 11, 16, 32
Verrucous cyst	60
Laryngeal papillomatosis	6, 11
Cancers	
Oropharyngeal cancer	16
Laryngeal or oesophageal carcinoma	16 and 18
Genital cancers: Seen in cervix, vulva, vagina, penis and anus.	Highest risk: 16, 18, 31, 45. Other high-risk: 33, 35, 39, 51, 52, 56, 58, 59. Probably high-risk: 26, 53, 66, 68, 73, 82

dose, 2nd dose should be given 6–12 months later. However, if first dose given after 15 years then total three doses should be given at 0, 2 and 6 months. Immune response is better, if given before 14 years. It provides long lasting immunity.

- **Quadrivalent:** It contains types 6, 11, 16 and 18. It is licensed in USA and recommended by CDC for administration to girls and young women from 9 to 26 years of age. Three doses are given by IM route: After the 1st dose, 2nd dose should be given 2 months later and 3rd dose should be given 6 months after the 1st dose (0, 2 and 6 months). Among contains of vaccine, types 16 and 18 are the most common types for cervical cancer worldwide with prevalence rate is >70%. Vaccine offered the protection against cervical cancer caused by type 16 and 18.
- **Bivalent:** It contains high risk type 16 and 18. It is given as single dose by IM route.

Treatment: HPV lesions generally heal spontaneously. **Topical agents** like podophyllum and imiquimod (immunotherapeutic agent induces IFN) are applied for genital warts. Recurrence and scarring is common following surgical excision of wart, so not recommended. **Cryosurgery and electrodessication** are useful. **Immunotherapy** includes preparation from two main oncogenes like E6 and E7 is useful especially in treatment of cancer but still under trial.

JC Virus

Meaning and history: John Cunningham virus, from the initial of patient's name. Patient was suffered with progressive multifocal leukoencephalopathy (PML) and virus was identified from the brain of this patient in 1965 by ZuRhein, Chou, Silverman and Rubinstan by EM. Later, it was isolated in culture.

Morphology: Described above.

Culture characteristics (C/Cs): It grows in human fetal glial cell culture.

Pathogenicity

- **Epidemiology:** It infects 70–90% of people. Antibodies are identified from human serum. It has **8 genotypes**. **(1) Types 1 and 4:** Both types are closely related. They found in Europe, Northern Japan, North-East Siberia and Northern Canada. **(2) Type 2:** It has several variants like 2A found in Japanese population and Native Americans, 2B found in Eurasians, 2D found in Indians and 2E found in Australians and Western Pacific populations. **(3) Types 3 and 6:** Both types are found in sub-Saharan Africa. Type 3 presents in Ethiopia, Tanzania and South Africa and type 6 found in Ghana. **(4) Type 5:** No more detail is known. **(5) Type 7:** It has several variants like 7A found in Southern China and South East Asia and Native Americans, 7B found in Northern China, Japan and Mongolia and 7C found in Northern and Southern China. **(6) Type 8:** Found in Papua New Guinea and Pacific Island. **Other subtypes are as per distribution** are described below.
  - **European types:** Like a, b and c.
  - **African types: Minor** called Af1 distributed in Central and West Africa and **major** called Af2 distributed throughout the Africa, West and South Asia.
  - **Asian types:** Like B1-a, B1-b, B1-d, B2, CY, MY and SC.
- **Reservoirs of infection:** Humans.
- **Sources of infection:** These are contaminated sewage water from urine of patients.
- **Modes of transmission:** Mostly by ingestion of contaminated water.
- **Portal of entry:** GIT.
- **Sites:** Brain and colon/rectum.
- **Precipitating factors: (1) Age:** Infection acquired in childhood and also by adolescent. **(2) IDDs:** It produces the PML and other lesions only in persons with AIDS, immunosuppressive therapy like rituximab and organ transplantation.

- **Pathogenesis:** Probable mechanisms are mentioned in **Flowchart 79.2**.
- **Clinical features:** Two types of latent forms occur in immunocompromized patients. **(1) Progressive multifocal leukoencephalopathy (PML):** It is the subacute demyelinating (degenerative) disease of older person whose immunity is suppressed and presents with deterioration of motor function, speech, vision and death occurs in 3–4 months. **(2) Malignancy:** Virus is malignant and produces malignant gliomas following intracerebral inoculation in new born hamsters. JC virus is found from the human colorectal cancer in immunodeficient subjects but still it is controversial to accept it as causative agent.

Diagnosis: Brain biopsy is tested by **EM, culture** (JC virus grows in human fetal glial cell culture) and **molecular method** like PCR for identification and to differentiate JC virus from BK virus. **Serological tests** for Abs detection are performed from serum.

Prevention: Avoid immunosuppressant like rituximab.

BK Virus

Meaning and history: Virus was isolated in 1971 from the urine of renal transplant patient. Name is given from the initial of patient.

Morphology: It is similar to JC virus, since their genomes share 75% sequence similarity. Both of these viruses can be identified and differentiated from each other by serological tests using specific antibodies or by using a PCR based genotyping approach. EM is also useful.

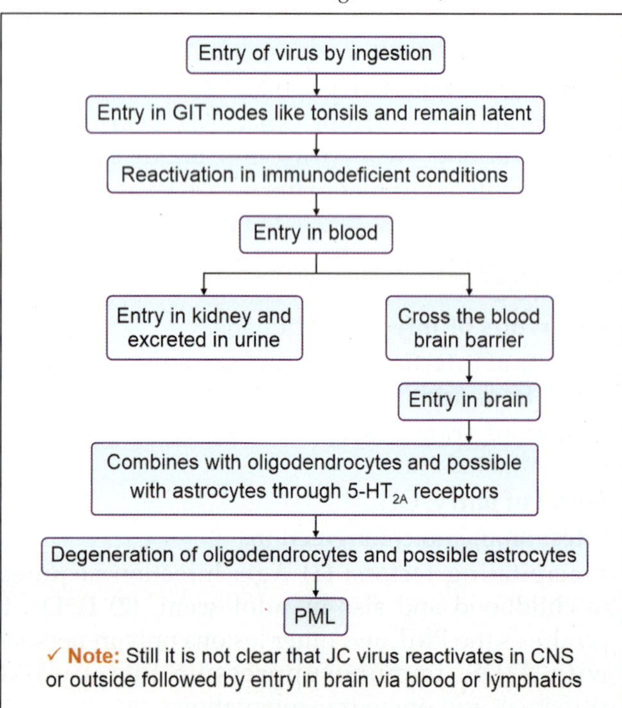

Flowchart 79.2: Pathogenesis of JC virus

Culture characteristics (C/Cs): It grows in primary and continuous cell cultures. Both JC and BK viruses agglutinate guinea pig and chicken RBCs.

Pathogenicity

- **Epidemiology:** It is distributed in 80% of population in asymptomatic form.
- **Reservoirs and sources of infection:** Virus excreted in respiratory droplets or in urine of humans that act as reservoirs and sources.
- **Modes of transmission:** Exactly not known but person to person transmission is possible from urine or respiratory droplets.
- **Portal of entry:** May be respiratory system.
- **Site:** Kidneys.
- **Precipitating factors:** Latent infection occurs in patient with immunocompromized conditions like renal or organs or bone marrow transplantation, and immunosuppressive therapy like tacrolimus, and mycophenolate mofetil.
- **Pathogenesis:** Virus remains in latent form in kidney for very long time and reactivated when immunity gets down produces renal dysfunction.
- **Clinical features**
  - Asymptomatic infection: It presents in 80% population.
  - Primary infection: It presents as mild infection, with fever or respiratory symptoms.
  - Latent infection: Virus reactivated in patient with renal or organ transplant leading to renal dysfunction, seen by a progressive rise in serum creatinine and an abnormal urinalysis revealing renal tubular cells and inflammatory cells. It is also associated with ureteral stenosis and interstitial nephritis. In bone marrow transplant recipient it is notable as a cause for hemorrhagic cystitis.

Diagnosis: Kidney biopsy is tested by **microscopy** (follow morphology), **culture** (follow C/Cs) and **molecular method** (PCR for identification and to differentiate JC virus from BK virus). **Serological tests** for Abs detection are performed from serum.

Prevention: Switch tacrolimus to cyclosporine.

Treatment: Cidofovir, Ig and fluoroquinolones are recommended. Ciprofloxacin lowers the viral load but no data on survival and graft loss exist.

PARVOVIRIDAE

Meaning

Family name given from word parvo (Latin) means small, because it is the smallest virus about 20 nm in size. It is as small as largest protein molecules like hemocyanin.

Classification

Follow **Flowchart 79.3**.

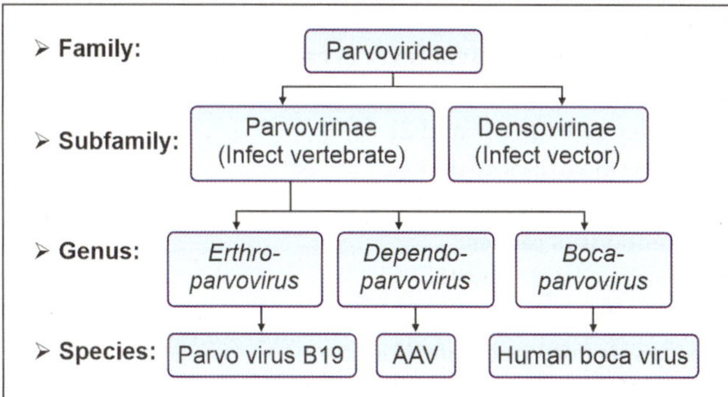

Flowchart labels:

Morphology

Size: It is smallest about 20 nm size nonenveloped (naked) virus. It is the only DNA type virus with ss-DNA. It is surrounded by icosahedral variety of capsid. Parvovirus detected from blood by EM.

Parvo Virus B19

Synonym: It also called B19 virus or erythro virus B19.

Meaning: It is able to invade the precursors of RBCs in bone marrow, hence the genus called *Erythroparvovirus.* It was identified by Australian virologist Yvonne Cossart in 1975 in the well number B-19 of larger size microtiter plate, hence the species called **B19.**

Morphology: Described above.

Genomic structure: Two types of genes like VP1 and VP2. Both are encoded for structure protein, where VP2 is the major capsid coding gene.

Culture characteristics (C/Cs): Difficult to grow it in culture.

Immunity: Past infection provides protection in half populations.

Pathogenicity

- **Disease name:** In children it produces the disease called **erythema infectiosum** or **slapped cheek disease** or **5th disease.**
- **Epidemiology:** It is distributed worldwide. It may produce the epidemic. Last epidemic was occurred in 1998. Outbreaks can arise especially in nurseries and schools. It has 3 genotypes.
- **Reservoirs and sources of infection:** Humans (cat and dog parvo viruses do not infect the humans).
- **Mode of transmission:** Primary mode of transmission is respiratory route. It also transmitted by blood and ≥30% chances of transplacental transmission.
- **Incubation period:** Symptoms appear after 6 days (average 16–17 days).
- **Portal of entry:** May be respiratory system.
- **Sites:** Skin and blood cells.
- **Precipitating factors:** It is common in children of 5–10 years and person with IDDs.
- **Pathogenesis:** Virus is able to invade the precursors of RBCs in bone marrow. It combines with specific

receptors present on RBCs surface called P antigens or globosides. It enters in RBCs and lyses the cells causing erythema infectiosum (slapped cheek disease or 5th disease). Sometimes, it is associated with aplastic anemia due to destruction of RBCs precursors.

- **Clinical features: In immunocompetent host** it causes many diseases. **(1) Respiratory infection:** It causes maculopapular rash and arthralgia. **(2) Erythema infectiosum or slapped cheek disease:** It is common in children of 5–10 years. It starts with erythema of cheeks (**Fig. 79.3**) and spread to trunk and limbs, followed by lymphadenopathy and arthralgia. It is called **5th disease,** as it was 5th in the list of old six exanthematous diseases. Brief prodromal phase presents with fever, headache, nausea and diarrhea. As the fever breaks, a red rash forms on the cheeks, with relative pallor around the mouth ("slapped cheek rash"), sparing the nasolabial folds, forehead, and mouth, "lace-like, (reticular)" red rash on trunk or extremities then follows the facial rash. Infection in adults usually involves the reticular rash, with multiple joint pain (arthralgia) predominating. Exacerbation of rash occurs by sunlight, heat and stress. Teenagers or young adults may develop pruritic erythematous papules with patechiae and edema on hands and feet, so-called **"papular purpuric gloves and socks syndrome".** **(3) Acute arthropathy:** It commonly presents in adults with arthralgia, arthritis and sometimes with rash. It involves the peripheral joints like wrists and knees. It is nondestructive and resolves by three weeks. **(4) Aplastic crisis:** "Aplastic crisis" means rapid and severe anemia due to temporary cessation of RBCs production (erythtopoiesis). It also called **reticulocytopenia** or **marrow failure.** It is most dangerous in children with pre-existing bone marrow stress, like hemolytic anemia, sickle

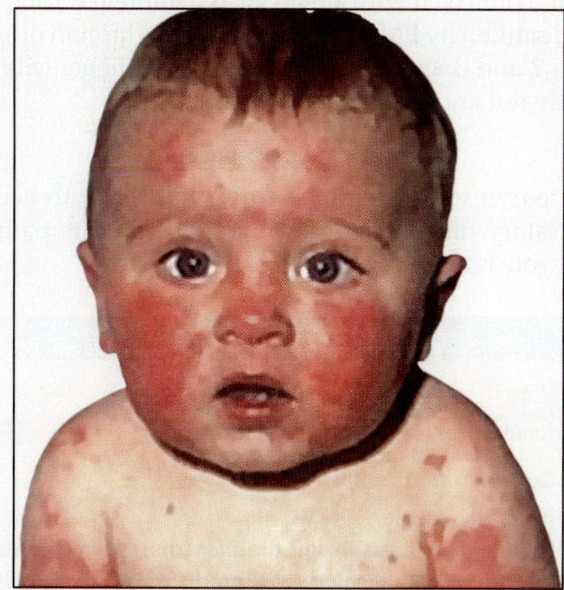

Fig. 79.3: Slapped cheek disease

cell disease or hereditary spherocytosis, Virus enters in the reticulocytes and destroy them, so reticulocyte count is decreased in blood (reticulocytopenia) followed by temporary cessation of RBCs production. Due to reduced RBCs, there is rapid and severe anemia (aplastic crisis) which is treated by blood transfusion. **(5) Hemophagocytic syndrome:** It also called **hemophagocytic lymphohistiocytosis.** It is a hematological disorder characterized by uncontrolled proliferation of lymphocytes and macrophages (histiocytes). Rarely it caused by parvovirus B19. It occurs more in children than adults. **In immunocompromised host** it causes persistent anemia. **In pregnancy** infection of 2nd and 3rd trimester causes nonimmune fetal hydrops and fetal loss (abortion).

Laboratory diagnosis: Blood/serum, tissues or respiratory specimens are tested by **microscopy** (follow morphology), **culture** (follow C/Cs) and **molecular method** (Real-time PCR, which is most sensitive and detects the viral load. In acute infection viral load is 10^{12} DNA copies/ml). **Serological test** like ELISA is performed for Abs detection against VP1 and VP2 Ags. IgM Ab is diagnosed in early stage (recent infection) and remains elevated for 2–3 months. IgG Ab is diagnosed in later stage (late infection) and persists for years. Ag is detected by immunohistochemistry method from bone marrow or fetal tissues.

Prevention and treatment: No effective vaccine or specific antiviral is available. Symptomatic treatment is given. Neutralizing Ig is also given.

AAV

It is the <u>A</u>deno <u>A</u>ssociated <u>V</u>irus (**AAV**) also called adeno satellite virus. It is a defective virus and remains dependent on helper virus for multiplication, hence genus name is *Dependoparvovirus*. It specifically depends on adeno virus so species called AAV. It multiplies only in cells coinfected with adeno virus. Different serotypes are identified by EM, IF test and CFT like human origin are 1, 2 and 3 and simian origin is 4. Pathogenicity is exactly not known.

Human Boca Virus

New parvo virus identified from children with acute respiratory disease and gastroenteritis, but its pathogenic role is not clear.

ACCESS YOURSELF

Short Note

1. Adenoviruses/HPV/Parvovirus B19.

Short Questions for Theory/Viva Questions

1. What is 5th disease?
2. What is aplastic crisis? Name the one virus responsible for it.
3. Write the one pathogenic disease produced by each RK virus and BK virus.

Adenoviridae

1. **Pharyngoconjunctival fever is caused hy:**
   a. Adeno viruses 3 and 7
   b. Adeno viruses 11 and 21
   c. Adeno viruses 40 and 41
   d. Adeno viruses 8 and 19
2. **Adeno virus causes:**
   a. Hemorrhagic cystitis
   b. Diarrhea
   c. Respiratory tract infection
   d. IMN

Papovaviridae

3. **Most common tumor caused by HPV is:**
   a. Warts
   b. Carcinoma of cervix
   c. Nasopharyngeal carcinoma
   d. Lymphoma
4. **Condyloma accuminata is caused by HPV types of:**
   a. 18, 31   b. 17, 12
   c. 6, 11   d. 16, 18
5. **Human papillomatsis is caused by:**
   a. HSV   b. HPV
   c. HIV   d. HBV
6. **Most common type of HPV associated with cervical cancer:**
   a. 6, 11   b. 5, 8
   c. 16, 18   d. 6, 8
7. **HPV vaccine is:**
   a. Monovalent   b. Bivalent
   c. Quadrivalent   d. Both bivalent and quadrivalent
8. **HPV vaccine is:**
   a. A live attenuated vaccine
   b. A killed vaccine
   c. A subunit vaccine
   d. A synthetic peptide vaccine
9. **Which of the following statement is correct?**
   a. Viral warts usually resolve spontaneously
   b. Plantar warts should not be excised
   c. Callosity is formed occupationally
   d. Corns are viral in etiology
   e. Plantar warts are painless
10. **All of the following are true about the papova viruses, *except*:**
    a. They are nonenveloped icosahedral viruses
    b. Produce papilloma
    c. RNA viruses
    d. SV-40 oncogenic
11. **Post transplant nephropathy after 1 month is most likely be due to:**
    a. Hepatitis C   b. HHV-6
    c. BK virus   d. Herpes simplex virus
12. **PML is caused by:**
    a. CMV   b. Papova virus
    c. HIV   d. Polio virus

Parvoviridae

13. **Infection with which of the following agents is particularly dangerous for anaemic patients?**
    a. Adeno virus   b. Cytomegalo virus
    c. Herpes simplex virus d. Parvo virus

14. **In parvovirus infection what is common in adult:**
    a. Bone marrow aplasia   b. PRCA
    c. Erythema infectiosum d. Arthropathy
15. **Parvo virus infection is associated with:**
    a. Hydrops fetalis
    b. Aplastic anemia
    c. Abortion
    d. Sixth disease
    e. Hemophagocytic syndrome
16. **All of the following statements about parvo virus B19 are true** *except*:
    a. DNA virus
    b. Crosses placenta in <10% cases
    c. Can cause severe anemia
    d. Can cause aplastic crisis
17. **All of the following statements about parvo virus B19 are true,** *except*:
    a. <10% spread by transplacental route
    b. Respiratory route is the primary mode of transmission
    c. It is a DNA virus
    d. Affects erythroid progenitors
18. **False regarding erythema infectiosum is:**
    a. Caused by HHV6
    b. Marked erythema of the cheeks or slapped cheek appearance often with relative circum oral pallor
    c. Infection during pregnancy can result in hydrops fetalis due to fetal anemia
    d. Arthritis is a complication

Answers and Explanation of MCQs

1. a
2. a, b, c
• Follow **Table 79.2** for explanation of answers of MCQs 1–2.
3. a
4. c
5. b
6. c
• Follow section, **Papovaviridae and Table 79.3** for explanation of answers of MCQs 3–6.
7. d
8. c
• Follow section, **Papovaviridae → HPV (immunoprophylaxis)** for explanation of answers of MCQs 7–8.
9. a, b and c
• Viral warts are mostly caused by HPV, that usually resolve spontaneously. Recurrence and scarring is common following surgical excision of warts (plantar warts), so not recommended. Callosity is formed occupationally, which is characterized by superficial circumscribed yellowish white flat thickened patch of hyperkerotic materials. Corn is the hyperkeratosis of skin at pressure points usually on sole, foot and toes. It is painful. Plantar wart causes pain on walking.
10. c
• Papova virus is ds-DNA virus. SV-40 causes the polyoma (malignant tumor) in mouse.
11. c
• Neuropathy following renal or organ transplantation is common by BK virus.
12. b
• Follow section, **Papovaviridae → JC virus (pathogenicity → clinical features)** for explanation.
13. d
14. d
15. a, b, c, e
16. b
17. a
18. a
• Follow section, **Parvoviridae → parvo virus B19** for explanation of answers of MCQs 13–18.

Infections of Picornaviridae

INTRODUCTION

Meaning

Picorna word derived from pico and RNA where pico means small size about 27–30 nm and RNA means type of genome in virus.

Classification

Family: Picornaviridae.

Genus: Two groups like human and animal pathogens.

A. Human pathogens: Following four genera.

1. ***Enterovirus* (EV):** Following are the species affecting the enteric tract.
   - **Polio virus:** 1–3.
   - **Coxsackie virus**
     - Coxsackie virus A: 1–24.
     - Coxsackie virus B: 1–6.
   - **ECHO virus:** 1–34.
   - **Other entero viruses 68–72**
     - EV-68: Pneumonia and bronchitis.
     - EV-69: No human disease.
     - EV-70: Acute hemorrhagic conjunctivitis.
     - EV-71: HFMD (2nd most common cause), meningitis and encephalitis.
     - EV-72: Reclassified as *Heparnavirus* (HAV).
2. ***Rhinovirus*:** More than 100 serotypes.
3. ***Heparnavirus* or *Hepatovirus*:** Hepa means liver and rna means RNA genome. It includes only one species like HAV (hepatitis A virus), which infects the liver. Previously, it was classified as EV 72. For more details of HAV follow **Ch. 88.**
4. ***Parechovirus*:** It includes 4 species like parecho virus A, B, C and D. They cause minor GIT and respiratory symptoms in humans; however, data for human infections are very limited.

B. Animal pathogens: Following two genera.

1. ***Aphthovirus*:** It causes foot and mouth disease in cattle.
2. ***Cardiovirus*:** It causes encephalo-myocarditis in mice.

Morphology

Shape and size: It is spherical, 27 nm in diameter and having 60 subunits, each consist four viral proteins (VP-1 to VP-4).

Genome and capsid: It is the naked ss-RNA (+) virus surrounded by icosahedral type of capsid (**Fig. 80.1**). It produces eosinophilic intranuclear Cowdry type-B inclusion bodies which may be seen in stained preparation.

Resistance

Sterilization: It is inactivated at 55°C for 30 minutes.

Disinfection: It is the nonenveloped virus resists to lipid solvents like ether, chloroform and bile salts. It also resists to proteolytic enzymes. It is killed by formaldehyde, oxidizing agents and chlorination. Echoviruses are stable over a wide range of pH (3–10).

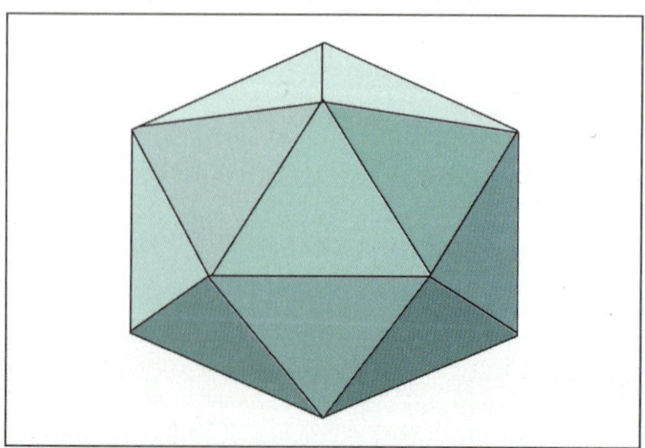

Fig. 80.1: Morphology of Picronaviridae (polio virus)

Polio Virus

History

It was first isolated in 1909 by Karl Landsteiner and Erwin Popper by inoculation of spinal cord and fecal extract in to monkey from the fatal case of polio. Virus infecting nerve system, but its non neural cultivation technique over human embryonic cell was developed by Nobel Laureate, Enders, Wellers and Robbins in 1949 with development of CPE.

Use

Polio virus is useful for understanding the biology of RNA viruses.

Morphology

It is described above (**Fig. 80.1**).

Culture Characteristics (C/Cs)

Egg culture: Old strain can grow in egg, not the fresh isolates.

Animal culture: Monkey is infected by intraspinal/ intracerebral mode. Chimpanzees or cynomolgus monkey may be infected orally.

Tissue culture: It grows only in human diploid (secondary) fibroblast cell line and primary monkey kidney cells. Growth is indicated by typical CPE in 2–3 days. Infected cells are round up, pyknotic, refractile, crenated or degenerated. Eosinophilic **intranuclear Cowdry type-B inclusion bodies** may be seen in stained preparation. Viral antigen from culture is identified by neutralization or IF test. Cell monolayer produce plaques with agar overlay. Virus is isolated from throat swab up to 3 weeks, from stool sample/rectal swab up to 12 weeks and more longer in persons with IDDs. From blood or CSF it is isolated very rarely. Specimen should be kept frozen during transportation.

Resistance

It is described above.

Antigenic Types

Two types of polio viruses like wild polio virus (WPV) and vaccine derived polio virus (VDPV).

WPV: It occurs naturally and has three strains identified by neutralization test like WPV1 (Brunhilde and Mahoney), WPV2 (Lencing and Mefi) and WPV3 (Lencing and Mefi). Pathologically, all strains are same produce same disease, but antigenically and genetically they are different. Antibody response is type-specific without cross reaction in between, hence to eradicate polio each strain is required to eradicate. WPV2 and WPV3 are globally eradicated in 1999 and 2019 respectively. WPV1 is the only strain to cause current polio cases, it is the common strain to cause polio till now.

VDVP: It is described later in this chapter.

Immunity

AMI: Abs are protective, but they are type specific. 1st infection produces type specific Abs while subsequent infection produces group specific Abs to all three strains.

CMI: Virus also induces the CMI but its role is uncertain.

Pathogenicity

Disease name: It called polio or poliomyelitis.

Epidemiology: On the bases of transmission countries are classified in to four categories. **(1) Endemic countries:** These countries have ongoing transmission of indigenous WPV. Currently WPV is endemic only in Pakistan (20 cases of WPV-1 in 2022) and Afghanistan (2 cases of WPV-1 in 2022). Nigeria was declared as polio free in August 2022. 8 cases of WPV-1 reported from Mozambique in 2022 after 30 years gap and several cVDPV cases were reported from nonendemic countries like Madagaskar and Nigeria. **(2) Outbreak countries:** These include total 26 countries where cases are not due to indigenous WPV, but due to imported WPV or VDPV from other country or by emergencies and circulation of VDPV. Implementation of outbreak guidelines are must to stop these outbreaks. **(3) Key at risk countries:** These include total 5 countries with low level of immunity and surveillance. **(4) Polio-free countries:** These include total 173 countries without a case of indigenous WPV in ≥3 years. Polio is eradicated from the India and WHO declared India as polio free in March 2014.

Reservoirs of infection: Human is the only natural reservoir of infection due to shedding of virus in **feces** for 3–4 months. There is no chronic carrier, but person with IDDs can shed it for longer time. It may be patient (case) or symptomless carrier. For every one case, about 1000 subclinical cases are present in children and 75 in adults that act as reservoirs.

Sources of infection: These are human feces, which contaminate the food or water and oropharyngeal secretion in early stage. Patients are infectious and shed viruses in stool from 7–10 days before appearance of symptoms and up to 2–3 weeks thereafter (**period of communicability**). Sometimes, virus sheds very long in stool up to 3–4 months.

Modes of transmission: It is transmitted by **feco-oral (ingestion)** route due to entry of food and water contaminated from human feces, **inhalation** or by **conjunctiva** from droplets of respiratory secretions- possible mode of entry in close contacts of patients in early stages of the disease. **Direct neural transmission** occurs in case of tonsillectomy.

Incubation period: 7–14 days on average (range is 4 days – 4 weeks).

Portal of entry: GIT.

Sites: Brain and spinal cord.

Precipitating factors (epidemiological determinants): **(1) Age:** Adults and older children are more susceptible than children. **(2) Sex:** It has been noted as 3 males to one female. **(3) Pregnancy:** It increases the risk due to hormonal changes. **(4) IDDs:** Polio is more severe and virus shedding is more prolonged in persons with IDDs. **((5) Tonsillectomy:** During the incubation period, it may predispose the bulbar poliomyelitis. It also precipitates the direct neural transmission. **(6) Heavy exercise:** Heavy exercise and increase muscular activities are increasing the risk of paralysis. **(7) Vaccination:** Injection of alum precipitated triple vaccine (DPT) may precipitate the paralysis in inoculated limbs. Mechanisms are uncertain. **(8) Trauma:** It may lead to the viral entry in to local nerve fibres or segment of spinal cord corresponding to the site may be more susceptible to viral damage due to reactive hyperemia. **(9) Cold environment:** Polio virus survives for long time in cold environment. **(10) Overcrowding and poor sanitation:** Also provide the opportunities for exposure.

Pathogenesis: Virus spreads by hematogenous (**Flowchart 80.1**) and neural routes. Neural transmission occurs following tonsillectomy.

Clinical features: In includes four stages. **(1) Inapparent infection:** About 90–95 % cases are asymptomatic. Only 5–10% produces the clinical illness. **(2) Minor illness or abortive polio:** It occurs in 4–8% of the infections. It is associated with primary viremia and production of minor symptoms like fever, headache, sore throat and malaise lasting for 1–5 days. It called the minor illness and in many cases this is the only manifestation. **(3) Non-paralytic polio (aseptic meningitis):** Sometimes, the infection will not progress beyond the stage of aseptic meningitis and there is no paralysis. Features of aseptic meningitis (stage of CNS invasion by virus) like fever (biphasic fever), headache, neck stiffness and rigidity are present. Aseptic meningitis occurs in 1% cases. **(4) Major illness or paralytic polio:** It presents with flaccid paralysis (**Fig. 80.2**) with following features.

- **Duration:** If the infection progresses, minor illness followed 3–4 days later by major illness.
- **Frequency:** Occurs in <1% of the infections.
- **Sites of paralysis:** Paralysis may be bulbar, spinal and bulbospinal. It affects the proximal muscles 1st followed by distal muscles. Quadriceps is the most common muscle affected in leg, while in hand most common affected muscle is oppenens pollicis. Tibialis anterior is the most common muscle undergoes to complete paralysis.
- **Tripod sign:** It also called **"Amoss's sign"** and first described by the American pathologist Harold Lindsay Amoss (1886-1956). Paralysis starts from the hip and proceeds towards extremities, which leads to typical 'tripod sign'. In tripod sign patient sits with flexed hip and both arms extended towards the back for support as shown in **Fig. 80.3a**.
  - Tripod sign is a useful sign of meningeal irritation. It is used for diagnosing conditions like meningitis, subarachnoid hemorrhage, and poliomyelitis.

Flowchart 80.1: Pathogenesis of polio

Entry of virus by feco-oral route. It attaches to host cells by CD155 receptors present on host cells

↓

Multiplication in intestinal epithelium and lymphatic tissues from tonsils to Peyer's patches

↓

Spreads to regional lymph nodes

↓

Viral multiplication in lymph nodes

↓

Released in to blood called primary viremia

↓

Entry in RE system

↓

Further multiplication

↓

Released in to blood called secondary viremia **leads to onset of symptoms (prodromal phase)**

↓

Entry in target organs like brain and spinal cord

↓

Virus multiplies in neurons and destroys them. Earliest change is degeneration of **Nissl bodies (chromatolysis)**. Nuclear changes also occur. Degenerated cells are then phagocytosed. Lesions mostly seen in anterior horns of spinal cord, but posterior horn and intermediate columns may also involve

↓

Development of clinical features of polio

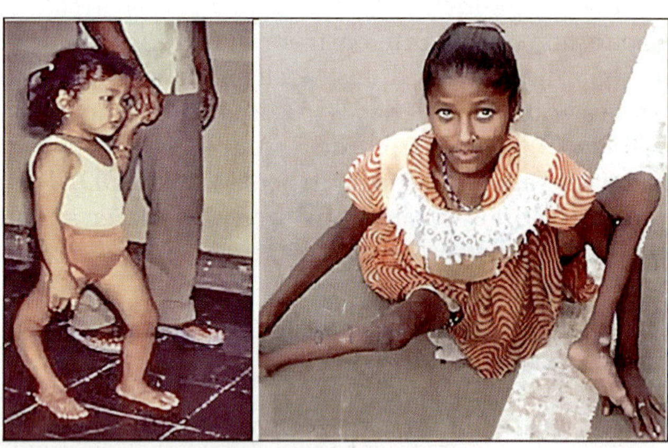

Fig. 80.2: Cases of polio

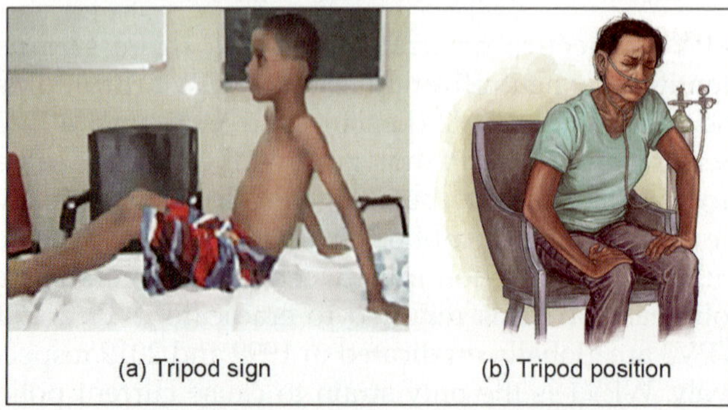

(a) Tripod sign   (b) Tripod position

Fig. 80.3: Tripod sign and tripod position

- **Other important points:** It is characterized by descending, asymmetric and acute flaccid paralysis. Cranial nerves are involved, but no sensory loss, no autonomic disturbance and nonprogressive. Lower motor neuron type and exaggerated tendon reflexes are absent.
- **Recovery:** Recovery of paralyzed muscles takes place in next 4–8 weeks and usually complete after 6 months, leaving behind the residual paralysis with variable degree.
- **Biphasic course:** Sometimes disease progression occurs in two phases. In 1st phase, aseptic meningitis occurs with full recovery. In 2nd phase, return of fever with paralysis in 1–2 days.

Complication: It includes postpolio muscle atrophy syndrome. Recrudescence of paralysis and muscle wasting has been occurred in individual, usually 20–40 years after the episode of paralytic polio. Mortality occurs in 5–10% cases due to respiratory failure.

> **Note: Tripod position**
> - **Features:** In tripod position, one sits or stands leaning forward and supporting the upper body with hands on the knees or on another surface as shown in **Fig. 80.3b**.
> - **Causes:** It mostly occurs when patient feels respiratory distress like in chronic obstructive pulmonary disease.

Laboratory Diagnosis

Diagnosis of Case

Specimens: Blood, throat swab, stool and CSF.

Testing methods

A. **Microscopy:** Follow morphology.

B. **Culture:** Follow C/Cs.

C. **Serological tests:** Rise in Ab (neutralizing Ab) titer in paired sera collected at 1–2 weeks interval is suggest polio. Neutralizing Abs are detected by neutralization test.

D. **Molecular methods:** RT-PCR is performed to differentiate between WPV and VDPV.

Diagnosis from Sewage

Detection of wild or vaccine virus from sewage is the tool of polio eradication program to know whether the transmission is being continued or stopped. Specimen should be kept frozen during transportation.

Prevention

General measures: These include improvement of food and water hygiene.

Immunoprophylaxis: It includes following two vaccines (both are effective).

1. Killed vaccine

- **Synonym:** It also called injectable polio vaccine or inactivated polio vaccine (IPV) or Salk vaccine.
- **History:** Developed by Salk and his team, in 1957 at the University of Pittsburgh.

- **Preparation:** It is prepared by using all three strains of virus by growing in a monkey kidney tissue culture (Vero cell line), which are then inactivated with formalin. One dose (0.5 ml) contains 40 units of type 1, 8 units of type 2 and 32 units of type 3 (total 80 units).
- **Administration and dose:** In 2015, IPV was included in national immunization schedule as single full dose of 0.5 ml given by IM route at 14th week along with bivalent OPV in thigh. **In 2017,** it is administered in fractional dose (f-IPV) of 0.1 ml by ID route in upper arm at 6th and 14th week along with bivalent OPV. Two doses f-IPV provide higher seroconversion than single full dose of IPV and also costs effective like 0.2 ml versus 0.5 ml.
- **Efficacy:** It provides 80–90% immunity after full course.
- **Mixed or combination:** In some states pentavalent vaccine **(Pentvac)** contains diphtheria, pertussis, tetanus, HBV, *H influenza* type b (Hib) is given at 6th week, 10th week and 14th week with OPV. Other same preparation is **Pentaxim** contains D, P, T, IPV and Hib given with OPV and HBV.
- **Advantages:** (1) IPV is more stable and safe than OPV, because the dead microbes cannot mutate back to their disease-producing state, so no disease to contact or recipient like vaccine associate paralytic polio (VAPP). (2) It does not require refrigeration, can be easily stored and transported at ambient temperature. (3) It is given in combination with polyvalent vaccine. (4) It is useful for person with IDDs, on steroid therapy and in pregnancy. (5) The IPV confers IgM and IgG-mediated immunity in the bloodstream, which prevents polio infection from progressing to viremia and protects the motor neurons, thus eliminating the risk of bulbar polio and post polio syndrome.
- **Disadvantages:** (1) IPV provides weaker immune response than OPV. So it would likely take several additional doses or booster doses to maintain a person's immunity. (2) Immunity lasts very short. (3) It provides only systemic, but not local immunity because it produces IgM and IgG not mucosal IgA. (4) It produces local reaction at the site of injection. (5) It cannot spread by feco-oral route, so it has no role in herd immunity. (6) IPV does not provide community protection; hence not useful in epidemic. Instead it precipitates the paralysis. (7) It is expensive than OPV.

2. Live 5vaccine

- **Synonym:** Oral polio vaccine (OPV) or Sabin vaccine.
- **History:** It was developed by Coprowsky, Cox and Sabin.
- **Preparation:** It is prepared by using all three strains. It is available as trivalent (type 1, 2 and 3), bivalent (type 1 and 3) and monovalent (type 1 only). Each 0.5 ml dose contains 3 lakhs TCID50 (tissue culture infective dose) of type 1, 1 lakh TCID50 of type 2

and 3 lakh TCID50 of type 3. Viruses are grown in a primary monkey kidney or in human diploid cell culture. Earlier trivalent OPV was used but type 2 was eradicated from world in 1999 and trivalent vaccine causes vaccine virus induced paralysis so it is replaced by bivalent OPV.

- **Administration and dose:** Two drops per dose are given by oral route; repeat if child spit out. OPV is included in national immunization schedule. Initial dose is given immediately after birth called **zero dose**. If missed then give within 15 days. First three doses are given at 6th, 10th and 14th week with f-IPV at 6th and 14th week. Booster dose given at 16–24 months. Total five doses should be completed before school age.
- **Efficacy:** It provides 90–100% immunity faster than IPV.
- **Reason for requirement of multiple doses:** Single dose is enough to provide sufficient immunity, but in practise multiple doses are required to ensure multiplication by all viruses after overcoming interference by themselves.
- **Advantages:** (1) OPV produces secretory IgA, which provides intestinal (local/mucosal) immunity, the primary site of wild polio virus entry, which helps to prevent infection with wild virus in areas, where the virus is endemic. (2) It also provides systemic immunity (IgG), so prevent paralysis. (3) The live virus used in the vaccine, sheds in the stool and can be spread to others within a community and provides herd or community immunity in addition to individual immunity. (4) It confers strong and long lasting immunity. (5) It is more economical than IPV. (6) It is given by oral route, so easy to administer. (6) It is useful during epidemic.
- **Disadvantages:** (1) **Temperature and pH sensitive:** OPV is heat labile and inactivated, if temperature (−20°C) is not maintained during storage and transportation. It makes thermostable by adding some salts like $MgCL_2$ and Na_2SO_4. These salts act at pH <7. The pH of vaccine (<7) is maintained by keeping the vial airtight. (2) It is safe but risky to give in person with IDDs, on steroid therapy and in pregnancy. (3) It causes Vaccine Associated Paralytic Polio (VAPP) and Vaccine Derived Polio Virus (VDPV).
- **Contraindications of OPV:** IDDs and pregnancy. In pregnancy OPV will be delayed unless immediate protection is required.
- **Effective factors on OPV failure: (1) Breast-milk:** Colostrum contains high level of IgA. OPV gets washed away in stool if given immediately before or after the breastfeeding. Avoid the breastfeeding 30 minutes before or after the OPV administration. **(2) Diarrhea:** OPV gets washed away in stool, if given in diarrheal patient. Dose given to child with diarrhea is not counted and repetition of dose is required after recovery. **(3) Preexisting viral disease:** *Enterovirus* may prevent the colonization by vaccine virus and effect over immunity. **Coxsackie B** may interfere,

while **coxsackie A** may be synergistic for polio virus. **(4) Improper cold-chain maintenance:** The shelf life of the vaccine is 4 months at 4–8°C and 2 years at −20°C. Improper storage and cold chain failure may be partly responsible for OPV failure. **(5) pH:** OPV best acts at pH below 7 which is maintained by keeping the vaccine air tight.

- **Criteria for attenuation of strain:** Vaccine virus (1) should not be neurovirulent as tested by intraspinal inoculation in monkey, (2) should be stable and should not acquire neurovirulence after serial enteric passage, (3) should be able to induce intestinal infection and immune response following feeding and (4) should possess genetic characteristics (marker) by which they can be differentiated from WPV.
- **Markers to differentiate WPV from attenuated strain: (1) D marker:** Wild strain will grow at low level of bicarbonate but the avirulent strain will not. **(2) ret 40:** Wild strain will grow well at 40°C but not the avirulent strain. **(3) MS:** Wild strain will grow in a stable cell line of monkey kidney, but the avirulent strain grows poorly. **(4) Mc Bride's intratypic antigenic marker:** Shown by rate of inactivation by specific antiserum. **(5) Other methods:** Above markers are not sufficient. Other methods used are molecular epidemiology by using monoclonal-Ab, oligonucleotide finger printing and nucleic acid sequencing.
- **Differences between IPV and OPV plus IPV and f-IPV:** Follow **Tables 80.1** and **80.2,** respectively.

> **Note: Vaccine originated polio**
> Two types like **VAPP and VDPV. (1) VAPP (Vaccine-associated paralytic polio):** OPV virus causes polio called VAPP. Virus shows the <1% genetic divergence from the parental strain called **OPV-like isolate.** It is ubiquitous at the place where OPV is given. It spread by feco-oral route to close contact. VAAP strain is not able to circulate so no chances of secondary case or outbreak. It is due to high immunity in community. It occurs in 1 case out of 2.7 million doses of OPV. It is more common in persons with IDDs (increased risk by 3000 fold) and following 1st dose than subsequent dose. It is most common in Sabin serotype 3 (60%) followed by Sabin type 2 (40%). **(2) VDPV (Vaccine derived polio virus):** OPV virus shows the genetic divergence from the parental strain and produces new strain called VDPV. Total 3 types of VDPV like circulating called cVDPV, immunodeficiency associated called iVDPV and ambiguous called aVDPV. Out of all these cVDPV circulates in community, produces same disease as like WPV, transmitted by feco-oral route, can cause outbreak and shows the >1% genetic divergence from the parental strain. It is most common in Sabin serotype 2 (90%) followed by Sabin type 1. It is more common in type 2 of vaccine because wild strain of type 2 is eradicated and not circulating in the community since 1999 (type 2 is eradicated in 1999). Since 2000 VDPV outbreaks have occured in several countries including nonendemic countries which are declared polio free. In 2013 over 700 strains and in 2019 about 365 cases of cVDPV have been isolated worldwide. To avoid chances of VDPV by type 2 it is eradicated from trivalent vaccine. Such trivalent vaccine is replaced by bivalent vaccine contains types 1 and 3.

TABLE 80.1: Differences between IPV and OPV

Features	IPV	OPV
Synonym	Killed or Salk vaccine	Live or Sabin vaccine
Preparation	Trivalent	Trivalent, bivalent and monovalent
Attenuation	By formalin	Live attenuated
Route	IM or SC	Oral
Dose	2 doses at 6th and 14th week	Total 5 doses
Contraindication	Not in IDDs	In IDDs
Cost	Viruses are 10,000 times higher than OPV, hence costlier	Cheap
Role	Prevent paralysis not reinfection	Prevent paralysis and reinfection
Efficacy	80–90%	90–100%
Immunity	Short lived	Long lived
Protection	Less and only paralysis	More including paralysis and reinfection
Ig produced	IgG and IgM	IgA, IgM and IgG
Local immunity	No	Yes
Herd immunity	No	Yes
In epidemic	Not useful	Useful
CMI	Poor/no	Yes
Reversion to virulence	No	Yes
Excretion of vaccine microbes and transmission to nonimmune contacts	No	Yes
Interference by other virus in host	No	Yes
Heat stability	Stable	Labile
Temp., maintenance during storage and transport	No	Yes
VAPP and VDPV	No chance	More chances

TABLE 80.2: Differences between IPV and f-IPV

Features	IPV	f-IPV
Administration	IM in thigh	ID in upper arm
Dose	0.5 ml	0.1 ml
Time	1 full dose at 14th weeks	2 fractional doses at 6th and 14th week
Syringe	0.5 ml AD	0.1 ml AD

Poliomyelitis eradication program

- **Strategies for eradication:** These are conduction of PPI including mopping up, high level of routine immunization, monitoring of vaccination, diagnosis of all cases of acute flaccid paralysis by polio or nonpolio, early diagnosis from stool and outbreak control of cases to stop transmission.

- **PPI (Pulse Polio Immunization) Program: Pulse means** mass immunization by bivalent OPV on a single day to all children between 0 and 5 years in the community regardless of previous or routine immunization. **Mopping up means** door-to door immunization in high risk district where wild virus is known or suspected to circulate. PPI was started globally, and in India, it is continued since 1995–96. Two rounds are scheduled every year in winter season.

- **Acute flaccid paralysis survey:** It is conducted to known the all remaining infected areas and to monitor progress toward eradication.

- **Endgame strategic plan (2022–26):** It aims to achieve polio free entire world. Strategies are (1) permanent stopping of transmission of all WPV in endemic countries like Pakistan and Afghanistan (2) stopping of transmission of cVDPV and (3) prevention of outbreaks in nonendemic countries.

World Polio Day

It was fixed on 24th Oct of every year by Rotary International over a decade ago to commemorate the birth of Jonas Salk, who led the first team to develop a vaccine against poliomyelitis.

Coxsackie Virus

Meaning: It called coxsackie, because 1st patients came from the Coxsackie village of New York.

History: It was first isolated in 1948 by Dalldorf and Sickles by inoculation of fecal extracts into suckling mouse from the pediatric case of paralytic polio, from which polio 1 was isolated. It caused paralysis in inoculated mouse.

Morphology: Described above (**Fig. 80.1**).

Culture characteristics (C/Cs): (1) Animal culture: It infects the suckling mouse but not the adult mouse by subcutaneous, intraperitoneal or intracerebral route. Identification is done by histopathology and by neutralization test. **Coxsackie A** produces generalized myositis, flaccid paralysis and death in a week. **Coxsackie B** produces patchy focal myositis, spastic paralysis, necrosis of brown fat, pancreatitis, hepatitis, myocarditis and encephalitis. **(2) Tissue culture:** A7, A9 and all coxcackie B can grow in monkey kidney tissue culture. A7 and A9 can grow in human diploid (secondary) fibroblast cell line. A21 grows in Hela cells. CPE will be produced in 5–14 days.

Resistance: Described above.

Antigenic structure and serotypes: Two species like Coxsackie A and Coxsackie B are identified by neutralization test and pathological changes in suckling mice. **Coxsackie A** has 1–24 serotypes (type 23 is same as echovirus-9 and type 24 is same as echovirus-34). **Coxsackie B** has 1–6 serotypes (all types share common complement fixing Ag).

 Immunity: Immunity is type specific.

Pathogenicity:

- **Epidemiology:** Coxsackie B produces the epidemic every 2–3 years. Different serotypes of coxsackie A and coxsackie B are distributed in different regions.
- **Reservoirs of infection:** Humans.
- **Sources of infection:** These are food and water contaminated from human feces.
- **Modes of transmission:** By feco-oral route.
- **Portal of entry:** GIT.
- **Sites:** Follow **Table 80.2.**
- **Precipitating factors:** Young infants are commonly infected.
- **Clinical features of coxsackie virus A: (1) Herpangina (vesicular pharyngitis):** It occurs in children. Local features are small painful vesicle on fauces and posterior pharyngeal wall, which break down to form ulcer. Systemic features fever, headache, vomiting and abdominal pain. **(2) Aseptic meningitis:** It is mostly caused by A7 and A9. It may be present with maculopapular rash. Type-7 produced outbreak of paralysis in Scotland, Russia and elsewhere, which was erroneously referred as type-4 polio virus. **(3) Exanthematous disease:** It is produced by A9, A16 and B1-3 in young children and presents with papulo-vesicular lesions on skin and oral mucosa, which resolved in 1–2 weeks. It occurs as sporadic and as an outbreak. **(4) Hand-foot-mouth disease (HFMD):** A16 is the most common cause, EV-71 is the second-most common cause, other causes are coxsackie virus B1-3, few strains of coxsackie virus and entero virus. It presents with fever, nausea, vomiting, and irritability in infants and toddlers. Rashes are presents on palms, soles, feet **(Fig. 80.4)**, buttocks and sometimes on the lips which may be flat or bumpy followed by vesicular sores with blisters. The rash is rarely itchy for children, but can be extremely itchy for adults. Painful facial ulcers, blisters, or lesions may also developed in or around the nose or mouth. HFMD usually resolves in 7–10 days. Most cases of the disease are relatively harmless, but complications including encephalitis, meningitis, and paralysis that mimic the neurological symptoms of polio. **(5) Minor respiratory infection:** Like common cold caused by A10, A21 and A24. **(6) Acute hemorrhagic conjunctivitis:** Rarely, it is caused by A24.

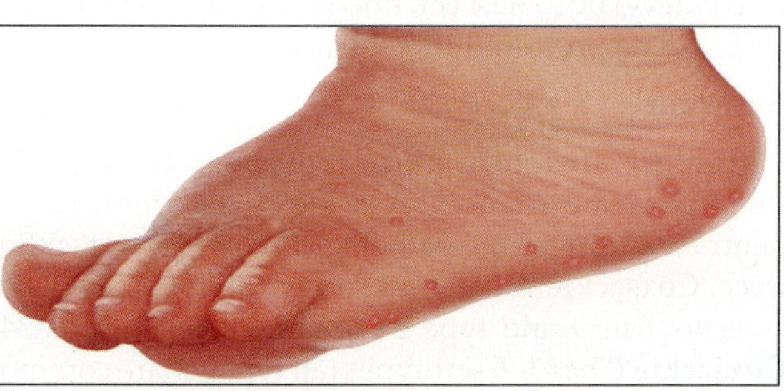

Fig. 80.4: Foot disease

- **Clinical features of coxsackie virus B: (1) Aseptic meningitis:** It is caused by all members of group B. **(2) HFMD:** It is caused by B1–3 viruses. **(3) Bornholm disease:** So called because it was 1st described in **Danish Island of Bornholm.** Disease have other names like Bamble disease, the devil's grip, devil's grippe, epidemic myalgia, epidemic pleurodynia, epidemic transient diaphragmatic spasm or The Grasp of the Phantom. It occurs as sporadic cases and as an epidemic. It presents with fever, headache and sever lower chest pain often on one side. The slightest movement of the rib cage causes a sharp increase of pain, which makes it very difficult to breathe, and an attack is therefore quite a frightening experience, although it generally passes off before any actual harm occurs. **(4) Myocarditis and pericarditis:** These occur in newborns, older children and adult with high fatality. **(5) Juvenile diabetes:** It is associated with B4, but causal role is still to prove. **(6) Orchitis:** It is an inflammation of testis. **(7) Minor respiratory infection:** Like common cold caused by B3. **(8) Transplacental and neonatal transmission:** It causes serious disseminated disease in newborns like meningitis, encephalitis, hepatitis and adrenocortical involvement. **(9) Post-viral fatigue syndrome:** It is associated with group B, but causal role is still to prove.

Differences between Coxsackie A and B: Follow Table 80.3.

Laboratory diagnosis: Stool, CSF, throat swab and sample from lesion are tested by microscopy (follow morphology) and culture (follow C/Cs). **Serological tests** are not useful because of many serotypes or useful to detect the neutralizing Abs. **Molecular method** like PCR is more useful as it is sensitive and serotype specific.

Prevention: Vaccination is not useful because of several serotypes and immunity is type specific.

Treatment: It is a self limited disease and symptomatic treatment is required.

ECHO Virus

Full name: Entero Cytopathogenic Human Orphan Viruses.

TABLE 80.3: Differences between coxsackie A and B		
Features	**Coxsackie A**	**Coxsackie B**
Intracerebral inoculation in mouse	Flaccid paralysis and systemic myositis	Spastic paralysis and focal myositis
Sites	More in skin and mucosa	More in heart, pancreas, pleura and liver
Clinical features	Follow text	Follow text
Serotypes	1–26	1–6
Receptors	Intercellular adhesion molecules	Coxsackie virus and adenovirus receptors

Meaning: It was not associated with any particular disease at the time of discovery hence called **orphan virus**. However, later, it is identified as a causative agent of many diseases, but the original name is still persist.

History: The first echovirus was isolated from the feces of asymptomatic children early in the 1950, just after cell culturing had been developed.

Morphology: Described above (**Fig. 80.1**).

Culture characteristics (C/Cs): (1) Animal culture: Animals are less susceptible. Monkey and newborn mouse produce paresis on inoculation. **(2) Tissue culture:** It grows only in human diploid (secondary) fibroblast cell line or in primary monkey kidney cells. Growth is indicated by typical CPE. Viral antigen from culture is identified by neutralization or hemagglutination test.

Resistance: Described above.

Antigenic structure and serotypes: ECHO virus is classified into 34 serotypes by neutralization test, but type 10 and 28 removed from the group and classified as reo virus-1 and rhino virus-1, respectively.

Pathogenicity

- **Epidemiology:** ECHO virus is distributed worldwide. Epidemics have been reported with significant morbidity and mortality. A Thai hospital reported the first nosocomial outbreak of HFMD due to ECHO virus type 11.
- **Reservoirs of infection:** Humans.
- **Sources of infection:** These are mostly contaminated food and water. Rarely, virus presents in saliva and swimming pool water.
- **Modes of transmission:** It is transmitted by feco-oral route, inhalation of oral secretions such as saliva, fomites, contaminated swimming water, or wading pools and by contaminated hands of hospital personnel.
- **Incubation period:** 4–6 days after the infection.
- **Portal of entry:** GIT.
- **Sites:** Meninges, GIT, liver, spleen, bone marrow, heart and lungs.
- **Pathogenesis:** Follow **Flowchart 80.2**.
- **Precipitating factors:** It is common in male children <15 years, summer season, overcrowding conditions with poor sanitation and in persons with IDDs.
- **Clinical features:** More than 90% infections are asymptomatic. Epidemic of fever, rash and aseptic meningitis is common by serotypes, 4, 6, 9, 16, 20, 28 and 30. Respiratory features are common by 1, 11, 19, 20 and 22, while gastroenteritis occurs due to type 18.
- **Complications:** Occasionally, it causes paralysis and hepatic necrosis.

Laboratory diagnosis: Throat swab, stool and CSF are tested by microscopy (follow morphology) and culture (follow C/Cs). **Serological tests** are not used routinely but useful in epidemic to identify causative serotype.

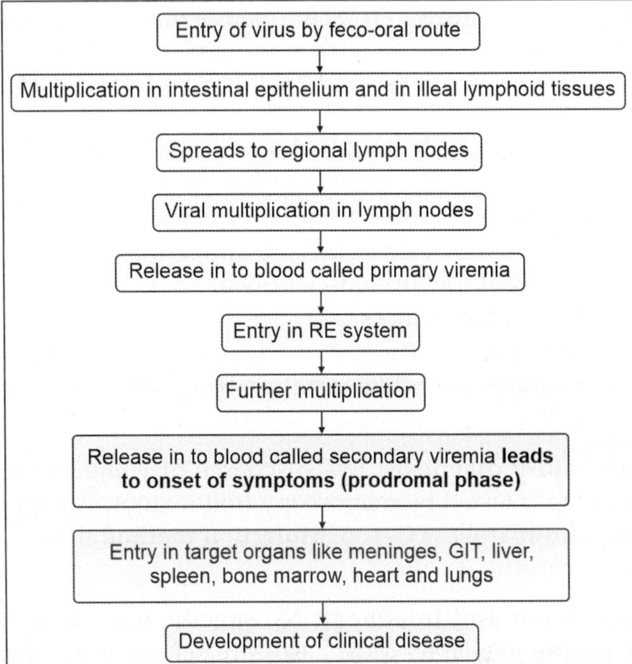

Flowchart 80.2: Pathogenesis of ECHO virus

Prevention and treatment: No specific vaccine and drugs are available. Symptomatic treatment is given. The antiviral drug pleconaril may be useful.

Entero Virus-70

Synonym: It also called **acute hemorrhagic conjunctivitis virus.**

Morphology: Described above (**Fig. 80.1**).

Culture characteristics (C/Cs): It grows in human embryonic kidney cells, Hela cells or monkey kidney cells.

Resistance: (1) Sterilization: It can be inactivated by extreme heat (>56°C), UV light and drying. It survives at least 24 hours on inanimate surfaces under high humidity. It is stable in liquid environments and survives for many weeks in water, body fluids and sewage. **(2) Disinfection:** Entero virus 70 is sensitive to a solution of 500 ppm sodium hypochlorite with a contact time of 2 minutes. It is resistant to phenyl mercuric borate and isopropyl alcohol.

Pathogenicity

- **Disease name:** It causes **acute hemorrhagic conjunctivitis.**
- **Epidemiology:** It is distributed in tropical countries. It causes epidemic, pandemic or large outbreaks.
- **Reservoirs of infection:** Humans.
- **Sources of infection:** These are eye secretions.
- **Modes of transmission:** Virus enters in eye through infected hands, fingers and fomites due to direct or indirect contact with eye secretions. No vector-borne transmission occurs.
- **Incubation period:** 24–48 hours.
- **Portal of entry:** Eyes.
- **Sites:** Ophthalmic and nonophthalmic sites.

Essentials of Medical Microbiology

- **Precipitating factors:** Health workers are at more risk.
- **Clinical features: (1) Acute hemorrhagic conjunctivitis:** It presents with eye pain, photophobia, swelling of eyelid and varying redness of the conjunctiva (from subconjunctival petechiae to hemorrhage). It resolved in 10 days. Transient corneal involvement may occur in some cases. Other virus produces same lesion is coxsackie type A-24. Both viruses show intratypic antigenic difference. **(2) Nonocular symptoms:** Like neurologic, respiratory and GIT disturbances.
- **Complications:** These are radiculomyelopathy and permanent polio-like paralysis (reported in some cases).

Laboratory diagnosis: Eye discharge or conjunctival scraping is tested by microscopy (follow morphology) and culture (follow C/Cs). **Molecular method** like RT-PCR is useful.

Prevention and treatment: No specific vaccine and drugs are available. Most cases are self-limiting and recover with symptomatic treatment.

RHINOVIRUS

Meaning: Rhino (Greek) means nose, as associated with nose infection.

History: Tyrrell and colleagues 1st isolated the virus in 1960 from specimen of an individual of common cold in to monkey kidney cell culture contains low bicarbonate followed by incubation at 33°C.

Morphology: Described above (**Fig. 80.1**).

Culture characteristics (C/Cs): It grows in human cell culture preferably in MRC-5 or WI-38 cells or in human diploid (secondary) fibroblast cell line. Growth is indicated by typical CPE in 2 weeks at 33°C (nasopharyngeal temperature), good O_2 tension and at neutral pH (around 7). According to growth in culture 3 groups like H, M and O are classified but these are not stable, so classification is no longer in use.

Resistance: (1) Sterilization: It proliferates between 30–35°C. Some serotypes are inactivated at 50°C. **(2) Disinfection:** It is acid labile and inactivated at pH <6.

Serotypes: More than 100 serotypes are identified by neutralization test.

Pathogenicity

- **Epidemiology:** It is distributed worldwide.
- **Reservoirs of infection:** Humans.
- **Sources of infection:** Nasal or respiratory secretions.
- **Modes of transmission:** It is transmitted by inhalation of droplets, hand to hand contact or by self inoculation of nasal secretion by contaminated hands or fingers.
- **Incubation period:** 12–72 hours, but may be longer.
- **Portal of entry:** Nose.

- **Site:** Most common in nose.
- **Precipitating factors:** It is common in children, smokers, overcrowding conditions, persons with poor sanitation and in the season of early fall usually September to November and again in the spring from March to May.
- **Clinical features: (1) Respiratory symptoms** are common cold (generally lasts for 7–11 days), nasal discharge, nasal congestion, sneezing and nasal obstruction, which interferes with sleep, feeding, smell and taste. About 30% patients develop a cough and 20% develop hoarseness, both of which may persist for up to 1 week. **(2) Other symptoms** are fever, malaise, headache, vomiting, irritability and restlessness.
- **Complications:** Secondary bacterial infections due to damage to cilia and epithelium cells, sinusitis, otitis and acute exacerbation of respiratory disease.

Laboratory diagnosis: Nasal swab or throat swab or nasal secretions are tested by microscopy (follow morphology) and culture (follow C/Cs). **Serological tests** are not useful, because of many serotypes. Antibody appears in nasal secretions. **Molecular method** like RT-PCR is useful.

Prevention and treatment: No specific vaccine (due to many serotypes) and drugs are available. Most cases are self-limiting and recover with symptomatic treatment.

ACCESS YOURSELF

Case Study

1. An 11-year-old male child is brought to the hospital with history of fever and sore throat before 5 days followed by flaccid paralysis in right leg. On examination, there is absence of deep tendon reflexes. In his other limb tone, movement, sensation and reflexes were normal. Fecal sample was collected and the reference laboratory identified ss-RNA. Identify the case and answer the following:
   a. Name the clinical condition and its etiological agent.
   b. Write pathogenicity of clinical condition.
   c. Write diagnosis of clinical condition.
   d. Write the immunoprophylaxis of given clinical condition.

Essay/Full Question

1. Discuss the etiopathogenicity, management and prevention of poliomyelitis in children.

Short Notes

1. Pathogenicity/laboratory diagnosis of poliovirus
2. Polio vaccine
3. Coxsackie virus/ECHO virus/*Rhinovirus.*

Short Questions for Theory/Viva Questions

1. Write the four differences between OPV and IPV.
2. Write the four differences between IPV and f-IPV.
3. What are tripod sign and tripod position?
4. Write the four differences between Coxsackie virus A and B.
5. What are herpengia and pleurodynia?

Comments on

1. Trivalent OPV has been replaced by bivalent OPV.
2. OPV is more useful than IPV.

MCQs for Chapter Review

Classification

1. All belong to picorna virus, *except*:
a. Entero virus 70   b. Coxsackie virus
c. *Rhinovirus*   d. HSV

2. Coxsackie virus is:
a. Herpes virus   b. Pox virus
c. *Enterovirus*   d. Myxo virus

3. Entero virus 72 is:
a. Hepatitis A   b. Hepatitis E
c. Hepatitis G   d. Hepatitis C

Polio Virus

4. All of the following statements are true regarding polio virus, *except*:
a. Transmitted by feco-oral route
b. Asymptomatic infections are common in children
c. Single serotype causes infection
d. Live attenuated vaccine produces herd immunity

5. Portal of entry of polio virus is mainly:
a. Gastrointestinal tract   b. Nasal mucosa
c. Lung   d. Skin

6. Most common mode of transmission of polio virus:
a. Droplet infection   b. Feco-oral route
c. Blood transfusion   d. Vertical transmission

7. False about polio:
a. Descending paralysis   b. Bilateral symmetrical
c. Nonprogressive   d. LMN type paralysis

8. True about polio:
a. Paralytic polio is most common
b. Spastic paralysis
c. IM injection and increased muscular activity lead to increased paralysis
d. Polio drops given only in <3 years

9. All are true about polio, *except*:
a. 99% nonparalytic
b. Flaccid paralysis
c. Exaggerated tendon reflex
d. Aseptic meningitis

10. Poliomyelitis involving the spinal cord, lesions are seen most common in:
a. Anterior horn cells   b. Posterior horn cells
c. Intermediate columns   d. Anterior and posterior horn cells

11. The most definitive method for laboratory diagnosis of poliomyelitis are A/E (all, *except*):
a. Virus isolation from blood
b. Virus isolation from CSF
c. Virus isolation from feces or throat
d. Serological diagnosis

12. Diagnosis of polio:
a. Detection of polio virus in stool
b. Serology
c. Limb wasting
d. AFP

13. Death in poliomyelitis is due to:
a. Infection   b. Neurological shock
c. Cardiac failure   d. Respiratory paralysis

14. All are false regarding polio virus *except*:
a. Most cases are symptomatic
b. Inactivated vaccine given by IM
c. Inactivated polio vaccines are given to child less than 3 years of age
d. Only one type exists

15. MgCL$_2$ is added to polio vaccine, because of the following:
a. Potentiate the vaccine
b. Vaccine can be kept at higher temperature
c. Preservative
d. None of the above

16. Common antibodies produced by IPV and OPV are:
a. IgA   b. IgM
c. IgG   d. b + c

Coxsackie Virus

17. Following virus can be grown in suckling mouse:
a. Coxsackie virus   b. *Rhinovirus*
c. ECHO virus   d. Polio virus

18. *Enterovirus* causes all *except*:
a. Hemorrhagic fever   b. Pleurodynia
c. Herpengina   d. Aseptic meningitis

19. Mode of spread of *Enterovirus*:
a. Vector mediated   b. Droplet infection
c. Feco-oral route   d. Skin contact

20. Coxsackie A virus does not causes:
a. Herpengina
b. Hand foot and mouth disease (HFMD)
c. Laryngotracheobronchitis
d. Aseptic meningitis

21. Coxsackie group A commonly causes:
a. Conjunctivitis   b. Aseptic meningitis
c. Hepatitis   d. Myocarditis

22. Herpengina is caused by:
a. Adeno virus   b. Entero virus-72
c. Coxsackie virus A   d. Coxsackie virus B

23. Coxsackie B virus causes all *except*:
a. Aseptic meningitis
b. Herpangina
c. Myocarditis
d. Bornholm disease

Entero Virus-70

24. Acute hemorrhagic conjunctivitis is caused by:
a. Entero virus 70   b. Adeno virus
c. Polio virus   d. Hepadna virus

25. Acute hemorrhagic conjunctivitis is caused by:
a. Entero virus 70   b. Adeno virus
c. Coxsackie A24   d. Hepadna virus

26. Entero viruses cause:
a. Acute hemorrhagic conjunctivitis
b. Acute follicular conjunctivitis
c. Posterior follicular conjunctivitis
d. Epidemic keratoconjunctivitis

Rhinovirus

27. No vaccine for *Rhinovirus* due to:
a. Cost
b. Multiple serotypes
c. Risk to deal with virus
d. None of above

28. **All are included in picorna group of virus,** *except*:
    a. Encephalomyocarditis
    b. HEV
    c. Foot and mouth virus
    d. Polio virus
29. **All of the following clinical features are associated with entero viruses,** *except*:
    a. Myocarditis   b. Pleurodynia
    c. Herpengina   d. Hemorrhagic fever
30. **True statements about entero viruses:**
    a. Composed of segmented genome
    b. Stable at pH 4
    c. Cause pleurodynia
    d. Cause encephalitis
    e. Cause meningitis

Answers and Explanation of MCQs

1. **d**
2. **c**
3. **a**
   - Follow section, **classification** for explanation of answers of MCQs 1–3.
4. **c**
5. **a**
6. **b**
7. **b**
8. **c**
9. **c**
10. **a**
11. **d**
12. **a**
13. **d**
14. **b**
15. **b**
16. **d**
   - Follow section, **polio virus** for explanation of answers of MCQs 4–16.
17. **a**
18. **a**
19. **c**
20. **c**
21. **b**
22. **c**
23. **b**
   - Follow section, **coxsackie virus** for explanation of answers of MCQs 17–23.
24. **a**
25. **a, c**
26. **a**
   - Follow section, **enterovirus-70 and coxsackie virus A** for explanation of answers of MCQs 24–26.
27. **b**
   - Follow section, *Rhinovirus* → **prevention** for explanation.
28. **b**
   - Viruses from Picornaviridae cause encephalo myocarditis (encephalitis by polio virus and myocarditis by coxsackie virus B) and HFMD (coxsackie virus A and B). It includes HAV (EV-72, *Hepatovirus*) not HEV (*Hepeviridae*).
29. **d**
   - Myocarditis and pleurodynia are caused by Coxsackie B (*Enterovirus*) while herpengina is caused by Coxsackie A (*Enterovirus*). Hemorrhagic fever is not caused by *Enterovirus*. Hemorrhagic fever viruses are described in **Ch. 89**.
30. **b, c, d and e**
   - **Entero viruses** are composed of nonsegmented genome, and stable at wide range of pH from 3–10. Pleurodynia, encephalitis and meningitis are caused by coxsackie viruses (*Enterovirus*).

CHAPTER 81

Infections of Orthomyxoviridae

Chapter Outline

INTRODUCTION

Myxo Virus

Commonly the virus is called **myxo virus,** because it has affinity for mucus (myxa means mucus) and includes two families like Orthomyxoviridae and Paramyxoviridae. Orthomyxoviridae is described here and for Paramyxoviridae follow **Ch. 82. Orthos (Greek) means straight** related to infection in vertebrate animals **and myxa (Greek) means mucus,** related to affinity of virus for mucin.

Classification of Orthomyxoviridae

Follow **Flowchart 81.1**.

INFLUENZA VIRUSES

History

Influenza virus was 1st isolated in 1933 by Smith, Andrews and Laidlaw.

Morphology

Shape and size: Influenza virus is spherical in shape and 80–120 nm in size. Pleomorphism is common. It is filaments in fresh isolates with several μm in length and visible under **DGI microscope**.

Genome: Influenza virus contains ss-RNA (–) with 8 pieces (**Fig. 81.1**) in type A and B. Type C contains 7 pieces of RNA lacking neuraminidase gene. It presents with RNA dependent RNA polymerase.

Capsid: NA is surrounded by helical variety of capsid.

Envelope: Nucleocapsid is surrounded by bilayered envelope which consist inner protein and outer lipid layer. Protein is virus-coded and lipid-derived from host cell membrane by budding method. Membrane protein also called **matrix** or **M protein** contains two components like M1 and M2.

Peplomeres: Viral envelop consists two types of projecting spikes called peplomeres like H and N.

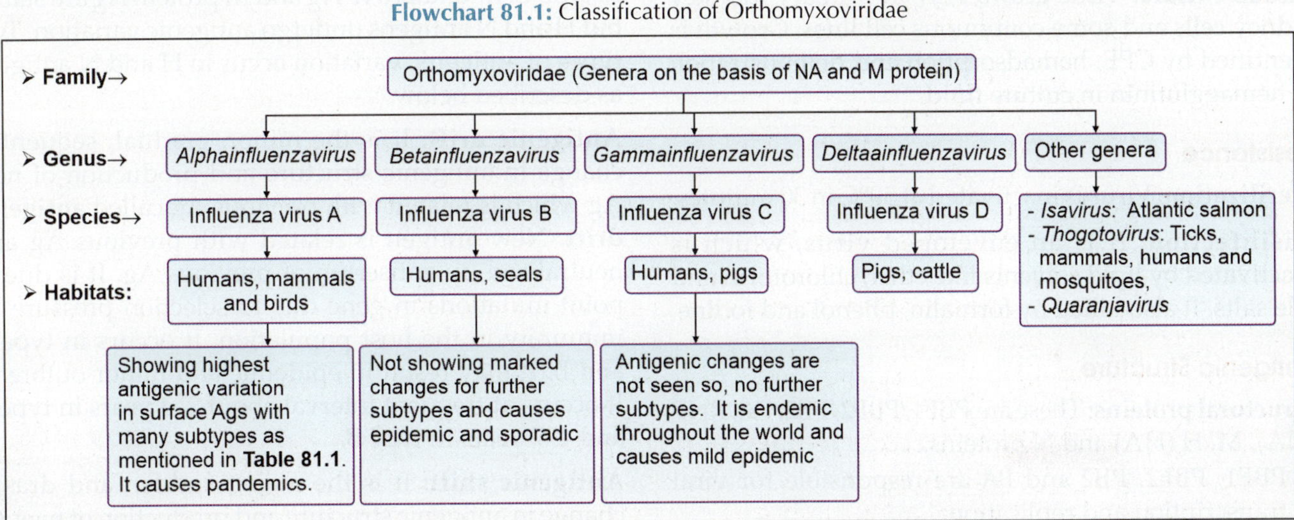

Flowchart 81.1: Classification of Orthomyxoviridae

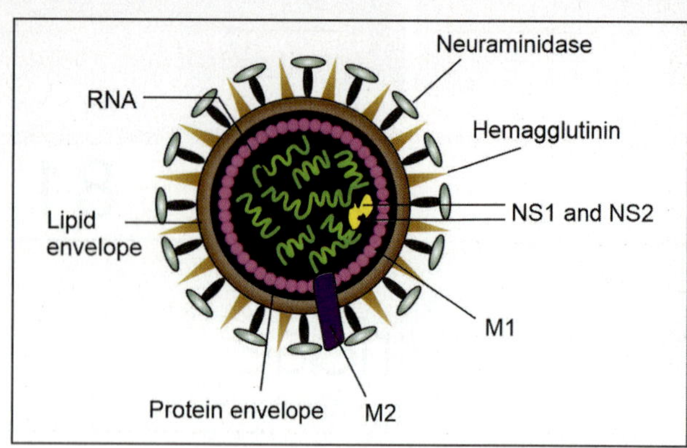

Fig. 81.1: Morphology of influenza A and B virus

Both are strain specific. **(1) Hemagglutinin (H/HA)** is triangular in cross section with 18 subtypes from H1-H18. **(2) Neuraminidase (N)** is mushroom with 11 subtypes from N1-N11. It is actually glycoprotein enzyme, which destroys the cell receptor by splitting of N-acetyl neuraminic acid from it. It destroys the hemagglutinin (H) receptors on RBCs hence called **receptor destroying enzyme (RDE)** and causes reversal of hemagglutination and releases the bounded viruses to infect the other RBCs called **elution.**

Synthesis: Virus synthesizes in nucleus.

Culture Characteristics (C/Cs)

Egg culture: Virus grows well in amniotic cavity of chick embryo. After a few egg passages, virus grows well in the allantoic cavity also, except type C. When passage serially in eggs, using undiluted infected allantoic fluid as inocula, the progeny viruses will show high hemagglutinin titer, but low infectivity called **von Magnus phenomenon,** and it is due to formation of incomplete viral particles lacking NA.

Animal culture: Intranasal instillation of virus in **ferrets** produces acute respiratory disease. Intranasal instillation of virus in **mouse** produces fatal respiratory disease while intracerebral inoculation of virus produces fatal encephalitis.

Tissue culture: Virus grows only in primary monkey kidney cells and some continuous cell lines. Growth is identified by CPE, hemadsorption and demonstration of hemagglutinin in culture fluid.

Resistance

Sterilization: Virus is inactivated at 50°C in 30 minutes.

Disinfection: It is an enveloped virus, which is inactivated by lipid solvents like ether, chloroform and bile salts. It also killed by formalin, phenol and iodine.

Antigenic Structure

Structural proteins: These are PBF1, PBF2, PB2, PA, RNP (NA), M, H (HA) and N proteins.

- PBF1, PBF2, PB2 and PA are responsible for viral transcription and replication.

- RNP (ribonucleoprotein)/NA (nucleic acid) occurs in supernatant in virus containing fluid when centrifuged. It also called internal Ag or soluble (S) Ag. It is not showing antigenic variation and detected by CFT and precipitation tests. Anti-RNP Ab develops after infection, but not after killed vaccine and useful to classify virus as A, B and C.

- M1 forms protein layer beneath the envelope and M2 forms the ion channel in envelope and it also helps in transport of molecule.

- H (HA) and N are the surface Ags. **(1) Hemagglutinin (H/HA): Biochemically,** it is glycoprotein and composed of two polypeptide chains HA1 and HA2. It **functions** as adhesion of virus with mucoprotein receptors on respiratory epithelium (adsorption) and on RBCs (hemadsoption) and causes hemagglutination of erythrocytes. It has total **18 subtypes** from H1-H18, only four have been found in human so far. **Anti-H-Ab** develops after infection and immunization, which prevents infection by preventing hemadsoption or adsorption. **(2) Neuraminidase (N): Biochemically** it is glycoprotein. It **functions** as elution (described above). It has total **11 subtypes** from N1-N11. **Anti-N-Ab** develops following infection and immunization. It is not protective as like antihemagglutinin-Ab. It prevents the release of progeny virus and thus prevents the spread and severity of infection. N Ag is a strain specific antigen and shows variation in antigenic structure, temperature and heat stability. It also presents in some bacteria like *V. cholerae* (red cell pretreated with *V. cholerae* culture suspension will resist hemagglutination by influenza virus due to destruction of receptors by RDE of *V. cholerae*).

Nonstructural proteins: These are NS1 (IFN antagonist and inhibits pre-mRNA splicing) and NS2 (exports molecules across the nucleus). Antibodies to these proteins are not protective.

Antigenic Variation in Influenza Virus A

It is highest in type A, less in type B and not demonstrated in type C. Internal RNP Ag and M protein Ag are stable, but H and N antigens undergo antigenic variation. Two types of antigenic variation occur in H and N antigens as described below.

Antigenic drift: It is the minor, gradual, sequential change in antigenic structure and production of new Ag which is related with previous Ag called **antigenic drift.** New antigen is related with previous Ag and neutralized by antiserum of previous Ag. It is due to point mutations in gene due to selection pressure by immunity in the host population. It occurs in type A and B. It may result in epidemic and minor outbreak. It occurs at frequent interval about 2–3 years in type A and 4–7 years in type B.

Antigenic shift: It is the major, sudden and drastic change in antigenic structure and production of new Ag

Essentials of Medical Microbiology

TABLE 81.1: Classification and nomenclature of strains of Influenza virus A

Old classification	Newer classification by WHO (1971)	Modified classification by WHO	Nomenclature of human strains
Asw/A swine	Hsw N1		A swine/Wisconsin/15/30 (H1N1)
A0	H0N1	H1N1	A/PR/8/34 (H1N1)
A1	H1N1		A/FM/1/47 (H1N1)
A2 (Asia)	H2N2	H2N2	A/Singapore/1/57 (H2N2)
A2 (Hong Kong)	H3N2	H3N2	A/Hong Kong/1/68 (H3N2)

which is unrelated with previous Ag called **antigenic shift**. New antigen is unrelated with previous Ag and not neutralized by antiserum of previous Ag. It is due to exchange of gene segments (genetic re-assortment). It occurs in type A only. It may result in major epidemic and pandemic. It occurs less frequently at 10–12 years interval.

Antigenic Classification and Nomenclature of Influenza Virus A Strains

Influenza virus A is further classified on the bases of variation H and N Ags as mentioned in **Table 81.1**.

Older classification: It was based only on H-Ag. Within each subtype strain shows the gradual antigenic drift. In 1930, earliest strain isolated from swine (pig) called **Asw (A swine)**. In 1934, earliest strain isolated from human called **A0**. In 1946, hemagglutinin undergoes changes and new strain identified called **A1 or A/ or A prime**. In 1957, new pandemic strain identified in Asia called **A2 (Asia)** specifically from Singapore. In 1968, new strain emerged in Hong Kong called **A2 (Hong Kong)**.

Newer classification by WHO-1971: It based on both H and N Ags. Based on H-Ag, earlier Asw, A0, A1, A2 (Asia) and A2 (Hong Kong) were designated as Hsw, H0, H1, H2 and H3, respectively. Based on N-Ag, earlier Asw, A0, A1 were classified in N1 category while A2 (Asia) and A2 (Hong Kong) were categorized as N2.

Modified classification by WHO: Earlier classification of WHO-1971 was again modified by grouping Hsw, H0 and H1 as H1.

Nomenclature of strain: Complete designation of strain will include the type of influenza virus, host only for nonhuman strain, place of origin, serial number and year of isolation followed by antigenic subtypes of H and N in brackets. For example, nonhuman strain is written as A/swine/Iowa/15/1930 (H1N1). Examples of human strains are mentioned **Table 81.1**.

Immunity

Humoral immunity: Single attack produces the effective antibodies provide protection for 1–2 years; however, problem is frequent antigenic variation by virus. Abs are subtype specific against structural and nonstructural proteins and their role is described above. IgA is synthesized locally by respiratory mucosa and provides local immunity. When individual repeatedly attacked

by different strain of influenza virus type A, antibodies are formed to earlier and also to newer strains, but the dominant antibodies are formed by earliest strain than the latest strain called **the doctrine of original antigenic sin**.

Cell-mediated immunity: Virus also induces the CMI, but its role is uncertain.

Pathogenicity

Disease name: Disease called **influenza.**

Epidemiology: Outbreak data before the isolation of virus (1933) are based on serological survey **(Seroarcheology)** of individuals alive during those years. Different influenza pandemics are mentioned in **Table 81.2**.

Reservoirs of infection: Humans, birds, animals, etc.

Sources of infection: Respiratory secretions.

Modes of transmission: It is transmitted by droplets/aerosols in early phase of disease and rare by contact.

Incubation period: 1–3 days.

Portal of entry: Respiratory system.

Sites: Pulmonary and extrapulmonary.

TABLE 81.2: Calendar of influenza pandemic

Years	Antigenic type	Nature	Common name
1889–90	H2N8	Severe pandemic	—
1900–03	H3N8	Moderate outbreak	—
1918–33	H1N1 (former HswN1/Asw)	Pandemic	Spanish flu (most sever with >50 million deaths)
1933–46	H1N1 (former H0N1/A0)	Epidemic	—
1946–57	H1N1 (former A1)	Epidemic	—
1957–68	H2N2 [former A2 (Asia)]	Pandemic	Asian flu
1968–69	H3N2 [former A2 (Hong Kong)]	Pandemic	Hong Kong flu
1977–78	H1N1 (former A1)	Pandemic	Russian or red flu
2009–10	H1N1	Pandemic	Swine flu
2023	H3N2	Outbreak	Seasonal flu

CHAPTER 83

Infections of Arboviruses and Roboviruses

Chapter Outline

- Arboviruses
  - Introduction
  - Togaviridae
  - Flaviridae
  - Bunyaviridae
  - Reoviridae
  - Rhabdoviridae
- Roboviruses
  - Introduction
  - Bunyaviridae
  - Arenaviridae

ARBOVIRUSES

Introduction

Meaning

Arboviruses means **Ar**thropod (vector) **bo**rne **viruses**. These are diverse group of RNA viruses belong to different families and transmitted by arthropods or vectors or insects.

Classification

- Viruses multiply inside the insects and establish a lifelong harmless infection in them; hence, viruses transmitted mechanically by vectors are not included in this group.
- Classification based on ecological and epidemiological, physical and chemical considerations. Families and respective genera are mentioned in **Table 83.1**.
- Species of each family and genus are further classified on the bases of clinical syndrome like fever and/or rash, fever and/or arthralgia group, encephalitis group, hemorrhagic fever group, multiple syndromes like dengue virus, etc.

Antigenic Properties

Three antigens are important in serological studies like hemagglutinin Ag, complement fixing Ag and neutralizing Ag. Antigenic cross reaction occurs in between the arboviruses. Plaque reduction neutralization test (PRNT) shows the greatest specificity for the identification of the arboviruses.

Arboviruses found in India

Arboviruses are distributed worldwide. They are more in tropical than temperate zone. Over 500 viruses have been listed, out of which about 100 can infect humans.

In India >40 viruses have been identified, of which more than 10 can infect humans.

Common arboviruses in India: Chikungunya virus, Dengue virus and Japanese encephalitis (JE) virus.

Rare in India: Kyasanur Forest disease (KFD) virus, West Nile virus, Sindbis virus, Crimean Congo hemorrhagic fever (CCHF) virus, Ganjam virus, Vellore virus, Chandipura virus, Bhanja virus, Umbre virus, Sathuperi virus, Chittor virus, Minnal virus, Venkatapuram virus, Dhori virus, Kaisodi virus and Sand fly fever virus, Zika virus.

Togaviridae

'Toga' means Roman mantle or cloak which refers to viral envelope. Different genera and species of Togaviridae are mentioned in **Table 83.1**. *Alphavirus* is the only arbovirus which includes about 32 species, of which 13 are known to infect the human. All are mosquitoes borne viruses. Cross reaction occurs in between the species. Human species of *Alphavirus* include two categories **encephalitis causing viruses and febrile illness plus arthritis causing viruses,** as described below.

Encephalitis Causing Viruses

They are transmitted by *Culex* and *Anopheles* mosquitoes.

Eastern Equine Encephalitis (EEE) Virus

It causes epidemic and sporadic cases of encephalitis in eastern Canada, USA and Caribbean. Formalin inactivated vaccine is developed.

Western Equine Encephalitis (WEE) Virus

It causes epidemic in America. Formalin inactivated vaccine is developed.

Family	Genus	Arbovirus species	
Togaviridae	*Rubivirus*	Rubella virus (It is not an arbovirus and described in **Ch. 86**)	
	Alphavirus	**Encephalitis causing viruses** • Eastern equine encephalitis (EEE) virus • Western equine encephalitis (WEE) virus • Venezuelan equine encephalitis (VEE) virus	**Febrile illness plus arthritis causing viruses** • Chikungunya virus • O'nyong-nyong virus • Semliki forest virus • Sindbis virus • Others: Mayaro virus and Ross River virus
Flaviridae	*Hepacivirus*	HCV (It is not an arbovirus and described in **Ch. 88**)	
	Pegivirus	GB (G Barker) virus C (not an arbovirus)	
	Pestivirus	Bovine viral diarrhea 1 (not an arbovirus)	
	Flavivirus	**Mosquito-borne species** **A. Encephalitis group** • Japanese encephalitis (JE) virus • St. Louis encephalitis virus • Ilheus virus • West Nile virus • Murray Valley encephalitis virus • Rocio encephalitis virus **B. Fever/hemorrhagic fever group** • Yellow fever virus • Dengue virus • Zika virus	**Tick-borne species** **A. Encephalitis group** • Russian spring summer encephalitis (RSSE) virus • Central European encephalitis virus • Western Siberian encephalitis virus • Powassan encephalitis virus • Louping ill virus **B. Hemorrhagic fever group** • Kyasanur forest disease (KFD) virus • Omsk hemorrhagic fever virus
Bunyaviridae	*Bunyavirus*	• California encephalitis virus • Oropouche virus	
	Phlebovirus	• Sand fly or *Phlebotomus* fever or *Pappataci* fever or three day fever virus • Rift Valley fever virus (hemorrhagic fever virus)	
	Nairovirus	• Nairobi sheep disease virus • Crimean Congo hemorrhagic fever (CCHF) virus • Hazara virus • Ganjamvirus	
	Hantavirus	It is not an arbovirus, but robo virus (**Table 83.2**) described below	
Reoviridae	*Rotavirus*	8 species from rota virus A to rota virus H (It is not an arbovirus and described in **Ch. 89**)	
	Orbivirus	• Orungo virus • Kemerovo virus	
	Coltivirus	Kolarado tick fever virus	
	Orthoreovirus	Not an arbovirus	
Rhabdoviridae	*Lyssavirus*	7 species like lyssa virus 1 to lyssa virus 7 (It is not an arbovirus and described in **Ch. 84**)	
	Vesiculovirus	• Vesicular stomatitis virus • Chandipura virus	

TABLE 83.1: Classification of arboviruses

Family	Genus	Robovirus species	
Bunyaviridae	*Hantavirus*	• Hantaan virus • Seoul virus • Puumala virus • Sin Nombre virus	
Arenaviridae	*Mammarenavirus*	**A. Old world viruses** • Lymphocytic choriomeningitis (LCM) virus • Lassa fever virus (hemorrhagic fever virus)	**B. New world viruses (hemorrhagic fever viruses)** • Junin virus • Machupo virus • Guanarito virus • Sabia virus • White water arroyo virus • Lujo virus

TABLE 83.2: Classification of roboviruses

It causes influenza like illness and encephalitis in Central and South America. Live attenuated vaccine has been developed.

Febrile illness plus arthritis causing viruses

Chikungunya Virus

History: Virus was 1st isolated from human case and mosquito in Tanzania in 1952.

Morphology (Fig. 83.1)
- Shape and size: Virus is spherical in shape and 50–70 nm in size.
- Genome: ss-RNA (+).
- Capsid: Icosahedral capsid.
- Envelope: Nucleocapsid is surrounded by envelop contains peplomeres like E1 and E2.
- Replication: It replicates in host cell cytoplasm and released by budding.

Culture characteristics (C/Cs): Virus is isolated in monkey cell lines in 1–2 weeks. It is useful for early diagnosis in 7–10 days of infection.

Resistance: Virus contains lipoprotein envelope and sensitive to lipid solvents like ether and chloroform.

Antigenic structure: Structural and nonstructural antigens are seen. **(1) Structural proteins:** These are proteins for capsid and envelope glycoproteins. Envelope glycoproteins are **E2 and E1.** E2 binds to host cell receptors. Virus enters in the host cell through endocytosis. **E1** is fusion peptide which, when exposed to the acidity of the endosome in eukaryotic cells, dissociates from E2 and initiates membrane fusion that allows the release of nucleocapsid in to the host cells. **(2) Nonstructural proteins:** Three in number like NS1, NS2 and NS3.

Pathogenicity
- **Disease name:** It called **chikungunya.**
- **Meaning:** Chikungunya word is derived from native word Kungunyala (kimakonde language) means

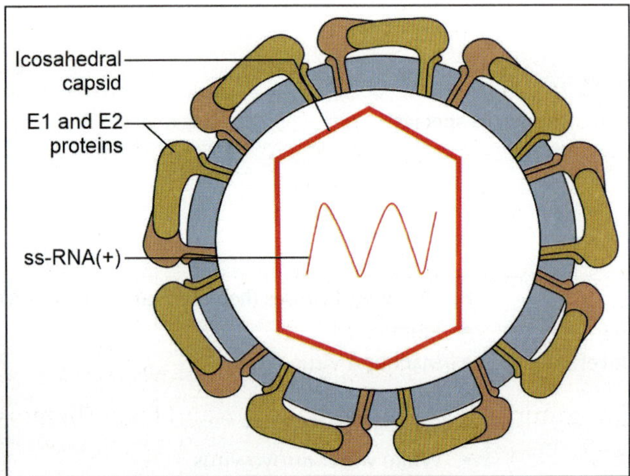

Fig. 83.1: Chikungunya virus

bending or folding, as the patient develops stooped or doubled up or contort posture in disease due to severe joint pain.

- **Epidemiology:** Three genotypes are seen according to its distribution like (1) West African genotype, (2) East/Central/South African genotype and (3) Asian genotype. Most recent epidemic was occurred in Colombia in 2014–15 with 82,977 cases. **In India** 1st outbreak was reported in 1963 in Kolkata, 2nd in 1964 in Pondicherry, Chennai-Vellore region and then in 1973 in Barsi, Maharashtra. Virus remained silent from 1973–2005 called **quiescent period.** Chikungunya **re-emerged** in 2005–06 from the Southern India (Andhra Pradesh, Tamil Nadu, Kerala, Karnataka) with 2,58,000 cases from entire country. Later it spread to other parts of India, Asia, Africa, Europe and America. This **novel virus** is due to two reasons. **(1) Mutation:** Where alanine in 226 position of E1 glycoprotein is replaced by valine. **(2) New vector:** Mutant virus is 100 times more infective to *Aedes albopticus* than *Aedes aegypticus.*
- **Reservoirs of infection:** Humans are the reservoir during epidemic or acute infection (**urban cycle**) because virus is available in high amount in blood during this stage. From human the virus has been taken up by mosquito and spreads to other humans. During other time (**sylvatic cycle**) monkeys, birds and other vertebrates are the reservoirs.
- **Source of infection:** Mosquitoes.
- **Mode of transmission:** It is transmitted by mosquito bite. Rarely it is transmitted by congenital route or by blood transfusion. Two types of cycles. **(1) Urban cycle** is maintained between human and *Aedes aegypti* (original virus and novel virus) or *Aedes albopticus* (novel virus). **(2) Sylvatic cycle** occurs in African forests. Cycle has been maintained between monkeys and forest mosquitoes like *A. furcifer, A. taylori, A. africanus* and *A. luteocephalus.*
- **Incubation period:** About 5 days (3–7 days).
- **Portal of entry:** Skin.
- **Sites:** Skin, conjunctiva, joints, lymph nodes and hemopoietic system.
- **Precipitating factors (epidemiological determinants):** Almost similar to other mosquito borne diseases as follows. **(1) Agent (virulence) factors** like novel virus which is 100 times more infective to *Aedes albopticus* than *Aedes aegypticus.* **(2) Vector factors** like density, life span, choice of host, resting habit, breeding habit, time of biting and resistance to insecticides. **(3) Host factors** like **age** (chronic chikungunya virus-induced arthralgia is common in old age group), **sex** (more in male than female because of outdoor visit and better covering of clothes in female, **arthritis** (joints with arthritis are more likely affected with more severe pain. It also favors the development of chronic chikungunya virus-induced arthralgia), **human habit** (sleeping outside and not using mosquito

repellents like net, cream, etc.), and **migration** (migration to endemic areas increases the spread). **(4) Environmental factors** like it is common after rainy season in July-November.

- **Pathogenesis:** It's not clearly understood, but after entry it can multiply in epithelial cells, endothelial cells, fibroblasts, monocytes, skeletal muscle progenitor cells and in myofibers.
- **Clinical features: (1) Chikungunya fever:** It presents with maculopapular rash (**Fig. 83.2**), sudden onset of fever, crippling joint pain, joint stiffness (predominantly in ankles and wrists), immobility, tenosynovitis, lymphadenopathy and conjunctivitis. Biphasic fever occurs with remission after 1–6 days. **(2) Chikungunya hemorrhagic fever:** Hemorrhagic tendency was present in epidemic of 1963.
- **Complications: (1) Chronic chikungunya virus-induced arthralgia:** It is more common in older age group and persons with arthritis. Arthralgia last for many years. Exact reason for this is not know but may be due to chronic persistence infection by virus. Virus or viral antigen can be detected from the synovial fluid. **(2) Neurological disorders:** These are reported very rarely like Guillian barre syndrome, palsies, meningoencaphalitis, flaccid paralysis and neuropathy.

Diagnosis
- **Specimens:** Blood/serum or synovial fluid.
- **Testing methods**
  1. **Microscopy:** Follow morphology.
  2. **Culture:** Follow C/Cs.
  3. **Serological tests:** IgM develops after 4 days of infection and lasts for 3 months. IgG develops after 2 weeks and lasts for year. Antibodies are diagnostically useful, so detection of IgM or four-fold rise in IgG titer is more significant. MAC ELISA has 95% sensitivity and 98% specificity. Rapid method for Ab detection is ICT which uses envelope antigens.
  4. **Molecular test:** Real time RT-PCR.

Prevention
- **General measures:** These are mainly mosquitoes controlling measures as described in **Ch. 110**.

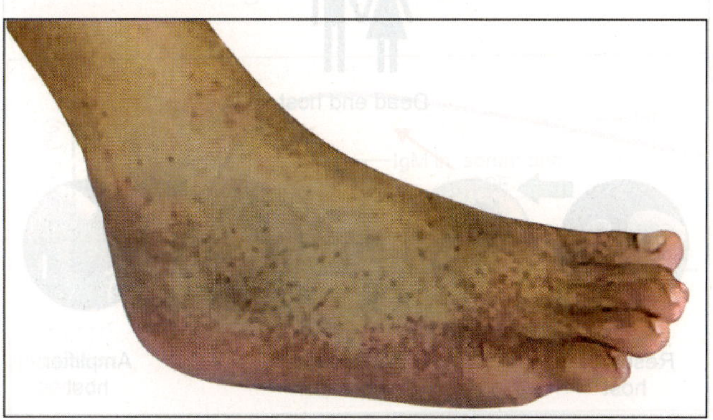

Fig. 83.2: Rashes of chikungunya on right feet

- **Immunoprophylaxis:** Currently no approved vaccine exists.

Treatment: No specific antiviral drug is available, symptomatic treatment is given as follows.

O'nyong-nyong Virus

Antigenically it is related to chikungunya virus and causes similar disease. Virus was 1st isolated in Uganda and confirmed to Africa. It is transmitted by *Anopheles*.

Semliki Forest Virus

Virus was 1st isolated in Uganda in 1942 from *Aedes* mosquitoes. It is not associated with human illness though neutralizing antibodies have been seen in Africa.

Sindbis Virus

Virus was 1st isolated in the Sindbis district of Egypt in 1952 from *Culex* mosquitoes. It is not associated with human illness though neutralizing antibodies have been seen in India.

Others Species of *Alphavirus*

These are Mayaro virus and Ross River virus.

Flaviridae

Flavi word derived from 'flavus' (Latin) means yellow, refer to the species, yellow fever virus of this family. Different genera and species of Flaviridae are mentioned in **Table 83.1**. *Flavivirus* is the only arbovirus and different species are described below in two categories like mosquito borne species and tick borne species.

> **Notes: Pegivirus**
>
> **Species:** GB (G Barker) virus C, because it was discovered in 1995 by G Barker. It is not an arbovirus. Formerly it called **Hepatitis G virus (HGV)**, but wrongly it was interpreted as HGV because neither it is hepatotropic nor causing hepatitis. In fact it multiplies in bone marrow and spleen and not known to cause any human disease called orphan virus.
>
> **Epidemiology:** It has six genotypes and each has its own geographical distribution.
>
> **Pathogenicity:** It is transmitted by blood/blood products and by sexual intercourse. It's RNA has been found from the patient of hepatitis, hemodialysis, IV drug addicts and blood donor, but it's role in hepatitis is still to be decided. It's genome resembles to HCV genome except, it lacks protein corresponding to the core protein of HCV that forms the nucleocapsid. HGV infection occurs independently and it does not require HCV. However its prevalence is higher in patient with HCV and HIV. It shows the 35% prevalence in HIV infected persons, but surprisingly this dual infection protects against HIV and patients survive longer.
>
> **Diagnosis:** HGV RNA can be detected by RT-PCR from the blood. It produces the Abs, which are protective but not useful in diagnosis.

Mosquito Borne Species of *Flavivirus*

Two groups like encephalitis group and fever/hemorrhagic fever group as follows.

A. Encephalitis group: It includes following five species.

652

Morphology: Almost same as JEV **(Fig. 83.3)**. Virus produces intra nuclear **Cowdry type-A** (acidophilic) inclusion body called **Torres body**. This is examined under light microscopy. All the necrosed cells of liver are coalesces to form multinucleated cell called **Councilman body**. Yellow fever antigen may be demonstrated in formalin fixed tissues of liver, kidney, and heart by immunocytochemical staining.

Culture characteristics (C/Cs): Virus is cultivated on Vero cells using serum free medium.

Resistance: It is enveloped virus sensitive to lipid solvents.

Antigenic structure: Antigenically homogenous and only one serotype is known to exist. Virus contains **structural antigens** of capsid, matrix and envelop. **Non-structural antigens** of virus are NS1, NS2a, NS2b, NS3, N4a, NS4b and NS5.

Immunity: Indian population is non-immune to yellow fever. Single infection confers lifelong immunity and second attack is uncommon. Infants born of immune mothers have passive immunity for 6 months. India is endemic for dengue and presence of dengue antibodies may provide little protection to yellow fever; however yellow fever vaccine is not providing any amount of protection to dengue.

Pathogenicity

- **Disease name:** Disease called **yellow fever**. The name "yellow fever" originated from its propensity to cause jaundice (present with yellowish discoloration of sclera) in victims. It also called **Yellow Jack** (familiar to the pirates of yellow jack).
- **Epidemiology:** It is endemic in West Africa and South plus Central America. It is not present in rest of the world including India. India is considered as receptive area because of availability of vector (*Aedes aegypti*) and non-immune human. Seven **genotypes** are identified based on genomic sequence, five of that are endemic in Africa and two in South America. 1st epidemic was reported in Kenya in 1992. WHO estimated about 200,000 cases and 30,000 deaths every year due to yellow fever with nearly 90% outbreaks are only from Africa.
- **Reservoirs of infection:** Humans are the reservoirs for urban infection while wild monkeys are the reservoirs for wild infection.
- **Sources of infection:** Mosquitoes.
- **Modes of transmission:** It is transmitted by bite of *Aedes aegypti* mosquito. Two types of cycle are observed like urban cycle which is maintained between human and *Aedes aegypti* while wild cycle maintained between wild monkeys and mosquitoes like *Haemagogus spegazzinii* in South America and *A. africanus* and *A. simpsoni* in Africa. Mosquito to mosquito spread is possible by transovarial route. Once it infected remains so for life.

- **Incubation period:** 3–6 days.
- **Portal of entry:** Skin.
- **Site:** Liver.
- **Precipitating factors (epidemiological determinants):** **(1) Age:** Less in below 6 months due to passive protection by maternal antibodies, otherwise all age groups are susceptible. **(2) Sex:** More in male due to outdoor visit. **(3) Occupation:** Like forest workers are at high risk. **(4) Migration of population:** From endemic to receptive areas increase the risk. **(5) Temperature and humidity:** Temperature >24°C and humidity >60% favor the multiplication of virus. **(6) Urbanization:** It brings population closer to the wild cycle.
- **Pathogenesis:** After transmission from a mosquito, the virus replicates in the lymph nodes and infects dendritic cells. From there, it reaches the liver and infects hepatocytes, which leads to eosinophilic degradation of cells and release of cytokines. Apoptotic masses called **Councilman bodies** appear in the cytoplasm of hepatocytes. Death occurs due to shock and multiple organs failure.
- **Clinical features: Prodromal phase** presents with sudden onset of high grade fever including chills, headache, nausea and vomiting. Pulse is slow despite high temperature. **Jaundice phase** presents with jaundice, albuminuria and hemorrhagic manifestation.
- **Complication:** Death due to renal or hepatic failure. Mortality is >20% in children and elders.

Diagnosis: Primary diagnosis is based on clinical parameters like signs, symptoms, travel history, activities, etc.

- **Specimens:** Body fluids, blood and tissues like liver, kidneys, heart, etc.
- **Testing methods**
  1. **Blood picture:** These are leukopenia, thrombocytopenia, abnormal liver function test, prolonged clotting time and abnormal electrolytes.
  2. **Urine picture:** Abnormal kidney function test. Urine tests may demonstrate elevated levels of urinary proteins and urobilinogen.
  3. **Microscopy:** Follow morphology.
  4. **Culture:** Follow C/Cs.
  5. **Serological tests:** MAC ELISA is positive within 5–7 days but cross-reacting antibodies to other species (like dengue virus, zika virus, etc.) of *Flavivirus* often complicate the diagnosis, hence positive results should be confirmed by PRNT or by HAI test. False positivity also occurs after immunization.
  6. **Molecular method:** Identification of viral RNA NS5 gene by RT–PCR is more sensitive and specific then MAC ELISA.

Prevention

- **General measures:** Mosquito control measures are described in **Ch. 110**. It does not exist in India and India is considered as '**receptive area**' because of

6

large numbers of availability of *Aedes aegypti*, non-immune humans and tropical climatic condition as like Africa. General measures taken at airport include **(1) Quarantine** of travelers for 6 days coming from endemic areas who are unimmunized. **(2) Strict vaccination:** Strict vigilance for vaccination for persons travel to or from endemic areas. **(3) International certificate of vaccination:** It is mandatory for persons travel to or from endemic areas. Certificate is issued 10 days after the date of vaccination and validate for 10 years. Validity is extends for 10 years more from the date of vaccination following revaccination. Vaccine is contraindicated in infants (<1 year) and even vaccine also induces yellow fever in <1 year, so decision about requirement of certificate for infant is rest on individual country. In this regard, India requires vaccination/certificate too. **(4) Breteau index or *Aedes aegypti* index:** It is defined as numbers of containers showing breeding of *Aedes aegypti* larvae/numbers of house surveyed × 1000. It should be <1, surrounding 400 meter of air port.

- **Immunoprophylaxis:** It's not occurs in India though vaccination is mandatory for travel to or from endemic areas. Vaccine starts protection after 10 days and continues for 10 years.
  1. **Killed/neurotropic/Dakar strain vaccine:** It was prepared from infected mouse brain. It was used in French West Africa. It produces encephalitis so not useful.
  2. **Live attenuated/non-neurotropic vaccine/17D/Asibi strain vaccine:** It is a live attenuated vaccine **prepared** by Theiler in 1937 in allantoic cavity of chick embryo from **Asibi strain**. In India it is prepared in Central Research Institute, Kasauli. During transportation it should be stored at temperature range from −30°C to + 5°C. It is available in lyophilized form and reconstituted before use with sterile saline. It should be used within 30 minutes following reconstitution. Single dose is **administered** by subcutaneous injection. **Efficacy** is very good, as it is safe, effective in 7–10 days and provides protection for 35 years. It is **contraindicated** in children <9 months (<6 months in endemic areas), pregnancy except during outbreak, person with IDDs like AIDS and people allergic to eggs proteins. Cholera and yellow fever vaccine interact with each other, hence should not be given together and 3 weeks gap should be maintained between two vaccines.

Treatment: No specific antiviral drugs are available for yellow fever. Only supportive treatment is given.

Dengue Virus

History: DEN 1 was isolated in Hawaii in 1944, DEN 2 (more dangerous/virulent than other types) from New Guinea in 1944, DEN 3 and 4 from Philippines in 1956. Recently DEN 5 is identified in 2013 from Malaysia.

Morphology (Fig. 83.3): Same as JE virus.

Culture characteristics (C/Cs): Virus isolation is best useful in initial 5 days. **Animal culture** includes use of suckling mouse, but less sensitive. **Tissue culture** includes use of Vero or BHK-cells or mosquito cells like C6/36 and AP61.

Resistance: It is an enveloped virus sensitive to lipid solvents.

Antigenic structure: Two types like **structural antigens** of capsid, matrix and envelop and **non-structural antigens** like NS1, NS2a, NS2b, NS3, N4a, NS4b and NS5. Spike protein is viral hemagglutinin which is a group and type-specific.

Immunity: Two types of antibodies are produced.
1. **Neutralizing antibodies:** These are antibodies against both infecting and non infecting serotypes. These are protective in nature and show the 60-80% cross reactivity with other serotypes of dengue virus. Antibodies induce long-life protection against the infecting serotype, but they gives short time cross protection against the other types.
2. **Non-neutralizing antibodies:** Non-neutralizing antibodies are heterotypic (because antibodies produce against non infecting serotypes but not against the infecting serotype) and last lifelong. The first (primary) dengue infection causes mostly minor disease, but secondary dengue infection has been reported to cause severe disease (DHF or DSS) in both children and adults. This phenomenon is called **antibody dependent enhancement**. Mechanisms of antibody dependent enhancement include formation of antibodies following 1st serotype are bound to 2nd serotype, but instead of neutralizing to 2nd serotype they protect the 2nd serotype against the action of host B-cells resulting more sever disease during secondary infection. It occurs most when serotype 1 infection is followed by serotype 2. This progresses to DHF and DSS.

Pathogenicity
- **Disease name:** Disease called **dengue** or **dengue fever**. Term dengue derived from the 'Ki denga pepo (Swahili language)' means sudden seizure by a demon.
- **Synonym:** Term **'break-borne fever'** was coined during the Philadelphia epidemic in 1780. It also called **'Dandy fever'** in West Indies. Word related with posture and gait in dengue fever in West Indies.
- **Epidemiology**
  - **Distribution:** It is widely distributed in tropics and subtropics. It occurs in both epidemic and endemic forms. It is endemic in >100 countries of world. Tropical countries of South East Asia and Western Pacific are at high risk. Epidemic starts in rainy season and usually explosive. Initial 4 types are present in India, but there is no indication of

the presence of DEN-5 in India. Seroprevalence varies between season and places but DEN-1 and DEN-2 are widespread. More than one type has been isolated from same patients.

  – **Genotypes:** Total 13 genotypes are identified by molecular typing like 3 for DEN 1, 2 for DEN 2 and 4 for each DEN 3 plus DEN 4.

- **Reservoirs of infection:** Only humans no animal reservoirs are known.
- **Sources of infection:** Mosquitoes.
- **Mode of transmission:** It is transmitted by bite of *Aedes aegypti* mosquito. Extrinsic incubation period is 8–10 days.
- **Incubation period:** 3–14 days.
- **Portal of entry:** Skin.
- **Sites:** Blood, lymph nodes, etc.
- **Precipitating factors (epidemiological determinants):** **(1) Age:** It is common at all ages. Children <12 years are more prone to develop DHF and DSS. **(2) Sex:** More in males due to outdoor visit. **(3) Immunity:** Secondary infection is less in persons having antibodies but more sever, if occurs. **(4) Environmental factors:** It is widely distributed throughout the tropics and subtropics. Epidemic occurs in rainy season and usually explosive.
- **Pathogenesis:** Viremia progress to fever, hemorrhage and shock. Initial dengue infection by any serotype called **primary dengue infection**. Subsequent dengue infection, months or year after the primary infection by other serotype, which is more severe, called **secondary dengue infection**. There is a hypersensitive or enhancement response due to infection by more than one dengue virus serotype (double infection) as described earlier. Serotype 1 followed by serotype 2 is more dangerous (it progress to DHF and DSS) than serotype 4 followed by serotype 2. Serotype 2 is considered as most virulent.
- **Clinical features:** Traditional (1997) WHO classification divide the dengue in to three categories (**Flowchart 83.1**). **(1) Dengue fever:** Sudden onset of fever which is biphasic (**saddle back**). It presents with chills, 39°–40°C temperature and subsides in 5–7 days. It rises again after 5–8 days hence called

biphasic or **saddle back fever.** Other features of dengue are headache, retro-orbital pain, conjunctival injection, pain in back and limbs (**break bone fever**), lymphadenopathy and maculopapular rashes appear in 80% cases during remission or during 2nd febrile phase. This stage lasts for 2 hours to several days. **(2) Dengue hemorrhagic fever (DHF):** This stage presents with high grade continuous fever, hepatomegaly, thrombocytopenia (count is ≤1,00,000/mm³ and hemorrhagic manifestations. Hemorrhage is detected by positive tourniquet test (>20 petechial spots per square inch area in cubital fossa) and spontaneous bleeding from skin, nose, mouth and gums. **(3) Dengue shock syndrome (DSS):** It includes shock in addition to all above features. Shock is characterized by circulatory failure with rapid and weak pulse, narrowing pulse pressure (20 mm Hg or less) or hypotension with cold clammy skin and restlessness.

- **2009 WHO classification:** It divides the dengue in following two categories based on the severity.
  – Dengue with or without warning sign.
  – Severe dengue.

Laboratory diagnosis: Diagnosis is done clinically on the bases of fever characteristics, blood picture, bleeding tendency, pulse, pressure, etc.

- **Specimens**
  – Whole blood/plasma/serum/CSF.
  – Autopsy materials like liver, spleen, lymph node, thymus, intestine, lungs, skin rashes materials, brain, kidneys, etc., are collected for virus isolation.
  – Virus also isolated from insect vectors or from animals/birds reservoirs.
  – Collect the blood from the heart in autopsy patients.
- **Testing methods**
  1. Blood picture:
  – Platelets count: ≤1,00,000/mm³ (thrombocytopenia).
  – Hematocrit: Increased by 20% or more of base line.

  2. Urine picture: Proteinuria, hematuria, etc.
  3. Microscopy: Follow morphology.
  4. Culture: Follow C/Cs.
  5. Serological tests:
  – Immunological reactions (**Fig. 83.6**): IgM and IgG antibodies are produced against the spike proteins. Immune response varies depending on whether the individual has a primary (1st dengue or other *Flavivirus*) v/s a secondary (had dengue or other *Flavivirus* infection in past) dengue infection. **Primary infection detected by** virus isolation or by identifying NS1Ag or by IgM detection followed by IgG detection in low titer. **Secondary infection is detected**

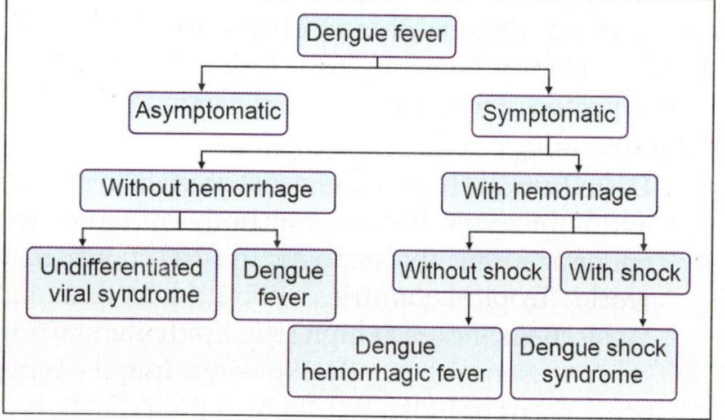

Flowchart 83.1: Clinical features of dengue

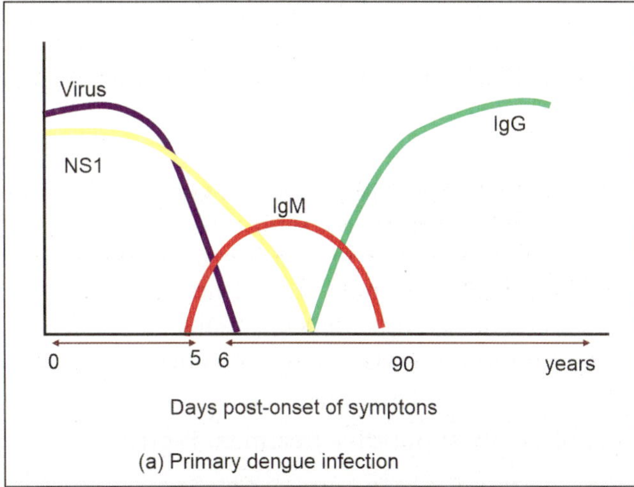

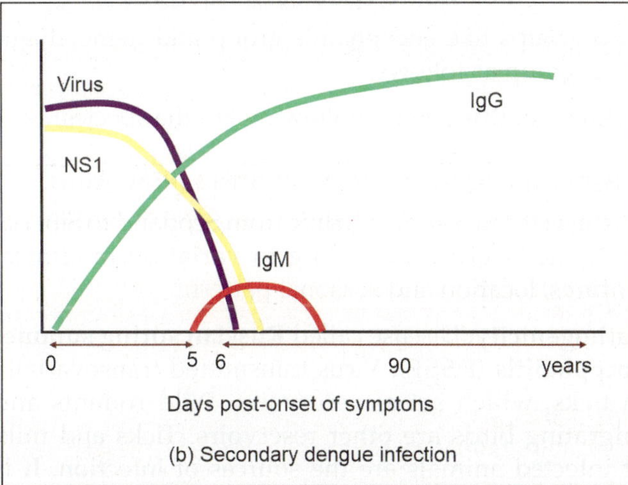

Fig. 83.6: Immunological reactions in dengue

by IgG detection. NS1 appears in serum from day 1, after the onset of fever and remains positive for 18 days. Its diagnosis differentiates between flavi viruses and also between different serotypes of dengue. IgM is produced after 4–5 days and persists for 90 days. IgG develops 6–15 days following illness and continue for long time even >60 years used to detect the **past infection** in absence of symptoms.

- Tests: **(1)** NS1 Ag is detected by ELISA and rapid test like ICT. **(2)** IgM is detected by MAC-ELISA or by rapid test like ICT. MAC-ELISA is useful for recent infection. It has 90% sensitivity and 98% specificity (cross reactivity with other flavi viruses is very low). **(3)** IgG is detected by ELISA. **(4)** PRNT is most sensitive and specific test, but expensive and labor intensive so routinely not used. **(5)** HAI is sensitive but less specific. The hemagglutinin antibodies last long, so useful for epidemiological study.
- Notification on rapid test like ICT: It provides result in few minutes, but less sensitive and less specific, hence Government of India issued a notification that dengue NS1 Ag or IgM positive by ICT is considered as probable diagnosis which should be confirmed by ELISA.

6. **Molecular methods:** Real-time RT-PCR (qRT-PCR) is more rapid and sensitive. It is positive from –1 to +5 days of beginning of clinical features. It is useful to detect the type of DEN and also for viral load.

Prevention

- **General measures:** Mosquito control measures are described in **Ch. 110.**
- **Immunoprophylaxis:** Live attenuated tetravalent vaccine based on Chimeric Yellow fever Dengue virus Tetravalent Dengue Vaccine (CYD-TDV) has been developed by Sanofi Pasteur Company with name of **dengvaxia.** It is found safe for human use and effective in Phase III clinical trial done in Latin America. WHO recommended its use for persons with 9–45 years after confirmed previous dengue infection. If it is given to previously uninfected person then subsequent infection may produces very severe disease. It is not available in India due to safety issues. It is available in lyophilized form and reconstituted before use with sterile saline. Three doses are **administered** by subcutaneous route in 0.5 ml amount at 6 months interval. **Efficacy** against hospitalized dengue is found around 80%. It is **contraindicated** in pregnancy, breastfeeding women, person with IDDs like AIDS and people allergic to vaccine.

Treatment: No specific antiviral drug for dengue virus only supportive treatment is given.

Zika Virus

Meaning: So called because virus was 1st identified from rhesus monkeys in zika forest, Uganda in 1947.

Morphology: Morphologically zika virus is similar to JE virus **(Fig. 83.3).** Flavi viruses generally replicate in cytoplasm, but zika multiplies in nucleus of host cells.

Culture characteristics (C/Cs): Zika has been successfully cultivated on Vero cells.

Resistance: Enveloped virus sensitive to lipid solvents.

Antigenic structure and serotypes: Include the three structural (capsid, matrix and envelope) and seven nonstructural antigens.

Pathogenicity

- **Disease name:** It called **zika fever** or **zika virus disease.**
- **Epidemiology**
  - **Genotypes (lineages):** There are two zika lineages like the African lineage and the Asian lineage. Phylogenetic studies indicate that the virus spreading in the America is 89% identical to African genotypes and also closely related to the Asian strain that circulated in French Polynesia during the 2013–2014 outbreaks.

- **Distribution:** Outbreaks have been recorded form Africa, Ameria and South East Asia. Largest outbreak occurred in Brazil in 2015–16. In India outbreaks were reported from Rajasthan, Madhya Pradesh and Gujarat in 2018.
- **Reservoirs of infection:** It is epizootic and monkeys are the reservoir hosts. Human is the accidental host.
- **Sources of infection:** Mosquitoes.
- **Mode of transmission:** It is transmitted by bite of *Aedes aegypti* mosquito, which is active mostly in the daytime. Extrinsic incubation period is about 10 days. Cycle is maintained between wild monkey and mosquitoes with occasional transmission to human. It also transmitted by sexual route including semen, blood transfusion, organ transplantation and transplacental with microcephaly in fetus.
- **Incubation period:** Not clear but may be few days.
- **Portal of entry:** Skin.
- **Sites:** Muscles, joints, lymph nodes, brain, etc.
- **Precipitating factors (epidemiological determinants): (1) Age:** No age bar. **(2) Sex:** More in males due to outdoor visit. **(3) Occupation:** Like forest workers are at high risk. **(4) Migration of population:** Migration in to endemic areas increases the risk. **(5) Pregnancy:** At most risk and development of microcephaly to fetus of infected mother. **(6) Urbanization:** It brings population closer to the wild cycle.
- **Pathogenesis:** After the mosquito bite virus enters in epidermal keratinocytes, fibroblasts and Langerhans cells of skin. Further progress is done through lymph nodes and blood stream to produce the viremia and other features. It multiplies in the nucleus of host cells unlike other flavi viruses.
- **Clinical features:** No symptoms or mild symptoms like fever, skin rash, conjunctivitis, muscle and joint pain, malaise or headache. These symptoms normally last for 2–7 days. Feature simulating the dengue fever.
- **Complications: (1) Fetal damages:** Zika can also spread from a pregnant woman to her fetus. This can results in microcephaly, severe brain malformations, and other birth defects. Incidence of fetal damage is 3% when infection occurs in 1st trimester and 1% when infection occurs in 2nd or 3rd trimester **(2) In adults:** Rarely Guillain–Barré syndrome.

Laboratory diagnosis

- **Specimens:** Blood, urine, semen, CSF or saliva.
- **Testing methods:**
1. **Microscopy:** Follow morphology.
2. **Culture:** Follow C/Cs.
3. **Serological tests:** PRNT detects the neutralizing antibodies. It is the confirmatory test. MAC-ELISA detects IgM, which appears after 1st week and remains raised up to 12 weeks. It is detected from CSF or serum. Cross reactivity may occur with other flavi viruses, which is cleared by PRNT.
4. **Molecular method:** Multiplex real-time RT-PCR is useful, which targets zika virus (NS5), chikungunya virus (NS4) and dengue 1–4 (3' untranslated region).

Prevention

- **General measures:** Mosquito control measures are described in **Ch. 110.** Other preventive measures are abstinence from sex with infected partner, no pregnancy planning by women living in affected areas, pregnant women should not travel to these areas and screening of blood donors.
- **Immunoprophylaxis:** There is no effective vaccine.

Treatment: No specific antiviral drugs for zika virus are available. Only supportive treatment is given.

Tick-borne Species of *Flavivirus*

Two groups like encephalitis group and hemorrhagic fever group as follows.

A. Encephalitis group: Following are the species.

Russian Spring Summer Encephalitis (RSSE) Virus

History: It causes encephalitis from Scotland to Siberia. Its name is given according to variation in clinical features, location and seasonal pattern.

Pathogenicity: Disease called **Russian spring summer encephalitis (RSSE).** Virus transmitted transovarially in ticks, which act as reservoirs. Wild rodents and migrating birds are other reservoirs. Ticks and milk of infected animals are the sources of infection. It is transmitted by bite of *Ixodes persulcatus* tick or by drinking the milk of infected goats. It presents with encephalitis with high fatality and permanent paralytic sequelae. Biphasic meningoencephalitis may occur.

Prevention: Avoiding ticks bite. Formalin inactivated vaccine is effective.

Other Tick-borne Species of *Flavivirus*

Central European encephalitis virus: It is transmitted by tick bite like *Ixodes ricinus.*

Western Siberian encephalitis virus: It is transmitted by tick bite like *Ixodes persulcatus.*

Powassan encephalitis virus: It is transmitted by tick bite like *Ixodes cookie.*

Louping ill virus: It is transmitted by tick bite like *Ixodes ricinus.* It causes acute viral encephalitis in sheep. Name given because infected sheep 'loup' (old Scottish word) or spring in the air. Human cases have been reported from Europe.

B. Hemorrhagic fever group: Following are two species.

Kyasanur Forest Disease (KFD) Virus

History and meaning: Virus was 1st isolated in Kyasanur forest, Shimoga district of Karnataka, India in 1957 from human and dead monkeys.

Morphology: Virus is belongs to the Russian spring summer encephalitis virus group. It is a spherical, enveloped virus of 45 nm in diameter and has a single-stranded, positive-sense RNA genome.

Culture characteristics (C/Cs): (1) Animal culture: Research using mouse model found that virus primarily replicated in the brain. Other research had identified the neurological changes in experimental mouse like gliosis, inflammation and cell death in the brain. **(2) Cell culture:** It produces the cytopathic effect on cell culture.

Resistance: It is an enveloped virus sensitive to lipid solvents.

Pathogenicity

- **Disease name:** It is a zoonosis. Transmission cycle involves monkeys and ticks. Human disease called **Kyasanur forest disease.** Animal disease called **monkey disease** or **monkey fever**.
- **Epidemiology:** Currently, it is endemic in 5 districts of Karnataka, India like Shimoga, North Kanara, South Kanara, Udupi and Chikkamagaluru.
- **Reservoirs of infection:** Monkeys, porcupines, rats, squirrels, mice, shrews and birds are the reservoirs. Virus transmitted transovarially in ticks, which act as reservoirs. Wild monkeys are the amplifier hosts. Man is the accidental and dead end. No human to human transmission can occur.
- **Sources of infection:** Ixodid or hard ticks.
- **Mode of transmission:** It is transmitted by ticks bite like *Haemaphysalis spinigera, H. turtura,* etc.
- **Incubation period:** 3–8 days.
- **Portal of entry:** Skin.
- **Sites:** Blood, conjunctiva, muscles, etc.
- **Precipitating factors (epidemiological determinants): (1) Age:** No age bar. **(2) Sex:** More in males due to outdoor visit. **(3) Occupation:** Like forest workers are at high risk. **(4) Migration of population:** Migration in to endemic areas increases the risk. **(4) Seasonal:** Dry season from November–June increase the risk, because of increase forest activities.
- **Clinical features: (1) Acute (prodromal) phase** presents with sudden onset of fever, headache, conjunctivitis, myalgia, severe prostration, etc. **(2) Hemorrhage phase** includes hemorrhage in skin, mucosa and inner viscera.
- **Complication:** Fatality rate of 3–10%.

Laboratory diagnosis

- **Specimens:** Blood/serum.
- **Testing methods:**
  1. **Microscopy:** Follow morphology.
  2. **Culture:** Follow C/Cs.
  3. **Serological tests:** IgM detected by ELISA
  4. **Molecular methods:** Nested RT-PCR and real-time RT-PCR are useful methods.

Prevention

- **General measures:** These are vector control, adequate clothing, avoiding sleeping on ground, etc.
- **Immunoprophylaxis:** A killed/formalin-inactivated vaccine is **prepared** from chick embryo by Haffkine institute, Mumbai. Two doses are **administered** at interval of 2 months followed by booster dose after 6–9 months and then every 5 years. The vaccine has a 62.4% **efficacy** for individual who receive two doses and 82.9% for those who receive an additional dose. It is **indicated** in villages, within 5 km range of endemic areas in Karnataka. Incidences are reduced after the vaccination.

Treatment: No specific antiviral drugs are available for KFD. Only supportive treatment is given.

Omsk Hemorrhagic Fever Virus

Clinically it is similar to KFD virus. It occurs in Russia and Romania and transmitted by *Dermacentor* ticks.

Bunyaviridae

Family includes four genera as described below.

Bunyavirus

It includes >150 species and only few can infect human.

California Encephalitis Virus

It is endemic in USA, transmitted by mosquito and causes encephalitis, aseptic meningitis and fever.

Oropouche Virus

It is transmitted by midges bite like *Culicoides paraensia* and causes aseptic meningitis and fever.

Phlebovirus

Following are two important human species.

Sand Fly Fever Virus

It also called *Phlebotomus* **fever virus** or *Pappataci* **fever virus** or **three day fever virus**. It is transmitted by sand fly like *Phlebotomus pappataci* and fever lasts for three days hence above mentioned names are given. It is endemic in Mediterranean coast and Central Asia and extended to East Pakistan to North West India. It has 20 antigenic types of which five can cause human disease like Chagres, Candiru, Naples, Silician and Punta Toro. Virus transmitted transovarially in sand flies which act as reservoirs. It causes self limited febrile illness in human.

Rift Valley Fever Virus

It named after Rift Valley, Kenya, where it was 1st recognized. It is transmitted by mosquitoe bite like *Culex tritaeniorhynchus* and *Aedes vexans* and also by direct contact with tissues of infected animals. It produces human disease resembling influenza. It causes fever or hemorrhagic fever.

Genus was named from its species. Following are two important human species.

Nairobi Sheep Disease Virus

It is transmitted by tick bite like *Rhipicephalus*. It produces disease in sheep and goats called **Nairobi sheep disease,** which is characterized by acute hemorrhagic gastroenteritis. Human infection presents with mild febrile illness.

Crimean Congo Hemorrhagic Fever (CCHF) Virus

Synonym: Commonly, it called Congo virus.

History: Virus was first identified in Crimea in 1945 and it was identical to the Congo fever virus identified in 1956 in Congo (Zaire), hence the name CCHF virus.

Morphology (Fig. 83.7): It is pleomorphic virus with 80–120 nm in size. It contains ss-RNA (–) with three segments like Small (S), Medium (M) and Large (L). L segment encodes for RNA polymerase, M for envelope proteins (Gc and Gn) and the S for nucleocapsid. NA is surrounded by helical capsid. It is an enveloped virus with 5–10 nm long projecting spikes.

Culture characteristics (C/Cs): Virus isolation is most effective in first 5 days. **Animal culture** is done in suckling mouse which is more sensitive. **Tissue culture** is done in SW-13, Vero cells, LLC-MK2 and BHK-21 cells.

Resistance: Virus can be inactivated by 1% hypochlorite solution, 2% glutaraldehyde solution and by boiling at 56°C.

Pathogenicity

- **Disease name:** It is a zoonosis, called **Crimean Congo Hemorrhagic fever (CCHF).**
- **Epidemiology**
  - **Distribution:** CCHF is found in Eastern Europe, in North-Western China, Central Asia, southern Europe, Africa and the Middle East. **In India** it was 1st identified in 2011.

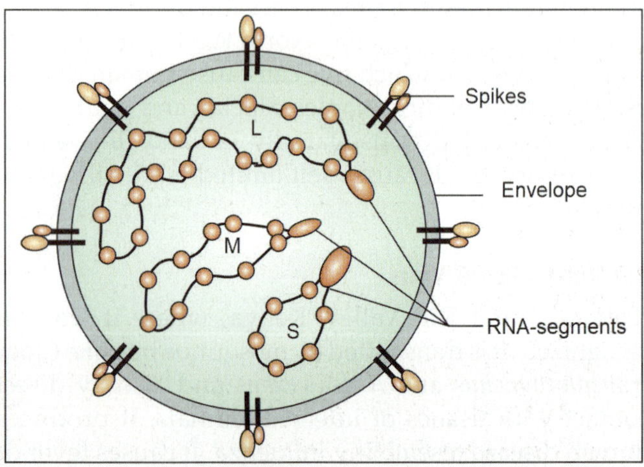

Fig. 83.7: Morphology of CCHF virus

  - **Genotypes:** Total **seven genotypes** are recognized as mentioned below.
    1. Africa 1: Senegal.
    2. Africa 2: Democratic Republic of the Congo and South Africa.
    3. Africa 3: Southern and western Africa.
    4. Europe 1: Albania, Bulgaria, Kosovo, Russia and Turkey.
    5. Europe 2: Greece.
    6. Asia 1: Middle East, Iran and Pakistan.
    7. Asia 2: China, Kazakistan, Tajikistan and Uzbekistan.
- **Reservoirs of infection:** Domestic animals like cattle, sheep, goats and others are the reservoirs.
- **Sources of infection:** Ixodid ticks (hard ticks).
- **Mode of transmission:** It is transmitted by bite of tick like *Hyalomma marginatum*, *H. anatolicum*, etc., direct contact with blood or infected tissues, aerosols, crushing of infected ticks with bare hands, drinking un-pasteurized milk from infected animals, vertically from mother to child or by fomites.
- **Incubation period:** Depends on the mode of acquisition of the virus like 1–3 days (maximum of 9 days) following tick bite and 5–6 days, (maximum of 13 days) following contact with infected blood or tissues.
- **Portal of entry:** Skin.
- **Sites:** Symptoms occurs in many systems.
- **Precipitating factors:** Agricultural workers, slaughterhouse workers, veterinarians and health care workers attending on suspected/probable/confirmed CCHF cases and not following contact precautions are at high risk.
- **Clinical features: (1) Prodromal (prehemorrhagic) phase:** It is characterized by sudden onset of fever (39–41°C, on an average, fever persists for 4–5 days) with chills, severe headache, photophobia, neck pain, dizziness, abdominal pain, diarrhea, vomiting, myalgia, hypotension, neuropsychiatric changes (early-mood swings, confusion, aggression. Later-sleepiness, depression and lassitude), tachycardia, lymphadenopathy, etc. **(2) Hemorrhagic phase:** It is short and lasts for 2–3 days. It begins on 3rd to 5th day and characterized by petechiae, ecchymosis, hematemesis, epistaxis, malena, leukopenia, thrombocytopenia, hematuria, etc. **(3) Convalescence phase:** Recovery starts by 9–10 days. Pronounced generalized weakness, weak pulse and sometimes complete loss of hair may be seen.
- **Complications:** These are hepatorenal failure, pulmonary failure and DIC. Mortality highest in 2nd week ranging from 5 to 50%.
- **Sequelae:** Polyneuritis, headache, nausea, poor vision, hearing loss and memory loss may be permanent. But they may persist for a year or more.

Laboratory diagnosis

- **Specimens:** These are blood plus tissue samples like liver, spleen, bone marrow, kidneys, lungs and brain. Handling of samples required bio-safety level 4 precautions.
- **Testing methods**
  1. **Microscopy:** Follow morphology.
  2. **Culture:** Follow C/Cs.
  3. **Serological tests:** Patients with fatal disease do not usually develop a measurable antibody response and in these individuals, as well as in patients in the first few days of illness, diagnosis is achieved by virus detection in blood or tissue samples.
     - **Ab detection tests:** IgM detectable after 6th day of illness and remains detectable for up to four months. It is detected by ELISA. **IgG** level declines but remains detectable for up to five years. It is detected by ELISA.
     - **Ag detection tests:** Viral antigens may sometimes be detected from tissues by immunofluorescence or ELISA.
  4. **Molecular method:** RT-PCR (rapid and sensitive).

Prevention

- **General measures:** (1) Intensive tick control measures have to be taken by spraying **acericide** drug on all animals in affected and neighborhood villages and repeated after one month. (2) Animal supervision animal husbandry department. (3) Insecticides have to be sprayed intensively in all breaks on floors and walls in cattle shades. (4) Surveillance among hospital contacts should be strengthened at hospital setting. (5) BMW at the hospitals should be strengthened. (6) Health education about causation, transmission and prevention of disease. (7) State wide sero-surveillance in animals to identify prevalence of disease. (8) Insect repellants containing DEET. (9) Wearing protective clothes. (10) Healthcare workers who had contact with tissue or blood from patients with suspected or confirmed CCHF should be followed up with daily temperature and symptoms monitoring for at least 14 days after the putative exposure.
- **Chemoprophylaxis:** Ribavirin is useful.
- **Immunoprophylaxis:** No specific safe and effective vaccine is available. An inactivated mouse brain derived vaccine against CCHF virus has been developed and used on a small scale.

Treatment

- Both oral and IV formulations of antiviral ribavirin seems to be effective.
- Other treatment includes fluid and electrolytes management.
- Passive immunotherapy: Hyperimmune serum has been tested in few cases. A new specific Ig CCHF-venin, has been prepared by using the plasma pool of boosted donors. It contains Ab to CCHF virus in a titer of 8.

Hazara Virus

Virus related to CCHF virus and isolated from Pakistan.

Ganjam Virus

Virus related to Nairobi sheep disease virus and isolated from ticks collected from sheep and goats in Orissa, India. It also isolated from human. Accidental transmission in laboratory causes mild febrile illness.

Hantavirus

It is not an arbovirus but robovirus and discussed below, under separate heading of **Roboviruses.**

Reoviridae

Meaning

It means **R**espiratory **E**nteric **O**rphan virus, because it was isolated from **r**espiratory and **e**nteric tract and **o**rphan means without causing any disease.

Morphology

Shape and size: Spherical in shape with 55–57 nm size.

Genome: ds-RNA (+/−) with 10–12 pieces as shown in **Fig. 83.8** (11 pieces in *Rotavirus*). It has three segments like large (L), medium (M) and small (S).

Capsid: It contains icosahedral type of capsid with two layers (3 layers in *Rotavirus*). Outer layer is fuzzy and indistinct while inner layer has 32 ring shaped capsomere.

No-envelope: It is nonenveloped or naked virus.

Classification

Follow **Flowchart 83.2**.

Orbivirus

Meaning: Orbi word derived orbi (Latin) means ring, because having 32 ring shaped structure in inner layer of icosahedral capsid.

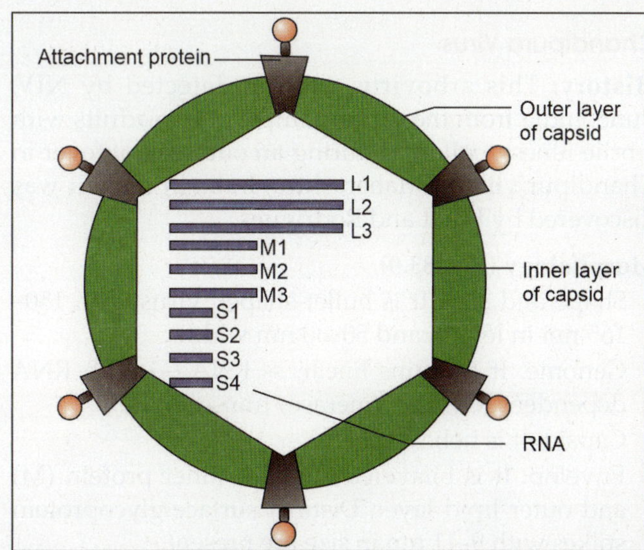

Fig. 83.8: Morphology of Reoviridae

Flowchart 83.2: Classification of Reoviridae

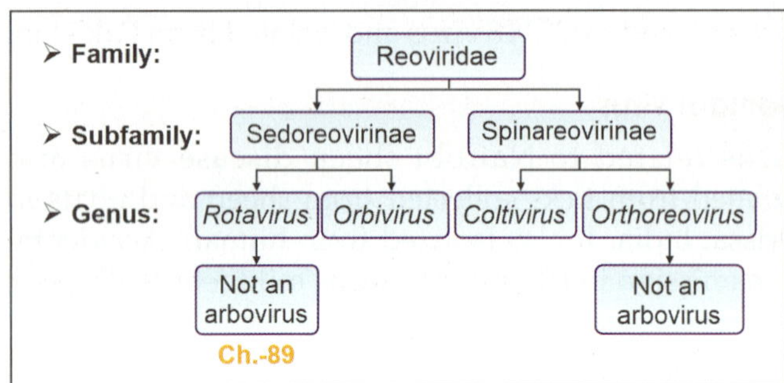

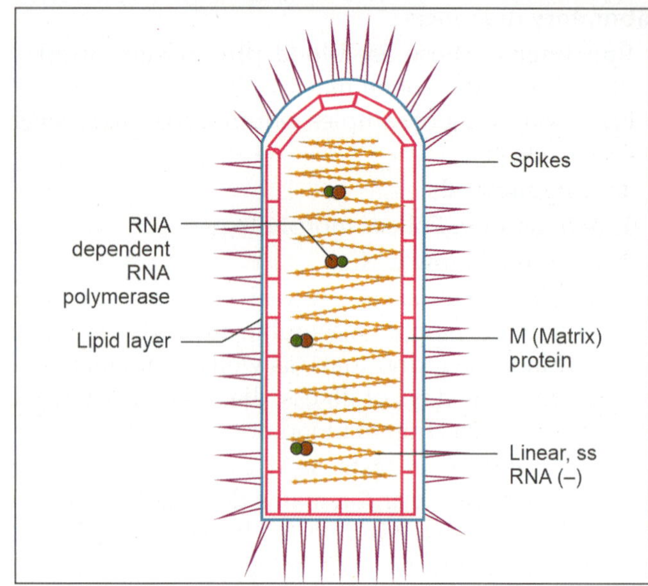

Fig. 83.9: Morphology of Chandipura virus

Species: Following are the different species.
1. **Orungo virus:** It is endemic in sub-Saharan Africa. It is either subclinical or causes acute febrile illness in human. It is transmitted by mosquito bite like *Aedes*.
2. **Kemerovo virus:** It causes febrile illness with involvement of meninges in human. Cases have been reported from Russia. It is transmitted by tick like *Ixodes persulcatus*.

Coltivirus

Kolarado tick fever virus: It is distributed in Western USA. It is transmitted by wood tick *Dermacentor andersoni*, which acts as vector and reservoir. It is a natural pathogen of rodents and in human produces self-limited mild fever without rash.

Rhabdoviridae

Word is derived from Rhabdos means rod shape, because of its shape like rod (bullet). Family Rhabdoviridae has two genera like *Lyssavirus* and *Vesiculovirus*. *Lyssavirus* is not an arbovirus and described in **Ch. 84.** Species of *Vesiculovirus* are described below.

Vesicular Stomatitis Virus

Virus causes oral mucosal vesicles and ulcers in cattle, horses and pigs as like hand-foot-mouth disease virus. Human infection is rare or presents with flu like illness. It is a zoonosis, transmitted by sand fly like *Lutzomyia shannoni*.

Chandipura Virus

History: This arbovirus was 1st detected by NIV, Pune, India from the serum sample of two adults with febrile illness, collected during an outbreak of fever in Chandipur village, Maharashtra, India in 1967. It was discovered by Bhatt and Rodrigues.

Morphology (Fig. 83.9)
- Shape and size: It is bullet-shaped virus with 150–165 nm in length and 50–60 nm width.
- Genome: It contains linear, ss RNA (–) with RNA dependent RNA polymerase/transcriptase.
- Capsid: It is helical type.
- Envelop: It is bilayered contains inner protein (M) and outer lipid layer. Distinct surface glycoprotein spikes with 9–11 nm in size are present.
- Replication: It occurs in cytoplasm.

Culture characteristics (C/Cs): Virus isolation is best useful in initial 5 days. **Animal culture** is done in infant mouse. **Tissue culture** is done over Vero and BHK-21 cells.

Pathogenicity
- **Epidemiology:** Since its discovery, it is limited in adjoining regions of Maharashtra, Central India and produced many unexplained outbreaks of encephalitis.
- **Reservoirs of infection:** Humans.
- **Sources of infection:** Sand flies.
- **Mode of transmission:** It is transmitted by sand flies like of *Phlebotomus* and *Sergentomyia*.
- **Incubation period**: 2–6 days.
- **Portal of entry:** Skin.
- **Site:** Brain.
- **Precipitating factors: (1) Age:** It is common between 9 months to 14 years. **(2) Sex:** Male: female ratio is 1:0.77. **(3) Seasonality:** It occurs in late summer and early monsoon (July-September). **(4) Rural Areas:** Mud, damp soil and other vegetative materials (high moisture areas) are increasing the risk. **(5) Socio-economic class:** It is common in lower class people.
- **Clinical features:** It produces mild symptoms like high grade fever of short duration and vomiting. Other features re altered sensorium, convulsions, decerebrate posture and acute encephalitis/ encephalopathy.
- **Complication:** Death within few hours to 48 hours of hospitalization.

Laboratory diagnosis
- **Specimens:** Blood/serum, throat swab, CSF and urine.
- **Testing methods:**
  1. **Microscopy:** Follow morphology.
  2. **Culture:** Follow C/Cs.

3. **Serological tests:**
   - **Ab detection tests:** IgM and IgG are detected by ELISA, neutralization test and HAI test.
   - **Antigen detection tests:** Viral antigens may sometimes be detected from tissues by immuno-fluorescence or ELISA.
4. **Molecular method:** RT-PCR is more rapid and sensitive method.

Prevention: There are no known efficient methods of controlling biting sand flies, but personal protection like wearing protective clothes (long sleeves/pants) will help in reducing exposure to their bites. Avoid localities, especially at dawn and dusk that are known to be frequented by biting sand flies.

Treatment: Specific antiviral drugs are not available. Only symptomatic treatment is given.

ROBOVIRUSES

Introduction

Meaning

It means __Ro__dent __B__orne __Viruses__. However, transmission by mites have been reported but not confirmed and in absence of proved arthropod-borne transmission it considered as **Robovirus**.

Classification

Follow **Table 83.2**.

Bunyaviridae

Genus *Hantavirus* of family Bunyaviridae infects human. Following four species infect human.

Hantaan Virus

Meaning: Virus is named for the Hantaan River which flows through the endemic region in Korea.

History: American tropical virologist, his colleagues and Korean virologist isolated Hantaan virus from the lungs of striped field mice from Korea.

Morphology: It is same as other virus of Bunyaviridae family which includes single-stranded, RNA (–) with three segments like large, medium, and small. It is an enveloped virus. It has three structural proteins like G1 (glycoproteins), G2 and N (nucleoprotein).

Pathogenicity

- **Disease name:** Disease called **hemorrhagic fever renal syndrome**. Clinically, resemble to typhoid, leptospirosis and scrub typhus.
- **Synonyms:** It also called Korean hemorrhagic fever, epidemic nephrosonephritis, endemic nephroso-nephritis, manchurian epidemic hemorrhagic fever, nephropathia epidemica and rodent borne nephropathy.
- **Epidemiology:** It is prevalent in Scandinavia, Far East, North Asia, Western Russia, China, South and Central America.

- **Reservoirs of infection:** Field mice (*Apodemus agrarius*) are the natural hosts.
- **Sources of infection:** These are aerosols from the excreta of rodents like urine, feces and saliva.
- **Modes of transmission:** It is transmitted by aerosols arise from excreta of rodents like urine, feces and saliva.
- **Incubation period:** 2–4 weeks.
- **Portal of entry:** Respiratory system.
- **Sites:** Kidneys.
- **Precipitating factors:** There are two seasonal peaks for almost all outbreaks of *Hantavirus* diseases. A small one in spring and a large one in fall. It is due to increase the activities in farms by farmers and exposed to rodents in the fields during planting and harvest periods. Unusually, high rainfall in dry parts of the country result in increased food sources for rodents and subsequently increased rodent populations.
- **Clinical features:** It presents in two forms like mild epidemic nephritis and serious epidemic hemorrhagic fever.
- **Complication:** Death occurs in 10–12% cases.

Diagnosis

- **Specimens:** Blood/serum and urine.
- **Testing methods**
  1. **Microscopy:** Follow morphology.
  2. **Serological tests:** These are IgM ELISA and IgG ELISA.
  3. **Molecular methods:** RT-PCR.

Prevention

- **General measures:** Rodent control measures.
- **Immunoprophylaxis:** Vaccine based on two glyco-proteins (G1 and G2) is more effective.

Treatment: Specific antiviral drugs are not available. Only symptomatic treatment is given.

Seoul Virus

Probably it presents world-wide. It produces mild disease. It is transmitted by rats (*Rattus rattus* and *R. norvegicus*).

Puumala Virus

It is responsible for nephropathia epidemica in northern and Eastern Europe. It is transmitted by rats (*Rattus rattus* and *R. norvegicus*).

Sin Nombre Virus

It was identified in South Western USA in 1993. It produces a disease called **hanta virus pulmonary syndrome.** It is transmitted by inhalation of aerosols arise from excreta of deer mouse and other rodents of Sigmoidontine family. It presents with sudden onset of fever, malaise, myalgia and GIT symptoms, which last for 3–4 days. Later, it causes pulmonary edema identified under radiological picture. In severe cases there is development of hypotension, tachycardia and tachypnoea leading to death.

Arenaviridae

Meaning

Word is came from arena (Latin) means land, resemble to grains of sand.

Morphology

Shape: It is the spherical **(Fig. 83.10)** or pleomorphic particles. Viral particles are actually ribosomes picked up from host cells during maturation and resemble to sand under EM. It is 80–300 nm in size.

Genome and capsid: It contains ss-RNA (–) in two segments with complex symmetry of capsid.

Envelope: Present.

Mammarenavirus

Genus *Mammarenavirus* of family Arenaviridae infects human. Two groups are identified like old world viruses and new world viruses, based on geographical and genetical properties.

Old world viruses: They found in the Eastern hemisphere in places such as Europe, Asia, and Africa. Following two are the old world viruses.

1. **Lymphocytic choriomeningitis (LCM) virus:** It exists in both world but classified as an old world virus. It is the natural parasite of rodents, which excrete the virus in urine and feces that act as **sources for infection**. It is **transmitted** from excreta of rodents. It is asymptomatic or **presents with** influenza like illness or meningitis. Death occurs due to deposition of immune complex in CNS or kidneys (glomerulonephritis). It is **diagnosed** from brain/ meninges biopsy by EM **(electron dense particles resemble to sand)** or inoculating in mouse.

2. **Lassa fever virus:** It was 1st noticed in 1969 in American Mission station in Lassa, Nigeria, hence the name. It is the natural parasite of rodents, which excrete the virus in urine and feces, which act as **sources for infection**. It is **transmitted** from excreta of rodents or by inhalation. **Incubation period** is 3–16

days. It **presents with** hemorrhagic fever called **lassa fever**. Case fatality is 35–70%. Virus is **diagnosed** from throat swab, blood and urine of patients. **Treatment** is given with ribavirin.

New world viruses: They found in the Western hemisphere, in places such as Argentina, Bolivia, Venezuela, Brazil, and the United States. Following six are the new world viruses.

1. **Junin virus:** It causes Argentinian hemorrhagic fever and transmitted from excreta of rodents.
2. **Machupo virus:** It causes Bolivian hemorrhagic fever and transmitted from excreta of rodents.
3. **Guanarito virus:** It causes hemorrhagic fever in Venezuala.
4. **Sabia virus:** It causes hemorrhagic fever in Brazil.
5. **White water arroyo virus:** It causes hemorrhagic fever in South Western USA.
6. **Lujo virus:** Name is given by using first two letters of the cities involved in the 2008 outbreak of the human viral hemorrhagic fever in **Lu**saka (Zambia) and **Jo**hannesburg (South Africa).

ACCESS YOURSELF

Case Studies

1. A 50-year-old male is brought to the hospital with stiffness in ankle since 6 months. He was suffered with fever before 6 months and from that time the neck stiffness is continue. Fluid is taken out from the joint and sent to the referral laboratory, which suggests the virus induced disease. Identify the case and answer the following:
   a. Name the clinical condition and its etiological agent.
   b. Write the morphology of etiological agent.
   c. Write pathogenicity and diagnosis of clinical condition.

2. A 25-year-old male is brought to the hospital with high fever and headache and retro-orbital pain. On clinical examination petechial rashes, hepatomegaly and positive tourniquet test were ruled out. Blood picture reported the thrombocytopenia with 30,000/mm^3 count. All lab., reports suggest, it could be virus induced condition. Identify the case and answer the following:
   a. Name the clinical condition and its etiological agent.
   b. Write the morphology of etiological agent.
   c. Write pathogenicity and diagnosis of clinical condition.

Essays/Full Questions

1. Arboviruses prevalent in India.
2. Roboviruses.

Short Notes

1. Chikungunya/Japanese encephalitis/Yellow fever/Dengue/Zika virus infection/CCHF.

Short Questions for Theory/Viva Questions

1. Name four arboviruses prevalent in India.
2. Name four mosquito borne viruses or flaviviruses or roboviruses.
3. What are primary and secondary dengue infections?
4. What is antibody dependent enhancement in dengue? Write its clinical significance.
5. Name the agent causing 3-day fever and 5-day fever.

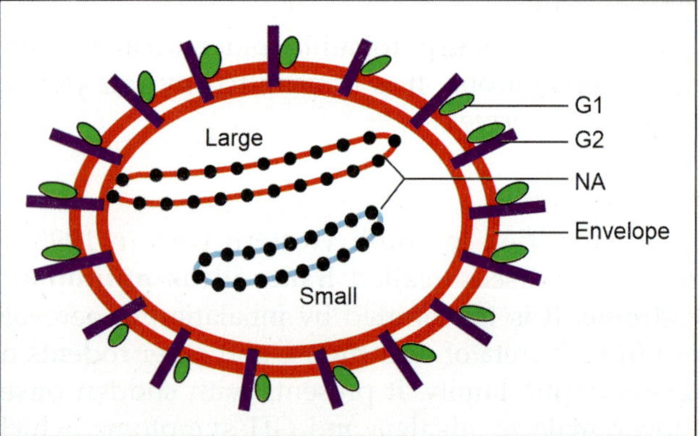

Fig. 83.10: Morphology of Arenaviridae

Comments on

1. India considered as receptive area for yellow fever.
2. In dengue neutralizing antibodies are protective, while non-neutralizing antibodies are dangerous or Secondary dengue infection is more severe than primary infection.
3. Dengue fever called biphasic or saddle back fever.

MCQs for Chapter Review

Arboviruses

Introduction

1. **Group B (flavi viruses) arboviruses is/are:**
   a. Dengue fever
   b. Rift valley fever
   c. Chikungunya
   d. JE
   e. Yellow fever

2. **Viruses cause hemorrhagic fever are:**
   a. Lassa fever virus
   b. Yellow fever virus
   c. West Nile virus
   d. Crimean Congo virus
   e. Ross river virus

3. **Fever and hemorrhagic rashes are seen in all *except*:**
   a. Dengue fever
   b. Lassa fever
   c. Rift valley fever
   d. Sand fly fever
   e. Yellow fever

4. **Hemorrhagic fever is not caused by:**
   a. Yellow fever
   b. KFD
   c. JE
   d. Dengue

5. **Viral infection is transmitted by tick?**
   a. JE
   b. Dengue
   c. KFD
   d. Yellow fever

6. **Zika virus belongs to which family?**
   a. Flaviviridae
   b. Togaviridae
   c. Bunyaviridae
   d. Reoviridae

7. **All of the following belong to Flaviviridae family *except*:**
   a. Yellow fever
   b. Chikungunya
   c. Dengue
   d. JE

8. **Which of the following is not common in India?**
   a. Japanese B encephalitis
   b. Lassa fever
   c. KFD
   d. Dengue

Togaviridae

9. **Chikungunya virus is transmitted by:**
   a. *Aedes aegypti*
   b. *Culex visnoi-complex*
   c. *Aedes albopticus*
   d. *Mansonia indiana*

Flaviridae

10. **Choose the false statement regarding hepatitis G virus:**
    a. Also called GB virus
    b. Blood borne RNA virus
    c. Mostly infected with HCV
    d. Respond to lamivudine

11. **Following is true for kyasanur forest disease, *except*:**
    a. Transmitted by soft ticks
    b. Caused by retro virus
    c. Incubation period is 3–8 days
    d. Killed vaccine available

12. **Which of the followings are true regarding KFD?**
    a. It is zoonosis
    b. Affects monkeys
    c. Caused by bacteria
    d. Caused by *Rickettsia*

13. **KFD is transmitted by:**
    a. Flea
    b. Mite
    c. Tick
    d. Mosquito

14. **Vector of Japanese B encephalitis virus is:**
    a. *Anopheles* mosquito
    b. *Culex* mosquito
    c. *Aedes* mosquito
    d. All of the above

15. **In Japanese encephalitis pig acts as:**
    a. Amplifier host
    b. Definitive host
    c. Intermediate host
    d. Any of the above

16. **In nature, JE virus life cycle runs between?**
    a. Pigs-mosquitoes
    b. Cattle-birds
    c. Pigs-humans
    d. Birds-pigs

17. **Iceberg phenomenon is feature of:**
    a. JE virus
    b. Dengue virus
    c. a + b
    d. None

18. **Certificate for yellow fever vaccination is valid for:**
    a. 1 year
    b. 5 years
    c. 10 years
    d. lifelong

19. **True about dengue fever:**
    a. Caused by 4 serotypes
    b. Effective vaccine available
    c. Presents with fever and joint pain
    d. Virus belongs to *Flavivirus* genus
    e. Contains segmented RNA

20. **Dengue is transmitted by:**
    a. *Mansonia annulifera*
    b. *Culex fatigans*
    c. *Anopheles punctulatus*
    d. *Aedes aegypti*

21. **Dengue hemorrhagic fever is caused by:**
    a. Type I dengue virus
    b. Reinfection with the same serotype of dengue virus
    c. Reinfection with the different serotype of dengue virus
    d. Infection is in the immunocompromized host

22. **Most virulent dengue fever strain is:**
    a. 1
    b. 2
    c. 3
    d. 4

23. **In India, human infections has been reported with dengue virus type:**
    a. Type 1 and 2
    b. Type 3 and 4
    c. Type 1 and 5
    d. Type 2 and 5

24. **Most sensitive diagnostic test for dengue is:**
    a. IgM ELISA
    b. Complement fixation test
    c. Neutralization test
    d. Electron microscopy

25. **Most specific for dengue diagnosis:**
    a. IgM ELISA
    b. Tissue culture
    c. CFT
    d. Electron microscopy

26. **CYD-TDV is a vaccine given for:**
    a. Yellow fever
    b. Chikungunya
    c. DHF
    d. JE

27. **The rate of brain or eye defects in fetuses is higher in which trimester of pregnancy with Zika virus disease?**
    a. First
    b. Second
    c. Third

Bunyaviridae

28. **CCHF virus belongs to:**
    a. Flaviviridae family
    b. Arenaviridae family
    c. Togaviridae family
    d. Bunyaviridae family

29. **True about CCHF:**
    a. Zoonosis
    b. Develop petechial rashes
    c. Transmitted by mites
    d. It has high fatality

Rhabdoviridae

30. **Chandipura virus is transmitted by:**
    a. Culex
    b. Tick
    c. Sand fly
    d. *Aedes*

31. All are true regarding *Hantavirus*, except:
   a. DNA virus
   b. Carried by rodents
   c. Causes recurrent respiratory infection
   d. Hemorrhagic manifestations may occur

32. True about *Hantavirus*:
   a. *Hantavirus* pulmonary syndrome
   b. Transmitted by arthropod
   c. Transmitted by rodents
   d. Hemorrhagic fever with renal failure
   e. *Hantavirus* pulmonary syndrome acquired from person to person

Mixed MCQs of Arboviruses

33. Which of the following statement is true regarding arboviruses?
   a. Yellow fever is endemic in India
   b. Dengue has only one serotype
   c. KFD is transmitted by ticks
   d. Mosquito of *Culex visnoi-complex* is the vector of dengue

Answers and Explanation of MCQs

1. a, d, e
2. a, b, d
3. d
4. c
5. c
6. a
7. b
- Follow section, **arboviruses → introduction (classification and Table 83.1) and Table 83.2** for explanation of answers of MCQs 1–7.
8. b
- Follow section, **arboviruses → introduction (arboviruses found in India)** for explanation.
- Lassa fever is robovirus and found in the Eastern hemisphere in places such as Europe, Asia, and Africa.
9. a and c
- Follow section, **arboviruses (Togaviridae → Chikungunya virus)** for explanation.
10. c and d

- Options a, b and c: Follow section, **arboviruses (Flaviviridae → *Pegivirus*)** for explanation. Lamivudine is reverse transcriptase inhibitor and useful in HBV and HIV.
11. a, b
12. a, b
13. c
- Follow section, **arboviruses → Kyasanur Forest Disease virus** for explanation of answers of MCQs 11–13.
14. b
15. a
16. a
17. a
- Follow section, **arboviruses → Japanese encephalitis virus** for explanation of answers of MCQs 14–17.
18. c
- Follow section, **arboviruses → yellow fever virus → prevention → international certificate of vaccination** for explanation.
19. c, d
20. d
21. c
22. b
23. a
24. a, c
25. a
26. c
- Follow section, **arboviruses → dengue virus** for explanation of answers of MCQs 19–26.
27. a
- Follow section, **arboviruses → zika virus → complications** for explanation.
28. d
29. a, b, d
- Follow section, **arboviruses → CCHF virus** for explanation.
30. c
- Follow section, **arboviruses → Chandipura virus → pathogenicity** for explanation.
31. a
32. a, c, d
- Follow section, **roboviruses (Hantaan virus and sin Nombre virus)** for explanation of answers of MCQs 31–32.
33. c
- Follow section, **arboviruses (Yellow fever virus, Dengue**

Infections of Rhabdoviridae

INTRODUCTION

Meaning

Word rhabdo derived from rhabdos (Greek) means rod, because of its shape like rod (bullet).

Classification

Family: Rhabdoviridae.

Genus: Two genera like *Vesiculovirus* (**Ch. 83**) and *Lyssavirus.* Lyssa (Greek) means rage (violent/anger), indicates characteristic feature of disease. *Lyssavirus* has 7 species or serotypes as shown in **Table 84.1**.

TABLE 84.1: Species or serotypes of *Lyssavirus*		
Species	**Common name**	**Isolated from**
Lyssa virus 1	Rabies virus	Warm blood animals
Lyssa virus 2	Lagos bat or Natal bat virus	Bat/cat
Lyssa virus 3	Mokola virus	Cat/dog/human
Lyssa virus 4	Duvenhage virus	Human/bat
Lyssa virus 5	European bat-I virus	Bat/human
Lyssa virus 6	European bat-II virus	Bat/human
Lyssa virus 7	Australian bat virus	Bat/human

LYSSA VIRUS 1

Common Name, Meaning and Use

Lyssa virus is **commonly known** as rabies virus. Rabies word came from rabidus (Latin) means mad (In Sanskrit Rabhas), indicating the features of disease produced by virus. Rabies virus is **used** in research to know viral synaptic connections and directionality of synaptic transmission.

Morphology

Shape (Figs 83.9 and 84.1) and size: It is bullet shaped with one round or conical end and other end is concave or planar. It is 100 nm × 70–75 nm in size.

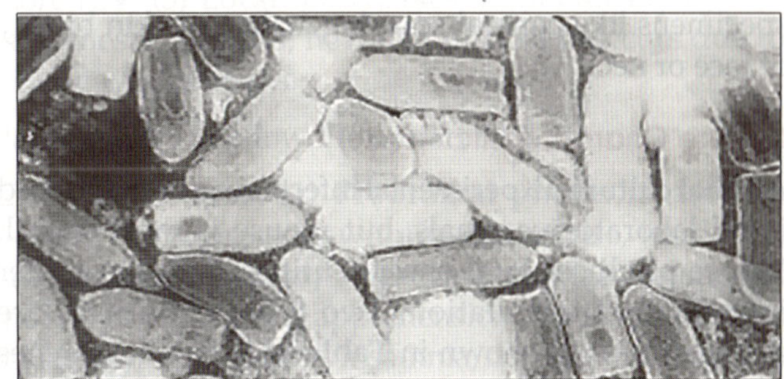

Fig. 84.1: Microscopic image of rabies virus

Genome and capsid: It contains single, linear ss-RNA (–) surrounded by helical capsid and RNA dependent RNA polymerase.

Envelope: Nucleocapsid is surrounded by bilayered envelop which consists inner protein called **matrix (M protein)** and outer lipid layer. M proteins are projecting outward at planar end and form blebs in some viruses. Envelop shows projecting glycoprotein spikes, which are absent at planar end.

Multiplication: It occurs in cytoplasm.

Inclusion body: Brain tissues (cerebellum and hippocampus) and neurons contain pathognomic intracytoplasmic inclusion bodies called **Negri bodies**. Name is given from the name of Adelchi Negri. It is eosinophilic (acidophilic) in nature. It presents in Purkinje cells of cerebellum and pyramidal neurons of hippocampus and rare in cortical and brainstem neurons. It is absent in 20% cases. Brain tissue is collected at post-mortem. Impression smear is stained with **Seller's technique** (basic fuchsin and methylene blue) and examined under light microscopy for presence of Negri bodies (**Fig. 84.2**). It also stained by **Giemsa or Mann's method.** Negri bodies are differentiated from other bodies like canine distemper virus inclusion bodies. Negri bodies are oval or spherical in shape,

Infections of Retroviridae

Chapter Outline
- Introduction
- HIV

INTRODUCTION

Meaning

Retro means reverse, because having enzyme reverse transcriptase (RNA dependent DNA polymerase).

Classification

Family: Retroviridae.

Genus: Important genera and respective species are mentioned below.

1. *Deltavirus:* Two species are HTLV-I (Human T-cell lymphotropic virus-I) and HTLV-II.
2. *Lentivirus:* Lenti means slow because causing slow virus diseases. Important species are mentioned below.
   - **Visna virus or visna-maedi virus or maedi-visna virus or ovine lenti virus:** It causes two types of disease in sheep like **visna** (demyelinating disease of sheep characterized by paresis and paralysis followed by death) and **maedi** (fatal hemorrhagic pneumonia in sheep).
   - **Simian immunodeficiency virus (SIV):** Causing Simian-AIDS (S-AIDS). It causes AIDS like disease in monkey.
   - **Human immunodeficiency virus (HIV):** Causing Human-AIDS (H-AIDS).

Note: HTLV-I

Synonym: Also called **human T-cell leukemia-lymphoma virus** because known to cause adult T cell leukemia/lymphoma.

History: HTLV-I was discovered by Robert Gallo and colleagues in 1980 from the cell culture of adult patient with cutaneous T cell lymphoma (mycosis fungoides) and leukemia (Sezwary syndrome) in the USA.

Morphology: It is spherical in shape with 100 nm in size. It contains two identical ss-RNAs (+) with enzyme reverse transcriptase. Genome is surrounded by double shell, icosahedral variety of capsid. Inner shell is cone shape and contains genome. It contains lipoprotein envelope.

Culture Characteristics (C/Cs): It is done by co-cultivation of normal lymphocytes with infected lymphocytes. Virus growth is detected in normal cells by identifying the antigen of HTLV-I.

Resistance: It is inactivated by heating 56°C in 30 minutes, mild acids, ether and formalin.

Pathogenicity
- **Epidemiology:** It is distributed worldwide, but disease presents only in endemic areas like Japan and Caribbean. Six **genotypes** are identified from A to F. Majority of infections are caused by genotype A. HTLV-1 and HTLV-2 both are involved in actively spreading epidemics.
- **Reservoirs of infection:** Humans.
- **Sources of infection:** Blood or breast milk.
- **Modes of transmission:** HTLV-I and HTLV-II can be transmitted by unprotected sex, blood/blood products or breast feeding.
- **Portal of entry:** Skin or mucosa.
- **Sites:** T cells (both CD4 and CD8), B cells, CNS, skin, etc.
- **Precipitating factors (epidemiological determinants):** It is common in adults and persons required multiple blood transfusion or taking IV drugs.
- **Pathogenesis:** It acts by inserting its genome in host cell DNA called **provirus** which continuously spreads from cell to cell. It has a tax (tex) gene which has oncogenic power.
- **Clinical features:** It causes adult T cell leukemia/lymphoma with rapid onset and fatal in nature. It also causes infectious dermatitis, cutaneous T cell lymphoma, tropical spastic paraparesis or demyelinating disease, uveitis, arthropathy and inflammatory disorders.
- **Complications:** Autoimmune disease and death.

Laboratory diagnosis: Blood/serum is tested by microscopy (follow morphology), culture (follow C/Cs), serological tests (ELISA, WB test and IF test) and molecular method (like RT-PCR).

Prevention: Preventive measures are same as like HIV-AIDs as described below.

Note: HTLV-II

History: It is closely related to HTLV-I and also discovered by Robert Gallo and colleagues from patient with adult T cell leukemia in Japan and Caribbean.

Pathogenicity: Transmission, replication and pathogenesis are similar to HTLV-I. HTLV-II is associated with T cell malignancies, mild neurologic diseases and chronic lung infections.

HIV

History and Synonym

- 1st indication was came in 1981, when there were outbreak of two unknown diseases like Kapsosi's sarcoma and *Pneumocystis jirovecii* pneumonia in young adults in Los Angeles and New York, USA who were homosexual and addicted to injected narcotics. They appeared to lose immune system rendering them vulnerable to secondary infections especially avirulent organisms and malignancies. This condition was called **AIDS**.
- In 1983, virus was isolated by Luc Montagnier and colleagues at Pasteur Institute, Paris from African patient of lymphadenopathy hence called **lymphadenopathy associated virus.**
- In 1984 it was isolated by Robert Gallo and colleagues at National Institute of Health (NIH), USA from AIDS patients and called **HTLV-III**, as HTLV-I and II were already described to cause human leukemia.
- Later, many similar isolates were identified from AIDS patient and in 1986 all such isolates were called **HIV** by International Committee on Virus Nomenclature.
- HIV was diagnosed on clinical background but antibodies detection method like ELISA was established in 1986.
- HIV-2 was discovered in 1986, on the bases of molecular and antigenic variation. It is endemic in West Africa. Differences between two viruses (or called serotypes) are mentioned in **Table 85.1**.
- **In India** 1st case of HIV infection has been found in female sex workers in Chennai in 1986 and 1st case of AIDS was found in Bombay in 1986.

Morphology

Shape and size: It is spherical in shape **(Fig. 85.1)** with 90–100 nm in size.

Genome and capsid: It has two identical ss-RNAs (+) with enzyme reverse transcriptase (RNA dependent DNA polymerase or DNA polymerase). Conversions of ss-RNA in to ss-DNA and then in to ds-DNA is take place by enzyme reverse transcriptase. DNA transcribed to mRNA, which is translated to protein. ds-DNA may integrated in to host cell chromosome after infection called provirus. Provirus remains in host cell for long time, influences the viral functions and also initiates the viral replication. Genome surrounded by double shell and icosahedral variety of capsid. Inner shell is cone shape and contains genome.

Envelope: It contains lipoprotein envelope. Lipids derived from the host cell plasma membrane by budding and proteins (glycoprotein-gp) are viral encoded. Protein presents as transmembrane pedicle (gp41) and projecting spike (gp120). Spike protein (gp120) helps in adsorption of virus with specific receptors present on CD4 cell surface, while gp41 helps in cell fusion.

Culture Characteristics (C/Cs)

Culture methods are of following two types.

Direct culture method: *In vitro* cultivation of patient cells in presence of phytohemagglutinin.

Cocultivation method: It is done between patient's lymphocytes and normal lymphocytes in presence of IL-2. Virus growth is detected by demonstrating Ags or reverse transcriptase enzyme in normal cells. Few **disadvantages** of this method are risk of HIV infection and it is available at higher center with adequate infection containment facilities.

TABLE 85.1: Differences between HIV-1 and HIV-2

Features	HIV-1	HIV-2
Origin	1st indication in 1981, USA	In 1986, West Africa
Distribution	World-wide and in India	West Africa and few report from western and southern India
Origin for human infection	Chimpanzees **(Pan troglodytes troglodytes)**	Sooty Mangabey monkeys **(Cercocebus atys)**
Virulence	More	Less
Spread	Rapid	Slow
Resembles to SIV	Less	More
Gene	vpu	vpx
Envelope Ags	gp120 and gp41	gp36
Genotypes (clades)	Total 11 from A-K	Total 8 from A-H
Similarities	Cross-reactivity occurs between nucleocapsid-Ags. HIV-2 has 40% genetic identity with HIV-1	

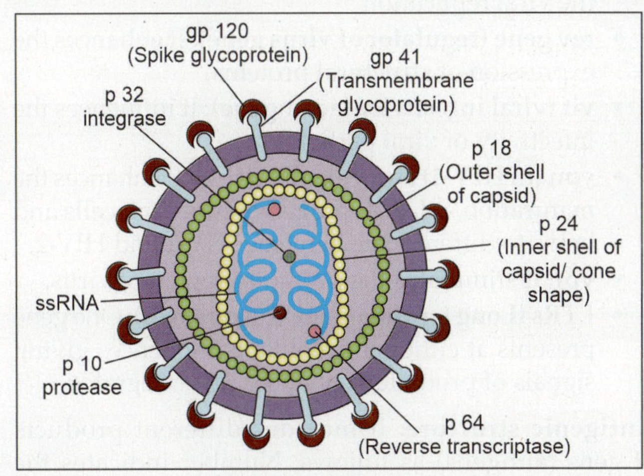

Fig. 85.1: Morphology of HIV

Sterilization (physical methods): HIV is heat labile (thermo labile) and inactivated by following methods

- Autoclaving at 121°C (15 lbs pressure) for 20 min.
- Dry heat 170°C for 1 hours.
- Boiling for 20 minutes.

Disinfection (chemical methods): Minimum 30 minutes contact time required to kill HIV with following agents.

- Extreme of pH: 1.0 or 13.0.
- Sodium hypochlorite (0.5–1%).
- Ethanol (70%) or isopropanol (35%).
- Lysol (0.5%) for 10 minutes.
- Providone iodine.
- Hydrogen peroxide (0.3%) for 10 minutes.
- Formalin (3–4%).
- Glutaraldehyde (2% called **Cidex**).
- House hold bleaching powder (10%) for 10 minutes, for infected needles or syringes.

Viral Genome and Antigenic Structure

Genes: Two types of genes as follows.

1. **Structural genes:** Following three types.
   - **gag gene:** It encodes for nucleocapsid antigens of virus and expressed as a precursor protein p18 (p17 by some authors) which is cleaved into 3 proteins like p15, p18 and p24.
   - **pol gene:** It encodes for reverse transcriptase and other viral enzymes like integrase (p32), protease (p10) and endonuclease. Reverse transcriptase is expressed as a precursor protein which is cleaved into p31, p51 and p66.
   - **env gene:** It encodes for envelope glycoproteins of virus. Precursor glycoprotein gp160 which is cleaved into 2 proteins like gp120 (forms the surface spikes) and gp41 (transmembrane anchoring protein).

2. **Non-structural genes:** Following are types.
   - **tat gene (transactivating gene):** It enhances the expression of all viral genes.
   - **nef gene (negative factor gene):** It down regulates the viral replication.
   - **rev gene (regulator of virus gene):** It enhances the expression of structural proteins.
   - **vif (viral infectivity factor gene):** It influences the infectivity of viral particles.
   - **vpu (in HIV-1) and vpx (in HIV-2):** It enhances the maturation, releases progeny virus from cells and helps to differentiate between HIV-1 and HIV-2.
   - **vpr:** It stimulates the promoter region of virus.
   - **LTRs (Long terminal repeat sequences):** One gene presents at either end, contains sequences giving signals of promoter, enhancer and integrator.

Antigenic structure: It includes different products of gens (antigens) as follows. Number indicates the molecular weight of antigen in kilodaltons.

1. **Major antigens (products of gene) of HIV**
   - **Nucleocapsid-Ags (gag gene)**
     - Principle nucleocapsid Ag (inner shell): p24.
     - Other nucleocapsid Ags (outer shell): p15, p18 and p55.
   - **Enzyme-Ags (pol gene)**
     - Reverse transcriptase: p31, p51 and p66.
     - Integrase: p32
     - Protease: p10
   - **Envelope-Ags (env gene)**
     - Projecting spikes glycoprotein: gp120 (principle envelopes Ag).
     - Trans-membrane pedicle glycoprotein: gp41.
     - gp36 for HIV-2.

2. **Diversity of HIV:** HIV shows frequent diversity in antigens (most in envelope-Ags and rare in other), nucleotide sequences, cell tropism, growth characteristics, cytopathology, place to place, person to person and different site of same person. It is due to error prone nature of reverse transcriptase. Antibodies are produced against env gene and mutation to env gene produced problem in vaccine production and in killing of HIV by body defense.

Immunity

CMI: After entry, virus binds by gp120 Ag to CD4 receptors of CD4[+] (helper or inducer) cells and resulting breakdown of cells with leukopenia and reversal of CD4:CD8 ratio. Decreasing CD4[+]cells is an indirect indicator for decreasing CMI, which is responsible for development of malignancies, opportunistic infections and inability to mount the DTH.

AMI: As the CMI decreasing, AMI is increased for the compensatory mechanisms resulting increasing the production of antibodies like IgA, IgM and IgG, but these are not protective. Decrease response to Ag and also production of type-III hypersensitivity and autoimmunity by useless antibodies. In end stage of AIDS there is complete paralysis of immune system and decreasing antibodies like anti-p24.

Pathogenicity

Disease name: It is called **AIDS**

Epidemiology

- **Genotypes/clades and distribution in HIV-1:** On the bases of sequence analysis of gag and env genes, HIV-1 is classified in to **three genotypes** like M (Major), N (New → HIV-1 strains isolated from Cameroon and Gabon) and O (Outlier → HIV-1 strains isolated from Cameroon which are different from M and O group) Recently **genotype P** is also added. **Genotype P** is defined for HIV like strain of gorilla SIV identified from woman of Cameroon in 2009. **M group** causes worldwide infection and it contains total 11 subtypes from A-K.

- A: It is most prevalent in West Africa.
- B: It presents in USA, Europe, Japan and Australia commonly spread by homosexual and blood route.
- C: It is distributed worldwide and common in East Africa, South Africa, China and **India**. Commonly spreads by heterosexual route.
- E: It is common in Thailand. This type is not considered as separate type, but recombinant type of A and E called **AE type** or called **CRFs AE (circulating recombinant forms AE)**.
- Common types in Africa: A, C and D.
- Common types in Asia: B, C and E.
- Same host has more than one genotype and/or CRFs which are collectively called **quasispecies**.
- **Genotypes in HIV-2:** It has total 8 from A-H. Type A is most common.

Reservoirs of infection: These are both human carrier and case. Actual origin for HIV is shown in **Table 85.1**.

Sources of infection

- These are body fluids like blood, semen, milk and vaginal secretions and cells/tissues, which are infectious.
- It presents in saliva, but no confirmed case of salivary transmission has been reported may be due to viral inactivation by salivary substances like secretory leukocyte protease, fibronectin and glycoproteins.
- It also presents in urine, CSF and tears, but not able to transmit the disease.

Modes of transmission: Following are the different modes for transmission of HIV. Percentage in brackets indicates the chances of infection per exposure.

1. Direct transmission

- **Direct contact**
  - Direct contact (splash) of infected body fluid (blood) with abraded skin/mucosa (in eyes, mouth and nose in hospital workers): Chances of risk are **0.09%**.
  - During sexual intercourse (0.1–1%): It includes oral, vaginal, (heterosexual), anal (homosexual) and digital sex. Male to female transmission by vaginal sex is approximately 8 times higher than female to male transmission. Heterosexual mode is the most common mode of transmission of HIV, especially in developing countries. Anal sex has more risk than vaginal intercourse.
- **Inoculation under skin/mucosa:** It occurs by contaminated needles or syringes in IV drug users or in hospital staff or surgical wound (**0.5–1%**). Percutaneous injury transmits HBV (**30%**) > HCV (**3%**) > HIV (**0.5–1%**). Lower segment cesarean section (LSCS) decreases the HIV transmission.
- **Perinatal/vertical/mother to baby (30%):** Risk is maximum when mother is recently infected or already developed AIDS. Range of vertical transmission is 15–25% in developed countries and 25–35%

in developing countries. This route transmits HIV (**30%**) > HBV (**20%**) > HCV (**6%**) > HDV (**±**) > HAV and HEV (**–**).
  - Antenatal (**23–30%**): By placenta.
  - Intranatal (**50–65%**): Contact of infected mother's blood to abraded skin/mucosa of baby. It has maximum risk than antenatal and postnatal transmission.
  - Postnatal (**12–20%**): By breast milk.
- **Indirect transmission:** It occurs by body fluids or by tissues.
  - Transfusion of blood or blood products (**>90%**).
  - Organ or semen transplantation (**50–90%**).
  - Virus load is maximum in blood, genital secretions and CSF. Variable in breast milk and saliva. Zero to minimal in other body fluids.

Exogenous and endogenous retro viruses

- **Exogenous retro viruses:** It includes the horizontal transmission from one host to another as mentioned earlier.
- **Endogenous retro viruses:** It includes the hereditary (vertical) transmission from parents to offspring, where virus is integrated into host cell genome, called provirus. Provirus behaves like host cell gene, regulated by host cell gene and spreads to offspring. However, provirus is usually silent, nonpathogenic and do not cause any disease or malignancy.

No transmission: HIV is not transmitted by arthropods like mosquitoes or others, by activities like shaking hands, hugging (body to body contact), dry kissing, putting cheeks together, eating in same dishes, by sharing common things like clothes, towels or bed sheets or by nonliving objects like air, food and water.

Incubation period: It is uncertain and from months to year or 10 years or more. Long incubation period presents, because of latent infection by provirus and time to time release of progeny virus.

Portal of entry: Skin/mucosa.

Sites: Almost all body organs/tissues/cells are susceptible to HIV infection.

Precipitating factors (epidemiological determinants): **(1) Age:** It occurs more in sexually active young adults. Children are infected in <3% cases. **(2) Sex:** HIV is more in men in America, Australia and Europe while in Africa ratio is equal. **(3) Occupation:** Health workers, drug addicts, patients required multiple blood transfusions like hemophiliacs and prostitutes are at increased risk. **(4) No use of condoms:** Sexual intercourse with infected partners without using condoms or suitable devices increases the risk. **(5) Iatrogenic:** Use of unsterile needle, syringes or surgical instruments by hospital staff may precipitate the infection. **(6) Nature of virus entry:** Infection chances are more due to entry of HIV-infected cells than entry of the cell free virus.

1. **Viral multiplication:** Follow **Flowchart 85.1.**
2. **Disease progression**
   - **HIV infects and destroys the CD4 cells resulting** decrease T4 numbers (selective T-cell deficiency), reversal of T4:T8 ration, reduction of lymphocytes in lymph node biopsy and also in peripheral blood (lymphopenia), decrease the release of IL-2, γ-IFN and other interleukins which reduce the CMI and decrease the DTH which gives false-negative skin tests.
   - **When CMI decrease, body increase AMI by compensatory mechanisms resulting** hyper-gammaglobulinemia (mostly useless antibodies) particularly raised IgA, IgG and in children raised IgM level, immune complex disease (type-III hypersensitivity) due to presence of useless antibodies, autoimmunity due to presence of autoantibodies and decrease response to Ags due to presence of useless Abs.
   - **HIV infects and destroys following cells**
     – Platelets: It produces thrombocytopenia.
     – Monocytes and microphage: Resulting decrease cytotoxic activity, chemotaxis, intracellular killing and antigen presentation.
     – Cells of CNS: Producing dementia (called **pre senile dementia**) and degenerative neurological lesions. HIV can cross the blood-brain barrier and neurological changes are due to direct HIV effects or by opportunistic infections or by malignancies.

Flowchart 85.1: Multiplication of HIV

Entry of virus by above mentioned routes
↓
Binding by gp120 Ag to CD4 receptors present on different cells like CD4$^+$ T cells (most commonly affected than other cells), B cells, keratinocytes and macrophages like monocytes, microglial brain cells, intestinal cells, PAM of lungs, follicular dendritic cells from tonsils, Langerhans cells in skin, etc.
↓
After binding **fusion** is required for infection which is possible by gp41 Ag with cooperation by other receptors called **co-receptors** like CXCR4 for T cell tropic HIV strain and CCR5 for macrophage (M) - tropic HIV strain
↓
After fusion entry of virus in cells
↓
Uncoating of viral genome and conversion of RNA to ds-DNA by reverse transcriptase
↓
ds-DNA called **provirus** which integrated in to genome of host cells with the help of viral enzyme integrase and produces **latent infection**
↓
This latency is different from other as it lyses the infected cells and releases the progeny virions that infect the other cells. During latent phase, patient is infective; virus is present in lymphoid cells (like CD4) and replicating very slowly and steadily with decrease in CD4 count like 50/ml/year. Abs tests are positive. Progress of disease directly depends on the HIV- RNA level in plasma. Patients with high level are faster progressors than with low level.

- **Clinical manifestations in HIV infection are not due to virus, but due to immunodeficiency resulting** secondary (opportunistic) infections by other microbes and malignancies.

Clinical features: Disease passes through following different stages. **(1) Acute HIV infection or acute HIV syndrome or acute retroviral syndrome or seroconversion illness:** Minor flu like or acute infectious mononucleosis (glandular fever) like symptoms may occur like fever, rash, headache, fatigue, malaise, lymphadenopathy, etc. Symptoms occur 6–12 weeks after infection. In this stage only p24 Ag is present while Abs are absent. Abs develop 4 weeks after infection called **window period** or in later part hence called **seroconversion illness.** These symptoms are due to effect by viral multiplication with high viral load and by immune complex. There is significant decrease in CD4 count in this stage. **(2) Asymptomatic or latent infection:** Virus remains latent in infected cells for very long time without any clinical features. Abs tests are positive. Progress of disease directly depends on the HIV-RNA level in plasma. Patients with high level are faster progressors than with low level. **(3) Persistent generalized lymphadenopathy:** It is defined as an enlarged lymph node with >1 cm size in two or more noncontagious sites that last for at least 3 months. It should be differentiated from other causes of lymphoma. **(4) ARC:** Patient presents with considerable immunodeficiency with fever >1 month, diarrhea >1 month, >10% of weight loss and mild opportunistic infections like oral candidiasis (thrush). **(5) AIDS:** It is an end stage of HIV infection. Patient presents with high viral load and total destruction of lymphoid tissues with fibrosis. After clinical infection, AIDS has been developing approximately in 10 years; however, about 5–10% patients escape the development of AIDS for long time around 15 years or more called **long-term survival** or **chronic nonprogressors. Clinically, AIDS is defined** as irreversible damage of immune system (CD4 count is <200) with opportunistic infection and/or malignancy.

Note: Opportunistic infections and malignancies

Bacterial infections

- Mycobacteria: *M. tuberculosis* and NTM like *M. avium intracellulare* are the common agents. Tuberculosis is the most common opportunistic infection in AIDS patients. In developing/tropical countries like India, *M. tuberculosis* is the most important pathogen with multidrug resistant strains. *M. avium intracellulare* is most common when CD4$^+$ count falls <50/mm^3.
- Other bacterial infections are salmonellosis, campylobacteriasis, nocardiasis, actinomycosis and legionellosis. Genital lesions in AIDS patients are also common by *H ducreyi*.

Viral infections

- Varicella Zoster virus (herpes zoster).
- Herpes simplex viruses: These are common pathogens causing genital lesions in AIDS patients.

(Contd…)

- **CMV:** Most common ocular features by CMV in AIDS patients are hemorrhagic retinitis and granular retinitis, when CD4$^+$ count falls <50/mm^3. These are usually bilateral and present with retinal exudates and perivascular hemorrhage. They complicated as retinal detachment, retinal atrophy or optic nerve disease.
- **Zoster infection by VZV:** Vey rarely in AIDS patients, when CD4$^+$ count falls <50/mm^3, it causes devastating necrotizing retinitis called **progressive outer retinal necrosis** characterized by bilateral rapid painless loss of vision.
- **EBV:** Oral hairy leukoplakia.
- **JC virus:** It produces the PML and other lesions only in immunodeficient persons.

Fungal infections

- These are candidiasis, aspergillosis, histoplasmosis, penicilliosis, cryptococcosis and pneumocystosis.
- Cryptococcosis: It is the most common cause of acute meningitis in AIDS patient when CD4 count falls <100/mm^3. Other opportunistic microbes affecting CNS in AIDS patient are *Toxoplasma gondii, M. tuberculosis* and herpes (herpetic encephalitis).
- Pneumocystosis: Caused by *Pneumocystis carinii*. Now species name changed to *Pneumocystis jiroveci*. It is the most common in USA and other Western countries. Pneumonia occurs in patient when CD4$^+$ count falls <200/mm^3 and required prophylaxis.

Parasitic infections

- Toxoplasmosis: It occurs when CD4$^+$ count falls <200/mm^3. Most common cause of seizure in AIDS patient. Toxoplasmosis may present with other features like hemiparesis, vomiting and headache.
- Diarrheal parasites: Mostly they cause persistent diarrhea. Most common type of diarrhea is cryptosporidiosis. Other types include cystoisosporiasis and microsporidiasis.
- Strongyloidiasis: It is the most common helminthic infection in AIDS patients.

Malignancies

- Kaposi's sarcoma: It is a most common tumor in AIDS patient.
- Lymphomas: These are Hodgkin's and non-Hodgkin's types. Most common lymphoma is immunoblastic lymphoma, while among the brain tumor the most common lymphoma is primary CNS lymphoma.
- Cervical cancer.

Others

- HIV encephalopathy.
- HIV associated nephropathy or cardiomyopathy.
- Illness progress inexorably and death of patients in months or years.

Clinical classification: CDC, USA classified HIV infection in to following different stages in 1993, based on clinical presentation and CD4 count as mentioned below. This classification was revised by WHO in 2007.

Stage-I: Acute HIV syndrome.

Stage-II: Asymptomatic infection.

Stage-III: Persistent generalized lymphadenopathy.

Stage-IV: Other diseases with following subgroups.
- **Substage A:** AIDS Related Complex.
- **Substage B:** Neurological diseases.

- **Substage C:** Secondary infectious diseases with following two subgroups. **(1) C1:** CMV or other herpes diseases, pneumocystosis, cryptococcosis, cryptosporidiosis, toxoplasmosis or strongyloidiasis. **(2) C2:** Oral hairy leukoplakia, salmonellosis, nocardiosis, tuberculosis, bacteremia or thrush.
- **Substage D:** Secondary cancer like Kaposi's sarcoma, lymphoma.
- **Substage E:** Other conditions.

Note: Exclusion of tests from the clinical definition of AIDS
Serological and molecular tests are excluded from the definition of clinical AIDS, because they also become positive in other stages of HIV infection like acute HIV syndrome and PGL.

Note: Correlation between CD4 count and opportunistic infections
- <50/mm^3: *M. avium intracellulare* and CMV retinitis.
- <100/mm^3: Cryptococcosis
- <200/mm^3: Pneumocystosis and toxoplasmosis
- 200–400/mm^3: *M. tuberculosis*, Varicella Zoster virus, Herpes simplex, candidiasis, oral hairy leukoplakia and Kaposi's sarcoma (KS).

Laboratory Diagnosis

Policy: In India, it is decided by NACO, Delhi. NACO provides guidelines for serological testing and interpretation of results. It also maintains the uniformity and quality in reporting the incidence and prevalence of disease. Testing policy includes **3Cs** that include **c**onsent, **c**ounseling and **c**onfidential testing.

Indications: (1) To diagnose, to treat and to prevent the HIV-AIDS. (2) To screen the blood before transfusion. (3) For antenatal screening of mother for **P**revention of **P**arent **T**o **C**hild **T**ransmission (PPTCT) or vertical transmission of HIV. (4) For postexposure management.

Steps

1. Suspect HIV infection (from history).
2. Pretest counseling at ICTC/PPTCT.
3. Informed consent should be taken in written format.
4. Sample collection.
5. Sample transport.
6. HIV testing and interpretation of results.
7. Confidential reporting is must. Patient name or HIV positive should not be written on report form.
8. Post-test counseling. It is done to motivate the patient to tell spouse or family and to do behavior change.
9. Refer HIV positive to ART center to assess clinical and immune status (CD4 count) and for ART.
10. Follow up counseling as required.

Specimens: Best sample is blood/serum, but can use CSF, saliva, cervical secretions, semen, tears or materials from organ biopsy.

Testing methods: All tests are categorized in three categories like nonspecific tests (tests to detect immune status or immunological tests), specific tests

 (tests to detect HIV-AIDS) and tests to diagnose the opportunistic infections.

Nonspecific Tests (Tests to Detect Immune Status or Immunological Tests)

- TLC: Leukopenia (<2000 mm^3).
- CD4 count: <200 mm^3. CD4 count is measured by flow cytometry method. It is very low in full blown AIDS (end stage) as suggested in **Fig. 85.2**. Normal CD4 range is 400–1600 cells/mm^3 and normal CD8 range is from 150–800 cells/mm^3.
- CD4:CD8 ratio: In normal adults and children CD4 to CD8 ratio is 2:1, it means CD4 constitutes 65% and CD8 constitutes 35% of total T cells, or it means that there are 2 CD4 cells for every CD8 cell out of total T cells. In AIDS ratio is <1 (reversed).
- Platelets: Thrombocytopenia.
- Antibodies status: Raised IgG and IgA.
- Diminished CMI: All skin tests are false-negative.
- Abnormal lymph node biopsy with fibrotic changes.

Specific Tests (Tests to Detect HIV-Aids)

A. **Microscopy:** Follow morphology.
B. **Culture:** Follow C/Cs.
C. **Serological tests:** Two types of tests like Ag and Ab detection tests.

1. **Ag detection tests**
   - **Synthesis of Ag:** P24 is the earliest Ag seen in blood after 2 weeks (**Fig. 85.2**) and lasts for 3–4 weeks. However, if the infecting dose is small like needle prick injury, appearances may be delayed. It again increased in late advance stage.
   - **Tests:** P24 Ag is detected by 4th generation **ELISA**. It has 30% sensitivity.
   - **Clinical significances of P24 Ag detection: (1) Diagnostic:** It is the earliest marker appears in blood and useful for diagnosis of HIV infection in **window period.** Newborns and children, where immune system is immature and Abs are not synthesized, disease has been diagnosed by p24. P24 will be detected from CSF also. **(2) Infectivity:** Virus and p24 Ag titer is high in acute phase, decline/absent in latent (asymptomatic) phase due to Ab formation (**seroconversion**

illness mostly after 4 weeks) and again high in the end, which shows maximum infectivity in initial part and end part while less in latent phase (**Fig. 85.2**). However, sometimes, p24 Ag which is already bound to Ab, gets dissociated from Ag-Ab complex and can be detected by capture ELISA in 30–50% cases. **(3) Prognostic:** It helps to monitor the disease progress.

2. **Ab detection tests**
   - **Synthesis of Ab:** IgM synthesized at 4–6 weeks followed by IgG. IgM disappears in 8–10 weeks while IgG persists throughout (**Fig. 85.2**). When clinical AIDS established with severe immunodeficiency some Ab components may disappear like anti-p24.
   - **Tests:** They detect HIV-1 and 2 either separately or together. These tests are recommended by NACO and they have high sensitive of $\geq99.5\%$ and high specificity of $\geq98\%$. Two types of Ab detection tests as mentioned in **Table 85.2** with name and differences. **Screening tests** include ELISA (3rd or 4th generation) and rapid tests like particle agglutination (latex, RBCs and gelatin) tests, immunoconcentration test (flow through/vertical/dot blot assays), ICT (lateral flow assay) and ELISA based rapid tests (immuno-comb HIV-1 and 2 bi-spot method). **Confirmatory tests** include WB test (**Ch. 122**), line immunoassay and IF test.
   - **Ag used in screening test kits**
     - HIV 1: p24, gp160, gp120 and gp41
     - HIV 2: gp36
   - **Clinical Significances**
     - **Screening:** It is defined as application of HIV testing whether voluntary or mandatory in entire population or target group; however, it is not possible in entire population. It is useful to prevent the transmission. It is possible in target groups like blood/organ/semen donors, IV drug abusers, antenatal women etc. Screening is best useful when done for P24 Ag detection.
     - **Diagnostic:** Most useful for diagnosis of HIV.
     - **Sero-epidemiological:** Ab survey is most useful in identifying the geographical extent of HIV infection and other epidemiological studies like spread of HIV from identified sources.

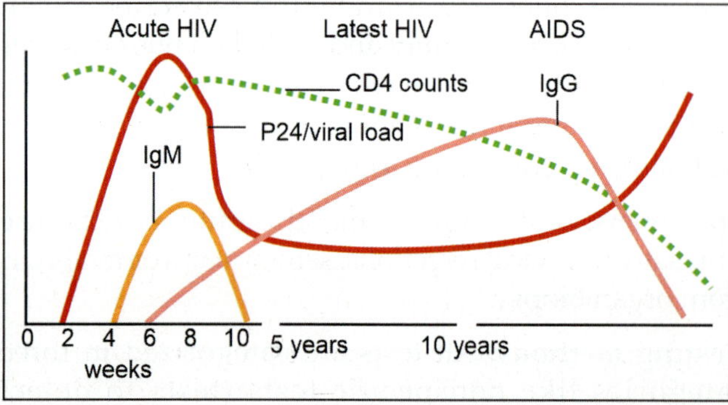

Fig. 85.2: HIV markers in different stage of HIV infection

TABLE 85.2: Differences between Ab detection tests	
Screening tests	**Confirmatory tests**
High sensitivity	High specificity
More false-positive	Few or no false-positive
Low cost	High cost
Easy to perform	Requires special instruments and expert person

- **Prognostic:** Loss of anti-p24 Ab from serum indicates end stage of disease and irreversible breakage of immunity. End stage also associated with high antigenic titer of p24 and high viral load in circulation.

- **Disadvantages:** Ab detection tests are not useful in **window period** as antibodies are not synthesized, in **end stage** of disease because of total immunodeficiency and when **infection occurs by other virus** like HIV-2 and testing is done by using HIV-1 antigens alone.

3. **NACO strategies of HIV testing by using serological tests:** By using screening tests like ELISA and rapid tests, total three strategies (I, II, III) are defined. In case of discordant/indeterminate result WB test/RT-PCR is used for confirmation. A1, A2, A3 represent test-1, test-2 and test-3 respectively. Serologically HIV-AIDS is declared positive when 3 tests are reactive with two different principles or Ags. Same kit should not be used again. Following are 3 strategies.

- **Strategy–I:** It is indicated for transfusion or transplantation. It is a single test strategy as shown in **Flowchart 85.2**.

- **Strategy–II**
  - Strategy–IIA: It is indicated for sentinel surveillance. It is a two tests strategy as shown in **Flowchart 85.3**.
  - Strategy–IIB: It is indicated for diagnosis of an individual with symptoms suggestive of AIDS. It is a three tests strategy as shown in **Flowchart 85.4**.

- **Strategy–III:** It is indicated for diagnosis of HIV infection in asymptomatic individual, antenatal patients and patients required surgeries. It is a 3 tests strategy as shown in **Flowchart 85.5**.

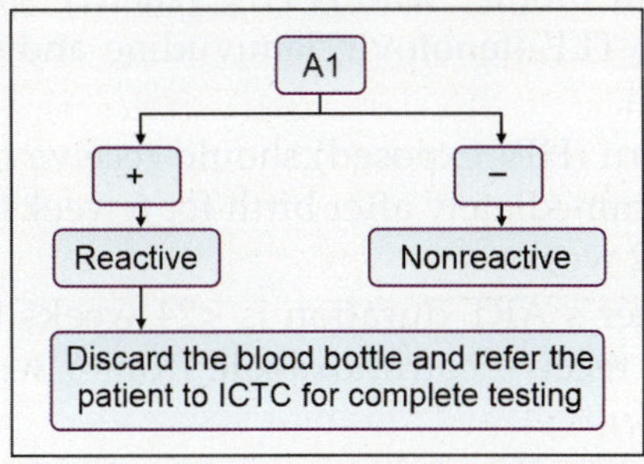

Flowchart 85.2: Strategy-I

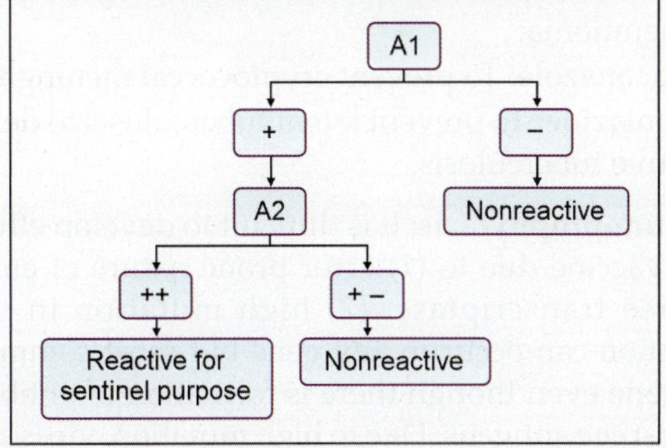

Flowchart 85.3: Strategy-IIA

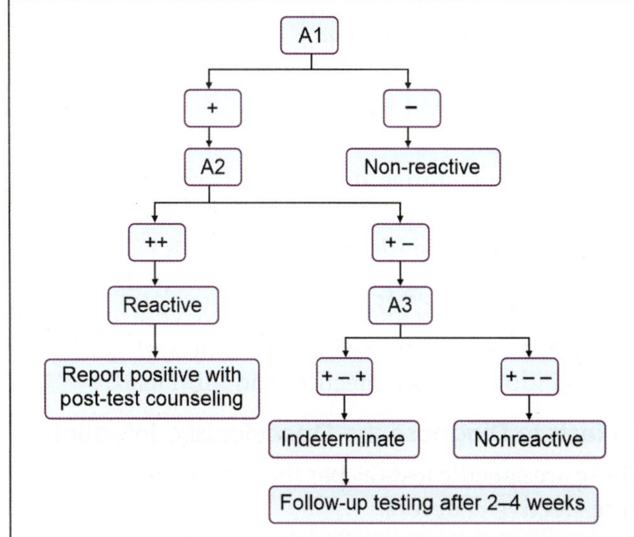

Flowchart 85.4: Strategy-IIB

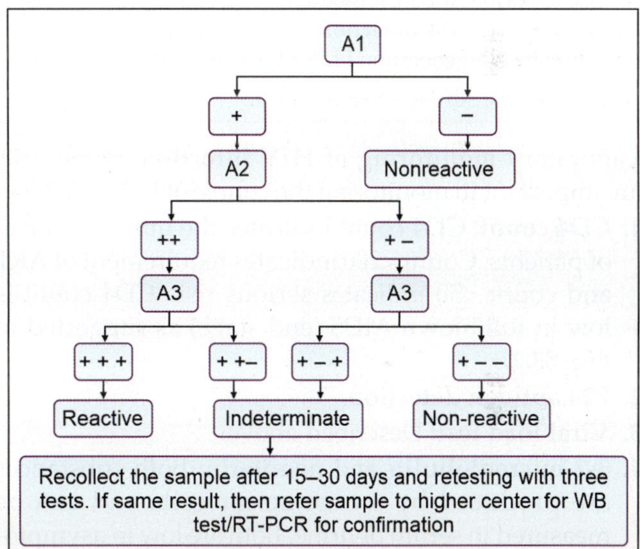

Flowchart 85.5: Strategy-III of HIV testing

D. Molecular methods

- **Different tests are:** DNA PCR, RT-PCR, real-time RT-PCR (qRT-PCR to detect the viral load), branched DNA assay and NASBA.

- **Clinical significances**
  - **Diagnostic:** Molecular method is highly sensitive and specific test, considered as gold standard for diagnosis and confirmation of HIV. It can detect very few copies of RNA from clinical samples. RT-PCR can detect as less as 40 copies/ml of HIV RNA, where DNA assay and NASBA can detect when concentration is >50 copies/ml and >80 copies/ml, respectively; hence, RT-PCR is considered most sensitive in compare to other molecular methods. It diagnoses HIV in window period (due to high viral load) within 12 days of exposure before the availability of p24 Ag in blood.
  - **Monitoring of therapy:** RT-PCR and branched DNA assay (bDNA) can measure the viral load and considered as best tools to monitor the response

Infections of Retroviridae

6

to ART, progress of disease and clinical outcome. Viral load is high in full blown AIDS (end stage) as suggested in **Fig. 85.2**.

- **Viral typing and genotyping:** It differentiates between HIV-1 and 2 and also detect the specific genotype.
- **Drug resistance detection:** It detects the drug resistance gene.
- **Clinical stage identification:** DNA PCR is useful to diagnose the pediatric HIV and to differentiate between latent infection and active viral replication. DNA PCR also used for detection of genotypes, window period infection and viral load.

Tests to Diagnose the Opportunistic Infections

These are specific tests as per the organism as described in respective chapters.

> **Note: Suggested test of HIV as per time after exposure**
>
> **Window period: RT-PCR** is best method and useful in 12 days after infection. **p24 detection is detected** after 2 weeks of infection by 4th generation ELISA but sensitivity is only 30%.
>
> **After 4 weeks:** Ab detection is useful.

Laboratory monitoring of HIV infection: Some tests are important in monitoring the course of HIV infection.

1. **CD4 count:** CD4 count indicates the immune status of patients. Count <200 indicates requirement of ART and count <50 indicates serious risk. CD4 count is low in full blown AIDS (end stage) as suggested in **Fig. 85.2**.
2. **P24 antigen detection.**
3. **Viral load test:** Described above.
4. **β-2-microglobulin and neopterin:** Both substances are produced by mononuclear cells and can be measured in serum or urine. Both are low in asymptomatic phase and high in end stage. Other useful substance is IL-2.

HIV-marker in different stages: Follow **Table 85.3 and Fig. 85.2**.

Pediatric AIDS

Modes of transmission: It is described above in the section → **Perinatal/vertical/mother to baby transmission**. It occurs in 30% cases from infected mothers.

Clinical features: Early deficiency of AMI leads recurrent bacterial infections. Other features are failure

to thrive, persistent oral candidiasis, tuberculosis, lymphadenopathy, recurrent diarrhea, abnormal neurological findings and lymphocytic interstitial pneumonia, Kaposi's sarcoma, cryptococcosis and toxoplasmosis are rare compared to adults.

Testing in children <18 months: Screening tests are not useful because child has immature immune system, so antibodies are not formed, interference by maternal antibodies especially by IgG and inability of screening tests to differentiate between child's IgG and maternal IgG. Maternal IgG disappears after 18 months. Following are useful tests.

- DNA PCR: It is the most useful method and detected by dry blood spot. If found positive then retest by DNA PCR and if again positive then refer child for lifelong ART.
- P24 Ag detection by ELISA.
- RT-PCR
- Viral culture method: Labor intensive, so less useful.

Testing in children >18 months: Screening tests detecting child's IgG are useful because of disappearance of maternal IgG after 18 months and there is no maternal IgG interference.

Prevention

General measures: These are improvement in health education, screening of blood, organs and tissues before of donation, safe sex, high-risk behavior modification like drug abuse, avoid direct contact with blood-body fluid and use of disposable sterile medical devices.

Prevention of sharp/splash injuries in HCWs: Ch. 29.

Preventions during surgeries: Ch. 29.

Preventions of neonatal transmission

- Pregnant mother who is HIV reactive should take lifelong TLE (tenofovir, lamivudine and efavirenz) regimen.
- Newborn (HIV exposed) should receive nevirapine syrup immediately after birth for 6 weeks or extend up to 12 weeks.
- If mother's ART duration is <24 weeks then baby should receive cotrimoxazole from 6 weeks to 18 months.

Chemoprophylaxis for opportunistic infections

- Cotrimoxazole: To prevent *Pneumocystis jiroveci* pneumonia.
- Fluconazole: To prevent cryptococcal meningitis.
- Isoniazide: To prevent latent tuberculosis to develop active tuberculosis.

Immunoprophylaxis: It is difficult to develop effective HIV vaccine due to (1) error prone nature of enzyme reverse transcriptase. (2) high mutation in virus. Mutation can occur in any gene but most common in *env* gene even though there is formation of antibodies against *env* antigens. Due to high mutation, virus shows

Stage	P24	IgM	IgG	Ab by WB test
TABLE 85.3: HIV-marker in different stages				
Acute	+	–/+	–	Partial p24 and/or gp120
Latent	–	+	–	Full pattern
ARC	–	–	+	Loss of p24
AIDS	+	–	–	Loss of p24

extensive genetic diversity and evades the host's defense and creates problems in development of effective HIV vaccine. (3) long latent period between exposure and appearance of symptoms. (4) lack of ideal animal model. (5) difficult to get human volunteers for vaccine trials. (6) viral genome get incorporated in to host cell genome very soon after exposure, hence it provides very short time to control the infection. (7) possible risk of viral reactivation. (8) HIV targets the immune system, so natural immunity is not able to control it. (9) incapability of HIV to induce the neutralizing antibodies. Following type of vaccine are developed but still under trial.

- **Killed/modified whole virus vaccine:** This trial is going in Canada since 2012 to till now. It includes genetically modified killed virus without *nef* and *vpu* genes. Results showed good efficacy and antibodies produced till 52 weeks.
- **Subunit vaccine:** Prepared from envelope gp120. This trial was done in 2003.
- **Vector vaccine:** Most of the vaccine trials are based on this principle. Different types vaccine trials have been done by inserting the immunogen of HIV in nonpathogenic bacteria or viruses. **(1) RV 144 trial (Thailand, 2003–09):** It was done by inserting gp120 in canarypox virus. Results showed 31.2% efficacy at 42 months. **(2) Step trial:** It was done by using adenovirus type 5 from 2005–07. **(3) Trials in India:** It used AAV and modified vaccinia ankara virus. Viruses were used by International AIDS Vaccine Initiative in collaboration with NACO, India. **(4) HVTM 505 trial:** It was done from 2009–13. It used the "prime and boost" strategy in which two or more different vaccines were used. It used vector virus to prime T cells with subunit vaccine or DNA vaccine as booster.
- **Target cell protection** by Anti-CD4-antibody or by genetically engineered CD4 cells.

Treatment

It includes management of normal person after exposure called **postexposure prophylaxis** and **management of case of HIV-AIDS**.

Postexposure Prophylaxis (PEP) of HIV

Definition: It is a short-term treatment to avoid the HIV infection after occupational exposure like skin injury by sharp items like needle, scalpel, etc., splash of potentially infectious materials **(PIMs)** over non-intact skin, splash of PIMs over intact skin for prolonged duration and splash of PIMs over mucosa (intact or abraded).

Infectious materials/specimens: PIMs are blood, semen, vaginal secretion, CSF, synovial fluid, pleural fluid, peritoneal fluid, pericardial fluid and amniotic fluid. **Non-infectious materials** (unless contaminated with blood) are feces, urine, sputum, saliva, nasal secretions, sweat, tears and vomitus.

1. **First aid**
   - **Do's**
     - If splash occurs then irrigates the site with water for 5 minutes.
     - If fluid entered in mouth then spit out immediately and rinse the mouth several times.
     - If wearing the contact lenses, then wash the eyes. Once eyes get clear remove the lenses and wash them with water.
   - **Don'ts**
     - Don't panic and put injured finger in mouth.
     - Don't squeeze blood from injured site.
     - Don't use antiseptics or detergents.
2. **Report to nodal officer of hospital:** Nodal officer should perform following functions from steps 3–8.
3. **Taking 1st dose of PEP for HIV:** It should be initiated within 2 hours but not >72 hours. If it started very soon it reduces the risk of HIV transmission up to 80%. NACO recommended **TLD regimen** (Tenofovir, Lamivudine and Dolutegravir). If the source is found HIV nonreactive on patient's case or any of hospital documentation system then PEP for HIV is not needed. If status of source is not available then immediately start the PEP for HIV without waiting for laboratory report.
4. **Viral rapid tests**
   - Do the viral rapid tests (results should be available in 1–2 hours) for anti-HIV antibody, anti-HBs antibody (for vaccinated person), anti-HCV antibody and HBsAg of source (patient) and HCW.
   - Risks of transmission of infections following needle prick/percutaneous injuries are HBV (**30%**) > HCV (**3%**) > HIV (**0.5–1%**).
   - Risk of transmission following splash of blood to mucosa of eyes, mouth and nose: HIV is 0.09%.
5. **Selection of PEP:** PEP regimen for HIV is given by NACO and for HBV given by CDC. Regimen depends on severity of exposure, type of exposure and HIV status of the source: Source includes the PIMs or instruments contaminated by these PIMs.
   - **HIV Exposure Code (EC):** Follow **Flowchart 85.6.**
   - **HIV Status Code (SC):** Follow **Flowchart 85.7.**
   - **Selection of regimens:** Follow **Table 85.4.**
   - **PEP not indicated**
     - Delay in reporting >72 hours.
     - Exposure is on intact skin.
     - SC unknown.
     - Source is HIV nonreactive.
     - Exposure of low risk specimens like feces, urine, sputum, saliva, nasal secretions, sweat, tears and vomitus.
     - EC-1,2,3 and SC-1.
   - **Problems with PEP**
     - High cost.

7. **The HIV can be destroyed *in vitro* by which of the following?**
   a. Boiling
   b. Ethanol
   c. Cidex
   d. All of above

8. **HIV gene is/are:**
   a. gp73
   b. p24
   c. gp120
   d. gp5
   e. None

9. **Nef gene in HIV is used for:**
   a. Enhancing the expression of genes
   b. Enhancing viral replication
   c. Decreasing viral replication
   d. Maturation

10. **Which of the following is not a structural gene of HIV?**
    a. Gag
    b. Pol
    c. Env
    d. Tat

11. **Which of the following genes are present in HIV genome?**
    a. Gag
    b. Tat
    c. p500
    d. Kinase
    e. p24

12. **Gene coded for core of HIV is:**
    a. Gag
    b. Pol
    c. Env
    d. Tat

13. **Which of the following gene is associated with encoding of reverse transcriptase?**
    a. Gag
    b. Pol
    c. Env
    d. Ltr

14. **Gag gene encodes for:**
    a. Reverse transcriptase
    b. Core antigen
    c. Envelope
    d. Gene activation

15. **Which of these gene is a negative regulatory gene for HIV virus replication?**
    a. vif gene
    b. tet gene
    c. nef gene
    d. rev gene

16. **Which of these nonstructural genes is absent in HIV-2?**
    a. tat gene
    b. rev gene
    c. vpu gene
    d. vpx gene

17. **A HIV mother delivers a baby. All are true, *except*:**
    a. Risk of HIV in the baby is up to 90%
    b. HIV infection cannot be diagnosed in the baby with available methods
    c. AIDS can be transmitted from mother to child during delivery
    d. Breast feeding can transmit AIDS

18. **In the heterosexual transmission of HIV:**
    a. There is a greater risk of transmission from man to woman
    b. There is a greater risk of transmission from woman to man
    c. Risk is equal in either ways
    d. HIV infection is not transmitted by heterosexual act.

19. **True about HIV:**
    a. Not transmitted through semen
    b. More chances of transmission during LSCS than normal labor
    c. More infectious than hepatitis B
    d. Male to female transmission > female to male

20. **Mother to child transmission of HIV:**
    a. 25%
    b. 50%
    c. 60%
    d. 75%

21. **The chances that a health worker gets HIV from an accidental needle prick is:**
    a. 1%
    b. 10%
    c. 95%
    d. 100%

22. **Most common mode of transmission of HIV worldwide is:**
    a. Heterosexual
    b. Homosexual
    c. IV drug abuse
    d. Contaminated blood products

23. **HIV is transmitted by following route, *except*:**
    a. Blood transfusion
    b. Sexual intercourse
    c. Saliva
    d. Organ transplantation

24. **Percentages of HIV infection by sexual intercourse are:**
    a. 0.1–1%
    b. 30%
    c. 50–90%
    d. More than 90%

25. **HIV:**
    a. Can cross blood-brain barrier
    b. Is RNA virus
    c. Inhibited by 0.3% H_2O_2
    d. Is thermostable

26. **Regarding HIV infection, not true is:**
    a. p24 is used for early diagnosis
    b. Lysis of infected CD4 cells is seen
    c. Dendritic cells do not support replication
    d. Macrophage is a reservoir for the virus

27. **Regarding HIV which of the following is not true?**
    a. It is a DNA retro virus
    b. Contains reverse transcriptase
    c. May infect host CD4 cells other than T lymphocytes
    d. Causes a reduction in host CD4 cells at late stage of disease

28. **HIV affects:**
    a. Only T helper cells
    b. T helper and macrophages
    c. NK cells
    d. B lymphocytes

29. **HIV infects most commonly:**
    a. $CD4^+$ helper cells
    b. $CD8^+$ cells
    c. Macrophage
    d. Neutrophil

30. **HIV can infect all *except*:**
    a. Circulating dendritic cells
    b. CD4 T lymphocytes
    c. Macrophages
    d. Cytotoxic T cells

31. **Receptors for HIV:**
    a. CD4
    b. CC3
    c. CD5
    d. CD56

32. **The receptor through which M tropic HIV strains bind?**
    a. CCR5
    b. CXR4
    c. CXCR5
    d. Any of the above

33. **Latent phase of HIV:**
    a. Viral replication
    b. Sequestered in lymphoid tissue
    c. Infective
    d. Non-infective
    e. Rapid decline of CD4 is common

34. **HIV infection is associated with:**
    a. A glandular fever like illness
    b. Generalized lymphadenopathy
    c. Gonococcal septicemia
    d. Sinus disease
    e. Pre-senile dementia

35. **Sero-conversion in HIV infection takes place in:**
    a. 2 weeks
    b. 4 weeks
    c. 9 weeks
    d. 12 weeks

36. **Most common opportunistic infections in AIDS in India:**
    a. Toxoplasmosis
    b. Cryptococcosis
    c. Cryptosporidium
    d. TB

37. **Most common agent causing tuberculosis in AIDS patient in tropical countries is:**
    a. *Mycobacterium tuberculosis*
    b. *Mycobacterium intracellulare*
    c. *Mycobacterium parvum*
    d. *Mycobacterium* atypical

38. **In India, most common cause of TB in HIV:**
    a. *M. tuberculosis*
    b. *M. avium intracellulare*
    c. *M. bovis*
    d. *M. scrofulaceum*

39. **CMV retinitis in HIV occurs when CD4 counts falls below:**
    a. 50
    b. 100
    c. 200
    d. 150

40. **An HIV patient complains of visual disturbances. Fundal examination shows bilateral retinal exudates and perivascular hemorrhage. Which of the following viruses are most likely to be responsible for this retinitis?**
    a. Herpes simplex
    b. Varicella zoster
    c. Cytomegalo virus
    d. EBV

41. **The most common organism amongst of the following that causes acute meningitis in AIDS patient is:**
    a. *Streptococcus pneumoniae*
    b. *Streptococcus agalactiae*
    c. *Cryptococcus neoformans*
    d. *Listeria monocytogenes*

42. **Common CNS lesion in HIV is caused by:**
    a. *Cryptococcus*
    b. *Toxoplasma*
    c. Neurocysticercosis
    d. Mucormycosis
    e. Lymphoma

43. **Important features of AIDS are:**
    a. Follicular tonsillitis
    b. Lichen panus
    c. Oral candidiasis
    d. Hairy leukoplakia
    e. Mitotic division of oral cavity

44. **In a patient having HIV infection, oral ulcer is most commonly due to:**
    a. *Candida*
    b. Cryptococcosis
    c. *Histoplasma*
    d. *Trychophyton*

45. **Persistent diarrhea in AIDS is caused by A/E:**
    a. *Microspora*
    b. *Cryptococcus*
    c. *Cryptosporidia*
    d. *Cystoisospora belli*
    e. *Giardia lamblia*

46. **Most common causative agent of diarrhea in AIDS:**
    a. Toxoplasmosis
    b. Cryptococcosis
    c. *Cryptosporidium*
    d. *Mycobacteria*

47. **Most common genital lesion in HIV patients is:**
    a. *Chlamydia*
    b. Herpes
    c. Syphilis
    d. *Candida*

48. **Commonest helminthic infection in AIDS is:**
    a. *Trichuris trichuria*
    b. *Strongyloides stercoralis*
    c. *Enterobius vermicularis*
    d. *Necator americanus*

49. **Which of the following lesion is not seen in HIV patient with CD4 count less than 100 per micro liter, who has non-productive cough?**
    a. *Mycobacterium tuberculosis*
    b. *Pneumocytis carinii*
    c. *Mycoplasma pneumoniae*
    d. Cryptococcal infection

50. **True about HIV all *except*:**
    a. PML caused by JC virus
    b. CNS lymphoma is the most common CNS tumor
    c. CMV is the most common cause of retinitis
    d. Most common cause of seizure is *Candida*

51. **All are true about AIDS, *except*:**
    a. Seen in heterosexual only
    b. Caused by retro virus
    c. Candidiasis is also common feature
    d. Retro virus is thermo labile

52. **During window period in HIV infection, the plasma viral load is:**
    a. Very high
    b. Low below detectable level
    c. Just around detectable level
    d. May be high or low

53. **HIV window period indicates:**
    a. Time period between infection and onset of first symptoms
    b. Time period between infection and detection of antibodies against HIV
    c. Time period between infection and minimum multiplication of the organism
    d. Time period between infection and maximum multiplication of the organism

54. **HIV can be detected and confirmed by:**
    a. Polymerase chain reaction (PCR)
    b. Reverse transcriptase-PCR
    c. Real time-PCR
    d. Mimic PCR

55. **When compared to Western Blot technique, ELISA test is:**
    a. Less sensitive, less specific
    b. More sensitive, more specific
    c. Less sensitive, more specific
    d. More sensitive, less specific

56. **More sensitive test for HIV infection:**
    a. Western blot
    b. ELISA
    c. Agglutination test
    d. CFT

57. **Screening test for AIDS:**
    a. ELISA
    b. PCR
    c. Western blot
    d. CD4 count

58. **More specific test for HIV infection:**
    a. Western blot
    b. ELISA
    c. Agglutination test
    d. CFT

59. **Confirmatory test for AIDS:**
    a. ELISA
    b. Rapid tests
    c. Western blot
    d. CD4 count

60. **All of the following methods are used for the diagnosis of HIV infection in a 2-month-old child, *except*:**
    a. DNA-PCR
    b. Viral culture
    c. HIV ELISA
    d. p24 antigen assay

61. **A patient comes to hospital with history of sore throat, diarrhea and sexual contact 2 weeks before. The best investigation to rule out HIV is:**
    a. p24 antigen assay
    b. ELISA
    c. Western blot
    d. Lymph node biopsy

62. **False about p24 is/are:**
    a. Developed after 3 weeks of infection
    b. Cant be seen in first week
    c. Cant be detected after sero-conversion
    d. Detected by ELISA

63. **Antenatal /maternal HIV diagnosis is of important:**
    a. To prevent vertical transmission
    b. To terminate
    c. To discharge
    d. To isolate the patient
64. **Which of the following is the most sensitive for diagnosis of HIV?**
    a. RT-PCR
    b. bDNA assay
    c. NASBA
    d. p24 detection
65. **Full blown immuno deficiency syndrome is:**
    a. High viral titers with low CD4 count
    b. Low viral titers with low CD4 count
    c. Low viral titers with high CD4 count
    d. High viral titers with high CD4 count
66. **Maternal to child transmission of HIV is prevented by:**
    a. Tenofovir   b. Lamivudine
    c. Efavirenz   d. Abacavir
67. **A resident doctor sustained a needle stick injury while sampling blood of a patient who is HIV positive, A decision is taken to offer him postexposure prophylaxis. Which one of the following would be the best recommendation?**
    a. Tenofovir + Lamivudine + Dolutegravir
    b. Zidovudine + Lamivudine + Nevirapine
    c. Zidovudine + Lamivudine + Indinavir
    d. Zidovudine + Stavudine + Nevirapine
68. **In HIV infection the term elite controllers refers to those cases where:**
    a. The virus load is always high.
    b. The response to ART is very good.
    c. AIDS does not develop for 10–30 years, without being on antiretroviral therapy.
    d. Complex combination of art drugs are required for their case management.

Answers and Explanation of MCQs

1. c
- HIV is a retro virus belongs to Retroviridae family.

2. a and c
- Follow section, **HTLV-I (pathogenicity → clinical features)** for explanation.

3. c
- Follow section, **HIV (Epidemiology → genotypes/clades in HIV 1)** for explanation.

4. c

5. b

6. a
- Follow section, **HIV (morphology)** for explanation of answers of MCQs 4–6.

7. d
- Follow section, **HIV (resistance)** for explanation.

8. e

9. c

10. d

11. a, b

12. a

13. b

14. b

15. c
- Follow section, **HIV (viral genome and antigenic structure)** for explanation of answers of MCQs 8–15.

16. c
- Follow section, **HIV (viral genome and antigenic structure) and Table 85.1** for explanation.

17. a and b
- Risk of HIV transmission from mother to baby is 15–25% in developed countries and 25–35% in developing countries.
- HIV infection can be diagnosed in the baby by P24 Ag detection by ELISA and viral nucleic acid by RT-PCR.
- AIDS can be transmitted from mother to child during delivery (intranatal) and by breast feeding (postnatal).

18. a

19. d

20. a

21. a

22. a

23. c

24. a
- Follow section, **HIV (pathogenicity → mode of transmission)** for explanation of answers of MCQs 18–24.

25. a, b, c
- Option a: Follow section, **HIV (pathogenicity → pathogenesis → HIV infects and destroys following cells)** for explanation.
- Option b: Follow section, **HIV (morphology)** for explanation.
- Option c and d: Follow section, **HIV (resistance)** for explanation.

26. c
- Option a: Follow section, **HIV (laboratory diagnosis → Ag detection tests)** for explanation.
- Option b, c and d: Follow section, **HIV [pathogenicity → pathogenesis and Flowchart 85.1)** for explanation.

27. a
- Option a and b: Follow section, **HIV (morphology)** for explanation.
- Option c and d: Follow section, **HIV [pathogenicity → pathogenesis (Flowchart 85.1)]** for explanation.

28. b, d

29. a

30. d

31. a

32. a
- Follow section, **HIV (pathogenicity → pathogenesis and Flowchart 85.1)** for explanation of answers of MCQs 28–32.

33. a, b, c, e

34. a, b, e
- Follow section, **HIV (pathogenicity → pathogenesis and Flowchart 85.1)** for explanation of answers of MCQs 33–34.

35. b
- Follow section, **HIV (pathogenicity → clinical features → acute HIV infection)** for explanation.

36. d

37. a

38. a

39. a

40. c

41. c

42. a, b

43. c, d

44. a

45. b, e

46. c

47. b

48. b

49. c

50. d

- Follow section, **HIV (pathogenicity → clinical features → AIDS → opportunistic infections and malignancies)** for explanation of answers of MCQs 36–50.

51. a

- AIDS is seen in both heterosexual and homosexual persons.

52. a

53. b

54. b

55. d

56. b

57. a

58. a

59. c

60. c

61. a

62. a, b, c

63. a

64. a

65. a

- Follow section, **HIV (laboratory diagnosis)** for explanation of answers of MCQs 52–65.

66. a, b and c

- Follow section, **HIV (prevention → Preventions of neonatal transmission)** for explanation.

67. a

- Follow section, **HIV (treatment → post exposure prophylaxis)** for explanation.

68. c

- Follow section, **HIV (elite controllers)** for explanation.

CHAPTER 86

Infections of Other RNA Viruses (Rubella Virus and Coronaviridae)

Chapter Outline
- Rubella Virus
- Coronaviridae
  - SARS-CoV2

RUBELLA VIRUS

Classification

Family: Togaviridae.

Genus: Following two genera.
1. *Alphavirus*: Ch. 83.
2. *Rubivirus*: Species is rubella virus

History

Teratogenic complication was diagnosed by Australian ophthalmologist Norman Gregg in 1941, with sudden increase of cataract in infants with rubella. Rubella virus was isolated in 1962 by tissue culture method.

Morphology

Shape and size: Spherical in shape (**Fig. 86.1**) with 50–70 nm in size.

Genome: It contains ss-RNA (+) with icosahedral capsid.

Enveloped: Lipoprotein envelope contains hemagglutinating spikes. They agglutinate the goose, pigeon, one-day-old chick and human RBCs at 4°C.

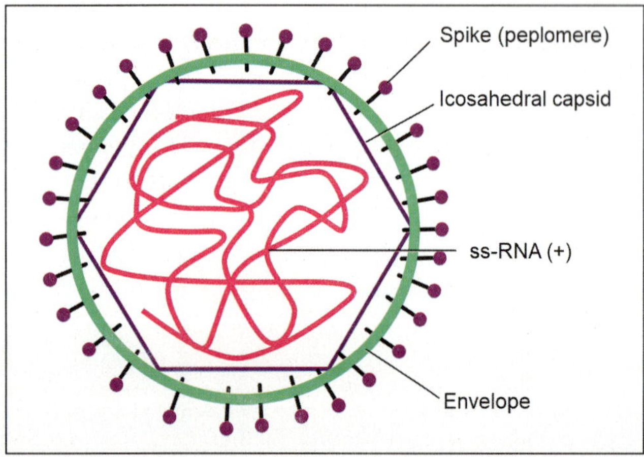

Fig. 86.1: Morphology of rubella virus

Culture Characteristics (C/Cs)

Animal culture: Experimental infection can be produced in monkeys. Pregnant rabbit can be used as laboratory model to study the congenital malformation.

Cell culture: It can grow in primary cell culture. It also grows in continuous cell lines like RK (Rabbit Kidney)-13 cells where growth is identified by CPE, BHK-21 and Vero cells where no CPE, but growth is identified by interference or by using challenger virus like ECHO-11. It is done from throat swab 6 days before and after onset of rashes and rarely from urine (up to 1 year) and CSF. Shell vial technique is used for rapid diagnosis.

Resistance

Sterilization: Virus is inactivated by heat (56°C).

Disinfection: Virus is inactivated by ether, chloroform, formaldehyde, BPL and desoxycholate.

Antigenic Structure and Serotypes

It includes structural antigens of capsid and envelope (E1 and E2) and nonstructural antigens like p150 and p90. It does not show any antigenic variation, only one sero types is known.

Immunity

Single attack provides the lifelong immunity and secondary attack is rare. Passive immunity occurs in infants of immune mother for 4–6 months following birth due to presence of passively transferred IgG.

Pathogenicity

Disease name: Disease called **rubella** or **german measles.**

Epidemiology

- **Distribution:** Disease occurs worldwide as epidemic and pandemic; however, due to use of effective vaccine epidemics are less encountered nowadays.
- **Genotypes:** On the basis of differences in the sequence of the E1 protein, two genotypes have been described, which differ by 8–10%. These have been subdivided into 13 recognized genotypes like 1A, 1B, 1C, 1D, 1E, 1F, 1G, 1H, 1I, 1J, 2A, 2B and 2C. 1A, 1E, 1F, 2A and 2B found in China, 1C found in Central and South America. 1E found in Africa, America, Asia and Europe. 1G found in Belarus, Cote d'Ivoire and Uganda. 1J found in Japan and Philippines. 2B found in South Africa and 2C found in Russia.

Reservoirs of infection: These are only human cases. Carrier stage does not exist.

Sources of infection: It presents in nasopharyngeal/throat secretion, blood, CSF and urine. **Period of infectivity (communicability)** is a week before the onset of symptoms (rashes) to a week after rashes appear.

Modes of transmission: It is transmitted by inhalation or by placenta.

Incubation period (IP): 2–3 weeks.

Portal of entry: Respiratory system.

Sites: Skin, CNS, bones, joints, fetal organs, etc.

Precipitating factors (epidemiological determinants): **(1) Age:** It is common in children between 3 and 10 years. Infants of immune mothers are protected up to 4–6 months. **(2) Pregnancy:** About 10–40% people can reach to adulthood without experiencing the rubella in the absence of immunization, thus child bearing age remains rubella susceptible. In India, 40% female are infected with rubella during child-bearing age. Percentages of fetal damage at different stages of pregnancy are mentioned in **Table 86.1**. **Early** (1st) trimester is the most dangerous because, organs are developing. It cause chromosomal breakages and inhibition of mitoses in infected embryonic cells resulting fetal death and spontaneous abortion. **Late** trimester results congenital malformations. **(3) Environmental factors:** It occurs throughout the years, with peak in the spring.

Pathogenesis: After the entry, virus replicates locally in the nasopharynx and then spread to the lymph nodes. Viremia develops after 7–9 days and lasts until 14th day by which time both antibodies and rashes appear simultaneously suggesting an immunological basis for the appearance of rash.

Clinical features: (1) Prodromal phase: It consists coryza (catarrhal inflammation of the mucous membrane in the nose, caused by a cold or by hay fever), low-grade-fever and sore throat. **(2) Lymphadenopathy:** Post-auricular and posterior cervical lymph nodes are enlarged; however, rubella cases without lymphadenopathy are also documented. **(3) Rashes:** Pinkish, minute, discrete, macular rashes typically appear on face and then extended to neck, trunk and extremities sparing the palms and soles.

Complications: These are arthralgia, arthritis, SSPE (very rare) and thrombocytopenic purpura (called **blue berry muffin syndrome**). In pregnancy it produces following effects:

- **1st trimester (before 12 weeks):** Fetal death and spontaneous abortion.
- **Later stage:** Congenital malformation called **congenital rubella syndrome** which has two types. **(1) Classical congenital rubella syndrome,** which is a triad of cataract, deafness and congenital cardiac defect (like ventricular septal defect, patent ductus arteriosus and pulmonary artery stenosis) in babies. **(2) Expanded congenital rubella syndrome** in which babies are present with extrafeatures like hepato-splenomegaly, thrombocytopenic purpura, myocarditis, bone lesions, glaucoma, retinopathy, mental, motor and growth retardation.

Laboratory Diagnosis

In Pregnancy or a Case Except Congenital Malformation

Specimens: Throat swab, blood, serum, etc.

Testing methods

1. **Microscopy:** Follow morphology.
2. **Culture:** Follow C/Cs.
3. **Serological tests: IgM** does not cross the placenta and synthesized after 20 weeks of gestational age, its presence before this time indicates intrauterine infection and its detection is useful in diagnosis of congenital rubella infection. **IgG** can cross the placenta and it can survive in infants for a period of 6 months. Persistent beyond 1 year in an unvaccinated child suggest the diagnosis of congenital rubella. IgM and IgG are detected by ELISA. **IgG avidity ELISA** is useful to differentiate between active infection or past infection or immunization.
4. **Molecular method:** RT-PCR is rapid and more sensitive method.

In Congenital Malformation from Fetus

Specimens: Virus is present in all parts of infected infants. Commonly used specimens are urine, CSF, throat swab, blood, etc.

Weeks of pregnancy	Fetal infection (%)	Fetal damage (%)	Overall fetal risk (%)
<11	90	100	90
11–16	55	37	20
17–26	33	0	0
>27	53	0	0

TABLE 86.1: Percentages of fetal risk at different stages of pregnancy

Testing methods

1. **Microscopy:** Follow morphology.
2. **Culture:** Follow C/Cs.
3. **Serological tests:** ELISA to detect IgM and IgG.
4. **Molecular method:** PCR is rapid and more sensitive method.

Prevention

General measures: These are airborne precaution.

Immunization: Live attenuated vaccine was **prepared** in 1979 by serial passage of virus **(RA 27/3 strain)** in human diploid cell culture. It is **administered** alone or in combination as MR or MMR or MMRV by SC route in 0.5 ml dose at 9–12 months and second dose at 16–24 months in selected states. Some **side effect**s like local reactions, fever, rash, arthralgia and lymphadenopathy may occur. **Efficacy (seroconversion)** is 90% and provides protection for 14–16 years or life-long. It is **indicated** to all women with child bearing age and to all children of 1–14 years. It is **contraindicated** in persons with IDDs like leukemia, AIDS, etc., in infants <1 year to avoid possible interference to maternal antibody and pregnancy. Pregnancy should be avoided for 28 days after vaccination because of teratogenic risk of vaccine; however, if woman conceives in <28 days of vaccination then do not terminate it, but wait and watch for development of any risk.

Treatment

Rubella is mild, self limited condition and no specific treatment is available.

CORONAVIRIDAE

Meaning

Name is given from the word **corona means solar**, because spherical shape and projecting spikes give solar like appearance.

Classification

Taxonomical classification

- **Family:** Coronaviridae.
- **Subfamilies:** Two subfamilies like Coronavirinae/ Orthocoronavirinae and Torovirinae. Torovirinae includes animal and avian pathogens. Coronavirinae includes following four genera with human species.
- **Genus and species**
  1. *Alphacoronavirus*
     - Human coronavirus 229E.
     - Human coronavirus NL63 (New heaven coronavirus).
  2. *Betacoronavirus*
     - Human coronavirus OC43.
     - Severe acute respiratory syndrome related coronavirus/-1 (SARS-CoV or SARS-CoV1).
     - Human coronavirus HKU1.

- Middle-East respiratory syndrome related coronavirus (MERS-CoV): Also called Novel coronavirus 2012 or human coronavirus-Erasmus Medical Centre (HCoV-EMC).
- Severe acute respiratory syndrome corona-virus-2 (SARS-CoV2): Initial name was given by WHO was 2019 Novel coronavirus (2019-nCoV). Also called Wuhan strain or Wuhan coronavirus or Wuhan seafood market pneumonia virus. Later, it renamed as SARS-CoV2 by International Committee on Taxonomy of Viruses.

Clinical classification

- **Mild disease:** Like mild URTI caused by 229E, NL63, OC43 and HKU1.
- **Severe disease:** Like severs respiratory tract infection caused by SARS-CoV1, MERS-CoV and SARS-CoV2.

History

Human coronaviruses were discovered in the 1960s. **Initial two** human coronaviruses HCoV-229E and HCoV-OC43 were identified from the nasal cavities of patients with the common cold. The discovery of SARS-CoV1 in 2002-03 added a **3rd** human coronavirus. By the end of 2004, three independent research laboratories reported the **4th** human coronavirus, NL63 or New Haven coronavirus. Early in 2005, a research team at the University of Hong Kong reported **5th** human coronavirus in two patients with pneumonia, named it human coronavirus HKU1. In September 2012, a **6th** new type of coronavirus, Novel coronavirus 2012 was identified and now officially labeled as Middle-East respiratory syndrome coronavirus (MERS-CoV). On 31st December 2019, a **7th** species of coronavirus, SARS-CoV2, was reported in Wuhan, China.

Morphology

Shape and size: It is spherical (**Fig. 86.2**) or pleomorphic in shape with 100 nm size.

Genome and capsid: It contains ss-RNA (+) with helical variety of capsid. RdRp (RNA dependent RNA polymerase) synthesizes RNA (–) from the positive-sense RNA for replication and transcription.

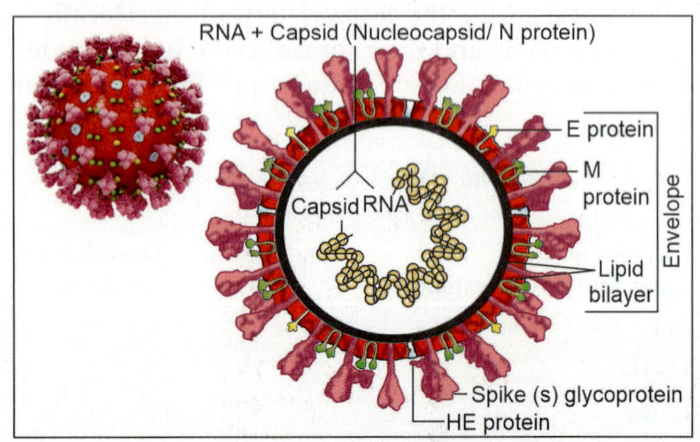

Fig. 86.2: Morphology of Coronaviridae

Envelope: It is lipoprotein in nature consists of M protein and E protein and lipid.

Peplomeres: It carries club or petal shaped glycoprotein spikes (S protein) on surface, giving solar like appearance. It also consists shorter spike (specifically the members of *Betacoronavirus* subgroup A) on surface called **hemagglutinin esterase (HE) protein**.

Note: SARS-CoV or SARS-CoV1

Full name: Described above.

History: Virus was isolated in February 2003 on Vero cells.

Morphology: Described above.

Culture characteristics (C/Cs): Described above.

Resistance: It is an enveloped virus susceptible to lipid solvents. Two groups are identified like **acid stable** causing gastroenteritis in humans and animals and **acid labile** causing common-cold like illness.

Immunity: Human coronavirus has many serotypes, resulting immunity is poor and reinfection can occur with same serotypes.

Pathogenicity

- **Disease name:** Disease called **severe acute respiratory syndrome (SARS).**
- **Epidemiology:** SARS-CoV1 is thought to be an animal virus with uncertain animal reservoir perhaps bat that spread virus to other animals (civet cats) and first human case was identified in the Guangdong province of southern China in 2002. This area is considered as a potential zone of re-emergence of SARS-CoV1. Later 2002–03 epidemic spread in almost 26 countries. **India** escaped the 2002–03 epidemics of SARS; however, few suspected cases were detected and quarantined.
- **Reservoirs of infection:** Humans and some animals like monkeys, civets, racoons, dogs, cats and rodents are the reservoirs.
- **Sources of infection:** Virus is excreted in respiratory secretions and stool from the patients, which carry the risk for transmission. Infectivity is more in 2nd week of illness.
- **Modes of transmission: (1) Direct entry by droplet nuclei:** Aerosols arise during coughing or sneezing carry the risk up to 3 feet. Virus deposited directly on mucosa of nose, mouth and eyes. **(2) Contact:** It can spread when a person touches a surface or object contaminated with infectious droplets and then touches his or her mouth, nose, or eyes. **(3)** It is possible that the SARS virus might spread more broadly by **inhalation** (airborne spread) or by other ways that are unknown.
- **Incubation period:** 2–10 days.
- **Portal of entry:** Respiratory system.
- **Sites:** Respiratory system and GIT.
- **Precipitating factors (epidemiological determinants): (1) Age:** The mortality rate was much higher in patients >50 years, in 2003 outbreak. **(2) Occupation:** Healthcare workers are at high risk. **(3) Close contact with the patients:** Like kissing or hugging, sharing, eating or drinking utensils, talking to someone within 3 feet and touching someone directly.
- **Clinical features: Initially in few days** flu-like symptoms like fever, malaise, myalgia, headache, diarrhea and shivering (rigors) occur. **In 1st/2nd week** cough (initially dry), shortness of breath and diarrhea are present.
- **Complications:** Like secondary bacterial infections, lymphopenia (low lymphocytes), liver or heart failure and respiratory distress/failure leading to death.

Laboratory diagnosis: Respiratory specimens like sputum, nasopharyngeal secretion and serum are tested by microscopy (follow morphology) and culture (follow C/Cs). **Serological tests** include antibody detection by HAI test or by ELISA. IgM peaks during the acute or early convalescent phase (week 3) and declines after 12th week. IgG Ab is produced later and peaks at 12th week. **Molecular method** like RT-PCR is more sensitive and specific test. **Chest X-ray** is done to confirm the pneumonia.

Prevention: Preventive measures are isolation of patients, avoid contact with patients, frequent hand washing and surface disinfection. No specific vaccine is available.

Treatment: No specific antiviral drugs are available, supportive measures are given. Antibiotics are given to prevent the secondary bacterial infections.

Note: MERS-CoV

Full name: Described above.

History and synonym

- Virus was isolated in 2012 from the sputum samples of a person who fell ill in a 2012 outbreak in Saudi Arabia.
- As it was diagnosed in 2012 and newer in the list of coronavirus so called **Novel coronavirus 2012.**
- After the Dutch Erasmus Medical Centre sequenced the virus in 2012, the virus called **human coronavirus-Erasmus Medical Centre 2012 (HCoV-EMC 2012)**.
- The official name for the virus is Middle-East respiratory syndrome-related coronavirus (MERS-CoV).

Morphology: Described above.

Culture characteristics (C/Cs): It produces ballooning and syncytium over Vero cells.

Resistance: It is sterilized by autoclave at 121°C for 15 minutes. It is susceptible to 70% alcohol or 0.5% sodium hypochlorite.

Pathogenicity

- **Disease name:** It called **MERS.** Virus was identified from 2012 outbreak in Saudi Arabia hence initially disease was known as **Saudi SARS.**
- **Epidemiology**
  - **Distribution:** The virus started the epidemic in Saudi Arabia in July 2012 and then it spread in >21 countries. It produced outbreak in Republic of Korea in 2015. It is not reported from **India** yet.
  - **Genotypes (clades):** Two clades like A and B were reported. The earliest cases of MERS-CoV were of clade A (EMC/2012 and Jordan-N3/2012) and newer cases were from clade B.
- **Reservoirs of infection:** These are unknown, but it may be acquired from camels or bats.
- **Sources of infection:** These are exactly unknown, but may be the respiratory droplets arise during sneezing and coughing from patients. Raw camel meat or milk may be the source for animal to human spread.
- **Modes of transmission: (1) Animal to human transmission:** The route of transmission from animals (camels) to humans is not fully understood. **(2) Human-to-human transmission:** The risk of sustained human to human transmission appears to be very low until there is as close contact with the patient especially in health care workers. MERS-CoV infects only 20% of respiratory epithelial cells, so a large numbers of virions are needed to be inhaled to cause infection.
- **Incubation period:** 2–14 days.
- **Portal of entry:** May be respiratory tract.

- **Sites:** Respiratory system, GIT and kidneys also.
- **Precipitating factors (epidemiological determinants):** Health care workers are at high risk.
- **Clinical features:** No symptoms or mild symptoms like fever, cough, shortness of breath, diarrhea and nausea/vomiting.
- **Complications:** Kidney failure or even death.

Laboratory diagnosis: Respiratory specimens like sputum, nasopharyngeal secretion and serum are tested by microscopy (follow morphology) and culture (follow C/Cs). **Serological tests** include ELISA, IF test or micro-neutralization test are useful. **Molecular method** like RT-PCR is most sensitive and specific method. **Chest X-ray:** It is done to confirm the pneumonia.

Prevention: Preventive measures are isolation of patients, avoid contact with patients, frequent hand washing, avoid visit of places where camels or their products like meat, milk, etc., are available, avoid contact with sick animals and surface disinfection. No specific vaccine is available.

Treatment: No specific antiviral drugs are available, supportive measures are given.

Culture Characteristics (C/Cs)

Animal culture: Also useful.

Organ/cell culture: Organ culture was done in past by using human embryonic trachea. **Cell culture** is done on Vero cells and human diploid embryonic fibroblast cells. SARS-CoV was isolated by using the Vero cells.

Antigenic Structure

Structural antigens are nucleocapsid (N) protein, envelope (E) protein, membrane (M) protein and spike (S) protein with S1 and S2 subunits. Single-stranded RNA genomes of SARS-CoV and MERS-CoV consist two large genes like ORF1a and ORF1b genes (ORF → Open Reading Frame), which are encode for **16 nonstructural antigens** (nsp1–nsp16) that are highly conserved throughout coronaviruses.

SARS-CoV2

Full form: Described above.

Synonym: 2019 Novel coronavirus (2019-nCoV), or Wuhan strain or Wuhan coronavirus or Wuhan seafood market pneumonia virus.

History: It was identified on 31st December 2019 from Wuhan, China. It is the genetically clusters with *Betacoronavirus* from group 2B. It is approximately 70% genetically similar to the SARS-CoV, 96% to a bat coronavirus RaTG13 and 92% to pangolin coronavirus. It is widely suspected that it is originated from bats (zoonotic) with evidences like (1) high sequence homology between SARS-CoV2 and bat coronavirus RaTG13, (2) both viruses have affinity for same receptors (ACE-2) and (3) both will infect the same cells.

Morphology: Described above.

Culture characteristics (C/Cs): It required BSL-3 facilities. Methods are described above.

Resistance: WHO recommended sodium hypochlorite, chlorine and bleach solution (in recommended dilution) as disinfectants for surfaces and objects. WHO recommended two types alcohol-based hand disinfectants against SARS-CoV2. **(1) Disinfectant one** consists of 80% ethanol, 1.45% glycerine and 0.125% hydrogen peroxide. **(2) Disinfectant two** consists of 75% isopropanol, 1.45% glycerine and 0.125% hydrogen peroxide.

Immunity

- **CMI:** CD8[+] T cells are the main inflammatory cells that play a vital role in viral clearance.
- **AMI:** Most patients display an antibody response not clear, how long immunity will lasts, **few researches shown that probably, it may last at least for 3 months or little longer.** But even if immunity wanes, reinfection would be less severe. People reinfected might not develop symptoms or might have a mild, cold-like infection with significantly lower levels of virus replication (low levels of SARS-2 viruses in their respiratory tracts) and onward transmission. This is due to presence of immunological memory cells. Children who first encountered the virus when they are very young may end up being infected several times over their lifetimes, but those later infections would not lead to severe illness, even when they are elderly. Antibody responses tended to be highest and durable in people with the most severe infection than those with mild infections.

Pathogenicity

- **Disease name:** Disease called **COVID-19 (Coronavirus disease-2019) or Wuhan pneumonia or China pneumonia.**
- **Epidemiology:** It is distributed worldwide. WHO declared the outbreak a public health emergency of international concern on 30th January 2020 and a pandemic on 11th March 2020. The 1st case of COVID-19 pandemic in India was reported on 30th January 2020.
- **Reservoirs of infection:** Mostly bats.
- **Sources of infection:** Virus is excreted in respiratory secretions and in stool also. Infectivity is highest in bronchoalveolar lavage (93%), followed by sputum (72%), nasal swab (63%), and pharyngeal swab (32%). Pharyngeal swabs may show positivity in the 1st week. Later, the virus can disappear in the throat but continue to multiply in the lungs.
- **Modes of transmission**
  - **Direct transmission: (1) Direct entry by droplet nuclei:** Aerosols arise during coughing or sneezing carry the risk up to 3 feet. Virus deposited directly on mucosa of nose, mouth and eyes. **(2) Contact:** It can spread when a person touches a surface or object contaminated with infectious droplets and then touches his or her mouth, nose, or eyes.

Note: SARS-CoV2 variants

Introduction: SARS-CoV-2 constantly undergoes to mutation to produce new virus called **variant**. When a new variant appears to be growing in a population, it called as an **emerging variant or new variant**. New variants which have particular importance **called notable variants**. The sequence WIV04/2019, belonging to the GISAID S clade/PANGOLIN A lineage/Nextstrain 19B clade, reflects the sequence of the original virus infecting humans **called "sequence zero"**, and is widely used as a reference sequence. **Lineage** is a group of closely related viruses with a common ancestor. SARS-CoV-2 has many lineages; all cause COVID-19.

Nomenclature system of variant: Three main, nomenclatures systems have been proposed.

1. As of January 2021, GISAID: Identified eight global clades (S, O, L, V, G, GH, GR, and GV).
2. In 2017, Hadfield et al. announced Nextstrain (real-time tracking of pathogen evolution): Nextstrain has later been used for tracking SARS-CoV-2, identifying 11 major clades (19A, 19B, and 20A–20I) as of January 2021.
3. In 2020, Rambaut et al. of the PANGOLIN: A dynamic nomenclature for SARS-CoV-2 lineages that focuses on actively circulating virus lineages and those that spread to new locations, as of February 2021, six major lineages (A, B, B.1, B.1.1, B.1.177, B.1.1.7) had been identified.

Problem with notable variants

- Evade natural immunity due to decreased susceptibility to neutralizing antibodies (e.g. causing reinfection).
- Increased transmissibility, morbidity and mortality.
- Affect diagnostic (evade detection), therapeutic (evade effect of antiviral drugs and convalescent plasma/monoclonal antibodies) and preventive measures (infect vaccinated individual).
- Increased risk of multisystem inflammatory syndrome.
- Increased affinity for particular demographic or clinical groups, such as children or persons with IDDs.
- Variants that appear to meet one or more of these criteria may be labeled as "variants of interest (VOI)" pending verification and validation of these properties. Once validated, "VOI" may be renamed as "Variants of concern (VOC)" by monitoring.

Committee for classification: The US Department of Health and Human Services established a SARS-CoV-2 Interagency Group (SIG) to improve coordination among the CDC, NIH, FDA, Biomedical Advanced Research and Development Authority (BARDA), and Department of Defence (DoD). This interagency group is focused on the rapid characterization of emerging variants and actively monitors their potential impact on diagnosis, prevention and treatment. The SIG meets regularly to evaluate the risk posed by new variants and to make recommendations about the classification of variants.

Classification: The WHO classified variants as alpha, beta, gamma, etc. Based on their attributes and prevalence in the United States, variants may be reclassified in four categories. (1) Variants being monitored (VBM). (2) Variants of interest (VOI). (3) Variants of concern (VOC). (4) Variants of high consequence (VOHC).

1. **Variants being monitored (VBM):** These include variants for which current medical measures are impacted or associated with more severe disease or increased transmission, but are not circulating or are circulating at very low levels in the community and do not pose a significant risk to public health. VBM are mentioned in **Table 86.2.**
2. **Variants of interest (VOI):** These variants have specific genetic markers that have been associated with changes to receptor binding, reduced neutralization by antibodies generated against previous infection or vaccination, reduced efficacy of treatments, potential diagnostic impact or predicted increase in transmissibility or disease severity. VOI are mentioned in **Table 86.3.**
3. **Variants of concern (VOC):** These variants are with increase in transmissibility, more severe disease (e.g. increased hospitalizations or deaths), significant reduction in neutralization by antibodies generated during previous infection or vaccination, reduced effectiveness of treatments or vaccines, or diagnostic detection failures. VOC are mentioned in **Table 86.4**.
4. **Variants of high consequence (VOHC):** These variants are with significantly reduced diagnostic, treatment and preventive measures and current vaccines do not offer protection. As of now, there are no any VOHC.

Infections of Other RNA Viruses (Rubella Virus and Coronaviridae)

TABLE 86.2: VBM						
WHO label	**Pango lineage**	**GISAID clade**	**Nextstrain clade**	**Additional amino acid changes monitored**	**Earliest documented samples**	**Date of designation**
Alpha	B.1.1.7 and Q lineage	GRY	20I (V1)	+S:484K +S:452R	United Kingdom, Sep-2020	18-Dec-2020
Beta	B.1.351 and descendent lineage	GH/501Y.V2	20H (V2)	+S:L18F	South Africa, May-2020	18-Dec-2020
Gamma	P.1 and descendent lineage	GR/501Y.V3	20J (V3)	+S:681H	Brazil, Nov-2020	11-Jan-2021
Delta	B.1.617.2 and AY.1 lineage or Delta plus variant	G/478K.V1	21A	+S:417N	India, Oct-2020	VOI: 4-Apr-2021 VOC: 11-May-2021
Epsilon	B.1.427, B.1.429					
Eta	B.1.525					
Iota	B.1.526					
Kappa	B.1.617.1					
N/A	B.1.617.3					
Zeta	P.2					

Note: Delta variant (B.1.617.2 has mutated further to form the AY.1 lineage or Delta plus variant. It has mutation K147N 1st identified in Europe, March 2021. In India it was 1st found in April, 2021.

TABLE 86.3: VOI

WHO label	Pango lineage	GISAID clade	Nextstrain clade	Earliest documented samples	Date of designation
Lambda	C.37	GR/452Q.V1	21G	Peru, Dec-2020	14-Jun-2021
Mu	B.1.621, B.1.621.1	GH	21H	Colombia, Jan-2021	30-Aug-2021

TABLE 86.4: VOC

WHO label	Pango lineage	GISAID clade	Nextstrain clade	Earliest documented samples	Date of designation
Omicron	B.1.1.529	GR/484A	21K	South Africa 24th Nov-2021	VUM: 24-Nov-2021 VOC: 26-Nov-2021

- **Indirect transmission:** In addition, it is possible that the SARS virus might spread more broadly through the air (airborne spread) or by other ways that are not now known.

- **Incubation period:** Typically around 5 days but may range from 1–14 days.

- **Portal of entry:** Respiratory system.

- **Sites:** Respiratory system and other organs.

- **Precipitating factors (epidemiological determinants): (1) Genetic factors, age, and sex:** ACE-2 expression depends on the genetic factors, age, and sex. Majority of the infected children are asymptomatic or experience mild symptoms like fever and cough during the COVID-19 pandemic. In children newly identified condition called **multiorgan inflammatory syndrome in children (MIS-C) or pediatric inflammatory multisystem syndrome (PIMS)**. Children present with fever, chills, pain in abdomen, rashes, myocarditis, vomiting and diarrhea. Diagnosis is done by detection of RNA or antibodies (IgM and IgG). In most of the cases CRP, LDH, troponin, D-dimer and proBNP are elevated. About 80% cases require hospitalization and over 10% require mechanical ventilation. Intravenous Ig and glucocorticoids are the treatment. Very low case fatality rate in pediatric patients compared to patients beyond 80 years of age (case fatality rate of 0% for individuals under 8 years of age versus 21.9% for patients above 80 years). Limited evidences suggest that ACE 2 expression is attenuated in females compared with the males which could justify the higher number of COVID-19 cases in men. **(2) Health conditions:** Risk is more in persons with IDDs. It is due to increase the ACE-2 expression in the presence of comorbid conditions. **(3) Occupation:** Healthcare workers are at high risk. **(4) Environmental factors:** Overcrowding and close contact with the patients increased the risk of transmission.

- **Pathogenesis:** It is discussed in 3 steps. **(1) Viral entry:** SARS-CoV2 attaches to ACE 2 receptors for S proteins and enters in human cells via endocytosis as shown in **Fig. 86.3**. ACE 2 is a membrane carboxypeptidase presents on type II alveolar cells (AT2), oral, esophageal, ileal epithelial cells, myocardial cells, proximal tubule cells of the kidneys, urothelial cells of the bladder, etc. Newly identified receptors for viral entry are CD-147 and glucose regulated protein 78. Other than ACE-2, the protease furin and serine protease, tansmembrane serine protease 2 (TMPRSS2) are essential for the viral entry. Furin breaks the S protein required for virus entry in cells. Activated S protein is primed by the TMPRSS2 and finally attaches ACE 2 receptors. **(2) Inflammatory phase:** The physiological role of ACE 2 is the degradation of angiotensin II to angiotensin, which has a fundamental role in inflammation and tissue injury. Following the viral replication in the host cells, down regulation of ACE 2 inhibits breakdown of angiotensin II into angiotensin that causes inflammation (vasoconstriction, electrolytes imbalance-hypokalemia, hypercoagulability, oxidation and fibrosis). Viral replication in cell leads cell death with release of immune markers that stimulates T cells to release cytokines like IL-2, IL-6, macrophage inflammatory proteins 1a and 1b, IP-10 (Interferon gamma-induced protein 10), MCP-1 (Monocyte chemoattractant protein-1), TNF-α, etc. **(3) Hyperinflammatory phase:** Further attraction of macrophages, monocytes and T cells at infective site called **hyperinflammation or cytokine storm,** which causes multi organs damage.

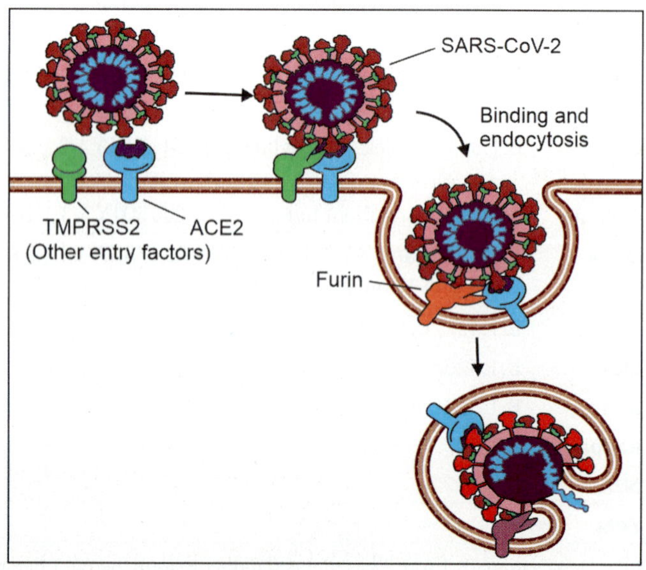

Fig. 86.3: Pathogenesis of SARS-CoV2

- **Clinical features:** Flu-like symptoms like fever, dry cough, fatigue, sore throat, headache, chills, vomiting and diarrhea. Other features are sputum production may contains blood, loss of the sense of smell, loss of taste, shortness of breath, myalgia, joint pain, rashes, persistent chest pain or pressure, sudden confusion, difficulty in walking, and bluish face or lips. Children might not develop fever and cough as frequently as adults. COVID-19 is divided in following 3 categories.

  - Mild disease: Mild URTI features like fever, dry cough, fatigue, sore throat, headache without shortness of breath or hypoxia (SpO_2 ≥94%).

  - Moderate disease (atypical pneumonia): Mild disease plus any one of respiratory rate ≥24/min, breathlessness or SpO_2 is 90–93% on room air.

  - Severe disease: Any one of respiratory rate >30/min, breathlessness or SpO_2 is <90% on room air.

- **Complications: (1) Respiratory:** Pulmonary fibrosis called **post COVID pulmonary fibrosis**/ARDS/respiratory failure leading to death. Average case fatality rate is 4%. **(2) Pancreas:** Both β cells and exocrine cells have ACE-2 receptors, hence direct damage to β cells cause hyperglycemia and ketoacidosis. **(3) CVS:** Myocarditis, cardiomyopathy and arrhythmia. **(4) Other systems:** Autopsies report suggested the presence of viral particles with damages in liver, kidneys, CNS (acute encephalomyelitis like picture), skin, bone marrow, lymph nodes, etc. **(5) Pregnancy:** It may develop preeclampsia and require hospitalization. **(6) Secondary bacterial infections:** It causes sepsis or septic shock. Common bacteria are GNB like *Klebsiella pneumoniae* and *Acinetobacter baumanii.* **(7) Secondary fungal infections:** This includes COVID associated mucormycosis and COVID associated aspergillosis.

Diagnosis

- **Specimens:** Two swabs like nasal (63% sensitivity)/oropharyngeal (32% sensitivity) and throat swabs (dacron or polyester flocked swabs) are collected in viral transport medium for RT-PCR. All samples are transported to laboratory by triple layer packing at 4°C. After arrival samples should be stored at 4°C for 72 hours, if delay in testing then stored the sample at ≤–70°C. Lower respiratory tract specimens such as broncho alveolar lavage (93% sensitivity), sputum (72% sensitivity) and endotracheal aspirate can be collected in patients with respiratory distress.

- **Testing methods**

  A. Biomarkers (Biochemical and hematological tests): (1) Lymphopenia, raised CRP and raised LDH, AST, ALT, amylase, alkaline phosphatase and skeletal muscles enzyme suggest severe disease and risk of mortality. **(2) Procalcitonin** helps to differentiate COVID-19 from sepsis. It is increased in sepsis and remains normal in COVID-19. It is the precursor of calcitonin and raised following tissue injury in bacterial infections. **(3) Coagulation marker like D-dimer:** It produced following breakdown of fibrin (clotted blood) by fibrinolysis. It increased in inflammation and hypoxia. Pneumonia increases the coagulation and fibrinolysis resulting raised level of D-dimer. It is the early marker to start anticoagulant therapy. **(4) Others:** Raised IL-2, IL-4, IL-6 and creatinine.

  B. Microscopy: Follow morphology.

  C. Culture: Follow C/Cs.

  D. Serological tests (Fig. 86.4): **(1) Antigen detection:** It is a point of care test. Nucleoprotein Ag is detected after 2–3 days from nasopharyngeal swab by ICT. Other Ag testing method is IF test. If antigen test is positive then report as true positive. However, if antigen test is negative and the patient is symptomatic then retest by RT-PCR. **(2) Antibody detection:** Antibodies require time to develop, so they are not the best marker of acute infection. Because they persist in blood for long, they are ideal for detecting past infections. Tests are used for surveillance purpose in high risk group and vulnerable group, but not for clinical diagnosis. IgM/IgG antibodies are detected by ELISA, ICT and chemiluminescence from blood, serum, or plasma. Antibodies after day 7 only indicate previous contact with the virus, but do not confirm the presence or shedding of the virus. **IgM** is detectable in 1st week with a peak in 2nd week. IgG detectable 10–14 days after the onset of infection with a peak around 28 days. The test may remain positive following several weeks after the infection. A positive test indicates exposure to SARS-CoV2, whereas a negative test does not rule out COVID-19 infection.

E. Molecular methods: (1) Conventional real-time reverse transcriptase PCR (qRT-PCR): It becomes positive in 3–4 days **(Fig. 86.4)**. It takes 4–6 hours. qRT-PCR protocol usually based on the detection of at least two genes. Detection of E gene as a screening tool followed by confirmation with one more gene out of RdRp gene/Orf1a or 1b gene/N gene/S gene/M gene. The E assay is specific for all SARS-CoV related coronaviruses (like SARS-CoV, SARS-CoV2 and related bat viruses). E gene PCR has higher sensitivity. **(2) Automated real-time PCR:** Like TRUENAAT and geneXpert are positive in 1–2 hrs. **(3) CRISPR (Cultured regularly interspaced short palindromic repeats):** It is a low cost, rapid and point of care test which identifies the viral RNA. **(4) Detection of omicron variant:** It is detected by real-time PCR by detecting specific S gene. **(5) Genomic sequencing:** All positive samples should be tested for genomic sequencing to detect the viral mutation. **(6) Others:** RT-LAMP, CBNAAT, etc.

F. Chest X-ray and high-resolution CT (HRCT) scan: These are done to confirm the pneumonia.

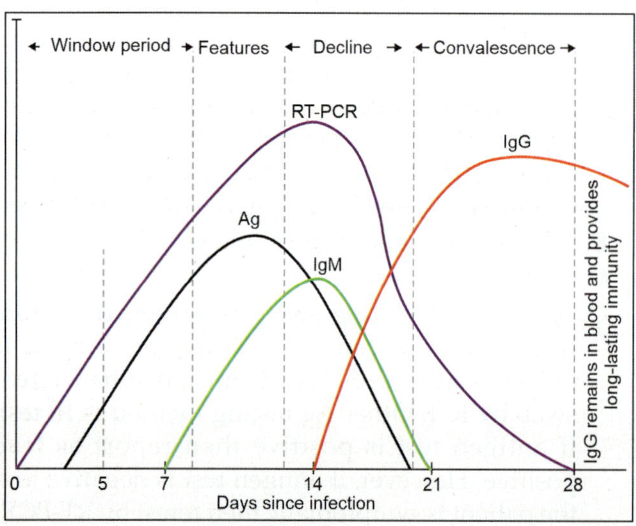

Fig. 86.4: Viral and serological markers in COVID-19

Prevention

- **General measures:** These are isolation of patients, avoid contact with patients, frequent hand washing, wearing of mask, avoid gathering of people, eat only well-cooked food, avoid spitting in public and surface disinfection. No specific vaccine is available.
- **Chemoprophylaxis:** No any specific drugs are available.
- **Immunoprophylaxis:** More than 20 vaccines are developed by different companies. Few are mentioned below.

Recombinant Vaccines

1. **Covishield**
   - **Manufacturer:** It is the Oxford-Astrazeneca, UK vaccine with local name as covishield, developed by serum institute of India.
   - **Preparation:** It is the **recombinant vaccine** prepared by using adeno virus as vector. It is made from a weakened version of adeno virus of chimpanzees. It has been modified to look more like coronavirus by introducing genes in it.
   - **Storage:** It can be stored at 2–8°C.
   - **Administration and dose:** Two doses of 0.5 ml are given by IM route in deltoid muscle in person ≥18 years. The 2nd dose should be administered between 4–6 weeks after the 1st dose. However, there is data available for administration of the 2nd dose up to 84 days (12 weeks) after the first dose.
   - **Efficacy:** It provides the protection against COVID-19 after 4 weeks of 2nd dose. The duration of protection against COVID-19 disease is currently unknown.
   - **Side effects:** Pain, itching, swelling at injection site, fever, decreased appetite, abdominal pain, nausea, vomiting, enlarged lymph nodes, excessive sweating, rash, dizziness, headache, joint pain or muscle ache.
   - **Contraindications: (1) Strict contraindications:** Allergic reaction to vaccine components or allergic reaction to 1st dose, pregnancy or not sure of pregnancy, breast feeding, individual <18 years of age and person with IDDs. **(2) Provisional or temporary contraindications:** Person with active SARS CoV2 infection, COVID-19 patients who took monoclonal antibodies or plasma therapy and unwell or hospitalized person due to any illness.

2. **Sputnik V vaccine**
   - **Manufacturer:** It is developed by Dr Reddy's Lab and Gamaleya National Centre in Russia.
   - **Preparation:** It is the **recombinant vaccine** prepared by using adeno virus serotypes 5 (Ad5) and 26 (Ad26) as vector with spike protein of SARS CoV 2. Use of two serotypes provides long-lasting immunity.
   - **Administration and dose:** Two doses are given by IM route in deltoid muscle in person ≥18 years.

The 2nd dose should be administered 3 weeks after the 1st dose.

- **Storage, side effects and contraindications:** Same as covishield.

3. **Janssen COVID-19 recombinant vaccine:** It prepared by using adeno virus serotypes 26 (Ad26) by Johnson and Johnson (J and J) in USA and Netherland.

Killed or Inactivated Vaccines

1. **Covaxin**
   - **Manufacturer:** Covaxin is the India's first indigenous COVID-19 vaccine developed by Bharat Biotech in collaboration with the ICMR, New Delhi and NIV, Pune.
   - **Preparation:** It is the whole virus killed Vero cell vaccine. It includes immune-potentiators, also called **adjuvants.**
   - **Storage:** It can be stored at 2–8°C.
   - **Administration and dose:** Two doses are given by IM route in deltoid muscle in person ≥15 years. The second dose should be administered 4 weeks after the 1st dose.
   - **Uses:** It is used in persons >15 years. As covaxin is killed vaccine it may not cause any safety risk in persons with IDDs.
   - **Side effects:** Same as covishield.
   - **Contraindications:** (1) Allergic reaction to any ingredients of vaccine or to 1st dose. (2) Individual <15 years of age. (3) Person with active SARS CoV2 infection or fever.

2. **Coronavac (sinovac):** It is developed by China.

mRNA Vaccines

1. **Pfizer BioNTech COVID-19 vaccine**
   - **Manufacturer:** Developed by Germany and USA.
   - **Preparation:** It is prepared from mRNA. Advantage of RNA vaccine is that RNA does not possess the risk of integrating in to host cells unlike DNA. Disadvantages of RNA vaccines are risk of disintegration (overcome by coformulation with lipid nanoparticles which protect the RNA and help in adsorption).
   - **Storage:** It can be stored at –20°C to –70°C.
   - **Administration and dose:** Two doses are given by IM route in deltoid muscle in person ≥16 years. The 2nd dose should be administered 3 weeks after the 1st dose.
   - **Uses:** It is used in person ≥16 years. The FDA authorized it for emergency use in children ages 5–15 years.
   - **Efficacy:** It has 95% efficacy.
   - **Side effects:** Same as covishield.
   - **Contraindications:** (1) Allergic reaction to any ingredients of vaccine or to 1st dose. (2) Individual <16 years of age. (3) Person with active SARS CoV2 infection or fever or received other COVID vaccine. (4) Pregnancy or not sure of pregnancy, breast feeding, and person with IDDs.

2. **Moderna COVID-19 vaccine:** Developed by USA from mRNA. It can be stored at –25°C to –15°C. Two doses are given by IM route in deltoid muscle in person ≥18 years. The second dose should be administered 4 weeks after the 1st dose. It has 94.1% efficacy. Side effects and contraindications are same as Pfizer BioNTech COVID-19 vaccine.

3. **HGCO19:** India's first mRNA vaccine made by Pune-based Genova in collaboration with Seattle-based HDT Biotech Corporation, using bits of genetic code.

DNA Vaccine

ZyCov-Di: It is the 1st DNA based vaccine developed by Ahmedabad, India-based Zydus-Cadila. The vaccine contains a DNA plasmid vector that carries the gene encoding the spike protein of SARS CoV2. As with other DNA vaccines, the recipient's cells then produce the spike protein, eliciting a protective immune response. The plasmid also contains unmethylated CpG motifs to enhance its immunostimulatory properties. The plasmid is produced by using *E. coli*. The vaccine is given ID route using a spring-powered jet injector. This is because successful transfection of DNA vaccine requires traveling across both the cell plasma membrane and the nuclear membrane and using a conventional needle gives poor results and leads to low immunogenicity.

Other Vaccines

1. **Novavax:** A 2nd vaccine being developed by Serum Institute of India and American vaccine development company.
2. **Nasal vaccine:** Developed by Bharat BioTech.
3. A vaccine being developed by Hyderabad-based Biological E, the first Indian private vaccine-making company, in collaboration with US-based Dynavax and Baylor College of Medicine.
4. **Vaccines for children:** Follow **Table 86.5.**

Treatment

- No specific antiviral drugs are available, symptomatic treatment, oxygen therapy, convalescent plasma therapy, ventilation, steroid, chloroquine/hydroxyl-chloroquine, ivermectin, interferon and tocilizumab (monoclonal antibody) are useful. Tocilizumab is an IL-6 inhibitor, is being tried targeting the cytokine storm that is the hallmark of this disease.
- Prophylactic anticoagulant is given like heparin or fractionated heparin.

TABLE 86.5: Vaccines for children		
Age	Pfizer BioNTech	Moderna
≤4	No	No
5–11	Yes	No
12–17	Yes	No

- Useful antiviral drugs are remdesivir and flavipiravir. These are not reducing the mortality.
  - Antibiotics like azithromycin, ceftriaxone or doxycycline are prescribed to prevent or to treat the secondary bacterial infections.

ACCESS YOURSELF

Case Study

1. A 30-year male visited the OPD with complain of dry cough, sore throat and fever. Nasopharyngeal swab was sent for RT-PCR which identified the E gene and orf gene. Identify the case and answer the following.
   a. Name the clinical condition and its most common etiological agent.
   b. Write the morphology of etiological agent.
   c. Write pathogenicity of clinical condition.
   d. Write preventive measures of clinical condition.

Essay/Full Question

1. SARS-CoV 2 (COVID-19).

Short Notes

1. Rubella virus.
2. SARS-CoV1/MERS-CoV.
3. Variants of SARS-CoV 2 (COVID-19).
4. COVID-associated mucormycosis.
5. SARS-CoV 2 (COVID-19) vaccines.

Short Questions for Theory/Viva Questions

1. Write the common manifestations (complications) caused by Rubella virus in pregnant women.
2. Name any four species of human coronavirus.
3. What is MIS-C?
4. Name the four biomarkers used to detect the COVID-19.

MCQs for Chapter Review

Rubella Virus

1. **Incubation period of rubella is:**
   a. 18–72 hours
   b. 2–3 weeks
   c. 1–3 months
   d. >1 year
2. **All of the following statements are true about congenital rubella** *except*:
   a. It is diagnosed when the infants has IgM antibodies at birth
   b. It is diagnosed when IgG antibodies persist for more than 6 months
   c. Most common congenital defects are deafness, cardiac malformations and cataract
   d. Infection after 16 weeks of gestation results in major congenital defects
3. **Risk of the damage to fetus by maternal rubella is maximum if mother gets infected in:**
   a. 6–12 weeks of pregnancy
   b. 20–24 weeks of pregnancy
   c. 24–28 weeks of pregnancy
   d. 32–36 weeks of pregnancy
4. **Classical triad of congenital rubella includes all,** *except*:
   a. Cataract
   b. Deafness

   c. Retinitis
   d. CHD (congenital heart disease)
5. **The congenital rubella syndrome:**
   a. May be prevented by vaccination in early pregnancy
   b. Causes IUGR
   c. Causes cataract
   d. Causes deafness only if acquired before 16 weeks of gestation
6. **Rubella vaccine is contraindicated in all,** *except*:
   a. Patient on immunosuppressant
   b. Girls with leukemia
   c. Girls between 11–14 years
   d. Pregnancy

Coronaviridae

7. **MERS was first reported from:**
   a. China
   b. Saudi Arabia
   c. India
   d. France
8. **Following is/are the human coronavirus (es):**
   a. 229E
   b. NL63
   c. OC43
   d. SV40
9. **SARS true are:**
   a. Severe acute respiratory syndrome
   b. Documented respiratory route spread
   c. Effective vaccine available
   d. Group coronavirus
10. **Which statement about COVID-19 is false?**
    a. Patients with comorbidities have a worse prognosis
    b. Morbidity and mortality increase significantly with age
    c. More than 80% of infected patients recover without intensive medical intervention
    d. None of the statements is false
11. **Children infected with SARS-CoV-2 and having pneumonia have increased levels of all of the following,** *except*:
    a. TNF-alpha
    b. Serum procalcitonin
    c. IL-10
    d. All of the answers are correct
    e. None of the answer is correct
12. **Among healthcare workers fully vaccinated against COVID-19 with the mRNA vaccine, levels of neutralizing antibodies decline faster in which group? Those with:**
    a. Previously infected with SARS-CoV-2
    b. No previous infection with SARS-CoV-2
    c. Prior infection does not matter
13. **The investigational antiviral remdesivir improves, which outcome in adults hospitalized with COVID-19?**
    a. Time to recovery
    b. Mortality

Answers and Explanation of MCQs

1. b
2. d
3. a
4. c
5. b, c
6. c

- Follow section, **rubella virus** for explanation of answers of MCQs 1–6.

7. b

- Follow section, **Coronaviridae (MERS-CoV)** for explanation.

8. a, b and c

- For explanation of options a, b and c, follow section, **Coronaviridae (classification)**. SV40 virus is belongs to Papovaviridae family. It is not a human pathogen. It causes the polyoma (malignant tumor) in mouse.

9. a, b and d

- Follow section, **Coronaviridae (SARS-CoV1)** for explanation.

10. d

- Follow section, **Coronaviridae (SARS-CoV2 → Pathogenicity → Precipitating factors)** for explanation.

11. e

- All are elevated; alkaline phosphatase, IL-2, IL-4 and IL-6.

12. b

- There was a substantial decline in neutralizing antibodies in those who had not previously been infected with the virus in a recent study.

13. a

- Mortality was numerically, but not significantly reduced with remdesivir.

Bacteriophages

BACTERIOPHAGES

Meaning

Bacteriophage word derived from two words like bacteria and phagein (Greek) means to devour or to eat bacteria, means virus which eats the bacterium.

Definition

Taxonomically, unrelated (either DNA or RNA) viruses, which eat the bacteria are called bacteriophages. Commonly, they are abbreviated as phages.

History

In 1917 Twort and d'Herelle observed the lysis of dysentery bacillus broth culture from stool filtrate of dysentery patients and gave the name as bacteriophage to such lytic agent.

Morphology

E. coli T-even phages (**Fig. 87.1**) like T2, T4, T6, etc., serve as model to describe the properties of bacteriophage.

Shape: T-even phage is tad pole in shape. Some phages are filamentous/spherical in shape.

Parts: Three parts (1) head part is 28–100 nm in size and hexagonal in shape. Some phages contain DNA, while some contain RNA, but never both. NA is may be ss or ds and surrounded by capsid. (2) Tail part is hollow core surrounded by contractile sheath. (3) Base plate contains prongs and tail fibers. Tail fibers help in adsorption to host cell.

Life Cycle

Follow **Ch. 8 (Fig. 8.8).**

Uses

1. **Therapeutic:** Initially, it was used in treatment of bacterial infection, but not satisfactory.

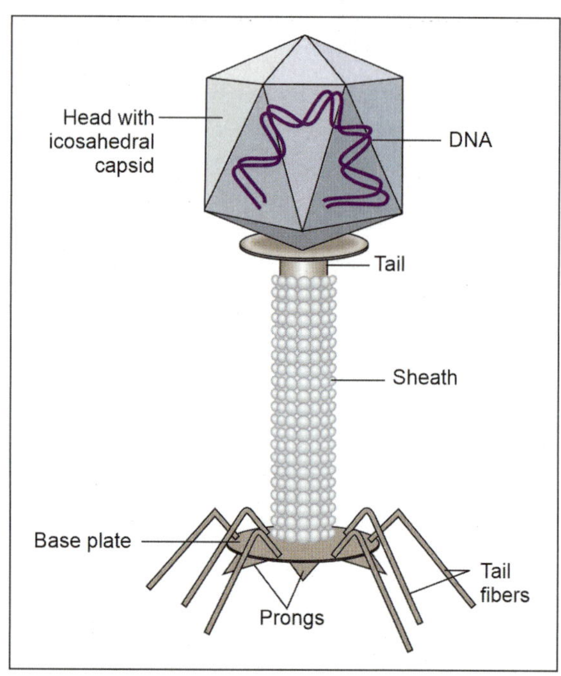

Fig. 87.1: *E. coli* phage (T-even phage)

2. **Transduction:** It transfers the genetic information between two cells. It transduced not only the chromosome, but also plasmid, episome and transposon. They are discussed in more detail in **Ch. 8 (Fig. 8.8).**

3. **Model:** To study the virus-host interaction.

4. **Cloning vector:** Used in genetic manipulation.

5. **Control of bacterial population:** Presences of 10^8 phages per ml of water suggest that they have a role in control of bacterial population.

6. **Diagnosis of bacteria and bacteriophage typing:** It is used for identification and epidemiological typing of bacteria which are difficult to distinguish by biochemical and serological methods. More detail is given in **Ch. 118.**

7. **Phage conversion (lysogenic conversion):** Presence of phage over bacteria confers certain new properties (like toxin production) in bacteria called **phage conversion**. Nontoxigenic strain of bacteria becomes toxigenic after infection by phage. Bacteriophage mediated exotoxins are written in **Ch. 9.**

8. **Alteration of antigenic properties of *Salmonella*:** Structural changes in O Ag of *Salmonella* can be induced by bacteriophage (lysogenization), which results in new serotypes of *S.* Anatum to *S.* Newington and later to *S.* Minneapolis. More detail is given in **Ch. 62 (Flowchart 62.1).**

Phage Assay

Definitions: Phage assay is one type of plaque assay, but when virus is bacteriophage called **phage assay**. When phage is applied to bacterial culture, it produces lytic zone called **plaque**.

Uses: It gives idea about numbers of phages in suspension. Size, shape and nature of plaque are give idea about type of phage.

Classification

Bacteriophage group contains either DNA or RNA viruses as shown in **Table 87.1.**

TABLE 87.1: Classification of bacteriophages

Naked or enveloped	Virus family	Genus/species	Nucleic acid type	Capsid and morphology
		DNA type bacteriophages		
Enveloped	Lipothrixviridae	Acidianus filamentous virus 1	Linear dsDNA	Rod-shaped
	Ampullaviridae		Linear dsDNA	Bottle-shaped
	Globuloviridae		Linear dsDNA	Isometric
	Plasmaviridae	MV-L2 phage	Circular dsDNA	Pleomorphic
Naked	Myoviridae	*E. coli* phage/coli phage (T even phage like T2, T4, T6 phage), Mu, PBSX, P1Puna-like, P2, I3, Bcep 1, Bcep 43, Bcep 78	Linear dsDNA	Icosahedral, contractile tail
	Sipho(stylo) viridae	*E. coli* phage (like T1, T5), λ phage, phi, C2, L5, HK97, N15	Linear dsDNA	Icosahedral, noncontractile tail (long)
	Podo(pedo)viridae	*E. coli* phage/coli phage (like T3, T7 phage), P22, P37	Linear dsDNA	Icosahedral, noncontractile tail (short)
	Rudiviridae	Sulfolobus islandicus virus 1	Linear dsDNA	Rod-shaped
	Tectiviridae	PRD1 phage	Linear dsDNA	Icosahedral, isometric
	Bicaudaviridae		Circular dsDNA	Lemon-shaped
	Clavaviridae		Circular dsDNA	Rod-shaped
	Corticoviridae	*Pseudomonas* phage MP2	Circular dsDNA	Icosahedral, isometric
	Fuselloviridae		Circular dsDNA	Lemon-shaped
	Guttaviridae		Circular dsDNA	Ovoid
	Inoviridae	M13, MV-L1, *E. coli* phage fd	Circular ssDNA	Helical capsid, filamentous
	Microviridae	*E. coli* phage (X174 phage)	Circular ssDNA	Icosahedral, isometric
		RNA type bacteriophages		
Enveloped	Cystoviridae	Phage φ6	Segmented dsRNA	Icosahedral, spherical
Naked	Leviviridae	Phage MS2, Qβ	Linear ssRNA	Icosahedral, isometric

ACCESS YOURSELF

Short Note

1. Bacteriophages.

Short Questions for Theory/Viva Questions

1. Write four uses of bacteriophage.
2. Define: Transduction and lysogenic conversion.
3. Draw the morphological labeled diagram of bacteriophage.

4. Define: Provirus and prophage.

5. Write four examples of bacteria producing toxin after phage conversion.

MCQs for Chapter Review

History

1. Bacteriophage was discovered by:
a. Edward Jenner
b. Twort and d' Herelle
c. Smith, Andrew and Laidlaw
d. Johnson and Goodpasture

Uses

2. Bacteriophages are mostly needed for:
a. Bacterial identification
b. Epidemiologically
c. As antibacterial agent
d. Conversion property in bacteria

3. All of the following statements are true about bacteriophages, *except*:
a. It is a virus that infects the bacteria.
b. It helps in transduction of bacteria.
c. It imparts toxigenicity to bacteria.
d. It transfers only chromosomal gene.

Answers and Explanation of MCQs

1. b
- Follow section, **introduction (history)** for explanation.

2. a, b, d

3. d
- Follow section, **introduction (uses)** for explanation of answers of MCQs 2–3.

CHAPTER 88

Infections of Hepatitis Viruses

Chapter Outline

INTRODUCTION

Meaning

From hepa (Greek) means liver and itis means inflammation, because viruses are infecting and causing inflammation of liver.

Definition

Group of heterogeneous or taxonomical unrelated viruses (HBV is DNA type and others are RNA type) causing inflammation of liver.

List of Viruses Causing Hepatitis

Hepatitis viruses group: HAV, HBV, HCV, HDV and HEV.

Others: They are not from hepatitis group, but causing inflammation of liver (hepatitis). For examples, yellow fever virus, lassa fever virus, marburg virus, CMV, HSV, VZV, rubella virus, measles virus and coxsackie virus.

Classification

Systemic/taxonomical classification: Follow **Flowchart 88.1**.

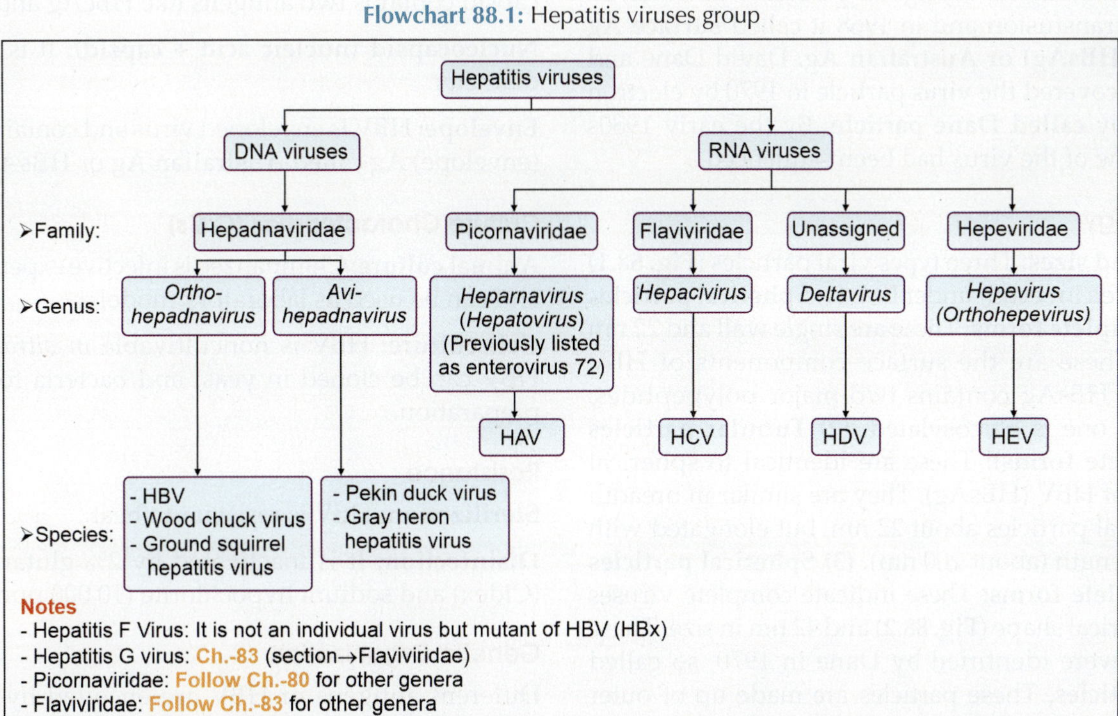

Flowchart 88.1: Hepatitis viruses group

Notes
- Hepatitis F Virus: It is not an individual virus but mutant of HBV (HBx)
- Hepatitis G virus: **Ch.-83** (section→Flaviviridae)
- Picornaviridae: **Follow Ch.-80** for other genera
- Flaviviridae: **Follow Ch.-83** for other genera

Based on modes of transmission

- **Parenteral transmission:** It includes route other than oral. For example Hepatitis B Virus (HBV), Hepatitis C Virus (HCV) and Hepatitis D Virus (HDV).
- **Non-parenteral (feco-oral) transmission:** It includes feco-oral route. For example Hepatitis A Virus (HAV) and hepatitis E Virus (HEV).

Epidemiological and clinical classification

- **Infectious/infective/type A hepatitis:** It occurs sporadically or produces epidemic, affects young adult or children and transmitted by feco-oral route, for example, HAV.
- **Transfusion/serum/type B hepatitis/homologous serum jaundice:** It is transmitted by transfusion of blood or blood products. For example HBV. As it is transmitted by homologous serum used for prevention or treatment, so called **homologous serum jaundice.**
- **Non-A Non-B hepatitis (NANB):** Includes all viruses causing hepatitis but not diagnosed as HAV or HBV like HCV, HDV and HEV.

World Hepatitis Day

It is celebrated on 28th July of every year. It aims to raise global awareness and to encourage the prevention, diagnosis and treatment of hepatitis A, B, C, D and E.

HBV

Synonym

Complete viral particle of HBV called **Dane particle.**

History

In 1965, it was identified by Baruch Blumberg in serum of Australian aboriginal person, who received multiple transfusion and in 1968 it called **surface Ag of HBV (HBsAg) or Australian Ag.** David Dane and others discovered the virus particle in 1970 by electron microscopy **called Dane particle**. By the early 1980s the genome of the virus had been sequenced.

Morphology

Shapes and sizes: Three types viral particles (**Fig. 88.1**) are observed in serum under EM. **(1) Spherical particles but incomplete forms:** These are single wall and 22 nm in size. These are the surface components of HBV (HBsAg). HBsAg contains two major polypeptides, of which, one is glycosylated. **(2) Tubular particles (incomplete forms):** These are identical to spherical particles of HBV (HBsAg). They are similar in breadth to spherical particles about 22 nm, but elongated with variable length (about 200 nm). **(3) Spherical particles but complete forms:** These indicate complete viruses with spherical shape (**Fig. 88.2**) and 42 nm in size. These particles were identified by Dane in 1970, so called **Dane particles.** These particles are made up of outer layer called envelope contains HBsAg and inner layer

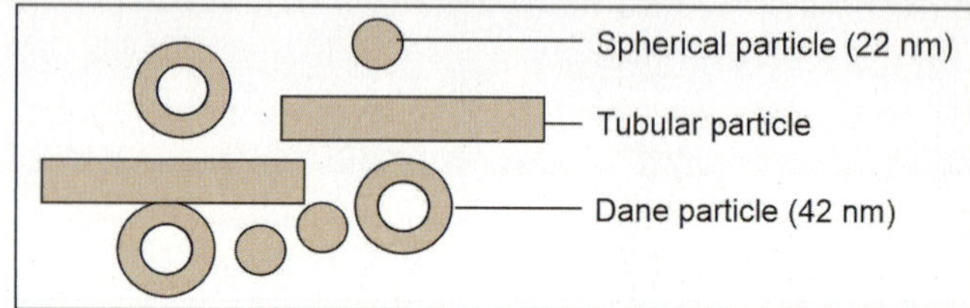

Fig. 88.1: HBV particles in serum

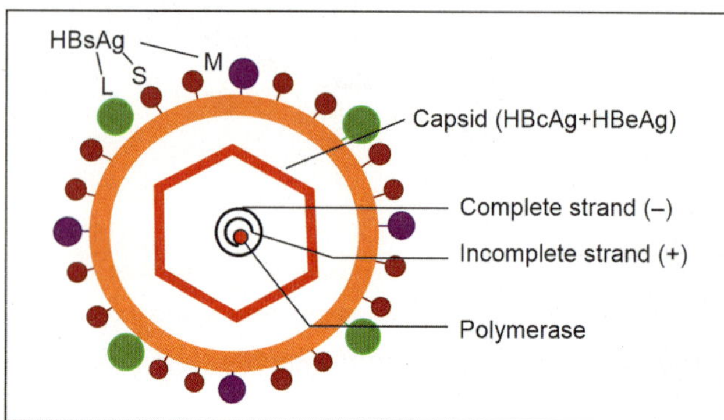

Fig. 88.2: Detail morphology of dane particle (complete form) of HBV

called nucleocapsid (27 nm) contains HBcAg, HBeAg and partially ds-DNA (core Ag).

Genome: HBV contains ds-DNA with one complete strand and other incomplete strand. Incomplete strand contains DNA polymerase (DNA dependent DNA polymerase), so called plus (+) stand and complete strand called minus (–) strand. DNA polymerase has two functions like DNA dependent DNA polymerase and RNA dependent reverse transcriptase. DNA polymerase repairs the gap in plus strand and makes the genome fully-ds.

Capsid: DNA surrounded by icosahedral variety of capsid contains two antigens like HBcAg and HBeAg.

Nucleocapsid (nucleic acid + capsid): It is 27 nm in diameter.

Envelope: HBV is enveloped virus and contains surface (envelope) Ag called **Australian Ag** or **HBsAg.**

Culture Characteristics (C/Cs)

Animal culture: Chimpanzee is infective experimentally and can be used as laboratory model.

Cell culture: HBV is noncultivable *in vitro.* DNA of HBV can be cloned in yeast and bacteria for vaccine preparation.

Resistance

Sterilization: HBV is resistant to heat.

Disinfection: It is inactivated by 2% glutaraldehyde (Cidex) and sodium hypochlorite (10,000 ppm).

Genetic Organization

Different antigens of HBV are encoded by different genes (**Fig. 88.3**).

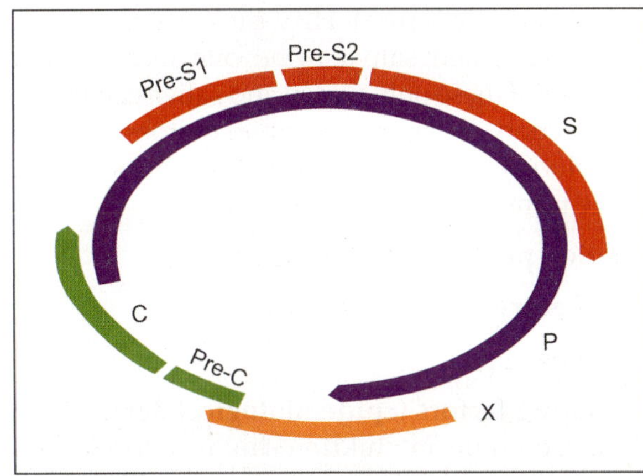

Fig. 88.3: HBV genome

1. **C-gene:** Two-regions.
   - Pre-C (core) region: Encodes for HBeAg.
   - C-region: Encodes for HBcAg.
2. **P-gene:** It is the largest (longest) gene and encodes for enzyme DNA polymerase.
3. **S-gene:** It encodes for HBsAg. Three regions.
   - Pre-S1: Encodes for large (L) proteins, not found free in circulation.
   - Pre-S2: Encodes for middle (M) protein.
   - S-region: Encodes for major (S) protein. M and S proteins are free in circulation as HBsAg particles.
4. **X-gene:** Encodes for non-particular protein called **HBxAg**. It has transactivating effect on viral and some cellular genes. It leads to replication of HBV and some other virus like HIV. HBxAg and its antibody are present in patient with severe chronic hepatitis and HCC.

Mutation in Genes of HBV

Mutation in genes of HBV leads to clinical emergency of mutant strain. Following three types of mutations are identified in genes of HBV.

Precore (pre-C) region mutation: Nonsense mutation in the pre-C region leads the formation of stop codons which lead the absence of HBeAg. It is distributed in Mediterranean countries and Europe. Patients infected with HBV carrying such mutants are diagnosed late and rapidly progress to chronic stage and cirrhosis. Diagnostically HBeAg is absent while other markers, are as usual.

P-gene mutation: It presents in **YMDD locus** of P-gene of HBV reverse transcriptase. It results in development of resistant to lamivudine.

S-gene mutation (in 'a' components of HBs Ag): Such mutant called **escape mutant.** It presents in 'a' components of HBsAg in S-gene. It is due to substitution of single amino acid from glycine to arginine at position 145. Presence of S-gene mutant may lack the HBsAg, which creates the **problems in diagnosis.** Escape mutant can cause infection in immunized person, because HBsAg antibodies present in immunized person cannot

neutralize the escape mutants resulting **vaccine failure.** It observed in following cases:
1. Infants born to HBeAg positive mothers.
2. Liver transplant recipient, who underwent the procedure for hepatitis B and who were treated with a high potency human monoclonal anti-HBs preparation.
3. In patients who were actively and passively immunized for HBV. In such cases antibody pressure may favors the evolutionary changes in 'a' component.

Antigenic Structure

Major antigens (products of gene) of HBV
1. **Capsid-Ags:** HBcAg and HBeAg.
2. **Envelope-Ag:** HBsAg.

Antigenic diversity of HBsAg: It contains group specific and type specific components as follows.
- **Group specific:** Group specific component is 'a'.
- **Type specific:** Type specific components are 'd-y' and 'r-w'. In addition to all these few other components like x, f, t, j, n, and g are also reported, but not been adequately studied.

Antigenic types: Virus is divided into four major serotypes like adr, adw, ayr and ayw based on epitope present on its envelope (HBsAg). These types are not important in immunity, because each type shows the dominant 'a' variant. Each type shows the distinct geographical importance as follows.
- adr: Common in South and East India and far East.
- adw: Common in Europe, Australia and America.
- ayr: Very rare prevalent.
- ayw: Common from West Asia through the Middle East, to West and Northern India.

Immunity

AMI: Antibodies form after onset of jaundice within week or two weeks. Order is 1st HBcAg then HBeAg and later HBsAg antibodies. Development of HBsAg antibodies indicates the recovery.

CMI: Development of CMI to HBsAg also noticed. Hepatocytes carry the antigens are subjected to kill by ADCC or by other cytotoxic T-cells. Absence of immune response may not causes hepatitis, but may produce carrier stage, hence infants and immune-deficient persons are become carrier than case.

Pathogenicity

Disease name: Disease called **hepatitis B.** It also called **transfusion/serum/homologous serum jaundice.**

Epidemiology
- **Distribution:** It is distributed worldwide and more than 1/3 of world's population is infected by HBV. Disease is usually sporadic and no seasonal outbreak occurs, though occasional outbreak occurs in hospitals, orphanages and asylum.

- **Genotypes:** There are eight major genotypes (A–H). Genotypes I and J are not universally accepted as of 2015. The genotypes have a distinct geographical distribution and are used in tracing the evolution and transmission of the virus. Differences between genotypes affect the disease severity, course, a complications, response to treatment and possibly vaccination. Genotypes differ by at least 8% of their sequence. Most genotypes are now divided into sub-genotypes with distinct properties, 1/4 of them becomes carrier and develop chronic liver diseases like chronic hepatitis, cirrhosis and hepatocellular carcinoma (HCC).

Reservoir of infection: Only humans (no animals) are reservoirs.

Source of infection

- Blood/blood products and body fluids are the main sources.
- Period of infectivity (communicability): Virus presents in blood/blood products and body fluid during incubation period and acute phase of disease. People infected with HBV remain carrier as long as they carry HBsAg in blood. Individual becomes non-infectious once HBsAg is replaced by anti-HBsAg Ab. Infectivity is determined by HBeAg. It is maximum when HBeAg is elevated.

Mode of transmission: Two modes.

1. Direct transmission

- **Direct contact**
  - Due to contact of contaminated devices like scalpels, scissors, endoscopes, etc., with abraded skin.
  - Sexual intercourse: It is more in male homosexual. Virus is presents in semen, so screening before donation of semen is obligatory.
- **Inoculation** under skin or mucosa by contaminated needles, syringes or by tattooing transmits the HBV. Percutaneous injury by infected needle or syringe transmits HBV (**30%**) > HCV (**3%**) > HIV (**0.5–1%**).
- **Perinatal/vertical/mother to baby:** It occurs from mother to infant. Infectivity is more if mother is HBeAg positive. Among the all hepatitis, HBV has maximum perinatal transmission. This route transmits HIV (**30%**) > HBV (**20%**) > HCV (**6%**) > HDV (±) > HAV and HEV (–).
  - Antenatal (transplacental): Low rate (20%) but higher than HCV (6%).
  - Intranatal: More common during birth by contact of maternal blood with skin or mucosa of fetus.
  - Postnatal: Infection by breast milk is possible but efficiency is very low. However, it is advisable to stop feeding if mother is infective and baby is ensured by other sources.

2. Indirect transmission: It occurs by contaminated blood or blood products or cryoprecipitates or by tissues (**vehicle borne**). HBV presents in other body fluids like semen, saliva, urine, bile, feces and breast milk, but of these only semen and saliva can transmits the infection. As it survives in mosquitoes for 2 weeks, after blood meal, but not able to multiply so **no vector borne** transmission.

Incubation period (IP): 1–6 months.

Portal of entry: Skin or mucosa.

Sites: Liver.

Precipitating factors (epidemiological determinants): (1) **Age:** Common in children due to perinatal transmission (babies of mothers with chronic HBV infection). (2) **Occupation:** HCWs, sex workers, homosexuals, drug addicts, person required blood transfusion or organ transplantation and barbers are considered as high risk groups. (3) **HIV-HBV coinfection:** About 10% of population infected with HIV is coinfected with HBV. HBV does not alter the progress of HIV, but coinfection with HIV will increases the chances of development of chronic hepatitis, cirrhosis and HCC by HBV.

Pathogenesis: It based on immune response. Hepatocytes carries the antigens are subjected to kill by ADCC or by other cytotoxic T-cells. Absence of immune response may not cause hepatitis but may produces carrier stage, hence infants and immune-deficient persons are become carrier than case. Mechanisms of hepatitis by HBV are mentioned in **Flowchart 88.2**.

Clinical features (C/Fs): In 90% patients, it presents as an acute hepatitis with following features.

- Insidious onset with minimal fever.
- Extrahepatic features are more like arthralgia, urticaria, glomerulonephritis.
- Complete recovery occurs in 90% cases in 1–2 months and virus eliminated from circulation in 6 months and patient remains immune thereafter.
- Simultaneous **HDV infection** leads to fatal fulminant hepatitis (poorest the prognosis).
- 1–10% (poor prognosis) cases develop carrier or chronic hepatitis.

Complications: In 1–10% patients it moves to chronic stage with following complications:

- Fulminant hepatitis: Instead of insidious onset, acute hepatitis sometimes presents with severe and sudden onset called **fulminant hepatitis.**

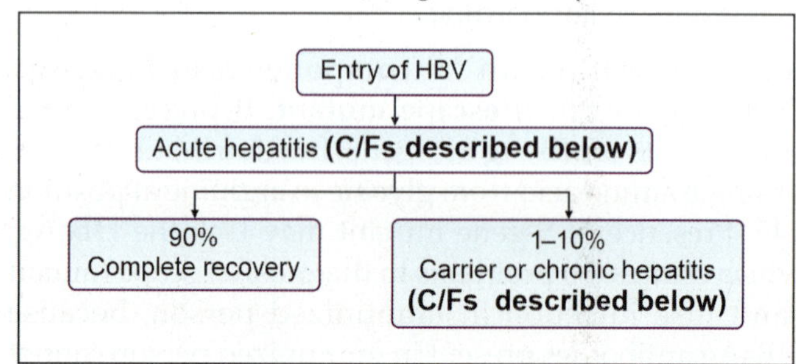

Flowchart 88.2: Pathogenesis of hepatitis B

- Chronic hepatitis, cirrhosis and HCC.
- Extrahepatic: It includes glomerulonephritis, popular acrodermatitis of child hood (Gianotti-crosti syndrome), poly arteritis nodosa, arthritis, angio-edema, etc. These are may be due to immune complex (type-III hypersensitivity) deposition.

Carrier

- **Definition:** It is defined as person carries HBsAg in blood for more than 6 months without any sign-symptoms.
- **Types:** Two types. **(1) Super carrier:** It has high titer of HBsAg with HBeAg, DNA polymerase and HBV in circulation. It has high infectivity and generally with raised transaminase. A quarter of carriers are HBeAg positive. It is considered as immunotolerant chronic HBV infection. **(2) Simple carrier:** It has low titer of HBsAg with negative HBeAg, DNA polymerase and HBV. It has low infectivity. Many super carriers become simple carrier in time. It is considered as chronic inactive HBV infection.
- **Carrier rate:** It is different in different countries depending on the living standard. **(1) Low endemicity:** Carrier rate is <2%. It is observed in America, Australia and European countries. Lowest is recorded as 0.01% in UK and Norway. **(2) Intermediate endemicity:** Carrier rate is 2–8%. India falls in intermediate group with high rate (2–7%) in South part and low in North part. China, Eastern Mediterranean and Southeast Asia are also fall in intermediate group. **(3) High endemicity:** Carrier rate is >8%. Few African and Western Pacific countries are fall in high group. In some countries rate is exceed >15% like South Sudan, Kiribati, Swaziland and Solomon Islands.
- **Other important aspects about carrier stage**
  - Following infection 90% neonates, 30% children and 5–10% adults are carrier.
  - There are 350 million carrier worldwide, of them 45 million are in India next to China.

Laboratory Diagnosis

Specimens: Blood/serum.

Testing methods

A. Microscopy: Follow morphology.

B. Culture: Follow C/Cs.

C. Serological tests

- **Antigens (Ags):** HBsAg and HBeAg are present free in blood. HBcAg is enclosed in capsid and not released free in circulation, hence not detectable from blood; however it is detectable from hepatocytes by IF test. Its antibody is detectable from blood.
- **Antibodies (Abs):** Anti-HBc, anti-HBs and anti-HBe.
- **Immunological reactions in HBV (Fig. 88.4)**
  1. **HBsAg and anti-HBs**
     - HBsAg is the 1st marker detectable after infection, even before elevation of liver enzymes (transaminase). It reaches peak at 8th–12th

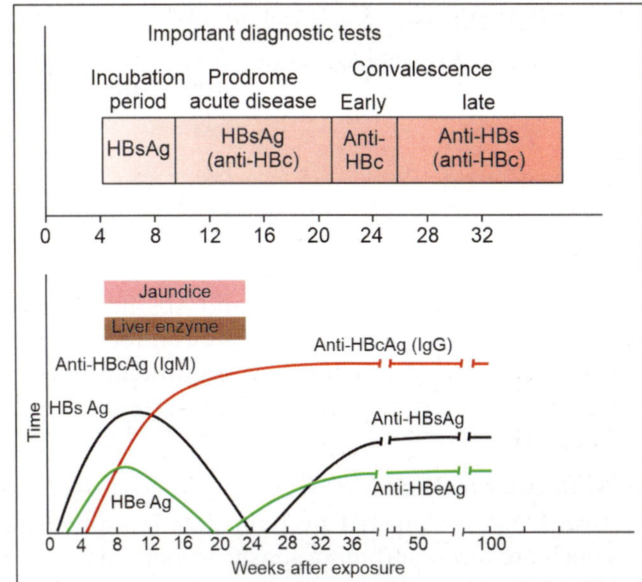

Fig. 88.4: Serological markers of HBV

weeks and disappears at 24th week (6 months). If lasts for more than 6 months **indicates carrier or chronic stage.** When it disappears, its antibody, anti-HBs starts to appear and remains for very long time.

- Anti-HBs is protective and provides immunity to HBV. Its presence without any serological marker (protective antibody) indicates immunization or recovery from infection.

2. **HBeAg and anti-HBe**
   - HBeAg develops almost same time or little later with HBsAg. Its presence indicates intrahepatic viral replication and **high infectivity.** DNA polymerase in blood also indicates **high infectivity.**
   - Disappearance of HBeAg is followed by anti-HBe.

3. **HBcAg and Anti-HBc**
   - HBcAg is enclosed in capsid and not releases in circulation, hence not detectable freely in blood. It remains in hepatocytes, from where it can be detected by immunohistochemical staining. Intrahepatic HBcAg indicates active viral replication.
   - Anti-HBc is detectable after 1st week. It is the earliest Ab seen in blood than other. It remains life-long and serves as useful indicators for HBV infection even in absence of others. It is IgM in nature in initial 6 months and indicates **recent infection** then after becomes IgG which indicates **remote infection** and used as **epidemiological markers.** Anti-HBc distinguishes between recent and remote infection.
   - **Acute stage markers:** HBsAg is the earliest marker. For diagnosis of acute stage HBsAg, HBV DNA and HBeAg are also useful, but IgM anti-HBcAg is most significant.

TABLE 88.1: Serological pattern in HBV infection

Stage	HBs Ag	HBe Ag	DNA	Anti-HBc	Anti-HBs	Anti-HBe
Acute	+	+	+	IgM	–	–
Chronic	+	+	+	IgG	IgG	–
Carrier	+	–	–	IgG	–	+/–
Immunity	–	–	–	–	+	–

- **Tests commonly used are:** Ags and Abs are detected by ELISA, ELFA, chemiluminescence and some rapid tests like ICT plus latex particle agglutination test.
- **Serological pattern in HBV infection:** Follow **Table 88.1**.

D. Molecular methods
- Viral DNA is detected by hybridization and PCR which are rapid and more sensitive methods.
- HBV-DNA: Like HBeAg, it indicates viral replication in liver and viral infectivity. Both these two present every time, so they cannot differentiate between acute and chronic stage. They help to assess the progress of infection to chronic stage. They are the quantitative markers for HBV replication. In a carrier HBV is actively multiplying and is highly infectious called **super carrier**.
- Level of HBV-DNA is correlates with the degree of liver damage indicated by raised ALT and AST.
- **Markers for active viral replication:** HBV-DNA, DNA polymerase, HBeAg, Intrahepatic HBcAg and liver enzymes.

E. Other supportive tests: Estimation of ALT, AST, bilirubin, etc.

Prevention
General measures
- Improvement in health education.
- Screening of blood, organs and tissues before donation.
- Safe sex like use of condom.
- High-risk behavior modification like drug abuse.
- Avoid direct contact with blood and body fluids.
- Use of disposable and sterile medical devices.

Prevention of sharp/splash injuries in HCWs: Follow **Ch. 29.**

Preventions during surgeries: Follow Ch. 29.

Preventions of neonatal transmission: Follow combined immunization as described above.

Immunoprophylaxis
A. Active immunization: Cloned/recombinant/genetic engineered/subunit vaccine is in use currently.
- **History:** It was prepared in 1986 and gradually replaced the plasma derived vaccine.
- **Preparation:** Prepared by cloning the HBsAg (non-glycosylated part alone) of HBV in **Baker's yeast**. It contains alum adjuvant It is available singly or in combination as **pentavac** which contains diphtheria, pertussis, tetanus, HBV, *H. influenza* type b (Hib).
- **Administration and dose:** Total three doses are given on 0, 1 and 6 months. It is given in the dose of 10–20 μg in adults and half in children <10 years by IM route in deltoid region or anterolateral aspect of the thigh in infants. Avoid the gluteal region, as it is less immunogenic because less numbers of APCs in gluteal fat delaying the presentation of vaccine antigens to T or B cells. Birth dose is given with BCG vaccine but at different site. It also included in national immunization schedule and given with DPT or as combined vaccine like **pentavac** which contains diphtheria, pertussis, tetanus, HBV, Hib at 6th, 10th and 14th week.
- **Efficacy:** It is highly immunogenic and has 95% efficacy including in infants and children. In persons of >65, efficacy is around 65–75%. It offered long lasting immunity up to 15 years; however, some research witnessed the lifelong immunity.
- **Disadvantage:** High cost.
- **Interruption of vaccine schedule:** Do not restart of schedule but complete the vaccine series regardless of when the last dose was taken. If primary series is missed after 1st dose then give 2nd dose immediately followed by 3rd dose after 4 weeks. If 3rd dose is interrupted, then give it immediately.
- **Testing for revaccination:** Effect of vaccine is tested by detecting anti-HBs titer. It should be checked 2 months after 3rd dose of vaccine and 6 months after HBVIg, otherwise it gives wrong results. It should also not check after 1–2 doses of vaccine. Recipients are considered as protective if titer is ≥10 mIU/ml. Start the 2nd series if titer falls below 10 mIU/ml; however few studies show that revaccination is not needed even if titer falls below 10 mIU/ml due to presence of immunological memory.
- **Contraindications:** It is contraindicated in persons allergic to vaccine components. Pregnancy and lactation are not the contraindications. HIV-AIDS, chronic liver disease, renal failure and diabetes shows the low immunogenicity.
- **Non/low responders:** About 5–10% recipients are not showing the seroconversion even after two series (total 6 doses) of vaccination called **non/low responders**. They do not need any further vaccination.

> **Note: Other HBV vaccines**
> These are plasma derived vaccine and synthetic peptide vaccine.

B. Passive immunization: HBVIg is **prepared** from human volunteers with high anti-HBs titer. It is **administered** in the dose of 0.05–0.07 ml/kg by IM route within hour of exposure but not after 7 days. It may not prevent the infection, but protects against illness and carrier state for 3–6 months. Following are the indications of HBVIg.

TABLE 88.2: PEP regimens for HBV by CDC

Vaccine status of HCW	If source is positive/unknown for HBsAg	If source is negative for HBsAg
Complete and protected (titer ≥10 mIU/ml)	No further treatment is required: • Regardless of the HBV status of source* • Regardless titer fall down later*	
Complete but not protected (titer <10 mIU/ml)	HBV Ig-1: Within hour of exposure but not after 7 days Vaccine: Start second series (3 doses)	Vaccine: Start second series (3 doses)
No or partial vaccine	HBVIg-1: Within hour of exposure but not after 7 days. Complete the remaining dose of vaccine (do not restart)	Vaccine: Complete the remaining dose of vaccine (do not restart)
Vaccinated for 2 series (6 doses), but not protective	HBV Ig: 2 doses at 1 month apart (0.06 ml/kg or 10–12 IU/kg)	Nothing is required

Note:
- HCW is said to be protected when titer raised (≥10 mIU/ml) after 3 or more doses. Rise of titer after 1–2 doses should not considered as protective.
- HBV Ig provides protection for 6 months.
- HBV Ig and vaccine should administer at different sites.
- Anti-HBs antibody titer should be checked after 2 months of last vaccine dose and 6 months after HBV Ig administration, otherwise it may give false results.
- Previous documented report of anti-HBs titer is accepted not verbal.

– Exposure to HBV positive blood in HCWs.
– Sexual contact with acute hepatitis B partner.
– Neonate born to HBV carrier mother.
– Post liver transplant patients who need protection against HBV infection.

C. Combined immunization: It includes the administration of vaccine plus HBVIg at different sites. It is indicated to neonate born to HBV carrier mother. 0.05–0.07 ml/kg HBVIg is given immediately after birth by IM route followed by full course of HBV vaccine. HBVIg gives protection for 6 months and it does not interfere with Ab to HBV vaccine.

National Viral Hepatitis Control Program (NVHCP): The program includes diagnostic, therapeutic and preventive measures for both HBV and HCV. Diagnostic part includes rapid tests of HBV and HCV followed by viral load.

Treatment

It includes management of normal person after exposure called **postexposure prophylaxis** and case of hepatitis B.

Postexposure Prophylaxis (PEP) of HBV

Definition: It is a short-term treatment to avoid the HBV infection after occupational exposure like skin injury by sharp items like needle, scalpel, etc., splash of PIMs over non-intact skin, splash of PIMs over intact skin for prolonged duration and splash of PIMs over mucosa (intact or abraded).

Infectious materials/specimens: Follow PEP of HIV described in **Ch. 85.**

Guidelines for PEP: Total 7 steps: (1) First aid. (2) Report to nodal officer of hospital. (3) PEP regimens for HBV (by CDC). (4) Viral rapid tests. (5) Informed consent and counseling. (6) Documentation and records of exposure. (7) Follow-up precautions. **For steps 1–2 and 4–7** follow PEP of HIV described in **Ch. 85. Step 3** is described in **Table 88.2** for HBV.

Management of Case of Hepatitis B

- New antiviral drugs like tenofovir and telbivudine are the choice of drugs currently.
- Tenofovir and emtricitabine are used in HIV-HBV coinfection.
- Interferon-alone is not use now, because of side effect.

HAV

History

It was demonstrated by Feinstone and colleagues in 1973 from the feces of experimentally infected human volunteers by immuno electron microscopy.

Morphology

Shape and size: It is spherical (**Fig. 88.5**) in shape and 27 nm in size.

Non enveloped: It is naked virus.

Genome and capsid: It contains ss-RNA (+) surrounded by icosahedral variety of capsid.

Immunoelectron microscopy: It used to identify the virus from stool, blood, bile and liver.

Culture Characteristics (C/Cs)

HAV is the only hepatitis virus which can be cultivated in cell culture (*in vitro*). It can grow in human and simian cell culture. Chimpanzees and marmosets can be infected experimentally. It has also been cloned.

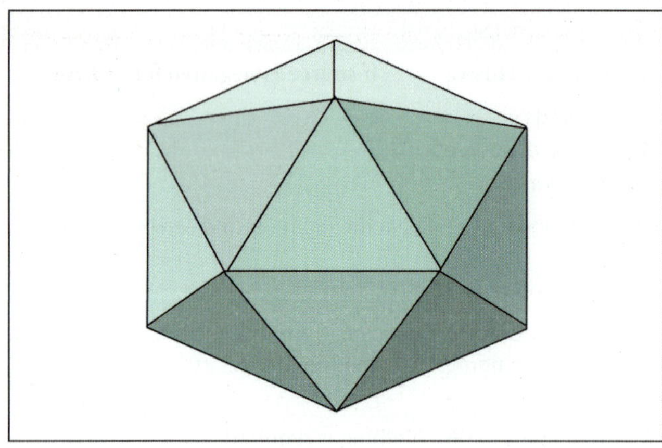

Fig. 88.5: Morphology of HAV

Resistance

Sterilization: It is resistant to heat at 60°C for 1 hr and pH at 3. It preserved at 4°C temperature or below.

Disinfection: It is resistant to ether, chloroform, bile salts and pH at 3. It is inactivated by formalin (1:4000), chlorine (1 ppm for 30 min.) and boiling for 1 minute.

Antigenic Structure

It has only one serotype.

Immunity

Primary infection produces the lifelong immunity; however, about 10–15% can get secondary infection. Antibodies are formed in 90% of population due to repeated exposure up to age of 10 years and incidences are declined after 10 years and in adults.

Pathogenicity

Disease name: Disease **called hepatitis** or **infectious hepatitis/infective hepatitis/type A hepatitis.**

Epidemiology

- **Distribution:** It is distributed worldwide. It occurs sporadically or in form of outbreak. Shellfish have been known to be responsible for outbreak.
- **Genotypes:** Seven genotypes (I–VII), including 4 humans and 3 simian have been described. Human genotypes are numbered as I–III and VII with subtypes like IA, IB, IIA, IIB, IIIA and IIIB. Genotype III has been isolated from both humans and owl monkeys. Commonest human isolate is IA. Simian genotypes are numbered as IV-VI.

Reservoirs of infection: Human are the only known reservoirs; however; chimpanzees can acquire the infection from human and can transmit it to human.

Sources of infection

- These are contaminated food, water or milk. It also presents in saliva and urine, but clinically, it is not significant. Virus present in blood for a very short time (short viremia) before 2 weeks of jaundice and after 1 week of jaundice, so no or very rare blood/serum transmission.

- **Period of infectivity (communicability):** Virus sheds in feces during incubation period (before 2 weeks of jaundice) and in prodromal phase (after 2 week of jaundice), but not in icteric phase.
- **Secondary attack rate:** Transmission from patient to case is 10–20%.

Modes of transmission: It is transmitted by feco-oral route (infected food handlers, raw shellfish). Vertical transmission is not possible.

Incubation period (IP): 2–6 weeks (½–1½ months).

Portal of entry: GIT.

Sites: Liver.

Precipitating factors (epidemiological determinants): **(1) Age:** Most common infection occurs in children and adolescent between 5–14 years, but illness is mild like mild anicteric gastroenteritis or subclinical. It is more icteric in adults and causes acute icteric febrile illness. Anicteric to icteric ratio in children is 12:1 while in adults is 1:3. Antibodies are developed after the age of 10 years in 90% of population due to repeated exposure, so remain protected. **(2) Overcrowding and poor sanitation:** Common in developing countries and persons with poor sanitation. Incidence is decline in persons with improved personal hygiene. Outbreaks are common in military camps, neonatal ICUs, day care centers, summer camps, families and institutes. **(3) Seasonal factor:** In India it is common in late rainy season and in winter.

Pathogenesis: Follow **Flowchart 88.3.**

Clinical features: Large numbers of cases are asymptomatic. Overt illness (acute hepatitis) occurs in 5% cases with following features. **Prodromal (pre-icteric phase)** presents with acute onset with fever, malaise, anorexia, nausea, vomiting and liver tenderness is seen in 5% of cases. Symptoms are subsides with onset of jaundice. **Icteric (jaundice) phase** presents with jaundice, which recovers in 99% cases over a period of 4–6 weeks.

Complications

- HAV and HEV are not showing carrier stage. Carrier stage is present in HBV, HCV and HDV.

Flowchart 88.3: Pathogenesis of hepatitis A

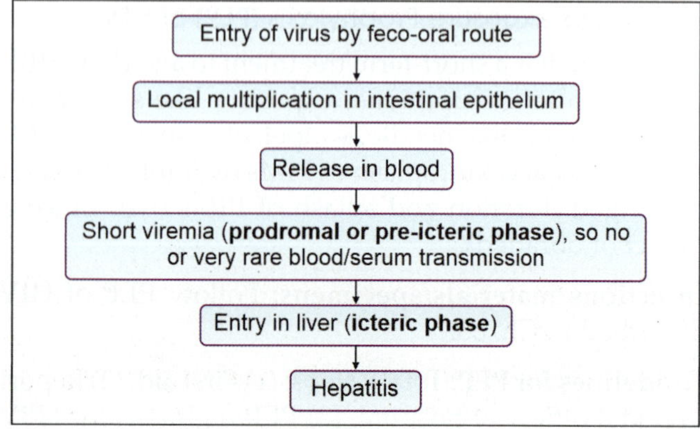

- HAV is also not responsible for chronic stage like cirrhosis, chronic hepatitis and HCC.
- Common complications of HAV are **fulminant hepatitis** (characterized by severe necrosis of hepatocytes), **relapsing hepatitis** (occurs weeks-months after recovery of acute stage) and **cholestatic hepatitis** (presents with protracted cholestatic jaundice and pruritus). Mortality rate of HAV is 0.1–1%.

Laboratory Diagnosis

Specimens: Feces and serum.

Testing methods:

A. **Microscopy:** Follow morphology.

B. **Culture:** Follow C/Cs.

C. **Serological tests:**
- **Antibody detection: IgM (Fig. 88.6)** appears in acute phase, reaches peak at 2nd week and disappears at 12th weeks. It indicates the recent or early infection. **IgG** appears a week after IgM, reaches peak at 7th–8th week and persists for long time or may be for life. It indicates the remote or late infection. IgM and IgG are detected by ELISA.
- **Antigen detection:** It is detected from stool (–2 to +2 weeks of jaundice) by ELISA.

D. **Molecular method:** RT-PCR is rapid and more sensitive method, performed from feces.

E. **Other supportive tests:** Estimation of ALT, bilirubin, etc.

Prevention

General measures
- Hand washing after toilet.
- Prevention of fecal contamination of food and water.
- Use of boiled (at least for 5 minutes) water during outbreak.

Immunoprophylaxis
1. **Active immunization:** Two types of vaccines. Both are free from side effects, highly immunogenic and provide long lasting immunity, possibly lifelong.
   - **Formalin inactivated human cell culture vaccine:** It is **prepared** by growing HAV on human fetal lung fibroblast cells like MRC-5 and WI-38 and inactivated by formalin. It is **indicated** for children >1 year. Single IM injection is **administered** followed by booster after 6–12 months. It has 94% efficacy.
   - **Live attenuated human diploid cell culture vaccine:** It is **prepared** by growing H2 and L-A-1 strains of HAV on human diploid cell culture in China. Single SC injection is **administered.**
2. **Passive immunisation:** HAVIg is prepared by using pooled normal human Ig. It is **administered** in dose of 0.02–0.12 ml/kg of body weight by IM route within 2 weeks of exposure. It gives protection for 1–2 months. It is **indicated** for intimate contacts of HAV patients like day care centers, house hold contacts, etc., and travelers. It is **not indicated** for vaccinated persons, olders (may be immuned) and casual contacts like office, school, hospital and factory.

Treatment

No specific anti-HAV drugs are available. Supportive measures are given.

HCV

History and Synonym

Virus was identified in 1889 and called **parenterally transmitted NANB hepatitis virus**. It is the major cause (about 80%) of parenterally transmitted NANB-hepatitis.

Morphology

Shape and size: It is spherical (**Fig. 88.7**) in shape and 50–60 nm in size.

Genome and capsid: It is linear, ss-RNA (+) surrounded by icosahedral variety of capsid.

Envelope: Envelope presents and carries glycoprotein spikes like E1 and E2.

Culture Characteristics (C/Cs)

Animal culture: Chimpanzee is infective *in vitro*.

Cell culture: HCV is non-cultivable *in vitro*. It can be cloned in *E. coli*.

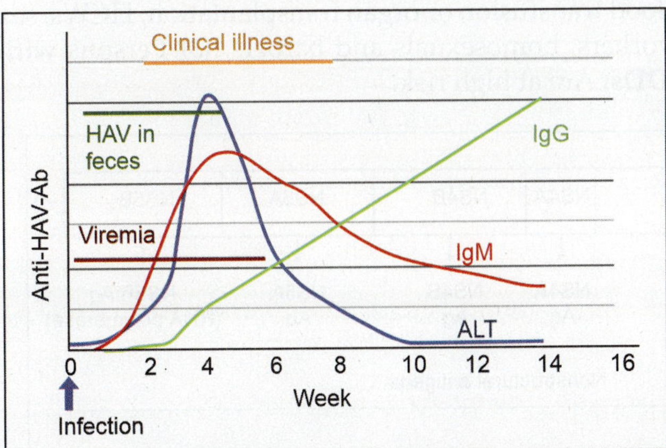

Fig. 88.6: Markers of HAV

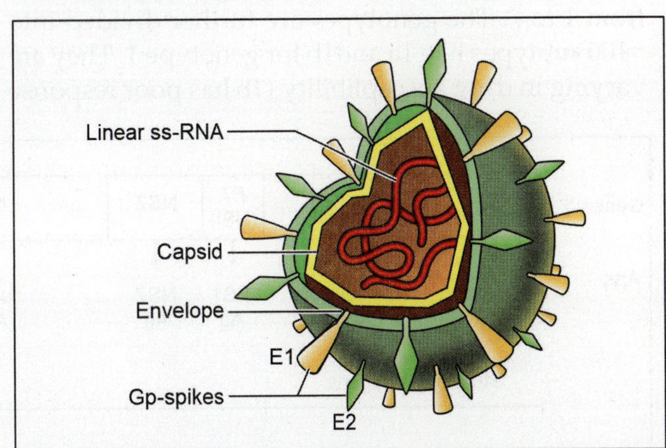

Fig. 88.7: Morphology of HCV

Resistance

HCV is an enveloped virus sensitive to lipid solvents.

Genetic Organisation and Antigenic Structure

Genes present in HCV encodes for structural and non-structural antigens as shown in **Fig. 88.8**. In HCV genes and antigens have almost same name.

- **Structural antigens:** It includes
  - Capsid antigens (encode by C gene): C22, C33 and C100.
  - Spikes antigens (encode by E1 and E2 genes): Glycoprotein antigens like E1 and E2.
- **Nonstructural antigens:** Encodes by NS genes and antigens are like NS1 (now known as p7 membrane protein which acts as an ion channel), NS2, NS3, NS4A, NS4B, NS5A and NS5B (function as RNA polymerase).

Mutation in Genes of HCV

The most variable gene in HCV is E2 gene followed by NS genes; hence, they are more prone to undergoes mutations. Mutation in E2 gene enables the emergency of mutant strain, which results in establishment of chronic infection and failure of development of effective vaccine.

Immunity

Because of genetic variation, there is less homologous or heterologous post-infection immunity. Protective (neutralizing) antibodies are produced against E2 antigens.

Pathogenicity

Disease name: Disease **called hepatitis C** or called **parenterally transmitted NANB hepatitis.**

Epidemiology

- **Distribution:** About 3% of people of the world population are living with chronic hepatitis C. High prevalence has been documented from Africa (10%), South America and Asia. In India prevalence is 1%. Prevalence is detected by detecting HCV-RNA.
- **Genotypes:** There are seven major genotypes of HCV from 1 to 7. The genotypes are further divided into >100 subtypes like 1a and1b for genotype 1. They are varying in drug susceptibility (1b has poor response to treatment than others), duration of treatment and in distribution as mentioned below.
  - In the United States about 70% of cases are caused by genotype 1, 20% by genotype 2 and about 1% by each of the other genotypes
  - Genotype 1: World wide
  - Genotype 4: Egypt
  - Genotype 5: South Africa
  - Genotype 6: Hong Kong
  - Genotype 1 and 3: India

Reservoirs of infection: Only human and no animal reservoirs.

Sources of infection: These are blood/blood products and body fluids.

Modes of transmission: Two modes like

1. **Direct transmission**
   - **Direct contact:** It is transmitted by contact of contaminated devices like scalpels, scissors, endoscopes etc., with abraded skin or during medical procedures.
   - **Inoculation** under skin or mucosa by contaminated needles, syringes in drug abusers or by tattooing transmits HCV. Percutaneous injury transmits HBV (**30%**) > HCV (**3%**) > HIV (**0.5–1%**).
   - **Sexual intercourse:** But less important.
   - **Perinatal (vertical):** This route transmits HIV (**30%**) > HBV (**20%**) > HCV (**6%**) > HDV(±) > HAV and HEV (–). It depends on degree of viremia which is detected by HCV RNA. It is not transmitted by breast feeding.
2. **Indirect transmission:** It occurs by transfusion of contaminated blood or blood products or by organ transplantation (**vehicle borne**).

Incubation period (IP): 15–160 days (½–5½ months).

Portal of entry: Skin or mucosa.

Sites: Liver.

Precipitating factors (epidemiological determinants): **(1) Age:** It is common in adults due to drug abuse or promiscuous sexual behavior. **(2) High-risk groups:** These are drug addicts, person required multiple blood transfusion or organ transplantation, HCWs, sex workers, homosexuals and barbers. **(3) Persons with IDDs:** Are at high risk.

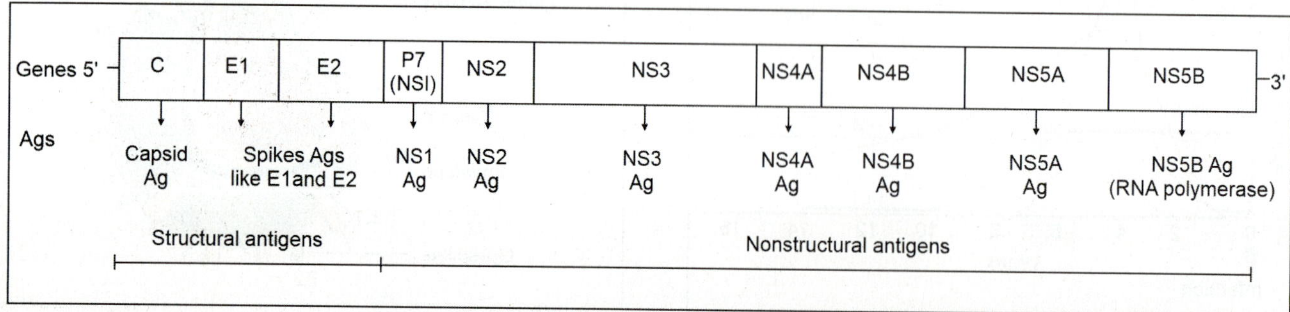

Fig. 88.8: Genetic organization and antigenic structure of HCV

Clinical features (C/Fs): In 20% patients it present as acute hepatitis with following features.

- Asymptomatic or anicteric: >75% cases are asymptomatic.
- Acute hepatitis: It occurs in around 20% patients. Symptoms are generally mild and vague includes decreased appetite, fatigue, nausea, muscle/joint pains and weight loss. Overt jaundice seen in 5% cases.

Complications: It is the most common virus among the hepatitis group which moves to chronic stage in 50–80% cases (hence called **notorious virus**) with following complications.

- Chronic hepatitis (60–70%).
- Cirrhosis (5–20%): HCV is most common indicator for liver transplantation following cirrhosis.
- Hepatocellular carcinoma (HCC): It occurs in 1–5% cases and it accounts about 25% of total liver cancers. It is next to HBV to produce HCC. Coinfection of HCV with HBV increases the risk for cirrhosis and HCC.
- Extrahepatic: It includes glomerulonephritis, arthritis, joint pain, mixed cryoglobulinemia, etc. These are may be due to immune complex (type-III hypersensitivity) deposition.

Carrier: In India carrier rate is 1–20%.

Laboratory Diagnosis

Specimens: Blood/serum.

Testing methods

A. **Microscopy:** Follow morphology.

B. **Culture:** Follow C/Cs.

C. **Serological tests**

- **Antigen detection:** ELISA detects capsid Ag, NS3Ag, NS4Ag and NS5Ag. It is useful to detect the acute stage and to monitor the therapeutic response. Other marker to detect the acute stage is HCV RNA detected by qRT-PCR.
- **Antibody detection:** Anti-HCV antibodies (mostly IgG) develop after 2–3 months (**Fig. 88.9**). In acute infection, antibodies present variably in 50–70% cases and in chronic they present in 95% cases. 3rd generation ELISA detects antibodies to capsid Ag, NS3Ag, NS4Ag and NS5Ag. It cannot differentiate between acute and chronic infection for which IgG avidity ELISA is useful.
- **Simultaneous detection of Ag and Ab:** ELISA detects capsid-Ag and antibodies to NS3Ag and NS4Ag.

D. **Molecular methods:** HCV RNA can be detected by qRT-PCR within few days after exposure, even before Ab production and elevation of enzymes. Molecular methods are not reliable to know the disease progression or prognosis, but help in identifying the acute stage, identifying the genotype/subtype and predicting the drug response. qRT-PCR is considered as a gold standard.

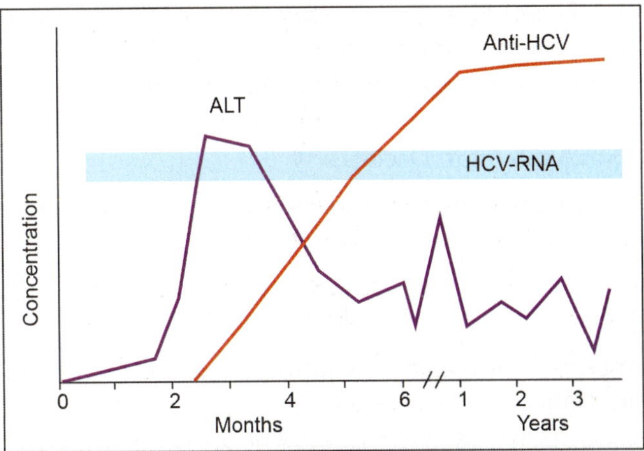

Fig. 88.9: Markers of HCV

E. **Liver biopsy:** It is a gold standard used in staging of hepatitis C. When biopsy is not available fibroscan/elastograpy is used to know the liver fibrosis.

F. **Testing sequence for HCV:** CDC recommended 2 steps testing as mentioned in **Flowchart 88.4**.

Prevention

General measures: These are same as HBV.

HCV screening

- Area with prevalence >0.1%: All individual with >18 years and all pregnant women in each pregnancy should go for HCV testing. Prevalence is detected by detecting HCV-RNA.
- High risk groups (mentioned above) regardless of prevalence: All should go for HCV testing.

Immunoprophylaxis: No specific active or passive prophylaxes are available.

National Viral Hepatitis Control Program (NVHCP): Described with HBV.

Treatment

Goal: HCV induces chronic infection in 50–80% of infected persons. Main goal of treatment is to arrest the development of chronic stage. In rare cases, infection can clear without treatment. Approximately 40–80% of cases required treatment.

Effective factors over treatment: Treatment depends on type of genotype and presence of cirrhosis.

Flowchart 88.4: Testing sequence for HCV

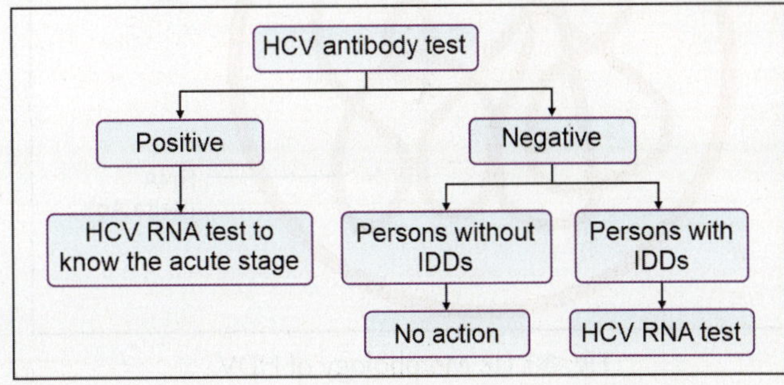

Directly acting antiviral drugs: Following drugs can clear about 90% of viral load.

- NS3/4A protease inhibitor: Grazoprevir.
- NS5B polymerase inhibitor: Dasabuvir, and sofosbuvir
- NS5A inhibitor: Daclatasvir.

Pegylated IFN and ribavirin: They can clear moderate degree of viral load and have high side effects.

Regimen: It includes directly acting antiviral drugs, pegylated IFN and ribavirin for 12–24 weeks.

Surgery: It is the most common indicator for liver transplant.

Prognosis: It is poor in genotype 4 and 1b>1a, presence of cirrhosis, coinfection with HBV-HIV, high viral load (>800000 IU/ml) and absence of IL28B. People contains IL28B (called CC genotype) induces strong Ir to HCV due to production of IFN-α by it.

HDV

Introduction

The delta agent (HDV) is a defective virus and always required helper virus like HBV, so it is dependent virus. It shows similarities with plants viruses like viroids or satellite viruses.

History

HDV was identified in Italy in 1977 by Rizzetto and colleagues as nuclear-Ag in nuclei of liver cells of patient infected with HBV. Nuclear Ag was initially considered as HBV Ag and called **the delta antigen**. Subsequent experiments in chimpanzees showed that the delta Ag was a separate pathogen that required HBV infection to replicate. The entire genome was cloned and sequenced in 1986. It was subsequently placed in its own genus *Deltavirus*.

Morphology

Shape and size: It is spherical (**Fig. 88.10**) in shape and 36 nm in size.

Genome and envelope: It has linear, ss-RNA (−) surrounded by HBsAg (envelope protein) of HBV.

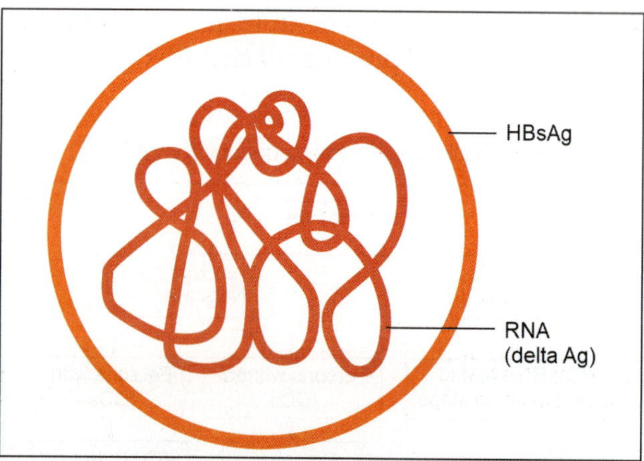

Fig. 88.10: Morphology of HDV

IF microscopy: Ag of HDV is detected from nuclei of liver cells by IF test.

Culture Characteristics (C/Cs)

Woodchuck (animal) is an ideal model for study of HDV infection.

Pathogenicity

Disease name: Disease called **hepatitis D** or called **hepatitis delta.**

Epidemiology

- **Distribution:** Globally 5% of HBV patients are co-infected with HDV. It is not prevalent in India in spite of maximum HBV carrier. In **endemic countries** like Mediterranean it is common in persons with close personal contact like sexual route. In **non-endemic countries** like North Europe or North America it is common in, who exposed with blood/blood products, drug abuses and hemophiliacs infected with HBV. It may cause **outbreak** in persons infected with HBV with high mortality.
- **Genotypes:** At least 8 genotypes of HDV are identified from 1 to 8. Phylogenetic studies suggest an African origin for this pathogen.

Reservoirs of infection: Only human and no animal reservoirs.

Sources of infection: Blood/blood products and body fluids.

Modes of transmission

1. **Percutaneous exposures:** Injecting drug, blood or by blood products.
2. **Permucosal exposures:** Sex contact.
3. **Vertical:** From mother to child.

Incubation period (IP): 1–6 months.

Portal of entry: Skin or mucosa.

Sites: Liver.

Precipitating factors (epidemiological determinants): Occupational factors are same as HBV described above.

Clinical features: Two types of infection according to transmission as mentioned in **Table 88.3.**

Laboratory Diagnosis

Specimens: Blood/serum and liver biopsy.

Testing methods

A. **Microscopy:** Follow morphology.
B. **Animal culture:** Follow C/Cs.
C. **Serological tests:** Follow **Table 88.3.**
D. **Molecular methods:** Follow **Table 88.3.**

Prevention

General measures: Same as HBV.

Immunoprophylaxis: HBV vaccine provides protection against HDV.

<table>
<tr><td colspan="3">TABLE 88.3: Differences between coinfection
and superinfection of HDV</td></tr>
</table>

Features	Coinfection	Superinfection
Definition	HDV and HBV transmitted simultaneously	Infection by HDV in a person already infected with HBV or carrier
Status	Healthy	Infected or carrier of HBV
Clinical pictures	Transient and self limited. Clinically indistinguishable from hepatitis B	It deteriorates the underlying HBV infection. **(1) Acute stage:** HDV replicates actively with high transaminase levels with suppression of HBV. **(2) Chronic stage:** HDV replication decrease, fluctuating transaminase levels with high HBV. Disease progresses to chronic hepatitis, cirrhosis and HCC in 5–20% cases. Mortality is >20%
Risk of development of		
Fulminant disease	> HBV alone	> Coinfection
Chronic hepatitis	Rare	High
Cirrhosis	Rare	More
HCC	Rare	More
Mortality	Rare	More
Diagnosis		
Ags/Abs	Anti-HDV IgM develops, 2–3 weeks after infection, immediately replaced by IgG. IgM also present in chronic stage. Anti-HBc IgM and anti-HDV IgM are elevated but anti-HDV IgM appears late and frequently short lived. IgM detected by ELISA	HBV is already present so, HBV markers already available are HBsAg, HBeAg and Anti-HBc (IgG). HDV markers are Anti-HDV IgM and IgG
Best diagnosis	Anti-HBc IgM + anti-HDV IgM	Anti-HBc IgG + mixture of anti-HDV IgM and IgG
Molecular method	HDV RNA is detectable from blood and liver in acute stage in both types of infections	

Treatment

Pegylated IFN reduces the viral load. More than 1 year of therapy may be necessary. Liver transplantation may be considered for cases of fulminant hepatitis and end-stage liver disease. Continue the same treatment of HBV.

HEV

Synonym

- Enterically transmitted NANB hepatitis virus.
- E-NANB hepatitis virus.
- Epidemic-NANB hepatitis virus (because causing epidemic).

History

Infection HEV was 1st documented in 1955 during an outbreak in New Delhi, India. HEV was discovered in 1990.

Morphology

Shape: It is spherical in shape (**Fig. 88.11**) with 32–34 nm in size. Virus has 32 cup shape depressions on capsid (surface), hence it was initially classified under the genus *Calcivirus* of family Calciviridae. **Calci** derived from **calyx** means **cup**. Now it is classified under the genus *Hepevirus* of family Hepeviridae.

Genome and capsid: It has ss-RNA (+) surrounded by icosahedral variety of capsid.

Envelope: It is non enveloped (naked) virus.

EM: Virus is detected from stool by EM.

Culture Characteristics (C/Cs)

Experimentally, it can be transmitted in primates. *In vitro* cultivation has not been successful so far. Virus genome has been cloned.

Resistance

It is naked virus resistant to lipid solvents.

Antigenic Structure

RNA antigen called **HEV-Ag.** It has only one serotype.

Immunity

Secondary attack rate is very low in house hold contact about 2–3% compare to 10–20% in HAV.

Pathogenicity

Disease name: Disease is zoonotic called **hepatitis E** or called **enterically transmitted NANB hepatitis** or **E-NANB hepatitis** or **Epidemic-NANB hepatitis virus.**

Epidemiology

- **Distribution:** HEV is the most common cause of viral hepatitis in India, Africa, and Central America. It also causes epidemic or sporadic cases. Two patterns are observed as shown in **Table 88.4**.

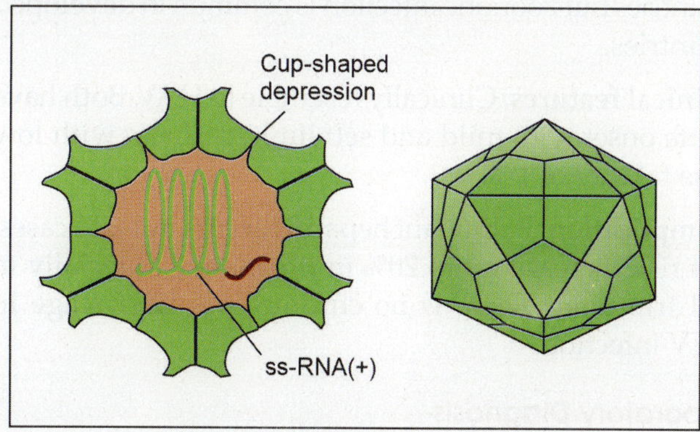

Fig. 88.11: Morphology of HEV

TABLE 88.4: Pattern of HEV		
Features	**Areas with poor water supply**	**Areas with safe water supply**
Reservoirs of infection	Humans	Animals like monkeys, cats, dogs, pigs, deer, etc.
Sources of infection	Water/food	Animal tissues (like pork in UK)
Modes of transmission	By ingestion of water/food	By ingestion of raw tissues of animal (liver)
Genotype(s)	1 and 2	3 and 4
Type of disease	Sporadic/epidemic	Sporadic
Countries	Developing	Developed

- **Genotypes:** Four genotypes are identified from 1 (5 subtypes), 2 (2 subtypes), 3 (10 subtypes) and 4 (7 subtypes) based on the nucleotide sequences. Genotypes 1 and 2 are restricted to humans and often associated with large outbreaks and epidemics in developing countries with poor sanitation. Genotypes 3 and 4 infect humans, pigs and other animal species and have been responsible for sporadic cases of HEV in both developing and industrial countries.

Reservoirs of infection: Follow **Table 88.4**.

Sources of infection: Follow **Table 88.4**. Secondary attack rate (transmission rate from patient to case) is 1–2%.

Modes of transmission: Follow **Table 88.4**.

Incubation period (IP): 15–60 days (½–2 months).

Portal of entry: GIT.

Sites: Liver.

Precipitating factors (epidemiological determinants): **(1) Age:** Common in young adults (15–40 years) unlike HAV which is common in children. **(2) Pregnancy:** It is more sever (fulminant hepatitis) and high fatality rate of 20–40% occurs in pregnancy especially in 3rd trimester. **(3) Immunocompromised hosts:** It develops chronic hepatitis with high mortality in persons with organ transplantation and immunosuppressive drugs. **(4) Seasonal factors:** Outbreaks most commonly occur after heavy rainfall and monsoon because of disruption of water supplies. **(5) Poor hygiene and poor sanitation:** These are the most favoring factors. **(6) Economical status:** It is common in developing countries but zoonotic infection is common in developed countries.

Clinical features: Clinically resemble to HAV. Both have acute onset with mild and self limited illness with low case fatality of 1%.

Complication: Fulminant hepatitis occurs in 1–2% cases, but risk increase up to 20% in pregnancy especially in 3rd trimester. There is no chronic or carrier stage in HEV infection.

Laboratory Diagnosis

Specimens: Feces, bile and serum.

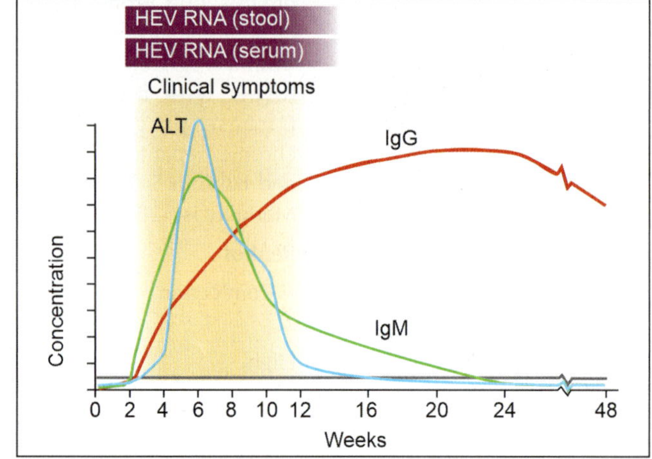

Fig. 88.12: Markers of HEV

Testing methods

A. **Microscopy:** Follow morphology.

B. **Culture:** Follow C/Cs.

C. **Serological tests (Fig. 88.12):** IgM appears early with ALT and indicates the acute infection. IgG replaces IgM after 2–4 weeks and lasts for years indicates the past infection. IgM and IgG Abs are diagnosed by ELISA.

D. **Molecular methods:** HEV RNA is detected by qRT-PCR from stool.

Prevention

General measures: Improvement of quality public water supplies and establishing proper disposal systems for human feces.

Immunoprophylaxis: China prepared the recombinant vaccine by using **HEV239** strain, but it is not available globally.

Treatment

It is self-limiting disease, hospitalization and treatment is generally not required. Ribavirin is given to people with fulminant hepatitis, symptomatic pregnant women and immunosuppressed people. In some specific situations, interferon has also been used successfully.

DIFFERENCES BETWEEN HEPATITIS VIRUSES

Follow **Table 88.5**.

TABLE 88.5: Differences between hepatitis viruses

Features	A	B	C	D	E
Synonym	Infective hepatitis	Serum hepatitis	NANB hepatitis	Delta agent	• NANB enterically transmitted hepatitis • E-NANB hepatitis virus • Epidemic-NANB hepatitis virus
Family	Picornaviridae	Hepadnaviridae	Flaviviridae	Unclassified like viroid	Hepeviridae
Genus	*Heparnavirus (Hepatovirus)*	*Orthohepadnavirus*	*Hepacivirus*	*Deltavirus*	*Hepevirus*
Type	RNA	DNA	RNA	RNA	RNA
Size	27 nm	42 nm	50–60 nm	36 nm	32–34 nm
Envelope	No	Yes	Yes	Yes	No
Mode of transmission	Feco-oral	Percutaneous, permucosal	Percutaneous, permucosal	Percutaneous, permucosal	Feco-oral
Incubation period	2–6 weeks (½–1½ months)	1–6 months	15–160 days (½–5½ months)	1–6 months	15–60 days (½–2 months)
Fulminant hepatitis	Rare (0.1%)	Rare (0.1–1%)	Rare (0.1%)	More (5–20%)	Rare (1–2%) but more in pregnancy (20%)
Chronic infection	No	Yes (1–10%)	Yes (50–80%/notorious virus)	Yes (5–20%)	No
Malignancy	No	Yes	Yes	No	No
Carrier	Nil	Yes (variable rate)	Yes (1–20%)	No	No
Prophylaxis	Ig and inactivated vaccine	Ig and recombinant vaccine	None	HBV vaccine	HEV239 vaccine in China
Prognosis	Excellent	Worse with age	Moderate	Acute–good Chronic–poor	Good

ACCESS YOURSELF

Case Study

1. A 35-year-old male, who received multiple transfusions of blood, admitted in hospital. He developed nausea, vomiting, loss of appetite, abdominal pain and passing yellowish dark colored urine after 6 months of transfusion. On clinical examination yellowish sclera was noticed. Blood is collected and sent for serological testing. Laboratory reported, positive test for surface antigen for one of the hepatitis virus. Identify the case and answer the following.
   a. Name the disease with clinical stage and its etiological agent.
   b. Write the morphology and genetic structure of etiological agent.
   c. Write pathogenicity and diagnosis of clinical condition.
   d. How you can prevent this clinical condition?

Essays/Full Questions

1. Hepatitis viruses.
2. Outline a diagnostic approach to liver disease based on hepatitis serology.

Short Notes

1. Pathogenicity/lab., diagnosis of HBV (Australian antigen)
2. HBV vaccine or discuss the prevention of hepatitis B infection.
3. Post-exposure prophylaxis of HBV.
4. HAV/HCV/HDV/HEV.

Short Questions for Theory/Viva Questions

1. Mention the immunological markers of HBV present in serum.
2. Write different antigenic types of HBV.
3. Write different modes of entry of HBV.
4. Mention the markers to detect active HBV replication.

Comments on

1. HBcAg of HBV is not detectable in blood.
2. HAV have no or very rare blood/serum transmission.
3. HCV called notorious virus among the all hepatitis group.

MCQs for Chapter Review

Classification

1. **Nonparenteral hepatitis is:**
   a. Hepatitis A   b. Hepatitis B
   c. Hepatitis C   d. Hepatitis D
2. **Hepatitis C virus resembles to which of the following virus group:**
   a. Picorna viruses   b. Herpes viruses
   c. Hepadna viruses   d. Flavi viruses

HBV

3. **All of the following are components of Dane particle, *except*:**
   a. Surface antigen   b. Core antigen
   c. C-antigen   d. Delta antigen
4. **Which of the following is not present in Dane particle?**
   a. Core antigen   b. Surface antigen
   c. p53   d. None of above
5. **Australian antigen is associated with:**
   a. HBsAg   b. HBcAg
   c. HBeAg   d. None of above

6. **Presence of x antigen in the serum of patient with hepatitis B virus infection is suggestive of:**
   a. Acute stage of hepatitis B infection
   b. Convalescent stage of hepatitis B infection
   c. Presence of protective immunity
   d. Infection going into chronicity

7. **The pre-core mutants in hepatitis B are characterized by notable absence of:**
   a. HBV DNA
   b. HBeAg
   c. HBcAg
   d. Anti HBeAg

8. **HBV presents in India is:**
   a. adw
   b. ayw
   c. adr
   d. ayr

9. **Reverse transcriptase of HBV is coded on following gene:**
   a. C gene
   b. S gene
   c. P gene
   d. X gene

10. **Hepatitis B is not transmitted by:**
    a. Blood transfusion
    b. Pasteurized albumin
    c. Cryoprecipitate
    d. Sexual contact

11. **HBV, all are true *except*:**
    a. DNA virus
    b. Spreads by blood transfusion
    c. HBsAg marker of infection
    d. Least chances of chronicity

12. **Presence of HBeAg in patients with hepatitis indicates:**
    a. Simple carrier
    b. Late convalescence
    c. High infectivity
    d. Carrier status

13. **HBV is associated with all of the following, *except*:**
    a. Hepatic cancer
    b. Chronic hepatitis
    c. Hepatic adenoma
    d. Cirrhosis

14. **Incubation period of HBV is:**
    a. 45–180 days
    b. 6–60 days
    c. 10 days
    d. 10 hours

15. **Hepatitis B vaccination is given to a patient. His serum will reveal:**
    a. HBsAg
    b. Anti-HBsAg
    c. IgM anti-HBcAg and HBsAg
    d. IgM and IgG anti-HBcAG

16. **Which of the following antigen is found within the nuclei of infected hepatocytes and not usually in the peripheral circulation in hepatitis B infection?**
    a. HBeAg
    b. HBcAG
    c. Anti-HBc
    d. HBsAg

17. **Serological testing of patients shows HBsAg, anti-HBc and HBeAg positive. The patient has:**
    a. Chronic hepatitis B with low infectivity
    b. Acute hepatitis B with high infectivity
    c. Chronic hepatitis B with high infectivity
    d. Acute on chronic hepatitis

18. **Acute hepatitis B can be earliest diagnosed by:**
    a. IgM anti-HBc Ab
    b. HBsAg
    c. IgG anti-HBc Ab
    d. Anti-HBsAg Ab

19. **In a patient of active chronic hepatitis B all are seen, *except*:**
    a. HBsAg
    b. IgM anti-HBcAb
    c. HBeAg
    d. Anti-HBsAg

20. **In a patient only anti HBsAg is positive in serum, all other markers are negative. This indicates:**
    a. Acute hepatitis
    b. Chronic active hepatitis
    c. Persistent carrier
    d. Hepatitis B vaccination

21. **First antibody to appears in HBV:**
    a. IgM anti-HBe
    b. IgG anti-HBe
    c. IgM anti-HBc
    d. IgM anti-HBs

22. **The best diagnostic test for recent hepatitis B is:**
    a. HBsAg
    b. IgM anti-HBc
    c. Anti-HBe
    d. Anti-HBs

23. **Best epidemiological tool for investigation of hepatitis B is:**
    a. Anti-HBsAg
    b. Anti-HBcAg
    c. Anti-HBeAg
    d. HBcAg

24. **Active replication in hepatitis B infection is indicated by?**
    a. HBeAg
    b. HBsAg
    c. Intra-hepatic HBcAg
    d. Anti-HBsAg

25. **Infectivity of HBsAg is best/commonly diagnosed by:**
    a. HBeAg
    b. HBsAg
    c. HBV DNA
    d. Anti-HBsAg

26. **Which of this is not a marker of active replicative phase of chronic hepatitis B?**
    a. HBV DNA
    b. HBV DNA polymerase
    c. Anti HBc
    d. AST and ALT

27. **Which of the following viral marker(s) is/are positive in simple carrier of Hepatitis B?**
    a. HBsAg
    b. HBeAg
    c. Both a and b
    d. None of the above

28. **Best means of giving hepatitis B vaccine is:**
    a. Subcutaneous
    b. Intradermal
    c. Intramuscular deltoid
    d. Intramuscular gluteal

29. **Hepatitis B virus vaccine contains:**
    a. HBsAg
    b. HBc
    c. HBe
    d. All of above

30. **A mother is HBsAg positive at 32 weeks of pregnancy. What should be given to newborn to prevent the neonatal infection?**
    a. Hepatitis B vaccine + immunoglobulin
    b. Immunoglobulin only
    c. Hepatitis B vaccine only
    d. Immunoglobulin followed by vaccine 1 month later.

HAV

31. **Transmission of hepatitis A virus occurs:**
    a. One week before and one week after the onset of symptoms
    b. 2 weeks before the onset of symptoms
    c. 2 weeks after the onset of symptoms
    d. 1 week after the onset of symptoms

32. **Hepatitis A is transmitted by:**
    a. Blood route
    b. Inhalation
    c. Feco-oral route
    d. All

33. **Age group affected by hepatitis A virus:**
    a. Children
    b. Adult
    c. Old age
    d. Any age

34. **About hepatitis A true is:**
    a. Causes mild illness in children
    b. 3% incidence of carrier state
    c. 10% transmission to HCC
    d. Vertical transmission never seen

35. **True about hepatitis A virus:**
    a. Causes cirrhosis
    b. Helps HDV replication
    c. Common cause of hepatitis in children
    d. Causes chronic hepatitis

36. **Hepatitis A virus is best diagnosed by:**
   a. IgM antibodies in serum
   b. Isolation from stool
   c. Culture from blood
   d. Isolation from bile

HCV

37. **HCV is:**
   a. Enveloped RNA virus
   b. Non-enveloped RNA virus
   c. Non-enveloped positive strand RNA virus
   d. DNA virus

38. **Maximum H transmission to fetus in pregnancy depends on:**
   a. Duration of illness   b. Time of infection
   c. Route of delivery   d. HIV infection
   e. High level of HCV infection

39. **Hepatitis C virus, true finding is:**
   a. Spreads by feco-oral route
   b. Antibody to HCV may not be seen in acute stage
   c. Does not cause chronic hepatitis
   d. It cannot be cultured.

40. **Which of the following statement about hepatitis C is true?**
   a. DNA virus
   b. Most common indication for liver transplant
   c. Does not cause liver cancer
   d. Does not cause coinfection with hepatitis B

41. **True about HCV includes all, *except*:**
   a. Highest rate of chronicity among all hepatitis viruses
   b. Can be cultured
   c. Diagnosed by detection of HCV RNA
   d. Transmitted through infected food

42. **HCV is associated with:**
   a. Anti-LKM-1 antibody b. Scleroderma
   c. Cryoglobulinemia   d. Polyarteritis

HDV

43. **HBV and HDV false is:**
   a. Both can infect simultaneously.
   b. HDV causes more serious infection due to superinfection.
   c. HDV cannot infect in absence of HBV.
   d. DNA viruses

44. **HDV is:**
   a. ss-RNA virus   b. ss-DNA virus
   c. ds-RNA virus   d. ds-DNA virus

HEV

45. **Most common route of spread of hepatitis E is:**
   a. Sex   b. Feco-oral route
   c. Blood transfusion   d. IV injections

46. **During epidemic of hepatitis E fatality is maximum in:**
   a. Pregnant woman   b. Infants
   c. Malnourished male   d. Adolscent

47. **Regarding hepatitis E, true is:**
   a. Occurs with hepatitis B
   b. Single stranded DNA virus
   c. Occurs along with HIV
   d. Mortality increased in pregnancy

Mixed MCQs for All Hepatitis Viruses

48. **Which of the following hepatitis virus is a DNA virus?**
   a. Hepatitis C virus   b. Hepatitis B virus
   c. Delta agent   d. Hepatitis E virus

49. **Which hepatitis virus had been called as enterovirus?**
   a. HAV   b. HBV
   c. HCV   d. HEV

50. **Which of the following hepatitis virus has significant perinatal transmission?**
   a. HEV   b. HCV
   c. HBV   d. HAV

51. **Which of the following hepatitis has a poor prognosis?**
   a. HAV   b. HBV
   c. NANB   d. HCV

52. **Cultivable (*in vitro*) hepatitis is:**
   a. Hepatitis A   b. Hepatitis B
   c. Hepatitis C   d. Hepatitis D

53. **Carrier state does not exist for:**
   a. Hepatitis B virus
   b. Hepatitis A virus
   c. Non A Non B hepatitis
   d. Delta agent

54. **Potent (prophylactic) vaccine is available for:**
   a. Hepatitis C   b. Hepatitis A
   c. Hepatitis D   d. Hepatitis E
   e. Hepatitis B

55. **Which hepatitis virus is notorious for causing chronic hepatitis evolving to cirrhosis?**
   a. HEV   b. HAV
   c. HBV   d. HCV

56. **Next to HBV, virus implicated in hepatocellular carcinoma is:**
   a. HCV   b. Herpes
   c. HAV   d. HEV

57. **Chronic liver disease is caused by:**
   a. Hepatitis B   b. Hepatitis A
   c. Hepatitis C   d. Hepatitis E

58. **The commonest cause of viral hepatitis in India:**
   a. Hepatitis A virus
   b. Enterically transmitted NANB hepatitis
   c. Hepatitis C virus
   d. Hepatitis B virus

59. **All the statements are correct about hepatitis viruses, *except*:**
   a. Maximum chances of chronic infection is HCV.
   b. Pregnant woman with HEV has 10–20% chances of mortality.
   c. Vaccine is available only against HBV.
   d. HBV and HCV has oncogenic potential.

60. **All of the following hepatitis viruses can be transmitted through blood *except*:**
   a. Hepatitis B   b. Hepatitis C
   c. Hepatitis D   d. Hepatitis E

61. **Which of the following is not matched correctly?**
   a. Hepatitis D – Defective virus
   b. Hepatitis C – Parenteral transmission
   c. Hepatitis B – RNA virus
   d. Hepatitis E – Feco-oral transmission

62. **Which of the following is a defective/dependent virus?**
   a. HAV   b. HBV
   c. HCV   d. HDV

63. **Which is ss-RNA un-enveloped virus?**
   a. HBV   b. HEV
   c. HCV   d. None

64. **With which of the following viral hepatitis infection in pregnancy, the maternal mortality is highest?**
   a. Hepatitis A   b. Hepatitis B
   c. Hepatitis C   d. Hepatitis E

Infections of Hepatitis Viruses

65. Hepatitis E clinically resembles:
   a. Hepatitis A
   b. Hepatitis B
   c. Hepatitis C
   d. Hepatitis D

Answers and Explanation of MCQs

1. a
2. d
- Follow section, **classification and Flowchart 88.1** for explanation of answers of MCQs 1–2.
3. d
4. c
5. a
6. d
7. d
8. b, c
9. c
10. c
11. d
12. c
13. c
14. a
15. b
16. b
17. b
18. a, b
19. b
20. d
21. c
22. b
23. b
24. c
25. a
26. c
27. a
28. c
29. a
30. a
- Follow section, **HBV** for explanation of answers of MCQs 3–30.
31. b, c
32. c
33. a
34. a, d
35. c
36. a
- Follow section, **HAV** for explanation of answers of MCQs 31–36.
37. a
38. e
39. b, d
40. b
41. b, d
- Follow section, **HCV** for explanation of answers of MCQs 37–41.
42. a and c
- Anti-LKM Ab is anti-liver kidney microsomal Ab. It is developed due to homology between viral particles and host tissues (cytochrome P450 system). Anti-LKM Ab has three types. (1) Anti-LKM Ab: HCV. (2) Anti-LKM Ab: Drug (tienilic acid) induced hepatitis. (3) Anti-LKM Ab: HDV and autoimmune polyendocrine syndrome.

43. d
44. a
- Follow section, **HDV** for explanation of answers of MCQs 43–44.
45. b
46. a
47. d
- Follow section, **HEV** for explanation of answers of MCQs 45–47.
48. b
49. a
- Follow **Flowchart 88.1** for explanation of answers of MCQs 48–49.
50. c
- Among the all hepatitis, HBV has maximum perinatal transmission. Risk includes HBV (+) > HCV and HDV (±) > HAV and HEV (–).
51. b
- HBV has poor prognosis with age, while HCV has moderate prognosis.
52. a
- HAV is the only hepatitis virus, which can be cultivated in cell culture.
53. b
- HAV and HEV are not showing carrier stage. Carrier stage presents in HBV, HCV and HDV.
54. b, e
- Among the hepatitis viruses potent vaccine available for HAV and HBV.
55. d
- HCV is the most common virus among the hepatitis group, which moves to chronic stage in 50–80% cases, hence called **notorious virus**.
56. a
- HCC occurs in 1–5% cases of HCV and it accounts about 25% of total liver patients. It is next to HBV to produce HCC. Co-infection of HCV with HBV increases the risk for cirrhosis and HCC.
57. a, c
- Chronicity of hepatitis viruses: HCV (50–80%, most notorious) > HDV (5–20%) > HBV (1–10%) > HAV = HEV (Almost none).
58. b
- HEV is the most common cause of viral hepatitis in India, Africa, and Central America.
59. c
- Potent vaccine is available against HBV and HAV.
60. d
- HEV transmitted by feco-oral route.
61. c
- HBV is DNA virus
62. d
- The HDV is a defective virus and always required helper virus like HBV, so it is dependent virus.
63. b
- HEV is nonenveloped (naked) virus, while HBV and HCV are enveloped viruses.
64. d
- Fulminant hepatitis occurs in 1–2% cases of HEV, but risk increase up to 20% in pregnancy especially in 3rd trimester.
65. a
- HEV and HAV have many similarities like mode of transmission, no chronic or carrier stage, acute onset with mild and self limited illness with low case fatality of 1%.

CHAPTER 89

Infections of Other Group of Viruses

Chapter Outline

- Hemorrhagic Fever Viruses
- Diarrheal Viruses
- Slow Disease Viruses
- Encephalitis Viruses
- UTI Causing Viruses

HEMORRHAGIC FEVER VIRUSES

These are taxonomically unrelated viruses causing vascular damage resulting symptomatic bleeding or hemorrhage. They are classified as follows.

Systemic/taxonomical classification

1. **DNA viruses**
   - **Poxviridae:** Pox viruses causing humna lesion like variola virus.
   - **Herpesviridae:** Like VZV.
2. **RNA viruses**
   - **Paramyxoviridae:** Measles virus.
   - **Togaviridae:** Chikungunya virus.
   - **Flaviridae:** Two groups like mosquito-borne and tick-borne as follows.
     - Mosquito-borne species: Yellow fever virus, dengue virus and zika virus.
     - Tick-borne species: KFD virus and Omsk hemorrhagic fever virus.
   - **Bunyaviridae**
     - *Nairovirus:* CCHF virus.
     - *Phlebovirus:* Rift Valley fever virus.
     - *Hantavirus:* Hantaan virus and puumala virus.
   - **Arenaviridae:** Lassa fever virus and all new world viruses are hemorrhagic viruses.
   - **Filoviridae:** Described below.

Clinical classification: It based on many features like clinical presentation, modes of transmission, etc.

1. **Exanthematous viruses with hemorrhagic property:** Smallpox, chickenpox, measles (all are described in respective chapters).
2. **Arboviruses**
   - Mosquitoe-borne: Yellow fever virus, dengue virus and chikungunya virus.
   - Tick-borne: KFD virus, Omsk hemorrhagic fever virus and CCHF virus.

3. **Roboviruses**
   - Hemorrhagic fever renal syndrome: Hantaan virus.
   - Hemorrhagic fever: Lassa fever virus and all new world viruses are the hemorrhagic species.
4. **Others:** Like Filoviridae (described below).

Filoviridae

Meaning

- **Name of family:** It is given from the word filum means thread, because thread like shape of virus.
- **Name of genera: (1)** *Ebolavirus*: Genus name is came from the Ebola river, because the 1st case was identified from the village Yambuku near to Ebola river in Zaire (now the Democratic Republic of the Congo) in 1976. **(2)** *Marburgvirus*: Genus name is derived from the Marburg, Germany, where the 1st infection was occurred in laboratory workers, who were working with tissue of African green monkeys (*Chlorocebus aethiops*).

Morphology

- **Shape and size:** It is elongated as like thread in shape **(Fig. 89.1)**, seen under EM. It is 80 nm in breath and 800–1000 (sometimes up to 14, 000) nm in length.

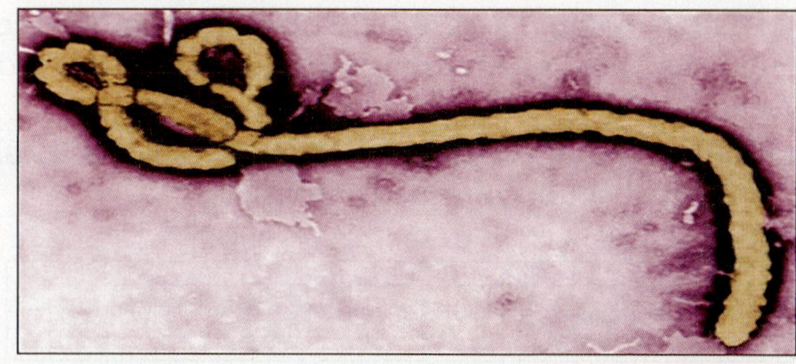

Fig. 89.1: Morphology of Filoviridae (*Ebolavirus*)

Flowchart 89.1: Classification of Filoviridae

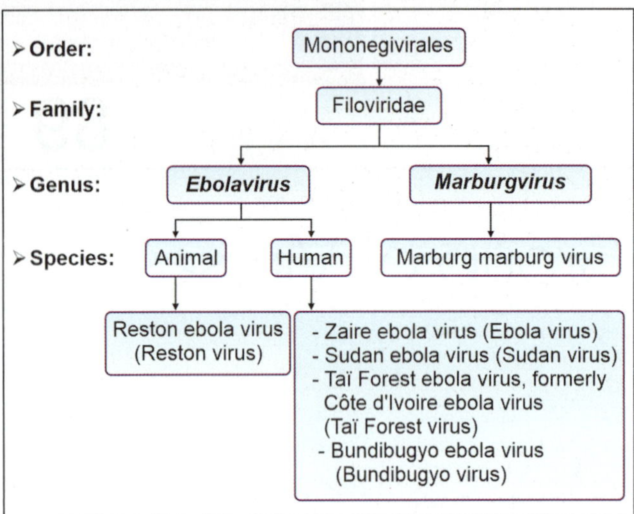

- **Genome and capsid:** It has ss-RNA (–) with helical variety of capsid.
- **Envelope:** Present.
- **Immunohistochemical staining and histopathology:** These are used to detect the viral antigens from tissues.

Classification: Follow **Flowchart 89.1**.

Zaire Ebola Virus

Common name: Ebola virus.

Abbreviation: ZEBOV or EBOV.

History: It was 1st isolated in 1976 from Yambuku, Zaire (now the Democratic Republic of the Congo).

Morphology: Described above (**Fig: 89.1**).

Antigenic structure: Virus has 7 structural Ags and one nonstructural Ags as shown in **Fig. 89.2**.

Culture characteristics (C/Cs): It is cultivated on Vero cells but culture method is not able to differentiate between the members of Filoviridae and also great risk of laboratory transmission (BSL-4).

Resistance: It is an enveloped virus susceptible to lipid solvents. It survives on dry objects for few hours and for a few days within body fluids outside of a person.

Immunity: People with ebola infection develop antibodies that last at least for 10 years.

Pathogenicity

- **Disease name:** Disease called **ebola virus disease** or called **ebola hemorrhagic fever** or simply called **ebola.**
- **Epidemiology:** It is distributed in African countries. India had not documented any case yet.

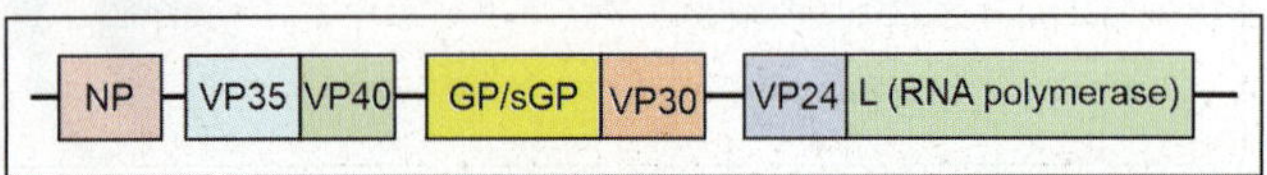

Fig. 89.2: Genomic structure of Ebola virus

- **Reservoirs of infection:** Fruit bats (*Hypsignathus monstrosus*, *Epomops franqueti* and *Myonycteris torquata*) are the main reservoir hosts. End hosts are humans and great apes.
- **Sources of infection:** These are body fluids (like blood, semen, breast milk, etc.) and tissues of infected animals or humans. Contaminated objects and water are also the sources of infection.
- **Modes of transmission:** Following are the possible cycles as shown in **Fig. 89.3**.
  1. Fruit bats to fruit bats.
  2. Fruit bats to animals: Bats drop the partially eaten fruits and pulp. Mammals such as gorillas, monkeys and duikers feed on these fallen fruits.
  3. Animals to animals: From one animal to other animal.
  4. Fruit bats to human: Exact route is unclear.
  5. Animals to human: Direct contact of abraded skin/mucosa with infected animals such as a gorillas or monkeys or their body fluid.
  6. Human to human
     - Direct contact of abraded skin/mucosa with blood and bodily fluids (including urine, saliva, feces, vomitus, breast milk and semen) from infected persons. It has not been reported to be transmitted through sweat.
     - Sexual transmission: Virus sheds in semen for >3 months to years after recovery, which leads the sexual infection.
     - Breast milk may be infectious in survivors for few months.
  7. Human infection from other sources
     - Contact with contaminated water.
     - Objects like clothes, bedding, needles, syringes/sharps or medical equipments that have been contaminated with the virus.
     - Airborne transmission has not been documented due to low levels of the virus in the lungs and other parts of the respiratory system in primates.

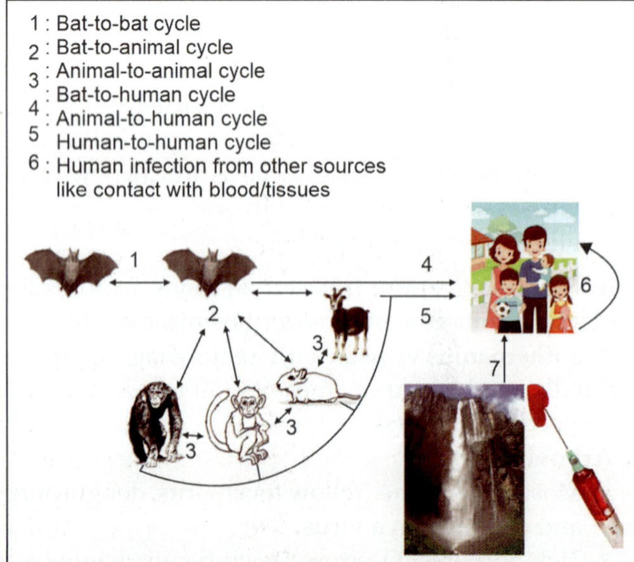

Fig. 89.3: Cycle of Ebola virus

- **Incubation period:** 2–21 days (average 8–10 days).
- **Portal of entry:** Skin or mucosa.
- **Sites:** Endothelial cells (cells lining the inside of blood vessels), liver cells, immune cells such as macrophages, monocytes, etc. and dendritic cells.
- **Precipitating factors (epidemiological determinants): (1) Occupation:** HCWs, forest workers and veterinarian are at high risk. **(2) Handler of dead bodies or animal tissues:** Because dead bodies are still infectious, the handling of the bodies of Ebola victims (during cremation or embalming) can only be done while observing proper barrier/separation procedures. **(3) Climate:** Ebola virus outbreak occurs when temperature is lower, and humidity is higher than the usual for Africa.
- **Pathogenesis:** Following entry in body, virus enters in endothelial cells lining the inside of blood vessels, liver cells, dendritic cells and immune cells such as macrophages, monocytes, etc. Breakdown of endothelial cells lead multiple bleeding points.
- **Clinical features:** It causes fever, severe headache, muscle pain, weakness, fatigue, diarrhea, vomiting, abdominal (stomach) pain and unexplained hemorrhage outside (bleeding) or inside the body (bruising) like from the gums, blood in the stools.
- **Complications:** High case fatality rate about 25–90% during outbreak. Normally, it is around 50%.

Diagnosis

- **Specimens:** Virus is diagnosed from blood and other body fluids of patients.
- **Testing methods**
  1. **Microscopy:** Follow morphology.
  2. **Culture:** Follow C/Cs.
  3. **Serological test**
     - **Antigen tests:** GP (glycoprotein), NP (nucleoprotein) and VP40 Ags are detected by capture ELISA.
     - **Antibody tests:** IgM and IgG Abs are developed against GP and NP Ags. IgM appears after 7 days of symptoms and lasts for 3–6 months. IgG appears after 2 weeks and lasts for 3–5 years. IgM and IgG are detected by ELISA (capture), IF test and antibody phage indicator assay.
  4. **Molecular method:** RT-PCR and qRT-PCR are used to detect the genes of NP and GP after 3 days of fever. They remain positive for 2–3 weeks.

Prevention

- **General measures**
  - Isolation of patients and avoid contact with relatives.
  - Avoid contact with bats and nonhuman primates or any secretion and raw meat prepared from these animals.
  - Avoid sharing objects used by patients like clothes, bedding, etc.
  - Avoid funeral or burial rituals that require handling the body of someone who has died from ebola.
  - Wearing personal protective equipments like gloves, etc.
  - Avoid to visit area with ebola outbreak.
- **Immunoprophylaxis:** No specific vaccine is available. STRIVE (**Si**erra leone **Tri**al to **I**ntroduce a **V**accine against **E**bola) prepared **r**ecombinant vaccine by using **V**esicular **S**tomatitis **V**irus and **Z**aire **eb**ola virus (rVSV-ZEBOV). It was used in 2018-19 outbreak of ebola in Democratic Republic of the Congo.

Note: Marburg marburg virus

Meaning: Described above.

Morphology: Same as other filoviruses as described above **(Fig. 89.1)**.

Pathogenicity: Disease called **marburg virus disease.** It is **transmitted** from person to person or by sexual route or as laboratory acquired infection. It causes hemorrhagic fever. Case fatality rate is 35%.

Diagnosis: Virus is diagnosed from tissue, blood, semen and anterior chamber of eye of patients by microscopical, culture, serological and molecular methods.

Treatment: No specific treatment is available. Supportive care like oral or intravenous fluids, blood products, immune therapies and drug therapies are given.

DIARRHEAL VIRUSES

These are taxonomically unrelated viruses causing gastroenteritis resulting diarrhea. They are classified as follows.

DNA viruses: It includes viruse of **Adenoviridae** family. Serotypes like 40 and 41 are producing diarrhea in children in summer season.

RNA viruses

1. **Picornaviridae:** *Enterovirus* like polio virus, coxsackie virus and ECHO virus.
2. **Orthomyxoviridae:** Influenza A/H5N1 and influenza A/H1N1.
3. **Retroviridae:** HIV-1 and 2.
4. **Coronaviridae:** SARS-CoV, MERS-CoV and SARS-CoV2.
5. **Reoviridae:** *Rotavirus* (rota virus A to rota virus H).
6. **Calciviridae:** *Norovirus* (norwalk virus) *Sapovirus* (sapporo virus).
7. **Astroviridae:** *Astrovirus* (human astro virus).

Reoviridae

Classification is given in **Flowchart 89.2.**

Rotavirus

Meaning: Word derived from rota (Latin) means wheel, because outer shell shows radiating spikes giving appearance like wheel.

History: Virus was identified by Ruth Bishop and colleagues in 1973, in Melbourne, Australia from

Flowchart 89.2: Classification of *Rotavirus*

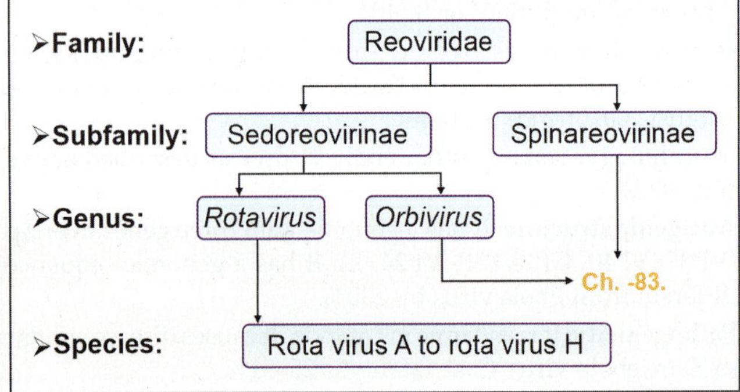

ultrathin section of biopsy under EM by using negative staining.

Classification: Total 8 species of *Rotavirus* (may be based on common group Ag present in inner shell) from rota virus A to rota virus H are identified on the bases of group specific VP6 antigen. Most common human species is A (about 90% of human infections are by species A) and other rare are by B and C. E and H infect pigs while D, F and G infect birds.

Morphology

- **Shape and size:** It is spherical in shape **(Fig. 89.4)** and about 60–80 nm in size.
- **Three types of viral particles in *Rotavirus*:** (1) Complete or double shelled particles are 70 nm, with smooth surface and radiating spikes giving wheel like appearance. (2) In-complete or single shelled particles are 60 nm, without outer shell and have rough surface. (3) Empty particles are without RNA core.
- **Genome:** It has ds-RNA (+/−) with 11 pieces in *Rotavirus* (10–12 pieces in other species of Reoviridae).
- **Capsid:** It has icosahedral type of capsid with three layers in *Rotavirus* (2 layers in other species of Reoviridae).
- **Nonenvelope:** Also called **naked virus.**
- **Immune Electron Microscopy (IEM):** It is expensive and required 10^6 particles/ml of stool for better result.

Antigenic structure, serotypes and genotypes

- **Antigens:** Structural antigens are VP1 (RNA polymerase), VP2 (inner capsid), VP3 (guanylyl transferase), VP4 (attachment protein), VP6 (middle capsid) and VP7 (outer capsid). Non-structural antigens are NSP1, NSP2, NSP3, NSP4, NSP5 and NSP6.
- **Serotypes and genotypes**
  - Within rota virus A there are different strains, called **serotypes**. Dual classification system is used which based on genetic reassortment of two proteins (as like influenza virus) on the surface of the virus. The glycoprotein VP7 defines the G serotype and the protease-sensitive protein VP4 defines P serotype.

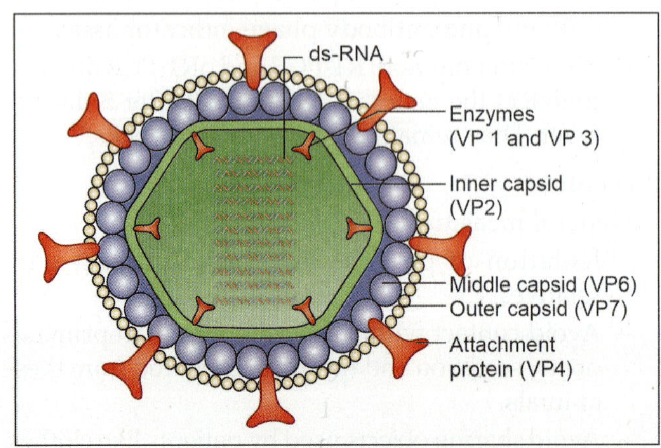

Fig. 89.4: Morphology of *Rotavirus*

– G serotype is further classified as G1, G2, G3 ... G19, while typing of P serotype is challenging. However genotyping of P is possible and labeled as P[1], P[2] P[3] .... P[28].

– Most widely used systems for expressing typing of rota virus is G serotype and P genotype based on neutralization test and RNA sequencing.

– Most common type presents in India and in world is G1P[8], which accounts about 70% of total isolates.

– Because the two genes that determine G-types and P-types can be passed separately to progeny viruses and different combinations (genetic reassortment) are found.

Culture characteristics (C/Cs): Human rota viruses are not growing on cell culture media except rota virus A. Growth of rota virus A is facilitated by adding the trypsin and rolling the tissue culture.

Resistance: Virus is stable in environment for long period. It is naked and resistant to lipid solvents.

Immunity: Children <6 months are protected by maternal Ab. Initial attack (clinical or subclinical infections) in child age provides immunity, so less in older children and adults. However, some strains produce diarrhea in older children and adults called **adult diarrhea rota virus (ADRV).**

Pathogenicity
- **Disease name:** Disease called **rota virus diarrhea.**
- **Epidemiology:** It is distributed worldwide. It produces the epidemic mostly in winter season.
  – Group A: It causes endemic diarrhea in infants and young children.
  – Group B: It causes outbreak of diarrhea in adults and children.
  – Group C: It causes sporadic and occasionally outbreak of diarrhea in children.
- **Reservoirs of infection:** Humans.
- **Source of infection:** The feces of an infected person is the source. Stool contains more than 10^{10} infectious particles per gram; fewer than 100 of these are required to transmit infection to another person.
- **Modes of transmission:** It occurs by feco-oral route, fomites-borne, contaminated hands, contact with infectious surface or rare by respiratory route.
- **Incubation period:** 2–3 days.
- **Portal of entry:** GIT. Virus also enters by respiratory tract.
- **Sites:** GIT.
- **Precipitating factors (epidemiological determinants):** **(1) Agent factors:** NSP4 (considered as enterotoxin) increases the vascular permeability and the secretion. **(2) Age:** It is common in infants and young children between 6–24 months. Children <6 months are protected by maternal Ab. Nearly every child in the world is infected with rota virus at least once by the age of five. **(3) Poor hygiene and low socio-economical status:** They precipitate the infection. **(4) IDDs:** Usually virus excreted for 2–12 days, but in HIV or malnourished person it prolonged. **(5) Season:** *Rotavirus* and other viral diarrhea are common in winter while of bacterial origin are common in summer. **(6) Climate:** In tropical countries *Rotavirus* occurs throughput the years but peak in cool seasons.

- **Pathogenesis:** After entry, virus damages the enterocytes (villous epithelium) in small intestine, resulting secretory diarrhea. It damages the small intestinal mucosa and sparing the gastric and colonic mucosa. NSP4 (act as enterotoxin) increases the vascular permeability and increase the secretion.

- **Clinical features:** Common features are acute watery diarrhea, abdominal pain, vomiting with or without fever and green color or pale (yellow) stool without blood or mucus. It is self limited and recovery occurs in 5–10 days. It has low mortality.

Diagnosis
- **Specimens:** Stool, mucosal biopsy and serum.
- **Testing methods**
  1. **Microscopy:** Follow morphology.
  2. **Culture:** Follow C/Cs.
  3. **Serological tests**
     – **Ag detection:** From the stool by ELISA, latex particle agglutination and IF test.
     – **Ab detection:** IgM and IgG are detected from serum by ELISA. It is used to detect the seroprevalence.
  4. **Molecular method:** RT-PCR is rapid and more sensitive method.
  5. **Serotyping and genotyping:** Described above.

Prevention
- **General measures:** These are sanitation, health education and frequent hand washing.
- **Immunoprophylaxis:** Following are the different types of vaccines. Intussusception or bowel obstruction is the disadvantage occurs when given to infants >12 weeks.
  1. **Monovalent human *Rotavirus* vaccine (Rotarix):** It is live attenuated vaccine **prepared** by using the type G1P[8]. It provides protection to type G1 and cross protection to G3, G4 and G9. Two doses are **administered** orally between 2 and 4 months, 1st at 6 weeks and 2nd at 10 weeks, not beyond 12 weeks. Interval between two doses should be at least 2 weeks. Two doses are completed by the age of 16 weeks, not after 24 weeks.
  2. **Rotavac:** Rotavac was licensed for use in India in 2014 and is manufactured by Bharat Biotech International Limited. It is a live attenuated, monovalent vaccine **prepared** from G9P[11] human strain isolated from an Indian child. It provides cross protection to many types including G1P[8]. Three doses (5 drops/dose) are **administered** orally at 6, 10 and 14 weeks. It has 55% **efficacy** in

Hemorrhagic Fever Viruses

1. Hemorrhagic fever is caused by:
a. West Nile fever   b. Sand fly fever
c. Ebola virus   d. All of above

2. Ebola virus is a:
a. Reo virus   b. Filo virus
c. Herpes virus   d. *Rotavirus*

3. Which of the following belongs to Filoviridae family?
a. Rota virus   b. Norwalk virus
c. Marburg virus   d. Adeno virus

Diarrheal Viruses

4. Which virus does not cause diarrhea?
a. *Rotavirus*   b. *Reovirus*
c. Adeno virus   d. Pox virus

5. Diarrhea is not a feature of:
a. *Rotavirus*   b. *Calcivirus*
c. *Enterovirus*   d. Rhabdo virus

6. Viral enterotoxin is detected as possible mechanism of pathogenesis in:
a. Adeno virus
b. *Rotavirus*
c. *Calcivirus*
d. *Astrovirus*

7. *Rotavirus* causes:
a. Acute non-bacterial gastroenteritis in adult
b. Infantile diarrhea
c. Teratogenic effects
d. Respiratory infection in immunocompromised

8. All are true about *Rotavirus* except:
a. Causes diarrhea in men and children
b. Rota B can be grown in culture
c. Rota C can cause diarrhea in children
d. Culture cannot be done

9. Which of the following is true about *Rotavirus*:
a. Commonly affects children
b. Double stranded DNA
c. Can be grown easily on cell culture
d. Egg shell appears under electron microscope

10. *Rotavirus* is diagnosed by:
a. IgM specific antibody in stool
b. ELISA which demonstrates antibody in stool
c. Immunofluorescence antigen in stool
d. Culture of *Rotavirus*

11. *Rotavirus* infection is diagnosed by the presence of:
a. Antigen in stool by ELISA
b. Virus in stool
c. Antigen in blood
d. Antibody in stool

12. Best vaccine for *Rotavirus* infection is:
a. Asymptomatic neonatal vaccine
b. DNA vaccine
c. Genetic reassortment
d. Capsular component vaccine

13. Genetic reassortment is seen with:
a. *Astrovirus*
b. Herpes virus
c. *Rotavirus*
d. *Hepadnavirus*

14. Genetic reassortment is seen with:
a. Influenza virus
b. Herpes virus
c. *Rotavirus*
d. *Hepadnavirus*

15. Vaccine causing intussusceptions is:
a. *Rotavirus*
b. Parvo virus
c. Inactivated polio
d. BCG
e. Measles

Slow Disease Viruses

16. Which of the following is not considered to be a slow virus disease?
a. Kuru   b. Scrapie
c. Maedi   d. Sarcoidosis

17. All of the following human diseases are caused by prion, *except*:
a. SSPE   b. Visna
c. CJD   d. Kuru

18. Creutzfeldt-Jacob disease is caused by:
a. Prion
b. JC virus
c. Genetic factors
d. Nutritional deficiency

Encephalitis Viruses

19. Encephalitis is caused by:
a. HSV-1
b. EBV
c. Infectious mononucleosis
d. CMV

20. Most common cause of sporadic viral encephalitis is:
a. Japanese encephalitis
b. Herpes simplex encephalitis
c. HIV encephalitis
d. Rubeola encephalitis

21. Not a cause of epidemic encephalitis:
a. Herpes simplex virus
b. Rabies
c. West Nile virus
d. Nipah virus
e. Japanese encephalitis virus

UTI Causing Viruses

22. Renal involvement is seen in which of the following infections?
a. Cytomegalovirus
b. Polyoma virus
c. Human papilloma virus
d. HIV
e. HBV

Answers and Explanation of MCQs

1. c
• Follow section, **hemorrhagic fever viruses (classification)** for explanation.

2. b

3. c
• Follow section, **hemorrhagic fever viruses (classification) and Filoviridae (Flowchart 89.1)** for explanation of answers of MCQs 2–3.

4. d

5. d

- Follow section, **diarrheal viruses (classification)** for explanation of answers of MCQs 4–5.

6. b

7. b

8. b

9. a

10. c

11. a, b

12. c

13. c

14. a, c

15. a

- Follow section, **diarrheal viruses (*Rotavirus*)** for explanation of answers of MCQs 6–15.

16. d

17. a

18. a

- Follow section, **slow disease viruses** for explanation of answers of MCQs 16–18.

19. a, b, c, d

20. b

21. a, b

- Follow section, **encephalitis viruses (classification)** for explanation of answers of MCQs 19–21.

22. a, d and e

- Follow section **UTI causing viruses (classification)** for explanation.

Fungal Infections

Superficial Mycoses

Fungal infections present in the skin and its appendages (like hair and nail) called **superficial mycoses** with two subtypes like surface mycoses and cutaneous mycoses.

SURFACE MYCOSES

Fungi affecting the outer most layer (dead layer or stratum corneum) of skin and its appendages like hairs and nails called **surface mycoses.** They do not produce the inflammatory reactions, but only produce the cosmetic effects. Examples of surface mycoses are *Malassezia* **infections, tinea nigra, piedra** and **onychomycoses.**

Malassezia Infections

Etiological agent: *Malassezia furfur.* It was previously known as *M. ovalis* or *Pityrosporum ovale* or *Pityrosporum malassezii.*

Pathogenicity

- **Reservoirs of infection, sources of infection, modes of transmission and portal of entry:** Fungus is the resident flora in skin of humans and animals. So humans and animals are the reservoirs and sources of infection and responsible for endogenous infection.
- **Sites:** Infection occurs in areas with excessive sweating. Lesions are found in neck, trunk, face, scalp, upper limbs and shoulders.
- **Precipitating factors (epidemiological determinants):** **(1) Age:** It is common in teenager. **(2) Other factors:** Hormonal factors, excessive sweating and immuno-suppressive status are responsible for recurrence.t
- **Pathogenesis:** It is a lipophilic fungus. It interferes with melanin production resulting hypo- or hyper-pigmented patches on skin.
- **Clinical features**
  1. **Ptyriasis versicolor/tinea:** Fungus interferes with melanin production resulting hypo- **(Fig. 90.1a)**

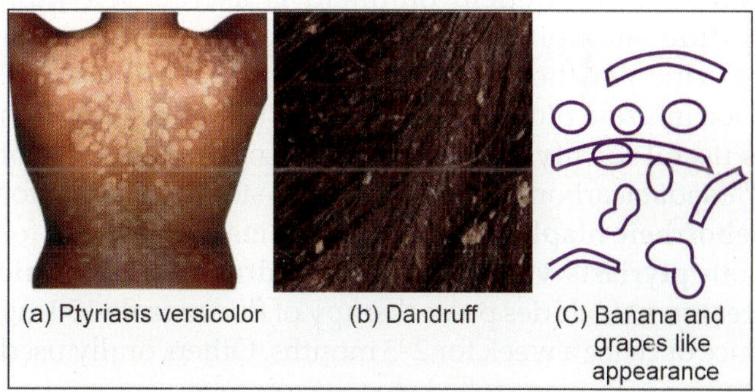

Fig. 90.1: Images of *M. furfur*

or hyperpigmented patches on skin and non-inflammatory macular lesions with fine scales.
  2. **Seborrheic dermatitis (dandruff or pityriasis capitis):** Sebo word derived from Sebum (Latin) means Grease and rrhea from rhea (Greek) means discharge/flow. It presents with whitish, dry and loose flakes on scalp **(Fig. 90.1b)**.
  3. **Seborrheic blapharitis (squamous blapharitis):** It is the chronic inflammation of lid margin usually associated with dandruff. It presents with deposition of whitish scales around the lid margin (like dandruff) with mild discomfort, irritation, watering and falling of eyelashes. In long-term cases it causes thickening of lid margin and rounding of posterior border leading to epiphora.
- **Complications:** Recurrence is common due to hormonal factors, excessive sweating and immuno-suppressive status.

Laboratory diagnosis

- **Wood's lamp examination:** Wood's lamp is made up with barium silicate and nickel oxide. It transmits the UV light which produces golden yellow fluorescence.
- **Specimens:** These are epidermal scales picked up by forceps or by using cellophane tap and skin scraping. All the samples are collected in black sterile envelope.

- **Testing**
  1. **10% KOH:** It shows the cluster of round yeast cells with 2–7 µm size and short, curved hyphae giving **banana and grapes like appearance** or **spaghetti and meatballs like appearance (Fig. 90.1c).**
  2. **Culture**
     - Medium: It is SDA contains antibiotics, tween-80 and olive oil. Fungus is lipophilic grows with difficulties on medium contains olive oil.
     - C/Cs on SDA: Small, creamy, yellowish and typically fried egg colonies appear.
     - Identification of growth by LCB/PHOL stain: Stains shows spherical yeast cells and short curved hyphae.
  3. **Urease test:** It gives urease positive.

Prevention: Removal of precipitating factors to prevent the recurrence.

Treatment: (1) Local application: 25% sodium thiosulfate or Whitefield ointment is useful. Selenium sulfide shampoo or 2% ketoconazole shampoo or 1% zinc pyrithone shampoo is useful for **dandruff**. Clotrimazole plus selenium sulfide (Candid TV) lotion is useful for **ptyriasis versicolor**. Local application of 3% sodabicarbonate, antifungal and steroid is useful for **seborrheic blapharitis**. Topical terbinafine is useful for both **ptyriasis versicolor and dandruff**. **(2) Systemic treatment** includes pulse therapy of fluconazole 150 mg once or twice a week for 2–3 months. Others orally used drugs are itraconazole, ketoconazole, etc.

Tinea Nigra

Meaning: Tinea means fungal infection and nigra means black.

Definition: Superficial fungal infection of palm or sole by black fungi called **tinea nigra.** It also called **superficial pheohyphomycosis**, because caused by black fungi.

Etiological agent: *Exophiala werneckii.* It also called *Hortae werneckii* or *Cladosporium werneckii* or *Phaeoannellomyces werneckii.*

Pathogenicity

- **Reservoirs and sources of infection:** Fungus presents in soil, sewage, decaying vegetation, wooden stick, salted fish and sea water. These act as sources and reservoirs for human infection. Fungus is halophilic and optimally can grow at 3–6% NaCl of sea water, but can tolerate up to 10% NaCl.
- **Modes of transmission:** Infection commonly acquired from sea water during bathing.
- **Incubation period:** 2–7 weeks.
- **Portal of entry:** Skin.
- **Sites:** These are palm and sole due to excessive sweating which precipitates the infection due to halophilic (required high salt for growth) nature of fungus.

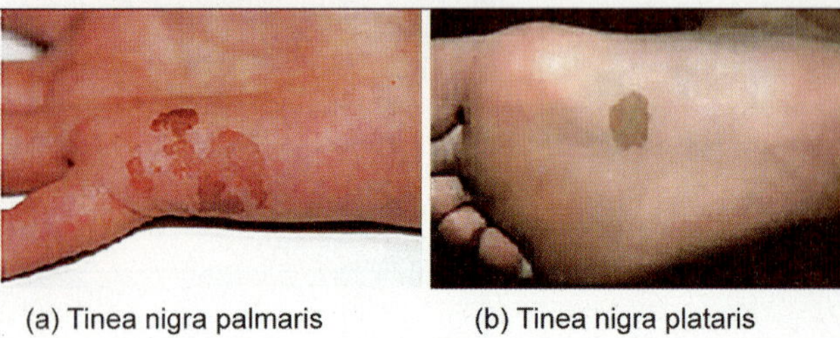

Fig. 90.2: Types of tinea nigra

- **Precipitating factors (epidemiological determinants): (1) Age:** More in teenager than <20 years. **(2) Sex:** Females are infected 3–5 times more than males. **(3) Hyperhydrosis:** It is common in patients with excessive sweating.
- **Pathogenesis and clinical features:** Fungus increases the melanin production. Presence of brown-black pigmented macular lesion on palm called **tinea nigra palmaris (Fig. 90.2a)** or sole called **tinea nigra plantaris (Fig. 90.2b).**

Laboratory diagnosis

- **Specimens:** Skin scraping collected in black envelope.
- **Testing**
  1. **10% KOH:** Cluster of round budding yeast cells with 2–8 µm size and brown septate branched hyphae are seen.
  2. **Culture**
     - Medium: It is SDA contains antibiotics.
     - C/Cs on SDA: Brown-black colonies appear.
     - Identification of growth by LCB/PHOL stain: Stains shows cluster of round budding yeast cells with 2–8 µm size and brown septate branched hyphae.

Prevention: No standard guidelines for prevention.

Treatment: (1) Local application of keratolytic agents like Whitefield ointment and 10% salicylic acid ointment. **(2) Systemic treatment:** Orally used drugs are itraconazole (200 mg daily for three week), griseofulvin, etc.

Piedra

Meaning: Word derived from piedra (Spanish) means stone, because of nodule formation within the hair shaft.

Definition: It is a fungal infection of hair characterized by nodule formation along the hair shaft.

Types: Following two types.

White Piedra

Synonym: It also called **trichomycosis nodularis** or **trichosporonis nodosa.**

Etiological agent: *Trichosporon beigelii.*

Pathogenicity

- **Reservoirs and sources of infection:** Fungus presents in soil and stagnant water which act as reservoirs and sources of infection.

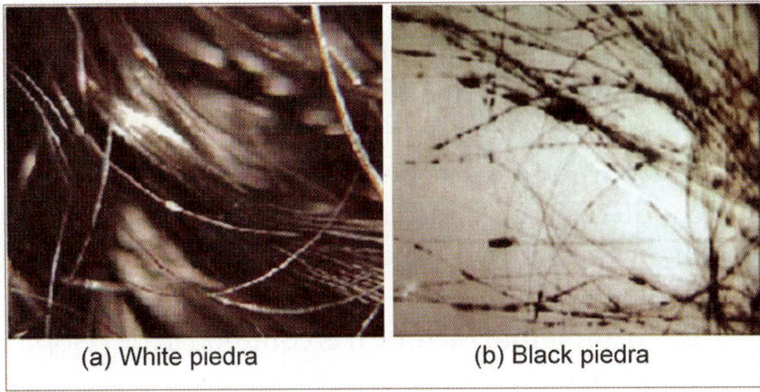

(a) White piedra   (b) Black piedra

Fig. 90.3: Types of piedra

- **Modes of transmission:** Exactly not known but may be due to contact with stagnant water of swimming pool. White piedra may occur by sexual intercourse.
- **Portal of entry:** Direct contact of hair with contaminated swimming pool stagnant water.
- **Sites:** Axillary, pubic, moustache and beard area.
- **Precipitating factors (epidemiological determinants):** It is common in persons with black complex.
- **Clinical features:** Grayish-white soft nodule occurs around the hair shaft as shown in **(Fig. 90.3a)**.
- **Complications:** Scalp colonization may lead persistent infection and recurrence.

Laboratory diagnosis: 10% KOH of hair shows hyaline septate hyphae with arthrospores. **Culture** is done on SDA contains antibiotics. It shows white creamy colonies. LCB/PHOL stain from culture growth shows hyaline septate hyphae.

Prevention: Good personal hygiene helps to prevent the infections.

Treatment: Removal of hairs, coating of hairs with azoles, use of antifungal shampoo or topical lotion or systemic use of azoles or AMB are effective.

Black Piedra

Synonym: Disease also called **tinea nodosa**.

Etiological agent: *Piedraia hortae*.

Pathogenicity

- **Reservoirs of infection:** Fungus presents in soil and infects monkey as well as humans.
- **Sources of infection:** These are contaminated combs, pillows and bed sheets.
- **Modes of transmission:** Common sharing or contact with infected items like combs, pillows and bed sheets.
- **Site:** Hair of scalp, moustache and beard area.
- **Precipitating factors (epidemiological determinants):** **(1) Sex:** It is more in girls than boys. **(2) Lifestyle:** Common in hostels girls due to sharing common combs, pillows and bed sheets.
- **Clinical features:** Presence of hard, black nodules along the hair shaft **(Fig. 90.3b)** produce metallic sound during combing of hair.

Laboratory diagnosis: 10% KOH of hair shows dark septate hyphae around hair shaft with asci containing ascospores. **Culture** is done on SDA contains antibiotics. It shows brown-black colonies. LCB/PHOL stain from culture growth shows dark septate hyphae.

Treatment: Systemic treatment includes terbinafine in a dose of 250 mg for 6 weeks. Other treatments are same as white piedra.

Onychomycoses

Meaning: Onycho word derived from Onyx (Greek) means nail and mycosis means fungal disease.

Definitions: Fungal infections of nail called **onychomycoses (Fig. 90.4a)**. Fungal infections of nail by dermatophytes are called **tinea unguium**.

Etiological agents

1. **Dermatophytes (tinea unguium)** *Trichophyton mentagrophytes, T. rubrum* and *Epidermophyton flocossum*.
2. **Nondermatophytes:** *Scopulariopsis brevicaulis, Scytalidium dimidiatum, Scytalidium hyalinum, Aspergillus flavus, Aspergillus fumigatus, Acremonium* species, *Fusarium oxysporum, Onychocola canadensis, Geotrichum candidum* and *C. albicans*.

More details about etiological agents: Follow respective chapter.

Prevention: Improvement in hygiene.

Treatment

1. **Treatment of dermatophytes:** Pulse therapy of itraconazole and terbinafine is given for 6–12 months. Clinical cure of onychomycoses is defined as disappearance of almost all lesions and residual lesions are not >10%. Mycological cure is defined as negative microscopy and culture.
2. **Treatment of nondermatophytes:** Itraconazole is effective for *S. brevicaulis, Aspergillus* spp., and *Fusarium oxysporum*.

CUTANEOUS MYCOSES

Fungal infections of skin layer beneath the dead layer called **cutaneous mycoses**. It produces the inflammatory and allergic reactions. Examples of cutaneous mycoses are **candidiasis** and **dermatophytoses**.

Candidiasis

Synonym

- **Moniliasis:** Because previously the genus was called *Monilia* or called *Odium* (species was referred as *Monilia albicans* or *Odium albicans*).
- **Candidosis:** Both candidosis and candidiasis are correct terminology for the disease. 'Osis' used as suffix in European countries while 'asis' in USA and rest of the world.

Definition: It is an infection of skin, its appendages, subcutaneous tissue and internal organs by *Candida* spp.

Etiological agents: It is caused by *Candida* species like *C. albicans* (*Monilia albicans*), *C. krusei*, *C. kefyr* (*C. pseudotropicalis*), *C. glabrata*, *C. guilliermondi*, *C. tropicalis*, *C. parapsilosis*, *C. lusitaniae*, *C. dubliniensis*, *C. visvanathii*, *C. pelliculosa* (*Pichia anomala*) and *C. auris*.

Pathogenicity

- **Reservoirs of infection, sources of infection, modes of transmission and portal of entry:** It is a normal flora of skin-mucosa and infection is endogenous.
- **Sites:** It infects many sites like skin, mucosa, meninges, GIT, etc.
- **Precipitating factors (epidemiological determinants):** These are agent factors (virulence factors), host factors and environmental factors as described below.
    - **Agent factors (virulence factors): (1) Adhesins:** Adhesins like mannose, C3d receptors, manno-protein, and saccharins presents on cell wall help to bind with fibronectin of host cells. **(2) Toxin:** It is a glycoprotein like bacterial endotoxin and produces pyogenic and anaphylactic reaction. **(3) Enzymes and cytokines:** *Candida* produces TNF and >14 enzymes like protease, enolase, aspartate proteinase, lipase, phospholipase, esterase, phosphatase, etc. They help in fungal invasion. **(4) Phenotyping switching:** It is an ability of fungus to adapt the different changing condition of host, which helps to evade the host defense mechanism and survival of fungus. **(5) Antigens:** They produce allergic reaction called **id reaction** or **candidid.**
    - **Host factors: (1) Age:** Common in old age. **(2) Trauma:** Local trauma or 3rd degree burn may precipitate the infection. **(3) Pregnancy:** During pregnancy, *Candida* is even more common. According to one study, about 20% of women have *Candida* in their vagina normally. That becomes 30% during pregnancy due to hormonal fluctuations. **(4) Absence of saliva:** Saliva contains antimicrobial proteins like lactoferrin, sialoperoxidase, lysozyme, histidine-rich polypeptides, and specific anti-*Candida* antibodies. They interact with the oral mucosa and prevent overgrowth of *Candida*. Absence of salivary secretion can precipitates the oral candidiasis. **(5) High carbohydrate diet:** It increases the growth and adherence of *Candida* to epithelial cells of oral mucosa. **(6) Denture:** Wearing of dentures produces a microenvironment favor the growth of oral candidiasis with low oxygen, low pH, and an anaerobic environment. This may be due to enhanced adherence of *Candida* spp., to acrylic, reduced saliva flow under the surfaces of the denture fittings, improperly fitted dentures, or poor oral hygiene. **(7) Immune status:** It causes infection in immunocompetent and immunosuppressive persons with long-term steroid therapy, antibiotic therapy (kills the bacteria and reduces the nutritional competition with

Candida spp.), contraceptive pills, intravenous drug abuse, malnutrition (due to deficiency of vitamins B6, folic acid, zinc, iron, etc.), DM, AIDS, CMI deficiency disease (like neutropenia, leukemia, DiGeorge syndrome, chronic mucocutaneous candidiasis, etc.) and combined immunodeficiency of AMI and CMI (like severe combined immuno deficiency).
    - **Environmental factors:** It is common in developing countries.
- **Clinical features:** Two groups like infectious diseases and allergic diseases.

Infectious Disease

1. **Cutaneous (skin + appendages) candidiasis**
   - Generalized skin infection.
   - Intertriginous: Skin folds infection.
   - Paronychia: Nail folds infection.
   - Onychomycosis **(Fig. 90.4a)**: Nail infection.
   - Diaper dermatitis [diaper rash or napkin candidiasis, **(Fig. 90.4b)**: It is the maculopapular skin rash in infants due to use of wet diaper which may form vesicle or pustule.
   - Candidal granuloma: It is the chronic granulomatous lesion contains giant cells over face, trunk, scalp or legs.

2. **Mucosal candidiasis**
   - Oral candidiasis: Thrush **(Fig. 90.4c)**, glossitis, cheilitis and stomatitis.
   - GIT candidiasis: Esophagitis and gastritis.
   - Genital candidiasis: Balanitis, balanoposthitis and vulvovaginitis.
   - Respiratory candidiasis: Tracheitis and bronchitis.

3. **Systemic candidiasis**
   - CNS candidiasis: Meningitis and encephalitis.
   - Respiratory candidiasis: Pneumonia.
   - Urinary tract/urogenital candidiasis: Urethritis.
   - CVS candidiasis: Endocarditis.
   - Bloodstream candidiasis: Candidemia.
   - Bones and joints candidiasis: Arthritis and osteomyelitis.
   - Ocular candidiasis: Keratitis, endophthalmitis, anterior uveitis, multifocal choreoretinitis, dacryocystitis and lesions in eyelashes.
   - Ears: Otomycocsis.
   - Nosocomial infection: *C. auris* is an emerging nosocomial pathogen. It causes asymptomatic

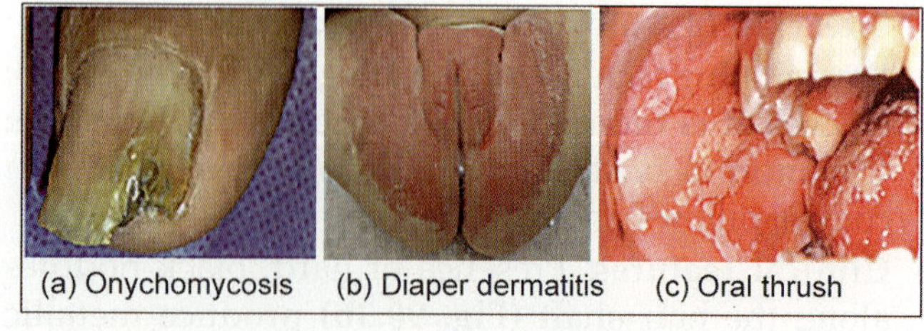

(a) Onychomycosis   (b) Diaper dermatitis   (c) Oral thrush

Fig. 90.4: Types of candidiasis

colonization to candidemia, abscess, ear infection, etc. It is difficult to diagnose *C. auris* by routine methods but best diagnosed by molecular test or by MALDI-TOF. It is prevented by taking appropriate infection control measures. It is multidrug resistant and caspofungin or AMB is the drug of choice.

Allergic Diseases

It called **'id' reaction** or **candidid.** It featured as gastritis, irritable bowel syndrome, urticaria, asthma or eczema.

Laboratory diagnosis

- **Specimens:** Urine, CSF, blood, discharges, mucosal biopsy, etc.
- **Testing**

A. Microscopy: (1) Direct 10% KOH wet mount: Round or oval budding yeast cells with pseudohyphae are seem. Schematic diagram is shown in **Fig. 90.5. (2) Gram's stain:** It looks gram-positive with presence of round or oval budding yeast cells with pseudohyphae **(Fig. 90.6)**.

B. Routine culture
- Medium: It is SDA contains antibiotics.
- Cultivation technique: **Ch. 120** (section → **Flowchart 120.1**).
- C/Cs on SDA: Creamy-white colonies **(Fig. 90.7)** appear.

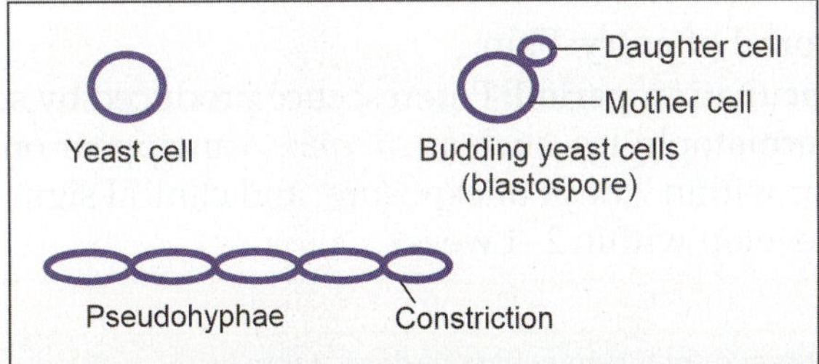

Fig. 90.5: Morphology of *Candida*

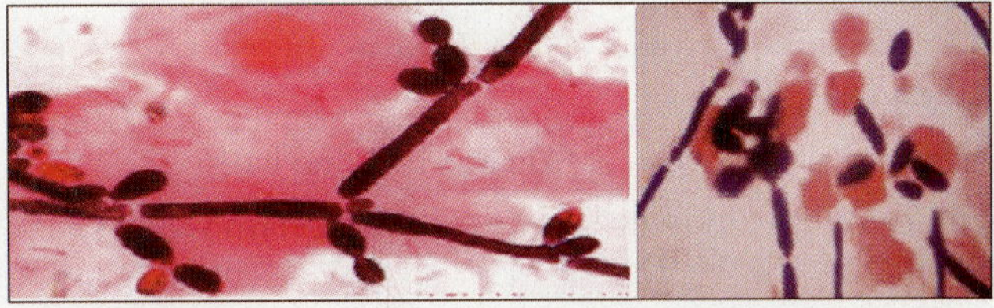

Fig. 90.6: *Candida* spp., (pseudohyphae) under Gram's stain

Fig. 90.7: *Candida* in SDA in bottle and Petri dish

- Identification of growth by LCB/PHOL stain: Round or oval budding yeast cells with pseudohyphae are seen.
- Growth is also identified by automated methods like VITEK and MALDI-TOF.
- Growth at 45°C: It is positive for *C. albicans*.

C. Culture on selective media: These are used to differentiate the *Candida* spp.

1. **Tetrazolium reduction medium:** Make it selective by adding neomycin. Color by different *Candida* spp., is mentioned below.
   - *C. albicans* and *C. glabrata*: Pale pink.
   - *C. tropicalis*: Orange pink.
   - *C. kefyr* (*C. pseudotropicalis*): Salmon pink.
   - *C. parapsilosis*: Rose pink.
   - *C. guilliermondi*: Pink and pasty.
   - *C. krusei*: Pink and dry.

2. **CHROM agar:** It detects the enzymatic activities of *Candida* spp., by adding fluorochrome substances. Advantages of using this medium are rapid isolation and identification of *Candida* spp., at 30°C in 48–72 hours.
   - *C albicans*: Light green.
   - *C. parapsilosis*: Cream colored and large colonies.
   - *C. tropicalis*: Blue with pink halo.
   - *C. krusei*: Pink.
   - *C. glabrata*: Purple.
   - *C. dubliniensis*: Dark green.
   - *Prototheca* spp.: Cream colored and small colonies.

D. Chlamydospores producing media: Deficiency of nutrition allows the production of spores. They are produced at 20°C. Following are chlamydospores producing media.
   - Cornmeal or cornmeal tween agar.
   - Rice starch agar.

E. Animal culture: Laboratory animals like rabbit, guinea pig, etc., are useful.

F. Sugar fermentation test and sugar assimilation test: They are useful to differentiate the *Candida* species and other yeast or yeast like fungi.

G. Serological tests: ELISA is useful to detect the cell wall mannan Ag plus cytoplasmic Ag and also to Ab against these Ags.

H. Enzymes detection: It includes the detection of enolase, aspartate proteinase, etc.

I. Molecular method: PCR detects the genes of various *Candida* spp.

J. Colorimetric EIA: It is used to detect the β-D-glucan. It increased in many invasive fungal diseases except zygomycosis, blastomycosis and cryptococcosis. It is a useful marker to monitor the therapy.

K. Gag liquid chromatography: To detect the fungal metabolites like D-arabinitol and D-mannose.

L. Special tests: These are teased mount, slide culture, cellophane tape mount, germ tube test and **dalmau plate culture as mentioned in Ch. 120.**

M. Antifungal sensitivity test: Report the result with antifungal sensitivity pattern.

Prevention: Removal of precipitating factors as mentioned earlier. Use appropriate antifungal drugs to prevent the deep seated infections.

Treatment: Three types of treatment.

1. **Local chemotherapy:** Local oral or mucocutaneous lesions are treated by applying 1% gentian violet, nystatin or azole cream.
2. **Systemic chemotherapy:** Intravenous AMB, fluconazole, voriconazole or caspofungin is prescribed.

Dermatophytoses

Synonym: It also called tinea or ring worm.

Definitions

- **Dematophytoses:** Superficial fungal infections of skin, hairs and nails caused by dermatophytes called **dermatophytoses.**
- **Dematomycoses:** Infections of skin due to different fungi like *Candida* and by cutaneous manifestations of deep mycoses called **dermatomycoses.**

Etiological agents: It is caused by dermatophytes spp., which are classified in different groups.

1. **Systemic/taxonomical classification:** Three genera and around 42 species are mentioned in **Table 90.1 and Fig. 90.8.**
2. **Ecological types:** These are based on habitat as mentioned in **Flowchart 90.1.**

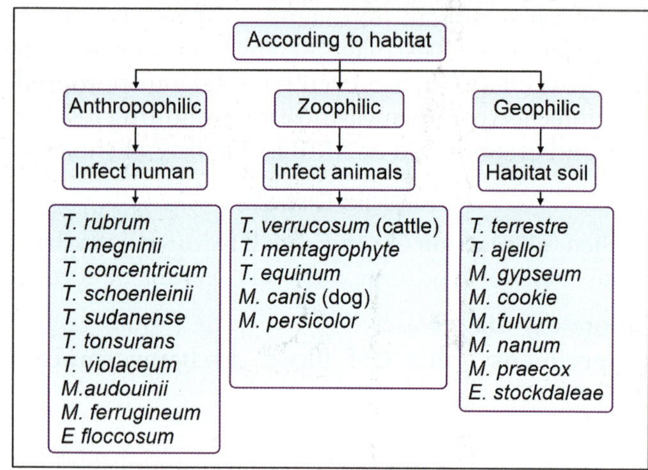

Flowchart 90.1: Ecological types of dermatophytes

Pathogenicity

- **Reservoirs and sources of infection:** Common reservoirs are living objects like humans and animals (zoonosis) and nonliving objects (soil, water, materials available at barber's shop, etc.) or fomites. All such objects are the sources of infection.
- **Modes of transmission:** It transmitted by direct contact with living and nonliving objects or by fomites. *T. capitis* transferred from barbershop by sharing the common articles like scissors, razors, blades and headrest. It is also transmitted by headrest of vehicle.
- **Portal of entry:** Skin.
- **Incubation period:** Fluorescence produced by some dermatophytes, such as *M. canis*, can appear on the fur within 7 days of exposure, and clinical signs can develop within 2–4 weeks.

Genus	*Trichophyton*	*Microsporum*	*Epidermophyton*
Species	T. mentagrophytes T. violaceum, T. schoenleinii T, verrucosum, T. rubrum	M. gypseum M. canis M. audounii, M.nanum	E. floccosum E. stockdaleae
Habitat	Skin, hairs and nails	Skin and hairs (no nails)	Skin and nails (no hairs)
C/Cs on SDA	Powdery, velvety or waxy with pigmentation	Cottony, powdery or velvety with white-brown pigments	Powdery and greenish yellow
Microconidia	Abudant and grapes like arrangement **(Fig. 90.8a)**	Scanty **(Fig. 90.8b)**	Absent
Macroconidia	Rare, thin-smooth wall, blunt end and pencil shape **(Fig. 90.8c)**	Numerous, thick-rough wall, pointed end and boat shape **(Fig. 90.8d)**	Numerous, smooth wall and club shape **(Fig. 90.8d)**

TABLE 90.1: Systemic/taxonomical classification and differences between genera of dermatophytes

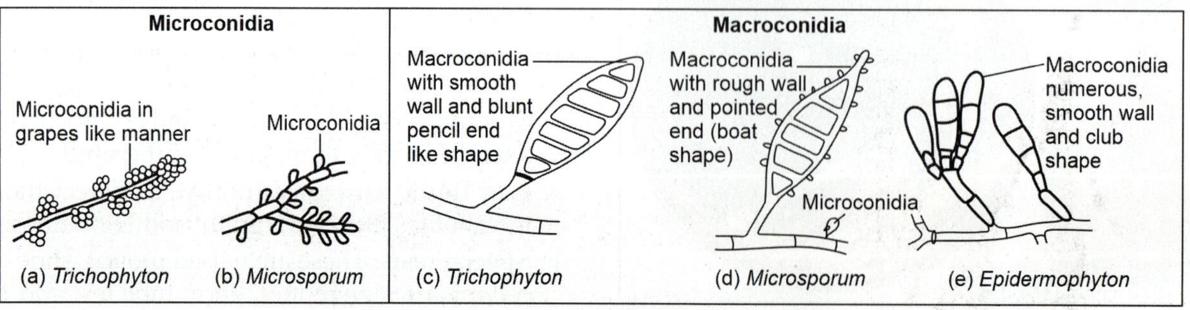

Fig. 90.8: Macroconidia and microconidia in dermatophytes

- **Sites:** It infects the skin of scalp, moustache area, beard area, chest, shoulder, neck, groin and perineum. Plantar aspect of feet, toes, interdigital web space, nails and hairs are also infected.
- **Precipitating factors (epidemiological determinants):** These are described below.
  - **Agent factors (virulence factors): (1) Keratophilic fungi:** It grows only in dead keratinized tissues. **(2) Enzymes:** Keratinases, elastase and collagenase help in fungal invasion. **(3) Fungal metabolites:** Fungal metabolites produce erythema, vesicles and pustules. When hyphae become old, they breakdown to release the arthrospores which clear the central ringworm lesions. **(4) Antigens:** They produce allergic reaction called **id reaction** or **dermatophytid.**
  - **Host factors:** These are excessive sweating, tight fitting garments like socks and underwear, poor hygiene, trauma, etc.
  - **Environmental factors:** As per different environmental conditions species of dermatophytes show the different geographical distribution. In **India** *T. rubrum, T. mentagrophytes* and *E. floccosum* are common.
- **Pathogenesis:** Fungal hyphae penetrate the stratum corneum by direct contact. Fungus grows only in dead keratinized tissues and releases fungal metabolites, which help in invasion and also for allergic reactions.
- **Clinical features:** Two types of diseases.

Infectious Diseases

1. **Tinea capitis:** Scalp infection with following variants **(Fig. 90.9).**
   - **Kerion:** Kerion (Greek) means honeycomb. It is an inflammatory boggy mass on scalp with multiple sinuses sometimes may discharge the mycetoma like grains. It mostly caused by *T. mentagrophytes* and *T. verrucosum.*
   - **Favus (tinea favosa):** Favus (Latin) means honeycomb. It is the honeycomb like fungal growth within hair follicles of scalp which may produce cup like crusts (scutulas), alopecia and scarring. It is mostly caused by *T. schoenleinii* and rarely by *T. violaceum* and *M. gypseum.*
   - **Black dot:** It is an endothrix type of invasion of fungus in hair shaft with abundant sporulation. It produces breakage of hairs from scalp surface and black dot formation over the surface. It mostly caused by *T. tonsurans* and *T. violaceum.*
   - **Ectothrix hair infection (Fig. 90.10a):** It shows the presence of arthrospores appear as mosaic or in chain outer to the hair shaft. It mostly caused by *T. mentagrophytes, T, verrucosum, M. gypseum, M. canis* and *M. audouinii.*
   - **Endothrix hair infection (Fig. 90.10b):** It shows the presence of arthrospores inside the hair shaft. It mostly caused by *T. schoenleinii, T. violaceum* and *T. tonsurans.*

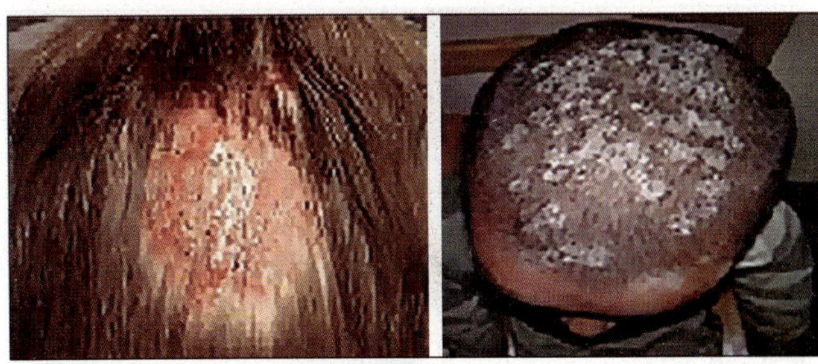

Fig. 90.9: Tinea capitis

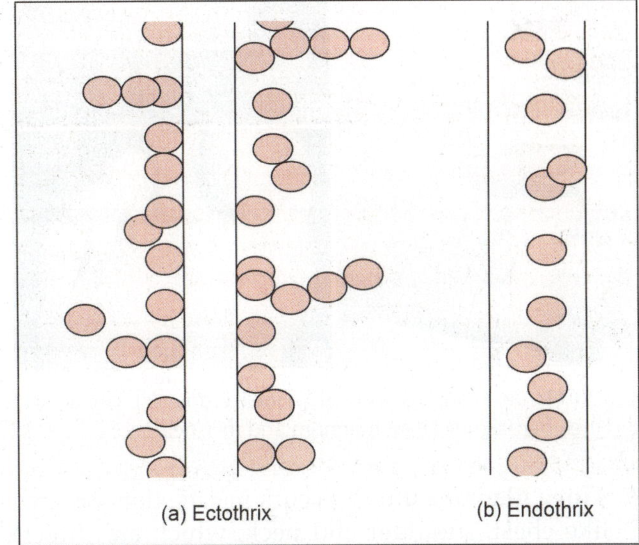

Fig. 90.10: Ectothrix and endothrix hair infection

2. **Tinea corporis (tinea glabrosa):** It is an infection in non-hairy skin like trunk and limbs **(Fig. 90.11a)** caused by *T. rubrum.*
3. **Tinea cruris (jock itch):** It is an infection of groin and perineum area **(Fig. 90.11b)** caused by *T. rubrum* and *E. floccosum.* Itching is predominant feature.
4. **Tinea faciei:** It is an infection of facial skin excluding beard area **(Fig. 90.11c).**
5. **Tinea barbe (barber's itch):** It is an infection of beard and moustache area **(Fig. 90.11d)** of skin mostly caused by *T. verrucosum, T. mentagrophytes* and rare by *M. canis* and *T. rubrum.*
6. **Tinea mannum:** It is an infection of palm **(Fig. 90.11e)** caused by *T. rubrum, T. mentagrophytes* and *E. floccosum.*
7. **Tinea pedis (athlet's foot):** It is an infection of plantar aspect of feet, toes and interdigital web space **(Fig. 90.11f).** It mostly occurs in athletes (like football player) and persons wearing shoes for long time. It is precipitated by warmth and moisture produced by shoes. It is caused by *T. rubrum, T. mentagrophytes* and *E. floccosum.*
8. **Tinea imbricata:** It is the concentric rings of papulo squamous scales produce intense pruritus which may lead to lechnification. It is caused by *T. concentricum.*

Essentials of Medical Microbiology

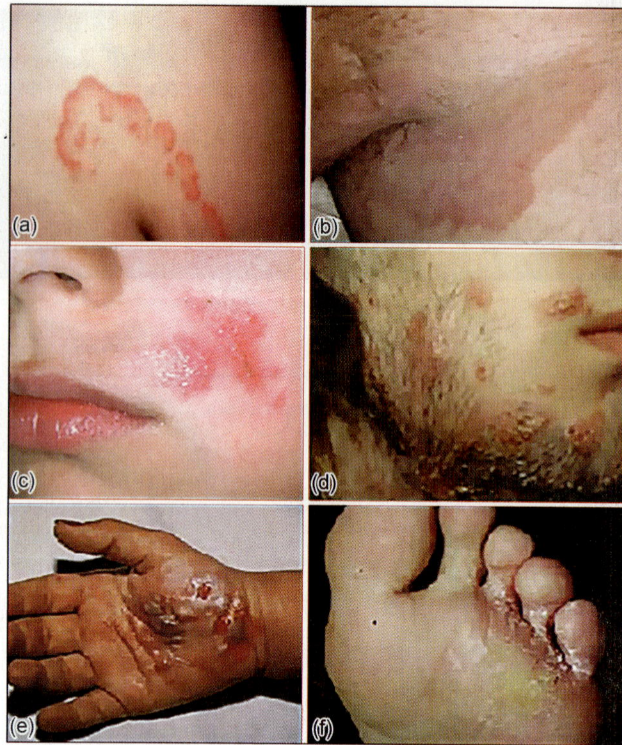

Fig. 90.11: (a) Tinea corporis; (b) Tinea cruris; (c) Tinea faciei; (d) Tinea barbae; (e) Tinea mannum and (f) Tinea pedis

9. **Tinea gladiatorum:** It occurs in wrestlers on areas like chest, shoulder and neck, which are directly coming in contact with infected wrestler. It is caused by *T. tonsurans*.

10. **Tinea incognito:** It occurs due to steroid therapy.

11. **Tinea unguium:** It is the infection of nail.

Allergic Diseases

It called **'id' reaction** or **dermatophytid**. It presents in two forms. (1) Lichen scrofulosorum-like. (2) Pompholyx-like.

Laboratory diagnosis

- **Wood's lamp examination:** It transmits the UV light which produces fluorescence by some fungi and by some bacteria.
  - Bright green: *M. audounii, M. ferruginium, M. canis* and *Pseudomonas* in burns wound.
  - Dull green: *T. schoenleinii*.
  - Dull yellow: *M. gypseum*.
  - Golden yellow: *Malassazia furfur*.
  - Coral-red: *Corynebacterium minutissimum* (erythrasma).
  - Brick-red: *Prevotella melaninogenica*.
- **Specimens:** Skin, hair and nail.
- **Testing**

A. **Microscopy: (1) Direct 10% KOH wet mount:** It shows hyaline septate hyphae with arthrospores **(Fig. 90.12). Slide method** is useful for thin samples like epidermal scales, skin scraping, hair, etc. **Tube method or 20–40% KOH** is useful for thick samples like nail. **(2) Histopathological stains** like PAS stain

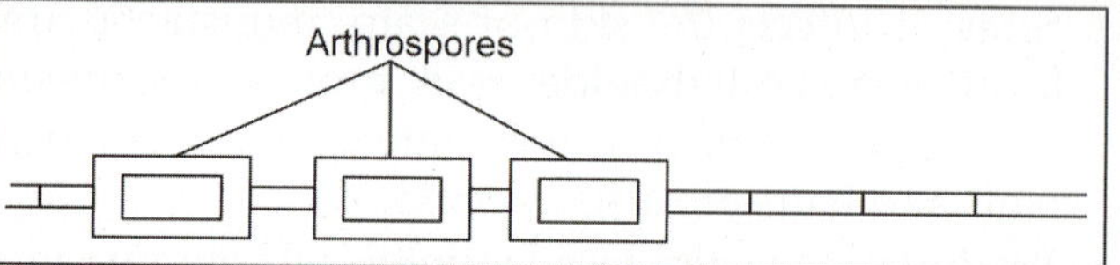

Fig. 90.12: Arthrospores

and GGMS stain are useful for nails. **(3) Fluorescent microscopy** by using calcofluor white stain is also useful.

B. **Routine culture**
- Medium: It is SDA contains antibiotics.
- Cultivation technique: **Ch. 120** (section → **Flowchart 120.1**).
- C/Cs on SDA: Follow **Table 90.1**.
- Identification of growth by LCB/PHOL stain: Macroconidia and microconidia are identified as described in **Table 90.1**. Also there is presence of racquet hypha, knot organ, spiral hypha, fevic chanderlier and pectinate body as described in **Ch. 13**.

C. **Culture on selective media:**
- Dermatophyte test medium: Red colony.
- Dermatophyte identification medium: Purple colony.

D. **Animal inoculation:** Guinea pig is useful.

E. **Urease test:** It is used to differentiate the *T. mentagrophytes* (positive) from *T. Rubrum* (negative).

F. **Serological tests:** Like immunodiffusion test.

G. **Molecular method:** PCR.

H. **Special tests:** These are teased mount, slide culture, cellophane tape mount, hair bait technique and hair perforation test **as mentioned in Ch. 120.**

Prevention: Improvement of hygiene.

Treatment: Local treatment includes Whitefield ointment, tolnaftate and azole cream. Terbinafine, itraconazole, 500 mg oral griseofulvin or ketoconazole is the drug of choice for systemic route.

ACCESS YOURSELF

Case Study

1. A 50-year-old HIV positive female visited the medical OPD with complain of ulcer on oral cavity. Oral examination rule out white patches over tongue and other part of oral mucosa. Tissue taken from ulcer and proceed by Gram's stain and culture on corn meal agar. Gram's stain revealed budding yeast cell with pseudohyphae and corn-meal agar produces the chlamydospores. Identify the organism and answer the following.
   a. Name clinical infection (disease) and the most common causative agent of it.
   b. Write two selective and two spore producing media of causative agent.
   c. Draw the labeled diagram of morphology of causative agent.
   d. Describe the virulence factors of causative agent.

e. Write in details about pathogenicity of causative agent in given case.

2. A 25-year-football player visited the surgical OPD with complains of ulcer on plantar aspect of foot. Tissue taken from ulcer and proceed with culture on Sabouraud's agar, which revealed smooth wall pencil shaped macroconidia and plenty of microconidia. Identify the organism and answer the following.
   a. Name clinical infection (disease) and the most common causative agent of it.
   b. Name four fungi causing infection in to toe nail.
   c. Write the differences between different genera of dermatophytes.
   d. Write in details about pathogenicity of causative agent.
   e. Write the laboratory diagnosis of causative agent.

Essays/Full Questions

1. Superficial mycoses.
2. Describe the etiology, microbiology, pathogenicity and diagnostic features of dermatophytes.

Short Notes

1. Surface mycoses.
2. Candidiasis.
3. Dermatophytoses.

Short Questions for Theory/Viva Questions

1. Write four examples of surface mycoses.
2. What is onychomycoses? Name two etiological agents.
3. Define: Dermatophytoses and dermatomycoses.
4. What is chlamydospore? Name two chlamydospores producing media.
5. Write four differences between *Trichophyton* and *Microsporum/ Trichophyton* and *Epidermophyton/Epidermophyton* and *Microsporum*.
6. What are clinical cure and mycological cure of onychomycoses?
7. Name the one example of each lipophilic, halophilic, keratophilic and hydrophilic fungus.

Comment on

1. *Candida* infection increases in pregnancy.

MCQs for Chapter Review

Surface Mycoses

1. **Ptyriasis versicolor is caused by:**
   a. *E. flocossum* b. *M. gypseum*
   c. *M. furfur* d. *T. tonsurans*
2. **Spaghetti and meatball appearance is seen in:**
   a. *Hortaea werneckii* b. *Trichosporon beigelii*
   c. *Piedraia hortae* d. *Malassezia furfur*
3. **Black piedra is produced by:**
   a. *Piedra hortae* b. *Trichsporon beigelii*
   c. *Hortae werneckii* d. *Malassezia furfur*
4. **White piedra is produced by:**
   a. *Piedra hortae*
   b. *Pityrosporum orbicularae*
   c. *Hortae werneckii*
   d. *Trichsporon beigelii*

Cutaneous Mycoses

5. ***Candida albicans* causes all of the followings, *except*:**
   a. Endocarditis b. Mycetoma
   c. Meningitis d. Oral thrush

6. **Most common fungal infection in febrile neutropenia is:**
   a. *Aspergillus niger* b. *Candida*
   c. Mucormycosis d. *Aspergillus fumigatus*
7. ***Candida albicans* infection is predisposed by all, *except*:**
   a. Menstruation b. Diabetes
   c. Minipill use d. Combined pill use
8. **Candidiasis is associated with all, *except*:**
   a. Oral contraceptive pill (OCP)
   b. IUCD user
   c. Diabetes
   d. Pregnancy
9. ***Candida* has predilection for all, *except*:**
   a. Diabetes b. Extreme of age
   c. Athletes d. Pregnant female
10. ***Candida* is most implicated in causation:**
    a. Conjunctivitis b. Tinea capitis
    c. Desert rheumatism d. Thrush
11. **Production of pink color on CHROMagar indicates:**
    a. *Candida albicans* b. *Candida krusei*
    c. *Candida tropicalis* d. None of the above
12. **Which of the following fungus does not infect hairs?**
    a. *Trichohyton* b. *Microsporum*
    c. *Epidermophyton* d. *Trichosporon*
13. **Dermatophyte not infecting nail is:**
    a. *Trichohyton* b. *Microsporum*
    c. *Epidermophyton* d. *Trichosporon*
14. **Dermatophytes infecting:**
    a. Subcutaneous tissue
    b. Systemic organs
    c. Nails, hair and skin
    d. Superficial skin and deeper tissue
15. **Which of the following infecting hair, skin and nail?**
    a. *Trichohyton* b. *Microsporum*
    c. *Epidermophyton* d. None of above
16. **Trichohyton species which is zoophilic?**
    a. *T. tonsurans* b. *T. violaceum*
    c. *T. mentagrophytes* d. *T. schoenleinii*
17. **Which of the following is spread from animals to man?**
    a. *T. rubrum* b. *T. tonsurans*
    c. *E. flocossum* d. *T. verrucosum*
18. **Which of the following does not produce dermatophytosis in India?**
    a. *T. rubrum*
    b. *T. mentagrophytes*
    c. *Microsporum distortum*
    d. *Epidermophyton flocossum*
19. **Most common cause of tinea capitis:**
    a. *M. canis* b. *Epidermophyton flocossum*
    c. *T. tonsurans* d. *M. distortum*
20. **Black dot ring worm is caused by:**
    a. *Microsporum* b. *Trichohyton*
    c. *Epidermophyton* d. *Candida*
21. **T capitis (endothrix) is caused by:**
    a. *Epidermophyton* b. *T. tonsurans*
    c. *T. violaceum* d. *Microsporum*
    e. *T. rubrum*
22. **Common causative agent of favus is:**
    a. *Trichophyton rubrum*
    b. *Epidermophyton floccosum*
    c. *Trichophyton mentagrophytes*
    d. *Trichophyton schoenleinii*

23. **Tinea cruris is caused by:**
   a. *Epidermophyton*   b. *Trichohyton*
   c. *Microsporum*   d. *Candida*

24. **Dermatophytids are:**
   a. Fungal hyphae in skin
   b. Vegetative fungal cells in keratinized tissue
   c. Cutaneous lesions secondary to hypersensitivity to fungal antigens
   d. Dead fungal tissue

Answers and Explanation of MCQs

1. c
- Ptyriasis versicolor is a type of surface mycosis caused by *M. furfur*, while *E. flocossum*, *M. gypseum* and *T. tonsurans* are the dermatophytes causing cutaneous mycosis.

2. d
- Follow section, **surface mycosis (*Malassezia* infections → diagnosis)** for explanation.

3. a

4. d
- Follow section, **surface mycoses (peidra)** for the explanation of answers of MCQs 3–4.

5. b

6. b

7. a

8. b

9. c

10. d

11. b
- Follow section, **candidiasis** for explanation of answers of MCQs 5–11.

12. c

13. b

14. c

15. a
- Follow **Table 90.1** for explanation of answers of MCQs 12–15.

16. c

17. d
- Follow **Flowchart 90.1** for explanation of answers of MCQs 16–17.

18. c
- Follow section, **cutaneous mycoses (dermatophytosis → pathogenicity → precipitating factors → environmental factors)** for explanation.

19. c

20. b

21. b, c

22. d

23. a, b

24. c
- Follow section, **cutaneous mycoses (dermatophytosis → pathogenicity → clinical feature)** for explanation of answers of MCQs 19–24.

Deep Mycoses

Fungal infections present in the tissues below the skin and in different systems called **deep mycoses** with three subtypes like subcutaneous mycoses, systemic mycoses and opportunistic mycoses.

SUBCUTANEOUS MYCOSES

These are the fungal infections of subcutaneous tissues mostly transmitted by trauma. Examples of subcutaneous mycoses are **mycetoma, sporotrichosis, chromomycosis, rhinosporidiosis plus oculosporidiosis, lobomycosis and subcutaneous zygomycosis.**

Mycetoma

Synonym: It also called **madurella mycosis** or **madura mycosis** (as it was 1st diagnosed in Madura district of Tamil Nadu, India by Gill in 1842) or **madura foot** (exclusive for foot lesion of mycetoma).

Definition: It is a chronic granulomatous lesion of subcutaneous and deeper tissues with multiple sinuses, discharging pus or granules.

Etiological agents: Two etiological types:

1. **Actinomycetoma:** Mycetoma caused by bacteria.
   - **Aerobic bacteria**
     – *Nocardia asteroids, N. brasiliensis, N. caviae, Streptomyces somaliensis* and *Actinomadura madurae:* They produce yellowish-white grains
     – *Actinomadura pelletieri:* It produces red grains. It is the most common cause of mycetoma in India in West coast.
     – *Nocardiopsis dassonvillei:* It produces cream color grains.
   - **Anaerobic bacteria:** These are *Actinomyces israelii* and *A bovis.* They produce yellowish grains called **sulfur granules.**
2. **Eumycetoma (mycotic mycetoma):** Mycetoma caused by following fungi.

- **Black grains fungi:** *Madurella mycetomatis, Madurella grisea, Exophiala* (formerly *Phialophora*) *jeanselmei* and *Cuvularia geniculata.*
- **White grains fungi:** *Pseudoallescheria boydii, Aspergillus nidulus, Acremonium falciforme* and *Fusarium graminiarum.*

Pathogenicity

- **Reservoirs and sources of infection:** Fungus present in soil or wooden sticks, which are the reservoirs and sources for infections.
- **Modes of transmission:** It is transmitted by trauma or thorn prick injury, which allows the entry of microbes.
- **Portal of entry:** Skin.
- **Sites:** It occurs mostly on foot **(Fig. 91.1a)**, but also presents on hands, shoulders, scalp or buttocks.
- **Precipitating factors (epidemiological determinants): (1) Occupation:** It is common in farmers, carpenters and field workers. **(2) Age:** It is common in adults between 20 and 40 years. **(3) Sex:** More in male. **(4) Areas:** More in rural areas than urban. **(5) Poor hygiene:** It also increases the risk.
- **Pathogenesis:** Mycetoma presents with granulomatous lesions and sinus that discharges the granules. Granules centrally contain organism (bacteria or fungi) which look gram-positive surrounded by deposition of Ag-Ab complexes in periphery due to host defense which look gram-negative.

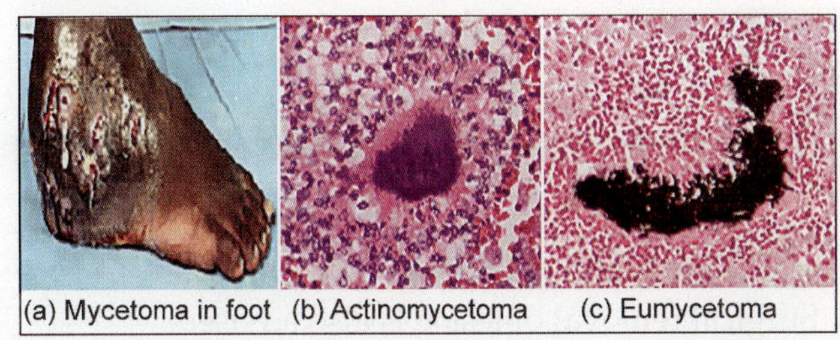

Fig. 91.1: Mycetoma

- **Clinical features:** Lesion is painless. It is a triad of granulomatous swelling (tumefaction), sinus and discharge of grains or granules. Grains are different in color according to causative agent as mentioned earlier. Pus discharge presents, due to secondary bacterial infections.
- **Complications:** (1) Secondary bacterial infections. (2) Local spread occurs in surrounding tissues to involve underlying fasciae and bones causing osteolytic or osteosclerotic bony lesions. No lymphatic or blood spread occurs in distant sites. (3) Recurrence occurs after treatment.

> **Note: Botryomycosis**
> It is a chronic granulomatous mycetoma like lesion caused by *Staph. aureus* and other pyrogenic bacteria. It resembles to mycetoma, so called **pseudomycosis.**

Laboratory diagnosis

- **Specimens:** Grains, discharge or tissue from lesion.
- **Testing**

A. Macroscopy of grains
- – Color of granules: Follow etiological agents.
- – Consistency and size of granules: Thin and <1 μm in actinomycetes while thick and larger in mycetoma.

B. Microscopy: Crush the granules between two slides and do the microscopy.
- – Actinomycetoma **(Fig. 91.1b)** is identified by Gram's stain. Granules contain centrally gram-positive filaments and peripherally gram-negative materials giving club shape or sun-ray appearance.
- – Eumycetoma **(Fig. 91.1c)** is identified by KOH, which shows hyaline septate hyphae in white grain fungi and black hyphae in black grain fungi.
- – Modified ZN stain is useful for *Nocardia* spp.
- – Histopathological stains are useful to know the granulomatous reaction.

C. Culture: It is done for bacteria and fungi.
- – Actinomycetoma: BHIA, LJ medium or blood agar
- – Eumycetoma: SDA with antibiotics. Cultivation technique is same as like other fungi.

D. Animal inoculation: Guinea pig and mouse are useful.

E. Special tests: Like slide culture, teased mount are useful for identification of particular fungi.

Prevention: Improvement of hygiene, avoiding traumatic injury, removal of risk factors, health education, etc., are the preventive measures.

Treatment

- Eumycetoma: Antifungal drugs like itraconazole or AMB are useful.
- Actinomycetoma is treated by Welsh regimen (co-trimoxazole and amikacin).
- Surgical removal of lesion is also advised.
- Recurrence is common after treatment.

Sporotrichosis

Synonym: Rose Gardner's disease or rose thorn disease, because it can be spread by rose.

Definition: It is a chronic pyogranulomatous lesion of skin and subcutaneous tissues, which may be fixed or spread to lymph nodes or other organs.

Etiological agent: It is caused by *Sporothrix schenckii*. It is a dimorphic fungus, which grows as yeast phase at 37°C in tissue and as mycelial phase at 22°C in environment.

Pathogenicity

- **Epidemiology:** It is reported from Central South America and South Africa. In India, it is endemic in sub-Himalayan holly areas of North-East states ranging from Himachal Pradesh to Assam. Other endemic areas are Northern Karnataka and Southern Maharashtra.
- **Reservoirs and sources of infection:** Fungus found as saprobe in soil, plants, wood, bark, leaves, straw and other dead materials, which are the reservoirs and sources for infection.
- **Modes of transmission:** It is introduced by trauma by wooden stick. Inhalation of conidia may elicit the lesion of lungs.
- **Incubation period:** 8–30 days (average 3 weeks).
- **Portal of entry:** Skin.
- **Sites:** Skin, subcutaneous tissue and lymphatics of feet and hands are common sites. It rarely occurs in lungs and in other organs.
- **Precipitating factors (epidemiological determinants):**
  - **Agent factors (virulence factors): (1) Fungal dimorphism:** As described in *H. capsulatum*. **(2) Enzymes:** Fungus releases the proteinase I (serine proteinase) and proteinase II (aspartic proteinase) that help in fungal invasion.
  - **Host factors: (1) Age:** It is common in young adults. **(2) Sex:** It is three times more in males than females due to outdoor job. **(3) Occupation:** Common in persons exposed to soil or plants like gardeners, farmers, carpenters, field workers, mine workers, florists and forest workers.
  - **Environmental factors:** Fungus is common in tropical and subtropical countries with high humidity (65%) and moderate temperature (25–28°C).
- **Pathogenesis:** Formation of granulomatous lesion with yeast cells in center surrounded by eosinophilic materials and Ag-Ab complex deposition called **asteroid body (Fig. 91.2b)**. Deposition of eosinophilic materials around yeast cells is due to immune phenomenon of body called **Splendore-Hoeppli phenomenon.**
- **Clinical features: (1) Fixed cutaneous sporotrichosis:** Granulomatous nodule seen at inoculation site in skin and subcutaneous tissue, which follows the ulcer and necrosis. It is the most common form of sporotrichosis.

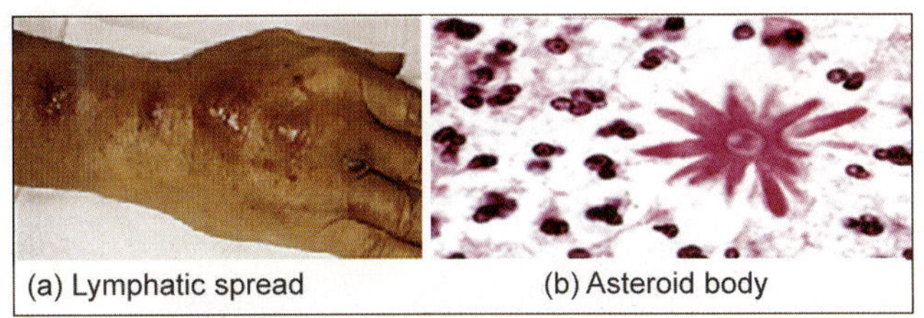

Fig. 91.2: Images of sporotrichosis

It remains fixed and no spread by lymph, hence called **'fixed cutaneous sporotrichosis'**. (2) **Lympho-cutaneous sporotrichosis:** Cutaneous lesion spreads along the course of lymphatic vessels to lymph nodes called **'lymphocutaneous sporotrichosis'**. Enlarged lymph nodes extend centripetally, like beaded chain **(Fig. 91.2a)**. (3) **Pulmonary sporotrichosis:** It may occur as primary infection due to inhalation of conidia.

- **Complications:** Very rarely nodules spread by hematogenous route to meninges, brain, bones, joints, kidneys, etc., called **'disseminated sporotrichosis'**.

Laboratory diagnosis

- **Specimens:** Skin scraping, discharge or tissue from granulomatous lesion.
- **Testing**

A. Microscopy: (1) Direct 10% KOH: It shows round, spherical yeast cells or cigar shaped body. Schematic diagram is shown in **Fig. 91.3a**. **(2) Gram's stain:** It shows gram-positive yeast cells. **(3) Histopathological stains:** H and E, PAS and GGMS stain show the asteroid body **(Fig. 91.2b)**.

B. Culture

- Medium: It is SDA contains antibiotics.
- Cultivation technique: **Ch. 120** (section → **Flowchart 120.1**).
- C/Cs on SDA: Creamy-white colonies appear in 2–5 days which may turn in to brown-black color in 10–14 days.
- Identification of growth by LCB/PHOL stain: **(1) From yeast phase plate:** Round-spherical or cigar shaped yeast cells are present. Schematic diagram is shown in **Fig. 91.3a**. **(2) From mycelia phase plate:** Shows hyaline septate hyphae and conidiophores with arrangement of conidia in chain giving **flower like pattern.** Schematic diagram is shown in **Fig. 91.3b**.

C. Animal inoculation: Mice and rats are susceptible.

D. Molecular methods: PCR.

E. Skin test: It is done by using sporotrichin-Ag of mycelial phase or peptido-rhamno-mannan Ag of yeast phase.

F. Special tests: Teased mount, slide culture cellophane tape mount and exo-antigen tests are useful.

Prevention: Preventive measures are same as mycetoma.

Treatment: Surgical removal of localized lesion. Itraconazole is the drug of choice except in disseminated lesion where AMB is the drug of choice. Fungus is sensitive to temperature above 39°C, so hyperthermic treatment is successful for disseminated lesions.

Chromomycosis

Meaning: Chromo means colored (black) and mycosis means fungal infection (black fungus).

Definition: It is a chronic granulomatous infection of skin and subcutaneous tissues caused by black fungi (phaeoid fungi or dematiaceous fungi).

Etiological agents: Two variants.

1. **Chromoblastomycosis (verrucous dermatitis):** It occurs due to yeast phase of fungus and characterized by presence of black yeast cells with dividing septum **called sclerotic bodies** or **muriform cells** or **medlar bodies** or **copper pennies bodies** as shown in **Fig. 91.4**. Sclerotic bodies are dark brown, about 5–12 µm in size with multiple internal transverse septa and chest nut in appearances. Sclerotic body is the pathognomic feature of chromoblastomycosis and not seen in any other phaeoid fugal infection. It is caused by following fungi.
   - *Fonsecaea compacta.*
   - *Fonsecaea pedrosoi.*
   - *Phialophora verrucosa.*
   - *Cladophialophora* (formerly called *Cladosporium*) *carrionii.*
   - *Rhinocladiella aquaspersa.*

2. **Phaeohyphomycosis:** It occurs due to mycelial (hyphae) phase of fungus and characterized by presence of black hyphae and no sclerotic bodies. It is caused by following fungi.
   - *Wangiella dermatitidis.*
   - *Exophiala jeanselmei.*
   - *Cladophialophora* (formerly *Cladosporium*) *bantiana.*
   - *Nattrassia magniferae.*

Deep Mycoses

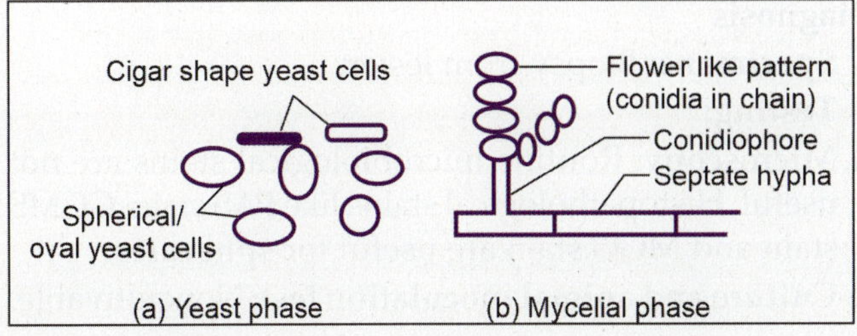

Fig. 91.3: Schematic diagram of *S. schenkii*

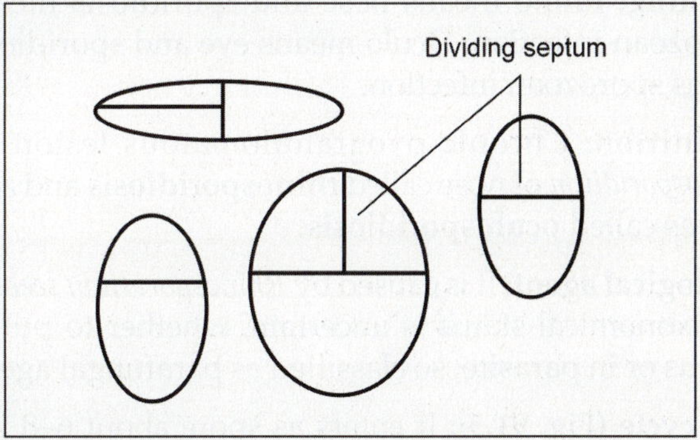

Fig. 91.4: Sclerotic bodies

Pathogenicity

- **Reservoirs and sources of infection:** All fungi found as saprobes in soil, plants, wood, barks, leaves, straws and other dead materials, which are the reservoirs and sources for infection.
- **Modes of transmission:** Fungus is introduced by wooden stick injury.
- **Portal of entry:** Skin.
- **Sites:** Skin and subcutaneous tissue of feet and hands are common sites.
- **Precipitating factors (epidemiological determinants):** **(1) Age:** It is more common in young adults. **(2) Sex:** It is more in males due to outdoor job. **(3) Occupation:** It is common in persons exposed to soil or plants like gardeners, farmers, carpenters and field workers, mine workers, florists and forest workers.
- **Pathogenesis, clinical features and complication**
  1. **Chromoblastomycosis:** It presents as nodular, verrucous, tumoral, plaques or cicatricial type skin lesion. It raised about 1–3 cm above the skin level with irregular surface and gives a cauliflower like appearance hence called **verrucous dermatitis.** It shows the presence of sclerotic bodies. No dissemination, but local spread may occur. Sometimes, it produces oral ulcer with local lymphadenopathy.
  2. **Pheohyphomycosis:** It presents as granulomatous lesion with abscess, giant cells, histiocytes and black hyphae. It disseminated to other organs (complication).

Diagnosis

- **Specimens:** Skin scraping, discharge or tissue from lesion.
- **Testing:** Microscopy revealed black hyphae and sclerotic bodies, while culture gives black colonies.

Prevention: Same as mycetoma.

Treatment: Surgical removal of localized lesion. Flucytosine alone or along with itraconazole or AMB are the drugs of choice. Fungi are sensitive to high temperature, so hyperthermic treatment is also an alternate. Local application of cryotherapy (liquid nitrogen) or laser therapy is useful.

Rhinosporidiosis and Oculosporidiosis

Meaning: Rhino means nose and sporidiosis means sporozoan infection. Oculo means eye and sporidiosis means sporozoan infection.

Definition: Chronic pyogranulomatous lesion by *Rhinosporidium* of nose called **rhinosporidiosis** and and of eyes called **oculosporidiosis.**

Etiological agent: It is caused by *Rhinosporidium seeberi.* Its taxonomical status is uncertain, whether to put in fungus or in parasite, so classified as parafungal agent.

Life cycle (Fig. 91.5): It enters as spore about 6–8 μm in size. Spore forms large, thick double wall structure

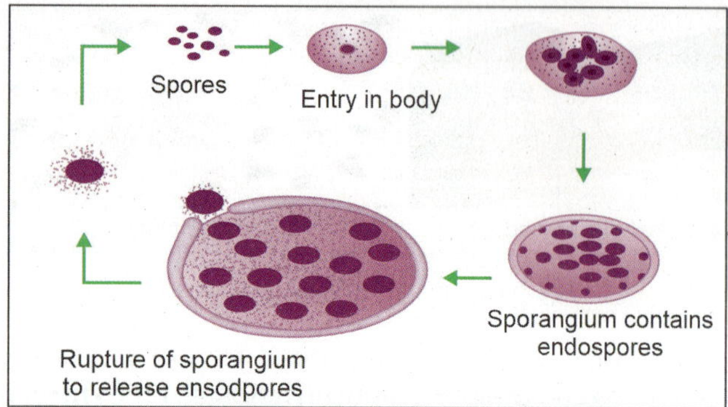

Fig. 91.5: Life cycle of *R. seeberi*

called **sporangium** (up to 580 μm) contains about 12,000–16,000 endospores (6–9 μm). Sporangium ruptures after maturation to release the endospores. Each endospore starts the new cycle.

Pathogenicity

- **Epidemiology:** It is endemic in South India (coastal areas like Tamil Nadu, Kerala, Andhra Pradesh and Pondicherry) and Sri Lanka. Isolated cases are reported from East Africa, South America, North America and Europe.
- **Reservoirs and sources of infection:** Water.
- **Modes of transmission:** It is a hydrophilic fungus habitats in fresh or stagnant water and introduced by direct contact with water.
- **Incubation period:** 1–10 months.
- **Portal of entry:** Nasal or ocular mucosa.
- **Sites:** Nose and eyes.
- **Precipitating factors (epidemiological determinants):** **(1) Age:** No age is exempted but it is more common between 2nd and 4th decades of life. **(2) Sex:** Male-female ratio is 3:1 to 4:1. **(3) IDDs:** It is reported from patient of AIDS.
- **Pathogenesis and clinical features: (1) Rhinosporidiosis:** It is chronic granulomatous lesion of nasal cavity characterized by development of polyp. Granuloma contains central sporangium surrounded by inflammatory cells called **spherules.** **(2) Oculosporidiosis:** It is chronic granulomatous lesion of eye (more in palpebral conjunctiva and rare in other parts) characterized by development of polyp. It also associated with sclera melting and staphyloma formation. Chronic dacryocystitis also presents sometimes.
- **Complications:** It spreads by blood to other organs like bone. Recurrence occurs after treatment.

Diagnosis

- **Specimens:** Biopsy from lesion.
- **Testing:**
A. **Microscopy:** Routine microbiological stains are not useful. Histopathological stains like PAS stain, GGMS stain and MGG stain are useful for **spherules.**
B. **Culture and animal inoculation test:** Noncultivable.

Prevention: Avoid contact with contaminated water.

Treatment: Surgical removal of lesion or electro-cauterization. Dapsone 100 mg is given orally for 1 year to prevent the recurrence. Antimony compound like neostibasan is also given. Despite all efforts recurrence is common.

Lobomycosis

Synonym: Jorge Lobo's disease or keloidal blastomycosis.

Definition: It is chronic granulomatous infection of skin and subcutaneous tissues without involvement of mucosa or deeper tissues.

Etiological agent: It is caused by *Loboa loboi* (*Lacazia loboi*).

Pathogenicity

- **Reservoir and source of infection:** Water.
- **Modes of transmission:** It is as hydrophilic fungus habitats in water and introduced by direct contact of abraded skin with infected water.
- **Incubation period:** Very long about 2.5 years.
- **Portal of entry:** Skin.
- **Sites:** Skin and subcutaneous tissues of lower limb, upper limb, face and ears.
- **Precipitating factors (epidemiological determinants):** **(1) Age:** It is more in young adults. **(2) Sex:** It is more in males due to farm activities. Male-to-female ratio is 20:1.
- **Pathogenesis:** Formation of granulomatous lesion with yeast cells in center surrounded by eosinophilic materials and deposition of Ag-Ab complexes called **asteroid body**. Deposition of eosinophilic material around yeast cells is due to immune phenomenon of body called **Splendore-Hoeppli phenomenon**.
- **Clinical features:** It presents as chronic granulomatous infection of skin and subcutaneous tissues without involvement of mucosa or deeper tissues, which may be keloid or verrucous type.
- **Complications:** Local and lymphatic spread occurs in surrounding tissues while disseminated lesion in other organs likely less. Recurrence occurs after treatment.

Diagnosis

- **Specimens:** Biopsy from lesion.
- **Testing**
- A. **Microscopy:** **(1) Direct 10% KOH:** It shows round, spherical, budding yeast cells. Yeast cells are also arranged in chain of 4–7 cells called **blastoconidia.** Schematic diagram is shown in **Fig. 91.6. (2) Gram's stain:** It shows gram-positive yeast cells. **(3) Histopathological stains:** Fungi contain melanin and stain by Masson-Fontana stain and shows the asteroid body.
- B. **Culture:** It is not cultivated on cell free media.
- C. **Animal inoculation:** Mouse is best useful.

Prevention: Avoid contact with contaminated water.

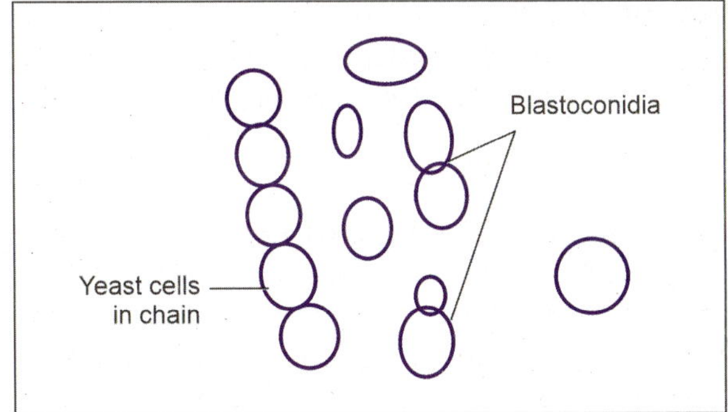

Fig. 91.6: Schematic diagram (yeast cells and blastoconidia) of *L. loboi*

Treatment: Surgical removal of localized lesion. Clofazimine produced good result in some patient either alone or with itraconazole.

Subcutaneous Zygomycosis

Synonym: Entomophthoromycosis or entomophthoramycosis, because all the fungi are belong to order Entomopthorales.

Meaning: Entomo (Greek) means **insect,** because primary pathogens of insects.

Etiological agents: It is caused by fungus from Zygomycetes group like *Conidiobolus coronatus* and *Basidiobolus ranarum*.

Pathogenicity: Written in **Table 91.1** with differences between two species.

Diagnosis

- **Specimens:** Biopsy from lesion.
- **Testing:** Difficult to isolate in culture. Histopathological stains show the nonseptate hyphae.

TABLE 91.1: Differences between *C. coronatus* and *B. ranarum*		
Features	***C. coronatus***	***B. ranarum***
Reservoir and source	Both are present on decaying vegetation, in soil and in insects	
Mode of transmission	Inhalation of spores	By insect bite
Portal of entry	Mucosa	Skin
Sites	Nose, paranasal sinuses, lips	Trunk and buttocks
Precipitating factors	Both are opportunistic fungi, cause infection only in immunodeficient persons	
Pathogenesis	Eosinophulic infiltration occurs around centraly located hyphae in both fungi called **Splendore-Hoeppli phenomenon**	
C/Fs	—	Painless, mucosal nodule, which may enlarge in size
Complication	Locally spread to involve entire face with disfigurement of face	Dissemination may occur in GIT, lungs and retroperitoneal tissues but very rare

Treatment: Oral potassium iodide alone or with surgical excision of lesion gives better result. Hyperbaric oxygen is believed to accelerate the healing process.

SYSTEMIC MYCOSES

These are the fungal infections of different systems. All the systemic fungi are dimorphic fungi. They are endemic in different regions, so called **endemic mycosis** (few authors defined only coccidioidomycosis as endemic mycosis). Examples of systemic mycoses are **histoplasmosis, blastomycosis, coccidioidomycosis** and **paracoccidioidomycosis.**

Histoplasmosis

Synonym: It also called Darling's disease (as disease was 1st described by Samuel Darling in 1905), Cave disease, Spelunker's lung, Caver's disease, Ohio Valley disease and reticuloendotheliosis.

Definition: It is a granulomatous infection of RE system.

Etiological agent: It is caused by *Histoplasma capsulatum* (species name is misnomer, infact fungus has no capsule). It is a dimorphic fungus with following three variants:

- *H. capsulatum* var. *capsulatum*: Causing classical histoplasmosis in humans.
- *H. capsulatum* var. *duboisi:* Causing African histoplasmosis in humans.
- *H. capsulatum* var. *farciminosum:* Causing histoplasmosis (epizootic) in animals.

Pathogenicity

- **Epidemiology:** Fungus is distributed worldwide. It is endemic in USA, particularly in states bordering the Ohio River Valley and the lower Mississippi river. In India it is endemic in West Bengal along the Ganga river.
- **Reservoirs and sources of infection:** It presents in soil with high nitrogen contents, especially in bird reservoirs (chicken coops, pigeon roosts) or bat dropping which are the reservoirs and sources for infection.
- **Mode of transmission:** It is transmitted by inhalation of spores. Person-to-person spread does not occur.
- **Incubation period:** 10–16 days.
- **Portal of entry:** Respiratory system.
- **Sites:** Pulmonary and extrapulmonary sites like skin, subcutaneous tissues, liver, spleen, nodes, etc.
- **Precipitating factors (epidemiological determinants)**
  - **Agent factors (virulence factors):** It is **dimorphic fungus** has two forms like mycelium and yeast phase, possesses different antigenic and surface features, and requires different host mechanisms to contain it.
  - **Host factors: (1) Age:** It is common in young adults. **(2) Sex:** It is more in males due to outdoor job. **(3) Occupation:** Accidental transmission

occurs in laboratory workers while handling the culture. **(4) IDDs:** It is common in persons with IDDs like AIDS.

- **Pathogenesis:** It produces acute as well as chronic inflammation. Inhaled spores are engulfed by macrophages, but evade the host defense due to dimorphism and change to yeast phase and elicit the immune response resulting formation of granuloma like tuberculosis contains yeast cells, giant cells, epithelioid cells with or without caseation necrosis. Granuloma heals with fibrosis and calcification. Unlike tuberculosis, histoplasmosis once heals rarely reactivates.
- **Clinical features**
  - **Classical histoplasmosis: (1) Acute pulmonary histoplasmosis:** Minor respiratory features like cough, fever and flu like illness occur. Chest X-ray findings are normal in 40–70% of cases. **(2) Chronic pulmonary histoplasmosis:** Granulomatous lesion in lungs called **histoplasmoma** and may simulates the tuberculosis. Chest X-ray suggest pulmonary infiltrate with hilar or mediastinal lymphadenopathy. **(3) Dissemination of infections:** It occurs in patient with low CMI in organs like skin, subcutaneous tissues, oral cavity (common in India) lymph nodes, liver, spleen, brain, etc.
  - **African histoplasmosis:** It is more in skin, subcutaneous tissues and bones.
- **Complications:** These are common in absence of proper treatment and especially in persons with IDDs. Common complications are recurrent pneumonia, respiratory failure, fibrosis, fibrosing mediastinitis (it can be fatal), superior vena cava syndrome, pulmonary vessels obstruction, progressive fibrosis of lymph, chronic cavitary histoplasmosis, hemoptysis (due to erosion of airway by hard calcified node), dissemination of infection in other organs, anemia and high fatality.

Diagnosis

- **Specimens:** BAL, sputum, biopsy from lesion or other suitable specimens.
- **Testing**

A. **Microscopy: (1) Direct 10% KOH:** In classical histoplasmosis round, spherical yeast cells about 2–4 µm are seen while African histoplasmosis they are larger about 7–15 µm inside the mononuclear or polymorphonuclear cells. Schematic diagram is shown in **Fig. 91.7a. (2) Gram's stain:** It shows gram-positive yeast cells. **(3) Histopathological stains:** Giemsa and PAS stains show the granulomatous lesion with intracellular fungus. **(4) Fluorescent stain:** It is also useful.

B. **Culture**
  - Medium: It is SDA contains antibiotics.
  - Cultivation technique: **Ch. 120** (section → **Flowchart 120.1**).

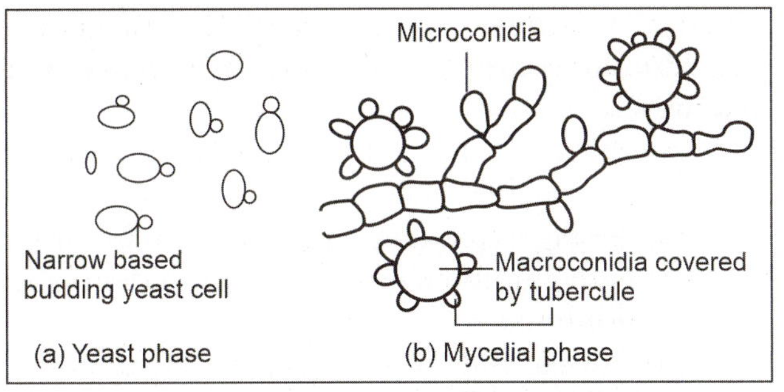

Fig. 91.7: Schematic diagram of *H. capsulatum*

- C/Cs on SDA: **(1) Yeast phase:** Mucoid cream color colonies are seen which may turn in to brown color in 10–14 days. **(2) Mycelia phase:** Initially white colonies, which later becomes brown.
- Identification of growth by LCB/PHOL stain: **(1) From yeast phase plate:** Round-spherical, narrow based budding yeast cells are present. Schematic diagram is shown in **Fig. 91.7a. (2) From mycelia phase plate:** It shows hyaline septate hyphae surrounded by microconidia and macroconidia. Macroconidia surrounded by a finger like projections or tubercles called **tuberculate spores**. Schematic diagram is shown in **Fig. 91.7b**.

C. Animal inoculation: Mouse is an ideal model.

D. Molecular methods: PCR.

E. Skin test: Detection of DTH by using **histoplasmin-Ag** of mycelial phase.

F. Special tests: Teased mount, slide culture, cellophane tape mount and exo-antigen tests are also useful.

Prevention: Avoid the visit of bird reservoirs and use of PPE especially by laboratory workers.

Treatment: Liposomal AMB is given for acute and disseminated lesions. Itraconazole is effective in chronic cavitary pulmonary lesions.

Blastomycosis

Synonym: It also called North American blastomycosis, as it is endemic in North American continents.

Definition: It is a chronic pyogranulomatous infection especially of lungs and skin.

Etiological agent: It is caused by *Blastomyces dermatitidis* (teleomorphic/sexual state is *Ajellomyces dermatitidis*).

Pathogenicity

- **Epidemiology:** Fungus is endemic in North American continents. In India, it is isolated from bronchial aspirates of a patient in Delhi and also from the lungs of insectivorous bats.
- **Reservoirs and sources of infection:** Exactly not known, but spores may present in moist soil.
- **Modes of transmission:** By inhalation of spores.
- **Incubation period:** 45 days.
- **Portal of entry:** Respiratory system.
- **Sites:** Pulmonary and extrapulmonary sites like skin, bone, genitourinary system, breast, etc.

- **Precipitating factors (epidemiological determinants)**
  - **Agent factors (virulence factors):** Fungal dimorphism as described with *H. capsulatum*
  - **Host factors: (1) Age:** It is common in young adults. **(2) Sex:** Male-to-female ratio is 6:1. **(3) IDDs:** It is common in persons with IDDs like AIDS, leukemia/lymphoma and steroid therapy.
- **Pathogenesis:** Inhaled spores, change to yeast phase and elicit the immune response resulting formation of pyogranulomatous lesion in lungs. Later it disseminates to other sites by blood.
- **Clinical features:** Pyogranulomatous lesion in lungs called **pulmonary blastomycosis.**
- **Complications:** Hematogenous spread occurs in other organs like skin (cutaneous blastomycosis), bones, genitourinary system, breasts, brain, etc. Relapse is common.

Diagnosis

- **Specimens:** BAL, sputum, biopsy from lesion or other suitable specimens.
- **Testing**

A. Microscopy: (1) Direct 10% KOH: It shows round, spherical, thick, double wall, broad-based budding yeast cells with 7–20 µm size and giving figure of 8 appearance. Schematic diagram is shown in **Fig. 91.8a. (2) Gram's stain:** It shows gram-positive yeast cells. **(3) Histopathological stains:** Giemsa and PAS stains show fungus surrounded by granuloma. **(4) Fluorescent stain:** It is also useful.

B. Culture

- Medium: It is SDA contains antibiotics.
- Cultivation technique: **Ch. 120 (section → Flowchart 120.1).**
- C/Cs on SDA: **(1) Yeast phase:** Cream color colonies are seen, which become tan color in old culture. **(2) Mycelia phase:** Same colonies as like yeast phase.
- Identification of growth by LCB/PHOL stain: **(1) From yeast phase plate:** Broad-based budding yeast cells are present. Schematic diagram is shown in **Fig. 91.8a. (2) From mycelia phase plate:** It shows hyaline septate hyphae with microconidia on tip of lateral branches. Schematic diagram is shown in **Fig. 91.8b**. It also shows the chlamydospores.

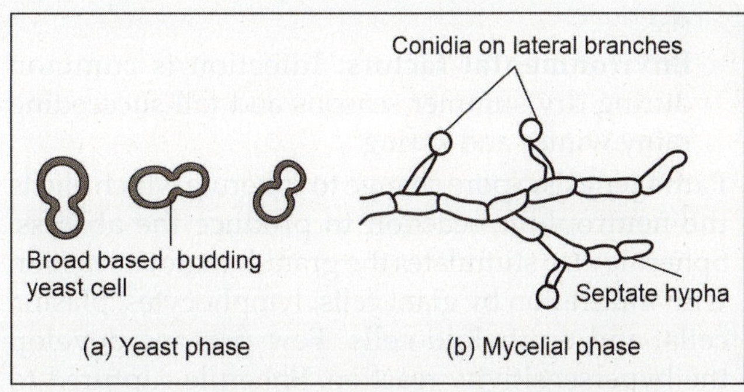

Fig. 91.8: Schematic diagram of *B. dermatitidis*

C. **Animal inoculation:** Mouse, rat, guinea pig, and hamster are susceptible.

D. **Molecular methods:** PCR.

E. **Skin test:** Detection of DTH by using **blastomycin-Ag** of mycelial phase.

F. **Special tests:** Teased mount, slide culture, cellophane tape mount and exo-antigen tests are also useful.

Prevention: Avoid the visit of suspected places. Live recombinant vaccine is developed by using surface antigen (called **BAD antigen → B. dermatitidis adhesin**), which is immunogenic in animals and under trial for human use.

Treatment: Liposomal AMB is given in all cases.

Coccidioidomycosis

Synonym: It also called San Joaquin Valley fever, Desert rheumatism, California fever or Kok-see in local language.

Definition: It is a granulomatous infection of respiratory system.

Etiological agent: It is caused by *Coccidioides immitis* (im means not and and mitis means mild, indicates not mild disease/serious nature of disease).

Pathogenicity

- **Epidemiology:** Fungus is endemic in Northern Mexico, Arizona, California, Nevada, Texas and Utah.
- **Reservoirs and sources of infection:** Spores are present in alkali soil of desert. Fungus produces the large numbers of arhthroconidia which are dispersed in air by recurring winds, which act as the reservoir and source for infection.
- **Modes of transmission:** It is transmitted by inhalation of spores. Person to spread is not reported.
- **Incubation period:** About 2 weeks.
- **Portal of entry:** Respiratory system.
- **Sites:** Pulmonary, extrapulmonary sites and almost all sites except GIT.
- **Precipitating factors (epidemiological determinants)**
  - **Agent factors (virulence factors): (1) Fungal di-morphism:** As described in *H. capsulatum*. **(2) Large size spherule:** Difficult to kill by host cells.
  - **Host factors: (1) Age:** It is common in young adults. **(2) Sex:** It is common in male than female. **(3) IDDs:** Immunodeficient persons are at high risk.
  - **Environmental factors:** Infection is common during dry summer seasons and fall succeeding rainy winter and spring.
- **Pathogenesis:** Spore change to spherule which elicits the neutrophilic reaction to produce the abscess. Spherule also stimulates the granulomatous reaction with infiltration by giant cells, lymphocytes, plasma cells, and epitheloid cells. Few persons develop the hypersensitivity reaction. Spherule ruptured to release the endospores.

- **Clinical features:** Lung lesion called **pulmonary coccidioidomycosis** which is present with following features.
  - Acute stage: Flu like symptoms like fever, cough, headache, malaise, etc.
  - Chronic stage: About 90% patients turn to chronic stage with granulomatous lesion formation called **coccidioidoma**.
- **Complications:** Calcification occurs in pulmonary lesion. Lung cavities may develop in 5% cases. Hematogenous spread occurs in almost all organs except GIT. Relapse is common.

Diagnosis

- **Specimens:** BAL, sputum, biopsy from lesion or other suitable specimens.
- **Testing**

A. **Microscopy:** Giemsa, GGMS and PAS stains show the spherules with granuloma formation. When spherule becomes mature it ruptures to release endospores. Spherule develops in tissue (lungs) not in culture. Barrel-shaped spores called **arthrospores** are also present in tissues.

B. **Culture**
  - Medium: It is SDA contains antibiotics.
  - Cultivation technique: **Ch. 120 (section → Flowchart 120.1).**
  - C/Cs on SDA: It grows as mold at 25°C and at 37°C. Spherule develops in tissue (lungs) not in culture. Mycelial phase produces white-tan colored colonies.
  - Identification of growth by LCB/PHOL stain: **(1) Yeast phase:** No yeast (spherule) phase develops in culture but it develops in body. Schematic diagram of spherule is shown in **Fig. 91.9a. (2) From mycelia phase plate:** It shows hyaline septate hyphae with arthrospores (barrel-shaped spores). Schematic diagram is shown in **Fig. 91.9b.** Rupture of hyphae release the arhthroconidia which are infectious.

C. **Animal inoculation:** Mouse, rat, guinea pig, and hamster are susceptible.

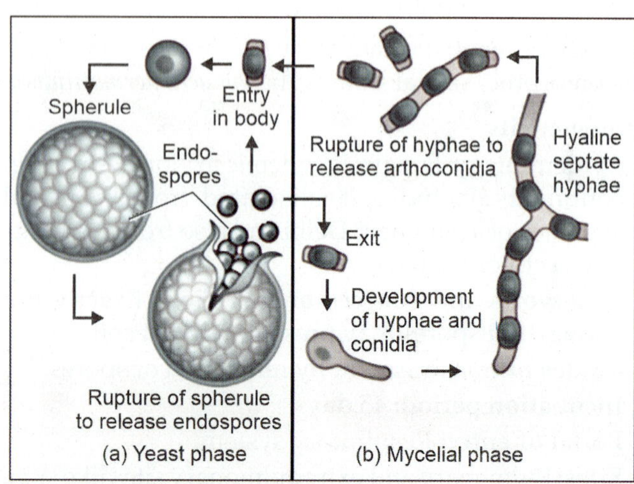

Fig. 91.9: Schematic diagram of *C. immitis*

D. Serological tests: Like agglutination test, CFT immunodiffusion test, etc.

E. Molecular methods: PCR.

F. Skin test: Detection of DTH by using **coccidiodin-Ag** of mycelial phase and **spherulin-Ag** from spherules.

G. Special tests: Teased mount, slide culture, cellophane tape mount and exo-antigen tests are also useful.

Prevention: Avoid visiting the suspected places. Immunity develops following primary infection provides the lifelong protection. Vaccine is developed, and it is useful for persons living in endemic areas.

Treatment: It is a self limiting condition. Itraconazole is effective in all cases except diffuse pneumonia, where AMB is recommended. Relapse is major issue following treatment.

Paracoccidioidomycosis

Synonym: It also called South American blastomycosis, because common in South American countries or Lutz Splendore de Almeida disease.

Definition: It is an acute or chronic granulomatous endemic systemic fungal infection primarily involving lungs and disseminated to other tissues.

Etiological agent: It is caused by *Paracoccidioides brasiliensis*.

Pathogenicity

- **Epidemiology:** It is Endemic in Brazil and other South American countries.
- **Reservoir and source of infection:** It is a saprophytic fungus and spores are present in soil.
- **Mode of transmission:** By inhalation of spores.
- **Incubation period:** Clinical illness develops after several years of exposure.
- **Portal of entry:** Respiratory system.
- **Sites:** Pulmonary and extrapulmonary.
- **Precipitating factors (epidemiological determinants)**
  - **Agent factors (virulence factors):** It is **fungal dimorphism** as described above in *H. capsulatum*.
  - **Host factors: (1) Age:** About 90% cases occur in adults between 20 and 50 years of age. **(2) Sex:** It is less in female due to inhibitory effect of estrogen for conversion of mycelium to yeast. **(3) IDDs:** Like AIDS may favor the infection.
  - **Environmental factors:** High humidity and temperature around 23°C favor growth of fungus.
- **Pathogenesis:** After entry spores changed to yeast cells to establish the infection. Acute stage is presents with nonspecific inflammatory response, while chronic stage presents with granulomatous reaction.
- **Clinical features:** Lung lesion called **pulmonary paracoccidioidomycosis,** which is present with following features.
  - Acute (juvenile) form: Pneumonia.
  - Chronic (adult) form: Granulomatous lesion.

- **Complications:** Infection spread from lungs via blood to other organs like skin, bones, genito-urinary system, breasts, CNS, etc. Disseminated coccidioidomycosis is difficult to treat and relapse is common.

Diagnosis

- **Specimens:** BAL, sputum, biopsy from lesion or other suitable specimens.
- **Testing**

A. Microscopy: (1) Direct 10% KOH: Round, spherical, multiple based budding yeast cells **like Mariner's wheel** or **Pilot's wheel** are present. Schematic diagram is shown in **Fig. 91.10a. (2) Gram's stain:** It shows gram-positive yeast cells. **(3) Histopathological stains:** H and E stain and GGMS stains show fungus with granulomatous lesion. **(4) Fluorescent stain:** It is also useful.

B. Culture
  - Medium: It is SDA contains antibiotics.
  - Cultivation technique: **Ch. 120** (section → **Flowchart 120.1**).
  - C/Cs on SDA: **(1) Yeast phase:** Colonies are off-white to cream and rough to pasty in appearance. **(2) Mycelia phase:** Colonies are white to tan and yellow-brown on reverse side.
  - Identification of growth by LCB/PHOL stain: **(1) From yeast phase plate:** Round, spherical, multiple based budding yeast cells **like Mariner's wheel or Pilot's wheel** are present. Schematic diagram is shown in **Fig. 91.10a. (2) From mycelia phase plate:** It shows hyaline septate hyphae with intercalary chlamydospores. Schematic diagram is shown in **Fig. 91.10b.**

C. Animal inoculation: Mouse, rat, guinea pig and hamster are susceptible. Nine banded armadillo (*Dasypus novemcinctus*) is the naturally infected animal and may provide clue to understand the pathophysiology of fungus.

D. Serological tests: Like agglutination test, CFT, immunodiffusion test, etc.

E. Molecular methods: PCR.

F. Skin test: Detection of DTH by using **paracoccidioidin-Ag**.

G. Special tests: Teased mount, slide culture, cellophane tape mount and exo-antigen tests are also useful.

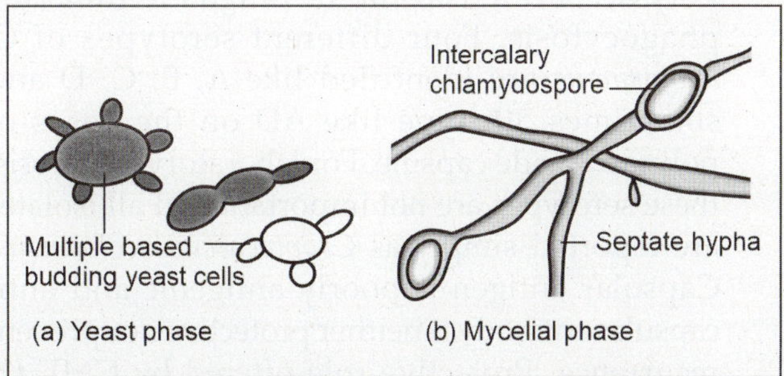

Fig. 91.10: Schematic diagram of *P. brasiliensis*

Prevention: Avoid visiting the suspected places. DNA based vaccine is developed but not useful for humans.

Treatment: Itraconazole is effective in all cases except seriously ill where AMB is recommended. Relapse is major issue following treatment.

OPPORTUNISTIC MYCOSES

These are the fungal infections occur in host with IDDs. Examples of systemic mycoses are **candidiasis, crypto-coccosis, pneumocystosis, penicilliosis, aspergillosis, systemic zygomycosis** and **others opportunistic mycoses.**

Candidiasis

Follow **Ch. 90.**

Cryptococcosis

Synonym: Previously, it was called **torulosis** or **European blastomycosis** (because originally reported from Europe).

Definition: It is an acute, subacute or chronic form of fungal infection primarily involving lungs, brain and disseminated to other tissues.

Etiological agents

- **Most common human pathogen:** It is *Cryptococcus neoformans* with three varieties like *C. neoformans* var. *neoformans, C. neoformans* var. *grubii* and *C. neoformans* var. *gatii.*
- **Other rare human pathogens:** These are *C. albidus* and *C. laurentii.*

Pathogenicity

- **Epidemiology:** Var. *neoformans* is common in Europe and var. *gattii* is common in Vancouver.
- **Reservoir and source of infection:** It is a saprophytic fungus found in house dust and soil. It is associated with Eucalyptus tree. It presents in pigeon's feces, because contains nitrogen which is essential growth factor for fungus. Pigeons are not suffered by disease.
- **Mode of transmission:** By inhalation of blastospores.
- **Incubation period:** 2–4 weeks.
- **Portal of entry:** Respiratory system.
- **Sites:** Pulmonary and extrapulmonary sites.
- **Precipitating factors (epidemiological determinants)**
  - **Agent factors (virulence factors): (1) Capsule:** Polysaccharide capsule of fungus inhibits the phagocytosis. Four different serotypes of *C. neoformans* are identified like A, B, C, D and sometimes 5th type like AD on the bases of polysaccharide capsule. For laboratory diagnosis, these serotypes are not important and all isolates are reported simply as *Cryptococcus neoformans.* Capsular antigen is poorly antigenic and anti-capsular antibody is neither protective nor prevent recurrence. Protective role offered by CMI. **(2) Ability to produce melanin:** Melanin protects the fungus from intracellular killing by phagocytes, high temperature, UV light, antibiotics like AMB and reduced state of iron. **(3) Ability to produce mannitol:** Mannitol protects the fungus from intracellular killing by phagocytes. **(4) Ability to survive at high temperature:** About 41–43°C. **(5) Urease production:** It converts urea in to ammonia, which acts as nitrogen source. **(6) Phospholipase production:** It helps in fungal invasion.
  - **Host factors: (1) Age:** It is common between 30 and 60 years and uncommon in children. **(2) Sex:** It is more in male. **(3) IDDs:** It is the 4th most common cause of CNS infection in AIDS patient. It may occur at any stage of AIDS but most common when CD4 count falls <100/mm^3. Others factors like long-term steroid/antibiotic/immunosuppressive therapy, DM, renal transplantation, malignancy, etc., also increase the risk of disease. **(4) Fungal predilection of CNS:** Because of presence of specific receptors of yeast cells on nerve cells, presence of nitrogen sources like asparigine and creatinine in CSF and absence of inhibitory factors (present in serum).
  - **Environmental factors:** Higher temperature about 41–43°C favors the growth of fungus.
- **Pathogenesis:** It is the only fungus having capsule. It primarily infects lung-lymph node complex. Later yeast cells spread to CNS by crossing the blood brain barrier either via direct migration through endothelium or carried inside the macrophages as **"Trojan horse".**
- **Clinical features:** Pulmonary cryptococcosis characterized by pneumonia.
- **Complications: (1) Meningitis or meningo-encephalitis:** Features include fever, headache, sensory loss, memory loss, cranial nerve paresis and loss of vision due to optic nerve involvement. Loss of vision may be rapid (sudden) or slow. Neck rigidity and stiffness are found in one third of patients. Kernig's and Brudzinski's sign are usually negative, but may be positive in few patients. Granulomatous lesion or ICSOL (intracranial space-occupying lesion) presents in brain called **cryptococcoma. (2) Generalized cryptococcosis:** Dissemination via blood to other organs like bones, prostate, eyes, skin (tumor like lesion), etc. Least invasion occurs in GIT, liver and kidneys. **(3) Relapse:** It is the common problem.

Diagnosis

- **Specimens:** Sputum, CSF or other suitable specimens.
- **Testing**

A. Microscopy: (1) Negative staining: Indian ink (60–70% sensitivity) or 10% Nigrosin shows round/oval budding yeast cells with capsule as shown in **Fig. 91.11.** Capsule is twice as thick as the diameter of yeast cells. **(2) Gram's stain:** It shows gram-positive yeast cells. **(3) Fluorescent stain:** It is also useful. **(4) Histopathological stains:** As mentioned below.

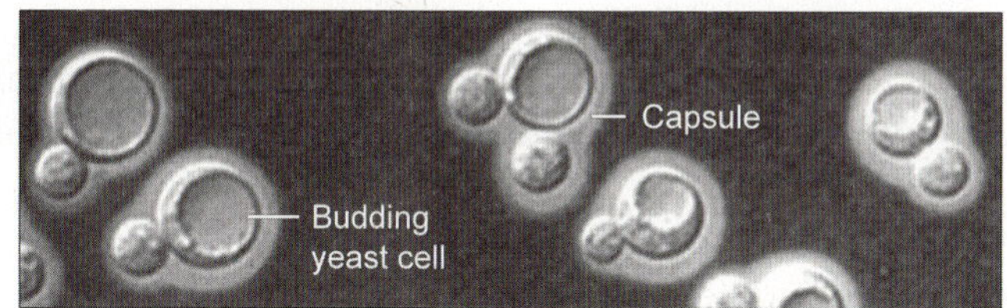

Fig. 91.11: Capsule of **C. neoformans** under Indian ink preparation

- HE stain, PAS and GGMS: They stain the fungus in tissue section.
- Mayer's mucicarmine stain: It stains capsule in deep rose in color.
- Masson Fontana Silver (MFS) stain: It demonstrates melanin production by fungus. Fungus stains black and nuclei in red.
- Alcian blue stain: It demonstrates the capsule of fungus.

B. Culture on routine media

- Medium: It is SDA contains antibiotics. Blood agar and chocolate agar are also useful media. CSF is inoculated followed by incubation at 37°C.
- C/Cs on SDA: Mucoid, cream to buff color colonies are developed. Mucoid colony is due to poly-saccharide capsule.
- Identification of growth by LCB/PHOL/negative stain: Round, budding yeast cells with capsule without hyphae (mycelium) or pseudohyphae.
- Growth is also identified by automated methods like MALDI-TOF.

C. Culture on selective media: These are bird seed/niger seed agar (brown color colonies) and sunflower seed agar.

D. Animal inoculation: Mouse is susceptible.

E. Biochemical tests: These are assimilation of inositol, urease positive and nitrate positive.

F. Serological tests: Capsular Ag is detected by latex agglutination test (95% sensitivity) from CSF and serum.

G. Molecular methods: PCR.

H. Antifungal sensitivity testing: Report the result with drug sensitivity pattern.

Prevention

- General measure: Avoid visiting the suspected places like bird reservoirs.
- Chemoprophylaxis: Use fluconazole in AIDS patient to avoid the risk of cryptococcosis.
- Immunoprophylaxis: Vaccine containing major capsular components like glucuronoxylomannan is developed. It provides protection in mice and also effective in humans.

Treatment: Initial therapy includes AMB and flucytosine. Maintenance therapy includes fluconazole to prevent the relapse.

Pneumocystosis

Meaning: Pneumo means air (suggest lungs) and cystis means cyst like structure.

Synonym: Pnemocystis pneumonia (PCP), *P. jirovecii* pneumonia (PJP).

Definition: It is an opportunistic fungal infection characterized by primary atypical (interstitial) pneumonia.

Etiological agent: It is caused by *Pneumocystis jirovecii* (previously it was known as *P. carinii*).

Taxonomy of *P jirovecii*

- **Initially protozoon:** Initially, it was classified as protozoon based on morphology, resistant to antifungal drugs (because no ergosterol in cell wall), sensitive to antiprotozoal drugs and no growth on fungal media, but in cell culture media.
- **Now fungus:** Now, it is classified as fungus as *Ascomycetes*, because (1) cell wall contains β-1,3 glucans, which is responsible for staining with histo-pathological stains as described below. (2) It contains chitins in cell wall and produces the enzyme chitinase as like other fungi. (3) rRNA study, protein synthesis elongation factor (EF3) and thymidylate synthase are homologous to the ascomycetes. (4) Sporozoites (intracystic bodies) of cyst structure are simulating the ascospores of ascomycetes. (5) Mitochondria are with laminar cristae as in ascospores of ascomycetes, while protozoal mitochondria are tubular and vesicular cristae. However, still it is controversial to classify it as fungus and better to classify as **atypical fungus** under the new phylum called **Protomycota**.

Nomenclature of *P. jirovecii*

- **Species name:** It was 1st identified from the lungs of guinea pig by Chagas in 1909 and from the lungs of rat by Antonio Carinii in 1910 (*Pneumocystis carinii* is considered as rat species) and after 2nd world war it was identified from the lung of infant in Central and Eastern Europe by Otto Jirovec and Joseph Vanek (*Pneumocystis jirovecii* is considered as human species).
- **Disease name:** When the name *P. jirovecii*, replaced the *P. carinii*, at first it was felt that *P. jirovecii* pneumonia (PJP) should replace *P. carinii* pneumonia (PCP). However, the term PCP remains continue with its full form as **P**neumo **C**ystis (jirovecii) **P**neumonia, because it was used by many physicians.

Life cycle of *P jirovecii*: Three stages as shown in **Fig. 91.12.**

- **Trophozoite:** It is thin wall, amoeboid-shaped and 1–5 μm in size. It contains tubular projection called **filopodia** (glycoprotein in nature) for attachment with host cells.
- **Precyst:** It is the intermediate stage with 5–8 μm length, oval shape and with few filopodia.
- **Mature cyst:** It is thick wall, 8 μm in size and spherical shaped. It contains 8 nuclei (intracystic bodies/sporozoites) and its rupture releases the sporozoites, which start the new cycle.

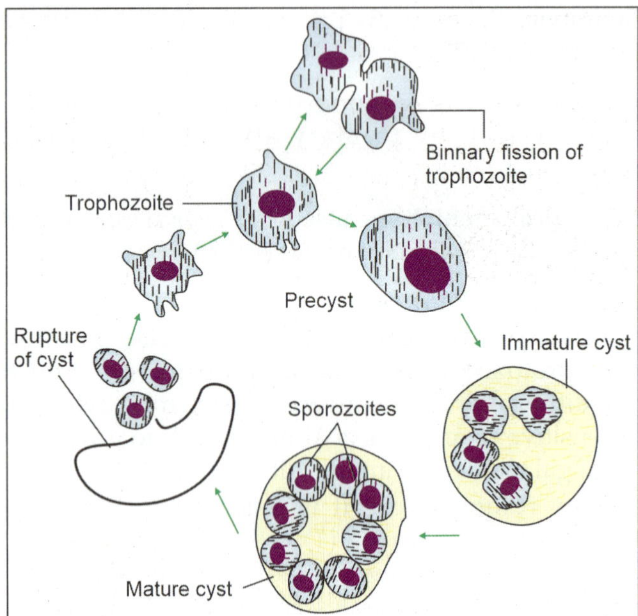

Fig. 91.12: Life cycle of *P. jirovecii*

Pathogenicity

- **Reservoirs and sources of infection:** Uncertainty about sources and reservoirs, but fungus presents in animals, in humans and free in the environment.
- **Modes of transmission:** It is transmitted by inhalation of trophozoites. Congenital transmission also reported.
- **Incubation period:** 4–8 weeks.
- **Portal of entry:** Respiratory system.
- **Sites:** Pulmonary and extrapulmonary sites.
- **Precipitating factors (epidemiological determinants): (1) Age:** It is common in old age **(2) Immunosuppressive disease or conditions:** Pneumonia occurs in patient when CD4+ count falls <200/mm³ in AIDS patients. Other risk factors are steroid/immunosuppressive therapy, DM, organ/bone marrow transplantation, leukemia, lymphoma, Cushing's syndrome, hypo-gammaglobulinemia, marasmus, etc.
- **Pathogenesis:** Inhaled trophozoites attach with alveolar epithelium and initiate the eosinophilic infiltration in alveoli or in interstitial space, called **primary atypical (interstitial) pneumonia.**
- **Clinical features:** Primary atypical (interstitial) pneumonia characterized by fever, nonproductive cough and shortness of breath.
- **Complications:** These are secondary bacterial infection in lungs, ocular pneumocystosis (it is characterized by conjunctivitis, retinitis with cotton wool spot in fundus and choroid lesions) and pneumocystosis of lymph nodes, spleen, etc.

Diagnosis

- **Key points for diagnosis of PCP**
  - CD4+ count <200/mm³.
  - Both trophozoite and cyst are the diagnostic features. Trophozoites are more in numbers than cysts with 10:1 ratio. Trophozoites are 1–5 µm in size while cysts are 8 µm in size.

  - Positivity of PCP in induced sputum is 60%.
  - Diagnosis also based on radiological study, clinical presentation and histopathological examination of BAL or lung biopsy.
- **Specimens:** Sputum, BAL, lung biopsy, etc.
- **Testing**

A. Microscopy

- **Histopathological stains for cyst wall:** These are best useful and stain the wall of cyst but neither stain the trophozoites nor the nuclei (sporozoites) of cysts. Cysts are appearing in cluster of 2–8 bodies. (1) Toluidine blue O stains blue colored cysts as shown in **Fig. 91.13a**. (2) Gram-Weigert stain gives characteristic appearance of cysts as shown in **Fig. 91.13b**. (3) Silver stains like GGMS showing black colored crushed cysts **(Fig. 91.14a)** with **ping-pong ball appearance** and silver Gram's stain **(Fig. 91.14b)** produces black colored cysts. (4) Other useful stain is Cresyl Echt violet stain.
- **Histopathological stains for trophozoites and sporozoites (internal nuclei) of cysts:** They do not stain the cyst wall but stain the sporozoites of cysts appear negatively stained. They also stain the trophozoites (1–5 µm). (1) Giemsa stain shows the nuclei in pink color and cytoplasm in blue color as shown in **Fig. 91.15a**. (2) Other useful stains are Trichrome stain, Papanicolaou stain and MGG stain.
- **Fluorescent stains:** Are also useful.

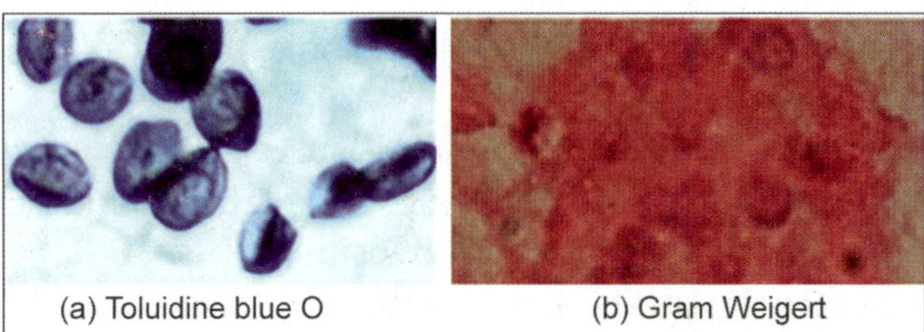

Fig. 91.13: Stains for cyst wall of PCP

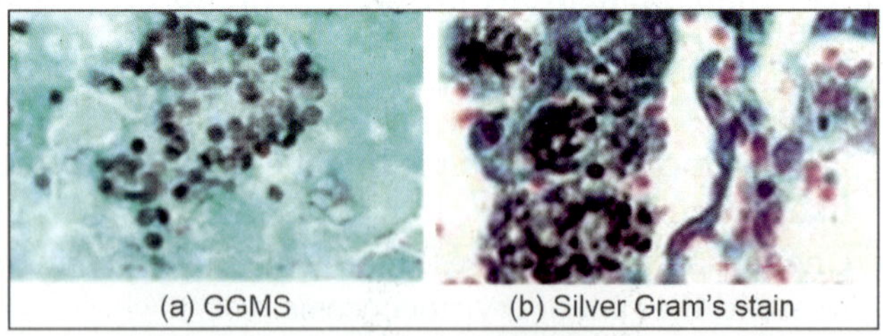

Fig. 91.14: Silver stains for cyst wall of PCP

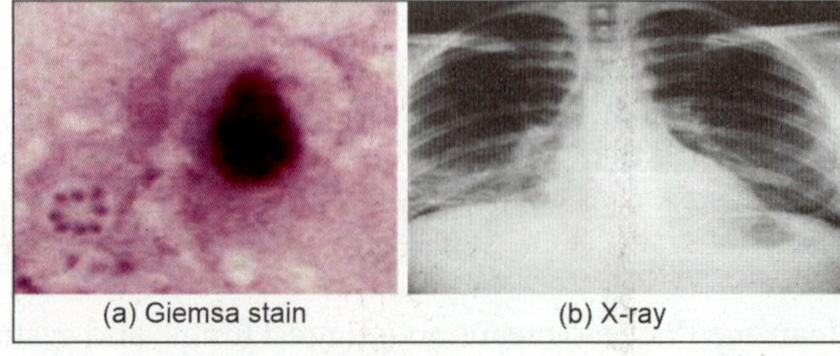

Fig. 91.15: Giemsa stains (a) and X-ray (B) findings for PCP

B. Culture: No growth occurs on media, but it can grow in several cell lines like A549 and WI-38.

C. Animal inoculation: Mouse is susceptible.

D. Molecular method: PCR.

E. Detection of fungal metabolites: Like 1,3 β-D glucan.

F. Pulmonary function test (PFT): It revealed the reduction in the vital capacity and total lung capacity.

G. Radiological findings: Chest X-ray shows wide spread pulmonary infiltration (mottling of lung fields) and opacification (whiteness) in lower lobes on both sides as shown in **Fig. 91.15b**. **CT of lungs** shows the ground glass opacities in early stage; however, atypical presentation like nodule or cavity formation may also noticed.

Prevention: Cotrimoxazole is given as **primary prophylaxis** (CD4+ count <200/mm^3 or with oral thrush or with unexplained fever >100F for >2 weeks but never had PCP) and **secondary prophylaxis** (HIV infected patients with history of PCP) in AIDS patient to prevent *Pneumocystis jiroveci* pneumonia. Lifelong secondary prophylaxis is given to the HIV patients.

Treatment: Cotrimoxazole is the drug of choice.

Penicilliosis

Etiological agent: It is caused by *Penicillium* with more than 250 species. *Penicillium marneffei* is the most common pathogenic species causing penicilliosis. It is the only dimorphic fungus; other species are not dimorphic and can grow at 25°C as mold. It multiplies by binary fission not by budding. It produces brick-red pigment. It is a common laboratory contaminant. It presents in environment and grows on various substances like bread, jam, fruit, vegetables, etc. **Other species** occur as laboratory contaminants and causing rare human diseases like otomycosis, keratitis, onychomycosis, endocarditis, endophthalmitis, allergic fungal disease (**Ch. 92**) and food poisoning (aflatoxin is produced by *P. puberulum*).

Pathogenicity

- **Epidemiology:** It is endemic in South East Asian countries like Thailand, Vietnam and India (Manipur).
- **Reservoirs and sources of infection:** Conidia (spores) are present in soil. Natural infection occurs in humans and bamboo rats [*Rhizomys sinensis* (Vietnam), *R. pruiosus* (China) *R. sumatrensis* and *Cannomys badius*].
- **Modes of transmission:** It is transmitted by inhalation of spores. In Southern China, it is acquired due to ingestion of infected rat/rat's meat. Direct transmission occurs by inoculation in skin.
- **Incubation period:** 4 weeks.
- **Portal of entry:** Respiratory system.
- **Sites:** Pulmonary and extrapulmonary sites.
- **Precipitating factors (epidemiological determinants):** **(1) Agent factors (virulence factors):** Fungal dimorphism as described in *H. capsulatum*. **(2) Host factors like IDDs:** Cases are increased in numbers following

increasing the cases of AIDS. It may causes fulminant disease in persons with IDDs. **(3) Environmental factors:** Disease is more likely to occur following the rainy season.

- **Pathogenesis:** After entry spores target the RE system, with formation of three types of lesions. **(1) Granulomatous lesion** which contains nonbudding yeast cells with transverse septa in center surrounded by inflammatory cells. **(2) Suppurative lesion** contains pus. **(3) Lesion with necrosis** contains little or no granuloma.
- **Clinical features:** Granulomatous and suppurative lesions occur in lungs.
- **Complications:** These are pericarditis, pericardial effusion, peritonitis, endophthalmitis, osteomyelitis, extrapulmonary lesions in lymph nodes, liver, spleen, etc., and relapse.

Diagnosis

- **Specimens:** Sputum, BAL or other suitable specimens.
- **Testing**

A. Microscopy

- **Histopathological stains:** H and E, PAS and GGMS stains are showing granulomatous lesion in tissue. Yeast cells are present in macrophages. Yeast cells are round, spherical, elongated with curved end and transverse septum, but without budding giving 'sausage like appearance'. Schematic diagram is shown in **Fig. 91.16a**.
- **Fluorescent stains:** Are also useful.

B. Culture

- Medium: It is SDA contains antibiotics.
- Cultivation technique: **Ch. 120 (section → Flowchart 120.1)**.
- C/Cs on SDA: **(1) Yeast phase:** Cerebriform, convoluted and pink-white colonies appear. **(2) Mycelia phase:** Grayish-white colonies appear. Colonies appear red on reverse side due to brick-red pigment production.
- Identification of growth by LCB/PHOL stain: **(1) From yeast phase plate:** Yeast cells are round, spherical, elongated with curved end and transverse septum giving 'sausage like appearance'. Schematic diagram is shown in **Fig. 91.16a**. **(2) From mycelia phase plate:** It shows hyaline

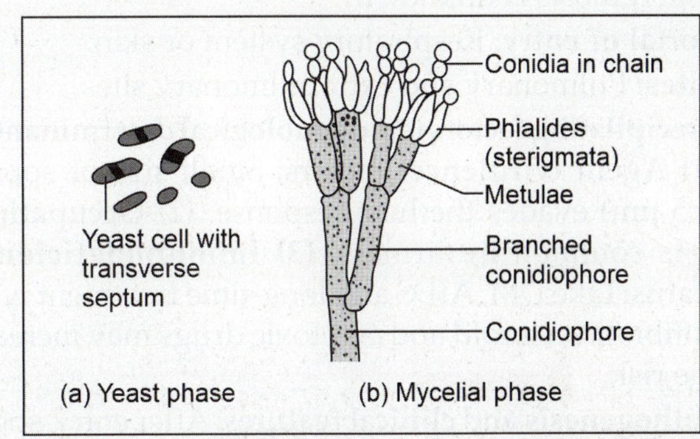

Fig. 91.16: Schematic diagram of *P. marneffei*

septate hyphae with conidiophores, verticilis (branched conidiophores) metulae, phialides and conidia, which are basipetal (youngest conidium at base) in nature and arranged in chain giving **brush like appearance.** Schematic diagram is shown in **Fig. 91.16b**.

C. **Animal inoculation:** Bamboo rat is susceptible.

D. **Serological tests:** Like agglutination test, CFT immunodiffusion test, etc.

E. **Molecular method:** PCR.

F. **Special tests:** Teased mount, slide culture, cellophane tape mount and exo-antigen tests are also useful.

Prevention: Antibiotics may prescribe to AIDS patient to prevent the infection.

Treatment: In sever case AMB is given followed by lifelong maintenance therapy with itraconazole to prevent the relapse. Mild cases are treated by itraconazole.

Aspergillosis

Meaning: *Aspergillus* word derived from aspergillum means perforated globe used to sprinkle the holy water, as chain like arrangement of conidia, radiating from central structure resemble to aspergillum.

Definition: It is the systemic fungal infection occurs in immunocompetent and immunodeficient individuals.

Etiological agents: It is caused by *Aspergillus* with more than 185 species. Around 35 species are pathogenic for humans and three are most pathogenic like *A. flavus*, *A. fumigatus* and *A. niger*. It is the common laboratory contaminant, and it is difficult to differentiate from true pathogen. It presents in environment and grows on various substances like bread, jam, fruit, vegetables, etc.

Pathogenicity

- **Reservoirs and sources of infection:** It is a saprophytic fungus and spores are present in soil, environment and on decaying vegetation.
- **Modes of transmission:** It is transmitted by inhalation of spores (conidia), specially working with decaying vegetation like moldy hay. Direct entry in traumatized skin may occur.
- **Incubation period:** Allergic illness develops after 6–9 months of exposure. Exact time for invasive aspergillosis is unknown.
- **Portal of entry:** Respiratory system or skin.
- **Sites:** Pulmonary and extrapulmonary sites.
- **Precipitating factors (epidemiological determinants):** **(1) Agent (virulence) factors:** Small size of spores (<5 µm) evades the host response. **(2) Occupation:** It is common in farmers. **(3) Immunodeficiency status:** Like DM, AIDS and long-time treatment with antibiotics, steroid and cytotoxic drugs may increase the risk.
- **Pathogenesis and clinical features:** After entry, spore establishes following types of diseases.

– **Pulmonary aspergillosis:** Three categories. **(1) Allergic:** It called **allergic bronchopulmonary aspergillosis.** It occurs by different mechanisms like type-I hypersensitivity presents with asthma (atopy type), type-III hypersensitivity presents with extrinsic alveolitis or combined type-I and III hypersensitivity. **(2) Noninvasive:** In the lungs fungal mass is surrounded by dense fibrous wall called **aspergilloma (fungus ball).** It presents with hemoptysis and common in patients with old tuberculosis or bronchiectasis. Surgical removal is necessary. **(3) Invasive:** At first, it causes the pneumonia and later disseminate to other organs like skin, CNS, paranasal sinuses, CVS, kidneys, heart, bones, etc.

– **PNS (paranasal sinuses) aspergillosis:** Most common agent is *A. flavus*. It includes nasal and paranasal sinuses with four categories. **(1) Allergic:** It presents with rhinitis, nasal discharge, nasal blockage, etc. It is mostly combined type-I and III hypersensitivity. **(2) Noninvasive:** It is characterized by aspergilloma (fungus ball) formation in sinuses. It presents as chronic sinusitis. Surgical removal is necessary. **(3) Invasive:** It presents as facial mass (like a neoplastic growth or polypoid growth) and proptosis. It disseminates through bone to orbit and finally to brain (called **rhinocerebral aspergillosis,** which may present with proptosis or polypoid lesion in nose). **(4) Fulminant:** It is angio-invasive, rapidly destructive and often fatal.

– **Food poisoning:** Preformed toxin production in food causes food poisoning.

– **Other disease:** In addition to above mentioned lesion *Aspergillus* spp., causes lesions in other organs or tissues either primarily or secondary due to distant spread. For examples, CNS aspergillosis, endocarditis (common following cardiac surgery), otomycosis, mastoiditis, mycotic keratitis, onychomycosis, cutaneous aspergillosis (cutaneous erythematous lesion progress to necrotic eschar mostly by *Aspergillus flavus*. It appears rarely as secondary infection or more as primarily in neutropenic patients at the site of insertion of IV catheter, following trauma or surgery, in burn patients) and osteomyelitis.

- **Complications:** Intracranial aneurysm and other invasive diseases are the major issues.

Diagnosis

- **Specimens:** BAL, sputum, CSF, biopsy from lesion or other suitable specimens.
- **Testing**

A. **Microscopy:** KOH (**Fig. 91.17a**) and histopathological stains show the hyaline septate hyphae. Fluorescent stain is also useful.

B. **Culture on routine media**

– Medium: It is SDA contains antibiotics.

– Cultivation technique: **Ch. 120** (section → **Flowchart 120.1**).

- C/Cs on SDA:
  1. *A. fumigatus*: Dark green colony **(Fig. 91.18a)**.
  2. *A. niger*: Black colony **(Fig. 91.18b)**.
  3. *A. flavus*: Yellow-green **(Fig. 91.18c)**.
- Identification of growth by LCB **(Fig. 91.17b)**/ PHOL stain: It shows hyaline septate hyphae with conidiophores, vesicle, phialides and chain like arrangement of conidia. Hyphae show dichotomous branches, where conidiophores and stems are equal in size (3–6 μm) and at acute/ narrow angle of 450 (V shaped). Follow **Fig. 91.19 and 91.20** for schematic diagram of *Aspergillus*.

B. Culture on selective medium: Czapek-Dox agar is used to differentiate the *Aspergillus* spp.

C. Animal inoculation: Rabbit and mouse are susceptible animals.

D. Serological tests: Early stage infection is diagnosed by detecting galactomannan antigen from serum and urine by ELISA. Invasive disease is diagnosed by detecting β-D glucan antigen.

E. Molecular method: PCR.

F. Skin test: To detect the type-I and type-III hypersensitivity.

G. Tests to detect fungal metabolites: Gas-liquid chromatography detects the fungal metabolites like D-mannitol. **G-test** detects β-1, 3-D-glucan.

H. Special tests: Slide culture is useful for better species identification.

I. Antifungal sensitivity test: Report the result with drug sensitivity pattern.

J. Prevention: Reduction of environmental exposure. Posaconazole is given as prophylaxis.

Treatment: It has high mortality rate. Surgery is performed to remove fungal ball. Azoles like voriconazole, itraconazole, etc., are useful drugs.

> **Note: Zygomycosis**
>
> **Definition:** It is an acute or chronic infection caused by several fungal agents belong to phylum Zygomycota and class Zygomycetes.
>
> **Types:** It has two main types **subcutaneous zygomycosis** (described above) and **systemic zygomycosis** (described below).

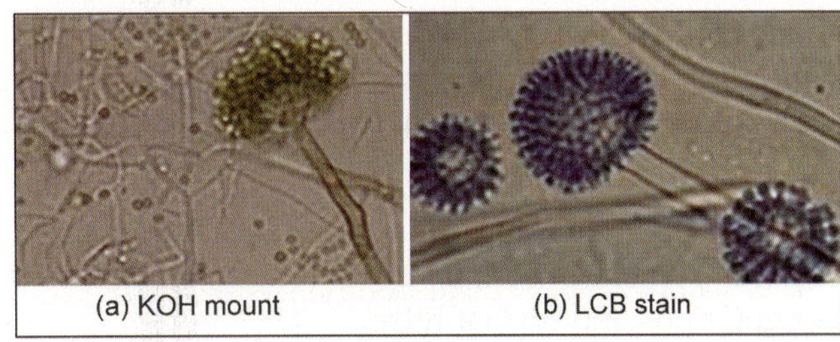

Fig. 91.17: KOH and LCB stains of *Aspergillus*

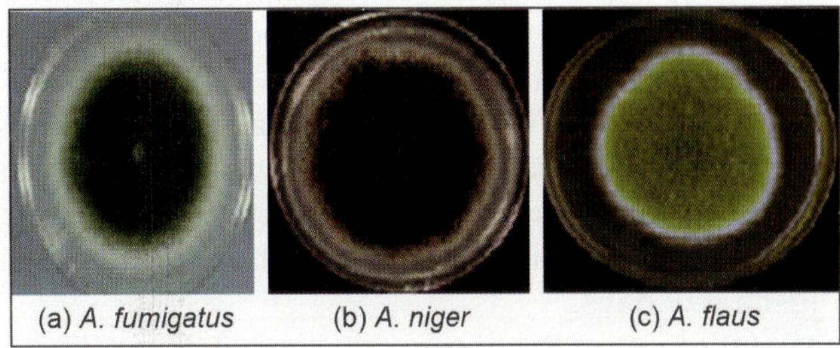

Fig. 91.18: C/Cs of *Aspergillus* spp. in SDA

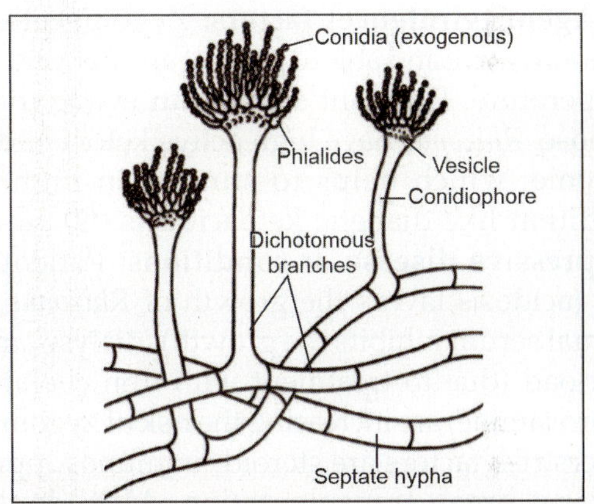

Fig. 91.19: Schematic diagram of *Aspergillus*

Systemic Zygomycosis

Synonym: It also called **mucormycosis,** because all the fungi are belong to order Mucorales or **phycomycosis** because all the fungi are belong to class Phacomysetes (Zygomycetes).

Etiological agents: It is caused by *Rhizopus arrhizus, Rhizopus microspores, Rhizomucor pusillus, Absidia*

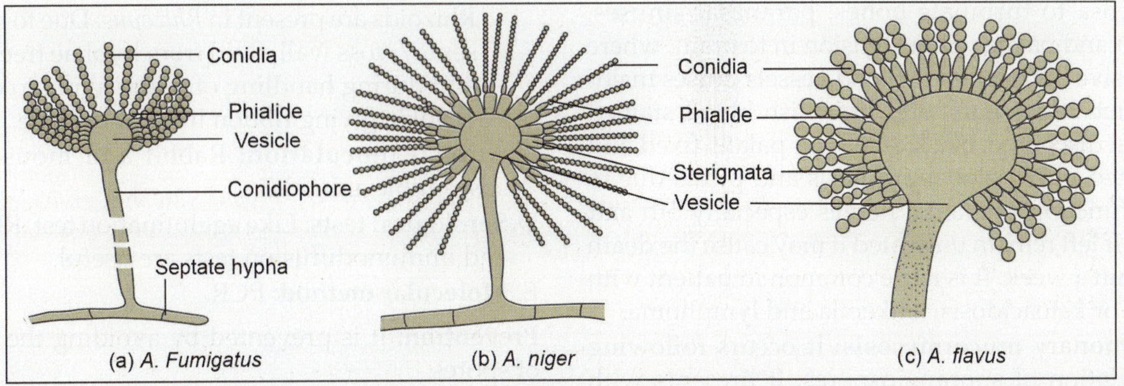

Fig. 91.20: Schematic diagram of different *Aspergillus* spp.

corymbifera, Apophysomyces elegans, Mucor racemosus, Cunninghamella bertholletiae, Saksenaea vasiformis, Cokeromyces recurvatus and *Syncephalastrum recemosum.*

Pathogenicity

- **Reservoirs and sources of infection:** Fungal spores present in soil, in environment, on decaying vegetations and on leaf littre.
- **Modes of transmission:** It is transmitted by inhalation of spores (conidia), especially working with decaying vegetations like moldy hay or by ingestion of spores. Direct infections may occur by percutaneous inoculation.
- **Incubation period:** The incubation period is measured in days. The clinical course can progress from normal to symptomatic in a week and from sinus opacification to uncal herniation and death in just a few days.
- **Portal of entry:** Respiratory system, skin or GIT.
- **Sites:** Pulmonary and extrapulmonary sites.
- **Precipitating factors (epidemiological determinants):** **(1) Agent (virulence) factors:** Zygomycetes is an angio-invasive and able to grow at and above the body temperature. Dormant spores can evade the body defense. *Rhizopus* have high active ketone reductase enzyme, which helps to survive in high acidic condition like diabetic ketoacidosis. **(2) Immuno-suppressive disease or conditions:** Patients with DM (acidosis favors the growth of *Rhizopus*, while normal serum inhibits the growth), dialysis and iron overload (due to treatment with iron chelator like deferoxamine) are increasing the risk of zygomycosis. Others risk factors are steroid/immunosuppressive therapy, organ transplantation, AIDS, leukemia, lymphoma, etc.
- **Pathogenesis:** Inhaled spores evade the host defense and change to mycelial form which have more predilection for elastic lamina of small and large arteries (angio-invasion) causing thrombosis, hemorrhage and infarction.
- **Clinical features:** Following are the primary lesions of mucormycosis.
  - **Rhinocerebral mucormycosis:** It is the most common and fulminant variety. Most common agent is *Rhizopus arrhizus.* It spreads from nasal mucosa to turbinate bones, paranasal sinuses, orbit and palate with extension in to brain, where massive invasion in to blood vessels causes major infarct. It presents with brownish blood stained nasal discharge, black eschar on palate, fixed and dilated pupil, global proptosis and ptosis due to dysfunction of cranial nerves especially 5th and 7th. If left remain untreated it may cause the death within a week. It is more common in patient with DM or ketoacidosis, leukemia and lymphoma.
  - **Pulmonary mucormycosis:** It occurs following inhalation of sporangiospores. It presents with chest pain, dyspnea and hemoptysis.

  - **Cutaneous mucormycosis:** It is primary due to direct inoculation of fungus in to skin or secondary due to blood spread from inner lesions. It mostly caused by *Rhizopus* and *Mucor* spp.
  - **Gastrointestinal mucorcosis:** It is primary due to direct ingestion of fungal spores via food or secondary due to blood spread.
- **Complications: (1) Local spread:** From nasal mucosa or sinus, fungus enters in orbit via cribriform plate called **orbital zygomycosis/orbital mucormycosis.** It presents with chemosis, periorbital cellulitis, ophthalmoplegia, proptosis, ptosis, abrupt visual loss, orbital pain and fascial hypoesthesia. Fungus may invade the blood vessels and causes vasculitis, thrombosis or ischemic necrosis. It also enters in brain to cause meningitis, brain abscess and patient may die within days or weeks. **(2) Hematogenous spread:** Fungus enters via blood to other organs like heart, bones, kidneys, eyes, etc.

Diagnosis: These are the common laboratory contaminants and also caused confusion in diagnosis of true infection or contaminants.

- **Specimens:** BAL, sputum, CSF, biopsy from lesion or other suitable specimens.
- **Testing:**

A. **Microscopy:** KOH and histopathological stain shows the hyaline nonseptate hyphae (co-enocytic hyphae). Fluorescent stain is also useful.

B. **Culture**

- Medium: It is SDA contains antibiotics.
- Cultivation technique: **Ch. 120** (section → **Flowchart 120.1**).
- C/Cs on SDA: Almost same colonies are produced by different species. Colonies are gray-white with thick-cottony fluffy surface.
- Identification of growth by LCB **(Fig. 91.17b)**/ PHOL stain: Shows irregular, broad (10–20 μm), hyaline, nonseptate (co-enocytic) or sparsely septate hyphae. Hyphae showing obtuse branches, where conidiophores and stems are at wide-angle or right angle of 900 (L-shaped) and at irregular interval with sporangiophores. Round sporangium contains endospores called **sporangiospores** is present. Schematic diagram is shown in **Fig. 91.21**. Rhizoids are present in *Rhizopus*. Due to absence of septa/cross wall, fluid from hyphae free to escape and during handling of tissue hyphae collapse or wrinkle giving ribbon like appearance.

C. **Animal inoculation:** Rabbit and mouse are the susceptible animals.

D. **Serological tests:** Like agglutination test, RIA, ELISA and immunodiffusion tests are useful.

E. **Molecular method:** PCR.

Prevention: It is prevented by avoiding the exposure of spores.

Treatment: Surgery and antifungal drugs are useful.

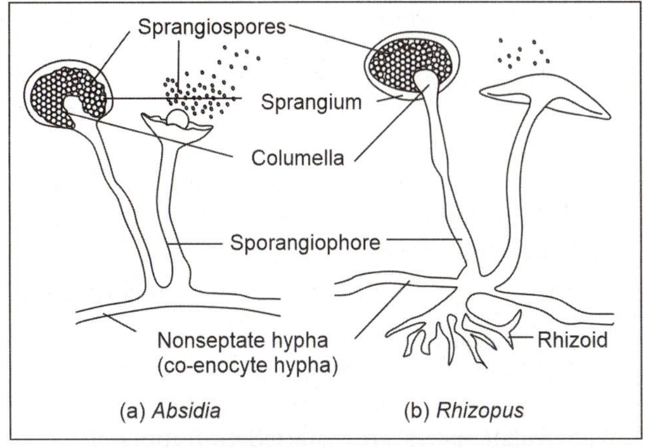

Fig. 91.21: Schematic diagram of Zygomycetes

Morphological differences between *Aspergillus* and Zygomycetes: Follow **Table 91.2**.

TABLE 91.2: Morphological differences between *Aspergillus* and Zygomycetes

Features	*Aspergillus*	Zygomycetes
Hyphae	Septate	Aseptate
Branches	Dichotomous branches, where conidiophore and stem are at acute/narrow angle of 45° (V-shaped)	Obtuse branches, where conidiophore and stem are at wide–angle or right angle of 90° (L-shaped)
Spores	Exogenous in chain	Endogenous in a sporangium
Rhizoides	Absent	Present in few species

Other Opportunistic Mycoses

These are *Trichosporon beigelii* (produces white piedra and other systemic lesions), *Fusarinm graminiarum*, *Geotrichum candidum*, *Blastoschizomyces capitatus*, *Saccharomyces cerevisiae* (Baker's yeast or Brewer's yeast) and *Rhodotorula* spp.

ACCESS YOURSELF

Case Studies

1. A 30-year-old male, resident of West Bengal, associated with chicken coops, presents with tuberculosis like features like chronic cough, hemoptysis and lymphadenopathy. Chest X-ray shows the pulmonary infiltrate. Lymph node aspirate cultured on SDA in two plates, one incubated at 25°C and other at 37°C. Microscopy of culture growth at 37°C revealed hyaline septate hyphae with microconidia and tuberculate macroconidia. Identify the case and answer the following.
   a. Name the clinical disease and causative agent in given case.
   b. Write the pathogenicity of causative agent.
   c. Write the laboratory diagnosis of causative agent.
2. A 22-year-old male, brought to emergency with headache, neck rigidity and sudden loss of vision. Lumber puncture is performed and CSF is collected. Microscopy performed with indian ink which revealed capsulated budding yeast cells. Identify the organism and answer the following.
   a. Name the clinical disease and causative agent in given case.
   b. Write the pathogenicity of causative agent.
   c. Write the laboratory diagnosis of causative agent.

Essays/Full Questions

1. Deep mycoses.
2. Opportunistic mycoses.

Short Notes

1. Mycetoma/sporotrichosis/histoplasmosis/cryptococcosis.
2. Aspergillosis (Aspergilloma).
3. Mucormycosis (systemic zygomycosis).

Short Questions for Theory/Viva Questions

1. Name the four fungi/bacteria causing mycetoma.
2. Write four examples of subcutaneous mycoses/systemic mycoses/opportunistic mycoses.
3. Write four examples of dimorphic fungi.
4. What are asteroid bodies and sclerotic bodies?
5. What is Splendore-Hoeppli phenomenon?
6. What are dichotomous branches of hyphae? Name the fungus with such haphae.
7. Write four morphological differences between *Aspergillus* and zygomycetes.

Comments On

1. Production of melanin by *Cryptococcus* offers a protective role to fungus.
2. *Cryptococcus* is carminophilic fungus.
3. *Pneumocystis* is classified as fungus.

MCQs for Chapter Review

Subcutaneous Mycoses

1. **Actinomycotic mycetoma is caused by:**
   a. *Actinomyces*
   b. *Nocardia*
   c. *Streptomyces*
   d. Madura mycosis
   e. *Staphylococcus*
2. **Mycetoma foot can be caused by following:**
   a. *Cladosporium*
   b. *Phialophora jeanselmei*
   c. *Madura mycetomatis*
   d. *Allesheria boydii*
3. **Color of granules in mycetoma caused by *Actinomyces* spp.**
   a. Black
   b. Yellow
   c. Red
   d. Brown
4. **Color of granules in mycetoma caused by *A. pelleiterri*:**
   a. Black
   b. Yellow
   c. Red
   d. Brown
5. **Yellow black granules are seen in which fungal infection?**
   a. Mucormycosis
   b. Mycetoma
   c. Aspergillosis
   d. Rhinosporidiosis
6. **Which of the following is the most predominant constituent of sulfur granules of actinomycosis?**
   a. Organisms
   b. Neutrophils and monocytes
   c. Monocytes and lymphocytes
   d. Eosinophils
7. **The granules discharged in mycetoma contains:**
   a. Bone specules
   b. Fungal colonies
   c. Pus cells
   d. Inflammatory cells
8. **True about mycetoma:**
   a. Commonly occurs on hands
   b. Commonly erodes bone
   c. Drain through lymphatics
   d. Inflammatory cells
9. **Which of the following is false about mycetoma?**
   a. Can affect lower and upper extremities
   b. Caused by actinomycetes and filamentous fungi

c. Diagnosed by examination of pus
d. Uncommon in India

10. **True about madura mycetoma:**
    a. Fungal infection
    b. Presented as nonpainful nodular lesion for many months
    c. Discharging sinus
    d. Bone involvement seen
    e. All of above

11. **A plant prick produces sporotrichosis. All are true statements about sporotrichosis, *except*:**
    a. Is a chronic mycotic disease that typically involves skin, subcutaneous tissue and regional lymphatics
    b. Most cases are acquired via cutaneous inoculation
    c. Enlarged lymph nodes extending centripetally as a beaded chain are a characteristic finding
    d. Is a occupational disease of butchers, doctors

12. **A gardner has multiple vesicles on hand and multiple eruption along the lymphatics. Most common fungus responsible is?**
    a. *Sprothrix schenckii*   b. *Cladosporium*
    c. *Histoplasma*   d. *Candida*

13. **Cigar body is seen in:**
    a. Cryptococcosis   b. Histoplasmosis
    c. Sprothrichosis   d. Aspergillosis

14. **Asteroid body and cigar-shaped globi may be produced by:**
    a. *Sporothrix*   b. *Phialophora*
    c. *Aspergillus*

15. **A series of ulcer in lower extremities in sub-Himalayan area is often caused by:**
    a. *Trichophyton rubrum*   b. *Pseudoallescheria boydii*
    c. *Cladosporium*   d. *Sporothrix schenckii*

16. **Causative organism of chromoblastomycosis:**
    a. *Cladosporium*   b. *Blastomyces*
    c. *Sporothrix*   d. *Histoplasma capsulatum*

17. **Sclerotic bodies 3–15 µm in size, multiseptate chest nut brown colour seen in:**
    a. Rhinosporidiosis   b. Chromoblastomycosis
    c. Phacohyphomycosis   d. Histoplasmosis

18. **Rhinosporidiosis is caused by:**
    a. Fungus   b. Bacteria
    c. Virus   d. Protozoan
    e. Parasite

19. **True about *Rhinosporidium seeberi*:**
    a. Virus
    b. Bacterium
    c. Ketoconazole – treatment
    d. Presents in coastal India

Systemic Mycoses

20. **The medium of choice for culturing yeast form of dimorphic fungi is:**
    a. Brain heart infusion
    b. Sabouraud's
    c. Sabouraud's plus antibiotics
    d. Any medium incubated at 35–37°C

21. **Systemic fungal infections can be caused by the following:**
    a. *Cryptococcus neoformans*
    b. *Histoplasma capsulatum*
    c. *Paracoccidioides brasiliensis*
    d. *Naegleria fowleri*
    e. *Cystoisospora belli*

22. **Endemic fungal infection is caused by all of the following *except*:**
    a. *Coccidioides immitis*   b. *Cryptococcus*
    c. *Histoplasma*   d. *Aspergillus*
    e. *Blastomyces*

23. **What is true about histoplasmosis?**
    a. In early stage it is indistinguishable from TB
    b. Culture is not diagnostic
    c. Hyphal forms are infectious forms
    d. Person-to-person spread occurs by droplet infection

24. **Darling disease is caused by:**
    a. *Histoplasma*   b. *Candida*
    c. *Cryptococcus*   d. *Rhizopus*

25. **'Tuberculate spores' are characteristic features of:**
    a. *Candida*   b. *Histoplasma*
    c. *Coccidioidomyces*   d. *Cryptococcus*

26. **Histoplasmosis is spread by:**
    a. Human   b. Water
    c. Soil   d. None

27. **_Blastomyces_ is characterized by all, *except*:**
    a. Yeast like fungus
    b. Commonly involves lung and skin
    c. Dimorphic fungus
    d. Common in South America

28. **Valley fever or Desert rheumatism is caused by:**
    a. *Coccidioides immitis*   b. *Cryptococcus*
    c. *Histoplasma*   d. *Aspergillus*

29. **In tissue *Coccidioides immitis* produces:**
    a. Spherules and endospores
    b. Encapsulated yeast cells
    c. Fine, delicate hyphae
    d. Coarse, septate hyphae

30. **_Coccidioides immitis_ is identified in tissues on the basis of the following:**
    a. Budding yeast cells with pseudohyphae
    b. Yeast like forms with very large capsules
    c. Arthospores
    d. Endosporulating spherules

31. **Barrel-shaped arthroconidia are seen in:**
    a. Histoplasmosis   b. Cryptococcosis
    c. Coccidioidomycosis   d. Paracoccidioidomycosis

32. **Following is not matched:**
    a. *H. capsulatum* – Narrow-based budding yeast cells
    b. *B. dermatitidis* – Broad-based budding yeast cells
    c. *H. capsulatum* – Yeast cells gives figure 8 appearance.
    d. *C. immitis* – Arthospores and endosporulating spherules
    e. *P. brazsliensis* – Multiple-based budding yeast cells

Opportunistic Mycoses

33. **_Cryptococcus neoformans_ is a:**
    a. Protozoon   b. Fungus
    c. Parasite   d. *Mycoplasma*

34. **_Eucalyptus camaldulensis_ is associated with the transmission of:**
    a. *Blastomyces dermatitidis* b. *Histoplasma*
    c. *Cryptococcus*   d. *Coccidioides immitis*

35. **Primary site of infection of cryptococcosis is:**
    a. Adrenal   b. Bone
    c. Central nerve system   d. Lung

36. **_Cryptococcus_ is least likely to cause infection of:**
    a. Skin   b. Bone
    c. Brain   d. Kidney

37. *Cryptococcus* has predilection for:
   a. Lungs
   b. Meninges
   c. Liver
   d. GIT

38. **Virulence factor of *C. neoformans* includes all, *except*:**
   a. Polysaccharide capsule
   b. Production of protease
   c. Ability to make melanin
   d. Urease production

39. **Phagocytosis of *C. neoformans* is inhibited by:**
   a. Cryptococcal capsular material
   b. The size of yeast cell
   c. The cell wall
   d. Toxins produced by the organism

40. **Cryptococcal meningitis is common in:**
   a. Renal transplant patient
   b. Agammaglobulinemia
   c. Neutropenia
   d. IgA deficiency

41. **The capsule of *Cryptococcus neoformans* in CSF sample is best seen by:**
   a. Gram's stain
   b. Indian ink preparation
   c. Giemsa stain
   d. Methenamine silver stain

42. *Cryptococcus* can be readily demonstrated by:
   a. Albert's stain
   b. Indian ink preparation
   c. Giemsa stain
   d. Gram's stain (e) ZN stain

43. **Latex agglutination study of the antigen in CSF helps in the diagnosis of:**
   a. Cryptococcosis
   b. Candidiasis
   c. Aspergillosis
   d. Histoplasmosis

44. **All are true regarding cryptococcal infection, *except*:**
   a. Occurs in immunodeficient states
   b. Capsular antigen in CSF is a rapid method of detection
   c. Anticapsular antibody is protective
   d. Urease +ve organism in tissue is diagnostic

45. **Which of the following features is used for identification of *Cryptococcus neoformans*?**
   a. Oxidase +ve
   b. Dextran fermentation
   c. Hydrolyse urea
   d. Growth at 42°C

46. **All are true about *Cryptococcus, except*:**
   a. Polysaccharide capsule
   b. Reproduced by budding
   c. Pseudohyphae
   d. Mycelium has a narrow base

47. **The technique used for staining capsule of *Cryptococcus neoformans* in biopsy material is:**
   a. Modified Ziehl-Neelsen
   b. India ink
   c. Mayer's mucicarmine staining
   d. Giemsa staining

48. **The important organism causing meningitis in immuno-compromised patient is:**
   a. *Histoplasma*
   b. *Cryptococcus*
   c. *Coccidioidomycosis*
   d. *Candida albicans*

49. *Pneumocystis carinii* infects:
   a. Human
   b. Monkeys
   c. Rat
   d. Cats

50. *Pneumocystis jirovecii* is a fungus because:
   a. rRNA, mitochondrial protein gene sequence and presence of thymidylate synthase
   b. Cell wall contains glucans
   c. Antifungals are effective
   d. Commonest infection in AIDS

51. *Pneumocystis jirovecii* is diagnosed by:
   a. Sputum examination for trophozoites and cyst under microscope
   b. Culture
   c. Positive serology
   d. Growth on artificial media

52. **The cysts resemble crushed ping-pong balls is seen with:**
   a. *Microsporum*
   b. *Pneumocystis jirovecii*
   c. *Epidermophyton*
   d. *Fusarium* species

53. **The pigment produced by *Penicilliun marneffei* in culture media is:**
   a. Blue
   b. Green
   c. Black
   d. Brick red

54. **Which is false about *Penicillium marneffei*:**
   a. Black colonies
   b. Dimorphic fungus
   c. Amphotericin B used for treatment
   d. Causes fulminant infection in immunocompromised patients

55. *Penicillium marneffei* is seen in:
   a. Tuberculosis
   b. AIDS
   c. Diabetes
   d. Kala azar

56. **Bronchopulmonary aspergillosis is mediated by:**
   a. Type I hypersensitivity
   b. Type III hypersensitivity
   c. Type II hypersensitivity
   d. Both a and b

57. **What is the most probable entry of *Aspergillus*:**
   a. Puncture wound
   b. Blood
   c. Lungs
   d. Gastrointestinal tract

58. *Aspergillus* causes all, *except*:
   a. Bronchopulmonary allergy
   b. Otomycosis
   c. Dermatophytosis
   d. Oculomycosis

59. **A diabetic patient presents with bloody nasal discharge, orbital swelling and pain. Culture of periorbital swelling showed branching septate hyphae. Which of the following is the most probable organism involved?**
   a. *Mucor*
   b. *Candida*
   c. *Aspergillus*
   d. *Rhizopus*

60. **Acute angled septate hyphae are seen in:**
   a. *Aspergillus*
   b. *Mucor*
   c. *Pencillium*
   d. *Candida*

61. **Aspergilloma has:**
   a. Septate hyphae
   b. Pseudohyphae
   c. Metachromatic hyphae
   d. No hyphae

62. **Aseptate hyphae are not seen in:**
   a. *Rhizopus*
   b. *Mucor*
   c. *Aspergillus*
   d. None

63. **Mucormycosis:**
   a. Angio-invasion
   b. Lymph invasion
   c. Septate hyphae
   d. Long-term deferoxamine is the precipitating factor
   e. It may lead to blindness.

64. **Orbital mucormycosis is a complication of:**
   a. AIDS
   b. Steroid therapy
   c. Cushing's disease
   d. Diabetic ketoacidosis

- **Ochratoxin:** It is produced by *Aspergillus ochraceous, Aspergillus niger* and *Penicillium verrucosum*. It is called ochratoxin, because originally it was identified from *Aspergillus ochraceous.* There are different types of ochratoxin, but type A is medically significant. It is ingested along with cereals (like wheat, oats), peanut, peas, meat, milky powder, hay, etc. It is identified as nephrotoxic (endemic nephropathy in different areas), mutagen, teratogen, carcinogen and immunosuppressant in all experimental animals. Ochratoxin A is identified as **urinary tract carcinogen** by IARC.
- **Cyclopiazonic acid:** It is produced by *Aspergillus flavus, Aspergillus oryzae* and also by *Penicillium cyclopium.* It is ingested along with groundnut and corn mostly as a contaminant with aflatoxin; however, it does not augment the action of other toxin. In humans, it causes symptoms of **"kodua poisoning"** which actually occurred due to consumption of kodo millet (*Paspulum scrobiculatum*) and characterized by giddiness, sleepiness and tremors recover after 1–3 days. It also produces other symptoms like anorexia, weight loss, dehydration, convulsion and death.
- **Fumonisins:** It is produced by *Fusarium* species especially by *Fusarium moneliforme*, which is the common contaminant in maize and maize-based products. It is responsible for major toxicological effects in animals like pigs (porcine edema), rats (hepatotoxicity) and horses (equine leuko-encephalomalacia). Fumonisins is classified as class 2B human carcinogen by IARC and responsible to produce the **esophageal cancer**, in endemic regions of South Africa, particularly in Transkei and also in parts of China. It also causes liver cancer in rat.
- **Zearalenone:** It is an estrogenic toxin produced by *Fusarium graminearum*. It is the common contaminant in maize, wheat and sorghum. It causes genital disorders in animals, and in humans, it causes precocious pubertal changes. It is having carcinogenic property to produce **cervical cancer** in human.
- **Trichothecene:** It is produced by *Fusarium* species like *F. graminearum* and rarely by *F. sporotrichioides.* It also produced by other genera like *Trichoderma, Trichothecium, Myrothecium* and *Acremonium (Cephalosporium).* There are four types of trichothecene, but type A is medically significant. It is ingested along with maize and soughum. Toxin is responsible for cardiomyopathy, alimentary toxic aleukia (necrotic lesion of oral cavity, esophagus and stomach with marked leukopenia) and immunodepression. It is not the carcinogen, but synergise the effect of aflatoxin. It was used in biological warfare in Laos called **yellow rain.**
- **Other toxins:** Like **patulin, nitropropionic acid,** etc., are able to produce the mycotoxicoses in humans and animals.

2. **Mycetism (mycetismus, muscarinism or mushroom poisoning):** Clinical illness due to ingestion of mycotoxin along with fungus called mycetism. Following are the examples.
   - *Claviceps* spp.: Ergot poisoning.
   - *Coprine* spp.: Coprine poisoning.
   - *Inocybe* spp.: Muscarine poisoning.

Diagnosis

Specimens: Toxins present in urine, bile, stool, milk and food stuff.

Testing methods

1. **Culture:** By using primary fetal bovine kidney cells.
2. **Serological tests:** ELISA, RIA and precipitation-based tests are useful.
3. **Molecular method:** DNA-DNA homology and RFLP.
4. **Animal inoculation:** Duckling, rat, rabbit and trout are useful animals.
5. **Chromatography:** GLC and HPLC.

Prevention

Decontamination: Removal of fungal growth from the stored grains by hand picking.

Detoxification: Conversion of aflatoxoin B1 to less active metabolites like aflatoxoin M1 by using physical, chemical or microbiological methods.

Legislation: Strict rule for overuse of mycotoxin in food above permitted level.

Treatment

Specific antidote for particular mycotoxin is not available, so symptomatic treatment is given.

ACCESS YOURSELF

Short Notes
1. Allergic fungal disease.
2. Fungal food poisoning.

Short Question for Theory/Viva Question
1. Define: Mycetism and mycotoxicoses.

MCQs for Chapter Review

Allergic Fungal Diseases
1. **Farmer's lung is caused by:**
   a. *Micromonospore faenia*   b. *Aspergillus*
   c. *Histoplasma capsulatum*   d. All of above

Fungal Food Poisoning
2. **Aflatoxins are produced by:**
   a. *Aspergillus flavus*   b. *Aspergillus niger*
   c. *Aspergillus fumigatus*   d. *Candida*

Answers and Explanation of MCQs

1. b
- Actual name of bacteria causing farmer's lung is *Micropolyspora faenia, Micromonospora candidus* and *Micromonospora vulgaris.* Fungi causing farmer's lung are mentioned in **Table 92.1.**
2. a
- Aflatoxin means <u>A</u>spergillus <u>*flavus*</u> <u>**toxin**</u>.

Parasitic Infections

Infections of Protozoa
93. Infections of Amoebae
94. Infections of Intestinal, Oral and Genital Flagellates
95. Infections of Blood and Tissue Flagellates
96. Infections of Ciliates
97. Infections of Blood Inhabiting Sporozoa
98. Infections of Gastrointestinal Sporozoa
99. Infections of Microspora and Blastocystis

Infections of Metazoa (Helminths/Worms)
Infections of cestodes
100. Cestodes: General Features and Classification
101. Infections of Cestodes: Pseudophyllidea
102. Infections of Cestodes: Cyclophyllidea

Infections of trematodes
103. Trematodes: General Features and Classification
104. Infections of Trematodes: Diecious Trematodes
105. Infections of Trematodes: Monoecious Trematodes

Infections of nematodes
106. Nematodes: General Features and Classification
107. Infections of Nematodes: Intestinal Nematodes
108. Infections of Nematodes: Somatic (Tissue) Nematodes

Infections of Miscellaneous Parasites
109. Miscellaneous Infections of Parasites

Infections of Amoebae

Chapter Outline

- Introduction
- *Entamoeba histolytica*
- Nonpathogenic Amoeba
- Free-living Amoeba

INTRODUCTION

Meaning

From **amoibe (Greek) means change**, as the organism have habit to change the shape constantly. Amoeba is singular and Amebas or Amoebas/amoebae is plural.

Classification

Systemic/taxonomical classification: Follow **Ch. 14 (Table 14.3)**.

Pathogenic classification: Two groups.

1. **Gastrointestinal amoebae:** Two subgroups according to site of infection.
   - **Pathogenic:** *Entamoeba histolytica* (large intestinal and extraintestinal).
   - **Nonpathogenic**
     - Mouth: *Entamoeba gingvalis.*
     - Intestinal: *E. coli, E. hartmanni, E. polecki, Endolimax nana, Iodamoeba butchlii* and *Dientamoeba fragilis.*
2. **Free leaving amoebae:** Following are **pathogenic** species.
   - *Naegleria:* Out of 18 species, only *N. fowleri* is pathogenic.
   - *Acanthamoeba: A. culbertsoni, A. castellanii, A. hatchetti, A. polyphaga, A. rhysodes, A. astronyxis, A. divionensis* and *A. healyi.*
   - *Balamuthia: B. mandrillaris.*

Entamoeba histolytica

History

E. histolytica was 1st described by **Losch** from stool of Russian suffering with dysentery in 1875. **Koch** reported the trophozoites in hepatic capillaries of patients of amoebic dysentery with hepatic abscess in 1887. **Boek** and **Drobhlav** described the polyaxenic medium for cultivation in 1925. **Diamond** described the axenic medium with successful *in vitro* cultivation in 1965.

Morphology

It is unicellular and single cell performs the all functions. It is studied in stained **[Iodine and Iron-Hematoxylin (I andH)]** and unstained **[Normal Saline (NS)]** preparations with three stages like trophozoite, precyst and cyst.

Trophozoite Stage

Synonym: It also called tropic stage, active stage, growing stage, feeding stage or vegetative stage.

Shape and size: Shape is not fixed, because of constantly changing position of pseudopodium. Average size is 18–40 μm.

Motility: Motility is due to pseudopodium. It shows crawling or gliding or jerky motility. Movement is jerky, and unidirectional due to formation of new pseudopodium at different site, when whole cytoplasm moves into the direction of new pseudopodium. In a culture test tube, trophozoites are crawl by side of tube. Pseudopodium is temporary. Motility and pseudopodium formation is inhibited at low temperature. Speed of motility is 0.2–3 μm per second. Pseudopodium also helps in ingestion of food.

Structure (Fig. 93.1a): It consist cytoplasm and nucleus.

1. **Cytoplasm:** It is divided in two portions.
   - **Endoplasm:** It is an internal, granular portion of cytoplasm. It **functions** for nutrition, excretion and reproduction. It **contains** RBCs called **erythrophagocytosis.** Only intestinal amoebae having RBCs in cytoplasm, and it is the diagnostic feature. It also contains WBCs, food debris, nucleus, but no bacteria.

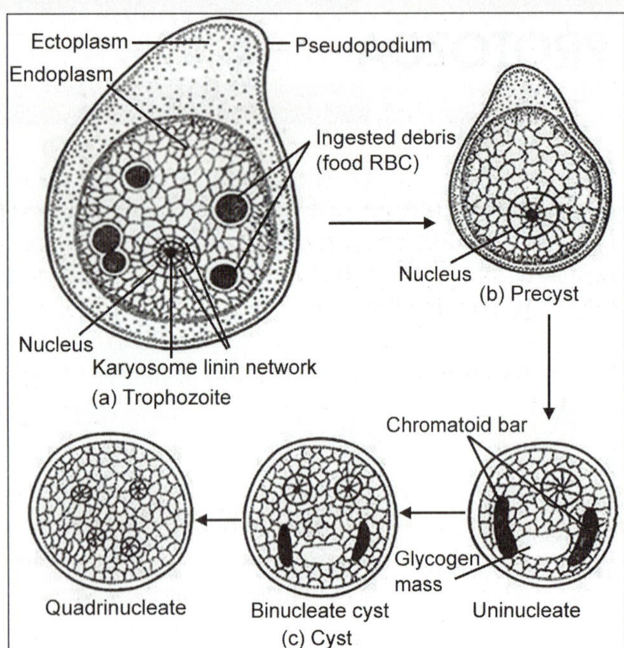

Fig. 93.1: Morphology of *E. histolytica*

- **Ectoplasm:** It is an external, clear, hyaline portion of cytoplasm. It **functions** for protection, locomotion and sensation. **Structures developed from ectoplasm** are pseudopodium and cyst-wall.
2. **Nucleus:** It is spherical in shape, 4–6 μm in size and single in number. **In unstained preparation** it is not visible due to motility, but when motility decreases, it observed with faint outline and eccentric position. **In stained preparation** it shows nucleolus (karyosome), linin network, nuclear membrane and nuclear sap. **Nucleolus** is a small dot like, concentrated chromatin material. It located either centrally or peripherally. Space between karyosome and nuclear membrane is filled by radially (spoke like) arranged fine threads called **linin network,** which gives cartwheel appearance. Nucleus bounded by thin and well defined **nuclear membrane** which is lined by fine chromatin granules. **Nuclear sap** is the fluid present in nucleus. It **functions** for controlling all activities of cell, growth, fertilization, replication, and hereditary transmission of genetic information.

Precyst Stage

It is oval in shape and 10–20 μm in size. It contains **(Fig. 93.1b)** cytoplasm and nucleus. **Cytoplasm** is divided into ectoplasm, which contains blunt pseudopodium and endoplasm, which is free of RBCs and other ingested food debris. **Nucleus** is spherical, single and smaller than trophozoite, but larger than cyst. Structure of nucleus is same as nucleus of trophozoite.

Cyst Stage

Synonym: It also called inactive stage, resistant stage or resting stage.

Shape and size: It is spherical in shape and variable in size from smaller about 6–9 μm to larger about 12–15 μm.

Structure (Fig. 93.1c)

1. **Structure of immature cyst**
   - **Cytoplasm:** It is clear, hyaline, free of RBCs and other ingested food debris. It **contains** chromatoid bars (chromodial bars) and glycogen mass. **Chromatoid bars** are 1–4 in number, variable in size from 1/2 to 2/3 of the size of cyst and disappear in mature cyst. In iodine, they remain unstained, in NS, they seen as rounded ends and in I and H stain, they seen in black color. **Glycogen mass** is single and disappears in mature cyst. In iodine, it stains in brown color; in NS, it is not seen and I and H stain, it remains unstained.
   - **Nucleus:** It is spherical in shape and size is smaller than the nucleus of trophozoite and precyst. Immature cyst is uninucleate, but nucleus divides by binary fission to develop in to binucleate and quadrinucleate bodies. Mature cyst is quadrinucleate in nature. **Structure** is same as trophozoites.
2. **Structure of mature cyst:** Immature cyst passed in feces and becomes mature over the soil.
   - **Cyst wall:** It develops from ectoplasm. It is thick, highly resistant and helps to survive in unfavorable condition.
   - **Cytoplasm:** It is clear, hyaline and free of RBCs, ingested food debris, chromatoid bars and glycogen mass.
   - **Nucleus:** It is spherical, quadrinucleate and smaller than trophozoite and precyst. Other structural features are same as trophozoites.

Culture

Culture on media: Three types of media.

1. **Axenic culture:** It is free of other microbial supplement. Examples of media are Nelson's medium, Craig's medium, Balamuth's medium, Robinson's medium, Cleverland and Sander's medium, St. Jhonson's medium and Diamond's medium.
2. **Monoaxenic culture:** It contains single microbial supplement.
   - **Philip's medium:** It contains thyoglycollate preparation, horse serum and *T. cruzi*.
   - **Shaffer and Frye's medium:** It contains thyoglycollate preparation and penicillin inhibited streptobacilli. Bacilli are living, but not multiplying.
3. **Polyaxenic culture:** It contains polymicrobial supplement. Examples of media are **Boeck and Drbohlav's medium** egg serum medium and NIH medium.

Culture in animals: Animals are used to study the pathogenicity and effect of amoebicidal drugs.
- Experimental amoebiasis can be produced in **cats, dogs and monkeys.**
- **Kittens:** It is fatal within 2–3 days due to extensive sloughing and colonic ulcer formation.

- **Pups (Dogs):** Lesion is similar to human amoebiasis, but animal lives longer than kittens.
- **Hamsters:** Production of hepatitis.
- **White rats:** Production of cecal ulcer after intra-cecal injection.
- **Guinea pigs:** Production of colonic ulcer after intra-ileal injection.

Resistance

Sterilization: Trophozoites are killed by drying and heating. Trophozoites survive for 5 hours at 37°C and even if enter from fresh stool; they are killed by stomach acidity; hence, infection is not transmitted by trophozoites.

Disinfection: Trophozoites are killed by disinfection.

Immunity

Specific antibodies are formed in invasive amoebiasis, which are detected by CFT, precipitation test, immobilization test and IF test. Antibodies are not protective against reinfection.

Life Cycle

Host: Human is the only host.

Methods of transformation (Flowchart 93.1):
1. **Excystation:** Transfer of cyst (inactive) to trophozoite (active) **called excystation.** It takes place inside the lumen of intestine of infected person. Mature cyst is quadrinucleate in nature. Each nucleus divides by binary fission, which gives birth to eight amoebulae. Each amoebula is capable to develop in to trophozoite. Finally, 1 cyst produces 8 trophozoites.

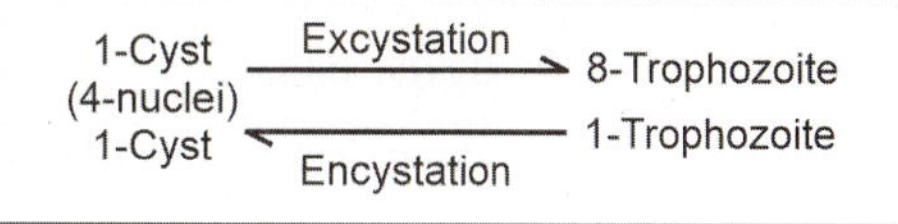

2. **Encystation:** Transfer of trophozoite (active) to cyst (inactive) **called encystation.** Encystation is not a reproductive process, but a process of protection. It takes place inside the lumen of intestine. It occurs in few hours and mature cyst lasts for 2 days inside the lumen. It does not occur in intestinal wall, liver, lungs, brain, etc. One trophozoite gives one cyst. Mature cyst is quadrinucleate in nature.

Methods of reproduction (multiplication): It occurs only in trophozoite stage. Trophozoite divides by binary fission every 8 hours. 1st nucleus divides by binary fission and then cytoplasmic body.

Cycles: Three stages of life cycle are described in **Fig. 93.2 and Flowchart 93.2.**

Summary: Follow **Flowchart 93.3.**

Pathogenicity

Disease name: It called **amoebiasis.**

Epidemiology: It is distributed world-wide.

Reservoirs of infection: Natural infection occurs in men and monkeys. Man is the commonest reservoir. In China, dog is the possible reservoir.

Sources of infection: These are contaminated food and water from feces. Cyst of *E. histolytica* also found in dropping of cockroaches, which also serve a source of infection.

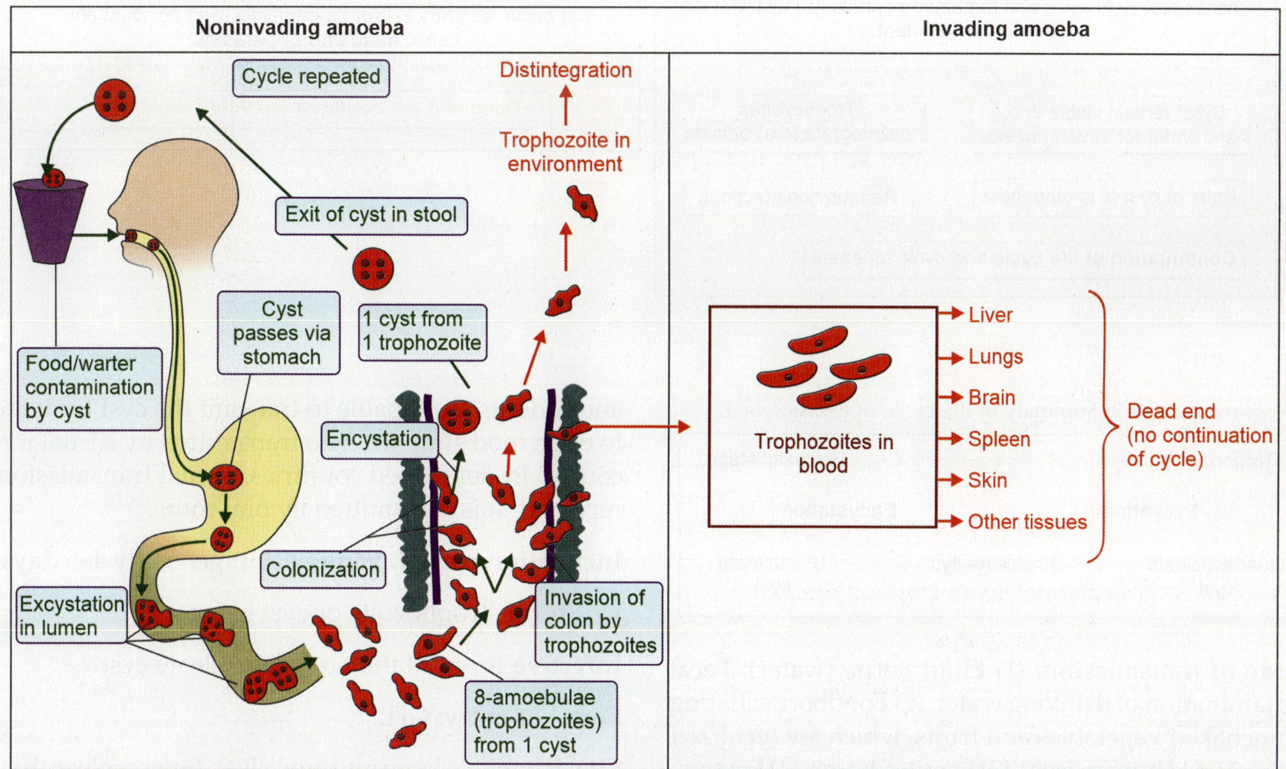

Fig. 93.2: Life cycle of *E. histolytica*

Infections of Amoebae

Flowchart 93.2: Life cycle of *E. histolytica*

Cystic stage
Mature quadrinucleate cyst enters in human by contaminated food and water
Resist the digestion by gastric juice and enters in to intestine. Digestion of cyst wall by action of trypsin in the intestine.
Entry into lower part of ileum (neutral or slightly alkaline medium) or cecum and excystation will take place.
Rupture of cyst wall and liberation of amoeba with four nuclei
Each nucleus divides by binary fission and finally releases the eight amoebulae.
Each amoebula gives the one trophozoite, so finally eight trophozoites are formed from one cyst.
Trophozoite stage
Trophozoites enter into the intestinal epithelium through crypt of Lieberkuhn
Trophozoites liberate the histolysin, which brings the destruction and coagulative necrosis of tissue and slough formation. Trophozoites receive the nourishment by absorption of dissolved tissue juices.
Slough fall down leaving behind an amoebic ulcer. Trophozoites continuously lyse the tissue till they reach the submucous coat as shown in Fig. 93.3.
Trophozoites then move in different directions
Intestinal/noninvasive amoebiasis
Noninvading trophozoites are remaining in lumen of intestine and cause an attack of acute of amoebic dysentery.
Precystic stage
After some time due to decrease in virulence of trophozoites and development of tolerance in host, they are not able to continue the life cycle, so some trophozoites are change to precystic stage followed by cystic stage. Few trophozoites remained unchanged. Both cysts and remained trophozoites exit with the stool in environment.
Cysts remain viable in soil and water for several weeks
Entry of cyst of in other host
Continuation of life cycle and cycle repeated
Trophozoites disintegrate (die) outside
Remain noninfectious
Extraintestinal/invasive amoebiasis
Tissue invading trophozoites move from dead tissues to the healthy margin
Enter into deeper layers and sometimes gain access to portal circulation
Carried to the liver where their further development is arrested. In the liver the trophic forms may multiply but encystation does not occur, so entry in liver by parasites is an accident and called dead end of parasite.
Dead end (no continuation of life cycle)

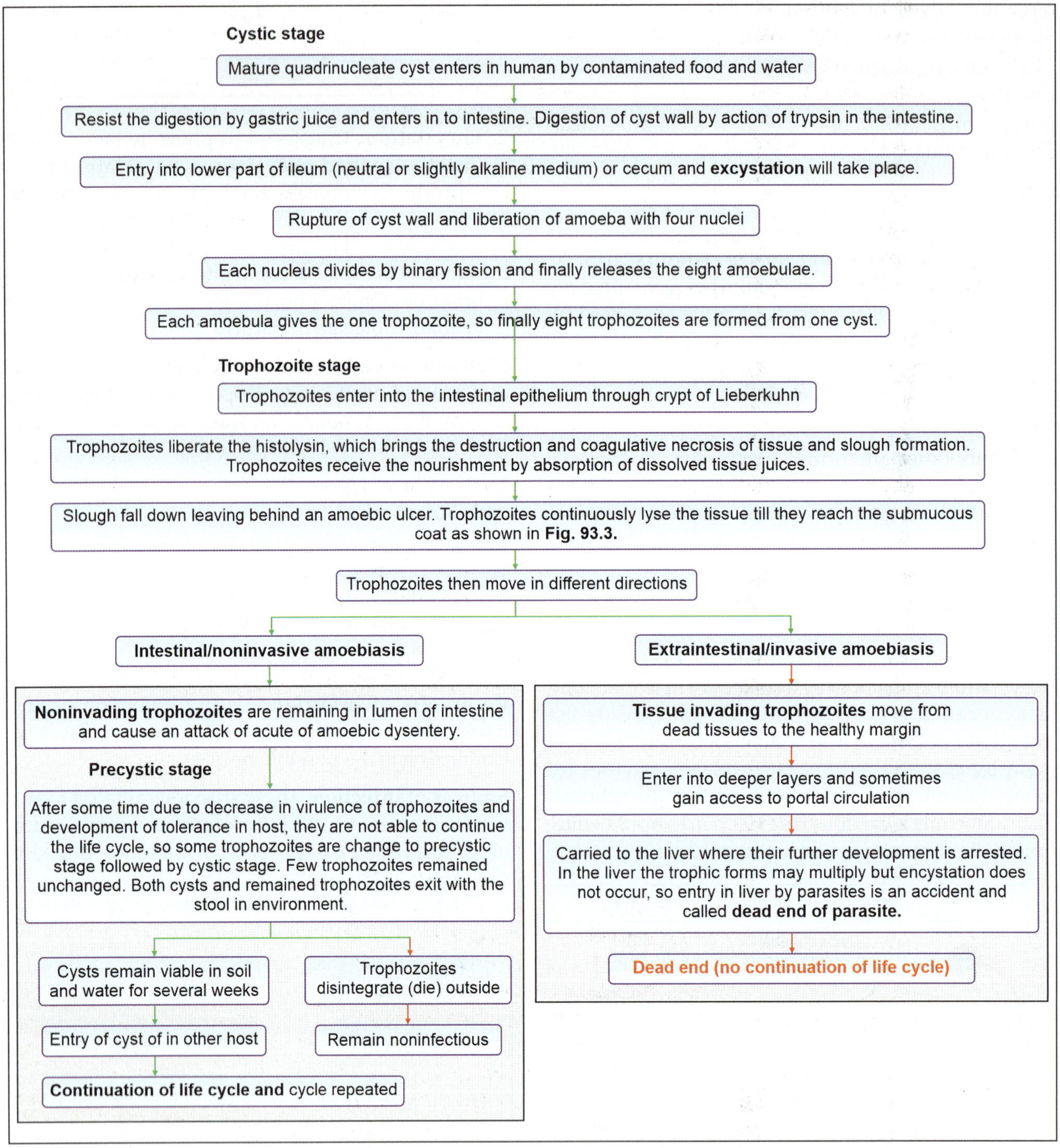

Flowchart 93.3: Summary of life cycle of *E. histolytica*

Trophozoites
Precystic stage
Excystation
Encystation
Quadrinucleate cyst
Binucleate cyst
Uninucleate cyst

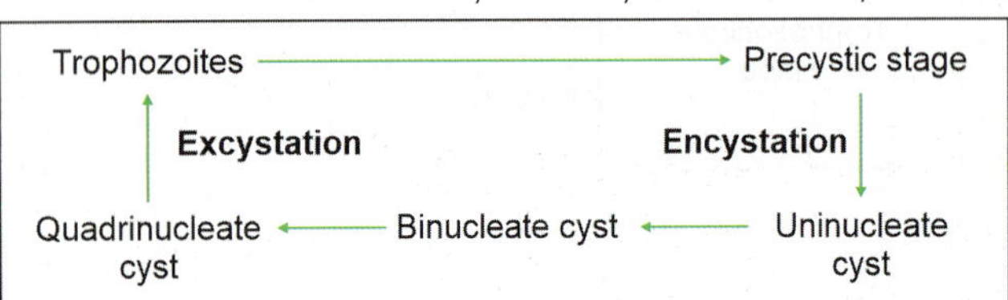

Mode of transmission: (1) Fluid borne (water): Fecal contamination of drinking water. (2) Foodborne: Eating of uncooked vegetables and fruits, which are fertilized with infected human feces. (3) Fomites-borne. (4) Fingers-borne. (5) Flies (vector-borne): House flies, cockroaches and rodents are capable to transmit the cyst from feces to open food stuff. (6) Also transmitted by sexual intercourse. In developed countries, sexual transmission is reported. It is transmitted by oral route.

Incubation period: Variable, but generally 4–5 days.

Exit form: Trophozoite or cyst in feces.

Infective form: Mature quadrinucleate cyst.

Portal of entry: GIT.

Site: It habitats large intestine, liver, lungs, spleen, brain, skin, etc.

Precipitating factors (epidemiological determinants)

- **Agent factors (virulence factors): (1) Lectin mediated adherence protein:** It helps in contact with target cells. **(2) Toxin:** Tissues damage by enterotoxin or cytotoxin has been studied in certain studies. **(3) Enzymes:** Histiolysin brings the destruction and necrosis of tissues and slough formation. Organisms receive the nourishment by absorption of dissolved tissue juice. Other enzymes produced by parasites are hyaluronidase, trypsin, pepsin, amylase, etc., may induce tissues destruction. **(4) Amoebapore:** It is a pore (hole) forming protein. It polymerize in target cell membrane → producing holes in cell membrane → cell lysis (cytolysis). **(5) Motility:** It is due to pseudopodium which helps in deeper penetration. **(6) Infective dose:** There is direct relation between the numbers of cyst ingested and disease development. **(7) Zymodeme pattern:** Isoenzyme pattern of *E. histolytica* called **zymodeme pattern**. It is useful to detect whether the strain is virulent or avirulent. *E. histolytica* has total 22 isoenzymes (zymodemes) of these 10 are invasive and 12 are noninvasive. Isoenzyme (zymodeme) is identified by electrophoretic mobility of 4 enzymes like L-malate (NADP + oxidoreductase), phosphoglucomutase (most important), glucose phosphate isomerise and hexokinase. Electrophoresis of phosphoglucomutase can show one or more of 4 bands such as α, β, γ and δ. Strain with absence of α-band with presence of β-band indicates the virulent strain.
- **Host factors: (1) Age:** It is more in adults than in children. **(2) Sex:** More in male than in female. **(3) Diet:** Low iron (restriction of trophozoite growth) and high protein restrict the disease. Vitamin C deficiency, high carbohydrate and high lipid precipitate the infection. **(4) Immune status:** Late pregnancy, DM, malignancy, steroid therapy, immunosuppressive drugs and antimetabolites increase the risk. **(5) Coinfection:** Coinfection by bacteria enhances the pathogenicity and evidenced by different ways. Better growth of amoebae in culture in presence of bacteria like streptobacilli (also in presence of parasite like *T. cruzi*). No growth of amoebae in germ-free animals. **(6) Resistant host:** Minimal injuries/superficial ulcers are produced, if parasite enters in to resistant host. **(7) Others are like:** Poor socioeconomic status, poor hygiene, etc., increase the risk.
- **Environmental factors:** It is more in tropics and sub-tropics than temperate zone.

Pathological types of amoebiasis:

Two types like intestinal/primary/noninvasive amoebiasis **and** extra-intestinal/secondary/invasive amoebiasis (metastatic amoebiasis).

Intestinal/Primary/Noninvasive Amoebiasis

Definition: Infection limited to large intestine called **intestinal (primary) amoebiasis.**

Pathological subtypes of intestinal amoebiasis: Two types like acute and chronic.

1. Macroscopic features of acute intestinal amoebiasis

- **Sites of ulcer: Generalized** ulcer is distributed through the whole length of large gut and up to internal anal sphincter. **Localized** ulcer has two sites. **(1) Ileocecal region:** It is the most common site. Ileocecal valve, cecum, ascending colon, appendix are involved. It is found twice than sigmoido-region. **(2) Sigmoido-region:** Sigmoido-colon and rectum are involved.
- **Size of ulcer:** Ulcer is varying from pin's head to one inch or greater in diameter.
- **Shape of ulcer:** It is round to oval or typically **flask shaped ulcer (water bottle or water decanter or undermined ulcer)** indicates the process of tissue necrosis as shown in **Fig. 93.3**.
- **Margin of ulcer:** Ragged and undermined, being formed by overhanging mucous membrane. On vertical section, it appears in flask-shape.
- **Base of   ulcer:** Formed by muscular coat.
- **Contains:** Ulcer contains yellowish-black slough (necrosed tissues and trophozoites).
- **Extension of ulcers: Superficial ulcer** does not extend beyond the muscularis mucosa. **Deep ulcer** does not extend beyond the submucous coat and extends laterally in surrounding tissues. However,

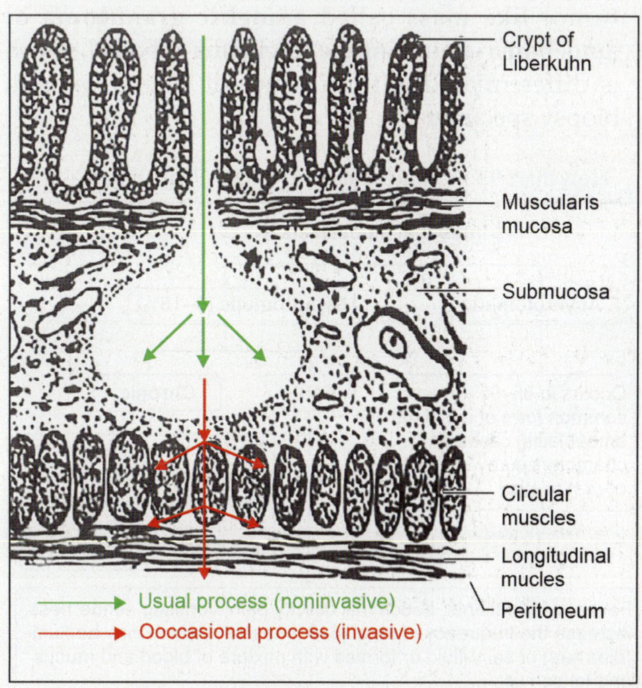

Fig. 93.3: Amoebic ulcer (flask shape)

sometimes, it extends beyond the submucous coat and even up to serous layer and produces complications which are described below.

- **Healing of ulcers: Superficial ulcer** heals without scar while **deep ulcer** heals with scar.

2. **Microscopic features of acute intestinal amoebiasis**
   - **Early ulcers:** Section in the middle of early ulcer shows centrally coagulative necrosis and trophozoites, either singly or in groups in periphery.
   - **Advanced ulcers:** Trophozoites migrate away from the site of actual ulcer invade the intermuscular spaces to reach the peritoneal coat. Parasites also invading the small venules, causing hyperplasia of endothelial cells and thrombosis of blood vessels.

3. **Chronic intestinal amoebiasis:** Chronic stage is different from acute and depends on immunity of host. It presents with small, superficial ulcer involving the mucosa and healed with marked scarring of intestinal wall.

Clinical features of intestinal amoebiasis: Follow **Flowchart 93.4**.

Complications of intestinal amoebiasis

- **Acute:** These are perforation, local peritonitis, generalized peritonitis, hemorrhage, pericecal or pericolic abscess, gangrene of large gut, amoebic appendicitis and excessive scarring of healed ulcers which leads to thickening of intestinal wall, stricture formation and partial intestinal obstruction.
- **Chronic: (1) Marked scarring** leads thinning, dilatation and sacculation of wall. **(2) Localized thickening** of intestinal wall leads to narrowing of the lumen of the gut. **(3) Generalized thickening** rendering the palpable gut. **(4)** Formations of granulomatous tumor like mass called **amoebic granuloma or amoeboma** simulating the carcinoma of bowel, which is differentiated with presence of trophozoites in biopsy specimens.

Flowchart 93.4: Clinical features of intestinal amoebiasis

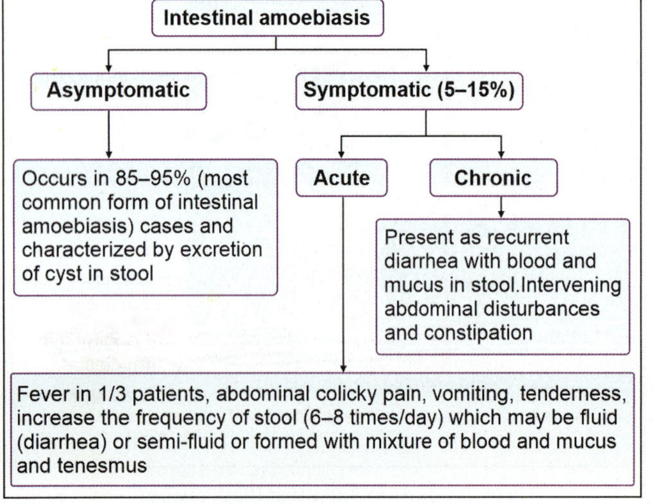

Extraintestinal/Secondary/Invasive Amoebiasis (Metastatic Amoebiasis)

Intestinal lesions in metastatic amoebiasis: Presence of small superficial ulcers with thickening of intestinal wall, latent ulcer in cecum and pigmented/non-pigmented scar, representing the sites of previous ulcer.

Sites of metastatic amoebiasis: It involves liver, lungs, brain, skin, spleen, urogenital, etc.

A. Hepatic amoebiasis (amoebic liver abscess)

- **Incidence:** It is the most common form of extraintestinal amoebiasis. Average 5% cases in *E. histolytica* infection produce the liver amoebiasis.
- **Pathogenesis:** Follow **Flowchart 93.5**.
- **Pathology:** Two subtypes. **Amoebic hepatitis** involves only inflammation of liver and no pus detected in liver. **Amoebic liver abscess** is not the pus formation (no suppuration), but a mixture of sloughed liver tissues and blood called **anchovy sauce pus**. It is bacteriologically sterile.

> **Note: Diffuse amoebic hepatitis**
> Nonspecific hepatomegaly present in patients of amoebic dysentery with enlarged tender liver, right upper quadrant pain, intermittent fever and leukocytosis called **diffuse amoebic hepatitis**. Liver biopsy does not reveal the presence of *E. histolytica*. Enlargement of liver is due to periportal inflammation and congestion by **bacteria** carried to liver by blood stream from ulcerated gut.

1. **Macroscopic features of liver abscess**
   - Sites: Most common site is posterosuperior surface of right lobe of liver. Rarely, it occurs in left lobe but more important because of more chances of rupture due to less liver mass.
   - Size: It is small in size of 10–15 cm. Sometimes, multiple small abscesses (miliary abscess) coalesce to form big sized abscess, which may hold liter of pus.
   - Other features: In >80% cases abscess is solitary, chocolate-brown in color, semifluid or grumous in consistency and with offensive smell.

Flowchart 93.5: Pathogenesis of hepatic amoebiasis

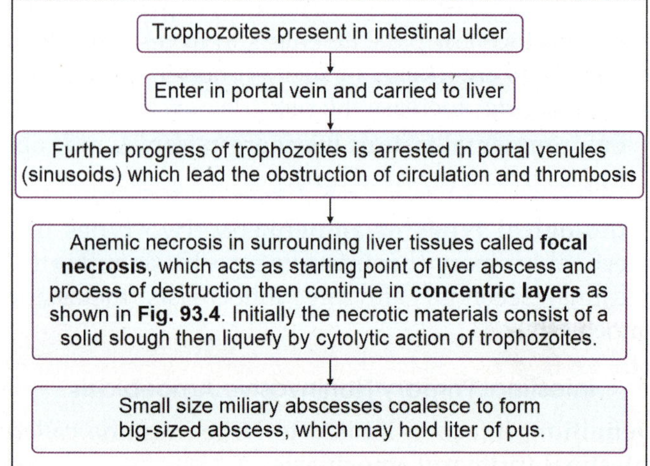

- Wall of abscess cavity: It is ragged, shaggy and formed by necrotic liver tissues which gradually merge in to healthy tissues with intervening zone of hyperemia.

2. **Microscopic features of liver abscess:** When a cut section made through the margin of liver abscess and examined microscopically, it shows three zones **(Fig. 93.4)**. **Central zone** contains cytolyzed granular materials and no amoebae. **Intermediate zone** contains RBCs, WBCs, degenerated liver cells and few trophozoites of *E. histolytica*. **Peripheral zone** contains congested capillaries, necrosed liver cells and amoebae (multiplying). In long-standing cases it may consists the lymphocytes, monocytes and connective tissues cells.

- **Clinical features of hepatic amoebiasis**
  - General features: High, continuous fever with chills, rigor and sweating. It can be remittent or intermittent or absent. Other symptoms include emaciation and shallowness of the skin.
  - GIT features: Sudden onset of abdominal pain and tenderness in right hypochondrium due to stretching of liver capsule. Referred pain present in right shoulder due to irritation of phrenic nerve which supplies the under surface of diaphragm. Also referred pain presents in lower abdomen and in right iliac region. There is no diarrhea/dysentery. Tenderness and thickening of bowel may elicited on abdominal palpation.
  - Liver features: Jaundice is rarely present and mild if present. Liver movement on right side of chest during respiration is decreased or absent. Liver is **tender and palpable** below the costal margin. **Dullness** of liver which may extend upwards. **Rigidity** on upper part of right rectus which may interfere with enlarged liver's palpation. **Compensatory hypertrophy** of left lobe of liver occurs in case of large abscess.
  - Lung features: Collapse of base of right lung seen due to growing liver abscess. Occasionally presence of right pleural effusion.

- **Complications (course and termination) of hepatic amoebiasis:** Complications are healing of lesion as encysted mass or with fibrosis/calcification, secondary bacterial infection occurs via bloodstream or during aspiration or during surgery or following spontaneous rupture, and hematogenous spread in distant organs like brain, kidney, etc. Sometimes lesion may terminate in neighboring tissues as described below **(Fig. 93.5)**.

1. **Rupture of right side liver abscess**
   - Into the lungs: Sputum shows the presence of anchovy-sauce pus with liver cells, trophozoites, viscid consistency and chocolate brown color (like hemoptysis).
   - Into the right pleural cavity: Leads to empyema thoracis. Pleuropulmonary termination is the most common complication of hepatic amoebiasis.
   - Below diaphragm: Producing subphrenic abscess
   - Externally into skin: Granulomatous lesions of skin due to secondary infection by trophozoites from ruptured liver abscess called granuloma cutis.
   - Into the peritoneal cavity: Producing generalized peritonitis.
   - Into the colon: Diarrhea and presence of anchovy-sauce pus in stool.

2. **Rupture of left side liver abscess**
   - In to the left pleural cavity: Leads empyema thoracis.

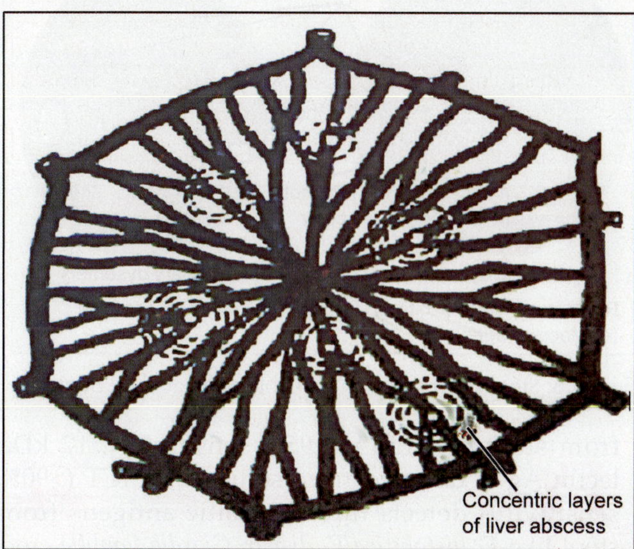

Fig. 93.4: Concentric layers of liver abscess

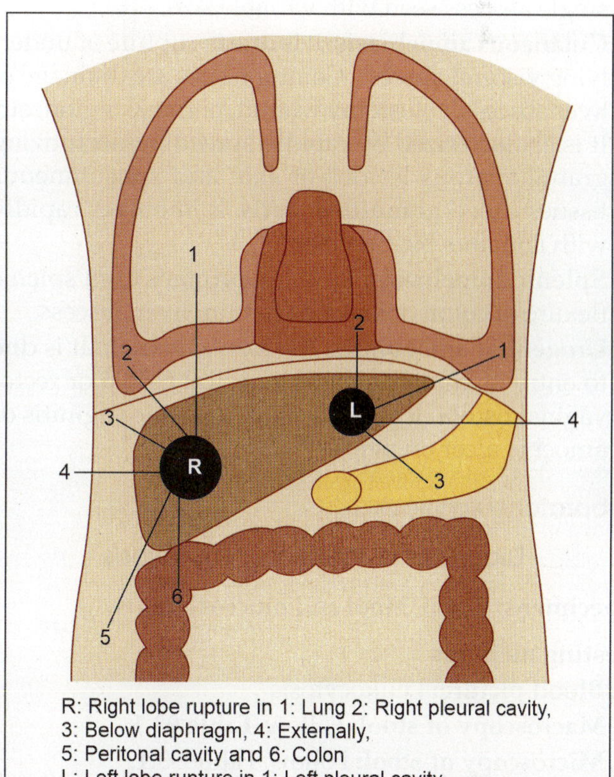

R: Right lobe rupture in 1: Lung 2: Right pleural cavity,
3: Below diaphragm, 4: Externally,
5: Peritonal cavity and 6: Colon
L: Left lobe rupture in 1: Left pleural cavity,
2: Pericardial cavity, 3: Stomach and 4: Externally

Fig. 93.5: Rupture of right (R) and Left (L) side liver abscess

– Into the pericardial cavity: Leads purulent pericarditis, which is fatal in nature.
  – Into the stomach: Vomitus shows the presence of anchovy-sauce pus with liver cells, trophozoites, viscid consistency and chocolate brown color (like hematemesis).
  – Externally: Through anterior abdominal wall in epigastric region.
3. **Rupture of inferior surface liver abscess**
  – Into the transverse colon/duodenum: Diarrhea and presence of anchovy-sauce pus in stool.
  – Into the peritoneal cavity: Causing a fatal peritonitis.
4. **Rupture of posterior surface liver abscess:** Into inferior vena cava, which is very are but fatal.
5. **Other sites of rupture:** These include common bile duct, pelvis of kidney and perinephric tissues of lumbar region.

B. **Pulmonary amoebiasis:** It may be primary or secondary.
  • **Primary:** It is very rare due to direct entry of trophozoites from gut wall by blood (portal circulation) in to the pulmonary capillaries. Single or multiple abscesses are present in lungs.
  • **Secondary:** It is very common due to rupture of liver abscess in to lungs. It affects the lower lobe of right lung. Mostly single abscess seen.

C. **Cerebral amoebiasis:** It is very common. It is due to complication of hepatic or lung amoebiasis or both. It affects the one of the cerebral hemisphere. Mostly single abscess seen with variable size.

D. **Cutaneous amoebiasis:** It is due to rupture of underlying visceral abscess. Common sites are drainage of liver abscess, colostomy wound, perianal region, etc. It is characterized by rapidly spreading necrotizing granulomatous lesions of skin and subcutaneous tissues called **granuloma cutis.** It improves rapidly with antiamoebic treatment.

E. **Splenic amoebiasis:** It occurs **primary** from splenic flexure of colon or **secondary** from liver abscess.

F. **Urogenital amoebiasis:** It is very common. It is due to entry of amoebae by rectovesical fistula or rectovaginal fistula. It presents with amoebic vaginitis or amoebic ulcer on penis.

Laboratory Diagnosis

Diagnosis of Intestinal Amoebiasis

Specimens: Blood, Stool and mucosal biopsy.

Testing methods

A. **Blood picture:** Leukocytosis
B. **Macroscopy of stool:** Follow **Table 93.1.**
C. **Microscopy of stool:** Follow **Table 93.1.**
D. **Culture:** Described above.
E. **Serological tests:** Ab appears in late stage so not useful in early stage. 17 kDa lectin Ag is detected

TABLE 93.1: Differences between amoebic and bacillary dysentery

Features	Amoebic dysentery	Bacillary dysentery
Causes	*E. histolytica*	*Shigella* spp., *Cl. difficile* EIEC, EHEC, *C. jejuni, Y. enterocolitica, V. parahaemolyticus, Aeromonas* spp., and may by few non-typhoidal *Salmonella* spp.
Macroscopy of stool		
Numbers	6–8 motions/day	>10 motions/day
Amount	Copious	Small
Odor	Offensive	Odorless
Color	Dark red	Bright red
Nature	Blood and mucus with feces	Blood and mucus, no feces
Reaction	Acid	Alkaline
Consistency	Not adherent to container	Adherent to the bottom of container
Microscopy of stool (in Fig. 93.6) by normal saline		
RBCs	In clumps, reddish-yellow in color	Discrete/in rouleaux, bright red in color
Pus cells	Scanty	Numerous
Macrophages	Very few	Large and numerous, contain RBCs, so mistaken as *E. histolytica*
Eosinophils	Present	Scarce
Cysts and motile trophozoites	Present	Absent
Ghost cells	Nil	Numerous
Pyknotic bodies	Very common	Nil
C L crystals	Present	Absent

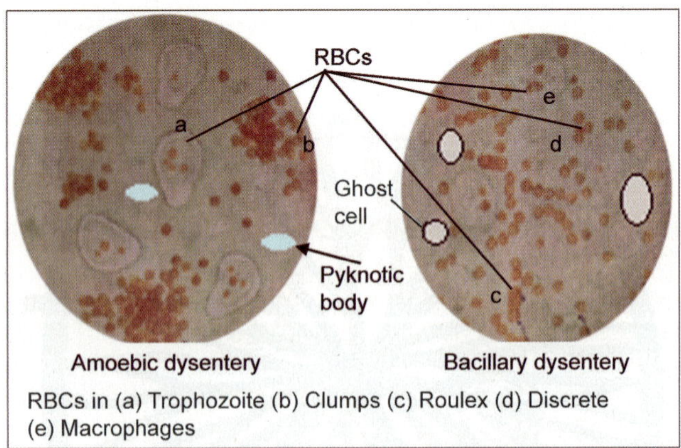

Fig. 93.6: Stool microscopy in amoebic and bacillary dysentery

from stool by ELISA (>95% sensitivity). 17 kDa lectin Ag is detected from serum also. ICT (>90% sensitivity) detects three parasitic antigens from stool like *E. histolytica/E. dispar, Giardia lamblia,* and *Cryptosporidium parvum.*

F. **Molecular method:** Nested multiplex and qPCR are useful. **BioFire Film Array** is an automated multiplex PCR which targets multiple bacteria, viruses and parasites causing diarrhea like *E. histolytica, Giardia, Cyclospora* and *Cryptosporidium parvum.*

G. **Zymodeme pattern (isoenzyme pattern):** Described above.

H. **Colonoscopy:** It detects the flask shaped ulcer.

I. **Amoebic dysentery should be differentiated from other conditions:** Like bacillary dysentery **(Table 93.1)**, parasitic dysentery (*B. coli, S. mansoni, S. japonicum, S. intercalatum, H. heterophyes, Trichuris trichuria, S. stercoralis,* etc.), diseases (psedomembranous colitis, ulcerative colitis, Crohn's disease, piles, malignancy, etc.) and nonpathogenic *Entamoeba coli* (described below).

Notes: Ghost cells, pyknotic bodies and CL crystals

Ghost cells: These are enlarged eosinophilic epithelial cells with eosinophilic cytoplasm without nucleus.

Pyknotic bodies: These are condensed nuclear/chromatin materials of epithelial cells, macrophages or pus cells in amoebic dysentery.

CL (Charcoat-Layden) crystals: These are breakdown products of eosinophils. Lysophospholipase, which is present in eosinophils, convert them into CL crystals. Their presence indicates that immune response has taken place which may be due to parasitic infections or allergic reactions. Their presence in stool may not correlate with eosinophilia in blood. They are slender, hexagonal or bipyramidal or diamond or whetstone shaped. They are different in size from 5–50 µm. They are **diagnosed** from stool in infection of *E. histolytica, A. duodenale, Trichuris trichuria,* etc., and from sputum in asthma and *A. lumbricoides* by NS preparation (diamond or whetstone shaped, clear and refractile as shown in **Fig. 93.7**) and trichrome stain (red-purple color as shown in **Fig. 93.8**). They are frequently present in exudates and granuloma containing eosinophils. Pineapple crystals, kiwis crystals and vegetables like aspargas sometimes looks like CL crystals and should be differentiated with history of ingestion of juice or vegetables.

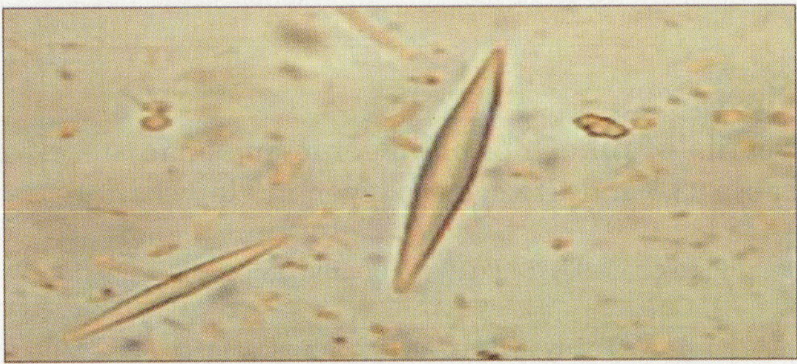

Fig. 93.7: CL crystals under NS preparation

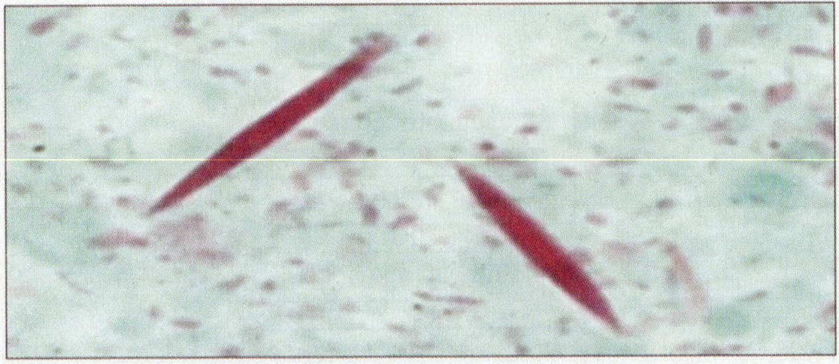

Fig. 93.8: CL crystals under trichrome stain

Specimens

- **Aspiration of liver pus or liver biopsy material:** For diagnosis of *E. histolytica* material should be taken from the abscess wall by scraping and not from the abscess contents.
- **Other specimens:** Blood and stool.

Testing methods

1. **Blood picture:** Leukocytosis (15,000–30,000 mm^3).
2. **Microscopy:** Motile trophozoites are detected from liver pus in 25% cases. Cysts are detected from stool by microscopy.
3. **Culture:** Follow C/Cs.
4. **Serological tests:** 17 kDa lectin Ag is detected from serum, saliva and liver pus by ELISA. Ab to 17 kDa lectin Ag is detected by ELISA. Inhibition of trophozoite's motility by specific antiserum by immobilization test.
5. **Molecular method:** Nested multiplex and qPCR are useful.
6. **Radiological test:** Like USG, CT scan and MRI are useful to know the location and extent of abscess. However, they cannot differentiate between amoebic abscess and pyogenic abscess.

Diagnosis of Pulmonary Amoebiasis

Specimens: Sputum, aspiration of pus from abscess, lung biopsy material, blood and stool.

Testing methods

1. **Microscopy:** Motile trophozoites are detected by microscopy from pus or biopsy material. Cysts are detected from stool by microscopy.
2. **Other methods:** Same as hepatic amoebiasis.

Carrier

Synonym: It also called **cyst passer.**

Types: There are two types of carrier. **(1) Healthy (contact) carrier** contains the organisms but never suffered from amoebic dysentery (health remains intact). **(2) Convalescent carrier,** who recovered from acute clinical attack of amoebic dysentery.

Diagnosis

- **Specimens:** Purged stool obtained after saline catheter or mucosal biopsy obtained by sigmoidoscope.
- **Testing methods**
  1. **Microscopy by concentration method:** Formol ether and zinc sulfate flotation techniques are useful. It increases the sensitivity of cyst identification from 30 to 70%. Repeated smear examination to identify the cyst.
  2. **Culture:** Follow C/Cs.
  3. **Serological tests:** Tests are positive after tissue invasion. IgG Ab will be detected against lectin Ag by ELISA.

Prevention

Preventive measures are prevention of contamination of food and water by human excreta, detection and treatment of carrier and their exclusion from food handling occupation and treatment of carriers by luminal amoebicides.

Treatment

- Oral rehydration and electrolyte replacement.
- Carrier is treated by luminal/intestinal amoebicides like iodoquinol (20 days) or paromomycin (10 days).
- Intestinal and extraintestinal amoebiasis: Treated by extraintestinal (tissue) amoebicides like metronidazole (5–10 days) or tinidazole (3 days) followed by luminal amoebicides.

NONPATHOGENIC AMOEBAE

Entamoeba coli

History: *E. coli* was 1st described by **Lewiss** in 1870 and **Cunnigham** in 1871 in Kolkatta. **Grassi** reported its presence in intestine of healthy person in 1878.

Morphology: It is unicellular and single cell performs the all functions. It has 3 morphological stages like trophozoite, precyst and cyst as shown in **Fig. 93.9 and Table 93.2**.

Life cycle: Octanucleated cyst ingested by food or water. Cyst produces eight trophozoites **(Flowchart 93.6)**. Trophozoites converted in to cyst, which pass through feces and cycle will repeat.

Pathogenicity: It is distributed world-wide, habitats large intestine, transmitted by feco-oral route and non-pathogenic.

Diagnosis: *E. coli* presents in dysenteric stool and differentiated morphologically from *E. histolytica* by **microscopic method** as shown in **Table 93.2**.

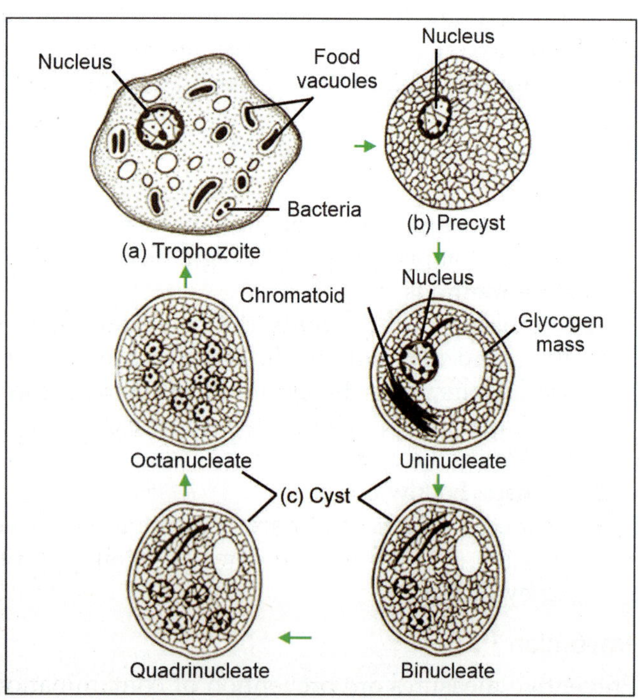

Fig. 93.9: Morphology of *E. coli*

TABLE 93.2: Differences between *E. histolytica* and *E. coli*

Features	E. histolytica	E. coli
Trophozoite		
NS preparation from fresh stool		
Size	18–40 µm	20–40 µm
Motility	Active	Sluggish
Cytoplasm	Clearly defined in to ectoplasm and endoplasm	Ill defined, ectoplasm scarcely seen
Cytoplasmic inclusion	• RBCs, WBCs and tissue debris • No bacteria	• Bacteria and other materials • No RBCs
Nucleus	Not visible in NS preparation	Visible in NS preparation
Iodine preparation (Nuclear character)		
Karyosome	Central	Eccentric
Linin network	Fine	Coarse
Nuclear membrane	Delicate	Thick
Chromatin granules	Fine and lined to nuclear membrane	Coarse and lined to nuclear membrane
Mature cyst		
NS preparation from fresh stool		
Size	6–15 µm	15–20 µm
Chromatoid bars	Rounded	Filamentous, thread like and pointed ends
Iodine preparation		
Nucleus	1–4	1–8
Glycogen mass	Visible in uni- and binucleate stage	Visible in bi- or quadricnucleate stage

Flowchart 93.6: Stages in life cycle of *E. coli*

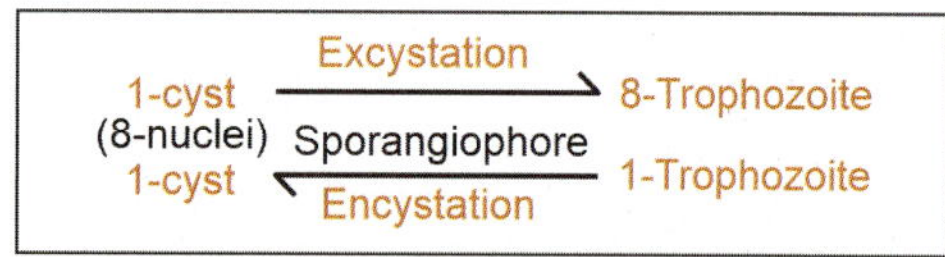

Others

Details of other nonpathogenic amoebae are given **Table 93.3** and **life cycle** is given in **Flowchart 93.7**.

Flowchart 93.7: Life cycle of nonpathogenic amoeba

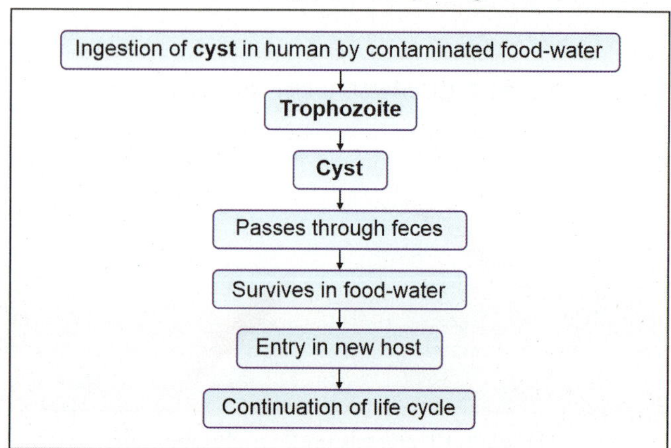

Essentials of Medical Microbiology

TABLE 93.3: Details of other nonpathogenic amoebae

Features	E. gingivalis	E. hartmanni	E. polecki	E. nana	I. butchlii	D. fragilis
Trophozoite						
History	Gross, 1849 Brumpt,1913	von Prowazek, 1912	von Prowazek, 1912	Wenyon and O,Conner, 1917 Brug, 1918	von Prowazek, 1912 Dobell, 1919	Jeps and Dobell, 1918
Figure 93.10	a	b	c	d	e	f
Size	10–20 µm	4–12 µm	8–25 µm	5–10 µm	8–20 µm	5–8 µm
Motility	Active	Sluggish	Sluggish	Sluggish	Sluggish	—
Cytoplasm	Divided in ectoplasm and endoplasm	Divided in ectoplasm and endoplasm	Divided in ectoplasm and endoplasm	Divided in ectoplasm and endoplasm		Fragile cytoplasm
Cytoplasmic inclusion	• Bacteria, food vacuoles • No RBCs	• Bacteria • No RBCs	• Bacteria • No RBCs	• Bacteria • No RBCs	• Bacteria, yeast and food particles • No RBCs	• Bacteria and food particles • No RBCs
Other features	Multiple pseudopodia	Same as like E. histolytica, but smaller in size, so called small race of E. histolytica		Trophozoite and cyst are in same size		So called, because it is binucleate and cytoplasm is fragile
Mature cyst						
Figure 93.11	No cyst stage	a	b	c	d	No cyst stage
Size		5–10 µm	10–20 µm	5–10 µm	5–10 µm	
Nuclei		1–4	1	4	1	
Chromatoid bars		Rounded ends	Pointed ends	Absent	Absent	
Glycogen mass		Diffuse	Diffuse	Absent	Rarely seen	
Nucleus						
Figure 93.12	a	b	c	d	e	f
Karyosome	Central	Central	Central	Large, irregular mass is connected with one small mass	Central in position and large mass surrounded by refractile globules	Karysome is divided in 6-chromatin granules
Chromatin granules	Coarse (Ring like deposit)	Fine	Fine, symmetrical	Not clearly visible	Not clearly visible	

They are also known as **opportunist amoebae** or **amphizoic amoebae, because** living free (exozoic) in fresh water, mud and moist soil and also in the body of host (endozoic). Classification is described above.

Naegleria fowleri

Synonym: Brain-eating amoeba.

Morphology

Trophozoite Stage

It has two types of trophic forms **(Fig. 93.13a)**.

1. **Amoeboid form:** It is elongated, about 10–35 μm × 7 μm in size and actively motile. Cytoplasm has pseudopodia called **lopopodia. Nucleus** is spherical with central karyosome (endosomes) and visible in stained preparation. It can multiply and converted in to cyst form.
2. **Flagellate form:** It is pear shape and motile with two flagella at anterior end. **Nucleus** is spherical with endosomes. It cannot multiply and comes back to amoeboid form to produce cyst stage.

Cystic Stage

It is spherical or oval in shape and 7–10 μm in size. **Nucleus** is spherical with central karyosome (endosomes) as shown **Fig. 93.13b**. It has smooth double layered cyst wall. It is not present in tissues.

Life cycle: Cyst stage never forms in human body. Entire life cycle completed in external environment and all life cycle stages are shown in **Fig. 93.13**. Human is infected accidently by amoeboid trophozoite form by contact of nasal mucosa with contaminated swimming pool water, which produces the disease in human with **dead end** of cycle.

Pathogenicity

- **Disease name:** It is called **primary amoebic meningo-encephalitis (PAM)** or **terramoebiasis.**
- **Modes of transmission:** It is transmitted by **direct contact** of nasal mucosa with contaminated swimming pool water or by **inhalation** of dust particles containing the cyst **or** air of air-cooler.

> **Note: Swimming pool originated microbes**
> - *Mycobacterium balnei* (*M. marinum*): Swimming pool granuloma on prominences like elbows, nose, etc.
> - *Chlamydia trachomatis:* Swimming pool conjunctivitis.
> - Adeno virus serotypes 3, 4, 7 and 11: Swimming pool (acute follicular) conjunctivitis.
> - *Rhinosporidium seeberi:* Rhinosporidiosis and oculosporidiosis.
> - *Trichsporon beigelii:* White piedra.
> - *Naegleris fowleri:* PAM.

Incubation period: 1–7 days.

Precipitating factors: It mostly occurs in children and young adults. Amoebas are common in air of air cooler and live symbiotically with *Legionella* or *Listeria*.

Sites: It habitats the CNS system.

Pathogenesis: Follow **Flowchart 93.8**.

Clinical features: Acute purulent PAM presents with severe headache, fever, vomiting, stiff neck, seizure and coma. It is most fatal and patient dies within a week (average in 5 days). **Allergic alveolitis** presents with humidifier fever due to inhalation of air from air cooler.

Complications: It causes cranial nerve palsies of 3rd, 4th and 6th nerve. Most patients die within a week.

Laboratory diagnosis

- **Specimens:** CSF and brain biopsy.
- **Testing methods**
1. **Microscopy by normal saline:** It is used to identify the motile trophozoites. Sugar and protein levels are higher than the bacterial meningitis in CSF. Cyst is not present in tissues.
2. **Culture:** It grows in proteose peptone glucose medium or non nutrient agar seeded with *Escherichia coli* as a food source. Both trophozoites and cyst can grow in culture.

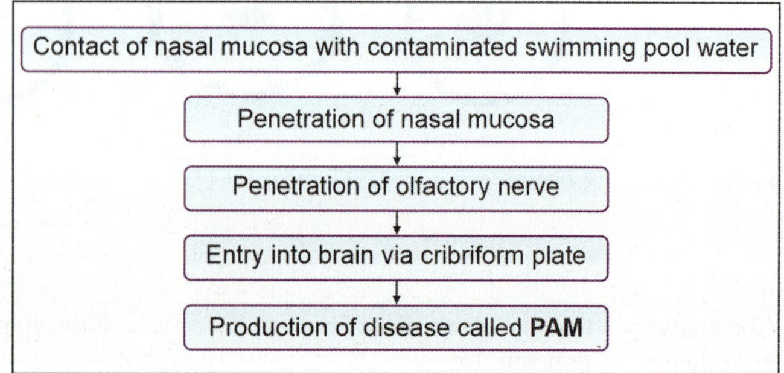

Fig. 93.13: Morphology and life cycle of *N. fowleri*

Flowchart 93.8: Pathogenesis of *N. fowleri*

Contact of nasal mucosa with contaminated swimming pool water
↓
Penetration of nasal mucosa
↓
Penetration of olfactory nerve
↓
Entry into brain via cribriform plate
↓
Production of disease called **PAM**

3. **Differential diagnosis from *Acanthamoeba*:** Follow **Table 93.4**.

Prevention: Avoiding contact with contaminated water.

Treatment: AMB and rifampicin are effective.

Acanthamoeba spp.

A. culbertsoni is most pathogenic for human.

Morphology: Follow **Fig. 93.14**.

Trophozoite Stage

It is oval in shape, 15–45 μm in size and sluggishly motile. Cytoplasm shows the multiple pointed pseudopodia called **spiny amoebae** or **acanthopodia** and having no flagellate form. **Nucleus** is same as like *N. fowleri*. It presents in tissues and can multiply and convert into cyst form.

Cystic Stage

It is spherical in shape and 8–25 μm in size. **Nucleus** is same as like *N. fowleri*. It has smooth double layered cyst wall. Inner wall is smooth and outer wrinkled. It presents in tissues.

Life cycle: Stages are shown in **Fig. 93.14**.

Pathogenicity

- **Mode of transmission:** It found freely in soil and lakes. It is transmitted by direct contact with contact lenses or to traumatized skin or by inhalation and ingestion.
- **Incubation period:** Few weeks to several months.
- **Precipitating factors:** Wearing contact lenses (especially soft one) or cleaning it with home-made saline or poor disinfectant.
- **Sites:** Eye and brain are natural habitat in human.

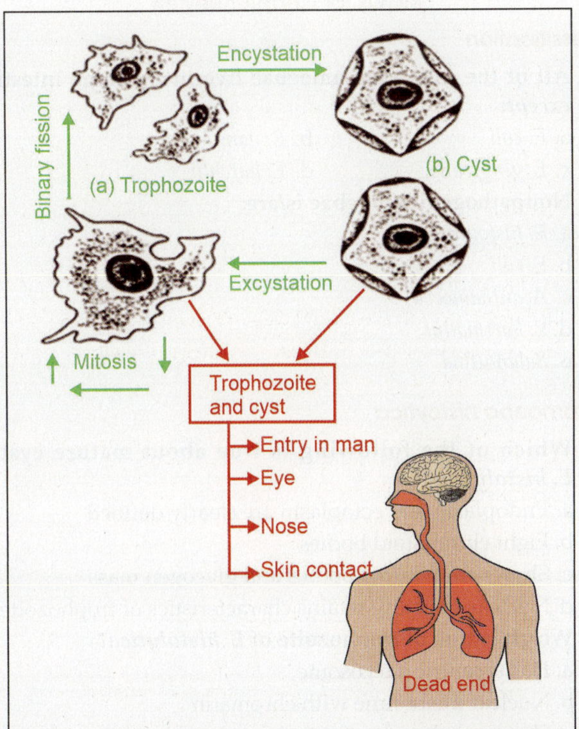

Fig. 93.14: Morphology and life cycle of *Acanthamoeba*

- **Clinical features:** It produces ulcerative keratitis called **chronic amoebic keratitis**, corneal ulcer, skin ulcer and opportunistic infection in sinuses and lungs in immunocompromised / AIDS patient.
- **Complications:** It causes GAM (granulomatous amoebic meningoencephalitis), unlike *Naegleria*, it is secondary by blood invasion of trophozoites from mucosa / ulcer of skin or from lower respiratory infection.

Laboratory diagnosis
- **Specimens:** CSF and corneal scraping
- **Testing methods**
1. **Microscopy by normal saline:** It is used to identify the motile trophozoites. Protein and PMNs counts are higher in CSF.
    1. **Culture in media:** Nutrient agar contains *Escherichia coli* and *K. aerogenes/Kleb pneumoniae* is inoculated with drop of CSF and incubated at 37°C or 42°C in an aerobic condition for 24 hours. Culture plate can be examined with dissecting microscope under 10 × magnification.
    2. **Animal culture:** Intranasal inoculation in mouse develops brain infection.
2. **Serological tests:** Follow **Table 93.4**.
3. **Molecular:** PCR.
4. **Differential diagnosis from *Naegleria*:** Follow **Table 93.4**.

TABLE 93.4: Differences between *Naegleria* and *Acanthamoeba*

Features	*Naegleria*	*Acanthamoeba*
Trophozoite		
• **Size**	• Smaller	• Larger
• **Psedopodia**	• Single and called **lopopodium**	• Multiple and called **spiny amoeba**
• **Flagellate**	• Present	• Absent
Cyst		
• **Size**	• Smaller	• Larger
• **Surface**	• Smooth	• Wrinkled
Nucleus	Nucleolus divides and nuclear membrane persist	Nuclear membrane dissolves
Pathogenicity		
CNS lesion	PAM	GAM
Portal of entry	Nose	Eye/respiratory tract
Precipitating factor	Swimming in contaminated water	Immunodeficiency disease/AIDS
Clinical course	Acute	Subacute/chronic
Pathology	Acute inflammation	Granulomatous inflammation
Laboratory diagnosis		
Tissues	Only trophozoite	Cyst and trophozoite
CSF	Predominantly PMNs	Predominantly leukocytes
Serological tests	Most species die rapidly so rare useful	Immunofluorescence or immunoperoxidase staining

Prevention: Improvement of hygiene specially of wearing contact lenses.

Treatment: No specific treatment is available for GAM.

Balamuthia mandrillaris
Morphology (Fig. 93.15)

Trophozoite Stage

It is irregular in shape, 12–60 μm in size and actively motile. **Cytoplasm** shows the broad pseudopodia. **Nucleus** is same as like *Acanthamoeba*. It presents in tissues, can multiply and convert into cyst form.

Cystic Stage

It is spherical in shape and 6–20 μm in size. **Nucleus** is same as like *Acanthamoeba*. It has smooth double layer cyst wall. Inner wall is smooth and outer irregular. It presents in tissues.

Life cycle (Fig. 93.15): Same as like *Acanthamoeba*.

Pathogenicity
- **Mode of transmission:** It found freely in soil and water. It is transmitted by **direct contact** with contact lenses or to traumatized skin.
- **Sites:** Eye and brain are natural habitat in human.
- **Clinical features:** Keratitis.
- **Complication:** GAM.

Laboratory diagnosis: Same as *Acanthamoeba*.

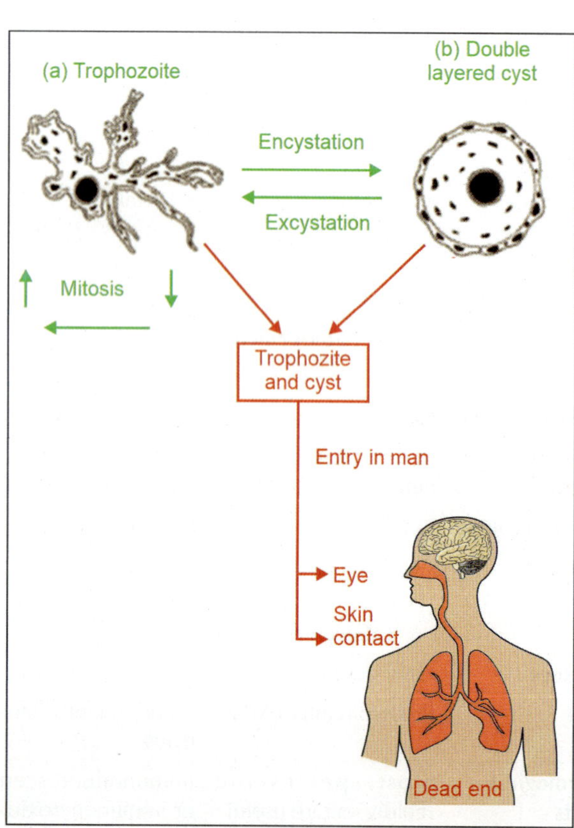

Fig. 93.15: Morphology and life cycle of *B. mandrillaris*

ACCESS YOURSELF

Case Study

1. A female came with complain of abdominal pain and diarrhea. Stool examination revealed presence of blood, mucus, motile trophozoites and RBCs in clumps. Identify the case and answer the following.
   a. Name and draw the properly labeled diagram of morphological stages of causative agent.
   b. Describe the life cycle of causative agent.
   c. Describe the pathogenicity of causative agent.

Essays/Full Questions

1. *E. histolytica*.
2. Free living amoebae.

Short Notes

1. Morphology/life cycle/pathogenicity/laboratory diagnosis of *E. histolytica*.
2. Hepatic amoebiasis.

Short Questions for Theory/Viva Questions

1. What is zymodeme pattern in *E. histolytica*?
2. What is granuloma cutis?
3. What is diffuse amoebic hepatitis?
4. What is Anchovy sauce pus?
5. What are pyknotic body and CL crystal?
6. Name the four parasites causing dysentery.
7. Name the free living amoebae.
8. Name the two microbes transmitted by contaminated swimming pool water.
9. What is spiny amoeba?

Comments on

1. Motility of *E. histolytica* is unidirectional.
2. Infection of *E. histolytica* is not transmitted by trophozoites.

MCQs for Chapter Review

Classification

1. **All of the following amoebae live in the large intestine, *except*:**
   a. *E. coli*   b. *E. nana*
   c. *E. gingivalis*   d. *E. butchlii*
2. **Nonpathogenic amoebae is/are:**
   a. *E. histolytica*
   b. *E. coli*
   c. *Acanthamoeba*
   d. *E. hartmanni*
   e. *Balamuthia*

Entamoeba histoytica

3. **Which of the following is true about mature cyst of *E. histolytica*?**
   a. Endoplasm and ectoplasm are clearly defined
   b. Eight chromatoid bodies
   c. Shows chromatoid bodies and glycogen mass
   d. Nuclear structure retains characteristics of trophozoite.
4. **Which is true of trophozoite of *E. histolytica*?**
   a. Has eccentric karyosome
   b. Nuclear membrane with chromatin
   c. Shows erythrophagocytosis
   d. Presence of bacteria inside cell

5. **Presence of RBCs in the endoplasm of trophozoite is a feature of:**
   a. *Entamoeba coli*
   b. *Entamoeba histolytica*
   c. *Entamoeba dispar*
   d. *Entamoeba hartmanni*

6. **All are true about *Entamoeba histolytica*, *except*:**
   a. Cysts are 8 nucleated
   b. Cysts are 4 nucleated
   c. Trophozoites colonize in the colon
   d. Chrmoatoid bars are stained by iodides

7. **Glycogen mass and chromidial bars are absent in the cyst of *Entamoeba histolytica* in:**
   a. Uninucleate cyst
   b. Binucleate cyst
   c. Quadrinucleate cyst
   d. Octanucleate cyst

8. **The main reservoir for *Entamoeba histolytica* is:**
   a. Man
   b. Dirty water
   c. Ponds

9. **Amoebiasis is transmitted by all, *except*:**
   a. Cockroaches
   b. Feco-oral
   c. Vertical transmission
   d. Oro-rectal

10. **Pathogenicity of *E. histolytica* is indicated by:**
    a. Zymodeme pattern
    b. Size
    c. Nuclear pattern
    d. ELISA

11. **True about amoebic colitis is/are:**
    a. Caused by *E. histolytica*
    b. Cyst contains 8 nuclei
    c. Flask shaped ulcers are present
    d. Cecum is most commonly affected (e) Is premalignant

12. **Diffuse amoebic hepatitis is caused by:**
    a. *E. histolytica*
    b. Bacteria
    c. a + b
    d. None of above

13. **Most common extrahepatic complication of amoebic hepatitis is:**
    a. Meningitis
    b. Lung abscess
    c. Nephritis
    d. Encephalitis

14. **Parasite having Charcoat-Layden crystals but no pus cells:**
    a. *Giardia*
    b. *Taenia*
    c. *E. histolytica*
    d. *Trichomonas*

15. **Charcoat-Layden crystals are derived from:**
    a. Macrophages
    b. Eosinophils
    c. Basophils
    d. Neutrophils

16. **Commonest site of extra intestinal amoebiasis is:**
    a. Brain
    b. Liver
    c. Spleen
    d. Lungs

17. **Regarding cutaneous amoebiasis which is not true?**
    a. It is a spreading necrotizing inflammation of the skin and subcutaneous tissue
    b. Rapid improvement occurs with antiamoebic treatment.
    c. Can occur in peri-anal region
    d. Infection reaches the skin through the bloodstream

18. **Invasive amoebiasis can be best diagnosed by:**
    a. ELISA
    b. Counter current immunoelectrophoresis
    c. Indirect hemagglutination test
    d. Complement fixation test

19. **Amoebic liver abscess can be diagnosed by demonstrating:**
    a. Cysts in sterile pus
    b. Trophozoite in the pus
    c. Cysts in the intestine
    d. Trophozoite in the feces

20. **A patient presents with lower gastrointestinal bleeding. Sigmoidoscopy shows ulcer in the sigmoido colon. Biopsy from area shows flask shaped ulcer. Which of the following is the most important treatment?**
    a. IV ceftriaxone
    b. IV metronidazole
    c. IV steroids and sulphasalazine
    d. Hydrocortison enemas

Nonpathogenic Amoebae

21. **Mature cyst of *Entamoeba histolytica* differs from *Entamoeba coli*, in the following, *except*:**
    a. Size is 6–15 microns
    b. Nuclei are 1–4 in number
    c. Karyosome is central in position
    d. Chromatoid bars seen

Free Living Amoebae

22. **A 30-year-old patient presented with features of acute meningoencephalitis in the casualty. His CSF on wet mount microscopy revealed motile unicellular microorganisms. The most likely organism is:**
    a. *Naegeria fowleri*
    b. *Acanthamoeba culbertsoni*
    c. *Entamoeba histolytica*
    d. *Trypanosoma cruzi*

23. **Acute primary meningoencephalitis true is:**
    a. Meningitis caused by *Acanthamoeba* species is acute in nature
    b. Diagnosis is done by demonstration of trophozoite in CSF
    c. Caused by feco-oral transmission
    d. More common in tropical climate

24. **Most fatal (fulminant) amoebic encephalitis is caused by:**
    a. *E. histolytica*
    b. *Naegleria*
    c. *E. dispar*
    d. *Acanthamoeba*

25. **Selective medium for *Naegleria fowleri* is:**
    a. Nutrient agar rich with *E. coli*
    b. NNN medium
    c. Non-nutrient agar rich with *E. coli*
    d. Diamond media

26. **Contamination of contact lenses can cause eye infection of:**
    a. *Acanthamoeba*
    b. *Entamoeba coli*
    c. *Onchocerca volvulus*
    d. *Toxocara canis*

Answers and Explanation of MCQs

1. c
2. b, d
- Follow section, **classification** for explanation of answers of MCQs 1–2.
3. d
4. b, c
5. b
6. a, d
7. c
- Follow section, *Entamoeba histolytica* (**morphology**) for explanation of answers of MCQs 3–7.
8. a
9. c
10. a
11. a, c, d
12. b
13. b
- Follow section, *Entamoeba histolytica* (**pathogenicity**) for explanation of answers of MCQs 8–13.
14. c
15. b
- Follow section, **CL (Charcoat-Layden) crystals** for explanation of answers of MCQs 14–15.

16. **b**
    - Follow section, **hepatic amoebiasis** → **incidence** for explanation.

17. **d**
    - Follow section, **extraintestinal amoebiasis** → **cutaneous amoebiasis** for explanation.

18. **a**

19. **b**
    - Follow section, *Entamoeba histolytica* (**diagnosis**) for explanation of answers of MCQs 18–19.

20. **b**
    - Follow section, *Entamoeba histolytica* (**treatment**) for explanation.

21. **d**
    - Follow section, **nonpathogenic amoebae (***Entamoeba coli***)** for explanation.

22. **a**

23. **b**

24. **b**

25. **c**
    - Follow section, *Naegeria fowleri* for explanation of answers of MCQs 22–25.

26. **a**
    - Follow section, *Acanthamoeba* spp., → **modes of transmission** for explanation.

CHAPTER 94

Infections of Intestinal, Oral and Genital Flagellates

INTRODUCTION

Definition

Parasites possess flagella as organs of locomotion and are called **flagellates.**

Classification

Systemic classification: Follow Ch. 14 (Table 14.3).

Pathogenic classification: Two groups.

1. **Intestinal**
   - **Pathogenic:** *Giardia lamblia* (duodenum and upper part of jejunum).
   - **Nonpathogenic**
     - *Trichomonas: T. hominis* (cecum)
     - *Chilomastix: C. mesnili* (cecum)
     - *Enteromonas: E. hominis* (colon)
     - *Retortamonas (Embadomonas): R. intestinalis* (colon)
2. **Oral (nonpathogenic):** *Trichomonas tenax* (mouth)
3. **Genital (pathogenic):** *Trichomonas vaginalis* (vagina and urethra).

Giardia lamblia

Synonym

It also called *Giardia intestinalis, Giardia duodenalis* or *Lamblia intestinalis.*

History

It was discovered by Leeuwenhoeck in 1681 in his own stool. Lamb (1859) and Alexeieff (1914) also contributed in discovery of this parasite.

Morphology

It exists in two morphological phase like trophozoite and cyst **(Fig. 94.1).**

Trophozoite Stage

Shape and size: In **front view (Fig. 94.1a)** it looks tennis/badminton racket or pear-shaped with broad anterior end and pointed posterior end. In **side view (Fig. 94.1b),** it looks longitudinally split pear. It is about 14 μm × 7 μm.

Motility: Resembles a falling leaf.

Structure: Its anterior end is broad, posterior end is pointed, dorsal surface is convex and ventral surface is concave with disc like structure called **sucking disc or ventral disc**, provides the organ of adhesion. It is bilaterally symmetrical and all the organs are paired. It has centrally two **axostyles,** which divide the trophozoite in two equal halves. It has two sausage shaped parabasal bodies (median bodies), four pairs of flagella and two nuclei with central karyosome (nucleolus), which give **monkey face appearance** to parasite under microscope.

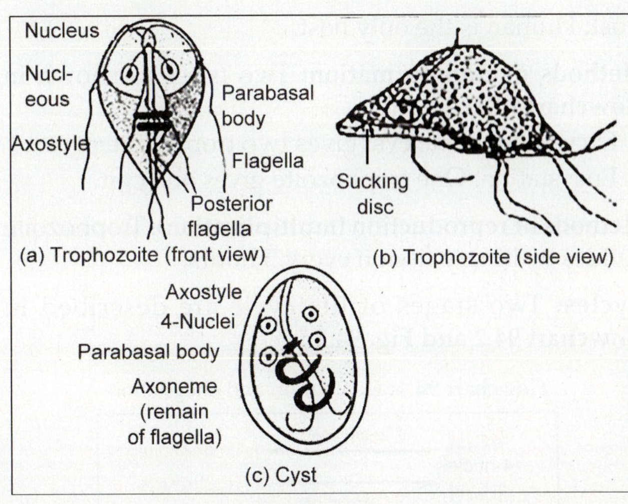

Fig. 94.1: Morphology of *G. lamblia*

Cyst Stage

Shape and size: It is oval in shape with 12 μm length and 7 μm breadth.

Structure (Fig. 94.1c): Axostyles lie more diagonally and form the dividing line. Remains of flagella called **axoneme** and sucking disc are seen inside the cytoplasm. It has two sausage-shaped parabasal bodies and four nuclei with central karyosome. Acid environment and wet condition favor the encystation.

Culture

Medium described by Karapetyan (1962) and Meyer (1970): It includes cultivation of *Giardia* with yeast, *Candida guilliermondi*, chick embryo extracts, human serum, Hottinger's digest (tryptic meat digest) and Hank's solution. It is very cumbersome to prepare and required human serum and chick embryo extracts.

Diamond's (TYI-S-33) medium described by Diamond et al., (1978): It developed to cultivate *Entamoeba* and *Trichomonas* species. Later Visvesvara (1980) noticed that "TYI-S-33" medium could support the axenic growth of *Giardia* trophozoites.

Modified Diamond's (TYI-S-33) medium: It was modified by Keister (1983). Modifications includes addition of bile, higher concentrations of cysteine, removal of the vitamin-tween 80 mixture, pH change to 7–7.2, and sterilization by filtration, rather than autoclaving. It is suitable for most *Giardia* strains/species and is the most commonly used medium for trophozoite culture.

Others: A comprehensive description of culture media is reviewed in Clark and Diamond (2002), including reference to a serum-free medium (Wieder et al., 1983).

Resistance

Sterilization: Cyst is removed from water by boiling and filtration by membrane filter.

Disinfection: Chlorination of water is ineffective to remove the cyst from water.

Life Cycle

Host: Human is the only host.

Methods of transformation: Two types as shown in Flowchart 94.1.
- Excystation: One cyst gives two trophozoites.
- Encystation: One trophozoite gives one cyst.

Methods of reproduction (multiplication): Trophozoite divides by binary fission every 5 hours.

Cycles: Two stages of life cycle are described in Flowchart 94.2 and Fig. 94.2.

Flowchart 94.1: Excystation and encystation

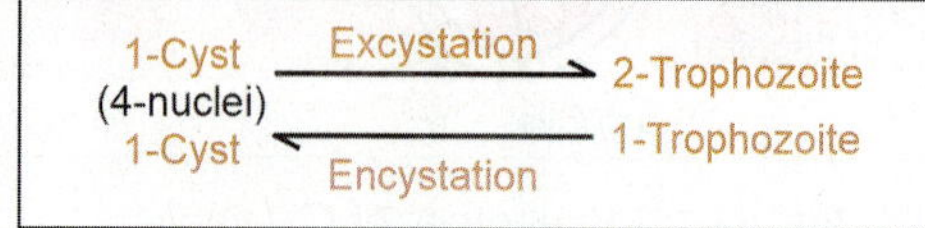

Flowchart 94.2: Life cycle of *G. lamblia*

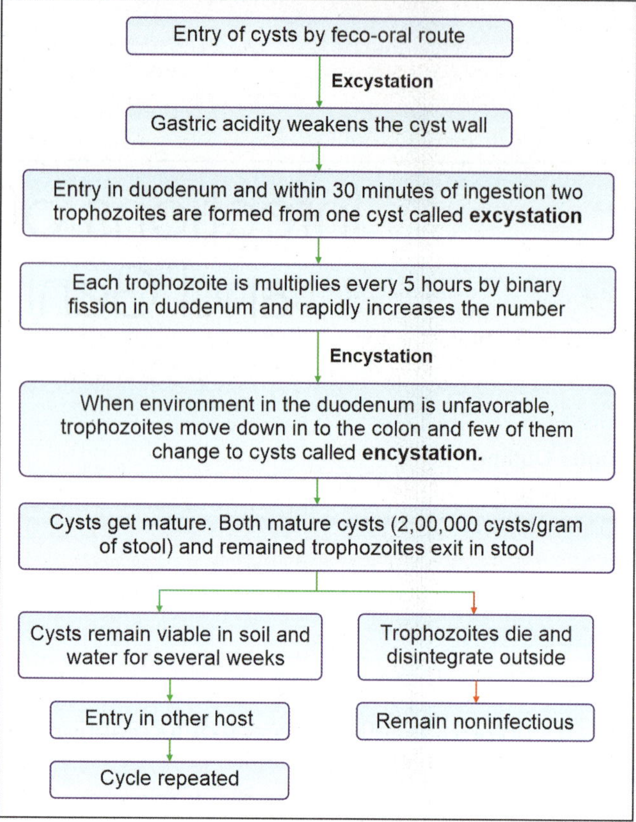

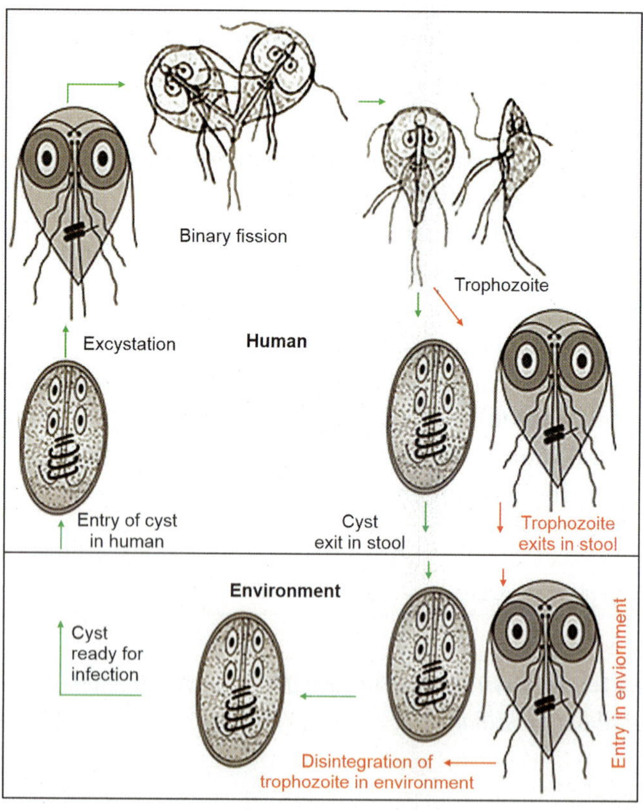

Fig. 94.2: Life cycle of *G. lamblia*

Pathogenicity

Disease name: It is called **giardiasis.**

Epidemiology: It is distributed worldwide. It is endemic in many areas and one of the agents responsible for traveler's diarrhea.

Reservoirs of infection: These are mainly humans plus wild and domestic animals.

Sources of infection: Food and water.

Modes of transmission: It is transmitted by **ingestion** of contaminated food and water. It is transmitted from man to man and also from animal to man. It also transmitted by **sexual intercourse** by male homosexual.

Incubation period: Average 10 days

Exit form: Trophozoite and cyst

Infective form: Cyst

Infective dose: Ingestion of up to 10 (range 10–100) cysts is enough to cause infection.

Portal of entry: GIT

Sites: Duodenum and upper part of jejunum

Precipitating factors (epidemiological determinants)

- **Agent factors (virulence factors): (1) Sucking disc:** It is an organ of adhesion. Parasite does not invade the tissue but remains tightly adhere to the intestinal epithelium. Parasite attaches to the convex surface of epithelium and causes intestinal disturbances leading to malabsorption of fat **(called steatorrhea)** in children and in adults. **(2) Toxin:** It causes allergic manifestations.
- **Host factors: (1) Age:** It is common in pediatric age. **(2) Blood group:** Group A is more prone to disease. **(3) Habit:** Common in cannabis users. **(4) Diseases:** Achlorhydria or hypochlorhydria, chronic pancreatitis (leads to deficiency of pancreatic enzymes) and malnutrition increase the risk. Recurrent giardiasis is associated with selective IgA deficiency, X-linked and autosomal recessive agammaglobulinemia, hypogammaglobulinemia, combined (common) variable immunodeficiency disease (CVIDD), AIDS, etc.

Pathogenesis: In majority of cases morphology of bowel remains unaltered, but in chronic cases flattened villi present with clinical picture resemble to tropical sprue and gluten sensitive enteropathy. Villous architecture may damage by cell apoptosis and increased lymphatic infiltration in lamina propria.

Clinical features: Follow **Flowchart 94.3**.

Laboratory Diagnosis

Specimens

- **Stool:** It is collected after defecation or by purgation. It contains both cysts and trophozoites.
- **Bile A:** Collected from duodenum by aspiration method or by using gelatin capsule called **entero test (string test)**.
- **Bile B:** Collected from biliary tract, because to avoid the acidity of duodenum, trophozoite often enters in to the biliary tract (gall bladder).
- **Mucosal biopsy:** From duodenum and proximal jejunum. Jejunal wash is not useful.

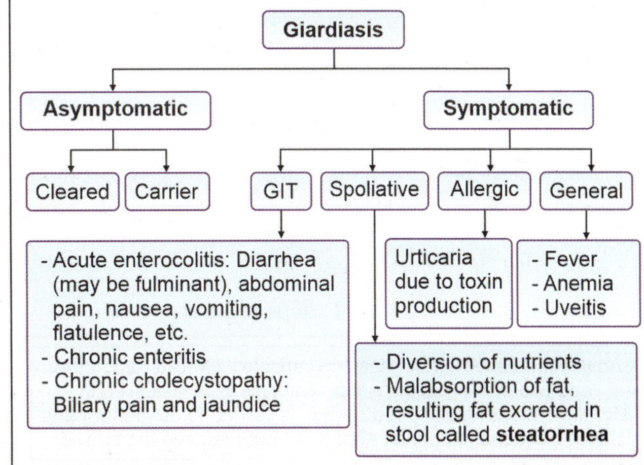

Testing methods

A. **Macroscopy of stool:** Pale or yellow colored stool due to presence of fat.

B. **Microscopy by normal saline preparation:** It is considered as gold standard for detection of trophozoites or cysts or both.
1. **For trophozoites:** Motile trophozoites will be detected under normal saline preparation with leaf like motility in fresh stool.
2. **For cysts:** Cysts shed intermittently in stool, so at least 3 specimens are collected at an alternate day and examined for cysts under normal saline mount. Sensitivity with one stool is 60–80% while with 3 stools it is >90%.
3. **No RBCs and pus cells:** *Giardia* do not cause dysentery, so RBCs and pus cells are absent.

C. **Histopathology:** Duodenal biopsy can be stained with Giemsa stain to demonstrate the trophozoites.

D. **Culture:** Described above.

E. **Serological tests:** Cyst wall protein **Ag** is detected from stool (coproantigen) by ELISA and direct IF test. It has 90–100% sensitivity and 99–100% specificity. **Ab** is detected from serum by ELISA and indirect IF test. Ab detection is used for epidemiological purpose.

F. **Molecular methods:** qPCR is useful. **BioFire Film Array** is an automated multiplex PCR which targets multiple bacteria, viruses and parasites causing diarrhea like *E. histolytica*, *Giardia*, *Cyclospora* and *Cryptosporidium parvum*.

G. **Radiological test:** X-ray after barium meal may reveal nonspecific mucosal thickening with large dilated loops of hypotonic bowel. It is positive in 20% cases.

> **Note: Entero test**
> - **Synonym:** String test or E-test in short.
> - **Apparatus:** Follow **Fig. 94.3**.
> - **Steps:** Follow **Flowchart 94.4**.
> - **Uses:** For detection of trophozoites of *G. lamblia,* larvae of *S. stercoralis*, eggs of liver flukes like *C. sinensis, O. felineus, F. hepatica* and oocysts of *C. bellei (Isospora bellei), C. parvum.*

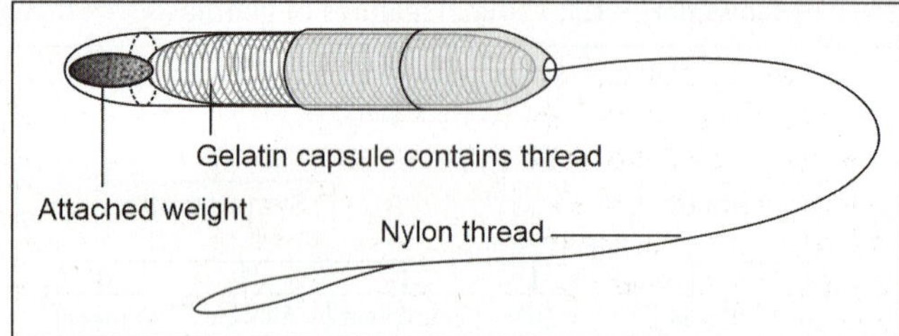

Fig. 94.3: Device of entero test

Flowchart 94.4: Steps of E-test

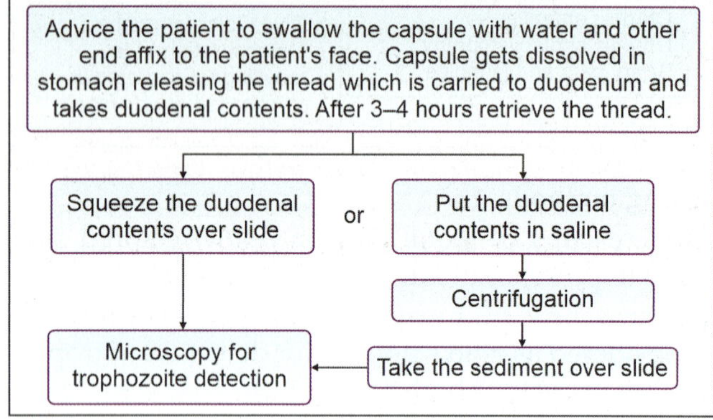

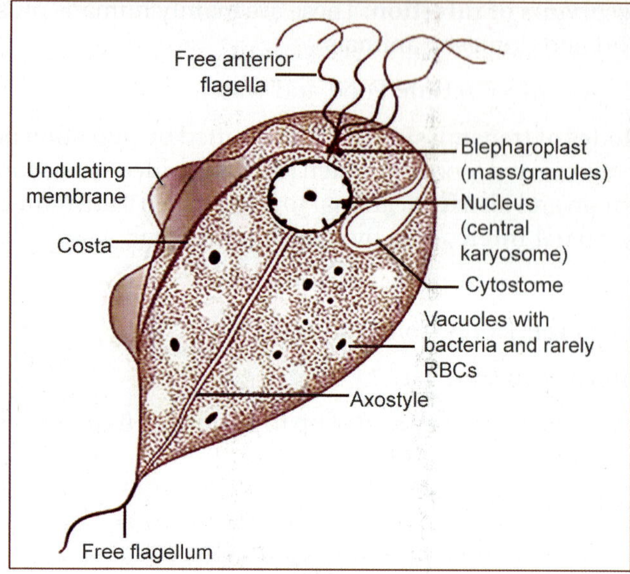

Fig. 94.4: Trophozoite of *T. vaginalis*

Prevention

It is prevented by prevention of contamination of food and water by human feces, frequent hand washing and boiling and filtration of water to remove the cysts

Treatment

Only symptomatic cases need treatment. Tinidazole (2 g single dose) and metronidazole (250 mg TDS for 5–7 days) are the drugs of choice. Paromomycin is used for pregnant woman.

Trichomonas vaginalis

History

It was 1st observed by Donne in 1836 in vaginal secretion.

Morphology

It has **no cystic stage,** only trophozoite stage occurs which has following properties **(Fig. 94.4).**

Shape and size: It is oval or pear shaped and 10–12 µm in length and 5–10 µm in breadth.

Motility: It has wobbling (side to side)/rotator or jerky movement.

Structure: Its anterior end is broad and posterior end is pointed. **Axostyles** run the middle of the body and end in posterior pointed end. It has 3–5 free anterior **flagella.** One thicker flagellum passes backwards along the body of parasite with formation of membrane like structure called **undulating membrane.** It covers the ½–⅓ of the body and becomes free at posterior end. Undulating membrane is supported at the base by thick rod like structure called **costa.** It has single nucleus, which is situated at the anterior end with central karyosome. It has a cleft like depression lies near by nucleus called **cytostome (mouth).**

Culture

It is the gold standard if microscopy is negative.

1. **Cell-free media:** It is most sensitive (>95%) if performed properly by using media like Bushly's medium, Feinberg-Whittington medium, Roiron's medium, Johnson-Trussel's medium, CPLM (cysteine peptone liver maltose) medium and plastic envelope medium.

2. **Tissue culture and egg culture media:** Also useful.

Life Cycle

Host: Human is an optimum host.

Methods of reproduction: Parasite does not have a cyst form only trophozoite form, which divides by binary fission.

Cycle: Trophozoites live in urogenital tract and multiply by binary fission as shown in **Fig. 94.5.** Cycle is maintained by sexual transmission of trophozoites from person to person. Few trophozoites are excreted with urine in external environment, but they do not survive.

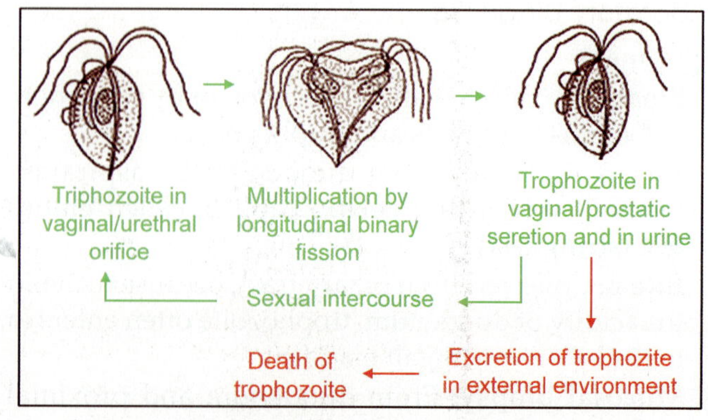

Fig. 94.5: Life cycle of *T. vaginalis*

Pathogenicity

Disease name: It called **trichomoniasis.**

Epidemiology: It is prevalent in the sexually active age groups in all climates and racial groups.

Reservoir of infection: Human (asymptomatic male) is the only known reservoir of infection.

Sources of infection: Genital materials.

Modes of transmission: It is transmitted by sexual intercourse (other parasites transmitted by sexual intercourse are *G. lamblia, E. histolytica, C. parvum*) or by fomites like toilet articles or clothes or vertically from infected birth canal to infant.

Incubation period: 4–28 days.

Infective and exit form: Trophozoite.

Portal of entry: Genital organs.

Sites: In female it presents in vagina, cervix, Bartholin's gland, urethra and in urinary bladder. **In male,** it presents mostly in the anterior urethra and rarely in prostate and preputial sac.

Precipitating factors (epidemiological determinants)
- **Agent factors (virulence factors): (1) Obligate parasite:** Parasite cannot live without close association with vaginal, urethral or prostatic tissues. **(2) Liberation of enzymes and other metabolites:** Like cysteine protease, lactic acid and acetic acid.
- **Host factors:** It is common in adults.

Pathology: It causes petechial hemorrhage (**strawberry mucosa**), metaplastic changes and desquamation of vaginal epithelium.

Clinical features: Follow **Flowchart 94.5.**

Complications: Follow **Flowchart 94.6.**

Laboratory Diagnosis

Specimens: (1) Vaginal discharge is collected from the posterior fornix. **(2) Urethral secretion** is collected in male from urethral orifice by prostatic massage. **(3) Urine** is collected in sterile test tube → centrifugation → use the sediment for testing.

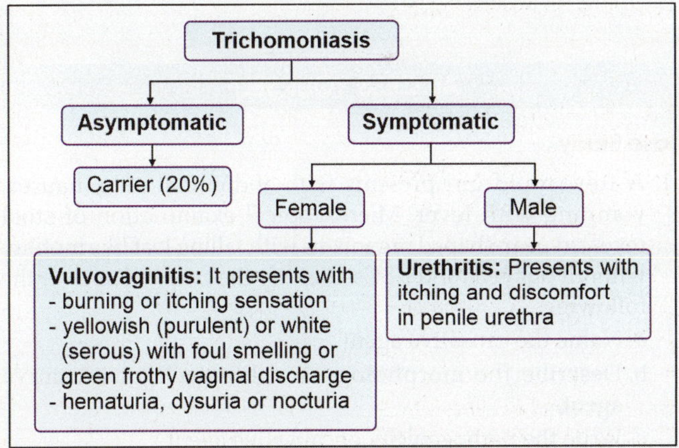

Flowchart 94.5: Clinical features of trichomoniasis

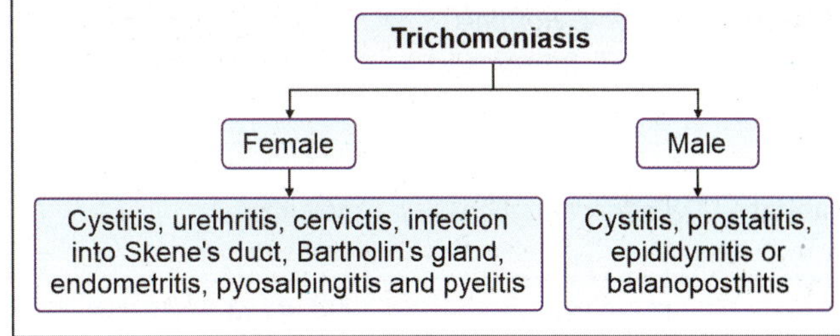

Flowchart 94.6: Complications of trichomoniasis

Testing methods

A. **Microscopy: (1) Wet mount by normal saline:** Prepare the smear from specimen by using NS and examine under the low power for motile trophozoite. **(2) Staining:** By using Papanicolaou stain, Giemsa stain, Gram's stain, periodic acid-Schiff stain, Leishman's stain, Diff-quick method or by Acridine-Orange stain.

B. **Culture:** Described above.

C. **Serological tests:** ELISA and ICT are useful to detect the 65 kDa surface polypeptide Ag of *T. vaginalis* from vaginal secretion.

D. **Molecular test:** PCR is useful.

E. **Differential diagnosis from other diseases:** Follow **Table 94.1.** Normal vaginal pH from puberty to menopause is 3.0–3.5.

	TABLE 94.1: D/D of *T. vaginalis*		
Features	*T. vaginalis*	*C. albicans*	*G. vaginalis*
Discharge	Yellowish-white	White	Gray and offensive
pH	>4.5	<4.5	>4.5

Prevention

Preventive measures are safe sex like wearing condom, sex education and testing plus treatment of partner for *T. vaginalis.*

Treatment

Metronidazole (2 g single dose or 500 mg BD for 5–7 days) or tinidazole is the drug of choice.

OTHER NONPATHOGENIC FLAGELLATES

Follow **Table 94.2 and Figs 94.6 to 94.9.**

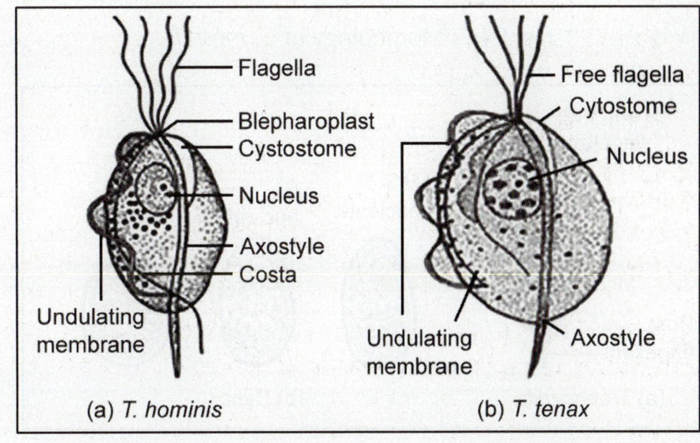

Fig. 94.6: Trophozoite of *T. hominis* and *T. tenax*

Essentials of Medical Microbiology

TABLE 94.2: Non-pathogenic flagellates

Features	T. hominis (T. intestinalis)	T. tenax (T. buccalis)	C. mesnili	E. hominis	R. intestinalis (E. intestinalis)
History	Davine, 1854		Wenyon, 1910 Alexeieff, 1912	Da Fonseca, 1915	Wenyon and O'Conor, 1917
Habitat	Ileocecal region	Mouth	Cecum	Colon	Colon
Trophozoite					
Figure	Fig. 94.6a	Fig. 94.6b	Fig. 94.7a	Fig. 94.8a	Fig. 94.9a
Size	5–15 µm × 7–10 µm	5–10 µm in length	10–15 µm × 5–6 µm	8 µm × 4 µm	5 µm × 3 µm
Shape	Pyriform	Pyriform	Pear	Pear	Oval
Flagella	3–5 anterior flagella. Parasite feeds the bacteria	4-anterior flagella	3-anterior flagella and 4th lies within cytostome	3-anterior flagella and 4th posterior adherent with body and becomes free in the end	1-anterior flagella and 2nd pass through cytostome, before becomes free
Nucleus	Single situated near anterior end				
Cytostome	Present	Present	Present	Absent	Present
Axostyle	Present	Present	Absent	Absent	Absent
Undulating membrane	Along the entire length of parasite		Absent	Absent	Absent
Motility	Jerky	Jerky	Rotatory	Jerky	Jerky
Cyst					
Figure			Fig. 94.7b	Fig. 94.8b	Fig. 94.9b
Size			7–10 µm × 4–6 µm	8 µm × 4 µm	4–5 µm × 4–6 µm
Shape			Lemon	Oval	Pear
Nucleus	No cyst stage	No cyst stage	One with central karyosome	1–4 nuclei situated at opposite pole	1 nuclei situated in the center
Other features			• Projection at anterior end called nipple • Remnant of cytostome present	It mimics a E. nana	Nucleus is surrounded by fibril called **bird's beak appearance**
Modes of transmission					
Route	Feco-oral	Kissing, salivary droplet	Feco-oral	Feco-oral	Feco-oral

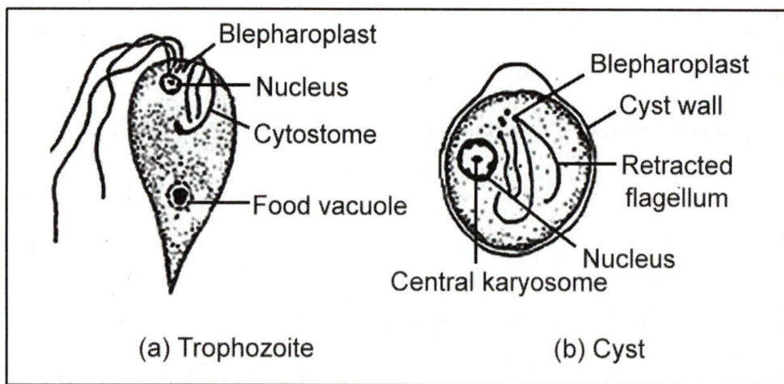

Fig. 94.7: Morphology of *C. mesnili*

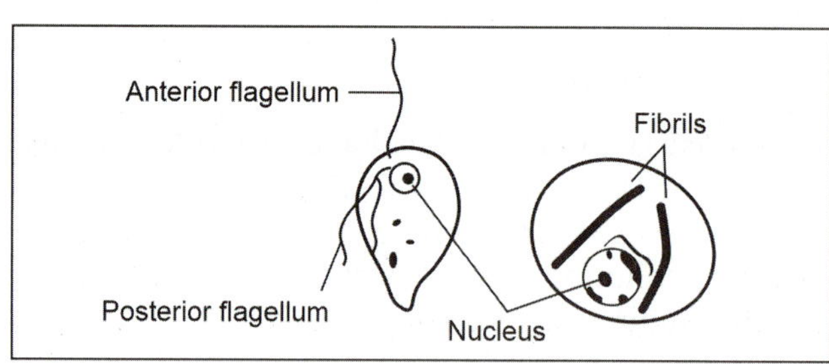

Fig. 94.9: Morphology of *E. intestinalis*

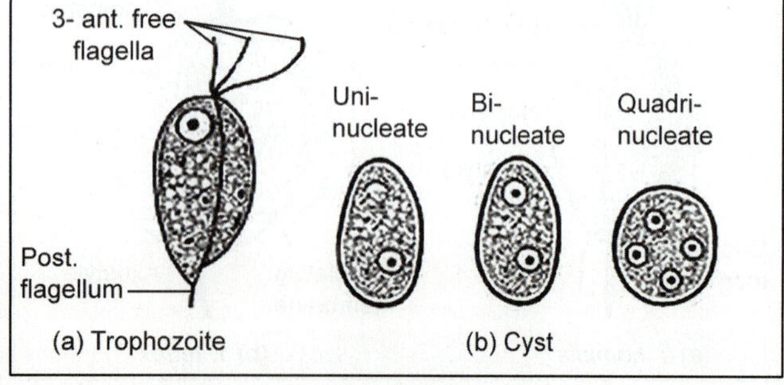

Fig. 94.8: Morphology of *E. hominis*

Case Study

1. A 10-year-old girl presents with abdominal pain, nausea, vomiting with fever. Microscopical examination of stool revealed pear-shaped organisms with falling leaf like motility without pus cells and RBCs. Identify the case and answer the following.
   a. Name the causative agent.
   b. Describe the morphology and life cycle of causative agent.
   c. Write the pathogenicity of causative agent.

Short Note

1. *Giardia lamblia/T. vaginalis.*

Short Questions for Theory/Viva Questions

1. What is E-test?
2. Name the organ of locomotion for following parasites: *B. coli, E. histolytica*, *T. cruzi* and *G. lamblia.*
3. Name the type of motility for following parasites: *B. coli, T. vaginalis, E. histolytica* and *G. lamblia.*

MCQs for Chapter Review

Giardia lamblia

1. **Normal habitat of *Giardia* is:**
   a. Duodenum and jejunum
   b. Stomach
   c. Cecum
   d. Ileum
2. **How many pairs of flagella does *Giardia lamblia* possess?**
   a. 1   b. 2
   c. 3   d. 4
3. **Recurrent giardiasis is associated with:**
   a. Severe combined immunodeficiency
   b. Common variable immunodeficiency
   c. DiGeorge syndrome
   d. C8 deficiency
4. **A case of giardiasis presents with:**
   a. Nausea and vomiting
   b. Abdominal pain
   c. Steatorrhea and flatulence
   d. All of the above
5. **True about giardiasis:**
   a. Only cyst is infective
   b. Reside in cecum
   c. Only man-to-man transmission
   d. Exist in one phases
6. **True about *Giardia* is:**
   a. May cause traveler's diarrhea
   b. *Giardia* inhabits ileum
   c. Trophozoite is an infective form
   d. None of the above
7. **Which of the following is true with *Giardia lamblia*?**
   a. Malabsorption commonly seen
   b. Trophozoite form is binucleate
   c. Diarrhea is seen
   d. Jejunal wash fluid is diagnostic
   e. Is a free living nematode
8. **Entero test is done to detect:**
   a. Larva of *Strongyloides*   b. Trophozoites of *Giardia*
   c. Oocyst of *Cystoisospora* d. All of the above

Trichomonas vaginalis

9. **Parasite cultured in Trussell and Johnson's medium is:**
   a. *Giardia lamblia*   b. *Trichomonas vaginalis*
   c. *Trypanosome cruzi*   d. *Leishmania dondovani*
10. **Following is the feature of *Trichomonas vaginalis*:**
    a. Trophozoite is the largest among all protozoa
    b. Cyst is absent
    c. Habitat in genital organs
    d. Transmitted sexually

Answers and Explanation of MCQs

1. a
2. d
3. b
4. d
5. a
6. a
7. b, c
- Follow section, *Giardia lamblia* for explanation of answers of MCQs 1–7.
8. a
- Follow section, **Entero test (uses)** for explanation.
9. b
10. b, c, d
- Largest protozoan is *Balantidium coli*. Follow section, *Trichomonas vaginalis* for explanation of other options.

Infections of Blood and Tissue Flagellates

Chapter Outline

INTRODUCTION

Synonym

Blood flagellates are also called **hemoflagellates.**

Classification

Systemic classification: Follow **Ch. 14 (Table 14.3)**.

Pathogenic classification: Two pathogenic groups like *Trypanosoma* and *Leishmania* are mentioned below.
- *Trypanosoma:* Follow **Flowchart 95.1**.
- *Leishmania: Leishmania donovani* causes kala azar or dum dum fever or visceral leishmaniasis. Diseases by other species are described later in this chapter.

Reproduction

Methods of reproduction: They multiply by longitudinal binary fission. **No sexual cycle** is known. It begins with division of kinetoplast, basal body and nucleus. Flagellum with undulating membrane remains with one half of basal body while new one develops from other. Cytoplasmic body then splits longitudinally from anterior end.

1. **Amastigote stage (Fig. 95.1a):** A means without and mastigote [from mastix (Greek)] means whip. It also called **aflagellar stage.** It is an oval or spherical in shape and contains nucleus, kinetoplast (**antenuclear**), basal body and axoneme, but no free flagellum. It occurs in *Leishmania* in human.
2. **Promastigote stage (Fig. 95.1b):** It is elongated in shape and contains nucleus, kinetoplast (**antenuclear**), basal body, axoneme and free flagellum, but no undulating membrane. It occurs in *Leishmania* in culture and vector.
3. **Epimastigote stage (Fig. 95.1c):** It is elongated in shape and contains nucleus, kinetoplast (near to the

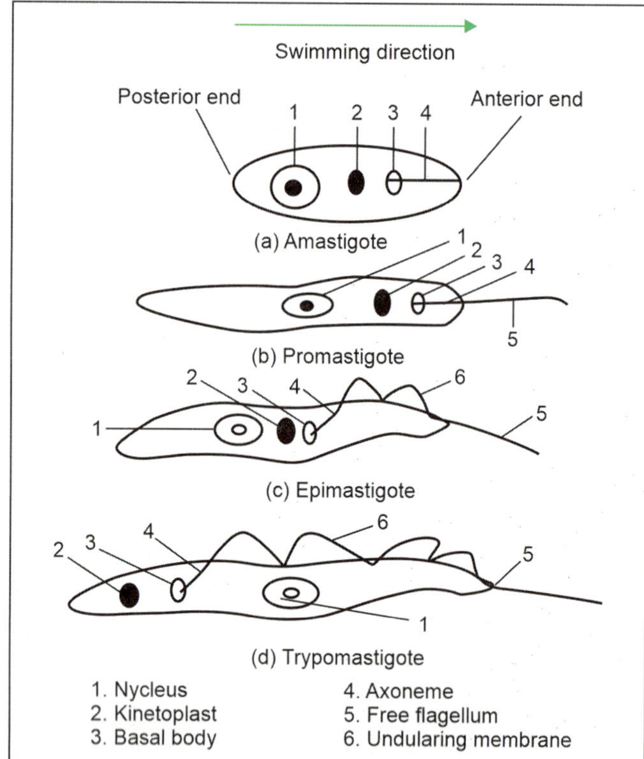

Fig. 95.1: Morphological parts and developmental stages of hemoflagellates

nucleus, so called **juxtanuclear kinetoplast**), basal body, axoneme, undulating membrane and free flagellum. It occurs in *Trypanosoma*.
4. **Trypomastigote stage (Fig. 95.1d):** It is elongated in shape with blunt posterior end and pointed anterior end. It contains nucleus, kinetoplast (**post-nuclear**), basal body, axoneme, undulating membrane and free flagellum. It occurs in *Trypanosoma*.

Trypanosomes infecting humans

- **Pathogenic species**
  - *T. brucei*
    - Strain/subspecies
      - Animal strain
        - *T. brucei brucei*
          - "**Nagana**" in wild (game) and domestic animals
      - Human strain
        - *T. brucei gambiense* Chronic
          - Gambian sleeping sickness *or* West African sleeping sickness *or* West African trypanosomiasis
        - *T. brucei rhodesiense* Acute
          - Rhodesian sleeping sickness *or* East African sleeping sickness *or* East African trypanosomiasis
        - *T. brucei zambiense* Chronic **or** acute
          - Zambian sleeping sickness
  - *T. cruzi (Schizotrypanum cruzi)*
    - Chaga's disease **or** South American trypanosomiasis
- **Nonpathogenic species**
  - *T. rangeli*
    - Found in human blood in Venezuela and Colombia

Trypanosomes infecting animals

- **Pathogenic species**
  - *T. evansi*
    - "**Surra**" in horses, camels, mules and elephants. Transmitted mechanically by tabanid
  - *T. equiperdum*
    - "**Stallion's disease**" (Venereal disease) in horses and asses
  - *T. equinum*
    - Causes disease in cattle and transmittted by tsetse flies
  - *T. vivax* and *T. congolense*
    - "**Mal de caderas**" in horses. Transmitted mechanically by tabanid
- **Nonpathogenic species**
  - *T. lewisi*
    - Found in rat all over the world. Transmitted by rat fleas

Mnemonic
- **BAfSTN:** **B**rucei, **Af**rican Trypanosomiasis, **S**leeping sickness, **T**se tse fly and **N**agana
- **CCB:** **C**ruzi, **C**haga's disease and **B**ug

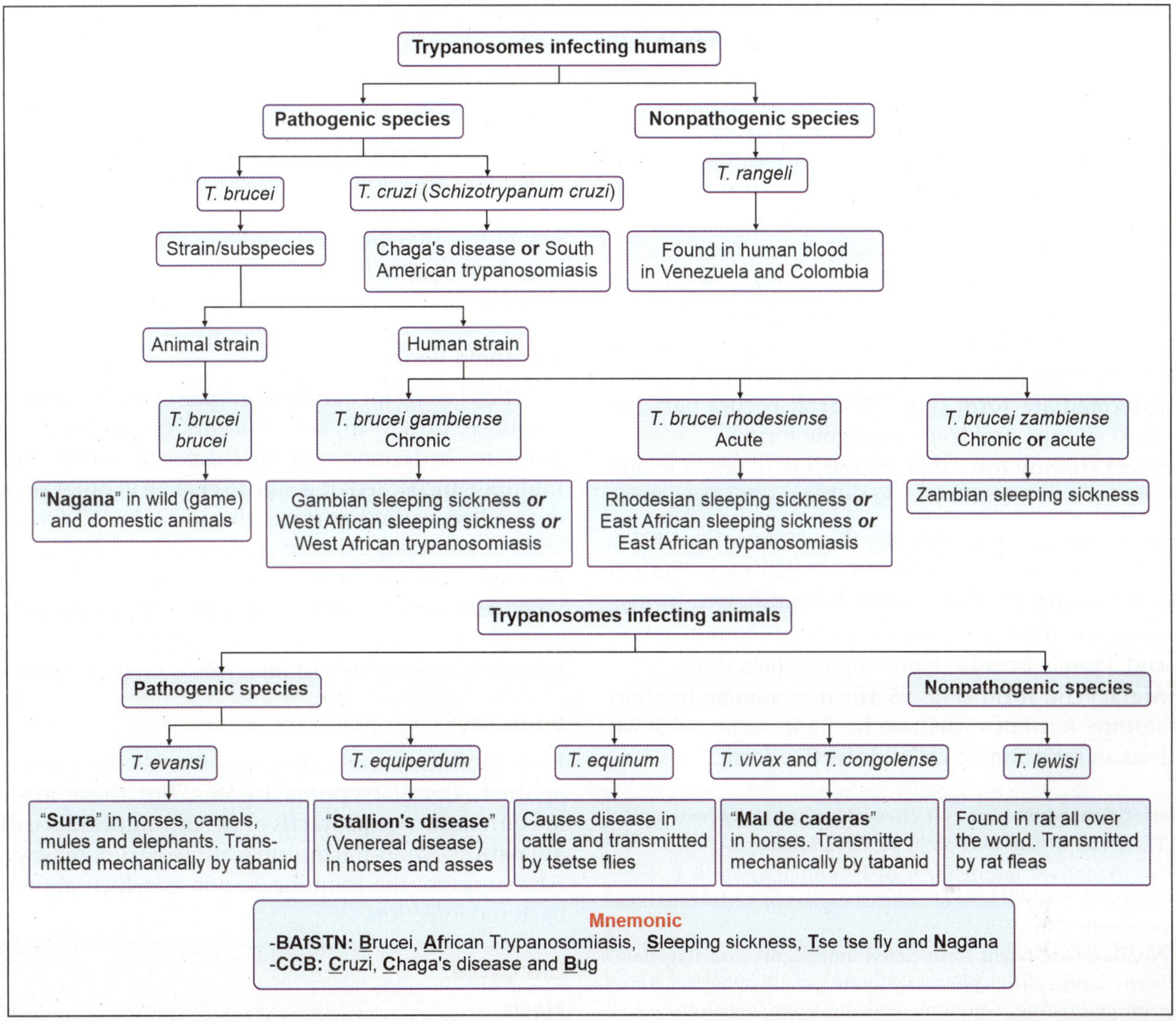

Reproductive stage: Nomenclature of stage is based on flagellar characteristic like starting point, its course and point of becoming free from body.

Staining Reaction

For body fluid: Thin and thick films are prepared by Romanowsky's stains like Giemsa, Leishman's and Wright's stain. Cytoplasm and undulating membrane stains blue, kinetoplast red, nucleus and flagellum appears pink (dark red) as shown in **Fig. 95.2.**

For tissues: Hematoxylin and eosin (H and E) stain is useful.

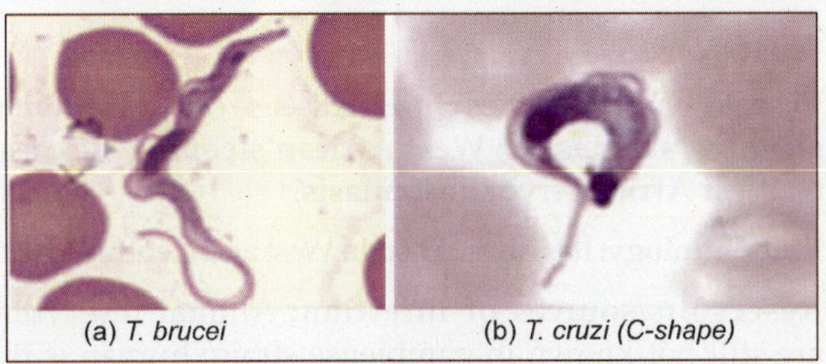

Fig. 95.2: Trypomastigotes in blood smear

Trypanosoma spp.

Trypanosoma brucei gambiense

Meaning of *Trypanosoma*

From **Greek** word **Trypanon means to bore** and **Soma means body.**

History

T. brucei was discovered by David Bruce in 1890 from cattle suffered with "**nagana**" in Zululand. Human strain was discovered in Gambia by Forde and Dutton in 1902.

Morphology

All members of trypanosome exist as trypomastigote in vertebrate hosts (humans, game and domestic animals), while some like *T. cruzi* assumes amastigote forms in vertebrate hosts. Trypomastigote of *T. brucei* exhibits variation in size and shape (called **pleomorphism**) at different stages of its life cycle with intermediate, short stumpy, long slender and metacyclic form.

Trypomastigote in vertebrate hosts (Fig. 95.1d/ Fig. 95.3a): It is elongated in shape with blunt posterior end and pointed anterior end. It is 16–42 µm in length and 1–3 µm in breadth and actively motile. It has large, oval and central nucleus. It has extranuclear DNA containing body having mitochondrial structure called kinetoplast which is situated near posterior end called postnuclear kinetoplast. Starting point of flagellum is indicated by basal body. Part of flagellum lying inside the body called axoneme or axial filament. There are 3–4 folds of flagellum present around the body of parasite called undulating membrane. Flagellum becomes free from anterior end called free flagellum. Trypomastigote shows **pleomorphism** with following different forms.

- **Intermediate form (Fig. 95.3b):** It occurs between short stumpy and trypomastigote forms.
- **Short stumpy form (Fig. 95.3c):** It is 10 µm in length and 5 µm in breadth with small or no free flagellum.

Epimastigote form in tsetse fly (Fig. 95.1c or Fig. 95.3e): It develops from long slender form in salivary gland. It shows **pleomorphism** with following different forms.

- **Long slender form (Fig. 95.3d):** It is 20 µm in length and 3 µm in breadth with long free flagellum.
- **Metacyclic form (Fig. 95.3f):** It is similar to short stumpy form of vertebrate host and responsible for human infection, so called **infective form**.

> **Note: Latent forms or occult visceral forms of trypomastigote**
> - These are present in choroid plexus and lungs.
> - Nonflagellate latent form of trypomastigote was first observed in capillaries of internal organs in vertebrate host by Fantham in 1911.
> - Nonflagellate latent form called **amastigote** and flagellate latent form called **spheromastigote** (small, spherical form with undulating membrane and short free flagellum) were found in choroid plexus of rat inoculated with trypomastigote of Botswana strain of *T. brucei* by Ormerod and Venkatesan in 1971.
> - **Clinical significance:** They are responsible for recurrence/ latent infection.

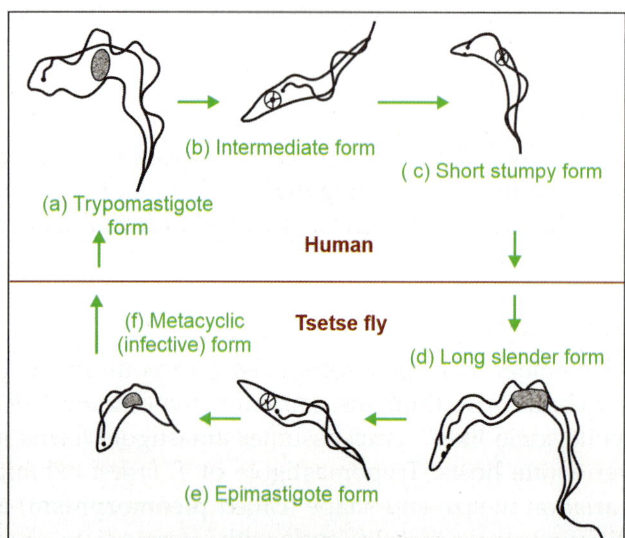

Fig. 95.3: Life cycle of *T. brucei*

Culture

Culture on media: Long slender form (mid-gut form of fly) grows over medium contains Ringer's solution, with sodium chloride, Tyrode's solution and citrated human blood.

Culture in animals: It is used in diagnosis when trypomastigote are scanty or difficult to find in blood smear and also to maintain the strain. It causes the overwhelming parasitemia in **rat, mouse** or **guinea pig** and kills the animal in few days. It produces chronic infection in **rabbit** and animal dies in 4–5 weeks.

Antigenic Variant

Infection lasting for long time in spite of strong antibodies response, due to mutation occurs every 3–4 days in parasite. It depends on host defense, when fails it multiply unchecked. It is due to change in outer protein coat called **variant surface glycoprotein (VSG).** Each VSG is immunogenic but antigenically different from previous VSG. Antibodies formed against each type, but when titer rises, homologous variant will disappear with emergence of new variant. Each variant is followed by fever and leukocytosis (monocytosis, no neutrophilia).

Immunity

Host produces strong Abs response like IgG and IgM against Ags of trypomastigotes, but these are not specific and nonprotective. Serum antibody called **ablastin** prevents the multiplication of trypomastigote. Abs suppress the immune response, which allow the bacterial infection.

Life Cycle

Hosts

- **Vertebrate hosts:** Human, game and domestic animals.
- **Insect host:** Tsetse fly like *Glossina palpalis, G. pallidipes* and *G. tachinoides*. Both male and female flies can transmit the infection in daylight usually in early morning and evening. They take 20 days to produce infective stage and once infected remain infective for rest of life (extending up to 185 days). No evidence of hereditary transmission in fly. Fly produces single larva and about 6–12 larvae in whole life, so reproduction is limited.

Method of reproduction: Longitudinal binary fission.

Cycles: Follow **Fig. 95.3 and Flowchart 95.2.**

Pathogenicity

Disease name: It causes chronic disease called **Gambian sleeping sickness** or **West African sleeping sickness** or **West African trypanosomiasis.**

Epidemiology: It is distributed in West and Central Africa.

Reservoirs/sources of infection: Animal reservoirs are still not known in gambiense strain; human is the reservoir for gambiense strain.

Flowchart 95.2: Life cycle of *T. brucei*

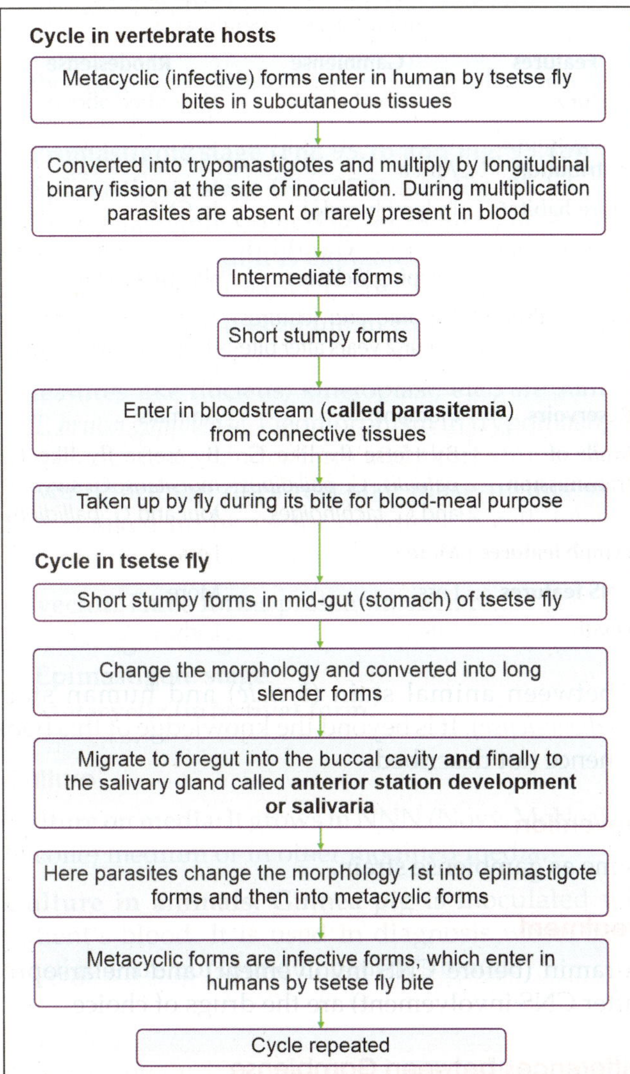

Modes of transmission: It is transmitted by bites of tsetse fly.

Exit form: Short stumpy form.

Infective form: Metacyclic form.

Portal of entry: Skin.

Sites: Parasite invades more in lymph nodes, less in the brain and consumes plenty of glucose from tissues. African sleeping sickness is a disease of CNS.

Pathogenesis: Pathogenic damages are not due to motility of organisms or by allergic reaction (as organism is nontoxic), but due to following reasons.

1. **Mutant of VSG:** Abs are formed, which destroy most of the trypomastigotes, but mutants of VSG escape this destruction and produce new wave of parasitemia characterized by fever.
2. **Autoimmune reaction:** Large numbers of non-specific Abs like IgM and IgG are formed, which are not able to sensitize the Ags of trypomastigote, but instead damage the normal tissues.
3. **Inflammatory reaction:** Follow **Flowchart 95.3**.

Flowchart 95.3: Inflammatory reaction

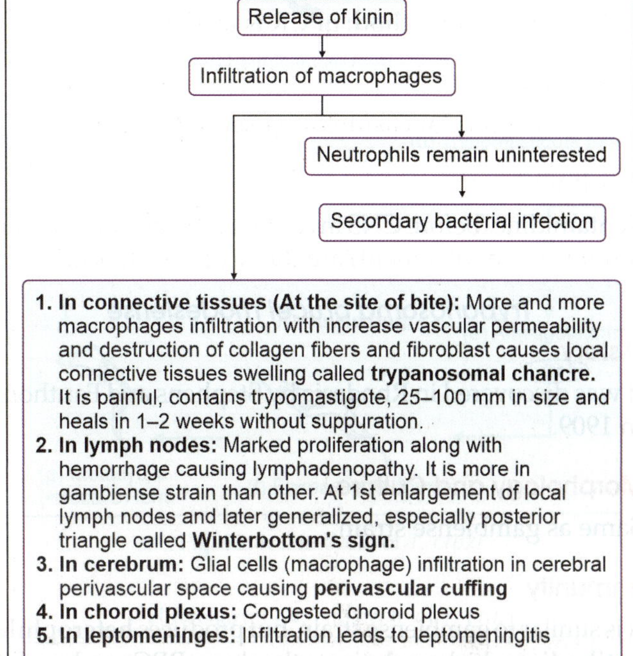

1. **In connective tissues (At the site of bite):** More and more macrophages infiltration with increase vascular permeability and destruction of collagen fibers and fibroblast causes local connective tissues swelling called **trypanosomal chancre.** It is painful, contains trypomastigote, 25–100 mm in size and heals in 1–2 weeks without suppuration.
2. **In lymph nodes:** Marked proliferation along with hemorrhage causing lymphadenopathy. It is more in gambiense strain than other. At 1st enlargement of local lymph nodes and later generalized, especially posterior triangle called **Winterbottom's sign.**
3. **In cerebrum:** Glial cells (macrophage) infiltration in cerebral perivascular space causing **perivascular cuffing**
4. **In choroid plexus:** Congested choroid plexus
5. **In leptomeninges:** Infiltration leads to leptomeningitis

Clinical features: In gambiense strain symptoms develop after 2 years of bite of testse fly.

1. **Trypanosomal chancre:** Follow **Flowchart 95.3**.
2. **Fever:** Intermittent fever with severe headache, loss of nocturnal sleep, feeling of oppression.
3. **Skin:** Fleeting, circinate erythematous rashes appear on chest and shoulders, which observe easily in person with fair skin and difficult in dark skin.
4. **Lymphadenopathy:** Follow **Flowchart 95.3**.
5. **CNS features:** It includes meningoencephalitis with classical sleeping sickness. Pressure on the palm or over the ulnar nerve may be followed by severe pain within short time after the pressure is released called **Keranadel's sign.** It is less in gambiense strain and more in rhodesiense.
6. **Final stage of disease:** Patient becomes thin and wasted accompany by various signs of malnutrition. Patient survives longer naturally unlike rhodesiense strain. Patient dies sooner or later, if remains untreated.

Laboratory Diagnosis

- **Specimens:** CSF and biopsies from chancre, lymph nodes and bone marrow are more useful. Blood (low parasitemia) is less useful.
- **Testing methods**

  A. **Blood picture:** Anemia, monocytosis and raised ESR due to increased IgM.

  B. **CSF picture:** Increased protein (100–150 mg%) and increased cell count (50–100/mm^3).

  C. **Microscopy of blood:** Motile trypomastigotes are seen in unstained of blood. Staining reactions are **mentioned in Fig. 95.2a.**

  D. **Culture:** Described above.

- Vector-borne: Metacyclic (infective) forms enter in human by rubbing the bug's feces against abraded skin or wound produced by bite of reduviid bug.
- Unclean hands and fingers: Rubbing the exposed mucosa or conjunctiva with contaminated fingers.

Incubation period: 7–14 days

Exit form: Trypomastigote form

Infective form: Metacyclic trypomastigote form.

Portal of entry: Skin or conjunctiva.

Sites: Two forms present in human. **Trypomastigote stage** presents in peripheral blood. **Amastigote stage** is an intracellular form presents in muscles (striated → heart and skeletal), nervous system and RE system.

Clinical features: Two clinical types.

1. **Acute:** It occurs in infants and children.
   - **At the site of entry:** Unilateral swelling of skin presents at the site of entry of parasite called **chagoma.** Sometimes chagoma presents on facial skin, because of feeding preference of reduviid bug called **kissing bug.** Swelling of conjunctiva at the site of entry called **Roman's sign.**
   - **General features:** Fever, enlargement of spleen and lymph nodes, anemia and lymphocytosis.
   - **Final stage:** It lasts for 20–30 days. Sometimes, patient died (5–10%) with symptoms of meningo-encephalitis or myocardial failure.
2. **Chronic:** It occurs in adults and adolescent.
   - **CVS features:** Heart block and Adams-Stokes syndrome.
   - **CNS features:** Spastic paralysis and psychological changes. It may last for 12 years.

Complications: Dilatation of organs occur due to damage to muscle and nerve that controls the tone, particularly in hollow organs as mentioned below.

- GIT: Megaesophagus and megacolon.
- Heart: Cardic myopathy and herniation of endo-cardium at apex of left ventricle
- Lungs: Dilatation of pulmonary conus.

> **Note: Congenital trypanosomiasis**
> - **Clinical features:** *T. cruzi* transmitted in acute and chronic stage producing low-birth weight, stillbirth, myocarditis and neurological complications.
> - **Diagnosis:** IgA-IgM antibodies (from serum) and antigens (from serum and urine) are detected by ELISA. PCR detects specific gene of kinetoplast and nuclear DNA.

Laboratory Diagnosis

- **Specimens:** Blood is collected for trypomastigote form. For amastigote form collect biopsies from swelling, lymph nodes, spleen, brain and muscles (calf/deltoid).
- **Testing methods**
  - A. **Blood picture:** Anemia and lymphocytosis.
  - B. **Microscopy of blood:** Motile trypomastigotes are seen in unstained film of blood. **C-shaped** trypomastigotes are seen from blood by Romnowsky stain as shown in **Fig. 95.2b.** Amastigotes are seen from tissues by H and E stain.
  - C. **Culture:** Described above.
  - D. **Serological tests:** Machado-Guerreiro test based on principle of CFT.
  - E. **Molecular method:** qPCR is useful.
  - F. **Skin test:** It based on DTH.
  - G. **Xenodiagnosis:** Allow the bug to bite infected person and detect the parasite from feces or intestinal contents of bug called xenodiagnosis.

Prevention

Preventive measures are control of reduviid bug (**Ch. 110**) by using insecticides and avoiding bug bite by using mosquito net or repellents.

Treatment

No specific and effective treatment is available. Nifutrimox and benzinidazole have been used for acute and chronic stages with some success.

Leishmania spp.

Leishmania donovani

History

It was discovered simultaneously by two persons. Sir William Leishman in May-1903 in London, from his name genus called *Leishmania* and Donovan in Chennai in July 1903, from his name species called *donovani. L infantum. L. chagasi, L. canis* were identified as separate species earlier, but later, it was found that all were identical to *L. donovani.*

Morphology

Amastigote/aflagellar (Fig. 95.1a and 95.5b) stage in vertebrate hosts: It presents inside RE cells called **LD (*Leishmania donovani*) body (Fig. 95.5).** Features are same as amastigote stage of *T cruzi* described above.

Promastigote/flagellar (Fig. 95.1b and 95.5f) stage in sand fly and in culture: It is an actively motile, elongated pear-shaped body with 5–10 μm in length and 2–3 μm in breadth. It has large, oval and central nucleus. Kinetoplast is situated near anterior end called **antetnuclear kinetoplast.** Starting point of flagellum called basal body. Part of flagellum lying inside the body called axoneme or axial filament. Light stained space lying alongside the axoneme called vacuole. Free flagellum present, but sometimes, it is absent. Undulating membrane is absent. It multiplies by binary fission.

Culture

Culture on media: Promastigote stage is developed on following culture media.

- **Monophasic media:** Schneider's insect tissue culture medium, Grace's insect medium, solid medium (prepared by J C Ray in 1932 contains agar, glucose, beef heart infusion, peptone, amino acid, glycerine and rabbit's blood) and liquid medium like HOMEM/Eagle's minimal essential medium **(transforms amastigote form of *Leishmania* to promastigote form in 48 hours in 60–80% cases).**
- **Biphasic media:** Brain heart infusion agar, Tobies medium and NNN [Novy, McNeal (1904) and Nicole (1908)] medium.

Specimens to be cultivated: Blood (collect 1–2 ml of blood and dilute with 10 ml citrated solution contains 0.85% saline with 2% sodium citrate. Either centrifuge or allow to settle at 22°C overnight. Cellular deposit is inoculated in liquid part of NNN medium and biopsy from spleen, bone marrow and lymph nodes in case of African and Chinese form.

Culture method and results: Follow **Flowchart 95.5**.

Culture in animals: It is used for diagnosis, research purpose and to test the efficacy of leishmanicidal drugs. Intraperitoneal inoculation kills the **hamster/small rodents** (*Cricetulus griseus*) in 2–3 months and parasite can be seen in organs like liver and spleen.

Immunity

Follow **Flowchart 95.6**.

Life Cycle

Hosts

- **Vertebrate hosts:** Human, dog and hamster.
- **Insect host:** It is transmitted by sand fly, *Phlebotomus* in old world and by *Lutzomyia* in new world. In India (old world), it is transmitted by *Phlebotomus argentipus* and by other species of *Phlebotomus* in other countries. In South America (new world) vector is *Lutzomyia longipalpis*.

Method of reproduction: Binary fission.

Cycles: Follow **Fig. 95.5 and Flowchart 95.7**.

Flowchart 95.5: Culture method and results

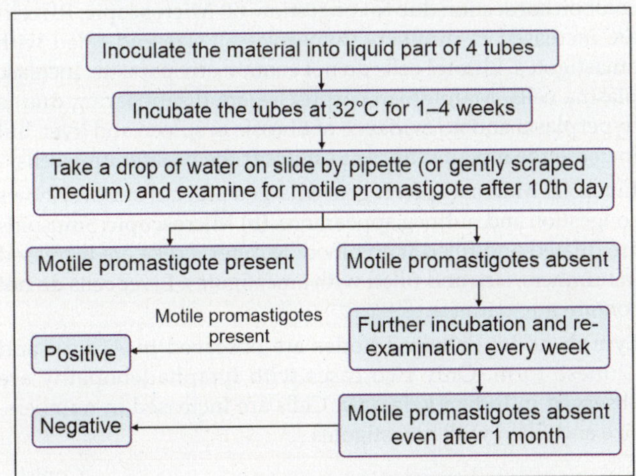

Flowchart 95.6: Immunity in *L. donovani*

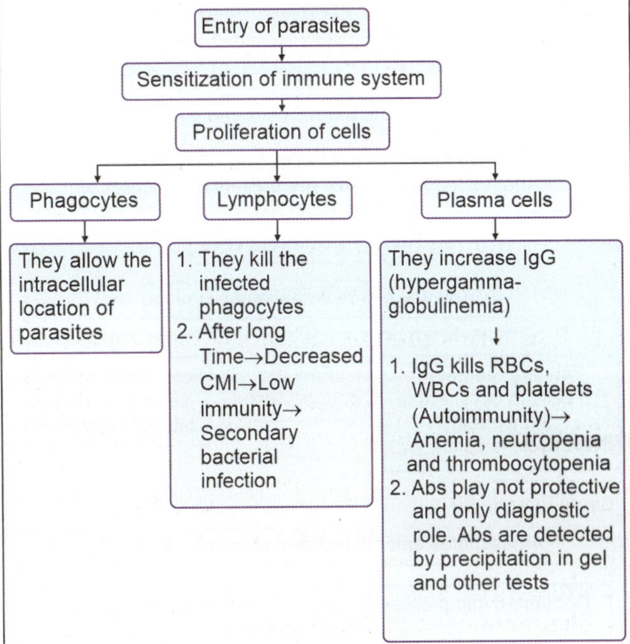

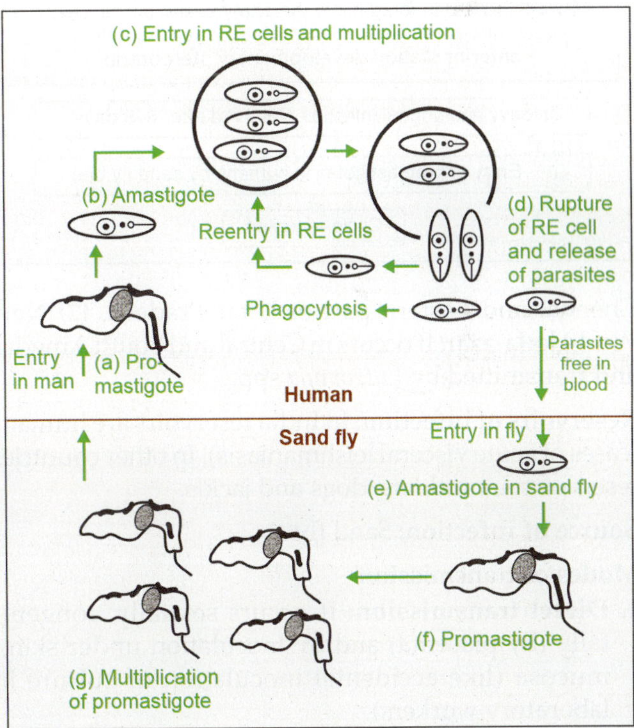

Fig. 95.5: Life cycle of *L. donovani*

Pathogenicity

Disease name: It called **visceral leishmaniasis** or called **kala-azar** or **dum dum fever.**

Meaning: Kala-azar is an Indian terminology. Here **Kala** means **black** and also **deadly**, while **Azar means illness**, which indicates blackening of skin and fatal nature of illness.

Epidemiology: (1) Old World kala-azar: It occurs in India, Africa and rare in Europe and transmitted by *Phlebotomus* spp. **In India,** it is endemic along the coastal area of Ganges and Brahmaputra and in Bihar, Odisha,

Flowchart 95.7: Life cycle of *L. donovani*

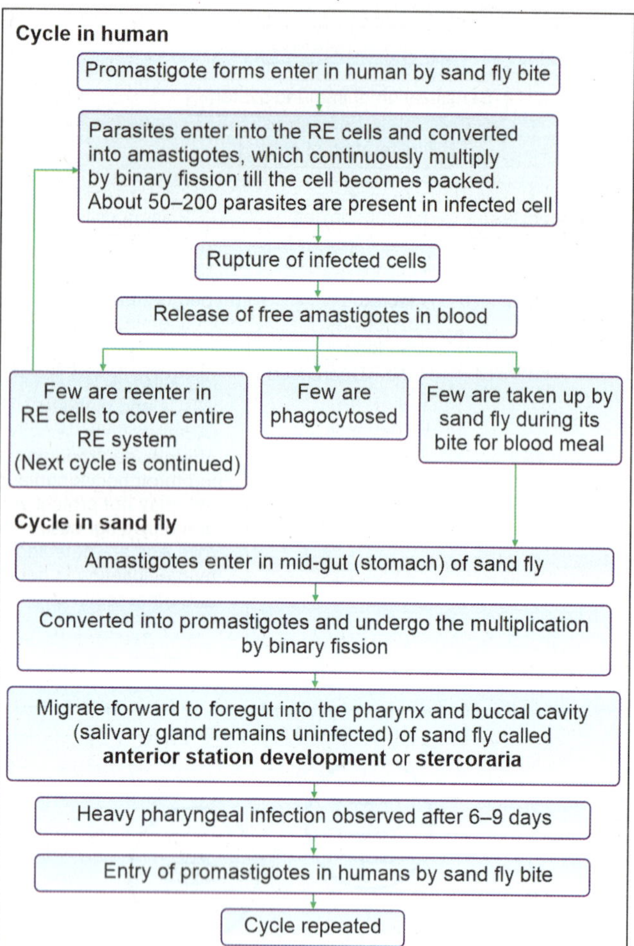

Chennai and Eastern part of Uttar Pradesh. **(2) New World kala-azar:** It occurs in Central and South America and transmitted by *Lutzomyia* spp.

Reservoirs of infection: In India reservoirs are humans (cases of acute visceral leishmaniasis). In other countries reservoirs are rodents, dogs and jackls.

Source of infection: Sand fly.

Modes of transmission

1. **Direct transmission:** It occurs sexually, congenitally (by placenta) and by inoculation under skin/mucosa (like accidental inoculation of culture in laboratory workers).

2. **Indirect (vehicle borne)**
   - Fluid-borne: Transfusion of blood/blood products.
   - Vector-borne: It occurs by bite of sand fly. Pharynx and buccal cavity of sand fly are completely blocked by presence of promastigote called **blocked sand fly**. Parasites are not present in salivary gland. Bites of such blocked sand fly transmit the infection in wound produced by proboscis. To take the blood meal fly has to empty the pharynx and buccal cavity.

Incubation period: It is generally 3–6 months, but sometimes exceed up to 1–2 years.

Exit form: Amastigote.

Infective form: Promastigote.

Portal of entry: Skin.

Sites: Amastigote stage present in RE system.

Clinical types: Early case of kala-azar of 2–6 weeks duration called **acute kala-azar** and after 2–6 weeks duration called **chronic kala-azar.**

Precipitating factors (epidemiological determinants): Coinfection of HIV and kala-azar has been reported from Ethiopia, Sudan, South Europe, Brazil and India (Bihar). In addition to features of kala-azar, it includes cutaneous/mucocutaneous leishmaniasis, diarrhea and pleural effusion. Abs tests are negative. It is diagnosed by detecting Ag from urine and amastigotes from BAL plus buffy coat smear of blood.

Clinical features

- **At the site of entry:** Nodular swelling developed in skin at the site of sand fly bite called **leishmanioma.** It is 1–1½ inch in diameter and lasts for 1–3 weeks. It observed in Sudan, Middle Asia and Kenya. It also observed after artificial inoculation of culture from human, gerbil or ground squirrel strain of *L. donovani.* It is not observed in India and South America. **Other part of skin** becomes dry, rough, harsh and pigmented (black color). In African kala-azar warty eruption of skin and mucocutaneous lesion may appear.
- **Hairs:** They become brittle and fall out.
- **Splenomegaly:** Spleen reaches below the coastal margin and often fills the entire abdomen.
- **Lymphadenopathy:** Observed in African and Chinese form.
- **Hepatomegaly:** Liver is enlarged, but not as much as spleen. No jaundice, unless liver is enlarged greatly.
- **Fever:** It may be continuous or remittent type, becoming remittent in later stage. Around 20% of cases show the double rise in 24 hours. Waves of pyrexia may be followed by apyrexia. Sometimes, fever, splenomegaly and hepatomegaly are considered as **visceral leishmaniasis triad**.

Note: Pathological changes in different systems

Spleen: (1) Macroscopic: It is the most commonly affected organ. It gets enlarged with thick capsule, soft consistency, easily friable by pressure with thumb and looks dull red or chocolate color on cut section due to congestion. **(2) Microscopic:** RE cells are increased in numbers (hyperplasia), size and filled with amastigotes. Littoral cells do not contain any parasite. Increase plasma cells, Malpighian corpuscles are disappearing due to hyperplasia and no evidence of fibrosis in spleen and liver, but increase the reticulin fibrils to support the proliferating cells.

Liver: (1) Macroscopic: It gets enlarged and cut surface shows congestion and nutmeg appearance. **(b) Microscopic:** Sinusoids are dilated and filled with blood. Kupffer's cells are increased in numbers, size and filled with amastigotes. Liver cells do not contain any parasite.

Lymph node: Enlarged nodes are observed in African and Chinese form. Only two cases with lymphadenopathy are observed in Indian kala-azar. Cells are increased in numbers, size and filled with amastigotes.

(Contd…)

(Contd...)

GIT: Intestinal ulcer occurs due to secondary bacterial infection but not due to *L. donovani*.

Bone marrow: RE cells are increased in numbers, size and filled with amastigotes. Increase plasma cells. Leukoblastic elements are increased, resulting leukopenia (neutropenia).

Blood picture: Pancytopenia is the feature.
- **Anemia:** 5–10% Hb.
- **RBCs count:** Decreased RBCs count due to autoantibodies, reduction of survival time of RBCs up to 50%, sequestration and destruction of RBC in spleen.
- **WBCs count:** Total count is 3000/mm^3 and sometimes reduced up to 1000/mm^3. **Neutropenia** occurs due to autoantibodies and increased leukoblastic elements in bone marrow. **Lymphocytosis and monocytosis** occur due to proliferation.
- **Platelets count:** Thrombocytopenia is due to autoantibodies.
- **WBC to RBC ratio:** Normal ratio is 1:750 and in **kala-azar** it is 1:1000 to 1:2000.

- **General features:** There is no malaise or apathy. Emaciation and patient remains unaware about fever. Epistaxis may present.
- **Final stage:** Around 75–95% chances of death if remains untreated due to following complications.

Complications: Immunosuppressive effect leads amoebic dysentery, bacillary dysentery, pneumonia, pulmonary tuberculosis, cancrum oris in case of severe neutropenia and other septic infections which may produce death.

Laboratory Diagnosis

Diagnosis of Chronic Kala-azar

Specimens
- Blood: It is useful in HIV positive patients.
- Biopsy from spleen: It is the most useful (98% sensitivity), but chances of bleeding, which may continue from soft and enlarged spleen leads death.
- Biopsy from bone marrow: It is collected from sternum or iliac crest. It is most useful (50–85% sensitivity) when spleen is not enlarged. It gives negative result, when parasites are scanty.
- Biopsy from lymph nodes: In case of African and Chinese form (98% sensitivity).
- Liver biopsy: Also useful

Testing methods

A. Blood picture: Described above.

B. Microscopy
- **Preparation of smear from blood**
  - Thin film (less useful because of scanty parasites) and thick film: As like *Plasmodium* spp.
  - Centrifuge the blood and prepare the smear from buffy coat area.

- **Preparation of smear from biopsies:** Rub the tissues over the slide and then fix with methanol.
- **Staining of smear:** Stain the smear by Giemsa, Leishman's stain or Wright's stain. Intracellular amastigotes called **LD bodies** are seen as shown in **Fig. 95.6a.** Cytoplasm appears blue, kinetoplast red and nucleus appears pink (violet) as shown in **Fig. 95.6b** in inset.
- **Grading of smear:** Follow **Table 95.2.**

C. Culture: Described above.

D. Serological tests
1. **Ag detection test:** Carbohydrate Ag is detected from urine by latex agglutination test. It is useful in coinfection with HIV, in active infection and as prognostic marker.
2. **Ab detection tests:** Ab detection is more sensitive but less specific, not useful in coinfection with HIV and unable to differentiate between past and current infections. Ab is detected by direct agglutination test, IHA, ICT, IF test and ELISA.
3. **Test to differentiate cutaneous (*L. tropica*) and mucocutaneous leishmaniasis (*L. brasiliensis*) from kala-azar:** It is done by **Adler's test** which based on inhibition of development of promastigotes in Locke's serum agar by using specific antiserum.
4. **Tests for detection of hypergamma globulin**
   - **Napier's aldehyde test:** It based on precipitation reaction. Mix 1–2 ml serum with 40% formalin in a test tube and allow to stand for 20 minutes. Jellification of milk-white opacity like a white of hard-boiled egg within 2–20 minutes called strong positive. It is positive in visceral leishmaniasis (after 3 months), African

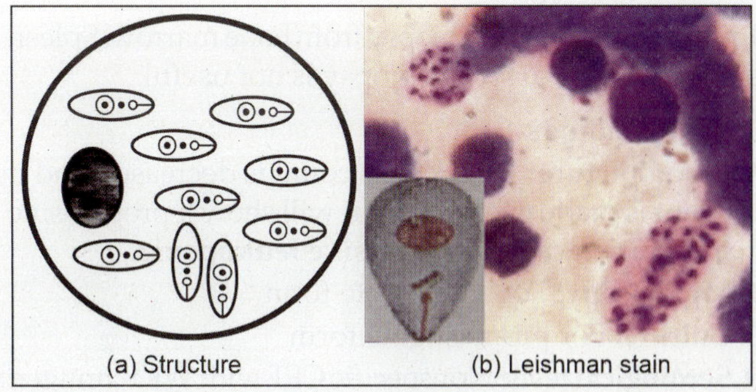

Fig. 95.6: LD body

TABLE 95.2: Grading of smear		
Grade	**No. of amastigotes**	**No. of fields**
0	0	1000
1	1–10	1000
2	1–10	100
3	1–10	10
4	1–10	1
5	>10–100	1
6	>100	1

trypanosomiasis (*T. brucei*), *S. japonicum*, *S. haematobium*, multiple myeloma and cirrhosis but negative in cutaneous leishmaniasis.

- **Chopra's antimony test:** It based on precipitation reaction. Mix 1–2 ml serum (whole serum or 1:10 diluted) with 4% urea stibamine solution in a test tube and allow to stand for few minutes. Formation of precipitate (flocculant) indicates positive test. It is positive in visceral leishmaniasis.
- **Nonspecific CFT with WKK antigen:** Sharing the common antigens by *Leishmania* and *M. tuberculosis* is the basis of this test. It is called nonspecific, as the antigen used in the test is not prepared from the *Leishmania* itself, but from human tubercle bacilli (*M. tuberculosis*) called **Kedrowsky's strain.** Antigen is prepared by **W**itebusky, **K**ligenstein and **K**uhn, hence called **WKK antigen.** WKK antigen detects the certain antibodies present in serum of kala-azar patients by using principle of CFT. It is positive in visceral leishmaniasis (in 3 weeks/acute infection), lepromatous leprosy, pulmonary tuberculosis and tropical pulmonary eosinophilia.

E. **Skin (allergic) test:** It is called **Leishmanin test** or **Montenegro reaction.** It is based on DTH. Inject 0.1–0.2 ml culture contains 6–10 million promastigote/ml (killed) by ID route. Development of induration after 72 hours indicates positive test. Test becomes positive in people with good CMI after 6–8 weeks of recovery of African kala-azar and oriental sore, but not in Indian kala-azar, Mediterranean kala-azar and PKDL.

Diagnosis of Acute Kala-azar

Specimens: Blood and biopsy from bone marrow. Spleen is not enlarged so spleen biopsy is not useful.

Testing methods

A. **Blood picture:** Total WBCs count is decreased and if count is made every week, it will show a progressive diminution called **progressive leukopenia.**

B. **Microscopy:** For amastigote form

C. **Culture:** For promastigote form

D. **Serological tests:** Nonspecific CFT with WKK antigen is positive and aldehyde test is negative.

Note: Post kala-azar dermal leishmaniasis (PKDL)
- **Synonym:** Dermal leishmanoid.
- **Definition:** Non-ulcerative cutaneous lesion occurs in kala-azar patients, 1–2 years after completion of treatment with antimonial for original disease, where visceral infection disappears and skin infection persists.
- **Epidemiology:** It seen in 2–50% cases of kala-azar. In India it seen in 2–20% cases after 2–10 years in any age group and persist for long period. It is rare in Chinese kala-azar and not in Mediterranean kala-azar.

- **Clinical features**
  - **Early lesions:** (1) **Depigmented patches:** They present on trunk, limbs and rare on face without complete loss of pigment. (2) **Butter fly erythematous patch:** Butterfly type erythematous patch presents on nose, cheeks and chin. It is photosensitive and becomes prominent in middle of the day.
  - **Late lesions: Yellowish pink nodules** replace the early lesions and occasionally appear from beginning. They present mostly on face and mucosa of tongue and eye; however, they can appear anywhere. They are soft and painless granulomatous ulcers, which differentiate these from espundia and oriental sore.
- **Complications:** Recurrence of kala-azar occurs in case associated with PKDL.
- **Differential diagnosis:** It is differentiated from leprosy, oriental sore and espundia.
- **Diagnosis:** Amastigotes are detected from nodular lesion (but not from depigmented lesion) by microscopy or by direct agglutination test. Abs are detected to rK39 Ags.
- **Treatment:** AMB (for 4 months) is the drug of choice.

Prevention

Preventive measures are control of sand fly (**Ch. 110**) by using insecticides, elimination of animal reservoirs and avoiding sand fly bite by using thick full sleeve clothes, mosquito net or repellents.

Treatment

Useful drugs are liposomal amphotericin B (current drug of choice), pentavalent antimonials (used in past bur resistant developed), oral miltefosine and paromomycin.

Old World and New World Cutaneous Leishmaniasis

Old World Cutaneous Leishmaniasis

Etiological agents: It also called **oriental sore.** It includes countries of Asia, Africa and Europe. Common species are *L. tropica, L. major* and *L. aethiopica* separately described in box.

Note: *L. tropica*

History: It was 1st defined by James Homer Wright. In 1914, it was divided into two subspecies, namely *L. tropica minor* and *L. tropica major*, based on the size of the parasites. Later, it was identified that *L. tropica minor* causes dry nodular lesions in urban areas, while *L. tropica major* causes wet ulcerating lesions in rural areas. Therefore In 1973 Bray et al described two separate species, *L. tropica major* became *L. major* and *L. tropica minor* became *L. tropica*.

Morphology: Same as *L. donovani*.

Culture: Promastigote stage can be developed in NNN medium.

Life cycle
- **Hosts: (1) Vertebrate hosts:** These are human, domestic dogs, red foxes, golden jackals, gundis and wild rodents. **(2) Insect host:** Sand fly species *Ph. sergenti* is the main vector. Other species are *Ph. arabicus, Ph. guggisbergi, Ph. chabaudi, Ph.rossi* and *Ph saevus*.
- **Cycle:** It is similar to *L. donovani*, except that amastigotes develop in RE cells of skin.

(Contd…)

- **Method of reproduction:** Amastigotes in humans and promastigotes in sand flies divide by binary fission.

Pathogenicity

- **Disease name:** Disease called **dry type cutaneous leishmaniasis** and its variant called **leishmaniasis recidivans** (also caused by *Leishmania brasiliensis*).
- **Epidemiology:** This parasite is distributed in Africa, Europe, Afghanistan, Iran, Syria, Yemen, Algeria, Morocco and Northern India.
- **Incubation period:** Generally few days to 6 months and in some cases, it may extend up to 1–2 years.
- **Clinical features: (1) Dry type cutaneous leishmaniasis:** It is characterized by nodular granulomatous swelling at the site of skin bite. It is single or 2–3 in numbers with one inch in size. In majority of cases it ulcerates with clear cut margin and raised indurated edge surrounded by red areola. It heals spontaneous or slowly in 6 months with production of white scar. Granulomatous lesion contains monocytes/macrophages with parasites inside, lymphocytes, plasma cells and fibroblast. **(2) Leishmaniasis recidivans:** It is DTH reaction (infective type) occurs after healing of dry lesion and characterized by erythematous nodular or papular skin lesion which looks like lupus vulgaris (cutaneous tuberculosis) and occurs in the center or periphery of previously healed lesion.

Note: *L. major*

Morphology and culture: Same as *L. donovani*.

Life cycle: Vertebrate hosts are rodents (main hosts), humans, dogs and cattle. **Insect host** is sand fly species like *Phlebotomus papatasi*. **Cycle and method of reproduction** are similar to *L. donovani* except that amastigotes develop in RE cells of skin.

Pathogenicity

- **Disease name:** Disease called **moist types cutaneous leishmaniasis**, zoonotic cutaneous leishmaniasis, Aleppo boil, Aleppo button, Aleppo evil, Baghdad boil, Bay sore, Biskra button, tropical sore, Chiclero's disease, Delhi boil, Kandahar sore, Lahore sore, Pian bois or Uta.
- **Epidemiology:** It presents in Northern Africa, Middle-East, North-Western China and North-Western India.
- **Incubation period:** 2–6 weeks.
- **Clinical features:** Nodular lesion develops at the site of bite, which later gets ulcerated. Lesion is moist and has raised outer borders, a granulating base, an overlying layer of white purulent exudate and have been described as **pizza like**.

Note: *L. aethiopica*

Morphology: Same as *L. donovani*.

Culture: Promastigote stage can be developed in NNN medium.

Life cycle

- **Hosts: (1) Vertebrate hosts:** These are humans. **(2) Insect host:** Sand fly species like *Ph. longipes*.
- **Cycle:** It is similar to *L. donovani* except that amastigotes develop in RE cells of skin.
- **Method of reproduction:** Amastigotes in humans and promastigotes in sand flies divide by binary fission.

Pathogenicity

- **Disease name:** Also called **leishmaniasis diffusa** or **diffuse cutaneous leishmaniasis.**
- **Epidemiology:** This parasite is distributed in Ethiopia, Uganda and Kenya.

- **Pathogenesis:** It is due to deficiency of CMI to leishmanin antigen and characterized by nodular lesion which looks like lepromatous leprosy.
- **Clinical features:** Nodular lesion which is neither destructive nor erosive and presents as single lesion on skin which spread to covers the face, ears, extremities, buttocks and entire body.

Laboratory diagnosis of old world cutaneous leishmaniasis: (1) Microscopy: Amastigotes are seen by Leishman or Giemsa stain of smear prepared from biopsies or tissues from the edge of lesion. **(2) Culture:** Described above. **(3) Leishmanin test/reaction:** Positive skin test after intracutaneous injection of suspension of promastigotes of *L tropica*. It detects the DTH. It is negative in diffuse cutaneous leishmaniasis due to absent or low CMI.

Prevention of old world cutaneous leishmaniasis: Same as kala-azar.

Treatment of old world cutaneous leishmaniasis: Local use of paromomycin ointment or intralesional antimonials. Systemic therapy required in case of disfiguring or scar formation.

New World Cutaneous Leishmaniasis

Etiological agents: It includes countries of Central and South America. Common species are **L. braziliensis, L. maxicana** and **L. amazonensis** separately described in box.

Laboratory diagnosis and prevention of new world cutaneous leishmaniasis: Same as old world cutaneous leishmaniasis.

Treatment of new world cutaneous leishmaniasis: Drugs are same as used in old world cutaneous leishmaniasis but given as local and systemic therapy.

Note: *L. braziliensis*

History: It was 1st observed in 1911 by Carini and defined as a new species by Vianna in 1911.

Morphology: Morphologically, it is similar to *L. donovani* and *L. tropica*.

Culture

- **Culture on media:** Promastigote stage can be developed in NNN medium.
- **Culture in animals:** Syrian hamster produces only local lesion and even intraperitoneal inoculation cannot produce visceral lesion.
- **Culture in chick embryo:** Also grow in chorio-allantoic membrane.

Immunity: CMI develops after metastasis in skin (clasmatocytes) and mucosa (monocytes).

Life cycle

- **Hosts: (1) Vertebrate hosts:** Humans. **(2) Insect host:** Sand fly like *Lutzomyia*.
- **Cycle:** It is similar to *L. donovani* except that amastigote form develops in skin and mucosa. Amastigote form in human and promastigote form in sand flies divide by binary fission.

(Contd…)

Infections of Blood and Tissue Flagellates

Essentials of Medical Microbiology

Pathogenicity
- **Epidemiology:** It presents in South and Central America.
- **Mode of transmission:** It is transmitted by bite of sand fly. Transmission by direct contact is possible.
- **Incubation period:** Few days to few weeks.
- **Reservoirs of infection:** Small forest rodents.
- **Sources of infection:** Sand flies.
- **Portal of entry:** Skin.
- **Sites:** Amastigote is present in RE cells of skin, nasal mucosa and buccal mucosa.
- **Clinical features: (1)** It causes **skin lesion** similar to that of oriental sore but more severe. **(2)** After healing of cutaneous lesions, patients remains asymptomatic for 3–20 years and during this time parasites spread to oro-nasal mucosa and causing mucocutaneous lesions called **espundia**. Espundia is destructive and mutilating erosion. Sometimes, it presents with destruction of nasal septum. It heals with scar producing typical tapir nose called **camel nose**. **(3) Leishmaniasis recidivans:** Described above with *L. tropica*.

Note: *L. maxicana*
- **Epidemiology:** Mexico and Central America.
- **Clinical features: (1) Cutaneous lesion:** It presents as an ulcer at the bite site. It develops in few days or several months after the initial bite. **(2) Diffuse cutaneous lesion:** It begins when the amastigote diffuses through the skin and metastasizes to other tissue. This type does not produce ulcers.

Note: *L. amazonensis*
- **Epidemiology:** It presents throughout the Brazilian Amazon region.
- **Clinical features: (1) Cutaneous lesion:** It is characterized with skin lesions, which can appear localized, or throughout the body. **(2) Mucocutaneous lesion:** It is characterized with ulcers around the skin, mouth, and nose. This form of leishmaniasis has also been known to spread by metastasis and can be deadly.

ACCESS YOURSELF

Case Study
1. A 55-year-old male presents with nodular swelling over feet. Examination revealed pallorness, splenomegaly and edema feet. Hemogram revealed 5% Hb, neutropenia and agranulocytosis. Blood microscopy shows an intracellular amastigote stage of parasite. Identify the case and answer the following.
   a. Name the causative agent.
   b. Describe the life cycle of causative agent.
   c. Write the pathogenicity of causative agent.
   d. Describe the lab., diagnosis of causative agent.

Essay/Full Question
1. *Trypanosoma/Leishmania.*

Short Notes
1. Morphology/life cycle/pathogenicity/Lab., diagnosis of *T. brucei* or *T. cruzi* or *L. donovani*.
2. Post kala-azar dermal leishmanoid.
3. Cutaneous leishmaniasis.

Short Questions for Theory/Viva Questions
1. Name two hemoflagellates and two intestinal flagellates.
2. Write the vector of *T. brucei* and *T. cruzi*.
3. What are Winterbottom's sign and Keranadel's sign?
4. What is Roman's sign?

MCQs for Chapter Review
Trypanosoma spp.

1. **Animal disease produced by *T. brucei* is known as:**
   a. Chaga's disease   b. Nagana
   c. Sleeping sickness   d. None of above
2. **Tsetse fly transmits:**
   a. *Trypanosoma brucei*   b. *T cruzi*
   c. Kala azar   d. Oriental sore
3. **Sleeping sickness is transmitted by:**
   a. Tsetse fly   b. House fly
   c. Sand fly   d. Simulium fly
4. **Winter bottom's sign in sleeping sickness refers to:**
   a. Unilateral conjunctivitis
   b. Posterior cervical lymphadenopathy
   c. Narcolepsy
   d. Transient erythema
5. **Roman's sign presents in:**
   a. *L. donovani*   b. *T. brucei*
   c. *L. infantum*   d. *T. cruzi*

Leishmania spp.

6. **Aflagellar stage of *L. donovani* is known as:**
   a. Amastigote stage   b. Promastigote stage
   c. a + b   d. None
7. **Amastigote form is seen in:**
   a. RE system   b. WBC
   c. RBC   d. Macrophage
8. **The most important reservoir of leishmaniasis in India is:**
   a. Dogs
   b. Rodents
   c. Acute visceral leishmaniasis
   d. Case of post kala azar dermal leishmaniasis
9. **Which of the following is most severely affected in kala-azar?**
   a. Spleen   b. Liver
   c. Adrenal gland   d. Bone marrow
10. **The followings tests help in laboratory diagnosis of kala-azar *except*:**
    a. Bone marrow examination
    b. Immobilization test
    c. Blood smear examination
    d. Aldehyde test
11. **In case of kala azar aldehydes test becomes positive after:**
    a. 2 weeks   b. 4 weeks
    c. 8 weeks   d. 12 weeks
12. **Visceral leishmaniasis:**
    a. Caused by *L. tropica*
    b. Post leishmaniasis dermatitis is common
    c. Antimonial are useful drugs
    d. Diagnosed by blood smear
    e. Vector is *Phlebotomus sergenti*
13. **The followings are true of kala-azar *except*:**
    a. Persistent hyper gamma globulinemia
    b. Pancytopenia
    c. Cancrum oris can occur
    d. Full treatment prevents post kala-azar dermal leishmaniasis

14. Mucocutaneous leishmaniasis is caused by:
 a. *L. braziliensis*   b. *L. tropica*
 c. *L. donovani*   d. *L. orientalis*

15. Oriental sore is caused by:
 a. *L. donovani*   b. *L. tropica*
 c. *L. braziliensis*   d. *L. infantum*

16. Nasopharyngeal leishmaniasis is due to:
 a. *Leishmania braziliensis*   b. *Leishmania tropica*
 c. *Leishmania chagasis*   d. *Leishmania donovani*

Answers and Explanation of MCQs

1. b
 • Follow section, **Flowchart 95.1** for explanation.

2. a

3. a
 • All the strains of *T. brucei* are transmitted by tsetse fly, but species are different.

4. b
 • Follow section, *Trypanosoma brucei gambiense* (**pathogenicity → clinical features**) for explanation.

5. d
 • Follow section, *Trypanosoma cruzi* (**pathogenicity → clinical features**) for explanation.

6. a

7. a
 • Follow section, *Leishmania donovani* (**morphology**) for explanation of answers of MCQs 6–7.

8. c
 • Follow section, *L donovani* (**pathogenicity → Reservoir of infection**) for explanation.

9. a
 • Follow section, *Leishmania donovani* (**pathogenicity → pathological changes in different organs**) for explanation.

10. b

11. d
 • Follow section, *Leishmania donovani* (**diagnosis**) for explanation of answers of MCQs 10–11.

12. c, d, e

13. d
 • Follow section, *Leishmania donovani* for explanation of answers of MCQs 12–13.

14. a

15. b

16. a
 • Follow section, **old world and new world cutaneous leishmaniasis** for explanation of answers of MCQs 14–16.

Infections of Ciliates

Chapter Outline
- Introduction
- *Balantidium coli*

INTRODUCTION

Definition

Thin, delicate and hair like processes cover the body of parasite called **cilia** and such parasites called **ciliates.**

Classification

Systemic classification: Follow Ch. 14 (Table 14.3).

Pathogenic classification: *Balantidium coli.*

Balantidium coli

History

It was 1st described by Malmsten in 1857 from the stool of two dysenteric patients. It was also discovered by Leuckart in 1861 and by Stein 1862, who classified the species in to the genus *Balantidium.*

Morphology

It exists in two forms **(Fig. 96.1).**

Trophozoite Stage

Shape and size: It found in dysenteric stool. It is oval in shape with 60–70 μm × 40–50 μm (largest protozoan) in size.

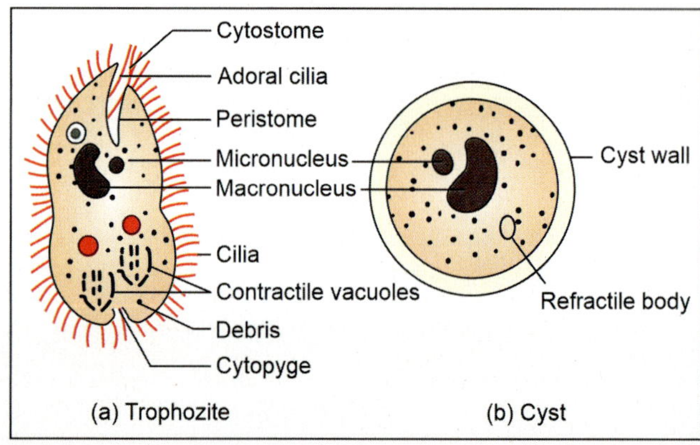

Fig. 96.1: Morphology *B. coli*

Motility: It resembles a 'thrown football'.

Structure (Fig. 96.1a): It has narrow anterior end and broad posterior end. Surface is covered with thin, delicate and hairs like processes called **cilia,** which are the organs of locomotion. They are shorter over entire body, but longer at mouth part. It has GIT, cytoplasm and nucleus.

- **GIT.** Anterior end presents with a groove like structure called **peristome.** Cilia present at the peristome **called adoral cilia.** It has mouth called **cytostome** and terminating into a short funnel-shaped gullet called **cytopharynx.** It extends up to one third of body. At the posterior end, there is an anus (but no intestine) called **cytopyge.** Parasite eats the intestinal bacteria, RBCs and epithelial cells.

- **Cytoplasm:** It is divided in two portions. **(1) Endoplasm:** It is an internal and granular portion. It functions for nutrition, excretion and reproduction. It contains RBCs, WBCs, food debris (eaten from host's gut), tissue debris, two nuclei and two contractile vacuoles (one near the posterior end and 2nd in middle of the body). **(2) Ectoplasm:** It presents as external thin layer.

- **Two nuclei:** Lager one is kidney-shaped and situated in the middle of the body called **macronucleus** and smaller is spherical and situated in the concavity of macronucleus called **micronucleus.**

Cyst Stage

Shape and size: It found in chronic case and carrier. It is spherical or oval in shape with 50–60 μm in size.

Structure (Fig. 96.1b): Cyst surrounded by thick transparent double layered wall. Cytoplasm is granular and contains the two nuclei and refractile body. Contractile vacuoles remain active for some time during encystment.

Flowchart 96.1: Life cycle of *B. coli*

Entry of cysts by feco-oral route
Excystation
Tough wall resists cysts from acid of stomach and alkali of small intestine.
Entry in large intestine and one trophozoite is formed from one cyst called excystation
Few trophozoites enter in mucosa and few remain in lumen of intestine
Asexual method
Sexual method
Trophozoites multiply by binary transverse fission
Few of them enclosed in a cyst like structure by conjugation to exchange the nuclear materials and after which they are separated.
Produce the daughter trophozoites
Produce the daughter trophozoites
Daughter trophozoites follow the successive divisions and rapidly increase in numbers. During the division 1st the micronucleus divides into two, then macronucleus and finally body by transverse septum. Daughter cells formed from anterior half, retain the cytostome of mother cells, but reconstruct the posterior end, while one formed from posterior end develops its own cytostome and other structure
Encystation
Encystation
Some trophozoites converted into the cysts in large intestine called encystation. Each trophozoite forms one cyst. Few trophozoites remained unchanged. Both cysts and remained trophozoites pass with the stool in environment.
Cyst remains viable in soil and water for several weeks
Trophozoites disintegrated outside
Entry in other host
Remain noninfectious
Continuation of life cycle and cycle repeated

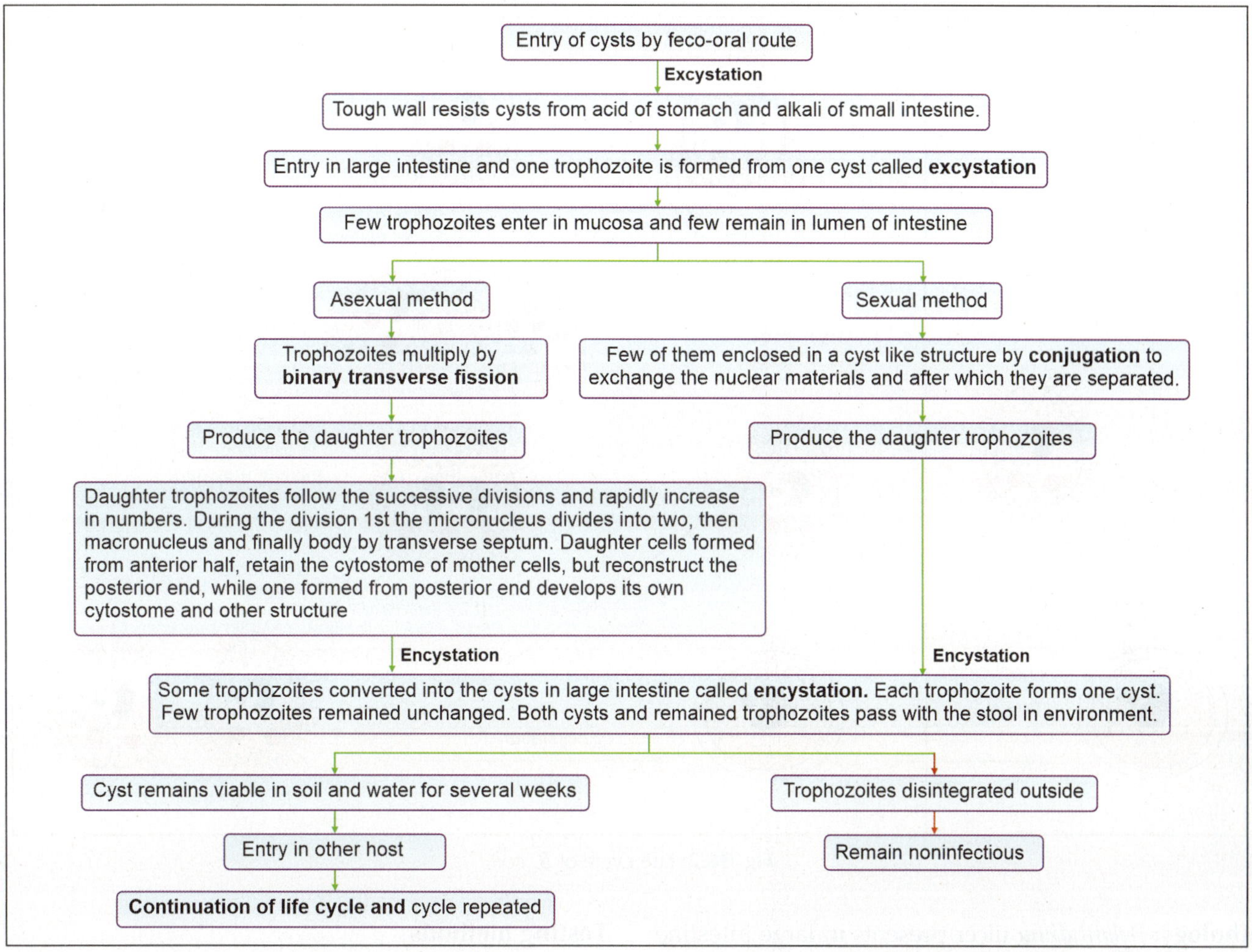

Culture

B. coli can be grown in media of *Entamoeba coli.*

Life Cycle

Host: Only one host is pig, while human is an incidental host; however, it has two stages of life cycle, of which trophozoite stage exists in host (in pig or in human) and cystic stage, in environment.

Methods of reproduction: Two types.

1. **Asexual method:** Binary transverse fission.
2. **Sexual method:** By conjugation.

Cycles: Follow **Flowcharts 96.1 and 96.2, and Fig. 96.2.**

Pathogenicity

Disease name: It called **balatidiasis** or **ciliate dysentery**.

Epidemiology: It is distributed worldwide.

Reservoirs of infection: Mainly pigs (parasite is non-pathogenic) and rarely monkeys and chimpanzees.

Flowchart 96.2: Excystation and encystation

Excystation
1-cyst
1-Trophozoite
(2-nuclei)
(2-nuclei)
1-cyst
1-Trophozoite
Encystation

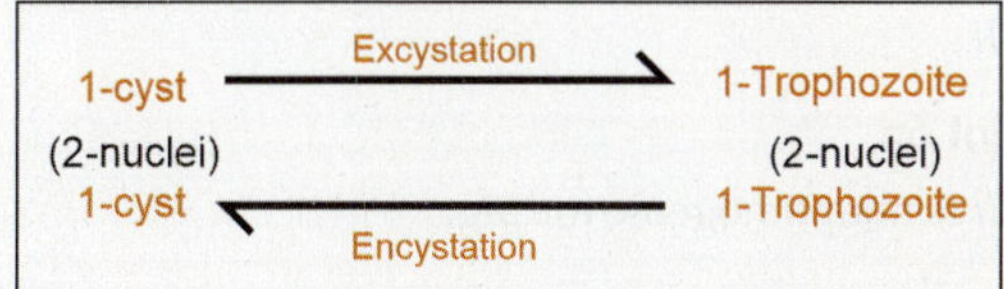

Sources of infection: Food and water.

Modes of transmission: Human transmission occurs by **ingestion** of contaminated food and water contain cysts from pig/human or by occupational **contact** with pigs. It transferred from pig to pig, pig to human, human to human or human to pig.

Incubation period: Few days to week.

Exit form: Trophozoite and cyst in feces.

Infective form: Cyst is an infective and the transmitting stage.

Portal of entry: GIT.

Sites: Large intestine (cecum and colon).

Precipitating factors (epidemiological determinants)

Agent factors (virulence factors): (1) Tough cyst wall: Protects the parasite from acid and alkali. **(2) Enzymes:** Parasite secretes hyaluronidase, which degrades the intestinal mucosa and facilitates penetration.

Host factors: (1) Concurrent infection: Presence of *T. trichuria* in intestine encourages the tissue invasion by *B. coli.* **(2) Nutritional:** Parasite thrives mainly on starchy food found in plenty in pig's intestine. **Hoare** stated that presence of abundant fat in pig's intestine renders it harmless, while scarcity in humans makes it pathogenic for human.

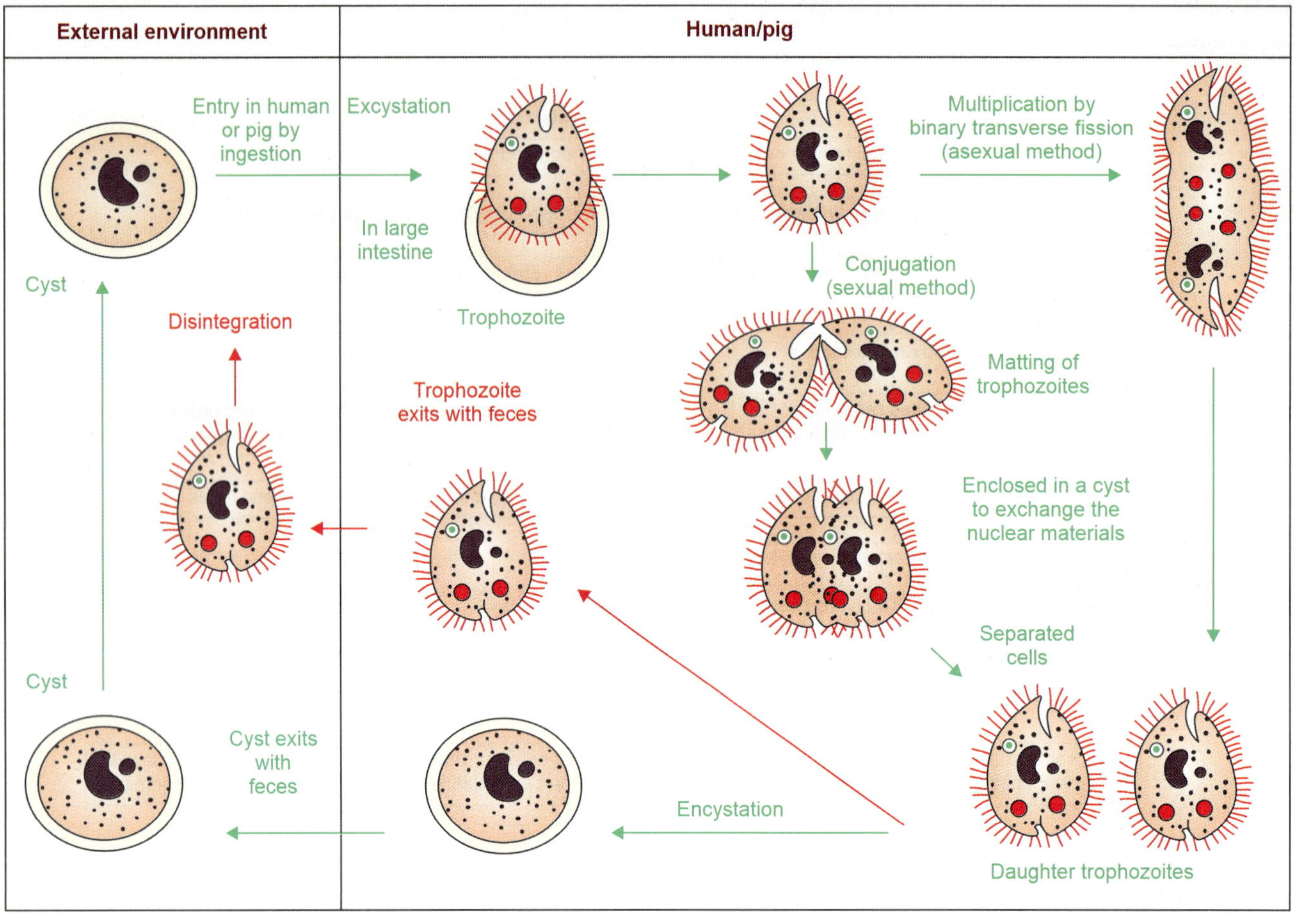

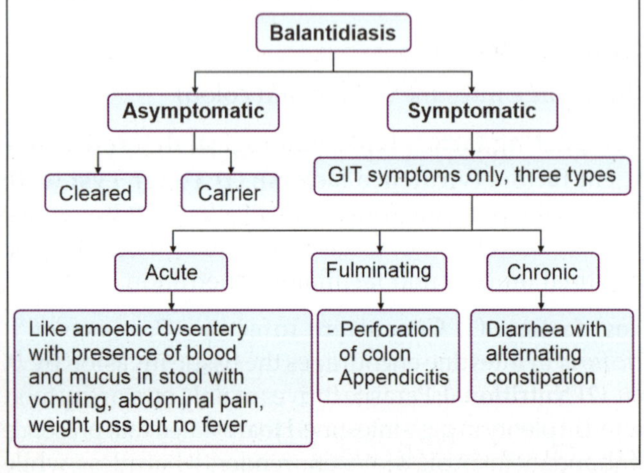

Fig. 96.2: Life cycle of *B. coli*

Pathology: *Balantidium* ulcer presents in large intestine (cecum or colon), varying in size from small to large, similar in shape to amoebic ulcer like **flask shape** and may extended to mucosa, sub mucosa and sometimes up to muscularis mucosae. Parasite is not able to enter the liver.

Clinical features: Follow **Flowchart 96.3.**

Laboratory Diagnosis

Specimens: Stool (contains both cysts and trophozoites) and **mucosal biopsy** (collected by sigmoidoscope)

Flowchart 96.3: Clinical features of balantidiasis

Testing methods

A. Macroscopy: Presence of blood and mucus in stool.

B. Microscopy: Motile trophozoites are detected under normal saline preparation in fresh stool. Major drawback of stool examination is that parasites shed intermittently and undergo the rapid degeneration. Trophozoites are not found in stool after 6 hours of passage of stool. Cysts are also present in stool. Rate of detection of trophozoites to cysts in stool is in ratio of 9:1.

C. Culture: Described above.

D. Molecular method: Like qPCR.

Prevention

Prevention of contamination of food and water and treatment of carrier are the useful measures.

Treatment

Tetracycline is the drug of choice.

Short Note

1. *B. coli.*

Comment on

1. *B. coli* is nonpathogenic for pig.

CHAPTER 97

Infections of Blood Inhabiting Sporozoa

Chapter Outline

- Introduction
- *Plasmodium* spp.
- *Babesia* spp.

INTRODUCTION

Reproduction

Reproduction in sporozoa is taken place by asexual method (called **schizogony**) and sexual method (called **sporogony**) called **alteration of generation.** Products of schizogony are called **merozoites** and those of sporogony are called **sporozoites.** Both these methods are taken place in different host called **alteration of host.**

Classification of Sporozoa

Systemic classification: Follow Ch. 14 (Table 14.3).

Pathogenic classification: Two groups like blood inhabiting sporozoa and GIT sporozoa. GIT sporozoa are mentioned in **Ch. 98.** Following are pathogenic species of blood inhabiting sporozoa.

A. *Plasmodium*: It includes around 172 species. Few pathogenic species are listed below.
- **Common species infecting man**
  - Benign tertian malaria: (1) *P. vivax* is described by Grassi and Feletti in 1890. (2) *P. ovale* is described by Stephens in 1922.
  - Malignant tertian malaria: *P. falciparum* is described by Welch in 1897.
  - Benign quartan malaria: *P. malariae* is described by Laveran in 1880 (Noble Prize in 1907) and Grassi and Feletti in 1890.
- **Rare species infecting human**
  - Benign tertian type: (1) *P. cynomolgi* is described by Mayer in 1907. (2) *P. cynomolgi bostianellii* is described by Garnham in 1959.
  - Ovale tertian type: *P. simium* is described by Fonseca in 1951.
  - Quartan type: (1) *P. inui* is described by Halberstadter and von Prowazek in 1907. (2) *P. brasilianum* is described by Gonder and von Berenberg-Gossler in 1908. (3) *P. shortii* is described by Bray (1963).
  - Quotidian type: *P. knowlesi* is described by Sinton and Mulligan in 1932. It is the simian parasite, which infects the human. *Anopheles leucosphyrus* is the vector. It presents in Malaysia, Thailand and Myanmar. In India, vector is found in coastal area of Kerala and Maharashtra. It is same as *P. ovale* except that fever occurs daily, because of short RBCs cycle and more severe because infects RBCs of all age. Early trophozoites are same as *P. falciparum*, while late trophozoites are same as *P. malariae*. Chloroquine or primaquine is the drug of choice.

B. *Babesia*: *B. divergens* and *B. microti*.

Plasmodium spp.

History

In 1880, In **Algeria Alpnonse Laveran**, a French army surgeon, discovered the malarial parasite in RBCs of a patient. Name of discoverers and year of discovery of important species of *Plasmodium* are described earlier with the pathogenic classification. In 1886 **Golgi** in Italy described the asexual development of the parasites in RBCs (erythrocytic schizogony); hence, it is also called **Golgi cycle.** In Russia, in 1891, **Romanowsky** developed the staining method of malarial parasite. In Secunderabad, India, in 1897, **Ronald Ross** identified the developing stage of malarial parasite in mosquito, and he described the mode of transmission. This led to the discovery of mosquito control measures to eradicate the malaria. Ronald Ross got Noble Prize in 1902.

Life Cycle

Hosts: *Plasmodium* parasites reproduced by asexual method (called **schizogony**) in intermediate host and by sexual method (called **sporogony**) in definitive host called **alteration of generation.** Products of schizogony called **merozoites** and products of sporogony called sporozoites. Both these methods are taken place in different host called **alteration of host.**

Types of hosts

- **Intermediate host:** It is human and cycle called **human/schizogony/asexual cycle.**
- **Definitive host:** It is female *Anopheles* mosquito and cycle called **mosquito/sporogony/sexual cycle.**

Methods of reproduction

- **Asexual method:** It occurs in human called **schizogony.**
- **Sexual method:** It occurs in mosquito (female anopheles) called **sporogony.**

Cycles: Two types of cycles as mentioned in **Flowchart 97.1 and Fig. 97.1.**

Flowchart 97.1: Life cycle of *Plasmodium* species

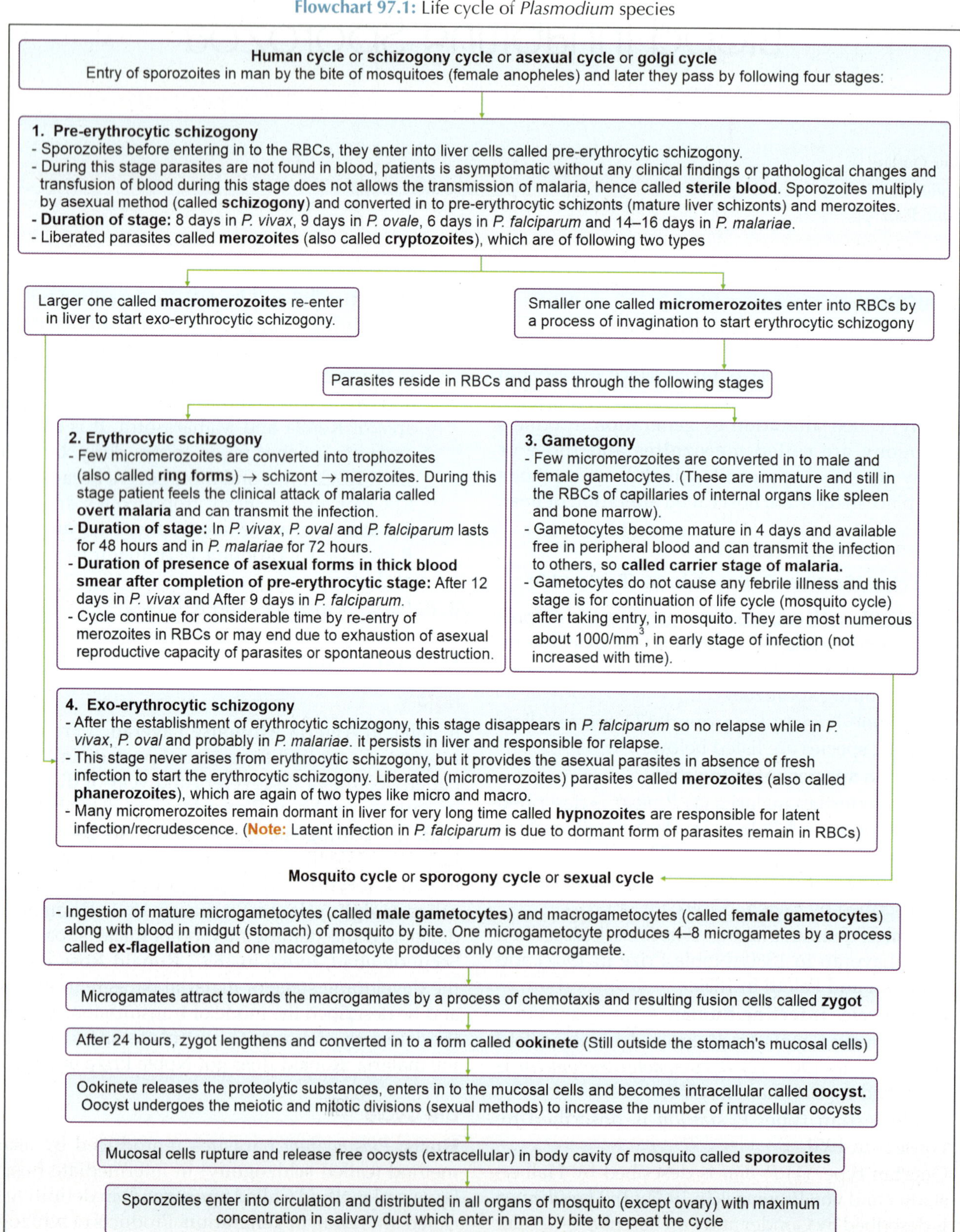

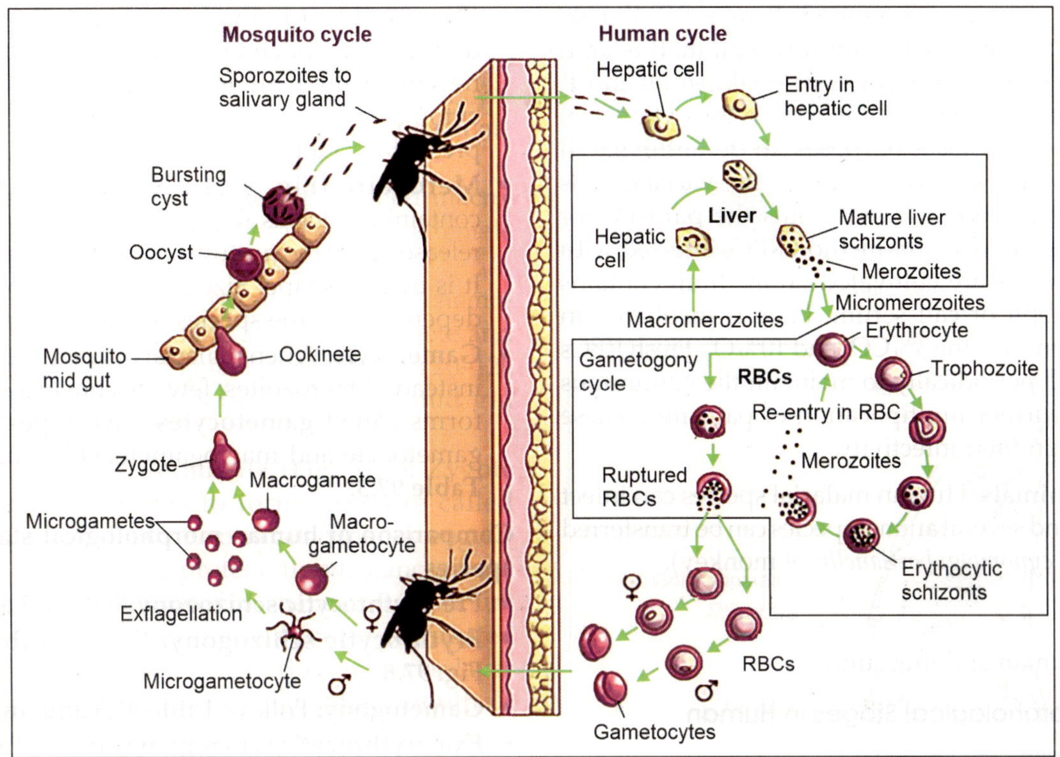

Fig. 97.1: Life cycle of *Plasmodium* species

Salient features of life cycle

A. Human cycle:

- **Phases:** Inefective form for human infection is sporozoite. There are mainly two phases of human cycle.
  - **Inside the liver: (1) Pre-erythrocytic schizogony:** Blood is sterile and there is no clinical attack of malaria. **(2) Exo-erythrocytic schizogony:** Blood is sterile and there is no clinical attack of malaria but it can cause relapse.
  - **Inside the RBCs: (1) Erythrocytic schizogony:** Inoculation is positive (case) and there is clinical attack of malaria (overt malaria). Method of entry of merozoites in to RBCs is not clear, because the envelope remains intact after the merozoites have entered. **(2) Gametogony:** Inoculation is positive (carrier), and it can infect the mosquito.
- **Summary of human cycle**
  - **For:** *P. vivax*, *P. malariae* and *P. ovale* **(Flowchart 97.2).**
  - **For:** *P. falciparum* **(Flowchart 97.3).**

B. Mosquito cycle

  - Male and female gametocytes are the infective for mosquito.
  - It starts in human with formation of gametocytes and enters in mosquito for further maturation.
  - Female *Anopheles* mosquito ingested by sexual and asexual forms, but only sexual forms follow the further development while rest will die immediately.
  - To infect the mosquito, human blood should contain at least 12 gametocytes per mm^3 of blood and female gametocytes must be in excess of the male gametocytes.

Flowchart 97.2: Summary of human cycle

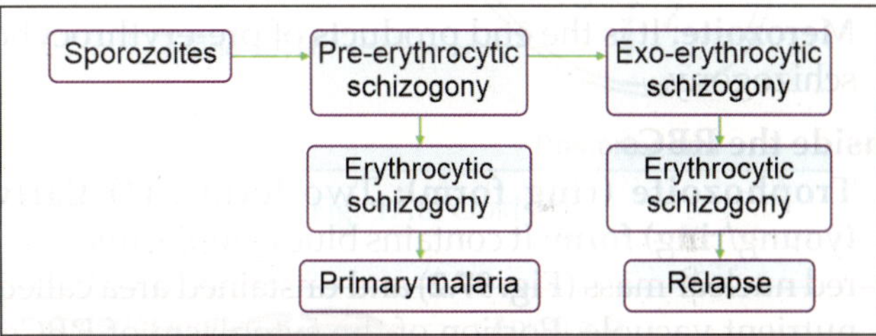

Flowchart 97.3: Summary of human cycle *P. falciparum*

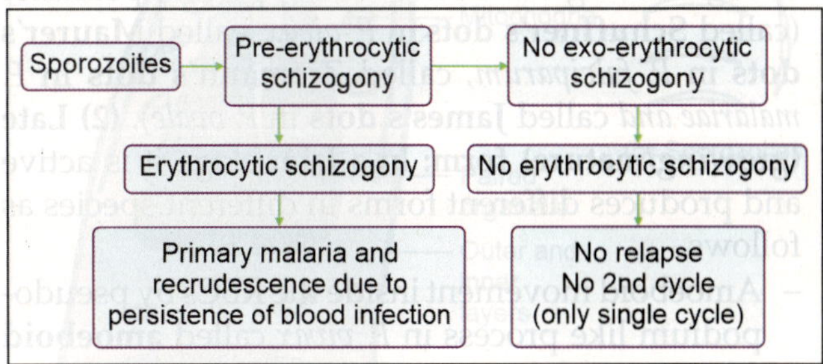

  - Presence of sporozoites in the salivary gland of mosquito is to be taken as a positive proof for epidemiological survey of malaria.
  - **Mixed infection:** Different malarial species can develop in same mosquito at a time and such mosquito can give rise mixed infection and it is common in *P. falciparum* and *P. vivax*.

Culture

Culture on media: (1) Brass and Jone's technique was invented in 1912 by using artificial medium, but it could not maintain the parasite for more than one cycle. **(2) Trager and Jensen's technique** was invented in 1976,

Epidemiology: It is distributed worldwide, mostly in tropical zone extending from 40°S to 60°N. **In India,** malaria continues to be major public health problem. In India, about 27% people lives in high transmission (>1 case/1000 population) and 58% people lives in low transmission (0–1 case/1000 population) area. **In India** about 50% cases by *P. falciparum,* 4–8% are mixed and rest by *P. vivax* but *P. ovale* is not reported in India. WHO classified the following types of endemicity based on spleen rate or parasite rate in children (2–9 years) and adults.

1. **Hypoendemic:** Transmission is low. Spleen or parasite rate is <10%.
2. **Mesoendemic:** Transmission is moderate. Spleen or parasite rate is 11–50%.
3. **Hyperendemic:** Transmission is intense but seasonal. Spleen or parasite rate is 51–75%.
4. **Holoendemic:** Transmission is of high intensity. Spleen or parasite rate is >75%.

Reservoirs of infection: Humans particularly children act as reservoirs for human infection. In some part of Africa, chimpanzees may act as reservoirs for *P. malariae.* Monkeys are the reservoirs for *P. knowlesi.*

Sources of infection: Mosquitoes.

Modes of transmission: It is discussed with natural and therapeutic types of malaria.

Natural Malaria

A. According to route of entry of parasites: Following four routes.

1. **Inoculation:** By use of contaminated (through blood) needles or syringes in IV drug users or by injection of secretion of salivary glands containing sporozoites.
2. **Transplacental:** Placental defect allows the transmission of malarial parasites from mother to baby called **congenital malaria.** Healthy placenta acts as a barrier for malarial parasites.
3. **Vector-borne**
   - **Species:** It is transmitted by bite of female *Anopheles* mosquito. More than 10 species of *Anopheles* mosquitoes are responsible for the transmission in India. In rural area, it is by *A. culcifacies* and in urban area it is by *A. stephensi.* Others responsible species are *A. philipinensis, A. fluviatilis, A. varuna, A. minimus, A. annularis, A. leucosphyrus (A. balabacensis), A. sundaicus* and *A. jeyporiensis candidiensis.*
   - **Method of introduction:** Female *Anopheles* mosquito act as a definitive host. It sucks the parasites from human blood during its bite for blood meal purpose. Parasite takes 8–10 days for development in mosquito, its presence in salivary gland and to be infective for man. This period called **extrinsic incubation period** (**definition:** Time taken by vector to become infective after receiving the microorganisms called **extrinsic incubation period**). It depends on environmental conditions. During the act of bite mosquito's proboscis pierce the skin and salivary secretion containing the sporozoites is introduced directly into blood through the puncture wound. However, parasites cannot be found in blood after about half an hour.

4. **Vehicle-borne:** Transfusion of blood containing asexual forms of erythrocytic schizogony called **trophozoite-induced malaia** or **merozoite-induced malaria.** Storage of blood may reduce the infectivity of malarial parasites. Infectivity is highest with *P. falciparum* and *P. malariae,* while not more than 2% in *P. vivax.*

B. According to entry of morphological stages of parasites: Two types like sporozoite and trophozoite induced malaria as differentiated in **Table 97.5.**

1. **Sporozoite induced malaria**
   - **Injection:** Injection of secretion of salivary glands containing sporozoites will also induce the infection.
   - **Vector-borne:** Transmission by mosquito bite.
2. **Trophozoite-induced malaria**
   - **Injection:** By contaminated (through blood) needle or syringe in IV drug users.
   - **Vertical or transplacental:** Placental defect allows the transmission of malarial parasites from mother to baby called **congenital malaria.**
   - **Vehicle-borne:** Transfusion of blood or blood products.

Therapeutic Malaria

It is an introduction of artificial malarial infection mostly by *P. vivax* in case of neurosyphilis (generalized paralysis of insane) caused by *T. pallidum.* It is introduced by following types of methods.

1. **Sporozoite-induced therapeutic malaria:** It is done by injection of secretion of salivary glands containing sporozoites or by allowing the laboratory bred infected mosquito bite to the recipient.
2. **Trophozoite-induced therapeutic malaria:** It is done by transfusion of blood or blood products of infected donor.

TABLE 97.5: Induction of malaria		
Features	**Sporozoite-induced malaria**	**Trophozoite-induced malaria**
Pre-erythrocytic (PE) stage	Present	Absent
Incubation period	Long	Short
Exo-erythrocytic (EE) stage	May be present	Absent
Relapse	May occur	No relapse
Schizonticidal drugs	No radical cure, because of presence of EE stage	Radical cure, because no EE stage

Prepatent period/biological incubation period: It is the time taken by microbes to appear first in blood after infection. Pre-patent period in different species is mentioned in **Table 97.3**. After entry in body parasites enter in liver and not detected in blood. Patient remains **asymptomatic** during this time. Parasites appear in blood after few days of infection. It is a part of intrinsic incubation period.

Incubation period (intrinsic/clinical incubation period)

- **Incubation period in different species**
  - In *P. vivax*: 8–17 (average 14) days.
  - In *P. falciparum* (shortest): 9–14 (average 12) days.
  - In *P. malariae* (longest): 18–40 (average 28) days to 6 weeks.
  - In *P. ovale*: 16–18 (average 17) days.
- **Effective factors:** It depends on many factors. **(1) Numbers of parasites injected:** Long incubation period about 38 weeks has been noticed in European stain of *P. vivax* and *P. falciparum*, which indicates that numbers of sporozoites injected may have relation with incubation period. **(2) Use of antimalarial drugs:** Long incubation period also noticed in person who used antimalarial drugs.

Exit form: Gametocyte.

Infective form: Sporozoite.

Portal of entry: Skin.

Site: First in liver cells then in RBCs and later in other organs by blood.

Precipitating factors (epidemiological determinants): Occurrence of malaria depends on chain of following factors. Breaking of this chain at any point can prevent the spread of malaria.

- **Agent (virulence) factors:** Parasite factors include numbers of sporozoites injected, malarial toxins, etc. Malarial toxins are described with pathogenesis in **Flowchart 97.4**.
- **Vector factors:** Mosquito factors include density, life span, choice of host, resting habit, breeding habit, time of biting and resistance to insecticides.
- **Host factors: (1) Age:** Common at all age but new born infants resist to *P. falciparum* due to high level of fetal hemoglobin. **(2) Sex:** More in male than female because of outdoor visit and better covering of clothes in female. **(3) Race:** Genetic resistance to *P. falciparum* occurs in some parts of Africa and Mediterranean coast. Genetic abnormalities in RBCs (sickle cell anemia) provide protection against malaria. **(4) Pregnancy:** It is at high risk in malaria and malaria in pregnancy may cause intrauterine death, premature labor or abortion. **(5) Immunity:** Infection immunity (premunition) may develop and reduce the risk of malaria. **(6) Human habit:** Malaria is common in person sleeping outside and not using mosquito repellents like net, cream, etc. **(7) Migration:** Migration to endemic areas increases the spread of malaria.

- **Environmental factors: (1) Rain fall:** Malaria is common after rainy season in July-November due to increase the breeding of mosquitoes. **(2) Altitude:** Mosquitoes are not found at altitude above 2000–2500 meters, due to unfavorable climatic conditions. **(3) Temperature:** Temperature between 20–30°C favors the breeding of mosquito. Mosquitoes are not surviving below and above this range. **(4) Humidity:** Humidity directly effect on mosquitoes not on parasites. Relative humidity for mosquitoes to live normal life is 60%. When relative humidity is high, it becomes more active and when decline, it may not live longer. **(5) Men made malaria:** Burrow pits, garden pools, irrigation channels, flower pots, building and construction work are allowing the mosquitoes to breed.

Pathogenesis: Follow **Flowchart 97.4**.

Clinical features (C/Fs): Clinical picture of malaria presents with triad of fever, anemia and splenomegaly.

Fever

Onset of fever: Generally it starts in the early afternoon, but may occur at any time.

Duration: Total duration of fever is 6–10 hours; however, variable with the species of *Plasmodium*.

Stages of fever: It presents with three stages. **(1) Cold stage:** It presents with chills and rigor and lasts for 20 minutes to an hour. **(2) Hot stage:** It lasts for 1–4 hours. **(3) Sweating stage:** It lasts for 2–3 hours.

Types of fever: Different types of fever are produced by each species on the basis of duration of erythrocytic schizogony as mentioned below.

- **Benign tertian malaria (Fig. 97.10):** It is less dangerous and fever recurs after 48 hours or 3rd day. It includes **vivax malaria** (produced by *P. vivax*) and **ovale malaria/ovale tertian type malaria** (produced by *P. ovale*). In *P. vivax* initial paroxysm could be continuous or remittent (or quotidian), and in later stage, it could be intermittent.
- **Malignant tertian malaria (Fig. 97.11):** It is more dangerous (most virulent) and fever recurs after 48 hours or 3rd day. It includes **falciparum malaria** (produced by *P. falciparum*). Fever may not show the three successive cold, hot and sweating stages. It could be continuous or remittent in nature.
- **Benign quartan malaria (Fig. 97.12):** It is less dangerous and fever recurs after 72 hours or 4th day. It includes **malariae malaria** (produced by *P. malariae*). Fever is intermittent in nature.
- **Quotidian malaria (Fig. 97.13):** Fever recurs after 24 hours or 2nd day. It occurs mainly in *P. knowlesi*. It has been observed in some cases of *P. vivax* and *P. malariae*. In *P. vivax* it is due to the maturation of two generation of tertian parasites on two successive days, so called **tertian duplex.** In *P. malariae* it is due to the maturation of three generation of quartan parasites on three successive days, so called **quartan triplex.**

Flowchart 97.4: Pathogenesis of malaria

Pathogenesis | **Clinical effects (complications)**

1. Fever
- Raised cholesterol during fever and normal during afebrile period
- In acute case raised WBCs count during fever and normal during afebrile period but leukopenia (neutropenia) in chronic case

2. Neo antigen or fibrin coat on infected RBCs
Formation of new antigen or fibrin coat on the surface of infected RBCs

Adherence of infected RBCs (cytoadherence / sequestration) to the receptors of cells of vascular endothelium in different organs and with other RBCs to form RBCs rosettes. → Anemia and raised ESR due to increased stickiness of RBCs

Occlusion of blood supply in distal portion of organs → Anoxic/hypoxic damage to the organs

3. Increase breakdown of infected RBCs
- Decreased O_2 carrying capacity
- Decreased RBCs count → Anemia (**Other mechanisms of anemia:** Described in C/Fs)
- Rise potassium (hyperkalemia) and decrease sodium

4. Proliferation of RE cells
- Blood: Monocytosis
- Spleen: Splenomegaly (most in *P. falciparum*).
- Liver: Enlarged liver → decreased liver function and albumin production → hypoalbuminemia, reversal of albumin:globulin ration and false-positive test of syphilis

5. Breakdown of Hb
- Release of globlin → Utilized by parasite as protein source called **parasitic protein**
- Release of heme (hematin) → Free hematin is toxic in nature and can binds to cell membrane, damaging cell structure and causing the lysis of the host erythrocytes → so the parasites convert it into an inert, nontoxic and insoluble crystalline form called **hemozoin**, which is called **malaria pigment**. It is taken up by RE cells; hence, organ becomes slaty gray to black colour. Hemozoin contains the iron, but stains black in color with potassium ferrocyanide instead of prussian blue color.

✓ **Note:**
- Heme will not take any other pathway in malaria
- Other pigments in malaria: Bile pigments, hemosiderin stercobilin and urobilin

6. Autoimmunization
Parasite utilised the glucose with help of G6PD enzyme called **glycolysis**
- Decrease blood sugar level (hypoglycemia)
- Persons deficient with this enzyme remains protected

Production of pyruvic acid and lactic acid these are called **malarial toxins** → Fall in blood pH

Toxin interfere with mitochondrial activity (Maegraith, 1967) → Decreased cellular respiration and metabolism

7. Immunosuppression → Secondary bacterial infections

8. Type-III Hypersensitivity:
Deposition of Ag +Ab complex in glomeruli → Malarial nephrosis

9. Autoimmunization:
Formation of autoantibody against RBCs in pregnancy → Autoimmune hemolytic anemia

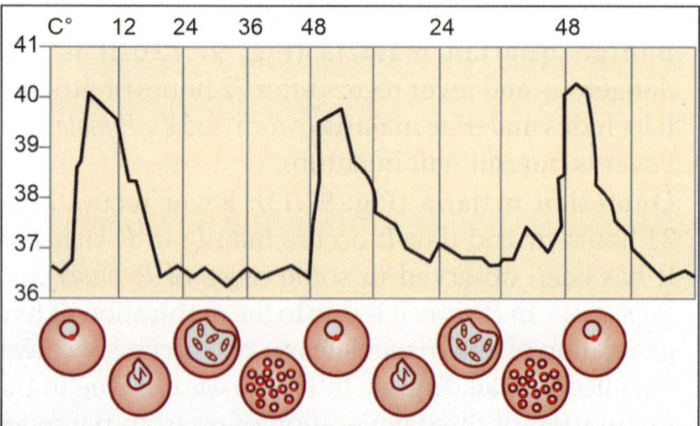
Fig. 97.10: Benign tertian fever

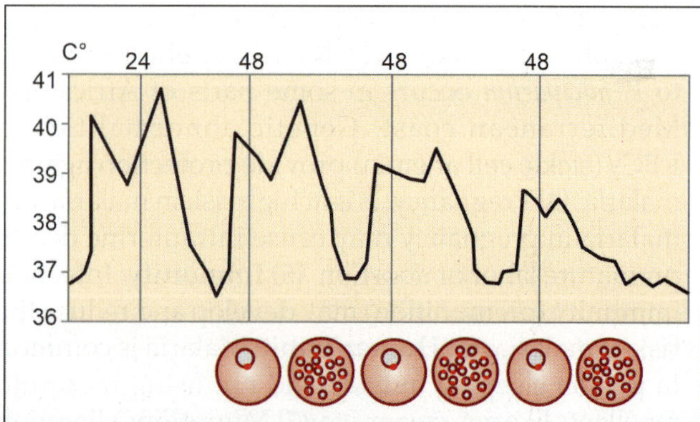
Fig. 97.11: Malignant tertian fever

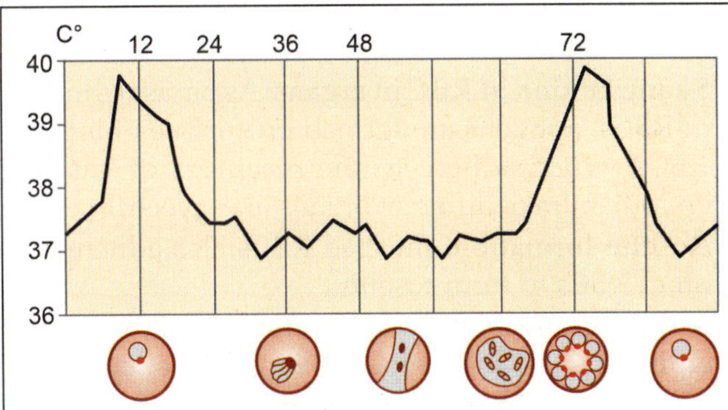

Fig. 97.12: Quartan fever

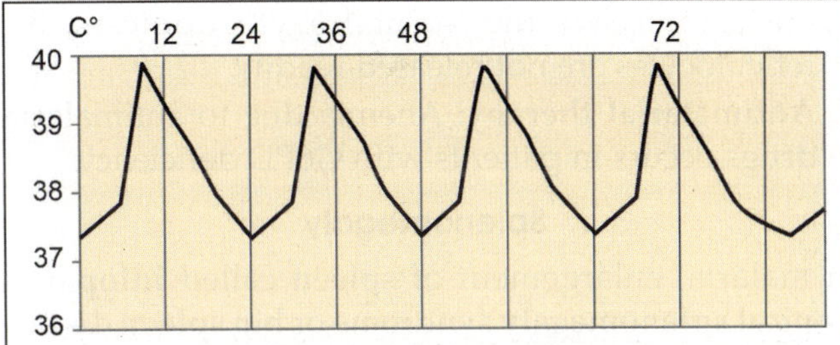

Fig. 97.13: Quatidian fever

Sometime in *P. malariae*, two generation of quartan parasites mature on successive days and fever occurs on two successive days followed by a day of apyrexia called **quartan duplex.**

Note: Pathological changes in different systems

- **Spleen:** Increase spleen functions in malaria with increasing macrophages in red pulp (Billroth cords) to phagocytose the parasites and malarial pigments like hemetin (hemozoin) and others like hemosiderin. **(1) Macroscopic:** Enlarged spleen, more compare to the other chronic conditions like kala azar. It is slate-gray or black in color due to deposition of pigments. Capsule becomes thin and stretched in acute case but in chronic case thickened capsule due to perisplenitis. In acute case it is soft and in chronic case it becomes firm called **ague cake.** Cut surface shows the homogenous black area with scattered white fibrous band (trabeculae), grayish white spots (Malpighian corpuscles) and hemorrhage under capsule and within the splenic tissue in acute falciparum malaria and in spontaneous rupture of organ. **(2) Microscopic:** Congestion presents in splenic sinusoids. Macrophages (RE cells) are increasing in numbers and contain pigments and parasites. Pigments are deposited in red pulp. White pulp (Malpighian corpuscles) does not contain any parasites or pigments. Parasites are seen as black dots in RBCs under hematoxylin and eosin stain.
- **Liver: (1) Macroscopic:** It gets enlarged and chocolate-red to slate-gray or black due to deposition of pigments and also due to congestion. Cut surface shows dilated lobular veins; where fatty changes are limited to the central zone of the lobule. Fatty changes are seen as yellowish area against a brownish red background. **(2) Microscopic:** Sinusoids are dilated and filled with infected and normal RBCs. Kupffer's cells are increased in numbers (hyperplasia) and filled with malarial pigments and parasites. Liver cells do not contain any parasite and show fatty changes, atrophy and necrosis. It is due to hypoxia from interference with the escape of venous blood. No evidence of fibrosis, if present indicates process of degeneration.

- **Bone marrow:** In acute cases long bones are showing very little changes. In chronic cases upper and lower thirds of long bones are showing changes. **(1) Macroscopic:** Slate-gray or black due to deposition of pigments. **(2) Microscopic:** Yellowish marrow is replaced by red formative marrow due to congestion. RE cells are increased in numbers (hyperplasia) and contains the malarial pigments. Increased parasitized RBCs. Bone marrow shows the erythroblastic reaction of the normoblastic type (increase the numbers of nucleated RBCs and reticulocytes). Depression of myeloblastic activity as evidenced by decrease granulocytes in blood.
- **Kidneys: (1) Malarial nephrosis** occurs due to hypersensitivity reaction (described above in immunology). **Renal anoxia syndrome (lower nephron nephrosis or renal tubular vascular syndrome)** presents in acute falciparum malaria and in black water fever. It is characterized by oliguria, anuria and acute uremia.
- **Blood picture**
  - **Substances released in blood following completion of erythrocytic schizogony and destruction of RBCs:** These are merozoites, malarial pigments, malarial toxins and unused portion of the cytoplasm of infected RBCs.
  - **Anemia:** 3–9% Hb. All the reasons for development of **anemia** in malaria are described with clinical features.
  - **RBCs count:** RBCs are more reduced in *P. falciparum* (1 million per mm^3) than *P. malariae* and *P. vivax* (2–3 million per mm^3) because this species infect young and mature (old) RBCs with heavy parasitemia.
  - **WBCs count:** Total count is increased (10,000–20,000/mm^3) during fever stage and returns to normal during apyrexial phase. In recurring case (chronic) total count reduced up to 3000–5000/mm^3 (leukopenia). **Neutropenia** occurs in recurring case (chronic). Neutrophils are reduced up to 50–70% and cells contain the malarial pigments. **Monocytosis** occurs with 10–20% increase in numbers, and they contain malarial pigments.
  - **Platelets count:** Thrombocytopenia is common in children and adults due to coagulation disturbances, splenomegaly and platelets destruction by macrophages, bone marrow alterations, antibody-mediated platelets destruction, oxidative stress and platelets aggregation.

Note: Terminology of fever

Continuous (Fig. 97.14a and 97.14b): Constant high temperature [>37.7°C (100°F)] throughout a day (24 hours) which does not fluctuate >1°C and never gets normal in-between.

Intermittent (Fig. 97.14c): High temperature for few hours with normal temperature in between.

Remittent (Fig. 97.14d): Constant high temperature throughout a day (24 hours) which may fluctuate >1°C and never gets normal in-between.

Note: Reading of fever

- **Normal body temperature:** 36.5°C (97.7°F) to 37.7°C (100°F) (Average = 37°C).
- **Fever:** >37.7°C (100°F).
- **Low-grade fever:** <38.3°C (101°F).
- **Mild-grade fever:** 38.3°C (101°F) to 40°C (104°F). A temperature above 38.3°C (101°F) in an infants younger than 3 months is considered high and should be treated as a medical emergency.
- **High-grade fever:** >40°C (104°F).

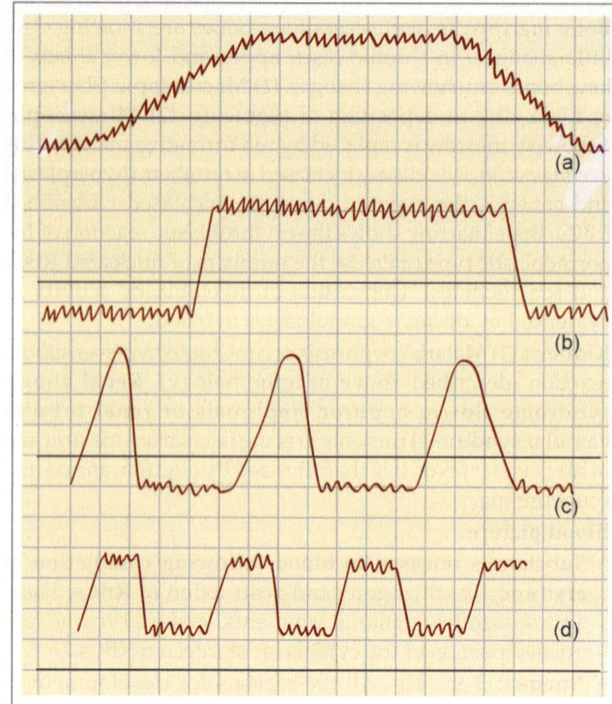

Fig. 97.14: (a) Continuous fever with gradual onset; (b) Continuous fever with sudden onset; (c) Intermittent fever; (d) Remittent fever

---

Note: Congenital malaria
- **Effect of malaria on pregnancy:** Placenta transmits trophozoites which cause miscarriage, abortion, low birth weight, stillbirth or complicate the pregnancy by causing the severe anemia. It is detected by thin and thick smears from cord blood, placental impression and HRP2 Ag.
- **Effect of pregnancy on malaria:** Pregnancy impairs the immunity and causing the relapse.

Note: Malaria in children
- **Severity:** Malaria presents in different forms and more marked in children than adults even with mild infection.
- **Features:** It presents with high fever, convulsion, dehydration due to vomiting and excessive sweating.

Note: Relapses in malaria
It includes following two types.
- **Recrudescence (latent infection):** It occurs due to latent (dormant) forms of parasites, which activate after 1 year or longer due to impairment of immunity or by any other stimuli. Types and reasons are mentioned below.
  - **Erythrocytic form in *P. falciparum*:** Latent form presents in RBCs and activates after 1 year.
  - **Liver form in *P. vivax, P. ovale* and *P. malariae*:** Latent form presents in liver called **hypnozoite,** which activated later to produce recrudescence. Term **"hypnozoite"** is derived from the **Greek** words **hypnos** means **sleep** and **zoon** means **animal**.
- **Recurrence (true relapse):** It is a repeated attack of malaria. It occurs in *P. vivax, P. ovale* and *P. malariae* due to presence of exoerythocytic stage but not in *P. falciparum* due to absence of exoerythocytic stage.

Anemia

1. **Hemolytic anemia:** Increase destruction of RBCs by parasites during its development. When RBCs becomes full with schizonts, cell wall is not able to

hold the weight and rupture of RBCs to release the mature schizonts.

2. **Sequestration of RBC in organs:** As parasites mature in RBCs, antigens are formed on surface of infected RBCs which adhere to the receptors of vascular endothelium of different organs like spleen, liver, etc.
3. **Rosette formation:** Infected RBCs also adhere with other RBCs to form rosette.
4. **Autoimmune hemolytic anemia:** Formation of auto antibodies to RBCs (especially in pregnancy).
5. **Normocytic/microcytic hypochromic anemia:** Decrease incorporation of iron into heme.
6. **Decreased production by bone marrow:** Bone marrow shows the normal erythropoiesis, but reticulocytes are not released readily.
7. **Antimalarial therapy:** Anemia due to antimalarial drugs occurs in patients with G6PD deficiency.

Splenomegaly

In malaria, enlargement of spleen called **idiopathic tropical splenomegaly syndrome** or **big spleen disease** or **hyperactive malaria splenomegaly syndrome.** Exact reason for spleen enlargement in malaria is not known, but may be due to increase spleen function to produce the IgM in response to malarial antigens and also the infiltration by RE cells.

Complications of malaria: In general complications by *Plamodium* spp., are mentioned in **Flowchart 97.4.** Major complications by *P. falciparum* like pernicious malaria and black water fever are described separately in box.

Note: Pernicious malaria
Definition: Series of phenomenon occurs in *P. falciparum* infection and if not effectively treated, threatens the life of patient within 1–3 days.

Pathogenesis: It is an anoxic damage in internal organs due to capillary blockage and decreased effective circulation by different mechanisms as shown in **Flowchart 97.5.**

Clinical types and pathological changes in different system
A. Cerebral malaria
- **Macroscopy of brain:** Cut section shows slate-gray or black color due to deposition of pigments. Multiple small hemorrhages are seen in white matter. Small infarct presents in brain substances.
- **Microscopy of brain: (1) Initial stage:** Dilatation and congestion of cerebral capillaries and plugged with parasitized RBCs. It is more in gray matter than white matter. Capillary blood contains parasites with erythrocytic (like trophozoites and schizonts) and gametogony stages. Hemorrhage presents around the plugged capillaries called **ring hemorrhage,** which contains non parasitized RBCs. There are scattered areas showing degeneration of nerve tissue called **soft areas. (2) Later stage (reparative stage):** Reparative work starts in soft areas with infiltration by glial cells and production of granuloma called **malarial granuloma of Durck.** All the glial cells are arranged radially around the occluded blood vessels. In advanced cases granuloma terminated in to multiple sclerosis and remains as sequel. Patients may present with psychotic syndrome after recovery from such illness.

(Contd...)

B. Algid type (algid means frozen): (1) GIT: Mucosa becomes slate gray color with hemorrhage without ulcer. Capillaries of mucosa and submucosa are packed with parasitized RBCs but capillaries of muscular and serous layer contains very few parasitized RBCs. **(2) Adrenal:** Necrosis of zona fasciculata and hemorrhage with congestion in zona reticulata. Capillaries are packed with parasitized RBCs and pigmented phagocytes.

C. Septicemia type: (1) Blood: Heavy parasitemia leads to the presence of erythrocytic schizogony and gametogony in blood and anemia. **(2) Heart:** Coronary blood vessels are packed with parasitized RBCs. Cardiac muscles show the fatty degeneration and necrosis. **(3) Lungs:** Cut section shows the evidence of hemorrhage and edema. Alveolar capillaries are congested and packed with parasitized RBCs. Alveoli contain extravasated RBCs and pigmented phagocytes. **(4) Eyes:** Vascular obstruction and hemolysis release the pigments causing eye damages.

Clinical features: (1) Cerebral type: It presents with hyperpyrexia, coma and paralysis. **(2) Algid type: GIT features include** cold clammy skin with vascular collapse leading to peripheral circulatory failure. Common features are vomiting, watery diarrhea and dysentery. **Adrenal features** are peripheral circulatory failure resulting death. It is due to adrenal damage or independent. **(3) Septicemic type:** It presents with high continuous typhoid like fever, anemia, cardiac syncope, pneumonia/pulmonary edema, nephritis/renal failure, retinal hemorrhage, optic neuritis, glaucoma, uveitis, oculomotor paralysis and cortical blindness.

Laboratory diagnosis: (1) Blood picture shows heavy parasitemia with presence of erythrocytic schizogony and gametogony in blood, anemia and mild leukopenia with total count is 5200/mm^3. **(2) Biopsy of internal organs** shows the capillary blockage by parasitized RBCs and other changes as described earlier.

Treatment: Start an intravenous therapy with quinine and antibiotics without delay. Blood transfusion including a erythrapheresis and plasmapheresis.

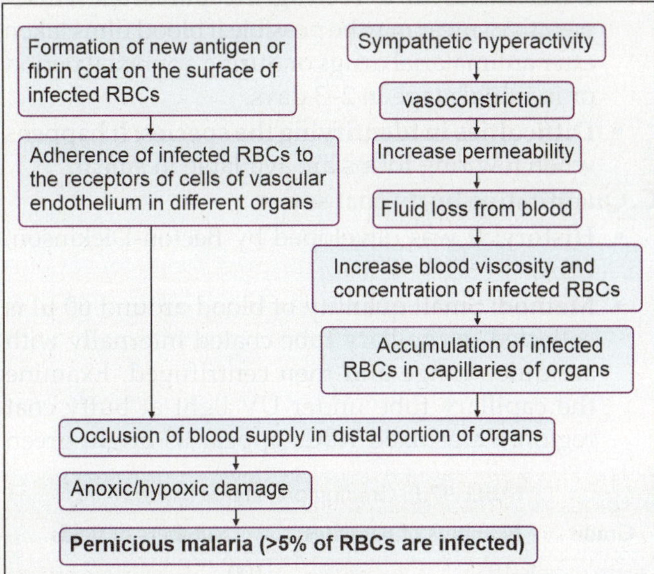

Flowchart 97.5: Pathogenesis of pernicious malaria

Laboratory Diagnosis of Malaria

Specimen and time of collection: Blood is an ideal sample. In *P. falciparum*, parasites present in blood (highest parasite density about 250000–300000 per ml than other species) during pyrexial period, so it is good to collect the blood at the height of fever in case of *P. falciparum*, while in case of *P. ovale, P. vivax* and *P. malariae* parasites can be demonstrated during febrile and afebrile period. It should be collected before taking antimalarial drugs. Blood is collected from ear lobe and finger prick in adults and older children while in infants it is collected from great toe.

Testing methods

A. Blood picture: Described above → pathological changes in different system.

> **Note: Black water fever**
>
> **Definition:** It is a manifestation of *P. falciparum* infection occurs in previously infected patients and characterized by sudden intravascular hemolysis (hemolytic crisis) followed by fever and hemoglobinuria.
>
> **Precipitating factors: (1) Inadequate quinine therapy:** Occurs in *P. falciparum* infection in non-immune (non-indigenous) individual who resided in malarious countries for 6 months to 1 year and have inadequate prophylactic plus therapeutic dose of quinine for repeated clinical attack. **(2) Others:** Cold, sun exposure, trauma, pregnancy, fatigue, parturition, X-ray treatment of spleen, G6PD deficiency, etc.
>
> **Pathogenesis:** Follow **Flowchart 97.6.**
>
> **Pathological changes in different systems**
> - **Kidneys: (1) Macroscopic:** It becomes large and dark in color due to congestion and pigmentation. **(2) Microscopic:** Degenerative changes in distal convoluted tubules, which are blocked with eosinophilic granules (hemoglobin cast). Parasitized RBCs may or may not be detected in renal capillaries.
> - **Liver: (1) Macroscopic:** It becomes large, yellow in color due to hemosiderin and with soft consistency. **(2) Microscopic:** Necrotic changes seen in central zone.
> - **Gallbladder:** Filled with dark green viscid bile.
> - **Spleen:** Larger in size and dark in color due to pigmentation.
> - **Blood picture**
>   - Parasites: Parasites do not present in blood during hemolytic crisis, but reappear within a week.
>   - Hb concentration: <10 g/dl.
>   - RBCs counts: 1–2 million/mm^3. Normocytic anemia. Polychromasia and basophilic stippling of RBCs.
>   - WBCs counts: Moderate degree of neutrophilia.
>   - Pigments: It includes methemalbumin, bilirubin (indirect van den Bergh positive) and absence of methemoglobin.
>   - Biochemical reactions: Increased blood urea, decreased blood cholesterol and reduced plasma hepatoglobin.
> - **Urine picture**
>   - Pigments: It includes hemoglobin (red color of urine), methemoglobin (dark brown/black color of urine), urobilin with positive testing and absence of methemalbumin.
>   - Biochemical reactions: Red to brown in color, acidic pH, amorphous deposit at bottom, when allow to settle, presence of hemoglobin cast and hemetin crystals.
>
> **Clinical features:** Fever with rigor, aching pain in the loin region, bilious vomiting (green color), black urine, icterus, circulatory collapse, acute renal failure (uremia) and acute liver failure.

Infections of Blood Inhabiting Sporozoa

(Contd...)

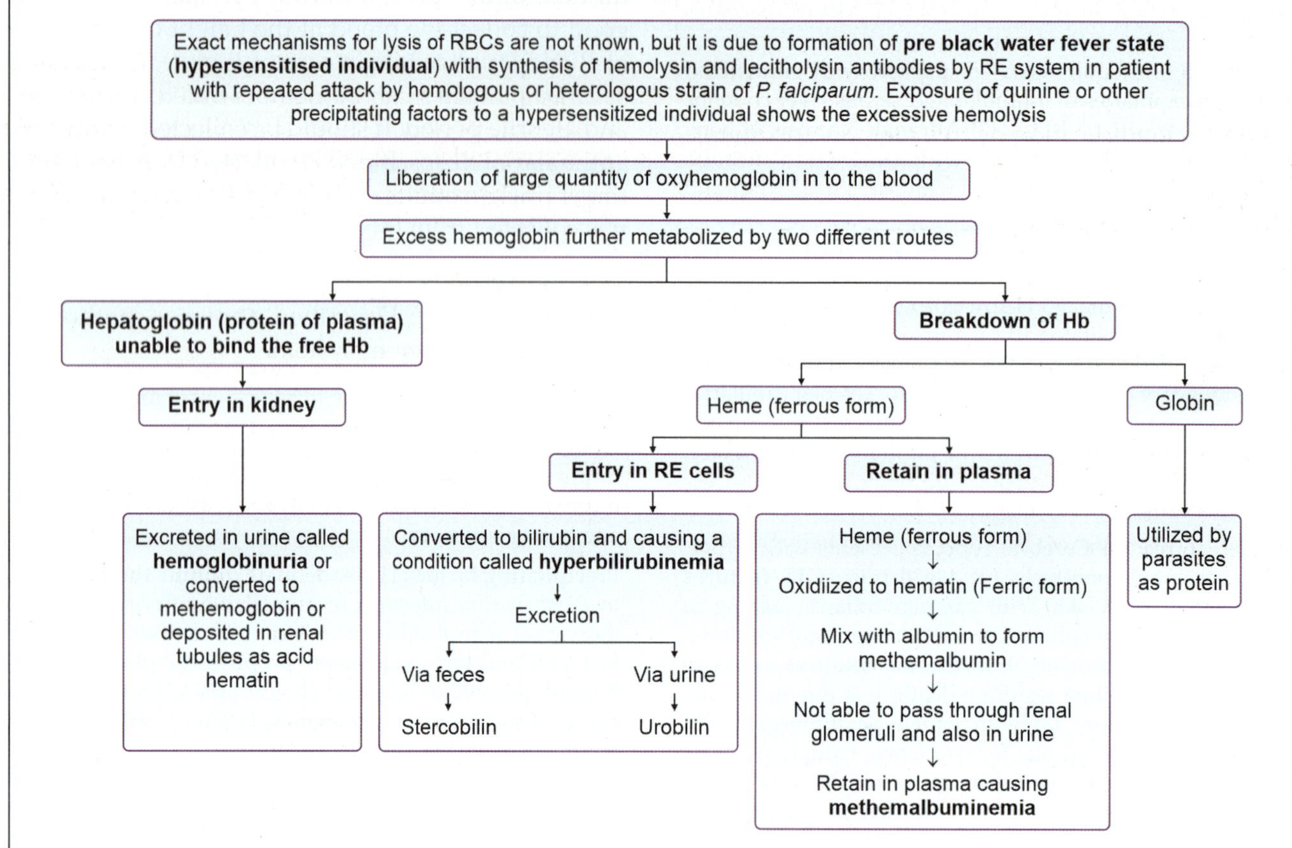

(Contd...)

Sequelae: Anemia and pigmented calculi.

Laboratory diagnosis: Blood picture and urine picture are described above.

Prevention: Avoid quinine and use other antimalarial drugs. Leave the endemic area and never return back.

Treatment: Start the chloroquine and in resistant area like South East Asia other drug is advisable.

B. Microscopy: It is the gold standard.

- **Types of smears:** Three types of smears like thin smear, thick smear and combined smear are described separately in boxes. Direct wet mount is also useful.
- **Rules to be adopted:** Follow **Flowchart 97.7**.

Flowchart 97.7: Rules for smear examination

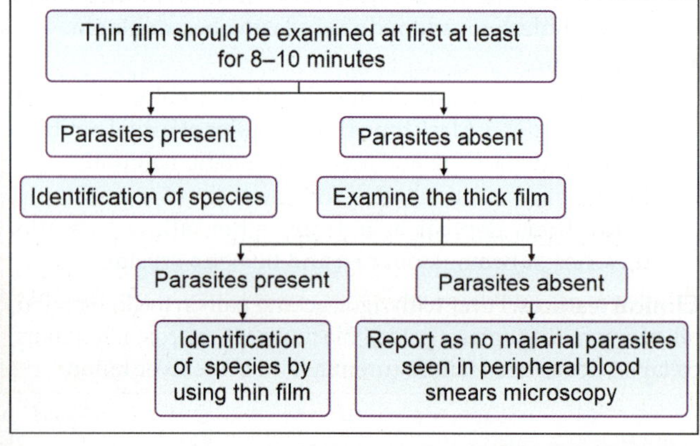

- **Grading of smear:** Grade the smear by counting the average numbers of parasites observed per thick film field (100X) as shown in **Table 97.6**.
- **Concentration technique:** It increases the chance of positivity rate. It changes the morphology of parasites. Blood is centrifuged at high speed after collection. Separate the sediment, prepare the smear and examine for parasites.
- **Difficulties in detecting the parasites:** False-negative smear may be possible if blood films taken after antimalarial drugs or during apyrexial period or in initial stage in 2–3 days.
- **Difficulties in identifying the species:** It happens when few ring forms are available in smear.

C. Quantitative buffy coat smear

- **History:** It was developed by Becton-Dickinson, USA.
- **Method:** Small quantity of blood around 60 µl is collected in capillary tube coated internally with acridine orange and then centrifuged. Examine the capillary tube under UV light at buffy coat region. Parasitized RBCs appear as bright green

TABLE 97.6: Grading of malarial smear		
Grade	**Numbers of parasites**	**Numbers of fields**
+	1–10	100
++	11–100	100
+++	1–10	1
++++	>10	1

dots. WBCs also take up the stain. Normal RBCs are non-nucleated and remain unstained.

- **Advantages:** It is more sensitive, rapid and quantitative method.
- **Disadvantage:** It is less specific, expensive and species identification by this method is difficult.

Note: Thin smear

Smear preparation (Fig. 97.15a): Prick the pulp of finger or ear lobe with sterile needle or lancet. Take a drop of blood on slide at a distance about half an inch from the right side. Hold the spreader (slide or cover slip) at 45° angle in contact with drop of blood and lower it to 30° angle. Push the spreader to the left till the blood is exhausted, which makes the tail of tongue-shaped smear in center of slide. Allow drying and labeled with pencil.

Properties of ideal smear: Even and uniform surface, tail end near about center of the slide, consists single layer of RBCs and margin of the smear do not extend to the sides of the slides.

Fixation: It is not required for Leishman's stain and Field's stain, because already alcohol present in stain, which acts as fixative, but required for Giemsa's and JSB stain. For Giemsa stain fix the smear with methyl alcohol or ethyl alcohol for 2–3 minutes and then dry it. For JSB stain fix the smear with methyl alcohol for 1–2 seconds and then dry it.

Staining: Stain the smear with one of the Romanowsky stain like Leishman's stain, Field's stain, Giemsa's, JSB (Jaswant Singh and Bhattcharji) stain or Wright's stain.

Advantages: It is useful to identify and to differentiate the species, because of clear morphology.

Disadvantages: Less sensitive compare to thick smear, because of few parasites.

Remarks on the smear examination: Few precautions are required before any slide declared as negative for malarial parasites by thin film. (1) Area to be examined is upper and lower margins of "tail end", because parasites are maximum in numbers at there. (2) Minimum 100 fields should be examined at least for 8–10 minutes. It is accepted that in nonimmune person, at least one parasite is present in 100 fields.

Appearance of parasites (Fig. 97.8)
- Cytoplasm: Blue in color.
- Nucleus: Red in color.
- Central unstained portion called vacuole.
- Pigment granules: Different in color.
- Portion of the cytoplasm of RBCs, unoccupied by the parasites gives dotted or stippled appearance as pink dots called **Schuffner's dots** in *P. vivax*, **Maurer's dots** in *P. falciparum*, **Ziemann's dots** in *P. malariae* and **James's dots** in *P. ovale*.
- Round gametocytes occur in all species except crescent/sickle shape in *P. falciparum*.
- RBCs are almost double in size in *P. vivax*.
- Multiple invasions of RBCs by ring forms or presence of ring forms on surface of RBCs called appliqué forms or accole forms in *P. falciparum*.

Note: Thick smear

Smear preparation (Fig. 97.15b): Take a one large drop of blood on slide. Spread it by using needle or corner of other slide to make an area of a half-inch. Labeled with pencil. Allow the smear to dry in horizontal position (at room temperature it takes 30 minutes for drying, which is accelerated by putting the slide in the incubator).

Properties of ideal smear: It allows the newsprint to be read or wrist watch to be seen through the dry smear. It contains 15–40 WBCs/oil immersion field.

Dehemoglobinization of smear: It means removal of coloring matter (Hb) from RBCs. Due to this thick smear consist of a thick layer of dehemoglobinized (lysed) RBCs which provide better opportunity to detect parasitic forms against a more transparent (colorless) background. It is not required for Field's stain. It is performed for Leishman's stain and Giemsa's stain by flooding the smear with the mixture of glacial acetic acid and tartaric acid. Drain off the fluid, as soon as the dehemoglobinization is over, which is indicated by the grayish white color of the film.

Fixation: Fix the smear by methyl alcohol for 3–5 minutes. Finally wash the slide by putting it in the vertical position in neutral or slightly alkaline distilled water for 5–10 minutes to remove the minor element of acid. Take out the slide, when it becomes the white, allow drying followed by staining.

Staining: Stain the smear with Leishman's stain, Field's stain, Giemsa's, JSB stain or Wright's stain.

Advantages: (1) It is more sensitive than thin smear, because of more parasites. One field is equal to 50 fields of thin film, hence it is a rapid method and used for mass survey. (2) It guides for treatment, to know the effect of drugs.

Disadvantages: (1) Species identification can be difficult in thick smear due to unclear morphology, unfamiliar setting of parasites, absence/destruction of RBCs or absence of the outlines of host-RBCs. (2) RBCs are destroyed (or remain unstained) and only elements seen in the smear are stained parasites and WBCs. So, relationship of RBCs with parasites or any morphological changes in RBCs are not observed.

Remarks on the smear examination: Thick film is neither a method of choice nor a substitute of thin film. It is a supplementary or actually a concentration method for detection of parasites, as it contains the large numbers of parasites in given field than thin film.

Appearance of parasites: Because of dehemoglobinization and lack of fixation the morphology of the parasite is altered.
- Trophozoites (early ring forms): Broken rings and detached nuclei are seen. Variable morphological pattern and certain terms are employed by Field and Fleming in 1938. (1) Comma shape: Curved cytoplasm with red nucleus. (2) Swallow or Gull's wing: Blue cytoplasm on either side of red nucleus like two wings. (3) Exclamatory mark (!): Blue cytoplasmic line with red nucleus at below.
- Schizonts and gametocytes: They retain the normal appearance; however, they appear smaller with irregular outlines.
- Pigments: Seen more clearly.
- RBCs: Absence of stippling appearance or all dots. Occasionally in *P. vivax*, outlines of enlarged RBCs (Ghost's cells) with Schuffner's dots are observed.
- WBCs: Tattered cytoplasm with deep purple nuclei.
- Platelets: Stain purple having woolly texture and outline.

Note: Combined smear

Smear preparation: Two drops of blood are taken on slide, one half an inch and second inch from the right end of slide. Draw the line with pencil inbetween. Make a thick smear from former and thin from later as shown in **Fig. 97.16**.

Staining: It is stained by Leishman's stain or Giemsa's stain or JSB stain.

Advantages: It is useful for surveillance work.

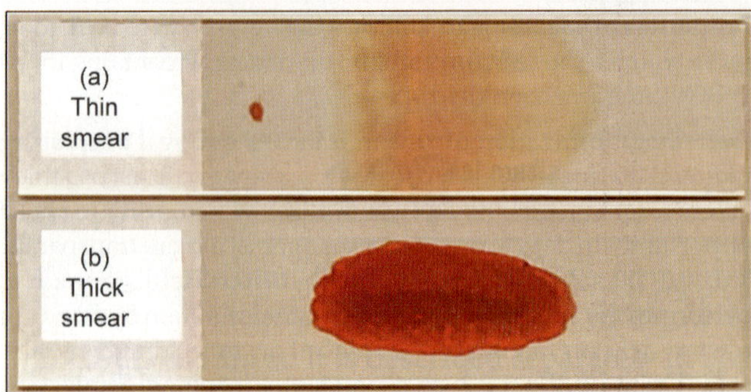

Fig. 97.15: Comparison of thin and thick smear

Fig. 97.16: Combined smear

D. Culture: Described above.

E. Serological tests

- ICT is performed to diagnose two different types of antigens. Specific Ags to *P. falciparum* like HRP-2 (Histidine Rich Protein-2) Ag and pfLDH (*P. falciparum* lactate de-hydrogenase) Ag and common *Plasmodium* spp., Ags like pan LDH and aldolase.
- Device is impregnated at three sites such as test site-1 with Abs to specific *P. falciparum* Ags (like HRP-2 and pfLDH), test site-2 with Abs to common *Plasmodium* spp., Ags (pan LDH and aldolase) and control site.
- Band at site-1 with control band: Indicates *P. falciparum.*
- Band at site-2 with control band: Indicates *Plasmodium* spp., other than *P. falciparum.*
- Band at site-1 and 2 with control band: Indicates mixed infections.
- Band at control site only: Indicates negative result.
- Band at site-1/2 without control band: Indicates invalid result.

F. Molecular methods

- PCR is useful to detect the drug resistant genes.
- qPCR is useful for quantification.
- Nested multiplex PCR is useful to diagnose the particular species.

Prevention

General measures: Like mosquito control measures are described in **Ch. 110.**

Chemoprophylaxis: It is given to travelers to endemic areas. For <6 weeks (short-term) doxycycline is prescribed. For >6 weeks (long-term) mefloquine is prescribed.

Immunoprophylaxis

- Pre-erythrocytic stage vaccine contains pf-CSP (*P. falciparum*-circum sporozoite protein) fused with HBsAg and chemical adjuvant like AS01 called RTSS/AS01 vaccine is approved by WHO, which is used in children of regions with moderate to high degree transmission like sub-Saharan Africa.
- Many vaccines are still under trial. **(1) Anti-sporozoite vaccine:** Prepared by using CSP, TRAP (thromboplastin-related anonymous protein), LSA-1/2 (Liver Stage Antigen-1/2) or SALSA (sporozoite and liver stage antigen). **(2) Anti-asexual blood stage vaccine:** Prepared by using the surface antigens of erythrocytic stage (trophozoites, schizonts and merozoites) like MSP-1/3 (merozoite surface protein-1/3), GLURP (glutamate rich protein), RAP 1/2 (rhoptry antigen), AMA-1 (apical membrane antigen-1), EBA-175 (erythrocyte-binding antigen-175), PfHRA-2 (*P. falciparum*-histidine rich protein-2), SERA (serine-rich antigen), pf 126/140/332 or pf EMP (*P. falciparum* erythrocyte membrane protein). **(3) Antigametocytes vaccine:** Prepared by using the pfs25 antigens. **(4) Multi-stage targeting vaccine:** Prepared by combining the 25 different antigens from *P. falciparum*.

Treatment

List of antimalarial drugs

- Quinoline group: Chloroquine, primaquine, quinine and mefloquine.
- Artemisinin derivatives: Artesunate, artemether and arteether.
- Hydroxynaphthoquinones: Atovaquone.
- Biguanide derivative: Proguanil.
- Diaminopyrimidines: Pyrimethamine.
- Sulfonamides: Sulfadiazine and sulfadoxine.
- Tetracycline: Tetracycline and doxycycline.

Regimen: It is given by NVBDCP, India as described below.

- Vivax malaria: Chloroquine plus primaquine.
- Falciparum malaria: (1) In North East states: Artemether–lumefantrene (for 3 days) and primaquine (single dose on 2nd day). (2) In other states: Artesunate (for 3 days) + pyrimethamine and sulfadoxine (single dose on 1st day) + primaquine (single dose on 2nd day).
- Severe malaria: IV use of quinine or artemisinin derivatives.

Drug resistance in *P. falciparum*: It occurs due to longer half-life of drugs, mutation in parasite gene encodes for chloroquine transport, irregular use of drugs and host immunity.

- Chloroquine resistance developed in India, maximum from North East states.
- Highest chloroquine and other drug resistance developed from Thailand, Cambodia and Myanmar.

World Malaria Day

25th April of every year is celebrated as world malaria day with an aim to control it.

Babesia spp.

Described separately in box.

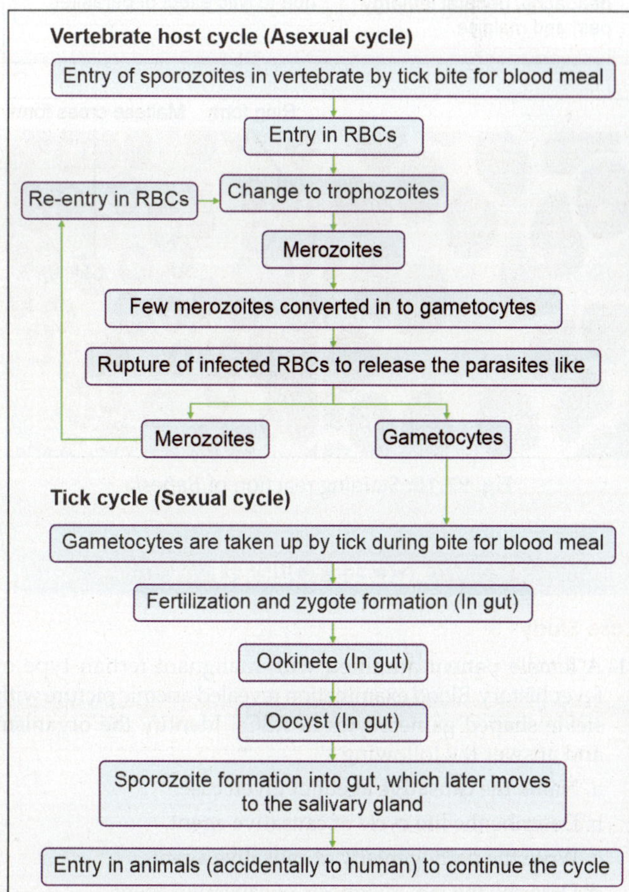

Flowchart 97.8: Life cycle of *Babesia*

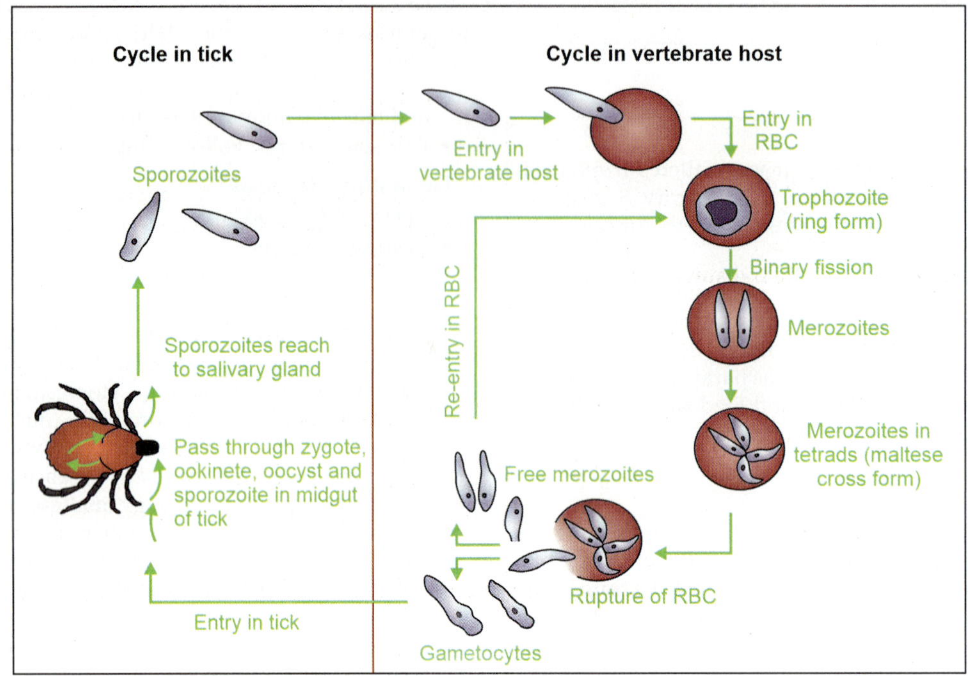

Fig. 97.17: Life cycle of *Babesia*

Flowchart 97.9: Clinical features of babesiosis

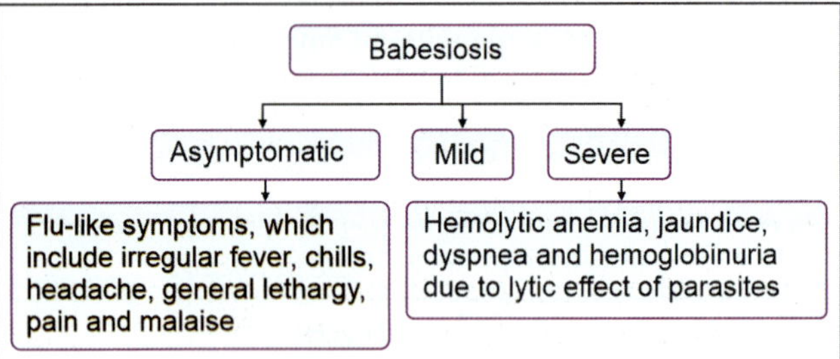

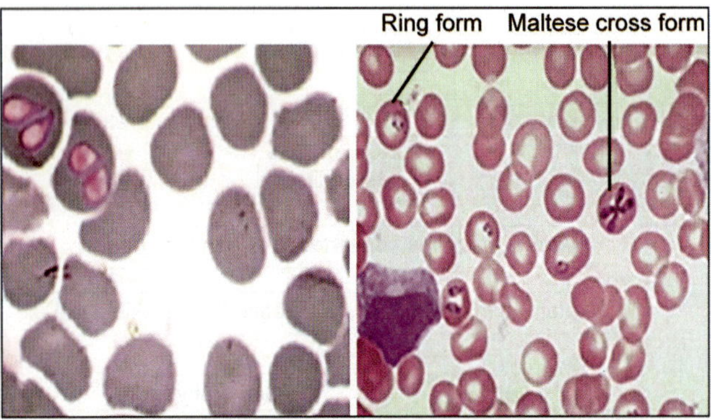

Fig. 97.18: Staining reaction of *Babesia*

ACCESS YOURSELF

Case Study

1. A female patient admitted with malignant tertian type of fever history. Blood examination revealed anemic picture with sickle shaped gametocytes in RBCs. Identify the organism and answer the following.
   a. Name the causative agent of given case.
   b. Describe the life cycle of causative agent.
   c. Write the pathogenicity of causative agent.
   d. Describe the lab., diagnosis of causative agent.

Essay/Full Question

1. Describe the pathogenicity of malaria.

Short Notes

1. Morphology/life cycle/pathogenicity/laboratory diagnosis of *P. vivax* or *P. falciparum*.
2. Immunity in malaria/black water fever/pernicious malaria.
3. Differences between *P. vivax* and *P. falciparum*.

Short Questions for Theory/Viva Questions

1. What are alteration of generation and alteration of host in sporozoa?
2. Name the end products of sporogony and schizogony in malarial parasites.
3. What is hypnozoite? Write its clinical significance.
4. What is "therapeutic malaria"?
5. Name the fever for following parasites.
   P. vivax, P. falciparum P. malariae and *P. ovale.*
6. Write four differences between trophozoite and sporozoite-induced malaria.
7. What is "maltese cross form"?
8. Name two blood inhabiting sporozoa.
9. Name the plasmodium species causing following type fever:
   • Benign tertian fever
   • Ovale tertian fever
   • Malignant tertian fever
   • Benign quartan fever
10. Define: Black water fever and red water fever.

Comments on

1. *P. falciparum* is not showing relapse.
2. In *P. falciparum* infection, peripheral blood smear shows only gametocytes and ring forms **or** In *P. falciparum*, late trophozoite stage and schizonts not identified in peripheral blood smear.
3. Anemia and splenomegaly are common in patient of malaria.

MCQs for Chapter Review

Plasmodium spp.

1. **Causative agent of malaria:**
   a. Protozoa
   b. Mosquito
   c. Bacteria
   d. Virus
2. **Malarial parasite was discovered by:**
   a. Ronald Ross
   b. Paul Muller
   c. Laveran
   d. Pampana
3. **The scientists who discovered the transmission of malaria by *Anopheles* mosquito:**
   a. Laveran
   b. Paul Muller
   c. Ronald Ross
   d. Pampana
4. **In malaria sexual cycle is:**
   a. Sprozoites to gametocytes
   b. Gametocytes to sporozoites
   c. Occurs in human
   d. Responsible for relapse
5. **In malaria, pre-erythrocytic schizogony occurs in:**
   a. Lungs
   b. Liver
   c. Spleen
   d. Kidney
6. **In transmission of malaria, mosquito bite transfers the infective form of malarial parasite for vertebrate host is? or which form of malarial parasite is present in saliva of infective mosquito?**
   a. Sporozoites
   b. Merozoites
   c. Hypnozoites
   d. Gametocytes
7. **Which is the infective stage for mosquito in case of *Plasmodium vivax*?**
   a. Gametocytes
   b. Sprozoites
   c. Zygotes
   d. Merozoites
8. **In which type of malarial parasite, exo-erythrocytic stage is absent?**
   a. *P. ovale*
   b. *P. vivax*
   c. *P. falciparum*
   d. *P. malariae*
9. **Malaria carrier contains:**
   a. Sporozoites
   b. Gametocytes
   c. Merozoites
   d. Trophozoites
10. **In *Plasmodium falciparum* the number of cycle the parasite undergoes in the liver is:**
    a. 0
    b. 1
    c. 2
    d. 3
    e. 5
11. **_Plasmodium vivax_ attacks:**
    a. Reticulocytes
    b. Young RBCs
    c. Old RBCs
    d. Dead RBCs
12. **True about *P falciparum* includes all, *except*:**
    a. Duration of erythrocytic cycle is 48 hours.
    b. Exo-erythrocytic phase is absent.
    c. Parasitic burden can be estimated by peripheral parasitemia.
    d. Causes rosette formation
13. **_Plasmodium falciparum_ infection of human is characterized by:**
    a. The erythrocytes are increased in size
    b. All stages of erythrocytic schizogony are seen in peripheral blood
    c. Multiple infections of erythrocytes are seen
    d. Each erythrocytic cycle lasts for 72 hours
14. **Stages seen in peripheral smear of *falciparum* malaria:**
    a. Schizonts
    b. Gametocytes
    c. Accole
    d. Ring form
    e. Trophozoite
15. **Stages of *P. falciparum* not seen in PBS is:**
    a. Schizonts
    b. Gametocytes
    c. Ring form
    d. Double ring
16. **Not seen in the peripheral smear in *Plasmodium falciparum* infection:**
    a. Accole
    b. Maurer's dots
    c. Schuffner's dots
    d. Schizonts
17. **Schizonts of *P. falciparum* are not seen in peripheral blood smear because:**
    a. Killed by antibodies
    b. Absent in life cycle
    c. Present in capillaries of internal organs
    d. None of above
18. **A patient diagnosed to have malaria, smear shows all stages of schizonts, 14–20 merozoites, yellowish-brown pigment. The type of malaria is:**
    a. *P. falciparum*
    b. *P. malariae*
    c. *P. vivax*
    d. *P. ovale*
19. **RBCs are enlarged in infection with:**
    a. *P. vivax*
    b. *P. malariae*
    c. *P. ovale*
    d. *P. falciparum*
20. **Band-shaped trophozoites are seen in:**
    a. *P. vivax*
    b. *P. malariae*
    c. *P. ovale*
    d. *P. falciparum*
21. **Multiple infections of RBCs are seen in:**
    a. *P. vivax*
    b. *P. malariae*
    c. *P. ovale*
    d. *P. falciparum*
22. **Infection with *Plasmodium falciparum* is suspected, if the infected RBCs are:**
    a. Normal in size
    b. Showing multiple infections
    c. a + b
    d. Showing motile form inside the RBCs
23. **Which of the following is true about malaria?**
    a. Size of RBC is enlarged in *vivax* infection
    b. Size of RBC is enlarged in *flaciparum* infection
    c. Schuffner's dots are seen in *malariae*
    d. Relapse is seen in *flaciparum* infection
24. **Most virulent *Plasmodium* species causing malaria is:**
    a. *P. ovale*
    b. *P. vivax*
    c. *P. falciparum*
    d. *P. malariae*
25. **The most common cause of malaria in India:**
    a. *P. ovale*
    b. *P. vivax*
    c. *P. falciparum*
    d. *P. malariae*
26. **In malaria reservoir, parasite remains as:**
    a. Merozoite
    b. Sporozoite
    c. Trophozoite
    d. None
27. **Most common *Anopheles* vector for transmission of malaria in urban area is:**
    a. *A. stephensi*
    b. *A. gambiae*
    c. Both
    d. None
28. **Malarial pigment is formed by:**
    a. Parasite
    b. Bilirubin
    c. Hemoglobin
    d. Any of the above
29. **Chronic complication of malaria:**
    a. Splenomegaly
    b. Nephrotic syndrome
    c. Pneumonia
    d. Hodgkin's disease
30. **Splenic rupture is most common in infection with:**
    a. *P. vivax*
    b. *P. malariae*
    c. *P. ovale*
    d. *P. falciparum*

31. **Complications in malaria are commonly with:**
    a. *P. vivax*   b. *P. malariae*
    c. *P. ovale*   d. *P. falciparum*

32. **Pernicious malaria is a complication seen in infection with:**
    a. *P. vivax*   b. *P. malariae*
    c. *P. ovale*   d. *P. falciparum*

33. **Cerebral malaria is caused by *Plasmodium*:**
    a. *falciparum*   b. *ovale*
    c. *malariae*   d. *vivax*

34. **Which of the following is true about *P. falciparum* malaria?**
    a. James dots are seen   b. Accole forms are seen
    c. Relapses are frequent   d. Longest incubation period

35. **Shortest incubation period in malaria:**
    a. *P. vivax*   b. *P. falciparum*
    c. *P. malariae*   d. *P. ovale*

36. **Parasitemia is highest in:**
    a. *P. vivax*   b. *P. falciparum*
    c. *P. malariae*   d. *P. ovale*

37. **Which of the following is detected by the antibody detection test used for the diagnosis of *P. falciparum* malaria?**
    a. Circum sporozoite protein
    b. Merozoite surface antigen
    c. Histidine Rich Protein-I (HRP-I)
    d. Histidine Rich Protein-II

38. **Thin blood smear of malaria is used to identify:**
    a. Schizont   b. Gametocyte
    c. *Plasmodium* parasite   d. Type of parasite

39. **JSB stain is used for which parasite?**
    a. Malaria   b. Filaria
    c. Kala azar   d. Sleeping sickness

40. **What is the treatment of choice for benign tertian malaria?**
    a. Sulphamethaxozole – pyrimethamine
    b. Quinine
    c. Mefloquine
    d. Choroquine

41. **Which is true of malaria?**
    a. Rods forms are seen in *P. malariae*
    b. RBCs size is more in *P. vivax*
    c. Relapse seen in *P. falciparum*
    d. Male and female mosquito transmit the disease

Babesia spp.

42. **True about babesiosis:**
    a. Caused by *B. microti*   b. Resides in RBCs
    c. Resides in WBCs   d. Chloroquine is the drug choice
    e. Filarial parasite

43. **Maltese cross seen on polarizing microscope in:**
    a. *Cryptococcus neoformans*
    b. *Penicillium marnefeii*
    c. *Blastomyces*
    d. *Candida albicans*
    e. *Babesia microti*

Answers and Explanation of MCQs

1. a
- Malaria is caused by *Plasmodium*, which is a protozoan and transmitted by mosquito.

2. c

3. c
- Follow section, ***Plasmodium* spp., (history)** for explanation of answers of MCQs 2–3.

4. b

5. b

6. a

7. a

8. c

9. b

10. b
- Follow section, ***Plasmodium* spp.,** (life cycle, Flowchart 97.1 and Flowchart 97.3) for explanation of answers of MCQs 4–10.

11. a, b

12. c

13. c

14. b, c, d

15. a

16. c, d

17. c

18. c

19. a, c

20. b

21. d

22. c
- Follow section, ***Plasmodium* spp.,** (morphology, Table 97.2 and note) for explanation of answers of MCQs 11–22.

23. a
- Option a, b and c: Follow section, ***Plasmodium* spp.,** (morphology and Table 97.2) for explanation.
- Option d: Follow section, ***Plasmodium* spp.,** (life cycle) for explanation.

24. c

25. c

26. d

27. a

28. c

29. a, b

30. d

31. d

32. d

33. a
- Follow section, ***Plasmodium* spp., (pathogenicity)** for explanation of answers of MCQs 24–33.

34. b

35. b
- *P. falciparum* contains Maurer's dots, has no exo-erythrocytic phase, hence no relapse and has shortest incubation period.

36. b
- Follow section, ***Plasmodium* spp., (laboratory diagnosis → Specimen and time of collection)** for explanation.

37. d

38. d

39. a
- Follow section, ***Plasmodium* spp., (laboratory diagnosis)** for explanation of answers of MCQs 37–39.

40. (d)
- Follow section, ***Plasmodium* spp., (treatment)** for explanation.

41. a, b
- Band forms are seen in *P. malariae*, RBC size is more in *P. vivax* and *P. ovale*, relapse not seen in *P. falciparum* and only female mosquito transmits the disease.

42. a, b

43. a, e
- Follow section, ***Babesia* spp.,** for explanation of answers of MCQs 42–43.

Infections of Gastrointestinal Sporozoa

INTRODUCTION

Systemic classification: Follow Ch. 14 (Table 14.3).

Pathogenic classification: Two groups like blood inhabiting sporozoa and GIT sporozoa. Blood inhabiting sporozoa are mentioned in **Ch. 97.** Following are pathogenic species of gastrointestinal sporozoa.

- *Cystoisospora*: Previously, it was called *Isospora*. Common species are *C. belli* (Wenyon-1923) and *C. natalensis* (rarely isolated).
- *Cryptosporidium:* So far, many species are identified, but *C. parvum* is the only human pathogen.
- *Toxoplasma: T. gondii*
- *Cyclospora: C. cayetanensis*
- *Sarcocystis:* **(1) Intestinal species:** *S. bovihominis* (in cattle also called *S. hominis*) and *S. suihominis* (in pig). **(2) Invasive or muscular species:** *S. lindemanni* in skeletal and cardiac muscles (in lymph nodes and vessels also).

Cystoisospora belli

Meaning

Word **belli** came from **bellum means war,** as it caused infection among troops in Middle East during 1st World War.

Morphology

Morphological stage called oocyst, which passes through the followings stages.

Immature (non-sporulated) oocyst (Fig. 98.2a): Zygote changes to oocyst, which contains single sporoblast called immature oocyst and passes through feces over soil. Sporoblast undergoes division in external environment to produce the two sporoblasts.

Mature (sporulated) oocyst (Fig. 98.2b): Immature oocyst undergoes further development in external environment. Each sporoblast undergoes division to produce four sporozoites, so finally 8-sporozoites are produced. Sporozoites are arranged in two groups, each contains four sporozoites in a specialized structure called **sporocyst (Greek → Sporos means seed)** within oocyst and such oocyst called mature oocyst. Mature oocyst is an infective form, elongated in shape, 20–30 μm × 10–20 μm in size and digested by digestive juices to release the eight sporozoites. Time required for maturation is 1–5 days.

Life Cycle

Hosts and methods of reproduction: Both asexual (schizogony) and sexual (gametogony followed by syngamy) stages occur in human.

Cycles: Two types **(Flowchart 98.1 and Fig. 98.1).**

Pathogenicity

Disease name: Called **cystoisosporiasis.**

Epidemiology: It is distributed in Central and South America, Africa and South-East Asia.

Reservoirs of infection: Human is the only reservoir and no any animal reservoirs are known.

Sources of infection: Food and water.

Modes of transmission: It is transmitted by **ingestion** of contaminated food and water contain oocyst or by oral-anal sex.

Incubation period: 1–4 days.

Exit form: Immature oocyst (contain 1 sporoblast).

Infective form: Mature oocyst.

Portal of entry: GIT.

Flowchart 98.1: Life cycle of *C. belli*

Cycle in animal/human host (Both sexual and asexual cycle)

Entry of oocysts in host by ingestion of contaminated food and water

Digestion of wall of oocysts (**excystation**) by digestive juices and release of 8- sporozoites

Re-entry in epithelial cells → Entry in epithelial cells

Change to trophozoites → Schizonts → Merozoites (**Schizogony**)

Rupture of epithelial cells to release the merozoites

Few merozoites re-enter in epithelial cells

Few merozoites converted into gametocytes (**Gametogony**)

Male gametocytes are smaller called **microgametocytes** and female gametocytes are larger called **macrogametocytes**

Male gametocyte penetrates the female gametocyte to form the structure called **zygote** (**Syngamy**)

Converted to an encysted structures called **oocysts** (**encystation**)

Each oocyst contains single sporoblast called **immature oocyst** and passes through feces over soil

Cycle outside the host / in external environment (Asexual cycle)

Oocysts contain single sporoblast which divides to form two sporoblasts

Each sporoblast divides to form four sausage shaped sporozoites. So, total 8-sporozoites and they are arranged in two groups called **sporocysts** within oocysts called **mature oocysts**

Oocysts enter in to other host via food and water to continue the life cycle

Note: Two modes of epithelial cells infection.
1. Fresh sporozoites from ruptured oocysts.
2. Infection by merozoites.

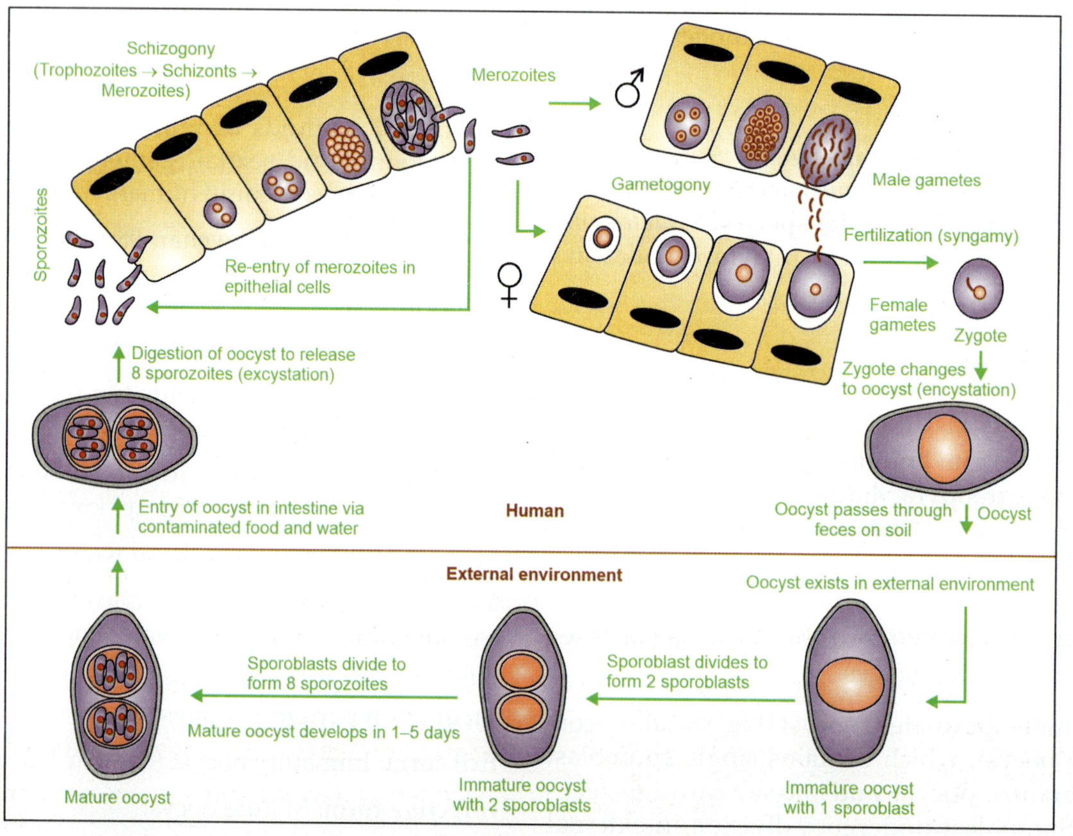

Fig. 98.1: Life cycle of *C. belli*

Site: Small intestine (lower part of ileum).

Precipitating factor: It is common in AIDS patient.

Clinical features: Follow **Flowchart 98.2**.

Laboratory Diagnosis

Specimens: (1) Stool: Preserve the stool in 2.5% potassium dichromate solution. It kills the bacteria and helps in clear identification of *Cystoisospora* after 24–48 hours. **(2) Muscle biopsy:** In case of extraintestinal infection. **(3) Others:** Like duodenal aspirate and mucosal biopsy.

Testing methods

A. **General picture: Blood** shows eosinophilia and **stool** shows the presence of high fat, fatty acid crystals and Charcot-Leyden (CL) crystals.

B. **Stool microscopy: (1) NS and iodine preparation:** In fresh stool, oocyst with nucleus can be identified. In old stool, unsporulated (immature) or sporulated (mature) oocyst can be identified as shown in **Fig. 98.2**. **(2) Other stains:** Modified ZN stain **(Fig 98.3)** and fluorescent microscopy (by using auramine dye) are useful.

Prevention and Treatment

It is prevented by improving the food hygiene. It causes self limited diarrhea. Treatment is required in AIDS patient with cotrimoxazole. Alternative drug is pytimethamine.

Flowchart 98.2: Clinical features of cytoisosporiasis

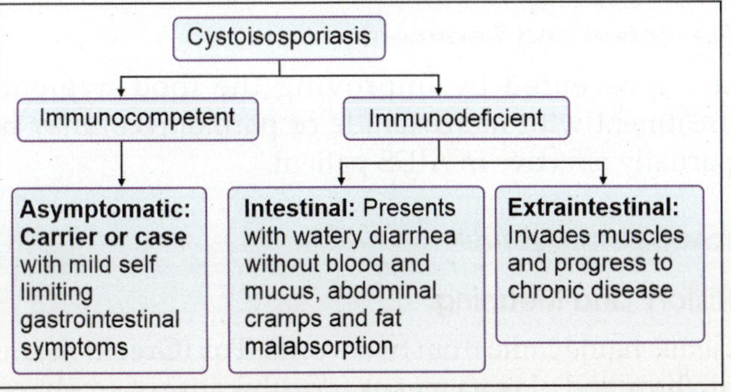

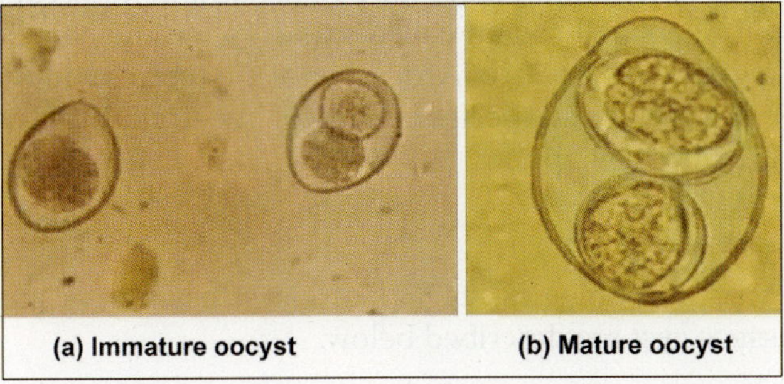

Fig. 98.2: Oocyst of *Cystoisospora* under NS and I preparation

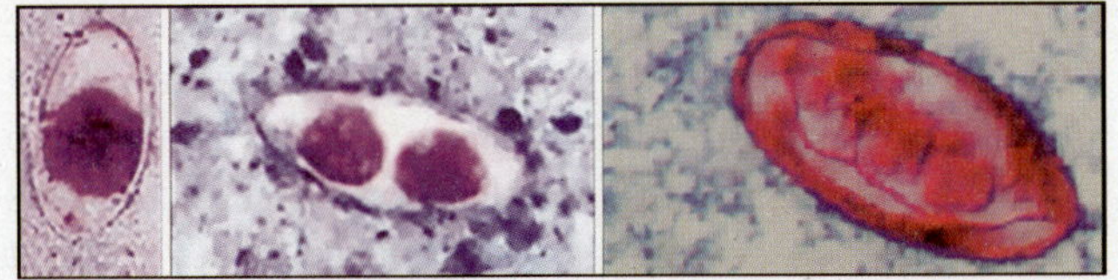

Fig. 98.3: Different types of oocysts of *Cystoisospora* under ZN stain

Cryptosporidium parvum

History

It was first reported by Tyzzer in 1907 in the gastric crypt of laboratory mouse. First human case was identified in children in 1976 causing self limited diarrhea.

Morphology

Morphological stage called oocyst has following two types.

Thin-walled oocyst (Fig. 98.4): It develops from zygote. It is not excreted into external environment and penetrates the epithelial cells to produce **autoinfection** (internal reinfection).

Thick-walled oocyst (Fig. 98.4): Initially, it is immature and excreted into external environment for further maturation. Oocyst undergoes further development in external environment and produces four sporozoites. Mature thick-walled oocyst is 4–5 μm in diameter.

Life Cycle

Hosts and methods of reproduction: Both asexual (schizogony) and sexual (gametogony followed by syngamy) stages occur in human.

Cycles: Two types **(Flowchart 98.3 and Fig. 98.4)**.

Pathogenicity

Disease name: Called **cryptosporidiosis.**

Epidemiology: It is an animal pathogen, distributed worldwide.

Reservoirs of infection: Humans and animals are the known reservoirs.

Sources of infection: Food and water.

Modes of transmission: It is transmitted by **ingestion** of contaminated food and water contain oocyst or by **oral-anal sex** or by **autoinfection (Flowchart 98.3)**.

Incubation period: 1–12 days.

Exit form: Thick-walled oocyst.

Infective form: Thin- and thick-walled oocyst.

Portal of entry: GIT.

Site: Small intestine (distal part of jejunum, ileum and colon).

Precipitating factors: Incidences are increased with AIDS. Animal handlers and homosexual persons are at more risk.

Clinical features: Follow **Flowchart 98.4**.

Laboratory Diagnosis

Specimens: Stool, intestinal, gallbladder or biliary tract biopsies or sputum (formalin fixed). Specimen is also collected by entero test **(follow Ch. 94)**.

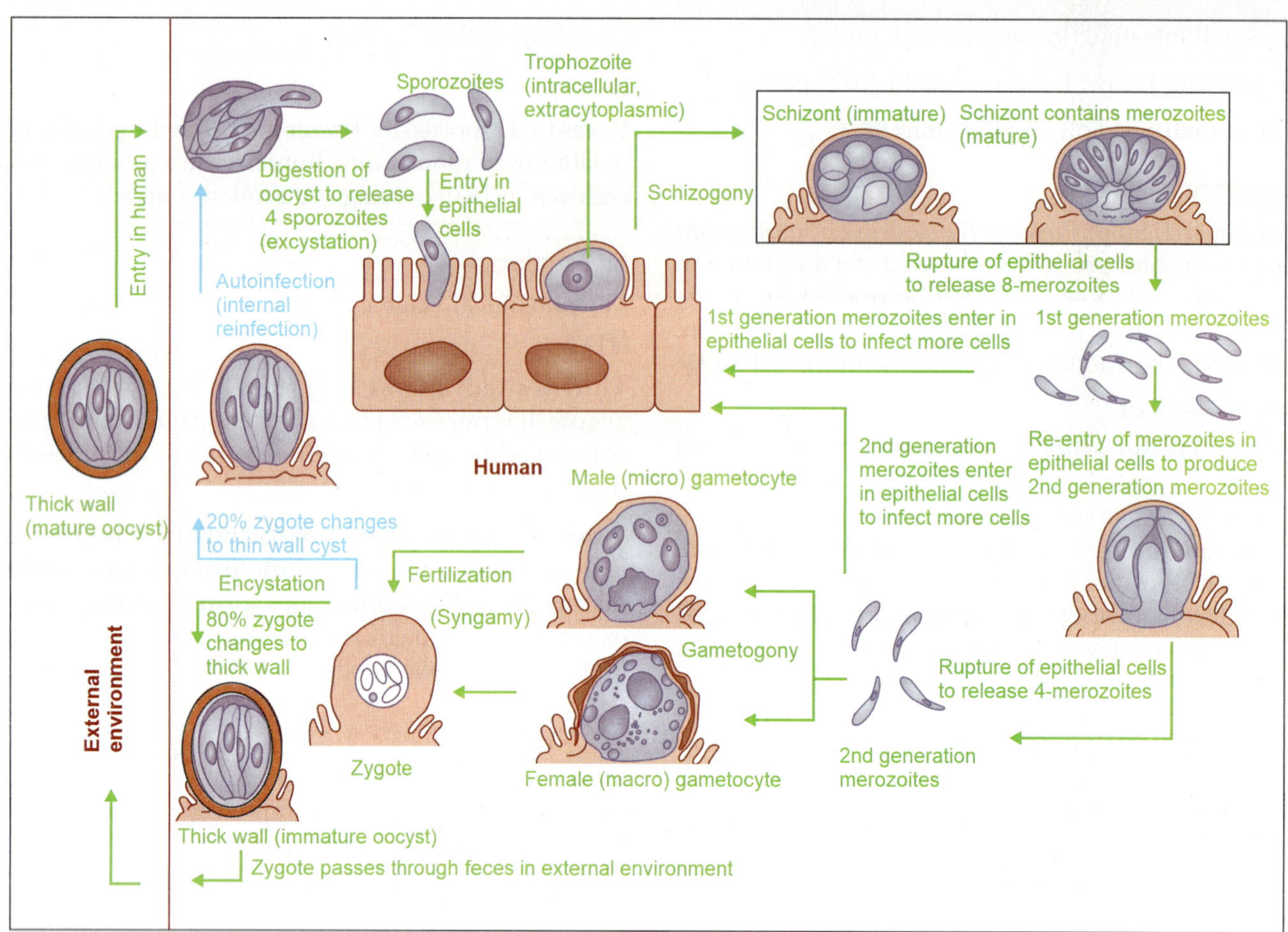

Fig. 98.4: Life cycle of *C. parvum*

Testing methods

A. Microscopy: (1) Wet mount and iodine preparation from fresh stool (Fig. 98.5): Oocysts with 4–5 µm in size can be identified. Numbers of oocysts are more in stool with water consistency. **(2) Stool concentration technique:** Sugar flotation (Sheather's) technique is useful. **(3) Modified ZN stain (Fig. 98.6):** Decolorizing agent is 3% HCL in 95% ethanol instead of 25% H_2SO_4. At least 5–6 smears should be examined before declaring the negative sample. **(4) Fluorescent microscopy:** Is also useful. **(5) Giemsa's stain:** For biopsy specimens. **(6) Diamidinophenyl indole (DAPI) stain:** It stains the nuclei of sporozoites in oocysts.

B. Serological tests

1. **Ag detection:** (1) Ag is detected from stool (coproantigen) by ELISA. (2) ICT diagnose the Ags of *Cryptosporidium*, *E. histolytica* and *Giardia*.

2. **Ab detection:** It is not useful in diagnosis, but for epidemiological purpose. Ab is detected against the oocyst Ag of *Cyclospora* and *Cryptosporidium parvum* by ELISA from serum.

C. Molecular method: qPCR is useful. **BioFire Film Array** is an automated multiplex PCR which targets multiple bacteria, viruses and parasites causing diarrhea like *E. histolytica*, *Giardia*, *Cyclospora* and *Cryptosporidium parvum*.

Prevention and Treatment

It is prevented by improving the food hygiene. Treatment with nitazoxanide or parmomycin may be partially effective in AIDS patient.

Toxoplasma gondii

History and Meaning

Genus name came from two words. **Tox (Greek) means arc/bow** and **plasma means form**, because of arc shape. It was first identified in 1908 by Nicolle and Manceaux from the small rodent called **gundi** (scientific name: *Ctenodactylus gundi*) of Africa, hence the species name is *gondii*. Only one species is known, but strain to strain variation may appear.

Morphology

Three morphological stages like oocyst, pseudocyst and tissues cyst are described below.

Oocyst

1. **Immature (nonsporulated) oocyst (Fig. 98.7):** It develops only in the definitive host of family **Felidae** (domestic/wild cat), but not in intermediate host (like human or animal or bird). Zygote changes to oocyst, which contains single sporoblast and passes through feces over soil. Freshly passed oocyst is not infective. Sporoblast undergoes division in external

Cycle in animal/human host (Both sexual and asexual cycle)

Entry of thick walled oocysts in host by ingestion of contaminated food and water

Digestion of wall of oocysts (**excystation**) by digestive juices and release of 4–sporozoites

Re-entry in epithelial cells → Entry in epithelial cells (Intracellular and extracytoplasmic in brush border/villi)

Change to trophozoites and each trophoozoite undergoes three successive nuclear divisions ⎫

Schizont ⎬ (**Schizogony**)

Finally 8-merozoites are produced by three successive nuclear divisions ⎭

Rupture of epithelial cells to release the 8-merozoites called **1st generation merozoites**

Few merozoites re-enter in epithelial cells to continue the schizogony

Few merozoites re-enter in epithelial cells and multiply by two successive nuclear divisions

Again 4 merozoites are produced by two successive nuclear divisions. Rupture of epithelial cells to release the 4-merozoites called **2nd generation merozoites**

Few merozoites re-enter in epithelial cells to continue the schizogony

Few merozoites re-enter in epithelial cells and converted into gametocytes (**gametogony**)

Male gametocytes (smaller called **microgametocytes**) penetrate the female gametocytes (larger called **macrogametocytes**) to form the structures called **zygotes (Syngamy)**

Converted into two types an encysted structures called **oocysts (encystation)** which are of two types

Thin-walled oocysts, not excreted in external environment

↓

Infect the epithelial cells, as like fresh infection called autoinfection

Thick-walled oocysts, excreted in external environment

Cycle outside the host/in external environment (Asexual cycle)

Immature oocystd undergo further development in external environment to produce mature oocysts (contains 4-sporozoites)

Mature oocysts enter in to other host via food and water to continue the life cycle

Note: Three modes of epithelial cells infection.
(1) Sporozoites from thick-walled oocysts: Fresh entry. (2) First generation merozoites.
(3) Second-generation merozoites. (4) Sporozoites from thin-walled oocysts: Autoinfection

Flowchart 98.4: Clinical features of cryptosporidiosis

Cryptosporidiosis

Immunocompetent | Immunodeficient

Asymptomatic: Carrier or case with self limiting diarrhea and gastrointestinal symptoms

Intestinal: Watery diarrhoea (17 liters/day), abdominal cramps, weight loss. It may be fatal. Traveller's diarrhoea or outbreak also reported

Extraintestinal: Respiratory illness

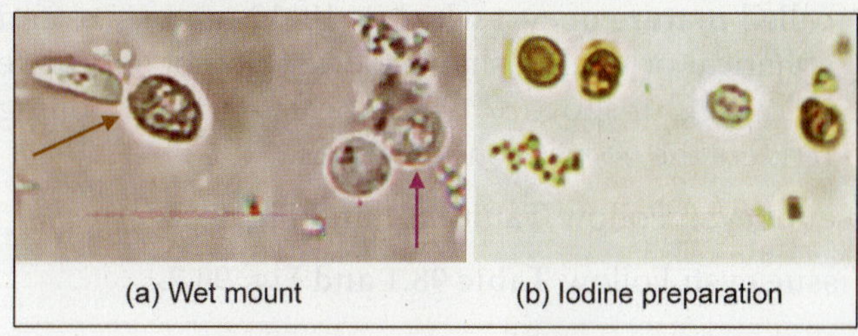

Fig. 98.5: Oocyst of *C. parvum* under wet mount and iodine preparation

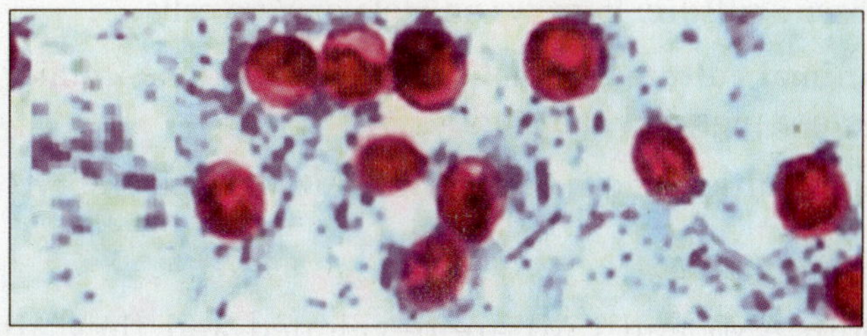

Fig. 98.6: Oocyst of *C. parvum* under ZN preparation

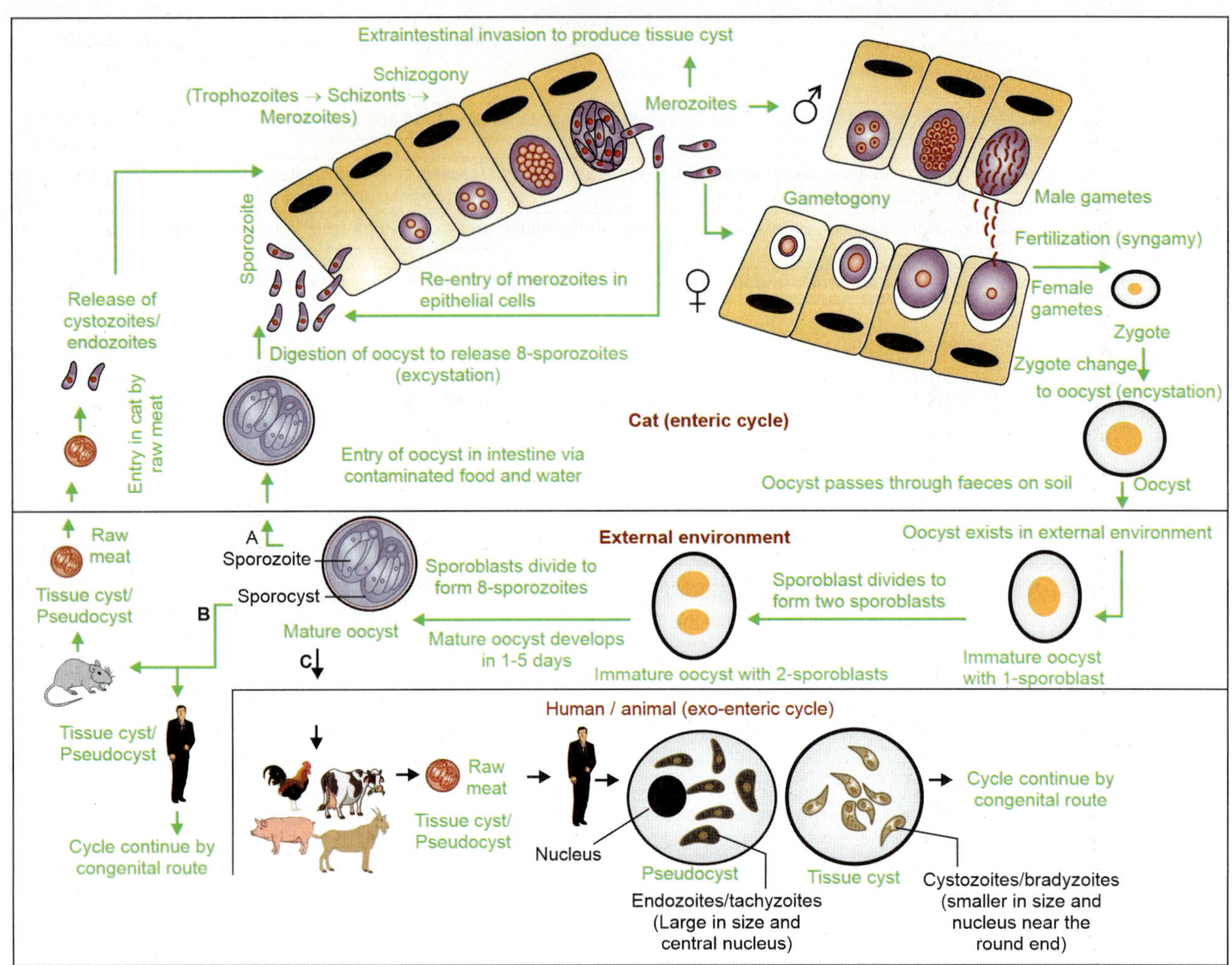

Fig. 98.7: Life cycle of *T. gondii*

A = Feline cycle, B = Feline–non feline cycle and C = Non feline–non feline cycle

environment to produce the two sporoblasts, such structure called **immature oocyst**.

2. **Mature (sporulated) oocyst (Fig. 98.7):** Each sporoblast follows the successive divisions to form four sausage shaped sporozoites. So, total 8-sporozoites and they are arranged in two groups in a specialized structure called **sporocyst** within oocyst. Such oocyst called **mature oocyst** which is 10–12 μm in size and spherical or oval in shape. Oocyst do not sporulate below 4°C or above 37°C. Dry heat (>66°C) or boiling renders the oocyst non-infectious.

Pseudocyst: Follow **Table 98.1** and **Fig. 98.7.**

Tissue cyst: Follow **Table 98.1** and **Fig. 98.7.**

Culture Characteristic (C/Cs)

Tissue culture: It grows on Hela or Vero cells.

Animal culture: Animal culture can be done in mice, guinea pigs and hamsters.

Eggs culture: It can be done in CAM of chick embryo.

Immunity

Both AMI and CMI are developed and they are protective.

Life Cycle

Types of host and types of reproductive method: Both are two types.

- **Definitive hosts:** Domestic or wild cat is the definitive host. Both asexual (schizogony) and sexual (gametogony followed by syngamy) stages occur in cat.
- **Intermediate or accidental hosts:** Other mammals like rodents, cattle, pigs and humans are the intermediate or accidental hosts. Endozoites of pseudocysts are multiplying by budding method called **endogeny.** Cystozoites of tissue cysts are multiplying asexually but by different method called **endopolygeny.**

Cycles: Two types.

- **Enteric cycle:** It occurs in two phases in definitive host and in external environment as shown in **Flowchart 98.5 and Fig. 98.7.**
- **Exo-enteric cycle:** It occurs only in one phase in intermediate host as shown in **Flowchart 98.6 and Fig. 98.7.**

> **Note: Few authors defined the life cycle by using 3 cycles** (indicated by **A, B** and **C** in **Fig. 98.7**) as below.
>
> **A. Feline cycle:** It is maintained between cat and external environment as follows. Cat → passes oocysts in feces → entry in cat by food and water.

TABLE 98.1: Differences between pseudocyst and tissue cyst

Features	Pseudocyst (Fig. 98.7)	Tissue cyst (Fig. 98.7)
Mnemonic	**PECT → P**seudocyst **E**ndozoite **C**entrally placed nucleus **T**achyzoite	**TCBS → T**issuecyst **C**ystozoite **B**radyzoite **S**hifted nucleus, means not in center and placed towards round end
Meaning	So, called because all the parasites are not enclosed by a specific epithelial membrane, but by a host cell wall	It is a true cyst and all the parasites are enclosed by a specific epithelial membrane
Sites	It develops inside the cells (intracellular) of RE system, placenta and almost all body cells except RBCs	It develops inside the muscles and CNS (inside the tissues but extracellular)
Size	10–200 µm in size	10–200 µm in size
Shape	Spherical or oval	Spherical or oval
Parasites Specific name	Called **endozoites** or **tachyzoites** (or **trophozoites** in the past)	Called **cystozoites** or **bradyzoites**
Meaning	From **tachos (Greek) means speed,** because quickly multiplying	From **brady (Greek) means slow,** because slowly multiplying
Numbers	50–100 parasites	Thousands in numbers
Shape	Rounded at one end and blunt pointed at other	Rounded at one end and pointed at other (crescent shape)
Size	3–7µm × 2–4 µm (larger)	7 µm × 1.5 µm (smaller)
Nucleus	Located centrally	Nucleus situated at round end
Reproductive method	Budding (asexual) method called **endogeny** or **endodyogeny**	Asexual method called **endopolygeny**
Clinical significances	1. It causes more extensive damage in tissues and responsible for clinical signs and symptoms 2. Pseudocyst is an infective form and is transmitted by using the raw meat of above tissues	1. It does not cause any damage in tissues and not associated with any signs and symptoms 2. Tissuecyst is also an infective form and is transmitted by using the raw meat of tissues 3. It is also a source of recrudescent (latent) infection in immunocompromised hosts

(Contd...)

B. **Feline–nonfeline cycle:** It is maintained between cat and intermediate hosts as follows. Cat → passes oocysts in feces → entry in other non-feline hosts like mouse, humans, etc., by food and water → development of pseudocysts/tissue cysts in mouse → entry in cat by raw meat of mouse.

C. **Nonfeline–nonfeline cycle:** It is maintained between two different intermediate hosts like animals to humans or one animal to other animals. Oocysts in external environment → entry in other animals by food-water → development of pseudocysts/tissue cysts → entry in human by raw meat → Cycle not continue from human to animals as human meat is not accessed by animals but it continue from human to human by congenital route.

Pathogenicity

Disease name: It is a zoonotic disease and transmitted from animals to humans. Disease called **toxoplasmosis.**

Epidemiology: It is distributed worldwide. About 30–40% population of different countries are sero-positive, but incidence is higher, about 75–80% in France and Scandinavia.

Reservoirs of infection: Both definitive host and intermediate host are the reservoirs. Sheep and swine are the sources of human infection. Dog does not play any role in human infection.

Sources of infection: Infected tissues or food-water.

Modes of transmission: Following are two modes.

1. **Direct transmission**
   - **Direct contact:** It enters through direct contact of abraded skin with infected tissues.
   - **Inoculation under skin:** By infected needles or syringes.
   - **Transplacental (vertical or congenital):** It is transmitted by placenta and also by mother's milk. One of the agent among ToRCH list (**Ch. 3**). Incidence of fetal transmission is less in 1st trimester, but once happened infection is more sever. Incidence is highest in 3rd trimester but fetal risk is not severe.

2. **Indirect transmission**
   - **Food and water borne (ingestion):** It enters by
     - ingestion of contaminated food and water contain oocysts from cat's feces. Freshly passed oocysts (contain sporoblast and called **immature oocysts**) are not infective, but become infective only after getting mature (contain 8 sporozoites and called **mature oocysts**) in external environment either in water or soil.
     - ingestion of raw meat or eggs (pseudocyst also found in ovary of hen) of infected animals contain pseudocysts or tissue cysts.
     - ingestion of milk from infected mother to children
   - **Inhalation:** Bronchial secretion also contains parasites and responsible for respiratory illness.

– Principle: It based on inhibition of alkaline methylene blue dye's (pH = 11.0) staining of tachyzoites by specific antiserum.

– Steps: Equal volume contains suspension of tachyzoites (act as Ags), accessory factor (human serum with complement) and patient's serum (contains *Toxoplasma* Abs) are mixed in a tube with serial dilution and incubated at 37°C in a water bath for one hour. Add one drop of alkaline methylene blue to each tube and re-incubated. One drop of suspension from each tube is examined under microscope.

– Examination and results: Tachyzoites are killed if patient's serum contains anti toxoplasma antibodies in presence of complement (complement mediated lysis). Dilution of the test serum at which 50% of tachyzoites are stained (hence not killed), is reported as significant titer. Significant titer is 1:4 and test becomes positive after 1–2 weeks

– Advantage: It is the gold standard test for Ab detection.

– Disadvantages: It is labored intensive and available only at reference centers. It cannot differentiate between IgM and IgG. False-positive test is reported in *T. vaginalis, Trypanosoma lewisii* and *Sarcocystis.*

4. **Interpretation of results of serological tests:** It is very difficult to interpret the serological result of *Toxoplasma*, because of high prevalence of *Toxoplasma* antibodies in population due to past and subclinical infection.

- **Toxoplasmosis in pregnant women:** Mother is mostly asymptomatic and IgM positive titer indicates recent infection and risk to fetus, while high IgG titer indicates infection of 4–8 weeks before conception.

- **Congenital toxoplasmosis**
  - **IgM:** In neonate IgM (does not cross the placenta) titer of ≥1:4 after 2 weeks indicate congenital infection.
  - **IgG:** Presence of IgG in neonate is either maternal or infection origin. Maternal IgG will disappear after 6–10 months and its persistence after this time indicates congenital infection. It also requires X-ray help.
  - **IgA:** It is more sensitive than IgM. IgA usually disappears within 10 days of infection. Its persistence after this time indicates postnatal infection.

F. Skin (Frenkel's) test: It based on DTH.

G. Molecular method: PCR detects parasite DNA from blood, CSF or amniotic fluid.

H. Radiological tests: Like CT, MRI or X-ray is useful. USG of fetus should be done at 20–24 weeks of pregnancy and then repeat every 2–4 weeks.

Prevention

General measures: These are improvement of food hygiene, avoid contact with cat or cat's feces, screening of all blood donors and women of child-bearing age for *Toxoplasma.*

Chemoprophylaxis: It required especially in AIDS patient with CD4 count <200/mm^3. Cotrimoxazole is the drug of choice if not tolerates then shifts to dapsone-pyrimethamine. Discontinue the prophylaxis if CD4 count is >200/mm^3.

Immunoprophylaxis: A genetically engineered vaccine is under trial.

Treatment

Treatment in immunocompetent: Presents with lymphadenopathy and not required treatment.

Treatment in patients with AIDS: Cotrimoxazole is the drug of choice if not tolerates then shifts to dapsone-pyrimethamine, atovaquone with or without pyrimethamine.

Treatment to pregnant woman: Spiramycin is the drug of choice.

Treatment in neonate: Oral pyrimethamine (1 mg/kg/day) plus sulfadiazine (100 mg/kg/day) with folinic acid for 1 year. Steroid is given to reduce the risk of chorioretinitis.

Cyclospora cayetanensis

History and Meaning

First human case was reported in 1985 in Peru. The species name refers to the **Cayetano Heredia University** in Lima, Peru, where early epidemiological and taxonomic work were done.

Morphology

Morphological stage called **oocyst,** which passes through the followings stages.

Immature (nonsporulated) oocyst (Fig. 98.10): Zygote changes to oocyst, which contains single sporoblast called **immature oocyst** and passes through feces over soil.

Mature (sporulated) oocyst (Fig. 98.10): Immature oocyst undergoes further development in external environment. Sporoblast undergoes successive division to produce four sporozoites which are arranged in two groups, in a specialized structure called **sporocyst** within oocyst and such oocyst called **mature oocyst.** Mature oocyst is spherical in shape, 8–10 µm in and gets mature in 7–15 days.

Life Cycle

It is similar to *Cystoisospora belli.*

Hosts and methods of reproduction: Both asexual (schizogony) and sexual (gametogony followed by syngamy) stages occur in human.

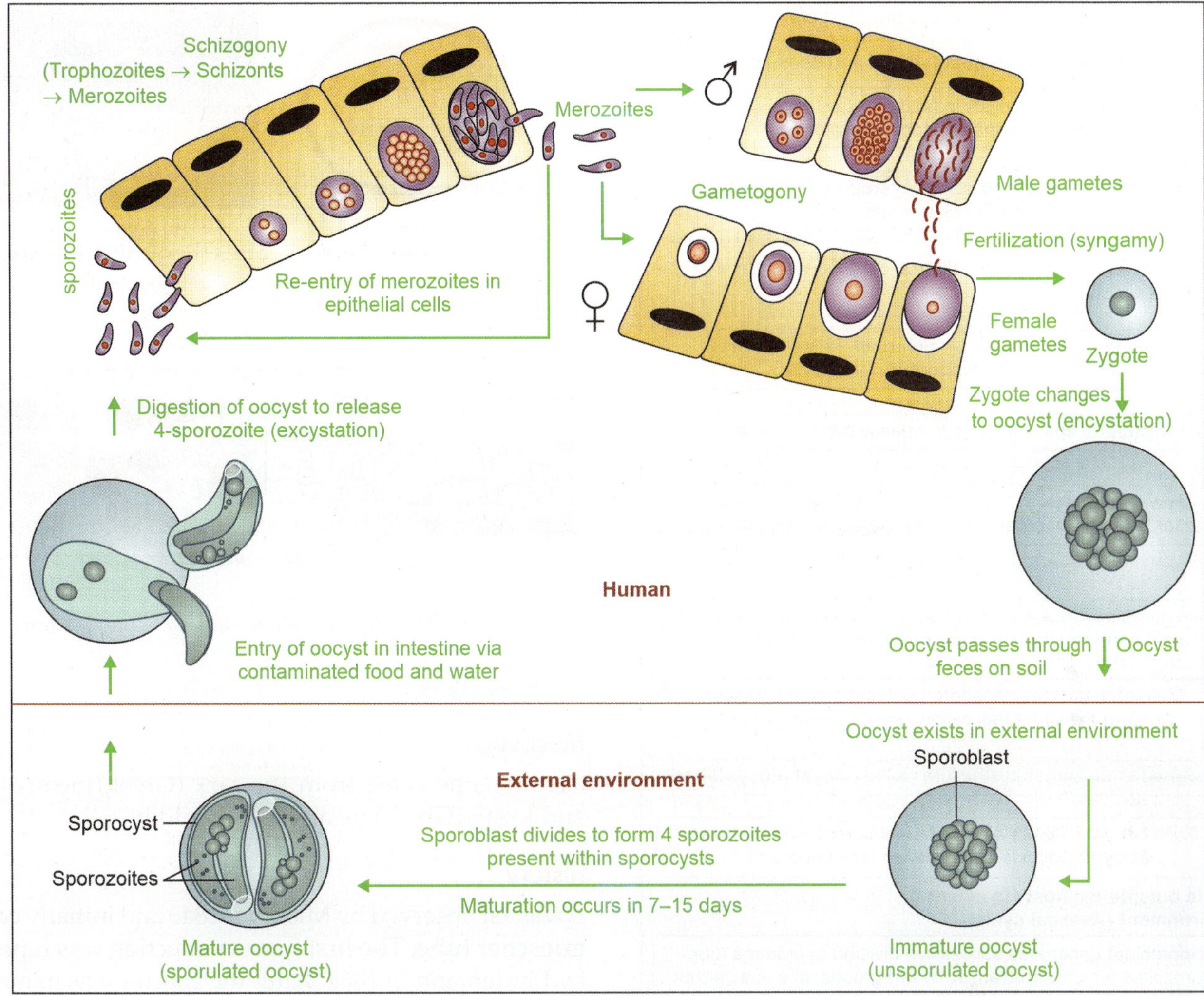

Fig. 98.10: Life cycle of *C. cayetanensis*

Cycles: Two types **(Flowchart 98.8 and Fig. 98.10).**

Pathogenicity

Disease name: Called **cyclosporiosis.**

Epidemiology: It is distributed worldwide in many animals like reptiles, birds and animals. It is reported from Nepal, India and South America.

Reservoirs of infection: Animals.

Sources of infection: Food-water.

Modes of transmission: By ingestion of contaminated food-water contain sporulated oocysts.

Incubation period: 2–11 days.

Portal of entry: GIT.

Site: Small intestine (duodenum and jejunum).

Infective form: Mature oocyst with 2 sporocysts (total 4 sporozoites).

Exit form: Oocyst with single sporoblast.

Precipitating factor: It is common in persons with IDDs/AIDS.

Clinical features: Follow **Flowchart 98.9.** In human, it is a cause of traveler's diarrhea.

Laboratory Diagnosis

Specimens: Stool.

Testing methods

A. Microscopy: (1) Wet mount preparation: Spherical oocyst is seen as shown in **Fig. 98.11a.** Oocyst is differentiated from *C. belli* and *C. parvum* as shown in **Table 98.2. (2) Modified ZN stain:** It used to differentiate from other acid fast oocyst as shown in **Fig. 98.12c. (3) Autofluorescence:** Cyclospora are fluorescent in nature under UV microscope called **autofluorescence** as shown in **Fig. 98.11b.** This nature could be taken in to consideration while performing immunofluorescent microscopy for *C. parvum* and *Giardia.* **(4) Stool concentration method:** Oocyst can be concentrated by modified zinc sulfate flotation technique.

B. Ab detection method and are molecular methods: Described above with *Cryptosporidium parvum.*

Prevention and Treatment

It is prevented by improving the food hygiene. Cotrimoxazole is the effective drug. Long-term suppressive therapy is required in AIDS patient.

Flowchart 98.8: Life cycle of *C. cayetanensis*

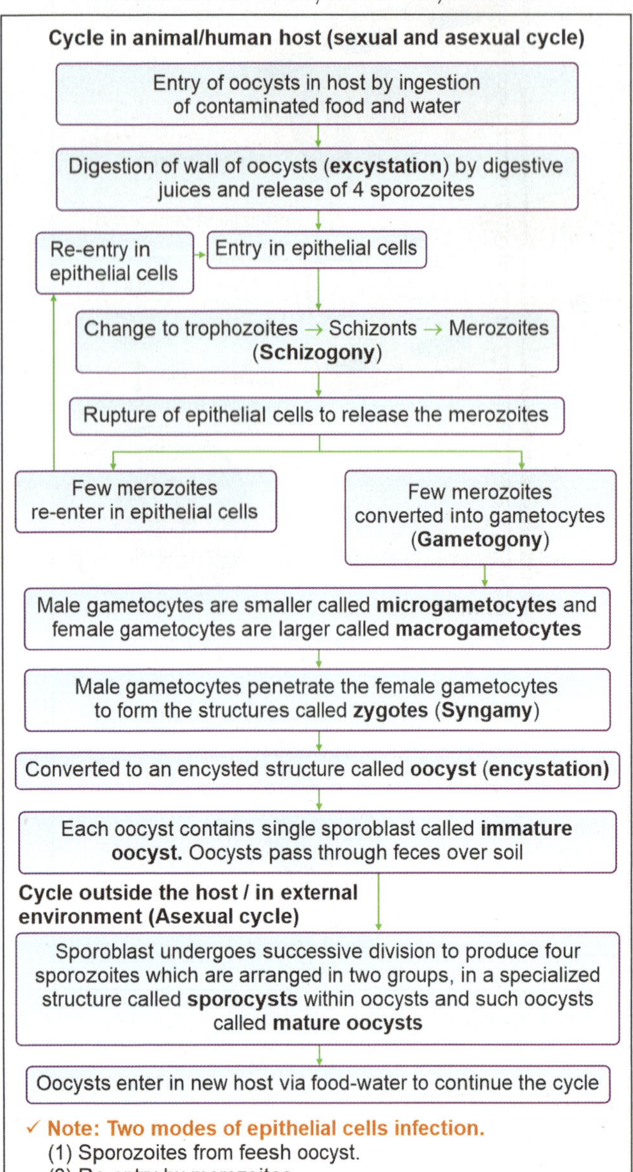

Flowchart 98.9: Clinical features of cyclosporiosis

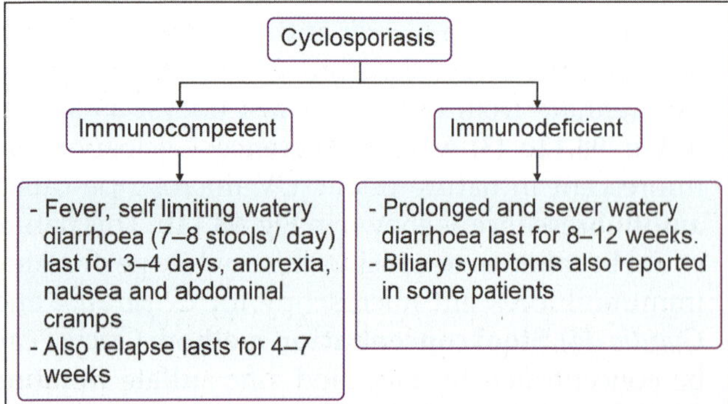

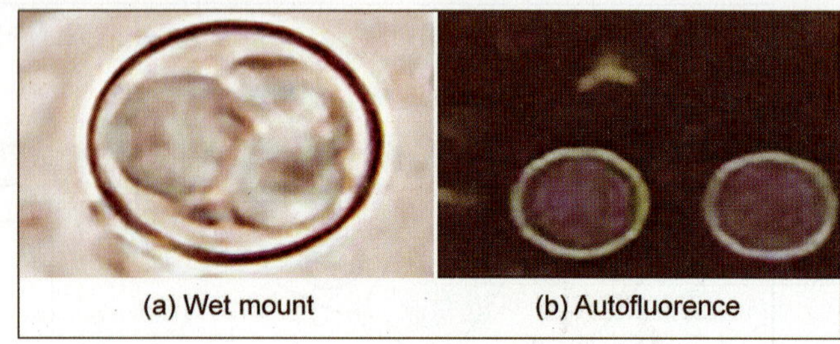

Fig. 98.11: Oocyst of *C. cayetanensis* (a) under wet mount and (b) autofluorescence

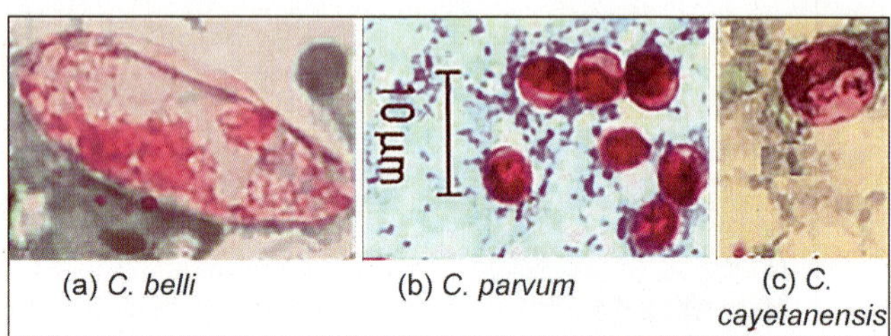

Fig. 98.12: Different oocyst under ZN preparation

Sarcocystis spp.

Meaning

Genus name came from the sarx (Greek) means meat and kystis (Greek) means cyst/bladder.

History

It was 1st observed by Miescher 1843 and initially **called miescher tube.** The first human infection was reported by Lindemann in 1868. Later the species was named as *S. lindemanni* by Rivolta.

Morphology

Three morphological stages are described below.

Oocyst (Fig. 98.13): Zygote differentiated to oocyst, which contains two sporocysts and four sporozoites in each sporocyst (total eight sporozoites). Oocyst excystes to release sporocysts. Sporocycts are excreted through feces over soil. Oocyst is oval in shape and 19 µm × 13 µm in diameter.

Sporocyst (Fig. 98.13): It is spherical or oval in shape, 13–16 µm × 8–10 µm in diameter and contains 4 banana shaped sporozoites.

Sarcocyst or muscular cyst (Fig. 98.13): It is a tissue cyst with thick wall, presents parallel to the long axis of muscle, contains parasites called **cystozoites or bradyzoites** (7–16 µm long), tongue or spindle shape and 300 µm × 100 µm in size.

Parasites	Size	No. of sporocysts	Sporozoites		Acid fast	Maturation time
			Per sporocyst	Total		
TABLE 98.2: Comparison of oocysts of different parasites						
C. belli	20–30 µm × 10–20 µm	2	4	8	Yes	1–5 days
C. parvum	4–5 µm	—	—	4	Yes	—
C. cayetanensis	8–10 µm	2	2	4	Variable	7–15 days

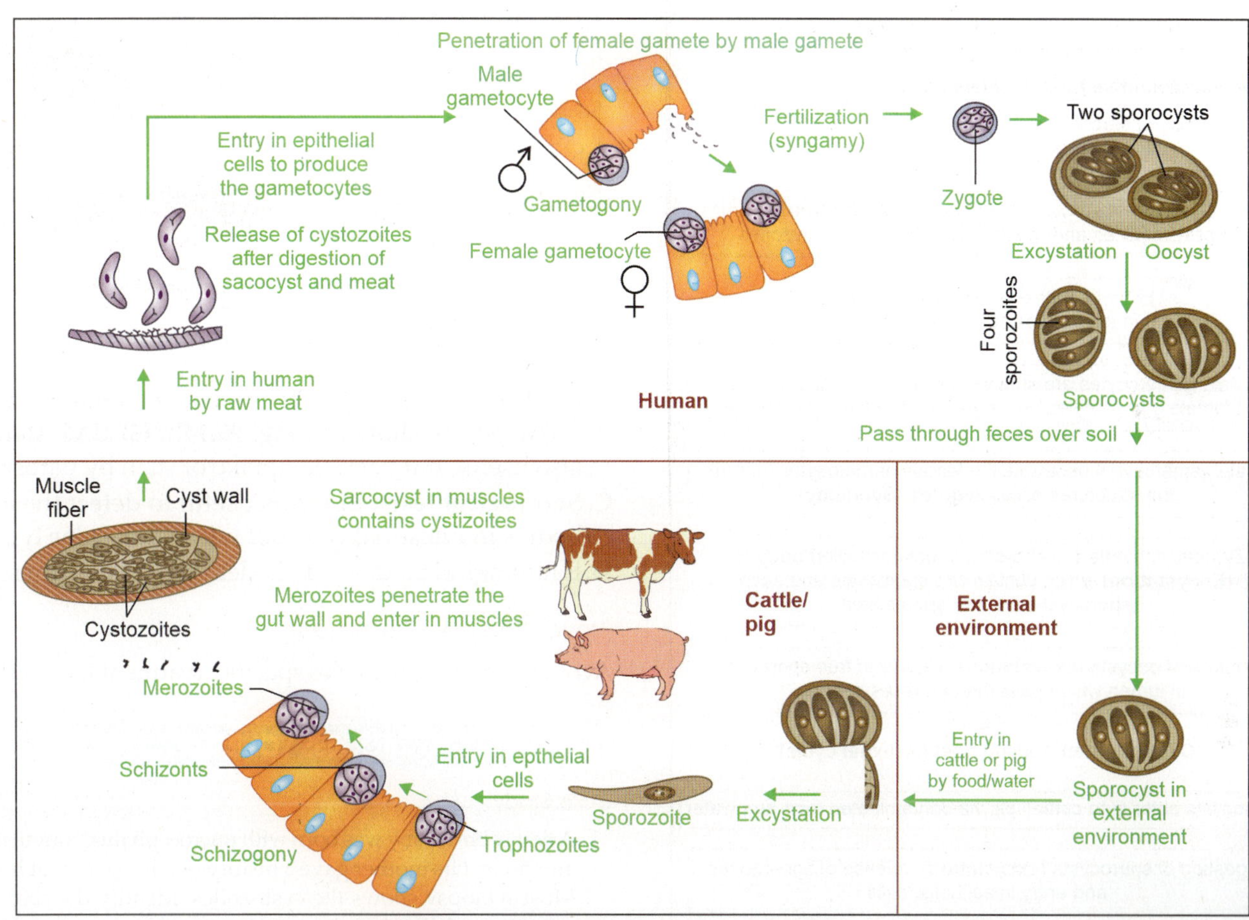

Fig. 98.13: Life cycle of intestinal *Sarcocystis*

Life Cycle

It is different in different species.

Intestinal (*S. bovihominis* and *S. suihominis*)

Hosts and methods of reproduction

- **Definitive host:** Human is the definitive host, which harbors the sexual (gametogony followed by syngamy) stage.
- **Intermediate host:** Cattle in case of *S. bovihominis* and pig in case of *S. suihominis* are the intermediate hosts, which harbor the asexual stage.

Cycles: Two types as mentioned in **Flowchart 98.10 and Fig. 98.13.**

Invasive or Muscular (*S. lindamanni*)

Hosts and methods of reproduction

- **Definitive host:** Cat or dog.
- **Intermediate host:** Human is the intermediate host, which harbors the asexual stage.

Cycles: Follow **Flowchart 98.11.**

Pathogenicity

Disease name: Called sarcocystosis.

Epidemiology: It is distributed worldwide in humans and animals like reptiles, birds, animals, etc.

Reservoirs of infection: Animals.

Sources of infection: Beef or pork.

Modes of transmission: It is transmitted by ingestion of raw beef or pork. Transmission by contaminated food and water is the only theoretical concept.

Incubation period: Symptoms appear after 3–6 hours of beef ingestion.

Exit form: Sporocyst with four sporozoites.

Infective form: Sarcocyst with cystozoites (intestinal type) and sporocyst with four sporozoites (invasive type).

Portal of entry: GIT.

Sites: Intestine and muscles (in lymph nodes and vessels also).

Clinical features: Intestinal sarcocystosis is asymptomatic in many cases but may present with diarrhea, anorexia, nausea and abdominal cramps. **Muscular sarcocystosis** is asymptomatic or presents with fever, painful muscular swelling, myonecrosis, myositis, perivascular interstitial inflammation, vasculitis and eosinophilic myositis (40%).

Laboratory Diagnosis

Specimens: Stool and muscle biopsy or autopsy specimens.

Flowchart 98.10: Life cycle of intestinal *Sarcocystis*

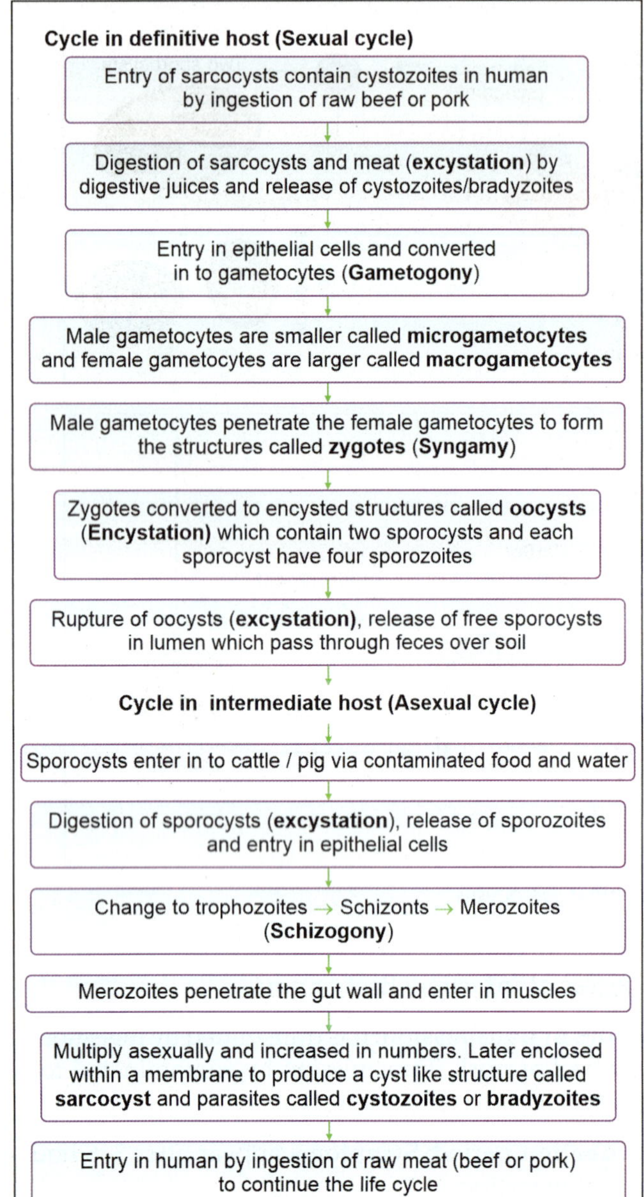

Flowchart 98.11: Life cycle of invasive *Sarcocystis*

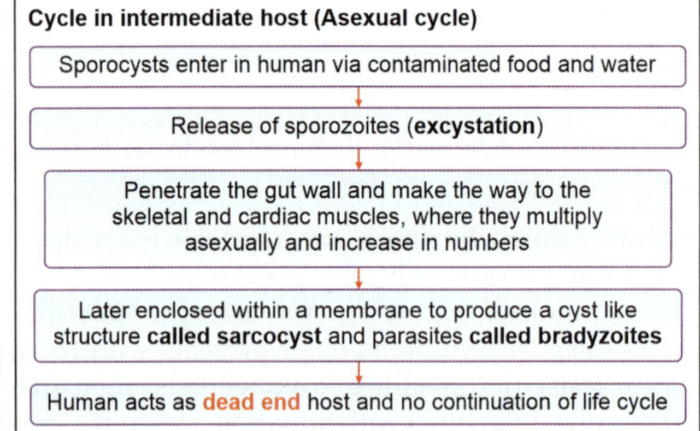

Testing methods

A. Blood picture: Eosinophilia (40%).

B. Macroscopy: (1) Wet mount preparation from stool shows the presence of sporocysts as shown in **Fig. 98.14a**.

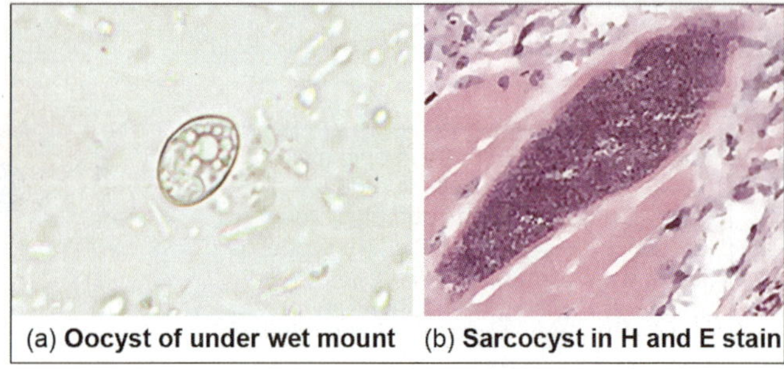

(a) **Oocyst of under wet mount**   (b) **Sarcocyst in H and E stain**

Fig. 98.14: Microscopic picture of *Sarcocystis*

(2) H and E stain show the presence of tongue shaped sarcocysts as shown in **Fig. 98.14b. (3) PAS stain** is also useful, but variable uptake of stain by parasites.

C. Serological tests: ELISA is useful to detect the antibodies in intestinal type but not in muscular type.

D. Radiological methods: They detect the cysts in muscles.

Prevention and Treatment

Avoid use of raw meat. No specific treatment is available.

ACCESS YOURSELF

Case Study

1. A newly born baby was born with microcephalus, convulsions and fever. His mother gave a history of having a cat at home. Muscle biopsy shows the cystozoites. Identify the case and answer the following.
   a. Name the causative agent and draw the diagram of morphological stages of causative agent.
   b. Describe the life cycle of causative agent.
   c. Write the pathogenicity of causative agent.

Essay/Full Question

1. *T. gondii.*

Short Notes

1. Congenital toxoplasmosis.
2. Differences between tissue cyst and pseudocyst of *T. gondii*.
3. Cystoisosporiasis/Cryptosporidiosis/Cyclosporiosis.
4. Intestinal sarcocystosis.

Short Questions for Theory/Viva Questions

1. Name the four gastrointestinal sporozoa.
2. What is autofluorescence?
3. Write four different mode of transmission of *T. gondii*.

MCQs for Chapter Review

Cystoisospora belli

1. **True about *Cystoisospora belii*:**
   a. Human is the definitive host
   b. Infective form is oocyst
   c. Transmitted by feco-oral route
   d. Oocyst has 4 sporocysts

Cryptosporidium parvum

2. **True about *Cryptosporidium* are all, *except*:**
   a. Oocyst is chlorine resistant
   b. Acid fast spores
   c. Oocyst >100 µm
   d. Enzyme immunoassay done

3. **True about *Cryptosporidium parvum*:**
   a. Affects only immunocompromised patient
   b. It is one type of the common opportunistic infection in AIDS
   c. Cyst size is 12–15 nm
   d. AFB +ve cyst
   e. Treatment is metronidazole
4. ***C parvum* cyst identified by which stain in stool sample?**
   a. PAS
   b. H and E
   c. Giemsa
   d. Acid fast stain
5. **In human cryptosporidiosis is present as:**
   a. Meningitis
   b. Diarrhea
   c. Pneumonia
   d. Hepatitis

Toxoplasma gondii

6. **Definitive host for *Toxoplasma*:**
   a. Man
   b. Dog
   c. Cat
   d. Rat
7. **Intermediate hosts for *Toxoplasma* are all, *except*:**
   a. Human
   b. Sheep
   c. Cat
   d. Pig
8. **Oocyst of *Toxoplasma* is found in:**
   a. Cat
   b. Dog
   c. Mosquito
   d. Cow
9. **Route of transmission of *Toxoplasma*:**
   a. Blood
   b. Feces
   c. Urine
   d. None
10. **Infective form of *Toxoplasma gondii* is:**
    a. Oocyst
    b. Bradyzoite
    c. Tachyzoite
    d. All of the above
11. **All of the following statements about toxoplasmosis are true *except*:**
    a. Oocyst in freshly passed cat's feces is infective
    b. May spread by organ transplantation
    c. Maternal infection acquired after 6 months has high risk of transmission
    d. Arthralgia, sore throat and abdominal pain are the most common manifestations
12. **True about toxoplasmosis:**
    a. Due to ingestion of sporocyst with meat
    b. Due to ingestion of oocyst from cat's feces
    c. Spiramycin is given pregnancy
    d. Due to bite of *Anopheles* mosquito
    e. Mostly symptomatic

Cyclospora cayetanensis

13. **Following is not the property of *Cyclospora cayetanensis*:**
    a. Autofluorescence
    b. Acid fastness
    c. Transmitted by feco-oral route
    d. Oocyst contains 4 sporocysts
14. **25-year-old male presented with diarrhea for 6 months. On examination the causative agent was found to be acid fast with 12 micrometers diameter. The most likely agent is:**
    a. *C parvum*
    b. *Cystoisospora*
    c. *Cyclospora*
    d. *Giardia*

Answers and Explanation of MCQs

1. **a, b, c**
- Oocyst of *Cystoisospora belli* contains 2 sporocysts and each sporocyst contains 4 sporozoites (total 8 sporozoites in oocyst).
2. **c**
3. **b, d**
4. **d**
5. **b**
- Follow section, *Cryptosporidium parvum* for explanation of answers of MCQs 2–5.
6. **c**
7. **c**
8. **a**
- Follow section, *Toxoplasma gondii* **(life cycle)** for explanation of answers of MCQs 6–8.
9. **a, b**
10. **d**
- Follow section, *Toxoplasma gondii* **(pathogenicity)** for explanation of answers of MCQs 9–10.
11. **a, d**
- Option a, b and c: Follow section, *Toxoplasma gondii* **(pathogenicity → mode of transmission and infective form)** for explanation. Most common manifestation of toxoplasmosis is posterior cervical lymphadenopathy.
12. **b**
- For options a, b and d follow section, *Toxoplasma gondii* **(modes of transmission)** and for option c **treatment.**
13. **d**
- Oocyst of *Cyclospora cayetanensis* contains 2 sporocysts and each sporocyst has two sporozoites (total 4 sporozoites in mature oocyst).
14. **c**
- All options except *Giardia* are acid fast. Oocyst is the diagnostic form present in stool. Its size helps to differentiate the option a, b and c as per **Table 98.2.**

CHAPTER 99

Infections of Microspora and *Blastocystis*

Chapter Outline

- Microspora
- *Blastocystis hominis*

MICROSPORA

Taxonomy

Taxonomic status is still unclear; few workers classified them as fungi, while protozoa by few. They are true eukaryotes, as they possess membrane bound nucleus and other cellular features, but they are unusual eukaryotes in that, they contains 70S ribosome, no mitochondria and Golgi membrane.

History

It was Pastuer who identify the Microspora (*Nosema bombycis*) as a causative agent of **pebrine (Silkworm disease)** in 1863 in France. Microspora had been known as animal pathogens for a very long time, their role as human pathogens were identified in 1980 and advancing with AIDS.

Classification

Systemic classification: Follow Ch. 14 (Table 14.3).

Pathogenic classification: Million of species are known under the title of Microspora, almost all are animal pathogens. Only 9 genera and 13 species are known as human pathogens especially in persons with IDDs called **opportunistic pathogens.** Few genera and species as human pathogens are mentioned in **Table 99.1.**

Morphology

Morphological stage called **spore,** as shown in **Fig. 99.1** with following features. It is gram-positive, acid fast in nature, up to 6 μm in diameter and variable in shape like spherical, oval or rod shaped. Structural parts are mentioned below.

- **Nucleus:** It is ninucleate or binucleate (diplokaryon).
- **Polar tube:** It is the tubular structure also called **polar filament** or **spiroplasm** which is 1.1–1.6 μm in length

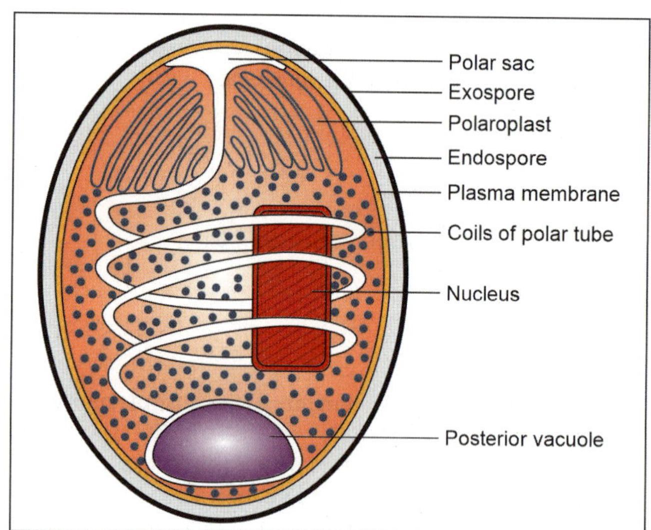

Fig. 99.1: Spore of Microspora

and 0.7–1.0 μm in breath. During infection spiroplasm is injected in to the host cells by extrusion mechanism. It is coiled up in the posterior half of the spore.

- **Polaroplast:** It presents at anterior part of the polar filament.
- **Vacuole:** It presents behind the polar filament.
- **Cell wall:** It consists three layers like **outer** which is electron-dense called exospores, **middle** which is wide, contain chitin and seems structure less called endospores and **inner** which is thin called plasma membrane.

Life Cycle

Methods of reproduction: Spores are produced by asexual method called **sporogony,** which may be either binary fission called **merogony** or multiple fissions called **schizogony.**

Cycles: Follow **Flowchart 99.1 and Fig. 99.2.**

TABLE 99.1: Classification, nonhuman host and clinical features of Microspora

Genus	Species	Nonhuman host	Clinical features in immunodeficient patients	Clinical features in immunocompetent patients
Common				
Enterocytozoon	E. bieneusi	Pigs, nonhuman primates	Chronic diarrhea, wasting syndrome cholangitis, AIDS cholengiopathy, cholecystitis, pneumonia, chronic sinusitis	Traveler's diarrhea
Encephalitozon	E. cuniculi (three strains I, II and III)	Mammals, rabbits, mice, blue foxes, dogs	Sinusitis, keratoconjunctivitis, bronchitis, pneumonia, hepatitis, encephalitis	Not described
	E. hllem	Parrots	Sinusitis, keratoconjunctivitis, cystitis, bronchitis, pneumonia, urethritis	Not described
	E. intestinalis	Goats, donkeys, pigs, cows, dogs	Sinusitis, bronchitis, pneumonia, nephritis, chronic diarrhea, cholangiopathy	Self limiting diarrhea, asymptomatic carrier
Uncommon				
Microsporidium	M. africanum	Not identified	Not described	Corneal ulcer, keratitis
	M. ceylonensis	,,	Not described	Corneal ulcer, keratitis
Nosema	N. connori	,,	Disseminated infection	Not described
	N. ocularum	,,	Not described	Keratitis
Pleistophora	P. ronneafier	,,	Myositis (skeletal)	Not described
Brachiola	B. vesicularum		Myositis (skeletal)	Not described
	B. conori		Myositis (smooth and cardiac)	Not described
Trachipleistophora	T. anthropophthera	,,	Disseminated infection (brain)	Not described
	T. hominis	,,	Sinusitis, keratoconjuctivitis, myositis	Not described
Vittaforma	V. corneae	,,	Keratitis	Not described

Flowchart 99.1: Life cycle of Microspora

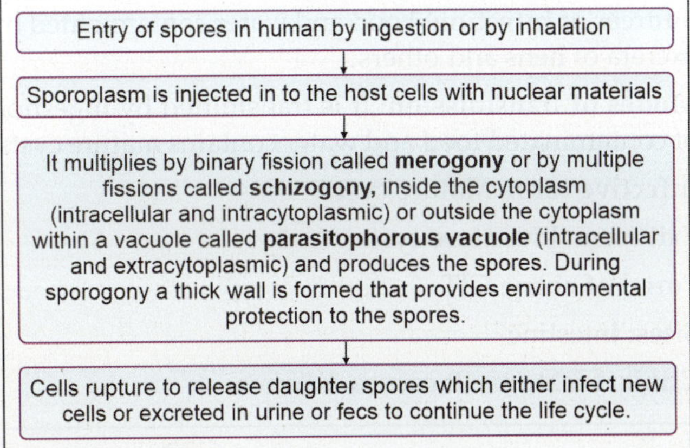

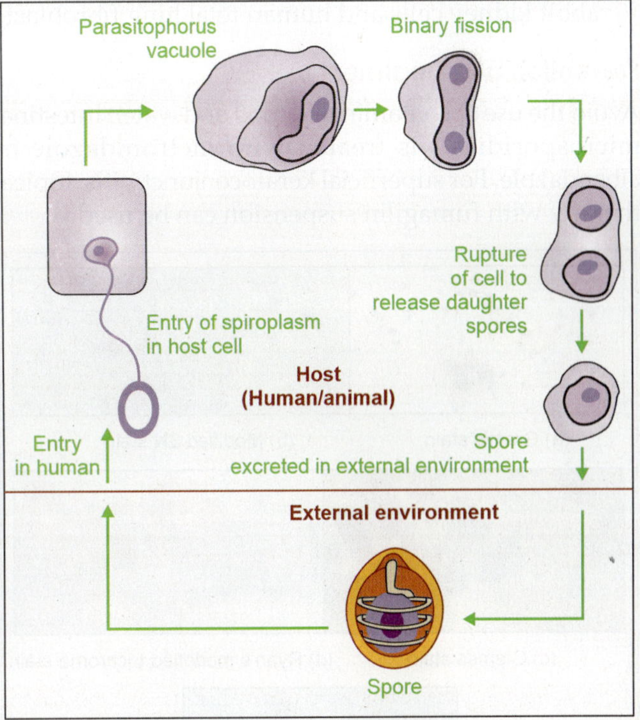

Fig. 99.2: Life cycle of Microspora

Pathogenicity

Disease name: Called **microsporidiosis.**

Epidemiology: Worldwide distribution.

Reservoirs of infection: Exactly unknown but many vertebrate and invertebrate animals are reservoirs.

Sources of infection: Food and water contaminated by excreta of vertebrate and invertebrate animals.

Modes of transmission: It is transmitted by **ingestion** of contaminated food and water contain spores that have been shed in the urine and feces of infected humans and animals. It also transmitted by **inhalation** route.

Infective and exit forms: Spores.

Portal of entry: GIT or respiratory tract.

Sites: Intestine and dissemination in other tissues as shown in **Table 99.1**. Obligate intracellular parasite present free in cytoplasm or extracytoplasmic within the parasitophorous vacuole. It survives outside the host in free environment.

Precipitating factor: It is most common in patient with AIDS and CMI defect.

Clinical features: Follow **Table 99.1**.

Laboratory Diagnosis

Specimens: Stool, urine, CSF, biopsy from infected tissues.

Testing methods

A. Microscopy

1. **Gram's stain:** Gram-positive spores appear as shown in **Fig. 99.3a**.
2. **Modified ZN stain:** Spores are acid fast (1% H_2SO_4) as shown in **Fig. 99.3b**. The acid-fast spores appeared bright red on a blue background, and a posterior vacuole and central diagonal strip within the spores were often visible. Bacteria and other cellular debris appear blue, owing to methylene blue counter stain.
3. **Giemsa stain:** The darkly stained spores could be identified as shown in **Fig. 99.3c**.
4. **Ryan's modified trichrome stain:** Parasite appears red in color as shown in **Fig. 99.3d**.
5. **Calco-fluor-white (CFW) stain:** Spores are seen as white oval bodies, often are clustered in groups, against a relatively dark background as shown in **Fig. 99.3e**. Spores displayed variable fluorescence intensities.

B. Culture: Microspora can be cultured in Vero cells, rabbit kidney cells and human fetal lung fibroblast.

Prevention and Treatment

Avoid the use of contaminated food and water. Intestinal microsporidiasis is treated with metronidazole or albendazole. For superficial keratoconjunctivitis, topical therapy with fumagilin suspension can be used.

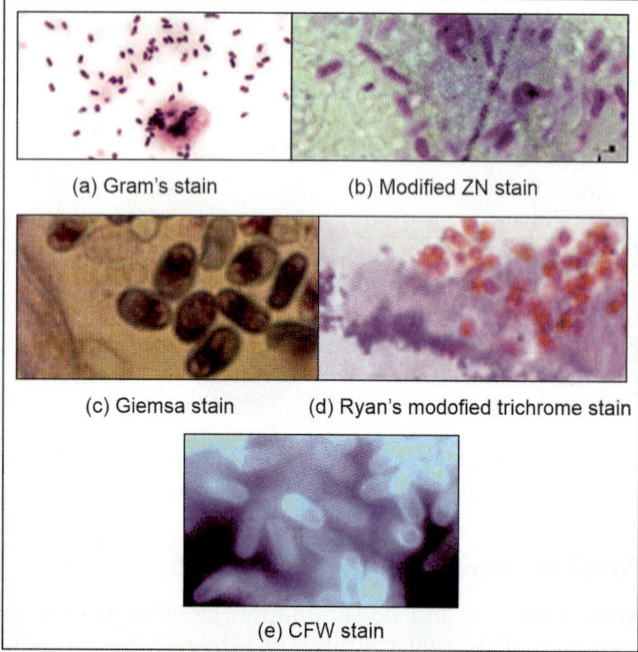

Fig. 99.3: Microspora under different stains

Blastocystis hominis

Taxonomy

Initially it was considered as an alga, then yeast and since 1982 a protozoan.

Morphology

Four morphological forms are seen as described below.

1. **Vacuolated form:** It is the most common form. About 8 µm (range 5–32 µm) in size. It has large central vacuole and thin rim of cytoplasm contains 2–4 nuclei. It multiplies by binary fission.
2. **Granular form:** It is the second most common form. It multiplies by endodyogeny and endosporulation.
3. **Amoeboid form:** It is non motile. A research study has reported that amoeboid forms are produced only in cultures taken from symptomatic individuals, with asymptomatic individuals producing exclusively vacuolar forms.
4. **Cyst form:** It is an infective form. Immature cyst released from host which gets mature in external environment and infects the other host.

Life Cycle

Follow **Fig. 99.4**

Pathogenicity

Disease name: Called **blastocytosis**

Reservoirs of infection: Like hens and other as shown in **Fig. 99.4**.

Sources of infection: Food and water contaminated by excreta of hens and others.

Modes of transmission: It is transmitted by ingestion of contaminated food and water contains mature cysts.

Infective form: Mature cyst

Exit form: Immature cyst

Portal of entry: GIT

Sites: Intestine

Clinical features: It causes diarrhea.

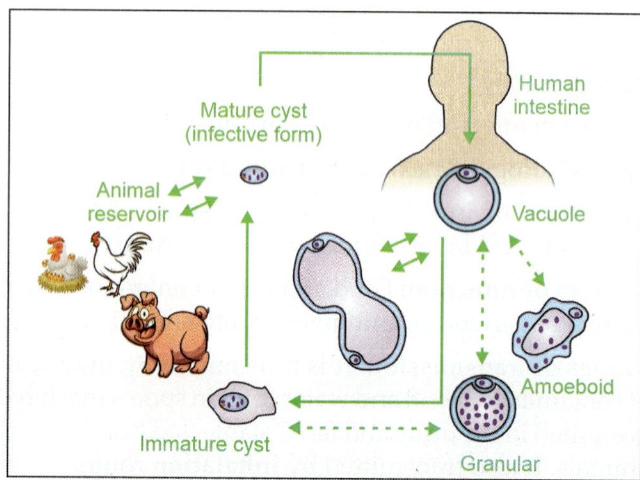

Fig. 99.4: Life cycle of *Blastocystis homonis*

Treatment

Iodoquinol and metronidazole are effective drugs.

ACCESS YOURSELF

MCQs for Chapter Review

Microspora

1. **Causative agent of silk worm disease was identified by:**
   a. Pasteur
   b. Jener
   c. Koch
   d. Philpis

2. **Which of the following is true with *Microsporidium*?**
   a. It is fungus
   b. It is protozoan
   c. It is bacteria
   d. It is associated with diarrhea in HIV

Answers and Explanation of MCQs

1. a
- Follow section, **Microspora (history)** for explanation.

2. b and d
- Follow section, **Microspora (taxonomy and pathogenicity).** for explanation.

Cestodes: General Features and Classification

Chapter Outline
- General Features
- Classification

GENERAL FEATURES

Introduction

Cestodes are commonly known as **tape worms** because dorso-ventrally flat and broad as like a measure tape. Word cestodes is originated from **Kestos (Greek) means ribbon** or **girdle and edios means appearance,** as the parasites are ribbon or measure tape in shape. Cestodes habitat the small intestine and the tissues.

Morphology

Adult Stage (Fig. 100.1a)

It is segmented and few millimeter to several meters in length. It has following three morphological parts.

- **Scolex (head):** It has different organs. **(1) Suckers:** These are organs for adhesion with host cells. Two slit like suckers (**Fig. 100.1b**) present in Pseudophyllidea (one on ventral and other on dorsal surface) called **bothria** or four cup (circular) like suckers (**Fig. 100.1c**) present in Cyclophyllidea called **acetabula. (2) Hooks:** These are also the organs for adhesion with host cells. Parasites with hooks called **armed tape worms** and which lack called **unarmed tape worms. (3) Rostellum:** It is a beak like projection on the head called **rostellum.** It presents in some tape worm like *Taenia solium*. It is surrounded by hooks. It is having characteristic shape in each family.

- **Neck:** It is the part behind head from which segments (proglottids) of the body are generated continuously.
- **Strobila (body or trunk):** Some authors used it for entire worm—head, neck and trunk. Nervous system, excretory system and reproductive system are present while alimentary canal and body cavity are absent. It is the chain of segments called **proglottids.** Proglottids are 3 types according to maturity with order from front to back.

> **Note: Reproductive system of cestodes (Figs 100.2 and 100.3)**
> **Definition:** In cestodes both male and female reproductive organs are present in single individual called **monoecious** or **hermaphrodite.**
>
> **Female reproductive organs:** They are present on ventral surface.
> - **Ovary:** It is bilobed, single or paired, lying behind equatorial plane of each segment and discharges the ova in the oviduct which connect the spermatic duct with the ootype.
> - **Vagina:** It extends from genital pore to the ootype and inner end contains the receptacle for storage of sperms called **seminal receptacle** followed by constricted tubule called **spermatic duct.** It provides the route for entrance of sperms.
> - **Uterus:** It is a straight tube arises from ootype. In Pseudo-phyllidea uterus is nonbranching and in Cyclophyllidea it may sometime have branches. It may be open in Pseudophyllidea (uterine pore present) or like blind sac as in Cyclophyllidea (uterine pore absent). It contains eggs when gravid.
> - **Ootype:** It is situated posteriorly in the middle of each segment. It provides platform for fertilization of eggs.
> - **Vitelline/yolk glands (vitellaria):** It scattered throughout the segment in Pseudophyllidea or remains as a single mass in Cyclophyllidea.
> - **Mehli's gland:** It also called **shell gland.** It is a very small organ surrounding the ootype and contains unicellular glands which open separately into the ootype.

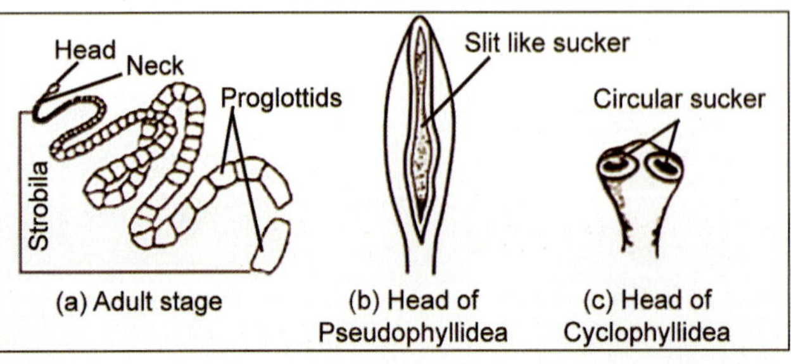

Fig. 100.1: Sucker of cestodes

(Contd…)

(Contd...)

Male reproductive organs: They are present on dorsal surface and mature prior to female system.

- **Testes:** It scattered throughout the segment as multiple follicles. It produces the sperms and each testis connected to vas deferens via small tubules called **vas efferentia.**
- **Vas deferens:** It is a convoluted tubule that begins from the center of each segment and ascends upward to open into the common genital pore in Pseudophyllidea or passes laterally to open into the genital atrium (common genital pore) in Cyclophyllidea. It enlarges for storage of sperms called **seminal vesicle.**
- **Cirrus sac and cirrus (penis):** It is a sac like structure presents at the end of vas deferens. It contains a muscular organ called **cirrus** or **penis.**

Genital atrium: Commonly it called **genital pore**. It presents on the mid-ventral surface in Pseudophyllidea and laterally in Cyclophyllidea. It provides place for common opening of penis and vagina.

Fertilization: It is the mating of sperm with ovum. Two types like self fertilization takes place in a same segment and cross fertilization takes place between the segments of same or other worm.

- **Immature:** Male and female organs are undifferentiated.
- **Mature:** Male and female organs are differentiated and male organs appear first.
- **Gravid:** Uterus is filled with eggs with disappearance or atrophy of other organs. Terminal segment is shrunken and empty due to constant discharging of eggs through uterine pore. Later dried segment break up from strobila in chain and passed in feces.

Egg Stage

Eggs are formed in ootype and surrounded by protective covering.

Eggs of Pseudophyllidea (Fig. 100.4)

- **Unembryonated egg:** Initially not containing the embryo and passed with feces in water for further maturation.
- **Embryonated egg:** It is oval, 70 μm × 45 μm in size, brown color (bile stained), has single cover called **egg shell,** operculated at one end, small knob at other end and contains yolk materials in abundant amount. **Embryo** is not present initially but formed later within 1–2 weeks of egg escape from host. Single egg gives the single embryo. Embryo is unsegmented **called coracidium.** Coracidium released free in water on hatching of egg and taken up by cyclops. Cilia around embryo are present.

Egg of Cyclophyllidea (Fig. 100.5): It is spherical, 31–43 μm in size, nonoperculated. It has two covers like thin outer cover called **egg shell** and thick inner cover called **embryophore.** Sometimes, egg shell ruptures before the eggs are excreted with feces. Embryophore becomes very thick after rupture of egg shell and radially striated to protect the inner oncosphere. Space between two covers is taken up by yolk material. Eggs contain embryo

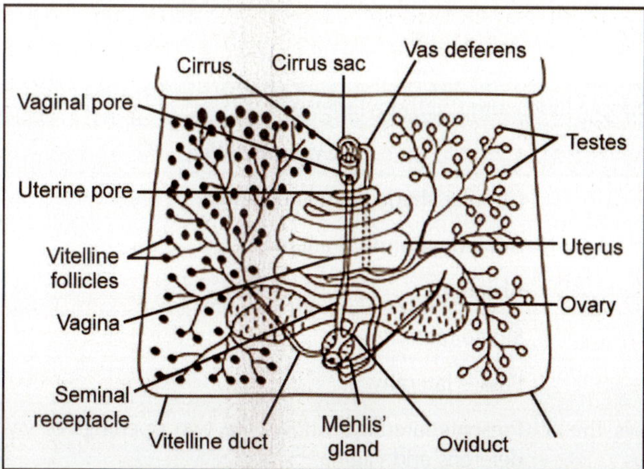

Fig. 100.2: Reproductive system of Pseudophyllidea

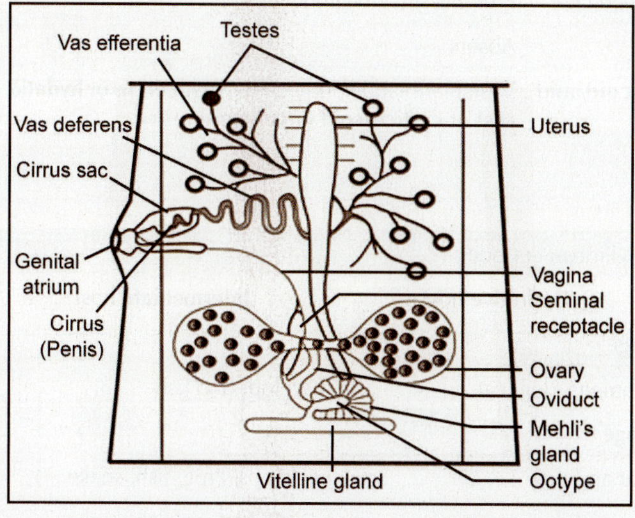

Fig. 100.3: Reproductive system of Cyclophyllidea

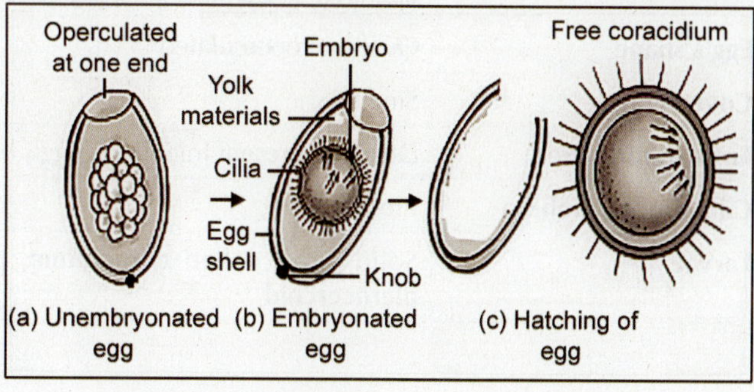

Fig. 100.4: Egg stage of Pseudophyllidea

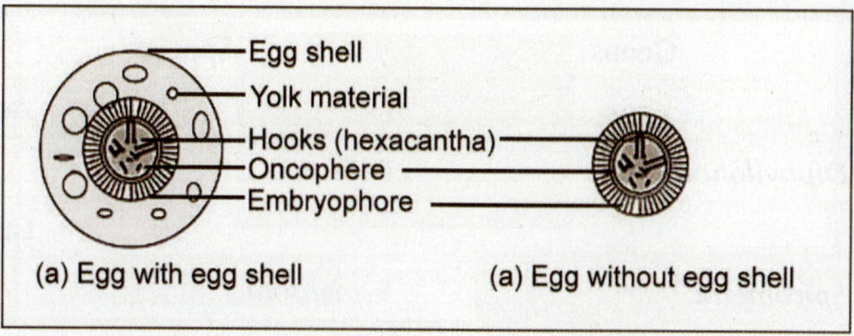

Fig. 100.5: Egg stage of Cyclophyllidea

from begining called **oncopshere** when laid down. It contains 6-hooklets (3-pairs) called **hexacanth.** Cilia around embryo are absent.

Larval Stage

Larva of Pseudophyllidea (Fig. 100.6): Solid forms of larvae are present in Pseudophyllidea. 1st stage develops free in water called **coracidium,** 2nd stage in cyclops (1s intermediate host) called **procercoid** and 3rd stage in fish/frog/snake (2nd intermediate host) called **plerocercoid** or **sparganum.**

Larva of Cyclophyllidea (Fig. 100.7): Vesicular/bladder forms of larvae are present in **Cyclophyllidea.** Larvae

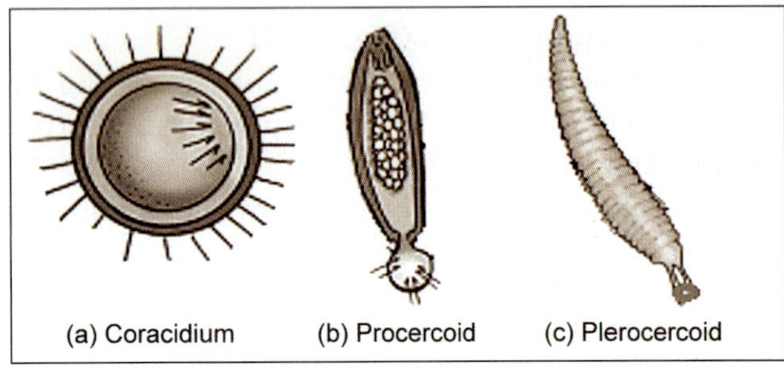
Fig. 100.6: Larval stage of Pseudophyllidea

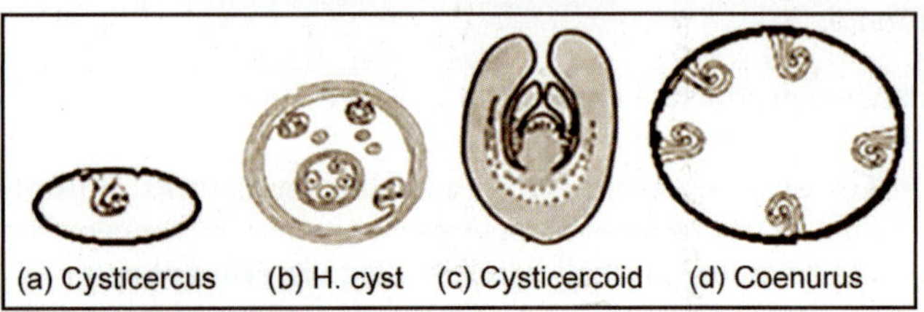
Fig. 100.7: Larval stage of Cyclophyllidea

called **cysticercus** (in *T. solium* and *T. saginata*) or **hydatid cyst** (in *E. granulosus*) or **cysticercoid** (in *H. nana* and in *D caninum*) or **coenurus** (in *M. multiceps*) discussed in respective chapters.

Differences between Pseudophyllidea and Cyclophyllidea

Follow **Table 100.1.**

CLASSIFICATION

Systemic classification: Follow Ch. 14 (Table 14.4).

Pathological classification: Pathogenic species of Pseudophyllidea and Cyclophyllidea are mentioned in **Tables 100.2** and **100.3,** respectively.

Features	Pseudophyllidea	Cyclophyllidea
TABLE 100.1: Differences between Pseudophyllidea and Cyclophyllidea		
Suckers	Slit like called **bothria**	Circular shape called **acetabula**
Uterus	No branching	Branching or no branching
Uterine pore (opening)	Present	Absent
Vitelline glands	Scattered	Single mass
Vas deferens	Ascends upward	Passes laterally
Genital atrium	Presents on mid ventral surface. Allows the 3 openings of vas deferens, vagina and uterus	Presents laterally. Allows the two openings of vas deferens and vagina
Egg's shape	Oval and operculated	Spherical and non-operculated
Covers of egg	Single	Two
Embryo formation	Does not present initially in eggs, but formed later	Presents since beginning
Cilia around embryo	Present	Absent
Larvae	Solid forms called **coracidium, procercoid** and **plerocercoid**	Vesicular/bladder form called **cysticercus** or **hydatid cyst** or **cysticercoid** or **coenurus**

TABLE 100.2: Pseudophyllidean cestodes			
Genus	**Species**	**Definitive host**	**Intermediate host**
Adult stage			
Diphyllobothrium	*D. latum*	Humans, dog, cat	Cyclops and fish
Larval stage			
Spirometra	*S. mansoni*	Cat and dog	Cyclops, Frog, fish, snake
	S. proliferum	Cat and dog	Cyclops, Frog, fish, snake

TABLE 100.3: Cyclophyllidean cestodes

Genus	Species	Definitive host	Intermediate host
Adult stage			
Taenia	*T. saginata*	Humans	Cow
	T. solium	Humans	Pig, humans
Hymenolepis	*H. nana*	Humans, rat	Rat flea, human flea
	H. diminuta	Rat, humans	Rat flea, beetle
Dypylidium	*D. caninum*	Dog, cat, humans	Dog flea, cat flea, human flea
Larval stage			
Echinicoccus	*E. granulosus*	Dog, wolf, jackal	Sheep, cattle, goat, humans
	E. multilocularis	Fox, dog, dingo, wolf	Humans, field mouse, tundra vole
Taenia	*Cysticercus cellulosae* of *T. solium*	Humans	Pig, humans
Multiceps	*Coenurus* of *M. multiceps*	Fox, dog, wolf	Sheep, cattle, goat, humans

ACCESS YOURSELF

Short Questions for Theory/Viva Questions

1. What are armed tape worm and unarmed tape worm?
2. Name the larval stage for following parasites.
   T. solium, T. saginata, E. granulosus, H. nana, and *M. multiceps*

MCQs for Chapter Review

General Features (Morphology)

1. **Acetabulum of cestodes refers to:**
   a. Hook
   b. Sucker
   c. Rostellum
   d. Proglotid
2. **Larval form of *Taenia* is referred to as:**
   a. Cysticerus
   b. Cysticercoid
   c. Echinococcosis
   d. Coenurus
3. ***Cysticerus cellulosae* seen in:**
   a. *T. saginata*
   b. *T. solium*
   c. *D. latum*
   d. *S. haematobium*

Classification

4. **Which of the following is not a cestodes?**
   a. *Diphyllobothrium latum* b. *Taenia saginata*
   c. *Schistosoma mansoni*   d. *Echinococcus granulosus*

Answers and Explanation of MCQs

1. **b**
   - Follow section, **General features → morphology (adult stage) and Table 100.1** for explanation.
2. **a, b and d**
   - Echinococcosis is the disease name caused by *E. granulosus*. Cysticercus is the larval stage in *T. solium* and *T. saginata*, cysticercoid is the larval stage in *H. nana (Taenia nana)* plus *D caninum* and coenurus is the larval stage in *M. multiceps (Taenia multiceps)*.
3. **b**
   - *Schistosoma haematobium* is trematode. Follow section, **larval stage → larvae of Cyclophyllidea → name)** for explanation of other options.
4. **c**
   - *Schistosoma mansoni* is trematode while all others are cestodes. Follow section, **classification (Table 100.2 and Table 100.3)** for explanation of other options.

A. **Blood picture:** Eosinophilia.
B. **Macroscopy of stool:** It shows the presence of adult stage or segments.
C. **Microscopy of stool**
  - **For eggs: (1) Wet mount by using normal saline and iodine:** It shows the bile stained (brown color) operculated egg with knob at the bottom as shown in **Fig. 101.2.** When operculum is not examined easily, tapped the cover slip of wet mount and pressure may open the operculum. **(2) Concentration technique:** In case of negative result by wet mount method.
  - **For segments:** A segment has been identified under the microscope by knowing the characters of uterus and position of uterine pore.

D. **Isozyme technique:** It is useful to identify the segment.
E. **Serological tests:** ELISA and latex agglutination are positive in late cases.

Prevention and Treatment

Preventive measures are proper cooking of fish, deep freezing (–10°C for 24–48 hours) of fish if it is consumed to be raw, periodical deworming of cats and dogs and avoiding the fecal contamination of water. Praziquantel (single dose 10 mg/kg) is effective. Parenteral B_{12} is given if deficiency is present.

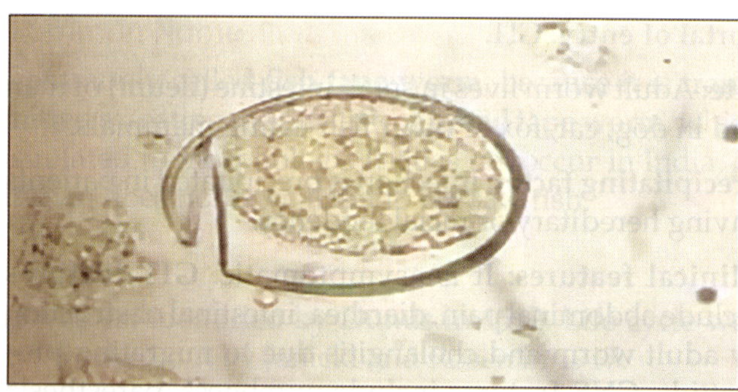

Fig. 101.2: Egg of *D. latum* under wet mount

Spirometra spp.

It is described separately in box.

Note: *Spirometra* spp.

Morphology: It is similar to *D. latum* with following important features.
- Adult stage does not occur in humans, but in dog or cat.
- Human is an accidental host and infected by 2nd or 3rd stage of larva.
- **Sparganum (3rd stage of larva):** It is creamy-white, 10–30 mm × 2–7 mm in size, elongated and flattened, motile and infective for human. Anterior end is without scolex (head) and marked by vertical groove. Body wall consists two layers of muscles and one layer of tegument. It can live for more than 9 years in human. Developmental variation occurs in *S. proliferum* when it founds in granulomatous nodule/cyst. It often proliferates to give supernumary buds which eventually detach themselves and develop in to larvae.

Life cycle
- **Types of host:** Three types.
  - **Definitive hosts:** Dog/cat.
  - **Intermediate hosts:** Larval stage develops in two different aquatic animals. **(1) 1st intermediate host:** Cyclops. **(2) 2nd intermediate host:** Frog/snakes/fish. Human is an accidental host.
- **Cycles:** Two main types of cycles as shown in **Flowchart 101.2 and Fig. 101.3. (1) Cycle between definitive and intermediate hosts:** It is continuing between dog/cat, cyclops and frog/snake/fish. **(2) Cycle in accidental host (cycle in humans):** It starts from definitive host (dog/cat), passes through intermediate hosts (cyclops and frog/snakes/fish) and ends in accidental host (humans).

Pathogenicity
- **Disease name:** Human is infected by second and third stage of larva, but disease is produced by 3rd stage of larva called sparganosis.
- **Epidemiology:** *S. mansoni* found in many parts of Asia like Japan, China, Vietnam, Taiwan, Korea and Thailand. *S. mansonoides* found in America. Few cases reported from India also.
- **Reservoirs of infection:** Dog/cat.
- **Sources of infection:** Cyclops/fish/frog.
- **Modes of transmission: (1) Direct contact:** By applying the raw flesh to skin sore or eye sore or vagina as poultices (in China as custom). Wounded skin allows the entry of parasites. **(2) Indirect:** By ingestion of drinking water contains infected cyclops with 2nd stage of larva or raw fish/frog contains 3rd stage of larva or pork (pig's meat) contains plerocercoid larvae (in case of *S. mansoni*).
- **Exit form:** Unembryonated egg exits from dog/cat.
- **Infective form:** 3rd stage of larva (2nd stage is only transmitting stage not infective).
- **Portal of entry:** GIT and skin.
- **Sites:** Adult worm naturally habitats in intestine of dog and cat. Human harbors the larval stage in peritoneum, subcutaneous tissue, eyes, muscles, brain and abdominal viscera.
- **Precipitating factor:** It is common in Chinese people due to customs of applying frog's flesh as poultices to skin sore or eye sore or vagina.
- **Pathogenesis:** It is characterized by granulomatous nodule with infiltration of lymphocytes, macrophages, eosinophils, neutrophils, plasma cells, giant cells. Charcoat-Layden crystals and multinucleated giant cells may be seen.
- **Clinical features: Granulomatous nodule/cyst** presents in peritoneum, subcutaneous tissue, muscles, eyes, brain and abdominal viscera. **Eye swelling** should be differentiated from Chaga's disease (Roman's sign).

Laboratory diagnosis
- **Specimens:** Tissue removal or biopsy from nodule.
- **Testing methods**
  A. **Blood picture:** Eosinophilia occurs in 30–40% cases.
  B. **Macroscopy:** To rule out the presence of sparganum. It looks like narrow tape worm and differentiated from adult stage of other tape worms by absence of suckers and hooklets.
  C. **Species identification:** It can be done by allowing the cat or dog to eat live sparganum and subsequent identification of adult worm.
  D. **Histopathological staining:** It is also useful.
  E. **Serological tests:** Agar gel diffusion and IHA tests are positive.

(Contd…)

Cycle between definitive and intermediate hosts

Cycle in definitive host (dog / cat): Infection to cat/dog by ingestion of fish or frog contains plerocercoid larvae. Entry in intestine and change to sexually mature adult stage which lays the umembryonated eggs, which passed with feces in water

Cycle in water: Change to embryonated eggs and embryo (larva) called **coracidium** or **1st stage of larva.** Release of coracidium free in water on hatching

Cycle in 1st intermediate host (cyclops): Coracidium taken up by cyclops and transformed to 2nd stage larva called **procercoid**

Cycle in accidental host (cycle in human): Accidental ingestion of **procercoid** in human by drinking the water contain the infected cyclops. Intestinal digestion of cyclops and release of procercoid. It transformed to 3rd stage larva called **plerocercoid/sparganum.**

It penetrates the gut wall and enters in peritoneum, subcutaneous tissue, muscles, brain and abdominal viscera.

Formation of granulomatous swelling called **sparganosis** and **dead end** of parasites (no continuation of cycle).

Direct entry in subcutaneous tissues

1. Applying the raw flesh of frog to skin sore or eye sore or vagina as poultices (In China as custom)

Accidental entry of **sparganum** in human by **1 and 2**

2. Ingestion of raw fish contains sparganum

Intestinal digestion of flesh and release of sparganum

Ingestion of infected cyclops in fish/snake/frog

Cycle in 2nd intermediate host (fish/snake/frog: Digestion of cyclops and release of procercoid free in intestine

Transformed to **sparganum**

Ingestion in dog/cat by eating the fish or frog

Cycle repeated

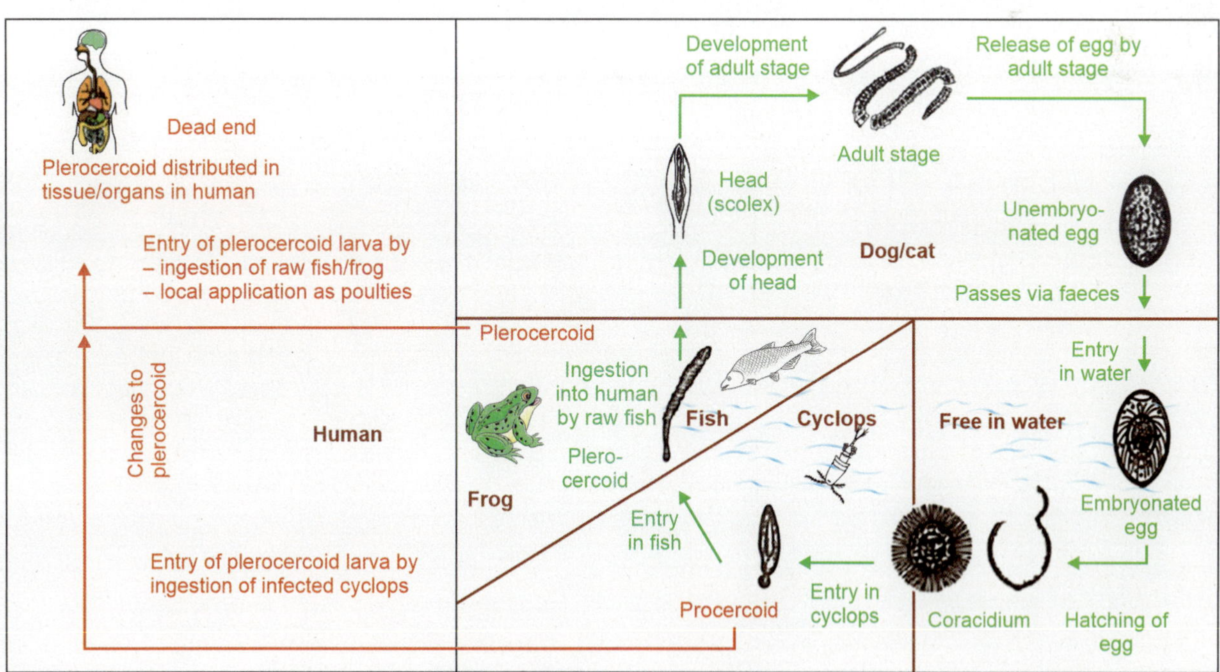

Fig. 101.3: Life cycle of *Spirometra*

(Contd...)

Prevention: Preventive measures are proper cooking of fish, proper boiling and filtering the drinking water, periodical deworming of cats and dogs and avoiding the fecal contamination of water from infected animals.

Treatment: Surgical removal of nodule.

Disease name: Human is infected by second and third stage of larva, but disease is produced by 3rd stage of larva called **sparganosis.**

Short Notes

1. Sparganosis.

Short Questions for Theory/Viva Questions

1. Name the four parasites in which cyclops acts as an intermediate host or cyclops required to complete the life cycle.
2. Write the name of 1st, 2nd and 3rd stages of larva of *D. latum*.

Comments On

1. *D. latum* causes pernicious and megaloblastic anemia.

MCQs for Chapter Review

Diphylobothrium latum

1. Second intermediate host of *Diphylobothrium latum* is:

a. Cyclops   b. Man
c. Snail   d. Fresh water fish

2. Vitamin deficiency due to *Diphylobothrium latum* infection:

a. B1
b. B2
c. B4
d. B12

3. *Diphylobothrium latum* is causative organism of:

a. Megaloblastic anemia
b. Iron deficiency anemia
c. Peptic ulcer
d. None

Answers and Explanation of MCQs

1. d
- Follow section, *D. latum* **(life cycle)** for explanation.

2. d

3. a
- Follow section, *D. latum* **(pathogenicity → clinical features)** for explanation of answers of MCQs 2–3.

CHAPTER 102

Infections of Cestodes: Cyclophyllidea

Chapter Outline

Taenia spp.

Two species like *Taenia saginata* and *Taenia solium*. *Taenia saginata* has two subspecies like *Taenia saginata saginata* and *T. saginata asiatica*.

Taenia saginata saginata

Meaning

Taenia (Greek) means **tape or band,** as it is dorsoventrally flat like measure tape.

Common Name

1. **Beef tape worm:** As it is transmitted by eating the raw beef (beef means cow's meat) contains *Cysticercus bovis* (larval stage). Such infected beef looks like human measles so called **measly beef** or **beef measles** (beef and larva means measly beef).
2. **Unarmed tape worm:** Word is related to morphology like unarmed means without hooklets while tape worm means dorsoventrally flat and broad adult stage as like a measure tape.

History

Tapeworm infection has been recorded in history from 1500 BC and has been recognized as one of the earliest human parasite. *T. saginata* was differentiated from *T. solium* infection by the late 1700s by Goeze. Exact life cycle of *T. saginata* was discovered around 1863 when the cattle were identified as the immediate host by Leuckart.

Morphology

Three morphological stages like adult, egg and larva are described below.

Adult Stage

Time of maturation: Larva transformed to sexually mature adult stage in 2–3 months in definitive host.

Color: White and semitransparent.

Size: 5–10 meters in length (may be up to 24 meters)

Shape: Like a measure tape

Life span: 10 years

Morphological parts: Three parts like head, neck and strobila (body/trunk) as shown in **Fig. 102.1a**.

- **Scolex/head (Fig. 102.1b):** It is quadrate in shape, larger about 1–2 mm in diameter, has four circular pigmented suckers called **acetabula** and without rostellum and hooklets.
- **Neck:** It is long, narrow about 0.5 mm wide and fragile.

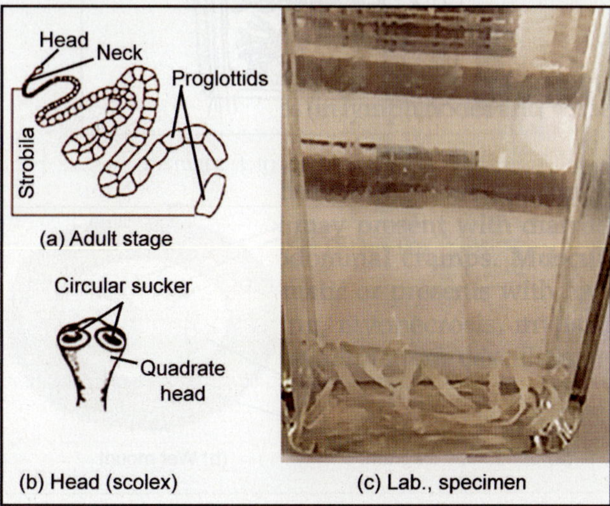

Fig. 102.1: Adult stage of *T. saginata*

858

- **Strobila (body or trunk):** It is composed of chain of segments called **proglottids or segments (Fig. 102.2).** They are 1,000–2,000 in numbers and 22 mm × 5 mm in size (gravid segment). Length is 3–4 times greater than breadth. They are of three types according to maturity and order from front to backwards is immature (male and female organs are not differentiated), mature (male and female organs are differentiated and male organs are appear first) and gravid (uterus contains eggs).
- **Expulsion of eggs:** They are expelled singly and may force anal sphincter. Free gravid segment while crawling out of anal orifice oviposits in the perianal skin, which are collected by NIH swab.
- **Reproductive system:** It is monoecious or hermaphrodite. Uterus has 15–30 lateral branches on each side, thin and dichotomous (sub-branches). Two ovaries (without accessory lobe) and vaginal sphincter are present. Testes have 300–400 follicles. Genital atrium presents near the posterior end of lateral margin of each segment and alternates irregularly between right and left margins. Features ootype, Mehli's gland, vas efferentia, vas deferens and cirrus (penis) are described in **Ch. 100.**

Egg Stage

Egg formed in ootype and surrounded by protective covering. Egg is brown color (bile stained), spherical **(Fig. 102.3)**, nonoperculated and 31–43 µm in diameter.

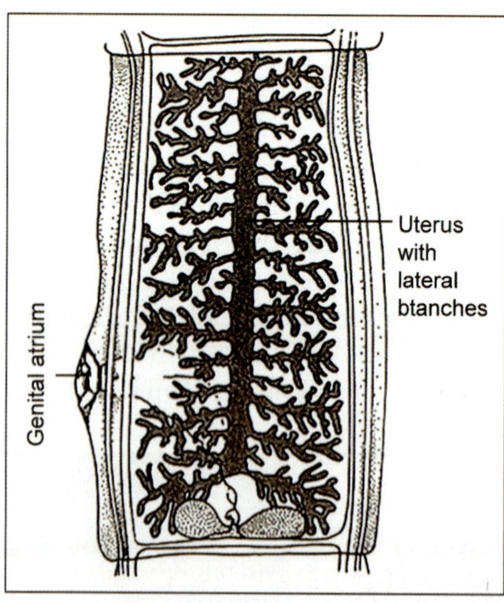

Fig. 102.2: Segment of *T. saginata*

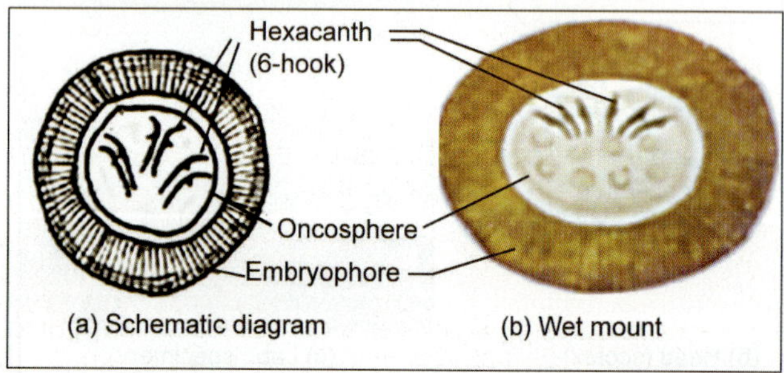

Fig. 102.3: Egg of *T. saginata* and *T. solium*

It lives about 8 weeks on field. Egg has two covers. **Outer cover** is thin called **egg shell.** Sometimes, it ruptures before the eggs are excreted, but when present it represents the remnant of yolk mass. It causes the eggs to clump together. **Inner cover** is thick called **embryophore.** It becomes very thick after rupture of egg shell and radially striated to protect the inner oncosphere. Egg contains embryo from beginning when laid down called **oncopshere** with 14–20 µm in size. It contains 6 hooklets (3 pairs) called **hexacanth,** which helps to penetrate the gut wall in intermediate host. There are no cilia around embryo. Egg does not float in saturated common salt solution and is infective only to cattle. It is acid fast under modified ZN stain which is the only way to differentiate it from the egg of *T. solium.*

Larval Stage

Larva develops in muscles of intermediate host (cow/buffalo) and matures further when ingested by human. It does not occur in human. It is oval in shape and 3–4 mm × 5–10 mm in size. Larva lives for 8 months in muscles of cattle. It called *Cysticercus bovis* (taxonomically not correct). It is without hooklets. Scolex (future head of adult worm) is invaginated from one side **(Fig. 102.4a),** later exvaginated **(Fig. 102.4b)** in definitive host. Egg develops the *Cysticercus* within 60–70 days.

Life Cycle

Types of hosts: Two types.
- **Definitive host:** Human (adult stage).
- **Intermediate host:** Cow/buffalo (larval stage).

Cycles: Two types as shown in **Flowchart 102.1 and Fig. 102.5.**

Pathogenicity

Disease name: Called **intestinal taeniasis.**

Epidemiology: It is distributed worldwide

Reservoir of infection: Human

Source of infection: Cattle (beef)

Mode of transmission: By ingestion of raw beef contains *C. bovis* (measly beef or beef measles)

Exit form: Embryonated egg/segment

Infective form: Larva called *Cysticercus bovis.*

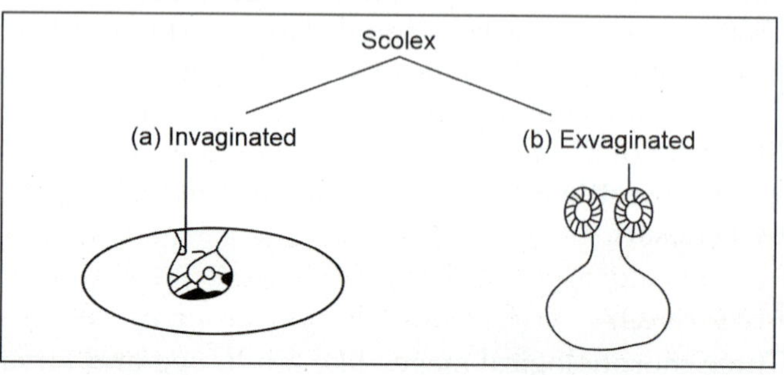

Fig. 102.4: Larva of *T. saginata* and *T. solium*

Flowchart 102.1: Life cycle of *T. saginata*

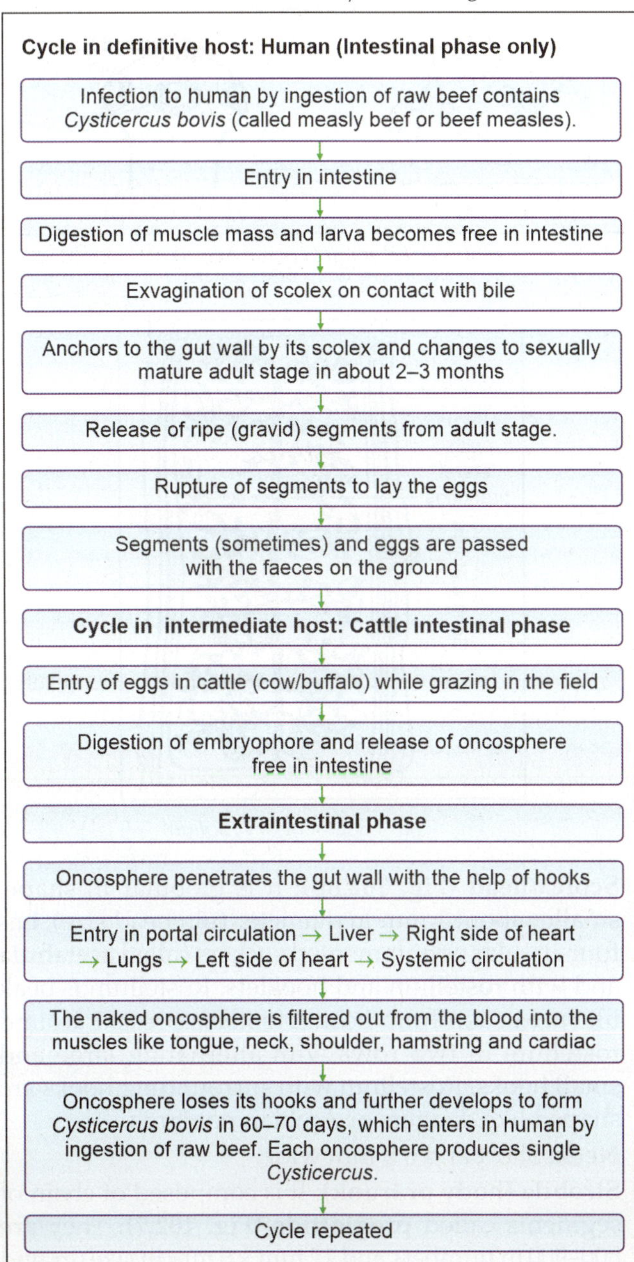

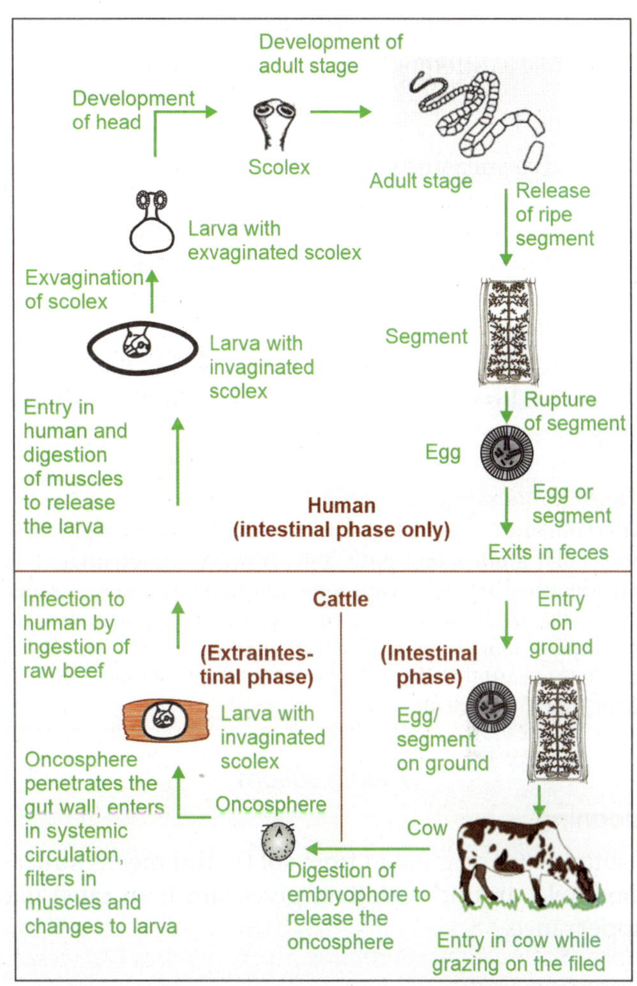

Fig. 102.5: Life cycle of *T. saginata*

Portal of entry: GIT

Site: Adult worm lives in small intestine (upper jejunum) of man and moves against peristalsis.

Clinical features: It is mostly asymptomatic, but sometimes features developed by adult stage like **GIT features,** which include abdominal discomfort like indigestion, diarrhea with alternating constipation, intestinal obstruction and appendicitis and anemia.

Laboratory Diagnosis

Specimens: (1) Stool contains segments of adult worm and eggs. **(2) Eggs from perianal region** are collected by NIH (cellophane) swab or by cellulose (scotch) tape method. **(3) Muscle biopsy** from slaughtered animal.

Testing methods

A. History: Of beef eating is present.

B. Macroscopic (naked eye)
- **For segments:** White segments present against yellowish stool. Sometimes, segments also present in clothes of patient.
- **For scolex:** Species differention can be made by macroscopic examination of scolex (if adult stage is available) after anthelmintic treatment.

C. Microscopy
- **For eggs: (1) Wet mount of NS and iodine** shows the bile stained (brown color) eggs as shown in **Fig. 102.3b. (2) Concentration technique** is useful in case of negative result by wet mount method. **(3) ZN stain** is useful to differentiate *T. saginata* from *T. solium*. Egg of *T. saginata* is acid fast, while of *T. solium* is non-acid fast.
- **For segments:** Species differention (as shown in **Table 102.1**) can be made by microscopic examination of segments. It can be pressed between two slides and examined by hand lens.

D. Isozyme technique: It is useful to identify the segment.

E. Serological tests: ELISA detects the Ag from stool. It is not able to differentiate *T. saginata* from *T. solium* and subspecies *T. saginata saginata* from *T. saginata asiatica*.

F. Molecular methods: PCR is useful to diagnose and to make a difference between species and subspecies.

Prevention

Preventive measures are periodical checking of beef for presence of larvae, proper cooking of beef (critical thermal point for larvae is 56°C for 5 min.) and avoiding the fecal contamination of soil, water and food.

Treatment

Praziquantel (single dose of 10 mg/kg) is the drug of choice. Niclosamide (2 g single dose) is another effective drug.

> **Note: *Taenia saginata asiatica***
> It is similar to *T. saginata saginata* on the basis of morphology, molecular studies and pathology; however differentiated as **(1)** identified from Taiwan, Korea, Phillipines and few other parts, **(2)** adult stage has two rows of rudimentary hooks, **(3)** smaller than *T. saginata saginata* and **(4)** intermediate host is pig (not cattle) harbors the larval (*Cysticercus bovis*) stage in liver (not in muscles).

Taenia solium

Meaning

Solium word originated from **sol (Latin)** means **sun,** as the hooks around rostellum gives **sun (sun rays)** like appearance.

Common Name

1. **Pork tape worm:** As it is transmitted by eating the raw pork (pork means pig's meat) contains *Cysticercus cellulosae* (larval stage). Such infected pork looks like human measles so called **measly pork** or **pork measles** (pork and larva means measly pork).
2. **Armed tape worm:** Word is related to morphology like armed means presence of hooklets on head while tape worm means dorsoventrally flat and broad adult stage as like a measure tape.

History

It was discovered by Linnaeus in 1758.

Morphology

It has three morphological stages like adult, egg and larva are described below.

Adult Stage

Time of maturation: Larva transformed to sexually mature adult stage in 2–3 months in definitive host.

Color: White and semitransparent

Size: 2–3 meters in length

Shape: Like a measure tape

Life span: 25 years.

Morphological parts: Three parts like head, neck and strobila (body/trunk) as shown in **Fig. 102.6.**

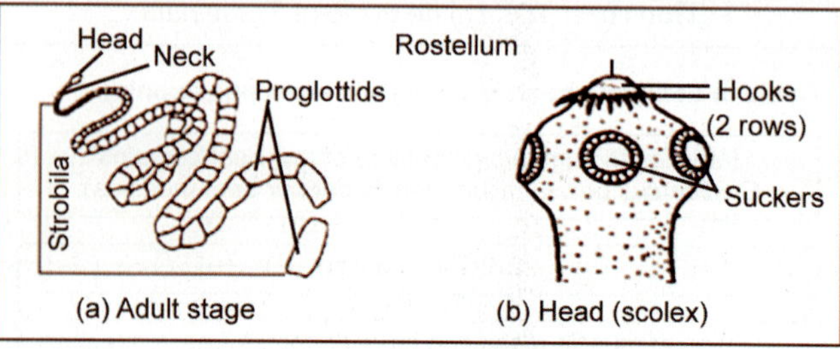

Fig. 102.6: Adult stage of *T. solium*

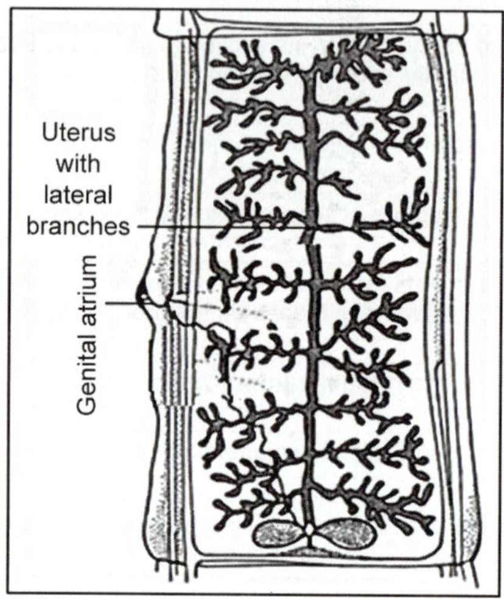

Fig. 102.7: Segment of *T. solium*

- **Scolex/head (Fig. 102.6b):** It is globular in shape, smaller about 1 mm in diameter (pin head size), has four circular nonpigmented suckers called **acetabula** and with rostellum and hooklets. Rostellum is beak like projection from head. Hooklets present around rostellum in two rows with alternating large and small hooks. Rostellum with surrounding hooks are shaped like daggers or Arabian poniards.
- **Neck:** Short and 0.5 mm wide.
- **Strobila (body or trunk):** It is composed of chain of segments called **proglottids (Fig. 102.7).** They are 800–900 in numbers and 12 mm × 6 mm in size (gravid segment). Length is twice than breadth. Proglottids are three types same as *T saginata saginata* described above.
- **Expulsion of eggs:** Expelled in chain of 5–6 numbers and passively.
- **Reproductive system:** It is monoecious or hermaphrodite. Uterus has 5–10 lateral branches on each side, thick and dendritic. Two ovaries (with an accessory lobe) are present and vaginal sphincter is absent. Testes have 150–200 follicles. Genital atrium presents near the middle of lateral margin of each segment and alternates irregularly between right and left margins. Features of ootype, Mehli's gland, vas efferentia, vas deferens and cirrus (penis) are described in **Ch. 100.**

Egg Stage

Eggs are same as *T. saginata* (**Fig. 102.3**) but non-acid fast.

Larval Stage

Larva develops in muscles of intermediate host (pig) and matures further when ingested by human. It also occurs in brain, eyes, subcutaneous tissues and muscles of human. It is oval in shape and 5 mm × 10 mm in size. Larva lives for 8 months in muscles of pig. It called *Cysticercus cellulosae* (taxonomically not correct). It is without hooklets. Scolex (future head of adult worm) is invaginated from one side **(Fig. 102.4a)**, later exvaginated **(Fig. 102.4b)** in definitive host. Egg develops the *Cysticercus* within 60–70 days. It contains fluid rich in salt and albuminous materials.

Life Cycle

Types of hosts: Two types
- **Definitive host:** Human (adult stage)
- **Intermediate host: (1) Pig:** Harbors the larval stage. **(2) Human:** Sometimes, human also act as an intermediate host and harbors the larval stage.

Cycles: Two types as shown in **Flowchart 102.2 and Fig. 102.8.**

Pathogenicity

Disease name: Disease produced by adult called **intestinal taeniasis** and by larvae called **cysticercosis.**

Flowchart 102.2: Life cycle of *T. solium*

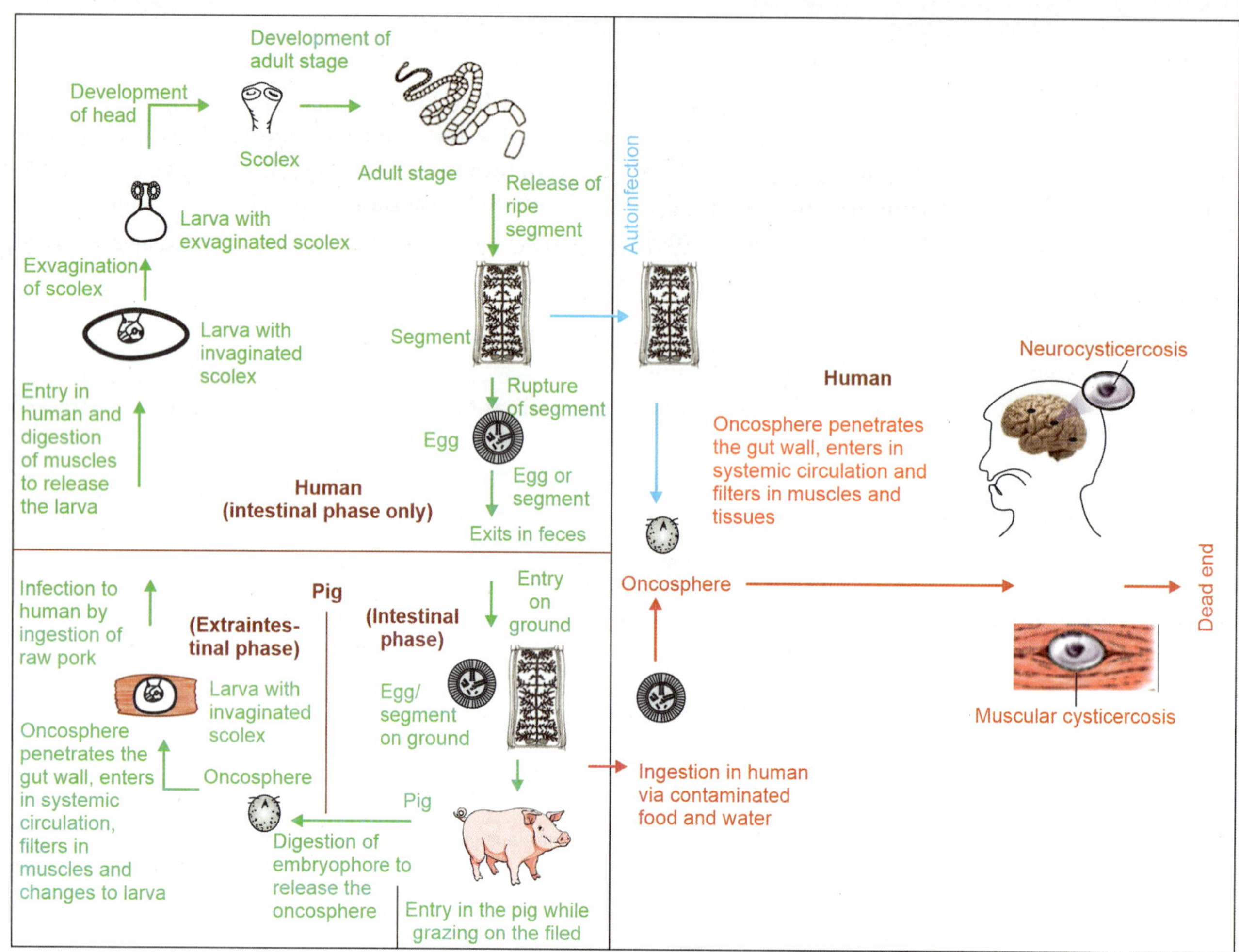

Fig. 102.8: Life cycle of *T. solium*

Epidemiology: It is distributed worldwide and common in pork eaters; however, it may occur in vegetarians due to consumption of contaminated food and water contain eggs.

Reservoir of infection: Human.

Source of infection: Pig (pork).

Modes of transmission:

- **Larva stage infection:** By **ingestion** of raw pork contains *C. cellulosae* (measly pork/pork measles).
- **Egg stage infection: (1) Autoinfection:** Entry of gravid segments in stomach from intestine by reversal of peristalsis (regurgitation) or by finger contamination with eggs (oral fecal route). **(2) Ingestion:** Ingestion of contaminated food and water contain eggs.

Exit form: Embryonated egg/segment.

Infective form: Human is infected by both larvae and eggs of *T. solium* and lesions are due to adult plus larval stages, while in *T. saginata* only egg is infective and lesions are only due to adult stage, hence *T. solium* is more dangerous than *T. saginata*.

Portal of entry: GIT.

Sites: Adult worm lives in small intestine (upper jejunum) of human while larva lives in brain, eyes, subcutaneous tissues and muscles.

Clinical types and features: Two clinical types like **adult stage disease** and **larval stage disease.** Features of adult stage disease are same as *T. saginata*. Larval stage disease called **cysticercosis** described below.

1. **Subcutaneous cysticercosis:** It infects mostly subcutaneous tissues and muscles. It presents as palpatory nodule or inspector swelling. It is mostly asymptomatic and should be excised.
2. **Muscular cysticercosis:** It may be present as an acute myositis.
3. **Ocular cysticercosis:** It is located in vitreous humor, subretinal space and conjunctiva. It produces blurred or loss of vision, iritis, uveitis and palpebral conjunctivitis. It is identified by fundoscopy.
4. **Cysticercosis of brain (neurocysticercosis):** Described separately in box.

Note: Neurocysticercosis

Epidemiology: It is the most common parasitic CNS lesion worldwide. It is the 2nd most common cause of ICSOL after tuberculosis in India. In India, it is reported from Bihar, Punjab, Uttar Pradesh and Odisha. It mostly infects adults of 30–40 years and risk increased due to AIDS.

Sites: Brain parenchyma is most common sites of neurocysticercosis. Extraparenchymal sites are meninges, ventricles, spinal cord and subarachnoid space.

Racemose cysticerci: It is due to *Cysticercus cellulosae*. Mode of entry in brain is described in **Flowchart 102.2** and **Fig. 102.8**. It is 5 mm in length but sometimes >20 cm called **racemose cysticerci.**

Symptoms: It is the most serious form. It is the causative agent of adult epilepsy in 70% cases. It also causes rise in intracranial tension, psychiatric disturbances, hydrocephalus, meningoencephalitis, transient paresis, behavior disorders, etc.

Complications: Fibrosis and calcification in 5–6 years.

Diagnosis
- **Specimens:** CSF and brain biopsy.
- **Testing methods: (1) Blood picture:** Eosinophilia. **(2) Ag is detected** from CSF or serum by ELISA. Ag disappears after treatment; hence it is useful to monitor the therapy. **(3) Ab is detected** from CSF or serum by ELISA and WB test. **(4) Radiological method** like CT scan or MRI detects the size, numbers, site, calcification or fibrosis of cyst. **(5) FNAC** of cyst of brain is performed by Giemsa stain. **(6) Revised Del Brutto's criteria** are used in endemic areas. They based on clinical, radiological, immunological and epidemiological parameters.

Treatment: Praziquantel 50 mg/kg in 3 divided doses for 20–30 days and albendazole, 400 mg BD for 30 days are given. Steroid is given to avoid the inflammatory reaction caused by dead larvae. Antileprotic drugs are advised until the brain reaction is subsided. Surgery is indicated in case of hydrocephalus.

Laboratory Diagnosis

Diagnosis of disease caused by adult stage: Stool (for segments of adult worm) and muscle biopsy (from slaughtered animals) are tested same as *T. saginata*.

Diagnosis of disease caused by larval stage: Follow cysticercosis.

Prevention

Preventive measures are periodical checking of pork for presence of larvae. Other measures are same as *T saginata*.

Treatment

Intestinal lesions: Same as *T saginata*.

Treatment of cysticercosis: Follow cysticercosis.

Differences between *T. saginata* and *T. solium*

Follow **Table 102.1**.

Echinococcus spp.

Echinococcus granulosus

Common Name

1. **Dog tape worm:** As dog is the definitive (optimum) host.
2. **Hydatid worm:** Name given from the larval stage name called **hydatid cyst** which causes the human disease.
3. **Hypertape worm.**

TABLE 102.1: Differences between *Taenia* spp.

Features	*T. saginata*	*T. solium*
Common name	• Beef tape worm • Unarmed tape worm	• Pork tape worm • Armed tape worm
Subspecies	Two like *saginata* and *asiatica*	No subspecies
Synonym	*Taeniarhynchus saginata*	*Taenia cucurbitiana*
Geographical distribution	In beef eater (non-vegetarian) and not in vegetarian	In pork eater (non-vegetarian) and in vegetarian
Habitat	Only intestine	Intestine and tissues
Size	5–10 meters	2–3 meters
Life span	10 years	25 years
Scolex (head)		
Figure	**Figure 102.1b**	**Figure 102.6b**
Shape	Quadrate	Globular
Size	Larger, 1–2 mm in diameter	Smaller, 1 mm in diameter
Suckers	Four circular suckers which may be pigmented	Four circular suckers which are not pigmented
Rostellum	Absent	Present
Hooklets	Absent	Present
Neck	Long, narrow	Short
Proglottids (gravid segments)		
Figure	**Figure 102.2**	**Figure 102.7**
Numbers	1,000–2,000	<1,000 (800–900)
Size	22 mm × 5 mm, length is 3–4 times more than breadth	12 mm × 6 mm Length is twice than breadth.
Expulsion of eggs	Singly and force to anal sphincter.	Expelled in chain of 5–6 and passively
Uterus	15–30 lateral branches on each side, thin and dichotomous	5–10 lateral branches on each side, thick and dendritic
Ovary	Two without any accessory lobe	Two with an accessory lobe
Vagina	Sphincter present	Sphincter absent
Testes	300–400 follicles.	150–200 follicles.
Genital atrium	Present near the posterior end of lateral margin	Present near the middle of lateral margin
Eggs (Figure 102.3)		
Stain	Acid fast	Non acid fast
Infective to human	No	Yes
Larvae (Figure 102.4)		
Name	C. bovis	C. cellulosae
Size	3–4 mm × 5–10 mm	5 mm × 10 mm
Host	Cattle only	Pig and human
Hooklets	Absent	Present
Hosts and other differences		
Definitive	Human	Human
Intermediate	Cattle only	Pig and human
Infection	Only by larvae	By larvae and eggs
Disease	Intestinal taeniasis	Intestinal taeniasis and cysticercosis

It is an ancient disease, known since the time of Hippocrates and other physician. Adult stage was described by Hartmann in 1695 from dog's intestine and larval stage was described by Goeze in 1782.

Morphology

Three morphological stages like adult, egg and larva are described below.

Adult Stage

Time of maturation: Larva transformed to sexually mature adult stage in 6–7 weeks in definitive host.

Color: White and semitransparent.

Size: 3–6 mm in length.

Shape: Like a measure tape.

Life span: 6–30 months in canine host.

Morphological parts: Three parts like head, neck and strobila (body/trunk) as shown in **Fig. 102.9**.
- **Scolex/head:** It is pyriform in shape with 4 circular suckers. Rostellum is prominent. Hooklets present around rostellum in 2 circular rows.
- **Neck:** It is shorter than rest of worms (3 mm × 6 mm).
- **Strobila (body or trunk):** It is composed of chain of segments called **proglottids.** They are 3-4 in numbers and 2–3 mm × 0.6 mm in size (gravid segment). Length is 3–4 times greater than breadth. Three segments are same as *T. saginata saginata* described above. **4th segment** present very rarely and gravid when present.
- **Reproductive system:** It is monoecious or hermaphrodite. Genital atrium presents on the lateral margin of each segment. All the female and male reproductive organs are shown in **Fig. 102.9** and also described in **Ch. 100**.

Egg Stage

Egg is brown color (bile stained), 32–36 μm × 25–32 μm in size and oval in shape. Other morphological features are same as *T. saginata* (**Fig. 102.3**).

Larval Stage

Development: It is a vesicular body with invaginated scolex develops in intermediate host like human and herbivorous animals like sheep, goat, cattle, pig and horse.

Size: Variable with 0.5–1 cm in diameter.

Shape: Oval or spherical.

Life span: It may continue to develop for many years.

Morphological features (Fig. 102.10): Larva is a vesicular body with invaginated scolex present in a cystic structure called **hydatid** or **hydatid cyst.** Word hydatid derived from hydatis (Greek) means drop of water, because containing fluid. Eggs develop the hydatid in about 6 months. Oncosphere secrets following two layer from inner to outer.

1. **Endocyst:** It also called **inner layer** or **germinal layer.** It is 22–25 μm thick, cellular and consists of a number of nuclei embedded in protoplasmic mass. It forms the ectocyst, secretes the hydatid fluid and gives brood capsules with scolices. Scolices are invaginated and present with hooks, which later exvaginated in definitive host to form adult stage.

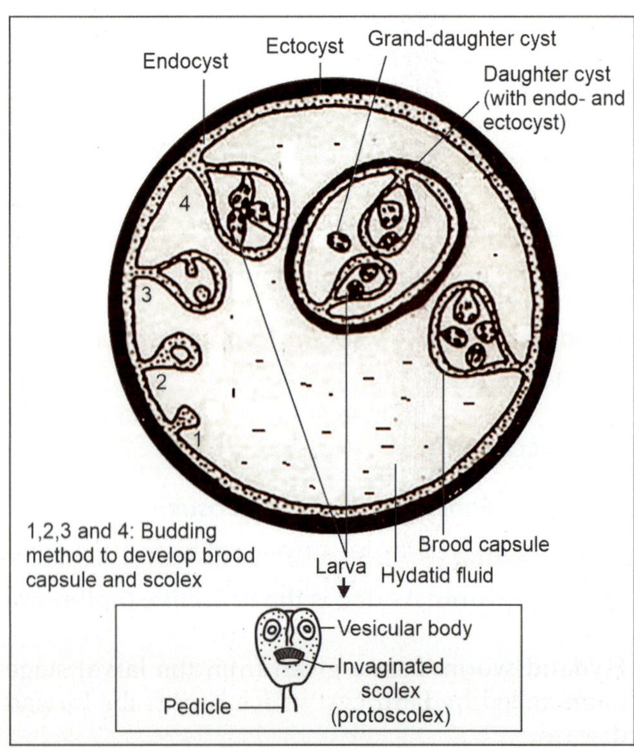

Fig. 102.9: Adult stage of *E. granulosus*

Fig. 102.10: Hydatid cyst

2. **Ectocyst:** It also called **outer layer** or **cuticular layer** or **laminated layer** or **hyaline layer.** It is 1 mm thick, elastic and appears like hard-boiled egg. It is protective and exposing the inner layer on incision or rupture.

Hydatid fluid: It may be clear and colorless or sometimes pale yellow, pH is 6.7 and slightly acidic and specific gravity is low from 1.005 to 1.010. It contains **biochemical substances** like sodium chloride, sodium sulfate, sodium phosphate and sodium and calcium salts of succinic acid (called **Fehling reducing substances**) and granular materials in form of liberated brood capsules, free scolices and loose hooks, present at the bottom of cyst called **hydatid sand.** Hydatid fluid is highly toxic and when absorbed in systemic circulation it gives allergic or anaphylaxis reactions. It is antigenic in nature and used for Casoni's test.

Development of brood capsule: Brood capsule sprouts from endocyst by budding method. Development starts with small spherical bud which becomes vacuolated and transformed in to vesicle.

Development of scolices: Scolices also developed by budding method from brood capsule. They are 5–20 in numbers. Newly formed scolices are called **protoscolices.** Scolex represents the development of future head of adult worm. It remains invaginated and presents with suckers and hooks. In growing hydatid cyst all stages of scolex are present from small bud to fully developed scolex with suckers and hooks. It is attached to the brood capsule by pedicle or remains free as grains of hydatid sand.

Life Cycle

Types of hosts

- **Definitive hosts:** Carnivorous animals like dog (optimal/ideal host), fox, jackal and wolf.
- **Intermediate hosts:** Herbivorous animals like sheep (optimal/ideal host), goat, cattle, pig and horse. Human (child) is an accidental (dead end) intermediate host.

Cycles: Two types as shown in **Flowchart 102.3 and Fig. 102.11.** Natural cycle continue between carnivorous animal and herbivorous animal.

Pathogenicity

Adult stage disease: Adult stage is found in small intestine (duodenum and jejunum) of carnivorous animals like dog, fox, jackal and wolf without any illness. 100–1000 parasites are present in dog that can be examined at postmortem examination. Eggs are passed in feces of dog.

Larval stage disease: Larval stage found in tissues of human and herbivorous animals like sheep, goat, cattle, pig and horse. Human infection is described below.

- **Disease name:** It is a zoonosis and human disease presents with unilocular cyst called **hydatid disease,** or **hydatidosis** or **echinococcal disease** or **echinococcosis.**
- **Epidemiology:** It is distributed worldwide. It is common in sheep and cattle raising countries like Africa, Australia, South America, Europe, China, Middle East and India.
- **Reservoir of infection:** Dog.
- **Source of infection:** Food is the source for human infection.
- **Modes of transmission:** Eggs enter in human by (1) ingestion of food contaminated with feces of infected canine. Eggs do not enter by water, as the heavy eggs sink at the bottom, (2) eating in a same dish with dog and (3) direct contact (may by finger contamination during handling or fondling of dog by children).
- **Exit form:** Human acts as dead end host and eggs exit from definitive host (dog).
- **Infective form:** Eggs.
- **Portal of entry:** GIT.
- **Sites:** (1) Liver (most common site and mostly in right lobe): 65–70%. (2) Lungs (2nd most common site after liver and mostly lower lobe of right lung): 20%. (3) Pelvic organs, muscles and other tissues: 1–3%. (4) Others: Brain (5%), kidneys (2%), spleen (2%) and bones (1%).
- **Precipitating factors:** Common in children, farmers, shepherds and shoe maker.
- **Pathology and pathogenesis:** It mostly presents as unilocular cyst. In man, it grows very slow and at the end of a year, it is approximately 4 cm in diameter and brood capsules and scolices begin to appear. H cyst shows variations in growth. **(1) Acephalocyst:** It is the sterile cyst, because it is without scolices. Large numbers of acephalocysts are found in cattle. Sometimes, brood capsule is not developed and if developed, cyst is without scolices called **acephalocyst. (2) Endogenous daughter cyst:** It develops after many years and particularly in man. It also presents with endocyst and ectocyst. It develops probably in mother cyst by detachment of fragment of germinal layer or by regressive changes or regressive metamorphosis of the young brood capsule and scolex. It is the characteristic of cestodes "to migrate against the law" with transformation of scolex in daughter cyst in a same host without passing to the adult stage. This phenomenon was observed very early by Naunyn in 1862 and enunciated by Van Beneden. Normally, fully developed scolex is the end product which changes to adult stage when entered in definitive host. It is not able to go regressive changes to produce daughter cyst. Experimental work had been confirmed by Deve in 1925, shows that *E. granulosus* does not follow the general biological rule of development and shows the regressive changes to produces the daughter cyst. **(3) Grand-daughter cyst:** It develops in daughter cyst. **(4) Pericyst:** It also called **adventitial layer.** Development of hydatid cyst from

Flowchart 102.3: Life cycle of *E. granulosus*

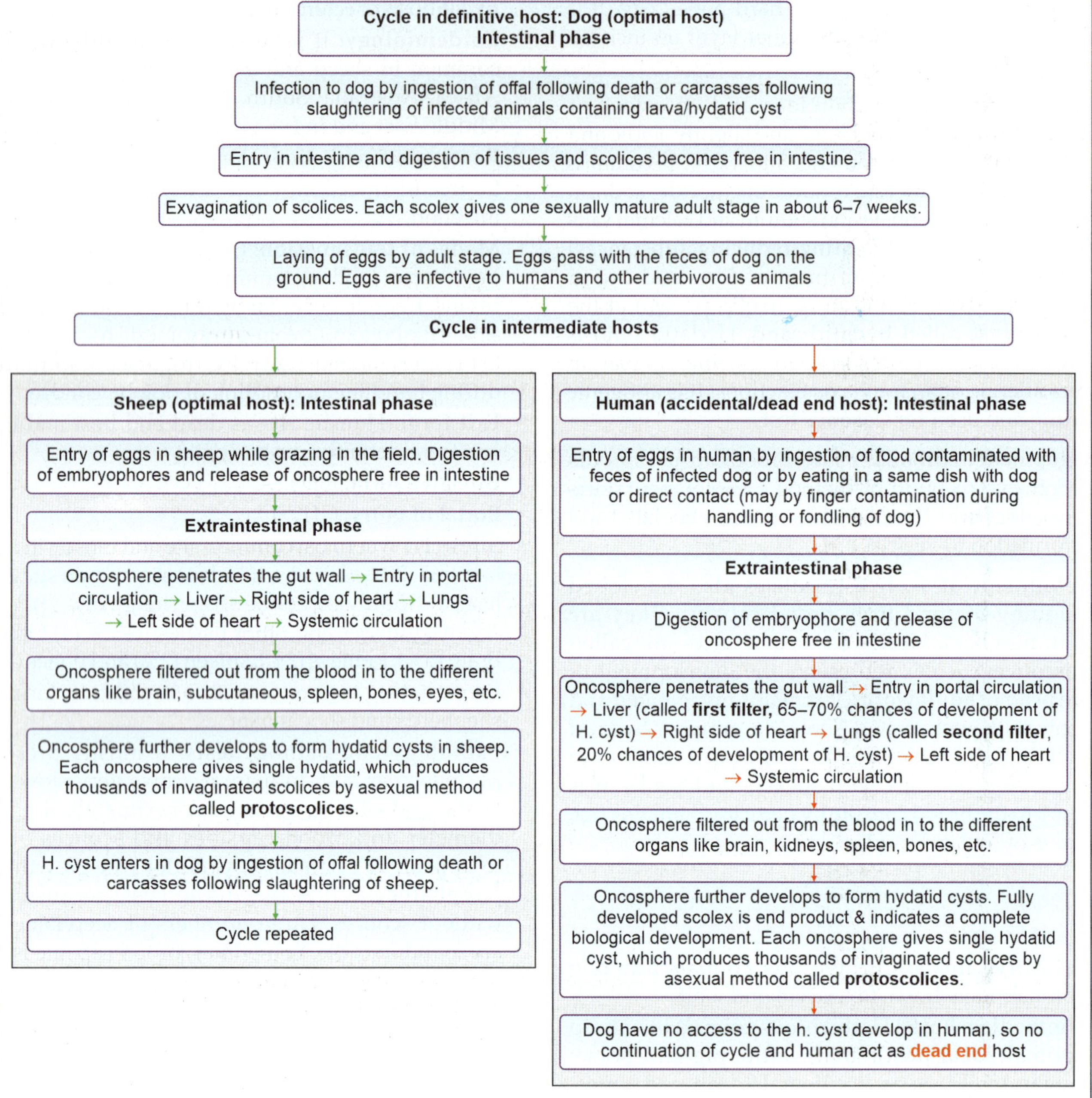

oncosphere is prevented by body defense mechanism with cellular infiltration of monocytes, lymphocytes and giant cells around it. Many of embryos have been destroyed, while those escapes will develop in to the hydatid cyst. Cellular infiltration is replaced by fibroblast and new blood vessels, which ultimately transferred in to new fibrous layer around cyst called **pericyst.** Pericyst is not the morphological part of hydatid cyst. It gradually merged in to surrounding healthy tissues and provides channel for nutrition of parasites. In older cyst, pericyst may be calcified or sclerosed and parasites within it may die or degenerate owing to lack of nutrition. **(5) Exogenous cyst:** It mostly found in bony hydatid. Mechanisms of exogenous cyst formation are mentioned in **Flowchart 102.4.**

- **Clinical features:** Cure or evacuation of cyst is due to inflammatory reaction. Clinical features depend on size, location and integrity of cyst.

  - **Size:** Smaller size is asymptomatic. Larger size produces pressure symptoms on surrounding organs.

  - **Location:** Superficial cyst is asymptomatic and presents as inspectory swelling. Deep cyst presents as palpatory nodule in organs like liver **(cut section in Fig. 102.12)**, lungs, spleen kidneys, etc., and detected at the time of autopsy **(Fig. 102.13)**. **Liver cyst** produces the hepatomegaly, pain in right hypochondrium and obstructive jaundices. **Lung cyst** produces the cough, chest pain, hemoptysis, dyspnea and pneumothorax.

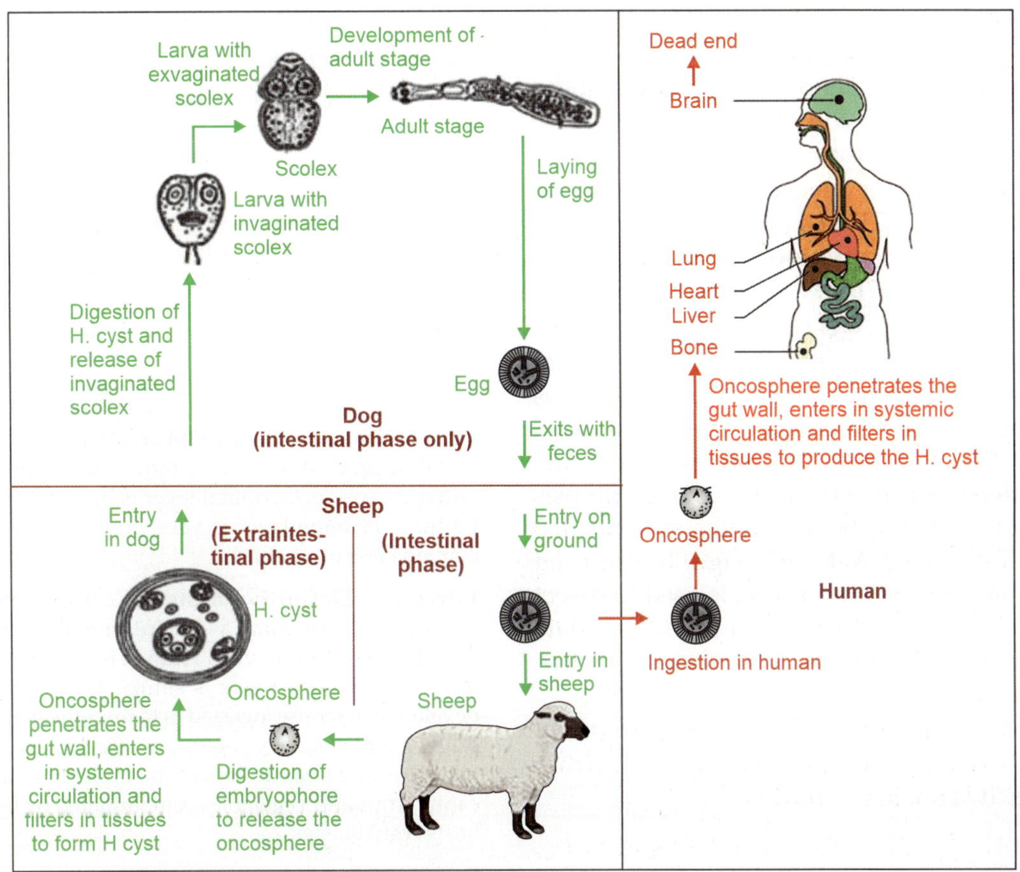

Fig. 102.11: Life cycle of *E. granulosus*

Flowchart 102.4: Mechanism of exogenous cyst

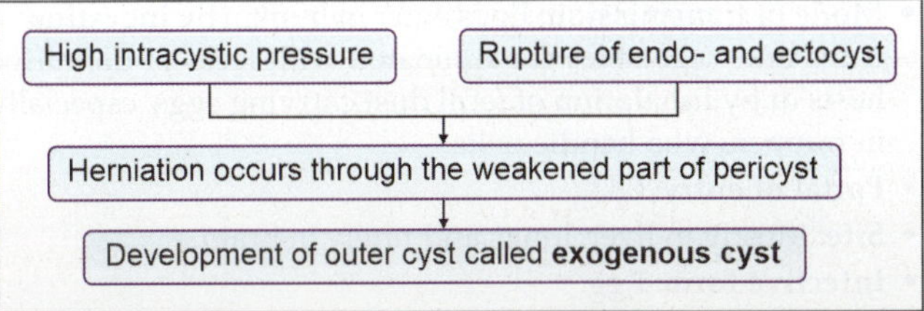

Fig. 102.13: H. cyst of liver after surgery

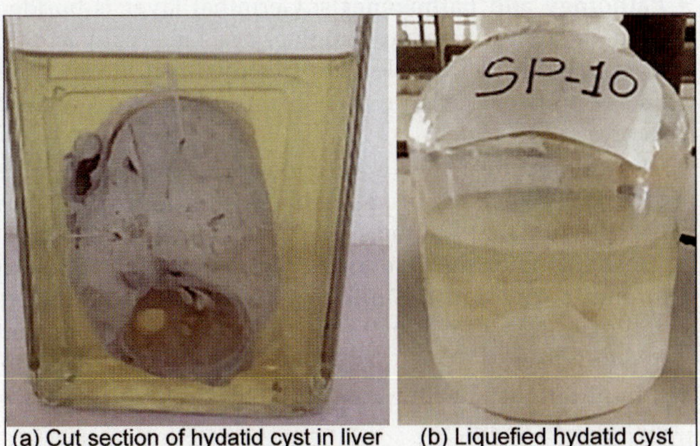

Fig. 102.12: Lab., specimens of hydatid cyst

- **Integrity:** Intact cyst is asymptomatic or shows pressure symptoms. Ruptured cyst produces allergic reactions like utricaria or anaphylaxis due to hydatid fluid.

- **Complications: (1) Rupture of cyst:** It ruptured spontaneously in surrounding organs and produces disseminated lesions. Like liver cyst rupture in lungs or body cavity. **(2) Suppuration:** Pus formation by secondary bacterial infections. **(3) Calcification:** Pulmonary hydatid cyst almost never calcified unlike mediastinal cyst. **(4) Sclerosis: (5) Death:** 10% of all diagnosed cases are fatal.

Laboratory Diagnosis

Specimens: (1) Hydatid fluid: Fluid aspiration is not advised generally, because spillage of fluid during aspiration may cause allergic reaction. Fluid is collected from surgically removed cyst. **(2) Urine and sputum:** Scolices and hooks are present when cyst is ruptured.

A. **Blood picture:** Eosinophilia (20–25%).

B. **Microscopy:** Scolices and hooks are identified under the microscope [by ZN, trichrome or lacto phenol blue (LCB) stain] in case of ruptured cyst.

C. **Serological tests:** Ab is detected by screening test like ELISA (by using B2T Ag) or by dot immunogold filtration assay which are more sensitive. Positive result is confirmed by WB test which is more specific.

D. **Molecular methods:** PCR is useful methods.

E. **Radiological: (1) X-ray:** Liver cyst and lung cyst are characterized by sharp outline and bone cyst gives mottled appearance. Calcified cyst is diagnosed easily. **(2) USG:** It detects the size, number, site, activity (active or dormant), calcification or fibrosis of cyst. Part of cyst may be detached and floats free within the fluid called **water lily sign**. It also used to monitor the therapy. **(3) CT scan** is used to detect the smaller cyst, extrahepatic cyst and to differentiate the h cyst from other cysts. **(4) MRI** is used for clear examination of cyst.

F. **Casoni's test:** It based on ITH. It is performed by using fresh hydatid fluid obtained by operation from human case. It is not in use now.

Prevention

Preventive measures are periodical deworming of pet dog, avoiding contact or kissing to infected dog and hand washing after touching the dog and ensuring that pet dog cannot eat the animal offal.

Treatment

PAIR: Basic steps are **P**uncture of cyst guided by ultrasound or CT, **A**spiration of cyst fluid, **I**njection of scolicidal (agent which kills the scolex/head of cyst) solution like 95% ethanol alternatively hypertonic saline and **R**e-aspiration of fluid after 5 minutes. It is **indicated** for single uncomplicated hepatic cyst. It is **contraindicated** when cyst is superficial, cysts has multiple thick internal septal divisions (honey-combing pattern), cyst communicated with biliary tree and extra hepatic cyst. **Advantages** of PAIR are less recurrence, less complications, less hospitalization and high cure rate.

Chemotherapy: Albendazole is the drug of choice to prevent the recurrence or to reduce the size before surgery/PAIR.

Surgery: It is **indicated** when PAIR is not possible, cyst communicated with biliary tree, secondary bacterial infection or advanced cyst. Pre- and postoperative albendazole for 2 years after surgery is advised. **Type of surgery** is laparoscopy for hepatic cyst and lobectomy or wedge resection for pulmonary cyst. Percutaneous thermal ablation of the germinal layer of cyst is performed by using radiofrequency ablation device. **Disadvantage** of surgery is recurrence.

Other *Echinococcus* spp.

E. multilocularis (*E. alveolaris*), *E. vogeli* and *E. oligarthus* are described separately in box.

Note: *Echinococcus multilocularis*

History: Multilocular disease was 1st recognized and differentiated from unilocular hydatid cyst by Virchow in 1855. Multilocular means cyst having multiple small chambers or cysts or daughter cysts. Leuckart discovered the causative agent in 1863. Life cycle was discovered by Thomas in 1954.

Morphology: (1) Adult stage: Same as *E. granulosus*, but smaller in size about 1.2–3.7 mm in length. **(2) Egg stage:** Same as other *Taenia* species, but more resistant to cold and other environmental conditions. **(3) Larval stage:** It is the multilocular larval stage. Cyst is sterile without scolices (protoscolices) and without capsule. Germinal layer is hyperplastic and folded or budded. Hyaline layer is very thin and less conspicuous. It contains less or no fluid.

Life cycle: Definitive hosts are carnivorous animals like fox (optimal), dog and wolf. **Intermediate hosts** are mouse (optimal) and tundra vole. Human is an accidental (dead end) intermediate host. **Cycle** is similar to *E. granulosus* and cyst developed in mouse ingested in to fox, which develops the adult stage. Adult stage lays eggs, pass through feces over soil and taken up by mouse to repeat the cycle. Human is infected by eating fruits and vegetables contaminated by feces of definitive hosts contain eggs.

Pathogenicity

- **Disease name:** It is a zoonosis and human disease presents as cyst with multiple locules called **multilocular hydatid cyst** or **alveolar hydatid cyst.**
- **Epidemiology:** Disease is prevalent in certain parts of Europe.
- **Mode of transmission:** Eggs enter in human by **ingestion** of fruits and vegetables contaminated with feces of definitive hosts or by **inhalation** of fecal dust carrying eggs, especially in trappers who handle pelts.
- **Portal of entry:** GIT.
- **Site:** Mostly in liver, lungs and rarely in brain.
- **Infective form:** Egg.
- **Exit form:** Embryonated egg.
- **Pathology and pathogenesis:** Germinal layer is budded (folded) and forms many daughter cysts internally as well as externally, hence called **multilocular cyst.** Eosinophils and endothelial cells produce persistent reaction around the cyst.
- **Clinical features:** Cyst is very slow growing and it takes around 30 years to produces the symptoms. Liver cyst may produce the symptoms like pain in right upper quadrant and hepatomegaly. Due to exogenous cyst formation, it mimics a carcinoma. Multilocular cyst mimics clinically and prognosis wise to malignancy, hence called **malignant hydatid disease**.
- **Complications:** Calcification, portal hypertension, cirrhosis and may be death.

Laboratory diagnosis: It is a sterile cyst and laboratory methods are not useful. It is better diagnosed by **serological** and **radiological** methods.

Prevention: Proper washing of vegetables before eating.

Treatment: Albendazole and surgical removal of cyst are recommended.

Hymenolepis spp.

Hymenolepis nana

Meaning

Hymeno word is originated from hymen (Greek) means membrane and lepis means shell or outer covering, as outer egg shell looks like membrane. Nana word is originated from nanus means dwarf or short.

Common Name

It also called **dwarf tape worm** because of short size.

History

It was first discovered by Bilharz in 1857.

Morphology

Three morphological stages like adult, egg and larva are described below.

Adult Stage

Time of maturation: Larva transformed to sexually mature adult stage in 1–2 weeks.

Color: White and semitransparent

Size: 1 to 4 cm in length × 1 mm in diameter

Shape: Like a measure tape

Life span: Very short about 2 weeks

Morphological parts: Three parts like head, neck and strobila (body/trunk) as shown in **Fig. 102.14a**.

- **Scolex/head (Fig. 102.14b):** It is globular in shape and 0.3 mm in diameter (1/3rd of *T. solium*). It has four circular suckers, rostellum and single row of hooklets around 20–30 in numbers. Rostellum with hooks gives tuning fork like appearance.
- **Neck:** It is long and slender.
- **Strobila (body or trunk):** It contains chain of segments called **proglottids**. They are **around** 200 in numbers and 0.3 mm × 0.9 mm in size (gravid segment). Breadth is 3–4 times greater than length (**Fig. 102.15**). Proglottids are three types same as *T saginata* described above.
- **Expulsion of eggs:** Released by disintegration of gravid segment.
- **Reproductive system:** It is monoecious or hermaphrodite. Uterus presents as transverse sac with lobulated wall. Gravid uterus filled with 80–100 eggs. It has bilobed coarsely granular ovary lies posteriorly between the testes. Testes are 3 in numbers. Genital atrium is marginal and presents on same side. Features of ootype, Mehli's gland, vas efferentia, vas deferens and cirrus (penis) are described in **Ch. 100.**

Egg Stage

Egg is colorless (non bile stained), 30–40 µm in diameter and spherical or oval in shape. It survives about 10 days in external environment. It has two covers (**Fig. 102.16**) outer cover is thin called **egg shell** and inner cover is thick called **embryophore.** Space between two covers is filled by yolk materials and polar filaments. Egg contains embryo called **oncopshere** with 6 hooklets (3 pairs) called **hexacanth.** It floats in saturated common salt solution.

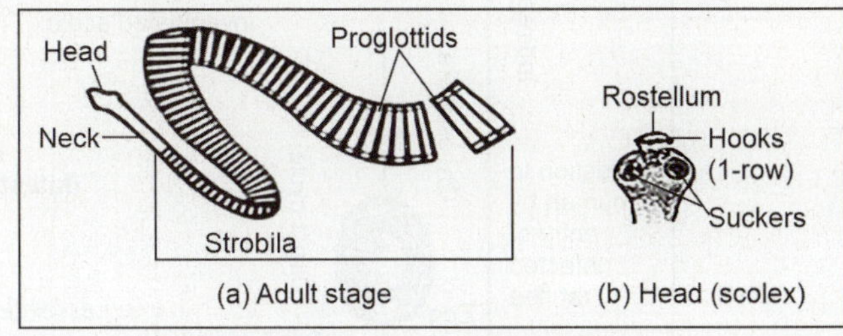

Fig. 102.14: Adult stage of *H. nana*

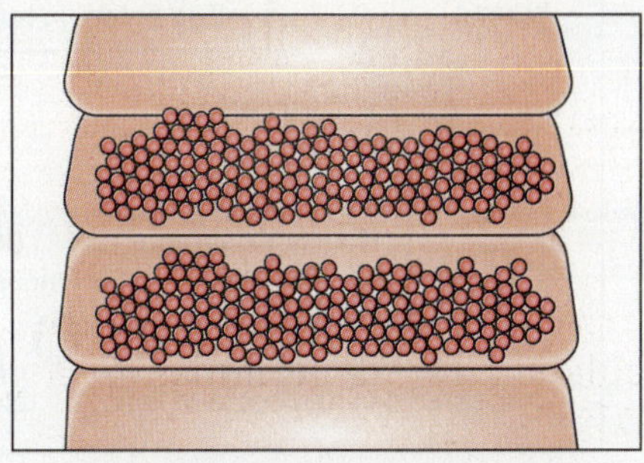

Fig. 102.15: Segment of *H. nana*

Infections of Cestodes: Cyclophyllidea

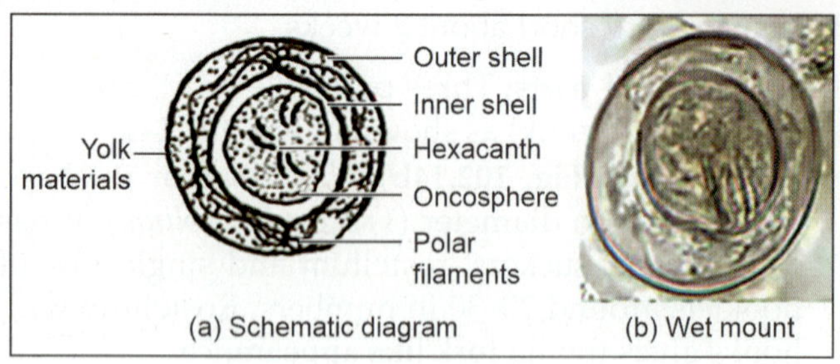

Fig. 102.16: Egg of *H. nana*

Laval Stage

Larva develops in small intestine of host in 4 days. It is pyriform in shape. It is known as **cysticercoid.** It contains hooklets and scolex is invaginated from one side **(Fig. 102.17)**. Egg develops the cysticercercoid in 4 days.

Life Cycle

Hosts: (1) Definitive host: Human or rat (adult stage). **(2) Intermediate host:** Rat flea (*Xenopsylla cheopis*) or beetle (larval stage).

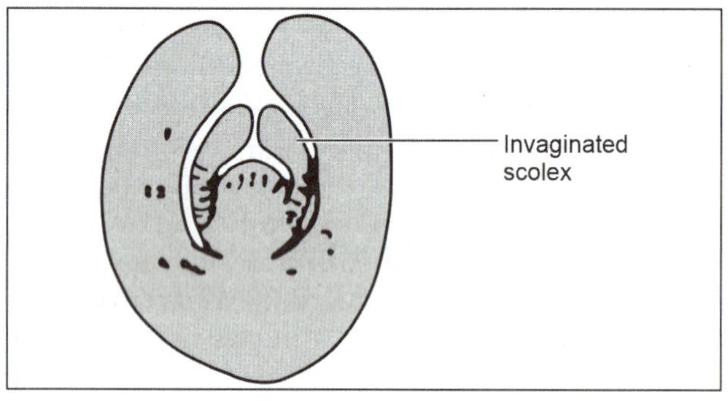

Fig. 102.17: Larva of *H. nana*

Cycle (Fig. 102.18): (1) Direct cycle: Human or rat acts as definitive host (adult stage) and intermediate host (larval stage) as shown in **Flowchart 102.5. (2) Indirect cycle:** As shown in **Flowchart 102.6**. This cycle is true in Argentina.

Strain of *H. nana*: In above cycles strain mentioned is able to infect human to rat and *vice versa*; however, few authors says that human strain can infect rat but not *vice versa*.

Pathogenicity

Disease name: Called **hymenolepiasis** or **hymeno-lepidosis** or **dwarf tape worm disease.**

Epidemiology: It is distributed worldwide. It is common in warm climates and in school children.

Reservoir and source of infection: Humans.

Mode of transmission: (1) Eggs enter in human by **ingestion** of contaminated food and water. (2) Internal reinfection occurs (**autoinfection**) due to persistence of few eggs in intestine rather exits. (3) Infection to human/rat occurs by **ingestion** of infected rat flea/bettle contain cysticercoids (larvae) in hemocele.

Exit form: Embryonated egg.

Infective form: Embryonated egg or larva.

Portal of entry: GIT.

Site: Adult worm lives in small intestine (proximal ileum) of man/rat.

Precipitating factors: It is common in children and in warm than cold climate. In person with immuno-suppression it can cause dissemination of infection.

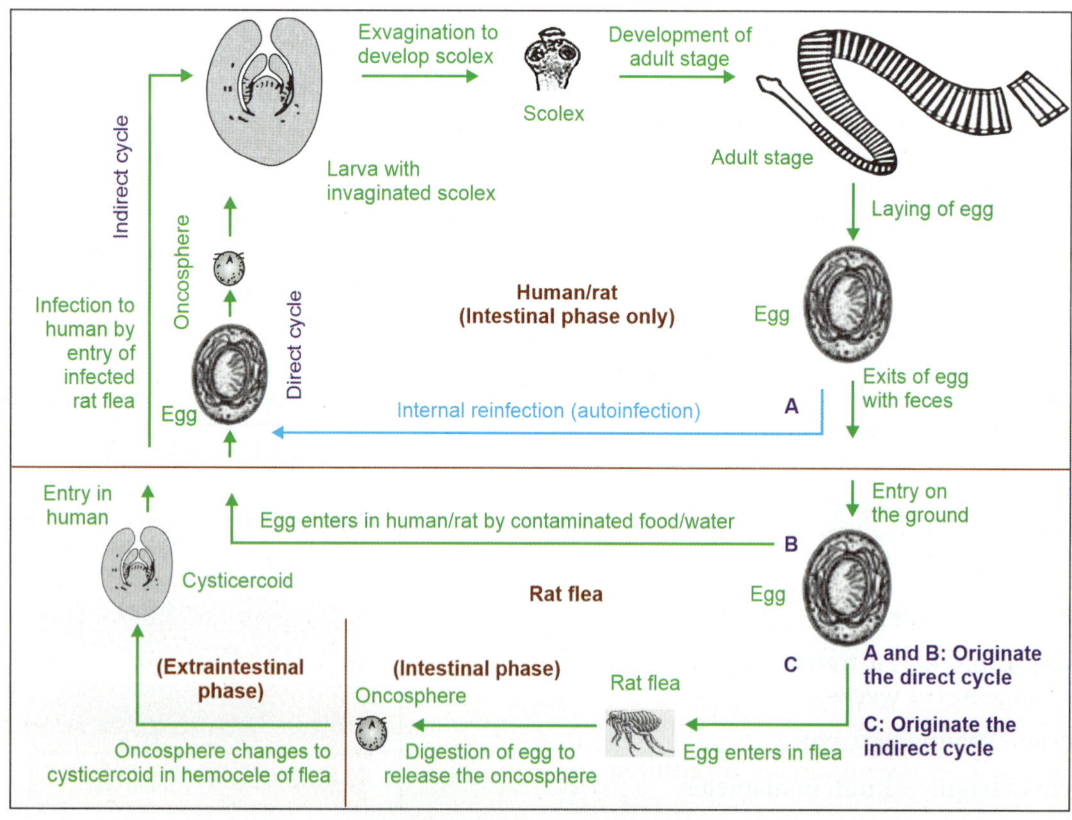

Fig. 102.18: Life cycle of *H. nana*

Flowchart 102.5: Direct cycle of *H. nana*

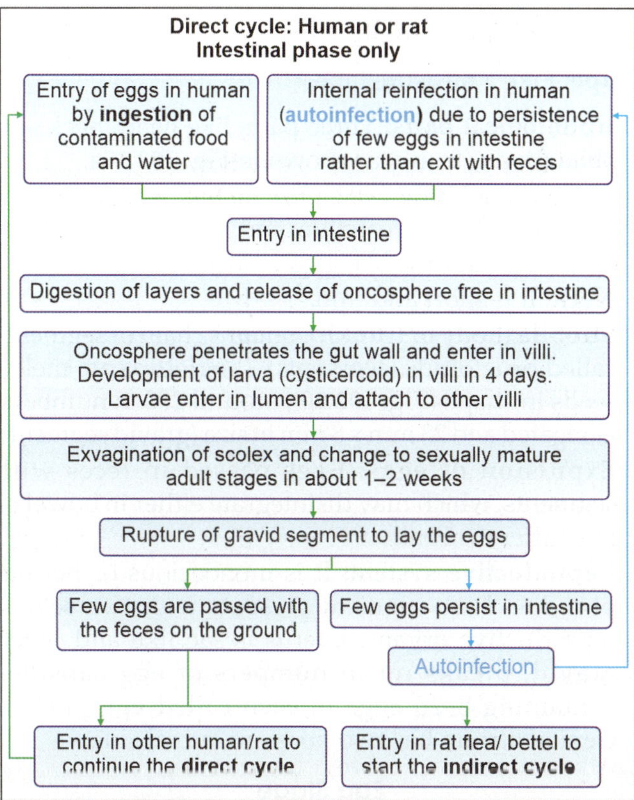

Flowchart 102.6: Indirect cycle of *H. nana*

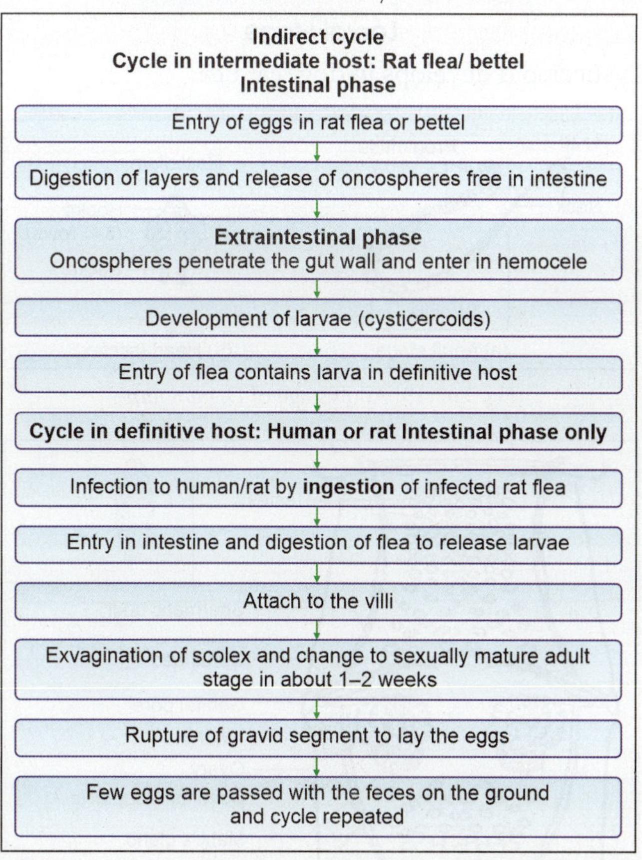

Clinical features: It is asymptomatic or adult stage produces minor symptoms like abdominal pain or diarrhea.

Laboratory Diagnosis

Specimen: Stool (eggs are infectious and unpreserved stool can be handled with care).

Testing methods

A. **Macroscopy:** For segments which are rarely or not available.

B. **Microscopy: (1) Wet mount by using NS and iodine:** It shows the nonbile stained (colorless) eggs as shown in **Fig. 102.16b. (2) Concentration technique:** In case of negative result by wet mount method.

C. **Serological test:** ELISA test has been developed with 80% sensitivity.

Prevention

Preventive measures are avoiding the use of contaminated food and water, rodent control and flea control.

Treatment

Praziquantel (single dose of 25 mg/kg) is the drug of choice for adult and larval stage. Nitazoxanide 500 mg BD is another effective drug.

> **Note:** *Hymenolepis diminuta*
>
> **Meaning:** Word diminuta is misnomer (inappropriate), as it is larger than *H. nana*.
>
> **Common name:** Rat tape worm.
>
> **History:** It was first discovered by Rudolph in 1819 and later by Blanchard in 1891.
>
> **Morphology**
> - **Adult stage** is white, semitransparent and 20–60 cm in length × 4 mm in breadth. It has three parts like head, neck and strobila (body/trunk). **(1) Scolex (head)** is 0.2–0.4 mm in size, contains 4 suckers and rostellum but no hooklets. **(2) Neck** is long and slender. **(3) Strobila (body or trunk)** contains chain of segments called proglottids. Proglottids (segments) are around 800–1000 in numbers, 0.75 mm × 2.5 mm in size (gravid segment). Breadth is greater than length. Other features are same as *H. nana*.
> - **Egg stage** is bile stained and larger than *H. nana* about 60–80 μm in diameter. It has outer thin and yellow cover. Inner cover is thick and with two knob like structure called **embryophore**. Space between two covers does not contain polar filaments. Eggs contain embryo called **oncopshere** with 6 hooklets (3 pairs) called **hexacanth**.
> - **Larval stage:** Cysticercoid develops in rat flea.
>
> **Life cycle**
> - **Hosts: (1) Definitive hosts:** Human or rat (adult stage). **(2) Intermediate hosts:** Rat flea (*Xenopsylla cheopis*) or flour beetle or ear wigs (larval stage).
> - **Cycles:** Two types of cycle as shown in **Flowchart 102.7**.
>
> **Pathogenicity**
> - **Disease name:** Called **hymenolepiasis diminuta**
> - **Epidemiology:** It is distributed worldwide
> - **Reservoir of infection:** Rat
> - **Source of infection:** Rat flea/bettle contain cysticercoids (larvae)
> - **Modes of transmission:** Accidental infection occurs in human by **ingestion** of infected rat flea/bettle.

(Contd...)

Essentials of Medical Microbiology

- **Exit form:** Embryonated egg
- **Infective form:** Larva
- **Portal of entry:** GIT
- **Site:** Adult worm lives in small intestine (upper 3/4 of ileum) of rat, mouse or human.
- **Precipitating factors:** It is common in children.
- **Clinical features:** Same as *H. nana*

Laboratory diagnosis: Wet mount prepared from stool by using NS and iodine shows the bile stained eggs without polar filaments. Concentration technique is useful in case of negative result by wet mount method.

Flowchart 102.7: Life cycle of *H. diminuta*

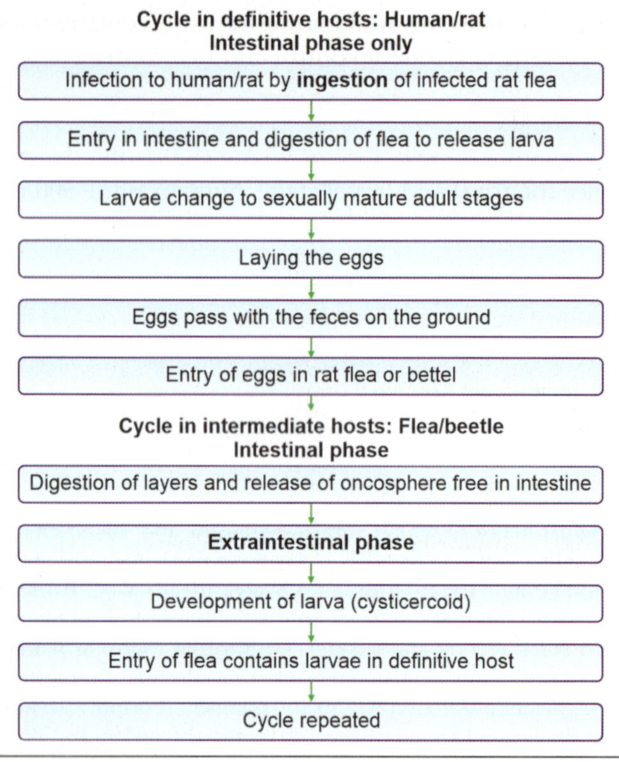

Dipylidium caninum

Meaning

Di word originated from **dis (Greek)** means **two and pylis means gate** or **entrance,** as parasite have bilateral genital pores in each segment (2 entrances). Caninum came from **canine** means **dog**, indicates dog species.

Common Name

Commonly it called double pored dog tape worm.

History

Linnaeus (1758) and Ralliet (1892) contributed in discovery of parasite.

Morphology

Adult Stage

Time of maturation: Larva change to sexually mature adult stage in 20 days in definitive host.

Color: White and semitransparent.

Size: 10–17 cm in length.

Shape: Like a measure tape.

Morphological parts: Three parts like head, neck and strobila (body/trunk) as shown in **Fig. 102.19a**.

- **Scolex/head (Fig. 102.19b):** It has four elliptical suckers, refractile rostellum and 3–4 rows of hooklets giving rose thorns like appearance.
- **Neck:** It is short and thin.
- **Strobila (body or trunk):** Contains chain of segments called proglottids (segments). Proglottids are melon seeds in shape **(Fig. 102.20),** around 200 in numbers, elongated and 23 mm × 8 mm in size (gravid segment).
- **Expulsion of eggs:** Eggs passed in feces with segments, which may disintegrate either in bowel or later after evacuation.
- **Reproductive system:** It is monoecious or herma-phrodite. Each segment presents with two sets of reproductive organs. Uterus is sac like and when gravid, breaks up in numbers of egg capsules containing 8–20 eggs in each called **egg pocket.** Genital atrium has two pores on lateral side.

Egg Stage

Egg is 20–35 µm in diameter. All the eggs are covered by capsule called egg pocket as shown in **Fig. 102.21.** Each capsule (pocket) contains 8–20 eggs.

Larval Stage

Cysticercoid develops in dog/cat flea.

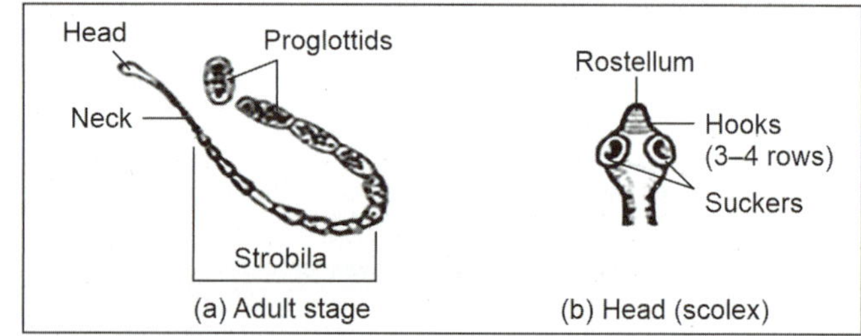

Fig. 102.19: Adult stage of *D. caninum*

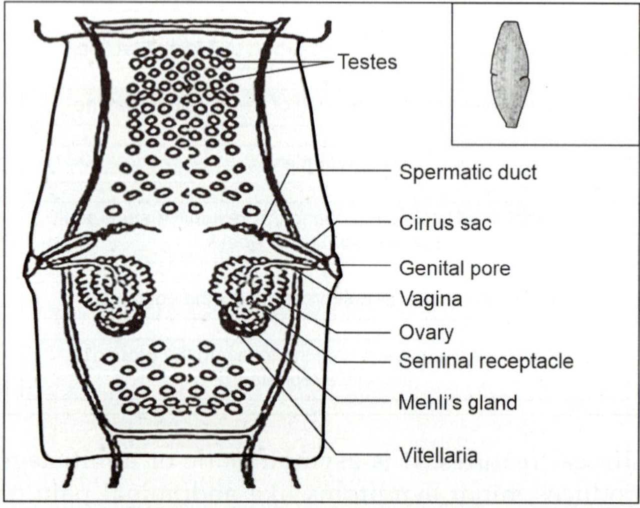

Fig. 102.20: Segment of *D. caninum*

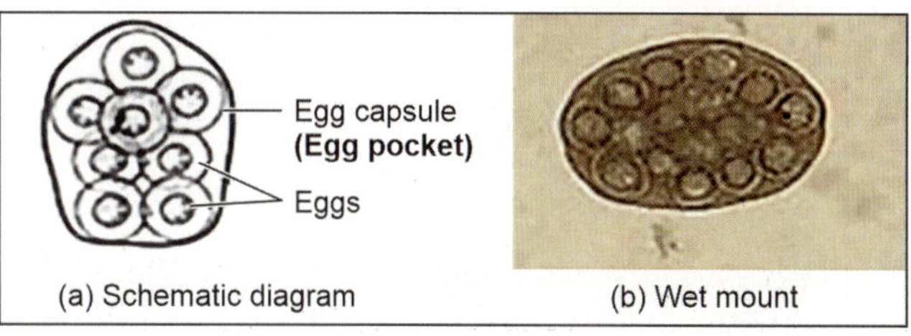

Fig. 102.21: Egg of *D. caninum*

Life Cycle

Hosts: (1) Definitive hosts: Human/dog/cat (adult stage). **(2) Intermediate host:** Dog flea (*Ctenocephalus canis*) or cat flea (*C. felis*) or human flea or dog louse (larval stage).

Cycles: Two types of cycle occur as shown in **Fig. 102.22** and **Flowchart 102.8**.

Pathogenicity

Disease name: Called **dipylidiasis.**

Epidemiology: Most of the cases occurred in European countries. Chaudhary and Bandyopadhyay from Kolkatta reported the cases from India.

Reservoirs of infection: Animals like dog or cat.

Sources of infection: Flea or louse.

Mode of transmission: Accidental infection to human occurs by **ingestion** of infected flea especially in children during fondling cat or dog.

Infective form: Larva.

Exit form: Proglottids or eggs.

Portal of entry: GIT.

Site: Adult worm lives in small intestine of human/dog/cat.

Precipitating factors: It is common in children.

Clinical features: Adult stage is asymptomatic or presents with minor symptoms like abdominal pain or diarrhea or pruritus ani (due to motile proglottids).

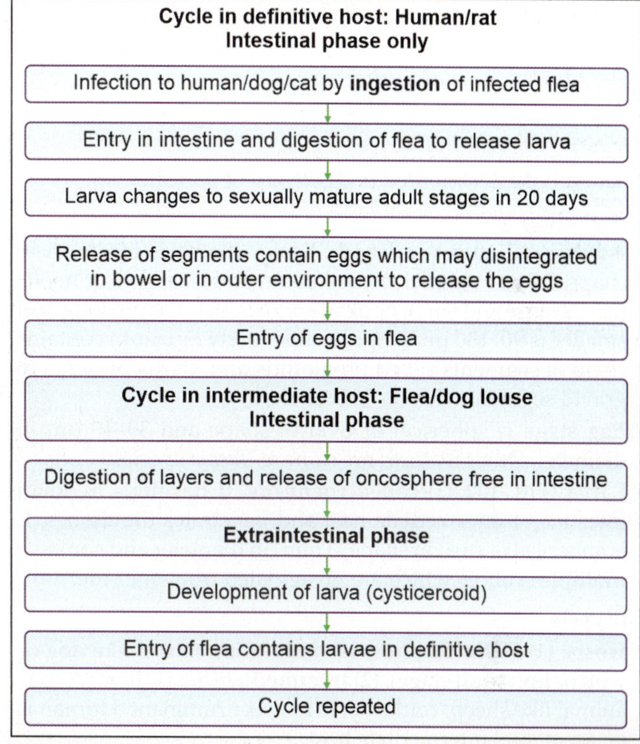

Laboratory Diagnosis

Macroscopy of stool is done for segments.

Microscopy: (1) Wet mount of stool by using NS and iodine shows the characteristic egg pocket as shown in **Fig. 102.21. (2) Concentration technique** is useful in case of negative result by wet mount method.

Prevention and Treatment

Flea control is useful measure. Praziquantel is the drug of choice.

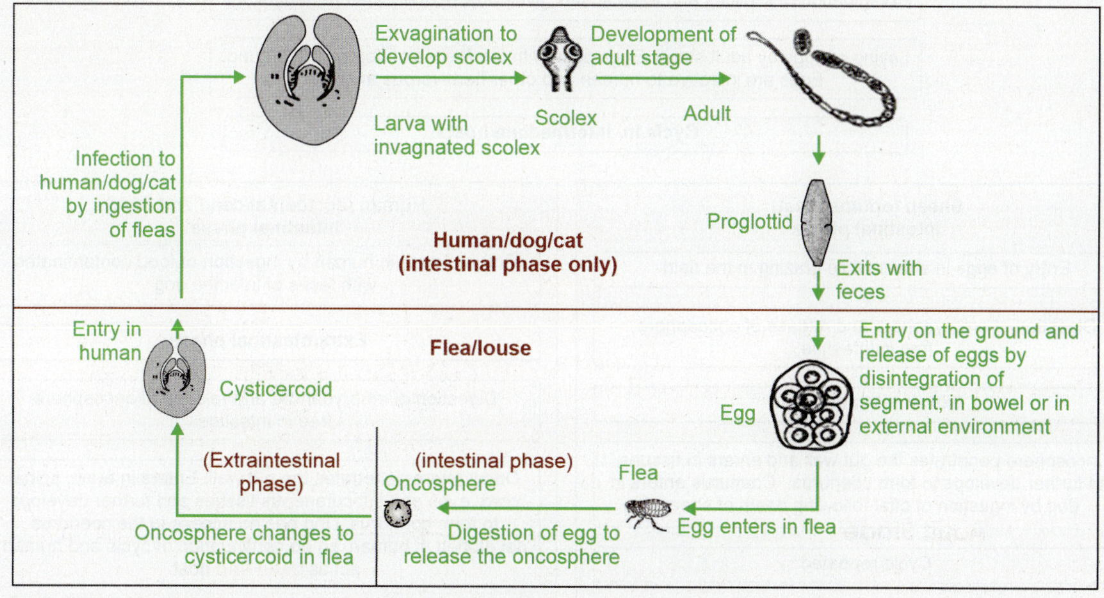

Fig. 102.22: Life cycle of *D. caninum*

It is described separately in box.

Note: *Multiceps multiceps*

Meaning: It contains multiple protoscolices hence the name *Multiceps*.

History: Luke contributed in discovery of parasite.

Morphology

- **Adult** is 40–60 cm in length. Scolex (head) is pyriform in shape. It has four suckers, rostellum and two rows of hooks (22–32 large and small hooks). Large is 150–170 µm long and smaller is 90–130 µm long. Strobila (body or trunk) contains chain of segments called proglottids and uterus presents in gravid segment with 18–26 branches.
- **Egg stage** is spherical or oval in shape and 30–40 µm in diameter. Other features are same as *Taenia* species.
- **Larva (Fig. 102.23)** called **coenurus**. It develops in small intestine of intermediate host and penetrates the tissues. It is spherical or oval in shape, 3 mm^3 in diameter and contains multiple scolices which are invaginated from one side.

Life cycle

- **Hosts: (1) Definitive host** is carnivorous animal like dog or wolf or fox (adult stage). **(2) Intermediate** host is herbivorous animal like sheep, cattle, horse or other ruminant. Human is an accidental intermediate host.
- **Cycles:** Natural cycle continues between carnivorous animal and herbivorous animal, while human act as dead end host as shown in **Flowchart 102.9**.

Pathogenicity

- **Disease name:** Called **coenurosis**.
- **Epidemiology:** Human coenurosis is most common in Africa, Europe and USA.
- **Reservoir of infection:** Dog.

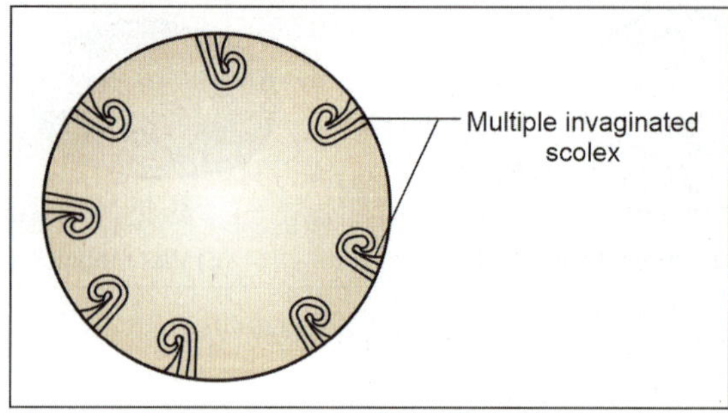

Fig. 102.23: Larva of *M. multiceps*

- **Sources of infection:** Food and water contaminated with faces of dogs.
- **Modes of transmission:** Accidental infection occurs to human by **ingestion** of food and water contaminated with feces of dog contains eggs.
- **Exit form:** Egg from definitive host (dog).
- **Infective form:** Larva.
- **Portal of entry:** GIT.
- **Sites:** Adult worm lives in small intestine (proximal ileum) of definitive hosts. Larva presents in brain, spinal cord and subcutaneous tissues in human and in other intermediate hosts.
- **Clinical features:** It occurs by larvae which produce the space occupying lesions. It is asymptomatic or produces pressure symptoms like headache, vomiting, paresis, seizures, etc.

Laboratory diagnosis

- **Specimen:** Biopsies from infected tissues.
- **Testing methods: (1) Microscopy:** Histological examination of biopsy material to identify the coenurus. **(2) Serological test:** Very useful methods. **(3) Radiological methods:** X-ray, CT-scan, MRI and USG are also useful methods.

Prevention: Avoid the fecal contamination of food.

Treatment: Surgical removal of space-occupying lesion.

Flowchart 102.9: Life cycle of *M. multiceps*

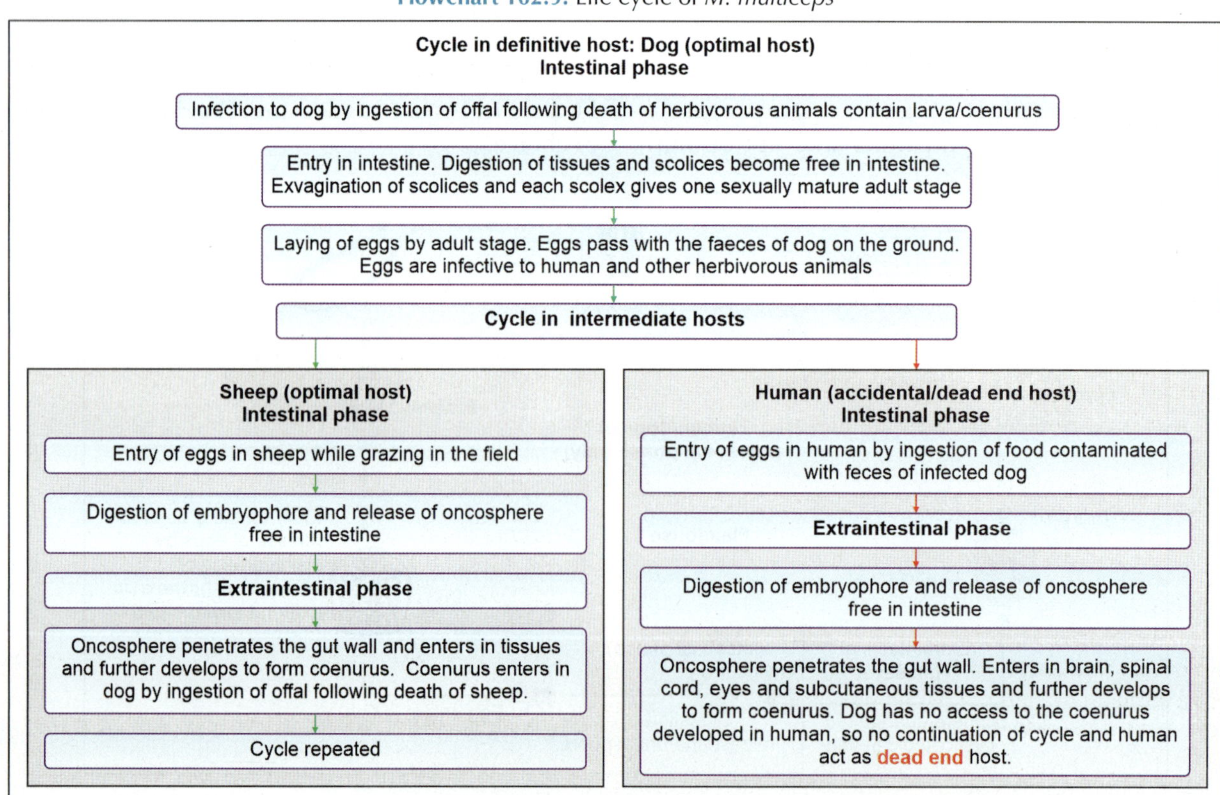

ACCESS YOURSELF

Case Studies

1. A 24-year-old male presents with abdominal discomfort since 10 days. He is giving history of beef ingestion. Stool examination revealed bile stained, spherical and hexacanth containing egg. Identify the case and answer the following.
   a. Name the causative agent.
   b. Describe the life cycle of causative agent.
   c. Write the pathogenicity of causative agent.
2. A 40-year-old patient admitted with complains of pain in right hypochondrium. Occupationally patient is farmer. USG reported the cyst in right lobe of liver. Identify the case and answer the following.
   a. Name the causative agent and draw the diagram of larval stage of causative agent.
   b. Describe the life cycle of causative agent.
   c. Write the pathogenicity of causative agent.
   d. Describe the lab., diagnosis of causative agent.

Essay/Full Question

1. *T. saginata* or *T. solium* or *E. granulosus*.

Short Notes

1. Morphology/life cycle/pathogenicity/laboratory diagnosis of *T. saginata* or *T. solium* or *E. granulosus*.
2. Differences between *T. saginata* and *T. solium*.
3. Cysticercosis/Neurocysticercosis.
4. Hydatid cyst/Multilocular hydatid cyst.
5. *H. nana*.

Short Questions for Theory/Viva Questions

1. What are measly beef (beef measles) and measly pork (pork measles)?
2. What are acephalocyst and pericyst in H. cyst?
3. Write two mechanisms of development of daughter cyst in H. cyst.
4. What are egg pocket and egg capsule?
5. What is malignant hydatid disease?
6. Name the causative agents of measles, german measles, beef measles (measly beef) and pork measles (measly pork).

Comments on

1. *T. solium* is more dangerous than *T. saginata*.
2. Diagnostic aspiration of hydatid fluid from hydatid cyst is generally not advised.

MCQs for Chapter Review

Taenia solium

1. Human is both definitive and intermediate host for:
   a. *T. solium*　　　　　b. *T. saginata*
   c. *E. granulosus*　　　d. *Dicroftis hominis*

2. Consumption of uncooked pork is likely to cause which of the following helminthic disease?
   a. *T. solium*
   b. *T. saginata*
   c. *H. nana*
   d. *H. cyst*

3. About neurocysticercosis, followings are true, except:
   a. Not acquired by eating contaminated vegetables
   b. Caused by regurgitation of larvae
   c. Acquired by orofecal route
   d. Acquired by eating pork

4. The most commonly affected tissue in cysticercosis is:
   a. Brain　　　　　b. Eye
   c. Muscles　　　　d. Liver

5. The followings regarding cysticercosis are true, *except*:
   a. Commonest sites are meninges and cerebral ventricle
   b. Calcification is common
   c. Causes focal neurological complication
   d. Found in subcutaneous tissues

6. Which of the following is the most common location of intracranial neurocysticercosis?
   a. Brain parenchyma　　b. Subarachnoid space
   c. Spinal cord　　　　　d. Orbit

7. Commonest parasite of CNS in India is:
   a. Schisosomiasis　　b. Cysticercosis
   c. *T. spiralis*　　　　d. H. cyst

8. Drug of choice in cerebral cysticercosis is:
   a. Piperizine　　　　b. Pyevinium
   c. Thiabendazole　　d. Mebendazole
   e. None

Echinococcus granulosus

9. Hydatid disease of liver is caused by:
   a. *Strongyloides*　　　b. *Echinococcus granulosus*
   c. *Taenia solium*　　　d. *Trichinella spiralis*
   e. *Echinococcus multilocularis*

10. Definitive host of hydatid disease is:
   a. Man　　　　b. Dog
   c. Horse　　　d. Sheep

11. Intermediate host of hydatid disease:
   a. Man　　　b. Dog
   c. Cat　　　d. Foxes

12. Most common site for hydatid cyst:
   a. Lung　　　b. Liver
   c. Brain　　　d. Kidney

13. Least common site for calcified hydatid cyst is:
   a. Lung　　　　　　　b. Mediastinum
   c. Extraperitoneal site　d. Liver

14. Which of the following shows regressive metamorphosis?
   a. Hydatid cyst　　　　b. Cysticercoid
   c. *Cysticercus bovis*　　d. *Cysticercus cellulosae*

Echinococcus multilocularis

15. The following infection resembles to malignancy:
   a. *E. granulosus*　　b. *E. multilocularis*
   c. *E. vogeli*　　　　d. *E. oligarthus*

16. What is malignant hydatid disease?
   a. Malignant change in to hydatid cyst
   b. Infection with *E. multilocularis*
   c. Hydatid disease in immunocompromised host
   d. None of the above

Hymenolepsi nana

17. Dwarf tape worm refers to:
   a. *Echinococcus*　　　　b. *Loa loa*
   c. *Hymenolepis nana*　　d. *Schistosoma haematobium*

18. Which of the following parasite's eggs consist polar filaments arising from either end of the embryophore?
   a. *Taenia saginata*
   b. *Taenia solium*
   c. *Echinococcus granulosus*
   d. *Hymenolepis nana*

876
19. The larva form of *Hymenolepsis nana* is called:
    a. Hydatid cyst
    b. Coenurus
    c. Cysticercus
    d. Cysticercoid

Answers and Explanation of MCQs

1. a
- Follow section, *Taenia solium* (**life cycle**) for explanation.

2. a
- Follow section, *Taenia solium* (**pathogenicity** → **mode of transmission**) for explanation.

3. b
- Follow section, *Taenia solium* (**life cycle** → **Flowchart 102.2 and pathogenicity** → **mode of transmission**) for explanation.

4. c

5. a

6. a

7. b

8. e
- Follow section, *Taenia solium* (**pathogenicity** → **larval stage disease**) for explanation of answers of MCQs 4–8.

9. b, e

10. b

11. a

12. b

13. a

14. a
- Follow section, *Echinococcus granulosus* for explanation of answers of MCQs 9–14.

15. b

16. b
- Follow section, *Echinococcus multilocularis* for explanation of answers of MCQs 15–16.

17. c

18. d

19. d
- Follow section, *Hymenolepis nana* for explanation of answers of MCQs 17–19.

CHAPTER 103

Trematodes: General Features and Classification

Chapter Outline
- General Features
- Classification

GENERAL FEATURES

Introduction

Trematodes are commonly known as **flukes,** word originated from **floc (Anglo-saxon)** means **flat fish,** because they are dorso-ventrally flat and unsegmented like flat fish or leaf of tree. Word trematodes originated from **trema (Greek)** means **hole and edios** means **appearance,** as the parasites of this group have large conspicuous sucker with hole in middle. Trematodes habitat in intestine, blood, liver and lungs.

Morphology

Adult Stage

Shape: Diecious (nonhermaphrodite) trematodes includes schistosomes, which are not flattened and leaf like, but cylindrical as like round worm as shown in **Fig. 103.1. Monoecious (hermaphrodite) trematodes** are dorsoventrally flat, like a leaf of tree or flatfish hence called flukes as shown in **Fig. 103.2.**

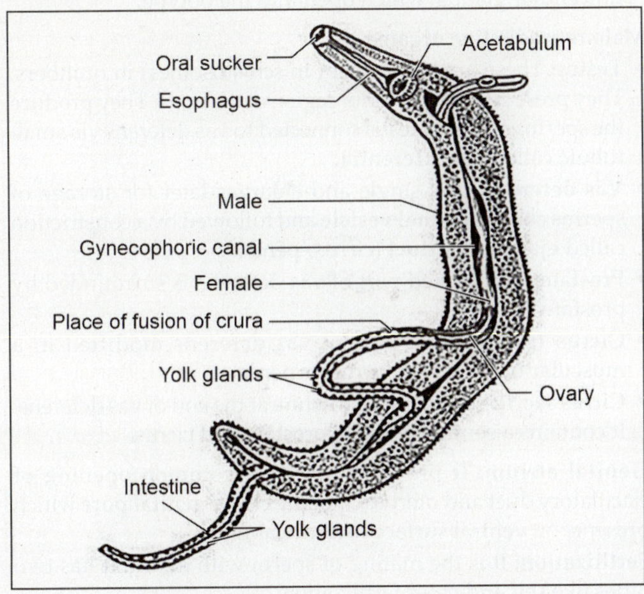

Fig. 103.1: Adult stage of diecious trematodes

Size: They are 1 mm to several centimeters in length.

Organs (Figs 103.1 and 103.2)

- **Tegument:** Body of parasite is covered with non-cellular layer called **tegument** or called **cuticle.** Under light microscope it looks as homogeneous layer about 7–16 μm in thickness, showing affinity for basic stains and may partially or completely covered with spines or tubercles or ridges. Under electron microscope two zones can be seen. **Outer zone** thrown in folds to form microvilli. **Inner zone** contains mitochondria, endoplasmic reticulum and glycogen granules. Tegument helps in absorption of carbohydrate and excretion of excessive mucus and others metabolites.

- **Suckers:** These are strong muscular cup shaped depression act as organs for adhesion. They are two in number, so species called **distomata (Greek word from di means two and stomata means mouth).** They are two types like oral sucker presents around the oral cavity and ventral sucker presents on ventral surface. It also called **acetabula** (acetabula is Latin word meaning a shallow vinegar vessel or cup).

- **Body cavity:** Absent.

- **Gastrointestinal system:** It presents, but without anus so called incomplete. Mouth is situated anteriorly in the center of oral sucker. Prepharynx is present, and pharynx is bulbous and muscular. Esophagus is bifurcated in front of ventral sucker into two tube-like structures called **intestinal ceca** or called **crura,** which may be simple (*C. sinensis*) or branched (*F. hepatica*) or curved in middle without branches (*F. buski*) or reunite to form the single cecum (*Schistosomes species*). Crura are lined by epithelial cells.

- **Excretory system:** It contains small tubules, which are united to form the collecting tubules. Tubules are lined with ciliated cells called **flame cells,** and they provide the basis for species identification. Tubules open posteriorly in excretory pore via excretory bladder.

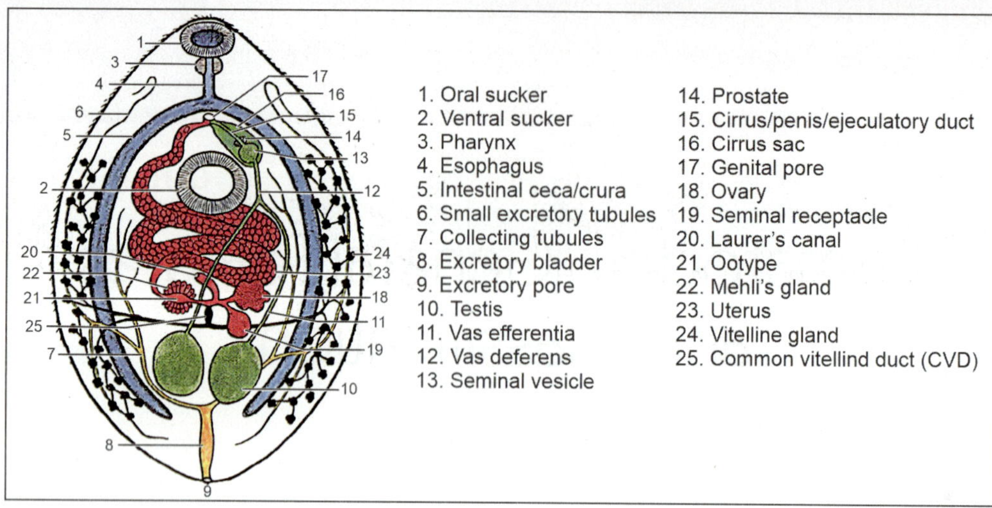

Fig. 103.2: Adult stage of monoecious trematodes

- **Nervous system:** It is primitive in nature and of low grade. It includes lateral ganglia connected by dorsal commissure. Sense organs are almost lacking.
- **Gynecophoric canal:** Lateral margin of males are folded ventrally with a canal like structure called **gynecophoric canal.** It presents only in diecious flukes and helps to hold the female during copulation. Both male and female are live together in gynecophoric canal called **coupled worm.**
- **Reproductive system:** It is highly developed and complete in each individual. Reproductive organs lie between the two branches of the esophagus. Female and male reproductive organs are described separately in box.

Egg Stage

Egg formed in ootype and surrounded by protective covering.

Egg of diecious trematodes (Fig. 103.3): It is oval and nonoperculated, 70–140 µm × 50–90 µm in size and surrounded by transparent and brownish yellow protective covering. Initially contains ciliated embryo when laid called **miracidium.** Miracidium released free in water on hatching of egg and taken up by intermediate host (snail). Spine (terminal/lateral) or lateral knob is the diagnostic feature of egg of the particular species. Yolk material presents in abundant amount.

Egg of monoecious trematodes (Fig. 103.4): It is oval and nonoperculated and surrounded by transparent and brownish yellow protective covering. In some eggs miracidium presents at the time of laying, while in some it develops after laying. Egg does not contain spine or knob except terminal hook in *C. sinensis.*

> **Note: Reproductive system of trematodes**
> **Definition:** In trematodes both male and female reproductive organs are present in single individual called **monoecious** or **hermaphrodite** except schistosomes in which separate male and female species are available called **diecious.**

Female reproductive organs
- **Ovary:** It is single, lies in front of testes and discharges the ova in the oviduct.
- **Laurer's canal (rudimentary vagina):** It represents the rudimentary vagina and inner end contains the receptacle for storage of sperms called **seminal receptacle.** It may not open or if opened then opening present on dorsal surface and acts as vagina for entrance of sperms and cross fertilization.
- **Uterus:** It is a tortuous tube emanate from the ootype and ends in genital atrium. Terminal part called **metaterm,** which in absence of Laurer's canal acts as vagina. It contains eggs when gravid.
- **Ootype:** It provides place for fertilization of eggs.
- **Vitelline/yolk glands (vitellaria):** It scattered lateral to intestinal crura and connected to vitelline ducts, which are join to form common vitelline duct (common yolk channel) which open in the ootype. It produces the shell material.
- **Mehli's gland:** It also called **shell gland** because material from the ovi duct, seminal receptacle and common vitelline duct comes in Mehli's gland where egg shell has been formed. It is a very small organ surrounding the ootype and contains unicellular glands, which open into the ootype.

Male reproductive organs
- **Testes:** These are two (except in schistosomes) in numbers. They present in the posterior region of the body. They produce the sperms and each testis connected to vas deferens via small tubule called **vas efferentia.**
- **Vas deferens:** It is single and enlarges later for storage of sperms called **seminal vesicle** and followed by a constriction called **ejaculatory duct** (cirrus/penis).
- **Prostate:** Constricted part of vas deferens is surrounded by prostate.
- **Cirrus (penis):** End part of vas deferens modified in a muscular organ called **cirrus** or **penis.**
- **Cirrus sac:** It is the sac like structure at the end of vas deferens. It contains a seminal vesicle, prostate and cirrus.

Genital atrium: It provides place for common opening of ejaculatory duct and uterus. Opening called **genital pore** which presents on ventral surface near acetabulum.

Fertilization: It is the mating of sperm with ovum. It has two types like self and cross fertilization.

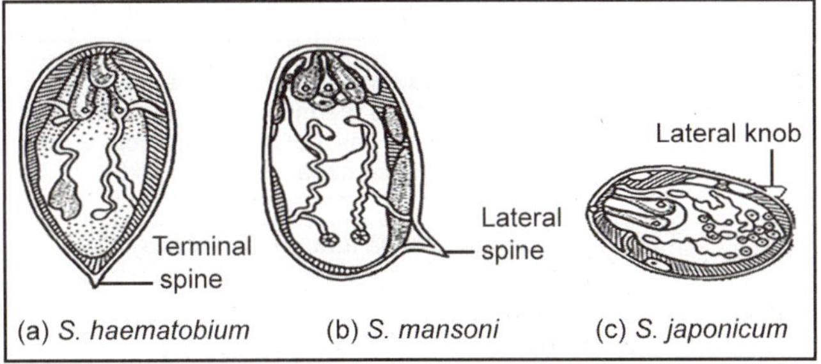

Fig. 103.3: Eggs of diecious trematodes (schistosomes)

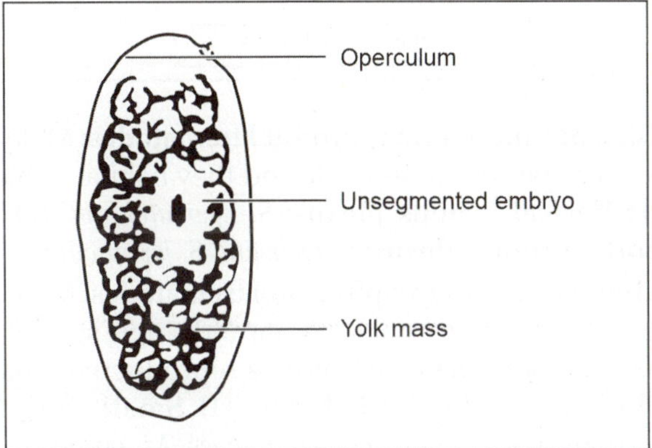

Fig. 103.4: Egg of monoecious trematodes (*F. buski*)

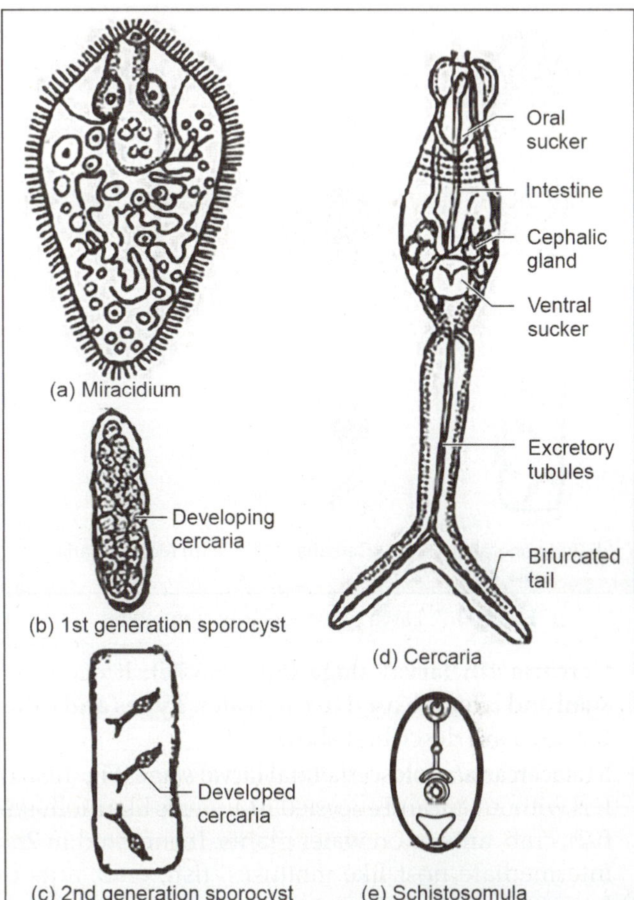

Fig. 103.5: Larva of diecious trematodes

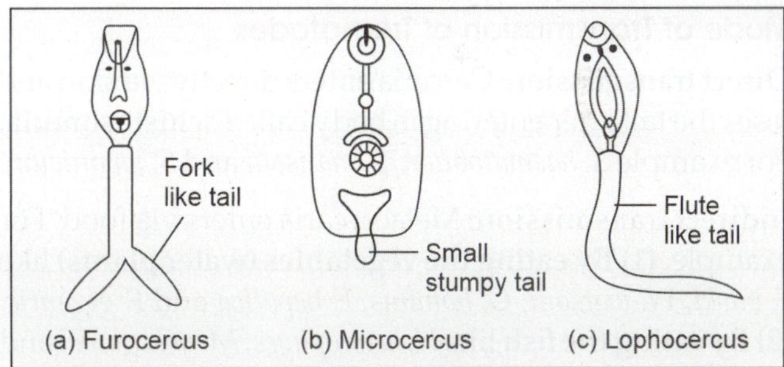

Fig. 103.6: Types of cercaria

Larval Stage

Larva of diecious trematodes: Solid form of larva is present with following stages, names and hosts.

- **Miracidium/1st larval stage (Fig. 103.5a):** Miracidium is Greek word meaning little boy. Eggs contain ciliated embryo called **miracidium** which releases free in water on hatching of egg and taken up by intermediate host (snail).
- **Sporocyst/2nd larval stage:** Sporos (Greek) means seed and kystis (Greek) means cyst or bladder. It occurs in molluscs or snail. Asexual multiplication takes place in this stage and it passes through 1st **(Fig. 103.5b)** and 2nd generation sporocysts **(Fig. 103.5c)**.
- **Cercaria/3rd larval stage (Fig. 103.5d):** Word derived from Kerkos (Greek) means tail. It occurs in molluscs/ snail and later released free in water. It possesses suckers, intestine like that of adult worm, body and tail. From single miracidium multiple cercariae can develop. According to nature of tail it is having following types as shown in **Fig. 103.6**. Types 2–3 belong to monoecious trematodes.
  1. **Furocercus cercaria:** Fork like/bifurcated tail (as in schistosomes).
  2. **Microcercus cercaria:** Small stumpy tail (as in *Paragonimus*).
  3. **Lophocercus cercaria:** Flute like tail (as in *Metagonimus, Clonorchis, Fasciolopsis, Fasciola* and *Heterophyes*).
  4. **Pleurolophocercus cercaria:** Long powerful tail with a pair of fin fold (as in *Opisthorchis*).
- **Schistosomula/final larval stage (Fig. 103.5e):** Cercaria penetrates the skin of definitive host and loses the tail called **schistosomula**. It develops in definitive host. It is same as cercaria, but without tail and infective to definitive hosts.

Larva of monoecious trematodes: It has following stages.
- **Miracidium:** Same as of diecious trematodes as mentioned above and shown in **Fig. 103.7a**.
- **Sporocyst:** Same as of diecious trematodes with few differences **(Fig. 103.7b)** like it occurs in snail and asexual multiplication does not take place in this stage so 1st and 2nd generation sporocysts are not formed.
- **Redia/3rd larval stage:** It named after Italian naturalist **Fransesco Redi**. It occurs in snail and asexual multiplication takes place in all trematodes and passes through 1st **(Fig. 103.7c)** and 2nd generation of redia **(Fig. 103.7d)**. It contains oral sucker, pharynx, a sac like intestine and birth pore through which next generation escape. It contains germ cells which develop into 2nd generation of redia (daughter redia) or into final larval stage called **cercaria**.

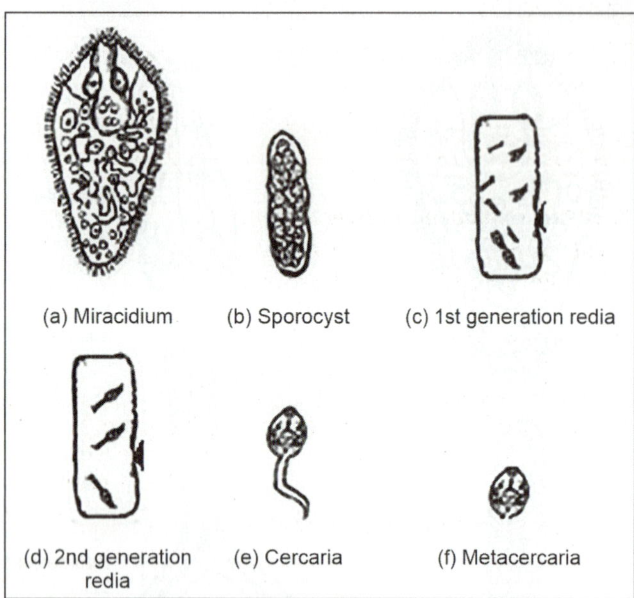

(a) Miracidium (b) Sporocyst (c) 1st generation redia

(d) 2nd generation redia (e) Cercaria (f) Metacercaria

Fig. 103.7: Larva of monoecious trematodes

- **Cercaria/4th larval stage (Fig. 103.7e):** It occurs in snail and later released free in water. **Types and other features** are described above.
- **Metacercaria/adolesceria/final larval stage (Fig. 103.7f):** It is without tail and encysted in animals like molluscs, fish, crab, ants or on water plants. It encysted in 2nd intermediate host like molluscs, fish, crab, ants or free on vegetables. Structure is same as cercaria, but without tail and infective to definitive hosts.

Mode of Transmission of Trematodes

Direct transmission: Cercaria enters directly via skin and loses the tail after entering in body called **schistosomula.** For example *S. haematobium, S. mansoni* and *S. japonicum.*

Indirect transmission: Metacercaria enters via food. For example. **(1) By eating the vegetables (water plants)** like *F. buski, W. watsoni, G. hominis, F. hepatica* and *F. gigantic.* **(2) By eating the fish** like *H. heterophyes, M. yokogawai* and *C. sinensi.* **(3) By eating the mollusks** like *Echinostoma* species. **(4) By eating the ants** like *D. dendriticum.* **(5) By eating the crab or crayfish** like *P. westermani.*

Life Cycle

Hosts: Worm passes its life cycle in two different hosts.
- **Definitive host:** Man who harbors the adult stage.
- **Intermediate hosts:** Larval stage developed in two different aquatic animals. **(1) 1st intermediate host** like snail or molluscs required for development. **(2) 2nd intermediate host** like water fish or crab or crayfish or ant required for encystment.

Cycles: Summarize in **Flowchart 103.1.**

CLASSIFICATION

Systemic classification: Follow Ch. 14 (Table 14.4).

Pathological classification: Pathogenic species are further classified on the basis of reproductive organs and habitat as below.

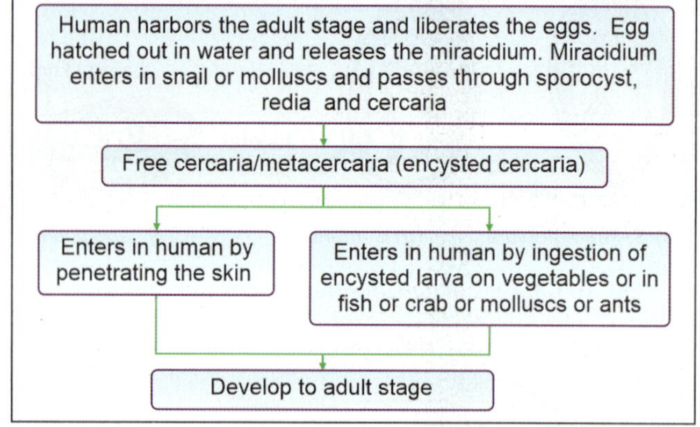

- **Diecious (nonhermaphrodite) trematodes:** Male and female species are separate and they habitat in blood. **(1) Vesical venous plexus:** *S. haematobium.* **(2) and portal venous plexus:** *S. mansoni, S. japonicum.*
- **Monoecious (hermaphrodite) trematodes:** Male and female species are not separate and single species can harbors both male and female reproductive organs. They have different habitats. **(1) Small intestinal trematodes:** *F. buski, H. heterophyes, M. yokogawai, E. ilocanum, E. malayanum, E. revolutum, P. supraryfex* and *W. watsoni.* **(2) Large intestinal trematodes:** *G. hominis.* **(3) Hepatic trematodes (liver flukes):** *C. sinensis, O. felineus, F. hepatica, F. gigantica, D. dendriticum* and *E. pancreaticum.* **(4) Pulmonary trematode (lung fluke):** *P. westermani* and other species.

Differences between Diecious and Monoecious Trematodes

Follow **Table 103.1.**

TABLE 103.1: Differences between diecious and monoecious trematodes		
Features	**Diecious**	**Monoecious**
Adult stage		
Sex	Separate	Nor separate
Shape	Cylindrical like round worm **(Fig. 103.1)**	Leaf like **(Fig. 103.2)**
Gyneco-phoric canal	Present	Absent
Egg stage		
Figure	Fig. 103.3	Fig. 103.4
Egg's shape	Non-operculated	Operculated
Spine/knob	Present	Absent
Embryo formation	Miracidium presents at the time of laying	In some eggs miracidium presents at the time of laying while in some it develops later
Larval stage		
Figure	Fig. 103.5	Fig. 103.7
Larva	No redia	Redia present
Infective stage	Schistosomula (tail less after infection)	Metacercaria (tail less before infection)

ACCESS YOURSELF

Short Note

1. Cercariae.

Short Questions for Theory/Viva Questions

1. What are metacercaria and schistosomulae?
2. Write the difference between diecious and monoecious trematodes.

MCQs for Chapter Review

General Features (Morphology)

1. Schistosomula of trematodes refers to:
   a. Cercaria without tail
   b. Cercaria with tail
   c. Cercaria with bifurcated tail
   d. Cercaria with short stumpy tail

Classification

2. Diecious trematode is:
   a. *S. haematobium*
   b. *C. sinensis*
   c. *F. hepatica*
   d. None

3. All are inhabitants of liver, *except*:
   a. *F. hepatica*   b. *F. buski*
   c. *C sinensis*   d. *s felineus*

Answers and Explanation of MCQs

1. a
- Follow section, **General features → morphology (larval stage → cercaria and schistosomula)** for explanation.
2. a
3. b
- Follow section, **classification** for explanation.

Infections of Trematodes: Diecious Trematodes

Chapter Outline

GENERAL FEATURES OF SCHISTOSOMES

Follow **Table 104.1**.

VESICAL PLEXUS TREMATODES

Schistosoma haematobium

Meaning

Schisto (Greek) means split and soma (Greek) means body; it looks vertically split, because of presence of gynecophoric canal in male.

Synonym and History

Initially genus was called *Bilharzia*, from the name of Theodor Bilharz, who 1st discovered the parasite in the mesenteric vein of Egyptian in 1851 from Cairo. Later name changed to *Schistosoma* with above mentioned meaning. Life cycle was 1st worked out by Leiper in Bulinus in Egypt in 1915.

Common Name

Follow **Table 104.1**.

Morphology

Three morphological stages like adult, egg and larva. Few important features are given below.

Adult Stage

Time of maturation: Schistosomulae transformed to sexually mature adult stage in 3 weeks in definitive host.

Size: Female is larger and stout than male.

Shape: Unlike other tramatodes, schistosomes are not flattened and leaf like, but cylindrical as like round worm as shown in **Fig. 103.1**.

Life span: 20–30 years.

Organs: Following are the important organs as shown in **Fig. 103.1**.
- **Tegument:** Follow **Ch. 103**.
- **Suckers:** Follow **Ch. 103**.
- **Body cavity:** Absent.
- **Gastrointestinal system:** It presents but without anus so called incomplete. Mouth is situated anteriorely in the center of oral sucker. Muscular pharynx is lacking. Esophagus is bifurcated in front of ventral sucker into two tube-like structures called **intestinal ceca** or **called crura** which reunite to form the single cecum. Length of reunited intestine is variable with different species.
- **Excretory and nervous systems:** Present.
- **Reproductive system:** Both male and female species are separate. Testes are 3–4 in numbers. Laurer's canal (rudimentary vagina) is absent. More details is given in **Ch. 103**.
- **Gynecophoric canal:** Follow **Ch. 103**.

Egg Stage

Morphology: Egg formed in ootype. Egg is oval and nonoperculated (**Fig. 103.3a**), 150 μm × 50 μm in size and surrounded by transparent and brownish yellow protective covering. Initially it contains ciliated embryo when laid called **miracidium.** Terminal spine is characteristic of the species and useful for identification.

Egg laying: Common sites of egg laying is small venules of vesical plexus and pelvic plexus. Ectopic sites are mesenteric portal system and pulmonary arterioles which are very rare. Mechanisms of egg laying are mentioned in **Flowchart 104.1**.

Mechanism of eggs expulsion: (1) From vesical plexus: Eggs laid in the vesical plexus make their way through

TABLE 104.1: General features and differences between major schistosomes

Features	*S. haematobium*	*S. mansoni*	*S. japonicum*
Common name	Vesical blood fluke	Manson's blood fluke	Oriental blood fluke
Geographical distribution	• Parts of Africa and middle east India: Few cases reported from Ratnagiri district of Maharashtra by Gadgil and Shah in 1951	• Various part of Africa and America	• Parasite of far East • Found in China, Japan, Philippines and Shah state of Burma
Habitat (sites) (Fig. 104.4)	• **Common site:** Adult worm lives in **vesical plexus (pelvic plexus)** supplying urinary bladder (inferior mesenteric vein) • **Ectopic sites:** Lungs, brain, spinal cord (due to outflow phenomenon in case of heavy infection)	• **Common site:** Adult worm lives in **plexus** supplying sigmoido–area (inferior mesenteric vein) • **Ectopic sites:** Liver (portal vein), lungs and spinal cord	• **Common sites: (1)** Superior mesenteric vein supplying ileocecal area (superior mesenteric vein), **(2) plexus** (hemorrhoides), **(3)** Liver **(portal vein/plexus)** • **Ectopic sites:** Lungs and brain
Adult stage (Fig. 103.1)			
Male			
Size	1–1.5 cm × 1 mm	1 cm × 1 mm	1.2–2 cm × 0.5 mm
Cuticula	Finely tuberculated	Grossly tuberculated	Nontuberculated
Testes	4–5 and in group	8–9 and in zig-zag row	6–7 and in single file
Female			
Size	2 cm × 0.25 mm	1.4 cm × 0.25 mm	2.6 cm × 0.3 mm
Ovary	Behind the middle of body	Anterior to the middle of body	In the middle of the body
Uterus	Contains 20–30 eggs	Contains 1–3 eggs	Contains ≥50 eggs
Egg stage (Fig. 103.3)			
Size	150 µm × 50 µm	150 µm × 60 µm	100 µm × 65 µm
Spine/knob	Terminal spine	Lateral spine	Lateral knob
ZN staining	Non-acid fast	Acid fast (egg shell)	Acid fast
Larval stage (Fig. 103.5)			
Cephalic gland	2 pairs oxyphilic and 3 pairs basophilic	2 pairs oxyphilic and 4 pairs basophilic	5 pairs oxyphilic and no basophilic
Hosts			
Intermediate	Snail (*Ferrissia*, etc.)	Snail (*Biomphalaria*, etc.)	Snail (*Oncomelania*)
Definitive	Humans	Humans	Humans and domestic animals

Flowchart 104.1: Mechanism of eggs laying

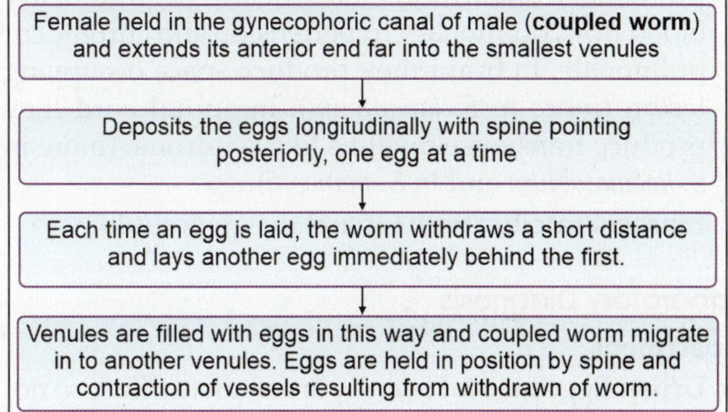

the vessels and mucosa of bladder. They entered in to the cavity and escape with urine usually at the end of micturition. They mostly expelled in mid day than other time with unknown reasons. **(2) From ectopic sites:** Eggs laid in the ectopic sites are generally died and evoke local tissue reactions. They are found in biopsy of local tissues and rarely expelled in local material like feces or sputum.

Larval Stage

One egg gives the one miracidium, which produces multiple cercariae. Larvae pass through the stages of miracidium, sporocysts (1st and 2nd generation sporocysts), cercaria and schistosomula as described in **Ch. 103 (Fig. 103.5).** There is no redia stage in schistosomes.

Immunity

Acquired immunity: Developing schistosomulae are more susceptible to the immune response and death occurs due to destruction of tegument. Adult worms are long-lasting worms and immunity develops in endemic areas after long time.

Premunition: It is an immunity to reinfection in presence of active adult infection. It was determined by Smithers and Terry in 1969.

Hypersensitivity: Toxic metabolites by schistosomulae produce the anaphylaxis (immediate type hypersensitivity). Toxic metabolites of eggs are coming out through the pores and elicit the cellular infiltration and granuloma formation (delayed type hypersensitivity).

Life Cycle

Types of hosts: Two types. **(1) Definitive host:** Human. **(2) Intermediate hosts:** Fresh water snail with different species like *Bulinus (Physopsis) truncates* in Africa, *Planorvarius metidjensis* in Morocco and Portugal and *Ferrissia tenuis* in India.

Cycles: As shown in **Flowchart 104.2** and **Fig. 104.1**.

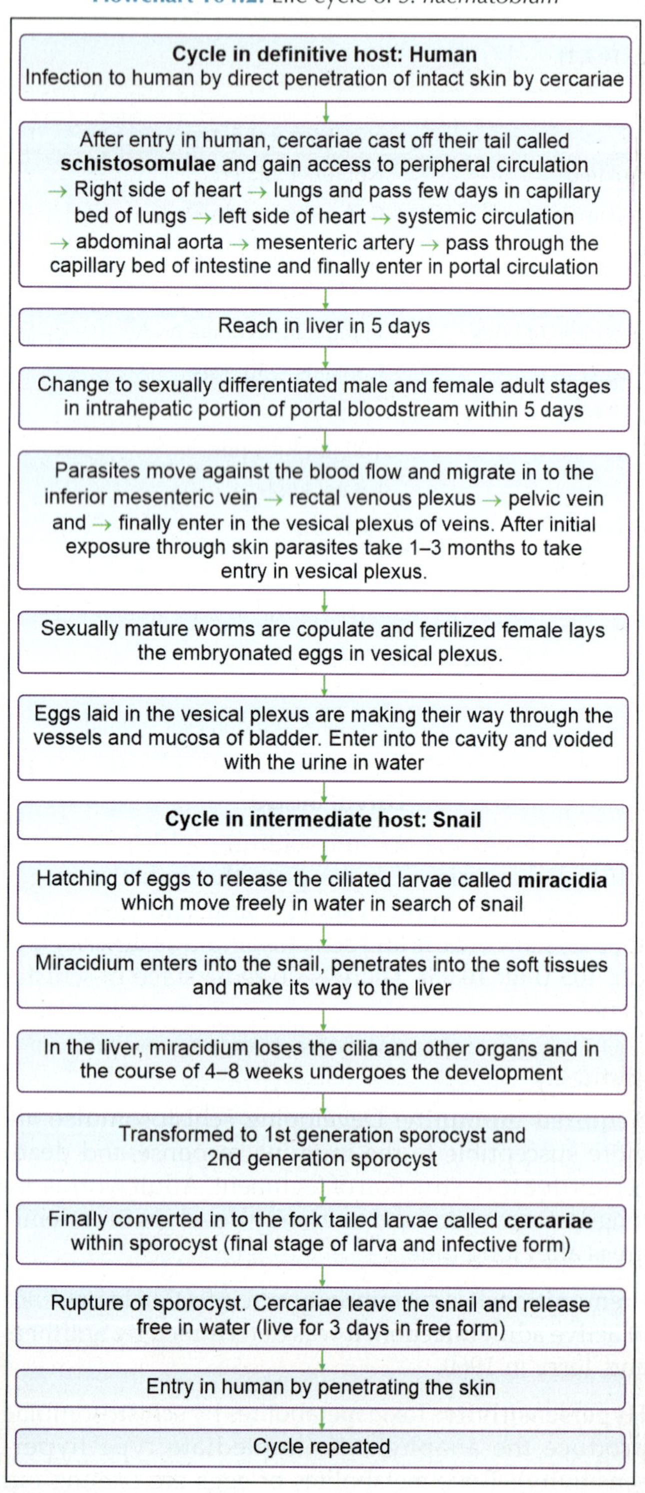

Flowchart 104.2: Life cycle of *S. haematobium*

Pathogenicity

Disease name: Called **schistosomiasis.**

Epidemiology: Geographical distribution is mentioned in **Table 104.1.**

Synonym: Also called **bilharziasis** or **urinary schistosomiasis** or **schistosomiasis hematobia** or **endemic hematuria.**

Virulence factors: (1) Cytolytic secretion by cercariae from cephalic gland causes local lesion called **swimmer's itch.** (2) Toxic metabolites by schistosomulae produce ITH. (3) Toxic metabolites by eggs are coming out through the pores and elicit the cellular infiltration and granuloma formation (DTH). (4) Mechanical factor erosion of blood vessels by spine of egg causes hemorrhages.

Reservoir of infection: Human

Source of infection: Water

Modes of transmission: Cercaria enters by penetrating the skin

Exit form: Embryonated egg

Infective form: Cercaria

Portal of entry: Skin

Sites: Follow **Table 104.1**

Pathogenesis and clinical features: Follow **Flowchart 104.3.**

Complications: They occur after very long time.
1. **Squamous cell carcinoma of bladder:** Close association between vesical schistosomiasis and cancer of urinary bladder has been observed in endemic areas.
2. **Ectopic lesions:** Due to out flow phenomenon in heavy infection. Eggs and worms escape in to pelvis vein and carried to lungs and sometimes from portal circulation to brain and spinal cord. In **lungs** they produce granulomatous lesion with fibrosis, pulmonary endarteritis, obstruction of pulmonary blood flow, pulmonary hypertension and chronic cor pulmonale. In **brain** they produce space occupying lesion (more in *S. japonicum*). In **spinal cord** they produce transverse myelitis like syndrome (more in *S. haematobium* and in *S. mansoni*).
3. **Fibrosis/calcification:** Occurs in egg granuloma.

Laboratory Diagnosis

Specimens

- **Urine:** Eggs present in end part of micturition, so not the mid stream but last few drops of urine should be collected. Miracidium can be found if urine is left to stand for few hours without preservatives. Urine can be preserved by using 10% formalin. For eggs counting 24 hours urine is collected to which add 10% of formalin (0.5 ml/100 ml urine).
- **Mucosal biopsy:** Piece of vesical mucosa is removed by cystoscope. Excised tissue is divided in two pieces. One piece is compressed between two slides and

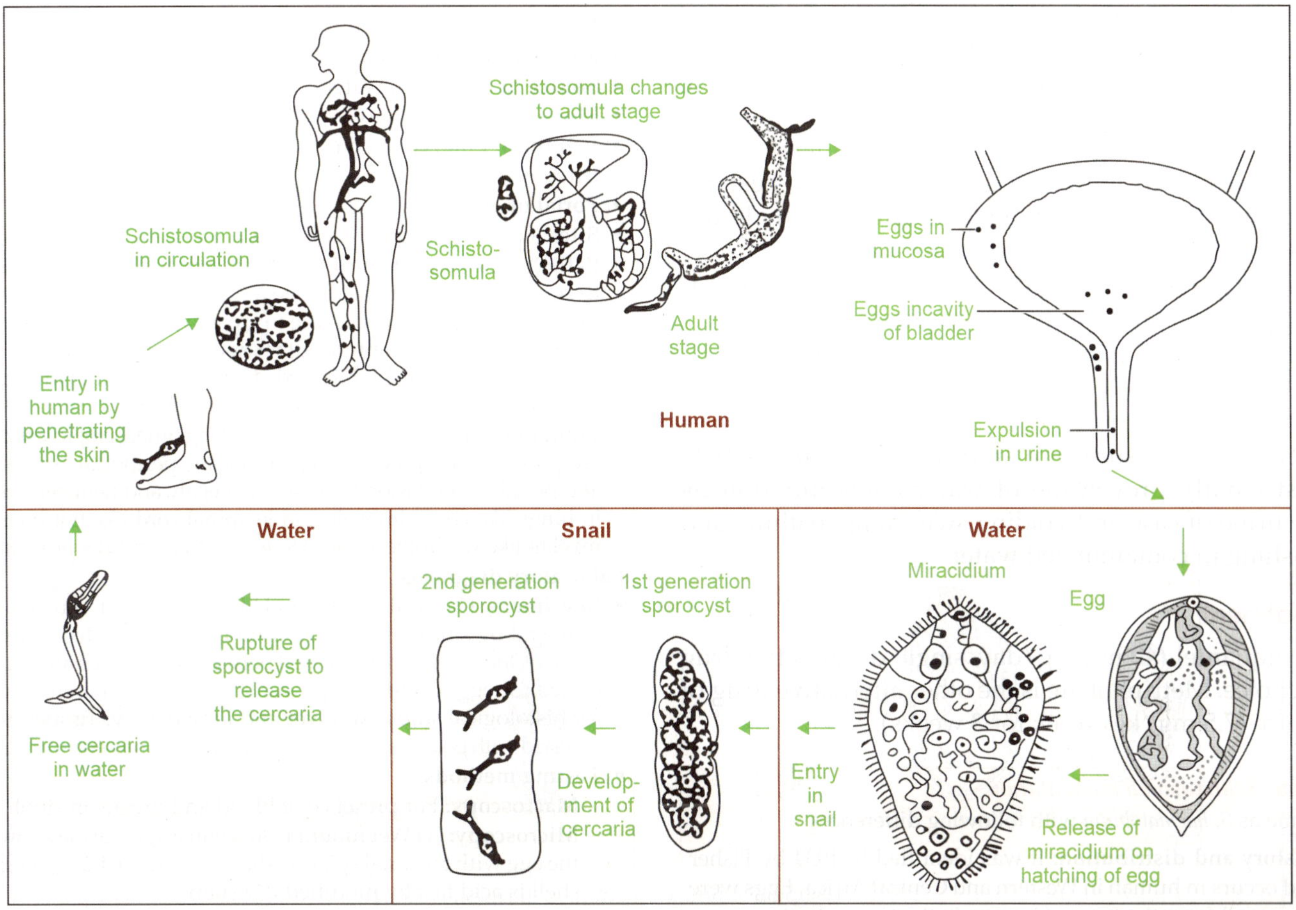

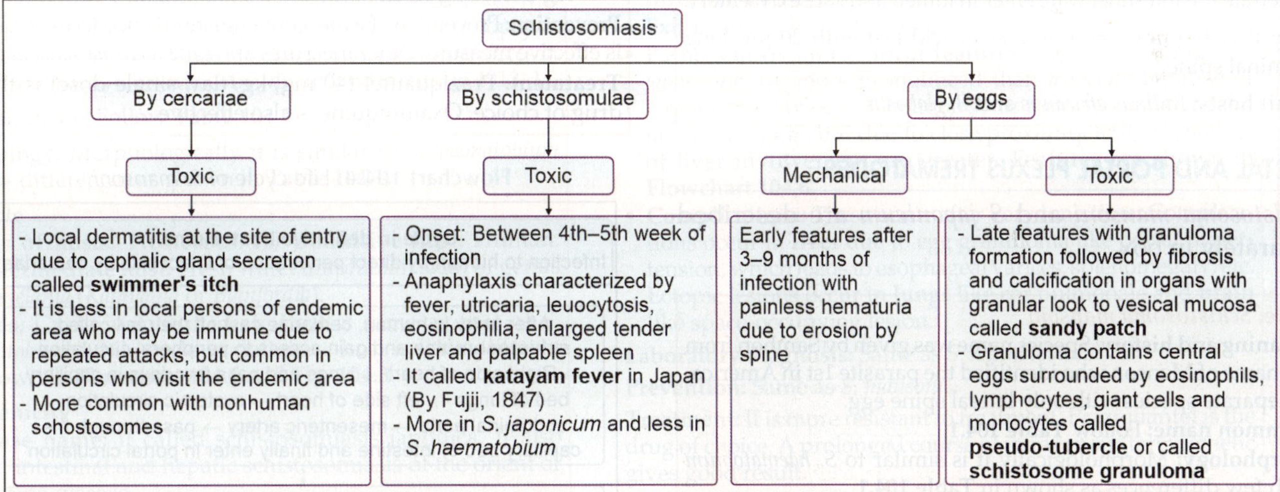

Fig. 104.1: Life cycle of *S. haematobium*

Flowchart 104.3: Clinical features and pathogenesis of *S. haematobium*

Schistosomiasis

By cercariae	By schistosomulae	By eggs	
Toxic	Toxic	Mechanical	Toxic
- Local dermatitis at the site of entry due to cephalic gland secretion called **swimmer's itch** - It is less in local persons of endemic area who may be immune due to repeated attacks, but common in persons who visit the endemic area - More common with nonhuman schostosomes	- Onset: Between 4th–5th week of infection - Anaphylaxis characterized by fever, utricaria, leuocytosis, eosinophilia, enlarged tender liver and palpable spleen - It called **katayam fever** in Japan (By Fujii, 1847) - More in *S. japonicum* and less in *S. haematobium*	Early features after 3–9 months of infection with painless hematuria due to erosion by spine	- Late features with granuloma formation followed by fibrosis and calcification in organs with granular look to vesical mucosa called **sandy patch** - Granuloma contains central eggs surrounded by eosinophils, lymphocytes, giant cells and monocytes called **pseudo-tubercle** or called **schistosome granuloma**

examined for eggs under low power. Put other piece in fixative for histological study.

Testing methods

A. Blood picture: Eosinophilia.

B. Macroscopy: Reddish or cloudy urine due to presence of blood.

C. Microscopy: Wet mount from centrifuged urine shows the egg with terminal spine as shown in **Fig. 104.2a**.

D. Serological tests: (1) Ag detection tests: Ag detection is useful in acute infection and to monitor the therapy. Ag concentration reduced after treatment. Circulating

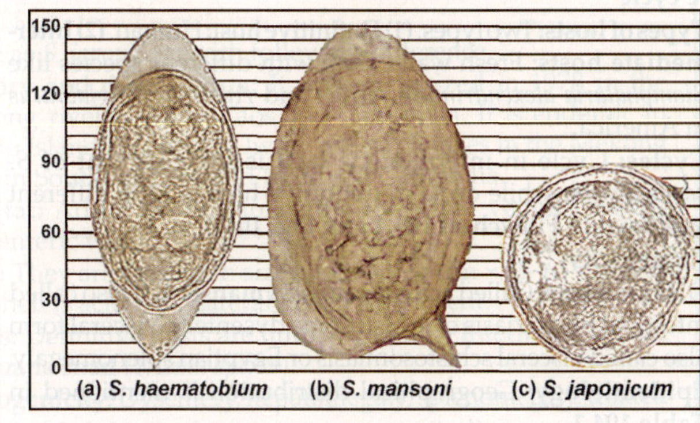

Fig. 104.2: Eggs of schistosomes under wet mount

8

Essentials of Medical Microbiology

Flowchart 104.5: Life cycle of *S. japonicum*

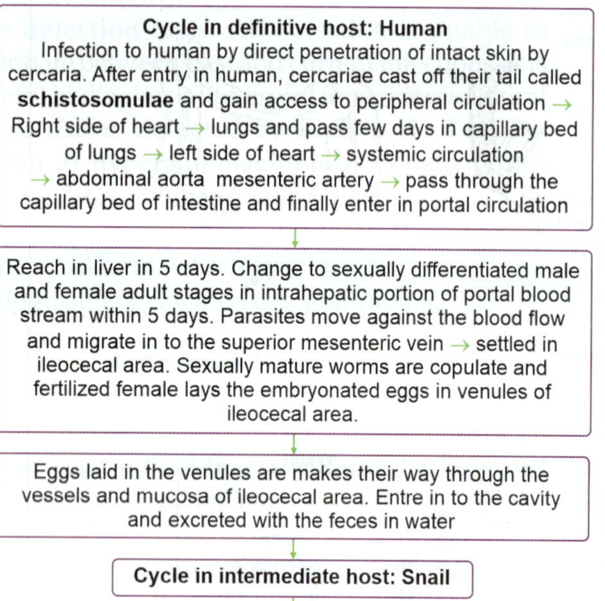

Flowchart 104.6: Clinical features of *S. japonicum*

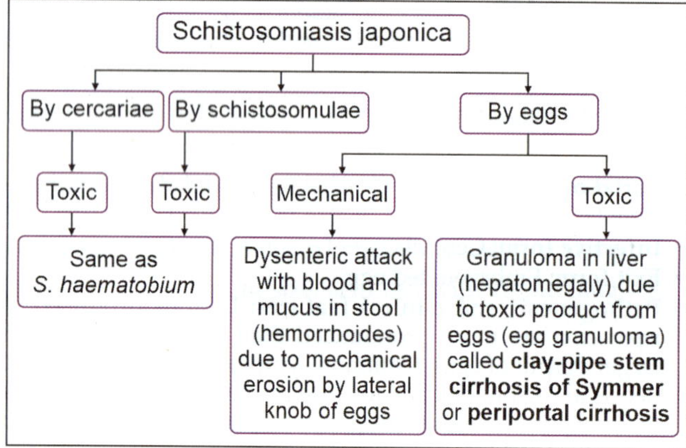

ANATOMICAL ASPECT OF PORTAL SYSTEM

Follow **Fig. 104.4.**

ACCESS YOURSELF

Case Study

1. A 35-year-man came with complain of local skin lesion and painless hematuria. He is giving history of swimming in water. Urine examination revealed non-acid fast egg with terminal spine. Identify the case and answer the following.
   a. Name and draw the properly labeled diagram of egg of causative agent.
   b. Describe the life cycle of causative agent.
   c. Describe the pathogenicity of causative agent.
   d. Describe the lab., diagnosis of causative agent.

Essay/Full Question

1. *Schistosoma haematobium.*

Short Note

1. Life cycle/pathogenicity/lab., diagnosis of *S haematobium*.

Short Questions for Theory/Viva Questions

1. What is gynaecophoric canal? Write its function.
2. Write the mechanisms of eggs laying and eggs expulsion in schistosomes.

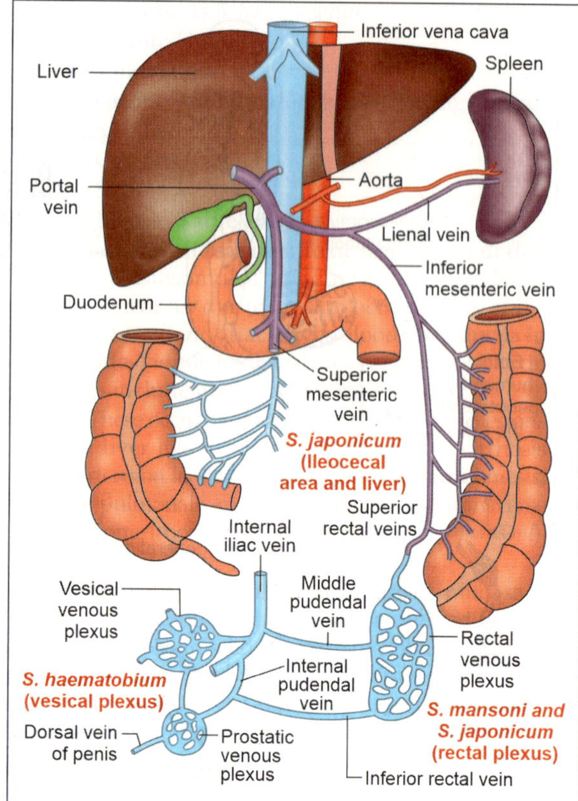

Fig. 104.4: Anatomical aspect and portal system

3. Name the four parasites causing dysentery.
4. Name the larval stage of schistosomes.
5. What is hatching test?

MCQs for Chapter Review

General Features

1. **Natural habitat of *Schistosoma* (blood flukes):**
   a. Veins of the urinary bladder
   b. Portal and pelvic veins
   c. Vesical plexus
   d. All of the above
2. **Terminal spine eggs are seen in:**
   a. *Schistosoma haematobium*   b. *Schistosoma mansoni*
   c. *Schistosoma japonicum*   d. *Chlonorchis sinensis*
3. **Water host required for schistosomiasis:**
   a. Fish   b. Cyclops
   c. Snails   d. Crabs

Schistosoma haematobium

4. **Painless hematuria is seen as one of the manifestation in the infection caused by:**
   a. *Schistosoma japonicum*   b. *Schistosoma mansoni*
   c. *Schistosoma haematobium* d. *Plasmodium falciparum*
5. **Katayam fever is caused by:**
   a. *F hepatica*   b. *C sinensis*
   c. *S. haematobium*   d. *A lumbricoides*

Answers and Explanation of MCQs

1. **d**
2. **a**
3. **c**
- Follow section, **general features (Table 104.1)** for explanation of answers of MCQs 1–3.
4. **c**
5. **c**
- Follow section, *Schistosoma haematobium* **(pathogenicity → clinical features → Flowchart 104.3)** for explanation of answers of MCQs 4–5.

CHAPTER 105

Infections of Trematodes: Monoecious Trematodes

Chapter Outline

- Small Intestinal Trematodes
- Large Intestinal Trematodes
- Hepatic Trematodes
- Pulmonary Trematodes

SMALL INTESTINAL TREMATODES

Fasciolopsis buski

Common name: (1) Giant or largest intestinal fluke: Because it is the largest intestinal fluke infecting human. **(2) Ginger worm:** This name is given, because it appears like a slice of ginger.

History: It was 1st discovered in the duodenum of East Indian sailor in 1843 by Busk.

Morphology

1. **Adult stage (Fig. 105.1)**
   - **Development:** Metacercaria changes to adult worm in 3 months time.
   - **Shape:** Elongated, flat and oval in shape. Anterior end is narrower than posterior.
   - **Size:** It is a largest trematode and it is 2–7.5 cm long, 8–20 mm in broad and 0.5–3 mm in thick.
   - **Life span:** Not more than 6 months.
   - **Organs: Tegument and suckers** are mentioned in **Ch. 103. Body cavity** is absent. **GIT** is present, but without anus so called incomplete. Esophagus is bifurcated into two tube like structures called **intestinal ceca** or called **crura.** Crura present with two characteristic curves in middle and do not bears any branches. **Excretory system** is present. **Nervous system** has no cephalic cone as like in *F. hepatica.*

2. **Egg stage (Fig. 105.1):** One adult worm lays 25,000 eggs/day. Egg is oval/spherical, with small operculum, 130–140 μm × 80–85 μm in size, yellowish-brown color (bile stained) and with single egg shell. Yolk material is present abundantly. Initially embryo is unsegmented, nonciliated and surrounded by yolk mass.

3. **Larval stage:** Solid form of larva is present. Different larval stages are mentioned in life cycle and shown in **Fig. 105.1.**

Life cycle

- **Hosts: (1) Definitive hosts** are human and pig. **(2) Intermediate host** is small flatty coiled aquatic snail of genus *Segmentina.*
- **Cycles:** As shown in **Flowchart 105.1 and Fig. 105.1.**

Pathogenicity

- **Disease name:** It called **fasciolopsiasis.**
- **Epidemiology:** This Asiatic trematode distributed in China, Thailand, Malaysia, Bengal, Assam and other oriental region.
- **Virulence factors:** Toxic metabolites by adult worms produce allergic reactions with edema and eosinophilia.
- **Reservoir of infection:** Pig
- **Source of infection:** Water
- **Modes of transmission:** It enters in human by **ingestion** of aquatic plant as raw food stuffs like water chestnut or water caltrop or **peeling** of water chestnut/water caltrop with teeth.
- **Exit form:** Unembryonated egg
- **Infective form:** Encysted cercaria called **metacercaria** or **adolesceria**
- **Portal of entry:** GIT
- **Site:** Adult stage lives in small intestine of human and pig.
- **Clinical features:** As shown in **Flowchart 105.2.**

Laboratory diagnosis

- **Specimen:** Stool
- **Testing methods**
  1. **Blood picture:** Eosinophilia in heavy infection
  2. **Macroscopy:** Adult worms can be demonstrated after purgatives
  3. **Microscopy:** Eggs are examined by NS wet mount. Eggs of *F. buski, Echinostoma* species, *G. hominis* and *F. hepatica* are indistinguishable.

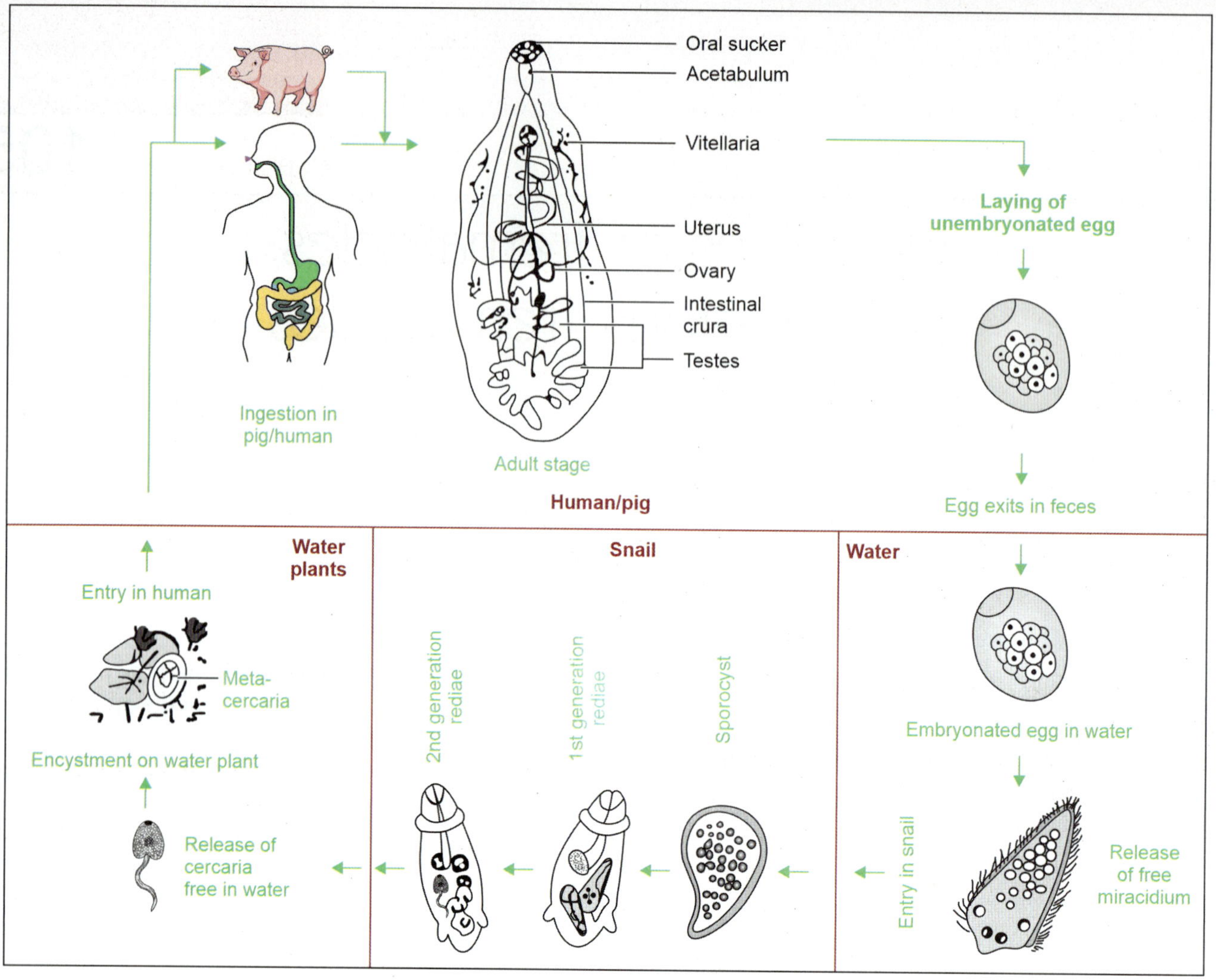

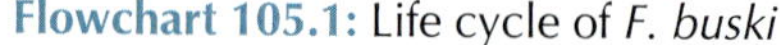

Fig. 105.1: Life cycle of *F. buski*

Flowchart 105.1: Life cycle of *F. buski*

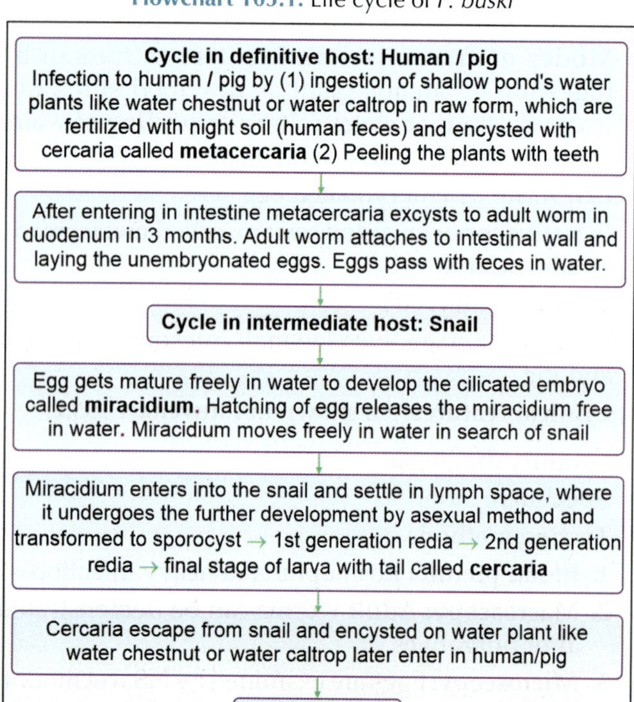

Cycle in definitive host: Human / pig
Infection to human / pig by (1) ingestion of shallow pond's water plants like water chestnut or water caltrop in raw form, which are fertilized with night soil (human feces) and encysted with cercaria called **metacercaria** (2) Peeling the plants with teeth

↓

After entering in intestine metacercaria excysts to adult worm in duodenum in 3 months. Adult worm attaches to intestinal wall and laying the unembryonated eggs. Eggs pass with feces in water.

↓

Cycle in intermediate host: Snail

↓

Egg gets mature freely in water to develop the cilicated embryo called **miracidium.** Hatching of egg releases the miracidium free in water. Miracidium moves freely in water in search of snail

↓

Miracidium enters into the snail and settle in lymph space, where it undergoes the further development by asexual method and transformed to sporocyst → 1st generation redia → 2nd generation redia → final stage of larva with tail called **cercaria**

↓

Cercaria escape from snail and encysted on water plant like water chestnut or water caltrop later enter in human/pig

↓

Cycle repeated

Flowchart 105.2: Clinical features of *F. buski*

- Fasciolopsiasis
  - Mild infection
    - General
      - Asthenia
      - Anemia
      - Malnutrition
    - GIT
      - Chronic diarrhea
      - Obstruction
  - Heavy infection
    - Severe anemia
    - Allergic reactions like edema and eosinophilia due to absorption of toxic products of adult worms

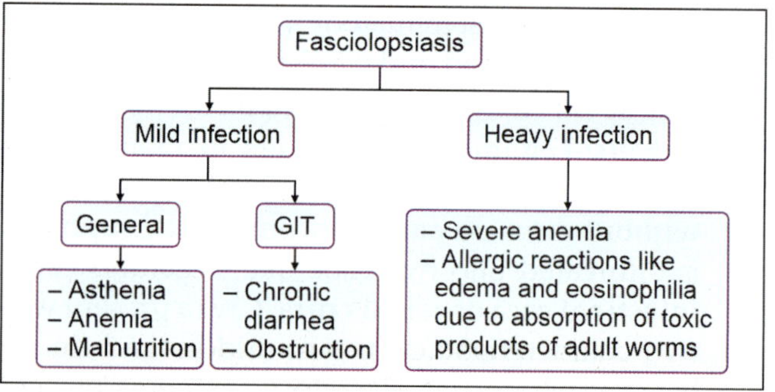

Prevention: Preventive measures are eradication of intermediate host (snail), prevention of water contamination by the feces of case and washing of food and vegetables before eating.

Treatment: Praziquantel is the drug of choice. Hexylresorcinol and tetrachlorethylene are also useful drugs.

Heterophyes heterophyes

Common name: Von Siebold's fluke, Dwarf fluke.

Meaning: It is the smallest fluke hence called **dwarf fluke.**

History: Infection was 1st reported by Bilharz in 1851.

Morphology

1. **Adult stage (Fig. 105.2)**
   - **Development:** Metacercaria changes to adult worm in 15–20 days
   - **Size:** 1.5 mm × 0.3 mm
   - **Shape:** Flattened dorso-ventrally with narrow anterior end and rounded posterior end
   - **Life span:** 2 months
   - **Organs:** For more details, **follow Ch. 103.**
2. **Egg stage (Fig. 105.2):** Egg is oval, operculated, 28–30 µm × 16–18 µm in size, yellowish-brown color (bile stained) and surrounded by transparent and brownish yellow protective covering. Ciliated embryo presents initially when laid called miracidium.
3. **Larval stage:** Solid form of larva is present. Single egg gives the single miracidium, which produces multiple cercariae. Different larval stages are mentioned in life cycle and shown in **Fig. 105.2.**

Life Cycle

- **Hosts: (1) Definitive hosts** are human, cat, dog, wolf, bird and fox. **(2) Intermediate hosts:** 1st intermediate host is marine and brackish water snail like *Pirenella conica* or *Ceithidea* while 2nd intermediate host is fish like mullet (*Mugil cephalus*) and tilapia.

- **Cycles:** As shown in **Flowchart 105.3** and **Fig. 105.2.** 891

Flowchart 105.3: Life cycle of *H. heterophyes*

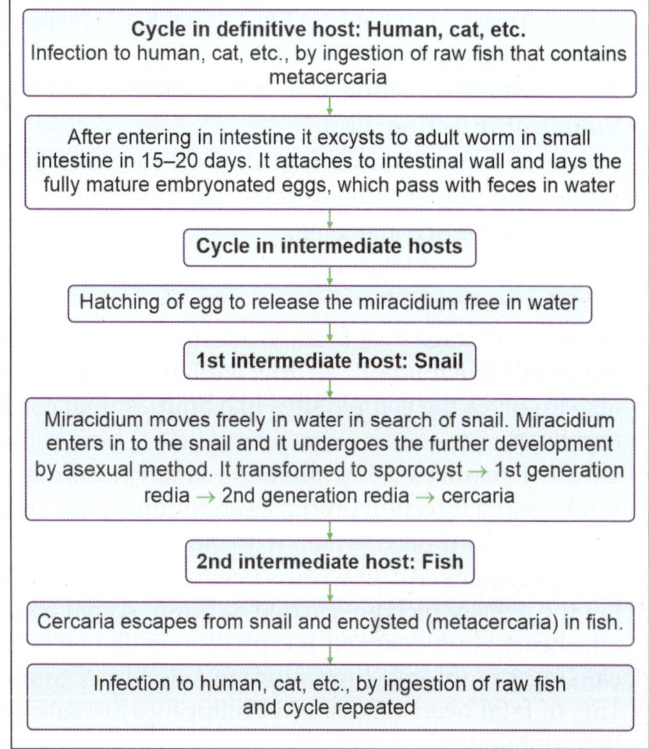

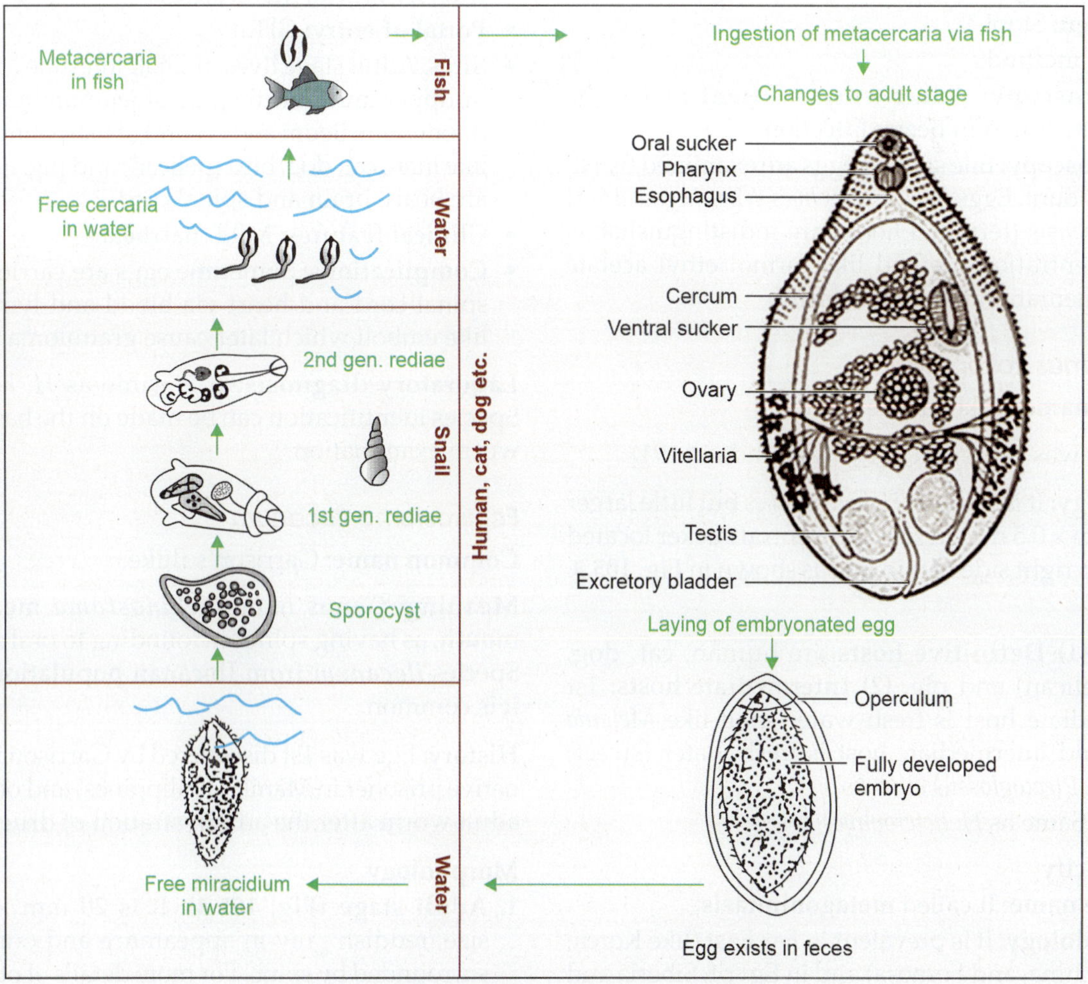

Fig. 105.2: Life cycle of *H. heterophyes*

Pathogenicity

- **Disease name:** It called heterophyiasis.
- **Epidemiology:** It is prevalent in Egypt, Iran, Israel, Sudan, Turkey and Far East like China, Korea, Japan, Taiwan, Philippines and India.
- **Reservoirs of infection:** Cat, dog, wolf, bird and fox
- **Source of infection:** Fish
- **Mode of transmission:** Ingestion of raw fish like mullet and tilapia contains metacercaria.
- **Exit form:** Embryonated egg
- **Infective form:** Metacercaria
- **Portal of entry:** GIT
- **Sites:** Adult stage lives in small intestine of fish eating mammals like human, cat, dog, wolf, bird and fox. It also involves the ectopic sites like brain, spinal cord and heart.
- **Clinical features:** Mild infection is asymptomatic while heavy infection produces abdominal pain and dysentery (chronic diarrhea with blood and mucus).
- **Complications:** Adult worms penetrate the gut and lay the eggs, which are carried to brain, spinal cord and heart via blood and lymphatics as like emboli causing granuloma formation and death. Around 15% of fatal heart diseases in Philippines are due to heterophyiasis.

Laboratory diagnosis

- **Specimen:** Stool
- **Testing methods**
  1. **Macroscopy:** Adult worms absent or can be demonstrated in heavy infection.
  2. **Microscopy:** Bile stained eggs are examined by NS wet mount. Eggs of *H. heterophyes*, *M. yokogawai* and *C. sinensis* (terminal hook) are indistinguishable. Concentration method like formol ethyl acetate sedimentation method is useful.

Metagonimus yokogawai

Common name: Yokogawa's fluke.

History: It was 1st reported by Katsurada in 1911.

Morphology: It is same as *H. heterophyes* but little larger about 2 mm × 0.5 mm in size and ventral sucker located laterally on right side of mid line as shown in **Fig. 105.3**.

Life cycle

- **Hosts: (1) Definitive hosts** are human, cat, dog, bird (pelican) and pig. **(2) Intermediate hosts:** 1st intermediate host is fresh water snail like *Melania* while 2nd intermediate host is fresh water (sweet) fish like *Plectoglossus altivelis*.
- **Cycles:** Same as *H. heterophyes*

Pathogenicity

- **Disease name:** It called **metagonimiasis.**
- **Epidemiology:** It is prevalent in Far East (like Korea, Japan, China, and Fomosa) and in Egypt, Siberia and Balkan states.

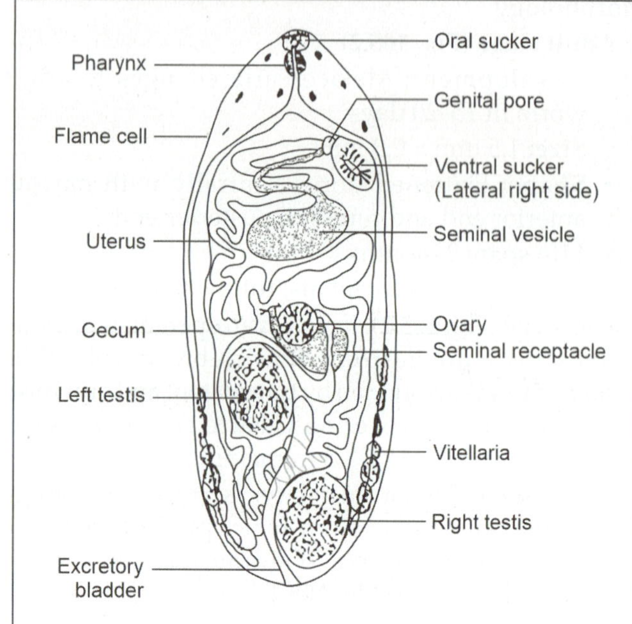

Fig. 105.3: Adult stage of *M. yokogawai*

- **Reservoirs of infection:** Cat, dog, bird (pelican) and pig
- **Source of infection:** Fish
- **Mode of transmission:** Ingestion of raw fish like *Plectoglossus altvalis* contains metacercaria.
- **Exit form:** Embryonated egg
- **Infective form:** Metacercaria
- **Portal of entry:** GIT.
- **Sites:** Adult stage lives in small intestine (particularly in upper and middle part of jejunum and rarely in duodenum, ileum and cecum) of fish eating mammals like man, cat, dog, bird (pelican) and pig. Ectopic sites are heart, brain and spinal cord.
- **Clinical features:** Mild diarrhea.
- **Complications:** Sometime eggs are carried to brain, spinal cord and heart via blood and lymphatics as like emboli which later cause granuloma formation.

Laboratory diagnosis: It is same as *H. heterophyes*. Species identification can be made on the basis of adult worm examination.

Echinostoma ilocanum

Common name: Garrison's fluke.

Meaning: Genus name *Echinostoma* **means spiny mouth,** as having spine surrounding to oral sucker and Species *ilocanum* **from Ilocanan population** in which it is common.

History: Egg was 1st discovered by Garrison in 1907 in a native prisoner in Manila (Philippines) and obtained the adult worm after the administration of drug filix mas.

Morphology

1. **Adult stage (Fig. 105.4):** It is 20 mm × 2 mm in size, reddish gray in appearance and oral sucker is surrounded by spine. For more details of other organs **follow Ch. 103.**

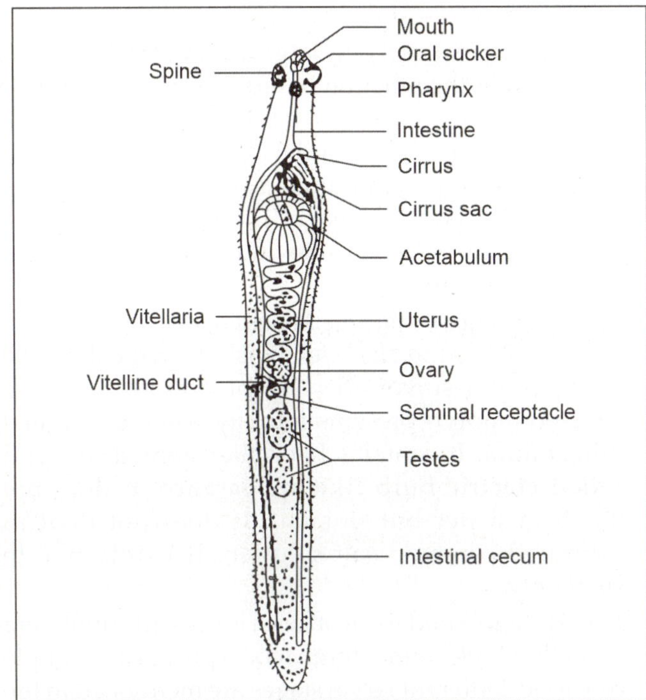

Fig. 105.4: Adult stage of *E. ilocanum*

2. **Egg stage:** Egg is oval or spherical with small operculum, 90–125 µm × 55–75 µm in size and yellowish-brown color (bile stained). It contains ciliated embryo when laid down called miracidium.
3. **Larval stage:** Solid form of larva is present. Single egg gives the single miracidium, which produces multiple cercariae. Different larval stages are same as *H. heterophyes* except that 2nd intermediate host is molluscs and no stage of sporocyst formation.

Life cycle

- **Hosts: (1) Definitive hosts** are human, rat and dog. **(2) 1st intermediate host** is fresh snail while **2nd intermediate host** is molluscs.
- **Cycles:** Same as *H. heterophyes* without sporocyst formation.

Pathogenicity

- **Disease name:** It is called echinostomiasis.
- **Epidemiology:** It is prevalent in **Ilocanan population** of Luzon province, the Manila-Philippines, Japan, China, Indonesia, Taiwan and India (Assam).
- **Reservoirs of infection:** Human, dog and rat
- **Source of infection:** Molluscs
- **Modes of transmission:** Ingestion of molluscs contains metacercaria
- **Exit form:** Embryonated egg
- **Infective form:** Metacercaria
- **Portal of entry:** GIT
- **Site:** Adult stage lives in small intestine
- **Clinical features:** Mild infection is asymptomatic. Heavy infection produces nausea, severe diarrhea, abdominal pain, distension, etc.

Laboratory diagnosis: It is same as *H. heterophyes*. Species identification can be made on the basis of adult worm examination.

> **Note:** *Echinostoma malayanum*
> It was discovered by Leiper in 1911. It was 1st reported from Singapore and Kuala Lumpur, hence commonly called **Malayan fluke.** It is the common intestinal parasite of tribes living in the Sino-Tibetian frontier (Bare, 1930).
>
> **Note:** *Echinostoma revolutum*
> It was discovered by Forhlichr in 1802. It is a parasite of duck, geese and fowl. Human infection is accidental.

Paryphostomum supraryfex

It is reported from India. It is a parasite of pig and clinically similar to *F. buski*.

Watsonius watsoni

Morphology: Adult stage is 8–10 mm × 4–5 mm in size, pear shaped and flat dorso-ventrally. Body is ventrally concave near the posterior sucker. Egg and larval stages are same as *F. buski*.

Life cycle: It has not been worked out yet.

Pathogenicity: It is prevalent in Africa, Japan and Malaysia. Adult stage lives in small intestine in human and monkey. It presents with severe diarrhea resulting death was found in Africa.

Laboratory diagnosis: Based on identification of eggs in stool.

LARGE INTESTINAL TREMATODES

Gastrodiscoides hominis

Synonym: *Amphostomum hominis.*

History: It was discovered in 1876, by Lewis and Mc Connell in the cecum of Indian patient.

Morphology

1. **Adult stage**
   Size: 5–10 mm × 4–6 mm (broad at the widest part)
   Shape: Pyriform shaped and flat dorso-ventrally.
   Organs (Fig. 105.5): Body is divided in two portions. Anterior part is conical. Posterior part is hemispherical and hollowed out ventrally to form a concave disc. Acetabulum (ventral sucker) is postero-terminal and situated ventrally. Notch presents at the posterior end. Other organs are same as other monoecious trematodes as mentioned in **Ch. 103.**
2. **Egg stage:** It is 130 µm × 60 µm in size and greenish gray. Other details are same as *F. buski*.
3. **Larval stage:** It is same as *F. buski*.

Life cycle: Definitive hosts are human, pig, monkey, deer and mouse. Intermediate host is planorbid snail like *Helicorbis coenosus* (larval stage). Cycle is same as *F. buski*.

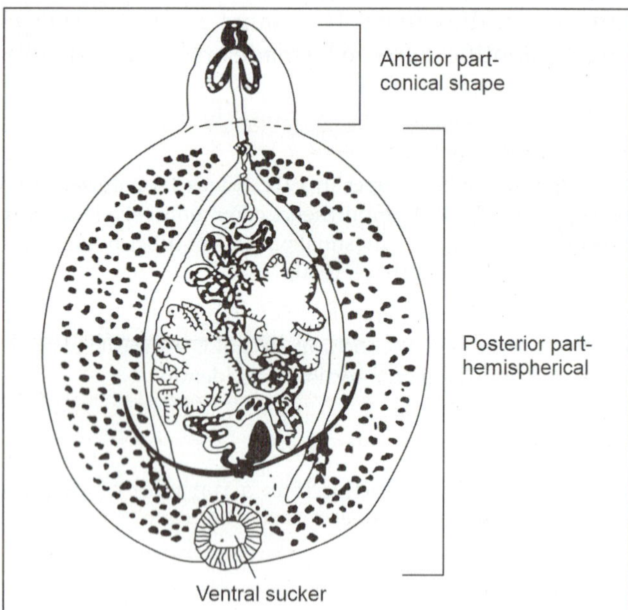

Fig. 105.5: Adult stage of *G. hominis*

Pathogenicity

- **Disease name:** It called **gastrodiscoidiosis**
- **Epidemiology:** It is prevalent in India (Bengal, Assam, Bihar, Orissa), Cochin-China Malaya and Indian immigrants to British Guiana.
- **Reservoir of infection:** Man, monkey, pig, deer and mouse
- **Source of infection:** Water vegetables
- **Mode of transmission:** Ingestion of raw vegetables like water chest-nut contains metacercaria
- **Exit form:** Unembryonated egg
- **Infective form:** Metacercaria
- **Portal of entry:** GIT
- **Site:** Adult stage lives in large intestine
- **Clinical features:** Mucoid diarrhea

Laboratory diagnosis: It is made by identifying the eggs in stool. Eggs are same as *F. buski* but narrower and greenish gray instead of yellowish brown.

HEPATIC TREMATODES

Clonorchis sinensis

Common name: Chinese liver fluke, Oriental liver fluke and Distoma of China.

History: Parasite was 1st discovered by McConnel in 1857 from the autopsy of a Chinese carpenter in the Calcutta Medical College Hospital. Life cycle was worked out by Faust and Khaw in 1927.

Morphology

1. **Adult stage (Fig. 105.6)**
   - **Development:** Metacercaria changes to adult worm in 1 month.
   - **Size:** Variable, 10–25 mm × 2–3 mm
   - **Shape:** Flattened dorso-ventrally with narrow anterior end and rounded posterior end.

- **Life span:** 20–30 years.
- **Organs:** Oral sucker is larger and ventral sucker is located at the junction of the anterior and middle third of the body. Blind intestinal ceca are simple and extend to the caudal region. Reproductive systems are two in numbers, in large size, deeply branched and located in the posterior third of the body, one behind the other. For more details, **follow Ch. 103.**

2. **Egg stage (Fig. 105.6):** Egg is flask shaped, operculated, 35 μm × 20 μm in size, yellowish-brown color (bile stained) and surrounded by protective covering. Ciliated embryo presents initially when laid called miracidium. It contains hook like spine at the end called **electric bulb like appearance**. It does not hatch in water but in snail. It does not float in saturated common salt solution. It is infective to snail only.

3. **Larval stage:** Solid form of larva is present. Single egg gives the single miracidium, which produces multiple cercariae. Different larval stages are mentioned in life cycle and shown in **Fig. 105.6.**

Life cycle

- **Hosts:** Worm passes its life cycle in two different hosts. **(1) Definitive hosts** are human, cat, dog, pig and rat. **(2) 1st intermediate host** is snail (*Bulimius suchsianus, Parafossarulus manchouricus* and *Alocina longicornis*), while **2nd intermediate host** is fresh water fish (*Cyprinoid*).
- **Cycles:** As shown in **Flowchart 105.4 and Fig. 105.6.**

Pathogenicity

- **Disease name:** It called clonorchiasis.

Flowchart 105.4: Life cycle of *C. sinensis*

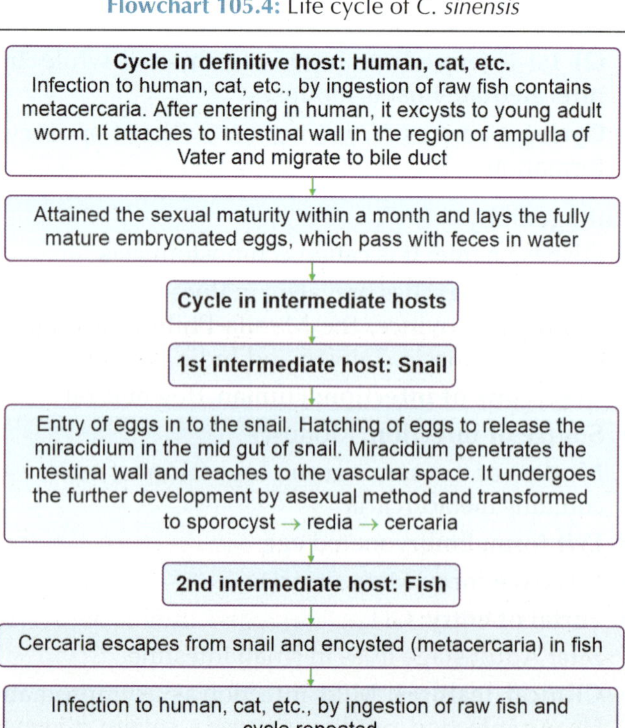

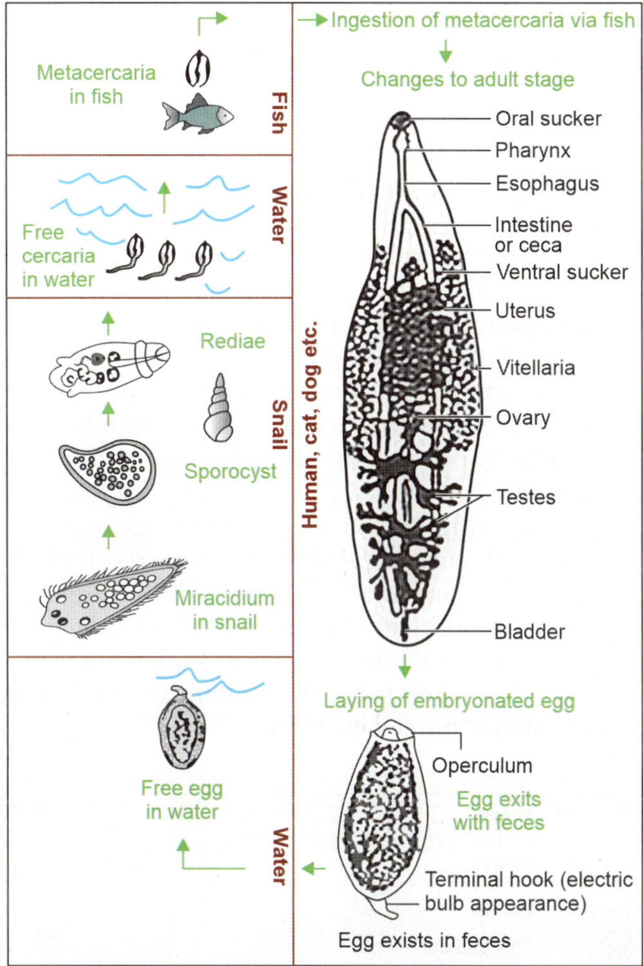

Fig. 105.6: Life cycle of *C. sinensis*

- **Epidemiology:** It is prevalent in Far East like China, Korea, Japan, South and North Vietnam and Formosa. It also prevalent in USA, France, Canada and Australia among Asian refugees from endemic area.
- **Virulence factors: (1) Mechanical** irritation to mucosa of biliary tract by adult worm → hyperplasia and obstruction of bile flow → jaundice and secondary infection. **(2) Toxic metabolites** of adult worm cause hyperplasia of mucosa.
- **Reservoir of infection:** Cat, dog, pig and rat.
- **Source of infection:** Fish.
- **Mode of transmission:** Ingestion of raw fish like *Cyprinoid* contains metacercaria.
- **Exit form:** Embryonated egg.
- **Infective form:** Metacercaria.
- **Portal of entry:** GIT.
- **Sites:** Adult stage lives in biliary tract and sometimes in pancreatic duct.
- **Clinical features:** Mild infection presents as mild cholangitis. Heavy infection presents as epithelial hyperplasia (adenoma formation) and congestion of peribiliary plexus, chronic diarrhea, hepatomegaly, recurring jaundice and blocking of duct.
- **Complications:** These occur after long-term (chronic infection) like.

1. **Secondary infection:** Hyperplasia of duct → narrowing of lumen → blockage of biliary tree → obstruction of bile flow → secondary bacterial infection (also by heavy infection).
2. **Stone:** Intrahepatic calculi formation.
3. **Cancer of bile duct:** Called cholangiocarcinoma which epidemiologically related to China.

Laboratory diagnosis

- **Specimen: (1) Stool:** Eggs are not present in case of biliary obstruction. **(2) Aspirated bile:** Collected by entero (string) test (**Ch. 94**).
- **Testing methods**
  - **Blood picture:** Leukocytosis with eosinophilia.
  - **Microscopy: (1) NS wet mount:** It shows bile stained eggs. Eggs of *H. heterophyes*, *M. yokogawai* and *C. sinensis* (terminal hook) are indistinguishable. **(2) Concentration method:** Eggs do not float in saturated common salt solution due to operculum. They are sediment with normal tap water concentration method.

Prevention: Preventive measures are eradication of intermediate hosts, prevention of water contamination by the feces of case and avoiding the use of raw fish.

Treatment: Praziquantel (25 mg/kg, 3 doses in 1 day) is the drug of choice. Surgical intervention is necessary in case of obstructive jaundice.

Opisthorchis felineus

Common name: Cat liver fluke

History: Rivolta 1884 and Blanchard 1895

Morphology: It is same as *C. sinensis*, but testes are not branched, eggs are small with 30 μm × 11 μm in size and cercariae are with long tail with pair of fin fold (pleurolophocercus type).

Life cycle: Definitive hosts are human, cat, dog, pig and fox. 1st intermediate host is snail (*Bithynia leachi*) and 2nd intermediate host is fresh water fish of genus *Cyprinoid*. Cycle is same as *C. sinensis*.

Pathogenicity: It has been reported from Prussia, Poland, Siberia and Japan. Chandler found that 60% of cat in Kolkata (India) were infected with the parasites and human cases were reported from India. Adult stage lives in biliary and pancreatic duct. Pathogenicity is same as *C. sinensis* but cholangio carcinoma is epidemiologically related to North-East Thailand.

Laboratory diagnosis: Same as *C. sinensis*.

> **Note:** *Opisthorchis viverini*
> **History:** Poirier, 1886; Stiles and Hasall, 1896.
>
> **Morphology:** (1) Differences from *C. sinensis*: Adult worm is smaller about 8–12 mm long, eggs are small with 25 μm × 15 μm in size and cercarias are with long tail with pair of fin fold (pleurolophocercus type). (2) Differences from *O. felineus*: Greater proximity of ovary and testes and few clusters of Vitellaria.

(Contd…)

Life cycle: Definitive hosts are human and civel cat (*Felis viverrus*). Intermediate hosts and cycle are same as *C. sinensis*.

Pathogenicity: It is prevalent in the Mekong river valley in Laos and Thailand. Adult stage lives in distal bile duct. Acute stage presents with diarrhea, pain in right hypochondrium and mild jaundice. Chronic stage presents with periportal fibrosis of liver, inflammation of biliary canaliculi, epithelial hyperplasia and cholangiocarcinoma of bile duct.

Laboratory diagnosis: Same as *C. sinensis*.

Note: *Opisthorchis noverca*
It was 1st reported from Pariah dog, pig and human in India.

Fasciola hepatica

Common name: Sheep liver fluke and common liver fluke.

History: It was 1st discovered by Jehan De Brie in 1379. Life cycle was worked out by Leuckart and Thomas in 1883, independently to one another.

Morphology
1. **Adult stage (Fig. 105.7)**

- **Development:** Metacercaria changes to adult worm in 3–4 months
- **Size:** 3 cm × 1.5 cm
- **Shape:** Leaf like
- **Color:** Brown to pale gray in color
- **Life span:** In sheep 5 years and in human 9–13 years
- **Organs:** Oral sucker is smaller and anterior end bearing the oral sucker form a conical projection (cephalic cone). Ventral sucker is located in line with the two shoulders formed by the broadening of the conical projection posteriorly. Both the intestinal ceca bear the numbers of lateral branches. Other features are described in Ch. 103.

2. **Egg stage (Fig. 105.7):** Egg is oval, operculated, 140 μm × 80 μm in size, yellowish-brown color (bile stained) and surrounded by protective covering. It contains unsegmented ovum surrounded by yolk mass initially and miracidium formed later in water. Egg hatch in water to release the miracidium. It does not float in saturated common salt solution.

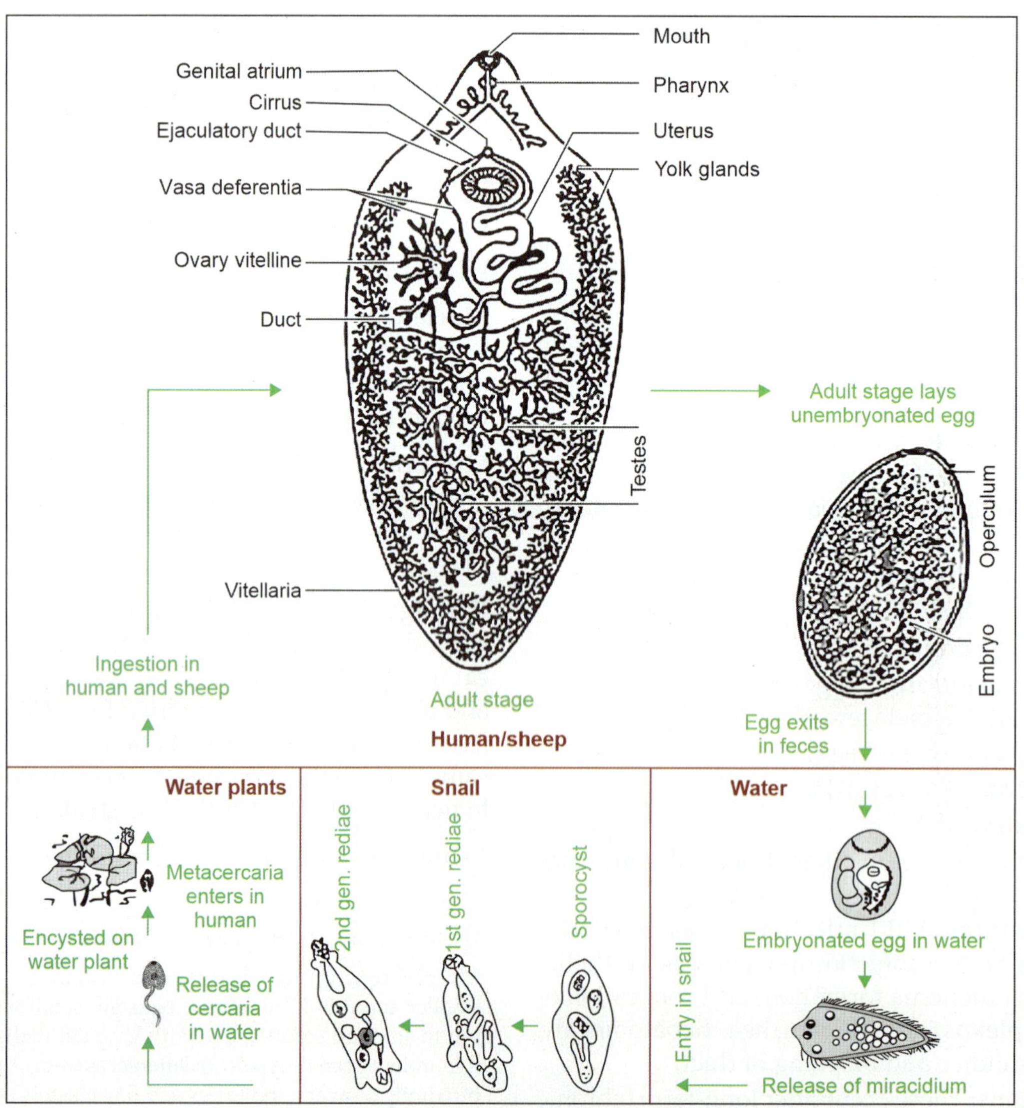

Fig. 105.7: Life cycle of *F. hepatica*

3. **Larval stage:** Solid form of larva is present. Single egg gives the single miracidium, which produces multiple cercariae. Different larval stages are mentioned in life cycle and shown in **Fig. 105.7**.

Life cycle

- **Hosts: (1) Definitive hosts** are human, sheep, goat and cattle. **(2) Intermediate host:** Amphibian snail (*Lymnaea truncatula*).
- **Cycles:** As shown in **Flowchart 105.5 and Fig. 105.7**.

Pathogenicity

- **Disease name:** It called fascioliasis.
- **Epidemiology:** It is cosmopolitan.
- **Reservoir of infection:** Sheep, goat and cattle
- **Source of infection:** Raw grass or water cress as salad encysted with cercaria called metacercaria
- **Mode of transmission:** Ingestion of raw grass or water cress as salad contains metacercaria
- **Exit form:** Embryonated egg
- **Infective form:** Metacercaria
- **Portal of entry:** GIT
- **Sites:** Adult stage lives in biliary tract. During migration larva may enter in lungs, cutaneous tissues, etc.
- **Clinical features: Acute infection** presents with vomiting, biliary colic, persistent diarrhea, tender hepatomegaly and eosinophilia (40–85%). **Chronic infection** presents with inflammation and obstruction of bile duct.
- **Complications:** During migration larva may enter in lungs and cutaneous tissues with production of abscess or fibrotic lesion.

Flowchart 105.5: Life cycle of *F. hepatica*

Cycle in definitive host: Human / sheep
Infection to human / sheep by ingestion of raw grass or water cress as salad encysted with cercaria called **metacercaria**. After entry in human, it excysts to adult worm in duodenum in 3–4 months

↓

Penetrates the intestinal wall → enters in peritoneal cavity and → eventually enters in biliary passage → Penetrates the liver capsule and traverses through parenchyma → Adult worm laying the unembryonated eggs, pass through bile in intestine

↓

Eggs exit with feces in water

Cycle in intermediate host: Snail

Egg gets mature freely in water to develop the cilicated embryo called **miracidium**. Hatching of egg to release the miracidium free in water. Miracidium moves freely in water in search of snail.

↓

Miracidium enters into the snail and settles in lymph space, where it undergoes the further development by asexual method

↓

Transformed to sporocyst → 1st generation redia → 2nd generation redia → final stage of larva with tail called **cercaria**. Cercaria escapes from snail and encysted in blades of grass or water cress called **metacercaria**.

↓

Entry in human/sheep and cycle repeated

Laboratory diagnosis

- **Specimens: (1) Stool:** Eggs are present. **(2) Bile:** Collected by entero (string) test (**Ch. 94**). **(3) Pus in case of ectopic lesion:** It may contain the worms.
- **Testing methods**
    A. **Blood picture:** Eosinophilia (40–85%).
    B. **Microscopy:** Same as *C. sinensis*.
    C. **Serological tests:** Antigen is detected from stool by ELISA. Ab is detected from serum by ELISA or WB test.
    D. **Molecular method:** PCR.
    E. **Radiological method:** CT and MRI are also useful.

Prevention: Preventive measures are eradication of intermediate hosts, prevention of water contamination by the feces of case and washing of vegetables before eating.

Treatment: Oral triclabendazole (10 mg/kg, single dose) is the drug of choice. Bithionol (30–50 mg for 10–15 days) is the other option.

Note: Halzoun

Meaning: Actual meaning is suffocation.

Cause: It is common in Syria (Lebanon) due to ingestion of raw liver of sacrificial goat. Initially, it was believed that it was due to invasion of immature *F. hepatica* and commonly called **pharyngeal fascioliasis**, but now it is considered due to nymph of *Linguatula serrata* (tongue shaped animal parasite belong to **pentastome** subclass).

Features: It presents with acute dysphagia and laryngeal obstruction.

Note: *Fasciola gigantica*

Meaning: Size of egg is larger than *F. hepatica* hence species called **gigantic.**

Morphology: It is same as *F. hepatica*, with few differences like lanceolate shape, less distinct cephalic cone and larger size egg about 160–190 μm × 70–90 μm.

Life cycle: Same as *F. hepatica*, but intermediate host is fully aquatic snail instead of amphibian snail.

Pathogenicity: Infection rate is high in China, Iraq and Northe East Thailand (50% cattle, 45% goats and 33% water buffalo). Sporadic cases have been reported from Zimbabwe, Uganda, Tashkent, Iraq, Vietnam and Hawai. It is common in herbivorous animals like cattle, goats, camels, water buffalo, etc. Other features are same as *F. hepatica*.

Laboratory diagnosis: Same as *F. hepatica*.

Dicrocoelium dendriticum

Morphology: Adult is lancet shape (so common name is **lancet fluke**), flat, transparent and 5–15 mm × 1.5–2.5 mm in size. Egg is thick shelled, operculated, about 38–45 μm × 22–30 μm in size and golden brown in color.

Life cycle: Definitive hosts are human and herbivorous animals like sheep, deer, etc. 1st intermediate host is land snail and 2nd intermediate host is ant of genus *Formica*. **Cycles** is shown in **Flowchart 105.6**.

Flowchart 105.6: Life cycle of *D. dendriticum*

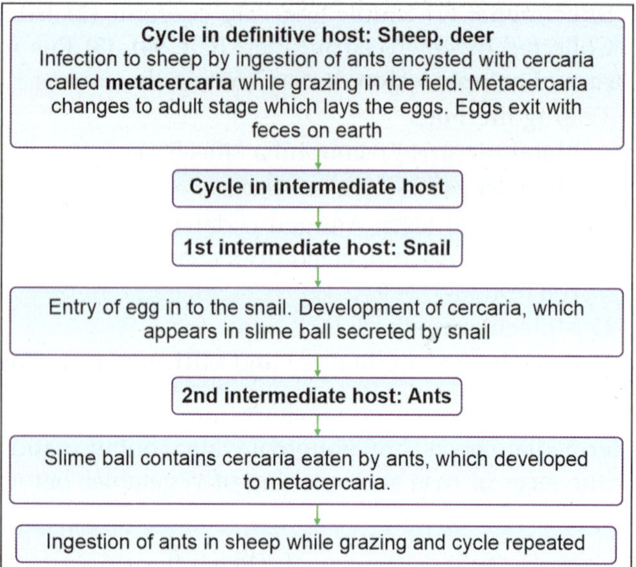

Pathogenicity: It is prevalent in Europe, North Africa, Northern Asia and Far east in sheep. It presents in biliary tract. Human infection is spurious and reported from Europe, Middle East and China.

Eurytrema pancreaticum

Adult is 8–16 mm × 5–9 mm in size. Egg is same as *D. dendriticum*. Human infection is noticed in China and Japan. It is common in pancreatic duct (so common name is **pancreatic fluke**) of cattle, sheep, camel and monkeys.

PULMONARY TREMATODES

Paragonimus westermani

Common name: Oriental lung fluke or lung distome.

Meaning: Species name given from the name Westerman, director of Amsterdam zoo by Kerbert in 1878.

History: Infection was 1st reported by Naterer in 1828 and later by Kerbert in 1878 from Bengal tiger that died in zoo of Amsterdam. Species name was given from Amsterdam.

Morphology

1. **Adult stage (Fig. 105.8)**
   - **Development:** Metacercaria changes to adult worm in 2–3 months.
   - **Size:** 8–12 mm in length, 4–6 mm in breadth and 3–5 mm in thickness.
   - **Shape:** Egg shaped with anterior end is broader than posterior end.
   - **Life span:** 6–7 years.
   - **Organs:** Tegument covers with spines. Oral sucker is situated anteriorly and ventral sucker is located in the middle of the body. Crura do not bears any branches and ends blindly in caudal (posterior) area. Excretory system/vesicle is large and extends from posterior to anterior region and divides the body into two equal halves. Other features are described in **Ch. 103.**

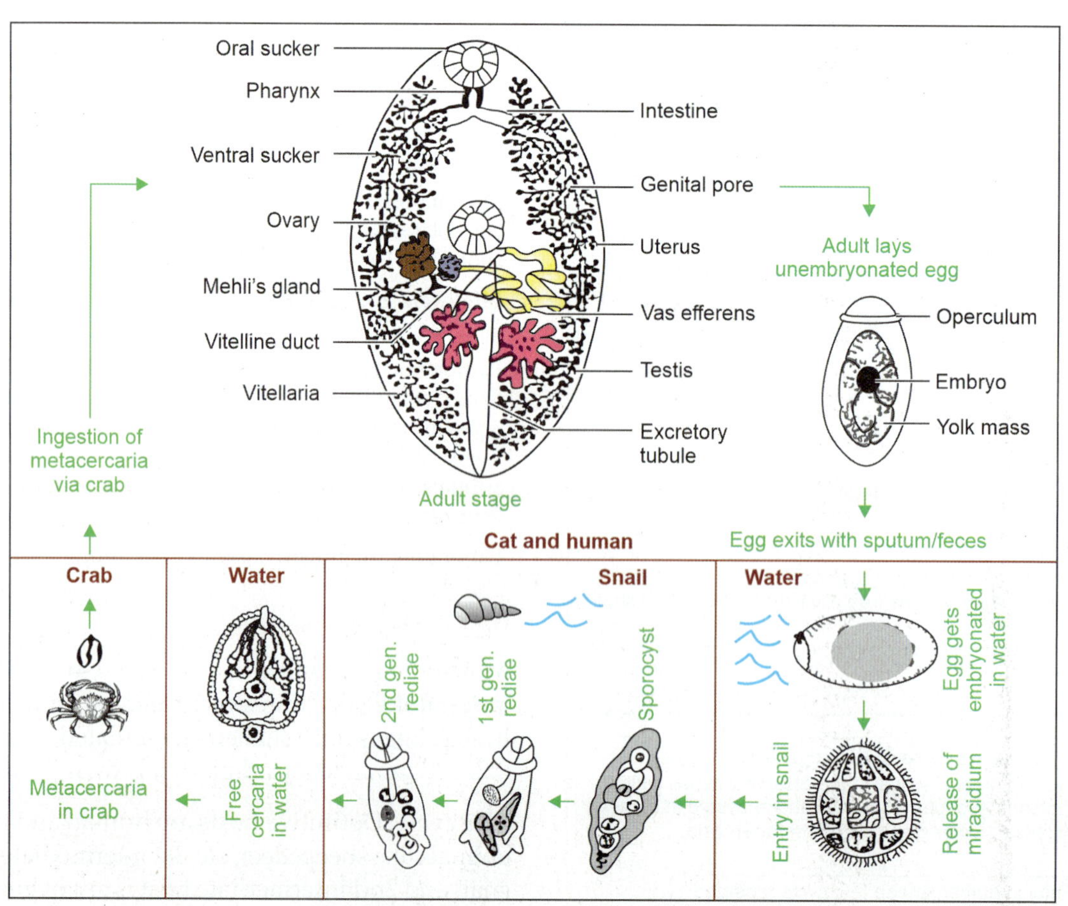

Fig. 105.8: Life cycle of *P. westermani*

2. **Egg stage (Fig. 105.8):** Egg is oval, flat, operculated, 80 μm × 55 μm in size, golden-brown color (bile stained) and surrounded by transparent and brownish yellow protective covering. It contains unsegmented embryo surrounded by yolk mass from initially (when laid). Miracidium develops later after releasing in water.

3. **Larval stage:** Solid form of larva is present. Single egg gives the single miracidium, which produces multiple cercariae. Different larval stages are mentioned in life cycle and shown in **Fig. 105.8.**

Life cycle

- **Hosts:** Worm passes its life cycle in two different hosts. **(1) Definitive hosts** are humans, cat, tiger and leopard. **(2) 1st intermediate host** is fresh water snail of genus *Melania liberta* (*Semisulcospira liberta*) or *Brotia*. While **2nd intermediate host** is crab or crayfish.
- **Cycles:** As shown in **Flowchart 105.7 and Fig. 105.8.**

Pathogenicity

- **Disease name:** It called paragonimiasis.
- **Epidemiology:** It is prevalent in Far East in China, Korea, Japan, Formosa and India (Bengal. Assam and South India). It also reported from Nepal, parts of Africa and South America.
- **Reservoir of infection:** Cat.
- **Source of infection:** Crab or crayfish.
- **Modes of transmission:** It is transmitted by ingestion of raw crab or crayfish contains metacercaria. Strips of raw crab meat soaked in alcohol called drunken crab a popular delicacy in China.
- **Incubation period:** Highly variable from 2–20 days too few months.

Flowchart 105.7: Life cycle of *P. westermani*

Cycle in definitive host: Human, cat, etc.
Infection to human, cat, etc., by ingestion of raw crab of crayfish contains metacercaria. After entry in human, it excysts to young adult worm in duodenum. Adult worm attaches to intestinal wall and enters in to the abdominal cavity. It pierces the diaphragm and two layers of pleura to enter into the lungs (sometimes, in extrapulmonary sites). Becomes fully mature in the lungs in 2–3 months and lays the unembryonated eggs in bronchioles. Eggs excreted with the sputum in water (few are also in feces due to swallowing)

↓

Cycle in intermediate hosts
Egg develops ciliated embryo in 2–3 weeks in water. Hatching of egg releases the miracidium free in water. Miracidium moves freely in water in search of snail

↓

1st intermediate host: Snail

↓

Miracidium enters in to the soft tissues of snail and undergoes the further development by asexual method. It transformed to sporocyst → 1st and 2nd generation redia → cercaria. Cercaria escapes from snail and encysted (metacercaria) in crab of crayfish

↓

2nd intermediate host: Crab or crayfish

↓

Entry in human, cat, etc., by ingestion of raw crab or crayfish and cycle repeated

---

- **Infective form:** Metacercaria.
- **Exit form:** Unembryonated egg.
- **Portal of entry:** GIT.
- **Sites:** Adult stage lives in lungs. Extrapulmonary sites are liver, intestine, peritoneum, brain, skin and lymph nodes.
- **Pathogenesis and virulence factors:** Follow **Flowchart 105.8.**
- **Clinical features:** It presents with fever, lymphadenitis, skin ulcer, abdominal pain and diarrhea. Pulmonary features include cough, dyspnea and hemoptysis simulating bronchiectasis or tuberculosis.
- **Complications:** These are ICSOL like brain tumor and Jacksonian type of epilepsy.

Laboratory diagnosis

- **Specimens:** Sputum and stool (contains eggs in case of swallowing of sputum).
- **Testing methods**

  A. **Blood picture:** Eosinophilia present constantly.

  B. **Macroscopy:** For presence of blood.

  C. **Microscopy: (1) NS wet mount** shows the bile stained eggs. **(2) Concentration method:** Mix the equal volume of sputum with 3–4% NaOH. Left the mixture for 10–15 minutes to dissolve the mucus. Centrifuge at 2000 rpm for 5 minutes. Use sediment for eggs examination.

  D. **Serological tests:** Ag is detected by ELISA which is useful in acute stage. Ab is detected by WB test which is useful for epidemiological purpose.

  E. **Radiological methods: (1) X-ray** reveals the abnormal nodular, cystic, ring or infiltrative type shadow. **(2) CT-scan** shows "soap bubble" like appearance in cerebral cyst.

Prevention: Preventive measures are eradication of intermediate hosts, prevention of water contamination by the sputum/feces of case and proper cooking of crab or crayfish.

Treatment: Praziquantal (25 mg/kg, TDS for 1–2 days) is the drug of choice. Bithinol and niclofolan are the other options.

Other *Paragonimus* spp.

P. mexicanus, P. africanus, P. uterobilateralis, P. heterotremus, P. kellicotti, P. miyazakii, P. philipinensis, P. skrjabini, P. hueitengenesis, etc.

Flowchart 105.8: Pathogenesis of *P. westermani*

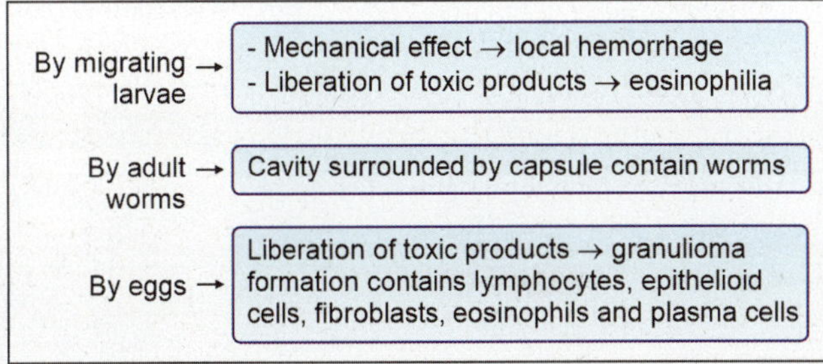

Case Study

1. A young male presents with history of cough, hemoptysis, fever and dyspnea. He took antituberculous treatment for 6 months without sign of improvement. Sputum microscopy revealed the bile stained eggs. Identify the case and answer the following.
   a. Name and draw the properly labeled diagram of egg of causative agent.
   b. Describe the life cycle of causative agent.
   c. Describe the pathogenicity of causative agent.

Essays/Full Questions

1. Small intestinal trematodes.
2. Hepatic trematodes.

Short Questions for Theory/Viva Questions

1. Write the scientific name of following parasites.
   a. Ginger worm
   b. Dwarf fluke
   c. Chinese liver fluke
   d. Cat liver fluke
   e. Sheep liver fluke
   f. Oriental lung fluke
2. What is halzoun?
3. What is drunken crab? Write its clinical significance.

MCQs for Chapter Review

Hepatic Trematodes

1. **Egg with electric bulb like appearance is:**
   a. *F buski*   b. *Fasciola*
   c. *Chlonorchis*   d. *Paragonimus*

Pulmonary Trematodes

2. **Which organism can be isolated from stool and sputum?**
   a. *Paragonimus*   b. *Fasciola*
   c. *Chlonorchis*   d. *F buski*

Answers of MCQs and Explanation

1. c
   - Follow **respective section,** for explanation.
2. a
   - Follow **respective section,** for explanation.

Nematodes: General Features and Classification

Chapter Outline

- General Features
- Classification

GENERAL FEATURES

Introduction

Nematodes are commonly known as **round worm,** as they are cylindrical like a tube. Nematodes word originated from nema (Greek) means thread and edios means appearance, as they are elongated, unsegmented, cylindrical and tapered at both the ends. They habitat in intestine and tissues.

Morphology

Intestinal nematodes have three morphological stages like adult, egg and larval stages are described below. **Somatic (tissues) nematodes** have two morphological stages like adult and larval stages are described in **Ch. 108.**

Adult Stage

Shape: They are elongated, unsegmented, cylindrical and tapered at both the ends. Posterior end of male species is curved or coiled ventrally.

Size: Great variation in size of nematodes from smallest (*T. spiralis* and *S. stercoralis*) less than 5 mm to longest (*D. medinensis*) about 1 meter in length. Male is smaller than female.

Organs

- **Cuticle:** Body of parasite is covered with acellular layer called cuticle. It may be smooth, bossed (rough), spiny, or striated. From outer to inner it has four layers **(Fig. 106.1)** like epicuticle, exocuticle, mesocuticle and endocuticle. It has protective and other metabolic role. Variations in cuticle include cervical alae (wing like expansion of the cuticula near the mouth) and cervical papillae (protuberance of the cuticula near esophagus).
- **Basal lamina (Fig. 106.1):** It separates the cuticle from underlying tissues.

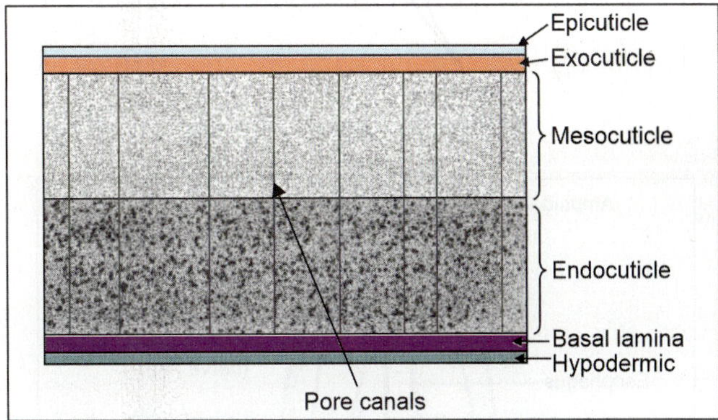

Fig. 106.1: Structure of cuticle

- **Body wall musculature:** All the muscles are spindle shaped and longitudinally oriented. Each cell is divided in to contractile and noncontractile portion. Myofibrils are located in the contractile portion, while mitochondria with glycogen granules, nucleus and lipids are located in the noncontractile portion.
- **Body cavity:** Present and contains all viscera.
- **GIT (Fig. 106.2):** It presents with anus so called complete. It presents with oral aperture, mouth cavity, esophagus, intestine and subterminal anus. Mouth cavity contains teeth or cutting plates. In case where mouth cavity is absent, the oral aperture is directly continuous with the esophagus.
- **Excretory system:** Present, but rudimentary
- **Nervous system:** Present, but rudimentary. It consists two parts. **(1) Commissures:** Two major nerve commissures are circumoesophageal commissure **(Fig. 106.3)** presents around esophagus and rectal commissure **(Fig. 106.4)** presents around rectum. **(2) Sensory papillae:** They are present on the cuticle and may be amphids **(Fig. 106.3)** or phasmids **(Fig. 106.4).**
- **Reproductive system:** In all nematodes both male and female species are separated so called diecious. Reproductive system is highly developed and

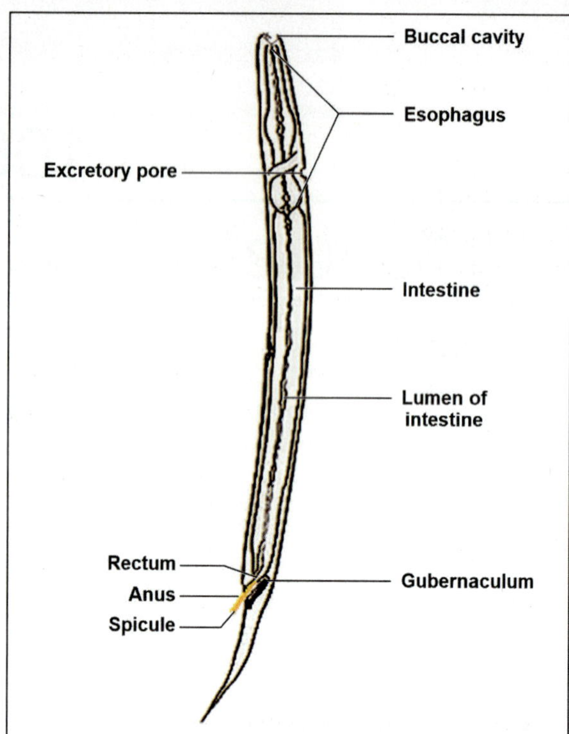

Fig. 106.2: GIT of nematodes

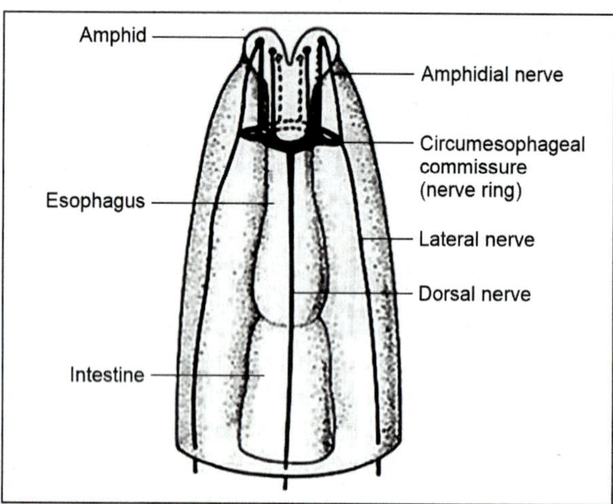

Fig. 106.3: Circumoesophageal commissure and amphid

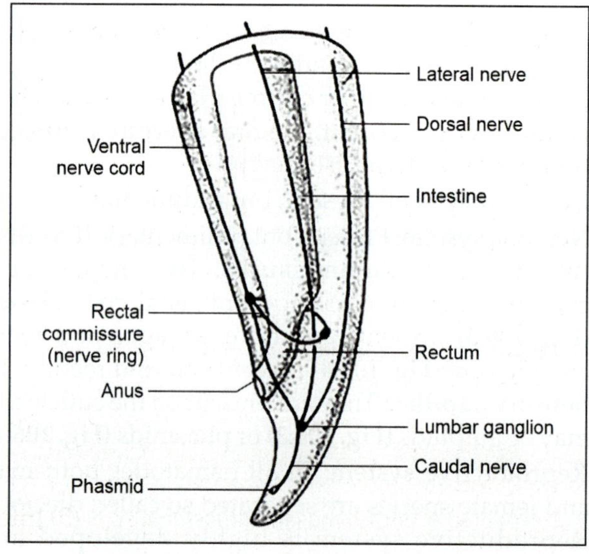

Fig. 106.4: Rectal commissure and phasmid

complete in each individual. Reproductive organs lie in body cavity. Reproductive organs are described separately in box.

> **Note: Reproductive organs of nematodes**
>
> **Female reproductive organs (Fig. 106.5a):** It is a single or double or multiple convoluted tube which consists organs like ovary, oviduct, seminal receptacle, uterus, vagina and vulva. Genital pore open in the middle of the body or near the mouth. When species have two or more tubes, they unite in a common duct before terminating into genital pore. Female nematodes are following 3 types.
>
> - **Oviparous:** Female lays eggs called oviparous. (1) Eggs contain unsegmented embryo: *A. lumbricoides* and *T. trichuria*. (2) Eggs contain segmented embryo: *A. duodenale* and *N. americanus*. (3) Egg contains larva: *E. vermicularis*.
> - **Viviparous (larviparous):** Female gives birth to larva called viviparous like *D. medinensis*, *W. bancrofti*, *B. malayi* and *T. spiralis*.
> - **Ovi-viviparous:** Female lays eggs contain rhabditiform larva which immediately hatch out called ovi-viviparous like *S. stercoralis*. Rhabditiform larva presents in freshly passed stool.
>
> **Male reproductive organs (Fig. 106.5b):** It is a single convoluted tube, which differentiated into testis, vas deferens, seminal vesicle and ejaculatory duct. Genital duct forms a common passage with intestine called cloaca. Accessory copulatory organs are **(1) Spicule:** It is rod like and protrucible. **(2) Gubernaculum:** An elevation on the dorsal wall of cloaca which guide the spicule during copulation. **(3) Copulatory bursa (Bursa copulatrix):** Umbrella like expansion of the cuticula surrounding the cloaca of the male nematode of certain species. It is supported by fleshy rays comparable with the ribs of umbrella.

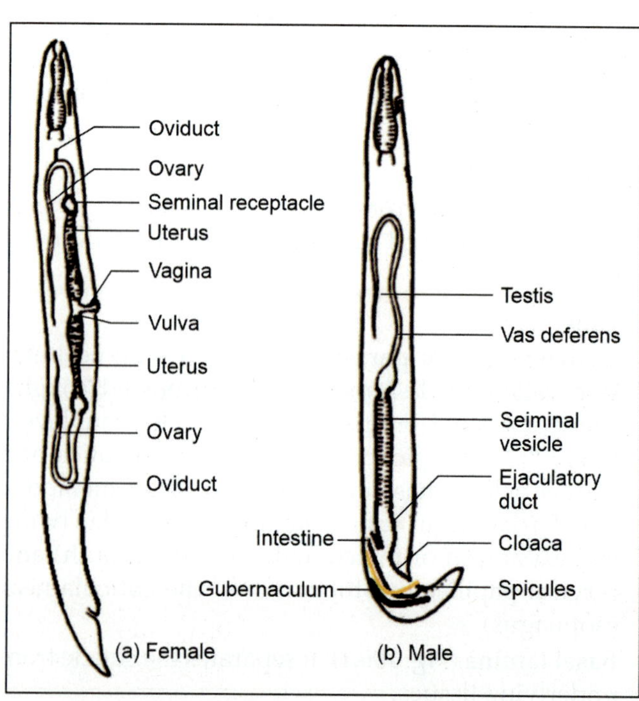

Fig. 106.5: Reproductive system of nematodes

Egg Stage

Two types eggs. **(1) Bile stained eggs of nematodes:** These are *A. lumbricoides* and *T. trichuria.* **(2) Non-bile stained eggs of nematodes:** It includes *A. duodenale, N. americanus* and *E. vermicularis.*

> **Note: Bile stained eggs**
>
> **Pseodophyllidean cestodes:** *D. latum.*
>
> **Psyclophyllidean cestodes:** *T. saginata, T. solium, E. granulosus* and *D. caninum.*
>
> **Only monoecious trematodes have bile stained eggs:** *F. buski, H. heterophyes, M. yokogawai, E. ilocanum, E. malayanum, E. revolutum, P. supraryfex, W. watsonis, G. hominis, C. sinensis, O. felineus, F. hepatica, F. gigantica, D. dendriticum* and *P. westermani* (other species also).
>
> **Nematodes:** Described above.

Larval Stage

Two types larvae. **(1) Rhabditiform larva:** Esophagus is short compared to the length of larva and posterior end is dilated like a bulb as shown in **Fig. 106.6a.** **(2) Filariform larva:** Esophagus is long compared to the length of larva and posterior end is not dilated (no bulb) as shown in **Fig. 106.6b.**

Modes of Transmission of Nematodes

Direct transmission: Filariform larva enters directly via penetrating the skin; for example, *A. duodenale, N. americanus* and *S. stercoralis.*

Indirect transmission

- **Ingestion:** (1) By ingestion of water or food contaminated with embryonated eggs; for example, *A. lumbricoides, T. trichuria* and *E. vermicularis.* (2) By ingestion of pig's flesh contains encysted embryo; for example, *T. spiralis.*
- **By inhalation (air-borne):** Inhalation of embryonated eggs present in dust; for example, *A. lumbricoides* and *E. vermicularis.*
- **Vector-borne:** (1) By mosquito bite like *W. Bancrofti* and *B. malayi* (superfamily Filarioidea). (2) Ingestion of cyclops contains larvae. For example *D. medinensis* (superfamily Dracunculoidea).

Life Cycle

Superfamily Filarioidea and Dracunculoidea: Worms pass their life cycle in two different hosts. **(1) Definitive hosts** are humans, who harbor the adult stage. **(2) Intermediate hosts** are arthropods like mosquito for Filarioidea and cyclops for Dracunculoidea who harbor the larval stage.

Other superfamily of nematodes: Worms pass their life cycle in one host with development in external environment. Human is an optimum host in all nematodes except *T. spiralis* where pig is an optimum host and human is an alternative.

CLASSIFICATION

Systemic classification: Follow **Ch. 14 (Table 14.4).**

Pathological classification: Following are the pathogenic species and further classified as per habitat.

1. **Intestinal nematodes**
   - **Small intestinal nematodes:** *T. spiralis, A. lumbricoides, A. duodenale, N. americanus, Trichostrongylus* species, *S. stercoralis* and *C. philippinensis.*
   - **Large intestinal nematodes:** *T. trichuria* and *E. vermicularis.*
   - **Arteries and arterioles of ileocecal part:** *Angiostrongylus costaricensis.*
2. **Somatic (tissues) nematodes:** Tissue nematodes causing following infections.
   - **Filariasis:** It is further classified on the basis of habitat by different species.
     - Lymphatic filariasis: *W. bancrofti, B. malayi* and *B. timori.*
     - Nonlymphatic filariasis: (1) Subcutaeous filariasis: *L. loa, O. volvulus* and *M. strptocerca.* (2) Subcutaneous and serous cavity filariasis: *M. ozzardi* and *M. perstans.* (3) Zoonotic filariasis: *B. pahangi, D. immitis, D. tenuis* and *D. repens.*
   - **Other infections:** Further classified on the basis of habitat by different species. (1) Infection in lungs: Larvae of *S. stercoralis, A. lumbricoides* and *A. duodenale.* (2) Infection in liver: *C. hepatica.* (3) Infection in brain: *A. cantonensis.* (4) Infection in brain and other tissues: *Gnathostoma* spp. (5) Infection in subcutaeous tissues: *D. medinensis.*

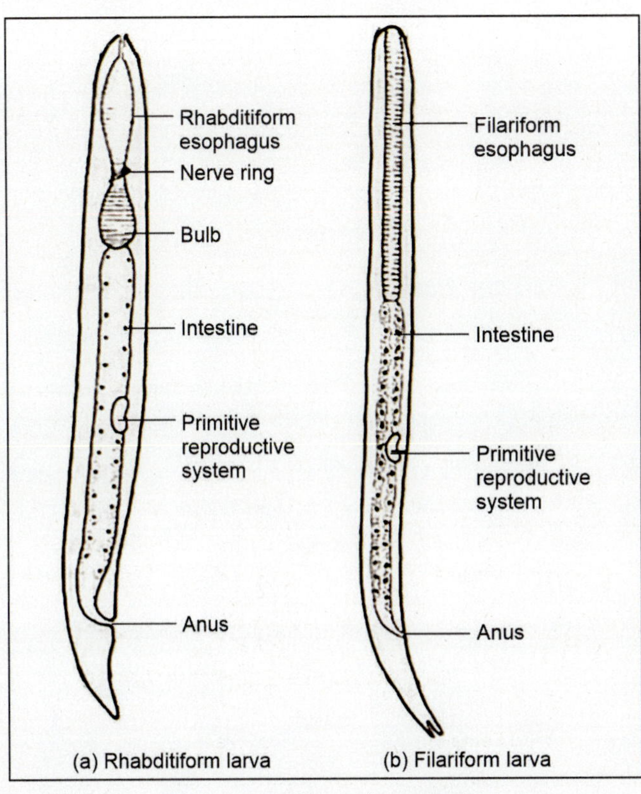

Fig. 106.6: Larvae of nematodes

ACCESS YOURSELF

Short Questions for Theory/Viva Questions

1. Name accessory copulatory organs in nematodes.
2. Write four names of bile stained eggs.
3. Write the two differences between filariform larvae and rhabditiform larvae.
4. Name the parasite transmitted by arthropod.

MCQs for Chapter Review

General Features (Morphology)

1. Which of the following is viviparous?
   a. *Strongyloides stercoralis*
   b. *Trichinella spiralis*
   c. *Enterobius*
   d. *Ascaris*

2. Ovi-viviparous nematode is:
   a. *Ascaris lumbricoides*   b. *Dracunculus medinensis*
   c. *Strongyloides stercoralis* d. *Enterobius vermicularis*

3. Rhabditiform larvae in freshly passed stool are seen in:
   a. *Ascaris lumbricoides*   b. *Toxoplasma*
   c. *Strongyloides stercoralis* d. *Enterobius vermicularis*

4. Which of the following produces bile stained eggs?
   a. *Ancyclostoma duodenale*
   b. *Taenia solium*
   c. *Enterobius vermicularis*
   d. *Necator americanus*

Introduction (Mode of Transmission of Nematodes)

5. Filariform larva is infective in:
   a. *E. vermicularis*   b. *A. lumbricoides*
   c. *N. americanus*   d. *T. trichuria*

Classification

6. Small intestinal helminths are:
   a. *Ascaris*   b. *Necator*
   c. *Trichuris*   d. *Enterobius*

7. Which of the following resides in cecum?
   a. *T trichuria*   b. *A lumbricoides*
   c. *Strongyloides*   d. *Ancyclostoma*

8. Lymphatic filariasis is caused by:
   a. *W. bancrofti*   b. *Brugia malayi*
   c. *Schistosoma*   d. *B. timori*

9. Which is nonlymphatic filariasis?
   a. *Loa loa*   b. *Wuchereria bancrofti*
   c. *Brugia malayi*   d. *Brugia timori*

Answers and Explanation of MCQs

1. b
2. c
3. c
- Follow section, **general features (adult stage → female reproductive organs)** for explanation of answers of MCQs 1–3.

4. b
- Follow section, **morphology (egg stage) and note (other bile stained eggs)** for explanation.

5. c
- Follow section, **mode of transmission of nematodes (direct transmission)** for explanation.

6. a, b
7. a
8. a, b
9. a
- Follow section, **classification** for explanation of answers of MCQs 6–9.

Infections of Nematodes: Intestinal Nematodes

SMALL INTESTINAL NEMATODES

Trichinella spiralis

Meaning

Trichos (Greek) means hair and ella (suffix) means spiral, indicates very thin parasite with spirally coiled larvae.

Common Name

Commonly it called trichina worm.

History

Tidemann, who 1st discovered the disease in Germany in 1821. Later disease has also been defined by Peacock and Own in 1825. Larval stage was 1st worked out by Herrick and Janewil in human's blood in 1909.

Morphology

Two morphological stages like adult and larva. Few important features of parasite are given below. For more details, follow **Ch. 106.**

Adult Stage

Time of maturation: Larva transformed to sexually mature adult stage in 2 days in definitive host after ingestion.

Size: Female is larger than male. Female is 3–6 mm × 0.06 mm in size and male is 1.4–1.6 mm × 0.04 mm in size.

Shape: It is elongated, unsegmented, and cylindrical as shown in **Fig. 107.1.**

Life span: Male dies after fertilizing the female about one week after infection. Female may live for 4 months.

Organs: Spicule and copulatory bursa are absent. Two conspicuous papillae present on either side near the tail end. Female is **viviparous** and laying the embryo.

Larval Stage

Form: Solid form of larva is present within a capsule in muscles called **Trichinella cyst** as shown in **Fig. 107.1**.

Host: Larvae are present in striated muscles (not smooth or cardiac muscles) like deltoid, biceps, gastrocnemius, diaphragm, intercostals and pectoralis major. Usually one larva presents per cyst, but thousands of larvae present per gram of muscles.

Development of capsule around larva: Encapsulation of larva begins on 21st day and completed in 3 months. A blunt ellipsoid lemon shaped sheath (0.4–0.25 mm) develops around tightly coiled larva. It is due to host reaction. Long axis of capsule is parallel to long axis of muscle fiber. Calcification starts after 6–18 months.

Development of larva within capsule: Initially, it is very small about 100 μm × 6 μm in size, continuously develops in host and achieves the maximum size about 1000 μm in 35 day. It presents in coiled form.

Immunity

Acquired immunity: Experimental removal of adult worm in mouse is possible by CMI. Antibodies (IgE) are increased, but not protective and useful for serological testing.

Hypersensitivity: ITH developed by Ags from larvae.

Life Cycle

Types of host: Whole cycle is possible in one host only like pig, rat or humans (both **definitive** and **intermediate**

and conspicuous buccal capsule is lined with hard substances and is provided with 6 teeth from that 4 are hook like on the ventral surface and 2 are knob like (triangular plates) on the dorsal surface. GIT contains five glands, one of them is esophageal gland, which secretes the anticoagulant substances. In **male genital pore** opens into the cloaca. In **female genital pore** opens at the junction of posterior and middle third of the body, owing to the position of genital pore worms assume the Y-shape appearance during copulation as shown in **Fig. 107.12**. **Copulatory bursa (Fig. 107.11b)** consists of three lobes and each lobe contains chitinous rays (total 13 rays). One dorsal lobe contains 3-rays like 1 dorsal and 2 externodorsal. Rays are present at tip and called tripartite. Two lateral lobes contain 10-rays like 3 lateral pairs and 2 ventral pairs. Rays present at base called bipartite. Female is an **oviparous** and lays the 25,000–30,000 eggs/day and 15–18 million eggs in whole life. Eggs contain segmented embryo.

Egg Stage (Fig. 107.13)

Egg is laid by fertilized female. It is elliptical or oval, 65 µm × 40 µm, colorless (nonbile stained) and covered

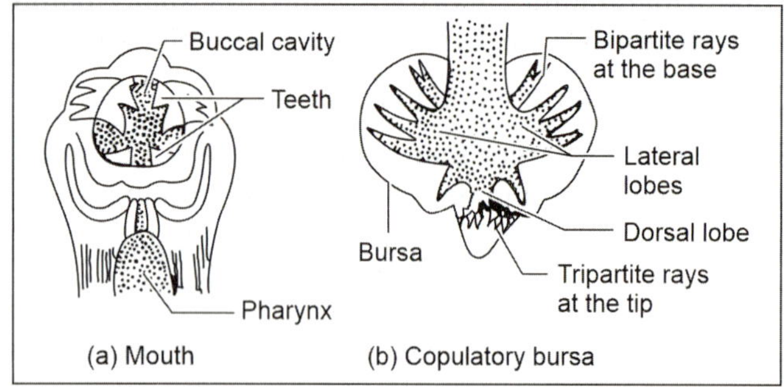

Fig. 107.11: Mouth and bursa of *A. duodenale*

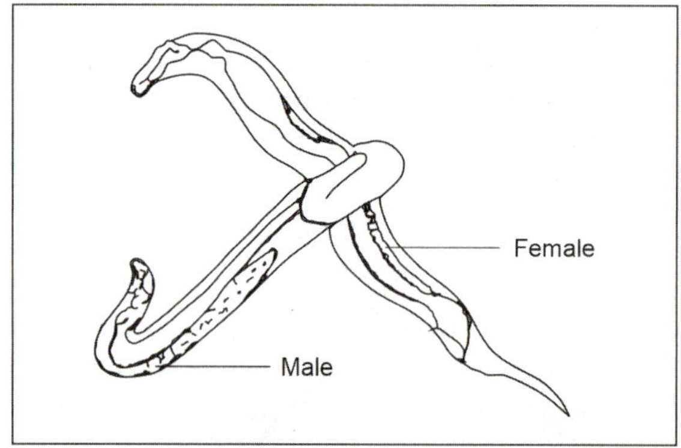

Fig. 107.12: Y-shape appearance during copulation

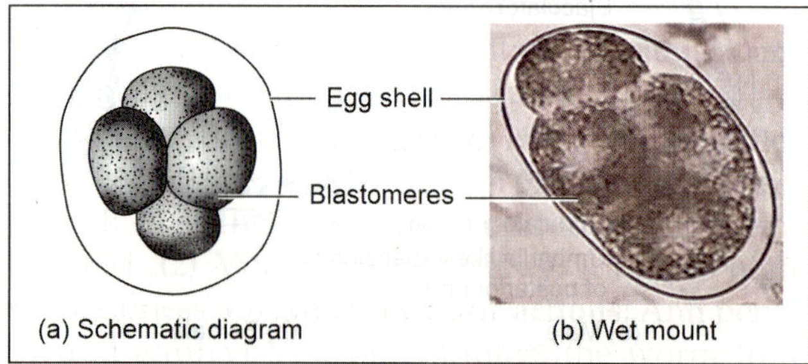

Fig. 107.13: Eggs of *A. duodenale*

by thin hyaline membrane called egg shell. Each egg contains embryo with four segments called blastomeres. Clear space presents between shell and blastomeres. Egg floats in saturated common salt solution.

Larval Stage

Two forms of larvae are present. **(1) Rhabditiform larva** developed from the segmented embryo within the egg shell in 2 days. It is 250 µm in length and releases over soil on hatching of egg. It moults twice on 3rd and 5th day and changes to filariform larva. It has bulbous esophagus. **(2) Filariform larva:** It is 500–600 µm in length with pointed tail. It is an infective stage and develops within 8–10 days. Esophagus is not bulbous. Both forms are developed over the soil. For more details follow **Ch. 106.**

Immunity

Small repeated natural infection in human may develop complete immunity. Asymptomatic patients may acts as carrier and may develop partial immunity. In endemic area patients with minimal intestinal infection without anemia may develop partial immunity. Patients with heavy intestinal infection with anemia may fail to develop resistance.

Life Cycle

Hosts

- **Definitive host:** Whole cycle is possible in one host only like human. Change of host (other human) is required for continuation of life cycle.
- **Intermediate host:** Not required.

Cycles: Following two types of cycles occur.
- **Cycle over soil:** ⎤ Follow **Flowchart 107.5**
- **Cycle in human:** ⎦ and **Fig. 107.14.**

Moulting: Larvae produce the total four moulting at following places.
- **1st moulting:** Outside the host over soil (3rd day).
- **2nd moulting:** Outside the host over soil (5th day).
- **3rd moulting:** Esophagus (around on 10th day after ingestion).
- **4th moulting:** Intestine (between 21st and 28th days after ingestion).

Pathogenicity

Disease name: It called ancyclostomiasis or uncinariasis or hook worm disease.

Epidemiology: It is widely distributed in all tropical and subtropical countries.

Reservoir of infection: Human

Source of infection: Soil

Modes of transmission
- **Direct transmission: (1) Direct contact of sole with contaminated soil:** Filariform larva enters in human by penetrating the skin when human walks with

Essentials of Medical Microbiology

Cycle over soil

Fertilized female lays the eggs. Egg contains segmented embryo called blastomeres, which passes through feces over soil. Rhabditiform larva developed from the segmented embryo within the egg shell. Egg hatched out over the soil within 48 hours to release the rhabditiform larva. Rhabditiform larva moults twice (**1st moulting** on 3rd day and **2nd moulting** on 5th day) and changes to filariform larva (**infective form**). Average time required to develop filariform larva from egg is around 8-10 days.

Cycle in human

Filariform larva casts of the sheath and enters in human by penetrating the skin when human walk with bare foot over the contaminated soil. Enters in subcutaneous tissues → Enters in lymphatics → Later enters in peripheral circulation → Moves in to the right side of heart from the circulation → Enters in lungs via pulmonary circulation

Breaches the capillary wall and enters in lung's alveoli. Larvae crawls up in bronchus and aided by ciliated epithelium → Trachea → Larynx → Pharynx → Swallowed → Passes through oesophagus (with **3rd moulting** around on 10th day after ingestion) and stomach → Enter in jejunum and shows the **4th moulting** (between 21st and 28th days after ingestion)

Larva change to sexually mature male and female adult stage within 3–4 weeks after injection. Male dies after fertilizing female. Fertilized female lays the eggs (6 weeks after ingestion) and clinical features start to begin.

Eggs pass through stool over soil and again development of rhabditiform larva from each egg. Entry in new human host and cycle repeated

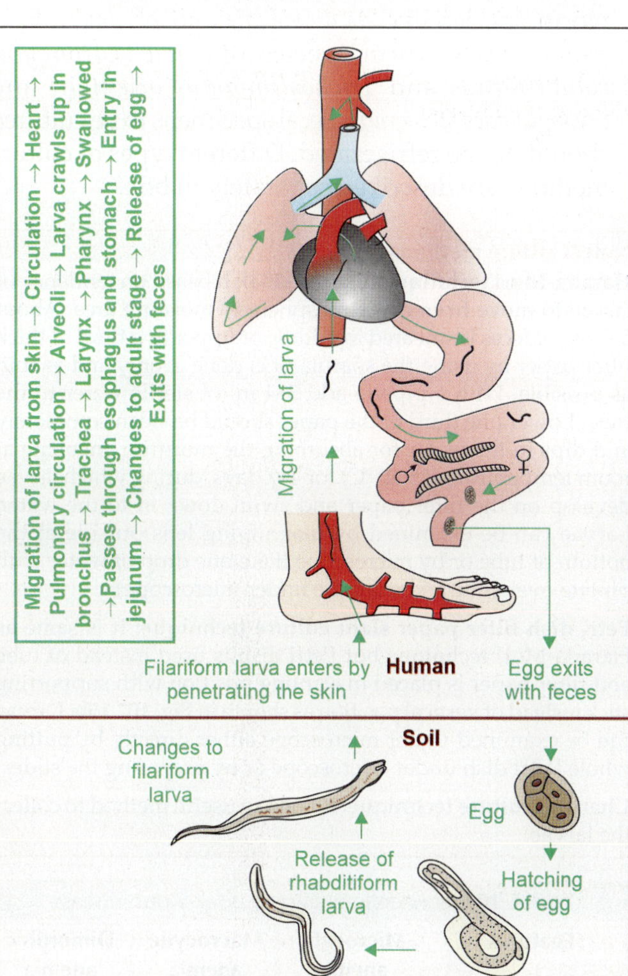

Fig. 107.14: Life cycle of *A. duodenale*

bare foot over the contaminated soil. Common entry points are thin skin between toes or dorsum of feet or inner side of soles. **(2) Direct contact of hands with contaminated soil:** It is possible in gardeners and miners. **(3) Direct contact of any part of skin with**

contaminated soil: Filariform larva can penetrate the any part of skin which is thin.

- **Indirect transmission:** Ingestion of water contaminated by filariform larvae can allow the infection. Infection by this route is very rare. Okamoto (1961) and Higo (1962) have demonstrated that filariform larva can develop in to adult stage without lung passage when swallowed.

Exit form: Egg contains blastomeres.

Infective form: Filariform larva.

Portal of entry: Skin.

Sites: Adult parasite habitat in jejunum, rare in duodenum and rarely in ileum.

Pathogenesis and clinical features: Divided in two categories.

- **Due to migrating larva**
  - **Skin features: (1) At the site of entry:** Lesion called ground itch or ancyclostome dermatitis or itchy dermatitis. It is characterized by severe itching with development of erythematous papular rash. It is a self limiting lesion lasting for 2–4 weeks. It is more common with *N. americanus* than *A. duodenale*. **(2) In between the layers of skin:** Condition called cutaneous **larva migrans** or **creeping eruption**. It is characterized by reddish itchy papule along the path traversed by larva. Larvae of animal species cannot penetrate below the level of stratum germinatum of skin, and they can migrate in between the layers of skin in aimless manner for several weeks or months or years (up to 2 years). They form the tunnel like structure where roof has been formed by stratum germinatum and floor by stratum cornium. Larvae migrate very slowly at speed of 1–2 cm/day. Human species

can enter in to the circulation, while animal species can die. Path left by larva becomes dry and crusty. It is more common with animal hook worm species than human hook species. Maplestone had observed such lesion in Indian tea-garden coolies due to larvae of *N. americanus* in 1933.

– **General features:** Marked eosinophilia.
– **Lungs features:** Bronchitis and bronchopneumonia may occur when larva enters in alveolar space after breaking the pulmonary capillaries.

- **Due to adult stage:** Worm attaches to the intestinal mucosa by powerful buccal capsule and produces the following effects.

  – **Acute infection:** It presents with dyspepsia, epigastric tenderness simulating duodenal ulcer, perverted taste for earth or mud or lime called **geophagy** or **pica,** constipation or fatty diarrhea (steatorrhea) with unknown pathogenicity, pigmentation of mucosa at the bleeding points after long time and achlorhydria or hypoacidity. Worms found to be attached to the mucosa, if intestinal examination done within 3 hours of death of patients and free in lumen, when examination delayed.

  – **Chronic infection:** It causes **anemia which is characterized** by low iron leads to achlorhydria or hypoacidity, shallow appearance of skin (light yellow), pale mucosa of tongue, lips and eyelids, edema of face (puffy face), lower eyelids, feet and ankle, spoon shaped nail called **koilonychia,** dry lustreless hair, protuberant abdomen, hyperkinetic circulation may lead circulatory failure, retardation of growth and development in children, fatty changes in liver, erythroblastic reaction in marrow and the yellow marrow transformed in to a red formative marrow. **Three types anemia are produced by parasite. (1) Iron deficiency (hypochromic microcytic) anemia:** It is either due to **spoliative,** means withdrawal of blood by worms for the purpose of diet (worm thrives on plasma which is the main source of nutrition and RBCs passes out from the worm practically unchanged in to the lumen of host's intestine) or **mechanical** means chronic blood loss from punctured sites. Blood loss is continuing due to secretion of anticoagulant substances by worms. Roche et al. estimated 0.2 ml blood loss per parasite per day by *A. duodenale* and 0.03 ml blood loss per parasite per day by *N. americanus*. It is estimated that 0.76 mg iron loss per day by *A. duodenale* and 0.45 mg iron loss per day by *N. americanus*. Darling estimated about 1% of hemoglobin loss by 12 worms. **(2) Megaloblastic (macrocytic) anemia:** It is due to deficiency of B_{12} or folic acid. **(3) Dimorphic anemia:** It is due to deficiency of iron and B_{12} or folic acid.

Laboratory Diagnosis

Specimens: (1) Stool: For presence of adult stage or eggs. **(2) Bile:** Obtained by duodenal intubation and useful for adult stage or egg stage. **(3) Blood and bone marrow biopsy:** To detect the effects of anemia.

Testing methods

A. **Blood picture:** It shows eosinophilia and anemic picture: Follow **Table 107.2**.

B. **Macroscopic examination:** For presence of adult stages and occult blood.

C. **C.L. crystals:** Present in stool.

D. **Microscopy: (1) Wet mount by NS and iodine:** It shows the colorless (nonbile stained) eggs as shown in **Fig. 107.8b. (2) Concentration technique:** It is useful in case of negative result by wet mount method. Eggs do not float in common salt solution.

E. **Culture/coproculture:** Culture technique from stool called **coproculture.** It is performed to obtain the larvae to diagnose or research or differentiate the species. It is useful to diagnose the larvae of hook worm species like *A. duodenale* and *N. americanus*, pseudohook worm species like *Trichostrongylus colubriformis* and *Trichostrongylus orientalis* and *Strongyloides stercoralis*. Fecal specimens to be cultured should not be refrigerated. Different types of culture methods are described separately in box.

Note: Culture methods

Harada-Mori technique (Fig. 107.15a): It based on tendency of larvae to move from dry atmosphere to moisture area. About 500 mg of feces is smeared in middle or upper two third of thick filter paper by using the spatula. It is done evenly and as thin as possible. With a pipette add 3–4 ml of sterile water to the tube. Lower one third of the paper should be clean completely and dips in the water for absorbing the moisture. Incubate at room temperature (25–28°C) for 4–7 days, during which larvae develop on the filter paper and swim down in to the water. Larvae can be examined by magnifying lens outside at the bottom of tube or by microscope (take one drop of water with pipette over slide and examine under microscope).

Petri dish filter paper slant culture technique: It is same as Harada-Mori technique but Petri dish is used instead of tube and filter paper is placed in slanting position with supporting stick instead of vertical position as shown in **Fig. 107.15b**. Larvae can be examined under microscope either directly by putting whole Petri dish under microscope or by preparing the slide.

Charcoal culture technique: It is also a useful method to collect the larvae.

TABLE 107.2: Anemic picture in hook worm disease			
Features	**Microcytic anemia**	**Macrocytic anemia**	**Dimorphic anemia**
RBC (million/mm³)	2.9	0.73	1.3
Hb (%)	4.64	3.48	3.77
PCV (%)	21	10	16
MCV in femtoliter	72	137	123
MCH in picogram	16	47	24
MCHC (%)	22	34	23

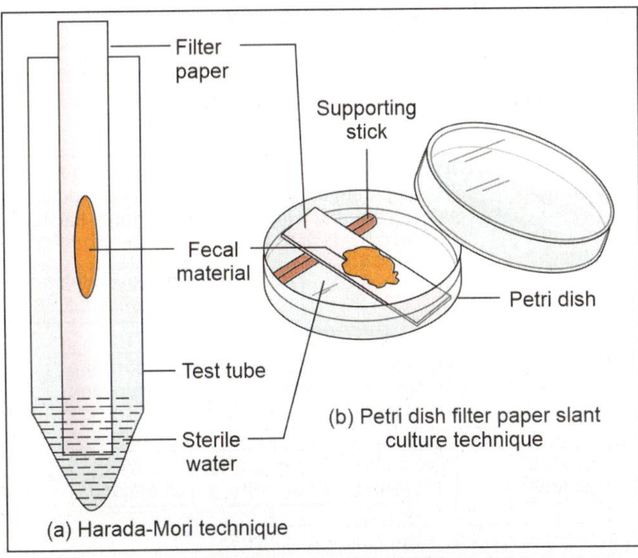

Fig. 107.15: Culture technique

F. Bone marrow biopsy: Erythroblastic reaction in marrow and the yellow marrow transformed in to a red formative marrow.

G. Molecular method: PCR has been developed which targets *Ancyclostoma* genes. It is useful to differentiate between hook worm species.

H. Radiological tests: Chest X-ray shows the pulmonary infiltration during migration of larvae.

I. Specific diagnosis

1. **Differentiation of** *A. duodenale* **from** *N. americanus*: Follow **Table 107.3**.
2. **Differentiation of larvae of hook worm species from** *Strongyloides*: If stool examination done after 24 hours then larva develop within egg and it indicates hook worm species, not the *Strongyloides*. Further differentiation can be done as per **Table 107.4** and **Fig. 107.17** after obtaining larvae by coproculture.

TABLE 107.3: Differences between *A. duodenale* and *N. americanus*

Features	*A. duodenale*	*N. americanus*
Meaning	**Ankylos (Greek) means hooked and stoma (Greek) means mouth** as anterior end of worm is bent slightly dorsally like hook, hence the name hook worm	**Necator (Latin) means murderer and americanus means America**
Common name	Old world hook worm	New world hook worm or American hook worm
Geographical distribution	Tropical and subtropical countries	It diagnosed from Texas in 1902 by Stiles, but origin was Africa and later spread to India, Sri Lanka, Far East, Australia and Africa
Adult stage		
Size	Large and thick	Small and thin
Life span	2–7 years	4–20 years
Anterior end	Bends in the same direction as the body curvature	Bends in the opposite direction as the body curvature
Buccal capsule	6 teeth and 4 hooks on ventral surface and 2 knobs on dorsal surface	4 chitinous plates, 2 on ventral surface and 2 on dorsal surface
Copulatory bursa	Total 13 rays and single dorsal ray, which divided at the tip	Total 14 rays and single dorsal ray, which divided at the base
Posterior end of female	Spine present	No spine
Vulval opening	Behind the middle of the body	In front of the middle of the body
Larval stage → Figs 107.16, Figs 107.17a and Figs 107.17b		
Head	Conical	Round
Buccal cavity • Length • Lumen • Chitinous wall • Oral depression	• Short, 10–10.5 μ • Larger • Thinner and bounded by one line dorsally and two less prominent lines ventrally converge anteriorly • Fine lines joins the oral depression and the anterior end of the buccal structure	• Long, 15–16 μ • Shorter • Thicker and dorsal plus ventral walls are of equal thickness; diverge anteriorly • No visible lines between the oral depression and the anterior end of the buccal structure
GIT • E-I junction • Lumen • Posterior end of intestine	• No gap at esophago-intestinal (E-I) junction • Anterior dilatation of intestinal lumen less prominent • Small refractive body (ampuliform dilatation) present	• An apparent gap at esophago-intestinal (E-I) junction • Anterior dilatation of intestinal lumen more prominent • Refractive body (ampuliform dilatation) absent
Genital pore	Behind the mid-point of end of esophagus and anus	In front of the mid-point of end of esophagus and anus
Cuticular striae	Less prominent	More prominent
Pathogenicity		
Risk	More dangerous (Reasons → described in text)	Less dangerous
Blood loss	0.2 ml per parasite per day	0.03 ml per parasite per day
Iron loss	0.76 mg per parasite per day	0.45 mg per parasite per day

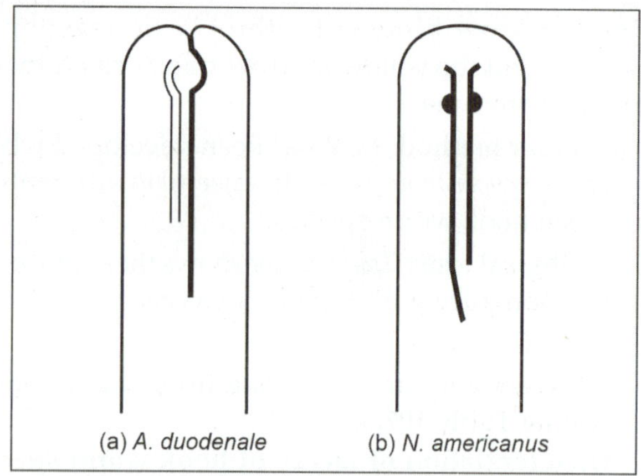

(a) *A. duodenale*   (b) *N. americanus*

Fig. 107.16: Buccal cavity of larvae of hook worm species

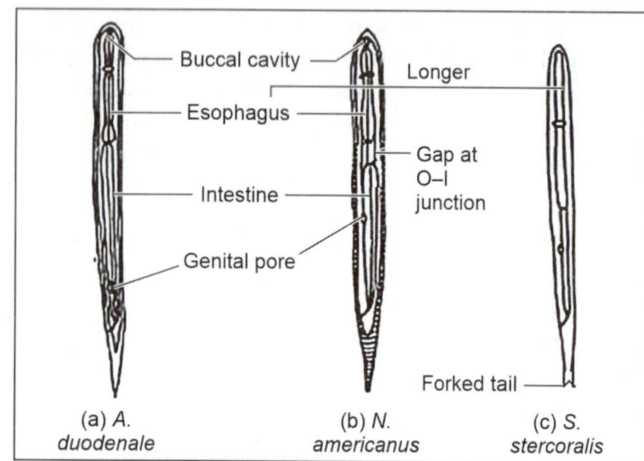

(a) *A.*   (b) *N.*   (c) *S.*
duodenale   americanus   stercoralis

Fig. 107.17: Filariform larvae of hook worm species and *Strongyloides*

TABLE 107.4: Differences between larvae of hook worm species and *Strongyloides*

Features	Hook worm	*Strongyloides*
Esophagus	Extended up to 25% of total body length	Extended up to 40% of total body length
Sheath	Present	Absent
Tail	Pointed	Forked

3. **Difference between hook worm species and other species that can be mistaken as hook worm:**
   - **Adult stage differentiation:** Follow **Flowchart 107.6**.
   - **Egg stage differentiation:** Following species can be differentiated.
     - *Trichostrongylus* **species (Fig. 107.18a):** Egg is elongated, pointed at one or both ends, longer and thinner than hook worm egg about 85–115 μm in length and contains segmented about with 16–32 blastomeres.
     - *Ternidens deminutus* **(Fig. 107.18b):** Egg is larger about 85 μm in length and contains segmented embryo with more numbers of blastomere.
     - *Strongyloides fuelleborni* **(Fig. 107.18c):** Egg is smaller than hook about 35–50 μm in length, oval in shape and may contain partially developed larva. Egg passes in feces rather than larvae.

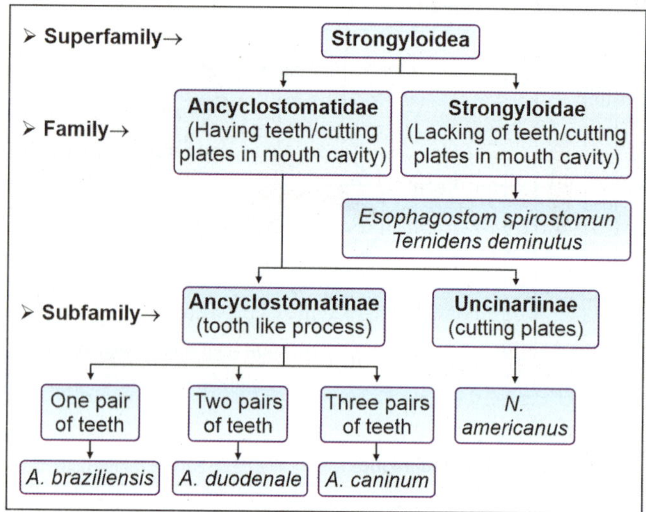

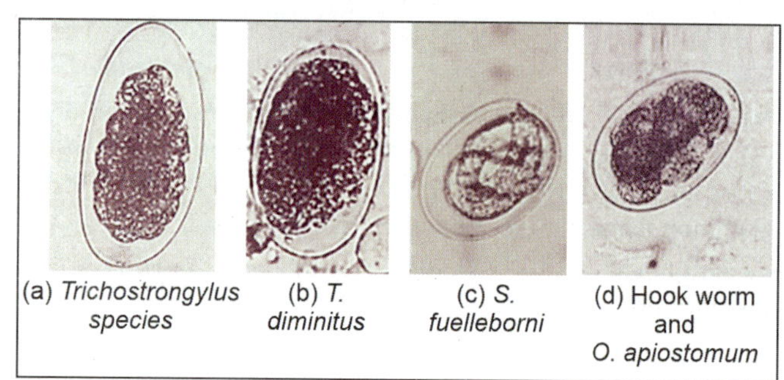

(a) *Trichostrongylus* species   (b) *T. diminitus*   (c) *S. fuelleborni*   (d) Hook worm and *O. apiostomum*

Fig. 107.18: Eggs can be mistaken as hook worm

- *Esophagostom spirostomum* **(Fig. 107.18d):** Egg is same as hook worm but passes in feces in advance stage of development.

Prevention

Preventive measures are prevention of fecal contamination of soil and wearing shoes at the time of walking and gloves while working in garden.

Treatment

Specific treatment of worm: Albendazole or mebendazole is the drug of choice. Pyrental pamoate is safe in pregnancy.

Treatment of anemia: Iron and folic acid given orally.

Other Hook Worm Species

Described separately in box.

Strongyloides spp.

Strongyloides stercoralis

Meaning

Strongylus (Greek) means round, edios means appearance (resembling) and stercoralis means feces, as this minute cylindrical (round) worm was first identified in diarrheic stool of French soldier in Cochin, China by Normand in 1876.

Essentials of Medical Microbiology

Note: Other hook worm species

A. braziliensis: It is a small adult worm. Buccal capsule has small orifice and ventral dental plate contains one pair of large teeth. **Life cycle** is same as *A. duodenale*. It is prevalent in Malaysia, Brazil and India. It is a parasite of dog and cat. In human it habitats only skin and no intestinal lesion. It wanders in between the layers of skin causing **creeping eruption** or **cutaneous larva migrans.**

A. ceylanicum: Morphology is same as *A. braziliensis* and sometimes considered as biological variant of *A. braziliensis* var *ceylanicum*. It presents in Sri Lanka. It is a parasite of civet cat. In human it habitats intestine and no skin lesion.

A. caninum: Buccal capsule has largest orifice and ventral plates contain 3 teeth, innermost being the smallest and outermost being the largest. **Life cycle** is same as *A. braziliensis*. It is cosmopolitan. It is a parasite of dog. In human, it habitats only skin and no intestinal lesion.

Necator americanus: General morphology, life cycle, pathogenicity and laboratory diagnosis are same as *A. duodenale*. Differences between *A. duodenale* and *N. americanus* are shown in **Table 107.3**. *A. duodenale* is more dangerous than *N. americanus*, because it causes more blood loss in intestinal infection due to larger size, armed with teeth and more migratory so leaving more bleeding points.

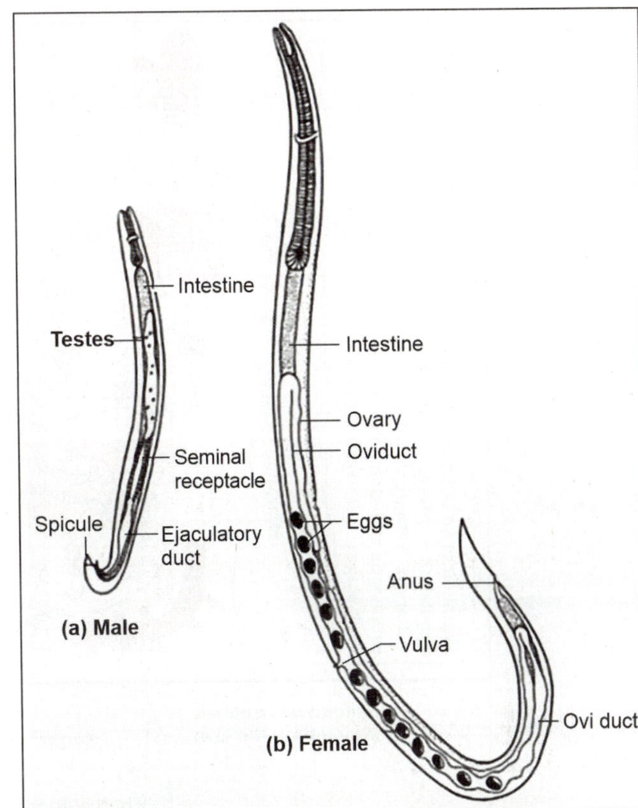

Fig. 107.19: Adult stage of *S. stercoralis*

History

Parasite was 1st discovered in diarrheic stool as mentioned above. Life cycle and pathogenesis were worked out by Stiles and Hassal in 1902.

Morphology

Three morphological stages like adult, egg and larva. Few important features of parasite are given below. For more details, follow **Ch. 106.**

Adult Stage (Fig. 107.19)

Time of maturation: Larva transformed to sexually mature adult stage in 15–20 days in definitive host after injection.

Size: In parasitic cycle females are readily identified but not the males. Male is 0.6–1 mm × 0.05 mm in size. Female is larger and narrower than male with 2.5 mm × 0.04–0.05 mm in size.

Shape: It is rounded, elongated, unsegmented, and cylindrical as shown in.

Life span: Adult worm lives about 3–4 months in human, but because it can cause autoinfection may survive longer up to 20–30 years.

Organs: Anterior end is rounded. Posterior (tail) end is pointed. Male has conspicuous buccal cavity **(mouth)**. Till date male was discovered by Kreis only in 1932 but other has failed. Muscular **esophagus** extends through the anterior third of body. **Intestine** extends through the posterior two thirds. Anus opens mid-ventrally, a short distance in front of the caudal tip. In **male genital pore** opens in to the cloaca. Copulatory spicules which penetrate the female during copulation are located on each side of the gubernaculum. Female contains paired uteri, vagina and vulva. **Female genital pore** opens at the junction of posterior and middle third of the body. Gravid uterus contains 8–10 eggs arranged antero-posteriorly. Female is an **ovi-viviparous** and lays eggs contain larva which immediately hatched out. It also called **parthenogenetic female,** as it lays eggs without fertilization by male.

Egg Stage (Fig. 107.20)

Fertilized female lays the egg. It is elliptical or oval, 55 µm × 30 µm in size and covered by thin hyaline membrane. Egg contains rhabditiform larva. Eggs are immediately hatched out to release larvae in the intestine; hence that are the larvae not the eggs present in stool.

Larval Stage

Two forms larvae are present.

Rhabditiform larva is 200–250 µm × 16 µm in size, has bulbous esophagus and moults twice on soil to change filariform larva. It is the exit form and present in stool. Rhabditiform larva develops in human intestine and also over soil from hatching of eggs.

Filariform larva is 500–600 µm × 20 µm in size, has non-bulbous esophagus and forked (notched) tail as shown in **Fig. 107.17c.** It is an infective stage and develops within 3–4 days. It develops from rhabditiform larva by following three modes as shown in **Flowchart 107.7.**

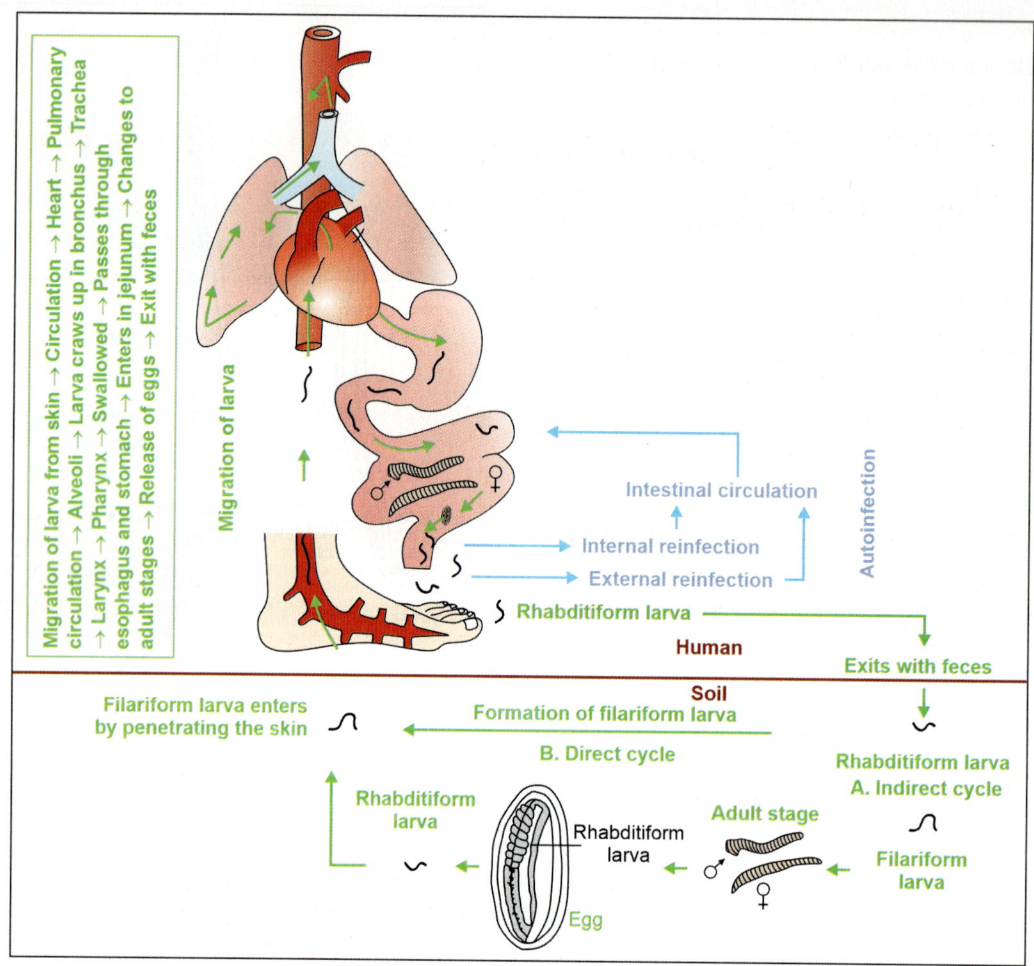

Fig. 107.20: Life cycle of *S. stercoralis*

Flowchart 107.7: Life cycle of *S. stercoralis*

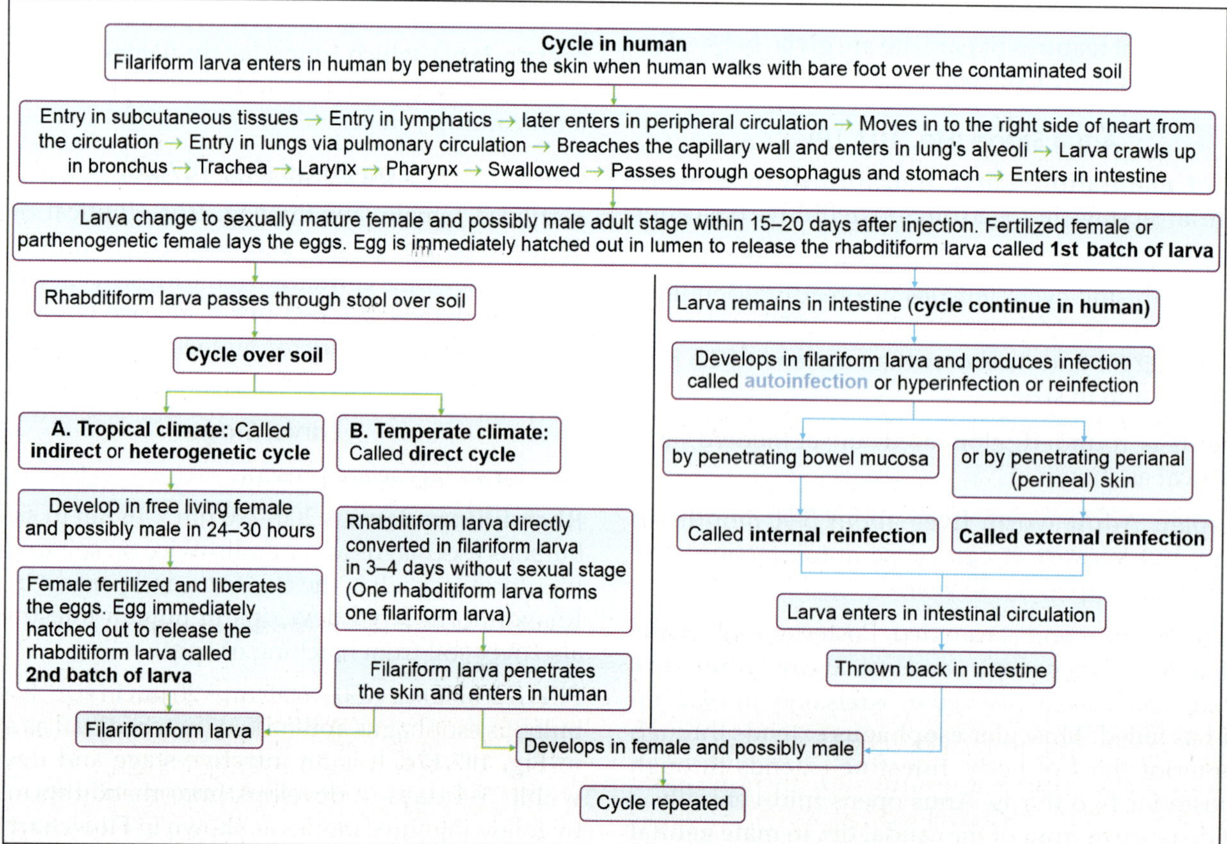

A. **Indirect cycle:** From the 2nd batch of rhabditiform larvae over the soil in tropical climate.

B. **Direct cycle:** From the 1st batch of rhabditiform larvae over the soil in temperate climate.

C. **Hyperinfection (autoinfection):** From the 1st batch of rhabditiform larvae in human.

Immunity

Acquired immunity: Immunity develops after primary infection in human which prevents the reinfection and also prevents tissue invasion by larvae and adult worms.

Hypersensitivity: It develops after reinfection and characterized by eosinophilia with utricaria.

Life Cycle

Hosts

- **Definitive host:** Whole cycle is possible in one host only like human. Change of host (other human) is not required for continuation of life cycle, as it undergoes the hyperinfective form **(autoinfection)**.
- **Intermediate host:** Not required.

Cycles: Following two types of cycle as shown in **Flowchart 107.7 and Fig. 107.20.**

- **Cycle in human**
- **Cycle over soil:** Two subtypes like
  A. **Indirect or heterogenetic cycle.**
  B. **Direct cycle.**

Pathogenicity

Disease name: It called strongyloidiasis.

Epidemiology: It is distributed world-wide, but common in Brazil, Far East (China Philippines) and Africa (tropical and temperate countries).

Reservoir of infection: Human.

Sources of infection: Soil and human itself in case of autoinfection.

Modes of transmission: (1) Filariform larva enters in human by penetrating the skin when human walks with bare foot over the contaminated soil. (2) Internal and external autoinfection can also occur as shown in **Flowchart 107.7.**

Exit form: Rhabditiform larva.

Infective form: Filariform larva.

Portal of entry: Skin.

Sites: Female parasite habitats in mucosa of duodenum and jejunum. It can be demonstrated postmortem by mucosal biopsy under low power microscopy. Male have no penetrating power and habitats in lumen only.

Precipitating factors: It is an opportunistic pathogen and adult stage causes autoinfection in immuno-compromised host with steroid therapy, AIDS or malignancy with extensive tissue damage.

Pathogenesis and clinical features: Divided in two categories as shown in **Flowchart 107.8.**

- **Due to migrating larva**
  - **Skin features: (1) At the site of entry:** Lesion called **ground itch** or **itchy dermatitis.** It presents with severe itching than *A. duodenale.* **(2) In between the layers of skin:** Normally, it completes the life cycle in human but due to repeated attacks host mount an immune response which prevents larva to complete the life cycle and larva limited only to skin. It traverses in between the layers of skin in serpigenous or linear fashion and produces lesion called **cutaneous larva migrans** or **creeping eruption** or **larva currens.** Currens (Latin) means running/race, as larva moves rapidly at speed of 3–4 cm/hour hence called **larva currens (racing larva).** This term was given in 1958. The track is started from anus and extended up to buttock, thigh or groin. It is characterized by reddish itchy papule along the path traversed by larva. Lesion lasts for few weeks. It also produced by other species of *Strongyloides.*
  - **General features:** Moderate eosinophilia (10–25%) in 60–80% of patients.
  - **Lungs features:** Migrating larva produces alveolar hemorrhage, bronchopneumonia and eosinophilic infiltration in lungs.
- **Due to adult stage:** Minimal invasion by adult stage which may be asymptomatic or present with mild diarrhea, abdominal pain (simulating peptic ulcer), gastrointestinal bleeding and mild chronic colitis.

Complication: Migrating larva carries bacteria along with, causing gram-negative bacteremia (Woodruff, 1968).

Laboratory Diagnosis

Specimens: Specimens can be examined to detect the rhabditiform larvae are stool, sputum, mucosal biopsy

Flowchart 107.8: Pathogenesis and clinical features of *S. stercoralis*

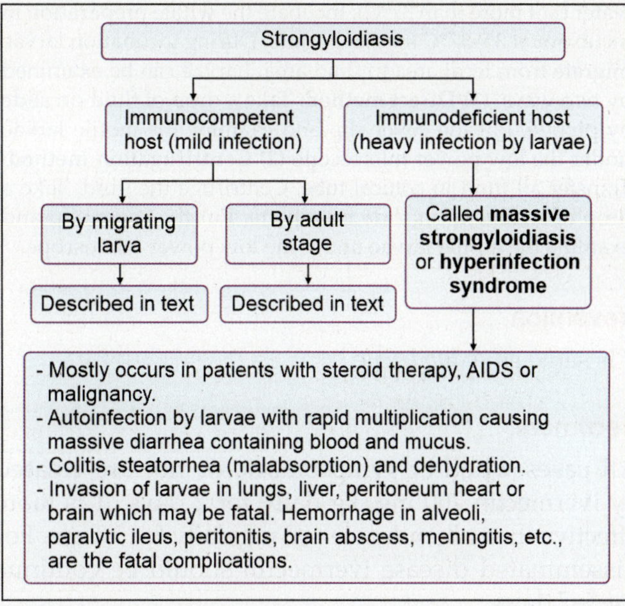

or bile/duodenal washing collected by E-test. CSF, ascetic fluid and urine are useful in immunodeficient patients to detect the larvae.

Testing Methods

A. **Blood picture:** Moderate eosinophilia (10–25%) in 60–80% of patients.

B. **Microscopy:** (1) Rhabditiform larva can be examined by **NS and iodine wet mount**. (2) Larva can be examined by **concentration techniques** as mentioned separately in box.

C. **Culture/coproculture:** Same as *A. duodenale*.

D. **Serological tests:** Serum antibodies are detected by ELISA. Ag is detected from stool by ELISA.

E. **Molecular method:** qPCR has been developed which targets *Strongyloides* genes from stool.

F. **Specific diagnosis:** To differentiate the larvae of hook worm species from *Strongyloides* as mentioned above with laboratory diagnosis of *A. duodenale*.

Note: Concentration techniques for *Strongyloides*

Formol-ether concentration technique: It is more sensitive than wet mount method. Larvae sediment at bottom of fluid.

Baermann's method (Baermannization): It based on tendency of larvae to move from colder to warmer area and most sensitive method. Baermann's apparatus is shown in **Fig. 107.21**. Take one funnel and fix on a stand. Funnel is about 10 cm in diameter with a short rubber tube which is screw clamped. Insert one round metal sieve in the funnel. Funnel is filled with warm water (42°C) up to sieve level. Appropriate amount of fresh feces is placed in cheese cloth, which is placed on the sieve. Incubate the whole preparation at warm temperature for 30 minutes. During incubation larvae migrate to the bottom of fluid. Loose the clamp, take a drop of fluid on slide, put the cover slip. Examine under the low power microscope for presence of motile larvae.

Water emergency semiconcentration technique: Principle is same as Baermannization. Transfer fresh feces (not more than 2 hours old) in a Petri dish. Make a central depression in fecal material by using a stick and fill the depression with warm water (not more than 37°C). Incubate the whole preparation in incubator at 35–37°C for 1½–3 hours. During incubation larvae migrate from fecal area to fluid area. Larvae can be examined by two ways. **(1) Direct method:** Take a drop of fluid on slide by pipette. Put the cover slip and examine the motile larvae under the low power microscope **(2) Centrifugation method:** Transfer all fluid in conical tube. Centrifuge the fluid. Take a drop of sediment on slide by pipette. Put the cover slip and examine the motile larvae under the low power microscope.

Prevention

It is same as *A duodenale*.

Treatment

All cases whether symptomatic or not are treated by ivermectin 200 mg/kg daily for 2 days. It is more effective than albendazole 400 mg daily for 3 days. For disseminated disease ivermectin should be continue for 5–7 days.

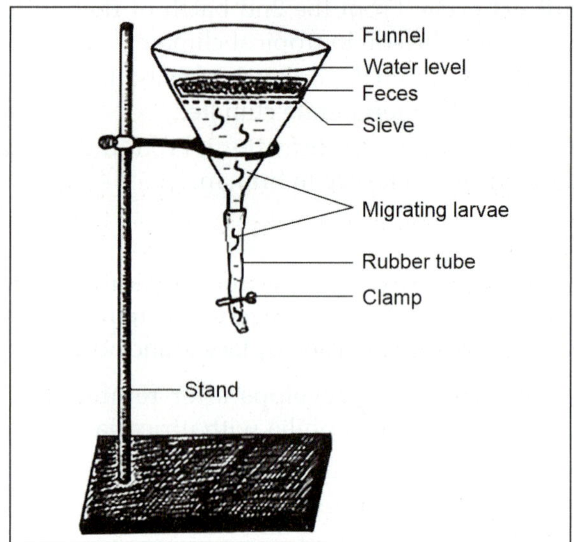

Fig. 107.21: Baermann's apparatus

Strongyloides fuelleborni

Same as *S. stercoralis* and differences are mentioned separately in box.

Note: *Strongyloides fuelleborni*

Morphology: Egg stage morphology **(Fig. 107.18c)** is described above with laboratory diagnosis of *A. duodenale*.

Pathogenicity: It produces disease in children called **swollen belly syndrome.** Two subspecies like *S. fuelleborni fuelleborni* is reported from Africa and *S. fuelleborni kellyi* is reported from Papua New Guinea. It is transmitted by breastfeeding from mother to child. It is a parasite of dog and monkey. It can cause intestinal infection in human specially babies as young as 2 months and young children. Infection is acute and fatal in nature. It presents with protein losing enteropathy, abdominal pain and diarrhea. Sometimes, coinfection occurs with *S. stercoralis* in same person.

Trichostrongylus species

Species are described separately in box.

Note: *Trichostrongylus* species

Species: Total 12 species, but species infecting human are *Trichostrongylus colubriformis* and *Trichostrongylus orientalis*.

Common name: Commonly it called pseudohook worm.

Morphology

- **Adult stage:** It is reddish colors, lacking buccal capsule, remains buried in mucosa and never found in stool. Male is 4–5 mm long while female is 4–6 mm long.
- **Egg stage:** Described above with laboratory diagnosis of *A. duodenale*.
- **Larval stage:** Eggs released through feces over soil and moults twice to form larvae (3rd stage). Larval stage is an infective stage and enters in human via contaminated vegetables or plants.

Life cycle: It completes its life cycle in single host. Larvae enter in human via contaminated vegetables or plants. On reaching the intestine it transformed to male and female adult stage after two more moultings. Female lays eggs, which pass through feces over soil and cycle gets repeated.

(Contd...)

Pathogenicity

- **Disease name:** It called pseudo hook worm disease.
- **Epidemiology:** It presents in Australia, Indonesia, Iran, Iraq, Japan and USSR. About 15% population of sheep and goat in Kashmir, India is infected by *Trichostrongylus* species.
- **Reservoir of infection:** Herbivorous animals like sheep, goat, cow, etc.
- **Source of infection:** Contaminated vegetables or plants.
- **Mode of transmission:** Larvae enter in human via contaminated vegetables or plants.
- **Exit form:** Egg.
- **Infective form:** Larva (3rd stage).
- **Portal of entry:** GIT.
- **Site:** Adult parasite habitats in small intestine in human.
- **Pathogenesis:** Mechanical damage occurs via penetration of anterior part (head) of adult stage into the mucosa causing traumatic damage and intestinal bleeding.
- **Clinical features:** Mild infection is mostly asymptomatic or rarely produces epigastric discomfort. Heavy infection presents with anemia, eosinophilia, prolonged diarrhea, emaciation and edema.

Laboratory diagnosis

- **Specimens:** Stool is an ideal sample. It contains egg, but not adult stage because it is embedded in mucosa and not excreted in stool.
- **Testing methods: (1) Microscopy:** It is performed for the presence of eggs, and it should be differentiated from hook worm eggs as described earlier. **(2) Culture/coproculture:** Harda-Mori technique is useful to obtain the larvae and to differentiate from related genera.

Capillaria philippinensis

Other species of genus is *C. hepatica* which is described in **Ch. 108**. *C. philippinensis* is described separately in box.

> **Note:** *Capillaria philippinensis*
>
> **Meaning and history:** Species name given because it was 1st discovered from a fatal case in Philippines in 1963 (contributed by Chitwood et al)
>
> **Morphology: (1) Adult stage:** It is same as *T. trichuria* but in smaller size. Female is 2.5–3.5 mm in length and male is 2.3–3.17 mm in length. **(2) Egg stage:** It is same as *T. trichuria* could be differentiated as per **Table 107.5 and Fig. 107.22**. **(3) Larval stage:** It presents in fresh or salt water fish.
>
> **Life cycle**
>
> **Hosts:** Definitive host is fish eating bird. Intermediate host is fresh or salt water fish.
>
> **Cycles: (1) Fish-bird-fish:** Cycle is not fully worked out. Cycle is continuing between fish-bird-fish. Fish is infected by egg which releases the larva. Such fish is eaten up by bird (accidentally by human), where larva changes to sexually mature male and female adult stage. Female lays eggs which pass through stool over soil to continue the cycle. **(2) Human cycle:** Human is an accidental host, infected by ingestion of partially cooked or pickled fish contains larvae and develops all stages in intestine. Larva changes to adult stage, which burrowed the jejunal mucosa. Female lays eggs which pass through stool over soil. **Internal autoinfection** occurs due to penetration of bowel mucosa by liberated larvae.

Pathogenicity

- **Disease name:** It called intestinal capillariasis.
- **Epidemiology:** Initially it was restricted to the Philippines and Thailand, later spread to Japan, Iran, Egypt, Taiwan, Indonesia and Columbia.
- **Reservoirs of infection:** Fish eating birds.
- **Modes of transmission:** Human is infected by ingestion of partially cooked or pickled fish contains larvae. Internal autoinfection can also occur.
- **Portal of entry:** GIT.
- **Site:** Adult parasite habitats the jejunum.
- **Exit form:** Egg.
- **Infective form:** Larva.
- **Pathogenesis and clinical features**
  - **Mild infection:** It presents with colicky abdominal pain, intestinal gurgling (borborygmi), chronic watery diarrhea with frequency of 8–10 stools per day, muscle wasting and edema.
  - **Heavy infection: (1) Spoliative action:** Heavy worm burden covers the intestinal mucosa, receive nutrition from intestinal contents, and derange intestinal function and produce protein loosing enteropathy, malabsorption of fat plus sugar and reduction of plasma level of potassium, sodium and calcium. **(2) Mechanical action:** Heavy worms cause lymphatic obstruction of intestinal wall and produce edema and even death also.

Laboratory diagnosis

- **Specimens:** Adult, larva and egg stage can be diagnosed from stool, bile or duodenal aspirates and mucosal (jejunal) biopsy.
- **Testing methods**
  - **Blood picture:** It shows eosinophilia and reduction of plasma level of potassium, sodium and calcium.
  - **Stool macroscopy:** For presence of adult stage.
  - **Microscopy of stool: (1)** Fat (fatty acid) and protein are present in stool. **(2)** Eggs are identified by **NS and iodine wet mount.** Eggs are same in appearance to *T. trichuria* and should be differentiated as per **Table 107.5 and Fig. 107.22**. **(3) Formol-ether concentration technique:** It is more sensitive than wet mount method. Larvae sediment at bottom of fluid.
  - **Serological tests:** Serum antibodies are developed, which are detected by ELISA.

Prevention: Avoid the use of raw fish meat.

Treatment: Mebendazole is effective drug.

TABLE 107.5: Differences between egg of *C. philippinensis* and *T. trichuria*		
Features	*C. philippinensis*	*T. trichuria*
Size	45 μm × 21 μm	50 μm × 25 μm
Shape	Barrel-shaped, but less elliptical	Barrel-shaped, but more elliptical
Covering	Striated	Smooth
Mucus plug	Bipolar and flattened	Bipolar and protuberant
Embryo	May or may not be embryonated, and if present, may be unsegmentd or with 2 segments	Contains unsegmented embryo

Prevention

Preventive measures are prevention of fecal contamination of soil and washing of vegetable before eating.

Treatment

Mebendazole (100 mg 12 hourly for 3–5 days) or albendazole (400 mg single dose) is effective with 70–90% cure rats.

Enterobius vermicularis

Meaning

Scientific name given from enteron (Greek) means intestine, bios means life and vermiculus means small worm like tail of hair.

Common Name

It called **thread worm** (because resembles to short piece of white thread), **pin worm** (because sharp tail like feature of female worm) or **seat worm**.

History

Eggs of *E. vermicularis* are detected in 10,000-year-old **coprolith** making it an oldest human parasite. **Coprolith** (fecalith or stercolith) means hard stony mass of dried feces in the intestine due to chronic constipation. Life cycle was 1st described by Leuckart in 1865. One common saying about this disease is "You had this infection as a child, you have it now or you will get it again when you have children".

Morphology

It has three morphological stages like adult, egg and larva. Few important features of parasite are given below. For more details, **follow Ch. 106.**

Adult Stage

Time of maturation: Egg transformed to sexually mature adult stage in 2 weeks–2 months in definitive host after ingestion.

Color: It is white in color.

Size: Female is larger than male as shown in **Fig. 107.26.** Male is 2–4 mm × 0.1–0.2 mm and Female is 8–12 mm × 0.3–0.5 mm in size.

Shape: It is spindle in shape. It looks like short white thread.

Life span: Male dies after fertilizing female, while female adult worm lives for 37–100 days.

Organs: Anterior end is expanded in both male and female like wings called **cervical alae.** In male posterior third of body is curved and sharply truncated. In female posterior end is straight, long, tapered and pointed which is nearly one third of the length of the worm. It has no buccal cavity. Posterior end of esophagus is dilated like a bulb called double bulb esophagus. Female is an **oviparous** and lays thousand of eggs/day. Egg contains tadpole-like larva.

Egg Stage (Fig. 107.27)

Egg is plano-convex (flat on ventral side and convex on dorsal side), 50–60 µm × 30 µm in size and colorless (nonbile stained). It is surrounded by transparent shell. It contains coiled tadpole-like larva. It floats in saturated common salt solution.

Larval Stage

Larva becomes free on digestion of egg shell in human intestine and develops adult stage.

Life Cycle

Hosts

- **Definitive host:** Whole cycle is possible in one host only like human. Change of host (other human) is not required for continuation of life cycle, due to autoinfection or retrograde infection.
- **Intermediate host:** Not required.

Cycles: Following two types of cycle.
- **Cycle in human:** Follow **Flowchart 107.10**
- **Cycle over soil:** and **Fig. 107.28.**

Pathogenicity

Disease name: It called enterobiasis or oxyuriasis or thread worm infection or pin worm infection or seat worm infection.

Epidemiology: It is distributed worldwide. It is the world's most common parasite specially infecting children. It is common in developed countries and in temperate climate than tropical countries.

Reservoir of infection: Human

Sources of infection: Food and water

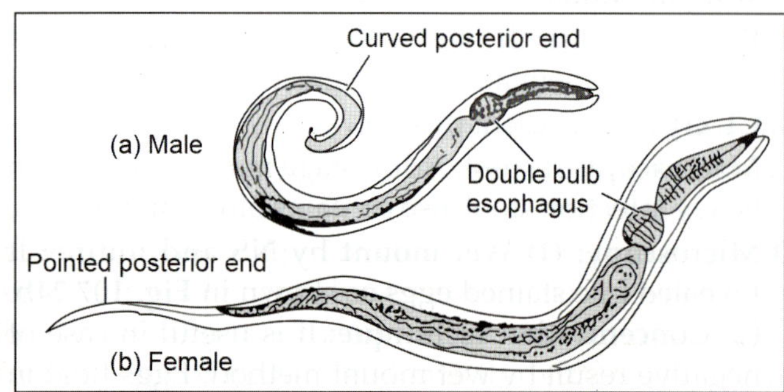

Fig. 107.26: Adult stage of *E. vermicularis*

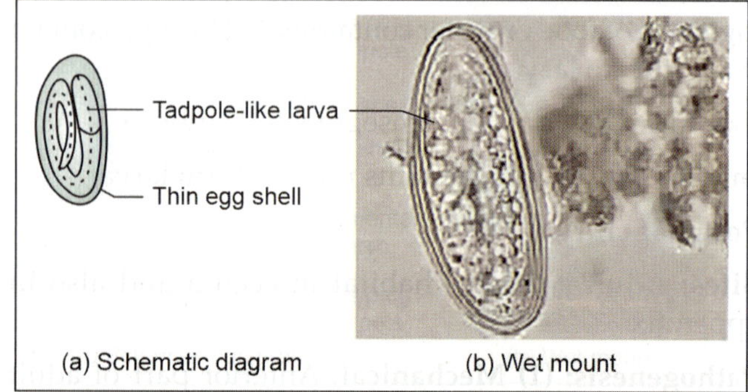

Fig. 107.27: Egg of *E. vermicularis*

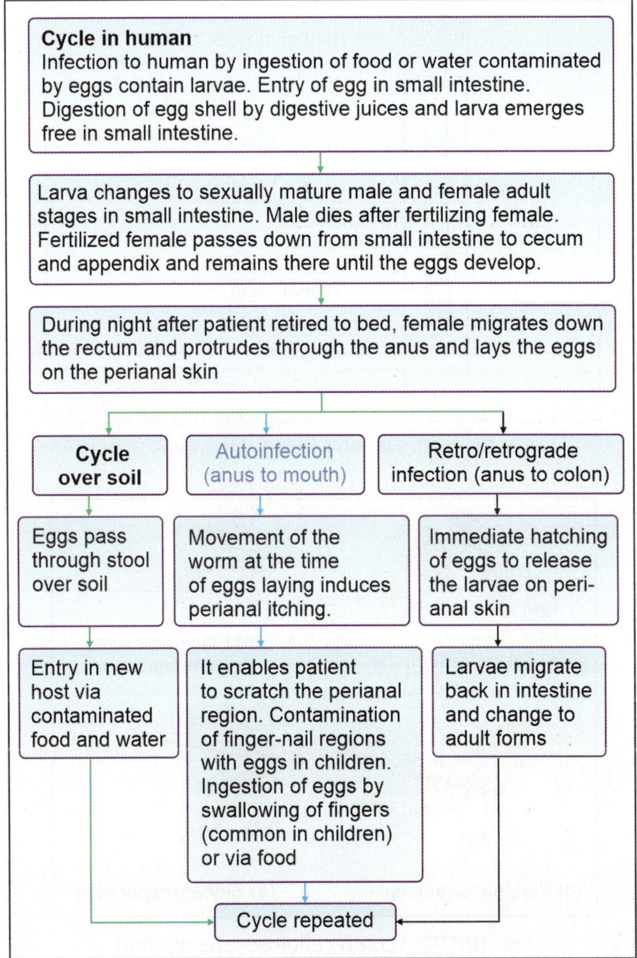

Modes of transmission: Four possible routes. (1) Human is infected by **ingestion** of contaminated food and water contain eggs. **(2) Inhalation** of dust particles of dried feces. **(3) Autoinfection (anus to mouth)** as mentioned

in **Flowchart 107.10. (4) Retro/retrograde infection (anus to colon)** as mentioned in **Flowchart 107.10.**

Exit form: Egg

Infective form: Egg

Portal of entry: GIT

Sites: Adult parasite habitat in cecum and also in appendix. It remains on the surface of mucosa and occasionally enters in submucosa.

Precipitating factors: It is common in children, persons with poor hygiene, in developed countries and in temperate climate than tropical countries.

Pathogenesis: It remains on the surface of mucosa and occasionally enters in submucosa. Mechanical irritation of anal mucosa by female worm at the time of laying of eggs is responsible for itching called **pruritus ani.**

Clinical features: It is mostly asymptomatic. It presents with itching and eczema around the skin of anus and perineum. In 2% cases, it produces an acute appendicitis. Nocturnal enuresis, abdominal pain and weight loss may occur following heavy infection.

Complications: Migrating female sometimes enters in female genital tract and causes urethritis, salpingitis and peritonitis after taking entry in peritoneal cavity by fallopian tube.

Laboratory Diagnosis

Adult stage diagnosis: (1) Macroscopic examination of stool: Stool is collected during normal defecation or by purgation with enema. Early morning time is preferable. Preserve the stool in alcohol or in 10% formaldehyde. Presence of white color adult worm in stool is examined by adult patients himself or by parents in case of

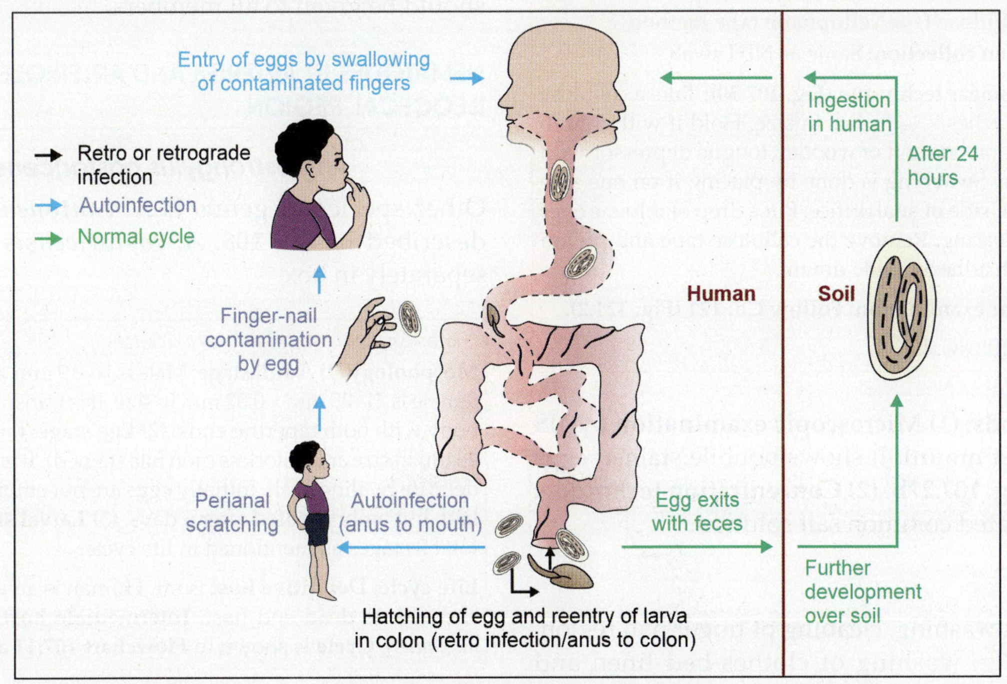

Fig. 107.28: Life cycle of *E. vermicularis*

pediatric patients. **(2) Direct examination of perianal region:** Migrating female can be noticed in perianal region at the time of laying of eggs.

Egg stage diagnosis

- **Specimens: (1) Stool:** Collection method, time of collection and preservation method are same as like adult stage **(2) Perianal swab:** It is collected by NIH swab or by scotch cellulose tape method as mentioned separately in box.

Note: National Institute of Health (NIH) swab

History: In 1937 Hall invented the cellophane anal swab commonly called NIH swab.

NIH swab's apparatus (Fig. 107.29): It constitutes a glass rod about 8–10 cm × 4 mm in size. One end of rod is covered with cellophane sprayer **(cellophane sprayer end)** about 1 inch square and held in place by rubber band. This end is inner end and used for swabbing. Outer end passes through rubber cork (stopper) with which test tube is closed.

Time of specimen collection: An early-morning sample, before the patient has bathed or used the toilet is optimal. Up to six successive days morning samples should be collected before a negative result is issued.

Swabbing technique: Remove the glass rod and rub the perianal region with cellophane sprayer end. After swabbing, rod with cellophane sprayer put inside the test tube and sent to the laboratory for smear preparation and testing.

Smear preparation technique: Put a drop of saline over glass slide and hold the cellophane end of rod over it. Push the rubber end up by using forceps until the cellophane is released. With the rod still held in position, the cellophane is spread out and smoothened in such a way that the materials adhering to the cellophane come in direct contact with glass slide. The eggs thus lie between the glass slide and cellophane.

Method of smear examination: Follow **Ch. 121 (Fig. 121.2).**

Uses: It is used to collect the peri-anal swab for *E. vermicularis*, *T. saginata* and *S. mansoni*.

Note: Scotch cellulose tape/cellophane tape method

Time of specimen collection: Same as NIH swab.

Swabbing and smear technique (Fig. 107.30): Take a cellulose tape about 3–4 inches × 3/4 inches in size. Hold it with thumb and index finger on one end of wooden tongue depressor with outer sticky side. Swabbing is done by placing it on one side and then on other side of anal orifice. Put a drop of toluene over glass slide for clearing. Remove the cellulose tape and put on a glass slide with adhesive side down.

Method of smear examination: Follow **Ch. 121 (Fig. 121.2).**

Uses: Same as NIH swab.

Testing methods: (1) Microscopic examination by NS and iodine wet mount: It shows nonbile stained eggs as shown in **Fig. 107.27b**. **(2) Concentration technique:** Floats in saturated common salt solution.

Prevention

Frequent hand washing, cleaning of finger nail region, regular bathing, washing of clothes-bed linen and avoiding the fecal contamination of soil.

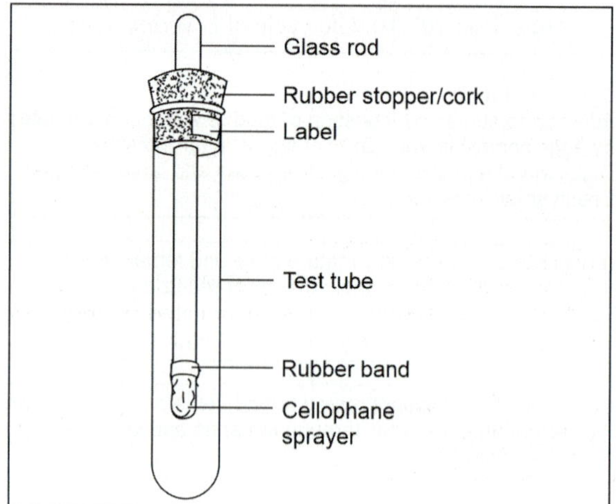

Fig. 107.29: NIH swab's apparatus

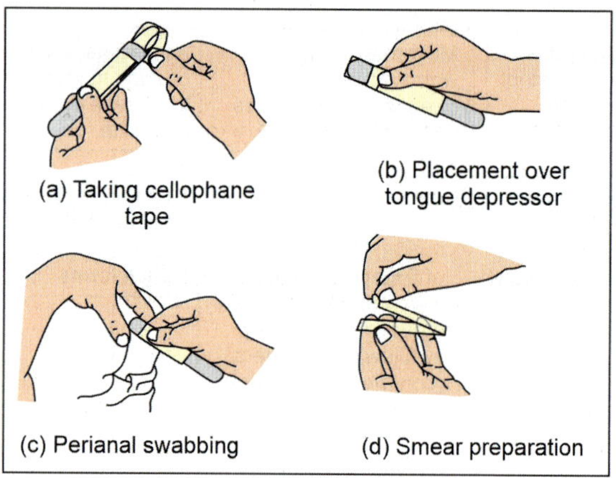

Fig. 107.30: Scotch cellulose tape method

Treatment

Pyrental pamoate (safe in pregnancy), albendazole or mebendazole is the drug of choice. Repeat the treatment after 2 weeks to prevent the autoinfection Usually pin worm affects the many members of family, so treatment should be given to all members.

NEMATODES IN ARTERIES AND ARTERIOLES OF ILEOCECAL REGION

Angiostrongylus costaricensis

Other species of genus is *A. cantonensis*, which is described in **Ch. 108.** *A. costaricensis* is described separately in box.

Note: *Angiostrongylus costaricensis*

Morphology (1) Adult stage: Male is 16–19 mm × 0.26 mm while female is 21–25 mm × 0.32 mm in size. It is transparent filariform body with both tapering ends. **(2) Egg stage:** It is oval, 72 µm × 46 µm in size and colorless (non bile stained). It is surrounded by delicate hyaline shell. Initially eggs are not embryonated when laid, but embryonated after 6 days. **(3) Larval stage:** Larva has total 3 stages as mentioned in life cycle.

Life cycle: Definitive host is rat. Human is an alternative host and acts as dead end host. **Intermediate host** is slug (*Limax maximus*). **Cycle** is shown in **Flowchart 107.11 and Fig. 107.31.**

(Contd…)

Pathogenicity

- **Disease name:** Human disease called abdominal angio-strongyliasis or eosinophilic gastroenteritis.
- **Epidemiology:** Human cases were reported from Costa Rica, Mexico and Central and South America.
- **Reservoir of infection:** Rat
- **Sources of infection:** Raw slug or food
- **Modes of transmission:** Human is infected by ingestion of raw slug or food contaminated with slug (contains 3rd stage of larva).
- **Exit form:** 1st stage of larva
- **Infective form:** 3rd stage of larva
- **Portal of entry:** GIT
- **Sites:** Adult parasite habitats in mesenteric artery.
- **Pathogenesis:** Pathology in human is due to both by adult and egg stages. Adult in the ileocecal arterioles causes an eosinophilic inflammation with thickening of walls. It affects mostly ileum, appendix and cecum.
- **Clinical features:** These are anorexia, diarrhea and vomiting. Abdominal pain simulates appendicitis.

Laboratory diagnosis:

- **Specimens:** Stool (for presence of eggs or larvae) and mucosal biopsy or local tissues.
- **Testing methods: (1) Blood picture:** Eosinophilia. **(2) Microscopy:** Wet mount of stool by NS-iodine and microscopical examination of tissues show the eggs.

Differences between *A. costaricensis* **and** *A. cantonensis***:** Follow **Table 107.6.**

Flowchart 107.11: Life cycle of *A. costaricensis*

Cycle in rat
Infection to rat by ingestion of slug contains 3rd stage larva. Larva enters in intestine. Larva changes to sexually mature male and female adult stages. Male dies after fertilizing female. Fertilized female lays the eggs in the intestine. Egg immediately hatched out and 1st stage larva passes with rat feces over soil.

Cycle in slug
1st stage larva taken up by slug, moults twice and changes to 2nd stage of larva and 3rd stage of larva (infective form)

Infective slug taken up by rat

Cycle repeated

Cycle in human
Infective slug taken up by human. Larva changes to sexually mature adult stage. Female lays the eggs. Adult and egg produce the eosinophilic gastroenteritis

No larval development from eggs in human intestine unlike rat, so human acts as **dead end** host

TABLE 107.6: Differences between *A. costaricensis* and *A. cantonensis*

Features	*A. costaricensis*	*A. cantonensis*
Habitat	Mesenteric artery	Pulmonary artery
Intermediate host	Slug	Snail
Pathogenic lesion	By adult and egg stage	By larval stage

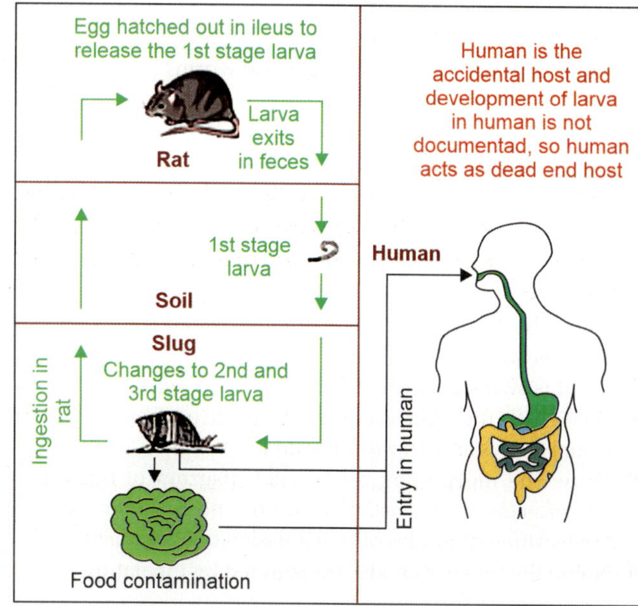

Fig. 107.31: Life cycle of *A. costaricensis*

Case Studies

1. A 40-year-female came with complain of fever, cough and dyspnea. Macroscopic stool examination shows the presence of adult worm with posterior end curved ventrally in the form of a hook having conical tip. Microscopic stool examination revealed bile stained egg with unsegmented embryo. Identify the case and answer the following.
   a. Name and draw the properly labeled diagram of egg of causative agent.
   b. Describe the life cycle of causative agent.
   c. Describe the pathogenicity of causative agent.
   d. Describe the lab., diagnosis of causative agent.
2. A 40-year-man came with complain of local skin lesion and pain in abdomen. Stool examination revealed nonbile stained egg with 4 segmented embryo. Blood examination shows the anaemic picture. Identify the case and answer the following.
   a. Name and draw the properly labeled diagram of egg of causative agent.
   b. Describe the life cycle of causative agent.
   c. Describe the pathogenicity of causative agent.
   d. Describe the lab., diagnosis of causative agent.

Essay/Full Question

1. Small intestinal nematodes.

Short Notes

1. Life cycle/pathogenicity/laboratory diagnosis of *Ascaris lumbricoides/Ancyclostoma duodenale/S. stercoralis/T. trichuria/E. vermicularis.*
2. Difference between *A. duodenale* and *N. americanus.*
3. NIH swab.

Short Questions for Theory/Viva Questions

1. What is trichinella cyst?
2. What is vulvar waist in *Ascaris lumbricoides*?
3. Write the four differences between fertilized and unfetilized eggs of *Ascaris lumbricoides.*
4. What is decorticated egg of *Ascaris lumbricoides*?
5. What is ripe egg of *Ascaris lumbricoides*?

6. What is Loeffler's syndrome?
7. What is ectopic ascariasis?
8. What is Y-shape appearance of hook worm?
9. What is blastomere?
10. What are coproculture and coprolith?
11. Define: Blastomere and blastospore.
12. Difference between larvae of hook worm and *Strongyloides*.
13. What is parthenogenetic parasite? Write one example.
14. Write different modes of formation of filariform larva in *Strongyloides stercoralis*.
15. Name the two nematodes transmitted by penetrating the skin.
16. What is larva currens?
17. What is swollen belly syndrome?
18. Write the four differences between eggs of *Capillaria philippinensis* and *Trichuris trichuria*.
19. Draw the morphological labeled disgram of eggs of *A. lumbricoides*, *A. duodenale*, *T. trichuria* and *E. vermicularis*.
20. Write different modes of transmission of *E. vermicularis*.
21. Name the two nematodes transmitted by inhalation route.

Comment on

1. Blood loss by *A. duodenale* is more than *N. americanus* in intestinal infection or *A. duodenale* is more dangerous than *N. americanus*.

MCQs for Chapter Review

Ascaris lumbricoides

1. **Ascariasis causes:**
   a. Appendicitis
   b. Intestinal obstruction
   c. Bile duct obstruction
   d. All of the above
2. **Specific diagnosis of ascariasis is made by:**
   a. Adult worm in stool
   b. Egg detection
   c. Antigen detection
   d. Antibody detection
   e. Skin test

Hook Worm spp.

3. **Habitat of hook worm:**
   a. Jejunum
   b. Ileum
   c. Colon
   d. Duodenum
4. ***Ancyclostoma* enters in human body by:**
   a. Ingestion
   b. Inhalation
   c. Penetration of skin
   d. Inoculation
5. **Infective stage of hook worm is:**
   a. Trophozoite form
   b. Filariform larva
   c. Cyst
   d. None
6. **Hook worm thrives on:**
   a. Whole blood
   b. Plasma
   c. Serum
   d. RBC

7. **The average blood loss in ancyclostomiasis per worm is:**
   a. 0.2 ml/day
   b. 2 ml/day
   c. 0.33 ml/day
   d. 1 ml/day

Trichuris trichuria

8. **Mucus plug at each pole of eggs is present in:**
   a. *T. spiralis*
   b. *A. lumbricoides*
   c. *T. trichuria*
   d. *E. vermicularis*
9. **Chronic dysentery, abdominal pain and rectal proplapse in children are caused by:**
   a. *Enterobius vermicularis*
   b. Ascariasis
   c. *Trichuris trichuria*
   d. *Trichinella spiralis*

Enterobius vermicularis

10. **Child having perianal pruritus is due to eggs of following:**
    a. *Enterobius vermicularis*
    b. *Ascaris*
    c. *Ancyclostoma duodenale*
    d. *Strongyloides stercoralis*

Answers and Explanation of MCQs

1. d
- Follow section, *Ascaris lumbricoides* (**pathogenicity → clinical features and Flowchart 107.4**) for explanation.

2. a and b
- Follow section, *Ascaris lumbricoides* (**laboratory diagnosis**) for explanation.

3. a, b and d
- Follow section, *Ancyclostoma duodenale* (**pathogenicity → sites**) for explanation.

4. c
- Follow section, *Ancyclostoma duodenale* (**pathogenicity → mode of transmission**) for explanation.

5. b
- Follow section, *Ancyclostoma duodenale* (**pathogenicity → infective form**) for explanation.

6. b

7. a
- Follow section, *Ancyclostoma duodenale* (**pathogenicity → pathogenesis and clinical features → due to adult stage**) for explanation of answers of MCQs 6–7.

8. c
- Follow section, *Trichuris trichuria* (**morphology → egg stage**) for explanation.

9. c
- From the given options only option 'a' is right. Follow section, *Trichuris trichuria* (**pathogenicity → clinical features**) for explanation.

10. a
- Follow section, *Enterobius vermicularis* (**pathogenicity → pathogenesis**) for explanation.

CHAPTER 108

Infections of Nematodes: Somatic (Tissue) Nematodes

Chapter Outline

FILARIASIS

General Aspects of Filariasis

Definition

Infection by any worm from the superfamily Filarioidea called **filariasis,** but it is limited to *W. bancrofti* and *B. malayi* (also newer species like *B. timori*).

Meaning

Filariasis word came from filum (Latin) means thread, because it is caused by slender thread like worms.

Taxonomy

Filarial parasites belong to the **super family filarioidea**, which has four families. Pathogenic spp., are belong to **Achantho-cheilonematidae family.**

Morphology

Filarial worm has two morphological stages like adult and larval stage as described below.

Adult Stage

Size: Adult worm is 80–100 mm × 0.25–0.30 mm in size, except female of *Onchocercus*. Female is larger than the male.

Shape: Adult worm is slender thread like.

Life span: Many years.

Organs: Worm possesses simple lipless mouth, cylindrical esophagus without bulb and simple intestine that may be atrophied at posteriorly. Tail of male worm has perianal papillae and unequal spicules but no caudal bursa.

Differential character of adult stages: Follow **Flowchart 108.1.**

Flowchart 108.1: Different features of adult worms

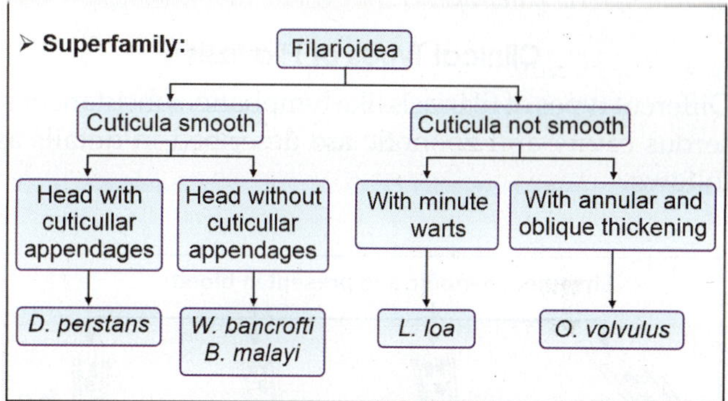

Larval Stage (Microfilaria)

Name of larvae: Female is viviparous giving birth to larva called microfilaria.

Size and shape: Microfilaria is about 290 μm × 6–7 μm in size and vermiform in appearance.

Life span: 3–36 months.

Organs: In some species, microfilariae retain their egg membrane, which develops into sheath called **sheathed microfilariae** and, which are lacking called **unsheathed microfilariae**. It is longer than the length of larva, so microfilariae can move to and fro within the sheath. **Nuclei** present along the length of microfilariae and interrupted by spaces or by special cells.

Periodicity: Depending on presence of microfilariae in blood, filarial worms can exhibit following types periodicity.

1. **Nocturnal periodicity:** Microfilariae present in blood during night between 10 pm and 4 am. Exact reason for nocturnal periodicity is not known but may be due to night feeding habit of intermediate host, e.g., *W. bancrofti.*

2. **Diurnal periodicity:** Microfilariae present in blood during day, e.g., *B. malayi.*

3. **Nonperiodic:** Microfilariae constantly present in blood during night and day, e.g., *O. volvulus.*

4. **Subperiodic or nocturnal subperiodic:** Microfilariae constantly present in blood during day but detected in late afternoon or during the night, e.g., *B. malayi.*

Hosts: Microfilariae complete their development in arthropod to achieve infective form for human.

Types: Microfilariae are differentiated on the basis of presence or absence of sheath, periodicity, numbers of nuclei at tail end and habitat as shown in **Fig. 108.1**.

Distribution: While examining blood films following thing should keep in mind.
- Africa: *Mf. bancrofti, Mf. perstans* and *Mf. loa.*
- South America: *Mf. bancrofti, Mf. perstans* and *Mf. ozzardi.*
- India: *Mf. bancrofti* and *Mf. malayi.*
- Brazil: *Mf. lewisi.*

Staining properties: Described below with morphology of *Wuchereria bancrofti.*

Clinical Types of Filariasis

Different types of filariasis like lymphatic, subcutaneous, serous cavity and zoonotic are described in details as follows.

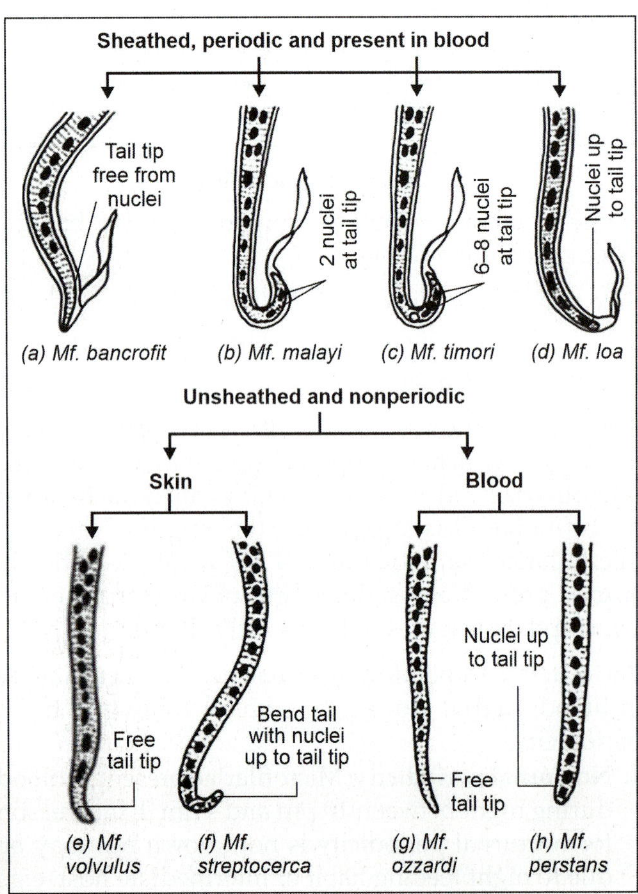

Fig. 108.1: Types of microfilariae

Lymphatic Filariasis: *Wuchereria bancrofti*

Common Name

Commonly it called Bancroft's filaria.

Meaning

Genus name was given from the **Wucherer** who found the microfilariae in chylous urine and **species name** came from the **Bancroft** who found the adult female.

History

Filariasis has been known since antiquity. **Elephantiasis** has been defined in India by Sushratha and in Persia by Rhazes and Avicenna. **"Malabar leg"** was applied to the condition by Clarke in Cochin in 1709. Larval stage called microfilaria was 1st identified by Demarquay in 1863 in the hydrocele fluid from Havana, Cuba. Wucher identified it in chylous urine in 1866 and Lewis identified it in blood in 1872 in Kolkata. Adult female was described by Bancroft in 1876 and the adult male was described by Bourne in 1888. Manson identified the *Culex* mosquito as vector in China in 1878. (1st human disease identified as vector borne). Manson also noticed the nocturnal periodicity of microfilariae in peripheral blood in 1879.

Immunology

Acquired immunity: It develops only against dead parasites, but not against live, and it is able to remove microfilariae from body not the adult worms. Both CMI and AMI response can develop. CMI plays dominant role than AMI. It lowers the microfilariae density in human in the age group of 15–20 years. In onchocerciasis microfilariae density does not decrease with age. High level of antibodies in occult filariasis prevents the entry of microfilariae (*W. bancrofti, B. malayi* and other animal parasites) in peripheral blood. A significant rise of serum IgM observed in Wuchereriasis and IgE in onchocerciasis. Antibodies are diagnosed by antigens prepared from *D. immitis.* Serological tests are positive in occult filariasis and Abs may cross react with *A. lumbricoides, S. stercoralis* and *Schistosoma* species.

Hypersensitivity: ITH develops and detected by intradermal test in onchocerciasis (called **Mazzotti's test**) and Wuchereriasis.

Morphology

Adult Stage (Fig. 108.3)

Time of maturation: Larva transformed to sexually mature adult stage in 5–18 months in definitive host after injection.

Size: Female is larger than male. Female is 8–10 cm × 0.2–0.3 mm and male is 2.5–4 cm × 0.1 mm in size.

Shape: Long thread or hair like body with both the ends are tapering.

Color: It is transparent and creamy-white in color.

Life span: Male dies after fertilizing the female about one week after infection. Female may live for 5–10 years.

Ends: The head end terminating in a slightly rounded swelling. In female tail end is narrow and abruptly pointed. In male tail end is curved ventrally and contains two spicules of unequal length. Female is viviparous and lays the larva.

Numbers: Male and female remains coiled together and separated with difficulties. Female's numbers are more than males.

Larval Stage/Microfilaria (Fig. 108.2)

Form: Solid form of larva is present **called microfilaria**.

Size: 290 µm × 6–7 µm.

Shape: Long body with blunt head and pointed tail end.

Color: Transparent and colorless.

Life span: In human body it can live for 70 days.

Host: It develops in human and later enters in mosquito.

Periodicity: (1) In India and China: Microfilaria shows **nocturnal periodicity** and present in blood during night between 10 pm and 4 am. During day time it presents in capillaries of lungs, kidneys, heart and artery such as carotid. Exact mechanisms for nocturnal periodicity are not known but may be due to night feeding habit of its intermediate host (*Culex fatigans*). **(2) In Pacific Island:** Microfilaria is **nonperiodic or subperiodic** and presents in blood during day plus night.

Density required to infect the mosquito: At least 15 microfilariae per drop of blood are required to infect the mosquito.

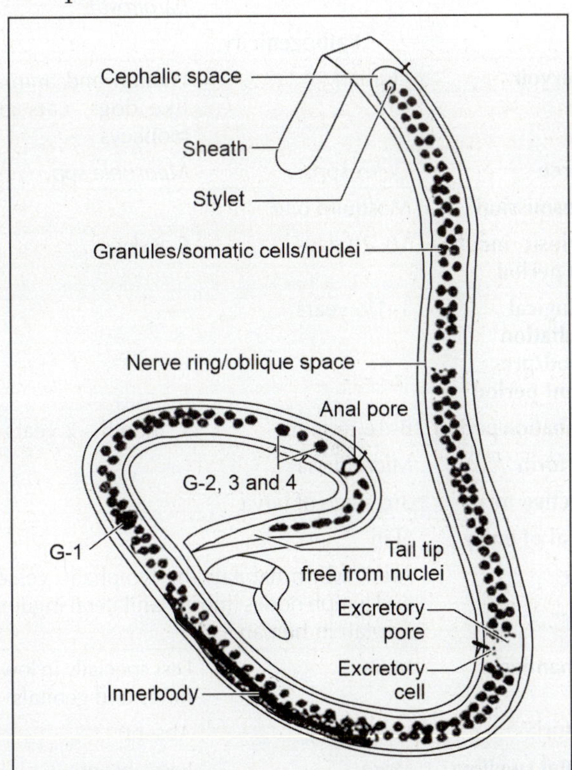

Fig. 108.2: Schematic diagram of microfilaria of *W. bancrofti*

Morphology and staining properties: (1) Wet mount from samples (unstained preparation): Actively motile microfilaria is seen under low or high power microscopy. In body it can move through and against the blood flow. **(2) Stained preparation:** Following morphological features are examined by Romanowsky's stain under oil immersion lens.

- **Sheath:** It is a structure less sac, seen best when projecting beyond the extremities of embryo. It represents the egg shell. It is longer (359 µm) than the length of larva, so larva can move to and fro within the sheath.
- **Cuticulla:** It is lined by subcuticullar cells.
- **Ends:** Head end is blunt and tail end is pointed.
- **Stylet:** It seen with vital stain.
- **Nuclei or granules or somatic cells:** They appear as granules in central axis of body and extend from the head to tail end except the terminal tip (terminal 5%). They are broken at certain places and making the landmark for identification of species as mentioned below.
  - Cephalic space: Nuclei are absent at head end called cephalic space.
  - Nerve ring: It is an oblique space.
  - Anterior V-spot/excretory pore: Represents the rudimentary excretory system.
  - Posterior V-spot or tail-spot: Represents the terminal part of GIT (anus/cloaca).
- **G-cells or genital cells:** G-2, 3 and 4 are situated just in front of anal pore while G-1 further front.
- **Inner body of Fulleborn or central body of Manson or internal body of Manson:** It extends from anterior V-spot to G-cells. It represents the rudimentary alimentary canal.

Life Cycle

- **Types of hosts: (1) Definitive host** is human. No animal host or reservoir is known. **(2) Intermediate host** is female mosquito (mosquito also act as an intermediate host in *Brugia* spp., *Mansonella* spp., and *Dirofilaria* spp.)
  - **In India:** *Culex fatigans* (*Culex quinquefasciatus*) night biter.
  - **In Polynesian Island:** *Aedes polynesiensis*, which can bite at any time.
  - **In Melanesian island:** *Anopheles punctulatus*, which can bite at any time.
- **Cycle:** It is possible between human and mosquito as shown in **Flowchart 108.2 and Fig. 108.3.**

Pathogenicity

Disease name: It called wuchereriasis or Bancroft's filariasis or lymphatic filariasis or commonly called filariasis.

Epidemiology: Parasite is confirmed to tropical and subtropical areas. **In world** it is common in West Indies, Puerto Rico, Africa, Southern China, Japan, Pacific

Flowchart 108.2: Life cycle of *W. bancrofti*

Cycle in human
Infection occurs to human by mosquito bite. Entry of third stage of larva in skin. Larva is not directly deposited in blood by mosquito, but enters via punctured skin or makes the way by its own.

↓

Entry in lymphatic vessels. Settle down in lymph nodes like inguinal, scrotal or abdominal. Larva changes to sexually mature male and female adult stages in 5–18 months in lymph nodes. Male dies after fertilizing female. Fertilized female gives birth to Microfilariae. Microfilariae enter in venous system via thoracic duct or via right lymphatic duct.

↓

Microfilaria migrates through pulmonary circulation and enters in peripheral blood and taken up by mosquito

↓

Cycle in mosquito
Microfilaria enters in anterior end of stomach, casts off the sheath, penetrates the gut wall and settled in thoracic muscles, where it can rest and can grow. After 2 days, it becomes short, thick, sausage shaped and 124–250 µm×10–17 µm in size called **1st stage larva**

↓

After 3–7 days it grows rapidly, moults twice and becomes 225–330 µm×15–30 µm in size called **2nd stage larva**

↓

After 10–11 days it becomes 1500–2000 µm×18–23 µm in size with complete development of all organs called **3rd stage larva**. It is the infective form and moves to the proboscis of mosquito. Extrinsic incubation period is 10–20 days.

↓

Enters in human during bite for blood meal purpose. There is no multiplication in mosquito and one microfilaria produces one infective form in proboscis. There are multiple microfilariae remaining coiled up and waiting for opportunity to infect human. Cycle repeated

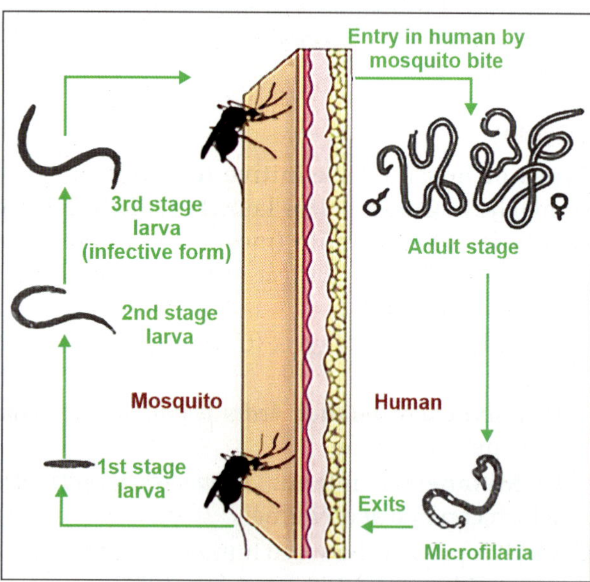

Fig. 108.3: Life cycle of *W. bancrofti*

Islands and South America. **In India,** it is prevalent along the coast of big rivers and also in Rajasthan, Bihar, Uttar Pradesh, Punjab and Delhi.

Reservoir of infection, source of infection, modes of transmission, incubation period, exit form, infective form, portal of entry and sites: Follow **Table 108.1.**

TABLE 108.1: Differences between *W. bancrofti* and *B. malayi*

Features	*W. bancrofti*	*B. malayi*
Adult worm		
Size	Larger	Smaller. Female is same, but male is different
Microfilaria		
Size	290 µm × 6–7 µm	230 µm × 6 µm
Shape	Long body with blunt head and pointed tail end	Long body with secondary curve and blunt head and bulb like tail end
Stylet	Single at anterior end	Double stylets at anterior end
Nuclei	Appear clean	Blurred hence counting is difficult
Cephalic space	Length and breadth are equal	Twice as long as broad
Tail tip	Free from nuclei	2 nuclei at tail tip with one at the end and second at midway between the tip and posterior column of nuclei
Periodicity	Nocturnal periodicity	• Nocturnal • Diurnal • Subperiodic
Development	Completed in 10–20 days	Completed in 6–8 days
Hosts		
Definitive	Only human	Human and animal
Intermediate (India)	*Culex fatigans*	*Mansonia annulifera, M. indiana, M. uniformis* and *Anopheles barbirostris*
Pathogenicity		
Reservoir	Human	Human and animals like dogs, cats and monkeys
Source	*Culex* spp.	*Mansonia* spp.
Transmission	Mosquito bite	
Extrinsic incubation period	10–20 days	6–8 days
Biological incubation period/pre-patent period	1–1½ years	—
Incubation period	8–16 months	1 month – 2 years
Exit form	Microfilaria	
Infective form	3rd stage of larva	
Portal of entry	Skin	
Sites	Adult worm habitats the lymphatic vessels and lymph nodes (mostly unilateral inguino-scrotal) in human	
Eliphantiasis	More	Less specially in lower limbs and genitals
Chyluria	More	Absent
Scrotal swelling (hydrocele)	More	Less/absent

Infective dose: It is not know, but many larvae fail to penetrate the skin by themselves and many are destroyed by immunological mechanisms. A very large number of infected mosquito bites are required to transmit the infection, perhaps about 15,000 infective bites per person.

Precipitating factors (epidemiological determinants): **(1) Agent (virulence) factors:** Like liberation of toxic products by live/dead adult stage and microfilariae. **(2) Vector factors:** Like density, life span, choice of host, resting habit, breeding habit, time of biting and resistance to insecticides. **(3) Host factors:** Morbidity increase with increasing age. It is more common in older age and male than female, because of outdoor visit and better covering of clothes in female. Sleeping outside and not using mosquito repellents like net, cream, etc., are increasing the risk. Migration to endemic areas increases the spread. **(4) Environmental factors:** Temperature between 22 and 38°C and humidity >70% increase the risk. Lack of town planning and bad drainage may aggravate the breeding of mosquitoes.

Clinical types: Clinical outcome of filariasis is varies in different person. **Person from endemic areas** is mostly asymptomatic, even with high microfilarial density in peripheral blood (20,000/ml). Such person can tolerate the microfilariae without any immune response. **Person from nonendemic areas** is progress to **classical and occult filariasis** as described separately in boxes.

Note: Classical filariasis

Definition: It is an acute inflammation of lymphatic vessels.

Pathogenesis, stage of disease and clinical features: Follow **Flowchart 108.3**.

Pathology: Different changes are seen in lymphatic system. Adult worms, which present in lymph nodes are aggravating the eosinophilic infiltration by releasing toxic metabolites, which reduce nutrition supply to the worm causing death. The tissue surrounding the dead worm undergoes necrosis and worm undergoes to degeneration or calcification. Dead worm (not the live worm) allows the response from RE system indicated by degeneration of dead worm by macrophages and giant cells. Eosinophilic infiltration which was appeared earlier starts to disappear and replaced by macrophages and giant cells. Fibroblasts begin to appear and form the concentric layers around the nuclei of dead worm, whose presence may not be determined later due to formation of scar tissue. New capillaries start to develop in fibrous layers. Such pathological layers formation around dead worm called **"parasitic onion"** as shown in **Fig. 108.4**. Besides the lymph nodes other organs of RE system like liver and spleen do not take any active part in wuchererial infection and they are spared to deal with disposal of microfilariae circulating in blood.

Complications

1. **Hydrocele:** It is defined as collection of fluid in scrotum. It is due to obstruction of para-aortic lymph nodes, which cause interference of fluid drainage from tunica vaginalis, epididymis and spermatic cord. Live microfilariae are present in hydrocele fluid, but not able to survive longer and they die in the fluid. It is common in East Africa, China, Japan but rare in Pacific Islands and India.

2. **Lymph varix or lymphangiovarix:** It is an abnormal irreversible dilatation of lymphatic vessels. Small vessels may rupture in large vessels and chyle [chyle word came from **chylous (Greek)** means **juice**. It is a milky body fluid consisting lymphs and lipids digested in intestine] may either excreted in urine/feces or collected in different body cavities. **(1)** Excretion of chyle in urine without blood called **chyluria.** It looks milky white **(Fig. 108.5)**. It is common in China, Japan and South India, but rare in Africa and Pacific Islands. **(2)** Excretion of chyle in urine with blood called **hematochyluria.** It contains fat cells (detected by fat solvents like ether, chloroform or xylol), albumin (precipitated on boiling), fibrinogen (coagulum formation when allowed to stands), RBCs, lymphocytes and microfilariae. **(3)** Excretion of chyle in feces called **chylous diarrhea. (4)** Collection of chyle in peritoneal cavity called **chylous ascites,** in pleural cavity called **chylothorax** and in any body cavity called **chylocele.**

3. **Elephantiasis:** It is defined as hyperplasia (solid edema) and hypertrophy of skin and subcutaneous tissues. It includes legs **(Fig. 108.6a)**, scrotum **(Fig. 108.6b)**, penis, labia, clitoris, vulva, breast **(Fig. 108.6c)** and forearms. It does not include the liver, lungs and spleen. Elephantiasis of leg and scrotum is common in China, India and Pacific, but rare in West Africa. It is due to presence of protein in the exudates which stimulate the connective tissues for excessive growth. Eliphantiasis of leg is due to obstruction of inguinal or iliac lymph nodes. Eliphantiasis of scrotum is due to obstruction of superficial inguinal lymph nodes. **(1) Macroscopic findings:** Overlying skin becomes rough, fissured and even papillomatous. Cut section of skin appears thickened, dense and fibrous. Subcutaneous tissues show a blubbery (edematous) appearance with presence of dilated and thickened lymphatics and veins. Underlying muscles and bones does not reveal any changes. Hair becomes rough and sparse. **(2) Microscopic findings:** Hyperplasia and hypertrophy of skin and subcutaneous tissues with granuloma formation (parasitic onion) including eosinophils, plasma cells, monocytes, giant cells and fibroblast around dead or clacified worms. Lymphatics vessel's walls show the inflammatory thickening and endothelial proliferation called **oblitrative endolymphangitis.** Thrombus formation also occurs in lymphatics. Blood microscopy shows no microfilariae in peripheral blood due to death of adult worms or failure of entry in blood due to lymphatic obstruction.

4. **Secondary bacterial infections:** By *Strept. pyogenes* and *Staph. aureus* cause septic lymphangitis, abscess and septicemia.

5. **Calcification of dead adult worms:** Examined in X-ray.

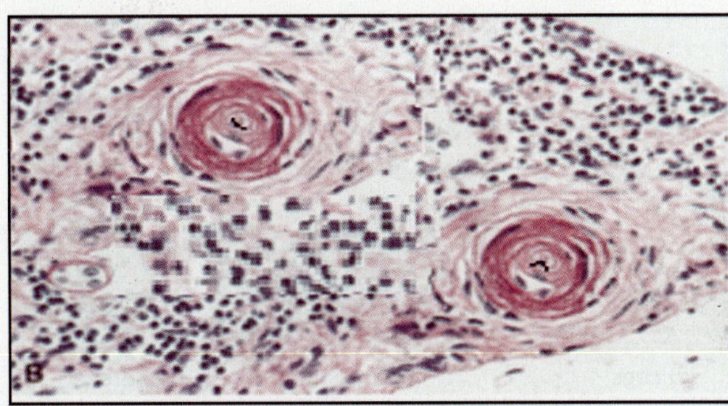

Fig. 108.4: "Parasitic onion" around dead worm

Flowchart 108.3: Pathogenesis, stage of disease, and clinical features of classical filariasis

Pathogenesis	Stages	Clinical features
Entry of 3rd stage of larva, and it migrates through lymphatic vessels with liberation of toxic products	**Allergic phase**	- Urticaria, fugitive swelling (painful, red, raised and tender area in limbs) and lymphoedema - Symptoms start with the growth of larva within 3 and half months
Larva settles in lymph nodes, changes to adult worm and produces lymphangitis and lymphadenitis by following mechanisms **1. Mechanical:** Mechanical irritation to vessel wall by moving adult worm. **2. Allergic** 　- Liberation of toxic metabolites by parturiting female. 　- Liberation of toxic metabolites from dead worm. **3. Infective:** Secondary infection by *Strept. pyogen* and *Staph. aureus*	**Inflammatory phase**	**Lymphangitis** - It involves the lymphatics of testis—epididymis (epididymo-orchitis), spermatic cord (funiculitis), abdomen (retroperitoneal lymphangitis), upper limbs and lower limbs. Lymphatic trunk appears as red streak in the superjacent skin. - Filarial fever (103°–104° F) which continues for 3–5 days and temperature falls down with profuse sweating. Fever is due to sensitization of host to metabolites released by parasites. **Lymphadenitis** - Site: Regional lymph nodes of groin and axilla. - Nodes appear as soft lobular masses. - Overlying skin is nonadherent, painless and nontender
Lymphatic obstruction by following mechanisms **1. Mechanical:** Mechanical blocking of lumen by live or dead adult worm. **2. Inflammatory:** Inflammatory thickening and endothelial proliferation of vessel's walls called **oblitrative endolymphangitis** **3. Fibrosis of vessel's walls:** Due to recurrent and repeated attack of lymphangitis. **4. Fibrosis of afferent lymph nodes** draining particular area	**Obstructive phase**	**Lymphatic obstruction takes long time around 20 years to develop and produces following complications** 1. **Hydrocele** 2. **Lymph varix** 3. **Elephantiasis** 4. **Secondary bacterial infection** ⎫ Described below 5. **Calcification of dead adult worm** ⎭ 6. **Occult filariasis:** Considered as separate type of filariasis and described below in detail.

Fig. 108.5: Chylous urine

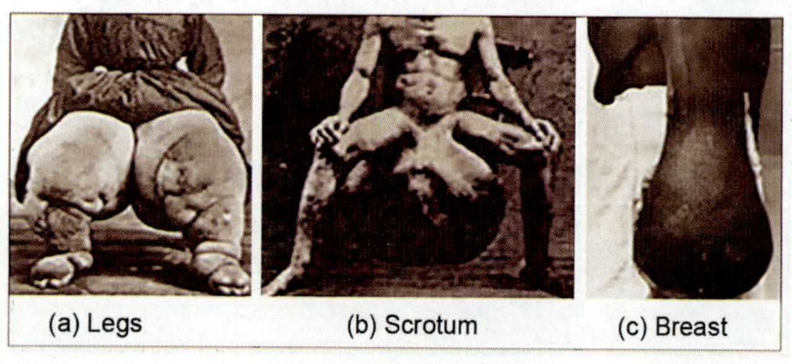

(a) Legs　　(b) Scrotum　　(c) Breast

Fig. 108.6: Elephantiasis

Note: Occult filariasis

Synonym: Also called **Meyers – Kowenaar syndrome.**

Definition: It is hypersensitive reaction characterized by massive eosinophilia (30–80%) around dead microfilariae or their products.

Geographical variation: Common in India, Sri Lanka, South East Asia, China, Philippines, Brazil and Africa.

Pathogenesis: Follow **Flowchart 108.4.**

Clinical features: Generalized lymphadenopathy, hepatosplenomegaly and tropical pulmonary eosinophilia (eosinophilic lungs or Weingarten's syndrome) of lungs which is characterized by low grade fever, loss of weight, paroxysmal (sudden and recurrent) cough with scanty sputum with or without blood and dyspnoea (not expiratory).

Pathological changes in tissues: It shows the eosinophilic granuloma around dead microfilariae or their products.

Laboratory diagnosis of occult filariasis: (1) Blood picture: It shows eosinophilia (30–80%) and 3000–5000 per mm^3 absolute count without microfilariae. **(2) Specific diagnosis of lung features:** Chest X-ray shows increased bronchovascular marking or diffuse miliary mottling in the lungs. Lung biopsy shows eosinophilic infiltration with microfilariae. Species identification is difficult. **(3) Serology:** Increased IgE and increased IgM.

Differences between classical filariasis and occult filariasis: Follow **Table 108.2.**

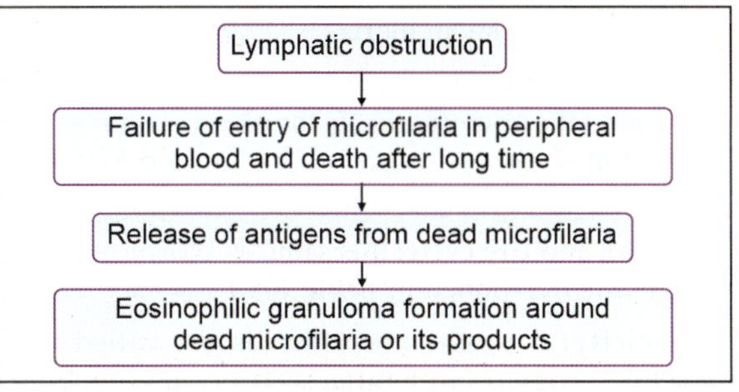

Flowchart 108.4: Pathogenesis of occult filariasis

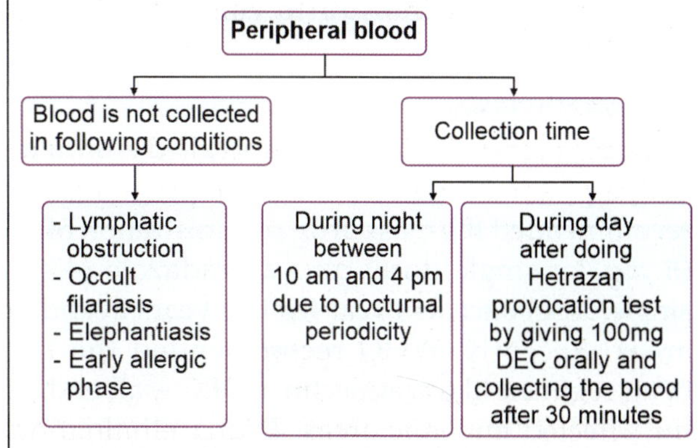

Flowchart 108.5: Collection of peripheral blood

TABLE 108.2: Differences between classical filariasis and occult filariasis

Features	Classical filariasis	Occult filariasis
Cause	Adult worm	Microfilariae
Mechanisms	Allergic, inflammatory and obstructive	Allergic
Organs involved	Lymphatic system (lymphatic vessels and lymph nodes)	Lymphatic system, liver, lungs, spleen
Microfilariae	In blood	In tissues not in blood

Laboratory Diagnosis of Classical Filariasis

Adult stage diagnosis: Adult stage presents in lymph nodes but chances of lymphatic obstruction, so not recommended.

Larval stage (microfilariae) diagnosis: It is performed either from mosquito (xenodiagnosis) or from human. Larval stage **diagnosis from mosquito (called xeno-diagnosis)** is performed by allowing the mosquito to bite the infected person followed by examining the microfilariae in the stomach blood of mosquito. **Diagnosis from human** is described below

A. **Blood picture:** Eosinophilia (5–51%) and increased IgE.

B. **Tests from peripheral blood:** It is the most commonly used specimen. It is useful for survey work, differentiation of species and next day examination because microfilariae remain viable for 1–2 days in blood. Time of blood collection depends on periodicity and stage of disease as shown in **Flowchart 108.5**. Blood is examined by different methods. **(1) Wet mount (unstained smear):** Take 2–3 drops of blood on clean glass slide. Add equal volume of water to lyse the RBCs. Put the cover slip and examine under the low power. Apply the vaseline to the rim of cover slip and put the preparation at room temperature for next day examination. Actively motile microfilariae are seen under low or high power microscopy. **(2) Thin and thick smears:** Details about thin and thick smear are given in **Ch. 97**. Appearance of microfilariae under Romanowsky's stain is shown in **Fig. 108.7**. **(3) Quantitative buffy coat smear:** Same as described in **Ch. 97**.

C. **Tests from other specimens:** Test like wet mount can be performed from chylous urine (10–20 ml of the

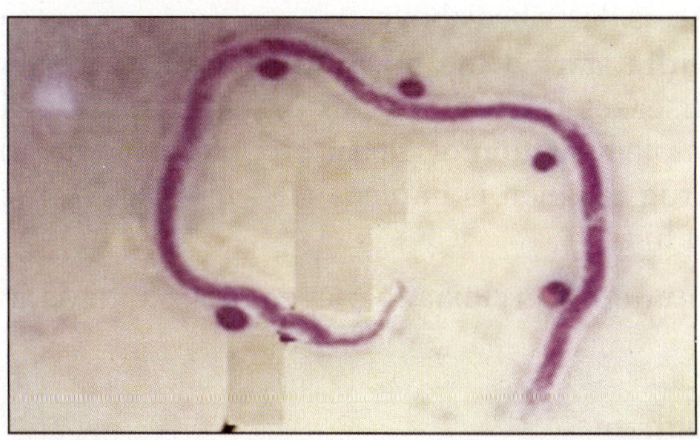

Fig. 108.7: Microfilaria under Romanowsky's stain

first early morning urine is collected), exudate from lymph varix or hydrocele fluid.

D. **Tests for microfilariae survey work:** It is done by thick film, or by millipore membrane/nucleopore filtration concentration technique. Later method is most sensitive. Blood is collected by venepuncture followed by filtration through membrane filter.

E. **Serological tests: (1) Antibody detection** is not useful in diagnosis but for epidemiological purpose in endemic areas. Flow through assay (immuno concentration test) or ELISA detects the IgG Ab to WbSXP-1 Ag. Luciferase immunoprecipitation test detects the Ab to Wb123 Ag. **(2) Antigen detection** is useful in diagnosis and to differentiate the current from past infection. Og4C3 and AD 12 antigens are detected from blood, serum or plasma by ELISA and ICT.

F. **Molecular tests:** PCR/qPCR provides sensitivity 10 folds greater than other methods and 100% specificity. These are also useful to monitor the therapy and to differentiate in between filarial species.

G. **Radiological tests:** More useful for adult stage. **(1) X-ray** shows the calcified worms. **(2) High frequency ultrasound and doppler within the scrotum, breast and spermatic cord** are showing motile adult parasites called **filaria dance sign**. Adult worms are examined in 80% cases in spermatic cord.

Laboratory Diagnosis of Occult Filariasis

It is described in box on page 934.

Essentials of Medical Microbiology

General measures (mosquito control measures): Follow Ch. 110.

Chemoprophylaxis

- **Global** program for filaria control called **elimination of lymphatic filariasis**. Under this program, WHO recommended the mass drug administration of DEC (6 mg/kg single dose) plus albendazole (400 mg single dose) once in a year for five years in endemic areas. Recently, WHO recommended the triple drug regimen like ivermectin + DEC + albendazole for selected endemic areas. Filaria elimination is validated by tests for survey work or by tests for Ag/Ab as described above.
- **India** started the **NVBDCP** for six vector-borne diseases, including filariasis. It supervizes the implementation of global program in India. Triple drug regimen is implemented in selected districts of India.

Immunoprophylaxis: No specific vaccine is available.

Treatment

Drugs: (1) DEC (6 mg/kg for 12 days) is the drug of choice for adult and larval stages. (2) Albendazole (400 mg two times in a day for 21 days) is effective for adult and larval stages. (3) Ivermectin (12 µg/kg) kills the microfilariae not adults. It is used in Africa not in India. Chances of recurrence are there.

Surgery: Required for elephantiasis and hydrocele.

Differences between *W. bancrofti* and *B. malayi*

Follow **Table 108.1**.

Lymphatic Filariasis: *Brugia* Species

Different species of *Brugia* are mentioned in **Flowchart 108.6**. Human species like *Brugia malayi* and *Brugia timori* are described separately in boxes.

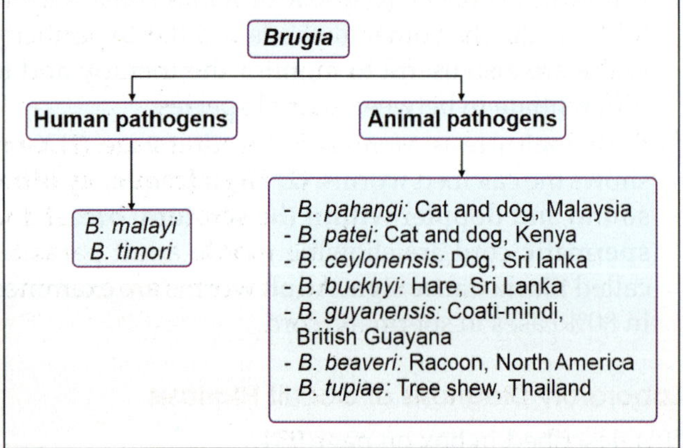

Flowchart 108.6: Brugia species

Note: *Brugia malayi*

Common name: Malayan filaria.

Meaning and history: Genus name was given from the Brug who 1st found the microfilariae in the blood of natives Sumatra in 1927. Adult worm was described by Rao and Maplestone in India in 1940.

Morphology and life cycle: It is same as *W. bancrofti* with few differences are mentioned in **Table 108.1**.

Pathogenicity: It causes zoonotic disease called malayan filariasis or brugian eliphantiasis. It occurs alone or with *W. bancrofti*. It is common in South Korea, China, Malaysia, Thailand, Vietnam, Indonesia, Borneo, Burma and Koshima Island (Japan). **In India** it is prevalent in Kerala, Orissa, Madhya Pradesh, West Bengal and Assam. Pathogenicity is same as *W. bancrofti* with few differences are mentioned in **Table 108.1**.

Laboratory diagnosis: It is same as *W. bancrofti* with characteristic identification of microfilariae (**Fig. 108.1b**).

Mosquito control measures: Certain plants like *Pistia*, in Sri Lanka and India are associated with breeding of mosquitoes. Removal of such plants can be effectively reduced the transmission. More details **follow Ch. 110**.

Chemoprophylaxis and treatment: Same as *W. bancrofti*.

Note: *Brugia timori*

Species name came from the Timor, Island from where it was 1st identified (1964). It is prevalent in Timor and Flores, Island of Eastern Indonesia. **Other details** are same as *B. malayi* with following differences.

Microfilariae (Fig. 108.1c): They are larger about 230 µm in length, 6–8 nuclei in tail tip and mostly showing nocturnal periodicity. Sheath fails to take Giemsa stain.

Life cycle: Definitive host is human, but no animal reservoirs. **Intermediate host** is female mosquito like *Anopheles barbirostris* which is breeds in rice fields and is a night feeder.

Pathogenicity: Lesions are milder than other. Lymphangitis and abscess formation occurs in lymph nodes and lymphatic vessels along the saphenous vein leading to scar formation. Lymphatic trunk becomes thickened like scar. Elephantiasis is minimal.

Subcutaneous Filariasis

Loa loa

Common name: African eye worm.

Meaning: Genus name was given from the local West African name for worm. **Common name** was given because of habit of adult worm to live in subconjunctival tissues of human.

History: It was 1st identified in eye of Negro girl in West Indies in 1770.

Morphology: Two morphological stages like adult and larva. (1) Adult stage: Larva transformed to sexually mature adult stage in 6–12 months in definitive host after injection. It is transparent. Female is 6 cm × 0.5 mm and male is 3 cm × 0.35 mm in size. Male dies after fertilizing the female. Gravid female may lives long for 4–17 years. Cuticula have numerous rounded protuberances called **cuticular bosses.** In female they are absent at head end, while in male they are absent at both ends. Female is **viviparous** and lays the embryo. **(2) Larval stage (microfilaria):** It is 300 μm × 7 μm in size, sheathed and not stained by Giemsa stain. Nuclei present up to tail tip **(Fig. 108.1d)**. It has diurnal periodicity by 12 pm–2 pm and presents in blood in day time in human.

Life cycle: Definitive host is human. **Intermediate host** is female mango or deer fly (*Chrysops silacea* and *C. dimidiate*). Both are canopy (cover) dwellers. Whole cycle is possible between human and fly as shown in **Flowchart 108.7**.

Pathogenicity

- **Disease name:** It called **loiasis**. It also called **calabar swelling** as the organism was first extracted from the eye of a woman of a West **Calabar** in Africa by Argyll Robertson in 1895 or **fugitive swelling** (edema of subcutaneous tissues), because it disappears quickly and reappears at somewhere else.
- **Epidemiology:** It is common in West and Central Africa.
- **Reservoir of infection:** Human. Parasite found in monkey, but it does not serve as human reservoir.
- **Mode of transmission:** By female mango or deer fly bite.
- **Incubation period:** Average 3–4 years, but may be long up to 17 years.
- **Exit form:** Microfilaria.
- **Infective form:** Third stage of larva.
- **Portal of entry:** Skin.
- **Site:** Subcutaneous tissues or often in subconjunctival tissues.
- **Pathogenesis:** Lesions are due to hypersensitivity reactions to the toxin produced by adult worms.
- **Clinical features:** These are due to adult worm. **(1) Lesions in subcutaneous tissue:** During migration adult produces the toxin gives fugitive swelling. Swelling is itchy, erythematous, 2–5 cm in size and mostly present on extremities. It disappears quickly

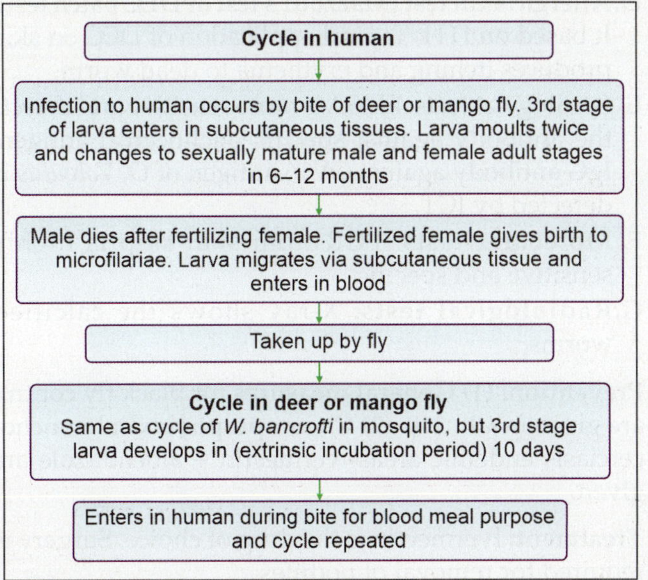

Flowchart 108.7: Life cycle of *L. loa*

(fugitive) in 2–3 days and reappears at somewhere else. **(2) Lesions in eyes:** Adult worm has special predilection for creeping in and around the eyes. It migrates from surrounding tissue and enters in subconjunctival area. Eye lesions include granuloma in bulbar conjunctiva, painless edema of the eyelids, proptosis, retionpathy, etc.

- **Complications:** Neuropathy, encephalopathy, meningitis, Jacksonian epilepsy, cardiomyopathy, renal diseases, etc., can occur but very rare.

Laboratory diagnosis: (1) Blood picture: Intense eosinophilia (30–40%). **(2) Adult stage diagnosis:** Adult stage can be detected from biopsy of subcutaneous lesion or eye lesion by surface bump. It can be extracted when it is crossing the conjunctiva. **(3) Larval stage (microfilariae) diagnosis:** It is detected from peripheral blood (12 pm–2 pm) and also from urine, CSF or sputum.

Prevention: (1) General measures like deerfly/mango fly control are given in **Ch. 110. (2) Chemoprophylaxis:** In loiasis endemic areas, albendazole is given twice in a year.

Treatment: DEC is the drug of choice for larval stage and adult stage. Multiple courses are required for complete cure of disease. Surgery is required for removal of adult worm.

Onchocercus volvulus

Common name: Commonly it called convoluted filaria.

Meaning: Genus name was given from the word Onkos (Greek) means hook and Kerkos (Greek) means tail because posterior end of parasite is hooked.

History: Microfilaria was 1st identified by O'Neil in 1875. Adult worm was 1st identified by Leuckart in 1893 in nodules removed from the chest and scalp of West African native.

Morphology: Two morphological stages like adult and larva. **(1) Adult stage:** Larva transformed to sexually mature adult stage in 12–15 months in definitive host after injection. Female is 50 cm × 0.4 mm in its greatest diameter and male is 3 cm × 0.13 mm in size. It is elongated and posterior end is turned like hook. It is whitish and has opalescent and transverse stria on the cuticula. Male dies after fertilizing the female. Gravid female may live long for 15 years. Cuticula is thick and raised in well marked annular and oblique thickenings. Female is **viviparous** and lays the embryo. **(2) Larval stage (microfilaria):** It is 300 μm × 6–8 μm in size and unsheathed. Tail tip is free from nuclei **(Fig. 108.1e)**. It is nonperiodic and presents in skin, subcutaneous lymphatics, conjunctiva and rarely in blood of human. It lives for about 1 year.

Life cycle: Definitive host is human. Animal reservoirs are not present. **Intermediate host** is day biting female black fly. Common species in Africa are *Simulium damnosum* (commonly called **buffalo gnat**) and *S. naevei*

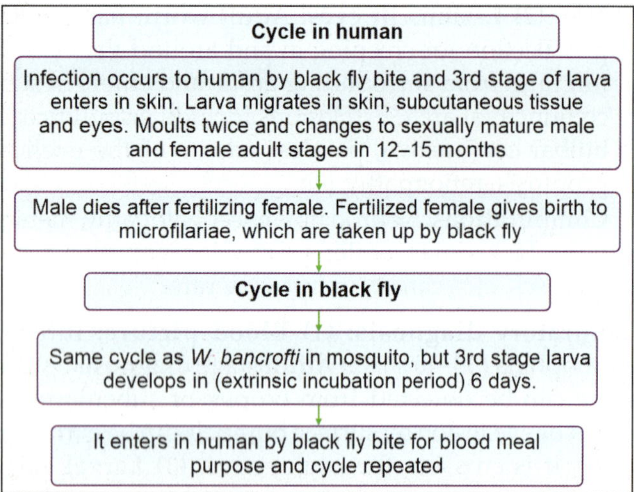

and in Central and South America are *S. ochraceum* (In Mexico and Guatemala) and *S. metallicum* (in Venezuela). Whole cycle is possible between human and female black fly as shown in **Flowchart 108.8**.

Pathogenicity

- **Disease name:** It called **onchocerciasis.** It also called **river blindness** as the vector breeds in fast flowing river and disease is common along the course of river or **blinding filaria** as it is the 2nd major cause of blindness in the world.
- **Epidemiology:** It is reported from Africa (Congo basin and Volta river basin), Central America (Guatemala, Venezuela and Mexico) and South Arabia.
- **Reservoir of infection:** Human only
- **Mode of transmission:** By female black fly bite
- **Incubation period:** One year
- **Exit form:** Microfilaria
- **Infective form:** Third stage of larva
- **Portal of entry:** Skin
- **Site:** Adult parasite habitats in subcutaneous connective tissues
- **Pathogenesis:** Lesions are due to hypersensitivity reactions to the dead adult worms or microfilariae.
- **Clinical features and pathology:** Two types
  - **Lesions due to adult worms: (1) Subcutaneous nodules:** Lesion called **onchocercoma.** It is few mm to cm (up to 6 cm) in size. It is single or average 3–6 in numbers but may be up to 15 in numbers. Site of lesion depends on habit of fly bite. **In Africa** it presents more on trunk and limbs than on the head as the fly bite commonly on trunk and limbs. **In Central and South America,** it presents over the scalp as the fly bite commonly on head and neck. Nodule raised above the skin surface, painless and non-suppurative. Nodule has honeycombed central area contains adult worms of both sexes. If worms are dead, they may initiate calcification and foreign body giant cell reaction. Peripheral part of nodule contains concentric mass of fibrous tissues. **(2) Skin lesions:** It is characterized by severe

dermatitis called **onchodermatitis/sowdah** with depigmentation and hyperpigmentation called **leopard skin** and pruritus, atrophy and fibrosis.
  - **Lesions due to microfilaria:** Mostly eyes are affected from surrounding lesion by wandering microfilaria. Eye lesion presents with simple conjunctivitis, small round opacities and pannus in the anterior quadrant of the cornea, keratitis, retinitis, iridocyclitis, secondary granuloma, optic atrophy and eventually blindness.
- **Complications:** These are hydrocele or lymph scrotum, elephantiasis (common in scrotum or legs), hanging groin (atrophied, inelastic and leopard like appearance of skin of groin which hangs down in a fold containing enlarged and sclerosed femoral or inguinal lymph nodes mostly in patients of Africa), calcification (of dead worms), abscess (due secondary bacterial infections) and dwarfism (due to invasion of pituitary by microfilariae).

Laboratory diagnosis

A. **Blood picture:** High eosinophilia.

B. **Adult stage diagnosis:** Adult stage can be detected by biopsy from subcutaneous lesion.

C. **Larval stage (microfilariae) diagnosis: (1) Skin snip preparation:** Maximum microfilarial density is present in iliac crest or trapezius (infrascapular) region. Sample is collected during day time by raising the skin with a needle and shaving the epidermal layer with razor. Skin snip is teased in water or saline and incubated for several hours. After sufficient incubation sample is examined by dissecting microscope for unsheathed microfilariae. **(2) Slip lamp examination of eye** revealed the microfilariae in anterior chamber. **(3) Conjunctival biopsy:** Unsheathed microfilariae are examined under microscope. **(4) Other specimens:** Unsheathed microfilariae are also present in urine, CSF, blood, sputum or lymph collected specially after DEC treatment.

D. **Allergic/skin test (Mazzotti's test or DEC patch test):** It based on ITH. Topical application of DEC on skin produces itching and erythema to dead worm.

E. **Serological tests:** ELISA is more sensitive and detects the antibody against specific onchocercal antigen. IgG antibody against OV16 antigen of *O. Volvulus* is detected by ICT.

F. **Molecular tests:** PCR from skin snip is highly sensitive and specific.

G. **Radiological tests:** X-ray shows the calcified worms.

Prevention: (1) General measures for black fly control are given in **Ch. 110. (2) Chemoprophylaxis:** In onchocerciasis endemic areas ivermectin + albendazole are given.

Treatment: Ivermectin is the drug of choice. Surgery is required for removal of nodules.

Mansonella streptocerca

Morphology: Microfilariae are 210 µm × 5–6 µm in size, unsheathed, contains nuclei up to tail tip **(Fig. 108.1f)**, hooked tail and nonperiodic.

Life cycle: Definitive host is chimpanzee. Human is also infected. **Intermediate host is** midges specially *Culicoides* spp., like *C. grahami* and *C. austeni.* Cycle is not fully worked out.

Pathogenicity: Disease called streptocerciasis. It is prevalent in West Africa, especially in Congo basin. Chimpanzees are the reservoirs of infection. It is transmitted by bite of midges. Adult parasite habitats the dermis and producing edema of skin, dermatitis with pruritus and hypopigmented macules.

Laboratory diagnosis: Microfilariae are identified from skin clipping or snips.

Prevention: General measures include midges control as described in **Ch. 110.**

Treatment: Ivermectin in single does is effective.

Subcutaeous (cutaneous) and serous cavity filariasis

Mansonella ozzardi

Common name: Commonly it called Ozzard's filaria.

History: Manson, 1897 and Faust, 1929 were contributed in discovery of parasite.

Morphology: (1) Adult stage: Male is inadequately known and a single incomplete male worm has been found. Female is 7 cm × 0.25 mm in size. Cuticula appears smooth. Posterior end is pointed and possesses a pair of flap like papillae. Female is viviparous and lays the embryo. **(2) Larval stage:** Microfilariae are 224 µm × 4–5 µm in size, unsheathed, tail tip is devoid of nuclei **(Fig. 108.1g)**, nonperiodic and present in the blood in human.

Life cycle: Definitive host is human. **Intermediate host** is midges specially *Culicoides furens.* Midges steps are same as other filaria.

Pathogenicity: It is prevalent in West Indies and parts of Central and South America. It is transmitted by bite of midges like *Culicoides furens.* Adult parasite habitats the mesenteric tissues. It is mostly asymptomatic or presents with utricaria, pruritus, lymph node swelling (lymphadenopathy) and occasionally hydrocele.

Laboratory diagnosis, prevention and treatment: Same as *Mansonella streptocerca.*

Mansonella perstans

History: Manson, 1891 and Railliet, Henry and Langeron, 1912 were contributed in discovery of parasite.

Morphology: (1) Adult stage: Female is 7 cm × 0.12 mm and male is 4 cm × 0.06 mm in size. Cuticula appears smooth. Anterior end has short cuticular shields called epaulettes. Posterior end is split and presents a cuticular thickening forming two triangular appendages. Female is viviparous and lays the embryo. **(2) Larval stage:** Microfilariae are 203 µm × 4–5 µm in size, unsheathed, nuclei present up to tail tip **(Fig. 108.1h)**, nonperiodic and present in the blood in human.

Life cycle: Definitive hosts are chimpanzee, gorilla and human. **Intermediate host** is midges specially *Culicoides grahami* and *C. austeni.* Midges steps are same as other filaria.

Pathogenicity: It is prevalent in South America and Africa. It is transmitted by bite of midges of *Culicoides* spp. Chimpanzee and gorilla are the reservoirs of infection. Adult parasite habitats the mesenteric tissues, pericardial and pleural cavities in humans. Mostly it is asymptomatic, but may cause lymphadenopathy, edema or allergic reaction in the skin. It may cause calabar swelling like *Loa loa.*

Laboratory diagnosis and prevention: Same as *Mansonella streptocerca.*

Treatment: Doxycycline (200 mg BID for 6 weeks) is the effective drug. It inhibits the endosymbiotic bacteria (*Wohlbachia* species) that are essential for the fertility of the worms.

Zoonotic Filariasis

Different species causing zoonotic filariasis are described separately in box.

> **Note: Zoonotic filariasis**
>
> **Introduction:** These are natural parasites of animals, but accidentally larvae enter in human by arthropod bite, which change to adult worms but not able to release the microfilariae. Later, adult worms die and inflammatory reaction around the adult worms gives all clinical manifestations.
>
> **Species: (1)** *Brugia pahangi*: It is a dog and cat parasite in Malaysia. In human it causes lymphangitis and lymphadenitis. **(2)** *Brugia beaveri*: It is a parasite of racoon in North America. In human it causes tender mass in cervical, axillary or inguinal region. **(3)** *Dirofilariae immitis*: Commonly it called **dog heart worm.** It is located in the right heart or branches of pulmonary artery in human. Dead parasite becomes embolus and blocks the small branches of pulmonary artery to produce infarct. Healed infarct appears as "coin lesion" under X-ray which can be mistaken as malignancy. Laboratory methods can't diagnose the parasite. It is diagnosed by serological tests and eosinophilia. X-ray method is insufficient. **(4)** *Dirofilaria repens*: It is a parasite of dog. In human it causes subcutaneous or subconjunctival nodules, so called *dirofilariae conjunctivae.* **(5) Other** *Dirofilaria* **spp., as human pathogen:** *D. tenuis, D. uris, D. striata* and *D. subdermata.*

OTHER INFECTIONS

Infections in Lungs

Lung infections are caused by larvae of *S. stercoralis, A. lumbricoides* and *A. duodenale.* All are described in **Ch. 107.**

Infection in Liver

Other species of genus is *Capillaria philippinensis* is described in **Ch. 107**. *Capillaria hepatica* is described separately in box.

Note: *Capillaria hepatica*

Meaning: It infects the liver capillaries, so called *hepatica*.

History: 1st case was reported in 1923 in a 20 years old British soldier in India at autopsy.

Morphology: Same as *C. philippinensis* **(Ch. 107)**

Life cycle: Definitive hosts are rat, squirrel, pig, chimpanzee, dog and monkey. **Intermediate host** does not required. Cycle continue between rat and other predators **(Flowchart 108.9)**. Human is an accidental host and acts as dead end host because no access of human tissues to animals.

Pathogenicity

- **Disease name:** It is called hepatic capillariasis.
- **Epidemiology:** Cases are reported from USA, England, Nigeria and South America.
- **Reservoirs of infection:** Rat and other animals
- **Sources of infection:** Food and water
- **Modes of transmission:** Human is infected by ingestion of food or water contains eggs.
- **Portal of entry:** GIT
- **Sites:** Eggs habitat in liver in human.
- **Exit and infective form:** Egg.
- **Pathogenesis:** Inflammation of liver (hepatitis) followed by fibrosis, cirrhosis and hepatomegaly.
- **Clinical features:** Hepatitis presents with abdominal pain in the liver area, weight loss, decreased appetite, fever and chills.
- **Complication:** Ascites and hepatolithiasis (gallstone in bile ducts).

Laboratory diagnosis: (1) Blood picture: Eosinophilia. **(2) Microscopy of liver biopsy** shows the eggs.

Flowchart 108.9: Life cycle of *C. hepatica*

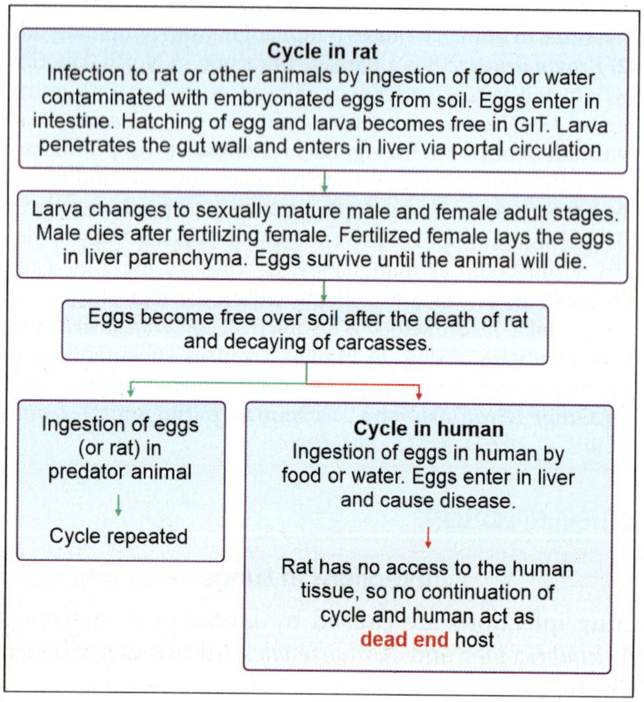

Cycle in rat
Infection to rat or other animals by ingestion of food or water contaminated with embryonated eggs from soil. Eggs enter in intestine. Hatching of egg and larva becomes free in GIT. Larva penetrates the gut wall and enters in liver via portal circulation

Larva changes to sexually mature male and female adult stages. Male dies after fertilizing female. Fertilized female lays the eggs in liver parenchyma. Eggs survive until the animal will die.

Eggs become free over soil after the death of rat and decaying of carcasses.

Ingestion of eggs (or rat) in predator animal

Cycle repeated

Cycle in human
Ingestion of eggs in human by food or water. Eggs enter in liver and cause disease.

Rat has no access to the human tissue, so no continuation of cycle and human act as **dead end** host

Infection in Brain

Other species of genus is *Angiostrongylus costaricensis* which is described in **Ch. 107**. *Angiostrongylus cantonensis* is described separately in box.

Note: *Angiostrongylus cantonensis*

Common name: Commonly it called rat lung worm.

History: Disease was 1st discovered from the CSF of a patient with eosinophilic meningitis by Nomura and Lim in Taiwan in 1944–45. Life cycle was worked out by Mackerass and Sanders in 1955.

Morphology: Adult stage and egg stages are same as *A. costaricensis* as described in **Ch. 107. Total 3 larval stages** are described along with life cycle in **Flowchart 108.10 and Fig. 108.8.**

Life cycle: Definitive hosts are brown rat (*Rattus norvegicus*) and the black rat (*Rattus rattus*). Human is an accidental host and acts as dead end host. **Intermediate host** is snail or mollusc (*Thelidomus aspera*). **Cycle** is shown in **Flowchart 108.10 and Fig. 108.8.**

Pathogenicity

- **Disease name:** It is a rat pathogen and human disease called cerebral angiostrogyliasis or eosinophilic meningoencephalitis.
- **Epidemiology:** Cases were reported from Taiwan, Thailand, Indonesia, Pacific Islands, India, Egypt, Cuba and USA.
- **Reservoir of infection:** Rat
- **Source of infection:** Raw slug or food
- **Mode of transmission:** Human is infected by ingestion of raw snail or food contaminated with snail/mollusc (3rd stage of larvae) or ingestion of crab/prawn contains 3rd stage of larvae.
- **Incubation period:** 2–3 weeks.
- **Exit form:** 1st stage of larva.
- **Infective form:** 3rd stage of larva.
- **Portal of entry:** GIT.
- **Sites:** Pathology in human is due to larval stage which habitats meninges and superficial plus deeper tissue of brain.
- **Pathogenesis:** Lesion occurs due to direct mechanical damage to neural tissue by migratory larva, toxic effect by products such as nitrogenous waste, release of antigens by dead and living parasite and inflammatory (eosinophilic) response in humans.
- **Clinical features:** Fever is often minor or absent, but the presence of high fever suggests severe disease. Bitemporal headache in frontal and occipital lobes, neck stiffness, photophobia, nausea with or without vomiting, paresthesia (tingling, prickling or numbing of skin), hyperesthesia (severe sensitivity to touch), bladder dysfunction with urinary retention, vertigo, blindness, local paralysis to one area, general paralysis (often ascending in nature starting with the feet and progressing upwards to involve the entire body), convulsion, coma and rarely death.
- **Complication:** Ocular damage is the possibility.

Laboratory diagnosis: (1) CSF picture: It looks like "aseptic meningitis" with slightly elevated protein level, normal or significantly decreased glucose is an indicator of severe meningoencephalitis and may indicate a poor prognosis. **(2) Blood picture:** Eosinophilia (90%). **(3) Brain biopsy:** Eosinophilia. **(4) Microscopic examination of CSF and brain biopsy:** Show the presence of larvae. **(5) Serological tests:** Ab is detected by ELISA against Ag of larva of *A. cantonensis*.

Prevention: Removal of intermediate host (snail or mollusc).

Treatment: Anthelmintics are not useful as the disease is caused by dead larvae. Drugs even may enhance the disease due to destruction of more larvae.

Flowchart 108.10: Life cycle of *A. cantonensis*

Cycle in rat
Infection to rat by ingestion of snail or mollusc or water contains 3rd stage larva. After entry in intestine larva released on digestion of snail or mollusc. Larva migrates through Portal circulation → Right heart → Pulmonary circulation Systemic circulation → Entry in brain and remains in subarachnoid space for 2 weeks → Entry in right ventricle and travels to the main branch of pulmonary artery where it lodges.

Changes to male and female adult stage. Male dies after fertilizing female. Fertilized female lays the eggs. Egg hatched out within 6 days to release 1st stage larva which breaches the alveoli and migrates through bronchus, trachea, larynx and finally enters in intestine

Larva passes with rat feces over soil
1st stage larva taken up by snail/mollusc

Cycle in snail/mollusc
1st stages taken up by snail/mollusc, it moults twice and changes to 2nd and 3rd stage of larva (infective form).

Infective snail/mollusc taken up by rat and cycle repeated

Cycle in human
Infection in human by ingestion of snail/mollusc or food contaminated with snail/mollusc (3rd stage larvae)

Larva penetrates the intestinal wall, enter in blood, carried to brain and produces the eosinophilic meningoencephalitis

Larva is not able to change to sexually mature adult stages in human unlike rat, so human acts as **dead end** host

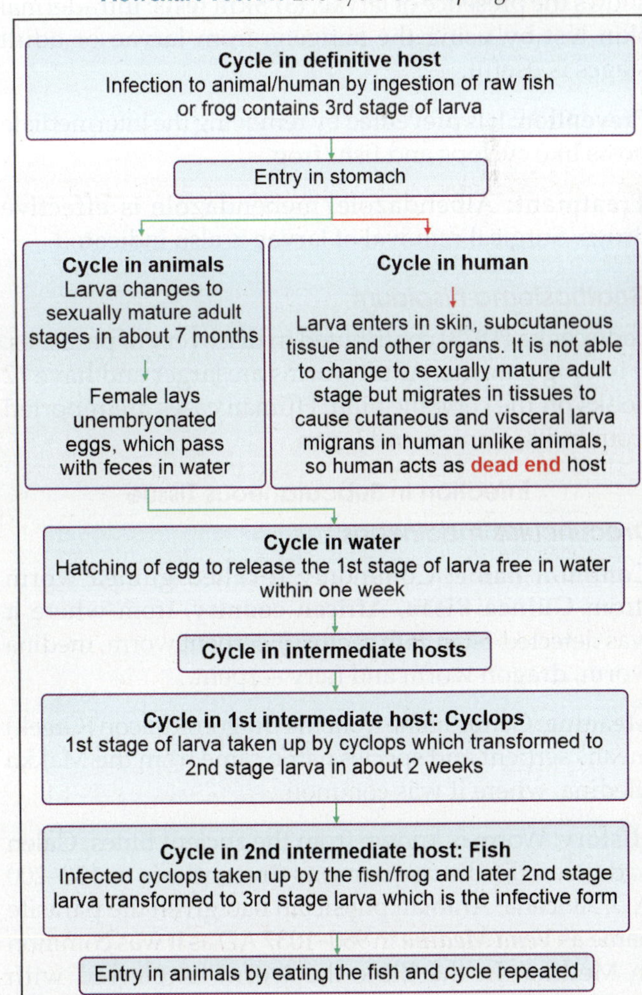

Fig. 108.8: Life cycle of *A. cantonensis*

Infection in Brain and Other Tissues

Gnathostoma spinigerum

Morphology: (1) Adult stage (Fig. 108.9a): Larva transformed to sexually mature adult stage in 18 months in animals. Female is 25–54 m × 3 mm and male is 21–25 mm × 2 mm in size. Parasite is curved ventrally at both the ends. Head end possesses 4–8 transverse rows of sharp hooklets on a distinct head bulb. Female contains single vulva situated in the middle of the body. Female is oviparous lays the eggs. **(2) Egg stage (Fig. 108.9b):** Egg is oval, transparent, 70 μm × 40 μm in size, brown color (bile stained) and contains mucous plug at one pole. Freshly passed eggs are not embryonated. **(3) Larval stage:** Total 3 larval stages are described along with life cycle in **Flowchart 108.11**.

Life cycle: Definitive hosts are domestic and wild animals. **1st intermediate host** is cyclops. **2nd intermediate host** is fresh water fish or frog. Whole cycle

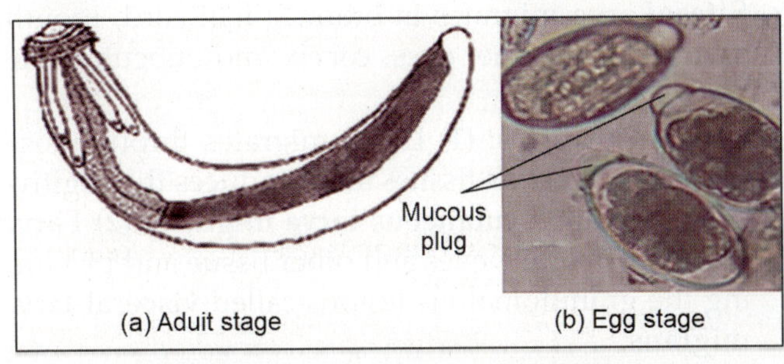

Fig. 108.9: Adult stage of *G. spinigerum*

Flowchart 108.11: Life cycle of *G. spinigerum*

Cycle in definitive host
Infection to animal/human by ingestion of raw fish or frog contains 3rd stage of larva

Entry in stomach

Cycle in animals
Larva changes to sexually mature adult stages in about 7 months

Female lays unembryonated eggs, which pass with feces in water

Cycle in human
Larva enters in skin, subcutaneous tissue and other organs. It is not able to change to sexually mature adult stage but migrates in tissues to cause cutaneous and visceral larva migrans in human unlike animals, so human acts as **dead end** host

Cycle in water
Hatching of egg to release the 1st stage of larva free in water within one week

Cycle in intermediate hosts

Cycle in 1st intermediate host: Cyclops
1st stage of larva taken up by cyclops which transformed to 2nd stage larva in about 2 weeks

Cycle in 2nd intermediate host: Fish
Infected cyclops taken up by the fish/frog and later 2nd stage larva transformed to 3rd stage larva which is the infective form

Entry in animals by eating the fish and cycle repeated

is possible between animals and intermediate host as shown in **Flowchart 108.11.** Human is an accidental host and acts as dead end host.

Pathogenicity

- **Disease name:** It called gnathostomiasis.
- **Epidemiology:** It is prevalent in Central East Asia, Japan, China, New Guinea and India.
- **Mode of transmission:** By ingestion of raw fish/frog contains third stage of larva.
- **Reservoir of infection:** Animals
- **Source of infection:** Fish/frog
- **Exit form:** Unembryonated egg
- **Infective form:** Third stage of larva
- **Portal of entry:** GIT
- **Sites:** Larva migrates in brain, spinal cord, mouth, pharynx, intestine, eyes, cervix and subcutaneous tissues.
- **Clinical features: (1)** Larva migrates through skin and subcutaneous tissues and produces the fugitive swelling called **cutaneous larva migrans. (2)** Larva migrates in brain, eyes and other tissue and producing the granulomatous lesions called **visceral larva migrans.**

Laboratory diagnosis: (1) Blood picture: Eosinophilia. **(2) Microscopical examination biopsy from lesions:** It shows the presence of larvae. **(3) Skin tests:** Intradermal skin test by using the antigens from larvae or adult stages is useful.

Prevention: It is prevented by removing the intermediate hosts like cyclops and fish/frog.

Treatment: Albendazole/mebendazole is effective drugs. Surgical removal of larvae is also indicated.

Gnathostoma hispidum

Fedtscenka (1873) contributed in discovery of parasites. It is a pig parasite. Adult worms are larger and have 12 hooks on the cephalic bulb. Human cases are reported from India, Canton and Japan.

Infection in Subcutaneous Tissue

Dracunculus medinensis

Common name: Commonly it called guinea worm (from **Guinea Bisau, African country,** from where it was detected 1st in 16th century), serpent worm, medina worm, dragon worm and fiery serpent.

Meaning: Genus name from the dragon/dracon (Greek) means serpent and species name came from the Makka Medina, where it was common.

History: Worm is known from the ancient times. Galen had given the disease name as **dracontiasis** in 130–200 AD. Avicena, Arabian physician had given the parasite name as *Vena Medina* in 980–1037 AD as it was common in Medina. In the Bible the worm is referred with common name as **"fiery serpent".**

Morphology: (1) Adult stage (Fig. 108.10): Larva transformed to sexually mature adult stage in 3–4 months in definitive host after ingestion. Female is 0.6–1 meter 1.5–1.7 mm (longest nematode) and male is 12–30 mm × 0.4 mm in size. It is cylindrical like a thread and milky-white in color. Male lives for not more than 6 months and dies after fertilizing the female. Male has been discovered from experimental animals but not from human except one case in India. Female may live for 1 year. The head end rounded and in female tail end is tapered and forms a hook. Female is **viviparous** and lays the embryo for a period of 3 weeks until the uterus of gravid female becomes empty. **(2) Larval stage:** Solid form of larva is present. It is 650–750 μm × 17–20 μm in size at the widest part, longer with blunt head and pointed tail, transparent and colorless. It released free in water by gravid female and lives for 4–5 days unless taken up by cyclops.

Life cycle

- **Hosts: Definitive host** is human. **Intermediate host** is cyclops. **In India** species is *Mesocyclops leuekarti.* **Other species** are *Mesocyclops hyalinus, Thermocyclops vermifer, Encyclops serrulatus, Tropocyclops multivolor,* etc.
- **Cycles:** As shown in **Flowchart 108.12** and **Fig. 108.11.**

Pathogenicity

- **Disease name: It called dracunculiasis** or **dracunculosis** or **dracontiasis** or **guinea worm disease.**
- **Epidemiology:** It is prevalent in African countries. It was eradicated from India and South East Asia region by 2000 and these countries are certified as guinea worm free by WHO.
- **Reservoir of infection:** Human
- **Source of infection:** Water contains infected cyclops
- **Mode of transmission:** By ingestion of contaminated water contains infected cyclops which is neither boiled nor filtered.
- **Exit form:** Coiled embryo
- **Infective form:** Larva
- **Portal of entry:** GIT

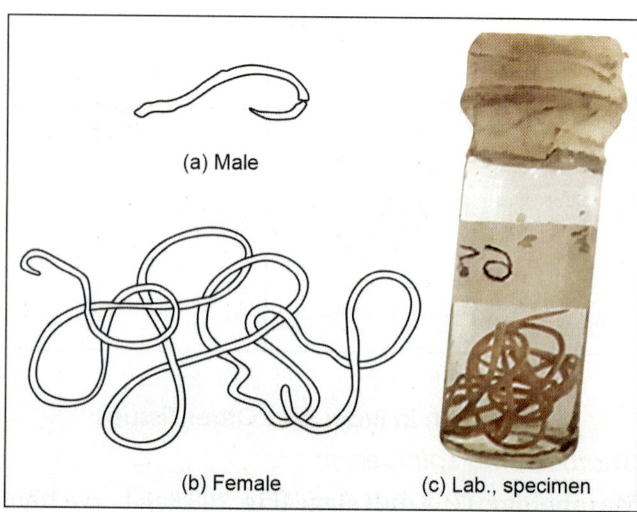

Fig. 108.10: Adult stage of *D. medinensis*

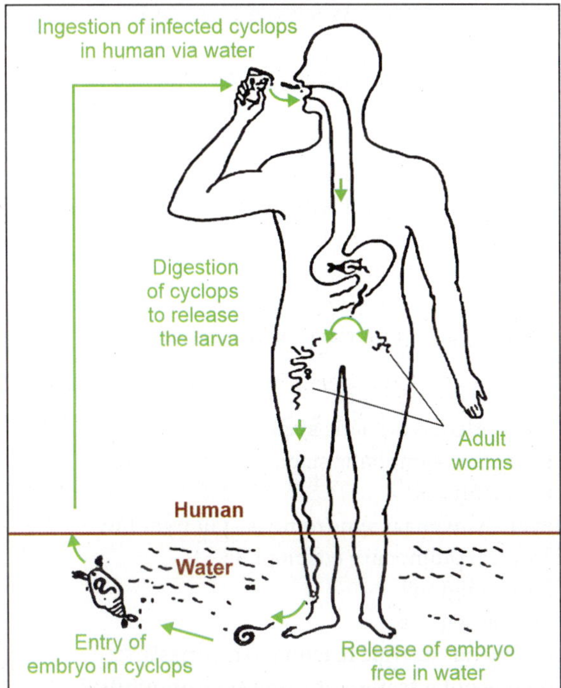

Fig. 108.11: Life cycle of *D. medinensis*

Flowchart 108.12: Life cycle of *D. medinensis*

Cycle in human
Infection to human by ingestion of water contains infected cyclops. Cyclops digested by gastric juices and larva become free in stomach. Larva penetrates the gut wall and enters in retroperitoneal connective tissues. Changes to sexually mature male and female adult stages in 5–18 months in lymph nodes

↓

Male dies after fertilizing female and disappears within 6 months. Fertilized female further grows and migrates through connective tissues and settles in to those body parts, which are repeatedly come in contact with water like legs, hands, back specially in water carrier (water carrier called **Bhistis** in India)

↓

Female liberates the toxic fluid which produces the blister. Blister ruptures to form the ulcer. When ulcer comes in contact with water, female protrudes through the center of ulcer and discharges the milky fluid containing coiled embryos (larvae), ejected from the prolapsed uterus.

↓

Released embryo (larva) enters in water and taken up by cyclops

Cycle in cyclops

Each cyclops is ingested by 15–20 embryos at a time. Normal life span of cyclops is 3 months but with such infection cyclops can live for 42 days while with heavy infection they do not live for more than 15 days

↓

Embryo penetrates the gut wall after 1–6 hours of ingestion and enters in body cavity of cyclops, where it undergoes further development with increase in size about 1 mm and within 1–2 weeks (extrinsic incubation period) it converted in to the infective form

↓

Entry of infected cyclops in human by ingestion of water and cycle repeated

- **Sites:** Adult female habitats the subcutaneous tissues of legs, arms and back.
- **Precipitating factors:** It is common in Bhistis.

- **Pathogenesis:** Symptoms are due to release of toxic fluid by female during parturition responsible for allergic manifestations and blister formation.
- **Clinical features: (1) Allergic reactions:** Burning or itching sensation at affected areas. **(2) Blister formation:** Initially lesion presents as red spot and later changes to bleb or blister. It presents on body parts which are repeatedly come in contact with water like legs (in between the metatarsal bones or on the ankles), hands and back especially in water carrier (Bhistis). **(3) Ulcer:** Rupture of blister produces the ulcer. Ulcer is 0.5–0.25 inches in diameter. In center of ulcer hole is present through which head of the worm will protrudes, when body parts come in contact with water. Toxic fluid present in ulcer is sterile, yellow in color and contains embryo, monocytes, eosinophils and neutrophils.
- **Complications: (1) Secondary bacterial infections in ulcer:** Sometimes, it may lead to tetanus. **(2) Worms travel to unusual sites:** Like in spinal canal, pericardium or in the eyes with serious effects. **(3) Calcification:** It presents in dead worm, which may cause arthritis, fibrosed joints or compression of spinal cord.

Laboratory diagnosis

A. **Blood picture:** Eosinophilia.

B. **Adult stage diagnosis:** It is examined direct over ulcer site and removed for diagnosis, prevention and treatment of disease. It is removed by **rolling over match stick (Fig. 108.12) or similar object.** There are chances of breaking of worm; so remove it very slowly, and roll it inch by inch daily. It takes about 15–20 days. Method is safe and effective. **Surgical removal under anesthesia** is also an useful method.

C. **Larval stage diagnosis:** Collect milky fluid from ulcer. If fluid is not present then bath the ulcer with sterile water which later collected as specimen. Larvae can be examined under microscope.

D. **X-ray:** It may reveal calcified worm.

Prevention: It is prevented by boiling or filtering the water by clothes (best way), destruction of cyclops by using abate (temephos) and cleaning the water reservoirs. Cyclops control measures are given in **Ch. 110.**

Treatment: Albendazole, niridazole and thiabendazole are effective in treatment.

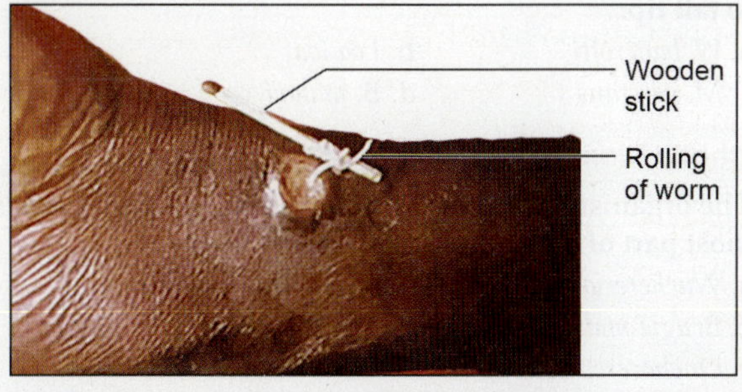

Fig. 108.12: Removal of adult worm by rolling over wooden stick

Case Study

1. A 55-year-old male presents with pain and swelling on left arm with fever. Examination revealed lymphangitis and enlargement of axillary lymph nodes. Blood picture shows the microfilariae with tail tip free from nuclei. Identify the case and answer the following.
   a. Name the causative agent and draw the diagram of microfilaria of causative agent.
   b. Describe the life cycle of causative agent.
   c. Write the pathogenicity of causative agent.

Essay/Full Question

1. Filariasis.

Short Notes

1. Microfilariae
2. Microfilariae of *W. bancrofti*
3. Life cycle/pathogenicity/lab., diagnosis of *W. bancrofti*
4. Classical filariasis/Occult filariasis
5. Darcunculiasis

Short Questions for Theory/Viva Questions

1. Name the four parasites in which microfilariae are present in peripheral blood.
2. Name the four parasites present in CSF.
3. Name the four parasites in which mosquito act as an intermediate host.
4. What is "parasitic onion" in wuchereriasis?
5. Difference between classical filariasis and occult filariasis.
6. What is hetrazan provocation test?
7. Name the four parasites in which fish act as a source of infection.

Comment on

1. Hetrazan provocation test is a useful tool in diagnosis of microfilariae of *W. bancrofti*.

MCQs for Chapter Review

General Aspect of Filariasis

1. Nocturnal periodicity of microfilaria is in relation with their occurrence in:
   a. Peripheral blood
   b. Urine
   c. Lymph
   d. Internal organs

2. Sheathed microfilariae is/are:
   a. *W. bancrofti*
   b. *Loa loa*
   c. *M. perstans*
   d. *B. malayi*

3. About *Wuchereria bancrofti* true is:
   a. Unsheathed
   b. Tail tip free from nuclei
   c. Non periodic
   d. All

4. One of the following microfilariae do not posses nuclei up to tail tip:
   a. *W. bancrofti*
   b. *Loa loa*
   c. *M. perstans*
   d. *B. malayi*

Lymphatic Filariasis

5. The organism most commonly causing genital filariasis in most part of Bihar and eastern Uttar Pradesh is:
   a. *Wuchereria bancrofti*
   b. *Brugia malayi*
   c. *Onchocerca vouvulus*
   d. *Dirofilaria*

6. Which is false about *Wuchereria bancrofti*?
   a. Causes filariasis
   b. Body is long and slender
   c. Terminal nuclei absent
   d. Human and *Anopheles* are hosts

7. False regarding filariasis is:
   a. Morbidity increased with age in endemic areas
   b. Humoral immunity plays dominant role
   c. Usually unilateral
   d. Human is the only host for filariasis.

8. In which stage of filariasis microfilariae seen in peripheral blood?
   a. Tropical eosinophilia
   b. Early adenolymphatic stage
   c. Late adenolymphatic stage
   d. Elephantiasis

9. Meyers Kowenaar syndrome is synonym for:
   a. Tropical pulmonary eosinophilia
   b. Larva migrans
   c. Occult filariasis
   d. Cutaneous allergic reaction to ascariasis

10. Which is not a feature of tropical eosinophilia?
   a. Eosinophilia more than $3000/mm^3$
   b. Microfilariae in tissues
   c. Microfilariae in blood
   d. Lymphadenopathy

11. Hydrocele and edema in foot occur in:
   a. *W. bancrofti*
   b. *B. malayi*
   c. *B. timori*
   d. *Onchocerca vouvulus*
   e. Guinea worm

12. All of the following are true about *Brugia malayi, except*:
   a. Intermediate host in India is *Mansonia*
   b. The tail is free from nuclei
   c. Nuclei are blurred, so counting is difficult
   d. Adult worm founds in lymphatic system

Subcutaneous Filariasis

13. River blindness is caused by:
   a. *Onchocerca*
   b. *Loa loa*
   c. *Ascaris*
   d. *B. malayi*

14. Kallu, a 30-year-old man, presented with subcutaneous itchy nodules over the left iliac crest. On examination they are firm, non-tender and mobile. Skin scraping contains microfilariae and adults worms of:
   a. *Loa loa*
   b. *Onchocerca vouvulus*
   c. *Brugia malayi*
   d. *Mansonella perstants*

15. Calabar swelling is produced by:
   a. *Onchocerca vouvulus*
   b. *Loa loa*
   c. *Brugia malayi*
   d. *Wuchereria bancrofti*

Dracunculus medinensis

16. Dracunculosis infection occurs through:
   a. Ingestion of water containing cyclops
   b. Ingestion of water containing parasite
   c. Ingestion of fish
   d. Penetration of skin

17. Definitive host for guinea worm is:
   a. Man
   b. Cyclops
   c. Snail
   d. Cyclops and human

Answers and Explanation of MCQs

1. a

2. a, b, d

3. b

4. a

- Follow section, **general aspect of filariasis (morphology → larval stage and Fig. 108.1)** for explanation of answers of MCQs 1–4.

5. a

6. b, d

7. b, d

- Follow section, **lymphatic filariasis (*Wuchereria bancrofti*)** for explanation of answers of MCQs 5–7.

8. b

- Tropical eosinophilia (feature of occult filariasis), late adenolymphatic stage and elephantiasis are features of obstructive phase **(Flowchart 108.3)**, where microfilariae are not able to come out from tissue in to the blood due to obstruction of lymphatic vessels and remain in tissues only.

9. c

10. c

- Follow section, **lymphatic filariasis (*Wuchereria bancrofti* → pathogenicity → occult filariasis)** for explanation of answers of MCQs 9–10.

11. a

- Hydrocele and edema in foot are the feature of *W. bancrofti* from the given options. Follow section, **lymphatic filariasis (Table 108.1)** for explanation.

12. b

- Follow section, **lymphatic filariasis (*Brugia malayi* → habitat and Table 108.1)** for explanation.

13. a

14. b

- Follow section, **subcutaneous filariasis (*Onchocercus volvulus*)** for explanation of answers of MCQs 13–14.

15. b

- Follow section, **subcutaneous filariasis (*Loa loa*)** for explanation.

16. a

17. a

- Follow section, ***Dracunculus medinensis*** for explanation of answers of MCQs 16–17.

C H A P T E R **109**

Miscellaneous Infections of Parasites

Chapter Outline
- Larva Migrans
- Parasites causing Autoinfection
- Parasites to Remember

LARVA MIGRANS

When certain helminths (mainly nonhuman like zoophilic helminths) enter in human, their further progress is arrested at certain stage, and they migrate in aimless manner in human body with production of clinical lesions called **larva migrans.** They are of two types according to mode of entrance like **cutaneous larva migrans** and **visceral larva migrans.** Differences between two are given in **Table 109.1.**

Cutaneous Larva Migrans

Definition: It is the migration of larvae in the skin.

Synonym: It also called **creeping eruption.**

Etiological agents
- **Rapidly moving larvae**: *Strongyloides stercoralis* (**Ch. 107**).
- **Slowly moving larvae:** *Ancyclostoma braziliensis* (**Ch. 107**), *Ancyclostoma caninum* (**Ch. 107**), *Gnathostoma spinigerum* (**Ch. 108**) and maggots (myiasis → described separately in box).

Common sites: Feet, legs, thighs, buttocks and back.

Clinical features: It presents with serpigenous tunnel formation at the path traversed by larvae. Intense irritation and eosinophilic infiltration are present in larval track. Mostly all larvae will die.

Diagnosis: It based on clinical ground. Rarely larvae found in local biopsy and also rarity of eosinophilia. Serological tests are not useful.

Prevention: Body is protected from larvae by using protective clothes and repellents.

Treatment: Local (15% cream) and oral thiabendazole is effective. Freezing of advancing part of the eruption with ethyle chloride is effective. Surgical removal of lesions is also advised.

Note: Myiasis

Definition: Human disease caused by migrating fly maggots called **myiasis.**

Agents: Female fly of genus *Hypoderma bovis* (cattle botfly), *Gastrophilus intestinalis* (horse botfly) and *Dermatobia hominis* (human botfly), *Chrysomyia* (Old World blow fly), *Chochiliomyia* (New World screw-worm fly), *Cordylobia* (mango fly), *Wohlfarhtia*, etc., deposit either eggs or larvae (maggots) directly in different body parts like nose, eye, ear, intestine, etc. Larvae settle on surface for easy oxygen exchange. Larvae produce the pathological lesions or secondary bacterial infection also.

Visceral Larva Migrans

Definition: It is the migration of larvae in the deeper tissues or organs.

Etiological agents: *Angiostrongylus costaricensis* (**Ch. 107**), *A. cantonensis* (**Ch. 108**), *Gnathostoma spinigerum* (**Ch. 108**), *Brugia patei* (**Ch. 108**), *Brugia pahangi* (**Ch. 108**), *Dirofilaria immitis* (**Ch. 108**), *Anisakis* (anisakiasis → described separately in box) spp., *Toxocara canis* and *Toxocara cati* (toxocariasis → described separately in box)

Note: Anisakiasis

Definition: Human disease caused by *Anisakis* species called **anisakiasis** or **herring worm.**

TABLE 109.1: Differences between CLM and VLM		
Features	**CLM**	**VLM**
Organs involved	Skin	Organs like liver, lungs, spleen, etc.
Portal of entry	Skin	Ingestion
Eosinophilia	Mild	High
Serology	Negative	Positive
Treatment	Thiabendazole	DEC and thiabendazole

(Contd…)

Agents: These are parasites of family Anisakidae like *Ascaris*. Human infection is from four genus *Anisakis, Pseudoterranova, Contracaecum* and *Phocanema*.

Hosts: Definitive host (Adult stage) for these parasites are sea mammals like dolphins, whales, etc. Shrimps is the 1st intermediate (2nd larval stage) host and fish is the 2nd intermediate (3rd larval stage) host.

Pathogenicity: It is prevalent in Japan, Netherland and USA. Human infection occurs due to ingestion of raw fish. On entering in human GIT, larvae penetrate the stomach wall and produce the eosinophilic granuloma which is characterized by epigastric pain, nausea and vomiting.

Diagnosis: Endoscopic biopsy is performed for presence of larvae. ELISA is also useful.

Treatment: Surgical removal of granuloma and thiabendazole are effective.

Note: Toxocariasis

Definition: Human disease caused by *Toxocara* species called toxocariasis.

Agents: These are parasites of dog (*Toxocara canis*) and cat (*Toxocara cati*).

Pathogenicity: Human infection occurs from ingestion of eggs contain 2nd stage of larvae. In intestine eggs are hatched to release the larvae which invade the intestinal wall and enter in liver, spleen, lungs, eyes, brain and in other organs. Further progress of larvae is arrested by formation of granuloma. Granuloma contains eosinophils, neutrophils, giant cell and macrophages. It presents with fever, hepatomegaly, spleno-megaly, pneumonitis, endophthamitis, space-occupying lesion in brain, convulsion, etc.

Diagnosis: (1) Biopsy of lesions shows the presence of larvae. (2) Clinical examination of eyes is also useful. (3) Serological tests are positive.

Treatment: DEC and thiabendazole are effective.

PARASITES CAUSING AUTOINFECTION

Synonym: Reinfection.

Definition: Reinfection by pathogen that is already present in body or infection caused by transfer of pathogen from one part of body to another.

Types: Two types. (1) Internal reinfection: Reinfection by parasites from inner side of body. (2) External reinfection: Reinfection by parasites from outer surface of body.

Examples: Parasites with mode of autoinfection are described below.

1. *Cryptosporidium parvum:* Thin walled oocyst, not excreted in external environment, but infects the epithelial cells, as like fresh infection.
2. *Taenia solium:* Entry of gravid segments in stomach from intestine by reversal of peristalsis or may be by contamination of fingers.
3. *Hymenolepis nana:* It is due to persistence of few eggs in intestine rather than exit in feces.

4. *Strongyloides stercoralis:* Occurs by two ways. **Internal reinfection** occurs by penetration of bowel mucosa by filariform larvae and **external reinfection** occurs by penetration of perianal (perineal) skin by filariform larvae.
5. *Enterobius vermicularis:* It is common in children by anus to mouth route. Scratching the perianal region may contaminate the fingers with eggs and swallowing of such fingers especially in children cause the autoinfection.
6. *Capillaria philippinensis:* **Internal reinfection** occurs due to penetration of bowel mucosa by liberated larvae.

Life cycle of parasite causing autoinfection: Life cycles of parasites in which autoinfection in human is possible are described with **sky blue arrow** (↓) in respective chapters.

Diagnosis and treatment: Follow respective chapters.

PARASITES TO REMEMBER

Largest protozoan: *Balantidium coli.*

Metazoa or helminths or worms:
- Largest/longest helminth or cestode: *T saginata.*
- Largest trematode: *F buski.*
- Largest or longest nematode: *D medinensis.*
- Largest liver fluke: *F. hepatica.*
- Smallest or shortest nematodes: *S. stercoralis* and *T. spiralis.*
- Smallest or shortest cestode: *H. nana.*

ACCESS YOURSELF

Short Notes

1. Larvae migrans.
2. Autoinfection by parasites.

Short Questions for Theory/Viva Questions

1. Name the four parasites causing cutaneous larva migrans.
2. Name the four parasites causing visceral larva migrans.
3. Name the four parasites causing autoinfections.

MCQs for Chapter Review

Larva Migrans

1. **Eosinophilic meningoencephalitis is caused by:**
   a. *Gnasthestoma spinigerum*
   b. *Naegleria*
   c. *Toxocara canis*
   d. *Angiostrongylous cantonensis*
2. **Cutaneous larva migrans is due to:**
   a. *A. braziliensis*　　　b. *W. bancrofti*
   c. *B. malayi*　　　d. *D. medinensis*
3. **Visceral larva migrans is associated with:**
   a. *Strongyloides stercoralis*
   b. *Ancyclostoma braziliensis*
   c. *Toxocara canis*
   d. *Visceral leishmaniasis*

4. **The cause of larva currens:**
   a. *Strongyloides stercoralis*
   b. *Necator americanus*
   c. *Ancyclostoma duodenale*
   d. *H. nana*

Parasites Causing Autoinfections

5. **Autoinfection is seen with:**
   a. *Cryptosporidium*   b. *Strongyloides*
   c. *Giardia*   d. *Ascaris*
6. **Autoinfection is a mode of transmission in:**
   a. *Trichinella*   b. Cysticercosis
   c. *Ancyclostoma*   d. *Ascaris*

Parasites to Remember

7. **Which of the following organism is biggest?**
   a. *Balantidium coli*   b. *Entamoeba coli*
   c. *Escherichia coli*   d. *Entamoeba histolytica*
8. **Largest intestinal protozoan is:**
   a. *E. coli*   b. *Balantidium coli*
   c. *Giardia*   d. *T. gondii*
9. **The largest trematodes infecting human is:**
   a. *F. hepatica*   b. *F. buski*
   c. *E. granulosus*   d. *C. sinensis*

10. **Which worm is longest?**
    a. *T. solium*
    b. *T. saginata*
    c. Hook worm
    d. *A. lumbricoides*

Answers and Explanation of MCQs

1. a, d
2. a
3. c
4. a
- Follow section, **larva migrans** for explanation of answers of MCQs 1–4.
5. a, b
6. b
- Follow section, **parasites causing autoinfection** for explanation of answers of MCQs 5–6.
7. a
8. b
9. b
10. b
- Follow section, **parasites to remember** for explanation of answers of MCQs 7–10.

Part III

Entomology, Practical Microbiology and MCQs

Entomology

Medical Entomology

Chapter Outline

INTRODUCTION

Definition

Study of the arthropods of medical importance is called **medical entomology.**

Clinical Significances

- They are human enemies and carry many microbes. They act as carriers or reservoirs or sources or vectors of microorganisms and transmit them to human.
- They also cause allergic reaction following bites and induce phobias.
- They help to fertilize the agricultural products.
- They may damage the agricultural products.

Classification of Arthropods

Morphological classification: There are 3 classes of arthropods of like **Insecta, Arachnida** and **Maxilopoda** as mentioned in **Table 110.1.**

TABLE 110.1: Morphological classification of arthropods			
Features	**Insecta**	**Arachnida**	**Maxilopoda**
Body parts	3 parts like • Head • Thorax • Abdomen	2 parts like • Cephalo-thorax • Abdomen	2 parts like • Cephalo-thorax • Abdomen
Antennae	1 pair	Absent	2 pairs
Legs	3 pairs	4 pairs	5 pairs
Wings and examples	• **1 pair:** Mosquitoes, flies and midges • **2 pairs:** Cockroach and reduviid bug • **Wingless:** Bed bug, louse and flea	Wingless like ticks and mites	Wingless like cyclops
Habitat	Soil	Soil	Water

Systemic classification: Follow **Table 110.2.**

Modes of Transmission of Arthropod (Vector) Borne Diseases

Follow **Ch. 47.**

MEDICALLY IMPORTANT VECTORS, MICROBES/DISEASES TRANSMITTED AND THEIR CONTROL MEASURES

Mosquito

Classification: Follow **Flowchart 110.1.**

Life cycle stages: Mosquito passes through the stage of egg, larva, pupa and adult stage.

Species and microbes/diseases transmitted by mosquitoes

- *Anopheles:* Total 45 species, of which only few are known to transmit the *Plasmodium* spp., (causing malaria) like *A. culcifacies* (In rural area, **Mnemonic: ll**) and *A. stephensi* (In urban area, **Mnemonic: nn**). Other species are *A. philipinensis, A. fluviatilis, A. varuna, A. minimus, A. annularis, A. leucosphyrus (A. balabacensis), A. sundaicus, A. jeyporiensis candidiensis* etc.
- *Aedes:* Arboviruses as described in **Ch. 83.**

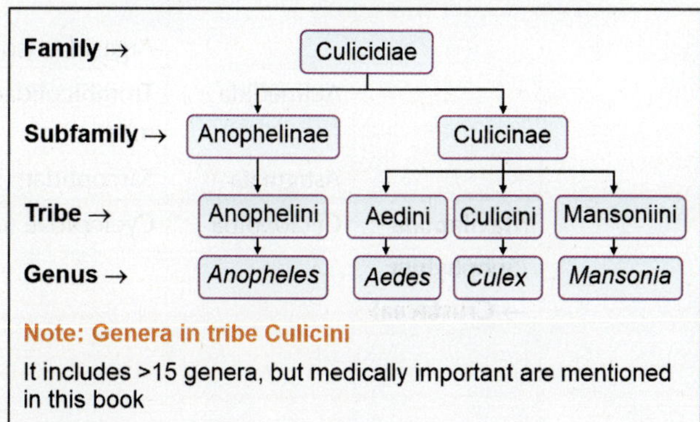

Flowchart 110.1: Classification of mosquito

Note: Genera in tribe Culicini

It includes >15 genera, but medically important are mentioned in this book

TABLE 110.2: Systemic classification of arthropods

Phylum	Class	Order	Family (common name)	Genus	Species
			Kingdom → Animalia		
Arthopoda	**Insecta**	Diptera	Culicidiae (mosquito)	*Anopheles*	All the species of arthropods, transmitting different microorganisms are mentioned with respective sections
				Aedes	
				Culex	
				Mansonia	
			Muscidae (house fly)	*Musca*	
			Glossinidae (tsetse fly)	*Glossina*	
			Psychodidae (sand fly)	*Phlebotomus*	
				Sergentomyia	
				Lutzomyia	
			Simuliidae (black fly)	*Simulium*	
			Tabanidae (deerfly, horse fly and yellow fly)	*Chrysops* (deerfly)	
				Tabanus (horse fly)	
				Diachlorus (yellow fly)	
			Calliphoridae (blow fly)	*Chrysomya*	
			Ceratopogonidae (midges)	*Culicoides*	
				Forcipomyia	
				Austroconops	
				Leptoconops	
		Blattaria	Blattidiae (cockroach)	*Blatta*	
		Hemiptera	Reduviidae (reduviid bug)	*Triatoma*	
			Cimicidae (bed bug)	*Cimex*	
		Phthiraptera	Pediculidae (head louse and body louse)	*Pediculushumanus capitis* (head)	
				Pediculushumanus corporis (head louse)	
			Phthiridae (pubic louse)	*Phthirus*	
		Siphonaptera	Pulicidae (flea)	*Xenopsylla*	
				Nosopsylla	
				Ctenocephalus	
				Pulex	
				Tunga	
	Arachnida	Ixodida	Ixodidae (hard tick)	*Dermacentor*	
				Haemaphysalis	
				Amblyoma	
				Ixodes	
				Rhipicephalus	
				Hyaloma	
			Argasidae (soft tick)	*Ornithodorus*	
		Actinedida	Trombiculidae (trombiculid mite)	*Trobicula*	
		Astigmata	Sarcoptidae (itch mite)	*Sarcoptes (Acarus)*	
	Maxilopoda (Subphylum → Crustacea)	Cyclopoida	Cyclopidae (cyclops)	*Cyclops*	
				Mesocyclops	
				Thermocyclops	
				Tropocyclops	
				Encyclops	

- *Culex* (nuisance mosquito)
  - *Culex fatigans* (*Culex quinquefasciatus*): *Wuchereria bancrofti* (Bancroft's filariasis) in India.
  - Arboviruses: **Follow Ch. 83.**
- *Mansonia:* *Mansonia annulifera, M. Indiana* and *M. uniformis* transmit *Burgia malayi* (malayan filariasis) in India.

Mosquito control measures
A. Anti-larval measures

- **Physical or environmental control:** Elimination mosquito breeding sites like filling, levelling and drainage of breeding places. Removal of aquatic plants to control the *Mansonia*.
- **Biological control:** (1) Use of fish like *Gambusia affinis, Lebister reticulates,* etc., which eats the larvae. (2) *Bacillus thuringenesis* subsp., *israelensis* (*Bti*) or S-methoprene are larvicides. (3) Commercial products "mosquito dunks" and "mosquito bits" containing *Bti* can be used at home.
- **Chemical control: (1) Mineral oils:** Like diesel, kerosene, crude oil etc., have direct toxic effect on larvae. All these oils can spread in water and make a thin film which cut the air supply to the larvae and pupae. Life cycle of mosquitoes is 8 days, so oil must to apply every week in all breeding places. It kills fish and also makes water unfit for drink. **(2) Paris green or copper acetoarsenite:** Paris green is stomach poison, so it should be ingested by larvae for desire effect. Good sample of paris green contains 50% of arsenious oxide. It is applied as 2% dust (20–25 microns particles) by using hand blower or rotary blower. 2% dust contains 2 kg paris green plus 98 kg diluents such as soapstone powder or slaked lime in a rotary mixture. Advised dose is 1 kg of actual paris green per hectare of water surface. It is a crystalline water insoluble powder and it kills the surface larvae like *Anopheles*, for bottom feeding larvae it should be applied in granular form. It does not ham fish, man and water quality. **(3) Organo phosphorous (OP) compounds:** They hydrolysed quickly in water. Few examples are abate (56–112 g/hectare), malathion (224–672 g/hectare), fenthion (22–112 g/hectare) and chlorpyrifors (11–16 g/hectare). Organochlorine compounds like DDT, HCH are not used as larvicides, because of long residual effect, water contamination and chances of resistance in the vector mosquitoes.

B. Anti-adult measures
- **Residual spray**
  - **DDT** is most effective and used as 1–2 g/square meter. Effect lasts for 6–12 months, so it is applied 1–3 times in a year to walls and other surfaces where mosquitoes are rest.
  - **DDT resistant areas:** DDT resistant areas should by sprayed by other insecticides. **(1) Malthion:** 2 g/square meter and effect lasts for 3 months. **(2) Gamma-HCH/lindane:** 0.5 g/square meter and effect lasts for 3 months. **(3) Propoxure/OMS33:** 2 g/square meter and effect lasts for 3 months.
- **Space spray:** It includes spraying of mist or fog to kill the insects. **(1) Pyrethrum extract (pyrethrin):** It is extracted from pyrethrum flowers. The active ingredient is pyrethrin. It is applied in the dose of 1 oz (contains 1% pyrethrin) of the spray solution/1000^3 feet area. Close door and windows for half an hour. For small scale, it is applied by hand gun with fine nozzle but for large scale aerosol dispenser or power sprayer is required. It is the nerve poison and kills insect on mere contact. It has no residual effect, so reinfestation from outside source will occur in short time. **(2) Residual insecticides:** Malathion and fenthion are useful. They are sprayed by ultra low volume sprayer.
- **Genetic control:** It includes genetic methods (under trial) of mosquito control like sterile male technique, cytoplasmic incompatibility, chromosomal translocations, sex distortion and gene replacement.

C. Protection against mosquito bite
- **Mosquito net:** Ordinary net is less effective. Chemically impregnated net with gamma-cyhalothrin is more effective. Selection criteria of nets are material should be in white color for easy detection of mosquito and the size of hole should not be >0.0475 inch in any diameter.
- **Mosquito repellents:** They are applied on a skin for short-term protection and prepared from different chemicals like diethyltoluamide/deet (outstanding in all purpose and effective for 18–20 hours against *Culex fatigans*), indalone, dimethyl phthalate, dimethyl carbate, ethyl hexanediol, etc.
- **Screening:** Screening of building with copper or bronze gauze having 16 meshes to the inch is recommended. The size of aperture should not be >0.0475 inch in any diameter. It is costly, but gives excellent result.

House Fly

Life cycle stages: Four stages like egg, larva, pupa and adult.

Species and microbes/diseases transmitted by house fly: Major species of flies available in India are *Musca domestica, M vicinia, M nebulo* and *M sorben*. House fly transmits the disease by different ways. **(1) Mechanical transmission:** Adult fly is restless insect and moves to and forth between foods; which allows the mechanical transmission (by wings or other body parts) of disease like enteric fever (typhoid and paratyphoid), cholera, shigellosis, *E. coli* diarrhea, amoebic dysentery, intestinal helminth diseases, conjunctivitis, trachoma, anthrax, yaws, hepatitis, etc. **(2) Defecation:** Fly is defecating frequently whole day, so depositing bacteria over food.

Medical Entomology

9

Species and microbes/diseases transmitted by reduviid bug: Reduviid bugs like *Panstrongylus megistus*, *Triatoma infestans*, *Triatoma dimidiata* and *Rhodnius prolix* are vectors for *Trypanosoma cruzi*, which causes Chagas disease/ South American trypanosomiasis (**Ch. 95**).

Reduviid bug control measures

- **Physical or environmental control:** (1) Elimination of bug's breeding places by improvement in housing quality and environment. (2) Bug bite is avoided by using mosquito net or repellents.
- **Chemical control:** Residual spray by using HCH ($0.5\ g/m^2$) or dieldrin ($1\ g/m^2$) is commonly used to control the bug.

Bedbug

Life cycle stages: Three stages like egg, nymph/larva and adult.

Species and microbes/diseases transmitted by bedbug

- *Cimex lactularius* and *Cimex hemipterus*: Are infected with bacteria and viruses. Hepatitis virus has been reported from West Africa, South America and Ethiopia with evidences.
- Bedbug bite produces inflammation and allergic reaction in host who are sensitive to bedbug saliva.

Bedbug control measures

- **Physical or environmental control:** Domestic cleaning and treating of mattresses an hour at a temperature of 45°C (113°F) or over, or two hours at less than –17°C (1°F) are effective.
- **Biological control:** The fungus *Beauveria bassiana* is able to control bedbugs.

Louse

Life cycle stages: Three stages like egg (nit), nymph (larva) and adult.

Types: Louse is the permanent ectoparasite. It is of three types (**Fig. 110.1**) like **head louse** (*Pediculus humanus capitis*), **body louse** (*Pediculus humanus corporis/Pediculus humanus humanus*) and **pubic louse/crab louse** (*Pthirus pubis*).

Head Louse

Species and microbes/diseases transmitted by head louse: Head louse itself acts as an ecto-parasite and causes a disease called **pediculosis** which is described separately in box. It also transmits *Rickettsia prowazekii* (epidemic typhus).

Head louse control measures

- **Physical control:** It includes avoiding contact with infested person or use of items used by infested person, regular checking of child's hairs, regular bathing with soap and water, washing all items in hot water, pressing with hot iron and improvement in health education.
- **Chemical control:** (1) **Lotion** prepared from malthion (0.5%) and keeping for 12–24 hours on hair followed by washing is effective. **(2) Louse dust:** Prepared from carbaryl is effective.

> **Note: Pediculosis**
>
> **Etiological agent:** Head louse (*Pediculus humanus capitis*) body louse (*Pediculus humanus corporis/Pediculus humanus humanus*)
>
> **Source and reservoir of infestation:** Only human host.
>
> **Precipitating factors:** (1) Age: Common in children. (2) Sex: Females are infested more often than males, probably due to long hairs. (3) Socioeconomic status: Body lice are usually seen in settings of poverty, war, overcrowding and homelessness.
>
> **Modes of transmission: (1)** Louse is transmitted by **direct head-to-head contact**. Contact is common during play (sports activities, playgrounds, at camp, and slumber parties) at school and at home. **(2) Indirect** transmission occurs by sharing common articles **(fomites borne)** like clothes, hats, scarves, sports uniforms, hair ribbons, combs, brushes, towels, bed, pillow, carpet, etc. It is less common with head louse and more common with body louse.
>
> **Clinical features:** The majority of head louse infestations are asymptomatic. When symptoms are noted, they may include a tickling feeling of something moving in the hair, itching, caused by an allergic reaction to louse saliva, irritability and dermatitis. People with long lasting disease have darkened and thickened skin called **Vagabond's disease.**
>
> **Complication:** Secondary bacterial infection.

Body Louse

Body louse is similar to head louse with following clinical differences.

- It habitats on body and on clothes and causes pediculosis on body parts with above mentioned features.
- Body louse can serve as vector for *Borrelia recurrentis* (louse-borne relapsing fever) and *Bartonella quintana* (trench fever).
- Powder containing 1% malthion is the treatment of choice. It applied on inner surface of clothes, socks and on body. About 50 g powder required for one person. After this treatment clothes are not removed. Single application eradicates the all lice, but sometimes 2nd application require after 7 days.
- Dust prepared from carbaryl is also effective.
- "Mass delousing" is carried out for large numbers of people with hand dusters.

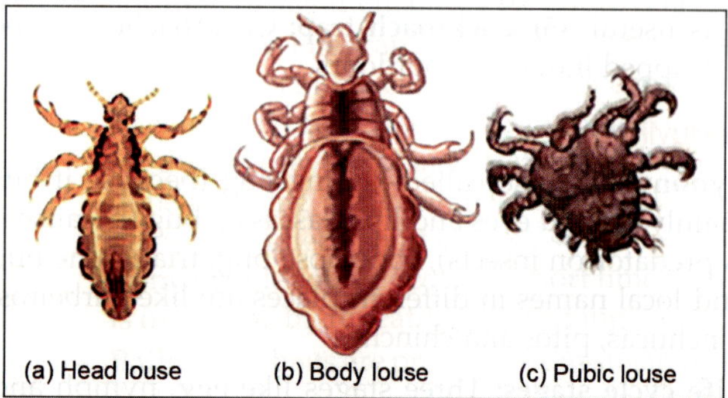

Fig. 110.1: Types of lice

Pubic Louse or Crab Louse

Crab louse is similar to head louse with few differences like it causes pediculosis on pubic parts (sexually transmitted), and it does not transmit any organisms. Autoclaving of clothes required to control body louse.

Fleas

Life cycle stages: Four stages like egg, larva, pupa and adult.

Types: Following are the different types of fleas.
- **Oriental (tropical) rat flea:** *Xenopsylla cheopis*, *X astia* and *X braziliensis*.
- **Northern (temperate) rat flea:** *Nosopsylla fasciatus*.
- **Cat flea:** *Ctenocephalides/Ctenocephalus felis*.
- **Dog flea:** *Ctenocephalides/Ctenocephalus canis*.
- **Human (house) flea:** *Pulex irritans*.
- **Sand flea (chigoe flea or jiggers):** *Tunga penetrans*.

Oriental (Tropical) Rat Flea

Species and microbes/diseases transmitted by Oriental (tropical) rat flea: *Xenopsylla cheopis* is a vector for *Yersina pestis* (plague), *Rickettsia typhi/Rickettsia mooseri* (endemic typhus/murine typhus/flea-borne typhus), *Hymenolepis nana* (hymenolepiasis) and *Hymenolepis diminuta* (hymenolepiasis diminuta).

Rat flea control measures: (1) Insecticidal spray or dust or powder: Dusting by 10% DDT or HCH or dieldrin is most widely used and cheapest way to control fleas. When rodents pass over the dust, they pick it up on the fur which kills fleas. Flea resistant to DDT or HCH or dieldrin, has been developed which is over come by using carbaryl, diazion (2%) and malthion (5%). It is applied to walls and floors up to a height of about 1 foot, over rat runs, rat burrows by using dust blower at about 30 g per burrow and cat plus dog resting places. **(2) Repellents:** Diethyltoluamide and benzyle benzoate are effective. Clothes impregnated with these repel fleas for >1 week.

Northern (Temperate) Rat Flea

Species and microbes/diseases transmitted: *Nosopsylla fasciatus* found on domestic rats (Norway rat *Rattus norvegicus*) and house mice. It has occasionally been observed feeding on humans and wild rodents. It is a minor vector for *Yersinia pestis* (plague), *Rickettsia typhi* (endemic typhus) and intermediate host of *Hymenolepis diminuta* (hymenolepiasis diminuta) in South America, Europe and Australia.

Cat Flea

Species and disease transmitted: *Ctenocephalides/Ctenocephalus felis* found on domestic cats. Also found on dogs and humans. It transmits *Rickettsia felis*/feline rickettsiae (endemic typhus/murine typhus/flea borne typhus) and *Dipylidium caninum* (dipylidiasis intermediate host). Very rare pathogens encountered by this flea are *Bartonella henselae* (cat-scratch disease) and *Borrelia burgdorferi* (agent of Lyme disease, but their ability to transmit the disease is unclear).

Dog Flea

Species and microbes/diseases transmitted: *Ctenocephalides/Ctenocephalus canis*, found on dogs. Also found on cats and humans. Dog fleas can transmit *Dipylidium caninum* (dipylidiasis → intermediate host).

Human (House) Flea

Species and microbes/diseases transmitted: *Pulex irritans* found on humans, pig, goat, sheep, cats, babbons, dogs, etc. Human flea transmits *Hymenolepis nana* (hymenolepiasis) and *Dipylidium caninum* (dipylidiasis → intermediate host).

Sand Flea (Chigoe Flea or Jiggers)

Synonym: Nigua, chica, pico, pique or suthi.

Species: *Tunga penetrans* (*Sarcopsylla penetrans* or *Pulex penetrates*) found on humans, cats, birds, monkeys, dogs, etc. It presents in tropical and subtropical climates. It has been reported from Western part of India. The fleas normally occur in sandy climates, including beaches, stables and farms. It flourishes in sandy soil.

Pathogenicity
- **Modes of transmission and disease name:** Both males and females feed intermittently on their host, but only mated females burrow by legs and proboscis into the skin (epidermis) of the host, where it causes a nodular swelling called **tungiasis (Fig. 110.2)**.
- **Clinical features**
  - Initial 1–2 days: Host may feel an itching or irritation but no pain.
  - Later stage: Inflammation and ulceration may become severe, and multiple lesions in the feet can lead to difficulty in walking. As the flea's abdomen swells with eggs later in the cycle, the pressure from the swelling may press neighboring nerves or blood vessels. Depending on the exact site, this can cause sensations ranging from mild irritation to serious discomfort.
- **Complication:** Secondary bacterial infections, including tetanus and gangrene.

Ticks

Life cycle stages: Four stages like egg, larva, nymph and adult.

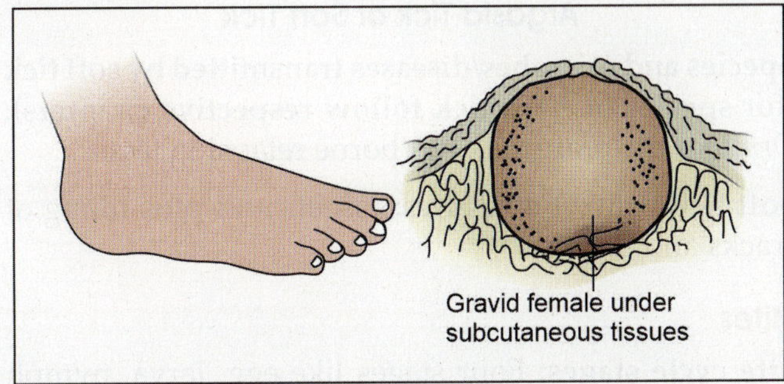

Fig. 110.2: Gravid female of sand flea under subcutaneous tissue

TABLE 110.3: Differences between hard tick and soft tick

Features	Hard tick	Soft tick
Morphology		
Scutum	Present	Absent
Head	Situated at anterior end	Lies ventrally, not seen from above
Life cycle		
Egg	100–1000 laid down at a time	20–100 over long time in different batches
Nymph	One stage	Five stage
Habits		
Breeding	Always on their host	Hide in crevices and cracks in day
Feeding	Feed in day and night. Cant withstand starvation	Feed in night. Can withstand starvation for years or more

Types: Two types like **ixodid tick or hard tick and argasid tick or soft tick.** Differences between two are mentioned in **Table 110.3.**

Ixodid Tick or Hard Tick

Species and microbes/diseases transmitted by hard tick (for species of hard tick follow respective chapters):

- *Dermacentor:* Rocky Mountain spotted fever, Colorado tick fever, tularemia, Siberian tick typhus, Omsk hemorrhagic fever and Central European tick-borne encephalitis.
- *Rhipicephalu:* Rocky Mountain spotted fever and boutonneuse fever.
- *Amblyomma:* Human monocytic ehrlichiosis, Rocky Mountain spotted fever and boutonneuse fever.
- *Ixodes:* Tularemia, Lyme disease, babesiosis, human granulocytic ehrlichiosis, and Russian spring-summer encephalitis.
- *Haemaphysalis:* Kyasanur forest disease.
- *Hyalomma:* Siberian tick typhus and Crimean-Congo hemorrhagic fever.

Hard tick control measures: (1) Insecticidal spray or dust: Dusting or spraying of DDT or dieldrin or lindane or malthion in animal places gives good result. **(2) Repellents:** Diethyltoluamide and benzyle benzoate or permethrin are effective. Clothes impregnated with these agents should be worn while working in tick infested areas.

Argasid Tick or Soft Tick

Species and microbes/diseases transmitted by soft tick (for species of hard tick follow respective chapters): *Ornithodorus* transmits tick borne relapsing fever.

Soft tick control measures: Insecticides plus filling of cracks and crevices

Mites

Life cycle stages: Four stages like egg, larva, nymph and adult.

Types: Two types **trombiculid mite** and **itch mite.**

Trombiculid Mite

Species and microbes/diseases transmitted by trombiculid mite: Species of trombiculid mite are *Leptotrombidium deliensis* (in India) and *Leptotrombidium akamushi* (in Japan). Larva (called **chigger**) of trombiculid mite is a vector of *Orientia tsutsugamushi* (scrub typhus or chigger borne typhus). It also transmits the *Rickettsia akari* (rickettsial pox).

Trombiculid mite control measures: These are same as hard tick.

Itch Mite

Species and disease by itch mite: *Acarus scabiei* (previously called *Sarcoptes scabiei*) itself acts as an ecto-parasite and causes a disease called **scabies,** which is described separately in box. It is not reported as vector for any microbes.

Note: Scabies

Etiological agent: *Acarus scabiei* (previously called *Sarcoptes scabiei*).

Synonym: Mange, crusted scabies, Norwegian scabies, seven-year itch.

Pathogenicity

- **Source and reservoir of infestation:** Only human host.
- **Precipitating factors:** It is common with overcrowding and poor hygiene.
- **Modes of transmission:** It is transmitted (1) by direct skin-to-skin contact with infested persons. Theoretically, touching an object that a mite is on is also a mode of transmission; however, this is not at all common. (2) rarely by sharing of clothes and bedding (fomites borne).
- **Incubation period:** Usually 2–6 weeks; however, in individuals with prior exposure to scabies, it is much shorter up to 1–4 days (Markell and Voge).
- **Sites:** Parasite burrows into the upper skin layer, never below the stratum corneum. It penetrates the thin skin with 63% in hands and wrists, 11% on elbows, 9% on feet and ankles, 12% in genital areas, and 2% in armpits. In women mites also affects the breast.
- **Pathogenesis:** Movement of mites produces an intense itch that may resemble to allergic reaction. A DTH reaction to the mites, their eggs, or scybala (packets of feces) occurs approximately 30 days after infestation. The presences of the eggs produce a massive allergic response that, in turn, produces more itching. Individuals who already are sensitized from a prior infestation can develop symptoms within hours.
- **Clinical features: (1) Common features** are superficial burrows **(Fig. 110.3a)**, intense pruritus (itching) and a generalized rash. **(2)** Acropustulosis, or blisters and pustules **(Fig. 110.3 b)** on the palms and soles of the feet, are characteristic symptoms of scabies in infants. **(3)** In immunocompromised, malnourished or elderly individuals, infestation can cause a more severe form of scabies called **crusted scabies or Norwegian scabies (Fig. 110.4)**. This syndrome is characterized by a scaly rash, slight itching and thickened crusts of skin contains thousands of mites. It is hardest to treat.
- **Complication:** Secondary bacterial infection.

(Contd…)

Diagnosis: It is best diagnosed clinically. Few laboratory tests are useful. **(1) Skin scraping and microscopy:** A drop of oil (enhance the stickiness of eggs to scalpel blade) or saline is placed on top of the affected skin area. A scalpel is then used to scrape the area of tissue samples, and the material is examined under the microscope for mites or eggs. **(2) Felt-tip marker test:** A washable felt-tip marker is drawn across the rash, followed by an alcohol wipe. This procedure identifies burrows because the ink penetrates deeply into the skin.

Treatment: (1) The topical medication: 5% Permethrin cream (elimite) or 10% crotamiton (eurax) cream is suggested for infants less than 2 months of age. 1% Gamma benzene hexachloride (lindane) causes neurotoxicity, especially in kids, so not recommended. 6% Sulfur ointment is effective. 5% Tetmosol applied 3 times in a day. **(2) Systemic:** Ivermectin (2 oral doses of 200 µg/kg is effective.

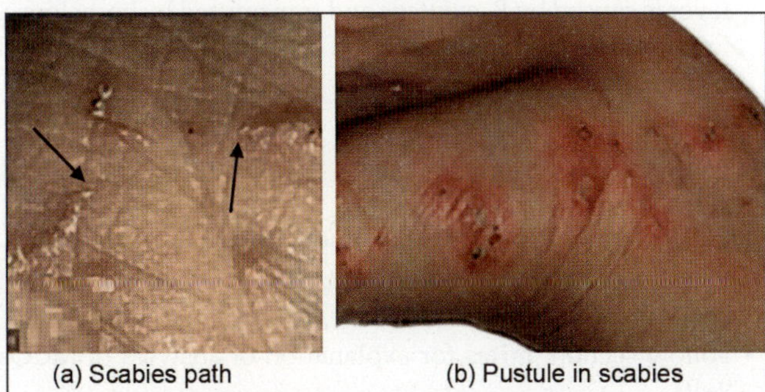

(a) Scabies path   (b) Pustule in scabies

Fig. 110.3: Clinical features of scabies

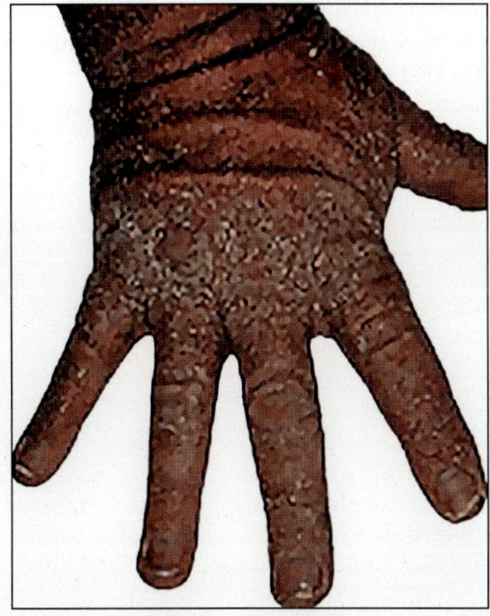

Fig. 110.4: Crusted scabies

Cyclops (Water flea)

Life cycle stages: Four stages like egg, larva or nauplii, copepodite and adult.

Species and microbes/diseases transmitted by cyclops

- **Species:** *Mesocyclops leukarti* (In India for *Dracunculus medinensis*), *Mesocyclops hyalinus*, *Thermocyclops vermifer*, *Encyclops serrulatus* and *Tropocyclops multivolor*.

- **Microbes/diseases transmitted:** Cyclops act as an intermediate (harbors larvae) host for (1) *Dyphylobothrium latum* (procercoid larvae). (2) *Spirometra* spp (procercoid larvae).(3) *Dracunculus medinensis* (larvae). (4) *Gnathostoma spinigerum* (3rd stage larvae).

Cyclops control measures

- **Physical or environmental control:** Filtering of water through fine cloth is enough to remove cyclops. Cyclops can easily killed by boiling water at 60°C. Other measures include provision of drinking water through piping water supply, use of tube wells and abolition of step wells.

- **Chemical control:** Chlorine destroys guinea worm larvae and cyclops in strength of 5 ppm. Drawbacks of this concentration of chlorine are bad odor and bad taste to water. Excess chlorine can be removed by dechlorination. Lime can be used in the dose of 4 grams/gallon of water. Temefos kills cyclops at concentration of 1 mg/liter.

- **Biological control:** Small fish like *Barbel* and *Gambusia* eat the cyclops.

ACCESS YOURSELF

Essay/Full Question

1. Role of insects in health and diseases.

Short Notes

1. Pediculosis
2. Scabies

Short Questions for Theory/Viva Questions

1. Write scientific name for jiggers and chiggers.
2. Name the four parasites in which cyclops act as an intermediate host.

MCQs for Chapter Review

Mosquito

1. *Anopheles* **is a vector of:**
   a. Relapsing fever   b. Malaria
   c. Babesiosis   d. Typhus fever

Sand Fly

2. **Vector of kala azar is:**
   a. Flea   b. Tsetse fly
   c. Sand fly   d. Mite

Reduviid bug

3. **Reduviid bug is vector for the transmission of:**
   a. Relapsing fever   b. Lyme disease
   c. Scrub typhus   d. Chaga's disease
4. **Vector of *T. cruzi*:**
   a. Reduviid bug   b. Tsetse fly
   c. Sand fly   d. Hard tick

Louse

5. **Lice are not vector for:**
   a. Relapsing fever   b. Q fever
   c. Trench fever   d. Epidemic typhus

6. **Babesiosis is transmitted by:**
   a. Tick
   b. Mites
   c. Flea
   d. Mosquito

7. **Soft tick transmits:**
   a. Relapsing fever
   b. KFD
   c. Tick typhus
   d. Tularemia

8. **Tick is a vector for:**
   a. Crimean Congo hemorrhagic fever
   b. Rocky mounted spotted fever
   c. Endemic typhus
   d. Scrub typhus

Mites

9. **Mite transmits:**
   a. Scrub typhus
   b. Trench fever
   c. Endemic typhus
   d. Epidemic typhus

10. **Scrub typhus is transmitted by:**
    a. Reduviid bug
    b. Tromiculid mite
    c. Enteric pathogen
    d. Cyclops

Cyclops

11. ***Dracunculus medinensis* is transmitted by:**
    a. Cyclops
    b. House fly
    c. Tick
    d. Flea

12. **Cyclops are intermediate host for:**
    a. Kala azar
    b. Schistosomiasis
    c. *Dracunculus medinensis*
    d. Taeniasis

13. **The slender larvae of which of the following helminths move about in water and are ingested by species of cyclops?**
    a. *D. latum*
    b. *D. medinensis*
    c. *W. bancrofti*
    d. *S. mansoni*

1. **b**
   - Relapsing fever is either tick borne or louse borne, babesiosis is tick borne and typhus fever has different types and transmitted by louse, flea, tick, or by mite.

2. **c**
   - Follow section, **sand fly** for explanation.

3. **d**
   - Relapsing fever is either tick borne or louse borne, Lyme disease is transmitted by *Ixodes tick* and scrub typhus is transmitted by bite of trombiculoid mite's larva called **chigger** (hence disease called **chigger-borne typhus**).

4. **a**
   - Remember mnemonic **CCB**: **C**haga's disease, **C**ruzi and **B**ug.

5. **b**
   - Lice are the vector for relapsing fever, caused by *B recurrnetis* (also can be tick borne). Q fever is caused by *Coxiella burnetii* and vector involved in transmission is *Ixodes tick*. Trench fever is caused by *B. quintana* and transmitted by body louse. Epidemic typhus is caused by *R. prowazekii* and transmitted by head louse.

6. **a**

7. **a**

8. **a, b**
   - Follow section, **ticks** for explanation of answers of MCQs 6–8.

9. **a**

10. **b**
    - Follow section, **mites** for explanation of answers of MCQs 9–10.

11. **a**

12. **c**

13. **b**
    - Follow section, **cyclops (water flea)** for explanation of answers of MCQs 11–13.

Practical Microbiology (DOAP/Skill Assessment)

CHAPTER 111

Visit to Microbiology Department

Chapter Outline

- Introduction of Staff
- Visit to All Sections
- Instructions to Students

INTRODUCTION OF STAFF

It includes the introduction between students, all faculties, resident doctors, technicians and sweepers/peons with full name, qualification and designation. It also includes the knowing the place of chamber/office of all faculties.

VISIT TO ALL SECTIONS

- It includes the visit to all laboratory sections like Bacteriology laboratory, Virology laboratory, Mycology laboratory, Parasitology laboratory, Mycobacteriology laboratory, Serology laboratory, Molecular laboratory, Autoclave room, Media preparation room, Washing room for Petri dishes, bottles, test tubes, etc., BMW collection room, Museum, Practical class, etc.
- It also includes the preliminary information (identification) of major equipments used in Microbiology like autoclave, hot air oven, biosafety cabinet, incubator, PCR machine, RNA extraction machine, etc.
- Distribution of students in batches with respective teachers (batch in-charge/batch teacher/table teacher)

for practical teaching, journal signature purpose and to contact for any queries.

INSTRUCTIONS TO STUDENTS

General instructions for students include followings.

- Wearing apron (white coat), gloves, etc.
- Coming regularly in class with journal/log book and other essential accessories.
- Proper nail and hair hygiene (girls should put hair inside the apron).
- Prohibition for drinking and eating in practical class.
- Instructions to handle, use and care of microscopes and other equipments.
- Staining of slides on the staining rod over wash basin.
- Do not overuse the reagents.
- After use cap all reagents bottles and put them back in rack.
- Contact to batch in-charge in case of accidental spillage or exposure to culture/specimen.
- Put microscope in cup board after its use.
- Switching off all lights, fans, water tapes, etc., and hand washing before leaving the class.

CHAPTER 112

Microscopy and Staining

Chapter Outline

MICROSCOPY

History

In 1674 Antony van Leeuwenhoek, the Dutchman was draper, prepares the lenses and observed the diverse materials through it. In 1678, Robert Koch developed a compound microscope and confirmed Leeuwenhoek's observation. In 1931, Ernst Ruska et al., invented the electron microscope for which he won the Nobel Prize in Physics in 1986.

Types of Microscope

Commonly used microscopes in microbiology are as follows.

- **Light microscope**
- **Dark-ground (field) Illumination Microscope (DGIM).**
- **Phase-contrast Microscope (PCM).**
- **Fluorescent Microscope (FM) and staining.**
- **Electron Microscope (EM).**
- **Other types.**

Light Microscope

Synonym: Simple/compound/optical/bright-field microscope.

Principle: It produces a dark image against a bright background. Follow **Fig. 112.1** for light transmission and image creation.

Parts: Two parts **(Fig. 112.2)**.

- **Optical parts:** Two types of lenses.
  - **Objective lenses:** They are attached to revolving nosepiece. They are of three types like low power (10×), high power (40×) and oil immersion lenses (100×).
  - **Ocular lenses (eye piece):** Eye piece have magnification of 10×, 12× or 15×.

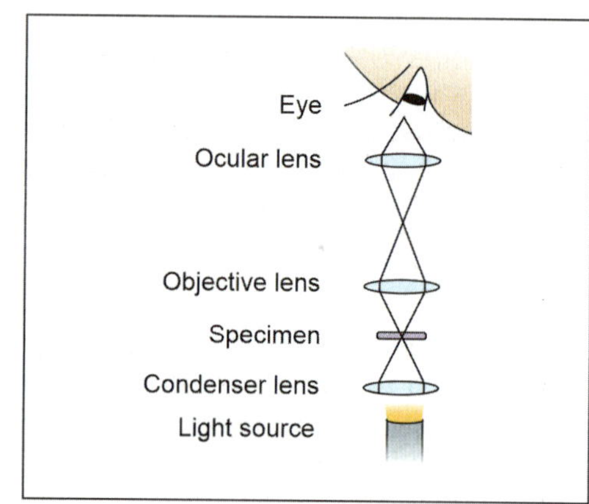

Fig. 112.1: Principle of light microscopy

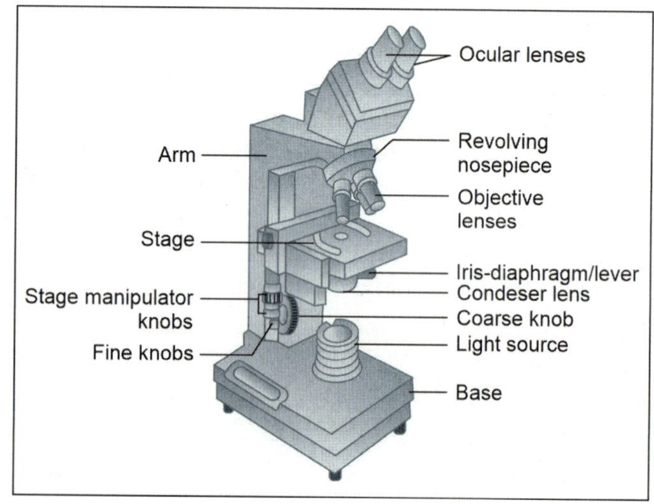

Fig. 112.2: Parts of light binocular microscope

- **Mechanical parts:** It includes base (stand), diaphragm, stage, etc., as shown in **Fig. 112.2**.

Magnification power of microscope: Degree of enlargement of image is called **magnification power of microscope**. For example eye lens (10×) × objective lens (10×) = Total magnification (100×).

Resolution power of microscope: Ability of a lens to distinguish two closely related structures called **resolution power of microscope**. It labeled as micron (µ) or micrometer (µm). Limit of resolution with unaided eye is 0.2 mm (200 µm or 2,00,000 nm), with light microscope is 0.0002 mm (0.2 µm or 200 nm) and with electron microscope is 0.0000005 mm (0.0005 µm or 0.5 nm). It depends on the refractive index of the medium. Oil has a higher refractive index than air, hence use of the oil enhances the resolution power of microscope.

Technique of use: The final magnification is the product of the magnification of the ocular and objective lenses. A slide is placed on the mechanical stage and is moved by rotating the stage control knobs. The height of the condenser may be varied (subjective variation) to give a bright, evenly illuminated field. Generally the condenser is used in its highest position for 100x and at lower level for 10x and 40x. A lever projects from the condenser and it is used to vary the opening of the condenser (or iris) diaphragm. For work with the scanning (4x) and low-power (10x) objectives, the condenser diaphragm should be wide open. For work with the high-dry (40x) and oil-immersion objectives (100x); however, the diaphragm should be closed slowly while looking at a sharply focused section until the level of illumination is just slightly reduced. This is the setting of the condenser diaphragm for optimum contrast and resolution (from a theoretical point of view this is not quite correct. The diaphragm should be adjusted for each magnification. In most instances; however, it is much less critical at the lower magnifications).

Uses

1. **Scanner (4×):** To view a large area at a glance.
2. **Low (10×) and high (40×) power:** For unstained preparations or wet mounts like KOH preparation, normal saline preparation, iodine preparation, hanging drop preparation, indian ink preparation, peripheral smear, etc.
3. **Oil immersion:** For stained preparations like Gram's stain, ZN stain, Albert stain, Giemsa stain, etc.

Care of microscope

- **Do's**
  - Keep at room temperature.
  - Keep it covered.
  - Clean frequently with soft camel hair brush, fine tissue paper, muslin silk or cotton cloth.
  - Clean the oil immersion lens with benzol/xylol.
- **Don'ts**
  - Use gauze piece and cotton for cleaning lenses.
  - Expose to direct sunlight.
  - Use alcohol or acetone for cleaning lens.

Dark-ground (Field) Illumination Microscope (DGIM)

Synonym: Dark-field microscope.

Principle: It produces a bright image of the object against a dark background. Dark-field condenser with a central circular stop illuminates the object with a cone of light, without allowing any light ray to fall directly on the objective lens. Light rays fall on the object are scattered or reflected on the objective lens, resulting illumination of object against dark background.

Uses: It is used to observe living organisms and parts of organisms in unstained preparations like (1) spirochetes, which are very thin and difficult to see under light microscope. They are better seen by DGIM. DGIM is also useful for serological test like TPI test, (2) flagella (motility of bacteria) and (3) sheathed microfilariae.

Phase Contrast Microscope (PCM)

Principle: It enhances the contrast between bacterial cells and surrounding medium with difference in refractive index. It also creates contrast between intracellular structures with slight differences in refractive index

Uses: It is used to study intracellular structures, bacterial components like endosome, inclusion body, etc., microbial motility, living cells and cell division.

Fluorescent Microscope (FM) and Staining

Introduction: Certain dyes called **fluors or fluorochrome** can be reached to high energy level from normal or low energy level after absorbing UV light. When dye back to normal or low energy level it release excess energy in the form of visible light. This process called **fluorescence.**

Types: Two types
A. Fluorochroming
- **Principle:** Dye combined with target bacteria and gives fluorescence **(Fig. 112.3a)**.
- **Advantages:** It increases the contrast. It is more sensitive (minimum concentration of organisms is 104/ml) than light microscopy (minimum concentration of organisms is 105/ml).
- **Dyes used are:** (1) **Acridine orange:** It stains the nuclei of all cells in smear. It does not discriminate between gram-positive and gram-negative cells. It also stains the host cells. **(2) Auramine/rhodamine:** It is used for *M. tuberculosis*. **(3) Calcofluor white stain:** It is used for fungus.

B. Immunofluorescence
- **Principle:** Dyes is conjugated with antibody. This conjugate will combine with antigen of target bacteria and gives fluorescence **(Fig. 112.3b)**.

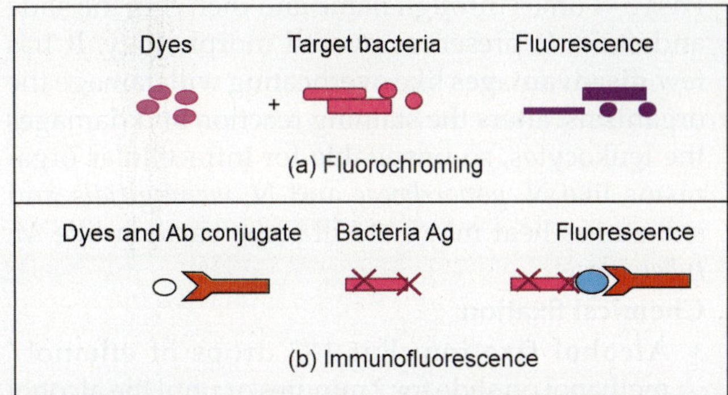

Fig. 112.3: Types and principle of fluorescent microscope (FM) and staining

- **Advantage:** More sensitive.
- **More detail: Ch. 20**

Electron Microscope (EM)

Principle: Beams of electrons are used instead of light (as used in light microscope) to produce images.

Types: Following two types

1. **Scanning Electron Microscope (SEM):** It uses the electrons reflected from the surface of a specimen to create the image. It produces a 3-dimensional image of specimen's surface features.
2. **Transmission Electron Microscope (TEM):** Electrons scatter when they pass through a specimen. Transmitted electrons (those that do not scatter) are used to produce image. Denser regions in specimen scatter more electrons and appear darker.

Uses: EM is used for diagnosis of virus and for detailed study of cell structure.

Other Types

These are interference microscope, polarization microscope, confocal microscopy, scanning probe microscopy, stereoscopic microscope, dissection microscope, etc.

STAINING

It includes three steps like smear preparation, fixation and staining, described below in detail.

Smear Preparation

Take one clean and dry slide. Label the slide with patient name and number. Take a drop of body fluid with sterile loop over slide and spread evenly covering an area about 15–20 mm. Hold the tissue with sterile forceps and rub over slide. Roll the swab on slide. Keep the slide on rack for drying. Avoid contact with dust, insect and sunlight during drying.

Fixation

It is a process by which organisms are killed and firmly attached to slide. It prevents the washing of microorganisms from smear during staining. Fixation of smear is done by following methods.

1. **Heat fixation:** Hold the smear with sterile forceps. Pass 3–4 times through flame and then cool the slide and stain. It preserves overall morphology. It has few **disadvantages** like overheating will damage the organisms, alters the staining reaction and damages the leukocytes, so unsuitable for intracellular organisms like *N. gonorrhoeae* and *N. meningitidis* and sometimes heat may not kill some bacteria like *M. tuberculosis*.
2. **Chemical fixation**
   - **Alcohol fixation:** Put 1–2 drops of ethanol/methanol on slide for 2 minutes or until the alcohol evaporates. It is less damageous to microorganisms and preserves morphology. It does not damage the leucocytes and therefore suitable for fixing smear contain intracellular organisms like *N. gonorrhoeae* and *N. meningitidis* (absolute alcohol is used). It kills bacteria like *M. tuberculosis* (70% alcohol is used).
   - **$KMnO_4$ (40 g/l):** For fixing *B. anthracis*.
   - **Formaldehyde vapor:** For fixing *M. tuberculosis*.

Staining

Two types like vital stains and supravital stains.

Vital stains: Stains which maintain the viability of organisms called **vital stains**. These are mostly unstained preparations as follows.

A. Required emulsification: By using NS.

- **Normal Saline (NS) wet mount from solid stool for parasites:** Take a drop of normal saline on slide. Add the stool particles. Mix with wooden stick. Put cover slip. Examine under low/high power. It is used for stool examination for parasites like *E. histolytica, T. vaginalis, G. lamblia*, etc.
- **Hanging drop preparation (Fig. 112.4b) from solid stool or culture:** Take a concavity slide (**Fig. 112.4a**) and clean it. Take a drop of normal saline on cover slip. Add the stool particles. Mix with wooden stick. Apply the adhesive jelly at all corner of cover slip. Invert the cover slip over concavity slide. Examine the edge of drop of NS under low power and than focus the motility under high power. It is used to detect the motility of bacteria and parasites in solid stool.

B. Not required emulsification

- **Direct wet mount:** Take a drop of liquid culture or body fluid (e.g. urine) on slide (called **direct wet mount**). Put a cover slip on it and examine under low/high power. It is used to examine inflammatory cells, deposits and motile organisms in body fluid.
- **Hanging drop preparation from liquid stool:** Take a drop of liquid stool over cover slip. Invert

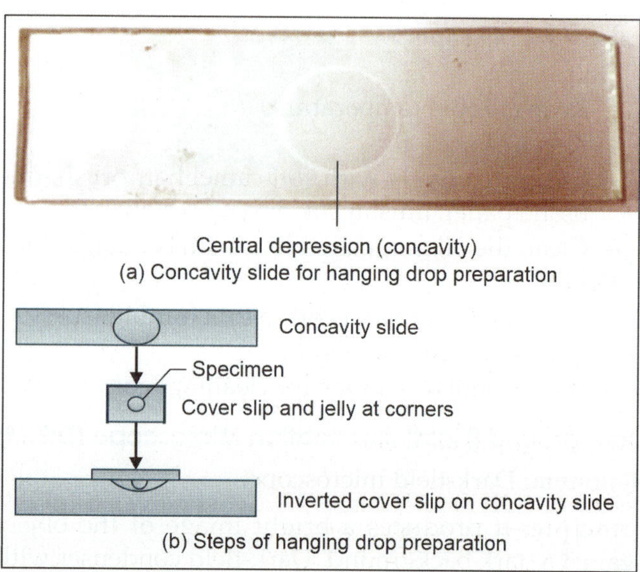

Fig. 112.4: Hanging drop preparation

the cover slip over concavity slide. Examine the edge of drop of liquid stool under low power and than focus the motility under high power. It is used to detect the motile bacteria in liquid stool like *V. cholerae*.

Supravital stains: Stains which do not maintain the viability of organisms called **supravital stains**. Different types are as follows.

A. **Simple staining (monochrome staining):** It is mostly one step staining. It stains all organisms in same color, however, provides color contrast between organisms and background. Following are the examples.
- **Wet mount stains:** Like iodine preparation for eggs of parasites from stool (**Ch. 121**), LCB stain and PHOL stain for fungi from culture.
- **Fix smear stains:** Methyle violet or methylene blue or basic fuchsin for bacteria.

B. **Negative staining (background staining):** Background is stained and structures to be demonstrated remain unstained. For example, Indian ink and nigrosin stains are useful for detection of bacterial capsule and spirochetes.

C. **Impregnation staining:** Very thin organisms are difficult to see under light microscope, and they are examined by increasing their thickness by applying the silver over surface. For example, Fontana's stain (contains silver nitrate) for spirochetes like *Treponema*, *Borrelia* and *Leptospira*.

D. **Fluorescent staining:** Described above.

E. **Differential staining:** It utilizes 2 different types of stains to distinguish between 2 different types of cells/ organisms or different parts of cells/organisms. For example, **histopathological stains** (for histological study of cells or tissues) like H and E stain, Giemsa stain, etc., and **microbiological stains** like Gram's stain (for gram-positive and gram-negative bacteria), acid fast stains (for acid fast microbes) and stains for polar bodies, fat globules, flagella and spores.

Gram's stain

History and meaning: So called because Hans Christian Gram, a histologist developed the technique to identify the bacteria in tissue in 1884.

Principle: Certain bacteria stained with basic dyes (like methyl violet/crystal violet/gentian violet for 5 minutes) followed by staining with mordant (like Gram's iodine for 1 minute), which fixes the dye in cell and prevents decolorization on subsequent treatment with decolorizing agent (like acetone for 1–2 seconds/alcohol for 20–30 seconds/acetone-alcohol for 10 seconds). Decolorized cells are finally stained by counter stain (like basic fuchsin/safranin).

Modifications: Following are the modifications in the original Gram's stain's method.

1. **KB (Kopeloff's and Beerman's) method/stain:** It uses the methyl violet as primary stain and basic fuchsin

as counter stain. It is best useful method nowaday for all bacteria and described separately in box.

2. **Jensen's modification:** It uses the absolute alcohol as decolorizer and neutral red as counter stain. It is used for *N. meningitidis* and *N. gonorrhoeae*.

3. **Weigert's modification:** It uses the aniline-xylol as decolorizer. It is useful for staining of tissue sections.

4. **Preston and Morrell's modification:** It uses the iodine-acetone as decolorizer. It is developed to overcome the irritating iodine in aerosols by reducing the iodine concentration to one-tenth and shortening the duration of decolorization to 10 seconds.

5. **Quick Gram's stain:** It includes the duration of 5 seconds for each primary stain, mordant and counter stain but 2 seconds for decolorizer. It is used for single slide.

6. **Gram's stain for multiple slides (10–15):** It includes the duration of 30 seconds for each step.

Note: KB (Kopeloff's and Beerman's) method/stain

History: Original Gram's stain was modified by Nicholas Kopeloff and Philip Beerman.

Principle: Same as Gram's stain described above.

Steps: Take clean and dry slide. Prepare the smear and fix with heat. Cover the smear part, not the entire slide with **primary stain** like methyl violet/gentian violet/crystal violet and keep for 5 min. Remove the primary stain and apply the iodine as **mordant** for 2 minutes. Wash and **decolorize** with 95–100% acetone (2–3 seconds) or ethanol (20–30 seconds) or acetone alcohol (10 seconds). Wash with tap water. Apply the basic fuchsin (or safranin) as **counter stain** and keep over smear for 30 seconds. Wash with tap water. Dry the slide. Add a drop of cedar wood oil and examine under oil immersion lens.

Action of ingredients: (1) **Methyl violet or crystal violet:** It is primary stain (basic dye). It stains gram-positive and gram-negative organisms in violet color. (2) **Iodine solution:** It acts as a mordant. It fixes the primary stain in the cell. (3) **Acetone or ethanol or acetone alcohol:** It is a decolorizing agent. It decolorizes only gram-negative organisms but not gram-positive organisms. (4) **Basic Fuchsin:** It acts as counter stain. Those organisms which are decolorized by acetone are counter stained by basic fuchsin.

Results: (1) **Gram-positive organisms:** Violet color (**Figs 112.5 and 112.6**). (2) **Gram-negative organisms:** Pink color (**Figs 112.7 and 112.8**).

Quality control: Always check new reagents with known gram-positive and gram-negative organisms.

Theories: Exact reasons for violet color in gram-positive and pink color in gram-negative organisms are not known, but few theories are responsible. (1) **pH Theory:** Gram-positive organisms have acidic pH and better stained with basic dye. Gram-negative organisms have basic pH and not stained by basic dye. (2) **Lipopolysaccharide (LPS):** LPS is present in gram-negative organisms. LPS is dissolved by lipid solvent like acetone, resulting damage to cell wall of gram-negative organisms and primary stain (violet color) will come out from cells, leave the gram-negative organisms colorless. Later gram-negative organisms will be stained in pink color by counter stain. **(3) Thick Peptidoglycan:** It is very thick in gram-positive organisms and cell wall remains unaffected to acetone and retains the original color of primary stain. It is very thin in gram-negative organisms and not able to retain the original color of primary stain. **(4) Magnesium ribonuclease theory:** Magnesium ribonuclease enzyme presents in gram-positive cells and makes dye-iodo-complex with dye and iodine. Complex is larger than the pores size of cell wall, hence not come out on treatment with acetone and maintains violet color in gram-positive cells. Enzyme is absent in gram-negative cells and no dye-iodo complex formation. Dye remains free in cytoplasm in gram-negative cells and easily come out on treatment with acetone.

Gram's variable reactions

1. **Gram-positive organisms may appear as gram-negative organisms** in following conditions
   - Cell wall damage due to antibiotic therapy or due to excessive heat fixation of smear.
   - Use of iodine solution which is too old (yellow instead of brown color, always store in brown glass bottle).
   - Over decolorization of smear.
   - Smear prepared from old culture like in *Cl. tetani.*
2. **Gram-negative organisms may appear as gram-positive organisms:** Thick smear gives incomplete decolorization of gram-negative organisms and retained the violet color.

Uses

1. **To classify the bacteria as gram-positive and gram-negative:** Follow **Table 112.1**.
2. **To identify the bacteria:** From culture growth.
3. **To identify the fungi:** Like *Candida* spp., or *Cryptococcus* spp.
4. **Early presumptive diagnosis:** It helps in early diagnosis of fastidious bacteria like *H. influenzae* which takes long time to grow on culture.
5. **To start the empirical therapy:** Broad spectrum antibiotic treatment can be started based on the shape, size and Gram's reaction of bacteria before the culture report will be available.
6. **Selection of anaerobic culture:** Preliminary diagnosis of bacteria like *Clostridium* spp., under Gram's stain gives ideas to perform anaerobic culture.

Advantages of Gram's stain: Cheap, fast and easy to perform.

Disadvantages (limitations) of Gram's stain

1. **Mycobacteria are not strict GPB:** In Gram's stain gram-positive bacteria give violet color following application of methyl violet, iodine, acetone and basic fuchsin. Mycobacteria resist decolorization even without use of iodine and stain in violet color. Hence mycobacteria are not following the complete principle of Gram's stain and thats why strictly it is not correct to classify mycobacteria as GPB.
2. **Bacteria difficult to examine under the Gram's stain**
   - *Treponema* and related genera like *Borrelia* and *Leptospira:* To thin to be seen.
   - *Mycobacterium:* High lipid content in the cell wall and detected by acid fast stain.
   - *Mycoplasma:* Cell wall deficient bacteria.
   - *Legionella:* Primarily intracellular.
   - *Rickettsia:* Intracellular and examined under Giemsa or other stains.
   - *Chlamydia:* Intracellular.

Mnemonic: <u>These Microbes May Lack Real Color</u>

Summary of Gram's stain: Follow **Table 112.2** and **Flowchart 112.1**.

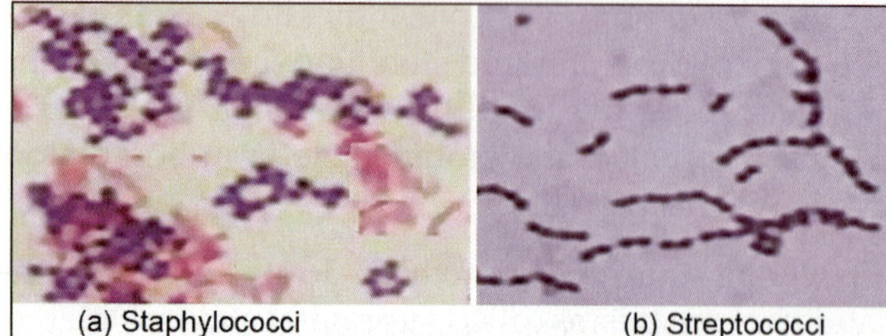

Fig. 112.5: Gram-positive cocci (GPC)

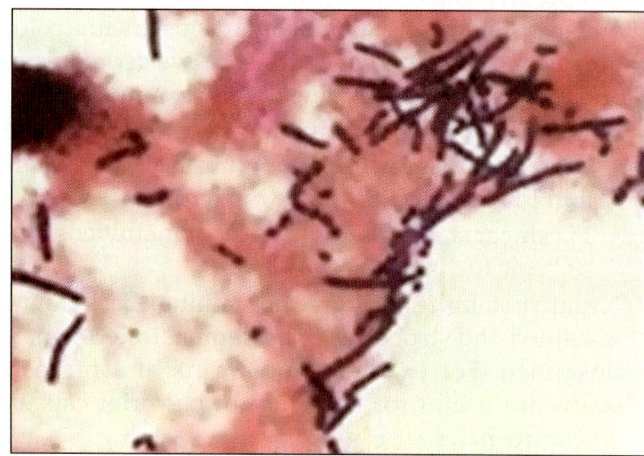

Fig. 112.6: Gram-positive bacilli (GPB)

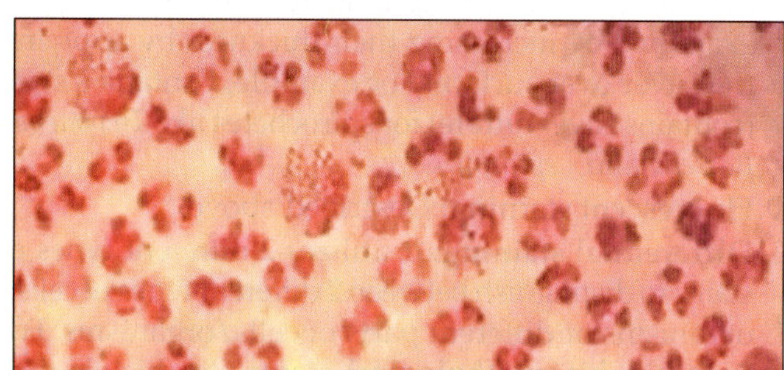

Fig. 112.7: Gram-negative cocci (GNC)

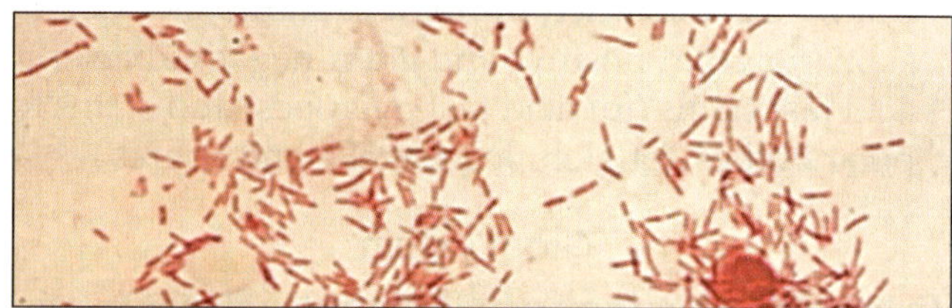

Fig. 112.8: Gram-negative bacilli (GNB)

Flowchart 112.1: Summary of Gram's stain

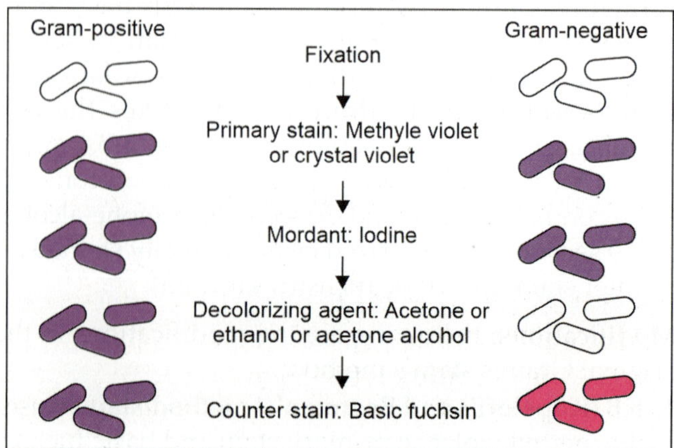

Essentials of Medical Microbiology

TABLE 112.1: Types of bacteria A/t Gram's stain

Types	Gram-positive	Gram–negative
Cocci	*Staphylococcus* *Streptococcus* *Strept. pneumoniae* *Stomatococcus* *Peptostreptococcus* *Peptococcus* *Enterococcus* *Micrococcus* **Mnemonic:** S_4P_2EM	*N. meningitidis* (**Men**ingococcus) *N. gonorrhoeae* (**Go**nococcus) *Mo*raxella *Ve*illonella **Mnemonic:** MenGo MoVe
Bacilli	*My*cobacterium *C*lostridium *Corynebacterium d*iphtheria *Erysipelothrix, Bacillius* (**O**thers) *N*ocardia, *A*ctinomyces *Listeria* *Lactobacillus* (**D**oderlein's bacilli) **Mnemonic:** MC DONALD	*Escherichia, K*lebsiella *Legionella, A*eromonas *Vibrio, Yersinia, Arcobacter, H*aemophilus *E*nterobacter, *S*almonella, *S*higella, *Proteus, Pseudomonas,* *Bordetella, Brucella* *A*naerobes (non-sporing) **Mnemonic:** EK LAVYA HE S_2P2B_2A (EK LAVYA HE with Sharp and Pointed Bow and Arrow

TABLE 112.2: Summary of Gram's stain

Reagents (mnemonic = MIAWBW)	Time	Action	Gram-positive color	Gram-negativee color
Methyl violet	5 minutes	Primary stain	Violet	Violet
Iodine solution	2 minutes	Mordant	Violet	Violet
Acetone (95–100%)	2–3 seconds	Decolorizer	Violet	Colorless
Water (tap water)		Washer		
Basic fuchsin	30 seconds	Counter stain	Violet	Pink/red
Water (tap water)		Washer		

Acid Fast Stains

History: Initially Ehrlich developed the acid fast stain in 1882 for tubercle bacilli with the use of aniline-gentian violet followed by nitric acid.

Meaning: Acid-fast bacteria once stain; resist decolorization even with the powerful solvent like acid or acid-alcohol so called **acid fast stain.**

Principle: Stain acid-fast bacteria with basic dye (like carbol fuchsin/aniline gentian violet). Fix the dye by using chemical (carbolic acid) or physical (heat) mordant followed by decolorization with acid (like nitric acid/ sulfuric acid) or acid-alcohol. Decolorized organisms are counter stained by methylene blue or malachite green.

Modifications: Called **modified acid fast stain.**

1. **ZN stain or hot stain:** Described separately in box.
2. **Kinyoun's stain or cold stain:** It avoids the use of heating, so called **cold stain.** It uses the carbol fuchsin as primary stain with higher concentration of carbolic acid than hot stain. Carbolic acid acts as chemical mordant. It takes more time than hot stain.
3. **Other modifications:** Like use of acid-alcohol as decolorizing agent and use of malachite green as counter stain.

Note: ZN (Ziehl-Neelsen) stain

Synonym: It also called hot stain because primary dye is fixed by using physical mordant like heat.

History: Initial Ehrlich technique was modified by Ziehl and Neelsen in 1882. This method is widely used for acid fast bacteria.

Principle: Stain acid-fast bacteria with basic dye like carbol fuchsin with simultaneous application of heat, which acts as physical mordant and fixes the dye in cell. Decolorization is done by 25% H_2SO_4. Decolorized organisms are counter stained by methylene blue.

Steps: Take one clean and dry slide. Prepare a thick smear from specimen. Air dry and heat fix it. Cover the entire smear with **basic dye** like carbol fuchsin (contains carbolic acid and basic fuchsin, so called **carbol fuchsin**). Simultaneously, apply the heat for 5 minutes. Cool and rinse with water. **Decolorize** by 25% H_2SO_4 by allowing it to sit for 15 seconds on smear. Wash the top and bottom of slide with water and clean the slide bottom well. **Counter-stain** with methylene blue for 30 seconds to 1 minute. Wash and blot the slide with absorbing paper. Add a drop of cedar wood oil. Examine under the oil immersion lens.

Precautions: (1) Apply the stain only on smear part, not on entire slide. (2) Carbol fuchsin is a carcinogen. Wear gloves when working with it. (3) Keep the stain steaming and do not boil. Add more stain if needed.

(Contd...)

Microscopy and Staining

Essentials of Medical Microbiology

Action of ingredients: (1) Carbol fuchsin: It acts as primary stain and stains all organisms in smear in pink color. **(2) Heat:** It acts as physical mordant and fixes the dye in cell. **(3) H_2SO_4 (25%):** It is decolorizer and it removes the color from organisms except acid fast. **(4) Methylene blue:** It acts as counter-stain and stains those organisms which are decolorized by 25% H_2SO_4.

Results: All acid fast organisms stain in pink color while non acid fast and background stain in blue color **(Fig. 112.9)**.

Reporting/grading of smear: Negative report should not be given before examining 100 fields or for 10 minutes of smear. Positive report can be given only if two or more bacilli have been seen. *M. tuberculosis* appears as long, curved, beaded or barred form and not uniformly stained red colored AFB. *M. bovis* appears as short, straight, stout and uniformly stained red colored AFB. Smear should be graded as per RNTCP, India as shown in **Table 112.3**. RNTCP grading is useful for monitoring the drug response, assessing the severity of the disease and assessing the infectiousness of the patient like higher the grade more is the infectiousness while smear negative ($<10^4$ bacilli/ml of sputum) are less infectious.

Quality control: Always check new reagents with known acid fast bacteria.

Acid fastness theory
- **Definition:** Bacteria once stain with primary dye resist decolorization even with the powerful solvent like acid-alcohol called **acid fastness.**
- **Reason for acid-fastness:** The acid fast genera have the unsaponifiable wax/lipid in their cell walls called **mycolic acid**. It is responsible for acid-fastness and prevents the decolorization by acid-alcohol. Peptidoglycan-polysaccharide-mycolic acids complex forms the skeleton in cell wall of mycobacteria, so acid fastness in mycobacteria is not the property of mycolic acid alone, but depends also on the integrity of cell wall.
- **Acid fastness increased by:** Growing organisms in presence of lipid.
- **Acid fastness decreased by:** Mechanical rupture of cell wall, autolysis, cell wall lysis by drug like isoniazide and cell wall lysis by fat solvents.

Further modifications in ZN stain: It called **modified ZN stain** and can be done by modifying the concentration of sulfuric acid like 0.25–0.5% for bacterial spore, 0.5–1% for sperm head, 1% for *M. smegmatism* (atypical mycobacteria), *Nocardia* spp., and for acid fast parasites and 5% for *M. leprae*.

Uses (acid-fast organisms)
- **Bacteria:** Follow **Table 112.4**.
- **Parasites:** Follow **Table 112.5**.
- **Human body cell:** Sperm head with 0.5–1% H_2SO_4.

Advantages of ZN stain: Cheap, fast, easy to perform, high sensitivity (>90%) and high specificity (98%).

Disadvantages (limitations) of ZN stain: (1) ZN stain is less sensitive than culture because at least 10,000 (10^4) bacilli should be present per ml of sputum for demonstration in direct smear, while culture can detect 10–100(10^1–10^2) bacilli per ml. (2) It is rare useful in young children who may not produce sputum. (3) Beaded or barred forms frequently seen in *M. tuberculosis* under ZN stain as shown in **Fig. 112.10**. *M. bovis* stains more uniformly under ZN stain.

Summary of ZN stain: Follow **Table 112.6 and Flowchart 112.2**.

10

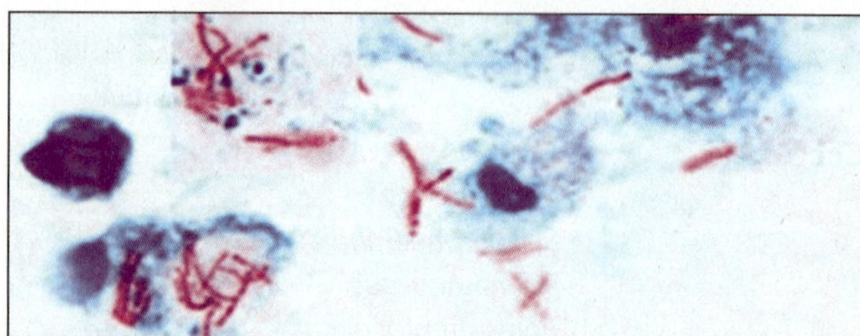

Fig. 112.9: Acid-fast bacilli under ZN stain

TABLE 112.3: Grading of smear as per RNTCP

No. of AFB/fields	Fields to be examine	Result	Grading
No bacilli/100	Entire smear/minimum 100 fields or 10 minutes	–ve	0
1–9/100	100	Scanty	Report with exact numbers of bacilli
10–99/100	100	+ve	1+
1–10 /1	50	+ve	2+
>10/1	20	+ve	3+

TABLE 112.4: Acid fast bacteria and concentration of H_2SO_4

Bacteria (Mnemonic: SAN MMRL)	Concentration of H_2SO_4
Bacterial spores	0.25–0.5%
Atypical mycobacteria *Nocardia asteroids, N. brasiliensis, N. caviae*	1%
M. leprae	5%
M. tuberculosis	25%
Rhodococcus	—
L. micdadei	—

TABLE 112.5: Acid-fast parasites

Parasites (Mnemonic: TC$_3$MS$_4$)	Diagnostic or acid fast stage (1% H_2SO_4)
Cystoisospora bellei (*Isospora bellei*)	Oocyst
Cryptosporidium parvum	Oocyst
Cyclospora cayetanensis	Oocyst
Microsporidium	Spore
Taenia saginatum	Egg
Schistosoma intercalatum	Egg
Schistosoma mansoni	Egg
Schistosoma japonicum	Egg
Schistosoma mekongi	Egg

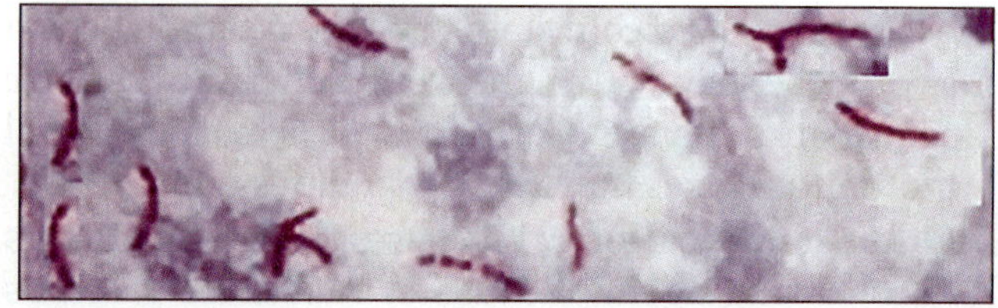

Fig. 112.10: Beaded forms under ZN stain

TABLE 112.6: Summary of ZN stain

Reagents (**Mnemonic = ChWHWMW**)	Time	Action	Acid-fast organism	Non-acid-fast organism
<u>C</u>arbol fuchsin and <u>h</u>eat	5 minutes	Phenol in carbolic acid acts as chemical mordant and heat as physical mordant. Basic dye stains all cells in pink color	Red/pink	Red/pink
<u>W</u>ater (tap water)		Washer		
<u>H</u>$_2$<u>S</u>O$_4$ (Acid/acid-alcohol)	15 seconds	Decolorizer	Red/pink	Colorless
<u>W</u>ater (tap water)		Washer		
<u>M</u>ethylene blue	1 minute	Counter-stain	Red/pink	Blue
<u>W</u>ater (tap water)		Washer		

Flowchart 112.2: Summary of ZN stain

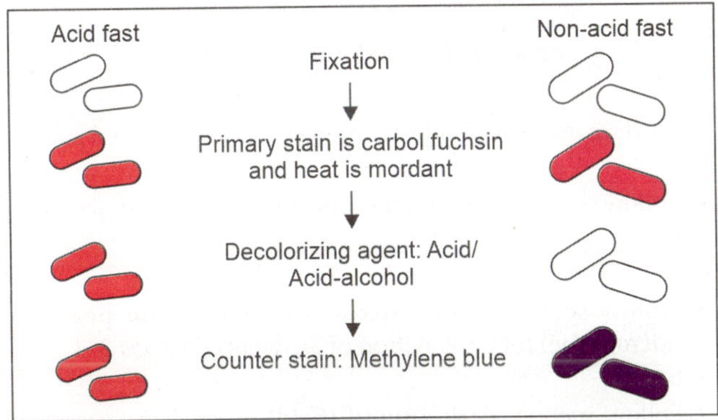

Albert Stain

History and meaning: So called because it was discovered by Albert (1878–1930), US Physician.

Principle: Albert-A stains the polar bodies and bacilli followed by mordant effect of Albert-B.

Reagent: Albert-A contains toludine blue and malachite green. Albert-B contains iodine.

Steps: Cover the smear with Albert-A for 5 minutes. Wash with water. Cover the smear with Albert-B for 1 minute. Wash and blot the slide with absorbing paper. Apply a drop of cedar wood oil. Focus under oil immersion lens.

Action of ingredients and result: Toluidine blue stains the granules in bluish black and malachite green stains the bacillary body in green color as shown in **Fig. 112.11**. Iodine acts as mordant.

Use: To stain the polar bodies (also called **metachromatic granules/volutin granules/Babes-Ernst granules**) in bacteria like *Corynebacterium diphtheriae*, *C. xerosis*, *M. leprae* (may be), *Bordetella pertusis*, *Gardnerella vaginalis*, and *Spirillum volutin*.

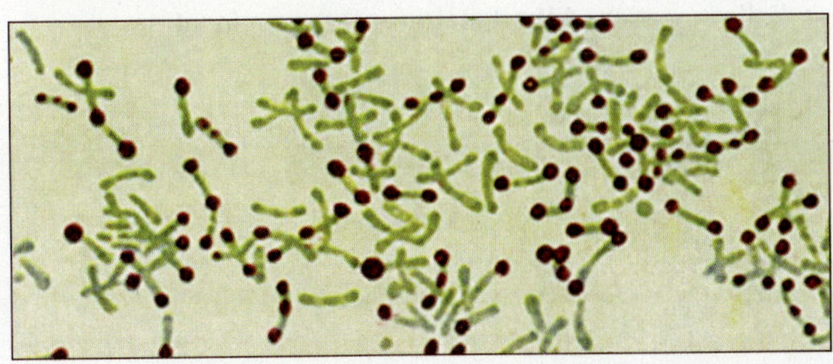
Fig. 112.11: Polar bodies under Albert's stain

ACCESS YOURSELF

Short Note

1. Fluorescent microscopy.

Short Questions for Theory/Viva Questions

1. Define: Magnification and resolution power of microscope.
2. Write the advantage of oil application on smear in microscopy.
3. What is fluorescence?
4. What are vital and supravital stains? Write one example of each.
5. Write four examples of gram-positive bacilli.
6. What is Gram's variable reaction?
7. Name the physical and chemical mordant of ZN stain.
8. What is acid fastness?
9. Name the four acid fast parasites.

Comments On

1. Heat fixation is not used in smear of intracellular organisms.
2. Alcohol fixation is used in smear of intracellular organisms.
3. In Gram's stain, gram-positive organisms stain in violet color and gram-negative organisms stain in pink color.
4. Strictly it is not correct to classify mycobacteria as GPB.
5. Carbol fuchsin so called carbol fuchsin.
6. In ZN stain, wearing of gloves is must during application of carbol fuchsin.
7. In ZN stain, acid fast organisms stain in pink color.
8. Acid fastness in mycobacteria is not only due to mycolic acid.
9. ZN stain is less sensitive than culture for detection of mycobacteria.
10. In albert stain, bacillary body stains in green color and polar bodies in bluish black color.

Question of Practical Exercise

1. Stain the given smear by Gram's stain and draw the labeled diagram of your findings. Report your findings to the examiner.
2. Stain the given smear by ZN stain and draw the labeled diagram of your findings. Report your findings to the examiner.

MCQs for Chapter Review

Microscopy

1. To see the bacteria, methods used are:
   a. Microscopy
   b. Stained preparation
   c. a + b
   d. None

Essentials of Medical Microbiology

2. Arrangement of lens from eye to source of light, in light microscope:
   a. Ocular lens: Subjective lens: Condensor lens
   b. Subjective lens: Ocular lens: Condensor lens
   c. Condensor lens: Subjective lens: Ocular lens
   d. Subjective lens: Condensor lens: Ocular lens

3. Limit of resolution with unaided eye is:
   a. 2000 microns   b. 200 microns
   c. 20 microns   d. 100 microns

4. Light microscopy resolution:
   a. 200 nm   b. 20 nm
   c. 0.2 nm   d. 300 nm

5. Dark ground microscopy is used to see:
   a. Refractile organisms   b. Flagella
   c. Capsule   d. Fimbriae

6. Dye used in fluorescent microscopy:
   a. Thioflavin T   b. Congo red
   c. Brilliant blue   d. Eosin
   e. Auramine

Staining

7. One of the following stain is an example of negative stain:
   a. Gram's stain   b. Fontana's stain
   c. Indian ink   d. ZN stain

8. In negative staining:
   a. The structure to be demonstrated is stained.
   b. The structure to be demonstrated is not stained.
   c. The background is not stained.
   d. The background and structure are stained.

9. Wet Indian ink preparation is used for demonstration of:
   a. Flagella   b. Capsule
   c. Spirochetes   d. Fimbriae

10. Sliver impregnation technique is used in the diagnosis of:
   a. Spirochetes   b. *Leptospira*
   c. *Borrelia*   d. All of above

11. Not used in Gram's staining:
   a. Methylene blue   b. Crystal violet
   c. Iodine   d. Safranin

12. Correct order of Gram's staining is:
   a. Methyl violet → Iodine → Acetone
   b. Methyl violet → Acetone → Iodine
   c. Methyl violet → Basic fuchsin → Iodine
   d. Methyl violet → Iodine → Basic fuchsin

13. Gram-negative stains in pink color because of:
   a. Polysaccharide   b. Lipopolysaccharide
   c. Techoic acid   d. None

14. Following are gram-negative cocci *except*:
   a. Pneumococci   b. Meningococci
   c. Gonococci   d. *Veillonella*

15. H_2SO_4 concentration to stain *M. leprae* is:
   a. 1%   b. 5%
   c. 10%   d. 25%

16. Mycobacteria are diagnosed on microscopy when counts:
   a. are 10,000 or more/ml   b. are 1, 00,000 or more/ml
   c. are 1000 or more/ml   d. are 10 or more/ml

17. The number of bacilli required in the sputum to yield a positive culture report for *Mycobacterium tuberculosis* is:
   a. 1–10 bacilli per ml
   b. 10–100 bacilli per ml
   c. 100–1000 bacilli per ml
   d. >1000 bacilli per ml

18. Acid fast structure is:
   a. *Vibrio*   b. *Nocardia*
   c. *E. coli*   d. *Bacillus anthracis*

19. Albert stain is used for:
   a. *Staph. aureus*   b. *C. diphtheriae*
   c. *Cl. perfringens*   d. *Cl. tetani*

Answers and Explanation of MCQs

1. c
- Bacteria are best visible by microscope with stained smear.

2. a
- Follow section, **light microscope (Fig. 112.1)** for explanation.

3. b

4. a
- Follow section, **light microscope (resolution power of microscope)** for explanation of answers of MCQs 3–4.

5. b
- Follow section, **dark ground (field) microscopy (uses)** for explanation.

6. e
- Follow section, **fluorescent microscopy** for explanation.

7. c

8. b

9. b, c
- Follow section, **staining (negative staining)** for explanation of answers of MCQs 7–9.

10. d
- Follow section, **staining (impregnation staining)** for explanation.

11. a

12. a

13. b

14. a
- Follow section, **Gram's stain and Table 112.1** for explanation of answers of MCQs 11–14.

15. b

16. a

17. b

18. b
- Follow section, **acid fast stains, Table 112.4 and Table 112.5** for explanation of answers of MCQs 15–18.

19. b
- Follow section, **Albert stain (use)** for explanation.

CHAPTER 113

Sample Collection, Transport and Storage

Chapter Outline
- Sample Collection
- Transportation and Storage

SAMPLE COLLECTION

Precautions

Following are the precautions to be taken before collecting the sample.
- Take all the aseptic precautions.
- Wearing the personal protective equipments (PPEs) like disposable gloves, eye shields, apron, cap, mask and shoes cover
- Use of disposable needle and syringe
- Container should be sterile, leak proof and securely fastened.
- Use appropriate transport medium whenever necessary.
- Selection of proper site, proper time (like for culture, obtain material during acute stage of illness or before administration of antimicrobial therapy), sufficient amount, numbers of specimens (like two swabs one for direct smear and other for culture), etc.
- For serological tests, collect paired serum samples one in acute stage and second 7–14 days later.
- Fill the request form with complete data.
- Wipe the spillage with 10% sodium hypochlorite.
- Proper waste disposal of hazardous materials.

Methods of Samples Collection

For bacterial infections: Follow **Table 113.1**.

For viral infections: Collect the samples in viral transport medium. Different types viral transport media (VTM) are described in **Ch. 114**. Collections of different samples according to system infected for viral diagnosis are shown in **Table 113.2**.

Fungal infections: Follow **Table 113.3**.

For parasitic infections: Samples collected are mentioned below with indications.

- **Blood:** *Trypanosoma* spp., *Leishmania* spp., in macrophages, *Plasmodium* spp., *Babesia* spp., *Toxoplasma gondii* and microfilariae.
- **Stool (GIT parasites): (1) Trophozoite or cyst stage** presents in *E. histolytica*, *E. coli*, *G. lamblia* and *B. coli*. **(2) Oocyst** presents in *Cystoisospora belli* and *Cryptopsoridium parvum*. **(3) Spores like yeast cells** present in *Microsporidium*. **(4) Adult stage of parasite** presents in common round worm, hook worm, thread worm and whip worm. **(5) Larval stage** presents in *S. stercoralis*. **(6) Egg stage** presents in common round worm **(bile stained)**, hook worm, thread worm, whip worm **(bile stained)**, *Taenia* spp., **(bile stained)**, *H. nana*, *Fasciolopsis buski*, *Fasciola* spp., *Schistosoma mansoni*, *S. japonicam*, *Dipylidium* spp., and *Diphylobothrium latum*. **(7) Segment of adult stage** presents in *Taenia* spp., *D. latum* and other tape worm species.
- **Urine:** It contains trophozoites of *T. vaginalis*, eggs of *S. haematobium* and microfilariae.
- **Sputum (pulmonary parasites/lung parasites):** It contains trophozoites of *E. histolytica*, scolex and hooklets of *E. granulosus* when hydatid cyst ruptures, eggs of *P. westermanii* and larvae of *Ascaris lumbricoides*, hook worm spp., *Toxocara canis*, *Toxocara cati* and *S. stercoralis*.
- **CSF:** It contains trophozoites of *E. histolytica*, trophozoites of *N. fowleri*, *Acanthamoeba* species, *T. brucei*, hydatid cyst and *A. cantonensis*.
- **Perianal swab:** It is collected by using **NIH swab** for *E. vermicularis*, *T. saginatum* and *S. mansoni*.
- **Aspiration of materials:** (1) H. cyst fluid to examine the scolex and hooklets. (2) Amoebic abscess to examine the trophozoites. (3) Hydrocele to examine the microfilariae.
- **Biopsy specimens:** (1) Spleen for *L. donovani*. (2) Bone marrow for *L. donovani*, *T. brucei*, etc. (3) Lymph nodes for *L. donovani*, *T. brucei* and microfilariae. (4) Muscles

Sample Collection

1. **Larval form in stool is found in:**
   a. *Strongyloides*
   b. *Ancyclostoma duodenale*
   c. *Ascaris lumbricoides*
   d. *Necator americanus*
   e. *Trichuris trichuria*

2. **Parasite/parasites present in sputum is/are:**
   a. *E. granulosus*
   b. *E. histolytica*
   c. None of above
   d. a + b

3. **Which of the following is best diagnosed by muscle biopsy?**
   a. *T. spiralis*
   b. *E. granulosus*
   c. *L. donovani*
   d. a + b + c

4. **Sputum examination is not useful in diagnosis of:**
   a. *T trichuria*
   b. *A. duodenale*
   c. *Paragonimus*
   d. *Strongyloides*

5. **Skin snip is used in the diagnosis of:**
   a. Trichinosis
   b. Strongyloidiasis
   c. Schistosomiasis
   d. Onchocerciasis

6. **Which of the following does not pass through the lungs?**
   a. Hook worm
   b. *Ascaris*
   c. *Strongyloides*
   d. *Enterobius vermicularis*

7. **Parasites causing lung infestation are:**
   a. *H. nana*
   b. *P. westermanii*
   c. *T. saginata*
   d. *E. granulosus*

Transportation

8. **In triple layer packing, the request form should be put in:**
   a. In primary container
   b. In secondary container
   c. Outer to secondary container
   d. In tertiary container

Answers and Explanation of MCQs

1. a
2. d
3. a
4. a
5. d
- Follow section, **methods of samples collection (for parasitic infections)** for explanation of answers of MCQs 1–5
6. d
7. b, d
- Sputum is the sample for diagnosis of hook worm, *Ascaris*, *P. westermanii*, *E. granulosus* and *Strongyloides*, means they can pass through the lungs but not for *T. saginata E. vermicularis*, *H. nana* and *E. multilocularis* it mean they does not pass through lungs.
8. c, d
- Primary container contains specimen while secondary container contains primary container surrounded by absorbent. Request form should be put outer to secondary container and inner to tertiary container.

Culture Media

INTRODUCTION

Definition

Media are artificial preparations of nutritional materials required for the growth of microorganisms.

History

Robert Koch invented the culture method for pure isolation of bacteria over solid media. Earliest solid medium was **cooked cut potato** used by him, later he used gelatin for solidification of media, but gelatin is not satisfactory as it has tendency to get liquefy at 24°C and by proteolytic bacteria. Use of agar in solidification of media was suggested by Frau Hesse, the wife of one investigator in Koch laboratory, who had seen her mother using agar to make jellies.

Ingredients/Components

Following materials are used for media preparation.

Protein sources

- **Peptone:** It is mixture of partially digested and uncoagulable proteins and available in powder form.
- **Meat extract:** It called **"lab lemco"**. It is semisolid in nature.
- **Neopeptone and proteose peptone:** It is the commercially available peptone brand.
- **Other proteins sources are:** Fetal calf serum (sterilized by filtration), albumin, etc.

Water: It is essential for the growth of all microorganisms. Deionized or distilled water is used to prepare the media. It is a source of H_2 and O_2.

Blood: It provides extranutrition to the fastidious organisms and used in enriched media. Sheep blood (5–10%) is ideal. Horse or ox blood is also useful. Human blood can be used but contains inhibitory substances. It is rendered noncoagulable by adding the anticoagulant (citrate or oxalate) or by defibrillation (by shaking the blood in a bottle contains sterile glass beads).

Agar/agar agar (Fig. 114.1a)

- **Properties:** It has no nutrient property, but provides solidification to media. It is a polysaccharide obtained from sea-weeds like red algae of species *Gelidium* and *Gracilaria*. Weeds being dried and finally supplied as dried strands or as a powder. On heating it becomes gelly at 45°C and liquefies at 98°C. Thus once gelled the medium remains solid throughout the temperature growth range of all bacteria. Blood, serum, etc., therefore can be added to nutrient agar at 55°C at which the medium is still fluid and the risk of denaturation of the serum proteins added is nil.
- **Concentration of agar in media**
  - Semi-solid media: 0.5%.
  - Solid media: They are different in properties from brand to brand like New Zealand agar gels at 1% and Japanese agar gels at 2%.
  - Solid media to inhibit the *Proteus* swarming: 3–4% or 6%.

Buffer: It contains carbonates and phosphates. It resist in pH change of medium.

Fig. 114.1: Agar and biphasic medium

Yeast extract: It is obtained from yeast cells like Baker's yeast. It contains proteins, amino acids, vitamins, carbohydrates, minerals (phosphate and potassium), etc.

Antibiotics: They prevent the growth of unwanted organisms.

Anticoagulants: In a blood culture media like taurocholate broth, blood contains inhibitory or bactericidal substance, which are obviated by fourfold dilution of blood or by adding liquoid/anticoagulant like SPS (sodium polyanethol sulfonate). Ratio of blood to medium is 1:10.

Other substances: These are electrolytes like sodium chloride, vitamin like vitamin K in thioglycollate broth and sugars like glucose, sucrose, lactose, maltose, mannitol, starch, etc.

CLASSIFICATION OF MEDIA

It is very difficult to classify a particular medium in particular group, because it has many properties and many uses; however, some types are described as below.

According to Physical form or Concentration or Consistency of Agar

Liquid media: They are without agar and simply called "broth" (e.g. nutrient broth). They are available in test-tubes, bottles or flasks. In liquid media bacterial growth produce uniform turbidity, surface pellicle (e.g. aerobic bacteria) or deposit at bottom.

Semisolid media: They contain 0.2–0.5% agar. They are available in test-tubes. They are soft, gelly like and flat surface at top. Their uses are given in **Table 114.1**.

Solid media: They contain 1–2% agar and simply called "agar" (e.g. nutrient agar). They are available in Petri-dish or as slant/slope in tubes and most commonly used for isolation of organisms.

Biphasic media (Fig. 114.1b): These media contain both liquid and solid phase are called **biphasic media**, e.g., Castaneda biphasic medium, Columbia biphasic medium, etc.

According to Nutritional Components

Simple/basal media: Follow **Table 114.1**.

Complex/special media: Exact nutritional components are difficult to estimate. All the media other than simple media are complex media, e.g., enriched media, selective media, transport media, enrichment media, differential media, sugar media and media for biochemical reactions.

According to Chemical Composition

Synthetic or defined media: They are prepared from pure chemical substances and the exact chemical composition (amount) is known. They are used for various special studies such as metabolic requirements, e.g., Dubo's medium with Tween 80.

Semisynthetic or semidefined media: Chemical composition is approximately known.

According to Oxygen Requirement

Aerobic media: Media used for growth of bacteria other than obligate (strict) anaerobes are called **aerobic media**, e.g., basal media, enriched media, etc.

Anaerobic media: Media used for growth of strict anaerobes or to create the anaerobiosis are called **anaerobic media**. For more details follow **Ch. 115**.

According to Uses or Applications

Simple/basal media: Follow **Table 114.1 and Fig. 114.2**.

Enriched media: Follow **Table 114.2 and Fig. 114.3**.

Enrichment media: Follow **Table 114.3 and Fig. 114.4**.

Transport (holding) media: Follow **Table 114.4 and Fig. 114.5**.

Selective media: Follow **Table 114.5 and Figs 114.6 and 114.7**.

TABLE 114.1: Simple/basal media

Definition: Media contain basic substances like meat extract, peptone, sodium chloride and water for growth of microorganisms are called **basal or simple media**. Exact nutritional components are known.

No.	Name of medium	Important ingredients	Action of important ingredients	Method of sterilization	Uses
1	Peptone Water (PW)	Peptone, salt, water	Basic requirement for growth of microbes	Autoclaving	(1) Used as a base for other media (2) Used for the growth of non-exacting microorganism
2	Nutrient Broth (NB)	Peptone water and meat extract	Do	Do	Do
3	Nutrient Agar (NA)	Nutrient broth and 1–2% agar	Do	Do	Do
4	Semisolid agar	Nutrient broth and 0.2–0.5% agar	Do	Do	(1) To detect the motility of bacteria. (2) For growth of anaerobic and microaerophilic organisms. (3) Hugh-Leifson's (**O**xidation – **F**ermentation/OF) test. (4) Gelatin liquefaction test

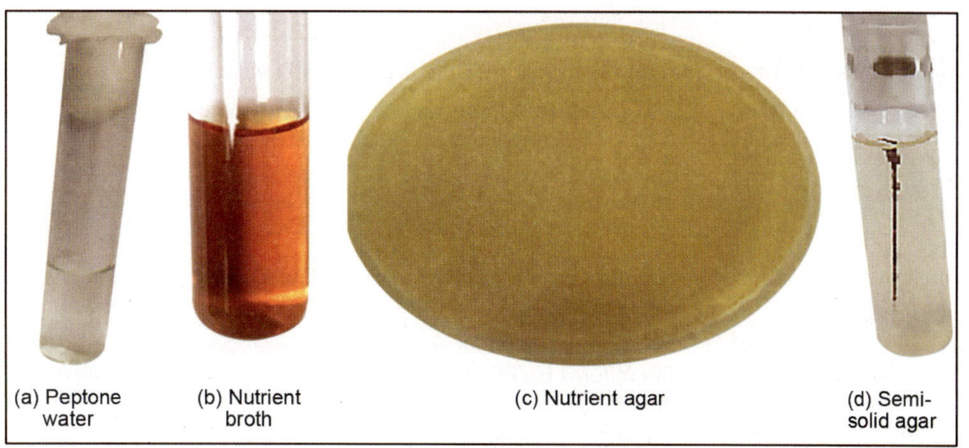

(a) Peptone water | (b) Nutrient broth | (c) Nutrient agar | (d) Semi-solid agar

Fig. 114.2: Simple/basal media

	TABLE 114.2: Enriched media				

Definition: Media enriched with blood, serum, eggs, hydrocele fluid in addition to basic substances for growth of fastidious organisms are called **enriched media**

No.	Name of medium	Important ingredients	Action of important ingredients	Method of sterilization	Uses
1	Brain Heart Infusion Broth (BHIB)/Agar BHIA)	NB/agar + glucose + calf brain and cow heart	**Glucose, calf brain and cow heart** are enriching the organisms	Autoclaving	For growth of fastidious organisms
2	Blood agar (BA)	NA + 5–10% sheep or horse blood	**Blood** is enriching the organisms	Autoclave the NA and add the blood when temp., comes down to 55°C	(1) To differentiate hemolytic from non-hemolytic organisms. (2) For growth of *Staphylococcus* spp., and *Neisseria* spp. (3) For antibiogram of *Streptococcus* and pneumococcus
3	Chocolate agar (CA)/heated blood agar)	BA + heating	When blood is heated, RBCs will lysed and release **X (Hemin) and V factors (Vitamins)** free in medium	Keep BA in oven at 55°C till become CA	To grow fastidious bacteria, like *H. influenzae*, *Staphylococcus* spp., *Neisseria* spp., *Streptococcus* spp., and pneumococcus
4	Loeffler's serum	NB + glucose + serum	**Serum and glucose** are enriching the medium	Inspissation	For growth of *C. diphtheriae*
5	Dorset's egg medium	NB + egg (white + yellow)	**Egg** is enriching the medium	Inspissation	For growth of *M. tuberculosis* (both human and bovine varieties)

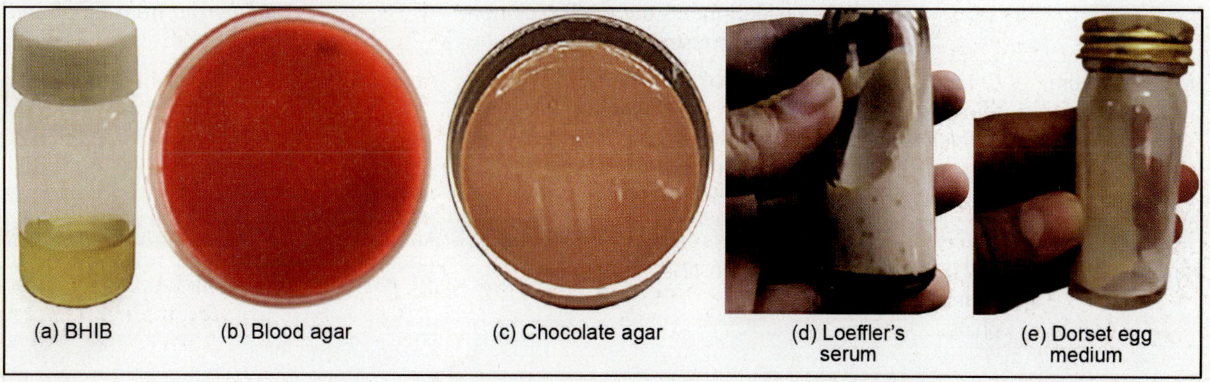

(a) BHIB | (b) Blood agar | (c) Chocolate agar | (d) Loeffler's serum | (e) Dorset egg medium

Fig. 114.3: Enriched media

TABLE 114.3: Enrichment media

Definition: Media that favor the growth of particular organisms are called **enrichment media**

No.	Name of medium	Important ingredients	Action of important ingredients	Method of sterilization	Uses
1	Glucose broth (GB)	NB + glucose	**Glucose** is enriching the organisms	Tyndallization	For pyogenic organisms and organisms causing subacute bacterial endocarditis **(SABE)**
2	Taurocholate broth	NB + sodium taurocholate	**Sodium taurocholate** is enriching the organisms	Autoclaving	For growth of *Salmonella* species from blood
3	Tetrathionate broth	Tetrathionate + potassium iodide + brilliant green	**Tetrathionate** inhibits the growth of coliform bacilli and favors the growth of *Salmonella*	Tyndallization	For growth of *Salmonella* species from feces
4	Selenite – F/broth	NB + lactose + selenite	**Selenite** is enriching the *Salmonella* species	Tyndallization	For growth of *Salmonella* (more) and *Shigella* (less) species from feces
5	Alkaline peptone water (APW)	Peptone water pH-8.6	**High pH** favors the *V. cholerae* growth	Autoclaving	For growth of *V. cholerae* from feces

Fig. 114.4: Enrichment media

TABLE 114.4: Transport or holding media

Definition: Media that maintain the viability of the microorganisms without allowing the multiplication in specimens during transport to the laboratory are called **transport** or **holding media**

No.	Name of medium	Important ingredients	Action of important ingredients	Method of sterilization	Uses
			Bacterial transport media (BTM)		
1	Pike's medium	Casein hydrosylate, tryptose, yeast extract, dextrose, crystal violet and sodium azide	Casein hydrosylate, tryptose and yeast extract serve as nitrogen source. Dextrose as energy source. Crystal violet inhibits the gram-positive bacteria. Sodium azide inhibits the GNB plus non-hemolytic streptococci	Autoclaving	(1) Selective, enrichment and transport medium for *Strept. pyogenes* and other hemolytic streptococci from throat swab. (2) It also preserves *Strept. pneumoniae* and *H. influenzae* from nasal and throat swabs
2	Stuart's medium	Charcoal, sodium thioglycolate (reducing agent), inorganic phosphates, buffer	**Charcoal** absorbs the inhibitory substances and **sodium thioglycolate** prevents oxidation during the transportation	Autoclaving	Provides viability of *Neisseria* spp.
3	Glycerol saline transport medium	Glycerol, sodium and potassium hydrogen phosphate	Prevents growth of intestinal commensals	Autoclaving	Preserves *Salmonella* and *Shigella* from feces

(Contd…)

Essentials of Medical Microbiology

TABLE 114.4: Transport or holding media (*Contd…*)

No.	Name of medium	Important ingredients	Action of important ingredients	Method of sterilization	Uses
	Bacterial transport media (BTM)				
4	Gram-negative broth	Tryptose, dextrose, mannitol, sodium citrate, sodium deoxycholate, monopotassium phosphate and dipotassium phosphate	Sodium citrate and sodium deoxycholate inhibit the growth of gram-positive bacteria and coliforms	Autoclaving	Transport and selective medium for *Salmonella* and *Shigella* from clinical (feces) and nonclinical specimens
5	VR medium	PW + Crude sea salt, PH 8.6	High pH preserves viability of *V. cholerae*	Autoclaving	For transportation of *V. cholerae*
6	Alkaline Peptone Water (APW)				For transportation of *V. cholerae*
7	Autoclaved sea water				For transportation of *V. cholerae*
8	Cary Blair transport medium				For transportation of EHEC (O157:H7), *Salmonella*, *Shigella*, *V. cholerae* and *C jejuni*
	Viral transport media (VTM)				
1	Stuart's medium, Amie's medium, Leibovitz-Emory medium, Hank's balanced salt solution (HBSS), Eagle's tissue culture medium	**Distilled water + protein sources** in this media are serum, albumin, veal infusion (calf meat) and gelatin + **antibiotics** like vancomycin, gentamicin and AMB	**Protein** stabilizes the virus. **Antibiotics** prevent bacterial and fungal contamination.	Filtration	For collection of throat and nasal swabs from **human**
2	Modified Stuart's medium, Modified Amie's medium and Leibovitz-Emory medium			Filtration	For collection of respiratory and stool specimens from **human**
3	Transport medium 199	0.5% bovine serum albumin (BSA) + **Antibiotics** like benzylpenicillin, streptomycin, polymyxin B, nystatin and gentamicin	**Antibiotics** prevent bacterial and fungal contamination	Filtration	For collection of clinical specimens from all **animals** for egg and tissue culture
4	PBS-Glycerol transport medium	Phosphate-buffered saline (PBS) + **antibiotics** as above	**Antibiotics** prevent the bacterial and fungal contamination	Autoclaving	For collection of specimens from all **animals** for egg culture but not for tissue culture

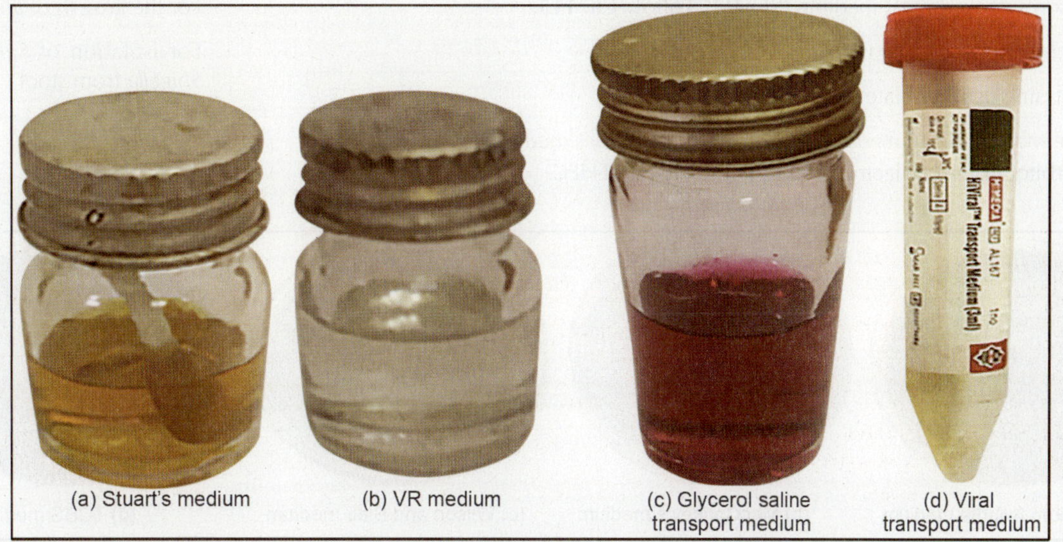

Fig. 114.5: Transport or holding media

TABLE 114.5: Selective media

Definition: Media contains inhibitory agents (e.g. antibiotics, chemicals) to prevent the growth of unwanted microbes called **selective media**.

No.	Name of medium	Important ingredients	Action of important ingredients	Method of sterilization	Uses
1	Potassium tellurite medium	NA + blood + potassium tellurite	**Blood** to enrich the medium. **Potassium tellurite** is less inhibitory to *C. diphtheriae* compared to other throat commensals	Autoclave	For the growth of *C. diphtheriae, Staph. aureus, E. faecalis, E. rhusiopathiae*. **Tellurite** is converted to metallic tellurium which gives black color colonies
2	Mac Conkey's agar	NA + sodium taurocholate + lactose + neutral red	**Sodium taurocholate** is inhibitory to nonintestinal organisms. **Lactose** will be fermented by coliform bacilli with acid production and reduction in pH. **Neutral red** will gives pink color at acidic pH	Autoclaving	(1) For differentiating lactose fermenting organisms from non-lactose fermenting organisms. (2) For growth of intestinal organisms
3	Wilson and Blair medium	NA + glucose phosphate + brilliant green + bismuth sulfite + ferric citrate	Brilliant green and bismuth sulfite inhibit intestinal organisms. Sulfite is converted to sulfide produces black colonies with ferric citrate	Autoclaving	For growth of *Salmonella*
4	TCBS (**T**hiosulfate **C**itrate **B**ile salt **S**ucrose) agar	Thiosulphate + citrate + bile salt + sucrose + Bromothymol Blue (BB)	**Sucrose** will be fermented by *V. cholerae* with acid production and reduction in pH of medium. **BB** gives yellow color at acidic pH and green color at alkaline pH (Sucrose nonfermenters)	Autoclaving	For growth of *V. cholerae*. It ferments sucrose produces yellow colonies while sucrose non-fermenters give green colonies
5	Lowen-stein Jensen (LJ) Medium	Egg + malachite green + mineral salts + glycerol	**Egg** enriches the medium. **Malachite green** is least inhibitory for *M. tuberculosis*. **Glycerol** allows the growth of human variety	Inspissation	For the growth of *M. tuberculosis*, human variety only
6	**S**abouraud's **D**extrose **A**gar (SDA)	Glucose + peptone + antibiotics like chloramphenicol, gentamicin and cycloheximide) + agar, pH: 5.4	**Acidic pH** and **high sugar** content make it selective for fungi. **Antibiotics** inhibit the bacterial growth	Autoclaving	For cultivation of fungi
7	Thayer-Martin medium	CA + vancomycin colistin + nystatin	**CA** to enrich the *Neisseria* spp., and **antibiotics** inhibit the other bacterial growth	Autoclaving	For the growth of the *Neisseria* spp.
8	Robertson's cooked meat medium (RCMM): Follow **Ch. 115**.				For the growth of anaerobes
9	Deoxycholate citrate agar (DCA)				For isolation of *Salmonella* and *Shigella* from stool
10	Xylose lysine deoxycholate (XLD) medium				

Note: Sorbitol Mac Conkey's medium/modified Mac Conkey's medium
It contains sorbitol instead of lactose and useful to identify EHEC.

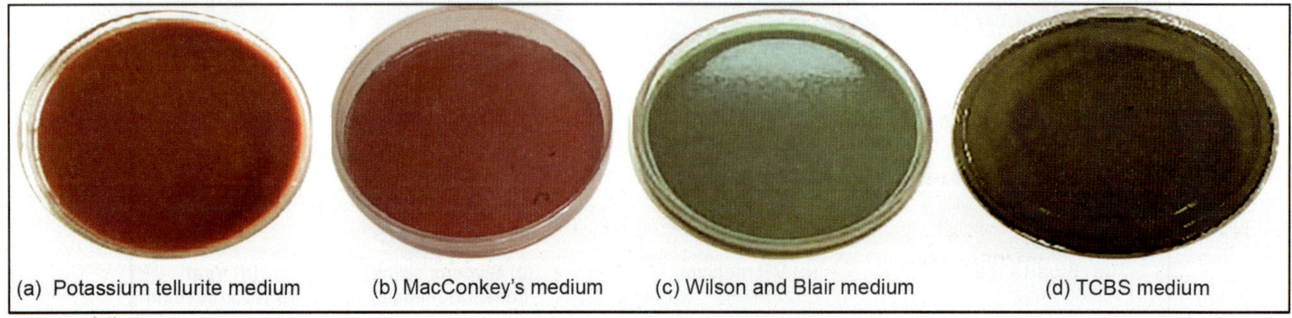

Fig. 114.6: Selective media

(a) LJ medium (b) SDA (c) RCMM

Fig. 114.7: Selective media

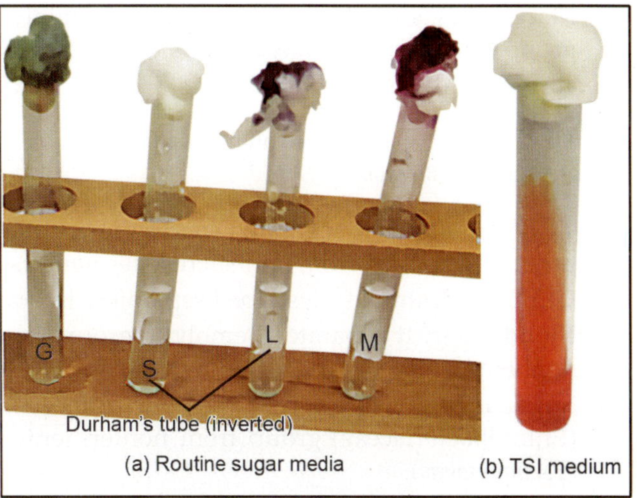

Fig. 114.8: Sugar media

Indicator media: Media contain substances (e.g., pH indicator) that indicate change in color of media, e.g., sugar media, TCBS medium, MacConkey's agar, etc.

Anaerobic media: For more details follow **Ch. 115.**

Sugar media: These media contain fermentable substances/sugars like **(1) Monosaccharides:** Pentose (xylose, arabinose) and hexose (glucose, mannose). **(2) Disaccharides:** Lactose and saccharose. **(3) Polysaccharides:** Starch and inulin. **(4) Trisaccharides:** Raffinose. **(5) Alcohol:** Sorbitol and glycerol. **(6) Glucosides:** Salicin and esculin. **(7) Noncarbohydrate substances:** Inositol. Different media are mentioned in **Fig. 114.8 and Table 114.6.**

Media for biochemical reactions

1. **Sugar fermentation test:** Sugar media are described above.
2. **Citrate test (Ch. 118):** It is performed in Koser's citrate broth or Simmon's citrate agar. Bromothymol blue is pH indicator.

3. **Urease test (Ch. 118):** It is performed in Stuart's urea broth or in Christinsen's urea agar. Phenol red is pH indicator.
4. **Gelatin liquefaction test:** Gelatin medium **(Table 114.7).**

Differential media

- **Synonym:** Sometimes differential media are defined as **indicator media.**
- **Definition:** These are media, which distinguish particular characteristics of bacterial group from other group.
- **Advantages of differential media**
  1. CLED medium is preferred over MacConkey's medium and blood agar for testing of urine, because it inhibits the *Proteus* swarming. It is less inhibitory than Mac Conkey's agar and supports the growth of microbes like *Candida* spp., and grampositive bacteria except β-hemolytic *Streptococcus*.

No.	Name of medium	Important ingredients	Action of important ingredients	Method of sterilization	Uses
	TABLE 114.6: Sugar media				
1	Routine sugar media are G, S, L and M	PW + 1 % sugar + Andrade's indicator + Durham's tube	Sugar fermentation produces **acid** which is indicated by pink color and **gas** indicated by bubbles formation in Durham's tube	Tyndallization	(1) For sugar fermentation tests. (2) For differentiation of sugar fermenters (Enterobacterales) and nonfermenters (*Pseudomonas*)
2	Hiss's serum	Sugar media + 3% serum	Do	Do	Sugar fermentation tests in *C. diphtheriae*, *Neisseria* spp., and pneumococcus
3	TSI agar/composite medium: Follow **Ch. 118.**				
4	Kligler Iron Agar (KIA): Follow **Ch. 118.**				

No.	Name of medium	Important ingredients	Action of important ingredients	Method of sterilization	Uses
	TABLE 114.7: Gelatin liquefaction medium				
1	Gelatin liquefaction medium	NB + Gelatin	Gelatin is liquefied by proteolytic organisms	Tyndallization	To detect the proteolytic activity of organisms

2. MacConkey's agar separates lactose fermenters group from nonlactose fermenters.

- **Examples of differential media**

> **Mnemonic: Differential media**
> TCBS + NBM (Nill By Mouth)

1. **TCBS medium:** It separates sucrose fermenters group (*V. cholerae*) from non-sucrose fermenters.
2. **CLED medium:** It is described separately in box.
3. **Blood Agar:** It separates hemolytic group from nonhemolytic.
4. **Sugar media:** They separates sugar fermenters (Enterobacteriaceae) group from nonfermenters (*Pseudomonas*).
5. **Nagler's medium:** It differentiates between lecithinase activities of different bacterial group.
6. **Bile-Esculin agar:** It separates *Enterococcus* (*Streptococcus* group D) group from non-*Enterococcus*.
7. **Mac Conkey's medium:** It separates lactose fermenters (*E. coli, K. pneumoniae*) group from non-lactose fermenter (*Salmonella, Shigella,* etc.).

Media for antibiogram: It is performed in Muller Hinton agar (MHA) plate **(Ch. 118)**.

Composite media: Follow **Ch. 118**.

> **Note: CLED medium**
>
> **Full name:** Cysteine Lactose Electrolyte Deficient medium.
>
> **History:** Previous medium used to inhibit the *Proteus* swarming was containing chloral hydrate, alcohol, sodium azide, surface-active agents, boric acid, and sulfonamides. In the mid 1960s, Sandys and Mackey reported the laboratory diagnosis of urinary pathogens by lacking the medium with electrolytes (NaCl), which helps to prevent the swarming of *Proteus* spp.
>
> **Ingredients and action:** Follow **Table 114.8**.
>
> **Use:** It is used to differentiate urinary LFs from NLFs, with controlling the *Proteus* swarming. It also supports the growth of many pathogens like Enterobacterales, *Pseudomonas, Staphylococcus, Enterococcus* and *Candida* and a number of contaminants such as diphtheroids, *Lactobacillus* and *Micrococcus*.
>
> **Result:** Follow **Table 114.9** and **Fig. 114.9**.

TABLE 114.8: Ingredients and action in CLED medium

Ingredients	Weight	Actions
Lactose	10.0 g	To separate LFs from NLFs
Pancreatic digest of gelatin	4.0 g	Nitrogen, vitamins and carbon will help in microbial growth
Pancreatic digest of casein	4.0 g	
Beef extract	3.0 g	
L-Cystine	0.128 g	To enhance the growth of cystine-dependent "dwarf colony" coliforms
Bromothymol blue	0.02 g	pH indicator
Agar	15.0 g	Solidifying agent
Final pH 7.3 +/– 0.2 at 25ºC		

TABLE 114.9: Result in CLED medium

Microbes	Culture characteristics
E. coli	Opaque yellow colonies with a slightly deeper yellow center
Klebsiella	Yellow to whitish-blue colonies, extremely mucoid
Proteus	Translucent blue colonies
P. aeruginosa	Green colonies with typical matted surface and rough periphery
Enterococcus	Small yellow colonies, about 0.5 mm in size
Staph. aureus	Deep yellow colonies, uniform in color
CoNS	Pale yellow colonies, more opaque than *Enterococcus faecalis*

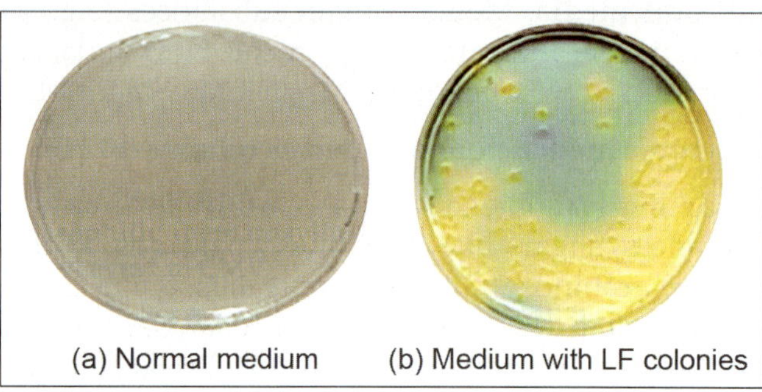

Fig. 114.9: Result in CLED medium

> **Note: Dehydrated media**
>
> **Preparation:** They are available commercially and prepared by different companies from dehydrated materials which are mentioned on the label of the container. They are reconstituted in distilled water and sterilized before use in laboratory.
>
> **Advantages:** They are useful in peripheral laboratory, time saving and less labor intensive for preparation.

According to Ingredients

Chemically defined media or cell free media or artificially prepared nutritional media

- **Preparation:** Prepared by using different biochemical or chemical substances like protein source, carbohydrate source, lipid source, vitamins, minerals, antibiotics, agar, etc.
- **Examples:** All earlier media discussed in this chapter are fall in this category.
- **Uses:** They are not useful for viruses, but useful for bacteria, fungi and few protozoa like *E. histolytica, Trepanosoma* spp., *Leishmania* spp., etc.
- **Limitations:** Few microbes are not growing in cell free media are mentioned in **Table 114.10**. Such microbes are called **viable but not cultivable.**

Cell culture or tissue culture media: They are prepared by using cells or tissues or bits of organ, e.g., uses and other details are mentioned in **Ch. 119**.

> **Note:** *Mycobacterium leprae,* pathogenic *Treponema pallidum* and *Lacazia loboi* are growing in animal culture.

TABLE 114.10: Microbes cannot grow in cell free media

Bacteria	
Treponema pallidum	*Rickettsia* spp.
Mycobacterium leprae	*Orientia tsutsugamushi*
Chlymydia spp.	*Ehrlichia* spp.
Fungi	
Loboa loboi	*Pneumocystis jiroveci*
Others	
Rhinosporidium seeberi, all viruses and metazoa	

ACCESS YOURSELF

Short Notes

1. Simple (basal) media/enriched media/enrichment media.
2. Transport (holding) media or viral transport media.
3. Selective media/sugar media/anaerobic media/differential media/composite media/cell culture (tissue culture) media.

Short Questions for Theory/Viva Questions

1. Define biphasic media. Write two examples.
2. What is agar? Write its role in preparation of media.
3. Name the one medium useful for following laboratory tests. citrate test, urease test, gelatin liquefaction test and anti-biogram.
4. Write four examples of selective media/differential media.
5. Name the pH indicator in following media.
6. Mac Conkey's medium, TCBS medium, sugar medium and Stuart's urea broth.
7. Name the two viral transport media.
8. What is modified Mac Conkey's medium? Write its use.
9. What are dehydrated media?
10. What are composite media? Write two examples.
11. Name four bacteria which cannot grow on cell free media.

Comment on

1. SPS is added in blood culture bottle.
2. Mac Conkey's medium is classified as both selective and differential medium.
3. Blood ager and TCBS are classified as differential media.
4. CLED medium is preferred over Mac Conkey's medium for testing of urine (UTI) sample.

Question of Practical Exercise

1. How you prepare a blood agar?

MCQs for Chapter Review

Introduction (History)

1. **Robert Koch was suggested to use agar in solidification of media by wife of one investigator of his laboratory, because:**
   a. Gelatin liquefies at 24°C.
   b. Gelatin liquefies by proteolytic bacteria.
   c. Agar is cheap.
   d. Gelatin is costly.
2. **Earliest solid medium used for bacterial cultivation was:**
   a. Cooked cut potato medium
   b. Blood agar
   c. Mac Conkey's medium
   d. Nutrient agar

Introduction (Ingredients/Components)

3. **Agar-agar is used in medium as:**
   a. Nutritional agent
   b. Solidifying agent
   c. Selective agent
   d. Preservative
4. **Which anticoagulant is used in blood for blood culture?**
   a. Sodium citrate
   b. EDTA
   c. Oxalate
   d. SPS
5. **In blood culture, the ratio of blood to reagent is:**
   a. 1:5
   b. 1:20
   c. 1:10
   d. 1:100

Classification of Media

6. **Which of the following is true?**
   a. Agar has nutrient property.
   b. Chocolate medium is selective medium.
   c. Addition of selective substance in a solid medium is called enrichment media.
   d. Nutrient broth is a basal medium.
7. **Chocolate agar is an example of?**
   a. Enriched medium
   b. Enrichment medium
   c. Selective medium
   d. Transport medium
8. **Blood agar is an example of?**
   a. Enriched medium
   b. Indicator medium
   c. Enrichment medium
   d. Selective medium
9. **Which is an enrichment medium?**
   a. Selenite F broth
   b. Chocolate medium
   c. Meat extract medium
   d. Egg medium
10. **Selenite F broth is an enrichment medium for?**
    a. *Salmonella*
    b. *Shigella*
    c. *E. coli*
    d. *Campylobacter*
11. **Lactose fermentation is seen in:**
    a. Blood agar
    b. Chocolate agar
    c. Mac Conkey's agar
    d. LJ medium
12. **MacConkey's agar is an example of?**
    a. Enriched medium
    b. Differential medium
    c. Enrichment medium
    d. Selective medium
13. **pH indicator in MacConkey's medium is:**
    a. Phenol red
    b. Andrade's indicator
    c. Neutral red
    d. Bromothymol blue
14. **Recommended transport medium for stool specimen suspected to contain enteric pathogen is:**
    a. Amies medium
    b. Buffered glycerol saline medium
    c. MacConkey's medium
    d. Stuart's medium
15. **pH of SDA is adjusted to:**
    a. 4–6
    b. 1-2
    c. 6–8
    d. 8–10
16. **In a patient with UTI, CLED medium is preferred over MacConkey's medium because:**
    a. It is a differential medium.
    b. It inhibits the swarming of *Proteus*.
    c. Promotes growth of *Pseudomonas*.
    d. Promotes growth of *Staph* and *Candida*
17. **Which organism cannot be cultured in cell free media?**
    a. *K rhinoscleromatis*
    b. *K ozaenae*
    c. *T pallidum*
    d. *Pneumocystis jiroveci*
    e. *Rhinosporidium seeberi*
18. **The term "viable but not cultivable" is used for:**
    a. *M leprae*
    b. *M tuberculosis*
    c. *T pallidum*
    d. *Salmonella*

1. **a, b**

2. **a**
- Follow section, **introduction (history** for explanation of answers of MCQs 1–2.

3. **b**

4. **d**

5. **c**
- Follow section, **introduction (ingredients/components)** for explanation of answers of MCQs 3–5.

6. **d**
- Agar has no nutrient property. Chocolate medium is enriched medium. Addition of selective substance in a solid medium is called selective medium.

7. **a**

8. **a**
- Follow section, **classification of media (enriched media → Table 114.2)** for explanation of answers of MCQs 7–8.

9. **a**

10. **a, b**
- Follow section, **classification of media (enrichment media → Table 114.3)** for explanation of answers of MCQs 9–10.

11. **c**
- MacConkey's agar is a selective and differential medium. It contains lactose, so it is useful to differentiate LF from NLF.

12. **b, d**

13. **c**
- Follow section, **classification of media (selective media → Table 114.5 and differential media)** for explanation of answers of MCQs 12–13.

14. **b**
- Follow section, **classification of media (transport media → Table 114.4)** for explanation.

15. **a**
- Follow section, **classification of media (selective media → Table 114.5)** for explanation.

16. **b**
- As far as CLED medium is concerned all options are right but here case is of UTI, where it is more useful to inhibit the swarming of *Proteus* by its electrolyte deficient nature.

17. **c, d, e**

18. **a, c**
- Follow **Table 114.10** for explanation of answers of MCQs 17–18.

Culture Methods

INTRODUCTION

Definitions

Culture method: Method used for growing (cultivating) microorganisms in media is called **culture method**.

Inoculum: Liquid culture or suspension of micro-organisms used for cultivation is called **inoculum**.

Aims of Culture

- For isolation of bacteria in pure culture: Isolation of bacteria is defined as separation of species/strain from clinical specimens/culture suspension, contain mixed population of microbes (identification of bacteria is defined as knowing the exact name of bacteria).
- For demonstration of bacterial properties
- For preparation of the antigens and vaccines
- For bacteriophage typing
- For bacteriocin typing
- For antibiotic sensitivity testing
- For estimation of viable counts
- For maintenance of stock cultures

Types

Two types like manual and automated culture methods described below.

MANUAL CULTURE METHODS

These include aerobic and anaerobic culture methods described below in details.

Aerobic Manual Culture Methods

Four types method like for solid media, for semi-solid media, for liquid media and for biphasic media

For Solid Media

1. **Streak culture (surface plating): Steps (Fig. 115.1)** include one loopful of culture is made as a primary inoculum and spread by streaking it with the nichrome loop (26 or 27 SWG) in a series of parallel lines in different segments of the plate. Loop flamed and cooled between the different sets of streaks. Incubate the plate for desired time and at desired temperature. On incubation growth may be confluent at the site of the original inoculation, but becomes progressively thinner and well separated colonies (pure culture) are obtained over the final series of streaks. It is **used** for routine isolation of bacteria and provides the isolated growth from the mixed culture.

2. **Lawn culture/carpet culture: Steps** include either flooding the surface of the plate with an inoculum or pipetting the surface of the plate with an inoculum or swabbing surface of the plate after soaking the swab in an inoculum. It is **used** for bacteriophage typing, antibiotic sensitivity testing (disc diffusion method), preparation of bacterial antigens and preparation of vaccines. It provides a uniform growth.

3. **Stroke culture (Fig. 115.2a): Steps** include streaking of agar slope/slant in test tubes by using nichrome loop **(Fig. 115.1a)** or nichrome straight wire **(Fig. 115.1b)**. It is **used** to identify biochemical properties like pigment production, citrate utilization, sugar fermentation in TSI medium etc. It provides a pure growth of the bacterium.

4. **Pour plate culture (Fig 115.3): Steps** include both agar and inoculum are plated simultaneously in Petri dish. Tube or flask contains 15 ml of the agar medium is melted and left to cool in a water bath at 45–50°C. Mix 1 ml of inoculum and melted agar in plate. Allow to settle. After incubation colonies will be developed throughout the depth of the medium

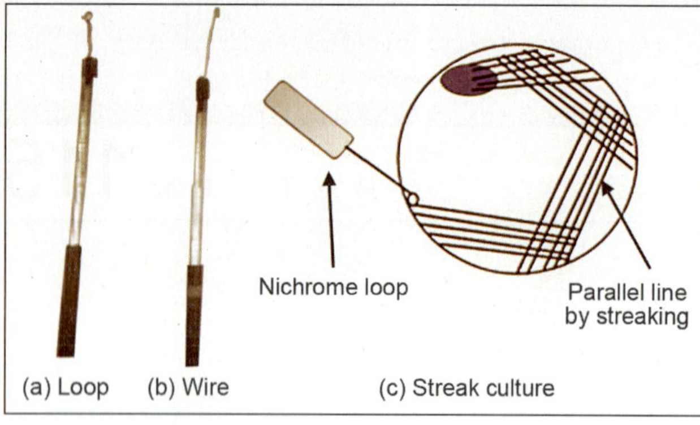

Fig. 115.1: Nichrome loop, wire and streak culture

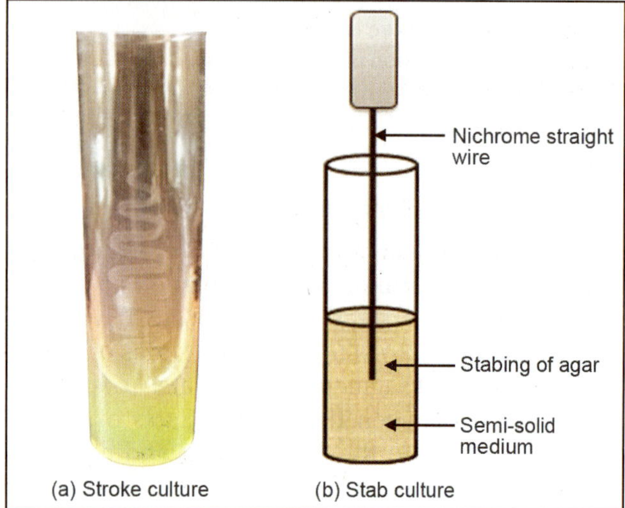

Fig. 115.2: Stroke culture and stab culture

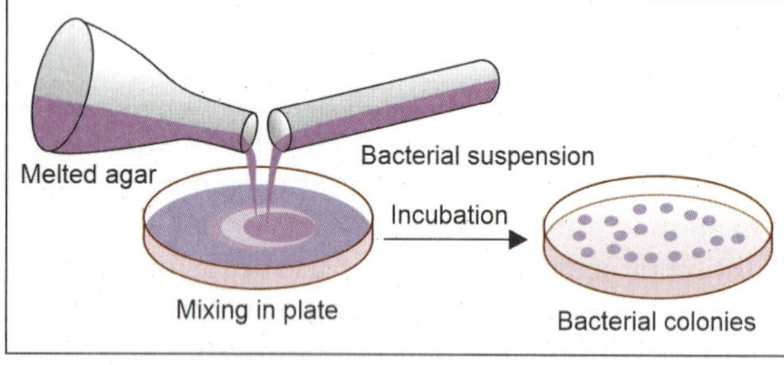

Fig. 115.3: Pour plate culture

which are enumerated using colony counters. It is **used** to estimate the viable bacterial count in a suspension and for quantitative urine culture.

For Semi-Solid Media

Method called **stab culture. Steps** include penetration of semi-solid agar medium (flat at top) in test tube by using nichrome straight wire (not nichrome loop) as shown in **Fig. 115.2b**. Stop once the wire reaches 0.5 inches away from the bottom of the stab media. It is **used** for motility detection, gelatin liquefaction test, O_2 requirement of the bacterium under research and to maintain the stock culture.

For Liquid Media

Method called **liquid culture method. Steps** include inoculation of liquid media either by touching with a loop dipped in inoculum or by adding the inoculum with Pasteur pipette or syringe. It is **used** for blood culture, other body fluid culture and antibiotic sensitivity testing (dilution method). It is preferred when large yields are desired. It does not provide pure growth from mixed culture.

For Biphasic Media

Method called **biphasic/diphasic culture method. Steps** include first inoculation of specimen in liquid part. Bottle is incubated at 37°C for 24 hours in upright position. Remove the bottle from the incubator. Tilt the bottle, so liquid part will run over solid part. Bottle is re-incubated at 37°C for 24 hours in upright position. Next day examine the solid phase for growth. It is **used** for isolation of bacteria like *N. meningitidis, S.* Typhi, *B. abortus*, etc. It eliminates the chances of aerial contamination and less labor intensive.

Anaerobic Manual Culture Methods

Synonym: Anaerobiosis

Definition: Method to create an anaerobic condition is called **anaerobiosis**

Use: For cultivation of anaerobic bacteria

Classification of anaerobic bacteria: Ch. 7.

Methods: These are physical methods, chemical methods, biological methods, method by using reducing agents in anaerobic media and method by using pre reduced system. All are described below.

Physical Methods

1. **Vacuum desiccators**
   Principle: Replacement of O_2 by vacuum.
   Disadvantages: Not useful because some O_2 left behind.
2. **Candle (CO_2) jar**
   Principle: Replacement of O_2 by CO_2 generated from candle. Candle will burn till O_2 is present in jar, then after it extinguished.
   Steps (Fig. 115.4): Put the culture plate in jar. Keep enlightened candle in jar. Put the lid and sealed the jar with vaseline.
   Advantage: It provides atmosphere with 3–5% CO_2 which stimulates the growth of most bacteria.
   Disadvantage: It is not useful because some O_2 left behind.

Chemical Methods

1. **McIntosh and Filde's anaerobic jar**
   Principle: It based on evacuation-replacement mechanism. Evacuation of O_2 from jar by vacuum pump and residual O_2 combines with H_2 to form water in presence of catalyst. Hydrogen pump acts as source of H_2.

Fig. 115.4: Candle jar

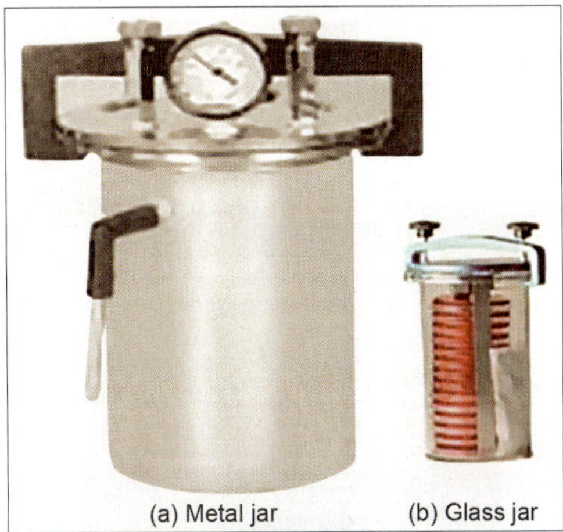

Fig. 115.5: McIntosh and Filde's jar

Structure: It consists metal (**Fig. 115.5a**) or stout glass jar (**Fig. 115.5b**) and metal lid. Lid has one inlet tube, one outlet tube, pressure gauge and electric terminal inner to which catalyst is attached.

Catalyst: It uses palladinized asbestos which acts after heating. Alumina pellets coated with palladium act at room temperature.

Steps: Put the inoculated test and control plates in a jar. Clamp the lid tightly. Evacuate the air from jar by attaching outlet with vacuum pump. Close the outlet and attach the inlet with H_2 pump. Close the inlet when jar is filled with H_2. Connect the electric terminal and start the electric supply. Palladinized asbestos catalyst is heated after electric supply and form water by combining the residual O_2 with H_2.

Advantage: It gives complete anaerobiosis.

Disadvantage: There are chances of explosion due to overheating if palladinized asbestos catalyst is used. That is overcome by using alumina pellets coated with palladium which acts at room temperature.

Controls

- **Chemical control:** Colorless methylene blue gives blue color if anaerobiosis is not maintained.
- **Biological control:**
    - Negative control: *P. aeruginosa* (aerobic bacteria) will grow if anaerobiosis is not maintained.
    - Positive control: *Cl. tetani* (anaerobic bacteria) will grow if anaerobiosis is maintained.

2. GasPak method

Principle: Different chemicals generate H_2 and CO_2 on addition of water. H_2 combines with O_2 to form water in presence of cold catalyst and CO_2 favors the bacterial growth as shown in **Fig. 115.6**.

Structure: It is a plastic jar with chemicals as shown in **Fig. 115.7**.

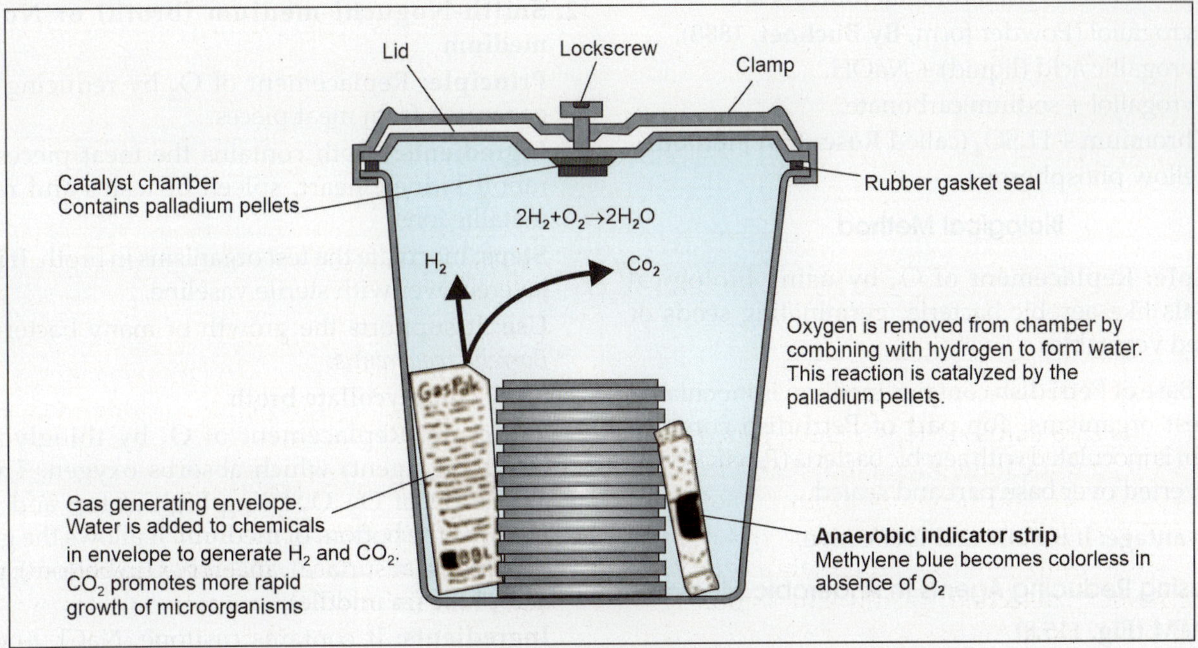

Fig. 115.6: Principle of GasPak method

Fig. 115.7: GasPak method

Fig. 115.8: RCMM

Catalyst: It includes alumina pellets coated with palladium called **cold catalyst.**

Steps: Put the inoculated test and control plates in a jar. Add the chemical (commercial available foil pack containing mixture of citric acid, sodium borohydrate and sodium bicarbonate). Add the water (about 10 ml) and screw tight the lid. H_2 and CO_2 are generated from chemicals after adding the water. H_2 combines with O_2 to form water in presence of cold catalyst and gives anaerobic condition.

Advantages: Method is simple, effective and commercial available as disposable envelope. There is no need of vacuum and H_2 pump. Some GasPak do not need water.

3. **Other chemicals used for anaerobiosis are**
   - Pyrogallol (Powder form, **By Buchner, 1888**).
   - Pyrogallic acid (liquid) + NaOH.
   - Pyrogallol + sodium carbonate.
   - Chromium + H_2SO_4 (called **Rosenthal method**).
   - Yellow phosphorous.

Biological Method

Principle: Replacement of O_2 by using biological materials like aerobic bacteria, germinating seeds or chopped vegetables.

Steps: Base of Petri dish contains medium is inoculated with test organisms. Top part of Petri dish contains medium is inoculated with aerobic bacteria (*Pseudomonas*) and inverted over base part and sealed.

Disadvantage: It is slow and ineffective.

By using Reducing Agents in Anaerobic Media

1. **RCMM (Fig. 115.8)**
   Full form: Robertson's Cooked Meat Medium.

Principle: Meat generates the **unsaturated fatty acids** and reducing substance (**glutathione**) which absorbed the O_2 and provides anaerobic condition (reduced Eh).

Ingredients: Nutrient broth and meat pieces.

Method of sterilisation: Autoclave.

Steps: Inoculate the test organism in RCMM. Proteolytic anaerobe like *Cl. tetani* attacks protein in meat and turns meat in black color with foul smelling. Saccharolytic anaerobe like *Cl. perfringens* attacks carbohydrates in meat and turns meat in pink/red color.

Uses: It is used to maintain the stock culture and for anaerobic culture.

2. **Smith-Noguchi medium (broth) or Noguchi medium**

Principle: Replacement of O_2 by reducing agent generated from meat pieces.

Ingredients: Broth contains the meat pieces from rabbit kidney, heart, spleen or testes and red hot metallic ions.

Steps: Inoculate the test organisms in broth. It is than layered over with sterile vaseline.

Use: It supports the growth of many bacteria like *Borrelia recurrentis*.

3. **0.1% Thioglycollate broth**

Principle: Replacement of O_2 by thioglycollate (reducing agent) which absorbs oxygen. There is gradation of O_2. O_2 being high at top and lower towards the bottom of medium. It allows the growth of aerobes (at surface), anaerobes (in bottom), micro-aerophilic (in middle).

Ingredients: It contains casitone, NaCl, l-cystine, thioglycollic acid, methylene blue, agar and pH 7.2.

Broth is enriched with hemin and vitamin K.

Steps: Boiled the broth to eliminate the O_2. Cool down and inoculate the test organism in broth. Bacterial growth is identified by turbidity.

Use: It supports the growth of anaerobes and micro-aerophilic bacteria.

Modification: Addition of small quantity of agar (semisolid) in broth prevents the diffusion of O_2 and increases the anaerobic capacity.

4. **Other reducing agents and anaerobic media:** Other reducing agents are 1% glucose, 0.1% ascorbic acid, 0.05% cysteine, etc. Other anaerobic media are anaerobic blood agar, brain-heart infusion agar with supplements like vitamin K and hemin, neomycin blood agar, egg yolk agar, phenyl ethyl agar, *Bacteroides* bile esculin agar, etc.

By Using Prereduced System

1. **Prereduced Anaerobic System (PRAS):** It is used for quantitative culture and for fastidious bacteria.
2. **Anaerobic chamber (glove box):** It is an airtight, glass fronted cabinet filled with inert-gas. It has entry locks for the introduction and removal of materials and gloves for the hands.

Aerobic Automated Culture Methods

Automated Systems (Instruments) for Culture from Clinical Specimen

Specimens: Blood, bone marrow and sterile body fluids like synovial fluid, CSF, plural fluid and peritoneal fluid.

Media: Tryptic soy broth and/or brain heart infusion broth with polymeric resin beads, which adsorbs and neutralizes the antimicrobials present in blood specimen.

Advantages: Methods are highly sensitive, less labor intensive and detect the growth very rapidly.

Disadvantages: Methods are costlier. Trained and highly qualified person is required. They use liquid medium so colony morphology is not observed.

Systems/instruments

1. **BacT/Alert 3D system:** It based on colorimetric detection of pH change due to CO_2 production by bacteria. When bacteria multiply they produce CO_2 which increases the pH which changes the color of a blue-green sensor present at the bottom of the bottle to yellow which is detected by colorimetric system.
2. **BacT/ALERT VIRTUO (bioMerieux):** It is an advanced form of BacT/ALERT 3D system with few advantages like rapid detection of growth, determine volume of blood present in bottle and automatic loading and unloading of bottles.
3. **BACTEC system:** It has different principle as per instrument/manufactures as follows
   - ^{14}C labeled palmitic acid is converted in to $^{14}CO_2$ by bacterial metabolism. $^{14}CO_2$ index is measured for interpretation of result.
   - Fluorescent based BACTEC: The **BD BACTEC** automated blood culture system utilizes fluorescent technology in detecting the growth of organisms in the blood culture bottles. When microbes present in the cultured vials, they metabolize nutrients in the culture medium, releasing CO_2 into the medium. A dye in the sensor at the bottom of the culture bottles will reacts with CO_2. This modulates the amount of light that is absorbed by a fluorescent material in the sensor. The instrument's photo detectors measure the level of fluorescence, which corresponds to the amount of CO_2 released by the organisms. Then the measurement is interpreted by the system according to preprogrammed positivity algorithms.
   - It also works a similar principle to MGIT (**Ch. 56**).

Automated Systems (Instruments) for Identification of Culture Growth from Petri Dish/Slant

Advantages: They are highly sensitive, produce faster result and identify the wide range of organisms like anaerobes which are difficult to identify by conventional biochemical tests.

Systems/instruments

1. **MALDI–TOF (Matrix Assisted Laser Desorption Ionization Time Of Flight):**

Types: Two systems are available commercially like VITEK MS (BioMérieux) and Biotyper (Bruker).

Uses: It is used to identify bacteria including myco-bacteria and fungi, with a turnaround time of few minutes and with absolute accuracy.

Principle: It examines the pattern of ribosomal proteins present in the organism.

Methods (Fig. 115.9)

- **Sample preparation:** Smeared the colony of an organism onto a well of the slide and add one drop of matrix solution contains cyano-hydroxy-cinnamic acid to the same well and mixed. Put the slide (load) in the system.
- **Steps after putting the slide in system**
  - Ionization chamber: Wells are exposed to laser beam. Matrix absorbs the laser and causes desorption and ionization of bacterial ribosomal proteins to generate the protonated ions singly.
  - Analyzer: Protonated ions are accelerated in to analyzer chamber via electric field. Analyzer separates the ions according to their time of flight in flight tube. Smaller ions travels faster followed by bigger as per mass of charge (m/z) ratio.
  - Detector: It converts the received ions in to electric signals/current which is amplified and digitized to form a characteristics graph (spectrum) which is unique for particular organism due to its unique ribosomal proteins. The test graph obtained is compared with known database.

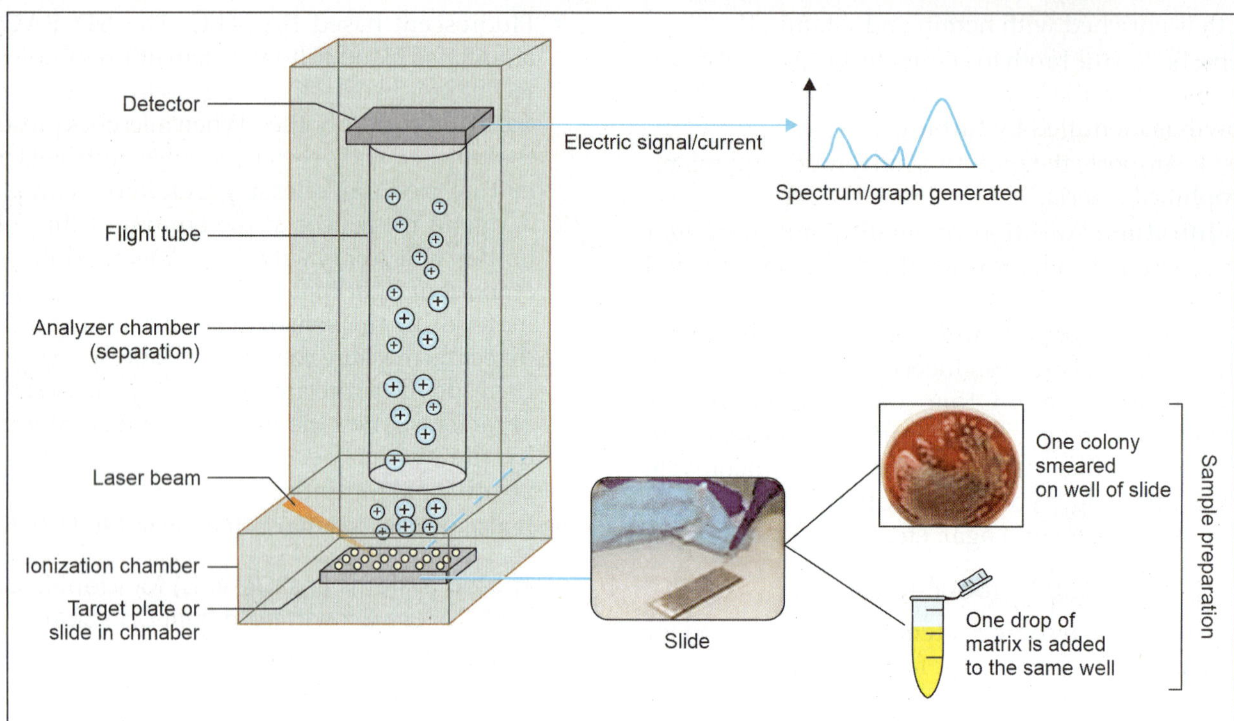

Fig. 115.9: Principle and steps of MALDI-TOF

Fig. 115.10: VITEK 2 compact

2. VITEK 2 compact (bioMérieux, Fig. 115.10):

Uses: It is used to identify and to do AST for both bacteria and yeast.

Principle: It uses colorimetric reagent card containing 64 wells (cards are incubated in the system at 35.5°C and reading is taken every 15 minutes by optical system), each well contains an individual test substrate. Separate cards are available for gram-positive, gram-negative and fastidious bacteria plus yeasts. Substrates in the well measure different metabolic activities like acidification, alkalinization, enzyme hydrolysis, etc., which reflected as colored product measured by optical system (colorimetry). All these help in identification of organisms. The test result obtained from test organism is compared and matched with known database to identify the species. Result of identification is usually available within 4–6 hours. Principle is illustrated in **Fig. 115.11**.

3. Phoenix (BD Diagnostics): It is used to identify and to do AST for both bacteria and yeast.

4. MicroScanWalkAway system (Beckman Coulter): It is used to identify and to do AST for both bacteria and yeast.

Automated Systems (Instruments) for AST

Follow **Ch. 116.**

Anaerobic Automated Culture Methods

System is **anoxomat** which based on **evacuation–replacement mechanism.** Evacuate O_2 from jar and

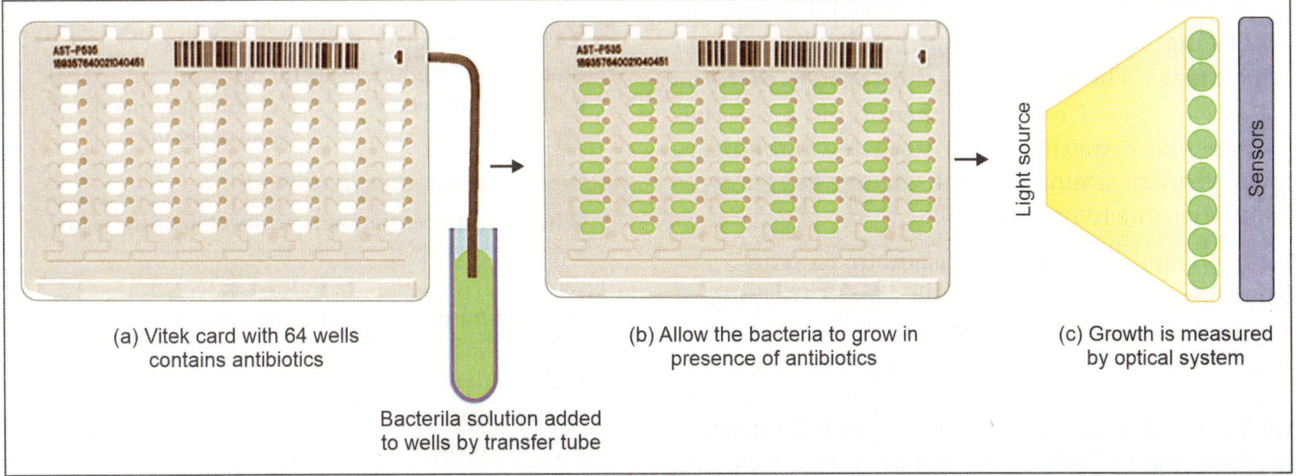

Fig. 115.11: Principle of VITEK 2 compact

replace with H_2 from cylinder in presence of catalyst same as McIntosh and Filde's jar. Hydrogen cylinder acts as source of H_2. It is more effective for anaerobiosis than McIntosh and Filde's jar.

METHODS OF PURE CULTURE ISOLATION

History

Robert Koch invented the culture method for pure isolation of bacteria over solid media.

Methods

1. **Surface plating:** To isolate pure culture from mixed growth.
2. **Enrichment, selective and indicator media:** To isolate pure culture from samples like feces, urine, etc., contain normal flora.
3. **Pretreatment of specimens with bactericidal agents:** Bactericidal agents like acid, alkali, etc., kill the unwanted organisms (e.g., commensals) from specimens. It is used to isolate *M. tuberculosis* from sputum and other samples.
4. **Temperature methods**
   - **Different temperature incubation methods:** Thermophilic can grow at 60°C. Mesophilic can grow at 37°C. Incubate a mixture of *N. meningitidis* and *N. catarrhalis* at 22°C, where later can grow.
   - **Heating method:** Heat a mixture of vegetative bacteria and spores 80°C, where later can grow.
5. **Separation of motile from non-motile bacteria**
   - **Craigie's tube method (Fig. 115.12):** Wide tube contains soft agar (0.2%) plus both ended open narrow and small inner tube. Inoculate the strain in inner tube and after sufficient incubation, subculture from agar surface outer to inner tube which yields the motile cells.
   - **U-tube method:** Inoculate the strain from one end and subculture from opposite end will yield motile cells.
6. **Laboratory animal inoculation method:** For example, separation of *Anthrax bacillus* from aerobic spore

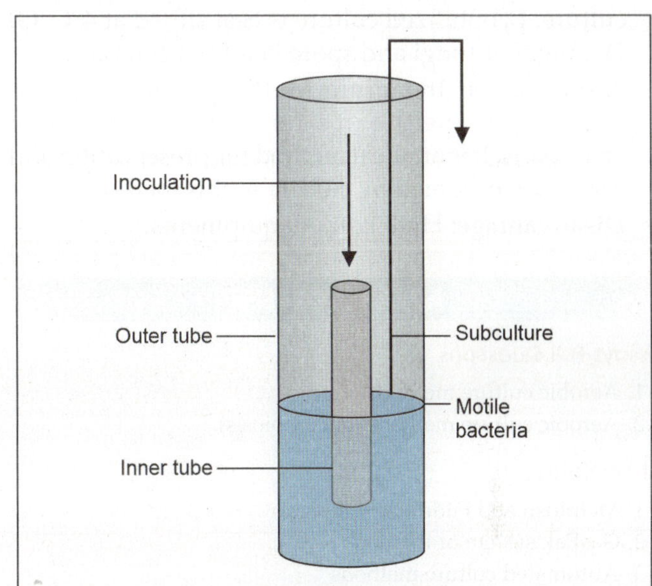

Fig. 115.12: Craigie's tube

bearer by inoculation in mice or guinea pigs. Produce fatal septicemia and *A. bacillus* will be cultured from heart blood.

7. **Filtration methods:** They separate the different size bacteria and also separate the viruses from bacteria.
8. **Anaerobiosis:** It separates the aerobes from anaerobes. Methods are described earlier.

METHODS OF PRESERVATION OF MICROORGANISMS

Aims of Preservation of Microorganisms

It required for academic work, research work and epidemiological purposes.

Methods

Short term methods: They are useful for weeks to months.
- **Methods: (1) Subculturing:** It is a routine method for preservation by using different liquid and solid media especially nutrient agar slant. RCMM is used for anaerobes. **(2) Freezing at −20°C (3) Immersion method:** Preservation by immersing the culture

in mineral oil, glycerol, sterile distilled water, etc.
(4) Drying: For fungi and spore bearing bacteria.

- **Advantages:** They are cheaper and easy to perform.
- **Disadvantages:** Change in phenotypic properties like loss of capsule due to repeated subculture make virulent strain avirulent. Change in genotypic properties due to mutation.

Long-term methods: They are useful up to years.

- **Methods: (1) Ultratemperature (deep) freezing:** It includes preservation of culture at −70°C with mixture of preservatives like glycerol, skimmed milk, sucrose, etc. Such preservatives are called **cryopreservatives. (2) Freeze drying:** It also called **lyophilization.** It involves the freezing of the liquid culture followed by dehydration to remove water from frozen bacterial suspension. It gives stable and readily dehydrated culture. Lyophilized culture is best stored at 4°C. **(3) Drying:** For fungi and spore bearing bacteria.
- **Advantages:** It maintains the phenotypic and genotypic properties. It maintains the viability of organisms. It is an ideal method for preservation and less space occupied by culture in this method.
- **Disadvantage:** High cost of equipments.

ACCESS YOURSELF

Essays/Full Questions

1. Aerobic culture methods.
2. Aerobic culture methods (anaerobiosis).

Short Notes

1. McIntosh and Filde's anaerobic jar
2. GasPak system or method
3. Automated culture methods
4. Methods to isolate the bacteria in pure culture
5. Methods for preservation of microorganisms

Short Questions for Theory/Viva Questions

1. Write the four uses of culture methods.
2. Write the principle of GasPak method/McIntosh and Filde's anaerobic jar method.
3. What is Craig's tube?
4. What is lyophilization?

Comment on

1. In McIntosh and Filde's jar anaerobiosis, alumina pellets coated with palladium acts as better catalyst than palladinized asbestos catalyst.

Question of Practical Exercise

1. How you do a streak culture or perform gas pack method?

MCQs for Chapter Review

Aerobic Manual Culture Methods

1. Lawn culture is employed for:
   a. Viable count
   b. Antibiotic sensitivity
   c. Motility of bacteria
   d. Primary inoculation of clinical specimen

2. Stab culture method is useful for:
   a. Motility detection   b. Gelatin liquefaction test
   c. a + b   d. None of above

Anaerobic Manual Culture Methods

3. What CO_2 concentration is achieved by candle jar?
   a. 1%   b. 3%
   c. 6%   d. 9%

4. Medium for growth of anaerobic bacteria is:
   a. Smith Noguchi medium
   b. LJ medium
   c. Blood agar
   d. Nutrient agar

Aerobic Automated Culture Methods

5. MALDI-TOF technology is used for:
   a. Blood culture
   b. Bacterial identification
   c. Antimicrobial susceptibility
   d. All of the above

Methods of Pure Culture Isolation

6. Who introduced methods of obtaining bacteria in pure culture using solid media?
   a. Robert Koch
   b. Joseph Lister
   c. Louis Pasteur
   d. Edward Jenner

7. Craigie's tube is used for:
   a. Monitoring autoclave procedures
   b. Antigen extraction
   c. Demonstration of motility
   d. Sub culture and stock cultures

Answers and Explanation of MCQs

1. b
- Follow section, **aerobic manual culture methods → lawn culture/carpet culture)** for explanation.

2. c
- Stab culture method is useful for motility detection and gelatin liquefaction test.

3. b
- Follow section, **anaerobic manual culture methods → candle (CO_2) jar** for more explanation.

4. a
- Follow section, **anaerobic manual culture methods** (by using reducing agents in anaerobic media) for more explanation. LJ medium is selective medium for *M. tuberculosis*. Blood agar is an enriched and nutrient agar is simple (basal) medium.

5. b
- Follow section, **aerobic automated culture methods (MALDI-TOF → uses)** for explanation.

6. a
- Follow section, **methods of pure culture isolation → history** for explanation.

7. c
- Follow section, **methods of pure culture isolation → Craigie's tube and Fig. 115.12** for explanation.

CHAPTER 116

Drug Resistance Mechanisms, Antibiotic Sensitivity Testing and Antimicrobial Stewardship Program

Chapter Outline

- Drug Resistance Mechanisms
- Antibiotic Sensitivity Testing
- Antimicrobial Stewardship Program (ASP)

DRUG RESISTANCE MECHANISMS

Definition

Irresponsiveness of microorganisms against antibiotics is called **antimicrobial drug resistance**.

Types

Two types like intrinsic and acquired as described below.

Intrinsic (Natural/Primary/Pretreatment/Initial) Resistance

Definition: Bacteria were resistant to the antibiotics even before the antibiotics were introduced.

Significance: It does not cause any problems in prevention or in treatment of diseases.

Mechanisms: Resistance occurs due to lack of metabolic process or target site required for drug action.

Examples: (1) GNB are normally unaffected by penicillin. (2) *M. tuberculosis* is insensitive to tetracycline.

Acquired (Secondary/Post-treatment) Resistance

Definition: Bacteria were previously sensitive to the antibiotics become resistant after the introduction of antibiotics.

Significance: It has great importance because it would result in treatment failure and transfer of resistance to other bacteria.

Mechanisms, subtypes and examples: Two subtypes.

A. **Physiological and biochemical mechanisms**

1. **Enzymatic inactivation of antibiotics:** Bacteria produce many enzymes which inactivate or destroy the drugs.
   - **Beta-lactamase (penicillinase):** It destroys the β-lactam ring and inactivates the β-lactam antibiotic (penicillin). Enzyme is inducible and it's production is controlled by plasmid. Such plasmid transmitted to other bacteria by conjugation or by transduction to make them resistant to penicillin. **Four types of penicillinase enzymes** are A to D. A is produced by hospital strains. **Examples of penicillinase producing bacteria** are *Staph. aureus, E. coli, K. pneumoniae, S. Typhi, P. mirabilis, N. gonorrhoeae, H. influenzae*, etc.
   - **Acetyl transferase:** It inactivates the chloramphenicol. **Examples of acetyl transferase producing bacteria** are *S.* Typhi, *Haemophilus* spp., and *E. coli*.
   - **Acetylase/adenylase/phophorylase:** It inactivates the aminoglycosides.

2. **Decrease in permeability of drugs:** Many hydrophilic antibiotics enter in bacteria by specific channels formed by proteins called **porin channel** or need specific transport channel. Losses of these channels produce the resistant. For examples, low degree penicillin-resistant in gonococci (gonococci are less permeable to penicillin) and aminoglycosides resistant in GNB.

3. **Development of alternate metabolic pathway to bypass the action of antibiotics:** Sulfonamide resistant bacteria like pneumococci, meningococci, gonococci, *E. coli, Staph. aureus, Shigella* spp., *Strept. pyogenes, Strept. viridans*, etc., utilize preformed folic acid from medium instead of synthesizing it.

4. **Active efflux (pump out) of antibiotics:** It occurs in tetracycline, cephalosporin, fluoroquinolones and in macrolides group of drugs.

5. **Cross resistance:** Acquisition of resistance to one drug confers resistance to other drug, to which organisms has not been exposed is called **cross resistance**. It is of following types.
   - **According to group of drugs**
     - **Between related drugs:** It may be complete or incomplete drug resistance. Resistance to one drug may extend to other drugs of same group

called **complete drug resistance**. For example (1) Resistance to one sulfonamide means resistance to all other drugs of sulfonamide group. (2) Resistance to one tetracycline means resistance to all other drugs of tetracycline group. Resistance to one drug may not extend to other drugs of same group called **incomplete drug resistance**. For example, resistance to one aminoglycoside may not extend to others (bacteria resistance to gentamicin may sensitive to amikacin).

– **Between unrelated drugs:** It includes only **incomplete drug resistance**. For example, (1) Tetracycline and chloramphenicol. (2) Erythromycin and lincomycin.

- **According to direction**
  – **Two ways:** Resistance to erythromycin extends to clindamycin and *vice versa*.
  – **One way:** Neomycin resistance to Enterobactericeae extends to streptomycin, but not *vice versa*.

B. Genetic mechanisms: It may be chromosome mediated or extrachromosomal.

1. **Chromosomal:** It occurs due to mutation. Some antibiotics work after binding with certain proteins. Mutation can alters such protein resulting in a protein with little or no affinity for the drug. Following are the types and examples of mutational resistant.

- **Step-wise mutations:** It is the high level of resistance achieved by a series of small-step mutations. For example **(1) Pn resistance in *Staph. aureus*:** Alteration in penicillin binding proteins (PBP2a) in cell wall of *Staph. aureus* cause unbinding or binding with low affinity to Pn (β-lactam) resulting Pn resistance. Such resistance extends to cover other β-lactam antibiotics like methicillin, cloxacillin or oxacillin called **methicillin resistance *Staph aureus* (MRSA). (2) Isoniazid and rifampin resistance in *M. tuberculosis*:** Mutations in genes are associated with isoniazid and rifampin resistance in *M. tuberculosis*.

- **One-step mutation:** It is the resistance in bacteria due to single mutation. For example, streptomycin resistance in *M. tuberculosis*. Initially bacilli are sensitive to streptomycin, but later resistant mutants develop, which multiply unchecked and replace the sensitive ones resulting treatment failure. Presence of other drug can kill such mutant bacilli. For this reason, multiple drugs are included in the treatment of tuberculosis. In addition to all this inadequate or irregular treatment can develop multiple drug resistance tuberculosis (MDR TB) and extended-drug resistance tuberculosis (XDT TB) strain.

2. **Extrachromosomal:** It occurs by different methods of gene transfer called **transferable or infectious drug resistance.**

- **Transformation:** It is the resistant transferred by agency of free DNA. Resistance transferred by transformation is demonstrated experimentally, but not useful clinically. For example, acquisition of altered PBP2a by pneumococci produces resistance against penicillin.

- **Transduction:** It is the resistant transferred by bacteriophage. Plasmid (R factor) is taken up by bacteriophage and transferred to other bacterium. For example, (1) Plasmid mediated penicillin resistance in *Staphylococcus*. (2) Plasmid mediated chloramphenicol resistance in *S. Typhi*. (3) Erythromycin resistance is also phage mediated.

- **Conjugation:** It is the resistant transferred by conjugation tube. Transfer of R factor plasmid by conjugation is the most important method of drug resistance. Acquisition of R factor confers resistance to several antibiotics. For example resistance to penicillin in gram-negative bacteria due to β-lactamase enzyme coded by plasmid.

Prevention of Drug Resistance

Elimination of R factor: By treating the bacteria with acridine dyes or ethidium bromide.

Preferring narrow spectrum drugs: Avoid the use of broad spectrum antibiotics.

Combination therapy: Avoid long-term treatment. This would minimize the selection pressure and time required for resistant strain to emerge. Use combined therapy whenever long-term treatment is required. For example in tuberculosis, SABE, etc.

Differences between Mutational and Transferable Drug Resistance

Follow **Table 116.1.**

TABLE 116.1: Differences between mutational and transferable drug resistance	
Mutational	**Transferable**
Mediated by chromosomal part	Mediated by extrachromosomal part
Low degree virulence	High degree virulence
One drug resistance	Multiple drug resistance
Treated by high dose	High dose is ineffective
Prevented by drug combination	Not prevented by drug combination
Mutants may be defective	Mutants may be not defective
Resistance does not spread	Resistance will spread to same or other species

ANTIBIOTIC SENSITIVITY TESTING

Synonym

It also called antibiotic/antimicrobial sensitivity/susceptibility testing (AST) or antibiogram or resistogram.

Definition

It is the detection of sensitivity of microorganisms to antimicrobial agents (antibiotics).

Uses

To select the antibiotics: AST is useful to choose the antibiotics to treat the patient.

To diagnose the organism: AST is useful to identify the organism and to differentiate one bacterium form other as mentioned below.

1. **Novobiocin** (5 mg) **sensitivity:** It is sensitive in *Staph. aureus* and in *Staph epidermidis* while resistant in *Staph. saprophyticus*, which helps in differentiation of these species.
2. **Bacitracin** (0.04 mg) **sensitivity:** It is sensitive in group A β-hemolytic streptococci and resistant in group B, which helps in differentiation of two β-hemolytic streptococci.
3. **Optochin** (ethyl hydrocuprein, 1/500,000) **sensitivity:** Pneumococcus is sensitive to optochin while viridans group is resistant, which helps in differentiation between all these bacteria.

Bacterial typing: It called **antibiogram typing** as like in vibrios and other bacteria.

Methods or Principle

Two types of methods or principles are in use as described below.

1. **By using liquid medium:** It called **dilution method** and it has following three subtypes.
   - Macrodilution method: It is performed by using liquid media in large quantity in test tubes.
   - Microdilution method: It is performed by using liquid media in very small quantity in microtiter wells.
   - Agar dilution method: It is performed by using Muller Hinton Agar (MHA).
2. **By using solid medium:** It called **disc diffusion method** and it has following for subtypes.
   - Rotary plate method: Discovered by Pearson-Whitefield, 1974. It is done by using control strain along with test strain.
   - Comparative disc diffusion method: Discovered by Stokes and Flemington, 1972. It is done by using control strain along with test strain.
   - Kirby-Bauer method: It is the most commonly used method and described separately in box.
   - Epsilomer test (E test): It is used to detect the MIC and described separately in box.
3. **Automated methods:** Following automated methods are used for identification of organisms and AST.
   - VITEK 2 (bioMérieux).
   - Phoenix (BD Diagnostics).
   - MicroScanWalkAway system (Beckman Coulter).

Materials required: (1) 5 ml Trypticase Soy Broth (TSB) or any other suitable broth: To prepare the inoculum or culture suspension of test organisms. (2) Antibiotic disc: It contains antibiotics for testing. (3) 0.5 Mac Farland standard inoculum: To compare the turbidity of broth contains test organisms. (4) Sterile forceps: To hold the disc. (5) Scale or calliper: To measure the size of inhibition zone. (6) MHA plate.

Steps: Inoculate few colonies from primary culture in TSB (called **inoculum**) and incubate at 37°C for few hours. (It called **direct AST** when performed directly from clinical specimen than using culture growth. It is performed in emergency.) Compare the TSB growth (inoculum) with 0.5 Mac Farland standard inoculums (1.5×108 cfu/ml). Take the MHA plate and transfer TSB growth in it by swabbing or flooding method. Allow the plate to dry for few minutes. Put the antibiotic disc by holding with sterile forceps. Incubate at 37°C for 24 hours and examine for sensitivity pattern.

Mechanisms during incubation: During the incubation drug released from disc and diffuse in medium along with the growth of microorganisms. At a particular point drug met to bacteria and if bacteria are sensitive then there is a production of zone of inhibition around the disc. If bacteria are resistant to drug then they grow up to the margin of disc without producing the zone of inhibition.

Results (Fig. 116.1): Measure the size of zone of inhibition by using the scale or calliper. (1) Sensitive (S) zone: Organisms are susceptible to drug. (2) Intermediate sensitive (IS) zone: Organisms are moderately susceptible to drug. (3) Resistance (R) zone: Organisms are not susceptible or resistant to drug

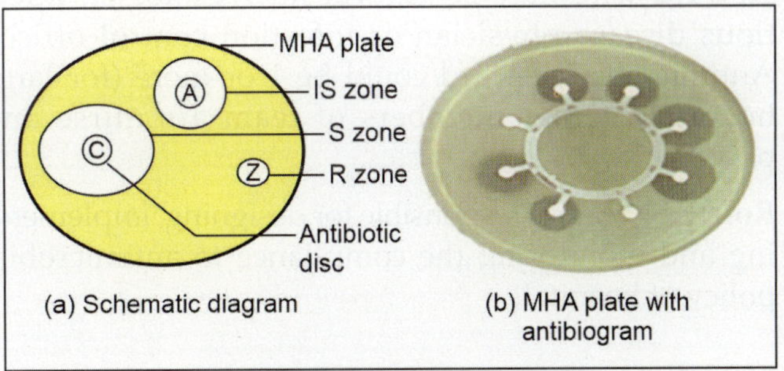

Fig. 116.1: Antibiogram

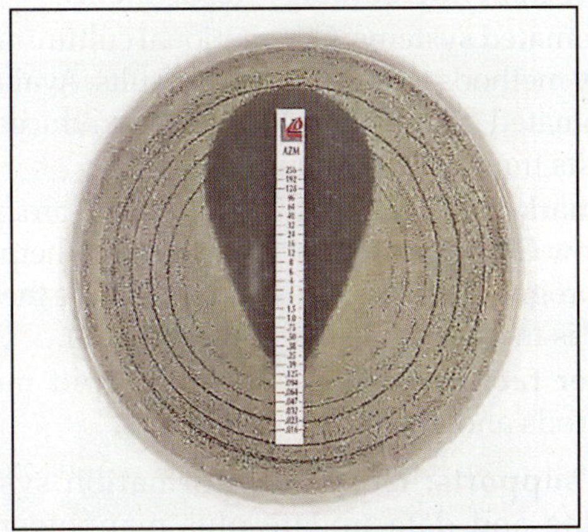

Fig. 116.2: Epsilomer test (E test)

Drug Resistance Mechanisms, Antibiotic Sensitivity Testing and Antimicrobial Stewardship Program

Synonym

It also called antimicrobial management program.

Definition

"Antimicrobial stewardship" is defined by CDC as the right antibiotic for the right patient, at the right time, with the right dose, by the right route, and causing the least harm to the patients and future patients.

Goals

Antimicrobial stewardship program (ASP) is required to prevent or to improve following problems.

Antimicrobial resistance: MDR organisms are available everywhere and resistance spread to other microbes creates problem in treatment.

Over use, misuse and widespread use of antimicrobials: Over use for self limiting or irrelevant infections like antibacterial for viral/fungal/parasitic infections. Misuse without need or prescription. Wide spread use for animal non therapeutic purposes (70%) followed by animal therapeutic purposes (15%), human therapeutic use (9%) and human nontherapeutic use (6%). A strong plan is required to control the wide spread use in animals.

Poor antimicrobial research: Research required to invent new drugs.

ASP Team

Team: Strong and committed leader called **antimicrobial steward** who may be clinical microbiologist, infectious disease physician or infection control officer. Antimicrobial steward could be 1 or more (for large hospital). Other members of team are nurse and pharmacist.

Role: ASP team is responsible for designing, implementing and monitoring the compliance to antimicrobial policy of hospital.

ASP Infrastructure

Support from microbiology department

- **Automated systems:** Conventional culture and sensitivity methods took 2–5 days of results. Availability of automated culture (**Ch. 115**) system reduces all over results time of 24–48 hours.
- **Biomarkers:** Availability of CRP and procalcitonin (follow **Ch. 51**) are the bacterial markers, help to know the prognosis of infections and also guide the therapy. CRP is more reliable than procalcitonin.
- **Other facilities:** These include rapid molecular methods and emergency laboratory.

Other supports: Hospital information system for reporting and data update plus manpower availability.

Designing Antimicrobial Policy

Every hospital has antimicrobial policy which is a pocket hand book prepared by ASP team after discussion with clinicians, microbiologist and administrators. It is prepared by considering all national and international antimicrobial guidelines and local AST reports. It provides information for drug uses and doses for particular infection.

Implementation of ASP

Strategies: Following two types.

- **Front end strategies (formulary restriction):** It divides the drugs in three categories with approval from ASP team for their uses as shown in **Table 116.2**. It is the strong strategy (**advantage**) with immediate effect and appears to be the most ideal way to achieve antimicrobial stewardship. It required approval from ASP team, so restrict the clinician's freedom (**disadvantage**) to choose empirical antibiotics especially in emergency time, so practically difficult to implement.
- **Back end strategies (prospective audit and feedback):** It includes audit and feedback between ASP team and clinicians. The ASP team goes for stewardship round during, which it discusses with the clinicians in detail about the compliance to the antimicrobial policy like proper use, dose and sensitivity report. Clinician should give justification about any noncompliance if occurred. This strategy is (**advantage**) more widely practiced, more easily accepted by clinicians, provides a higher opportunity for educating and training the health care professionals and impact is delayed but sustainable. It is more labor-intensive (**disadvantage**).

Education and training: It includes continuous education, training, motivation and assessment of the health care providers as like other programs.

Monitoring the Compliance

Policy adherence (process) indicators: It is achieved by conducting audit as described earlier and calculating of prescription and administrative compliance.

- **Indicators of prescription compliance**
  - Percentage of time the empirical antibiotic given, is it as per infectious disease suspected?

TABLE 116.2: Front end strategies		
Restricted drugs	**Semirestricted drugs**	**Unrestricted drugs**
Pharmacy supply of >1 days requires prior approval from ASP team	Pharmacy supply of >3 days requires prior approval from ASP team	Pharmacy supply does not require approval from ASP team
Colistin, carbapenem and tigecycline	Vancomycin, teicoplanin, daptomycin, linezolid and 3rd–4th generation cephalosporins	1st and 2nd generation cephalosporins, azithromycin, clarithromycin, cotrimoxazole and fluoroquinolones

- Percentage of time the empirical antibiotic is modified as per AST report.
- Percentages of time samples are taken for cultures before the start of antibiotics.
- Percentage of time the choice of surgical antimicrobial prophylaxis given is according to the hospital antibiotic policy.

- **Indicators of administrative compliance:** Percentage of time the antibiotic and surgical antimicrobial prophylaxis is given in correct dose, correct time, correct frequency and by correct route.

Antibiotic usage outcome indicators: Two indicators like daily defined dose and days of therapy are discussed below.

- **Daily defined dose:** It is the average maintenance of dose of drug per day for its indication in adults. It does not include the use of therapeutic dose as it varies from person to person depending on weight, infection type and renal function. Therefore, daily defined dose is a better indicator to calculate the antimicrobial consumption. For calculation of daily defined dose, one can use WHO website. Calculation of daily defined dose is as follows. For example, levofloxacin is given as 750 mg per oral daily for 7 days. The WHO assigned daily defined dose for levofloxacin is 0.5 g. Therefore, the number of daily defined dose is calculated as → 0.75 g dose × 7 days/0.5 g daily defined dose = 10.5 daily defined dose.

  No. of daily defined doses =

$$\frac{\text{Therapeutic dose (Number of tablet/vial used} \times \text{gram per tablet/vial)}}{\text{WHO defined daily defined dose of the antimicrobial agent}}$$

- **Days of therapy:** Days of therapy of an antibiotic are the numbers of days that patient receives at least one dose of that antibiotic. It can be used for estimating antibiotic consumption in paediatric patients and renal failure patients; hence it is chosen before daily defined dose. Examples include (1) A patient has received meropenem 1 g, twice daily for 3 days; the days of therapy are 3. (2) A patient has received meropenem 0.5 g, thrice daily for 3 days; the days of therapy are 3. (3) A patient has received meropenem 1 g, twice daily and vancomycin 1 g thrice daily for 3 days; the days of therapy are 3 + 3 = 6.

Antimicrobial resistance outcome indicator: It includes antimicrobial resistance surveillance.

Clinical outcome indicators: These include morbidity and mortality.

Financial outcome indicators: It includes cost of drug per day or per year or per admission.

Rational Use of Antimicrobials

Prescribe antimicrobial only when needed: Antibiotics are not required routinely for treatment or for prophylaxis in some conditions like diarrhea (ORS is enough before final diagnosis), sore throat (mostly by viral except diagnosis done for *Strept. pyogenes* or *C. diphtheriae*) or when alternative diagnosis like dengue or chikungunya is suspected.

Cultures sampling before antibiotics are given: Sample should be collected from specific site before antibiotics administration to avoid the possibility of false-negative result.

Empirical versus targeted therapy: Empirical therapy should be given after knowing type of infection, common etiological agents and local hospital antibiotic policy. Empirical therapy should be modified subsequently to targeted or pathogen-directed therapy, as per AST report (escalation or de-escalation).

Escalation versus de-escalation approach

- It is designed to know the rank of antibiotic prescription. It based on spectrum of activity and local antibiotic policy. For example; Antibiotics given for *Escherichia coli* are ranked according to decreasing order of susceptibility. Order is colistin (rank-1) → tigecycline → carbapenems → piperacillin-tazobactam → cefoperazone sulbactam → amikacin → cefepime → ceftazidime → cotrimoxazole → ceftazidime → ciprofloxacin → ceftriaxone (lowest rank).
- Escalation approach: It is selected when patient is stable and/or drug resistance is unlikely. Start empirical therapy with low ranked drug (like ceftriaxone for *E. coli*) and if AST report shows the resistance then escalate to higher ranked drug (like piperacillin-tazobactam for *E. coli*).
- De-escalation approach: It is selected when patient is unstable and/or drug resistance is likely to be high. Start empirical therapy with high ranked drug (like piperacillin-tazobactam for *E. coli*) if AST report shows the sensitivity then de-escalate to lower ranked drug (like ceftriaxone for *E. coli*).
- Few higher ranked/restricted drugs (like colistin, tigecycline, etc.) are not given as empirical therapy even under de-escalation approach. They are given only when AST report shows the resistance to all other drugs.

Site-specific antimicrobials approach: Few antibiotics are not active at particular sites hence they should not be included in therapy. Following are the examples.

- **CSF:** Any oral antibiotic, 1st and 2nd generation cephalosporins, tetracyclines, macrolides, quinolones and clindamycin are not active in CSF.
- **Urine:** Chloramphenicol, macrolide and clindamycin do not achieve adequate urinary concentrations hence they should not be included in UTI treatment.
- **Lungs:** Daptomycin is not active at respiratory site because it is inactivated by pulmonary surfactants.

No administration errors: Antimicrobials should be given at the correct dose (as per the age/body weight), frequency and duration of therapy.

- Loading dose: It requires in concentration dependent drugs like aminoglycoside, vancomycin and colistin.
- Infusion: Vancomycin gives good effect when mixed with saline and given as an IV infusion.
- Renal adjustment: Dosage of the nephrotoxic drugs like aminoglycoside, vancomycin, and colistin should be adjusted as per creatinine clearance.

MIC-guided therapy: MIC defined as minimum concentration of drug requires to inhibit the bacterial growth. It is determined by AST methods like epsilomer test or others. When MIC is low therapeutic efficacy is better. In certain infections antibiotics are decided by MIC like endocarditis, pneumococcal meningitis/pneumonia. Vancomycin should be avoided for *Staph. aureus* if MIC is >1 μg/Ml. If >1 antimicrobials are susceptible then antibiotic having lowest MIC (when compared with the susceptibility breakpoint) should be chosen for therapy. However, it is better to use therapeutic index than using only (absolute) MIC.

Therapeutic index: It is the ratio of susceptibility breakpoint divided by MIC of the test isolate (Therapeutic index = Susceptible breakpoint/MIC of the test isolate). Higher the therapeutic index better is the efficacy. Therapeutic index is more useful than absolute MIC. For example, if a clinical isolate of *E. coli* is susceptible to both meropenem (MIC 1 μg/ml) and amikacin (MIC 8 μg/ml), then meropenem falsely appears to be more efficacious as its absolute MIC value is lower than amikacin. However, the MIC value compared with the standard susceptibility breakpoint (and not the absolute MIC value) to determine therapeutic efficacy. Susceptible breakpoints of meropenem and amikacin for *E. coli* are 1 μg/ml and 16 μg/ml respectively, according to CLSI guideline 2020. Therapeutic index of meropenem is (1/1) 1. Therapeutic index of meropenem (16/8) is 2. Therefore, in this case amikacin is superior than meropenem.

Therapeutic drug monitoring: Therapeutic efficacy is dependent on *in vitro* (MIC) susceptibility and *in vivo* activity. *In vivo* depends on pharmacokinetic and pharmacodynemic parameters of the antimicrobial agent. Therapeutic drug monitoring is necessary to find out how the drug behaves *in vivo* like vancomycin, amikacin and colistin. Depending on pharmacokinetic and pharmacodynemic parameters, antibiotics are classified as **(1) Concentration dependent:** For example, aminoglycosides (better if the drug concentration in serum is much higher than the MIC of the drug). **(2) Time-dependent:** For example, beta-lactams (efficacy of the drug is dependent upon how much time the drug concentration remains higher in serum than the MIC. To keep serum level high it is given frequently like thrice daily).

Timely stoppage of antimicrobial: It should be stopped at appropriate time, which is determined by clinical improvement, negative culture report or by using of biomarkers.

Misuse of antibiotics: (1) Avoid overlapping spectra: Meropenem and piperacillin-tazobactam are belong to beta-lactam group and share common antibacterial spectra hence combination therapy of these two should be avoided. **(2) Combined effect (redundant antibiotic):** Meropenem is active against anaerobes and gram-negative bacteria while metronidazole is active against anaerobes. Hence metronidazole is excluded when combined action required for suspected GN and anaerobic sepsis. **(3) Ineffective antibiotic:** Cloxacillin is not effective in MRSA; hence it should be excluded. Vancomycin is the drug of choice for MRSA. **(4) Inferior antibiotic:** Vancomycin is an inferior cell wall acting agent compared to cloxacillin for MSSA.

Hospital Antibiogram

It is the overall susceptibility profile of antimicrobial agents available in hospital to a specific microorganism as shown in **Table 116.3**. Antimicrobial agents available in hospital called **antimicrobial battery**. It is prepared by Microbiology department and should be shared with clinicians. **Uses** of antibiogram are (1) it guides the clinicians to choose the best empirical therapy when culture and sensitivity reports are pending. (2) it is useful for detection and monitoring antimicrobial resistance within the hospital. (3) it compares susceptibility rates across institutions and track resistance trends and thereby contributing to national AMR surveillance database.

Chennai Declaration, 2012

Definition: It is a document, prepared by representatives of various medical societies and eminent experts in India, to control the antimicrobial resistance from an Indian perspective in August 2012.

TABLE 116.3: Profile of hospital antibiogram.

Microbes \ Drugs	Ciprofloxacin	Amikacin	Ceftazidime	Piperacillin-tazobactam	Meropenem	Colistin
E. coli	25	85	55	75	82	99
Klebsiella	15	75	35	65	70	85
Pseudomonas	32	80	60	80	80	99
Acinetobacter	29	70	10	18	20	99

Background: It was published in December 2012; although Ministry of Health, India had initiated efforts to control antimicrobial resistance in the country way back in 2010. Chennai declaration created widespread awareness about the antimicrobial resistance among the Indian medical communities. It created great impact at national and international levels.

Preparation: It is prepared by representatives from various fields like representatives from most medical societies in India, eminent policy makers from both central and state governments, representatives of World Health Organization, National Accreditation Board of Hospitals, Medical Council of India, Drug Controller General of India, and Indian Council of Medical Research along with well-known dignitaries in the Indian medical field.

Strategies: It is prepared 5 years plan to control antimicrobial resistance at 1st year, 2nd year and 5th year. It based on step-by-step implementation of various components in developing countries. Detail description is beyond the knowledge of this book hence not given.

Uses: It was used at different level in country with following progress report.
1. Document was reviewed by hospitals, international journals, international academics, health policy related conferences, Indian Ministry of Health, media, etc.
2. It created international awareness regarding the ground reality in developing countries and how a policy has to be tailored as per local requirement.
3. Discussed in Indian parliament to prepare the new over-the-counter rule to put many antibiotics in restricted category.
4. Ministry of Health launched public education initiatives via print media and radio channels.

Red Line Campaign and Red Line Medicines

Antimicrobial resistance awareness campaign by union health ministry to urge the people not to use medicines marked with a red vertical line, including antibiotics, without a doctor's prescription called red **line campaign**. Medicines marked with a red vertical line are called **'red line medicines'**.

ACCESS YOURSELF

Short Notes
1. Drug resistance mechanisms.
2. Antibiogram.
3. Describe and discuss the rational use of antimicrobials including antibiotic stewardship program.
4. Chennai declaration, 2012.

Short Questions for Theory/Viva Questions
1. Write four differences between mutational and transferable drug resistance.
2. Write two uses of antibiogram.
3. What is E test?
4. How Microbiology department can help to antibiotic stewardship program?
5. What is Chennai declaration? Write its importance in Microbiology.
6. Define: Red line campaign and red line medicines.

Comments on
1. Multiple drugs are included in treatment of tuberculosis.
2. Removal/loss of porin channels in bacteria produces drug resistance.
3. Alteration in PBP2a produces drug resistance in *Staph. aureus*.
4. Front end strategy (formulary restriction) is practically difficult to implement.
5. Therapeutic index is more useful than absolute MIC/MIC.

MCQs for Chapter Review

Drug Resistance Mechanisms
1. Multiple drug resistance is spread by:
   a. Transformation
   b. Conjugation
   c. Transduction
   d. Mutation

Antibiotic Sensitivity Testing
2. Which of these technique for antimicrobial sensitivity testing, has a built in control strain along with the test strain?
   a. Kirby Bauer disc diffusion technique
   b. Stokes disc diffusion technique
   c. E test
   d. Broth dilution technique

3. Kirby-Bauer method is based on principle of:
   a. Dilution method
   b. Disc diffusion method
   c. a + b
   d. None

4. For antibiotic sensitivity test, the organism broth prepared should match with:
   a. Mc Farland standard 0.5
   b. Mc Farland standard 1
   c. Mc Farland standard 2
   d. Mc Farland standard

5. For antibiotic sensitivity test, the organism in 0.5 Mc Farland standard are:
   a. 1.5×10^8 cfu/ml
   b. 1.5×10^7 cfu/ml
   c. 1.4×10^8 cfu/ml
   d. 1.4×10^7 cfu/ml

Antimicrobial Stewardship Program
6. Maximum consumption of antibiotics occurs for:
   a. Animal non-therapeutic use
   b. Animal therapeutic use
   c. Human therapeutic use
   d. Human non-therapeutic use

7. Who can act as antimicrobial steward?
   a. Infectious disease physician
   b. Clinical microbiologists
   c. Infection control officer
   d. Any of the above

8. Chennai declaration is published in?
   a. 2012
   b. 2013
   c. 2014
   d. 2010

1. a, b, c

- Multiple drug resistance is mediated by extrachromosomal part which is spread by transformation, conjugation and transduction while single drug resistance is mediated by chromosomal part which is spread by mutation.

2. b

3. b

4. a

5. a

- Follow section, **antibiotic sensitivity testing** for explanation of answers of MCQs 2–5.

6. a

- Follow section, **Antimicrobial stewardship program** → **goals** for explanation.

7. d

- Follow section, **Antimicrobial stewardship program** → **ASP team** for explanation.

8. a

- Follow section, **Antimicrobial stewardship program** → **Chennai declaration, 2012** → **Definition** for explanation.

Laboratory Animals

INTRODUCTION

Advantages of Animals

1. Isolation of microorganisms.
2. Identification of microorganisms.
3. Preparation of immune sera/therapeutic sera diagnostic sera/vaccines (antigens).

Disadvantages of Animals

1. Ethical issues regarding uses of animal. Required to take permission from the animal ethical committee of institute.
2. Time consuming about 6–9 months.
3. May not give desired results due to interference by animal immunity.

SPECIFIC ANIMAL AND USES

Nine-banded Armadillo

Synonym: Long nosed armadillo.

Scientific name: *Dasypus novemcinctus.*

Identification features (Fig. 117.1): There are nine (sometimes fewer) narrow, jointed armor bands on its midsection that let it bend.

Uses: (1) It develops generalized lepromatous leprosy following inoculation of *M. leprae*. (2) It acts as reservoir of infection in *T. cruzi*. (3) It naturally infected in *P. brasiliensis* and may provide clue to understand the pathophysiology of fungus.

Rabbit

Scientific name: It is shown in **Fig. 117.2**. There are many genera and species of rabbits. One Europian rabbit species is *Oryctolagus cuniculus*.

Uses: (1) Pathogenic *Treponema* does not grow in artificial culture media, and it maintained for many decades by serial testicular passage in rabbits. (2) It differentiate between human and bovine variety of tuberculosis, where animal is susceptible to later. (3) It is used to prepare the diagnostic sera. (4) It also useful in virus isolation and identification.

Fig. 117.2: Rabbit

Guinea Pig

Scientific name: It is shown in **Fig. 117.3**. Many species like *Cavia porcelus*.

Uses: It is used: (1) To isolate human and bovine variety of tuberculosis and animal is susceptible to both. (2) To do virulence test in diphtheria and tetanus. (3) To obtain the complement for CFT. (4) To investigate the typhus fever. (5) To investigate the Entero invasive *Escherichia*

Fig. 117.1: Nine-banded armadillo

Fig. 117.3: Guinea pig

coli (EICE) or *Shigella* called **Sereny test** or *Listeria* called **Anton test.** (6) To isolate and to identify the viruses. (7) To diagnose the anaphylaxis. (8) It was used to establish many theories like germ theory by Pasteur, Koch's phenomenon, Pfiffer's phenomenon and Theobald Smith phenomenon (anaphylaxis).

Rat

Scientific name: Different types.

- **Black rat (Fig. 117.4):** *Rattus rattus.*
- **Brown rat (Fig. 117.5):** *Rattus norvegicus.*
- **White rat or albino rat (Fig. 117.6):** *Rattus albus.*

Identification features: These are larger than mouse, rough fur and different colors as mentioned above.

Uses: (1) It is useful to differentiate between *Yersinia pseudotuberculosis* and *Yersinia pestis.* (2) It also associated with other zoonotic pathogens like *Leptospira*, *T. gondii*, *C. jejuni*, etc.

Fig. 117.4: Black rat

Fig. 117.5: Brown rat

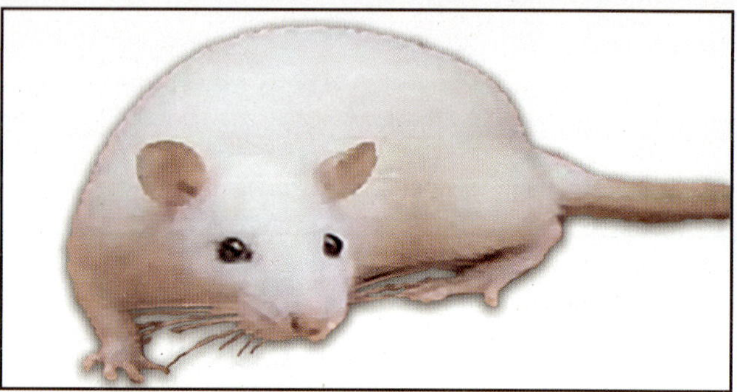

Fig. 117.6: White rat

Mouse (Singular)/Mice (Plural)

Scientific name: *Mus musculus.*

Identification features (Fig. 117.7): It is smaller than rat and has smooth fur.

Uses: (1) It is very susceptible to pneumococcus and useful to isolate it from the infective materials. (2) Foot pad of mouse is useful to cultivate the *M. leprae.* (3) It is used to demonstrate the ascending and descending tetanus. (4) It is used to demonstrate the influenza virus and *Chlamydia.* (5) Infant suckling mouse, but not the adult mouse is useful in virus isolation and identification like coxsackie viruses. (6) It is used to isolate the causative organisms of relapsing fever, rat bite fever and trypanosomiasis.

Sheep

Uses: (1) Sheep blood is useful to prepare the blood agar. (2) RBCs of sheep are useful in CFT, Paul Bunnel test, SRBC (sheep red blood cells) rosette or E (erythrocyte) rosette formation with CD2 cells and EAC (erythrocyte antibody complement) rosette formation with CR2 (complement receptor-2).

Monkey

Use: Monkey kidney cells called **Vero cells** are useful for viral culture.

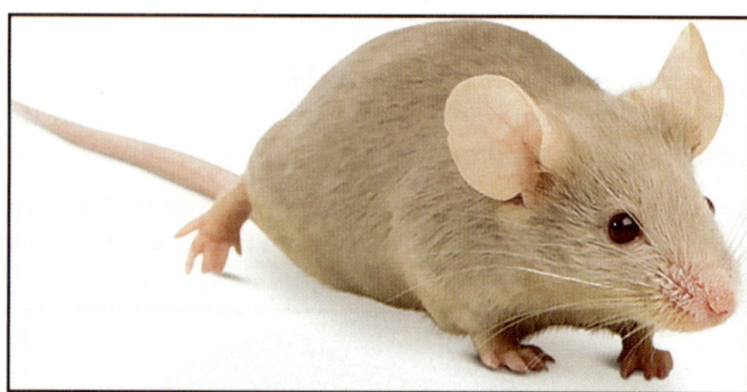

Fig. 117.7: Mouse

Short Questions for Theory/Viva Questions

1. Write the two uses of nine-banded armadillo/rabbit/guinea pig.

2. Name the parasite in which armadillo acts as reservoir of infection.
3. Write the scientific name of following animals: Nine-banded armadillo, rabbit, guinea pig and black rat.

MCQs for Chapter Review

Animal and Specific Uses

1. **Scientific name of nine banded armadillo**
   a. *Oryctolagus cuniculus*
   b. *Dasypus novemcinctus*
   c. *Cavia porcelus*
   d. *Rattus rattus*
2. **Animal used to demonstrate the anaphylaxis in the lab is:**
   a. Rabbit
   b. Adult mouse
   c. Monkey
   d. Guinea pig

3. **Suckling mouse is used for isolation of:**
   a. Coxsackie virus
   b. Pox virus
   c. Herpes virus
   d. Adeno virus

Answers of MCQs and Explanation

1. **b**
 • Scientific name of all animals are described in the text.
2. **d**
 • Guinea pig is the highly susceptible animal to diagnose the anaphylaxis.
3. **a**
 • Follow section (**animal and specific uses → mouse/mice),** for explanation.

Laboratory Diagnosis of Bacterial Infections

Chapter Outline

- Sample Rejection Criteria
- Pretreatment of Sample
- Testing Methods
- Reference Centers

Following are different steps employed for diagnosis of bacterial infections.

Sample collection, transportation and storage: Described in **Ch. 113**.

Sample rejection criteria: Described below.

Pretreatment of sample: Described below.

Testing methods: Described below.

SAMPLE REJECTION CRITERIA

After receiving the sample it should rejected or disallowed for testing if anyone from following is noticed.

- Unsterile or leakage or breakage of container.
- Patient's details on specimen are not matched with request form or unlabeled specimen.
- Dry swab, saliva mixed with sputum, urine mixed with feces, sample mixed with disinfectant or contaminated by other materials.

PRETREATMENT OF SAMPLE

CSF and urine: Centrifuged at 1st and use sediment for testing.

Sputum: It required digestion and decontamination, especially in case of tuberculosis.

TESTING METHODS

Microscopy and Staining

Light microscopy

1. **Wet mount:** It is useful for identification of capsule (Indian ink mount), motility (hanging drop preparation), etc.
2. **Fixed smear and staining**
   - **Differential stain:** Gram's stain and acid fast stains as described in **Ch. 112**.
   - **Special stains for flagella, polar bodies, spores and fat globules:** Follow **Ch. 6**.
   - **Special stains for spirochetes:** These are silver impregnation stains like Fontana stain for culture and Levaditi stain for tissue section.
   - **Histopathological stain:** Like Giemsa stain, Castaneda stains, Gimenez and Machiavello stain for *Rickettsia* and other bacteria which are poorly stained by Gram's stain.

DGIM and PCM: These are useful for thin structures like flagella and spirochetes.

EM: Useful for detail study of bacteria.

Fluorescent Microscopy (FM) and staining: Follow **Ch. 112**.

Culture

Culture in artificial prepared nutritional (cell free) media: Bacterial culture includes following three steps.

- **Inoculation:** Different types bacterial media are described in **Ch. 114**. Methods for inoculation of these media are called **culture methods** which are described in **Ch. 115**.
- **Incubation:** Different effective factors for bacterial incubation are temperature (normally it is 37°C for all bacteria), presence (aerobic incubation) or absence of O_2 (anaerobic incubation), 5–10% CO_2 for capnophilic bacteria use of, 5% O_2 + 10% CO_2 + 85% nitrogen for *C. jejuni* and *H. pylori*.
- **Identification of growth**
  - **Liquid media: (1) Growth properties:** Bacterial growth in liquid media is identified by surface pellicle formation (aerobic bacteria), deposit at bottom (anaerobic bacteria), uniform turbidity (facultative aerobes), color changes (like pink by *Cl. perfringens* due to saccharolytic properties in RCMM and black by *Cl. tetani* due to proteolytic properties in RCMM) and presence of typical odor (foul smell by *Cl. tetani* due to proteolytic properties in RCMM). **(2) Disadvantage:** No pure growth.

– **Semi-solid media:** Motility detected by fine lines around the streaking line.
– **Solid media: (1) Cultural characteristics (C/Cs):** Size, shape, elevation, margins, surface, edge, color or pigment production, hemolysis, consistency, emulsification, etc. **(2) Advantage:** It gives pure growth.
– **Biphasic media: (1) Growth properties:** Growth is examined over solid phase. **(2) Advantage:** Less chances of contamination and less labor intensive.

Automated culture method: Follow **Ch. 115**

Egg culture: Inoculation in yolk sac or CAM (chorio allantoic membrane).

Animal culture: Laboratory animals used for bacterial cultivation are guinea pig, rabbit, rat, nine-banded armadillo, mouse, etc.

Biochemical Reactions

1. Sugar fermentation test (Fig. 118.1)

Commonly used sugars: Are glucose, sucrose, lactose and mannitol abbreviated as GSLM respectively.

Other rarely used sugars: Maltose, rhamnose, etc.

Reagents: Medium contains peptone water (PW) + 1% sugar + Andrade's indicator (NaOH and acid fuchsin) + Durham's tube.

Principle: Bacteria ferment sugar with production of only acid (A) or acid with gas (AG). pH (Andrade's) indicator converted in to pink color at acidic pH and gas collected in Durham's tube.

Method: Inoculate a loopful culture suspension in medium and incubate at 37°C for 24 hours. Examine for result after 24 hours.

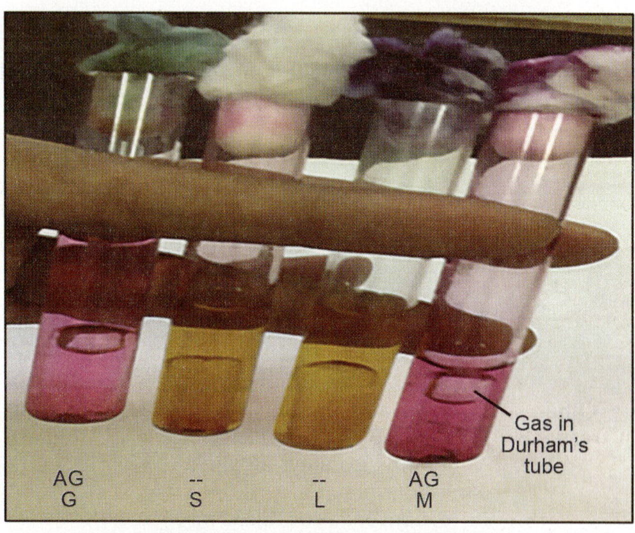

Fig. 118.1: Sugar fermentation test

Results: (1) Positive result: If only acid (A) production is indicated by pink color medium. Acid with gas (AG) production is indicated by pink color medium and gas collected in Durham's tube in form of bubbles. **(2) Negative result:** No color change of medium.

2. IMViC tests: Indole test, methyle red (MR) test, Voges-Proskauer (VP/Vi) test and citrate test are commonly abbreviated as **IMViC tests** and described in details in **Table 118.1**.

3. Oxidase test

Principle: Cytochrome oxidase enzyme presents in aerobic bacteria. It takes part in aerobic respiration (oxidation) by transferring electron (H_2) to O_2 to form H_2O. Cytochrome oxidase acts as substitute (artificial electron acceptor) for O_2 and oxidizes the dye (artificial electron donor) in purple/violet color.

Laboratory Diagnosis of Bacterial Infections

	TABLE 118.1: IMViC tests			
Particulars	**Indole test**	**MR test**	**VP test**	**Citrate test**
Principle	Indole produces from the degradation of tryptophan by enzyme tryptophanase **(Flowchart 118.1)**. When indole mixed with reagent it gives red color ring at top due to reaction of indole with aldehydes group of PDAB.	Production of pyruvic acid from glucose by glycolysis **(E-M pathway)** which further metabolized to produce lactic acid, acetic acid and formic acid **(mixed acid pathway)**. MR is a pH indicator which is yellow at pH 6.0 and red at acidic pH (pH = 4.4) as shown in **Flowchart 118.2**.	Production of pyruvic acid from glucose by glycolysis **(E-M pathway)** which further metabolized by mixed acid pathway to produce lactic acid, acetic acid and formic acid **(mixed acid pathway)**. Sometimes, instead of mixed acid pathway it passes by acetoin pathway **(butylene glycol pathway)** to produce acetoin (also called acetyl methyl carbinol). Acetoin oxidized by KOH and environmental O_2 to produce diacetyl which gives red color on contact with α-naphthol **(Flowchart 118.2)**.	Citrate is salt of citric acid produced during Krebs' cycle in the form of citric acid. Citrate is utilized by many bacteria as source of carbon to produce NH_3 which increases the pH of medium up to 7.6. Bromothymol blue is a pH indicator which is green initially converted in blue color at high pH and gives positive test.

(Contd...)

10

Essentials of Medical Microbiology

TABLE 118.1: IMViC tests (*Contd...*)

Particulars	Indole test	MR test	VP test	Citrate test	
Reagents • **Medium**	• PW rich in tryptophan.	• GPB contains glucose, phosphate and peptone also called MR broth or VP broth.	• Same as MR test.	Citrate acts as sole source of energy. Bromothymol blue is pH indicator mixed in medium. Two types of media as follows 1. **Solid medium:** Called Simmon's citrate agar which is available in test tube in slant form 2. **Liquid medium:** Called **Koser's citrate broth**	
• **Other reagents**	**1. Kovac's reagent** • PDAB: 10 g • Amyl/iso amyl alcohol: 150 ml • Concentrated HCl: 50 ml **2. Ehrlich's reagent** • PDAB: 2 g • Absolute ethyl alcohol: 190 ml • Concentrated HCl: 40 ml	• MR reagent (pH indicator).	• 40% KOH + α-naphthol.		
Quality control • +ve control • –ve control	• *E. coli* • *K. pneumoniae*	• *E. coli* • *K. aerogenes*	• *K aerogenes* • *E. coli*	• *K aerogenes* • *E. coli*	
Methods	• **Test tube method:** Inoculate a loopful culture suspension in medium and incubate at 37°C for 24 hours. After 24 hours remove medium and add the reagent. Examine for the results. • **Filter paper strip method impregnated with PDAB** • **In clinical practise indole test is used in combination with other tests:** Like (1) SIM (sulfide indole motility) test: More sensitive than TSI for H_2S detection in *S.* Typhi. (2) MIO (motile indole ornithine) test: (3) IN (indole nitrate/ indoso nitrate) test: For *V. cholerae*	Inoculate a loopful culture suspension in medium and incubate at 37°C in incubator for 24 hrs. After 24 hrs remove medium from incubator and add the MR reagent. Examine for the results.	Inoculate a loopful culture suspension in medium and incubate at 37°C for 24 hrs. After 24 hrs remove medium and add the reagent. Examine for the results.	Pick the colony from primary culture by nichrome loop/wire and inoculate the medium. Incubate at 37°C for 24 hours and read the results.	
Results (Fig. 118.2) • +ve test	• Red/pink color ring at top of medium	• Red/pink color (mostly upper half) in medium	• Red/pink color in whole medium	• **Solid [Fig. 118.2d] medium:** Blue color in medium due to alkaline pH. Test also considered positive with growth without blue color. • No change or original green color of medium.	• **Liquid medium:** Turbidity production • No change
• –ve test	• No color change	• No color change	• No color change		

Reagent: Tetra methyl para phenyl di amine Hydro-chloride (TPDH) is colorless in reduced state and purple/violet color in oxidized state.

Quality control: (1) Positive control: *P. aeruginosa*. (2) Negative control: *E. coli*.

Methods: Two type of methods.

1. **Direct method/Petri dish method:** Apply 2–3 drops of reagent directly on colonies in Petri dish.
2. **Indirect method/Kovac's method/filter paper strip method:** Take clean-sterile slide. Add few drops of sterile saline on slide. Emulsify the reagent in saline. Take a piece of filter paper and put over reagent containing slide (filter paper suck the reagent). Add colonies over reagent containing filter paper and then examine for the results.

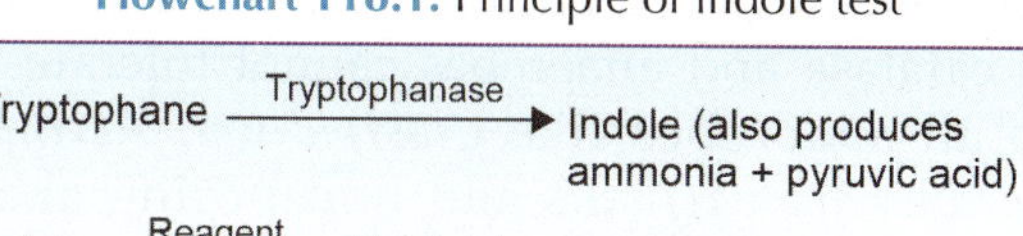

Flowchart 118.1: Principle of indole test

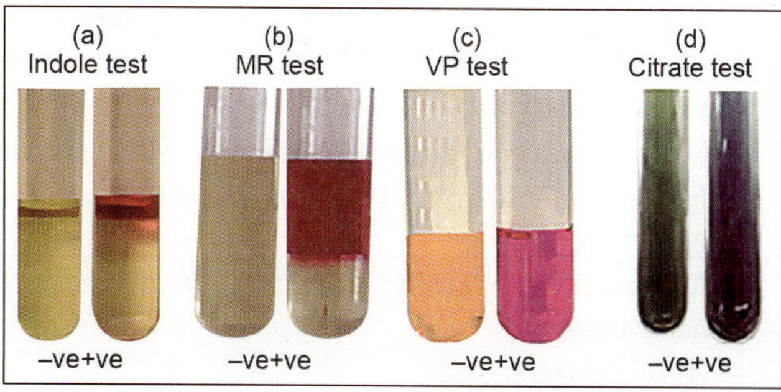

Fig. 118.2: IMViC tests

Flowchart 118.2: Principle of MR and VP test

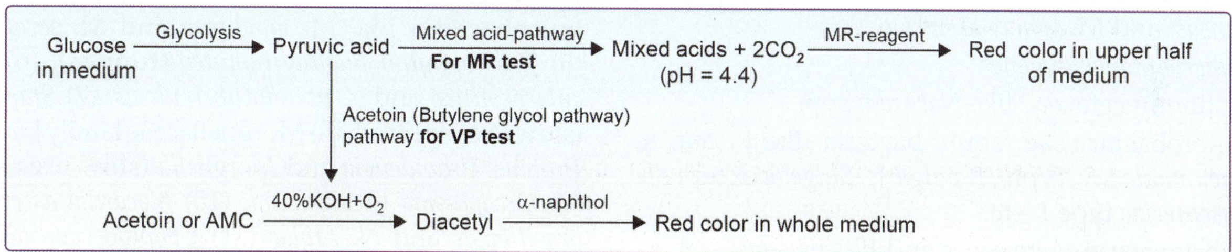

Results (Fig. 118.3a): (1) Positive result: Purple color development in 10 seconds. **(2) Negative result:** No color change. **(3) False-positive result:** Use of nichrome loop or wire may also give false-positive reaction.

Note: Kovac's reagent
It is used in indole test.

Uses
- To identify following oxidase positive bacteria.
  - Strict aerobic: *Micrococcus, Neisseria meningitidis, Moraxella lacunata, Acinetobacter baumanii, V. cholerae, P. aeruginosa, Bordetella pertusis, Brucella abortus* and *Alkaligenes faecalis*.
  - Facultative anaerobes: *Neisseria gonorrheae, H. influenzae* and *Pasteurella multocida*.
  - Microaerophilic category: *Campylobacter jejuni* and *Helicobacter pylori*.
  - *Flavobacterium meningosepticum*.
  - *Moraxella catarrhalis*.
- To differentiate oxidase negative bacteria like Enterobacterales from aerobic bacteria.

Mnemonic
- **Oxidase +ve bacteria:** M₃N₂ A₂B₂ P₂H₂ CVF
- **Catalase +ve bacteria:** C₄B₃N₃P₃ L₂ VF SAME
- **Urease +ve bacteria and fungi:** KRETY'S PUNCH Stop Motorcycle, Car, Bus and Truck

4. Catalase test
Principle: $2H_2O_2 \xrightarrow{Catalase} 2H_2O + O_2$ (produces bubbles)

Reagent: 3% H_2O_2

Quality control: (1) Positive control: *Staph. aureus*. (2) Negative control: *Streptococcus pyogenes*.

Method: Pick up the colonies by using wooden stick and apply in 3% H_2O_2 solution either on slide or in tube. Examine for results.

Results (Fig. 118.3b): (1) Positive result: It produces bubbles. Few bacteria poses enzymes which can degrade the H_2O_2 and produce bubbles after 20–30 seconds but it is not considered as significant. **(2) Negative result:** No bubbles. **(3) False-positive result:** It occurs when test performed from blood agar, because RBCs contain catalase enzymes.

Uses: It is used to differentiate catalase-positive and catalase-negative bacteria from related families.

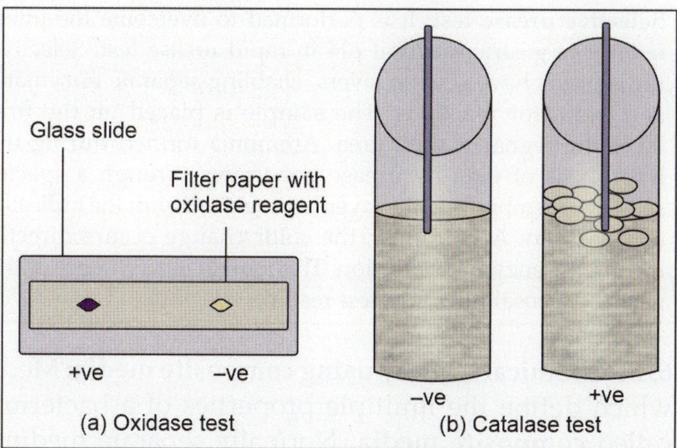

Fig. 118.3: (a) Oxidase test; (b) Catalase test

- **Catalase-negative bacteria:** O_2 is the end product of catalase and anaerobes cannot tolerate the O_2, so all anaerobes like *Clostridium*, *Actinomyces*, *Gardnerella vaginalis* and nonsporing anaerobes are catalase-negative. However, it is not absolute because some other members of different families like Streptococcaceae (*Streptococcus pyogenes*) and *Erysipelothrix rhusiopathiae* are also catalase negative.
- **Catalase-positive bacteria**
  - Micrococcaceae family bacteria like <u>S</u>*taph. aureus.*
  - <u>N</u>eisseriae family like *N. meningitidis* and *N. gonorrhoeae.*
  - <u>M</u>oraxellaceae family like *Moraxella catarrhalis* and *M. lcunata.*
  - <u>A</u>*cinetobacter* spp.
  - <u>C</u>*orynebacterium diphtheriae.*
  - <u>B</u>acillaceae like *B. anthracis.*
  - <u>N</u>on-tuberculous mycobacteria (NTM) like *M. kansasi* and *M. xenopi* at 68°C.
  - <u>L</u>*isteria monocytogenes.*
  - Actinomycetaceae like <u>N</u>*ocardia* spp.
  - <u>E</u>nterobactariceae family bacteria like *E. coli, K. pneumoniae, S.* Typhi, *P. vulgaris, Shigella* except *Sh. dysenteriae* type I, etc.
  - <u>V</u>ibrionaceae, *Aeromonas* and *Plesiomonas.*
  - <u>C</u>ampylobacterales like *C. jejuni.*
  - Nonfermenters like <u>P</u>*. aeruginosa.*
  - Pasteurellales like <u>H</u>*. influenza* and <u>P</u>*. multocida.*
  - <u>B</u>*ordetella* spp.
  - <u>B</u>*rucella* spp.
  - Miscellaneous GNB like <u>F</u>*rancisella tularensis, Chromobacterium violaceum, Capnicytophaga canimorsu* and <u>L</u> *pneumophila.*

5. Urease test

Principle: Urea is converted by enzyme urease into NH_3 which increases the pH of medium >8.1. Phenol red (pH indicator) is converted from colorless (at pH <8.1) form to red/pink color (at pH >8.1).

$$\text{Urea} \xrightarrow{\text{Urease}} NH_3 \xrightarrow{\text{Phenol red}} \text{red/pink color}$$
$$\text{(Alkaline pH)}$$

Reagents: Urea and phenol red (pH indicator) are mixed in medium. There are two types of media like **(1) solid medium** called **Christensen's urea agar** which is available in test tube in slant form **(Fig. 118.4a)** and **(2) liquid medium** called **Stuart's urea broth (Fig. 118.4b).**

Quality control: (1) Positive control: *Proteus* spp. (2) Negative control: *E. coli.*

Method: Pick the colony from primary culture by nichrome loop/wire and inoculate the medium. Incubate at 37°C for 4 hours and than after 24 hours.

Results (Fig. 118.4): (1) Positive result: Red/pink color in whole medium due to alkaline pH. **(2) Negative result:** No change in original yellow color of medium.

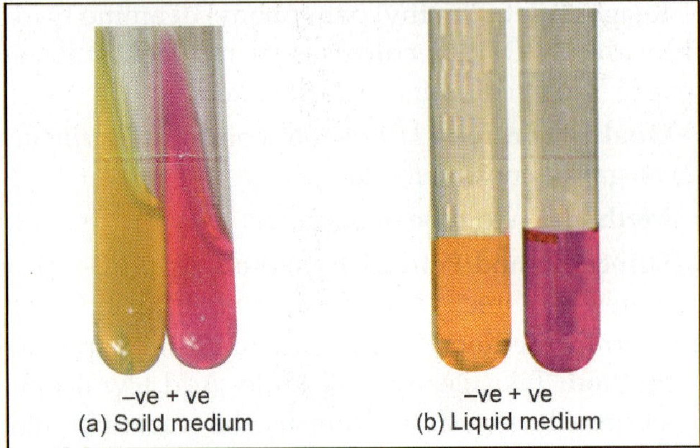

Fig. 118.4: Urease test

Uses: It is used to identify urease positive bacteria and fungi. (1) <u>K</u>*lebsiella pneumoniae* **(slow urea splitter).** (2) <u>M</u>*alassezia furfur* **(fungus).** (3) <u>R</u>*hodotorula mucinaginosa* **(fungus).** (4) <u>E</u>nvironmental (Nontuberculous/atypical) mycobacteria like *M. marinum* and *M. scrofulaceum.* (5) <u>T</u>*richophyton mentagrophytes* **(fungus).** (6) <u>Y</u>*ersinia enterocolitica* and *Y. pseudotuberculosis.* (7) <u>S</u>*taph. aureus* **(slow urea splitter).** (8) <u>M</u>orgnellaceae family bacteria like *Proteus, Providencia* and *Morgnella* **(slow urea splitter).** (9) <u>U</u>*reaplasma urealyticum.* (10) <u>N</u>*ocardia asteroides, N. brasiliensis* and *N. caviae* (11) <u>C</u>*ryptococcus neoformans* **(fungus).** (12) <u>H</u>*elobacter pylori* **(rapid urea splitter).** (13) <u>S</u>ome strain of *V. parahaemolyticus.* (14) <u>T</u>*richosporon beigelii* **(fungus).** (15) *Brucella abortus.* (16) <u>C</u>orynebacterium spp., like *C. ulcerans, C. psudodiphthericum* and *C. pseudo-tuberculosis.*

> **Note: Rapid urease test and selective urease test**
>
> **Rapid urease test:** Also called **CLO test** (*Campylobacter*-like organism test). It called **rapid** because result is available in 1–3 hours. It is useful to detect *H. pylori* infection. Urease produced by *H. pylori* is 100 times more active than produced by *Pr. vulgaris.* Maximum rapid urease positive is given by *H. pylori.* It is performed by placing gastric mucosa (antrum part collected by gastroscopy) in urease medium. It is performed as liquid-based medium, gel-based medium, and a special design type with single-layer sensitive element. Red/pink color indicates positive test. Major drawback of this test is interference of gastric mucosal pH with pH of medium, which is overcome by selective urease test.
>
> **Selective urease test:** It is performed to overcome the interference of gastric mucosal pH in rapid urease test. Selective urease tests have several layers, enabling separate enzymatic and indicator reactions. The sample is placed on the first layer impregnated with urea. Ammonia formed during the hydrolysis of urea by urease penetrates through a special selective membrane to the layer impregnated with the indicator composition. Accordingly, the color change occurs directly during the enzymatic reaction. This avoids the influence of the gastric mucosal pH to the test result.

6. Biochemical tests by using composite media: Media which define the multiple properties of a bacterium called **composite media.** Normally separate medium is required to define every property of bacterium, but composite media define many properties of bacteria. Their use is increased now days for identification of bacteria. Different types composite media/tests are TSI test, KIA test and bile esculin agar.

TSI Test

Full form: Triple Sugar Iron test, because contains three sugars (glucose, sucrose and lactose) and iron.

Advantages: Test is most convenient and economical. It defines many properties of bacterium like fermentation of glucose, fermentation of lactose, fermentation of sucrose, gas production and H_2S production.

Disadvantage: It cannot differentiate between lactose fermentation and sucrose fermentation, which is overcome by using KIA test.

Principle: Slant portion, constantly exposed to air, is aerobic and butt (deep) portion relatively less exposed to air, is anaerobic. Four types of reaction occur as follows.

- **Nonfermenters:** Nonfermenters are not able to utilize the carbohydrates, so they utilize the peptone and produce NH_3, which gives alkaline pH (red color) to medium.
- **Glucose fermenters but NLFs and NSFs:** Bacteria utilize the glucose to produce acid [acidic (A) pH → yellow color] but when glucose is exhausted they utilize peptone to produce NH_3 [alkaline (K) pH → red color].
- **Fermenters:** Bacteria ferment all sugars to produce large amount of acid (acidic pH → yellow color) in slant and in butt. Sometimes, gas produced which present as bubbles in bottom or in between the medium.
- **Glucose fermenters with H_2S production:** Bacteria utilize the glucose to produce acid (acidic pH → yellow color) but when glucose is exhausted they utilize peptone to produce NH_3 (alkaline pH → red color). Acid is required for H_2S production and due to presence of acid some bacteria utilize sodium thiosulfate to produce H_2S which react with ferrous sulfate and convert it in to ferrous sulfide to gives black color.

Reagents: Medium called **TSI agar.** It has two parts like slant and butt (butt also called **deep**). TSI contains three sugars like G (0.1%), L (1%) and S (1%). G to L/S ratio is 1:10. Other ingredients of TSI are four protein derivatives (beef extract, yeast extract, peptone and proteose peptone), pH indicator (phenol red which is yellow at acidic pH and red/pink at alkaline pH), sodium thiosulfate (source of sulfur atoms for H_2S production), ferrous sulfate (H_2S detector which gives black color on combination with H_2S if produced), 2% agar and pH is around 7.2. TSI is sterilized by autoclave.

Method: Pick the colonies from primary culture by nichrome wire and inoculate the medium up to the butt (deep) and than streaking over the slant. Incubate at 37°C for 24 hours and then examine for results.

Results: End products of test are acid (A) → yellow color, alkaline (K) → red or pink color, gas (G) → bubbles and H_2S → black color. Following four types of results are observed

1. **Nonfermenters or no reaction**
   - Red slant/red butt: K/K or K/Nil **(Fig. 118.5a)**.
   - Organism: *Pseudomonas aeruginosa.*
2. **Glucose fermenters but NLFs and NSFs:**
   - Red slant/yellow butt: K/A **(Fig. 118.5b).**
   - Organism: *Shigella dysenteriae.*
3. **Fermenters**
   - Yellow slant and yellow butt with gas (A/AG) or without gas A/A **(Figs 118.5c** and **118.5d,** respectively).
   - Organisms: Coliform bacilli like *E. coli, K. pneumoniae,* etc.
4. **Glucose fermenters with H_2S production**
   - Red slant/black butt: K/A + H_2S **(Fig. 118.5e)**.
   - Organisms: These are glucose fermenters and H_2S producing bacteria, but nonlactose/nonsucrose fermenters like *Salmonella* Typhi, *Proteus* spp., *Citrobacter frundi,* etc.
   - Acid is required for H_2S production and H_2S is present where acid is present, so yellow color of acid is masked by black color of H_2S, but it read as acid (yellow color).

Uses: To determine the carbohydrate fermenting and H_2S producing bacteria as described above.

KIA Test

Full form: Kligler Iron Agar test.

Reagents: Similar medium to TSI but contains only glucose and lactose.

Uses: It is used to differentiate *Y. enterocolitica* (NLF but sucrose fermenter gives A/A in TSI, K/A in KIA) from *E. coli* (LF and sucrose variable gives A/A in both), so by some microbiologist, KIA is preferred over TSI for fecal samples.

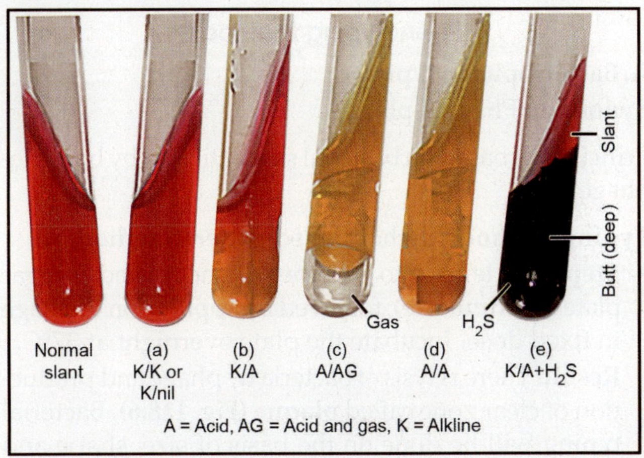

Fig. 118.5: TSI test

Bile Esculin Agar

- It is used to differentiate *Enterococcous* (group D) group of Streptococcaceae from non-*Enterococcus* (*Strept. mitis*, not group D) group.
- It also used to know the growth of bacterium in presence of 40% bile, ability to hydrolyse the esculin to esculetin and glucose. Esculetin combines with ferric citrate to form dark brown or black colonies.

7. Other biochemical tests

- Nitrate reduction test
- Phenyl pyruvic acid (PPA) test
- Amino acid (lysine, arginine and ornithine) decarboxylation test
- Coagulase test and phosphatase test: **Ch. 49**
- Hugh-Leifson's (**oxidation–fermentation/OF**) test
- Bile solubility test: **Ch. 51**
- O-nitro phenyl- β-D-galactosidase (ONPG) test.

Serological Tests

Agglutination based tests, precipitation based tests, Toxin-antitoxin neutralization tests, ELISA, radio-immunoassay (RIA), etc.

Allergic (Skin) Tests

Tuberculin tests, Lepromin test, Frei's test, etc.

Molecular Methods

Amplification methods: PCR, real time/quantitative PCR (qPCR), LAMP, Automated Biofire film array, CBNAAT, etc.

Non-amplification method: Hybridization like LPA, etc.

Physiological Methods

Like GLC and HPLC.

Test to Detect Metabolic Activities

1. Detection of tuberculo-stearic acid and adenosine deaminase in *M. tuberculosis*.
2. Pigment production by *Staphylococcus* spp., *Pseudomonas aeruginosa* and other bacteria.

Microbial Typing

Phenotyping Methods

A. Bacteriophage typing

Synonym: Phage typing.

Principle: It based on bacterial susceptibility by bacteriophage.

Typing method: Method called **pattern method.**

- **Steps:** Bacterium to be typed is inoculated in agar plate (lawn culture), followed by application of phage in fixed dose. Incubate the plate overnight at 37°C.
- **Result:** There is lysis of bacteria by phage and production of clear zone called **plaque (Fig. 118.6).** Bacterial typing will be done on the basis of size, shape and nature of plaque.

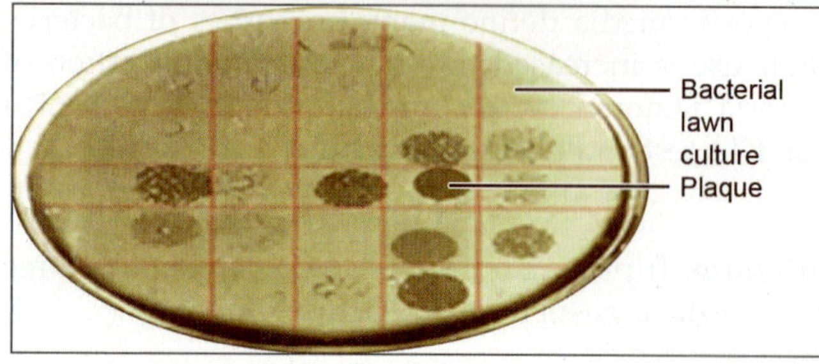

Fig. 118.6: Phage typing

Properties: (1) Genus specific: Phage affects all the genera of bacterium like in *Salmonella*. **(2) Species specific:** Phage affects all the species of genus like in *B. anthracis*. **(3) Biotype or subspecies specific:** Phage affects all the strains or subspecies like Mukharjee phage–IV which lyses all strains of Classical *Vibrio*, but not El Tor *Vibrio*.

Uses: It is used for identification and epidemiological typing of bacteria, which are difficult to distinguish by biochemical and serological methods. It is useful for following bacteria.

1. *Staph. aureus:* If strain of *Staph aureus* is lysed by phages 52, 79 and 80 it is called **phage type 52/79/80.** Phage type 80/81 is associated with outbreak in hospital and called **epidemic strain of *Staph. aureus.*** International basic set of phages for typing of *Staph aureus* of human origin is mentioned in **Table 118.2.**
2. *C. diphtheriae:* Bacteriophage infecting *C. diphtheriae* is called **corynephage.** Around 15 bacteriophage types are identified. Type 7 is avirulent, 1 and 3 are **mitis**, type 4 and 6 are **intermedius**, and remainders virulent are **gravis.**
3. *M. tuberculosis:* Bacteriophage infecting *M. tuberculosis* is called **mycobacteriophage.** Viral chromosome instead of being integrated in bacterial chromosome remains free like plasmid called **pseudolysogeny.** It includes four types of *M. tuberculosis* like A, B, C, I (I = Intermediate between A and B, common in India).
4. *Salmonella*
   - **O-Ag variation:** Structural changes in O Ag of *Salmonella* can be induced by bacteriophage (lysogenization) which results in new serotypes as shown in **Flowchart 118.3.**
   - **Vi phage type-II:** It acts on Vi Ag of typhoid bacilli hence called **Vi phage type II.** It was developed by Craigie and Yen in 1937. It is useful for intraspecies

TABLE 118.2: Phage typing of *Staph aureus*	
Lytic group of the phages	**Designation of phages**
Group I	29, 52, 52A, 79, 80
Group II	3A, 3C, 55, 71
Group III	6, 42E, 47, 53, 54, 75, 77, 83A, 84, 85
Group V	94, 96
Unclassified	81, 95

Flowchart 118.3: O-Ag variation by phage

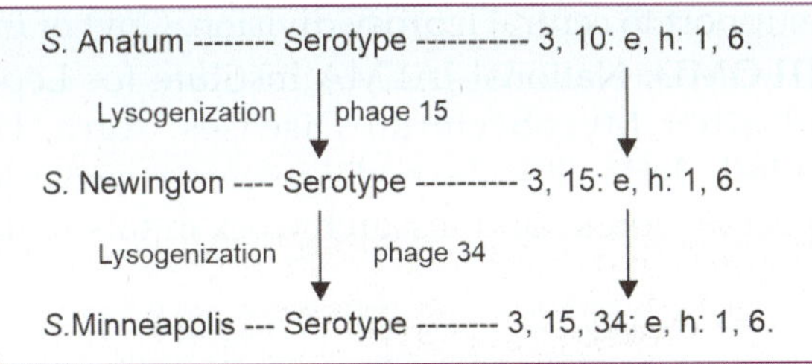

classification of *Salmonella.* It also provides the information regarding trend and pattern of typhoid at local, national and international levels. **Common Salmonella phages in India are** phage A and phage E1 of *S.* Typhi and type 1 and type 2 of *S.* Paratyphi A. Typing by Vi phage has few **limitations** like strain without Vi Ag are untypable and not able to discriminate between the other species.

- **Nicole's complementary phage typing:** In some region one or more phages are available in such cases typing is done through Nicole's complementary phage like typing of type A strain in to 10 types.

5. *Vibrio cholerae:* Bacteriophage called **vibriophage.** Typing is useful in epidemiological studies and to differentiate Classical *Vibrio* from El Tor *Vibrio.* Following are the different phage typing of *Vibrio cholerae.*
   - **Classical *Vibrio* typing:** *V. cholerae* is classified in to five types by using Mukherjee's 4 phages (I-IV). Type IV lyses the all strains of Classical *Vibrio* but none of the strain of El Tor *Vibrio.*
   - **El Tor *Vibrio* typing:** *V. cholerae* is classified in to six types by using Basu and Mukherjee's phage 5. Type V lyses the all strains of El Tor *Vibrio* but none of the strain of Classical *Vibrio.*
   - *Vibrio cholerae* **O1 typing:** Typing is done by Lee and Basu by using 14 phages.

6. *Brucells* **spp.**
   - **Tblisi phage (Tb phage):** Tb phage was isolated in former Soviet Union in 1950. It is designated as reference phage. *B. abortus* lysed at routine test dose (RTD) while *B. suis* lysed at 10, 000 RTD and *B melitensis* not lysed at all. It is less applicable because all the strains are not sensitive to Tb phage, so not identify all the strains.
   - **Newer phages for typing:** Numbers of newer phages are identified in last decades with wide host range. Corbel defined 6 groups of *Brucella* on the bases of this newer phage in 1987.

7. **Other bacteria:** Phage typing is done for few other bacteria like *P. aeruginosa* (less useful because less discriminatory power), *Proteus* spp., and *Anthrax bacillus* but less useful.

Bacteriophage typing centers for different bacteria: Follow section, reference centers (Described below).

B. Bacteriocin typing: Ch. 10.

C. Biotyping: It based on different biochemical properties of the organisms. More detail of biotypes of each bacterium is given in respective chapter. Biotyping is useful for following bacteria.
1. *C. diphtheriae:* It classifies the bacterium into three biotypes like gravis, intermedius and mitis.
2. *S.* Typhi: Four subgenera from I-IV are defined on the basis of biochemical properties.
3. *Shigella* spp.: Four species on the basis of biochemical and serological properties like *Shigella dysenteriae, Sh boydii, Sh flexneri* and *Sh sonnei.*
4. *Vibrio cholerae:* It classifies the bacterium into two biotypes like classical and El Tor.
5. *Y. pestis:* It classifies the bacterium into three biotypes like *Y. pestis* var *antigua, Y. pestis* var *medievalis* and *Y. pestis* var *orientalis.*
6. *H. influenzae:* It classifies the bacterium into eight biotypes like from I to VIII.
7. *Brucella* spp.: Several biotypes are defined in *brucella* spp.

D. Sero typing: It based on the serological properties of the organisms. It is useful for following bacteria.
1. *Streptococcus* spp.: Lancefield grouping based on Carbohydrate antigens of cell wall.
2. *E. coli, Shigell* spp., *Salmonella* spp., and *V. cholerae:* Grouping based on cell wall antigens.
3. *Strept pneumoniae, H. influenzae, N. meningitidis:* Grouping based on capsular Ag.

E. Antibiogram typing: It based on the susceptibility of the organisms to different antibiotics. It is done with antibiotic sensitivity testing for many bacteria and gives clue about occurrence of outbreak of resistant strain like MRSA, VISA and VRSA in *Staph. aureus,* MDR-TB and XDR-TB in *M. tuberculosis,* MDR strains in *Pseudomonas* and *Acinetobacter,* VRE, PPNG, etc.

F. Auxo typing: It based on the nutritional requirement of the organisms. It is useful for *N. gonorrhoeae.*

G. Morpho typing: It based on the cultural characteristics of the organisms on different media. It is useful in few bacteria like *P. aeruginosa.*

Genotyping Methods

Different genotyping methods are RFLP, PFGE, sequence based methods, plasmid profile analysis, ribotyping and AFLP.

> **Note: Genotyping methods**
>
> **RFLP (Restriction Fragment Length Polymorphism):** It based on southern blot technique. DNA is cleaved into small fragments by using 2 restriction enzymes. DNA fragments are separated by electrophoresis and transferred on nitrocellulose membrane which are detected by DNA probe.
>
> **PFGE (Pulse Field Gel Electrophoresis):** Lyse the organism to release the free DNA. Free DNA is cleaved in to lager size and less numbered DNA fragments. Such lager size DNA

(Contd...)

fragments are then subject to PFGE. PFGE gives mobility to DNA fragments of various strains obtained during an outbreak are compared manually or by computer soft ware. It is a gold standard (but required expertise and also labor-intensive) hence sequence based methods are more useful.

Sequence based methods: It can be done by specific equipment called **sequencer**. Sequencing is useful to detect the relatedness of bacteria. Sequencing methods are following two types.

- **Nucleotide sequencing:** Sequencing is performed by sanger sequencer. It can be done at one or multiple genes. **(1) Single gene sequencing:** It is used for identification of polymorphism within the gene (called **single nucleotide polymorphism/ SNP analysis**) and identification of organisms. **(2) Multiple gene sequencing:** It called **MLST (multi locus sequence typing).**
- **Whole genome sequencing:** Sequencing is performed by next generation sequencer. It determines the complete DNA sequence of an organism's genome at a single time. Identification of whole genome may useful to treat the infections.

REFERENCE CENTERS

Following are the reference centers in India for further diagnosis of bacteria up to strain/subsp./type level.

Staphylococcus aureus: Bacteriophage typing center of *Staph. aureus* is available at Maulana Azad Medical College, New Delhi, India.

Streptococcus **spp.:** Reference laboratories for streptococcal Lancefield grouping are Lady Harding Medical College, New Delhi, India and Christian Medical College, Vellore, India. Data from reference laboratories showed that about 45% of hemolytic streptococcal isolates are from group A, 10–15% from group B and C each, 25% from group G while 5% from group F.

Salmonella **Typhi:** (1) National *Salmonella* Vi **phage typing** center is available at Lady Harding Medical College, New Delhi, India. (2) National *Salmonella* **sero typing** reference centers are Central Research Institute, Kasauli, India for human origin and Indian Veterinary Research Institute, Izzatnagar, India for animal origin.

Vibrio cholerae: International reference center for **vibriophage typing** is National Institute of Cholera and Enteric Disease (NICED), Kolkata, India.

Brucella: **Biotyping** center is Central Veterinary Laboratory, New Haw, UK.

Mycobacteria

1. **CLTRI:** Central Leprosy Teaching and Research Institute, Chengalpattu, Chennai, Tamil Nadu, India. Followings are the aims and objectives of CLTRI.
   - Research into the basic problems relating to the inception and spread of leprosy
   - Field studies for controlling leprosy
   - Training of leprosy workers
   - To give technical guidance and advice for the promotion of anti leprosy work

- Monitoring and evaluation of NLEP and providing support to central leprosy division, Govt of India.
2. **NJILOMD:** National JALMA Institute for Leprosy and other Mycobacterial Diseases, Agra, Uttar Pradesh, India. It is the nodal reference center in all aspect of leprosy and research work in tuberculosis.

ACCESS YOURSELF

Essay/Full Question

1. Laboratory diagnosis of bacterial infections.

Short Notes

1. IMViC tests/catalase test/oxidase test/urease test/TSI test.
2. Composite media.
3. Microbial typing/bacteriophage typing.

Short Questions for Theory/Viva Questions

1. Describe in short: Rapid urease test and selective urease test.
2. What are composite media? Write two examples.
3. Name the bacteriophage typing center for following organisms: *Staph. aureus* and *S.* Typhi.
4. Name the sugar present in TSI medium and KIA medium.
5. Write the principle of MR test and VP test.
6. Write the clinical use of Kovac's reagent and Kovac's method.

Comments on

1. Performance of catalase test from blood agar can gives false-positive result.
2. Selective urease test is more advantageous than rapid urease test to detect *H. pylori.*

MCQs for Chapter Review

Sample Rejection Criteria

1. **Specific reason to disallow the sample for culture:**
   a. Sample brought within 2 hours of collection
   b. Sample brought in sterile plastic container
   c. Sample brought in formalin
   d. Sample obtained after cleaning the collection site

Testing Methods (Biochemical Reactions)

2. **Reagent(s) used in indole test is/are:**
   a. Kovac's reagent b. Ehrlich's reagent
   c. a + b d. None
3. **Medium (media) useful for citrate test is/are:**
   a. Koser's broth b. Simmon 's agar
   c. a + b d. None
4. **Oxidase positive bacteria are:**
   a. *Vibrio* b. *Pseudomonas*
   c. *Clostridium* d. *E. coli*
5. **Medium (media) useful for urease test is/are:**
   a. Stuart's broth b. Christensen 's agar
   c. a + b d. None
6. **Maximum urease positive is produced by:**
   a. *H. pylori* b. *P. mirabilis*
   c. *K. rhinoscleromatis* d. *Ureaplasma*
7. **Urease-negative is:**
   a. *E. coli* b. *Proteus*
   c. *Klebsiella* d. *Staphylococcus*
8. **All of the following bacteria test "urease positive",** *except***:**
   a. *E. coli* b. *Proteus*
   c. *Klebsiella* d. *Staphylococcus*

9. **Triple Sugar Iron agar (TSI) medium contain all the following carbohydrates, *except*:**
   a. Glucose
   b. Lactose
   c. Sucrose
   d. Mannitol

Testing Methods (Microbial Typing)

10. **Lysis of bacterial colony in culture is seen by which virus?**
    a. Pox
    b. HSV
    c. Bacteriophage
    d. CMV

11. **Phage typing is useful as epidemiological tool in all, *except*:**
    a. *Salmonella*
    b. *Staph. aureus*
    c. *V. cholerae*
    d. *Shigella dysenteriae*

12. **Phage typing can be done for:**
    a. *Salmonella*
    b. *Streptococcus*
    c. *Shigella*
    d. *Pseudomonas*

13. **Phage typing is widely used for the intraspecies classification of one of the following bacteria:**
    a. *Staph. aureus*
    b. *E. coli*
    c. *K. pneumoniae*
    d. *P. aeruginosa*

Answers and Explanation of MCQs

1. **c**
   - Reason for disallowing or rejecting sample for culture from given option is presence of disinfectant (formalin).

2. **c**
   - Follow section, **Table 118.1 (indole test)** for explanation.

3. **c**
   - Follow section, **Table 118.1 (citrate test)** for explanation.

4. **a and b**
   - Follow section, **oxidase test (uses)** for explanation.

5. **c**
   - Follow section, **urease test (reagents)** for explanation.

6. **a**
   - Urease produced by *H. pylori* is 100 times more active than produced by *Pr. vulgaris*. Maximum (rapid) urease positive given by *H. pylori*.

7. **a**

8. **a**
   - Follow section, **urease test (uses)** for explanation of answers of MCQs 7–8.

9. **d**
   - Follow section, **TSI test (full form)** for explanation.

10. **c**

11. **d**

12. **a, d**

13. **a**
    - Follow section, **microbial typing (bacteriophage typing)** for explanation of answers of MCQs 10–13.

Laboratory Diagnosis of Viral Infections

Chapter Outline

□ Testing Methods

TESTING METHODS

Following are different steps employed for diagnosis of viral infections.

Sample collection, transportation and storage: Described in **Ch. 113**.

Sample rejection criteria: Described in **Ch. 118**.

Testing methods: Described below.

Microscopy

Light (simple) microscopy: It is useful to detect the following things.

- **Viruses:** Viruses are very small in size and difficult to see under the light microscope; however, some viruses **like pox virus** can be examined under light microscope.
- **Viral antigens:** They are detected by immuno-peroxidase staining. Tissue sections or cells coated with viral antigens are stained by antibodies fixed with horseradish peroxidase enzyme, followed by adding of hydrogene peroxide and colored reagent (benzidine derivative). Colored complex formed can be viewed under light microscope.
- **Inclusion body:** It is visible under light microscope. Described separately in box.

Note: Inclusion body

Definition: It is an intracellular aggregate of viral proteins or antigens.

Properties: It is variable in size and shape. It includes protein from microbes and rarely from host cells. It may be intranuclear, intra-cytoplasmic or mixed.

Staining reactions: It is variable in staining reactions.
- **Acidophilic (eosinophilic):** It stains pink with Giemsa or eosin methylene blue, e.g. All the viral inclusion bodies are acidophilic like Cowdry type A, polio, etc., except adeno virus inclusion body and "owl eyes" in CMV.

- **Basophilic (hematoxylin):** For example, adeno virus and "owl eyes" in CMV.

Mechanism of formation: When genes from one organism are expressed in another the resulting proteins sometimes form inclusion bodies.

Etiological classification: Two main types.

1. Infectious origin

- **Bacterial (basophilic):** Like HP body (Halberstaedter Prowazeki body) of *C. trachomatis* and LCL (Levinthal-Cole-Lillie) body of *C. psittaci*.
- **Parasitic:** *Leishmania* spp., *Plasmodium* spp., *Babesia* spp.
- **Viral:** Three subtypes as per location in cell.
  - **Intracytoplasmic (Fig. 119.1)/acidophilic: (1)** Poxviridae family: Paschen body in variola virus (smallpox), Guarnieri body in vaccinia virus, Bollinger body in fow lpox virus and Henderson-Peterson body or molluscum body (larger size 20–30 µm) in molluscum contagiosum virus. **(2)** Rhabdoviridae family: Negri body in infected brain cell by rabies virus.
  - **Intranuclear:** It also called **Cowdry type inclusion body.** It has two subtypes. **(1) Cowdry type A (acidophilic):** It is variable in size and with granular appearance. For example Lipschutz body of herpes simplex virus and varicella zoster virus plus Torres body in yellow fever virus. **(2) Cowdry type B:** It is variable in numbers and more circumscribed. For example polio virus (acidophilic), adeno virus (basophilic) and "owl's eye" in CMV (basophilic).
  - **Both intranuclear and intracytoplasmic:** It seen in measles.

2. Noninfectious origin: Inclusion body in RBCs, presents in many noninfective conditions like as Hb precipitate, lead poisoning, anemia, etc.

Clinical significance of inclusion body: It is easily examined under light microscope and helps in diagnosis of viral infection.

Note: Pseudo-inclusions body

It is an invaginations of the cytoplasm into the cell nucleus, which may give the appearance of intranuclear inclusions body. It may appear in papillary thyroid carcinoma.

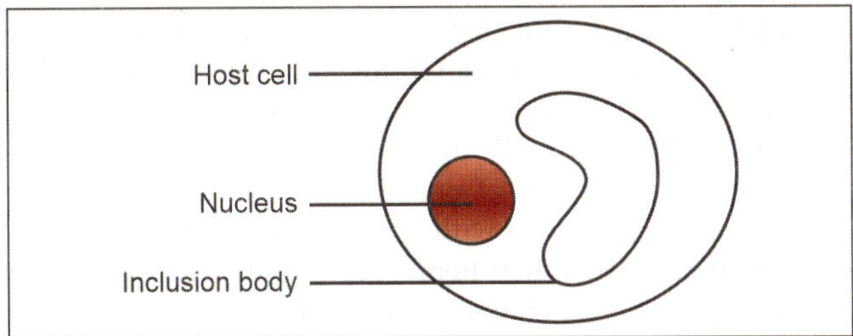

Fig. 119.1: Intracytoplasmic inclusion body

Electron microscopy: Viruses are very small, so simple microscopy is not useful and best examined by electron microscopy.

Immuno electron microscopy: Specific Ab is added to react with viral Ag and then examined under electron microscopy.

Immunofluorescent (IF) microscopy: Both direct and indirect are useful.

Viral Culture

Virus cannot grow over artificially prepared nutritional (chemically defined or cell free) media. Three techniques are employed for viral cultivation.

Animal Culture

Indications: Because of legal issues animal culture is restricted. It is used for **research purpose**s (to study the immune response, virus-host interaction, pathogenesis, epidemiology and oncogenesis) and in **diagnosis** of few viruses like coxsackie viruses and arboviruses.

Commonly used animals: Infant suckling mice (for coxsackie and arbo viruses), guinea pigs, rabbits, ferrets, etc.

Growth detection in animals: By death, disease or visible lesion production in animals.

Disadvantages: Animal immunity and latent microbes in animal are interfere with viral growth.

Embryonated Egg Culture

History: It was developed by Good pasture in 1931. With the development of new diagnostic methods use of egg culture is start to decrease.

Indications

- **Virus isolation:** Different parts (**Fig. 119.2**) of embryonated hen's egg are to be inoculated for particular virus. **(1) Chorioallantoic membrane (CAM):** It is used for vaccinia or variola virus, HSV, pox virus and Rous sarcoma virus cultivation. Visible lesion called **pock** with different morphology by different virus. Number of pocks will estimate the numbers of particles in suspension. **(2) Allantoic cavity:** It is used for orthomyxo virus (influenza) or paramyxo virus (mumps virus) cultivation. **(3) Amniotic cavity:** It is used for isolation influenza virus, mumps virus, avian adeno virus, new castle disease virus, etc. **(4) Yolk sac:** Used for some viruses (may like herpes virus),

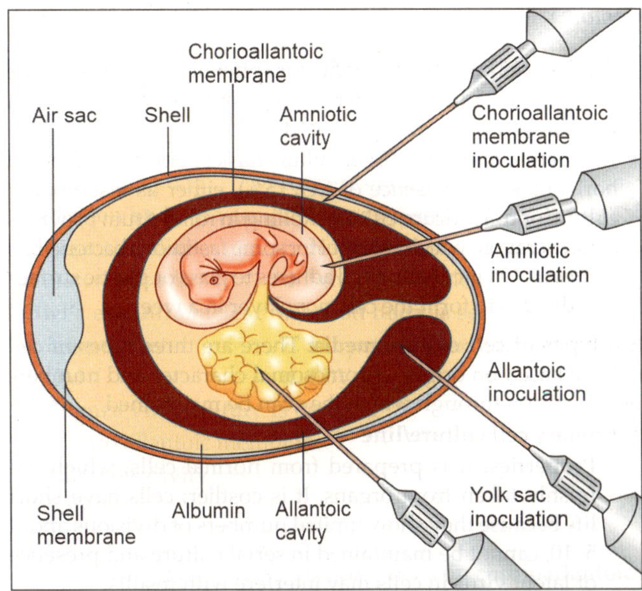

Fig. 119.2: Different parts of hen's egg

Chlamydia spp., *Haemophilus ducreyi* and *Rickettsia* spp.

- **Vaccine preparation: (1) Hen's eggs:** Allantoic cavity is inoculated to prepare the vaccine of influenza virus, yellow fever virus (17D) and rabies virus (Flury strain). **(2) Duck's eggs:** To prepare the inactivated duck egg (non-neural) vaccine for rabies virus.

Cell or Tissue Culture

History: Steinhardt was the first to use tissue culture in virology in 1913. He maintained the vaccinia virus in fragments of rabbit cornea. Enders, Robbins and Weller were made the method most useful for virus cultivation since 1949 by cultivating neurotropic virus like polio virus in non-neural tissue.

Types: Three types of tissue culture.

1. **Organ culture:** It is done by using bits of organs with preservation of normal Anatomy and Physiology. Useful for viruses which are special pathogens for certain organs, for example, tracheal ring for corona virus.
2. **Explant culture:** It is done by mincing the tissue called **explant culture.** For example, mincing of adenoid tissue for adeno virus.
3. **Cell culture:** It is described separately in box.

> **Note: Cell culture**
> **Preparation of cell culture media (monolayer)**
> - **Pretreatment:** It is done by dissociating the tissue in to cells by trypsin (proteolytic enzyme) or by mechanical shaking.
> - **Preparation of media:** Such cells are washed, counted and suspended in containers like bottles, Petri dish or test tubes or flask to prepare the cell culture media.
> - **Adding of different ingredients in media:** Media are enriched by adding vitamins, amino acids, salts, glucose, 5–10% foetal-calf serum. Antibiotics are added to prevent the bacterial contamination. Set the appropriate pH about

(Contd…)

20. **One of the following virus is not a cultivable virus:**
    a. Papova
    b. Parvovirus B-19
    c. Herpes
    d. Adenovirus
21. **All are cultivable viruses, *except*:**
    a. *Rotavirus*   b. *Enterovirus*
    c. ECHO virus   d. Coxsackie virus

Answers and Explanation of MCQs

1. **b**
  • *Rickettsia*, *Mycoplasma* and *H. pylori* are not producing any type of inclusion body.
2. **c**
3. **c**
4. **a**
5. **a**
6. **a**
7. **c**
  • Follow section, **inclusion body** for explanation of answers of MCQs 2–7.
8. **b**
9. **c**

10. **c**
  • Follow section, **embryonated egg culture** for explanation of answers of MCQs 8–10.
11. **d**
12. **a**
13. **b**
14. **c**
15. **c**
16. **c**
  • Follow section, **cell or tissue culture** for explanation of answers of MCQs 11–16.
17. **b**
  • Budding is the reproductive method of fungi and few bacteria like *Mycoplasma pneumoniae* and *Francisella tularensis*.
18. **b, c**
  • Follow section, **cell or tissue culture (methods for detection of viral growth in tissue culture → hemadsorption)** for explanation
19. **a**
  • Follow section, **cell or tissue culture → uses** for explanation.
20. **b**
21. **a**
  • Follow section, **cell or tissue culture → limitations and Table 119.2** for explanation of answers of MCQs 20–21.

CHAPTER 120

Laboratory Diagnosis of Fungal Infections

Chapter Outline
- Diagnosis
- Laboratory Fungal Contaminants

DIAGNOSIS

Following are different steps employed for diagnosis of fungal infections

Sample collection, transportation and storage: Described in **Ch. 113**.

Sample rejection criteria: Described in **Ch. 118**.

Testing methods: Described below.

Microscopy and Staining

Wet mount preparations

- **Direct KOH**
  - **Actions:** About 10% KOH will liquefy the keratin presents in samples and makes the clear fungal examination as shown in **Fig. 120.1a**. It called direct because performed directly from sample.
  - **Subtypes: (1) Slide method:** It is useful for thin samples like epidermal scales, skin scraping, etc. **(2) Tube method:** It is useful for thick samples like biopsy or tissues. It requires overnight incubation.
  - About 20–40% KOH is required for specimen like nail. Dimethyl sulfoxide can be added to improve the digestion.
  - Glycerol can be added to prevent the drying.
- **Negative staining:** Indian ink and 10% nigrosin are used to examine the capsule of *C. neoformans*.

- **Lactophenol cotton blue (LCB) stain:** It is used to identify the morphological parts of fungi from culture as shown in **Fig. 120.1b**. It contains lactic acid, which preserves the fungal morphology, phenol as disinfectant, cotton blue which stains the fungal elements in blue and glycerol to prevent the drying.
- **Pal, Hasegawa, Ono and Lee (PHOL) stain:** Same use like LCB stain.
- **Others:** Neutral red stain and Diazonium blue B stains are also useful.

Gram's stain: All fungi are gram-positive as shown in **Fig. 120.1c**.

ZN stain: It differentiates acid fast bacteria from mycetoma causing fungi.

Histopathological stains: These stains are useful to identify the fungus from tissues/biopsy materials.

1. **HE (Hematoxylin and eosin) stain**
   Ingredients: Stain contains hematoxylin dyes, eosin and ammonium or potassium alum (metal) as mordant effect. The hematoxylin dye could be of Ehrlich's hematoxylin, Mayer's hematoxylin, Harris's hematoxylin, Coles' hematoxylin, Caeazzi's hematoxylin and Gill's hematoxylin.
   Principle: Hematoxylin is oxidized to hematin on air or light exposure (or by adding oxidizing agents in stain) which stains nuclei in blue with mordant effect by metal while eosin stains cytoplasm and tissue fibers in pink color.

2. **PAS (Periodic-acid Schiff) stain**
   Ingredients: Stain contains periodic acid and Schiff's reagent (basic fuchsin, distilled water, potassium/sodium metabisulfite, HCl and decoloring charcoal).
   Principle: It based on the principle of Feulgen reaction where oxidization of polysaccharide of fungi and bacteria by periodic acid releases aldehydes

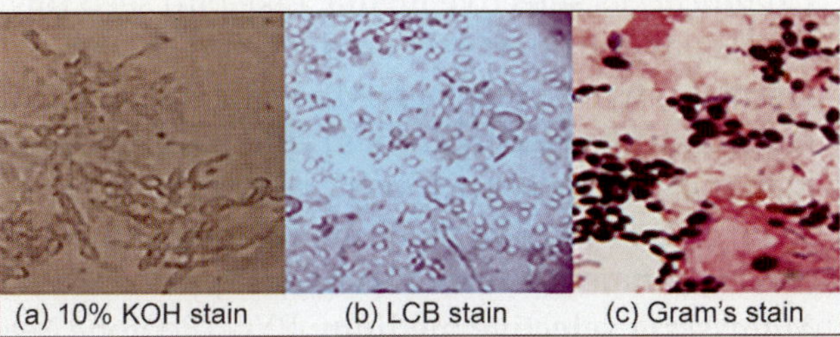

Fig. 120.1: Fungal stains

that yield red color compound with basic fuchsin of Schiff's reagent. Nuclei stain blue, fungi magenta/red and background as light green. Protein and nucleic acids remain unstained.

Uses: Stain is used for diagnosing fungi in tissues that are stained darker than surrounding tissues.

Disadvantages: It also stains the tissue contains carbohydrate making fungus demonstration more difficult. Few microbes like actinomycetes (*Nocardia*) are not stained by PAS stain. It stains only live not the dead fungi.

3. **Gridley's fungal stain:** Almost parallel to PAS stain.

 Ingredients: Stain contains chromic acid and Schiff's reagent (aldehyde fuchsin, distilled water, sodium metabisulfite, HCL and tartrazine).

 Principle: Oxidization of polysaccharide of fungi and bacteria by chromic acid releases aldehyde that yield purple color compound with aldehyde fuchsin of Schiff's reagent. Fungi (mycelia and yeast cells), elastic tissues and mucin are stained in purple and background in yellow color.

 Uses: Stain is used for diagnosing fungi in tissues with purple color and background in yellow color.

 Disadvantages: It stains the mucin and elastic fibers present in tissue in purple color, making fungus demonstration more difficult.

4. **GGMS (Grocott Gomori's methenamine silver) stain**

 Ingredients: Stain contains chromic acid, methenamine silver nitrate, sodium metabisulfite, sodium thiosulfate, etc.

 Principle: Oxidization of polysaccharide of fungi and bacteria by chromic acid releases aldehyde which reduces methenamine silver nitrate to metallic silver. Metallic silver deposited at the place where aldehydes is available yielding brown-black color to the fungal cell wall and bacteria, tissue pale green, cytoplasm old rose and mucopolysaccharide dark gray.

 Uses: (1) It stains both live and dead fungi compared to PAS stain which stain only live fungi. (2) It also stains the actinomycetes (like *Nocardia, Actinomyces, Actinomadura* and *Streptomyces*) which are not stained by PAS stain. (3) For *Pneumocystis jirovecii*, it stains the cyst wall in black color.

5. **Mayer's mucicarmine stain**

 Ingredients: Stain contains carmine (red color) dye, aluminum hydroxide and aluminum chloride. Aluminum hydroxide and aluminum chloride have mordant effect.

 Principle: It based on carminophilic (ability of microbes to eat carmine/red color dye) property of microbe especially due to presence of mucin (glycoprotein) in cell wall or polysaccharide capsule. Here mucin part of cell (like cancer cells and *R. seeberi*) stains in red, capsule in deep rose color while nuclei in blue color.

 Uses: (1) *R. seeberi* (mucin in cell wall): Sporangium and endospores stain in red color while nuclei in blue color. (2) *C. neoformans*: Capsule stains deep rose in color. (3) Cancer cells: Because malignant cells contain mucin.

6. **Masson Fontana silver (MFS) stain**

 Ingredients: Stain contains silver nitrate (of Fontans's silver stain used for *T. pallidum*).

 Principle: Melanin produced by fungi stains black in color by silver nitrate.

 Uses: It is used for melanin producing fungi like *C. neoformans* and black (phaeoid) fungi. Fungi stain black and nuclei in red.

7. **Giemsa stain:** For trophozoites and sporozoites (internal nuclei) of cyst of *Pneumocystis jirovecii*, where nuclei are stain in pink color and cytoplasm in blue color.

8. **Alcian blue stain:** For capsulated fungus like *C. neoformans*.

Fluorescent Microscopy (FM) and staining

1. **Fluorochrome staining:** It is performed by using dye like calco-fluor-white (CFW) stain. It directly binds with fungal cell wall presents in samples and enhances the fungal visualization.

2. **Immunofluorescent staining:** Detection of antigen or antibody by using a dye like FITC (fluorescein isothiocyanate).

Culture

Culture media

1. **Routine fungal culture media:** It includes **Sabouraud dextrose agar (SDA)** with antibiotics like cycloheximide which prevents the contamination by saprophytic fungi and gentamicin plus chloramphenicol that prevent the bacterial growth and make the medium selective for fungus. **Two sets** of media are used. 1st set is incubated at room temperature (20–25°C) for mycelial phase. 2nd set is incubated at 37°C for yeast phase.

2. **Nutritionally deficient (spore producing) media:** Deficiency of nutrition allows the spore production. **Chlamydospores producing media** are cornmeal or cornmeal tween agar and rice starch agar.

3. **Selective media**

 Candida **selective media:** They are used to differentiate *Candida* spp. For examples, tetrazolium reduction medium and chrom agar.

 Cryptococcus **selective media:** Like bird seed (niger seed) agar and sunflower seed agar.

 Dermatophyte selective media: Like dermatophyte test medium (red colonies) and dermatophyte identification medium (purple colonies).

 Aspergillus **selective medium:** Czapek-Dox agar is used to differentiate the *Aspergillus* spp.

 Malassezia **selective medium:** PYG (Peptone Yeast extract Glucose) agar.

Culture technique and identification of growth: Follow Flowchart 120.1.

Automated culture method: Like VITEK and MALDI-TOF are used in diagnosis of yeasts and to some extent for molds.

Biochemical Reactions (B/Rs)

Sugar fermentation test and sugar assimilation test: To differentiate the *Candida* species and other yeast or yeast like fungi.

Urease test: It is positive in fungi like *Malassezia furfur*, *Cryptococcus neoformans*, *Trichosporon beigelii*, *Trichophyton mentagrophytes* and *Rhodotorula mucinaginosa*.

Serological Tests

Like ELISA, precipitation based tests, RIA, CFT, agglutination tests, etc., are useful to detect the fungal infections.

Molecular Tests

PCR, real time/quantitative PCR (qPCR), nested PCR, multiplex PCR and DNA sequencing methods are developed to identify the fungi from specimens or from culture.

Other Tests

Skin test: Based on DTH.

Animal inoculation: Laboratory animals like rabbit, guinea pig, etc., are useful.

Gag Liquid Chromatography: To detect the fungal metabolites like D-arabinitol, D-mannose.

Wood's lamp examination: It also called black light lamp or UV-A light lamp or ultraviolet light lamp. It was devised by Robert William Wood in 1903 using "Wood's glass". It is made up with barium silicate and nickel oxide. It transmits the UV light (at a wavelength of approximately 365 nanometers) which produces different fluorescence as mentioned below on exposure to skin.

1. **Infectious conditions**
   - **Bright green:** *Microscporum audounii, M. canis, M. ferruginium* and *Pseudomonas* in burns wound.
   - **Dull green:** *Trichophyton schoenleinii.*
   - **Dull yellow:** *Microscporum gypseum.*
   - **Golden yellow:** *Malassazia furfur* (causing ptyriasis versicolor).
   - **Coral-red:** *Corynebacterium minutissimum* (causing erythrasma).
   - **Brick-red:** *Prevotella melaninogenica.*
   - **Orange glow:** *Propionibacterium acne.*
2. **Noninfectious conditions**
   - **Ethylene glycol poisoning:** Manufacturers of ethylene glycol-containing antifreezes commonly add fluorescein, which causes the patient's urine to fluoresce under Wood's lamp.
   - **Vitiligo:** Wood's lamps have also been used to differentiate hypopigmentation from depigmentation such as with vitiligo. A vitiligo patient's skin will appear yellow-green or blue under the Wood's lamp.
   - **Porphyria cutanea tarda:** In this condition urine turns pink upon illumination with Wood's lamp.
   - **Phototherapy:** Bili light a type of phototheraphy that uses blue light with a range of 420–470 nm, used to treat neonatal jaundice.
   - **Others:** Tuberous sclerosis, melanoma and freckles.

Special tests

1. **Teased mount for fungi:** Teased out the beats of colony over glass slide, put LCB stain and examine under microscope. It does not give *in situ* appearance but separate the fungus from mass and makes the easy examination.

Flowchart 120.1: Fungal culture technique and identification of growth

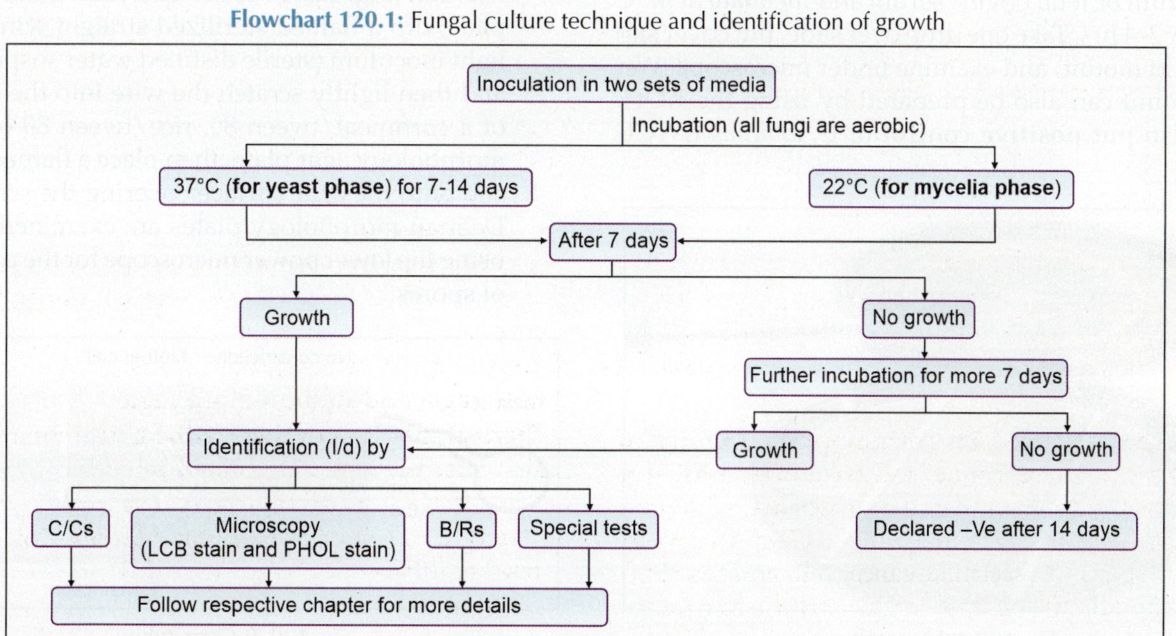

2. **Slide culture for fungi (Fig. 120.2):** It preserves the *in situ* (undisturbed) fungal growth under microscope. Place a sterile slide in Petri dish and place a 1 cm SDA block on slide. Inoculate the test species at four sides of agar block. Cover the agar block with cover slip. Incubate at 25°C for 48 hours. Take new slide and add a drop of LCB stain. Transfer the cover slip from Petri dish slide to new slide and examine under the microscope.

3. **Cellophane tape mount:** Touch the cellophane to the colony in media and fix on slide with LCB stain. Examine the *in situ* appearance under microscope.

4. **Exo-antigens tests:** They detect free antigens of fungus produced during growth in culture broth.

5. **Germ tube test**
   - **Synonym and history:** Also called **Reynolds-Braude phenomenon**, because it was first reported by Reynolds and Braude in 1956.
   - **Indications:** It is used for rapid identification of *Candida* species like **(1) *Candida albicans* (+):** Half rate of positivity than *Candida dubliniensis* Approximately 95–97% of *C. albicans* isolates develop germ tubes when incubated in a proteinaceous media. **(2) *Candida dubliniensis* (++):** Double rate of positivity than *Candida albicans*.
   - **Principle:** Formation of germ tube is associated with increased synthesis of protein and ribonucleic acid. Germ tube solutions contain serum, essential nutrients for protein synthesis. It is lyophilized for stability.
   - **Material required**
     - Culture: Suspected *Candida* culture plate.
     - Reagent: Serum (human, sheep, fetal bovine) or other commercially produced media.
     - Others: Test tubes, loop or wooden applicator, Pasteur pipette, slide, cover slip and microscope.
   - **Method:** Mix fungal culture with sheep or human serum or fetal bovine serum and incubate at 37°C for 2–4 hrs. Take one drop over slide, put cover slip (wet mount) and examine under microscope. Wet mount can also be prepared by using the KOH. Also put **positive control** of *C. albicans* (ATCC

10231) and **negative control** of *C. tropicalis* (ATCC 13803)/*C. glabrata* (ATCC 2001).

- **Results: (1) Positive test:** A short hyphal (filamentous) extension arising laterally from a yeast cell, with no constriction at the point of origin (**Fig. 120.3**). Examples include *Candida albicans* and *C. dubliniensis*. **(2) Negative test:** No hyphal (filamentous) extension arising from a yeast cell or a short hyphal extension constricted at the point of origin. Examples include *C. tropicalis*, *C. glabrata* and other yeasts.
- **Features of germ tube:** Germ tube is half the width and 3–4 times the length of the yeast cell. There is no presence of nucleus. It is nonseptate and without constriction at the point of origin.
- **Advantages:** (1) It is a rapid test for diagnosis and differentiation of *Candida albicans* from other *Candida* isolates. It can be performed in less than three hours. (2) It is easy to perform.
- **Disadvantages (limitations):** (1) Germ tube is confused with pseudohyphae or buds, as there is no constriction at the site of origin of germ tube from yeast cells. (2) *Candida tropicalis* may produce pseudo-germ tubes after 3 hours of incubation but they show constriction at the point of origin. (3) Too heavy inoculum will inhibit germ tube formation. (4) This test is only part of the overall scheme for identification of yeasts. Further testing is required for definite identification.
- **Other methods to detect the germ tube:** These are milk medium, amino acid synthetic medium and Muller Hinton agar.
- **Clinical significance of germ tube:** Germ tube is one of the virulence factors of *Candida albicans*.

6. **Dalmau plate culture**
   - **Indication:** For *in situ* identification of chlamydospores.
   - **Method (Fig. 120.4):** To set up a yeast morphology plate, dip a flamed sterilized straight wire into a light inoculum (sterile distilled water suspension) and then lightly scratch the wire into the surface of a cornmeal/tween 80, rice/tween 80 or yeast morphology agar plate, then place a flamed cover slip onto the agar surface covering the scratches. Dalmau morphology plates are examined *in situ* using the lower power microscope for the presence of spores.

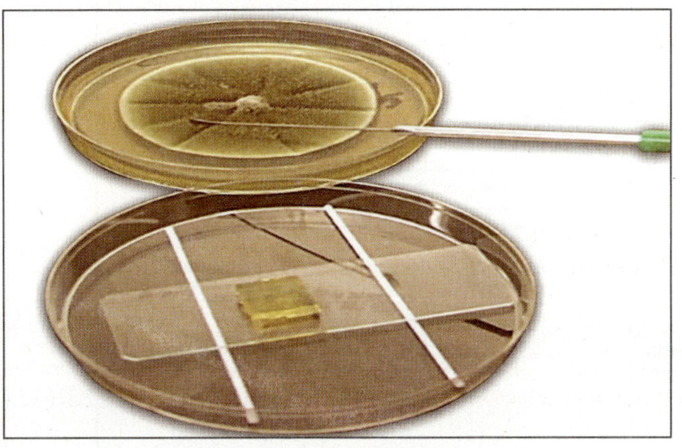

Fig. 120.2: Slide culture

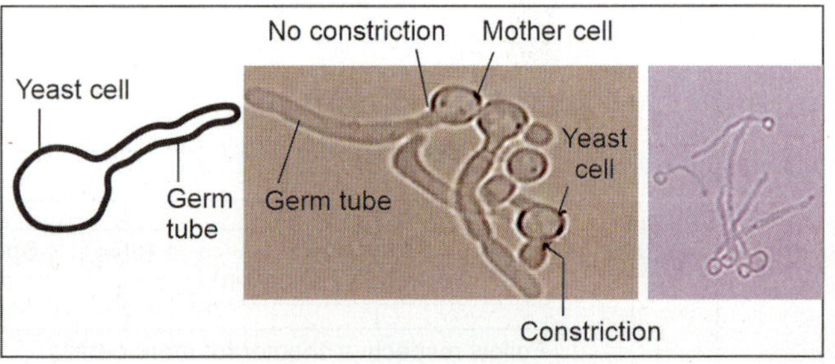

Fig. 120.3: Germ tube

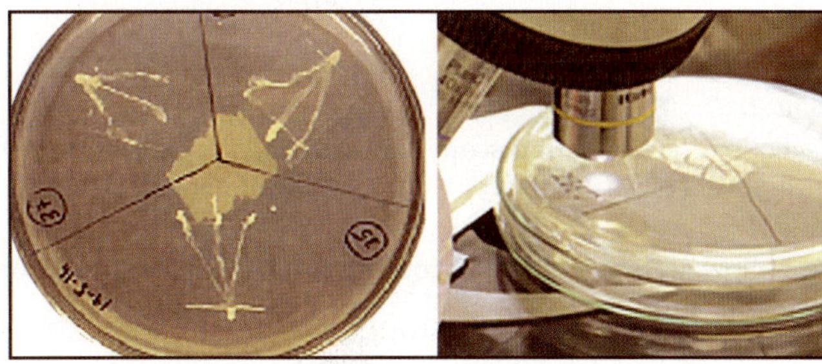

Fig. 120.4: Dalmau plate culture

7. Hair baits technique

- **Indication:** For identification of keratinophilic fungi from soil.
- **Principle:** It based on utilization of keratin by fungus.
- **Method:** Place soil samples in sterile Petri dish. Place around 30 pieces of sterile (by autoclave) horse's hairs on top of soil. Wet the soil surface by using sterile distilled water (do not flood it). Add more water and incubate at 30°C. When white to brown fungal growth appear around the hairs, transfer them with fungal growth on SDA plate in scattered manner. Buried one end of hair under agar surface. Take hair with fungal growth for microscopic observation under LCB mount. When fungal colonies appear on SDA plate, make the LCB mount for identification of fungus under microscope. Fungi which grow on SDA plate should be transferred to other plate for further confirmation.
- **Hair perforation technique:** Place filter paper strip in sterile Petri dish and cover the strip by using sterile distilled water. Add small portion of sterilized prepubertal or infant's hair in to water. Add 5–6 drops of 10% yeast extracts to speed up the reaction. Inoculate the colony directly on hair and incubate at 25°C for 4 weeks. Take the hair on slide, prepare the LCB mount and examine under microscope for perforation of hair shaft as shown in **Fig. 120.5**. Test is positive in *Trichophyton mentagrophytes* and *Microsporum canis* and negative in *Trichophyton rubrum* and *Microsporum equinum*.

Antifungal sensitivity test: Report the result with antifungal sensitivity pattern.

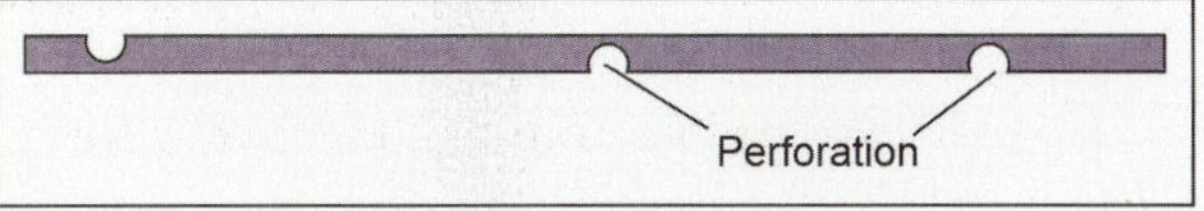

Fig. 120.5: Hair perforation test

LABORATORY FUNGAL CONTAMINANTS

Note: Laboratory fungal contaminant

Introduction: Mycology laboratory encounters more contaminants than actual pathogens. They always create the confusion in the diagnosis of actual pathogen. It is very difficult to draw a line between contaminants and actual pathogens; however, contaminants can be differentiated by (1) repeated growth in all clinical specimens, (2) presence of growth in un-inoculated areas (pathogens can grow only at the inoculated site, but not anywhere) and (3) may be absent in direct smear examination.

Classification: Almost many are described in respective chapters. All they are summarised in following five groups.

- **Zygomycets:** *Absidia* species, *Mucor* species, *Rhizopus* species, *Syncephalustrum* species and *Cunninghamella* species.
- **Hyaline hyphomycetes:** *Acremonium* species, *Aspergillus* species, *Chrysosporium* species, *Fusarium* species, *Penicillium* species, *Paecilomyces* species, *Scopulariopsis* species, *Sepedonium* species, *Trichoderma* species, *Trichothecium* species and *Verticillium* species.
- **Phaeoid fungi:** *Alternaria* species, *Aureobasidium* species, *Bipolaris* species, *Chaetomium* species, *Cladosporium* species, *Curvularia* species, *Helminthosporium* species, *Nigrospora* species and *Phoma* species.
- **Yeast and yeast like:** *Geotrichum* species, *Rhodotorula* species, *Saccharomyces* species, *Trichosporon* species and *Ustilago* species.
- **Look alike fungi:** *Sepedonium* species, *Malbranchia* species, *Chrysosporium* species and *Acrodonium* species.

ACCESS YOURSELF

Essay/Full Question

1. Laboratory diagnosis of fungal infections.

Short Notes

1. Germ tube test.
2. Fungal laboratory contaminants.

Short Questions for Theory/Viva Questions

1. Name two selective culture media for each, *Candida* spp., and *C. neoformans*.
2. Name four fungi giving urease test positive.
3. What are hair bait technique and hair perforation technique?

Comment on

1. Thick tissue or skin scraping sample is treated with KOH before examination.

MCQs for Chapter Review

Diagnosis

1. **KOH wet mount can be prepared for:**
   a. Bacteria   b. Virus
   c. Fungus   d. Parasite
2. **Role of KOH in fungal wet mount is:**
   a. To preserve the fungus
   b. To digest the keratin
   c. TO stabilizes the morphology
   d. To adjust the ph
3. **Which dye is most suitable for fungus demonstration in biopsy?**
   a. Alizarin red   b. Veirhoff dye
   c. Manson's trichome   d. PAS
4. **Stain used for staining fungal elements:**
   a. Acid fast stain   b. Mucicarmine
   c. Methenamine silver   d. Gram's stain
5. **Fungal staining is done by:**
   a. Calcofluor white   b. Leishman's stain
   c. ZN stain   d. None
6. **Which of the following is added in SDA to suppress the growth of contaminating (saprophytic) fungi and bacteria?**
   a. Cycloheximide   b. Chloramphenicol
   c. Gentamicin   d. All of the above

7. Wood lamp can be used for evaluation of:
   a. Tinea capitis   b. Freckles
   c. Vitiligo   d. Tuberous sclerosis

8. Reynold-Braude phenomenon, is seen in:
   a. *C. albicans*   b. *C. psitasi*
   c. *Histoplasma*   d. *Cryptococcus*

9. Germ tube test is used in diagnosis for:
   a. *Candida albicans*   b. *Cryptococcus*
   c. *Histoplasma*   d. Coccidioidomycosis

10. Hair perforation test is positive in infection with:
   a. *Trichohyton*   b. *Microsporum*
   c. *Epidermophyton*   d. All of above

Answers and Explanation of MCQs

1. c

2. b
- KOH will liquefy the keratin or thick tissue presents in samples and makes the clear fungal examination. It can be done by slide or tube method.

3. d

4. b, c, d

5. a
- Follow section, **microscopy and staining** for explanation of answers of MCQs 3–5.

6. d
- Follow section, **culture → routine fungal culture media** for explanation.

7. a, b, c, d
- Tinea capitis is caused by *Microscporum* spp., or *Trichophyton* spp. (dermatophyte category), which fluoresces in different color under UV light. Follow section, **Wood's lamp examination → uses** for explanation of other options.

8. a

9. a
- Follow section, **germ tube test** for more explanation of answers of MCQs 8–9.

10. a, b
- Follow section, **hair perforation technique** for more explanation.

Laboratory Diagnosis of Parasitic Infections

Chapter Outline

- Testing Methods

TESTING METHODS

Following are different steps employed for diagnosis of parasitic infections.

Sample collection, transportation and storage: Described in **Ch. 113**.

Sample rejection criteria: Described in **Ch. 118**.

Testing methods: Described below.

Macroscopic Examination

Stool is examined macroscopically for adult stage or segments of adult stage.

Microscopic Examination

Microscopic Examination of Stool

Collection of stool: Advice the patient to collect the stool (about 20–40 grams or 5–6 table spoonfuls of watery stool) directly in wide mouth container.

Precautions during collection of stool: Do not mix it with urine. Container does not contain or pretreated with antiseptics. Oil or barium should not be given to the patient.

Preservation of stool

1. **5% fomalin:** For preservation of protozoan cysts.
2. **10% fomalin:** For preservation of helminthic eggs and larvae. Around 1–2 g stool can be preserved with 8–10 ml of 10% formalin and thoroughly shakes before use.
3. **Merthiolate iodine formalin (MIF):** About 2.35 ml of stock MF (merthiolate-formaldehyde) is mixed with 0.15 ml of stock iodine (Lugol's iodine). Feces with pea size added to this mixture which acts as both fixative and stain. Permanent slide cannot be made from this mixture.
4. **Schaudin's fixative:** It contains mercuric chloride.

5. **Polyvinyl alcohol (PVA) or modified Schaudin's fixative:** It is used for preservation of protozoan cysts and trophozoites and to prepare the permanent slide. PVA has long shelf life (6–12 months). For preservation, 3 parts of PVA are added to one part of feces. Mercuric chloride has been replaced with cupric sulfate.
6. **Sodium acetate acetic acid formalin (SAF):** Less toxic than PVA.

Transportation of stool: Transport the stool within 30 minutes in triple layer packing, as motile protozoa can die and unidentifiable after that.

Examination of stool

A. NS and 1% iodine wet mount

- **Preparation of iodine solution:** Follow box.
- **Steps of preparation of NS and iodine smear (Fig. 121.1):** Take one clean glass slide. Put one drop of NS on one side and iodine on other. Add one drop of stool material in NS and iodine. Mix with wooden stick and remove large undigested food particles. Put cover slip and examine under the microscope.

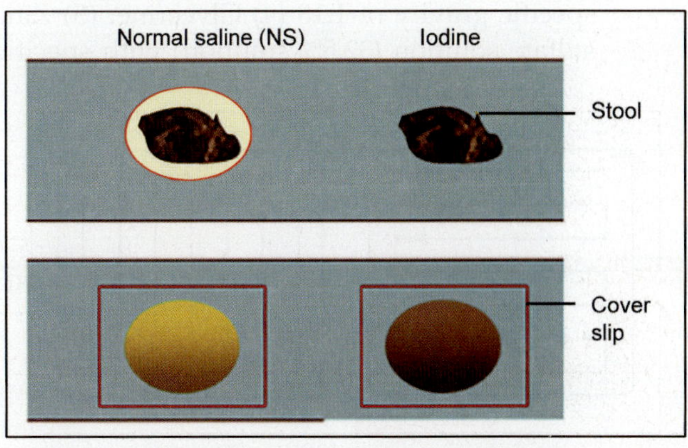

Fig. 121.1: NS and iodine preparation

- **Method (direction) of smear examination:** Start from low power and when egg found move to high power for details microscopic study. Start from one side and move to opposite side as shown in **Fig. 121.2**.

Note: Preparation of iodine solution

Preparation of Lugol's iodine (5%)
- **Reagents**
  - Iodine crystal/powder: 5 g
  - Potassium iodide: 10 g
  - Distilled water: 100 ml
- **Steps:** Dissolve the potassium iodide with distilled water. Add the iodine crystal slowly. Filtered the solution and stored in amber color bottle.

Preparation of 1% iodine (weak iodine solution): For parasitic examination 1% iodine is recommended by Dobell and O'Connor, which is prepared by diluting Lugol's iodine about five times with distilled water. It deteriorates very quickly, hence it should be prepared every two weeks.

B. Buffered Methylene Blue (BMB) preparation: It is indicated for amoebic trophozoites, but not for cyst from fresh samples not on preserved samples.

C. Concentration methods
- **Indication:** Due to less numbers of parasites in stool, routine methods like NS and 1% iodine preparation may give false-negative result. To overcome this drawback concentration methods are indicated. They selectively concentrate the protozoan cysts, heliminthic eggs and larvae, thus increasing the chances of positive result.
- **Types:** Three types.
  1. **Flotation methods**
     - **Principle:** Stool is suspended in a solution of high specific gravity than of parasitic eggs, so that parasitic eggs and cysts float up and get concentrated at surface.
     - **Common floatation fluids/solutions:** (1) Saturated solution of common salt (400 g NaCl/1000 ml water) with specific gravity of 1.18–1.20. (2) Saturated sugar/sucrose solution (500 g sugar/1000 ml water) with specific gravity of 1.25. (3) Saturated solution of sodium nitrate (400 g sodium nitrate/1000 ml water) with specific gravity of 1.18 (4) Glycerine. (5) Zinc sulfate solution (32.5% solution) with specific

gravity of 1.18 (6) A.E.X. (Acid/HCl of with specific gravity of 1.03 + Ether + Xylol equal parts) solution. (7) Magnesium sulfate solution.
  - **Uses and limitations:** All the eggs are float except unfertilized eggs of *A. lumbricoides*, larvae of *S. stercoralis*, eggs of *T. saginata*, *T. solium* and intestinal flukes.
  - **Methods:** Follow box.

Note: Flotation methods

Willi's (Leviation) simple floatation technique by using saturated salt solution
- **Steps:** About 2 ml of saturated salt is taken in flat bottom tube (or penicillin tube). Add 1 g of stool. Fill the tube completely with salt solution up to brim. Care should be taken to see that not a drop of the contents overflows. Apply a slide over the tube, so that it is in direct contact with the surface of solution without any intervening bubbles. Allow it to stand for 10–15 minutes by which time all the eggs would have floated up. Remove the slide without jerking and reversed to bring the wet surface on top, put cover slip and then examine under the low power for eggs.
- **Advantages:** Light infections are detected by this technique invariably. Only eggs of few nematodes like common round worm, hook worm, whip worm and thread worm are clearly visible without the hindrance of fecal and fibers materials.
- **Disadvantages:** (1) This method is not useful for protozoan cysts, eggs of cestodes, eggs of trematodes and unfertilized eggs of *A. lumbricoides*. (2) Different solution with different specific gravity is needed to float the eggs of different kinds of parasites. (3) The eggs may distort if kept in a floatation solution for long time. (4) Any delay in examination may cause salt crystal to develop, interfering in examination.

Zinc sulfate centrifuge flotation technique
- **History:** Discovered by Faust et al., in 1939.
- **Steps:** A fine suspension is made by adding 1 g of freshly passed stool to the 10 ml of luke warm distilled water. Strain through wire gauze (40 meshes to an inch) to remove coarse faecal materials. Filtrate is collected in Wassermann tube and centrifuged at the rate of 2500 rpm for one minute. Poured off the supernatant fluid and add the water to the sediment. It is shaken well, centrifuged and repeat the process for 2–3 times till the supernatant fluid become clear, which is then poured off. Add the 3–4 ml zinc sulfate solution to the sediment, sediment is stirred and further zinc sulfate solution is added to fill the tube up to the top. Again centrifuge the tube at the rate of 2500 rpm/minute for one minute. Remove the surface film by a platinum wire loop of 5 mm size, on to a clean glass slide, put cover slip and then examined under the low power of the microscope for eggs. Add the one drop of iodine before putting the cover slip for protozoan cysts.
- **Advantage:** It is used for protozoan cysts, eggs of nematodes and small cestodes.
- **Disadvantage:** It is not useful for large cestodes, eggs of most trematodes and unfertilized eggs of *A. lumbricoides* and larvae of nematodes.

Sheather's sugar floatation method: This method is useful for oocyst of *Cryptosporidium* spp., and *Cystoisospora belli*.

Lane's direct centrifugal floatation method: It is used to concentrate the eggs (quantitative measurement).

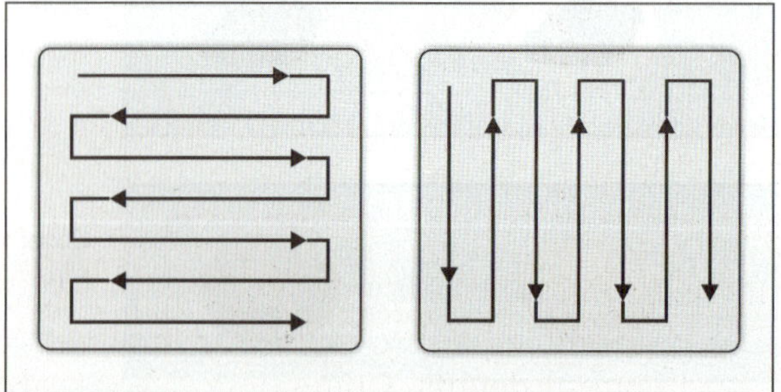

Fig. 121.2: Direction of smear examination

2. Sedimentation methods

- **Principle:** Stool is suspended in a solution of low specific gravity than of parasitic eggs, so that parasitic eggs and cysts sediment and get concentrated at bottom either spontaneously or after centrifugation.
- **Common floatation fluids/solutions: (1)** Simple tap water. **(2)** Formol ether (formalin-ethyl aceatate).
- **Methods:** Follow box.

3. Other concentration methods

- Baermann's method.
- Water emergency semi-concentration technique. } **Ch. 107**
- Kato-Katz preparation.

Note: Sedimentation methods

Simple sedimentation method with tap water

- **Steps:** Add about 2 g feces in a suitable container. Add water about 3/4th in to the fecal material. Thoroughly emulsify the fecal material. Strain the emulsion through a sieve (tea strainer) to remove all the coarser particles in a cone-shaped tube (urine analysis flask). 0.5% glycerine in water can be used to emulsify the feces instead of plain water. Centrifuge at 2,000 rpm for 2 minutes. All the eggs get packed at the bottom of the tube along with the sediment. The supernatant fluid is then poured off. A drop of the sediment is placed on the slide, covered with a thin cover slip and examine under low power microscope.
- **Advantages:** This method is the most reliable to recover the eggs of cestodes, trematodes and nematodes.
- **Disadvantages:** Presence of too many fecal materials and fibers may hinder the examination of eggs.

Formol ether sedimentation technique

- **Synonym:** Formalin-ethyl acetate sedimentation method.
- **Steps (Fig. 121.3):** Add about 2 g feces in a suitable container. Add about 10 ml of water in to the fecal material. Thoroughly emulsify the fecal material. Strain the emulsion through a sieve (tea strainer) to remove all the coarser particles in a cone shaped tube. Centrifuge at 2,000 rpm for not less than 10 minutes. All the eggs get packed at the bottom of the tube along with the sediment. The supernatant fluid is then poured off and wash the sediment with 10 ml of saline. Centrifuge again and repeat washing until supernatant is clear. After the last wash, decant the supernatant and add 10 ml of 10% formalin to the sediment. Mix and let stand for 5 minutes to effect fixation. Add 1–2 ml of ethyl acetate, Stopper the tube and shake vigorously. Centrifuge at 1500 rpm for 10 minutes. Four layers should result like a top layer of ether (ethyl acetate), plug of debris, layer of formalin (formal water) and sediment. Detach the plug of debris from the side of the tube by an applicator stick. Carefully decant the top three layers. With a pipette, mix the remaining sediment which contains eggs, with the small amount or remaining fluid and transfer one drop, each to a drop of saline and iodine on a glass slide. Apply the cover slip and examine under the low power microscope.
- **Advantages:** Formalin fixes the helminthic eggs, larvae, protozoan cysts, oocysts, and spores, so that they are no longer infectious, as well as preserves their morphology.
- **Disadvantage:** Ether is explosive and inflammable, which is over come by ethyl acetate.

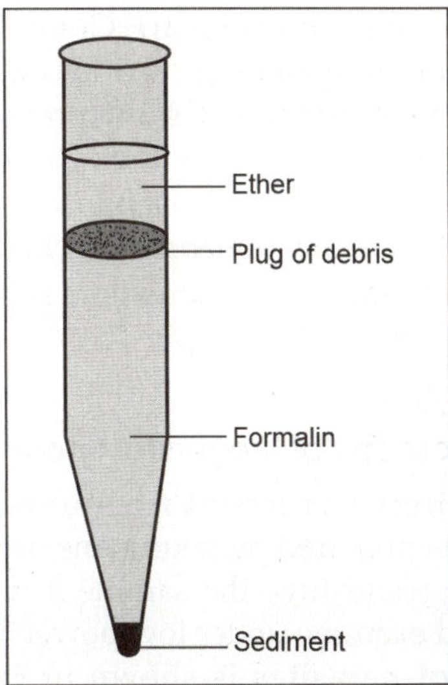

Fig. 121.3: Formol ether sedimentation technique

Microscopic appearance of parasites under normal saline and iodine preparation present in stool: Follow Fig. 93.1, Fig. 94.1, Fig. 102.3, Fig. 102.16, Fig. 107.5, Fig. 107.6, Fig. 107.7, Fig. 107.8, Fig. 107.13, Fig. 107.24 and Fig. 107.27.

Artefacts in stool: Materials present in stool which look like parasites called **artefacts**. For example yeasts, air bubbles, Charcot-Layden crystals, plants, cells like leukocytes, RBCs, muscle cells, epithelial cells, pus cells, fat cells, etc., undigested food materials and hairs or other fibers.

D. Eggs counting technique in stool

- **Indications:** (1) For epidemiological survey. (2) For monitoring of therapy by estimating numbers of worm infecting patients.
- **Techniques:** (1) Stoll's method: It is most popular method described in box. (2) Direct smear technique: It is rarely useful. (3) Kato-Katz preparation:

Note: Stoll's method

Transfer 4 g feces to Stoll's flask containing 60 ml, 0.1N NaOH. Add the several glass beads to make a uniform suspension. Close the flask with rubber stopper and shake vigorously. Take exactly 0.15 ml suspension by pipette over slide. Put cover slip over it and examine under microscope. Count the total numbers of eggs in entire smear. Number of eggs per gram of feces is obtained by multiplying the count of two such preparations by 100. It is roughly estimated that 100 eggs can be passed by one female per gram of feces per day. Exact numbers depend on consistency of stool and correction factors (CF) applied as follows.

Mushy formed stool: CF is 1.5.

Mushy stool: CF is 2.

Mushy diarrheic stool: CF is 3.

Franky diarrheic stool: CF is 4.

Watery stool: CF is 5.

E. Permanent stained smear of stool

1. **Histopathological stains:** They are useful for cytological details. Commonly used fixatives are Schaudin's fixatives (best fixative) and PVA.

Essentials of Medical Microbiology

Commonly used stains are Gomori's trichome stain (stains nucleus in red and cytoplasm in bluish green and H and E stain (provides excellent contrast between nucleus and cytoplasm).

2. **Modified ZN satin:** It is useful for acid-fast parasites which are mentioned in **Ch. 112 (Table 112.5).**
3. **Trichrome stain:** For Microspora spp.
4. **Fluorescence microscopy:** For *Cyclospora* spp., *Toxoplasma gondii*, etc.

Microscopy of Urogenital Specimen

It includes **direct wet mount** from urine or urethral discharge. It is prepared by taking one drop of sample on clean side (centrifuge the sample if required). Put cover slip and examine under low power. **Microscopic appearance of parasites is** shown in **Fig. 94.4 and Fig. 104.2.**

Microscopic Examination of Peripheral Blood

Following parasites are present in blood.
- For *Trypanosoma* spp., **Ch. 95.**
- For *Leishmania* spp., in macrophages: **Ch. 95.**
- For *Plasmodium* spp., and *Babesia* spp., **Ch. 97.**
- For *Toxoplasma gondii:* **Ch. 98.**
- For microfilariae: **Ch. 108.**

Microscopic Examination of Sputum

It is examined for CL crystals, eggs or larvae.

Culture

Cell free media/chemically defined media: Some protozoa can grow over cell free media. Culture methods in details for each parasite are described in **respective chapters.** Few important media or culture techniques with parasites are listed in **Table 121.1.**

Animal culture: It is useful in *E histolytica*, *Trypanosoma* spp., *Plasmodium* spp., *Babesia spp.*, and *T gondii*. Detail description of animal culture is given in respective chapter.

TABLE 121.1: Culture media of parasites

Parasite	Culture medium
E. histolytica	Different axenix, polyaxenic and monoaxenic media
G. lamblia	• Medium described by Karapetyan and Meyer • Diamond's (TYI-S-33) medium • Modified Diamond's (TYI-S-33) medium
T. vaginalis	Bushly's medium, Feinberg-Whittington medium, Roiron's medium, Johnson-Trussel's medium, CPLM (Cysteine Peptone Liver Maltose) medium and plastic envelope medium
Leishmania spp., and T. cruzi	Novy, McNeal and Nicolle (NNN) medium
B. coli	Same as *Entamoeba coli*
Plasmodium spp.	RPMI1640 (Rose Park Memorial Institute) medium

Cell or tissue culture: It is useful in *T. vaginalis*, *T. gondii* and Microspora spp.

Egg culture: It is useful in *T. vaginalis*.

Coproculture: Useful for hook worm spp., pseudohook worm spp., and *S. stercoralis*.

Xenodiagnosis

It based on diagnosis of parasites from the body of vectors by allowing vector to bite the infected person in *T. cruzi* and *W. bancrofti*.

Serological Tests

They are used for detection of Ag or Ab. Different tests are ELISA, ICT, etc. Particular test for particular parasite is described in respective chapter.

Molecular Methods

These are PCR, real-time PCR (qPCR), LAMP, (developed for kala azar and malaria) and automated Biofire film array (automated nested multiplex PCR which targets multiple bacteria, viruses and parasites causing diarrhea like *E. histolytica*, *Giardia*, *Cyclospora* and *Cryptosporidium parvum*).

Skin Tests

Tests based on immediate type of hypersensitivity: (1) Casoni's test in *H. cyst*. (2) Skin test for *T. spiralis*. (3) Skin test for wuchereriasis. (4) Mazzotti's test for *O. volvulus*.

Tests based on delayed type of hypersensitivity: (1) Skin test for trypanosomes. (2) Leishmanin test: Positive in African kala azar. (3) Frenkel's test: For *T. gondii*. (4) Scratch test: For *A. lumbricoides*.

Cytological Examination of Body Fluids

Eosinophilia and cell counting is done from different body fluids like blood, CSF, etc.

Other Indications of Stool Examination

For CL crystals: In *E. histolytica*, *A. duodenale*, *Trichuris trichuria*.

For ghost cells and pyknotic bodies: In *E. histolytica*.

Radiological Methods

Like X-ray, USG, CT-scan and MRI are useful.

ACCESS YOURSELF

Essay/Full Question

1. Laboratory diagnosis of parasitic infections.

Short Notes

1. Stool examination in parasitology.
2. Blood examination in parasitology.
3. Parasitic culture.

Short Questions for Theory/Viva Questions

1. Name the four parasites present in peripheral blood/CSF.
2. Name the four protozoa present in stool.
3. Name the culture media for following parasites: *G. lamblia*, *T. vaginalis*, *Leishmania* spp., and *Plasmodium* spp.

Question of Practical Exercise

1. How you examine the stool sample for trophozoite, cyst or ova? Draw the labeled diagram of your findings and report to the examiner.

MCQs for Chapter Review

Diagnosis (Testing Methods)

1. **Protozoan cysts are stored in:**
   a. Saline
   b. Phenol
   c. Sodium hypochlorite
   d. Formalin
2. **On microscopic examination eggs are seen, but on saturation with salt solution, no eggs are seen. The eggs are likely to be of:**
   a. *Trichuris trichuria*
   b. *Taenia solium*
   c. *Ascaris lumbricoides*
   d. *Ancyclostoma duodenale*
3. **All float in a saturated salt solution, *except*:**
   a. *Chlonerchis sinensis*
   b. Fertilized eggs of *Ascaris*
   c. Larvae of *Strongyloides*
   d. *Trichuris trichuria*

4. **The egg of which helminth, can be concentrated in saturated salt solution:**
   a. *Taenia saginata*
   b. *Taenia solium*
   c. Unfertilized egg of *Ascaris*
   d. *Ancyclostoma duodenale*
5. **In formol ether concentration technique, which layer contain parasites?**
   a. Ether
   b. Fecal debris
   c. Formal water
   d. Sediment

Answers of MCQs and Explanation

1. **d**
- Follow section, **microscopic examination of stool → preservation of stool)** for explanation.
2. **b**
3. **a, c,**
4. **d**
- Follow section, **concentration methods → Willi's (Leviation) simple floatation technique by using saturated salt solution)** for explanation of answers of MCQs 2–4.
5. **d**
- Follow section, **concentration methods → formol ether sedimentation technique)** for explanation.

Molecular Methods

Chapter Outline

Development of molecular genetic has provided basis in diagnosis, prevention and treatment of disease over the conventional methods. Molecular methods include DNA cloning and NA diagnostic techniques.

DNA CLONING

Synonym

It also called DNA recombinant (rDNA) technology or genetic engineering or genetic modification or genetic manipulation.

Definition

It is an artificial manipulation of DNA to identify and to obtain the gene or gene products.

Steps

Developments of DNA fragment (Fig. 122.1a): It is developed with the help of restriction endonuclease enzyme which cleaves the DNA in to nucleotide sequence specific fragment.

Cloning vector (Fig. 122.1b): It is DNA molecule in which DNA fragment is inserted. Commonly used vectors are plasmid (like R factor), temperate bacteriophage, cosmid (hybrid vector containing both plasmid and phage), etc.

Ligation (Fig. 122.1c): Combined DNA molecule (vector) and DNA fragment called **recombinant DNA**. This method is called **ligation** and it is cleaved by DNA ligase.

Selection of host cell (Fig. 122.1d): Commonly used host cells are bacteria like *E. coli*, fungi like yeast cells or viruses like vaccinia virus.

Introduction of recombinant DNA into host cell (Fig. 122.1e): Recombinant DNA is introduced into host cell by transformation.

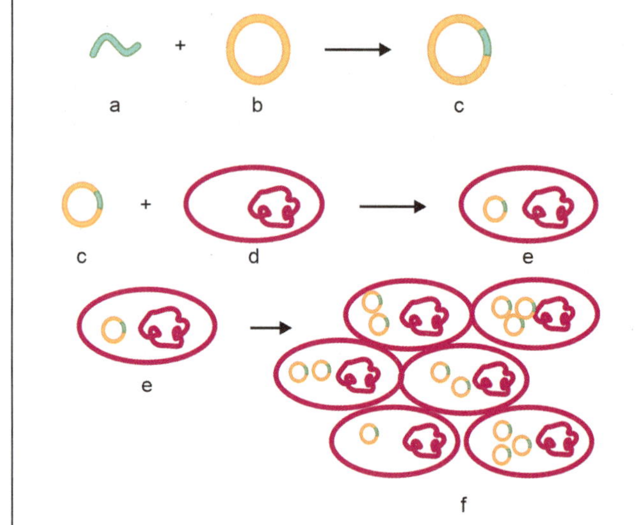

Fig. 122.1: Steps or DNA cloning

Multiplication and obtaining desired product (Fig. 122.1f): Host cell containing recombinant DNA is allowed to multiply in suitable medium to obtain desired protein or gene in large quantity.

Significances

1. **Production of hormones:** Like growth hormone, somatostatin, insulin, thymosin (used for lung and brain cancer), etc.
2. **Production of proteins:** Like enzymes like urokinase (used to dissolve the blood clots), interleukin, TNF, etc.
3. **Preparation of vaccines:** Like HBV vaccine, rabies vaccine, malaria vaccine (under trial), HIV capsid protein (under trial), etc.
4. **Gene therapy:** To treat inherited disease like hemophilia, thalassemia, sickle cell anemia, etc.
5. **Other uses:** Like production of interferons, antibiotics and some extramedical uses.

NUCLEIC ACID (NA) DIAGNOSTIC TECHNIQUES

Advantages over conventional methods: Molecular methods have few advantages than conventional (microscopy, culture, etc.) methods like rapid result, less labor intensive, highly sensitive and highly specific.

Disadvantages: They are with high cost and required special instruments-infrastructure, so difficult to establish in peripheral set-up.

Uses

- To diagnose different bacterial, viral, fungal and parasitic infections
- To diagnose antimicrobial drug resistance
- To diagnose neoplastic diseases
- Forensic investigations
- Examination of phylogenetic relationship in evolution.
- Archaeological studies

NA diagnostic techniques: Generally three methods are useful like blotting techniques, DNA probe-hybridization and NA amplification techniques.

> **Note: Nucleic acid diagnostic techniques**
>
> **Nonamplification techniques:** These are southern blotting, northern blotting and hybridization. Western blotting and Eastern blotting are not for NA detection; however, for better understanding purpose they are described under the heading of blotting techniques.
>
> **Amplification techniques:** These are PCR with its different types and LAMP.

Blotting Techniques

These are methods of separation of proteins (like DNA, RNA, Ag or Ab) by gel electrophoresis and transferring them on a carrier (like nylon or cellulose membrane) so, they can be identified later. Blotting techniques are following four types.

Southern Blotting

History: It was developed by EM Southern, hence the name.

Use: It is used for identification of DNA. One example of this method is RFLP.

Steps

1. **Fragmentation:** DNA molecule is fragmented by restriction endonuclease enzyme.
2. **Separation:** DNA fragment is separated by gel electrophoresis.
3. **Blotting or fixation:** Transfer separated piece on to nylon membrane or cellulose membrane and fixed.
4. **Denaturation:** DNA bound to membrane is denatured in to ss DNA.
5. **Hybridization:** Membrane bound DNA is hybridized with radio labeled ss DNA probe.
6. **Identification:** Hybridization is detected on X-ray films.

Northern Blotting

It is opposite (for RNA) to Southern blotting (for DNA) hence the name. It is used for identification of RNA. RNA is separated by gel electrophoresis, blotted and identified by labeled probe.

Western (Immuno) Blotting

Steps: These are described below with example of HIV (Fig. 122.2).

1. **Separation of Ags:** Antigens (Ags) are separated by SDS-PAGE (sodium dodecyl sulfate-poly acrylamide gel electrophoresis). Commonly used Ags of HIV are p18, p24, p40 and p55 of gag gene, p31, p51/55 and p65/66 of pol gene and gp160, gp120 and gp41 of env gene.
2. **Blotting or fixation:** Transfer separated Ags on to nylon membrane or cellulose membrane strip and fixed.
3. **Reaction with test serum:** Strip is reacted with sera contains Abs (antibodies) and forms Ag-Ab complexes.
4. **Conjugate step:** Strip then reacts with radio-labeled or enzyme labeled Abs (antihuman globulins).
5. **Substrate:** Add the substrate in last, which produces color band at Ag-Ab complex sites.

Interpretation of positive results: (1) WHO criteria: Presence of at least two env bands out of gp160, gp120 and gp41 with or without gag or pol bands. **(2) CDC criteria:** Presence of any two bands out of p24, gp160, gp120 and gp41.

Advantages: (1) It is more specific test than screening tests considered as 'gold standard' for HIV diagnosis. (2) It confirms the discordant result obtained by screening tests. (3) It differentiates between HIV-1 and 2.

Disadvantage: It is less sensitive than screening tests.

Eastern Blotting

It is the modification of Western blotting for protein analysis for post-translation modification by using probes to detect the lipids, carbohydrate and protein.

DNA Probe and Hybridization

DNA Probe

Definition: It is a biotinylated or radiolabeled or chemiluminscent piece of ss DNA for detection of homologous DNA.

Production: In the past DNA probe was prepared by DNA recombinant or DNA cloning. Nowadays, it is prepared by specific instrumentation.

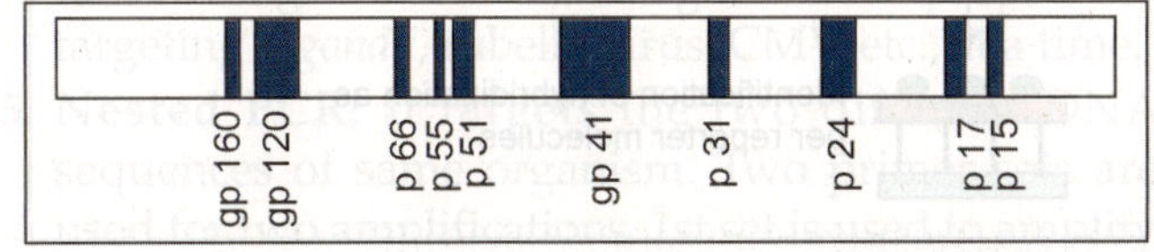

Fig. 122.2: Western blot test for HIV-Ab

Molecular Methods

10

target nucleic acid for second amplification. It is highly sensitive because two primers produce more yields of DNA. It is highly specific because two primers targets two different DNA of same organism. It is used for detection of *M tuberculosis* which targets IS 6110 gene.

6. **Automated multiplex nested PCR or Biofire film array (biomerieux):** It includes all steps automatically from sample preparation to amplification, detection and analysis. Four panels are designed which target around 20–25 common respiratory, gastrointestinal, meningoencephalitis or blood stream pathogens. It gives result around in one hour. It is very costly.

7. **Automated real-time PCR (automated qPCR):** Like CBNAAT and chip-based nucleic acid amplification test used for tuberculosis. For more details follow **Ch 56**.

Disadvantages of PCR

1. **False-positive:** Contamination by environmental DNA leading to false-positive results. It is for this reason that sample preparation, running PCR and post-amplification detection must be carried out in separate rooms.

2. **False-negative:** PCR inhibitors present in some samples (like blood, feces, etc.) may inhibit the amplification of target DNA leading to false-negative results.

3. **Concentration of Mg:** Is very crucial as low Mg^{2+} leads to low yields (or no yield) and high Mg^{2+} leads to accumulation of nonspecific products.

4. **High cost:** Reagents and equipments are costly, hence cannot be afforded by small laboratories.

5. **Viability:** It cannot differentiate between viable or non-viable organisms.

Advantages of PCR

1. It is having more sensitivity and specificity.
2. It can be done from clinical specimens and also from culture growth.
3. It can detect the drug resistance gene like mec A gene in MRSA.
4. It is used for fastidious organisms or noncultivable organisms.
5. Used in diagnosis of genetic diseases like thalassemia, sickle cell anemia, etc.

LAMP (Loop Mediated Isothermal Amplification)

- It is an isothermal nucleic acid amplification method where amplification is done at constant temperature of 60–65°C unlike alternating temperature cycles in PCR.
- It uses four primers targeting different regions on the target nucleic acid.
- It uses specific DNA polymerase enzyme (unlike taq polymerase in PCR) derived from *Bacillus stearo-thermophilus*. It has strand displacement capacity.
- Amplicon is present in form of turbidity or visual fluorescence which is detected by naked eyes.
- Use: It is useful for tuberculosis.
- Advantage: Very cheap because does not require thermocycler or gel electrophoresis and easy to perform.

- Disadvantage: High false-positive results due to cross contamination between the reaction tubes.

ACCESS YOURSELF

Short Notes

1. Blotting technique.
2. PCR.

Short Questions for Theory/Viva Questions

1. Write four uses of genetic engineering.
2. How you interpret the positive result for HIV by WB test?

Comment on

1. Taq polymerase is used for extension of primer–target duplex in PCR.

MCQs for Chapter Review

Nucleic Acid (NA) Diagnostic Techniques

1. **DNA is detected by:**
   a. Southern blot
   b. Northern blot
   c. Western blot
   d. Eastern blot

2. **Cy bromide green dye is used for:**
   a. HLPR
   b. PCR
   c. ELISA
   d. Immunofluorescence

3. **PCR was discovered by:**
   a. Robert Koch
   b. Kerry Mullis
   c. Edward Jenner
   d. Alexander Fleming

4. **Taq polymerase is obtained from:**
   a. *E coli*
   b. *Bacillus subtilis*
   c. *Bacillus stearothermophilus*
   d. *Thermus aquaticus*

5. **In PCR, denaturation occurs at:**
   a. 55°C
   b. 72°C
   c. 95°C
   d. None of the above

6. **Real-time PCR is used for:**
   a. Multiplication of RNA
   b. Multiplication of specific segment of DNA
   c. Multiplication of proteins
   d. To know how much amplification of DNA has occurred

7. **Reverse transcriptase polymerase chain reactions can be aided in diagnosis of all of the following viral infections,** *except*:
   a. Adeno virus
   b. *Astrovirus*
   c. *Rotavirus*
   d. Polio virus

Answers and Explanation of MCQs

1. a
- Southern blot is used for DNA, Northern blot is used for RNA and Western (immuno) blot is for Ag/Ab detection. Follow section, **blotting techniques** for more details.

2. b
- Follow section, **DNA probe → labeling** for explanation.

3. b

4. d

5. c

6. d
- Follow section, **PCR** for explanation of answers of MCQs 3–6.

7. a
- Reverse transcriptase polymerase chain reaction is useful for RNA viruses. Adeno virus is DNA virus while all other options are RNA viruses.

CHAPTER 123

Specimens for Spotting and Exercises for Practical Examination

Chapter Outline

- Specimens for spotting
- Exercises

SPECIMENS FOR SPOTTING

Following specimens can be kept in practical examination for spotting and viva.

Media

- All media described in **Ch. 114** can be kept in practical examination for spotting and viva.
- Blood agar preparation could be part of exercise.

Biochemical Tests

- All biochemical tests (listed below) described in **Ch. 49, Ch. 61 and Ch. 118,** can be kept in practical examination for spotting and viva. Performing biochemical test could be part of exercise.
  1. Sugar fermentation tests, IMViC tests, oxidase test, catalase test, urease test and TSI test: Follow **Ch. 118.**
  2. Coagulase test and phosphatase test: **Ch. 49.**
  3. PPA test: **Ch. 61.**

Serological Tests

- All serological tests (listed below) described in **Ch. 20** can be kept in practical examination as specimen for spotting and viva. Performing serological test could be part of exercise.
  1. VDRL tile for VDRL test
  2. RPR card for RPR test
  3. Slide for latex particle agglutination tests like RA/RF test and ASO test
  4. Widal test
  5. RPHA plate with fix microtiter well for RPHA test
  6. ELISA plate with free microtiter well for ELISA
  7. Immuno comb HIV 1 and 2 bispot test (rapid ELISA)
  8. Immunochromatographic test: Like strip form (Virucheck) for HBsAg and biscuits (card) form (SD Bioline) for HBsAg
  9. Immunoconcentration test: Like biscuit/card form (HIV-Tridot) for HIV and comb form (HIV Comb AIDS-RS) for HIV.

Laboratory Animals

All laboratory animals described in **Ch. 117** can be kept in practical examination as specimen for spotting and viva.

Parasite's Specimens

1. Tape worm: **Fig. 102.1c**
2. Cut section of hydatid cyst in liver: **Fig. 102.12a**
3. Liquefied hydatid cyst: **Fig. 102.12b**
4. Common round worm (*Ascaris lumbricoides*): **Fig. 107.2c**
5. NIH swab: **Fig. 107.29**
6. Chylous urine: **Fig. 108.5**
7. Guinea worm (*Dracunculous medinensis*): **Fig. 108.10c**

Graphs or Charts or Figures or Pictures

Following graphs or charts or figures or pictures can be kept in practical examination as specimen for spotting and viva.

- Chart of bacterial growth curve: **Fig. 7.1**
- Healthcare-related symbols: **Figs 5.1 to 5.7**
- Different color containers/bags for BMW disposal: **Fig. 31.1**
- Figure of trophozoite/cyst of *E. histolytica*: **Fig. 93.1**
- Figure of trophozoite/cyst of *G. lamblia*: **Fig. 94.1**
- Figure of trophozoite of *T. vaginalis*: **Fig. 94.4**
- Egg(s) of *T. saginata* and *T. solium*: **Fig. 102.3**
- Figure of hydatid cyst: **Fig. 102.10**
- Egg of *H. nana*: **Fig. 102.16**
- Figure of eggs of diecious trematodes (schistosomes): **Fig. 103.3**

- Figure of fertilized, unfertilized, fertilized decorticated and unfertilized decorticated eggs of *A. lumbricoides*: **Figs 107.5 to 107.8.**
- Figure of egg of *A. duodenale*: **Fig. 107.13**
- Figure of egg of *T. trichuria*: **Fig. 107.24**
- Figure of egg of *E. vermicularis*: **Fig. 107.27**
- Figure of microfilariae of *W. bancrofti*: **Fig. 108.2**
- Craigie's tube: **Fig. 115.12**
- Egg (hen's egg) culture in viruses: **Fig. 119.2**

Other Specimens

Petri dish

- **Introduction:** It is named after German Bacteriologist Julius Richard Petri. It also called **Petri plate** or **culture dish.**
- **Variants:** It is available in different diameters ranging from 30–200 mm. Different variants **(Fig. 123.1)** are described below.
  - **As per structure: (1) Glass (reusable):** It is recycled and used repeatedly. **(2) Plastic (disposable):** It is used single time.
  - **As per shape:** Round shape and square shape.
- **Sterilization:** Glass Petri dishes are sterilized by hot air oven while plastic plates are sterilized by gamma (ionising) radiation or by Ethylene oxide (EO) before use and after use they are discarded following autoclave.
- **Uses: (1) Routine cultivation:** Petri dish is widely used for routine cultivation of bacteria, parasites and fungi. **(2) Cell culture:** It also used for cell culture of viruses and few other microbes. **(3) Germination in plants:** Petri dish may be used to observe the early stages germination of plants. **(4) Entomology:** It is the convenient enclosures to study the behaviors of insects and other small animals. **(5) Storage:** Petri dish also makes convenient temporary storage for samples, especially liquid, granular, or powdered ones, and small objects such as insects or seeds. Its transparency and flat profile allow the contents to be inspected with the naked eye, magnifying lens, or low-power microscope without removing the lid. **(6) Use in chemistry:** Due to its large open surface, Petri dish is effective container to evaporate the solvents and dry out precipitates, either at room temperature or in ovens and desiccators.

Nichrome loop (Fig. 123.2a) and nichrome straight wire (Fig. 123.2b)

- **Introduction:** It is named as nichrome, because it is prepared from two metals like nickel and chromium. Advantages of nichrome are rapid cooling, not too rigid and less expensive than platinum loop. Ideal size of wire is 26 or 27 SWG. Length of wire from loop holder to loop is 6 cm and different sizes of loop are 1.3 mm, 2 mm and 4 mm. Loop should be completely closed to avoid the aerosols.
- **Sterilization:** Flaming (red hot/red heat).
- **Uses: (1) Nichrome loop:** It is used to pick up colonies from culture media/clinical samples for streak culture/surface plating, to make smears and to make bacterial suspensions. **(2) Nichrome straight wire:** It is used to pick up colonies from culture media/ clinical samples for stab culture in semisolid media for motility detection test, gelatin liquefaction test or test for anaerobic organism detection (stop once the wire reaches 0.5 inches away from the bottom of the stab media), to make smears and to make bacterial suspensions.

> **Note: Plastic (disposable) loop**
> It is prepared from plastic materials **(Fig. 123.2c)**, but expensive and also cause pollution, as it is not a biodegradable product.

Pasteur pipette

- **Introduction:** The Pasteur pipette name came from the French scientist Louis Pasteur, who used a variant of it extensively during his research. It also called **dropper.** For use in eyedrops applications it called **eyedropper.**
- **Variants: (1) Plastic pipette:** It is available as a single piece as shown in **Fig. 123.3a.** It also available as a tube with narrow point and fitted with a rubber bulb at the top **(eyedropper)** as shown in **Fig. 123.3b.**

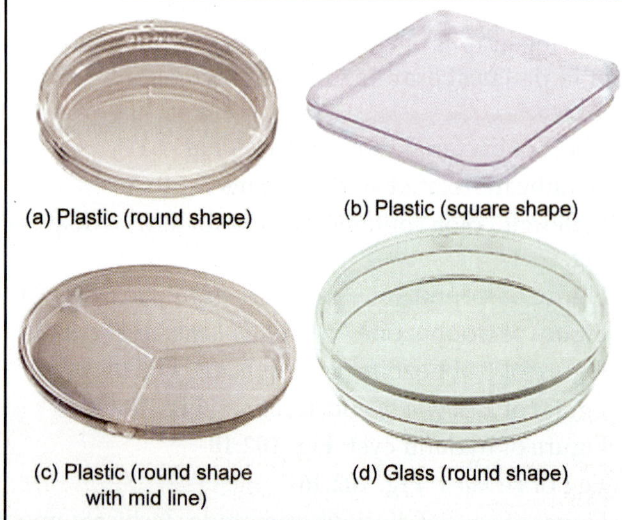

Fig. 123.1: Different variants of Petri dishes

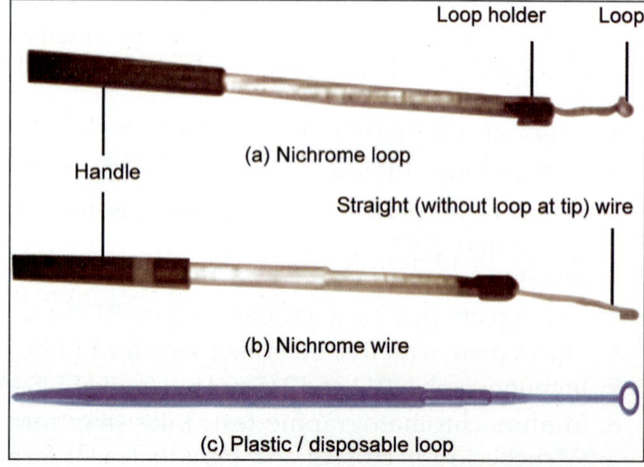

Fig. 123.2: Nichrome loop and nichrome straight wire

Essentials of Medical Microbiology

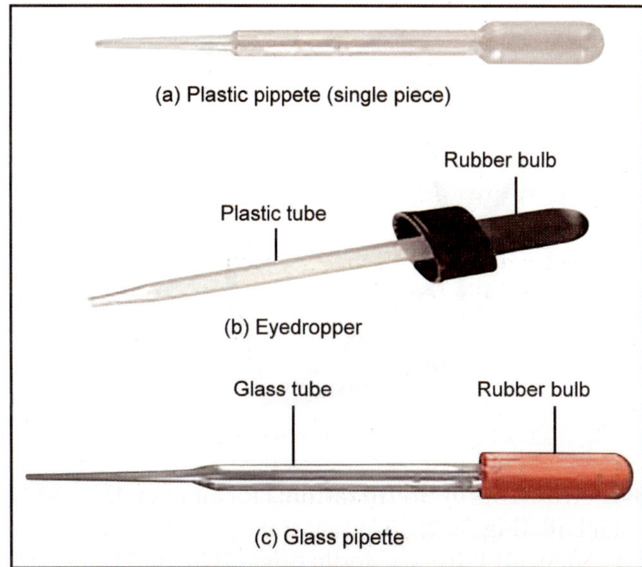

Fig. 123.3: Variants of Pasteur pipette

(2) Glass pipette: The glass tube tapered to a narrow point and fitted with a rubber bulb at the top as shown in **Fig. 123.3c**. It also called **teat pipette**.

- **Sterilization:** Glass tube is sterilized by hot air oven while bulb and plastic pipette are sterilized by gamma (ionizing) radiation or by EO.
- **Uses:** It is used to transfer small quantities of liquids in the laboratory and also to dispense small amounts of liquid medicines like eye drops. It may or may not be calibrated for any particular volume.

Incubator

- **Definitions:** The length of time in which humidity, temperature and other environmental factors are utilized to achieve the required growth and development of microbial culture is called **incubation**. Instrument used for the incubation process is called **incubator**.
- **Principle:** It acts by maintaining the temperature, humidity and other environmental factors like CO_2 and O_2 artificially inside the instrument for optimal growth of microbes in culture.
- **Uses and variants:** Incubators are in different types as per uses.
  - **Medical incubator:** Two subtypes. **(1) Microbiological/bacteriological/laboratory incubator (Fig. 123.4):** Used in microbial culture. It maintains the temperature around 37°C. **(2) Infant incubator:** Used to provide controlled environmental conditions for the care and protection of premature or sick babies.
  - **Poultry incubator:** It acts as a substitute for hen, results in higher hatch rates due to the ability to control both temperature and humidity.

Note: BOD incubator
Full name: Biological (Biochemical) Oxygen Demand incubator.
Difference from normal incubator: It is the low temperature incubator which maintains the low temperature up to 20°C.

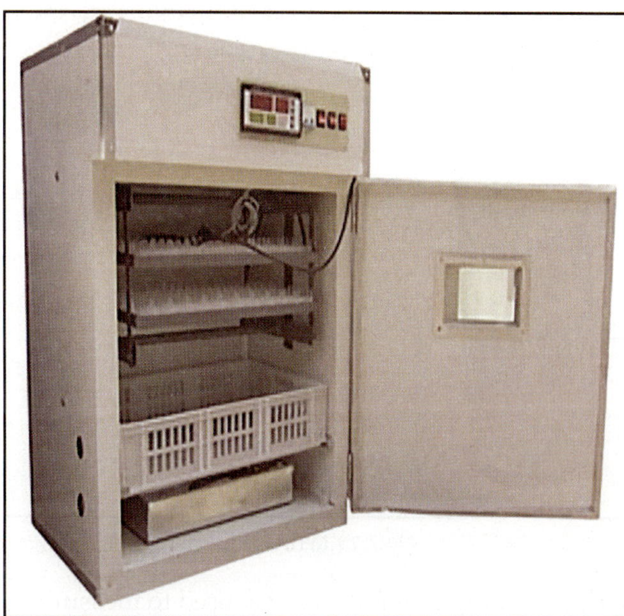

Fig. 123.4: Laboratory incubator

Sterile cotton swab

- **Introduction:** It consists small wads of cotton wrapped around one (**Fig. 123.5**) end of a short rod made of wood, rolled paper or plastic. Due to pollution issues some countries banning the plastic-stemmed versions in favor of the biodegradable alternatives.
- **Sterilization:** It is sterilized by gamma (ionizing) radiation or by EO.
- **Uses:** (1) To collect the specimen like pus or discharge followed by rubbing onto culture media, where bacteria from the swab may grow. (2) For preparing lawn culture for antibiogram. (3) To take DNA samples, most commonly by scraping cells from the inner cheek in the case of humans. (4) For applying and removing cosmetics materials, as well as for household uses such as cleaning, arts and crafts. (5) To apply medicines (like betadine) to a targeted area and to selectively remove substances from a targeted area.

West's post nasal swab: Follow **Ch. 52. (Fig. 52.3 and 52.4)**.

pH strip

- **Introduction:** It is a piece of paper (**Fig. 123.6**) that changes color depending on the pH of a liquid.
- **Uses:** It is a cheap and relatively accurate way of measuring the pH of any liquid like broth, chemical in laboratory, water, urine and saliva.

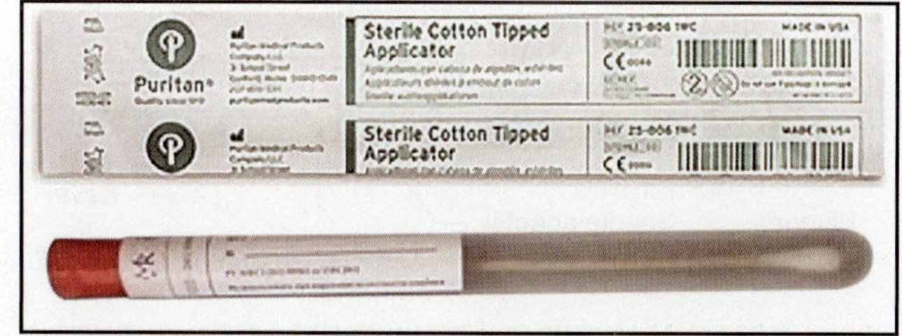

Fig. 123.5: Sterile cotton swabs

Essentials of Medical Microbiology

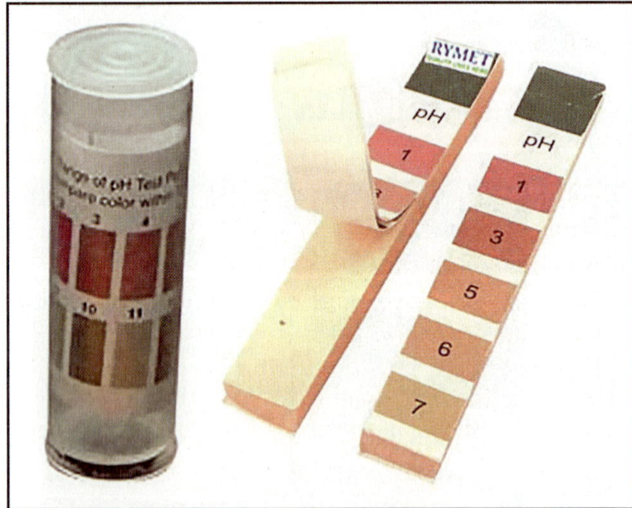

Fig. 123.6: pH strip

Lovibond comparator: It was also used to measure the pH but currently not useful.

Disposable syringe

- **Introduction:** The word "syringe" is derived from syrinx (Greek) meaning "pan flute" or "tube". It is made up of plastic. Its available in different sizes like 2 ml, 5 ml, etc. Image of disposable syringe and its different parts are shown in **Fig. 123.7 and Fig. 123.8,** respectively.
- **Sterilization:** Sterilized by gamma (ionizing) radiation or by EO.
- **Uses:** (1) To collect the blood sample by venepuncture. (2) To collect the fluid or pus (if in large amount). (3) For injecting medicines.

Icosahedral model of virus: Fig. 123.9.

Others: (1) Tuberculin syringe: Fig. 56.5. **(2)** Concavity slide and cover slip for motility testing: Fig. 112.4.

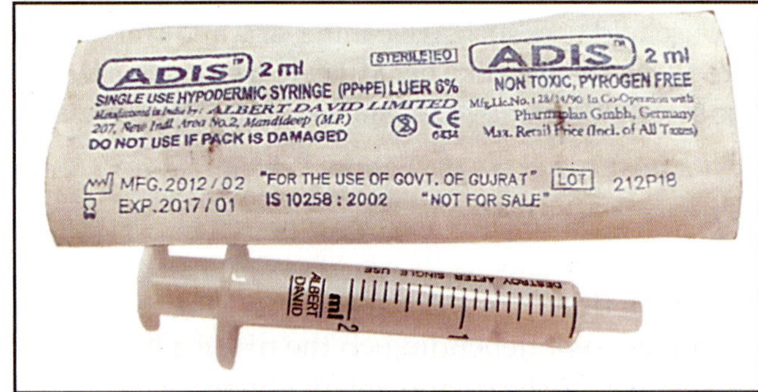

Fig. 123.7: Disposable syringe

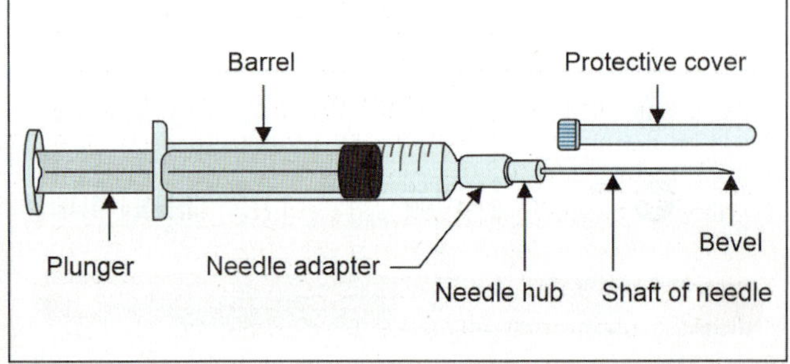

Fig. 123.8: Parts of disposable syringe and needle

Fig. 123.9: Lab., specimen of icosahedral model of virus

(3) Instruments of Sterilization: Hot air oven **(Fig. 30.1)**, water bath **(Fig. 30.2)**, inspissator **(Fig. 30.3)** and autoclave **(Fig. 30.5)**. **(4) Filters:** Candle filters **(Fig. 30.7)**, asbestos filter **(Fig. 30.8)** and sintered glass filter **(Fig. 30.9)**. **(5) Commonly used disinfectants in laboratory:** Fig. 30.10. **(6) Instruments and media of anaerobiosis:** Candle jar **(Fig. 115.4)**, McIntosh and Filde's anaerobic jar **(Fig. 115.5),** Gas Pack system **(Fig. 115.7)** and RCMM **(Fig. 115.8)**. **(7) Antibiotic sensitivity test:** Kirby – Bauer method **(Fig. 116.1)** and E test **(Fig. 116.2)**. **(8) Vaccines.**

EXERCISES

Staining

- Gram's stain/ZN stain: **Ch. 112.**
- NS and iodine preparation of stool for parasites: **Ch. 121.**

Sample Collection, Storage and Transport Including Triple Layer Packing

- Urine/blood/stool/CSF/sterile body fluids/pus or pus swab/sputum/anaerobic sample collection for microscopy and culture.
- Nasopharyngeal/throat swab collection for molecular tests.
- Triple layer packing: Follow **Ch. 113 (Fig. 113.1).**

Exercises of HAIs

- Different types of hand hygiene methods: Follow **Ch. 29 (Fig. 29.3 to Fig. 29.5).**
- PPE: Follow **Ch. 29** (donning and doffing → **Fig. 29.6 to Fig. 29.13).**
- BMW disposal and bio hazard symbol: Follow **Ch. 31 (Fig. 31.1 to Fig. 31.2).**

Culture Exercises

Bacterial culture

- Pigment production by *Staphylococcus* spp., on nutrient agar slant: **Flowchart 49.1 and Fig. 49.1.**
- Hemolytic colony on blood agar by streptococci: **Fig. 50.1.**
- Black colony by *C. diphtheriae* in PT medium: **Fig. 53.2.**

- Pink spreading colony of *E. coli* on MacConkey's agar: **Fig. 61.2a**.
- Pink mucoid colony of *K. pneumoniae* on MacConkey's agar: **Fig. 61.2b**.
- Pale/colorless colony of non-lactose fermenters on MacConkey's agar: **Fig. 61.9b**.
- Black colony by *Salmonella* spp., in Wilson and Blair (W and B) agar: **Fig. 62.2b**.
- Yellow colony by *V. cholerae* in TCBS medium: **Fig. 65.2b**.
- Pigments production by *P. aeruginosa* in nutrient agar plate/slant: **Fig. 67.2**.

Fungal culture
- White colony by *Candida* in SDA bottle and Petri dish: **Fig. 90.7**.
- Different colonies by *Aspergillus* spp., in SDA: **Fig. 91.18**.

Antibiotic sensitivity testing plates: Kirby – Bauer method **(Fig. 116.1)** and E test **(Fig. 116.2)**.

Microscopic Slides
Following microscopic slides could be part of exercise or spotting.
1. Gram's stain smears
   - Gram-positive cocci: **Fig. 112.5**.
   - Gram-positive bacilli: **Fig. 112.6**.
   - Gram-negative cocci: **Fig. 112.7**.
   - Gram-negative bacilli: **Fig. 112.8**.
2. Acid fast bacilli under ZN stain: **Fig. 112.9 and Fig. 112.10**.
3. Polar (*C. diphtheriae*) bodies under Albert stain: **Fig. 112.11**.
4. *T. pallidum* under Fontana's stain: **Fig. 73.5**.
5. *Candida* spp., (pseudohyphae) under Gram's stain: **Fig. 90.6**.
6. Filamentous fungi (*Aspergillus*) under KOH/LCB mount: **Fig. 91.17**.
7. LD body (amastigote form) of *L. donovani*: **Fig. 95.7b**.
8. Malarial parasites: Mostly
   - Ring stage of all species: **Fig. 97.8a**.
   - Gametocytes (Crescent) of all species: **Fig. 97.9**.
9. Egg of *T. saginata* and *T. solium*: **Fig. 102.3b**.
10. Egg of *H. nana*: **Fig. 102.16b**.
11. Eggs of *A. lumbricoides*:
    - Fertilized eggs: **Fig. 107.5b**.
    - Unfertilized eggs: **Fig. 107.6b**.
12. Egg of *A. duodenale*: **Fig. 107.13b**.
13. Egg of *T. trichuria*: **Fig. 107.24b**.
14. Egg of *E. vermicularis*: **Fig. 107.27b**.
15. Microfilariae under Romanowsky's stain: **Fig. 108.7**.

Other Exercises
1. Demonstration of light microscope.
2. Preparation of blood agar.
3. Performance of any biochemical tests.
4. Any method(s) of aerobic or anaerobic culture.
5. AST (Kirby-Bauer) method.
6. Methods to disinfect the spillage of infectious fluid.
7. Performance of serological test(s).
8. Triple layer packing
9. Albert staining for *C. diphtheriae*.
10. Hanging drop preparation for motility detection/ Hanging drop preparation for darting motility of *V. cholerae*.
11. Satellitism.
12. Demonstration of DGI microscopy for *T. pallidum*.
13. Demonstration of icosahedral model of virus.
14. Inclusion body detection in virus.
15. Egg/cell culture in virus.
16. Tzanck smear in herpes virus.
17. PEP for HBV and HIV.
18. Demonstration of bacteriophage model.
19. KOH/LCB mount for fungi.
20. Indian ink for *C. neoformans*.
21. Teased mount for fungi.
22. Slide culture for fungi.
23. Cellophane tape mounts for fungi.
24. Germ tube preparation.
25. Dalmau plate culture
26. Hair baits technique for fungi.
27. Hair perforation technique for fungi.
28. Identification of laboratory fungal contaminants.
29. Giemsa stain for PCP.
30. Concentration methods for parasites.
31. Microscopic examination of peripheral blood smear (PBS) for following parasites:
    - For *Trypanosoma* spp., **Ch. 95**.
    - For *Leishmania* spp., in macrophages: **Ch. 95**.
    - For *Plasmodium* spp., and *Babesia* spp., **Ch. 97**.
    - For *Toxoplasma gondii*: **Ch. 98**.
    - For microfilariae: **Ch. 108**.
32. NIH swabbing.
33. Scotch cellulose tape method.
34. AETCOM in Microbiology.

ACCESS YOURSELF

Spotting Questions
1. Identify the medium and write its use.
2. Identify the medium and write its method of sterilization.
3. Identify the medium and write its type.
4. Identify the biochemical test and name the two bacteria/ pathogens giving the test positive.
5. Identify the serological test and write the principle.
6. Identify the serological test and write its use.
7. Identify the animal and write its use.
8. Identify the animal and write its scientific name.
9. Identify the pathogen in smear.

10. Identify the parasite in specimen and write the name of host (hosts).
11. Identify the parasite in specimen and write its mode of transmission.
12. Identify the pathogen in graph/figure/picture.
13. Identify the specimen and answer the given question.

Note: MCQs and the short questions given at the end of each chapter are also the part of spotting exercise.

Question of Practical Exercise

1. Stain the given smear by Gram's stain and draw the labeled diagram of your findings. Report your findings to the examiner.
2. Stain the given smear by ZN stain and draw the labeled diagram of your findings. Report your findings to the examiner.
3. How you examine the stool sample for trophozoite, cyst or ova? Draw the labeled diagram of your findings and report to the examiner.
4. How you examine the blood sample for *Plasmodium* spp.? Draw the labeled diagram of your findings and report to the examiner.
5. Identify the culture medium given to you. Describe the growth/cultural characteristic, identify the pathogen. Draw the labeled diagram of your findings and report to the examiner.
6. Perform the AST (Kirby-Bauer) method for given culture suspension.
7. Read the case and perform the serological test for given serum.
8. Perform the slide/tube coagulase test.
9. Perform the hanging drop preparation for darting motility

CHAPTER 124

IMG, AETCOM and Competencies in Microbiology

Chapter Outline
- IMG (Indian Medical Graduate)
- AETCOM in Microbiology
- Competencies in Microbiology

IMG (INDIAN MEDICAL GRADUATE)

Definition

"Indian Medical Graduate" (IMG) possessing requisite knowledge, skills, attitudes, values and responsiveness, so that he or she may function appropriately and effectively as a physician of first contact of the community while being globally relevant.

Role of IMG

In order to fulfill this goal, the IMG must be able to function in the following roles appropriately and effectively.

1. **Clinician** who understands and provides preventive, promotive, curative, palliative and holistic care with compassion.
2. **Leader and member of the healthcare team and system** with capabilities to collect, analyze, synthesize and communicate health data appropriately.
3. **Communicator** with patients, families, colleagues and community.
4. **Lifelong learner** committed to continuous improvement of skills and knowledge.
5. **Professional**, who is committed to excellence, is ethical, responsive and accountable to patients, community and profession.

AETCOM IN MICROBIOLOGY

Meaning

AETCOM refers to **a**ttitude, **et**hics and **com**munication for performing medical practices.

Definitions

Attitude means thinking or approach toward something. **Ethics** means moral principle of medical practice. **Communication** means sharing or exchanging knowledge or information or skill by writing, speaking, or by other way.

Important

It is important (1) to remove medical errors like wrong diagnosis, poor behavior with patient, wrong or improper treatment, compromised patient safety, etc. (2) for addressing some medical problems like violence on doctor, litigation on doctors, distrust on doctors or health care facilities. (3) for conducting routine health care activities in hospital/laboratory.

AETCOM Modules in Microbiology

Followings are the different AETCOM modules assigned to microbiology.

Module 1: Healthcare as a right

- **Importance:** (1) To educate people about disease prevention and treatment. (2) To know the rights as patients. (3) To maintain the uniform and quality medical services at primary, secondary and tertiary level. (4) To improve the living standard of persons with poor socioeconomic status. (5) To notify health care facility about any outbreak or epidemic.
- **Problems in implementation:** It is difficult to implement in rural areas. (2) Public healthcare facility is not the 1st choice for affordable person.

Module 2: Working in a healthcare team

- **Aims to work in a health care team:** (1) To improve the confidence of patient in clinicians and treatment given. (2) To know the patient satisfaction. (3) To improve the health education. (4) To clear the doubt of patients. (5) To monitor the patients about treatment response. (6) To avoid unwanted hospitalization, testing and treatment.
- **Problems to work effectively in health care team:** (1) Changing role. (2) Un cooperation by staff and patients.

Module 3: What does it mean to be family member of a sick patient?

- **Spillover effects:** Having an ill or disabled person in family imposes a burden on family caregivers and non-care giving family members. These called "spillover effects". It affects quality of life and well-being of family member. Sick person itself acts as source of infection and also put family on medical risk.
- **Domains affected by spillover**
  1. **Physical:** Increase the risk of disease to family member. Other effects include decreased attention to self-care, sexual intimacy, etc.
  2. **Mental:** These are sleep disturbance, anxiety, loss of appetite, fatigue, change in behavior, frustration/depression, etc.
  3. **Social:** Tough to spare the time for any social function, child education or relation.
  4. **Economical:** Costing behind diagnosis, treatment and also not allotting time behind profession affects the financial burden of family.

COMPETENCIES IN MICROBIOLOGY

Followings are the different competencies assigned to microbiology.

Understanding of role of microbial agents in health and disease

Human health is continuously targeted by different types of microbial agents present in environment, water, food, soil, and in living body of animal, vectors or human itself. Microbial agents cause different types of diseases in humans like infectious diseases, intoxications, hypersensitive (allergic) diseases, immunodeficiency diseases, autoimmune diseases and malignancies. Detail description of all such diseases is given in respective chapters.

Understanding of the immunological mechanisms in health and disease

Types of Ag-Ab reactions are mentioned in **Flowchart 20.1.** Useful effect includes immunity which is described in **Ch. 16**. Harmful effects include hypersensitivity (allergy), immunodeficiency and autoimmunity which are described in **Ch. 24, Ch. 25** and **Ch. 26** respectively.

Ability to correlate the natural history, mechanisms and clinical manifestations of infectious diseases as they relate to the properties of microbial agents

Natural history means epidemiology, mechanisms means pathogenesis and clinical manifestations means clinical features plus complications of different microbial agents are described in respective chapters as per organisms.

Knowledge of the principles and application of infection control measures

Infection control measures include general measures, chemoprophylaxis, immunoprophylaxis and vector control. Principles and applications of all such measures are given in respective chapters as per organisms.

Understanding the basis of choice of laboratory diagnostic tests and their interpretations, antimicrobial therapy, control and prevention of infectious diseases

Principle, method, result, interpretation and all other details of different types of laboratory tests for bacteria, viruses, fungi and parasites are described in **Ch. 118, Ch. 119, Ch. 120,** and **Ch. 121,** respectively.

Specific laboratory tests for particular organism are described in concerned chapters.

Antimicrobial therapy for particular organism is described in concerned chapter under the section of treatment.

Knowledge regarding control and prevention of infectious diseases is already explained in earlier competency.

Knowledge of outbreak investigation and its control

Knowledge regarding outbreak management is almost same as pandemic management as described in **Ch. 48.**

Demonstration of confidentiality pertaining to patient's identity on laboratory results

Domain: Attitude.

Background knowledge
- Principle of medical ethics.
- Medicolegal aspects of confidentiality.
- Confidentiality and communication of laboratory results.
- Modes of transmission, diagnosis, post exposure prophylaxis and social issues related to some sensitive diseases, like HIV-AIDS, HBV, COVID-19, etc.

Case scenarios
1. **Case 1 (disclosing HIV result):** A 30-year-old female suffered with accident with multiple injuries. She admitted in hospital and then died. On enquiry she was identified as HIV positive and residents put "retro positive" label on her case file. Her family members asked for "retro positive" label. They became stressed out and shouted at the patient. Later all they except her mother stop to visit the lady.
2. **Case 2 (social stigma in COVID-19):** A 35-year-old male come in hospital with influenza like illness and he confirmed COVID-19 positive. He wants treatment, but does not want to be quarantined for fear of losing job. He is the only earning member in family.

Teaching learning methods: (1) Small group discussion. (2) Self directed learning. (3) Writing. (4) Interaction with laboratory technicians and counsellors of ICTC (Integrated Counselling Testing Centre) concerned with counseling of patients and laboratory tests of HIV. (5) Interaction with laboratory technicians concerned with COVID-19 testing. (6) Interaction with other required authority (like infection control officer) or person.

Assessments: (1) Formative assessment: It is the day-to-day assessment like small group discussion, MCQs, etc. **(2) Summative assessment:** It is the qualifying examination which decides pass or fail.

- Short notes and short questions.
- Practical (OSPE → Objective Structured Practical Examination).

Demonstration of respect for patient samples sent to Microbiology laboratory for performance of tests to detect the microbial agents causing infectious diseases

Domain: Attitude.

Background knowledge

- Appropriate sample for the test: Sample type, amount, collection method, preservative (if used), container type, transportation and storage.
- Proper labeling.
- Completely filled requisition form with all details.
- Possible issues (medicolegal, social, etc.) following misidentification of sample, improper storage or transportation.

Case scenarios

1. **Case 1 (rejection due to improper container):** Sequestrum is collected for culture from a case of osteomyelitis in formalin container, hence it is rejected. There is no more sample available for culture.
2. **Case 2 (misguided report due to inadequate information in requisition form):** A urine sample was received in laboratory for culture and sensitivity. Requisition form received in laboratory not contains complete details of patient, hence it is rejected. There is no more sample available for culture.

Teaching learning methods: (1) Small group discussion. (2) Self directed learning. (3) Writing. (4) Interaction with laboratory staff and discussion plus demonstration of methods for sample collection, storage, transport and filling requisition form.

Assessment: (1) Formative assessment: It is the day to day assessment like small group discussion, MCQs, etc. **(2) Summative assessment:** It is the qualifying examination, which decides pass or fail.

- Short notes and short questions.
- Practical (OSPE → Objective Structured Practical Examination): Asking to students to perform or to demonstrate methods for sample collection, storage, transport and filling requisition form.

MCQs

CHAPTER 125

MCQs for Section Review

Note: This chapter includes MCQs which required knowledge of many chapters of different sections/systems.

General Microbiology

1. **Lipopolysaccharide is an outer membrane seen with:**
   a. Gram-positive bacteria b. Gram-negative bacteria
   c. a + b
   d. None of above
2. **Following is mismatched:**
   a. Lysozyme – Gram +ve bacteria – Protoplast
   b. Taq polymerase – *Thermus aquaticus* – PCR
   c. Recombinase – Rearrangement at DNA level – Genes of Ig
   d. *Listeria* – IgA protease – Loss of local immunity
3. **S. Typhi is the causative agent of typhoid fever. The infective does of *S. Typhi*:**
   a. 1 bacillus
   b. 10^3–10^6 bacilli
   c. 10^8–10^{10} bacilli
   d. 1–10 bacilli
4. **Arena viruses contain:**
   a. 3 segments of ssRNA b. 2 segments of ssRNA
   c. 3 segments of dsRNA d. 2 segments of dsRNA

Answers and Explanation of MCQs

1. b
- Lipopolysaccharide presents in gram-negative bacteria.

2. d
- Lysozyme acts on cell wall of gram-positive bacteria to form protoplast. Taq polymerase is derived from *Thermus aquaticus* useful in PCR. Recombinase is useful in rearrangement at DNA level for genes of Ig. IgA protease produce by *N. meningitidis*, *N. gonorrhoeae*, *H. influenzae* and *Strept. pneumoniae*, but not by *Listeria*. It destroys the IgA resulting loss of local immunity.

3. b
- Follow **Table 9.1** for explanation.

4. b
- Remember the **mnemonic ROBRA**. Follow **Ch. 11** for explanation.

Immunology

1. **The following statements are true about DPT vaccine except:**
   a. Aluminum salt has an adjuvant effect.
   b. Whole killed bacteria of *Bordetella pertussis* has an adjuvant effect
   c. Presence of acellular pertussis components increase its immunogenicity
   d. Presence of *H. influenzae* type B components increase its immunogenicity.

2. **Following is/are true:**
   a. All immunogens are antigens, but all antigens are not immunogens.
   b. All antibodies are immunoglobulins, but all immuno-globulins may not be antibodies.
   c. All antigens are immunogens, but all immunogens are not antigens.
   d. All immunoglobulins are antibodies, but all antibodies may not be immunoglobulins.
3. **Prozone phenomenon is a feature of:**
   a. Tularemia
   b. Legionnaires' disease
   c. Plague
   d. Brucellosis
4. **All statements are correct about immunogenic technique, *except*:**
   a. ELISA can detects both antigens and antibodies.
   b. Immunofluorescence test uses fluorescein and rhodamine.
   c. Radioimmunoassay used to quntitate antigens and haptens.
   d. Immunoblotting also called northern blotting
5. **Antigen-antibody complexes are detected by:**
   a. Western blot
   b. Southern blot
   c. Northern blot
   d. ELISA
6. **Paul Bunnel reaction is a type of:**
   a. Agglutination
   b. CFT
   c. Precipitation
   d. Flocculation test
7. **Test based on agglutination principle is/are:**
   a. Widal test
   b. Weil-Felix test
   c. Blood grouping
   d. a + b + c
8. **Nagler reaction is type of:**
   a. Neutralization
   b. CFT
   c. Precipitation
   d. Agglutination
9. **Skin test based on ITH:**
   a. Casoni test
   b. Lepromin test
   c. Tuberculin test
   d. Schick test
10. **Frenkel's skin test is positive in:**
    a. Spinal cord compression
    b. Toxoplasmosis
    c. Pemphigus
    d. Pemphigoid
11. **Fairley's test is done for:**
    a. Cysticercosis
    b. LGV
    c. Schistosomiasis
    d. Filariasis
12. **Following is the helper or inducer cell:**
    a. CD1
    b. CD4
    c. CD2
    d. CD8
13. **Following is the cytotoxic or suppressor cell:**
    a. CD1
    b. CD4
    c. CD2
    d. CD8

14. All of the following are glycoprotein, *except*:
   a. Blood antigen
   b. Albumin
   c. Immunoglobulin
   d. hCG

15. Virus infected cells are killed by:
   a. Interferon
   b. Macrophages
   c. Neutrophils
   d. Autolysis
   e. None of above

16. Function (s) of T lymphocytes is/are:
   a. Production of interferon
   b. Lymphokine production
   c. Rosette formation
   d. All of above

17. Chromosomal order for location of genes for H chain, KL chain, γL chain and MHC is:
   a. 14, 2, 22, and 6
   b. 14, 22, 2, and 6
   c. 14, 22, 22, and 6
   d. 6, 2, 22, and 14

Answers and Explanation of MCQs

1. d
- DPT vaccine does not contain *H. influenzae* type B.

2. a and b
- For explanation, Follow **chapters 18 and 19.**

3. d
- Prozone phenomenon means Ab excess, while postzone phenomenon means Ag excess. Excess antibodies in serum are present in brucellosis and in secondary syphilis/syphilis with HIV, which interfere with diagnosis. It is overcome by dilution of serum.

4. d
- ELISA can detect both antigens and antibodies. Immuno-fluorescence test uses fluorescein isothiocyanate (FITC) (Blue-green fluorescence) or lissamine rhodamine (orange-red fluorescence). Radioimmunoassay used in quantitation of hormones, drugs, tumor markers, IgE and viral antigens. Immunoblotting also called **Western blotting.**

5. a and d
- Southern blot is used for DNA, Northern blot is used for RNA and Western (immuno) blot is for Ag/Ab detection. ELISA is also used for Ag/Ab detection.

6. a

7. d

8. a
- Follow **Table 125.1** for explanation of answers of MCQs 6–8.

9. a
- Casoni test is based on ITH, while lepromin test and tuberculin test are based on DTH. Schick test based on neutralization test.

10. b

11. c
- Follow **Table 125.2** for explanation of answers of MCQs 10–11.

12. b

13. d
- CD4 is the helper/inducer cell while CD8 is cytotoxic/suppressor cell.

14. b
- Different glycoproteins are Ig, mucin, lectin, IFN β and γ, glycophorin, alkaline phosphatase, collagen, enzymes, selectin, blood group antigens, transferring, ceruloplasmin, HLA class I and II, hCG (human chorionic gonadotropin), TSH and plasma protein except albumin.

TABLE 125.1: List of serological tests

Serological test	Use	Agent
Precipitation		
Ascoli's thermo-precipitin test.	Anthrax	*B. anthracis*
Lancefield grouping	Streptococcal disease	Streptococci
Elek's gel precipitation test	Toxigenicity test	*C. diphtheriae*
Flocculation		
VDRL test	Syphilis	*T. pallidum*
RPR test		
Kahn test		
Agglutination		
Widal test	Enteric fever	*S.* Typhi
Weil Felix reaction	Typhus fever	*Rickettsia*
Paul Bunnel test	Infectious mono-nucleosis	EBV
Cold agglutinin test	Primary atypical pneumonia	*M. pneumoniae*
Streptococci MG test		
Blood grouping/cross matching	Blood transfusion	—
RA factor test	Rheumatoid Arthritis	—
ASO test	Streptolysin –O detection	*Streptococcus*
SAT (Standard Agglutination Test)	Human brucellosis	*Brucella* spp.
Coombs test	Hemolytic anemia	—
CFT		
Wassermann reaction	Syphilis	*T. pallidum*
Neutralization		
Schick test	Susceptibility and hypersensitivity to diphtheria toxin	*C. diphtheriae*
Nagler reaction	Gas gangrene	*Cl. welchii*

TABLE 125.2: List of allergic tests

Allergic/skin test	Use	Agent
Immediate type hypersensitivity (ITH)		
Casoni test	H cyst	*E. granulosus*
Mazzotti's test	Onchocerciasis	*O. volvulus*
Delayed type hypersensitivity (DTH)		
Tuberculin test	Tuberculosis	*M. tuberculosis*
Lepromin test	Leprosy	*M. leprae*
Mallein test	Glanders (malleus)	*Bukholderia mallei*
Brucellin test	Human brucellosis	*Brucella* spp.
Frei's test	LGV	*C. trachomatis*
Leishmanin test	African kala azar	*L. donovanii*
Frenkel's test	Toxoplasmosis	*T. gondii*
Fairley's test	Schistosomiasis japonica (chronic)	*S. japonicum*
Scratch test	Ascariasis	*A. lumbricoides*

15. b

- Virus infected cells are killed by macrophages, Tc cells and NK cells. Interferon has no direct action on virus, but acts on other cells of same species, rendering them refractory to viral infections. IFN inhibits the viral transcription in infected cell.

16. d

- T lymphocytes produce lymphokines (lymphotoxin) and interferon. T cells bind with sheep RBC (Erythrocytes = E) to form SRBC or E-rosette by CD2 receptor while B cells binds with sheep Erythrocytes (E) coated with Ab (A) and Complement (C) to form EAC-rosette due to C3 receptors (CR2) on B cell surface.

17. a

- Chromosomal orders for location of genes for H chain, KL chain, γL chain and MHC is 14, 2, 22, and 6.

Healthcare-associated Infections (HAIs)

1. **All are true about *Coxiella burnetii, except*:**
   a. It is an obligate intracellular.
   b. It causes 'Q fever'.
   c. It is extremely fastidious.
   d. It is killed by Holder's method of pasteurization.
2. **CURB-65 score is used for prediction of prognosis in:**
   a. Ventilator-associated pneumonia
   b. Upper respiratory tract infections
   c. Community-associated pneumonia
   d. Tuberculosis

Answers and Explanation of MCQs

1. d

- *Coxiella burnetii* is not killed by Holder's method of pasteurization but by Flash (HTST: high temperature and short time) method.

2. c

- Clinical pulmonary infection score (CPIS) is useful for ventilator-associated pneumonia as described in **Ch. 29 (Table 29.1)**.
- CURB-65, also called **CURB criterion**, is a clinical prediction rule that has been validated for predicting mortality in community-acquired pneumonia and infection of any site. CRB-65 is a modified version of CURB-65 tool for assessing severity of community-acquired pneumonia, and determining whether the patient requires inpatient or outpatient treatment. CURB-65 is the acronym for each of the risk factors measured, as mentioned below. Each risk factor scores one point, for a maximum score of 5.
  - **C**onfusion of new onset (defined as an AMTS ≤8).
  - Blood **U**rea nitrogen >7 mmol/L (19 mg/dl).
  - **R**espiratory rate of ≥30 breaths per minute.
  - **B**lood pressure: Systolic <90 mm Hg or diastolic ≤60 mm Hg.
  - Age **65**.

Infective Syndrome

1. **A patient has prosthetic valve replacement and he develops endocarditis 8 months later. Organism responsible is:**
   a. *Staphylococcus aureus*
   b. *Streptococcus* viridans
   c. *Staphylococcus epidermidis*
   d. HACEK
2. **Which of the following sexually transmitted infection produces painful genital ulcers and painful lymph nodes?**
   a. Syphilis
   b. Chancroid
   c. Gonorrhea
   d. Donovanosis

1. c

- Early onset prosthetic-valve endocarditis (<1 year of valvular surgery) is due to *Staph. epidermidis* as it is capable of growing as a biofilm on plastic surfaces. Late onset prosthetic-valve endocarditis (>1 year following valvular surgery) is due to community acquired microorganisms mainly by viridans group of bacteria.

2. b

- In chancroid genital ulcers and enlarged lymph nodes are painful, while in syphilis and donovanosis they are painless. In gonorrhea there is urethritis and cervictis rather than presence of genital ulcers and enlarged lymph nodes.

Bacterial Infections

1. **False about gram-positive cocci is:**
   a. *Staph. saprophyticus* cause UTI in female
   b. Most enterococci are sensitive to penicillin
   c. Nonpathogenic strains are coagulase negative
   d. Neonatal meningitis causing streptococci hydrolyses hippurate
2. **Dental caries is caused by:**
   a. *Streptococcus pyogenes*   b. *Streptococcus mutans*
   c. *Enterococcus*   d. *H. influenzae*
3. **In a patient of orbital cellulitis, microorganism on culture shows the greenish colonies and optochin sensitivity. The most likely organism is:**
   a. *Strept. viridans*   b. *Staphylococcus*
   c. *Pseudomonas*   d. Pneumococcus
4. **Which of the following regarding bacterial drug resistance is not true?**
   a. Most bacterial drug resistance is due to production of penicillin degrading enzyme.
   b. Plasmid mediated resistance is always transmitted vertically
   c. Complete elimination of target produce resistance to vancomycin
   d. Alteration of target leads to pneumococcal resistance
5. **A man presented with 3 days H/o of lacrimation, redness and discharge from left eye. Later on he developed perforation. Discharge from his eye demonstrated gram-negative cocci, which were oxidase-positive. Which of the following can be the probable organism?**
   a. *Pseudomonas*   b. *Acinetobacter*
   c. *Neisseria gonorrhoeae*   d. *Moraxella catarrhalis*
6. **Gastrointestinal enteritis necroticans is caused by:**
   a. *Clostridium difficile*   b. *Clostridium perfringens*
   c. Botulinum   d. *C. jejuni*
   e. *Pseudomonas*
7. **Swarming growth on culture is a feature of following gram-positive bacterium:**
   a. *Clostridium tetani*   b. *Clostridium perfringens*
   c. *Proteus mirabilis*   d. *Bacillus cereus*
8. **Heating and subsequent plating is a method used for isolating:**
   a. *Corynebacterium*   b. *Vibrio*
   c. *Salmonella*   d. *Clostridium*
9. **A patient of acute lymphocytic leukemia with fever and neutropenia develops diarrhea after administration of amoxicillin therapy. Which of the following organism is most likely to be the causative agent?**
   a. *Salmonella* Typhi   b. *Clostridium difficile*
   c. *Clostridium perfringens*   d. *Shigella flexneri*

10. **Which of the following statement is true regarding pathogenicity of *Mycobacterium* species?**
    a. *M. tuberculosis* is more pathogenic than *M. bovis* to the human
    b. *M. kansasi* can cause a disease indistinguishable from tuberculosis
    c. *M. africanum* infection is acquired from the environmental sources
    d. *M. marinum* is responsible for tubercular lymphadenopathy

11. **Antibodies against PGL-1 are seen in:**
    a. *M. leprae*
    b. *M. tuberculosis*
    c. *Borrelia*
    d. *Brucella*

12. **Buruli ulcer is caused by:**
    a. *M. tuberculosis*
    b. *M. ulcerans*
    c. *M. buruli*
    d. *M. kansasi*

13. **Motility difference according to temperature change is shown by:**
    a. *L. monocytogenes*
    b. *P vulgaris*
    c. *Vibrio*
    d. *Leptospira*

14. **Microorganism motility at 25°C, but not motile at 37°C:**
    a. *Listeria monocytogenes*
    b. *Campylobacter*
    c. *Yersinia pestis*
    d. *Streptococcus agalactiae*
    e. *E. coli*

15. **A 28-year-lady presented with headache, Kernig's sign positive. Culture showed gram-positive bacilli, most probable organism is:**
    a. *L monocytogenes*
    b. *H. influenzae*
    c. Meningococci
    d. *Streptococcus pneumoniae*

16. **All of the following are true *except*:**
    a. *E. coli* is an aerobe and facultative anaerobe
    b. *Proteus* forms uric acid stone
    c. *E. coli* is motile by peritrichate flagella
    d. *Proteus* causes de-amination of phenyl alanine to phenyl pyruvic acid

17. **The bacteria producing a picture resembling "pseudo-hemoptysis" in a sputum sample due to prodigiosin production is:**
    a. *Serratia marcesens*
    b. *Erwinia herbicola*
    c. *Ehrlichia seenetsu*
    d. *Legionella pneumophila*

18. **Which of the following toxin inhibits the protein synthesis?**
    a. Cholera toxin
    b. Shiga toxin
    c. Pertusis toxin
    d. LT of enterotoxigenic *E. coli*

19. **A person returns to Delhi from Bangladesh after 2 days and has diarrhea. Stool examination shows RBCs in stool. The likely organism causing is:**
    a. Enteropathogenic *E. coli*
    b. Enterotoxigenic *E. coli*
    c. *Salmonella* Typhi
    d. *Shigella dysenteriae*

20. **Swarming growth on culture is characteristic of which gram-negative organism:**
    a. *Clostridium welchii*
    b. *Clostridium tetani*
    c. *Bacillus cereus*
    d. *Proteus mirabilis*

21. **Stalactite growth in ghee broth is due to:**
    a. *H. influenzae*
    b. *C. diphtheriae*
    c. *Y. pestis*
    d. *T. pallidum*

22. **Causative agent of plague:**
    a. *Y. pestis*
    b. *Y. enterocolitica*
    c. *Y. pseudotuberculosis*
    d. *Pasteurella septica*

23. **Appendicitis like syndrome is caused by all *except*:**
    a. *Yersinia enterocolitica*
    b. *Yersinia pseudotuberculosis*
    c. *Pasteurella septic*
    d. *Yersinia pestis*

24. **Which of the following bacterium acts by increasing cAMP?**
    a. *Vibrio cholerae*
    b. *Staphylococcus aureus*
    c. *E. coli* heat stable toxin
    d. *Salmonella*

25. **Selective medium for *Vibrio*:**
    a. TCBS
    b. Stuart
    c. Skirrows
    d. MYPA

26. **Which organism grows in alkaline pH?**
    a. *Vibrio cholerae*
    b. *E. coli*
    c. *Pseudomonas aeruginosa*
    d. *Salmonella* Typhi

27. **Swarm of gnats is a feature of:**
    a. *V. cholerae*
    b. *S.* Typhi
    c. *P. aeruginosa*
    d. *E. coli*

28. **Invasive infection is caused by all, *except*:**
    a. *V. cholerae*
    b. *Neisseria*
    c. *Streptococcus*
    d. *H. influenzae*

29. **A 32-year-old male, Kallu, who recently visited a sea coast developed ulcer over the left leg. The probable causes is:**
    a. *Pasteurella multocida*
    b. *Micrococcus halophilus*
    c. *V. fluvialius*
    d. *Neisseria gonorrhoeae*

30. **Which of the following has shortest incubation period?**
    a. Plague
    b. Cholera
    c. Measles
    d. Typhoid

31. **Pontiac fever is caused by:**
    a. *Legionella*
    b. *Listeria*
    c. Scrub typhus
    d. *Leptospira*
    e. *Rickettsia*

32. **Thumb print appearance in culture film smear is seen in:**
    a. *B. anthracis*
    b. *Brucella* species
    c. *B. pertusis*
    d. *Cl. welchii*

33. **Whooping cough is produced by:**
    a. *Y. pestis*
    b. *B. pertussis*
    c. *H. influenzae*
    d. *P. multocida*

34. **Milk ring test is seen in:**
    a. Brucellosis
    b. *Bacteroides*
    c. Tuberculosis
    d. Salmonellosis

35. **Positive cold agglutination test is seen in infection with:**
    a. *Mycoplasma*
    b. *Chlamydia*
    c. Infectious mononucleosis
    d. Varicella

36. **Which of the following is an obligate parasite?**
    a. *Mycoplasma*
    b. *Chlamydia trachomatis*
    c. Gram –ve bacilli
    d. Gram +ve cocci

37. **A 23-year-old male have unprotected sexual intercourse with a commercial sex worker. After 2 weeks he developed a painless indurated ulcer on the glans that exuded clear serum on pressure. Inguinal lymph nodes on both groin were enlarged and not tender. The most appropriate diagnostic test is:**
    a. Gram's stain of ulcer discharge
    b. Dark field microscopy of ulcer discharge
    c. Giemsa stain of lymph node aspirate
    d. ELISA for HIV infection

38. **Lyme disease caused by:**
    a. *Leptospira*
    b. *Borrelia*
    c. *Treponema*
    d. *Bordetella*
    e. Arbovirus

39. **Query fever is caused by:**
    a. *Pseudomonas*
    b. *Francisella*
    c. *Coxiella burnetii*
    d. *R. typhi*

40. Following is growing in cell free medium, *except*:
    a. *Rickettsia*   b. *M. leprae*
    c. *Bartonella*   d. Syphilis

41. **Granuloma with stellate abscess seen in:**
    a. Tuberculosis   b. Tularemia
    c. Sarcoidosis   d. *Staphylococcus*

42. **Patoc 1 strain is belong to:**
    a. *L. biflexa*   b. *T pallidum*
    c. *S. Typhi*   d. *M. bovis*

43. **Malta fever is caused by:**
    a. *Treponema pallidum*   b. *Borrelia burgdorferi*
    c. *Brucella melitensis* d. *Pseudomonas aeruginosa*

44. **Katayama fever is caused by:**
    a. *Schistosoma japonicum* b. *Fasiola hepatica*
    c. *Clonorchis sinensis*   d. *Fasiola buski*

45. **A child with fever and pharyngitis, which of the following investigation should not be done?**
    a. Widal test   b. ASO
    c. Throat swab culture   d. Chest X-ray

46. **Acute intravascular hemolysis can be caused by infection due to all of the following organisms, *except*:**
    a. *Clostridium tetani*
    b. *Bartonella bacilliformis*
    c. *Plasmodium falciparum*
    d. *Babesia microti*

47. **Correct combination of incubation period is:**
    a. Syphilis: 9–90 days   b. Herpes genitalis: 4–5 weeks
    c. LGV: 3–6 weeks   d. Donovanosis: 1–4 weeks
    e. Chancroid: 2–3 weeks

48. **Draining sinuses are seen in:**
    a. Mycetoma   b. Scrofula
    c. Lupug vulgaris   d. Pediculosis

49. **Pus is aspirated from a patient with frontal abscess. Pus shows red fluorescence on ultraviolet examination. The most likely organism causing frontal abscess is:**
    a. *Bacteroides*   b. *Peptostreptococcus*
    c. *Pseudomonas*   d. *Acanthamoeba*

Answers and Explanation of MCQs

1. b
- *Staph. saprophyticus* causes UTI in sexually active young females. Most enterococci are intrinsically resistant to penicillin. CoNS are also pathogenic but less virulent than CoPS. Neonatal meningitis causing streptococci are group B, β-hemolytic streptococci (*Strept. agalactiae*), giving hippurate hydrolyses test positive.

2. b
- *Streptococcus mutans* belongs to viridans group is breaking the dietary fibers to produces the dextran which helps in adhesion of bacteria, mucus, food debris and epithelial cells with teeth to produces the dental caries.

3. d
- A greenish colony (on blood agar) means alpha hemolysis which is given by viridans group (*Strept mutans, Strept. sanguis, Strept. mitis, Strept. salivarius, Strept. mitior*, etc.) and pneumoniae group (*Streptococcus pneumoniae*). Differences between two are given in **Table 50.2**. Organism is pneumococcus has following diagnostic features.
  – Arranged in pair with lancet/flame in shape.
  – α-hemolytic colonies on blood agar.
  – Catalase test: Negative.
  – Bile solubility test: Positive.

4. b
- Most bacterial drug resistance to penicillin is due to production of penicillinase (beta-lactamase). Mostly plasmid transmitted horizontally. Vancomycin resistant to *Enterococcus* is due to alteration or elimination of target site like D-alanyl-D-alanine chain. Pneumococcal resistance to Pn is due to alteration in penicillin binding proteins.

5. c
- From the given options gram-negative coccus and oxidase positive is *Neisseria gonorrhoeae*. Ocular manifestations of *Neisseria gonorrhoeae* are explained in **Ch. 52** in section complications (autoinoculation).

6. b
- Necrotizing enteritis or enteritis necroticans or pigbel is caused by *Clostridium perfringens* due to ingestion of pig meat with trypsin inhibitor like sweet potatoes.

7. a
- *Clostridium tetani* is gram-positive bacillus, and it produces swarming growth on opposite half of plate after 1–2 days in anaerobic incubation. *Clostridium perfringens* is nonmotile and not able to produce the swarming growth. *Proteus mirabilis* can produce swarming growth but it is gram-negative bacillus. From the genus *Bacillus, Bacillus subtilis* can produce the swarming growth but not the *Bacillus cereus*.

8. d
- In *Clostridium tetani* samples are collected in three bottles of RCMM. (1) 1st bottle: Heated at 80°C for 15 min. (2) 2nd bottle: Heated at 80°C for 5 min. (3) 3rd bottle: Left unheated.
- Heating kills vegetative forms and leaves spores undamaged. Incubated at 37°C for 24–48 hours followed by subculture on blood agar daily for 4 days and observe for growth.

9. b
- Given case is of antibiotic associated diarrhea caused by *Clostridium difficile* where acute lymphocytic leukemia and amoxicillin therapy are the precipitating factors.

10. b
- *M. tuberculosis* and *M. bovis* are equally pathogenic for humans. *M. kansasi* can cause chronic pulmonary disease resembling tuberculosis. *M. africanum* is listed in the category of *Mycobacterium tuberculosis* complex (MTC) causing tuberculosis in West Africans by airborne route but pathogenicity is lower than *M. tuberculosis*. Environment acts as a source for atypical group not for MTC. *M. marinum* is responsible for swimming pool (fish tank) granuloma.

11. a
- PGL (Phenolicglycolipid)-1 antigen is secreted by *M. leprae*, so antibody is against the *M. leprae*. Sensitivity is 95% in untreated LL and 60% in TT cases. Titer decrease with effective therapy. It also presents in non leprosy cases, so rarely useful.

12. b and c
- *M. ulcerans* and *M. buruli* are synonym for each other, causing buruli ulcer.

13. a

14. a
- Tumbling motility is the feature of *L. monocytogenes* which occurs at 25°C, but not at 37°C. Other bacteria given in other options are motile with different types of motility and not changed by temperature effect.

15. a
- All options are known to cause the meningitis, but gram-positive bacillus is *L. monocytogenes*.

16. b

- *Proteus* forms the enzyme urease which cleaves the urea to ammonia and contributes to the alkalinity. Alkalinity causes the necrosis of renal tubular epithelium and precipitates the phosphate stone. Uric acid stone forms in acidic urine, mostly in gout or Lesch-Nyhan syndrome.

17. a

- *Serratia marcesens* produces the red pigment called **prodigiosin** and its presence in sputum simulating the presence of blood in sputum called **pseudohemoptysis** (hemoptysis means blood in sputum).

18. b

- Diphtheria toxin and *Pseudomonas* toxin inhibit the protein synthesis by inactivation of EF-2. Shiga like toxin (SLT) and shiga toxin inhibit the protein synthesis by inactivation of 60S ribosome. Cholera toxin and LT of enterotoxigenic *E. coli* act through activation of cAMP. Follow **Ch. 69** for actions of pertusis toxin.

19. d

- Given case is of traveler's diarrhea. Enterotoxigenic *E. coli* (ETEC) is also known to cause traveler's diarrhea from the given option. Stool contains blood (RBCs) in given case which is not the feature of ETEC. EPEC causes watery diarrhea without blood (RBCs). Some author claimed that *Salmonella* Typhi causing diarrhea contains blood (RBCs) with 7–14 days incubation period, but incubation period in given case is 2 days. *Shigella dysenteriae* causing diarrhea contains blood (RBCs) and mucus with 1–7 days incubation period, which fulfil all features of given case.

20. d

- Gram-positive bacteria showing swarming are *Clostridium tetani* and *Bacillus subtilis*. Gram-negative bacteria showing swarming are: *Proteus vulgaris*, *Proteus mirabilis* and *Pseudomonas aeruginosa*.

21. c

- A characteristic growth occurs by *Y. pestis* in oil/ghee broth in flask (contains oil/ghee at top), which hangs down into broth from the surface called **'stalactite growth'** or called **'stalactites'**.

22. a

23. a, b and c

- *Y. pestis* causes plague. *Y. enterocolitica* and *Y. pseudotuberculosis* cause yersiniosis characterized by acute/subacute appendicitis (called **pseudoappedicular syndrome**). *Pasteurella septica* causes lesions like appendicitis, meningitis, abscess, pneumonia, etc.

24. a

- Cholera toxin or heat labile toxin (LT) of *Vibrio cholerae* acts by c-AMP. *Staphylococcus aureus* produces many toxins, but none of them act through cAMP. Heat labile toxin (LT) of *E. coli* acts by cAMP while heat stable toxin acts by cGMP. *Salmonella* does not produce toxin.

25. a

- TCBS is selective medium for *Vibrio cholerae*. Stuart is not selective but transport (holding) medium. Skirrows is selective medium for *C. jejuni*. MYPA (Baker's MYPA medium) is selective medium for *Bacillus cereus*.

26. a

- *Vibrio cholerae* can grow at high (alkaline) pH about 6.4–9.6 and salt concentration about 0.5–1%.

27. a

- Microscopy of young culture or acute cholera stool reveals actively motile *Vibrio cholerae* suggests **'swarm of gnats'**.

28. a

- *Vibrio cholerae* can cause noninvasive infection like watery/secretory diarrhea called **cholera.** *Neisseria, Streptococcus* and *H. influenzae* are causing invasive infection like meningitis.

29. c

- *Pasteurella multocida* is transmitted by dog bite or contact with infected animals. It also present in human respiratory tract and in nasal sinus as commensals, which carries to traumatic site by blood. *Micrococcus halophilus* is nonpathogenic, related to *Staphylococcus* and no relation with sea coast. *V. fluvialius* can cause wound infection on exposure to sea water. *Neisseria gonorrhoeae* is transmitted sexually and no relation with sea coast.

30. b

- Incubation period for cholera is <24 hrs to 5 days, bubonic plague is 2–7 days, pneumonic plague is 1–3 days, septicemic plague is 2–7 days, measles is 10–14 days (fever occurs after 10 day and rashes occur after 14 day) and typhoid is 7–14 days.

31. a

- *Legionella pneumophila* causes a disease called **legionellosis** having two forms like mild influenza like symptoms called **pontiac fever** while pneumonia called **Legionnaires' disease**. *Listeria* causes a disease called **listeriosis**. Scrub typhus is caused by *Orientia tsutsugamushi*. *Leptospira* causes a disease called **leptospirosis** with different clinical types. *Rickettsia* causes a disease classified in two groups like typhus fever group and spotted fever group.

32. c

- Culture smear of *B pertussis* shows loose clumps of bacilli with space in between known as **'thumb print appearance'**.

33. b

- *Y. pestis* causes plague, *B. pertussis* causes whooping cough, *H. influenzae* causes meningitis, respiratory infection, etc., and *P. multocida* causes local (abscess, cellulitis, etc.) and systemic lesions (like meningitis, respiratory infection, appendicitis, etc.).

34. a

- Milk ring test is performed to diagnose the animal brucellosis from milk.

35. a

- Antigens from *Mycoplasma* are producing the antibodies that agglutinate human O RBCs at 4°C called **cold agglutination.**

36. b

- Mycoplasmas, gram –ve bacilli and gram +ve cocci are living free in the environment or present as commensals in humans or animals, while *Chlamydia trachomatis* is an obligate intracellular organism in cells of humans, animals or birds with tropism for squamous epithelial cells, macrophages of GIT and respiratory system.

37. b

- Painless indurated ulcer on the glans with bilateral inguinal lymphadenopathy suggests syphilis due to *T. pallidum*, which is best diagnosed by DGI microscopy from the given options.

38. b

- Causative agent of Lyme disease was identified in1985 by Bergdorfer in USA hence called ***B. bergdorferi***. *B. mayonii* is also a causative agent of Lyme disease in USA. In Europe and Asia, *B. afzelii* and *B. garinii* are the causative agents of Lyme disease.

39. c

- Query fever is caused by *Coxiella burnetii*. *Pseudomonas* causes sanghai fever. *Francisella tularensis* causes tularemia. *R typhi* causes endemic typhus.

40. a, b and d
- *Rickettsiae, M. leprae* and syphilis (*T. pallidum*) cannot grow in cell free media while *Bartonella* can grow.

41. a, b and c
- Lymph node contains granulomatous lesions with central star shaped necrosis surrounded by histiocytes, B cells and giant cells called **stellate abscess.** It presents in chronic granulomatous disease (tuberculosis and sarcoidosis), tularemia, actinomycosis, nocardiosis, cat scratch disease, candidiasis and in leishmaniasis.

42. a
- Follow **Table 125.3** for explanation.

43. c
- *Treponema pallidum* causes syphilis. *Borrelia burgdorferi* causes lyme disease. *Pseudomonas aeruginosa* causes Sanghai fever. Follow **Table 125.4** for other common fever and etiological agents.

44. a
- Follow **Table 125.4** for other common fever and etiological agents.

TABLE 125.3: Common strain, related agent and its application		
Strain	**Agent**	**Application**
	Bacterial strains	
Cowan I strain	*Staph. aureus*	Co-agglutination test
Streptococcus MG (MG = Minute Group) strain	*Strept. angiosus*	Used for diagnosis of primary atypical pneumonia caused by *Mycoplasma pneumoniae.*
Por 1A	*N. gonorrhoeae*	It causes disseminated infection of *N. gonorrhoeae* to other organs by blood called **disseminated gonococcal infections** and produces arthritis, ulcerative endocarditis and rarely meningitis.
Park Williams 8 Strain	*C. diphtheriae*	Widely used for toxin production
Sterne strain	*B. anthrcis*	Used for prepare the Sterne vaccine
Carbazzo strain	*B. anthrcis*	Used for prepare the Mazzucchi vaccine
Kedrowsky's strain	*M. tuberculosis*	Used to prepare the WKK antigen for non-specific CFT of *Leishmania*
Danish 1331	*M. bovis*	In India, WHO recommended strain of BCG for preparation of vaccine. It is prepared in BCG laboratory, Guindy, Chennai
Onco TICE	*M. bovis* (BCG)	Multiple injection of BCG vaccine has been tried in bladder cancer and in other diseases as an adjuvant therapy
O157:H7	*E. coli*	All strains of *E. coli* are sorbitol fermenters while O157:H7 is sorbitol non-fermenter. It diagnosed by using Sorbitol MacConkey's medium (also called **modified Mac Conkey's medium**
K 12 strain	*E. coli*	Conjugation was 1st discovered by Joshua Lederberg and Tatum in 1946 in *E. coli* K 12 strain
X strain (OX19, OX2) OXK)	*Proteus* spp.	Nonmotile strains, OX2 and 0X19 of *Proteus vulgaris* and OXK of *Proteus mirabilis* Used in Weil-Felix reaction for diagnosis of some rickettsial infections
Bhatanagar strain	*Citrobacter freundii*	It shares common Ag **(like Vi Ag)** with *S.* Typhi and *S.* Paratyphi C and confusing the diagnosis
S. Typhi 901 H and O strain	*S.* Typhi	For preparation of H and O antigen of widal test respectively
Ty21a	*S.* Typhi	Preparation of Typhoral (Live Typhoid oral) vaccine of *S.* Typhi
Ty2	*S.* Typhi	Preparation of Vi vaccine (typhim-Vi) of *S.* Typhi
DT 104	*S.* Typhimurium	Multidrug resistant strain emerged in early 1990s and associated with an increased risk of blood stream infection and hospitalization. Acquired by raw or partially cooked meat products
Bengal strain (O139)	*V. cholerae*	Occurred in October 1992 in Madras (Chennai), India and later similar outbreaks occurred in other parts of India
American strain (H$_2$S +Ve) and Danish strain (H$_2$S –Ve)	*Brucella suis*	Biotypes of *Brucella suis* on the basis of H$_2$S production
Nichol's strain	*Treponema pallidum*	• Do not grow in artificial culture media. It maintained for many decades by serial testicular passage in rabbits • Nichol's strain is useful to prepare the antigen for species specific treponemal tests
Reiter's strain, Kazan's strain	*T. phagedenis*	• Grow in artificial culture medium like **thioglycollate medium containing serum** • Reiter's strain is useful to prepare the antigen for group specific treponemal tests
Noguchi's strain	*T. refringens*	Grow in artificial culture medium like **thioglycollate medium containing serum**
T strain or T form	*U urealyticum*	Produce very tiny colonies on H agar hence the name

(Contd...)

TABLE 125.3: Common strain, related agent and its application (*Contd...*)		
Strain	**Agent**	**Application**
Patoc 1 strain	*L. biflexa*	Antigen for genus specific (screening) tests of *Leptospira* is prepared from *L. biflexa* (nonpathogenic)
TWAR strain	*Chlamydia pneumoniae*	It was 1st reported by Grayston et al. in 1986 from the adult patient with acute respiratory disease in Taiwan
Viral strains		
Oka strain	HHV-3 (VZV)	To prepare the live attenuated vaccine of VZV
Towne 125 and AD 169 strain	HHV-5 (CMV)	To prepare the live attenuated vaccine
Brunhilde and Mahoney	Polio virus 1	
Lencing and Mefi	Polio virus 2	
Leon and Sauket	Polio virus 3	
H1N1	Influenza virus A	Causing swine flu (pandemic 2009)
H5N1		Causing bird (avian) flu
Jeryl-Lynn strain	Mumps virus	To prepare the live attenuated vaccine of mumps
Schwartz strain	Measles virus	To prepare the live attenuated (injectable) vaccine of measles
Edmonston and Zagreb strain		
Moraten strain		
SA 14-14-2 strain	Japanese Encephalitis virus	To prepare the live vaccine of Japanese encephalitis virus
Nakayama strain		To prepare the killed vaccine of Japanese encephalitis virus
Beijing strain		
Beijing P3 strain		
Dakar strain	Yellow fever virus	To prepare the killed vaccine/Dakar vaccine of yellow fever
Asibi strain		To prepare the live attenuated/non- neurotropic vaccine/17D vaccine of yellow fever
Pitman Moore strain	Rabies virus	To prepare the human diploid cell culture vaccine (HDCV) of rabies
RA 27/3 strain	Rubella virus	To prepare the live attenuated vaccine of rubella
HEV 239	HEV	Used to prepare recombinant HEV vaccine (China), but not available globally
Parasitic strains		
Leishmania **strain**	*Leishmania* spp.	To prepare the vaccine of cutaneous leishmaniasis

TABLE 125.4: Common fever and etiological agents	
Fever	**Agent**
Scarlet fever	*Strept pyogenes*
Enteric fever	*Salmonella* Typhi and *Salmonella* Paratyphi
Typhoid fever (step-ladder pyrexia)	*Salmonella* Typhi
Paratyphoid fever	*Salmonella* Paratyphi
Shanghai fever	*Pseudomonas aeruginosa*
Pontiac fever	*Legionella pneumophilla*
Mediterranean fever/Malta fever/Undulant fever	*Brucella melitensis*
Lemming fever (in Norway)	*Francisella tularensis*
Rat bite fever (RBF)	• *Streptobacillus moniliformis*: Called **haverhill fever or erythema arthriticum epidemicum or streptobacillary RBF** • *Spirillum minus*: Called **sodoku or spirillary RBF**
Relapsing fever	In USA: *B. bergdorferi* and *B. mayonii* In Europe and Asia: *B. afzelii* and *B. garinii*
Break bone fever	Dengue virus
Katayam fever	More in *S. japonicum* and less in *Schistosoma haematobium*

45. a
- Given case is of respiratory infection, by organism belong to family Streptococcaceae. Widal test is used for intestinal disease caused by *S.* Typhi.

46. a

47. a, c, d

48. a, b
- Follow **respective chapters** for explanation.

49. a
- All organisms are able to produce the brain abscess, but with red fluorescence is *P. melaninigenica* which was initially classified as a species of *Bacteroides*. Other microbes which get fluorescence under UV light are mentioned in **Ch. 90**.

Viral Infections

1. Paul Bunnell test is for:
- a. Malta fever
- b. Typhus fever
- c. Enteric fever
- d. Infectious mononucleosis

2. Parvo virus causes:
- a. Aplastic anemia
- b. Erythematosum infectiosum
- c. Roseola infantum
- d. Arthritis

3. Parvo virus B19 does not cause:
- a. Roseola infantum
- b. Aplastic anemia
- c. Fetal hydrops

4. Genital herpes simplex is diagnosed by:
- a. Gram's stain
- b. KOH preparation
- c. Tzanck smear
- d. Acid fast stain

5. Genital infection by HPV is diagnosed by:
- a. Gram's stain
- b. KOH preparation
- c. Tzanck smear
- d. Pap smear

6. Property of elution is found in:
- a. Myxo virus
- b. Toga virus
- c. Parvo virus
- d. Adeno virus

7. Which of the following pair is/are correct?
- a. RSV – Bronchiolitis
- b. HHV5 – Infectious mononucleosis
- c. Parvo virus – Exanthma subitum
- d. HHV6 – Kaposi sarcoma
- e. VZV – Chickenpox

8. Most common viral disease affecting parotid gland:
- a. Mumps
- b. Measles
- c. Rubella
- d. Varicella

9. Most common agent responsible for bronchiolitis:
- a. RSV
- b. Adeno virus
- c. Herpes virus
- d. Influenza virus

10. Most common cause of viral pneumonia in infant
- a. *Rhinovirus*
- b. RSV
- c. Reo virus
- d. CMV

11. Which of the following pair is true?
- a. *Hantavirus* pulmonary syndrome is caused by inhalation of rodent's urine and feces
- b. KFD is caused by bite of wild animal
- c. *Lyssavirus* is transmitted by tick
- d. Chikungunya is caused by *Anopheles*

12. Active immunization following exposure is given most commonly for:
- a. Rabies
- b. Polio
- c. Plague
- d. Measles

13. Postexposure immunization is done for:
- a. Measles
- b. Polio
- c. Rabies
- d. Chickenpox

14. Croup is most commonly due to:
- a. Respiratory syncytial virus
- b. Parainfluenza virus
- c. Adeno virus
- d. Coronavirus

15. T4/T8 ratio reversal is seen in:
- a. T cell lymphoma
- b. Hairy cell leukemia
- c. AIDS
- d. Infectious mononucleosis

16. Choose the correct matches:
- a. Mumps – RA 27/3 strain
- b. Rubella – Jeryl-Lynn strain
- c. Measles – Edmonston-Zagreb strain
- d. BCG – Danish 1331 strain

17. Nakayama strain is used to prepare which vaccine?
- a. Typhoid
- b. Chicken pox
- c. Japanese encephalitis
- d. Yellow fever

18. Latency seen in viral infections by:
- a. HSV-II
- b. CMV
- c. *Rotavirus*
- d. HIV
- e. EBV

19. Human to human transmission not seen in:
- a. SARS
- b. Japanese B encephalitis
- c. Bird's flu
- d. Chickenpox
- e. Rabies

20. Which of the following pair is correct?
- a. RSV – Bronchiolitis
- b. Orf – Viral infection is transmitted from sheep
- c. Parvo virus B19 – Exanthema subitum
- d. HHV6 – Kaposi sarcoma

21. Which pathogen adhere to respiratory epithelium?
- a. RSV
- b. Influenza
- c. Parainfluenza
- d. HBV
- e. Picorna virus

22. Which of the following contain(s) a RNA dependent DNA polymerase as a structural component of the virion?
- a. Adeno viruses
- b. Orthomyxo viruses
- c. Rhabdo viruses
- d. Retro viruses

23. Break bone fever is caused by:
- a. Variola
- b. Coxsackie
- c. Dengue
- d. Adeno virus

24. Man is the only reservoir of:
- a. Rabies
- b. Influenzae
- c. Typhoid
- d. Japanese B encephalitis

25. Hypoplasia of limb and scarring are caused by:
- a. Varicella
- b. Herpes simplex
- c. Rubella
- d. *Toxoplasma*

26. Which virus reactivates and involves the eye?
- a. Herpes zoster
- b. CMV
- c. EB virus
- d. Entero virus 70

27. Lymphocytosis with atypical lymphocytes are seen in infection with:
- a. HSV
- b. HBV
- c. EBV
- d. RSV

28. Immunocompromised patient due to transplantation is suffered with pyrexia and neutropenia. Most likely cause is:
- a. HSV
- b. CMV
- c. Gram-negative organism
- d. Gram-positive organism

29. Virus causing hemorrhagic cystitis, diarrhea and conjunctivitis is:
- a. RSV
- b. *Rhinovirus*
- c. Adeno virus
- d. Rota virus

30. **Condyloma accuminata is caused by:**
    a. HSV
    b. HPV
    c. HIV
    d. VZV

31. **The virus which causes aplastic anemia in chronic hemolytic disease is:**
    a. Adeno virus
    b. Hepatitis virus
    c. EB virus
    d. Parvo virus

32. **Slapped cheek sign is seen in:**
    a. Parvo virus B19
    b. JC virus
    c. *Rotavirus*
    d. Mumps

33. **Which viral disease is transmitted by orofecal route?**
    a. Dengue
    b. Polio virus
    c. Hepatitis B
    d. Influenza virus

34. **Virus which spreads by hematogenous and neural route is?**
    a. Rabies virus
    b. Varicella zoster virus
    c. Polio virus
    d. EB virus

35. **Herpengina is caused by:**
    a. *Enterovirus*
    b. *Rhinovirus*
    c. Myxo virus
    d. Rabies virus

36. **Antigenic variation is seen in which of the following:**
    a. Influenza virus
    b. Hepatitis virus
    c. Yellow fever virus
    d. *Leptospira*

37. **Subacute sclerosing pan encephalitis is the delayed manifestation of:**
    a. Influenzae
    b. Measles
    c. Mumps
    d. Polio

38. **Giant cell (Hecht's) pneumonia is due:**
    a. CMV
    b. Measles
    c. Malaria
    d. *P. carinii*

39. **Following are arboviral diseases/viruses:**
    a. KFD
    b. West Nile fever
    c. Ganjam virus
    d. RSV
    e. Puumala virus

40. **Which of the following is/are arboviral diseas(es)?**
    a. Japanese encephalitis
    b. Dengue
    c. Yellow fever
    d. Hand foot mouth disease
    e. Rocky mountain spotted fever

41. **Which of the following virus is composed of two distinct capsids enclosing the double stranded RNA?**
    a. Adeno virus
    b. Reo virus
    c. Herpes virus
    d. Myxo virus

42. **Causative agent of SARS:**
    a. H1N1
    b. Corona virus
    c. *Rotavirus*
    d. RSV

43. **Which enzyme is not found inside the core of the virus?**
    a. DNA polymerase
    b. Reverse transcriptase
    c. Neuraminidase
    d. RNA polymerase

Answers and Explanation of MCQs

1. **d**
   • Malta fever is caused by *Brucella* spp., and detected by SAT and few other serological tests. Typhus fever is caused by *Rickettsia* spp., and detected by Weil-Felix reaction. Enteric fever is caused by *Salmonella* spp., and detected by Widal test. Infectious mononucleosis is caused by EBV and detected by Paul Bunnell test.

2. **a, b, d**

3. **a**
   • Parvo virus B19 causes aplastic anemia, erythematosum infectiosum (5th disease or slapped cheek disease), fetal hydrops and arthritis. HHV-6B causes sixth disease or roseola infantum or exanthema subitum.

4. **c**
   • Gram's stain is useful for bacteria. KOH preparation is for fungi. Tzanck smear is used for Herpesviridae. Acid fast stain is for acid fast organisms.

5. **d**
   • Gram's stain is useful for bacteria. KOH preparation is for fungi. Tzanck smear is for Herpesviridae. Pap smear is for cervical cancer caused by HPV.

6. **a**
   • Myxo virus includes influenza virus which has H and N peplomeres. H is responsible to make complex with hemagglutinin (H) receptors on host cells (RBCs) and causing hemagglutination. N is called **neuraminidase**. It destroy the H receptors on host cells hence called **receptor destroying enzyme (RDE)** and causes reversal of hemagglutination and release the bounded virus to infect the other cells called **elution.** This property present only in Myxo virus not in other virus.

7. **a and e**
   • HHV4 (EBV) causes infectious mononucleosis. Parvo virus B19 causes aplastic anemia, erythematosum infectiosum (5th disease or slapped cheek disease), fetal hydrops and arthritis. HHV6 causes exanthea subitum and HHV causes Kaposi sarcoma.

8. **a**
   • Measles, rubella and varicella are the exanthematous diseases of skin.

9. **a**

10. **b**
    • Adeno virus, herpes virus, influenza virus, *Rhinovirus* and reo virus are also a respiratory pathogens but RSV out rank the other pathogens and most common cause of lower respiratory tract infections like bronchitis, bronchiolitis and pneumonia in infants (<1 year children) and young children. CMV is not the respiratory pathogen.

11. **a**
    • *Hantavirus* pulmonary syndrome is caused by inhalation of aerosols arise from excreta of deer mouse and other rodents of *Sigmoidontine* family. KFD is caused by ticks bite (Ixodid or hard tick) like *Haemaphysalis spinigera, H. turtura,* etc. *Lyssavirus* is a genus of Rhabdoviridae family. It has 7 species or serotypes like lyssa virus-1 as rabies virus, which is transmitted by dog bite. *Lyssavirus* is not transmitted by vector. Chikungunya is caused by *Aedes.*

12. **a**
    • For polio and plague pre-exposure prophylaxis is useful. For measles and rabies post-exposure prophylaxis is available, but question is for most common, so answer is rabies.

13. **a and c**
    • Post-exposure immunization is done for measles, rabies and tetanus.

14. **b**
    • Parainfluenza virus-2 (human rubula virus 2) causes croup (acute laryngotracheobronchitis). Follow respective chapters for lesions produced by respiratory syncytial virus, adeno virus and corona virus.

15. **c**
    • In AIDS, HIV attacks over CD4 cells, resulting decrease CD4 and there is reversal of T4/T8 ratio.

16. **c, d**

17. c
- Follow **Table 125.3** for explanation.

18. a, b, d and e
- Follow **Table 125.5** for explanation.

19. b and c
- Follow respective chapters for explanation of options.

20. a and b
- Parvo virus B19 causes erythema infectiosum or slapped cheek disease or 5th disease. HHV-6 causes sixth disease or roseola infantum or exanthema subitum. HHV-8 causes Kaposi sarcoma.

21. a, b, c and e
- Follow respective chapters for explanation of options.

22. d
- RNA dependent DNA polymerase is the reverse transcriptase, which present in viruses of family Retroviridae and Hepadnaviridae.

23. c
- Follow **Table 125.4** for explanation.

24. b and c
- Reservoirs in rabies are **domestic animals** like dogs, cats and cattle and **wild animals** like jackals, wolves, foxes, bats, mongooses and skunks. Reservoirs in influenza are humans, birds, animals, etc. Reservoirs in *Salmonella* Typhi are the humans. Cases or carriers of *Salmonella* are the reservoir hosts. Reservoirs in Japanese B encephalitis are *Ardeid* birds (herons and egrets), while pigs are the amplifier hosts.

25. a
- Hypoplasia of limb and scarring are features of fetal varicella syndrome caused by VZV.

26. a
- After primary infection, herpes zoster virus (actual name VZV) remains latent in trigeminal (semilunar/Gasserian or Gasser's) ganglion for several years and reactivated in old age around 50 years when immunity waned due to disease like AIDS, age factor or by other stimuli and it causes disease called **zoster** which involves nerves, skin, ears and eyes.

27. c
- EBV causes formation of Ag (neoantigen) on B-cell surface, which induces blast-transformation in T-cell. Multiple atypical T-cells (lymphoblasts) are examined in peripheral smear in EBV infection called **infectious mononucleosis**.

28. b
- CMV is the most common agent causes infection following organ transplantation.

TABLE 125.5: Agents causing latent infection	
Bacteria	
T. pallidum	*R. prowazekii*
B. bergdorferi	*B. bacilliformis*
	B. (Rochalimaea) quintana
Viruses	
DNA viruses	**RNA viruses**
Some pox viruses	Measles virus
HHV 1–8	Rabies virus
HPV	HIV
Parasites	
T. brucei (non-flagellate latent form and flagellate latent form)	*T. gondii* (tissue cyst)

29. c
- Adeno virus serotypes 11 and 21 are causing hemorrhagic cystitis, 40 and 41 causing diarrhea and serotypes 3, 4, 7, 11, 8, 19 and 37 are causing conjunctivitis.

30. b
- Condyloma accuminata is genital wart caused by HPV.

31. d
- Aplastic anemia or aplastic crisis is the severe anemia due to temporary cessation of RBCs production by parvovirus B 19.

32. a
- Erythema infectiosum or slapped cheek disease or 5th disease is caused by parvovirus B19.

33. b
- Dengue is transmitted by bite of *Aedes aegypti*, hepatitis B by blood and influenza virus by inhalation route.

34. c
- Rabies virus spreads by neural route. Varicella zoster virus and EB virus spread through blood route. Polio virus spreads by blood and also by neural route following tonsillectomy.

35. a
- Herpengia is caused by coxsackie virus A which is belongs to genus *Enterovirus*.

36. a
- Antigenic variations (like antigenic shift and drift) are the most common features of influenza virus.

37. b
- SSPE is very rare and delayed sequelae of measles infection occur several years after the primary infection. Rarely it also occurs by rubella infection.

38. b
- Giant cells called **Warthin-Finkeldey cells** occur only in measles not in CMV, malaria and *P carinii*.

39. a, b and c
- RSV is paramyxo virus. Puumala virus is robovirus.

40. a, b and c
- Hand foot mouth disease is caused by coxsackie virus. Rocky mountain spotted fever is caused by *Rickettsia rickettsii*.

41. b
- Adeno virus and herpes virus are DNA viruses with icosahedral capsid but with one layer. Reo virus is RNA type with icosahedral capsid but with two layers. Myxo virus is RNA type with helical capsid.

42. b
- H1N1 causes swine flu, coronavirus causes respiratory disease called **SARS**, *Rotavirus* causes diarrhea and RSV causes minor respiratory symptoms.

43. c
- Neuraminidase is the receptor destroying enzyme found in H1N1. It present as spike on envelope.

Fungal Infections

1. Fungal growth is diagnosed with:
   a. Giemsa stain
   b. KOH
   c. Foot pad culture
   d. Albert stain

2. Germ tube test is:
   a. Reynolds Braude phenomenon
   b. Pfiffer's phenomenon
   c. Theobald smith phenomenon
   d. Splendore-Hoeppli phenomenon

3. Which of the following organism cannot be cultured?
   a. *K. rhinoscleromatis*   b. *K. ozaenae*
   c. *K. granulomatis*   d. *Pneumocystis jiroveci*
   e. *Rhinosporidium seeberi*

4. Fungus not cultivable usually:
   a. *Rhinosporidium*   b. *Cryptococcus*
   c. Dermatophytes   d. *Histoplasma*

5. The most common organism amongst the following that causes acute meningitis in an AIDS patient is:
   a. *Streptococcus pneumoniae*
   b. *Streptococcus agalactiae*
   c. *Cryptococcus neoformans*
   d. *Listeria monocytogenes*

6. Which of the following is the most common fungal infection in immunocompetent patient?
   a. *Candia*   b. *Aspergillus*
   c. *Cryptococcus*   d. *Penicillium*

7. Neurotropic fungus is/are:
   a. *Cryptococcus neoformans*
   b. *Histoplasma*
   c. *Trichophyton*
   d. *Candia*
   e. Aspergillosis

8. Most fulminant fungal meningitis is caused by:
   a. *Coccidioides*   b. *Histoplasma*
   c. *Cryptococcus*   d. Mucormycosis

9. Maltese cross form seen under polarizing microscopy in:
   a. *Cryptococcus neoformans*
   b. *Penicillium marneffi*
   c. *Blastomyces*
   d. *Candida albicans*

10. Culture medium for fungus is:
   a. Tellurite medium   b. NNN medium
   c. Chocolate medium   d. Sabouraud's medium

11. Correctly matched stain:
   a. Mucicarmine – *Cryptococcus*
   b. Giemsa – *Candida*
   c. Methanamine silver – *Histoplasma*
   d. Gram's – *Pneumocystis carinii*

12. Which of the following is difficult to isolate by culture?
   a. *Candida*   b. Dermatophytes
   c. *Cryptococcus*   d. *Malassezia furfur*
   e. Coccidioidomycosis

13. The following is not true of *Candida albicans*:
   a. Yeast like fungus
   b. Forms chlamydospores
   c. Blastomere is seen in isolates
   d. Causes meningitis in immunocompromised

14. Which of the following disease is endogenous in origin?
   a. Aspergillosis   b. Candidiasis
   c. Phycomycosis   d. All of the above

15. Kerion is caused by:
   a. *Candida*   b. *Streptococcus*
   c. Dermatophytes   d. Herpes

16. The most common cause of mycetoma in India:
   a. *Nocardia brasiliensis*   b. *Actinomedua maduraei*
   c. Piedra   d. Tinea cruris

17. Discharging sinus is seen in:
   a. Sprotrichosis   b. Cryptococcosis
   c. Histoplasmosis   d. Mycetoma

18. A farmer from the sub Himalayan region presents with multiple leg ulcers. The most likely causative agent is:
   a. *Trichophyton rubrum*   b. *Cladosporium* species
   c. *Sprothrix schenckii*   d. *Aspergillus*

19. Which is not a fungal disease?
   a. Rhinosporidiosis   b. Sporotrichosis
   c. Torulosis   d. Candidiasis

20. True about *Histoplasma capsulatum*:
   a. Dimorphic fungus
   b. Causative organisms of moniliasis
   c. Causative organism of valley fever
   d. Capsulated

21. Which of the following is the most common etiological agent in paranasal sinuses?
   a. *Aspergillus* spp.   b. *Histoplasma*
   c. *Conidiobolus coronatus*   d. *Candida albicans*

Answers and Explanation of MCQs

1. a and b
- Giemsa stain is useful for histopathological study of fungi. KOH is useful for fungi. Foot pad culture is useful in *M. leprae*. Albert stain is useful for *C. diphtheriae*.

2. a
- Pfiffer's phenomenon is associated with complement system. Theobald smith phenomenon is related with anaphylaxis. Deposition of eosinophilic material around yeast cells in tissues is due to immune phenomenon of body called **Splendore-Hoeppli phenomenon.**

3. d, e

4. a
- Several attempts has been made for cultivation of *Rhinosporidium seeberi* in cell free media, cell containing media and animal culture, but successes are not achieved.

5. c
- All the microbes given in options are able to produce the meningitis, but in AIDS patients meningitis is caused by *C. neofromanswhen* when CD4 count is <100/mm^3, while others cause meningitis in immunocompetent individuals.

6. a
- *Candida* causes infection in immunocompetent and also in individual with immunosuppressive conditions while *Aspergillus*, *Cryptococcus* and *Penicillium* cause infection in person with IDDs.

7. a, b, d and e
- Neurotropic fungi are *Candida* spp., *Aspergillus* spp., *C. neoformans*, *H. capsulatum*, *B. dermatitids* and *S. schenckii*.

8. c
- Among the given options *Cryptococcus* is the most common fungus to cause meningitis, especially in AIDS patients when CD4 count fall below 100/mm^3.

9. a
- Maltese cross form is seen in infectious diseases like *Babesia* spp., *Cryptococcus neoformans*, *P. brasiliensis* and *M. furfur* and non infectious diseases like (in urine) nephritic syndrome, eclampsia, renal toxicity, fat embolism, crush injury, Fabry's disease (due to aggregates of glycophospholipids) and in arthroscopic fluid after local trauma.

10. d
- Tellurite medium used for *C. diphtheriae*, *Staph. aureus*, *E. faecalis*, *E. rhusiopathiae*. NNN medium is used for parasite like *Leishmania* spp., and *T. cruzi*. Chocolate medium is used for fastidious bacteria, like *H. influenzae*, *Staphylococcus* spp., *Neisseria* spp., *Streptococcus* spp., and pneumococcus. Sabouraud's medium is used for fungi.

11. a
- Correct matchings are Giemsa – *Pneumocystis carinii.*, Methanamine silver – *Pneumocystis carinii* and actinomycetes and Gram's stain – *Candida* and all fungi.

12. d
- *Malassezia furfur* is lipophilic fungus grows with difficulties on medium after adding the olive oil. Other fungi given in options can grow very easily on media.

13. c
- *Candida* produces blastospores. Blastomere is the segmented embryo present in the eggs of hook worm parasites like *Ancyclostoma duodenale* and *Nectar americanus.*

14. b
- Aspergillosis and phycomycosis have exogenous origin, while candidiasis has endogenous origin.

15. c
- Inflammatory boggy mass on scalp with multiple sinuses sometimes may discharge the mycetoma like grains. Mostly caused by dermatophytes like *T. mentagrophytes* and *T. verrucosum.*

16. b
- *Nocardia brasiliensis* also causes mycetoma, but less than *Actinomedua maduraei*. Piedra and tinea cruris are the cutaneous mycoses but different from mycetoma.

17. d
- Mycetoma presents with chronic granulomatous lesion with sinus, which discharges the granules, but no grains in lesions of sprotrichosis, cryptococcosis and histoplasmosis.

18. c
- *Sprothrix schenckii* is common in sub Himalayan region but not the other fungi given in options.

19. a
- Torulosis is the former name of cryptococcosis, which is a fungal disease. Sporotrichosis and candidiasis are fungal diseases. Rhisporidiosis is not fungal disease, but disease of parafungal agent.

20. a
- Moniliasis is caused by *Candida* spp. Valley fever is caused by *C immitis* while *Histoplasma* causes valley disease. Species name is *capsulatum*, but infact it is without capsule and only capsulated fungus is *C. neoformans*.

21. a
- *Aspergillus* is the most common etiological agent of infection in paranasal sinuses, while others are rare.

Parasitic Infections

1. Culture medium used for *Entamoeba histolytica* is:
   a. Blood agar   b. Philip's medium
   c. CLED medium   d. Trypticase serum

2. Primary amoebic meningoencephalitis (PAM) is caused by:
   a. *E. histolytica*   b. *N. fowleri*
   c. *B. coli*   d. *G. lamblia*

3. Cyst form is found in:
   a. *E. fragilis*   b. *B. coli*
   c. *T. vaginalis*   d. *T. intestinalis*

4. The cystic form of all are seen in man, *except*:
   a. *E. histolytica*   b. *Giardia*
   c. *Trichomonas*   d. *Toxoplasma*

5. Falling leaf like motility is the characteristic of:
   a. *E. histolytica*   b. *T. vaginalis*
   c. *G. lamblia*   d. All of above

6. A patient present with diarrhea. Analysis of stool on wet mount shows mobile protozoan without RBCs and pus cells. The diagnosis is:
   a. *B. coli*   b. Giradiasis
   c. *T. hominis*   d. *E. histolytica*

7. Protozoan associated with megaesophagus:
   a. Trypanosome   b. Amoeba
   c. *Giardia*   d. *Gnathostoma*

8. Parasitic disease/parasite affecting intestine is:
   a. Chaga's disease   b. Malaria
   c. Kala azar   d. *Fasciola*

9. Leishmania is cultured in .... medium:
   a. Chocolate agar   b. NNN
   c. Tellurite   d. Sabourauds

10. A 40-year-old male from Bihar complains of pain in abdomen, having hepatosplenomegaly, peripheral smear on stain shows:
   a. *Plasmodium vivax*   b. *Leishmania*
   c. Microfilaria   d. None

11. Cysticercosis is caused by:
   a. *T. solium*   b. *T. saginata*
   c. *A. duodenale*   d. *E. granulosus*
   e. *E. multilocularis*

12. 24-year-old AIDS patient develops chronic abdominal pain, low-grade fever, diarrhea and malabsorption. Oocyst demonstrated in stool. Likely cause of his diarrhea is:
   a. *E. histolytica*   b. *G. lamblia*
   c. Microspora   d. *Cystoisospora belli*

13. HIV patient with malabsorption, fever, chronic diarrhea with acid fast organism. What is the causative agent?
   a. *Giardia*   b. Microspora
   c. *Cystoisospora*   d. *E. histolytica*

14. Larva found in muscles is:
   a. *Trichinella spiralis*   b. *Ancyclostoma duodenale*
   c. *Trichuris trichuria*   d. *Enterobius vermicularis*

15. Larval form of which parasite resides in muscle:
   a. *Taenia saginata*   b. *Echinococcus*
   c. *Trichinella*   d. All of above

16. Dogs are responsible for transmission of all the following, *except*:
   a. Hydatid disease   b. Toxoplasmosis
   c. Kala-azar   d. *Toxocara canis*

17. Abdominal pain, fat malabsorption and frothy stool suggest:
   a. Amoebiasis   b. Bacillary dysentery
   c. Giardiasis   d. Pancreatic enzyme deficiency

18. Pigs are reservoir for:
   a. *T. solium*   b. *T. saginata*
   c. *Trichinella spiralis*   d. *Ancyclostoma*

19. All diseases are acquired by ingestion, *except*:
   a. Taeniasis   b. Guinea worm
   c. Toxoplasmosis   d. Leishmaniasis

20. Which of the following is transmitted by egg ingestion?
   a. Taeniasis   b. Trichinosis
   c. Hydatidosis   d. Strongyloidosis

21. Cercarias are infective form of:
   a. *S. haematobium*   b. *P. westermanii*
   c. *F. hepatica*   d. *T. solium*

22. All cause brain lesion, *except*:
   a. Giardiasis   b. Tuberculosis
   c. Cysticercosis   d. *Bacteroides*

23. Enteropathogenic organisms are:
   a. *Cryptococcus*   b. *B coli*
   c. *Microsporidium* species d. *E. dispar*
   e. *G. intestinalis*

24. Respiratory symptoms are associated with:
   a. Rocky mounted spotted fever
   b. *Strongyloides*
   c. *T. solium*
   d. *Onchocerca*

25. Which of the following congenital infection leads to maximum CNS damage in the fetus?
   a. Rubella and CMV
   b. Rubella and toxoplasmosis
   c. CMV and toxoplasmosis
   d. HIV and CMV

26. Human is secondary host for:
   a. Malaria   b. Tuberculosis
   c. Filariasis   d. Relapsing fever

27. Human is definitive host for:
   a. Echinococcosis   b. Malaria
   c. Filariasis   d. Rabies
   e. Leishmaniasis

28. Amastigote forms are seen in:
   a. *Leishmania donovani*   b. *Toxoplasma gondii*
   c. *Leishmania major*   d. *Entamoeba*

29. Double rise of temperature in 24 hours is seen in case of:
   a. Kala-azar   b. Malaria
   c. TB   d. Hodgkin's lymphoma
   e. Brucellosis

30. Numbers of nuclei in the cyst of *B. coli*:
   a. 2   b. 4
   c. 6   d. 8

31. Sabin Feldman dye test is used to demonstrate infection with:
   a. Filaria   b. *Toxoplasma*
   c. *Histoplasma*   d. *Ascaris*

32. Tachyzoites are seen in:
   a. *Toxoplasma*
   b. *Toxocara*
   c. Pulmonary eosinophilia
   d. *Ascaris*

33. Acid fast organism with oocyst of size 5 μm on stool examination, causing diarrhea in HIV positive patient:
   a. *Cryptosporidium*   b. *Cystoisospora belli*
   c. *Microsporidium*   d. *Blastocystis hominis*

34. The following eggs have hexacanth embryo, *except*:
   a. *T. solium*   b. *T. saginata*
   c. *Chlonorchis*   d. *H. nana*

35. Which is not a liver fluke?
   a. *Paragonimus*   b. Whip worm
   c. *Chlonorchis sinensis*   d. *Gnathestoma spinigerum*
   e. *Opisthorchis*

36. Liver is a target organ for:
   a. *Fasciola buski*   b. *Paragonimus westermanii*
   c. *Chlonorchis sinensis*   d. *Schistosoma haematobium*

37. A man on return a country complains of pain in abdomen, jaundice, with increased alkaline phosphatase and conjugated hyperbilirubinemia. USG shows blockage in biliary tree. What could be the cause?
   a. *Fasciola buski*   b. *Chlonorchis sinensis*
   c. *Strongyloides*   d. *Ancyclostoma*

38. Seat worm is:
   a. *Enterobius*   b. *Dracunculus*
   c. *Ancyclostoma*   d. *Necator*

39. Dragon or serpent worm is:
   a. *Enterobius*   b. *Trichuris*
   c. *Dracunculus*   d. *T. solium*

Answers and Explanation of MCQs

1. b
- Blood agar, CLED medium and trypticase serum are the media for bacteria.

2. b
- *E. histolytica* causes cerebral amoebiasis, but it is secondary due to intestinal lesion, not primary. *B. coli* and *G. lamblia* cause intestinal lesions.

3. b

4. c
- Parasites without cystic stage are *Entamoeba gingvalis*, *Dientamoeba fragilis*, *Trichomonas vaginalis*, *Trichomonas hominis* (*T intestialis*) and *Trichomonas tenax* (*T buccalis*).
- *Gardnerella vaginalis* is bacterium and having no cystic stage.

5. c
- *E. histolytica* shows crawling or gliding motility due to pseudopodium. Movement is jerky and unidirectional. *T. vaginalis* shows wobbling (side to side)/rotatory or jerky movement.

6. b
- *Balanidium coli* and *Entamoeba histolytica* cause dysentery, so RBCs and pus cells are present in stool. Giradiasis is a case of diarrhea, so RBCs and pus cells are not present in stool. *Trichomonas hominis* is nonpathogenic.

7. a
- *T. cruzi* causes Chaga's disease, which causes complications in GIT (megaoesophagus and megacolon), heart (cardic myopathy and herniation of endocardium at apex of left ventricle) and lungs (dilatation of pulmonary conus). Amoeba causes dysentery and *Giardia* causes diarrhea but no megaesophagus or megacolon. *Gnathostoma* is known to produce larva migrans.

8. a
- *T. cruzi* causes Chaga's disease which causes complications in GIT like megaesophagus and megacolon. Malaria affects RBCs and spleen. Kala azar affects skin and lymph nodes. *Fasciola hepatica* affects liver.

9. b
- Chocolate agar and tellurite are bacterial media. Sabouraud's is fungal medium.

10. b
- All features and location (Bihar) of patient are suggesting the *Leishmania*.

11. a
- Follow **respective chapter** for more explanation and disease produced by all given options.

12. d
- From the given options agents causing diarrhea in AIDS patient are Microspora and *Cystoisospora belli*. Oocyst present only in *Cystoisospora belli*.

13. c
- From the given options agents causing diarrhea in AIDS patient with acid fast natures are Microspora and *Cystoisospora belli*, but malabsorption is a feature of *Cystoisospora belli*. For acid fast parasites follow **Table 112.5**.

14. a

15. d
- Parasitic larvae present in human are mentioned in **Table 125.6**.

TABLE 125.6: Parasites with larval stage in human
Cestodes
Spirometra spp. (In human, in muscles)
T. saginata (Not in human but in animal, in muscles, *C. bovis*)
Taenia solium (In human and animal, in muscles and tissues, *C. cellulosae*)
Echinococcus granulosus (In human and animal, in muscles and tissues, H cyst)
Nematodes
Trichinella spiralis (In human and animal in muscles)
Gnathostoma spinigerum (In human and animal in tissues)
Filarial parasites (Microfilariae in blood)

16. b
- In hydatid disease dog is the definitive host. In kala-azar canine reservoirs present in Mediterranean or infantile or Chinese or South American kala-azar. *Toxocara canis* is a parasite of dog, causing visceral larva migrans in human called **toxocariasis.** Toxoplasmosis is transmitted by cat.

17. c and d
- Abdominal pain, fat malabsorption and frothy stool are not the features of amoebiasis and bacillary dysentery but features of giardiasis and pancreatic enzyme deficiency.

18. a, c

19. d

20. a, c

21. a

22. a

23. b, c, d, e
- Follow **respective chapters** for more explanation.

24. a and b
- Rocky mounted spotted fever is caused by *R. rickettsia*. It infects lungs in 7–16% cases. *Strongyloides* larvae are infecting lungs.

25. c
- Among the all options, rubella associated with cardiac anomalies, while CMV and *Toxoplasma* are associated with brain damage.

26. a
- In malaria, human acts as an intermediate (secondary) host, while in filariasis, human acts as definitive (primary) host. In relapsing fever and tuberculosis, there are no requirements of primary or secondary hosts.

27. c
- In echinococcosis and malaria, human is an intermediate host. In filariasis, human allows the sexual stage of life cycle hence it is a definitive host. Rabies is viral disease, does not required like definitive and intermediate host. In leishmaniasis human (also sand fly) allows the asexual stage of life cycle, hence it is not a definitive host.

28. a and c
- Morphological forms present in *Leishmania* spp., are amastigote and promastigote. Morphological forms present in *Toxoplasma gondii* are oocyst, pseudocyst and tissue cyst. Morphological forms present in *Entamoeba* are trophozoite and cyst.

29. a
- Fever in kala-azar may be continuous or remittent type, becoming remittent in later stage. In 20% of cases fever shows the double rise in 24 hrs. Waves of pyrexia may be followed by apyrexia. In malaria fever types depended on type of species. In brucellosis fever is irregular (undulating/on-off fever).

30. a
- Two nuclei are present in cyst of *B. coli*. 4 nuclei are present in cyst of *E. histolytica* (mature cyst) and *G. lamblia*. 8 nuclei are present in cyst of *E. coli* (mature cyst). No any parasite with 6 nucleated cyst.

31. b
- Sabin Feldman dye test is based on inhibition of alkaline methylene blue dye's staining of tachyzoites of *T. gondii* by specific antiserum finally examined under microscope. There is no such test for filarial parasite, *Histoplasma* and *Ascaris*.

32. a
- Tachyzoite is the morphological stage of pseudocyst of *Toxoplasma* not the feature of *Toxocara*, pulmonary eosinophilia (due to filarial parasite) and *Ascaris*.

33. a
- Acid fast oocyst is the morphological form of *Cryptosporidium* and *Cystoisospora belli*. *Microsporidium* and *Blastocystis hominis* are not containing oocyst. Size of oocyst is 4–5 μm in *Cryptosporidium* and 20–30 μm × 10–20 μm in *C belli*.

34. c
- Hexacanth embryo is the feature of eggs of Cyclophyllidean cestodes which include *T solium, T saginata* and *H nana*. *Chlonorchis* is the monoecious trematode.

35. a, b and d
- *Paragonimus* is lung fluke while whip worm and *Gnathestoma spinigerum* are nematodes.

36. c
- *F. buski* targets intestine (intestinal fluke) and *P. westermanii* targets lungs (lung fluke). *Schistosoma haematobium* is targeting the vesical plexus.

37. b
- Among the all options *Chlonorchis sinensis* is the liver fluke especially infects biliary tract and rarely pancreatic duct.

38. a

39. c
- Common name for *Dracunculus* is dragon worm, for *Ancyclostoma* is old word hook worm and for *Necator* is new word/American hook worm, for *Enterobius* is thread/seat/pin worm, for *Trichuris* is whip worm and for *T. solium* is pork tape worm.

Entomology

1. ***Aedes aegypticus* transmits:**
   a. JE
   b. KFD
   c. Yellow fever
   d. Filaria
   e. Dengue

2. ***Culex* mosquito is associated with the transmission of:**
   a. Malaria
   b. Filariasis
   c. Dengue
   d. Japanese encephalitis
   e. Typhus

Answers and Explanation of MCQs

1. c and e
- JE: *Culex tritaeniorhynchus* is the worldwide vector including India while in India it also transmitted by *Culex vishnui complex.*
- KFD: Transmitted by ticks bite (Ixodid or hard tick) like *Haemaphysalis spinigera, H. turtura.*
- Filaria: Different types of filaria have different vectors. Not transmitted by *Aedes.*

2. b and d

- Malaria is transmitted by *Anopheles* spp. Dengue is transmitted by *Aedes aegypticus*. Different types of typhus are caused by different species of *Rickettsia* which are transmitted by louse, flea, tick, or by mite.

Practical Microbiology

1. Microscope used in microbiology:
a. Light microscope
b. Phase contrast microscope
c. Fluorescent microscope
d. All

2. Culture media are sterilized by:
a. Autoclave
b. Tyndallization
c. Hot air oven
d. Radiation

3. Electron microscopy is used for following, *except*:
a. To differentiate T and B lymphocytes
b. IgG deposits in kidney
c. TPI
d. Flagella

4. Smith Noguchi medium is used for:
a. *Salmonella*
b. *Klebsiella*
c. Spirochetes
d. *Bacillus*

Answers and Explanation of MCQs

1. d

- Microscope used in microbiology are mentioned in **Ch. 112**.

2. a and b

- Almost all media are sterilized by autoclave while media contain gelatin and sugars are sterilized by tyndallization.

3. c

- In TPI (*Treponema pallidum* immobilization) test, *T. pallidum* is immobilized by specific antiserum in presence of complement, which is examined under dark ground illumination microscope, but not under electron microscope.

4. c

- Smith Noguchi medium is used for anaerobic bacteria and from the given options anaerobes are spirochetes.

Image-based MCQs

1. Identify the scientist given in below image who was belonged to Germany:
   a. Louis Pasteur
   b. Robert Koch
   c. Edward Jener
   d. Landsteiner

2. Name the phase in bacterial growth curve marked by question mark (?) in below image:
   a. Lag
   b. Log
   c. Stationary
   d. Decline

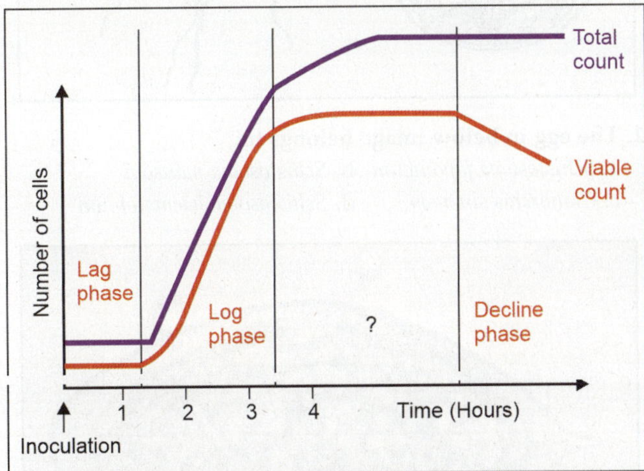

3. Name the site marked by arrow in below image:
   a. Chromosome
   b. Episome
   c. Transposon
   d. Plasmid

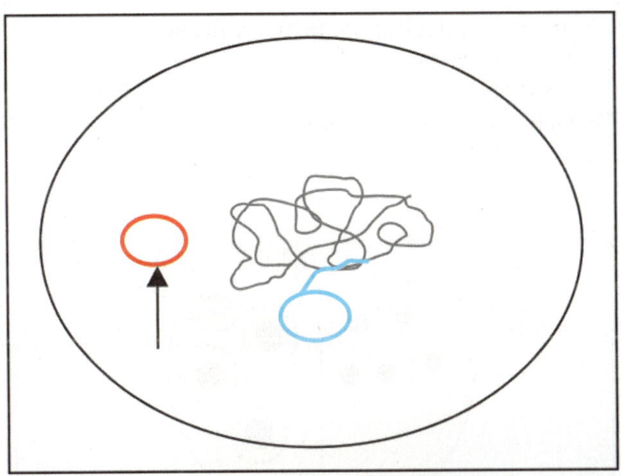

4. Waste collected in below color-coded bag is:
   a. Body organs
   b. Needle
   c. Glass items
   d. Food and papers

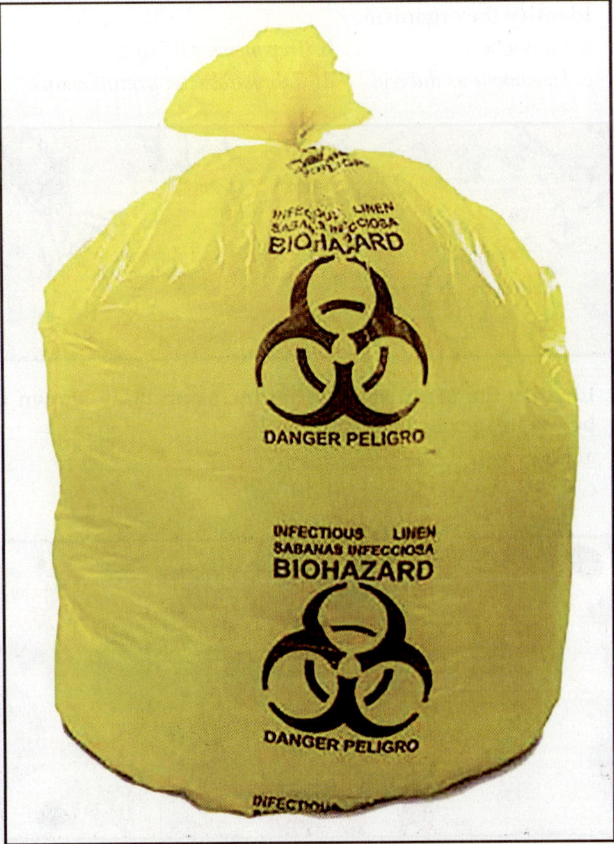

5. Identify the disease having "butter fly rash" appearance in below image:

a. SLE
b. Anthrax
c. Impetigo
d. Psoriasis

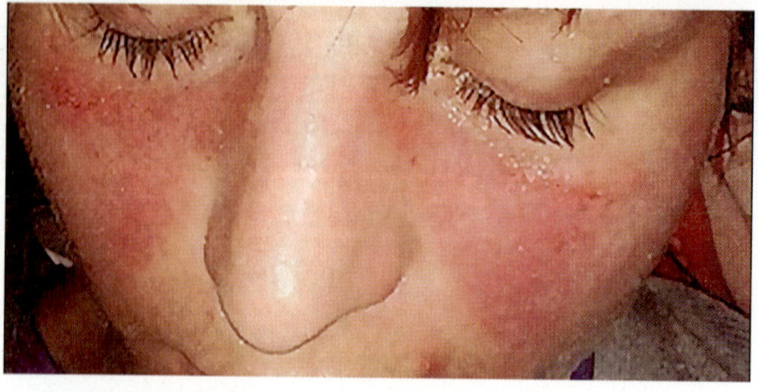

6. Name the serological test in below image:

a. Nagler reaction
b. CAMP test
c. Elek's gel precipitation test
d. Neutralization-based test

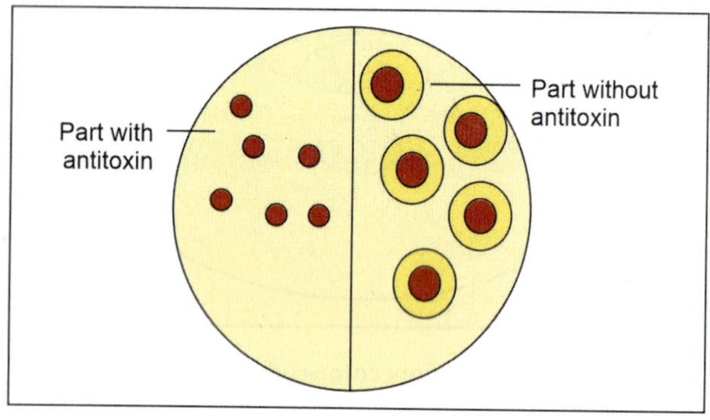

7. The microorganism circled in the given image is the causative agent of disease that is transmitted sexually. Identify the organism.

a. Gonococci
b. *Treponema pallidum*
c. *Haemophilus ducreyi*
d. *Calymatobacter granulomatis*

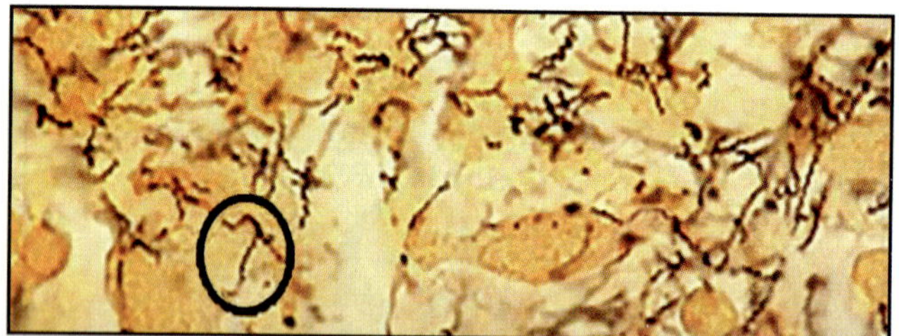

8. Identify the organism causing inclusion body shown in below image:

a. HSV
b. VZV
c. CMV
d. EBV

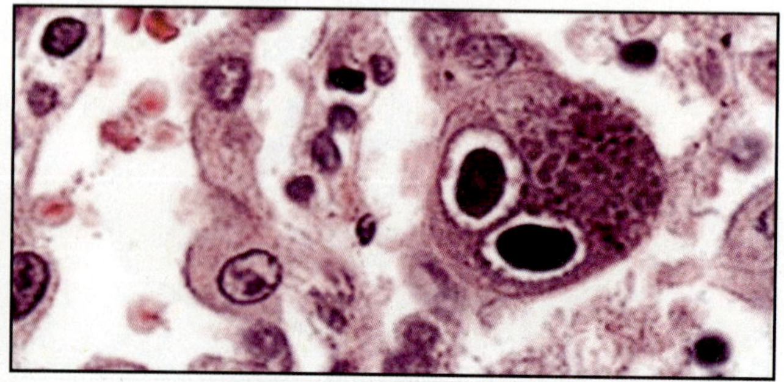

9. Identify the virus shown in below image:

a. Adeno virus
b. *Rotavirus*
c. Rabies virus
d. EBV

10. A patient is having TB like clinical picture, and peripheral smear with lacto phenol cotton blue stain is showing below image. Diagnosis is:

a. *Mycoplasma*
b. *Aspergillus*
c. *Cryptococcus*
d. *Candida*

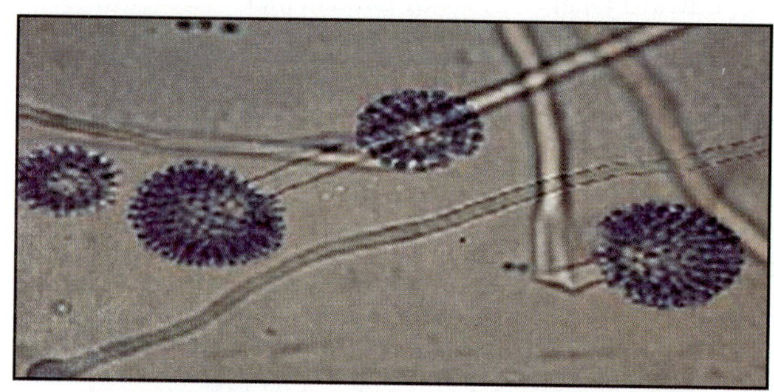

11. An anxious mother brought her 4 year old daughter to the pediatrician. The girl was passing loose bulky stools for the past 20 days. This was often associated with pain in abdomen. The pediatrician ordered the stool examination, which showed the organism with following image. Identify the organism.

a. *Entamoeba histolytica*
b. *Giardia lamblia*
c. *Cryptospiridium*
d. *E. coli*

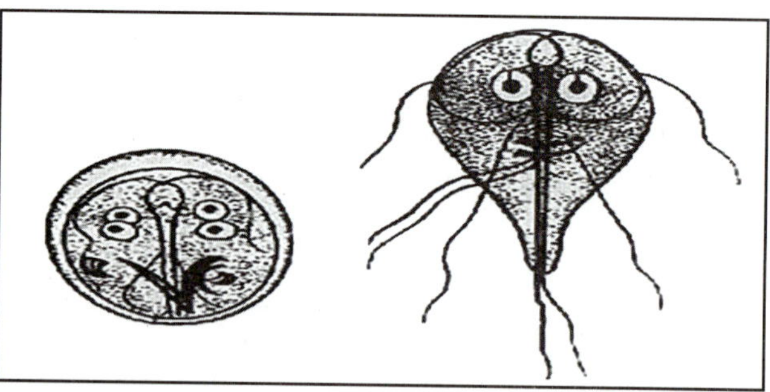

12. The egg in below image belongs to:

a. *Schistosoma japonicum*
b. *Schistosoma mansoni*
c. *Clonorchis sinensis*
d. *Schistosoma haematobium*

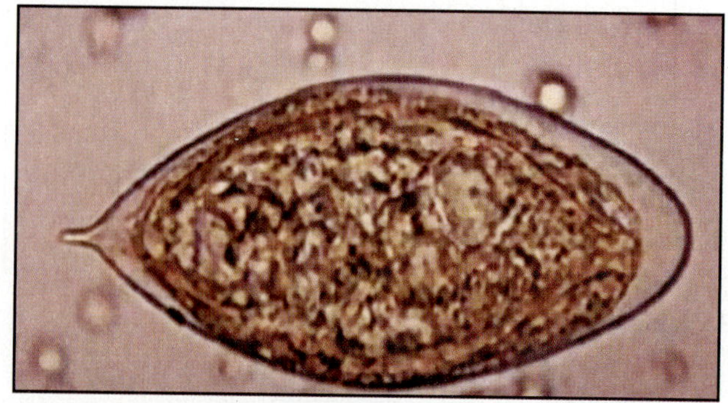

Essentials of Medical Microbiology

13. **Image shows the debridement of necrosed tissues in and around the external genitals. Most common clinical entity is:**
   a. Fournier gangrene
   b. Gas gangrene
   c. Pyoderma
   d. Mycetoma

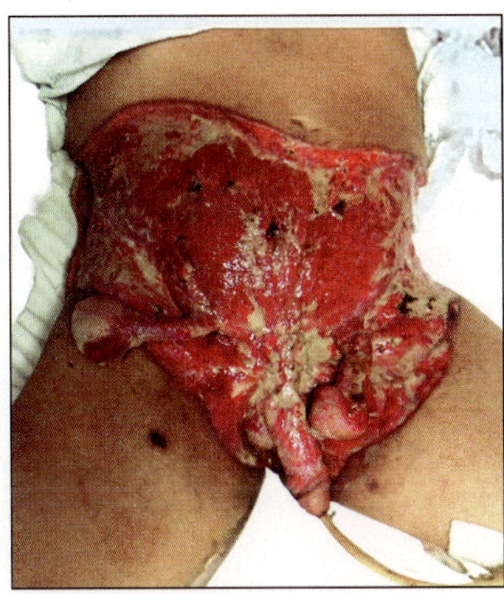

Answers of MCQs and Explanation

1. b
- Follow **Ch. 1** for images of important scientists.

2. c
- Follow **Ch. 7** for explanation.

3. d
- Follow **Ch. 8** for explanation.

4. a
- Follow **Ch. 31** for explanation.

5. a
- Follow **Ch. 26** for explanation.

6. a and d
- Follow **Ch. 55** for explanation.

7. b
- Follow **Ch. 73** for explanation.

8. c
- Given image is the owl's eye inclusion body produced by CMV. Follow **Ch. 78** for more explanation.

9. c
- Given image shows bullet shape of virus indicating rabies virus. Follow **Ch. 84** for more explanation.

10. b
- Image shows fungus with hyaline septate hyphae with exogenous conidia arranged in chain, suggestive of *Aspergillus*
- Follow **Ch. 91** for explanation.

11. b
- Follow **Ch. 94** for explanation.

12. d
- Egg in given image has terminal spine, so it is an egg of *Schistosoma haematobium.*

13. a
- Follow **Ch. 39** for explanation.